Thieme

Imaging Anatomy

Text and Atlas Volume 2
Abdomen and Pelvis

Farhood Saremi, MD
Professor of Radiology and Medicine
Department of Radiology
University of Southern California
Keck Medicine of USC
Los Angeles, California, USA

Associate Editors:

Damián Sánchez-Quintana, MD, PhD
Professor of Human Anatomy
Faculty of Medicine
Department of Anatomy and Cell Biology
University of Extremadura
Badajoz, Spain

Hiro Kiyosue, MD, PhD
Professor of Diagnostic Imaging and Analysis
Faculty of Life Sciences
Kumamoto University
Kumamoto, Japan

Dakshesh B. Patel, MD
Associate Professor of Clinical Radiology
Department of Radiology
University of Southern California
Keck Medicine of USC
Los Angeles, California, USA

Meng Law, MD
Professor of Radiology, Neurology, and Neurological Surgery
University of Southern California;
Biomedical Engineering
Viterbi School of Engineering;
Director of Neuroradiology and the Neuroradiology Fellowship Program
Keck Medicine of USC
Los Angeles, California, USA

R. Shane Tubbs, MD
Professor, Chief Scientific Officer, and Vice President
Seattle Science Foundation
Seattle, Washington, USA

1612 illustrations

Thieme
New York • Stuttgart • Delhi • Rio de Janeiro

Library of Congress Cataloging-in-Publication Data
is available from the publisher

Important note: Medicine is an ever-changing science undergoing continual development. Research and clinical experience are continually expanding our knowledge, in particular our knowledge of proper treatment and drug therapy. Insofar as this book mentions any dosage or application, readers may rest assured that the authors, editors, and publishers have made every effort to ensure that such references are in accordance with **the state of knowledge at the time of production of the book.**

Nevertheless, this does not involve, imply, or express any guarantee or responsibility on the part of the publishers in respect to any dosage instructions and forms of applications stated in the book. **Every user is requested to examine carefully** the manufacturers' leaflets accompanying each drug and to check, if necessary in consultation with a physician or specialist, whether the dosage schedules mentioned therein or the contraindications stated by the manufacturers differ from the statements made in the present book. Such examination is particularly important with drugs that are either rarely used or havebeennewly released on the market. Every dosage schedule or every form of application used is entirely at the user's ownrisk and responsibility. The authors and publishers request every user to report to the publishers any discrepancies or inaccuracies noticed. If errors in this work are found after publication, errata will be posted at www.thieme.com on the product description page.

Some of the product names, patents, and registered designs referred to in this book are in fact registered trademarks or proprietary names even though specific reference to this fact is not always made in the text. Therefore, the appearance of a name without designation as proprietary is not to be construed as a representation by the publisher that it is in the public domain.

Thieme addresses people of all gender identities equally. We encourage our authors to use gender-neutral or gender-equal expressions wherever the context allows.

Thieme Medical Publishers, Inc.
333 Seventh Avenue, 18th Floor,
New York, NY 10001, USA
www.thieme.com
+1 800 782 3488, customerservice@thieme.com

Cover design: © Thieme
Cover images source: © Thieme
Typesetting by Thomson Digital, India

FSC
www.fsc.org
MIX
Papier aus verantwortungsvollen Quellen
FSC® C089473

Printed in Germany by Beltz Grafische Betriebe 5 4 3 2 1

ISBN: 978-1-62623-982-1

Also available as an e-book:
eISBN (PDF): 978-1-62623-983-8
eISBN (epub): 978-1-63853-611-6

Contents

Preface

In the preface of his anatomy text collection *de Humani Corporis Fabrica* published in 1543, Andreas Vesalius wrote "... anatomy should rightly be regarded as the strong foundation of the whole art of medicine ...," acknowledging the fundamental role the study of anatomy has in the practice of medicine. Furthermore, one of the major lasting successes of his influential anatomy series was the ability to reproduce many superb illustrations of human anatomy dissection which was made possible by technical developments during his time.

Likewise, recent advancement in imaging technology and the development of newer postprocessing software have increased our ability to image and show anatomic detail of the human organs. At the same time, over the past decade, the literature is filled with many precious articles on anatomy and related topics. Unfortunately, there is an increasing gap between our fundamental core anatomy reference texts and our understanding and depiction of human anatomy using modern imaging and postprocessing methods. The idea for this project arose from the necessity for a text to fill this gap and describe anatomy in the context of current advances in imaging technology and science.

Computed tomography (CT) and magnetic resonance (MR) have been the traditional methods in performing noninvasive studies, and both have immensely contributed to our ability to deliver accurate diagnostic information. Current generation MR and CT as well as dedicated peripheral and intraluminal ultrasound provide the spatial and contrast resolutions required to demonstrate anatomic details with unprecedented accuracy.

When it comes to assessment of details, there is no imaging modality that can compete with the speed, accuracy, and spatial resolution of new CT scanners. Volume rendering and mutliplanar reformations facilitate understanding of the complex structures such as heart, vessels, and bones. With current scanners, the entire anatomic span of the major organs can be covered in a few seconds with spatial resolution of less than 0.5 mm^3. In this series, this technology along with the state-of-the-art postprocessing methods have been used to create volumetric color-coded images.

On the other hand, MRI provides high levels of contrast resolution that would be difficult to obtain with CT. High-field MRI provides superb resolution of brain anatomy. Taking advantage of a 7 Tesla MR scanner at our facility, high-quality images were obtained to complete the chapters of the neuroanatomy volume (volume 4). Using MR tractography, it is possible to depict the anatomic course of the white tracts in detail, and with new postprocessing software it is further possible to map areas of the brain cortex and subfield regions of small structures such as hippocampus. These technologies have ushered in a new era for investigation of brain anatomy and are being used for the planning of sophisticated brain surgeries and for localization of small regions of interest.

This imaging anatomy series is a reference atlas and text with a practical discussion of specific topics along with examples of anatomical variants. Divided into four major volumes, each volume takes the instructive format of a classic text but is infused with a large number of images and illustrations to teach readers anatomy via state-of-the-art cross-sectional and volumetric imaging. The volumes are divided into lung, mediastinum, and heart; abdomen and pelvis; musculoskeletal; and head, neck, and spine imaging. Peripheral vasculature and nerves have been included in the musculoskeletal volume.

The length of each chapter and number of images vary in accordance with the complexity of the topic. In all instances, effort has been made to provide a concise yet comprehensive review of the topic. Each chapter contains an in-depth review of the anatomy and anatomical variants. Pertinent embryology, microanatomy, and a brief description of physiology are discussed in each chapter. Postsurgical anatomy and important gross and surgical pathology images are included.

The text is supported by high-quality cross-sectional images with correlative three-dimensional and color-coded CT and MR views. Up-to-date references are used to support the text. Many new topics in radiology and surgery have been gathered from the recent 10-year literature. Surgical and clinical applications in each anatomy topic are presented with relevant images. In order to make each topic understandable, difficult anatomical concepts are supported by sketches as well as cross-sectional and topographic cadaveric views provided by internationally known anatomists. Superb cadaveric views have been provided by Professors Damián Sánchez-Quintana, R. Shane Tubbs, and the late Professor Albert L. Rhoton Jr. Images of high-resolution axial cadaveric cuts have been provided by the University of Auckland, New Zealand, thanks to efforts of Professor Ali Mirjalili.

As the author and chief editor of this textbook, I was fortunate to have assistance of other editors who have shared their experiences for specific parts of this compilation. Countless contributors from high-ranking academic institutions, with extensive experience in imaging, surgery, and human anatomy and embryology, have also provided substantial work and shared their expertise. Together over more than 5 years of constant work, we have created a comprehensive textbook and atlas on imaging and surgical anatomy that aims to serve a broad spectrum of the medical community including the medical students, radiologists, surgeons, clinicians, and research scholars.

Finally, this collaborative effort would have remained unfinished without the unreserved assistance and expert guidance of the product development editors of Thieme Publishers, William Lamsback (Executive Editor), Brenda Bunch (Production Editor), Shipra Sehgal (Managing Editor), Mary Wilson (Testing Specialist), and their team of professional medical illustrators.

Farhood Saremi
Los Angeles, California

Contributors

Frederic J. Bertino, MD
Department of Radiology and Imaging Sciences
Emory University
Atlanta, Georgia, USA

Alessandro Campari, MD
Scuola di Specializzazione in Radiodiagnostica
Università degli Studi di Milano, A.O. San Paolo
Milano, Italy

Frank Chen, MD
Assistant Professor
Department of Radiology
Mayo Clinic
Jacksonville, Florida, USA

Gianpaolo Cornalba, MD
Dipartimento di Scienze della Salute
Università degli Studi di Milano
A.O. San Paolo, Milano, Italy

Corinne Deurdulian, MD
Associate Professor of Radiology
Department of Radiology
University of Southern California
USC University Hospital
Los Angeles, California, USA

Jean H.D. Fasel, MD
Full Professor
Head of Clinical Anatomy Research Group
Departments of Cell Physiology and Metabolism, and Surgery
University Medical Centre and Hospitals
Geneva, Switzerland

Nicola Flor, MD
U.O. Radiologia Diagnostic ed. Interventistica
A.O. San Paolo, Milano, Italy

Ioannis Georgopoulos, MD
Department of Pediatric Surgery
Agia Sofia Children's Hospital
Athens, Greece

Mittul Gulati, MD
Associate Professor of Radiology
Department of Radiology
University of Southern California
USC University Hospital
Los Angeles, California, USA

Norio Hongo, MD
Department of Radiology
Oita University Faculty of Medicine
Oita, Japan

Kevin G. King, MD
Assistant Professor of Clinical Radiology
Department of Radiology
Keck Medicine of USC
USC University of Southern California
Los Angeles, California, USA

Maki Kiyonaga, MD
Department of Radiology
Oita University Hospital
Hasama, Yufu, Oita, Japan

Hiro Kiyosue, MD, PhD
Professor of Diagnostic Imaging and Analysis
Faculty of Life Sciences
Kumamoto University
Kumamoto, Japan

Ana Maliglig, MD
Assistant Professor
Department of Radiology
University of Southern California
Keck Medical Center of USC
Los Angeles, California, USA

Takashi Matsubara MD, PhD
Department of Radiology
Kanazawa University Hospital
Takaramachi Kanazawa City, Ishikawa, Japan

Shunro Matsumoto, MD
Department of Radiology
Faculty of Medicine
Oita University
Hasama-machi, Yufu, Oita, Japan

Aramsadat Meraji, MD
Department of Radiology
University of South California
USC University Hospital
Los Angeles, California, USA

Petros Mirilas, MD, PhD
Clinical Professor of Surgical Anatomy
University School of Medicine
Emory University
Atlanta, Georgia, USA

Ali Mirjalili, MD, PhD
Associate Professor
Anatomy and Medical Imaging Department
Faculty of Medical and Health Sciences
University of Otago
Grafton, Auckland, New Zealand

Navid Moradshahi, MD
Department of Radiology
University of Southern California
USC University Hospital
Los Angeles, California, USA

Hiromu Mori, MD
Department of Radiology
Oita University Hospital
Hasama, Yufu, Oita, Japan

Mojgan Motevali MD
Radiologist
University of Southern California
USC University Hospital
Los Angeles, California, USA

Eleni Paschalidou, MD
Department of Gynecology
G. Gennimatas General Hospital of Athens
Athens, Greece

Ali Rivandi, MD
Radiologist
Department of Radiology
University of Southern California
USC University Hospital
Los Angeles, California, USA

Damián Sánchez-Quintana, MD, PhD
Professor of Human Anatomy
Department of Anatomy and Cell Biology
Faculty of Medicine
University of Extremadura
Badajoz, Spain

Farhood Saremi, MD
Professor of Radiology and Medicine
Department of Radiology
University of Southern California
USC University Hospital
Los Angeles, California, USA

Andrea Schenk, MD
Member of Management Board
Fraunhofer Institute for Medical Imaging Computing MEVIS
Bremen, Germany

Scott A. Sylvester, MD
Department of Radiology
University of Florida – Jacksonville
Jacksonville, Florida, USA

Tapas K. Tejura, MD
Associate Professor
Department of Radiology
University of Southern California
Keck Medical Center of USC
Los Angeles, California, USA

R. Shane Tubbs, MD
Professor
Chief Scientific Officer and Vice President
Seattle Science Foundation
Seattle, Washington, USA

Daiji Uchiyama, MD
Department of Radiology
Tobata Kyoritsu Hospital
Kitakyushu, Fukuoka, Japan

1 Abdominopelvic Wall

Farhood Saremi

Introduction

The upper border of the abdominal cavity is delimited by the thoracoabdominal line, a horizontal plane passing through the base of the xiphoid appendix and the spinous process of T12.[1,2] The upper border of the anterior abdominal wall is defined by the "costal line," a line passing along the lowest edge of the costal arch which corresponds to the anatomic location of the diaphragm (▶Fig. 1.1). The lower limit of the abdominal cavity is bordered by the pubic symphysis, inguinal arc, and iliac crest anteriorly and laterally, and by the spinous process of the L5 posteriorly. Using horizontal and vertical umbilical planes, the anterior abdomen/pelvis is subdivided into four quadrants extending between the costal line and inguinal ligaments. In French literature, nine regions are defined, namely, the epigastrium, mesogastrium (umbilical), and hypogastrium in the middle and the hypochondrium, flanks, and iliac on the sides (▶Fig. 1.1).

Using axillary lines (anterior, mid, and posterior), the abdomen is divided into four walls: anterior, posterior, right, and left lateral.

These surface landmarks have been historically used to address pathologies in certain parts of the abdominal cavity in clinical examinations. Obviously, clinically relevant limits are usually different from the anatomical limits. On the other hand, the internal border between the abdomen and pelvis is not clear and both cavities can be discussed under abdominal or abdominopelvic cavity which extends between the diaphragm and pelvic floor. This delimitation better suits radiologic descriptions of this part of the body.

Layers of the abdominal wall include the skin, superficial fascia, muscles with their aponeurosis, investing muscle fasciae, transversalis fascia, and peritoneum. Although each layer is a distinct anatomic structure, they function together as a solitary unit.

Fascial System of the Abdominal Wall

Before reviewing the anatomy of the abdominal muscles and associated structures, it is necessary to address the definition of fascia, aponeurosis, and retinaculum. Embryonic mesenchyme is the source of many structures including hematopoietic tissue, bones, cartilage, lymphoid, and fascial system. Federative International Programme on Anatomical Terminologies (FIPAT) has defined the fascia as aggregated sheets of connective tissue beneath the skin that attach, enclose, and separate muscles and other internal organs. The Fascia Nomenclature Committee introduced a

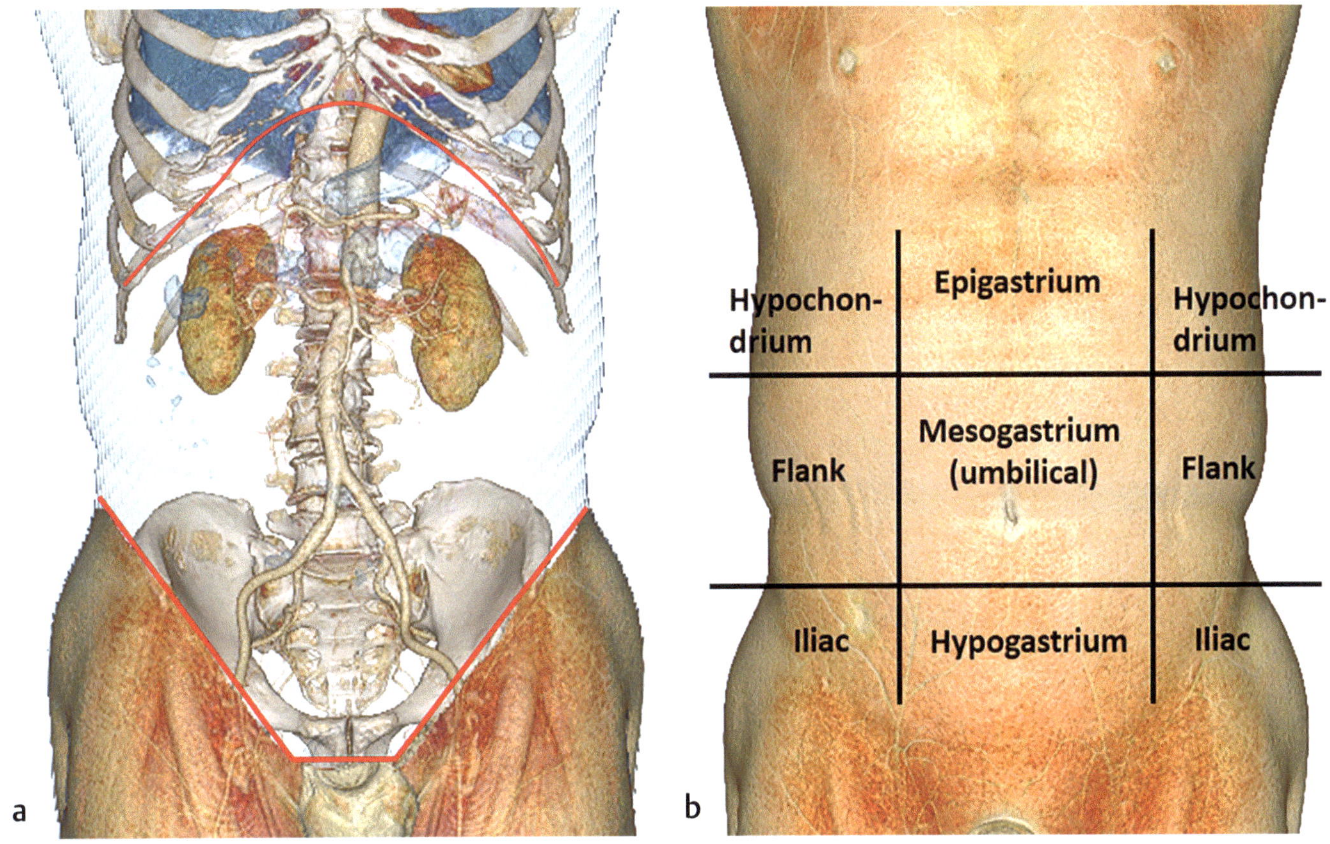

Fig. 1.1 (a) The upper border of the anterior abdominal wall is defined by the costal line which corresponds to the anatomic location of the diaphragm. The lower limit of the abdominal cavity is bordered by the pubic symphysis, inguinal arc, and iliac crest. **(b)** Clinical divisions of the abdominal cavity.

very broad definition for the fascial layers. In their definition, the fascial system consists of the three-dimensional continuum of collagen-containing loose and dense fibrous connective tissues that include many structures in the body such as adipose tissue, adventitia and neurovascular sheaths, aponeuroses, deep and superficial fasciae, epineurium, joint capsules, ligaments, membranes, meninges, myofascial expansions, periosteum, retinacula, tendons, visceral fasciae, and all intramuscular septa.

The fascia is composed of irregularly arranged collagen fibers. Conversely, tendons, ligaments, retinaculum, and aponeuroses are formed by the collagen fibers that are regularly arranged.[3,4] The irregular arrangement of collagen fibers in fascia increases the tensile strength of the fascia and allows it to resist tensional forces in every direction. An aponeurosis is a flat fibrous sheet that connects a muscle to a fixed point. Aponeurosis is covered with fascia.

The fascial system of the abdominal wall can be divided into two groups: (1) superficial or pannicular and (2) deep or investing. The superficial fascia refers to the fascia of the subcutaneous tissue. There has been considerable controversy with respect to the organization of superficial fascia in the human body. According to Lockwood, all the connective tissue between the dermis and muscle fascia in the body should be considered as a superficial fascial system which has the basic function of supporting the skin and fat of the body.[4] This fascial system consists of one to several thin, horizontal membranous layers separated by varying amounts of fat with interconnecting vertical or oblique fibrous septa (▶Fig. 1.2, ▶Fig. 1.3). In cross-sectional imaging this membranous layer is seen as a thin, 1 to 2 mm band surrounded on both sides by fat (▶Fig. 1.2).[5,6] In the abdomen this membranous layer is called the Scarpa's fascia and in the perineum, the Colles' fascia. The fatty layer superficial to the membranous layer was previously named the Camper's fascia. Below the inguinal ligament, the Scarpa's fascia blends with the fascia lata, the deep fascia of the thigh. The arrangement and thickness of this membranous layer varies according to the sex and location of the body. The membranous layer may be double or even triple layer on the posterior trunk, thigh, and arm. It is thicker in the posterior abdominal wall (▶Fig. 1.2). Functionally, the membranous superficial fascia plays a role in the integrity of the skin and support for subcutaneous structures.[7]

The deep fascia invests all bones, cartilages, muscles, tendons, ligaments, and aponeuroses, and blends into the periosteum of bone, epimysium of skeletal muscle, and peritenon of tendons and ligaments.[8] The investing fascia of the muscles that are anterior to the transverse process of the vertebrae is called hypaxial fascia (▶Fig. 1.2, ▶Fig. 1.3, ▶Fig. 1.4). These muscles are innervated by the anterior ramus. The fascia investing the muscles posterior to the transverse process (paraspinal muscles) is called epaxial fascia and the inside muscles are innervated by the posterior ramus.[8] The hypaxial and epaxial fasciae together fuse to the transverse process of the vertebrae, creating an intermuscular septum (▶Fig. 1.2, ▶Fig. 1.3).

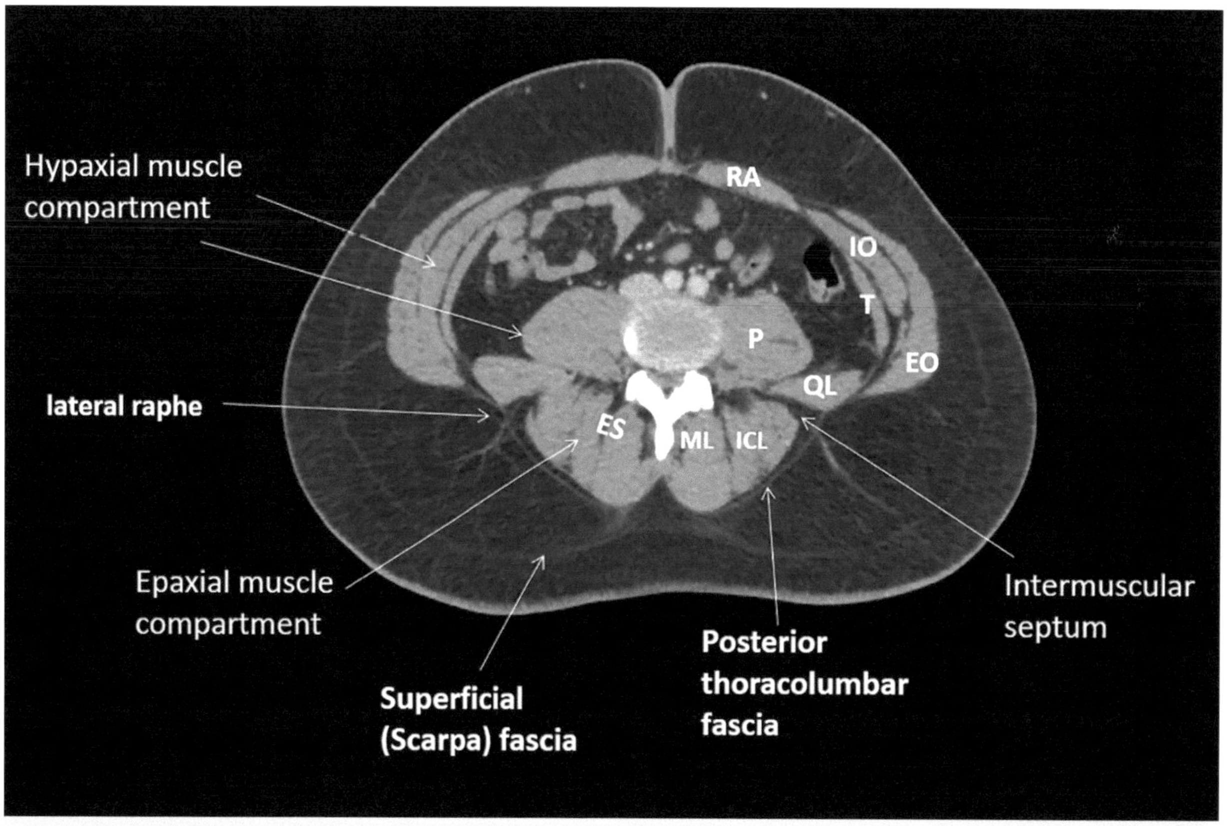

Fig. 1.2 Myofascial compartments on computed tomography (CT) scan. The hypaxial myofascial compartment is located anteriorly surrounding the body cavity and the epaxial myofascial compartment is located posteriorly. The epaxial compartment is divided into two subcompartments by the spinous process of the vertebra. The hypaxial and epaxial compartments are separated by an intermuscular septum that medially attaches to the transverse processes of the vertebra. In the lumbar region, this septum forms the middle layer of the thoracolumbar fascia. The membranous layer of the superficial fascia is shown. It is thicker in the posterior abdominal wall. EO, external oblique; ES, erector spinae muscles including medially positioned multifidus lumborum (ML) and laterally located iliocostalis lumborum (ICL); IO, internal oblique; LD, latissimus dorsi; P, psoas; QL, quadratus lumborum; RA, rectus abdominis; T, transverse muscle.

Abdominal wall muscles including the internal oblique, external oblique, and transversus muscles are covered on both sides by the deep fascia. This investing fascia originates from the embryonic parietal fascia and is divided into three layers, namely, deep, intermediate, and superficial.[9–11] The superficial layer is called the Gallaudet's fascia that covers the outer surface of the external oblique muscle. The deep layer is called the transversalis fascia that covers the inner surface of the transverse abdominal muscle.

All the fascial layers require hyaluronic acid for normal viscoelasticity. Decrease in quantity of hyaluronic acid compromises the sliding ability of the connective tissues and they become adhesive, stiff, and painful. Fibrosis, scar tissue, and calcification of connective tissue are the result of proliferation of fibroblasts following chronic inflammation, nonphysiological mechanical stress, or dysmotility.

Transversalis Fascia

The transversalis fascia or endoabdominal fascia is the investing fascia of the abdominal muscles. It is continuous with the

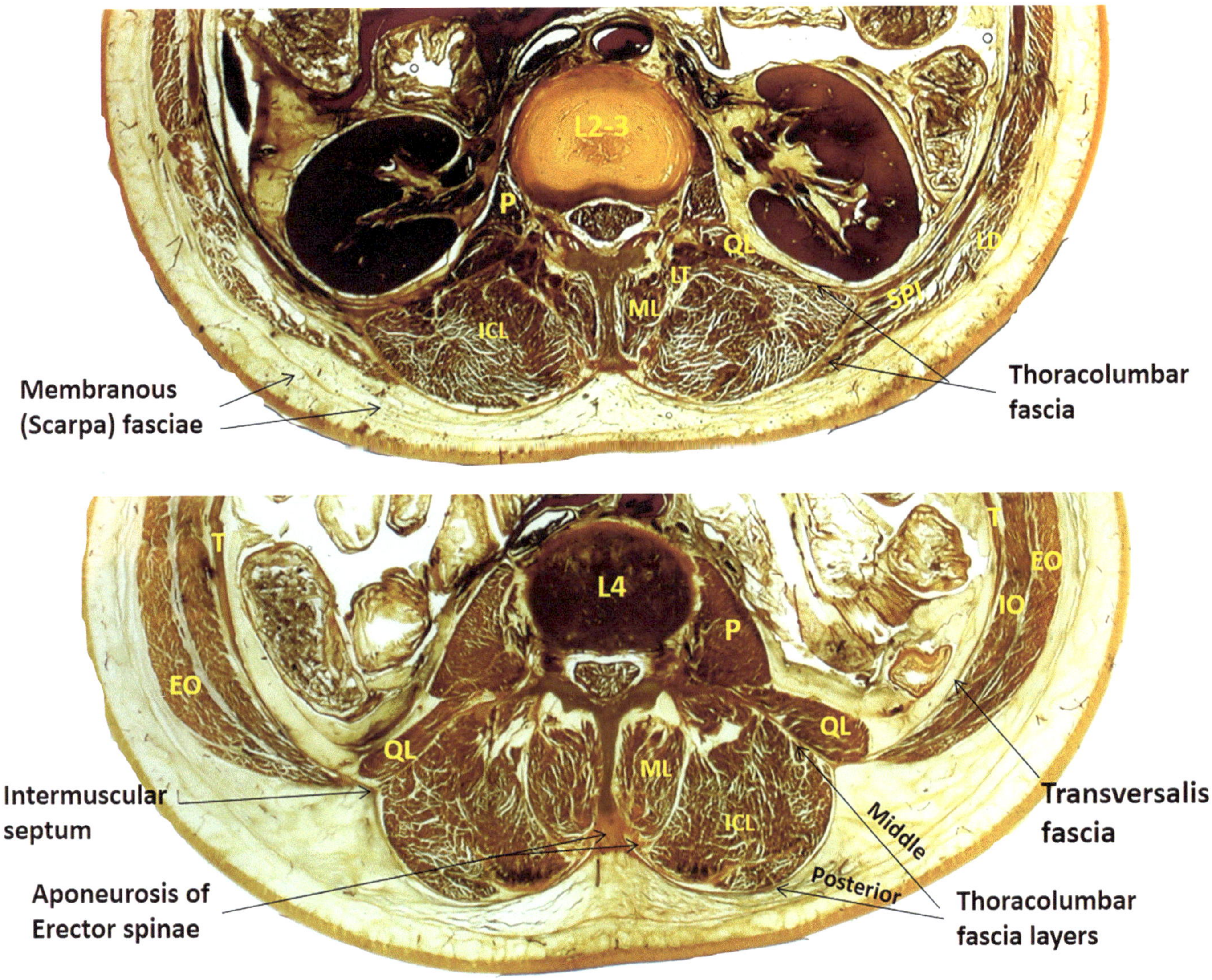

Fig. 1.3 The hypaxial and epaxial compartments are separated by an intermuscular septum that medially attaches to the transverse processes of the vertebra. In the lumbar region, this septum forms the middle layer of the thoracolumbar fascia. The posterior layer of the thoracolumbar fascia covers the posterior surface of the erector spinae. Latissimus dorsi (LD) aponeurosis forms the superficial lamina of the posterior layer of the thoracolumbar fascia. The deeper lamina of the posterior layer forms an encapsulating retinacular sheath around the paraspinal muscles. The aponeurosis of the erector spinae muscles and the overlying thoracolumbar fascia are separate, but they fuse at or slightly above the level of the posterior superior iliac spine to form the thoracolumbar composite. The transversalis fascia continues medially covering the anterior side of the investing fascia of the quadratus lumborum (QL) and also fuses with the psoas muscle (P) fascia. The internal (IO) and external obliques (EO) are seen external to the transverse abdominal muscle. Erector spinae muscles are shown including medially positioned multifidus lumborum (ML), laterally located iliocostalis lumborum (ICL), and deeply located longissimus thoracis (LT). The serratus posterior inferior (SPI) is often not present caudal to the L3 level. EO, external oblique; IO, internal oblique; LD, latissimus dorsi; P, psoas; QL, quadratus lumborum; RA, rectus abdominis; SPI, serratus posterior inferior; T, transverse muscle.

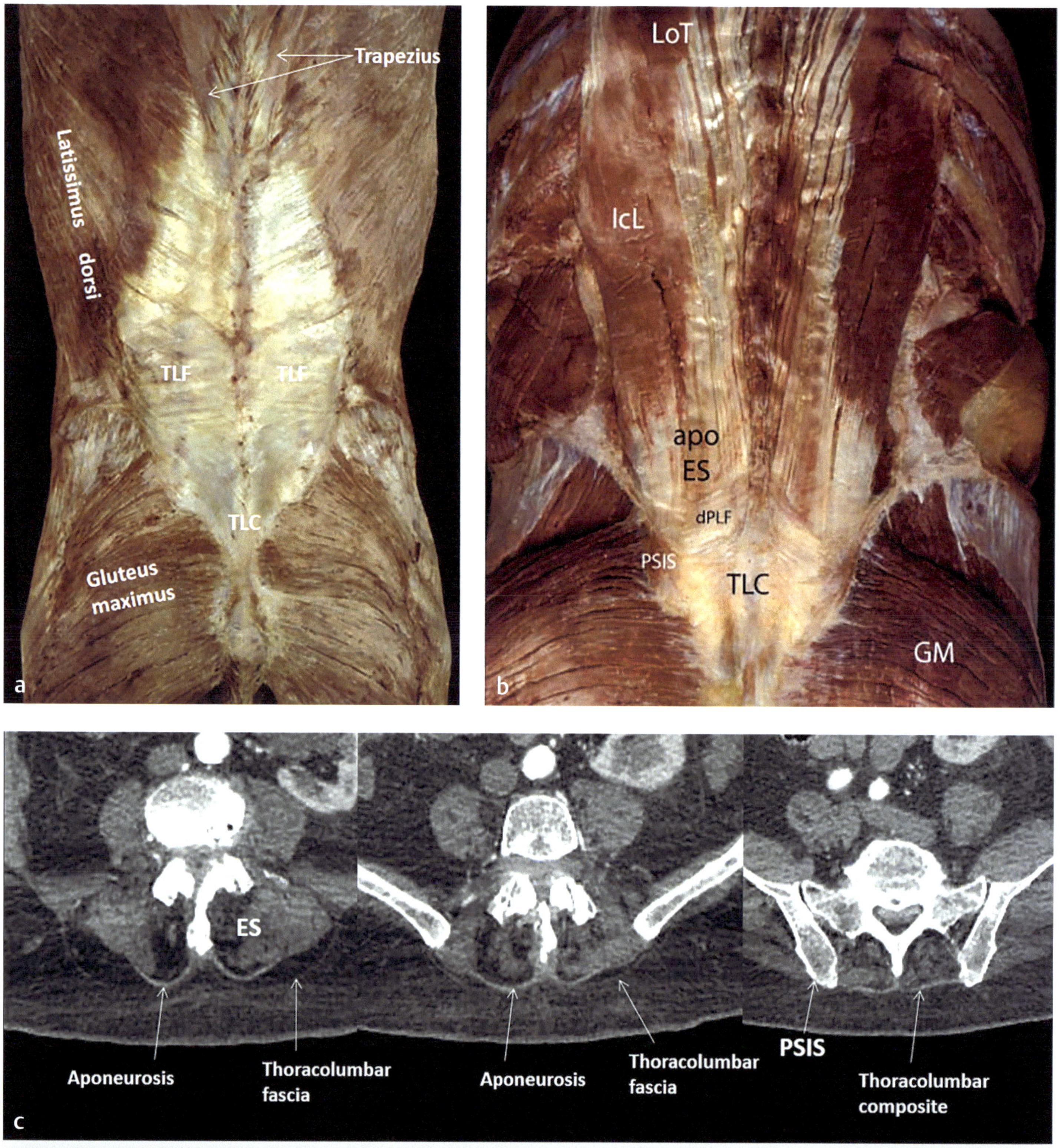

Fig. 1.4 (a) Posterior view of the back illustrating the attachments of the latissimus dorsi, trapezius, and gluteus maximus to the thoracolumbar fascia (TLF) and thoracolumbar composite (TLC). **(b)** Posterior view of the paraspinal muscles in the lumbosacral region. The posterior layer of the thoracolumbar fascia (PLF) has been removed to expose the iliocostalis lumborum (IcL) and the longissimus thoracis (LoT), as well as the aponeurosis of these two muscles (apo ES). A narrow rim of the deep lamina (dPLF) is seen at the point where the apo ES and the overlying TLF fuse to form the TLC. This fusion occurs at or slightly above the level of the posterior superior iliac spine (PSIS). (Reproduced with permission from Willard et al.[17]) **(c)** Axial computed tomography (CT) images showing the aponeurosis of the erector spinae (ES) muscles separate from the overlying TLF in the first two images. These two fasciae fuse at or slightly above the level of the posterior superior iliac spine (PSIS) to form the TLC.

endopelvic fascia that covers the internal surface of the pelvic muscles. The deep layer of the parietal fascia has been regionally named for the muscles that it covers. Similar to the transversalis fascia that is named for the transverse muscle, the fasciae of the abdominal diaphragm, iliopsoas, quadratus lumborum, obturator internus, and pelvic diaphragm are named after their parent muscles.

In the abdomen, the transversalis fascia adheres to the epimysium of the transverse muscles and their aponeurosis and can be easily separated from the peritoneum (except at the exit points such as the umbilicus). Medially, the transversalis fascia is continuous with the fascia covering the anterior side of the quadratus lumborum. The transversalis fascia is better developed in the lower abdomen compared to the upper abdomen (above the umbilicus). The transversalis fascia forms the posterior lamina of the rectus sheath below the arcuate line on which the muscle can slide freely. Compared with the abdominal wall muscles, there is no investing fascia in the thorax to separate the intercostal muscles from the parietal pleura.[10]

The transversalis fascia and peritoneum are separated by the preperitoneal connective tissue containing varying amount of fat. The preperitoneal space, the prevesical space of Retzius, and the inguinal space of Bogros are regional parts of the extraperitoneal space of the abdominal wall.[12] In the inguinal region the fat in this preperitoneal layer may protrude through the abdominal inguinal ring to form fatty inguinal hernia which is commonly seen especially in individual with abundant extraperitoneal fat.

The transversalis fascia is considered a single-layer fascia by most investigators. In the region of the internal inguinal canal, the transversalis fascia forms two laminae and the inferior epigastric vessels pass in between them.[11] At this level, the transversalis fascia also leaves an opening through which the spermatic cord or round ligament of the uterus passes. The transversalis fascia bridges the myopectineal orifice and forms a sling (sling of Henle) at the internal inguinal canal to prevent herniation when intra-abdominal pressure is raised (e.g., coughing). Previous surgeries may disturb innervation and function of this sphincter-like mechanism.[13,14] The transversalis fascia continues in the inguinal canal to form the internal spermatic fascia around the spermatic cord. It also participates in the formation of the iliopubic tract, a focal fascial thickening running deep and parallel to the inguinal ligament.[15]

Thoracolumbar Fascia

The thoracolumbar fascia is a complex of several membranous layers that separate the paraspinal muscles from the muscles of the posterior abdominal wall, quadratus lumborum, and psoas major.[16,17] This complex structure wraps around the paraspinal muscles of the lower back and sacral region (▶Fig. 1.3, ▶Fig. 1.4, ▶Fig. 1.5). It is continuous with paraspinal fascia in the thoracic

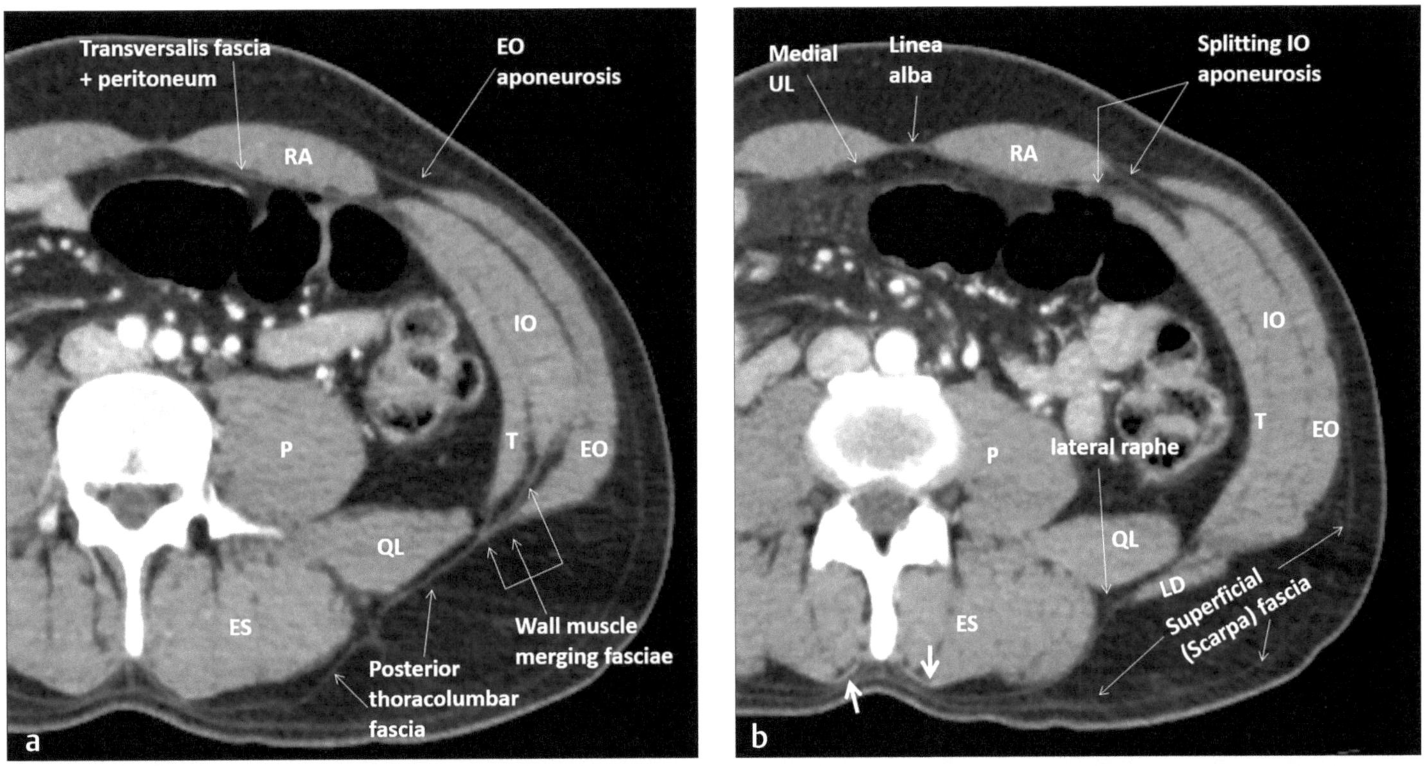

Fig. 1.5 (a, b) Axial computed tomography (CT) images. The posterior layer of the thoracolumbar fascia is continuous with the aponeuroses of the latissimus dorsi and serratus posterior inferior muscles. Anterolaterally, it attaches to the transverse processes and aponeurosis of the transversus abdominis. This anterolateral junction is termed the lateral raphe, which is an intermuscular septum separating the hypaxial and epaxial muscles. Medially, the thoracolumbar fascia attaches to the spinous processes in midline. The latissimus dorsi (LD) is seen lying on the external wall of the hypaxial compartment and extending over the epaxial compartment to reach its attachments on the midline. In doing so, the aponeurosis of the LD contributes to the superficial lamina of the posterior layer of the thoracolumbar fascia (PLF). *Large arrows* point to the aponeurosis of the erector spinae muscle. EO, external oblique; ES, erector spinae muscles; IO, internal oblique; LT, latissimus dorsi; P, psoas; QL, quadratus lumborum; RA, rectus abdominis; T, transverse muscle; UL, umbilical ligament.

and cervical regions, eventually fusing to the cranial base. It is especially prominent at lower lumbar spine where the thoracolumbar fascia and aponeurosis of paraspinal muscles blend to form a thickened brace between the two posterior superior iliac spines which extends caudally to the level of ischial tuberosities (▶Fig. 1.4, ▶Fig. 1.5). The posterior layer of the thoracolumbar fascia is continuous with the aponeuroses of the latissimus dorsi and serratus posterior inferior muscles. Medially, the thoracolumbar fascia attaches to the spinous processes in midline. Anterolaterally, it attaches to the transverse processes and aponeurosis of the transversus abdominis. This anterolateral junction is termed the lateral raphe which is an intermuscular septum separating the hypaxial and epaxial muscles (▶Fig. 1.2, ▶Fig. 1.3). The thoracolumbar fascia plays an important role in posture, load transfer, and respiration.

◆ Anterior Abdominal Wall Ligaments

Peritoneal folds on the inner surface of the anterior abdominal wall form three ligaments below the umbilicus and one above it. The inferior epigastric artery courses for a distance in the preperitoneal space before entering the rectus sheath just below the arcuate line. The deep inferior epigastric arteries and veins, with their peritoneal covering, form the lateral umbilical ligaments. Other structures in the infraumbilical preperitoneal space include the urachus, which forms the median umbilical ligament, and the obliterated umbilical arteries, which form the medial umbilical ligaments[18] (▶Fig. 1.6, ▶Fig. 1.7). Depending on the pattern of insertion and relationship, the median and medial umbilical ligaments are classified into four types. In the most common type (40%), the median umbilical ligament terminates to join one or both medial umbilical ligaments.[19] These ligaments form three infraumbilical peritoneal fossae. The fossa between the median and medial umbilical ligaments is the supravesical fossa. The fossa between the medial and lateral inguinal ligaments is the medial inguinal fossa, and the shallow fossa lateral to the lateral umbilical ligament is the lateral inguinal fossa (▶Fig. 1.6).

Above the umbilicus another peritoneal fold is formed by the teres ligament, the remnant of the umbilical vein. The umbilical vein lumen is generally obliterated by subendothelial connective tissue; however, a small part of the lumen may remain patent. This residual lumen transmits blood to the portal system from the paraumbilical and systemic sources.

The ligamentum teres hepatis (teres ligament) frequently divides into several branches cranial to the umbilical ring. It may fuse with the umbilical ring or connect to the umbilical ligaments.[19]

The umbilical vein is accompanied by the paraumbilical veins between the layers of the falciform ligament (▶Fig. 1.8). Together they constitute an accessory portal channel, which is in communication with the veins of the anterior abdominal wall.[20] These

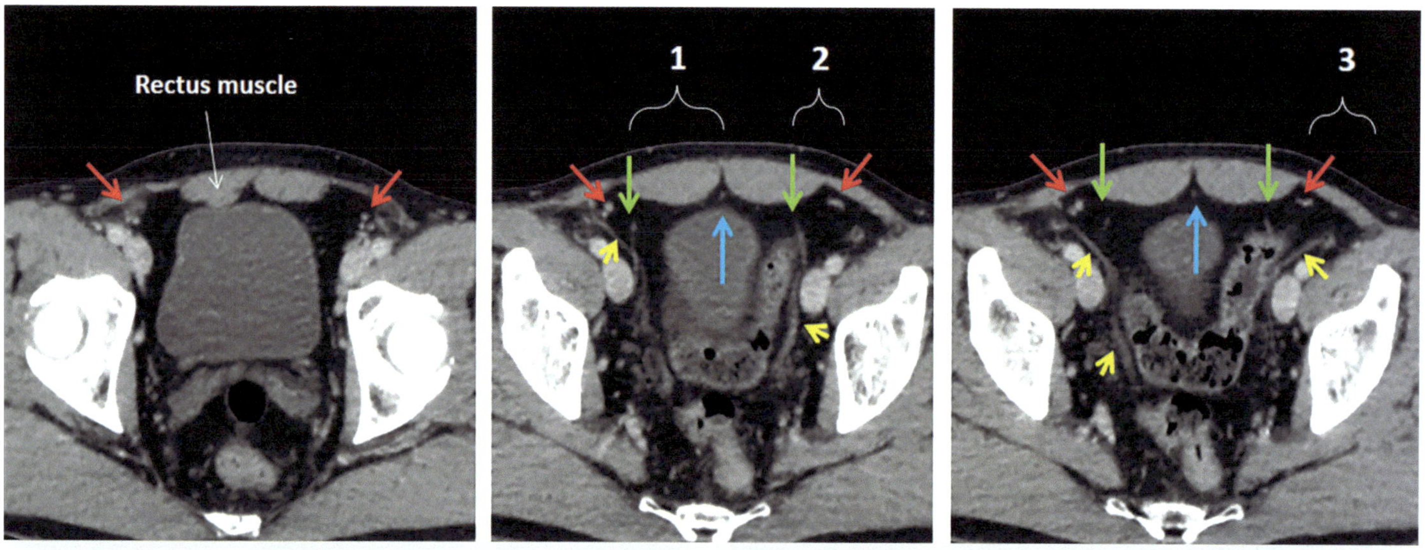

Fig. 1.6 Axial computed tomography (CT) images showing the relationship of the remnants of the urachus (median umbilical ligament) (*blue arrows*), umbilical arteries (medial umbilical ligaments) (*green arrows*), and inferior epigastric vessels (*red arrows*). The ductus deferens in men or round ligament of the uterus (*yellow arrows*) cross the median umbilical ligaments en route to the deep inguinal ring. The left and right deep inferior epigastric arteries and veins, with their peritoneal coverings, form the lateral umbilical ligaments. These ligaments form three peritoneal fossae, which could be shallow or deep, and can be observed within the peritoneal cavity: (1) supravesical fossa, (2) medial inguinal fossa, and (3) lateral umbilical ligament. Each fossa may be the potential site of an inguinal hernia.

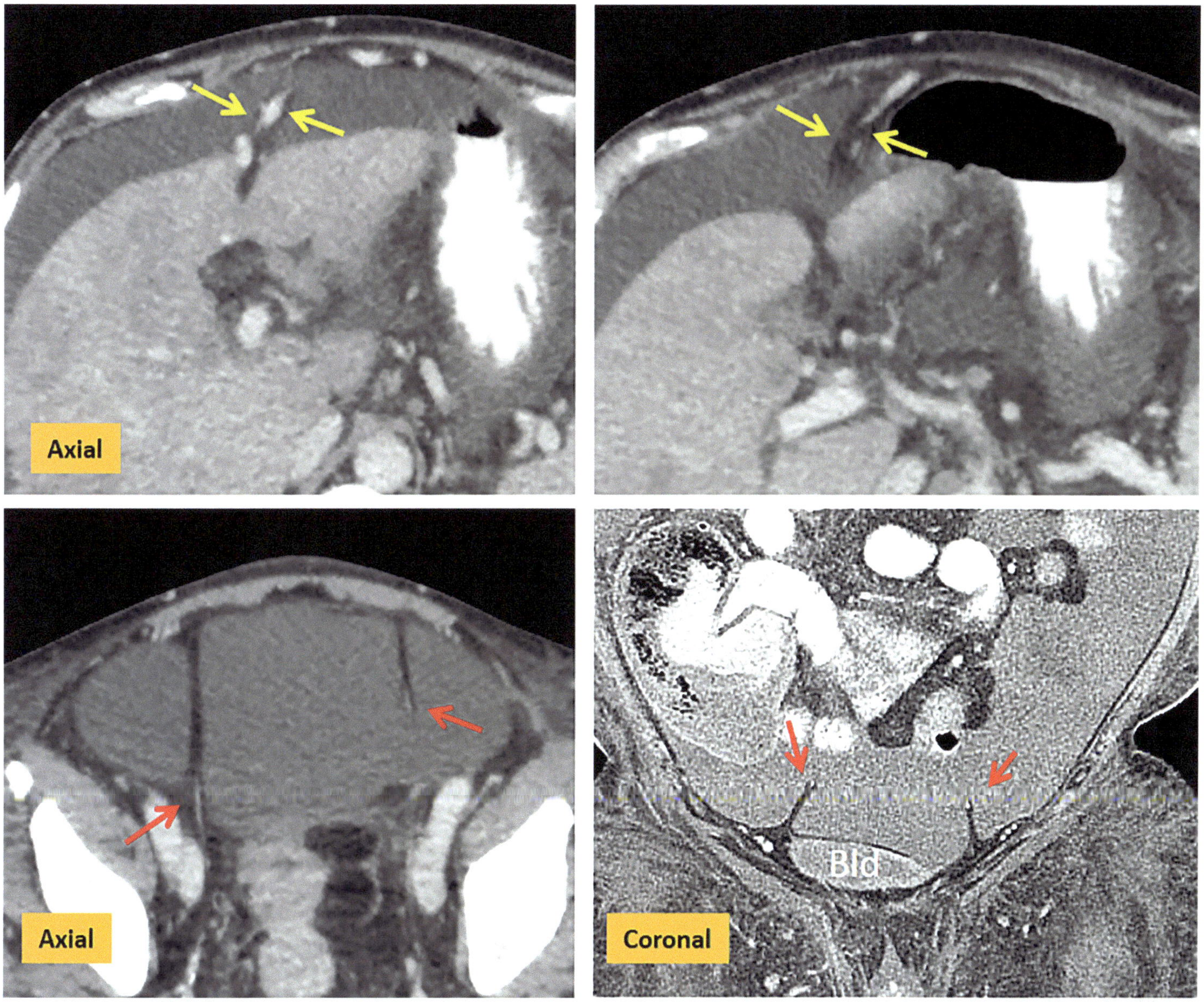

Fig. 1.7 Peritoneal folds (*arrows*) formed by the paraumbilical vein (*upper row*) and umbilical arteries (*lower row*) are shown in this patient with diffuse ascites. Bld, bladder.

veins are one of important portosystemic collateral pathways in patients with portal hypertension[21] (▶Fig. 1.8).

◆ Abdominal Wall Muscles

Muscle's layout is used to divide each side of the anterior abdominal wall into two subdivisions: Anterolateral and paramedian. The anterolateral portion is composed of the external oblique (obliquus externus abdominis), internal oblique (obliquus internus abdominis), and transverse abdomen (transversus abdominis) muscles.[13] The paramedian portion is composed of the rectus abdominis and pyramidal muscles (▶Fig. 1.9, ▶Fig. 1.10, ▶Fig. 1.11, ▶Fig. 1.12). The posterior abdominal wall is limited medially by the lateral margin of the paraspinal muscles corresponding to the posterior axillary lines. The posterolateral is also formed by the same muscles of the anterolateral aspect of the abdomen and is partly covered by the latissimus dorsi and serratus posterior inferior muscles at thoracoabdominal junction. The abdominal wall muscles play an important role in stabilizing the trunk, respiration, and postural adjustment.[22]

External Oblique (Obliquus Externus Abdominis)

The external oblique muscle arises from the lower margin of the 5th to 12th anterior ribs. The upper half of the muscle interdigitates

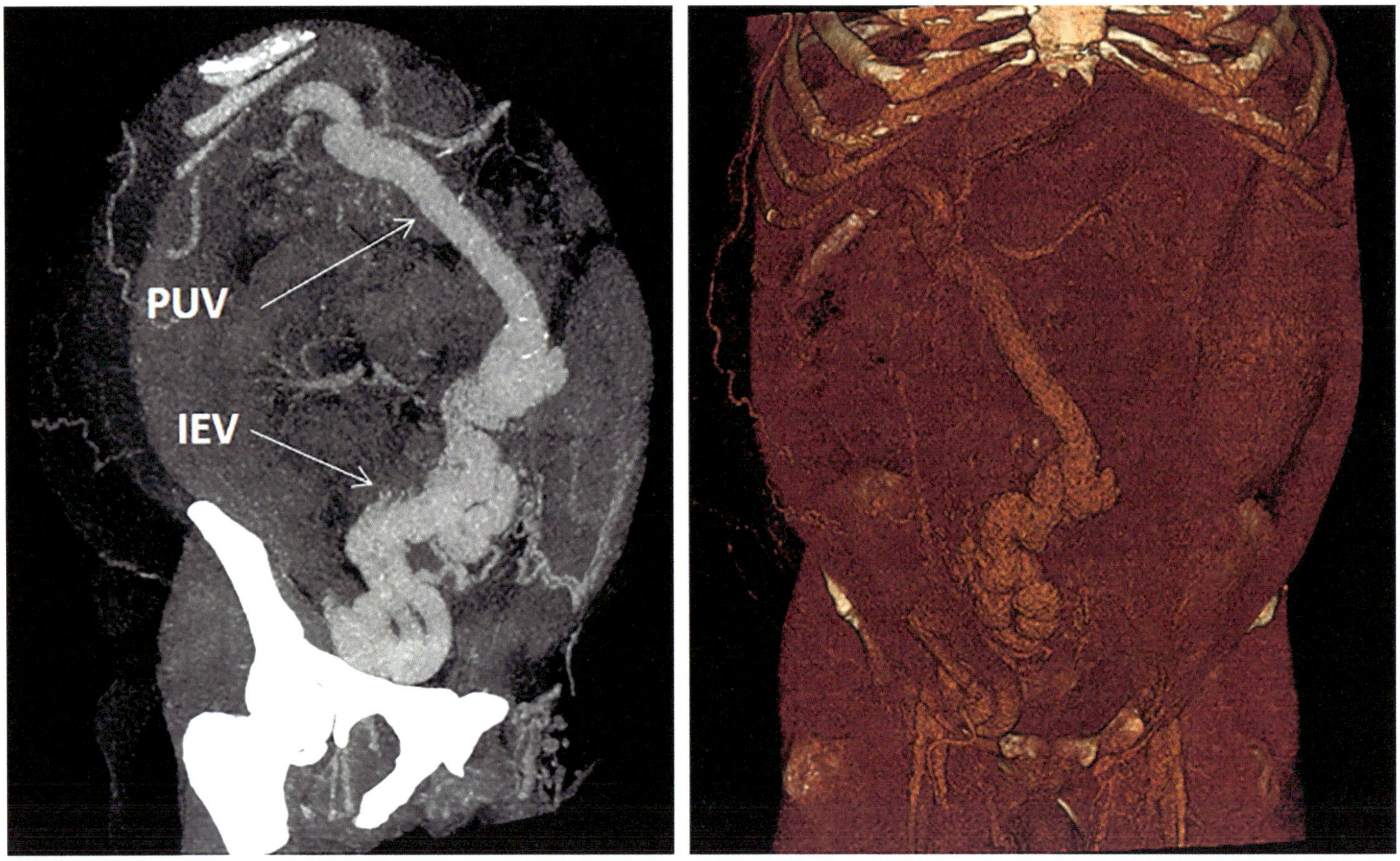

Fig. 1.8 Coronal computed tomography (CT) and volume-rendered images showing anterior abdominal wall varices due to portosystemic shunting of blood between the paraumbilical vein (PUV) (portal system) and the inferior epigastric vein (IEV) (systemic circulation).

with the serratus anterior and the lower half with the latissimus dorsi at the costal origin (▶Fig. 1.9, ▶Fig. 1.10, ▶Fig. 1.11, ▶Fig. 1.12). The muscle fibers extend anteriorly and inferiorly before ending in the extensive ventral tendinous aponeurosis that connects to the anterior lamina of the rectus sheath and linea alba anteriorly, and into the anterolateral half of the iliac crest, anterior superior iliac spine, and pubic tubercle inferiorly. The muscle aponeurosis participates in the formation of the inguinal ligament of Poupart (▶Fig. 1.12).

The upper fibers of the muscle show an almost horizontal direction and the lower fibers move obliquely toward the inguinal ligament. This fiber orientation allows some relaxation of the fascial structures in the inguinal region with muscle contraction.

The arterial supply to the external oblique muscle is from the lower six posterior intercostal arteries. The caudal part of the muscle is supplied by branches of the deep circumflex iliac artery.[23] The external oblique muscle is innervated by the ventral rami of the lower six thoracic spinal nerves and the iliohypogastric nerve.[23,24]

Internal Oblique (Obliquus Internus Abdominis)

The internal oblique muscle is located deep to the external oblique muscle (▶Fig. 1.11, ▶Fig. 1.12, ▶Fig. 1.13). The fibers of the internal oblique muscle fan out from their origin and run almost perpendicular to the external oblique. The posterior fibers connect to the thoracolumbar fascia and the inferolateral fibers arise from the middle part of the iliac crest and outer inguinal ligament. Fibers travel medially and superiorly to end in the rectus sheath through the muscle aponeurosis. The very superior fibers insert into the inferior border of the last three to four ribs. The caudal muscle fibers end as roof of the inguinal canal. Contraction of the internal oblique muscle results in relaxation of the transversalis fascia in the inguinal region.

The internal oblique muscle is vascularized by the ascending branch of the deep circumflex iliac artery, the lower six posterior intercostal arteries, and the lateral branches of the deep inferior epigastric artery.[24] The ascending branch of the deep circumflex iliac artery represents the dominant arterial supply. The muscle is innervated by the ventral rami of the lower six or seven thoracic spinal nerves and the first lumbar spinal nerve.

Transverse Abdominal (Transversus Abdominis)

The transverse abdominal muscle is the innermost muscle of the abdominal wall and, as implied by its name, contain horizontally oriented muscle fibers. It originates from the five lower ribs, thoracolumbar fascia, iliac crest, and the outer third of the inguinal ligament (▶Fig. 1.11, ▶Fig. 1.13). The muscle inserts medially into the rectus sheath through a broad aponeurosis. The inferior rim of the aponeurosis of transverse abdominal and internal oblique

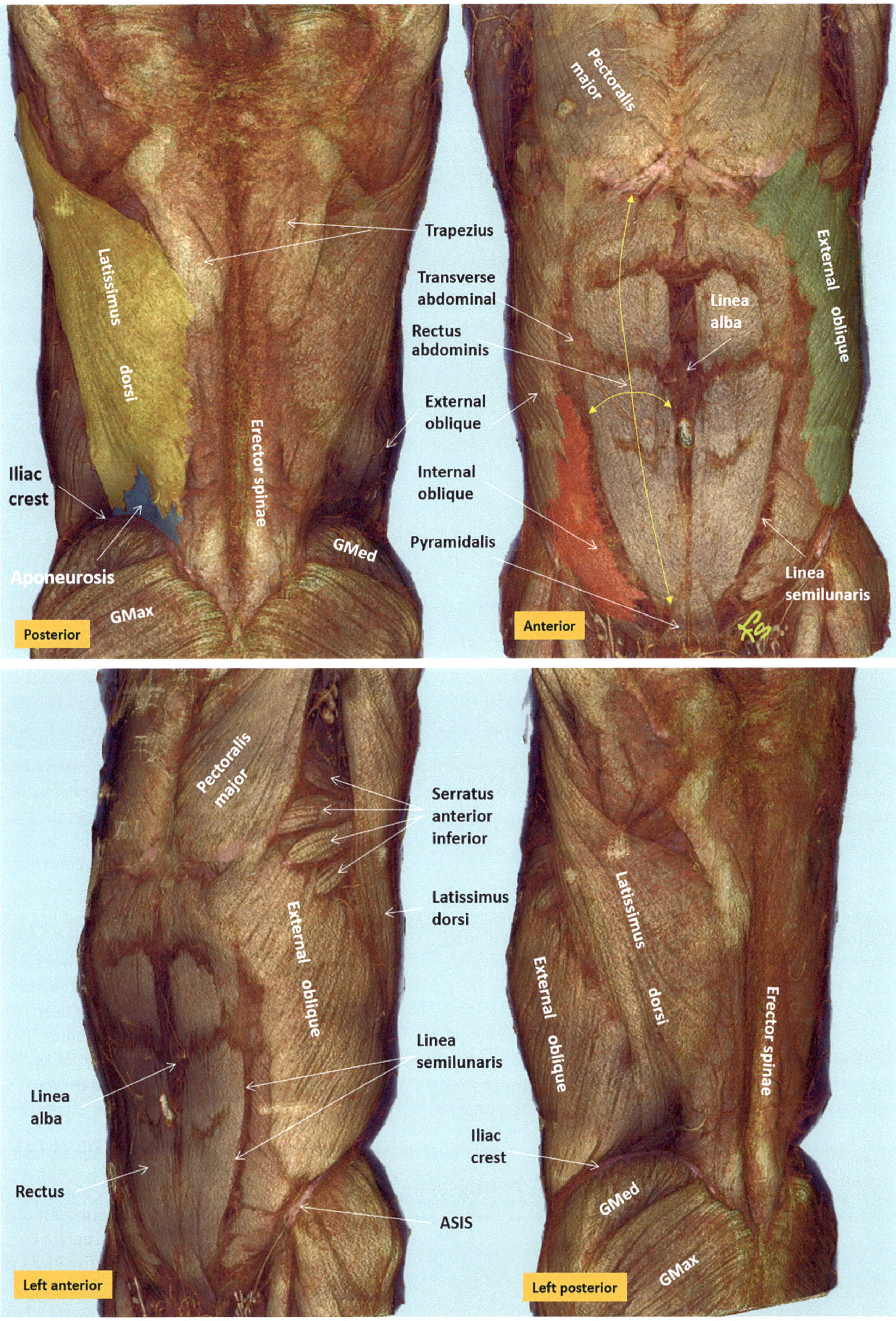

Fig. 1.9 Volume-rendered, color-coded computed tomography (CT) views of the abdomen show the layout of the abdominal wall muscles in different projections. *Yellow arrows* show the nice boundaries of the rectus abdominis muscle. ASIS, anterior superior iliac spine; GMax, gluteus maximus; GMed, gluteus medius.

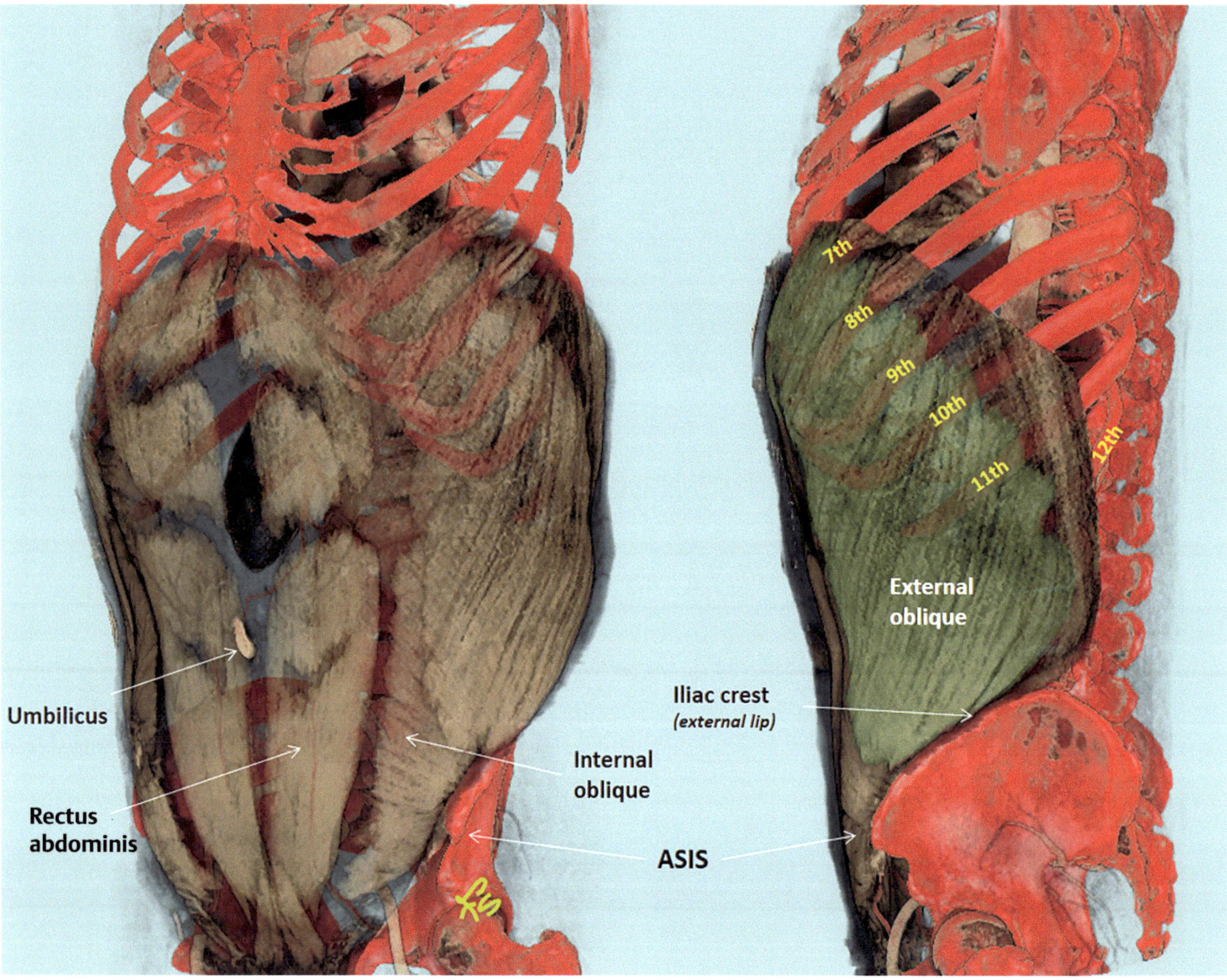

Fig. 1.10 Anterior abdominal wall muscles shown in relation to the bones. ASIS, anterior superior iliac spine.

curve inferomedially to form the conjoint tendon (falx inguinalis) and attach to the crest and pecten of the pubis. Inferior muscle fibers of the transverse abdominal and internal oblique muscles mix together and form the cremaster muscle of the spermatic cord (detail can be found below: inguinal canal).

The transverse abdominal muscles play a major role in stabilizing the lumbopelvic region together with the multifidus muscles. One important action of the transverse abdominal is to "draw-in" the abdominal wall. Weakening of these muscles may play a role in the development of low-back pain. Drawing in the abdominal wall is a specific exercise that appear to strengthen (like a corset) the transverse abdominal muscle (in cocontraction with the multifidus), and is used in the treatment of back pain and to improve the stabilization of the lumbopelvic region.[25–27]

Rectus Abdominis (Abdominal)

The rectus abdominis muscle originates from the fifth to seventh costal cartilages and the xiphoid process and inserts at the upper edge of the pubic symphysis, the superior ramus of the pubic bone, and the pubic tubercle (▶Fig. 1.9, ▶Fig. 1.10, ▶Fig. 1.11, ▶Fig. 1.12, ▶Fig. 1.13, ▶Fig. 1.14, ▶Fig. 1.15).

The muscle is covered by the rectus sheath except at the posterior wall below the arcuate line. The arcuate line (linea semicircularis) is a fibrous union of the rectus sheath with the aponeurosis of the transverse and oblique abdominal. It has a variable position that changes between the umbilicus and pubic line and is usually found approximately 5 cm below the umbilicus (▶Fig. 1.13). Above the arcuate line, the rectus sheath is formed by a

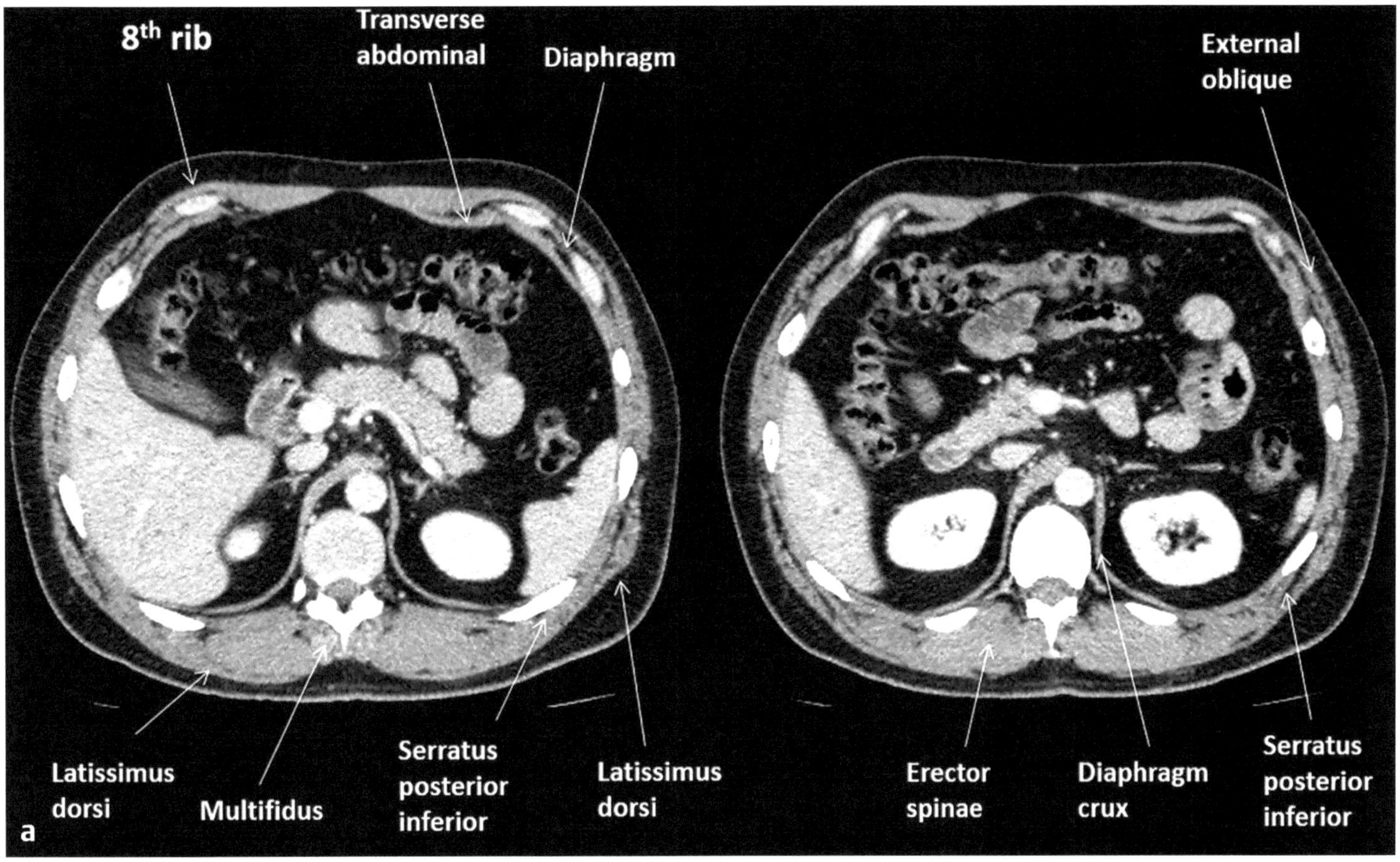

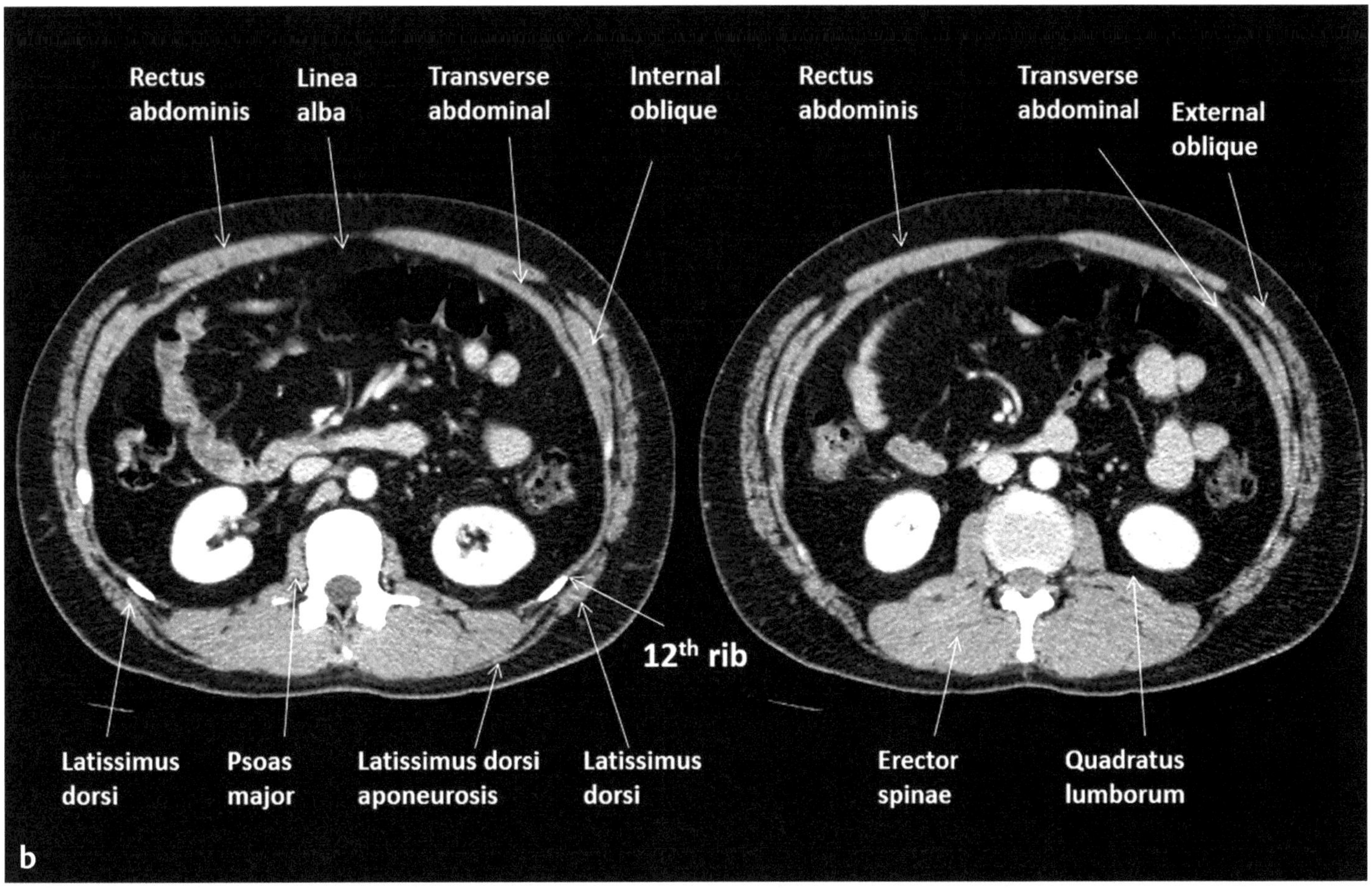

Fig. 1.11 (a–e) Cross-sectional views of the abdomen from top to bottom showing muscles of the abdominal wall.

(Continued)

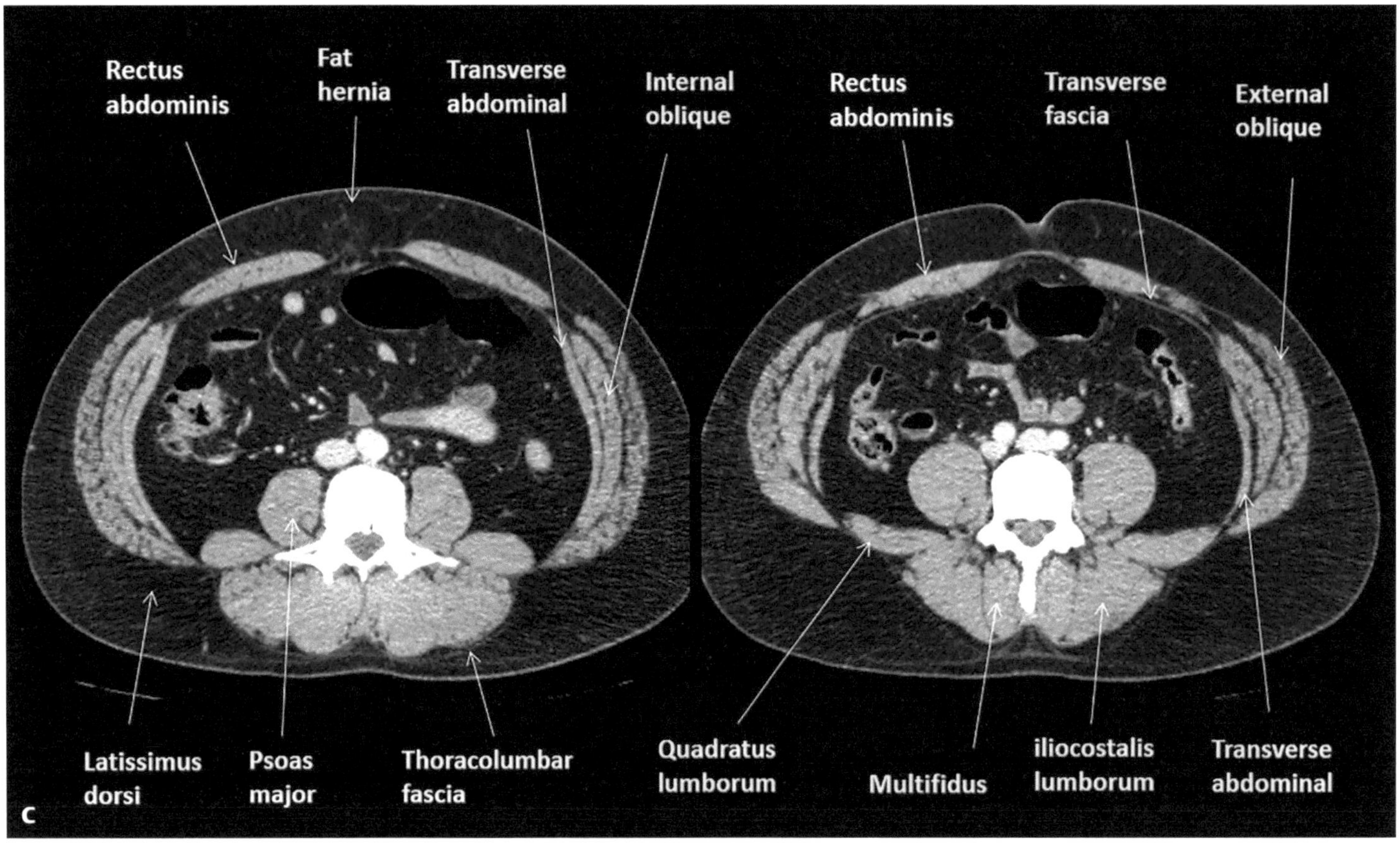

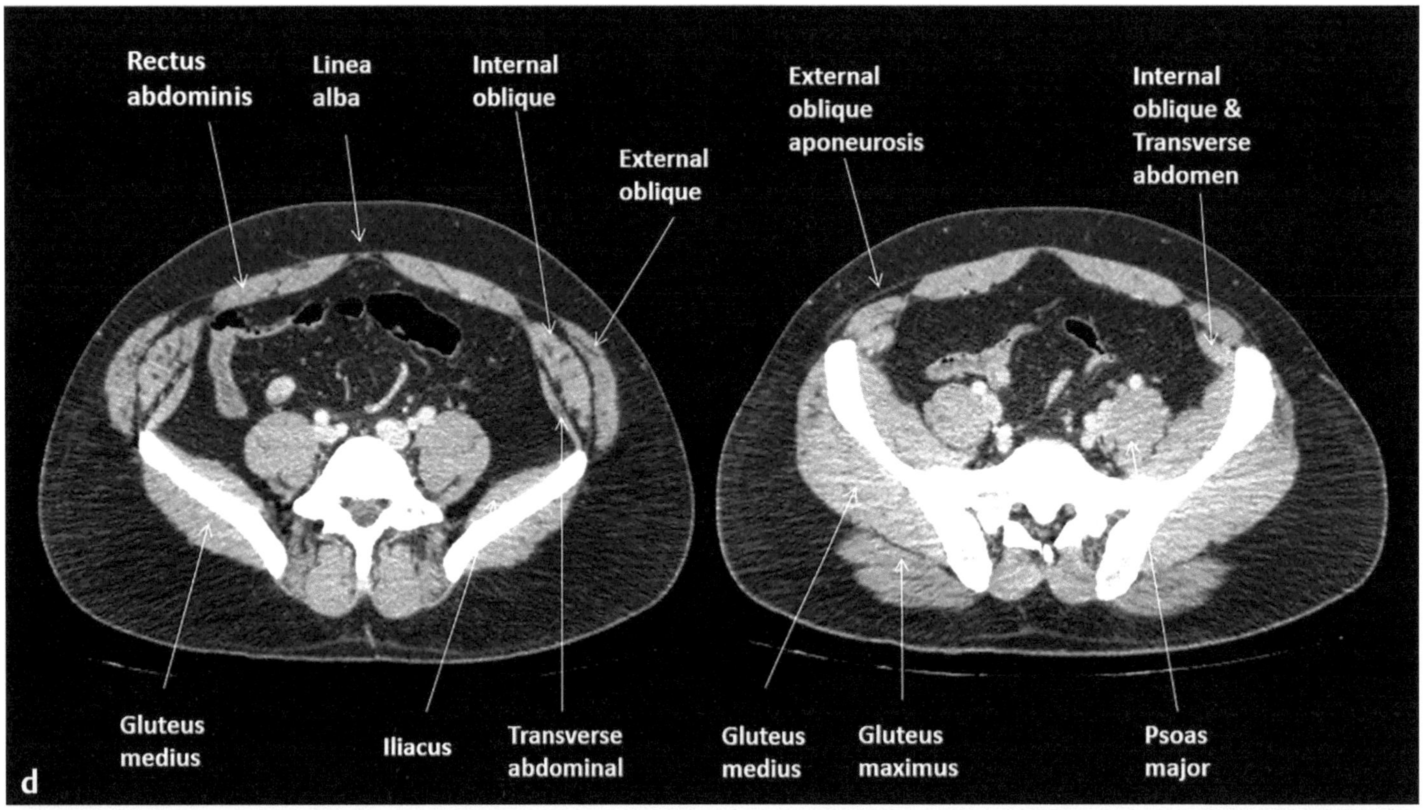

Fig. 1.11 *(Continued)*

(Continued)

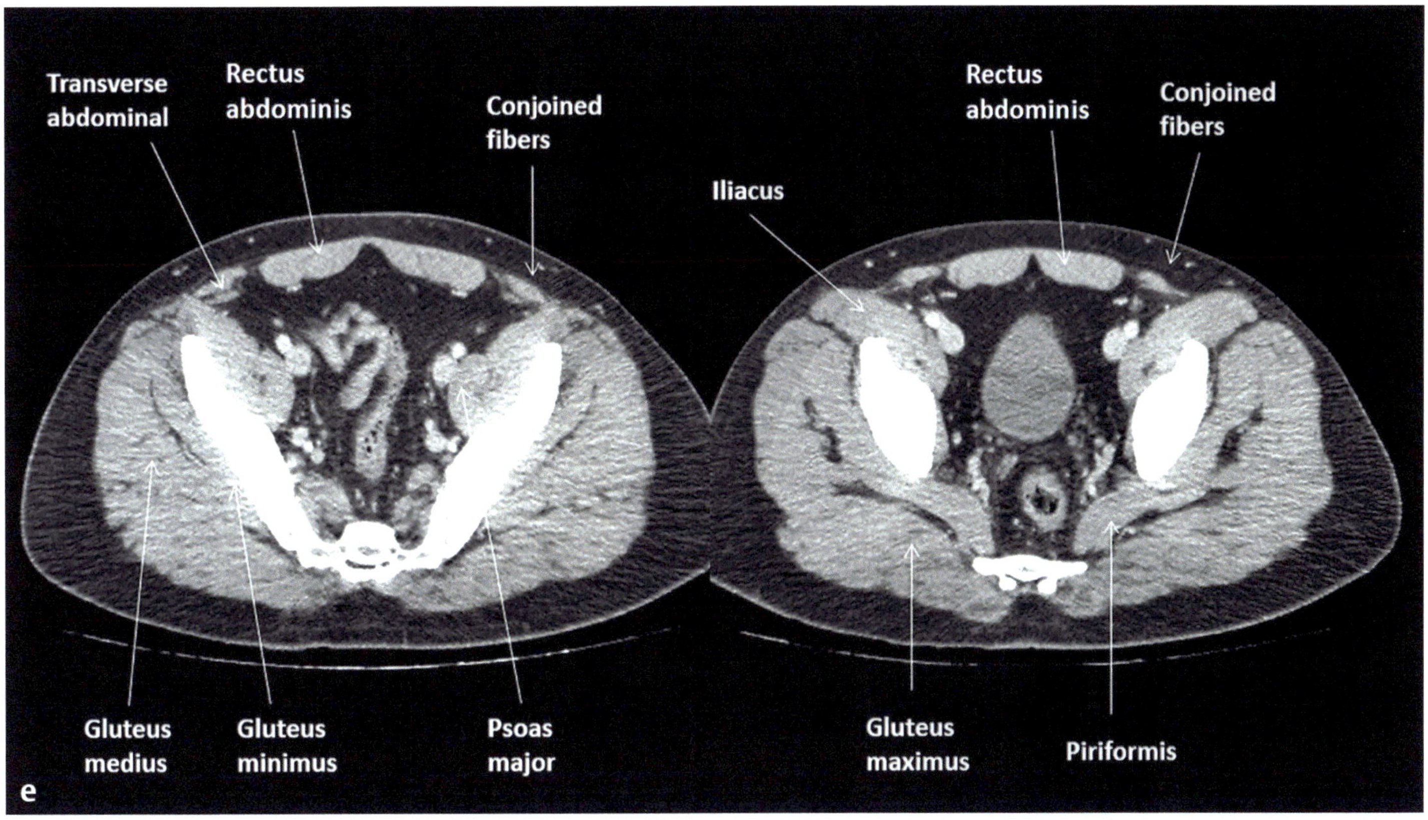

Fig. 1.11 (*Continued*)

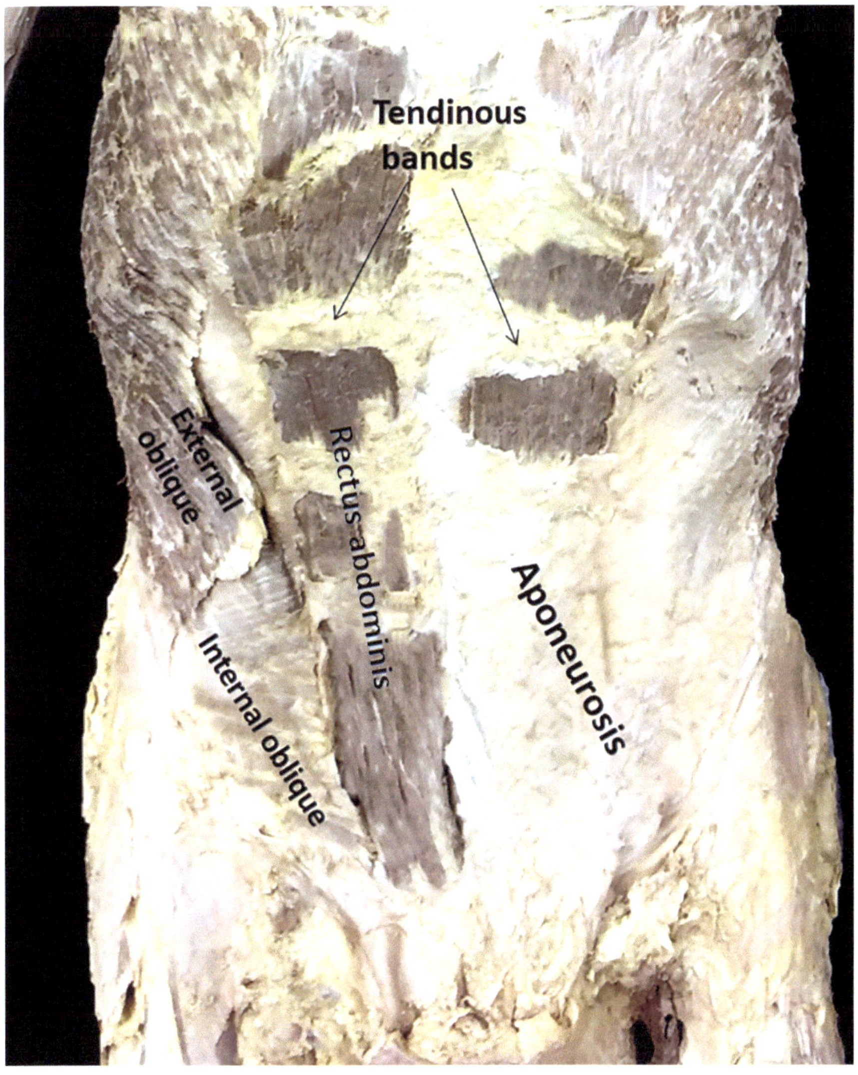

Fig. 1.12 Cadaveric view of the anterior abdominal wall before (*left side*) and after (*right side*) removal of the aponeurosis of the external and internal oblique muscles.

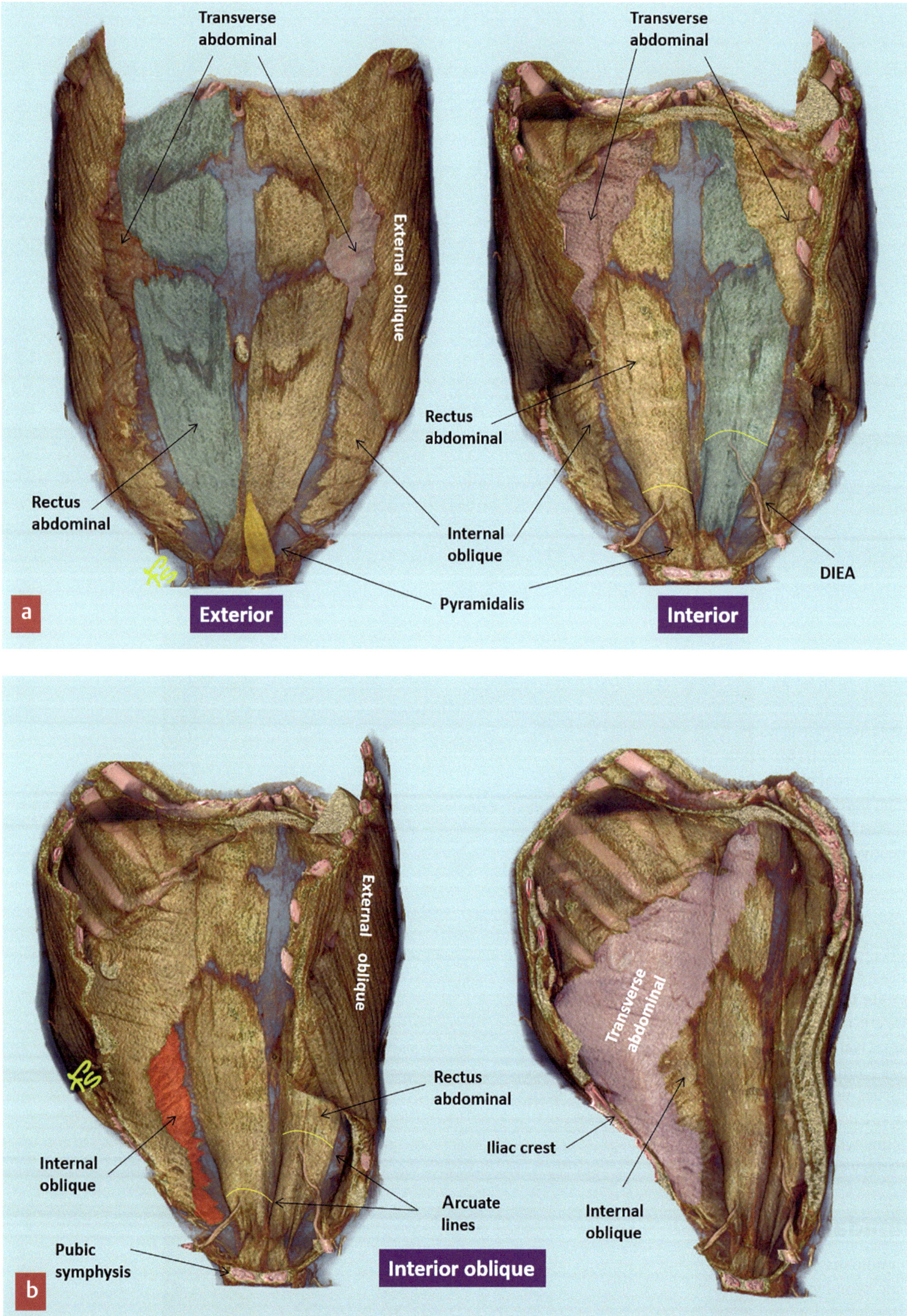

Fig. 1.13 (a, b) Volume-rendered, color-coded computed tomography (CT) views of the abdomen show the layout of the anterior abdominal wall muscles in different projections. Note opposed oblique orientation of the oblique muscle fibers and transverse orientation of the transverse abdominal muscle fibers. DIEA, deep inferior epigastric artery.

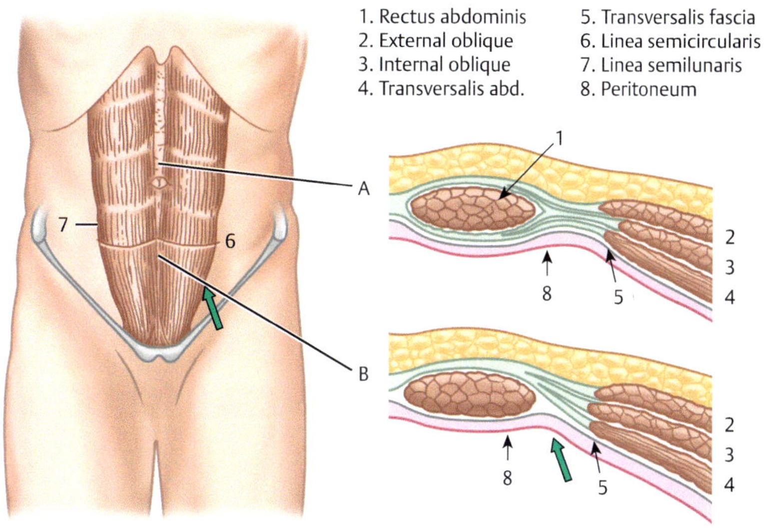

Fig. 1.14 Schematic anatomy of the anterior abdominal wall muscle aponeurosis.

central splitting of the aponeurosis of the internal oblique muscle enclosing the rectus abdominis muscle (▸Fig. 1.4, ▸Fig. 1.13). The rectus sheath is covered by the aponeurosis of external oblique muscle anteriorly and by the aponeurosis of transverse abdominal muscle and transversalis fascia posteriorly. Below the arcuate line, all three musculoaponeurotic layers pass anterior to the rectus muscle and the fascia transversalis remains the only structure covering the posterior surface of the rectus abdominis (▸Fig. 1.13, ▸Fig. 1.14, ▸Fig. 1.15). In many cases, the aponeurosis of the internal oblique and transverse muscles fuse medial to the lateral border of rectus muscle.

The rectus abdominis muscles are separated by a midline band of connective tissue called the linea alba. Tendinous bands of connective tissue also divide each muscle horizontally into three to four smaller muscle bellies (▸Fig. 1.11). Main arterial blood supply is provided by the inferior and superior epigastric arteries and veins. Multiple small branches from the lower six intercostal arteries also play a role is vascular supply to the muscle. The muscles are innervated by thoracoabdominal nerves. The rectus muscles are commonly used for reconstructive surgery of the abdominal wall and breast after mastectomy. Following abdominoplasty (tummy tuck surgery), the rectus muscles and overlying tissues will be tightened in midline with internal sutures (▸Fig. 1.15). Rectus sheath hematomas usually develop below the arcuate line where perforating branches of the deep inferior epigastric artery enter the muscle.

Pyramidalis

The pyramidalis is a small and triangular muscle just above the pubic symphysis and anterior to the rectus muscles (▸Fig. 1.13). It is located within the rectus sheath and attaches to the linea alba. The base of muscle arises from the pubic symphysis and pubic crest. The precise function of pyramidalis muscles is unclear, but together the muscles are thought to tense the linea alba. The muscles are present in 30 to 90% of individuals in one or two sides.[28] Occasionally, it is double on one side, and the muscles of the two sides are sometimes of unequal size. It may also extend higher than the usual level.

Iliopsoas Unit

The iliopsoas musculotendinous unit include the psoas major, psoas minor, and iliacus muscles. These muscles form part of the posterolateral abdominoplevic wall muscles.[29–32]

The psoas major arises from the transverse processes, intervertebral disks, and adjacent vertebral margins of T12 to L5. The muscle passes deep to the medial arcuate ligament of the diaphragm and moves along the posteromedial side of the abdomen (▸Fig. 1.16, ▸Fig. 1.17). The iliacus muscle arises from the superior two-thirds of the iliac fossa. Both muscle passes beneath the inguinal ligament to insert on the lesser trochanter of the femur via the iliopsoas tendon. In this region the iliopsoas forms part of the floor of the femoral triangle. A large bursa separates the tendon from the pubis and the hip capsule. The psoas minor tendon originates from the vertebral bodies of T12 and L1 and is found in 10 to 65% of individuals. The psoas minor tendon is a long slender muscle located anterior to the psoas major and inserts into the iliopectineal eminence. The fascia covering the iliopsoas and quadratus lumborum muscles is continuous with the endoabdominal fascia (transversalis fascia). Although the iliopsoas compartment is retrofascial and posterior to the retroperitoneal structures, it can be violated by infection, hemorrhage, and neoplasm (▸Fig. 1.18).

The blood supply to psoas major is derived from the four lumber vessels. Iliacus receives its arterial supply from the medial circumflex femoral artery and the iliac branch of the iliolumbar artery, the first posterior branch of the internal iliac artery. The psoas major muscle is innervated by the roots of L1–L3 and iliacus

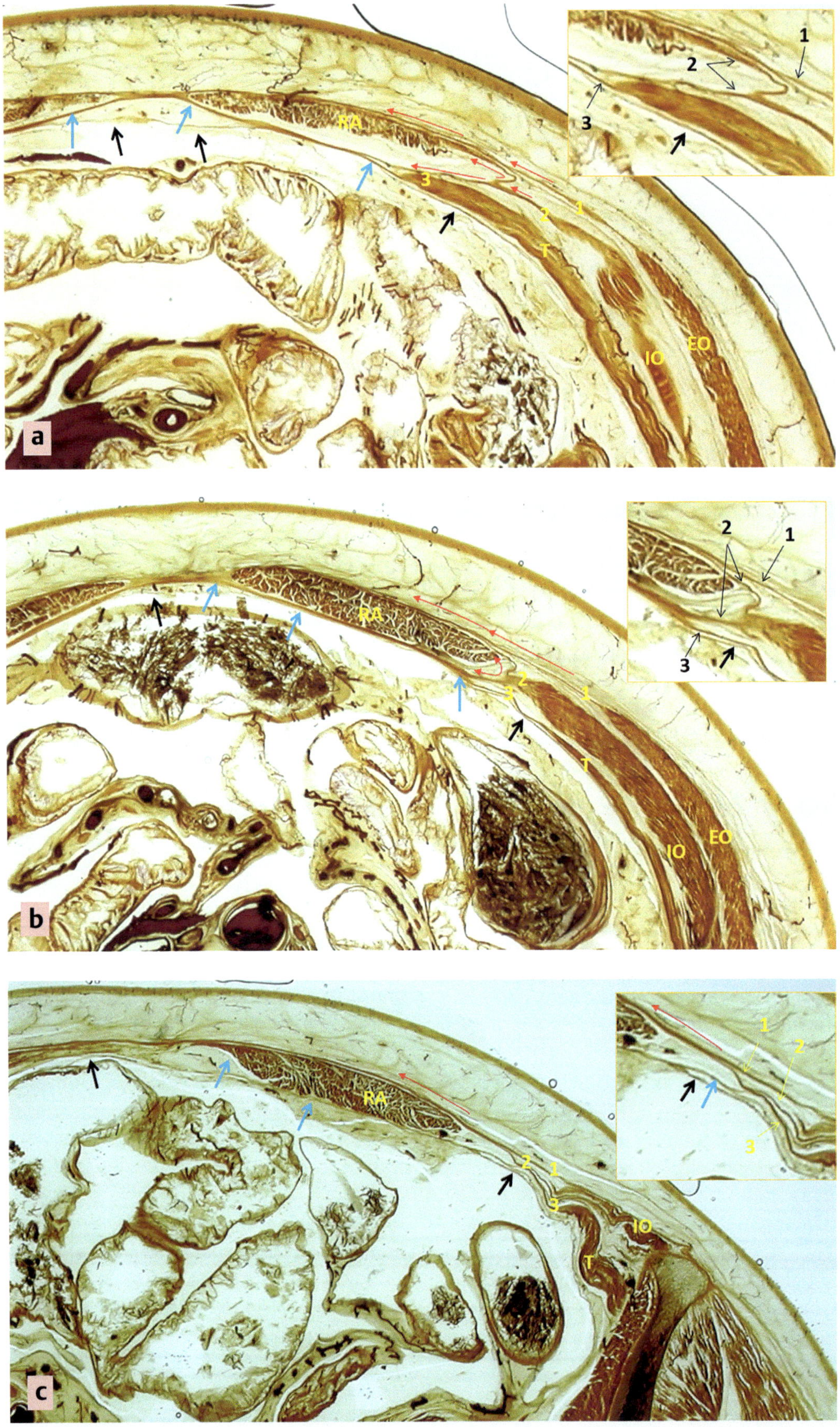

Fig. 1.15 Rectus sheath and anterior muscle aponeurosis (*red arrows*) at different levels. Above the arcuate line **(a, b)** the rectus abdominis (RA) sheath is formed by a central splitting of the aponeurosis of the internal oblique (IO) muscle (2) investing the rectus abdominis muscle. The rectus sheath is covered by the aponeurosis of external oblique (EO) muscle (1) anteriorly and by the aponeurosis of transverse (T) abdominal muscle (3) and transversalis fascia (*blue arrows*) posteriorly. Below the arcuate line **(c)**, all three musculoaponeurotic layers pass anterior to the rectus muscle (*red arrows*) and the fascia transversalis (*blue arrows*) remains the only structure covering the posterior surface of the rectus abdominis. In many cases, the aponeurosis of the internal oblique and transverse muscles fuse medial to the lateral border of rectus muscle. Note transversalis fascia moving toward the linea alba.

(*Continued*)

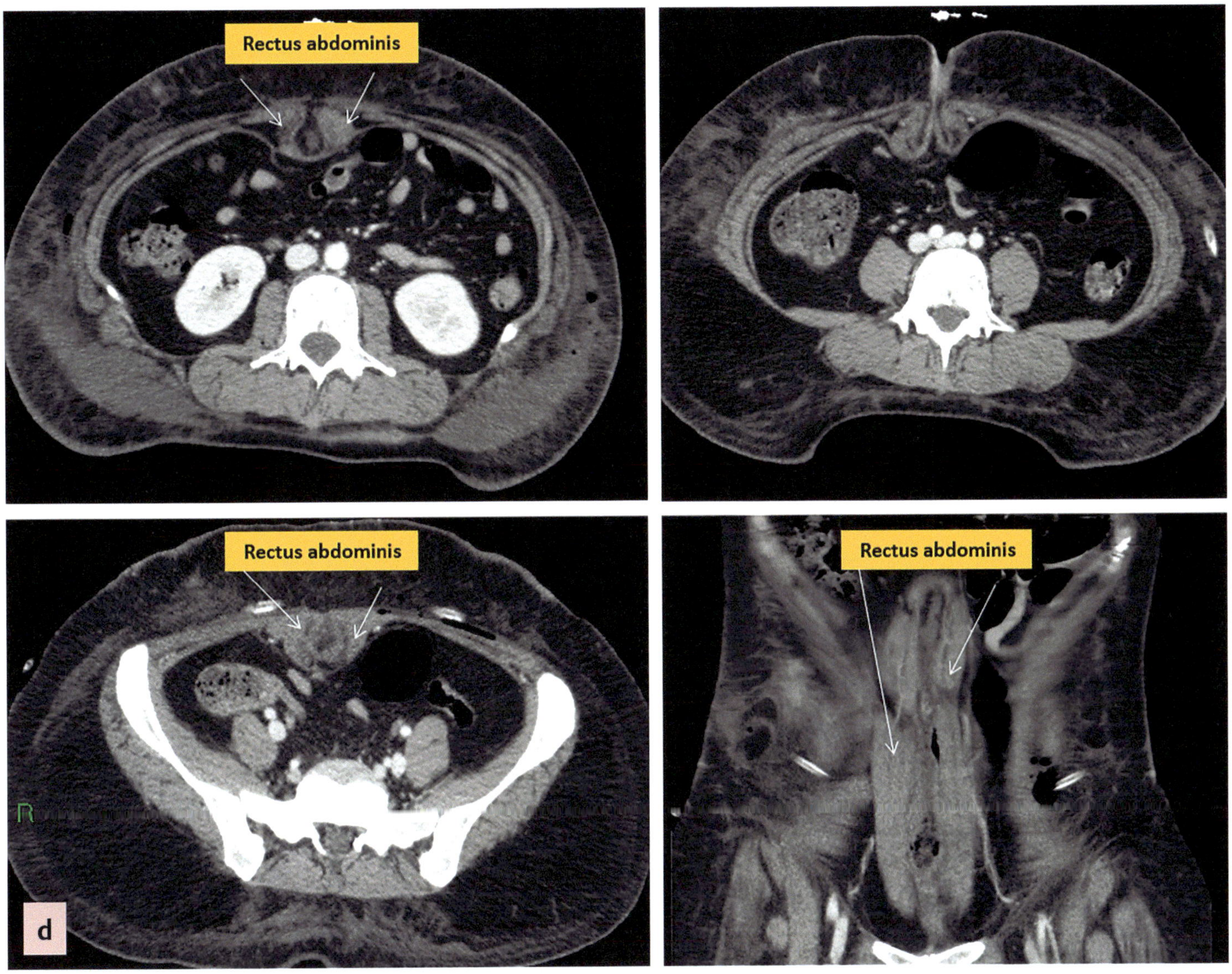

Fig. 1.15 (*Continued*) **(d)** Abdominoplasty (tummy tuck surgery). Image showing the midline incision. Following removal of excess skin and fat, any muscles that are loosened or torn will be repaired and the rectus muscles tightened in midline with internal sutures. Note tightened rectus muscles and overlying tissues in midline. Extensive edema of the subcutaneous tissue is seen due to recent operation.

by the L2 and L3 roots. The lumbar neural plexus is enclosed within psoas major with branches exiting from the muscle. This explains the fact that mass lesions of the psoas may cause muscle weakness or neurological pain.

The major action of the psoas and iliacus muscles is to flex the thigh on the pelvis as well as to maintain the erect position by supporting the spine and pelvis. The iliopsoas unit has close relationship with the diaphragm, spine, kidneys, ureters, aorta, inferior vena cava, sigmoid colon, and pelvic bone. It is, therefore, a potential conduit for the spread of pathology from the chest, abdomen, or retroperitoneum into the lower extremity.

Quadratus Lumborum

The quadratus lumborum muscle lies lateral to the psoas muscle and connects the ilium to the 12th rib and to the lumbar transverse processes.[33] The muscle passes proximally beneath the lateral arcuate ligament of the diaphragm (▸Fig. 1.16, ▸Fig. 1.17). At the level of L4 transverse process, the muscle occupies an area of 5 to 7 cm on the iliac crest.

The function of the quadratus lumborum muscle is not completely known. Because of its costal attachment, it may have a role in respiration and because of its vertebral attachments, some believe it causes lateral lumbar flexion and probably helps to extend the lumbar spine.[33] Atrophy and volume loss of this muscle can predispose to lumbar spine stress fracture and back pain.[34]

Latissimus Dorsi

The latissimus dorsi is a large, flat, dorsolateral muscle on the trunk which extends posterior to the arm (▸Fig. 1.19). Medially, it is partly covered by the trapezius (▸Fig. 1.9). The latissimus dorsi originates from the spinous processes of thoracic T7–T12, thoracolumbar fascia, iliac crest, inferior three or four ribs, and inferior angle of scapula. The tendon of this muscle inserts on floor of the

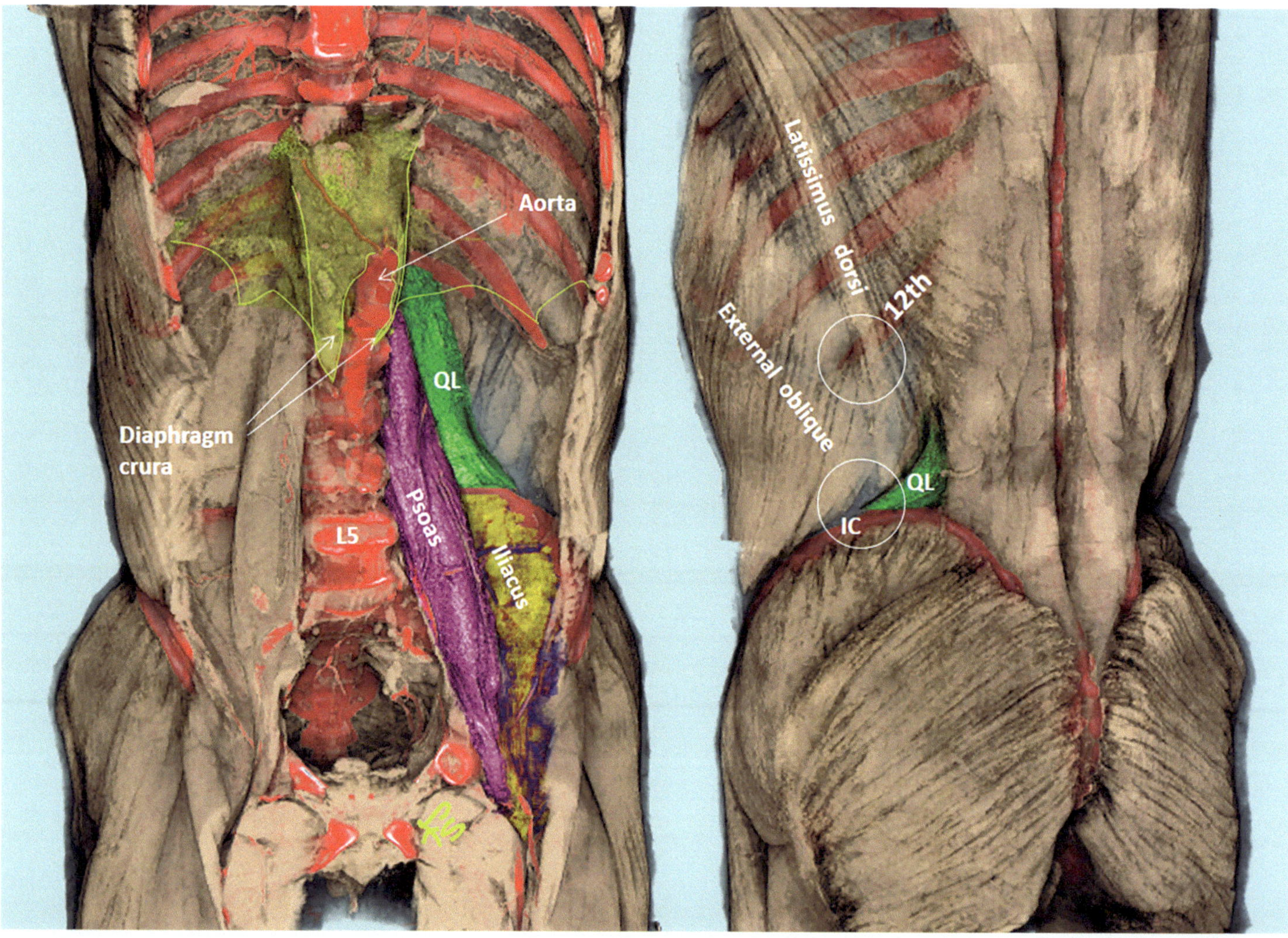

Fig. 1.16 Posterior abdominal wall muscles and lumbar triangles. The psoas major arises from the transverse processes and adjacent vertebral margins of T12 to L5 and inserts on the lesser trochanter of the femur. The quadratus lumborum (QL) muscle lies lateral to the psoas muscle and connects the ilium to the 12th rib. Approximate posterior attachment of the diaphragm is shown by *green line*. The superior lumbar triangle (*upper circle*) is an inverted triangle bordered superiorly by the 12th rib, medially by the QL muscle, and anterolaterally by the external oblique muscle and superficially covered by the latissimus dorsi. Its floor is formed by the transversalis fascia and other fasciae. The inferior lumbar triangle (*lower circle*) is bordered inferiorly by the iliac crest (IC), laterally by the external oblique muscle, and medially by the QL. It is not covered with muscles and floor is formed by the thoracolumbar fascia and other fasciae.

intertubercular groove of the humerus. In the posterior abdomen, the aponeurosis of the latissimus dorsi blends with the thoracolumbar fascia (▶Fig. 1.4). The inferolateral margin of the latissimus dorsi right above the iliac crest is separated from the external abdominis by the lumbar triangle of Petit (▶Fig. 1.16). The latissimus dorsi is supplied by the sixth, seventh, and eighth cervical nerves through the thoracodorsal (long scapular) nerve. The latissimus dorsi is involved in movements of the shoulder joint and extension and lateral flexion of the lumbar spine. Tight latissimus dorsi may be one cause of chronic shoulder and back pain.

Serratus Posterior Inferior Muscle

The serratus posterior inferior muscle is a relatively small, comb-like muscle covering a small portion of the posterior superior abdominal wall (▶Fig. 1.9). It is overlaid by the latissimus dorsi muscle and extends from the spinous processes of T11–L2 to insert on to the lower border of the last four ribs. Medially, its aponeurosis blends with the thoracolumbar fascia (▶Fig. 1.11). The serratus posterior inferior muscle contributes to forced expiration and extension/rotation of the trunk.

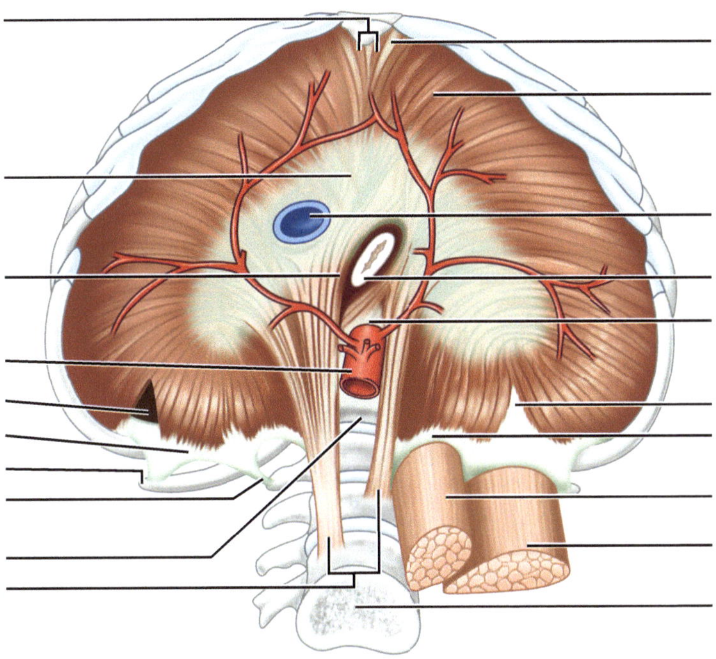

Fig. 1.17 The cranial aspect of the psoas major passes dorsal to the diaphragm behind the medial arcuate ligament. The quadratus lumborum passes beneath the lateral arcuate ligament. IVC, inferior vena cava.

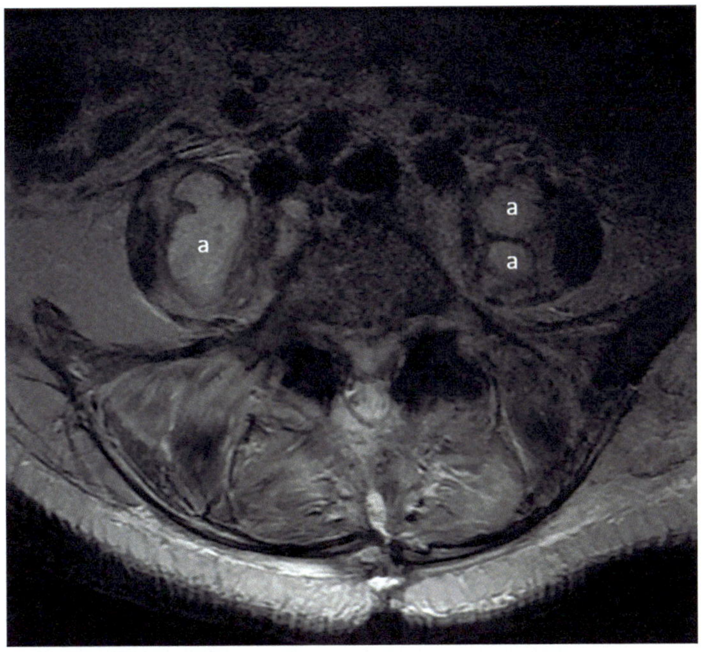

Fig. 1.18 Psoas muscle abscesses (a) shown by T2-weighted magnetic resonance imaging (MRI).

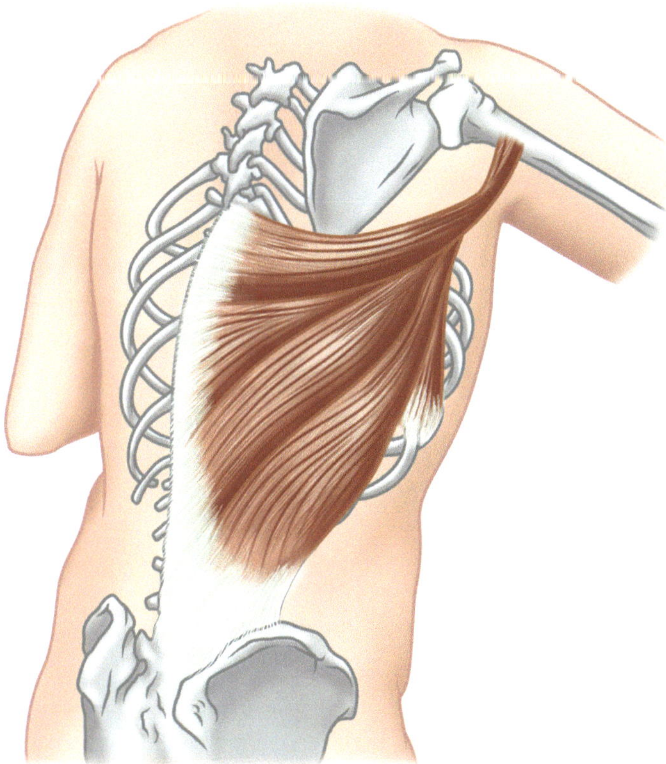

Fig. 1.19 Latissimus dorsi muscle.

1

◆ Pelvic Floor Muscles, Fascia, and Ligaments

The pelvic diaphragm, perineal membrane (inferior fascia of the urogenital diaphragm), and deep perineal pouch (urogenital diaphragm) form the pelvic floor. The pelvic diaphragm consists of the levator ani and ischiococcygeus muscles of the inferior pelvic floor. The piriformis and internal obturator muscles form posterolateral part of the pelvic floor on each side but functionally are categorized as muscles of the lower limb. The urogenital hiatus (for vagina, prostate, and urethra) and the rectal hiatus are formed in midline by slings of the anterior levator ani muscle fibers.

Obturator Internus

The obturator internus or internal obturator muscle is a hip external rotator arising from the internal surface of the obturator membrane and the obturator foramen (▶Fig. 1.20, ▶Fig. 1.21, ▶Fig. 1.22). It passes through the lesser sciatic foramen, underneath the sacrotuberous ligament, to exit the pelvis.[35] The obturator internus covers the internal aspect of the obturator foramen. The obturator internus, along with the gemellus superior and inferior muscles, usually forms a conjoined tendon to attach anteriorly to the medial aspect of the greater trochanter. Medial thickening of the obturator fascia provides support for the attachment of the levator ani and vagina. The arcus tendinous fascia pelvis provides attachment to the paravaginal connective tissue, and arcus tendinous levator ani connects the obturator fascia to the levator ani muscle.

Obturator Foramen and Canal

The obturator foramen is the largest foramen in the body located on the anterolateral pelvic wall. It is formed by the pubic rami anteriorly and medially and by the ischial ramus and body inferiorly[36,37] (▶Fig. 1.21, ▶Fig. 1.22, ▶Fig. 1.23). The obturator membrane covers most of this foramen. The membrane is attached to the periosteum and is continuous with the tendinous attachments of the obturator muscles. There is a small opening at the anterosuperior of the obturator foramen that is not covered by the membrane and forms the opening of the obturator canal.[37]

The obturator canal is a tunnel 2- to 3-cm long and 1-cm wide in the obturator region of the thigh. It is bordered by bone superiorly and laterally, and by the obturator muscles inferiorly and medially. The obturator canal is the passage for the obturator

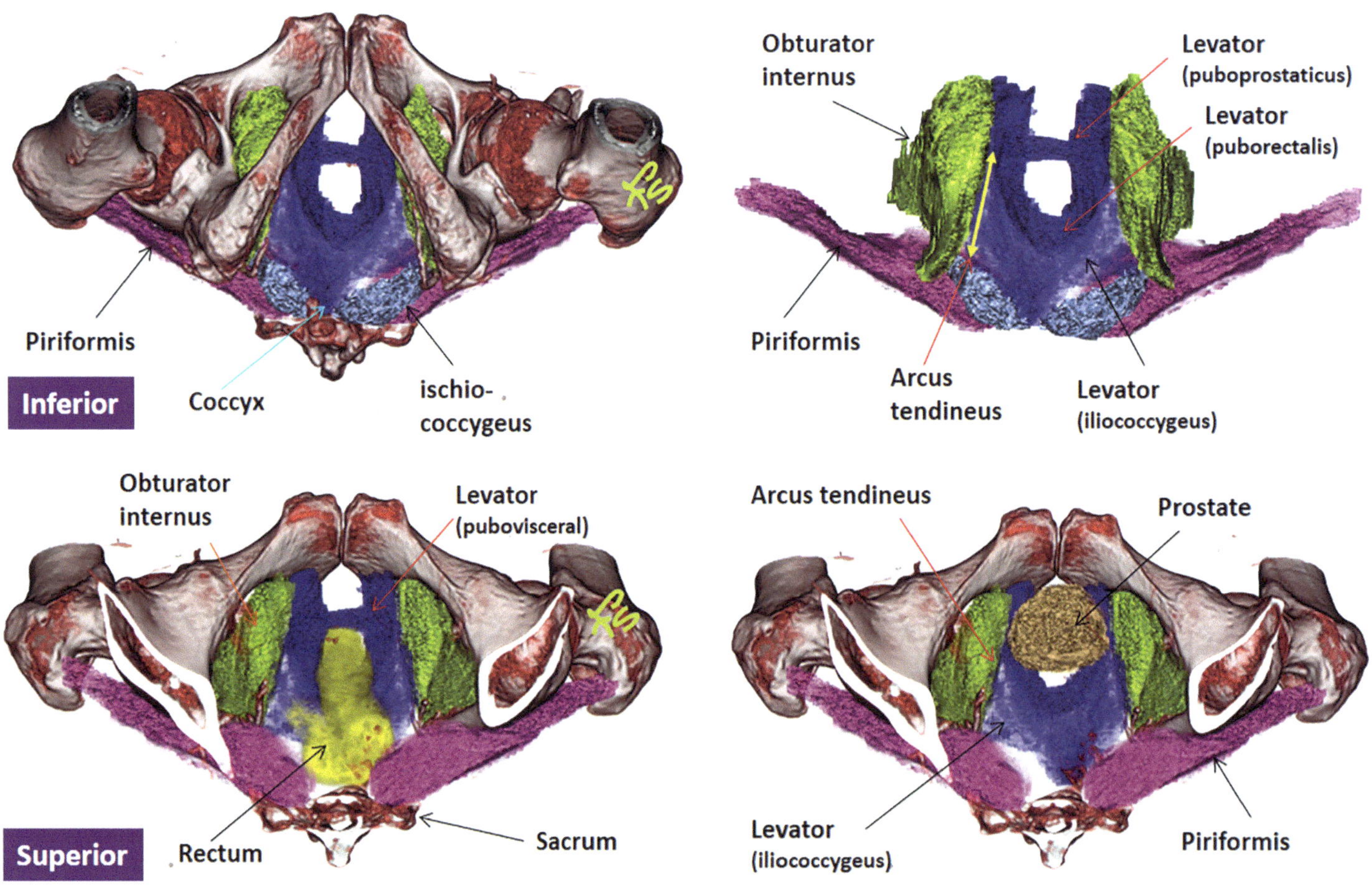

Fig. 1.20 Volume-rendered, color-coded computed tomography (CT) images showing inferior (*upper row*) and superior (*lower row*) views of the muscles of the pelvis.

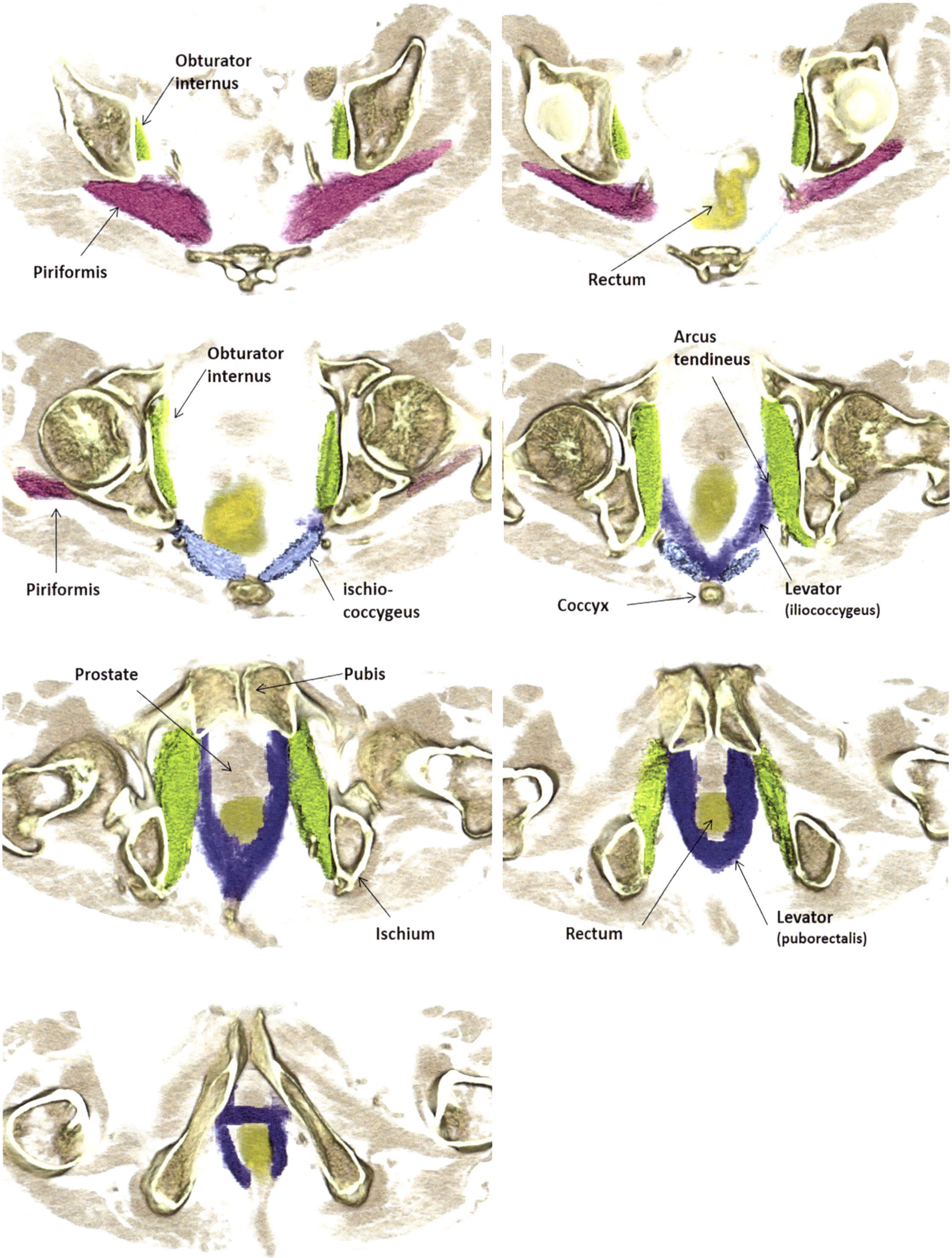

Fig. 1.21 Axial, color-coded computed tomography (CT) images from top to bottom showing muscles of the pelvic floor.

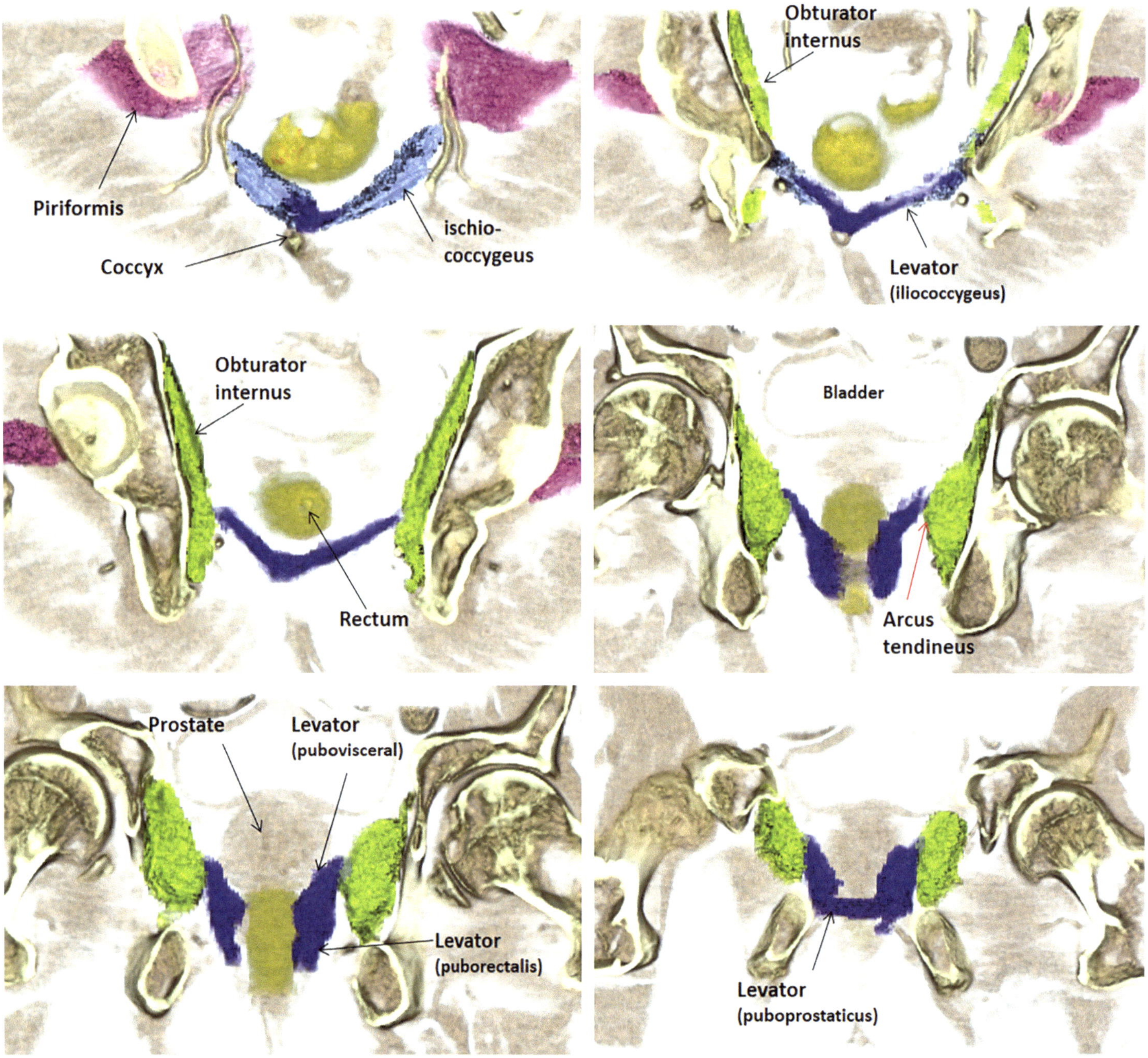

Fig. 1.22 Oblique, coronal, color-coded computed tomography (CT) images from posterior to anterior showing muscles of the pelvic floor.

nerve, artery, and vein and is filled with fat.[37] The obturator nerve arises from the ventral rami L3–L4 of the lumbar plexus.[38] It emerges from the medial border of the psoas major muscle and usually enters the obturator canal superior to the artery and vein. Outside the obturator canal, the obturator nerve divides into anterior and posterior divisions to supply the adductor muscles of the thigh, the gracilis muscle, and provides articular branches to the knee and hip joints.[37,38]

Piriformis

The piriformis originates from the anterior sacrum (▶Fig. 1.20, ▶Fig. 1.21, ▶Fig. 1.22). It is mainly located in the greater sciatic foramen and is closely related to branches of the sacral plexus and gluteal branches of the internal iliac vessels.[35] The piriformis muscle has two muscle bellies, superior and inferior. Outside the greater sciatic foramen, the muscle fibers converge to form a long

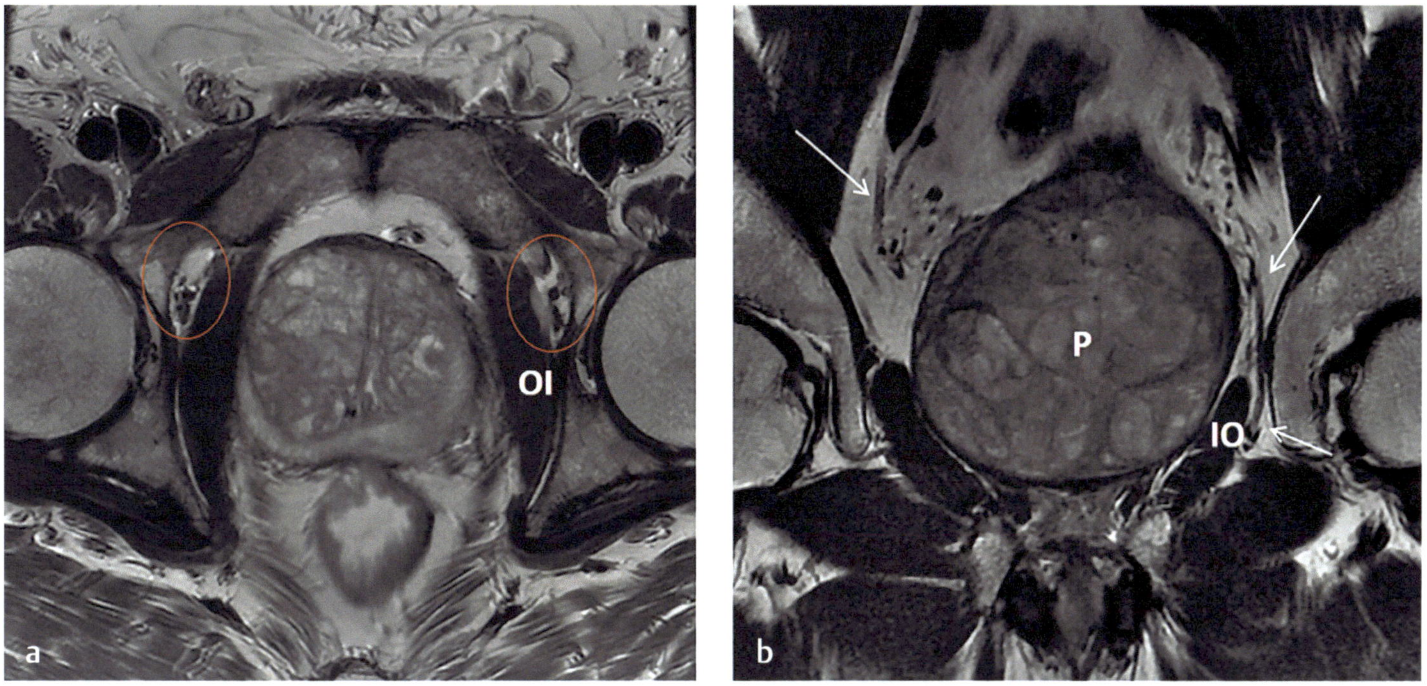

Fig. 1.23 Obturator foramen and canal. **(a)** Axial and **(b)** coronal magnetic resonance (MR) images showing the obturator foramen and canal located on the anterolateral sides of pelvis (*red circles*) between the pelvic bone and the obturator internus (OI) muscle. The obturator nerves are best shown on the coronal image (*arrows*). The obturator canal is the passage for the obturator nerve, artery, and vein and is filled with fat. A large prostate (P) is seen in this patient.

tendon that attaches to the superior surface of the greater trochanter. The piriformis is innervated by the sacral plexus. In 80 to 94% of cases, the sciatic nerve passes anterior to the pirifomis muscle belly without branching. Compression of the sciatic nerve by the piriformis, also known as piriformis syndrome, is identical in clinical presentation to low back pain caused by L5 and S1 radiculopathy.

Coccygeus Muscle

The coccygeus or ischiococcygeus is a triangular shaped muscle in the pelvic floor extending between the ischial spine and the margin of the distal sacrum.[39] It is located superior and posterior to the levator ani, inferior to the piriformis, and anterior to the gluteus maximus (▶ Fig. 1.20, ▶ Fig. 1.21, ▶ Fig. 1.22). The sacrospinous ligament is a triangular ligament behind the coccygeus (▶ Fig. 1.24, ▶ Fig. 1.25). This ligament forms the greater sciatic foramen above and the lesser sciatic foramen below it. The pudendal nerve is a branch of the sacral plexus (S2–S4) passing anterior to the piriformis and then between the sacrospinous and sacrotuberous ligaments via the lesser sciatic foramen before entering the pudendal (Alcock) canal (along with pudendal vessels) formed by a fascial duplication of the obturator internus fascia (▶ Fig. 1.24, ▶ Fig. 1.25, ▶ Fig. 1.27). The pudendal nerve gives off three branches, namely, the inferior rectal, perineal, and dorsal nerve of the penis or clitoris, to innervate the external anal sphincter, external urethral sphincter, perineal musculature, clitoris, and skin.[40,41] This nerve is at risk for compression or injury by tumors, pelvic surgery, and trauma.

Levator Ani

The levator ani is a broad flat thin muscle situated on the side of the pelvis (▶ Fig. 1.20, ▶ Fig. 1.21, ▶ Fig. 1.22). It provides main support for the urethra, prostate, vagina, and rectum. Traditionally, it is divided into three major components, the pubococcygeus, puborectalis, and iliococcygeus muscles.[42,43] The term pubococcygeus muscle incorrectly implies a connection between the pubis and coccyx while this muscle, in fact, inserts into the walls of the vagina and anorectum. Therefore, the pubococcygeus muscle has been recently renamed as the pubovisceral and is divided into three parts: the puboperineal, puboanalis, and pubovaginalis (female) or puboprostaticus (male) (▶ Fig. 1.26, ▶ Fig. 1.27, ▶ Fig. 1.28, ▶ Fig. 1.29, ▶ Fig. 1.30).

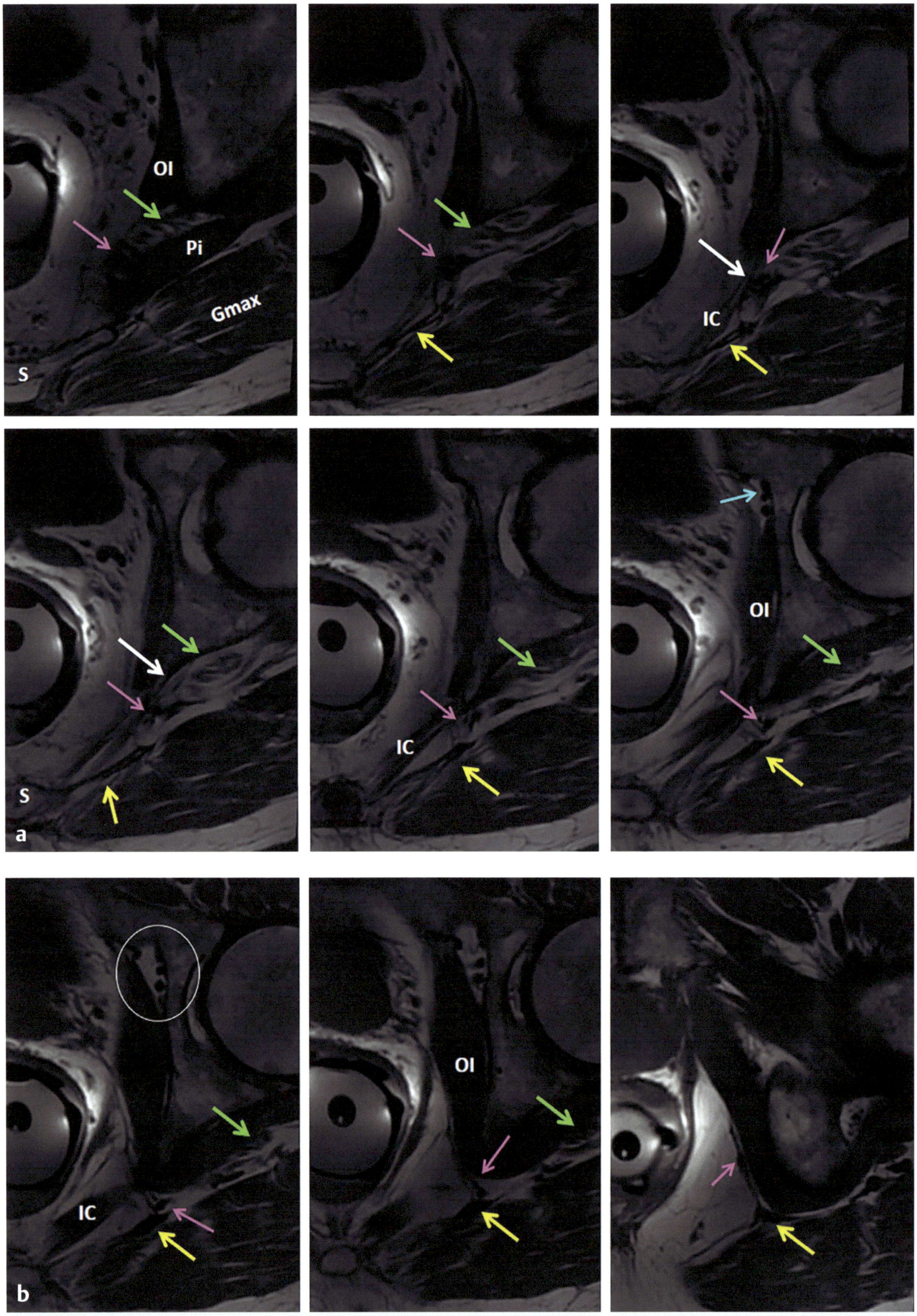

Fig. 1.24 (a, b) Pudendal nerve and associated vessels shown by axial, T1-weighted magnetic resonance imaging (MRI) from superior to inferior. The pudendal nerve (*pink arrows*) passes anterior to the piriformis (Pi) and then anterior to sacrotuberous (*yellow arrow*) ligaments before entering the pudendal (Alcock) canal. The sacrospinous ligament (*white arrow*) passes behind the ischiococcygeus (IC) muscle. *Blue arrows*, obturator nerve; *green arrows*, sciatic nerve; OI, obturator internus muscle; S, sacrum; *white circle*, contents of the obturator canal.

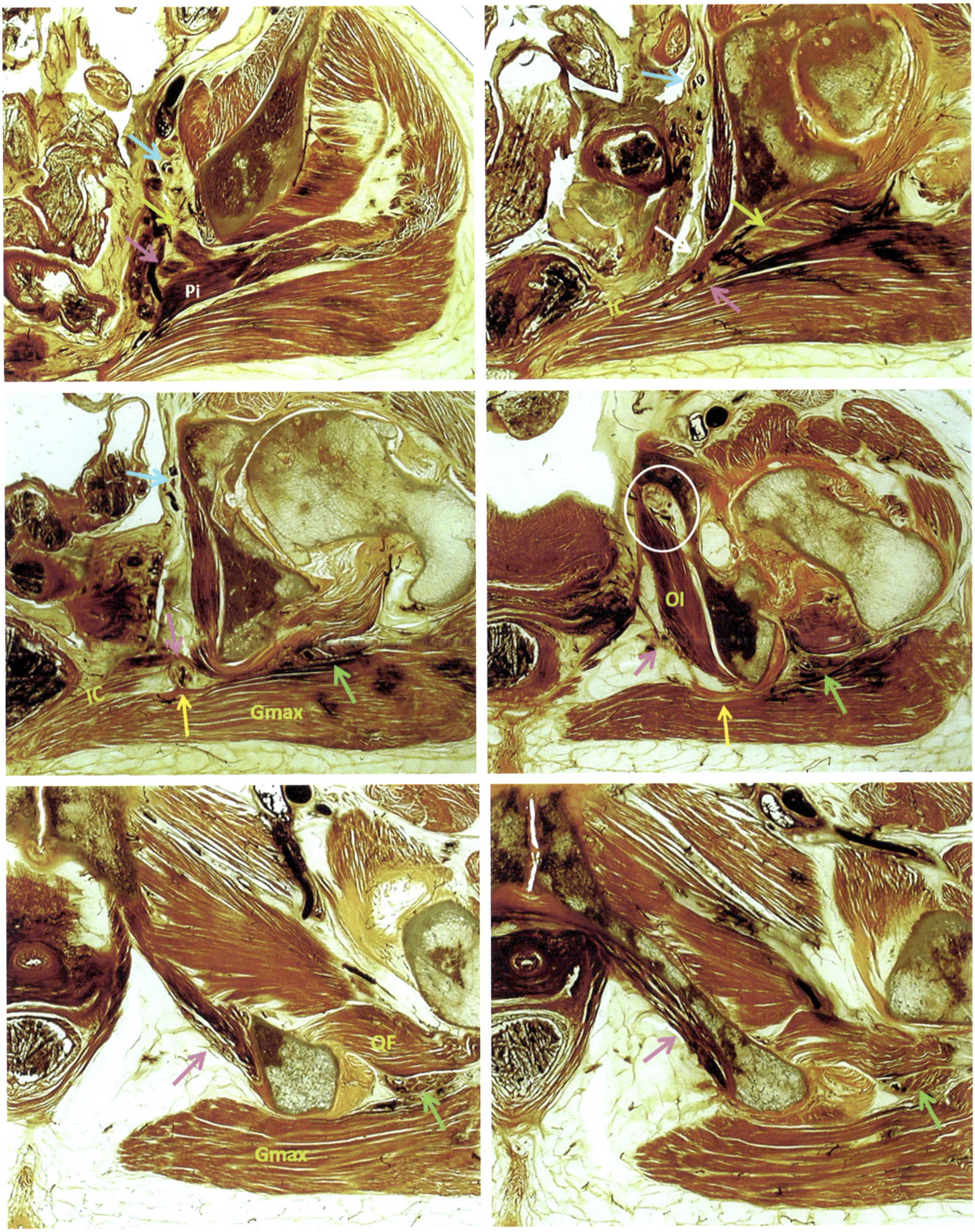

Fig. 1.25 Pudendal and obturator canals shown by axial cadaveric cuts from top to bottom. The pudendal nerve (*pink arrows*) is a branch of the sacral plexus (S2–S4) passing anterior to the piriformis (Pi) and then between the sacrospinous (*white arrow*) and sacrotuberous (*yellow arrow*) ligaments before entering the pudendal (Alcock) canal. The sacrospinous ligament (*white arrow*) passes behind the ischiococcygeus (IC) muscle. The pudendal nerve gives off three branches: the inferior rectal, perineal, and dorsal nerve of the penis or clitoris. Gmax, gluteus maximus; QF, quadratus femoris.

1

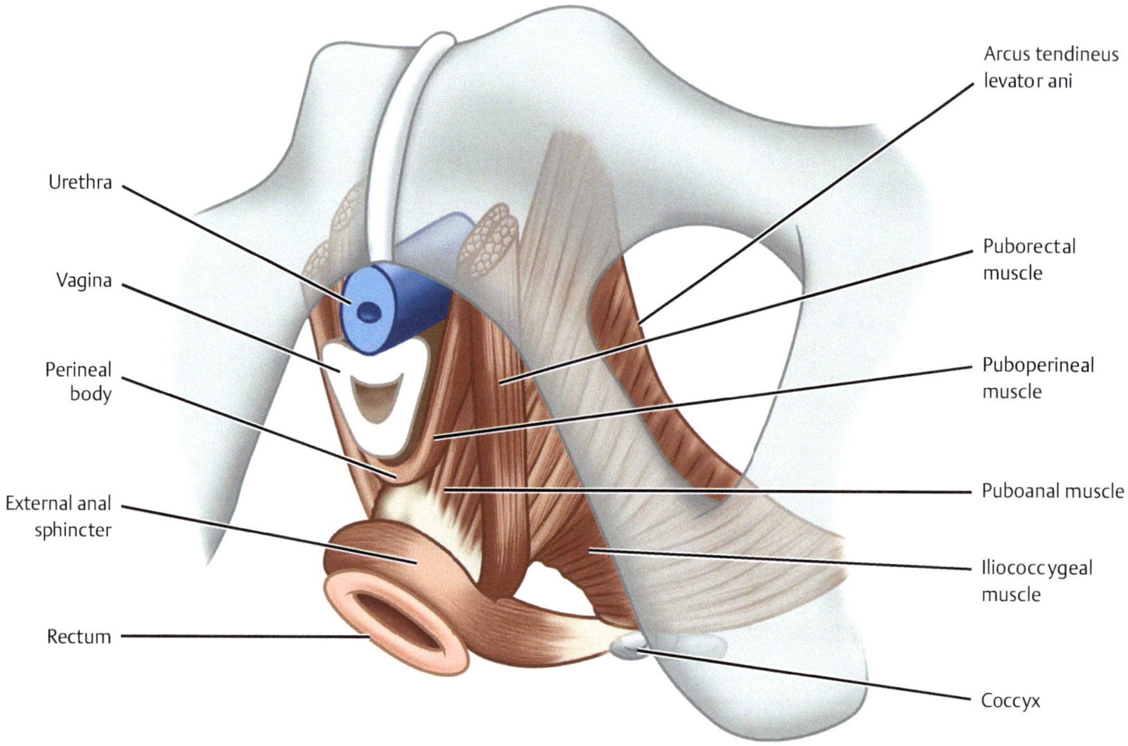

Fig. 1.26 Schematic view of the levator ani muscles from below after the vulvar structures and perineal membrane have been removed showing the arcus tendineus levator ani, external anal sphincter, puboanal muscle, perineal body uniting the two ends of the puboperineal muscle, iliococcygeal muscle, and puborectal muscle. Note that the urethra and vagina have been transected just above the hymenal ring. (Modified from Kearney et al.[43])

Laterally, the levator muscle complex arises from the posterior surface of the superior pubic ramus and the ischial spine on either end. Between these two bony attachments, the levator muscles fuse to the inner fascia of the obturator by an aponeurotic extension known as "arcus tendineus" (▶Fig. 1.27, ▶Fig. 1.29). Medially, the muscle fibers unite with the contralateral side by a fibrous band known as median (anococcygeal) raphé. However, the most posterior fibers connect medially to the side of the distal coccyx. Middle and anterior fibers form the pubovaginalis (or puboprostaticus in male) and puborectalis (and puboanalis) components that originate from the inner side of the pubic bone on each side anteriorly and insert into the perineal body, vaginal wall (at the level of mid urethra), prostate, and intersphincteric groove between the internal and external anal sphincters in the anus. Their fibers sling around these organs. For example, the puborectal muscle forms a sling around the rectum to support the angle between the anus and rectum, and the puboanalis subdivision forms a sling around the anus that helps to elevate it. The fascia of the levator ani invests the muscle and separates it from the pelvic organs. At the level of the anus, the fibers of the puboanalis and the deep portion of the external sphincter of the anus are continuous but in cross-sectional imaging they may be distinguished due to their different origins and attachments.[44]

The levator ani muscles are supplied by branches of the inferior gluteal artery, inferior vesical artery, and pudendal artery. The levator ani muscle complex is mainly innervated by the sacral nerve roots (S3–S5) that travel on the superior surface of the pelvic floor (levator ani nerve).[45] The levator ani nerve and pudendal nerves are so close at the level of the ischial spine that a transvaginal "pudendal nerve blockade" would block both nerves simultaneously.[46]

The knowledge of the anatomic location of the components of the levator muscles is important to localize the site of injury including muscle tears or avulsion during vaginal child birth, or after surgery and trauma. Major levator ani defects can cause pelvic organ prolapse or pelvic floor dysfunction. In case of damage of the puboanalis muscle, the elevation of the anus could be lost and the genital hiatus widened. The pubovaginalis component supports the urethral supports and its injury might affect urethral function.

Disruptions to the puboanalis or puborectalis muscles would be expected to result in different patterns of pelvic floor dysfunction. Vaginal delivery injuries are probably the most common cause of levator injury that occurs in a characteristic "injury zone" on the inner surface of the pubic bone at the origin of the pubovaginalis muscle but do not involve the puborectal muscle.[47–49] These defects can be detected by magnetic resonance imaging (MRI), computed tomography (CT), and ultrasound.[50–52] MRI studies of levator ani muscle injuries have revealed two patterns: (1) gross tear and architectural distortion of the anatomy and (2) muscle atrophy. In imaging study of the pelvic floor, attention to the size and shape of the bony pelvis is important since deformities or small dimensions of the pelvis may influence soft tissue injuries (LI). Atrophy of muscle can also be secondary to nerve injury (▶Fig. 1.28).

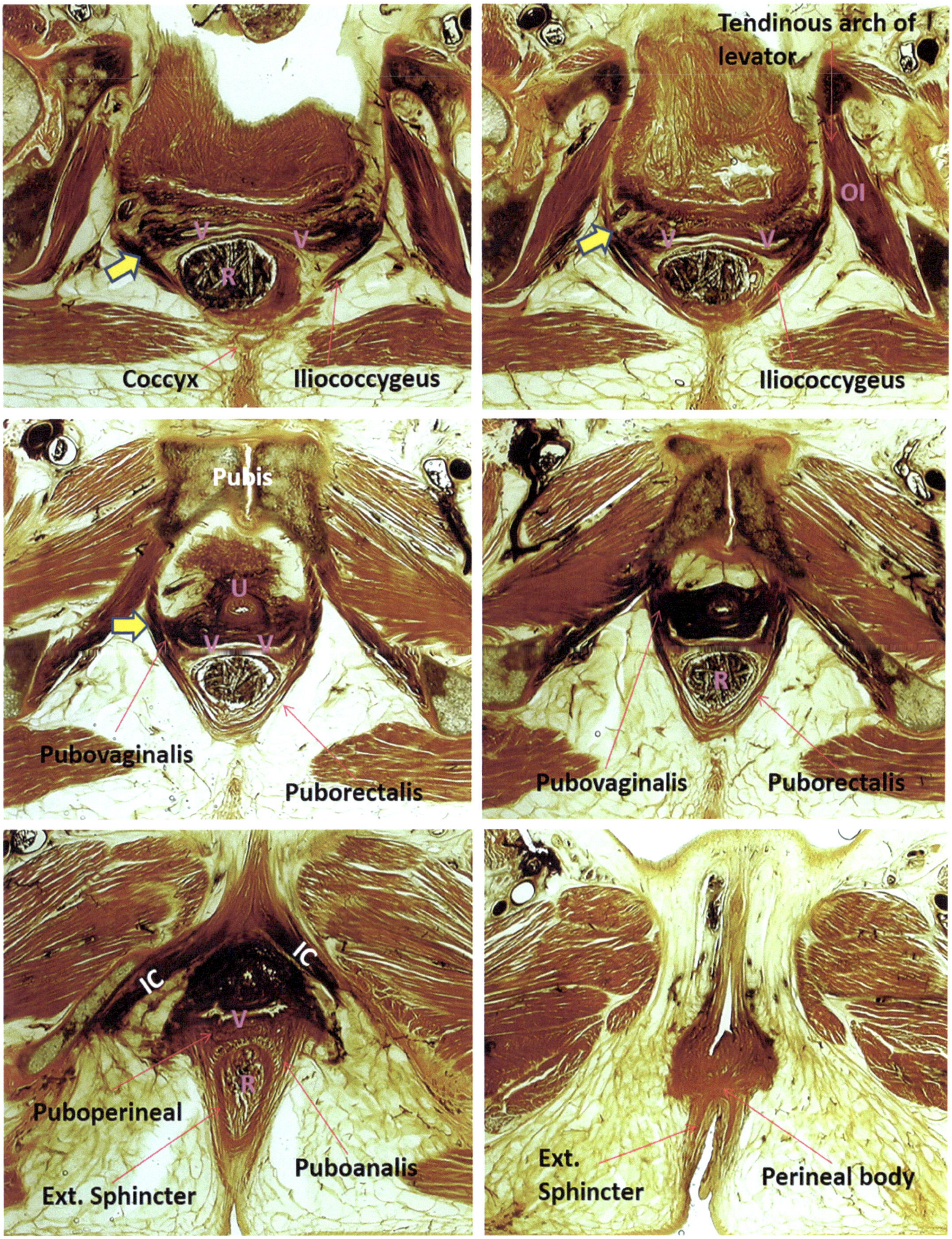

Fig. 1.27 Components of the levator ani in a female. Axial cadaveric cuts. Iliococcygeus component of the levator muscle arises from the inner aspect of the obturator internus (OI) fascia and unites with the contralateral side by a fibrous median raphé. However, the most posterior fibers of the iliococcygeus connect medially to the side of the distal coccyx. Middle and anterior fibers of the levator form the pubovisceral (rectoperineal, puboanalis, and pubovaginalis) and puborectalis components that originate from the inner side of the pubic bone on each side anteriorly and insert into the perineal body, vaginal wall (at the level of mid urethra), and intersphincteric groove of the anus. The fibers of the puboanalis and the deep portion of the external (Ext.) sphincter of the anus will merge. The tendinous arch of the pelvic fascia (*yellow arrows*) is where the lateral ligaments of the bladder and the transverse cervical ligaments fuse with the parietal fascia. IC, ischiocavernosus muscle; R, rectum; U, urethra; V, vagina.

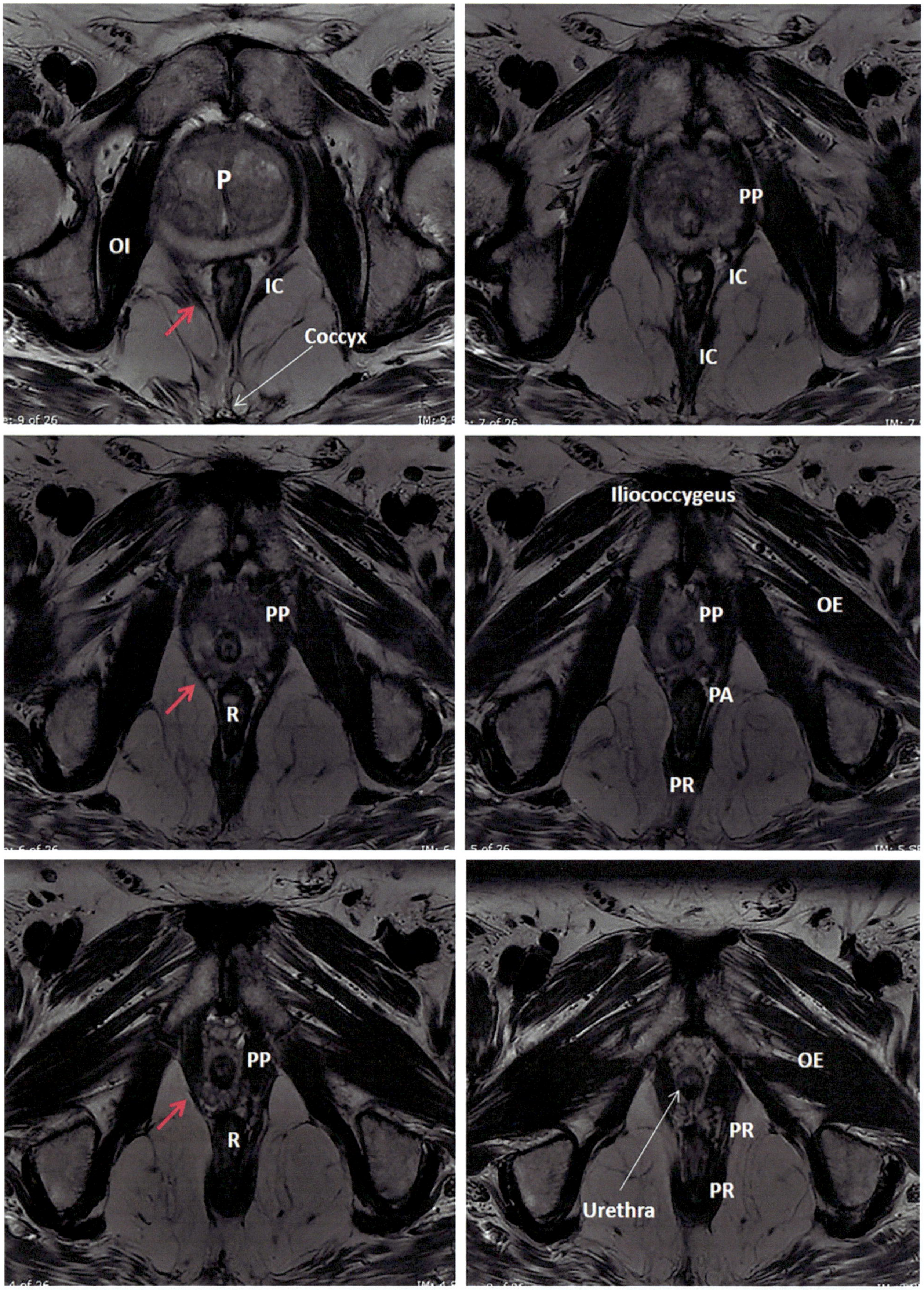

Fig. 1.28 Levator muscle components in a male patient with enlarged prostate (P). Axial magnetic resonance imaging (MRI) from top to bottom. Note atrophy of the levator on the right side (*red arrows*). The levator ani is divided into three major components, the pubococcygeus and puborectalis (PR), and iliococcygeus (IC) muscles. The pubococcygeus muscle has been renamed to pubovisceral and is subdivided into three parts: the rectoperineal, puboanalis (PA), and puboprostaticus (PP). Middle and anterior fibers of the levator form the puboprostaticus and puborectalis components that originate from the inner side of the pubic bone on each side anteriorly and insert into the perineal body, vaginal wall (at the level of mid urethra), prostate, and intersphincteric groove between the internal and external anal sphincters in the anus. OI, obturator internus; OE, obturator externus.

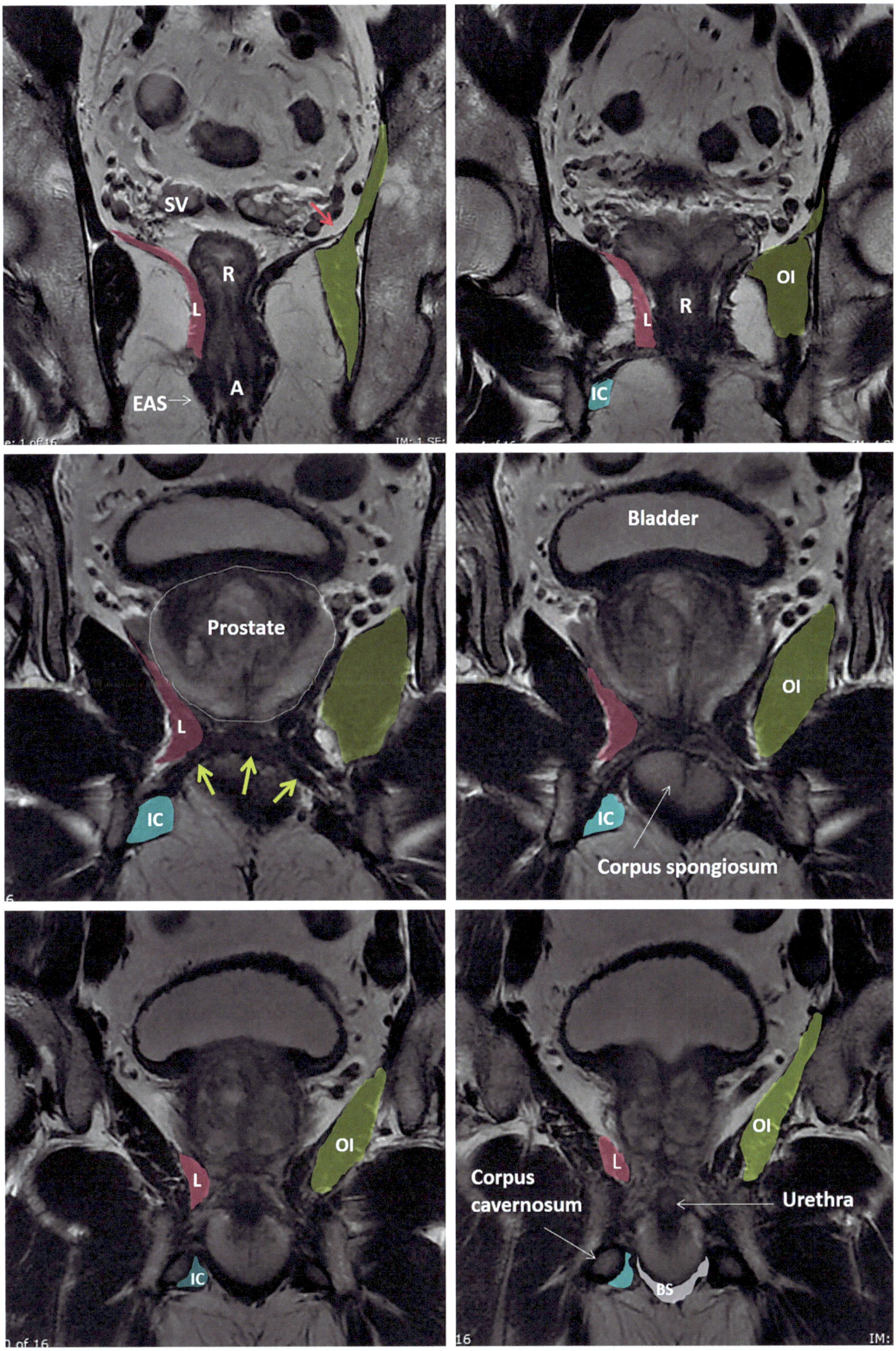

Fig. 1.29 Coronal magnetic resonance (MR) scan in a patient with enlarged prostate from posterior to anterior. Middle images show the deep transverse perineal muscle (*green arrows*). The levator muscle (L) complex is shown in *red*. The iliococcygeus part of the levator complex connects to the obturator internus by thin fibrous band (*red arrow*) known as "arcus tendineus." The ischiocavernosus (IC) muscle is shown in *blue* with the corpus cavernosum next to it. The bulbospongiosus (BS) muscle is shown in *light burgundy* with the corpus spongiosum anterior to it. A, anus; EAS, external anal sphincter; OI, obturator internus; R, rectum; SV, seminal vesicle.

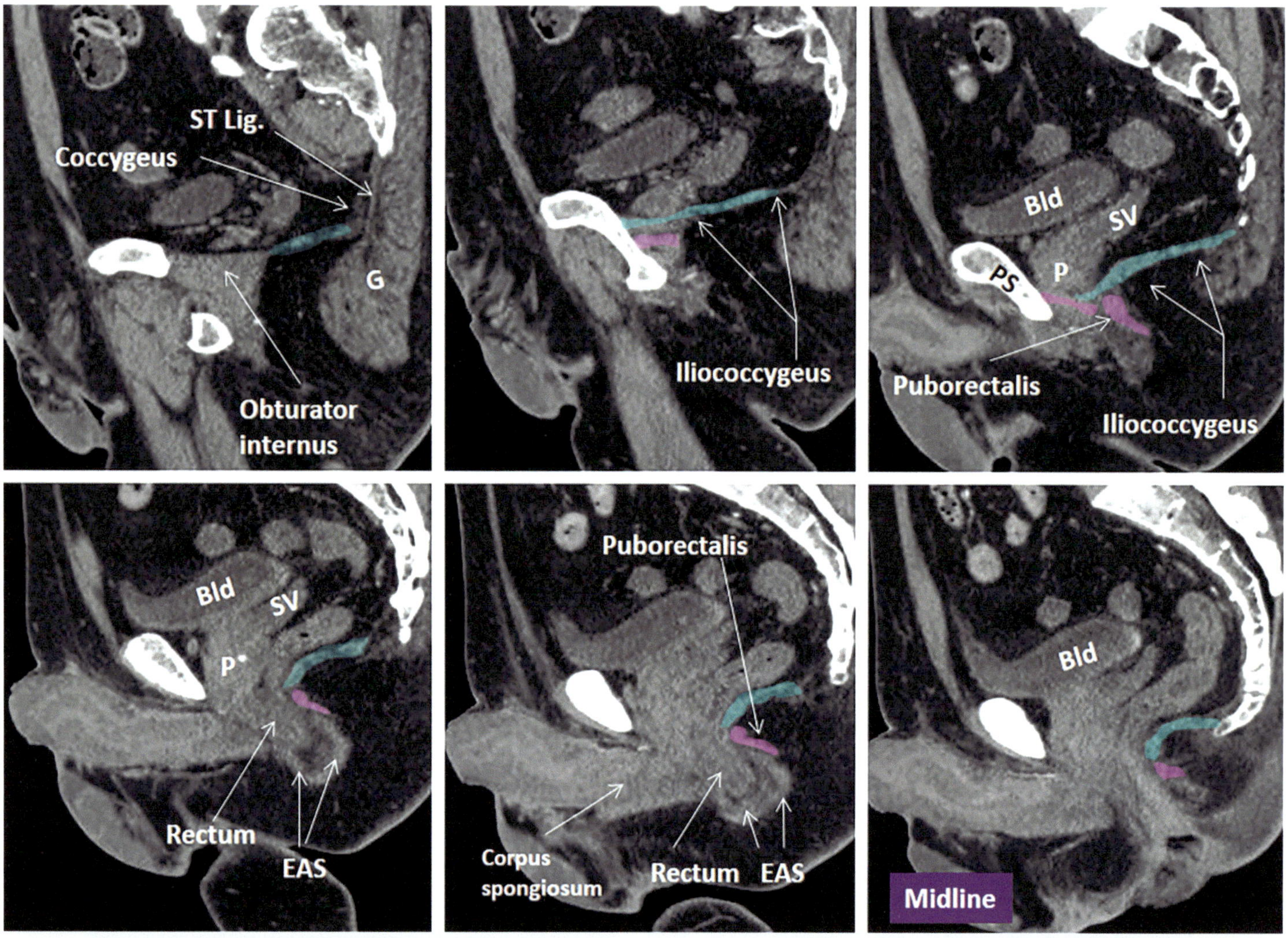

Fig. 1.30 Sagittal computed tomography (CT) images from lateral to midline showing layout of the levator muscles. The puborectalis muscle (*in red*) slings behind the rectum and attaches to the inner side of the pubic bone. The iliococcygeus (*in green*) extends between the coccyx and pubis and connects to the obturator internus by an aponeurotic extension known as "arcus tendineus." Bld, bladder; EAS, external anal sphincter; G, gluteus muscles; P, prostate; PS, pubic symphysis; ST lig., sacrotuberous ligament; SV, seminal vesicle.

Perineum and Perineal Membrane

The perineum is a diamond-shaped area lying superficial to the perineal membrane and is bounded by the pubic symphysis, ischiopubic rami, sacrotuberous ligaments, and coccyx.[53–55] The perineum is divided into two anal and urogenital triangles by a line passing between the two ischial tuberosities. The anal triangle contains the anal orifice. The urogenital triangle includes the external genitalia (▶ Fig. 1.31). The perineal membrane covers the urogenital triangle and divides it into superficial and deep perineal spaces. The superficial perineal space contains the bulbocavernosus (bulbospongiosus), ischiocavernosus, and superficial transverse perineal muscles, and vestibular bulbs.[56] The deep perineal space is bounded between the endopelvic fascia and perineal membrane. It contains sphincter urethrae, urethrovaginal sphincter (in females), compressor urethrae (in females), and deep transverse perineal muscle. The levator ani muscle is attached to the superior surface of the perineal membrane and to the perineal body. The transverse perineal muscle originates from the ischial tuberosity on each side and inserts on the perineal body (▶ Fig. 1.29).

The perineal body is the meeting point of the perineal structures including bulbospongiosus, external anal sphincter, superficial

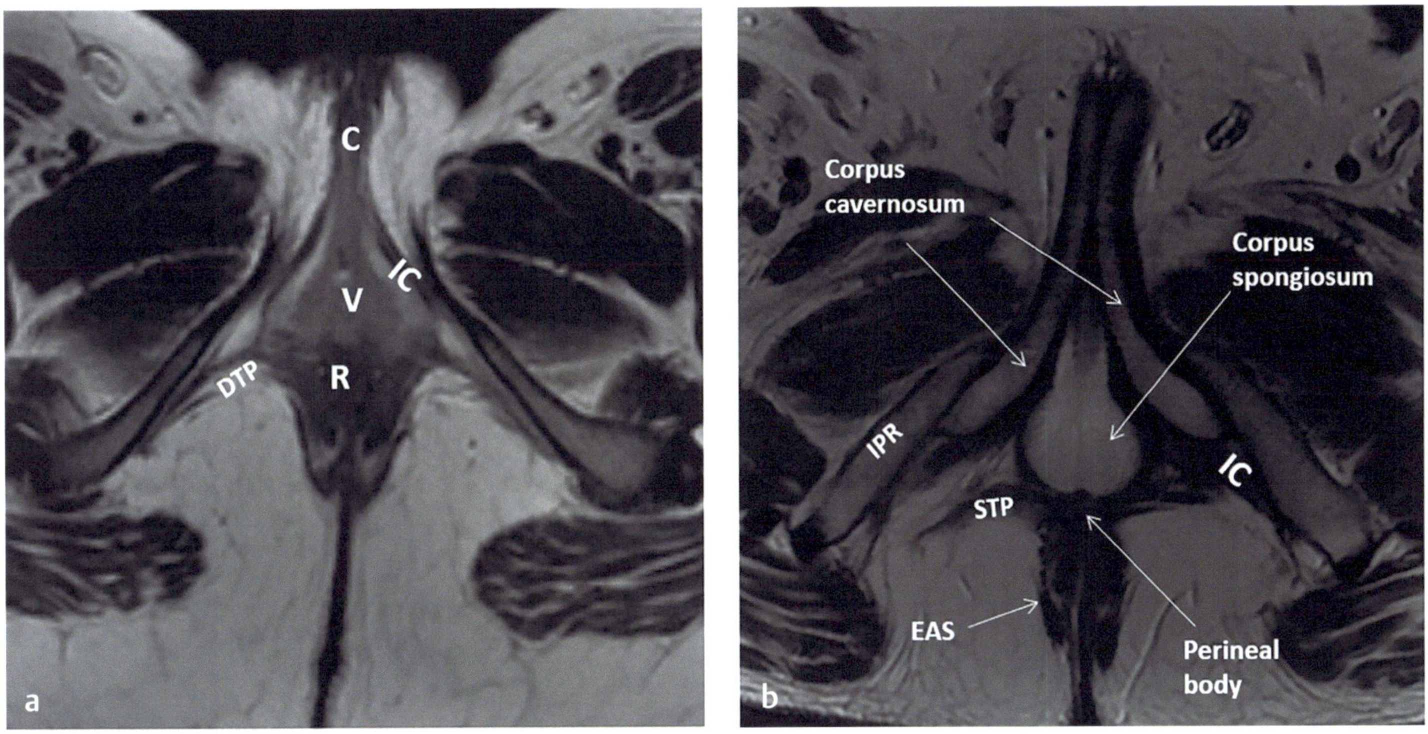

Fig. 1.31 Axial magnetic resonance (MR) images in **(a)** female and **(b)** male showing the deep transverse perineal (DTP) and ischiocavernosus (IC) muscles. The perineum is divided into the anal and urogenital triangles by a line passing between the two ischial tuberosities. The anal triangle contains the anal orifice. The urogenital triangle includes the external genitalia. The perineal body is the meeting point of the perineal structures. C, body of clitoris; EAS, external anal sphincter; IPR, inferior pubic ramus; R, rectum; V, vagina.

transverse perineal muscle, perineal membrane, and pubovisceral muscle, bringing stability to the region[57] (▸Fig. 1.27, ▸Fig. 1.31). The superficial Colles' fascia of the urogenital triangle is attached to the margins of the ischiopubic rami and is continuous with the Scarpa's fascia. The innervation to the perineum is provided by perineal branch of the pudendal nerve. Vascular supply is provided by the branches of pudendal and inferior rectal vessels.

◆ Neurovasculature of the Abdominal Wall

Vascular and neurologic injuries of the abdominal wall are frequent complications after trauma, surgery, and interventional procedures. As a greater number of surgical and interventional procedures are performed by instruments without opening of the abdominal wall, it remains increasingly important to know the anatomic course of vessels and nerves in order to minimize injuries to the nerves and vessels of the anterior abdominal wall. Demonstration of the vessels with CT or conventional angiography is easy, but it is still very difficult to show the nerves of the abdominal wall using imaging methods.

The lower six dorsal and the first lumbar nerves innervate the upper abdominal wall. The abdominal cutaneous branches of the lumbar plexus supplying the lower abdomen include the iliohypogastric, ilioinguinal, genitofemoral, lateral femoral-cutaneous, and obturator nerves. The iliohypogastric nerve originates from the T12 and L1 interspaces. The ilioinguinal nerve originates from L1. The genitofemoral nerve arises from L2 and L3 nerves. The nerves travel between the transversus abdominus and internal oblique muscles. The ilioinguinal and iliohypogastric nerves are probably the most important sensory nerves of the lower abdominal wall. Injury of these nerves has been reported in up to 3.7% of surgical procedures due to direct nerve trauma and neuroma formation from transection or entrapment of these nerves.[58,59] A classic manifestation of ilioinguinal/iliohypogastric neuropathies includes burning pain in the groin region that radiates to the genitalia. Nerve injury may also contribute to subsequent formation of direct inguinal hernias.[60] The ilioinguinal and iliohypogastric nerves pass on the posterior abdominal wall and between the anterior abdominal wall layers. These nerves may be seen as they emerge from the lateral border of the upper psoas muscles. They course laterally over the anterior surface of the quadratus lumborum toward the iliac crest. Near the iliac crest, both nerves

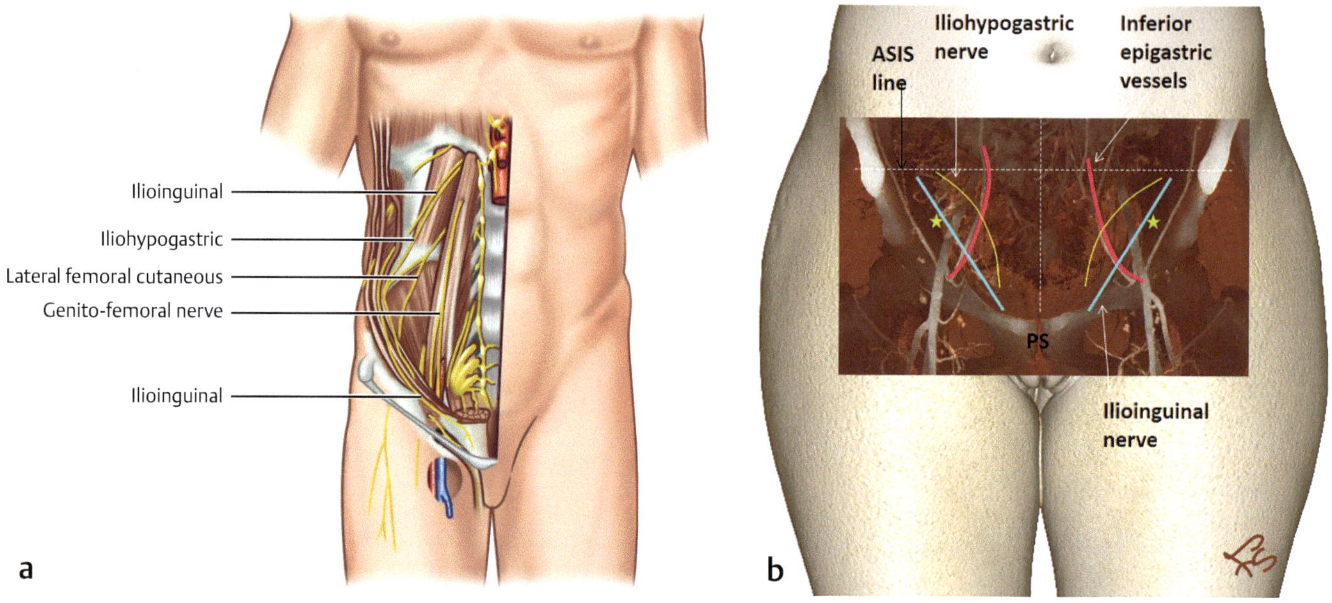

Fig. 1.32 (a) Ilioinguinal and iliohypogastric nerves. **(b)** Mean location of ilioinguinal and iliohypogastric nerves and inferior epigastric vessels are shown relative to anterior superior iliac spine (ASIS). Surface landmarks are often employed to localize the nerves. One such landmark is the point 5 cm superior to the pubic symphysis (PS) and 8 cm from the midline which is a common point for accessory trocar placement. Point 2 cm superior to the PS is common starting site for Pfannenstiel (bikini line) incision.[61] The surface landmarks of the inferior epigastric vessels are 4 cm from midline at the level of the ASIS and approximately 2 cm superior to the PS and 6 cm from the midline (lateral margin of the rectus muscles).

pierce the posterior surface of the transversus abdominis muscle and then course in the plane between the transversus abdominis and internal oblique muscles. Next, the nerves pierce the internal oblique muscles, and travel between the aponeuroses of the internal and external oblique muscles (▶ Fig. 1.32). The ilioinguinal nerve courses medially and inferiorly and enters the inguinal canal at variable distances from the pubic tubercles. Surface landmarks are often employed to localize the nerves. One such landmark is the point 5 cm superior to the pubic symphysis and 8 cm from the midline[61] (▶ Fig. 1.32).

Superficial vessels of the anterior abdominal wall are shown in ▶ Fig. 1.33. The anterior abdominal wall muscles are supplied by the lower intercostal, deep circumflex iliac, superior epigastric, and inferior epigastric arteries. The posterior abdominal wall muscles are mainly supplied by the posterior intercostal, lumbar, and iliolumbar arteries.

The superior epigastric artery originates from the internal mammary artery at the level of the seventh rib (▶ Fig. 1.34, ▶ Fig. 1.35). It descends between the slips of the diaphragm, anterior to the transverse thoracic and transverse abdominal muscles. The artery penetrates the rectus sheath of the rectus abdominis muscle and gives off several musculocutaneous perforators that anastomose with the perforators of the deep inferior epigastric branches.[62]

The inferior epigastric arteries move obliquely and superiorly toward the umbilicus. Near the origin from the external iliac artery, the inferior epigastric arteries are seen just medial to the spermatic cord (or round ligament in female). The inferior epigastric arteries enter the rectus sheath at the arcuate line and gives off perforating branches. The surface landmarks of the inferior epigastric vessels are 4 (2.5 –5.5) cm from midline at the level of the anterior superior iliac spine and approximately 2 cm superior to the pubic symphysis and 6 cm from midline (lateral margin of the rectus muscles)[58,59] (▶ Fig. 1.33, ▶ Fig. 1.34). Communication between the superior and inferior epigastric arteries occurs within rectus abdominis muscle, at a level above the umbilicus. These epigastric arteries also anastomose laterally with the intercostal, subcostal, and the ascending branch of the deep circumflex iliac artery.

The deep inferior epigastric artery perforator flap has been used for breast reconstruction after mastectomy. It involves the transfer of the patient's own skin and subcutaneous tissues from the lower abdominal wall to the chest to form the breast mound. Knowing the branching pattern with imaging permits the preoperative selection and localization of the optimal perforator branch to include with the vascular pedicle of the flap.[63,64] There are three general branching patterns of the inferior epigastric artery: Type 1, a single vessel (29%); Type 2, a bifurcated vessel (57%); and Type 3, a trifurcated vessel (14%). These branching patterns can be easily seen with CT angiography[62] (▶ Fig. 1.33, ▶ Fig. 1.34). Perforator with a short intramuscular course (<4 cm) through the rectus muscle are the easiest and safest to dissect for flap preparation.[64]

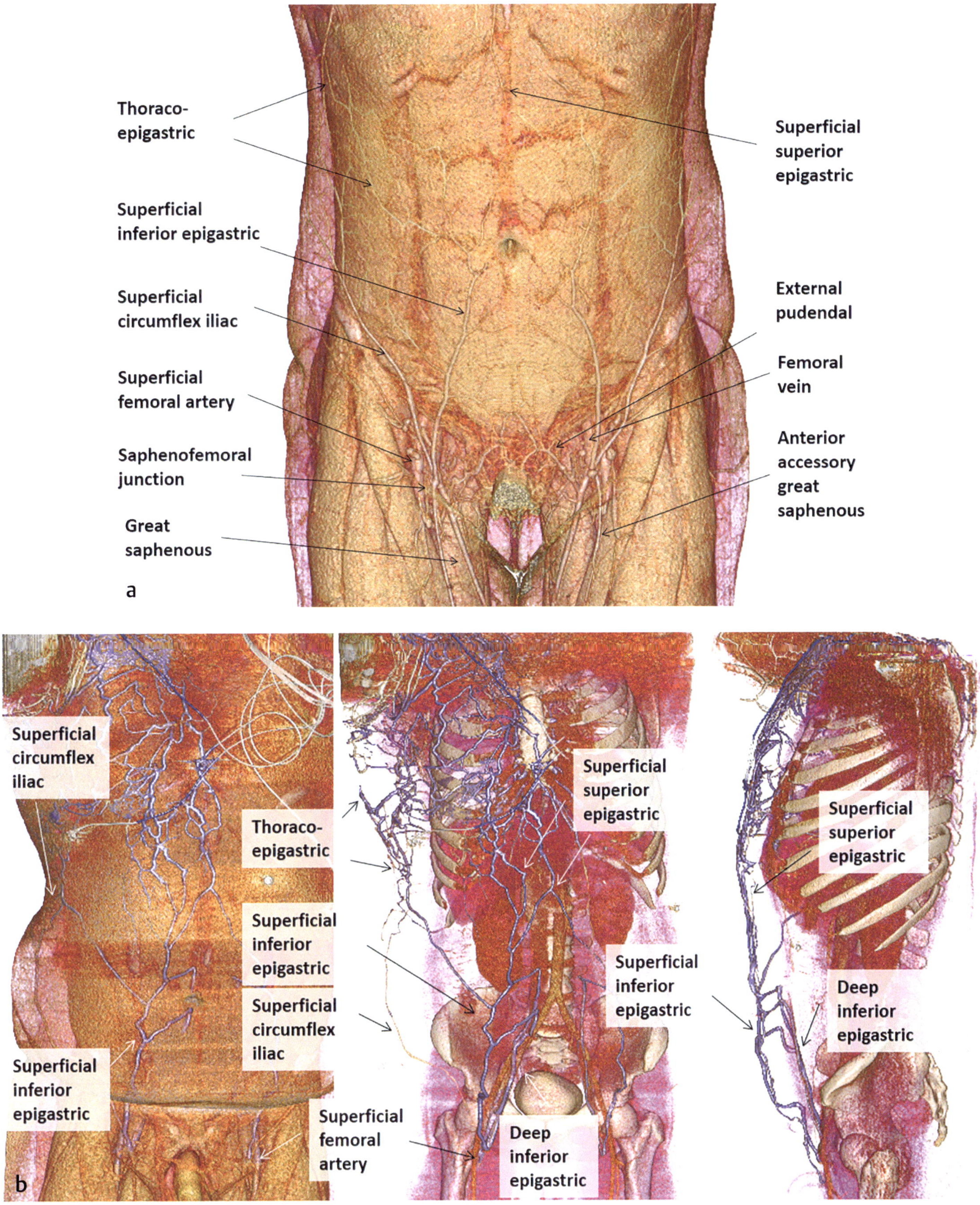

Fig. 1.33 (a) Classic pattern of the superficial veins of the abdominal wall. Veins are accompanied by arteries (not shown). The great saphenous vein (GSV) ascends posteromedial to the knee and medial to the thigh to drain at the confluence of the superficial inguinal veins, also known saphenofemoral junction (SFJ) approximately 3 to 4 cm inferior and lateral to the pubic tubercle. The confluence of superficial inguinal veins is formed by the GSV, the superficial circumflex iliac vein, superficial epigastric vein, and the external pudendal vein. The GSV also receives blood from the anterior and posterior accessory great saphenous veins. **(b)** Extensive collateral pathways of superficial thoracic and abdominal wall venous system developed between the right subclavian vein and the femoral veins after occlusion of the superior vena cava.

◆ Pelvic Wall Arteries

The external iliac artery has a superficial course along the medial aspect of the psoas muscle. It passes under the iliopubic tract and inguinal ligament to become the femoral artery. The external iliac artery gives off to the deep inferior epigastric and deep circumflex arteries which supply lower abdominal wall muscles (▶ Fig. 1.34). In the inguinal area, the inferior epigastric artery gives origin to the cremasteric artery and a pubic branch. In their upward course, the inferior epigastric arteries (and veins) travel between the transversalis fascia and the peritoneum. At the arcuate line, the artery enters the posterior rectus sheath. The cremasteric artery enters the deep inguinal ring along with the spermatic cord (or the round ligament in female).

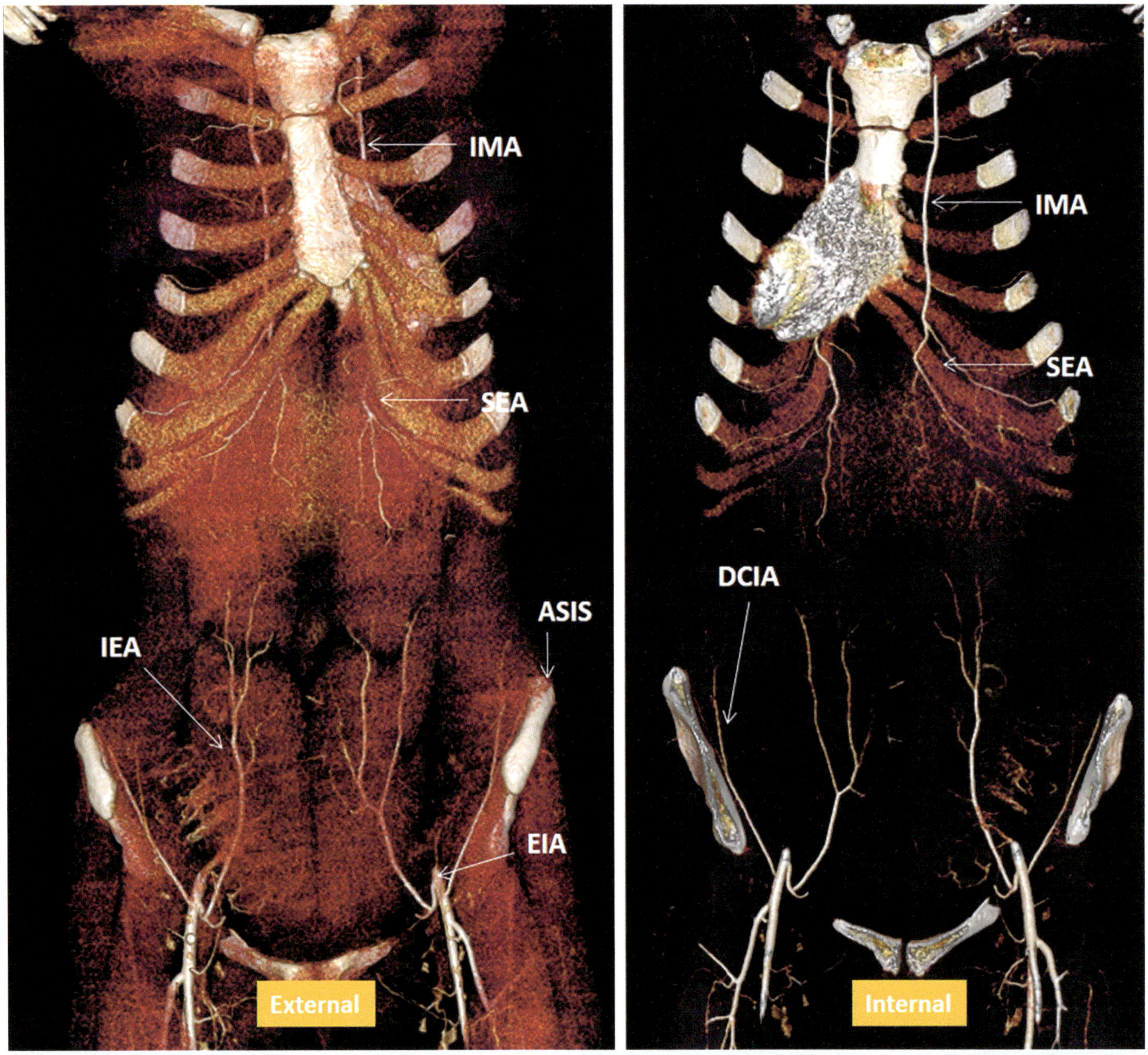

Fig. 1.34 Volume-rendered computed tomography (CT) angiography showing external and internal views of the anterior abdominal wall. Deep vessels including the superior epigastric and inferior epigastric arteries and their branching pattern are shown. The superior epigastric artery (SEA) originates from the internal mammary artery (IMA) at the level of the seventh rib. The inferior epigastric artery (IEA) originates from the external iliac artery (EIA). Communication between the superior and inferior epigastric artery perforators occurs within rectus abdominis muscle above the umbilicus. The IEA usually divides into two branches above the arcuate line. DCIA, deep circumflex iliac artery.

The deep circumflex artery passes deep to the iliopubic tract and then between the internal oblique and transverse abdominal. Near the anterior superior iliac spine, the deep circumflex artery gives origin to an ascending branch to supply the anterior abdominal wall (▶Fig. 1.35). Staples placed here laterally during laparoscopic herniorrhaphy may result in significant hematoma. The deep circumflex iliac artery anastomoses with the iliolumbar arteries.

The internal iliac artery divides into the anterior and posterior trunks. The posterior trunk branches include the iliolumbar, lateral sacral, and superior gluteal arteries. The anterior trunk branches include the obturator, umbilical, inferior gluteal, inferior vesical, uterine, vaginal, middle rectal, and internal pudendal (▶Fig. 1.36). The obturator artery usually originates from the main or posterior branch of the internal iliac artery. In some cases, it may arise from the inferior epigastric or external iliac arteries.

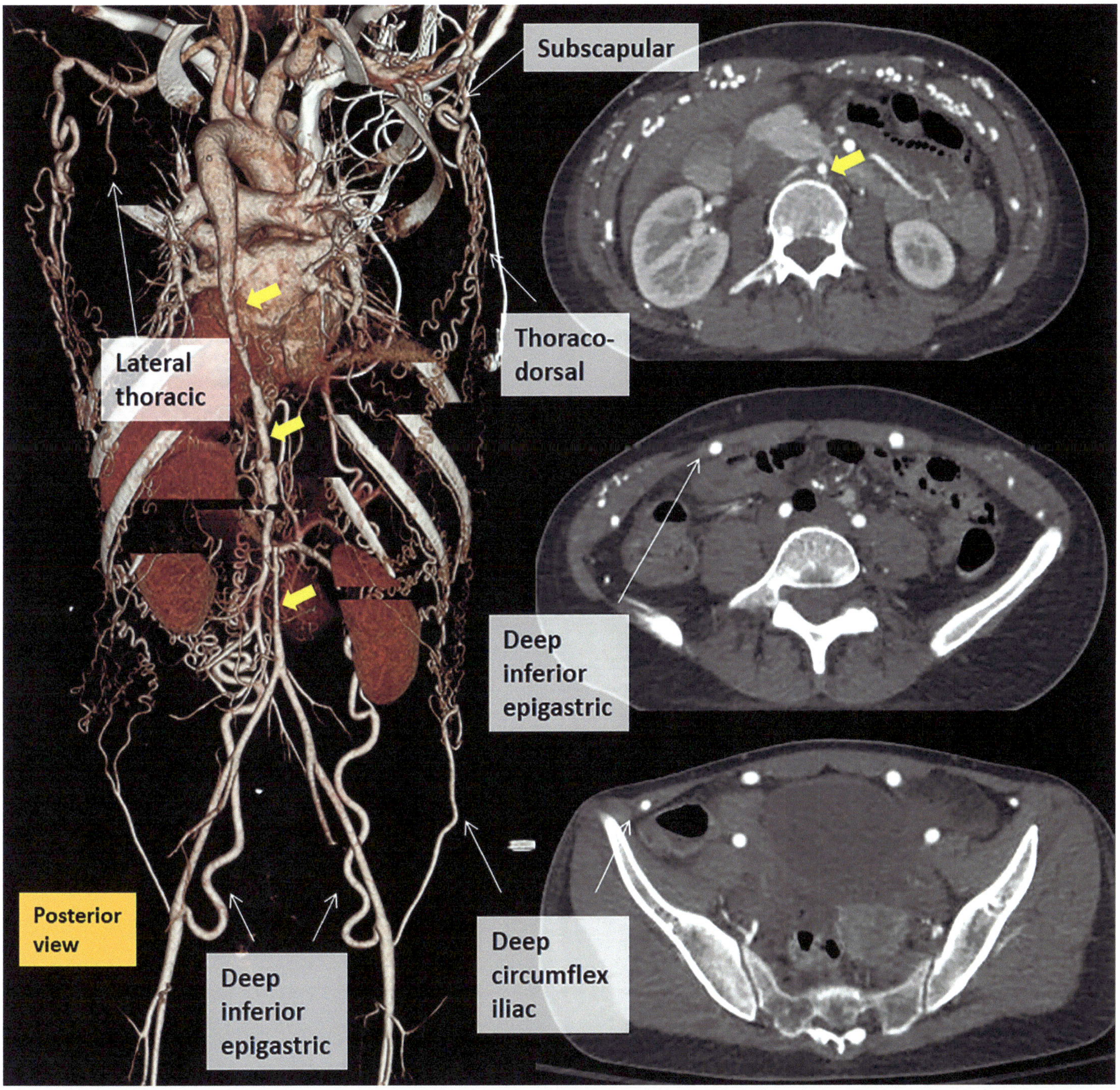

Fig. 1.35 Extensive collateral arterial pathway in the thoracic and abdominal walls due to chronic stenosis of the thoracoabdominal aorta (*yellow arrows*) in a patient with Takayasu's arteritis.

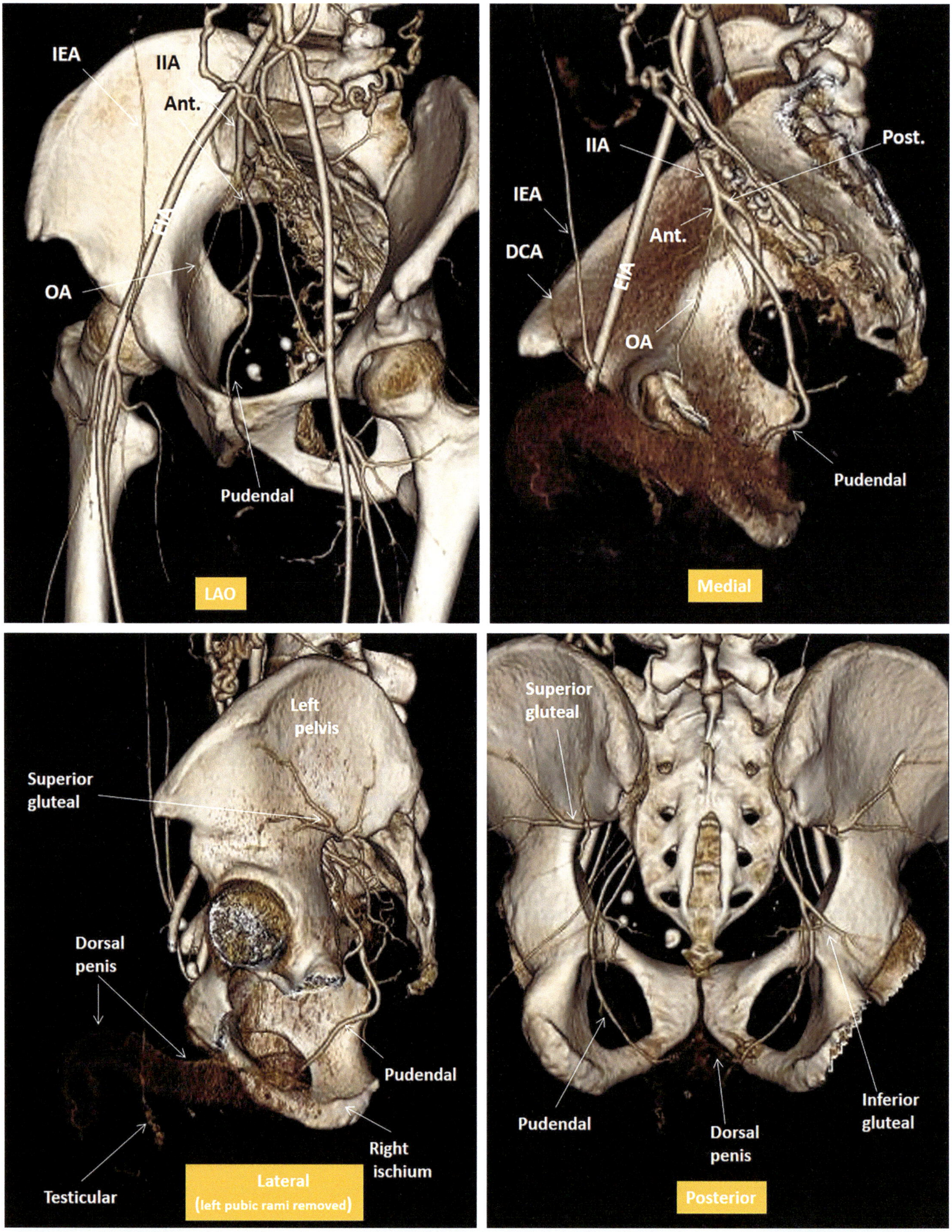

Fig. 1.36 Arterial branches to the pelvic floor. Arterial supply to the pelvic floor is from branches of the anterior division of the internal iliac artery. The levator ani muscles are supplied by branches of the inferior gluteal artery, inferior vesical artery, and pudendal artery. The pudendal artery branches include dorsal and deep arteries of penis (clitoris in female), perineal artery, scrotal (labial in female), and inferior rectal. Ant., anterior trunk of IIA; DCA, deep circumflex artery; EIA, external iliac artery; IEA, inferior epigastric artery; IIA, internal iliac artery; post., posterior trunk of IIA.

The pudendal artery is continuation of the posterior branch. It exits the pelvis by passing through the lesser sciatic foramen (between the sacrospinous and sacrotuberous ligaments) and then in the pudendal canal before giving off branches to supply muscles of the pelvic floor, external genital, perineal, and anorectal region (▶ Fig. 1.37). Pudendal artery branches include the dorsal and deep arteries of penis (clitoris in female), perineal artery, scrotal (labial in female), and inferior rectal.[36]

◆ Inguinal Region

Inguinal Canal

The inguinal canal is a tunnel formed by the aponeurosis of the three anterior abdominal wall muscles. Its inner and outer openings are called deep and superficial rings, respectively. To understand the anatomy of the inguinal canal, it is worth describing several anatomic landmarks.

- ***Inguinal ligament***: It is formed by the aponeurosis of the external oblique. It extends between the anterior superior iliac spine and the pubic tubercle.[65] At the pubic tubercle, the ligament fibers fan out to become the lacunar ligament. The spermatic cord passes superior to the inguinal ligament and the femoral vessels inferior to it (▶ Fig. 1.38).
- ***External inguinal ring***: It is formed by the fibers of the external oblique aponeurosis.
- ***Conjoined tendon***: Also known as inguinal falx, it is a tendinous structure formed by fusion of the internal oblique and transverse abdominal muscles aponeurosis that reaches the pubic tubercle. This tendon exists in only 3 to 6% of patients.
- ***Myopectineal orifice* (Fruchaud orifice)**: The anatomic orifice connecting the false pelvis to the lower extremity measuring

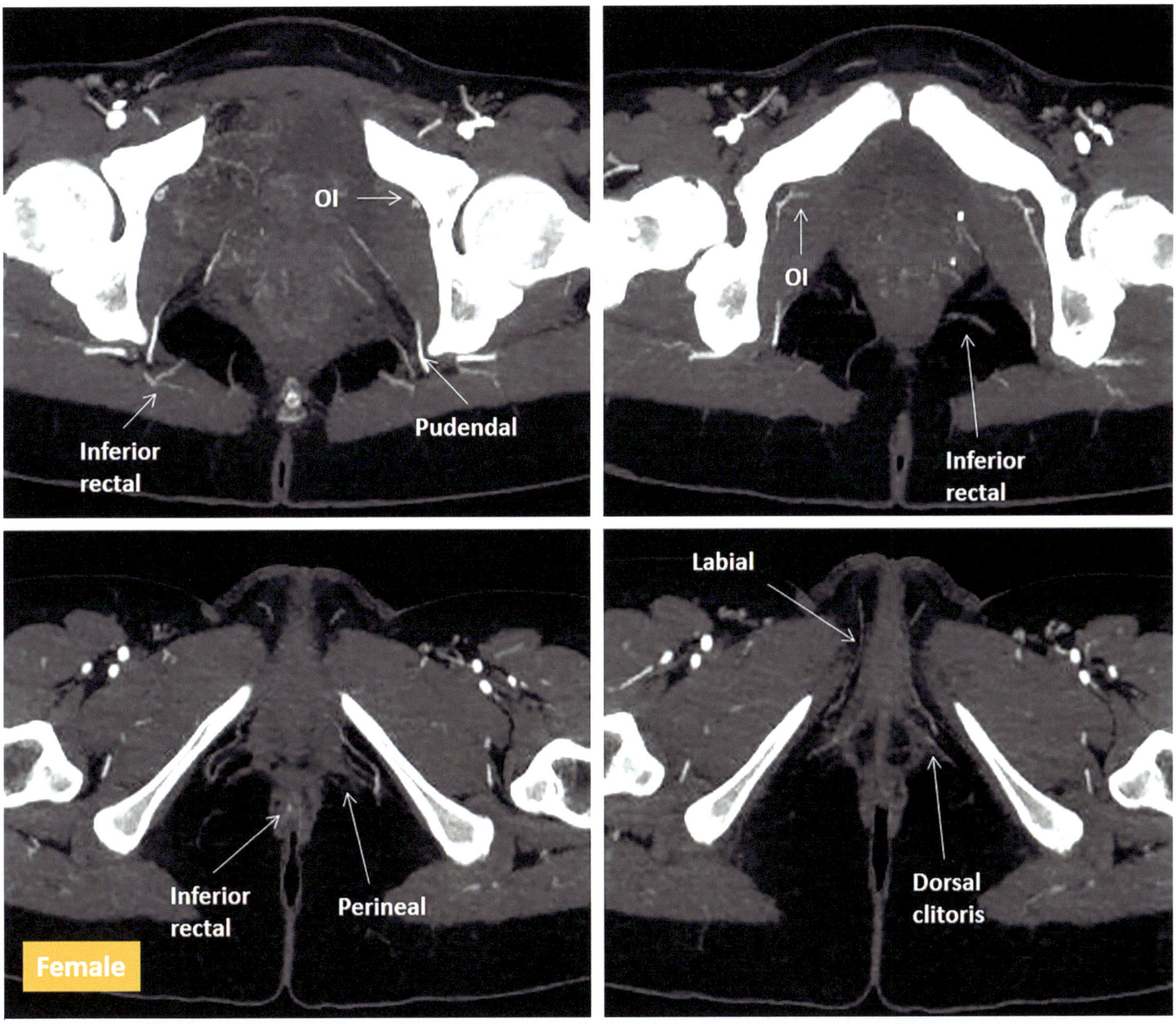

Fig. 1.37 Pudendal artery branches to the pelvic floor in female and male patients as shown by axial computed tomography (CT) scan (from top to bottom). The pudendal artery branches include the dorsal and deep arteries of penis (clitoris in female), perineal artery, scrotal (labial in female), and inferior rectal. OI, obturator internus artery.

(Continued)

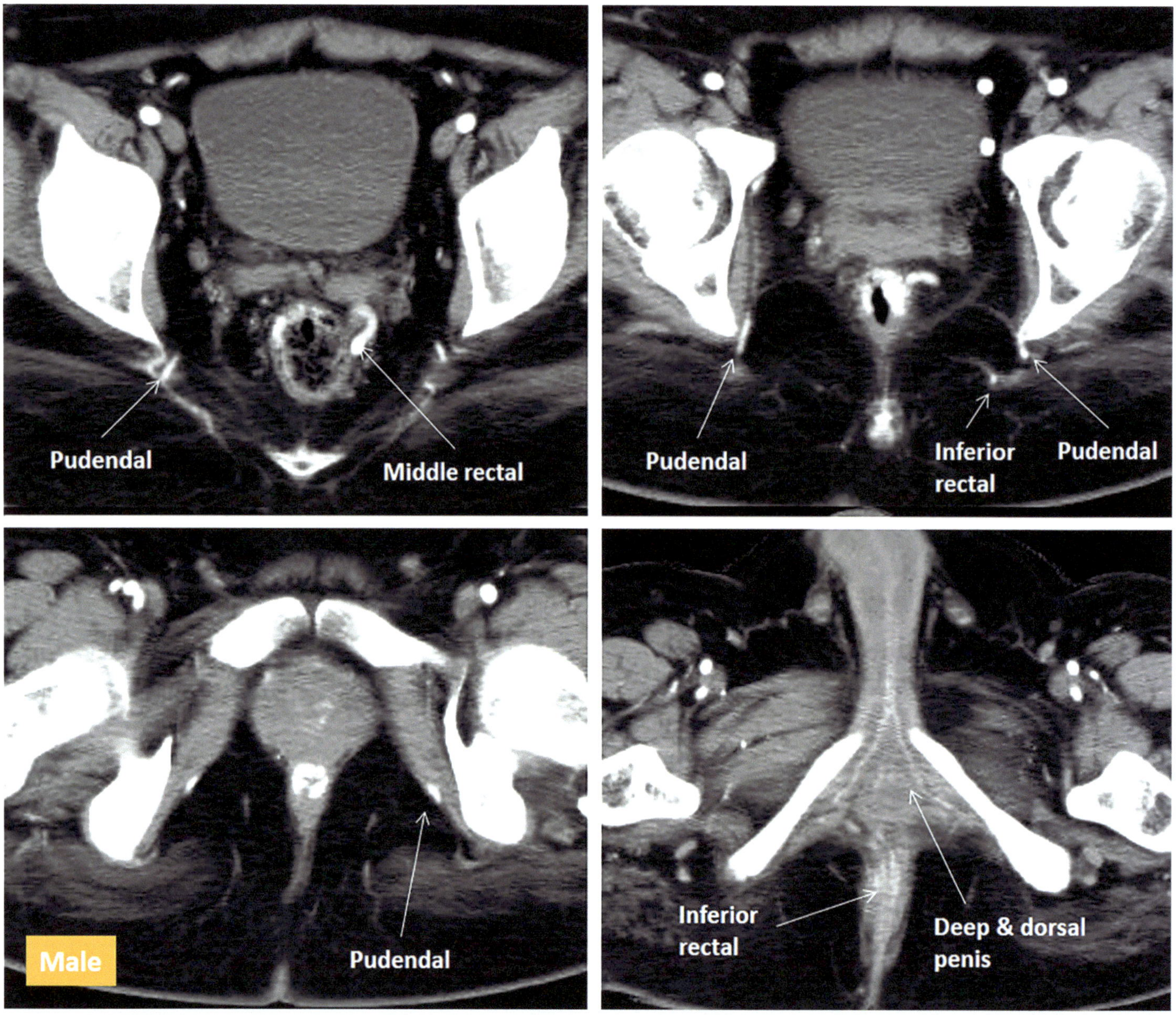

Fig. 1.37 *(Continued)*

approximately 8 cm in width and 6.5 cm in height.[66,67] It is bounded by the arching fiber of the internal oblique and transverse muscles superiorly, rectus abdominis muscle medially, anterior borders of the iliac bone (Cooper's ligament) inferiorly, and iliopsoas and iliopectineal arch laterally. The Cooper's ligament is the thickened periosteum of the superior pubic ramus.

The myopectineal orifice is divided by the inguinal ligament into two compartments, which are potential sites of indirect inguinal, direct inguinal, femoral, and some interstitial hernias. The superior compartment is subdivided by the deep inferior epigastric vessels into two triangles (▶Fig. 1.39). The medial triangle, also known as the triangle of Hesselbach, is bordered by the rectus abdominis muscle medially and the inguinal ligament inferiorly (▶Fig. 1.40). The floor of the triangle of Hesselbach is supported only by the transversalis fascia. It is the site of direct inguinal herniation. The lateral triangle is defined by the deep epigastric vessels medially, the inguinal ligament inferiorly, and a line drawn from the lateral end of the inguinal ligament to the intersection of the deep epigastric vessels with the arching fibers of the internal oblique. The internal inguinal ring is located in this triangle where the spermatic cord structures pass. The inferior compartment is known as the femoral triangle. It is bordered the inguinal ligament and iliopubic tract superiorly, iliopsoas muscle laterally, and pectineus muscle medially.

Tissue Layers of the Inguinal Canal Region

There are several important anatomic layers at the level of inguinal canal[65,68] (▶Fig. 1.41).

The first anatomic layer is the Scarpa's fascia. The Scarpa's fascia is a dense membranous layer of the superficial fascia within the subcutaneous tissue.

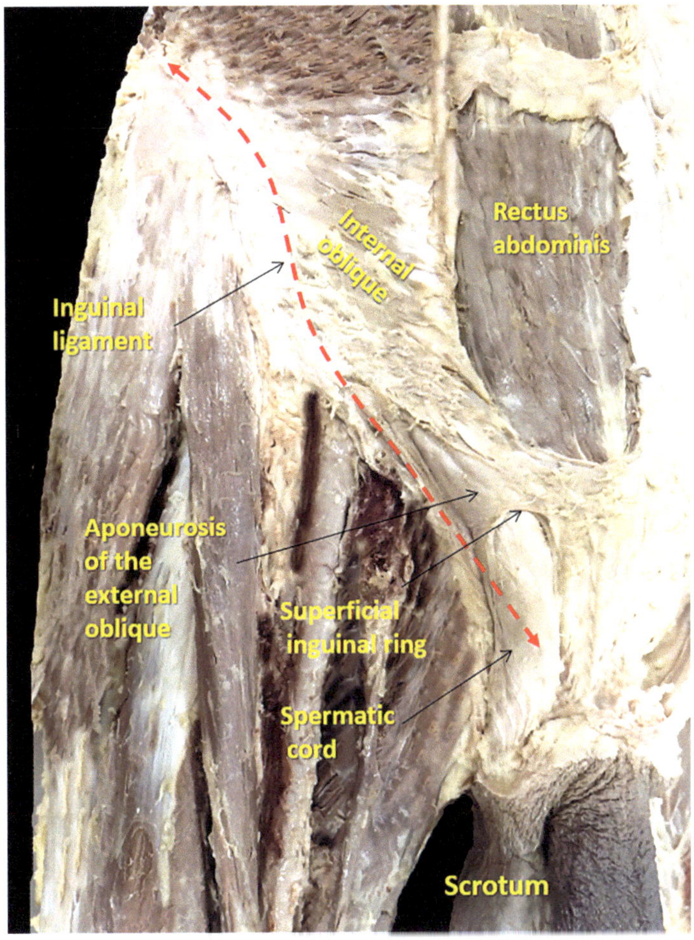

Fig. 1.38 External inguinal ring. The external oblique aponeurosis is the most superficial structure and forms the external inguinal ring. It continues over the spermatic cord as the external spermatic cord fascia.

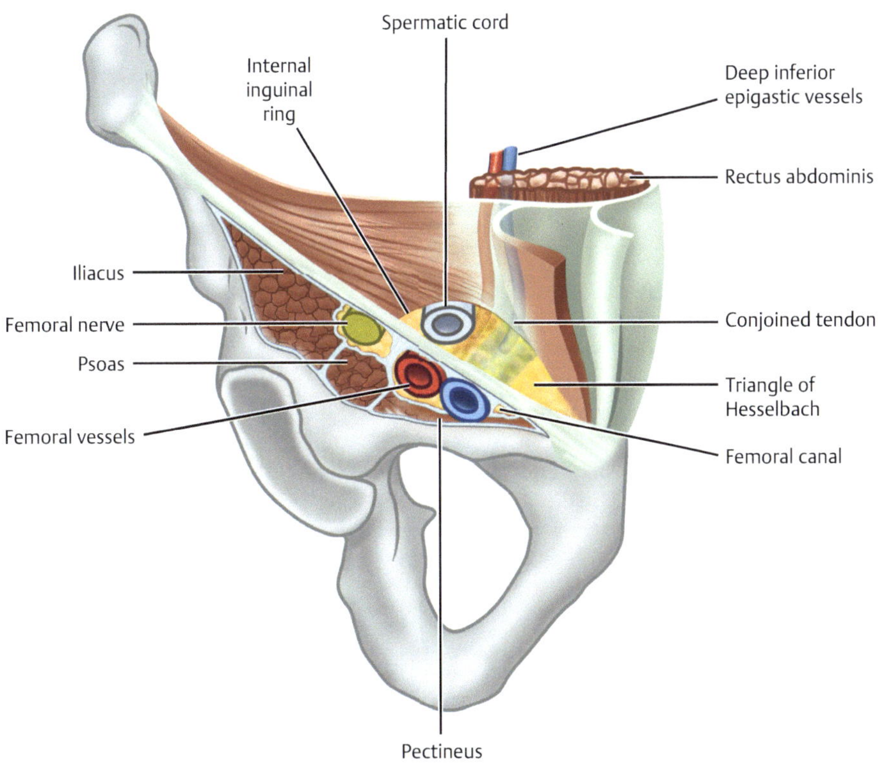

Fig. 1.39 Myopectineal orifice demonstrating medial, lateral, and femoral triangles. It is bounded by the arching fiber of the internal oblique and transverse muscles superiorly, rectus abdominis muscle medially, anterior borders of the superior pubic ramus (Cooper's ligament) inferiorly, and iliopsoas laterally. The myopectineal orifice is divided by the inguinal ligament into two compartments. The superior compartment is subdivided by the deep inferior epigastric vessels into the triangle of Hesselbach medially and the internal inguinal ring laterally. The inferior compartment is known as the femoral triangle.[65]

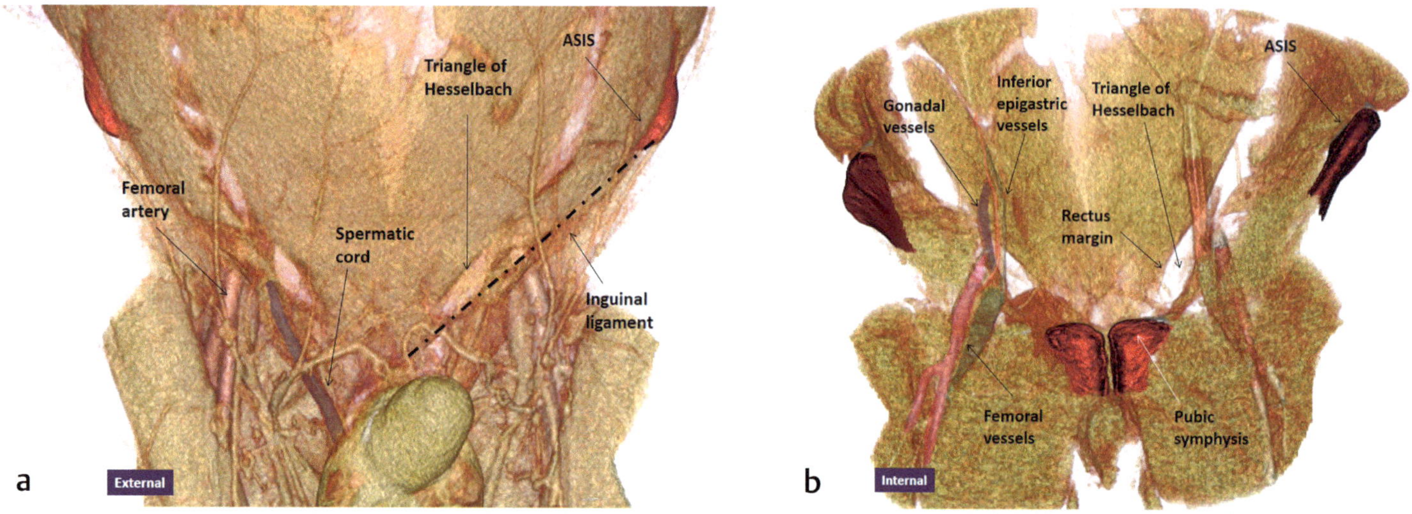

Fig. 1.40 Computed tomography (CT) of the inguinal region. **(a)** External view. **(b)** Internal view. The triangle of Hesselbach is bordered by the rectus abdominis muscle medially, inferior epigastric vessels laterally, and the inguinal ligament inferiorly. The floor of the triangle of Hesselbach is supported only by the transversalis fascia, the site of direct inguinal herniation. The inguinal ligament extends between the anterior superior iliac spine (ASIS) and the pubic tubercle.

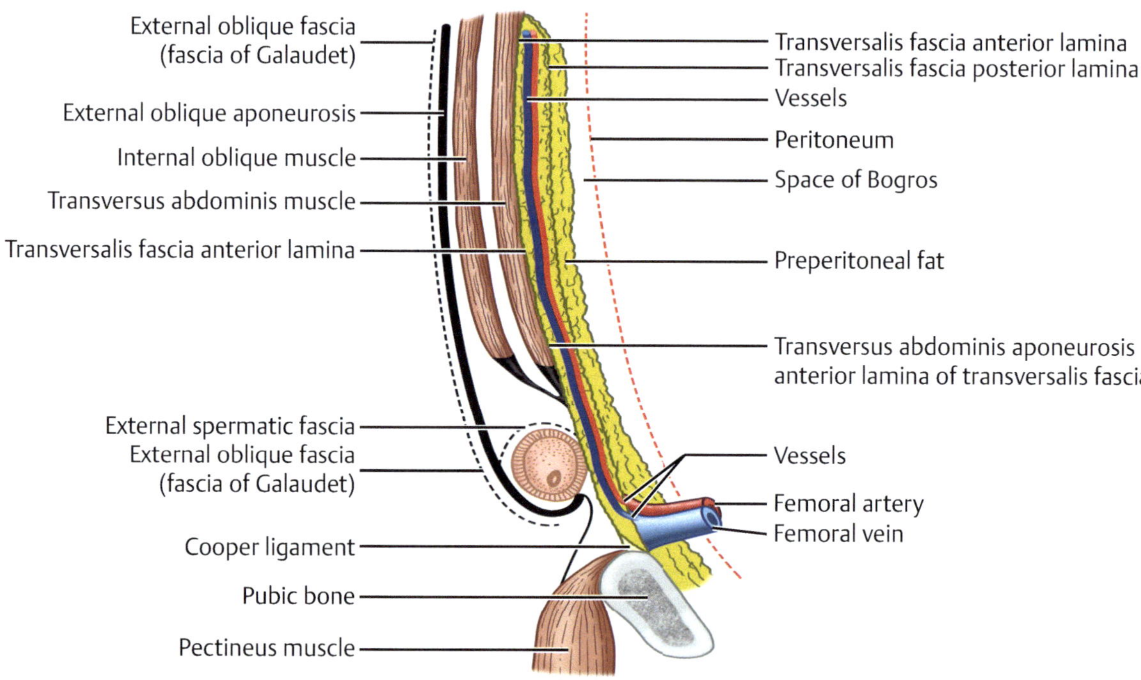

Fig. 1.41 Sagittal cross section of the lower abdominal wall at the inguinal region showing anatomic layers at the level of inguinal canal. 1, external oblique fascia (fascia of Gallaudet); 2, external oblique aponeurosis; 3, internal oblique muscle; 4, transversus abdominis muscle; 5, transversalis fascia anterior; 6, external spermatic fascia; 7, Cooper's ligament; 8, pubic bone; 9, pectineus muscle; 10, transversalis fascia; 11, transversalis fascia posterior lamina; 12, vessels; 13, peritoneum; 14, space of Bogros; 15, preperitoneal fat; 16, transversus abdominis aponeurosis and anterior lamina of transversalis fascia; 17, femoral artery; and 18, femoral vein. (Modified from Skandalakis et al.[68])

The next layer is the aponeurosis of the external oblique. The external oblique aponeurosis gives rise to the external (superficial) inguinal ring and inguinal ligament. It continues over the spermatic cord as the external spermatic cord fascia (▶ Fig. 1.38, ▶ Fig. 1.42). In contrast to the external oblique, the internal oblique and transverse muscles are composed of both a muscular portion and their aponeurosis within the region of the myopectineal orifice. These muscles support the lateral wall of the inguinal floor. Their muscle fibers travel medially and arch cephalad to the inguinal ligament, thereby producing a potentially weak area, variable in size in different individuals (▶ Fig. 1.42, ▶ Fig. 1.43, ▶ Fig. 1.44). Muscle fibers extending from these muscles into the inguinal canal form the cremasteric muscle (▶ Fig. 1.42, ▶ Fig. 1.43).

The lateral posterior floor of the inguinal canal is relatively strong as it is supported by all three muscles of the lateral abdominal wall. On the other hand, the medial inguinal floor is weak as it is covered only by the aponeurosis of the external oblique externally (external inguinal ring) and the transversalis fascia

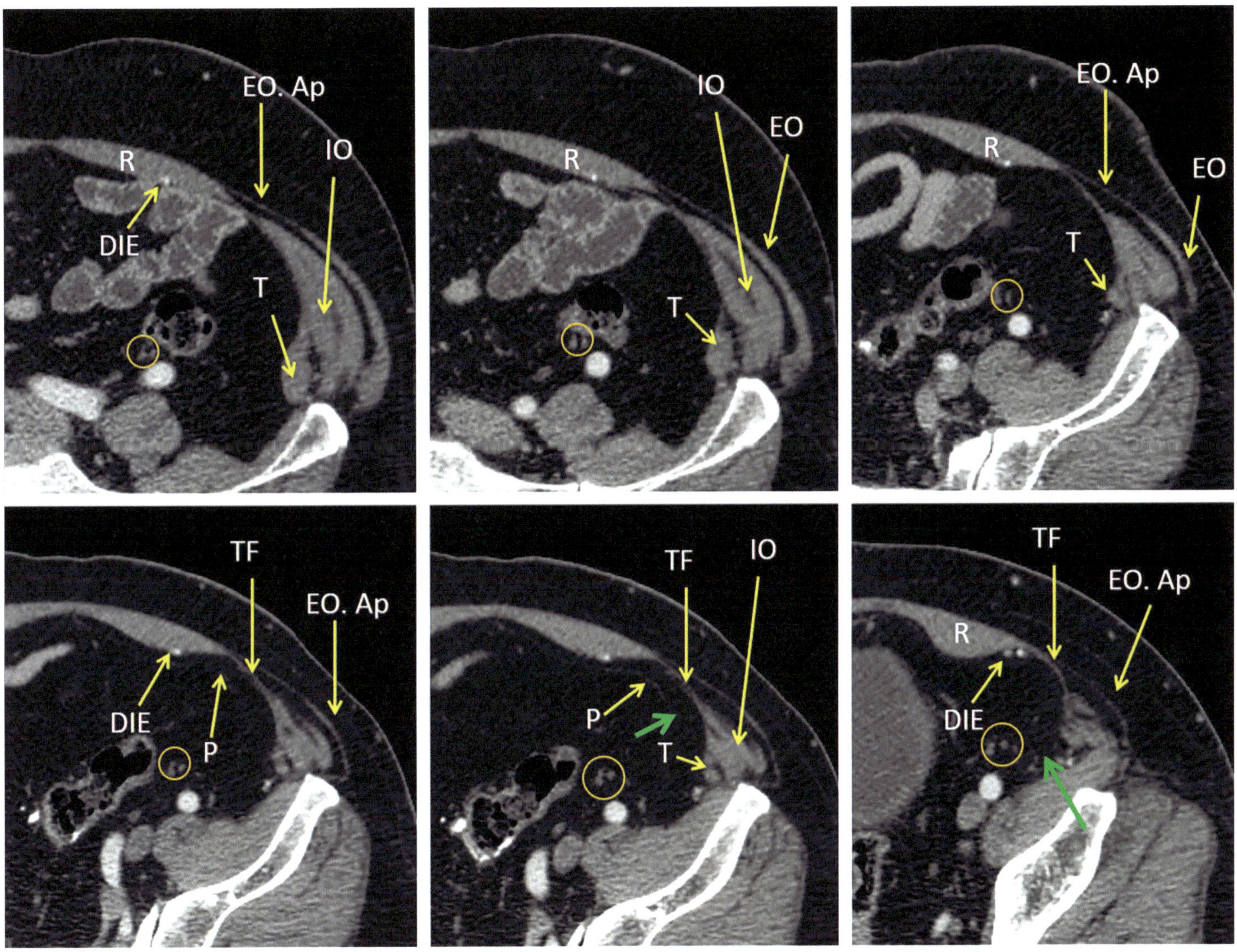

Fig. 1.42 Axial computed tomography (CT) images showing anatomic structures of the left lower abdomen and the inguinal region. The external oblique aponeurosis (EO. Ap.) is the most superficial structure at the external inguinal ring. Above the inguinal ligament and anterior to the deep inferior epigastric (DIE) vessels, the transversalis fascia (TF) and external oblique aponeurosis are the only structures covering the abdominal wall. This portion is called the triangle of Hesselbach (*red arrow*). The internal oblique (IO) and transverse (T) muscles travel medially and arch cephalad to the inguinal ligament and their aponeurosis join together to form the conjoined tendon (Con. T.). Muscle fibers extending from these muscles into the inguinal canal form the cremasteric muscle (Cr. m). The preperitoneal space between the fascia transversalis and the peritoneum above the inguinal canal is called the space of Bogros (*green arrow*). The vas deferens and round ligament of the uterus pass in this space before entering the inguinal canal. Anatomic course of the gonadal vessels is shown in the *circles*. EO, external oblique; IC, inguinal canal; IO, internal oblique; IIR, internal inguinal ring; P, peritoneum; Pec, pectineus; R, rectus abdominis; T, transverse muscle.

(*Continued*)

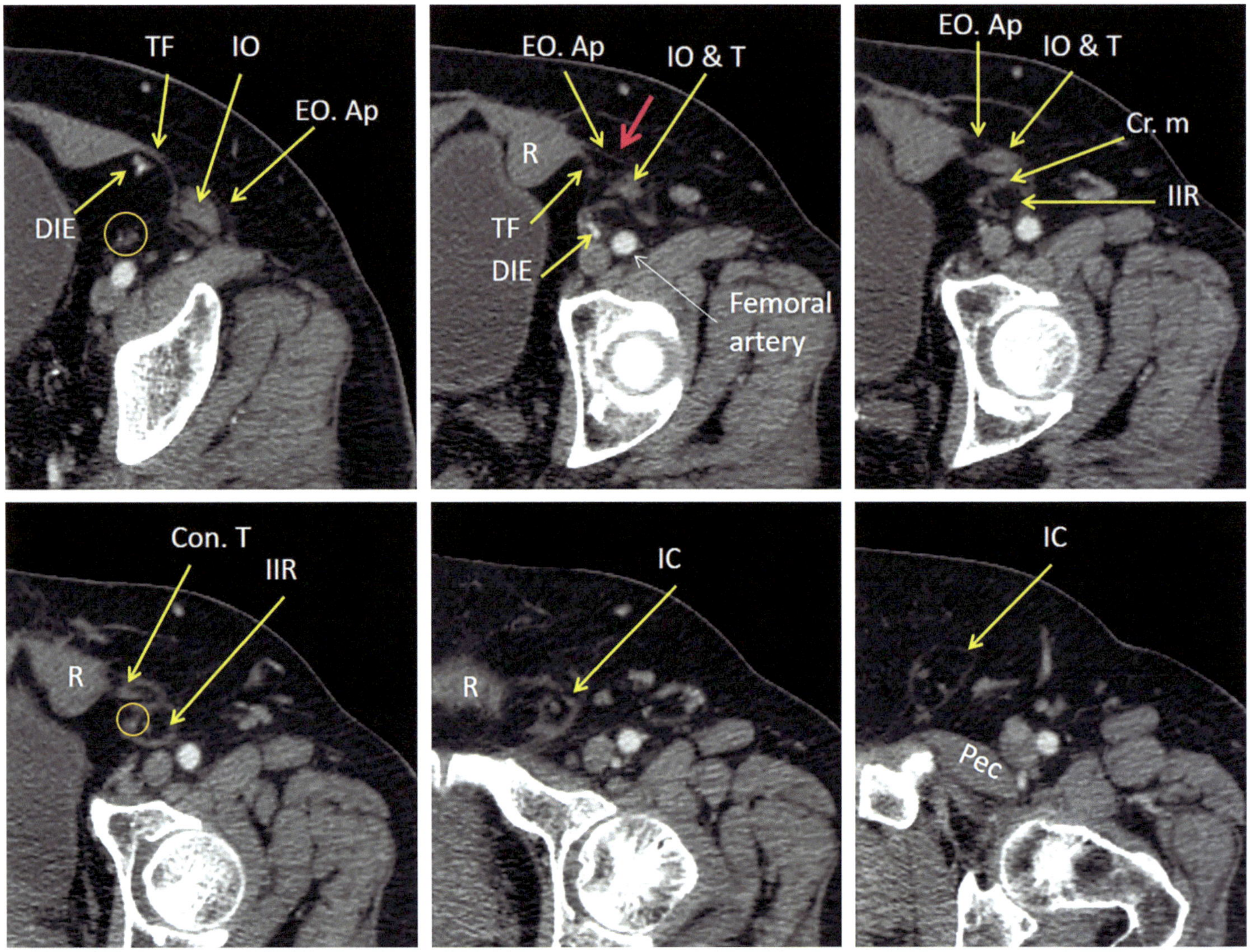

Fig. 1.42 (*Continued*)

internally. The transversalis fascia forms the posterior wall of the inguinal canal and continues outward around the spermatic cord to form the internal spermatic fascia. The transversalis fascia also participates to form the iliopubic tract, lacunar ligament, anterior and posterior crura of the internal ring, and the anterior portion of the femoral sheath.[65]

The final and an important tissue layer or space of the inguinal region is the preperitoneal space, located between the transversalis fascia and the peritoneum, also known as the retroinguinal space of Bogros[68,69] (▶Fig. 1.41, ▶Fig. 1.42). This space is occupied by fat and connective tissue and appears to be continuous superiorly with the posterior pararenal space and anteromedially with the space of Retzius, with a partial border between the two formed by the inferior epigastric vessels. The preperitoneal space is commonly used increasingly during both laparoscopic and open inguinal herniorrhaphy procedures. The vas deferens and round ligament of the uterus pass in this space before entering the inguinal canal. The vas deferens meets the testicular vessels at the internal inguinal ring to participate in formation of the spermatic cord. The testicular arteries usually arise from the anterior aspect of the abdominal aorta just inferior to the renal arteries at the level of the second lumbar vertebra. The left testicular vein drains into the left renal vein and the right testicular vein drains into the inferior vena cava. The relationship of the testicular arteries and veins in the abdomen is variable.[70]

Femoral Canal

The femoral sheath (crural sheath) is formed by continuation of the deep parietal fascia (transversalis and iliac fasciae) around the femoral vessels underneath the inguinal ligament. It is

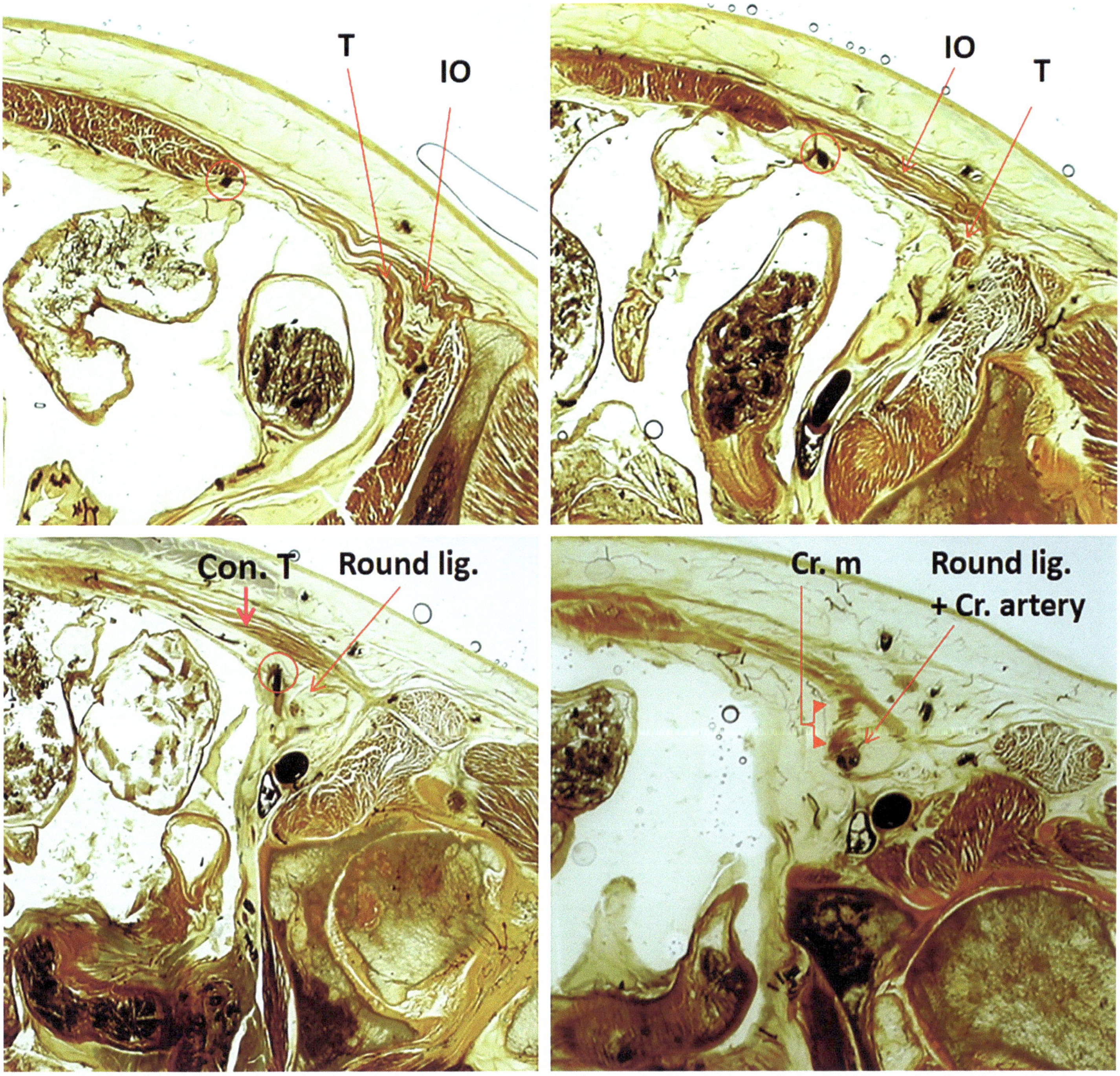

Fig. 1.43 Inguinal canal is shown by transaxial cuts in a female cadaver. The internal oblique (IO) and transverse (T) muscles travel medially and arch cephalad to the inguinal ligament and their aponeurosis join together to form the conjoined tendon (Con. T.). The cremastric muscle (Cr. m) is formed by fibers arising from the IO and T muscles and extends into the inguinal canal. The cremastric (Cr.) artery arises from the deep inferior epigastric artery (*red circles*) and joins the round ligament (or spermatic cord in male) within the inguinal canal.

1

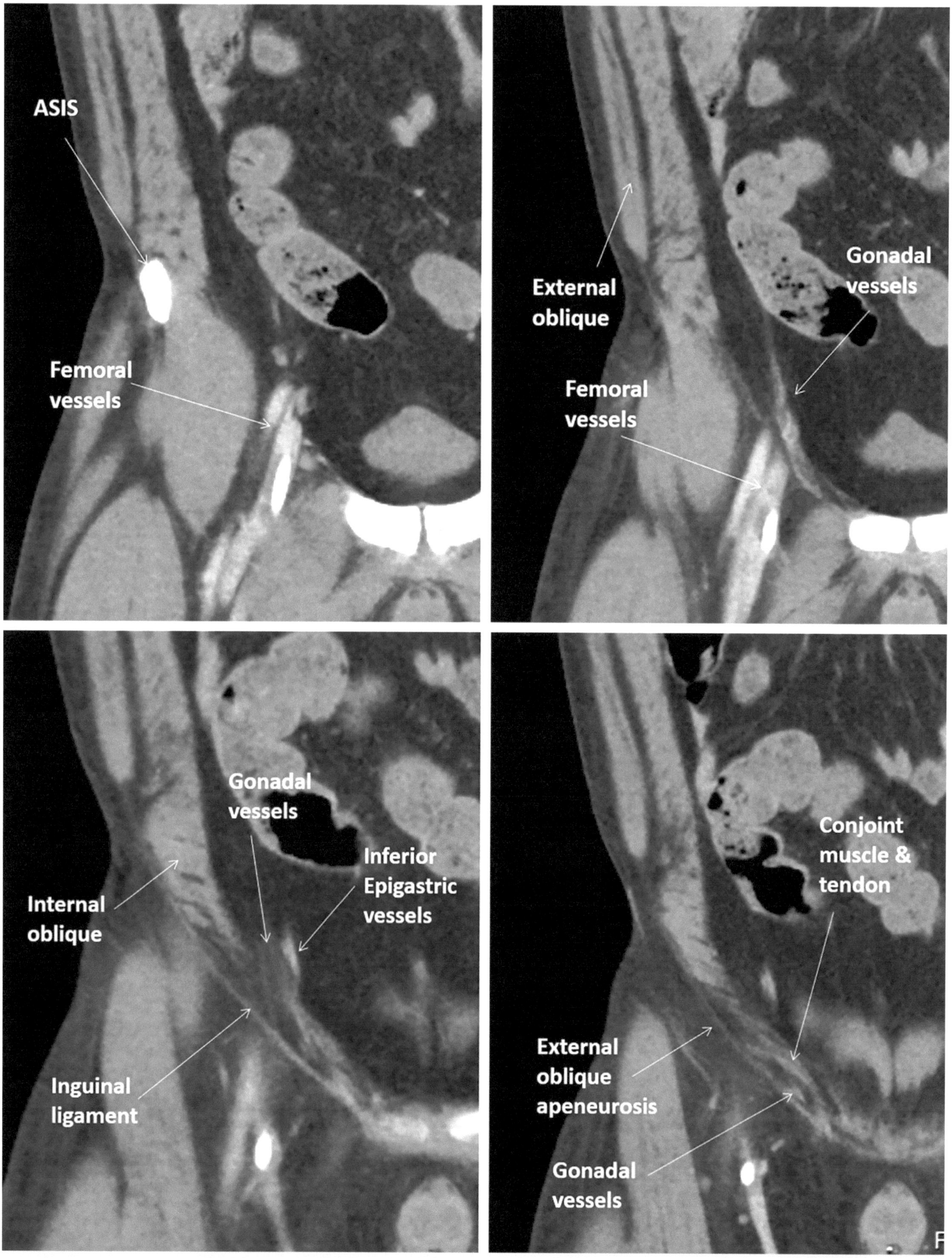

Fig. 1.44 Coronal computed tomography (CT) images, from posterior to anterior showing anatomic structures of the right lower abdomen and the inguinal region. The external oblique aponeurosis is the most superficial structure and forms the inguinal ligament and external inguinal ring. The internal oblique and transverse muscles travel medially and arch cephalad to the inguinal ligament and their aponeurosis join together to form the conjoined tendon. The preperitoneal space, between the transversalis fascia and the peritoneum above the inguinal canal, is called the space of Bogros. ASIS, anterior superior iliac spine.

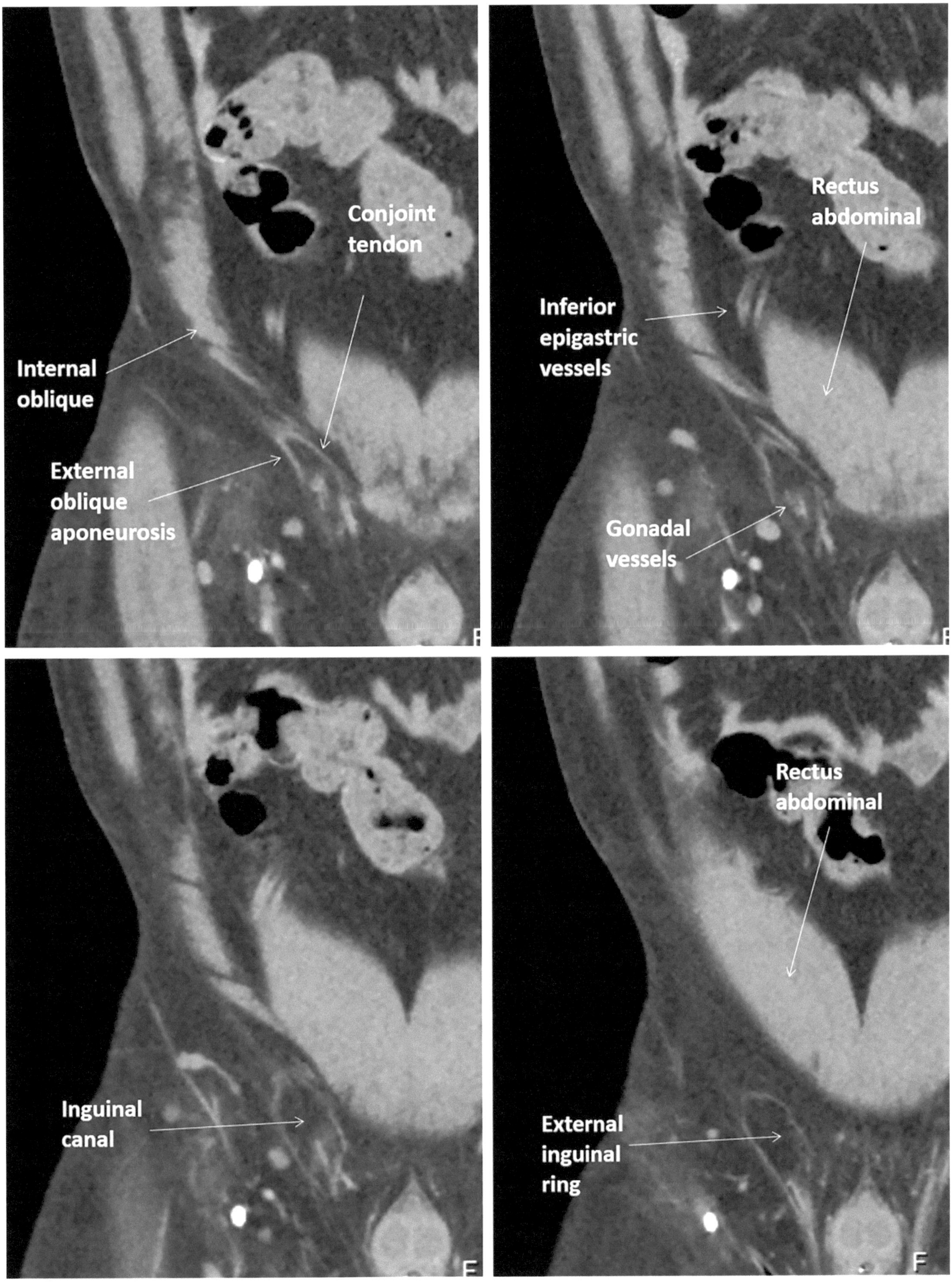

Fig. 1.44 *(Continued)*

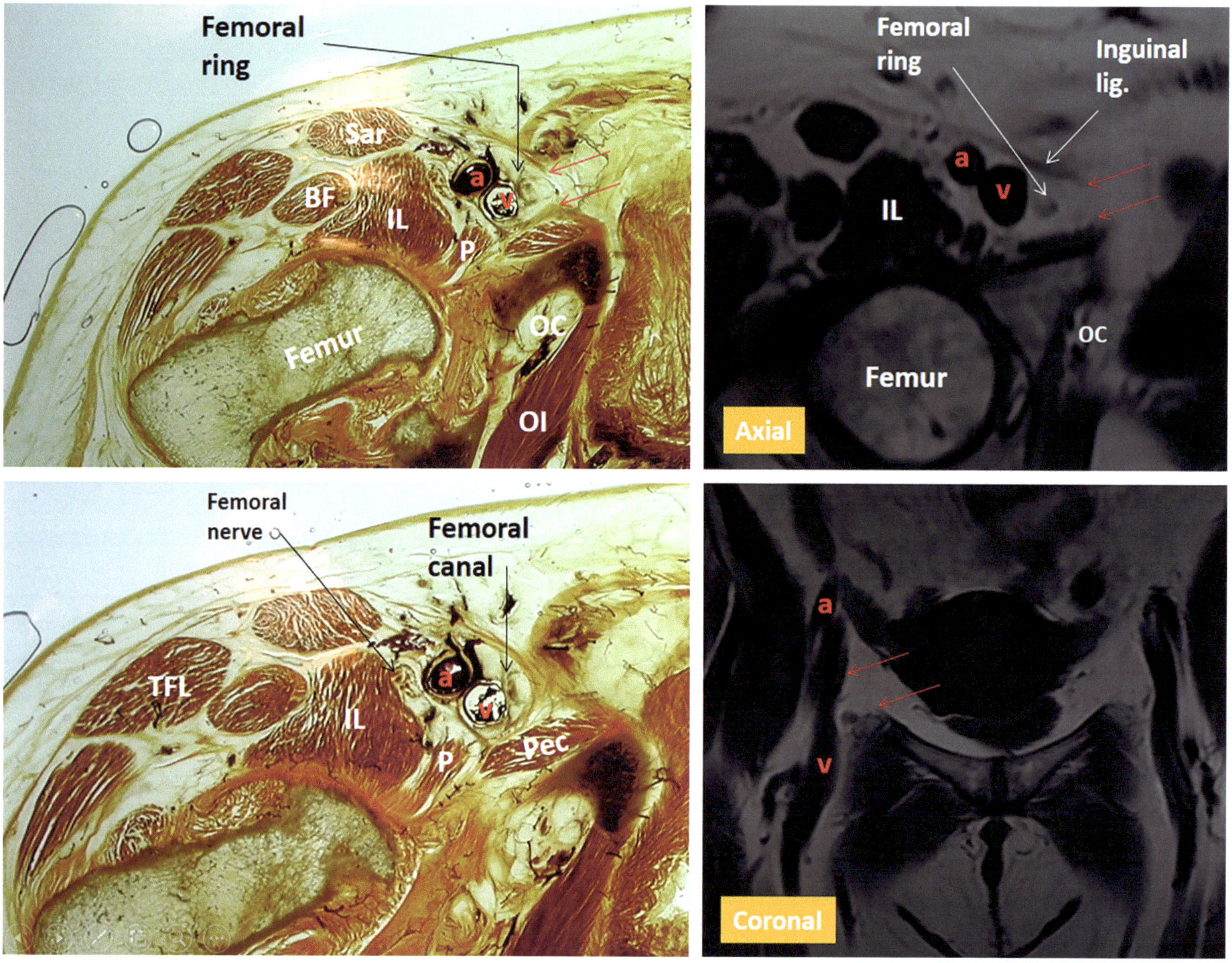

Fig. 1.45 Femoral canal. The femoral sheath (crural sheath) is formed by continuation of the deep parietal fascia (transversalis and iliac fascia) around the femoral vessels underneath the inguinal ligament. It is partitioned into three compartments for the femoral artery (a), femoral vein (v), and femoral canal. The femoral canal is medially located containing fat. The opening of the canal (*double red arrows*) is called the femoral ring. BF, biceps femoris; IL, iliacus; OC, obturator canal; OI, obturator internus; P, psoas; Pec, pectineus; Sar, sartorius; TFL, tensor fascia lata.

partitioned into three compartments for the femoral artery, femoral vein, and femoral canal. The femoral canal is medially located containing fat and lymph nodes (▶Fig. 1.45). The base of femoral canal is named the femoral ring. The pectineus is located posterior and the inguinal ligament anterior to it. Large femoral ring predisposes to femoral hernia.

◆ Weak Zones of the Abdominopelvic Wall

In the anterior abdominal wall, several weak zones exist that are prone to spontaneous hernia. The incidence of spontaneous abdominal wall hernias is approximately 5%. Approximately 75% of these are inguinal, 5% femoral, and the remaining 20% comprise incisional, umbilical, Spigelian, and other types.[71–74] Umbilical hernias are the most common type of anterior upper ventral hernia. They are usually small, asymptomatic containing omental fat. Most hernias contain omental fat or the entire wall circumference of a loop or segment of the bowel. Richter hernia refers to herniation of only the antimesenteric wall of the bowel. It most frequently occurs in association with femoral hernias but can be seen with umbilical or incisional hernia as well (▶Fig. 1.46).

Spigelian Hernia

The linea semilunaris is a weak level for development of abdominal wall hernia. It is located at the margin of the rectus muscle below the level of the arcuate line (▶Fig. 1.9). At this level, the internal oblique and transverse muscles constitute thin muscle and aponeurotic bands that run parallel, a situation prone to development of interparietal or interstitial herniations. The most common form of interstitial hernias is the

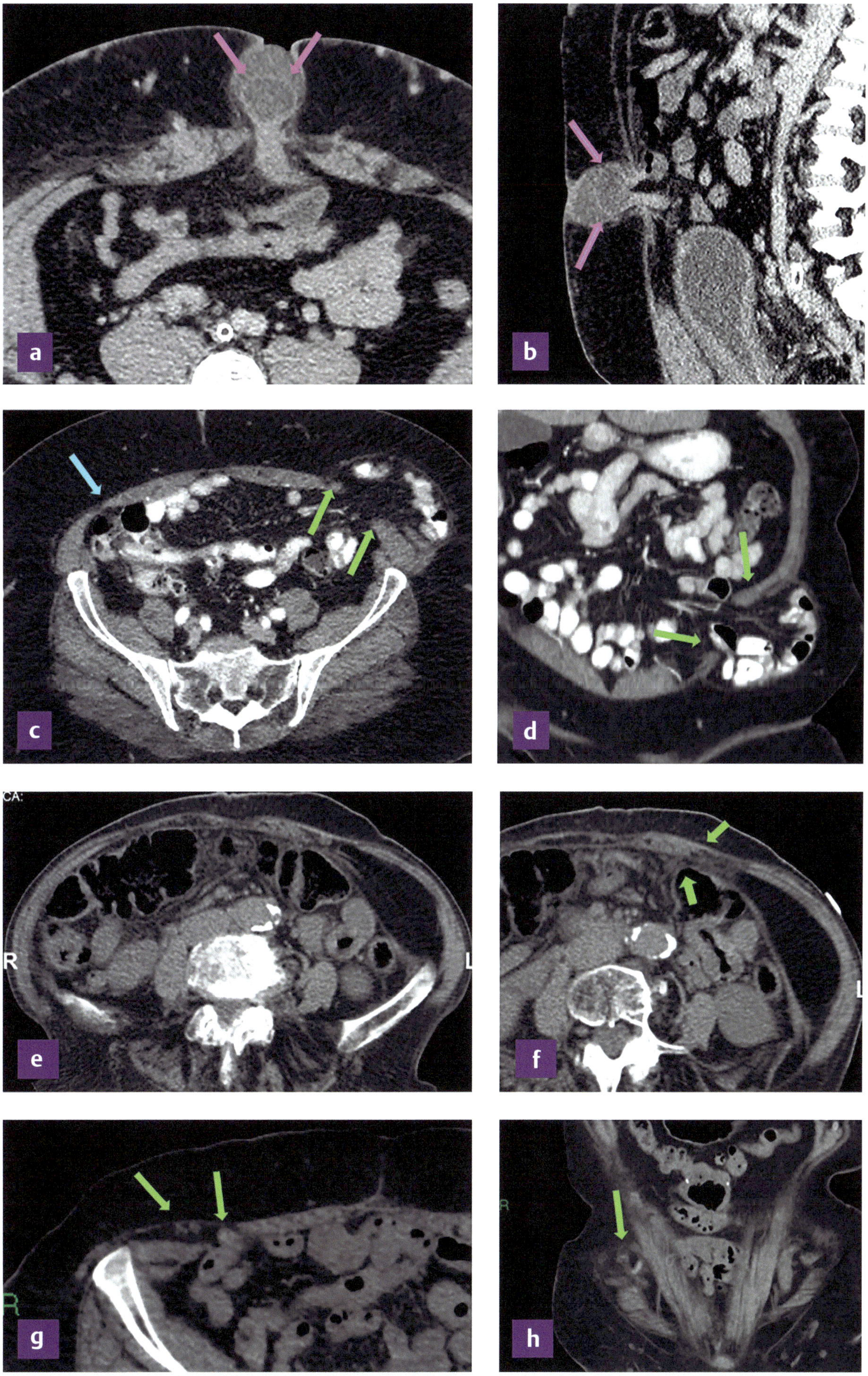

Fig. 1.46 (a) Axial and **(b)** sagittal computed tomography (CT) showing a Richter type of umbilical hernia (*arrows*). Richter hernia refers to herniation of only the antimesenteric wall of the bowel. **(c–f)** Spigelian hernia. **(c)** Axial and **(d)** coronal CT images. Spigelian hernia occurs through the weakening of the linea semilunaris (*blue arrow*) at the margin of the rectus muscle. It penetrates between the muscles of the abdominal wall; therefore, it is a form of interstitial hernia. Spigelian hernias are usually small and may not be detectable clinically. **(e, f)** Elderly woman (different patient) with abdominal pain. Axial CT images showing herniation of fat between transverse and oblique muscles of the left lower abdomen. The opening of the hernia sac is shown by *green arrows* in both cases. **(g, h)** A small Spigelian hernia containing the appendix (*arrows*).

1

Spigelian hernia seen mainly in elderly with muscle wasting. It usually consists of an asymptomatic ascending protrusion of fat tissue and/or intestinal loops[74,75] (▶ Fig. 1.46). Other interstitial hernias are rare and remain between the layers of the abdominal wall musculature or between a muscle aponeurosis and transversalis fascia.

Lumbar Hernia

In the posterolateral (flank or lumbar) abdominal wall two weak zones are particularly predisposed to the formation of lumbar hernia. These are the superior (Grynfelt-Lesshaft) and inferior (Petit) lumbar triangles (▶ Fig. 1.16). The superior lumbar triangle is an inverted triangle bordered superiorly by the 12th rib and the serratus posterior inferior muscle, medially by the erector spinae and quadratus lumborum muscles, and anterolaterally by the external oblique muscle. Its floor is formed by the transversalis fascia and other fasciae. The superior lumbar triangle is superficially covered by the latissimus dorsi aponeurosis.[76–78]

The inferior lumbar triangle is smaller. It is bordered inferiorly by the iliac crest, laterally by the external oblique muscle, and medially by the quadratus lumborum. Its floor is formed by the thoracolumbar fascia and other fasciae. The inferior lumbar triangle is not covered by muscles. The size and location of the inferior triangle is variable.[77] Surface area of the triangle may be as large as 15 cm^2. In some cases, the inferior triangle may reach in the vicinity of the midaxillary line. Lumbar hernia can be congenital or acquired (pressure, surgery, trauma, idiopathic). Hernia of the superior triangle is commonly spontaneous but the inferior lumbar triangle hernias are frequently traumatic (▶ Fig. 1.47). Postsurgical hernias can occur at any level (▶ Fig. 1.47).

Inguinal Hernia

Inguinal hernias are the most common type of abdominal wall hernia and traverse the inguinal canal.[71] In normal inguinal canal, a small amount of fat exists, associated with the gonadal vessels. This normal fat may be abundant in patients with large amount of extraperitoneal fat and should not be mistaken for the hernia of peritoneal or extraperitoneal fat into the inguinal canal (▶ Fig. 1.48). The deep inferior epigastric vessels are an important landmark to localize the inguinal hernia. The deep inguinal ring, which is the entry route for indirect inguinal hernias, lies immediately lateral to the deep inferior epigastric vessels (▶ Fig. 1.49, ▶ Fig. 1.50, ▶ Fig. 1.51). Bladder hernia is a rare form of indirect hernia (▶ Fig. 1.49). This hernia is usually associated with herniation of fatty contents of the prevesical

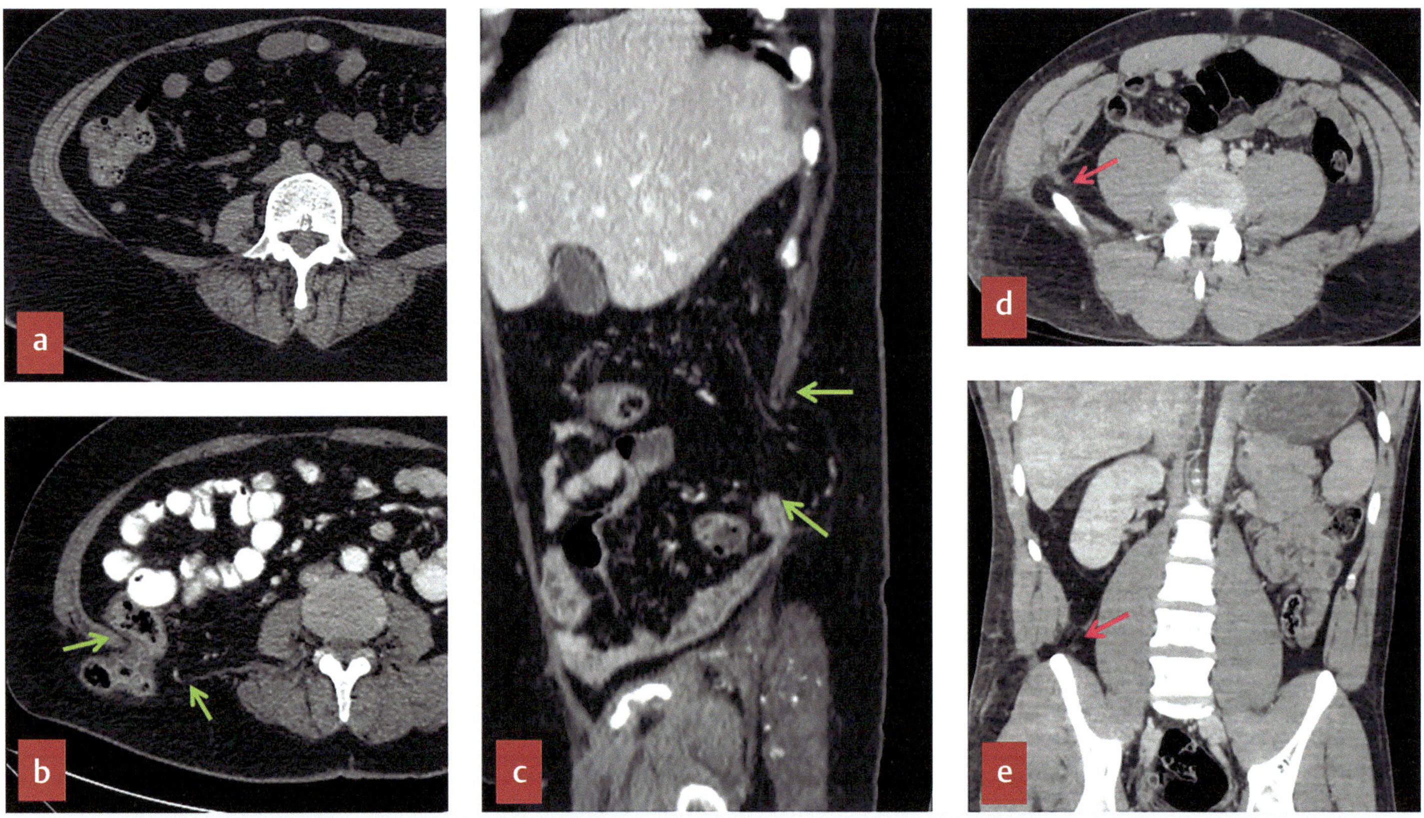

Fig. 1.47 (a–c) Postsurgical lumbar hernia. **(a)** Axial computed tomography (CT) before surgery. **(b)** Axial and **(c)** sagittal CT scans after right flank surgery demonstrating lumbar hernia containing fat and a portion of the ascending colon. The opening of the hernia sac is shown by *green arrows*. **(d)** Axial and **(e)** coronal CT in a patient with a small posttraumatic right hernia through the inferior (Petit) lumbar triangle (*red arrows*).

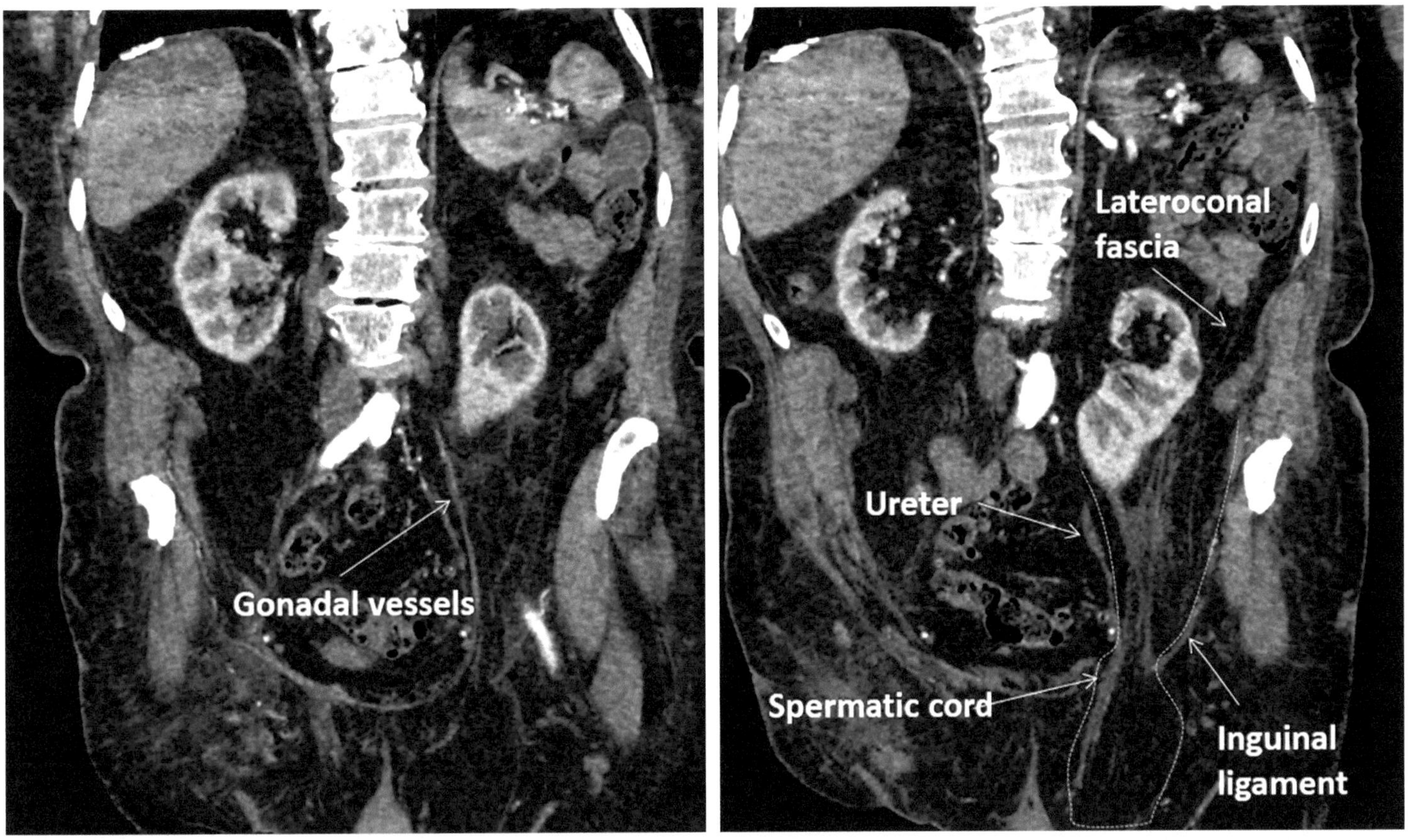

Fig. 1.48 Coronal computed tomography (CT) images showing large amount of extraperitoneal fat protruding into the left inguinal canal indirectly through a patulous inguinal canal. The left kidney is pulled inferiorly. Note the orifice of the sac is located lateral to the left inferior epigastric artery.

space of Retzius and preperitoneal space in patients with large amount of extraperitoneal fat (▶Fig. 1.50).

The direct inguinal hernia neck is located medial and superior to the deep inferior epigastric vessels and protrude through a weak Hesselbach's triangle.[72] In direct inguinal hernia, the fatty inguinal canal is usually compressed against the hernia sac forming a lateral crescent (▶Fig. 1.52). This is used as a sign to diagnose the direct hernia, the so-called lateral fat crescent sign.[79,80]

The sac of inguinal hernia usually contains a portion of peritoneal fat. The bowel, appendix, and rarely Meckel's diverticulum (Littre hernia) or bladder are seen in the hernia sac.

Femoral Hernia

Femoral hernias are uncommon and usually seen in females.[71] They occur medial to the femoral vein and inferior to the inguinal ligament, usually on the right side. The hernia sac is usually localized lateral to pubic tubercle, as opposed to inguinal hernia that lies medially (▶Fig. 1.53).

Obturator Hernia

Other rare hernias include obturator, sciatic, and perineal hernias (▶Fig. 1.51, ▶Fig. 1.54). The obturator canal is a potential pathway for the development of obturator hernias especially in underweight individual with reduction of body fat (▶Fig. 1.29). Obturator hernia occurs rarely predominantly in elderly females characterized by protrusion of the intra-abdominal viscera into the obturator foramen causing bowel obstruction and strangulation of the herniated loop.[81] There are three pathways an obturator hernia can follow. The most frequent type occurs along the course of anterior division of the obturator nerve (▶Fig. 1.51). In this case, the hernia sac lies between the external obturator and pectineus muscles.[81] Sciatic hernias are rare. Most of these hernias occur through the greater sciatic foramen and can present with sciatica.[82] Conditions causing atrophy of the piriformis muscle have been suggested as a predisposing factor (▶Fig. 1.52). Perineal hernias are uncommon and usually seen in older women. These hernias usually occur through a defect in the urogenital diaphragm or levator ani muscle.[83]

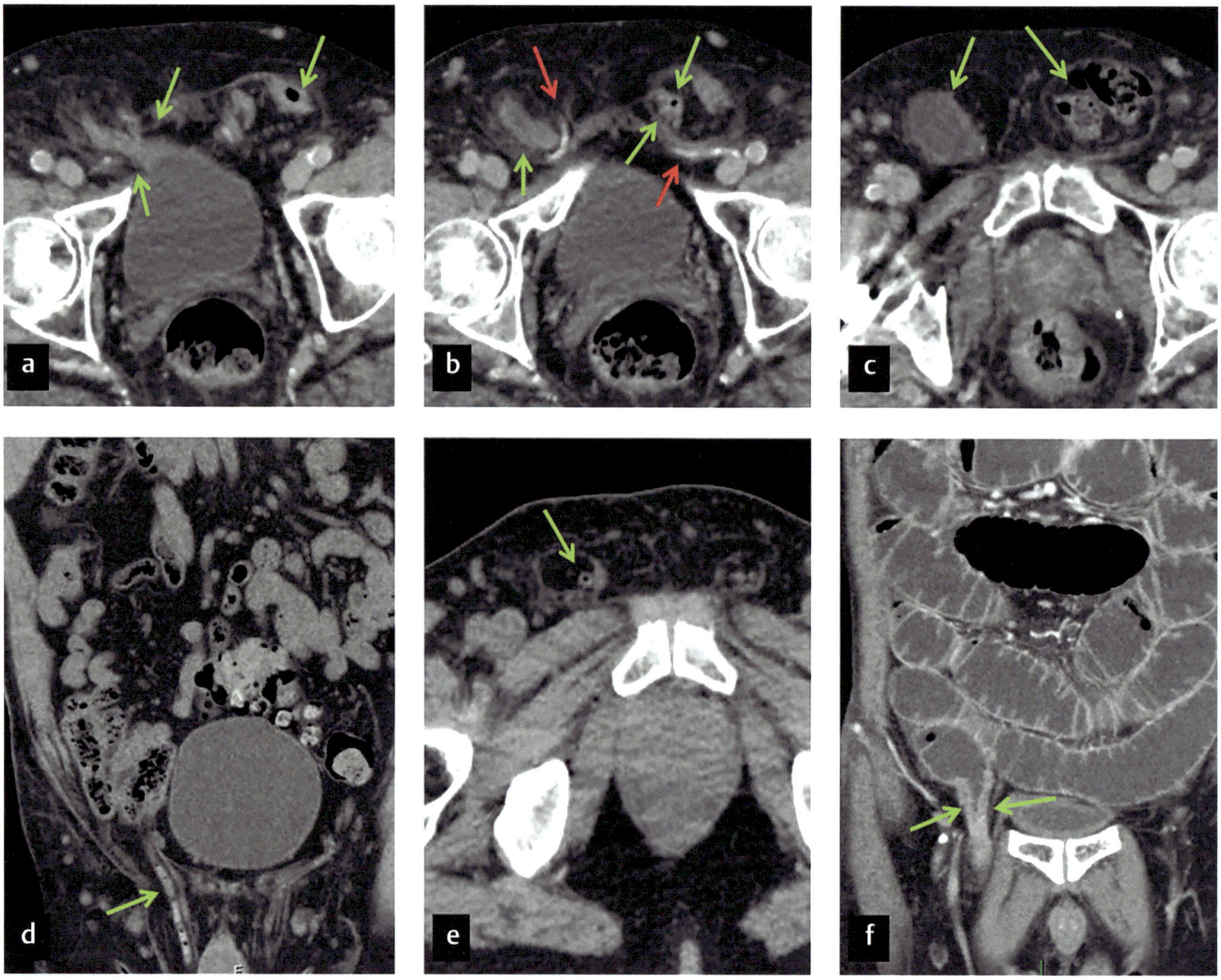

Fig. 1.49 Different forms of indirect visceral hernia into the inguinal canal. **(a–c)** Axial computed tomography (CT) images from superior to inferior. Bilateral indirect hernias containing part of the bladder on the right side and the sigmoid on the left side. Hernia sacs are shown by *green arrows*. Note the orifice of the sac is located lateral to the inferior epigastric arteries (*red arrows*). **(d)** Coronal and **(e)** axial CT showing right indirect hernia containing the appendix (coronal). **(f)** Coronal CT showing small bowel obstruction due to right indirect inguinal hernia containing a loop of small bowel (*green arrow*).

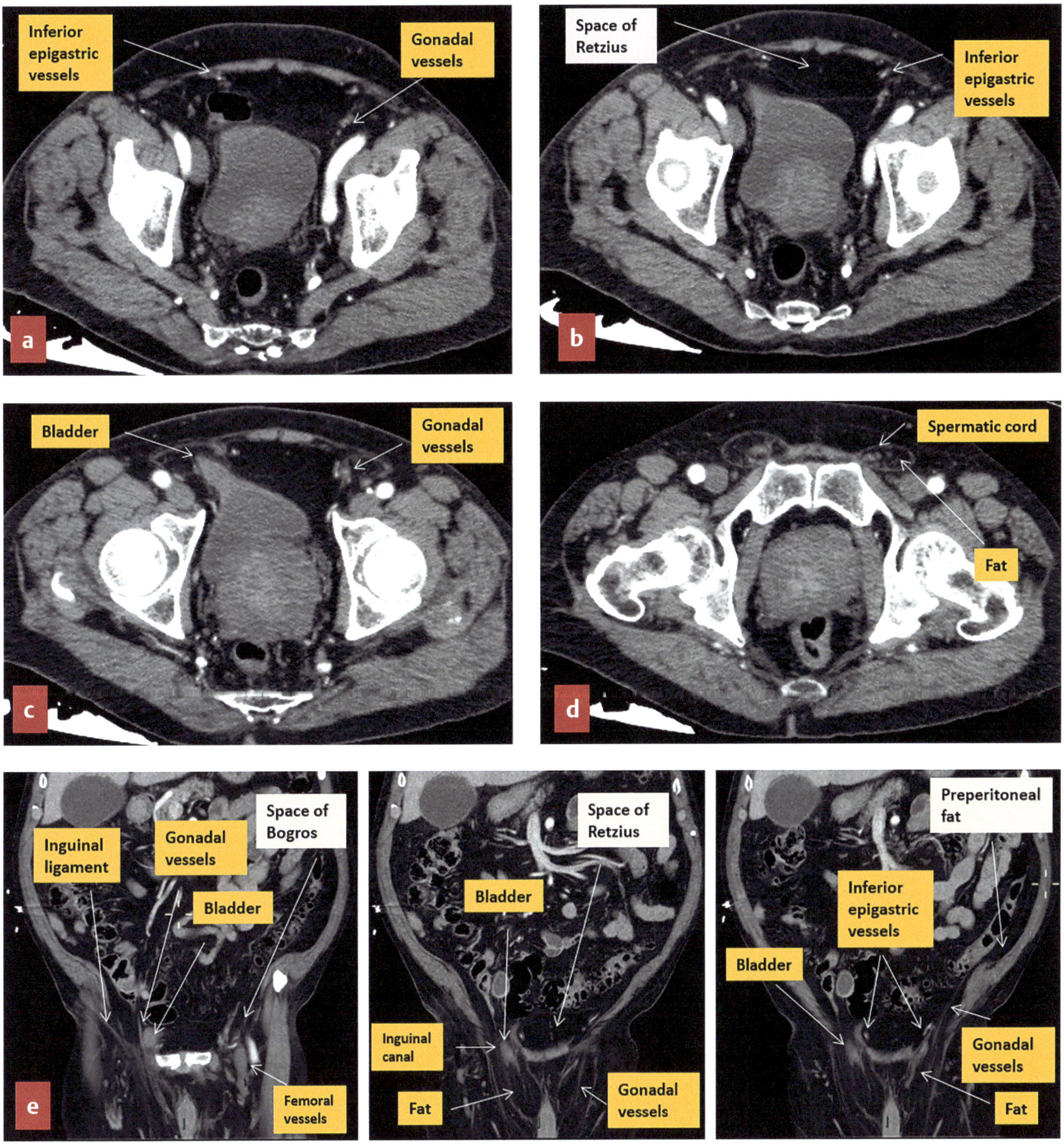

Fig. 1.50 Axial computed tomography (CT) images **(a–d)** from superior to inferior and coronal images **(e)** from posterior to anterior showing an indirect hernia containing part of the bladder on the right side. On the left side, normal contents of the inguinal canal are shown surrounded by large amount of extraperitoneal fat carried into the inguinal canal by the gonadal vessels and vas deferens. Normal fat in the canal is in continuation with the preperitoneal (posterior pararenal space) fat which is also in connection with the retroinguinal space of Bogros and the space of Retzius. Therefore, pathologies along with air, hematoma, and inflammatory tissues can extend from these spaces into the inguinal canal. Herniation of the bladder is an indirect form of hernia that along with the contents in the space of Retzius herniates into the inguinal canal, lateral to the inferior epigastric vessels.

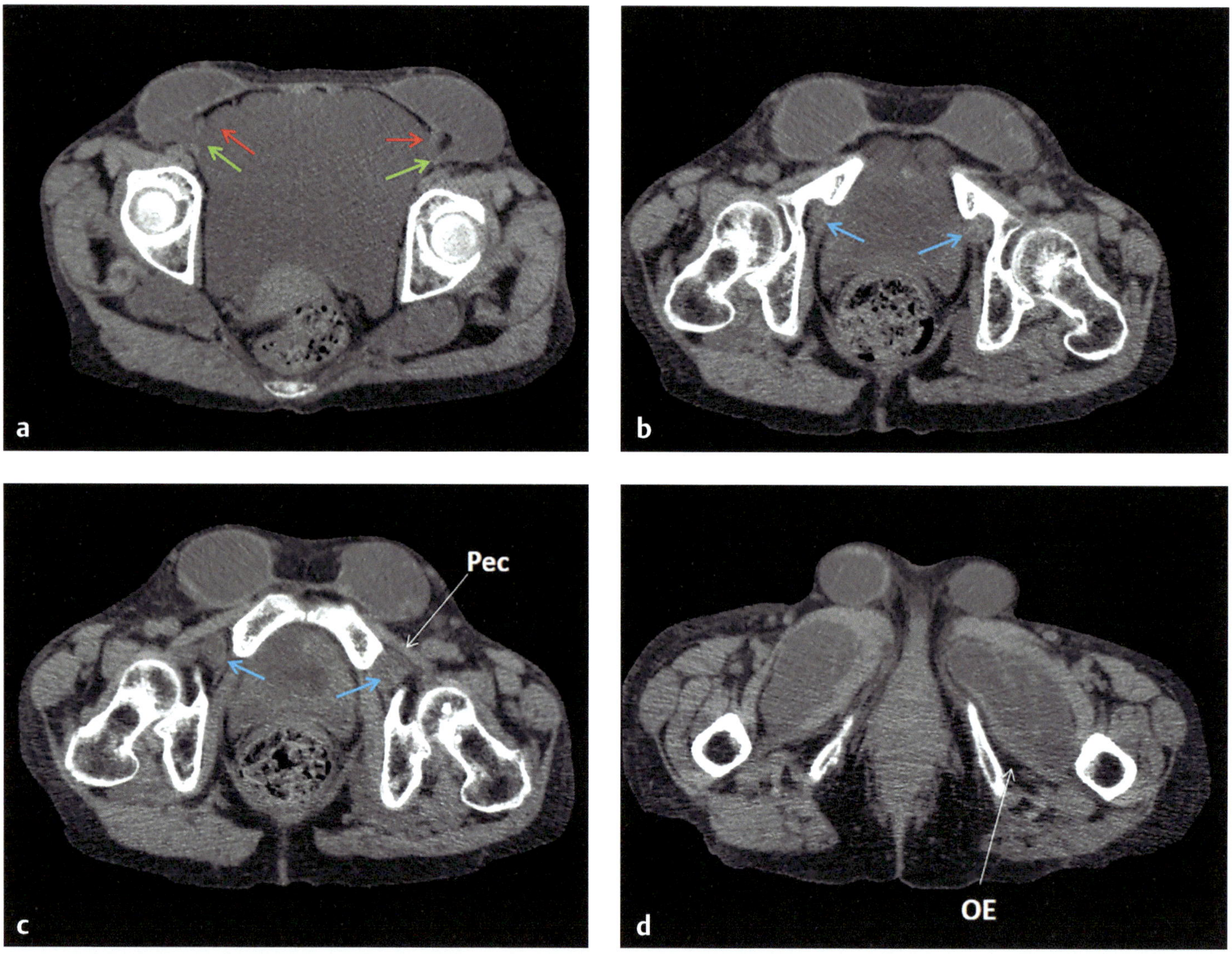

Fig. 1.51 (a–d) Inguinal and obturator hernias containing ascitic fluid in a patient with history of cirrhosis. Axial computed tomography (CT) images from top to bottom. The inguinal hernias are indirect and protrude (*green arrows*) lateral to the inferior epigastric vessels (*red arrows*). The orifice of the obturator hernias is at the obturator canal (*blue arrows*). Hernia sac in obturator hernia lies between the obturator externus (OE) and pectineus (Pec) muscles.

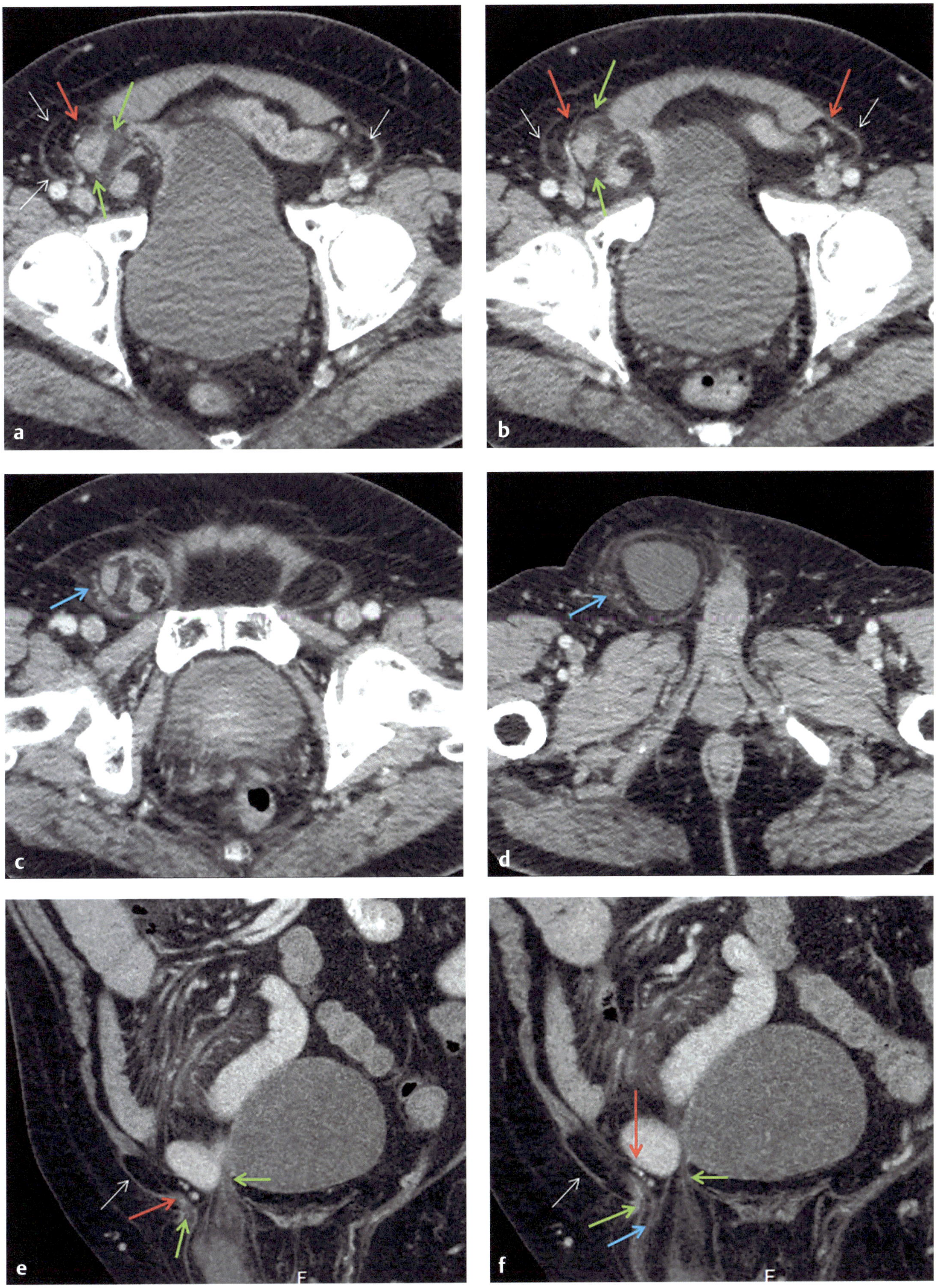

Fig. 1.52 Direct inguinal hernia containing a loop of small bowel and peritoneal fluid. **(a–d)** Axial and **(e, f)** coronal computed tomography (CT) scans. The orifice of the hernia sac (*green arrows*) is medial to the deep inferior epigastric artery (*red arrows*) and hernia sac extends into the scrotum medial to the spermatic cord (*blue arrows*). Lateral fat crescent is seen lateral to the hernia sac. *White arrows* denote the aponeurosis of the external oblique muscle.

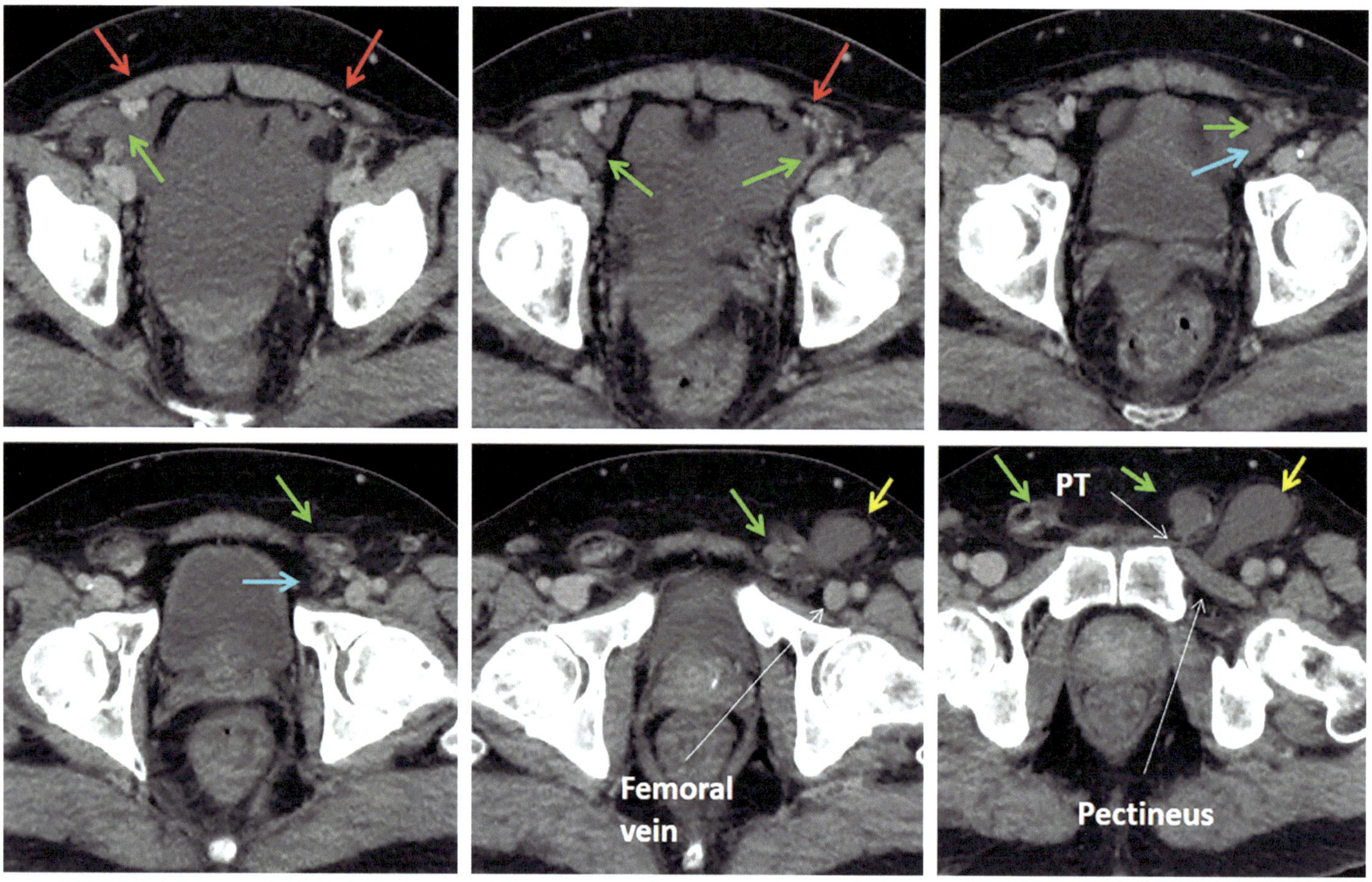

Fig. 1.53 Left femoral and bilateral indirect inguinal hernias. Axial computed tomography (CT) scans from top to bottom. Small bilateral inguinal hernias are shown with the hernia sac (*green arrows*) lateral to the inferior epigastric vessels (*red arrows*) consistent with an indirect hernia. On the left side the peritoneal sac herniates into both the inguinal and femoral canals. The femoral hernia containing peritoneal fluid (*yellow arrows*) is shown on the left side medial to the femoral vessels. The opening of the femoral canal is shown by *blue arrows*. The hernia sac (*yellow arrows*) locates medial to the femoral vein, inferior to the inguinal ligament, anterior to the pectineus muscle, and lateral to the pubic tubercle (PT).

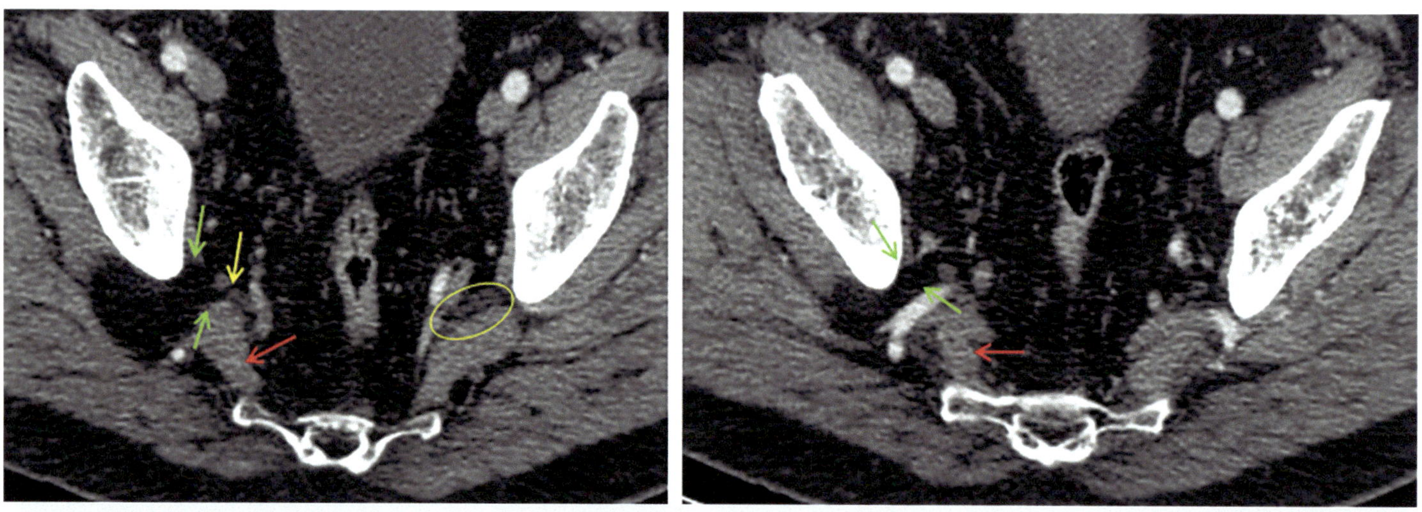

Fig. 1.54 Sciatic hernia containing fat. Opening of hernia through greater sciatic foramen is shown by *green arrows*. It may cause pain by compression of the sciatic nerve (*yellow arrow*). Note atrophy of the piriformis muscle on the side of hernia (*red arrow*). Left sciatic nerve bundles are shown on the left side within the *yellow circle*.

References

1. Cirocchi R, Boselli C, Renzi C, et al. The surface landmarks of the abdominal wall: a plea for standardization. Arch Med Sci 2014;10(3):566–569
2. Skandalakis J. The Embryologic and Anatomic Basis of Modern Surgery. New York: McGraw-Hill Professional Publishing; 2004
3. Chopra J, Rani A, Rani A, Srivastava AK, Sharma PK. Re-evaluation of superficial fascia of anterior abdominal wall: a computed tomographic study. Surg Radiol Anat 2011;33(10):843–849
4. Lockwood TE. Superficial fascial system (SFS) of the trunk and extremities: a new concept. Plast Reconstr Surg 1991;87(6):1009–1018
5. Johnson D, Dixon AK, Abrahams PH. The abdominal subcutaneous tissue: computed tomographic, magnetic resonance, and anatomical observations. Clin Anat 1996;9(1):19–24
6. Markman B, Barton FE Jr. Anatomy of the subcutaneous tissue of the trunk and lower extremity. Plast Reconstr Surg 1987;80(2):248–254
7. Abu-Hijleh MF, Roshier AL, Al-Shboul Q, Dharap AS, Harris PF. The membranous layer of superficial fascia: evidence for its widespread distribution in the body. Surg Radiol Anat 2006;28(6):606–619
8. Lancerotto L, Stecco C, Macchi V, Porzionato A, Stecco A, De Caro R. Layers of the abdominal wall: anatomical investigation of subcutaneous tissue and superficial fascia. Surg Radiol Anat 2011;33(10):835–842
9. Mirilas P, Skandalakis JE. Surgical anatomy of the retroperitoneal spaces—part I: embryogenesis and anatomy. Am Surg 2009;75(11):1091–1097
10. Skandalakis PN, Zoras O, Skandalakis JE, Mirilas P. Transversalis, endoabdominal, endothoracic fascia: who's who? Am Surg 2006;72(1):16–18
11. Mirilas P, Colborn GL, McClusky DA III, Skandalakis LJ, Skandalakis PN, Skandalakis JE. The history of anatomy and surgery of the preperitoneal space. Arch Surg 2005;140(1):90–94
12. Skandalakis JE, Gray SW, Skandalakis LJ, Colborn GL, Pemberton LB. Surgical anatomy of the inguinal area. World J Surg 1989;13(5):490–498
13. Peiper C, Junge K, Prescher A, Stumpf M, Schumpelick V. Abdominal musculature and the transversalis fascia: an anatomical viewpoint. Hernia 2004;8(4):376–380
14. Arnbjörnsson E. A neuromuscular basis for the development of right inguinal hernia after appendectomy. Am J Surg 1982;143(3):367–369
15. Teoh LS, Hingston G, Al-Ali S, Dawson B, Windsor JA. The iliopubic tract: an important anatomical landmark in surgery. J Anat 1999;194(Pt 1):137–141
16. Barker PJ, Urquhart DM, Story IH, Fahrer M, Briggs CA. The middle layer of lumbar fascia and attachments to lumbar transverse processes: implications for segmental control and fracture. Eur Spine J 2007;16(12):2232–2237
17. Willard FH, Vleeming A, Schuenke MD, Danneels L, Schleip R. The thoracolumbar fascia: anatomy, function and clinical considerations. J Anat 2012;221(6): 507–536
18. Oh CS, Won HS, Kwon CH, Chung IH. Morphologic variations of the umbilical ring, umbilical ligaments and ligamentum teres hepatis. Yonsei Med J 2008; 49(6):1004–1007
19. Hammond G, Yglesias L, Davis JE. The urachus, its anatomy and associated fasciae. Anat Rec 1941;80(3):271–287
20. Martin BF, Tudor RG. The umbilical and paraumbilical veins of man. J Anat 1980;130(Pt 2):305–322
21. Ohhira M, Ohta H, Ohhira M, et al. Altered intrahepatic pathway of para-umbilical vein in portal hypertension. J Gastroenterol Hepatol 1998;13(7):691–695
22. van Geffen HJ, Simmermacher RK, Bosscha K, van der Werken C, Hillen B. Anatomical considerations for surgery of the anterolateral abdominal wall. Hernia 2004;8(2):93–97
23. Schlenz I, Burggasser G, Kuzbari R, Eichberger H, Gruber H, Holle J. External oblique abdominal muscle: a new look on its blood supply and innervation. Anat Rec 1999;255(4):388–395
24. Yang D, Morris SF, Geddes CR, Tang M. Neurovascular territories of the external and internal oblique muscles. Plast Reconstr Surg 2003;112(6):1591–1595
25. Lee JS, Kim TH, Kim DY, Shim JH, Lim JY. Effects of selective exercise for the deep abdominal muscles and lumbar stabilization exercise on the thickness of the transversus abdominis and postural maintenance. J Phys Ther Sci 2015;27(2):367–370
26. Hides J, Wilson S, Stanton W, et al. An MRI investigation into the function of the transversus abdominis muscle during "drawing-in" of the abdominal wall. Spine 2006;31(6):E175–E178
27. Richardson CA, Snijders CJ, Hides JA, Damen L, Pas MS, Storm J. The relation between the transversus abdominis muscles, sacroiliac joint mechanics, and low back pain. Spine 2002;27(4):399–405
28. Lovering RM, Anderson LD. Architecture and fiber type of the pyramidalis muscle. Anat Sci Int 2008;83(4):294–297
29. Van Dyke JA, Holley HC, Anderson SD. Review of iliopsoas anatomy and pathology. Radiographics 1987;7(1):53–84
30. Polster JM, Elgabaly M, Lee H, Klika A, Drake R, Barsoum W. MRI and gross anatomy of the iliopsoas tendon complex. Skeletal Radiol 2008;37(1):55–58
31. Cronin CG, Lohan DG, Meehan CP, et al. Anatomy, pathology, imaging and intervention of the iliopsoas muscle revisited. Emerg Radiol 2008;15(5):295–310
32. Blankenbaker DG, Tuite MJ. Iliopsoas musculotendinous unit. Semin Musculoskelet Radiol 2008;12(1):13–27
33. Phillips S, Mercer S, Bogduk N. Anatomy and biomechanics of quadratus lumborum. Proc Inst Mech Eng H 2008;222(2):151–159
34. Engstrom CM, Walker DG, Kippers V, Mehnert AJ. Quadratus lumborum asymmetry and L4 pars injury in fast bowlers: a prospective MR study. Med Sci Sports Exerc 2007;39(6):910–917
35. Solomon LB, Lee YC, Callary SA, Beck M, Howie DW. Anatomy of piriformis, obturator internus and obturator externus: implications for the posterior surgical approach to the hip. J Bone Joint Surg Br 2010;92(9):1317–1324
36. Pine J, Binns M, Wright P, Soames R. Piriformis and obturator internus morphology: a cadaveric study. Clin Anat 2011;24(1):70–76
37. Windisch G, Braun EM, Anderhuber F. Piriformis muscle: clinical anatomy and consideration of the piriformis syndrome. Surg Radiol Anat 2007;29(1):37–45
38. Pokorný D, Jahoda D, Veigl D, Pinskerová V, Sosna A. Topographic variations of the relationship of the sciatic nerve and the piriformis muscle and its relevance to palsy after total hip arthroplasty. Surg Radiol Anat 2006;28(1):88–91
39. Niikura H, Jin ZW, Cho BH, et al. Human fetal anatomy of the coccygeal attachments of the levator ani muscle. Clin Anat 2010;23(5):566–574
40. Furtmüller GJ, McKenna CA, Ebmer J, Dellon AL. Pudendal nerve 3-dimensional illustration gives insight into surgical approaches. Ann Plast Surg 2014;73(6):670–678
41. Khoder W, Hale D. Pudendal neuralgia. Obstet Gynecol Clin North Am 2014; 41(3):443–452
42. Lawson JO. Pelvic anatomy. I. Pelvic floor muscles. Ann R Coll Surg Engl 1974; 54(5):244–252
43. Kearney R, Sawhney R, DeLancey JO. Levator ani muscle anatomy evaluated by origin-insertion pairs. Obstet Gynecol 2004;104(1):168–173
44. Fröhlich B, Hötzinger H, Fritsch H. Tomographical anatomy of the pelvis, pelvic floor, and related structures. Clin Anat 1997;10(4):223–230
45. Barber MD, Bremer RE, Thor KB, Dolber PC, Kuehl TJ, Coates KW. Innervation of the female levator ani muscles. Am J Obstet Gynecol 2002;187(1):64–71
46. Wallner C, Maas CP, Dabhoiwala NF, Lamers WH, DeRuiter MC. Innervation of the pelvic floor muscles: a reappraisal for the levator ani nerve. Obstet Gynecol 2006;108(3 Pt 1):529–534
47. Kim J, Betschart C, Ramanah R, Ashton-Miller JA, DeLancey JO. Anatomy of the pubovisceral muscle origin: macroscopic and microscopic findings within the injury zone. Neurourol Urodyn 2015;34(8):774–780
48. Ren S, Xie B, Wang J, Rong Q. Three-dimensional modeling of the pelvic floor support systems of subjects with and without pelvic organ prolapse. BioMed Res Int 2015;2015:845985
49. Nyangoh Timoh K, Bessede T, Zaitouna M, et al. [Anatomy of the levator ani muscle and implications for obstetrics and gynaecology]. Gynécol Obstét Fertil 2015;43(1):84–90
50. Hoyte L, Schierlitz L, Zou K, Flesh G, Fielding JR. Two- and 3-dimensional MRI comparison of levator ani structure, volume, and integrity in women with stress incontinence and prolapse. Am J Obstet Gynecol 2001;185(1):11–19
51. Margulies RU, Hsu Y, Kearney R, Stein T, Umek WH, DeLancey JO. Appearance of the levator ani muscle subdivisions in magnetic resonance images. Obstet Gynecol 2006;107(5):1064–1069
52. Li R, Song Y, Ma M. Relationship between levator ani and bony pelvis morphology and clinical grade of prolapse in women. Clin Anat 2015;28(6):813–819
53. Brandon CJ, Lewicky-Gaupp C, Larson KA, Delancey JO. Anatomy of the perineal membrane as seen in magnetic resonance images of nulliparous women. Am J Obstet Gynecol 2009;200(5):583.e1–583.e6

54. Yavagal S, de Farias TF, Medina CA, Takacs P. Normal vulvovaginal, perineal, and pelvic anatomy with reconstructive considerations. Semin Plast Surg 2011;25(2):121–129
55. Hsu Y, Lewicky-Gaupp C, DeLancey JOL. Posterior compartment anatomy as seen in magnetic resonance imaging and 3-dimensional reconstruction from asymptomatic nulliparas. Am J Obstet Gynecol 2008;198(6):651.e1–651.e7
56. Oelrich TM. The striated urogenital sphincter muscle in the female. Anat Rec 1983;205(2):223–232
57. Shafik A, Sibai OE, Shafik AA, Shafik IA. A novel concept for the surgical anatomy of the perineal body. Dis Colon Rectum 2007;50(12):2120–2125
58. Hurd WW, Bude RO, DeLancey JOL, Newman JS. The location of abdominal wall blood vessels in relationship to abdominal landmarks apparent at laparoscopy. Am J Obstet Gynecol 1994;171(3):642–646
59. Saber AA, Meslemani AM, Davis R, Pimentel R. Safety zones for anterior abdominal wall entry during laparoscopy: a CT scan mapping of epigastric vessels. Ann Surg 2004;239(2):182–185
60. Stark E, Oestreich K, Wendl K, Rumstadt B, Hagmüller E. Nerve irritation after laparoscopic hernia repair. Surg Endosc 1999;13(9):878–881
61. Rahn DD, Phelan JN, Roshanravan SM, White AB, Corton MM. Anterior abdominal wall nerve and vessel anatomy: clinical implications for gynecologic surgery. Am J Obstet Gynecol 2010;202(3):234.e1–234.e5
62. Rozen WM, Ashton MW, Taylor GI. Reviewing the vascular supply of the anterior abdominal wall: redefining anatomy for increasingly refined surgery. Clin Anat 2008;21(2):89–98
63. Lam DL, Mitsumori LM, Neligan PC, Warren BH, Shuman WP, Dubinsky TJ. Preoperative CT angiography and three-dimensional image post processing for deep inferior epigastric perforator flap breast reconstructive surgery. Br J Radiol 2012;85(1020):e1293–e1297
64. Scott JR, Liu D, Said H, Neligan PC, Mathes DW. Computed tomographic angiography in planning abdomen-based microsurgical breast reconstruction: a comparison with color duplex ultrasound. Plast Reconstr Surg 2010;125(2): 446–453
65. Fagan SP, Awad SS. Abdominal wall anatomy: the key to a successful inguinal hernia repair. Am J Surg 2004;188(6A, Suppl):3S–8S
66. Wolloscheck T, Konerding MA. Dimensions of the myopectineal orifice: a human cadaver study. Hernia 2009;13(6):639–642
67. Annibali R, Fitzgibbons R, Quinn T. Surgical anatomy of the inguinal region. In: Bendavid R, ed. Abdominal Wall Hernias, Principles and Management. New York: Springer; 2001:72–85
68. Skandalakis JE, Colborn GL, Androulakis JA, Skandalakis LJ, Pemberton LB. Embryologic and anatomic basis of inguinal herniorrhaphy. Surg Clin North Am 1993;73(4):799–836
69. Bendavid R. The space of Bogros and the deep inguinal venous circulation. Surg Gynecol Obstet 1992;174(5):355–358
70. Ozdemir MB, Celik HH, Aldur MM. Altered course of the right testicular artery. Clin Anat 2004;17(1):67–69
71. Trainer V, Leung C, Owen RE, Venkatanarasimha N. External anterior abdominal wall and pelvic hernias with emphasis on the key diagnostic features on MDCT. Clin Radiol 2013;68(4):388–396
72. Aguirre DA, Santosa AC, Casola G, Sirlin CB. Abdominal wall hernias: imaging features, complications, and diagnostic pitfalls at multi-detector row CT. Radiographics 2005;25(6):1501–1520
73. Felfel M, El Khoury M, Marboeuf Y, Strohl D, Menu Y. Incarcerated hernia through the posterior rectus sheath. AJR Am J Roentgenol 2005;185(5):1185–1186
74. Coulier B. Multidetector computed tomography features of linea arcuata (arcuate-line of Douglas) and linea arcuata hernias. Surg Radiol Anat 2007;29(5):397–403
75. Martin M, Paquette B, Badet N, Sheppard F, Aubry S, Delabrousse E. Spigelian hernia: CT findings and clinical relevance. Abdom Imaging 2013;38(2):260–264
76. Martín J, Mellado JM, Solanas S, Yanguas N, Salceda J, Cozcolluela MR. MDCT of abdominal wall lumbar hernias: anatomical review, pathologic findings and differential diagnosis. Surg Radiol Anat 2012;34(5):455–463
77. Loukas M, El-Zammar D, Shoja MM, et al. The clinical anatomy of the triangle of Grynfeltt. Hernia 2008;12(3):227–231
78. Jankovic ZB, du Feu FM, McConnell P. An anatomical study of the transversus abdominis plane block: location of the lumbar triangle of Petit and adjacent nerves. Anesth Analg 2009;109(3):981–985
79. Burkhardt JH, Arshanskiy Y, Munson JL, Scholz FJ. Diagnosis of inguinal region hernias with axial CT: the lateral crescent sign and other key findings. Radiographics 2011;31(2):E1–E12
80. Cherian PT, Parnell AP. Radiologic anatomy of the inguinofemoral region: insights from MDCT. AJR Am J Roentgenol 2007;189(4):W177-83
81. Losanoff JE, Richman BW, Jones JW. Obturator hernia. J Am Coll Surg 2002; 194(5):657–663
82. Miklos JR, O'Reilly MJ, Saye WB. Sciatic hernia as a cause of chronic pelvic pain in women. Obstet Gynecol 1998;91(6):998–1001
83. Lubat E, Gordon RB, Birnbaum BA, Megibow AJ. CT diagnosis of posterior perineal hernia. AJR Am J Roentgenol 1990;154(4):761–762

2 Esophagus

Farhood Saremi

Introduction

The esophagus is a muscular tube connecting the pharynx to the stomach. It serves as a conduit for passage of food between the mouth and the stomach. Topographically, it is subdivided into three distinct regions: cervical, thoracic, and abdominal. The esophageal phase of swallowing consists of highly coordinated control mechanism steps involving the central nervous system, enteric nervous system, and esophageal musculature.

Foregut Development

After the second week, the embryonic disc is formed by three layers: the ectoderm, the endoderm, and the mesoderms on each side. The endoderm forms the lining of the yolk sac. After the third week, along with elongation of the embryo, the mesoderms grow and form longitudinal masses called the paraxial mesoderm which later fragment into 33 to 35 cube-like somites.[1] The ectoderm gives rise to the epidermis and the neural plates which later evolve into neural tube and neural crest cells (▶ Fig. 2.1). The neural tube forms the spinal cord and brain and the neural crest cells are the precursor for the peripheral nervous system. Both the endoderm and mesoderm participate in the formation of the digestive tract. The endoderm produces the epithelium and submucosal glands. The mesoderm is substrate for the connective tissue, vessels, smooth muscle, and serosal layers of the gut. During the fourth week, along with lateral folding and cylindrical transformation of the embryo, the dorsal part of the yolk sac will be incorporated into the embryo dividing the yolk sac into intraembryonic and extraembryonic parts. The extraembryonic part regresses but the intraembryonic part forms the digestive tube.

Early digestive tube is divided into foregut, midgut, and hindgut (▶ Fig. 2.2). Both the esophagus and trachea arise from the foregut which is a derivative of endoderm. The primitive respiratory tract develops resembling appearance of the tracheobronchial diverticulum from the anterior aspect of the foregut after the fourth week. Separation of the trachea and esophagus occurs before the fifth week of gestation. Elongation of esophagus occurs by cranial growth of embryonic body. The smooth muscles of the muscularis propria arise from the mesenchyme of the somites. The striated muscles of the upper part of the esophagus are derived from the mesenchyme of the branchial arches, which also give origin to the vagal and recurrent laryngeal nerves. The circular muscle coat develops early in the 6th week, the longitudinal muscle fibers at approximately between the 9th and 12th weeks, and the muscularis mucosa at approximately the fourth month of gestation.[1] The neural crest cells migrate within the digestive tract, from top to bottom, between 4 and 7 weeks of gestation and form the ganglion cells of the myenteric (Auerbach) and submucosal (Meissner) plexuses.[2,3,4] This process is coordinated by several genes and its disruptions could result in Hirschsprung's disease. Interstitial cells of Cajal arise from the gut mesenchyme around the myenteric plexus.[2] The Cajal cells are pacemakers of the gut and generate slow wave contractions. Neurodegeneration with aging causes loss of excitatory (e.g., cholinergic) enteric neurons and interstitial cells of Cajal resulting in esophageal dysfunction, constipation, and evacuation disorders.[5] At 7 or 8 weeks of gestation, the luminal epithelium proliferates and partially occludes the foregut. Within 2 weeks, vacuoles appear in the luminal cells of the foregut and coalesce to form a single esophageal lumen lined with a layer of ciliated epithelial cells. These epithelial cells are stratified columnar in early life and later become stratified columnar which will be gradually replaced by stratified squamous starting at the fourth month that continues until birth. Blood vessels enter the esophageal wall during the week 7 of gestation, and lymph capillaries between the third and fourth months of life. The embryologic origin of the gastroesophageal junction is still controversial, but gastric rotation together with augmentation of the fundus of the stomach is believed to determine its formation. The esophagus reaches a length of 8 to 10 cm at birth. The esophagus is capable of peristalsis in the first to second trimester. In the first month after birth the lower esophageal sphincter (LES) is still immature, resulting in frequent regurgitation of food.

Anatomy

The esophagus is a 20 to 25 cm muscular tube extending between inferior margin of the cricoid cartilage (C5–C6 vertebral interspace) and the diaphragmatic hiatus (T10 vertebral level) (▶ Fig. 2.3, ▶ Fig. 2.4). The length of esophagus varies with age and sex but it is not directly related to height. Topographically, there are three distinct regions: cervical, thoracic, and abdominal. The cervical esophagus is 4- to 5-cm long extending from the cricoid cartilage to the suprasternal notch. The thoracic esophagus extends to the diaphragmatic hiatus. The abdominal esophagus refers to a short segment (2–3 cm) below the diaphragm. Functionally, the esophagus has three regions, the upper and lower esophageal sphincters, and the esophageal body. The upper 3 to 4 cm of the esophagus—roughly the level of the aortic arch—consists of striated muscle. The middle 35 to 40%, also known as "transition zone," contains both striated and smooth muscles. The distal 10 to 12 cm of the esophageal body is entirely smooth muscle. At each level, the esophagus is covered with an inner circular and an outer longitudinal muscle layer. A neural network (myenteric plexus), which provides terminal motor innervation to the smooth muscle of the esophagus, is sandwiched between the muscular layers.[1]

The esophagus meets several structures along its course (▶ Fig. 2.4, ▶ Fig. 2.5). In the upper part, the esophagus deviates slightly to the left and in the lower part it deviates to the right. The midline crossing point is approximately at the level of T4.

The cervical esophagus is located between the trachea anteriorly and the alar fascia, prevertebral fascia, longus colli muscles,

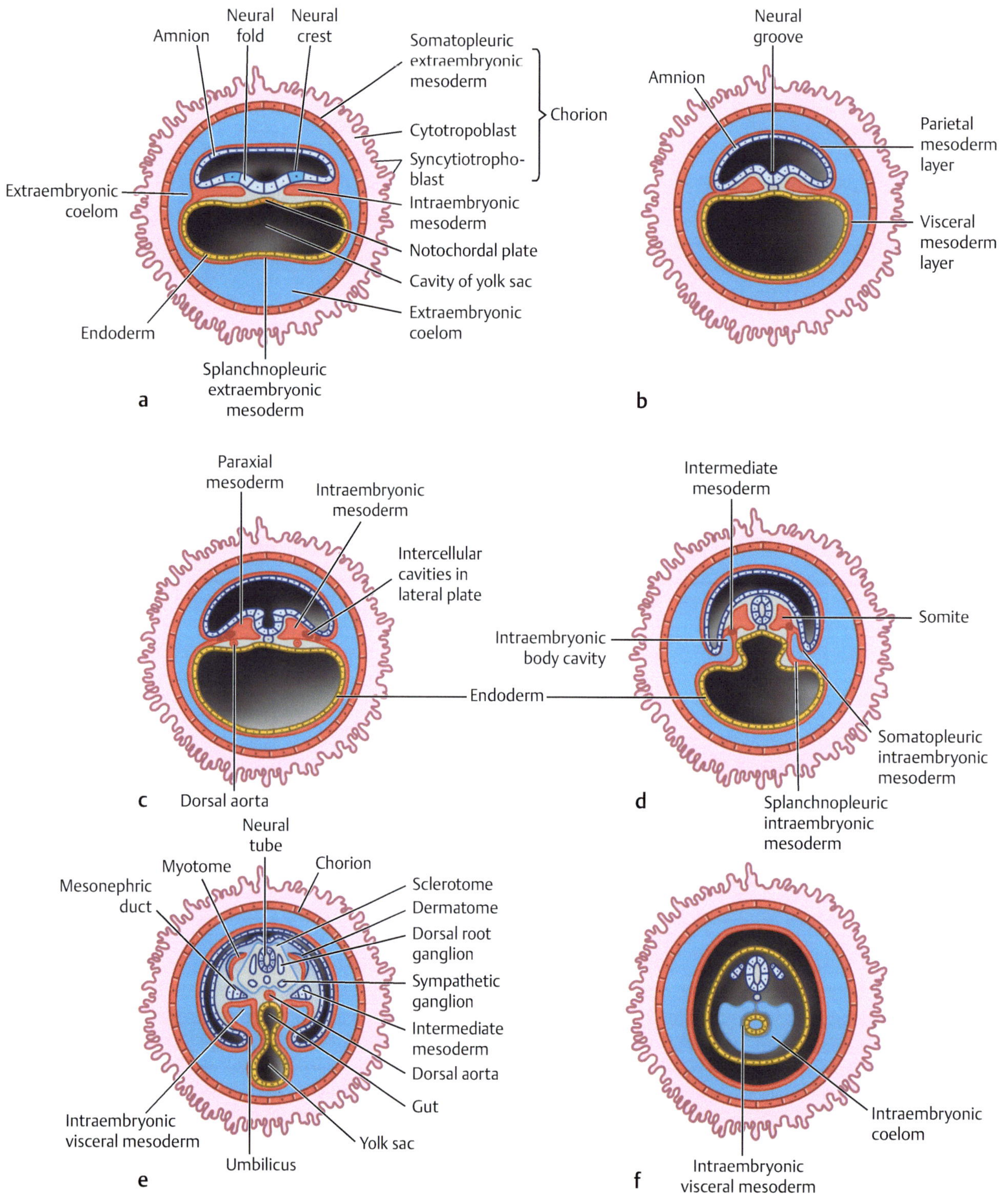

Fig. 2.1 (a–f) Early embryonic development of the foregut. The dorsal part of the yolk sac, composed of endoderm, is compressed by the lateral folding of the embryo and is incorporated as a rim during the fourth week. Therefore, the yolk sac is divided into intraembryonic and extraembryonic components. The intraembryonic component is the origin of digestive tube and its accessory glands. The extraembryonic part regresses and disappears around week 12. At this point, the early digestive system divides into foregut, midgut, and hindgut.

and vertebral column posteriorly. Between the alar and prevertebral fasciae is the retrovisceral space, also known as "danger space." The retrovisceral space extends from the skull base to the diaphragm.[1] It is attached to the trachea by loose areolar tissue and muscle strands. The recurrent laryngeal nerves pass in the groove between the trachea and esophagus in 50% of cases. The next common location for the recurrent laryngeal nerve is anterior to the groove or in a paratracheal position. The thyroid lobe may extend posteriorly to cover the lateral surfaces of the cervical esophagus.

In its course, the thoracic esophagus passes posterior to the trachea, tracheal bifurcation, left main stem bronchus, and left atrium (▶Fig. 2.4, ▶Fig. 2.5). The left subclavian artery is very close to the proximal thoracic esophagus. The thoracic esophagus

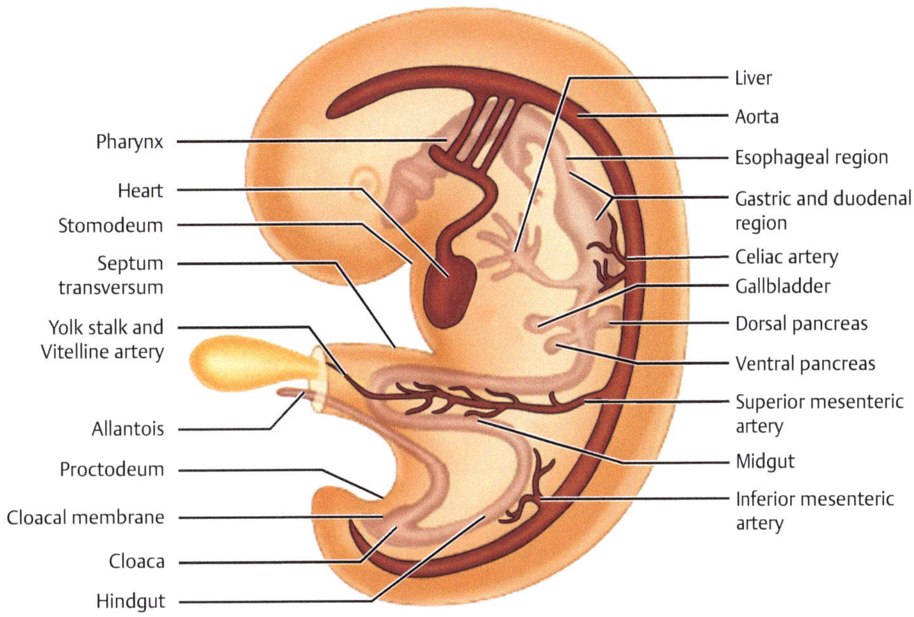

Fig. 2.2 Drawing of median section of the embryo showing the early digestive system and its blood supply. The primordial gut is a long tube extending throughout the length of the embryo. Its blood vessels are derived from the vessels that supplied the yolk sac. The early digestive system is divided into foregut, midgut, and hindgut.

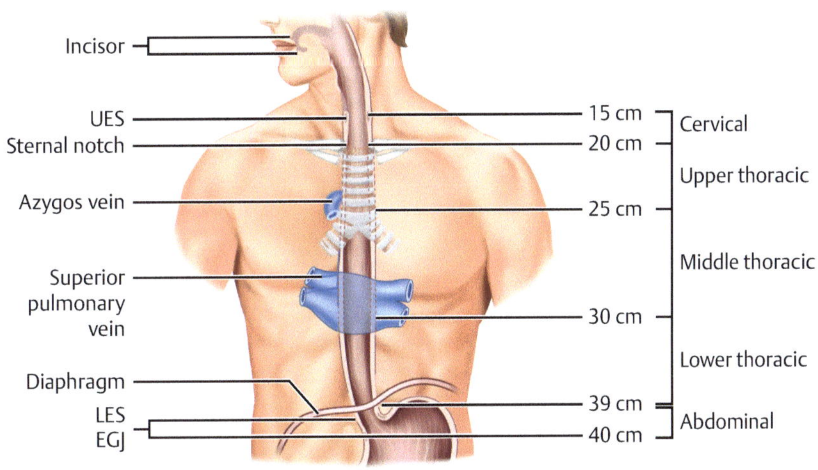

Fig. 2.3 Anatomic levels of the esophagus. EGJ, esophagogastric junction; LES, lower esophageal sphincter; UES, upper esophageal sphincter.

passes the left main stem bronchus at T5 level. The trachea and the esophagus are closely associated and carcinoma arising from the esophagus or the trachea and bronchi may invade the other. The esophagus lies anterolateral to the descending aorta, and distally it moves to the right to stay medial to the aorta at above the diaphragm. The azygos vein passes along the right posterolateral margin of the esophagus at midthoracic level. Division of the arch of the azygos vein provides excellent access to the upper thoracic esophagus during right surgical approaches. The thoracic duct lies behind the esophagus before it crosses midline at T5 level to stay along the left side of esophagus. The anterior vagal trunk and esophageal plexus are anterior to the lower thoracic esophagus. The esophageal plexus of the vagus nerve below the tracheal bifurcation and sometimes the posterior vagal trunk pass posterior to the lower thoracic esophagus. The entire length of the thoracic esophagus is directly related to the right mediastinal pleura except where the arch of the azygos vein crosses. Large posterior mediastinal fat may separate lung from the distal thoracic esophagus. On the left side, the left lung can reach the esophagus above the diaphragm (posterior junctional zone). The posterior convergence of the right and left pleurae between the esophagus and aorta forms the so-called “mesoesophagus” (▶Fig. 2.6). Below the diaphragm, the abdominal esophagus lies in the esophageal groove on the posterior surface of the left liver lobe.

A triangular surgical access area to the upper esophagus exists in the left upper mediastinum known as “esophageal triangle” (▶Fig. 2.5). It is formed by the descending arch of the aorta, left subclavian artery, and vertebral column. Because of the aortic

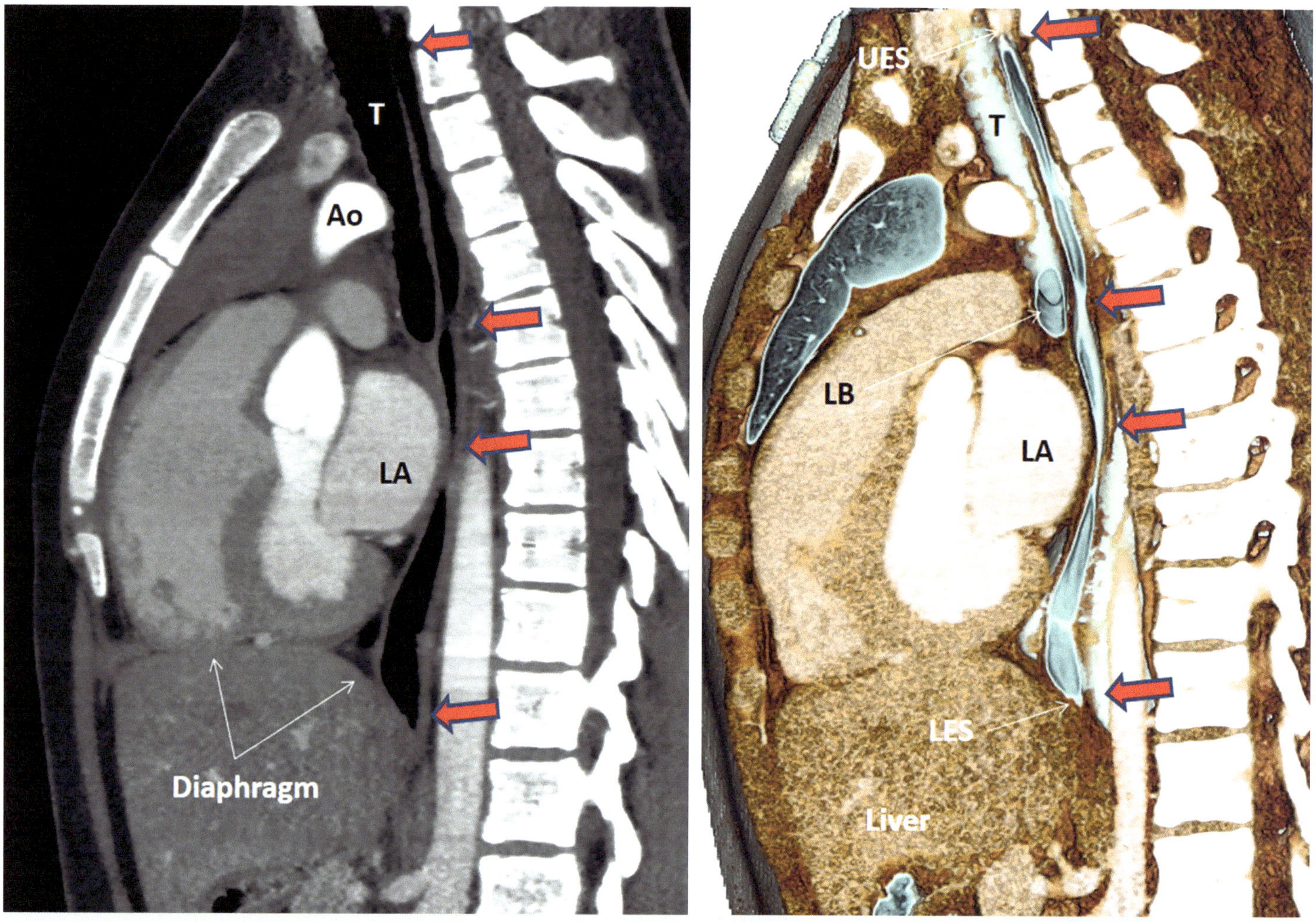

Fig. 2.4 Sagittal computed tomography (CT) showing normal narrowings of the esophagus from top to bottom (*red arrows*) at the upper esophageal sphincter (UES), aortic arch (Ao), left bronchus (LB), left atrium (LA), and lower esophageal sphincter (LES). T, trachea.

arch, generally lesions of the upper half of the thoracic esophagus are approached through a right thoracotomy to avoid technical problems. On the other hand, lesions of the distal thoracic esophagus can be explored through a left or a right thoracotomy. To expose the distal esophagus by a left thoracic approach (triangle of Truesdale), the left pulmonary ligament must be divided[1] (▸Fig. 2.5). This ligament extends between the inferior pulmonary vein and the diaphragm and is related to the anterior esophageal wall (▸Fig. 2.7).

The esophagus is normally collapsed at rest. In elderly, it may retain air and stay patulous (▸Fig. 2.6). The esophagus can dilate 2 to 3 cm during swallowing. The normal esophagus shows several benign narrowing best observed during barium esophagram. Normal esophageal narrowings along its course are related to the cricopharyngeus muscle, distal aortic arch, left main stem bronchus, and LES (▸Fig. 2.4).

Esophagus Wall Layers

The esophagus is made up of mucosa, muscularis mucosa, submucosa, muscularis propria, intermuscular septum, and adventitia (▸Fig. 2.8). In the esophagus, the mucosa and submucosa together form a thick strong layer of tissue.[1] The mucous membrane is a nonkeratinized squamous epithelium. It covers the entire inner surface of the esophagus down to the level of LES where both squamous and columnar epithelium coexists. The muscularis mucosa is a thin layer of two to three longitudinally arranged muscle cells. The submucosa contains the Meissner's plexus, vascular Heller's plexus, and mucous glands. The muscularis propria is approximately 0.75-mm thick. Intermuscular septum between the two layers contains myenteric plexus. The adventitia is composed of loose connective tissue, small vessels, and lymphatic channels. Absence of the serosal covering in the esophagus (in contrast to other parts of the gastrointestinal tract) allows faster spread of esophageal tumors and would make surgical repair of esophageal perforation difficult.

Muscles of the Esophagus

The muscularis propria is organized into outer longitudinal muscle fibers and inner circular muscle layer. In the stomach, there is an additional oblique muscle layer between the two layers. In the colon, the longitudinal muscle is organized into three tenia coli. Muscles forming the upper esophagus mainly consists of striated muscle and the lower esophagus consists of smooth muscles. The border between the two groups is formed by a mix of striated and smooth muscles.

The inner circular muscle fibers extend between the inferior pharyngeal constrictor muscle proximally and the LES distally. At

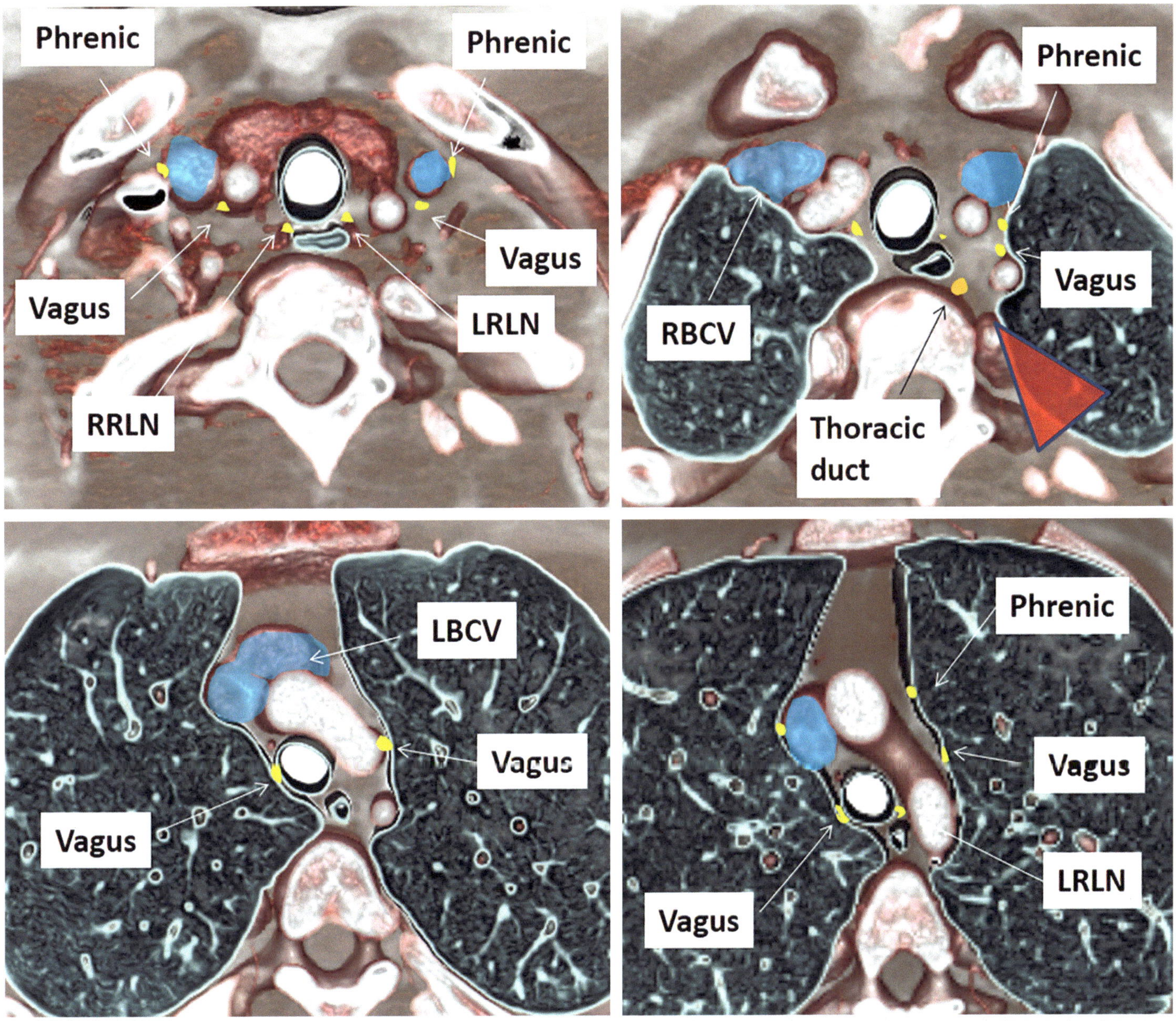

Fig. 2.5 Anatomic course of esophagus and its relationship with adjacent structures are shown using axial computed tomography (CT) scans. Major veins are color coded in *blue*, nerves in *yellow*, and thoracic duct in *orange*. The entire course of the esophagus meets the right pleura except where the azygos vein enters the superior vena cava. The esophagus can be accessed easily by right thoracotomy. A triangular surgical access area to the upper esophagus exists in the upper left mediastinum known as "esophageal triangle" (*red triangle*). It is formed by the descending arch of the aorta, left subclavian artery, and vertebral column. Another left side access to the esophagus is through the triangle of Truesdale (*yellow triangle*). The triangle is formed by the diaphragm below, pericardium anteriorly, and descending aorta posteriorly. LBCV, left brachiocephalic vein; LRLN, left recurrent laryngeal nerve; RBCV, right brachiocephalic vein; RRLN, right recurrent laryngeal nerve; SVC, superior vena cava.

(*Continued*)

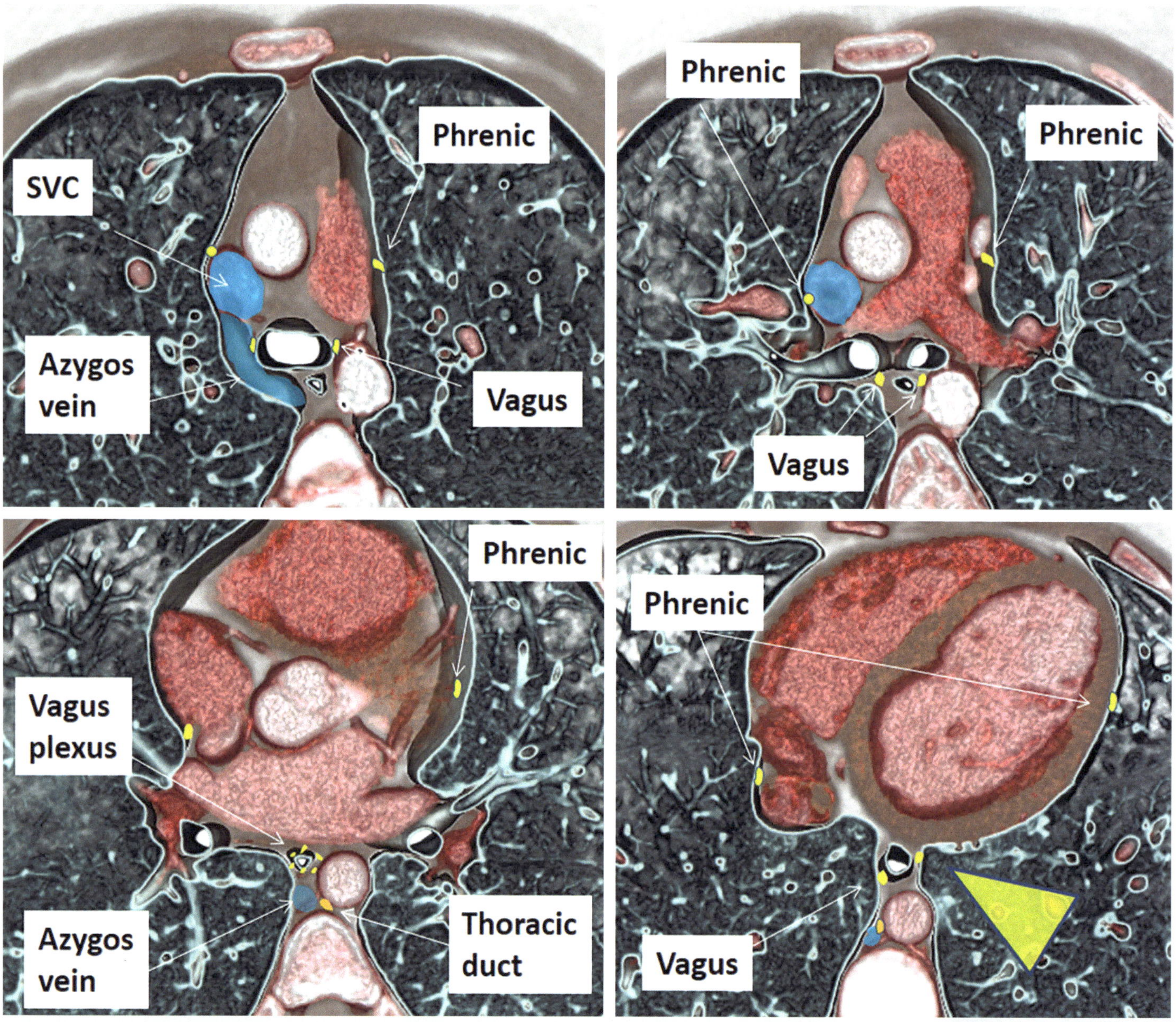

Fig. 2.5 (*Continued*)

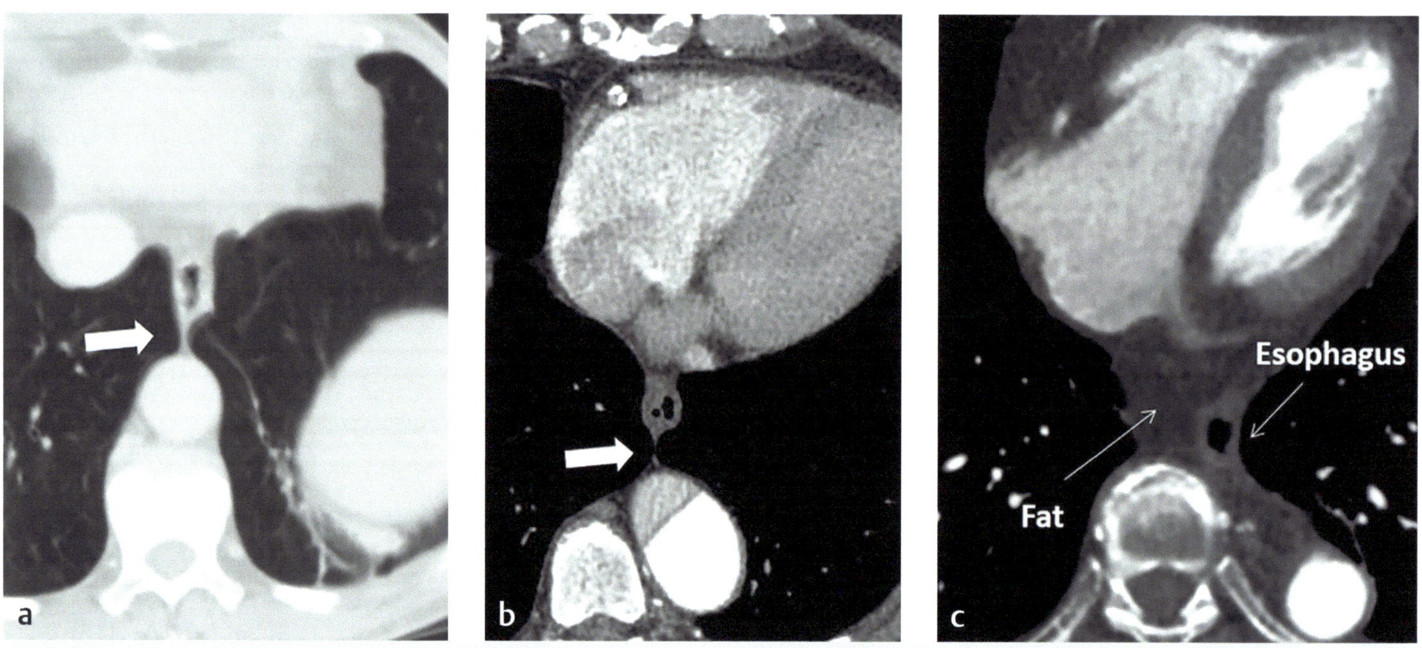

Fig. 2.6 Posterior junctional zone. **(a, b)** The posterior convergence of the right and left pleurae (*arrow*) between the esophagus and aorta above the diaphragm is a common occurrence. **(c)** Excess posterior mediastinal fat and herniation of fatty abdominal content can separate the two pleural lines in this region.

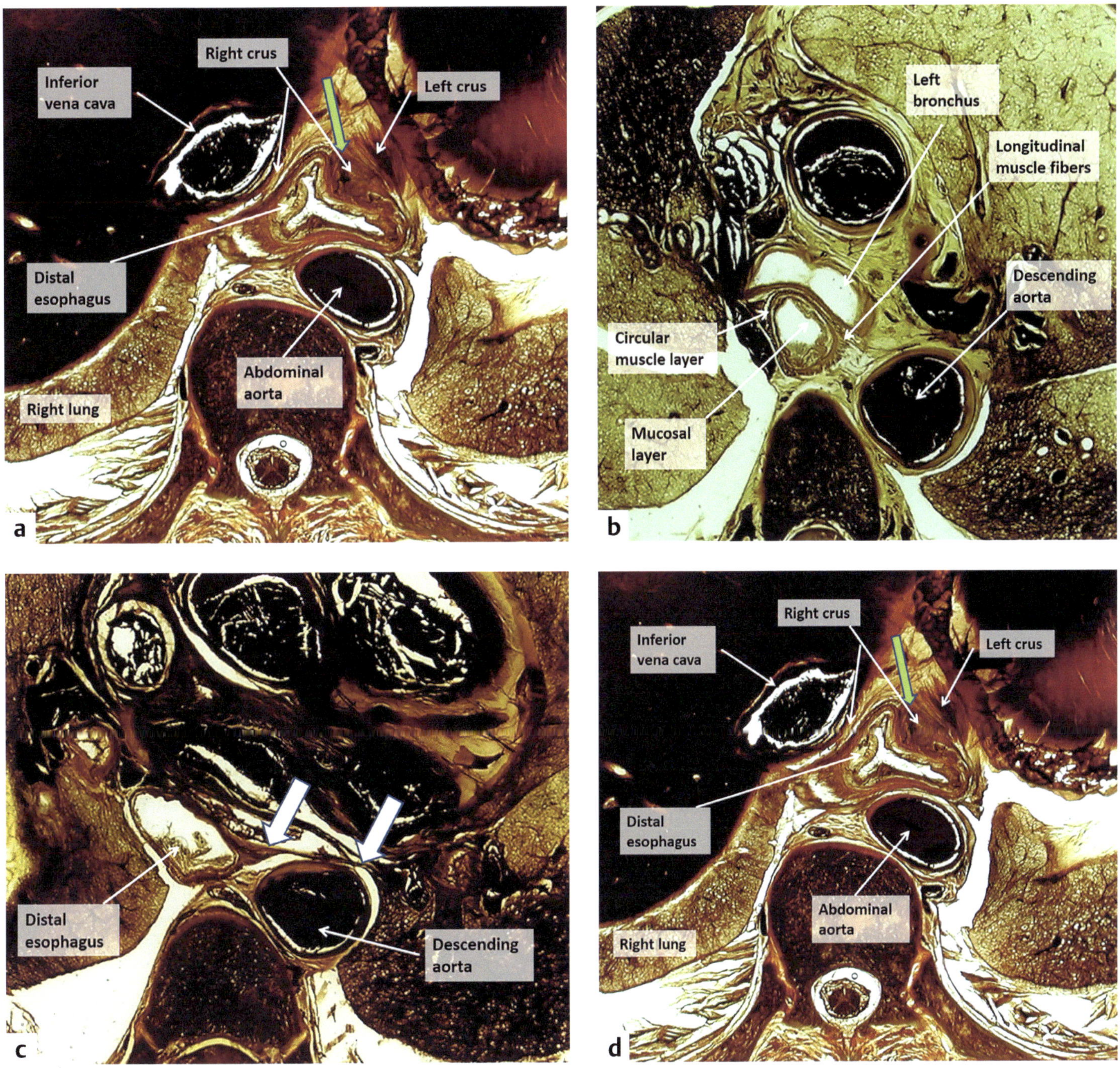

Fig. 2.7 (a–f) Anatomic course of mid to distal esophagus. Longitudinal muscle layer is thicker than the circular muscle in proximal part of the esophagus, but the two layers are equal in thickness in their distal extent. The strongest layer is the mucosa/submucosa. The inferior pulmonary ligament (*white arrows*) is seen extending from the medial margin of the left lower lung to the anterior esophageal wall. The inferior pulmonary vein rests at the upper part of the pulmonary ligament. Also shown in this series is the course of the right diaphragmatic crus (*yellow arrows*) passing along the anterior margin of the esophagus from right to left at the lower esophageal sphincter. The left part of the right diaphragmatic crus is shown by *green arrow*.

(*Continued*)

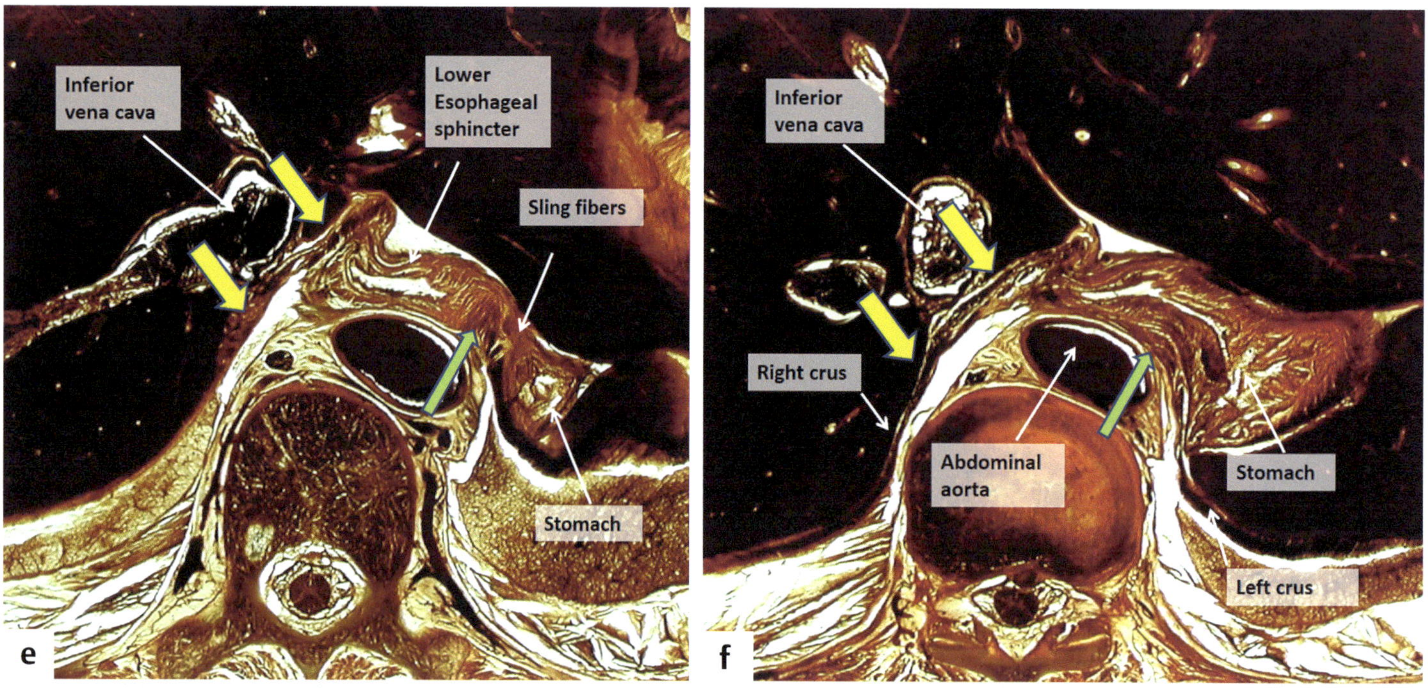

Fig. 2.7 (*Continued*)

the two ends of the esophagus, the muscle fibers run transversely and blend with the inferior constrictor muscle fibers proximally and the LES distally. In the remainder of the esophagus, the circular muscle fibers run obliquely and orient in a helical fashion especially in the upper part of esophagus.

The outer longitudinal muscle fibers connect proximally to the posterior and lateral surfaces of the cricoid cartilage and the cricopharyngeus muscle. They continue distally in the stomach terminating along the anterior and posterior gastric walls. Some accessory posterior muscle bundles of the proximal end of the longitudinal muscles move anterolaterally to connect with the posterolateral surface of the lamina of the cricoid cartilage. These accessory bundles are also called "cricoesophageal tendon." Separation of these muscle bundles behind the esophagus creates a weak area of sparse longitudinal muscle fibers below the cricopharyngeus muscle. This weak area, which is called V-shaped area of the Laimer or the Killian's triangle, is a potential site for development of a diverticulum.

Pharyngeal Muscle

The pharynx is a tube-like structure that is composed of three external circular and three internal longitudinal muscles. The external pharyngeal wall musculature is composed of three overlapping muscles arranged like a three-part telescopic plastic cup. These muscles are the superior, middle, and inferior pharyngeal constrictors. The three internal longitudinal muscles include the stylopharyngeus, palatopharyngeus, and salpingopharyngeus muscles. They merge with each other to form the internal wall of the pharynx and act in coordination to shorten the pharynx and elevate the larynx.[7]

The cricopharyngeus muscle fibers blend superiorly with the inferior constrictor muscle and inferiorly with the circular muscle of the cervical esophagus. It forms a C-shaped transversely oriented muscular band attached to the cricoid cartilage (▶Fig. 2.9, ▶Fig. 2.10). The cricopharyngeus muscle is formed by circular fibers that produce maximum tension in the anteroposterior direction and less tension in the lateral direction. Weak areas exist at several places of the hypopharynx, including above and below the cricopharyngeus muscle, that can become the site for development of pulsion diverticulum and inadvertent instrumental perforations.[8] Diverticulum formed above the cricopharyngeus muscle is called Zenker's diverticulum and the one below it is named Laimer's diverticulum.[1] These diverticula usually pass to the left and occupy the space between the prevertebral fascia and the pretracheal fascia.

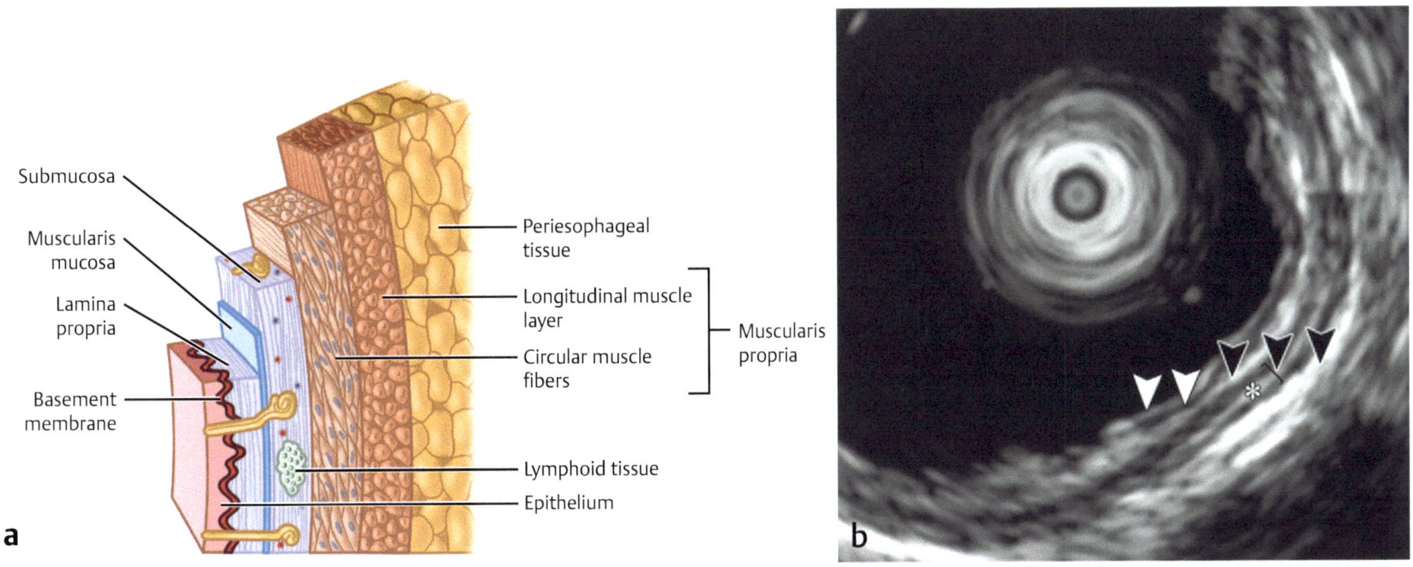

Fig. 2.8 (a) Esophagus wall layers. The esophagus is made up of mucosa, muscularis mucosa, submucosa, muscularis propria, intermuscular septum, and adventitia. **(b)** Endoscopic ultrasound (US) image shows the normal esophageal wall, with five alternating hyper- and hypoechoic layers (*arrowheads*). The hyperechoic layer between the hypoechoic inner and outer muscularis propria (*) is the inner muscular connective tissue layer and in some cases appear prominent. (Adapted with permission from Hong et al.[42])

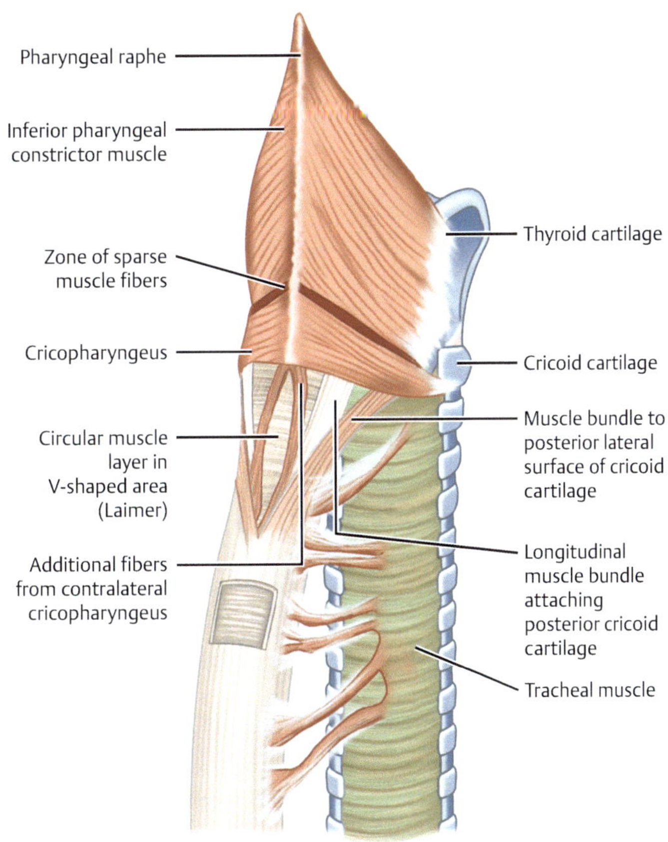

Fig. 2.9 Upper esophageal sphincter. The esophageal longitudinal muscle starts proximally as three bundles: a main longitudinal muscle bundle that originates from the posterior surface of the cricoid cartilage, an accessory muscle bundle from the posterolateral surface of the cricoid cartilage, and an accessory muscle bundle from the contralateral surface of the cricopharyngeus muscle. Muscle bundles from the two sides quickly fan out to surround the entire circumference of the esophagus, leaving only a triangular space, the area of Laimer, uncovered by longitudinal muscle at the most cranial end of the esophagus.

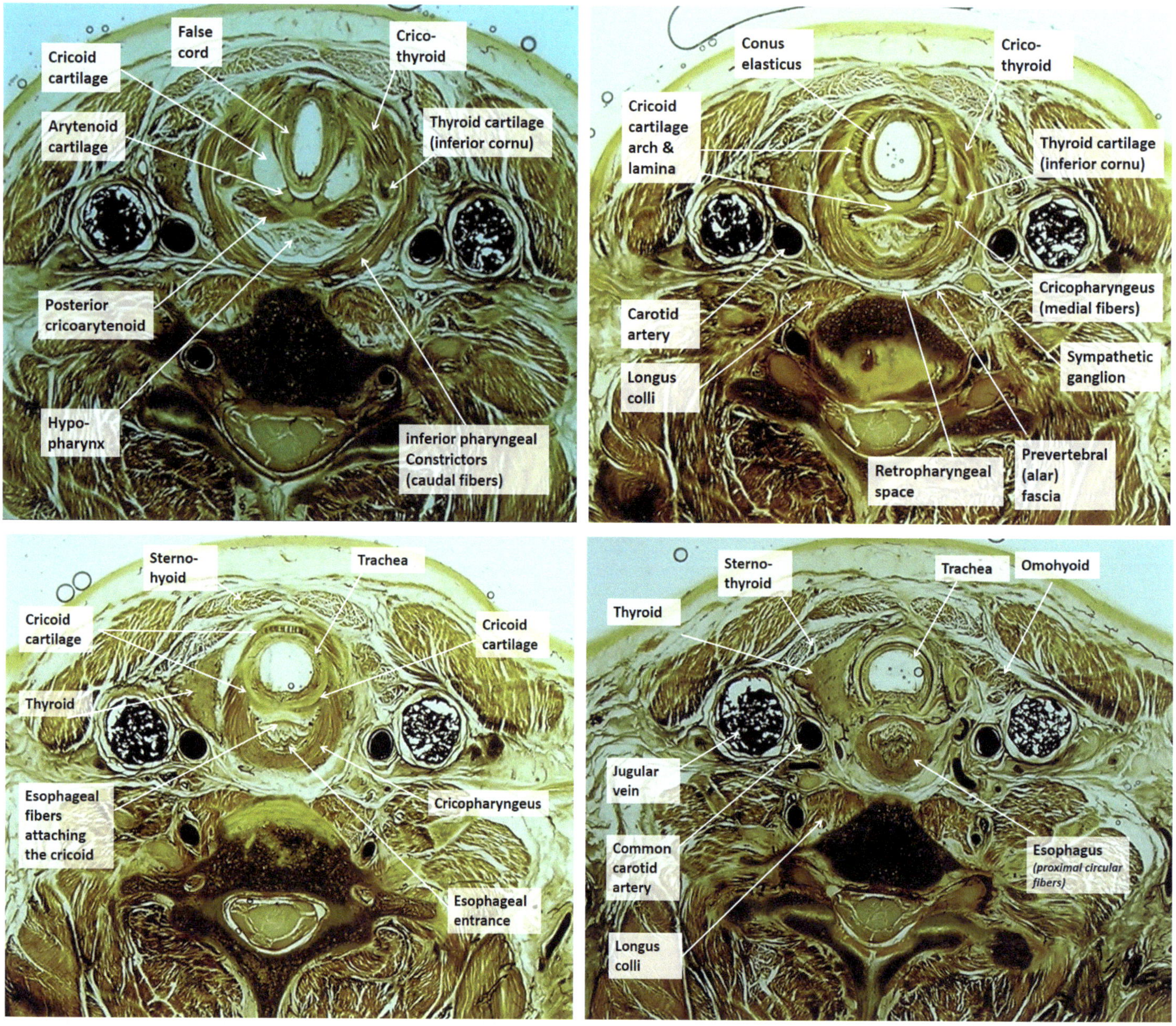

Fig. 2.10 Upper esophageal sphincter. The upper esophageal sphincter (UES) is composed of three striated circular muscles: the caudal fibers of the inferior pharyngeal constrictor (thyropharyngeus), cricopharyngeus, and proximal circular muscle of the esophagus. The UES is supported anteriorly by the posterior surface of the thyroid and cricoid cartilages. The outer longitudinal muscle fibers of the esophagus connect proximally to the posterior and lateral surfaces of the cricoid cartilage and the cricopharyngeus muscle.

◆ Function

The physiology of normal swallowing involves three organized phases: the oral phase, the pharyngeal phase, and the esophageal phase.[6] The oral phase of swallowing is mostly under voluntary muscle control and involves transit of a prepared bolus beyond the oral cavity. During the pharyngeal phase the superior, middle, and inferior pharyngeal constrictor muscles contract in peristaltic series while the intrinsic laryngeal muscles activate to approximate the epiglottis and arytenoids to protect against laryngeal aspiration. Simultaneously, the suprahyoid and pharyngeal elevator muscles contract to move the larynx anterosuperiorly, further shielding it from aspiration.[8,9] The esophageal phase begins with relaxation of the cricopharyngeus muscle.

The esophageal phase of deglutition consists of highly coordinated events from integration of numerous control mechanisms involving the central nervous system, enteric nervous system, and esophageal musculature.[10] Cortical input from the region of the internal capsule regulates the brainstem swallowing center located in the brainstem at the junction of the medulla and pons.[9]

Muscle contractions and relaxations can be easily recorded using intraluminal pressure probes. Longitudinal muscle contraction causes axial shortening of the esophagus, and it would be difficult to measure the contraction and relaxation in the

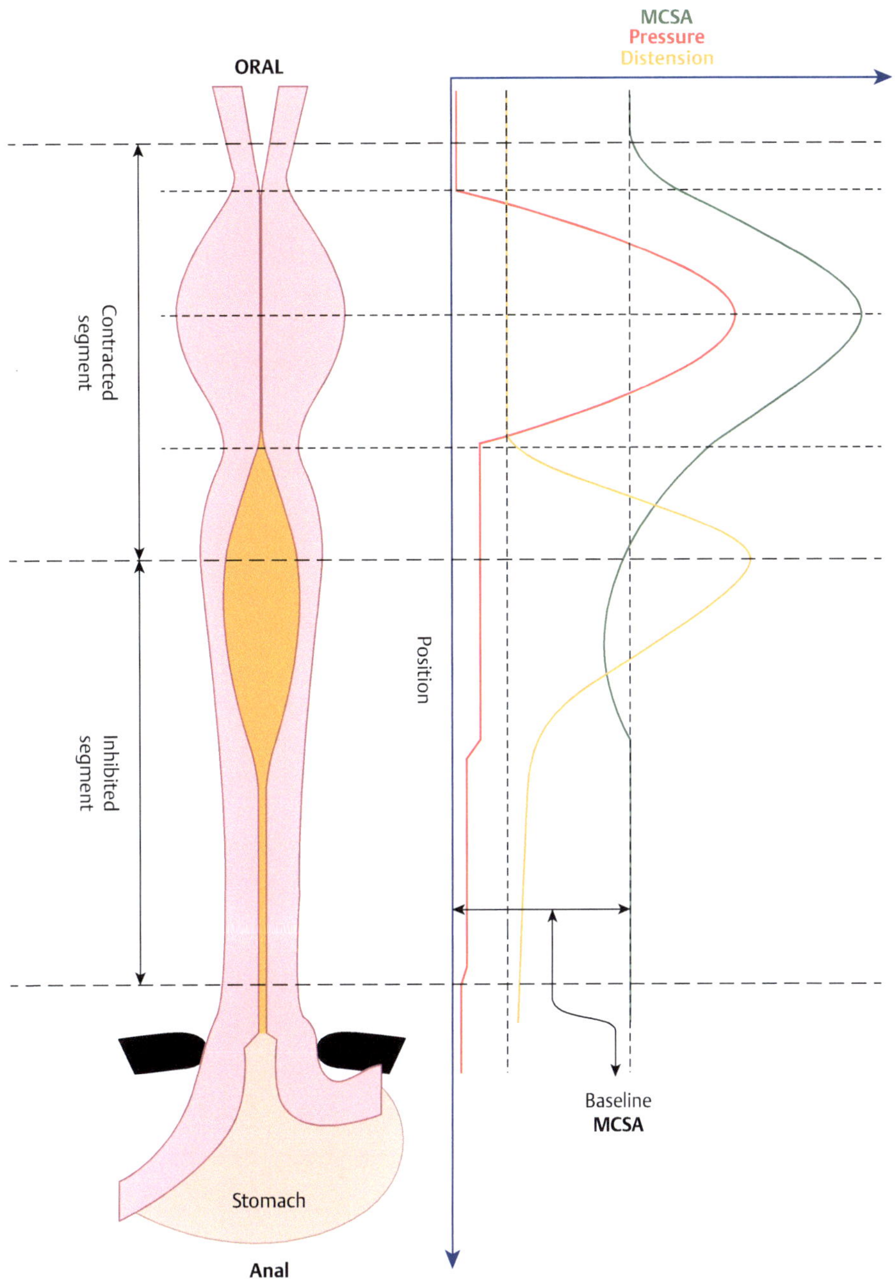

Fig. 2.11 Schematic of contraction and distension during swallow-induced peristalsis. Note that the pressure and muscle cross-sectional area (MCSA), surrogate markers of circular and longitudinal muscle contractions, respectively, precede distension that marches distally in front of the onset of contraction wave in a peristaltic fashion. (Modified from Mittal[11] and Abrahao et al.[12])

longitudinal muscle layer by routine esophageal manometry recordings.[11] Instead high-resolution ultrasound probes have been used to study contraction and relaxation of the longitudinal muscle layer. When the esophagus shortens in length due to longitudinal muscle contraction, it increases in thickness which can be easily measured by ultrasound imaging. More recently, a sensor that utilizes magnet and electromagnetic fields (Hall's effect) has been used to study axial motions of the squamocolumnar junction at the gastroesophageal junction, as an indirect marker of longitudinal muscle contraction of the esophagus.[11]

During normal primary esophageal peristalsis mediated by the vagus nerve, sequential contractions of a segment of the circular muscle start and move from the proximal to the distal end of the esophagus.[35] The length of this contracted segment increases as it moves distally. The longitudinal muscles also contract simultaneously and in perfect harmony with the circular muscle contractions. Pressure curve in the contracted segment shows a bell-shaped appearance, with the peak pressure in its middle[10] (▶Fig. 2.11). Secondary peristalsis occurs by distension of the esophagus (i.e., presence of food or fluid in the esophageal lumen) in a pattern similar to primary peristalsis. Transient relaxation of the LES has a different pattern in which the longitudinal muscle contracts independent of the circular muscle, starting at the caudal end of the esophagus.[9] Swallowing induces a peristaltic wave of distension in front of the peristaltic wave of contraction.[12]

Reflux of gastric acid in the esophagus can cause longitudinal muscle contraction. In patients with nutcracker esophagus and early achalasia, discoordination of contraction between the circular and longitudinal muscle layers occurs that also interferes with normal caudal relaxation of the esophagus leading to improper propulsion of the bolus.[13] Nutcracker esophagus is characterized by high-amplitude peristaltic esophageal contractions, and these patients often present with symptoms of "angina-like" or noncardiac chest pain.[14]

Peristaltic failure of the skeletal muscle of proximal esophagus is rare and can occur in skeletal muscle myopathy (e.g., scleroderma and dermatomyositis) or neurologic dysfunction (e.g., amyotrophic lateral sclerosis, myasthenia gravis, and cerebrovascular accident). Peristaltic failure of the distal esophagus smooth muscle occurs in certain central nervous system disorders such as Parkinson's disease. Animal studies have demonstrated that both excitatory cholinergic and inhibitory nonadrenergic noncholinergic (NANC) neurons are present in the smooth muscle of the esophagus. Cholinergic neurons have a greater influence on the proximal smooth muscle and NANC neurons influence predominates in the distal smooth muscle and in the LES.[10,11] In achalasia, the prototypic example of inhibitory failure, NANC inhibitory ganglion cells are diminished, causing incomplete relation of the LES.[36]

Upper Esophageal Sphincter

The upper esophageal sphincter (UES) is composed of three striated circular muscles: the caudal fibers of the inferior pharyngeal constrictor (thyropharyngeus), the cricopharyngeus, and the proximal circular muscle of the esophagus (▶Fig. 2.9, ▶Fig. 2.10, ▶Fig. 2.12). This region is supported anteriorly by the posterior surface of the thyroid and cricoid cartilages.

The UES is a high-pressure zone which remains tonically contracted between swallows to protect the upper airway from material refluxed from the esophagus and stomach.[10,15] It also prevents breathed-in air from entering the esophagus. This high-pressure zone spans 2 to 4 cm in length and is created primarily by the cricopharyngeus and to a lesser degree by the thyropharyngeus muscle (▶Fig. 2.13).

Cricopharyngeal dysfunction leads to an obstruction of bolus passage, oropharyngeal dysphagia, and later, development of Zenker's diverticulum (▶Fig. 2.14). Oropharyngeal swallowing dysfunction usually occurs in the elderly population where neuromuscular incoordination and weakness are more prevalent. Reliable interpretation of UES motor patterns is mostly based on pressure measurement (manometry).[16] Imaging evaluation of patients with cricopharyngeal dysfunction demonstrates stenosis

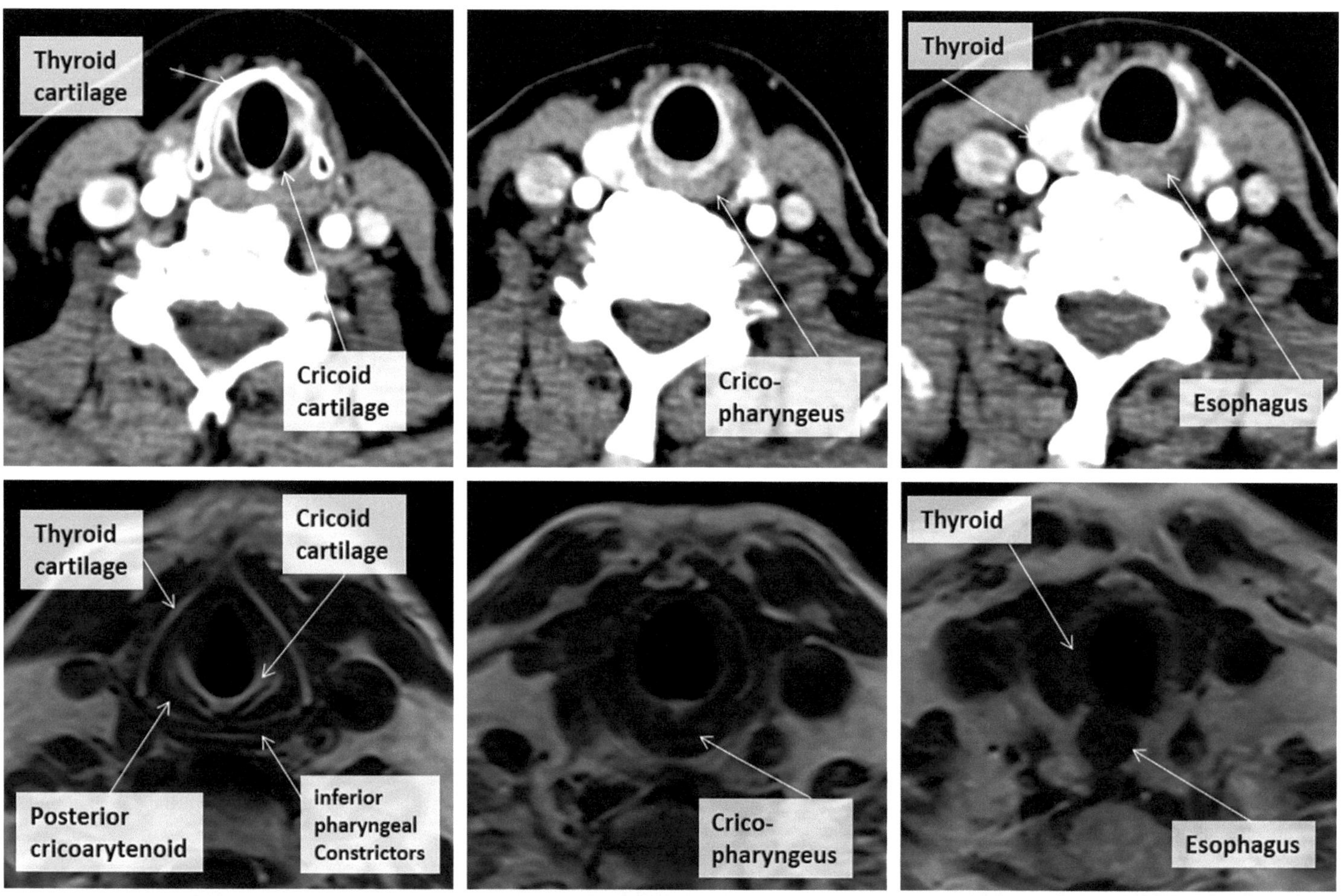

Fig. 2.12 Upper esophageal sphincter region shown by computed tomography (CT) (*upper row*) and magnetic resonance imaging (MRI) (*lower row*). This region is supported anteriorly by the posterior surface of the thyroid and cricoid cartilages.

of the UES which is believed to be secondary to cricopharyngeal hypertrophy or spasm. It creates a stable cricopharyngeal bar on barium swallow videopharyngogram or computed tomography (CT) studies of this region. The size of UES opening in patients with cricopharyngeal dysfunction is reported to be 57% of the size in normal controls.[8] Surgical myotomy is currently used to treat cricopharyngeal dysfunction. Aside from cricopharyngeal dysfunction, other possibilities for stenosis in this region include mucosal webs, fibrosis, and neoplasm.

Lower Esophageal Sphincter

The LES is a complex functional unit at the esophagogastric junction (EGJ) composed of 2- to 4-cm segment of asymmetrically thickened circular muscle fibers creating a high-pressure zone at the distal end of the esophagus.[17,18,19,20] It is called a physiologic sphincter because there is no anatomical structure, such as the pylorus. Because of the resting tone, the LES appears thicker in live situation but in autopsy specimen the LES appears similar to the rest of the esophagus. The upper part of the LES normally lies in the diaphragmatic hiatus, while the lower section is intra-abdominal.

The LES is formed by two components: intrinsic and extrinsic. The intrinsic component is composed of the gastric sling fibers on the left and the distal esophageal clasp fibers on the right (▶Fig. 2.15). The extrinsic component is formed by the crura of the diaphragm wrapping the proximal 2 cm of the sphincter and augments pressure during inspiration. The phrenoesophageal ligament is a fibrous layer of connective tissue arising from the crura, which keeps the LES within the abdominal cavity. Malfunction of any of the above components may result in gastroesophageal reflux.

Anatomical studies have shown that circular muscles at the LES are actually C shaped, not completely circular. On the left side, the open ends of "C-shaped" fibers (clasp fibers) intermingle

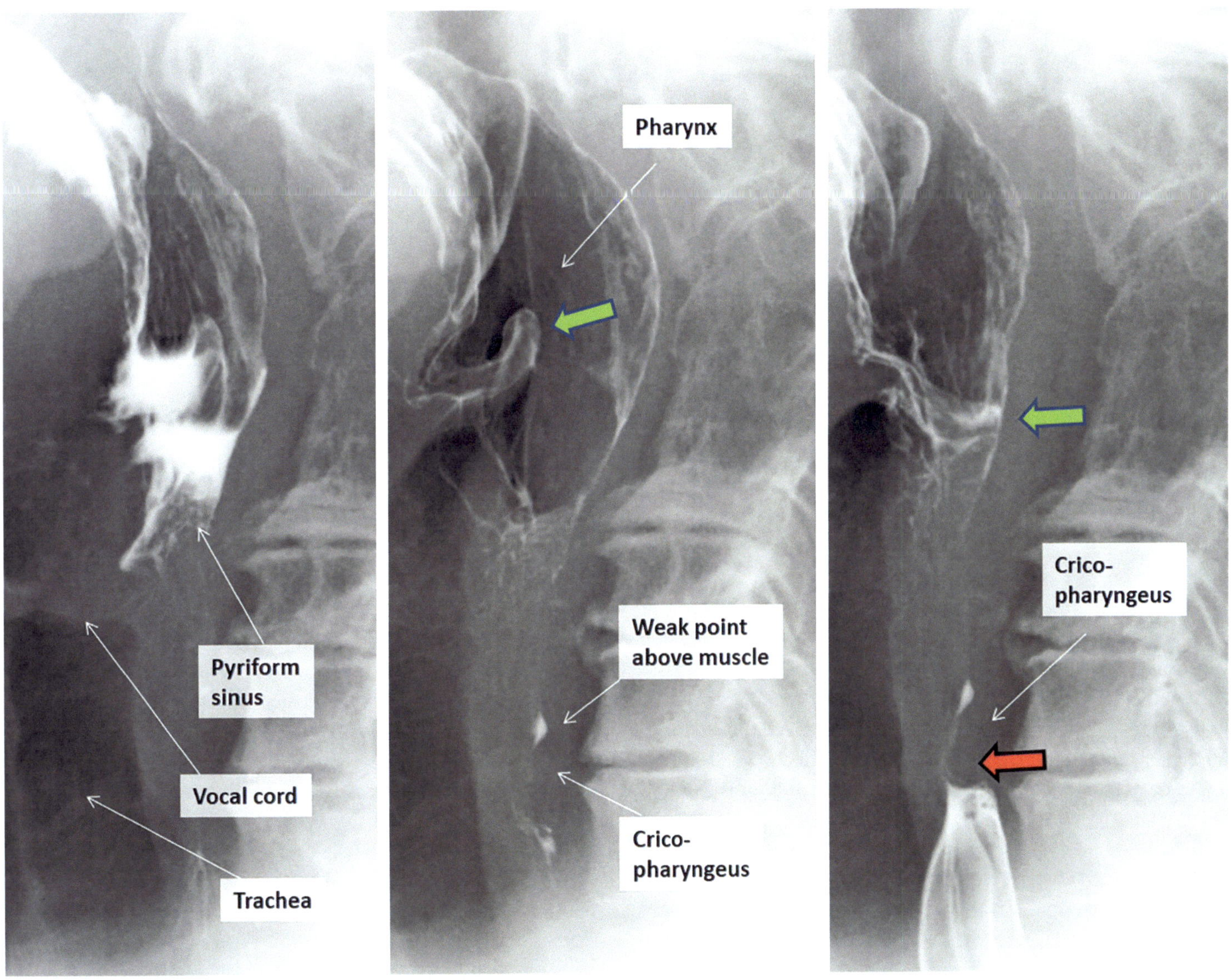

Fig. 2.13 Upper esophageal sphincter (UES). Contraction of the cricopharyngeus muscle (*red arrow*) is shown as the bolus passes the UES creating a transient bar. Weak areas exist at several places of the hypopharynx including above and below the cricopharyngeus muscle that can become the site for development of pulsion diverticulum. Note approximation of the epiglottis (*green arrows*) to the arytenoids and elevation of the larynx to protect against laryngeal aspiration.

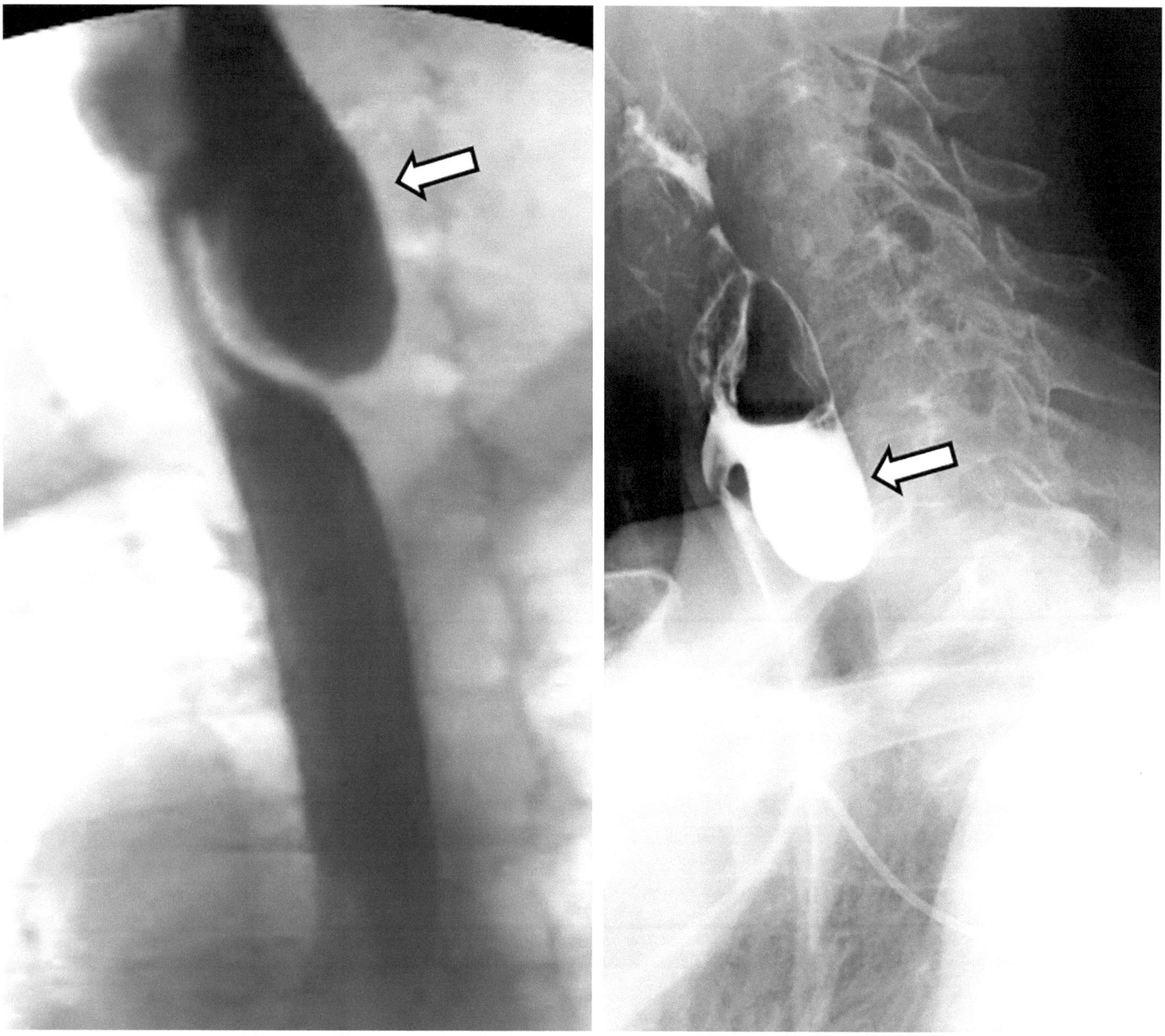

Fig. 2.14 Examples of Zenker's diverticulum (*arrows*).

with the gastric sling fibers (▶Fig. 2.15, ▶Fig. 2.16). The gastric sling fibers, first described by Liebermann-Meffert et al, are actually medial fibers of the oblique muscle of the stomach which loop around the EGJ on the left side, and its two limbs continue on the anterior and posterior walls of the stomach, parallel to the lesser curvature of stomach, to end in the vicinity of the gastric antrum.[21,22]

The diaphragm consists of two components, namely, the crural and costal, with different embryonic origin. Myoblasts of the body wall invade two pleuroperitoneal membranes and form the costal diaphragm. However, the two crura develop in the mesoesophagus. The costal diaphragm is the respiratory muscle, but the crural diaphragm has both respiratory and gastrointestinal functions.[11,22] The diaphragmatic sphincter of LES is a skeletal muscle and formed primarily by the right crus, but the left crus also contributes partially to its formation (Fig. 2.17). The inner fibers of the right crux are oriented in circumferential direction, covering the anterior aspect of the LES, and outer fibers are directed in a craniocaudal direction. The unique arrangement of muscle fibers results in two different types of actions of the diaphragmatic sphincter on the esophagus when it contracts: a craniocaudal motion and a circumferential squeeze.[18,20] Phrenic nerves formed by the branches of C5, C6, and C7 nerve roots provide both motor and sensory innervation to the crural and costal diaphragm.[23]

The LES and crural diaphragm are anchored to each other by the phrenoesophageal ligament. The phrenoesophageal ligament is formed from condensation of the loose areolar tissue below the diaphragm which is continuous with the transversalis fascia. It anchors the distal esophagus to the undersurface of the diaphragmatic sphincter (▶Fig. 2.15). It forms two leaflets near the esophageal wall. The ascending leaflet inserts above the squamocolumnar junction and the descending leaflet is attached

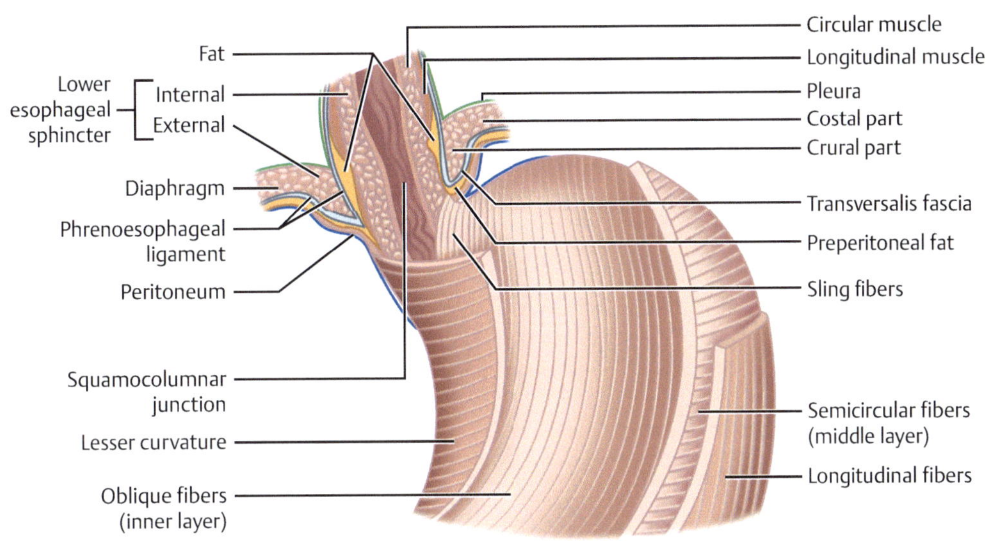

Fig. 2.15 Anatomy of lower esophageal sphincter (LES, internal sphincter) and the crural diaphragm (external sphincter). The two sphincters are anatomically superimposed on each other and are anchored by the phrenoesophageal ligament. The LES internal sphincter consists of clasp and sling fibers. Clasp fibers are continuous with the circular muscles of the lower esophagus. They are semicircular and C shaped at the level of LES with their open ends mixing with the sling fibers. The sling fibers are thickened muscle layer that form a collar at the left lower end of the gastroesophageal junction that extends caudally and toward the lesser curvature and remain continuous with the oblique muscles of stomach. The sling fibers form the angle of His. Oblique fibers of the stomach wall form a collar around the left lower end of the esophagus that extends caudally and toward the lesser curvature. (Adapted with permission from Liebermann-Meffert et al[21] and Mittal and Balaban.[22])

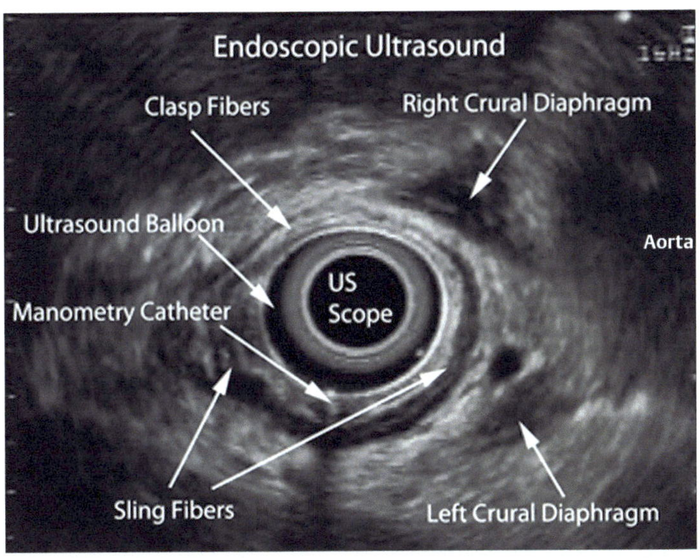

Fig. 2.16 Endoscopic ultrasound image at the level of the lower esophageal sphincter (LES) organized as clasp and sling fibers. Also shown are the left and right crural diaphragm. (Adapted with permission from Rampado et al.[40])

several centimeters below the ascending leaflet around the cardia. Between the phrenoesophageal ligament and the wall of esophagus is a ring of fat. Because of this specific morphology and internal elastic property, the phrenoesophageal ligament allows the distal esophagus to slide in and out of the hiatus several by centimeters. Weakening of the phrenoesophageal ligament, as may occur with age, results in a sliding hiatal hernia, which may be reducible or nonreducible[24] (▸ Fig. 2.18).

Normal LES relaxation occurs during three main major events: (1) triggered by pharyngeal stimulation during a swallow; (2) triggered by esophageal distension with secondary peristalsis; and (3) triggered by gastric distension particularly of the cardia.

The clasp and sling fibers show marked differences in their physiological properties and have been referred to as two distinct lower esophageal sphincters. Exact mechanism of circumferential squeeze by the clasp and sling fibers is not clear. The LES remains tonically contracted at rest from cholinergic innervation mediated by acetylcholine (neurogenic tone) and intracellular calcium influx (myogenic tone).[23,25] The clasp-like semicircular fibers have significant myogenic tone at rest but are not very responsive to cholinergic stimulation, whereas the gastric oblique sling fibers have little resting tone but contract vigorously to cholinergic stimulation. The gastric sling plays a role in the formation and modulation of angle of His. The angle of His acts as a flap or pinchcock in preventing reflux of gastric contents into the esophagus. This angle can be compromised by either a sliding or paraesophageal hernia, leading to decreased LES pressures. The crural muscles also act as an adjunctive extrinsic component raising intraesophageal pressure during inspiration. In general, end-expiratory pressure at the lower end of the esophagus is due to tonic contraction of the LES, and the increase in pressure with inspiration is due to the contribution from the diaphragmatic sphincter. With forceful inspiration the LES pressure can reach 100 to 150 mmHg.[17] The phrenoesophageal ligament anchors the lower end of the esophagus to the diaphragmatic sphincter and plays a key role in the maintenance of an intact antireflux barrier.

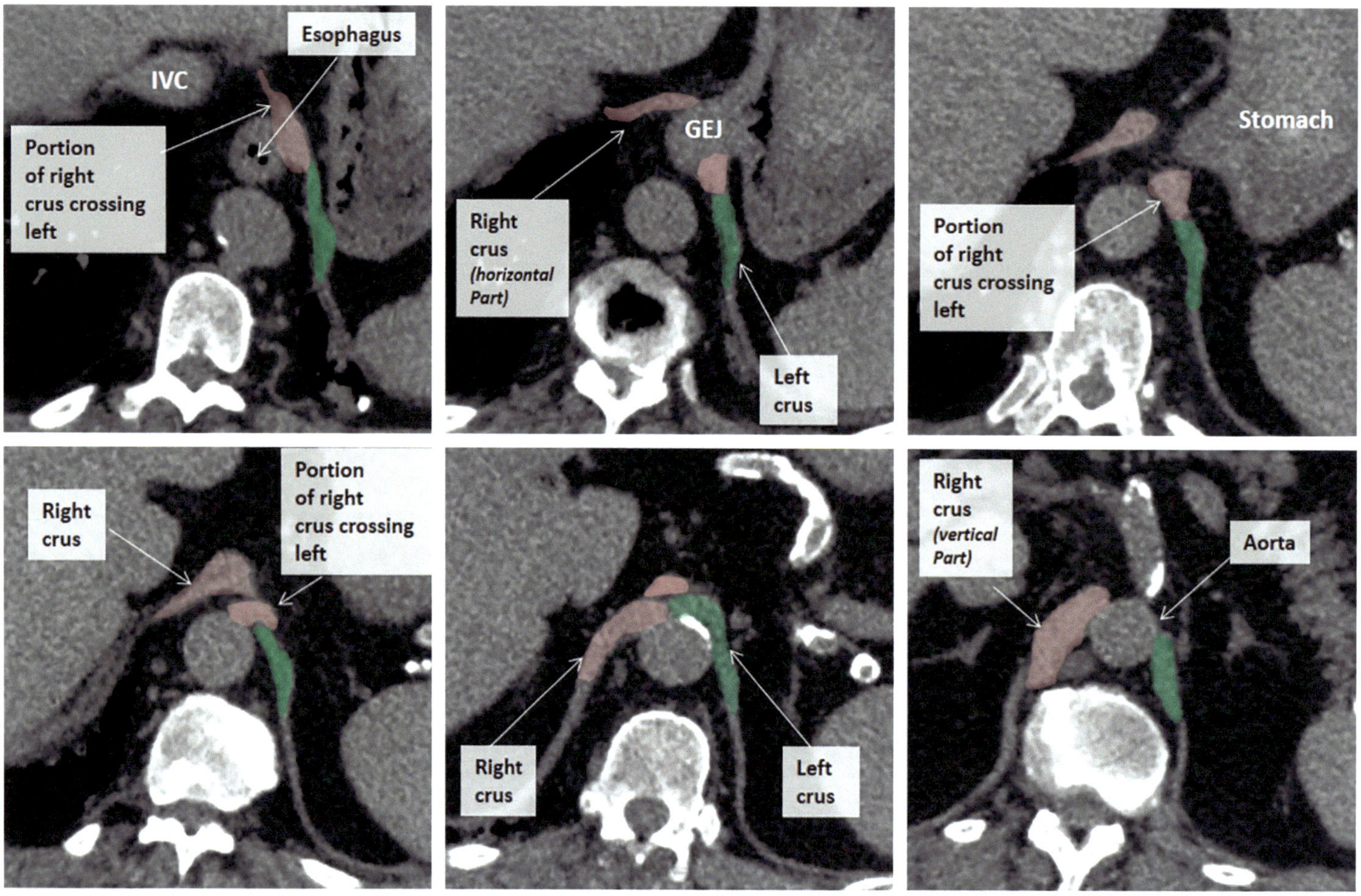

Fig. 2.17 Extrinsic component of the lower esophageal sphincter (LES) mechanism axial CT images from top to bottom. The extrinsic component is formed by the crura of the diaphragm wrapping the proximal 2 cm of the sphincter and augments pressure during inspiration. The crural sphincter of LES is a skeletal muscle and is formed primarily by the right crus (*red*), but the left crus (*green*) also contributes partially to its formation. The inner fibers of the right crux are oriented in circumferential direction covering the anterior and left aspects of the LES. GEJ, gastroesophageal junction; IVC, inferior vena cava.

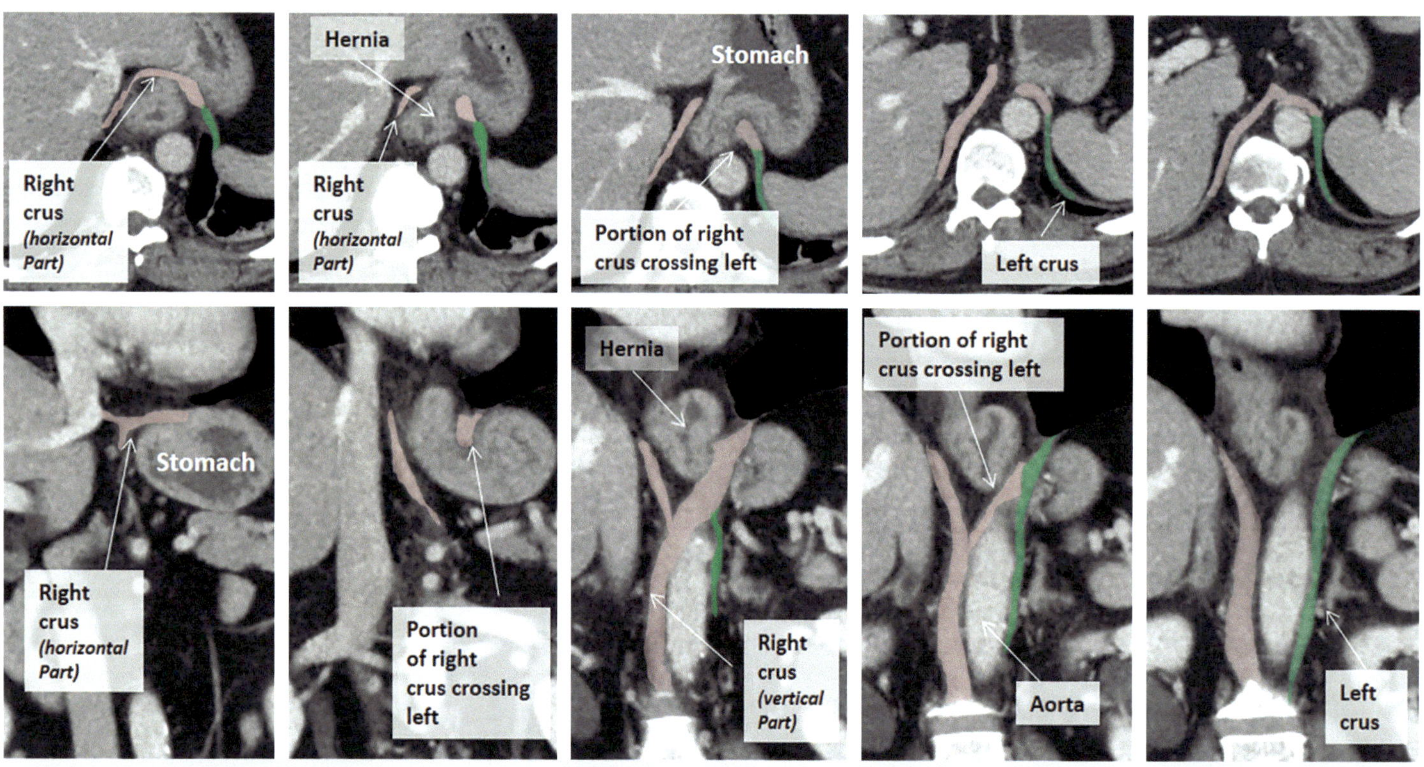

Fig. 2.18 Mildly dilated crural sphincter of the lower esophageal sphincter causing mild hernia of the stomach axial CT images (*top*) and coronal views (*bottom*). The crural component is formed primarily by the right crus (*red*), but the left crus (*green*) also contributes partially to its formation. Weakening of the phrenoesophageal ligament occurs with age and can result in a sliding hiatal hernia.

Disruption of the aligned intrinsic and extrinsic musculature caused by increased abdominal pressure from obesity, accumulation of fat in the hiatus, or a large sliding hiatal hernia may play a key role in the pathophysiology of gastroesophageal reflux. The asymmetry in structure of the gastroesophageal junction results in highest radial pressures on the left posterior side where the intrinsic LES and extrinsic crural muscle fibers coexist. This may explain, in part, why esophageal erosions or tears tend to occur on the right side of the wall.

Endoscopic localization of the LES is different from the manometric localization.[26,27,28] Endoscopically, the LES is defined as the junction of transition from nonstratified squamous esophageal epithelium to the gastric mucosa. This mucosal junction is known as the Z-line. The location of the functional LES is wider and extends 3 cm distal to the Z-line. The distal esophagus has been called the phrenic ampulla because of its bulbar shape on barium swallow. The phrenic ampulla most likely corresponds to the anatomical structures of the LES, along with a part of the proximal stomach. Radiologically, the LES is defined as an area of luminal dilation or "vestibule" (upper part of ampulla), as visualized during swallows when this area is relaxed (▶Fig. 2.19). The proximal end of the vestibule is sometimes marked by an indentation called "A" muscular or "A" ring. The distal end of the vestibule in some patients may be marked by another ring-like structure known as "B" (Schatzki's) ring. The latter generally corresponds to the squamocolumnar junction.

The central control mechanism for the LES is located in the brainstem. The rostral cells of the dorsal motor nucleus of the vagus nerve are involved in the excitatory innervation to the LES and its caudal cells are involved in the inhibitory innervation.[22]

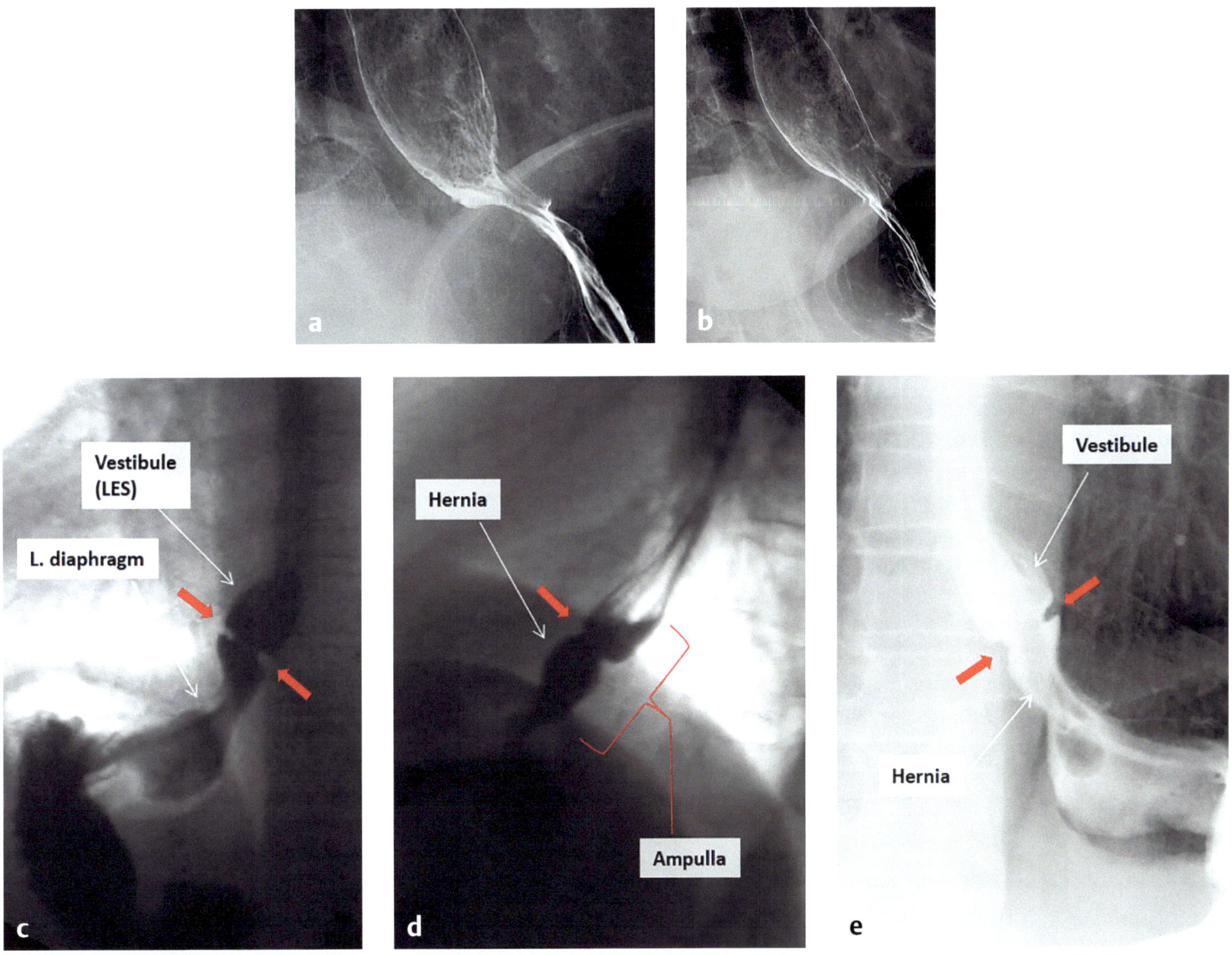

Fig. 2.19 (a, b) Normal opening of the lower esophageal sphincter (LES). **(c–e)** Schatzki's ring. Radiologically, the LES is defined as an area of luminal dilation or "vestibule" as visualized during swallows when this area is relaxed. The proximal end of the vestibule is sometimes marked by an indentation called "A" (muscular) ring. In some cases, the distal end may be marked by another ring, which is due to thickened mucosal/submucosal layers and called Schatzki's or B ring (*red arrows*). The latter generally corresponds to the squamocolumnar junction. The Schatzki's ring is almost always associated with a small sliding hernia.

The resting tone of the LES will be decreased with the act of swallowing, and with distension of the esophageal body or the gastric fundus.[27] Relaxation of the LES in the absence of swallow is known as transient LES relaxation (TLESR).[28] TLESR allows retrograde transport of gastric contents into the esophagus during belching and vomiting especially in sitting position. TLESR is shown to be the major mechanism of gastroesophageal reflux in normal subjects and patients with gastroesophageal reflux disease (GERD). Gastric distension (i.e., 750 mL) is a potent stimulus for TLESR. Pharyngeal intubation can also increase the rate of TLESRs.[27,28] Both the LES and diaphragmatic sphincter distensibility is increased in reflux disease compared to normal subjects.[26] H2 antagonists and Baclofen (a GABA-B agonist) seem to reduce the frequency of TLESRs.[29] Fundoplication operation may reduce TLESR frequency by 50%.[30]

◆ Neurovasculature

Arteries and Veins

The branches of the inferior thyroid artery provide arterial blood supply to the UES and cervical esophagus. Accessory branches may arise from branches of the subclavian vertebral or carotid arteries. The terminal branches of bronchial arteries especially the left bronchial supply the upper thoracic esophagus. The right intercostal arteries, especially the fifth branch, occasionally provide branches to the esophagus. The lower thoracic esophagus is supplied by the two esophageal branches directly arising from the descending aorta at T6–T8 level. These two branches move behind the esophagus and connect with other branches. The left gastric artery, short gastric artery, and a recurrent branch of the left inferior phrenic artery supply the most distal segment of the esophagus.

Venous drainage of the proximal esophagus is into the inferior thyroid vein which connects with the brachiocephalic veins. Occasionally, drainage occurs directly into the brachiocephalic or vertebral veins. The veins of the proximal and distal esophagus primarily drain into the azygos on the right and hemiazygos/accessory hemiazygos on the left. Venous drainage of the abdominal esophagus occurs into the left gastric vein, a branch of the portal vein, or in some cases directly into the left phrenic vein which drains into the systemic veins. At submucosal level connections exist between the portal and systemic venous systems in the distal esophagus. These veins will be enlarged and become varicoid in cirrhosis with portal hypertension (▶Fig. 2.20).

Lymphatics

The lymphatics of the upper esophagus drain into the deep cervical lymph nodes and the lymphatics of the midthoracic esophagus drain into the superior and posterior mediastinal nodes. Lymphatics of the distal esophagus follow the left gastric artery to the gastric and celiac lymph nodes. There is extensive submucosal lymphatic network of the esophagus which allows considerable interconnections among drainage sites of esophagus and bidirectional lymph flow. This network increases the likelihood of early regional lymph node metastases from esophageal cancer and longitudinal spread along the esophageal wall.[31] Consequently, cancer of the upper or lower esophagus can metastasize to involve the nodes on the other end (▶Fig. 2.21). At CT, normal lymph nodes are usually smaller than 1 cm in short-axis diameter and have a smooth, well-defined border; a uniform, homogeneous attenuation; and a central fatty hilum (▶Fig. 2.22). Detection of pathologic lymph nodes on CT depends primarily on size criteria. Intrathoracic and abdominal lymph nodes larger than 1 cm and supraclavicular lymph nodes larger than 5 mm in short-axis diameter are considered abnormal lymph nodes. However, normal-sized lymph nodes that contain microscopic metastatic foci cannot be differentiated from non-metastatic lymph nodes on CT, which can lead to understaging. Furthermore, mildly enlarged, benign, reactive and inflammatory lymph nodes seen on CT may lead to overstaging.[31] Therefore, CT is a nonsensitive test for depicting metastatic spread to regional lymph nodes in which tumor cells tend to involve nodes of normal size (▶Fig. 2.23). Differentiation between benign and malignant lymph nodes is generally difficult with either CT or fluorodeoxyglucose (FDG)-PET. Endoscopic ultrasound appears to have higher accuracy rates (72–80%) for assessing lymph node metastases than does CT (46–58%). Nodes that are round, have a hypoechoic central echo pattern, and have sharply demarcated borders are more likely to represent malignancy. PET-CT is also superior to CT for detecting lymph node metastases and can depict metastases in normal-sized lymph nodes through the uptake of FDG.[31,32]

Nerves

The motor innervation of the esophagus is predominantly via the vagus nerve.[34,35,36] Vagal nerve fibers originate from four medullary nuclei, namely, the dorsal nucleus of the vagal nerve (responsible for parasympathetic output to viscera), nucleus ambiguus (origin of preganglionic parasympathetic to the viscera), solitary nucleus (receives afferents from viscera), and the spinal trigeminal nucleus (receives sensory input from the dura and posterior cranial fossa).[33,37]

The innervation of the striated muscle of the pharynx and upper esophagus originates from nucleus ambiguus in the brainstem and travels along the vagus to synapse directly on motor end plates.[37] The vagal efferents (motor neurons) of the distal esophagus and LES originate in the dorsal motor nucleus of the vagus and terminates in ganglia in the myenteric plexus. The vagal afferents (sensory neurons) from the LES travel to the solitary nucleus.[22]

The vagus nerves enter the thorax behind the common carotid arteries. On the right it passes posterior to the innominate artery and superior vena cava and lateral to the trachea (▶Fig. 2.5). The right recurrent laryngeal nerve originates from the right vagal nerve and turns around the right subclavian artery and ascends to the neck. The left vagus passes anterior to the left subclavian artery. The left vagal nerve passes lateral to the aortic arch and then behind the left main stem bronchus. The left vagal nerve gives rise to the left recurrent laryngeal nerve, which returns around the aortic arch and travels back to the neck. Both vagal nerves descend posterior to the main stem bronchi and form the

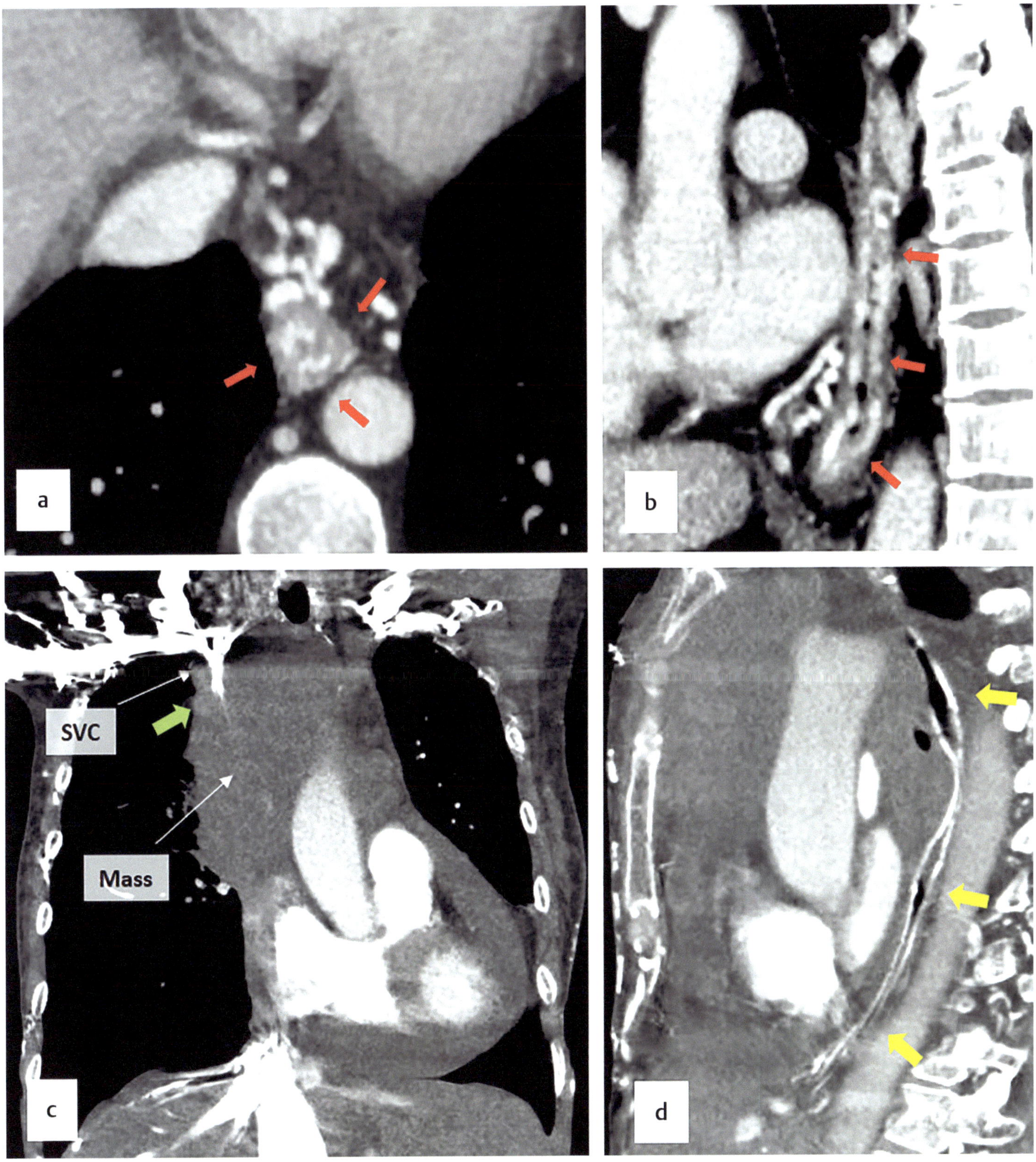

Fig. 2.20 Esophageal varices. **(a)** Axial and **(b)** sagittal computed tomography (CT) showing distal esophageal varices (*red arrows*) in a patient with cirrhosis and portal hypertension. **(c)** Coronal and **(d)** sagittal CT showing extensive esophageal varices (*yellow arrows*) due to extensive mediastinal collateral venous channel formation after obstruction (*green arrow*) of the superior vena cava (SVC) by large mediastinal mass.

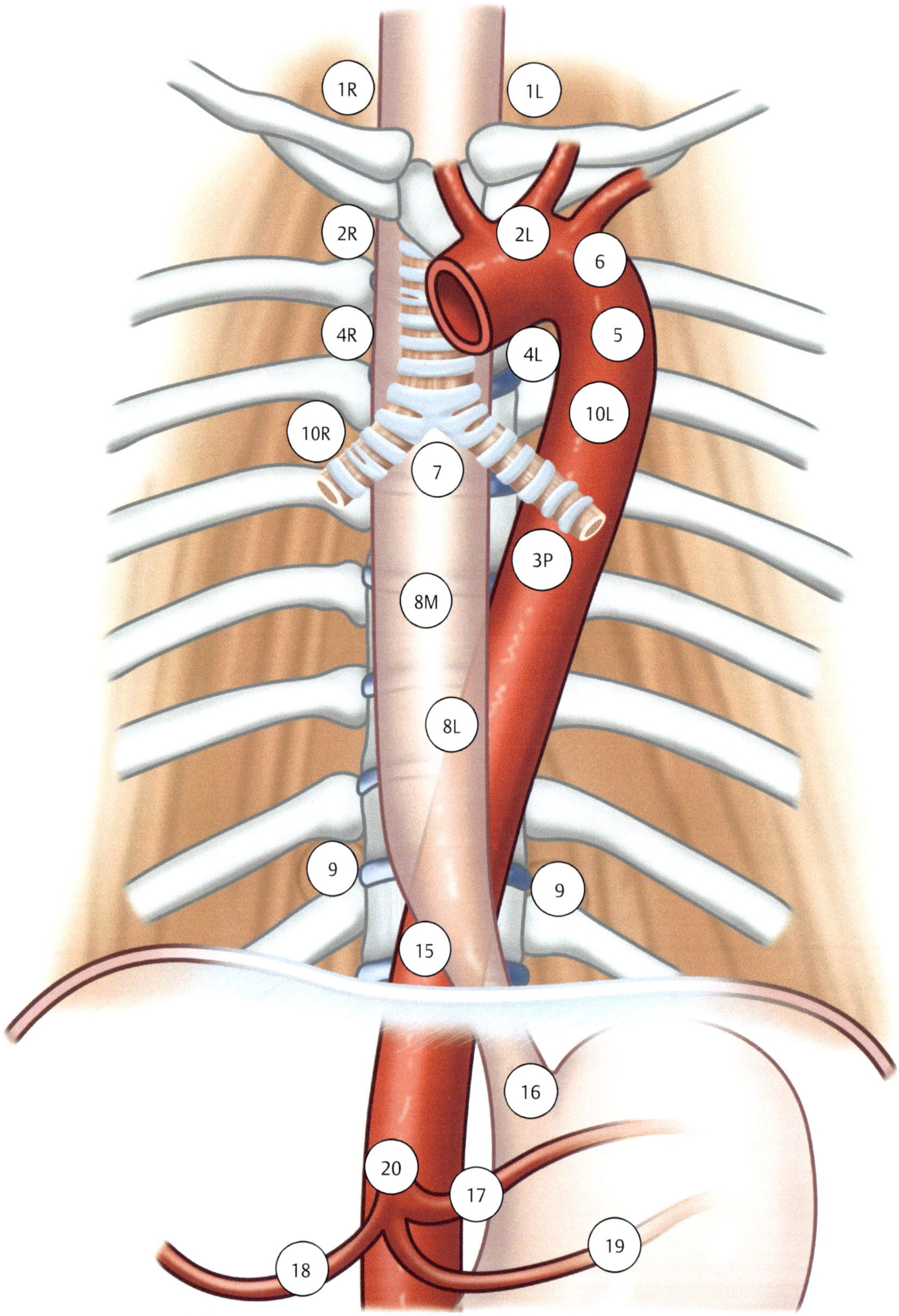

Fig. 2.21 Regional lymph nodes according to the seventh edition of the staging manual for esophageal cancer. 1L, left supraclavicular; 1R, right supraclavicular; 2L, left upper paratracheal; 2R, right upper paratracheal; 4L, left lower paratracheal; 4R, right lower paratracheal; 5, aortopulmonary; 6, anterior mediastinal; 7, subcarinal; 8L, lower paraesophageal; 8M, middle paraesophageal; 9, pulmonary ligament; 10L, left tracheobronchial; 10R, right tracheobronchial; 15, diaphragmatic; 16, paracardial; 17, left gastric; 18, common hepatic; 19, splenic; 20, celiac. The posterior mediastinal lymph node (3P) is not shown. (Modified with permission from Hong et al.[42])

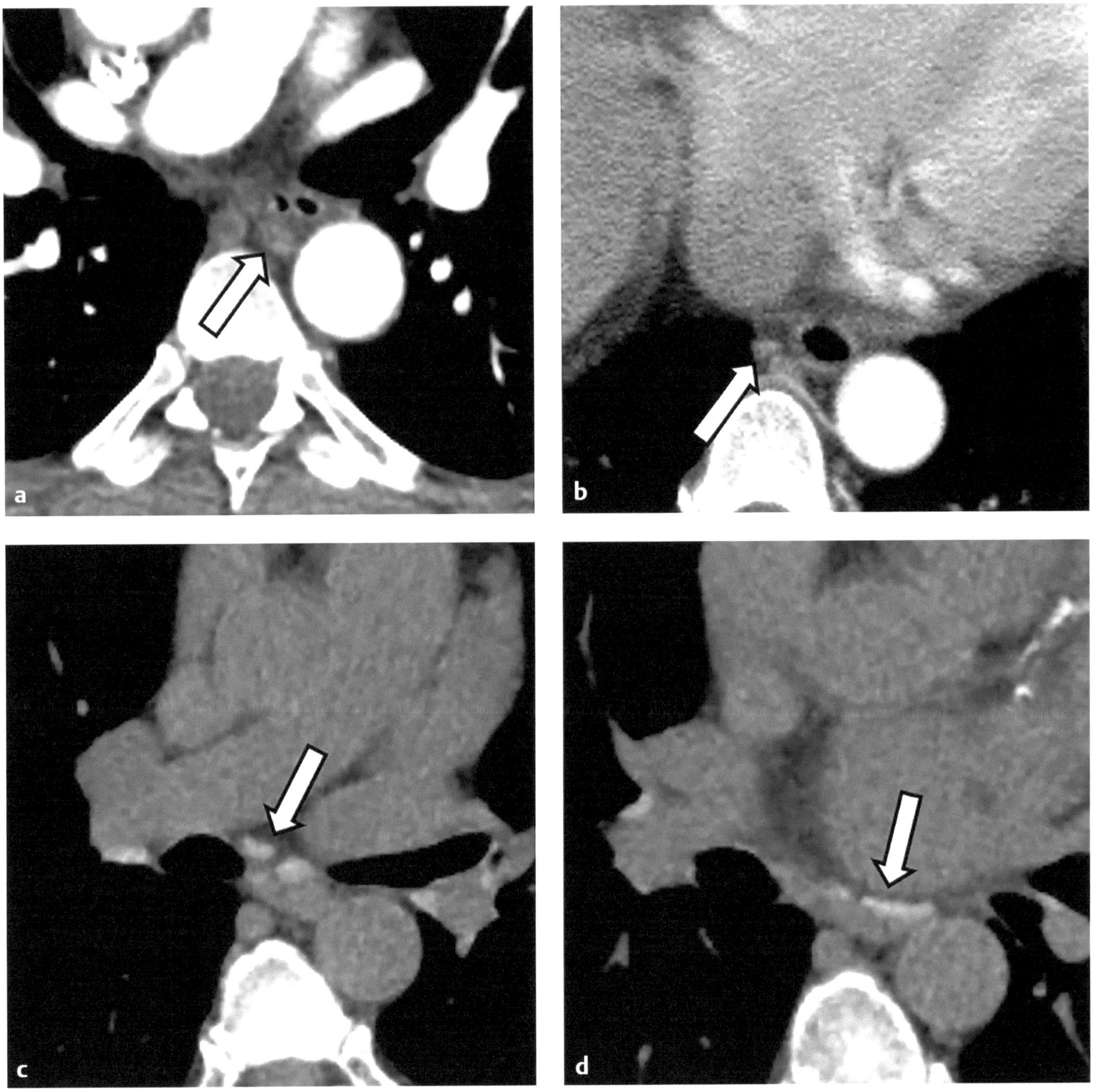

Fig. 2.22 Normal posterior mediastinal node shown with computed tomography (CT) (*arrows*). **(a)** Posterior to the esophagus and between the descending thoracic aorta and the azygos vein. **(b)** Normal juxtaesophageal node. **(c, d)** Calcified, normal-sized nodes in subcarinal and juxtaesophageal regions.

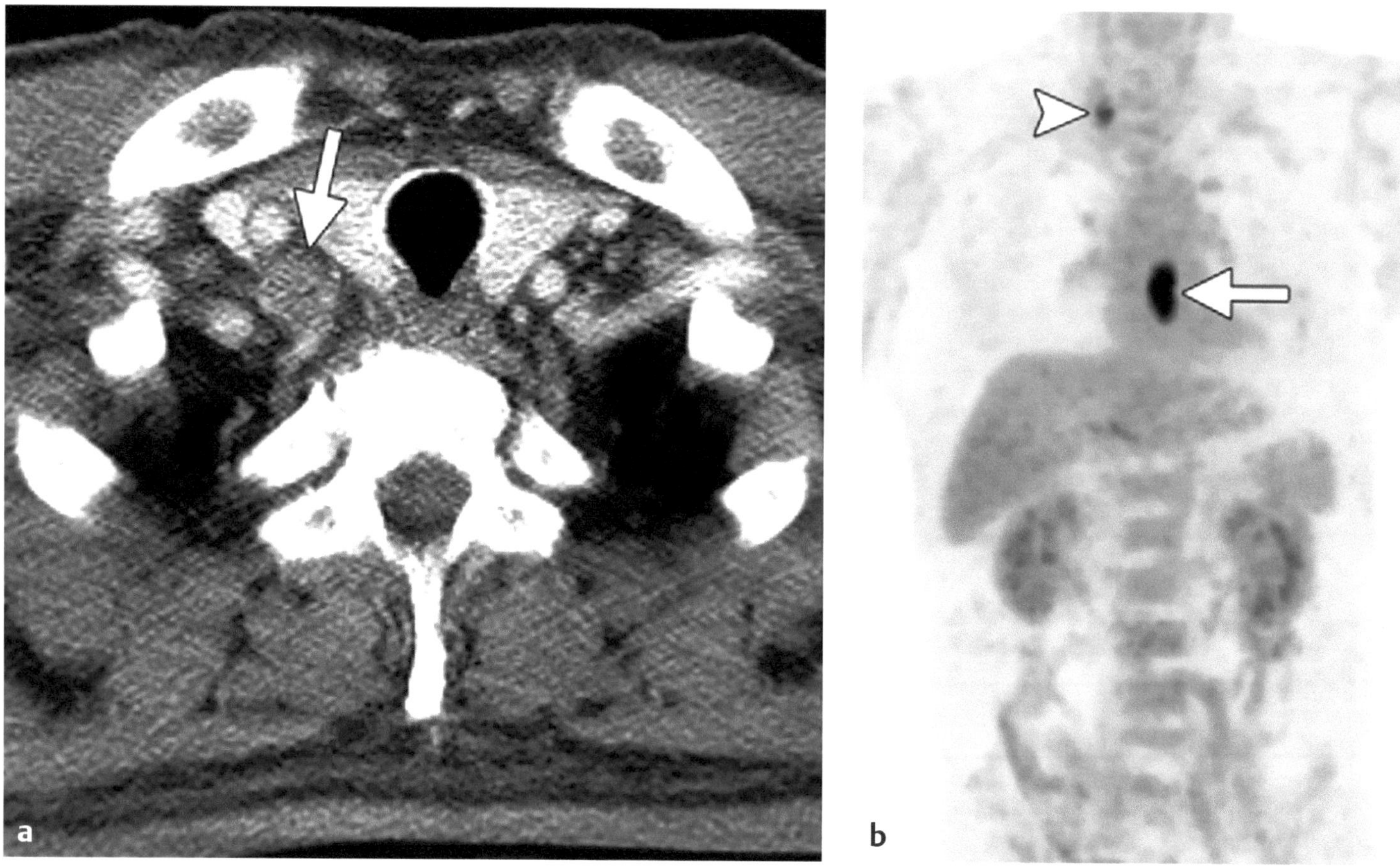

Fig. 2.23 N1 disease. **(a)** Axial, contrast-enhanced computed tomography (CT) image shows an enlarged right supraclavicular lymph node (*arrow*) measuring 11 mm in short diameter. **(b)** Coronal positron emission tomography (PET) image shows fluorodeoxyglucose (FDG) uptake in the primary tumor in the midesophagus (*arrow*) and in the right supraclavicular lymph node (*arrowhead*). Surgery confirmed squamous cell carcinoma with a malignant lymph node (N1 disease), which was considered M1b disease in the sixth edition of the TNM staging system. (Adapted with permission from Hong et al.[42])

cardiac, pulmonary, and esophageal plexuses. Then the vagus nerves form an extensive plexus around the thoracic esophagus (▶Fig. 2.24). Above the diaphragm they coalesce once more into two trunks.[37] The left trunk courses anteriorly and the right trunk posteriorly as they pass through the esophageal hiatus. The anterior vagus gives off the hepatic branch and the anterior nerve of Latarjet and the posterior vagus gives off the celiac branch and the posterior nerve of Latarjet.

Sympathetic innervation is provided by branches of the superior and inferior cervical ganglia in the neck and the splanchnic nerves and the celiac plexus in the chest and abdomen. The LES is innervated by both parasympathetic (vagal) and sympathetic (mainly splanchnic) nerves.

The intramural innervation of the esophagus is provided by the myenteric and the submucosal plexuses. The myenteric or Auerbach's plexus lies between the longitudinal and circular muscles, the submucous or Meissner's plexus lies in the submucosa. The myenteric plexus regulates contraction of the outer muscle layers, whereas submucosal plexus regulates secretion and the peristaltic contractions of the muscularis mucosae.

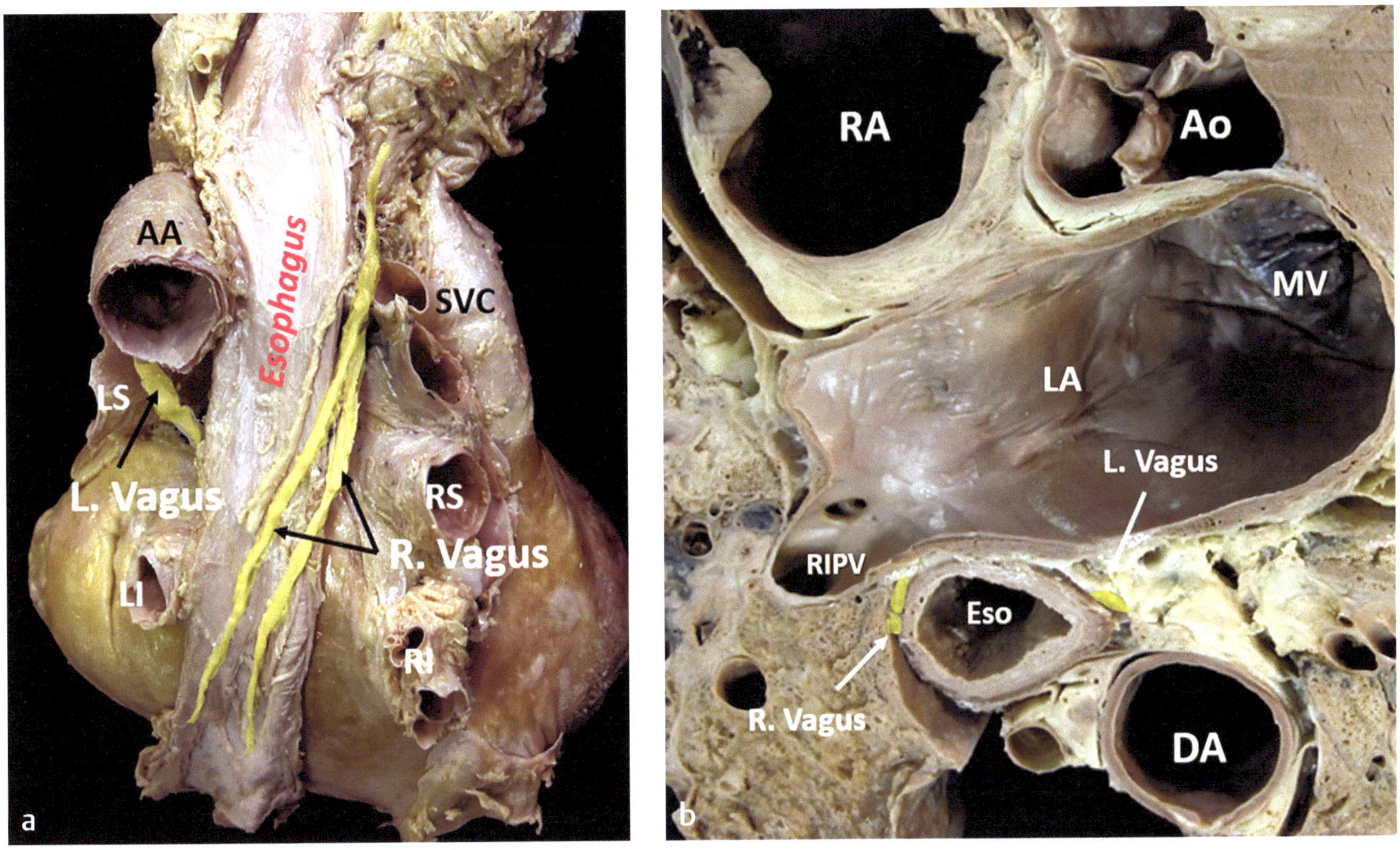

Fig. 2.24 (a) Thoracic organs viewed from the back after removal of the descending thoracic aorta to show the course of the esophagus in situ and its relationship to the right (R) and left (L) vagus nerves. The right vagus nerves move from the left side of the esophagus to its posterior aspect as descending toward the diaphragm. **(b)** Transthoracic section through a cadaver showing the locations of the esophagus, the descending aorta, and the vagus nerves (in *yellow*). Ablation procedures of the left atrium can damage the esophagus and vagal nerves. Pulmonary veins: AA, ascending aorta; Ao, aorta; DA, descending aorta; Eso, esophagus; LA, left atrium; LI, left inferior; LS, left superior; MV, mitral valve; RA, right atrium; RI, right inferior; RIPV, right inferior pulmonary vein; RS, right superior; SVC, superior vena cava.

◆ Diagnosis of Esophageal Diseases

Historically, barium esophagram has been used as the primary method to evaluate esophageal structure and function. Since the introduction of endoscopy and manometry, barium esophagram use is limited as a complementary technique to identify structural abnormalities and give a clear picture of solid and liquid bolus transit. Nowadays, endoscopy is the tool of choice for evaluating structural abnormalities primarily because it allows biopsy and treatment of anatomical abnormalities. It is not a sensitive test for identifying esophageal motor dysfunction.

Transluminal ultrasound using high-frequency transducers combined with endoscopy yields remarkable findings for research purposes in the evaluation of the anatomy, physiology, sphincteric dysfunctions, early cancer staging, and regional lymph nodes[39,40,41] (▸Fig. 2.8). CT and MRI are complementary tests to evaluate the extent of disease beyond esophageal margin and in postoperative cases to study complications or recurrence of disease.[42]

When esophageal motor function is normal, the esophagus clears the contrast bolus in <15 s. This normal transit time is delayed in neuromuscular diseases, achalasia, scleroderma, and structural stenosis. Abnormal bolus transit is common and could be seen in two-thirds of patients with dysphagia without anatomical obstruction or manometric abnormality. In normal barium esophagrams, delicate longitudinal folds with clear margins are seen. Folds are better discerned when the esophagus is collapsed but disappear easily with distension in double contrast studies (▸Fig. 2.25). In CT scan, normal esophageal wall is less than 3 mm (▸Fig. 2.5). Any thickness above 5 mm is abnormal.

Transverse esophageal folds also known as feline esophagus were originally described by Bremner et al.[43] The feline esophagus appears as transient 1- to 2-mm wide, closely spaced transverse folds crossing the entire circumference of the middle and distal thirds of the esophagus at about 5 to 10% double-contrast esophagography (▸Fig. 2.25d). There is an association between reflux disease and hiatal hernia since gastroesophageal reflux is detected on 50% barium studies. The differential diagnosis of a feline esophagus includes idiopathic eosinophilic esophagitis and reflux esophagitis.[44]

As described above, the LES at its full distension appears as a distended ampulla on barium studies of the esophagus (▸Fig. 2.19). Type A ring is located at the proximal end of the LES. A Type B or Schatzki's ring is a 1- to 2-mm thick circumferential mucosal/submucosal ring, 2 cm below the ring A, that protrudes into the

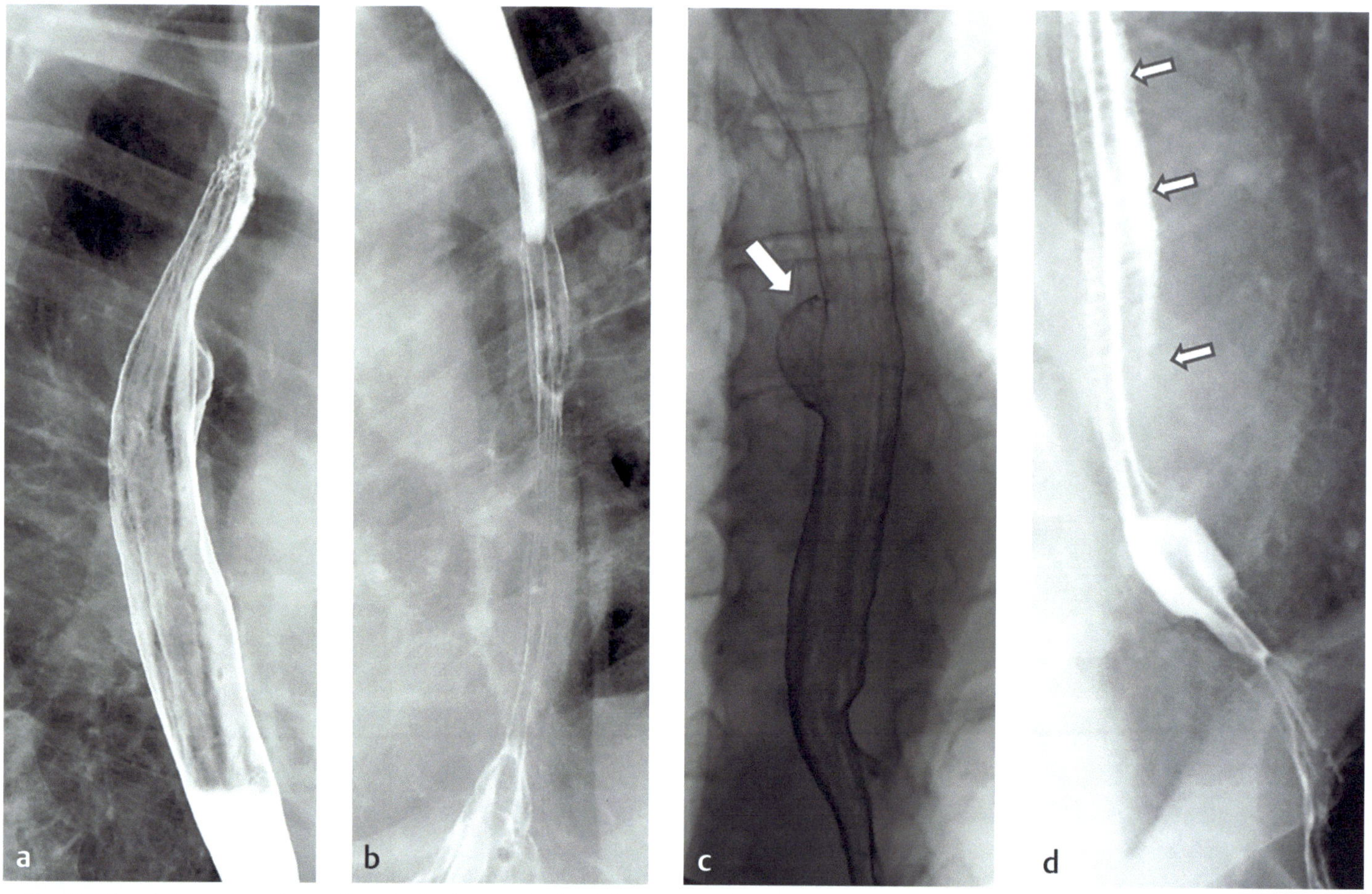

Fig. 2.25 (a–c) Normal esophageal longitudinal mucosal folds. **(c)** A traction diverticulum at the level of the tracheal bifurcation. **(d)** Feline esophagus. Right posterior oblique view from double-contrast esophagram shows moderate transient transverse folds (*arrows*) in the middle to distal esophagus. A small hiatal hernia also is present.

lumen of the distal end of the LES ampulla where gastroesophageal junction is located (▶Fig. 2.19). First described by Richard Schatzki, it is found in 6 to 14% of routine barium radiographs in both sexes.[45,46] It is often asymptomatic but the large ones can cause intermittent dysphagia to solid foods especially when the patent lumen diameter at the level of ring is less than 13 mm. The etiology of a Schatzki's ring is not completely understood and could be related to repeated injury to the lamina propria and an edematous esophageal mucosa from acid reflux. Association with hiatal hernia is very common. The type C esophageal ring is a transient narrowing where the esophagus passes the diaphragmatic crura.

◆ Pathologies

Common congenital esophageal diseases include esophageal rings/webs, esophageal atria with various forms of tracheoesophageal fistula, stenosis, and duplication cysts. Tracheoesophageal fistula results from defects in the separation of the respiratory tract from the foregut. In adult, esophageal rings can be found incidentally within the esophagus. Webs and rings can be mucosal, submucosal, or muscular. Congenital webs are most frequently found in the cervical esophagus.[47] Duplication cysts arising from the foregut include esophageal cysts (tubular duplications) and bronchogenic cysts. Bronchogenic cysts are more common and usually found incidentally at the level of the tracheal bifurcation.

Extrinsic vascular compression of the esophagus is common. The most common vascular compressions are at the level of the aortic arch and left atrium. In pediatric population, double aortic arch, other vascular rings, and aberrant right subclavian artery are common anomalies causing esophageal narrowing. In adults, aberrant right subclavian artery is common causing vascular compression below the aortic arch. It may rarely cause dysphagia lusoria and stenosis (▶Fig. 2.26).

Common acquired esophageal diseases include esophagitis, diverticulum, and cancer. Esophagitis is usually secondary to gastroesophageal reflux. Rarely, it is due to bacterial and fungal infections especially in immunocompromised patients. Esophagitis appears as thickened mucosa, erosions, and ulcers on imaging studies. Perforation of the esophagus due to instrumentation and foreign bodies are not uncommon and CT is the best imaging technique to show the complication.

Barrett's esophagus is an acquired disorder that predisposes to esophageal adenocarcinoma.[48] Barrett's esophagus probably

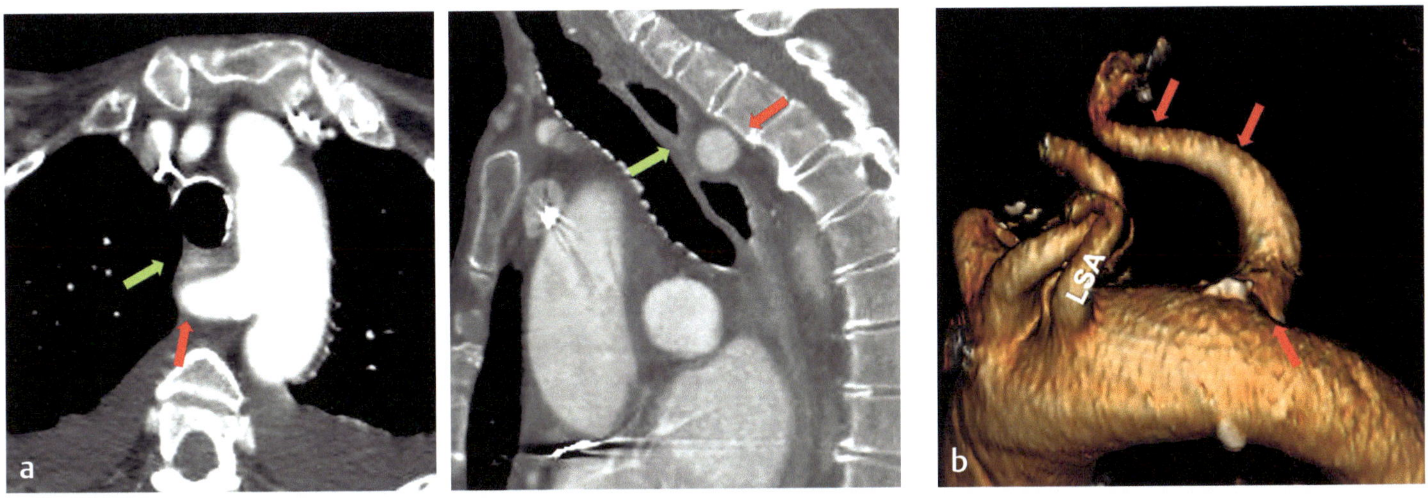

Fig. 2.26 Aberrant right subclavian artery (ARSA) compressing the upper esophagus causing dysphasia. **(a)** Sagittal computed tomography (CT) and **(b)** superior, volume-rendered view of the aortic arch demonstrate ARSA (*red arrows*) coursing behind the upper esophagus (*green arrow*) and compressing it. LSA = left subclavian artery.

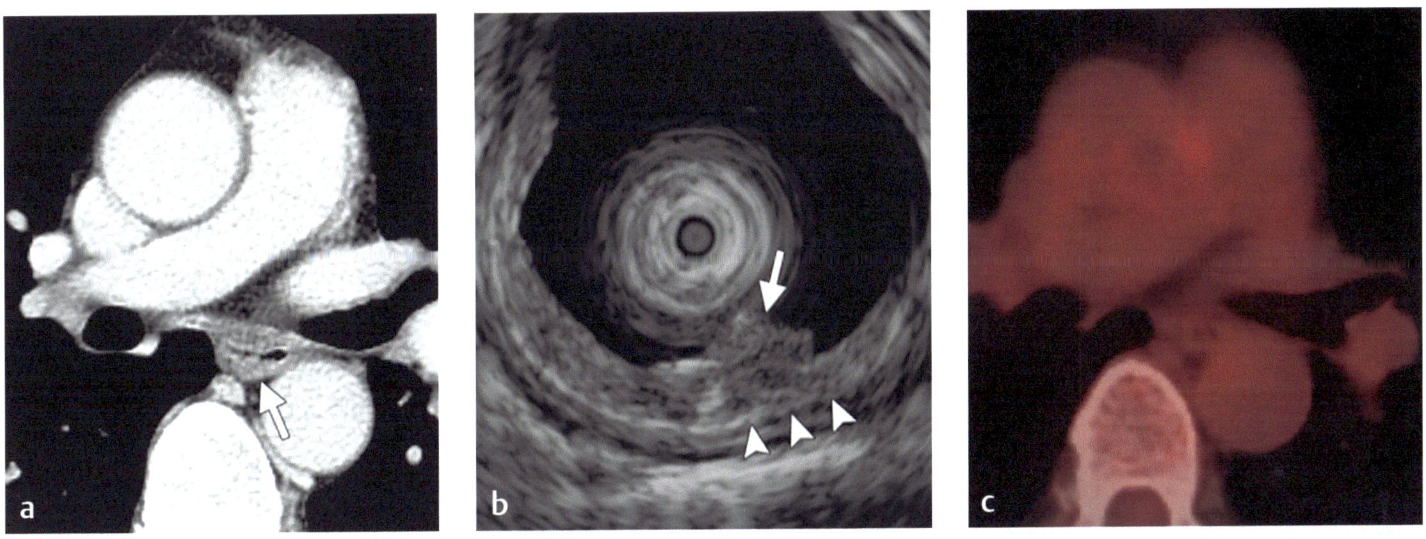

Fig. 2.27 T1a esophageal squamous cell carcinoma. **(a)** Axial, contrast-enhanced computed tomography (CT) image at the level of the right pulmonary artery shows a suspicious small nodular lesion in the midesophagus (*arrow*), a finding that is not easy to detect without endoscopy. **(b)** Endoscopic ultrasound (US) image shows an irregularly shaped nodule (*arrow*) and preservation of the hyperechoic third layer (submucosa) (*arrowheads*). **(c)** Axial positron emission tomography–computed tomography (PET-CT) image at the same level shows no definite fluorodeoxyglucose (FDG) uptake in the primary tumor. (Adapted with permission from Hong et al.[42])

results from reflux-induced damage to the normal squamous epithelium, with subsequent replacement by columnar epithelium. Most patients with Barrett's esophagus have a hiatal hernia. Diagnosis is usually made with endoscopy. Endoscopic ultrasound is superior to CT scan for detecting depth of invasion and nodal involvement in patients with early Barrett's neoplasia.[49,51]

Esophageal cancer (squamous cell carcinoma) is the fifth most common cause of cancer-related deaths in men with a low (<20%) 5-year relative survival rate.[42,50] Precise pretreatment staging of esophageal cancer is important for assessing operability and finding a suitable treatment plan. Endoscopy combined with CT scan staging are common studies before surgery. Esophagogastroduodenoscopy with endoscopic ultrasound is claimed to be more accurate than a CT scan alone for diagnosis and staging of the esophageal cancer (▸Fig. 2.27). Cancer is curative only in case of no lymph node involvement.[40] Esophageal cancer can be seen at CT as regional wall thickening or intraluminal mass causing obstruction. CT is limited to detect the exact depth of early tumor invasion into the esophageal wall (▸Fig. 2.27).

◆ Postsurgery Esophagus

Various surgical techniques are used for esophagectomy of benign or malignant lesions depending on location and extent of the abnormality. Esophagectomy is currently the primary treatment for local and locally advanced esophageal cancer.

Transthoracic esophagectomy is usually performed through a right or left thoracotomy (▸Fig. 2.28, ▸Fig. 2.29). Right thoracotomy for esophagectomy is preferred for lesions of the upper two-thirds of the esophagus. Left thoracotomy is usually performed

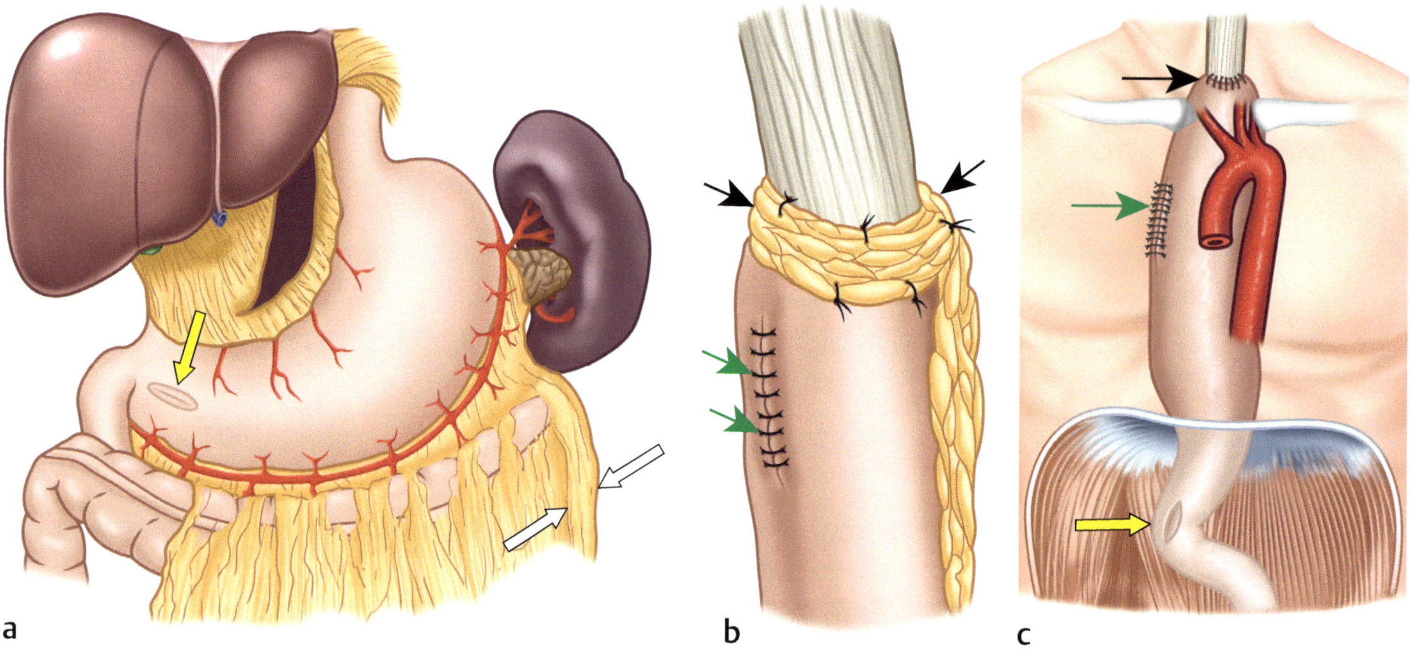

Fig. 2.28 Ivor Lewis esophagectomy. **(a)** First the stomach is elevated by grasping its greater curve and the short gastric arteries are tied. The greater omentum and gastrohepatic ligament are carefully dissected to avoid injury to gastric vessels. The esophageal hiatus and distal esophagus are dissected free, and a pyloromyotomy (*yellow arrow*) is performed. Note the tongue of the omentum (*white arrows*) attached to the greater curvature of the stomach; it will be used to wrap the anastomosis. **(b, c)** Following right thoracotomy, the stomach is pulled up into the chest and divided. An anastomosis is created between the end of the esophagus and the fundus of the stomach. The omentum that was preserved with the stomach is wrapped around the anastomosis (*black arrows*). *Green arrows* indicate the resected site of the cardioesophageal junction. (Modified from Kim et al[52] and Flanagan et al.[53])

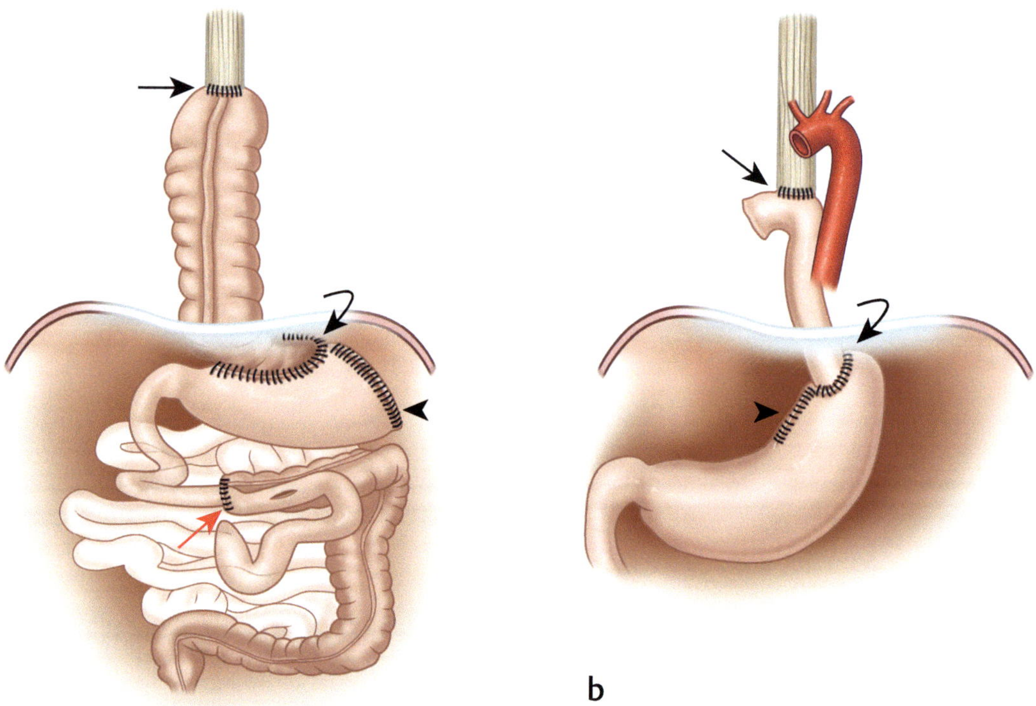

Fig. 2.29 (a) Retrosternal colonic interposition after esophagectomy shows the esophagocolonic anastomosis (*straight black arrow*), cologastric anastomosis (*curved black arrow*), gastric staple line (*arrowhead*), and ileocolonic anastomosis (*red arrow*). **(b)** Jejunal interposition after esophagectomy shows the intrathoracic esophagojejunal anastomosis (*straight arrow*), jejunogastric anastomosis (*curved arrow*), and gastric staple line (*arrowhead*). (Modified from Kim et al[52] and Flanagan et al.[53])

for distal esophageal and gastroesophageal junction.[52,53] After esophagectomy, stomach is usually used to replace the resected esophagus in the posterior mediastinal. Colonic interposition is commonly used in patients with benign disease and a long life expectancy (i.e., caustic or peptic strictures). After esophagectomy, the colon is transected and elevated to the neck through one of three subcutaneous, substernal, or posterior mediastinal approaches. When the stomach and colon are unavailable, the jejunum can be a suitable choice for esophageal reconstruction (▶Fig. 2.29).

Ivor Lewis (separate laparotomy and right thoracotomy incisions) is the most common type of esophagectomy (▶Fig. 2.28). The stomach is divided at its cardiac portion and pulled up into the thorax through the esophageal hiatus. An anastomosis between the distal end of the remaining esophagus and the fundus of the stomach is performed. In McKeown procedure a left neck incision is also performed which allows an easier resection of more proximal esophageal tumors and provides a direct access for a cervical esophageal anastomosis.

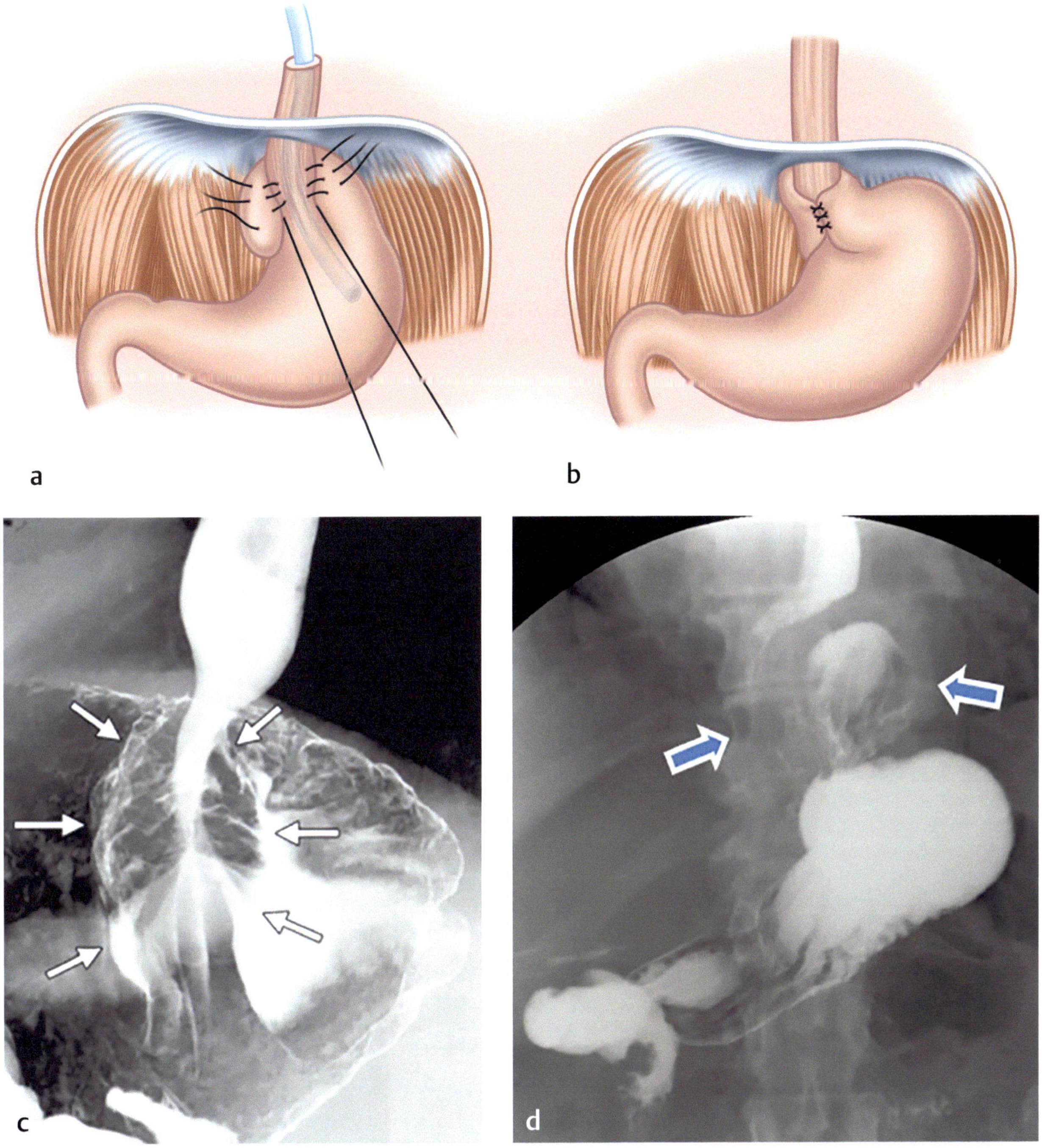

Fig. 2.30 Nissen fundoplication technique. **(a)** A bougie dilator is inserted into the esophageal lumen and sutures positioned between the gastric wrap and esophageal wall. **(b)** Final wrapping of the stomach around the distal esophagus. **(c)** Normal appearance of the Nissen fundoplication (*arrows*) on barium study. (Reproduced with permission from Carbo et al.[54]) **(d)** Transhiatal herniation of the Nissen fundoplication (*arrows*).

In transhiatal esophagectomy no thoracotomy is necessary. The esophagus is transected at the neck and at the EGJ and then removed through the hiatus of diaphragm. The stomach is brought into the neck through the posterior mediastinum for a cervical esophagogastrostomy.

Potentially curative surgical resection may be performed in only 25% of patients with esophageal carcinoma. In cases of unresectable esophageal cancer, a palliative substernal or rarely subcutaneous bypass surgery is performed using ascending colon or transverse colon interposition between stomach and cervical esophagus.

Antireflux surgeries are performed through a thoracotomy (Belsey Mark IV operation) or laparotomy (Nissen, Hill, and Guarner procedures) to restore and maintain the function of the LES (▶Fig. 2.30). Antireflux repairs may control reflux in up to 85% of cases.[54]

References

1. Skandalakis JE, Ellis H. Embryologic and anatomic basis of esophageal surgery. Surg Clin North Am 2000;80(1):85–155, x
2. Wallace AS, Burns AJ. Development of the enteric nervous system, smooth muscle and interstitial cells of Cajal in the human gastrointestinal tract. Cell Tissue Res 2005;319(3):367–382
3. Fu M, Tam PK, Sham MH, Lui VC. Embryonic development of the ganglion plexuses and the concentric layer structure of human gut: a topographical study. Anat Embryol (Berl) 2004;208(1):33–41
4. Lake JI, Heuckeroth RO. Enteric nervous system development: migration, differentiation, and disease. Am J Physiol Gastrointest Liver Physiol 2013;305(1): G1–G24
5. Camilleri M, Cowen T, Koch TR. Enteric neurodegeneration in ageing. Neurogastroenterol Motil 2008;20(4):418–429
6. Goyal RK, Chaudhury A. Physiology of normal esophageal motility. J Clin Gastroenterol 2008;42(5):610–619
7. Choi DY, Bae JH, Youn KH, Kim HJ, Hu KS. Anatomical considerations of the longitudinal pharyngeal muscles in relation to their function on the internal surface of pharynx. Dysphagia 2014;29(6):722–730
8. Yip HT, Leonard R, Kendall KA. Cricopharyngeal myotomy normalizes the opening size of the upper esophageal sphincter in cricopharyngeal dysfunction. Laryngoscope 2006;116(1):93–96
9. Gyawali CP, Bredenoord AJ, Conklin JL, et al. Evaluation of esophageal motor function in clinical practice. Neurogastroenterol Motil 2013;25(2):99–133
10. Mittal RK, Padda B, Bhalla V, Bhargava V, Liu J. Synchrony between circular and longitudinal muscle contractions during peristalsis in normal subjects. Am J Physiol Gastrointest Liver Physiol 2006;290(3):G431–G438
11. Mittal RK. Longitudinal muscle of the esophagus: its role in esophageal health and disease. Curr Opin Gastroenterol 2013;29(4):421–430
12. Abrahao L Jr, Bhargava V, Babaei A, Ho A, Mittal RK. Swallow induces a peristaltic wave of distension that marches in front of the peristaltic wave of contraction. Neurogastroenterol Motil 2011;23(3):201–207, e110
13. Jung HY, Puckett JL, Bhalla V, et al. Asynchrony between the circular and the longitudinal muscle contraction in patients with nutcracker esophagus. Gastroenterology 2005;128(5):1179–1186
14. Jiang Y, Mittal RK. Low esophageal mucosal blood flow in patients with nutcracker esophagus. Am J Physiol Gastrointest Liver Physiol 2016;310(6):G410–G416
15. Ahuja NK, Chan WW. Assessing upper esophageal sphincter function in clinical practice: a primer. Curr Gastroenterol Rep 2016;18(2):7
16. Sivarao DV, Goyal RK. Functional anatomy and physiology of the upper esophageal sphincter. Am J Med 2000;108(Suppl 4a):27S–37S
17. Mittal RK, Rochester DF, McCallum RW. Electrical and mechanical activity in the human lower esophageal sphincter during diaphragmatic contraction. J Clin Invest 1988;81(4):1182–1189
18. Hershcovici T, Mashimo H, Fass R. The lower esophageal sphincter. Neurogastroenterol Motil 2011;23(9):819–830
19. Pickering M, Jones JF. The diaphragm: two physiological muscles in one. J Anat 2002;201(4):305–312
20. Delattre JF, Palot JP, Ducasse A, Flament JB, Hureau J. The crura of the diaphragm and diaphragmatic passage. Applications to gastroesophageal reflux, its investigation and treatment. Anat Clin 1985;7(4):271–283
21. Liebermann-Meffert D, Allgöwer M, Schmid P, Blum AL. Muscular equivalent of the lower esophageal sphincter. Gastroenterology 1979;76(1):31–38
22. Mittal RK, Balaban DH. The esophagogastric junction. N Engl J Med 1997; 336(13):924–932
23. Mittal RK, Goyal RK. Sphincter mechanisms at the lower end of the esophagus. In: Goyal R, Shaker R, eds. GI Motility Online. PART 1 Oral Cavity, Pharynx and Esophagus. doi:10.1038/gimo14, May 2006
24. Shi G, Pandolfino JE, Joehl RJ, Brasseur JG, Kahrilas PJ. Distinct patterns of oesophageal shortening during primary peristalsis, secondary peristalsis and transient lower oesophageal sphincter relaxation. Neurogastroenterol Motil 2002;14(5):505–512
25. Delattre JF, Avisse C, Marcus C, Flament JB. Functional anatomy of the gastroesophageal junction. Surg Clin North Am 2000;80(1):241–260
26. Pandolfino JE, Shi G, Curry J, Joehl RJ, Brasseur JG, Kahrilas PJ. Esophagogastric junction distensibility: a factor contributing to sphincter incompetence. Am J Physiol Gastrointest Liver Physiol 2002;282(6):G1052–G1058
27. Mittal RK, Stewart WR, Schirmer BD. Effect of a catheter in the pharynx on the frequency of transient lower esophageal sphincter relaxations. Gastroenterology 1992;103(4):1236–1240
28. Leslie E, Bhargava V, Mittal RK. A novel pattern of longitudinal muscle contraction with subthreshold pharyngeal stimulus: a possible mechanism of lower esophageal sphincter relaxation. Am J Physiol Gastrointest Liver Physiol 2012;302(5):G542–G547
29. Zhang Q, Lehmann A, Rigda R, Dent J, Holloway RH. Control of transient lower oesophageal sphincter relaxations and reflux by the GABA(B) agonist baclofen in patients with gastro-oesophageal reflux disease. Gut 2002;50(1):19–24
30. Scheffer RC, Tatum RP, Shi G, Akkermans LM, Joehl RJ, Kahrilas PJ. Reduced tLESR elicitation in response to gastric distension in fundoplication patients. Am J Physiol Gastrointest Liver Physiol 2003;284(5):G815–G820
31. Yoon YC, Lee KS, Shim YM, Kim BT, Kim K, Kim TS. Metastasis to regional lymph nodes in patients with esophageal squamous cell carcinoma: CT versus FDG PET for presurgical detection prospective study. Radiology 2003;227(3):764–770
32. Kumbasar B. Carcinoma of esophagus: radiologic diagnosis and staging. Eur J Radiol 2002;42(3):170–180
33. Finsterer J, Grisold W. Disorders of the lower cranial nerves. J Neurosci Rural Pract 2015;6(3):377–391
34. Brock C, Brokjaer A, Drewes AM, et al. Neurophysiology of the esophagus. Ann N Y Acad Sci 2014;1325:57–68
35. Park H, Conklin JL. Neuromuscular control of esophageal peristalsis. Curr Gastroenterol Rep 1999;1(3):186–197
36. Pandolfino JE, Gawron AJ. Achalasia: a systematic review. 2015;313(18): 1841–1852
37. Patti MG, Gantert W, Way LW. Surgery of the esophagus. Anatomy and physiology. Surg Clin North Am 1997;77(5):959–970
39. Miller L, Dai Q, Korimilli A, Levitt B, Ramzan Z, Brasseur J. Use of endoluminal ultrasound to evaluate gastrointestinal motility. Dig Dis 2006;24(3-4):319–341
40. Rampado S, Bocus P, Battaglia G, Ruol A, Portale G, Ancona E. Endoscopic ultrasound: accuracy in staging superficial carcinomas of the esophagus. Ann Thorac Surg 2008;85(1):251–256
41. Vegesna AK, Sloan JA, Singh B, et al. Characterization of the distal esophagus high-pressure zone with manometry, ultrasound and micro-computed tomography. Neurogastroenterol Motil 2013;25(1):53–60.e6
42. Hong SJ, Kim TJ, Nam KB, et al. New TNM staging system for esophageal cancer: what chest radiologists need to know. Radiographics 2014;34(6):1722–1740
43. Bremner CG, Shorter RG, Ellis FH Jr. Anatomy of feline esophagus with special reference to its muscular wall and phrenoesophageal membrane. J Surg Res 1970;10(7):327–331
44. Samadi F, Levine MS, Rubesin SE, Katzka DA, Laufer I. Feline esophagus and gastroesophageal reflux. AJR Am J Roentgenol 2010;194(4):972–976
45. Schatzki R, Gary JE. The lower esophageal ring. Am J Roentgenol Radium Ther Nucl Med 1956;75(2):246–261
46. Patel B, Han E, Swan K. Richard Schatzki: a familiar ring. AJR Am J Roentgenol 2013;201(5):W678-82

47. Berrocal T, Torres I, Gutiérrez J, Prieto C, del Hoyo ML, Lamas M. Congenital anomalies of the upper gastrointestinal tract. Radiographics 1999;19(4):855–872
48. Cameron AJ. Barrett's esophagus: prevalence and size of hiatal hernia. Am J Gastroenterol 1999;94(8):2054–2059
49. Thomas T, Gilbert D, Kaye PV, Penman I, Aithal GP, Ragunath K. High-resolution endoscopy and endoscopic ultrasound for evaluation of early neoplasia in Barrett's esophagus. Surg Endosc 2010;24(5):1110–1116
50. Siegel R, Naishadham D, Jemal A. Cancer statistics, 2012. CA Cancer J Clin 2012;62(1):10–29
51. Pech O, May A, Günter E, Gossner L, Ell C. The impact of endoscopic ultrasound and computed tomography on the TNM staging of early cancer in Barrett's esophagus. Am J Gastroenterol 2006;101(10):2223–2229
52. Kim SH, Lee KS, Shim YM, Kim K, Yang PS, Kim TS. Esophageal resection: indications, techniques, and radiologic assessment. Radiographics 2001;21(5):1119–1137, discussion 1138–1140
53. Flanagan JC, Batz R, Saboo SS, et al. Esophagectomy and gastric pull-through procedures: surgical techniques, imaging features, and potential complications. Radiographics 2016;36(1):107–121
54. Carbo AI, Kim RH, Gates T, D'Agostino HR. Imaging findings of successful and failed fundoplication. Radiographics 2014;34(7):1873–1884

3 Stomach

Farhood Saremi and Damián Sánchez-Quintana

3

◆ Introduction

The stomach is a J-shaped reservoir in upper abdomen surrounded by the liver and gallbladder on the right, the spleen on the left, the diaphragm superiorly, the transverse colon inferiorly, and pancreas and spine posteriorly. It is subdivided into four parts based on physiologic characteristics and distribution of the gastric mucosa: the cardia, the fundus, the body (corpus), and the pylorus.

◆ Embryology

The part of foregut that forms the stomach and is suspended in the dorsal and ventral mesentery starts to enlarge at week 4 of gestation. As the stomach region dilates, it rotates 90 degrees so that the posterior margin which is attached to the dorsal mesogastrium moves to the left to form the greater curvature and anterior margin attached to the ventral mesentery moves to the right and posteriorly to become the lesser curvature (▶Fig. 3.1). There is a gradual caudal descent of

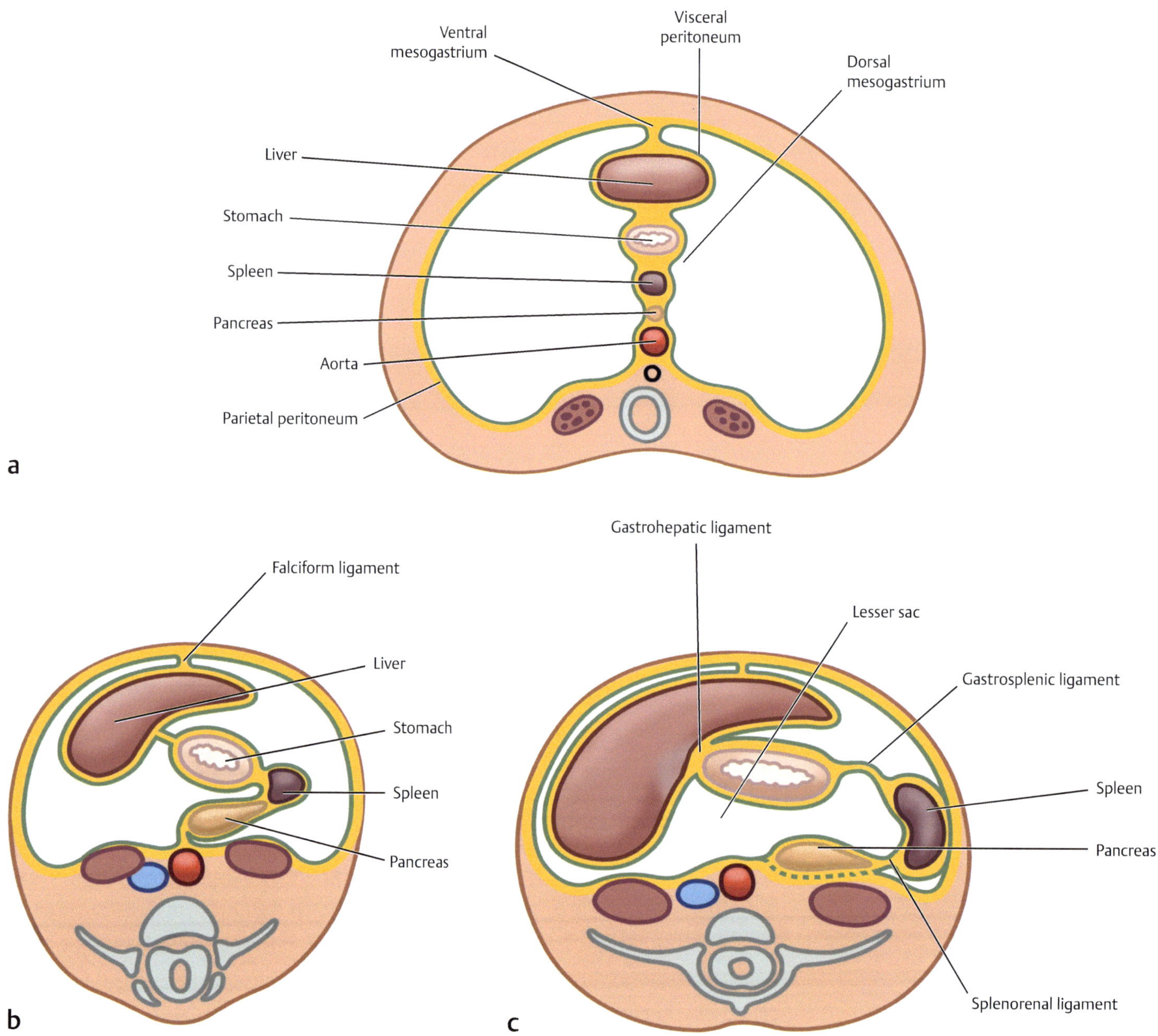

Fig. 3.1 (a–c) Development of stomach. As the stomach region dilates, it rotates 90 degrees so that the posterior margin which is attached to the dorsal mesogastrium moves to the left to form the greater curvature and anterior margin attached to the ventral mesentery moves to the right and posteriorly to become the lesser curvature. The dorsal mesogastrium elongates to form the fibrofatty greater omentum and the gastrosplenic ligament. The ventral mesogastrium remains as the lesser omentum or gastrohepatic ligament.

the greater curvature as the duodenum advances to the right. The vagal nerves originally arranged in a left–right orientation of the esophagus will be moved to lie in an anterior–posterior position as the stomach rotation is complete. Therefore, the anterior gastric wall will be supplied by the left vagus and the posterior gastric wall by the right vagus. Dilatation of the stomach into a final orientation will be completed by 7 weeks. At this stage, the dorsal mesogastrium has elongated sufficiently to form the fibrofatty greater omentum, a double layer of visceral peritoneum, suspended from the greater curvature and transverse colon. The greater omentum also forms the inferior recess of the lesser sac (bursa omentalis) between its two frontal and two dorsal sheets. The lesser sac opens medially into the peritoneal cavity through the foramen of Winslow. It is bordered anteriorly by the stomach and posteriorly by the pancreas and dorsal parietal peritoneum. The spleen and dorsal pancreas, which originally lied in the dorsal mesogastrium, move to the left and find their find positions. The gastrocolic ligament connects the stomach and colon, and the gastrosplenic ligament forms between the stomach and spleen. These ligaments also arise from the dorsal mesogastrium and contain vessels and lymphatics. The ventral mesogastrium which connects the anterior margin of the stomach to the anterior abdominal wall will be pushed away by the enlarging liver and remains as the lesser omentum or gastrohepatic ligament extending between the stomach and porta hepatis and the falciform ligament between the anterior margin of the liver and abdominal wall. The inferior edge of the lesser omentum is thicker and forms the hepatoduodenal ligament which connects the duodenum to the liver. The hepatoduodenal ligament forms the roof of the foramen of Winslow. It contains the portal vein, hepatic vein, and common bile duct.

◆ Gross Anatomy

The stomach has two surfaces, anterior and posterior, and two borders, the lesser curvature and the greater curvature. The lesser curvature forms an indentation inferiorly called "incisura angularis." The stomach is divided into four parts: the cardia, the fundus, the body (corpus), and the pylorus (▸Fig. 3.2, ▸Fig. 3.3).

The cardia is defined as the region distal to the squamocolumnar border of the gastroesophageal junction. The two stomach curvatures meet at the cardia. The fundus is formed primarily by the superior curvature of the stomach and defined as area above the horizontal line drawn from the gastroesophageal junction to the greater curvature. The body is the central part of stomach. The pylorus is divided into the pyloric antrum (vestibule) and the pyloric canal. The antrum starts from the "incisura angularis" of the lesser curvature. The pyloric canal is the sphincteric region of the stomach that ends the pyloric valve at the junction with the duodenum. The pyloric canal is usually closed.

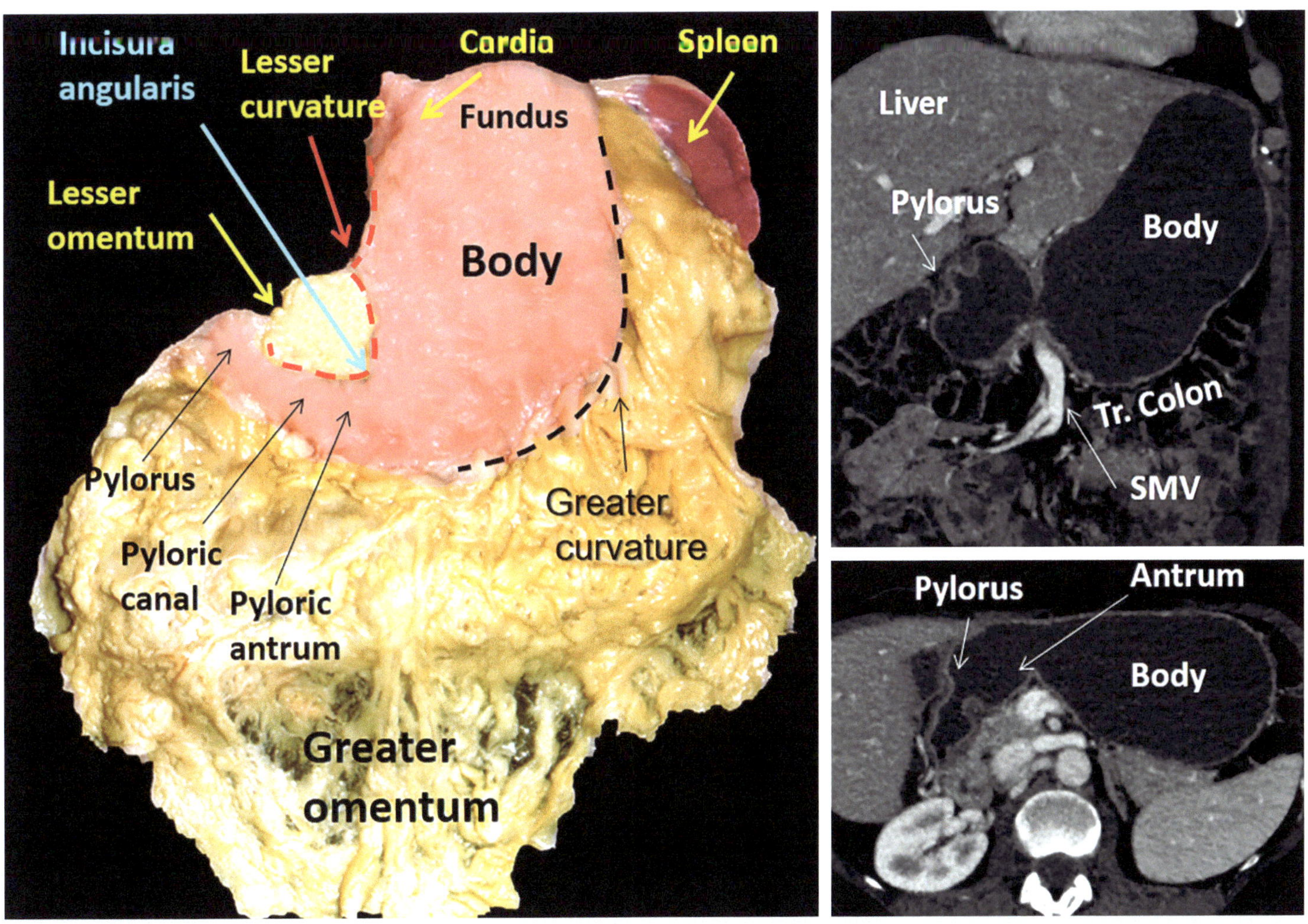

Fig. 3.2 Anatomic divisions of the stomach. The stomach is divided into four parts: the cardia, the fundus, the body (corpus), and the pylorus. The pylorus is divided into the pyloric antrum (vestibule) and the pyloric canal. The antrum is distal to the "incisura angularis" of the lesser curvature. A distended stomach shows a thin wall and the rugal pattern disappears. SMV, superior mesenteric vein; Tr., transverse.

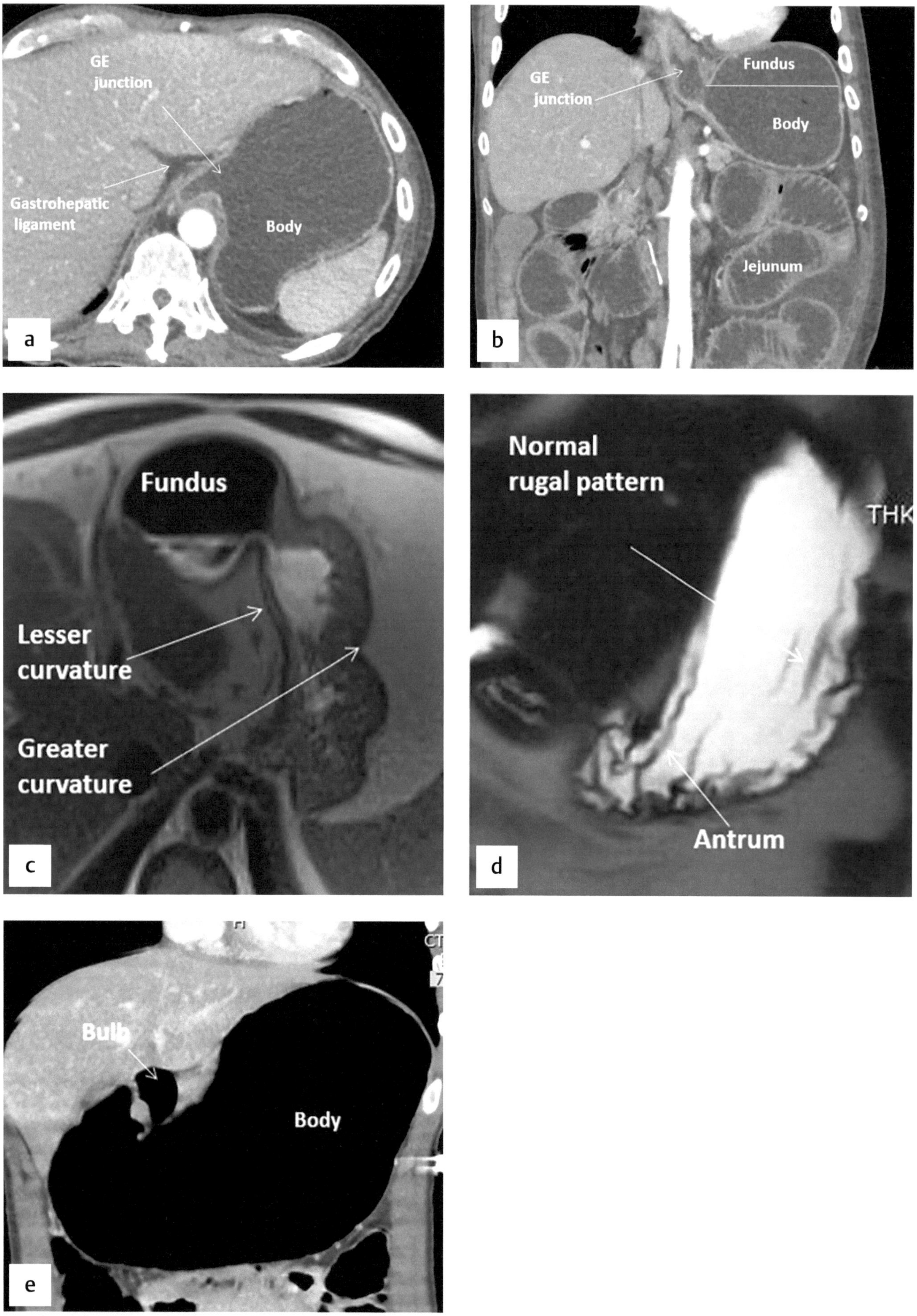

Fig. 3.3 Normal stomach morphology. Upper row. Axial **(a)** and coronal **(b)** computed tomography (CT) images. Lower row: Axial **(c)** and coronal **(d)** magnetic resonance (MR) and coronal CT **(e)**. The cardia is defined as the region distal to the gastroesophageal (GE) junction. The fundus is the area above the horizontal line drawn from the GE junction to the greater curvature. In a distended stomach the wall is thin and the rugal pattern disappears. Note that loss of mucosal pattern does not occur in a distended jejunum.

Topographic divisions of stomach are somewhat based on physiologic characteristics and distribution of the gastric mucosa. For example, the antrum contains no acid secreting parietal cells whereas the body has a large amount of parietal cells. The antrum has largest amount of gastrin cells whereas the fundus and corpus have few. Overall, there is no gross anatomic delineation between parts of stomach and histologic boundaries are also ill-defined.[1]

Musculature

The musculature of the stomach is composed of three layers: (1) innermost layer is formed by oblique fibers, (2) the middle layer is formed by circular muscles, and (3) the outer layer is formed by longitudinal muscle fibers.

The innermost muscular layer is more concentrated in the fundus than in the antrum and its muscle fibers run obliquely from the fundus to the greater curvature. At the esophagogastric junction the inner fibers join the circular muscular layer of the lower esophageal sphincter to form the angle of His (see Chapter 2: Esophagus). The inner oblique fibers blend with those of the middle circular fibers along the greater curvature. The outer longitudinal muscle layer of the stomach is continuous with the longitudinal muscle fibers of the esophagus and duodenum. Its fibers are more concentrated along the lesser and greater curvatures of the stomach. Therefore, most of the anterior and posterior walls of the stomach is relatively devoid of longitudinal muscle fibers. The middle layer is the only complete muscular layer covering the entire stomach. Its fibers become more concentrated in the antrum and play an important role in gastric emptying. This antral pump mechanism along with the pyloric sphincter function determines the rate of gastric pumping, particularly for solid foods. In nonselective vagotomy, the antral pump will be disrupted causing incomplete gastric emptying. In contrast, in superselective parietal cell vagotomy, the antral function remains intact.

◆ Gastroduodenal Junction

The pyloric sphincter is a thick band of circular muscle that separates the stomach from the duodenum (▶Fig. 3.4). Its fibers appear to be continuous with the middle layer of the stomach but is

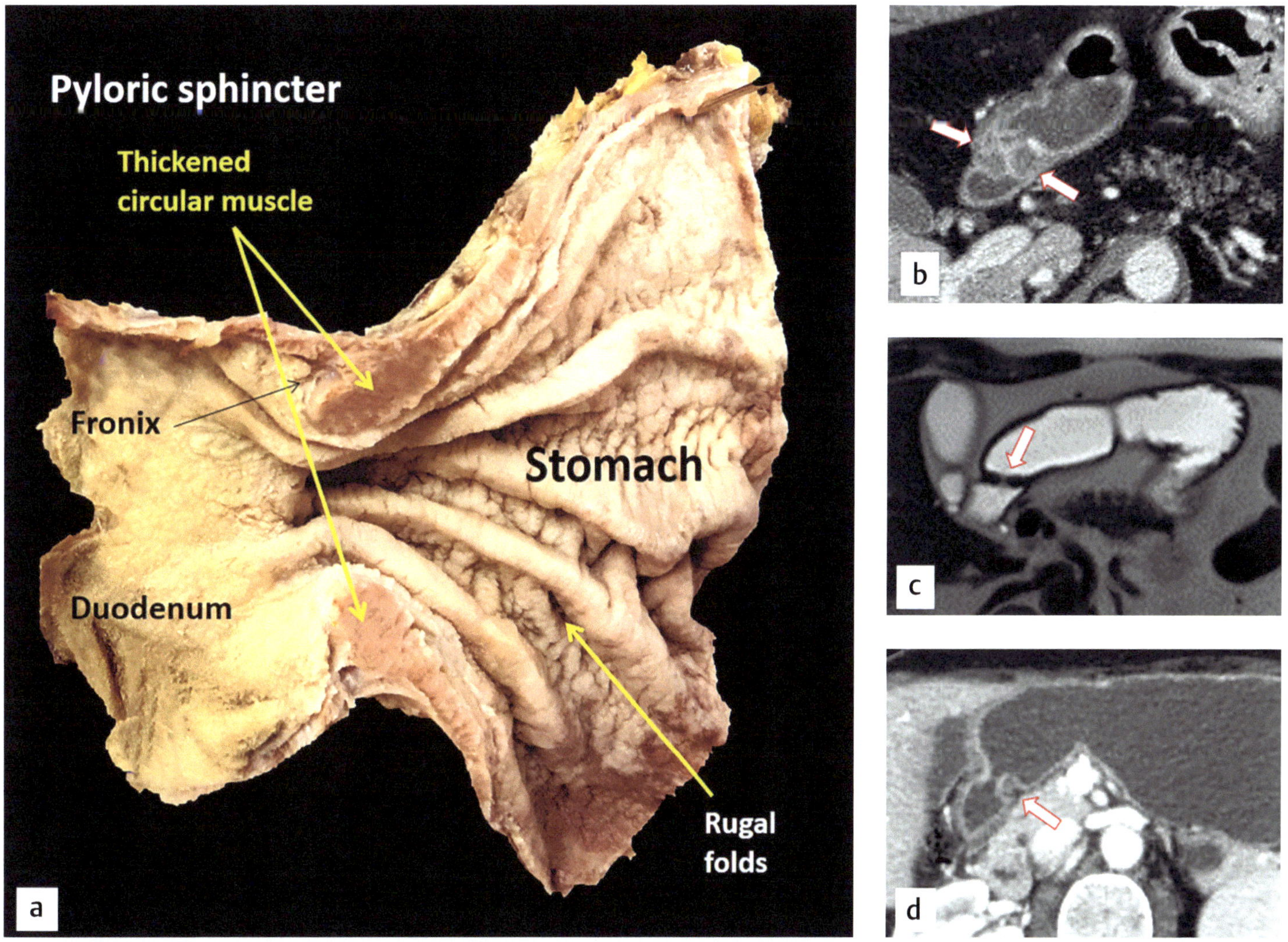

Fig. 3.4 Pyloric sphincter. **(a)** The pyloric sphincter is a thick band of circular muscle that separates the stomach from the duodenum. The thickness of the circular muscle is significantly decreased distal to the sphincter to form the relatively thin-walled duodenum. **(b–d)** Axial images show the pyloric sphincter (*arrows*) with different thicknesses. The muscle appears relaxed when the channel is open **(d)**.

actually separated from the middle layer of the duodenum by fibroconnective tissue. At the gastroduodenal junction, the continuity of the circular muscle of the pyloric sphincter is interrupted by a ring-shaped septum of connective tissue derived from the submucosa to isolate the pyloric sphincter. Distal to this fibroconnective ring, the thickness of the circular muscle is significantly decreased to form the relatively thin-walled duodenum. This decrease in thickness of the duodenal fornix can result in inadvertent entrance into the duodenum when pyloromyotomy is performed.

◆ Gastric Wall

The serosal surface of stomach is covered by visceral peritoneum except for small areas along the lesser and greater curvatures where the serosal layers continue on to the anterior and posterior leaves of the lesser and greater omentum (▶Fig. 3.5). The submucosa is primarily made of connective tissue collagen containing networks of vessels and the plexus of Meissner. An artery enters the stomach wall from the serosal surface and passes the muscle layers without branching until it reaches the submucosa. The submucosa appears to be the strongest part of the stomach and should be stitched separately when a gastric anastomosis is performed. The gastric mucosa consists of epithelium, lamina propria, and muscularis mucosae. The muscularis mucosae is a thin smooth muscle layer between the mucosa and submucosa. Its muscle tone is important in maintaining normal rugal fold of the stomach mucosa and increases the absorption surface area which may be increase almost fourfold with complete stomach distension. Rugal folds are most marked along the greater curvature (▶Fig. 3.2, ▶Fig. 3.3). The lamina propria is a vascular connective tissue located between the gastric epithelium and the muscularis mucosa.

The stomach is covered by a single layer of columnar epithelium. At the gastroesophageal junction, the "squamocolumnar junction" known as the "Z line" exists where sudden transition from the squamous epithelium of the esophagus to the gastric columnar epithelium occurs. The Z line determines the lower margin of the high-pressure zone but its location is quite variable. The types and concentration of the gastric mucosal cells are quite variable both anatomically and physiologically. There are three types of glands in the stomach mucosa:

- The cardiac region is mainly formed by branching glands containing mucus-secreting cells. The parietal and chief cells are sparse.
- The fundus and body mucosa are composed of long tubular glands with little branching. Parietal cells which secrete hydrochloric acid are concentrated in the body. The chief cells are located in the body and fundus and produce pepsinogen and histamine. The parietal and chief cells are relatively absent in the antrum. The parietal cells are concentrated more frequently in the isthmus and neck than the base of the gland. The chief cells are located at the base of glands.
- The antral and pyloric regions contain gastrin-producing G cells which stimulates acid production from the parietal cells. G cells are also found in the duodenum and small bowel.

◆ Neurovasculature

Arteries

The arterial blood supply to the distal esophagus, stomach, and proximal duodenum is quite variable and largely through branches of the celiac axis. The celiac artery trunk/axis usually gives origin to three branches: the left gastric, splenic, and common hepatic arteries in 55% of the population. The celiac artery may not give off the left gastric in 3.5%, the hepatic in 5.5%, or to the splenic artery in 1.5%. Instead, these branches may arise directly from the aorta or superior mesenteric artery (SMA). Blood supply to the stomach and distal esophagus is provided by six primary arteries. Collateral pathways are common and develop once the celiac or gastric arteries are occluded or surgically ligated. The primary arteries include: the left gastric, right gastric, gastroduodenal, right gastroepiploic, left gastroepiploic, and short gastrics (▶Fig. 3.6, ▶Fig. 3.7). The collateral branches will be supplied by the superior pancreaticoduodenal, supraduodenal, retroduodenal, transverse pancreatic, dorsal pancreatic, and left inferior phrenic arteries. The right and left gastric vessels travel in the lesser omentum.

The left gastric artery is the first and the smallest branch of the celiac axis. It ascends along the lesser curvature through the lesser omentum to reach the gastroesophageal junction (▶Fig. 3.5). It provides branches to the distal esophagus and cardia of the stomach. Its ascending branch connects with the branches of the inferior phrenic artery and short gastrics. The inferior phrenic artery may be the main provider to the proximal gastric pouch after a subtotal gastrectomy where the left gastric and short gastric arteries are ligated. The descending branch of the left gastric artery gives off branches to the lesser curvature of the stomach and anastomoses with the right gastric artery. The left hepatic artery may be a branch of the left gastric artery in 25% of cases. Identifying this variant is important during gastrectomies. In 12% of cases, the left hepatic artery is the only blood supply to the lateral segment of the left liver lobe and its ligation results in hepatic necrosis.[2,3] The right gastric artery is a small branch of the hepatic artery distal to the gastroduodenal artery (GDA) (▶Fig. 3.7). It descends in the hepatoduodenal ligament above the superior margin of the pylorus, gives off branches to the pylorus and antrum, and connects to the left gastric artery. The right gastric artery is a major source of blood supply to the remaining distal stomach following esophagogastrectomy.

The GDA is the first branch of the common hepatic artery and provides branches to the duodenum, stomach, pancreas, and gallbladder. The GDA gives off the superior pancreaticoduodenal arteries which connect with the inferior pancreaticoduodenal

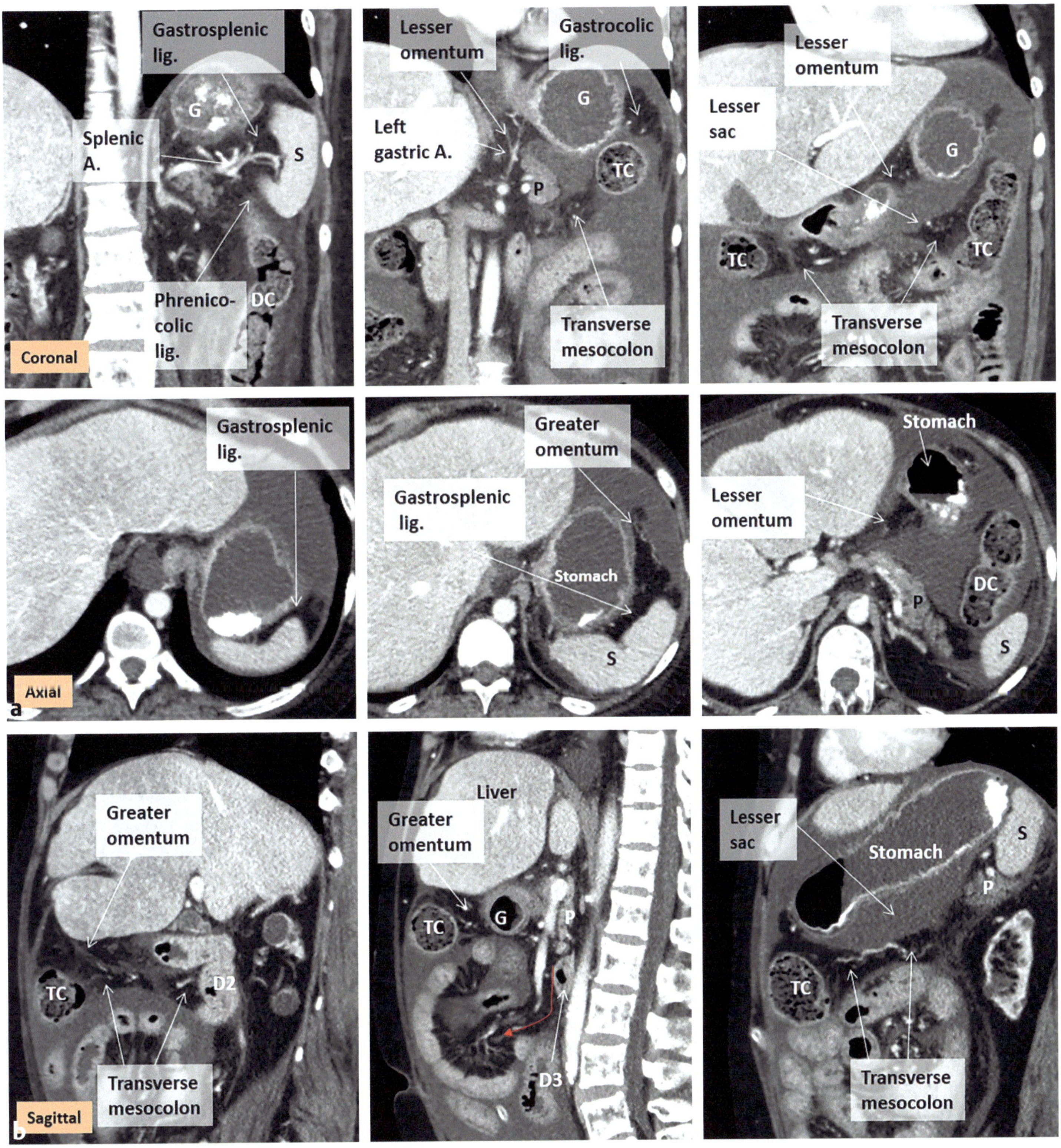

Fig. 3.5 (a, b) Ligaments supporting the stomach in a patient with ascites. The gastrosplenic ligament (gastrolienal ligament) and the gastrocolic ligament are parts of the greater omentum and extend between the greater curvature of the stomach and the spleen and the transverse colon respectively. The lesser omentum extends from the liver to the stomach (hepatogastric ligament) and the first part of the duodenum (hepatoduodenal ligament). The phrenicocolic ligament is between the left colic flexure and the diaphragm and below the spleen. Also note the line of attachment of the transverse mesocolon on the second duodenum (D2) on sagittal view. The root of small bowel mesentery including the mesenteric vessels (*red arrow*) pass anterior to the third part of duodenum (D3). DC, descending colon; G, gastric lumen; P, pancreas; S, spleen; TC, transverse colon.

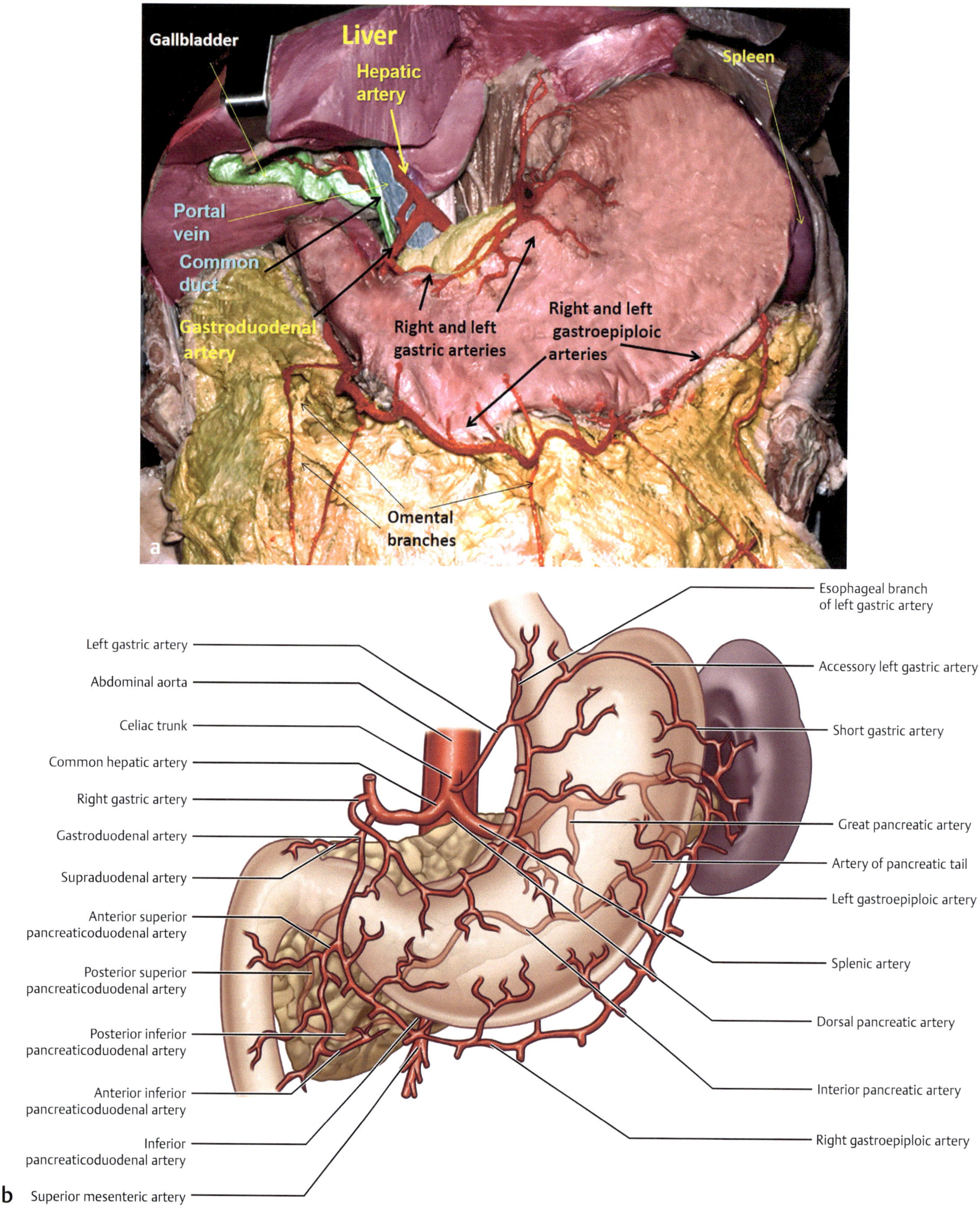

Fig. 3.6 (a, b) The arterial blood supply to the stomach. **(a)** The primary arteries include: the left gastric, right gastric, gastroduodenal, right gastroepiploic, left gastroepiploic, and short gastrics. The portal vein, common duct, and hepatic artery pass in the hepatoduodenal ligament. The right and left gastric vessels pass in the lesser omentum and the gastroepiploic vessels pass in the greater omentum. **(b)** Diagram of the blood supply to the pancreas, stomach, duodenum, and spleen.

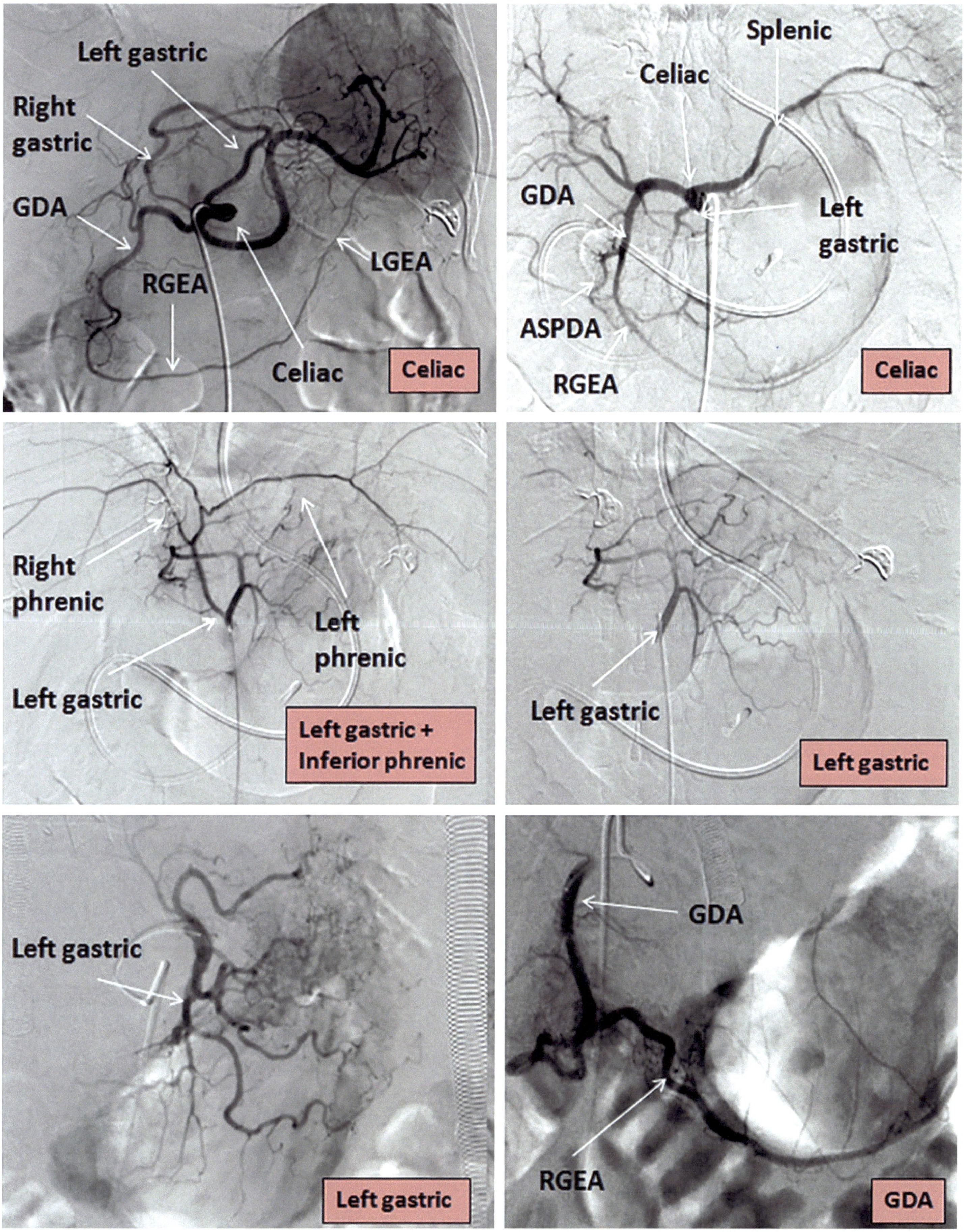

Fig. 3.7 Arterial blood supply to the stomach shown by celiac catheter angiograms. The left gastric artery is the first and the smallest branch of the celiac axis moving along the lesser curvature. It may have a common stem with the inferior phrenic artery. The descending branch of the left gastric artery gives off branches to the lesser curvature of the stomach and anastomoses with the right gastric artery. The termination of the gastroduodenal artery (GDA) is the right gastroepiploic artery (RGEA). This vessel courses along the greater curvature of the stomach giving off branches to the pyloric antrum, the greater curvature of the stomach, and the greater omentum before anastomosing with the left gastroepiploic artery (LGEA). ASPDA, anterior superior pancreaticoduodenal artery.

arteries of the superior mesenteric artery to form the pancreaticoduodenal arcade. This arcade is an important source of collateral blood flow to the stomach if the celiac trunk is occluded. The termination of the GDA is the right gastroepiploic artery. This vessel courses along the greater curvature of the stomach giving off branches to the pyloric antrum and greater curvature of the stomach and greater omentum before anastomosing with the left gastroepiploic artery. The left gastroepiploic artery arises from the distal or, less commonly, mid splenic artery at the splenic hilum and gives off branches to the greater curvature of the stomach and greater omentum. It reaches the stomach below the fundus by means of the phrenicopancreaticosplenic ligament.

In more than 60% of cases, great anastomoses exist between the right and left gastroepiploic arteries.[4] In 10% of cases, the left and right gastroepiploic arteries do not anastomose. Absence of enough collateral between the two arteries may increase the rate of ischemia and secondary fistula at upper anastomosis after gastric pull-up surgery and esophagoplasty.[4,5] Omental branch of the left gastroepiploic artery that descends in the anterior layer of the greater omentum continues into the posterior layer of the omentum below the transverse colon and connects with the omental branch of the right gastroepiploic artery or the transverse pancreatic artery through the posterior omental arch of Barkow (▶Fig. 3.6). The short gastric arteries are 2 to 10 branches of the splenic artery and are divided into upper, middle, and lower groups. They may originate from the superior polar or left gastroepiploic arteries. Leaving the phrenicolienal ligament, the short gastric arteries pass into the gastrolienal ligament to reach the fundus and anastomose with branches of the left gastric artery, left gastroepiploic artery, and left inferior phrenic artery. Given the existence of all the collaterals described above, infarction of the stomach is very unusual. In case of spleen after a spleen-preserving distal pancreatectomy with resection of the splenic vessels, the left gastroepiploic artery appears to be the main collateral vessel to maintain blood splenic flow, not the short gastric.[6]

Veins

There are five veins draining the stomach: the left gastric or coronary vein, the right gastric vein with its pyloric tributaries, the right and left gastroepiploic veins, and the short gastric veins (▶Fig. 3.5, ▶Fig. 3.8, ▶Fig. 3.9). The left and right gastric veins are in the lesser omentum and drain into the portal vein. The right and left gastroepiploic veins anastomose with each other in the greater omentum and together bridge between the splenic vein

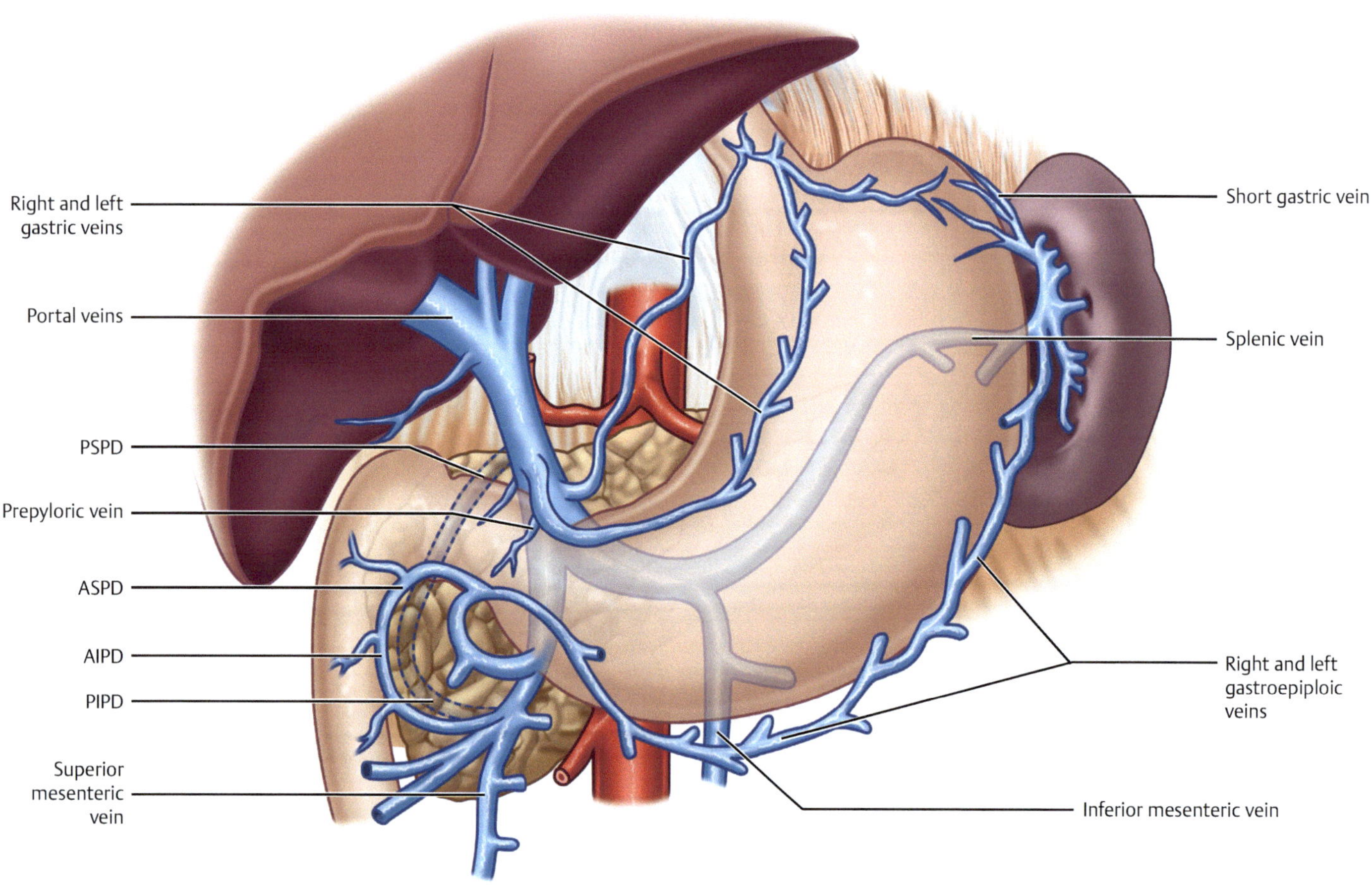

Fig. 3.8 Gastric and duodenal veins. There are five veins that drain the stomach: the left gastric or coronary vein, the right gastric vein with its pyloric tributaries, the right and left gastroepiploic veins, and the short gastric veins. AIPD, posterior inferior pancreaticoduodenal vein; ASPD, posterior anterior pancreaticoduodenal vein; PIPD, posterior inferior pancreaticoduodenal vein; PSPD, posterior superior pancreaticoduodenal vein.

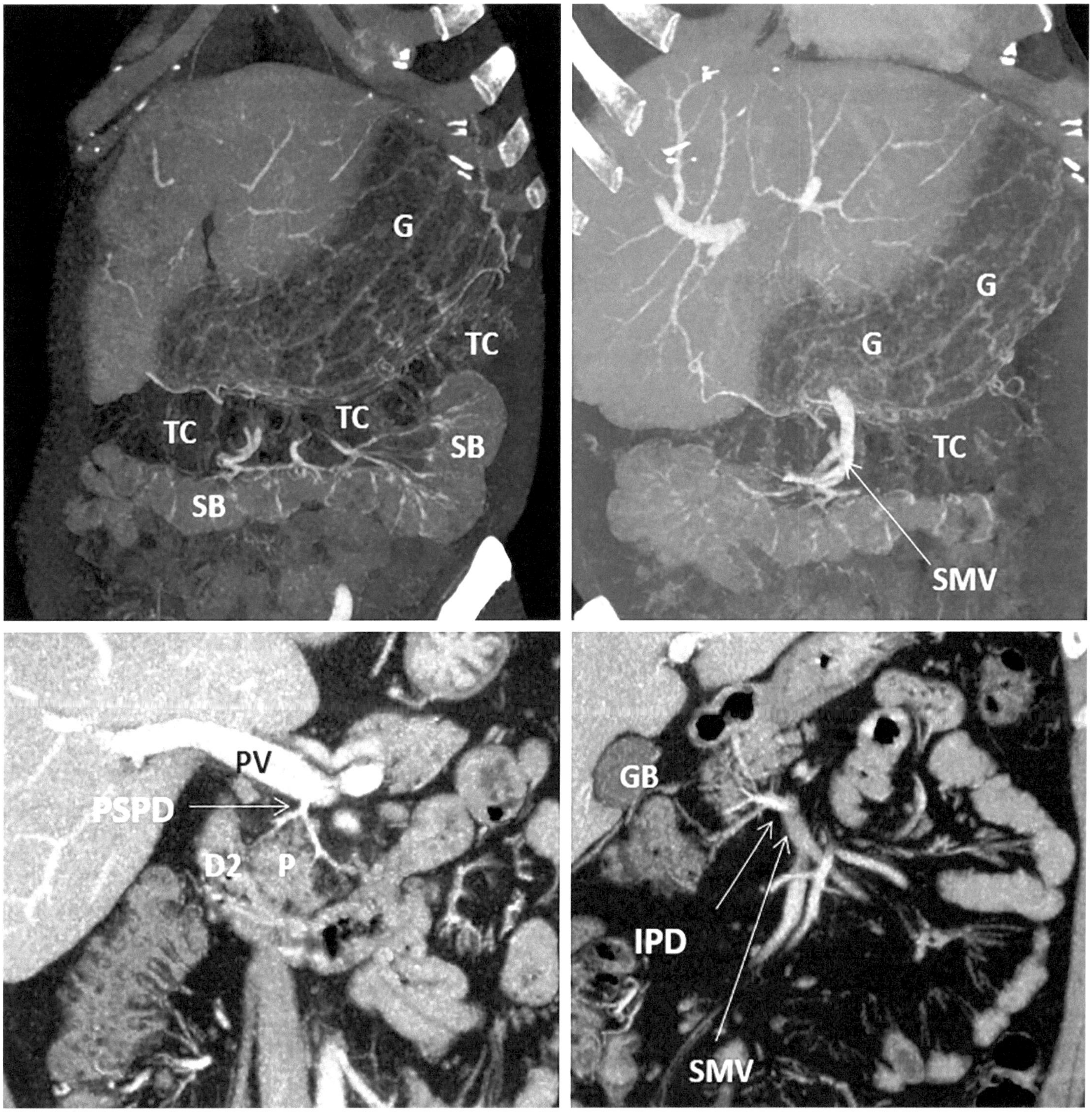

Fig. 3.9 Gastric and duodenal veins. Upper row: Coronal maximum intensity projection computed tomography (MIP CT) angiographies at the venous phase showing extensive venous branches on the surface of the stomach. Lower row: Axial and coronal CT images showing pancreaticoduodenal veins. The PSPD vein usually passes behind the common bile duct to enter the portal vein. D2, second duodenum; G, stomach; GB, gallbladder; IPD, inferior pancreaticoduodenal vein; P, pancreas; PSPD, posterior superior pancreaticoduodenal vein; PV, portal vein; SB, small bowel; SMV, superior mesenteric vein; TC, transverse colon.

3

and the superior mesenteric vein near the confluence with the portal vein. Prepyloric vein of Mayo is a tributary of the right gastric vein which is used as an anatomic landmark since it passes directly anterior to the pylorus (▶ Fig. 3.8).

The distal esophagus, part of lesser curvature, cardia, and fundus of the stomach blood supply drains into the portal vein via the coronary vein. The clinical significance of the gastric veins is primarily related to portal hypertension and secondary varices. Although the risk of bleeding from gastric varices is relatively low (10–36%), gastric variceal rupture carries a high mortality rate (14–45%).[7,8]

Esophageal and gastric varices usually involve the coronary and short gastric veins as they communicate with each other (▶ Fig. 3.10). Gastric varices of the left gastric vein are frequently located at the cardia and sometimes extend cephalad to coalesce with the paraesophageal varices. Gastric varices of the posterior gastric and short gastric veins are frequently located at the fundus. The majority of gastric varices located at the fundus drain into the inferior phrenic vein, which later joins with the left renal vein to form the gastrorenal shunt.[8] In splenic vein thrombosis, the short gastric veins drain splenic venous blood into the cardia and upper gastric veins causing varices of the submucosal venous saccules of the cardia and fundus of the stomach but in rare occasion intrapancreatic veins will be enlarged (▶ Fig. 3.10). In distal splenorenal shunt surgery for decompression of the esophageal and gastric varices in portal hypertension, the coronary vein is ligated and the distal end of the splenic vein is anastomosed to the renal vein.

Lymphatics

Lymph from the gastric wall is collected into a dense subperitoneal lymphatic network which is drained into the first group of lymph nodes along the lesser and greater curvature of the stomach.

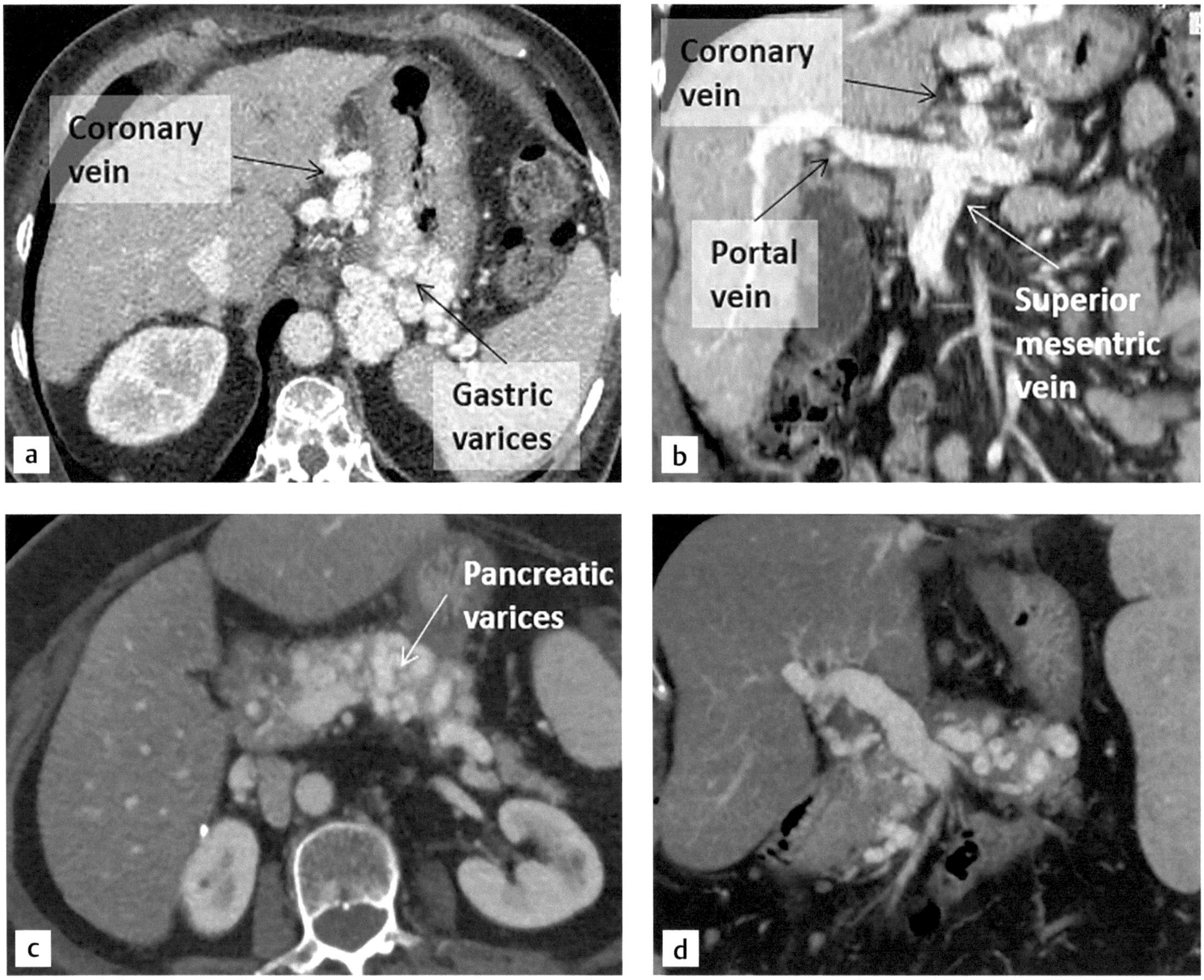

Fig. 3.10 Portal system varices. **(a)** Axial and **(b)** coronal computed tomography (CT) in a patient with cirrhosis and portal hypertension. Gastric fundus varices are seen. The coronary vein (left gastric) is enlarged and appears varicoid. **(c)** Axial and **(d)** coronal CT images show pancreatic varices in a patient with splenic vein thrombosis.

The stomach can be divided into four zones that drain into specific lymphatic node groups:

- The distal esophagus, cardia, and lesser curvature down to incisura angularis drain into the left gastric nodes along the left gastric artery.
- The lesser curvature portion of the antrum and pyloric canal drain along the right gastric artery into the suprapyloric and right suprapancreatic nodes.
- Lymphatics of the proximal greater curvature drain via the left gastroepiploic nodes into the left suprapancreatic and splenic lymph nodes.
- Lymphatics of the distal greater curvature down to the pylorus are drained by the right gastroepiploic lymph nodes into the subpyloric lymph nodes which lie anterior to the head of the pancreas. All of the lymph from the stomach eventually drain into the celiac nodes and the thoracic duct. Knowing the anatomy of lymphatics can help in predicting the root of lymph node metastasis from gastric carcinoma. However, due to abundant submucosal lymphatics, carcinoma of the stomach can spread to other parts of the stomach and involve any gastric or remote lymph nodes (▸Fig. 3.11). The thoracic duct drains into the venous system in the neck at the junction of the left internal jugular vein and the left subclavian vein. Many abdominal malignancies especially from the stomach can present with the left supraclavicular lymph node metastasis (see Chapter 21 for lymphatics of the abdomen and pelvis).

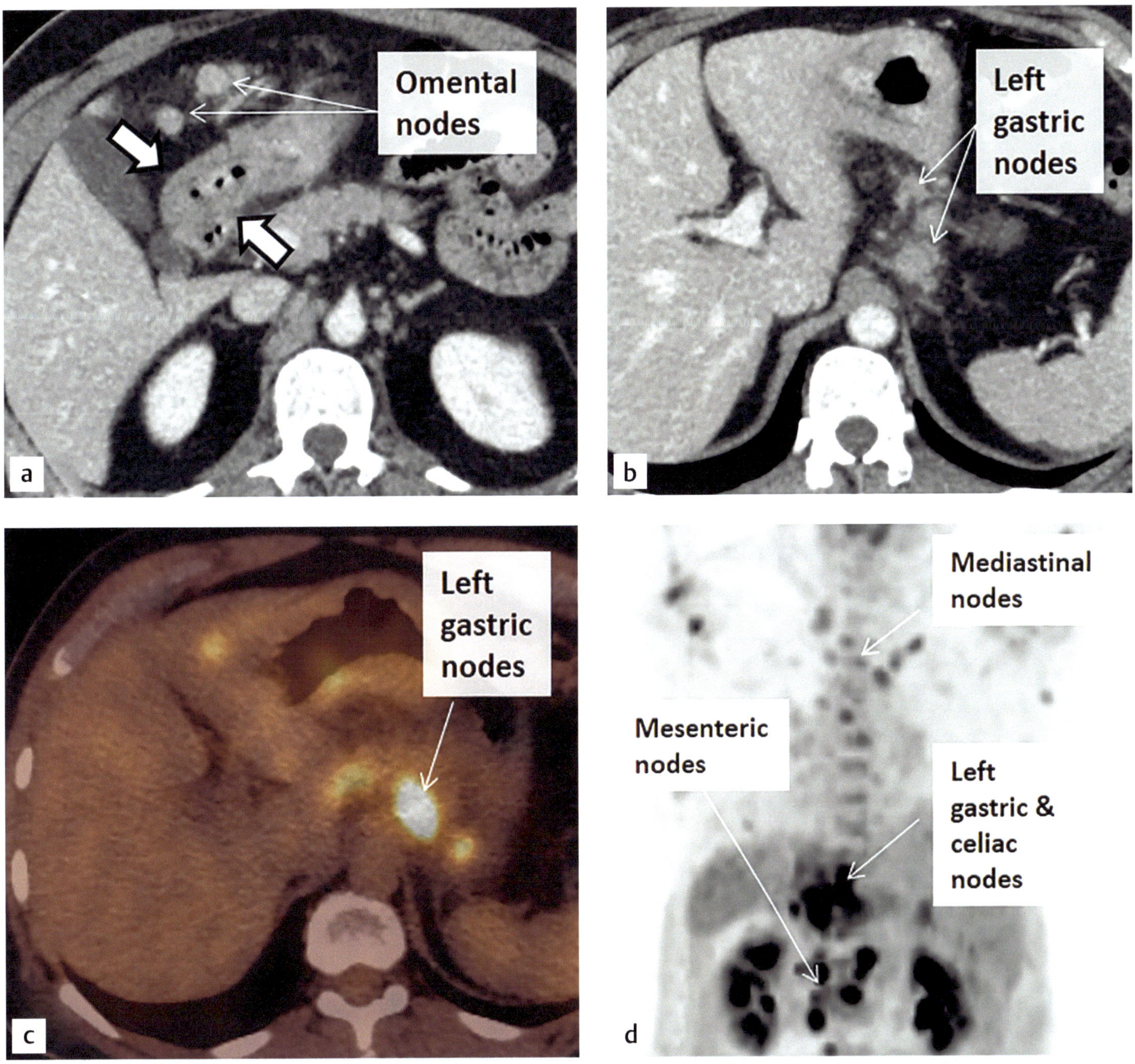

Fig. 3.11 Axial CT images showing **a** and **b**. Gastric antral cancer (*arrows* in **a**) with regional and remote nodal metastasis confirmed by positron emission tomography–computed tomography (PET-CT) (**c** and **d**).

3

Innervation

Since both the esophagus and vagal nerves develop together, the vagal nerves enter the abdomen through the esophageal hiatus. Above the diaphragm the vagus nerves of the periesophageal plexus coalesce into two trunks.[9] It is not uncommon for the vagal nerves to form up to five branches as they enter the abdomen. This variation can cause problem for a successful vagotomy. The left trunk courses anteriorly and the right trunk posteriorly as they pass through the esophageal hiatus. The anterior vagus gives off the hepatic branch and the anterior nerve of Latarjet along the lesser curvature and the posterior vagus gives off the celiac branch and the posterior nerve of Latarjet which is the largest branch (▶ Fig. 3.12). The anterior nerve of Latarjet innervates the anterior wall of the stomach and the first portion of the duodenum and the posterior nerve of Latarjet innervates the posterior gastric wall.

Several types of vagotomies have been developed. In truncal vagotomy, basically all vagal innervation of the abdominal viscera is transected (▶ Fig. 3.12). Selective vagotomy only denervates the stomach from the gastroesophageal junction to the pylorus. In highly selective vagotomy only the body of the stomach from the gastroesophageal junction to the antrum will be denervated.

The sympathetic innervation of the stomach is provided by preganglionic greater splanchnic nerves which synapse in the celiac plexus. Postganglionic fibers from the celiac ganglia accompany the arterial supply to innervate the stomach.

The pain that originates from the stomach, duodenum, and pancreas is carried by special visceral afferent fibers that relay through the splanchnic nerves and celiac plexus. Surgical resection or alcohol ablation of the celiac ganglia have been performed to alleviate pain of pancreatic and gastric cancers invading bone and retroperitoneal tissue.[10,11] Using a posterior approach and under computed tomography (CT) guidance, a 22-G needle is advanced to the antecrural space and neurolysis is achieved by injecting a volume of 20 to 50 mL of ethanol together with a local anesthetic and contrast medium.

Celiac Plexus

The celiac plexus is the largest visceral plexus and is located deep in the retroperitoneum along the anterolateral aspect of the aorta and close to the origins of the celiac axis and the superior mesenteric artery.[12] Posteriorly, it is separated from the vertebral column by the diaphragmatic crura, which are important anatomic barriers between the celiac plexus, situated in the antecrural space, and the splanchnic nerves, situated posterior to the crus, in the retrocrural space.[10] The celiac plexus consists of several ganglia including the two celiac ganglia, one superior mesenteric ganglion, and two aorticorenal ganglia, which form a dense network of interconnecting nerve fibers (▶ Fig. 3.13). In 94% of cases, the celiac ganglia are located at the level of T12 or L1.[13] The celiac artery is a reliable landmark for localizing the celiac plexus. The left celiac ganglion is slightly more caudal than the right, and both are located on the sides of the celiac trunk. The mean size of the celiac ganglion is about 2.7 cm (from 0.5 to 4.5 cm) and can be localized on CT images (▶ Fig. 3.13). The ganglia have the same CT attenuation as the adjacent adrenal glands on nonenhanced and portal, venous-phase images.[13]

The celiac plexus is composed of the preganglionic sympathetic efferent nerve fibers, which derive from the greater splanchnic

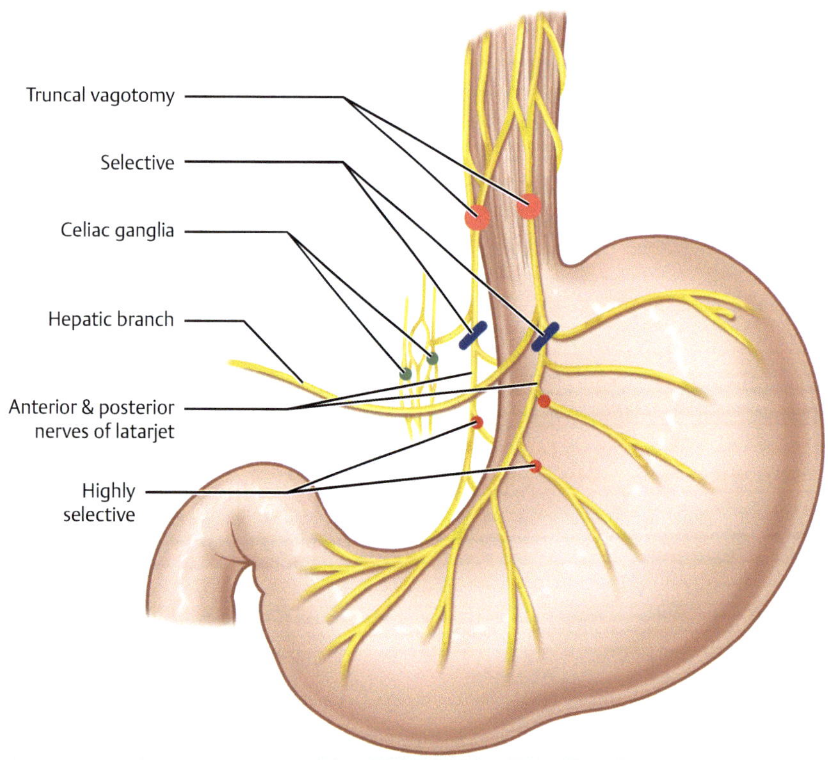

Fig. 3.12 Gastric vagal nerve branches and types of vagotomy. The anterior vagus gives off the hepatic branch and the anterior nerve of Latarjet along the lesser curvature and the posterior vagus gives off the celiac branch and the posterior nerve of Latarjet which is the largest branch.

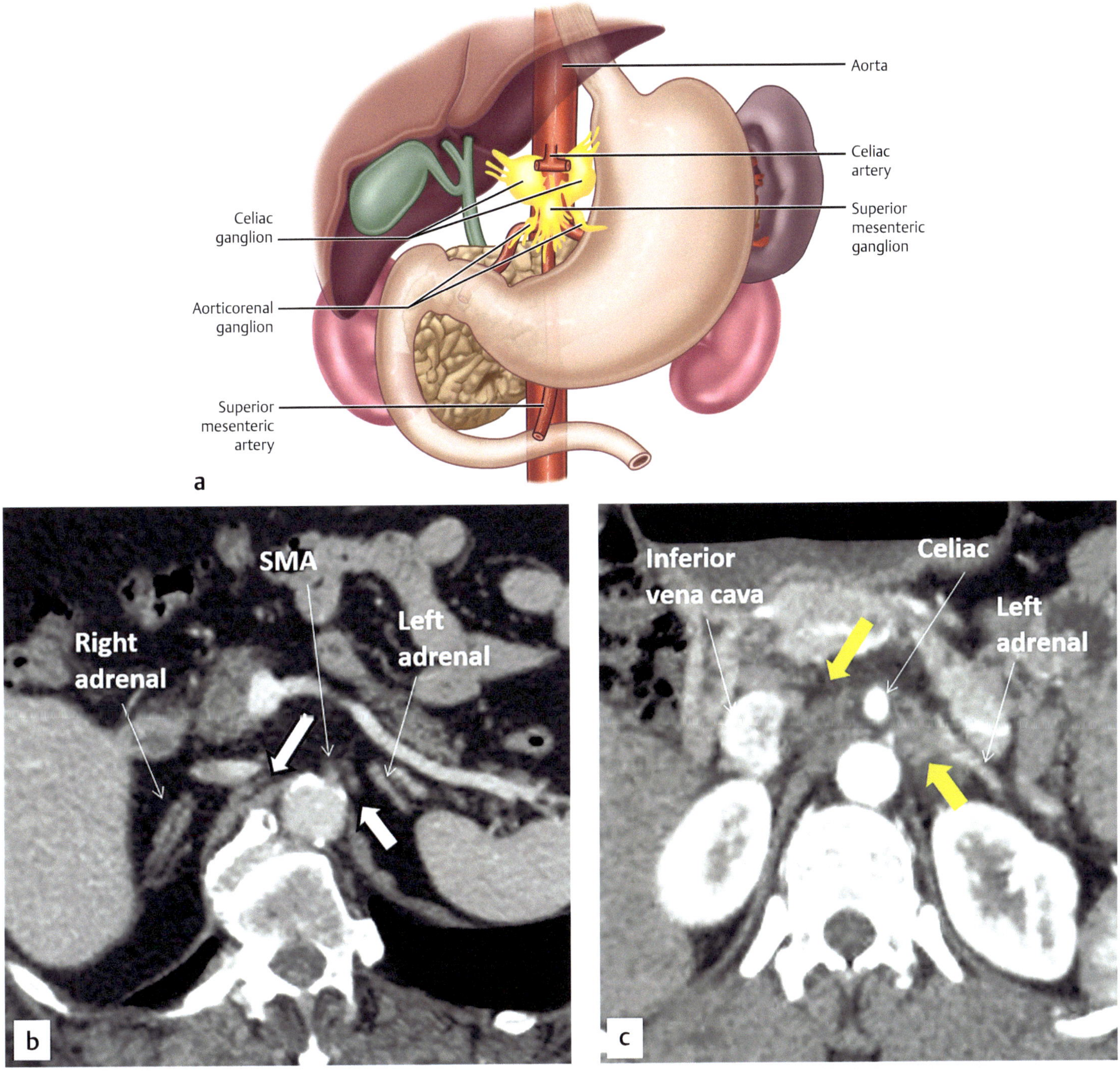

Fig. 3.13 (a) Relationship of the celiac plexus with the aorta and the celiac artery. The celiac plexus consists of two celiac ganglia, one superior mesenteric ganglion, and two aorticorenal ganglia, which form a dense network of interconnecting nerve fibers. **(b)** Axial, contrast-enhanced computed tomography (CT) scan just above the origin of the superior mesenteric artery (SMA) showing the left and right celiac ganglia (*arrows*) medial to the adrenal glands. **(c)** Metastatic lesions (*arrows*) to both celiac ganglia in a patient with gastric cancer.

(T5 through T9), lesser splanchnic (T10 through T11), and least splanchnic (T12) nerves. The preganglionic parasympathetic efferent fibers from the posterior trunk of the vagus nerve also contribute to the celiac plexus. Furthermore, the visceral afferent fibers that carry nociceptive stimuli from the upper abdominal viscera (the distal esophagus to the transverse colon) pass through the celiac plexus before terminating in the spinal cord. The celiac plexus innervates the liver, gallbladder, biliary tract, pancreas, spleen, adrenal glands, kidneys, mesentery, and the small and large bowel proximal to the transverse colon.

The celiac plexus represents the main target point of pain transmission to the upper abdominal organs. Therefore, neurolysis of the celiac plexus is an effective method of controlling pain that originates from these organs.

◆ Gastric Function

Gastric function and feeding process are under the control of the vagus and splanchnic nerves. Peristalsis starts in the antrum as shallow synchronized indentations in the incisura angularis. In a few minutes, the antral and later body contractions increase in strength and frequency as they move progressively toward the pylorus (▶ Fig. 3.14). During this early phase of digestion, the pylorus only opens incompletely and most of the material reaching

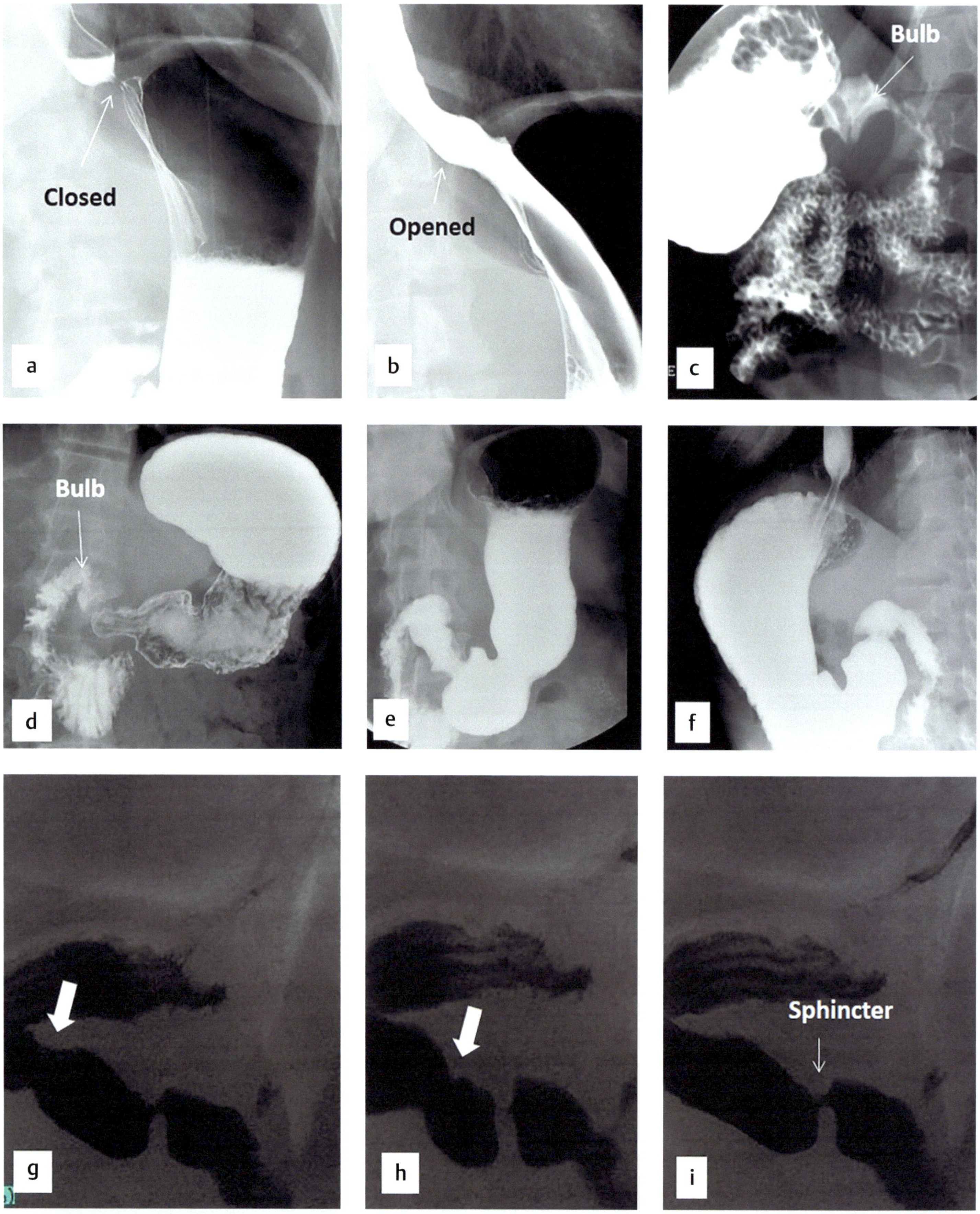

Fig. 3.14 Upper gastrointestinal (GI) barium study of the stomach and duodenum. Typical positions and image findings are shown. **(a, b)** Upright images showing closed and opened gastroesophageal junction. **(c, d)** Prone and supine images showing the bulb. **(e, f)** Upright and prone with full contrast showing gastric contour. **(g–i)** Fluoroscopy views showing circumferential peristalsis (*arrows*) starting in proximal antrum and extending to the pylorus but not opening the pyloric sphincter.

the pyloric portion is pushed back into the fundus. These contractions push the food particles and juice forward against a closed sphincter muscle and grind them into small particles. Once the food content consistency is reduced to semifluid, it passes through the pylorus into the duodenum. The tone of the pyloric sphincter and evacuation through the pylorus are regulated by the influence of gastrointestinal hormones secreted by the stomach and duodenum and nerve stimuli. Ghrelin is a peptide hormone, produced by enteroendocrine cells in the gastric mucosa, which has been shown to affect gastric motility in humans.[14] Hypertonicity, large amount of food, and fatty food content will all slow gastric motility and emptying. In addition to regulation of gastric emptying, the pylorus helps to prevent duodenal reflux.

The electrophysiologic pattern of gastric motor activity is based on a constant slow wave pattern at approximately 3 cycles per minute. This slow wave pattern originates at the gastric pacemaker somewhere in the middle of the greater curvature but never passes the pylorus. Muscle activity of the fundus is separate from muscle activity of the antrum and the pylorus.[15] Slow wave pattern electrical impulses also exist in the duodenum but it is totally separate from the stomach and has a faster rate of approximately 11 to 12 cycles per minute. In contrast to stomach where contractions propagate in any direction, in the duodenum and small bowel, contractions are progressive and move forward only. The lag phase before gastric emptying for solid food lasts about 60 minutes, whereas for the fluid it is faster. Emptying usually lasts 3 to 4 hours. Once emptying is complete, electrophysiologic activity of the stomach returns to the basic migrating motor complex. Sporadic waves of strong contractions (housekeeper waves) can occur between meals to clean stomach from any indigestible particles.

◆ Diagnosis of the Functional Disorders

Functional gastroduodenal disorders are common and classified into four groups: functional dyspepsia, belching (i.e., aerophagia) disorders, nausea and vomiting disorders, and rumination syndrome.[15,16] A number of tests to evaluate gastric neuromuscular function are available, including measures of gastric emptying, contractility, electrical activity, regional motility of fundus, antrum or pylorus, and tests of gastric sensitivity and compliance.[15] Wireless capsules are developed that records luminal pH and pressure during transit through the gastrointestinal tract, and provide data on gastric emptying time, small bowel transit time, and whole gut transit time.[17] Functional dyspepsia is persistent or recurrent pain or discomfort of the upper abdomen (e.g., postprandial fullness and bloating) in the absence of an organic disease. A negative endoscopy is also necessary for diagnosis. It is recommended that biopsies be routinely obtained at the time of endoscopy to detect *H. pylori* infection. A barium swallow study is less sensitive and specific than upper endoscopy, and hence it is not generally recommended. It is shown that less than 25% of patients with functional dyspepsia have delayed gastric emptying.[16] In delayed gastric emptying, the pyloric sphincter may not open efficiently or at the right times to allow food to pass through. Delayed gastric emptying is a frequent finding in diabetic patients likely related to disruption of the interstitial cells of Cajal (ICC) network.[18] Although gastric emptying study (e.g., scintigraphy) is not currently recommended as a routine clinical test to evaluate gastric dysfunction, it could help to exclude gastroparesis. Gastric emptying scintigraphy involves eating food that has added radioactive substance so that the rate of emptying of the stomach can be measured.

Gastroesophageal reflux disease (GERD) is a common, chronic gastrointestinal disorder. It is reported that approximately 20% of the population have heartburn weekly.[19] The diagnosis of GERD is made by barium fluoroscopy, endoscopy, catheters, and capsules measuring intraesophageal pH and empiric treatment trials. Esophageal pH monitoring is the gold standard to diagnose GERD. A sudden decrease in intraesophageal pH to below 4.0 is the accepted criterion to identify a gastroesophageal reflux during.[20]

◆ Imaging of the Stomach

Common imaging studies of the stomach and duodenum include single- and double-contrast gastrography and CT scan (▸Fig. 3.14). Air-contrast, high-resolution CT has been performed to enhance detection of gastric lesions including ulcers and carcinoma.[21–23] CT virtual endoscopy is capable of simulating gastric mucosal changes within the lumen as seen at conventional endoscopy. Transluminal ultrasound at the time of endoscopy is another technique that provides high-resolution images of the stomach. Although upper gastrointestinal barium study can reveal mucosal lesions, it has been largely replaced by endoscopy methods which allow direct visualization of the esophagus and stomach. Esophagogastroduodenoscopy (EGD) is superior to upper barium study for diagnosis and grading of erosive esophagitis. Still barium study remains an important tool in diagnosis of gastric hiatal hernia, GERD, postoperation complications, strictures, perforation, and fistula. With regards to GERD, presence or absence of gastroesophageal reflux during fluoroscopy should be interpreted cautiously given the limited sensitivity and specificity of this finding. Limited amounts of gastroesophageal reflux can occur even in normal healthy volunteers, and absence of reflux in the short time of study does not rule it out.

◆ Postoperative Stomach

Common gastric resection surgeries include Billroth I gastroduodenostomy, Billroth II gastrojejunostomy with a Roux-en-Y or loop-type gastrojejunostomy, and total gastrectomy with esophagojejunostomy. These techniques are commonly used for the treatment of malignancy (▸Fig. 3.15). Billroth I gastroduodenostomy

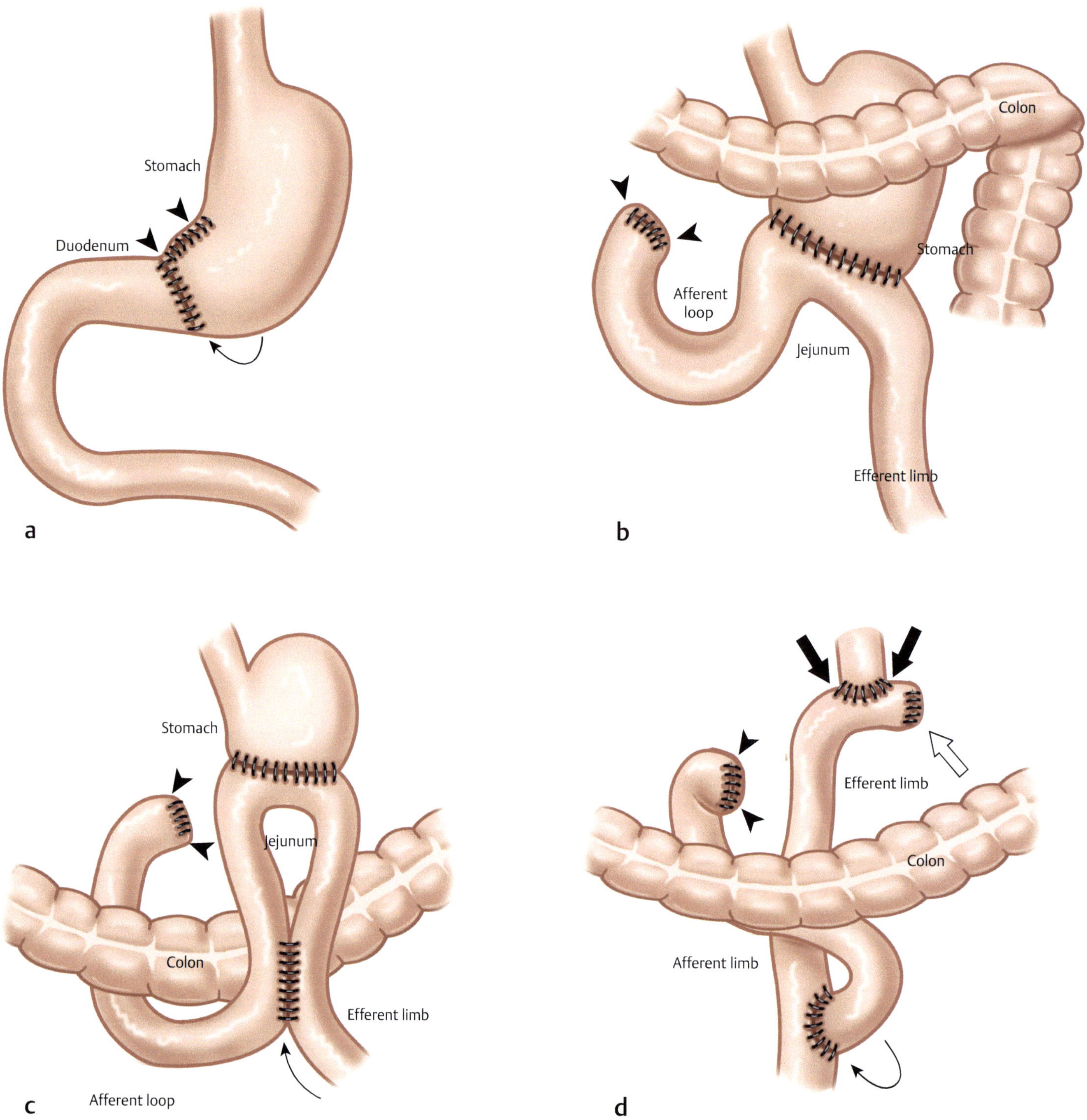

Fig. 3.15 Gastrectomy types. **(a)** Billroth I gastroduodenostomy include antrectomy and end-to-end anastomosis between the remnant stomach and the duodenum. In Hofmeister modification, usually the lesser curvature aspect of the cut end of the stomach is partially closed (*arrowheads*) and an anastomosis is created on the greater curvature side (*arrow*). **(b, c)** In Billroth II gastrojejunostomy, the duodenal stump (*arrowheads*) is closed and an anastomosis is made between the stomach and the jejunum. The afferent loop is shorter with retrocolic gastrojejunostomy **(b)** than with antecolic gastrojejunostomy **(c)**. In some instances, a side-to-side anastomosis is also created between the two jejunal limbs (*arrow* in **c**). **(d)** *A Roux-en-Y* procedure is performed following total gastrectomy. End-to-side esophagojejunostomy (*straight solid arrows*) with the blind end of the proximal jejunal limb (*straight arrow*) and end-to-side jejunojejunostomy (*curved arrow*) are performed. The closure of the duodenal stump (*arrowheads*) must be followed up.

is defined as antrectomy with gastroduodenal end-to-end anastomosis.[24] The entire incised end of the stomach may be used for the anastomosis (Polya procedure), or only a portion of the lumen may be used and the remainder closed (Hofmeister modification). In Billroth II gastrojejunostomy, the duodenal stump is closed and an anastomosis is created between the remnant stomach and the jejunum. The type of anastomosis is either Polya or Hofmeister techniques, the type of gastrojejunostomy is either Roux-en-Y or loop type (the latter in either an isoperistaltic or antiperistaltic configuration), and the placement of the anastomotic loop is either retrocolic or antecolic position. The antecolic gastrojejunostomy is frequently used in advanced gastric cancer to keep the anastomosis out of the lesser sac where recurrence usually occurs first in preference. However, retrocolic gastrojejunostomy creates a shorter afferent (blind) loop that results in improved nutritional status and less likelihood of retrostomal herniation. Roux-en-Y gastrojejunostomy with end-to-side anastomosis is common nowadays.

There are three types of bariatric surgical procedures for morbid obesity: restrictive, malabsorptive, and combined. Roux-en-Y gastric bypass combines restrictive and malabsorptive properties by creating a small gastric pouch and a Roux limb. Restrictive-only types create a small gastric pouch to induce weight loss which include vertical banded gastroplasty and adjustable gastric banding.[25,26] The Roux-en-Y gastric bypass can be done after laparotomy or through laparoscopic techniques.[25] Current standard laparoscopic Roux-en-Y gastric bypass procedure involves the following: construction of a pouch, Roux limb, and enteroenteric anastomosis; placement of the Roux limb; and creation of the gastrojejunal anastomosis (▶ Fig. 3.16). Laparoscopic adjustable gastric banding (LAGB) is one of the most frequently used and least invasive surgical treatments for morbid obesity and involves laparoscopic placement of the band around the proximal part of the stomach to create a small pouch (▶ Fig. 3.17).

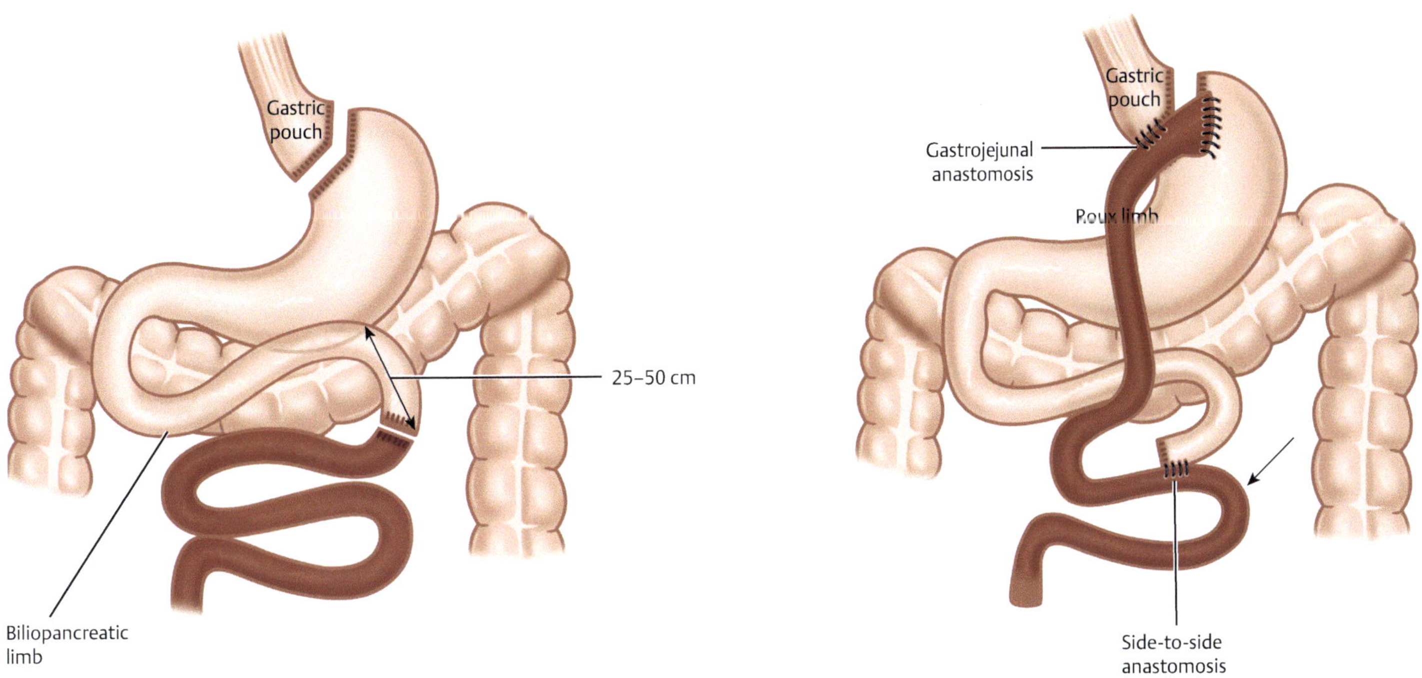

Fig. 3.16 Roux-en-Y gastric bypass. A gastric pouch is created by dividing the stomach primarily along the lesser curvature. The jejunum is divided 25 to 50 cm distal to the ligament of Treitz. The biliopancreatic limb is attached to the *Roux limb with a side-to-side anastomosis.* The Roux limb is brought anterior to the transverse colon and stomach and anastomosed to the gastric pouch. *Arrow* demarcates the common channel.

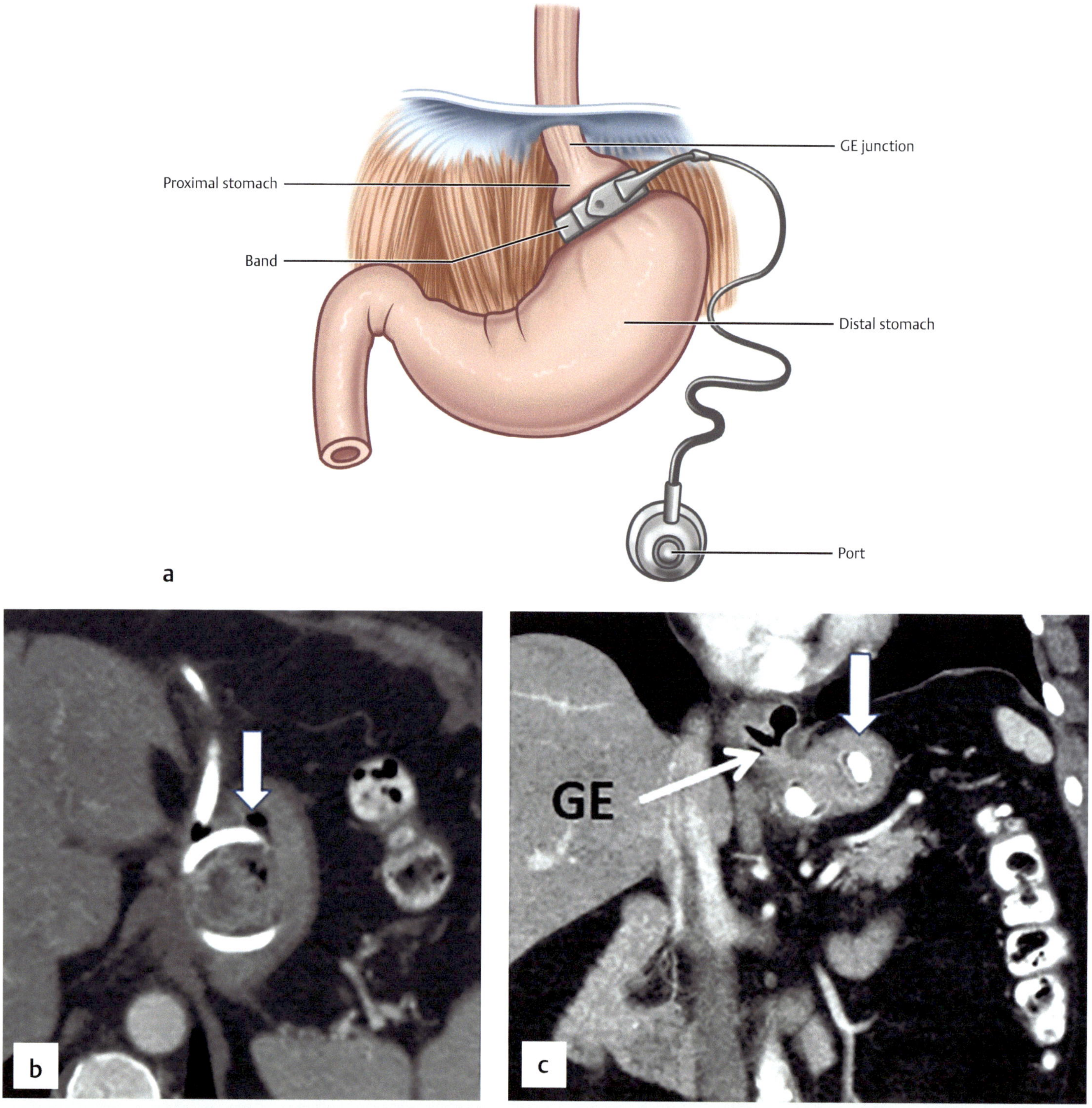

Fig. 3.17 (a) Drawing depicts the altered gastric anatomy after a laparoscopic adjustable gastric banding (LAGB) procedure. **(b)** Axial and **(c)** coronal computed tomography (CT) images. GE, gastroesophageal. Axial and coronal CT images show an LAGB device with a radiopaque silicone band (*arrow*).

◆ Gastric Pathology

Among developmental anomalies of stomach, hypertrophic pyloric stenosis is a serious gastric disorder seen frequently in infancy. The predominant symptom is regurgitation of bile-free vomitus within the first few hours after birth. Abdominal radiography shows distension of the stomach proximal to the obstruction and the absence of gas in the small bowel and colon, resulting in a "single bubble" appearance. Ectopic pancreatic tissue may be found at various sites in the gastrointestinal tract, most commonly in the gastric antrum. It is seen as a dome-shaped filling defect 1 to 3 cm in diameter on the larger curvature. Duplications of the stomach account for about 7% of gastrointestinal tract duplications. Most are noncommunicating.[27]

Gastric diverticulum is relatively common and usually seen in the fundus posteriorly (▶Fig. 3.18). Adult form of hypertrophic

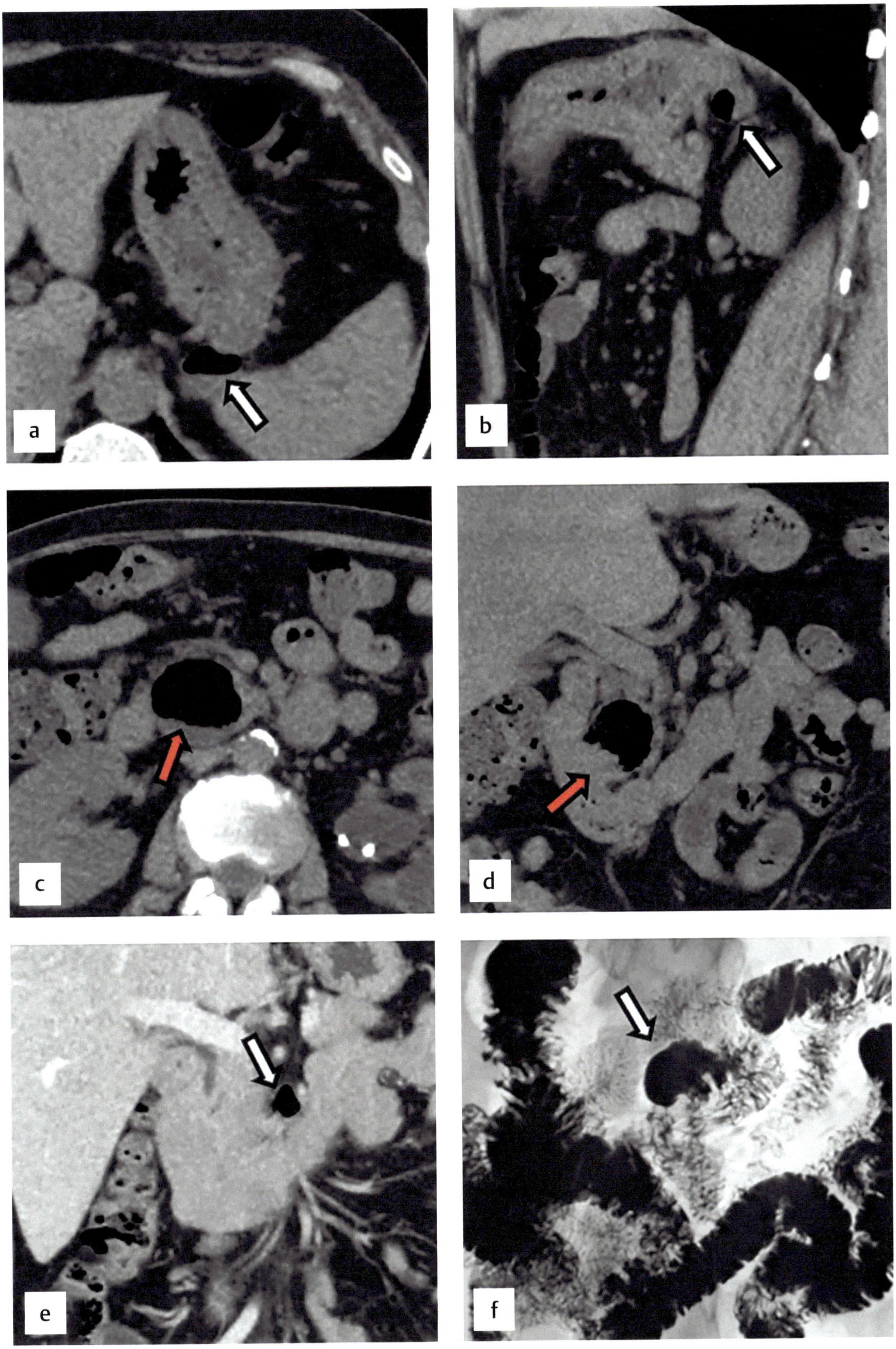

Fig. 3.18 Gastric and duodenal diverticulum (*arrows*). **(a)** Axial and **(b)** coronal computed tomography (CT) images showing a gastric diverticulum in the posterior aspect of the fundus. **(c)** Axial (*arrows*) and **(d)** coronal CT images show a relatively large diverticulum (*arrows*) of the second duodenum projecting toward the pancreas. This type of diverticulum may cause compressive obstruction of the distal common bile duct. **(e)** Coronal and **(f)** barium study showing diverticulum of the third duodenum (*arrows*).

pyloric stenosis is rare and is usually associated with gastric ulcer and gastritis. Gastric ulcer and gastritis are common gastric pathologies that are usually diagnosed with endoscopy but contrast fluoroscopy and CT of stomach may add further information (▶ Fig. 3.19). Stomach cancer is the second most common cancer worldwide and probably the most prevalent malignant neoplasm in Far East countries. CT with or without fluorodeoxyglucose (FDG)-positron emission tomography (PET) has been the modality of choice for preoperative evaluation and staging in patients with gastric carcinoma (▶ Fig. 3.11). In addition, CT is the primary tool for evaluation of recurrent tumors and response to chemotherapy. Early gastric cancer (EGC) is defined as a carcinoma in which invasion is limited to the mucosa and submucosa. Radiological diagnosis at this stage is usually difficult and endoscopy is the modality of choice. Recently, CT scan with virtual endoscopy following air distension of the stomach is used for the evaluation of gastric endoluminal pathologies including early gastric cancer[22] (▶ Fig. 3.20).

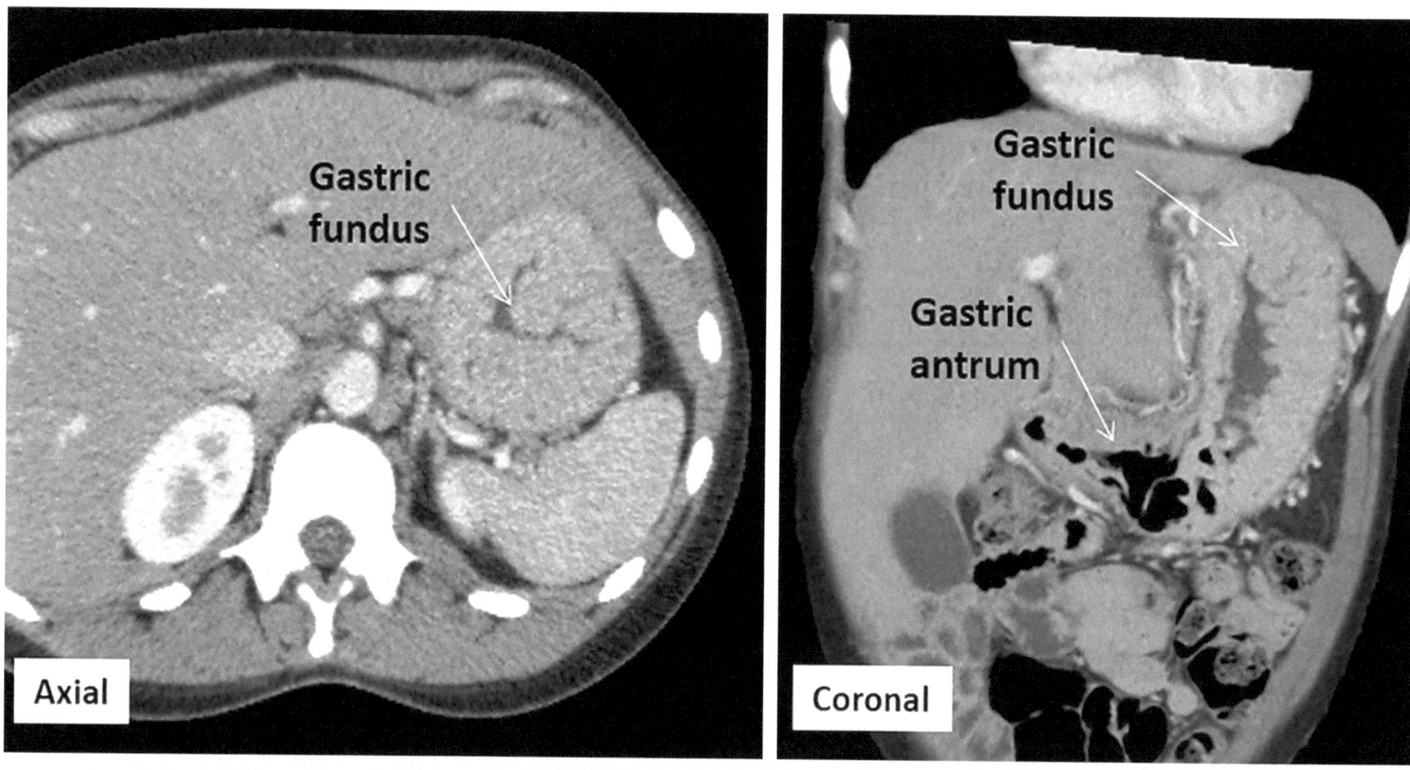

Fig. 3.19 Hypertrophic gastritis involving the body and fundus of stomach (Ménétrier disease). Severe thickening of the gastric mucosa is evident.

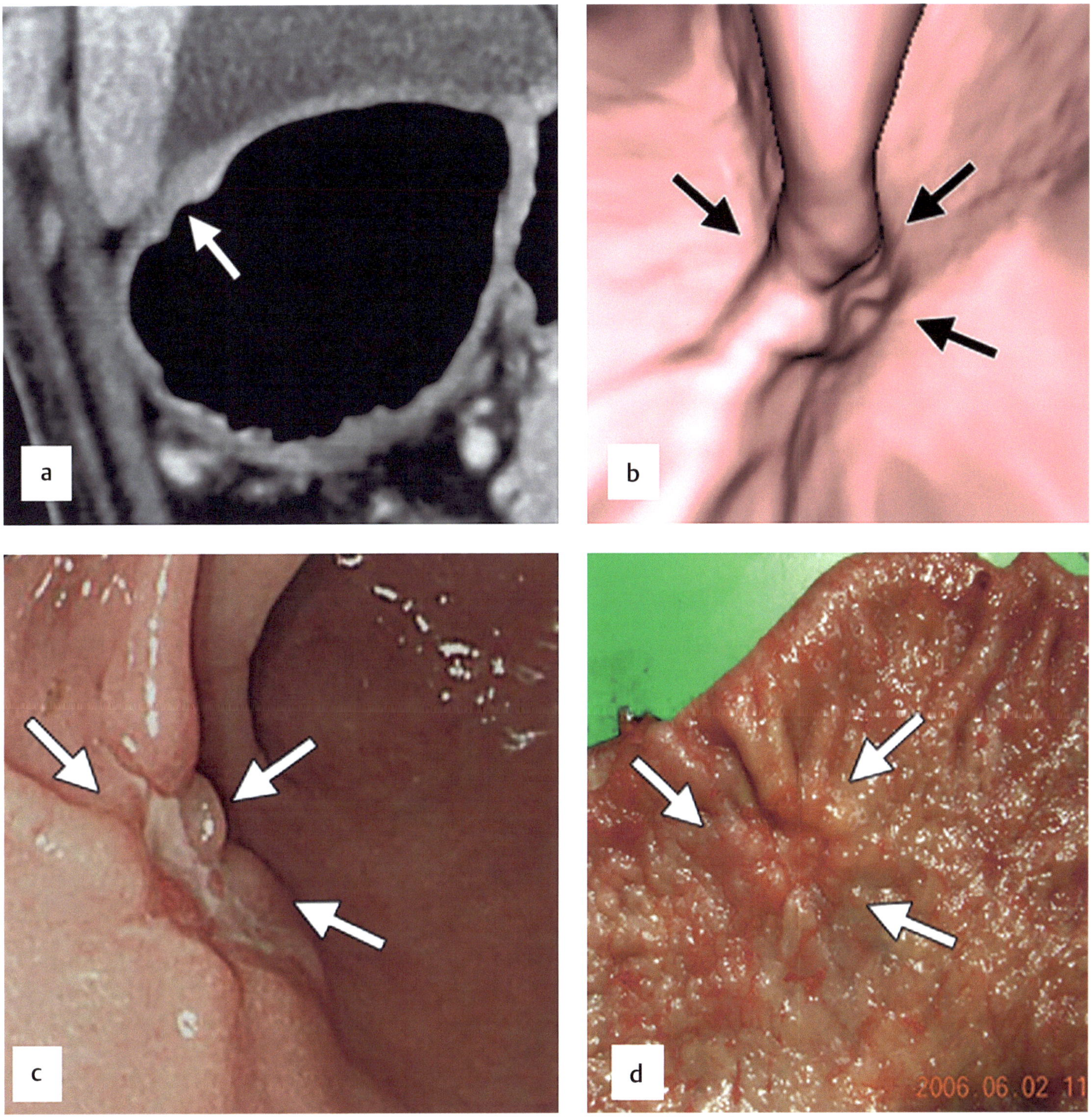

Fig. 3.20 Type IIa+IIc gastric cancer in a 63-year-old man. **(a)** Sagittal computed tomography (CT) and **(b)** virtual endoscopic image show a superficial elevated-depressed lesion (*arrows*) in the antrum of the stomach. **(c)** Conventional endoscopic image and **(d)** photograph of the gross specimen show the lesion (*arrows*). (Reproduced with permission from Shen et al.[22])

References

1. Flotch MH. Anatomy of stomach. In: Flotch MH, ed. Netter's Gastroenterology. 2nd ed. Philadelphia: Saunders-Elsevier; 2010
2. el-Eishi HI, Ayoub SF, el-Khalek MA. The arterial supply of the human stomach. Acta Anat (Basel) 1973;86(3):565–580
3. Vandamme JP, Bonte J. The blood supply of the stomach. Acta Anat (Basel) 1988;131(2):89–96
4. Koskas F, Gayet B. Anatomical study of retrosternal gastric esophagoplasties. Anat Clin 1985;7(4):237–256
5. Ndoye JM, Dia A, Ndiaye A, et al. Arteriography of three models of gastric oesophagoplasty: the whole stomach, a wide gastric tube and a narrow gastric tube. Surg Radiol Anat 2006;28(5):429–437
6. Egorov VI, Yashina NI, Zhurenkova TV, et al. Spleen-preserving distal pancreatectomy with resection of the splenic vessels. Should one rely on the short gastric arteries? JOP 2011;12(5):445–457
7. Sarin SK, Lahoti D, Saxena SP, Murthy NS, Makwana UK. Prevalence, classification and natural history of gastric varices: a long-term follow-up study in 568 portal hypertension patients. Hepatology 1992;16(6):1343–1349
8. Kiyosue H, Mori H, Matsumoto S, Yamada Y, Hori Y, Okino Y. Transcatheter obliteration of gastric varices. Part 1. Anatomic classification. Radiographics 2003;23(4):911–920
9. Androulakis J, Colborn GL, Skandalakis PN, Skandalakis LJ, Skandalakis JE. Embryologic and anatomic basis of duodenal surgery. Surg Clin North Am 2000;80(1):171–199
10. Kambadakone A, Thabet A, Gervais DA, Mueller PR, Arellano RS. CT-guided celiac plexus neurolysis: a review of anatomy, indications, technique, and tips for successful treatment. Radiographics 2011;31(6):1599–1621
11. Wang ZJ, Webb EM, Westphalen AC, Coakley FV, Yeh BM. Multi-detector row computed tomographic appearance of celiac ganglia. J Comput Assist Tomogr 2010;34(3):343–347
12. Loukas M, Klaassen Z, Merbs W, Tubbs RS, Gielecki J, Zurada A. A review of the thoracic splanchnic nerves and celiac ganglia. Clin Anat 2010;23(5):512–522
13. Zhang XM, Zhao QH, Zeng NL, et al. The celiac ganglia: anatomic study using MRI in cadavers. AJR Am J Roentgenol 2006;186(6):1520–1523
14. Lee KJ, Cha DY, Cheon SJ, Yeo M, Cho SW. Plasma ghrelin levels and their relationship with gastric emptying in patients with dysmotility-like functional dyspepsia. Digestion 2009;80(1):58–63
15. Tack J, Janssen P. Gastroduodenal motility. Curr Opin Gastroenterol 2010;26(6): 647–655
16. Tack J, Talley NJ, Camilleri M, et al. Functional gastroduodenal disorders. Gastroenterology 2006;130(5):1466–1479
17. Maqbool S, Parkman HP, Friedenberg FK. Wireless capsule motility: comparison of the SmartPill GI monitoring system with scintigraphy for measuring whole gut transit. Dig Dis Sci 2009;54(10):2167–2174
18. Camilleri M. Functional dyspepsia and gastroparesis. Dig Dis 2016;34(5): 491–499
19. Locke GR III, Talley NJ, Fett SL, Zinsmeister AR, Melton LJ III. Prevalence and clinical spectrum of gastroesophageal reflux: a population-based study in Olmsted County, Minnesota. Gastroenterology 1997;112(5):1448–1456
20. Tutuian R. Update in the diagnosis of gastroesophageal reflux disease. J Gastrointestin Liver Dis 2006;15(3):243–247
21. Kim JW, Shin SS, Heo SH, et al. Diagnostic performance of 64-section CT using CT gastrography in preoperative T staging of gastric cancer according to 7th edition of AJCC cancer staging manual. Eur Radiol 2012;22(3):654–662
22. Shen Y, Kang HK, Jeong YY, et al. Evaluation of early gastric cancer at multidetector CT with multiplanar reformation and virtual endoscopy. Radiographics 2011;31(1):189–199
23. Komori M, Kawanami S, Tsurumaru D, et al. Contrast-enhanced MDCT gastrography for detection of early gastric cancer: Initial assessment of "wall-carving image," a novel volume rendering technique. Eur J Radiol 2012;81(8): 1695–1701
24. Kim KW, Choi BI, Han JK, et al. Postoperative anatomic and pathologic findings at CT following gastrectomy. Radiographics 2002;22(2):323–336
25. Scheirey CD, Scholz FJ, Shah PC, Brams DM, Wong BB, Pedrosa M. Radiology of the laparoscopic Roux-en-Y gastric bypass procedure: conceptualization and precise interpretation of results. Radiographics 2006;26(5):1355–1371
26. Sonavane SK, Menias CO, Kantawala KP, et al. Laparoscopic adjustable gastric banding: what radiologists need to know. Radiographics 2012;32(4): 1161–1178
27. Berrocal T, Torres I, Gutiérrez J, Prieto C, del Hoyo ML, Lamas M. Congenital anomalies of the upper gastrointestinal tract. Radiographics 1999;19(4):855–872

4 Small Intestine

Farhood Saremi and Damián Sánchez-Quintana

◆ Introduction

The small intestine is a part of gastrointestinal tract where digestion and absorption of most proteins, lipids, and carbohydrates take place. It extends between the pylorus sphincter and the ileocecal valve. It is divided into the duodenum, jejunum, and ileum. Food breakdown occurs mainly in the duodenum. Digestive enzymes secreted by the pancreas and liver enter into the duodenum by the pancreatic duct and common bile duct, respectively. The alkaline secretions of the duodenal Brunner's glands help to protect duodenal wall against the acidic content of chyme. The jejunum is where the products of digestion will be absorbed. The absorption is determined by the surface area available which is highest in the jejunum. The ileum absorbs mainly vitamin B12. Bile acids are also absorbed in the ileum. Absorbed nutrients move to the liver by the portal venous capillaries.

◆ Embryology

The proximal duodenum is formed by the distal part of the foregut and the distal duodenum after ampulla is formed from the proximal part of the midgut. The jejunum, ileum, and ascending colon are formed by the midgut. As a result, the duodenum has a dual blood supply from the celiac trunk and superior mesenteric artery, whereas the rest of small intestine is supplied by the superior mesenteric artery.[1,2] The liver and pancreas are also formed by the foregut. Around the end of third week of gestation, a ventral diverticulum arises from the duodenum which by the end of fourth week differentiates into the primitive liver, the gallbladder and cystic duct, and the ventral pancreatic duct. The dorsal pancreatic primordium arises from the dorsal duodenum by the end of fourth week.

Early in embryonic life, before approximately 4 weeks of gestation, the gut is a short, straight, and featureless tube. Between 4 and 8 weeks, because of rapid elongation, most of the embryonic bowel protrudes outside the abdomen into the yolk sac, but between 10 and 12 weeks, it returns to the abdominal cavity. During the period of elongation, the duodenojejunal junction and colon undergo a simultaneous, two-phase "rotation" of 270 degrees counterclockwise around the mesenteric vessels' axis along with posterior fixation to assume their normal anatomic locations (▸Fig. 4.1). In the first stage, the duodenum and colon rotate 90 degrees around the omphalomesenteric vessels, which brings them from a superior–inferior position to a right–left orientation relative to the superior mesenteric artery. In the second stage, the duodenum rotates another 90 degrees. As a result, the duodenojejunal junction passes behind the superior mesenteric artery and later becomes fixed to the retroperitoneum on the left of spine. On the other hand, the colon rotates 180 degrees until the cecum is located in the right lower quadrant of the abdomen.[2] After 12 weeks until birth, the cecum continues to descend from the right upper to the right lower abdomen (third stage). At the completion of rotation, the mesentery becomes fixed to the retroperitoneum in a broad band extending from the duodenojejunal junction (ligament of Treitz) to the ileocecal junction.

In anomalies of rotation and fixation, various degrees of malrotation of the small or large bowel may occur and is classified into nonrotation, reversed rotation, and incomplete rotation.[3,4] In nonrotation, after rotating only 180 degrees, the postarterial limb of the midgut reenters the abdomen first instead of last. Therefore, the small intestine resides on the right side of the abdomen, and the colon and cecum on the left side. Upper gastrointestinal (UGI) series show an abnormal position of the duodenojejunal junction. In reversed rotation, the cecum returns to the abdomen first and passes to the right behind the superior mesenteric artery. In incomplete rotation prearterial segment of the midgut returns to the abdomen first; therefore, it is located in correct position within the abdomen. However, location of the cecum is abnormal. The cecum may be on the left side, higher than normal on the right side, or somewhere in the middle (▸Fig. 4.2). The mesentery in malrotation still attaches to the retroperitoneum but the attachment is very narrow. This configuration predisposes postnatal rotation around the narrow vascular mesenteric axis, leading to midgut volvulus with subsequent intestinal ischemia and necrosis.

◆ Anatomy

Duodenum

The duodenum is a C-shaped loop that extends between pyloric sphincter and jejunum. At the gastroduodenal junction, pyloric sphincter is surrounded by the duodenal fornix. The attachment of the suspensory ligament of Treitz marks the end of duodenum.[5] Histologically, the appearance of the submucosal glands of Brunner marks the beginning of the duodenum and their disappearance marks the duodenojejunal junction. The submucosal glands may extend a few centimeters into the pylorus.

A major part of the duodenum is located retroperitoneally (▸Fig. 4.3). Only the proximal half of the first part (duodenal bulb) of the duodenum is completely covered by the peritoneum. Bare areas of the duodenum, where no peritoneal coverage exists, include the medial margin of the second and third parts facing the head of the pancreas, and a small part of the anterior surface of the second part of the duodenum where the transverse colon is attached, and where the mesenteric root covers the third part of the duodenum (▸Fig. 4.4).[5] The superior mesenteric vessels pass anterior to the third part within the mesenteric root. Paraduodenal fossae are two pockets of peritoneum close to the third part of duodenum. It is formed because of incomplete fusion of the dorsal mesocolon with the parietal peritoneum. These pockets are potential sites of paraduodenal hernia which usually occurs toward the right-side mesentery.

The duodenum is divided into four parts (▸Fig. 4.5).

4

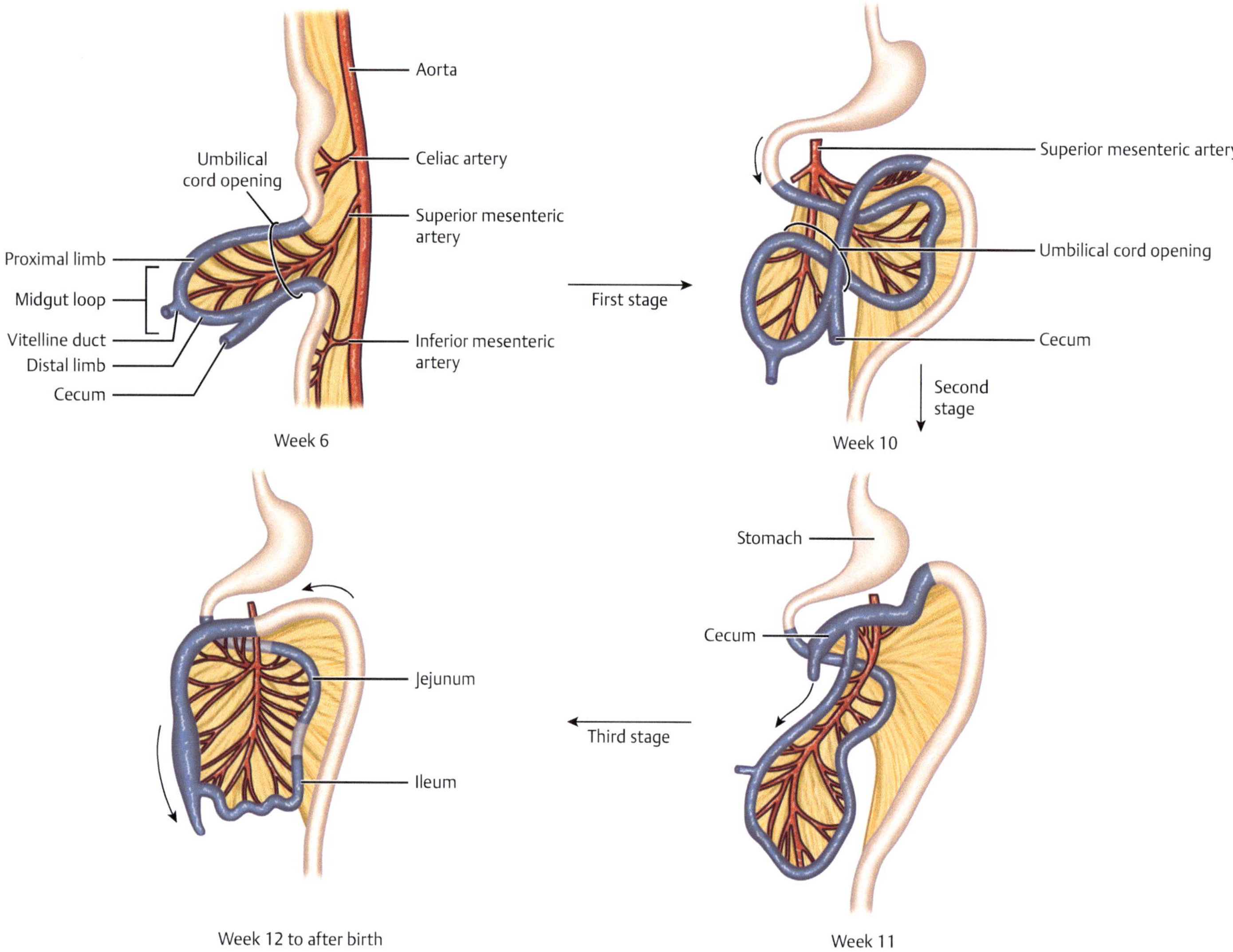

Fig. 4.1 Midgut rotation and fixation. During embryonic development, in the first two stages, the duodenojejunal and ileocolic segments of the primitive digestive tube rotate 270 degrees counterclockwise about the omphalomesenteric vessels (the future superior mesenteric vessels) to reside in their final normal locations. First, the duodenum and cecum rotate 90 degrees counterclockwise around the superior mesenteric artery so that the duodenum lies to the right and cecum lies to the left of the mesenteric vessels. Later, the duodenum rotates another 90 degrees counterclockwise so that it moves posterior to the superior mesenteric artery; the rest of the midgut is in the umbilical cord. By the 10th to 12th weeks of gestation, the intestine slides back into the peritoneal cavity, where the final 180 degrees rotation of the colon occurs allowing the cecum to move to the right and descend into the right lower quadrant. This rotation is followed in the third stage by peritoneal fixation of the bowel.

First Part

The first (or ascending) part of the duodenum is 5 cm in length. Its proximal half is mobile and intraperitoneal but the distal part is extraperitoneal (▸Fig. 4.5). The common bile duct, portal vein, gastroduodenal artery, and inferior vena cava pass posterior to the distal half of the first part. The quadrate lobe of the liver and the gallbladder are located anteriorly and the epiploic foramen superiorly. The hepatoduodenal portion of the lesser omentum attaches to the superior border of the duodenum and forms the roof of the epiploic foramen. The greater omentum attaches to its inferior border. A duodenal sphincter is described at the distal end of the duodenal bulb that is perhaps responsible for cases with bulbar enlargement (megabulb).[5]

Second Part

The second (or descending) part of the duodenum is 7.5 cm in length. It extends between the first and second duodenal flexures and ends approximately at the level of the superior border of L4 (▸Fig. 4.3). It is crossed by the transverse mesocolon dividing the second part into the supramesocolic and inframesocolic segments. The area between these two segments is not covered with the peritoneum (▸Fig. 4.4), whereas the rest of second part is covered with the visceral peritoneum.

The anatomic structures posterior to the second part include the right kidney, right ureter, right renal vessels, right psoas major, and edge of the inferior vena cava (▸Fig. 4.3, ▸Fig. 4.5). The common bile duct passes between the pancreatic head and the

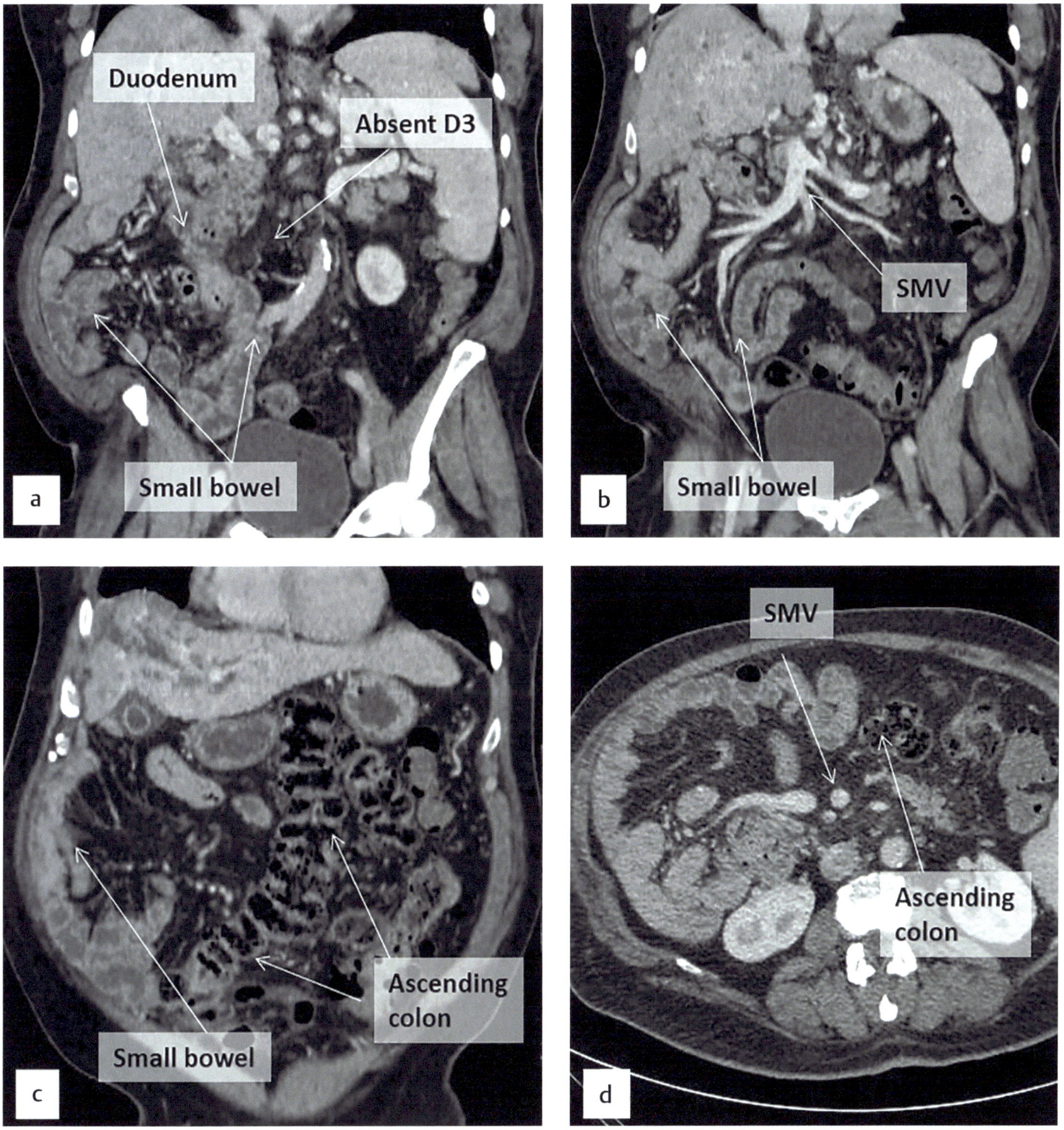

Fig. 4.2 Malrotation of the bowel in an adult. **(a–c)** Coronal images from posterior to anterior and **(d)** axial image showing abnormal position of the small bowel in the right side of abdomen and ascending colon in midline. The duodenojejunal junction and third duodenum (D3) are not formed. The mesenteric vessels are free and do not pass anterior to the D3 as expected. SMV, superior mesenteric vein.

second duodenum. The ascending colon is on the right side of the second part. The structures anterior to the second part are the right lobe of the liver, transverse colon, and jejunum. The ampulla of Vater opens into the midpoint of the second part of the duodenum on its posteromedial side.

Third Part

The third (or horizontal) part of the duodenum is 10 cm in length. It extends from about 5 cm of midline at the lower margin of L3 to the left side of the aorta where it curves superiorly to form the third flexure of the duodenum before connecting the

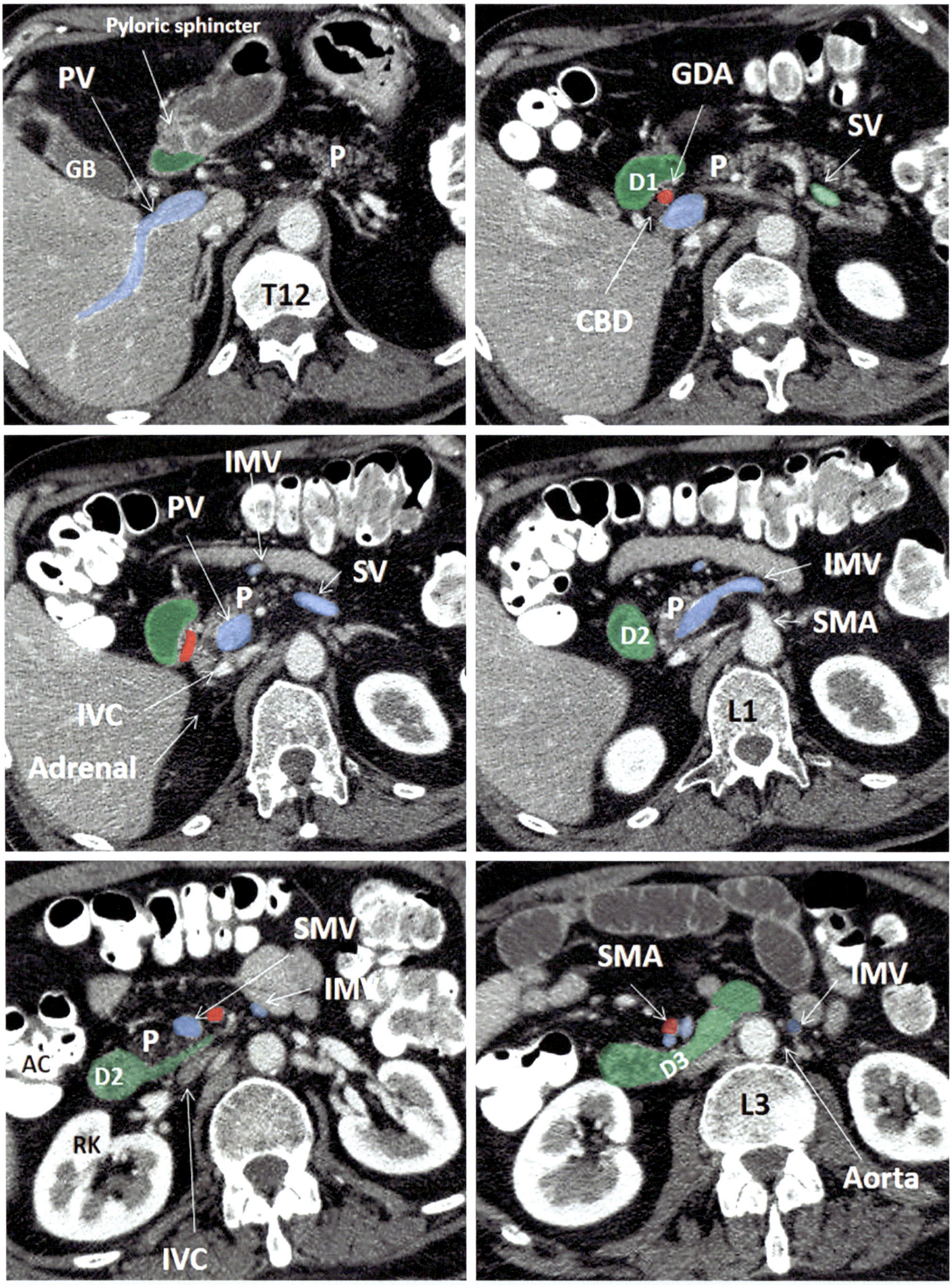

Fig. 4.3 Axial computed tomography (CT) scan images (from superior to inferior) showing the anatomic structures around the duodenum (*colored in green*). Portal veins are colored in *blue* and arteries in *red*. A major part of the duodenum is located retroperitoneally. Only the proximal half of the first part (D1) of the duodenum is completely covered by the peritoneum. The common bile duct (CBD), portal vein (PV), gastroduodenal artery (GDA), and inferior vena cava (IVC) pass posterior to the distal half of the D1. The gallbladder is located laterally and anteriorly. The anatomic structures posterior to the second part (D2) are the right kidney (RK) and its renal vessels. The common bile duct passes between the pancreatic head (P) and the second part of the duodenum (D2), and the ascending colon (AC) is on the right side of the D2. The third part (D3) passes anterior to the IVC, aorta, and right psoas muscle. It is crossed anteriorly by the superior mesenteric vessels and the root of the mesentery of the small intestine. The head of the pancreas (P) is located within the C loop of the duodenum. GB, gallbladder; IMV, inferior mesenteric vein; SMA, superior mesenteric artery; SMV, superior mesenteric vein; SV, splenic vein.

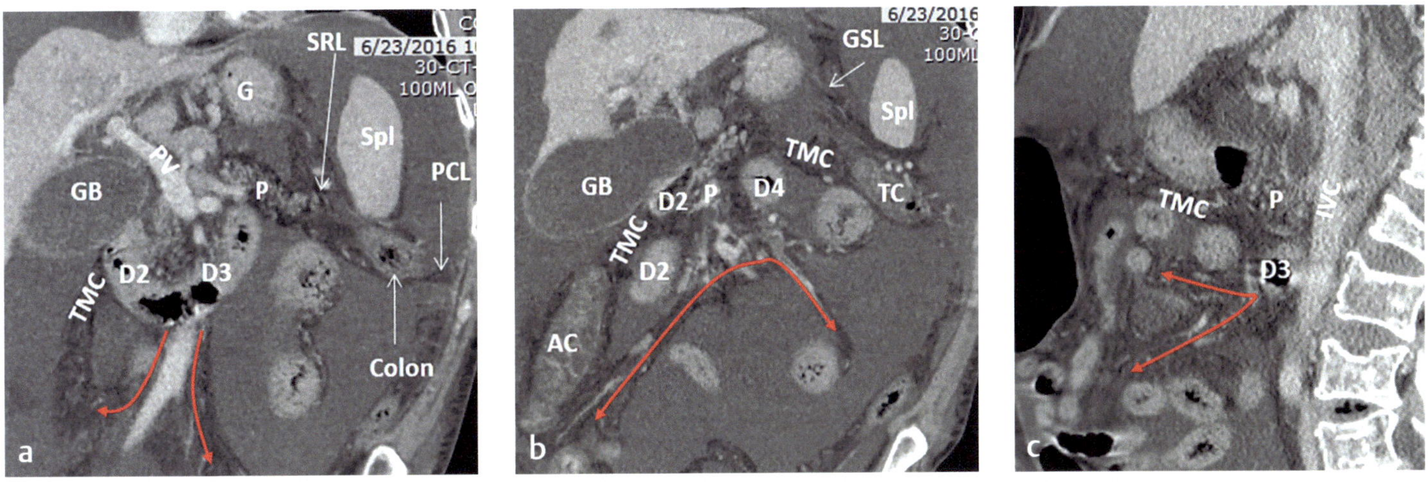

Fig. 4.4 Bare areas of the duodenum in a patient with severe ascites. **(a, b)** Coronal computed tomography (CT) scans from posterior to anterior and **(c)** sagittal CT at the level of the inferior vena cava (IVC). Most of the duodenum is covered by the peritoneum. Bare areas of the duodenum, where no peritoneal coverage exists, include the medial margin of the second (D2) and third (D3) parts facing the head of the pancreas (P), posterior wall facing retroperitoneum, a small part of the anterior surface of the second part of the duodenum (D2) where the transverse mesocolon (TMC) is attached, and where the mesenteric root (*red arrows*) covers the third part (D3) of the duodenum. The transverse mesocolon divides the D2 into the supramesocolic and inframesocolic segments. D3 is crossed anteriorly by the superior mesenteric vessels and the root of the mesentery of the small intestine. AC, ascending colon; G, stomach; GB, gallbladder; GSL, gastrosplenic ligament; PCL, phrenicocolic ligament; PV, portal vein; Spl, spleen; SRL, splenorenal ligament; TC, transverse colon.

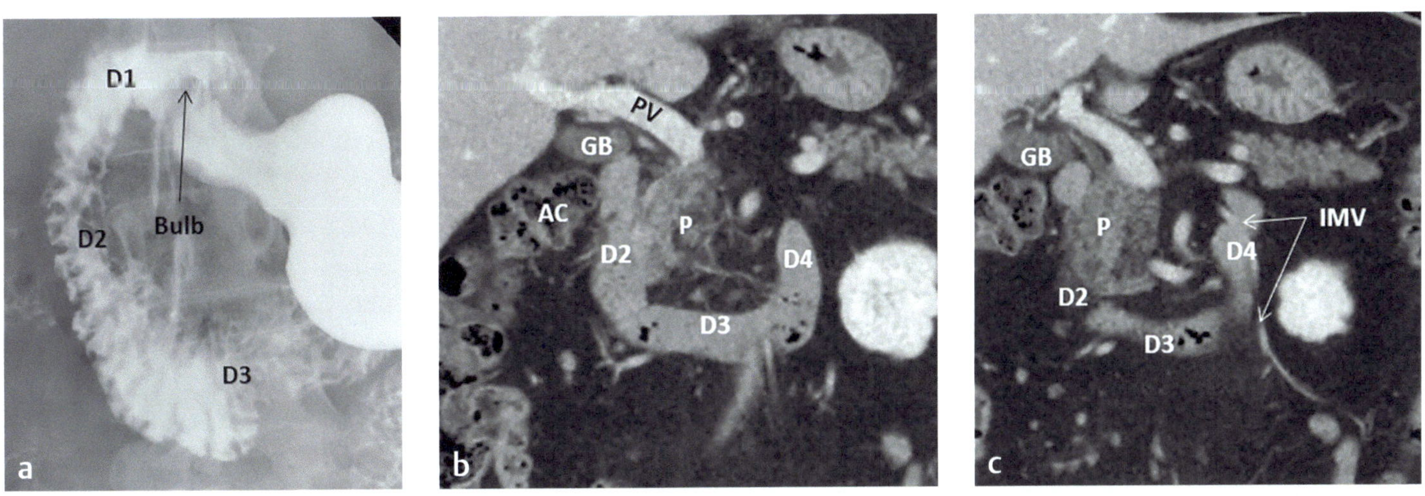

Fig. 4.5 Parts of the duodenum. **(a)** Barium study. **(b, c)** Coronal computed tomography (CT) from posterior to anterior. The duodenum is divided into four parts. The first (D1 or ascending) part of the duodenum is 5 cm in length and includes the bulb. The second (D2 or descending) part of the duodenum is 7.5 cm in length extending between the first and second duodenal flexures. The third (D3 or horizontal) part of the duodenum is 10 cm in length. The third part is related superiorly to the head and uncinate process of the pancreas (p). The fourth (D4 or ascending) part of the duodenum is 2.5 cm in length and extends from the left side of the aorta to the duodenojejunal flexure. Its termination is close to the terminal part of the inferior mesenteric vein (IMV). AC, ascending colon; GB, gallbladder; PV, portal vein.

fourth part of the duodenum. The third part passes anterior to the right ureter/renal hilum, and right gonadal vessels, inferior vena cava, aorta, and psoas muscle. It is crossed anteriorly by the superior mesenteric vessels and the root of the mesentery of the small intestine (▸Fig. 4.3, ▸Fig. 4.5). The third part is related superiorly to the head and uncinate process of the pancreas. The inferior pancreaticoduodenal artery and vein lie in a groove between the pancreas and duodenum (▸Fig. 4.5). The duodenum is separated from the inferior vena cava by a small amount of connective tissue.

Fourth Part

The fourth (or ascending) part of the duodenum is 2.5 cm in length and extends from the left side of the aorta to the duodenojejunal flexure below the root of the transverse mesocolon at the left side of L2. The fourth part is related posteriorly to the left sympathetic trunk, psoas muscle, and left renal and gonadal vessels. Its termination is close to the terminal part of the inferior mesenteric vein (▸Fig. 4.5). The duodenojejunal junction is suspended by the ligament of Treitz.

Suspensory Ligament of Treitz

The suspensory muscle or ligament is a remnant of the dorsal mesentery. It presents as a fibromuscular band that extends between the right crus of the diaphragm or the celiac axis and the upper margin of the duodenojejunal flexure.[6] It may also insert on the third and fourth portions of the duodenum. The ligament will blend with the fibers of the longitudinal duodenal muscle. It passes next to the celiac axis, posterior to the pancreas and splenic vein, and anterior to the left renal vein (▶ Fig. 4.6). Demonstration of this ligament with CT scan is very difficult.[6]

Periampullary Structures

The structures around pancreato-bilio-duodenal junction include: the fourth part (intramural) of the common bile duct (CBD), the main pancreatic duct (Wirsung), the major and minor papillae, the ampulla of Vater, and the sphincteric mechanism of Boyden (see Chapter 9 "Pancreas").

The terminal portion of the CBD passing through the duodenal wall is approximately 1.5 cm in length and usually lies side by side with the pancreatic duct in a common adventitia for several millimeters. The major duodenal papilla lies approximately 7 to 10 cm from the pylorus at L2 or L3 level. The orifice of the papilla may be obscured by a fold of duodenal mucosa. A longitudinal fold (the plica longitudinalis) may extend inferiorly from the inferior margin of the papilla (see Chapter 9 "Pancreas"). The orifice of the papilla may be filled by villous-like projections called valvulae.[7] A medially located duodenal diverticulum near the papilla is not an uncommon finding and may cause difficulty in cannulation of the pancreatic duct or in rare occasion obstruct the lumen of the CBD (▶ Fig. 3.18d; Chapter 3 "Stomach").

The ampulla of Vater is a prominence of the common pancreaticobiliary channel within the papilla and below the junction of the two ducts. It is present in 60 to 85% of cases with an average length of 5 mm and may not be ampullary-shaped (dilated) in 25%. In cases with separate channels of the CBD and pancreatic duct, the ampulla does not exist (10–15%).[5] A complex sphincteric mechanism composed of circular or spiral smooth muscle fibers is seen around the intramural part of the CBD, the main pancreatic duct, and the ampulla. This sphincteric complex is also known as the sphincter of Boyden.[5]

The minor duodenal papilla, through which the accessory pancreatic duct of Santorini opens, is approximately 2 cm proximal and slightly anterior to the major papilla. The gastroduodenal artery passes behind it and can be used as a localizing landmark during surgery.[5] It does not have a sphincter but muscular and elastic fibers in special arrangement exist at its orifice that allow opening of the duct orifice after contraction.[8,9] In 10% of individuals, the duct of Santorini is the only duct draining most of the pancreas.

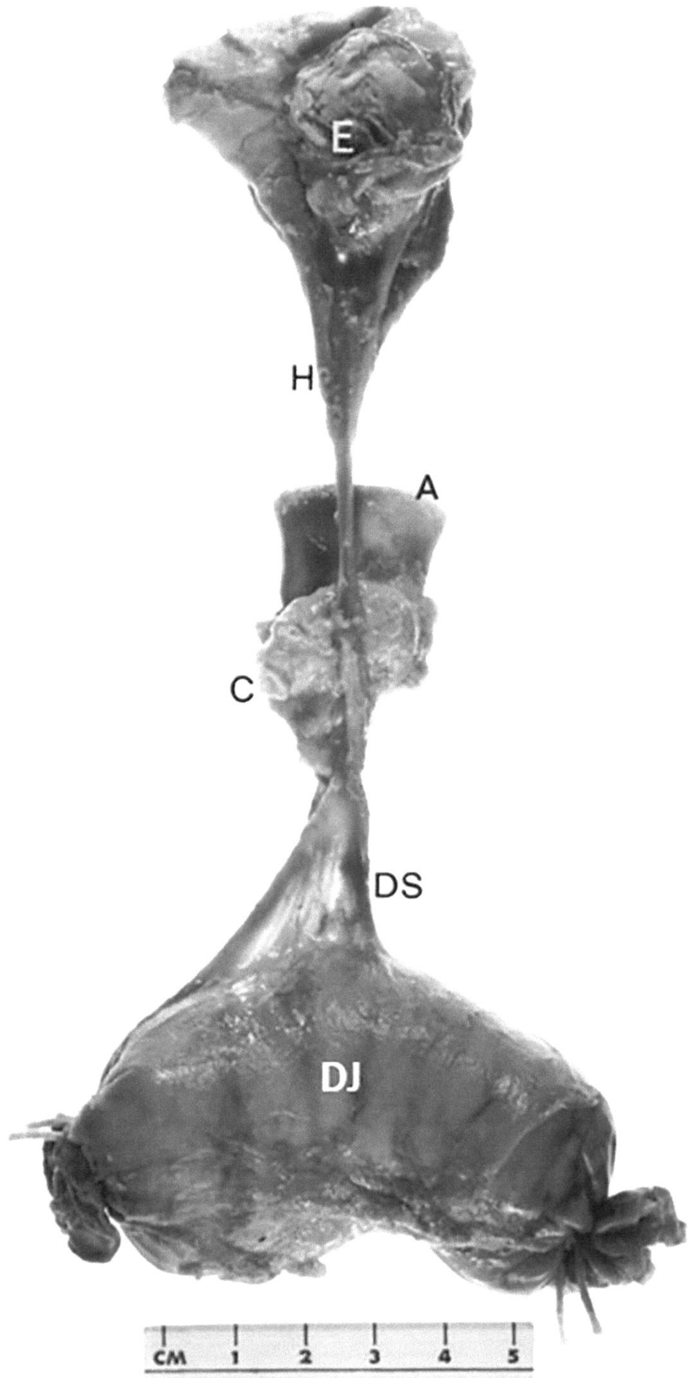

Fig. 4.6 Suspensory ligament of Treitz. Two parts of the ligament of Treitz are shown. The upper muscular part also known as Hilfsmuskel (H) arises from the esophageal (E) diaphragmatic crus and inserts to the celiac axis (C) at the aorta (A). The upper part is absent in many cases. The lower membranous part known as the duodenal suspensory ligament (DS) starts from the celiac axis (C), passes behind the pancreas and in front of the left renal vein, and finally attaches to the third part of duodenum or the duodenojejunal junction (DJ). (Reproduced with permission from reference from Kim et al.[6])

Jejunum and Ileum

The average length of the jejunum and ileum in adult human is usually 5-m long and about 40% or 2 m of the small bowel forms the jejunum. The jejunum is located in the left upper abdomen and the ileum in the right lower abdomen but their loops move freely. The division between the two parts is not anatomically distinct. The average diameter of the jejunum is 3.5 cm and the ileum is 2.5 cm or less (▶ Fig. 4.7, ▶ Fig. 4.8, ▶ Fig. 4.9). The jejunum

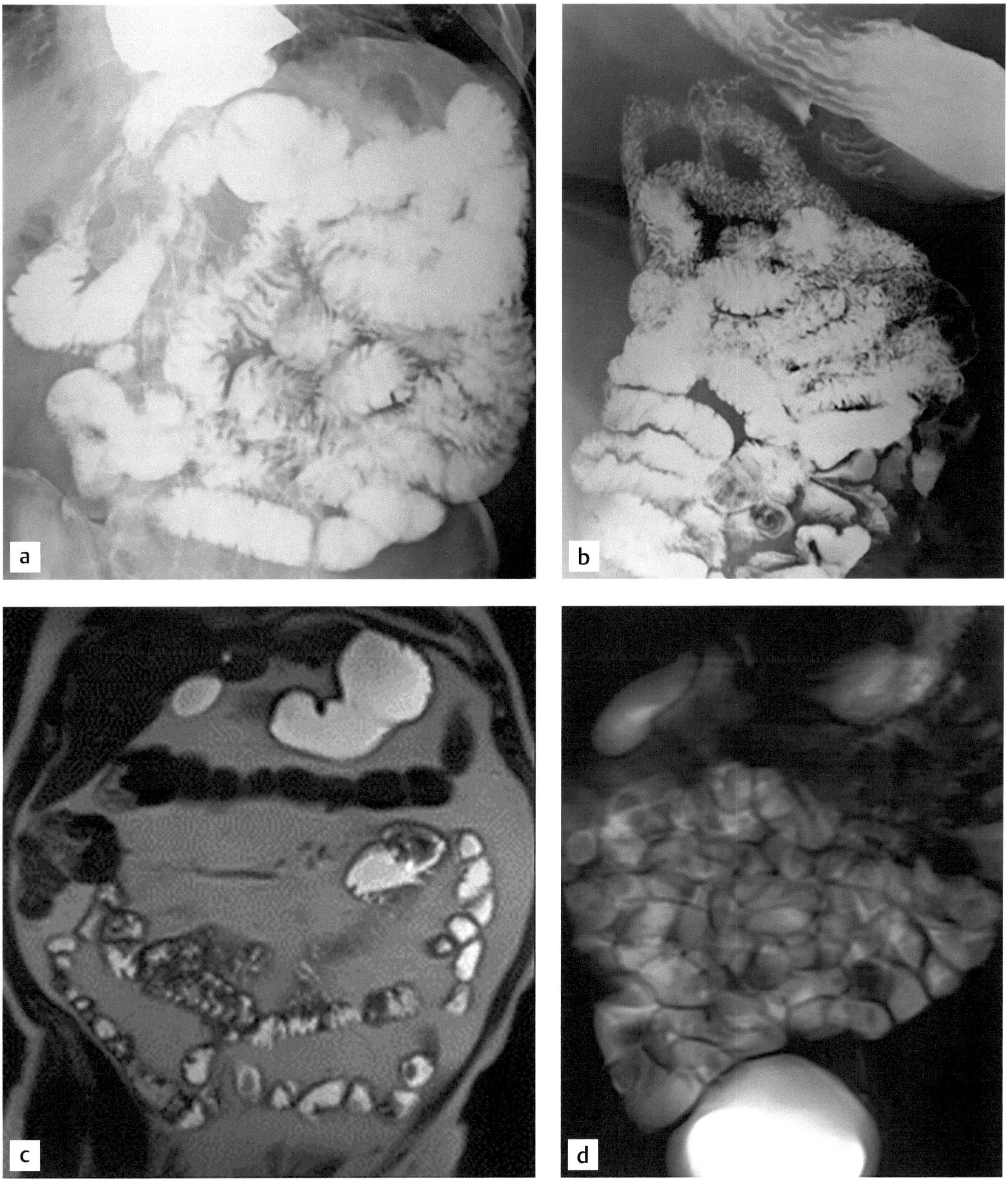

Fig. 4.7 Small bowel shown with barium follow-through examination **(a, b)**, magnetic resonance imaging (MRI) **(c)**, and computed tomography (CT) **(d)**.

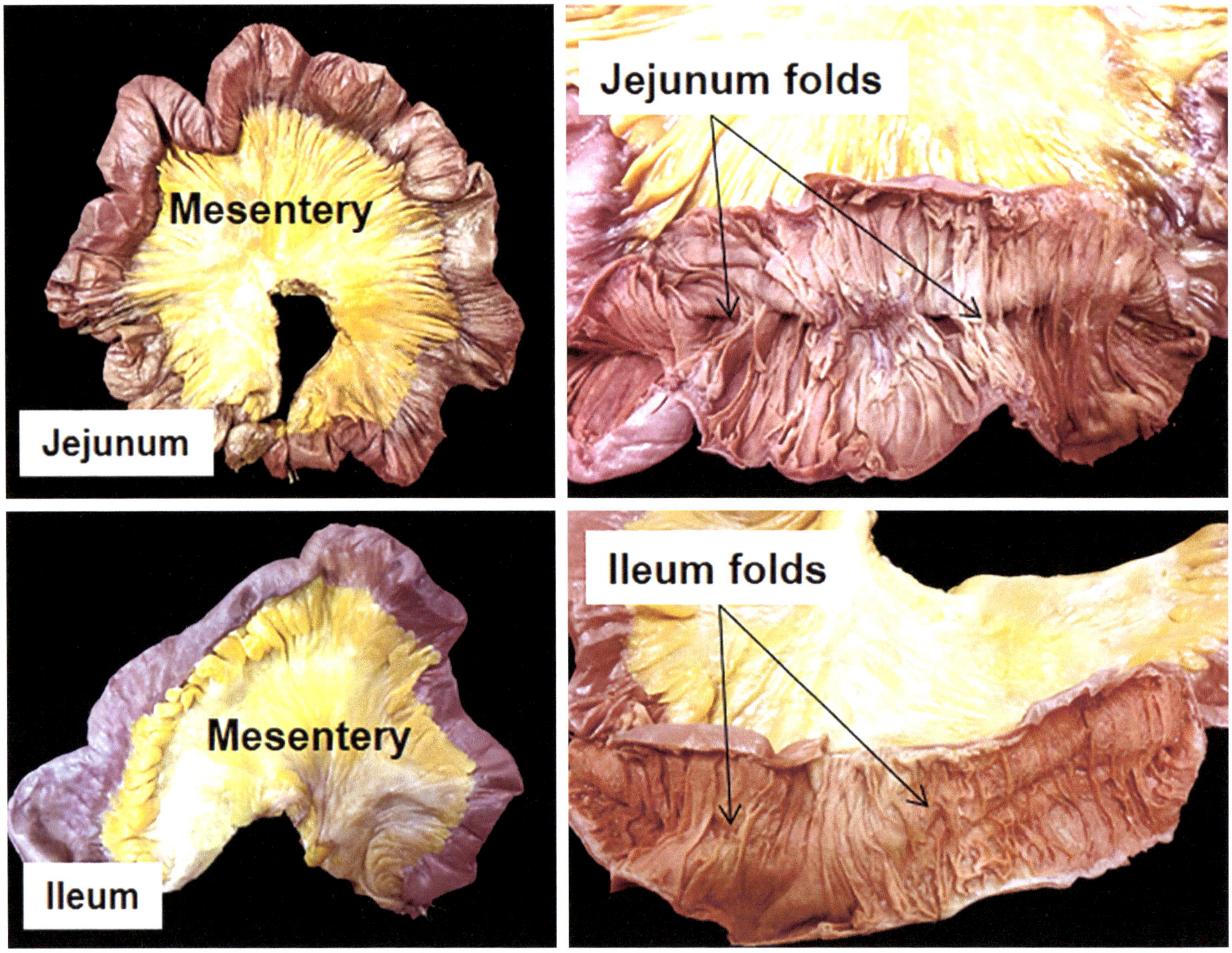

Fig. 4.8 Outer and inner views of the jejunum and ileum.

begins at the duodenojejunal flexure and the ileum ends at the ileocecal junction. The duodenojejunal flexure is usually below the transverse mesocolon but may be partially covered by it. From this point, the mesenteric root starts, crosses the midline, and ends at the ileocecal junction in the right lower quadrant of the abdomen (▸Fig. 4.4, ▸Fig. 4.10). The mesenteric root is only 15 to 20 cm in length at its base but fans out toward the intestine. The mesentery is formed by two layers of the visceral peritoneum. It contains vessels, lymphatics, lymph nodes, nerves, and varying amounts of connective tissues and fat (▸Fig. 4.10). The ileum has more fat inside the mesentery than the jejunum.

As with the entire gastrointestinal tract, the walls of the jejunum and ileum consistent of five layers: mucosa, submucosa, circular muscle, longitudinal muscle, and subserosa/serosa (▸Fig. 4.11). The mucosa is divided into three parts: muscularis mucosae, lamina propria, and epithelium. The mucous membrane is heavily plicated by convoluted folds known as valvulae conniventes (Kerckring folds or plicae circulares). These folds are permanent markings of the small bowel mucosa and, in contrast to rugal pattern of stomach, would not disappear with bowel distension (▸Fig. 4.8, ▸Fig. 4.9). The folds contain many small finger-like projection or villi (▸Fig. 4.11). This morphology is essentially designed to increase absorptive surface area of the small bowel. The folds are oriented in a circular, semicircular, or spiral patterns perpendicular to the lumen's longitudinal axis. The height of folds varies between 3 and 10 mm. The entire mucosal surface of the small intestine is covered by a single line of columnar epithelial cells. Each epithelial cell is covered by approximately 1,000 microvilli which further increase the surface area. The epithelia of the villi extend down into the lamina propria where they form crypts of Lieberkühn. The columnar epithelial cells are interspersed with three other types of cells in the crypts of Lieberkühn: goblet cells, Paneth cells, and enterochromaffin cells. The goblet cells are found in the crypts of Lieberkühn or lower part of villi and secrete an alkaline mucous fluid. The Paneth cells are found in the floor of the crypts and secrete antimicrobial

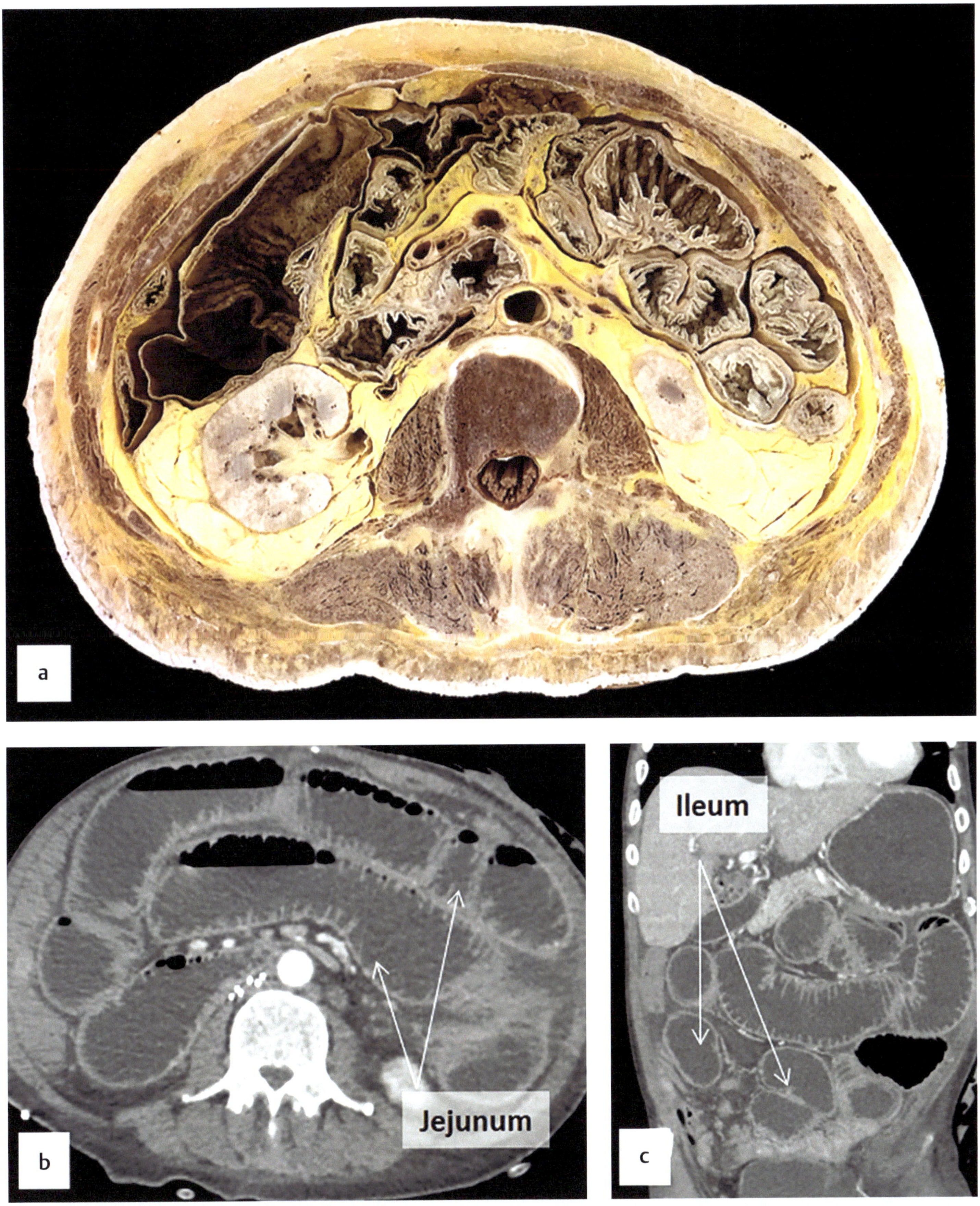

Fig. 4.9 (a) Cadaveric axial view showing the jejunal loops. **(b)** Axial and **(c)** coronal computed tomography (CT) images show cross-sectional appearance of dilated jejunal and ileal mucosal folds. The valvulae conniventes folds are permanent markings of the small bowel mucosa and, in contrast to rugal pattern of stomach, would not disappear with bowel distension.

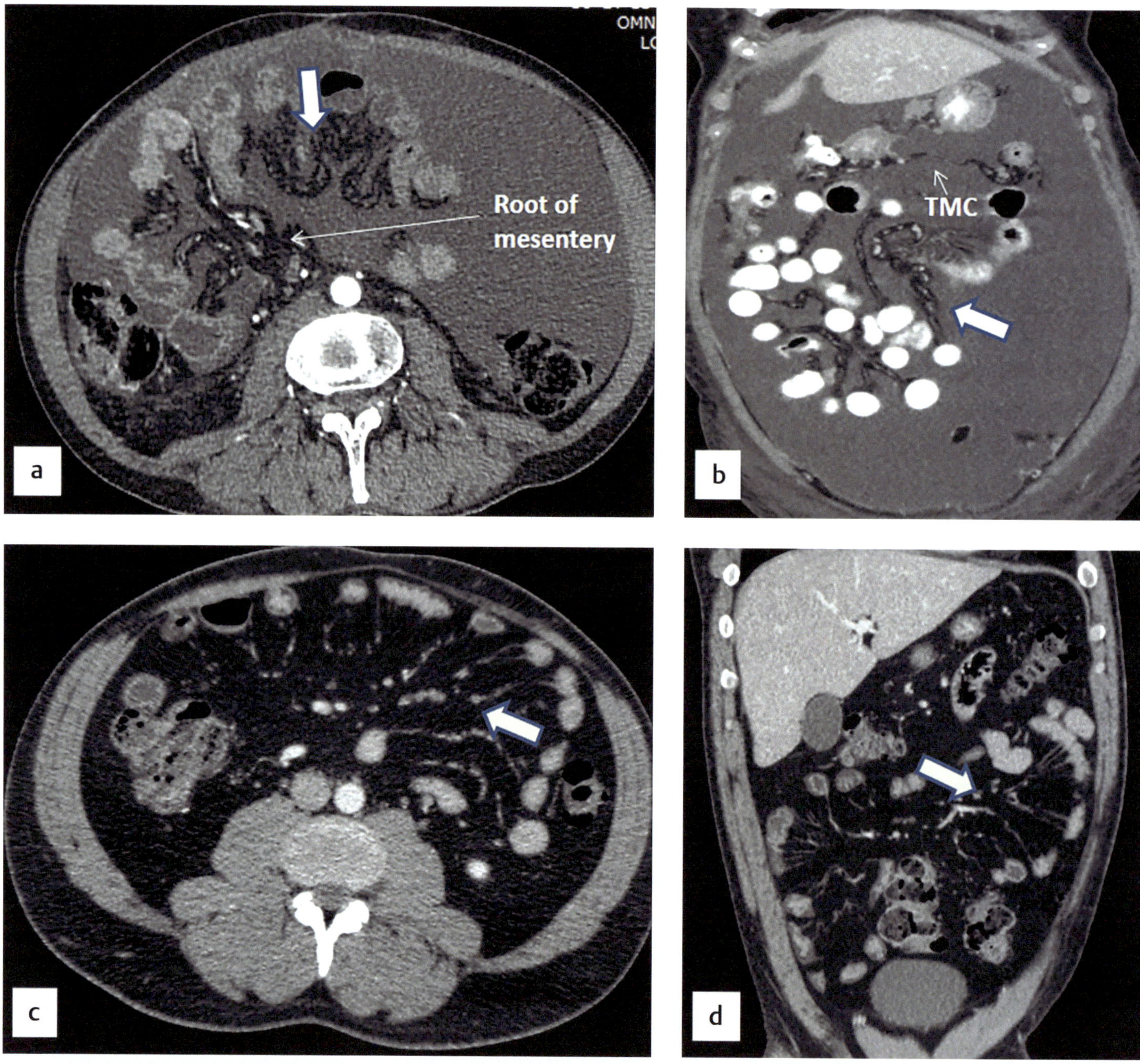

Fig. 4.10 Small bowel mesentery is shown by axial (left) and coronal (right) computed tomography (CT) images. **(a, b)** A patient with large ascites and **(c, d)** a patient with large visceral fat. Note all vessels traveling in the mesentery (*arrows*). TMC, transverse mesocolon.

peptides. The enterochromaffin cells are also located within the crypts and have neuroendocrine function. These cells are not present in the colon.

The lamina propria is a layer of reticular connective tissue with lymph nodes below the epithelial surface of the mucosa. Terminal vessels, lymphatics, and nerves extend into each villus. The lamina propria also contains fibrils of the smooth muscle arising from the muscularis mucosa into the villi. Contraction of these muscle fibers transforms the surface appearance of the villa from a more relaxed into jagged and serrated. The folds and villi become smaller and decreased in number in the ileum and are only seen sporadically in the terminal ileum.

The submucosa consists of a mesh-like network of collagenous connective tissue adaptive to constant changes in diameter and length of the intestinal lumen. The submucosa also contains numerous vessels and lymphatics and the submucous nerve plexus of Meissner. Lymphatics are more numerous and more pronounced in the ileum and the large intestine. In the jejunum, lymphatic tissue is encountered only in the form of solitary nodules whereas in the ileum aggregates of lymph nodules (20–30 in number) form the Peyer's patches, each measuring 2 to 10 cm in length and 1 to 1.5 cm in width. Therefore, jejunum contains very few Brunner's glands that are found in the duodenum or Peyer's patches found in the ileum. Peyer's patches are an important part of the local immune system of the intestinal tract.

The muscularis mucosa is between the submucosa layer and the serosa. It is composed of an inner circular muscle layer and an outer longitudinal muscle layer. The myenteric plexus of

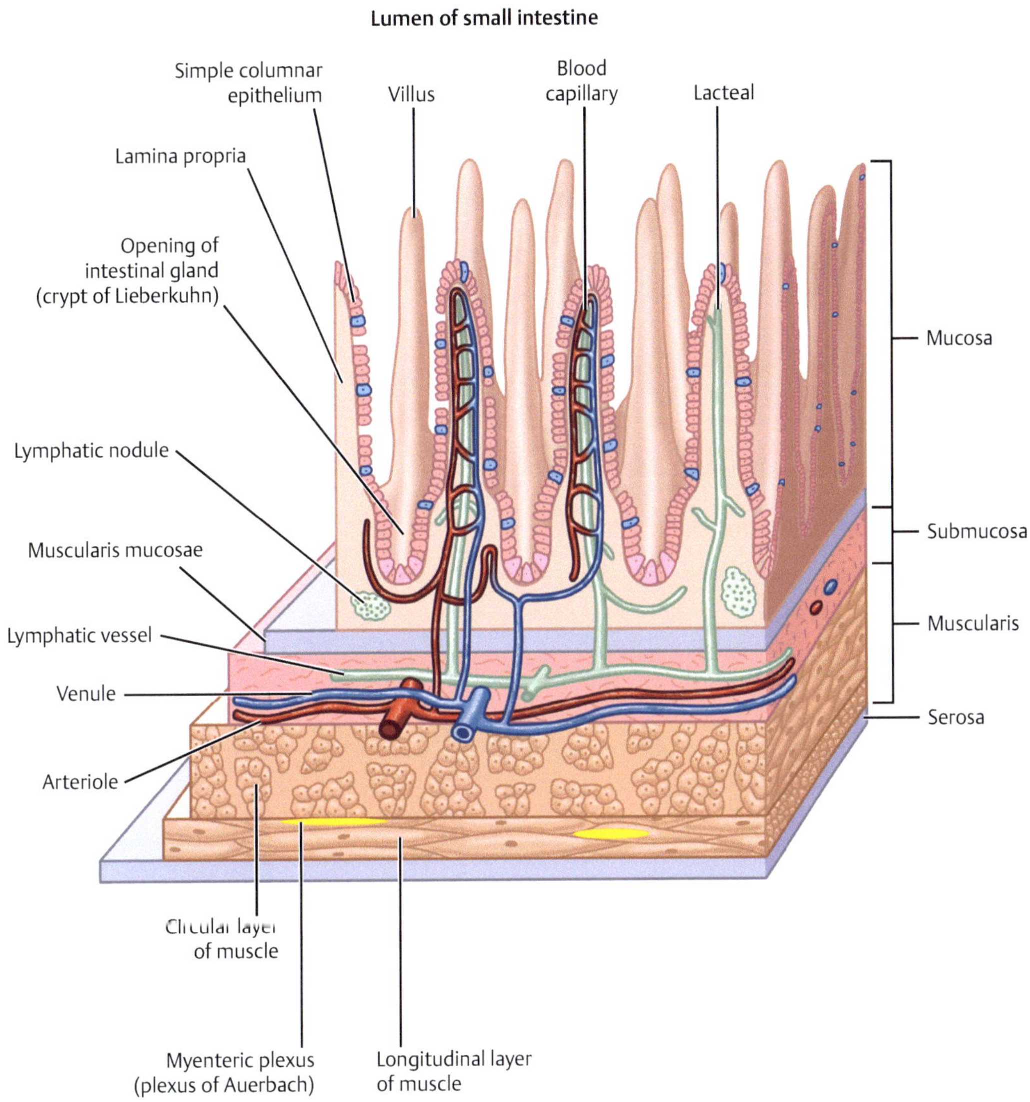

Fig. 4.11 The wall of the gastrointestinal track. The walls of the jejunum and ileum consistent of five layers: mucosa, submucosa, circular muscle, longitudinal muscle, and subserosa/serosa.

Auerbach is located between the circular and longitudinal muscle layers. The serosa is composed of a layer of flat epithelia continuous with the visceral peritoneum and a subserosa of loose connective tissue. It covers the entire circumference of the intestinal tube except where the mesentery is attached.

Terminal Ileum

The terminal ileum is the most caudal part of the small intestine, usually located in the right iliac fossa (▶Fig. 4.12). The terminal ileum joins the cecum at an acute angle and the ileal mucosa, submucosa, and muscular layer continue through the wall of the colon and combine with corresponding layers in the cecum inferiorly and the colon superiorly to form the ileocecal sphincter. The ileal papilla is the preferred term used by the *Terminologia Anatomica* to describe the ileocecal junction. The ileocecal valve is a true sphincter muscle valve containing thickened circular muscle layer of the ileum. Manometric studies have demonstrated a high-pressure zone in this region and suggest that the term ileocecal junction is a more accurate description than ileocecal valve.[10–13] Microscopic studies have shown progressive thickening of the ileal and large bowel muscle layer over a distance of about 1 cm leading up to the base of the ileal papilla. From this point, the muscle gradually tapered toward the tip of the papilla. The superficial bundles of the longitudinal muscle of the ileum continued with the large intestine but the deep bundles penetrated into the papilla. The degree of the submucosal vascularity or neuronal density in the ileocecal junction is not different from the rest of the ileum.

From a colonic perspective, the ileal aperture is bounded by the two folds referred to as the upper and lower lips of the valve. The lips coalesce to form two mucosal ridges at each end (see Chapter 5 "Colon"). These crescentic horizontally oriented ridges or "frenulum" demarcate the border between the cecum and the ascending colon (▶Fig. 4.12). Radiologists and endoscopists have reported a spasmodic circumferential or semilunar contraction (Busi's sphincter) and a prominent ileocecal fold at the border area between the cecum and ascending colon, 2 to 2.5 cm distal

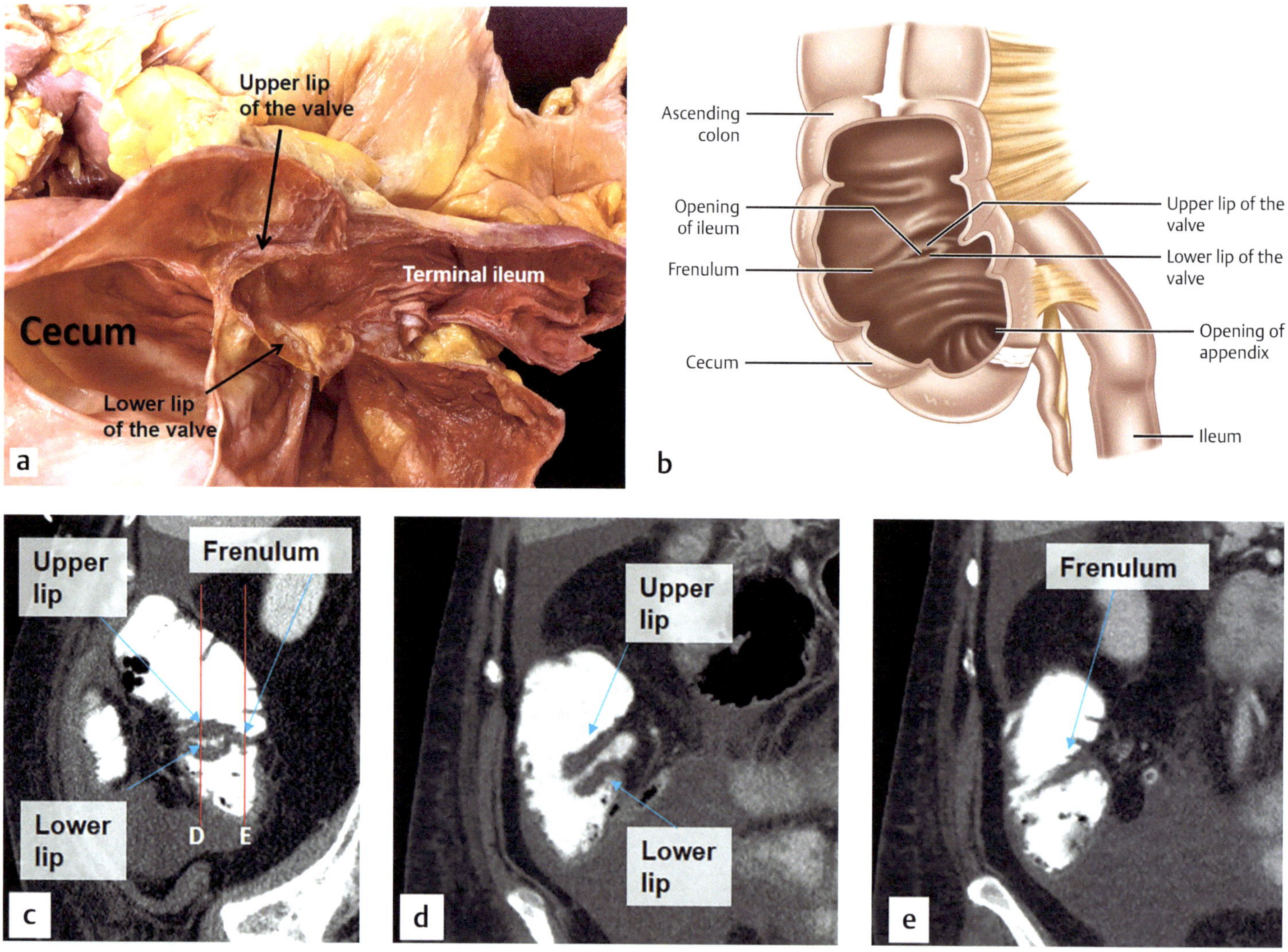

Fig. 4.12 (a, b) Ileocecal junction. **(c–e)** Computed tomography (CT) images showing normal ileocecal valve. **(c)** Sagittal view through the cecum (marked by the *red lines*) and **(d, e)** coronal images through the lips and the frenulum.

to the ileocecal papilla.[12] Microscopically, this cecocolonic fold consists of mucosa, submucosa, and muscle. The circular muscle layer is thicker than that of cecum or ascending colon. Fatty infiltration of the ileocecal is common and in some cases gives the false appearance of a mass. It also predisposes the junction to intussusceptions.[14]

Major function of the high pressure or valve-like mechanism of the ileocecal junction is to prevent coloileal reflux and therefore colonization of the ileum by colonic microflora. The percentage of time for phasic valve opening increases after food intake.[15] This is related to the "gastroileal" reflex causing an increased motor activity of the distal ileum after a meal. Colonic distension is followed consistently by contraction of the sphincter in humans. Backward flow from the cecum to the ileum can occur but it is episodic and minimal. In barium studies mild reflux of contrast from colon into the ileum is seen in up to 50% of studies.[16] Surgical removal of the ileocecal sphincter (i.e., ileocolic anastomosis) maximizes reflux, ultimately leading to bacterial overgrowth.

Various peritoneal folds and recesses may be formed around the cecum and terminal ileum during its descent from right upper abdomen to the right iliac fossa. Four types of peritoneal recesses are found in the pericecal region, namely, the superior ileocecal recess, inferior ileocecal recess, retrocecal recess, and paracolic sulcus[17] (▶Fig. 4.13). The superior ileocecal fold is above the terminal ileum.[1] This peritoneal fold arises from the ileomesentery, crosses the ileocecal junction, and attaches to the anterior wall of the cecum. It contains the anterior cecal artery. The superior ileocecal recess is located between the fold and the ileomesentery anterior to the terminal ileum. The inferior ileocecal fold is below the terminal ileum. It extends between the terminal ileum and the cecum and forms a recess anterior to the mesoappendix. The inferior ileocecal fold is also called the ligament, veil, or bloodless fold of Treves because it contains no artery. Paracecal recesses may be a potential site for internal hernia. Internal abdominal herniation is defined as the protrusion of an abdominal organ through a normal or abnormal mesenteric, epiploic, or peritoneal aperture. Internal herniation of the ileal

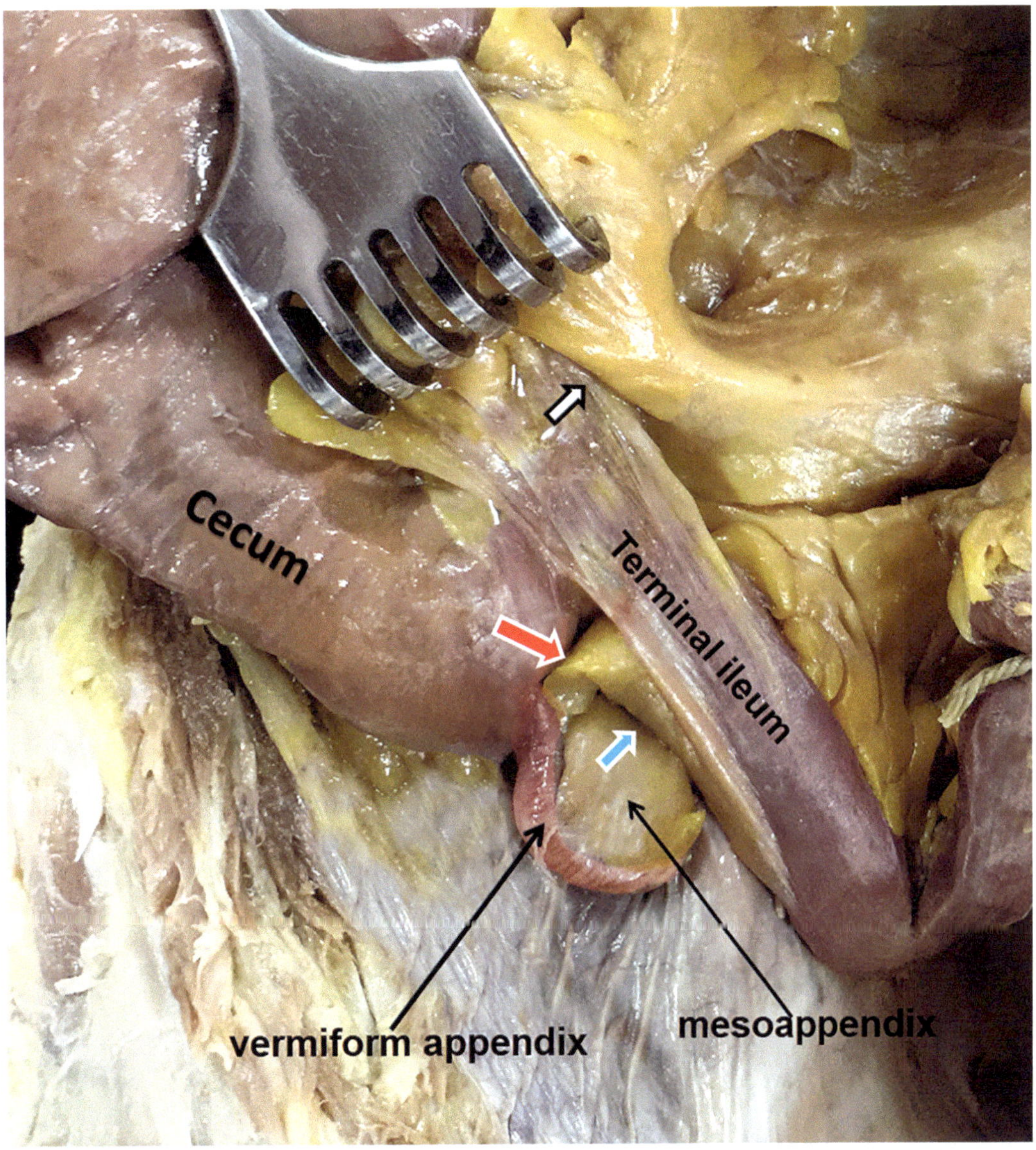

Fig. 4.13 Right lower quadrant peritoneal folds and recesses. Four types of peritoneal recesses are formed in the pericecal region by the peritoneal folds, namely, the superior ileocecal recess, inferior ileocecal recess, retrocecal recess, and paracolic sulcus. The superior ileocecal fold is above the terminal ileum and extends from the ileomesentery to the anterior wall of the cecum forming the superior ileocecal recess behind it (*white arrow*). The inferior ileocecal fold (*red arrow*) extends between the terminal ileum and the cecum and forms the inferior ileocecal recess (*blue arrow*) anterior to the mesoappendix.

loops into the inferior ileocecal recess is rare and present with bowel obstruction.[18]

◆ Vessels of Duodenum

Arteries

The blood supply of the duodenum is complex because of its origin, distribution, and individual variations. In general, the duodenum is supplied by the gastroduodenal artery and the inferior pancreaticoduodenal artery (IPDA). The common hepatic artery, a branch of the celiac trunk, courses toward the right near the pancreatic neck and head and divides into the proper hepatic artery and gastroduodenal artery. The gastroduodenal artery gives origin to the posterior superior pancreaticoduodenal artery (PSPDA) or retroduodenal artery and supraduodenal artery (► Fig. 4.6; Chapter 3 "Stomach").[14] The gastroduodenal artery terminates by dividing into the right gastroepiploic artery and anterior superior pancreaticoduodenal artery (ASPDA). The superior mesenteric artery (SMA) usually gives rise to the IPDA which subsequently divides into posterior and anterior branches. Small arteries supplying the duodenum are called "arteriae rectae," which form a large plexus in the submucosa and another one just beneath the epithelium of the villi.

4

The first part of the duodenum is usually supplied by the supraduodenal artery and the PSPDA. The supraduodenal artery may extend distally to supply the proximal duodenum. In 50% of cases, the artery anastomoses with the right gastric artery and PSPDA.[5] The supraduodenal artery may arise from the right gastric artery in less than 10% of cases. The PSPDA usually passes in front of the common bile duct and descends on the posterior surface of the pancreas to join the posterior branch of the IPDA. In many individuals, the first 1 cm is also supplied by branches of the right gastric artery. The gastroduodenal artery descends between the first part of the duodenum and the head of the pancreas and terminates by dividing into the right gastroepiploic artery and ASPDA, both supplying the proximal part of the duodenum. The remaining three parts of the duodenum are supplied by an anterior and a posterior arcade. Four arteries contribute to the pancreaticoduodenal vascular arcades, namely, the ASPDA, PSPDA, and anterior and posterior branches of the IPDA.

Veins

The upper first part of the duodenum is drained by subpyloric and suprapyloric veins which drain into the right gastroepiploic, portal vein, or posterior superior pancreaticoduodenal vein. Anastomoses between the subpyloric and suprapyloric veins pass around the duodenum and form the prepyloric vein of Mayo (▶Fig. 3.8, Chapter 3 "Stomach"). It is not a great indicator of the location of the pylorus.

The pancreaticoduodenal venous arcades draining the duodenum follow the arterial arcades and tend to lie superficial to them. The anterosuperior vein drains into the right gastroepiploic vein, whereas the posterosuperior vein usually passes behind the common bile duct to enter the portal vein (▶Fig. 3.9, Chapter 3 "Stomach"). Several tributaries of veins draining the second and third parts of the duodenum drain into dorsal pancreatic vein in 30% of cases. The inferior duodenal veins can enter the superior mesenteric or other tributaries of the portal system.

Lymphatics

The duodenum is richly supplied with lymphatics which originate as blind ending central lacteals in the mucosa of each villus and form a plexus in the lamina propria and a second one in the submucosa. Collecting trunks pass toward the lesser curvature of duodenum to enter the anterior and posterior pancreaticoduodenal lymph nodes. The duodenal lymphatics have two pathways: superior ones connect to the pyloric (suprapyloric, retropyloric, and infrapyloric) and hepatic nodes and inferior ones to the nodes in the vicinity of the origin of the superior mesenteric artery. The valves of lymphatic vessels connecting the duodenal wall with the head of the pancreas are arranged so that normal lymph flow is from pancreas to duodenum and not the reverse.

◆ Vessels of Jejunum and Ileum

Arteries

Arterial supply of the jejunum and ileum are provided by branches arising from the left side of superior mesenteric artery. Those arising proximal to the ileocolic artery of SMA (3–7 branches) are all jejunal and those distal to the ileocolic artery (8–16 branches) supply distal jejunum and entire ileum (▶Fig. 4.14, ▶Fig. 4.15, ▶Fig. 4.16). Iliac branch of the ileocolic artery connects to the distal SMA after giving off branches to the appendix and terminal ileum.[19]

The jejunal and ileal branches travel straight for variable distance before branching. The branches of these stem arteries unite with the branches of the adjacent stem vessels to form the primary arcuate vessels in the mesentery. Branches of these primary arcuate arteries form secondary arcuate arteries (▶Fig. 4.17). The vasa recta arise from the last set of arcuate arteries. In general, the jejunal arteries are longer than the ileal arteries. The first jejunal branch is generally large (5 mm) but may share a common origin with a small IPDA.

The vasa recta (straight arteries) are seen in both small intestine and colon. However, there are major differences in vessel branching and the number of vasa recta between the two organs.[20]

The vasa recta of the small intestine become mural trunks and anastomose freely with one another.[21] This network is more elaborate than that found in the colon. Indeed, while the upper jejunum has simple arcuate vessels, the more distal portion of the small intestine is complex and has an increased number of arcuate vessels. The arcuate vessels of the terminal ileum also have quaternary and even quinary loops.[20] This suggests that the small intestine in the ileal side potentially has a better collateral blood supply than the jejunum and colon and appears to be more resistant to ischemia and reperfusion injury. It is also shown that higher levels of nitric oxide in the ileum than jejunum has a protective effect on intestinal perfusion.[22]

The clinical relevance of these finding is in the rate of development of ischemia in case an artery is occluded or embolized and for the selection of best free graft for transplantation purposes (▶Fig. 4.18). Embolization of the vasa recta has been successfully performed to stop lower gastrointestinal tract bleeding, without sacrificing blood flow around the embolized vasa recta branches.[23,24]

In both the small intestine and colon, the vasa recta consist of a long branch and a short branch (▶Fig. 4.17). There are rich anastomoses between the long and short branches (see Chapter 5 "Colon"). Therefore, blood flow may be preserved owing to anastomosis from the surrounding vasa recta in case one is occluded.

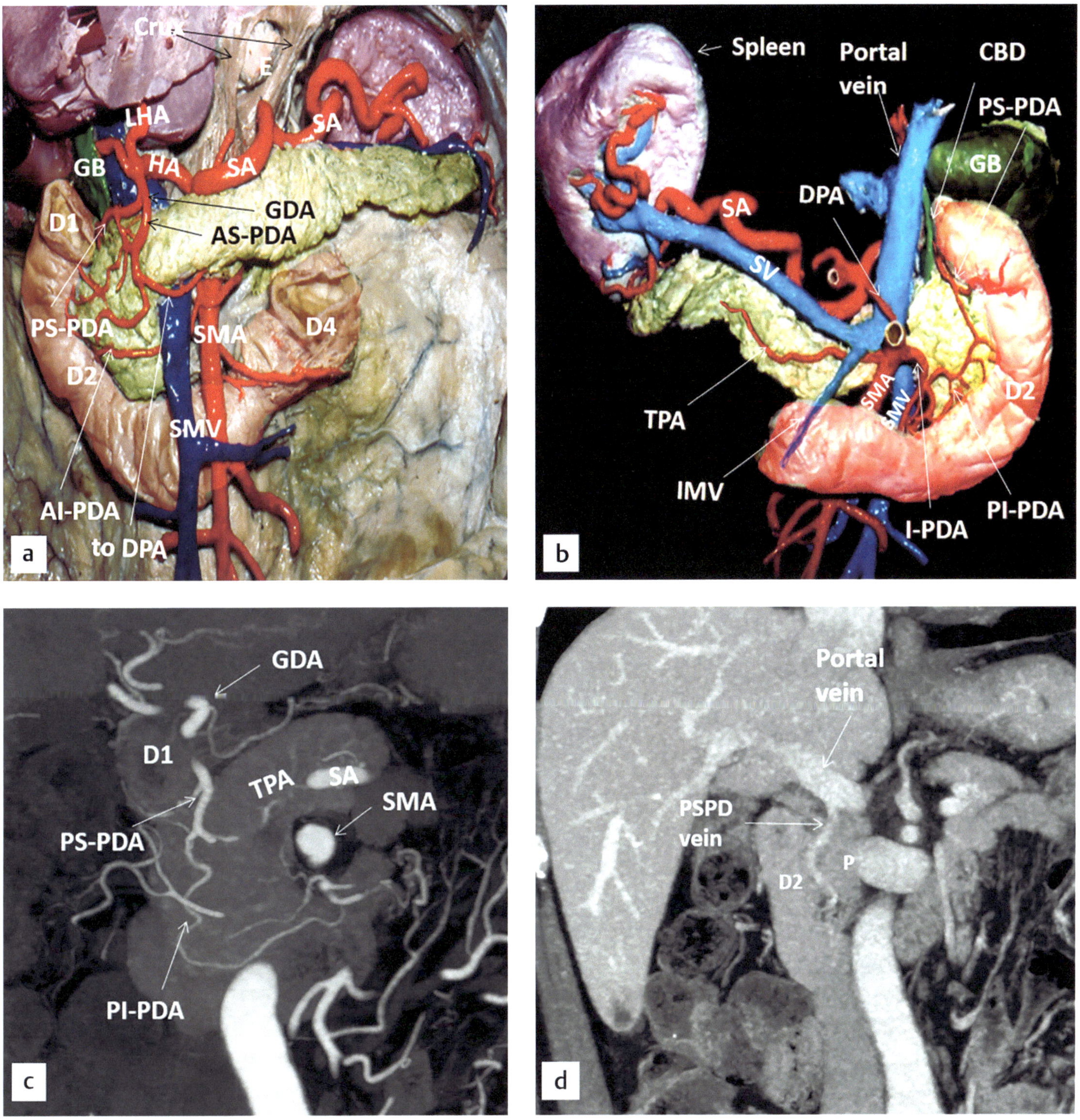

Fig. 4.14 (a–c) Arterial supply to the pancreatic head and duodenum by an anterior and a posterior arcade. In general, the duodenum is supplied by the gastroduodenal artery and the inferior pancreaticoduodenal artery (IPDA). The gastroduodenal artery (GDA) is a branch of the hepatic artery (HA) which gives origin to the posterior superior pancreaticoduodenal artery (PS-PDA), retroduodenal artery, and supraduodenal artery. The GDA terminates by dividing into the right gastroepiploic and anterior superior pancreaticoduodenal arteries (AS-PDA). The superior mesenteric artery (SMA) usually gives rise to the I-PDA which subsequently divides into posterior (PI-PDA) and anterior (AI-PDA) branches. The transverse pancreatic artery (TPA) arises from the splenic artery (SA) or GDA and supplies body and tail. **(d)** Coronal computed tomography (CT) shows some of the duodenal veins. CBD, common bile duct; D1/D2/D3, first, second, and third portions of the duodenum; DPA, dorsal pancreatic artery; E, esophagus; GB, gallbladder; LHA, left hepatic artery; P, pancreas; PSPD, posterior superior pancreaticoduodenal; SA, splenic artery; SMV, superior mesenteric vein; SV, splenic vein.

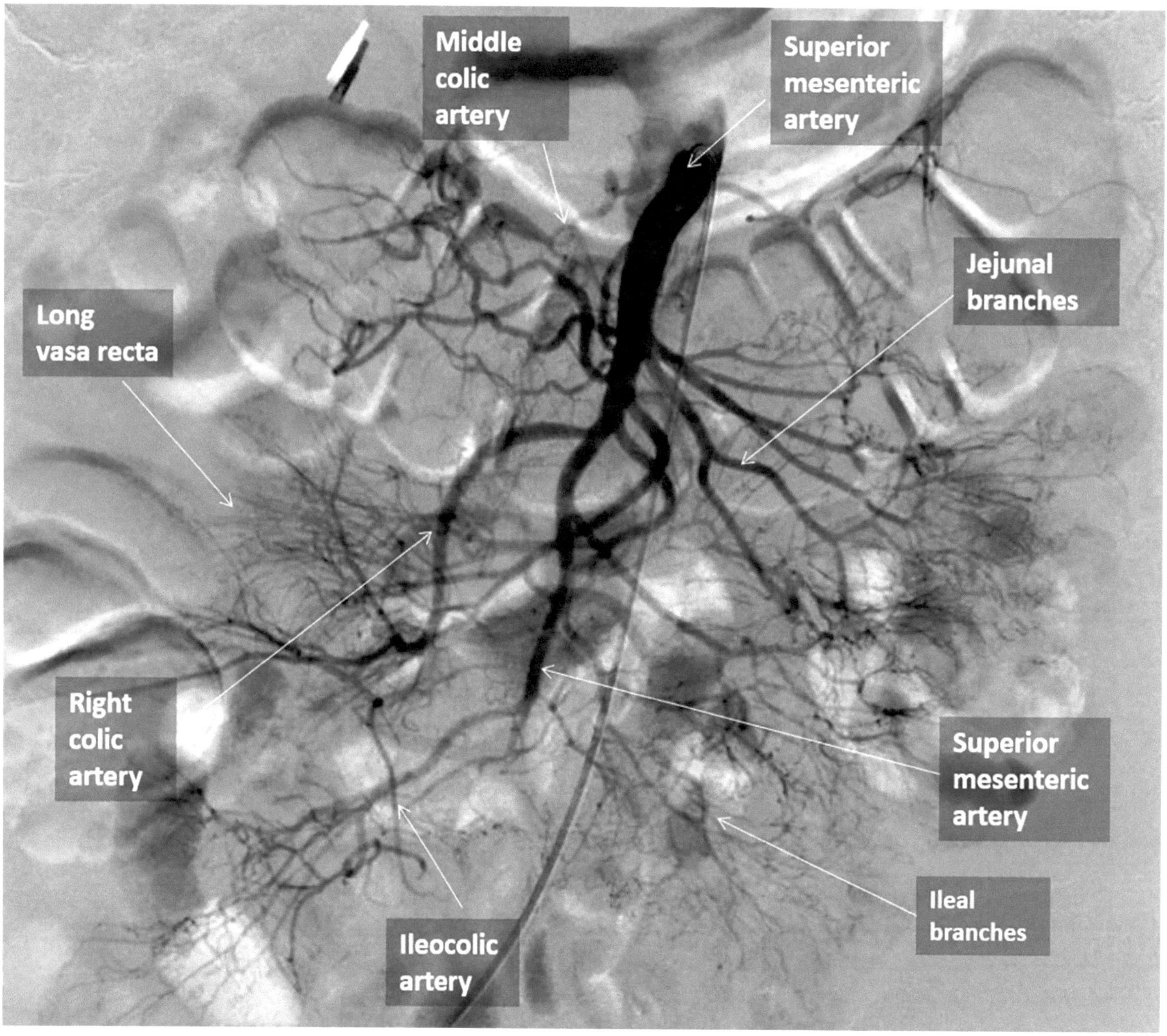

Fig. 4.15 Jejunal and ileal branches of superior mesenteric artery are shown in this selective angiogram.

Veins

Veins of the jejunum and ileum follow the same pattern of branching and distribution as the corresponding arteries and finally drain into the superior mesenteric vein (SMV) (▸Fig. 4.16). The SMV finally connects to the portal vein. The inferior mesenteric vein drains into the SMV in 30% of cases or at the confluence of the SMV/portal vein in 30%. In the remainder, the SMV drains into the splenic vein. Thrombosis of these veins in pregnancy and hypercoagulable states can occur leading to extensive mesenteric ischemia (▸Fig. 4.18). It is also an uncommon complication of appendicitis and diverticulitis.[25] Diagnosis is usually made with CT angiography.

Lymphatics

Intramural lymphatics of the small bowel starts with small central lacteals in the villi. These lacteals join together to form networks

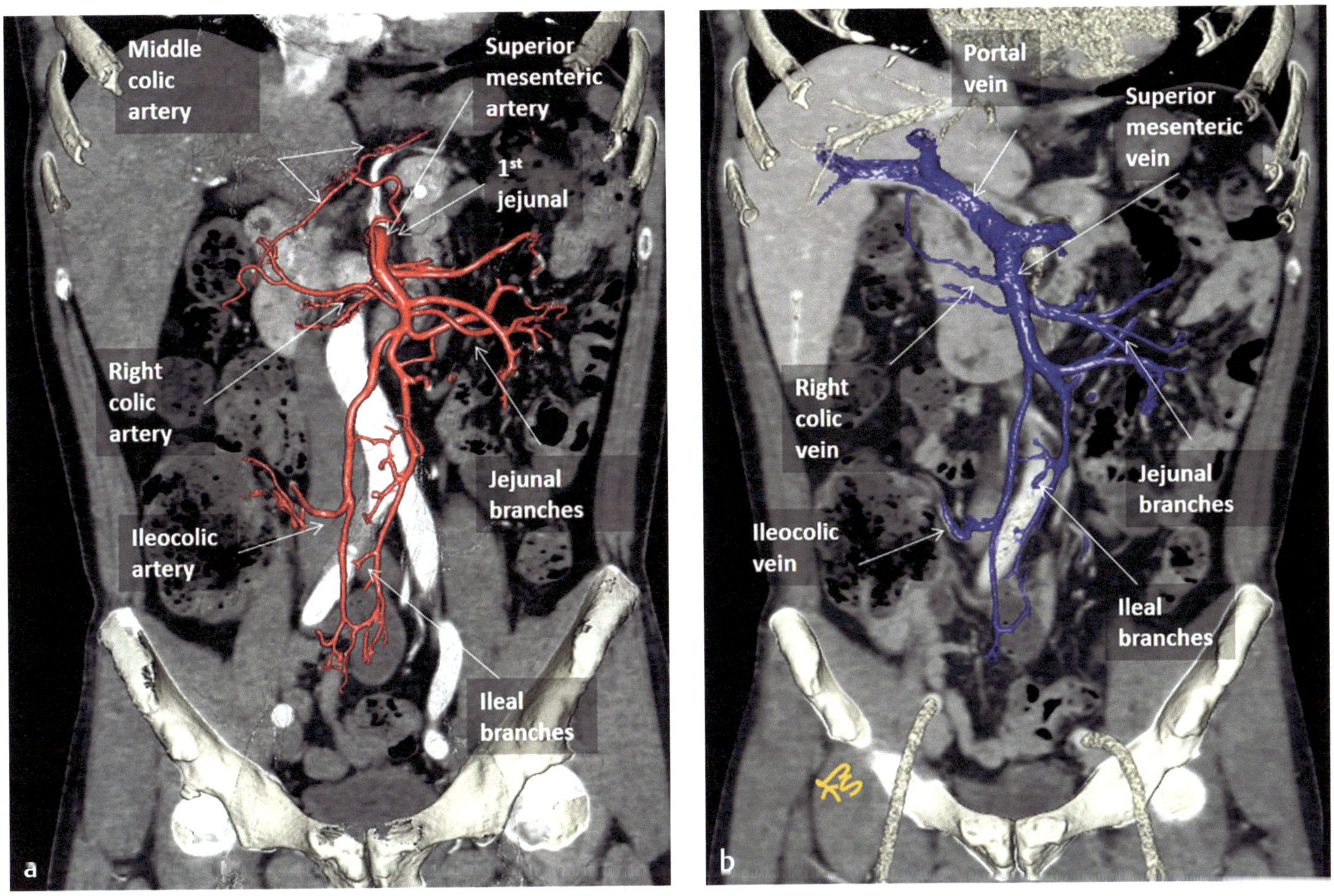

Fig. 4.16 (a, b) Jejunal and ileal superior mesenteric arterial and venous branches are shown in this computed tomography (CT) angiogram.

of lymph capillaries in the lamina propria and submucosa. Large lymph vessels containing valves emanate from the submucosal network and leave the intestinal wall, together with the veins and arteries, to enter the mesentery. Lymphatics of the jejunum and ileum drain into mesenteric nodes, in large numbers, and finally into the superior mesenteric nodes (▶ Fig. 4.19). It is common to find prominent lymph nodes around the distal ileum in CT or ultrasound imaging studies. Reactive mesenteric nodal enlargement is common in inflammation of the small bowel (i.e., ileitis, enteritis).

Innervation

There are two plexuses in the wall. The plexus of Meissner is in the submucosa, and the plexus of Auerbach is in the connective tissue between circular and longitudinal layers of muscularis externa (▶ Fig. 4.11). These plexuses contain preganglionic parasympathetic fibers carried initially by the vagus nerves, neuronal cell bodies of the postganglionic parasympathetic fibers, and sympathetic postganglionic fibers from cell bodies located in the celiac and superior mesenteric plexuses.

4

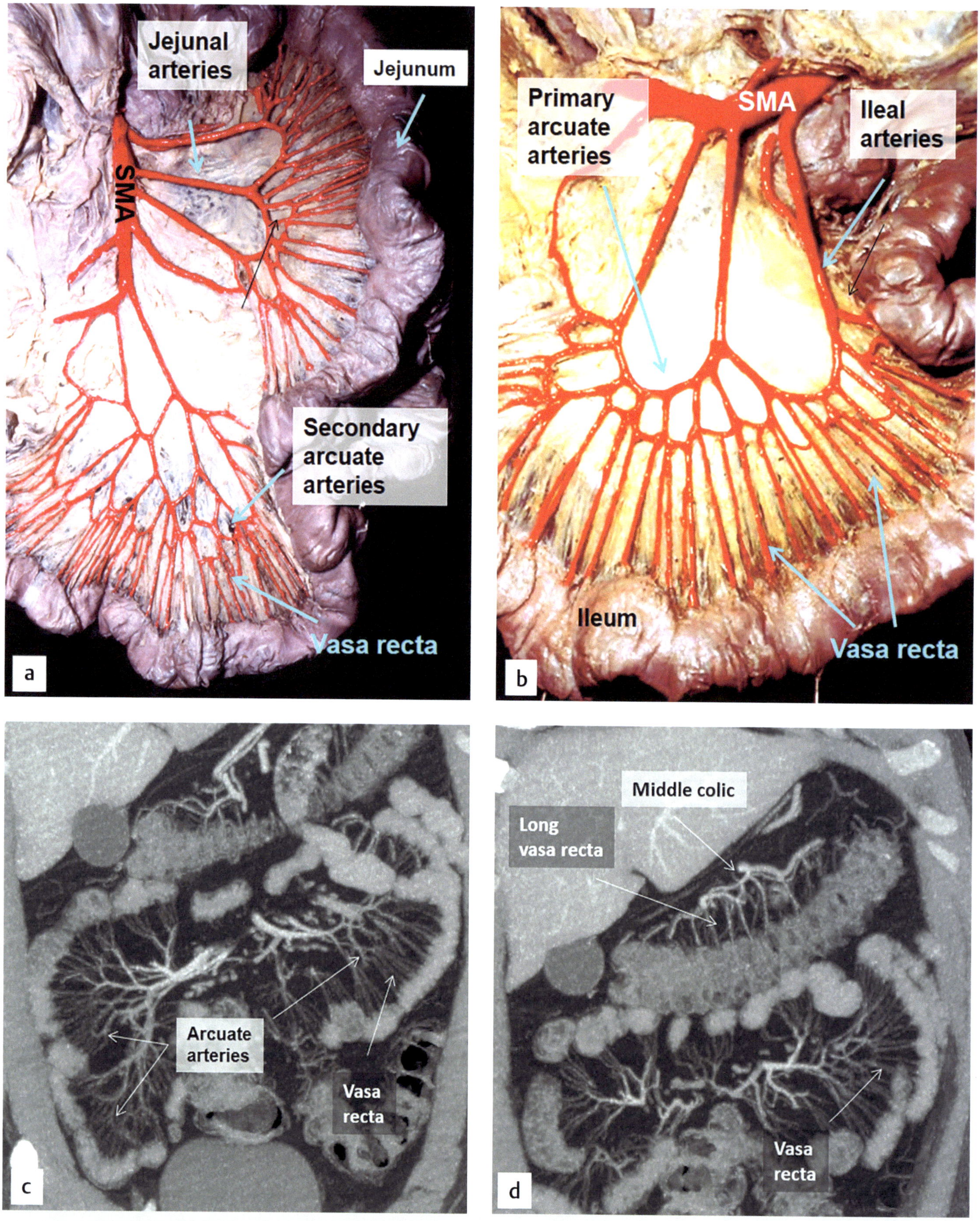

Fig. 4.17 (a, b) Cadaveric views showing primary and secondary arcuate vessels and vasa recta in the mesentery. **(c, d)** Computed tomography (CT) scan showing both artery and venous branches. SMA, superior mesenteric artery.

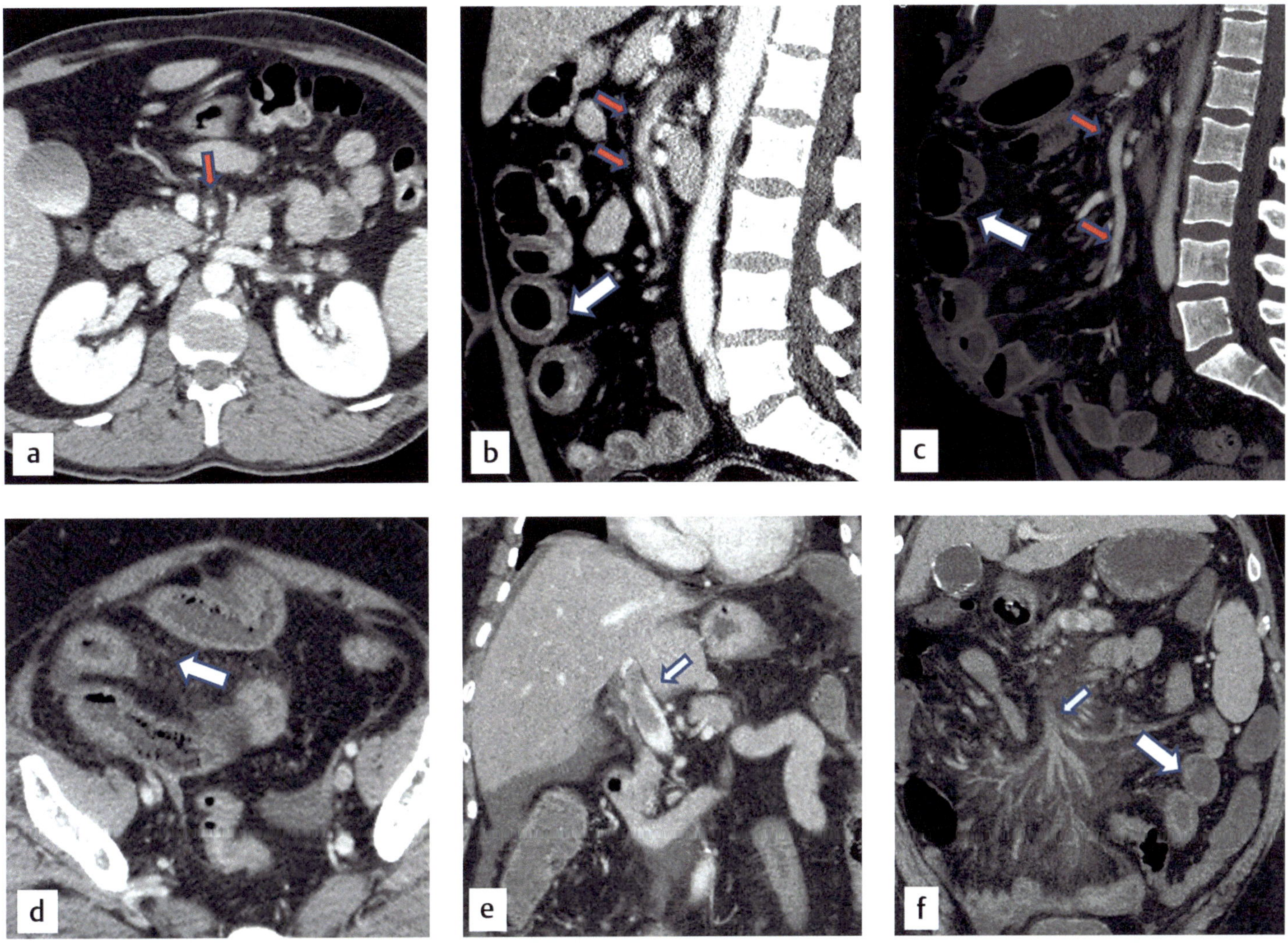

Fig. 4.18 Upper row. **(a)** Axial and **(b)** sagittal computed tomography (CT) angiography showing extensive thrombosis of the superior mesenteric artery (*red arrows*). Loops of jejunum are thickened (*white arrow*). **(c)** Patent superior mesenteric artery (*red arrows*) and improved jejunal wall thickening (*white arrow*) after operation. Lower row: **(d–f)** Mesenteric and portal vein thrombosis (*blue arrows*). Again note thickened loops of jejunum and ileum (*white arrows*). Edema of the mesentery is shown in **f**.

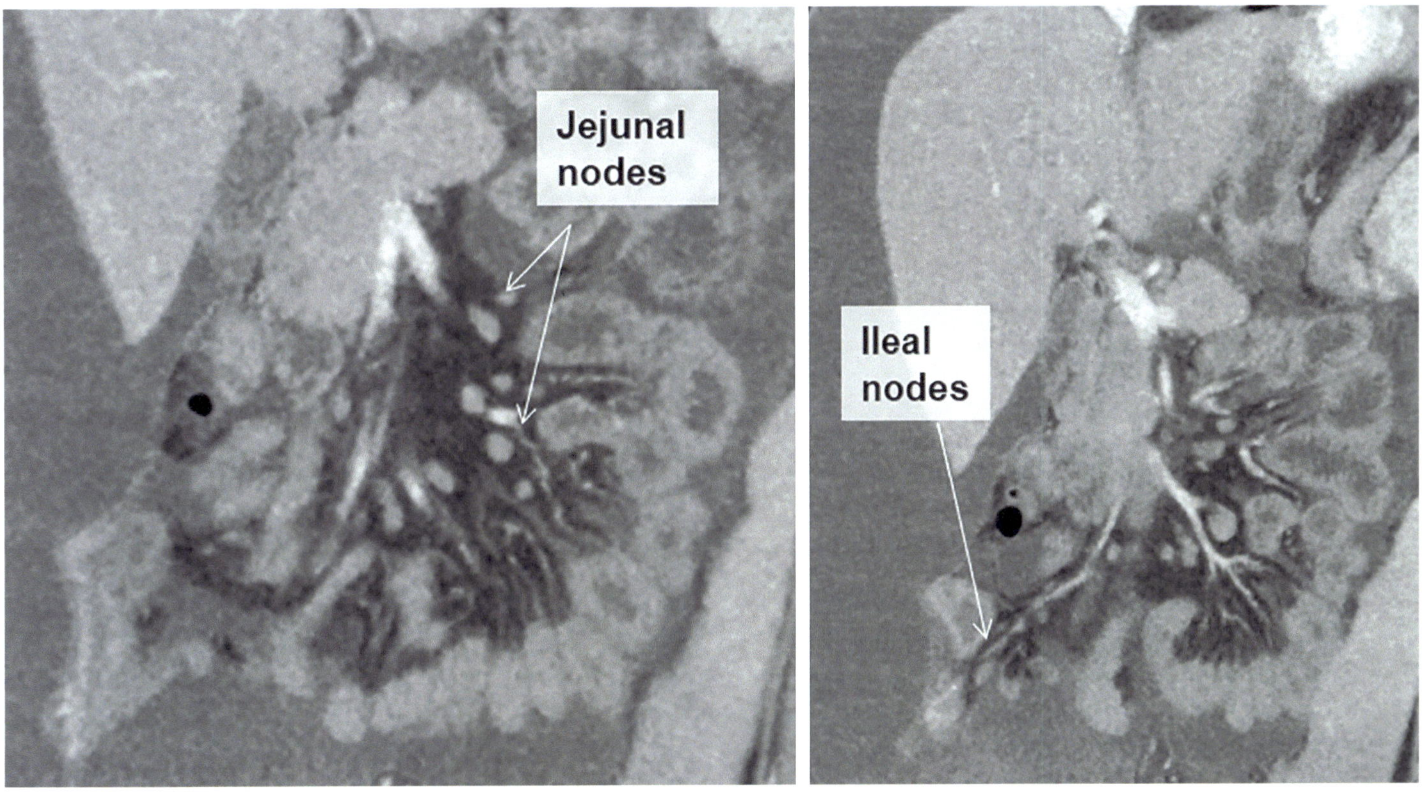

Fig. 4.19 Prominent mesenteric nodes are shown on coronal computed tomography (CT) images.

◆ Pathologies

Congenital

Congenital defects of the duodenum and small bowel are relatively common. Approximately 75% of all intestinal stenoses and 40% of intestinal atresias are found in the duodenum.[5] Duodenal obstruction is usually caused by duodenal atresia, annular pancreas, and midgut volvulus. Less frequently, it may be secondary to duodenal web, Ladd's bands, or preduodenal portal vein. Duodenal atresia presents with severe vomiting which is usually delayed until after the first feeding but increases progressively thereafter. The classic radiographic finding is the so-called double bubble sign related to dilated gas-filled stomach and proximal duodenum. Duodenal duplication cysts may be seen incidentally on CT scans (▶ Fig. 4.20).

Malrotation is defined as congenital abnormal position of the bowel within the peritoneal cavity and usually involves both the small and large bowel. Midgut volvulus is the most dramatic consequence of intestinal malrotation. It usually presents with partial obstruction of the duodenum; therefore, symptoms occur in the immediate postnatal period.[26] Evaluation for malrotation should be a part of every upper gastrointestinal tract examination in pediatric patients.

In cases that are diagnosed later in life, CT scan of the abdomen has a special place (▶ Fig. 4.2).

Meckel's diverticulum is a congenital diverticulum seen in 2% of the population, approximately 60 cm from the ileocecal valve on the antimesenteric side of the ileum. It is a vestigial remnant of the vitelline (omphalomesenteric) duct. It is usually found incidentally during CT scan of the abdomen or causes complications including bowel obstruction, diverticulitis, or hemorrhage due to ectopic gastric mucosa (▶ Fig. 4.21).

Acquired

Common primary lesions of the small intestine include ulcer, infection, diverticulum, ischemia, Crohn's disease, carcinoma, and lymphoma. Diverticulum is a very common incidental finding in the duodenum. Crohn's disease and lymphoma usually involve the terminal ileum (▶ Fig. 4.22). Ulcers are common in the bulb of duodenum but relatively uncommon in the rest of small intestine (▶ Fig. 4.23). Pain, hemorrhage, diarrhea, malabsorption, and obstruction are common presentations. Acute small intestinal bleeding is a rare disorder, typically caused by ulcers, diverticular disease, vascular malformation, and neoplasm. Chronic intestinal ischemia is characterized by postprandial abdominal pain and profound weight loss in the old patients and is usually secondary to atherosclerotic narrowing or occlusion of the mesenteric and celiac arteries. Vascular compression of the third part of the duodenum against the aorta and vertebral

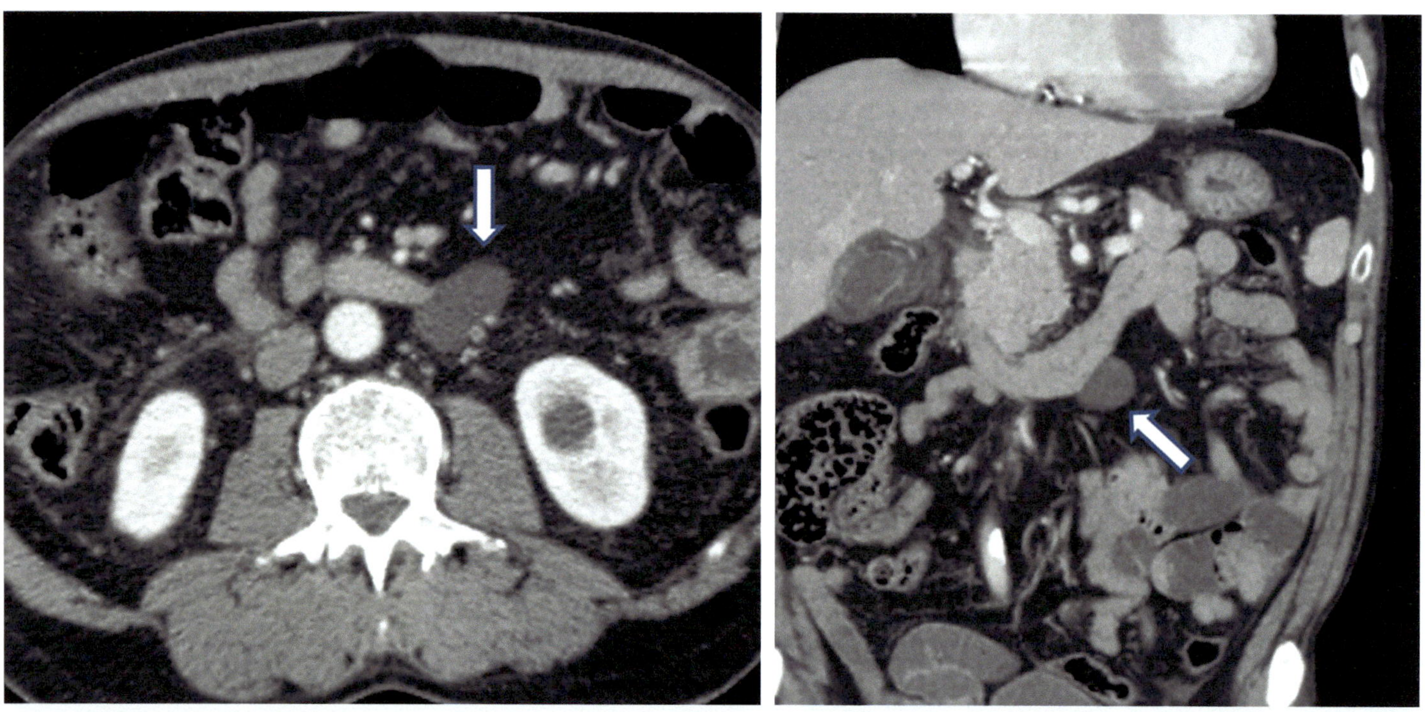

Fig. 4.20 Axial and coronal computed tomography (CT) images showing small duodenal duplication cyst (*arrows*) at the junction of the third part and fourth part of duodenum.

column by the SMA at the level of L2 or L3 is known as SMA syndrome.[27] It is usually asymptomatic but in severe cases acute duodenal obstruction may be seen (▶Fig. 4.24). Normal angle between the ascending aorta and the SMA is typically greater than 25 degrees, but it can be narrower in thin patients with little amount of mesenteric fat. Angles less than 10 degrees can cause compression of the left renal vein and the third portion of the duodenum.

Diagnostic imaging tools for evaluation of the small intestine are barium follow-through examination, CT, MR, and ultrasound. Small intestinal bleeding is usually diagnosed by capsule endoscopy. Selective angiography is a great tool for diagnosis and treatment of the bleeding sites. Transcatheter arterial embolization is recommended as the first-line treatment for acute small intestinal bleeding (especially distal to the ligament of Treitz) given the endoscopic approach limitations.

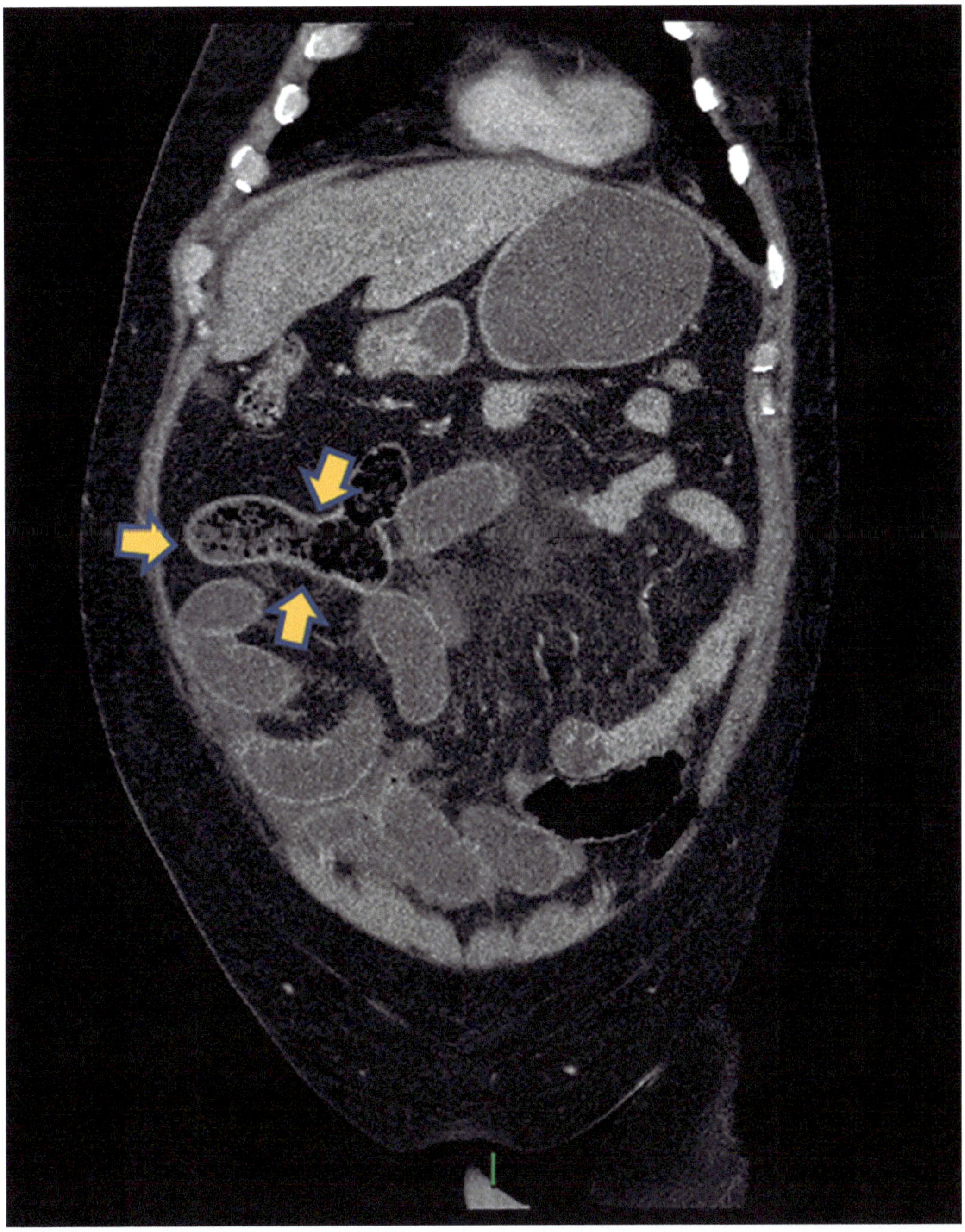

Fig. 4.21 Meckel's diverticulum is seen as a blind end outpouching of the distal ileum in this coronal computed tomography (CT). Retained ingested material is seen in the diverticulum, which has partially obstructed the ileum. The diverticulum is located on the antimesenteric side of the ileum (*arrows*). It shows mild thickening of the wall due to diverticulitis.

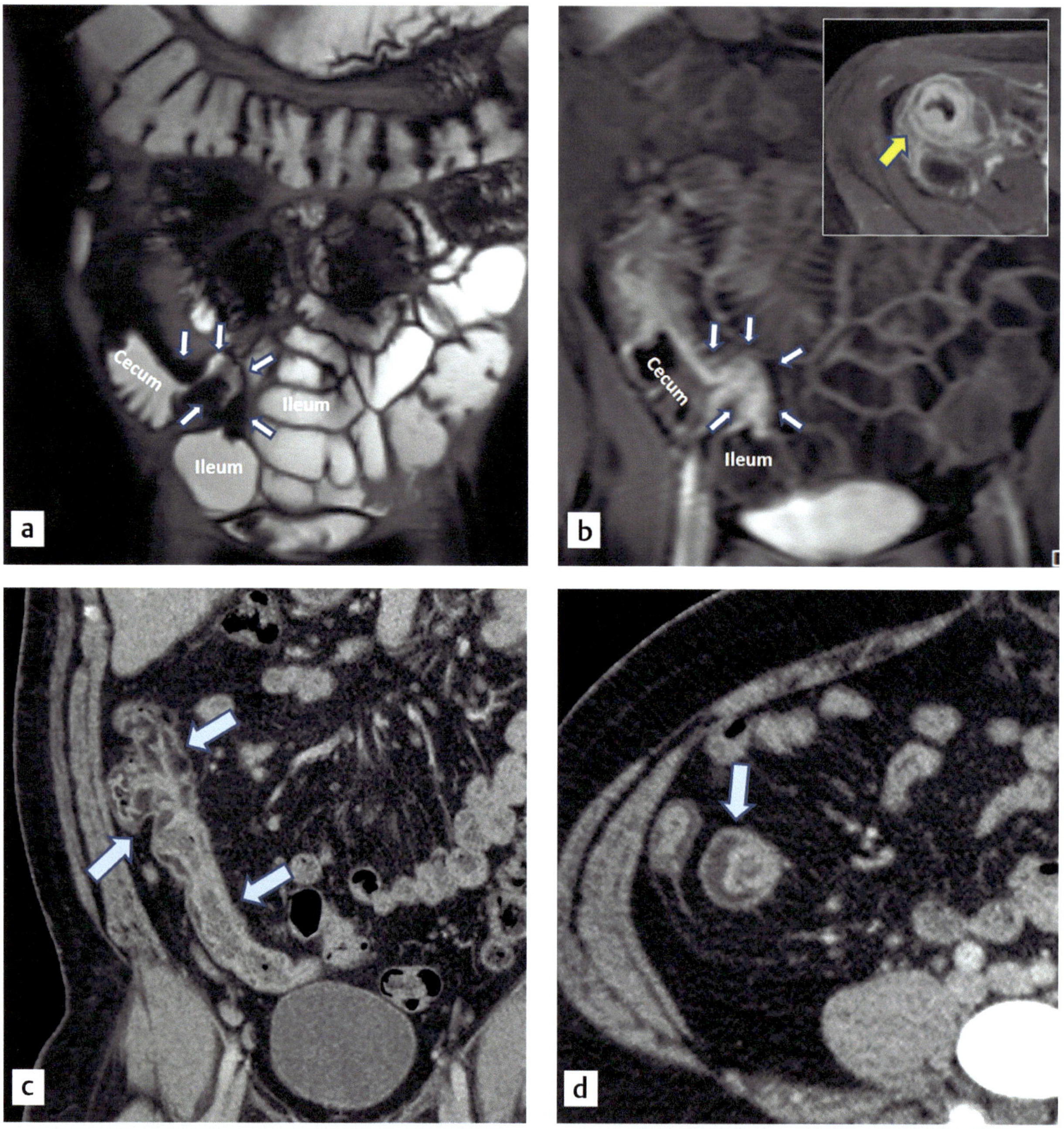

Fig. 4.22 Crohn's disease involving terminal ileum (*arrows*). **(a, b)** Acute phase of Crohn's disease. **(a)** T2-weighted coronal image showing thickened wall terminal ileum (thick dark band) causing partial obstruction. Note dilated fluid-filled loops of the small bowel. **(b)** Postcontrast coronal image showing severe enhancement of the terminal ileum wall. Inlay axial image showing thickening of the mucosa and enhancement of the mesentery (*arrow*). Muscularis layer is shown as a dark ring. **(c, d)** Chronic phase of Crohn's disease. **(c)** Coronal and **(d)** axial computed tomography (CT) showing extensive thickening of the distal ileum and the ileocecal valve (*arrows*). There is fatty infiltration of the involved bowel.

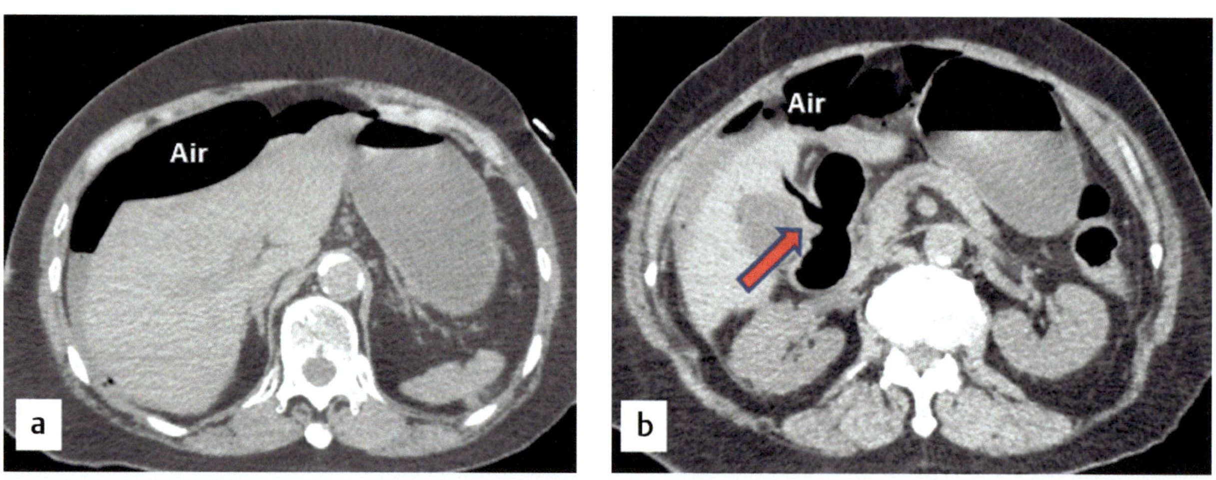

Fig. 4.23 (a, b) Axial computed tomography (CT) image of the upper abdomen in a patient with severe abdominal pain showing extensive pneumoperitoneum due to perforation of the duodenal bulb (*red arrow*).

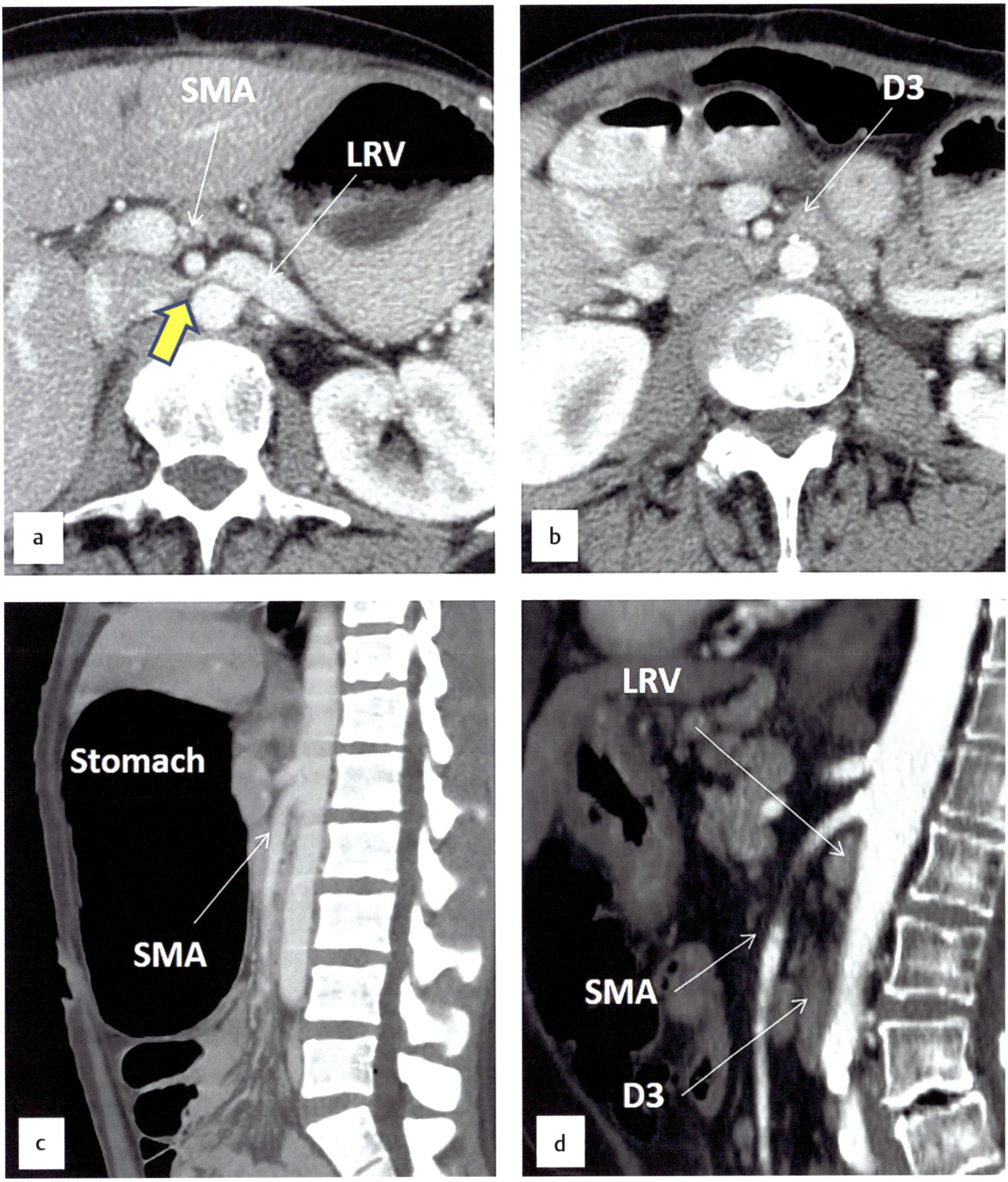

Fig. 4.24 Superior mesenteric artery (SMA) syndrome. **(a, b)** Axial computed tomography (CT) showing severe compression of the left renal vein (LRV) (*arrow*) and third duodenum (D3) between the aorta and SMA. **(c)** Sagittal CT showing close anatomic course of the SMA and the aorta causing stomach dilatation due to obstruction of the third duodenum. **(d)** Sagittal CT showing normal anatomic course of the SMA in relation to the aorta.

References

1. Snyder WH Jr, Chaffin L. Embryology and pathology of the intestinal tract: presentation of 40 cases of malrotation. Ann Surg 1954;140(3):368–379
2. Applegate KE, Anderson JM, Klatte EC. Intestinal malrotation in children: a problem-solving approach to the upper gastrointestinal series. Radiographics 2006;26(5):1485–1500
3. Dott NM. Anomalies of intestinal rotation: their embryology and surgical aspects—with report of five cases. Br J Surg 1923;11(42):251–286
4. Ford EG, Senac MO Jr, Srikanth MS, Weitzman JJ. Malrotation of the intestine in children. Ann Surg 1992;215(2):172–178
5. Androulakis J, Colborn GL, Skandalakis PN, Skandalakis LJ, Skandalakis JE. Embryologic and anatomic basis of duodenal surgery. Surg Clin North Am 2000;80(1):171–199
6. Kim SK, Cho CD, Wojtowycz AR. The ligament of Treitz (the suspensory ligament of the duodenum): anatomic and radiographic correlation. Abdom Imaging 2008;33(4):395–397
7. Dardinski VJ. The anatomy of the major duodenal papilla of man, with special reference to its musculature. J Anat 1935;69(Pt 4):469–478.1
8. Stimec B, Alempijevic T, Micev M, et al. Relationship between papillary fibrosis and patency of the accessory pancreatic duct. Ann Anat 2005;187(2):99–103
9. Valverde Barbato de Prates NE, Smanio T, De Maio Domingos M, Ferraz de Carvalho CA. "Sphincter" of the minor papilla of the human duodenum. Clin Anat 1996;9(1):34–40
10. Shafik AA, Shafik A, Asaad S, Wahdan M. A study of an anatomic-physiological cecocolonic sphincter in humans. Clin Anat 2010;23(7):851–861
11. DiDio LJA, Geocze S, Ferrari Júnior AP. Endoscopic study of the ileal papilla in a normal case and in patients under several pathological conditions. Ann Anat 1994;176(3):263–268
12. Faussone Pellegrini MS, Manneschi LI, Manneschi L. The caecocolonic junction in humans has a sphincteric anatomy and function. Gut 1995;37(4):493–498
13. Pollard MF, Thompson-Fawcett MW, Stringer MD. The human ileocaecal junction: anatomical evidence of a sphincter. Surg Radiol Anat 2012;34(1):21–29
14. Katsinelos P, Paroutoglou G, Jouvaras G, et al. Lipohyperplasia of the ileocecal valve as a cause of intussusception. Acta Gastroenterol Belg 2005;68(2):280–282
15. Spiller RC, Brown ML, Phillips SF. Emptying of the terminal ileum in intact humans. Influence of meal residue and ileal motility. Gastroenterology 1987;92(3):724–729
16. Machado WM, Morceli J. Prevalence, classification and characteristics of the coloileal reflux diagnosed by barium enema. Radiol Bras 2006;39(2):107–111
17. Rivkind AI, Shiloni E, Muggia-Sullam M, Weiss Y, Lax E, Freund HR. Paracecal hernia: a cause of intestinal obstruction. Dis Colon Rectum 1986;29(11):752–754
18. Fu CY, Chang WC, Lu HE, Su CJ, Tan KH. Pericecal hernia of the inferior ileocecal recess: CT findings. Abdom Imaging 2007;32(1):81–83
19. Johanowicz-Wiśniewska H, Ratajczyk-Pakalska E. Arteries of the jejunum and ileum. Folia Morphol (Warsz) 1976;35(2):159–168
20. Noer RJ, Derr JW, Johnston CG. The circulation of the small intestine: an evaluation of its revascularizing potential. Ann Surg 1949;130(4):608–621
21. Kachlik D, Baca V, Stingl J. The spatial arrangement of the human large intestinal wall blood circulation. J Anat 2010;216(3):335–343
22. Chan KL, Zhang XH, Fung PC, Guo WH, Tam PK. Role of nitric oxide in intestinal ischaemia-reperfusion injury studied using electron paramagnetic resonance. Br J Surg 1999;86(11):1427–1432
23. Marion Y, Lebreton G, Le Pennec V, Hourna E, Viennot S, Alves A. The management of lower gastrointestinal bleeding. J Visc Surg 2014;151(3):191–201
24. Koganemaru M, Nonoshita M, Iwamoto R, et al. Ultraselective embolization using a 1.7-Fr catheter and soft bare coil for small intestinal bleeding. Minim Invasive Ther Allied Technol 2016;25(6):345–350
25. Pinto S, Lerner T, Lingamaneni G, Richards K. Superior mesenteric vein thrombosis as a complication of cecal diverticulitis: a case report. Int J Surg Case Rep 2016;25:71–74
26. Berrocal T, Torres I, Gutiérrez J, Prieto C, del Hoyo ML, Lamas M. Congenital anomalies of the upper gastrointestinal tract. Radiographics 1999;19(4):855–872
27. Biank V, Werlin S. Superior mesenteric artery syndrome in children: a 20-year experience. J Pediatr Gastroenterol Nutr 2006;42(5):522–525

5 Colon

Frank Chen, Aramsadat Meraji, and Farhood Saremi

◆ Introduction

The large intestine, the last portion of the gastrointestinal tract, is a saccular tubular organ that begins at the ileocecal valve and extends to the anal canal. It is subdivided into the colon and the rectum, with the colon further subdivided into the cecum and appendix vermiformis, ascending colon, hepatic flexure, transverse colon, splenic flexure, descending colon, and sigmoid colon for descriptive purposes (▶Fig. 5.1).[1] The primary functions of the large intestine include the conservation of water and electrolytes and the storage and controlled evacuation of feces.[2]

5

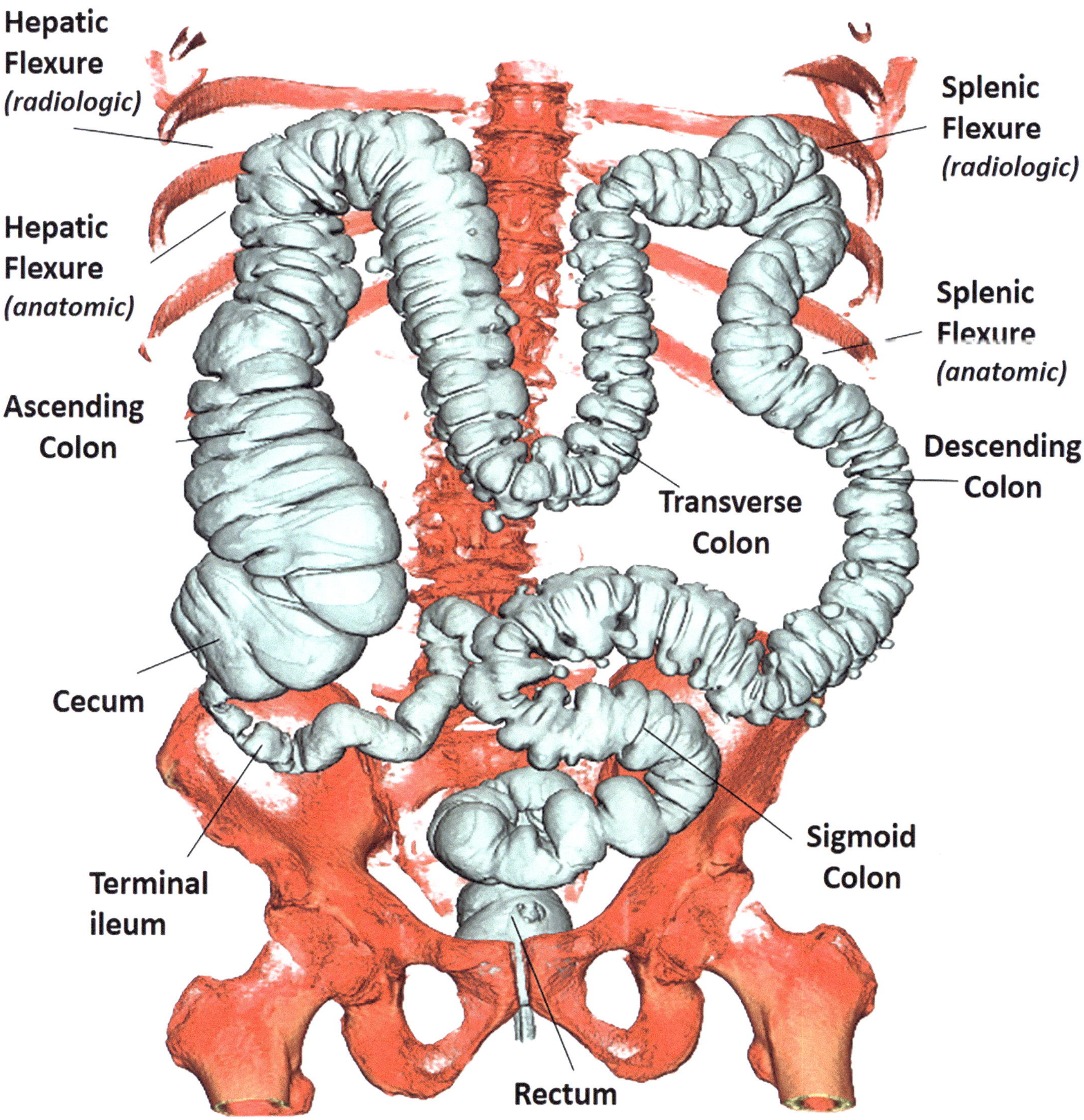

Fig. 5.1 Colorectal segments.

5

◆ Embryology

An intraembryonic endoderm results secondary to lateral and longitudinal folding of the embryo, and it forms the primitive gut tube. During the third week of gestation, the primitive gut tube is divided into the primitive foregut, midgut, and hindgut (▶ Fig. 5.2a). The midgut becomes the duodenum beyond the duodenal ampulla, jejunum, ileum, ascending colon, and proximal two-third of the colon, which are supplied by the superior mesenteric artery. The hindgut becomes the distal third of the transverse colon, sigmoid colon, rectum, and anal canal above the dentate line, which are supplied by the inferior mesenteric artery. The lower third of the anal canal below the dentate line originates from the ectoderm and is supplied by the internal pudendal artery.

Normal development of the colon is related to complete midgut development. At the end of the sixth week of gestation, the midgut, which is supplied by the superior mesenteric artery, rapidly elongates. Because the peritoneal cavity is too small to accommodate the elongating midgut, it bulges out of the abdominal cavity into the umbilical cord as a physiologic umbilical hernia (▶ Fig. 5.2b). As it herniates, the midgut revolves 90 degrees counterclockwise around the superior mesenteric artery. During the 10th week of gestation, the midgut returns into the peritoneal cavity as it revolves 180 degrees counterclockwise around the superior mesenteric artery and the physiological umbilical hernia closes. The intestine and its mesentery merge with the primitive posterior parietal peritoneum. The posterior fusion of the colon first involves the right colic angle. Later, migration of the cecum–appendix to the right and its descent from right upper abdomen toward the iliac fossa along with fusion of the right colon to the primitive posterior parietal peritoneum occur toward the end of the fetal life. The ascending colon develops through this descent. Abnormal migration of the cecum results in incomplete fixation of the ascending colon which explains floating cecum and right colon seen in some individuals (▶ Fig. 5.2).[3–5]

After birth, the ascending colon and the descending colon are fixed in their permanent retroperitoneal position. The cecum, transverse colon, and sigmoid colon are completely covered with peritoneum and have some freedom to move. Transverse mesocolon extends from right colic flexure to left colic flexure. The

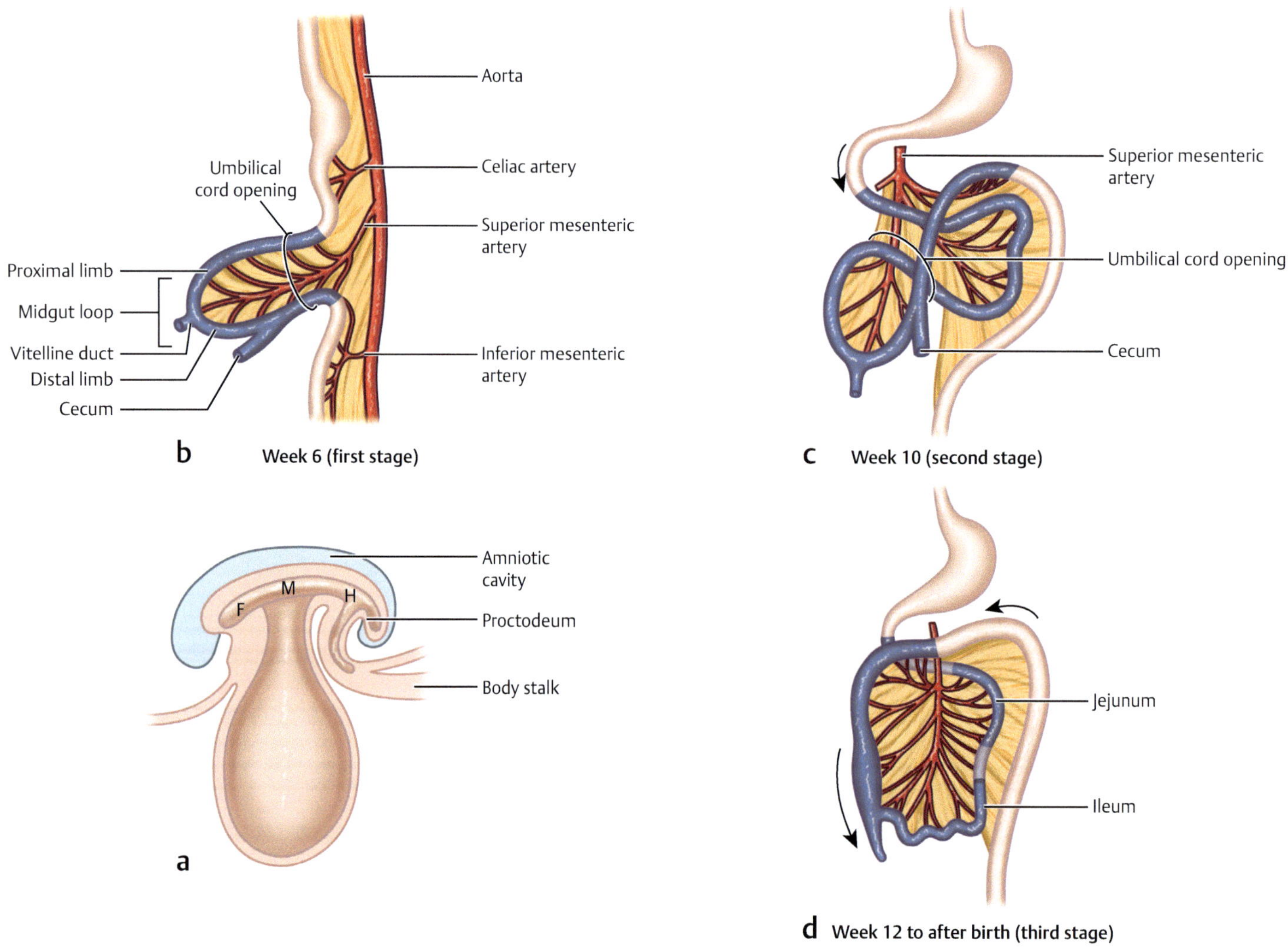

Fig. 5.2 Embryology of the colon and rectum. **(a)** At the third week of gestation, the primitive gut tube is divided into the foregut (F), midgut (M), and hindgut (H). **(b)** At the end of the sixth week of gestation, the midgut herniates out of the abdominal cavity. **(c)** During the 10th week of gestation, the midgut returns into the abdominal cavity. **(d)** The bowel becomes fixated.

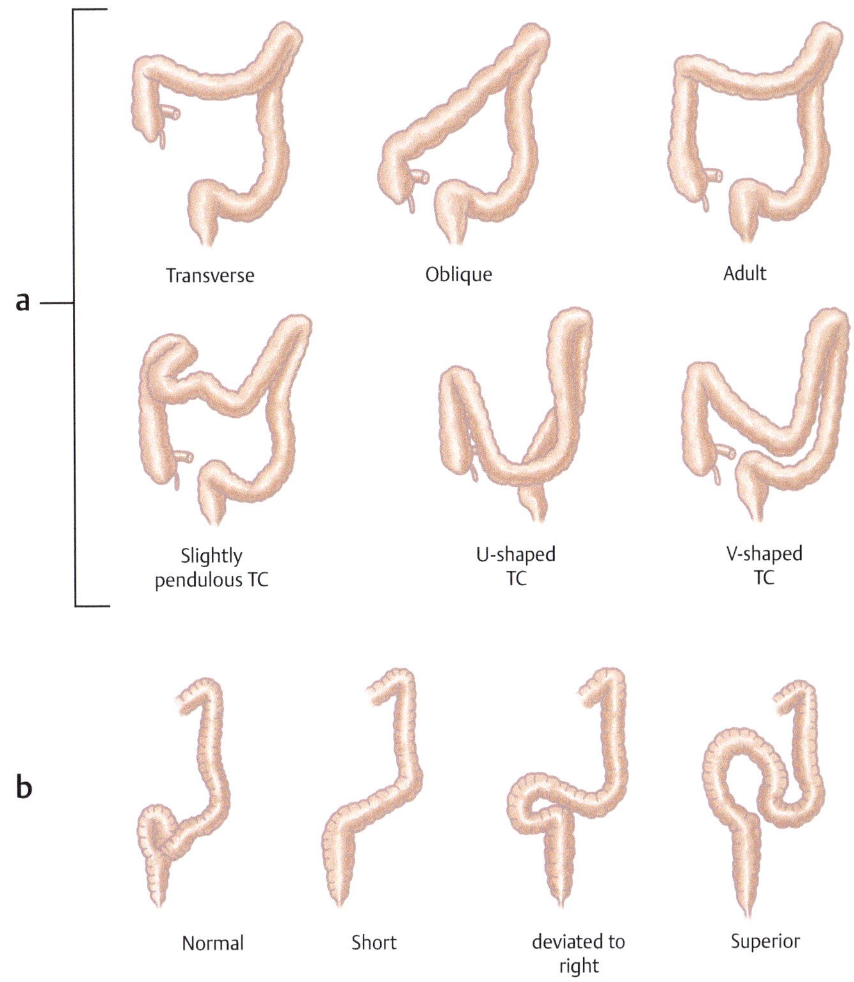

Fig. 5.3 (a) Types of colon parts between the ileal orifice and the sigmoid colon. TC, transverse colon. **(b)** Types of sigmoid colon. Normal sigmoid colon type is the most common type during the fetal period.

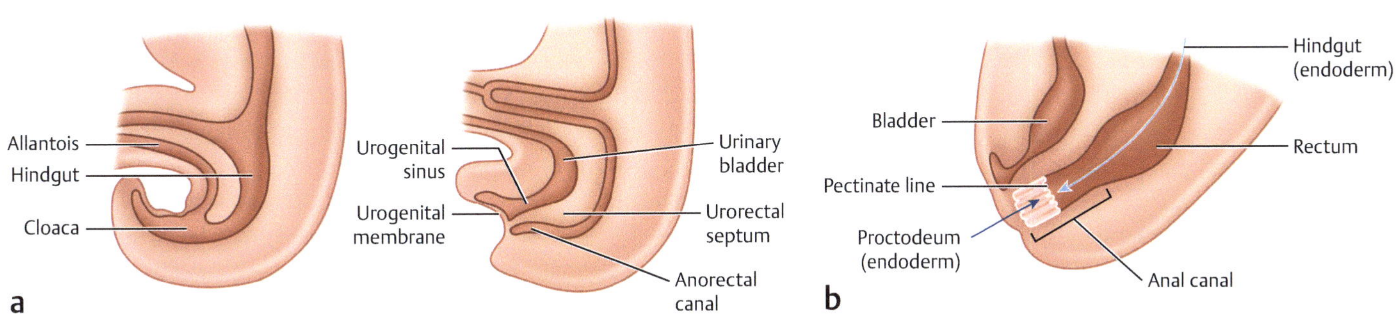

Fig. 5.4 (a) Embryonic development of the anorectal region between 4 and 7 weeks. **(b)** Components of the anorectal canal.

haustra and taenia coli are well developed in full-term fetus. Different morphological patterns of the colon and sigmoid colon exist that differ before and after birth (▶Fig. 5.3).[3]

Embryonic development of the anorectal region is different. At approximately the sixth week of gestation, the cloaca, which is located at the distal portion of the hindgut, is separated into the ventral urogenital sinus and the dorsal rectum by the caudal migration of the urogenital septum (▶Fig. 5.4a). The cloaca, which is endodermal in origin and becomes the upper two-thirds of the anal canal, fuses with proctodeum, which is ectodermal in origin and becomes the lower third of the anal canal (▶Fig. 5.4b). Incomplete fusion causes anorectal malformation including imperforate anus. The dentate line marks the transition from endoderm to ectoderm. The external sphincter originates from the posterior cloaca and the internal sphincter originates from the enlarging rectal circular muscles during the 10th and

12th weeks of gestation, respectively. During development, the external sphincter migrates cephalad, and the internal sphincter migrates caudally.[6,7]

◆ Colon

The colon extends from the ileocecal valve to the proximal rectum and varies considerably in its size, measuring 90 to 150 cm in length. The variation in length is usually related to the length of the rectosigmoid segment. Colonic caliber measures up to 7.5 cm at the cecum and progressively decreases to 2.5 cm at the sigmoid colon but may be quite variable based on degree of distension (▶ Fig. 5.1). Besides its length (approximately a quarter of the small bowel), location, degree of fixation (greater than the small bowel), and caliber (greater than the small bowel), the colon can be distinguished from the small bowel based on the presence of three distinct features: taeniae coli, haustra, and epiploic appendices. Taeniae coli are three equidistant, equally divided bands (taenia libera, taenia mesocolica, and taenia omentalis) of thickened longitudinal muscle, measuring approximately 0.5 cm wide, which are seen from the base of the appendix to the rectosigmoid junction (▶ Fig. 5.5). Haustra are the colonic saccular outpouchings that result secondary to the disproportionate length of the colon in relation to the taeniae coli (▶ Fig. 5.5). The taeniae coli are approximately one-sixth the length of the colon. The disproportionate length of the colon and taeniae coli also results in fixed transverse mucosal folds named semilunar folds. Epiploic appendices are outpouchings of peritoneal fat seen along the serosa which are most concentrated along the distal descending and sigmoid colon but absent along the appendix, cecum, and rectum[8] (▶ Fig. 5.4, ▶ Fig. 5.6). Inflammation of these appendices may occur and cause vague abdominal pain (▶ Fig. 5.6).

Cecum

The cecum is a blind-ending sacculated pouch of the large intestine that is normally situated in the right iliac fossa. It extends 6 to 8 cm below the ileocecal valve and measures 7.5 to 8.5 cm in diameter. At least the lower half of the cecum, if not the entire cecum, is invested by peritoneum forming a retrocecal space in the peritoneal cavity.

The terminal ileum empties into the posteromedial aspect of the cecum, forming a nipple-like invagination, the ileocecal valve or valve de Bauhin, which originates from a thickened circular muscular layer of the terminal ileum (▶ Fig. 5.7). Plump fatty ileocecal valve due to increased submucosal fatty tissue (lipohyperplasia) is common and should not be mistaken for lipoma or polyp of this region.

Vermiform Appendix

The appendix is a blind-ending, worm-like tube that extends from the posteromedial aspect of the cecum, at the fusion of the taenia coli, approximately 2 to 3 cm below the ileocecal valve (▶ Fig. 5.8). It measures approximately 8 to 10 cm in length and 3 to 8 mm in diameter. The appendix grows from the cecal bud and becomes visible during the eighth week of gestation. As the right haustrum of the cecum grows, the appendix is displaced, from an inferior position, upward and medially. Lymphatic nodules appear during the fourth and fifth months and continue to grow postpartum up to puberty. Agenesis of the appendix is very rare. The appendix has a distinct, twisted or curled appearance because the mesoappendix is shorter than the appendix (▶ Fig. 5.8a). Given its great mobility, the location of the appendix can vary in relation to the cecum and ileum—retrocecal position (65%) is the most common, followed by pelvic (31%), subcecal (2.3%), preileal (1.0%), and retroileal (0.4%) positions.[9,10] The pelvic location is the classic location, medial to the cecum at the brim of the pelvis over the psoas muscle (▶ Fig. 5.8a). In retrocecal variant, the appendix is usually intraperitoneal and behind or lateral to the cecum and may be free or attached to the posterior wall of the cecum. The retroperitoneal variant of the retrocecal appendix is relatively rare (2.5%). The retroperitoneal variant is usually medial to the cecum and behind it. In cases with a movable cecum, the appendix may be found in the pelvis, mid abdomen, or right upper quadrant (▶ Fig. 5.8c). The position of the appendix is determined mainly by the changes in the position and shape of the cecum that occur during organ development[10] (▶ Fig. 5.8e–g). Awareness of the appendiceal location, including the extraperitoneal variant, helps the surgeon avoid aggressive and time-consuming search for the inflamed appendix. The appendicular artery, a branch of ileocolic or cecal artery, provides blood supply to the appendix. Lymphatics of the cecum and appendix drain to the lymph nodes of the mesoappendix and follow the ileocolic artery to reach the superior mesenteric lymph node.

Ascending Colon

The ascending colon extends retroperitoneally from the ileocecal valve to the right colonic or anatomic hepatic flexure, underneath the right hepatic lobe. It measures approximately 15 cm in length and between 4 and 7 cm in diameter (▶ Fig. 5.9, ▶ Fig. 5.10). The anterior, medial, and lateral aspects of the ascending colon are covered by peritoneum. The posterior aspect of the ascending colon is fixed against the retroperitoneum and is devoid of peritoneum. Instead, the posterior aspect of the ascending colon is covered by areolar tissue, the fascia of Toldt. A mobile portion in the ascending colon due a short mesocolon is seen in two-thirds of the individuals (▶ Fig. 5.10) causing medial displacement of the ascending colon and lateral interposition of the small bowel that should not be mistaken with internal hernia.

Transverse Colon

The transverse colon, the longest segment of large bowel measuring approximately 45 cm in length, is located intraperitoneally, hanging between the splenic and hepatic flexures, which are relatively fixed structures (▶ Fig. 5.5, ▶ Fig. 5.9). It suspends loosely by the transverse mesocolon, allowing variable mobility and length, and is almost entirely enveloped in visceral peritoneum. The greater omentum, which consists of a fused double layer of visceral and parietal peritoneum, is attached

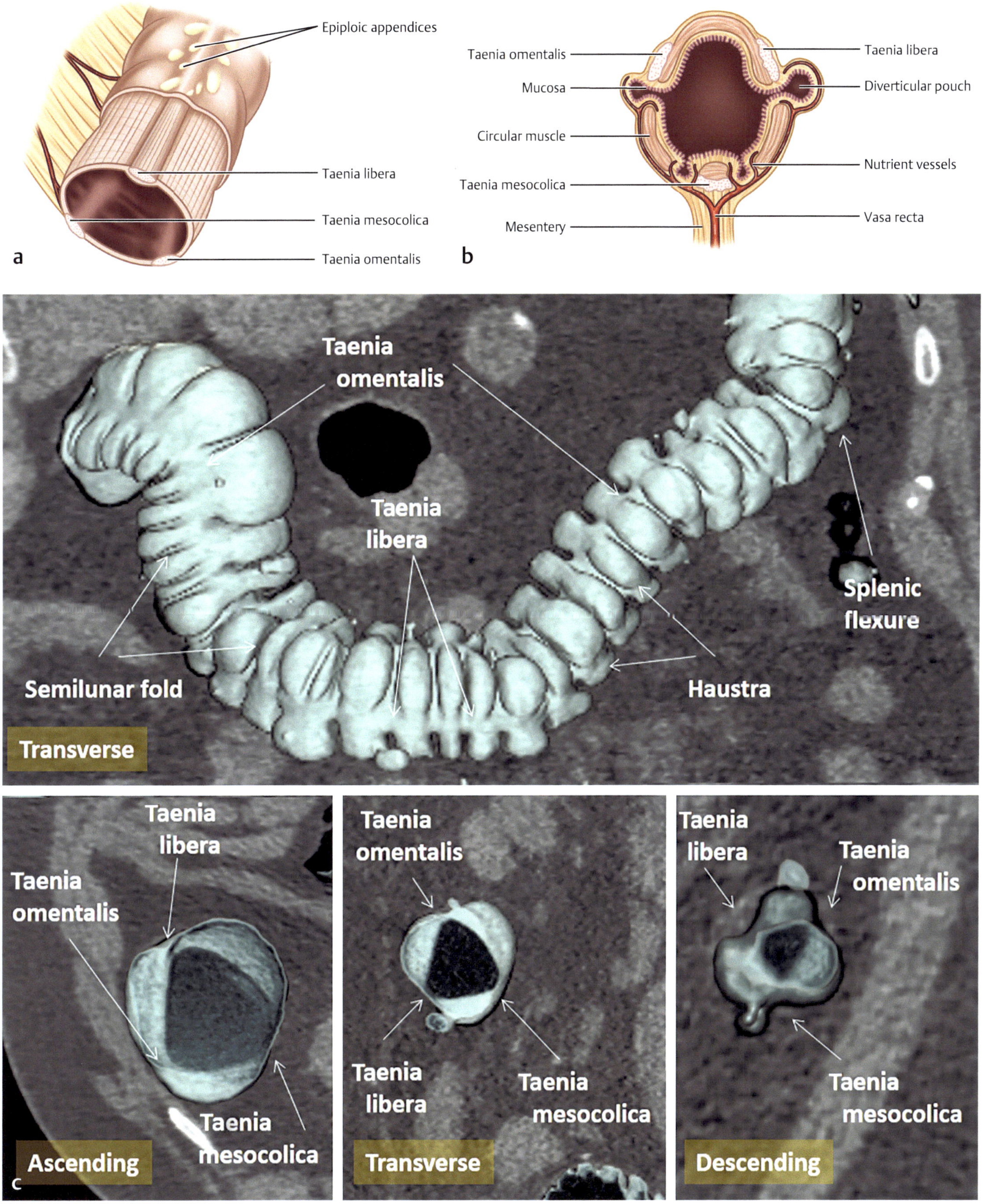

Fig. 5.5 (a) Cross section of a colon segment showing the taeniae coli and epiploic appendices. **(b)** Cross section of the colon showing the relationship of the penetrating arteries with the taeniae coli and pathogenesis of the diverticula development. Diverticula are typically found adjacent to the taeniae coli at the weak points. **(c)** Colon haustral pattern and taeniae coli. Taeniae coli are three bands (taenia libera, taenia mesocolica, and taenia omentalis) of thickened longitudinal muscle, measuring approximately 0.5 cm wide, which are seen from the base of the appendix to the rectosigmoid junction. Note diverticular arrangement adjacent to the taeniae coli.

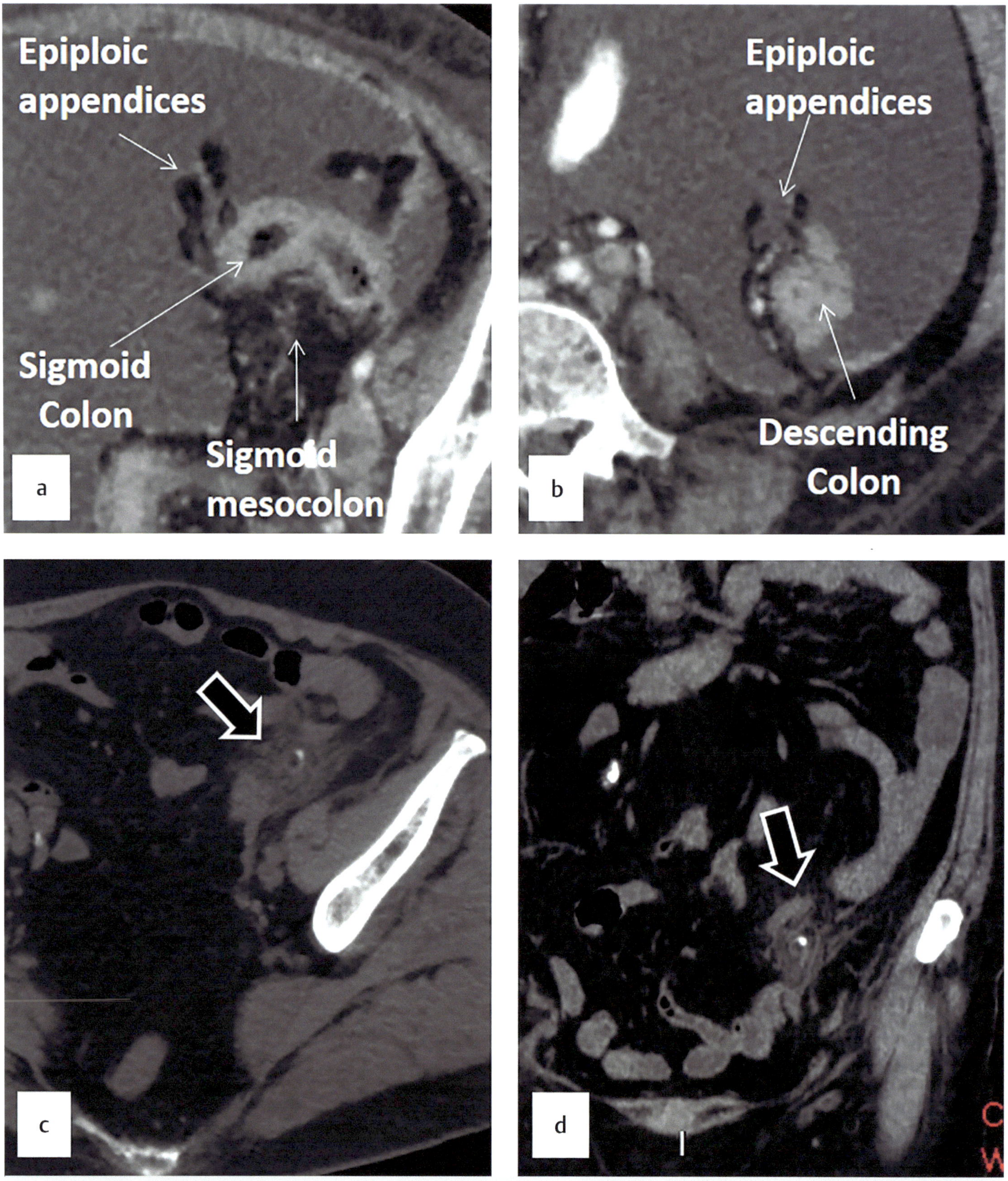

Fig. 5.6 Epiploic appendices. Epiploic appendices are outpouchings of peritoneal fat seen along the serosa which are most concentrated along the sigmoid and distal descending colon **(a, b)**. **(c)** Axial and **(d)** coronal computed tomography (CT) images showing inflammation of the epiploic appendix (*arrows*) causing abdominal pain.

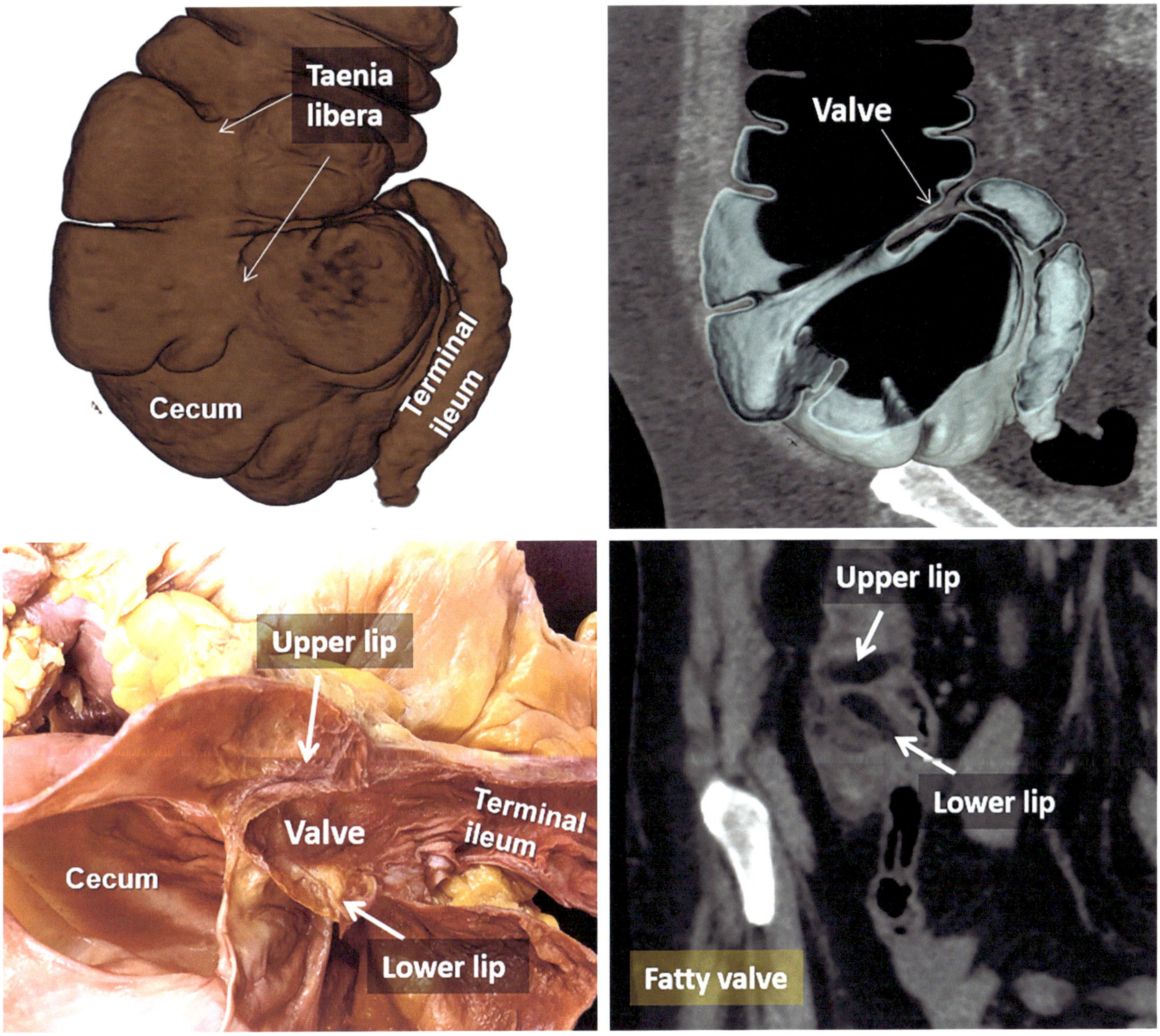

Fig. 5.7 Cecum and ileocecal valve. Computed tomography (CT) and cadaveric views are shown. A fatty valve is shown in the coronal CT view.

along the anterosuperior aspect of the transverse colon to the taenia omentalis. The hepatic flexure is fixed by the right renocolic, hepatocolic, and right phrenicocolic ligaments, while the splenic flexure is fixed by the splenocolic, left renocolic, and left phrenicocolic ligaments. An elongated right transverse colon may interpose between the liver and the right hemidiaphragm (▸Fig. 5.11).

Descending Colon

The descending colon extends retroperitoneally within the left anterior pararenal space from the splenic flexure to the pelvic brim and is located more posterior than the ascending colon (▸Fig. 5.9, ▸Fig. 5.10). It measures approximately 25 cm in length and between 3 and 4 cm in diameter. The anterior and lateral aspects of descending colon are covered by peritoneum. Similar to the ascending colon, the descending colon may have a short mesentery that makes it partly mobile (▸Fig. 5.10).

Sigmoid Colon

The sigmoid colon is an omega-shaped, small caliber, muscular tube that varies in length from 15 to 50 cm (average 40 cm).[11,12] It extends from the pelvic brim to the upper rectum, at approximately the level of the third sacral vertebrae. It is entirely invested by peritoneum and freely suspends by the sigmoid mesocolon, making it very mobile and variable in position (▸Fig. 5.1, ▸Fig. 5.6). The sigmoid mesocolon is longer in men.

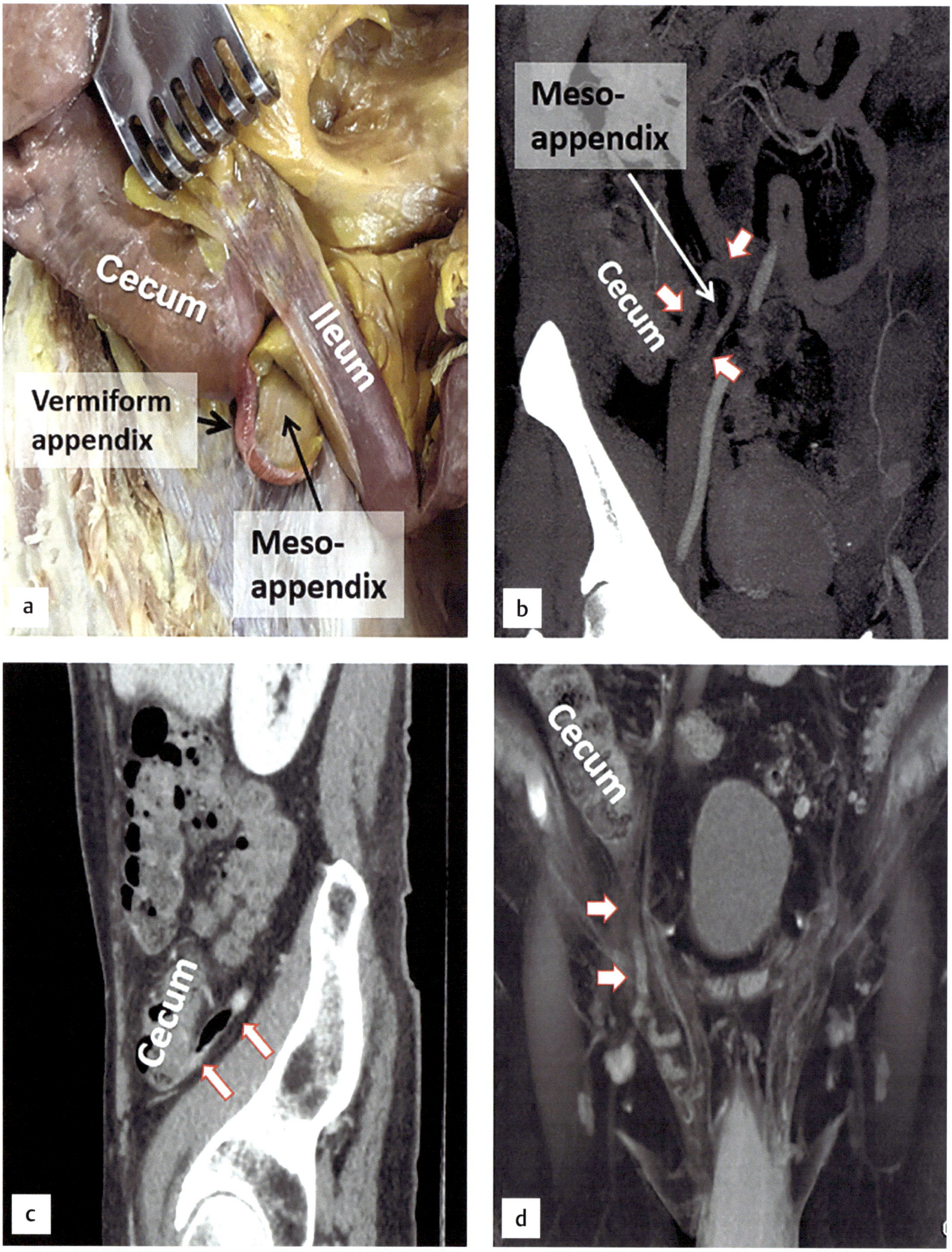

Fig. 5.8 Variations of appendix. **(a)** Cadaveric view of a typical vermiform appendix arising from the inferomedial margin of the cecum. **(b)** Coronal computed tomography (CT) showing an inverted U-shaped appendix (*arrows*) with typical location in the medial side of the cecum. **(c)** Sagittal CT shows a retrocecal appendix (*arrows*), the most common location of a normal appendix. **(d)** Coronal CT showing the appendix (*arrows*) extending into a right inguinal hernia. **(e)** Axial and **(f, g)** coronal CT images showing high position of a retrocecal appendix (*blue arrows*) in the right subhepatic region. Note that the cecum is also high in position indicating incomplete descent of the cecum. *White arrow* shows the location of the ileocecal valve.

(*Continued*)

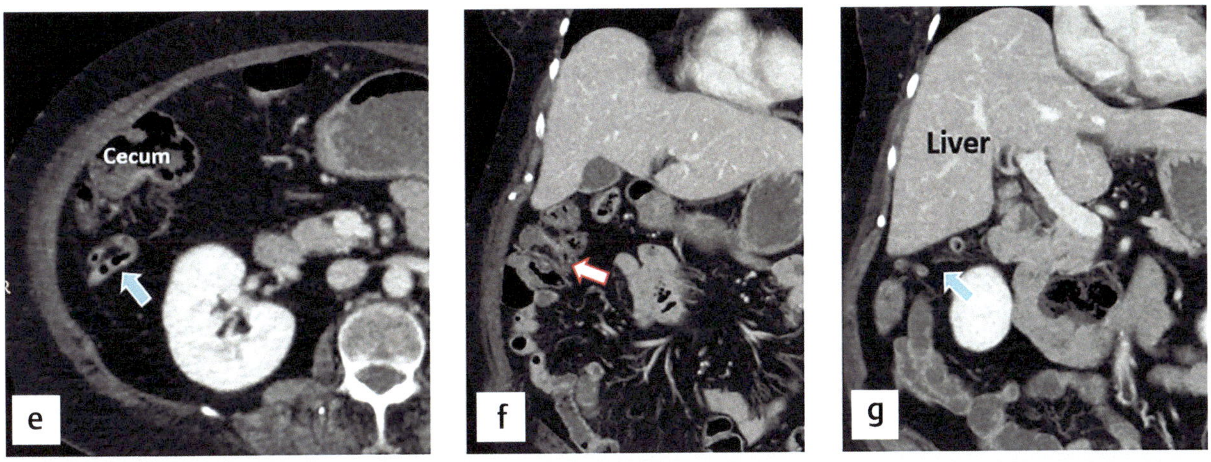

Fig. 5.8 *(Continued)* **(e-g)**

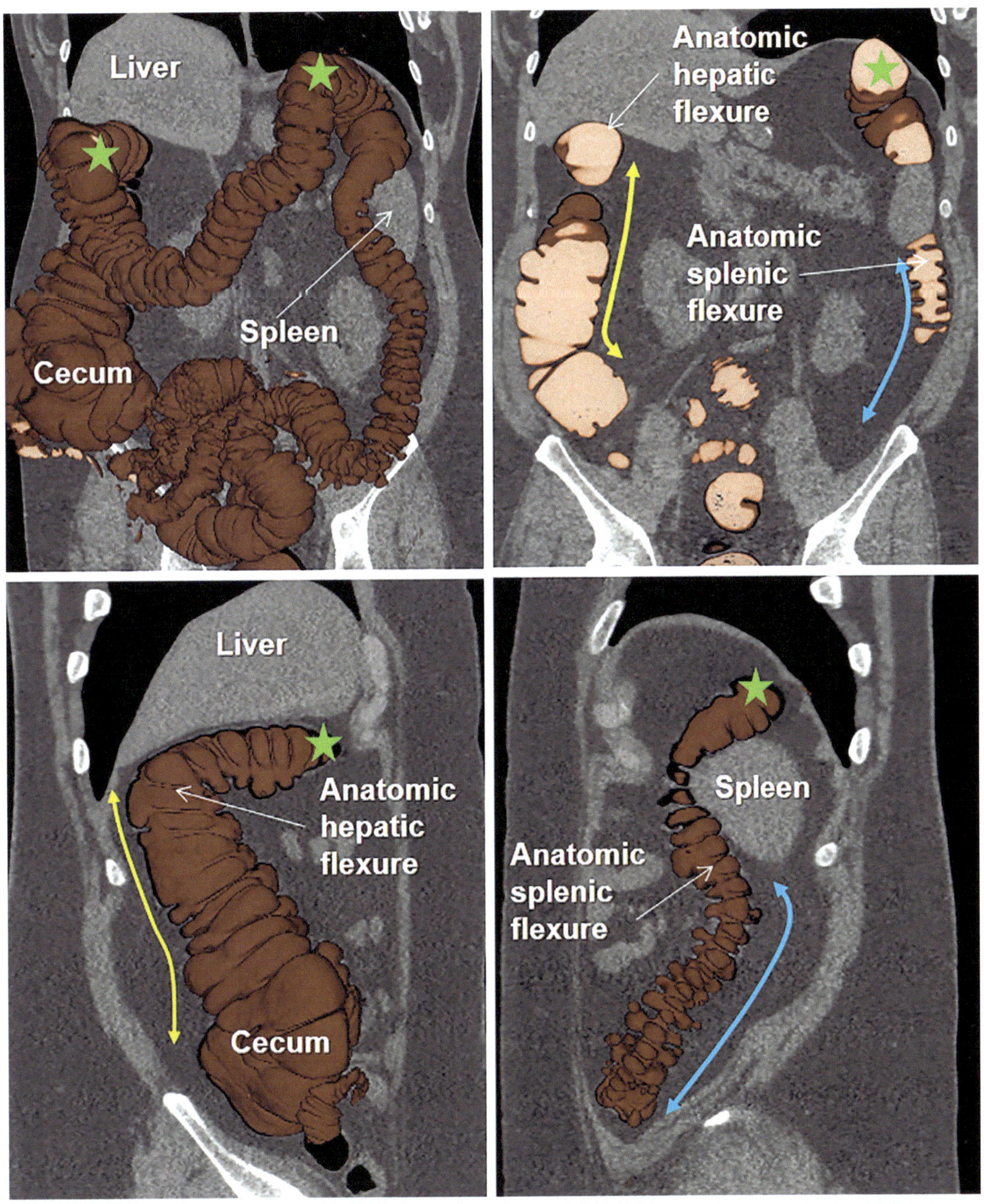

Fig. 5.9 Ascending and descending colon. Coronal (*upper row*) and sagittal (*lower row*) views are shown. The ascending colon is retroperitoneal, extending between the ileocecal valve and the anatomic hepatic flexure (*double-headed yellow arrows*). The descending colon is also retroperitoneal, extending between the sigmoid and the anatomic splenic flexure (*double-headed blue arrows*). Radiological flexures (*green stars*) are intraperitoneal and may extend high depending on the colon's length.

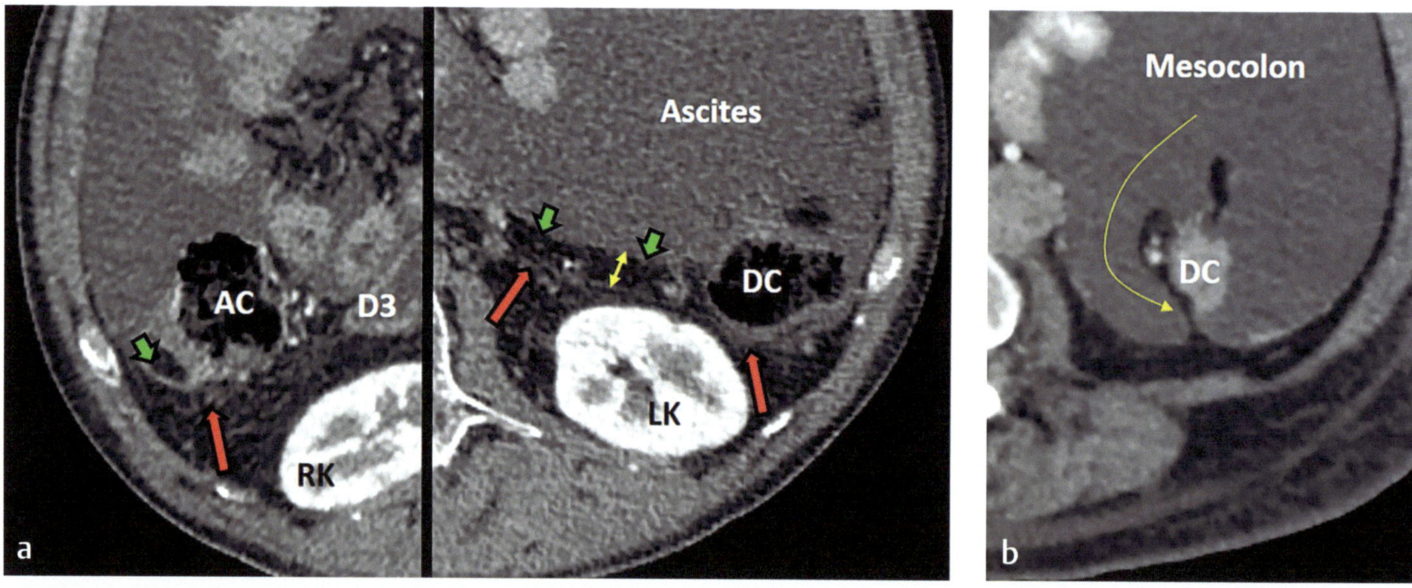

Fig. 5.10 (a, b) Axial computed tomography (CT) images. The posterior aspect of the ascending and descending colon is fixed against the anterior pararenal space of the retroperitoneum. The anterior pararenal space where the pancreas, second duodenum, third duodenum (D3), ascending colon (AC), and descending colon (DC) are located is a small potential space filled with fat (*double-headed yellow arrows*). It is anterior to the anterior renal fascia (*red arrow*). The anterior border of this space is shown by *green arrows*. **(b)** In some cases, due to a short mesentery, a portion of the ascending or descending colon may move freely. Note medially displaced descending colon. Relative position of the ascending and descending colon to the kidneys varies depending on the amount of perirenal fat. LK, left kidney; RK, right kidney.

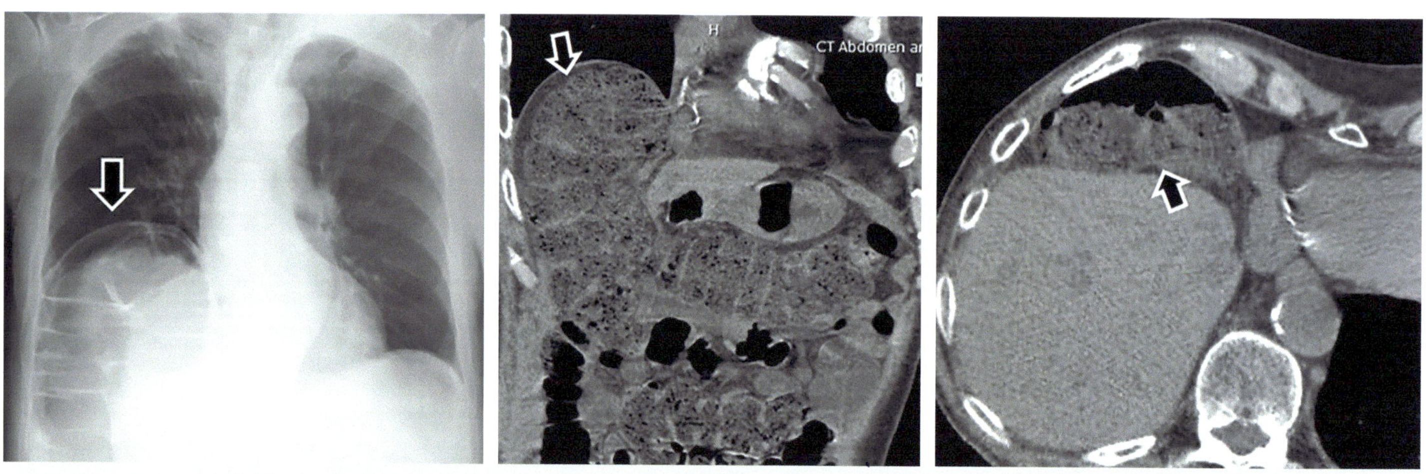

Fig. 5.11 Colon interposition (*arrows*). Chilaiditi's syndrome or hepatodiaphragmatic interposition of the large intestine is a relatively common finding. It is typically asymptomatic but it can present with intermittent mild abdominal pain or constipation and rarely acute intestinal obstruction.

◆ Colon Neurovasculature

Vessels

The cecum, appendix, ascending colon, and proximal two-thirds of the transverse colon receive their blood supply from the superior mesenteric artery (▶Fig. 5.12a). Following the takeoff of the ileal arteries, the superior mesenteric artery continues as the ileocolic artery, which branches off into a superior branch and an inferior branch. The inferior branch divides into anterior and posterior cecal arteries and the appendiceal arteries, which supply the cecum and appendix respectively. The right colic artery, which may arise from the superior mesenteric artery, the ileocolonic artery, or the middle colic artery, supplies the ascending colon and the hepatic flexure. It may be absent in 2 to 18% of specimens. In these cases, the right colon is supplied by colic artery of the ileocolonic artery and the right branch of the middle colic artery. The middle colic artery arises from the superior mesenteric artery near the inferior margin of the pancreas and bifurcates into a right branch, which supplies the hepatic flexure and proximal transverse colon, and a left branch, which supplies the distal transverse colon (▶Fig. 5.12b). Anatomic variation in the origin of the middle colic artery is common and absence of it can occur in a significant percentage of individuals. In the

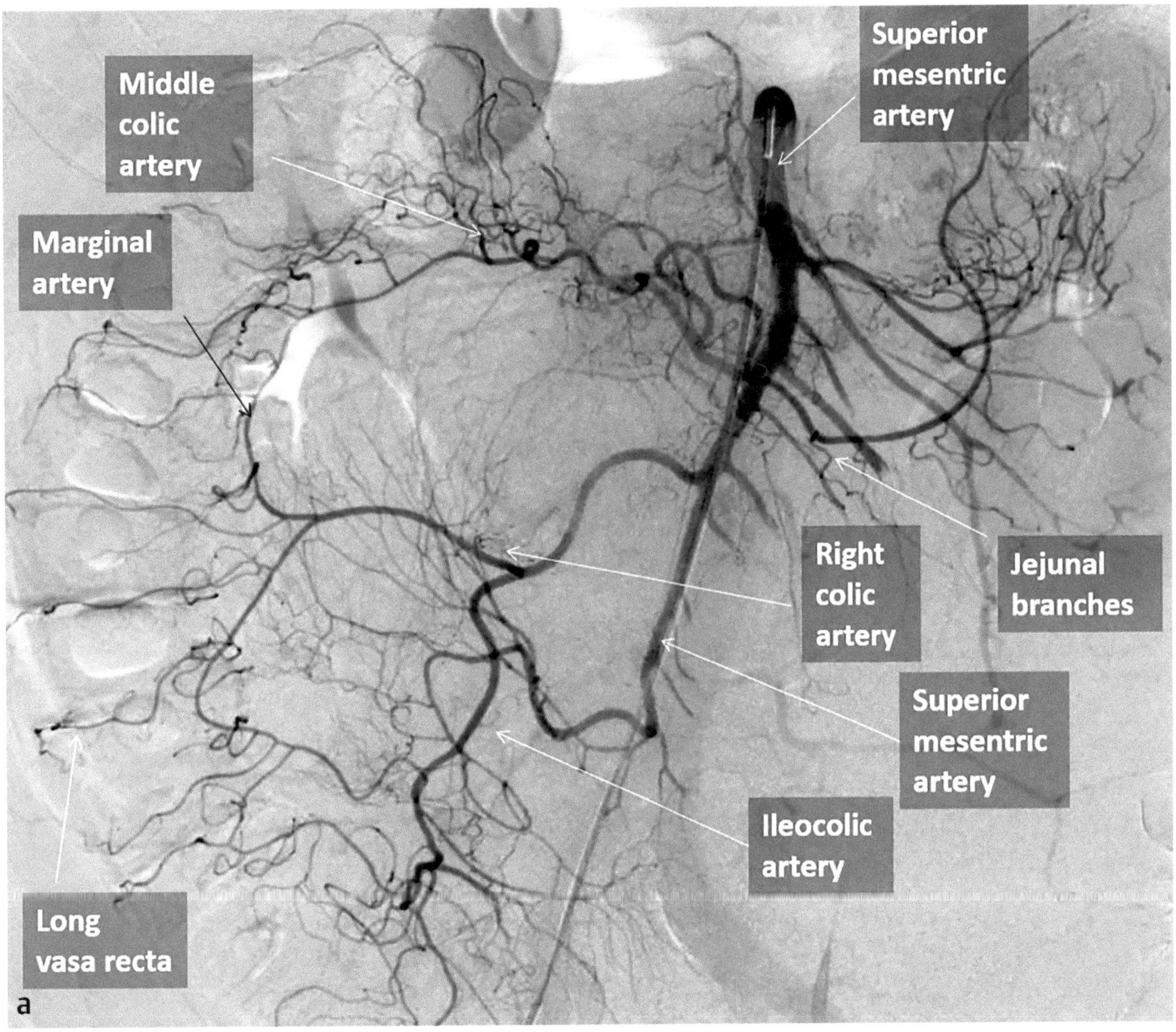

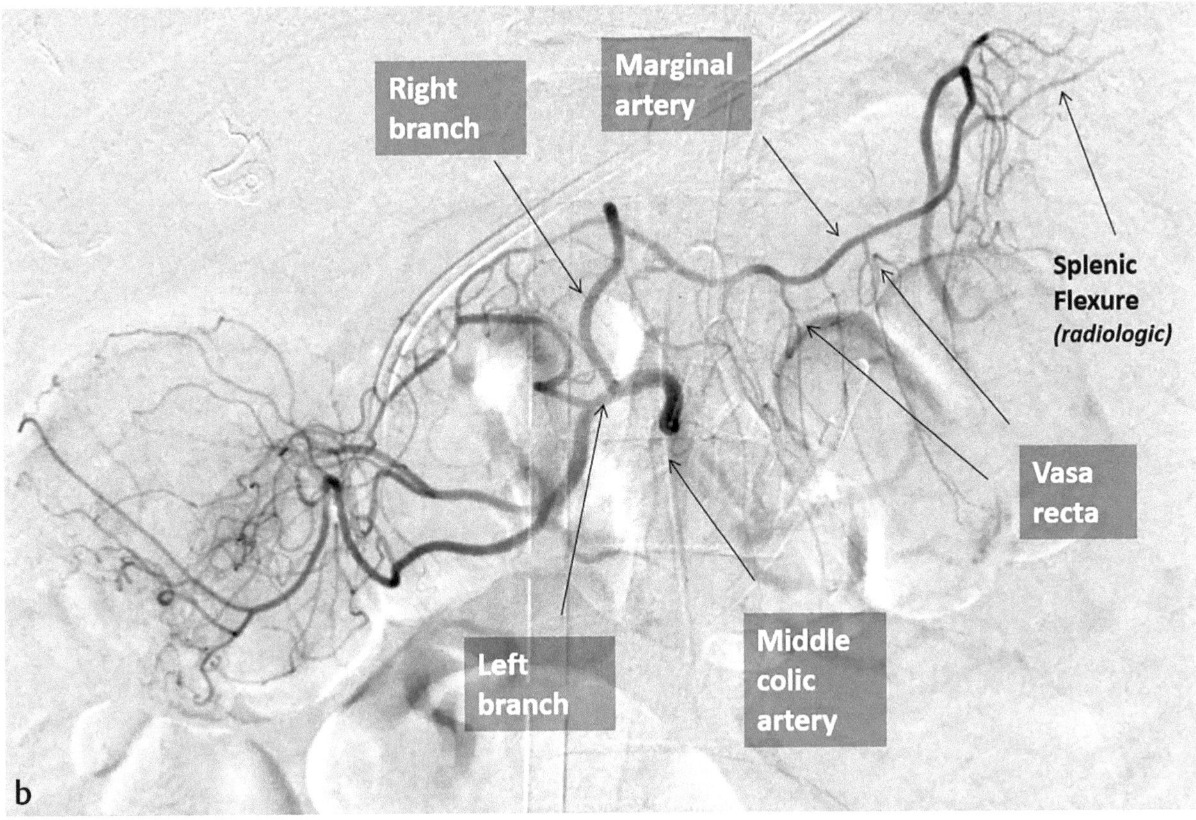

Fig. 5.12 (a) Superior mesenteric artery branches are shown in this selective angiogram. **(b)** Middle colic artery injection showing the extent of the marginal artery on the left to the splenic flexure. The middle colic artery is the first branch of the superior mesenteric artery near the inferior margin of the pancreas and bifurcates into the right and left marginal branches, which supply the transverse colon. The blood supply to the anatomic splenic flexure is provided by the inferior mesenteric artery in 90% of individuals, in the rest it is supplied by the middle colic artery.

absence of the middle colic artery, the right colic artery appears large.

The distal third of the transverse colon, the descending colon, the sigmoid colon, and the majority of the rectum receive their blood supply from the inferior mesenteric artery (▶ Fig. 5.13). The inferior mesenteric artery gives rise to the left colic artery and the sigmoid arteries, which are between two and five in number. The left colic artery divides into an ascending branch, which provides blood supply to the splenic flexure, and a descending branch, which provides blood supply to the descending colon (▶ Fig. 5.13).

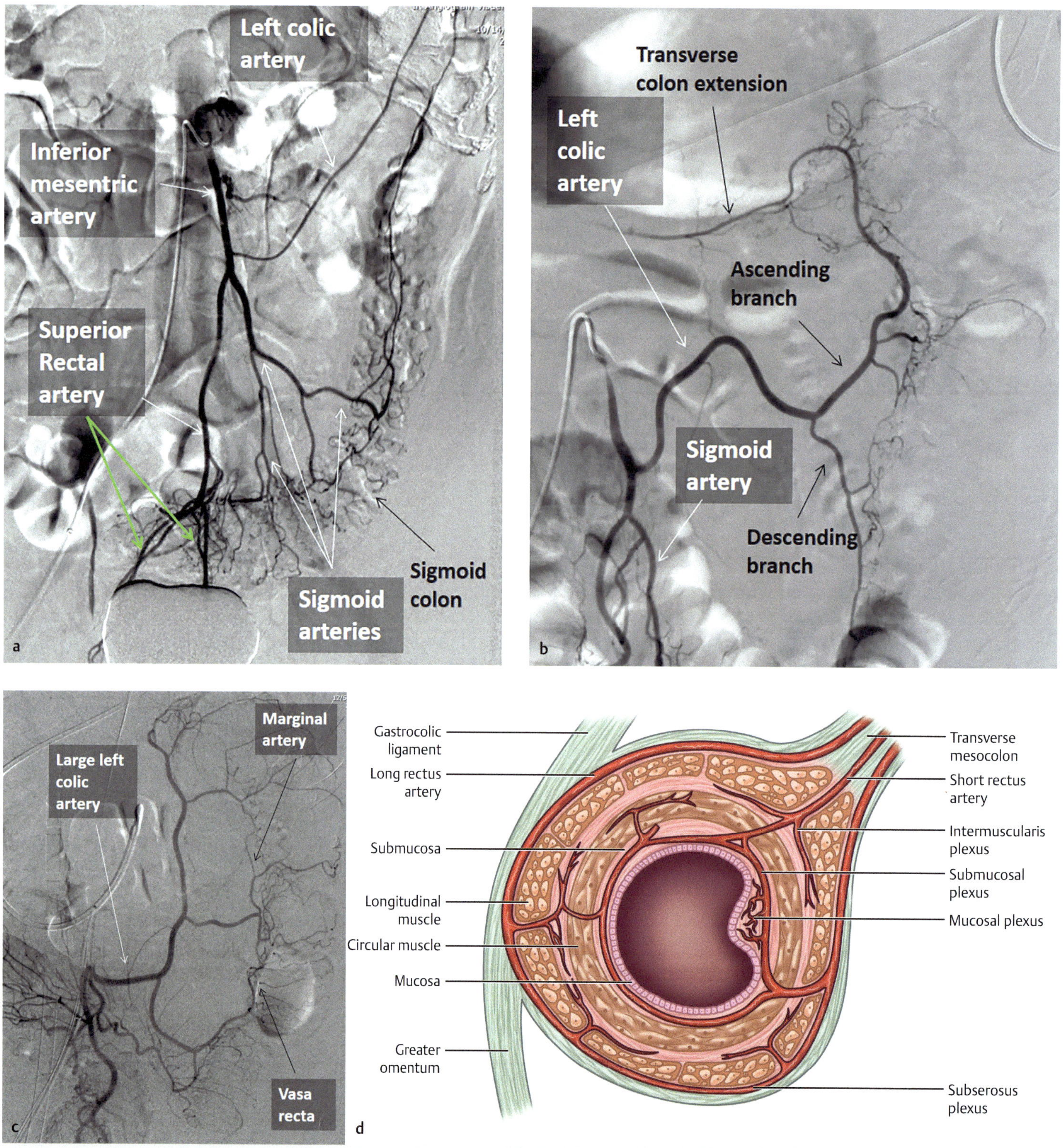

Fig. 5.13 (a–c) Inferior mesenteric artery (IMA). The IMA is smaller than superior mesenteric artery and originates from the anterior aorta 4 cm above its bifurcation and below the lower border of the third part of the duodenum. The anatomic course of the IMA is close to the left ureter and left gonadal vessels. The left colic artery arises as a separate branch or a common trunk with the sigmoid artery. Variation of the left colic artery is common. Typically, the ascending branch of the left colic artery supplies the distal third of the transverse colon and the splenic flexure **(b)**. In **a**, the descending branch is continuous with the sigmoid artery. In **c**, the left colic artery is large. The superior rectal artery with anterior and posterior branches (*green arrows*) are shown in **a**. **(d)** Intramural arterial plexuses in cross section of the transverse colon wall. (Modified with permission from Kachlik et al.[13])

The terminal branches of the superior mesenteric artery and inferior mesenteric artery anastomose to form an arcade along the mesenteric border of the colon named the marginal artery of Drummond. The arc of Riolan, otherwise known as the meandering mesenteric artery, is another collateral blood circulation that connects the left colic branch of the inferior mesenteric artery and the middle colic branch of the superior mesenteric artery. The marginal artery of Drummond and the arc of Riolan are important collateral pathways in the setting of arterial occlusion or significant stenosis. The splenic flexure, which is at the junction of the superior mesenteric artery and inferior mesenteric artery territory, is often referred to as a watershed area given its susceptibility to ischemia in the setting of hypotension. The arterial collateralization in this region is often poor or absent, and as a result, the marginal artery is small or discontinuous.[13]

The microcirculatory bed of the colon plays a central role in the absorptive, secretory, and protective functions of the colon. The vasa recta are branches from the marginal artery of Drummond or from the last peripheral arcades of the colic arteries. There are two groups of vasa recta, short and long, with rich anastomoses between them. The short vasa recta only supply the area bordering the attachment of the mesocolon to the intestinal wall and the taenia mesocolica. The long vasa recta travel in the subserosa and have a characteristic sinusoid course. Once the antimesocolic border is reached, they can take an oblique or parallel turn to supply other taenia before penetrating the intestinal wall or turn back to the mesocolic border.

The long vasa recta finally enter the serosa and form a fine network called the subserous plexus. The subserous plexus sends minor branches to feed the muscular layer of the intestine. The feeding vessels of the long vasa recta reach submucosa to connect to the submucous plexus which is composed of large caliber vessels running parallel to the longitudinal axis of the intestine. Mucosal branches arise from the submucous plexus to form the mucous plexus.[13]

Distally, the venous drainage of the colon directly mirrors its corresponding colic arteries. The superior mesenteric vein drains the cecum, ascending colon, and transverse colon (▶ Fig. 5.14). It then ascends and joins with the splenic vein posterior to the pancreatic neck to become the portal vein. The inferior mesenteric vein drains the descending colon, sigmoid colon, and upper rectum. It then ascends retroperitoneally to empty into the splenic vein posterior to the pancreatic body (▶ Fig. 5.14).[14,15]

Lymphatics

A rich network of intramural lymphatic plexuses is seen within the colonic wall, which drains into an extramural lymphatic system that parallels its arterial supply. The extramural lymphatic system can be divided into four lymph node groups: epiploic lymph nodes, paracolic lymph nodes, intermediate lymph nodes, and main (principal) lymph nodes. The epiploic lymph nodes are found along the colonic serosal surface, deep to the peritoneum, and within the epiploic appendages. The paracolic lymph nodes are located adjacent to the colonic wall, along the marginal arteries and arcades. The intermediate lymph nodes are seen along the primary colic vessels, and the main lymph nodes are seen along the superior and inferior mesenteric vessels. The main lymph nodes drain into the para-aortic nodal chain, which drains into the cisterna chili.[16] Inflammatory and malignant pathologies of the colon usually involve regional lymph nodes first before reaching the main nodes. The lymphatics of the transverse colon also communicate with those of the greater omentum and the hilum of the spleen.

Innervation

The sympathetic and parasympathetic nerves to the colon follow the vascular supply. The sympathetic nerves that innervate the right colon and proximal two-thirds of the transverse colon originate from the 6th to 12th thoracic segments. They then travel within the thoracic splanchnic nerves to the celiac plexus, followed by the superior mesenteric plexus, and enter the colonic wall. The sympathetic nerves that innervate the distal transverse colon and left colon originate from the first three lumbar segments. They then travel to the inferior mesenteric plexus, to the superior and inferior hypogastric plexus, and enter the colonic wall. The sympathetic nerves of the colon are responsible for peristaltic inhibition, colonic wall relaxation, contraction of the ileocecal valve and vascular musculature, and visceral pain sensation.

The parasympathetic nerves that innervate the right colon and proximal two-thirds of the transverse colon originate from the right vagus nerve. They travel within the celiac and superior mesenteric plexus, and into the colonic wall. The parasympathetic nerves that innervate the distal transverse colon and left colon originate from the second to fourth sacral segments. They then travel to the pelvic splanchnic nerves, to the superior and inferior hypogastric plexus, and into the colonic wall. The parasympathetic nerves of the colon are responsible for stimulating peristalsis and relaxation of the internal anal sphincter.[17]

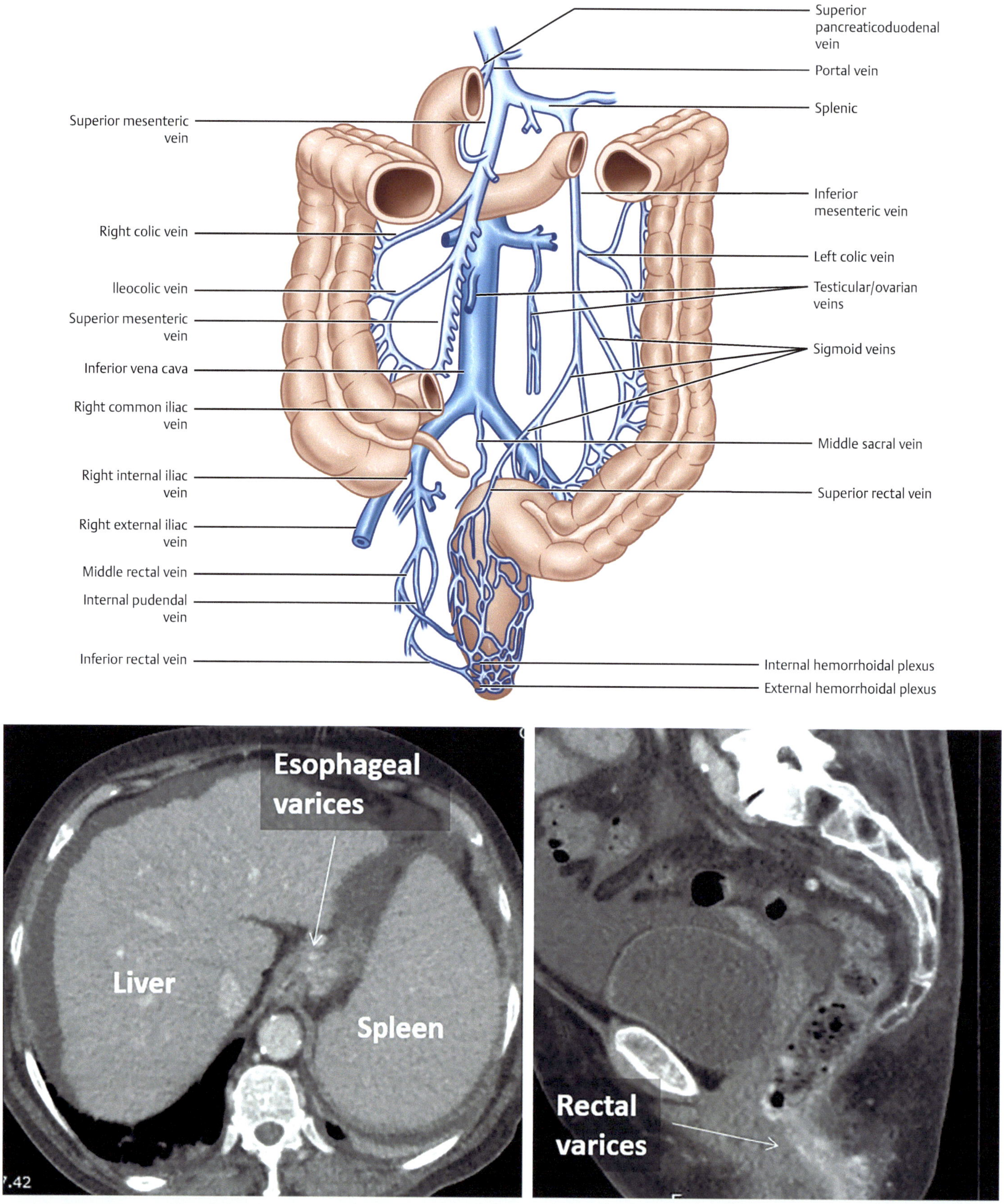

Fig. 5.14 Venous anatomy of the colon.

◆ Rectum and Anal Canal

The anorectum is the final portion of the large intestine and serves a dual purpose: reservoir for feces and controlled expulsion of feces. The rectum is located in the pelvis, along the curvature of the sacrum, and is 12 to 16 cm in length (▶Fig. 5.15). There is controversy in regards to the proximal and distal margins of the rectum. Anatomists consider the rectosigmoid junction to be at the third sacral vertebra, and surgeons consider it to be at the sacral promontory. The distal extent of the rectum is considered to be at the dentate line (2 cm from the anal verge) by anatomists but at the muscular anorectal ring by the surgeons. The rectum is anatomically different from the colon in several aspects. At the rectosigmoid junction, the taeniae coli coalesce into a continuous outer longitudinal muscle layer. Unlike the colon, the rectum lacks epiploic appendices, haustra, and a well-defined mesentery. It is characterized by the presence of two or three endoluminal submucosal folds, which are called valves of Houston. The peritoneum invests the proximal one-third of the rectum anteriorly and laterally and the middle third of the rectum only anteriorly (▶Fig. 5.16). The distal third of the rectum, on the other hand, is entirely extraperitoneal in location. The reflection of the peritoneum usually occurs 6 to 8 cm above the anal verge. Below the peritoneal reflection, the rectum is enveloped by the endopelvic fascia.

The anal canal, which is an entirely extraperitoneal structure, is 2.5 to 5 cm in length. The surgical anal canal extends from the anal verge, which is the junction between the anal epithelium and perianal skin, to the anorectal ring, which is the upper margin of the

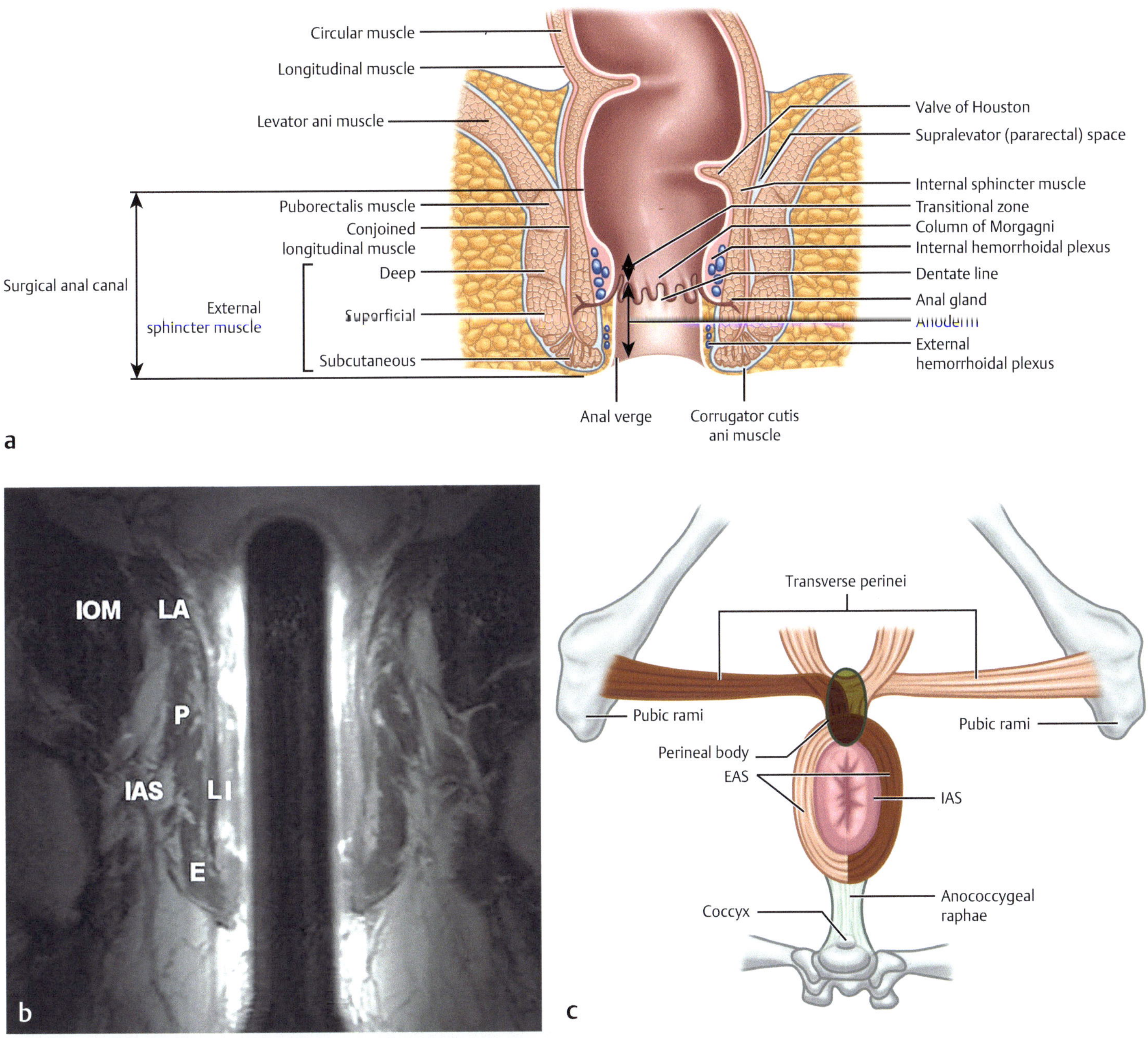

Fig. 5.15 (a) Rectoanal canal and internal and external anal sphincters. **(b)** Coronal oblique, endoanal, T2-weighted magnetic resonance (MR) demonstrates the internal sphincter (I), longitudinal layer (L), external sphincter (E), puborectalis (P), levator ani (LA), ischioanal space (IAS), and internal obturator muscle (IOM). Some subdivision of the external sphincter is visible. (Reproduced with permission from Stoker.[18]) **(c)** Anteroinferior fibers of the external anal sphincter is continuous with transverse perineal muscles at the perineal body. EAS, external anal sphincter; IAS, internal anal sphincter.

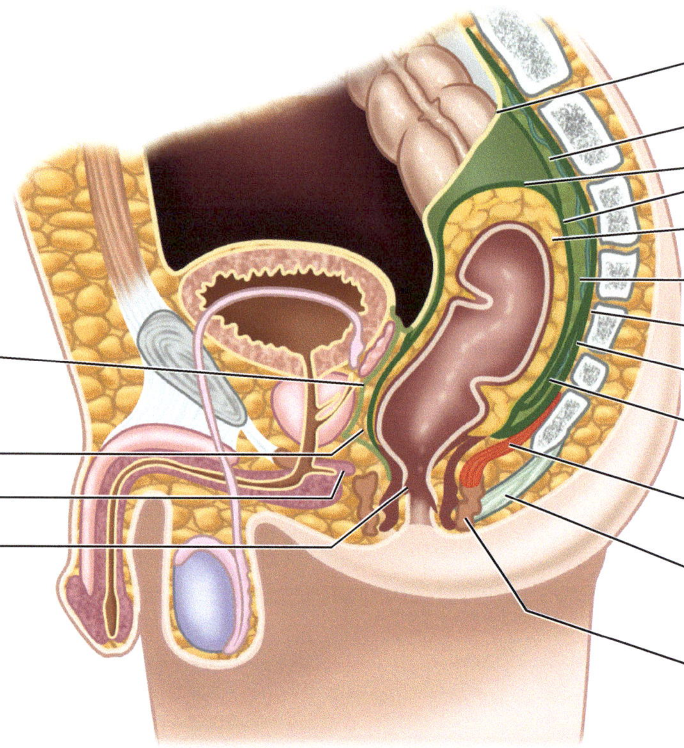

Fig. 5.16 Midsagittal view. Rectum and pelvic fascia are shown. The deep postanal space (space of Courtney) is located above the anococcygeal body (ligament). The visceral layer of the endopelvic fascia (fascia propria of the rectum) lines the rectum, and the parietal layer of the endopelvic fascia (presacral fascia) covers the sacrum. The two layers fuse a few centimeters above the coccyx to form Waldeyer's fascia. Anteriorly, Denonvilliers' fascia is interposed between the bladder and rectum. (Modified with permission from Kaiser and Ortega[19] and Barleben and Mills.[20])

anal sphincter complex. The anatomical anal canal extends from the anal verge to the dentate line, which embryologically marks the transition from endoderm to ectoderm. Just beyond the anal verge is the anus, otherwise known as the anal orifice. The levator ani muscles, external anal sphincter, and internal anal sphincter surround the anal canal. The anal canal consists of three different types of epithelia depending on the level. It begins as a columnar epithelium. Approximately 6 to 12 mm above the dentate line is a transitional zone where the columnar epithelium becomes cuboidal, and then squamous epithelium. The squamous epithelial portion of the anal canal differs from the perianal skin in that it lacks hair follicles, sebaceous glands, and sweat glands. Extending above the dentate line are the columns of Morgagni, which are mucosal folds that parallel the length of the anal canal. The columns of Morgagni are 8 to 14 in number and are surrounded by anal crypts, which drain several anal glands. The anal glands are normally located within the intersphincteric space and traverse the internal anal sphincter, sparing the external anal sphincter (▶ Fig. 5.15).[19,20]

◆ Anorectal Neurovasculature

Vessels

The inferior mesenteric artery continues distally as the superior rectal or hemorrhoidal artery. The superior rectal (hemorrhoidal) artery is a single branch that travels within the sigmoid mesocolon, bifurcates at the S3 level, and then provides blood supply to the rectum, mesorectum, and proximal anus to the dentate line (▶ Fig. 5.17, ▶ Fig. 5.18, ▶ Fig. 5.19). It is responsible for approximately 80% of rectal blood supply. The superior rectal artery gives off two branches, with the right branch supplying the posterior and lateral sides of the rectum and the anterior branch supplying the anterior surface of the rectum. The middle rectal artery has variable course, number, origin, and size, and supplies the distal rectum and proximal anal canal (▶ Fig. 5.18, ▶ Fig. 5.19). It typically arises from the anterior division of the internal iliac artery or rarely the inferior vesicle artery, runs in the extraperitoneal pelvic tissue, and gives off branches to the prostate and seminal vesicles in men, or vagina in female. The inferior rectal artery, which arises from the internal pudendal artery below the levator muscle, supplies the distal rectum, anal canal, and anal sphincters. Anastomoses exist between the superior and inferior rectal artery at the level of the dentate line and between the two inferior rectal arteries of the opposite side, which helps to explain the low incidence of rectal ischemia. However, in dorsocaudal area of the rectal ampulla vessel anastomosis with the opposite side is minimal, which explains higher rate of anastomotic leak after surgery in this watershed area. The median sacral artery arises from back of the aorta above its bifurcation and gives off branches to the posterior rectum.

The venous drainage of the rectum mirrors the arterial anatomy. The superior rectal vein drains the majority of the rectum

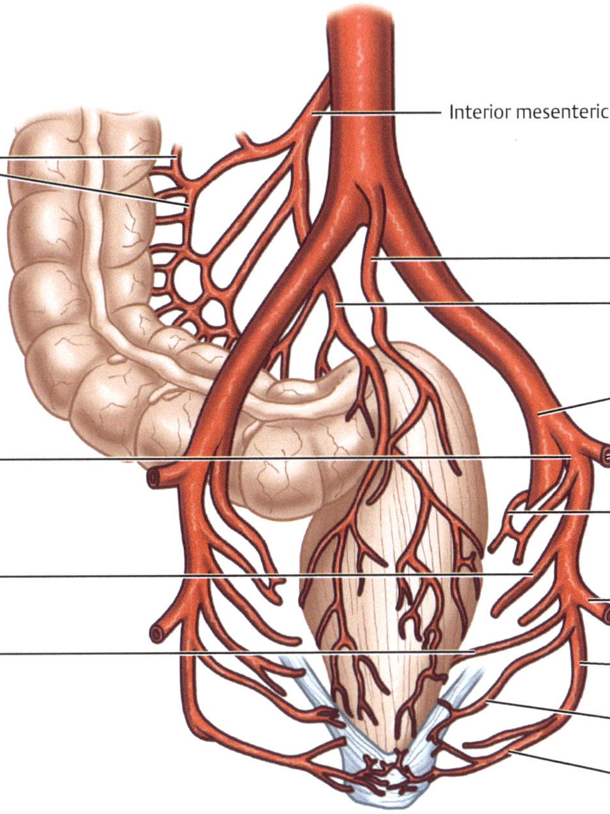

Fig. 5.17 Rectal arteries (posterior view). A single superior rectal artery divides into the anterior and posterior branches and supply major portion of the rectum. They communicate with branches from the inferior rectal and middle rectal arteries of both sides, leaving a potentially watershed area in the dorsocaudal aspect of the rectal ampulla. The middle sacral artery may give small branches to the superior rectum.

and empties into the portal circulation via the inferior mesenteric vein. The middle and inferior rectal veins drain the distal rectum and anal canal and empty into the internal iliac vein (systemic venous circulation) directly or via the internal pudendal vein.[18,21] This pattern of drainage explains common occurrence of pulmonary metastasis (systemic circulation) without involvement of the liver (portal venous system) in cancer of the anorectal region.

Lymphatics

The lymphatic drainage of the rectum mirrors the arterial anatomy (Chapter 21 "Lymphatics of the Abdomen, Pelvis, and Perineum"). The proximal two-thirds of the rectum drains solely to the inferior mesenteric lymph nodes and then to the para-aortic lymph nodes. The distal third of the rectum not only drains to the inferior mesenteric and then to the para-aortic lymph nodes; it can also drain along the middle rectal artery to the internal iliac lymph nodes. Above the dentate line (mucocutaneous junction), the anal canal drains to the inferior mesenteric lymph nodes and then to the para-aortic lymph nodes. Below the dentate line, the anal canal drains to the superficial inguinal lymph nodes and then to the superficial inguinal nodes. Regional lymph node enlargement in the perirectal space is a common finding in inflammatory and neoplastic diseases of the rectum. The external hemorrhoidal venous plexus is located in the perianal space (▸Fig. 5.6) and communicates with the internal hemorrhoidal plexus.

Innervation

The rectum and upper anal canal are innervated by the sympathetic and parasympathetic nerves. The external anal sphincter and the lining of the anal canal are supplied by somatic nerves. The sympathetic nerves that innervate the rectum arise from the first three lumbar segments. They travel along the inferior mesenteric and superior rectal arteries to form the inferior mesenteric plexus and superior hypogastric plexus. The right and left hypogastric nerves then continue caudally to supply the rectum. The parasympathetic nerves that innervate the rectum and upper anal canal arise from the second through fourth sacral roots and then combine with the hypogastric nerve to form the pelvic plexi before supplying the rectum and proximal anal canal.[22] The distal anal canal, below the dentate line, is innervated by somatic nerves. These somatic nerves are inferior rectal and perineal branches of the pudendal nerve and provide the sensations of pressure, pain, touch, and temperature to this region.

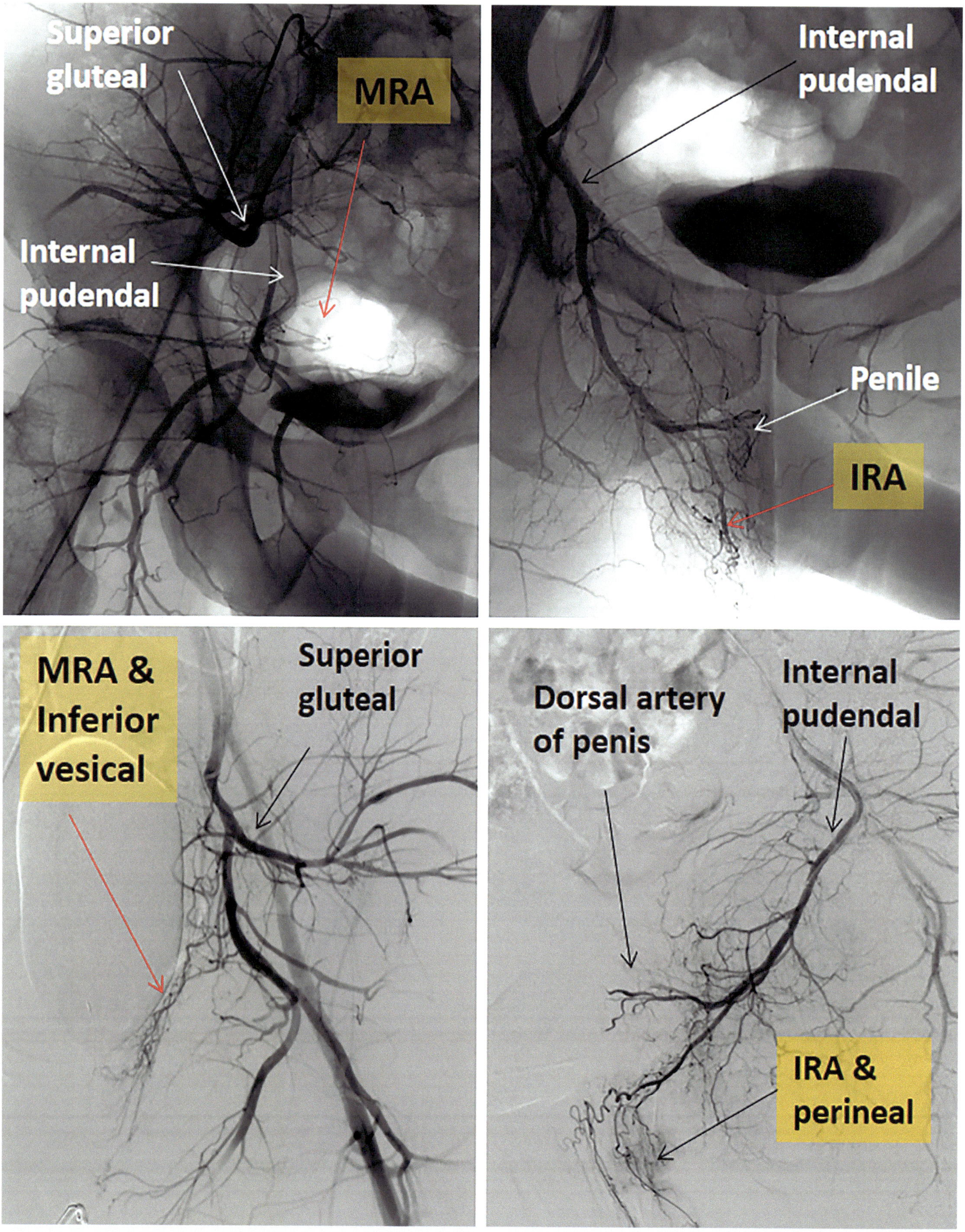

Fig. 5.18 Catheter angiograms of the internal iliac artery and selective internal pudendal artery injections showing the middle rectal artery (MRA) arising from the internal iliac artery (IIC) and the inferior rectal artery (IRA) arising from the internal pudendal artery.

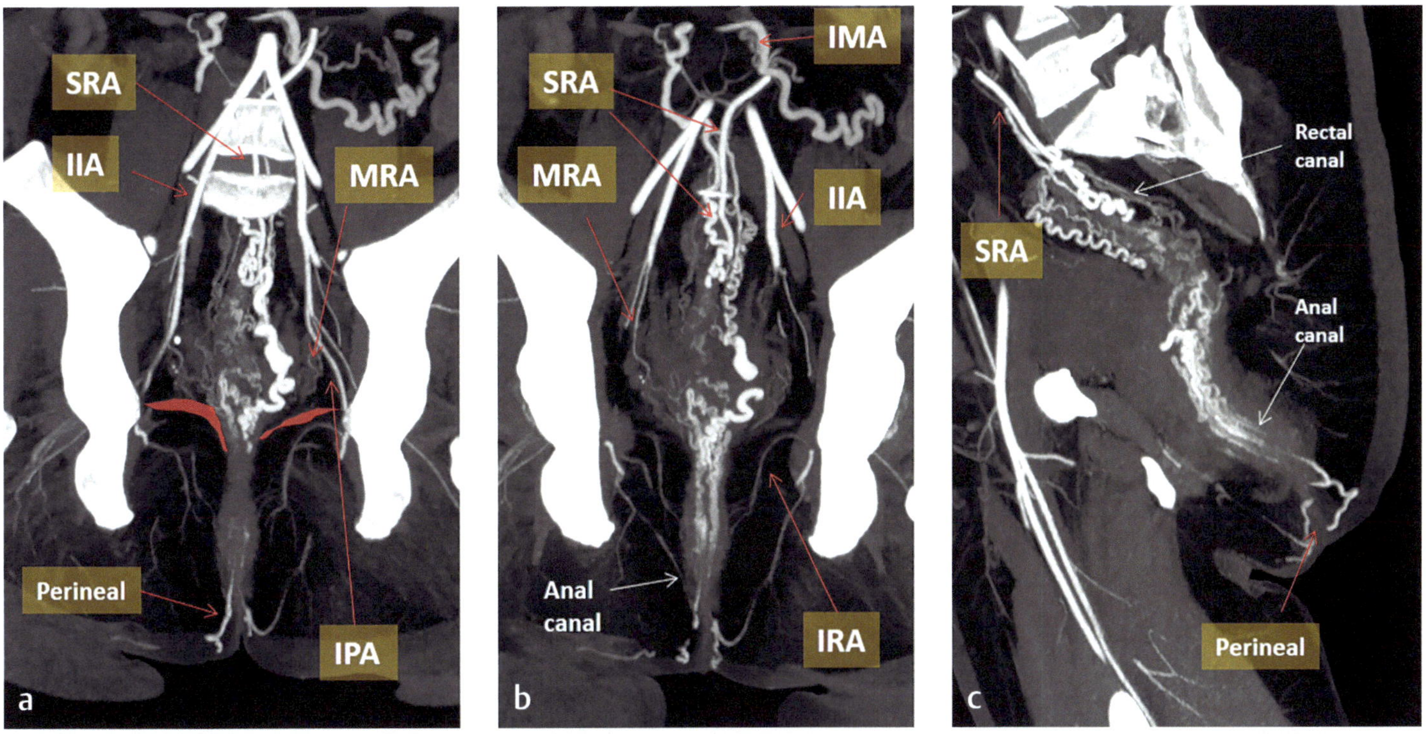

Fig. 5.19 Rectal artery branches in a patient with enlarged inferior mesenteric artery (IMA) and superior rectal artery (SRA). **(a, b)** Coronal and **(c)** sagittal computed tomography (CT) images showing extensive submucosal collateral vessels in the wall of the rectum and anus. Middle rectal artery (MRA) is shown arising from the internal iliac artery (IIA). The inferior rectal artery (IRA) is shown arising from the internal pudendal artery (IPA) below the levator muscle (*in red*).

◆ Sphincters

Internal Anal Sphincter

The internal anal sphincter consists of condensed, smooth muscle bundles that are the caudal extension of the rectal inner circular muscles (▸Fig. 5.15). It is 5- to 8-mm thick and 2.5 to 4.0 cm in length, extending from the anorectal ring to approximately 1 cm above the lower margin of the anal sphincter complex. Therefore, the lowermost portion of the anal sphincter complex consists of only the external anal sph incter. Superiorly, the internal anal sphincter is wrapped by the levator ani muscle. The internal anal sphincter encircles the anal canal and is in a constant state of contraction, accounting for approximately 50 to 85% resting anal tone.

Conjoined Longitudinal Muscle

The conjoined longitudinal muscle is located within the intersphincteric space, which is a thin, fat-containing space between the internal and external anal sphincters, and consists of diverging, smooth muscle bundles of the rectal outer longitudinal muscles combined with the skeletal muscles of the pelvic floor, in particular the puboanalis muscle (▸Fig. 5.15, ▸Fig. 5.20). Proximally, the muscle is predominantly muscular but becomes increasingly more fibroelastic distally. These fibroelastic fibers extend to the perianal skin to form the corrugator cutis ani (▸Fig. 5.15). The function of the conjoined longitudinal muscle is not entirely clear. It may serve to attach the anal canal to the pelvis and act as a framework to support and hold the internal and external anal sphincters together.

External Anal Sphincter

The external anal sphincter is an elliptical cylinder of striated muscles that surround the intersphincteric space and consist predominantly of slow twitch muscle fibers, enabling to maintain prolonged tonic contraction. There are also fast twitch muscles responsible for rapid contraction during voluntary squeeze. The external anal sphincter is 2.7 cm in length and may be shorter anteriorly specially in women, approximately 1.5 cm. The thickness measures 4 mm on magnetic resonance (MR) or computed tomography (CT) imaging (▸Fig. 5.20, ▸Fig. 5.21). The sphincter is separated into three portions: subcutaneous, superficial, and deep (▸Fig. 5.15, ▸Fig. 5.20, ▸Fig. 5.21). The subcutaneous portion extends 1 cm below the inferior margin of the internal anal sphincter and encompasses the anal orifice. The superficial portion encircles the lower internal anal sphincter. The deep portion has a sling-shaped morphology and forms the more cranial internal anal sphincter. It is inseparable from the puborectalis muscle.[23,24] That is why some believe that the external anal sphincter is a two-part structure; the deep part is most likely the puborectalis muscle.[23] The puborectalis muscle which is superior to external anal sphincter forms the top of the anorectal ring and can be palpated digitally. Voluntary contraction of the proximal and

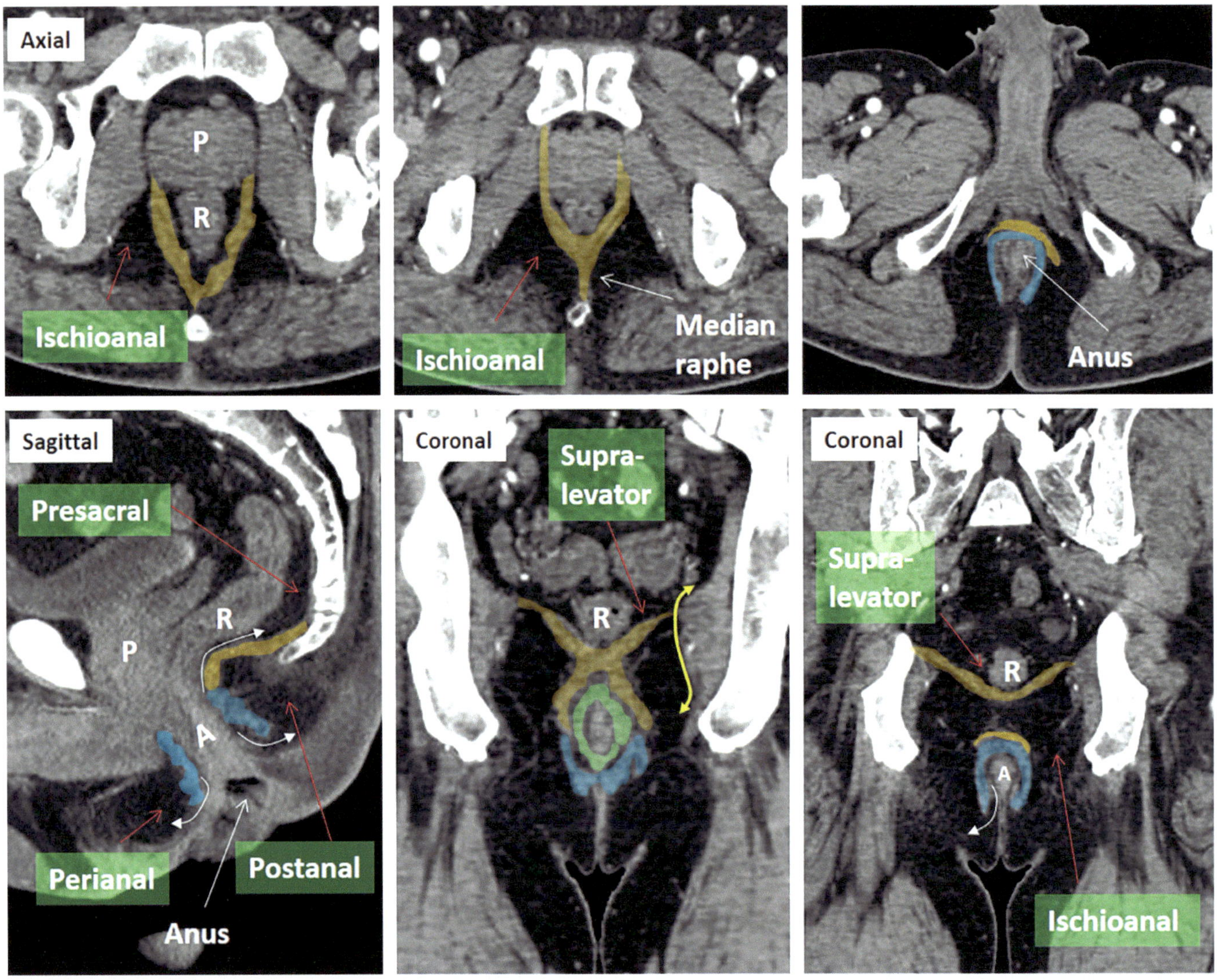

Fig. 5.20 Computed tomography (CT) images in different projections at the lower pelvis showing the spaces around the rectal (R) and anal (A) canals. Levator muscle (*yellow*), and external (*blue*) and internal (*green*) anal sphincters are shown in color. The intersphincteric space is shown between the two anal sphincters as a thin line of fat that communicates with the perianal space and supralevator space (*curved white arrows*). Supralevator space may communicate with the ischioanal space through the obturator fasciae (*yellow arrow*). Note, the most inferior part of the external sphincter does not have a continuous ring. P, prostate.

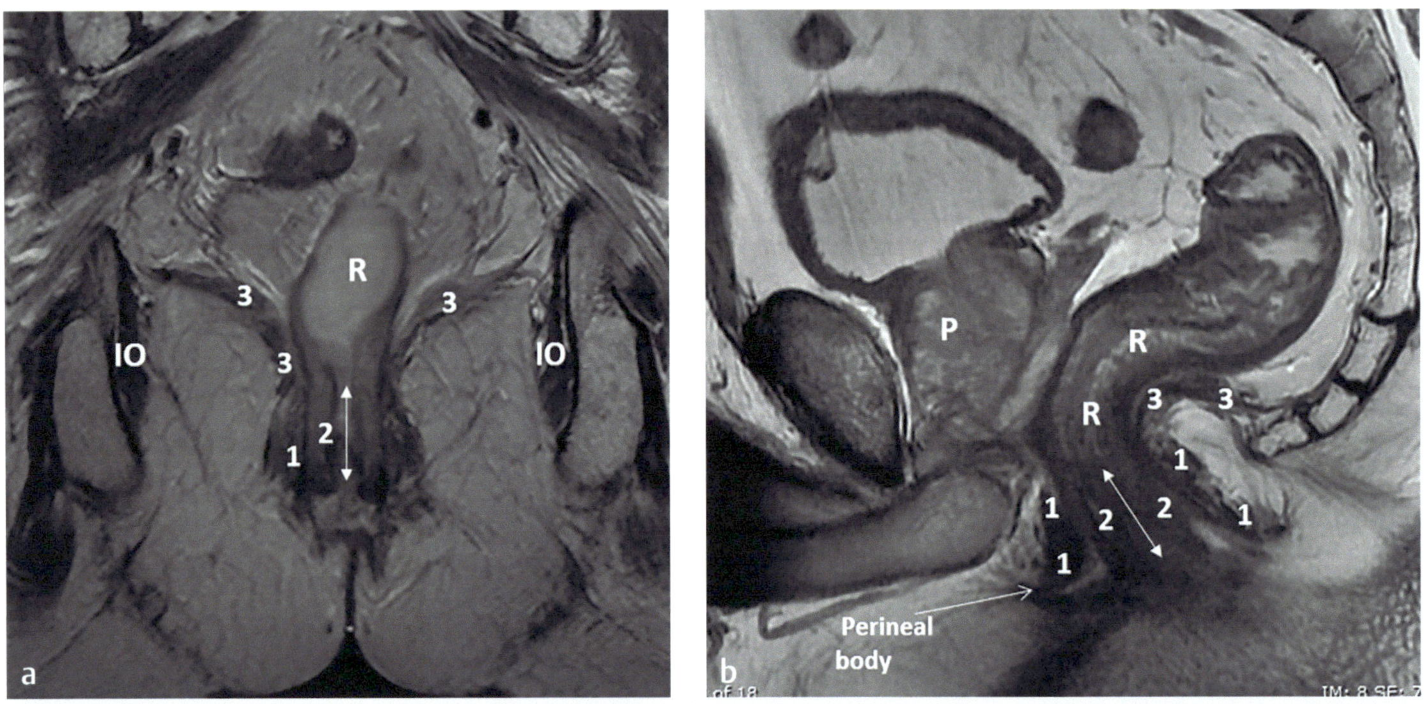

Fig. 5.21 Normal perianal anatomy on coronal **(a)** and sagittal **(b)** magnetic resonance imaging (MRI). 1, external anal sphincter; 2, internal anal sphincter/rectal wall; 3, levator ani; IO, internal obturator muscle; *double-headed arrow*, anal canal; P, prostate; R, rectum.

distal halves of the anal canal is more related to contraction of the puborectalis muscle and external anal sphincter respectively.[25] The anococcygeal body/ligament is a complex musculotendinous structure situated between the anterior aspect of the coccyx and the posterior wall of the anus. It is formed by the posterior extension of the superficial fibers of the external anal sphincter. Superior to that in midline is the median tendinous raphe of the levator muscle. The anterior margin of the external anal sphincter is attached to the perineal body and its fibers cross the perineal body and continue in the contralateral transverse perineal and bulbospongiosus muscles forming a "purse-string" morphology as recently suggested[25] (▶Fig. 5.14c). The external anal sphincter may be injured in up to one-third of women during the vaginal delivery and episiotomy that can cause anal incontinence.[25]

◆ Pelvic Floor Musculature

The pelvic floor musculature serves two major functions: provides support for the pelvic organs and aids in coordinating micturition, defecation, and parturition. It is predominantly comprised of the levator ani muscle or pelvic diaphragm (Chapter 1 "Abdominopelvic Wall"). The levator ani muscle is a broad, thin sheet of paired striated muscles that consist of the iliococcygeus, pubococcygeus, and puborectalis muscles. Two midline ventral defects are present in the levator ani muscle, the urogenital hiatus and anal hiatus, which allow passage of the urethra/vagina and the anal canal through each respectively. The iliococcygeus muscle originates from the ischial spine and the posterior tendinous arc formed from the condensation of the obturator fascia and inserts into the coccyx and last two sacral vertebrae. The right and left iliococcygeus muscles fuse at midline, contributing to the anococcygeal raphe, which extends from the coccyx to the anorectal junction. The pubococcygeus muscle originates from the anterior tendinous arc and the dorsal aspect of the pubic bone and insert into the coccyx and anococcygeal raphe. The puborectalis muscle, the most prominent muscle of the levator ani muscle, originates from the dorsal aspect of the pubic bone, where it is inseparable from the pubococcygeus muscle. It then fuses with the contralateral puborectalis muscle to form a U-shaped sling around the rectum, at the anorectal ring, and is thought to be responsible for creating the anorectal angle (▶Fig. 5.21). The puborectalis muscle is in a perpetual state of contraction that plays a major role in fecal continence. When the puborectalis muscle relaxes, the anorectal angle widens, allowing for fecal evacuation.[26]

Perianal and Perirectal Spaces

The rectum and anal canal are surrounded by spaces that normally contain loose areolar tissue or fat. The pelvis is invested by the endopelvic fascia (▶Fig. 5.16). The parietal layer covers the bones and muscles and the visceral layer covers the viscera including the rectum. The rectal (visceral) fascia is condensation fascia that surrounds the rectum and forms the perirectal space containing fibrofatty mesorectum (▶Fig. 5.22). Outside the rectal fascia is the pararectal space which contains loose areolar fatty tissue. The posterior part of the pararectal space is the presacral space and the lateral pararectal space is continuous with the paravesical space. The levator ani divides the pelvis into two separate spaces, namely, the supralevator space and the extrapelvic infralevator space. The supralevator space which is located between the levator and peritoneal reflection is continuous with the intersphincteric space (▶Fig. 5.20, ▶Fig. 5.22). The intersphincteric space is of significance because most of the anal glands are found in this location and can be the source of abscess formation (▶Fig. 5.15). The supralevator space can also communicate with the ischioanal space via the fascial planes such as the obturator internus fascia and through the pudendal canal. These connections are potential pathways for spread of infection. The ischioanal (formerly ischiorectal fossa) space is a triangular space bordered medially by the external sphincter, superiorly by the levator muscle, laterally by the obturator fascia, and anteriorly by the transverse perineal muscle. Inferiorly, a thin transverse fascia separates the ischioanal fossa from the perianal space. The postanal space is the space behind the rectum and below the levator (▶Fig. 5.16). It is continuous with the each ischioanal space anterolaterally. The postanal space is divided by the anococcygeal ligament (body) into superficial and deep compartments (▶Fig. 5.23).

◆ Postoperative Colon

Colon surgery is a common procedure for several benign and malignant colon diseases. Surgical treatment of colorectal disease requires resection of a segment or entire colon and rectum with or without colostomy.[27–29]

Several standardized types of colonic resections are available in the clinical practice (▶Fig. 5.24):

- Segmentectomy: Most commonly involves the sigmoid. The intestinal continuity is established with an immediate or delayed end-to-end or end-to-side anastomosis. In some cases (i.e., perforated colon in diverticulitis), after sigmoid resection, it may be necessary to postpone the anastomosis. In these cases, a temporary colostomy is created and the distal rectal pouch is closed off (Hartmann's pouch).
- Ileocecal resection: Resection of the terminal ileum and cecum, with an ileocolic anastomosis, for example, Crohn's ileitis.
- Right hemicolectomy: Resection of the terminal ileum, cecum, ascending colon, and proximal transverse colon with an ileo-transverse colon anastomosis.
- Left hemicolectomy: Resection of the splenic flexure, descending colon, and proximal sigmoid with colosigmoid anastomosis.
- Total colectomy or proctocolectomy followed by ileostomy or more commonly ileoanal anastomosis, after establishment of an ileal pouch (ileal pouch anal anastomosis or IPAA). It is indicated for diffuse colorectal diseases, such as ulcerative colitis and familial polyposis. In IPAA, a large pouch is created from the distal end of ileum by anastomosing two ileal loops and removing the wall between them. The pouch will be

5

5

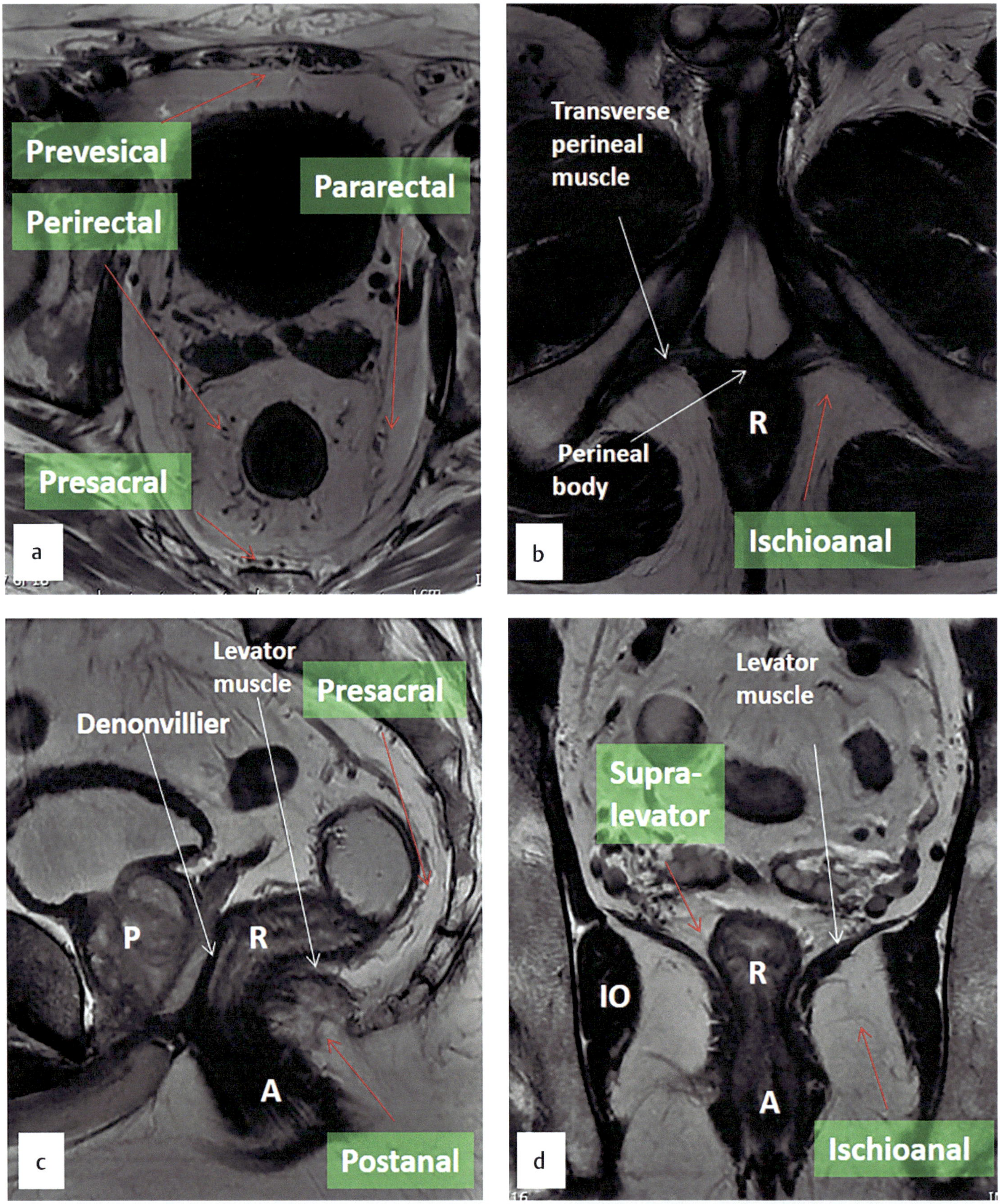

Fig. 5.22 Pelvic spaces. **(a, b)** Axial, **(c)** sagittal, and **(d)** coronal magnetic resonance (MR) images. **(a)** The perirectal space separated from the pararectal and presacral spaces by the rectal fascia. The ischioanal (also known as ischiorectal fossa) space is a triangular space bordered medially by the external sphincter/levator muscle, superiorly by the levator muscle, laterally by the internal obturator muscle, and anteriorly by the urogenital diaphragm as well as the transverse perineal muscle. Supralevator space can communicate with the ischioanal space through the obturator fasciae. A, anal canal; IO, internal obturator muscle; P, prostate; R, rectum.

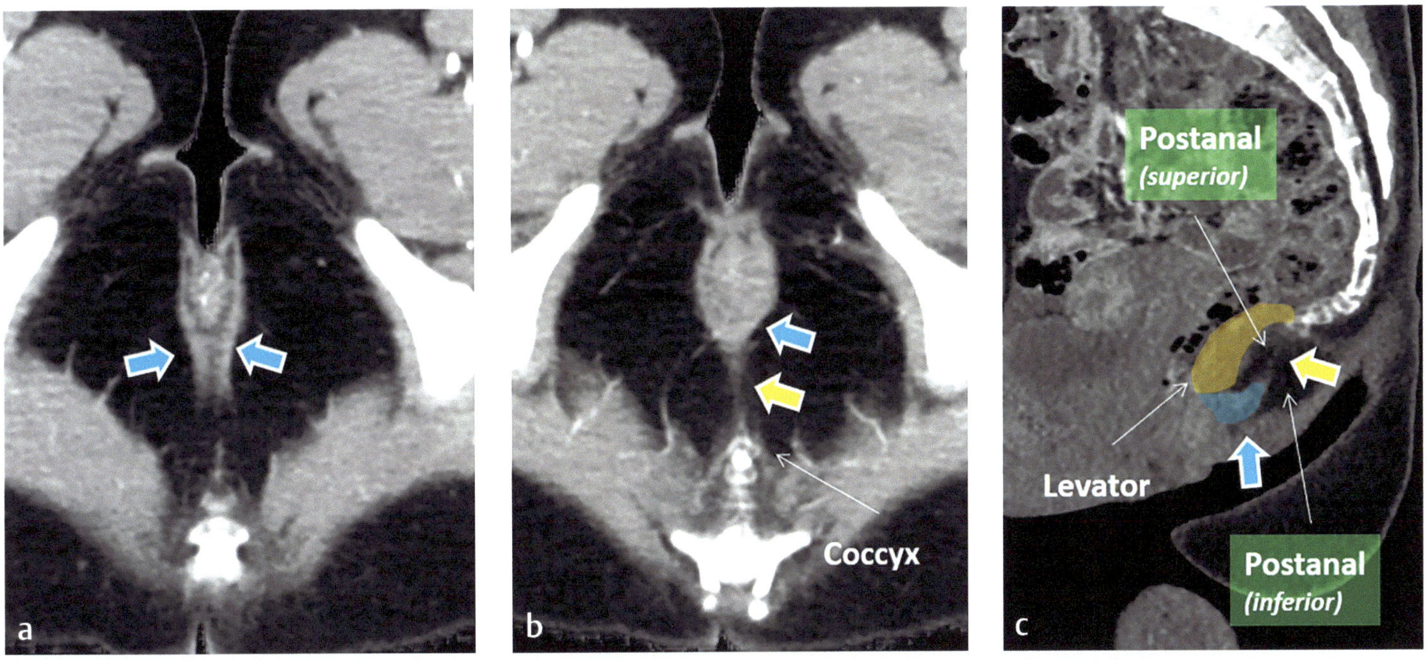

Fig. 5.23 Postanal space. **(a, b)** Axial and **(c)** sagittal computed tomography (CT) images. The postanal space is divided by the anococcygeal ligament (*yellow arrow*) into the superficial and deep compartments (image **c**). The posterior extension of the superficial fibers of the external anal sphincter (*blue arrows*) is shown in *blue color* **(c)**. The levator muscle is shown in *orange color*. The anococcygeal body/ligament is a complex musculotendinous structure extending between the anterior aspect of the coccyx and the posterior margin of the external anal sphincter. It is formed by the posterior extension of the superficial fibers of the external anal sphincter. Superior to that in midline is the median tendinous raphe of the levator muscle. These structures are seen in some high-quality CT scans.

anastomosed to the anus. A temporary loop ileostomy is made which is usually reversed 6 to 12 weeks later. Multiple configurations exist for the ileal pouch, with the J pouch being the most common.

- Anterior resection of rectum: Resection of distal portion of left colon, sigmoid, and distal rectum with a low rectal anastomosis. It is used for lesions of the rectosigmoid junction and proximal rectum over 5 cm from the anal border.
- Abdominoperineal resection: Resection of distal descending colon, sigmoid, rectum, and entire anus followed by a permanent colostomy. It is performed for lesions of lower 5 cm of rectum.

In some colon operations, a temporary or permanent colostomy is created. In this procedure, the colon is brought through the abdominal wall and stitched to the skin. Three types of colostomies exist: (1) end colostomy, (2) loop colostomy, and (3) double barrel colostomy (▶Fig. 5.25). In loop colostomy, the colon is sutured to the abdominal wall from its side; therefore, two openings are created in one stoma. In double barrel colostomy both ends are brought out onto the abdomen but only the proximal stoma is functioning.

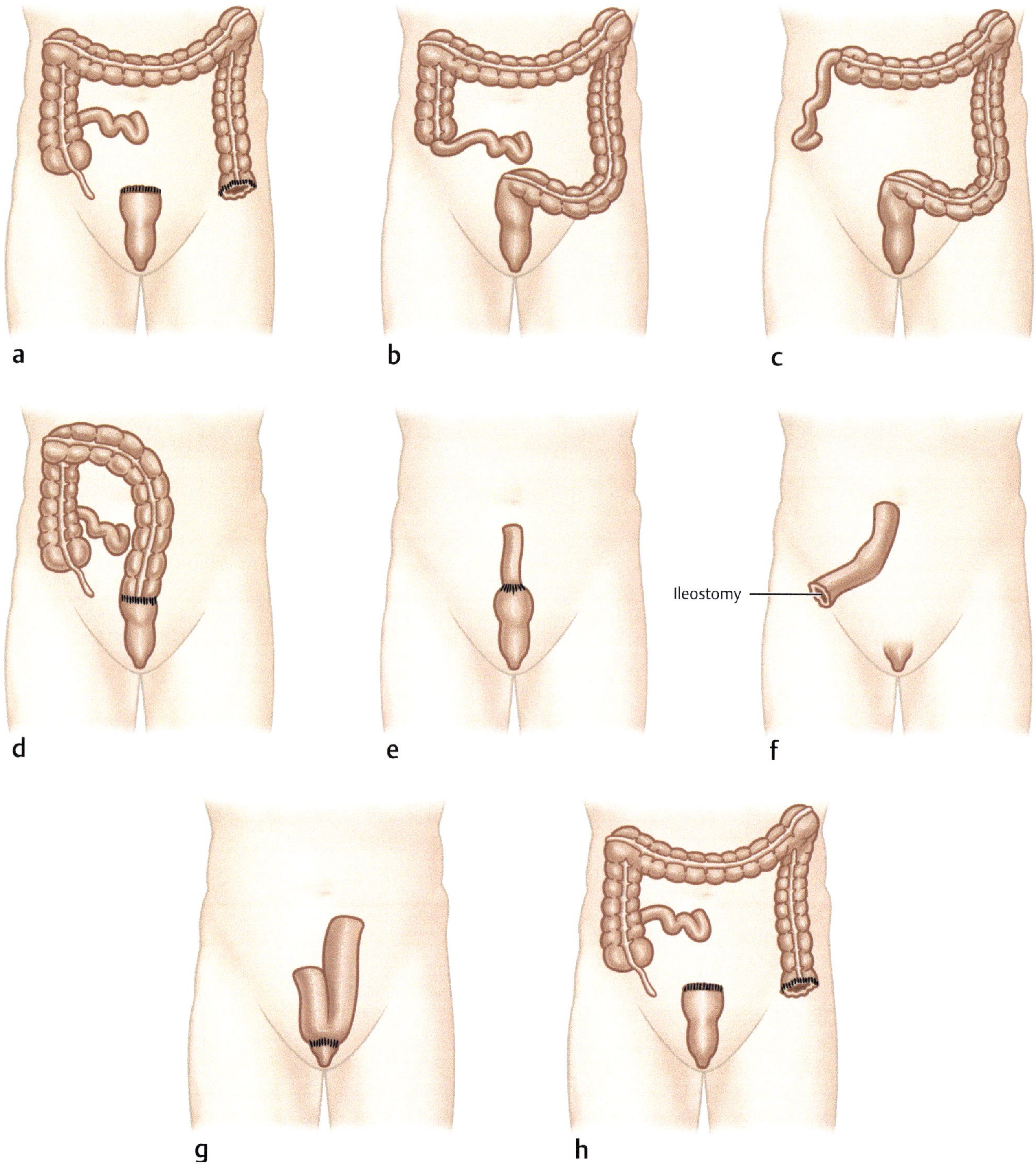

Fig. 5.24 (a) Normal colon. Different types of colonic resection. **(a)** Hartmann's procedure, **(b)** ileocecal resection, **(c)** right hemicolectomy, **(d)** left hemicolectomy, **(e)** total colectomy, **(f)** total proctocolectomy with ileostomy, **(g)** ileal pouch anal anastomosis (IPAA), and **(h)** abdominoperineal resection.

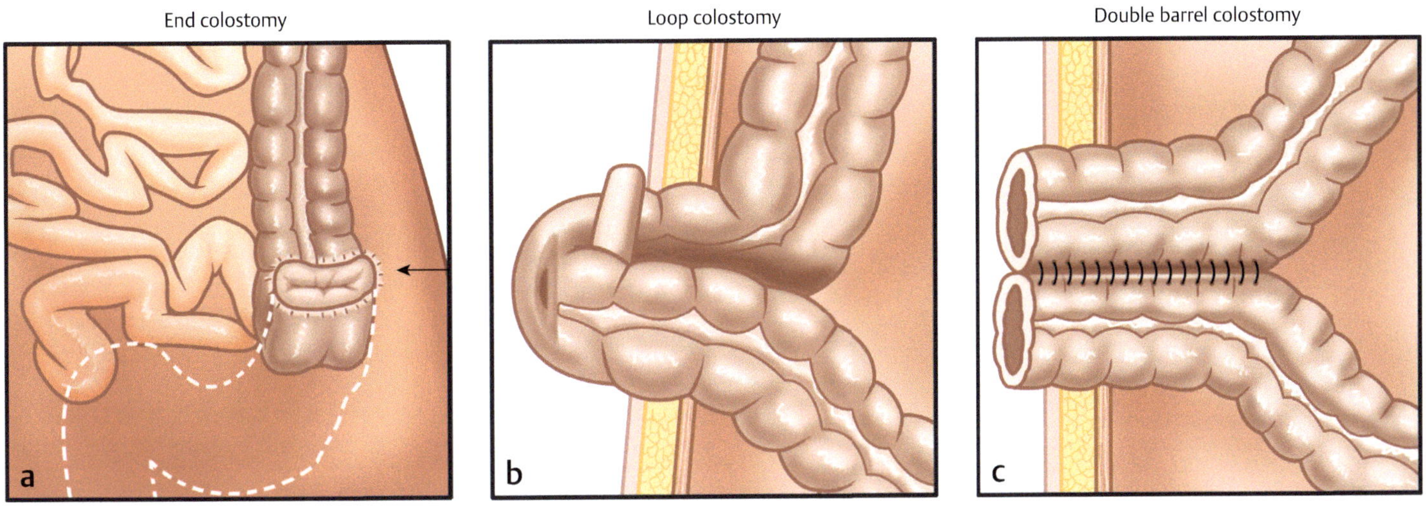

Fig. 5.25 Types of colostomies: **(a)** end colostomy, **(b)** loop colostomy, and **(c)** double barrel colostomy.

◆ Common Diseases

Congenital Abnormalities

Failure of the midgut to return to the peritoneal cavity between the 10th to 12th weeks of gestation may result in large umbilical herniation of the bowel and other intra-abdominal viscera known as omphalocele. The abdominal wall defect is variable in size, ranging from 2 to 10 cm.[30] Hirschsprung's disease is a developmental motor disorder of the large bowel characterized by incomplete craniocaudal migration of neural crest cells during the fourth to seventh weeks of gestation, resulting in absence of ganglion cells in the myenteric and submucosal plexuses. The anus is regularly involved, and the aganglionosis extends proximally for a variable distance. Due to aganglionosis, the diseased segment of colon fails to relax resulting in a functional distal bowel obstruction, chronic constipation, and failure to thrive.[31]

Anorectal malformations consist of a wide spectrum of developmental defects involving the lower intestinal and urogenital tracts. They are believed to occur due to the incomplete caudal descent of the urogenital septum, and range from an isolated imperforate anus to a persistent cloaca. The malformations are classified as high or low anorectal malformations based on the relation of the terminal colon and the levator sling and also based on the presence or absence of fistulas. The terminal colon is located above the levator sling in patients with high lesions and is located below the levator sling in patients with low lesion.[32]

Acquired Pathologies

Colorectal Cancer

Adenocarcinoma of the colon and rectum is the third most commonly diagnosed cancer and the fourth leading cause of cancer-related deaths in the world. The distribution of colorectal cancers is as follows: 30% ascending colon, 10% transverse colon, 15% descending colon, 25% sigmoid colon, and 20% rectum. Patients with colorectal cancer are often asymptomatic and are diagnosed during routine screening. Diagnosis of colorectal cancer is made during colonoscopy with tissue biopsy, but CT and MRI are commonly used for staging (▶ Fig. 5.26). The most common metastatic sites are regional lymph nodes, liver, lung, and peritoneum.[33]

Appendicitis

Appendicitis is one of the most common indications for emergent abdominal surgery. The lifetime incidence of appendicitis is 8.6% in men and 6.7% in women. It is believed that appendicitis is secondary to appendiceal obstruction by fecaliths, lymphoid followed by bacterial overgrowth, secondary infection, ischemia, perforation, and regional abscess.[34] The classic presentation of appendicitis is right lower quadrant pain, nausea, vomiting, and fever. In retroperitoneal variant of the retrocecal appendix, the usual signs or symptoms of appendicitis due to peritoneal irritation may not be present. The diagnosis of appendicitis is based on a combination of physical examination, laboratory findings, and radiologic imaging (▶ Fig. 5.27). CT is a great modality for diagnosis. In children ultrasound and in pregnant women MRI are commonly used for diagnosis. Inflammation of retrocecal intraperitoneal appendix may trigger inflammatory process in the extraperitoneal soft tissue that should not be mistaken for extraperitoneally located appendix (▶ Fig. 5.27c). In extraperitoneal variant, separation of the interfascial planes of the extraperitoneal space (split interfascial plane sign) by the inflammatory process behind and medial to the cecum helps in finding the location of the appendix.[35] Abscess formation after perforation is commonly intraperitoneal in the right lower quadrant or within the pelvis but in retroperitoneal variant the abscess may extend to the mesenteric root, posterior pararenal space, or presacral space.

Diverticular Disease

Diverticular disease is a common incidental finding in imaging studies of middle-aged individuals. A colonic diverticulum develops after a saccular herniation of the mucosa through muscular layer of the colonic wall. In Western countries, diverticulosis

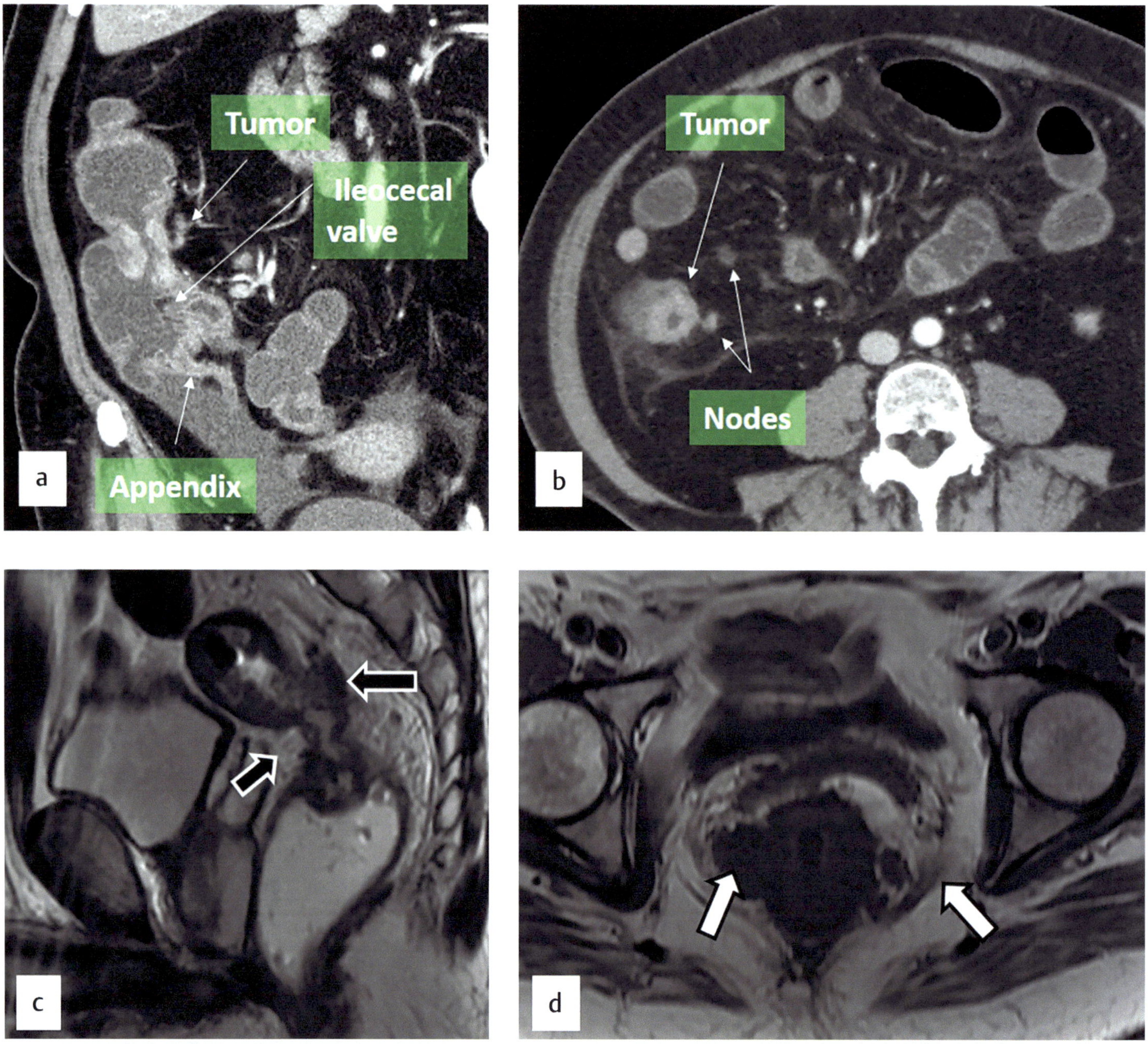

Fig. 5.26 (a, b) Coronal computed tomography (CT) showing annular narrowing with thickening wall of the ascending colon approximately 2 cm above the ileocecal valve due to colon cancer. **(b)** Axial image showing the annular thickening area and mildly enlarged regional lymph nodes. **(c, d)** Rectal cancer. **(c)** Sagittal view showing tubular, mass-like thickening of the rectosigmoid junction (*black arrow*), consistent with rectal cancer, with invasion into the perirectal fat. **(d)** Axial CT showing thickening of the rectal wall due to cancer with regional lymph node enlargement (*arrows*) in the perirectal space.

is more commonly seen in the sigmoid colon, but in Asia, it is more commonly seen in the right colon.[36] High luminal colonic pressures predisposes to herniation of the mucosa through the muscular layer in the weakened regions where the blood vessels penetrate the mucosa and submucosa. As a result, diverticula are typically found in parallel rows next to the taeniae coli (▶ Fig. 5.5). The penetrating vessels drape over the dome of the diverticulum and is only covered by mucosa. Inflammation of a diverticulum (diverticulitis) is a common cause of lower abdominal pain and fever. CT scan is the best imaging way of diagnosis, showing thickened colonic wall, mesenteric inflammation, regional abscesses, and fistula (▶ Fig. 5.28).

Inflammation and Ischemia of Colon

Colon is a common site of acute, subacute, and chronic inflammation (▶ Fig. 5.29). Common imaging findings are segmental or diffuse thickening of the wall of the colon/rectum, mesenteric or pericolonic edema, and mildly enlarged regional lymph nodes. Ulcerative colitis and Crohn's disease are uncommon idiopathic inflammatory processes of the gastrointestinal tract showing variable image findings depending on the stage of the disease (▶ Fig. 5.29).[37]

Colonic ischemia is the most common type of ischemic injury to the gastrointestinal tract. The three main mechanisms which result in colonic ischemia are: (1) Nonocclusive colonic ischemia

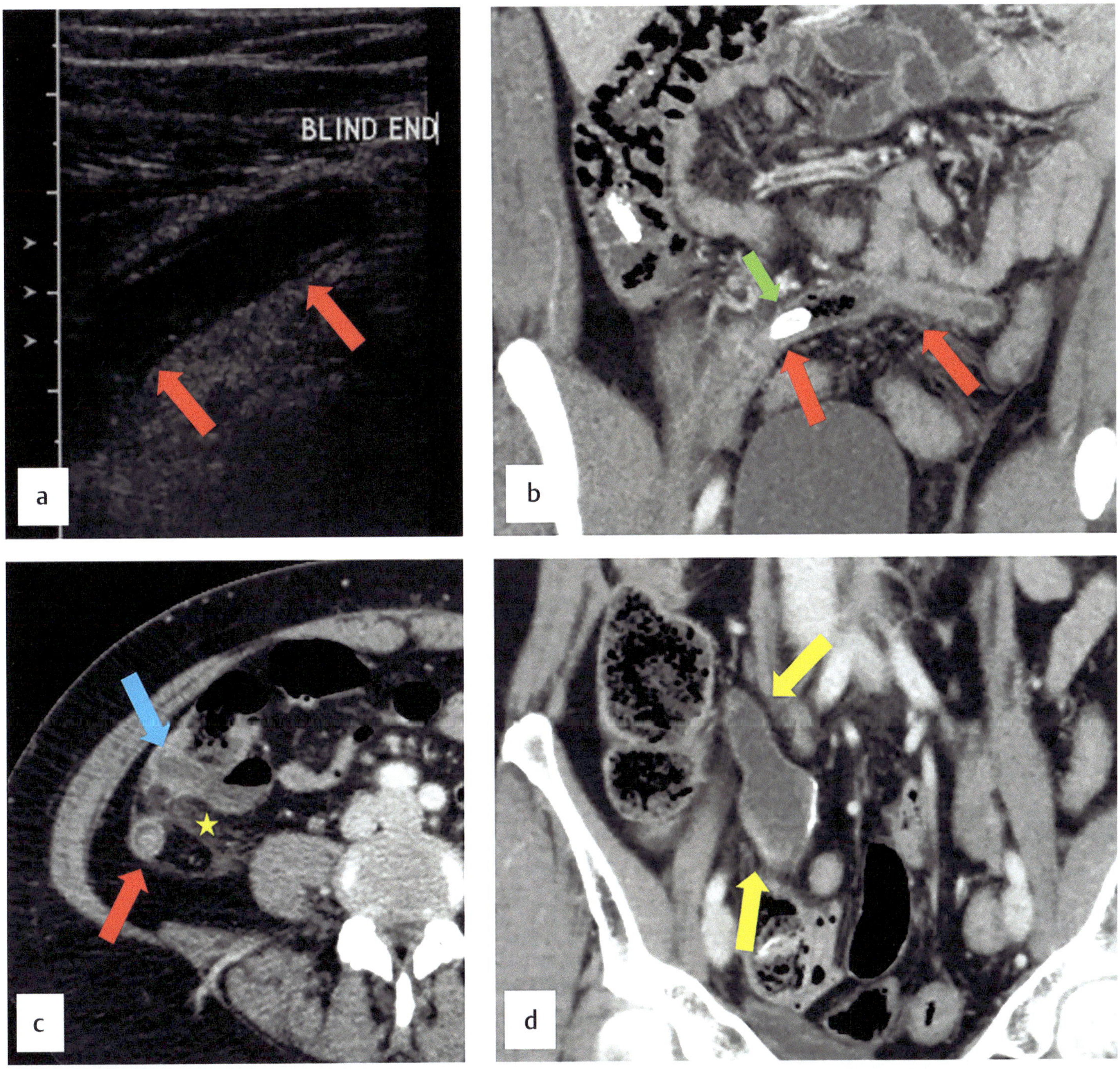

Fig. 5.27 (a–c) Acute appendicitis: **(a)** Ultrasound, **(b)** coronal computed tomography (CT), and **(c)** axial CT. Thick-walled tubular structure consistent with an inflamed appendix (*red arrows*) is seen in different patients. The cecum is shown by *blue arrow*. In **b,** the appendix is intraperitoneal and extends medially. An appendicolith is seen (*green arrow*) at the base of the inflamed appendix. In **c,** the appendix is retrocecal but intraperitoneal. There is associated inflammatory changes in the surrounding. **(d)** Mucocele of the appendix (*yellow arrows*). Coronal CT shows a bottle-shaped, fluid-filled appendix, an incidental finding in this patient.

due to hypoperfusion and reperfusion injury, which accounts for 95% of cases of colonic ischemia and most commonly affects the watershed areas of the colon, such as the splenic flexure and rectosigmoid colon; (2) embolic and thrombotic arterial occlusion; and (3) mesenteric vein thrombosis, which is a rare form of colonic ischemia that most commonly involves the distal small bowel and proximal colon.[38] Image findings include edema of the colonic wall, pericolonic inflammation, lack of enhancement of the involved segment after contrast administration, intramural gas, and perforation.

Anorectal Fistula

Perianal fistula is a common disease of middle-aged males causing perianal discharge, fistula tracts, and abscess.[39,40] Most of the perianal fistulas (almost 90%) are related to the infection of the anal glands. Treatment of most of these fistulas is surgery.

Fistulas can be unilateral, bilateral, and single or multiple. Horseshoe fistula is a pattern of extension when the extension occurs in the horizontal plane on either side of the midline. The anatomical course of the fistula tracts in relation to the anal

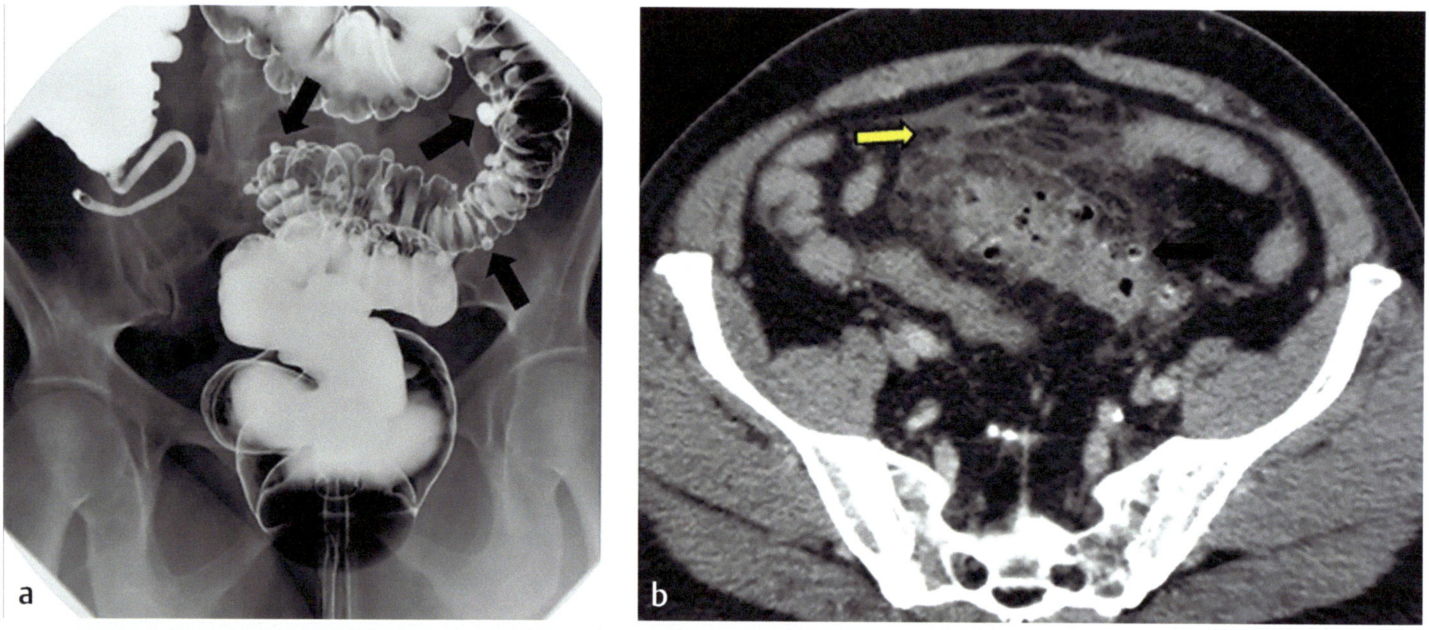

Fig. 5.28 Diverticular disease. **(a)** Focal outpouchings (*black arrows*) arising from the sigmoid colon, consistent with diverticulosis. Note a normal-appearing appendix. **(b)** Focal outpouchings arising from the sigmoid colon (*black arrow*) with thickening of the sigmoid colon and surrounding inflammatory changes and fluid (*yellow arrow*), consistent with acute diverticulitis.

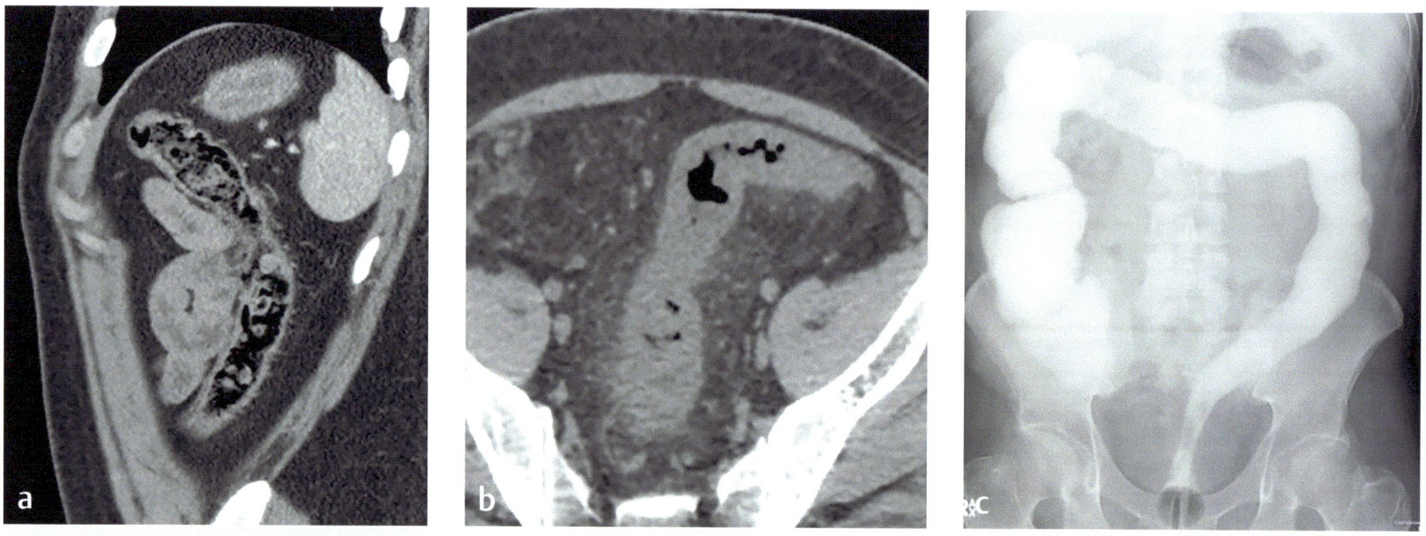

Fig. 5.29 (a) Sagittal computed tomography (CT) showing mild colonic wall thickening due to acute colitis involving the descending colon. **(b)** Axial CT showing severe colonic wall thickening, edema around the colon, and multiple enlarged lymph nodes due to acute colitis involving the sigmoid colon. **(c)** Ulcerative colitis. Barium enema showing featureless colon with loss of normal haustral markings, luminal narrowing, and bowel shortening, so-called lead pipe sign, which are classic imaging appearance of chronic ulcerative colitis.

sphincters and levator muscle is classified as follows (▸Fig. 5.30). In intersphincteric fistulas, the fistula traverses caudally within the intersphincteric space to open in the perianal skin. The external sphincter and ischiorectal remain unaffected (▸Fig. 5.31). In trans-sphincteric fistula the fistula tract extends from the anal canal into the internal and external sphincters and traverses through the ischioanal fossa to reach the perianal skin (▸Fig. 5.32). Translevatoric or supralevatoric fistulas are rare wherein a suprasphincteric fistula tract passes upward in the intersphincteric plane over the top of the puborectal muscle and then descends through the levator plate to the ischioanal fossa and finally to the skin. MRI provides valuable information about the morphological features of the fistula tract in relation to the anal sphincter complex, extent of fistula, multiplicity, and complications such as abscess formation (▸Fig. 5.30, ▸Fig. 5.31).

Vascular Abnormalities

A common finding in endoscopic and selective angiography of the colon is dilated, tortuous, and usually small (<10 mm) blood vessels within the mucosal and submucosal layers of the gut known as angiodysplasia. It is a cause of gastrointestinal bleeding in patients older than 60 years of age (▸Fig. 5.33). The colon is the most frequent site of angiodysplasia in the

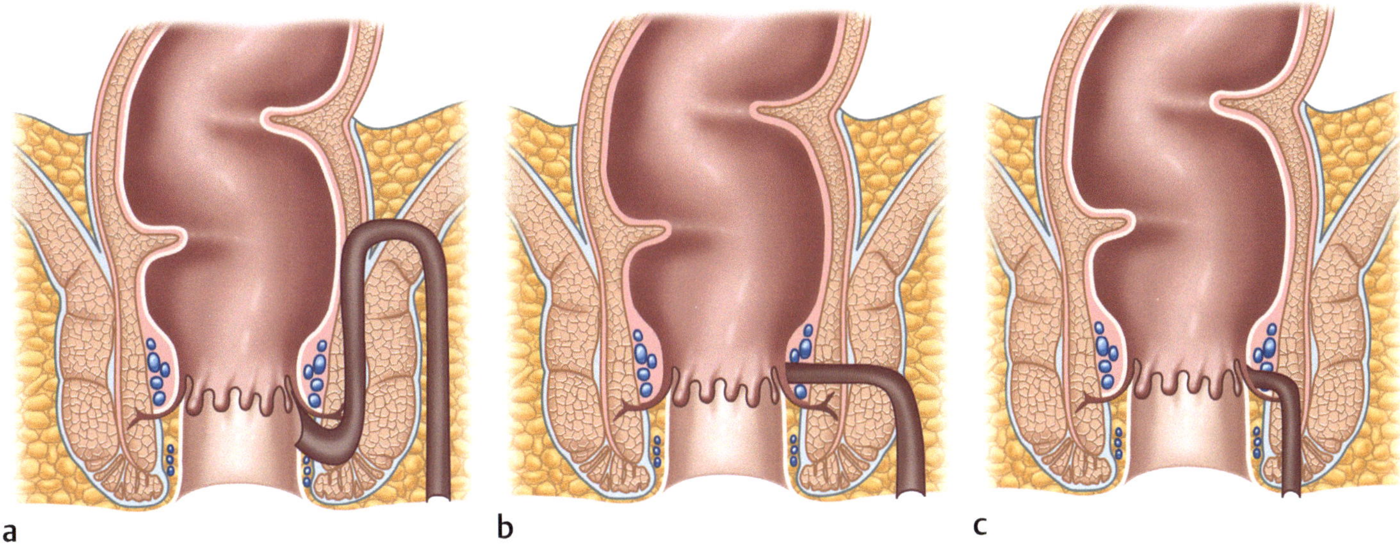

Fig. 5.30 Different types of perianal fistula are presented. **(a)** Supralevator perianal fistula. **(b)** Simple trans-sphincteric perianal fistula. **(c)** Simple intersphincteric perianal fistula.

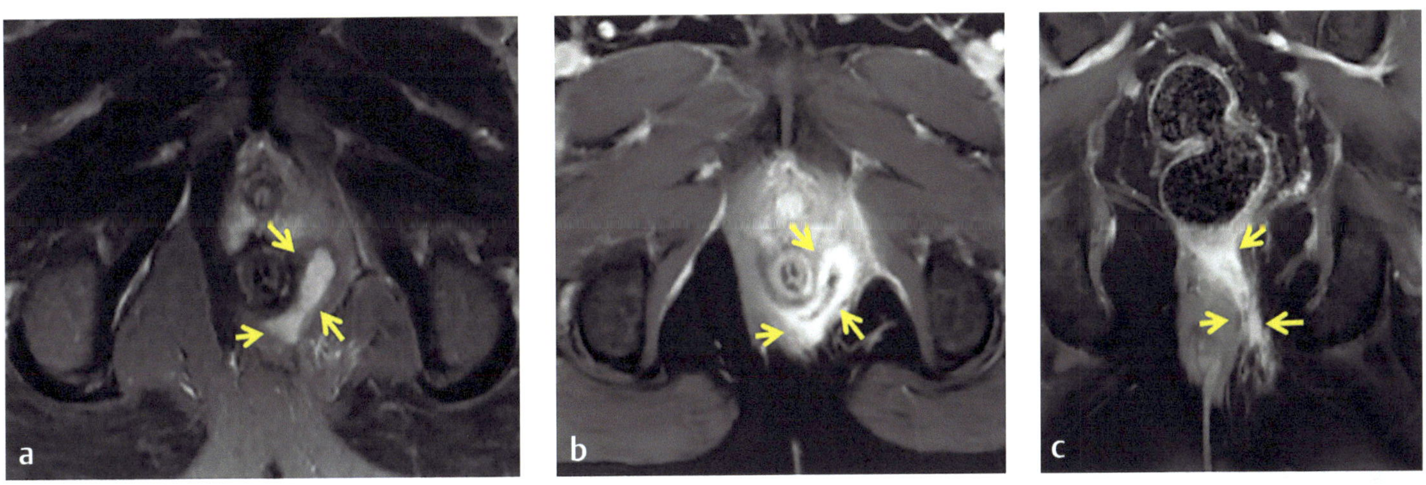

Fig. 5.31 (a) Axial T2-weighted, **(b)** axial, postcontrast, and **(c)** coronal, postcontrast images showing an intersphincteric fistula on the left side extending into the left ischioanal fossa (*arrows*). Note severe enhancement of the wall of fistula.

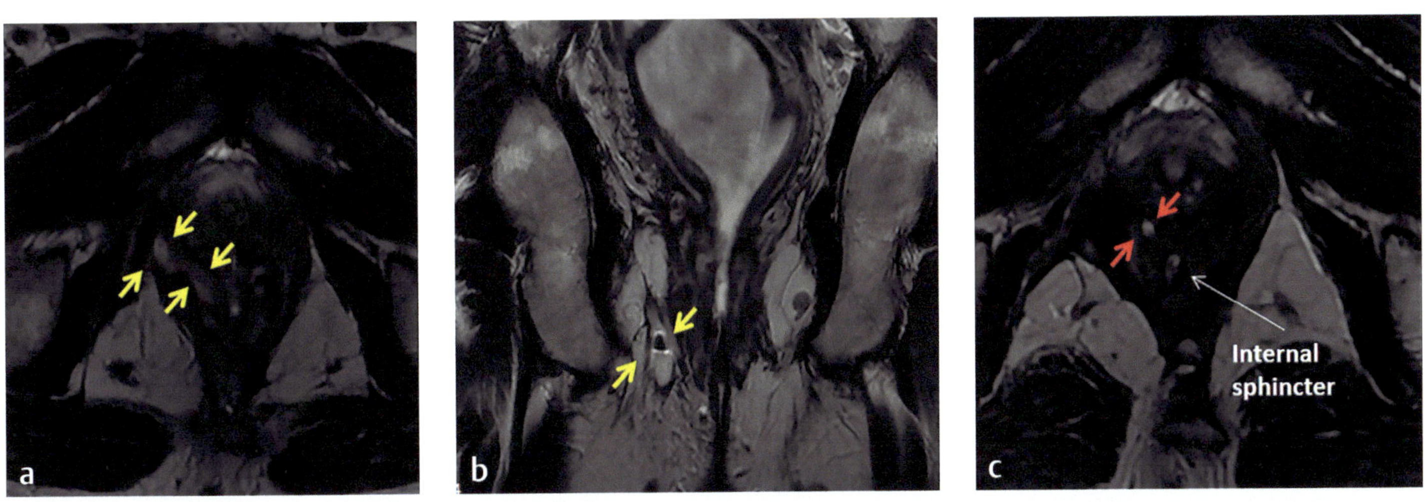

Fig. 5.32 (a) Axial and **(b)** coronal T2-weighted magnetic resonance imaging (MRI) images show a trans-sphincteric fistula below the levator muscle on the right side (*arrows*). Following passage through the sphincters, it has extended in the right ischioanal fossa toward the perineum (*arrows* in **b**). **(c)** After 1 year of treatment, axial MR shows reduction in size of fistula, now intersphincteric in location (*red arrows*). Note normal high signal of the internal sphincter.

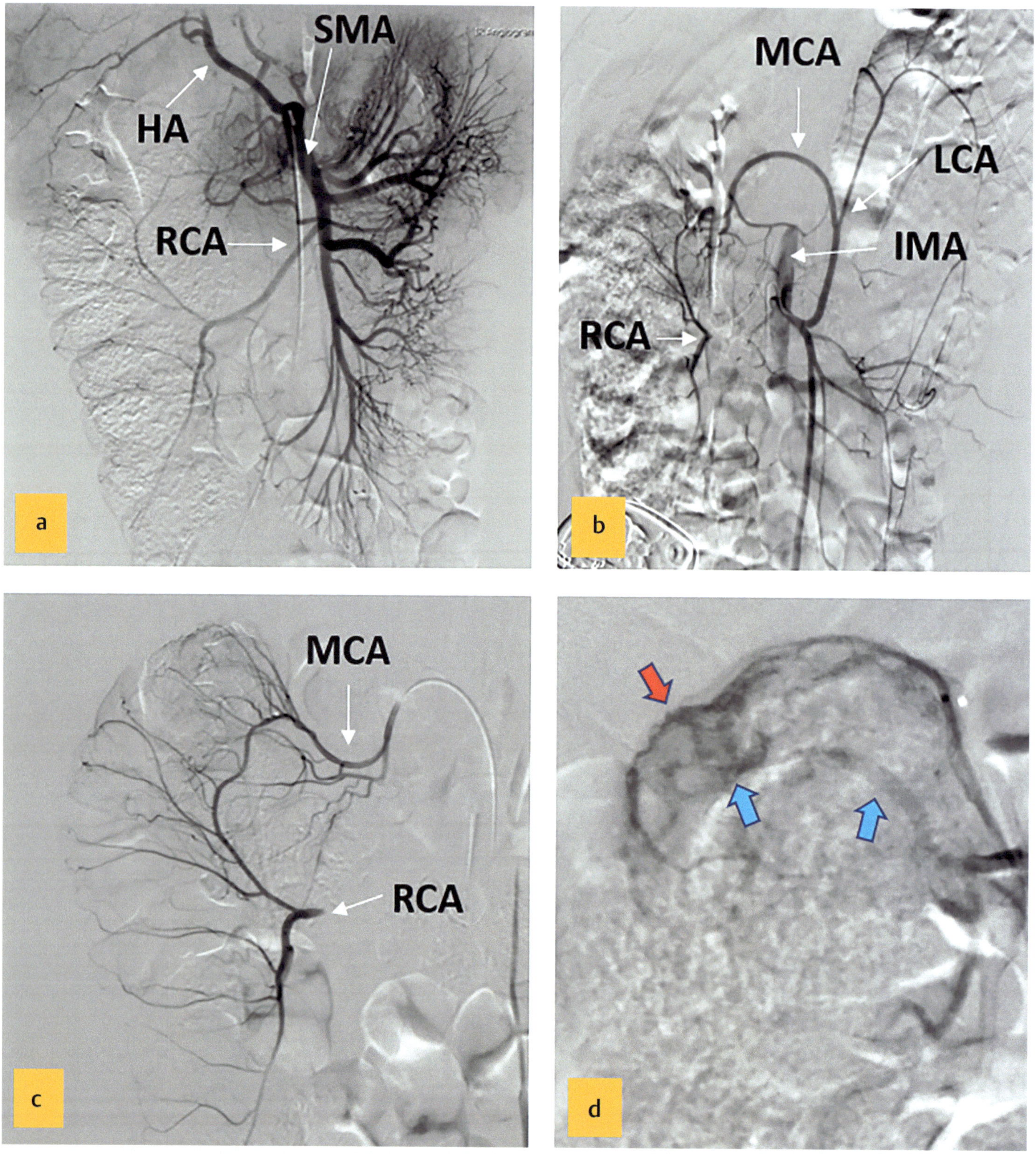

Fig. 5.33 Angiodysplasia of the hepatic flexure of colon in a patient with gastrointestinal bleed without evidence of a clear source. **(a)** Superior mesenteric artery (SMA) angiogram demonstrates unconventional vascular anatomy. There is a replaced right hepatic artery (HA) off the SMA. The middle colic artery (MCA) is not visualized to originate from the SMA. **(b)** Inferior mesenteric artery (IMA) demonstrates unconventional vascular anatomy with the MCA arising in a common trunk with the left colic artery (LCA). Cross-filling is noted from the left middle colic artery to the left colic artery at the splenic flexure. **(c)** Selective MCA injection showing cross-filling from the right middle colic artery to the right colic artery (RCA) at the hepatic flexure. **(d)** Hypervascularity is noted in the hepatic flexure (*red arrow*) with enlarged draining veins (*blue arrows*). Coil embolization was performed in one of the tertiary branches of the right middle colic artery at the hepatic flexure (not shown).

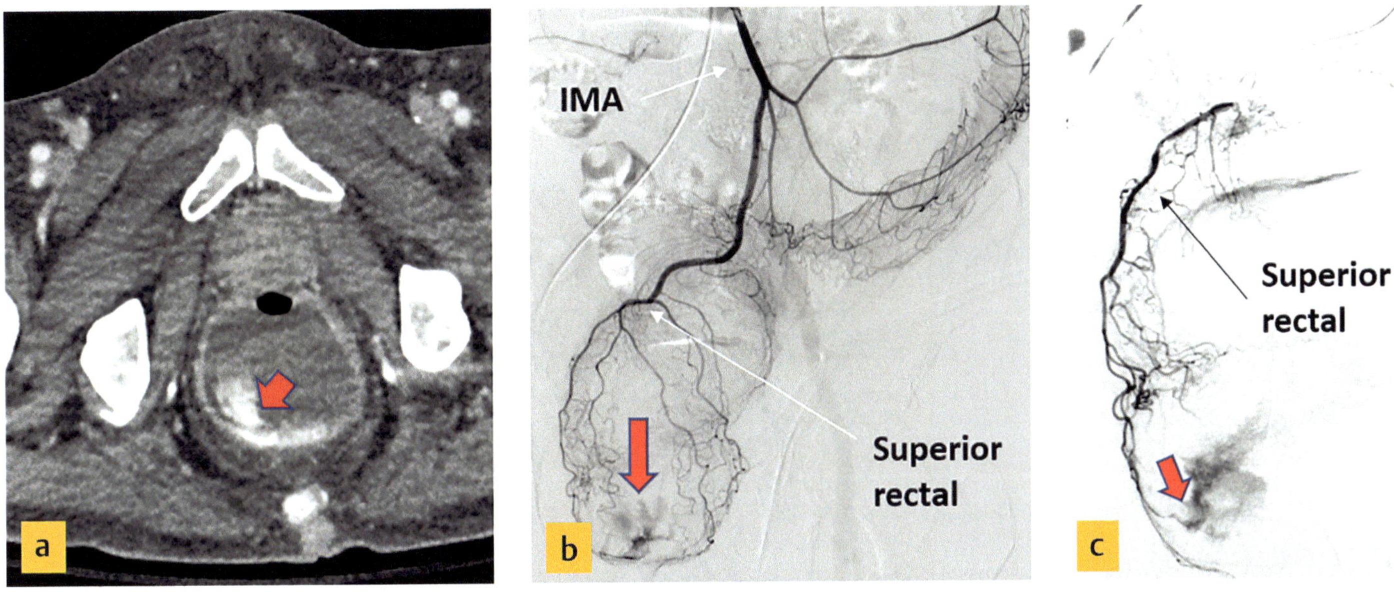

Fig. 5.34 (a) Axial computed tomography (CT) angiogram at the level of the rectum showing active bleed from the posterolateral wall of the rectum (*arrow*). **(b)** Inferior mesenteric artery (IMA) angiogram and **(c)** selective distal superior mesenteric artery (SMA) injection demonstrate the bleeding site (*arrows*) from the superior rectal artery.

gastrointestinal tract, predominantly involving the cecum and ascending colon. Chronic increased pressures during contractions on the right colon results in compression of the venules, dilated tortuous submucosal veins, and loss of precapillary sphincters competency, resulting in development of small arteriovenous collaterals.

It is well known that in patients with a large lower abdominal aneurysm, the inferior mesenteric artery is often occluded or severely stenosed due to severe atheromatous disease. Inferior mesenteric artery is also commonly occluded during aortic stent graft placement or ligated during repair of the aortic aneurysm or colon resection for the rectosigmoid cancer.

Bleeding from the colonic wall can occur following biopsy and in colitis, diverticulitis, polyps, malignancies, and vascular malformations. The bleeding vessels can be embolized (▶ Fig. 5.34).

References

1. Ellis H, Mahadevan V. Anatomy of the caecum, appendix and colon. Surgery 2014;32:155–158
2. Irving MH, Catchpole B. ABC of colorectal diseases. Anatomy and physiology of the colon, rectum, and anus. BMJ 1992;304(6834):1106–1108
3. Malas MA, Aslankoç R, Ungör B, Sulak O, Candir O. The development of large intestine during the fetal period. Early Hum Dev 2004;78(1):1–13 PubMed
4. Hollinshead WH. Embryology and surgical anatomy of the colon. Dis Colon Rectum 1962;5:23–27
5. Rigoard P, Haustein SV, Doucet C, Scepi M, Richer JP, Faure JP. Development of the right colon and the peritoneal surface during the human fetal period: human ontogeny of the right colon. Surg Radiol Anat 2009;31(8):585–589
6. Kluth D, Fiegel HC, Metzger R. Embryology of the hindgut. Semin Pediatr Surg 2011;20(3):152–160
7. Mitchell B, Sharma R. The digestive system. In: Mitchell B, Sharma R, eds. Embryology: An Illustrated Colour Text. 2nd ed. United Kingdom: Churchill Livingstone; 2009:50–57
8. Garza A, Beart R. Anatomy and embryology of the anus, rectum, and colon. In: Corman M, ed. Corman's Colon and Rectal Surgery. 2nd ed. Philadelphia: Lippincott Williams and Wilkins; 2013
9. Barlow A, Muhleman M, Gielecki J, Matusz P, Tubbs RS, Loukas M. The vermiform appendix: a review. Clin Anat 2013;26(7):833–842
10. Wakeley CP. The position of the vermiform appendix as ascertained by an analysis of 10,000 cases. J Anat 1933;67(Pt 2):277–283
11. Fry RD, Mahmoud NN, Maron DJ, Bleier J. Colon and rectum. In: Townsend CM, Beauchamp RD, Evers BM, Mattox KL, eds. Sabiston Textbook of Surgery: The Biological Basis of Modern Surgical Practice. 19th ed. Philadelphia: Elsevier Saunders; 2012:1294–1380
12. Wedel T. Colon, rectum, anus, pelvic floor. In: Herold A, Lehur PA, Matzel KE, O'Connell PR, eds. Coloproctology. Berlin: Springer-Verlag; 2008:5–20
13. Kachlik D, Baca V, Stingl J. The spatial arrangement of the human large intestinal wall blood circulation. J Anat 2010;216(3):335–343
14. Ogino T, Takemasa I, Horitsugi G, et al. Preoperative evaluation of venous anatomy in laparoscopic complete mesocolic excision for right colon cancer. Ann Surg Oncol 2014;21(Suppl 3):S429–S435
15. Yamaguchi S, Kuroyanagi H, Milsom JW, Sim R, Shimada H. Venous anatomy of the right colon: precise structure of the major veins and gastrocolic trunk in 58 cadavers. Dis Colon Rectum 2002;45(10):1337–1340
16. Cohn SM, Birnbaum EH, Friel CM. Colon: anatomy and structural anomalies. In: Yamada T, ed. Textbook of Gastroenterology. Oxford: Blackwell Publishing Ltd; 2008:1370–1385
17. Welton ML, Shelton AA, Chang GJ, Varma MG. Colon, rectum, and anus. In: Norton JA, Barie PS, Bollinger RR, et al, eds. Surgery: Basic Science and Clinical Evidence. 2nd ed. New York: Springer; 2008:1101–1110
18. Stoker J. Anorectal and pelvic floor anatomy. Best Pract Res Clin Gastroenterol 2009;23(4):463–475
19. Kaiser AM, Ortega AE. Anorectal anatomy. Surg Clin North Am 2002;82(6): 1125–1138, v
20. Barleben A, Mills S. Anorectal anatomy and physiology. Surg Clin North Am 2010;90(1):1–15
21. Sakorafas GH, Zouros E, Peros G. Applied vascular anatomy of the colon and rectum: clinical implications for the surgical oncologist. Surg Oncol 2006;15(4):243–255
22. Davies MR. Anatomy of the nerve supply of the rectum, bladder, and internal genitalia in anorectal dysgenesis in the male. J Pediatr Surg 1997;32(4):536–541
23. Hussain SM, Stoker J, Laméris JS. Anal sphincter complex: endoanal MR imaging of normal anatomy. Radiology 1995;197(3):671–677
24. Guo M, Gao C, Li D, et al. MRI anatomy of the anal region. Dis Colon Rectum 2010;53(11):1542–1548

25. Mittal RK, Bhargava V, Sheean G, Ledgerwood M, Sinha S. Purse-string morphology of external anal sphincter revealed by novel imaging techniques. Am J Physiol Gastrointest Liver Physiol 2014;306(6):G505–G514
26. Raizada V, Mittal RK. Pelvic floor anatomy and applied physiology. Gastroenterol Clin North Am 2008;37(3):493–509, vii
27. Scardapane A, Brindicci D, Fracella MR, Angelelli G. Post colon surgery complications: imaging findings. Eur J Radiol 2005;53(3):397–409
28. Broder JC, Tkacz JN, Anderson SW, Soto JA, Gupta A. Ileal pouch-anal anastomosis surgery: imaging and intervention for post-operative complications. Radiographics 2010;30(1):221–233
29. Cherukuri R, Levine MS, Maki DD, Rubesin SE, Laufer I, Rosato EF. Hartmann's pouch: radiographic evaluation of postoperative findings. AJR Am J Roentgenol 1998;171(6):1577–1582
30. Blazer S, Zimmer EZ, Gover A, Bronshtein M. Fetal omphalocele detected early in pregnancy: associated anomalies and outcomes. Radiology 2004;232(1): 191–195
31. Kim HJ, Kim AY, Lee CW, et al. Hirschsprung disease and hypoganglionosis in adults: radiologic findings and differentiation. Radiology 2008;247(2):428–434
32. Alamo L, Meyrat BJ, Meuwly JY, Meuli RA, Gudinchet F. Anorectal malformations: finding the pathway out of the labyrinth. Radiographics 2013;33(2):491–512
33. Johnson CM, Wei C, Ensor JE, et al. Meta-analyses of colorectal cancer risk factors. Cancer Causes Control 2013;24(6):1207–1222
34. Birnbaum BA, Wilson SR. Appendicitis at the millennium. Radiology 2000;215(2):337–348
35. Ku YM, Lee SL, Kim Y, Won Y. Extraperitoneal ascending appendicitis: usefulness of the split interfascial plane sign on MDCT. Diagn Interv Imaging 2016; 97(6):667–672
36. Gaglia A, Probert C. Diverticular disease. Medicine (Baltimore) 2015;43(6): 320–323
37. Nikolaus S, Schreiber S. Diagnostics of inflammatory bowel disease. Gastroenterology 2007;133(5):1670–1689
38. Brandt LJ, Feuerstadt P, Longstreth GF, Boley SJ; American College of Gastroenterology. ACG clinical guideline: epidemiology, risk factors, patterns of presentation, diagnosis, and management of colon ischemia (CI). Am J Gastroenterol 2015;110(1):18–44, quiz 45
39. Vanbeckevoort D, Bielen D, Vanslembrouck R, Van Assche G. Magnetic resonance imaging of perianal fistulas. Magn Reson Imaging Clin N Am 2014;22(1):113–123
40. Kumar N, Agarwal Y, Chawla AS, Jain R, Thukral BB. MRI of perianal fistulae: a pictorial kaleidoscope. Clin Radiol 2015;70(12):1451–1461

6 Liver

Farhood Saremi, Andrea Schenk, and Jean H.D. Fasel

◆ Introduction

Liver and spleen are common sites of pathologies and probably the most common reason for the indication for abdominal imaging studies after trauma. Knowledge of liver anatomy and related vasculature is important for interpretation of lesions. Another common application of assessment of liver anatomy is for patients who require partial hepatic resection and candidates of split liver transplantation. Liver and spleen can be examined by ultrasound, computed tomography (CT), magnetic resonance imaging (MRI), or catheter angiogram. Anatomical variant of liver vasculature and biliary system is very common and potentially can be problematic during surgery. It is crucial to learn about liver anatomy, anatomical variants, microcirculation, and function.

◆ Embryology

The endoderm covers the primitive gut and gives rise to the epithelium of the gastrointestinal tract and several organs such as the thyroid, the liver, and the pancreas. The embryonic live bud arises from the duodenal region of the primitive gut[1,2] (see Chapter 8 "Biliary System" and Chapter 9 "Pancreas"). The cranial portion of the endodermal diverticulum gives rise to the liver, which includes the intrahepatic biliary network, and its caudal portion is at the origin of the extrahepatic biliary tract.[3] The liver diverticulum is lined by endoderm cells called hepatoblasts which express typical hepatic genes, such as albumin, alpha-fetoprotein, and liver-specific transcription factors. The hepatoblasts then leave the endoderm epithelium, invade the septum transversum, and continue proliferating to expand the liver. The hepatoblasts are the precursors of the hepatocytes and cholangiocytes. The hepatocytes organize as cords and become tightly associated with the sinusoids. Hepatocyte maturation is not complete at birth. The cholangiocytes delineate the intrahepatic bile ducts. The intrahepatic ducts development is linked to development of the portal vein and the hepatic artery to constitute the portal triad.[2,3] The extrahepatic bile ducts, peribiliary glands, and ventral pancreatic bud are formed directly from an endodermal outpouching located inferior to the liver.

Formation of intrahepatic vessels is very complicated. It starts with sinusoids. The sinusoidal capillaries including the endothelial and stellate cells develop by angiogenesis from the vessels in the septum transversum mesenchyme. The sinusoidal endothelial cells have important roles in the transfer or clearance of molecules between serum and hepatocytes. The stellate cells are in the space of Disse between the sinusoidal endothelial cells and the basal surface of the hepatocytes. They help to modulate hepatic microcirculation. Chronic activation of stellate cells under pathological conditions leads to liver fibrosis.

The fetal liver is in contact with two major venous systems, the umbilical veins and the vitelline veins. The left intrahepatic part of the umbilical vein, in conjunction with the portal sinus, right and left portal veins, ductus venosus, and hepatic veins, is embryologically derived from the vitelline veins, while the intra-abdominal part of the umbilical vein is derived from the left umbilical vein[4] (see Chapter 11 "Portal Venous System"). The umbilical vein is the major afferent vessel in the fetal liver, but it disappears after birth and will be substituted by the portal vein. In fetus, the portal sinus (umbilical sinus) is wide L-shaped venous confluence at the terminal end of the umbilical vein (▸Fig. 6.1). The extrahepatic portal vein drains into the portal sinus near the origin of the right intrahepatic portal vein. The ductus venosus is a straight vessel extending between the portal sinus and inferior vena cava (IVC). During fetal life, the ductus venosus is responsible for shunting most blood flow of the umbilical vein directly into the IVC, transporting oxygenated blood from the placenta to the fetus. After birth, the umbilical vein closes and remains as the ligamentum teres.

Hepatic artery development occurs later than venous development and is driven by ductal plates which also participate in formation of the biliary ducts. It starts within the parenchyma and gradually extends toward the periphery.[1,2]

◆ Anatomy

The liver is in the right upper quadrant between the diaphragm, abdominal/chest wall, stomach, and transverse colon. It is the largest internal organ in the body, accounting for approximately 2 to 5% of the total body weight.[4,5]

Surface Landmarks

In coronal plane, the normal liver appears triangular with superior, right lateral, and inferior surfaces (▸Fig. 6.2, ▸Fig. 6.3). In this plane, the superior surface is facing the diaphragm, the right lateral surface is facing the lower chest wall, and the inferomedial surface is facing the gallbladder, transverse colon, and stomach anteriorly and the right kidney, right adrenal, and duodenum posteriorly (▸Fig. 6.3). The porta hepatis is located in the middle of the inferomedial surface.

In sagittal plane, three surfaces for the liver are recognizable: anterosuperior, posterior, and inferior surfaces. The posterior surface is retroperitoneal (including the bare area) facing the retrohepatic part of the IVC, the right adrenal gland, and the superior pole of the right kidney. The anterosuperior surface is related to the diaphragmatic dome. This anterosuperior surface is covered by the peritoneum except for the attachment of the falciform ligament and where the anterosuperior reflection of the coronary ligament exists. The inferior surface is the visceral hepatic surface facing several abdominal organs.

In axial plane, there are three surfaces including right lateral surface facing the chest wall, anterior surface facing the abdominal

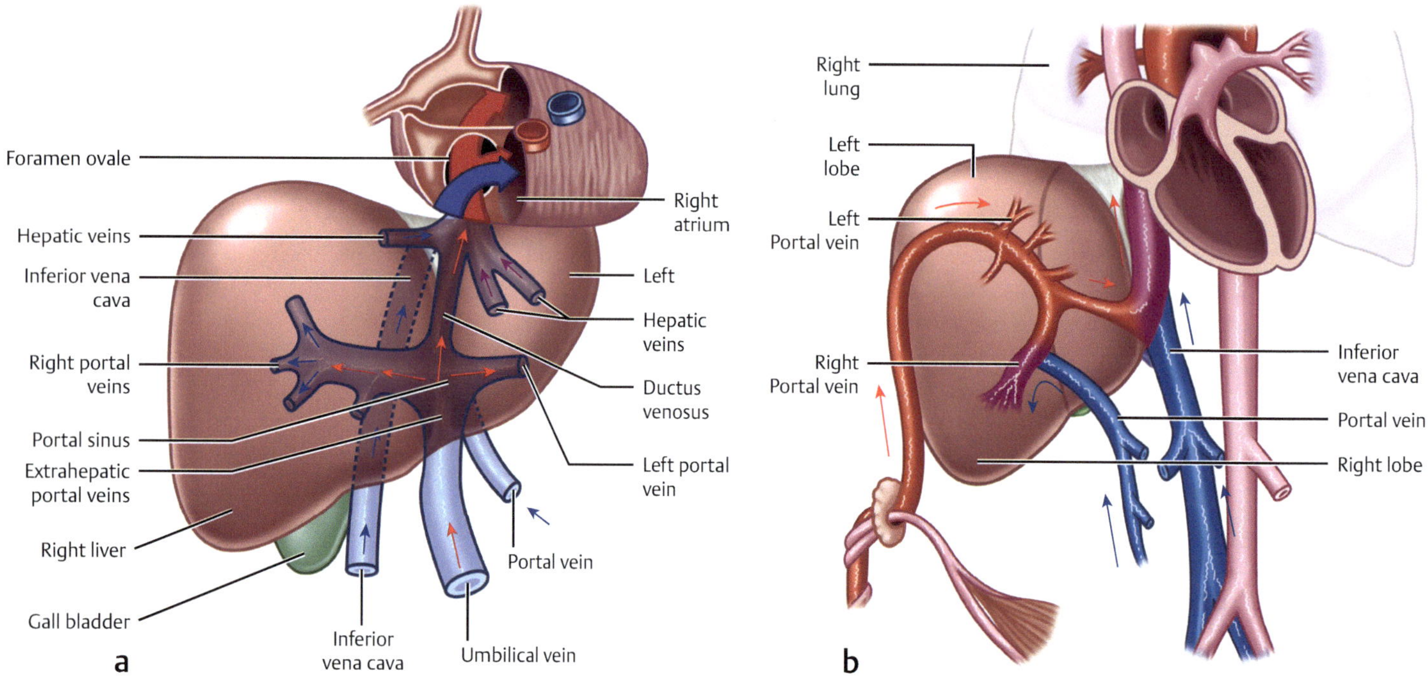

Fig. 6.1 Schematic representation of the fetal umbilical, portal and hepatic venous circulations. **(a)** Frontal view. **(b)** Lateral view. The arrows indicate the direction of blood flow and the color indicates the degree of oxygen content (*red* = high, *purple* = medium, *blue* = low). (Modified with permission from Mavrides et al.[4])

wall, and the medial surface facing the gastric antrum and gallbladder anteriorly and the right kidney and adrenal posteriorly.

The space under the right lobe is the subhepatic space of Morison and the space under the left lobe consists of the left subhepatic space and the lesser sac. Looking at the inferior surface of the liver, a H configuration can be recognized by the gallbladder and IVC on the right and the round ligament and ligamentum venosum on the left (▸Fig. 6.2). The cross bar of the H is the porta hepatis (hilum of the liver). Within the porta hepatis are the hepatic artery, the hepatic duct, the branches of the portal vein, nerves, and lymphatics (▸Fig. 6.2).

Hepatic Fissures

On the frontal view, only one fissure is visible which is related to portoumbilical fissure where the falciform ligament is attached[5] (▸Fig. 6.2). As the name implies, it contains the ligamentum teres and the left portal vein (▸Fig. 6.4, ▸Fig. 6.5). It also contains a major tributary of the left hepatic vein called umbilical fissure vein. Occasionally, a small umbilical artery is seen arising from the left hepatic artery and ending on the capsular surface of the left lobe (▸Fig. 6.5). The ligamentum venosum fissure is in the inferior surface of the liver between the caudate lobe posteriorly and the left lobe anteriorly (▸Fig. 6.5). It is invested by the peritoneal folds of the lesser omentum (gastrohepatic ligament). In isolated origin of the left hepatic artery, it passes in the ligamentum venosum fissure (▸Fig. 6.5).

Other fissures (incisura) are not real and only are named for classification of surgical segmental resections based of the anatomic plane of the hepatic veins (▸Fig. 6.6). The right portal fissure is along the right hepatic vein. Superiorly, the right fissure follows the attachment of the right anterosuperior coronary ligament and ends at inferior liver margin midway between the gallbladder fossa and the right liver.[5] The median (main) portal fissure moves along the plane of the middle hepatic vein between the gallbladder fossa and the left margin of the IVC. The left portal fissure runs from the left side of the IVC to a point between the dorsal one-third and ventral two-thirds of the left margin of the liver.

Hepatic Ligaments; Falciform, Coronary, Triangular, and Round

The liver is covered with the capsule of Glisson which is continuation of visceral peritoneum over the liver. At liver hilum, the capsule of Glisson envelops the hepatic artery, portal vein, and bile duct and follows these structures in the liver. Peritoneal attachments of the liver to the diaphragm include the falciform ligament, and the coronary ligaments with the right and left triangular ligaments (▸Fig. 6.7, ▸Fig. 6.8). These structures are not true ligaments and only represent peritoneal reflections. Membranous extensions of the Glisson's capsule around the IVC form the caval ligaments which closely attach the vein to the caudate lobe and right hepatic lobe.

The liver is suspended posterosuperiorly from the diaphragm by the hepatic veins at the confluence with the IVC. This region along with the part of liver facing the right adrenal gland is called bare area because it is not covered by the peritoneum as the rest of liver (▸Fig. 6.8, ▸Fig. 6.9). The anterior and posterior limits of the bare area are determined by the coronary ligaments (▸Fig. 6.7, ▸Fig. 6.8, ▸Fig. 6.9). Embryologically, these ligaments reflect the remnants of the developmental pathways of the liver in the septum transversum, and in the mature organism they only represent

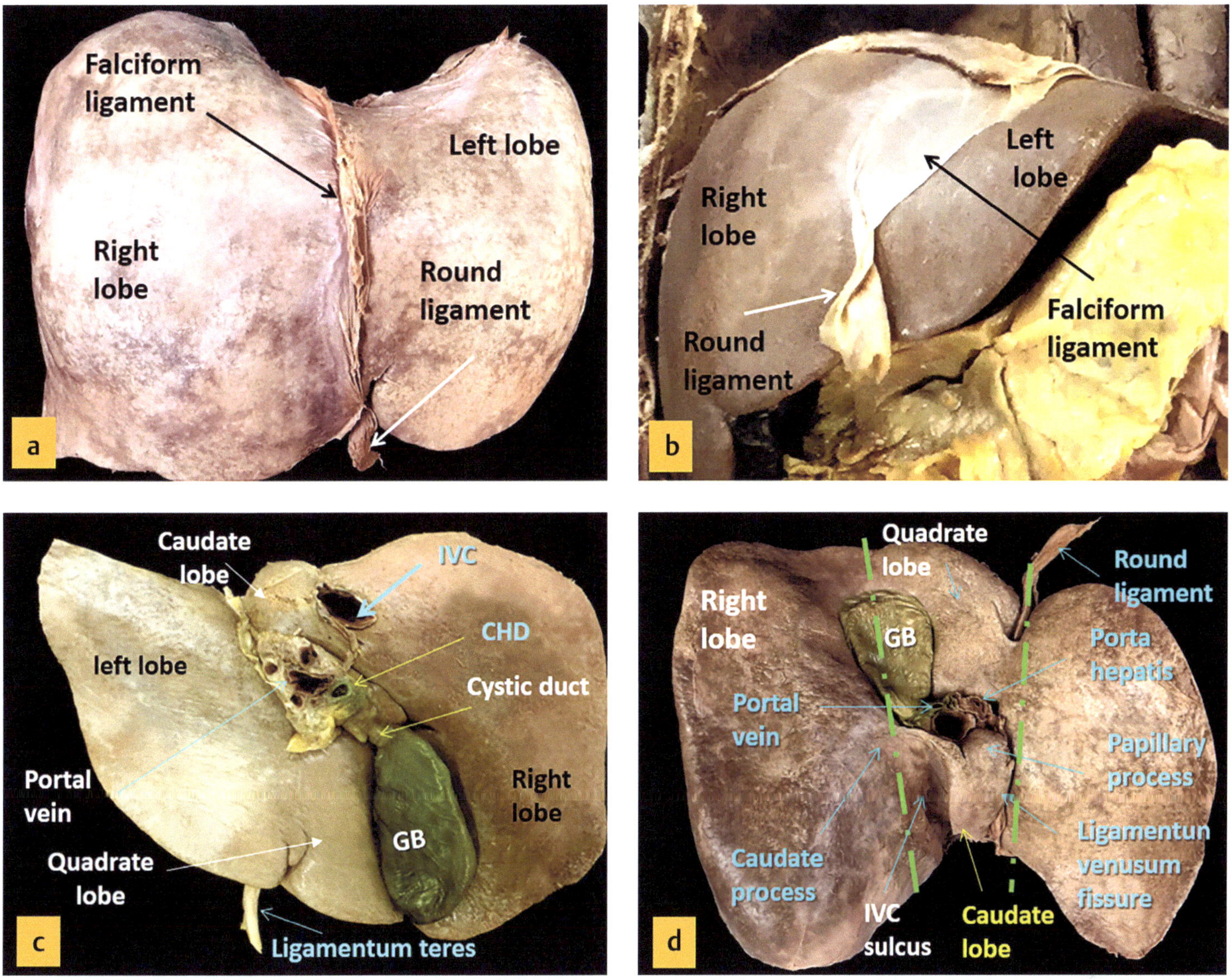

Fig. 6.2 Liver surface. **(a, b)** Anterior views. **(c, d)** Posteroinferior (visceral) views. On the frontal view, the only visible fissure is the umbilical fissure. Typical orientation of the H configuration (*green lines*) of the portal structures are shown on the visceral surface. The falciform ligament extends to the anterior surface of the liver at the umbilical fissure blending into the hepatic peritoneal covering (Glisson's capsule). The leaves of the peritoneal fold of the falciform ligament separate over the liver and blend with the anterosuperior leaf of the coronary ligament. The round ligament travels in the free edge of the falciform ligament. CHD, common hepatic duct; GB, gallbladder; IVC, inferior vena cava.

peritoneal reflections between the diaphragm and superior surface of the liver. Therefore, the coronary ligaments are not true ligaments.[6] On the right side, the distance between the anterior (anterosuperior) and posterior leaves of the coronary ligament, which determines the boundaries of the bare area, is larger compared to the left side. Enclosing the bare area on the right, the coronary ligaments unite laterally to form the right triangular ligament. On the left side, the distance between the anterior and posterior coronary ligaments is negligible and the two ligaments are in close apposition to each other and together form the left triangular ligament (▶ Fig. 6.7). Mobilization of the liver during surgery requires division of these ligamentous attachments to the diaphragm.

The posterior leaves of the right and left coronary ligaments continue to meet at the anterior wall of abdominal esophagus down to the gastroesophageal junction, where the esophagus is totally enveloped by the continuation of the lesser omentum (▶ Fig. 6.7b). The posterior wall of the abdominal esophagus is a bare area on the diaphragm and is not covered by the peritoneum (▶ Fig. 6.9).

The falciform ligament is a remnant of the primitive ventral mesentery. It originates from the anterior abdominal wall (between the diaphragm and the umbilicus) and extends to the anterior surface of the liver at the umbilical fissure blending into the hepatic peritoneal covering (Glisson's capsule). At the base of the falciform ligament along the liver, the hepatic veins drain into the IVC. The leaves of the peritoneal fold of the falciform ligament separate over the liver and blend with the anterosuperior leaf of the coronary ligament on the right and left sides of the diaphragmatic surface of the liver (▶ Fig. 6.8, ▶ Fig. 6.9).

The round ligament of the liver, which represents the obliterated left umbilical vein, travels in the free edge of the falciform ligament between the left portal vein and umbilicus (▶ Fig. 6.9).

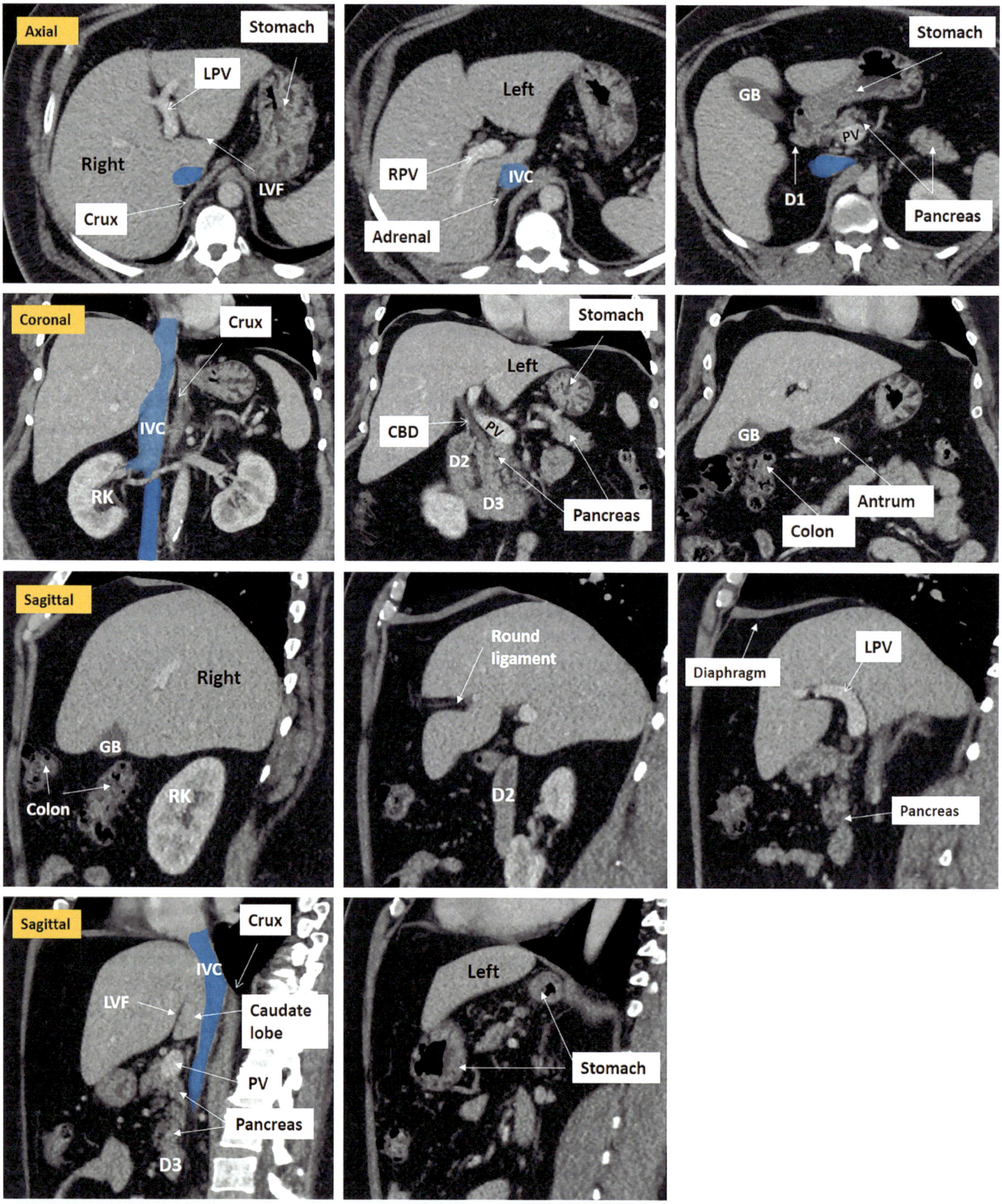

Fig. 6.3 Anatomical relationship of the liver with adjacent structures. The inferomedial (visceral) surface is facing the gallbladder (GB), transverse colon, and stomach anteriorly and the right kidney, right adrenal, and duodenum posteriorly. The porta hepatis is located in the middle of the inferomedial surface. The posterior surface is retroperitoneal (including the bare area) facing the retrohepatic part of the IVC, the right adrenal gland, and the superior pole of the right kidney. CBD, common bile duct; D, duodenum; D1, first part of duodenum; D2, second part of duodenum; D3, third part of duodenum; IVC, inferior vena cava; LPV, left portal vein; LVF, ligamentum venosum fissure; PV, portal vein; RK, right kidney; RPV, right portal vein.

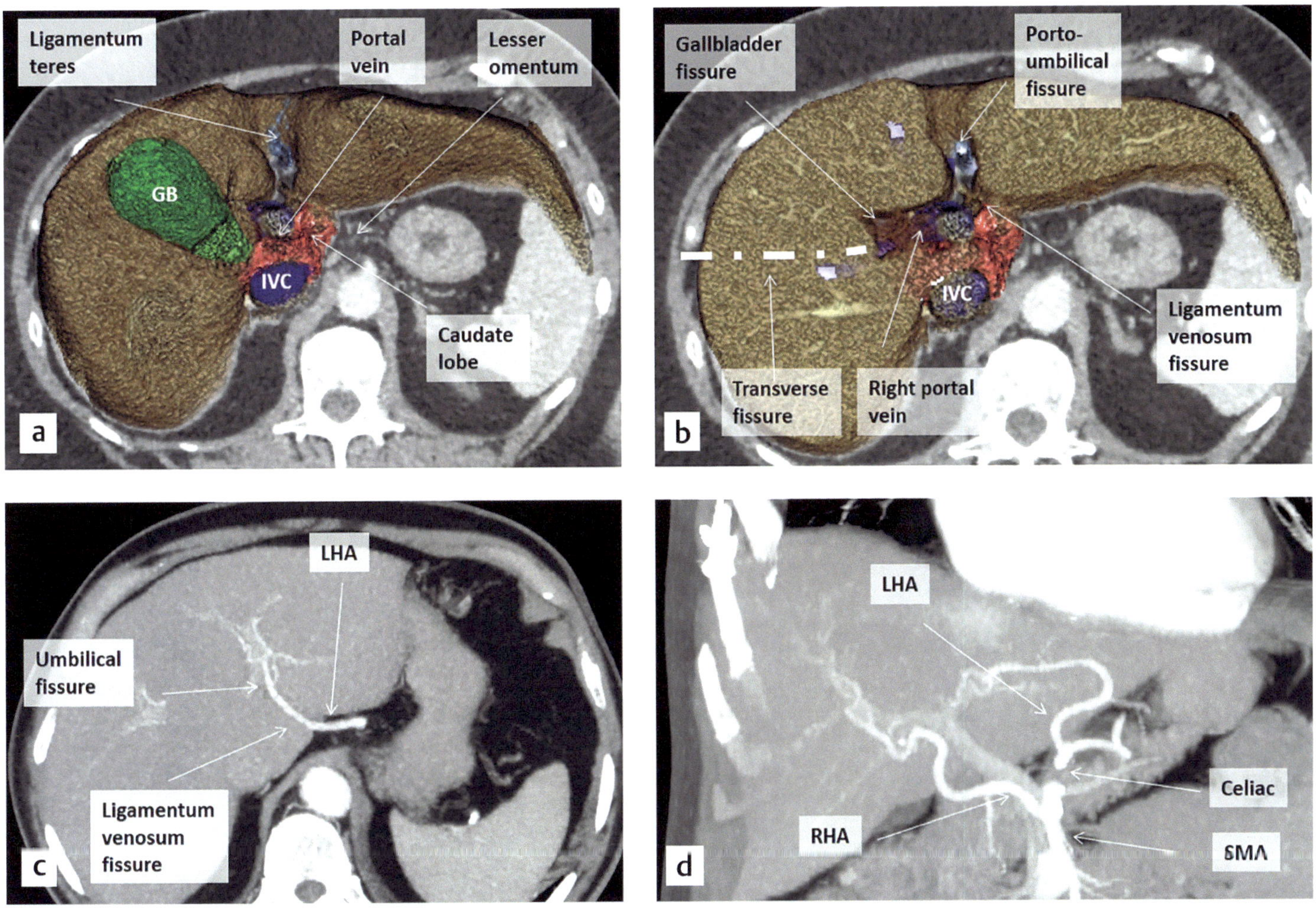

Fig. 6.4 Fissures, inferior views. **(a)** Inferior view of the liver. **(b)** Axial cut through the porta hepatis. Caudate lobe is shown in *red*. The ligamentum venosum fissure is between the caudate lobe posteriorly and the left lobe anteriorly. It is invested by the peritoneal folds of the lesser omentum (gastrohepatic ligament). **(c)** Axial and **(d)** coronal computed tomography (CT) angiography showing isolated origin of the left hepatic artery (LHA) from the celiac axis along with aberrant origin of the right hepatic artery (RHA) from the superior mesenteric artery. The LHA passes in the ligamentum venosum fissure before entering into the umbilical fissure. GB, gallbladder; IVC, inferior vena cava; SMA, superior mesenteric artery.

6

Recanalization and dilation of the umbilical vein along with formation of varicoid paraumbilical veins are common in patients with long-standing portal hypertension and eventually communicate with the systemic epigastric and diaphragmatic veins creating varicoid portosystemic shunts in the anterior abdominal wall known as caput medusae (▶Fig. 6.9, ▶Fig. 6.10).

Architectural Unit of the Liver

The basic architectural unit of the liver is the liver lobule. The "classic lobule," comprising several cone-shaped primary lobules, is a polygonal structure with ill-defined margin separated from adjacent lobules by variable amount of connective tissue (▶Fig. 6.11, ▶Fig. 6.12). A hepatic venule is seen in the center of a classic lobule and portal tracts are distributed at each of the six corners distinguished by the presence of the portal triad of vessels consisting of a portal vein, bile duct, and hepatic artery. Blood in the portal vein and hepatic artery flows through a network of sinusoidal capillaries before leaving the lobule through the central vein. Additionally, intersinusoidal anastomoses exist between adjacent lobules. The hepatic "acinus" is formed by the port tract in the middle and the neighboring central hepatic venules in the periphery. Three zones exist within the acinus depending on the levels of oxygenation and metabolic function.

6

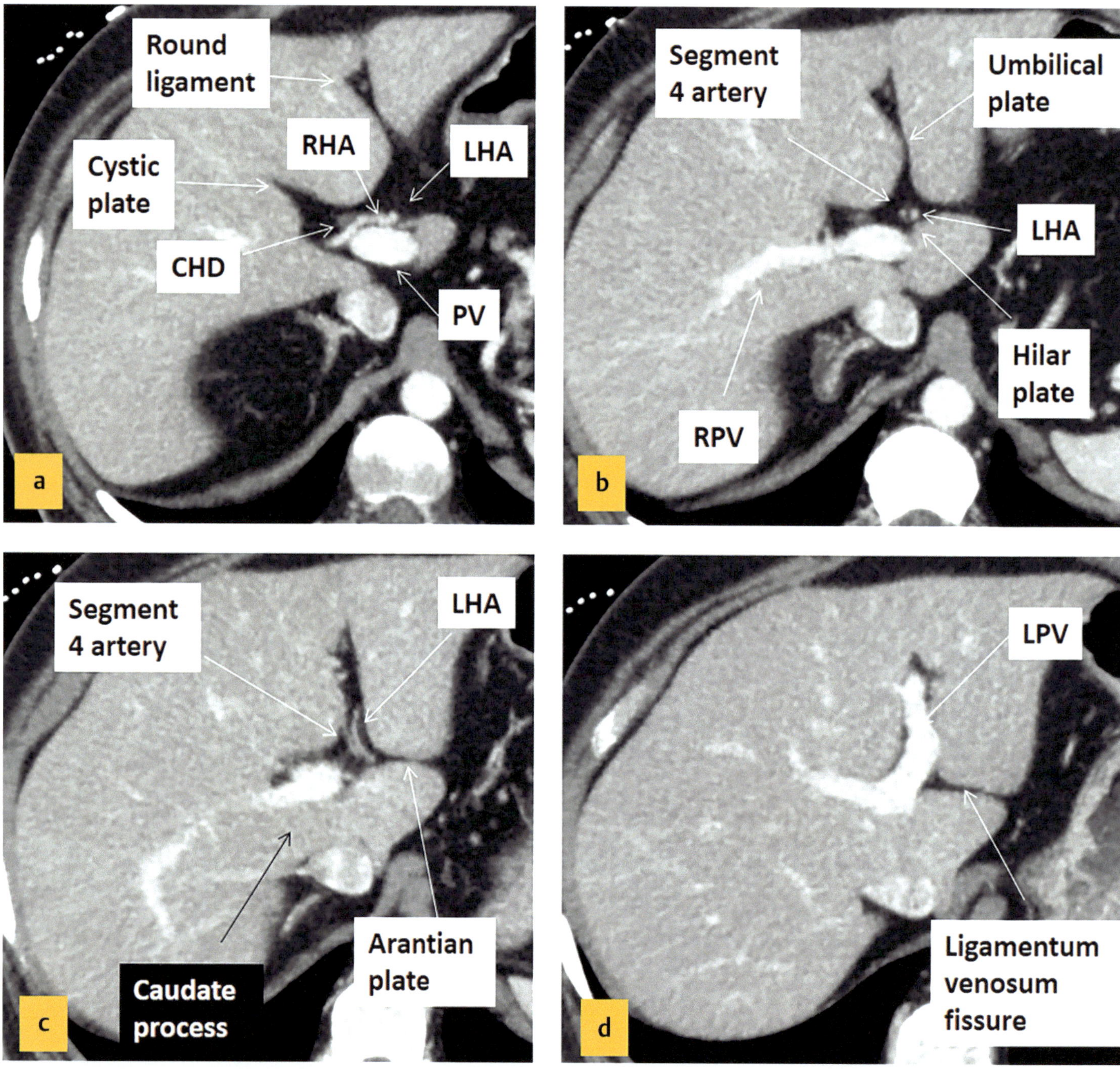

Fig. 6.5 (a–d) Structures in the portoumbilical fissure shown by axial computed tomography (CT) from inferior to superior. **(e, f)** Umbilical artery. It contains the round ligament, the left portal vein (LPV), left hepatic artery (LHA), and a major tributary of the left hepatic vein called umbilical fissure vein. Occasionally a small umbilical artery is seen arising from the left hepatic artery and ending on the capsular surface of the left liver. Approximate location of the plate system is shown in **a–d** comprising the hilar plate in the hepatic hilum, the umbilical plate enveloping the umbilical portion of the portal vein, the cystic plate in the gallbladder bed, and the Arantian plate covering the ligamentum venosum. CHD, common hepatic duct; PV, portal vein; RHA, right hepatic artery; RPV, right portal vein.

(*Continued*)

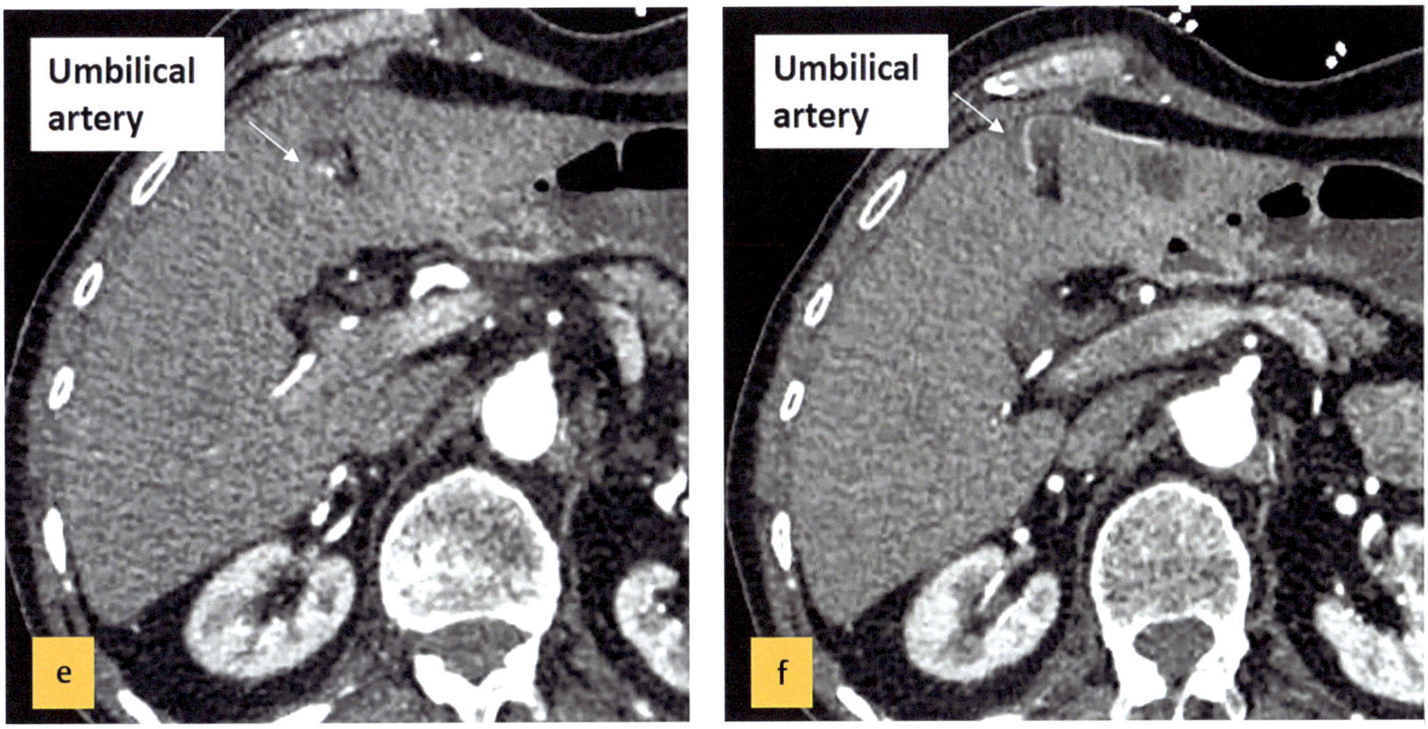

Fig. 6.5 (*Continued*) **(e, f)**

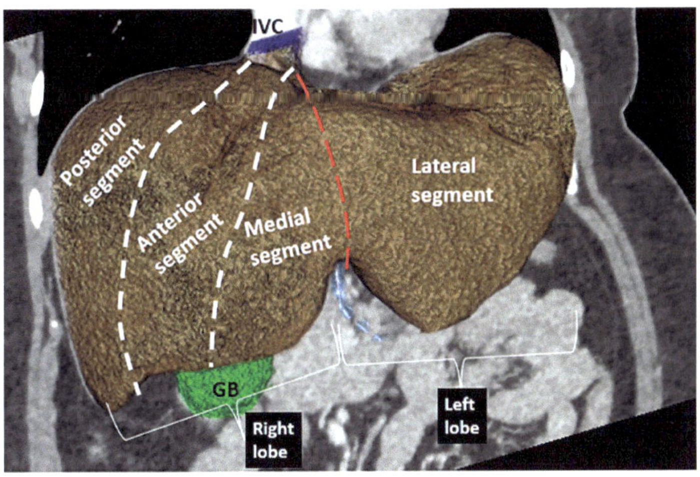

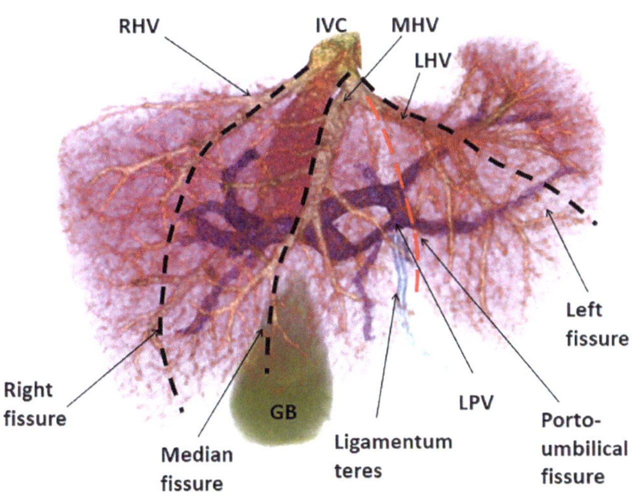

Fig. 6.6 The four fissures of liver surface on the frontal view. The portoumbilical fissure is the only real fissure where the falciform ligament is attached. It contains the ligamentum teres and vertical part of the left portal vein (LPV). The plane of Cantlie's line passes through the gallbladder and IVC bed and corresponds to the median fissure. GB, gallbladder; IVC, inferior vena cava; LHV, left hepatic vein; MHV, middle hepatic vein; RHV, right hepatic vein.

The term "portal lobules" is defined as a unit with the portal tract in the center and the central veins around it (▶Fig. 6.11). Hepatocytes account for 80% of liver cells. Other parenchymal cells include cholangiocytes (biliary epithelial cells), sinusoidal endothelial cells, Kupffer cells (resident liver macrophages), pit cells (natural killer cells), and hepatic stellate cells (▶Fig. 6.11). The hepatocytes are arranged as cords. The basolateral surfaces of the hepatocyte face the fenestrated sinusoidal endothelial cells. Tight junctions between neighboring hepatocytes form a canaliculus that surrounds each hepatocyte. Bile secreted by the cholangiocytes is carried by the canaliculi to the bile ductules within the portal triad.

Hepatic Lobes and Segments

Preoperative analysis of individual liver segments in terms of volume, shape, and location has become a common practice before surgical resection of liver tumors. Using preoperative CT or MRI, the location of a tumor, the course of the intrahepatic vessels, and the liver tissue to be resected can be calculated.

The portoumbilical fissure is considered to divide the liver into right and left hemilivers.[5] Traditionally, the liver is divided into four lobes based on external landmarks: the right, left, quadrate, and caudate lobes with the caudate lobe anatomically classified as a part of the right lobe (▶Fig. 6.2). On the visceral surface of the liver, the quadrate lobe is delimited by the gallbladder fossa, the porta hepatis, and the portoumbilical fissure (▶Fig. 6.2). The caudate lobe is demarcated by the groove for the IVC, the porta hepatis, and the fissure of the ligamentum venosum.

In surgical and radiological practices, the above subdivision is largely replaced by classifications based on internal hepatic architecture and functional features. Functional subdivision of the liver is performed based on arterial blood supply, portal venous blood supply, biliary drainage, and hepatic venous drainage. Using these

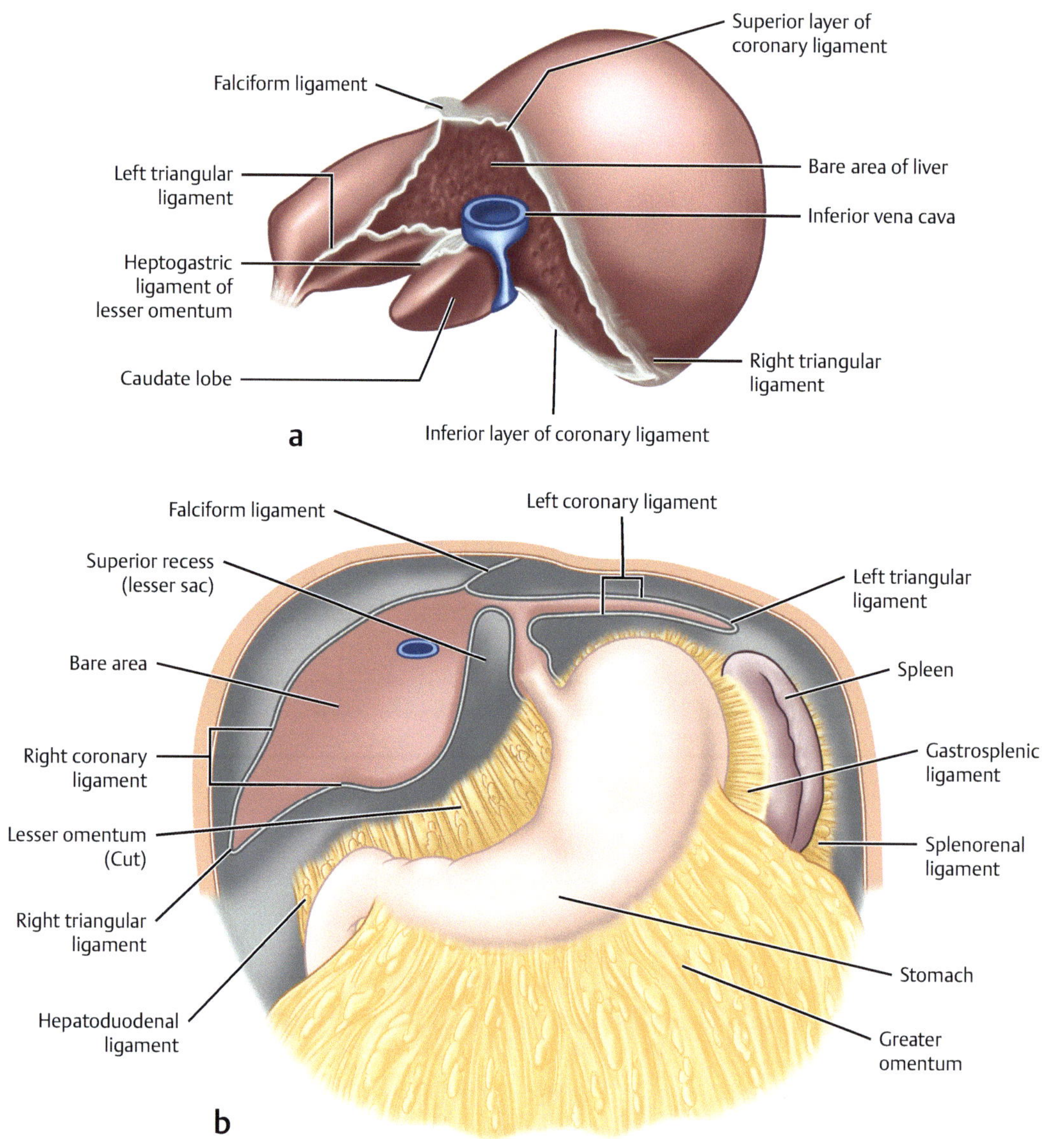

Fig. 6.7 Peritoneal attachments and bare area of the liver. **(a)** Visceral attachments. **(b)** Diaphragmatic attachments. Peritoneal attachments of the liver to the diaphragm include the falciform ligament, and the coronary ligaments with the right and left triangular ligaments. The anterior and posterior limits of the bare area are determined by the coronary ligaments. The inferior layer of the right triangular ligament meets the right margin of the lesser omentum anterior to the caudate lobe. The inferior layer of the left triangular ligament is continuous with the left margin of the lesser omentum.

vascular landmarks, the location of a focal lesion can be determined using cross-sectional imaging methods.

Healey and Schroy (1953) and Couinaud (1957) introduced the segmental terminology that is widely used[7–9] (▶Table 6.1). In most classifications, the liver is divided into two hemilivers by Cantlie's line (median fissure; ▶Fig. 6.6). The plane of Cantlie's line regularly passes through the gallbladder and IVC bed. It is located on the left side of the middle hepatic vein (MHV) and passes through the medial side of Couinaud's segment VIII in 60 to 70% of cases.[10] There is no standard definition for liver subunits including sectors, segments, and subsegments. Based on Glissonean division of portoductal branches, the main right and left trunks are first-order branches (usually 2), sectoral divisions (usually 4) are second-order branches, and segmental levels are considered third-order ducts (usually 8–20).

In most classifications, segment I is formed by the liver tissue that lies between the posterior aspect of the portal bifurcation and the vena cava. Cantlie's line divides the caudate lobe into right and left parts (▶Fig. 6.13, ▶Fig. 6.14, ▶Fig. 6.15). The left part of segment I corresponds to the Spiegel's portion of the caudate lobe. The caudate process of the caudate lobe is continuous with the right lobe and forms the superior boundary of the epiploic foramen. Its bile ducts, arteries, and portal veins arise from both right and left main branches. The caudate lobe is usually drained by two small hepatic veins that enter the left side of the IVC.

Healey and Schroy divided the human liver into eight subsegments based on the distribution of bile ducts and portal vein branches (▶Table 6.1). Four segments are initially made by the right, middle, and umbilical fissures. The right and middle fissures were defined by following the right and middle hepatic

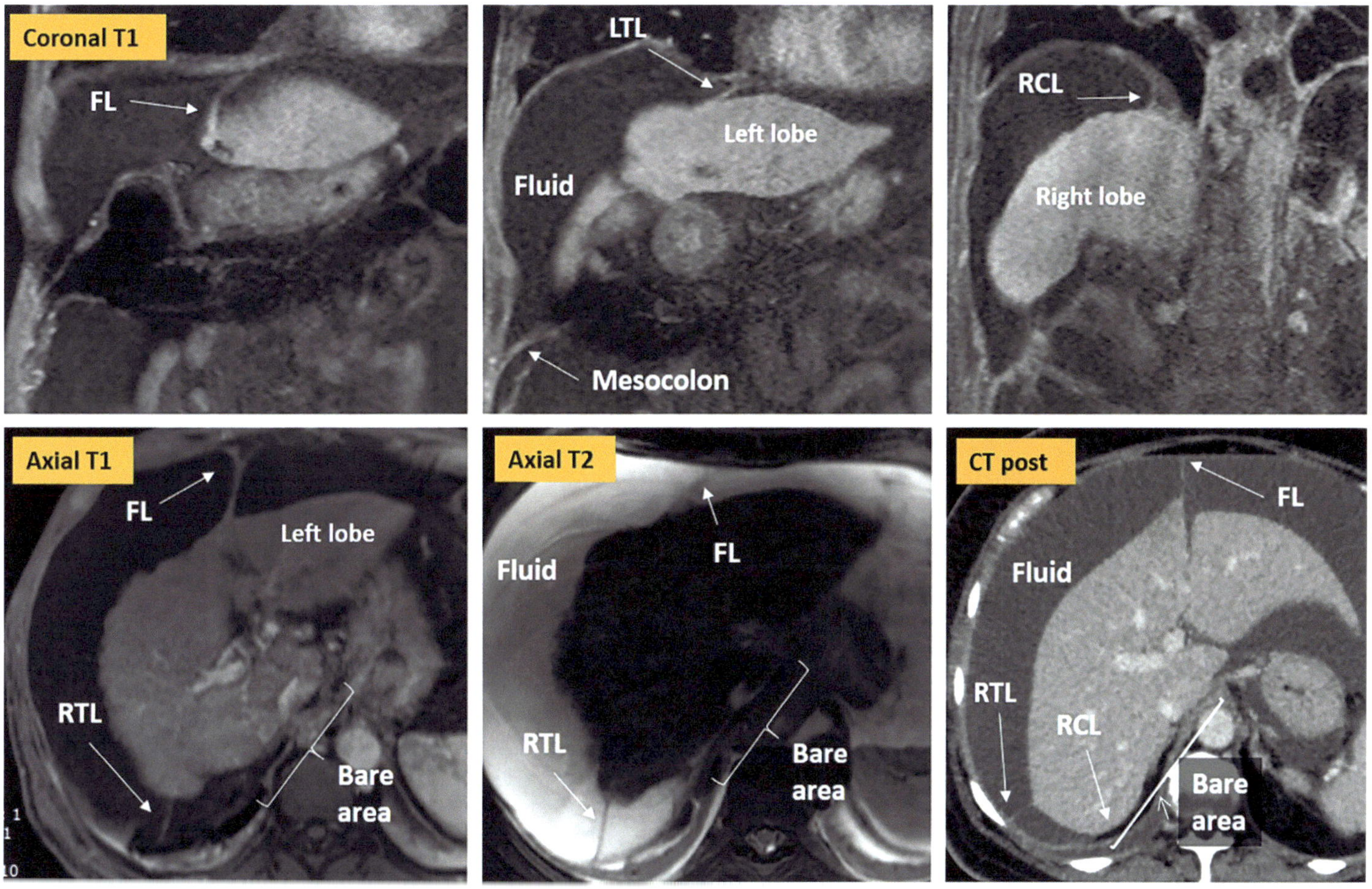

Fig. 6.8 Peritoneal attachments of the liver to the diaphragm include the falciform ligament, and the coronary ligaments with the right and left triangular ligaments. In computed tomography (CT) or magnetic resonance imaging (MRI) these structures are only seen when surrounded by air of ascetic fluid as shown in these patients with cirrhosis. Upper row shows T1-weighted coronal images from anterior to posterior. Lower low shows axial T1, axial T2, and axial contrast-enhanced CT. FL, falciform ligament; LTL, left triangular ligament; RCL, right coronary ligament; RTL, right triangular ligament.

veins, respectively. Each of these segments was then divided into two subsegments by a transverse line drawn through the right and left portal branches. Most of the caudate lobe was placed in the medial segment of the left lobe.[5,10–12]

Couinaud divides the liver into segments I to VIII in clockwise order based on the distribution of the portal and hepatic veins. Four sectors were created by the right, middle, and left fissures (▶Fig. 6.16). He separated the caudate lobe as segment I. The lateral segment of the left liver was divided by the left fissure (left hepatic vein) into segments II and III. The umbilical fissure separates segment IV from segment III. The right lobe was divided into paramedian (anterior) and lateral (posterior) sectors, being nourished by the right anterior (segments V and VIII) and right posterior (segments VI and VII) portal branches, respectively. Each right sector was divided into two segments by an imaginary transverse line drawn through the right portal branch. As per Couinaud's classification, second-order division of the portal vein is called "sector," which corresponds to segment as defined by Healey and Schroy, and the third-order branching is named "segment" not "subsegment" of Healey and Schroy. In this classification, subdivision of segment IV and the caudate lobe into two parts is not justified. Segment VIII is the largest, accounting for a median of 25% of total liver volume, whereas segments II and III each represents <10% of the liver volume.[13]

Bismuth's liver segmentation also used the three fissures (scissurae) hosting the hepatic veins to divide the liver into four sectors and a transverse plane passing through the right and left portal branches to form the segments.[14] Each sector is supplied by a portal pedicle. The left portal scissura divides the left liver into two sectors: anterior and posterior. The anterior sector is divided

6

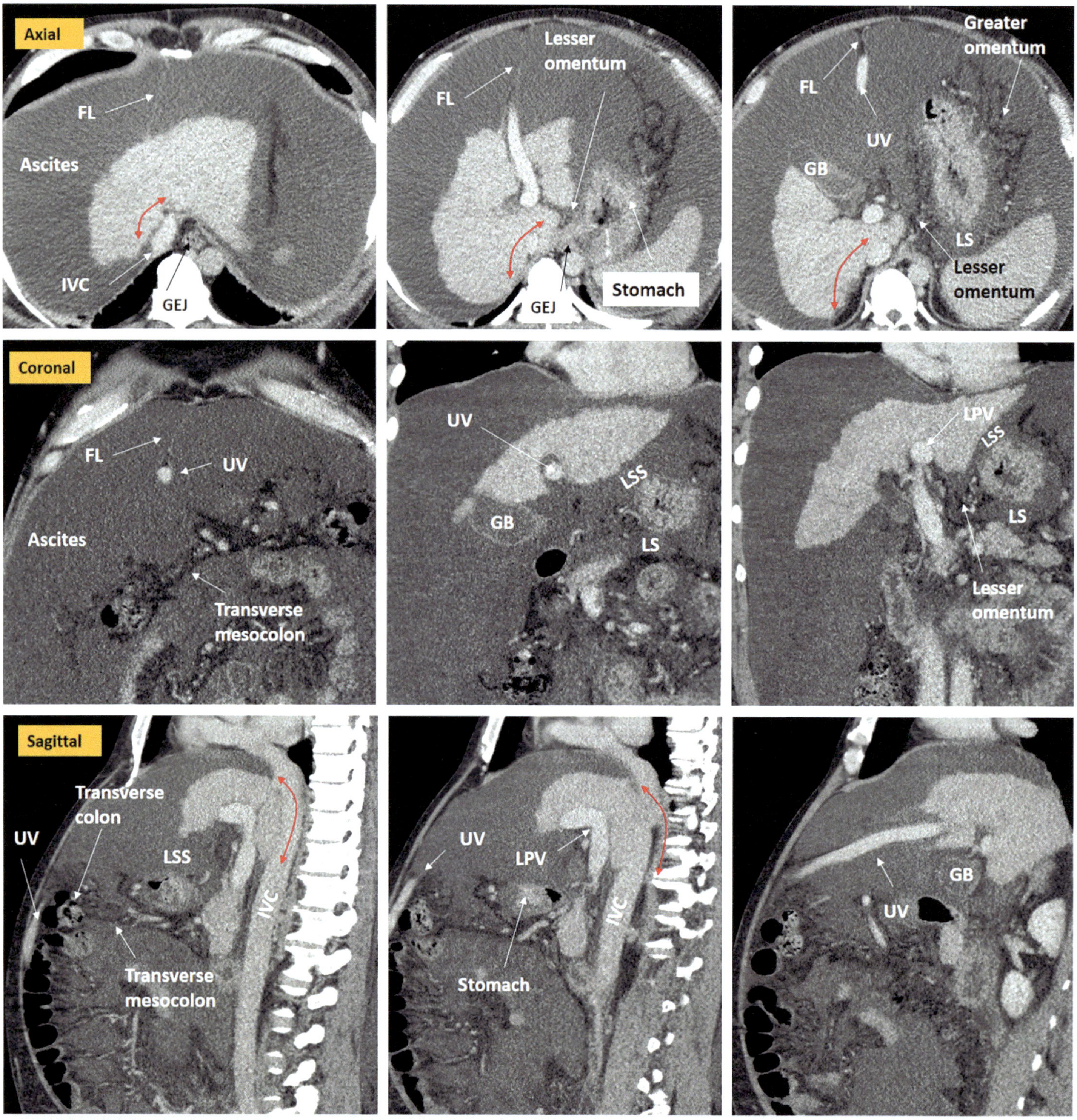

Fig. 6.9 Falciform ligament (FL) and recanalized umbilical vein (UV) are shown by computed tomography (CT) in a patient with cirrhosis and large ascites. The extent of bare area is shown by *double-headed red arrows*. The falciform ligament originates from the anterior abdominal wall (between the diaphragm and the umbilicus) and extends to the anterior surface of the liver at the umbilical fissure. The leaves of the peritoneal fold of the falciform ligament separate over the liver. The round ligament (umbilical vein) travels between the two leaves of the peritoneal fold (containing a little fat in this case) at the inferior edge falciform ligament. The space under the right lobe is the subhepatic space of Morison and the space under the left lobe consists of the left subhepatic (LSH) and the lesser sac (LS) spaces; these two spaces are separated by the lesser omentum. GB, gallbladder; GEJ, gastroesophageal junction; IVC, inferior vena cava; LPV, left portal vein; LSS, lesser sac splenic recess.

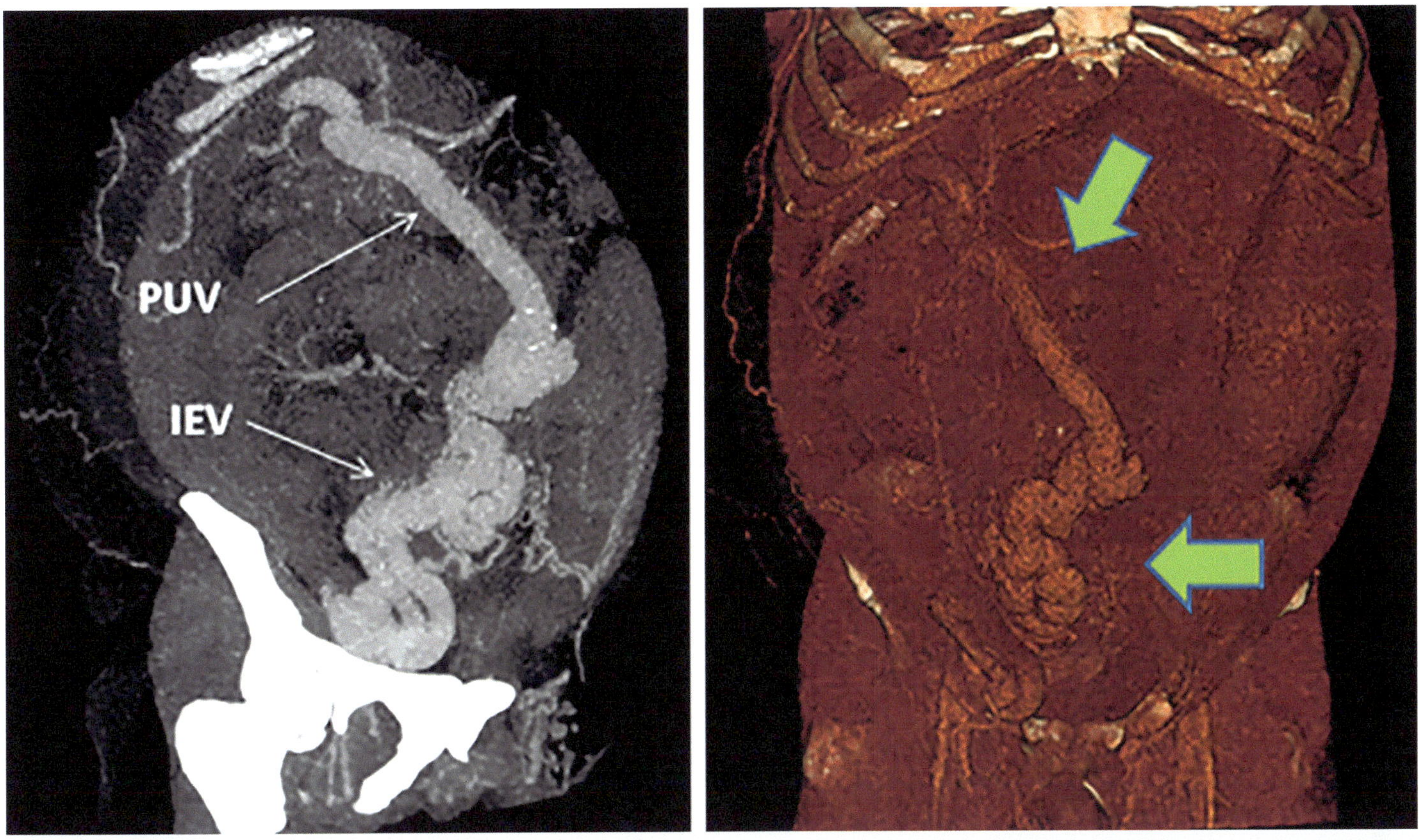

Fig. 6.10 Recanalization and dilation of the umbilical vein along with formation of varicoid paraumbilical veins (PUV) are common in patients with long-standing portal hypertension and eventually communicate with the system epigastric and diaphragmatic veins creating varicoid portosystemic shunts in the anterior abdominal wall known as caput medusa (*arrows*). IEV, inferior epigastric vein.

into segments III and IV by the umbilical fissure. The caudate is called segment I and segment IV is subdivided into IVa and IVb subsegments.

Federative Committee on Anatomical Terminology (FCAT) combined different classifications and divided liver based on distributions of the portal vein, hepatic arteries, and biliary ducts.[11]

Many surgeons prefer to use the classification of liver anatomy and resections defined by Terminology Committee of the International Hepato-Pancreatico-Biliary Association (IHPBA) held in Brisbane 2000 in which differences between biliary ducts, hepatic artery, and portal vein branches in segments and sectors are amplified.[15] They suggest calling the "section" of liver based on bile ducts and hepatic artery (similar to Healey and Schroy) and the "sector" based on portal vein (similar to Couinaud). Based on the ramification of the hepatic artery and bile duct, the medial (paramedian) sector/section of the left liver lobe is formed by segment IV only and based on portal branches it is formed by segments III and IV (▸Table 6.1).

Recently, Hjortsjo's anatomical model (1951), proposing six segments (three in each lobe) in the liver, has regained interest, which involves a partition plane in the conventional anterior (paramedian) sector.[14,16] In this model, the anterior sector (between the right and middle hepatic veins) is divided into two areas with a vertical plane (anterior segment fissure) passing along a hepatic venous tributary of the middle or the right hepatic vein or an independent hepatic vein. Therefore, the right hemiliver has three segments. This anatomy has surgical relevance for subsegmentectomy of the anterior sector.[12]

Limitations of Couinaud's Classification

Surgeons and radiologists commonly encounter liver segmentations that are inconsistent with those expected based on Couinaud's classification.[17] Its major limitation is because the volume, position, and shape of liver segments and their borders show significant variability. This is specially the case when a specific segment is being resected. However, for large resections such as hemihepatectomy, the segmental anatomy of the hepatic veins follows Couinaud's segmental anatomy.[17]

Using corrosion cast models of intrahepatic portal branches, it has been shown that portal veins give off much more branches than the eight suggested by Couinaud.[18–20] The human liver has an average of 20 (and not 8) segmental portal venous branches, ranging from 9 to 44.[18,19] These 20 territories can be grouped together

6

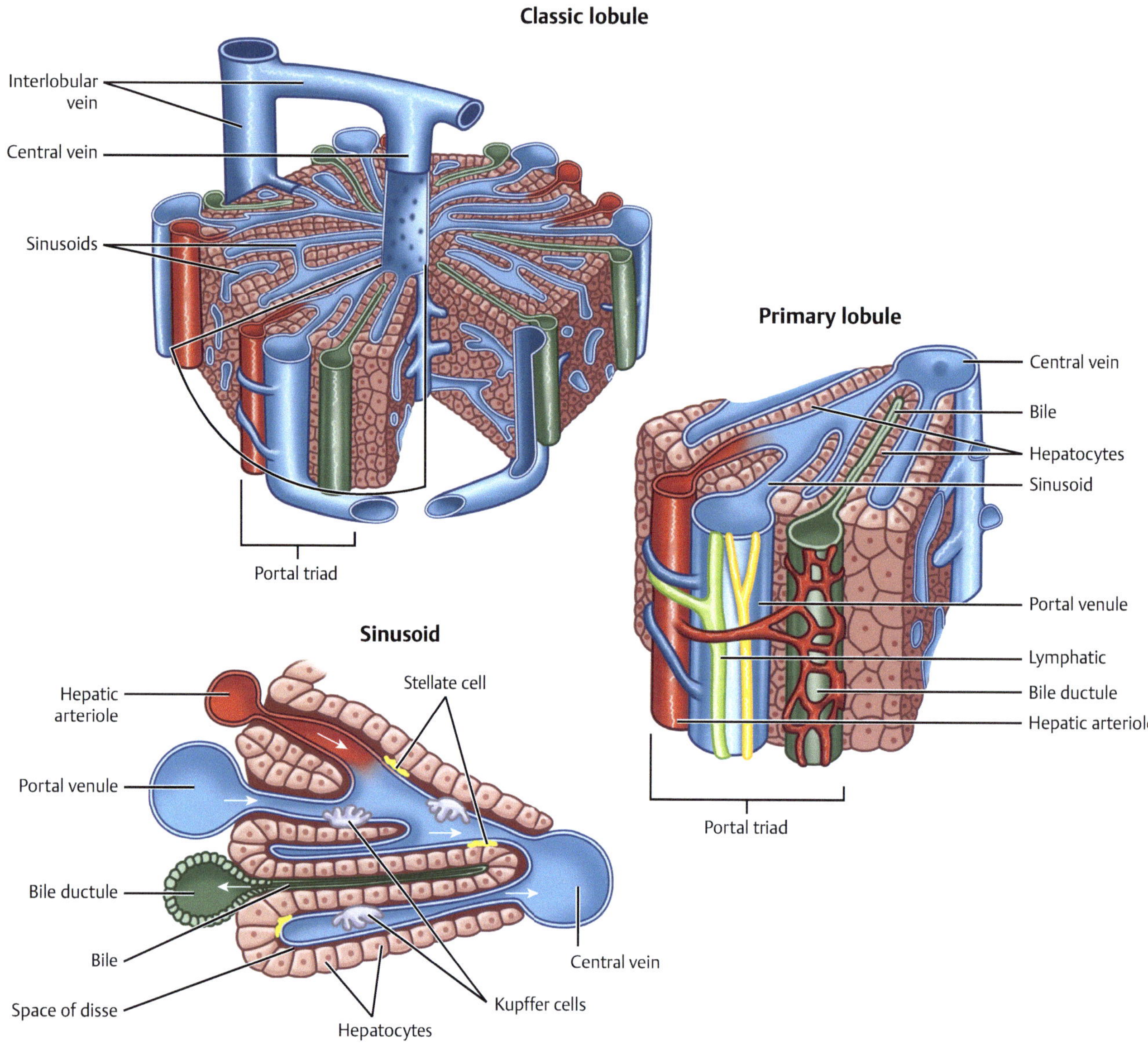

Fig. 6.11 Liver microvascular anatomy and vascular communications around the portal triad. *Arrows* in sinusoids indicate direction of flow.

based on the three main hepatic veins and the portal bifurcation to fit the eight segments of Couinaud's scheme (▶ Fig. 6.17). For example, three or four subsegmental vessels commonly visible on the right side of the umbilical portion of the left portal vein can be grouped together in order to obtain a single segment IV (▶ Fig. 6.17, ▶ Fig. 6.18).

Despite all these variations, it is important to know that the traditional eight-segment model of Couinaud still serves as the most acceptable model to address the location of focal hepatic lesions. However, the real branching of the major portal pedicles and of the hepatic veins is taken into account when a segment of the liver is to be resected. Respecting the segmental territorial portal branching is important in segmental liver resection to decrease complications such as hemorrhage, ischemia, congestion, or biliary leaks.[21]

Glisson's Capsule and Hilar Plate

Glisson's capsule wraps the liver and continues in the hepatoduodenal ligament around the portal vein, hepatic artery, and bile duct.[22–24] This capsule does not exist around the hepatic veins. The combination of this capsule with sheaths around hilar vessels and bile ducts is called Glissonean system forming two Glissonean pedicles which wrap the first-order ducts and vessels and continue as far as second- or even third-order branches.[22,25] Externally, Glisson's capsule continues into the subserosa of the gastrohepatic and hepatoduodenal ligaments. This anatomic continuity at subserosal/subperitoneal space provides a path for the spread of disease including edema, inflammation, and cancer between the intraperitoneal and extraperitoneal organs. Best clinical example is seen in acute pancreatitis in which edema and

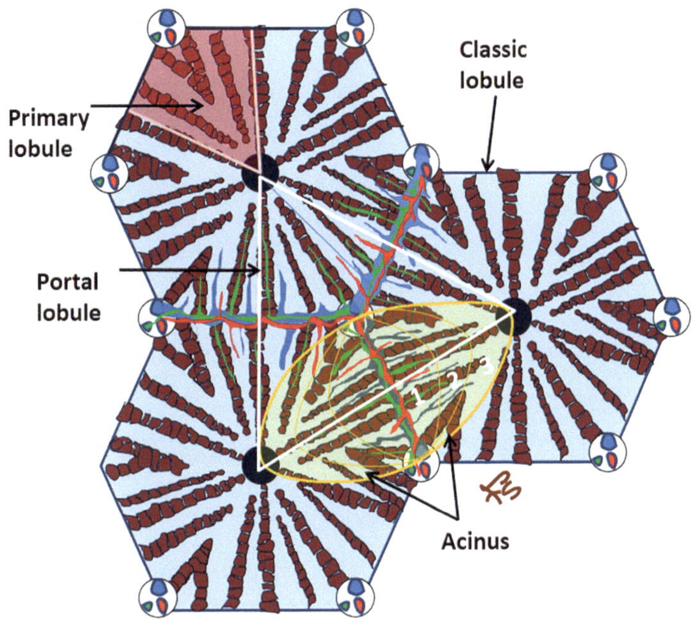

Fig. 6.12 Functional units of the liver. The "primary lobule" represents a slice (*red region*) of the hexagonal "classic lobule" with the central hepatic vein at its apex and portal triad vessels at the outer margin. The "acinus" borders (*oval-shaped yellow line*) are formed by the penetrating vessels at the bases where inflow blood from the portal vein and hepatic artery enters sinusoids and ends in the central hepatic vein. Three zones (1, 2, and 3) of the acinus represent metabolic regions that are increasingly distant from the inflow vessels. The *white triangle* represents the "portal lobule."

Table 6.1 Classification of liver segments.

Part	Classification									
	J. E. Healey and P. C. Schroy (1953)		C. Couinaud (1957)		H. Bismuth (1982)		FACT (1998)		Brisbane 2000 system	
	Segment	Subsegment	Sector	Segment	Sector	Segment	Sector	Segment	Section	Segment
Dorsal	Caudate	Right	Caudate	I	Caudate	I	Caudate	Posterior, I	Caudate	I
		Left	lobe		lobe		lobe			
Left	Lateral	Superior	Lateral	II	Posterior	II	Lateral	Posterior, II	Lateral	II
		Inferior	Paramedian	III	Anterior	III		Anterior, III		III
	Medial	Superior		IV		IVa, IVb	Medial	Medial, IV	Medial	IV
		Inferior								
Right	Anterior	Inferior	Paramedian	V	Anteriomedial	V	Medial	Anterior, V	Anterior	V
		Superior		VIII		VIII		Posterior, VIII		VIII
	Posterior	Inferior	Lateral	VI	Posterolateral	VI	Lateral	Anterior, VI	Posterior	VI
		Superior		VII		VII		Posterior, VII		VII
	A+a.d.		**PV+HV**		**PV+HV+A+ad**				**A+b.d.**	

Source: Adapted with permission from Rutkauskas et al[11] and Shindoh et al.[12]

Healey and Schroy classification is based on the distribution of bile ducts and portal vein branches. The Couinaud's segmentation system is based on the distribution of both the portal vein and the hepatic veins. Federative Committee on Anatomical Terminology (FCAT) combined different classifications and divided liver based on distributions of the portal vein, hepatic arteries, and biliary ducts. The umbilical fissure is the division plane between segments III and IV. Based on the ramification of the hepatic artery and bile duct, the medial/paramedian section/sector of the left liver lobe is formed by segment IV only and based on portal branches it is formed by segments III and IV.

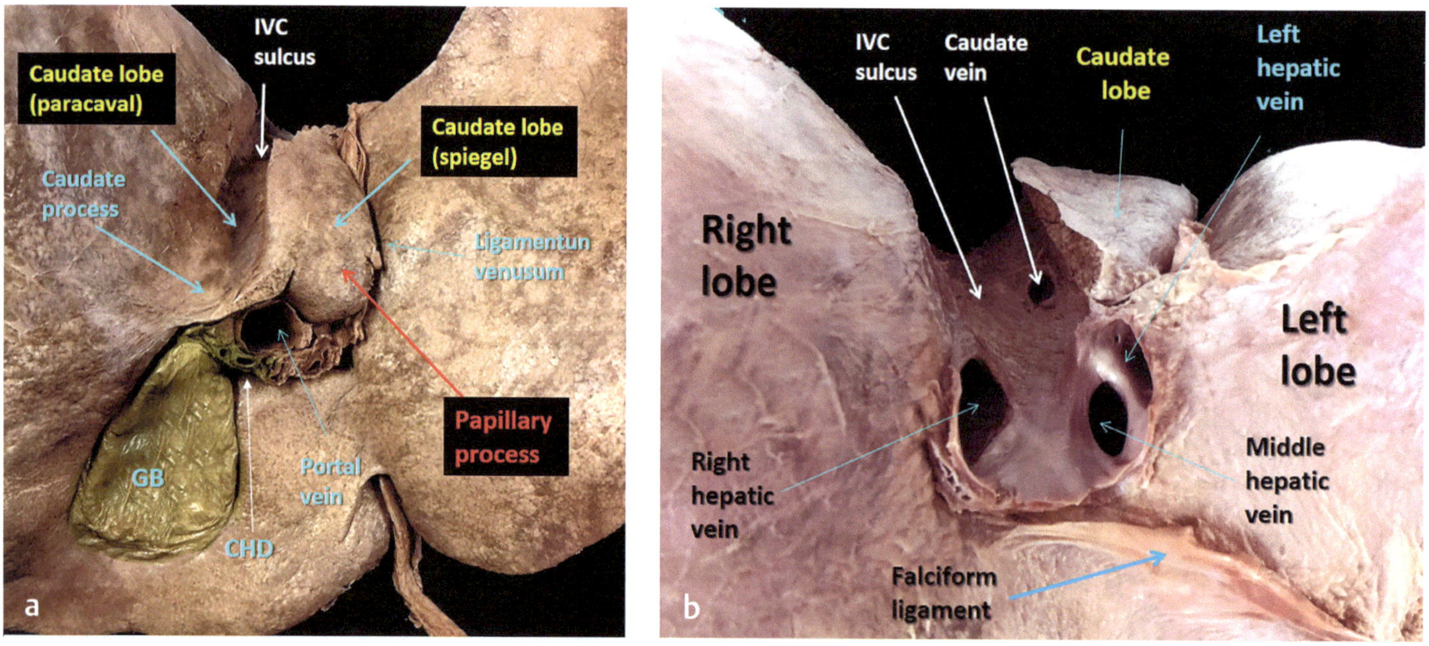

Fig. 6.13 Caudate lobe and porta hepatis. **(a)** Inferior view. **(b)** Diaphragmatic view. The Cantlie's line divides the caudate lobe two parts. The left part of segment I corresponds to the Spiegel's portion of the caudate lobe. The caudate process of the caudate lobe is continuous with the right lobe and forms the superior boundary of the epiploic foramen. CHD, common hepatic duct; GB, gallbladder; IVC, inferior vena cava.

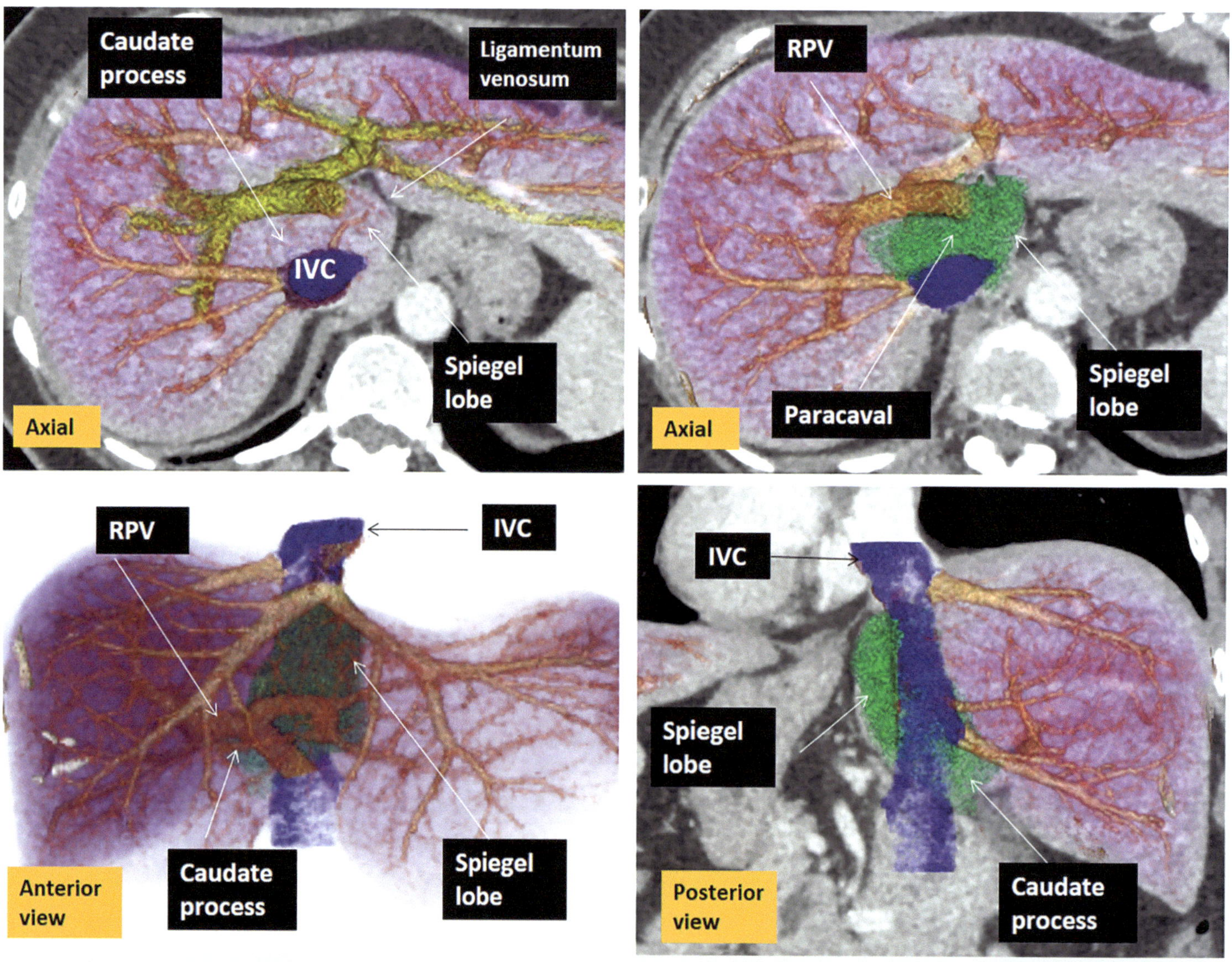

Fig. 6.14 Caudate lobe (segment I) shown in *green*. Based on the portal blood supply the caudate lobe is subdivided into the Spiegel's lobe (occupying the left side of the umbilical fissure), the paracaval portion (in front of the IVC), and the caudate process (on the right inferior side of the RPV). The Spiegel's lobe and the paracaval part together is known as the caudate proper. The caudate process connects the caudate lobe to the right hepatic lobe. Resection of the caudate lobe is technically challenging since this segment is located deep within the liver and is close to hepatic hilar structures. IVC, inferior vena cava; RPV, right portal vein.

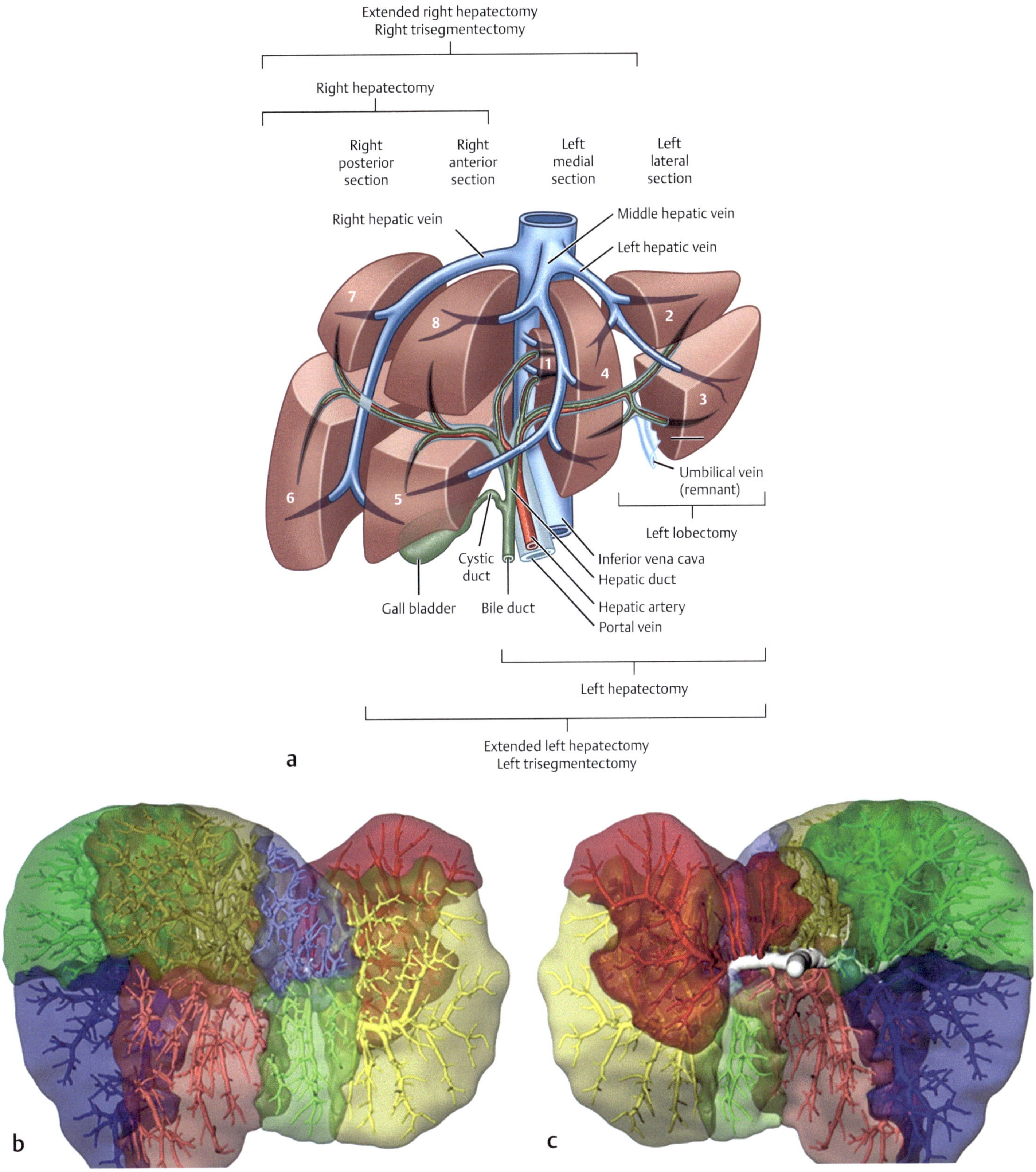

Fig. 6.15 (a) Liver segments based on Couinaud's method. **(b)** Anterior and **(c)** inferior views of the computed tomography (CT)-based reconstruction of the portal venous branching pattern in an anatomical corrosion cast.

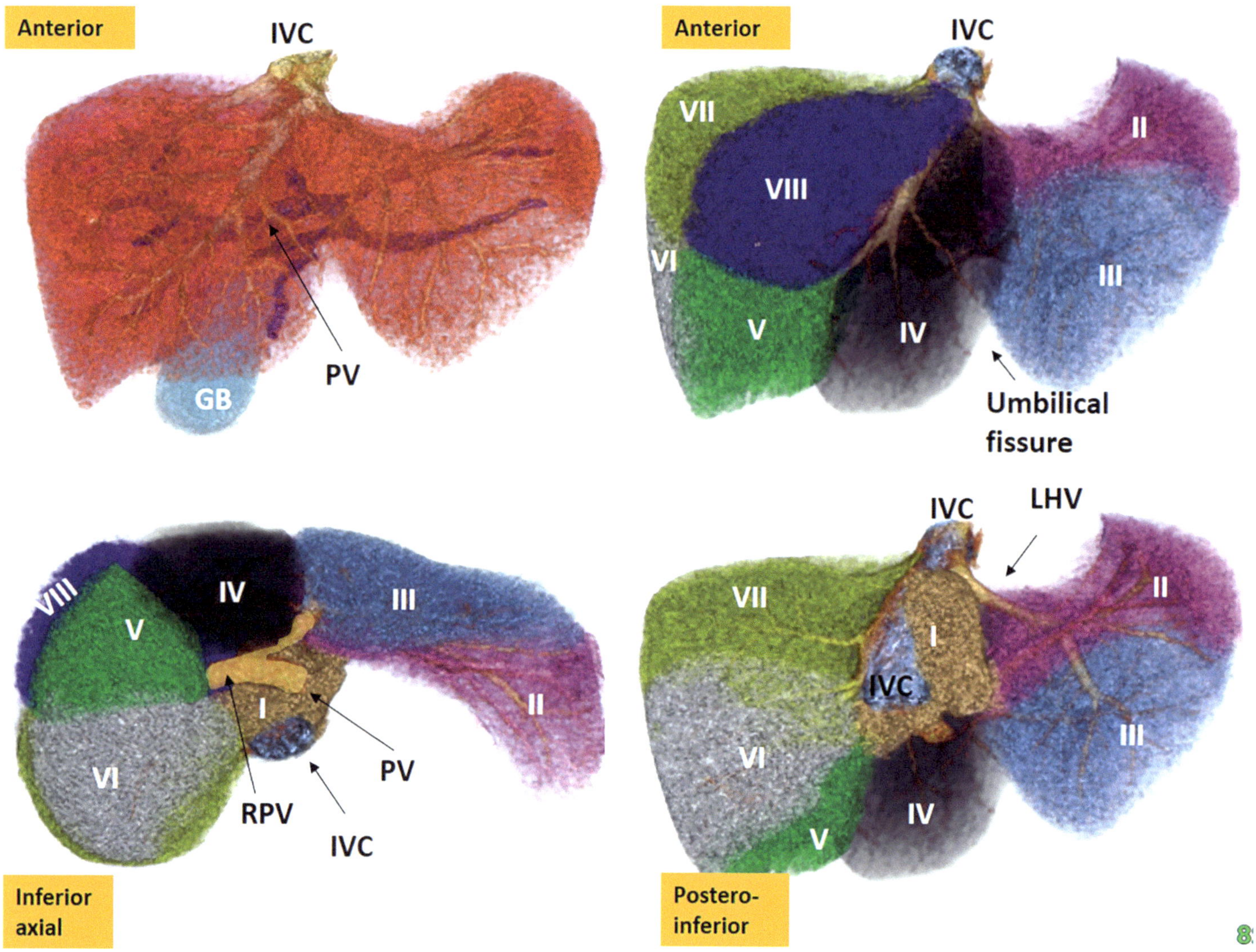

Fig. 6.16 Segments according to Couinaud shown by color-coded computed tomography (CT). The umbilical fissure separates segment IV from segment III. The lateral segment of the left liver was divided by the left fissure (LHV) into segments II and III. GB, gallbladder; IVC, inferior vena cava; LHV, left hepatic vein; PV, portal vein; RPV, right portal vein.

inflammation in the retroperitoneum can spread along subserosal planes to remote regions such as the portal tracks, gallbladder, and even abdominal wall.

Glisson's capsule also forms a thick fibrous connective tissue known as "plate system" at the inferior part of the liver, comprised of the hilar plate in the hepatic hilum, the umbilical plate enveloping the umbilical portion of the portal vein, the cystic plate in the gallbladder bed, and the Arantian plate covering the ligamentum venosum[26] (▶Fig. 6.19). The hilar plate can be detached from the liver when required to resect the liver to access the right and left Glissonean pedicles whereas Glisson's pedicle is difficult to remove from the hepatic parenchyma because of the presence of small branches. The left hepatic duct runs at the base of the segment IV and is covered by the hilar plate.

Recent progress in surgical liver resections of hilar region and caudate lobe for carcinoma requires detailed anatomical knowledge of the plate system especially the anatomy of bile ducts in this region[24,27] (see Chapter 8 "Biliary System"). The bile ducts of the caudate lobe are very small in diameter and run on the dorsal side of the hilar plate. Cancer involvement of the bile duct in the plate system is resectable without the need to resect liver parenchyma; however, when the tumor involves the Glisson's pedicle, it would be very difficult to resect the invaded bile duct alone.

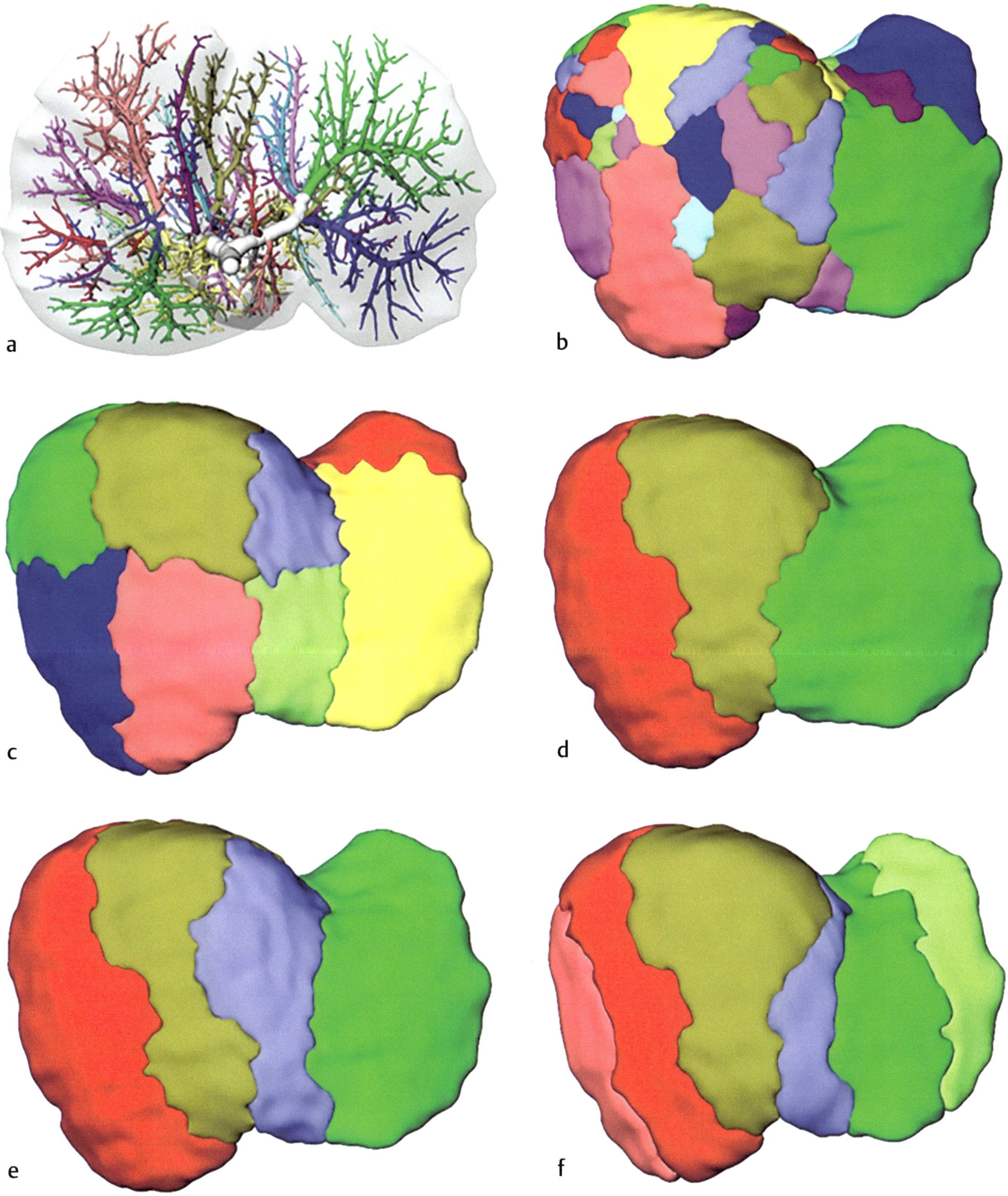

Fig. 6.17 (a) Inferior view of the computed tomography (CT)-based reconstruction of the portal venous branching pattern in an anatomical corrosion cast. **(b–f)** Images showing reconstruction of seemingly contradictory portal venous segmentation models of the same liver. **(c–f)** Obtained by different gatherings of the smaller territories presented in **b**, corresponding to the branching pattern in **a**. **(b)** The portal venous territories, in an anterior view, attributed to each of the branches colored differently in **a**. **(c)** A presentation of eight segments credited to Couinaud (1957). **(d)** The tripartition advocated by Takasaki (2007). **(e)** Echos Goldsmith's and Woodburne's (1957) four sectors. **(f)** Three territories both in the right and left hemilivers, according to Hjortsjo (1951). (Adapted with permission from Couinaud,[9] Hjortsjo,[16] Fasel and Schenk,[18] Fasel,[19] Fasel, Majno, and Peitgen,[20] and Takasaki.[22])

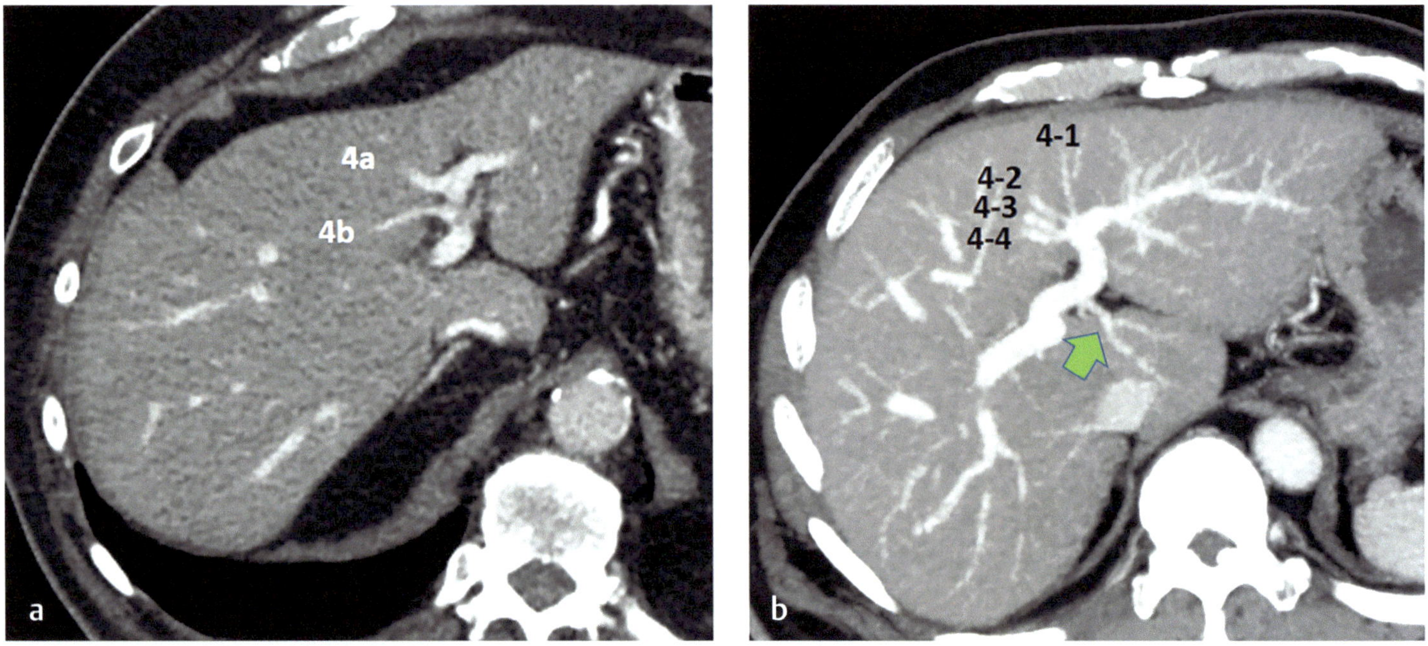

Fig. 6.18 Segment 4 portal supply. **(a)** Classic portal venous supply of the segment IV by two branches arising from the right side of the left portal vein. **(b)** Four branches are visible. Also, note a large single vein supplying the caudate lobe (*arrow*).

Surgical Implication of the Segmental Anatomy

Modern liver resections and transplantation have reached a level of complexity in which the traditional eight-segment model is not sufficient. As discussed previously, segments of the conventional Couinaud's scheme could be fed by third-order (segmental) branches originating from a different second-order branch. Surgeons have difficulties in performing the subsegmental resection of the anterior segment using Couinaud's segmental model because the border between segments V and VIII is only defined by an imaginary transverse line through the right portal vein.

For living-donor transplantations knowledge of the actual vascular anatomy is important where anatomical variation does not allow to simply resect the right or the left liver lobe. Knowing the portal branching territory is even more important before segmental or subsegmental liver resections because there are no significant vasobiliary anastomosis between adjacent segments. Therefore, in segmental liver resections, removal of the vascular pedicle and all biliary radicles is the goal to avoid postoperative hemorrhage and bile leak.

During liver resections, different technical approaches allow recognizing the portal territories. In extrahepatic dissection technique, dissection of intra-Glissonean portal and arterial branches within the liver hilum is performed without liver dissection followed by selective clamping to see a plane of ischemic demarcation between the right and the left liver lobes.[22,23] Using intra-Glissonean approach three segments of the liver can be reached as per the ramification of the second-order Glissonean pedicles (Takasaki's classification).[22]

Intra-Glissonean approach cannot be used to show segmental territories of the right liver. For segmental levels, other techniques will be used. In one technique, ultrasound-guided balloon occlusion of a portal branch followed by indigo carmine blue injection is used to identify second- and third-order branch territories. It can be used for resections of small carcinomas or monosegmental transplantation. In another technique, the hilar plate is exposed followed by an intrahepatic/extra-Glissonean dissection to ligate second-order branches feeding the territory that has to be removed.[23] With extra-Glissonean approach, any form of segmental territorial vascular pedicle can be identified and resected leaving a clean resection bed devoid of the portal veins, and the biliary radicles of the liver remaining in place. Glissonean approaches give access to left portal vein beyond the round ligament in split liver procedures.

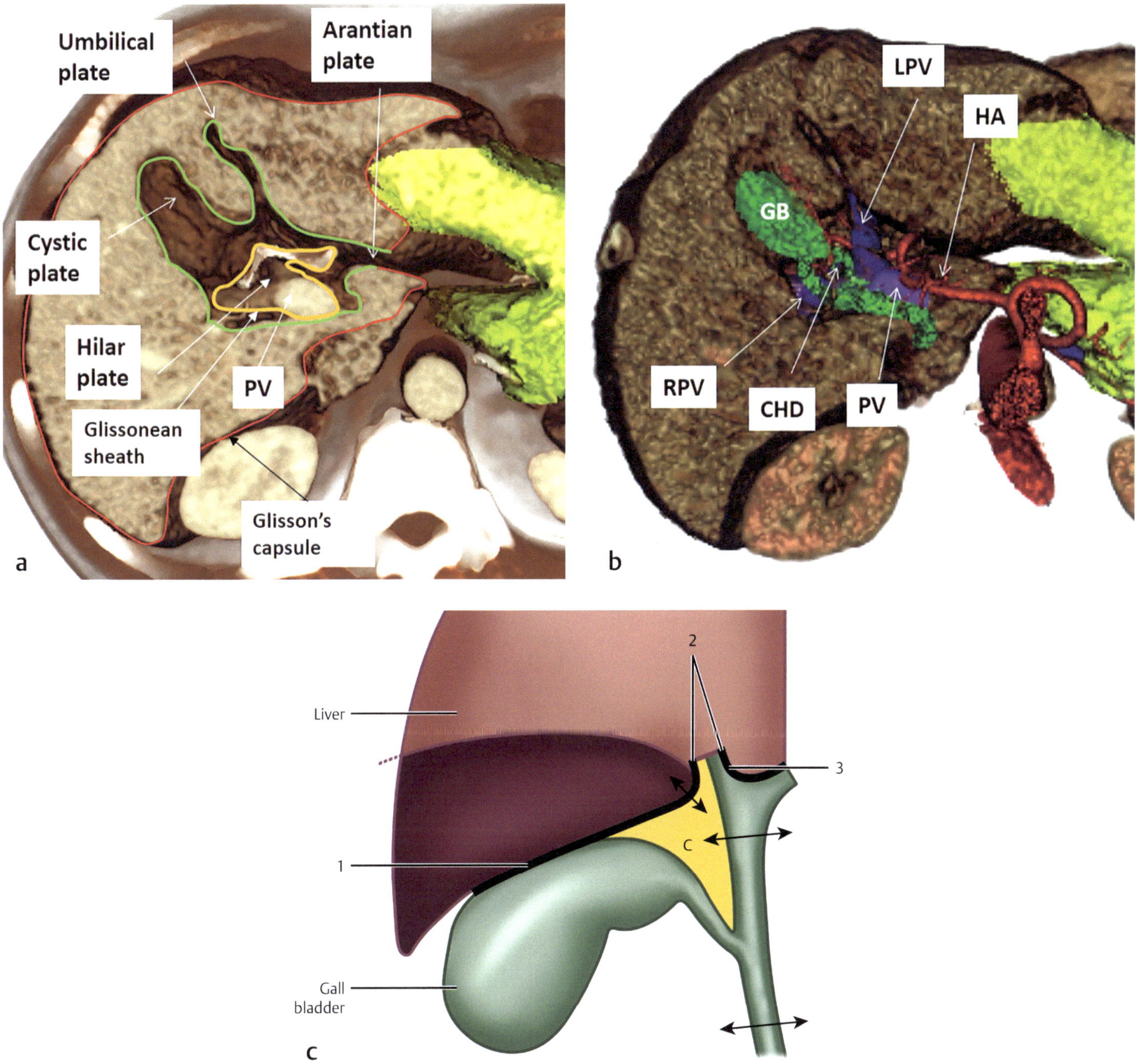

Fig. 6.19 (a–c) Plate system. Glisson's capsule forms a thick fibrous connective tissue known as "plate system" at the inferior part of the liver. It is comprised of the hilar plate in the hepatic hilum, the umbilical plate enveloping the left portal vein, the cystic plate in the gallbladder bed, and the Arantian plate covering the ligamentum venosum. Glisson's capsule also continues in the hepatoduodenal ligament to form Glissonean system which wraps the portal vein, hepatic artery, and bile duct. 1, cystic plate; 2, the dense fibrous sheath encircling the right portal pedicle; 3, hilar plate; C, the triangle of Calot; CHD, common hepatic duct; GB, gallbladder; HA, hepatic artery; LPV, left portal vein; PV, portal vein; RPV, right portal vein.

◆ Vasculature

The most common indication for hepatic vessel study is for assessment of perfusion in tumoral processes. Another common application is for patients who require partial hepatic resection or candidates of liver transplantation. Common problems in imaging assessment of the liver vasculature and perfusion include dual blood supply, parasitic and aberrant vascular supply, and arteriovenous shunting. One important question that may be raised in perfusion imaging of the liver is whether the detected perfusion abnormality is real or simply a pseudolesion due to vascular territorial variants in tissue perfusion. Knowing the anatomy can answer these questions.

Macrovasculature and Anatomical Variants

The study of hepatic vasculature is important in several clinical situations: before or after liver surgeries such as liver transplantation and segmentectomy procedures and in chemotherapy treatment of primary or secondary tumors of liver. Knowledge of the complex vascular anatomy of the liver is also important in perfusion studies since perfusion heterogeneities may be related to anatomical variants in liver vasculature and small arterioportal shuntings and not just diffuse parenchymal disease. The liver has a dual blood supply, approximately 25% comes from the hepatic artery and the remaining 75% from the portal vein.[28] The drainage of the liver is mainly provided by the hepatic veins. The hepatic artery is a vessel of resistance, whereas the portal and hepatic veins are vessels of capacitance.

Arteries

The classification of Michels is the most frequently employed system to describe hepatic artery variants[29] (▶ Fig. 6.20). Conventional anatomy (Type 1) is seen in 60 to 80% of CT studies consisting of a main hepatic artery arising from the celiac trunk.[30–34] The main

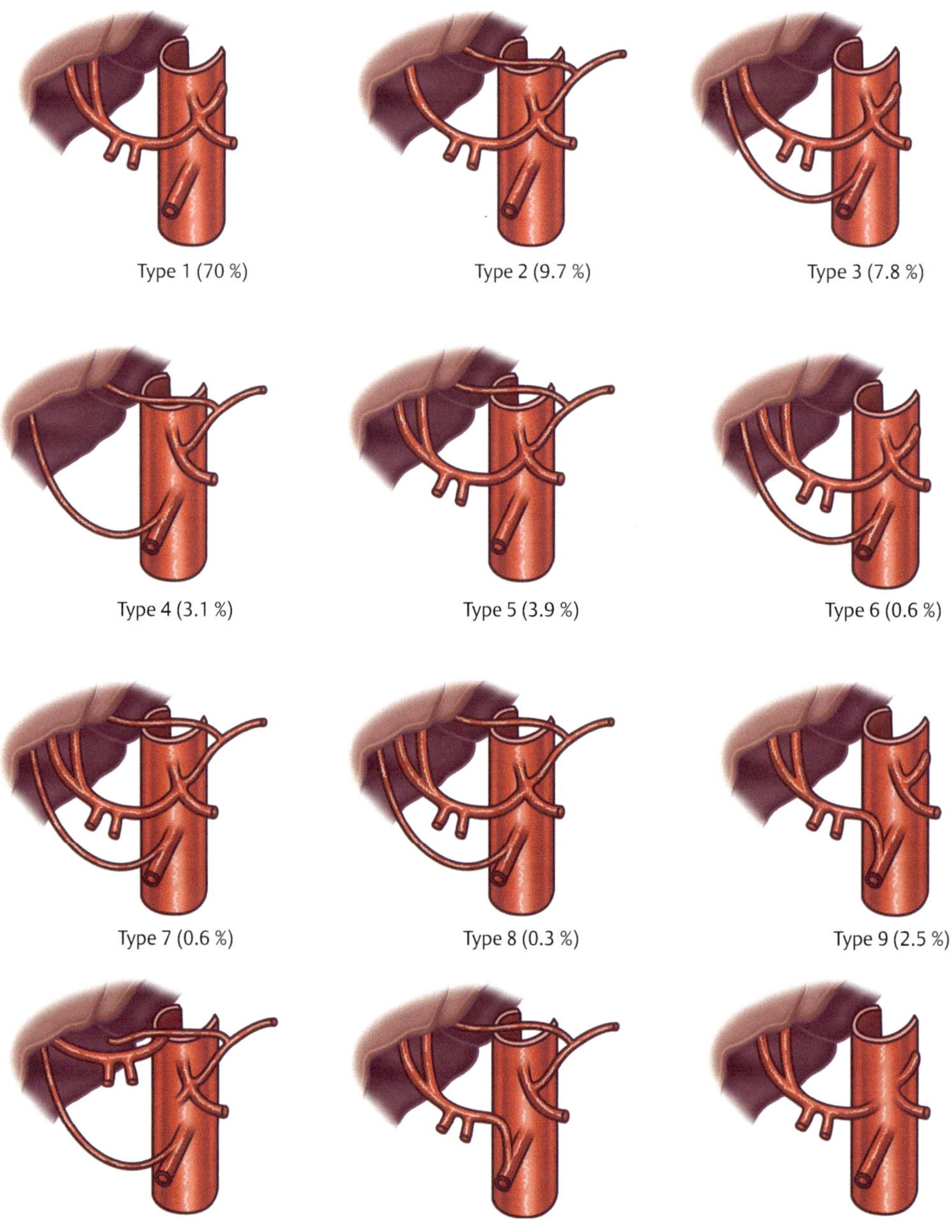

Fig. 6.20 Hepatic arterial anatomy in 1,081 cases according to Michels' classification.[29] Percentages corresponding to each anatomical type are shown.

hepatic artery gives off the gastroduodenal artery and continues as the proper hepatic artery. The proper hepatic artery continues as the right hepatic artery after giving off the left hepatic artery and then the right hepatic artery splits into its anterior and posterior sectoral branches. The left hepatic artery splits into branches that supply segments II and III.[34,35]

The commonest variant is replaced left hepatic artery arising from the left gastric artery in 2.5 to 14% of cases and a replaced right hepatic artery from the superior mesenteric artery in 8 to 11% of cases.[30–32] Direct origin of the right or proper hepatic arteries from the aorta is rarely seen (▶Fig. 6.21). From a surgical point of view, replaced arteries must be always preserved in reconstructive surgeries. In contrast, accessory arteries do not necessarily need to be reconstructed if adequate flow is established after the anastomosis of the major branches.[36] The replaced left hepatic artery from the left gastric artery typically runs to the right within the superior part of the lesser omentum, entering the liver through the fissure of the venous ligament. Yet, the abnormal artery may take a different path relative to hilar structures. While passing the lesser omentum, a replaced left hepatic artery could be damaged during gastrectomy and hernia repairs. The replaced right hepatic artery typically arises from the superior mesenteric artery, runs upward behind the pancreas and portal vein, and anterior to the IVC into the hepatic pedicle (▶Fig. 6.22). The replaced right hepatic artery is the source of the cystic artery, and because of its low run, it could be ligated during cholecystectomy by mistake (see Chapter 8 "Biliary System"). As the replaced right hepatic artery passes through the pancreatic head, it could be damaged in pancreatic resections.[32] Some anatomic variations can be of advantage in liver surgery. For example, a replaced left hepatic artery may serve as an ideal artery for anastomosis in left lobe liver transplantation by allowing rapid dissection of the porta hepatis.

The right hepatic artery divides into anterior and posterior sectional divisions (▶Fig. 6.23). The anterior division travels straight right and upward and the posterior branch turns proximally. Arterial supply for segment VI (A6) from the left posterior

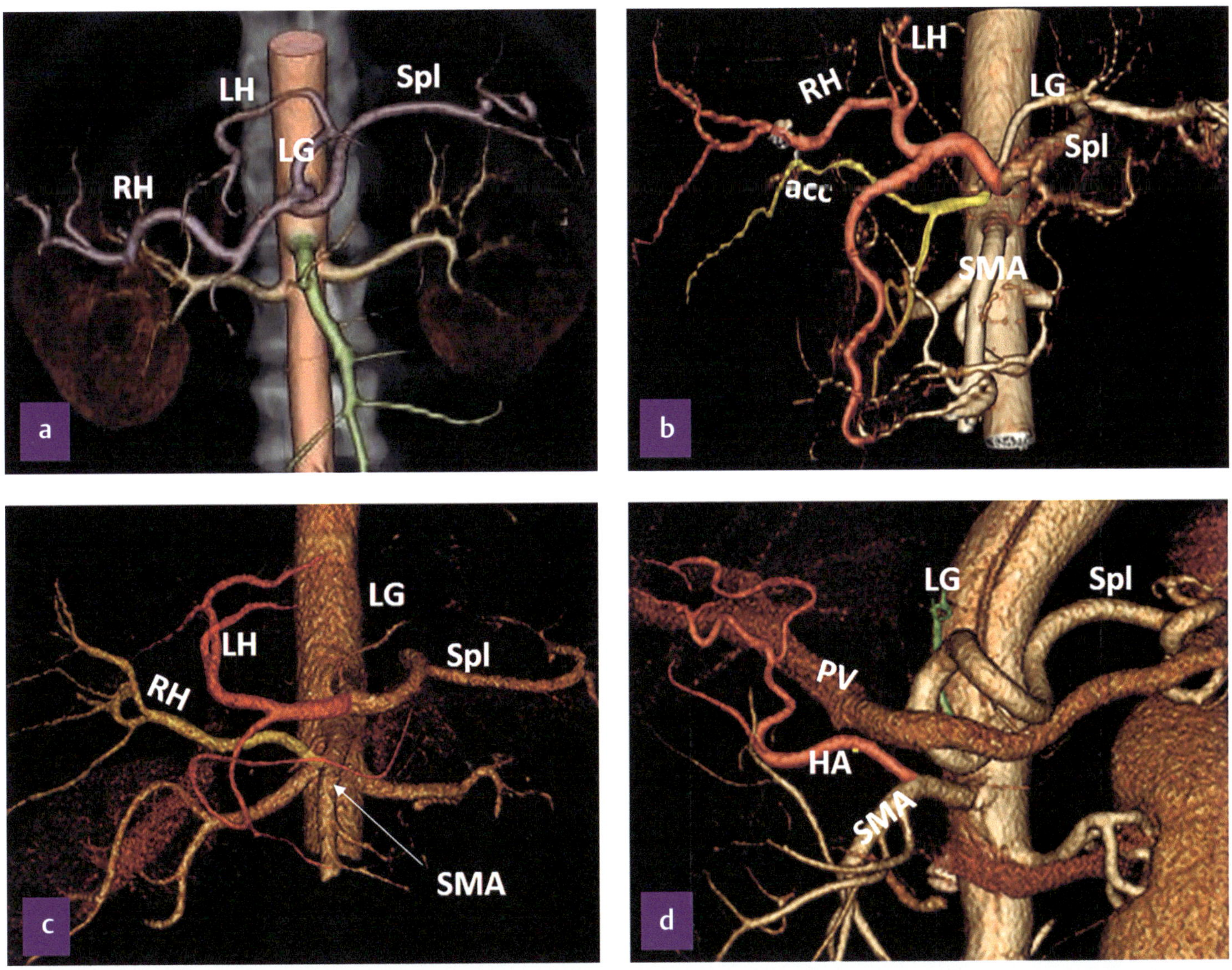

Fig. 6.21 Anatomical variants of the hepatic arteries. **(a)** Left hepatic (LH) artery arising from the left gastric (LG) artery is seen in 2.5 to 14%. **(b)** Accessory (acc) right hepatic (RH) artery arising directly from the aorta. Accessory right hepatic artery is seen in 8%, supplying posterior right liver, and usually arises from the superior mesenteric artery (SMA) but rarely from the aorta. **(c)** Replaced right hepatic artery from the SMA is seen in 8 to 11%. **(d)** Replaced hepatic artery from the SMA is seen in 2.5% based on Michels.[29] PV, portal vein; Spl, splenic artery.

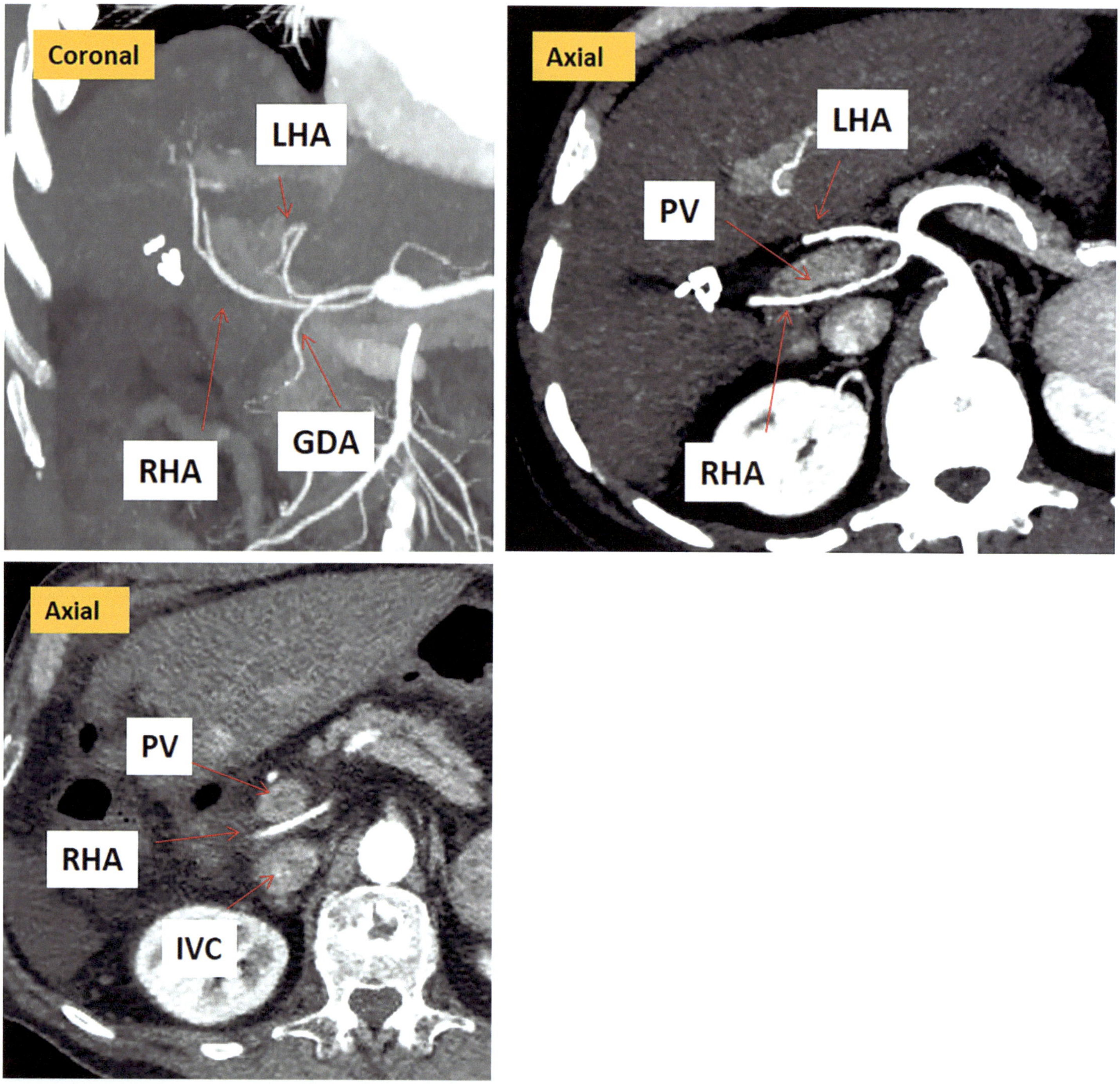

Fig. 6.22 Replaced right hepatic artery (RHA) from the celiac is seen passing between the portal vein (PV) and the inferior vena cava (IVC). The left hepatic artery (LHA) also arises from the celiac and gives rise to the gastroduodenal artery (GDA). Note in this example the replaced RHA does not pass behind the pancreatic head. If the replaced RHA originated from the superior mesenteric artery it will pass behind the pancreatic head.

division is often the longest and most prominent of the branches that runs inferolaterally[31] (▶Fig. 6.24, ▶Fig. 6.25). The territories of the segmental branches of the anterior division usually course upward and often do not correspond to segments V and VIII of the Couinaud's classification. The anterior division usually splits into an A8 dorsal and an A8 ventral. A5 is usually not a single vessel branching from the anterior division.[35,36] It represents a group of vessels that branch and course inferiorly from the A8 division (▶Fig. 6.24, ▶Fig. 6.25).

In pretransplantation evaluation of the donor liver, it is important to describe the origin and location of the artery supplying segment IV to prevent parenchymal damage to the medial segment of the liver. Arterial supply for segment IV may be from one to three arteries. Usually, the segment IV artery originates from the left hepatic artery in 65 to 75% of patients and from the right hepatic or common hepatic artery in 25%[37–39] (▶Fig. 6.26).

Portal Veins

The portal vein is a valveless afferent vessel that drains the blood from the capillary system of the intestine, spleen, pancreas, omentum, and gallbladder (see Chapter 11 "Portal Venous System"). It is in the hepatoduodenal ligament, behind the extrahepatic bile ducts and hepatic artery (▶Fig. 6.2, ▶Fig. 6.3). The portal vein is

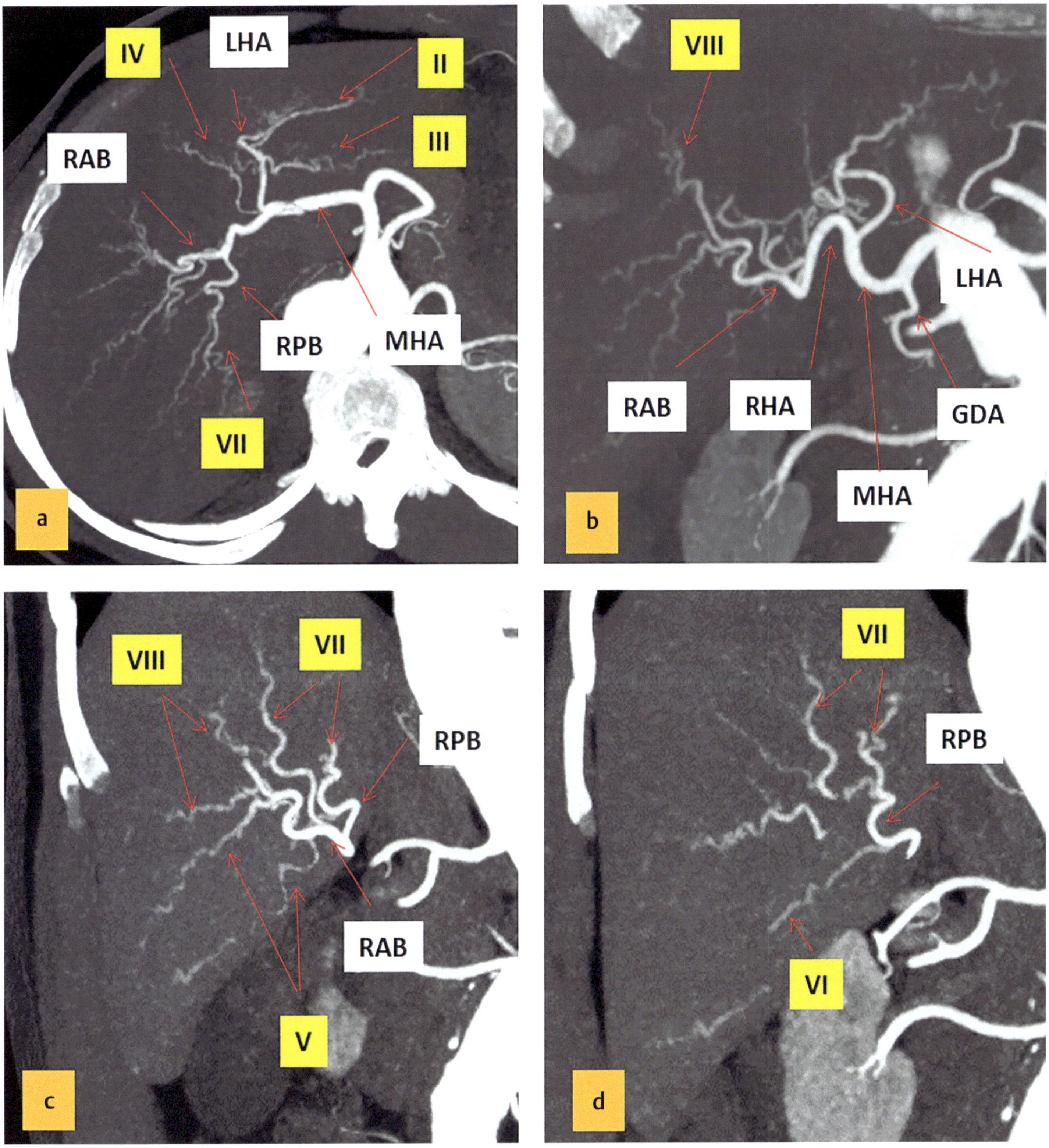

Fig. 6.23 Computed tomography (CT) images in axial **(a, b)** and coronal **(c, d)** planes show typical right hepatic artery branching pattern in a patient with cirrhosis. Arteries are enlarged and show a cork screw appearance. Segment IV is arising from the right hepatic artery (RHA). **(c)** Image is obtained anteriorly through the right anterior branch (RAB) giving off to VIII and V branches. **(d)** Image is obtained more posteriorly through the right posterior sectional branch (RPB) giving off to VII and VI segmental branches. GDA, gastroduodenal artery; LHA, left hepatic artery; MHA, main hepatic artery.

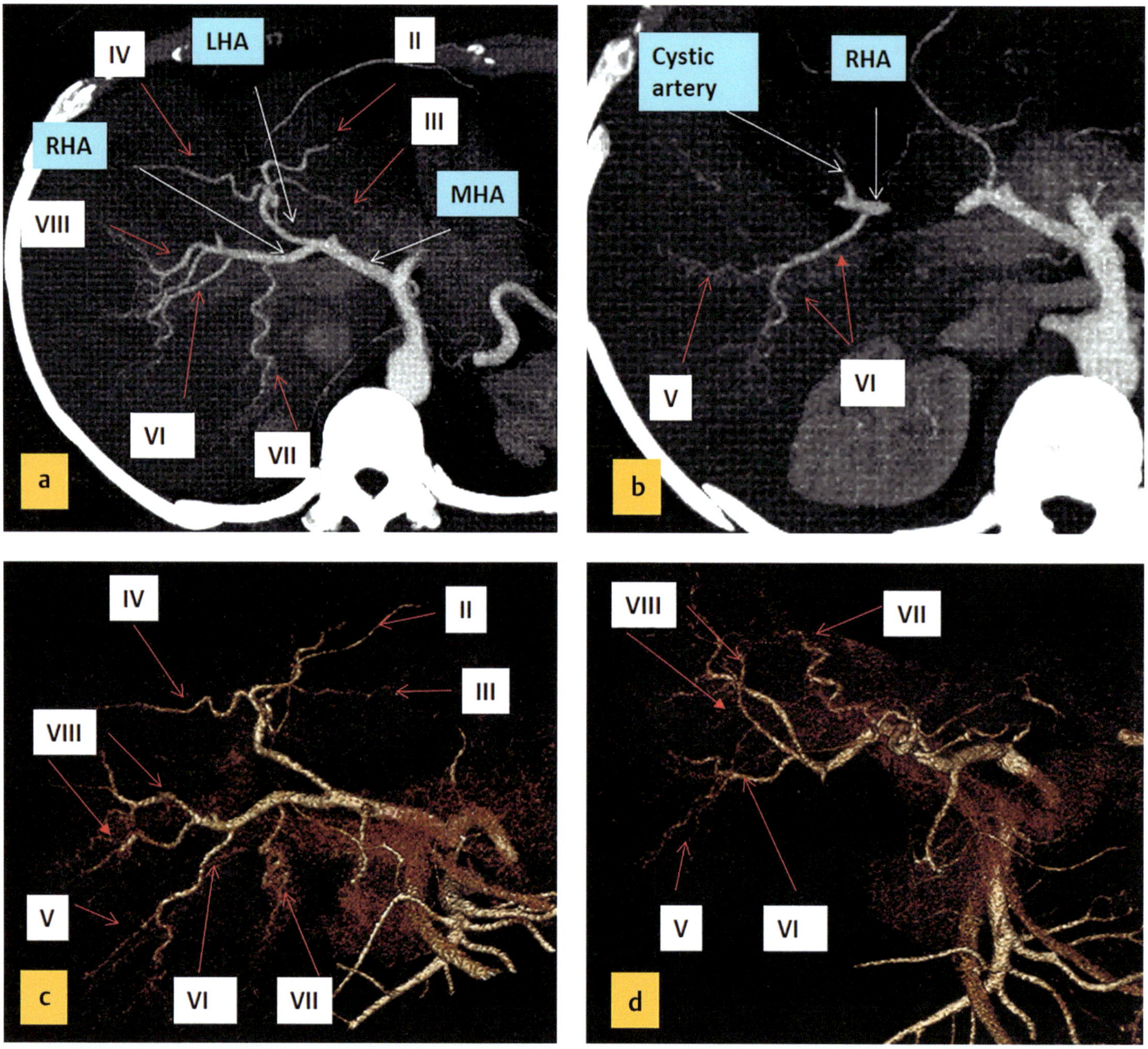

Fig. 6.24 Computed tomography (CT) images in axial **(a–c)** and coronal **(d)** planes show typical variant of right hepatic artery (RHA) branching pattern with segment VI arising as a single branch from the RHA followed by segment VI branch. Segments VI and VII both give rise to the segment V. Also note two branches moving superiorly to supply segment VIII. LHA, left hepatic artery; MHA, main hepatic artery.

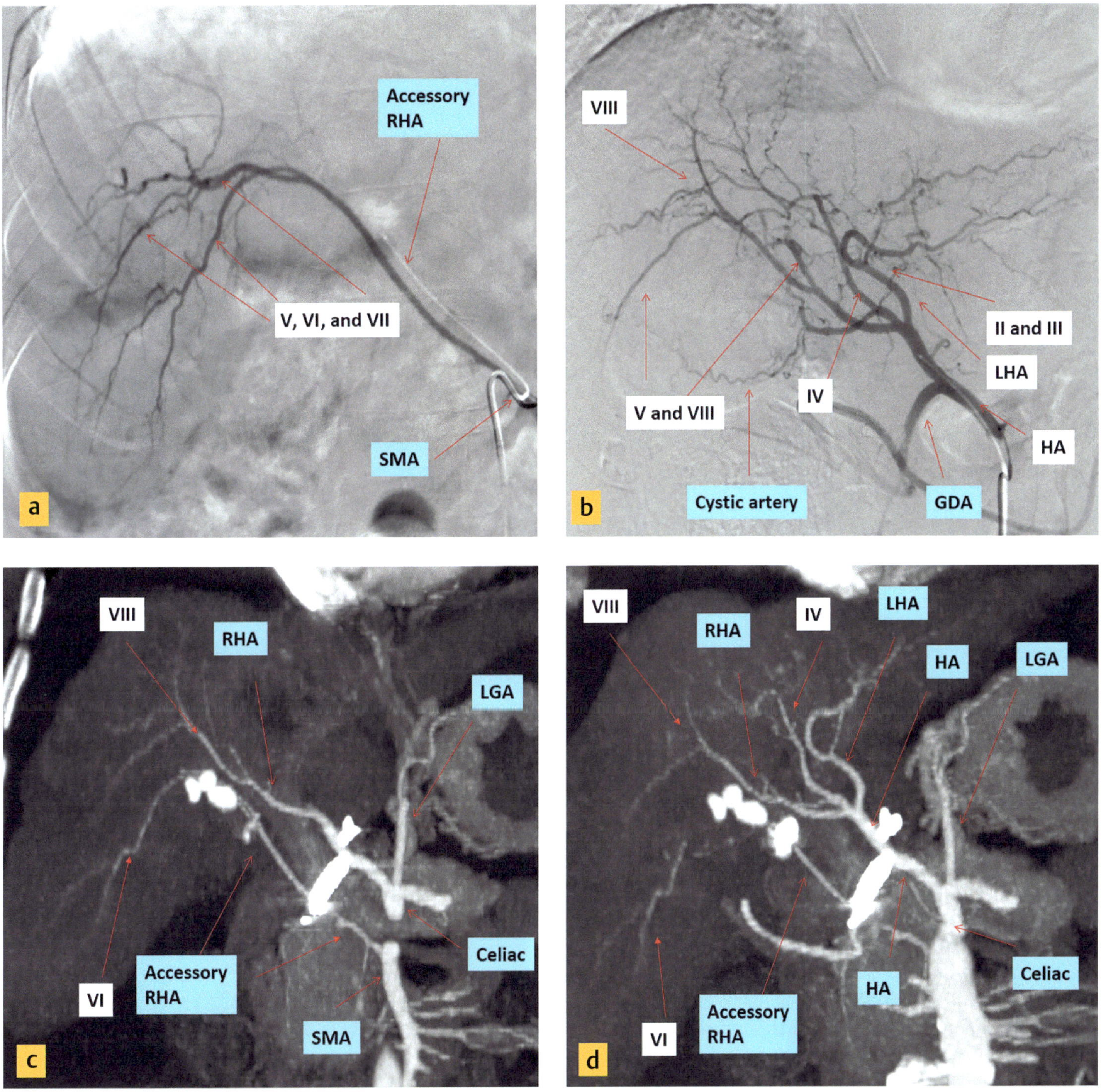

Fig. 6.25 Catheter angiogram **(a, b)** and coronal computed tomography (CT) **(c, d)** show an accessory right hepatic artery (RHA) from the superior mesenteric artery (SMA) supplying segments V, VI, and VII. The hepatic artery (HA) arises from the celiac trunk and trifurcates into the left hepatic artery (LHA) and segment V and VIII arteries. GDA, gastroduodenal artery; LGA, left gastric artery.

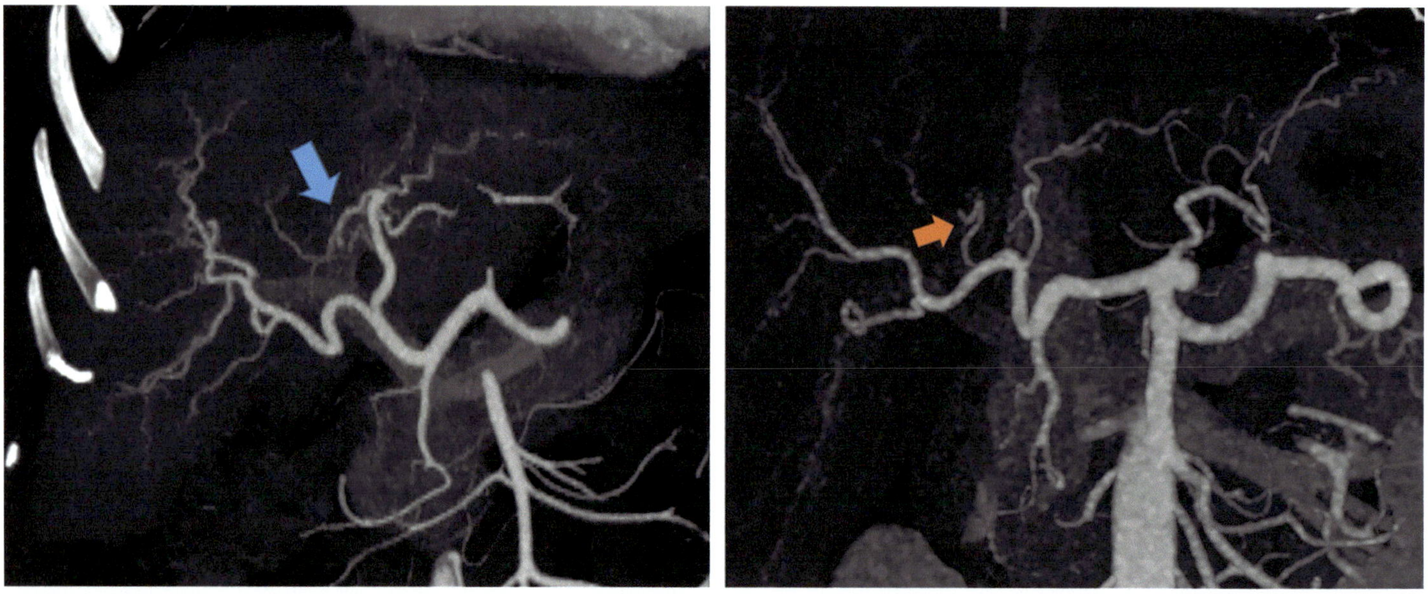

Fig. 6.26 Arterial supply for segment IV may be from one to three arteries. Usually, the segment IV artery originates from the left hepatic artery (*blue arrow*) in 65 to 75% of patients and from the right hepatic or common hepatic artery in 25% (*orange arrow*). Attention to arterial branches that feed segment IV may be of interest for pretransplantation evaluation of the liver because of its crucial position in surgical procedures.

7- to 10-cm long with a diameter of less than 1.5 cm. It is surrounded by lymphatics and soft tissue. Variations in portal vein branching and configuration are common.[40–42] The conventional pattern is for the main portal vein to divide into the left and right portal branches.[42] The portal bifurcation may be extrahepatic (50% of cases), intrahepatic (25%), or located right at the entrance of the liver (25%)[43] (▶ Fig. 6.27).

The right portal vein is located anterior to the caudate lobe and gives off a branch to the caudate before dividing into the anterior and posterior sectoral branches. Each anterior or posterior sectoral branch divides into superior and inferior segmental branches. This conventional pattern is present in 64 to 80% of the patients. Venous anomalies in the form of early connection of the right posterior branch or trifurcation of the portal vein can make surgery difficult, as more vascular anastomosis is then required with the recipient portal vein. Trifurcation of the main portal vein into the left portal, right anterior, and right posterior branches is seen in 9 to 11% of cases (▶ Fig. 6.28). Split, living-donor related liver transplantation of segments V to VIII can be challenging when portal trifurcation exists. In 10 to 23% of the patients, the main portal vein is formed by the right posterior portal vein as first branch which then connects to a common trunk of left portal and right anterior portal veins (▶ Fig. 6.27). The later variant may have important surgical implications. For example, if the left portal vein is resected at the bifurcation, the right anterior portal vein will be sacrificed resulting in devascularization of segments V and VIII. Therefore, the resection has to be distal to the origin of the right anterior portal vein. In case of liver transplant, if a portal vein branch is too short for anastomosis to the recipient portal vein, an interposed graft of greater saphenous vein may be required. Another common variant is the absence of right posterior branch due to immediate bifurcation in two branches for segments VI and VII. A further variation is the origin of the segment V branch from the right posterior portal vein or directly from the right portal vein. Rare portal vein variants include quadrification and absence of portal vein bifurcation. In the absence of portal vein bifurcation, the portal vein gives only a single right portal branch in the liver hilum but the origin of the left portal branch is absent. Instead, a large left portal branch arises from the distal right anterior branch, crosses segments VIII and VI, and ends in the umbilical portion of the portal vein to give rise to the segment II, III, and IV branches[40] (▶ Fig. 6.28).

The left portal vein tends to have a long transverse extrahepatic course in the porta hepatis. It gives off a caudate branch at this level and then moves vertically in the umbilical fissure toward the round ligament (▶ Fig. 6.29). The segment II and III branches arise from the left margin of the umbilical (vertical) segment of the left portal vein, separated by the left hepatic vein. Three to five branches arise from the right aspect of this segment to supply segment IV (▶ Fig. 6.29). The left hepatic artery follows the umbilical segment of the left portal vein not the porta hepatic segment (▶ Fig. 6.29).

It is shown by CT scan that portal vein variants are associated with a significantly higher number of biliary hilar variation. Conventional biliary branching pattern with a right posterior biliary duct joining the right anterior branch to form the right

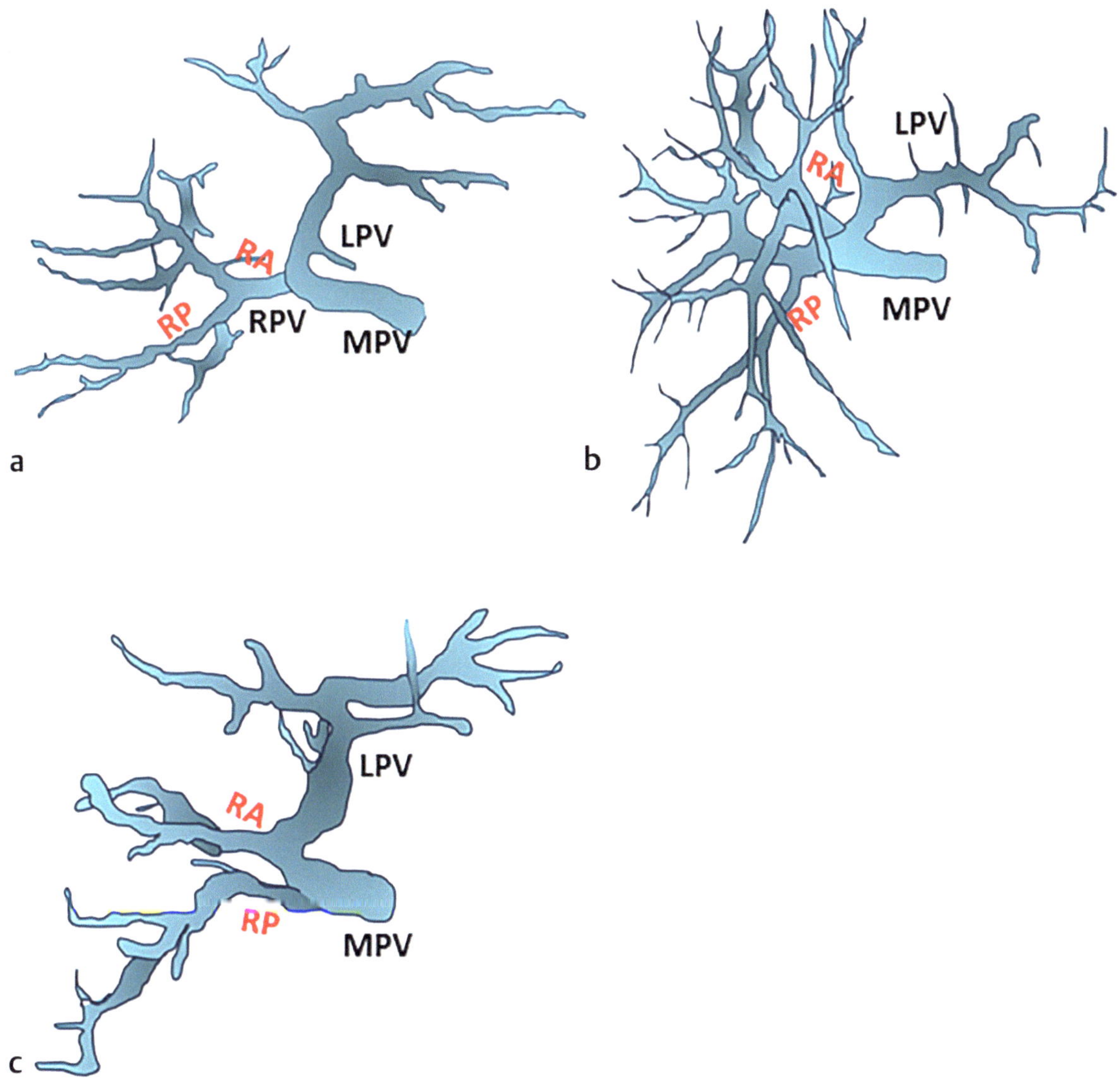

Fig. 6.27 Main types of portal vein division. **(a)** Conventional pattern is defined by bifurcation of the main portal vein (MPV) into the left portal vein (LPV) and the right portal vein (RPV) branches. The RPV is formed by the right anterior (RA) and the right posterior (RP) branches. **(b)** Anterior view of the MPV with trifurcation that is formed directly by three branches, the LPV, RA, and RP. **(c)** Anterior view of a portal system with the early origin of the RP branch from the MPV.

hepatic duct is seen in 74% of cases of patients without portal variant and in only 32% of cases with portal vein variation.[44]

Hepatic Veins

Knowing the anatomy of the hepatic veins is important for many radiological interventional procedures (i.e., transjugular intrahepatic portosystemic shunt [TIPS]). Preoperative mapping of the hepatic venous system is also important prior to liver transplantation surgery, as the transection plane is determined by the anatomic distribution of the hepatic veins.[45–47] Analysis of patency of all hepatic veins is also helpful before interpretation of liver perfusion imaging since absence or thrombosis may affect the results.

The hepatic venous system usually consists of three major hepatic veins which travel between lobes and segments (intersegmental planes) forming liver fissures (right, left, and middle scissura) (▶Fig. 6.30, ▶Fig. 6.31). Veins of the caudate lobe are directly connected to the IVC by small veins (▶Fig. 6.13).

The middle and left hepatic veins form a common trunk in 65 to 85% and are separate in 15 to 35% of cases[46] (▶Fig. 6.30). Selective clamping of the left hepatic vein can be difficult due to common occurrence of a common trunk with the middle hepatic vein.

The right hepatic vein is larger, with a short extrahepatic course (<1.5 cm). A caliber equal to or greater than 0.5 cm is considered "substantial."[48] Small caliber is usually associated with the presence of one or two inferior right accessory veins (▶Fig. 6.31e, f).

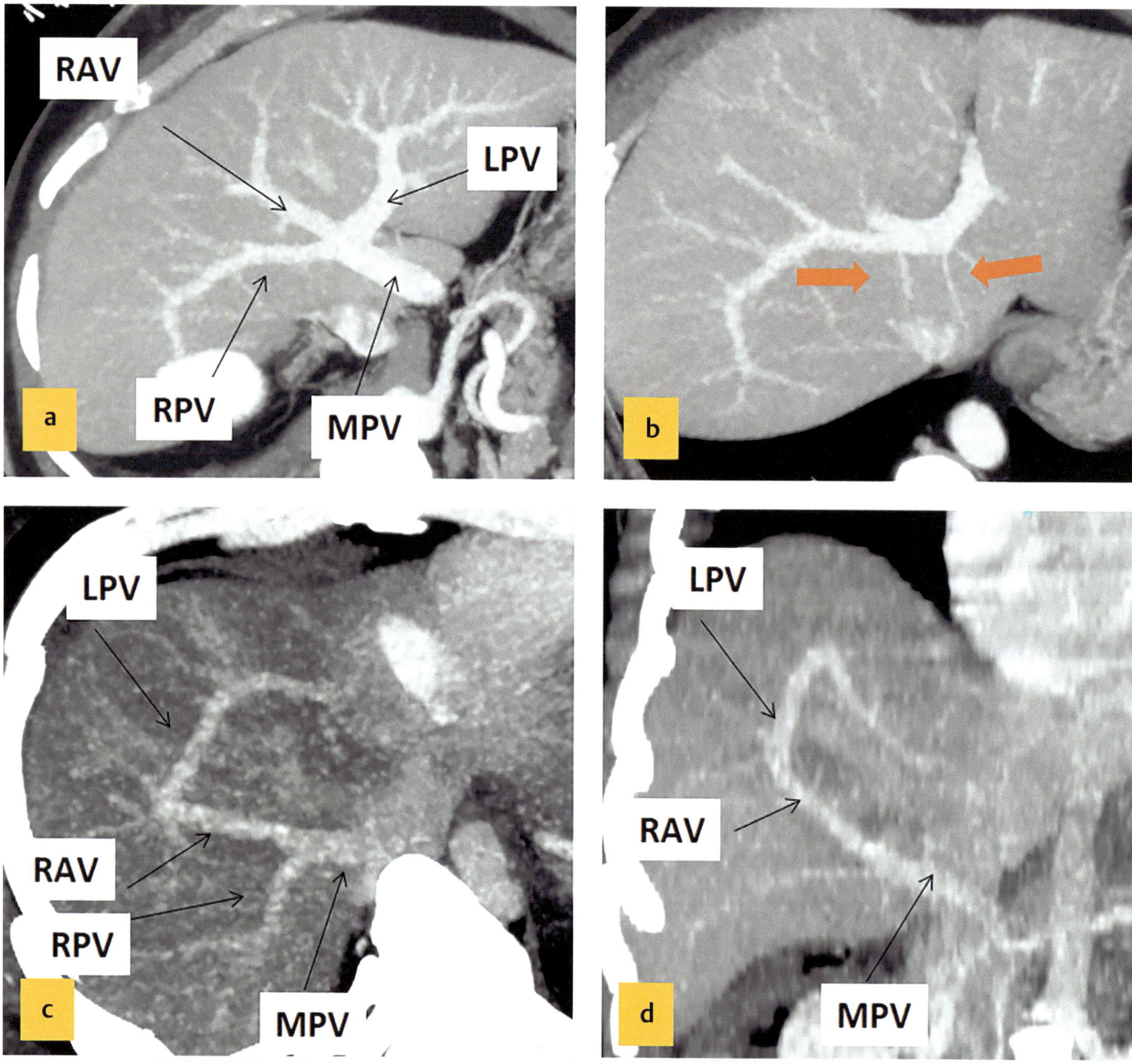

Fig. 6.28 (a) Trifurcation of portal vein. **(b)** Classic portal venous supply of the caudate lobe, one from right and one from left, is shown (*arrow*). **(c)** Axial and **(d)** coronal computed tomography (CT) showing absence of the left portal vein. The left portal vein supply is seen as anomalous vessels arising from the distal anterior right sectoral branch (RAV), running parallel to the liver surface, and traveling from segment VIII to segments IV, II, and III. LPV, left portal vein; MPV, main portal vein; RAV, right anterior vein; RPV, right posterior vein.

The right hepatic vein with its two or three main tributaries predominantly drains segments V, VI, and VII and part of segment VIII (▶ Fig. 6.31). Depending on distal right hepatic vein confluence pattern, surgeons may be able to perform a lateral (posterior) or medial (anterior) sectionectomy rather than a right hepatectomy.[45]

The middle hepatic vein drains the segments IVa/IVb, V, and part of VIII, thereby forming functional anastomosis between the left and right hepatic veins (▶ Fig. 6.31). Ligation or thrombosis of the middle hepatic vein is associated with congestion of the "medial zones" in both hemilivers.[45,46] A large volumetric variance is seen in the right subsegment of the middle hepatic vein territory. When the right territory (segments V and VIII) of middle hepatic vein is large, the inferior territory (segment VI) of the right hepatic vein is small (▶ Fig. 6.31d). The middle hepatic vein could be replaced by two veins, one draining segments V and VIII and second draining segment IV. These variants are important to consider when calculating liver volumes before surgical lobar resections. The left hepatic vein drains the left lateral sector (segments II and III) in up to 95% of all livers and part of segment IV.[49,50]

The right hepatic vein covers 47% of total liver volume followed by the middle hepatic vein (32%) and the left hepatic vein (21%).[51,52]

Additional liver drainage commonly occurs via several short retrohepatic veins (e.g., for caudate) and inferior right accessory hepatic veins. The venous drainage of the caudate lobe and its

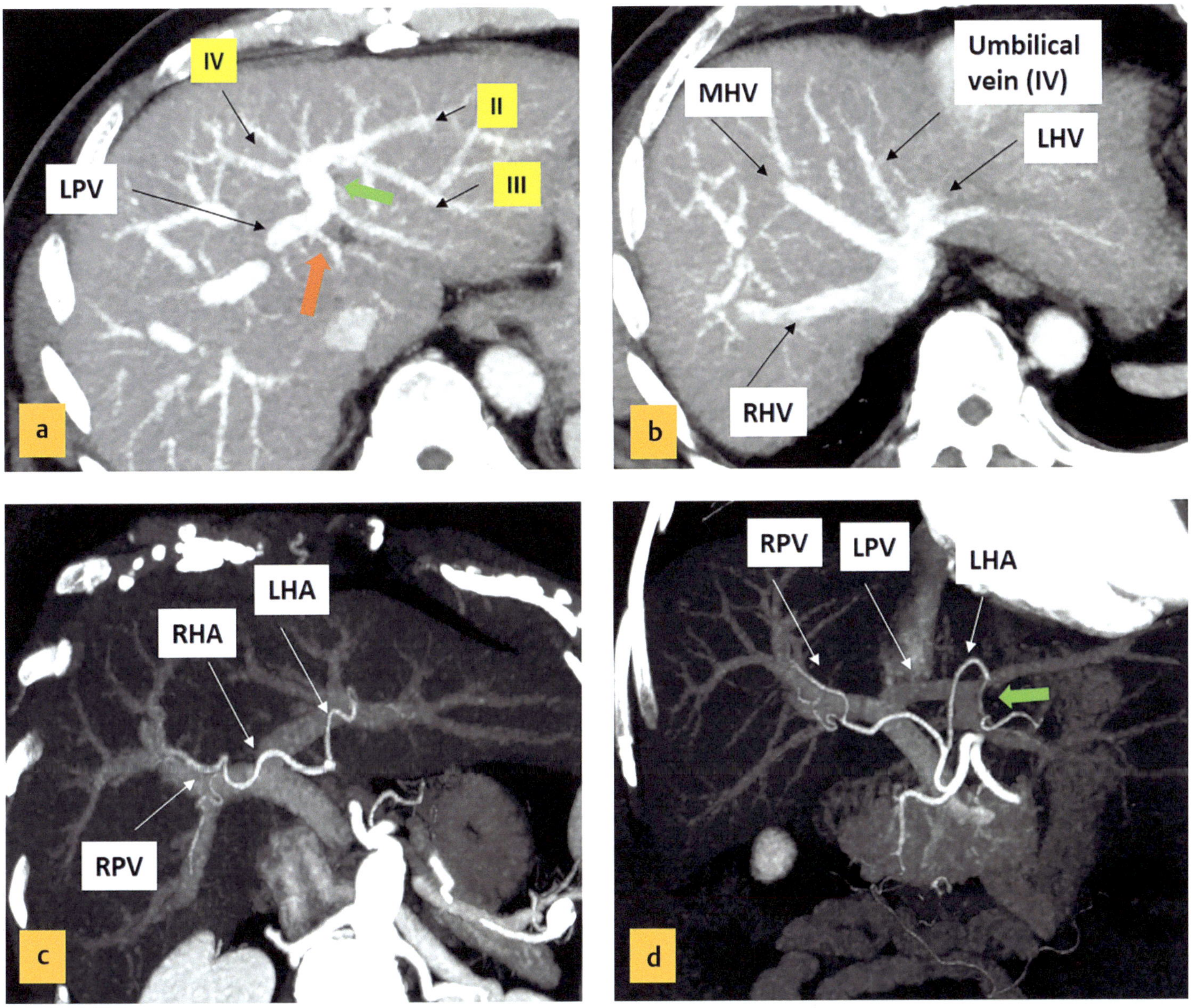

Fig. 6.29 Left portal vein (LPV) and umbilical hepatic vein. Axial computed tomography (CT) **(a–c)** and coronal CT images are shown. **(a)** The left portal vein tends to have a longer transverse extrahepatic course in the porta hepatis, giving rise to a caudate branch at this level (*orange arrow*). The segment II and III branches arise from the left margin of the umbilical (vertical) segment (*green arrow*) of the left portal vein. Three to five branches arise from the right aspect of this segment to supply segment IV. **(b)** Umbilical hepatic vein is seen connecting the left hepatic vein (LHV). **(c, d)** The left hepatic artery (LHA) follows the umbilical segment of the left portal vein (*green arrow*), not the porta hepatic segment. The right hepatic artery (RHA), on the other hand, follows the right portal vein (RPV). MHV, main hepatic vein; RHV, right hepatic vein.

papillary process is usually provided by the middle caudate vein which could be single or through a superior or inferior caudate hepatic veins[53,54] (▸Fig. 6.13). The draining veins of segment IV are highly variable. Drainage of segment IV into the middle hepatic vein occurs in 40% and into the left hepatic vein in 60% of individuals.[47] In 50% of cases with drainage of the segment IV into the left hepatic vein, an accessory umbilical (scissural) hepatic vein is seen in the umbilical fissure draining a portion of segment IV (▸Fig. 6.29, ▸Fig. 6.30). Ligation of the umbilical hepatic vein should be avoided as it could result in severe backflow bleeding.

In many cases, one or two accessory/replaced inferior right hepatic vein is seen draining the posterior aspect of segment VI or VII of the liver into the lower portion of the IVC[53] (▸Fig. 6.31). Information on the location of accessory hepatic veins and distances from the main hepatic veins is crucial before surgery as their inadvertent transection can cause bleeding, and their ligation results in atrophy of the supplied segment of the liver.

Intrahepatic Circulation

The liver receives up to 25% of the cardiac output and its oxygen consumption accounts for 20% of total body oxygen consumption. Approximately 80% of the blood entering the liver is poorly oxygenated venous blood supplied by the portal vein, whereas the remainder is well oxygenated and supplied by the hepatic artery. Yet 50% of the hepatic oxygen requirement is provided by portal venous blood due to its high flow rate.

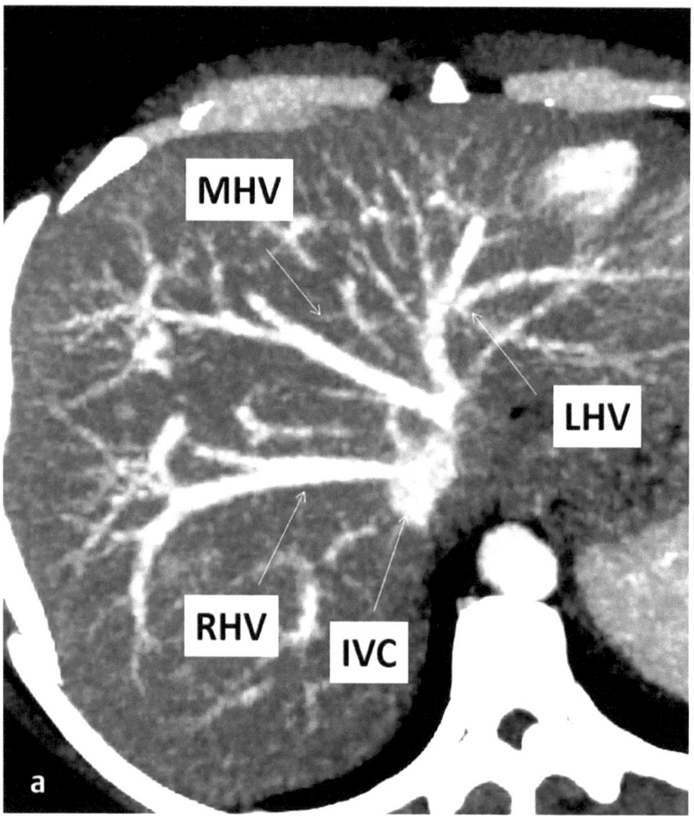

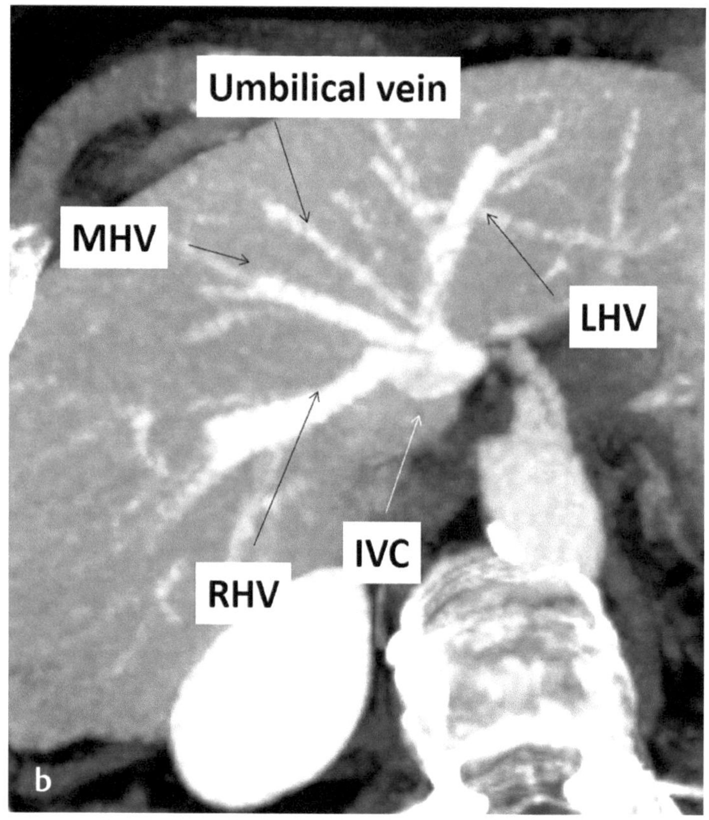

Fig. 6.30 Typical confluence pattern of the hepatic veins into the inferior vena cava (IVC). The middle and left hepatic veins form a common trunk in 65 to 85% **(a)** and are separate in 15 to 35% **(b)** of cases. An umbilical hepatic vein branch is seen in **b**. LHV, left hepatic vein; MHV, main hepatic vein; RHV, right hepatic vein.

Intrahepatic afferent branches of the portal vein and hepatic artery run in parallel (▶Fig. 6.29). In the liver, there is no hepatic capillary network. Instead, terminal branches of the afferent vasculature (portal venules and hepatic arterioles) supply blood to the fenestrated sinusoids contained within hepatic lobules (▶Fig. 6.11). The hepatic sinusoids represent the segment of the microcirculation in which supply of nutrients and removal of metabolic products take place. Sinusoids are lined by unique endothelial cells fenestrated by clusters of pores known as "sieve plates." The diameter of sinusoids increases in the pericentral area. Sinusoids lining cells are separated from hepatocytes by the space of Disse.[55] Sinusoid is also composed of nonparenchymal cells, such as hepatic stellate cells and Kupffer cells, which are involved in the regulation of blood flow through sinusoids. Both afferent vessels interconnect with each other through sinusoidal spaces. Flowing blood in the sinusoids collects in the central venules, which in turn drain into hepatic veins. Other forms of anastomosis between afferent vessels are transvasal (arterioportal) and through the peribiliary plexus.[56] Branches of hepatic arterioles also supply blood to the liver capsule and the capillaries of the peribiliary plexus, which finally drain into the sinusoids. Hepatic arterioles also send branches to the portal venules causing direct arterio-portal anastomoses.

The dual blood supply of the liver is a unique feature of the hepatic vasculature and there is an intimate relationship between the two vascular systems termed the "hepatic arterial buffer response" (HABR).[55] HABR describes the ability of the hepatic artery to produce compensatory flow changes in response to changes in portal venous flow. With this adenosine-mediated arterial autoregulation, hepatic arterial blood flow can increase diameter to buffer approximately 25 to 60% of the decreased portal flow. Enlargement of the hepatic arteries is common in severe liver cirrhosis with hepatofugal portal flow (▶Fig. 6.23). However, there is no reciprocity of the HABR and the portal vein cannot control its blood flow. In other words, loss of hepatic arterial perfusion does not induce compensatory increase in the portal flow.

Several models of the organization of the liver into structural or functional units have been proposed. These include the classic lobule, the acinus, and the portal lobule (▶Fig. 6.11). Overall, the hepatic microcirculatory subunit is the "primary lobule," also known as "hepatic microvascular subunits (HMS)." It is a conical-shaped area consisting of the two terminal afferent vessels, the associated cholangioles and canaliculi, the network of sinusoids, and the efferent terminal hepatic venule.[55] Decreased perfusion of sinusoids is a major factor in the pathogenesis of liver parenchymal diseases. Loss of balance of vasomotor control due to injury results in contraction of sinusoidal endothelium, which reduces the diameters of the hepatic sinusoids and increases vascular resistance and decreases blood flow. The development of liver fibrosis and cirrhosis is associated with vasomotor disturbances, increased intrahepatic resistance, and microarchitectural perfusion changes causing portal hypertension.

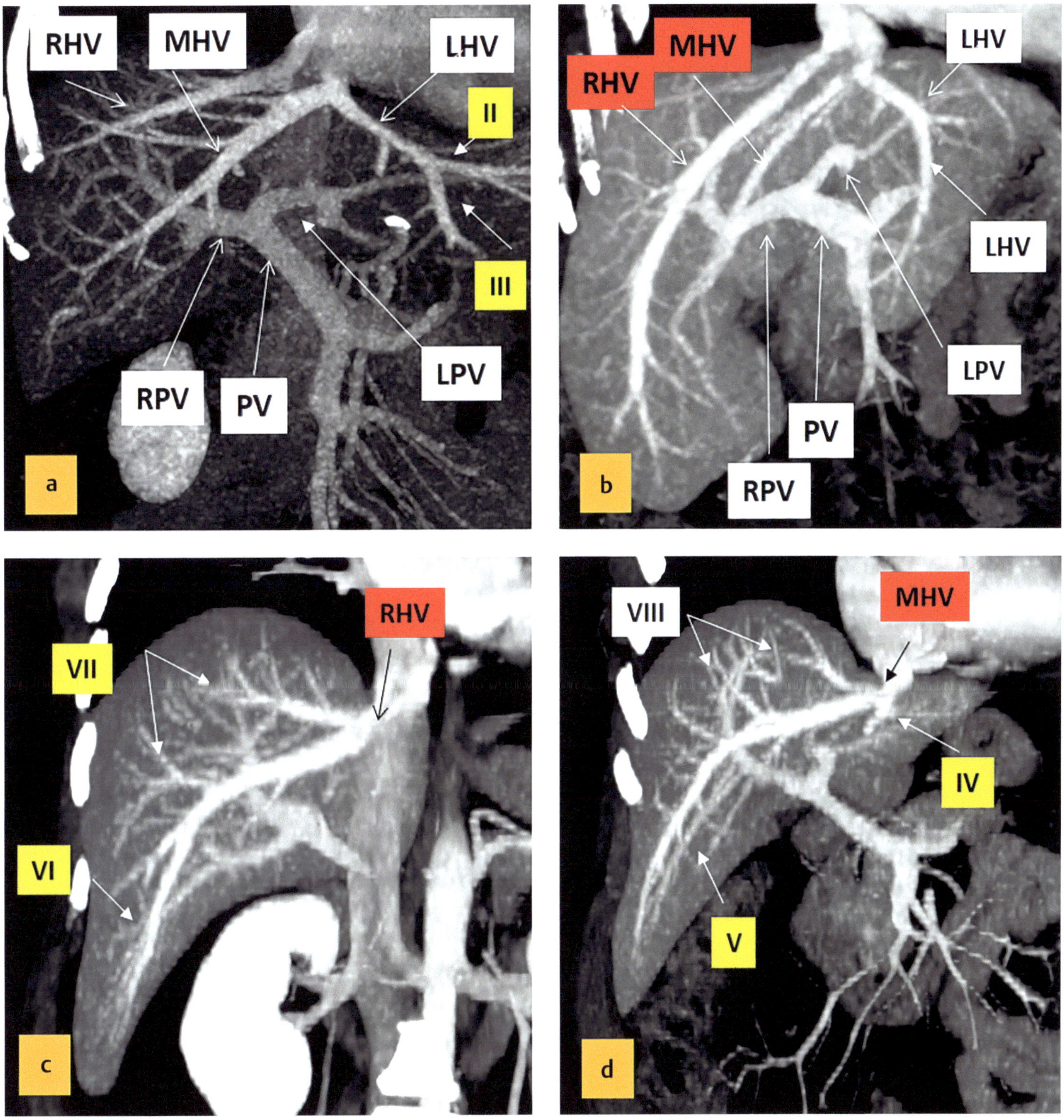

Fig. 6.31 Confluence pattern of the hepatic veins into the inferior vena cava (IVC) and relationship with the portal branches. **(a, b)** The middle and left hepatic veins form a common trunk in 65 to 85% of cases. **(c, d)** Territorial distribution of the right and middle hepatic veins are shown. **(e)** Hepatic veins in axial view. **(f)** Relationship of hepatic veins with the portal branches in axial view. Acc., accessory; LHV, left hepatic vein; LPV, left portal vein; MHV, middle hepatic vein; PV, portal vein; RHV, right hepatic vein; RPV, right portal vein.

(Continued)

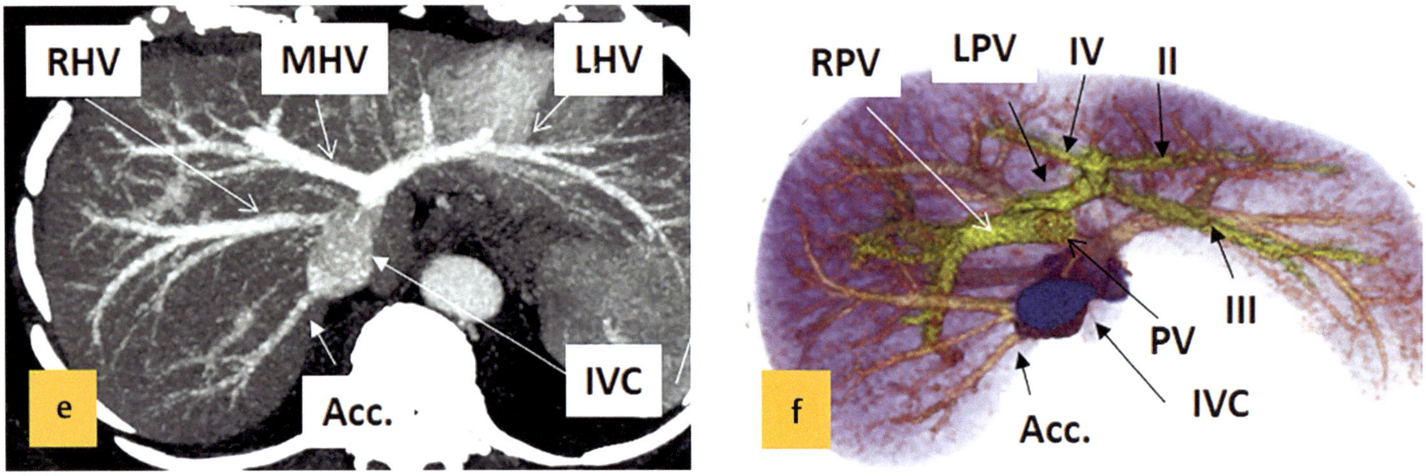

Fig. 6.31 (*Continued*) (**e, f**)

6

Lymphatics

The liver produces 25 to 50% of lymph flowing through the thoracic duct.[57] The hepatic lymph originates from the hepatic sinusoids, and flows into the space of Disse before entering interstitial space of the portal tract, sublobular veins, or liver capsule when it enters the lymphatic vessels. Lymphatic vessels originate as blind-ending capillaries in the portal tracts. Approximately 80% of hepatic lymph drains into portal tract lymphatic vessels, while the remainder drains through sublobular and subcapsular lymphatic vessels. Sublobular lymphatic vessels lead into lymphatic vessels running along the inferior vena cava. Some subcapsular lymphatic vessels run in the falciform ligament and coronary ligament toward the diaphragm, while others pass downward into the lymph nodes of the porta hepatis. Lymphatic vessels have many valves that allow unidirectional lymph flow. The lymph functions as a tissue drainage system and helps to control immunological system. Increase in lymphatic vessels occurs in liver cirrhosis and portal hypertension and that by itself can cause further reduction in portal blood flow. Benign periportal and portal hepatis edema is a common finding in acute hepatitis, pancreatic diseases, after liver trauma, and following liver transplant and may be related to interstitial edema and increase in lymphatic flow (▶ Fig. 6.32). The superior surface of the liver drains to the diaphragmatic nodes. The deep hepatic vessels that drain the liver parenchyma largely drain to the hilar, hepatic, and portal nodal groups (see Chapter 21 "Lymphatics of the Abdomen, Pelvis, and Perineum").

Aberrant Blood Supply and Arterioportal Shunting

In dynamic contrast-enhanced study of the liver and especially during arterial phase it is common to find areas of hyperenhancement in different locations with different configurations in normal and abnormal liver tissue. These pseudolesions are probably secondary to aberrant venous drainage into systemic veins or intraparenchymal microscopic arteriovenous shunting.[58,59] Extrahepatic venous channels can drain into the liver sinusoids through small veins of the liver capsule, accessory cystic vein, or aberrant right gastric vein (▶ Fig. 6.33). One characteristic example of heterogeneous liver enhancement is seen in patients with chronic obstruction or severe narrowing of the superior vena cava or inferior vena cava causing abnormal transcapsular connection of the tributaries of the diaphragmatic, epigastric, and abdominal wall veins with the liver sinusoids around the falciform ligament and within the quadrate lobe of the liver (▶ Fig. 6.34). Focal enhancing areas are seen on arterial phase images due to the early outflow of contrast material into these veins compared with the inflow into the surrounding parenchyma provided by the portal system branches. These benign enhancing foci characteristically disappear on the late venous phase, which differentiates them from pathological states such as portal venous thrombosis which usually remain abnormal on the venous phase (▶ Fig. 6.34). Common locations are around the gallbladder and falciform ligament, and in the subcapsular region. Pseudolesions at the periphery of the left

6

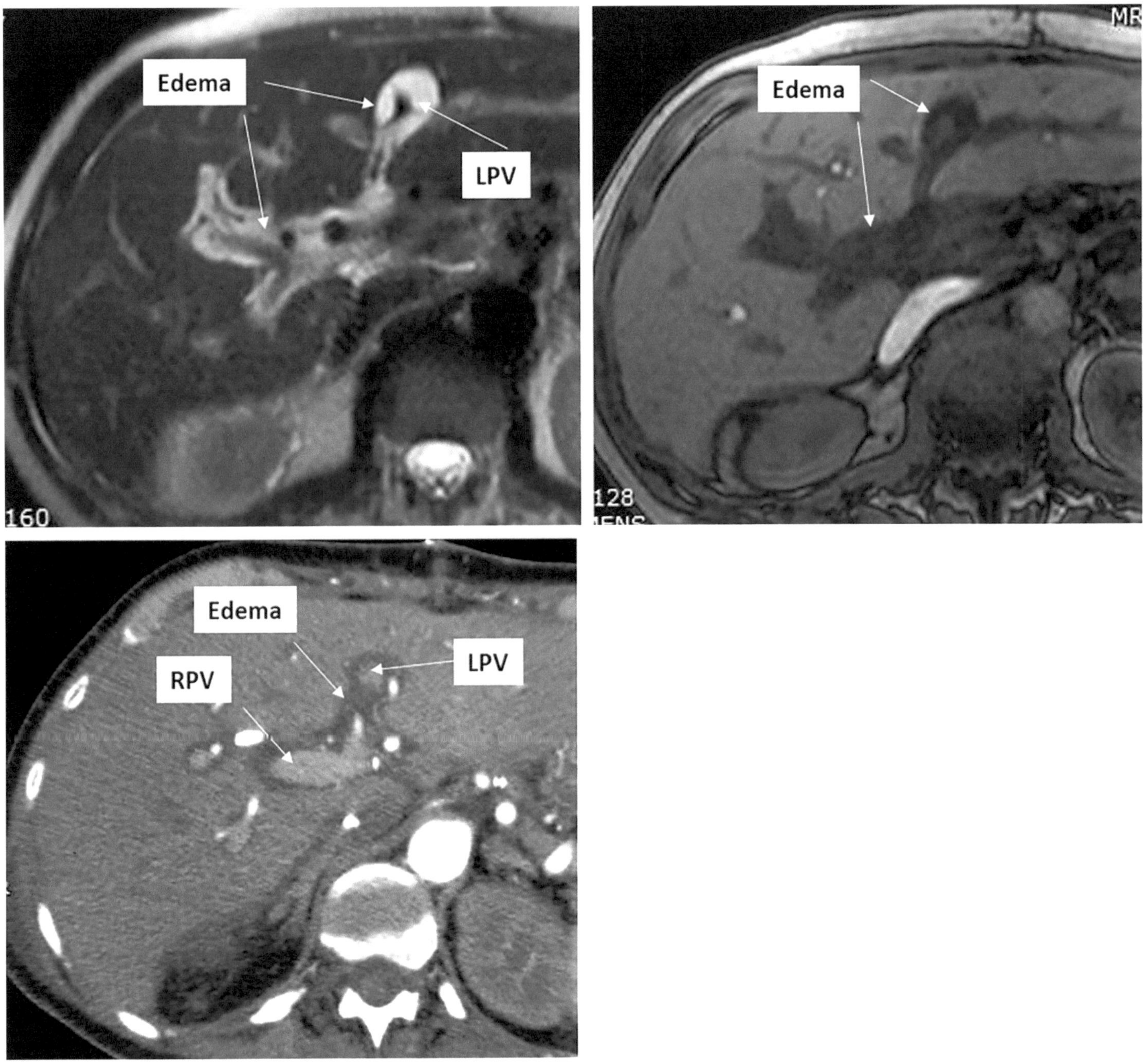

Fig. 6.32 Periportal edema in a patient after liver transplant. Edema appears bright on T2-weighted images and low density on computed tomography (CT) and low signal on T1-weighted. Benign periportal and portal hepatitis edema is a common finding in acute hepatitis, pancreatic diseases, after liver trauma, and following liver transplant and may be related to interstitial edema and increase in lymphatic flow. LPV, left portal vein; RPV, right portal vein.

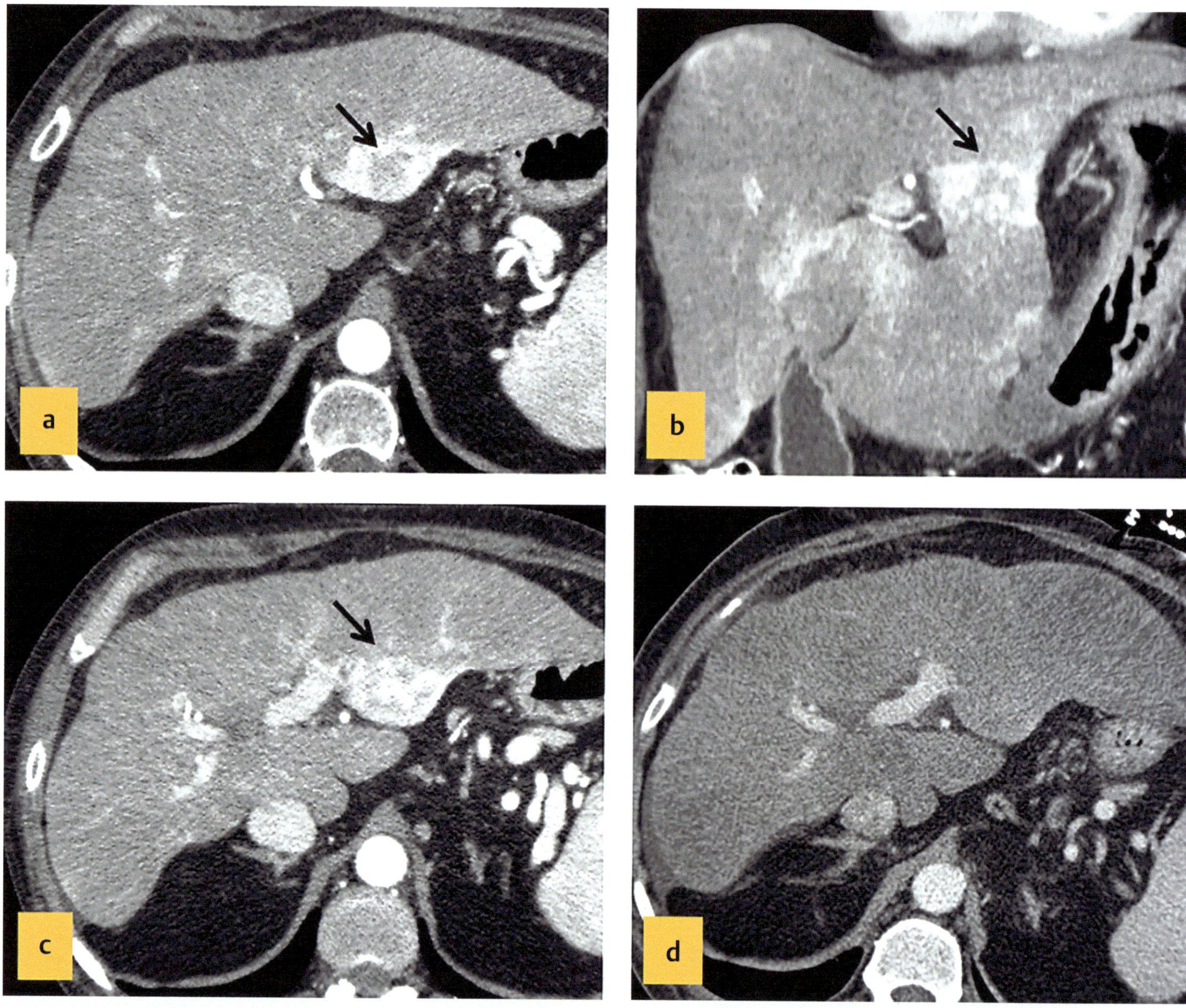

Fig. 6.33 Axial and coronal images **(a, b)** in early arterial phase and axial late arterial phase **(c)** in a 52-year-old female with cirrhosis show transient hepatic enhancement (*arrows*) in posterior margin of segment III which disappears in axial portal phase image **(d)**. This perfusion disorder probably corresponds to the third hepatic inflow tracts (aberrant gastric venous drainage or parabiliary venous system).

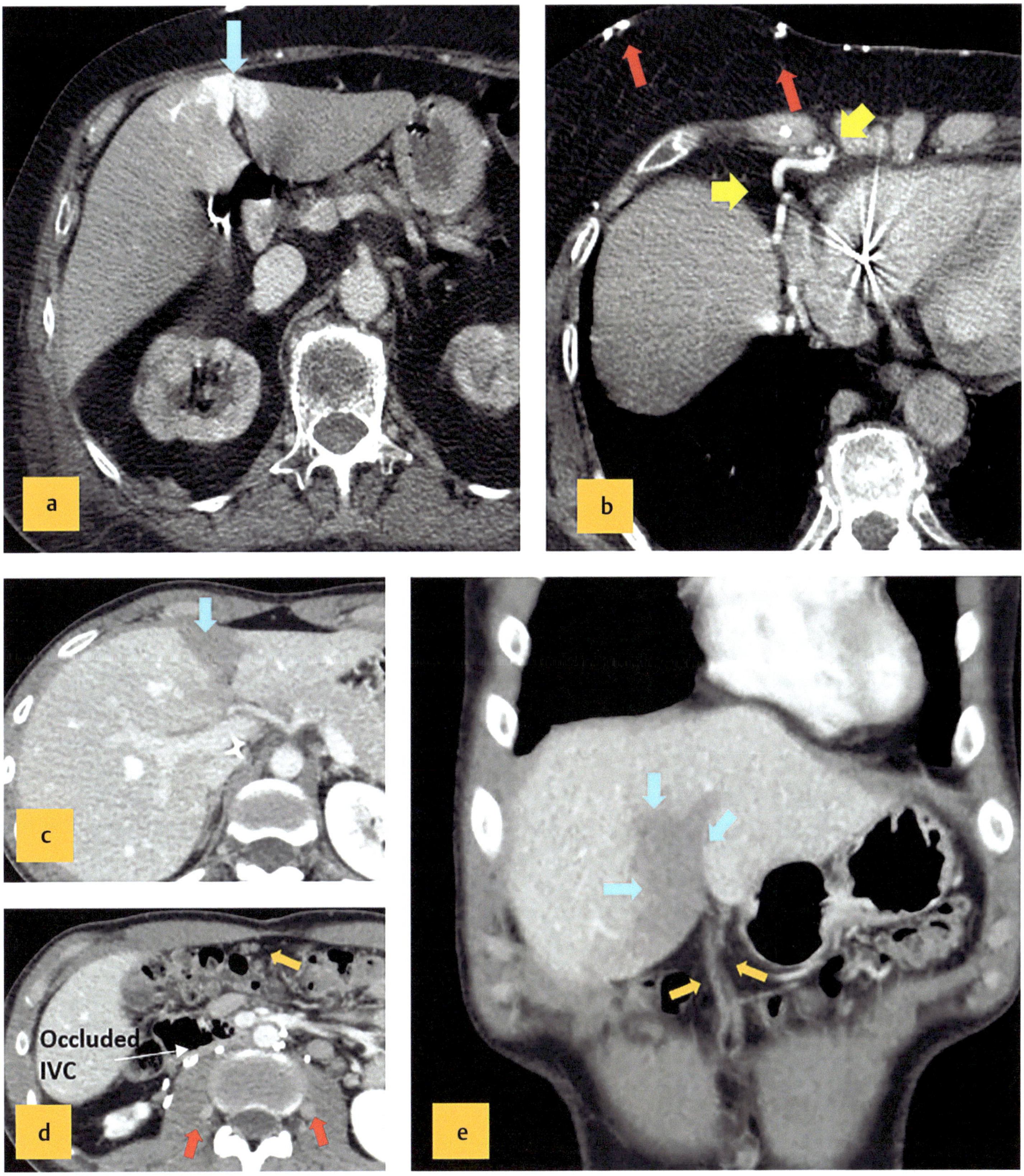

Fig. 6.34 (a, b) Axial computed tomography (CT) image in venous phase over the liver showing unusual enhancement of the liver sinusoids around the falciform ligament (*blue arrow*) due to reverse transcapsular flow of the tributaries of the epigastric veins (*yellow arrows*) caused by obstruction of the superior vena cava. Note venous collateral of the skin (*red arrows*). **(c–e)** Early venous phase axial **(c, d)** and coronal **(e)** CT images of the liver showing no enhancement of the liver around the falciform ligament (*blue arrows*) due to reverse transcapsular flow of the tributaries of the epigastric veins (*orange arrows*) caused by obstruction of the inferior vena cava (IVC). Note paravertebral venous collaterals (*red arrows*). **(f, g)** CT scan at portal phase shows a wedge-shaped area of enhancement (*red arrows*) in a patient following liver transplant due to portal vein thrombosis (*green arrow*). In contrast to this patient, benign enhancing foci characteristically disappear on the hepatic venous phase **(a–e)**.

(*Continued*)

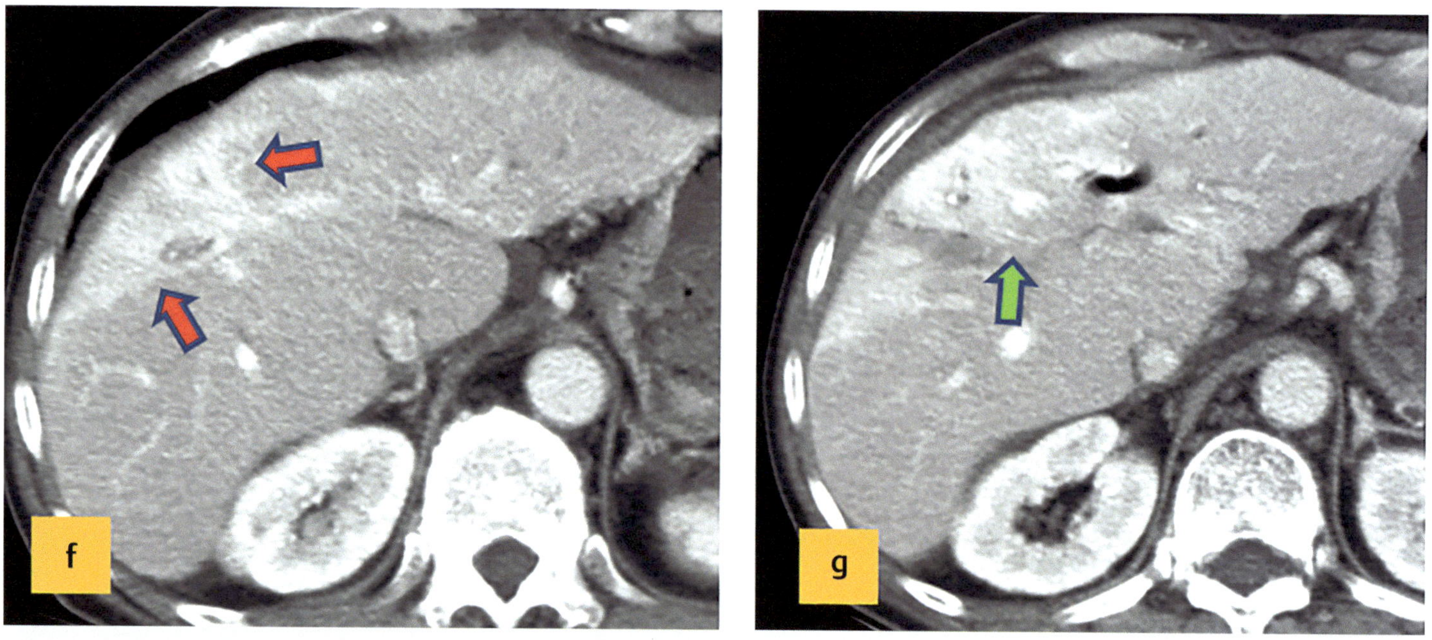

Fig. 6.34 (*Continued*) (f, g)

liver lobe could be secondary to aberrant right gastric antrum venous drainage to the liver.[60]

Hepatic arterial supply derived from the cystic artery branches can also produce peripheral rim of hyperattenuation around the gallbladder fossa. Another reason for benign focal enhancing areas in the liver is transsinusoidal or peribiliary plexi communications between microscopic interlobular arterioles and portal venules, resulting in redistribution of arterial flow into a focal region of portal venous flow. Cirrhosis is a known cause of transsinusoidal arterioportal shunting and is believed to be secondary to thrombosis of the small hepatic venules and retrograde hepatofugal flow filling of the small portal vein branches.[61] Image findings include transient, wedge-shaped peripheral parenchymal enhancement and early visualization of the peripheral portal vein branches. Pseudolesions are sometimes seen on nonenhanced CT scans as focal spared areas in a fatty liver or as areas of focal fatty infiltration. Due to arterioportal shunting, the hepatocyte tissues in that region are subject to less portal and more arterial blood and therefore are protected from fatty changes compared to the rest of the liver.

Typical perifalciform perfusion defect at portal venous phase is related to early drainage by paraumbilical venous system which also connects with the internal thoracic and inferior epigastric veins through veins of Sappey. These veins become more apparent in the setting of portal hypertension or superior vena cava obstruction.[61] Parabiliary venous plexus early drainage also causes small defects in the portal venous perfusion of the perihilar liver and defects frequently seen at the dorsal aspect of segment IV and around the gallbladder fossa.

◆ Presurgical Liver Assessment

Liver Resections

As described in previous sections, mapping of the liver vasculature and biliary tract facilitates liver surgery and decreases the rate of vascular and biliary complications (▸Fig. 6.15, ▸Fig. 6.35, ▸Fig. 6.36). Anomalies in the branching pattern of portal vein may at times complicate liver resection. For example, anomalous origin of the anterior branch of the right portal vein from the left portal vein could increase the risk of ischemia of the segments V and VIII in left lobe resections.

The course of the middle hepatic vein determines the cut plane in lobar resections (▸Fig. 6.37). The right middle hepatic vein territory drains the paramedian sectors of both left and right liver lobes. In lobectomies of the liver, ligation of the middle hepatic vein can cause ischemia of the residual lobe. The extent of ischemia depends on the territorial coverage of the middle hepatic vein in the remaining lobe. Presence of several accessory inferior hepatic veins could be another potential problem in liver resections.

Pretransplantation Donor

In most adult-to-adult living donor liver transplantation (LDLT), it is the donor's right hepatic lobe which is used to transplant into the recipient[62,63] (▸Fig. 6.37, ▸Fig. 6.38). Surgeons usually leave 30 to 40% of the liver behind to meet donor's metabolic demands during liver regeneration. A graft that provides 40% of the

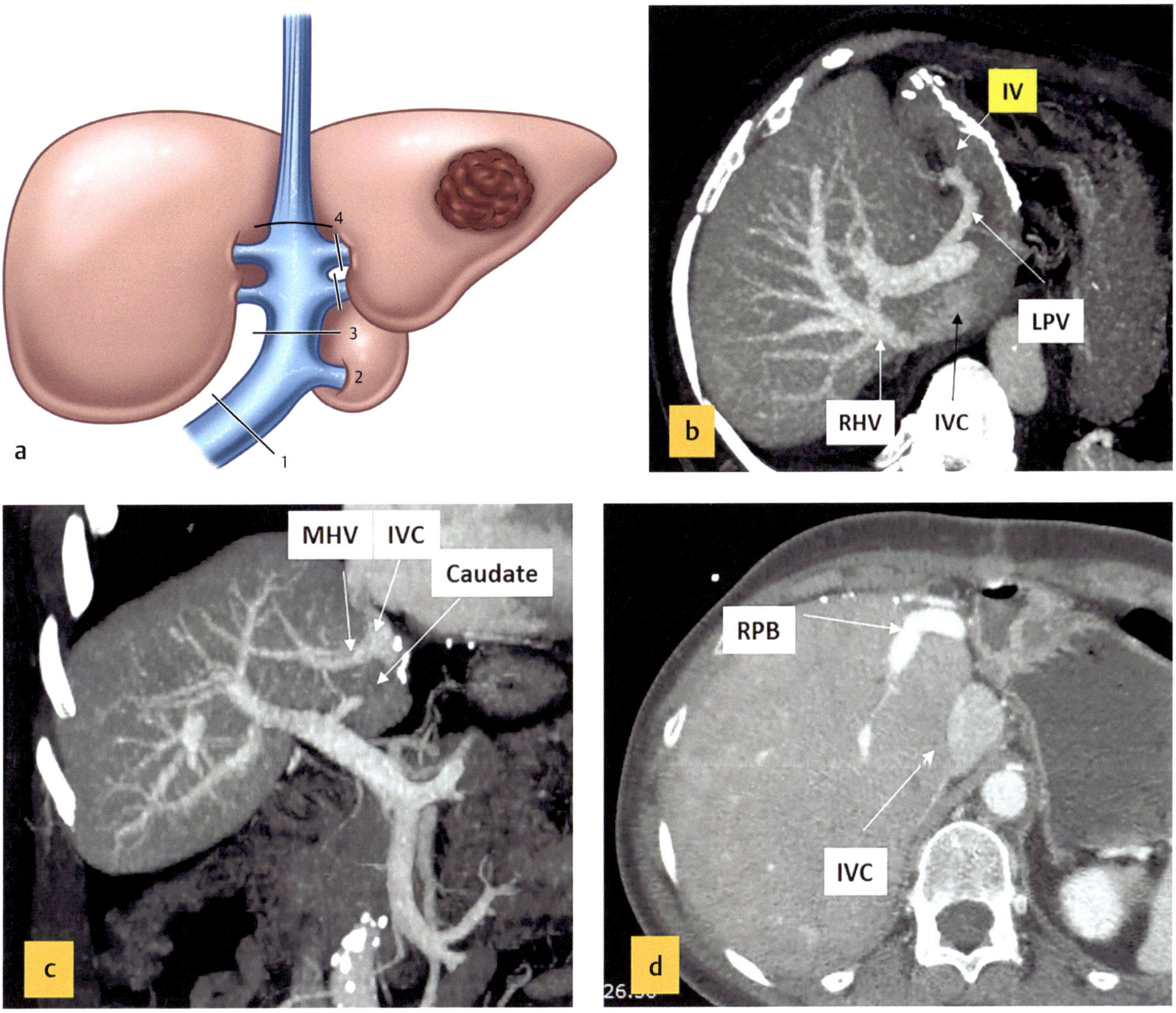

Fig. 6.35 Left lobe resections. **(a)** Various levels of portal vein transection. Level #1 is for a left hepatectomy and caudate lobectomy. Level #2 is a left hepatectomy. Level #3 is for a segment II resection. Level #4 is for a segment III resection. **(b)** Axial and **(c)** coronal computed tomography (CT) after left lobectomy (segments II and III are removed). Caudate and segment IV remain. Left hepatic vein is resected. **(d)** Axial CT after extended left hepatectomy. The only segments left include VI and VII supplied by the right posterior branch of the right portal vein. IVC, inferior vena cava; LPV, left portal vein part remained; MHV, middle hepatic vein; RHV, right hepatic vein; RPB, right posterior branch.

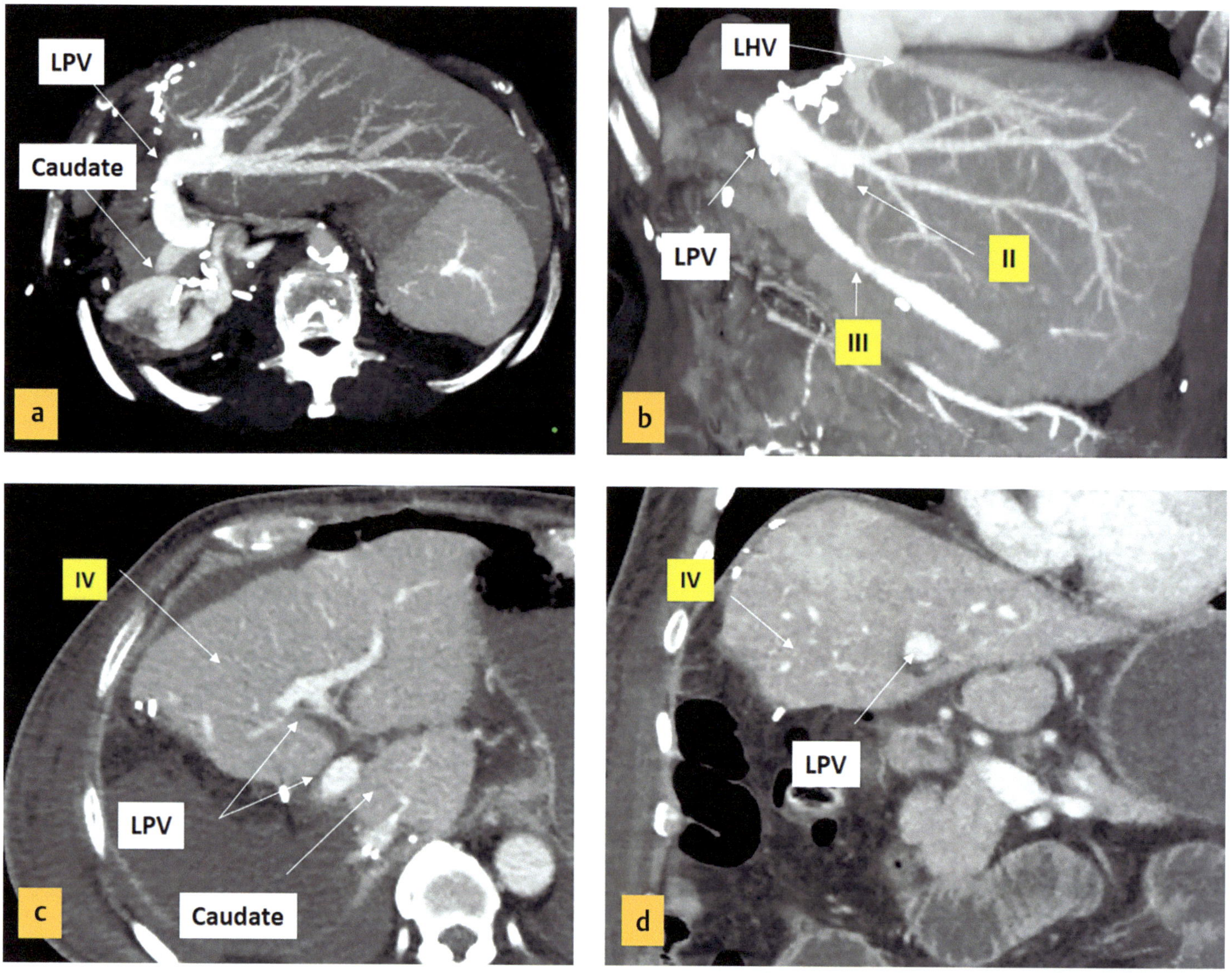

Fig. 6.36 Variants of right hepatectomy. **(a)** Axial and **(b)** coronal computed tomography (CT) after extended right hepatectomy. Segment IV is removed. LPV remained intact. **(c)** Axial and **(d)** coronal CT after right hepatectomy. Segment IV is not removed. LHV, left hepatic vein; LPV, left portal vein.

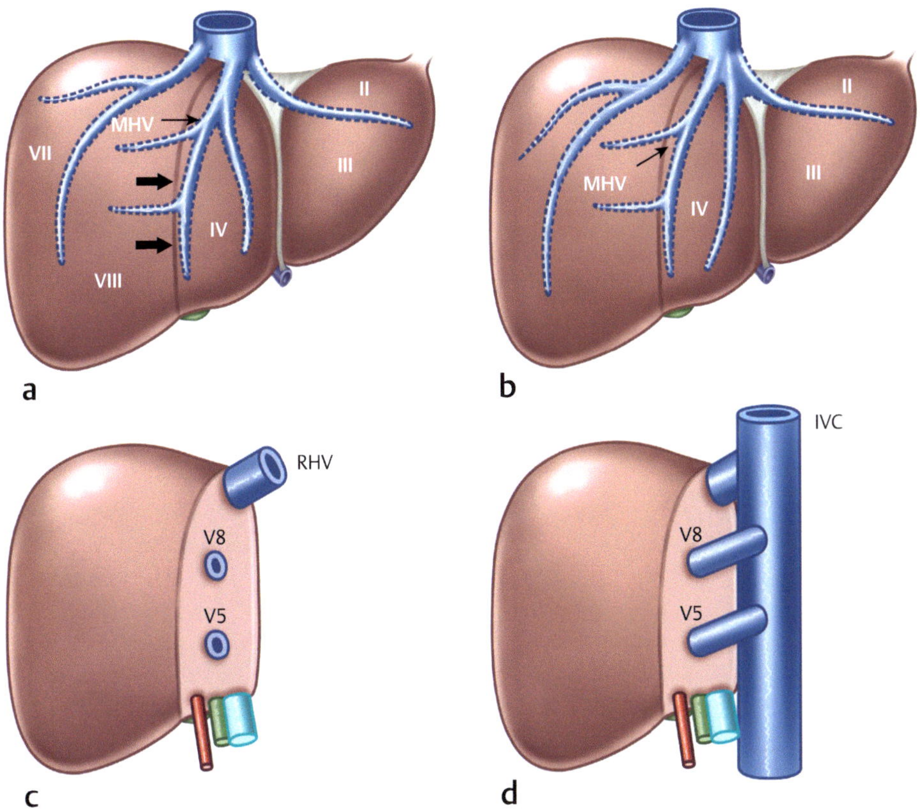

Fig. 6.37 Right hepatic transplant. **(a)** Transection plane (*arrows*) used on the right of the middle hepatic vein (MHV) to preserve the MHV when it drains segment IV. **(b)** Transection plane (*arrows*) remains on the left of the MHV when segment IV drains into the left hepatic vein. **(c, d)** When transection plane is on the right of the MHV, the right hepatic vein (RHV), V8, and V5 are anastomosed to the inferior vena cava (IVC) directly or by a saphenous vein graft. (Modified from Moon and Lee.[63])

estimated standard liver volume is necessary for the recipient's recovery. If the recipient is an adolescent or much smaller than the donor, the left hepatic lobe can be used. For babies, the left lateral segment, comprising segments II and III (20% of hepatic mass), is used.

Using CT, a liver donor candidate will be assessed for steatosis, a potential cause of donor liver dysfunction.[64,65] Identification of the origin of the segment IV artery from left or right hepatic arteries is very important in liver donor to prevent its ligation in candidates for left or right harvesting.[64] In right liver transplant, if the artery takes off from the right hepatic artery the distance between its orifice and the origin of the right hepatic artery from the proper hepatic artery should be measured. If the right portal vein is absent, two portal vein anastomoses for each anterior and posterior sectors may have to be performed and this increases the risk of postoperative portal vein thrombosis. As described above, anatomic variation of origin of the anterior branch of the right portal vein from the left is important in right liver harvesting.

Early bifurcation of the middle hepatic vein is important to identify preoperatively because it may alter the plane of resection. Presence of a large vein (>5 mm) draining segment VIII to the middle hepatic vein should be reported. If the distance of an accessory inferior right hepatic vein to the right hepatic vein exceeds 4 cm, it may be difficult to implant both veins with a single clamp on the recipient's IVC. A large (>3 mm) accessory inferior right hepatic vein should be not reconstructed to prevent graft dysfunction due to ischemic changes in its territory.[65,66]

During donor right hepatectomy, a cholecystectomy is performed and biliary anatomy is assessed using intraoperative cholangiogram.[62,63] Intraoperative ultrasound is used to evaluate hilar structures. In most cases, the liver is divided to the right of the middle hepatic vein. A graft without the middle hepatic vein is prone to congestion with cholestasis and portal hypertension early in the recipient, but obviously it is safer for the donor. In the recipient, the biliary anastomosis is created between the donor's right hepatic duct and the recipient's common bile duct remnant (choledochocholedochostomy) or the donor's right hepatic duct with the recipient's jejunum (choledochojejunostomy and a Roux-en-Y anastomosis). Biliary complications are most challenging in LDLT. Vascular complications in the recipient include hepatic artery thrombosis and hepatic venous congestion and are the most common cause of graft loss.

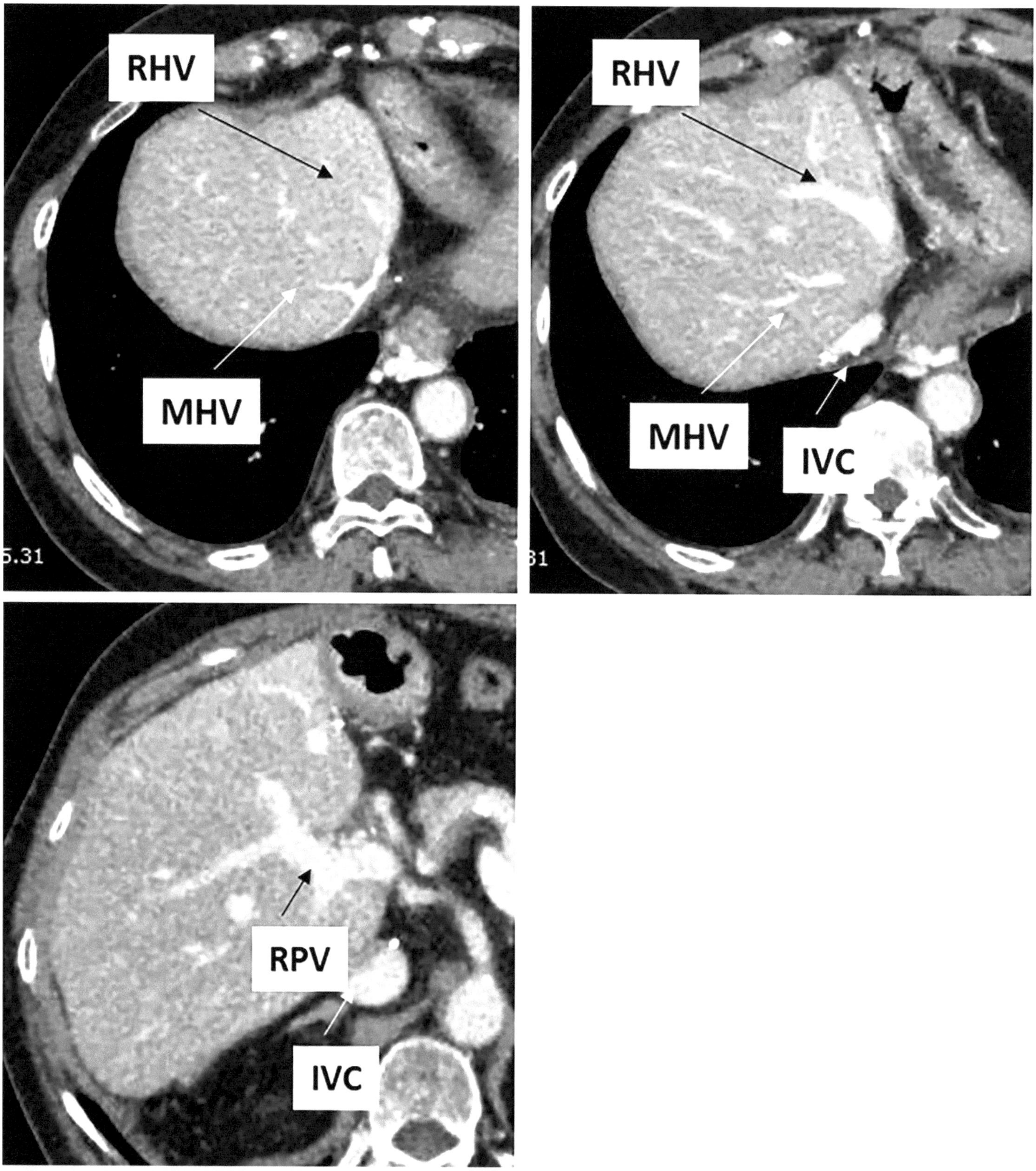

Fig. 6.38 Right liver transplant shown by axial computed tomography (CT) images. IVC, inferior vena cava; MHV, middle hepatic vein; RHV, right hepatic vein; RPV, right portal vein.

◆ Common Pathologies

Liver morphology varies in different individuals (▶ Fig. 6.39). Left liver lobe may be large extending to the left subdiaphragmatic region close to the spleen or it may be hypoplastic or rarely absent. Abnormal morphology and small size of the right liver lobe are unusual and, in most cases, indicate intrinsic liver pathology such as cirrhosis. Accessory lobe of the liver is a relatively rare congenital ectopic tissue that is seen as an accessory tissue attached to the liver or completely separate from the liver. It should not be mistaken with splenosis (▶ Fig. 6.39g,h) or other liver lesions. Congenital malformations of the liver are rare. Liver cysts are very common benign process. Among other benign liver lesions, cavernous angioma, focal nodular hyperplasia, and adenoma can be named. The most common primary liver malignancy is hepatocellular carcinoma followed by cholangiocarcinoma. Liver is a common site of metastasis from remote malignancies. Fatty liver, hepatitis, and cirrhosis are common parenchymal diseases. Liver and spleen are the most common intra-abdominal solid organ site of injuries.

Fatty Liver

Fatty liver infiltration is a common abnormality detected by imaging. It is usually diffuse but could involve part of the liver or be focal or multifocal.[67,68] The two most common conditions associated with fatty liver are alcoholic liver disease due to excess alcohol consumption and nonalcoholic fatty liver disease due to diabetes and metabolic syndrome.

On ultrasound, normal liver echogenicity is almost similar to the renal cortex or spleen. In fatty infiltration the liver echogenicity exceeds that of renal cortex and spleen and there is poor delineation of the intrahepatic architecture.

On unenhanced CT, the normal liver has slightly greater attenuation than the spleen and blood (▶ Fig. 6.40). Fatty liver is diagnosed when the attenuation of the liver is at least 10 HU less than spleen or if the attenuation of the liver is <40 HU. In contrast-enhanced CT, the comparison of liver and spleen attenuation values is not as reliable for the diagnosis of fatty liver (▶ Fig. 6.3). On portal phase contrast-enhanced CT, a density difference greater than 25 HU between spleen and liver suggests fatty liver deposition. On MRI, increased fat content of the liver is associated with increased signal on T1-weighted images. On chemical shift gradient-echo (GRE) imaging, the signal intensity of the normal liver parenchyma is similar on both in-phase and opposed-phase images. On the other hand, signal loss on opposed-phase images in comparison with in-phase images indicates fatty liver.

Focal fat deposition is less common than diffuse fatty liver and should not be mistaken for other focal lesions. Focal fat deposition or focal fat sparing is usually seen in subcapsular regions near the falciform ligament, ligamentum venosum, or gallbladder fossa (▶ Fig. 6.40). Focal fat accumulation tends to show a wedge-shaped appearance with no mass effect or disturbance of normal vessels distribution. Focal fat deposition or sparing may be linked to local hemodynamic disturbances due to unusual systemic venous flow involving the capsular, peribiliary, or cystic veins without normal portal venous flow. Multifocal fat deposition is uncommon and usually seen in patients with chronic liver parenchymal diseases.

Liver Angioma

Hemangioma is the most common benign hepatic tumor. Angiomas of the liver in adults are mostly cavernous angioma. These benign lesions are echogenic in ultrasound and hypoattenuating lesion on nonenhanced images. At MRI, hemangiomas are characterized by well-defined margins and high signal intensity on T2-weighted images.[69] After intravenous administration of contrast material, they demonstrate a characteristic early peripheral enhancement with progressive and persistent centripetal filling of the lesion on delayed CT or MR images (▶ Fig. 6.41).

Liver Cirrhosis and Hepatocellular Carcinoma

Hepatic cirrhosis is the pathological sequela of all chronic liver diseases. The most common causes of hepatic cirrhosis are alcoholic fatty liver disease, nonalcoholic fatty liver disease, and viral hepatitis. The disease process is characterized by right lobe atrophy and left and caudate lobes enlargement. Imaging findings include a nodular hepatic surface, parenchymal abnormalities, dilated collateral veins, splenomegaly, and ascites[70] (▶ Fig. 6.8, ▶ Fig. 6.9).

Hepatocellular carcinoma (HCC) is the most common primary liver cancer. Approximately 80% of all HCCs arise from an underlying cirrhotic liver.[71,72] Early diagnosis of HCC is important as several potentially curative treatment options are available that can then lead to an improved outcome. The disease can be studied by various imaging modalities. The presence of arterial enhancement of a nodule 2 cm or more in size with subsequent washout on portal or delayed phases are the definitive imaging features of HCC. Resection and liver transplantation are common surgical treatment methods. Image-guided percutaneous ablation therapies have been widely performed on patients with small HCCs.

6

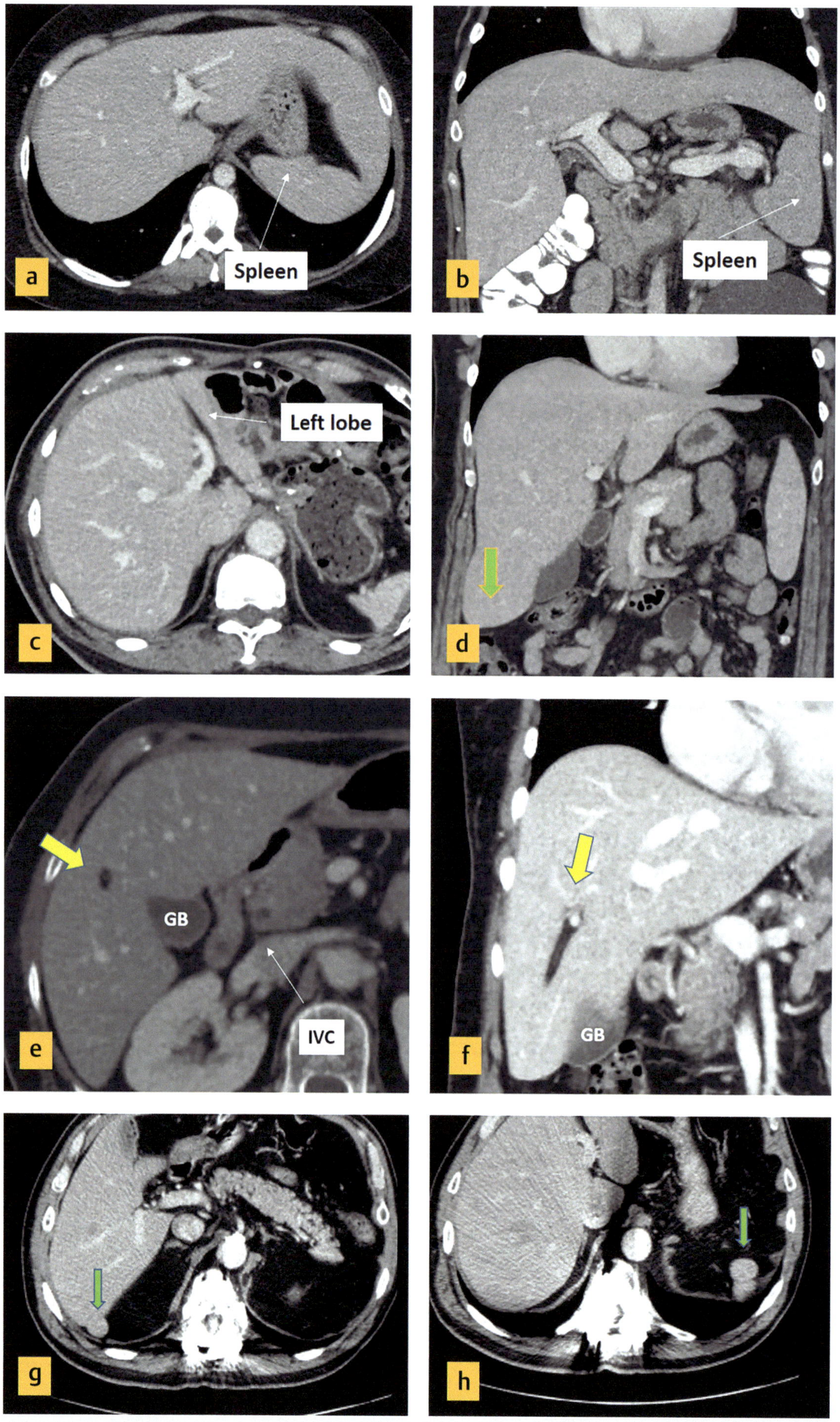

Fig. 6.39 Variants of liver morphology. **(a, b)** Long left liver lobe. **(c)** Small hypoplastic left liver lobe. **(d)** Long inferior tip of the right liver lobe (*arrow*). The later variant is called Riedel's lobe and relates to elongation of the anterior edge of the right lobe of the liver. **(e)** Axial and **(f)** coronal computed tomography (CT) showing unusual location of the umbilical fissure and falciform ligament (*arrows*) at the anatomic plane of Cantlie's line passing through the gallbladder and inferior vena cava (IVC) bed. (**g**,**h**) Axial CT images showing splenosis (*arrows*) following splenic injury and splenectomy. Splenosis is usually seen at the splenectomy bed, but infrequently is seen near pancreas, liver, stomach, or even intrathoracic. Those near liver may be mistaken with an accessory liver or liver lesions. GB, gallbladder.

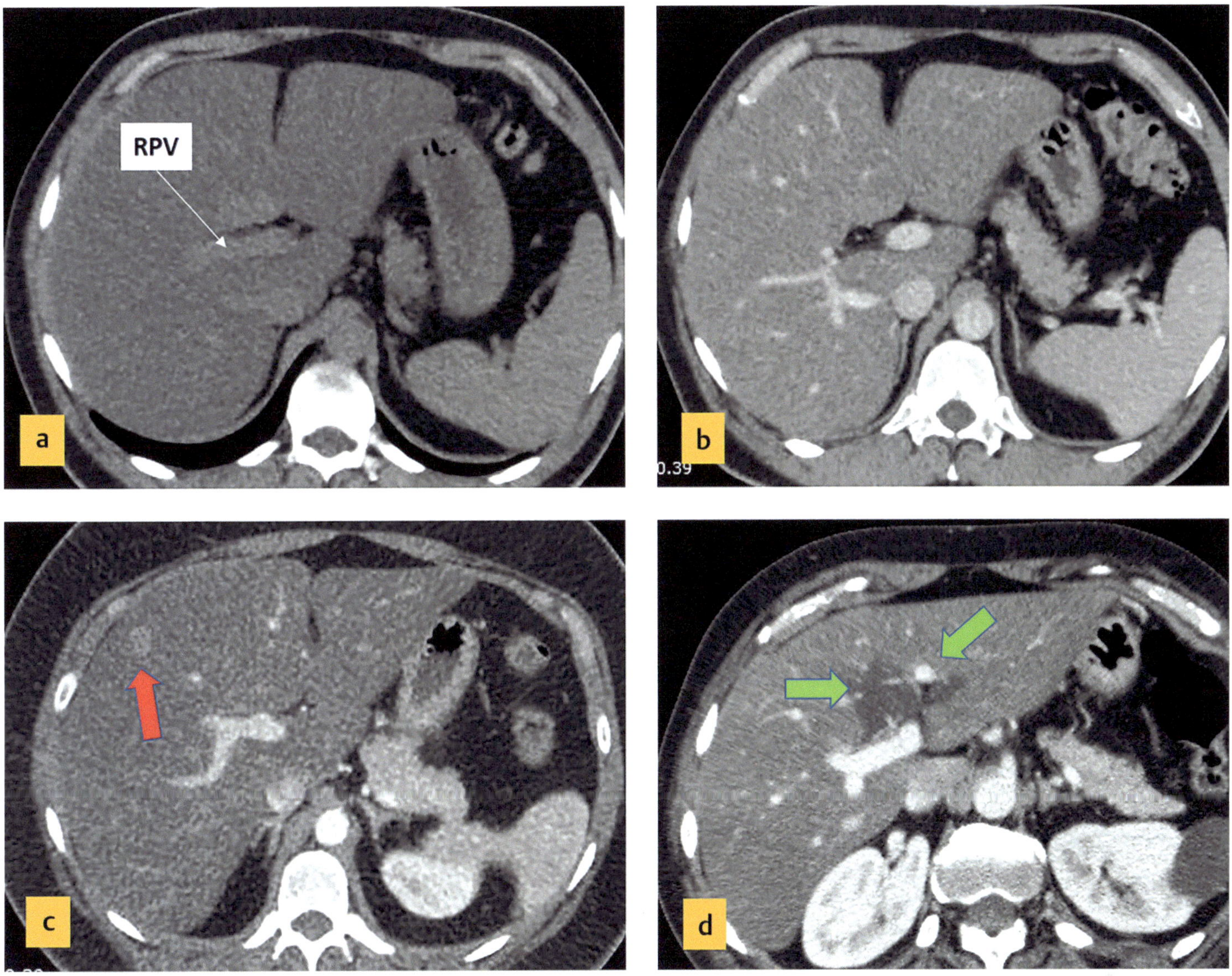

Fig. 6.40 Fatty liver. **(a)** Precontrast and **(b)** portal phase postcontrast images showing mild diffuse fatty liver. On unenhanced computed tomography (CT), the normal liver has slightly greater attenuation than the spleen and blood. **(c)** Fatty liver with a nodular focal sparing (*arrow*). **(d)** Focal fat deposition involving segment IV. Focal fat deposition or focal fat sparing is usually seen in subcapsular regions near the falciform ligament, ligamentum venosum, or gallbladder fossa. RPV, right portal vein.

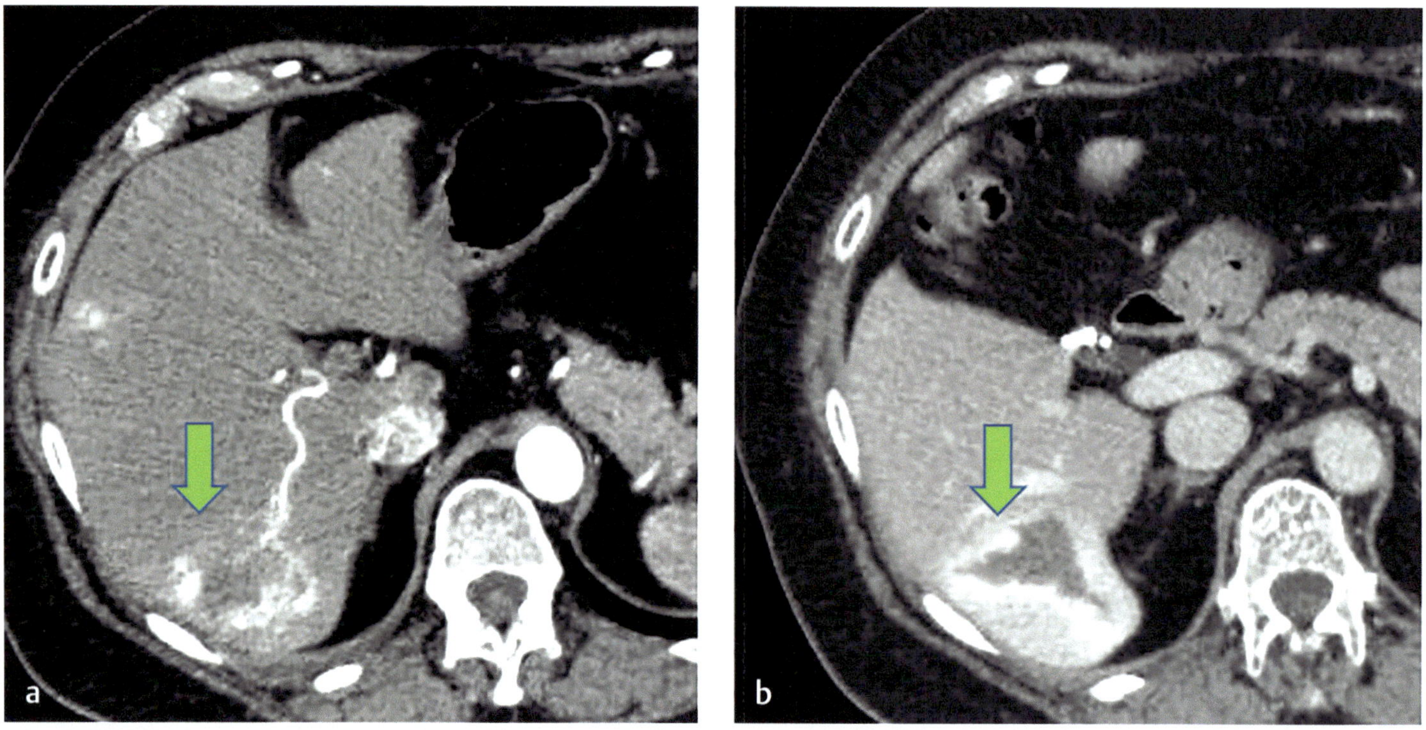

Fig. 6.41 Liver hemangioma (*arrows*). **(a)** Arterial phase and **(b)** portal phase axial computed tomography (CT) showing a characteristic early arterial peripheral enhancement with progressive and persistent centripetal filling of the lesion on delayed image.

References

1. Lemaigre FP. Mechanisms of liver development: concepts for understanding liver disorders and design of novel therapies. Gastroenterology 2009;137(1):62–79
2. Raynaud P, Carpentier R, Antoniou A, Lemaigre FP. Biliary differentiation and bile duct morphogenesis in development and disease. Int J Biochem Cell Biol 2011;43(2):245–256
3. Si-Tayeb K, Lemaigre FP, Duncan SA. Organogenesis and development of the liver. Dev Cell 2010;18(2):175–189
4. Mavrides E, Moscoso G, Carvalho JS, Campbell S, Thilaganathan B. The anatomy of the umbilical, portal and hepatic venous systems in the human fetus at 14–19 weeks of gestation. Ultrasound Obstet Gynecol 2001;18(6):598–604
5. Skandalakis JE, Skandalakis LJ, Skandalakis PN, Mirilas P. Hepatic surgical anatomy. Surg Clin North Am 2004;84(2):413–435, viii
6. Mirilas P, Skandalakis JE. Benign anatomical mistakes: right and left coronary ligaments. Am Surg 2002;68(9):832–835
7. Dodd GD III. An American's guide to Couinaud's numbering system. AJR Am J Roentgenol 1993;161(3):574–575
8. Healey JE Jr, Schroy PC. Anatomy of the biliary ducts within the human liver; analysis of the prevailing pattern of branchings and the major variations of the biliary ducts. AMA Arch Surg 1953;66(5):599–616
9. Couinaud C. Le foie, etudes anatomiques et chirurgicales. Paris: Masson & Cie; 1957
10. Ko S, Murakami G, Kanamura T, Sato TJ, Nakajima Y. Cantlie's plane in major variations of the primary portal vein ramification at the porta hepatis: cutting experiment using cadaveric livers. World J Surg 2004;28(1):13–18
11. Rutkauskas S, Gedrimas V, Pundzius J, Barauskas G, Basevicius A. Clinical and anatomical basis for the classification of the structural parts of liver. Medicina (Kaunas) 2006;42(2):98–106
12. Shindoh J, Satou S, Aoki T, et al. Hidden symmetry in asymmetric morphology: significance of Hjortsjo's anatomical model in liver surgery. Hepatogastroenterology 2012;59(114):519–525
13. Mise Y, Satou S, Shindoh J, et al. Three-dimensional volumetry in 107 normal livers reveals clinically relevant inter-segment variation in size. HPB (Oxford) 2014;16(5):439–447
14. Kogure K, Kuwano H, Fujimaki N, Ishikawa H, Takada K. Reproposal for Hjortsjo's segmental anatomy on the anterior segment in human liver. Arch Surg 2002; 137(10):1118–1124
15. Strasberg SM. Nomenclature of hepatic anatomy and resections: a review of the Brisbane 2000 system. J Hepatobiliary Pancreat Surg 2005;12(5):351–355
16. Hjortsjo CH. The topography of the intrahepatic duct systems. Acta Anat (Basel) 1951;11(4):599–615
17. Fischer L, Cardenas C, Thorn M, et al. Limits of Couinaud's liver segment classification: a quantitative computer-based three-dimensional analysis. J Comput Assist Tomogr 2002;26(6):962–967
18. Fasel JH, Schenk A. Concepts for liver segment classification: neither old ones nor new ones, but a comprehensive one. J Clin Imaging Sci 2013;3:48
19. Fasel JH. Portal venous territories within the human liver: an anatomical reappraisal. Anat Rec (Hoboken) 2008;291(6):636–642
20. Fasel JH, Majno PE, Peitgen HO. Liver segments: an anatomical rationale for explaining inconsistencies with Couinaud's eight-segment concept. Surg Radiol Anat 2010;32(8):761–765
21. Majno P, Mentha G, Toso C, Morel P, Peitgen HO, Fasel JH. Anatomy of the liver: an outline with three levels of complexity—a further step towards tailored territorial liver resections. J Hepatol 2014;60(3):654–662
22. Takasaki K. Glissonean pedicle transection method for hepatic resection: a new concept of liver segmentation. J Hepatobiliary Pancreat Surg 1998;5(3):286–291
23. Launois B, Jamieson GG. The importance of Glisson's capsule and its sheaths in the intrahepatic approach to resection of the liver. Surg Gynecol Obstet 1992; 174(1):7–10
24. Masunari H, Shimada H, Endo I, et al. Surgical anatomy of hepatic hilum with special reference of the plate system and extrahepatic duct. J Gastrointest Surg 2008;12(6):1047–1053
25. Couinaud C. The paracaval segments of the liver. J Hepatobiliary Pancreat Surg 1994;1(2):145–151
26. Kawarada Y, Das BC, Taoka H. Anatomy of the hepatic hilar area: the plate system. J Hepatobiliary Pancreat Surg 2000;7(6):580–586
27. Katagiri S, Ariizumi S, Kotera Y, Takahashi Y, Yamamoto M. Right hepatectomy using Glissonean pedicle transection method with anterior approach (with video). J Hepatobiliary Pancreat Sci 2012;19(1):25–29

28. Ronot M, Lambert S, Daire JL, et al. Can we justify not doing liver perfusion imaging in 2013? Diagn Interv Imaging 2013;94(12):1323–1336
29. Michels NA. Newer anatomy of the liver and its variant blood supply and collateral circulation. Am J Surg 1966;112(3):337–347
30. De Cecco CN, Ferrari R, Rengo M, Paolantonio P, Vecchietti F, Laghi A. Anatomic variations of the hepatic arteries in 250 patients studied with 64-row CT angiography. Eur Radiol 2009;19(11):2765–2770
31. Covey AM, Brody LA, Maluccio MA, Getrajdman GI, Brown KT. Variant hepatic arterial anatomy revisited: digital subtraction angiography performed in 600 patients. Radiology 2002;224(2):542–547
32. Koops A, Wojciechowski B, Broering DC, Adam G, Krupski-Berdien G. Anatomic variations of the hepatic arteries in 604 selective celiac and superior mesenteric angiographies. Surg Radiol Anat 2004;26(3):239–244
33. Winston CB, Lee NA, Jarnagin WR, et al. CT angiography for delineation of celiac and superior mesenteric artery variants in patients undergoing hepatobiliary and pancreatic surgery. AJR Am J Roentgenol 2007;189(1):W13-9
34. Onishi H, Kawarada Y, Das BC, et al. Surgical anatomy of the medial segment (S4) of the liver with special reference to bile ducts and vessels. Hepatogastroenterology 2000;47(31):143–150
35. Cho A, Okazumi S, Takayama W, et al. Anatomy of the right anterosuperior area (segment 8) of the liver: evaluation with helical CT during arterial portography. Radiology 2000;214(2):491–495
36. Furuta T, Maeda E, Akai H, et al. Hepatic segments and vasculature: projecting CT anatomy onto angiograms. Radiographics 2009;29(7):1–22
37. Sahani D, D'souza R, Kadavigere R, et al. Evaluation of living liver transplant donors: method for precise anatomic definition by using a dedicated contrast-enhanced MR imaging protocol. Radiographics 2004;24(4):957–967
38. López-Andújar R, Moya A, Montalvá E, et al. Lessons learned from anatomic variants of the hepatic artery in 1,081 transplanted livers. Liver Transpl 2007; 13(10):1401–1404
39. Saylisoy S, Atasoy C, Ersöz S, Karayalçin K, Akyar S. Multislice CT angiography in the evaluation of hepatic vascular anatomy in potential right lobe donors. Diagn Interv Radiol 2005;11(1):51–59
40. Schmidt S, Demartines N, Soler L, Schnyder P, Denys A. Portal vein normal anatomy and variants: implication for liver surgery and portal vein embolization. Semin Intervent Radiol 2008;25(2):86–91
41. Koç Z, Oğuzkurt L, Ulusan S. Portal vein variations: clinical implications and frequencies in routine abdominal multidetector CT. Diagn Interv Radiol 2007;13(2):75–80
42. Atasoy C, Ozyürek E. Prevalence and types of main and right portal vein branching variations on MDCT. AJR Am J Roentgenol 2006;187(3):676–681
43. Madoff DC, Hicks ME, Vauthey JN, et al. Transhepatic portal vein embolization: anatomy, indications, and technical considerations. Radiographics 2002;22(5):1063–1076
44. Kitami M, Takase K, Murakami G, et al. Types and frequencies of biliary tract variations associated with a major portal venous anomaly: analysis with multi-detector row CT cholangiography. Radiology 2006;238(1):156–166
45. Barbaro B, Soglia G, Alvaro G, et al. Hepatic veins in presurgical planning of hepatic resection: what a radiologist should know. Abdom Imaging 2013; 38(3):442–460
46. Cheng YF, Huang TL, Chen CL, et al. Variations of the middle and inferior right hepatic vein: application in hepatectomy. J Clin Ultrasound 1997;25(4):175–182
47. Scatton O, Plasse M, Dondero F, Vilgrain V, Sauvanet A, Belghiti J. Impact of localized congestion related to venous deprivation after hepatectomy. Surgery 2008;143(4):483–489
48. De Cecchis L, Hribernik M, Ravnik D, Gadzijev EM. Anatomical variations in the pattern of the right hepatic veins: possibilities for type classification. J Anat 2000;197(Pt 3):487–493
49. Gupta SC, Gupta CD. The hepatic veins—a radiographic and corrosion cast study. Indian J Med Res 1979;70:333–344
50. Masselot R, Leborgne J. Anatomical study of the hepatic veins. Anat Clin 1978;1:109–125
51. Kogure K, Kuwano H, Yorifuji H, Ishikawa H, Takata K, Makuuchi M. The caudate processus hepatic vein: a boundary hepatic vein between the caudate lobe and the right liver. Ann Surg 2008;247(2):288–293
52. Fischer L, Thorn M, Neumann JO, et al. The segments of the hepatic veins—is there a spatial correlation to the Couinaud liver segments? Eur J Radiol 2005;53(2):245–255
53. Buhe S, Miyaki T, Saito T, et al. A study of the accessory hepatic vein to segments VI and VII with a morphological reconsideration of the human liver. Surg Radiol Anat 2008;30(3):201–207
54. Ortale JR, Borges Keiralla LC. Anatomy of the portal branches and the hepatic veins in the caudate lobe of the liver. Surg Radiol Anat 2004;26(5):384–391
55. Vollmar B, Menger MD. The hepatic microcirculation: mechanistic contributions and therapeutic targets in liver injury and repair. Physiol Rev 2009;89(4):1269–1339
56. McCuskey RS. The hepatic microvascular system in health and its response to toxicants. Anat Rec (Hoboken) 2008;291(6):661–671
57. Ohtani O, Ohtani Y. Lymph circulation in the liver. Anat Rec (Hoboken) 2008;291(6):643–652
58. Quiroga S, Sebastià C, Pallisa E, Castellà E, Pérez-Lafuente M, Alvarez-Castells A. Improved diagnosis of hepatic perfusion disorders: value of hepatic arterial phase imaging during helical CT. Radiographics 2001;21(1):65–81, 288–294
59. Kim TK, Choi BI, Han JK, Chung JW, Park JH, Han MC. Nontumorous arterioportal shunt mimicking hypervascular tumor in cirrhotic liver: two-phase spiral CT findings. Radiology 1998;208(3):597–603
60. Matsui O, Takahashi S, Kadoya M, et al. Pseudolesion in segment IV of the liver at CT during arterial portography: correlation with aberrant gastric venous drainage. Radiology 1994;193(1):31–35
61. Breen DJ, Rutherford EE, Stedman B, Lee-Elliott C, Hacking CN. Intrahepatic arterioportal shunting and anomalous venous drainage: understanding the CT features in the liver. Eur Radiol 2004;14(12):2249–2260
62. Brown RS Jr. Live donors in liver transplantation. Gastroenterology 2008; 134(6):1802–1813
63. Moon DB, Lee SG. Liver transplantation. Gut Liver 2009;3(3):145–165
64. Kamel IR, Kruskal JB, Pomfret EA, Keogan MT, Warmbrand G, Raptopoulos V. Impact of multidetector CT on donor selection and surgical planning before living adult right lobe liver transplantation. AJR Am J Roentgenol 2001;176(1): 193–200
65. Atasoy C, Akyar S. Multidetector CT: contributions in liver imaging. Eur J Radiol 2004;52(1):2–17
66. Togo S, Shizawa R, Kanemura E, et al. Usefulness of 3-dimensional computed tomography for caudate lobectomy by transhepatic anterior approach. Hepatogastroenterology 2002;49(44):461–466
67. Hamer OW, Aguirre DA, Casola G, Lavine JE, Woenckhaus M, Sirlin CB. Fatty liver: imaging patterns and pitfalls. Radiographics 2006;26(6):1637–1653
68. Décarie PO, Lepanto L, Billiard JS, et al. Fatty liver deposition and sparing: a pictorial review. Insights Imaging 2011;2(5):533–538
69. Vilgrain V, Boulos L, Vullierme MP, Denys A, Terris B, Menu Y. Imaging of atypical hemangiomas of the liver with pathologic correlation. Radiographics 2000;20(2):379–397
70. Ito K, Mitchell DG, Gabata T, Hussain SM. Expanded gallbladder fossa: simple MR imaging sign of cirrhosis. Radiology 1999;211(3):723–726
71. Hennedige T, Venkatesh SK. Imaging of hepatocellular carcinoma: diagnosis, staging and treatment monitoring. Cancer Imaging 2013;12(3):530–547
72. Bruix J, Sherman M; Practice Guidelines Committee, American Association for the Study of Liver Diseases. Management of hepatocellular carcinoma. Hepatology 2005;42(5):1208–1236

7 Spleen

Farhood Saremi

◆ Introduction

The spleen is the largest secondary lymphatic organ in the body containing about one-fourth of the body's lymphocytes. The spleen is a highly vascular organ that filters blood.[1] Major functions of the spleen include red blood cell breakdown and immunologic surveillance. It is a crucial site of early exposure to encapsulated bacteria. It is believed that splenic contraction after hemorrhage and blood loss helps to augment blood volume. A wide range of pathologic conditions can affect the spleen.

◆ Anatomy

The splenic anlage is derived from the splanchnic mesodermal plate (SMP) in the dorsal mesogastrium around embryonic day 12.[1] Along with rotation of the stomach, the spleen will be moved to the left upper abdomen in its final position which will be completed by 7 weeks (see Chapter 3 "Stomach"). The first cells that colonize the spleen are progenitors of the erythroid and myeloid lineages at day 14.5.[1] With migration of the spleen from midline to the left, the dorsal mesentery (the dorsal mesogastrium) will be elongated sufficiently to form the fibrofatty gastrosplenic and phrenicosplenic ligament.

The spleen is a crescent-shaped intraperitoneal organ in the left upper quadrant of the abdomen below the left hemidiaphragm and anterior to the 10th and 11th ribs. It is attached to the retroperitoneum by peritoneal ligaments that also provide the splenic vascular supply. The gross appearance and size of the spleen are variable, depending on the species and the degree of distension. The spleen reaches peak development at puberty followed by gradual involution and decrease in lymphocyte numbers with age.

The spleen has two major surfaces named according to their locations, namely, the diaphragmatic (phrenic) and visceral surfaces. The visceral surface has two impressions; gastric impression is located anteriorly and the left renal impression is seen posteriorly (▶Fig. 7.1, ▶Fig. 7.2, ▶Fig. 7.3). The splenic hilum, where the splenic vessels emerge, is directed anteromedially and is close to the tail of the pancreas (▶Fig. 7.2). Inferior margin of the spleen approaches the splenic flexure of the colon.

Microscopic Anatomy

The spleen is surrounded by a fibrous capsule of connective tissue which is firmly attached to the visceral peritoneum (▶Fig. 7.4). Multiple trabeculae containing smooth muscles and fibroelastic tissue stem from the capsule into the splenic parenchyma which supports the larger blood and lymph vessels and nerves. There are no afferent lymphatic vessels in the spleen. The lymph vessels are primarily efferent-only vessels through which lymphocytes migrate to the splenic lymph nodes. In some species (i.e., dogs), the splenic capsule and trabeculae contain more smooth muscles that help spleen to contract.[2]

The spleen is divided into two functionally and morphologically distinct compartments, namely, the red and white pulps, separated by the marginal zone (▶Fig. 7.4). The red pulp is composed of a meshwork of splenic cords and venous sinuses. The red pulp macrophages remove foreign material, bacteria, and old erythrocytes. It is also a storage site for iron pigments (i.e., hemosiderin and ferritin), erythrocytes, and platelets. Extramedullary hematopoiesis occurs in the red pulp in pathological states. Hemosiderin deposits in the cytoplasm of macrophages in the red pulp.

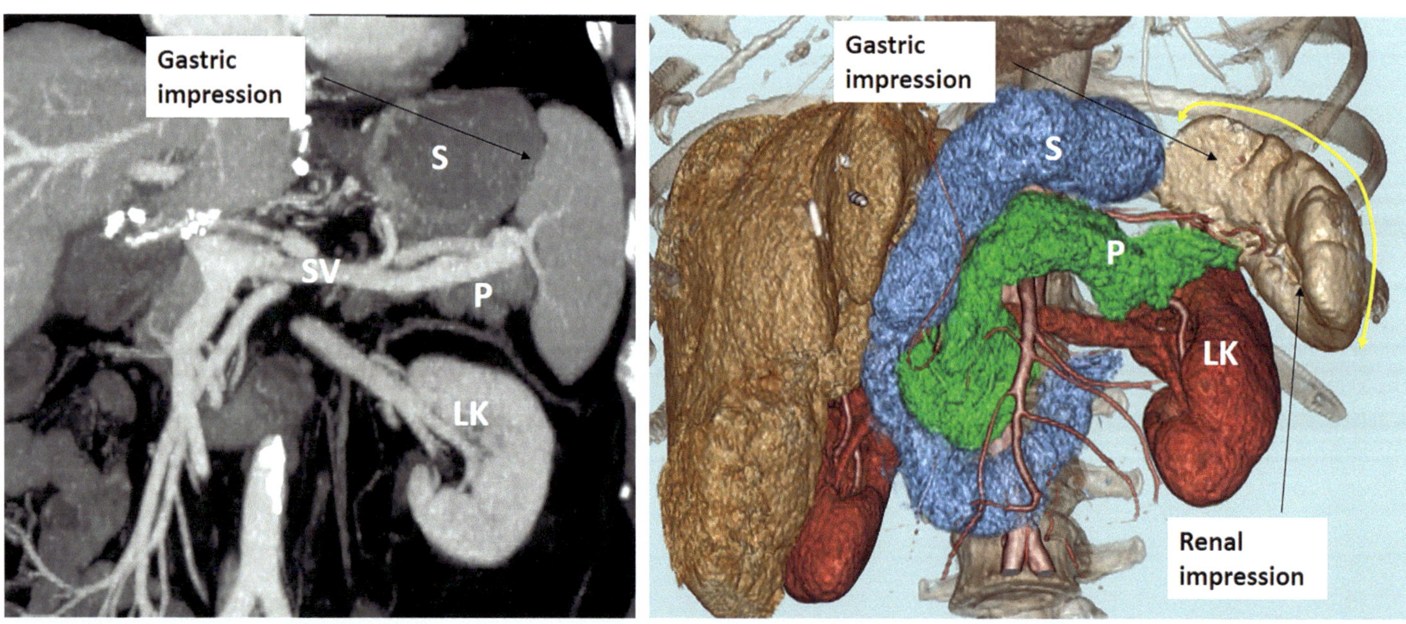

Fig. 7.1 Anatomic surfaces and relationships of the spleen. Coronal views. The diaphragmatic (phrenic) surface is shown by *double-headed arrow*. The visceral surface has two impressions: gastric and left renal. LK, left kidney; P, pancreas; S, stomach; SA, splenic artery; SV, splenic vein.

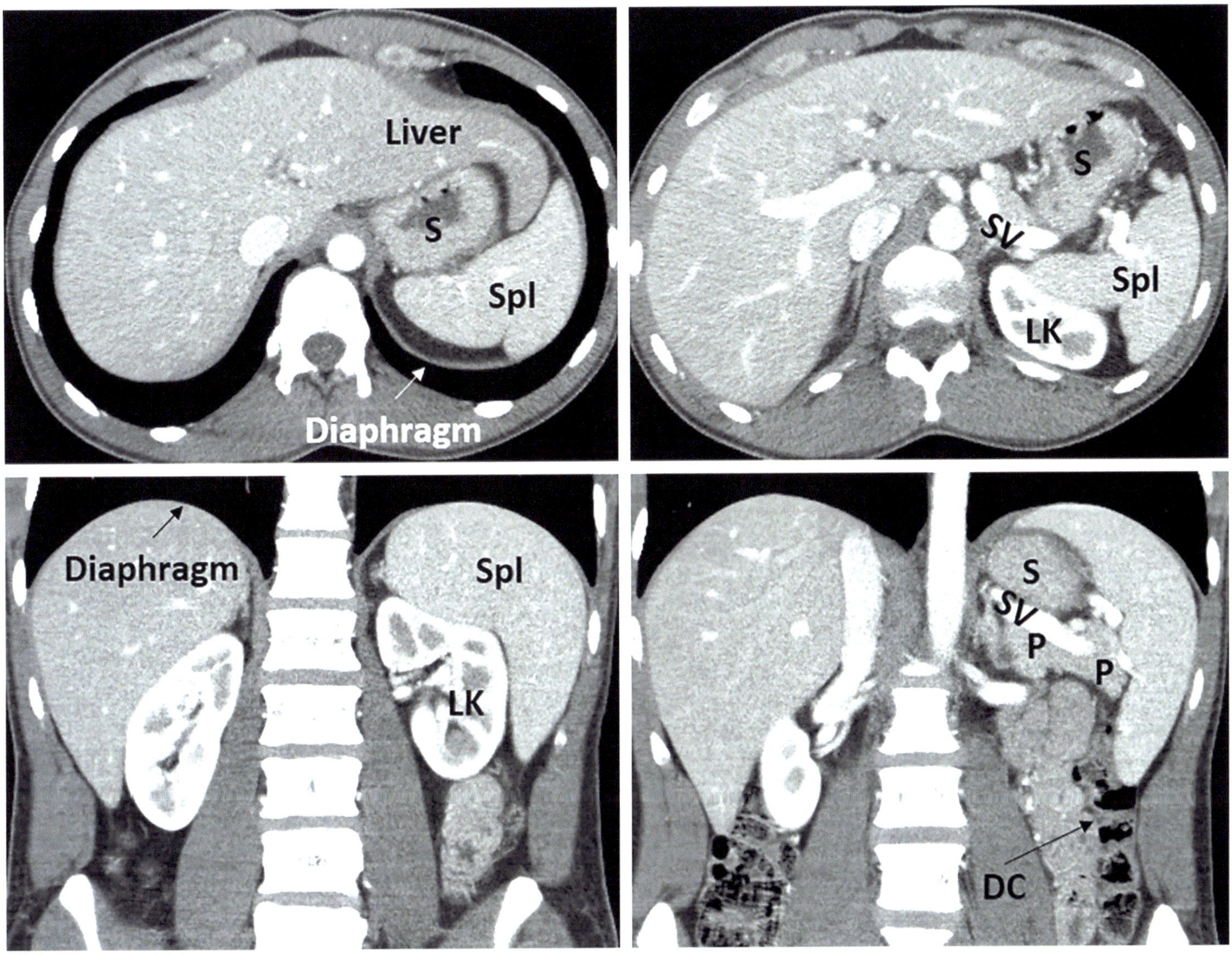

Fig. 7.2 Anatomic relationships of the spleen (Spl) shown by axial (*upper row*) and coronal (*lower row*) computed tomography (CT) views. The gastric impression is located anteriorly and the left kidney impression is seen posteriorly. Inferior margin of the spleen approaches the splenic flexure of the colon. DC, descending colon; LK, left kidney; P, pancreas; S, stomach; SV, splenic vein.

The white pulp is located centrally around the central arterioles. It contains T and B lymphocytes to initiate immune responses to blood-borne antigens.[3] The white pulp is composed of three compartments: the periarteriolar lymphoid sheath (PALS) containing T cells, the follicles containing B cells, and the marginal zone (▶Fig. 7.4). The marginal zone is located at the interface of the red and white pulps and blends into the red pulp. It plays an important role in antigen processing and clearance of microorganisms.[1]

◆ Vessels

The splenic artery is a branch of the celiac trunk. On rare occasions, it arises directly from the aorta. It supplies the spleen and parts of the stomach and pancreas. The splenic artery has a serpiginous course along the superior margin of the pancreas. At this level, it is slightly superior to the splenic vein. At or near the splenic hilum, the artery divides into two (superior and inferior) terminal branches in 70% of cases and into three (superior, middle, and inferior) in 20%[4] (▶Fig. 7.5, ▶Fig. 7.6, ▶Fig. 7.7). Each branch further subdivides into four to six segmental intrasplenic branches. The superior terminal branches are usually longer than the inferior terminal branches and provide the major splenic arterial supply[4] (▶Fig. 7.5).

The splenic artery provides several branches that supply the pancreatic body and tail. The first large branch is called the dorsal pancreatic artery, and the second large branch is the greater pancreatic artery (or arteria pancreatica magna) which connects to the transverse pancreatic artery (see Chapter 9 "Pancreas") (▶Fig. 7.6). When transcatheter embolization of the splenic artery is planned, the catheter should be placed distal to the pancreatic arteries to reduce the risk of unintended embolization.[5] Splenic arterial embolization is often performed to treat post-traumatic splenic injuries and in patients with hypersplenism.

In 25% of individuals a superior polar artery arises from the distal splenic artery or the superior terminal branch (▶Fig. 7.7). The inferior polar artery is seen in 50% of cases.[4] The left gastroepiploic artery usually arises from the inferior polar artery, but the latter may also originate from the distal splenic or terminal inferior branch artery.[5]

The left gastroepiploic artery is an important branch that supplies blood to the greater curvature of the stomach and the

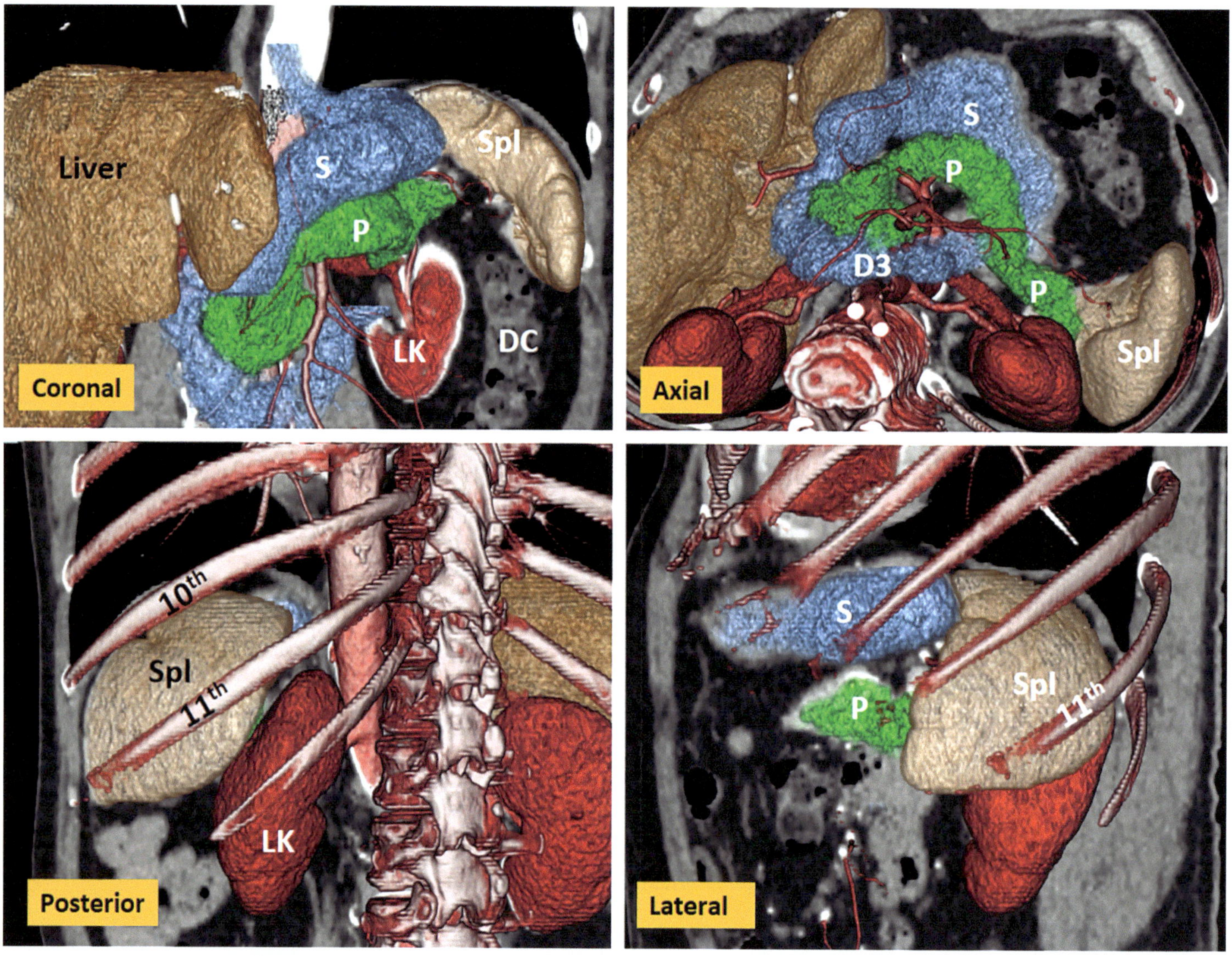

Fig. 7.3 Anatomic relationships of the spleen (Spl) by computed tomography (CT) views. The spleen is located in the left upper quadrant of the abdomen below the left hemidiaphragm and anterior to the 10th and 11th ribs. Inferior margin of the spleen approaches the splenic flexure of the colon. D3, third portion of the duodenum; DC, descending colon; LK, left kidney; P, pancreas; S, stomach.

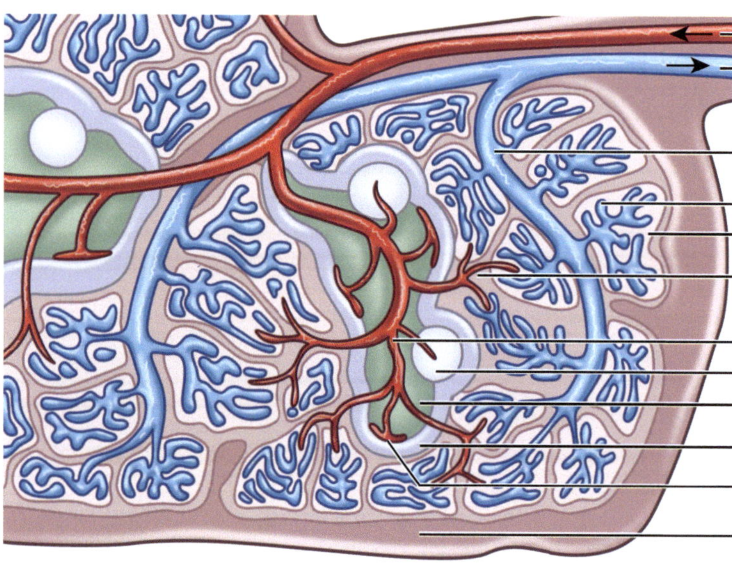

Fig. 7.4 Structure of the spleen. The afferent splenic artery branches into central arterioles, which are sheathed by white pulp areas. The white pulp areas consist of the T-cell zone (also known as the periarteriolar lymphoid sheath, PALS), arterioles, and B-cell follicles. The arterioles end in cords in the red pulp, from where the blood runs into venous sinuses, which collect into the efferent splenic vein. The larger arteries and veins run together in connective-tissue trabeculae, which are continuous with the capsule that surrounds the spleen. (Modified with permission from Mebius and Kraal.[1])

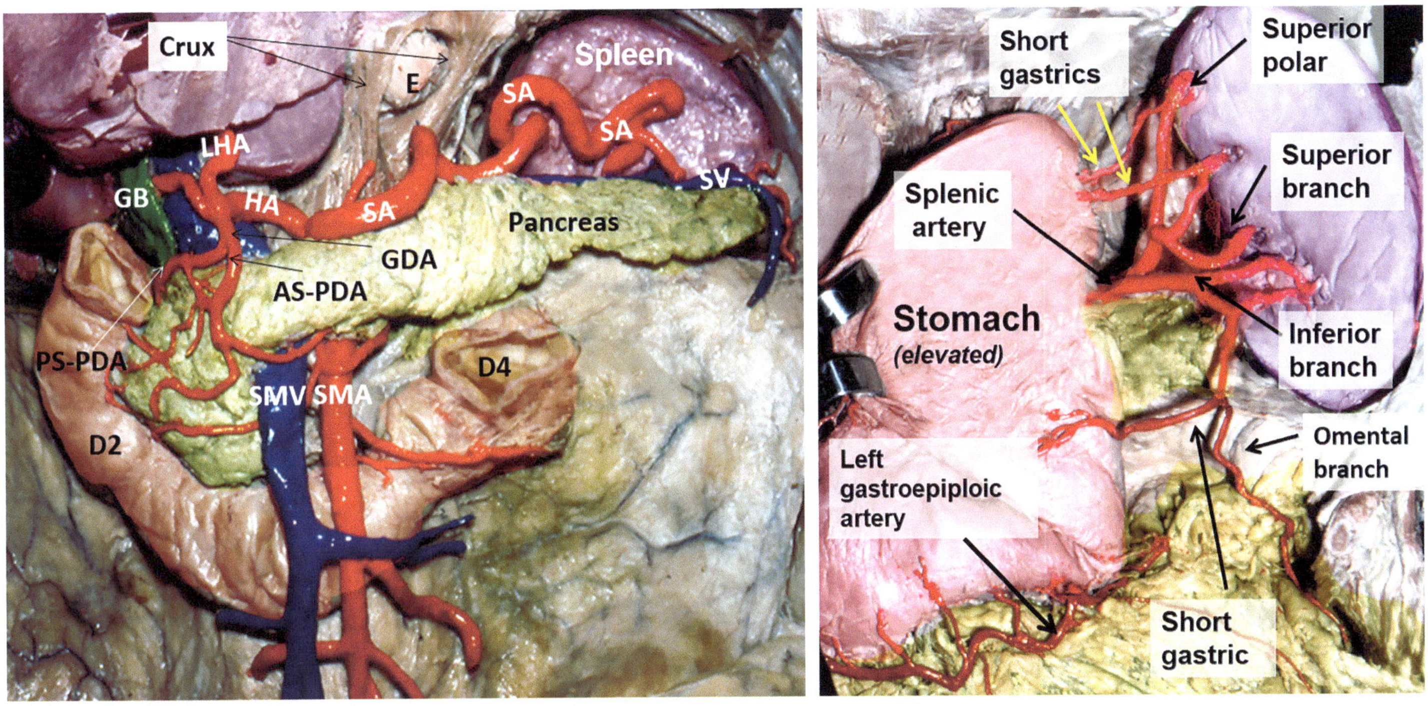

Fig. 7.5 Arterial supply to the spleen. The splenic artery (SA) and hepatic artery (HA) are branches of the celiac trunk. At or near the splenic hilum, the SA divides into two (superior and inferior) terminal branches. In 25% of individuals a superior polar artery arises from the distal splenic artery or the superior terminal branch. Multiple short gastric branches arise from the terminal splenic branches. The gastroduodenal artery (GDA) is a branch of the hepatic artery (HA) which gives origin to the posterior superior pancreaticoduodenal artery (PSPDA), right gastroepiploic and anterior superior pancreaticoduodenal arteries (ASPDA). D2/D4, second and fourth portions of the duodenum; E, esophagus; GB, gallbladder; LHA, left hepatic artery; SMA, superior mesenteric artery; SMV, superior mesenteric vein; SV, splenic vein.

7

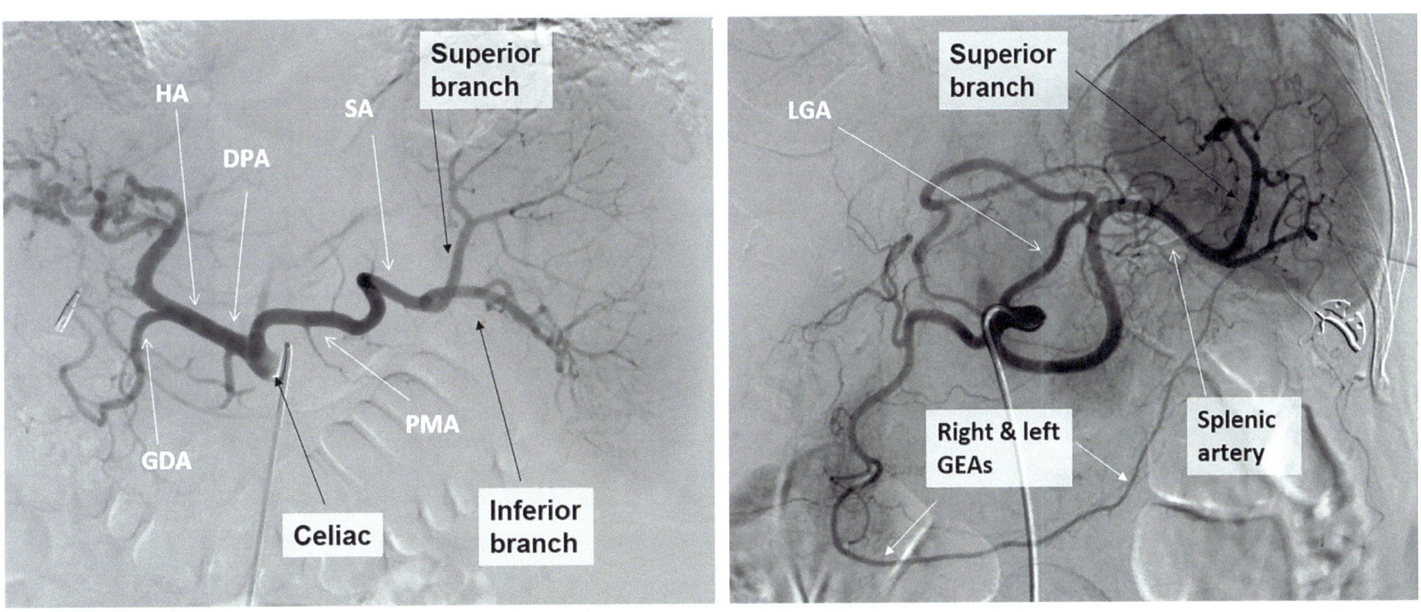

Fig. 7.6 Catheter angiogram of the celiac artery showing typical branches of the splenic artery (SA). At or near the splenic hilum, the SA divides into two (superior and inferior) terminal branches. The SA gives rise to the dorsal pancreatic artery (DPA) and the pancreatic magna artery (PMA). The DPA is the largest vessel arising from the splenic artery. The DPA also commonly arises from the gastroduodenal artery (GDA). GEA, gastroepiploic artery; HA, hepatic artery; LGA, left gastric artery.

greater omentum. Multiple short gastric branches arise from the terminal splenic branches or left gastroepiploic artery to supply the gastric cardia and fundus (▶Fig. 7.5).

Within the spleen, the terminal branches of the splenic artery divide into trabecular arteries which course within the trabeculae toward the white pulp. Small arterioles of the white pulp are called central arterioles as they are ensheathed by the lymphoid tissue. Some of these arterioles terminate in the marginal sinus at the junction of the white pulp and the marginal zone; a few end in the perifollicular zone outside the marginal zone, and some pass the marginal zone to terminate in the red pulp.[1,3] In the red pulp the central arteries are named the penicillar arteries.

Once through the marginal zone, the blood flows directly into the venous sinuses lined by Stave cells whose open ends are continuous with the marginal zone, the so-called "fast pathway."

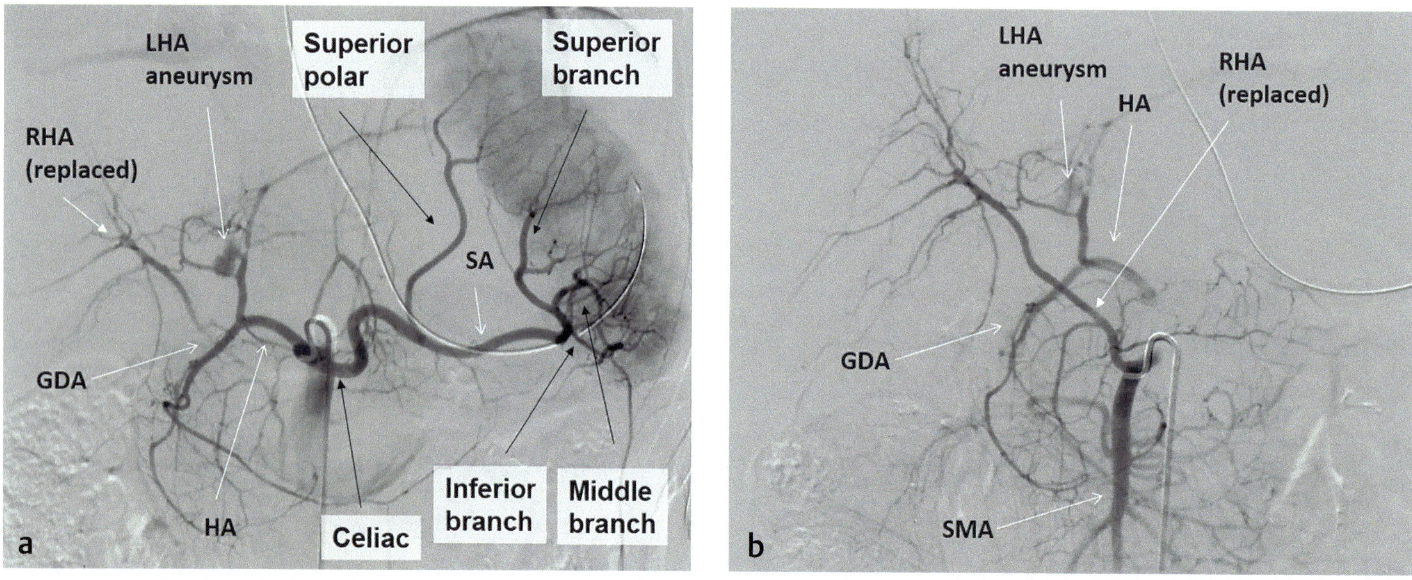

Fig. 7.7 Catheter angiogram of the celiac artery **(a)** and superior mesenteric artery (SMA) **(b)** showing unusual branches of the splenic artery (SA) into three terminal branches and an early superior polar branch. **(b)** Replaced right hepatic artery (RHA) is shown. Patient has a posttraumatic pseudoaneurysm arising from the left hepatic artery (LHA) which was embolized. GDA, gastroduodenal artery; HA, hepatic artery.

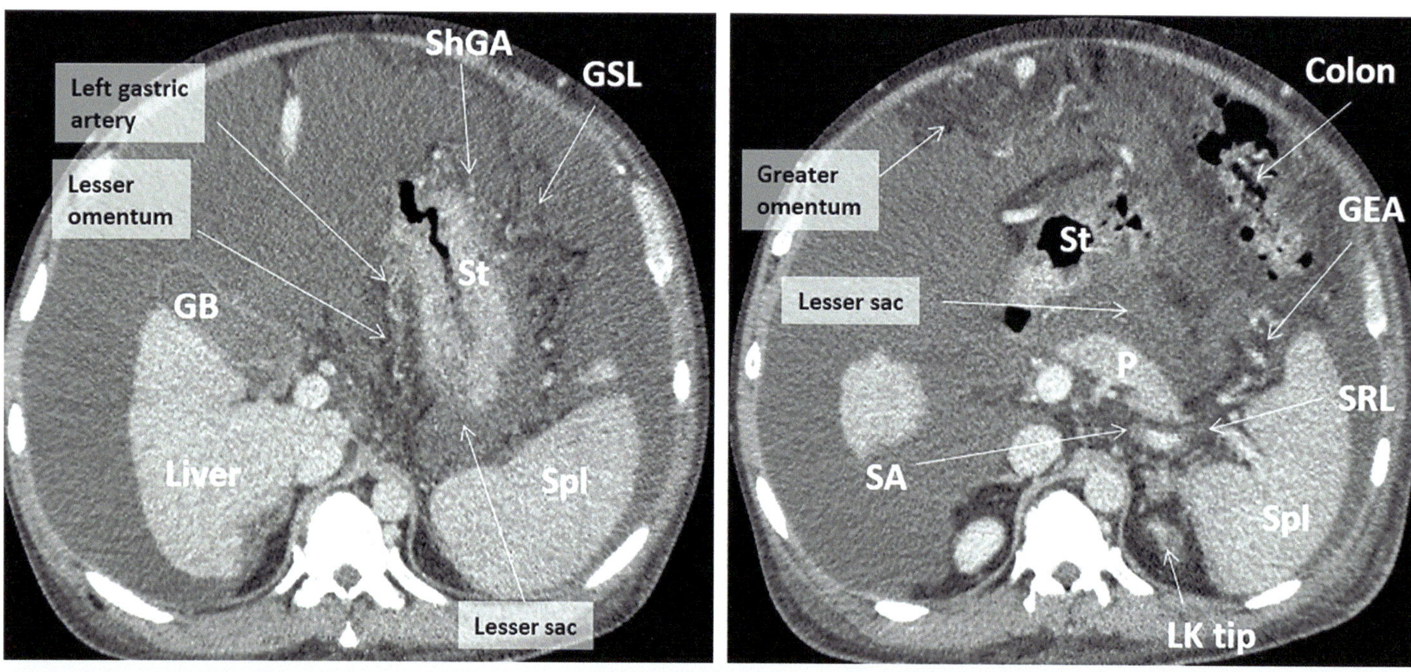

Fig. 7.8 Axial computed tomography (CT) images in a patient with large ascites showing the spleen (Spl) ligaments. The gastrosplenic ligament (GSL) is located between the greater curvature of the stomach (St) and splenic hilum containing short gastric arteries (ShGA) and left gastroepiploic artery (GEA). The splenorenal ligament (SRL) is the site of passage of the splenic artery (SA) and vein toward the splenic hilum between the left kidney (LK) and the spleen at the level of the pancreas (p) tail. GB, gallbladder; St, stomach.

Within the red pulp, in 90% of cases, the arteriolar flow enters a cellular reticular meshwork without an endothelial lining known as "cords." This open circulation is the "slow pathway" for the arterial flow before entering the venous sinusoids (▸ Fig. 7.4). In 10% of cases, the blood flows directly through a "fast pathway," into adjacent venous sinuses.[1,2] The venous sinus blood enters the trabeculae and merge into the trabecular veins (▸ Fig. 7.4). The trabecular veins then converge at the hilum to form the splenic vein which drains into the portal vein.

◆ Ligaments

There are four principal ligaments of the spleen: the gastrosplenic ligament, the colicosplenic ligament, the phrenicocolic ligament, and the splenorenal (also known as phrenicosplenic or phrenicolienal) ligament[6,7] (▸ Fig. 7.8, ▸ Fig. 7.9, ▸ Fig. 7.10).

The gastrosplenic ligament is located between the greater curvature of the stomach and splenic hilum. It is part of the greater omentum. Short gastric, superior polar, and left gastroepiploic

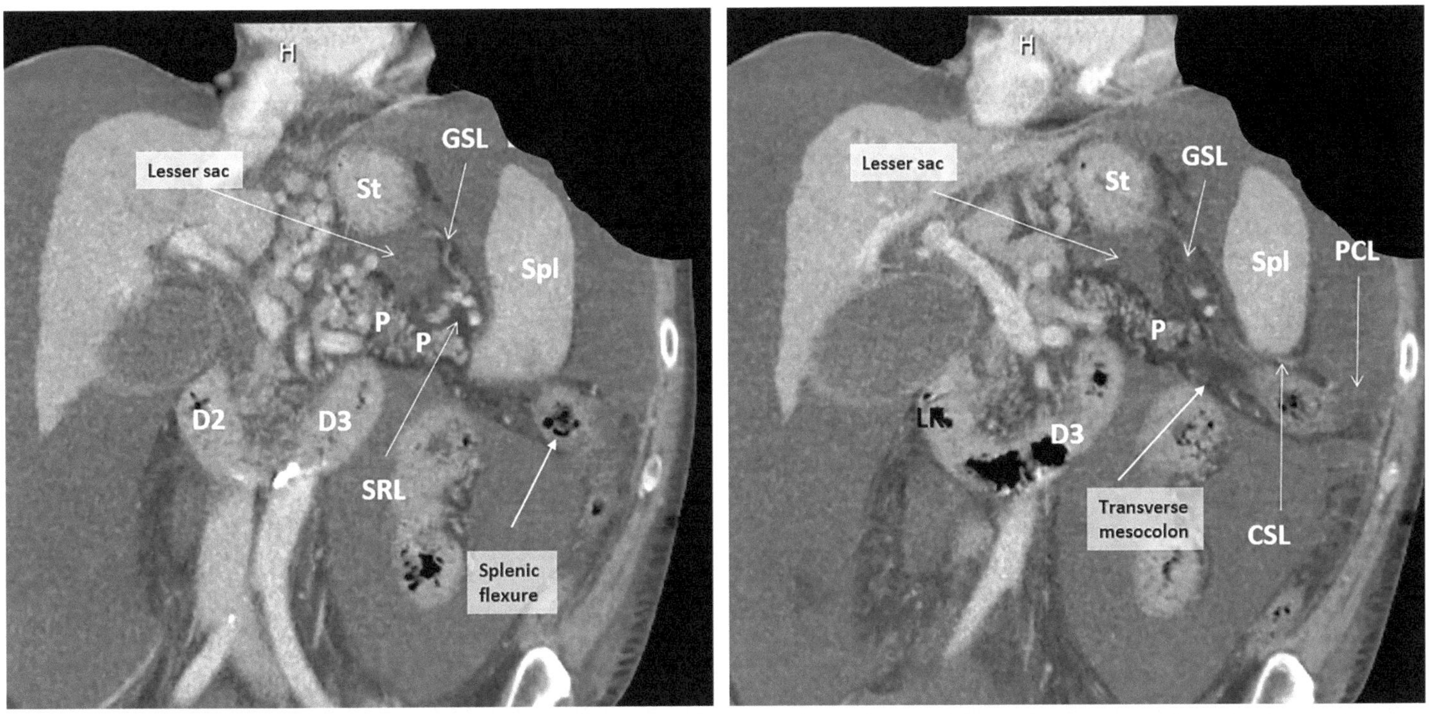

Fig. 7.9 Coronal computed tomography (CT) images at the level of the splenic (Spl) hilum in a patient with cirrhosis and severe ascites showing confluence of the splenic ligaments at the splenic hilum including gastrosplenic (GSL), splenorenal (SRL), and colicosplenic (CSL) ligaments. Note vessels traveling in the ligaments. The splenorenal ligament is the site of passage of the splenic artery and vein where tail of pancreas (p) is located. The gastrosplenic ligament is the anatomic site of the short gastric and left gastroepiploic vessels. The phrenicocolic ligament (PCL) is located below the spleen. D2/D3, second and third portions of the duodenum; H, heart; St, stomach.

vessels pass in the gastrosplenic ligament to reach the splenic arteries and veins at the splenic hilum[6] (▶Fig. 7.8). In case of thrombosis of the splenic vein, collateral venous vascularization develops in this ligament (▶Fig. 7.11).

The phrenicocolic ligament is located below the spleen between the diaphragm and the splenic flexure of colon and supports the spleen in its position.

The colicosplenic and splenorenal ligaments connect the inferior pole of the spleen to posterior abdominal wall. The colicosplenic ligament receives little branches from the left gastroepiploic vessels.

The splenorenal ligament is the site of passage of the splenic artery and vein between the left kidney and the spleen at the level of the pancreatic tail. The term splenorenal ligament is vague. At this place the general peritoneal cavity comes close to the lesser sac especially when the amount of the retroperitoneal fat is small (▶Fig. 7.8). In individuals with large amount of retroperitoneal fat, the ligament is filled with fat and the distance between the left kidney and spleen is large, interposed by the pararenal and perirenal fat (▶Fig. 7.10).

Wandering spleen is a rare anatomic variant of spleen due to redundancy or absence of the splenic ligaments with potentially serious clinical implications such as splenic torsion (▶Fig. 7.12). To fix the problem, splenopexy is performed by suturing the left colophrenic ligament to the lateral diaphragm, and the gastrocolic ligament to the anterior diaphragm.

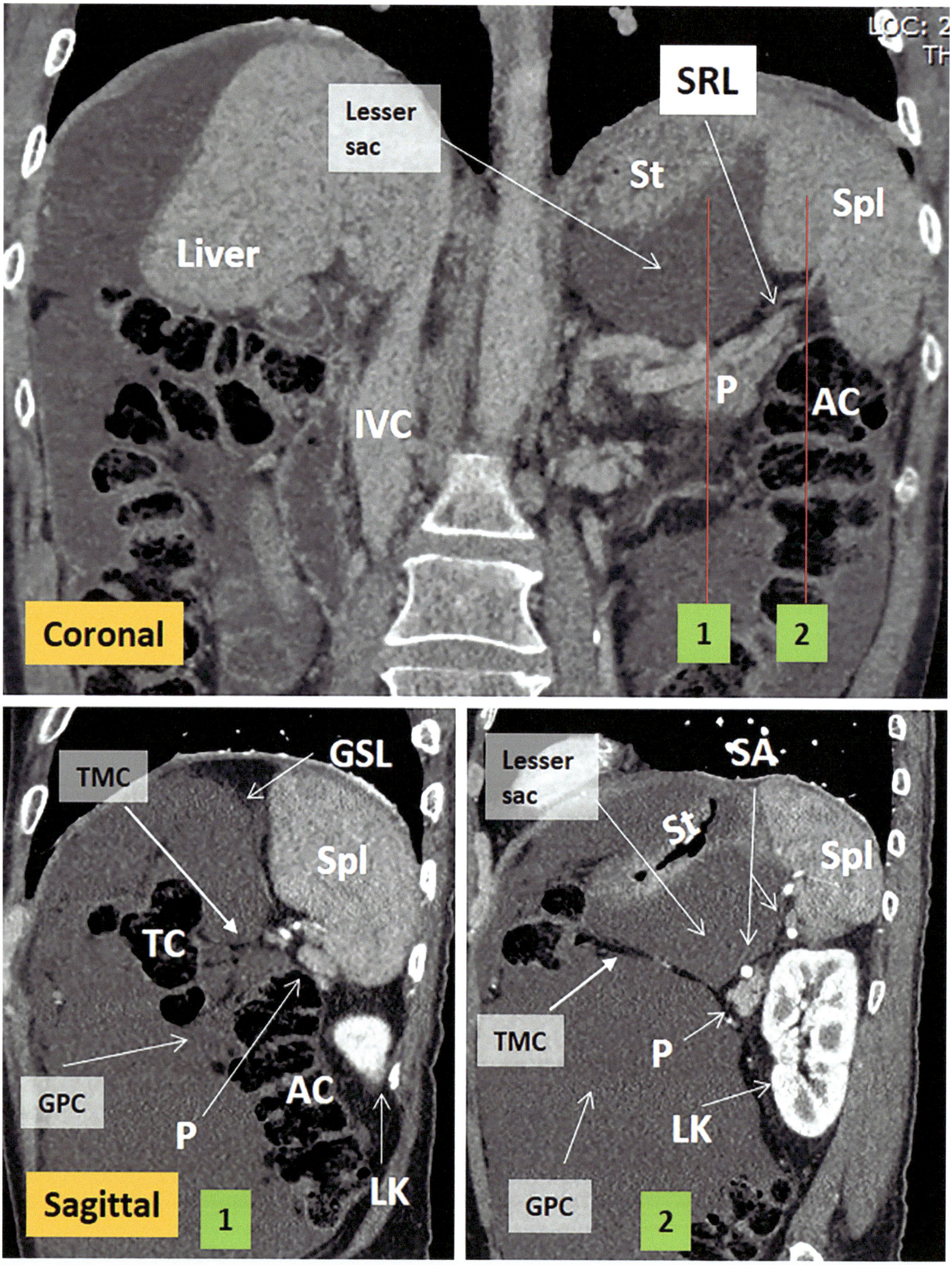

Fig. 7.10 The splenorenal ligament (SRL) is shown on sagittal cuts through the splenic hilum using a coronal image as the reference for the two slices. The patient has large ascites which outlines the spleen and associated structures. The splenorenal ligament is a controversial ligament. It is the site of passage of the splenic artery (SA) and vein between the left kidney (LK) and the spleen (Spl) at the level of the pancreatic tail. At this place the general peritoneal cavity (GPC) comes close to the lesser sac. However, in many individuals the ligament is filled with fat and the distance between the left kidney and spleen is also interposed by the pararenal and perirenal fat. AC, ascending colon; GSL, gastrosplenic ligament; IVC, inferior vena cava; St, stomach; TC, transverse colon; TMC, transverse mesocolon.

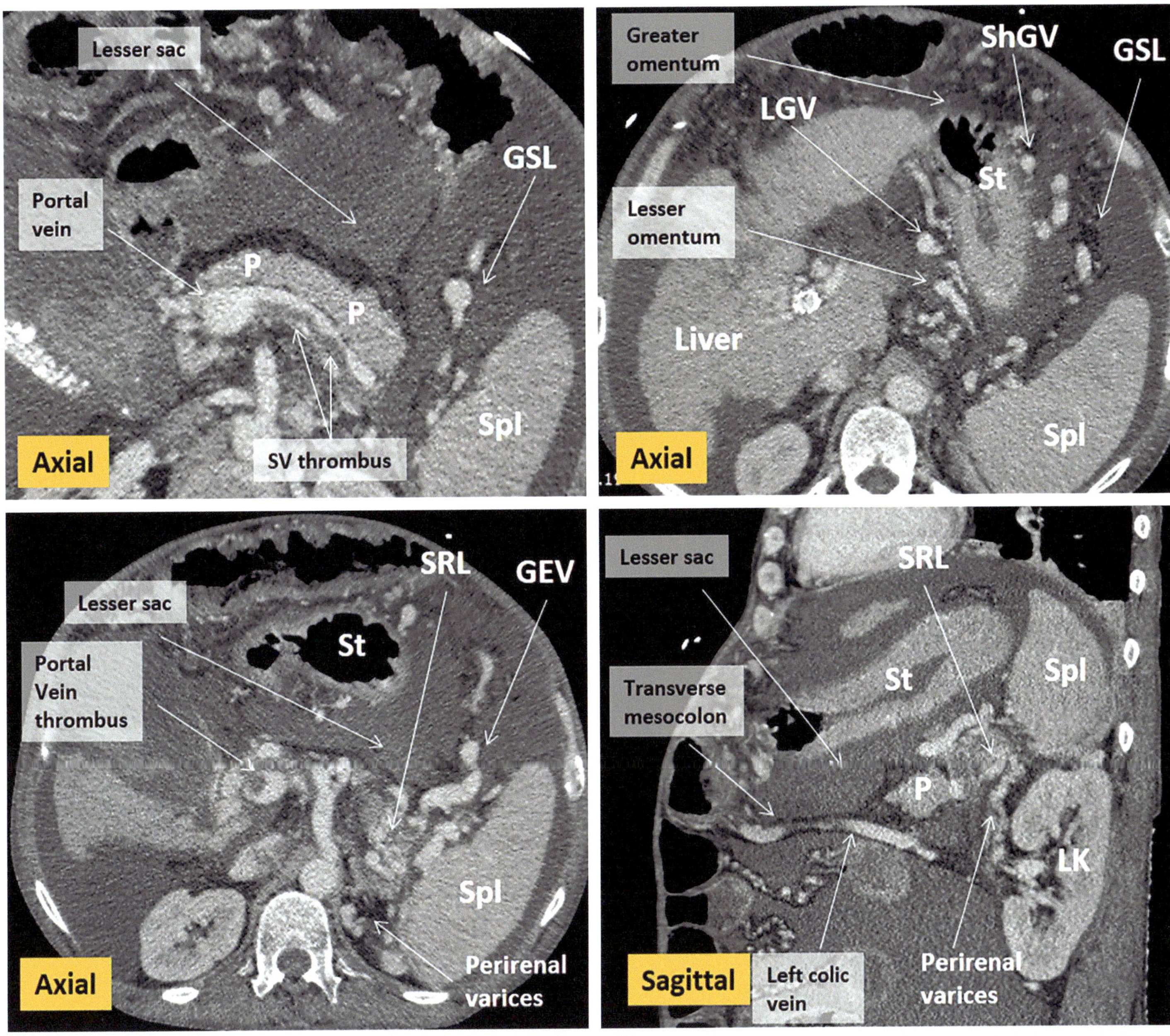

Fig. 7.11 Portal vein and splenic vein (SV) thrombosis in a patient with cirrhosis and ascites. Enlarged varicoid collateral short gastric veins (ShGV) and left gastroepiploic vein (GEV) are seen as they pass in the gastrosplenic ligament (GSL). Also shown are varicoid collateral veins passing through the splenorenal ligament (SRL) and connecting splenic hilum veins with the left renal vein to cause a splenorenal shunt. LGV, left gastric (coronary) vein; LK, left kidney; P, pancreas; Spl, spleen; St, stomach.

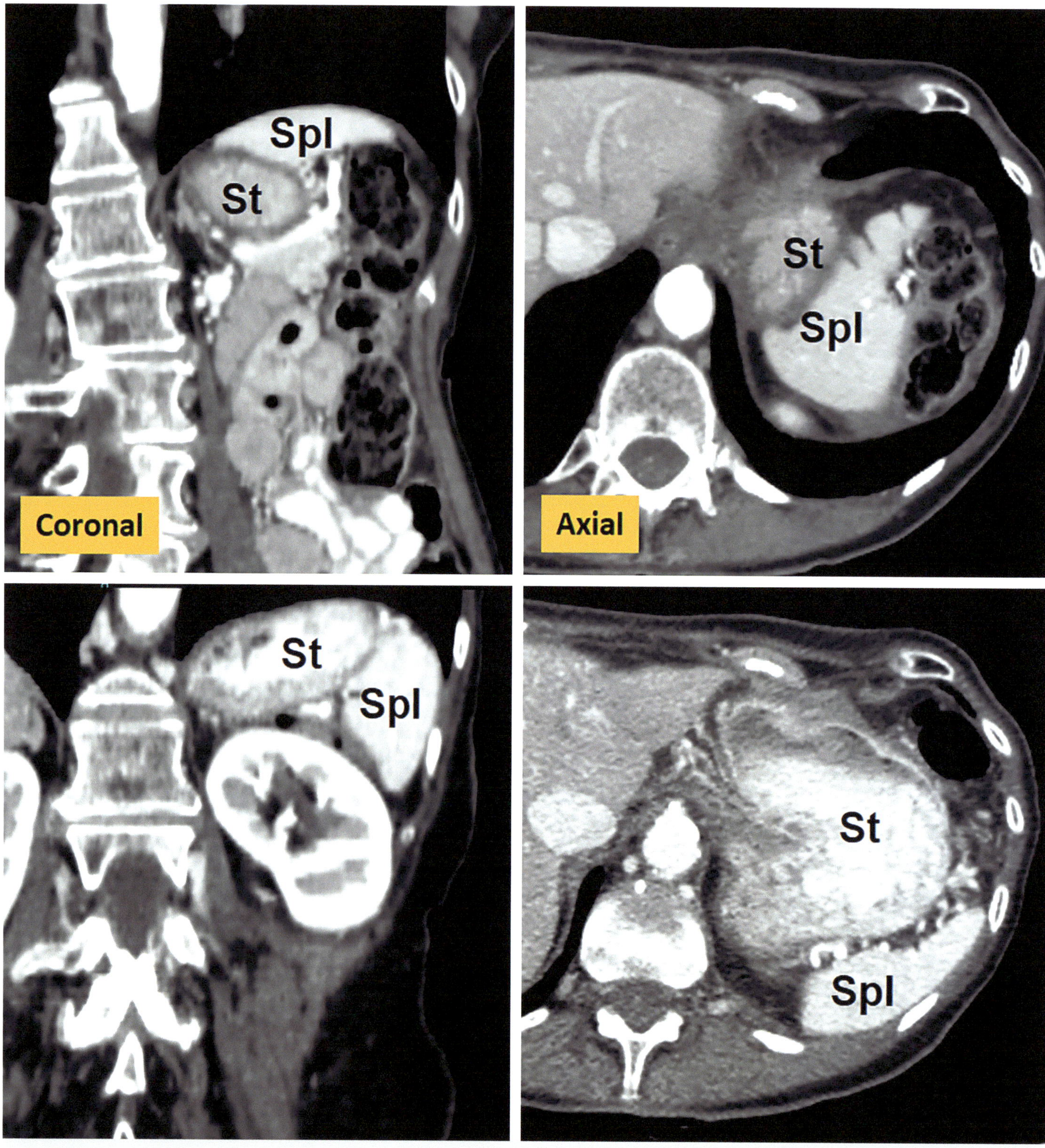

Fig. 7.12 Wandering spleen due to laxity of supporting ligaments is seen. Computed tomography (CT) images show superior location of the spleen (Spl) with a nondistended stomach (St) in the *upper row* images. In *lower row* images, with stomach distension, the spleen rotates laterally and posteriorly.

◆ Imaging

On ultrasound, the spleen appears homogeneous with mid- to low-level echotexture.

On noncontrast computed tomography (CT), the spleen has a density of around 45 Hounsfield Units (HU) and appears isodense to slightly hyperdense compared to liver (▶ Fig. 7.13). An increase in diffuse splenic density can be observed in patients with hemosiderosis, sickle-cell disease, and lymphoma.

On magnetic resonance imaging (MRI), the spleen has low signal intensity on T1-weighted imaging and high signal intensity on T2-weighted and diffusion-weighted imaging.[8,9] A decrease in diffuse splenic intensity is seen in patients with hemosiderosis or iron overload due to frequent blood transfusions (i.e., thalassemia) (▶ Fig. 7.14).

A similar enhancement pattern is seen on MRI as on CT (▶ Fig. 7.13, ▶ Fig. 7.14). During arterial phase scanning, alternating high and low attenuation regions within the spleen results in a serpentine or zebra pattern due to variable flow rates of the splenic red pulp. This pattern disappears on late portal phase which is approximately 60 to 90 seconds after contrast material administration, and the spleen becomes homogeneous, once all venous sinusoids are filled with contrast (▶ Fig. 7.13).

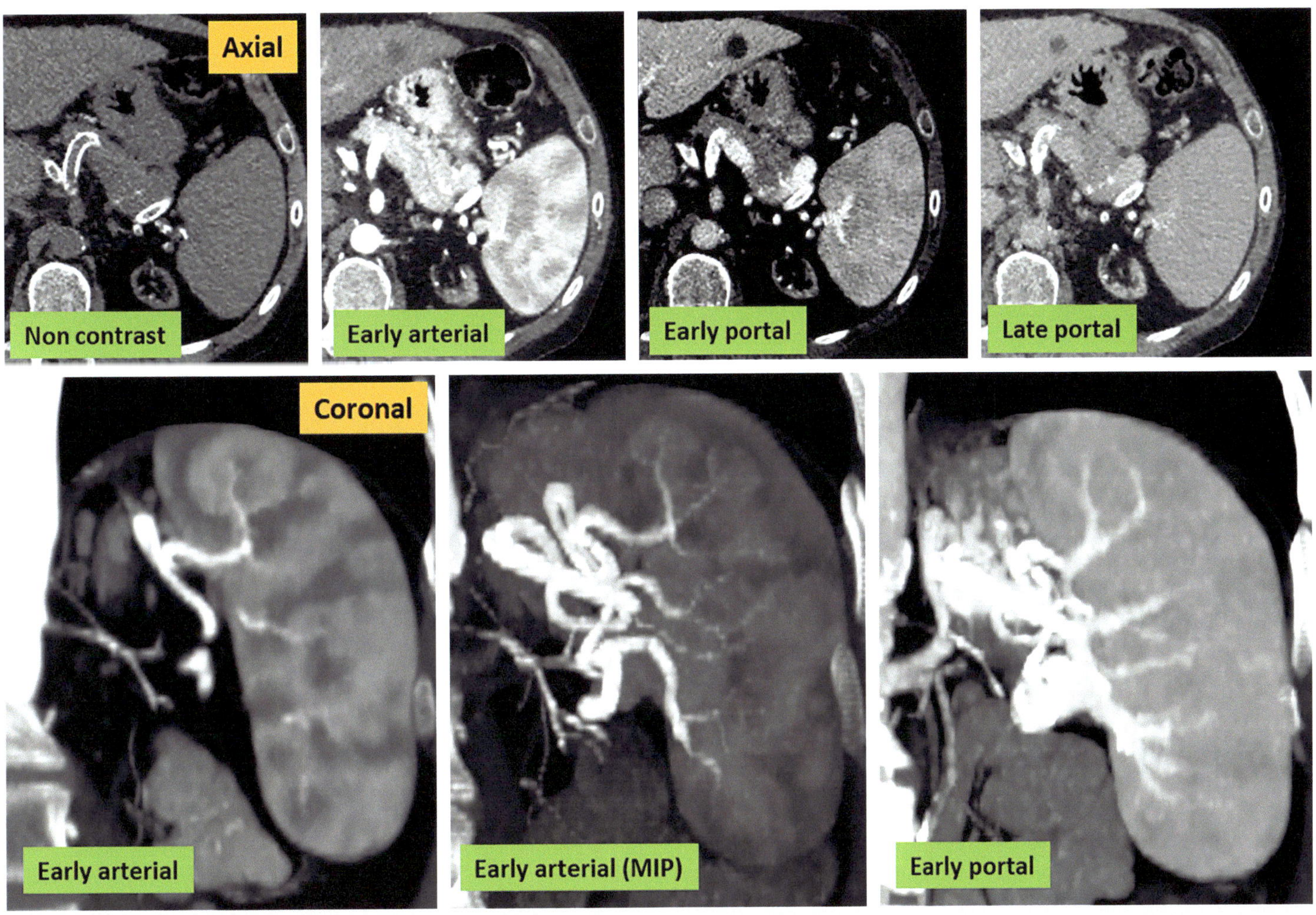

Fig. 7.13 Normal computed tomography (CT) appearance of the spleen on precontrast and dynamic postcontrast images. The trabecular arterial branches of the spleen move at the margin of the white and red pulps. Enhancement of the venous sinuses occur in a centripetal fashion from the periphery toward the center of the sinusoids giving a serpentine appearance to the spleen on early phase images that will become homogenous on late phase images. MIP, maximal intensity pixel.

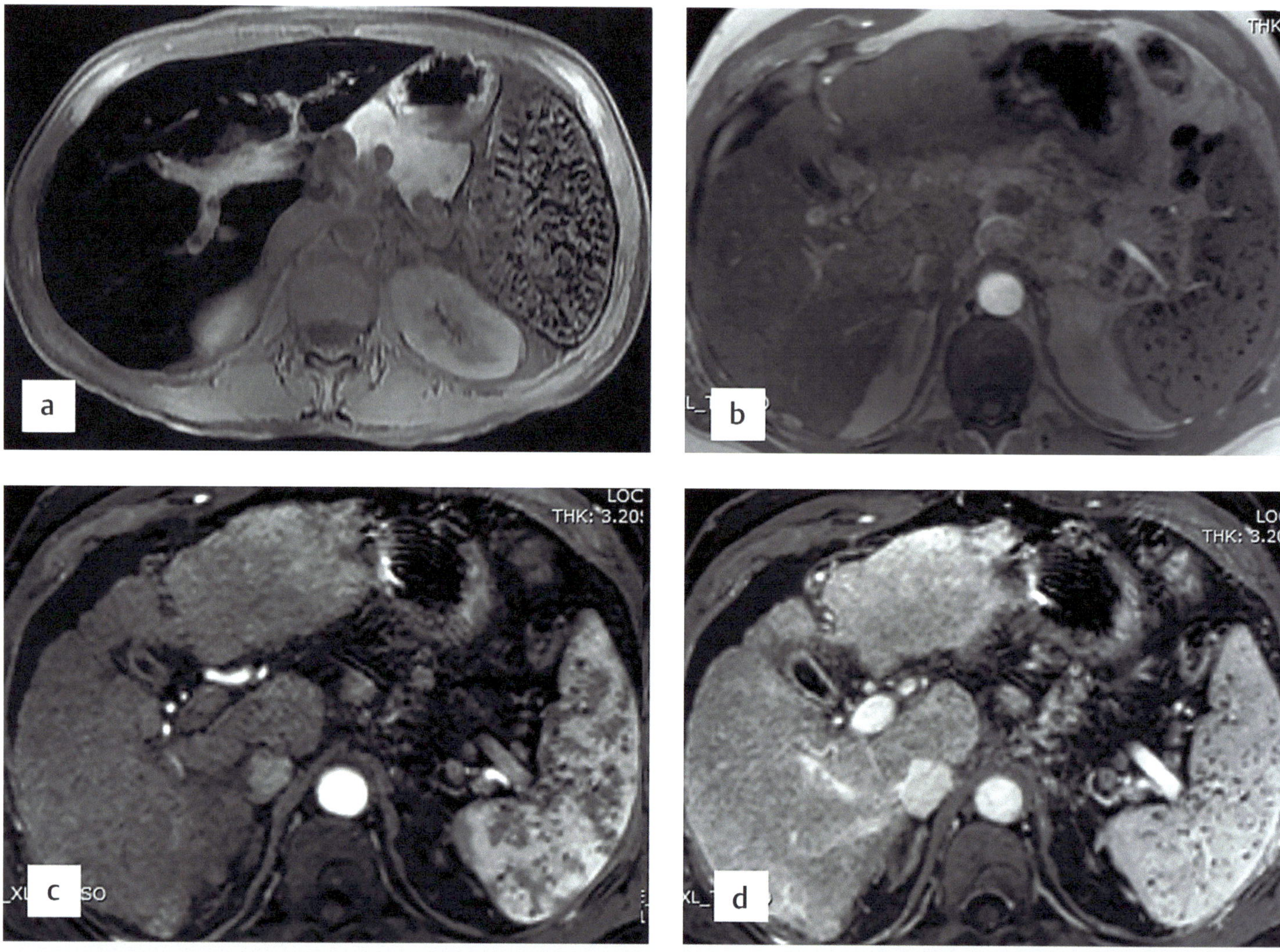

Fig. 7.14 (a) Postcontrast magnetic resonance imaging (MRI) shows diffusely decreased signal intensity of the liver due to secondary hemosiderosis in thalassemia. Linear and nodular signal void foci are seen in the spleen due to hemosiderin deposits. **(b)** T1-weighted, **(c)** arterial postcontrast, and **(d)** portal phase postcontrast MR images in a patient with cirrhosis showing Gamna-Gandy bodies in the spleen. These foci are seen due to hemosiderin deposits in the cytoplasm of macrophages in the red pulp. The red pulp containing large venous sinusoids are best shown in arterial phase image before complete enhancement.

◆ Variants of Morphology

The shape and size of spleen vary (▶Fig. 7.15). On CT images, the spleen usually measures 10 cm × 6 cm in axial diameters. In adults, the craniocaudal diameter usually does not exceed 15 cm.[8]

Solitary or multiple accessory spleens are seen in approximately 7 to 15% of the population.[10] They are usually less than 2 cm in size near the splenic hilum (▶Fig. 7.15). Image characteristics of accessory spleen including enhancement patterns are similar to normal spleen. Rare locations of accessory spleen include the greater omentum, within the pancreatic tail, or in the liver.

Bilobed spleen is not a rare image finding[11] (▶Fig. 7.16). Polysplenia and asplenia are seen in two spectra of patients with heterotaxy syndromes (▶Fig. 7.15, ▶Fig. 7.16). Polysplenia syndrome is classified as one of the situs ambiguous or heterotaxy syndromes and is classically termed left isomerism or bilateral left-sidedness.[12] However, isolated polyspenia without or with azygos continuation of the inferior vena cava is probably the most common form of polysplenia syndrome incidentally found on imaging studies of the adults (▶Fig. 7.16).

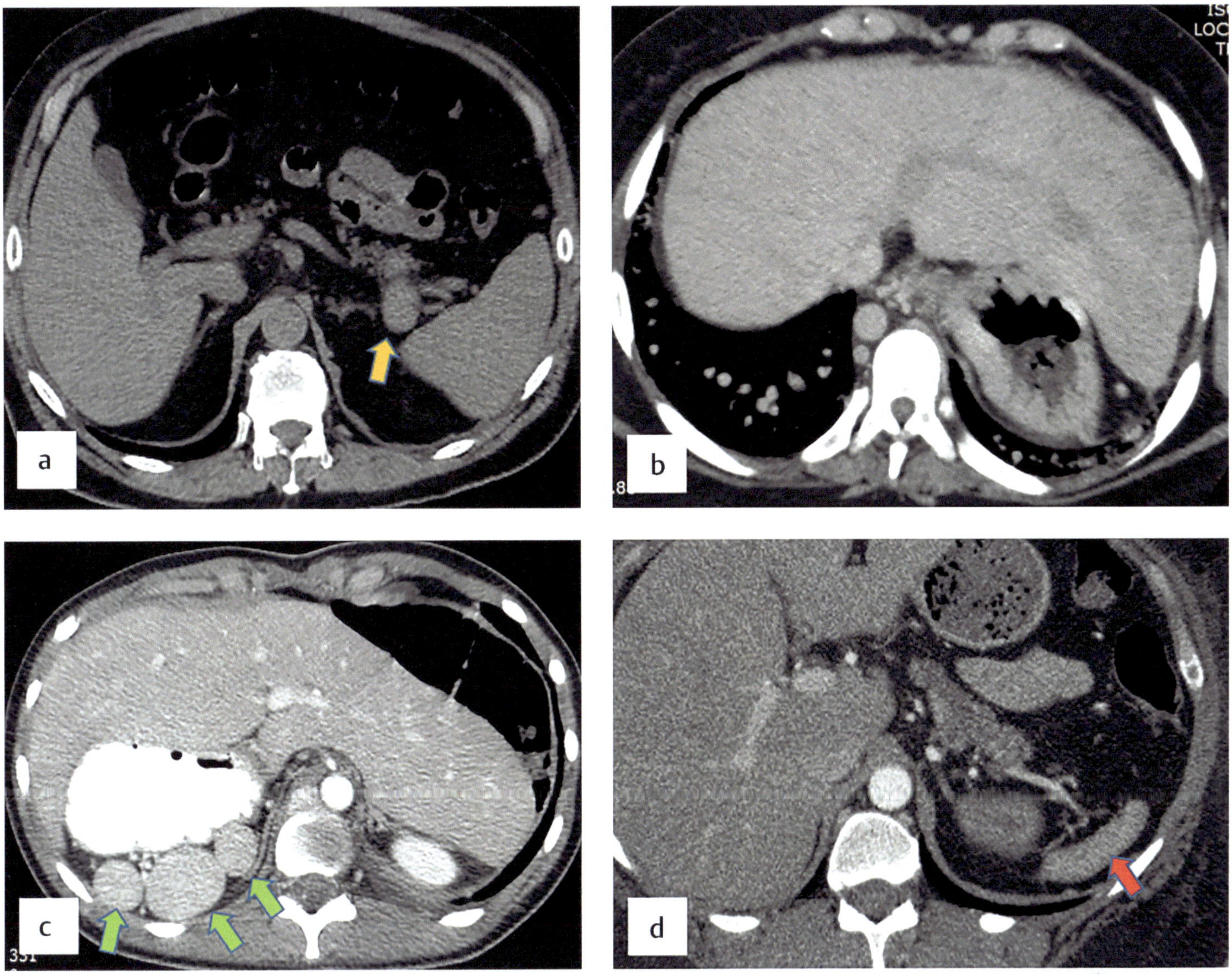

Fig. 7.15 Different morphologies of the spleen. **(a)** Splenule (*arrow*) or accessory spleens are common and usually seen near the tail of pancreas. **(b)** Asplenia. **(c)** Absent spleen on left with polysplenia (*arrows*) on the right behind a right-sided stomach and liver. **(d)** Small spleen (*arrow*) in sickle cell disease.

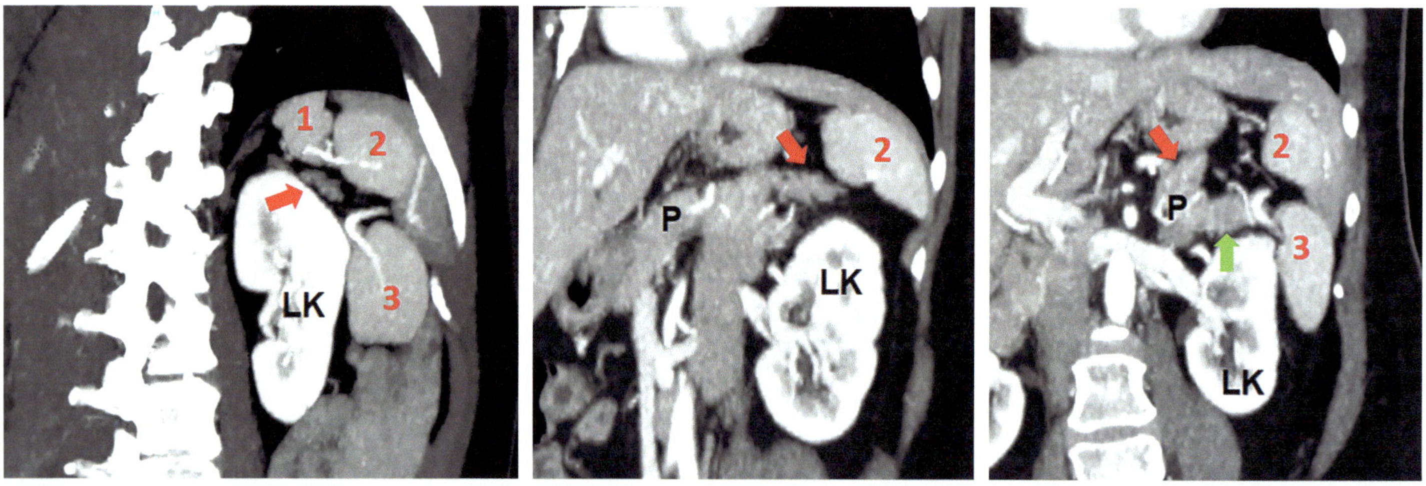

Fig. 7.16 Isolated polysplenia with a bifurcating pancreatic tail. The hilum of subdiaphragmatic part of the spleen, which is bilobed (#1+2), faces the upper pancreatic tail (*red arrow*) and the lower part of the spleen (#3), which is lateral to the left kidney (LK), faces the lower pancreatic tail (*green arrow*). P, pancreas.

◆ Common Pathologies

Parenchymal

The spleen is the most common solid organ injury after blunt trauma due to its complex ligamentous attachments and spongy parenchymal consistency.[8] Following splenic injury and splenectomy, seeding of splenic fragments onto serosal surfaces can result in regrown of splenic tissue. This process is called splenosis. These small spleens follow image characteristic of a normal spleen that can be used as diagnostic tool in difficult cases. Splenosis is usually seen at the splenectomy bed, but infrequently is seen near the pancreas, liver, stomach, or within thorax.

Benign lesions such as cysts and angiomas are common in the spleen. Primary splenic cysts are usually epithelial cell lined cysts, as opposed to pseudocysts. Hemangioma is the most common primary benign neoplasm of the spleen and is composed of endothelium-lined vascular channels filled with blood. Most hemangiomas are hypointense to the spleen on T1-weighted images and hyperintense on T2-weighted images. After contrast material administration, they usually demonstrate uniform enhancement (▶Fig. 7.17). Characteristic peripheral nodular enhancement, like that seen in hepatic hemangiomas, is not often seen in splenic angiomas.

Lymphoma is the commonest malignant tumor of the spleen and usually seen as simple splenomegaly without or with single or multiple nodular lesions. Splenic metastases are relatively uncommon.

Portal hypertension is the most common cause of splenomegaly in the United States. Foci of hemosiderin deposition are seen in about 9 to 12% of patients with portal hypertension. These foci are called Gamna-Gandy bodies (▶Fig. 7.14).

Splenic infarct is a common image finding due to embolic clots arising from major vessels or within the left heart chambers. The spleen is the organ most commonly involved by sickle cell disease. Autosplenectomy is often found in patients with homozygous sickle cell disease due to frequent splenic infarct (▶Fig. 7.15). In patients with sickle cell disease and thalassemia, the spleen appears dark on MRI due to iron deposition from blood transfusion (▶Fig. 7.14).

Vascular

Splenic artery occlusion is rare. It could be performed as a part spleen-preserving distal pancreatectomy. The splenic perfusion will be provided mainly by the gastroepiploic arcade and through the short gastric arteries via collateral blood supply from the left gastric and phrenic arteries. The left gastroepiploic artery is probably the major spleen feeding collateral after occlusion or transection of the splenic artery.[13,14] Compensatory dilation of short arteries and veins of the gastric body and fundus after occlusion of the splenic artery or vein increases the risk of gastric bleeding.

Splenic artery aneurysms are the most common visceral artery aneurysms caused by medial degeneration and superimposed atherosclerosis. Most aneurysms are small (less than 2 cm diameter), saccular, and located at a bifurcation of distal segment of the splenic artery. Splenic artery aneurysms are multiple in 20% of cases.[5] An aneurysm larger than 2.5 cm should be treated with surgery or coil embolization (▶Fig. 7.18). Splenic pseudoaneurysm is a known complication of pancreatitis due to enzymatic injury of the arterial wall (▶Fig. 7.19).

Splenic vein thrombosis has multiple causes but it is most commonly secondary to pancreatitis. It has been reported in at least 20% of patients with chronic pancreatitis (▶Fig. 7.11).

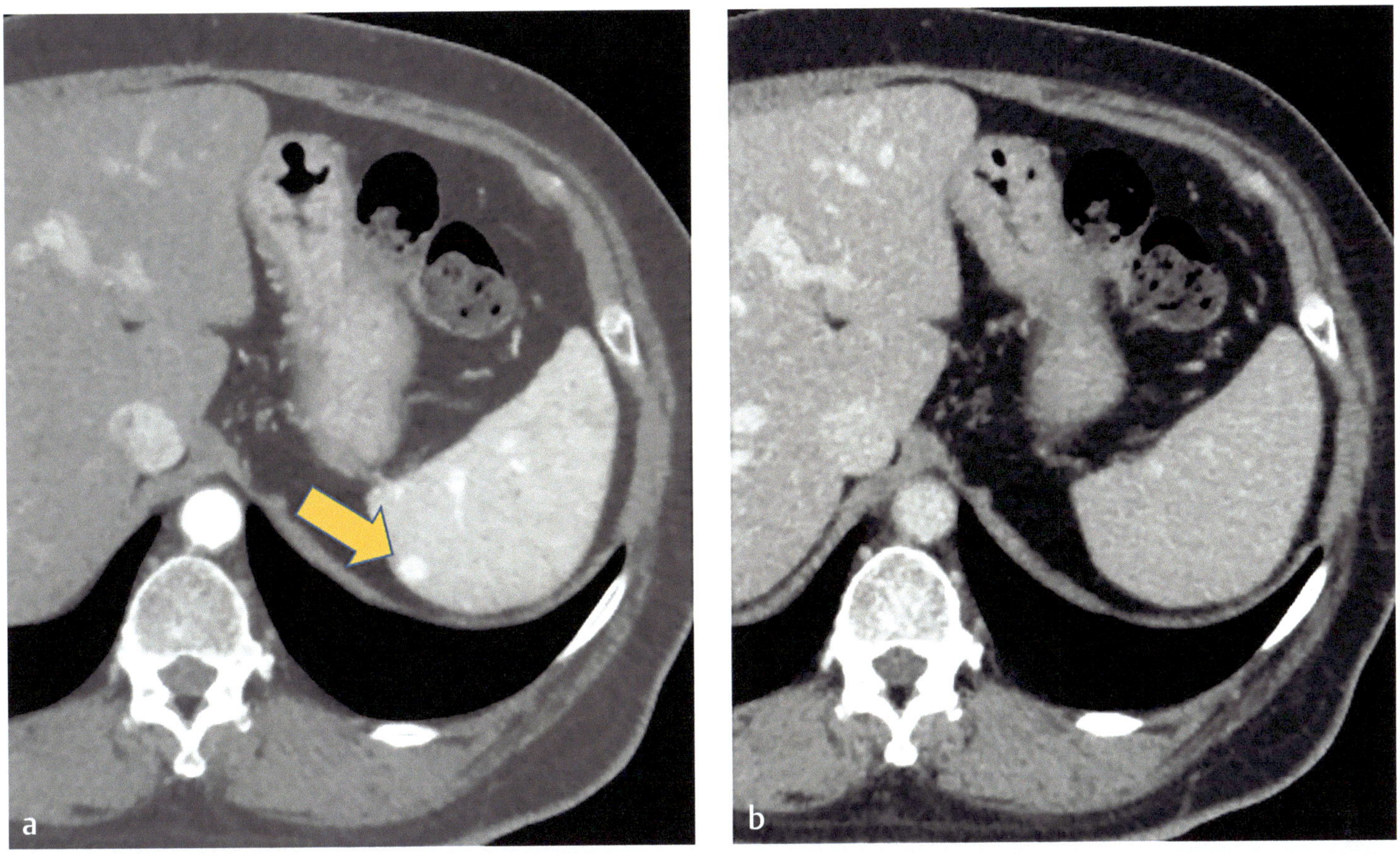

Fig. 7.17 (a) Axial early and **(b)** late postcontrast images showing a small enhancing splenic angioma (*arrow*).

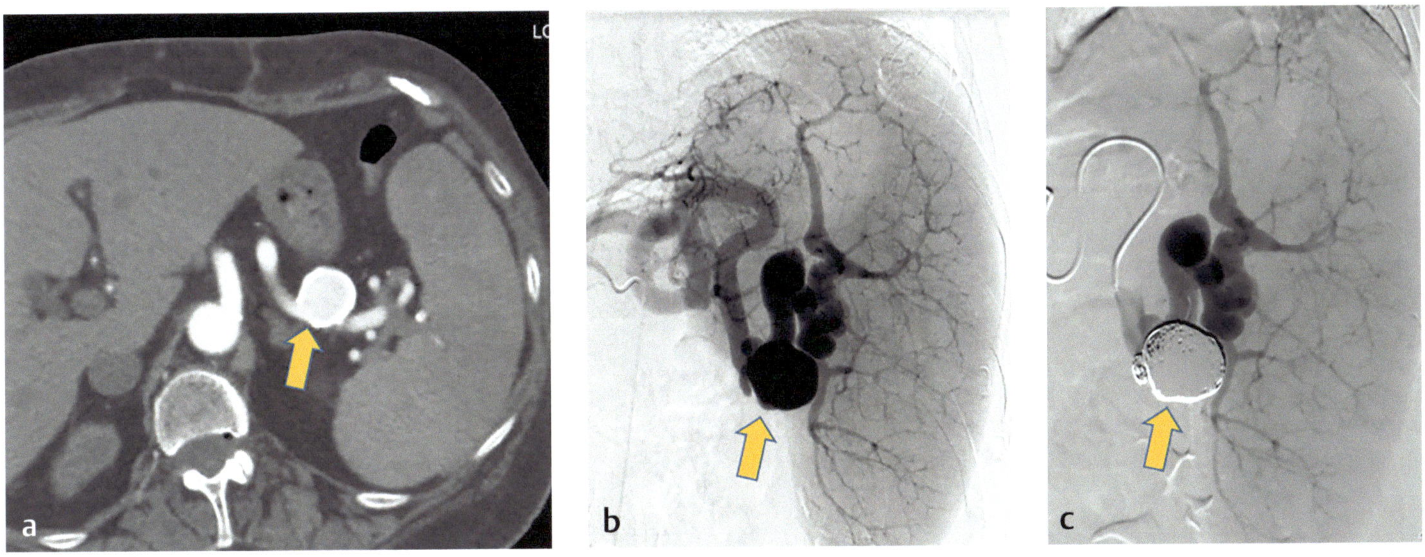

Fig. 7.18 (a) Axial postcontrast computed tomography (CT) showing a peripherally calcified splenic aneurysm measuring 2 cm. **(b)** Splenic artery angiogram showing the aneurysm (*arrow*) which was embolized with coils **(c)**.

7

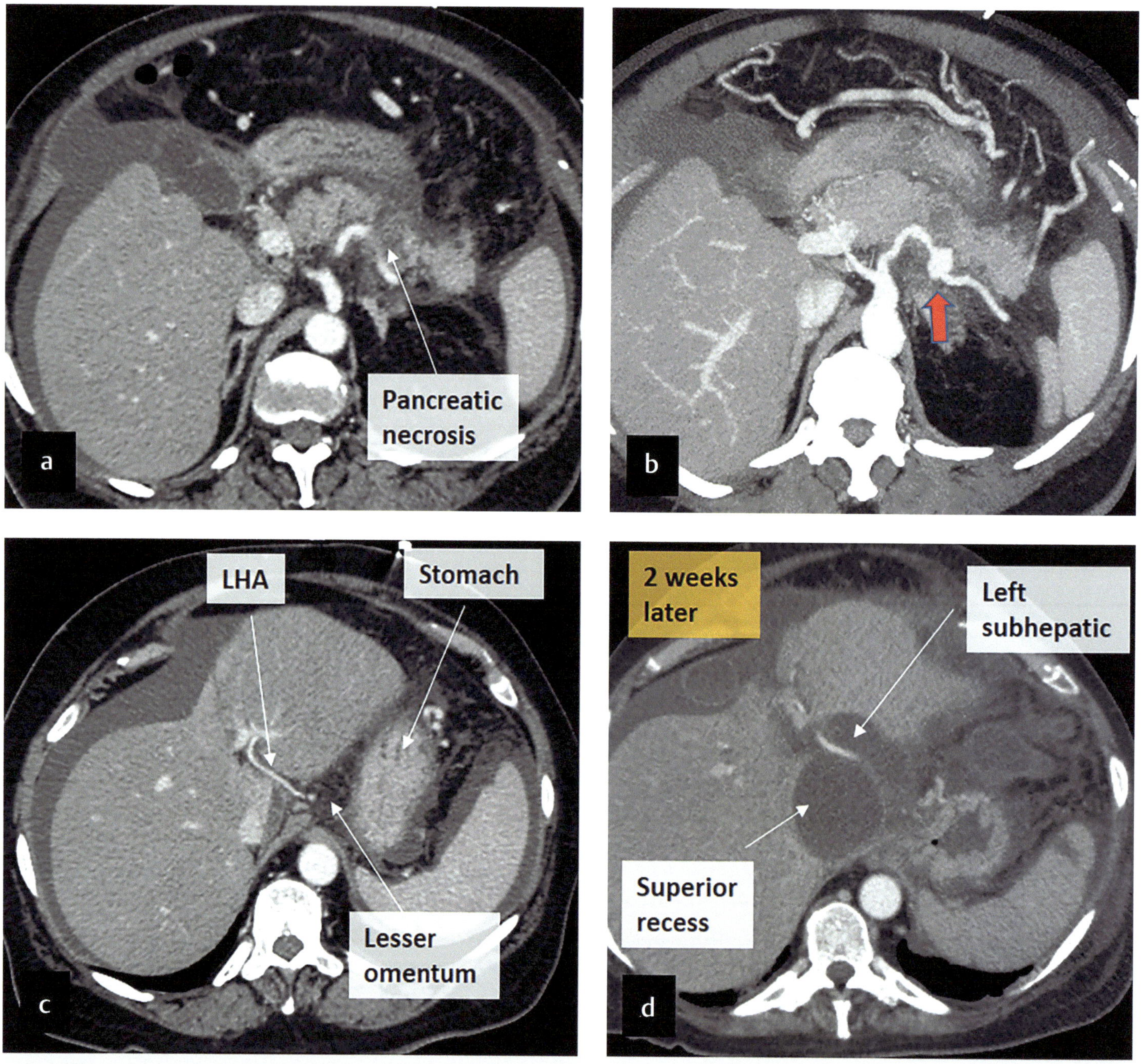

Fig. 7.19 Acute pancreatitis causing splenic artery pseudoaneurysm and lesser sac fluid collections. **(a–c)** Axial computed tomography (CT) images showing pancreatic necrosis with splenic artery pseudoaneurysm (*red arrow*). **(c)** Anatomic course of the replaced left hepatic artery (LHA) from the left gastric artery in the lesser omentum and the ligamentum venosum fissure. **(d)** Progression of disease 2 weeks later has caused several fluid collections in the lesser sac including one in the superior recess. Note the LHA separating the later collection from a smaller one in the left subhepatic space.

References

1. Mebius RE, Kraal G. Structure and function of the spleen. Nat Rev Immunol 2005;5(8):606–616
2. Cesta MF. Normal structure, function, and histology of the spleen. Toxicol Pathol 2006;34(5):455–465
3. Balogh P, Horváth G, Szakal AK. Immunoarchitecture of distinct reticular fibroblastic domains in the white pulp of mouse spleen. J Histochem Cytochem 2004;52(10):1287–1298
4. Mikhail Y, Kamel R, Nawar NN, Rafla MF. Observations on the mode of termination and parenchymal distribution of the splenic artery with evidence of splenic lobation and segmentation. J Anat 1979;128(Pt 2):253–258
5. Madoff DC, Denys A, Wallace MJ, et al. Splenic arterial interventions: anatomy, indications, technical considerations, and potential complications. Radiographics 2005;25(Suppl 1):S191–S211
6. Ostermann PA, Schreiber HW, Lierse W. Der Bandapparat der Milz und seine Bedeutung bei chirurgischen Eingriffen. [The ligament system of the spleen and its significance for surgical interventions]. Langenbecks Arch Chir 1987; 371(3):207–216
7. Zhao Z, Liu S, Li Z, et al. Sectional anatomy of the peritoneal reflections of the upper abdomen in the coronal plane. J Comput Assist Tomogr 2005;29(4):430–437
8. Elsayes KM, Narra VR, Mukundan G, Lewis JS Jr, Menias CO, Heiken JP. MR imaging of the spleen: spectrum of abnormalities. Radiographics 2005;25(4):967–982
9. Karlo CA, Stolzmann P, Do RK, Alkadhi H. Computed tomography of the spleen: how to interpret the hypodense lesion. Insights Imaging 2013;4(1):65–76
10. Gill N, Nasir A, Douglin J, et al. Accessory spleen in the greater omentum: embryology and revisited prevalence rates. Cells Tissues Organs 2017;203(6):374–378
11. Menezes RG, Murlimanju BV, Kharoshah MA, Madadin M, Naik R, Senthilkumaran S. Lobed spleen: an autopsy case report with emphasis on surgical anatomy. Clin Ter 2014;165(3):143–145
12. Tawfik AM, Batouty NM, Zaky MM, Eladalany MA, Elmokadem AH. Polysplenia syndrome: a review of the relationship with viscero-atrial situs and the spectrum of extra-cardiac anomalies. Surg Radiol Anat 2013;35(8):647–653
13. Romero-Torres R. The true splenic blood supply and its surgical applications. Hepatogastroenterology 1998;45(21):885–888
14. Ndoye JM, Savadogo J, Ndiaye A, et al. L'artère gastro-omentale gauche : une origine splénique mais un niveau de naissance variable. [The left gastroepiploic artery: a splenic origin but a variable birthplace]. Morphologie 2008; 92(296):11–15

8 Biliary System

Farhood Saremi, Takashi Matsubara, and Ana Maliglig

◆ Introduction

The biliary system or tract is composed of intrahepatic and extrahepatic ducts. The role of the biliary system is to transfer bile from canaliculi into the duodenum and concentrate it in the gallbladder. The bile, composed of cholesterol, bile acids, pigments, and electrolytes, is produced by the hepatocytes. From hepatocytes, bile will be secreted into the bile canaliculi at the apical pole of the hepatocytes. Bile flows from the canaliculi toward the center of the hepatic cords and drains into hepatic ductules. The ductules then coalesce and drain into the network of intrahepatic bile ducts. The bile flows from the intrahepatic ducts to the hilar hepatic, common, and cystic ducts to be stored in the gallbladder. With contraction of the gallbladder bile is excreted in the gut via the common duct. The lumen of biliary tract is lined by specialized cells called biliary epithelial cells or cholangiocytes.

◆ Embryology

Development of the biliary tract is a complicated process that is controlled by several molecular regulators. Around the third to fourth weeks of gestation (3–4 mm embryo) a hepatic diverticulum (liver bud) originates from the "anterior intestinal portal," an area in the ventral aspect of the foregut where the future duodenum forms.[1–4] This endodermal hepatic diverticulum, which forms the liver and biliary ducts, is divided into cranial and caudal parts.

The cranial portion (pars hepatica) gives rise to the liver and the intrahepatic bile ducts. Its caudal portion (pars cystica) forms the extrahepatic bile ducts and the ventral pancreas (▸Fig. 9.1, Chapter 9 "Pancreas"). The endodermal cells that form the cranial part of the liver diverticulum (hepatoblasts) express typical hepatic genes, such as albumin, alpha-fetoprotein, and liver-specific transcription factors.[1] The hepatoblasts are the precursors of the hepatocytes and cholangiocytes. Therefore, the cholangiocytes that form the biliary tract have two different origins. The cholangiocytes lining the extrahepatic bile ducts primarily originate directly from the endoderm whereas the cholangiocytes of the intrahepatic bile ducts derive from hepatoblasts.

◆ Extrahepatic Ducts

The caudal part of the hepatic diverticulum develops to form the gallbladder, the cystic duct, the common bile duct (CBD, ductus choledochus), and the ventral pancreas. The gallbladder anlage is first visible along the distal half of the hepatic diverticulum at day 29 after fertilization. Later, gallbladder myoblasts develop from condensing mesenchyme and the cystic duct can be seen at day 34.

The growth and lengthening of the extrahepatic biliary tree continue up to 8 weeks of gestation, and the extrahepatic biliary tree remains patent with the developing liver at all stages. Therefore, the concept that extrahepatic bile duct atresia may be caused by failure of recanalization of the CBD proved to be wrong by recent investigations.[5–7] The ventral pancreas arises from a portion of the hepatic diverticulum between the CBD and the gut (▸Fig. 9.1, Chapter 9 "Pancreas"). At the fifth week of gestation, the duodenum rotates and displaces the common bile duct from anterior to its definitive position on the posterior side of the duodenum.

During the fifth week, a rapid endodermal proliferation takes place above the junction of CBD and the cystic duct to form the distal right and left hepatic ducts at the porta hepatis.[3] However, the proximal part of the two ducts derives from the first intrahepatic ductal plates. This remodeling explains the existence of the several normal anatomic variants in the arrangement of the right and left hepatic ducts. The right and left hepatic ducts are clearly defined tubular structures by 12 weeks of gestation. Bile formation in the liver begins in the 12th week of gestation.

◆ Intrahepatic Ducts

Multiple sprouts of endodermal cells originate from the cranial part of the hepatic diverticulum and extend into the mesenchymal septum transversum (an area between the peritoneal and pericardial cavities) to form the liver, vessels, and intrahepatic bile ducts. The intrahepatic bile duct system develops from the hepatoblasts after 7 weeks of embryonic life.[7] This process starts with transformation of hepatoblasts to the cholangiocytes (ductogenesis) and continues with three-dimensional formation of the ductal plate and then with successive development of segmental ducts, interlobular ducts, and bile ductules (cholangiogenesis). The development and branching pattern of the intrahepatic bile ducts and the portal veins are closely associated, starting from the hilum of the liver and extending to the periphery. Therefore, anomalies of the biliary and portal system frequently occur together.

Ductal Plates

At about the eighth week of gestation, the primitive hepatoblasts surrounding the largest hilar portal vein branches form a continuous single-layered ring around the portal mesenchyme called the "ductal plate."[4] During the following weeks, ductal plates also appear around the smaller portal vein, become bilayered, and enter a phase of remodeling to form the tubular intrahepatic bile ducts (IHBD)[1] (▸Fig. 8.1).

The parts of the ductal plate not involved in the formation of the ducts progressively regress. In the final stage, which starts around birth, the ducts are incorporated into the portal mesenchyme and form the future portal tract, the intrahepatic branches of the hepatic artery, and the portal ducts that remain connected

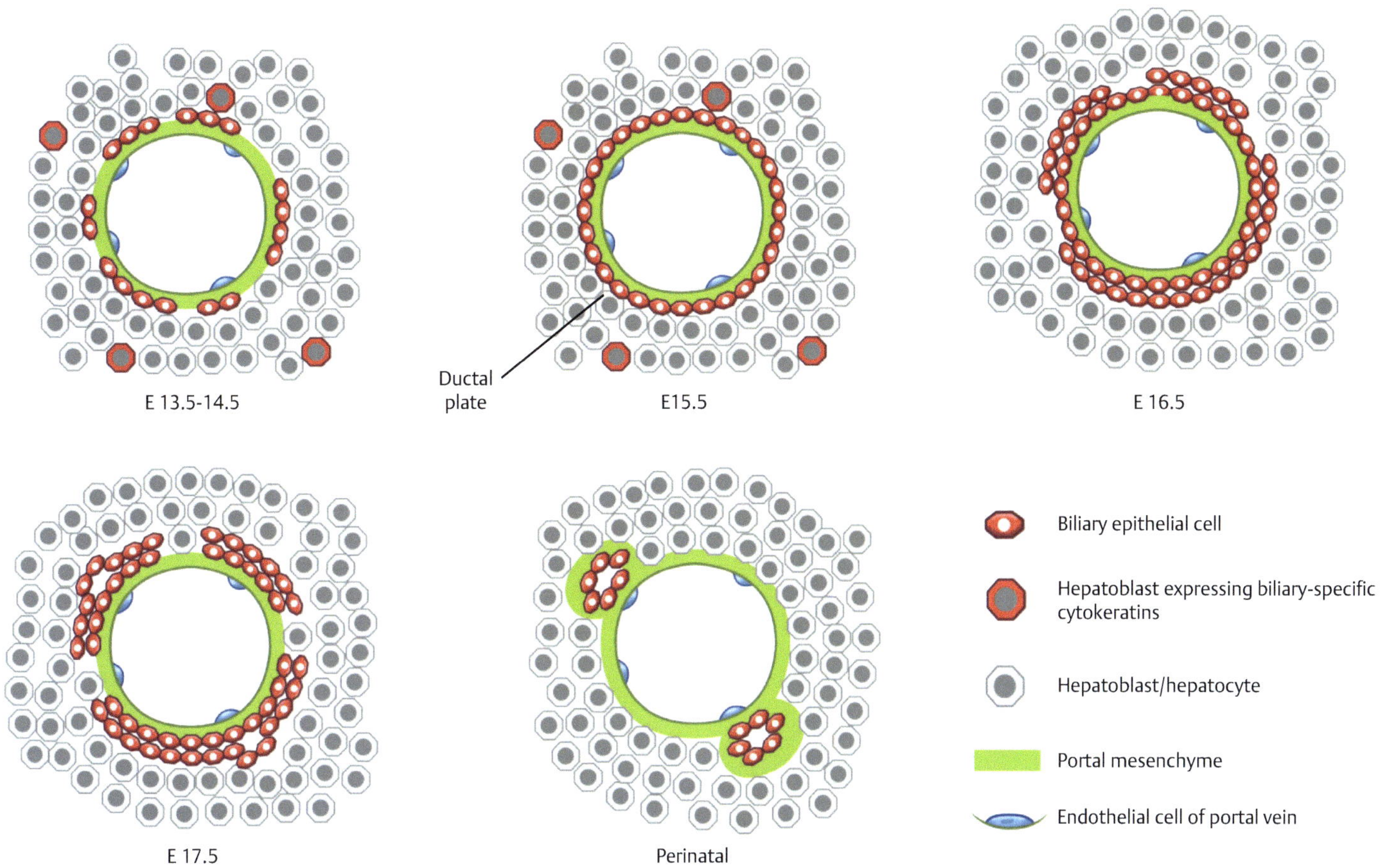

Fig. 8.1 Morphogenesis of the intrahepatic bile ducts in the mouse. At E13.5–14.5, the biliary precursor cells, which express biliary-specific cytokeratins, are located close to the portal mesenchyme, and some hepatoblasts expressing low levels of biliary-specific cytokeratins are found throughout the parenchyma. Around E15.5, the biliary precursor cells form a continuous single-layered ring called ductal plate. The ductal plate becomes partly bilayered around E16.5. One day later, around E17.5, focal dilations appear between the two cell layers. These dilations give rise to the bile ducts, while the rest of the ductal plate progressively regresses. Around birth, the ducts become incorporated into the portal mesenchyme. The ducts develop from the hilum to the periphery of the liver lobes. (Modified with permission from Lemaigre.[1])

to the ductal plate and its adjacent parenchyma by thin epithelial channels.

The primitive intrahepatic bile ducts are discontinuous but eventually they coalesce and connect with the extrahepatic bile ducts. The ramification of the biliary tree toward the liver periphery and remodeling of the ductal plates to small ductules continue throughout fetal life and until 4 weeks after birth. This finding explains some degree of immaturity and incompleteness of the intrahepatic bile duct system in the neonate and premature infants.[5]

◆ Imaging of Biliary System

Noninvasive imaging of the biliary ducts can be performed by ultrasound, computed tomography (CT), magnetic resonance imaging (MRI), intravenous CT cholangiography, and nuclear imaging. Ultrasound is the least expensive method and is used as the first step for imaging analysis of the biliary system. Biliary duct can also be opacified after contrast injection by invasive techniques such as percutaneous transhepatic cholangiography (PTC) and endoscopic retrograde cholangiopancreatography (ERCP). ERCP is an invasive technique and is usually reserved for confirmation and interventions such as stone extraction or biliary stent placement. In the adult liver, the intrahepatic bile ducts from the hepatic ducts to the smallest ductules can be best shown with intraductal administration of contrast. However, even best cholangiographic techniques can only show 11 to 12 orders of branching (Ludwig) and finer ramification of terminal bile ducts which can extend to 18 to 20 orders of branches would be hard to demonstrate.

◆ Extrahepatic Bile System

The extrahepatic bile system includes the gallbladder, the cystic duct, the common duct, and the hepatic ducts.

Gallbladder

The gallbladder is a contractile muscular sac in the right lower quadrant of the abdomen beneath the liver between the segments IV and V (▶Fig. 8.2, ▶Fig. 8.3, ▶Fig. 8.4). The gallbladder is 7- to 10-cm long, 3 to 5 cm in transverse diameter, and its wall is 2- to 3-mm thick in adults. It can accumulate bile because the resting

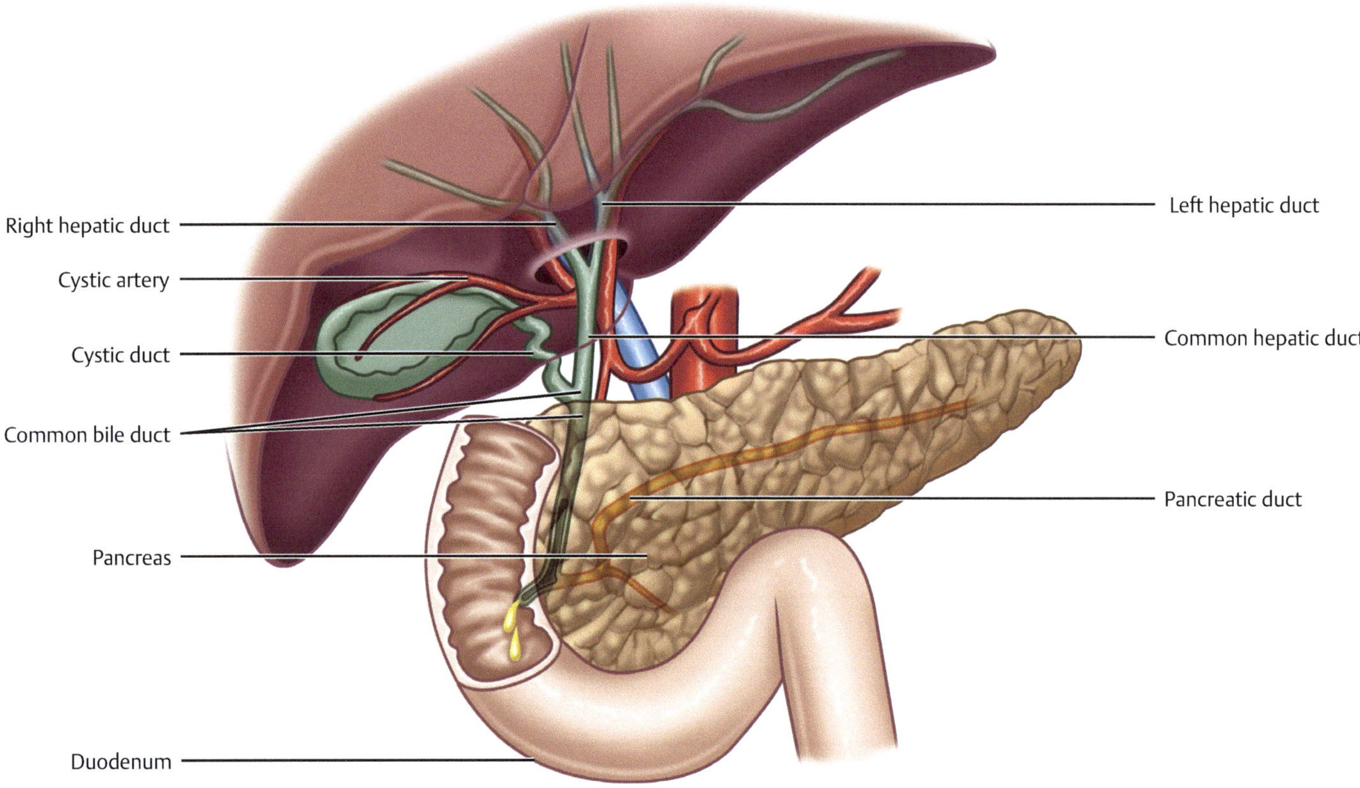

Fig. 8.2 Biliary tract anatomy.

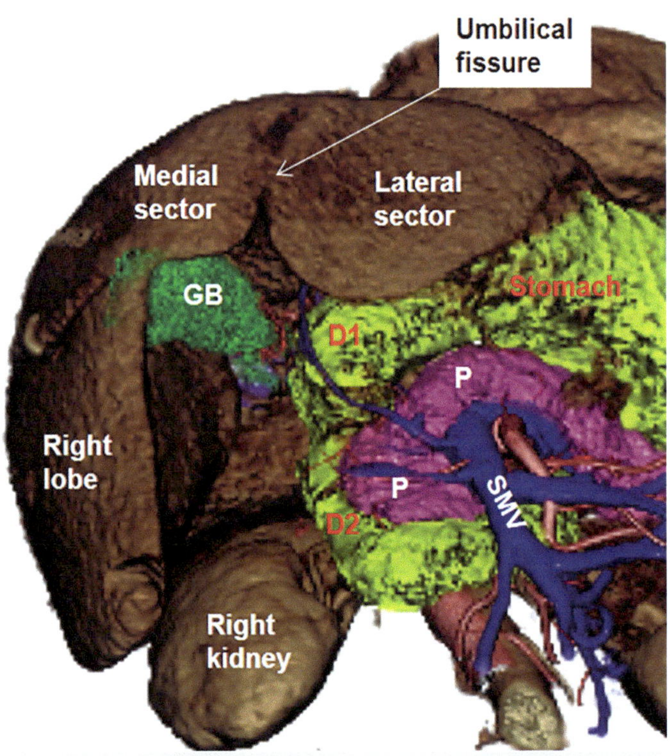

Fig. 8.3 Color-coded computed tomography (CT) showing inferior view of the gallbladder (GB). The GB is beneath the liver between the segments IV and V. The border between the right and left lobes of the liver is determined by the Cantlie's line, which corresponds to an oblique line through the GB fossa and the fossa of the inferior vena cava. The GB is closely associated with the first part of duodenum (D1). D, duodenum; D2, the second part of duodenum; P, pancreas; SMV, superior mesenteric vein.

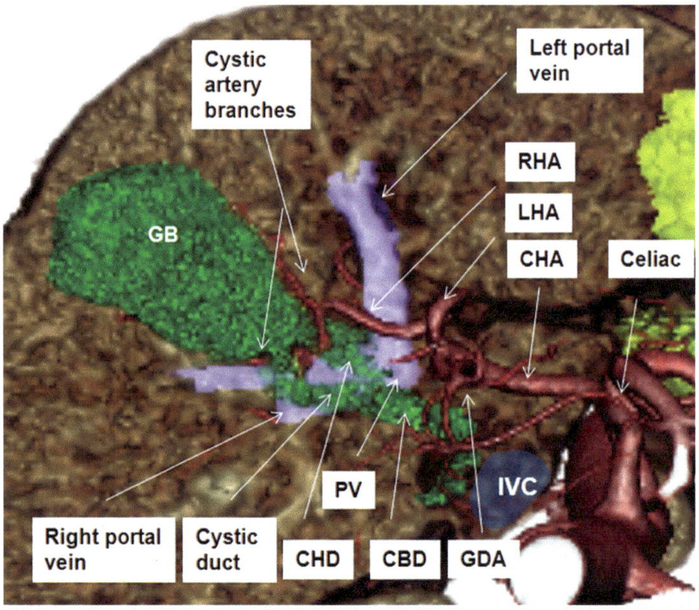

Fig. 8.4 Anatomic structures in porta hepatis are shown in this inferior view. The portal vein (PV) is located behind the common hepatic duct (CHD) and the hepatic arteries in porta hepatis. CBD, common bile duct; CHA, common hepatic duct; GB, gallbladder; GDA, gastroduodenal artery; IVC, inferior vena cava; LHA, left hepatic artery; RHA, right hepatic artery; SMV, superior mesenteric vein.

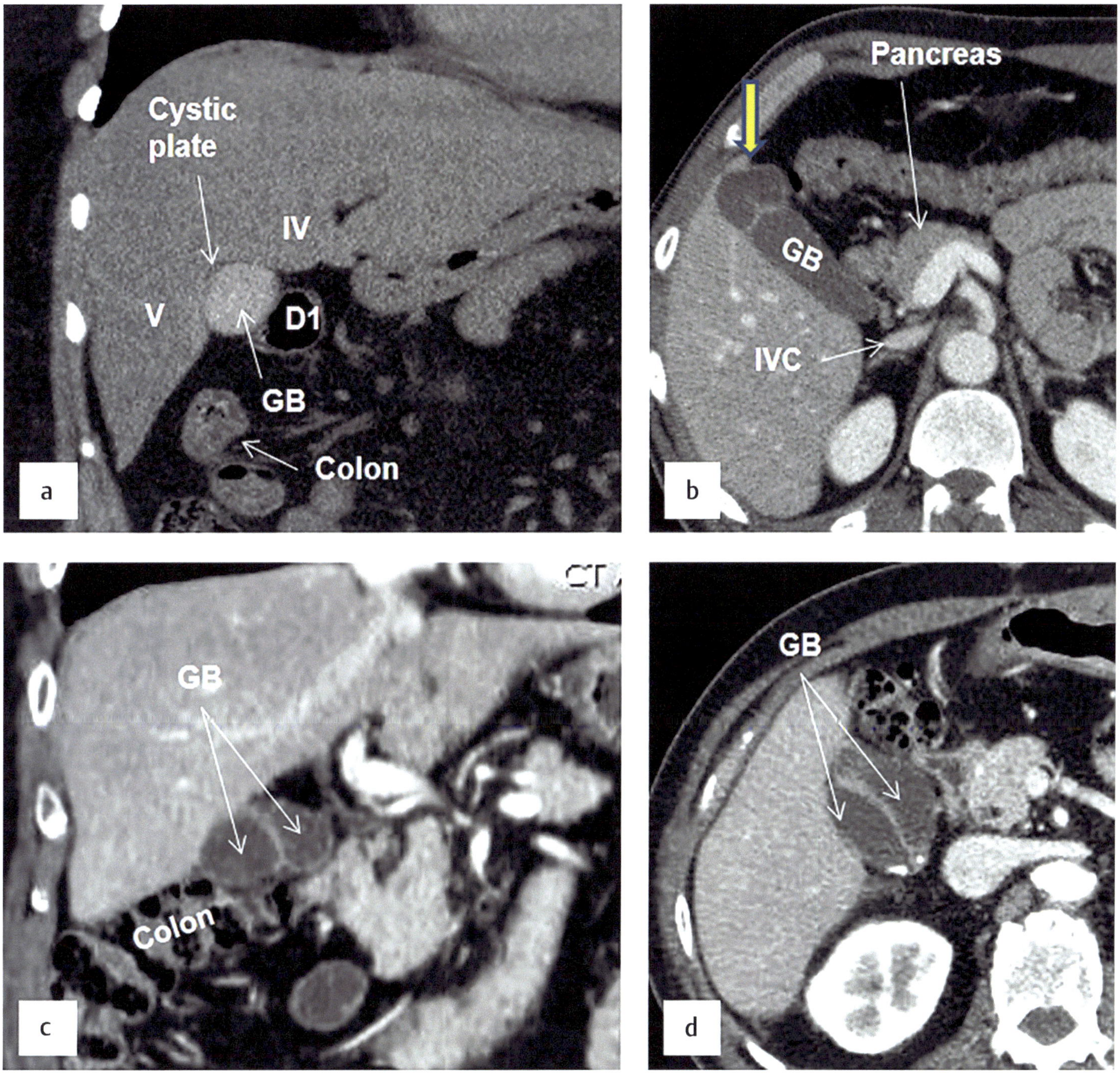

Fig. 8.5 Computed tomography (CT) images in coronal **(a, c)** and axial **(b, d)** views. **(a)** Noncontrast CT showing normal gallbladder (GB) with vicarious excretion of intravenous contrast material injected on prior day. Segments V and IV are seen. Note close relationship of the GB with the first part of duodenum (D1). **(b)** Phrygian cap of the gallbladder is shown (*arrow*). **(c, d)** Bilobed gallbladder containing stone. The fundus of the gallbladder may reside on the hepatic flexure of the colon. The body of the GB is very close to the duodenal bulb (D1). The GB is separated from the liver parenchyma by the cystic plate. IVC, inferior vena cava.

tone of the sphincter of Oddi prevents constant flow of bile into the duodenum.[8] The average gallbladder capacity is about 30 to 50 mL but under gradual increase pressure (i.e., bile obstruction) it can accommodate up to 300 mL of fluid.

The gallbladder during embryonic period is intrahepatic and then migrates to extrahepatic position. The gallbladder is covered by the visceral peritoneum on its free site where it is not in direct contact with the liver. It is separated from the liver parenchyma by the cystic plate, which is in continuity with the hilar plate (▶ Fig. 8.5). The hilar plate system is formed by the Glisson's sheath and the connective tissue surrounding the bile ducts and blood vessels extending to the hilar region.

The wall of the gallbladder is composed of the subserosa, muscular layer, lamina propria, and columnar epithelium. The gallbladder epithelium concentrates bile by active transport of sodium chloride and passive transport of water through its cell membrane channels.

Anatomically, the gallbladder is divided into the fundus, body, and neck. In some gallbladders, proximal of the neck, a focal distension or recess is seen which is called the infundibulum or the

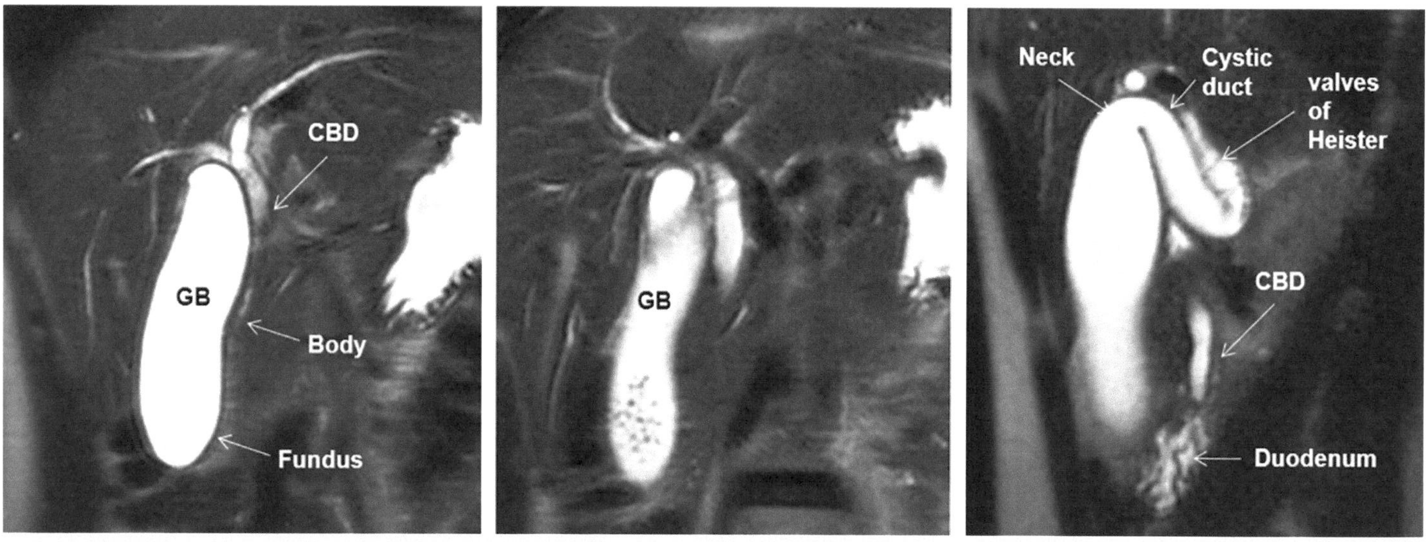

Fig. 8.6 Gallbladder (GB) shown by magnetic resonance cholangiopancreatography (MRCP) containing small stones. The lumen of the cystic duct is characterized by spiral mucosal folds called the valves of Heister. CBD, common bile duct.

Hartmann's pouch (▸Fig. 8.6). The neck connects the gallbladder to the cystic duct.

Rarely, the gallbladder is embedded deeply in the liver or it may be completely intraperitoneal by having a mesentery. The latter is called floating gallbladder. The gallbladder with a mobile mesentery is hypermobile and susceptible to torsion and volvulus.[9] Congenital absence of the gallbladder has been reported and should not be mistaken with hidden intrahepatic ones.[10] Other rare anomalies of the gallbladder and cystic duct include duplication of the gallbladder (two cystic ducts), bilobed gallbladder (single cystic duct) (▸Fig. 8.5), and left-sided gallbladder draining into the left hepatic duct. Ectopic position in the suprarenal, suprahepatic, or along the anterior abdominal wall could be seen.

Contraction of the gallbladder is under the regulatory control of multiple signals. Major positive mediators of contraction are cholecystokinin (CCK) and parasympathetic innervation. CCK secretion from the duodenum results in postprandial gallbladder contraction. Lack of gallbladder contraction results in stasis of bile which contributes to the formation of gallstones.

Cystic Duct

The mean diameter of cystic duct is 4 mm with an average length of 30 mm (range, 4 to 65 mm).[11] The lumen of the cystic duct is characterized by 5 to 12 spiral mucosal folds called the valves of Heister (▸Fig. 8.6). In more than two-third of individuals, the cystic duct joins the right wall of the mid segment of the common hepatic duct (CHD), approximately 2 cm inferior to the hepatic ducts confluence. In 20% of individuals, the cystic duct travels parallel to the CBD usually in a common fibrous sheath. In 10%, the cystic duct joins to anterior or posterior wall of the CBD to the confluence of the hepatic ducts creating a trifurcation, and rarely draining into the right hepatic duct[12] (▸Fig. 8.7). High insertion, short length, and parallel course of the cystic duct increase the risk of complications during or after cholecystectomy.

After cholecystectomy, the importance of the residual cystic duct is in postcholecystectomy syndrome which is described as persistent abdominal symptoms after surgery (e.g., abdominal pain and diarrhea). The length of its remnant is usually between 1 and 2 cm but may be up to 6 cm, especially in parallel course of the cystic duct with CBD (▸Fig. 8.7). The cystic duct remnant could be the site of retained calculi or bile leak. Calculi in the CBD or cystic duct remnant are generally the most common radiological findings of postcholecystectomy syndrome.

Blood Supply and Innervation

The gallbladder and cystic duct are supplied by the cystic artery, which is usually a branch of the right hepatic artery after crossing posterior to the CHD. It is in the hepatobiliary triangle of Calot in 85% of cases. The Calot's triangle is an anatomical landmark of special value in cholecystectomy. It is formed by inferior surface of the liver in the superior part, the CHD in the medial, and the cystic duct in the inferior wall (▸Fig. 8.8, ▸Fig. 8.9, ▸Fig. 8.10). Normal content of the triangle includes the right hepatic artery and branches, the cystic artery, and the sentinel cystic node.

The cystic artery is the second most common arterial variation of the hepatic pedicle after the right hepatic artery. The cystic artery moves anteromedial to the cystic duct, near the sentinel cystic lymph in 75% of cases.[13,14] In 5 to 13%, the cystic artery moves outside the Calot's triangle, inferior to the cystic duct.[13,15] The cystic artery may arise from an accessory of replaced right hepatic artery. Rarely, the cystic artery originates from the left hepatic artery, hepatic artery proper, gastroduodenal artery, or a replaced right hepatic artery (▸Fig. 8.10). The cystic artery is usually single but may be duplicated (7%) or multiple or rarely absent congenitally. Frequent anatomic variation of the cystic artery is important during cholecystectomy. When cystic artery runs parallel to the cystic duct, it should not be confused with right hepatic artery. In cases where cystic artery arises from unusual locations, damage to the duct may cause bile leakage

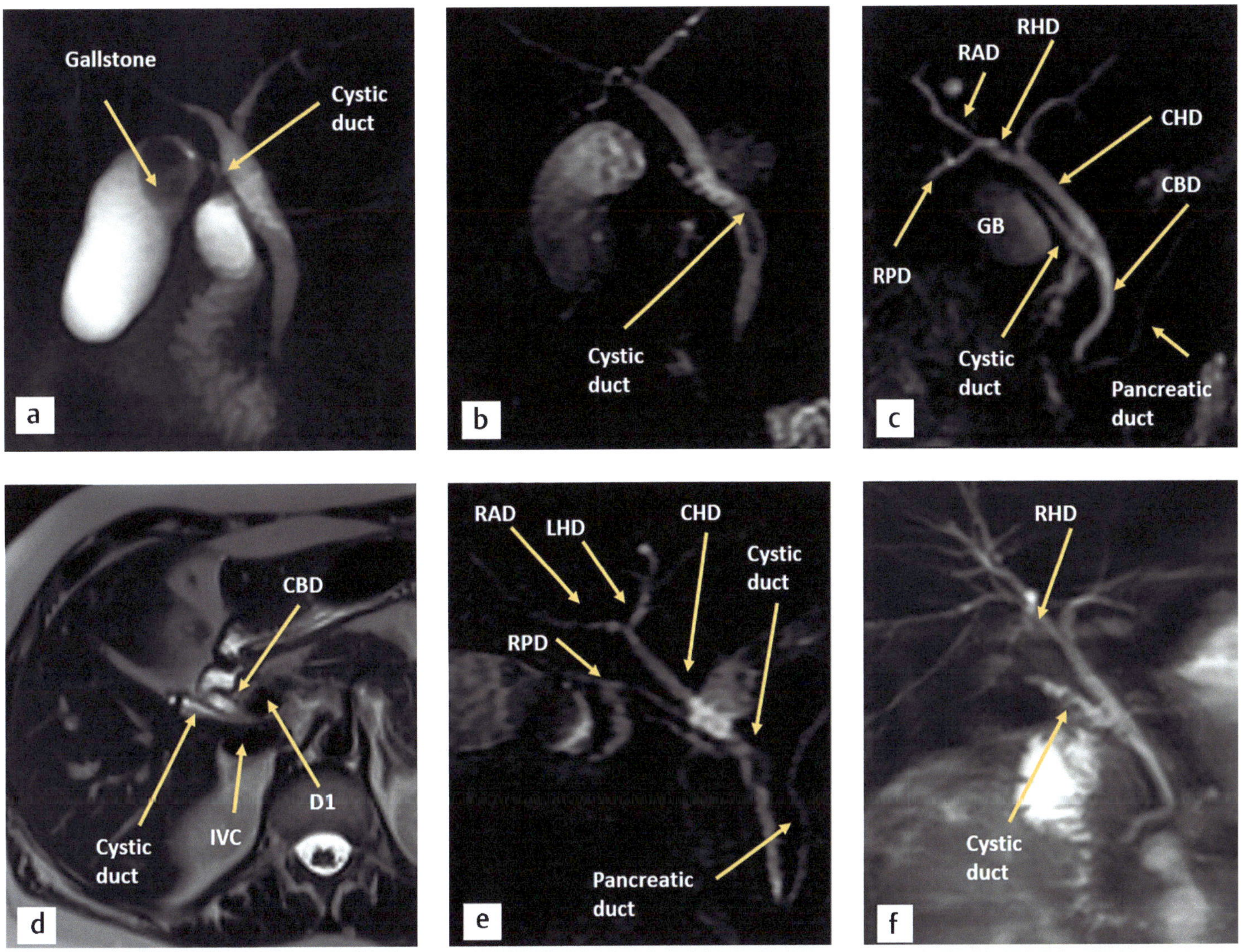

Fig. 8.7 Normal cystic duct anatomy. Magnetic resonance cholangiopancreatography (MRCP) images demonstrating conventional biliary anatomy. In two-thirds of individuals, the cystic duct joins the right wall of the mid segment of the common duct, approximately 2 cm inferior to the hepatic ducts confluence. The cystic duct insertion site defines the border between the common hepatic duct (CHD) proximally and common bile duct (CBD) distally. **(a)** Conventional lateral but high insertion. **(b)** Low medial insertion. **(c)** Parallel course of the cystic duct with the common duct is seen in 20% of individuals. **(d)** Axial T2-weighted magnetic resonance imaging (MRI) showing parallel course of the cystic duct with the common duct. **(e)** Spiral course of the cystic duct with medial insertion is seen in 5%. **(f)** Cystic duct remnant. D, duodenum; GB, gallbladder; IVC, inferior vena cava; LHD, left hepatic duct; RAD, right anterior duct; RHD, right hepatic duct; RPD, right posterior duct.

post cholecystectomy (▶Fig. 8.11). Injury of the vessels or the CBD during laparoscopic cholecystectomy can be avoided by careful dissection of Calot's triangle and the hepatoduodenal ligament.

Venous drainage of the gallbladder includes veins that drain into the portal system or directly from the gallbladder into the liver. Venous branches of the gallbladder are difficult to be seen in cross-sectional imaging unless they are dilated such as in patient with portal vein hypertension or thrombosis (▶Fig. 8.12). The hepatic branch of the vagus nerve provides parasympathetic innervation, which also promotes contraction. The gallbladder receives innervation from the sympathetic system via the celiac plexus which promotes relaxation of the smooth muscle of the gallbladder.

Lymphatic vessels in the gallbladder are located in the subserosal layer and drain to the Calot's node and lymph nodes along the porta hepatis. Lymphatic drainage can also course directly into the liver along segments V and IVb before reaching the lymph nodes within the hepatoduodenal ligament. Edema of the pericholecystic tissue and subserosal plane of the gallbladder is a common and nonspecific radiologic finding and should not be confused with real gallbladder wall thickening (i.e., cholecystitis) (▶Fig. 8.13). Pericholecystic tissue edema with or without periportal edema is seen in pancreatitis, liver trauma, volume overload, and lymphatic obstruction.

Gallbladder cancer has a high tendency to lymphatic invasion with a poor prognosis; therefore, an adequate lymphadenectomy is required to obtain a proper tumor staging.[16] Radical cholecystectomy is a common technique for surgery of gallbladder cancer and includes resection for the complete excision of the liver cystic plate or segment IVb–V bisegmentectomy, and resection of the suprapancreatic segment of the extrahepatic bile duct with regional lymphadenectomy.[17] The cystic duct and pericholedochal node groups have been regarded as the first-echelon nodes of the gallbladder.

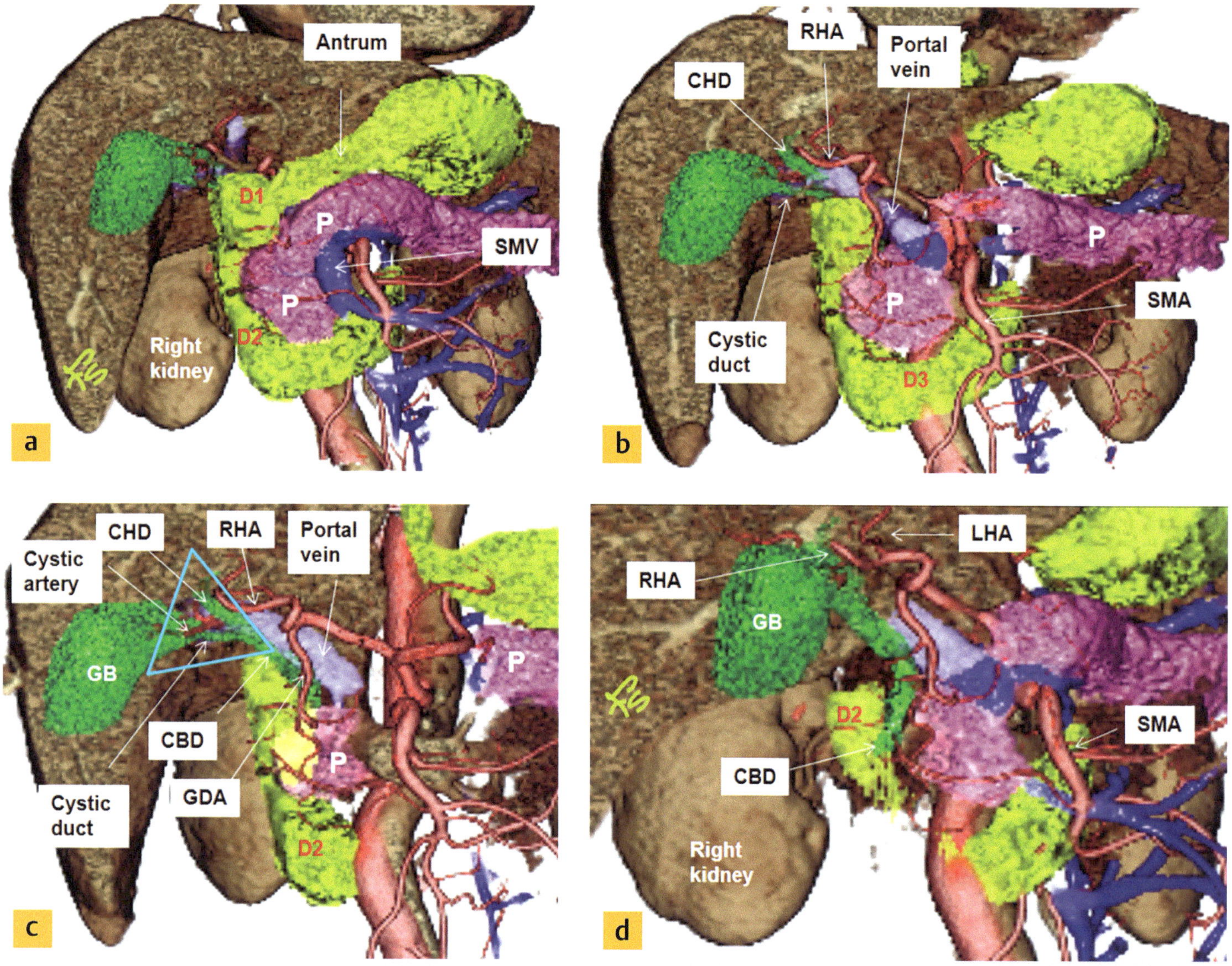

Fig. 8.8 Anatomic structures of the porta hepatis shown by color-coded, volume-rendered computed tomography (CT), frontal views. **(a)** Before removal of the antrum. **(b)** The antrum is removed. **(c)** The first part of duodenum (D1) (bulb) is removed. **(d)** Upper pancreatic head is removed. The Calot's triangle (*blue*) is formed by inferior surface of the liver superiorly, the common hepatic duct (CHD) medially, and the cystic duct inferiorly. Normal content of the triangle includes the right hepatic artery and branches, the cystic artery, and the sentinel cystic node. The cystic artery usually moves anteromedial to the cystic duct. CBD, common bile duct; D, duodenum; GB, gall bladder; GDA, gastroduodenal artery; LHA, left hepatic artery; P, pancreas; RHA, right hepatic artery; SMA, superior mesenteric artery; SMV, superior mesenteric vein.

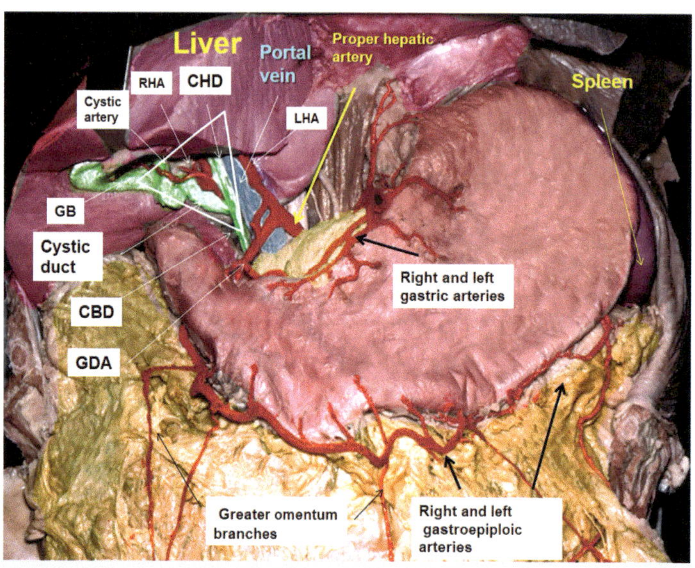

Fig. 8.9 Anatomic structures of the porta hepatis. The Calot's triangle is shown. It is an anatomical landmark of special value in cholecystectomy. CBD, common bile duct; CHD, common hepatic duct; GB, gallbladder; GDA, gastroduodenal artery; LHA, left hepatic artery; RHA, right hepatic artery.

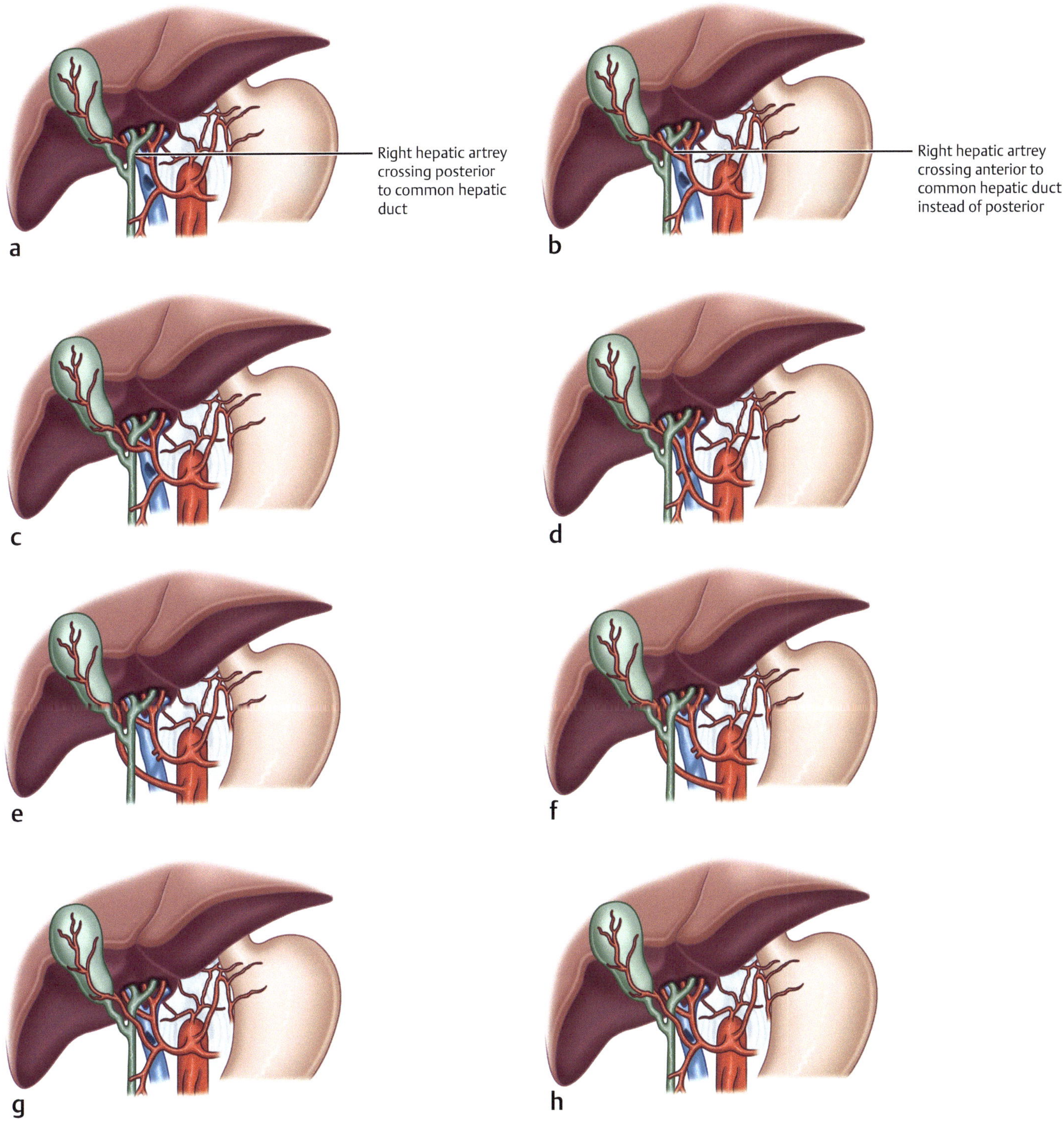

Fig. 8.10 Variations in hepatic arterial anatomy. **(a)** The most common type. The right hepatic artery crossing posterior to the common hepatic duct with the cystic artery arising from it. **(b)** Right hepatic artery crossing anterior to the common hepatic duct with the cystic artery arising from it. **(c)** The cystic artery passing anterior to the common hepatic duct. **(d)** Replaced right hepatic artery with the cystic artery arising from it. **(e)** Accessory right hepatic artery with the cystic artery arising from it. **(f)** Double cystic arteries. **(g)** The cystic artery arising from the gastroduodenal artery. **(h)** The cystic artery arising from the left hepatic artery.

8

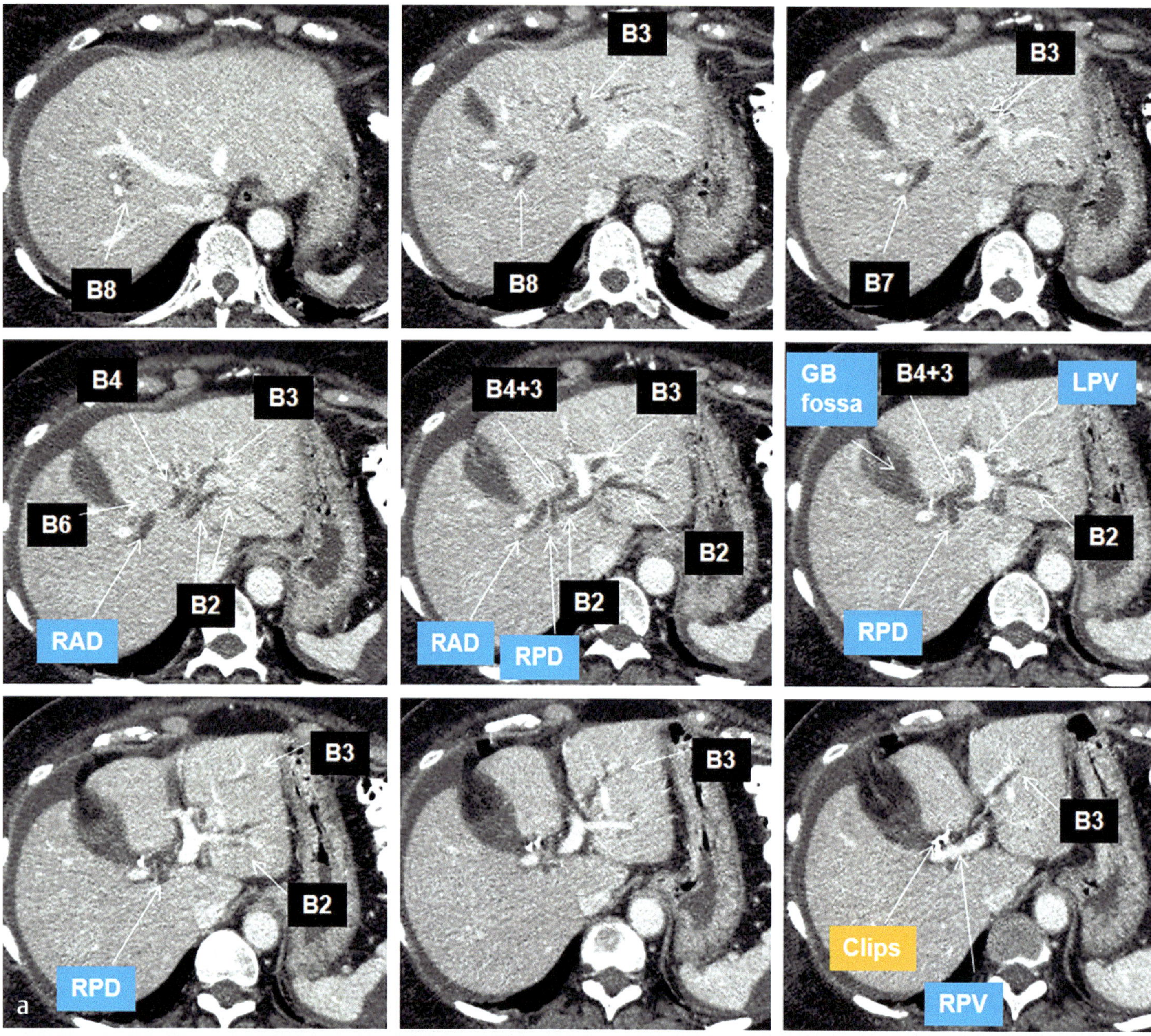

Fig. 8.11 Status post cholecystectomy. Inadvertent clipping of the common hepatic duct (CHD) at the time of surgery causing biliary obstruction. **(a)** Axial postcontrast computed tomography (CT) scan from superior to inferior. **(b)** Coronal postcontrast CT scan from posterior to anterior. Intrahepatic bile dilatation due to occlusion (clipping) of the CHD is seen. Anatomic variant of the confluence of the B2 and B3 bile ducts distal to the left portal vein (LPV), and umbilical fissure is also seen. Also evident is the unusual configuration of the confluence of the B2 and B3 with right posterior duct and right anterior duct at a single point, best shown on axial CT. This unusual configuration with a short CHD increases the risk of inadvertent clipping as happened in this case. Also note the aberrant right hepatic artery (RHA) arising from superior mesenteric artery (SMA) passing anterior to the portal vein (PV). The right hepatic artery is also ligated anterior to the portal vein. The aberrant RHA was the source cystic artery origin, where both were ligated. Concomitant injury to right hepatic artery can compromise liver perfusion and bile duct repair. B, Brisbane's classification of second order bile ducts; CBD, common bile duct; GB, gallbladder; RAD, right anterior duct; RAPV, right anterior portal vein; RPD, right posterior duct; RPPV, right posterior portal vein; RPV, right portal vein.

(Continued)

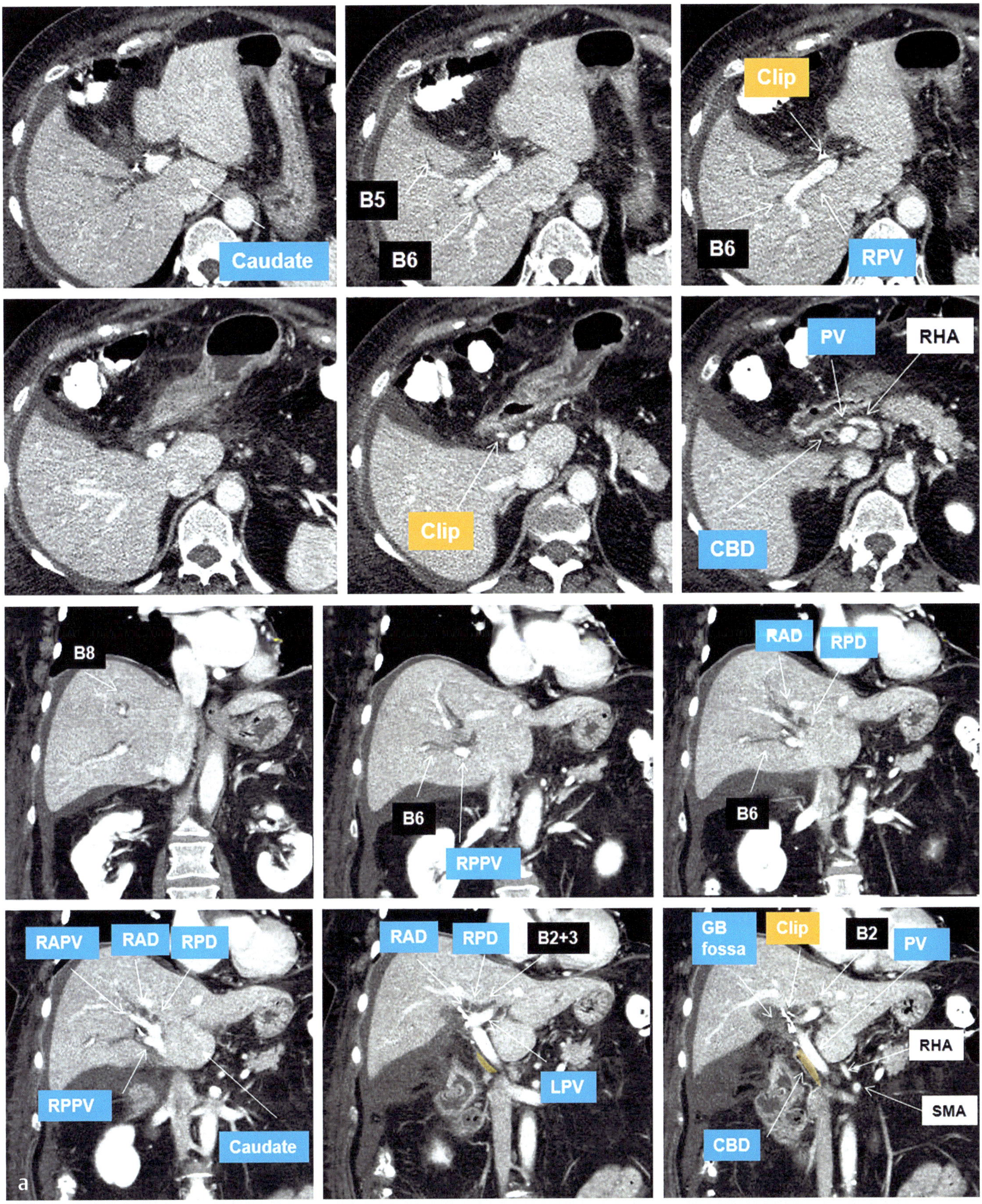

Fig. 8.11 (*Continued*)

(*Continued*)

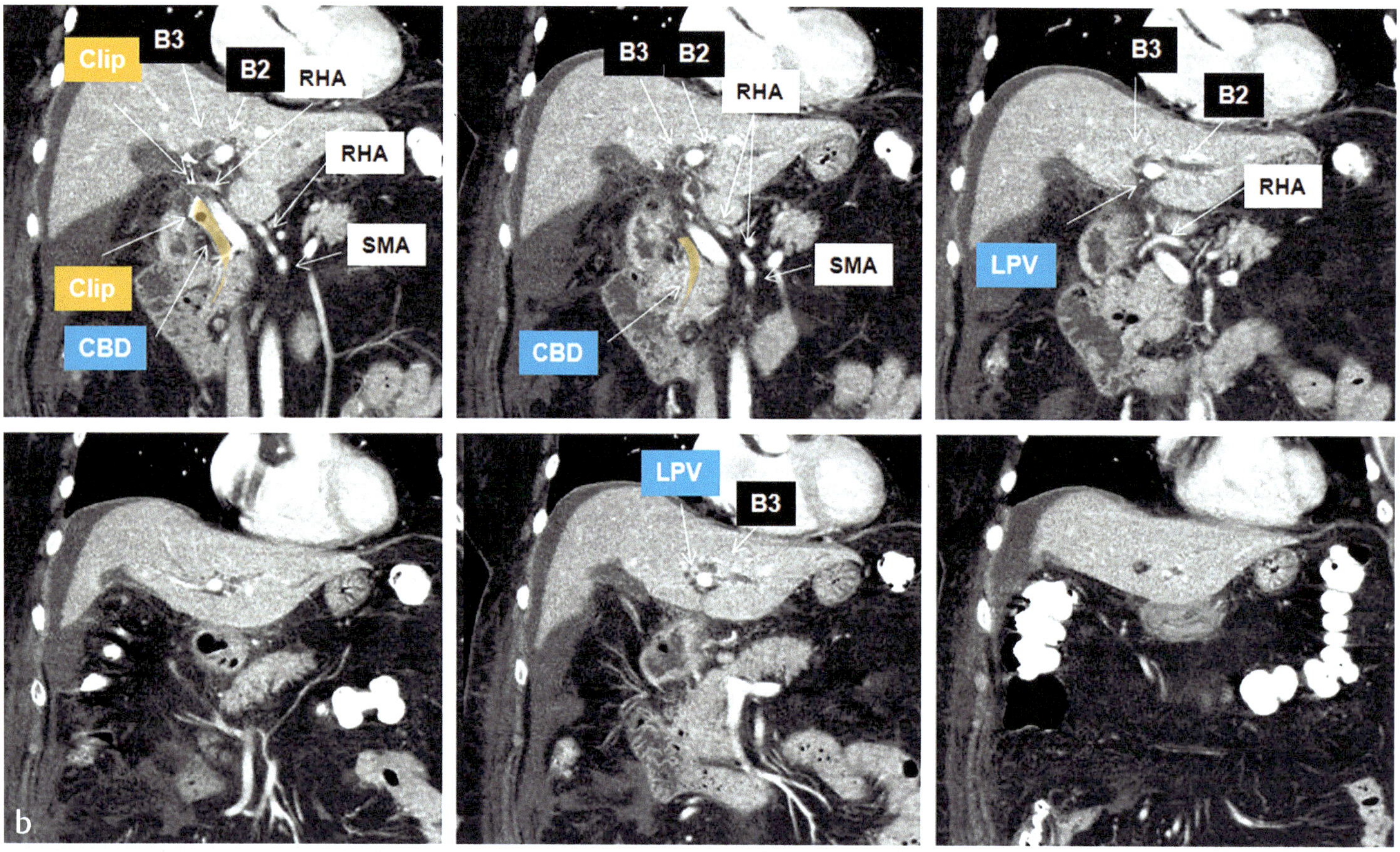

Fig. 8.11 (*Continued*)

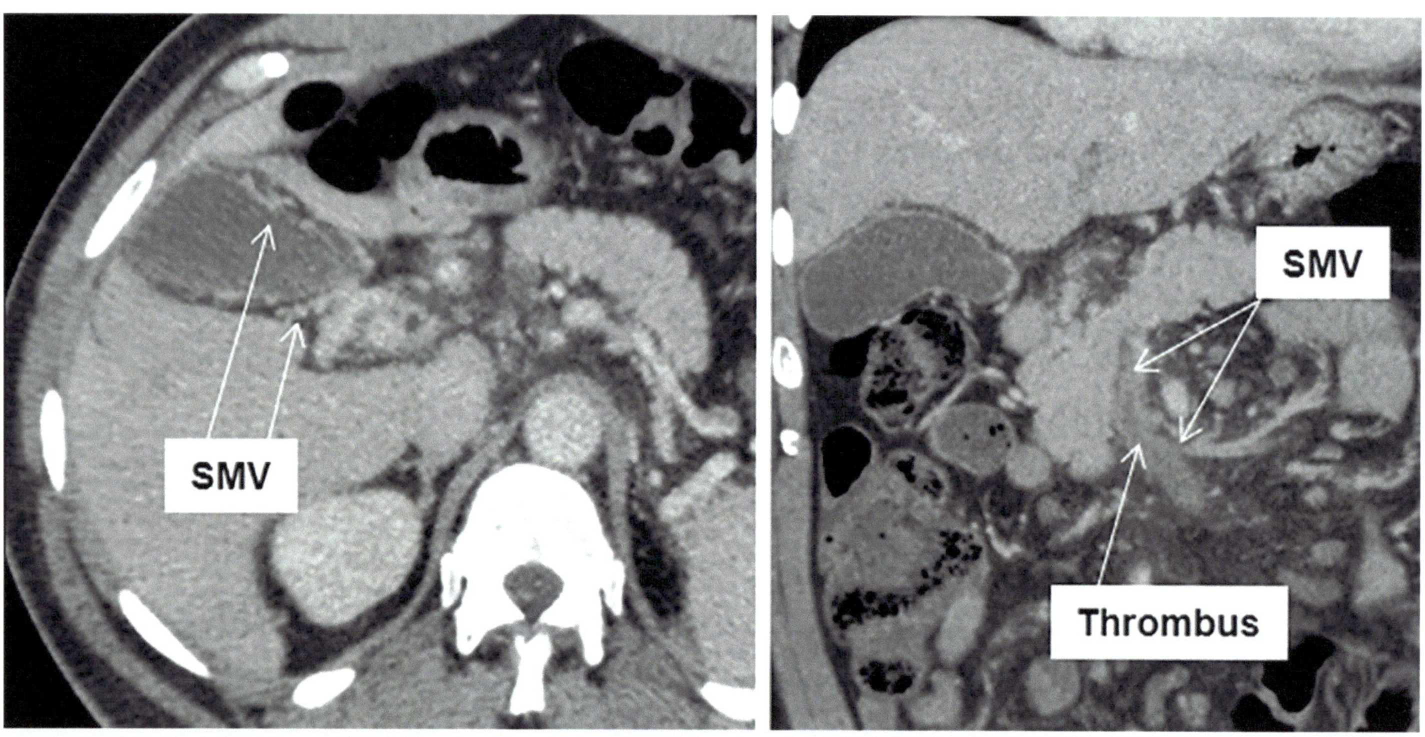

Fig. 8.12 Pericholecystic collateral veins in a patient with superior mesenteric vein (SMV) thrombosis.

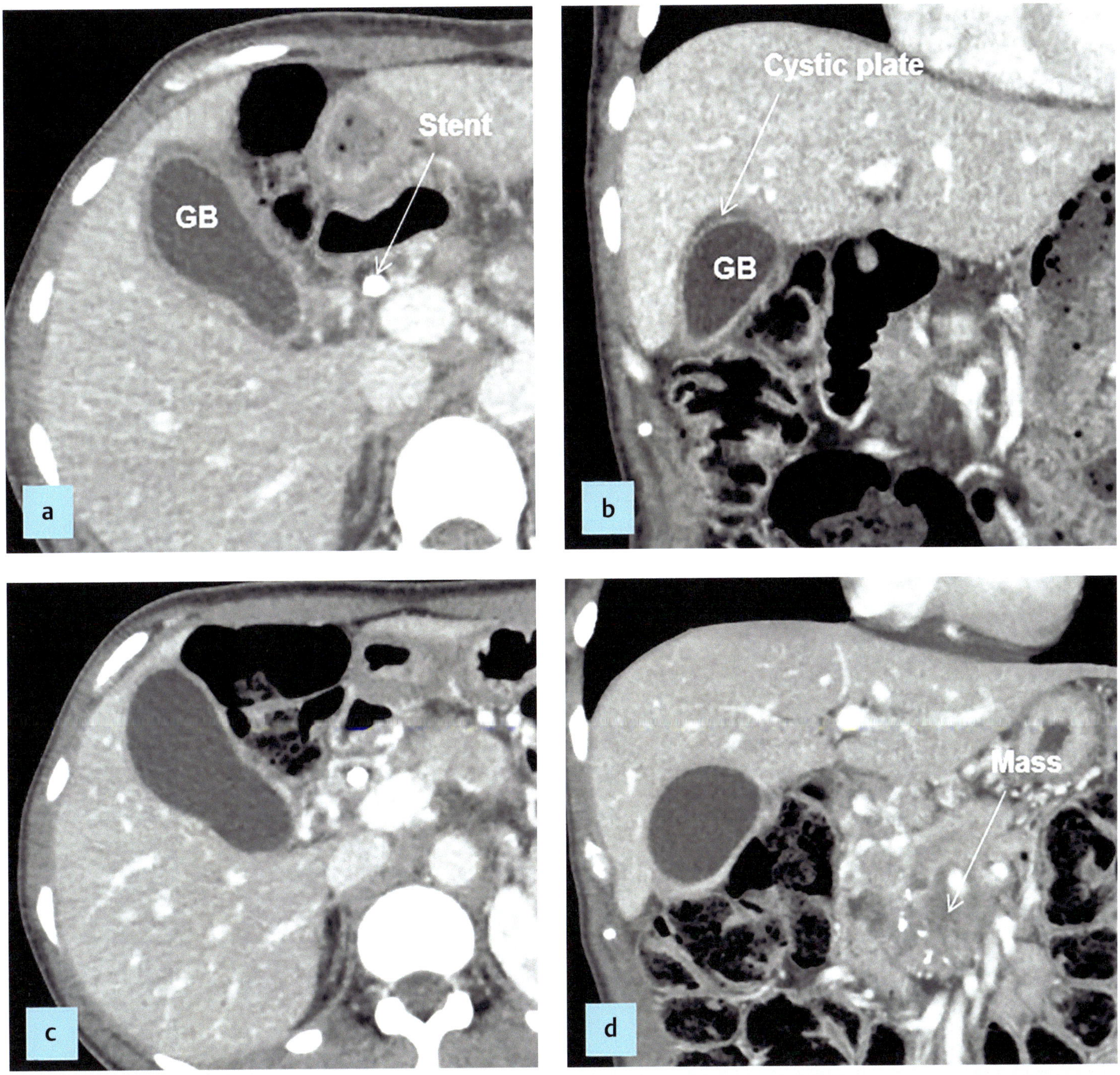

Fig. 8.13 Pericholecystic tissue edema **(a, b)** due to sepsis and extracellular fluid retention which resolved after a few days **(c, d)** in this patient with pancreatic cancer and bile obstruction. A stent is placed in the common bile duct. Edema of the pericholecystic tissue and subserosal plane of the gallbladder (GB) is a common and nonspecific radiologic finding and should not be confused with real GB wall thickening (i.e., cholecystitis).

Common Duct

The common duct measures between 7 and 11 cm in length and its internal diameter measures up to 2 to 8 mm at normal physiologic pressures. The cystic duct insertion site defines the border between the CHD proximally and CBD distally[8] (▶Fig. 8.7). The common duct travels in the hepatoduodenal ligament lateral to the hepatic artery. The portal vein is located behind the CBD and the hepatic artery in porta hepatis (▶Fig. 8.8, ▶Fig. 8.9). It then travels posterior to the first portion of the duodenum. At the pancreatic level, the CBD moves posterior to the pancreas in a groove or a tunnel made by a thin layer of pancreas. Before ending into the second portion of the duodenum, the CBD may join the pancreatic duct to form a common channel (▶Fig. 8.14), 1 to 8 mm in length, known as ampulla of Vater in 60% of cases (see Chapter 4 "Small Intestine" and Chapter 9 "Pancreas"). The ampulla of Vater consists of the ampullary pancreatic duct, the ampullary CBD, and the major duodenal papilla surrounded by the sphincter of Oddi. A double-barreled opening of the pancreatic and common ducts is seen in 38% of cases and separate duodenal openings for the two canals in 2% of cases.

The common duct lumen is lined by a columnar epithelium that contains mucus-secreting cells.

The main arteries supply of the CBD originates from the gastroduodenal and right hepatic arteries. Injury of these arteries during surgery can lead to ischemia of the CBD wall. In general, arterial supply of the proximal extrahepatic ducts is less than pancreatic portion of the CBD. The hepatic artery proper ascends in the hepatoduodenal ligament and in 70% of cases bifurcates into the left and right hepatic arteries (▶Fig. 8.8, ▶Fig. 8.9, ▶Fig. 8.10). The left hepatic artery then continues along the medial aspect of the hepatoduodenal ligament to enter the left liver through the umbilical fissure. The right hepatic artery crosses the hepatoduodenal ligament, typically passing behind the CHD to enter the right liver. The cystic artery usually arises from the right hepatic artery. The right hepatic artery usually courses posterior to the common duct before entering the liver. In approximately 25% of patients, the right hepatic artery will lie anterior to the duct (▶Fig. 8.15). In 10%, the right hepatic artery will cross posterior to the portal vein. In 20%, an accessory or a replaced right hepatic artery is seen, usually arising from the superior mesenteric artery and passing behind the portal vein (▶Fig. 8.10, ▶Fig. 8.11). The cystic artery usually arises from a normal or replaced right hepatic artery. If a replaced right hepatic artery is sacrificed during surgery, a cholecystectomy is commonly performed because flow to the cystic artery is likely compromised (▶Fig. 8.11).

Right and Left Hepatic Ducts

The right and left hepatic ducts run a short course outside of the liver parenchyma before forming the common duct. The right hepatic duct length is variable from 0.2 to 2.5 cm. Absence of the left hepatic vein is rare but more common in patient with portal vein anomaly.[18] The average length of the left hepatic duct is 3.2 cm with a range of 0.5 to 5.7 cm. The left hepatic duct arises from the base of the umbilical fissure along the inferior border of segment IV of the left lobe. The two ducts meet just below the infundibulum of the gallbladder (▶Fig. 8.7). Anatomical variations of the confluence of the bile ducts are not uncommon (▶Fig. 8.16). The "normal" Y-shaped confluence of hepatic ducts with the CBD is found in less than 60% of adults. Atypical branching patterns of both the right and left hepatic ducts are seen in 14 and 8%, respectively.[19] Rarely, the right and left ducts join within the liver. Because the left hepatic duct is longer than the right hepatic duct, it is a better choice for operative biliary decompression or bypass and reconstruction surgeries.

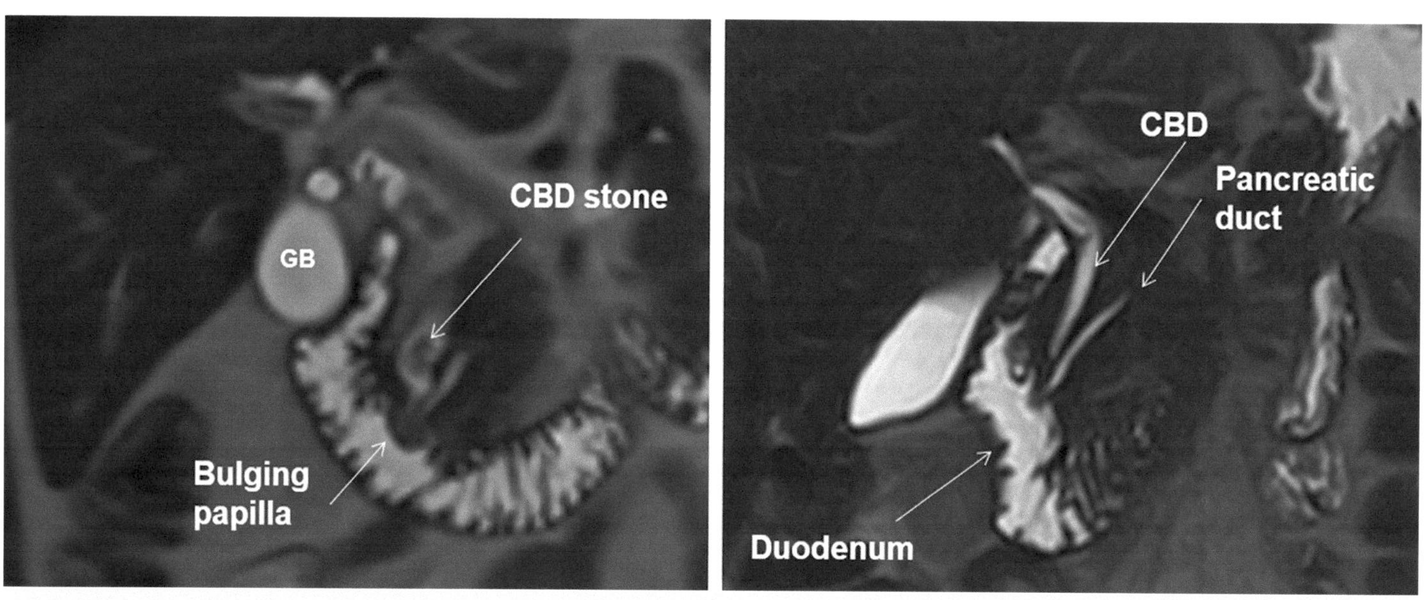

Fig. 8.14 Ampulla of Vater shown by T2-weighted magnetic resonance imaging (MRI). Before ending into the second portion of the duodenum, the common bile duct (CBD) joins the pancreatic duct to form a common channel known as ampulla of Vater in 60% of cases. A double-barreled opening of the pancreatic and common ducts is seen in 38% of cases. A bulging papilla greater than 10 mm in diameter usually indicates pathologies. GB, gallbladder.

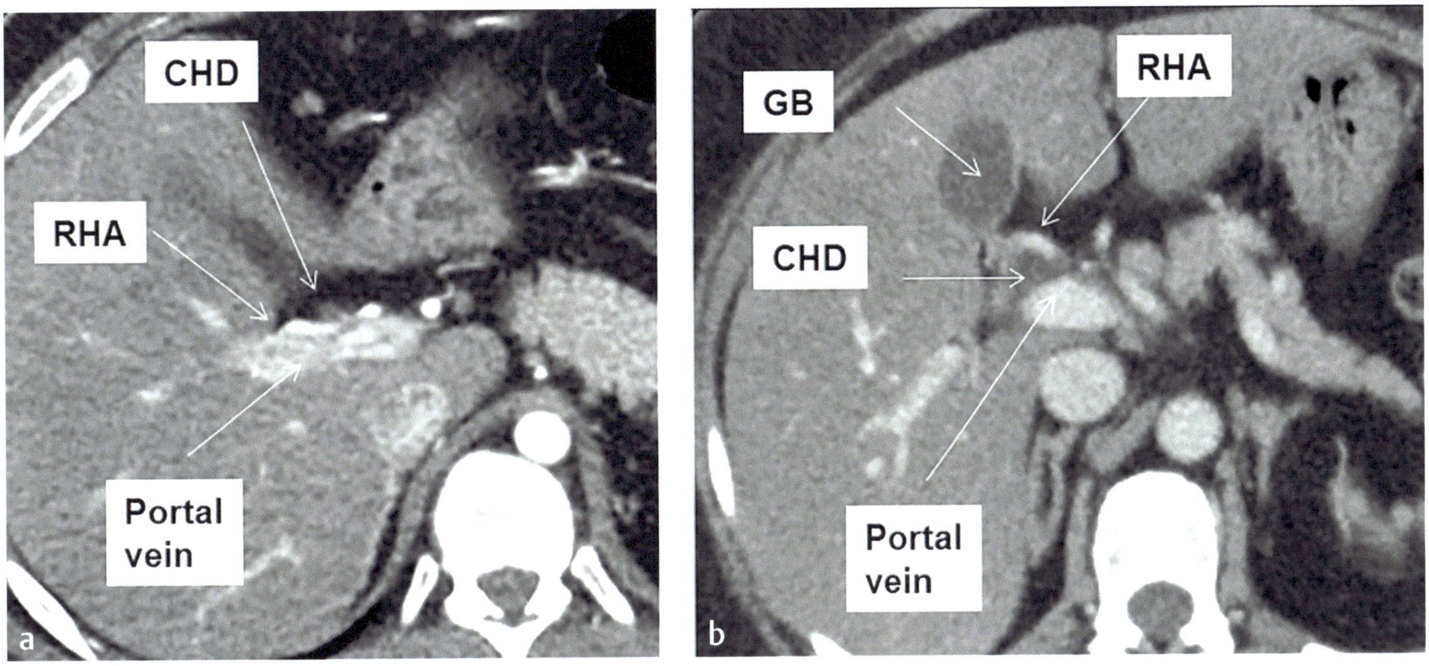

Fig. 8.15 Porta hepatis anatomy. Axial postcontrast computed tomography (CT) images are shown. The common hepatic duct (CHD) is invariably located slightly to the right and anterior to the portal vein. **(a)** The right hepatic artery (RHA) usually courses posterior to the CHD. **(b)** In 25% of cases, it is situated anterior to the CHD. GB, gallbladder.

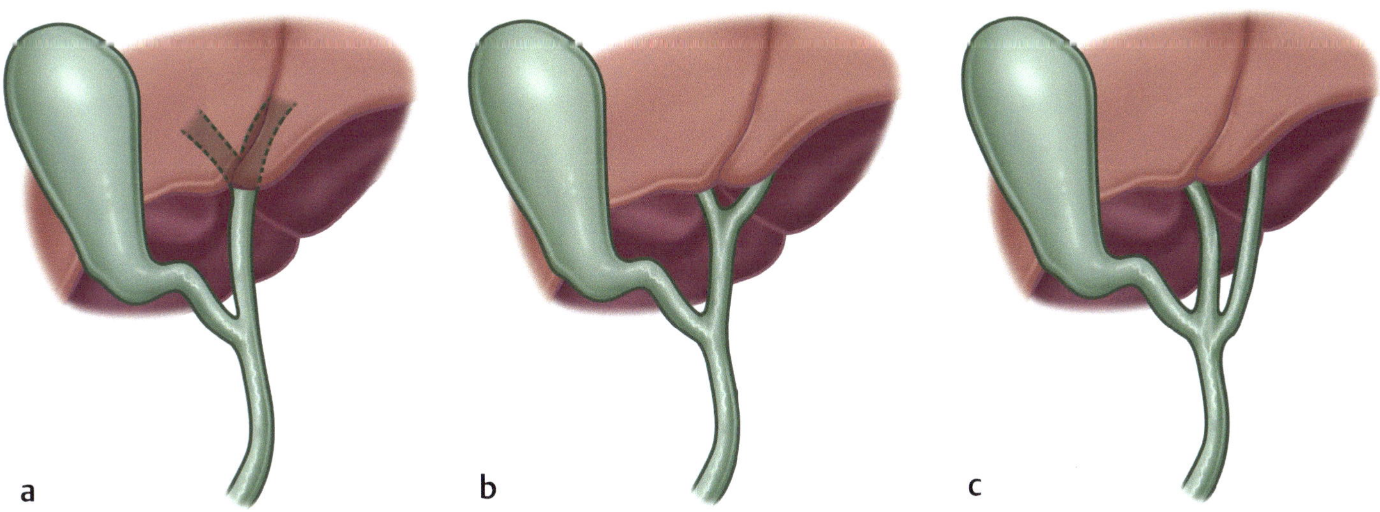

Fig. 8.16 Variation in confluence of right and left hepatic ducts. **(a)** Intrahepatic, **(b)** extrahepatic/ typical, and **(c)** low confluence of the right and left hepatic ducts.

◆ Intrahepatic Bile System

Intrahepatic biliary tree has a high frequency of anatomical variations. Variation is more common in patients with major portal venous anomaly.[8,20] Anomalous drainage of the segmental biliary ducts can lead to difficulties in opacification or drainage of the entire ductal system during biliary system percutaneous intervention. Furthermore, with increasing frequency and complexity of surgical procedures such as liver resection and partial living donor liver transplantation (LDLT), an accurate knowledge of the anatomy of intrahepatic bile ducts is critical to avoid biliary tract complications which may occur in 4 to 10% of operations.[21] Laparoscopic cholecystectomy is a less invasive procedure, but the limited visual field and biliary variants occasionally result in biliary complications such as bile leakage and vascular injury.

Classification

The border between the right and left lobes of the liver is determined by the Cantlie's line corresponding to an oblique line through the gallbladder fossa and the fossa of the inferior vena cava[22] (▶ Fig. 8.4, ▶ Fig. 8.5, ▶ Fig. 8.17).

Traditionally, nomenclature of the intrahepatic bile ducts is based on Couinaud's liver segment classification[23]

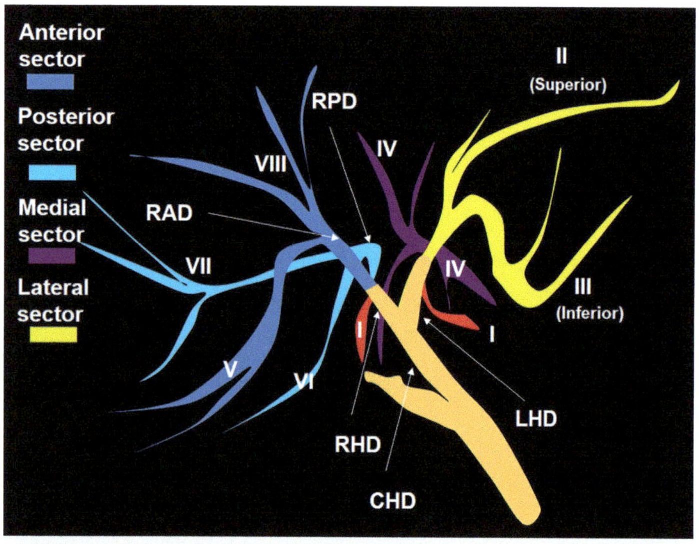

Fig. 8.17 Intrahepatic biliary tract branching classification (Brisbane's terminology). Segmental bile ducts are considered third-order ducts, sectoral ducts second-order ducts, and the right and left hepatic bile ducts first-order ducts. The right lobe is divided into anterior (paramedian, segments V and VIII) and posterior (posterolateral, segments VI and VII) sectors. The left lobe is divided into medial (segment IV) and lateral (segments II and III or superior and inferior segments respectively) sectors, which are separated by the umbilical fissure. The caudate lobe is classified separately (segments I and IX in Couinaud's classification). CHD, common hepatic duct; LHD, left hepatic duct; RAD, right anterior duct; RHD, right hepatic duct; RPD, right posterior duct.

8

(see Chapter 6 "Liver"). Many surgeons prefer to use Brisbane's terminology for intrahepatic bile duct classification and segmental liver resections[24] (▶Table 8.1).

The right lobe is divided into anterior/paramedian (segments V and VIII) and posterior/lateral (segments VI and VII) sectors which will be drained by the anterior and posterior sectional ducts, respectively[25] (▶Fig. 8.11, ▶Fig. 8.17, ▶Fig. 8.18). The location of segments in these two sectors/sections are classified into superior (VII and VIII) and inferior (V and VI) segments. The right anterior duct drains Couinaud's segments V and VIII, and the right posterior duct drains segments VI and VII.

Based on bile duct distribution, the left lobe is divided into medial (segment IV) and lateral (segments II and III or superior and inferior segments respectively) sections, which are separated by the umbilical fissure (▶Fig. 8.17, ▶Fig. 8.18). Therefore, left hepatic duct drains segments II, III, and IV. Segmental bile ducts are considered third-order ducts and sectoral bile ducts are second-order ducts.[25] The right and left hepatic bile ducts are first-order ducts (▶Fig. 8.17, ▶Fig. 8.18).

Right Intrahepatic Ducts

Various reports have described the branching pattern of the right hepatic duct.[19,26] In general, greater variability is seen in the branching pattern of the right hepatic duct (40%) compared with the left hepatic duct.[27,28] Therefore, in a case where the left lobe

Table 8.1 IHPBA Brisbane 2000 terminology of liver anatomy

Section	Second-order division (Couinaud's segments)		Sector
	Based on bile ducts and hepatic artery	Based on portal vein	
Left lateral	II, III	II	Left lateral or left posterior
Left medial	IV	III, IV	Left medial or left paramedian
Right anterior	V, VIII		Right anterior or right paramedian
Right posterior	VI, VII		Right posterior or right lateral

of the liver is large enough, it should be the preferred lobe for a transplantation when both the donor and recipient are alive. The right posterior duct is longer with a horizontal and more superior direction as opposed to the right anterior duct, which has an oblique orientation (▶Fig. 8.17). The right posterior duct moves posterior to the right portal vein and joins the anterior duct superior to the right portal vein.

Classification according to the site of insertion of the right posterior duct is probably the most practical way to show anatomical variants of the right hepatic duct branches (▶Fig. 8.19).

The most common form is defined as the right posterior sectoral duct joining the right anterior sectoral duct to form a right hepatic duct. This type is seen in 60 to 65% of cases.[28,29] The second most common form (10–17%) is trifurcation pattern of the right posterior, right anterior, and left hepatic bile ducts in which the three ducts join at the same point to form the CHD (▶Fig. 8.20).[30] Other less common types include right posterior duct draining into the left bile duct (6%), CHD, or cystic duct (2%) (▶Fig. 8.21).

Drainage of the right anterior duct into the left bile duct is seen in 10% of cases. Drainage of the right hepatic duct into the cystic duct is rare and an accessory duct leading to the CHD or right hepatic duct in seen 5%.[28] In 20%, segment VIII duct (anterosuperior) connects the right posterior sectoral duct not the right anterior duct.[22] Rarely, segment V duct (anteroinferior) drains into the right posterior duct, right hepatic duct, or CBD.[25]

Trifurcation pattern can cause difficulty for biliary-enteric anastomotic surgery and for a safe right hepatectomy in liver donor candidates (▶Fig. 8.20, ▶Fig. 8.21, ▶Fig. 8.22). Other potential causes of surgical complications include an infraportal course of the right posterior duct, a right posterior duct draining into the left hepatic duct beyond the hepatic hilum, and a right posterior duct draining into the CHD near the confluence of the cystic duct (▶Fig. 8.21). The right posterior duct draining into the left bile duct is considered a potential contraindication for both right and left liver donation (▶Fig. 8.21). Inadvertent ligation of a right posterior duct draining into the cystic duct may cause cholestasis in the sectoral liver from which the duct originates.

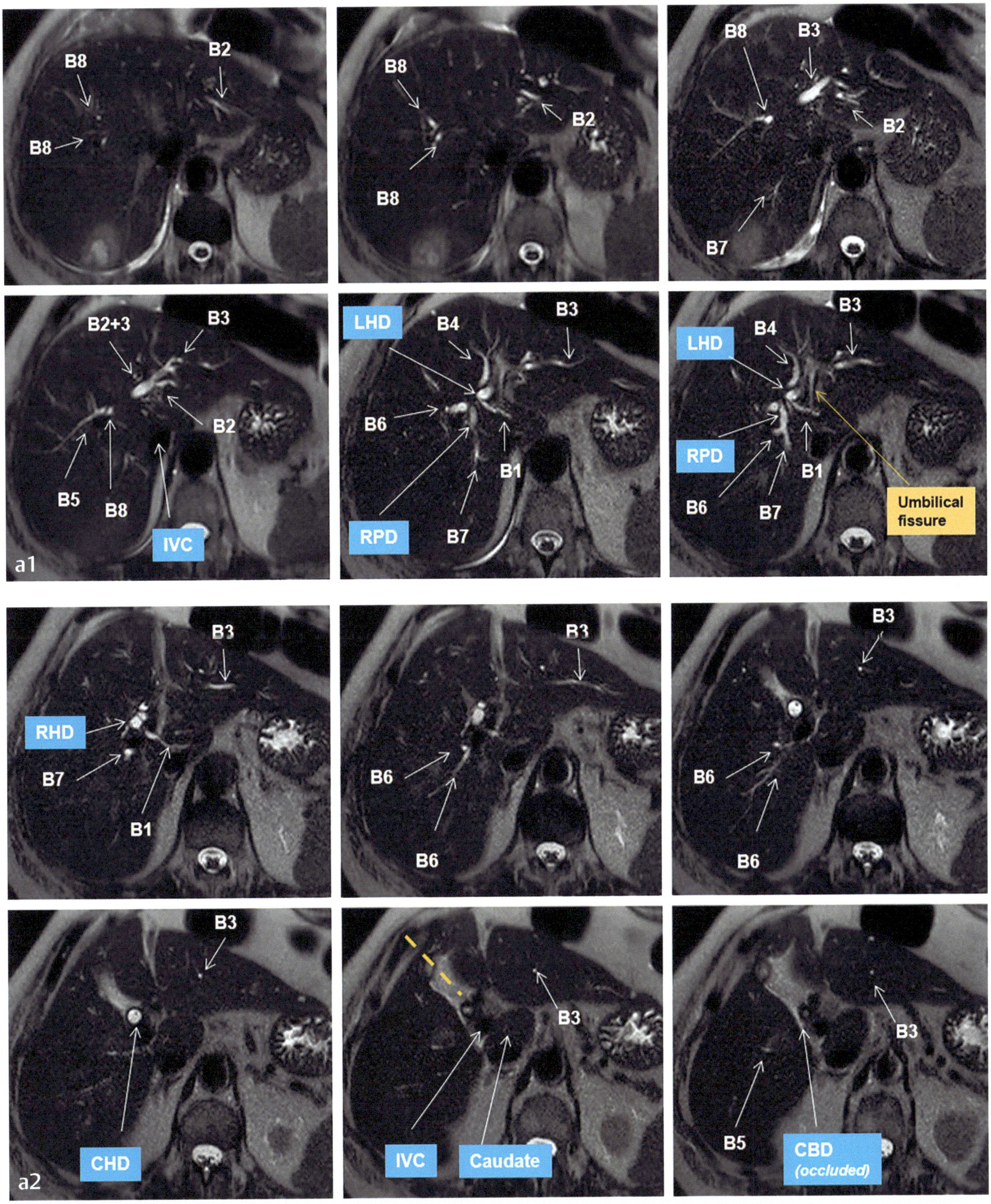

Fig. 8.18 Intrahepatic biliary tract anatomy shown by T2-weighted magnetic resonance imaging (MRI). Status post cholecystectomy. Intrahepatic bile dilatation due to inadvertent clipping of the common hepatic duct (CHD) is seen. **(a)** Axial images from superior to inferior. **(b)** Coronal images from anterior to posterior. **(c)** Sagittal images from lateral to medial. The border between the right and left lobes of the liver is determined by the Cantlie's line (*broken orange line*), which corresponds to an oblique line through the gallbladder fossa and the fossa of the inferior vena cava (IVC). The left lobe is divided into medial (B4) and lateral (B2 and B3) sectors, which are separated by the umbilical fissure. CBD, common bile duct; D, duodenum; GB, gallbladder; LHD, left hepatic duct; LPV, left portal vein; LV, ligamentum venosum; PD, pancreatic duct; RAD, right anterior duct; RHD, right hepatic duct; RPD, right posterior duct.

(*Continued*)

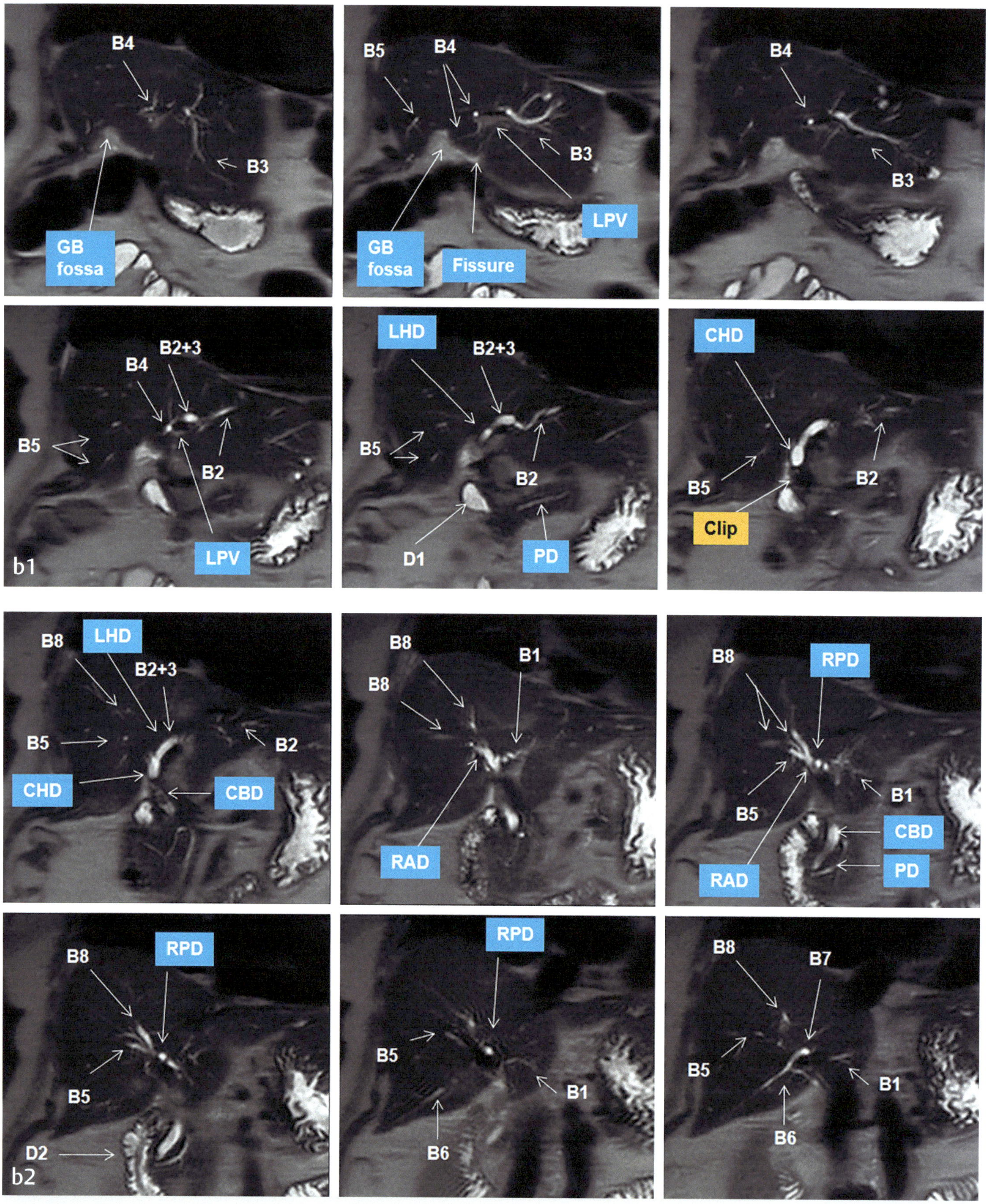

Fig. 8.18 *(Contined)*

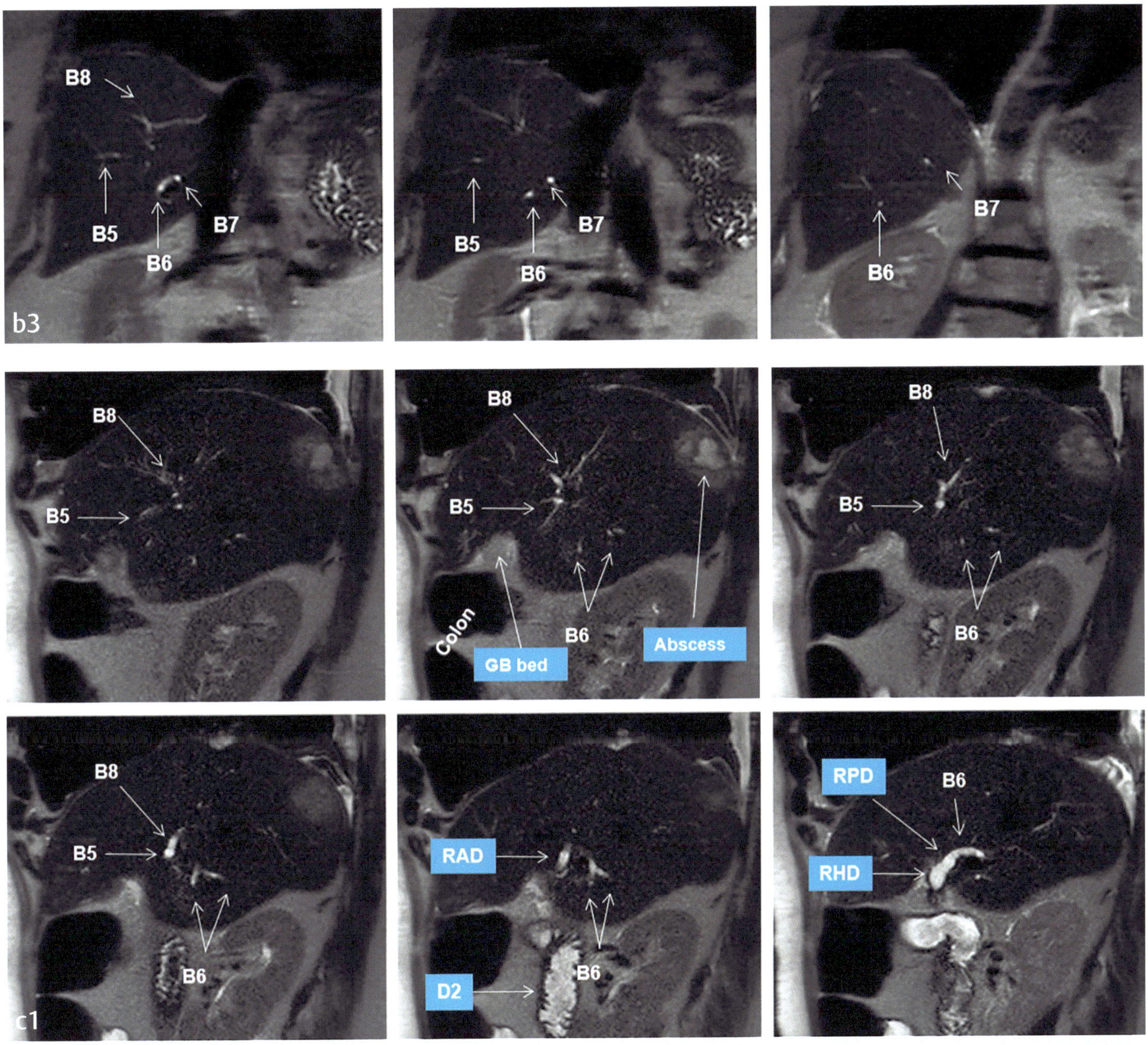

Fig. 8.18 (*Contined*)

(*Continued*)

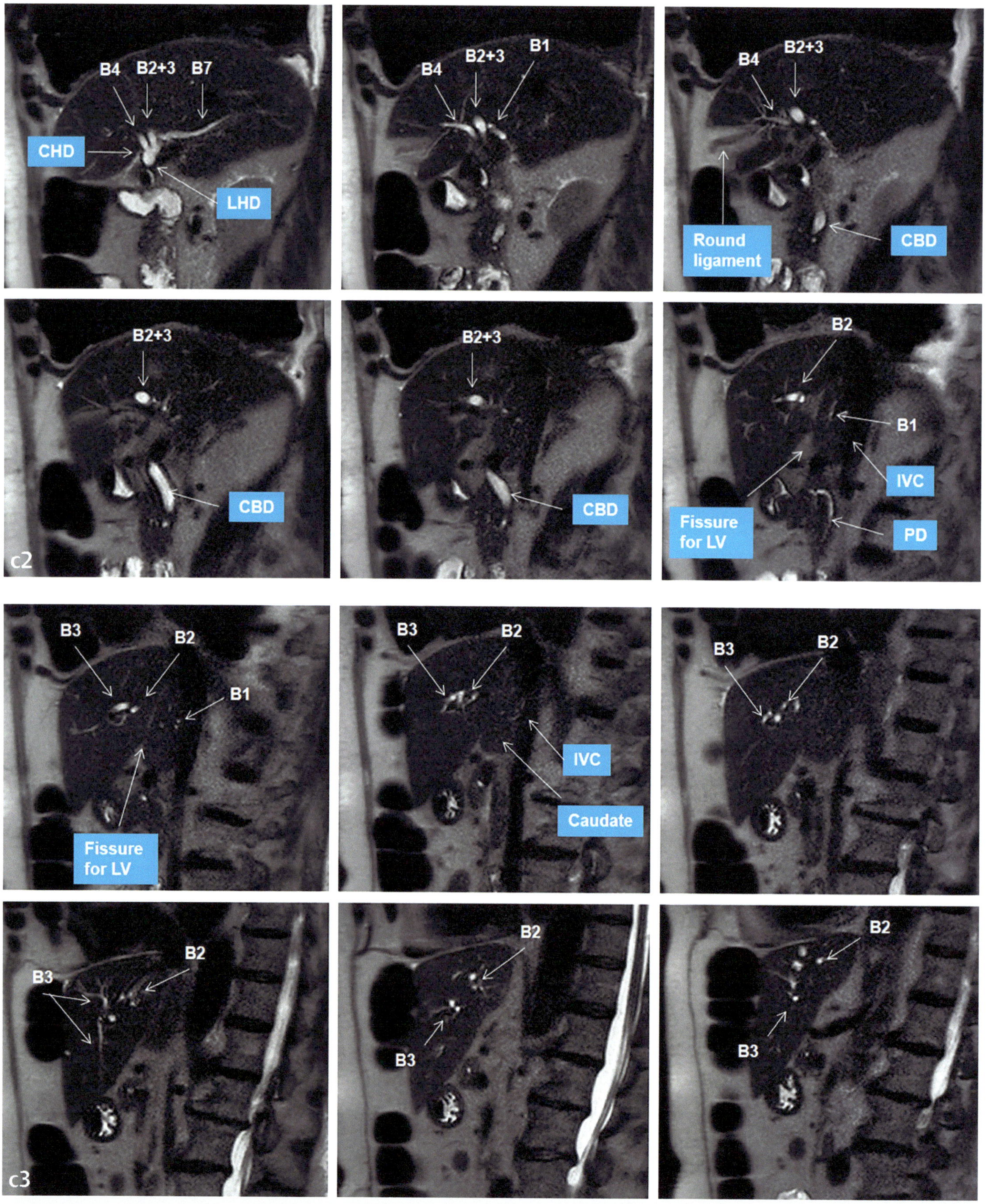

Fig. 8.18 *(Contined)*

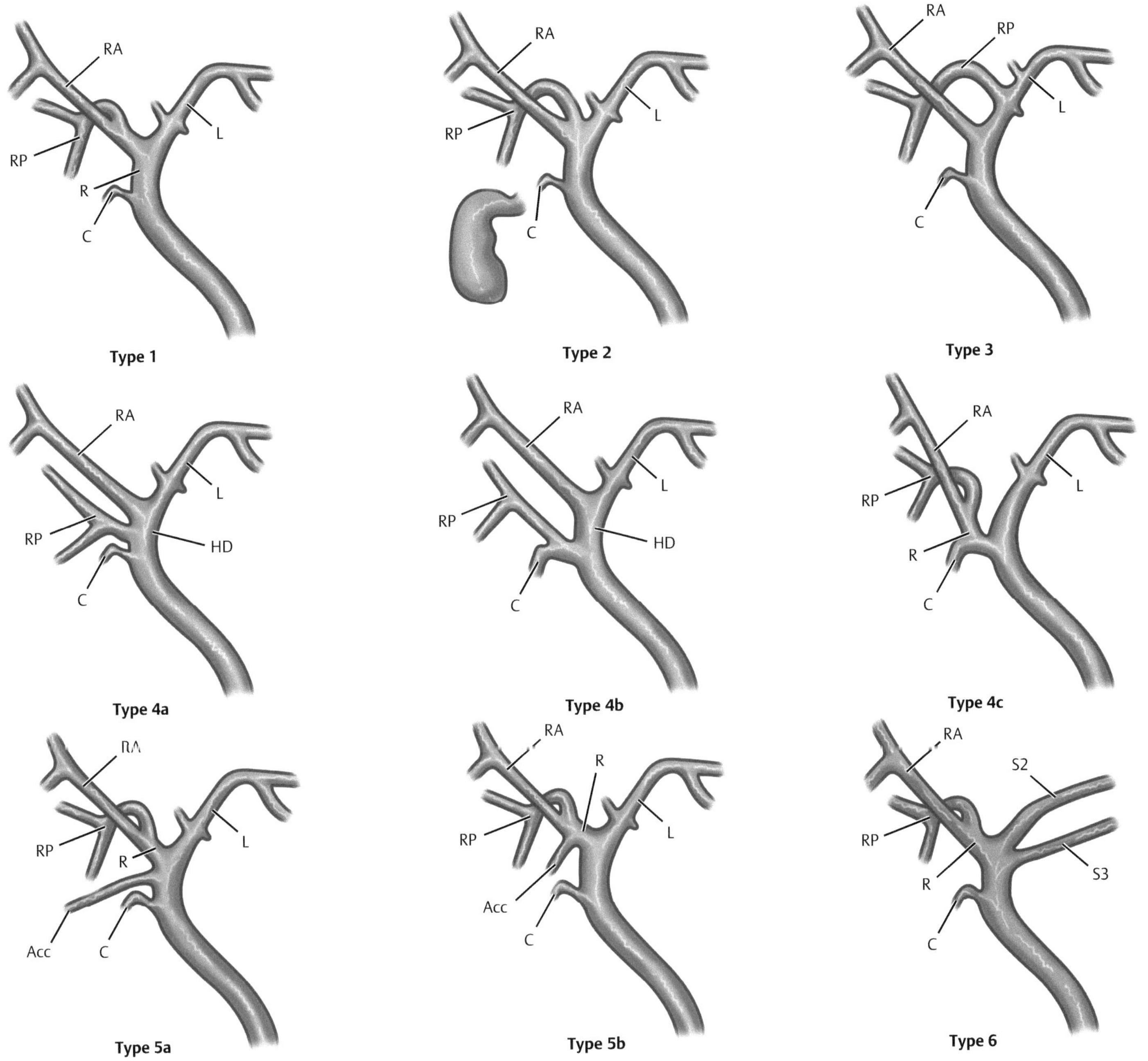

Fig. 8.19 Variation of the right biliary tree anatomy. The most common form is type 1 where the right posterior (RP) sectoral duct joins the right anterior (RA) sectoral duct to form a right hepatic (R) duct. This type is seen in 60 to 65% of cases. Type 2 is the second most common form (10–17%) showing a trifurcation pattern. In type 3, the RP drains in the left (L) duct. In type 4, anomalous insertion of RP duct occurs into hepatic duct (HD) or the cystic (C) duct. In type 5, an accessory (Acc) duct is present, and in type 6, quadrifurcation occurs. (Modified from Choi et al.[28])

8

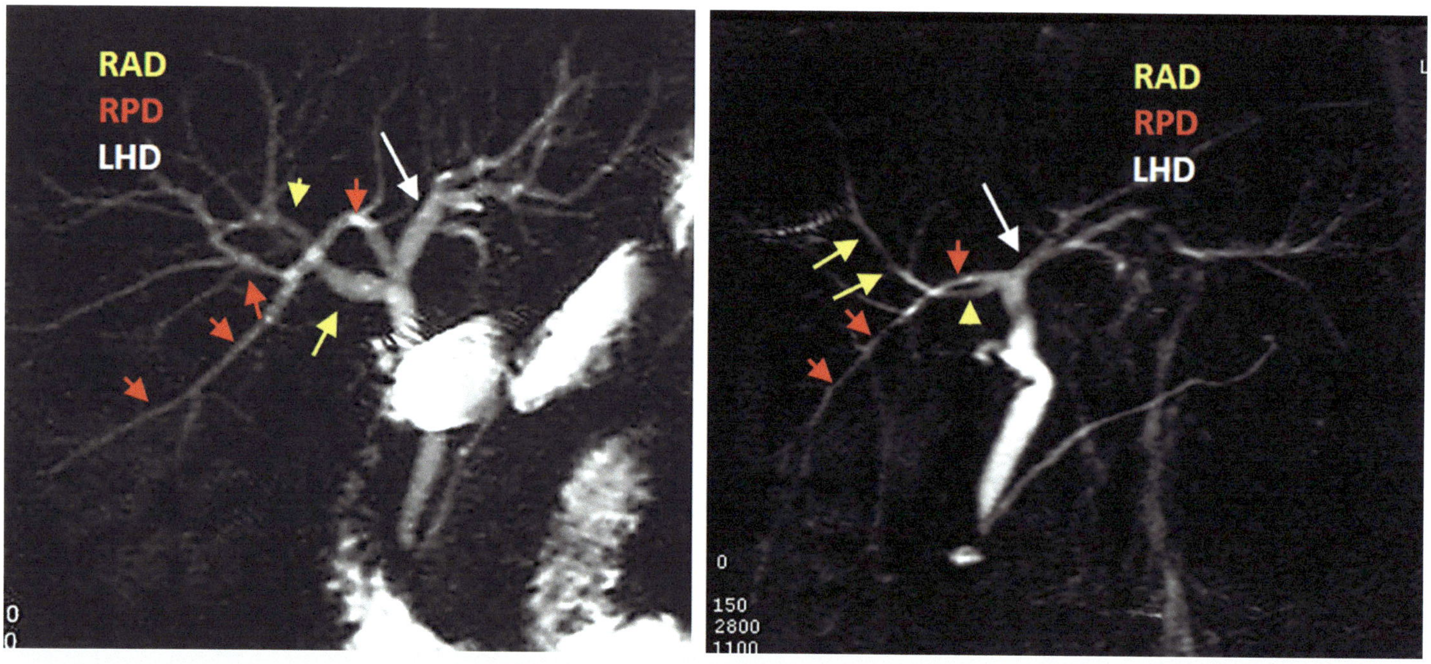

Fig. 8.20 Right bile variant; triple confluence. Triple confluence may be a concern during right hepatic transplantation (*purple area*). Due to the absence of the right hepatic duct, following liver resection, this variant may require hepaticojejunostomy to multiple ducts. LHD, left hepatic duct; RAD, right anterior duct; RPD, right posterior duct.

Left Hepatic Bile Duct Branches

The left hepatic duct runs superior and parallel to the left portal vein and joins the right hepatic duct anterior to the portal vein bifurcation. It is formed by confluence of the segments II, III, and IV near the umbilical fissure (▶Fig. 8.17, ▶Fig. 8.18). It shows less anatomical variation than the right (▶Fig. 8.23).

Segment III is usually larger with an anteroinferior course and appears more curvilinear than the segment II duct.[29,30] The most constant anatomical pattern is the confluence of segments II and III at (50%) or to the left (30%) of the ligamentum venosum[27] which makes it a good choice for biliary anastomosis in transplantation of the left liver lobe (▶Fig. 8.24).

Variation in the anatomy of the bile ducts draining the left medial sector (segment IV) is common.[19] The left medial sector is divided into superior (IVa) and inferior (IVb) segments.[25] The ducts of the left medial sector join to form a single duct in 60% of cases[22] (▶Fig. 8.23). This single duct then joins the common trunk of the segments II and III. However, in some cases a trifurcation is formed by the segments II, III, and IV. It is not uncommon (up to 19%) for the segment IV to drain into the segment III duct. The segment IV duct sometimes drains into the segment II duct, the CHD, the right anterior duct, or the right posterior duct.[18,19,31,32]

In 65% of individuals, the segment IV duct joins the left hepatic duct some distance away from the confluence of the main hepatic ducts. However, in 30% it inserts to the left hepatic duct near the liver hilum which predisposes the segment IV duct to being involved by hilar duct carcinoma requiring resection of that segment.[27,33]

Caudate Lobe Ducts

The caudate lobe is defined anatomically as the area supplied by the first branches of the left or right portal veins. The caudate lobe (segments I and IX in Couinaud's classification) is divided into a caudate lobe proper (segment I), which is located between the inferior vena cava and the umbilical fissure, and the caudate process (segment IX), which connects the caudate lobe to the right hepatic lobe[25,34] (▶Fig. 8.11, ▶Fig. 8.18).

Resection of the caudate lobe is technically challenging, since this segment is located deep within the liver and is close to hepatic hilar structures. Based on the portal blood supply three parts are recognized: the Spiegel's lobe (left) which is the main part of the lobe, the paracaval portion (right) which is the part of liver in front of the inferior vena cava, and the caudate process.[34–36] Based on the biliary drainage, the caudate lobe cannot be designated to a particular lobe. Because the caudate lobe is located adjacent to the porta hepatis, there are usually two or three biliary ducts which join the hepatic ducts.[26] Some believe that in caudate lobe anatomical dissociation can exists between the portal vein and biliary systems. For example, the right hepatic duct contributes to the drainage of the left caudate (Spiegel's lobe) in 40% of cases[37] and the left hepatic duct is likely to drain the paracaval portion. The caudate process duct usually drains into the right hepatic duct.

Aberrant and Accessory Bile Ducts

An aberrant bile duct is an anomalous duct and the only bile duct draining a segment of the liver (▶Fig. 8.21). Ligation, dissection, or ischemia of an aberrant duct can result in cholangitis, focal

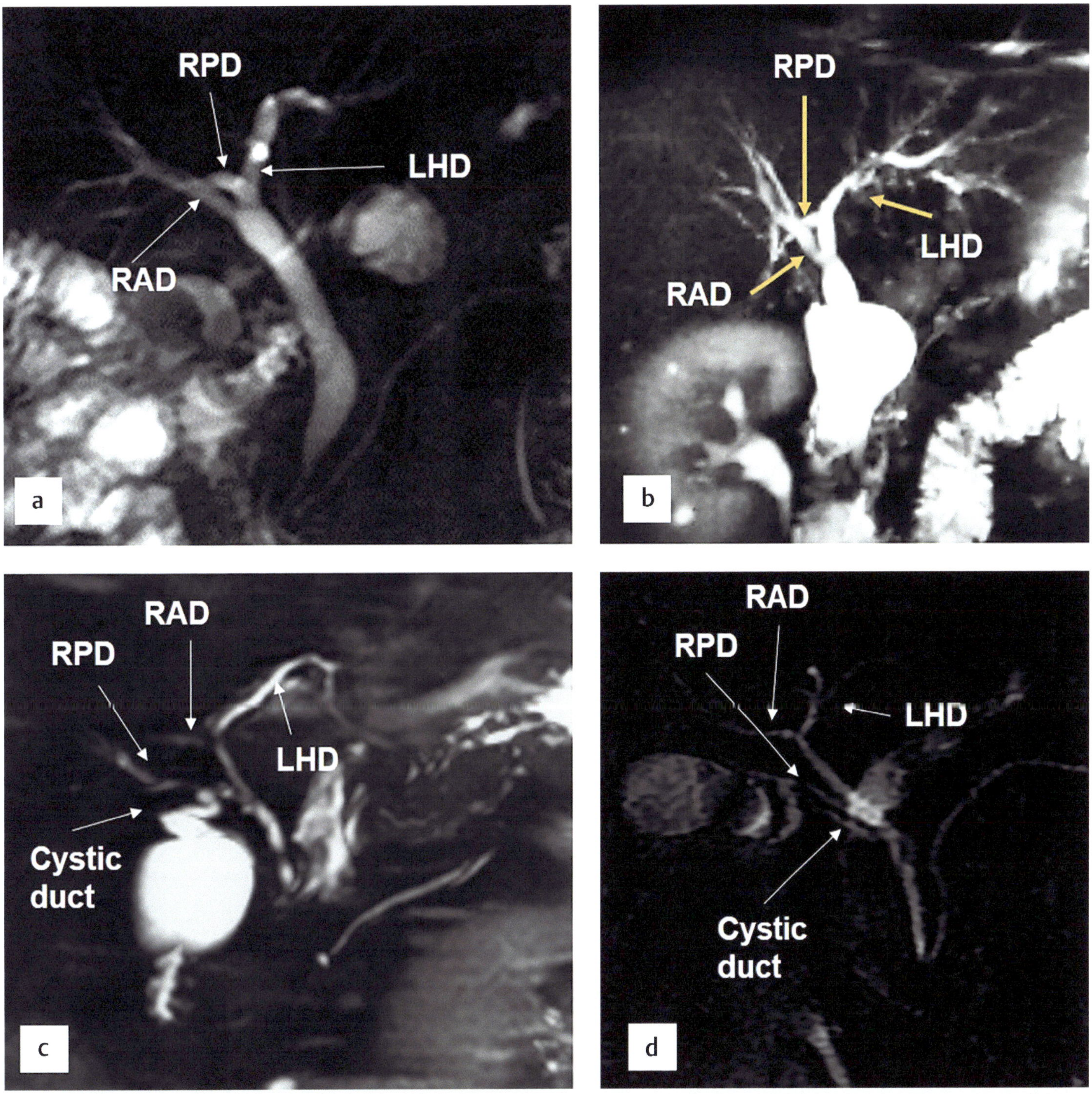

Fig. 8.21 (a, b) Aberrant insertion of the right posterior duct. The right posterior duct (RPD) drains directly into the left hepatic duct (LHD). Awareness of this variant is necessary to ensure complete biliary drainage following liver resections. **(c, d)** Infraportal aberrant course of the right posterior duct. **(c)** Right posterior duct insertion into the cystic duct may require open approach for cholecystectomy. Awareness of potential anomalies is important during cholecystectomy to prevent bile damage and bile leak. **(d)** Aberrant insertion of the RPD into the common duct. RAD, right anterior duct,

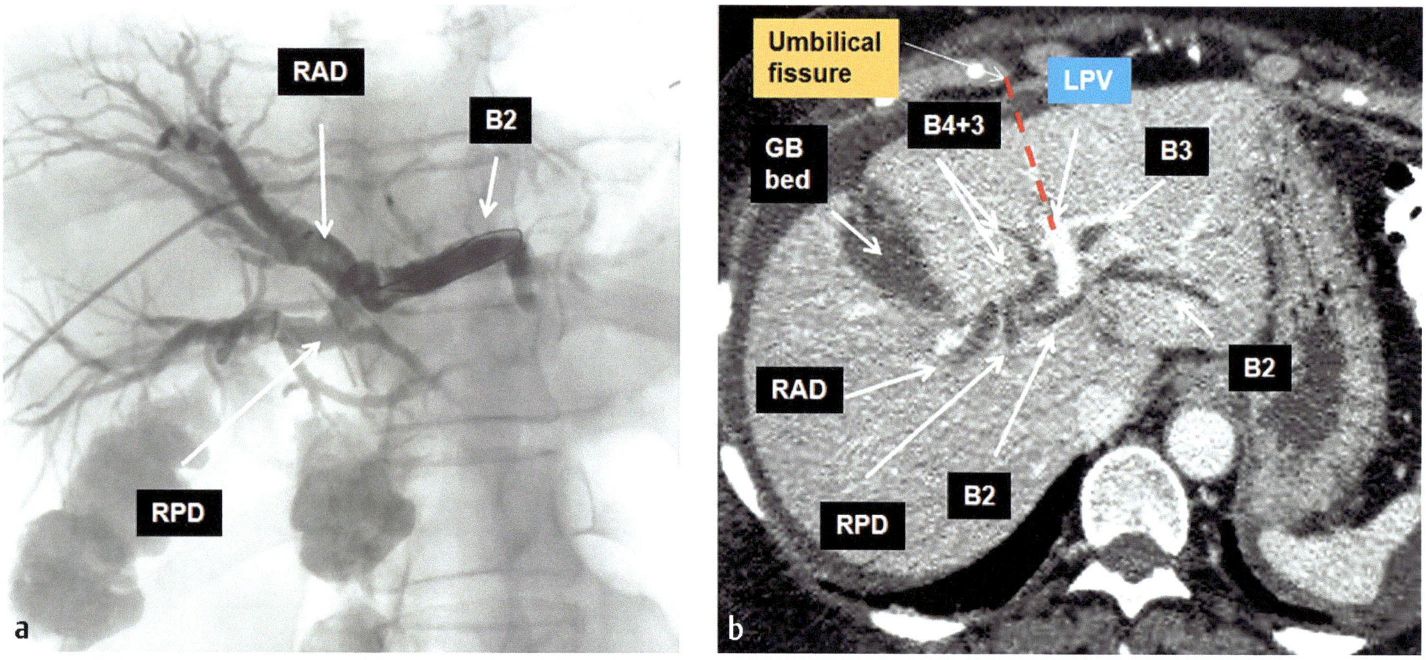

Fig. 8.22 Status post cholecystectomy. Inadvertent clipping of the common hepatic duct (CHD) at the time of surgery causing biliary obstruction. **(a)** Transhepatic cholangiogram. **(b)** Axial postcontrast computed tomography (CT) scan. Intrahepatic bile dilatation due to occlusion (clipping) of the CHD is seen. Anatomic variant of the confluence of the P2 and P3 bile ducts distal to the left portal vein (LPV) and umbilical fissure. Also evident is the unusual confluence of the P2 and P3 with right posterior duct and right anterior duct at one point (type 6), best shown on CT. This unusual configuration predisposes for a short CHD which increases risk of inadvertent clipping as happened in this case. Note that the right posterior duct (RPD) is longer with a horizontal direction as opposed to the right anterior duct (RAD), which has an oblique orientation. GB, gallbladder; LPV, left portal vein; RAD, right anterior duct; RPD, right posterior duct.

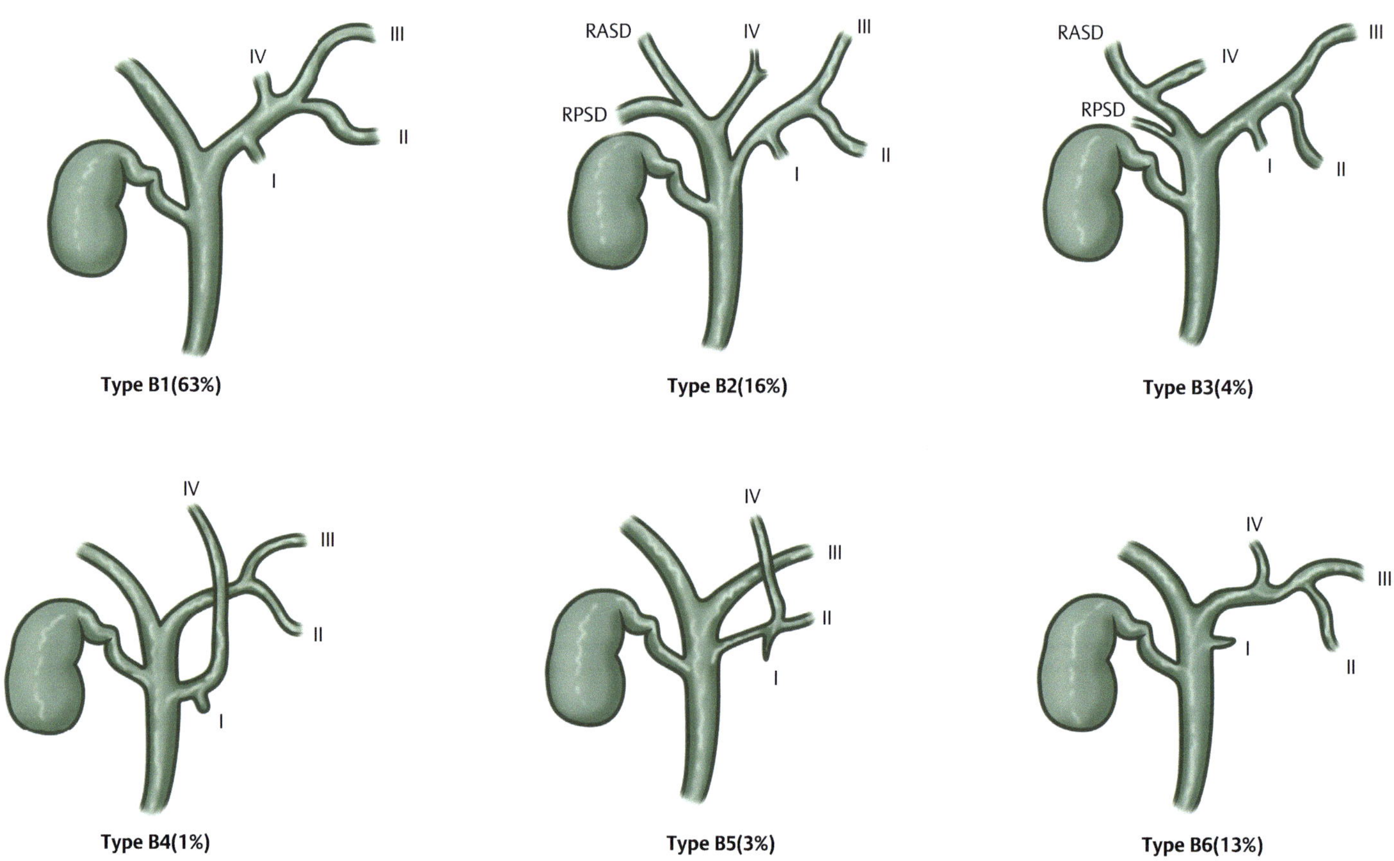

Fig. 8.23 Variation of the left biliary tree anatomy. Left hepatic duct is divided into six types according to segmental bile drainage and its respective frequencies. RASD, right anterior sectoral duct; RPSD, right posterior sectoral duct. (Modified with permission from Chaib et al.[19])

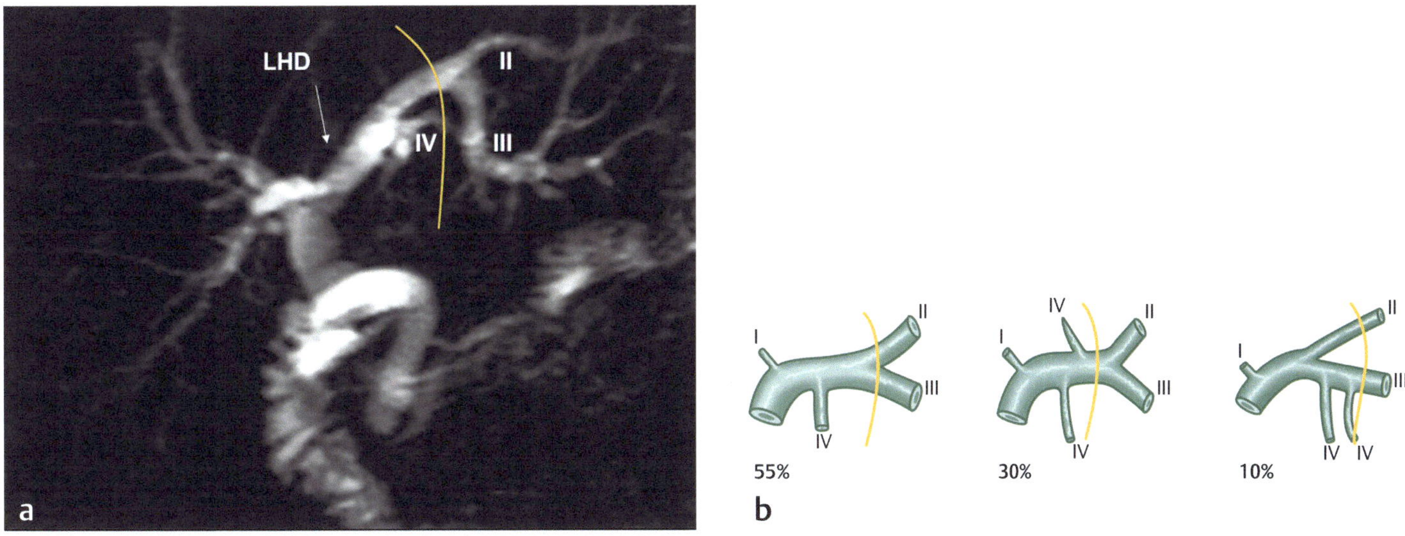

Fig. 8.24 (a, b) Patterns of drainage of the left hepatic duct (LHD) according to the confluence of segment II and segment III in relation to the location of the ligamentum venosum. The *orange line* represents approximate location of the ligamentum venosum. (Modified with permission from Reichert et al.[29])

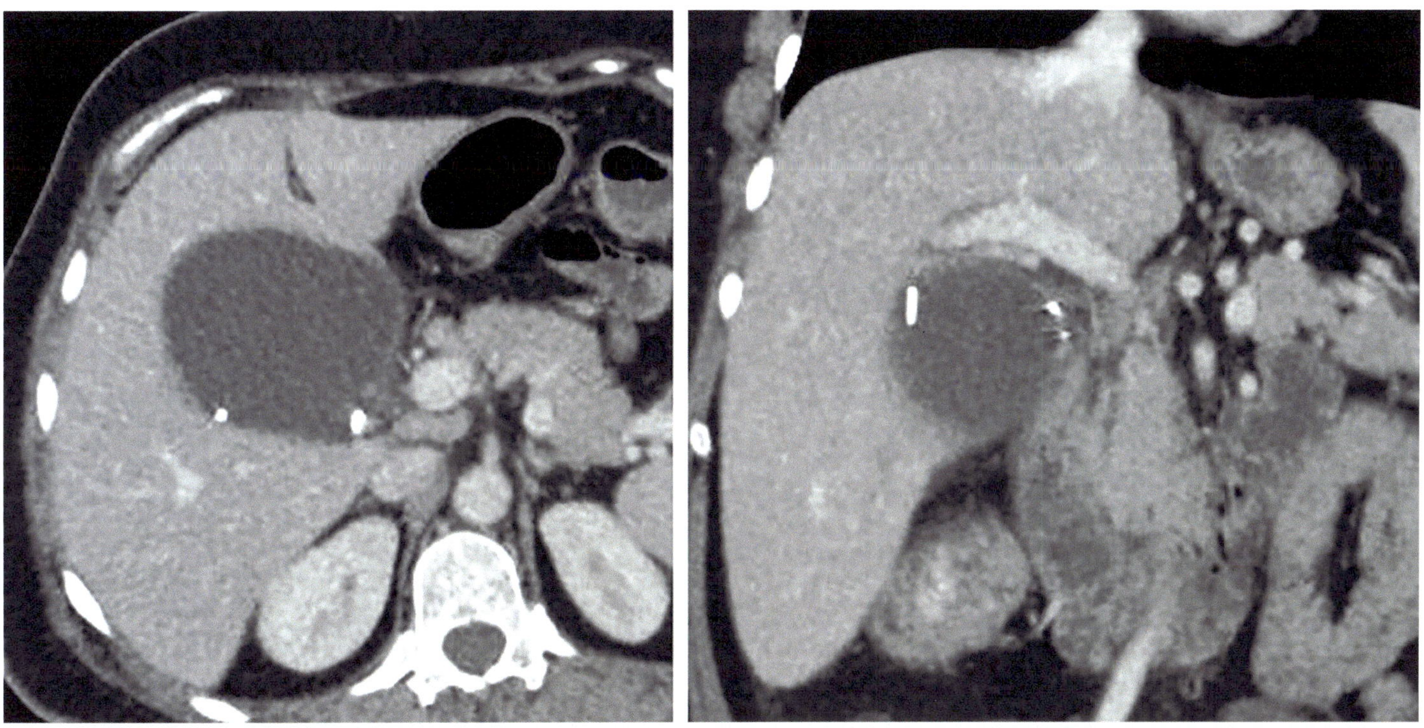

Fig. 8.25 Biloma after cholecystectomy shown by computed tomography (CT). Subvesical ducts may be the source of postcholecystectomy bile leaks and biloma formation.

fibrosis of the liver, or bile leak. On the other hand, an accessory bile duct is a supernumerary duct draining a segment of the liver. Ligation of an accessory bile duct usually does not cause much problem.[38,39]

The bile ducts of Luschka, or subvesical ducts, are small accessory or aberrant ducts which originate from the right hepatic lobe. These ducts exit the liver, course along the gallbladder fossa, and usually drain in the extrahepatic bile ducts in 6 to 30% of studies.[22,28] The mean diameter of the subvesical ducts is 2 mm (range 1–18 mm) and they are not accompanied by a portal vein branch[38]; therefore, it would be difficult to show them on cross-sectional imaging. The subvesical ducts may be the source of postcholecystectomy bile leaks and biloma formation[39] (▶Fig. 8.25).

◆ Pathologies

Congenital Malformations

Incomplete vacuolation causes gallbladder deformities such as a fundus septum (Phrygian cap), which is a common anatomical variant and abnormal development causing duplication or triplication (▸ Fig. 8.5). Malformations of the extrahepatic biliary ducts result in atresia and choledochal cysts.[40] Congenital diseases of the small- to medium-sized biliary tract are usually related to abnormal development of the ductal plates. These diseases are collectively termed "ductal plate malformations" (▸ Fig. 8.26). In ductal plate malformations, abnormal biliary ducts are often associated with anomalies of the portal mesenchyme and vessels. Ductal plate malformations involve a complex group of pathological abnormalities. Anomalies of the small-sized intrahepatic ducts result in biliary hamartomas and congenital hepatic fibrosis. Anomalies of the medium-sized intrahepatic duct results in biliary atresia, Caroli's disease, and recessive and dominant forms of polycystic kidney disease. Several polycystic diseases affecting the kidneys also display biliary anomalies that are referred to as "hepatorenal fibrocystic syndromes."[40–42]

Biliary Hamartoma

Biliary hamartomas, also known as Von Meyenburg complexes, are characterized by innumerable tiny cystic dilation of small intrahepatic bile ducts embedded within a dense fibrous stroma.[43,44] Cysts are less than 10 mm and distributed uniformly throughout the liver (▸ Fig. 8.26). Biliary hamartomas are usually discovered incidentally and are considered benign conditions. Like other ductal plate malformations, associations between Von Meyenburg complexes and cholangiocarcinoma have been described.

Caroli's Disease

Caroli's disease is characterized by saccular or tubular dilatation of the segmental ducts without any other associated anomaly (▸ Fig. 8.26). Caroli's syndrome which is a more common type of anomaly is characterized by congenital hepatic fibrosis, biliary dysgenesis, and intrahepatic biliary cysts.[41,46] Caroli's syndrome may be seen with autosomal recessive polycystic kidney disease (see section Peribiliary Glands). Imaging demonstrates intrahepatic, saccular or fusiform cystic areas of varying sizes measuring up to 5 cm, often containing calculi and sludge and communicating with the bile ducts.

Choledochal Cyst

Choledochal cysts are uncommon anomalies. They are characterized by dilatation of intrahepatic or extrahepatic biliary ducts or both[41,45,46] (▸ Fig. 8.26). Pathophysiology is either the result of ductal plate malformation or related to pancreatic enzymes reflux and secondary inflammation by an anomalous proximal junction of the CBD with the pancreatic duct. Classic presentation triad of abdominal pain, right upper quadrant mass, and jaundice is seen in 30% of patients. Lesions appear as fusiform or cystic dilations of the CBD/CHD and/or bile ducts communicating with the biliary tree (▸ Fig. 8.26). The most common type (80–90%) is fusiform dilation of the extrahepatic ducts, which may involve the cystic duct (type 1).

Common Pathologies

Abnormal gallbladder contraction results in stasis of bile which contributes to the formation of gallstones. Most gallstones are not calcified and will be best shown by ultrasound or MRI (▸ Fig. 8.27). Acute cholecystitis is the most common cause of acute pain in the right upper quadrant and occurs in approximately one-third of patients with gallstones.[47] Ultrasound is the modality of choice for the initial investigation of stones and gallbladder inflammation. CT or MR is reserved for complicated cases (▸ Fig. 8.28). Adenomyomatosis and cholesterol polyps are also common pathologies of the gallbladder wall (▸ Fig. 8.27). Adenomyomatosis is as a well-formed enhancing polypoid mass or circumferential wall thickening and intramural microcystic diverticula known the Rokitansky-Aschoff sinuses (▸ Fig. 8.27e). Gallbladder carcinoma is rare and invades the liver and surrounding structures. The main modes of hepatic spread from gallbladder cancer involve both direct invasion and portal tract invasion via intrahepatic lymphatic system. That is why surgery of gallbladder cancer involves both the gallbladder and tissues around it.

Prominent intrahepatic ducts can be found without any underlying pathology. However, most of the time dilated intrahepatic ducts indicate a pathologic obstruction. Biliary stones are seen as well-defined, sometimes movable, filling defects, without or with calcifications (▸ Fig. 8.29). Malignant strictures usually manifest as areas of narrowing or polypoid mass with proximal bile duct dilatation (▸ Fig. 8.29).

Cholangiocarcinomas are the most common malignant tumors of biliary ducts. Peripheral cholangiocarcinomas manifest as low-attenuation intrahepatic tumors with mild enhancement on contrast-enhanced CT or MRI.[41] Segmental dilatation and nonunion of the right and left hepatic ducts at the porta hepatis usually indicate a malignant etiology at hilar level (▸ Fig. 8.30). This hilar cholangiocarcinoma, also called Klatskin tumor, has a poor prognosis. Biliary stents are usually placed for palliation. Dilatation of both biliary or pancreatic ducts down to the level of the ampulla usually indicates an obstruction at the level of ampulla which can be caused by ampullary carcinomas, stone, papillitis, and sphincter of Oddi dysfunction.[47,48] Periampullary tumors are neoplasms that arise within 2 cm of the major papilla in the duodenum. They can originate from the pancreas, duodenum, distal CBD, or the ampullary complex. At the level of ampulla, a bulging papilla greater than 10 mm in diameter usually indicates pathologies at or near the ampulla of Vater[49] (▸ Fig. 8.14). CBD stone dilates the CBD or may result in a bulging papilla but it is unusual for it to dilate the pancreatic duct. Sometimes pancreatitis is a complication of acute CBD obstruction by a stone, a condition called gallstone pancreatitis (▸ Fig. 8.28).

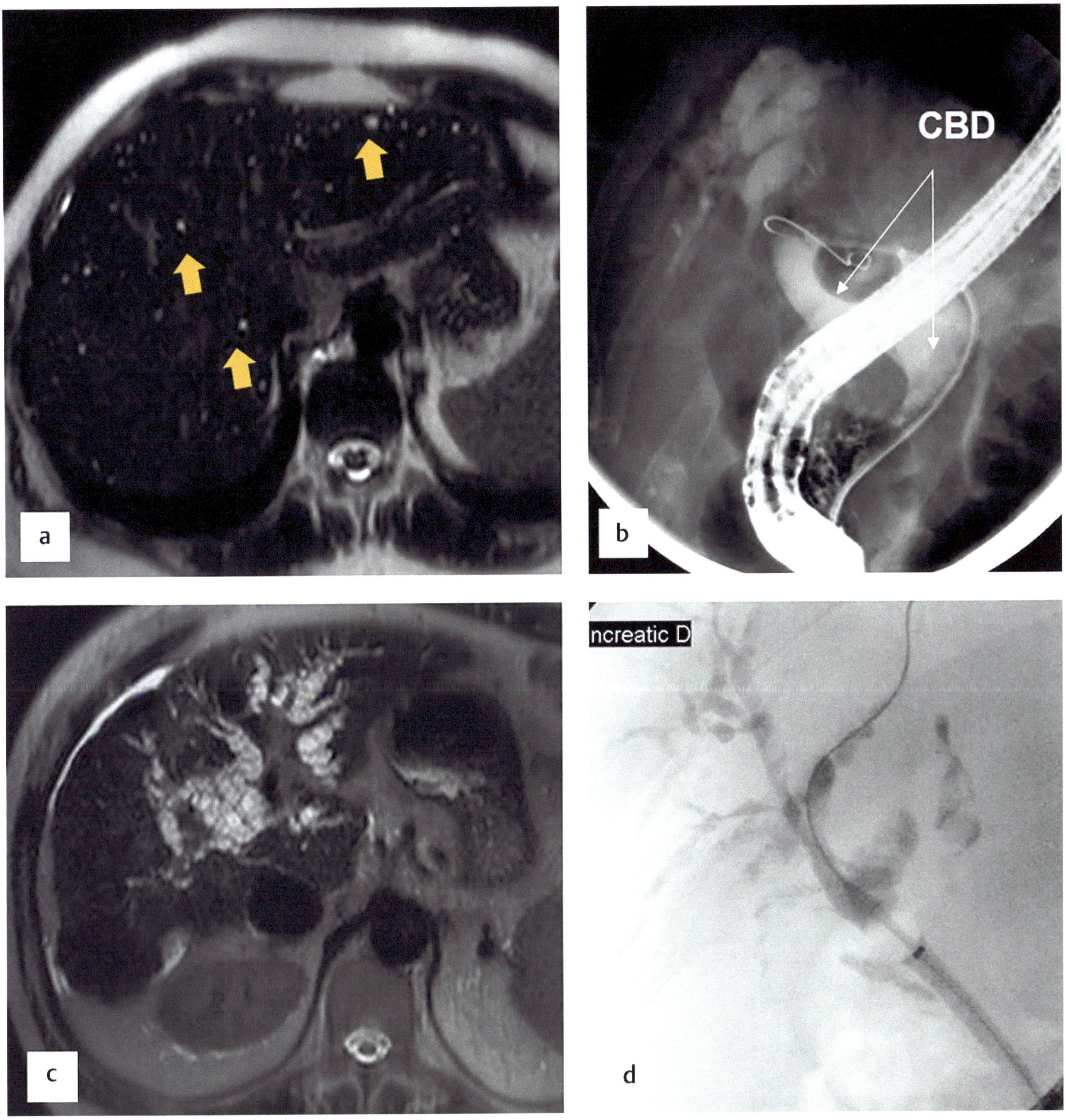

Fig. 8.26 Ductal plate malformations. **(a)** Biliary hamartomas. Axial T2-weighted magnetic resonance (MR) image shows multiple bright signal microcysts (*arrows*) in the liver. Intrahepatic bile ducts look normal. **(b)** Choledochal cyst. Endoscopic retrograde cholangiopancreatography (ERCP) shows cystic dilatation of the common bile duct (CBD) and some of the intrahepatic ducts. **(c, d)** Caroli's disease. **(c)** Axial T2-weighted MR image shows tiny parabiliary cysts around multiple intrahepatic bile ducts. **(d)** ERCP shows multiple saccular dilatations of the intrahepatic bile ducts consistent with Caroli's disease.

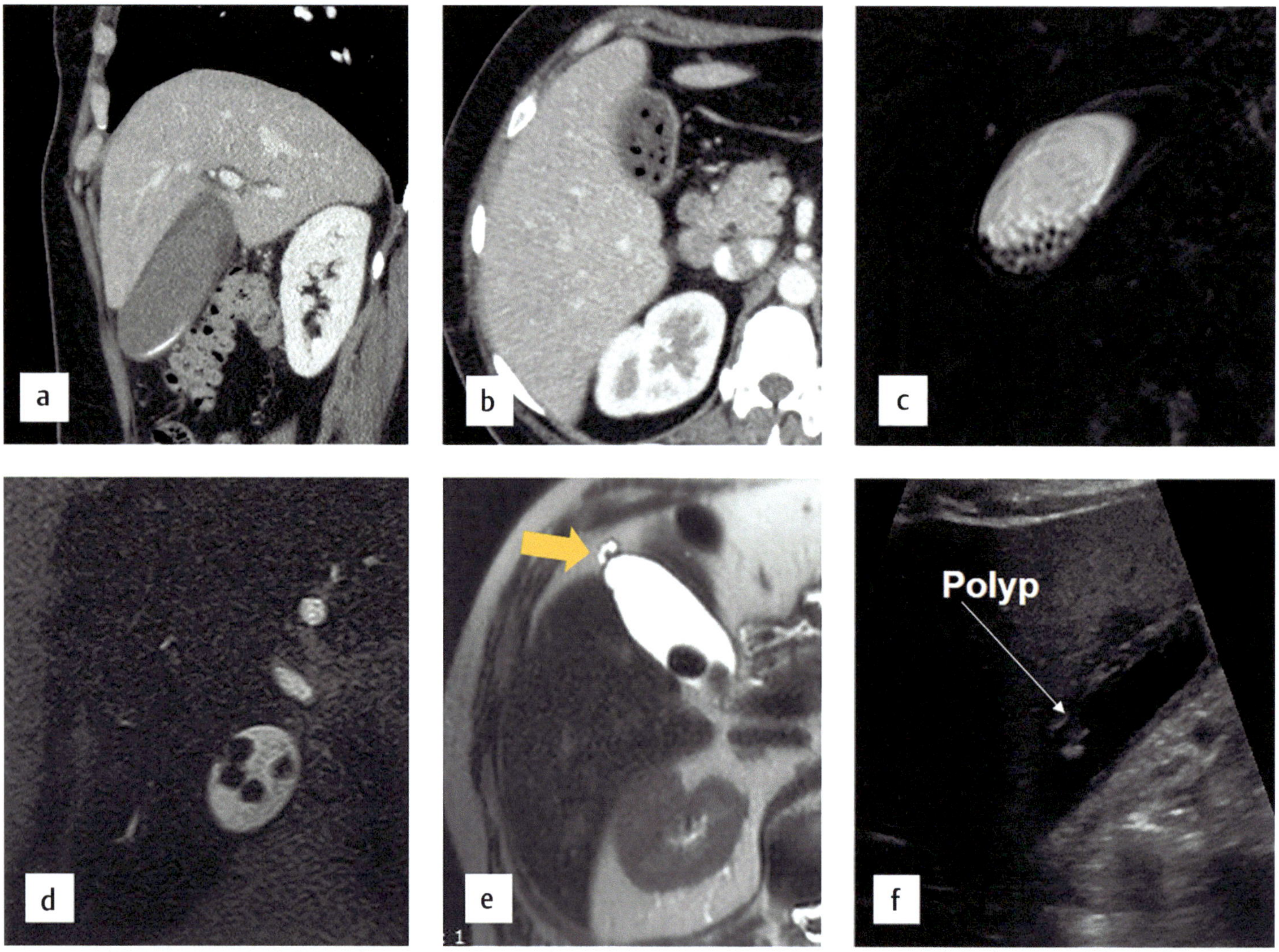

Fig. 8.27 Different morphologies of gallstones in different modalities. **(a)** Computed tomography (CT) shows small stones layering in the dependent aspect of the gallbladder fundus. **(b)** CT shows noncalcified stones containing gas. **(c)** Magnetic resonance cholangiopancreatography (MRCP) shows gallbladder filled with tiny stones. **(d)** MRI showing multiple dark signal stones. **(e)** T2-weighted MR showing a single gallstone. Also note a small adenomyomatosis (*arrow*). **(f)** Nondependent polyps are shown by ultrasound.

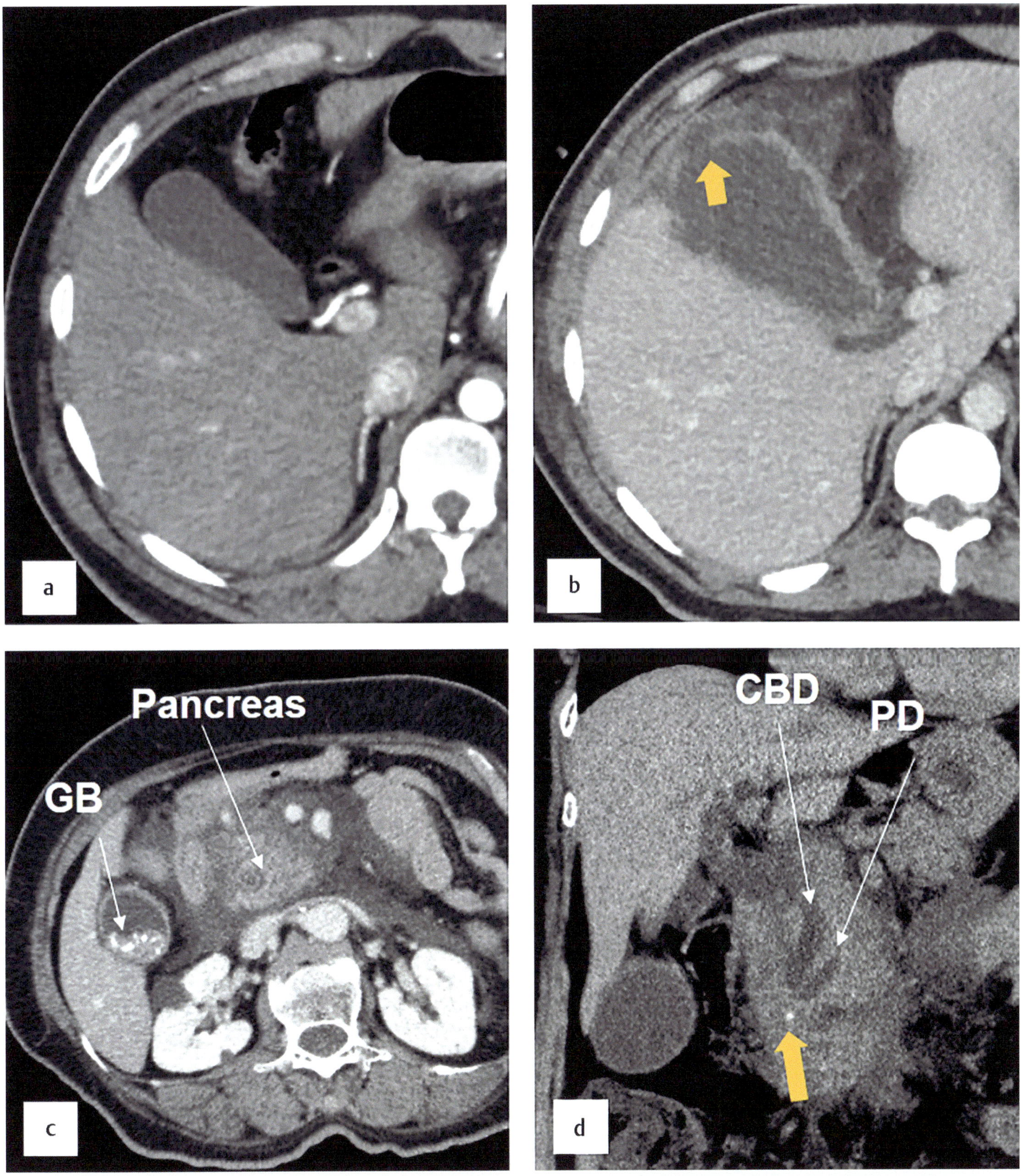

Fig. 8.28 (a, b) Axial computed tomography (CT) images showing a normal gallbladder (GB) **(a)** and following development of acute cholecystitis a few weeks later **(b)** causing gallbladder wall thickening, perforation (*arrow*), and pericholecystic fluid. **(c)** Axial and **(d)** coronal CT showing gallstone pancreatitis due to a small stone lodged in the ampulla (*arrow*). Gallstones, peripancreatic inflammation, and dilation of the common bile duct (CBD) and pancreatic duct (PD) are shown.

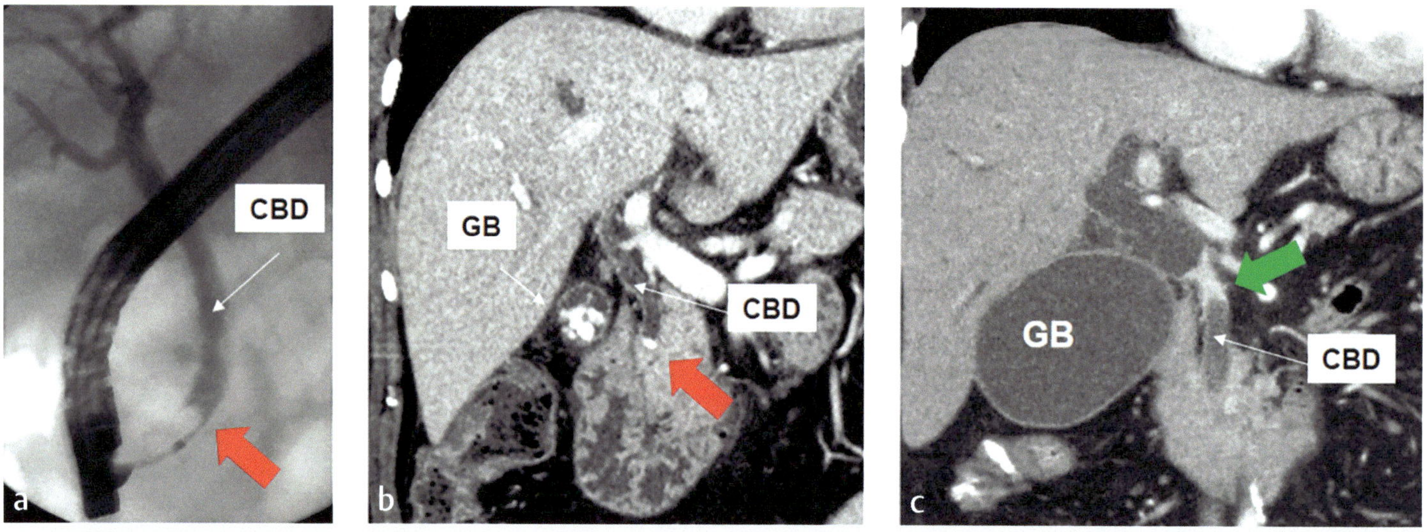

Fig. 8.29 (a) Endoscopic retrograde cholangiopancreatography (ERCP) and **(b)** coronal computed tomography (CT) showing an obstructing stone (*arrows*) in the distal common bile duct (CBD). **(c)** Bile duct cholangiocarcinoma is shown by coronal CT as a stenosing enhancing lesion involving the mid CBD (*arrow*). GB, gallbladder.

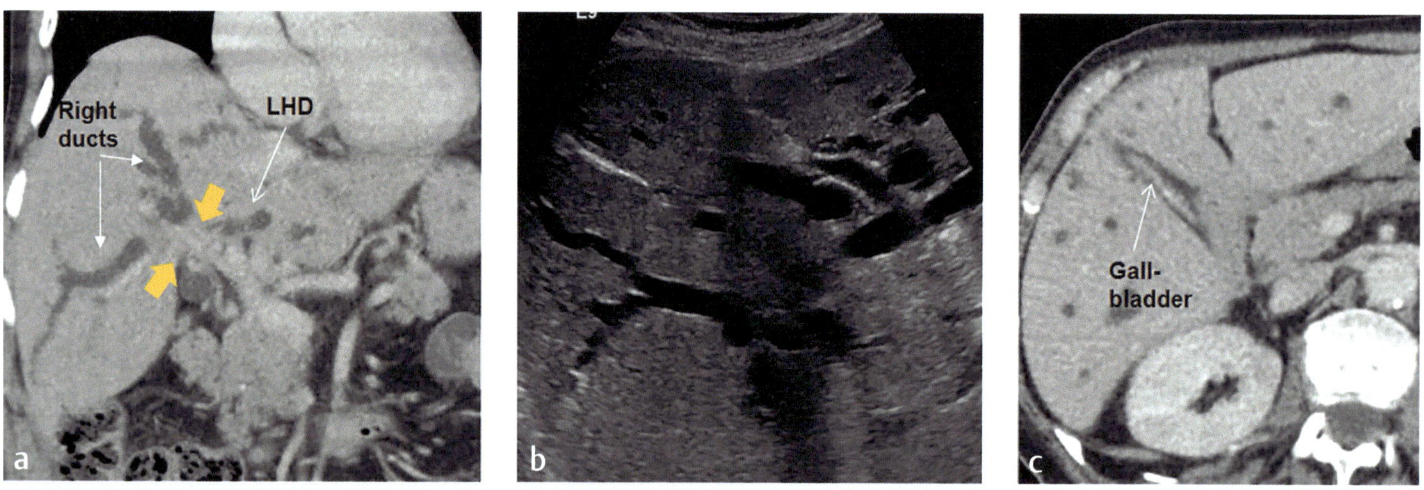

Fig. 8.30 Klatskin tumor. Segmental dilatation and nonunion of the right and left hepatic ducts at the porta hepatis usually indicate a malignant etiology at hilar level. **(a)** Coronal computed tomography (CT) image showing an infiltrating hilar cholangiocarcinoma causing obstruction of all ductal branches. **(b)** Dilated ducts are shown by ultrasound. **(c)** Axial CT at the level of the gallbladder shows decompressed gallbladder due to obstruction of normal bile flow into the cystic duct. LHD, left hepatic duct.

◆ Postoperative Imaging

Cholecystectomy is the most commonly performed operation in surgical practice.[50] Laparoscopic approach is the preferred technique for cholecystectomy because of its less painful postoperative period and a more rapid return to normal activities. Surgical clips placed at the cholecystectomy site are frequent findings on imaging studies (▶Fig. 8.25). The diameter of the CBD may increase after cholecystectomy, but it should not exceed 13 mm. After cholecystectomy, a variable length of the cystic duct is left in place. The cystic duct remnant is usually 1 to 2 cm long, but remnants of up to 6 cm in length may be seen (▶Fig. 8.7, ▶Fig. 8.31).

Major biliary complications of laparoscopic cholecystectomy may be prevented by an understanding of extrahepatic biliary ductal and arterial anatomic relationships.[50–52] A common cause of biliary tract injury is failure to recognize the anatomic variant of the triangle of Calot (▶Fig. 8.32, ▶Fig. 8.33).

Bile leak occurs in up to 2% of cases after cholecystectomy and could be a significant cause of increased morbidity for patients undergoing this procedure. The more common location for bile leaks after cholecystectomy is the cystic duct stump (▶Fig. 8.34). Bile leakage from the cystic duct remnant can be embolized percutaneously (▶Fig. 8.34).

Injury of subvesical ducts is the second most common cause of postcholecystectomy bile leaks. Leakage from these small ducts is usually self-limiting. Bile duct leaks can be detected by drip-infusion CT cholangiography, nuclear hepatobiliary (HIDA) scan, and ERCP preoperatively (▶Fig. 8.34). Routine intraoperative

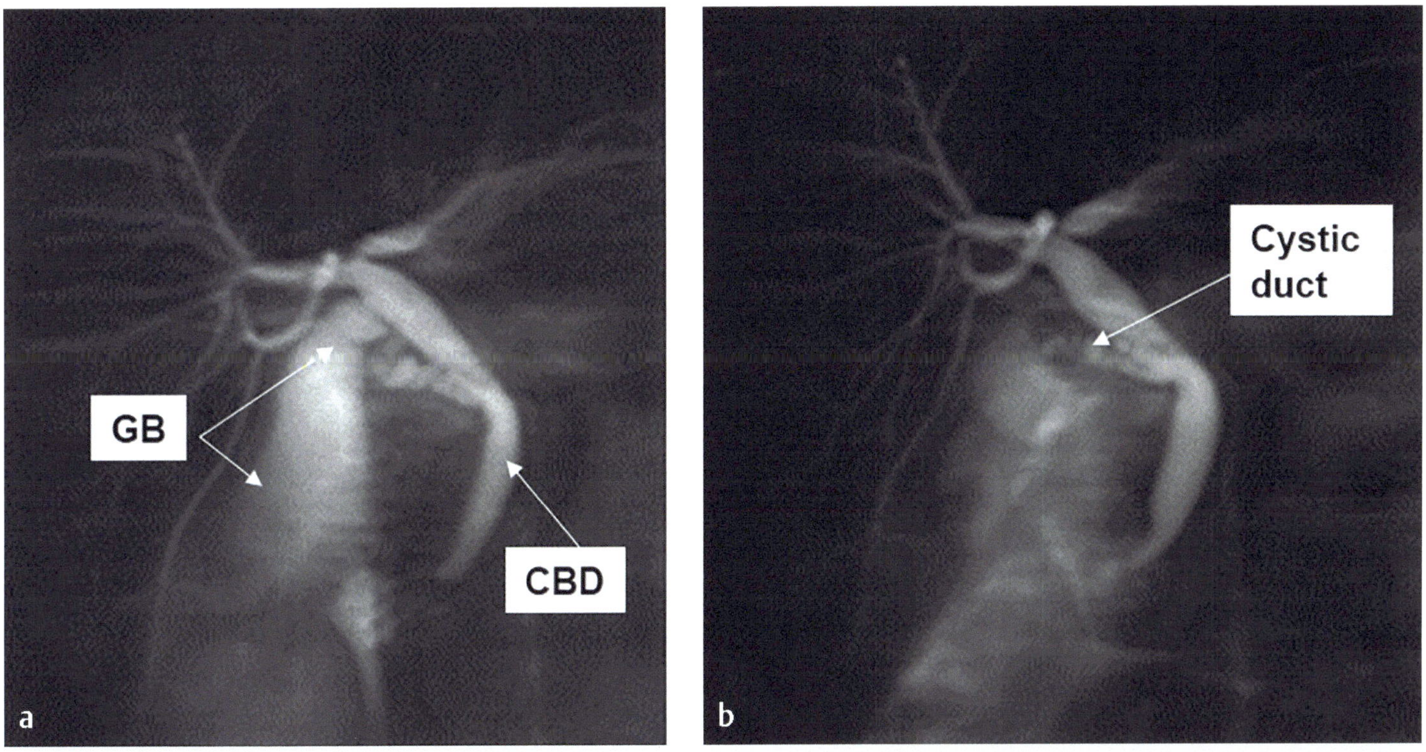

Fig. 8.31 Magnetic resonance cholangiopancreatography (MRCP) images **(a)** before and **(b)** after cholecystectomy showing mild dilatation of the common bile duct (CBD) in a few months. The diameter of the CBD may increase after cholecystectomy, but it should not exceed 13 mm. GB, gallbladder.

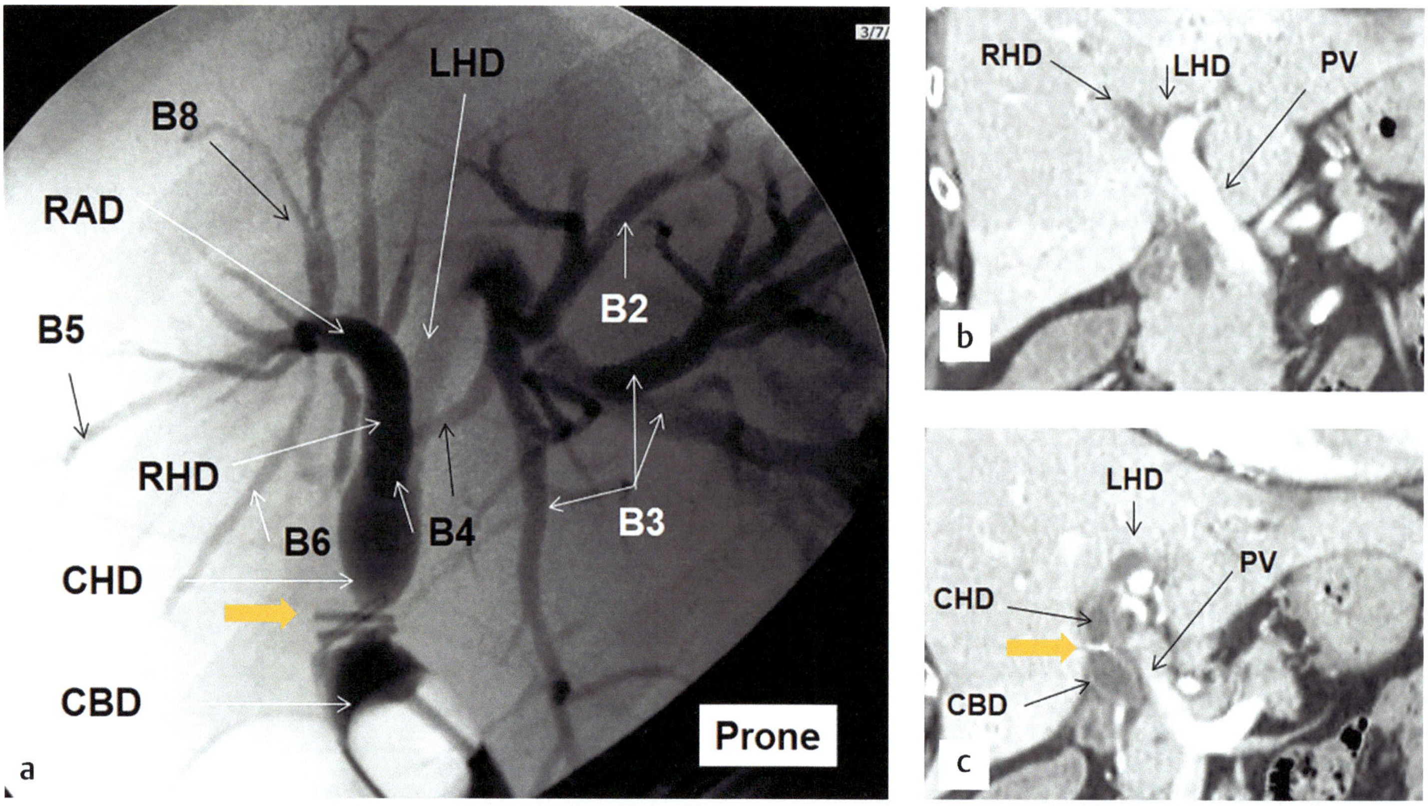

Fig. 8.32 Inadvertent clipping of the common hepatic duct (CHD) at the time of surgery causing biliary obstruction. **(a)** Endoscopic retrograde cholangiography (ERCP) shows severe stenosis of the CHD (*arrow*) at its junction with the common bile duct (CBD) due to inadvertent clipping during cholecystectomy. Status post cholecystectomy. **(b, c)** Coronal postcontrast computed tomography (CT) scan showing intrahepatic bile dilatation due to clipping (*arrow*) of the CHD. LHD, left hepatic duct; PV, portal vein; RAD, right anterior duct; RHD, right hepatic duct.

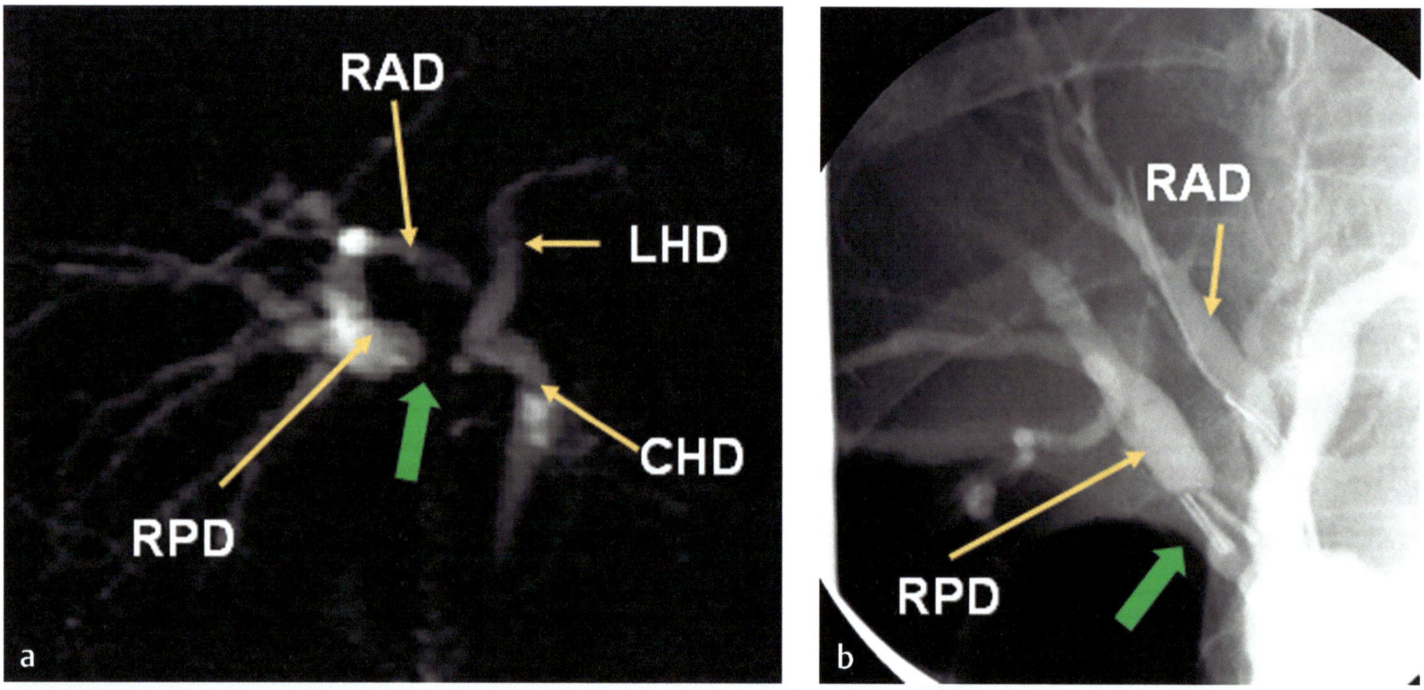

Fig. 8.33 Inadvertent clipping (*green arrows*) of a subportal right posterior duct (RPD) during cholecystectomy shown by magnetic resonance cholangiopancreatography (MRCP) **(a)** and endoscopic retrograde cholangiopancreatography (ERCP) **(b)**. There is aberrant low insertion of the RPD into the common hepatic duct (CHD). Stenosis is present proximal to the confluence with the CHD. Stent could not be placed across the stricture. LHD, left hepatic duct; RAD, right anterior duct.

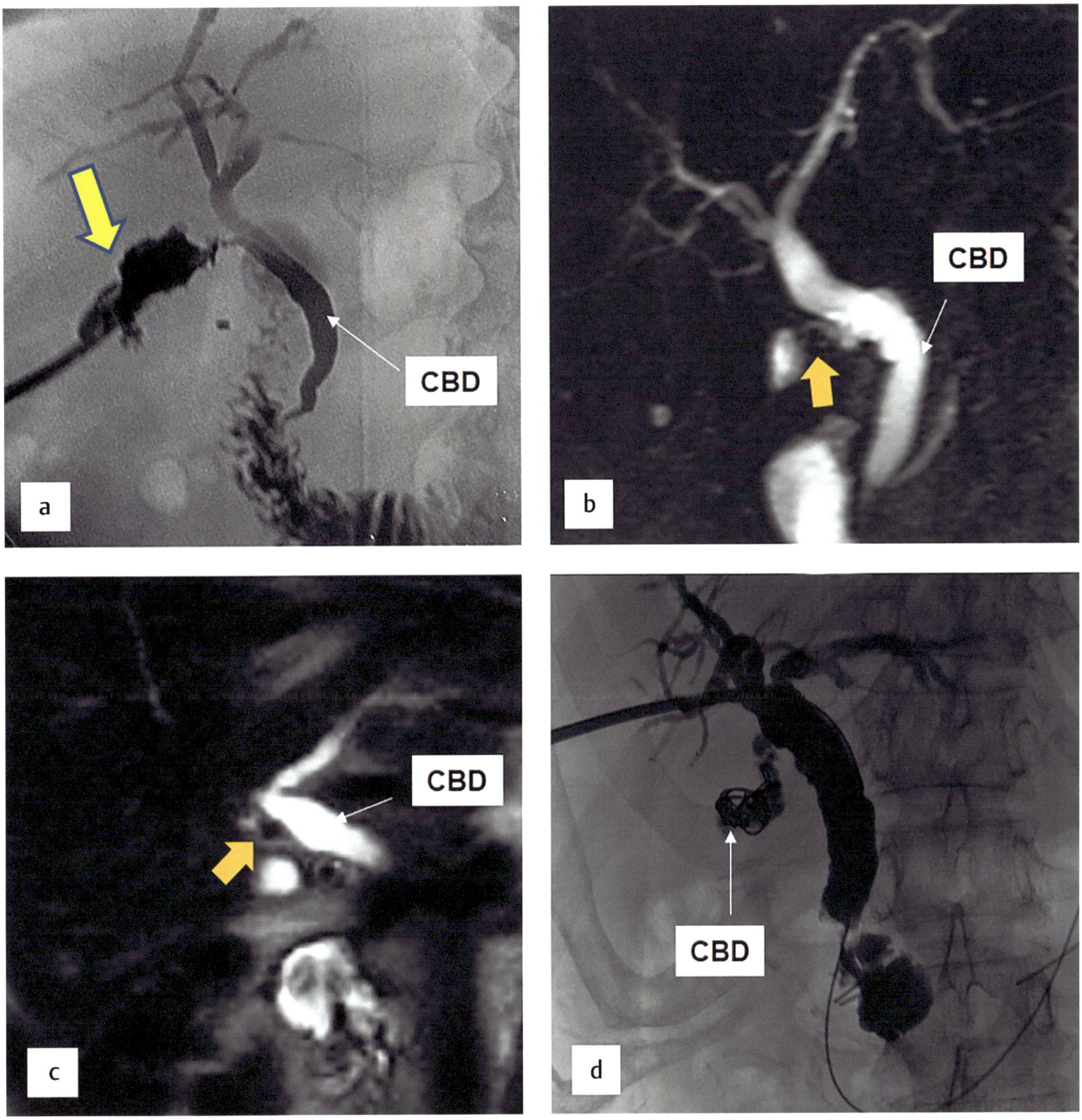

Fig. 8.34 Postcholecystectomy leak treated with cystic duct coil embolization. **(a)** T-tube cholangiogram shows leakage of contrast around cystic duct remnant (*arrow*). **(b)** Magnetic resonance cholangiopancreatography (MRCP) and **(c)** coronal T2-weighted magnetic resonance (MR) also show retained calculi in the cystic duct remnant (*arrows*). **(d)** Transcatheter cholangiogram after coil embolization of the cystic duct is shown. The contrast extravasation has stopped.

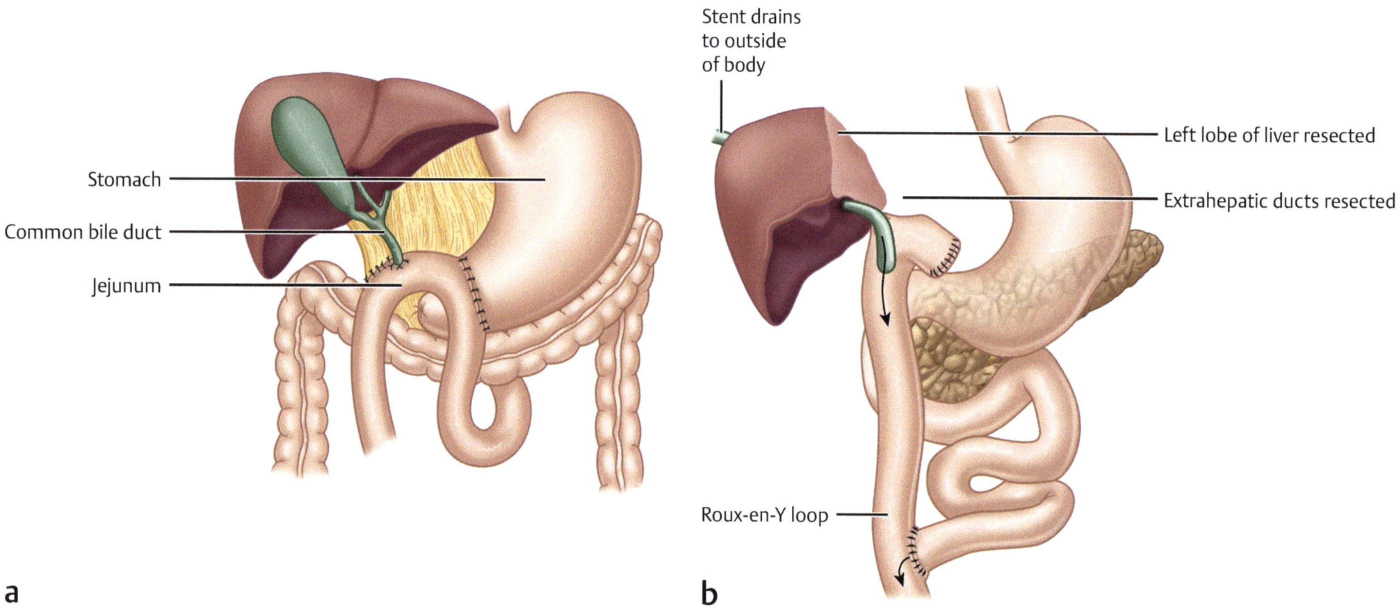

Fig. 8.35 Biliary-enteric anastomosis. In biliary-enteric anastomosis an anastomosis is made between a normal bile duct segment and a segment of the gastrointestinal tract, commonly a Roux-en-Y loop of the jejunum. **(a)** Choledochojejunostomy with gastrojejunostomy following resection of the duodenum. **(b)** Hepaticojejunostomy. It is performed after left hepatectomy, biliary stricture at porta hepatis, or following right liver transplant.

cholangiography is associated with a lower incidence and early recognition of bile duct injury.[50,53]

Common biliary duct surgeries include bile duct anastomosis after liver transplantation, biliary-enteric anastomosis (hepaticojejunostomy), ligation of hepatic ducts during partial hepatectomy or segmentectomy, and choledochoduodenostomy. In biliary-enteric anastomosis an anastomosis is made between a normal bile duct segment and a segment of the gastrointestinal tract, commonly a Roux-en-Y loop of the jejunum (▶Fig. 8.35). Common surgical procedures with a biliary-enteric anastomosis include the Whipple procedure (choledochojejunostomy), hepatic transplantation, and surgical repair of a bile and cancer of bile ducts or gallbladder.[53]

A choledochocholedochostomy and hepaticocholedochostomy are performed during hepatic transplantation or surgical treatment of biliary tract injury (▶Fig. 8.36). A T-tube or straight biliary tube is placed in the CBD or in the donor cystic duct remnant for a few weeks after transplantation.

On posttransplant imaging, the CBD of donor is usually thin and smooth but the CBD of recipient is slightly enlarged. Slight narrowing or kinking at the anastomosis is common and should not be mistaken with stenosis. Since the gallbladders of both the donor and recipient are removed prior to transplantation, two cystic duct remnants may be observed after surgery.

Presence of air in the biliary tract is a common finding after biliary-enteric anastomotic surgery, papillotomy, or less likely sphincter of Oddi dysfunction. Other common imaging findings in postoperative cases are T-tube, surgical drains, and biliary stents (▶Fig. 8.34, ▶Fig. 8.36).

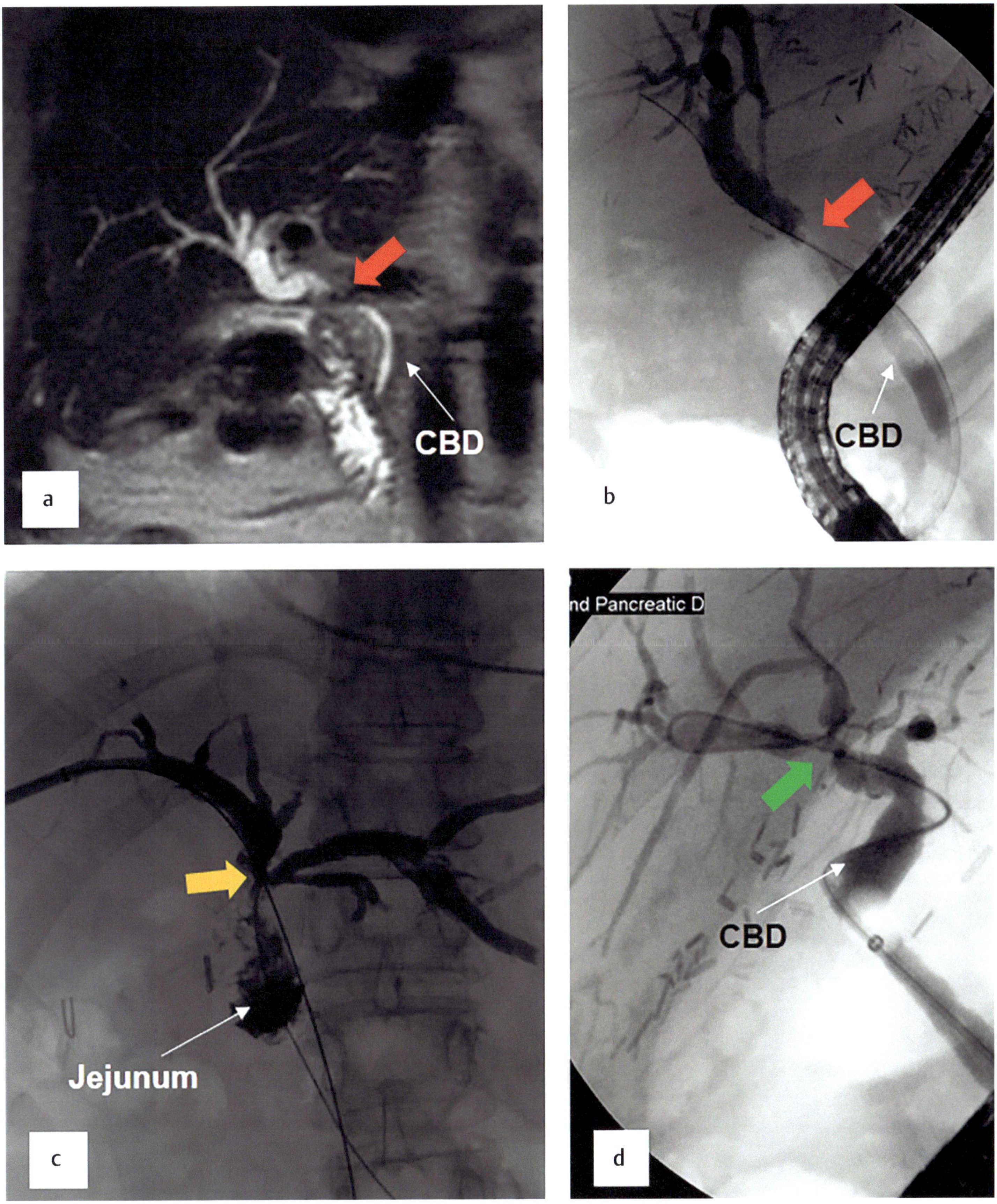

Fig. 8.36 Biliary stenosis at the anastomosis after liver transplant. **(a, b)** Coronal T2-weighted magnetic resonance imaging (MRI) and endoscopic retrograde cholangiopancreatography (ERCP) images showing stricture at choledochocholedochostomy (*arrows*). **(c)** Percutaneous transhepatic cholangiogram showing stricture (*arrow*) at hepaticojejunostomy anastomosis. **(d)** Stricture (*arrow*) at hepaticocholedochostomy anastomosis following right liver transplant. CBD, common bile duct.

◆ Peribiliary Glands

The extrahepatic and large intrahepatic bile ducts are surrounded by peribiliary glands[54–56] (▶Fig. 8.37). The peribiliary glands are classified into intramural type which are located within the bile duct wall and extramural type located within the periductal connective tissue. The extramural peribiliary glands are branched seromucous glands which appear larger than the intramural ones and may be seen with cross-sectional imaging when pathologically enlarged and become cystic (▶Fig. 8.38). The functions of the peribiliary glands are to lubricate the bile by secreting mucus and protect the mucosal surface by secreting immunoglobulins.[54]

These glands are very similar to pancreatic periductal glands and share similar pathologies.

For example, patients with chronic alcoholism can develop peribiliary cysts and liver cirrhosis as well as chronic pancreatitis and pancreatic fibrosis. Formation of peribiliary and intrapancreatic cysts are probably secondary to alcoholic injuries resulting in adenitis and cyst formation.[56] Immunoglobulin G4 (IgG4)–related sclerosing cholangitis is the biliary disease that can be associated with type 1 autoimmune pancreatitis. Also, intraductal papillary neoplasm of the bile duct and pancreatic intraductal papillary mucinous neoplasm (IPMN) share some clinicopathological characteristics.[57] In radiology, cystic dilation of the peribiliary glands is usually seen as a cluster of microcysts in the liver near the hilum in patients with cirrhosis but has also been described in normal livers[58] (▶Fig. 8.39).

8

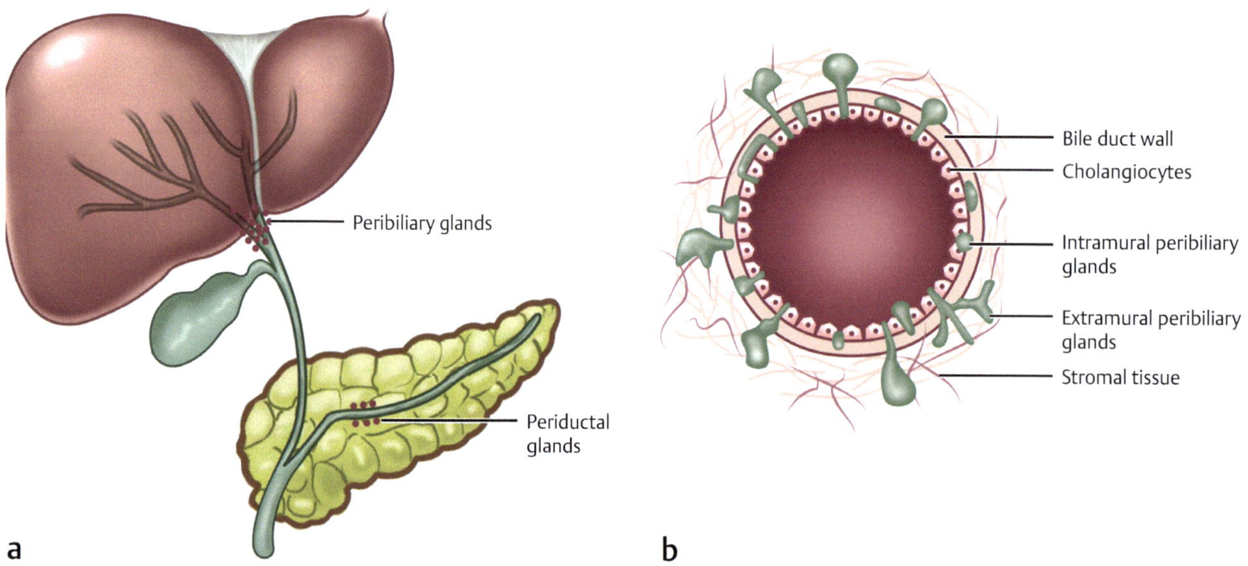

Fig. 8.37 (a) Common location of peribiliary glands. The peribiliary glands usually surround the extrahepatic and large intrahepatic bile ducts at porta hepatis. Also shown are the pancreatic periductal glands. **(b)** Cross section of an extrahepatic bile duct containing both intra- and extramural peribiliary glands. Cystic dilation of the peribiliary glands is usually seen as a cluster of small cysts in the liver near the hilum in cirrhotic patients. It is also seen in normal livers.

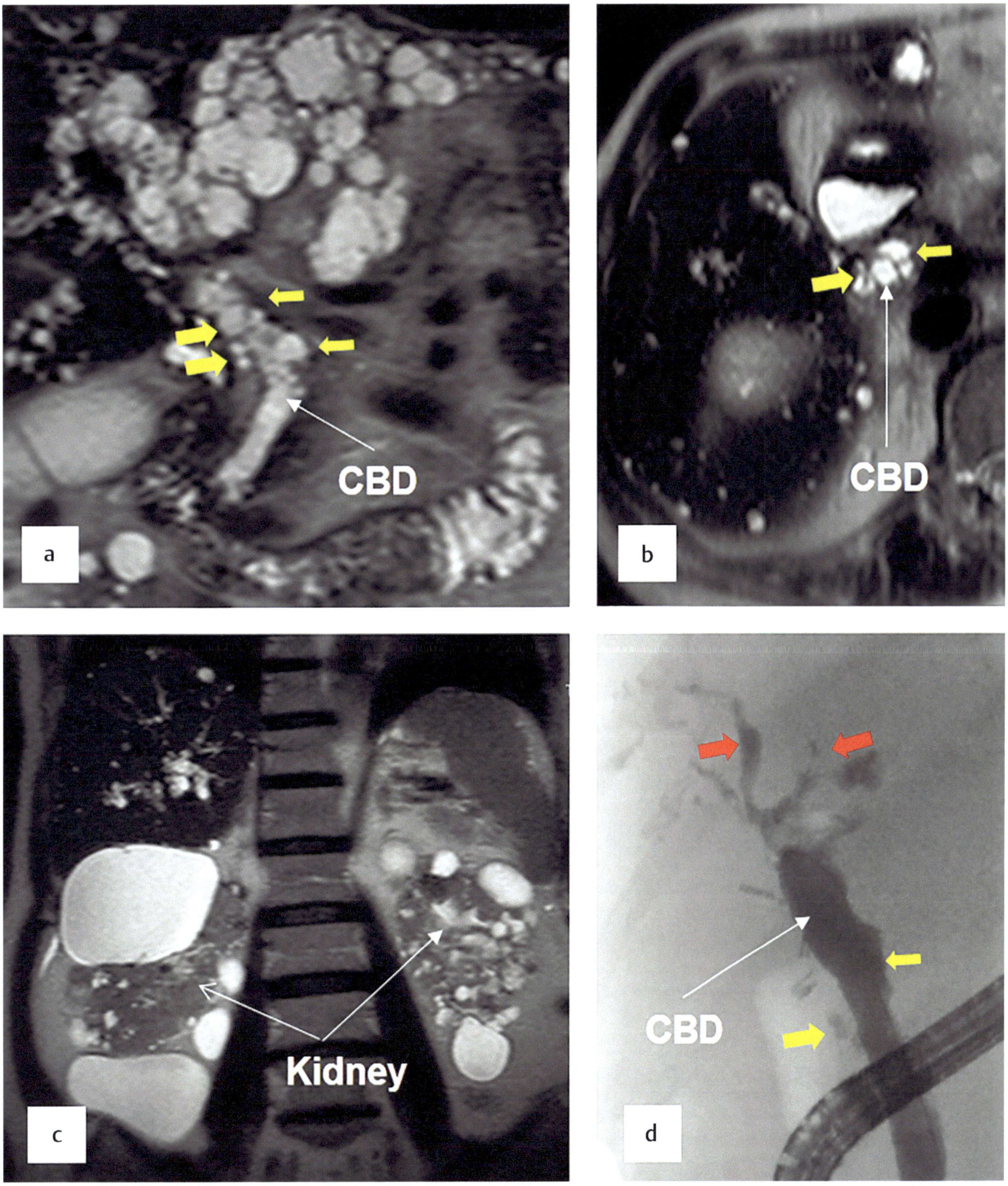

Fig. 8.38 (a–c) Magnetic resonance imaging (MRI) and **(d)** endoscopic retrograde cholangiopancreatography (ERCP) show cystic dilatation of the peribiliary glands around common bile duct (CBD) and major biliary ducts (*yellow arrows*) along with multiple intrahepatic cysts, intrarenal cysts, and biliary stenosis due to sclerosing cholangitis (*red arrows*) consistent with Caroli's syndrome.

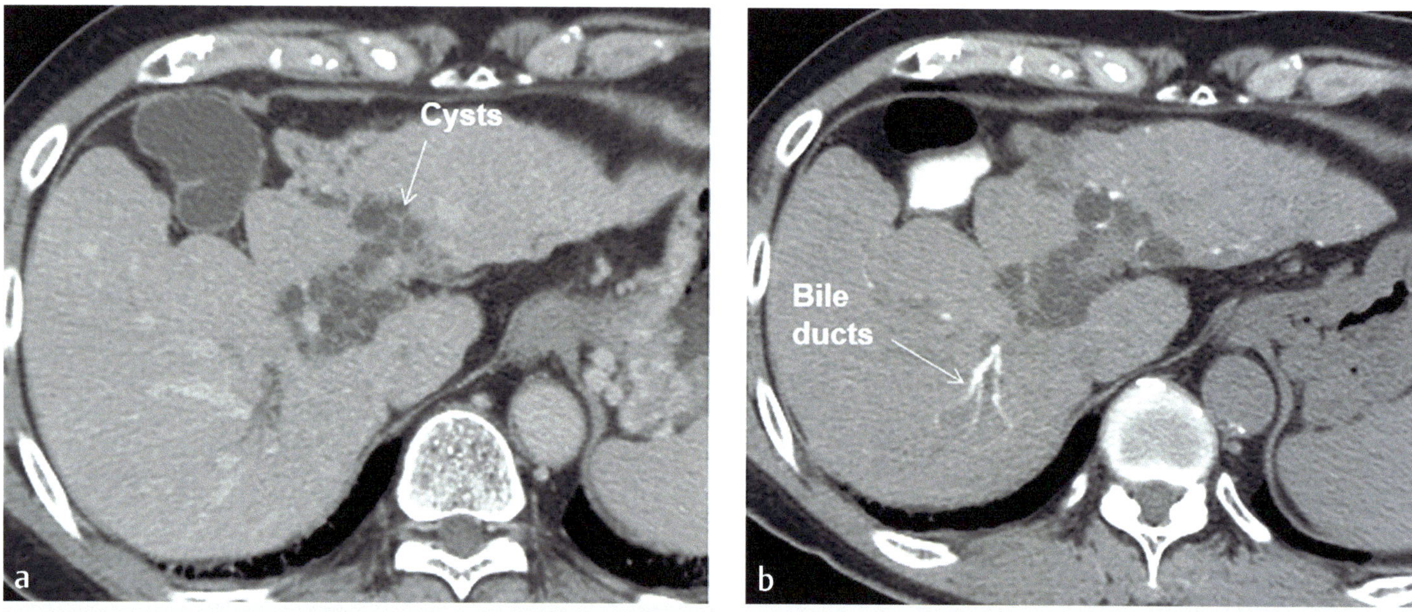

8

Fig. 8.39 Peribiliary cysts in a patient with alcoholic cirrhosis. **(a)** Postcontrast computed tomography (CT). There are multiple cysts around the portal veins in hilar to perihilar region. **(b)** Drip infusion cholangiographic CT shows contrast enhancement of intrahepatic bile ducts and no enhancement of multiple hilar to perihilar cysts indicating no bile duct communication.

References

1. Lemaigre FP. Development of the biliary tract. Mech Dev 2003;120(1):81–87
2. Lindner HH, Peña VA, Ruggeri RA. A clinical and anatomical study of anomalous terminations of the common bile duct into the duodenum. Ann Surg 1976;184(5):626–632
3. Tan J, Hytiroglou P, Wieczorek R, et al. Immunohistochemical evidence for hepatic progenitor cells in liver diseases. Liver 2002;22(5):365–373
4. Van Eyken P, Sciot R, Callea F, Van der Steen K, Moerman P, Desmet VJ. The development of the intrahepatic bile ducts in man: a keratin-immunohistochemical study. Hepatology 1988;8(6):1586–1595
5. Kahn E, Markowitz J, Aiges H, Daum F. Human ontogeny of the bile duct to portal space ratio. Hepatology 1989;10(1):21–23
6. Libbrecht L, Cassiman D, Desmet V, Roskams T. The correlation between portal myofibroblasts and development of intrahepatic bile ducts and arterial branches in human liver. Liver 2002;22(3):252–258
7. Ruebner BH, Blankenberg TA, Burrows DA, SooHoo W, Lund JK. Development and transformation of the ductal plate in the developing human liver. Pediatr Pathol 1990;10(1-2):55–68
8. Keplinger KM, Bloomston M. Anatomy and embryology of the biliary tract. Surg Clin North Am 2014;94(2):203–217
9. Shaikh AA, Charles A, Domingo S, Schaub G. Gallbladder volvulus: report of two original cases and review of the literature. Am Surg 2005;71(1):87–89
10. Kasi PM, Ramirez R, Rogal SS, Littleton K, Fasanella KE. Gallbladder agenesis. Case Rep Gastroenterol 2011;5(3):654–662
11. Moosman DA, Collier FA. Prevention of traumatic injury to the bile ducts; a study of the structures of the cystohepatic angle encountered in cholecystectomy and supraduodenal choledochostomy. Am J Surg 1951;82(1):132–143
12. D'Angelo T, Racchiusa S, Mazziotti S, Cicero G. Magnetic resonance (MR) cholangiopancreatography demonstration of the cystic duct entering the right hepatic duct. Am J Case Rep 2017;18:242–245
13. Andall RG, Matusz P, du Plessis M, Ward R, Tubbs RS, Loukas M. The clinical anatomy of cystic artery variations: a review of over 9800 cases. Surg Radiol Anat 2016;38(5):529–539
14. Suzuki M, Akaishi S, Rikiyama T, Naitoh T, Rahman MM, Matsuno S. Laparoscopic cholecystectomy, Calot's triangle, and variations in cystic arterial supply. Surg Endosc 2000;14(2):141–144
15. Ding YM, Wang B, Wang WX, Wang P, Yan JS. New classification of the anatomic variations of cystic artery during laparoscopic cholecystectomy. World J Gastroenterol 2007;13(42):5629–5634
16. Zimmitti G, Manzoni A, Guerini F, et al. Current role of minimally invasive radical cholecystectomy for gallbladder cancer. Gastroenterol Res Pract 2016;2016:7684915
17. Shirai Y, Sakata J, Wakai T, Ohashi T, Hatakeyama K. "Extended" radical cholecystectomy for gallbladder cancer: long-term outcomes, indications and limitations. World J Gastroenterol 2012;18(34):4736–4743
18. Furusawa N, Kobayashi A, Yokoyama T, et al. Biliary tract variations of the left liver with special reference to the left medial sectional bile duct in 500 patients. Am J Surg 2015;210(2):351–356
19. Chaib E, Kanas AF, Galvão FH, D'Albuquerque LA. Bile duct confluence: anatomic variations and its classification. Surg Radiol Anat 2014;36(2):105–109
20. Kitami M, Takase K, Murakami G, et al. Types and frequencies of biliary tract variations associated with a major portal venous anomaly: analysis with multi-detector row CT cholangiography. Radiology 2006;238(1):156–166
21. Taketomi A, Morita K, Toshima T, et al. Living donor hepatectomies with procedures to prevent biliary complications. J Am Coll Surg 2010;211(4):456–464
22. Healey JE Jr, Schroy PC. Anatomy of the biliary ducts within the human liver; analysis of the prevailing pattern of branchings and the major variations of the biliary ducts. AMA Arch Surg 1953;66(5):599–616
23. Couinaud C. Le foie, etudes anatomiques et chirurgicales. Paris: Masson & Cie; 1957
24. Strasberg SM, Belghiti J, Clavien PA, et al. The Brisbane 2000 terminology of liver anatomy and resections. HPB (Oxford) 2000;2:333–339
25. Vakili K, Pomfret EA. Biliary anatomy and embryology. Surg Clin North Am 2008;88(6):1159–1174, vii
26. Hyodo T, Kumano S, Kushihata F, et al. CT and MR cholangiography: advantages and pitfalls in perioperative evaluation of biliary tree. Br J Radiol 2012; 85(1015):887–896
27. Farias F, Vincente Bigolin A, Totti Cavazzola L, Pereira da Costa Filho O, Gonçalves da Costa R, Nocchi Kalil A. Anatomical study of the intrahepatic biliary ducts. Parameters that guide the surgical approach in transplanting the left lobe of the liver. G Chir 2013;34(7-8):210–215
28. Choi JW, Kim TK, Kim KW, et al. Anatomic variation in intrahepatic bile ducts: an analysis of intraoperative cholangiograms in 300 consecutive donors for living donor liver transplantation. Korean J Radiol 2003;4(2):85–90

29. Reichert PR, Renz JF, D'Albuquerque LA, et al. Surgical anatomy of the left lateral segment as applied to living-donor and split-liver transplantation: a clinicopathologic study. Ann Surg 2000;232(5):658–664
30. Yoshida J, Chijiiwa K, Yamaguchi K, Yokohata K, Tanaka M. Practical classification of the branching types of the biliary tree: an analysis of 1,094 consecutive direct cholangiograms. J Am Coll Surg 1996;182(1):37–40
31. Onishi H, Kawarada Y, Das BC, et al. Surgical anatomy of the medial segment (S4) of the liver with special reference to bile ducts and vessels. Hepatogastroenterology 2000;47(31):143–150
32. Ohkubo M, Nagino M, Kamiya J, et al. Surgical anatomy of the bile ducts at the hepatic hilum as applied to living donor liver transplantation. Ann Surg 2004;239(1):82–86
33. Kawarada Y, Das BC, Taoka H. Anatomy of the hepatic hilar area: the plate system. J Hepatobiliary Pancreat Surg 2000;7(6):580–586
34. Murakami G, Hata F. Human liver caudate lobe and liver segment. Anat Sci Int 2002;77(4):211–224
35. Kumon M. Anatomy of the caudate lobe with special reference to portal vein and bile duct. Acta Hepatol Jpn 1985;26:1193–1199
36. Furukawa H, Sano K, Kosuge T, et al. Analysis of biliary drainage in the caudate lobe of the liver: comparison of three-dimensional CT cholangiography and rotating cine cholangiography. Radiology 1997;204(1):113–117
37. Kamiya J, Nimura Y, Hayakawa N, Kondo S, Nagino M, Kanai M. Preoperative cholangiography of the caudate lobe: surgical anatomy and staging for biliary carcinoma. J Hep Bil Pancr Surg 1994;4:385–389
38. Schnelldorfer T, Sarr MG, Adams DB. What is the duct of Luschka?—A systematic review. J Gastrointest Surg 2012;16(3):656–662
39. Spanos CP, Syrakos T. Bile leaks from the duct of Luschka (subvesical duct): a review. Langenbecks Arch Surg 2006;391(5):441–447
40. Raynaud P, Carpentier R, Antoniou A, Lemaigre FP. Biliary differentiation and bile duct morphogenesis in development and disease. Int J Biochem Cell Biol 2011;43(2):245–256
41. Hashimoto M, Itoh K, Takeda K, et al. Evaluation of biliary abnormalities with 64-channel multidetector CT. Radiographics 2008;28(1):119–134
42. Venkatanarasimha N, Thomas R, Armstrong EM, Shirley JF, Fox BM, Jackson SA. Imaging features of ductal plate malformations in adults. Clin Radiol 2011;66(11):1086–1093
43. Mortelé B, Mortelé K, Seynaeve P, Vandevelde D, Kunnen M, Ros PR. Hepatic bile duct hamartomas (von Meyenburg complexes): MR and MR cholangiography findings. J Comput Assist Tomogr 2002;26(3):438–443
44. Pech L, Favelier S, Falcoz MT, Loffroy R, Krause D, Cercueil JP. Imaging of Von Meyenburg complexes. Diagn Interv Imaging 2016;97(4):401–409
45. Todani T, Watanabe Y, Fujii T, Toki A, Uemura S, Koike Y. Congenital choledochal cyst with intrahepatic involvement. Arch Surg 1984;119(9):1038–1043
46. Söreide K, Körner H, Havnen J, Söreide JA. Bile duct cysts in adults. Br J Surg 2004;91(12):1538–1548
47. Kim TU, Kim S, Lee JW, et al. Ampulla of Vater: comprehensive anatomy, MR imaging of pathologic conditions, and correlation with endoscopy. Eur J Radiol 2008;66(1):48–64
48. Avisse C, Flament JB, Delattre JF. Ampulla of Vater. Anatomic, embryologic, and surgical aspects. Surg Clin North Am 2000;80(1):201–212
49. Nikolaidis P, Hammond NA, Day K, et al. Imaging features of benign and malignant ampullary and periampullary lesions. Radiographics 2014;34(3):624–641
50. Connor S, Garden OJ. Bile duct injury in the era of laparoscopic cholecystectomy. Br J Surg 2006;93(2):158–168
51. Adams DB. The importance of extrahepatic biliary anatomy in preventing complications at laparoscopic cholecystectomy. Surg Clin North Am 1993;73(4): 861–871
52. Parmeggiani D, Cimmino G, Cerbone D, et al. Biliary tract injuries during laparoscopic cholecystectomy: three case reports and literature review. G Chir 2010;31(1-2):16–19
53. Hoeffel C, Azizi L, Lewin M, et al. Normal and pathologic features of the postoperative biliary tract at 3D MR cholangiopancreatography and MR imaging. Radiographics 2006;26(6):1603–1620
54. Nakanuma Y. A novel approach to biliary tract pathology based on similarities to pancreatic counterparts: is the biliary tract an incomplete pancreas? Pathol Int 2010;60(6):419–429
55. Katabathina VS, Flaherty EM, Dasyam AK, et al. "Biliary diseases with pancreatic counterparts": cross-sectional imaging findings. Radiographics 2016;36(2): 374–392
56. Matsubara T, Sato Y, Igarashi S, Matsui O, Gabata T, Nakanuma Y. Alcohol-related injury to peribiliary glands is a cause of peribiliary cysts: based on analysis of clinical and autopsy cases. J Clin Gastroenterol 2014;48(2):153–159
57. Uchida T, Yamamoto Y, Ito T, et al. Cystic micropapillary neoplasm of peribiliary glands with concomitant perihilar cholangiocarcinoma. World J Gastroenterol 2016;22(7):2391–2397
58. Seguchi T, Akiyama Y, Itoh H, et al. Multiple hepatic peribiliary cysts with cirrhosis. J Gastroenterol 2004;39(4):384–390

8

9 Pancreas

Tapas K. Tejura and Farhood Saremi

◆ Introduction

The pancreas is a mixed exocrine and endocrine accessory gland, the function of which is critical to normal digestion and metabolism. The pancreatic acinar cells produce pancreatic juice, an exocrine secretion that enters into the duodenum to assist in digestion. The pancreatic islet cells (of Langerhans) produce endocrine secretions that are released as hormones into the blood, including insulin, glucagon, and somatostatin.

The history of the pancreas is complex and has been closely scrutinized. Given its relatively hidden location, the pancreas has been referred to as a "hermit organ."[1] The first description of the pancreas is attributed to Herophilus of Chalcedon (circa 300 BC), considered to be the Father of Scientific Anatomy.[2] Rufus of Ephesus (circa 100 AD), a Greek physician, is generally credited for naming the organ pancreas (*pan*, all; *kreas*, flesh).[2] It took more than 100 years before anatomic descriptions of the pancreas were completed.

9

Knowledge of pancreatic anatomy has become increasingly clinically important. The modern usage of imaging allows for correlation of classical anatomical descriptions of the pancreas and diagnosis of pancreatic diseases. Thin-slice multidetector computed tomography (CT) and magnetic resonance imaging (MRI) can provide detailed evaluation of the pancreas and are the primary imaging modalities used to diagnose a variety of congenital, neoplastic, inflammatory, and traumatic processes of the pancreas. CT angiography and conventional catheter angiography are useful in providing detailed pancreatic vascular anatomy and variants. Conventional ultrasound (US) is a noninvasive, nonradiating imaging modality that is often used for the initial evaluation of the pancreas. Endoscopic ultrasound (EUS) has an increasing role in evaluating the pancreas with increased spatial resolution than conventional US, and allowing for targeted intervention of the pancreas and adjacent structures. Endoscopic retrograde cholangiopancreatography (ERCP) and magnetic resonance cholangiopancreatography (MRCP) are techniques that are commonly used to evaluate the pancreatic duct and its side branches.

In this chapter, the embryological development of the pancreas will be reviewed. Examples from a wide variety of imaging modalities will be used to help delineate pancreatic anatomy and its ductal system, with a focus on CT and MRI. In addition, common anatomical variants of the pancreas and expected postoperative pancreatic anatomy will be discussed.

◆ Embryology

Familiarity with the embryology and development of the pancreas is essential to recognize anatomical variants and congenital anomalies. The pancreas is formed from two separate embryonic buds in the primitive foregut: dorsal bud and ventral bud (▸Fig. 9.1a–d). At the end of the fourth gestational week of development, the dorsal bud arises from the endodermal lining of the duodenum slightly inferior to the hepatic diverticulum. The ventral bud is close to the bile duct and arises from the base of the hepatic diverticulum. The ventral bud may be divided into the left and right portions. The left portion of the ventral bud usually regresses, while the right portion rotates from the right to the left around the posterior aspect of the duodenum to lie inferior and posterior to the dorsal bud. The ventral bud differentiates into part of the pancreatic head (primarily posterior) and uncinate process, while the dorsal bud becomes the pancreatic tail, body, and remaining portions of the head (primarily anterior).[3]

During the seventh gestational week of development, the two pancreatic buds usually fuse in the region of the neck. Following fusion of the pancreatic buds, the ductal systems anastomose. The dorsal duct in the body and tail of the pancreas prior to the site of fusion is commonly referred to as the "main pancreatic duct." The portion of the ventral duct between the dorsal–ventral duct fusion point and the major papilla is referred to as the duct of Wirsung. The duct of Wirsung, along with the common bile duct, drains into the major papilla (papilla of Vater), a prominence located on the posteromedial wall of the second portion of the duodenum. If a segment of the dorsal duct persists distal to the dorsal–ventral duct fusion point, it is referred to as the accessory pancreatic duct (of Santorini), which drains through the minor papilla into the duodenum[3] (▸Fig. 9.1e).

As the fetus matures, there is fusion of the dorsal mesogastrium and posterior parietal layer of the peritoneum which becomes the posterior wall of the lesser sac. An outpouching of the dorsal mesogastrium between the stomach and pancreas forms the omentum. The posterior leaf of this extension fuses with the mesentery that attaches to the migrated midgut to form the transverse mesocolon over the pancreas. This fusion fixes the pancreas to the retroperitoneum. The exception is the lateral portion of the mesogastrium, which contains the distal-most portion of the pancreatic tail and becomes the splenorenal ligament. Because

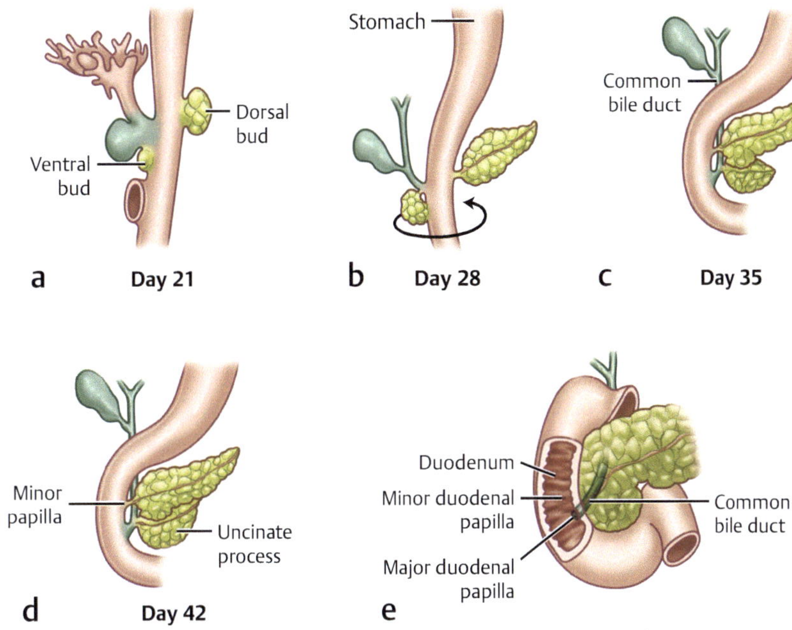

Fig. 9.1 (a–d) Normal development of the pancreas. **(a)** The ventral pancreatic bud arises from hepatic diverticulum, and the dorsal pancreatic bud arises from dorsal mesogastrium. **(b–d)** There is clockwise rotation of the ventral pancreatic bud behind the duodenum which fuses with the dorsal pancreatic bud. The ventral pancreatic bud differentiates into the posterior head and uncinate process of the pancreas. The dorsal pancreatic bud becomes the pancreatic tail, body, and anterior portion of the head. **(e)** Normal pancreatic duct anatomy. The "main pancreatic duct" is the portion of the main duct in the pancreatic body and tail. The "duct of Wirsung" exits into the ampulla of Vater at the major duodenal papilla. The "duct of Santorini" drains into the minor duodenal papilla, located approximately 2 cm superior to the major duodenal papilla.

9

of this unique embryological development, the pancreas can be considered as a secondary retroperitoneal organ.[4]

◆ Anatomy

The pancreas is an unpaired, predominantly retroperitoneal organ situated in the anterior pararenal space, partially surrounded by the duodenum and located behind the stomach (▶Fig. 9.2, ▶Fig. 9.3). The anterior pararenal space is bounded anteriorly by the posterior parietal peritoneum, posteriorly by the anterior leaf of Gerota's (renal) fascia, and laterally by the lateroconal fascia (▶Fig. 9.4, ▶Fig. 9.5). Thickening of the anterior renal fascia is commonly seen on CT in patients with acute pancreatitis, with the inflammatory process often extending to involve the anterior pararenal space[5] (▶Fig. 9.4).

In addition to the pancreas, the anterior pararenal space contains the duodenum, the descending and ascending colon, and the peripancreatic blood vessels. The anterior pararenal space extends superiorly to the subdiaphragmatic area and inferiorly to the pelvis. The two layers of the transverse mesocolon, which are reflections of the posterior parietal peritoneum, split near the anterior surface of the pancreas. The anterior pararenal space communicates with the transverse mesocolon and small bowel mesentery, which originates inferior to the pancreatic body. Therefore, pancreatic diseases can involve the stomach and duodenum by direct spread and potentially involve the small bowel and colon via the small bowel mesentery and transverse mesocolon, respectively[5,6] (▶Fig. 9.6, ▶Fig. 9.7).

Pancreatic Gland

The pancreas is a soft, pale yellow, lobulated organ that measures approximately 12 to 20 cm in length, 3 to 5 cm in height, and 1 to 3 cm in thickness and weighs 60 to 125 g in adults.[7] The pancreas has characteristic appearances on CT, MRI, and US (▶Fig. 9.8, ▶Fig. 9.9, ▶Fig. 9.10). Macroscopically, the pancreas has a granular surface. Its contour is smooth in about 80% and lobulated in 20%.[8] The lobulated contour of the pancreas is a characteristic

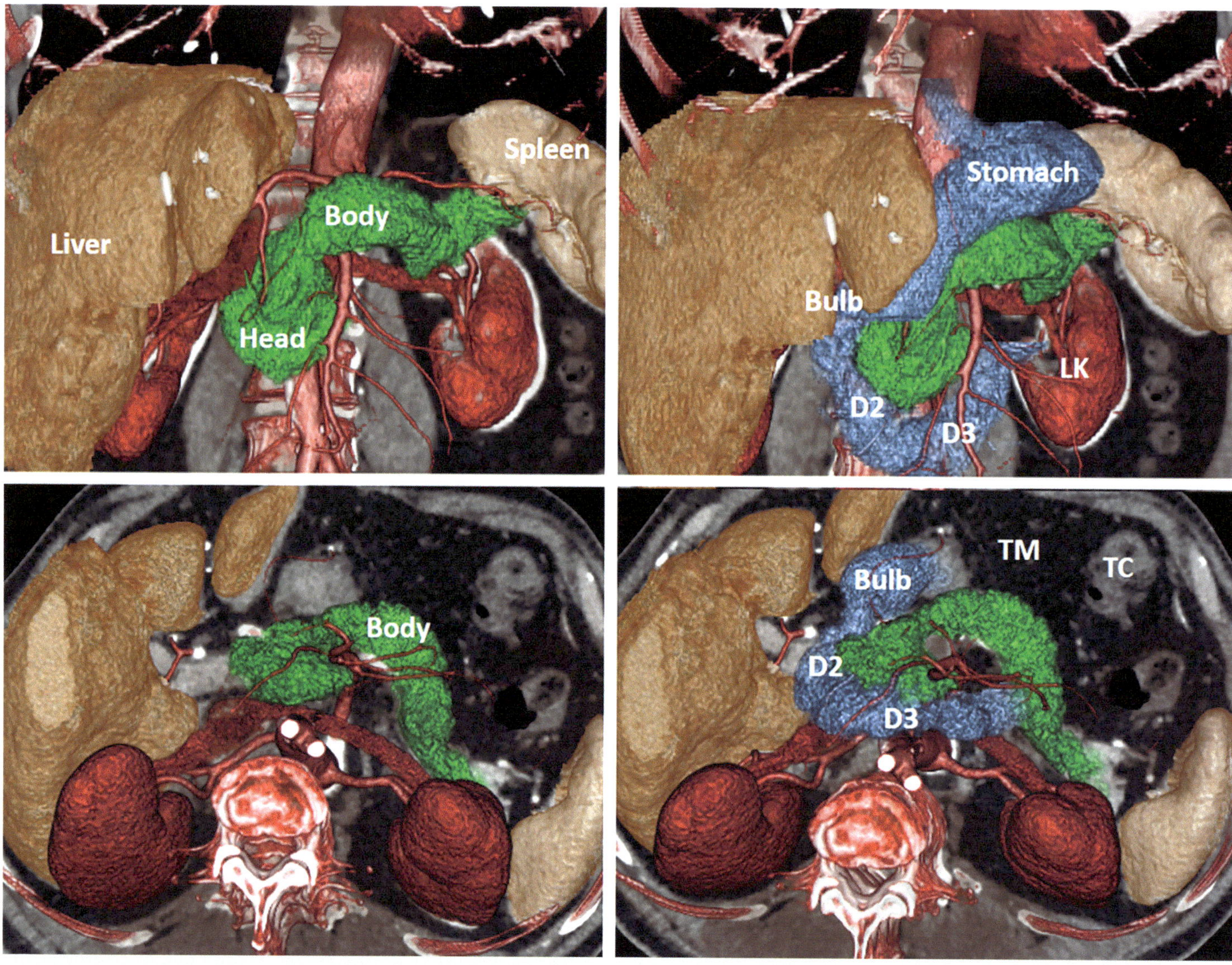

Fig. 9.2 Normal pancreatic orientation and relationship to surrounding structures. Coronal (*upper row*) and axial (*lower row*) computed tomography (CT) images are shown. The pancreas lies transversely along its long axis and has an oblong shape, slightly oblique in orientation. D2/D3, second and third portions of the duodenum; LK, left kidney; TC, transverse colon; TM, transverse mesocolon.

feature that may be lost in certain pathological conditions, particularly in pancreatitis (▸Fig. 9.11).

The shape, position, axis, and fatty content of the pancreas are variable and are influenced by many factors, including age, body habitus, and organomegaly. A gradual decrease in the size of the glandular pancreas is seen with age. The reported anteroposterior dimensions of the pancreas in people of 71 to 80 years of age are 21.2 mm (±4.3 mm) for the head, 14.4 mm (±2.7 mm) for the body, and 13.0 mm (±2.1 mm) for the tail. The reported anteroposterior dimensions of the pancreas in people of 20 to 30 years of age are 28.6 mm (±3.8 mm) for the head, 19.1 mm (±2.1 mm) for the body, and 18.0 mm (±1.6 mm) for the tail.[9]

The pancreas lies transversely along its long axis and has an oblong shape, slightly oblique in orientation to the left and superiorly (▸Fig. 9.2). The longitudinal axis of the pancreas is approximately 20 degrees with respect to the transverse plane and can be shaped like an L, S, or inverted V.[10] The pancreas is divided into the head, neck, body, and tail (▸Fig. 9.2, ▸Fig. 9.12). The divisions of the pancreas represent arbitrary boundaries and do not represent anatomically or functionally distinct units. The pancreatic head, neck, and body are fixed in the retroperitoneal space. The pancreatic tail, after extending across the left kidney, is an intraperitoneal structure incorporated within the splenorenal ligament.[5]

The uncinate ("hook like") process is the triangular, most inferior extension of the head of the pancreas and lies dorsal to the superior mesenteric vessels. The uncinate process may be absent or may completely encircle the superior mesenteric vessels (▸Fig. 9.13a–c). The left renal vein lies superior and the third portion of the duodenum lies inferior to the uncinate process. The anterior and posteromedial borders of the uncinate process are usually straight or concave. A convex appearance of the uncinate process border may suggest the possibility of underlying disease.[8]

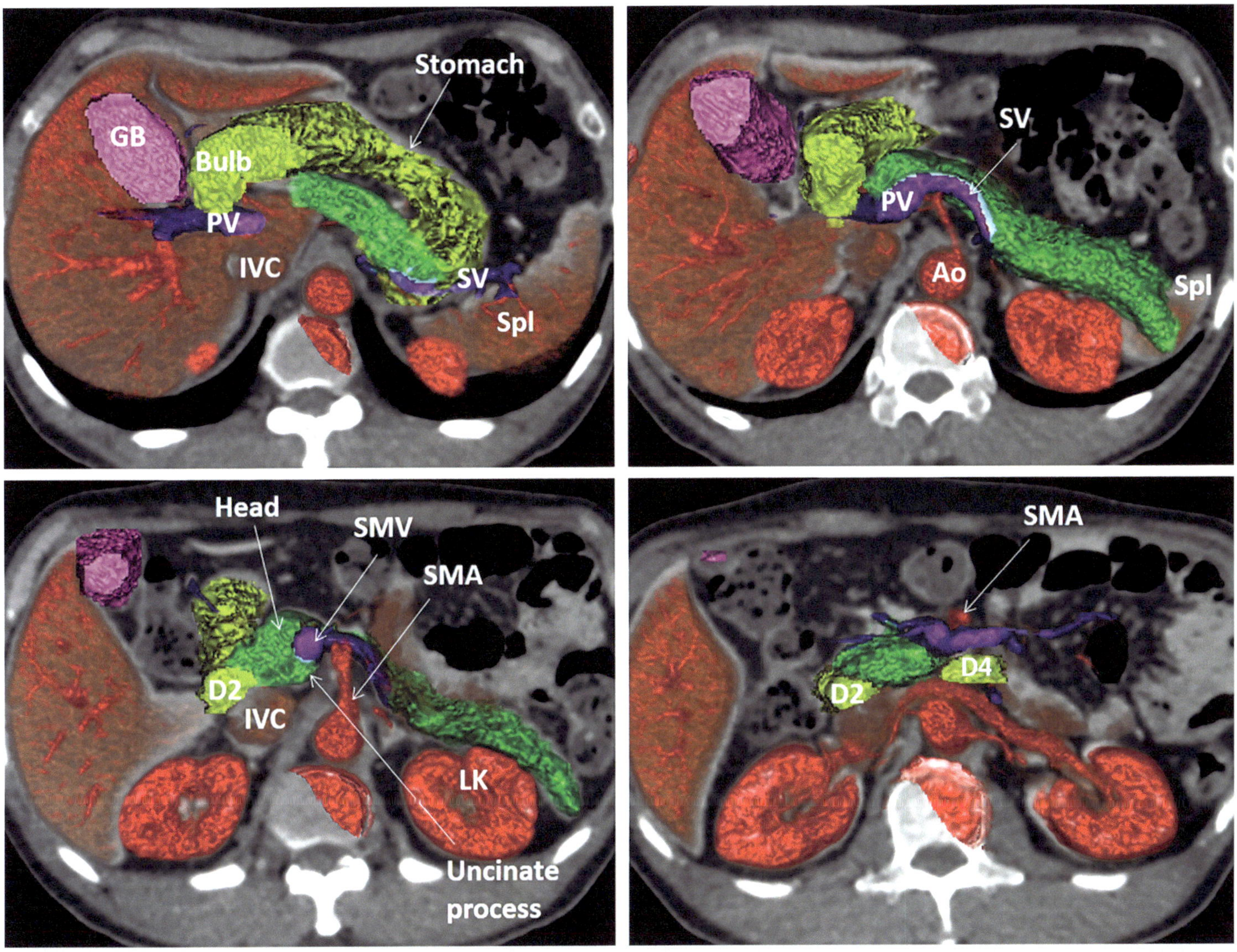

Fig. 9.3 Axial computed tomography (CT) images of the upper abdomen. The pancreas is shown in *green*. The relationship of the pancreas with the mesenteric vessels, duodenum (D), inferior vena cava (IVC), spleen (Spl), and left kidney (LK) is shown. Ao, aorta; D2/D4, second and fourth portions of the duodenum; GB, gallbladder; PV, portal vein; SMA, superior mesenteric artery; SMV, superior mesenteric vein; SV, splenic vein.

The pancreatic head usually lies at the level of L1–L2 and crosses the midline of the spine at L1. The head is the thickest portion of the pancreas and may occupy 60 to 70% of total pancreatic volume, especially in elderly[11] (▶Fig. 9.2). The anterior surface of the pancreatic head is covered by the gastric pylorus superiorly and the transverse colon inferiorly. Posterior to the pancreatic head are the renal hilum and vessels, the inferior vena cava, left renal vein, and right gonadal vein. The pancreatic head lies within the "C" loop of the second portion of the duodenum and to the right of the portal and superior mesenteric veins (▶Fig. 9.2, ▶Fig. 9.12). The lateral surface of the pancreatic head abuts the serosa of the duodenum and is adherent to the duodenum by a narrow soft tissue plane, referred to as the pancreatoduodenal groove. This is clinically important in inflammatory processes, such as groove pancreatitis, and for staging of pancreatic and ampullary tumors (▶Fig. 9.14).

The neck of the pancreas is a short thin transition between the head and body, measuring 1 to 2 cm in anteroposterior dimension[11] (▶Fig. 9.12). The pancreatic neck is located immediately anterior to the superior mesenteric vessels, which form a groove in its posterior aspect. The pancreatic neck is covered anteriorly by the gastric pylorus and lesser sac. The superior mesenteric vein (SMV) joins the splenic vein posterior to the pancreatic neck to form the portal vein (▶Fig. 9.12). The superior mesenteric artery (SMA) courses inferior to the pancreatic neck. The celiac artery is usually located at the superior margin of the pancreatic neck.

The body and tail of the pancreas extend to the left toward the splenic hilum, with the transverse mesocolon attaching to the inferior portion (▶Fig. 9.7). The body and tail of the pancreas together measure up to 13 cm in length and usually no more than 3 cm in maximum anteroposterior dimension.[11] The lesser sac separates the anterior surface of the pancreatic body from the stomach (▶Fig. 9.7). The inferior aspect of the body is covered by the peritoneum of the transverse mesocolon and lies adjacent to the fourth portion of the duodenum and the ligament of Treitz.

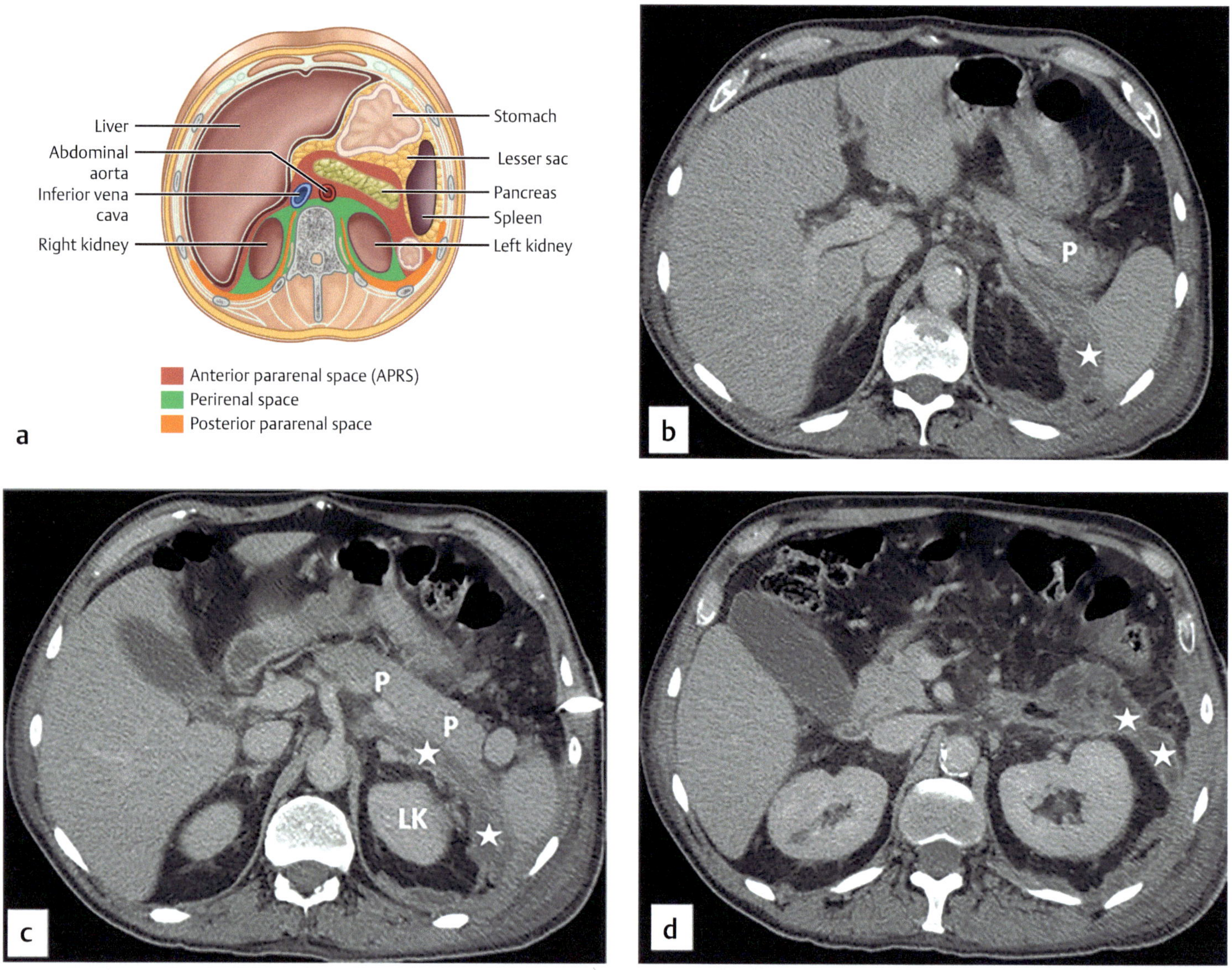

Fig. 9.4 (a) Retroperitoneal compartments. The pancreas (P) is located in the anterior pararenal space which is bounded anteriorly by the posterior parietal peritoneum, posteriorly by the anterior leaf of Gerota's fascia, and laterally by the lateroconal fascia. **(b–d)** Pancreatitis extending through the left anterior pararenal space. Axial computed tomography (CT) images demonstrate thickening of the left anterior renal fascia with peripancreatic infiltration and fluid (*stars*) extending from the body and tail of the pancreas (P) through the left anterior pararenal space. LK, left kidney.

The posterior surface of the pancreatic body opposes the aorta, SMA, left diaphragmatic crus, left adrenal gland, left kidney, and left renal vessels (▶Fig. 9.3, ▶Fig. 9.5). Depending on its size, the left lobe of the liver may lie anterior to the pancreatic body. The more superior portion of the body is usually in close contact with the celiac trunk. The splenic vein courses along the posterior surface of the pancreatic body and is fixed to it by short tributaries[12] (▶Fig. 9.12). The splenic artery is usually located posterior and superior to the pancreatic body, although this varies depending on its tortuosity.

The border between the pancreatic body and tail is not clearly defined, although it can be determined using one-half of the distance between the neck and the end of the pancreas.[13] The pancreatic tail is covered by the peritoneum and is relatively mobile,

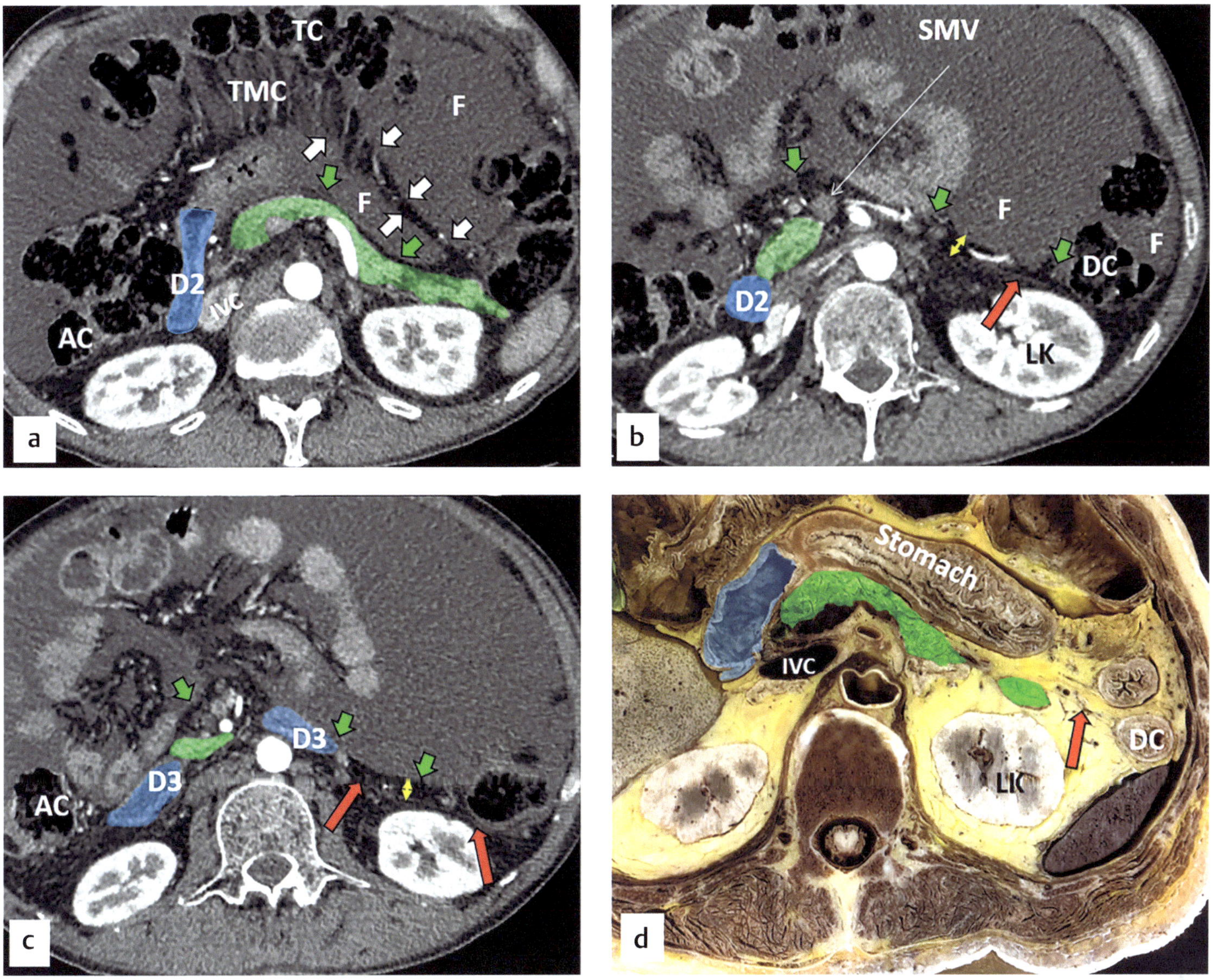

Fig. 9.5 Anterior pararenal space. Axial computed tomography (CT) images **(a–c)** in a patient with large ascites and a cadaveric cut **(d)** through the pancreas (in *green*) demonstrate the anatomic position of the pancreas in the anterior pararenal space. The anterior margin of the pancreas is well seen because the peritoneal cavity is filled with ascitic fluid (F). Attachment of the transverse mesocolon (TMC, *white arrows*) to the anterior surface of the pancreas can be seen in figure **a**. The anterior pararenal space where the pancreas, second portion of the duodenum (D2), third portion of the duodenum (D3), ascending colon (AC), and descending colon (DC) are located is a small potential space filled with fat (*double-headed yellow arrows*). The anterior pararenal space is located anterior to the anterior renal fascia (*red arrow*). The anterior border of this space is demarcated by *green arrows*. IVC, inferior vena cava; LK, left kidney; SMV, superior mesenteric vein; TC, transverse colon.

with the tip usually blunted or oriented superiorly in the region of the splenic hilum. The pancreatic tail is relatively mobile and can have varying relations with the splenic hilum[14] (▶Fig. 9.13d). In patients with a surgically or congenitally absent left kidney, the pancreatic tail extends into the empty renal fossa along with bowel and spleen situated dorsomedial to the spine[6] (▶Fig. 9.15).

Histology

The pancreas has a lobulated architecture, with the lobules varying in size (1–10 mm) and shape. The pancreas is surrounded by areolar tissue and lacks a distinct capsule. The pancreatic parenchyma consists of the islet cells (of Langerhans) and the acinar cells. The islet cells form small networks of cells that secrete

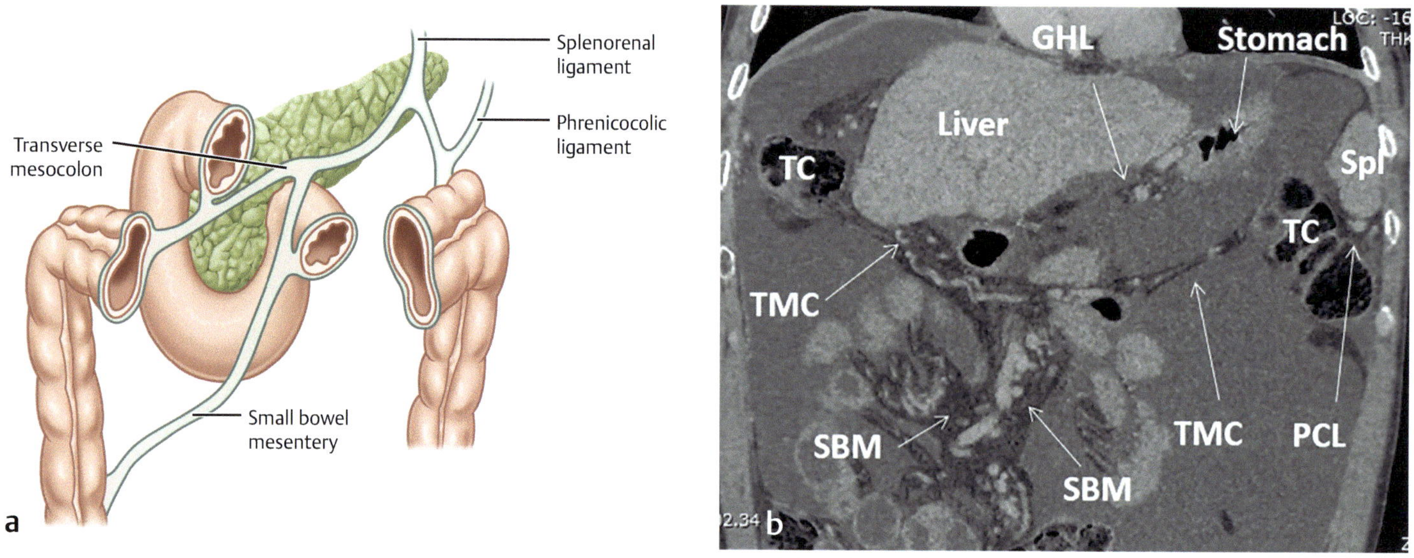

Fig. 9.6 (a) Frontal representation of the mesenteric relationships of the pancreas. **(b)** Coronal computed tomography (CT) in a patient with large ascites. GHL, gastrohepatic ligament; PCL, phrenicocolic ligament; SBM, small bowel mesentery; Spl, spleen; SRL, splenorenal ligament; TC, transverse colon; TMC, transverse mesocolon.

9

hormones that regulate blood glucose levels and constitute only 2% of the pancreatic mass.[11] The majority of the islet cells are present in the distal pancreas. The acinar cells constitute the majority of the pancreatic mass and are responsible for the secretion of enzymes of digestion, pancreatic fluids, and electrolytes. All of the secretions of the acini drain into the duodenum through the major and minor duodenal papillae. The acinar lobules are separated by fibrous septa containing arteries, veins, and variable amounts of fat. Fat-laden interlobular septa are more common in obese individuals and in conditions with glandular atrophy.[8]

Pancreatic Duct

The main channel of drainage of pancreatic exocrine secretions through the main pancreatic duct (MPD) begins in the tail. The MPD has a gradually enlarging caliber as it extends to the ampulla. Through the tail and body of the pancreas, the duct lies midway between the superior and inferior margins, slightly more posterior than anterior.[15] The diameter of the MPD is up to 3.5 mm in the head, 2.5 mm in the body, and 1.5 mm in the tail, and the length of the MPD varies from 9.5 to 25 cm.[13] The diameter of the pancreatic duct can increase with respiration.[16] The MPD usually receives short tributaries (side branches) entering at right angles to the long axis as it courses toward the head. When describing MPD anatomy, the pancreatic tail is anatomically proximal ("upstream"), while the portion of the duct toward the ampulla is distal ("downstream").

The pancreatic duct can have a variety of configurations. A straight descending course is most common (50%), although other studies show different results.[17] Other variants include a vertical course, a sigmoid course, and a looped appearance (▶ Fig. 9.16). Partial duplication variant of the MPD is relatively common (▶ Fig. 9.17). Partial duplication of the pancreatic parenchyma, however, is very rare.

Pancreatic ductal dilatation can indicate obstruction of the normal flow of pancreatic secretions. This can be due to a variety of etiologies, including chronic pancreatitis or pancreatic cancer (▶ Fig. 9.18, ▶ Fig. 9.19). There is also suggestion that dilatation of the MPD without obvious cause is an independent predictor for subsequent development of pancreatic cancer.[18]

Pancreaticobiliary Union

The MPD usually bifurcates into the duct of Wirsung and the accessory duct (of Santorini).[19–23] The duct of Wirsung joins with the distal common bile duct and drains into the duodenum at the major duodenal papilla (▶ Fig. 9.1e). The ampulla of Vater (biliaropancreatic ampulla) is strictly defined as a slightly dilated conduit that results from the union of the bile duct and pancreatic

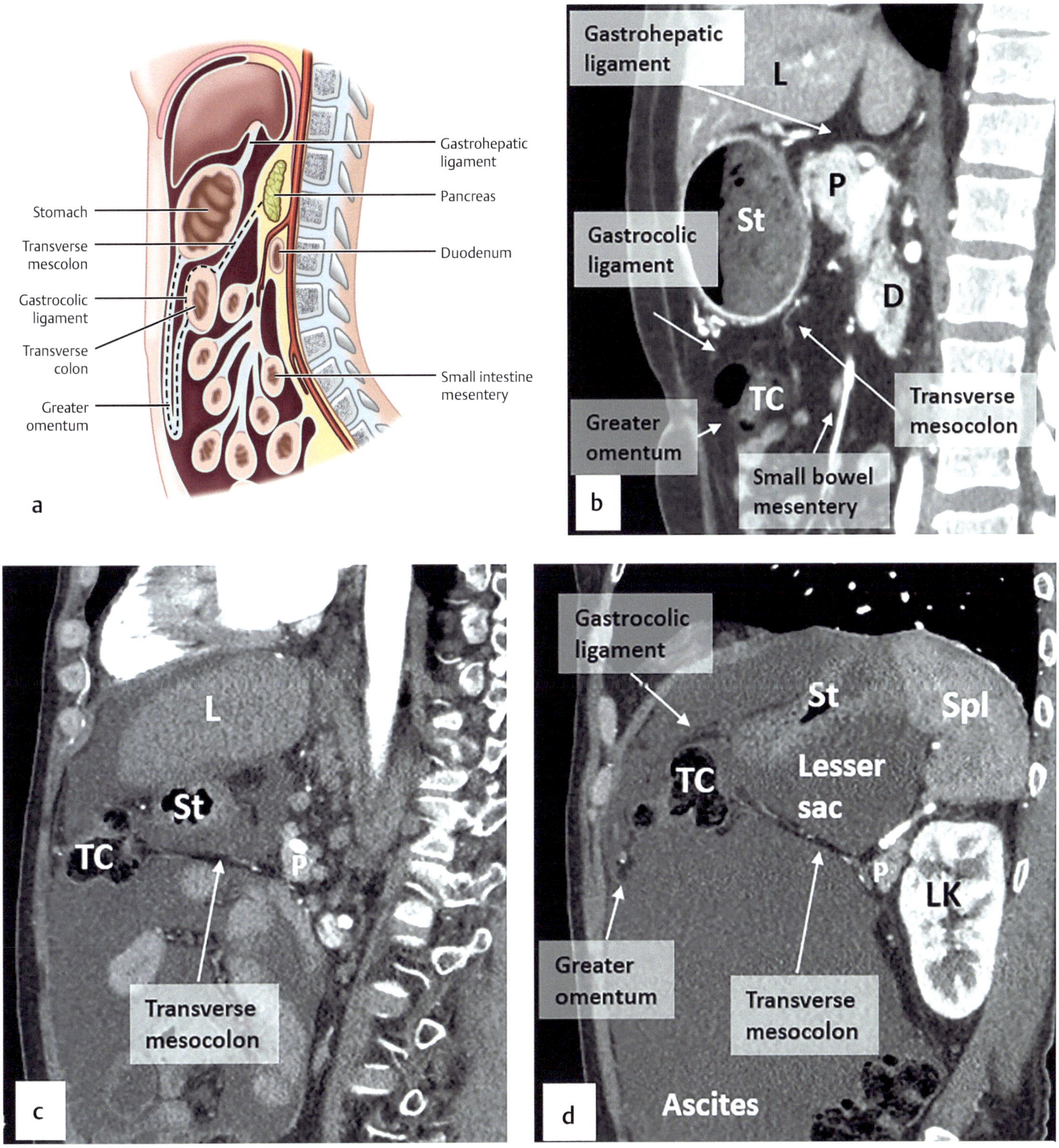

Fig. 9.7 (a) Sagittal drawing and computed tomography (CT) representation of the mesenteric relationships of the pancreas. Patients **(b)** without ascites and **(c, d)** with large ascites. The posterior leaf of the dorsal meosgastrium fuses with the mesentery that attaches to the migrated midgut to form the transverse mesocolon, thus creating a direct contiguity between the transverse mesocolon and the greater omentum. The lesser sac separates the anterior surface of the pancreatic body from the stomach. D, third part of duodenum; L, liver; LK, left kidney; P, pancreas; Spl, spleen; St, stomach; TC, transverse colon.

9

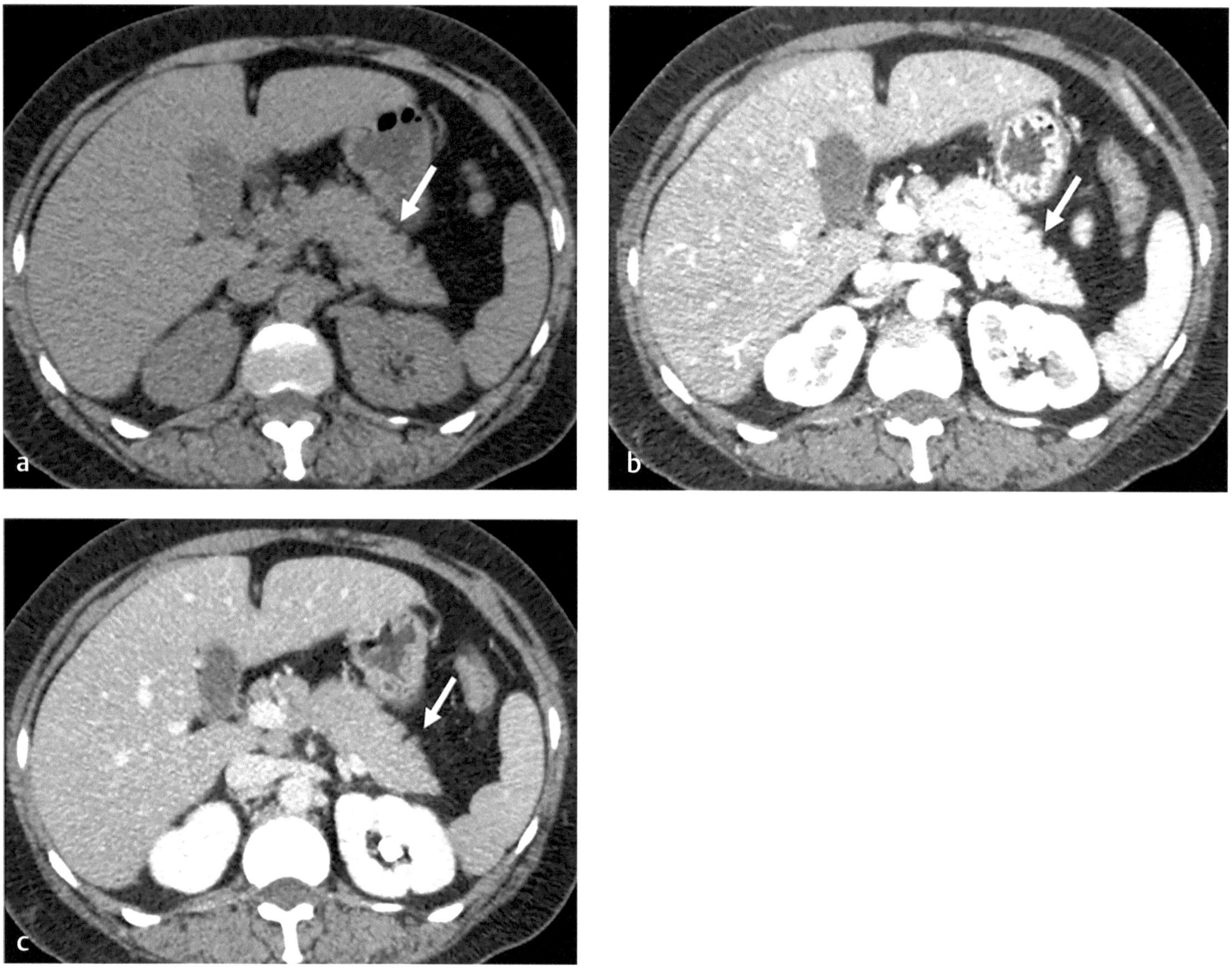

Fig. 9.8 Computed tomography (CT) showing anatomy of the pancreas. **(a)** Precontrast axial CT image of the pancreas demonstrates normal glandular contour and thickness (*arrow*). **(b)** Axial CT image in the arterial phase demonstrates pancreas with intense, homogenous enhancement similar to the renal cortex (*arrow*). **(c)** Axial CT image in the portal venous phase demonstrates homogenous washout of the pancreas (*arrow*).

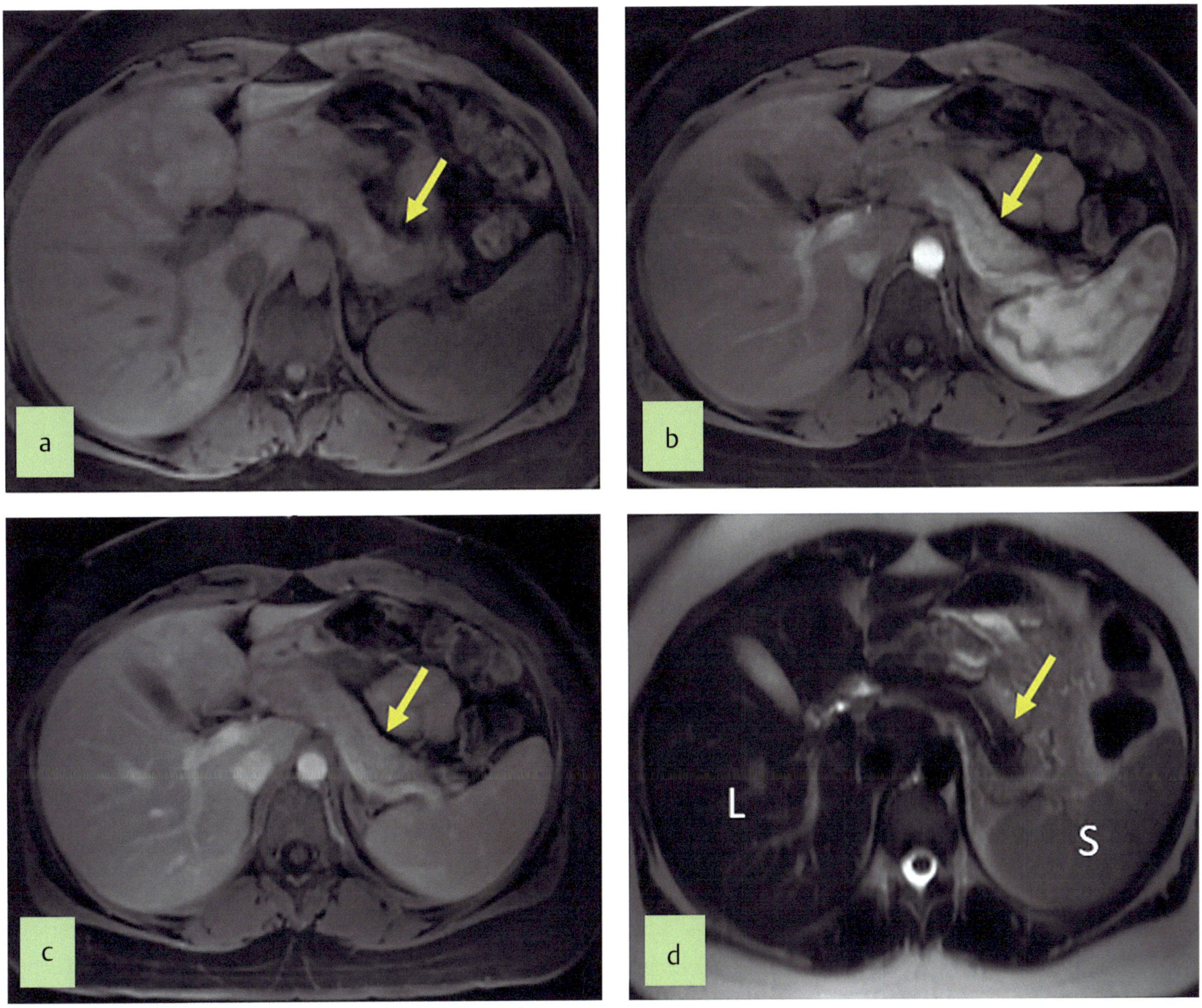

Fig. 9.9 Magnetic resonance imaging (MRI) showing anatomy of the pancreas. **(a)** Axial, T1-weighted, gradient-echo image of the pancreas (*arrow*) before the administration of gadolinium chelates. **(b, c)** Axial, T1-weighted, gradient-echo images of the pancreas (*arrow*) during the arterial **(b)** and portal venous **(c)** phases of the dynamic contrast-enhanced study. **(d)** Axial T2-weighted image demonstrates normal signal of the pancreas (*arrow*). Note that signal intensity of the pancreas is lower compared with the liver (L) and spleen (S).

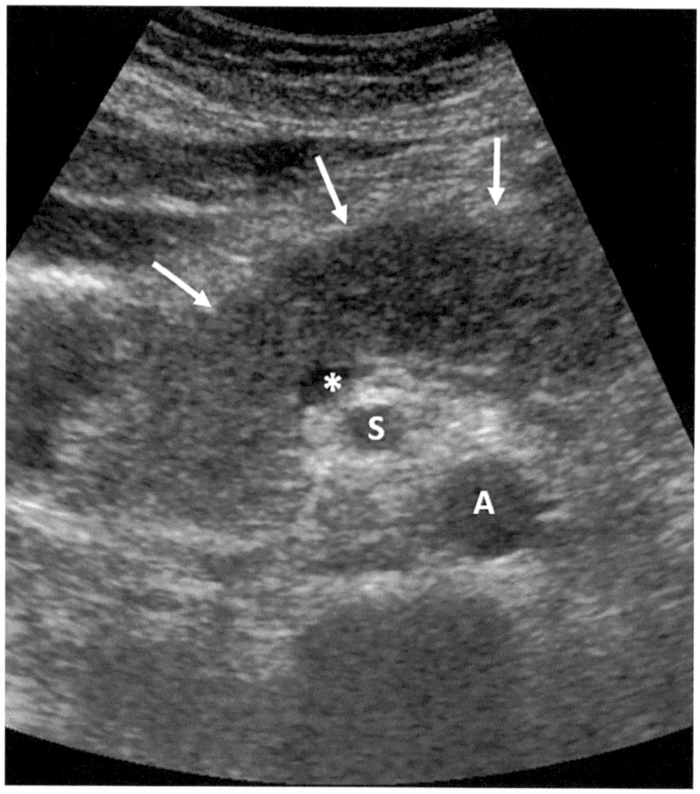

Fig. 9.10 Ultrasound of the pancreas. Transverse grayscale ultrasound image demonstrates hypoechoic echotexture of the pancreas (*arrows*). Note that the pancreas is located anterior to the splenic vein (*). A, abdominal aorta; S, superior mesenteric artery.

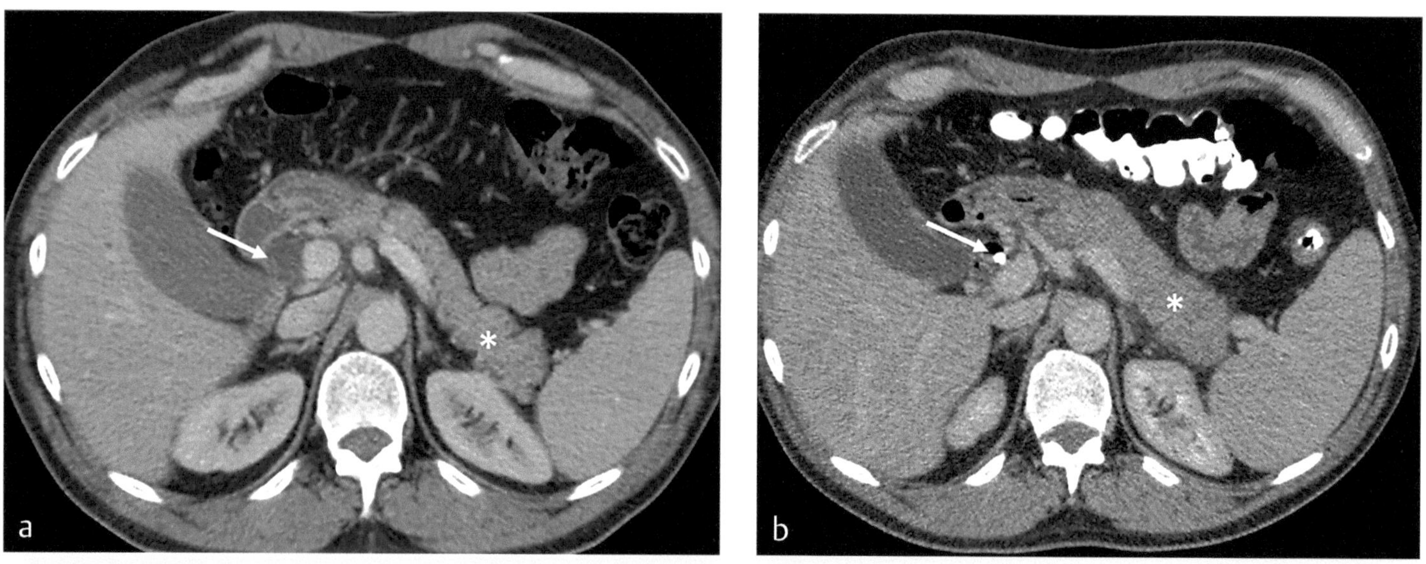

Fig. 9.11 Contour of the pancreas. **(a)** Axial computed tomography (CT) image demonstrates normal lobulated contour of the pancreas (*). Note dilatation of the extrahepatic duct (*arrow*). **(b)** Subsequent axial CT image from the same patient following extrahepatic duct stent placement (*arrow*) demonstrates edematous enlargement of the pancreas with loss of the lobulated contour (*), compatible with mild acute pancreatitis.

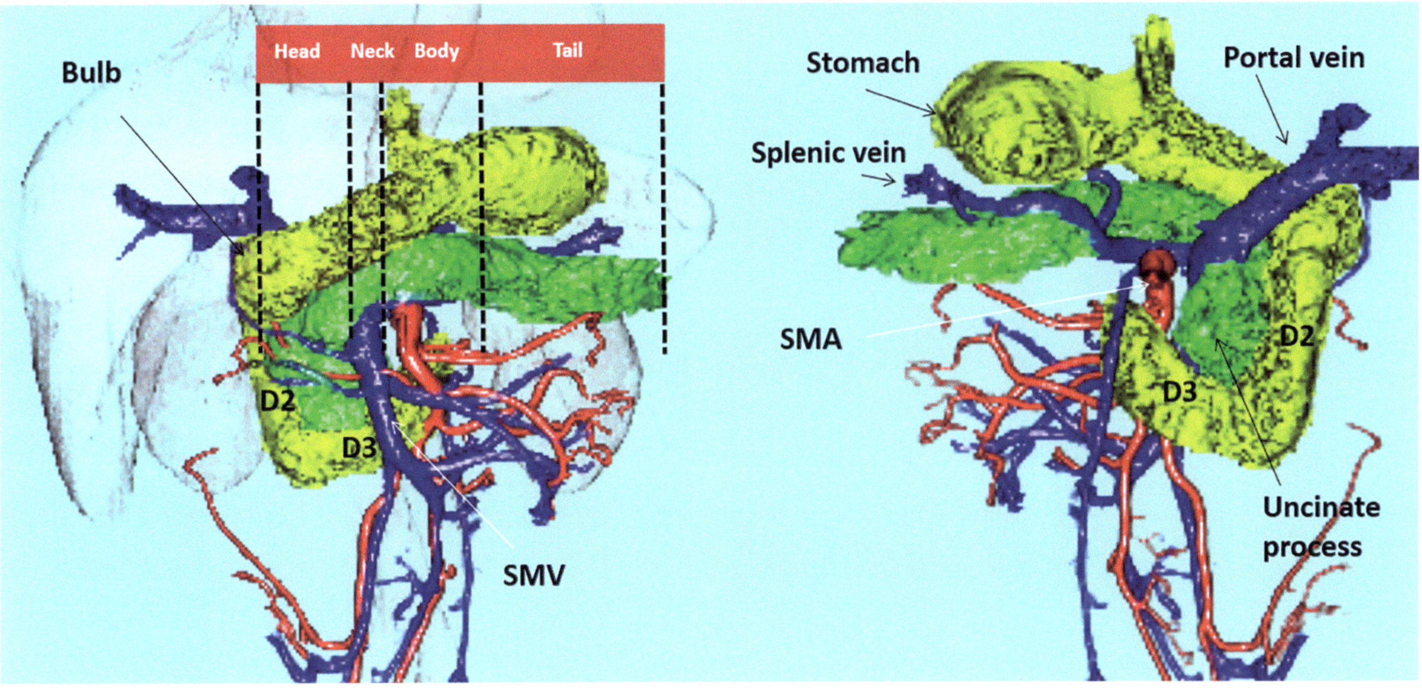

Fig. 9.12 Divisions of the pancreas (in *green*) into the head, neck, body, and tail. The pancreatic head is located to the right of the superior mesenteric vein (SMV). The pancreatic neck is located anterior to the SMV. The border between the body and tail is not clearly defined, although it can be determined using one-half of the distance between the neck and the end of the pancreas. Note the uncinate process extending behind the mesenteric vessels. The superior mesenteric artery (SMA) originates inferior to the pancreatic neck. The pancreatic neck is located immediately anterior to the superior mesenteric vein. D2/D3, second and third parts of the duodenum.

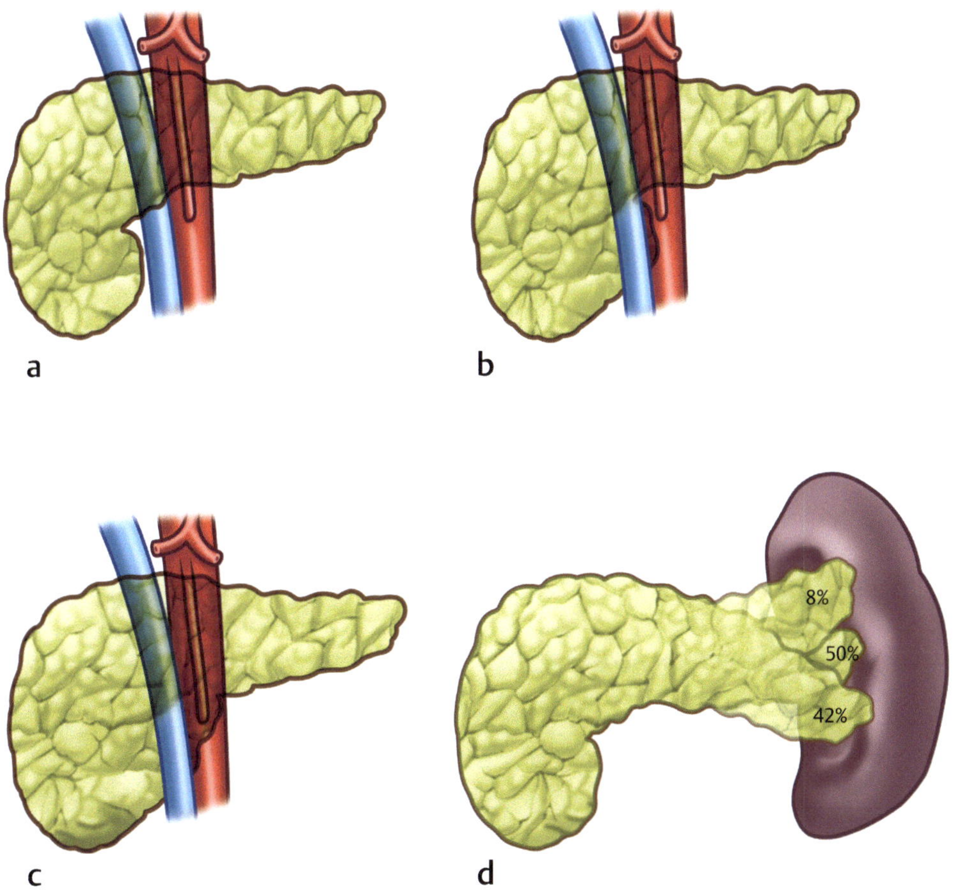

Fig. 9.13 (a–c) Uncinate process of the pancreas. The prominence of the uncinate process varies. It may not reach the superior mesenteric vessels **(a)**, it may extend behind the superior mesenteric vein almost to the superior mesenteric artery **(b)**, or it may extend behind the superior mesenteric vein and artery **(c)**. **(d)** Relations of the pancreatic tail with the splenic hilum. In most cases (50%), the tail of the pancreas extends to the midportion of the splenic hilum, followed by the lower third portion (42%).

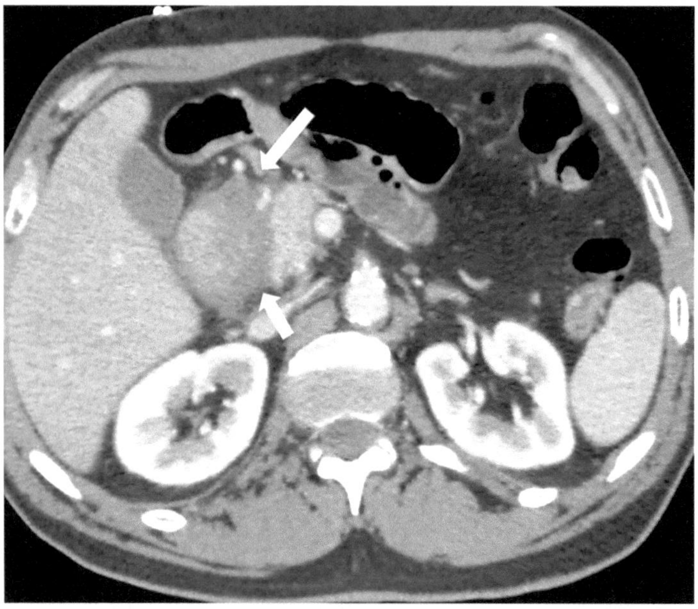

Fig. 9.14 Groove pancreatitis. Axial computed tomography (CT) image demonstrates soft tissue infiltration in the pancreaticoduodenal groove (*arrows*), with suggestion of a discrete duodenal mass on endoscopic ultrasound. This process was found to be groove pancreatitis after pancreaticoduodenectomy.

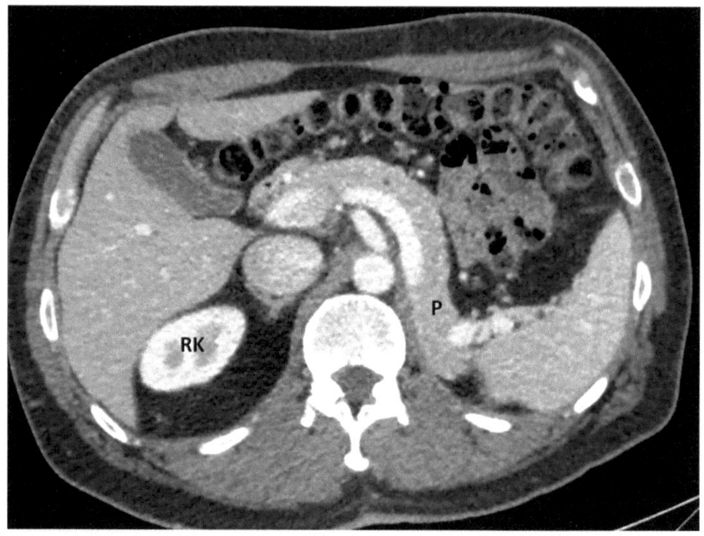

Fig. 9.15 Displacement of the pancreas. Axial computed tomography (CT) image demonstrates displacement of the pancreatic body and tail (P) into an empty left renal fossa in this patient who underwent left nephrectomy for renal cell carcinoma. RK, right kidney.

9

duct.[19,23] As the distal bile duct and pancreatic duct approach the duodenal wall, they become invested by three separate smooth muscle fibers referred to as the sphincter of Oddi, which regulates the passage of bile and pancreatic secretions into the duodenum. The sphincter of Oddi has an average length of 14 mm.[20] The common bile duct and duct of Wirsung unite within this sphincteric segment forming a common channel in the majority of cases, with an average length of 5 mm.[21] This common channel may be Y-shaped if the ducts form a relatively long common channel (70%), it may be V-shaped with a short channel (20%), or the ducts may open separately at closely adjacent sites in the papilla, forming a U-shaped configuration[22] (▶ Fig. 9.20).

The major duodenal papilla is a mucosal covered oval protuberance in the posteromedial wall of the second portion of the duodenum through which the distal end of the ampulla of Vater passes into (▶ Fig. 9.21). The major duodenal papilla is found at the intersection of a transverse mucosal fold that partially covers the papilla and a longitudinal mucosal fold that emanates from its inferior aspect, thus forming a "T" configuration. The location, size, and shape of the major duodenal papilla are variable. The location of the papilla is in the middle third portion of the descending duodenum in 75% of cases or in the horizontal portion of the duodenum in 25% of cases.[23]

The common bile duct enters the head of the pancreas after it passes the first portion of the duodenum in the hepatoduodenal ligament, coursing inferiorly and dorsally to join the duct of Wirsung. There are three major anatomic relationships between the common bile duct and head of the pancreas[1]: Most commonly, there is partial coverage of the duct by pancreatic tissue posteriorly (51.5%),[2] totally covered (30%),[3] and no coverage at all, with the duct running externally in a groove on the superior aspect of the gland (16.5%). The common bile duct may also occasionally course lateral to the pancreatic head[14,24] (▶ Fig. 9.22). The common bile duct is 7 cm long, with an average diameter of 7.4 mm.[25]

The accessory duct (of Santorini) drains the anterior and superior portion of the head into the minor duodenal papilla. This smaller papilla is located approximately 2 cm proximal (superior) and slightly anterior to the major duodenal papilla. The gastroduodenal artery (GDA) serves as an anatomical landmark, positioned anterior to the accessory duct (of Santorini) and the minor duodenal papilla[15] (▶ Fig. 9.22).

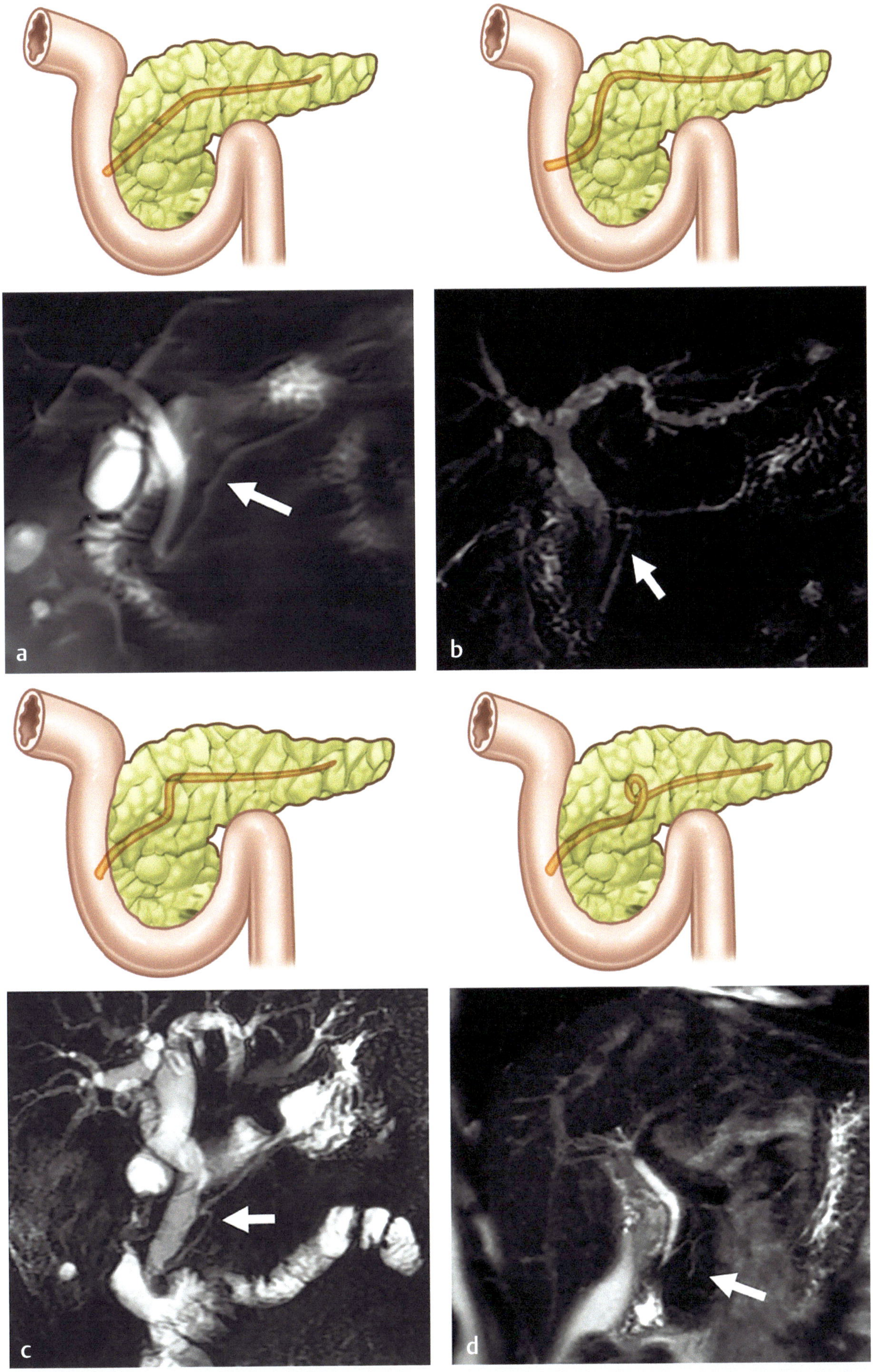

Fig. 9.16 Variations in the course of the pancreatic duct. Schematic drawings (*upper row*) and magnetic resonance (MR) cholangiograms (*lower row*) are shown. The most common course of the pancreatic duct is the descending course (*arrow*) **(a)**. Other variations (*arrows*) include a vertical course **(b)**, a sigmoid course **(c)**, or a loop-shaped course **(d)**.

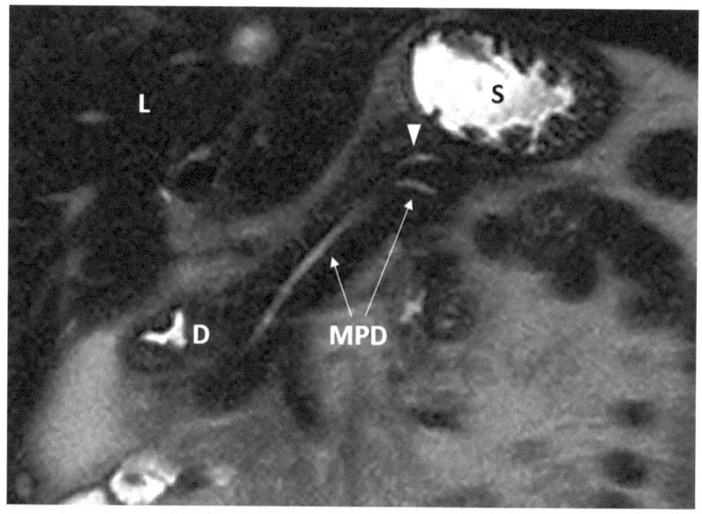

Fig. 9.17 Focal duplication of the pancreatic duct. Coronal magnetic resonance image demonstrates focal duplication of the pancreatic duct in the body and tail (*arrowhead*). D, duodenum; L, liver; MPD (*arrows*), main pancreatic duct; S, stomach.

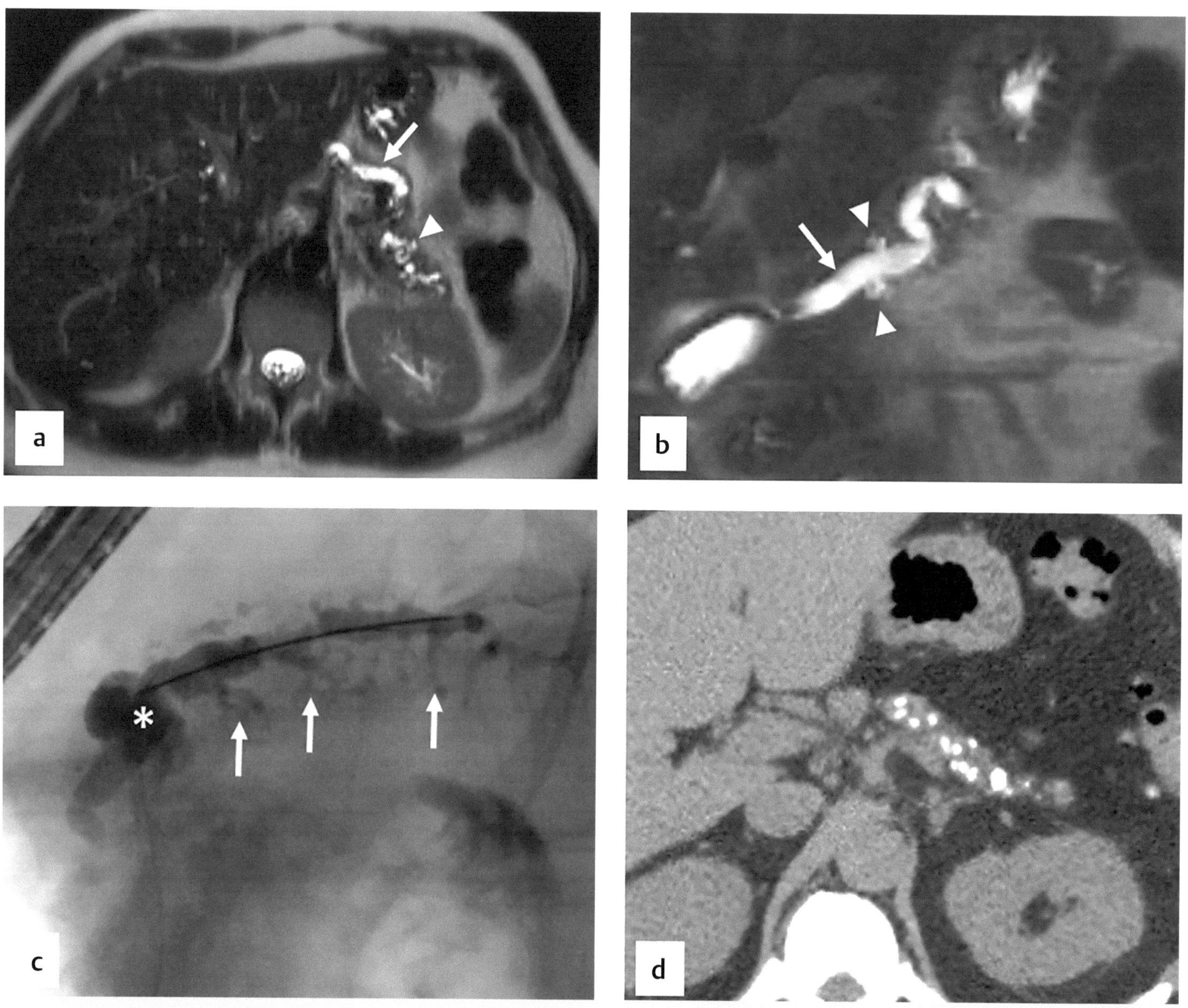

Fig. 9.18 Chronic pancreatitis. **(a, b)** Axial and coronal T2-weighted magnetic resonance (MR) images demonstrate dilatation and tortuosity of the main pancreatic duct (*arrow*) and its side branches (*arrowheads*). Note that the pancreatic parenchyma is diffusely atrophic. **(c)** Endoscopic retrograde cholangiopancreatography (ERCP) image demonstrates segmental areas of irregular dilatation (*) and strictures of the main pancreatic duct. The side branches are also dilated and tortuous (*arrows*). **(d)** Axial, noncontrast computed tomography (CT) from a different patient with chronic pancreatitis demonstrates diffuse pancreatic nodular calcifications.

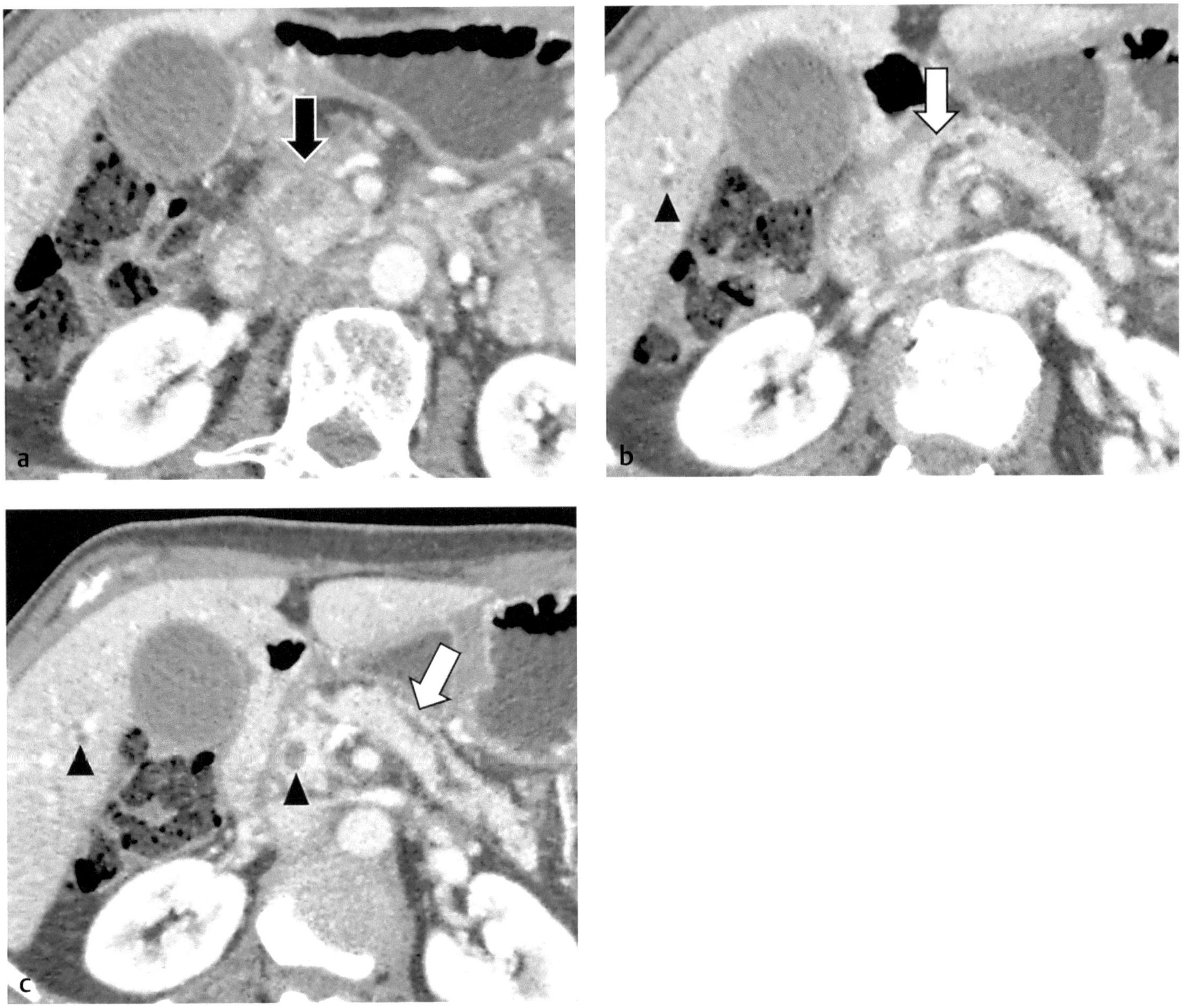

Fig. 9.19 Pancreatic adenocarcinoma. **(a)** Axial computed tomography (CT) image demonstrates an ill-defined hypodense mass in the pancreatic head, compatible with adenocarcinoma (*black arrow*). **(b, c)** Axial CT images demonstrate upstream dilatation of the main pancreatic duct (*white arrows*). There is also dilatation of the intrahepatic and extrahepatic bile ducts (*arrowheads*).

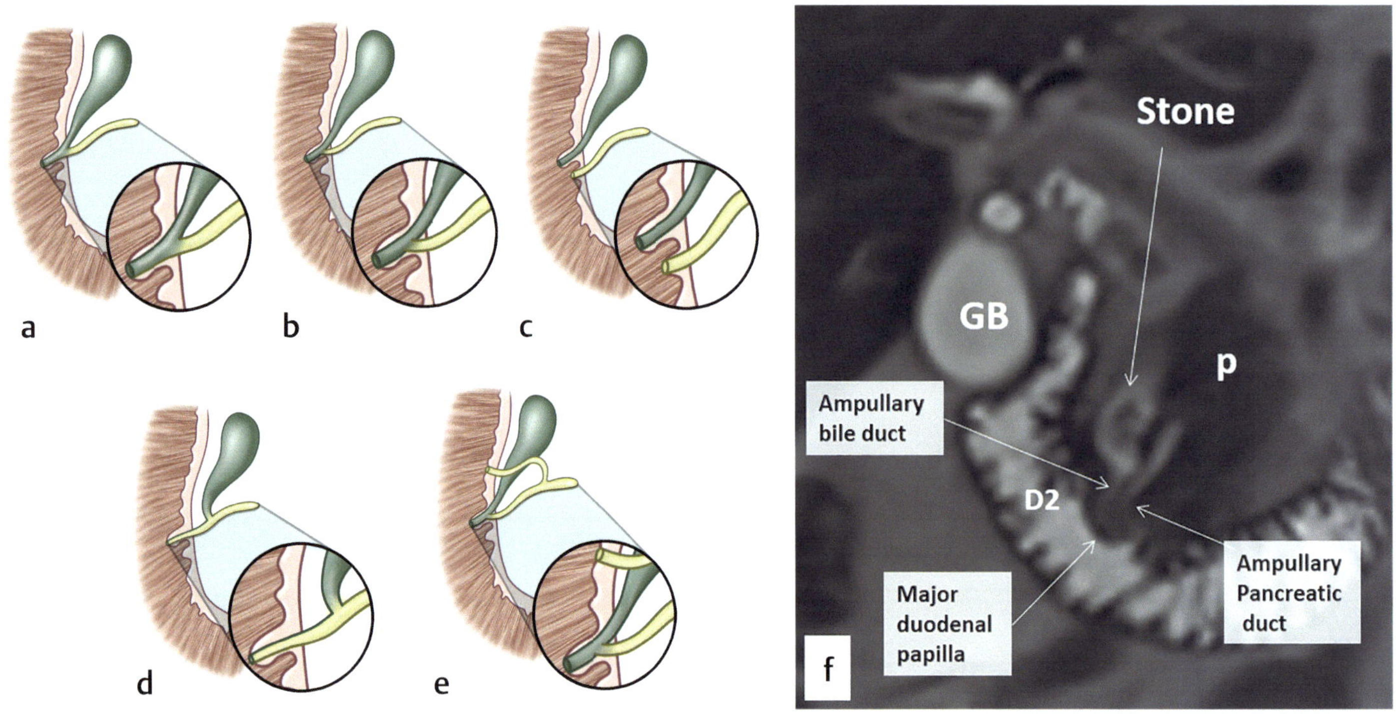

Fig. 9.20 (a–d) Variations in the relationship of the common bile duct and main pancreatic duct at the duodenal papilla. **(a)** Y-shaped. Common channel with near-complete coverage of the duodenal wall. This variant is seen in 70% of patients. **(b)** V-shaped. Double-barreled opening with partial absorption of the common channel into the duodenal wall. **(c)** Separate duodenal openings for the two ducts. **(d)** Pancreaticobiliary maljunction. Abnormal fusion of the pancreatic and common bile duct outside the duodenal wall. **(e)** Diagram of the normal ampulla of Vater. The ampulla of Vater consists of the ampullary portion (ampullary bile duct, ampullary pancreatic duct) and the papillary portion (major duodenal papilla) surrounded by the sphincter of Oddi. The *dashed white circle* indicates the periampullary region located within 2 cm of the major duodenal papilla. **(f)** Magnetic resonance (MR) cholangiogram demonstrating components of the ampulla of Vater. The distal common bile duct is dilated and contains a stone (*arrow*). D2, second part of the duodenum; GB, gallbladder; P, pancreatic head.

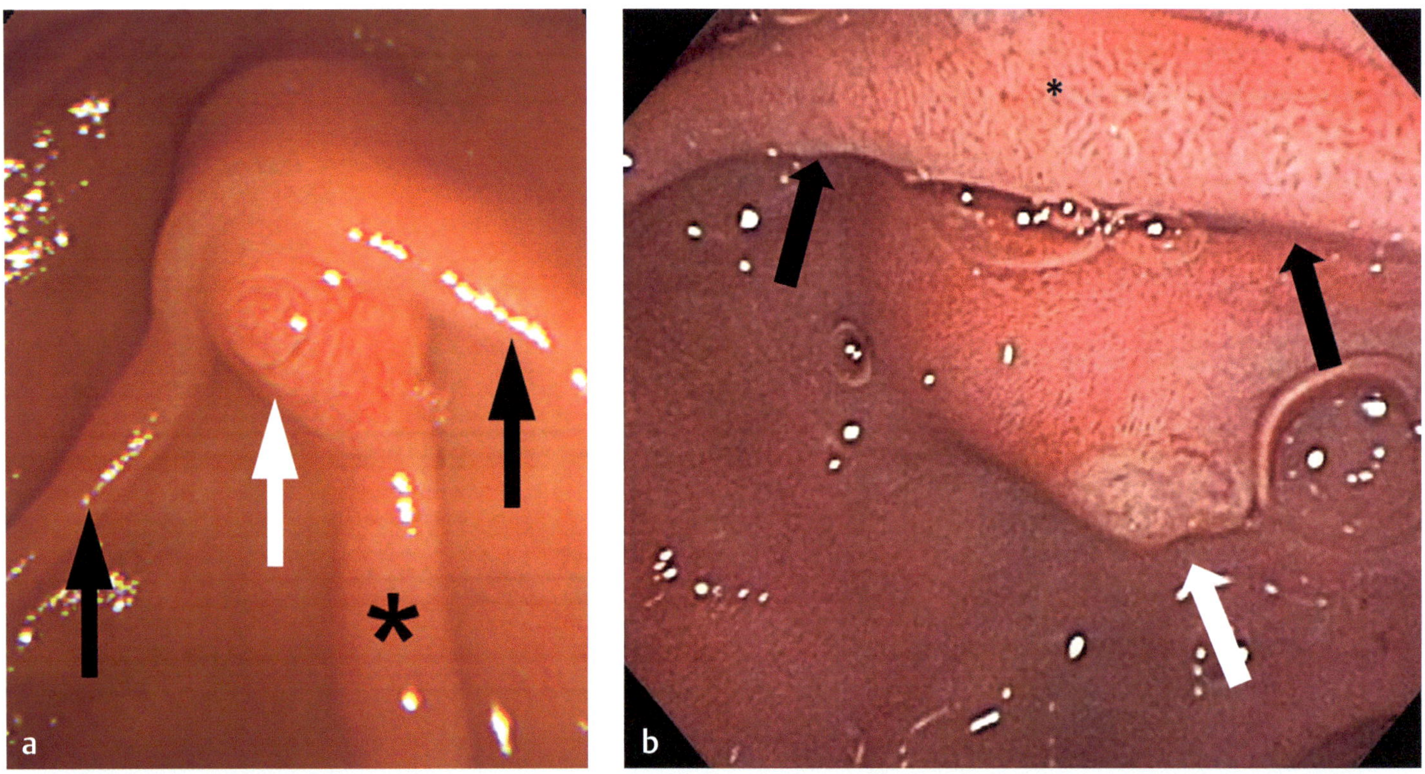

Fig. 9.21 Endoscopic images of the major duodenal papilla. **(a)** The major duodenal papilla (*white arrow*) is found at the intersection of a transverse mucosal fold (*black arrows*) that partially covers the papilla and a longitudinal mucosal fold (*asterisk*) that emanates from its inferior aspect, thus forming a "T" configuration. (Reproduced with permission from Kim et al.[21]) **(b)** The major duodenal papilla (*white arrow*) is seen as an oval protuberance in the posteromedial wall of the second portion of the duodenum. Note the overlying transverse mucosal fold (*black arrows*).

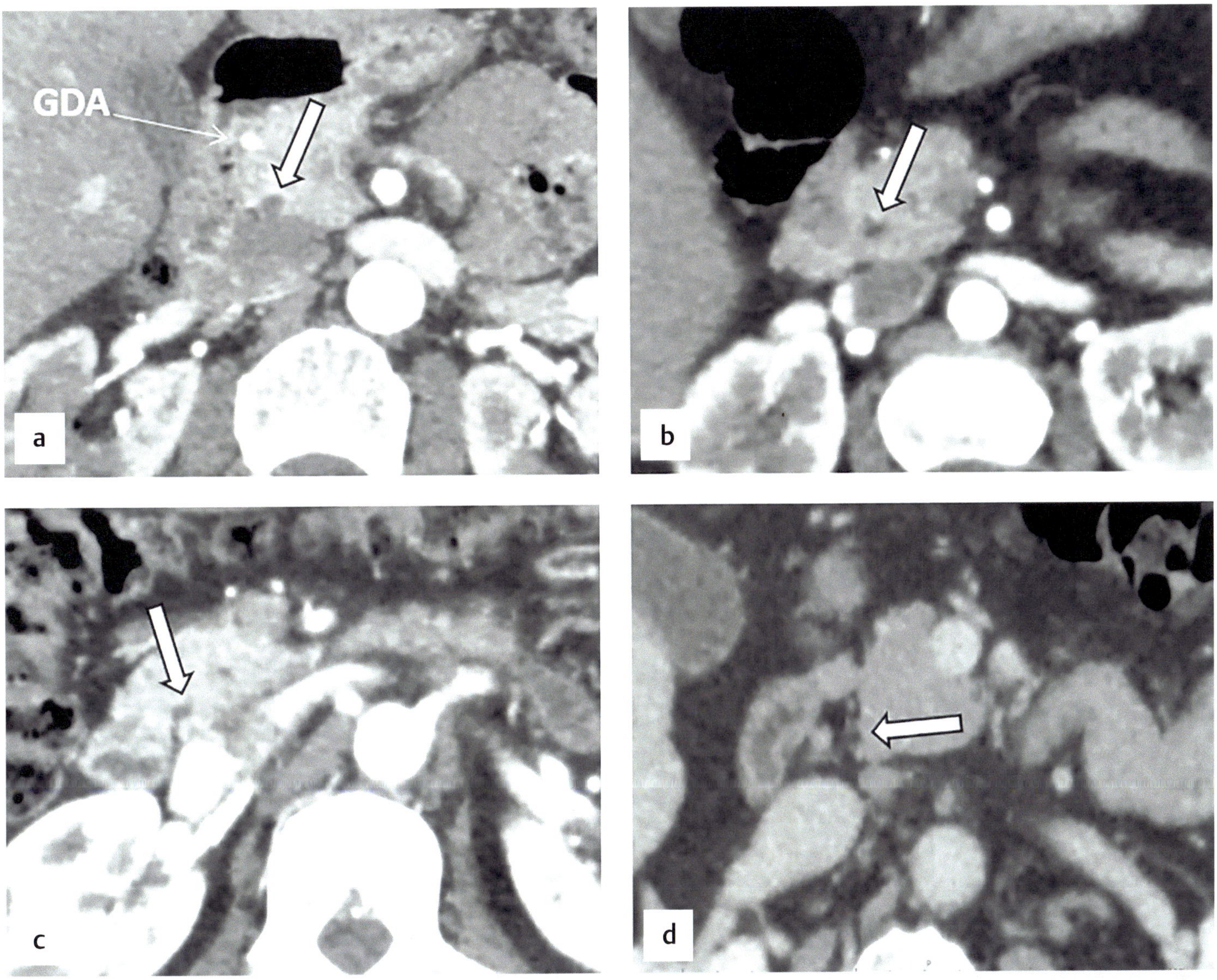

Fig. 9.22 Anatomic relationship between common bile duct (CBD) and pancreas on axial computed tomography (CT) images. Partial **(a)**, total **(b)**, and no **(c)** coverage of the CBD (*arrow*) by pancreatic tissue. Occasionally, the CBD may course lateral to the pancreatic head **(d)**. The gastroduodenal artery (GDA) serves as an anatomical landmark, positioned anterior to the accessory duct (of Santorini) and the minor duodenal papilla.

◆ Variations in Pancreatic Duct Anatomy

Communication between the Dorsal and Ventral Pancreatic Ducts

Due to the developmental origin of the two pancreatic ducts, several anatomical pancreatic ductal variants can be seen. Most commonly (60%), the downstream configuration of the MPD presents as a bifid configuration, with communicating dual drainage of the MPD formed by the duct of Wirsung and duct of Santorini. Less common configurations include a rudimentary duct of Santorini (30%) and a dominant duct of Santorini (1%)[13] (▸Fig. 9.23).

Another variant is "ansa pancreatica" in which there is caudally looping communication between the ducts of Santorini and Wirsung that connects the main duct to the minor papilla (▸Fig. 9.23). However, the entrance of the ansa pancreatica into the duodenum is not patent in the majority of cases.[26]

Pancreas Divisum

Pancreas divisum is the most common congenital anomaly of the pancreatic ductal system seen in 4 to 10% of people.[27] It is usually asymptomatic, although it has been implicated in the development of acute and chronic pancreatitis.[28] Pancreas divisum is characterized by failure of complete fusion of the dorsal duct (of Santorini) and ventral duct (of Wirsung) draining the pancreas. In pancreatic divisum, the duct of Santorini drains the body and tail of the pancreas (majority of the gland) into the duodenum via the minor papilla, while the duct of Wirsung drains the caudal portion of the head and uncinate process (minority of the gland) into the duodenum via the major papilla. Pancreatic divisum can be divided into three types: type I, or the classic form,

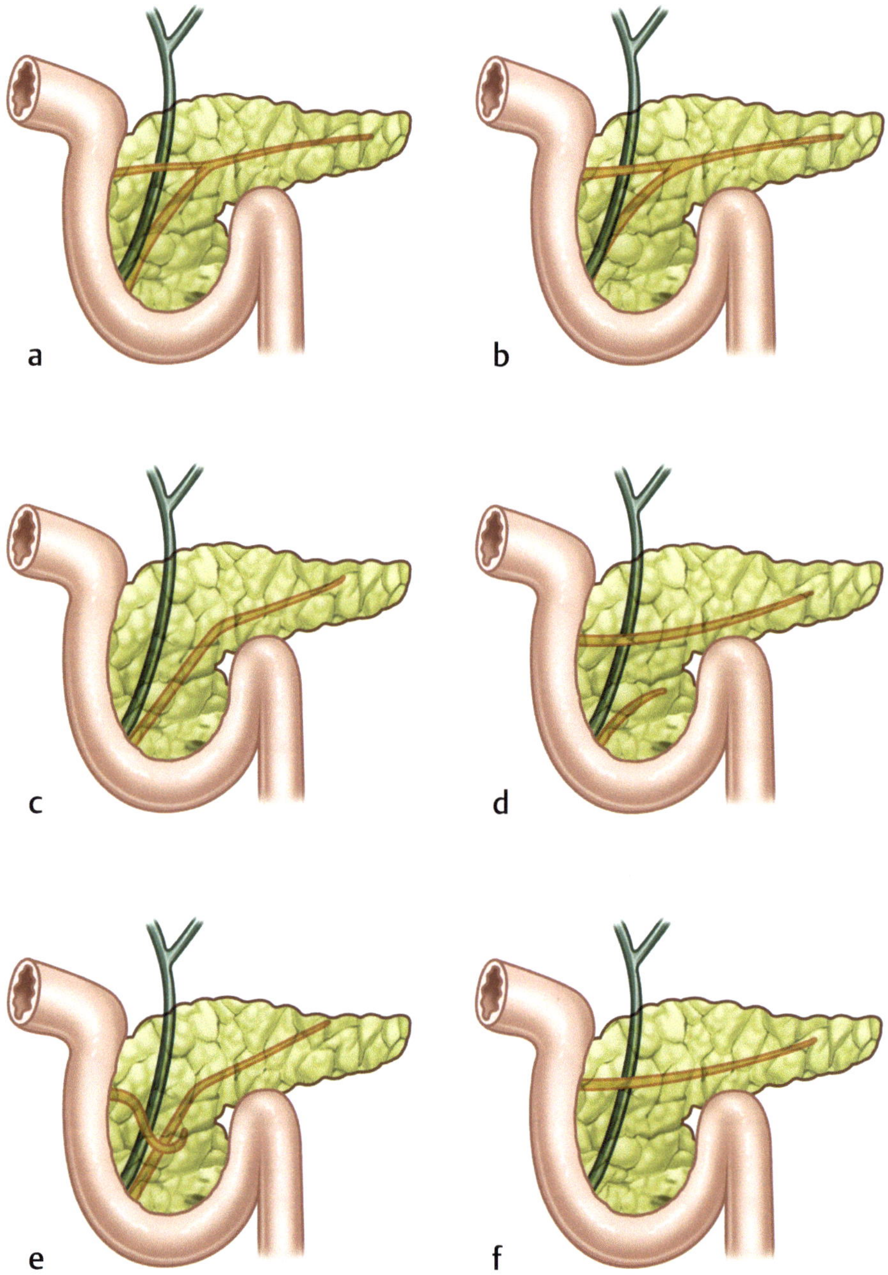

Fig. 9.23 Variations in configuration of the pancreatic duct. **(a)** Bifid configuration, with dominant duct of Wirsung drainage. **(b)** Bifid configuration with dominant duct of Santorini drainage (type III pancreatic divisum). **(c)** Rudimentary duct of Santorini. **(d)** Type I pancreatic divisum. **(e)** Ansa pancreatica. **(f)** Dominant duct of Santorini drainage and absent Wirsung (type II pancreatic divisum). (Modified from Türkvatan et al.[26])

9

with complete lack of fusion between the dorsal and ventral duct systems; type II, where the duct of Wirsung is absent; and type III, or the partial form, in which there is filamentous communication between the dorsal and ventral duct system, with the majority of the pancreas being drained through the dorsal system[29] (▶ Fig. 9.23, ▶ Fig. 9.24).

In pancreatic divisum, the ventral ducts are typically short, but do vary in size, ranging from barely visible to several centimeters long.[30] The ducts usually arborize and gradually decrease in diameter. The dorsal duct is typically larger in caliber than the ventral duct.[31] Cystic focal dilatation of the terminal portion of the dorsal pancreatic duct, known as santorinicele, is thought to be associated with pancreas divisum and relative obstruction at the minor duodenal papilla.[32]

Anomalous Pancreaticobiliary Junction

In 60 to 80% of people, the pancreatic and bile ducts unite to form a common channel 2 to 15 mm long, averaging 5 mm.[33] An anomalous pancreaticobiliary junction (also known as pancreaticobiliary maljunction) is characterized by abnormal fusion of the pancreatic and common bile duct outside the duodenal wall and beyond the influence of the sphincter of Oddi, with formation of a long common channel (>15 mm) (▶ Fig. 9.20d). The normal angle between the common bile duct and pancreatic duct is 5 to 30 degrees but can be as much as 90 degrees.[34] This anomaly is seen in less than 10% of individuals. The presence of a long common channel is clinically important, as it can predispose to intrabiliary reflux of pancreatic

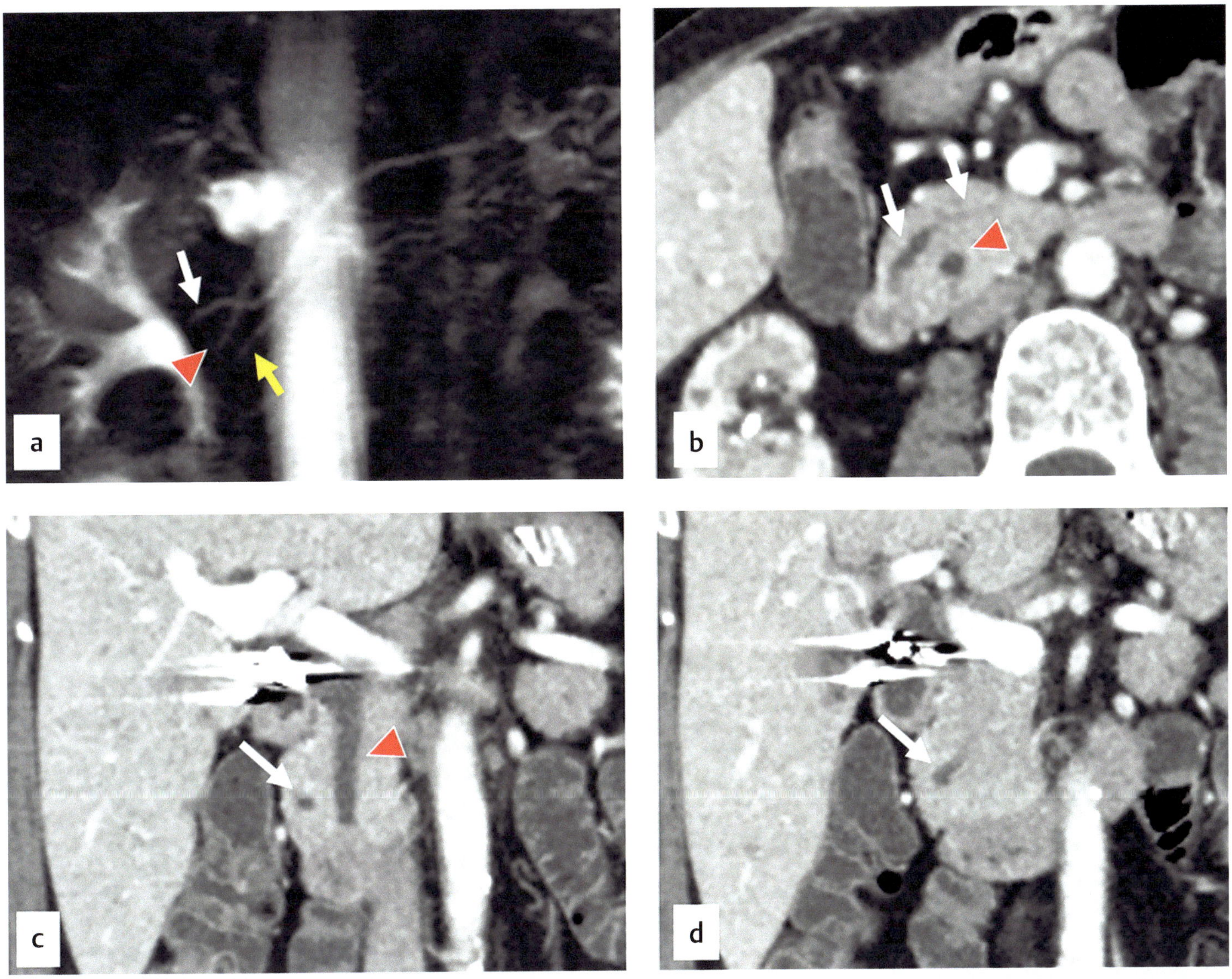

Fig. 9.24 Pancreatic divisum. **(a)** Coronal magnetic resonance pancreatography image demonstrates classic form (type I) of pancreatic divisum where there is complete lack of fusion between the duct of Santorini (*white arrow*) and duct of Wirsung (*yellow arrow*). The common bile duct (*red arrowhead*) and the Wirsung duct drain into the major duodenal papilla. **(b–d)** Axial **(b)** and coronal **(c, d)** images demonstrate type II pancreatic divisum, with dominant duct of Santorini (*white arrow*) and absent duct of Wirsung. The common bile duct is shown by *red arrowheads*.

secretions and formation, possibly contributing to development of some types of choledochal cysts and pancreatitis.[33] Anomalous pancreaticobiliary junction is also reported to be associated with cholangiocarcinoma and gallbladder carcinoma.[35,36]

◆ Developmental Anomalies

Annular Pancreas

Annular pancreas is a rare congenital anomaly in which pancreatic tissue partially or completely encircles the duodenum, usually in its second portion, almost always at or above the major papilla (▶ Fig. 9.25a). The annular tissue may consist of a fibrous band or glandular pancreatic tissue. The etiology of annular pancreas is not entirely clear, although it is likely related to failure of normal migration. The most widely accepted theory (Lecco's theory) is that the tip of the right ventral bud adheres to the duodenal wall and stretches to form a ring during normal rotation.[37] Baldwin's theory suggests the persistence of the left ventral bud.[38] A third theory suggests that the tip of the left ventral bud adheres to the duodenum and stretches to form a ring during development. In this theory, several arrangements of the annular ducts can be created depending on whether the tip is proximal or distal to the bile ducts.[39]

Several anatomic classifications of annular pancreas have been described, one of which has two subtypes: intramural and extramural.[40] In the intramural type, the pancreatic tissue is intermingled with muscle fibers in the duodenal wall, with small ducts draining directly into the duodenum. In the extramural type, the ventral pancreatic duct encircles the duodenum to join the MPD. The prevalence of annular pancreas is 5 to 15 cases per 100,000 adults.[41] The clinical presentation in adults differs from that in children. Children usually present because of upper gastrointestinal obstruction while adults less often present due to symptoms of upper gastrointestinal obstruction, peptic ulcer disease, or pancreatitis.[42] Of note, pancreas divisum has been reported in up to 36% of cases in patients with annular pancreas.[43]

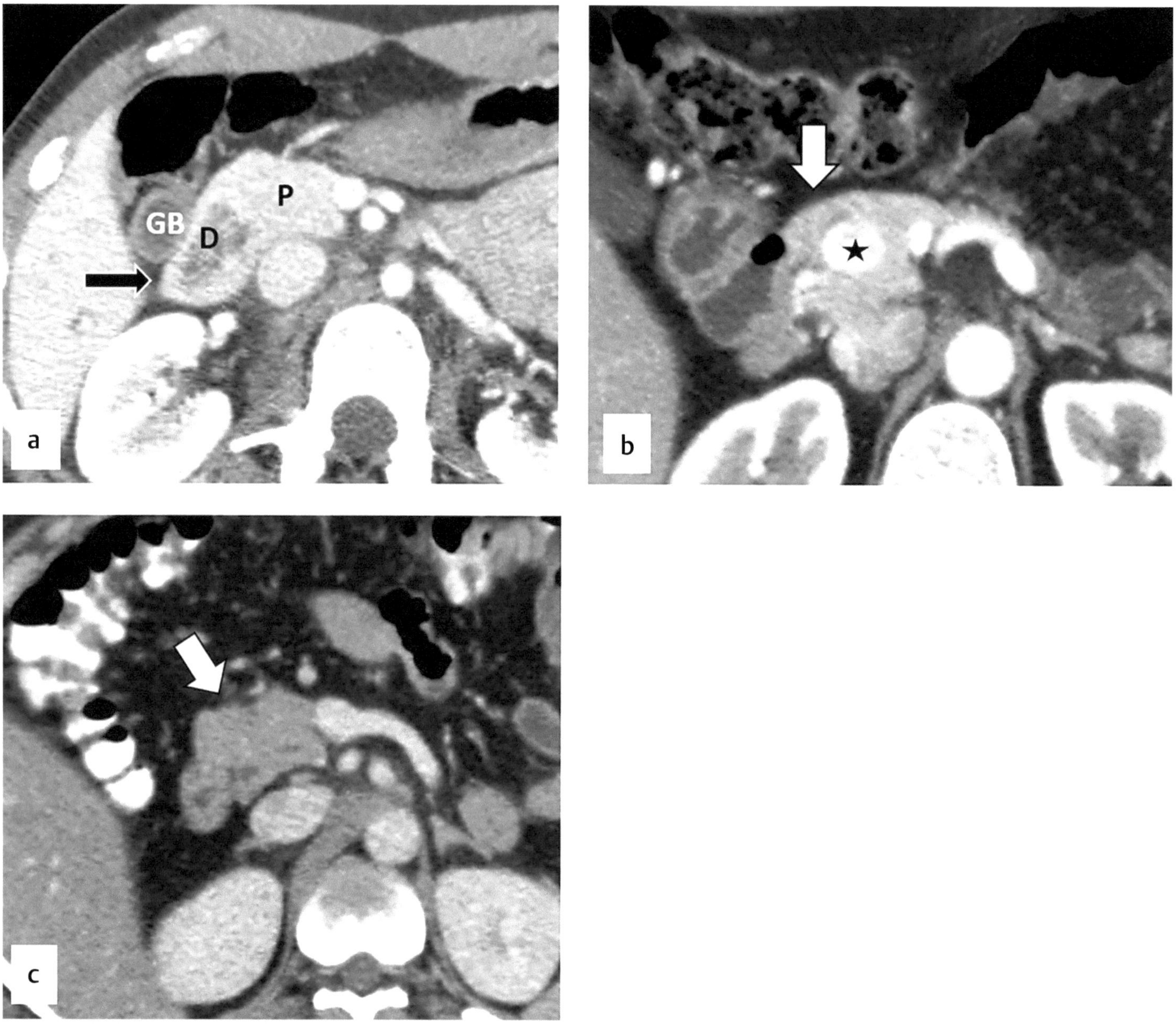

Fig. 9.25 (a) Annular pancreas. Axial computed tomography (CT) image demonstrates the pancreas (P) encircling the descending portion of the duodenum (D) (*arrow*). GB, gallbladder. **(b)** Portal annular pancreas. Axial CT image demonstrates the pancreas (*arrow*) encircling the portal vein (*star*). **(c)** Dorsal agenesis of the pancreas. Axial CT image demonstrates the pancreatic head (*arrow*) and absence of the pancreatic neck, body, and tail.

Very rarely, the uncinate process of the pancreas fuses to the dorsal surface of the body of the pancreas by encircling the portal vein, referred to as portal annular pancreas. Although generally asymptomatic, this anomaly is important to recognize when pancreatic surgery is being contemplated[44] (▶ Fig. 9.25b).

Pancreas Agenesis

Complete agenesis of both the dorsal and ventral pancreas is extremely rare and is incompatible with life. Partial agenesis of the pancreas may involve either the ventral or dorsal component.[45] Agenesis of the dorsal component is much more common than the ventral component.[46] In addition, agenesis of the dorsal or ventral component may be complete or incomplete. In the complete form of agenesis of the dorsal pancreas, only the pancreatic head and uncinate process develop. The pancreatic body, tail, and dorsal pancreatic duct system, including the accessory duct (of Santorini), are completely absent. In the partial form, which is relatively more common, the terminal end of the accessory duct, minor papilla, and the neck and proximal body of the pancreas are present[47] (▶ Fig. 9.25c). Dorsal agenesis of the pancreas may be asymptomatic or associated with recurrent pancreatitis and diabetes.[48] In addition, dorsal agenesis of the pancreas can be associated with other congenital abnormalities, including polysplenia and malrotation.

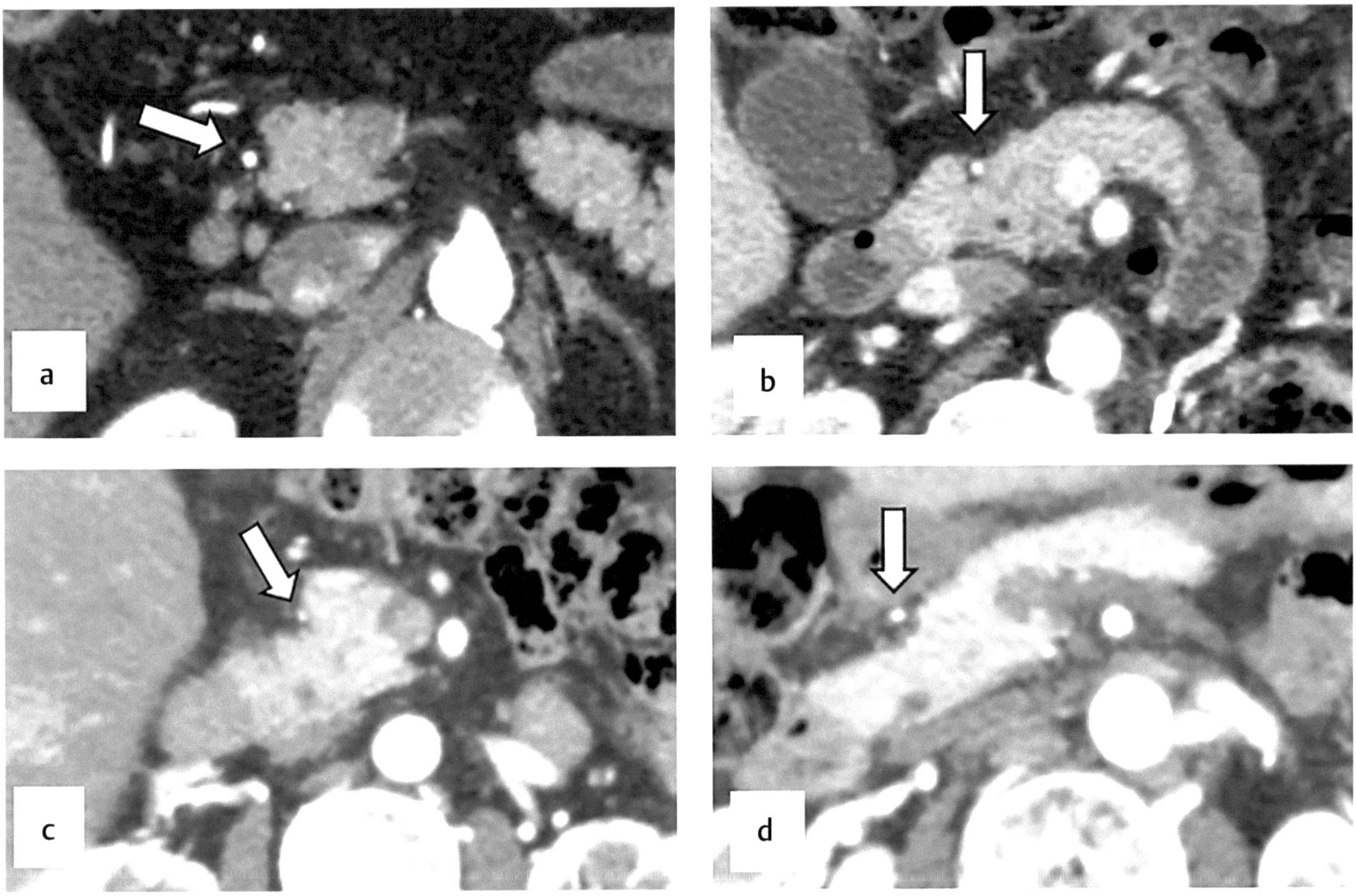

Fig. 9.26 (a–d) Pancreatic head configurations relative to the anterior superior pancreaticoduodenal artery (*arrows*). Axial computed tomography (CT) scan demonstrates normal position **(a)**, anterior position **(b)**, posterior position **(c)**, and horizontal position **(d)**.

Pancreatic Head Lobulations

Pancreatic head and neck lobulations have been reported in 35% of the general population and are defined as outpouchings of the gland more than 1.0 cm beyond the anterior superior pancreaticoduodenal artery (ASPDA). These are usually classified into three types: in type 1, the lobulation is anterior to the artery and is seen in 10%; in type 2, the lobulation is oriented posteriorly and is seen in 19%; and in type 3, the lobulation is oriented horizontally and laterally and is seen in 5%[49] (▶ Fig. 9.26).

Pancreatic Lipomatosis

Fatty replacement of the pancreas represents the most common histologic change in the adult pancreas. This is usually age related and involves the whole organ homogeneously. However, uneven pancreatic lipomatosis can occur and is thought to be due to the different histologic composition of the ventral and dorsal ducts.[50] The peribiliary region of the pancreatic head is usually spared. Fatty infiltration may be limited to the head (50%) or extend to involve body and tail.[50] Uneven pancreatic lipomatosis should not be mistaken for neoplasm (▶ Fig. 9.27). Total fatty replacement of the pancreas is a common finding in cystic fibrosis.

Ectopic Pancreas

Ectopic pancreas is defined as pancreatic tissue located outside the normal confines of the pancreas and lacking any vascular connection. On histologic examination, the ectopic pancreatic tissue may have all of the elements of the normal pancreas, which include pancreatic acini, pancreatic ducts, and islets of Langerhans. Ectopic pancreas presumably occurs as a mishap in the embryological development of the pancreas. The incidence ranges from 0.6 to 13.7%.[51] Most commonly, ectopic pancreatic tissue is seen in the stomach (26%), duodenum,[28] and jejunum (16%)[52] (▶ Fig. 9.28). Most ectopic pancreatic tissue is functional and may be affected by inflammatory or neoplastic disease.[53]

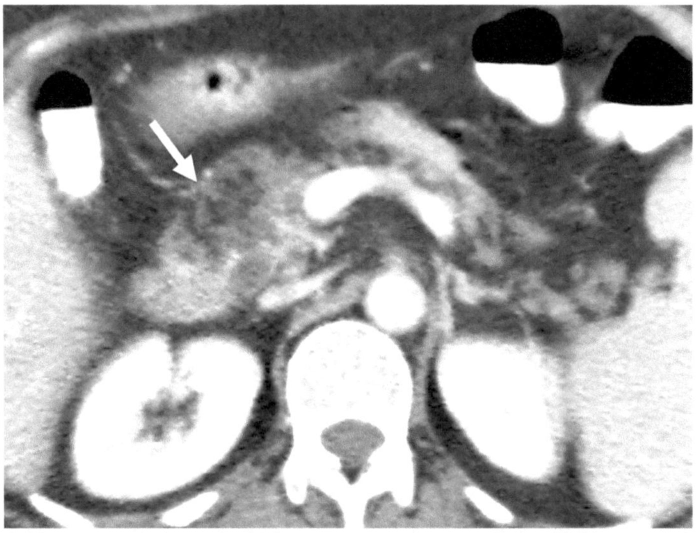

Fig. 9.27 Uneven pancreatic lipomatosis. Axial computed tomography (CT) demonstrates focal hypoattenuation in the pancreatic head that corresponds with fatty replacement (*arrow*). This should not be mistaken for potential neoplasm. Note the posterior part of the pancreatic head, including the peribiliary tissue, and uncinate process are spared.

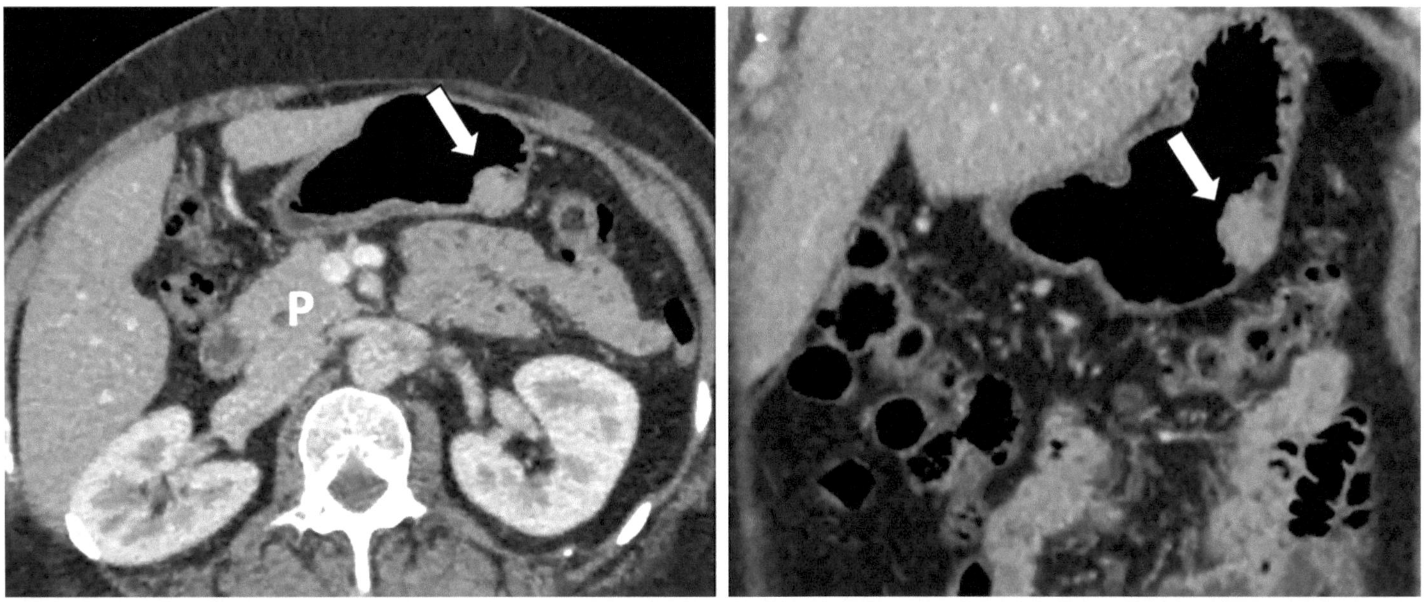

Fig. 9.28 Ectopic pancreas in a 53-year-old female. Axial and coronal computed tomography (CT) images demonstrate an ovoid submucosal lesion in the wall of the gastric body along the greater curvature (*arrows*). Note the enhancement is similar to that of normal pancreas (P). The lesion was confirmed as ectopic pancreas after surgical excision.

◆ Neurovasculature

Arteries

The arterial blood supply of the pancreas is complex, arising from branches of the celiac artery and the superior mesenteric arteries (▶ Fig. 9.29, ▶ Fig. 9.30, ▶ Fig. 9.31, ▶ Fig. 9.32). Knowledge of pancreatic arterial anatomy is particularly important in planning pancreatic intervention and interpretation of pancreatic imaging.[54]

The common hepatic artery (CHA) courses toward the right near the pancreatic neck and head. At this point, the CHA usually divides into the proper hepatic artery and GDA. The splenic artery directly gives rise to several important vessels that supply the pancreatic body and tail, including the dorsal pancreatic artery (DPA), the pancreatic magna artery (PMA), and the caudal pancreatic artery (CPA) (▶ Fig. 9.30).

The SMA arises anteriorly from the abdominal aorta approximately 1 to 2 cm distal to the origin of the celiac artery. The origin of the SMA is usually on the left side of the SMV. The SMA courses posterior and inferior to the neck of the pancreas and anterior to the uncinate process. The SMA usually gives rise to the inferior pancreaticoduodenal artery (IPDA) which subsequently divides into posterior and anterior branches.

The head of the pancreas, which shares its arterial supply with the adjacent duodenum, is supplied mainly by the GDA and IPDA. The GDA gives origin to the posterior superior pancreaticoduodenal artery (PSPDA), retroduodenal artery, and supraduodenal artery. The GDA terminates by dividing into the right gastroepiploic artery and ASPDA (▶ Fig. 9.29, ▶ Fig. 9.30). The ASPDA and PSPDA supply the head of the pancreas, sometimes through a common superior pancreaticoduodenal artery (SPDA).[55]

The extensive collateral network in the pancreas is important, particularly in cases of celiac, superior mesenteric, or splenic artery occlusion. A posterior arcade is formed by anastomoses between the posterior branch of the IPDA and the PSPDA. An anterior arcade is formed by anastomoses between the ASPDA and anterior branch of the IPDA.

The neck and body of the pancreas is mainly supplied by branches of the splenic artery, DPA, PMA, and transverse pancreatic artery (TPA) (▶ Fig. 9.32). The DPA supplies the head and neck of the pancreas and most commonly arises from the proximal splenic artery.[56] The DPA is considered to be the largest vessel arising from the splenic artery, sometimes as large as one-third of the splenic artery[57] (▶ Fig. 9.32). The right terminal branch of the DPA commonly communicates with the anterior pancreatic arcade or other GDA branch. The left terminal branch of the DPA is known as the TPA, which courses along the inferior surface of the pancreatic body and tail.[58] The DPA also commonly arises from the GDA. The PMA may be single, but is usually a group of comb-shaped branches that are perpendicular to the splenic artery.[58] The PMA courses along the superior margin of the pancreatic body and tail and usually anastomoses with the TPA in majority of cases, and less frequently with the DPA and CPA.[59]

The tail of the pancreas is supplied by the CPA and may be single or represented by multiple independent branches which originate from the splenic artery or right gastroepiploic artery. These branches may anastomose with the TPA and PMA.[58]

Variations of the peripancreatic arteries are common.[57] The GDA rarely arises from left hepatic artery. Anastomotic branch between the ASPDA and the DPA is seen in 15% (▶ Fig. 9.29). The PSPDA may directly originate from the hepatic artery. The IPDA arises from the SMA in 64%. Another origin site of the IPDA is

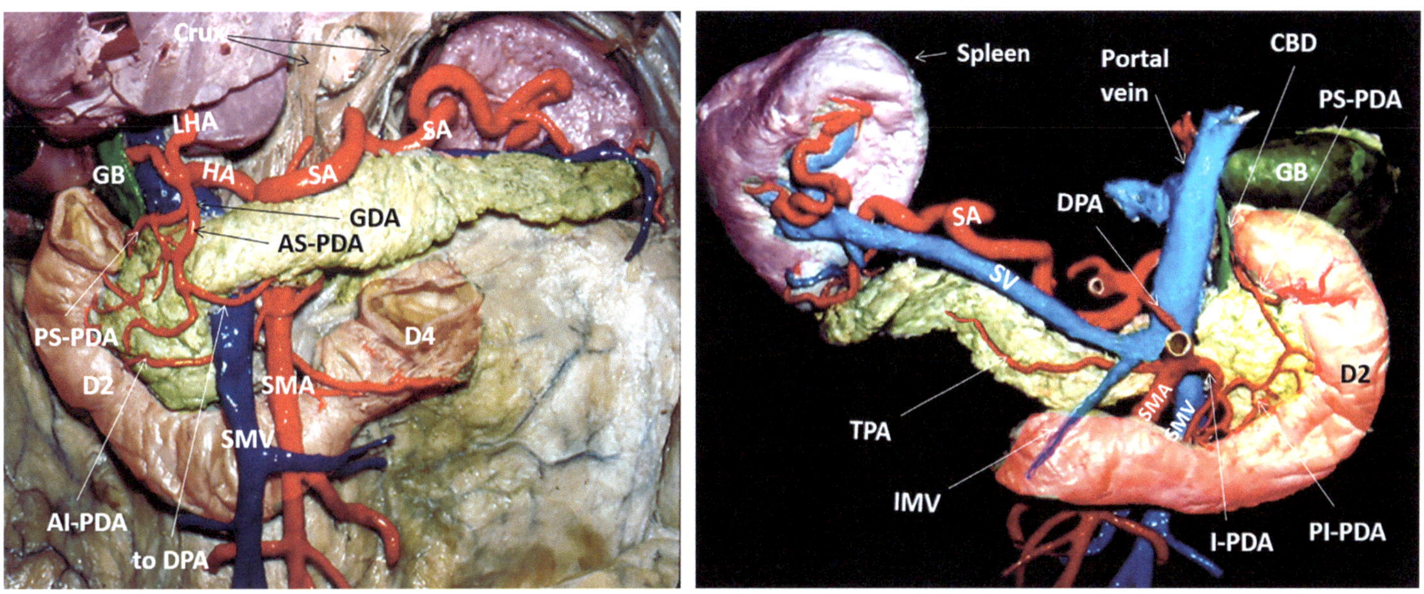

Fig. 9.29 Arterial supply to the pancreatic head. The gastroduodenal artery (GDA) is a branch of the hepatic artery (HA) which gives origin to the posterior superior pancreaticoduodenal artery (PSPDA), retroduodenal artery, and supraduodenal artery. The GDA terminates by dividing into the right gastroepiploic artery and anterior superior pancreaticoduodenal artery (ASPDA). The superior mesenteric artery (SMA) usually gives rise to the inferior pancreaticoduodenal artery (IPDA) which subsequently divides into posterior (PIPDA) and anterior (AIPDA) branches. The transverse pancreatic artery (TPA) arise from the splenic artery (SA) or GDA and supplies the body and tail. CBD, common bile duct; D2/D4, second and fourth portions of the duodenum; DPA, dorsal pancreatic artery; E, esophagus; GB, gallbladder; IMV, inferior mesenteric vein; LHA, left hepatic artery; SMV, superior mesenteric vein; SV, splenic vein.

9

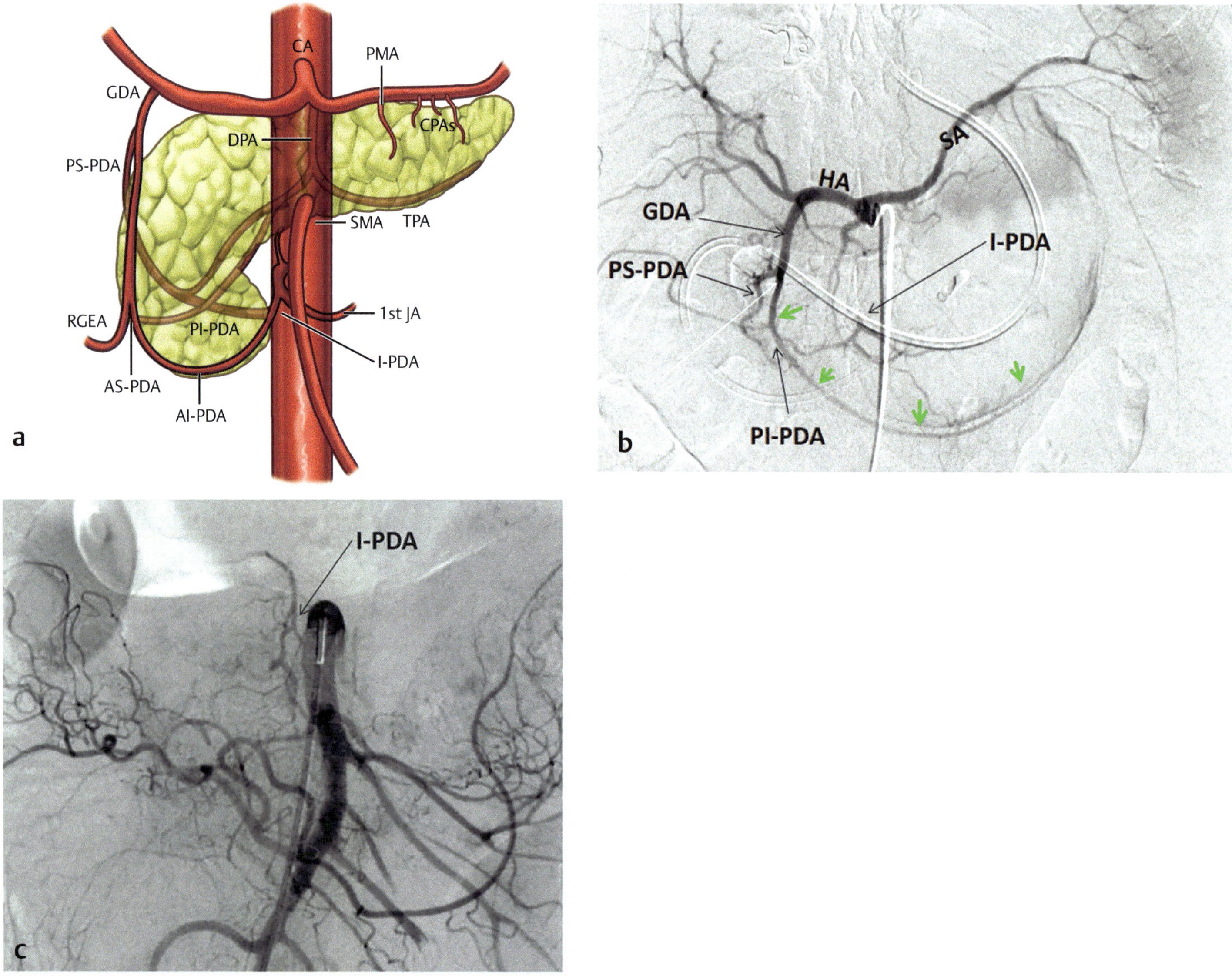

Fig. 9.30 (a) Arterial supply of the pancreas. The arterial supply of the pancreas is complex, arising from branches of the celiac artery and the superior mesenteric arteries. **(b)** Catheter angiogram of the celiac artery showing branches of the gastroduodenal artery (GDA). **(c)** Catheter angiogram of the superior mesenteric artery (SMA). The SMA gives rise to the inferior pancreaticoduodenal artery (I-PDA) which subsequently divides into posterior and anterior branches. The GDA terminates by dividing into the right gastroepiploic (*arrows*) and anterior superior pancreaticoduodenal arteries (AS-PDAs). I-PDA is filled in retrograde in figure **b**. AI-PDA, anterior inferior pancreaticoduodenal artery; CA, celiac artery; CPA, caudal pancreatic artery; DPA, dorsal pancreatic artery; RGEA, right gastroepiploic artery; HA, hepatic artery; JA, jejunal artery; PI-PDA, posterior inferior pancreaticoduodenal artery; PMA, pancreatic magna artery; PS-PDA, posterior superior pancreaticoduodenal artery; SA, splenic artery; TPA, transverse pancreatic artery.

the first jejunal branch artery (34%), and rarely from a replaced right hepatic artery or a common trunk with the DPA from the SMA. The DPA usually arises from the splenic artery. Other origins of DPA include the IPDA (33%), replaced SMA, the GDA, and first jejunal artery. The TPA arises from the DPA (80%) or IPDA, ASPDA, and SMA.

Pancreatoduodenal arcade branches are common sites for interventional procedures, such as embolization to stop a bleeding site or treat pseudoaneurysm (i.e., trauma, pancreatitis) (▶ Fig. 9.33).

Veins

The venous drainage of the pancreatic head initially courses with the arteries and can be divided mainly into anterior and posterior arcades[60] (▶ Fig. 9.34). The anterior pancreaticoduodenal venous arcade is formed by the ASPD and AIPD veins. The ASPD vein empties into the right gastroepiploic vein, which then drains into the SMV through the gastrocolic trunk.[60] The AIPD vein runs transversely between the third portion of the duodenum and the

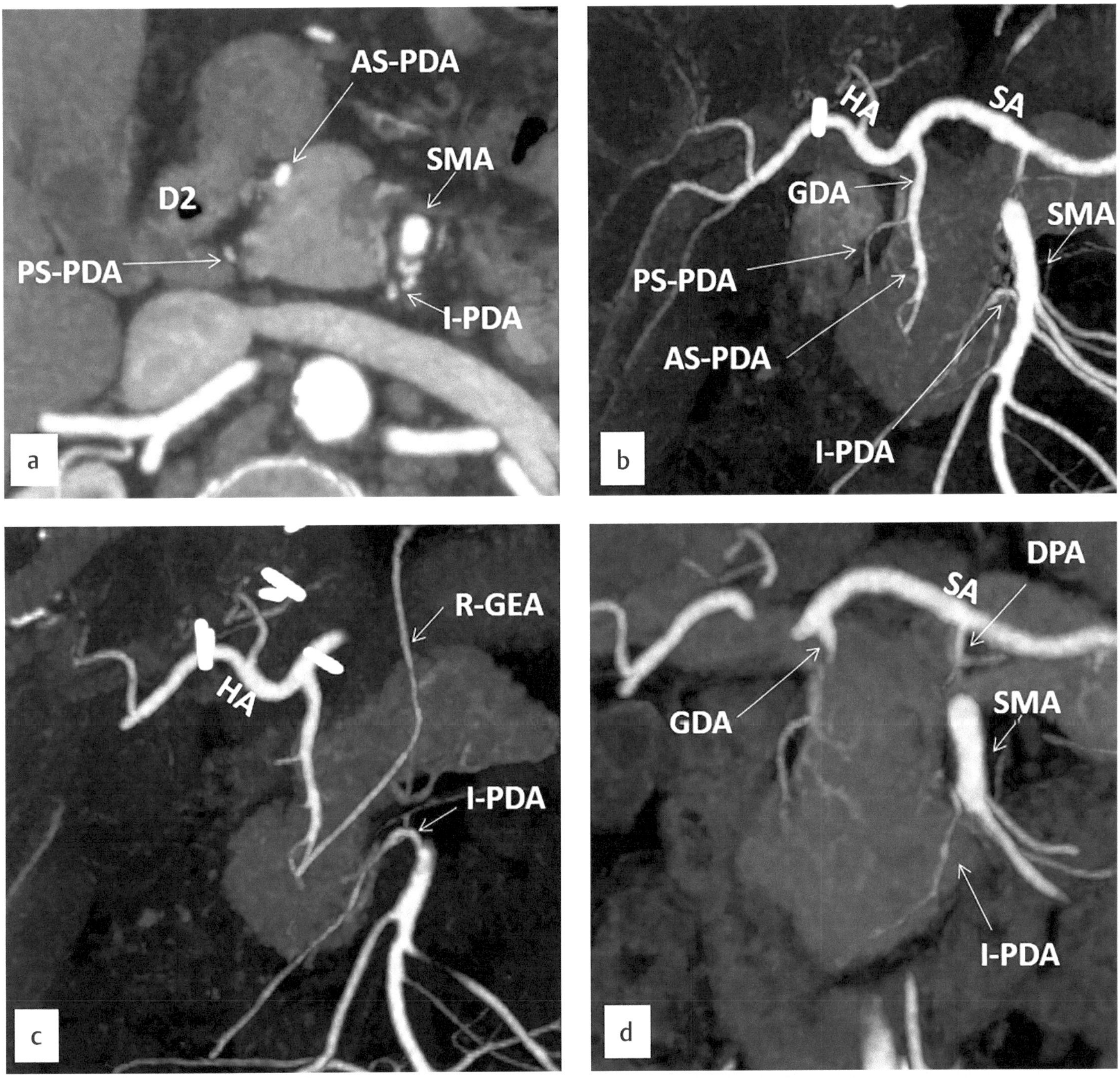

Fig. 9.31 Axial **(a, b)** and coronal **(c, d)** computed tomography (CT) angiography images showing branches of the pancreatic head. AS-PDA, anterior superior pancreaticoduodenal artery; D2, second duodenum; DPA, dorsal pancreatic artery; GDA, gastroduodenal artery; HA, hepatic artery; I-PDA, inferior pancreaticoduodenal artery; PS-PDA, posterior superior pancreaticoduodenal artery; R-GEA, right gastroepiploic artery; SA, splenic artery; SMA, superior mesenteric artery.

uncinate process and joins the uppermost jejunal vein, usually via common trunk with the PIPD vein or less frequently drains directly into the SMV.[61] The posterior pancreaticoduodenal venous arcade is formed by the PSPD and PIPD veins. The PSPD vein is the largest draining vein of the pancreatic head, running upward between the common bile duct and the second portion of the duodenum and joining into the portal vein. The PIPD vein passes transversely to the posterior aspect of the pancreatic head and may join the AIPD vein or may directly drain into the SMV.[61,62] The ASPD vein runs transversely at the anterior surface of the pancreatic head and drains into the SMV. Two other vessels may also contribute to the venous drainage in the pancreatic head, including an inferior

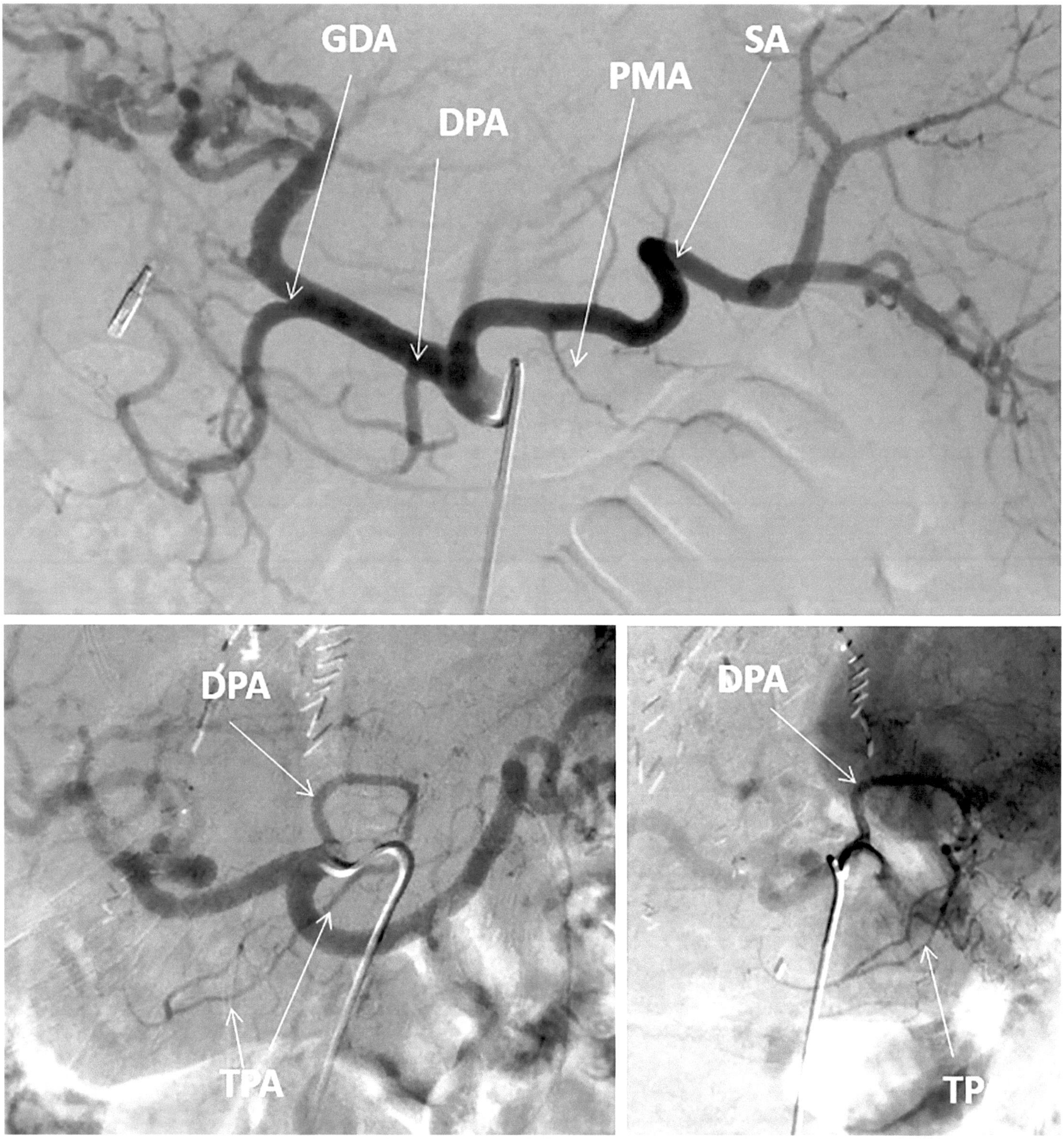

Fig. 9.32 Catheter angiogram of the celiac artery showing branches of the body and tail of pancreas. The splenic artery (SA) gives rise to the dorsal pancreatic artery (DPA) and the pancreatic magna artery (PMA). The DPA is the largest vessel arising from the SA. The left terminal branch of the DPA is known as the transverse pancreatic artery (TPA), which courses along the inferior surface of the pancreatic body and tail. The DPA also commonly arises from the gastroduodenal artery (GDA).

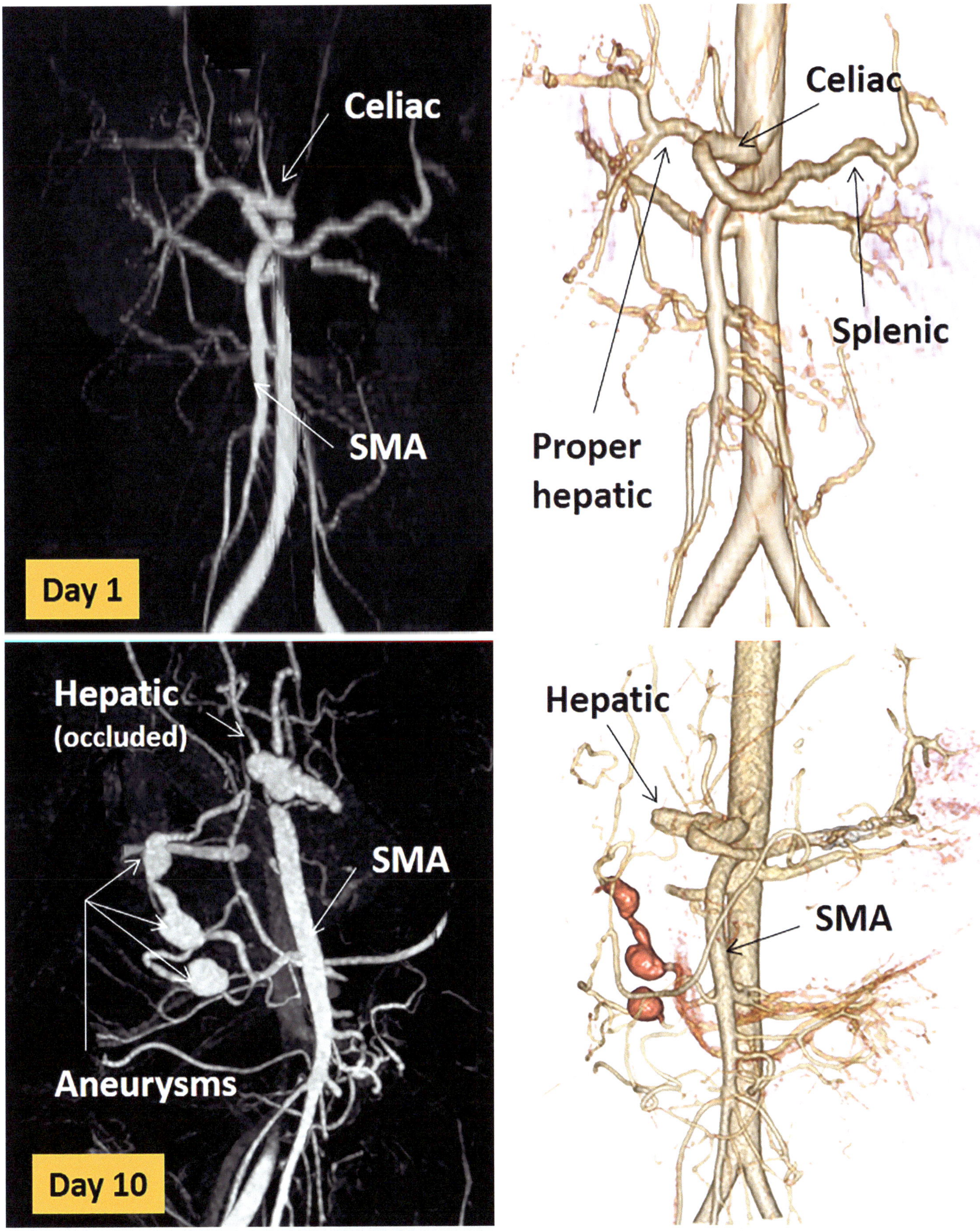

Fig. 9.33 Computed tomography (CT) angiograms at day 1 and day 10 after trauma demonstrate occlusion of the hepatic artery after dissection of the celiac trunk indicated by injury and secondary development of multiple pseudoaneurysms in the pancreatoduodenal arcade due to blood redistribution. No pancreatitis was shown on CT. Lesions were finally embolized successfully. SMA, superior mesenteric artery.

9

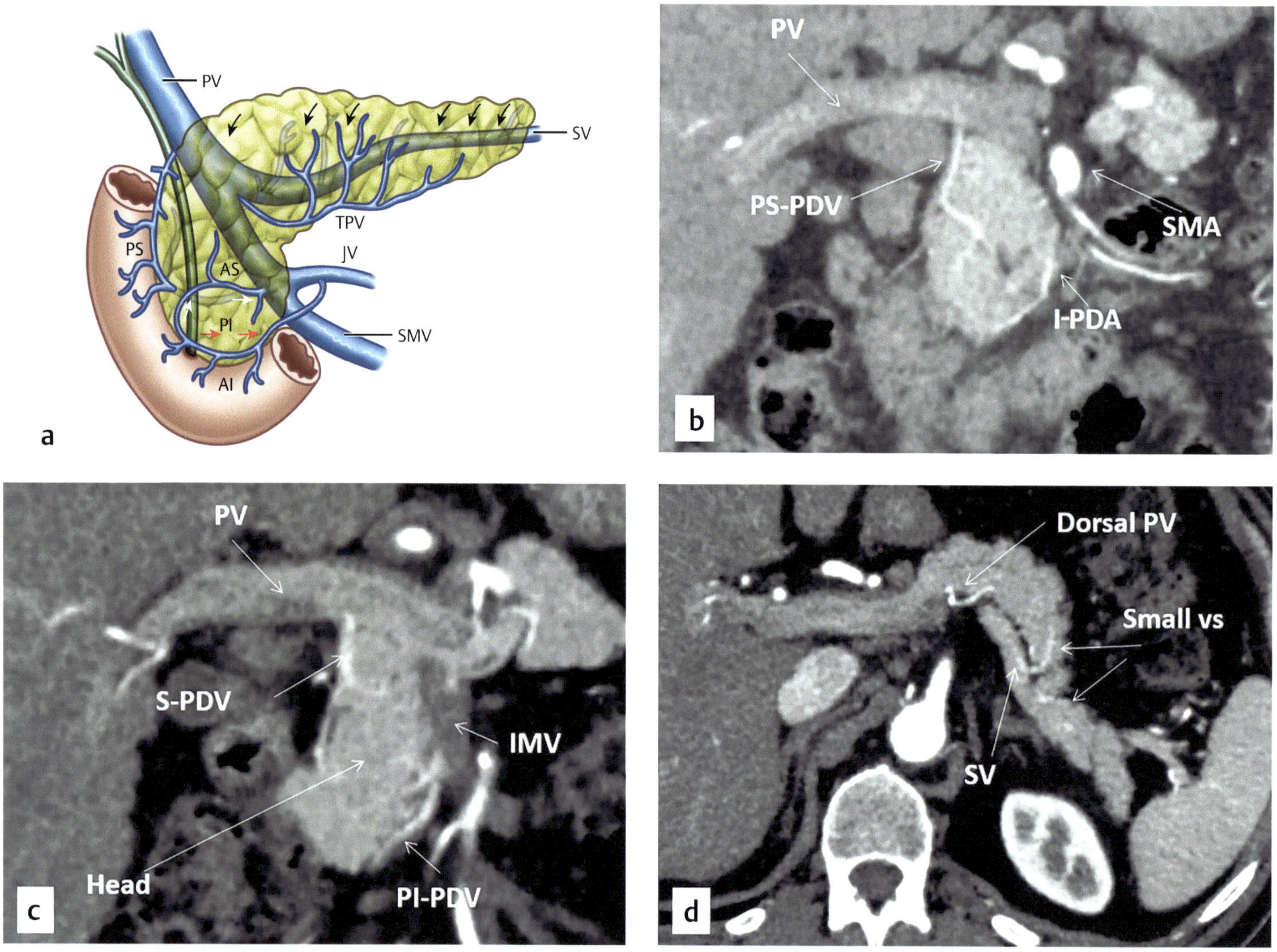

Fig. 9.34 Venous drainage of the pancreas. (Modified from Hongo et al.[62]) **(a)** Illustration of the main venous drainage of the pancreas. Coronal **(b, c)** and axial **(d)** computed tomography (CT) images. The anterior and posterior arcades drain the head. The anterior pancreaticoduodenal venous arcade is formed by the anterior superior pancreaticoduodenal (AS-PD) and anterior inferior pancreaticoduodenal (AI-PD) veins. The posterior pancreaticoduodenal venous arcade is formed by the posterior superior pancreaticoduodenal (PS-PD) and posterior inferior pancreaticoduodenal (PI-PD) veins. The PS-PD vein is the largest draining vein of the pancreatic head. AI, anterioinferior pancreaticoduodenal; AS, anterosuperior pancreaticoduodenal; IMV, inferior mesenteric vein; I-PDA, inferior pancreaticoduodenal artery; JV, jejunal vein; PI, inferior pancreaticoduodenal; PS, posterosuperior pancreaticoduodenal; PV, portal vein; SMV, superior mesenteric vein; SV, splenic vein; TPV, transverse pancreatic vein.

venous arcade which joins with the IPD veins and an anteromedial pancreaticoduodenal vein which empties into the SMV or right gastroepiploic vein.[62,63] The uncinate vein is seen in 30% and runs upward behind the medial side of the pancreatic head and drains into the posterior wall of the portal vein.[62]

The venous drainage of the pancreatic neck, body, and tail is from veins that usually follow the same course as the homonymous arteries. A system of small superior pancreatic veins (ranging from 3 to 13) drains into the splenic vein.[60,63] The transverse pancreatic vein (TPV) or inferior pancreatic vein (IPV) is comprised of one or several veins that pass transversely at the inferior surface of the pancreatic body.[62] The IPV drains the anterior and inferior parts of the pancreatic body into the splenic vein, SMV, or inferior mesenteric vein (IMV) at the level of the pancreatic neck (▶ Fig. 9.34). The dorsal pancreatic vein is a short vein that runs downward from the posterior aspect of the pancreatic neck and drains into the splenic vein or SMV.

Knowledge of the pancreatic venous anatomy is important for precise venous sampling for localization of functioning neuroendocrine tumors of the pancreas. Dilated intrapancreatic veins are often observed as drainage vessels around endocrine tumors of the pancreas. Chronic thrombosis of portal venous system can lead to dilatation of the intrapancreatic and peripancreatic veins.

Lymphatics

The lymphatic vessels of the pancreas arise in the interlobular tissue. Lymph capillaries form an anastomosing network to form large lymph ducts which reach the surface of the pancreas in the

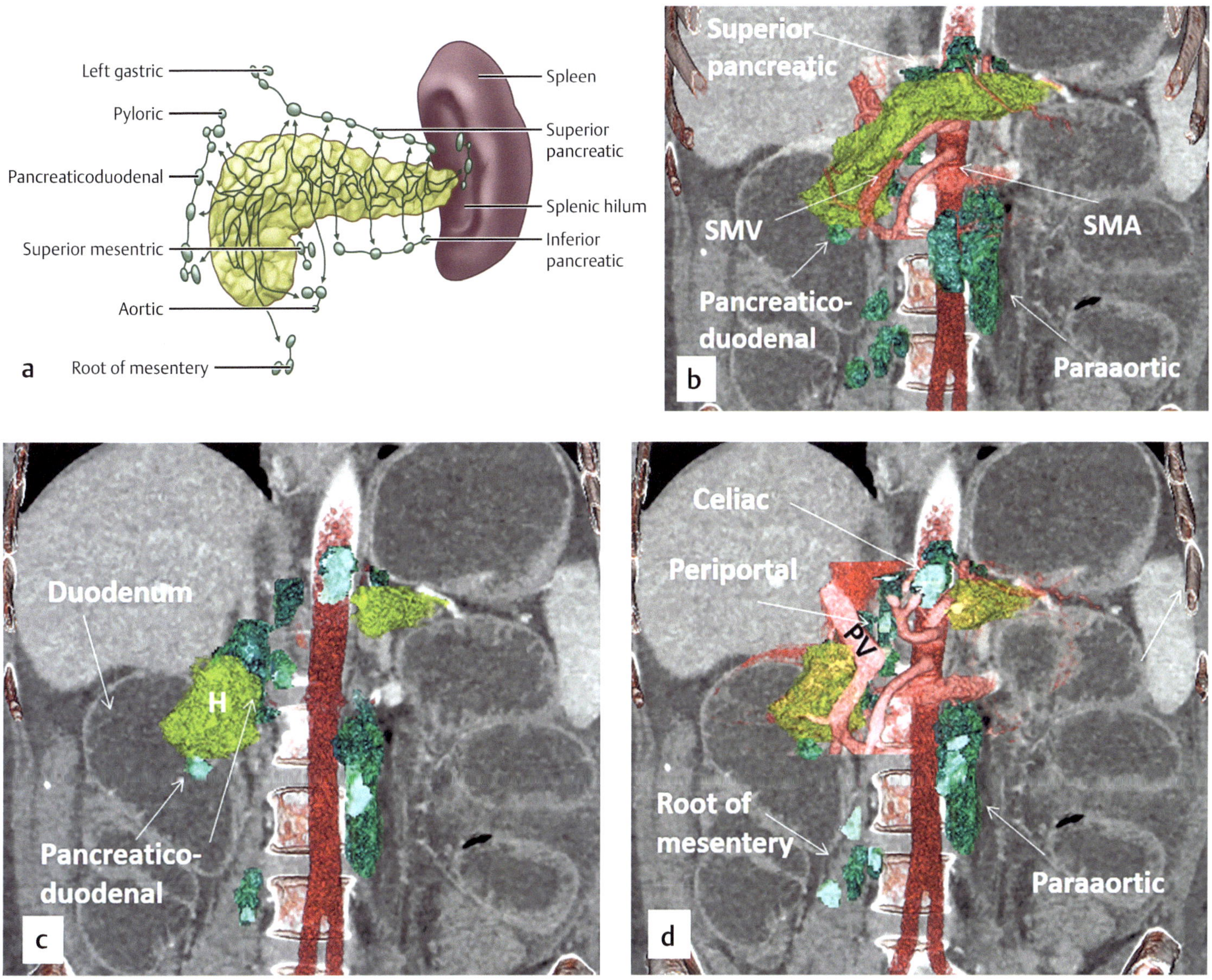

Fig. 9.35 Lymphatic system of the pancreas. **(a)** Illustration of the lymphatic system of the pancreas. **(b–d)** Coronal computed tomography (CT) images showing enlarged upper abdominal lymph nodes. H, head of pancreas; PV, portal vein; SMA, superior mesenteric artery; SMV, superior mesenteric vein.

interlobular spaces with the blood vessels. The main lymphatic flow of the pancreas is to five main collecting trunks and lymph node groups: Superior, inferior, anterior, posterior, and splenic nodes[64] (▸Fig. 9.35). The superior collecting trunks arise from the anterior and posterior upper half of the pancreas, with most ending in the lymph nodes located along the superior border of the pancreas. Some of these lymph nodes may terminate in the nodes of the gastropancreatic fold or hepatic chain.[64] The inferior collecting trunks drain the anterior and posterior lower halves of the head and body of the pancreas, with most ending along the inferior border of the head and body of the pancreas. Some of these lymph nodes may extend into the superior mesenteric and left para-aortic lymph nodes.[64]

The anterior collecting trunks drain the anterior surface of the superior and inferior portions of the head of the pancreas, extending to the infrapyloric and anterior pancreaticoduodenal lymph nodes. Some of these lymph nodes may extend to the root of the mesentery.[64]

The posterior collecting trunks drain the posterior surface of the superior and inferior portions of the head of the pancreas. They drain into the posterior pancreaticoduodenal lymph nodes and common bile duct lymph nodes, right para-aortic lymph nodes, and some nodes near the origin of the SMA. Most of the lymphatics of the common bile duct and ampulla of Vater also terminate in the posterior pancreaticoduodenal lymph node group.[64] The splenic collecting trunks drain the tail of the pancreas into the nodal groups at the hilum of the spleen, splenophrenic ligament, and at the inferior and superior lymph nodes of the tail of the pancreas. Some of these lymph nodes terminate superior and inferior to the body of the pancreas.[64]

Innervation

The pancreas receives sympathetic and parasympathetic innervation. The sympathetic innervations occur through the greater and lesser thoracic splanchnic nerves and the parasympathetic division occurs through the celiac division of the posterior vagus nerve. The nerves pierce the diaphragm to enter the celiac plexus and ganglion that are located close to the celiac trunk and the superior mesenteric plexus and ganglion that surround the SMA.[6]

◆ Postoperative Anatomy

Partial pancreaticoduodenectomy, also known as "Whipple" procedure, is the most common surgical procedure for resection of lesions in the pancreatic head, uncinate process, and neck in addition to lesions in the ampulla, common bile duct, and duodenum. The classic Whipple procedure includes en bloc resection of the pancreatic head, distal common bile duct, gallbladder, entire duodenum, pylorus, distal stomach, and a short segment of jejunum.[65] This is followed by reconstruction of the gastrointestinal tract by pancreaticojejunostomy, hepaticojejunostomy, and gastrojejunostomy (▶ Fig. 9.36). The pylorus-sparing Whipple procedure retains the gastric antrum and a 2-cm cuff of the duodenum, with subsequent gastrointestinal tract reconstruction as a duodenojejunostomy.[66] This was originally proposed to reduce bile reflux, although studies have shown no significant difference in perioperative morbidity and mortality compared to the classic Whipple procedure.[67]

Distal pancreatectomy is reserved for lesions involving the distal neck, body, and tail of the pancreas. A splenectomy is also performed in the setting of malignancy, except for select cases of cystic lesions and low-grade neoplasms[68] (▶ Fig. 9.37). In cases of celiac artery involvement, some pancreatic body lesions can be

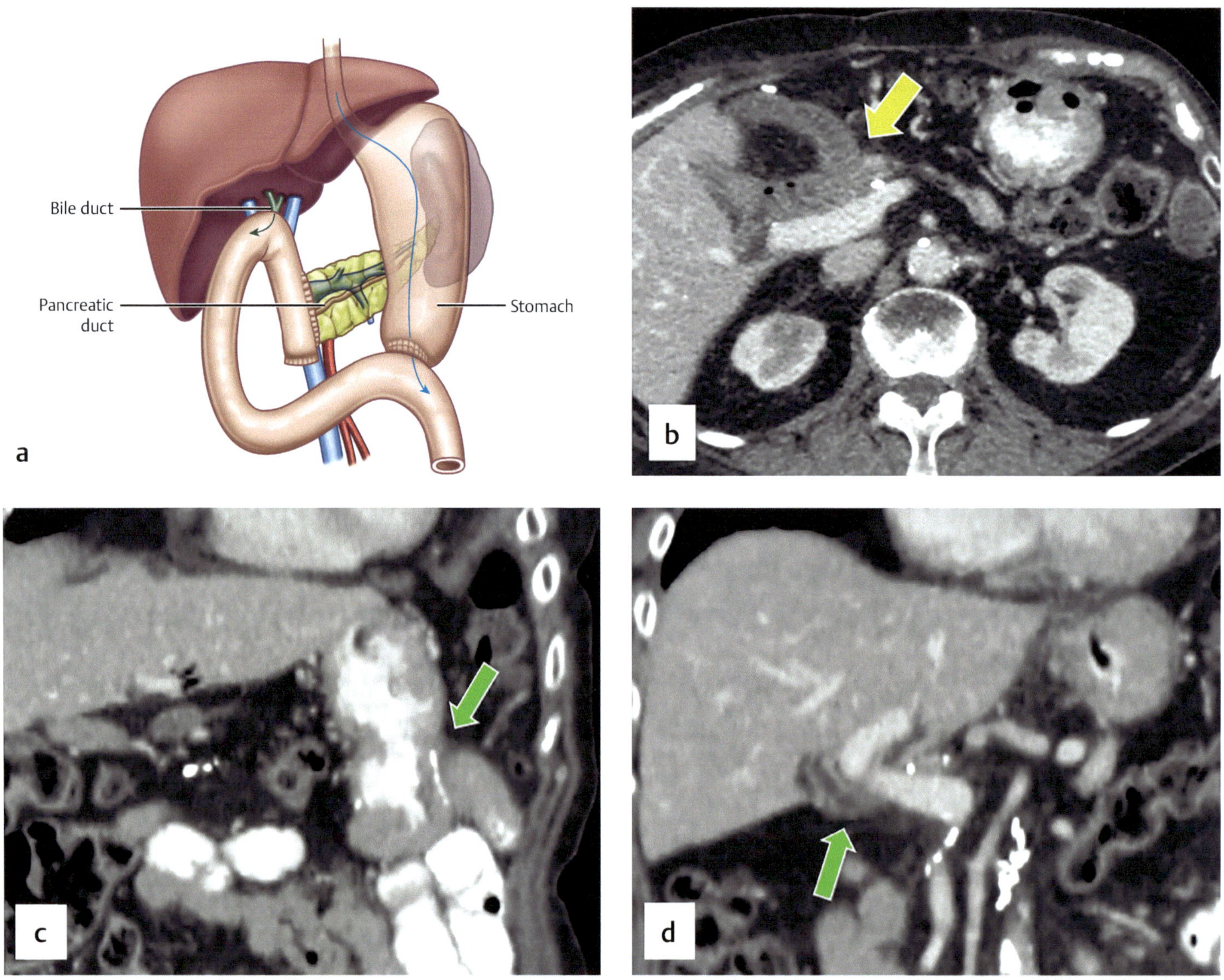

Fig. 9.36 (a) Classic Whipple procedure reconstruction includes resection of the pancreatic head, distal common bile duct, duodenum, pylorus, distal stomach, and a short segment of jejunum with reconstruction of the gastrointestinal tract by pancreaticojejunostomy, hepaticojejunostomy, and gastrojejunostomy. **(b–d)** Computed tomography (CT) demonstration of Whipple procedure reconstruction. **(b)** Axial CT image demonstrates the pancreaticojejunostomy (*arrow*). Note atrophy of the remaining pancreas. **(c)** Coronal CT image demonstrates the gastrojejunostomy (*arrow*). **(d)** Coronal CT image demonstrates the hepaticojejunostomy (*arrow*).

resected through distal pancreatectomy with en bloc resection of the celiac axis (Appleby procedure).[69]

The Puestow procedure (lateral side-to-side pancreaticojejunostomy) is the surgical procedure of choice for relieving abdominal pain related to chronic pancreatitis and pancreatic duct dilatation associated with obstruction.[70] In this procedure, the dilated pancreatic duct is longitudinally opened from the uncinate process to the tail and directly anastomosed in a side-to-side manner to a longitudinally opened Roux-en-Y segment of the jejunum anterior to the pancreas (▶Fig. 9.38).

Pancreas Transplant

A pancreatic transplant is indicated in patients with insulin-dependent or poorly controlled severe diabetes mellitus. The pancreas is usually transplanted along with a renal transplant from the same donor (simultaneous pancreas–kidney transplant), unless the patient already has a functioning renal transplant or in those patients who do not have gross nephropathy. Occasionally, the pancreas is transplanted after a kidney transplantation, although the 1-year survival rates for pancreas grafts are lower

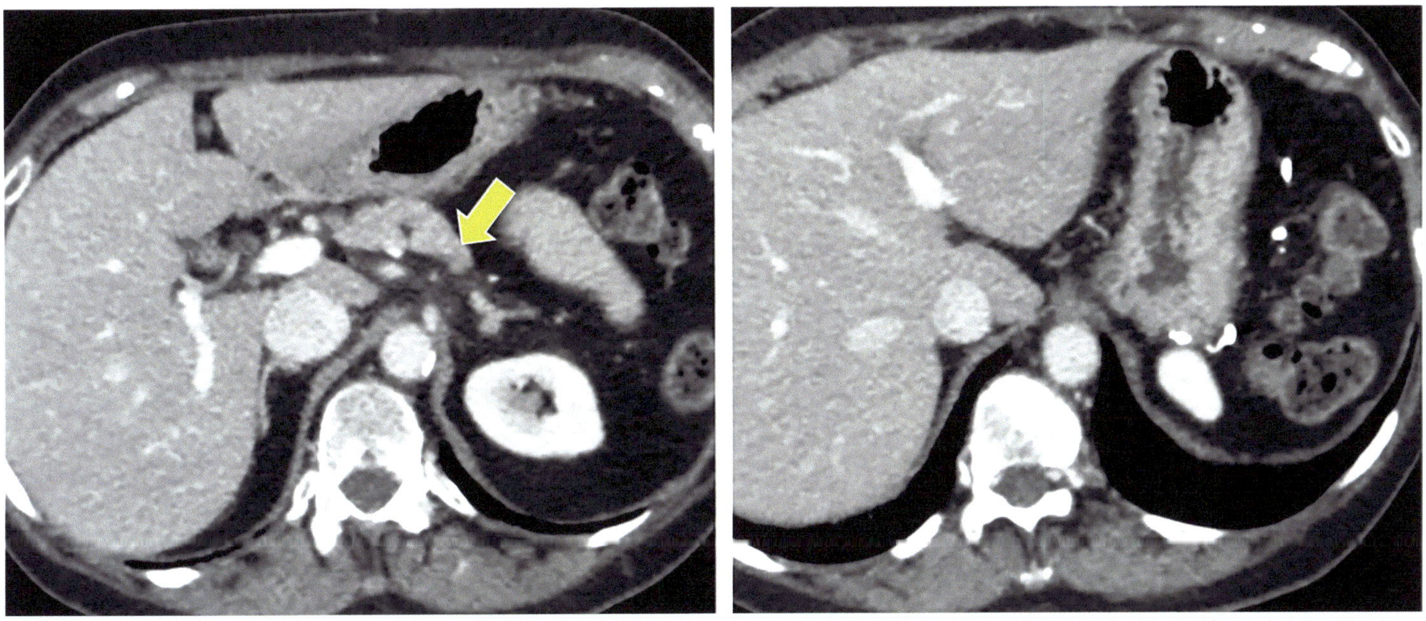

Fig. 9.37 Distal pancreatectomy. This procedure involves surgical resection of the distal pancreas and usually spleen in the setting of malignancy. Axial computed tomography (CT) images demonstrate expected findings after distal pancreatectomy (*arrow*), with absence of the distal pancreatic body, tail, and spleen.

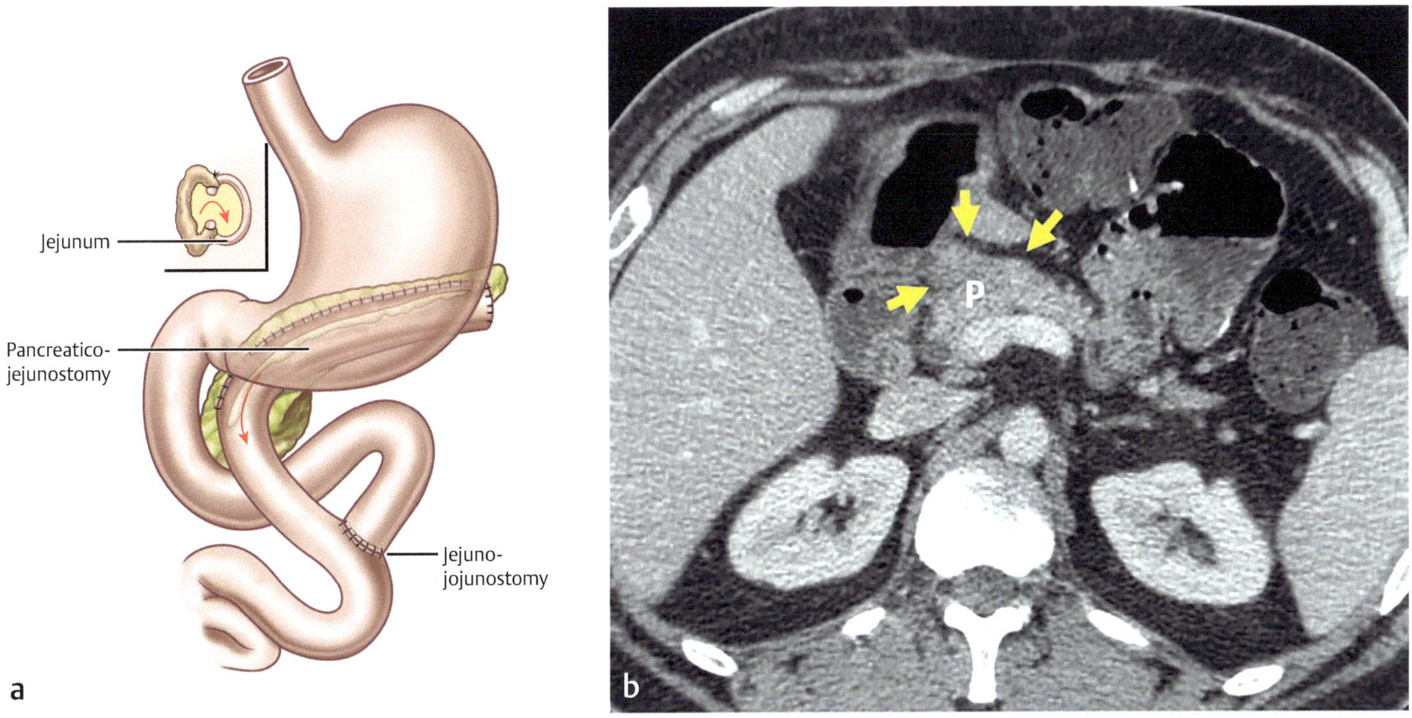

Fig. 9.38 Puestow procedure. **(a)** The Puestow procedure consists of longitudinal anastomosis of the pancreatic duct with a Roux-en-Y segment of the jejunum anterior to the pancreas. **(b)** Axial computed tomography (CT) demonstrates expected findings after Puestow procedure. The collapsed pancreaticojejunostomy anastomosis (*yellow arrows*) is immediately anterior to the pancreatic neck and body (P).

with pancreas transplant after kidney transplantation than with simultaneous pancreas and kidney transplants.[71]

The transplanted pancreas with attached duodenum is usually placed in the right iliac fossa, with a side-to-side duodenovesical or duodenojejunal anastomosis. A simultaneously transplanted kidney is placed in the left iliac fossa. The arterial supply of the transplanted pancreas is from the donor SMA and the donor splenic artery. The donor common, internal, and external iliac arteries are attached to the donor SMA and splenic arteries, forming a Y-graft (▶ Fig. 9.39). The common iliac artery portion of the Y-graft is then anastomosed with the recipient common iliac or external iliac artery.[71] The venous drainage may be via the recipient portal or systemic venous system. Portal venous drainage involves an anastomosis between the donor portal vein, which serves as the main graft vein, and recipient external iliac vein or SMV.[71] The current technique involves an anastomosis between the donor portal vein and the recipient SMV (▶ Fig. 9.39).

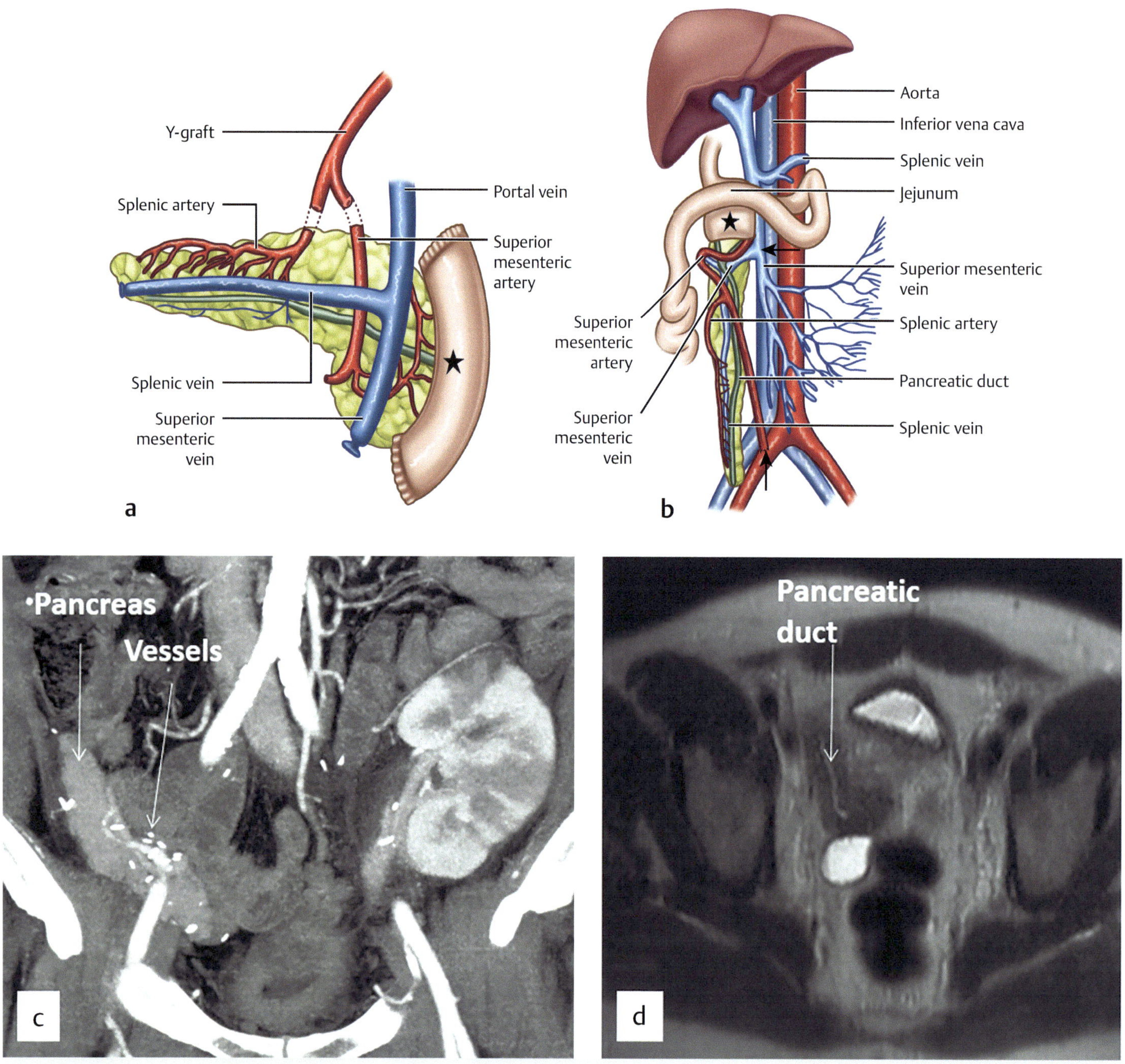

Fig. 9.39 Vascular reconstruction of pancreas transplant. **(a)** The donor iliac Y-graft is attached to the donor splenic artery and superior mesenteric artery. The donor portal vein (PV) serves as the main graft vein and drains the donor splenic vein (SV) and donor superior mesenteric veins (SMV). The donor duodenal stump (*) is also transplanted along with the pancreas. **(b)** In portal venous and enteric exocrine drainage, the graft vein (PV) is attached to the recipient SMV (*arrow*) for portal venous drainage and the duodenal stump (*) is anastomosed to the jejunum. The donor iliac Y-graft is anastomosed to the recipient common iliac artery. **(c)** Coronal computed tomography (CT) shows the transplant pancreas in the right lower quadrant with its vessels. A left lower quadrant transplant kidney is also shown. **(d)** Axial magnetic resonance imaging (MRI) showing transplant pancreas in the right lower quadrant with normal pancreatic duct.

References

1. Skandalakis JE, Gray SW, Rowe JS, et al. Anatomical complications of pancreatic surgery. Part 1. Contemp Surg 1979;15:17
2. McClusky DA III, Skandalakis LJ, Colborn GL, Skandalakis JE. Harbinger or hermit? Pancreatic anatomy and surgery through the ages—part 1. World J Surg 2002;26(9):1175–1185
3. Schulte SJ. Embryology, normal variation, and congenital anomalies of the pancreas. In: Freeny PC, Stevenson GW, eds. Margulis' and Burhenne's Alimentary Tract Radiology. 5th ed. St. Louis: Mosby; 1994:1039–1051
4. Rubenstein WA, Auh YH, Zirinsky K, Kneeland JB, Whalen JP, Kazam E. Posterior peritoneal recesses: assessment using CT. Radiology 1985;156(2):461–468
5. Myers M, Charnsangavej C, Oliphant M. Meyers' Dynamic Radiology of the Abdomen, Normal and Pathologic Anatomy. 6th ed. Springer; 2010:333–396
6. Hoff FL, Gabriel H, Hammond NA, Gore RM. Pancreas: Normal Anatomy and Examination Techniques. In: Gore RM, Levine MS, eds. Textbook of Gastrointestinal Radiology. Philadelphia: Saunders Elsevier; 2008:1839–1853
7. Quinlan RM. Anatomy and embryology of the pancreas. In: Zuidema GD, ed. Schackelford's Surgery of the Alimentary Tract. Vol. III. 3rd ed. Philadelphia: Saunders; 1991:3–18
8. Van Hoe L, Claikens B. The pancreas: normal radiological anatomy and variants. In: Baert AL, Delorme G, Hoe LV, eds. Radiology of the Pancreas. Springer Science & Business Media; 1999:19–68
9. Heuck A, Maubach PA, Reiser M, et al. Age-related morphology of the normal pancreas on computed tomography. Gastrointest Radiol 1987;12(1):18–22
10. Zylak CJ, Pallie W. Correlative anatomy and computed tomography: a module on the pancreas and posterior abdominal wall. Radiographics 1981;1:61–84
11. Campbell F, Verbeke CS. Pathology of the Pancreas, A Practical Approach. Springer; 2013:3–20
12. Trede M. Embryology and surgical anatomy of the pancreas. In: Trede M, Carter DC, eds. Surgery of the Pancreas. Churchill Livingstone; 1997:17–27
13. Mortelé KJ, Rocha TC, Streeter JL, Taylor AJ. Multimodality imaging of pancreatic and biliary congenital anomalies. Radiographics 2006;26(3):715–731
14. Skandalakis LJ, Rowe JS Jr, Gray SW, Skandalakis JE. Surgical embryology and anatomy of the pancreas. Surg Clin North Am 1993;73(4):661–697
15. Skandalakis JE, Skandalakis LJ, Kingsnorth AN, Colborn GL, Weidman TA, Skandalakis PN. Pancreas. In: Skandalakis JE, Colborn GL, et al, eds. Surgical Anatomy: The Embryologic and Anatomic Basis of Modern Surgery. Athens, Greece: Paschalidis Medical Publications; 2004
16. Wachsberg RH. Respiratory variation of the diameter of the pancreatic duct on sonography. AJR Am J Roentgenol 2000;175(5):1459–1461
17. Varley PF, Rohrmann CA Jr, Silvis SE, Vennes JA. The normal endoscopic pancreatogram. Radiology 1976;118(2):295–300
18. Tanaka S, Nakao M, Ioka T, et al. Slight dilatation of the main pancreatic duct and presence of pancreatic cysts as predictive signs of pancreatic cancer: a prospective study. Radiology 2010;254(3):965–972
19. Avisse C, Flament JB, Delattre JF. Ampulla of Vater. Anatomic, embryologic, and surgical aspects. Surg Clin North Am 2000;80(1):201–212
20. Teilum D. In vitro measurement of the length of the sphincter of Oddi. Endoscopy 1991;23(3):114–116
21. Kim TU, Kim S, Lee JW, et al. Ampulla of Vater: comprehensive anatomy, MR imaging of pathologic conditions, and correlation with endoscopy. Eur J Radiol 2008;66(1):48–64
22. Wiedmeyer DA, Stewart ET, Taylor AJ. Radiologic evaluation of structure and function of the sphincter of Oddi. Gastrointest Endosc Clin N Am 1993;3:13–40
23. Mirilas P, Colborn GL, Skandalakis LJ, Skandalakis PN, Zoras O, Skandalakis JE. Benign anatomical mistakes: "ampulla of Vater" and "papilla of Vater." Am Surg 2005;71(3):269–274
24. Smanio T. Varying relations of the common bile duct with the posterior face of the pancreatic head in Negroes and white persons. J Int Coll Surg 1954;22(2 1):150–173
25. Mahour GH, Wakim KG, Ferris DO. The common bile duct in man: its diameter and circumference. Ann Surg 1967;165(3):415–419
26. Türkvatan A, Erden A, Türkoğlu MA, Yener Ö. Congenital variants and anomalies of the pancreas and pancreatic duct: imaging by magnetic resonance cholangiopancreaticography and multidetector computed tomography. Korean J Radiol 2013;14(6):905–913
27. Agha FP, Williams KD. Pancreas divisum: incidence, detection, and clinical significance. Am J Gastroenterol 1987;82(4):315–320
28. Klein SD, Affronti JP. Pancreas divisum, an evidence-based review: part I, pathophysiology. Gastrointest Endosc 2004;60(3):419–425
29. Warshaw AL, Simeone JF, Schapiro RH, Flavin-Warshaw B. Evaluation and treatment of the dominant dorsal duct syndrome (pancreas divisum redefined). Am J Surg 1990;159(1):59–64, discussion 64–66
30. Belber JP, Bill K. Fusion anomalies of the pancreatic ductal system differentiation from pathologic states. Radiology 1977;122(3):637–642
31. Soto JA, Lucey BC, Stuhlfaut JW. Pancreas divisum: depiction with multi-detector row CT. Radiology 2005;235(2):503–508
32. Eisen G, Schutz S, Metzler D, Baillie J, Cotton PB. Santorinicele: new evidence for obstruction in pancreas divisum. Gastrointest Endosc 1994;40(1):73–76
33. Kochhar R, Nagi B, Chawla S, et al. The clinical spectrum of anomalous pancreatobiliary junction. Surg Endosc 1989;3(2):83–86
34. Dowdy GS, Waldron GW, Brown WG. Surgical anatomy of the pancreatobiliary ductal system. Observations. Arch Surg 1962;84:229–246
35. Yamauchi S, Koga A, Matsumoto S, Tanaka M, Nakayama F. Anomalous junction of pancreaticobiliary duct without congenital choledochal cyst: a possible risk factor for gallbladder cancer. Am J Gastroenterol 1987;82(1):20–24
36. Unozawa T, Kimura K, Ohto M, et al. [Clinical study on relationship between an anomalous pancreatico-biliary ductal union and carcinoma of the gallbladder]. Nihon Shokakibyo Gakkai Zasshi 1985;82(3):473–482
37. Lecco TM. Zur Morphologie des Pankreas annulare. Sitzungs berichte der Akademie der Wissenschaften Mathematischnatur Wien 1910;119:391–406
38. Baldwin WM. A specimen of annular pancreas. Anat Rec 1910;4:299–304
39. Kamisawa T, Yuyang T, Egawa N, Ishiwata J, Okamoto A. A new embryologic hypothesis of annular pancreas. Hepatogastroenterology 2001;48(37):277–278
40. Johnston DW. Annular pancreas: a new classification and clinical observations. Can J Surg 1978;21(3):241–244
41. Ravitch MM, Woods AC Jr. Annular pancreas. Ann Surg 1950;132(6):1116–1127
42. Lloyd-Jones W, Mountain JC, Warren KW. Annular pancreas in the adult. Ann Surg 1972;176(2):163–170
43. Lehman GA, O'Connor KW. Coexistence of annular pancreas and pancreas divisum—ERCP diagnosis. Gastrointest Endosc 1985;31(1):25–28
44. Karasaki H, Mizukami Y, Ishizaki A, et al. Portal annular pancreas, a notable pancreatic malformation: frequency, morphology, and implications for pancreatic surgery. Surgery 2009;146(3):515–518
45. Lechner GW, Read RC. Agenesis of the dorsal pancreas in an adult diabetic presenting with duodenal ileus. Ann Surg 1966;163(2):311–314
46. Wang JT, Lin JT, Chuang CN, et al. Complete agenesis of the dorsal pancreas—a case report and review of the literature. Pancreas 1990;5(4):493–497
47. Schnedl WJ, Reisinger EC, Schreiber F, Pieber TR, Lipp RW, Krejs GJ. Complete and partial agenesis of the dorsal pancreas within one family. Gastrointest Endosc 1995;42(5):485–487
48. Schnedl WJ, Piswanger-Soelkner C, Wallner SJ, et al. Agenesis of the dorsal pancreas and associated diseases. Dig Dis Sci 2009;54(3):481–487
49. Ross BA, Jeffrey RB Jr, Mindelzun RE. Normal variations in the lateral contour of the head and neck of the pancreas mimicking neoplasm: evaluation with dual-phase helical CT. AJR Am J Roentgenol 1996;166(4):799–801
50. Matsumoto S, Mori H, Miyake H, et al. Uneven fatty replacement of the pancreas: evaluation with CT. Radiology 1995;194(2):453–458
51. Cho JS, Shin KS, Kwon ST, et al. Heterotopic pancreas in the stomach: CT findings. Radiology 2000;217(1):139–144
52. Wei R, Wang QB, Chen QH, Liu JS, Zhang B. Upper gastrointestinal tract heterotopic pancreas: findings from CT and endoscopic imaging with histopathologic correlation. Clin Imaging 2011;35(5):353–359
53. Fékété F, Noun R, Sauvanet A, Fléjou JF, Bernades P, Belghiti J. Pseudotumor developing in heterotopic pancreas. World J Surg 1996;20(3):295–298
54. Covey AM, Brody LA, Maluccio MA, Getrajdman GI, Brown KT. Variant hepatic arterial anatomy revisited: digital subtraction angiography performed in 600 patients. Radiology 2002;224(2):542–547
55. Bertelli E, Di Gregorio F, Bertelli L, Mosca S. The arterial blood supply of the pancreas: a review. I. The superior pancreaticoduodenal and the anterior superior

9

pancreaticoduodenal arteries. An anatomical and radiological study. Surg Radiol Anat 1995;17(2):97–106, 1–3
56. Bertelli E, Di Gregorio F, Mosca S, Bastianini A. The arterial blood supply of the pancreas: a review. V. The dorsal pancreatic artery. An anatomic review and a radiologic study. Surg Radiol Anat 1998;20(6):445–452
57. Okahara M, Mori H, Kiyosue H, Yamada Y, Sagara Y, Matsumoto S. Arterial supply to the pancreas; variations and cross-sectional anatomy. Abdom Imaging 2010;35(2):134–142
58. Uflacker R. Atlas of Vascular Anatomy, An Angiographic Approach. Lippincott Williams and Wilkins; 2007:464
59. Toni R, Favero L, Bolzani R, Roversi R, Vezzadini P. Further observations on the anatomical variation in the arteries of the human pancreas. IRCS Med Sci 1985;13:605–606
60. Skandalakis JE, Gray SW, Rowe JS, Skandalakis LJ. Surgical anatomy of the pancreas. Contemp Surg 1979;15:17–40
61. Falconer CWA, Griffiths E. The anatomy of the blood-vessels in the region of the pancreas. Br J Surg 1950;37(147):334–344
62. Hongo N, Mori H, Matsumoto S, Okino Y, Ueda S, Shuto R. Anatomical variations of peripancreatic veins and their intrapancreatic tributaries: multidetector-row CT scanning. Abdom Imaging 2010;35(2):143–153
63. Peri G, Veralli E, Trivellini G. La vascolarizzazione del pancreas [Vascularization of the pancreas]. Arch Ital Chir 1969;95(2):287–300
64. Cubilla AL, Fortner J, Fitzgerald PJ. Lymph node involvement in carcinoma of the head of the pancreas area. Cancer 1978;41(3):880–887
65. Whipple AO, Parsons WB, Mullins CR. Treatment of carcinoma of the ampulla of Vater. Ann Surg 1935;102(4):763–779
66. Traverso LW, Longmire WP Jr. Preservation of the pylorus in pancreaticoduodenectomy a follow-up evaluation. Ann Surg 1980;192(3):306–310
67. Wolfgang CL, Corl F, Johnson PT, et al. Pancreatic surgery for the radiologist, 2011: an illustrated review of classic and newer surgical techniques for pancreatic tumor resection. AJR Am J Roentgenol 2011;197(6):1343–1350
68. Bruzoni M, Sasson AR. Open and laparoscopic spleen-preserving, splenic vessel-preserving distal pancreatectomy: indications and outcomes. J Gastrointest Surg 2008;12(7):1202–1206
69. Kondo S, Katoh H, Hirano S, et al. Results of radical distal pancreatectomy with en bloc resection of the celiac artery for locally advanced cancer of the pancreatic body. Langenbecks Arch Surg 2003;388(2):101–106
70. Freed KS, Paulson EK, Frederick MG, Keogan MT, Pappas TN. Abdomen after a Puestow procedure: postoperative CT appearance, complications, and potential pitfalls. Radiology 1997;203(3):790–794
71. Vandermeer FQ, Manning MA, Frazier AA, Wong-You-Cheong JJ. Imaging of whole-organ pancreas transplants. Radiographics 2012;32(2):411–435

10 Mesenteric Vasculature of Lower Gastrointestinal System

Alessandro Campari, Nicola Flor, and Gianpaolo Cornalba

◆ Introduction

The extramural and intramural vascular system of the gastrointestinal system has been studied extensively because a thorough knowledge of vessel anatomy is fundamental to the practice of surgery and radiology. The blood supply to the intestinal tract is derived from the three major anterior branches of the abdominal aorta: the celiac trunk, which feeds the upper gastrointestinal viscera, the superior mesenteric artery (SMA), which supplies the small bowel and the proximal half of the large bowel, and the inferior mesenteric artery (IMA), which feeds the distal half of the large bowel. Although this subdivision is found in almost every individual, many variations can be encountered in the branching pattern of the celiac and mesenteric arteries; therefore, only about half of the population fits into the textbook pattern of anatomy.[1] High variability also exists in the number and course of the many collateral pathways between the celiac trunk and both mesenteric arteries; thus, there is still confusion about the number of collateral vessels and their descriptive terminology. In the surgical practice, collateral pathways between splanchnic arteries are routinely employed to provide alternative blood supply after surgical ligation. Moreover, both surgical procedures and lymph node dissection are defined according to the structure of vessels, using arteries as landmarks; therefore, preoperative assessment of mesenteric artery anatomy is often required. The role of imaging in the evaluation of the gastrointestinal vasculature is to provide anatomic mapping for preoperative planning, with particular regard to variants and collateral pathways, and to diagnose acute and chronic conditions affecting arteries and veins. The leading imaging modality is multidetector computed tomography (MDCT). Magnetic resonance (MR) and conventional angiography are other useful available tools.

◆ Embryology

Knowledge of the embryologic development of the mesenteric arterial circulation enables to better understand the anatomy of the frequently found anatomic variations, the many arterial anastomoses, and the potential collateral circulation in the case of obstruction of one or more arteries. Studies of 5- to 17-mm human embryo show the main development stages of the bowel vascularization. Tandler first described the embryology of the celiac and mesenteric vessels in the early 1900. In 5-mm embryo, the descending aorta gives several primitive ventral segmental arteries that are connected by the ventral longitudinal anastomosis (Tandler's anastomosis) running anteriorly and parallel to the aorta. The ventral longitudinal anastomosis runs close to the primitive digestive tube and gives several ventral splanchnic arteries. As ordinary development proceeds, the digestive tube lengthens and the complex arterial network is simplified with the regression of a great number of segmental arteries and segments of the ventral anastomosis. At 17-mm embryo stage, the only remaining arteries are the precursors to the three major mesenteric vessels: the 10th segmental artery forms the celiac trunk that supplies the foregut (from the esophagus to the distal duodenum), the 13th segmental artery forms the SMA that supplies the midgut (from the proximal duodenum to the midtransverse colon), and the 21st segmental artery forms the IMA that supplies the hindgut (▶ Fig. 10.1). The presence of Tandler's longitudinal anastomosis connecting all primitive arteries allows the development of the numerous collateral channels among the three main digestive trunks. Furthermore, the high variability in mesenteric arteries branching pattern can be explained by the occurrence of different ontogenic developments, with either persistence or regression of primitive arteries and segments of the ventral anastomosis.[2–5] The arc of Bühler is present when the ventral anastomosis between the 10th and 13th segmental arteries (CT and SMA) persists. Regression of the 10th segmental artery with persistence of the ventral anastomosis leads to the development of a single celiomesenteric trunk (▶ Fig. 10.2). The position of some vessels is determined by the rotation of the gut around the SMA. If the rotation occurs downstream to the origin of the inferior pancreaticoduodenal artery, it will arise from the left side of the SMA, whereas if it occurs upstream to the origin, the inferior pancreaticoduodenal artery will originate from the right side of the SMA. Relationship between the SMA and the superior mesenteric vein (SMV) depends on the degree of rotation of the gut (see section Variants in Relationship between Arteries and Veins).[6]

◆ Arteries

The arterial circulation of the bowel is characterized by a complex redundant interconnection of branches from major abdominal vessels forming collateral arcades. The small bowel receives its blood supply almost exclusively from the SMA. Jejunal and ileal arteries divide and anastomose several times in the mesentery, forming multiple orders of arcades (up to five). The distal order of arteries is the marginal arcade of Dwight and Drummond, defined as the closest to, and parallel with, the mesenteric border of the intestine. Numerous short, straight branches known as vasa recta arise perpendicularly from the marginal arcade and penetrate the external muscle layer of the bowel wall, reaching the submucosal arterial plexus. In the upper jejunum there are one or two arcades and the vasa recta are 3 to 5 cm long. In the ileum the number of arcades increase up to five and the vasa recta become shorter. The SMA is also the primary blood source for cecum, and ascending and proximal/middle transverse colon. The distal transverse, descending, and sigmoid colon are supplied by the IMA (Fig. 10.3). The colonic vasculature architecture can be simplified to a square-shaped model (▶ Fig. 10.4), with four

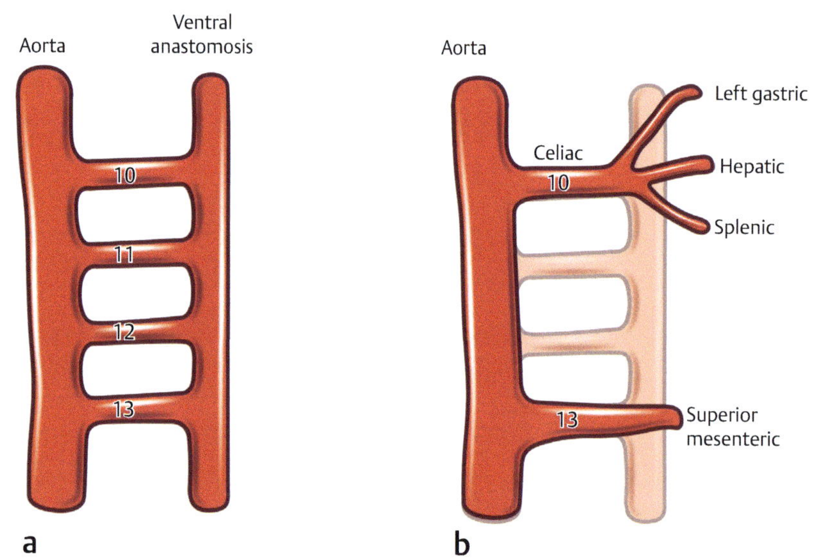

Fig. 10.1 Most common embryologic development. **(a)** The descending aorta gives several primitive segmental arteries that are connected by Tandler's ventral longitudinal anastomosis. **(b)** Regression of 11th and 12th segmental arteries and persistence of 10th and 13th segmental arteries that form the celiac trunk and the superior mesenteric artery. (Image modified with permission from Walker.[3])

main arteries arising from the SMA and IMA forming the sides of the square: ileocolic artery, middle colic artery, left colic artery, and sigmoid artery.[7] Branches from these vessels form arcades that are less numerous and less complex than those in the small bowel. A single, large anastomotic channel, known as marginal artery of Drummond, runs within 2 cm from the colon, and short vasa recta with few or no anastomoses enter the colonic wall (▶Fig. 10.5). The rectum and anal canal are supplied by a rich anastomosing system from branches of the IMA and internal iliac arteries. In contrast to the small and large bowel arteries, branches of the rectal arteries do not form arcades but penetrate the external and internal anal sphincters and enter the gut wall independently, joining in the submucosa.[8,9]

10

Superior Mesenteric Artery

The **SMA** emerges anteriorly from the aorta at the level of the first lumbar vertebra, behind the body of the pancreas, 1 to 2 cm below the celiac trunk (▶Fig. 10.6). It descends crossing the third portion of the duodenum on the left side of the SMV and enters the mesentery. Right-sided branches of the SMA are the inferior pancreaticoduodenal artery, the middle colic artery, the right colic artery, and the ileocolic artery (▶Fig. 10.7). Left-sided branches are ileal and jejunal arteries. Within the mesentery, the SMA courses to the right iliac fossa and ends forming an anastomosis with the ileocolic artery.[3,8,10] The **inferior pancreaticoduodenal artery** is the first right-sided branch of SMA. It runs upward and to the right to anastomose with the superior pancreaticoduodenal artery (from the hepatic artery), forming the pancreaticoduodenal arcade. It may arise on the left side of the SMA from a common origin with a proximal jejunal artery.[1] The **middle colic artery** is a constant branch, usually arising from the anterior/right side of the SMA, under the lower border of the pancreatic uncinate process before it enters the small bowel mesentery. This pattern has been described in about 50% of cases, otherwise the middle colic artery can derive from a common trunk with the right colic artery or ileocolic artery (▶Fig. 10.8), or, occasionally, originate from the celiac trunk, gastroduodenal artery, or IMA (see Chapter 5 "Colon"). An accessory middle colic artery is a common branch arising from the SMA in up to 40% of cases or rarely from the IMA, and usually supplies the left colon. The middle colic artery divides into two branches for the ascending and transverse colon, communicating with branches from the right colic artery, ileocolic artery, and left colic artery (▶Fig. 10.8). An additional inconstant large branch arising from the middle colic artery and running parallel to the transverse colon is described as the artery or arc of Riolan.[7,8] The **right colic artery** is an inconstant right-sided branch of the SMA that supplies the ascending colon. It has the greatest variations among colic arteries, being described in 10 to 40% of cadavers[7] (▶Fig. 10.9). The right colic artery may arise directly from the SMA between middle colic artery and ileocolic artery or from a common trunk with one of them. It usually divides into two branches anastomosing with corresponding branches from the middle colic artery and ileocolic artery. The **ileocolic artery** is the last and most constant tributary leaving the SMA from the right side. It provides multiple collaterals to the terminal ileum, appendix, cecum, and proximal right colon (▶Fig. 10.10). It gives two major branches that anastomose with each other forming an "ileocolic loop." Although the branching pattern is variable, an appendicular artery and two (anterior and posterior) colic (cecum) arteries usually arise from the ileocolic loop. Distally, the ascending (or colic) branch anastomoses with

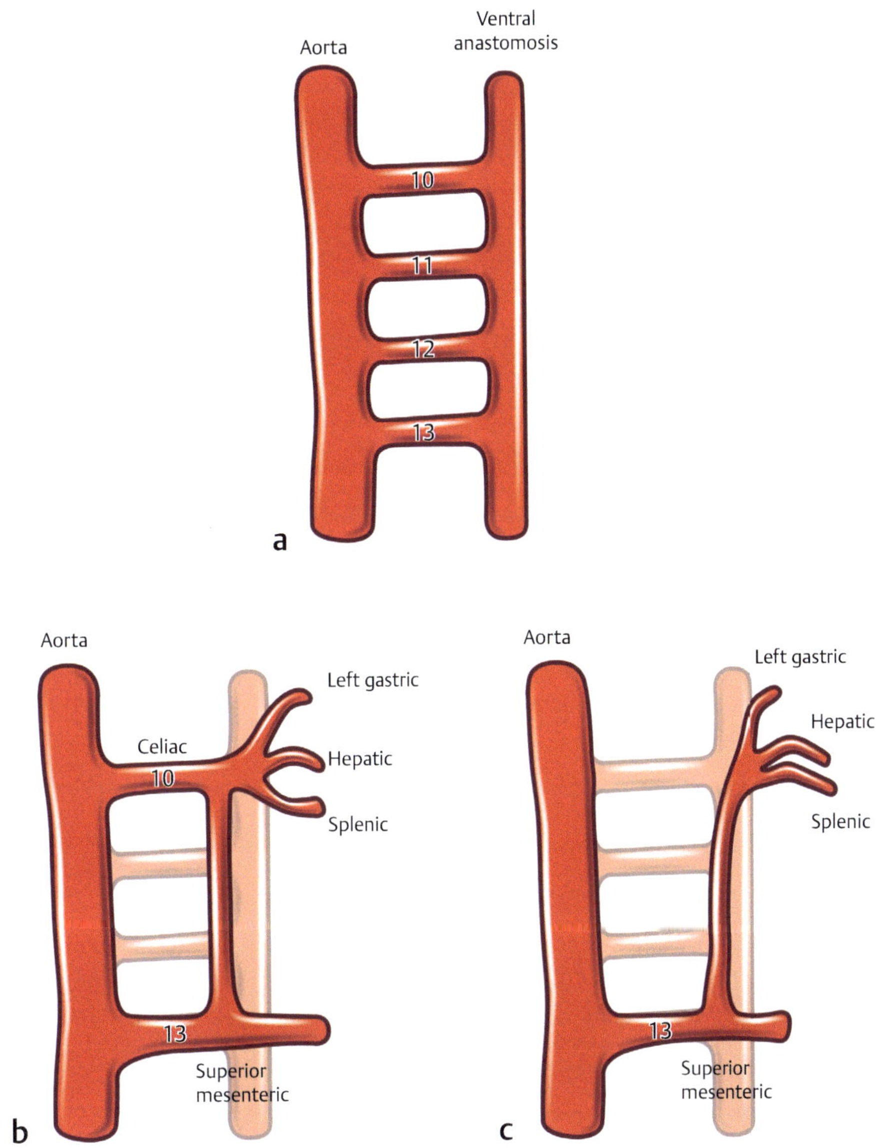

Fig. 10.2 Variations in embryologic development. **(a)** The descending aorta gives several primitive segmental arteries that are connected by Tandler's ventral longitudinal anastomosis. **(b)** Persistent ventral anastomosis between 10th and 13th segmental arteries forms the arc of Bühler. **(c)** Development of a celiomesenteric trunk with regression of the 10th segmental artery and persistence of the ventral anastomosis. (Reproduced with permission from Walker.[3])

the descending branch of right colic artery or middle colic artery, while the ileal branch forms an arch joining with the distal end of the main stem of the SMA (▶Fig. 10.10).[10] In 70 to 90% of cadavers, two arteries arise from the SMA to supply the right colon, while 10 to 30% have three arteries; therefore, in the majority of cases only the middle colic artery and the ileocolic artery are present.[7,11] Left-sided branches of SMA supply the small bowel. The ileocolic artery marks the transition from jejunal to ileal arteries (▶Fig. 10.11): 4 to 6 **jejunal arteries** arise cranially to its origin and 11 to 14 **ileal arteries** arise caudally.[9] Several branches may originate together as trunks (▶Fig. 10.7). Each artery gives two branches in the mesentery which anastomose with adjacent ones forming the multiple orders of arcades previously described.

Inferior Mesenteric Artery

The origin of the **IMA** is located on the anterior aspect of the aorta at the level of L3–L4, about 6 to 7 cm above the SMA, and is frequently overlaid by the third part of the duodenum (▶Fig. 10.6). The IMA courses caudally to the left, giving all its branches from the left side. Collaterals of the IMA include the left colic artery, the sigmoid arteries, and the superior rectal artery, with a highly variable branching pattern: in up to 85% of cases the left colic artery originates directly from IMA, while in 15% of cases the left colic artery and one or more sigmoid arteries arise from a common trunk (▶Fig. 10.12).[7] The **left colic artery** is the first branch of the IMA, which runs upward to the left and gives two branches

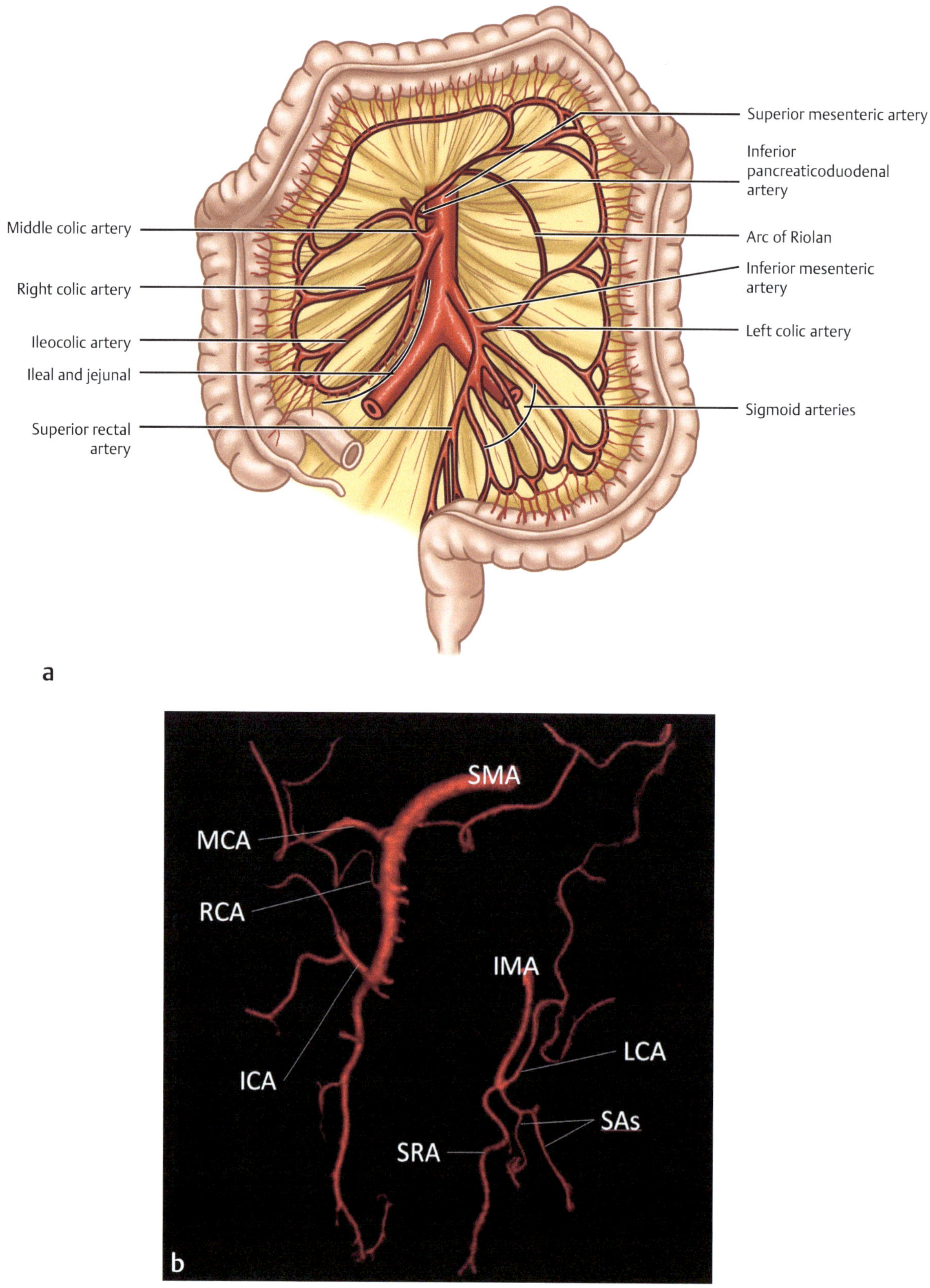

Fig. 10.3 Common mesenteric arteries anatomy. Comparison of artwork **(a)** and 3D volume rendered (VR) **(b)** reformatted computed tomography (CT) image. ICA, ileocolic artery; IMA, inferior mesenteric artery; LCA, left colic artery; MCA, middle colic artery; RCA, right colic artery; SAs, sigmoid arteries; SMA, superior mesenteric artery; SRA, superior rectal artery.

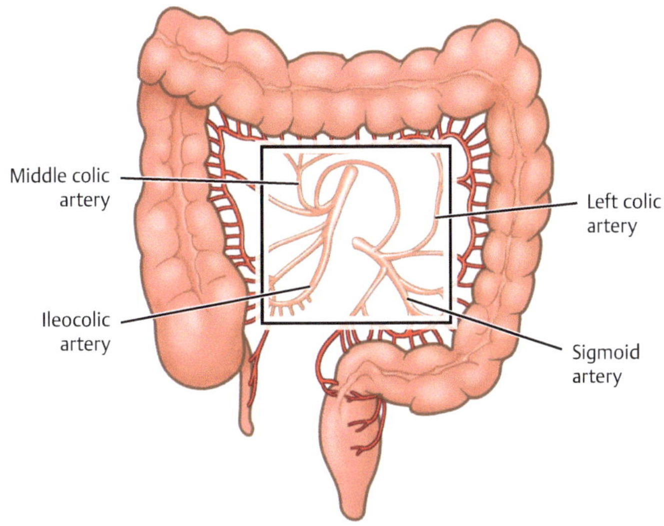

Fig. 10.4 Blood supply of the colonic vasculature shown by a square-shaped model. Four main arteries arising from the superior and inferior mesenteric arteries including ileocolic, middle colic, left colic, and sigmoid arteries.

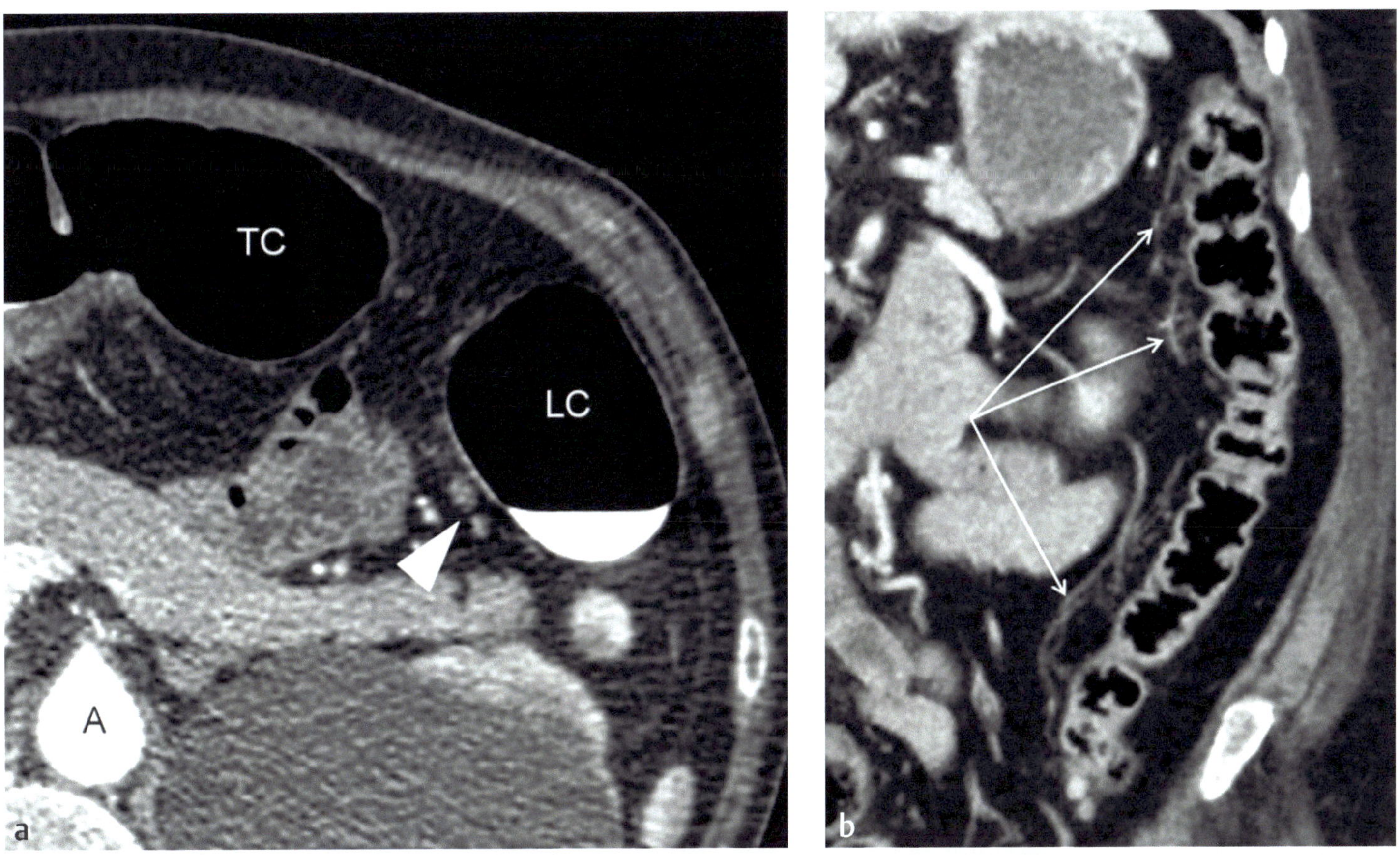

Fig. 10.5 (a) Axial and **(b)** coronal images from contrast-enhanced computed tomography (CT) colonography demonstrate the marginal artery of Drummond running parallel to the left colon (*arrowhead* and *arrows*). A venous vessel runs close to the artery. A, aorta; LC, left colon; TC, transverse colon.

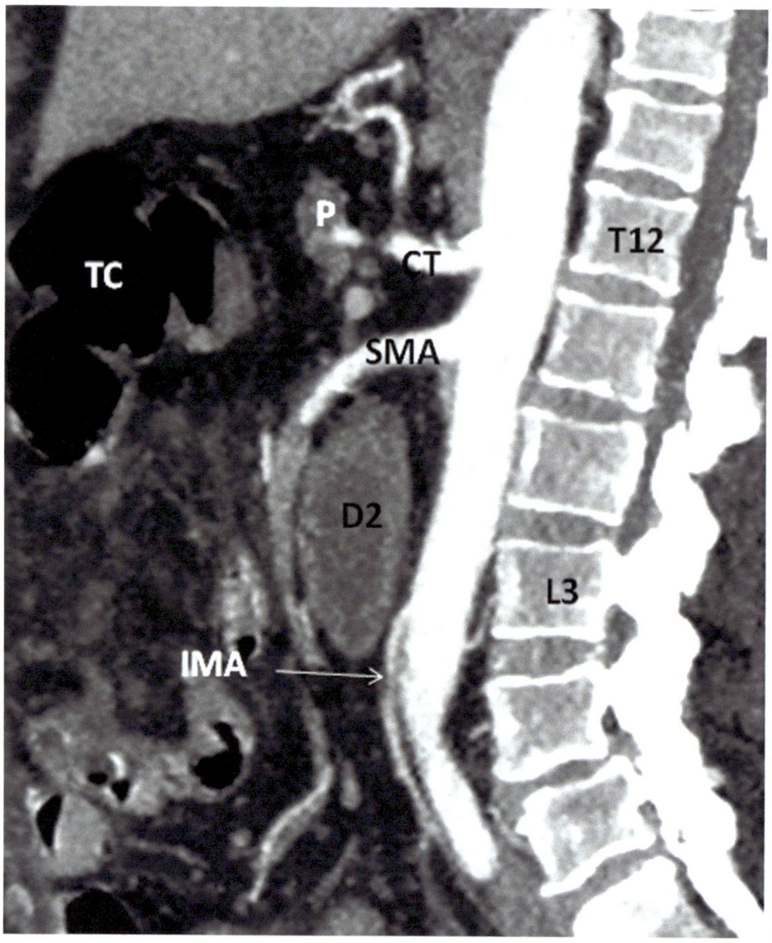

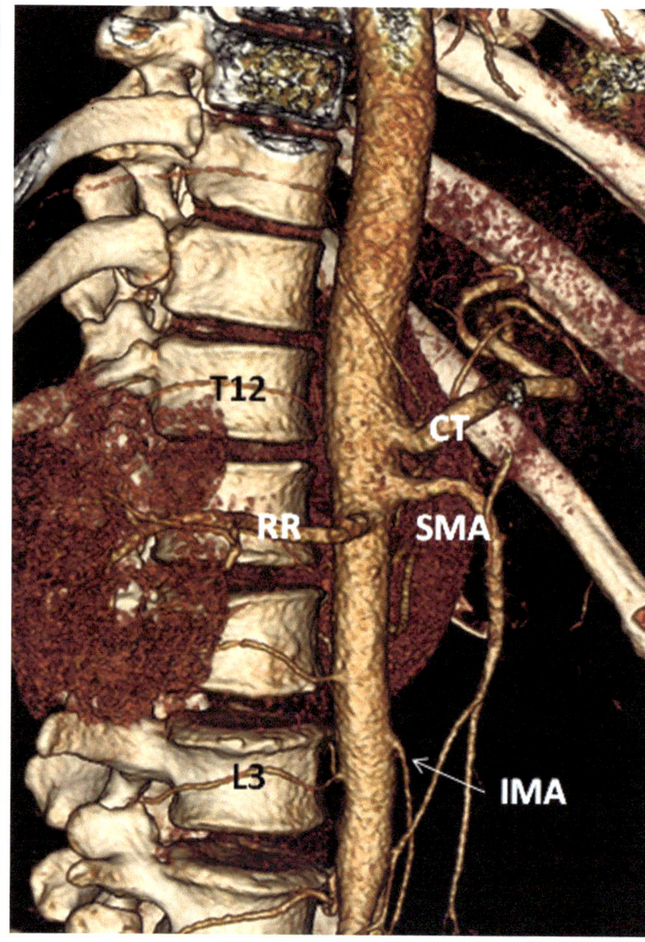

Fig. 10.6 Sagittal two-dimensional and three-dimensional images from computed tomography angiography show origins of the splanchnic arteries from the aorta. CT, celiac trunk; D2, second duodenum; IMA, inferior mesenteric artery; P, pancreas; RR, right renal artery; SMA, superior mesenteric artery; TC, transverse colon.

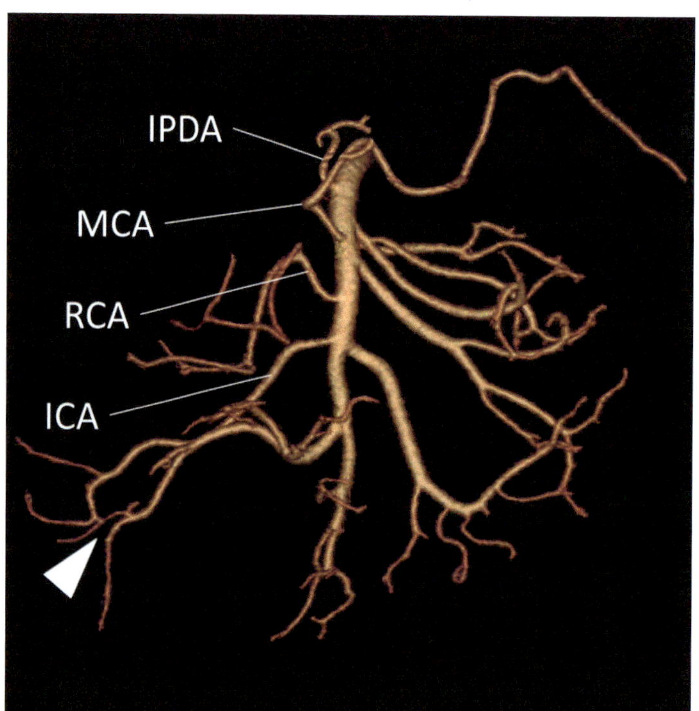

Fig. 10.7 3D volume rendered (VR) image from computed tomography (CT) angiography shows the branches of the superior mesenteric artery (SMA). The ileocolic artery anastomoses with the distal end of the main stem of the SMA (*arrowhead*). Left-sided branches are ileal and jejunal arteries. ICA, ileocolic artery; IPDA, inferior pancreaticoduodenal artery; MCA, middle colic artery; RCA, right colic artery.

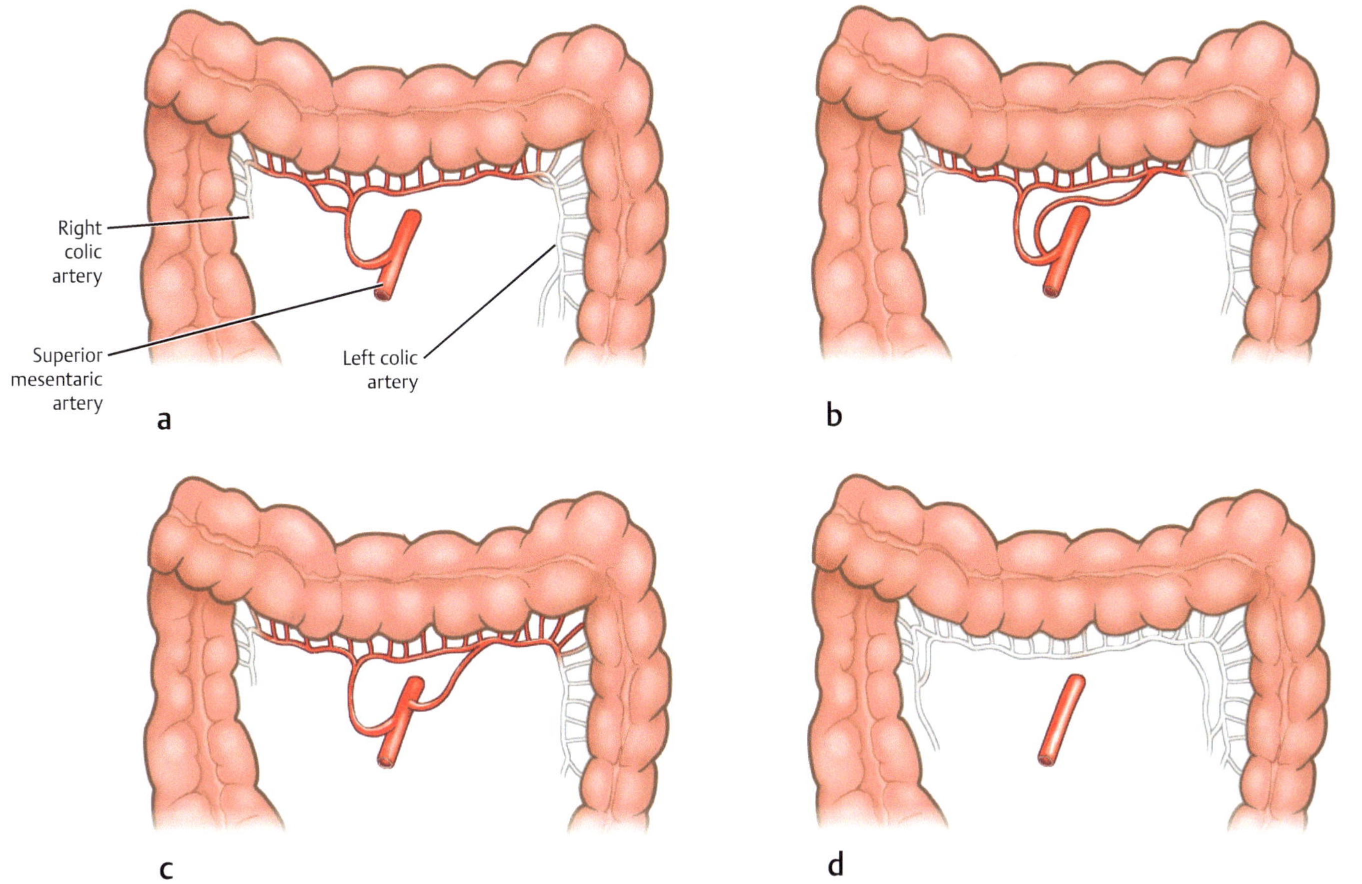

Fig. 10.8 Variants of the middle colic artery (shown in *red*) blood supply. **(a)** Middle colic artery extending to near hepatic and splenic flexures. **(b)** Early branching of the middle colic artery that reaches the splenic flexure. **(c)** Accessory middle colic artery supplying the splenic flexure and **(d)** absence of middle colic artery.

10

(▶Fig. 10.13, ▶Fig. 10.14). The ascending branch feeds the splenic flexure and anastomoses with the left-sided branch of the middle colic artery. The descending branch supplies the descending and sigmoid colon joining with the sigmoid arteries (▶Fig. 10.13). An inconstant branch of the left colic artery is the artery of Riolan, connecting left and middle colic arteries. One or more sigmoid arteries can arise from the left colic artery (▶Fig. 10.15). An anomalous origin of the left colic artery from the SMA occurs at a frequency of 1%.[12,13] The number of **sigmoid arteries** varies from 1 to 13 and their pattern is very changeable.[14] They can arise independently from the IMA or from common trunks. It is not uncommon for one or more sigmoid arteries to originate from the left colic artery or from a common trunk with it. They anastomose with each other and with the descending branch of the left colic artery, supplying the descending and sigmoid colon. The **superior rectal** (or **superior hemorrhoidal**) **artery** is a terminal branch of the IMA that descends in the sigmoid mesocolon and bifurcates into branches supplying the upper rectum (▶Fig. 10.16). It is commonly defined as the distal end of the IMA after the diverging of the last sigmoid artery: this point is known as Sudeck's point (see section Intermesenteric Collateral Circulation).[7] Some authors define the superior rectal artery as the portion of the IMA distal to its intersection with the left common iliac artery.[10] Although the superior rectal artery provides the main arterial supply to the rectum, the rectum is also supplied by the **middle sacral artery** and by the bilateral **middle rectal arteries** and **inferior rectal arteries**. The middle sacral artery originates from the posterior aspect of the aorta, above its bifurcation, and reaches the posterior wall of the rectum in the retroperitoneum. Middle and inferior rectal arteries derive indirectly from the internal iliac arteries below the pelvic peritoneum. The middle rectal arteries may be absent.[10]

Collateral Arterial Circulation

Adequate perfusion to the entire bowel is ensured by the presence of extensive collateral pathways. This sophisticated network allows normal digestive process and protects the bowel from potential ischemia due to poor blood supply or vessel occlusion.[3] Normally, most collateral pathways are not visible on imaging, but these vessels increase in size in order to respond to an ischemic stimulus.[15] Connections can be subdivided into three categories: between the SMA and the celiac trunk (celiacomesenteric), between the SMA and the IMA (intermesenteric), and between the IMA and collaterals from the aorta and internal iliac

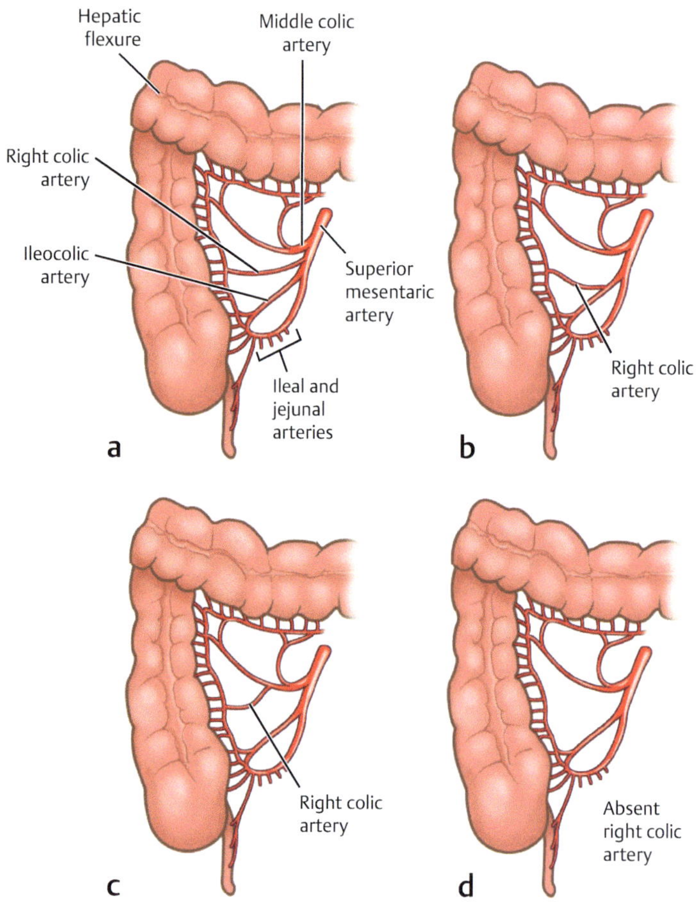

Fig. 10.9 Variants of blood supply to the cecum and right colon. **(a)** Origin of the right colic artery (in *red*) from the superior mesenteric artery. **(b)** Origin of the right colic artery from the ileocolic. **(c)** Origin of the right colic artery from the middle colic. **(d)** Absence of right colic artery. In **(b)** and **(d)** variants, the hepatic flexure is supplied by the middle colic artery.

10

arteries (systemic-mesenteric).[16] Those communications may be absent when variants in arterial branching occur.

Celiacomesenteric Collateral Circulation

The SMA and the CT communicate principally through the **pancreaticoduodenal arcade** (▸Fig. 10.17, ▸Fig. 10.18, ▸Fig. 10.19), a relatively constant collateral which supply the pancreatic head and the second portion of the duodenum. It is made by the anastomosis between the superior pancreaticoduodenal artery, originating from the gastroduodenal artery (from the common hepatic artery), and the inferior pancreaticoduodenal artery, first right-sided branch of the SMA. The right and left epiploic arteries are tributaries of the ipsilateral gastroepiploic arteries, derived from the gastroduodenal and splenic arteries, respectively. Their anastomosis is situated between the layers of the greater omentum and constitutes the **arc of Barkow**, or greater omental arcade, which provides blood supply to the greater omentum and collateral vascularization to adjacent organs. Numerous thin branches from the arc of Barkow join with collaterals from the middle colic artery and the left colic artery, increasing the vascular network of the transverse colon.[5,8,10] The **arc of Bühler** (▸Fig.10.20) is a remnant of the ventral anastomosis of the primitive intestinal vessels, which persist in less than 2% of patients. This rare but well-described vessel runs vertically behind the pancreatic body and creates a direct connection between the SMA or one of its branches and the splenic or the common hepatic artery.[2,17] Another rare (less than 1% described) additional collateral pathway is present when the middle colic artery originates from the celiac trunk or from one of its branches.[3]

Intermesenteric Collateral Circulation

Each colic artery arising from the SMA and IMA gives two or more branches that join with the others creating a single anastomotic bridging running within the mesentery, parallel to the colonic wall, along the entire length of the colon. This arcade, known as **marginal artery of Drummond** (▸Fig. 10.21, ▸Fig. 10.22, ▸Fig. 10.23, ▸Fig. 10.24), is the major collateral between the ileocolic, right colic, middle colic, left colic, and sigmoid arteries, and gives numerous short vasa recta that enter the visceral wall. It is small in caliber, but it can enlarge markedly in case of mesenteric arterial occlusion.[18] The marginal artery can provide blood supply to the entire left colon in case of occlusion of the IMA: during abdominal aortic aneurysm repair procedures, for example, the IMA is generally ligated or covered with an endoprosthesis without ischemic injury to the colon. Given a certain degree of variability regarding the marginal artery, two points of weakness have been described: the Griffiths' point at the splenic flexure and the Sudeck's point at the rectosigmoid junction. The Griffiths' point (▸Fig. 10.23) is defined as the site of communication of branches of the left colic artery with branches of the middle colic artery and represents a potential watershed area in the collateral circulation between the SMA and the IMA. Absence or insufficiency of this anastomosis has been reported in up to 50% of patients, who are more prone to ischemic damage of the splenic flecture.[17,19] The Sudeck's point (▸Fig. 10.15) is defined as the point of origin of the last sigmoid artery from the IMA, so it can also be considered as the point of origin of the superior rectal artery. In the majority of patients there are no anastomoses between the last sigmoid artery and the superior rectal artery distal to this point; therefore, the Sudeck's point is an area of critical weakness in the rectal blood supply after sigmoid colon resection. Ligation of the IMA proximal or distal to the Sudeck's point is crucial in terms of postoperative blood supply to the caudal stump. When a distal anastomosis between the last sigmoid artery and the superior rectal artery is present, the marginal artery can provide collateral circulation to the distal rectum.[20] The **arc of Riolan** (▸Fig. 10.24) is an inconstant collateral branch that connects the middle colic artery and the left colic artery, located centrally within the mesentery and parallel to the marginal artery. It has been described in 7 to 10% of population, but some authors suggest this is an enlargement of a constant collateral vessel which is normally too thin to be depicted.[3,17] The arc of Riolan is also known as the accessory left colic artery.[21] The **meandering artery of Moskowitz** has been described as an inconstant anastomotic channel running along the base of the mesocolon, connecting the middle colic artery and the left colic

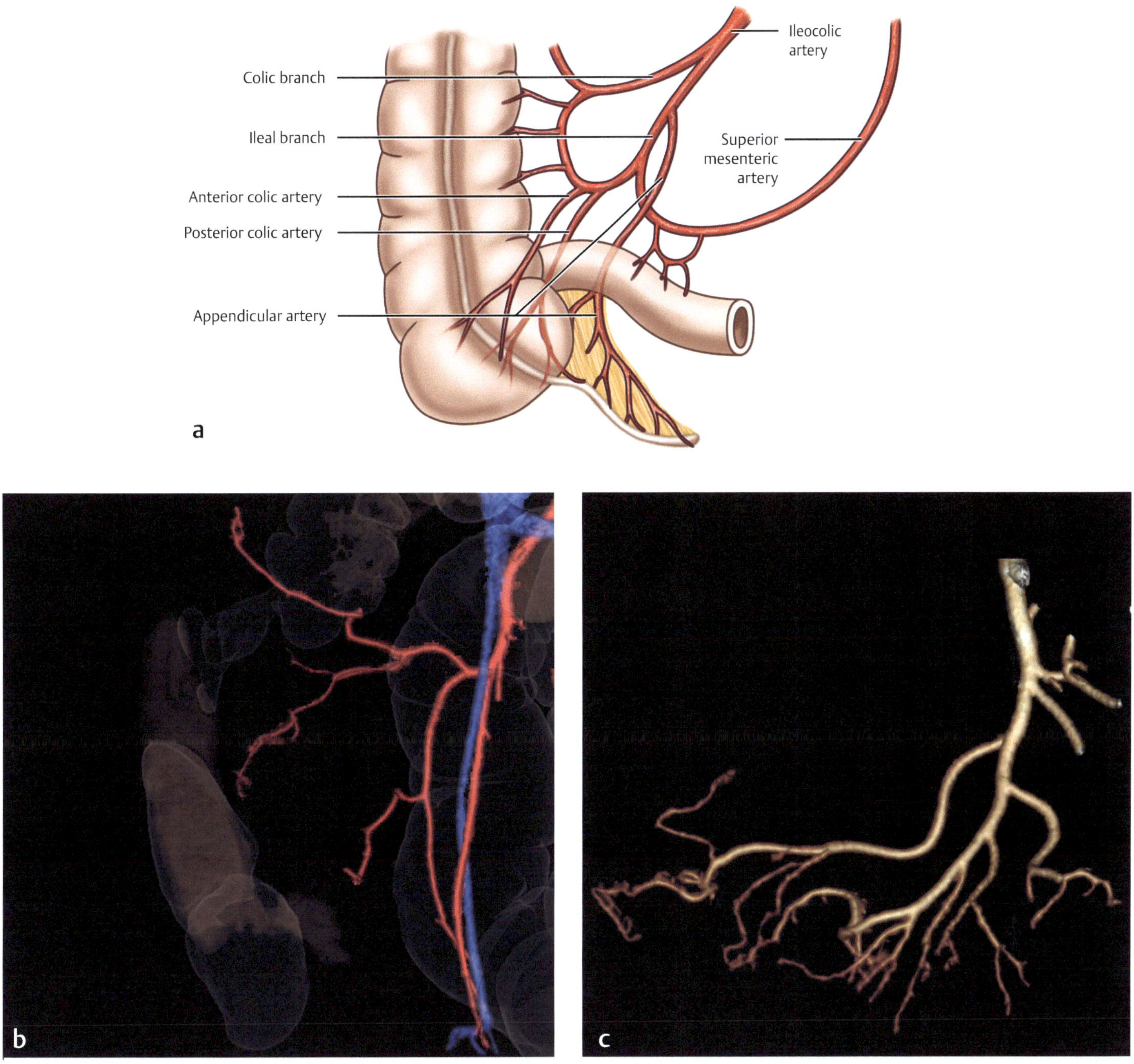

Fig. 10.10 (a) The "ileocolic loop" formed by the anastomosis of the two major branches of the ileocolic artery. An appendicular artery and two (anterior and posterior) colic arteries usually arise from the ileocolic loop. The ileal branch of the ileocolic artery (ICA) forms an arch joining with the distal end of the main stem of the superior mesenteric artery (SMA). **(b, c)** 3D volume rendered (VR) image from contrast-enhanced computed tomography (CT) colonography. **(b)** Common origin of the middle colic artery and the ileocolic artery. **(c)** Detail of the anatomy of the ileocolic artery with anastomosis between its ileal branch and the distal end of the main stem of the SMA. **(d)** Common branching pattern of the ICA in a patient with severe bleeding (*yellow arrows*) from branches of the right colic artery (RCA). In this example, selective angiogram of the SMA and ICA before embolization (pre) showing the RCA arising from the ICA. The ICA then divides into the cecal and ileal branches. The distal trunk of the ICA was embolized. The area of embolization is shown by *green arrows*. Postembolization angiograms of the SMA and middle colic artery (MCA) show occlusion of the ICA with distal runoff to the cecal and ileal branches provided by the marginal arteries (MA) and the distal SMA (*long curved yellow lines*).

(*Continued*)

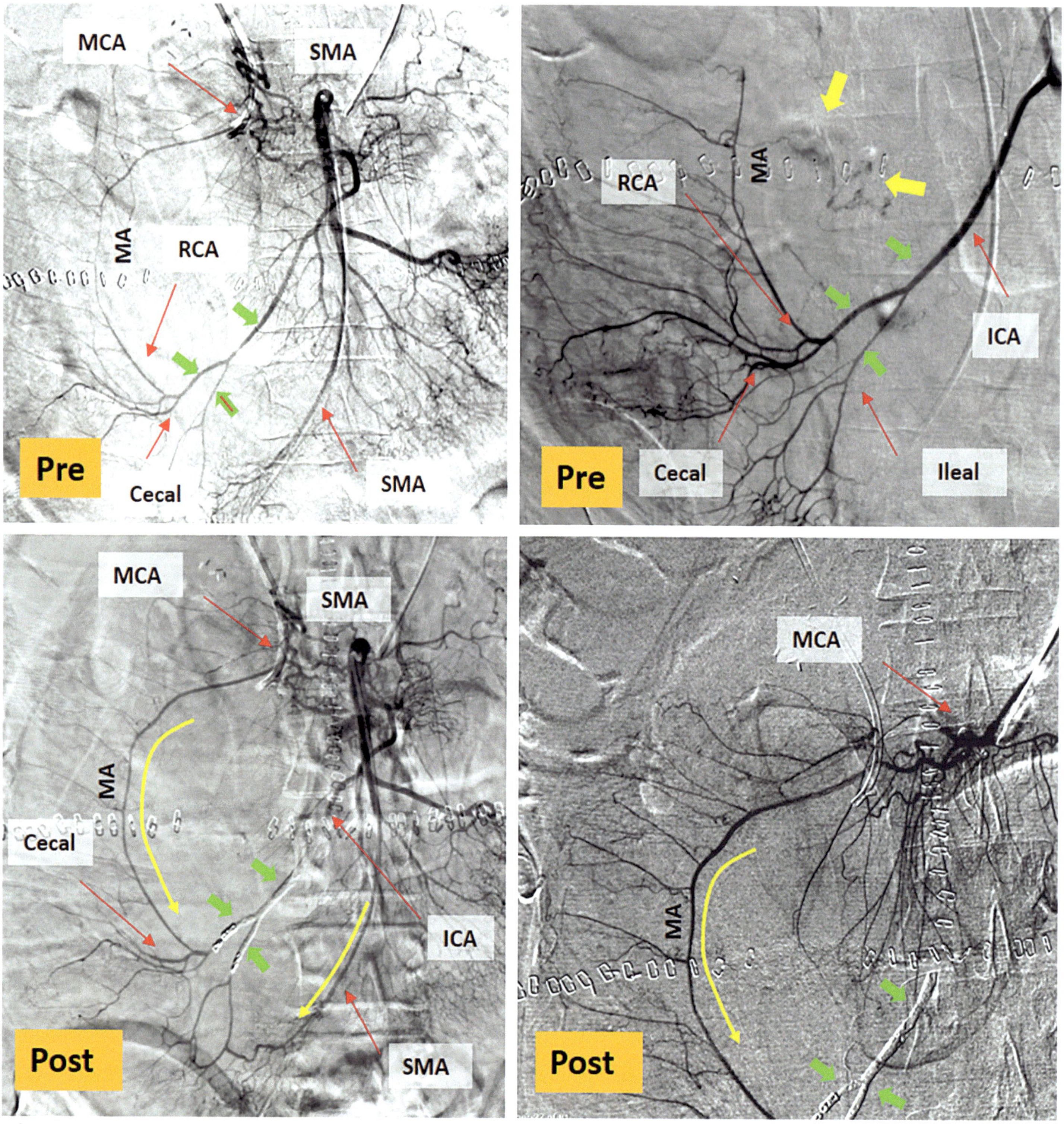

d

Fig. 10.10 (*Continued*) (**d**)

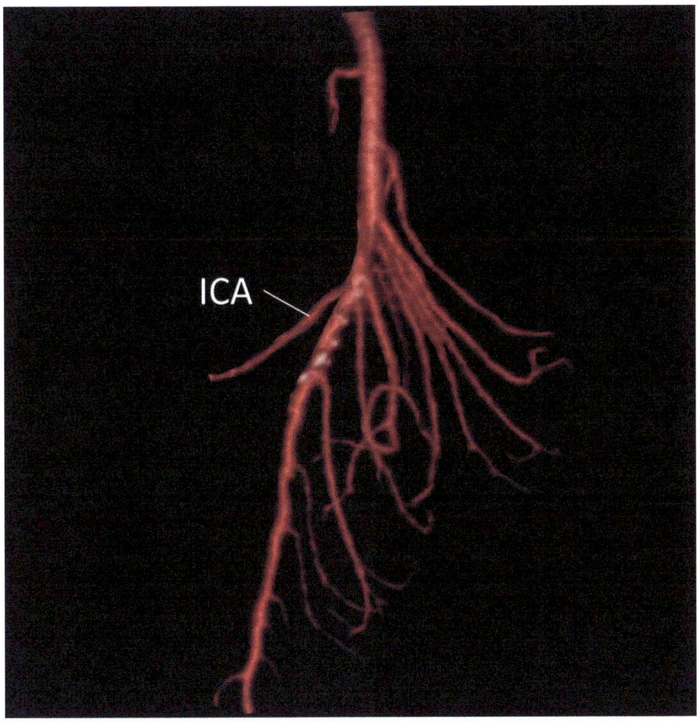

Fig. 10.11 3D volume rendered (VR) image from computed tomography (CT) angiography showing origins of jejunal and ileal branches from the superior mesenteric artery (SMA). Jejunal arteries arise cranially to the origin of the ileocolic artery, while ileal arteries arise caudally. ICA, ileocolic artery.

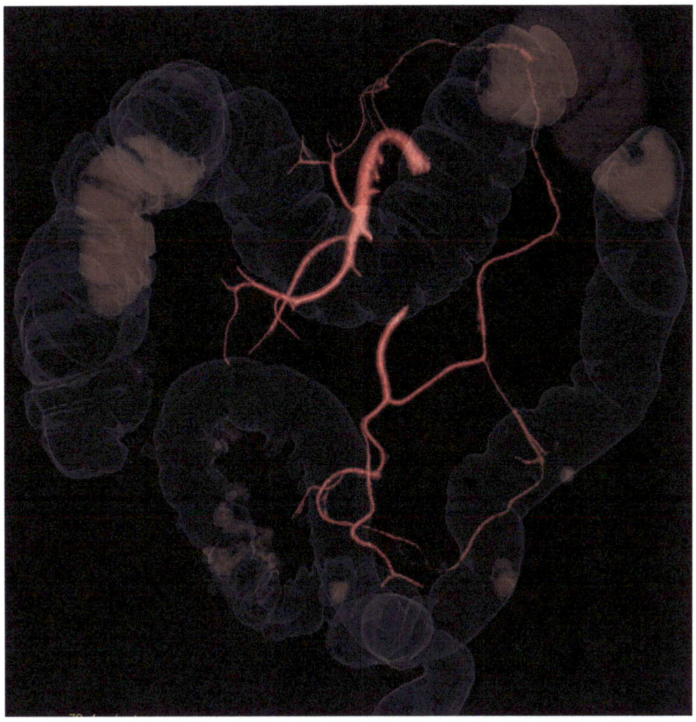

Fig. 10.13 3D volume rendered (VR) image from contrast-enhanced computed tomography (CT) colonography demonstrates anatomy of the inferior mesenteric artery (IMA) and connections between its branches and the superior mesenteric artery (SMA). The middle and left colic arteries are connected by the marginal artery of Drummond, which also anastomoses with sigmoid arteries. (Courtesy of Department of Diagnostic and Interventional Radiology, Ospedale S. Paolo, Milano, Italy.)

10

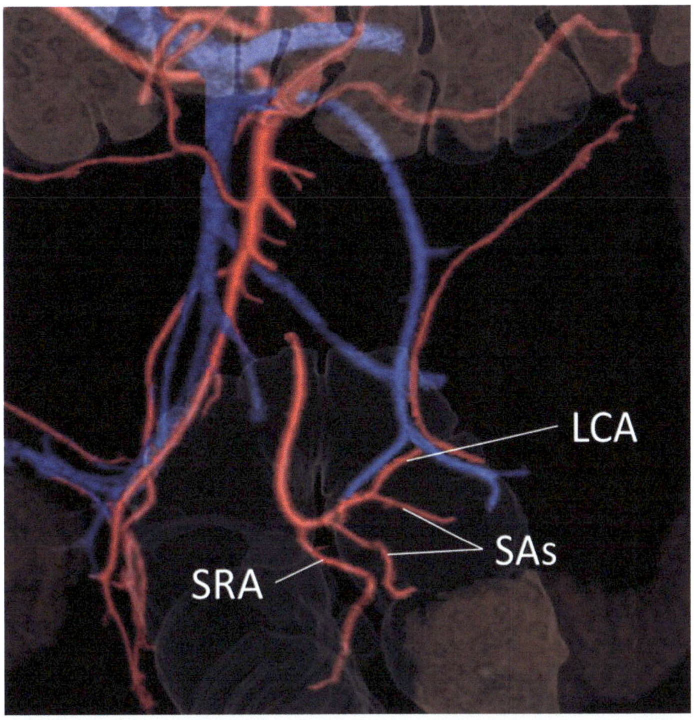

Fig. 10.12 3D volume rendered (VR) image from contrast-enhanced computed tomography (CT) colonography depicting the anatomy of the inferior mesenteric artery (IMA). In this patient the left colic artery (LCA) and the sigmoid arteries (SAs) have a common origin. The terminal branch of the IMA is the superior rectal artery (SRA). The ascending branch of the left colic artery feeds the splenic flexure and anastomoses with the left-sided branch of the middle colic artery. Relationship between IMA, superior mesenteric artery (SMA), and mesenteric veins are also depicted: both ileocolic and left colic artery cross posteriorly to the superior and inferior mesenteric veins.

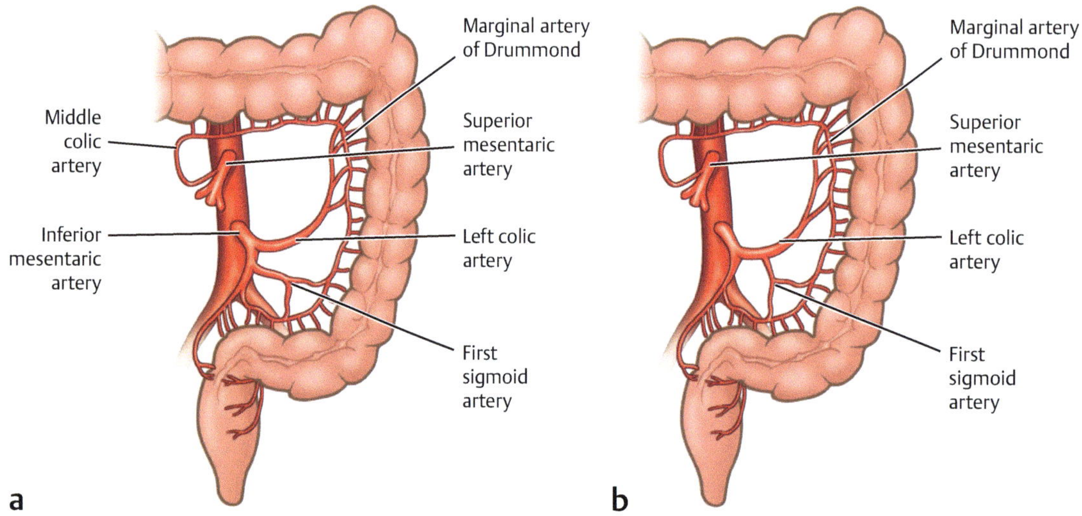

Fig. 10.14 Branching patterns of the left colic artery. **(a)** In 50 to 60% of individuals, the left colic artery arises from the inferior mesenteric artery. **(b)** In the remainder, left colic artery and the first sigmoidal artery form a short or long common trunk.

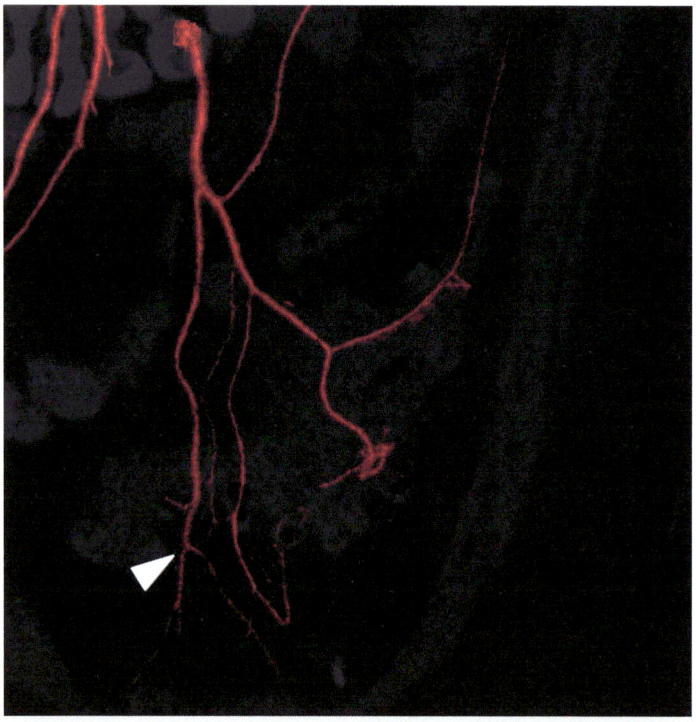

Fig. 10.15 3D volume rendered (VR) image from computed tomography (CT) angiography. Another variant in the origin of the inferior mesenteric artery (IMA) branches. Some sigmoid arteries arise from the left colic arteries, while the last one originates from the IMA at the Sudeck's point (*arrowhead*). This patient had acute diverticulitis. CT demonstrated distribution of mesenteric branches to the affected colic tract.

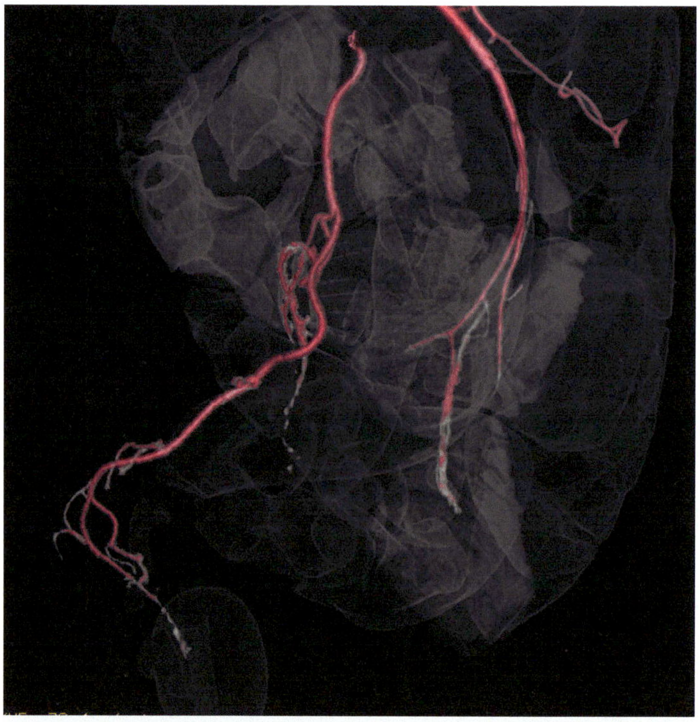

Fig. 10.16 3D volume rendered (VR) image from contrast-enhanced computed tomography (CT) colonography. Lateral view depicts anatomy of the superior rectal artery, terminal branch of the inferior mesenteric artery (IMA). The vessel runs parallel to the posterior aspect of the rectum.

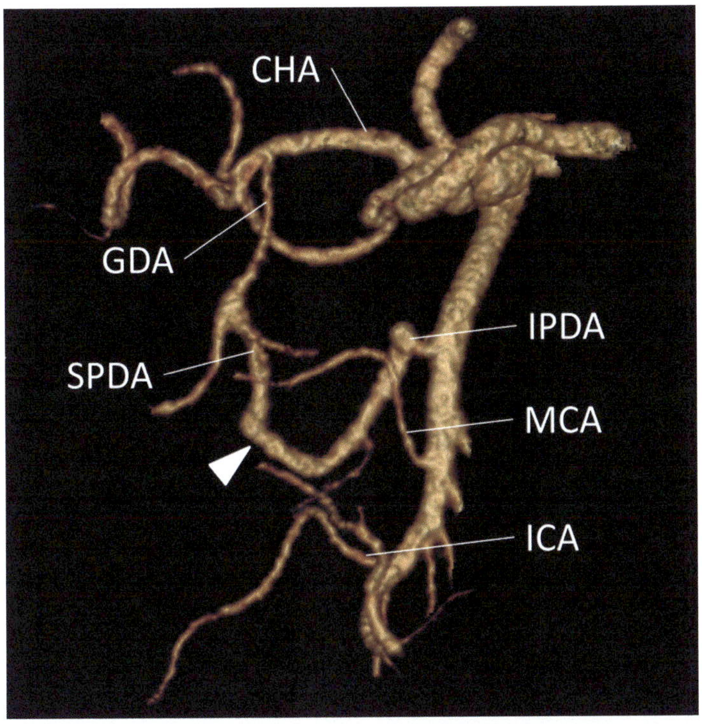

Fig. 10.17 3D volume rendered (VR) image from computed tomography (CT) angiography. Dilated pancreaticoduodenal arcade (*arrowhead*) in a patient with celiac trunk dissection. CHA, common hepatic artery; GDA, gastroduodenal artery; ICA, ileocolic artery; IPDA, inferior pancreaticoduodenal artery; MCA, middle colic artery; SPDA, superior pancreaticoduodenal artery.

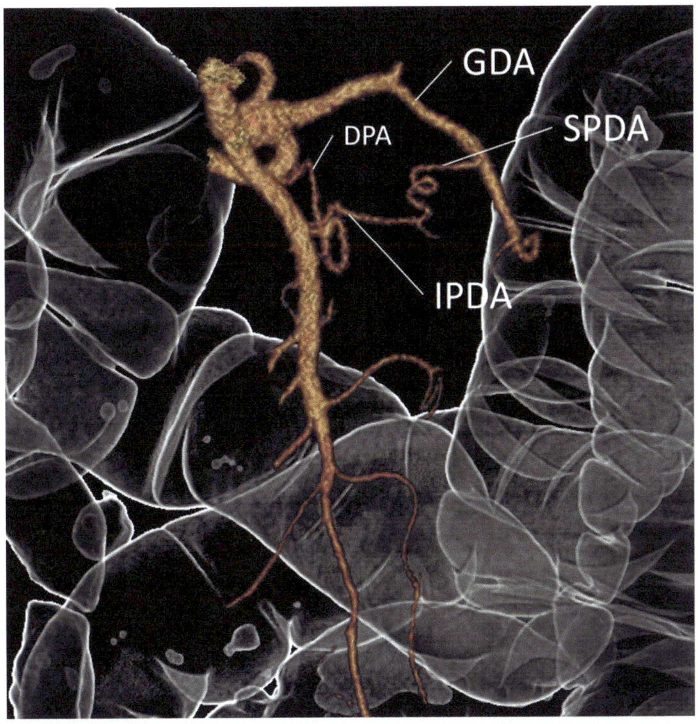

Fig. 10.18 3D volume rendered (VR) image from contrast-enhanced computed tomography (CT) colonography. Posterior view. Pancreaticoduodenal arcade. The dorsal pancreatic artery (DPA) is also depicted. GDA, gastroduodenal artery; IPDA, inferior pancreaticoduodenal artery; SPDA, superior pancreaticoduodenal artery.

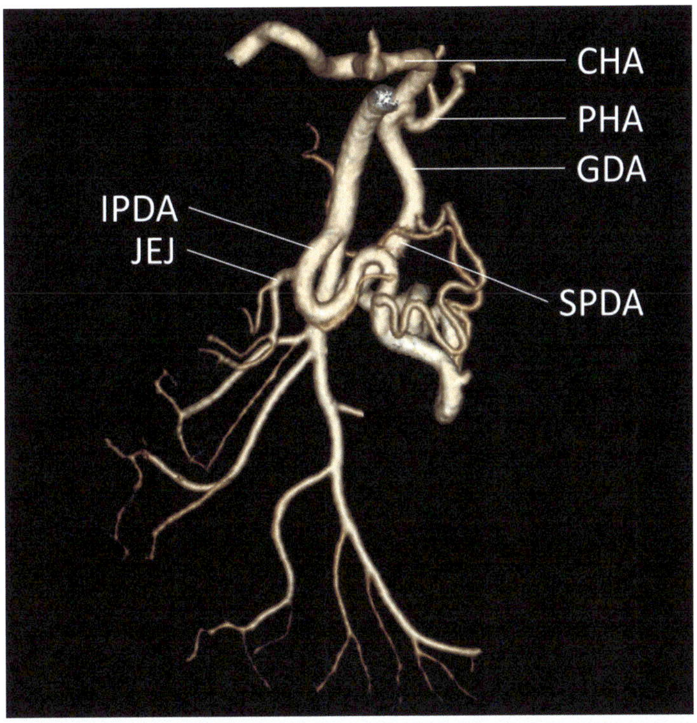

Fig. 10.19 3D volume rendered (VR) image from computed tomography (CT) angiography, posterior view. Patient with proximal celiac trunk occlusion. Celiacomesenteric collateral circulation is granted by a dilated pancreaticoduodenal arcade. Tortuous, hypertrophic gastroepiploic arcade is also depicted. The first jejunal artery arises from the IPDA. CHA, common hepatic artery; GDA, gastroduodenal artery; IPDA, inferior pancreaticoduodenal artery; JEJ, first jejunal artery; PHA, proper hepatic artery; SPDA, superior pancreaticoduodenal artery.

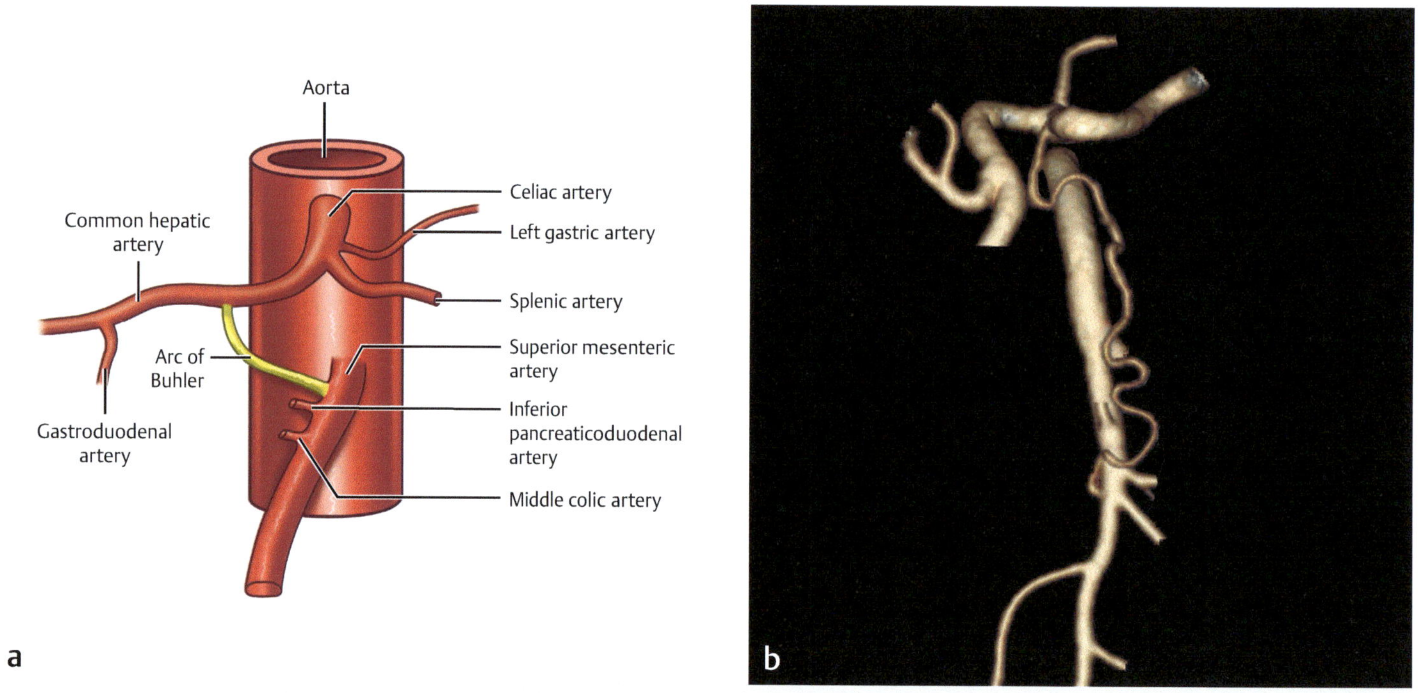

Fig. 10.20 (a) The arc of Bühler is shown, connecting the common hepatic artery and the superior mesenteric artery. This connection provides a direct collateral circulation when stenosis of the celiac artery or superior mesenteric artery is present. In some cases, the arc of Bühler may connect to another celiac branch or superior mesenteric artery branches. **(b)** 3D volume rendered (VR) image from computed tomography (CT) angiography, anterior view. A patient with proximal CT occlusion (same as Fig. 10.19). A tortuous arc of Bühler provides collateral circulation (*arrows*) between the SMA and the celiac trunk.

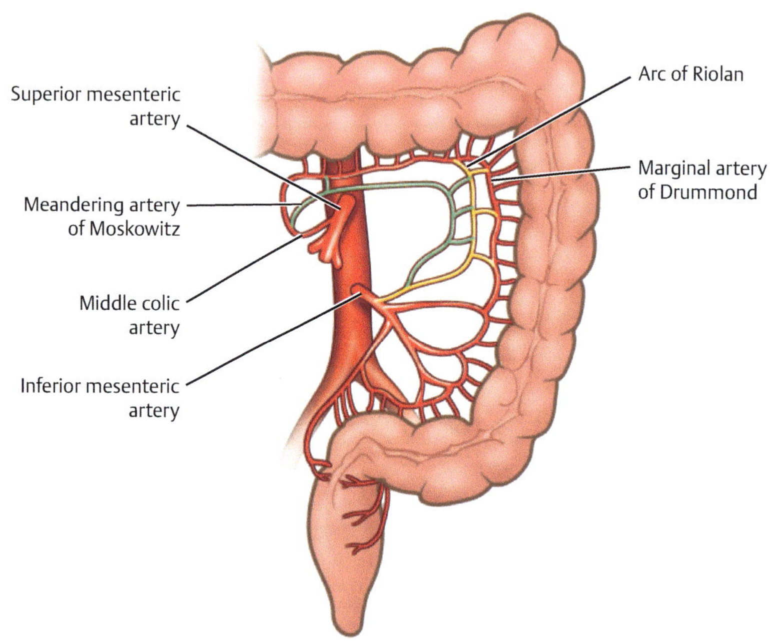

Fig. 10.21 Collateral arteries between the superior and inferior mesenteric arteries: marginal artery of Drummond, the arc of Riolan (*yellow*), and the meandering artery of Moskowitz (*green*) are shown.

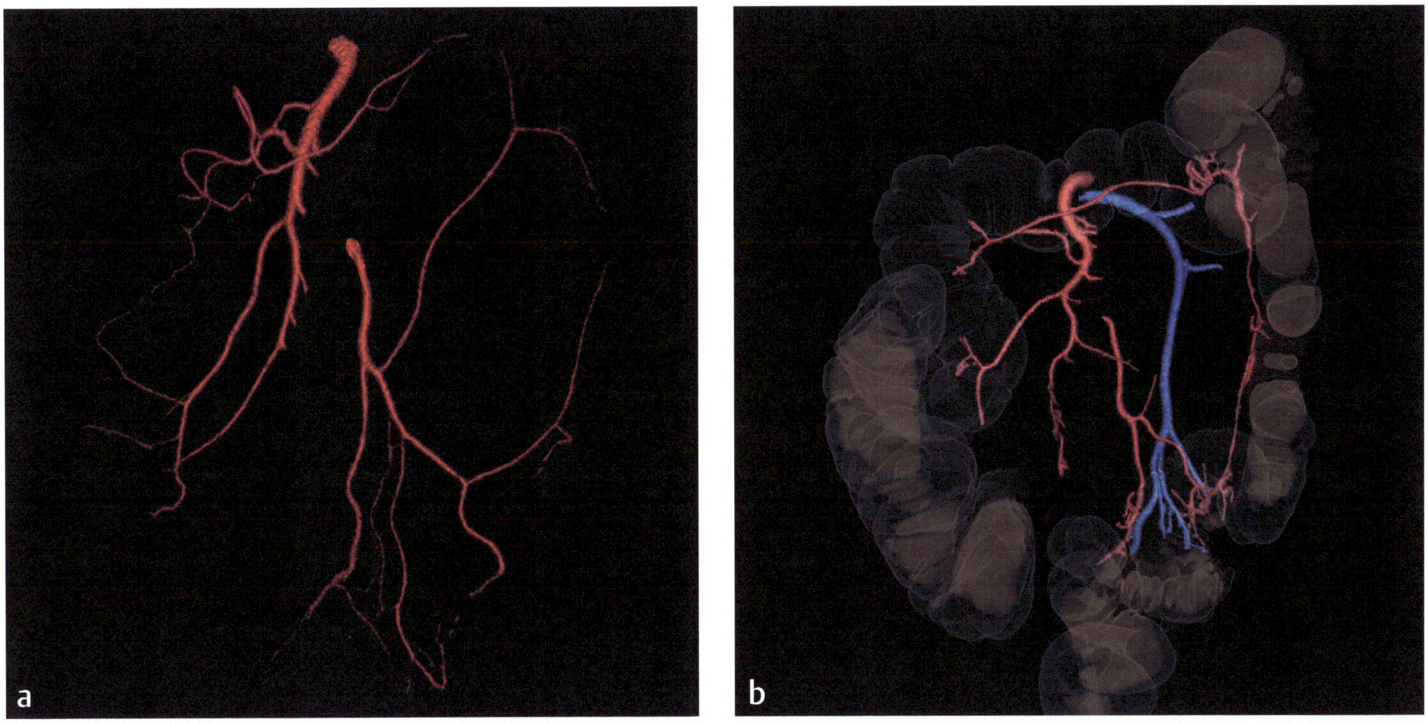

Fig. 10.22 (a) 3D volume rendered (VR) image from computed tomography (CT) angiography (same as **Fig. 10.15**) shows the entire extent of the marginal artery of Drummond connecting all branches of superior mesenteric artery (SMA) and inferior mesenteric artery (IMA). **(b)** 3D VR image from contrast-enhanced CT colonography shows the course of the marginal artery of Drummond, which follows the transverse and left colon, and gives small visceral branches.

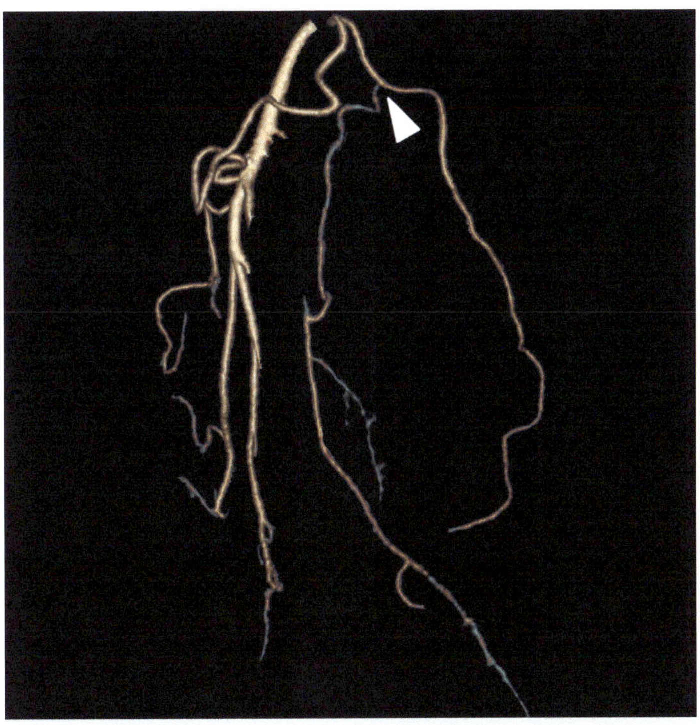

Fig. 10.23 3D volume rendered (VR) image from computed tomography (CT) angiography. Patient with inferior mesenteric artery (IMA) occlusion. Blood supply to the left colon is provided by a hypertrophic marginal artery. The Griffiths' point at the splenic flexure (*arrowhead*) is the site of anastomosis between the ascending branch of the left colic artery with branches of the middle colic artery.

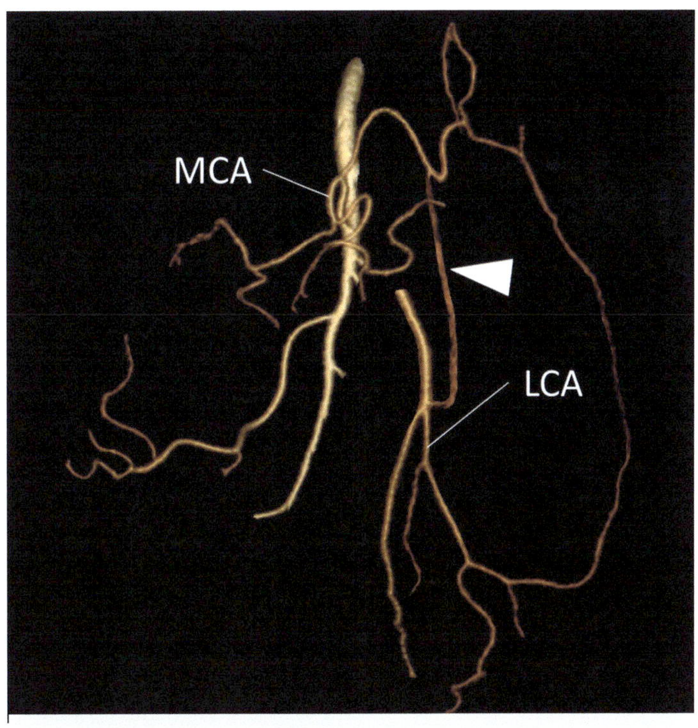

Fig. 10.24 3D volume rendered (VR) image from computed tomography (CT) angiography depicts the arc of Riolan (*arrowhead*) connecting the left branch of the middle colic artery with the inferior mesenteric artery (IMA) at the origin of the left colic artery. It provides additional intermesenteric collateral circulation in addition to the marginal artery. LCA, left colic artery; MCA, middle colic artery.

artery. Controversy exists regarding the presence of this vessel: some authors state that it represents a dilated the arc of Riolan, while others believe that it is a separate anastomotic channel.[22]

Systemic-mesenteric Collateral Circulation

Although the vascularization of the rectum lacks a system of arterial arcades as the rest of the bowel, there is a collateral network between the IMA with the aorta and internal iliac arteries in the pelvis. The superior rectal artery, middle rectal arteries, inferior rectal arteries, and middle sacral artery reach the rectal wall and divide into several branches that independently penetrate the anal sphincters and enter the gut wall, merging into a submucosal anastomosing plexus. Furthermore, additional anastomosis can be present between branches of the IMA and branches of the aorta such as lumbar arteries and femoral arteries and their branches.[10]

Variations in Major Arterial Branching

Variations in vessel branching patterns from the splanchnic arteries occur frequently and the increasing use of MDCT allows their diagnosis in more and more patients. Moreover, in daily radiological practice it is easy to collect large sample of data about the angiographic anatomy of the abdomen. Therefore, many case reports describing patients with unusual vessel anatomy and their prevalence have been recently published. Different classification systems for the branching patterns have been proposed.[2,4,23,24]

10

The **hepatic arteries** have a very variable anatomy: the normal origin of the common hepatic artery from the celiac trunk is reported in only 55% of individuals. Accessory or replaced hepatic arteries arise from the SMA in 20% of individuals. An "accessory" hepatic artery is a vessel with an aberrant origin that supplies an isolated area of a particular hepatic segment, whereas a "replaced" hepatic artery supplies an entire lobe. The most common anomaly is the origin of a replaced right hepatic artery from the SMA, described in up to 15% of patients (▶ Fig. 10.25).[5,25–27] Arteries arising from the SMA and IMA have a wide spectrum of variations in number, site of origin, and branching pattern: most common mesenteric variants have been previously described (see section Arteries). Anomalies in the origin of the CT, the SMA, and the IMA from the aorta are much less common. One of the most striking of these is the common origin of the CT and the SMA in a single **celiomesenteric trunk** (▶ Fig. 10.26). In this situation, described in about 1% of the population, a single artery is the only source of vascularization of the supramesocolic organs and significant collateral flow is only possible from the IMA.[4,14,25,27–29] Separate direct origin of the hepatic and splenic arteries from the aorta is not rare. Even less common is the anomalous **origin of the IMA from the SMA**: an artery with all the characteristics of the IMA, arising from the SMA instead of the aorta, has been reported in 10 published cases. In the majority of these cases, the IMA diverged as the first branch of the SMA.[12,30] The congenital lack of any identifiable SMA in newborns has been associated with congenital duodenal-jejunal atresia, whereas the **absence of the SMA** in a healthy adult has been observed in just one published report to date, in which the investigators found compensatory dilatation of the IMA.[24] Separate origin from the aorta of gastric, splenic, and hepatic arteries and middle colic artery has been also described.[31]

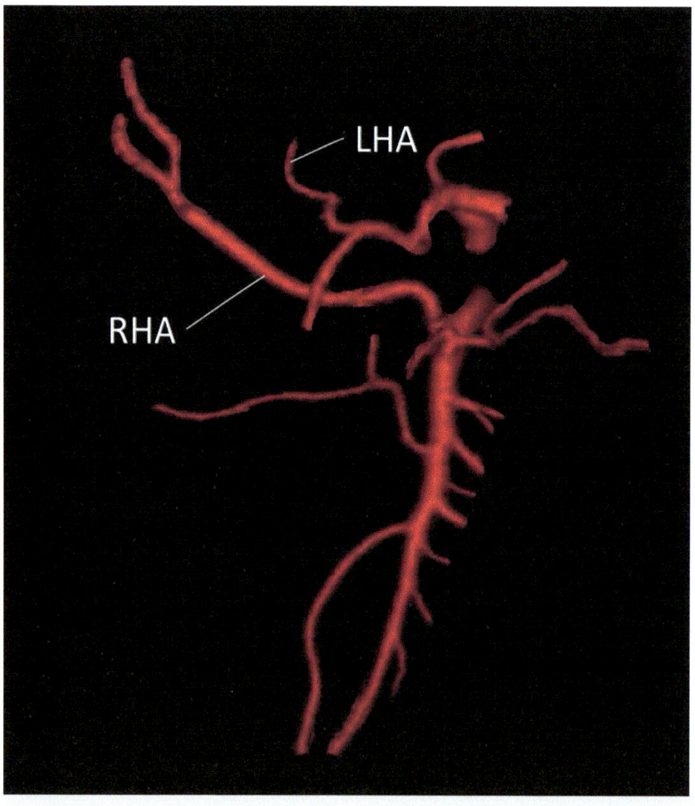

Fig. 10.25 3D volume rendered (VR) image from computed tomography (CT) angiography shows a replaced right hepatic artery (RHA) originating from the superior mesenteric artery (SMA). The common hepatic artery bifurcates into left hepatic artery (LHA) and the gastroduodenal artery.

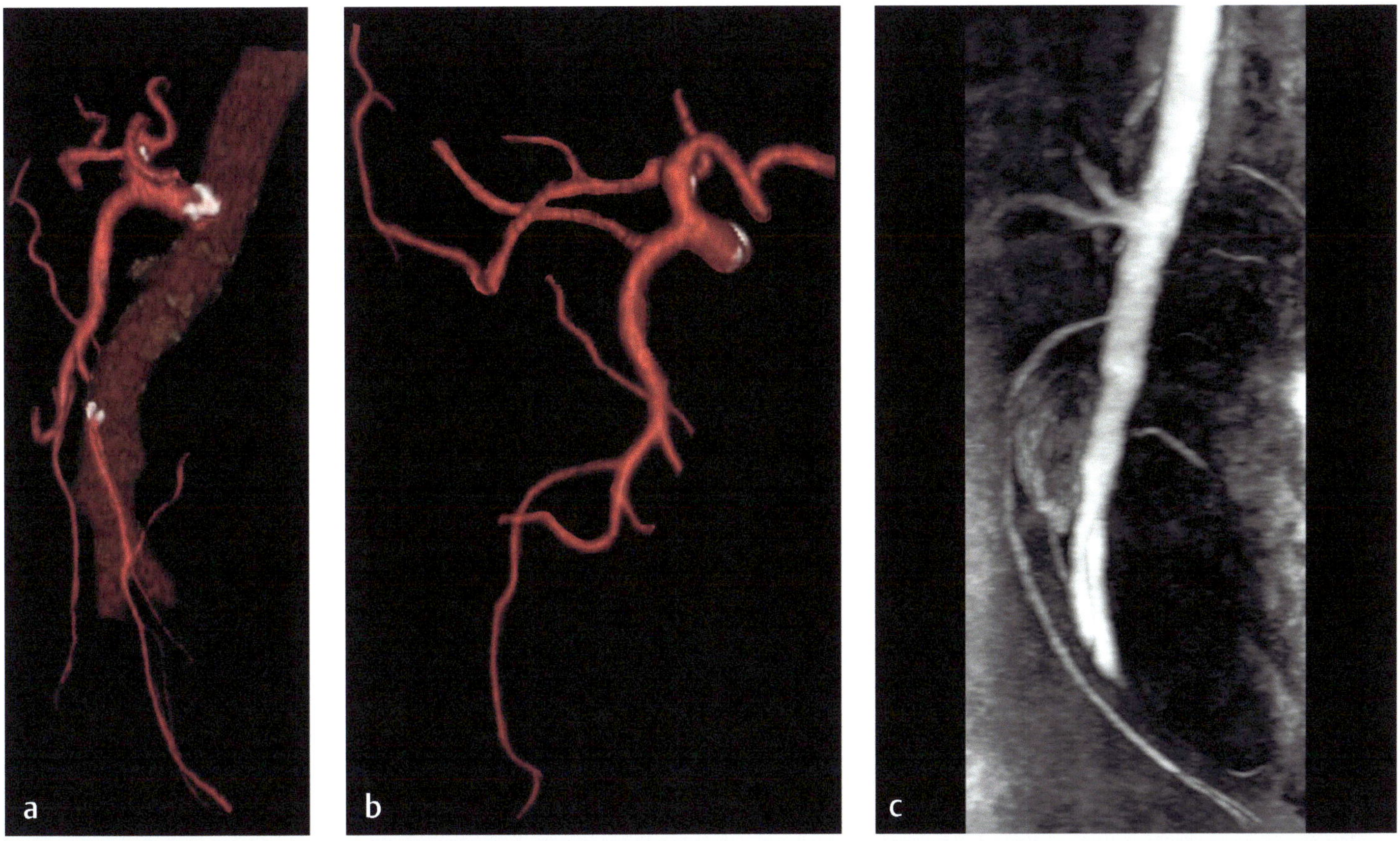

Fig. 10.26 (a, b) 3D volume rendered (VR) image from computed tomography (CT) angiography. Common origin of the celiac trunk and the superior mesenteric artery (SMA) from a celiomesenteric trunk. **(a)** Origin of the celiomesenteric trunk and the inferior mesenteric artery (IMA) from the aorta. **(b)** Branching pattern of the CT and the SMA. The right hepatic artery arises from the SMA. **(c)** Sagittal maximum intensity projection (MIP) image from magnetic resonance (MR) angiography shows common origin of the CT and the SMA from a celiomesenteric trunk and the origin of the IMA.

◆ Veins

Venous drainage from the bowel is provided by vessels that closely accompany the corresponding arteries (▶Fig. 10.27). The blood reaches the liver through the portal venous system by two main tributaries. Veins from the small bowel and right colon open into the SMV, which converges with the splenic vein to form the portal vein behind the neck of the pancreas. From the left colon and upper rectum, veins drain into the inferior mesenteric vein (IMV), which inconstantly joins the splenic vein. Venous blood from the mid and lower rectum drains into the systemic venous circulation via the internal iliac veins.[10] The splenic vein courses behind the pancreas below the splenic artery and conveys blood from the stomach, pancreas, and portions of the large intestine through the short gastric, left gastroepiploic, pancreatic, and inferior mesenteric veins.[8] The SMV receives venous blood from the stomach, the distal duodenum, the jejunum, the ileum, and the right and part of the transverse colon. In the majority of individuals, it is a vessel of variable length (5–50 mm) formed by two large intestinal branches (right and left). In some patients the main trunk is absent and there are large right and left mesenteric branches, which together join the splenic vein.[32] The right gastroepiploic, pancreaticoduodenal, and right colonic veins frequently form a common gastrocolic trunk that enters the SMA or its right branch.[3] The IMV drains the upper rectum, and the left and part of the transverse colon receiving blood from the superior rectal, sigmoidal, and left colic veins. The superior rectal vein and sigmoid vein usually join into a common trunk before uniting with the left colic vein.[9] Termination of the IMV is variable (▶Fig. 10.28): in the majority of cases the IMV drains into the splenic vein posterior to the body of the pancreas, but it may terminate at the angle of the confluence of splenic vein and SMV or, less frequently, drain into the SMV. Absence of IMV has been described.[32,33]

Variations in Relationship between Arteries and Veins

The majority of mesenteric veins run adjacent to the corresponding arteries and they cross each other along their course. Variability exists in the relationships between arteries and veins (▶Fig. 10.29, ▶Fig. 10.30, ▶Fig. 10.31). Surgeons are showing increasing interest in the preoperative evaluation of the vascular anatomy while planning for bowel resections. The laparoscopic approach to abdominal surgery is widely used because of its multiple advantages when compared to conventional laparotomy. It allows smaller surgical incisions, minor intraoperative blood loss, faster recovery of normal bowel function, and shorter hospitalization. Nevertheless, it suffers from a restricted two-dimensional operative field of view and the inability to manipulate tissues, leading to potential

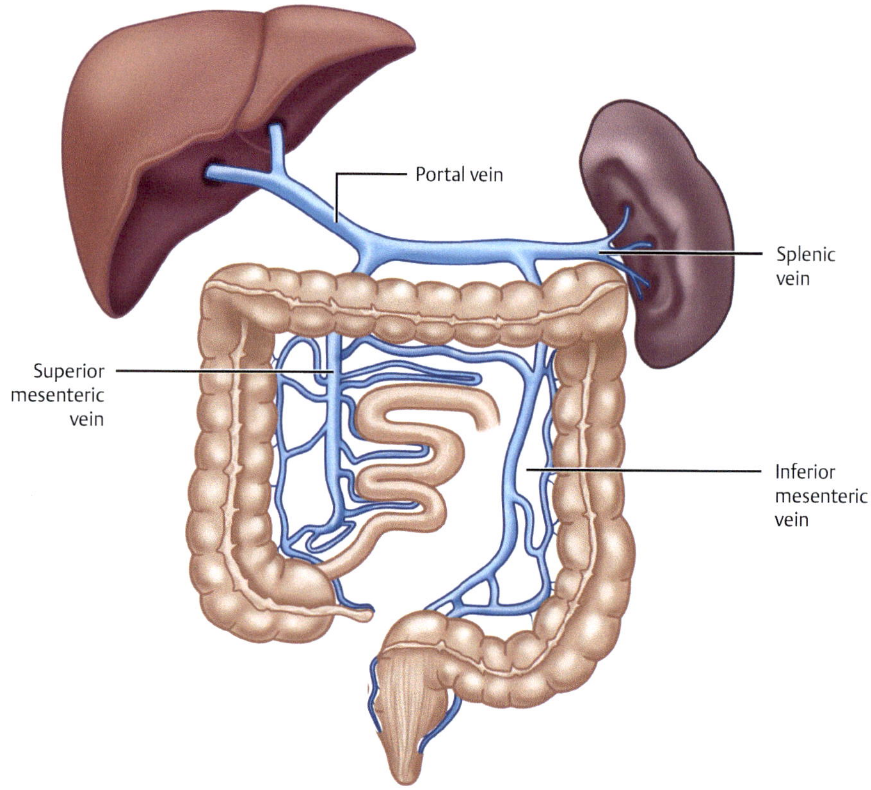

Fig. 10.27 Anatomy of common mesenteric veins.

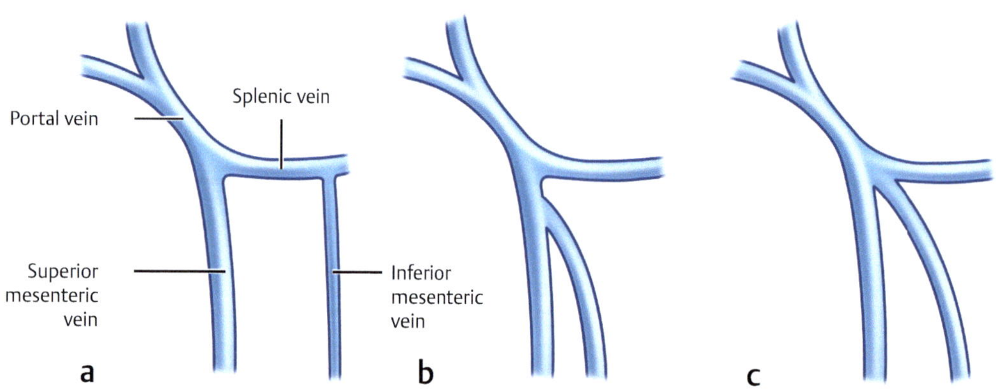

Fig. 10.28 Variations in inferior mesenteric vein (IMV) termination: in the majority of cases the IMV drains into the splenic vein **(a)**, but it may also terminate into the superior mesenteric vein (SMV) **(b)** or at the angle of the confluence of splenic vein and SMV **(c)**.

inaccurate localization of vessels and difficulties in ligation, hemorrhage control, and lymph nodes dissection. Therefore, knowledge of variants in vascular anatomy and crossing pattern between mesenteric arteries and veins is mandatory.[11,13,21,34–36] The SMV runs to the right side of the SMA as it crosses over the third portion of the duodenum and lies anteriorly where they enter the mesentery (▶ Fig. 10.30). This position is variable, especially in case of bowel rotation anomalies. In these patients there can be a variable degree of malrotation, depending on the degree of rotation of the bowel around the SMA during embryogenesis; therefore, different positions of the SMV can be observed. In patients with bowel nonrotation (complete malrotation) the small bowel is situated on the right side of the abdomen, the duodenum does not cross the spine, and the whole colon is on the left side. In these cases, the relationship between the SMA and SMV is reversed, and the SMV lies posteriorly to the SMA.[6,9] One of the most significant variants

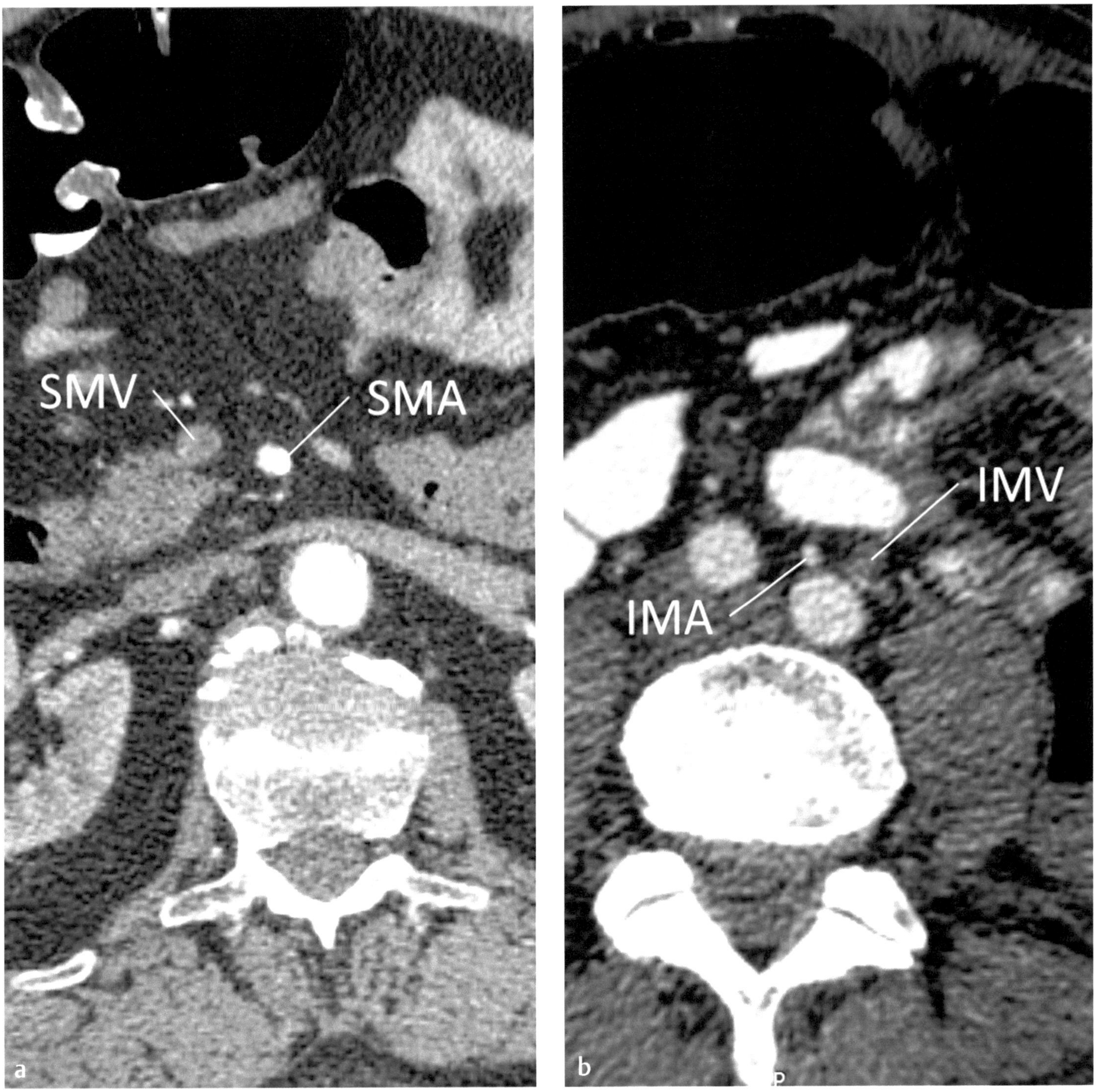

Fig. 10.29 Axial images from different contrast-enhanced computed tomography (CT) colonography. Relationship between mesenteric arteries and veins. **(a)** The superior mesenteric vein (SMV) running on the right side of the superior mesenteric artery (SMA). **(b)** The inferior mesenteric vein (IMV) running on the left side of the inferior mesenteric artery (IMA). Arteries arising from SMA and IMA cross the SMV and the IMV to reach the right and left colon.

to consider during right colon surgery affects the relationship between colic arteries and the SMV. Distally in the mesentery, the SMV runs on the right side of the SMA, and colic arteries need to cross it to reach the right colon. In the majority of cases, the middle colic artery and the right colic artery run anterior to the SMV, whereas the ileocolic artery runs posterior to the SMV in 67 to 79% of cases. Posterior course is described in less than 40% of cases for right colic artery and none for the middle colic artery.[11,37–39] Relationship between the IMV and arteries arising from the IMA can also vary: the left colic artery and the sigmoid arteries can have either an anterior or a posterior course to the IMV; however, no prevalence data are available in the literature.

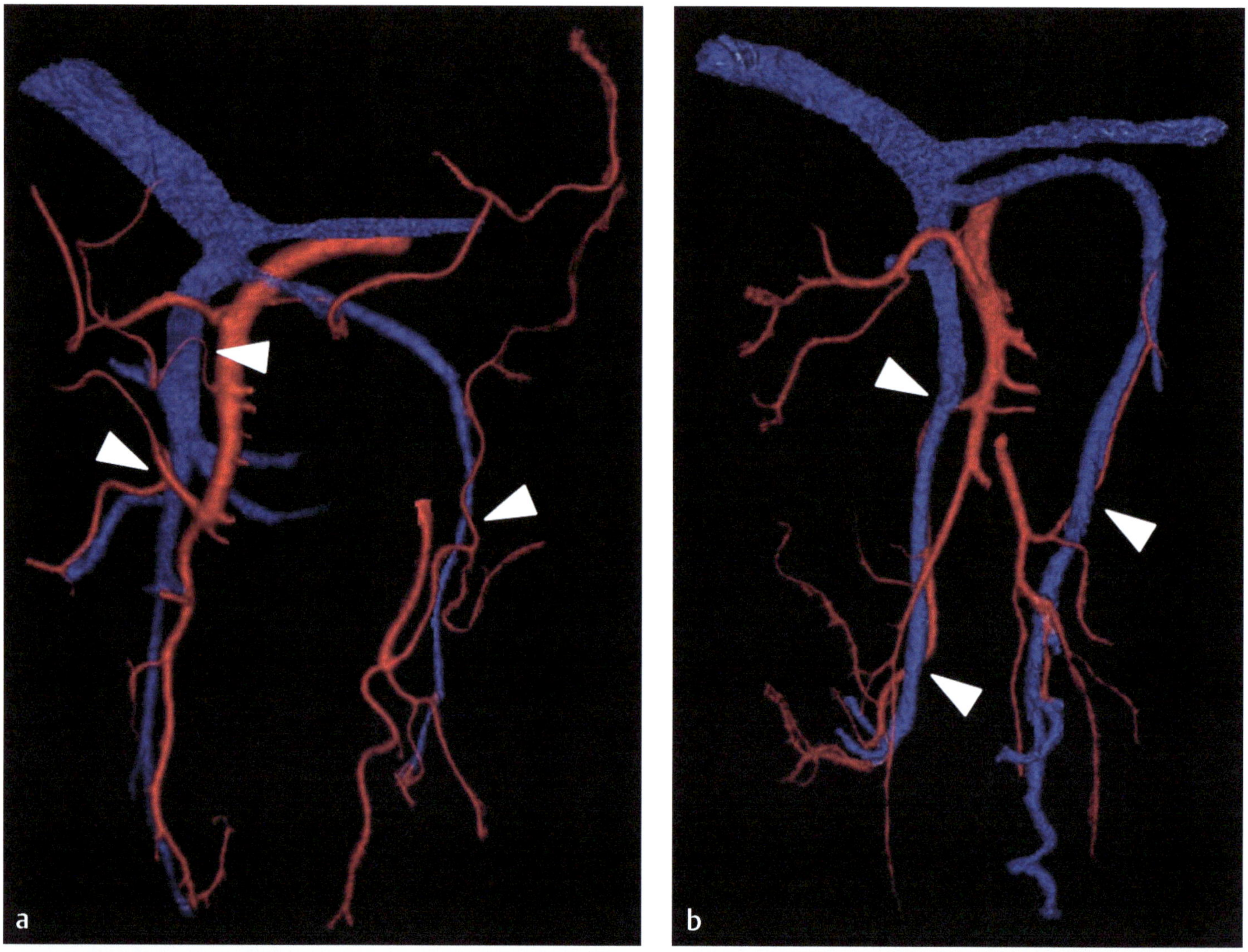

Fig. 10.30 3D volume rendered (VR) images from different computed tomography (CT) angiography showing variations in relationship between arteries and veins. **(a)** All arterial branches from the superior mesenteric artery (SMA) and the inferior mesenteric artery (IMA) cross anteriorly to the superior mesenteric vein (SMV) and inferior mesenteric vein (IMV) (*arrowheads*). **(b)** The middle colic artery crosses anteriorly to the SMV, while the ileocolic artery crosses posteriorly to the SMV (*arrowheads*); the ascending branch of the left colic artery crosses posteriorly to the IMV (*arrowhead*).

◆ Imaging

Conventional angiography has been for years the only radiological modality allowing the visualization of vessels; therefore, it was traditionally considered the gold standard diagnostic test for evaluation of mesenteric circulation. Its main drawbacks are that it is invasive, time consuming, and not available in all centers, particularly after routine working hours.[40] In the era of modern cross-sectional imaging modalities, conventional angiography represents a second-line diagnostic test and should only be performed when endovascular intervention is required. MDCT is today the best imaging modality for the assessment of vascular anatomy and pathology. Computed tomography angiography (CTA) is noninvasive, readily available, and fast; when performed with state-of-the-art equipment and settings, it is highly accurate in the depiction of vessels, with submillimeter spatial resolution achievable. CTA demonstrated 100% accuracy in identifying celiac and mesenteric arterial anatomy variants, stenosis, obstructions, and aneurysms of the celiac and mesenteric branches.[28,41] The American College of Radiology Appropriateness Criteria recommends CTA as the first-line imaging modality to evaluate both acute and chronic mesenteric ischemia.[40] Arterial phase scans easily allow visualization of the entire abdominal arterial tree, from the aorta to the submillimeter distal branches.[42] Proper portal venous phase images, with maximum opacification of SMV, IMV, and portal vein can be achieved with specific scan timing (ranging from 70 to 90 seconds after contrast material injection). Although most radiologists depend on axial source data when interpreting CT scans, the evaluation of small vessels and their relationship to surrounding structures and disease processes is easier in a three-dimensional manner.[32] Multiplanar reformatted (MPR), maximum intensity projection (MIP), and volume

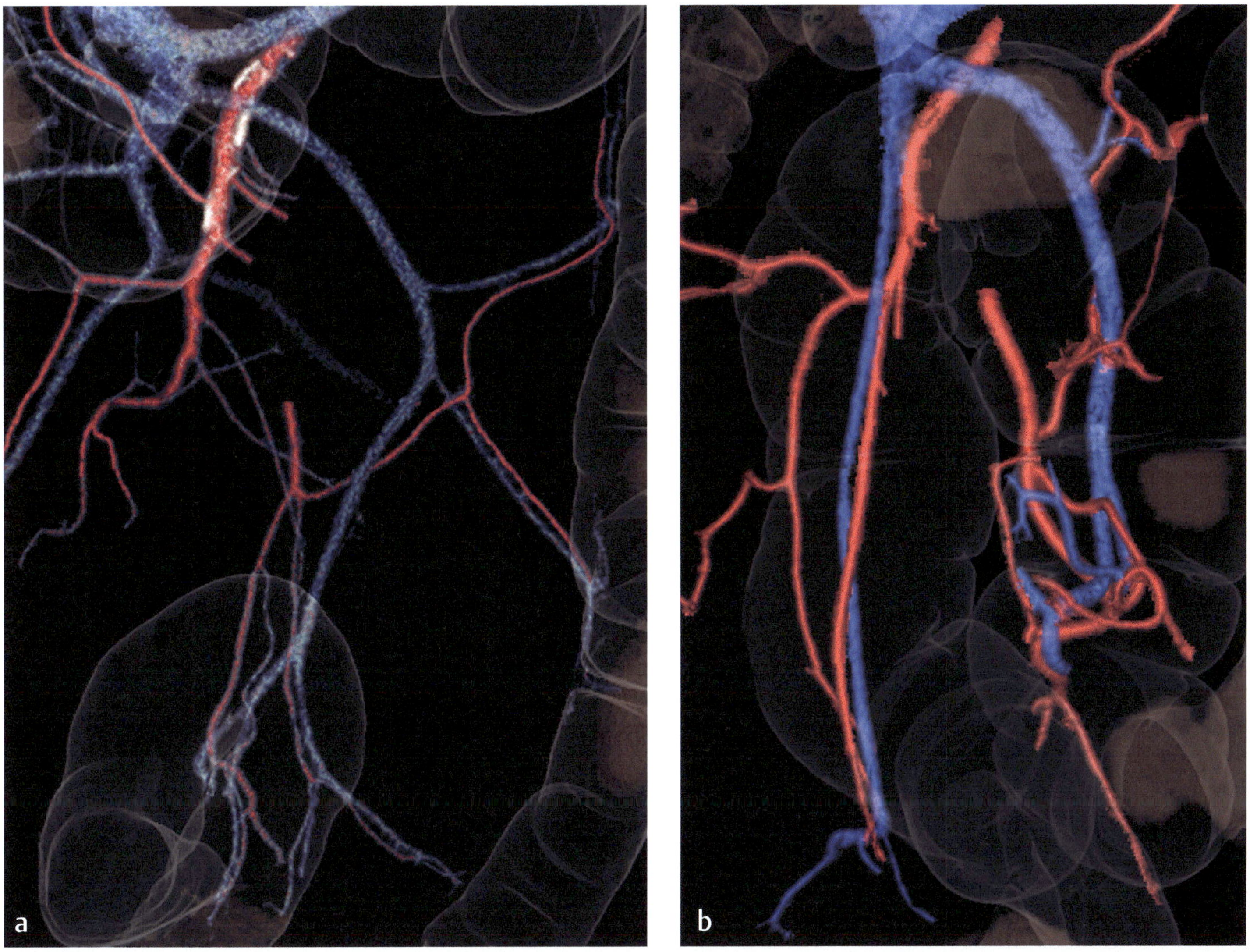

Fig. 10.31 3D volume rendered (VR) images from different contrast-enhanced computed tomography (CT) colonography showing variations in relationship between arteries and veins. **(a)** Branches of the superior mesenteric artery (SMA) crosses anteriorly to the superior mesenteric vein (SMV). **(b)** The middle colic and the ileocolic arteries arise from a common trunk, which crosses posteriorly to the SMV. In both images, the left colic artery crosses posteriorly to the inferior mesenteric vein (IMV).

rendered (VR) images are indispensable tools for the evaluation of complex vascular three-dimensional anatomy, particularly for inexperienced readers. There is legitimate concern about radiation exposure provided by the increasing number of CT studies. Pursuing ways to reduce radiation dose while maintaining image quality, CT vendors developed dose modulation techniques and iterative reconstruction algorithms that should always be employed when available. Moreover, the use of low tube voltage parameters (80–100 kV) enables to achieve higher contrast between opacified vessels and surrounding structures with a significant reduction of the dose to the patient.

Magnetic resonance (MR) is a valid alternative to CT for the evaluation of mesenteric vessels, with the advantage of not using ionizing radiation and iodinated contrast agents. Three-dimensional contrast-enhanced MR angiography (CEMRA) is the most widely used technique for the examination of the aorta and the mesenteric vasculature. High-quality mesenteric MRA achieves datasets of submillimeter, uninterpolated, isotropic voxels that can be post-processed with MPR, MIP, and VR algorithms (▶ Fig. 10.32). However, both spatial and time resolution are inferior to CTA, and CEMRA images still suffers from motion artifacts due to bowel peristalsis and breathing. It is therefore difficult to visualize thin and peripheral vessels. There is increasing interest in noncontrast MRA because of concerns about nephrogenic systemic fibrosis caused by gadolinium-based contrast agents. Time-of-flight sequences have been used in the past but ultimately have been abandoned with the advent of newer steady-state free precession (SSFE) sequences. This MRI technique is however not as robust as the CEMRA. In terms of spatial resolution, all described MRA sequences are reliable in imaging aorta, CT, SMA, SMV, and portal vein, but unsatisfactory image quality is usually achieved for SMA branches as well as the IMA and its branches.[42,43]

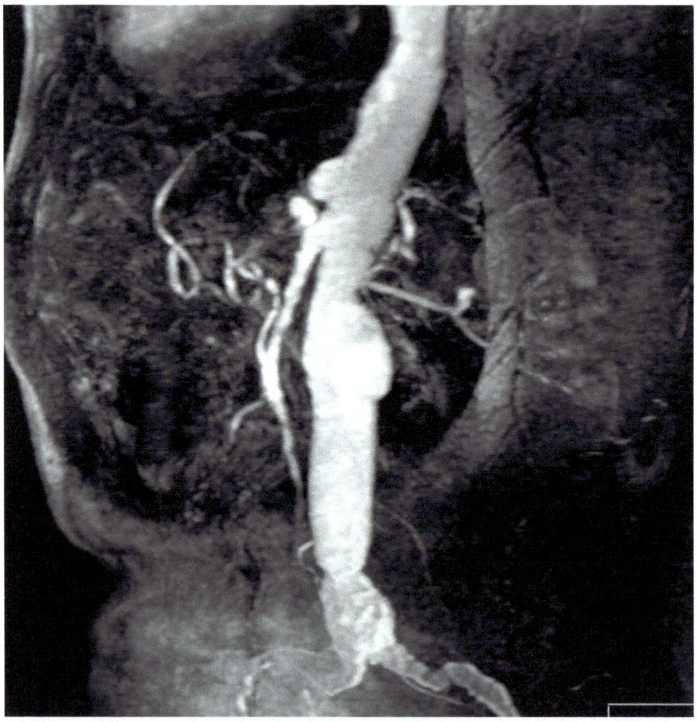

Fig. 10.32 Sagittal maximum intensity projection (MIP) image from contrast-enhanced magnetic resonance (MR) angiography shows the superior mesenteric artery and its branches in a patient with abdominal aorta aneurysm.

◆ Acknowledgments

The authors would like to thank Dr. Daniele Andreini, Dr. Andrea Annoni, Dr. Alberto Formenti, Dr. Enrica Nobili, and Dr. Gianluca Pontone from the Department of Cardiovascular Imaging, Centro Cardiologico Fondazione Monzino, Milano, Italy for providing state-of-the-art CT and MR images.

References

1. Reuter SR, Redman HC. Gastrointestinal Angiography. W.B. Saunders Company; 1972
2. Douard R, Chevallier JM, Delmas V, Cugnenc PH. Clinical interest of digestive arterial trunk anastomoses. Surg Radiol Anat 2006;28(3):219–227
3. Walker TG. Mesenteric vasculature and collateral pathways. Semin Intervent Radiol 2009;26(3):167–174
4. Wang Y, Cheng C, Wang L, Li R, Chen JH, Gong SG. Anatomical variations in the origins of the celiac axis and the superior mesenteric artery: MDCT angiographic findings and their probable embryological mechanisms. Eur Radiol 2014;24(8):1777–1784
5. White RD, Weir-McCall JR, Sullivan CM, et al. The celiac axis revisited: anatomic variants, pathologic features, and implications for modern endovascular management. Radiographics 2015;35(3):879–898
6. Sodhi KS, Bhatia A, Saxena AK, Rao KL, Menon P, Khandelwal N. Anticlockwise swirl of mesenteric vessels: a normal CT appearance, retrospective analysis of 200 pediatric patients. Eur J Radiol 2014;83(4):710–714
7. Mike M, Kano N. Reappraisal of the vascular anatomy of the colon and consequences for the definition of surgical resection. Dig Surg 2013;30(4-6):383–392
8. Geboes K, Geboes KP, Maleux G. Vascular anatomy of the gastrointestinal tract. Best Pract Res Clin Gastroenterol 2001;15(1):1–14
9. Horton KM, Fishman EK. Volume-rendered 3D CT of the mesenteric vasculature: normal anatomy, anatomic variants, and pathologic conditions. Radiographics 2002;22(1):161–172
10. Sakorafas GH, Zouros E, Peros G. Applied vascular anatomy of the colon and rectum: clinical implications for the surgical oncologist. Surg Oncol 2006;15(4):243–255
11. Tajima Y, Ishida H, Ohsawa T, et al. Three-dimensional vascular anatomy relevant to oncologic resection of right colon cancer. Int Surg 2011;96(4):300–304
12. Yoo SJ, Ku MJ, Cho SS, Yoon SP. A case of the inferior mesenteric artery arising from the superior mesenteric artery in a Korean woman. J Korean Med Sci 2011;26(10):1382–1385
13. Kobayashi M, Morishita S, Okabayashi T, et al. Preoperative assessment of vascular anatomy of inferior mesenteric artery by volume-rendered 3D-CT for laparoscopic lymph node dissection with left colic artery preservation in lower sigmoid and rectal cancer. World J Gastroenterol 2006;12(4):553–555
14. Ferrari R, De Cecco CN, Iafrate F, Paolantonio P, Rengo M, Laghi A. Anatomical variations of the coeliac trunk and the mesenteric arteries evaluated with 64-row CT angiography. Radiol Med (Torino) 2007;112(7):988–998
15. van Petersen AS, Kolkman JJ, Meerwaldt R, et al. Mesenteric stenosis, collaterals, and compensatory blood flow. J Vasc Surg 2014;60(1):111–119, 119.e1–119.e2
16. Tendler D. Overview of intestinal ischemia in adults; In: UpToDate. http://www.uptodate.com/. 2015. Accessed February 24, 2015
17. DiPoce J, Jimenez G, Weintraub J. Historical perspective: eponyms of vascular radiology. Radiographics 2014;34(4):1120–1140
18. Shah T, Singh M, Bhuriya R, Kovacs D, Khosla S. Rare case of "wandering artery of Drummond" as a result of chronic triple mesenteric vessel occlusion treated by isolated angioplasty and stenting of the inferior mesenteric artery. Int J Angiol 2013;22(4):245–250
19. Meyers MA. Griffiths' point: critical anastomosis at the splenic flexure. Significance in ischemia of the colon. AJR Am J Roentgenol 1976;126(1):77–94
20. van Tonder JJ, Boon JM, Becker JH, van Schoor AN. Anatomical considerations on Sudeck's critical point and its relevance to colorectal surgery. Clin Anat 2007;20(4):424–427
21. Matsuki M, Okuda J, Kanazawa S, et al. Virtual CT colectomy by three-dimensional imaging using multidetector-row CT for laparoscopic colorectal surgery. Abdom Imaging 2005;30(6):698–708 PubMed
22. Gourley EJ, Gering SA. The meandering mesenteric artery: a historic review and surgical implications. Dis Colon Rectum 2005;48(5):996–1000
23. Maldjian PD, Chorney MA. Celiomesenteric and hepatosplenomesenteric trunks: characterization of two rare vascular anomalies with CT. Abdom Imaging 2015;40(6):1800–1807
24. Wu Y, Peng W, Wu H, Chen G, Zhu J, Xing C. Absence of the superior mesenteric artery in an adult and a new classification method for superior-inferior mesenteric arterial variations. Surg Radiol Anat 2014;36(5):511–515
25. Singh B, Anand M, Gupta S. A rare variant angioarchitecture of upper abdomen. Anat Cell Biol 2014;47(1):73–76
26. Lee SW, Shinohara H, Matsuki M, et al. Preoperative simulation of vascular anatomy by three-dimensional computed tomography imaging in laparoscopic gastric cancer surgery. J Am Coll Surg 2003;197(6):927–936
27. Winston CB, Lee NA, Jarnagin WR, et al. CT angiography for delineation of celiac and superior mesenteric artery variants in patients undergoing hepatobiliary and pancreatic surgery. AJR Am J Roentgenol 2007;189(1):W13-9
28. Egorov VI, Yashina NI, Fedorov AV, Karmazanovsky GG, Vishnevsky VA, Shevchenko TV. Celiaco-mesenterial arterial aberrations in patients undergoing extended pancreatic resections: correlation of CT angiography with findings at surgery. JOP 2010;11(4):348–357
29. Yan J, Nagasawa Y, Nakano M, Hitomi J. Origin of the celiac and superior mesenteric arteries in a common trunk: description of a rare vessel variation of the celiacomesenteric trunk with a literature review. Okajimas Folia Anat Jpn 2014;91(2):45–48
30. Osawa T, Feng XY, Sasaki N, et al. Rare case of the inferior mesenteric artery and the common hepatic artery arising from the superior mesenteric artery. Clin Anat 2004;17(6):518–521

31. Yoshida T, Suzuki S, Sato T. Middle mesenteric artery: an anomalous origin of a middle colic artery. Surg Radiol Anat 1993;15(4):361–363
32. Graf O, Boland GW, Kaufman JA, Warshaw AL, Fernandez del Castillo C, Mueller PR. Anatomic variants of mesenteric veins: depiction with helical CT venography. AJR Am J Roentgenol 1997;168(5):1209–1213
33. Akpinar E, Turkbey B, Karcaaltincaba M, Karaosmanoglu D, Akata D. MDCT of inferior mesenteric vein: normal anatomy and pathology. Clin Radiol 2008;63(7):819–823
34. Mari FS, Nigri G, Pancaldi A, et al. Role of CT angiography with three-dimensional reconstruction of mesenteric vessels in laparoscopic colorectal resections: a randomized controlled trial. Surg Endosc 2013;27(6):2058–2067
35. Kanamoto T, Matsuki M, Okuda J, et al. Preoperative evaluation of local invasion and metastatic lymph nodes of colorectal cancer and mesenteric vascular variations using multidetector-row computed tomography before laparoscopic surgery. J Comput Assist Tomogr 2007;31(6):831–839
36. Hirai K, Yoshinari D, Ogawa H, et al. Three-dimensional computed tomography for analyzing the vascular anatomy in laparoscopic surgery for right-sided colon cancer. Surg Laparosc Endosc Percutan Tech 2013;23(6):536–539
37. Ignjatovic D, Sund S, Stimec B, Bergamaschi R. Vascular relationships in right colectomy for cancer: clinical implications. Tech Coloproctol 2007;11(3):247–250
38. Shatari T, Fujita M, Nozawa K, et al. Vascular anatomy for right colon lymphadenectomy. Surg Radiol Anat 2003;25(2):86–88
39. Spasojevic M, Stimec BV, Fasel JF, Terraz S, Ignjatovic D. 3D relations between right colon arteries and the superior mesenteric vein: a preliminary study with multidetector computed tomography. Surg Endosc 2011;25(6):1883–1886
40. Costa AF, Chidambaram V, Lee JJ, Asquith J, Skaff ER, Thipphavong S. Multidetector computed tomography of mesenteric ischaemia. Insights Imaging 2014;5(6):657–666
41. Heller MT, Shah A, Furlan A. MDCT of acute conditions affecting the mesenteric vasculature. Clin Radiol 2014;69(7):765–772
42. PubHagspiel KD, Flors L, Hanley M, Norton PT. Computed tomography angiography and magnetic resonance angiography imaging of the mesenteric vasculature. Tech Vasc Interv Radiol 2015 Mar;18(1):2-13
43. Runge VM, et al. The physics of clinical MR taught through images. 3rd ed. Thieme; 2014

11 Portal Venous System

Shunro Matsumoto, Hiro Kiyosue, Maki Kiyonaga, and Hiromu Mori

◆ Introduction

Portal vein is formed by the confluence of the splenic and mesenteric veins and their tributaries. The portal venous system shows different anatomical variants and congenital malformations and can be involved in multiple acquired diseases. State-of-the-art imaging modalities, including color Doppler ultrasonography (US), contrast-enhanced multidetector row computed tomography (MDCT), and magnetic resonance imaging (MRI), provide us with high-resolution vascular anatomy, which are essential contributors for imaging diagnosis of a variety of diseases and successful gastrointestinal interventions including new interventional procedures such as balloon-occluded retrograde transvenous obliteration (BRTO) for gastric varices. Knowledge of embryology, normal anatomy, anatomical variants, and common diseases of the portal venous system is important for stratifying treatment strategies. This chapter focuses on the embryology and normal anatomy, with emphasis on the radiological imaging of common anatomical variants and diseases of the portal venous system.

◆ Embryology

11

Pairs of symmetrical vitelline and umbilical veins are primarily responsible for development of the portal venous system. Initially, the vitelline veins arise from the capillary plexuses of the yolk sac and are followed by the appearance of the umbilical veins during the development of placenta. Both vitelline veins and umbilical veins open into the sinus venosus just below the heart (▶Fig. 11.1). Right and left vitelline plexuses (vitelline sinusoid) also develop in the septum transversum and connect to the vitelline veins. Simultaneously, liver cords develop in the septum transversum and form the liver sinusoids. The vitelline plexuses will be surrounded by the liver sinusoids, and together form a dense network of anastomosing venous spaces (▶Fig. 11.1a). As the heart develops, the distal end of the left vitelline vein diminishes and its connection to the left common cardinal vein disappears by 12 weeks in the embryo. As a consequence, all the umbilical veins' blood drains into the vitelline sinusoid (▶Fig. 11.1b). Subsequently, the right umbilical vein diminishes and disappears completely. These changes lead to the formation of ductus venosus and the entire placenta blood will drain into right sinus venosus via the left umbilical vein (▶Fig. 11.1c). The hepatic vitelline sinusoid become interrupted by the development of liver cords and is finally divided into the portal and hepatic vein components in the liver (▶Fig. 11.1d). The left umbilical vein and ductus venous regress after birth and consequently will be obliterated and remain as ligamentum teres and ligamentum venosus, respectively.[1–3]

The process of development of the extrahepatic portal venous system is complicated. It is believed that the right and left vitelline veins lie on the sides of the duodenum and are connected together with three anastomotic branches. During 6th to 10th weeks of embryo, both vitelline veins and their anastomoses partially develop or regress and consequently form the portal vein. Of these anastomoses, the cranial and caudal anastomoses run anterior to the duodenum and the middle anastomosis runs posterior to the duodenum.[1–3] The reason why and how the vitelline veins and their anastomoses develop is not clear. It is presumed that the formation of the portal venous system is closely linked with the intestinal rotation/fixation and pancreato-biliary system formation.[1–4] It is also believed that the ductus venosus connecting the left umbilical vein with the right sinus venosus would be an important structure for the development of the extrahepatic portal venous system. In fact, this venous pathway causes regression of the cranial to the mid-portion of left vitelline vein in association with the development of the duodenal bulb, and also assists in regression of caudal to the mid-portion of right vitelline vein along with the intestinal rotation. As a result, the portal venous trunk of sigmoid configuration forms behind the duodenum. Furthermore, it appears that the rotation of the ventral bud of the pancreas dissects the right vitelline vein.

◆ Anatomy

Intrahepatic Portal Vein

The intrahepatic vessels consist of the hepatic arteries, the hepatic veins, and the portal veins (▶Fig. 11.2). Of these, the portal veins have less variants compared to hepatic arteries, and therefore, it is a good landmark for localization of the hepatic segments. Regarding the hepatic segments, Couinaud's division (▶Fig. 11.3) is commonly used. It divides the liver into eight (subsequently nine) functional segments, based on the distribution of portal venous branches and the location of the hepatic veins in the hepatic parenchyma.[5,6] The portal vein is usually divided into the right and left branches at the hepatic hilum, but the portal veins do not always repeat bifurcating in the liver. The portal vein has two common branching patterns. The most common pattern is trifurcation into the right anterior, right posterior, and left branches. Another common pattern is that the right posterior branch initially diverges from the portal venous trunk, and then the right anterior branch and the left branch arise from a common trunk. Initially, the left branch of portal vein runs transversely at the hepatic hilum; it then curves sharply to the ventral site. This sharp curve is known as the umbilical portion, which serves a good landmark for localizing the left liver segments. On the other hand, the "P-point" is used to localize the right posterior branch where it diverges sharply from the right main branch and travels posteriorly.

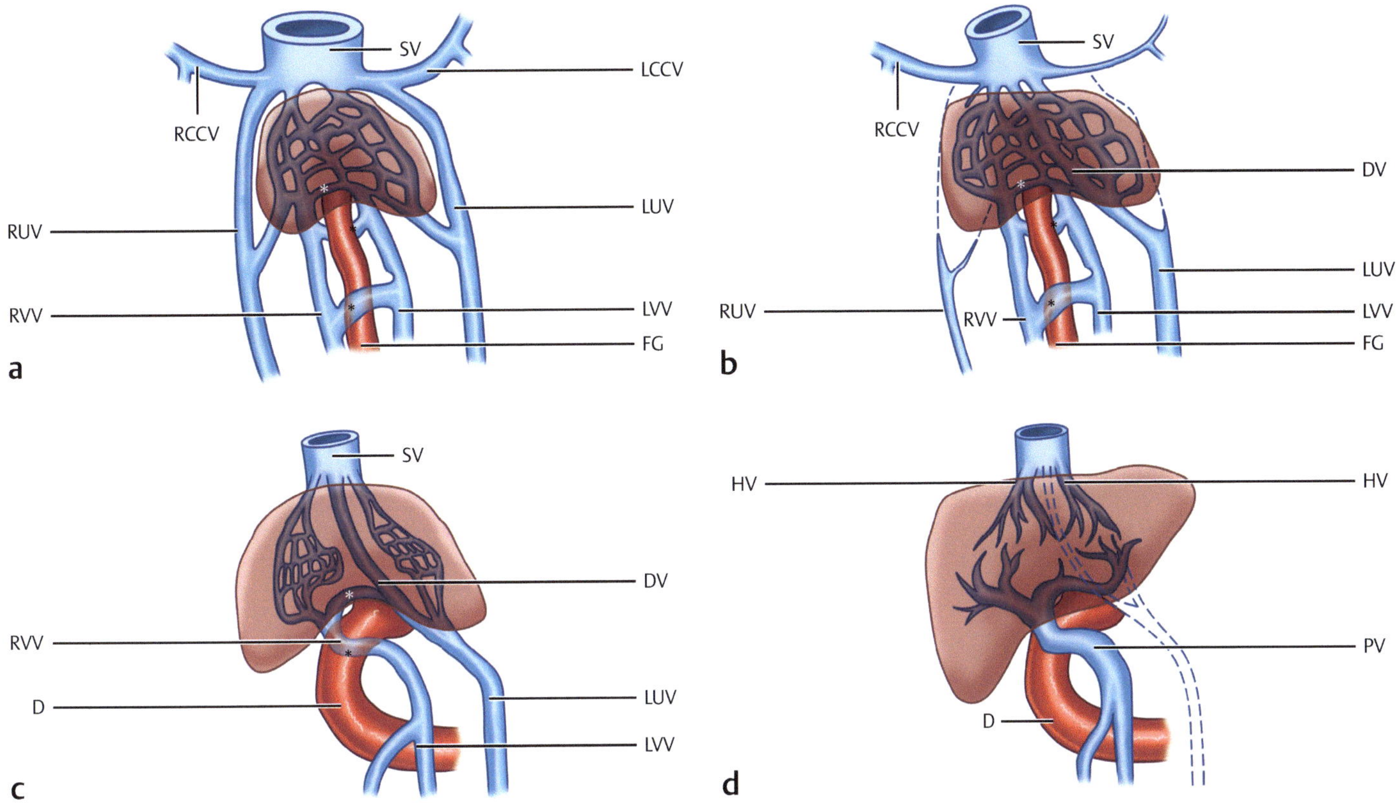

Fig. 11.1 (a–d) Development of portal venous system. A pair of vitelline veins appears on both sides of the foregut. After that, a pair of umbilical veins (UV) develops and drains into the sinus venosus (SV). Liver cords develop in the septum transversum with incorporating the vitelline sinusoid. During the development of the heart, the left vitelline vein (LVV) diminishes, and the pathway of right vitelline vein (RVV) becomes predominant. Finally, all umbilical vein blood drains into the vitelline sinusoids of the liver. Subsequently, the right umbilical vein (RUV) and the left umbilical vein (LUV) at the hepatic segment disappear and finally, almost all of the blood flow from the placenta drains into the right SV via the LUV. Hepatic vitelline sinusoids are interrupted due to the development of liver cords and then undergo differentiation into the portal veins (PV) and the hepatic veins (HV) in the liver. The cranial and caudal vitelline anastomoses run anterior to the duodenum and the middle anastomosis runs posterior to the duodenum (*stars*). D, duodenum; DV, ducts venosus; FG, foregut; HV, hepatic vein; LCCV, left common cardinal vein; LUV, left umbilical vein; PV, portal vein; RCCV, right common cardinal vein; RUV, right umbilical vein; RVV, right vitelline vein; SV, sinus venosus.

11

Extrahepatic Portal Vein

Portal Venous Trunk

The portal venous trunk (PVT) is formed behind the neck of the pancreas by the confluence of the superior mesenteric vein (SMV) and the splenic vein (SV) (▶ Fig. 11.2). There are three types of confluence of the inferior mesenteric vein (IMV) (▶ Fig. 11.4).[7] Type 1 shows the IMV draining into the SV. Type 2 shows the IMV draining into the confluence point of the SMV and SV. Type 3 shows the IMV draining into the SMV. These types are seen at the same rate in Western society, while type 1 is most frequent in Japanese. The PVT usually originates in the right paramedian region at the level of the first to second lumber vertebra, and ascends anteriorly to the hepatic hilum. The PVT runs behind the head of the pancreas and the descending portion of the duodenum, and passes through the hepatoduodenal ligament, just on the left side of the common bile duct. Proper hepatic artery ascends anterior to the PVT and divides into the right and left hepatic arteries. Common hepatic bile duct is located anterior to the PVT in the hepatic hilum.

Splenic Vein

The splenic vein (SV) runs inferior to the splenic artery and posterior to the pancreas. The SV is classified into three types according to the root of veins derived from the splenic hilum. In type 1, the SV is formed by two roots, almost equal in size. In type 2, the SV is formed by three venous roots. In type 3, the SV is formed by multiple roots of the veins. The short gastric veins, left gastroepiploic vein, posterior gastric vein, pancreatic veins, and IMV merge into the SV.[8,9]

Superior Mesenteric Vein

The SMV receives blood flow from the small bowel and right colon. It is formed by the union of the right colic vein and the ileocolic vein with the tributaries from the ileum. The SMV initially runs just left to the superior mesenteric artery (SMA), but gradually crosses the SMA and ascends on the right of the SMA. The SMV ascends anterior to the uncinate process of the pancreas and distal portion of the duodenum, and posterior to the body of the

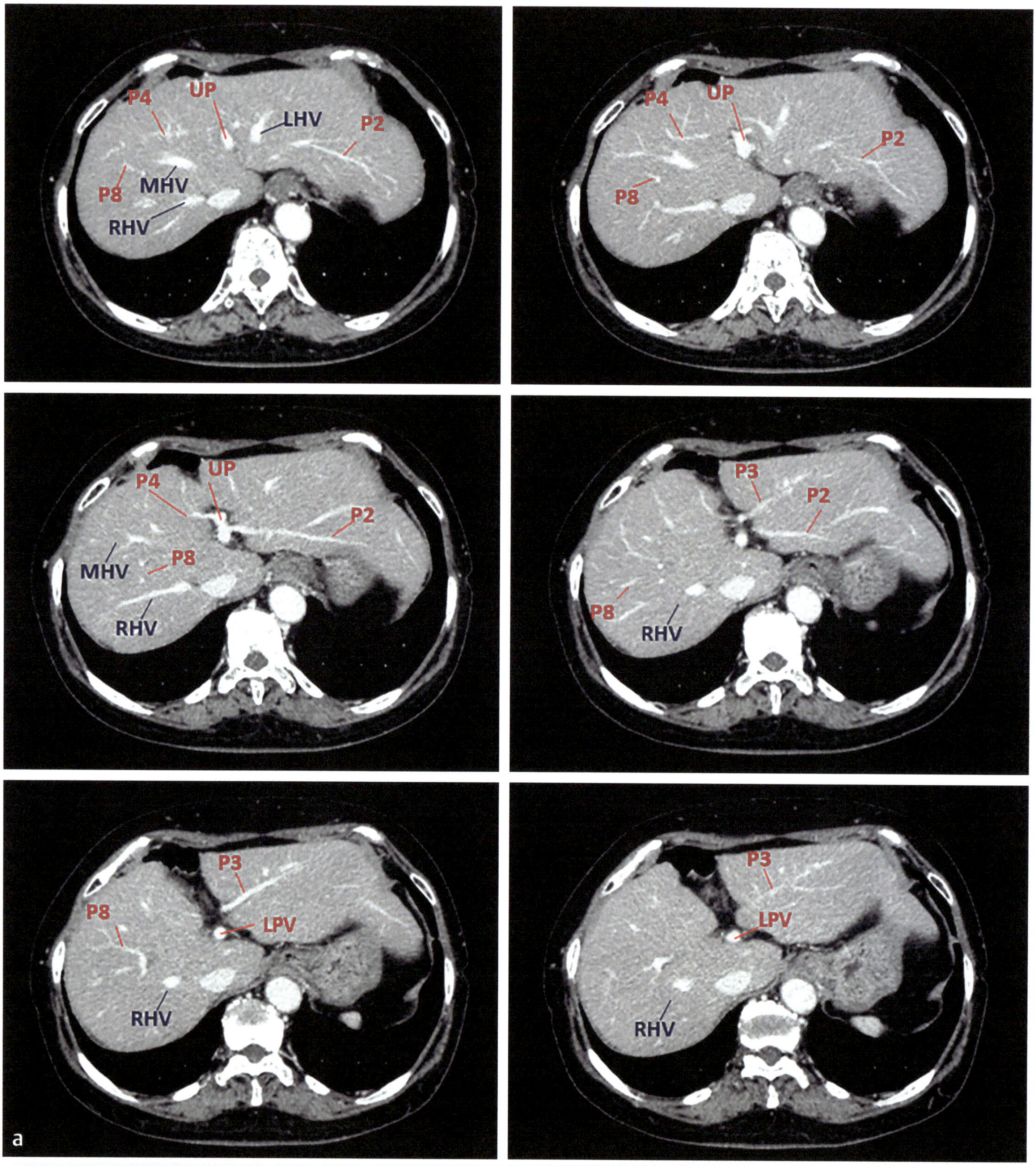

Fig. 11.2 (a–d) Sequential axial multidetector row computed tomography (MDCT) image after contrast-material injection. GCT, gastrocolic trunk; IMV, inferior mesenteric vein; LHV, left hepatic vein; MHV, middle hepatic vein; P2, lateral superior branch of the left portal vein (LPV); P3, lateral inferior branch of the left PV; P4, medial branch of the left PV; P5, anterior inferior branch of the right PV (RPV); P6, posterior inferior branch of the right PV; P7, posterior superior branch of the right PV; P8, anterior superior branch of the right PV; PSPDV, posterior superior pancreatoduodenal vein; PVT, portal venous trunk; RGEV, right gastric epiploic vein; RHV, right hepatic vein; SMV, superior mesenteric vein; UP, umbilical portion of the left PV.

(Continued)

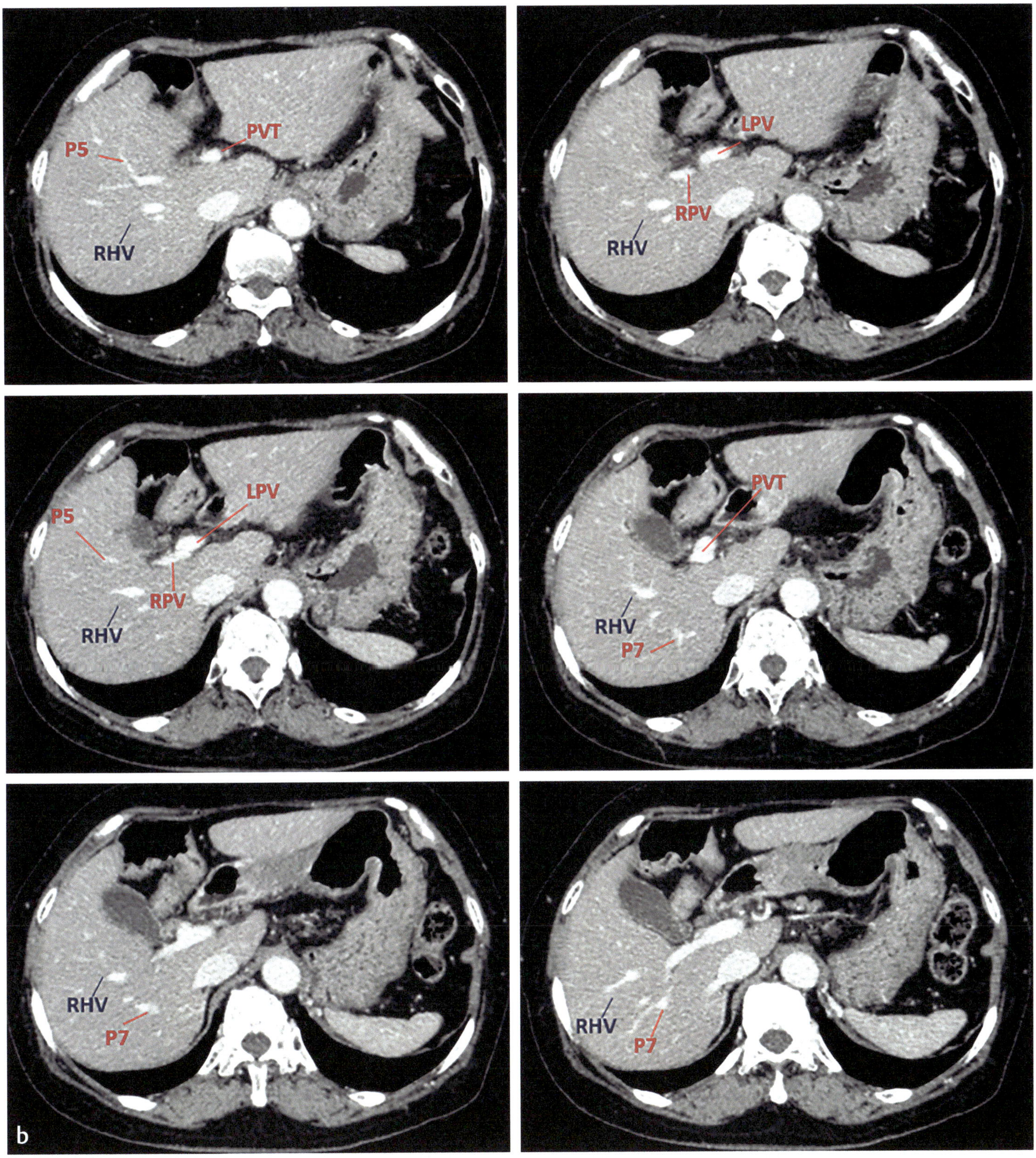

Fig. 11.2 (b) *(Continued)*

(Continued)

11

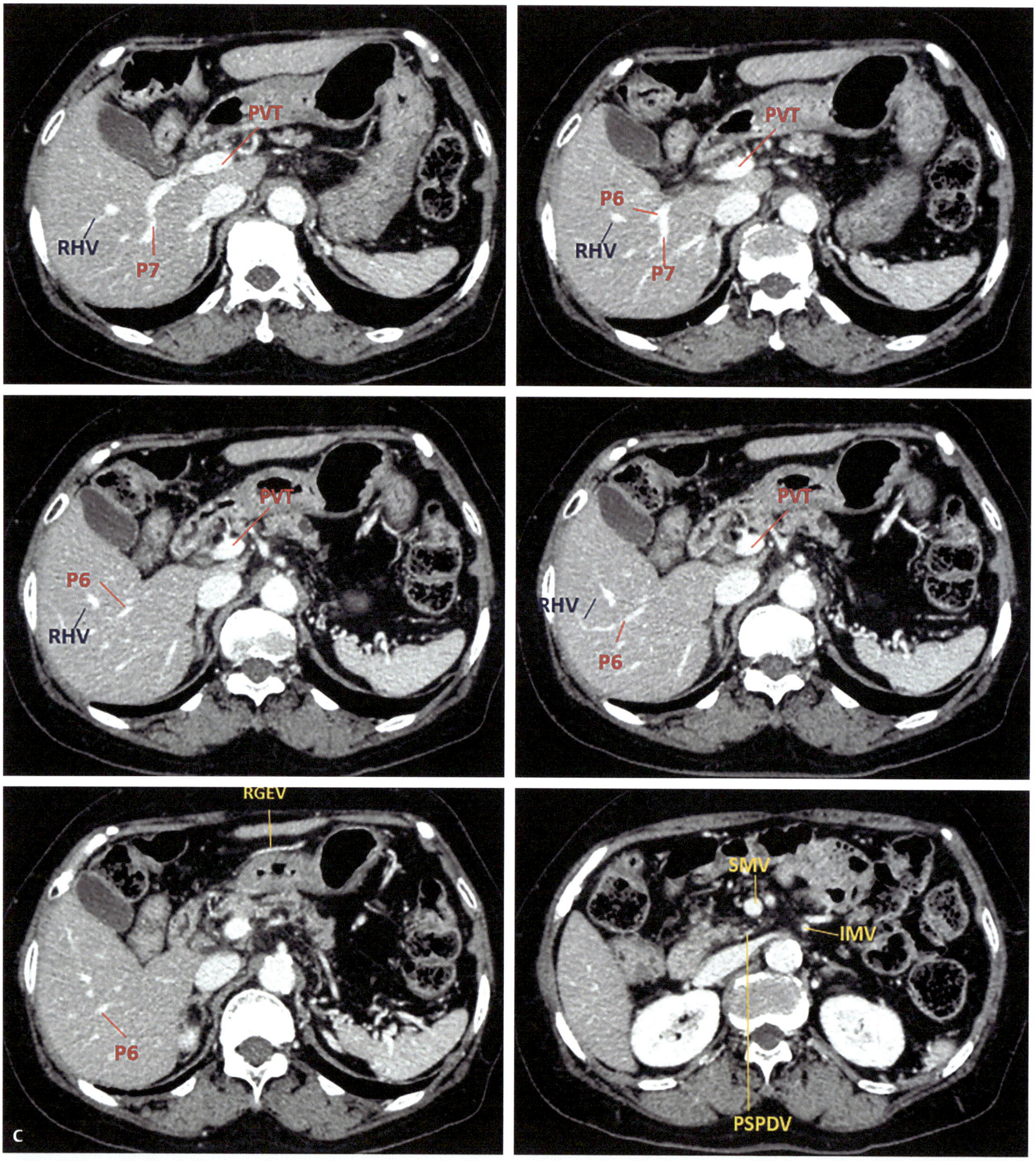

Fig. 11.2 (c) *(Continued)*

(Continued)

11

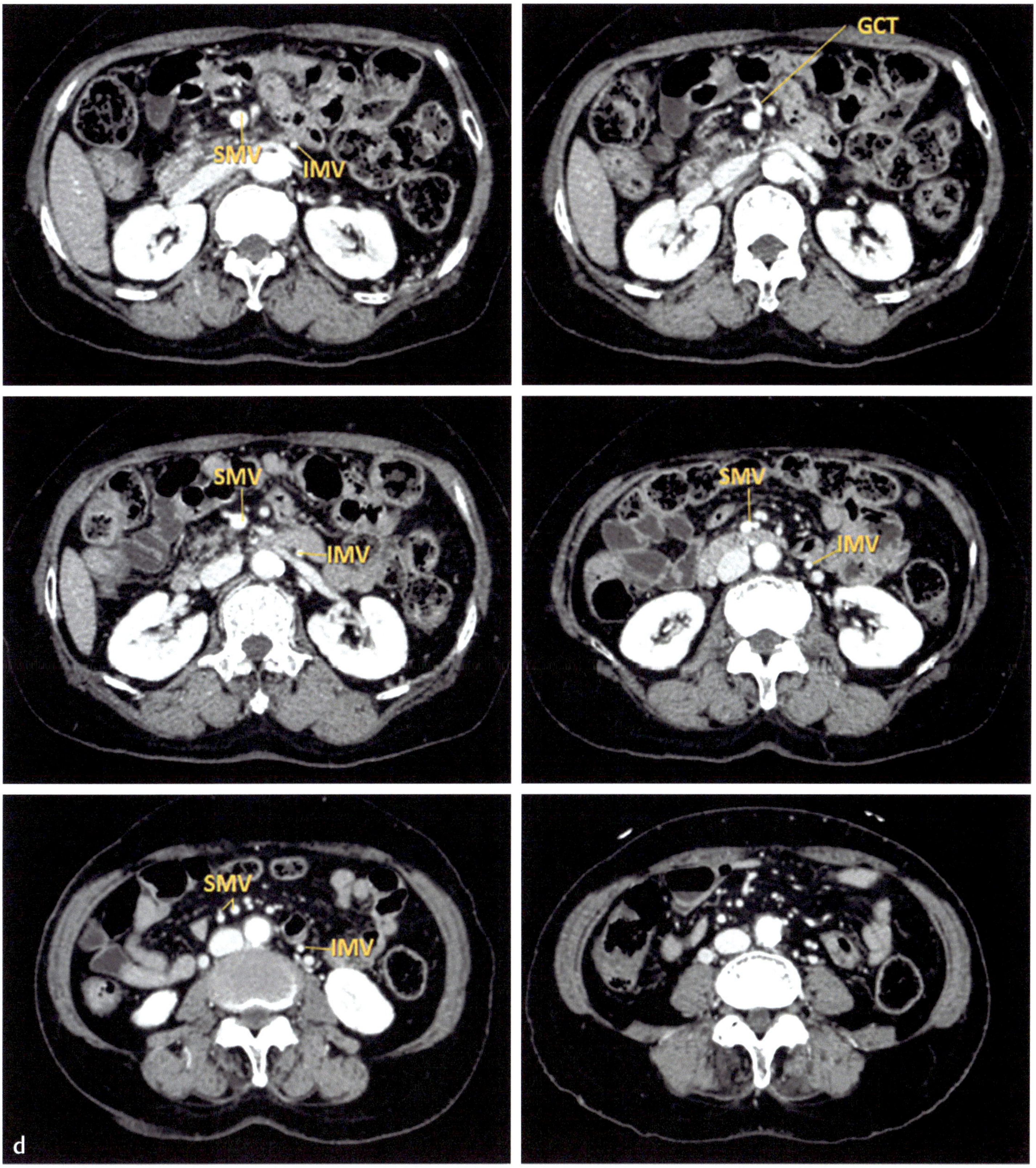

Fig. 11.2 (d) *(Continued)*

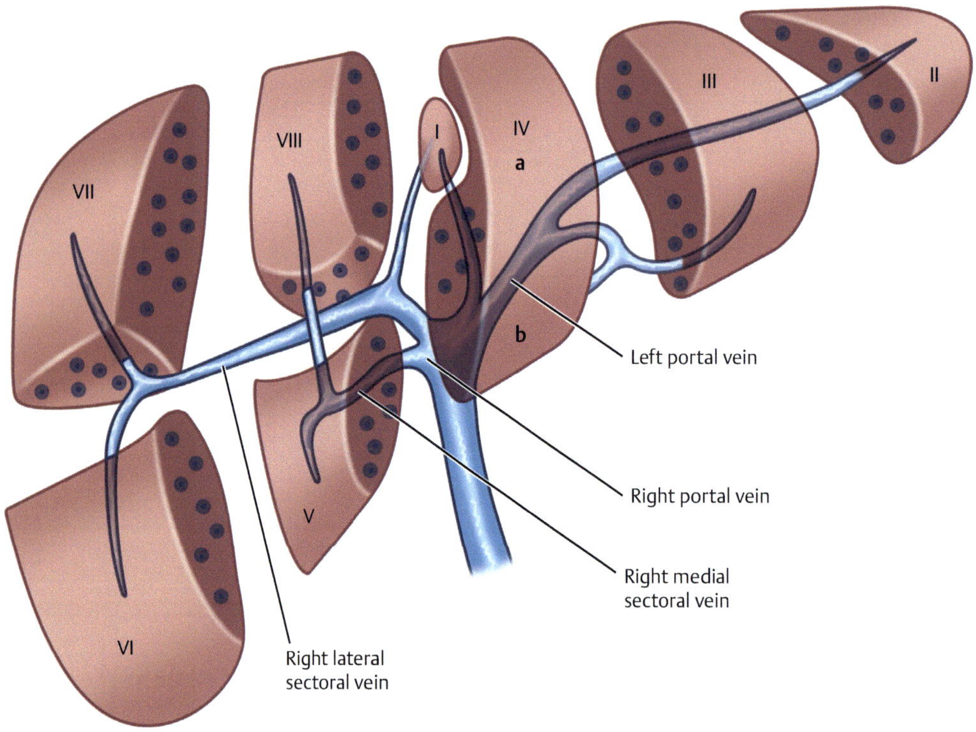

Fig. 11.3 Couinaud's division by portal vein branches. I, caudate; II, superior (medial) subsegment of the lateral segment; III, inferior (lateral) subsegment of the lateral segment; IVa, superior subsegment of the medial segment; IVb, inferior subsegment of the medial segment; V, inferior subsegment of the anterior segment; VI, inferior subsegment of the posterior segment; VII, superior subsegment of the posterior segment; VIII, superior subsegment of the anterior segment.

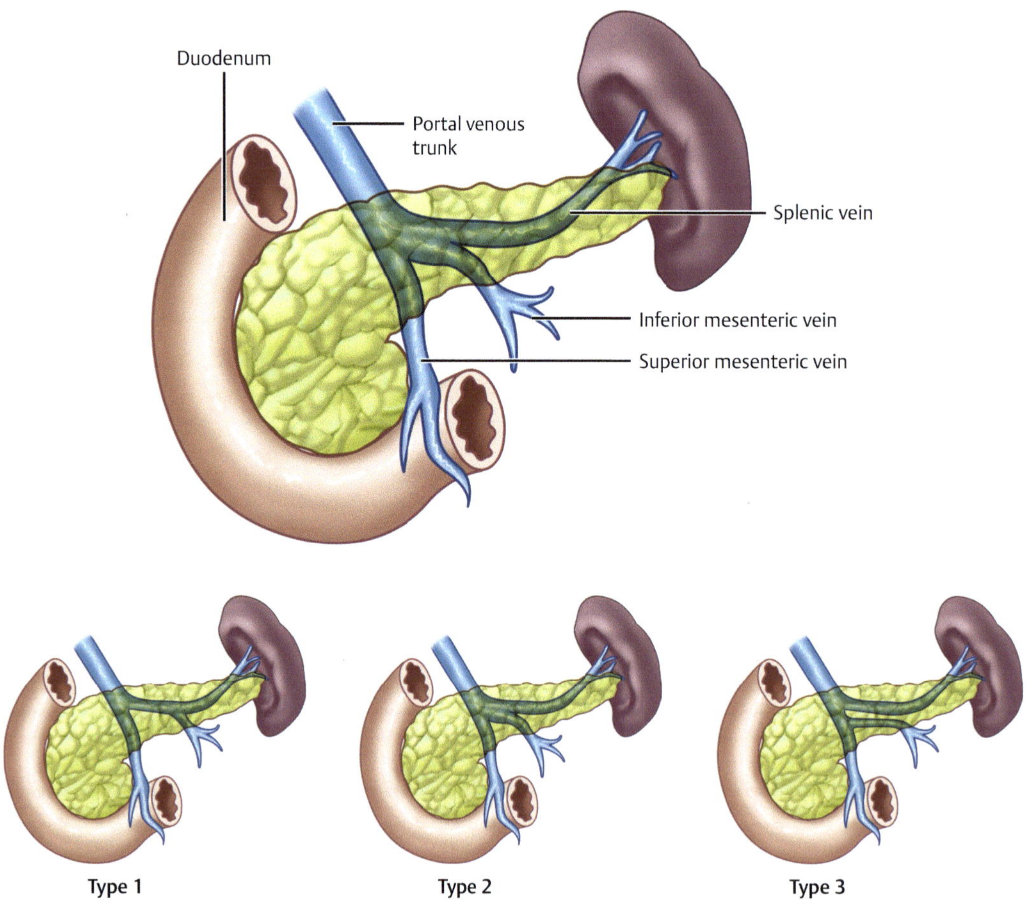

Fig. 11.4 Confluence patterns of the major three portal branches. Type 1: Inferior mesenteric vein drains into the splenic vein. Type 2: Inferior mesenteric vein drains into the junction of the superior mesenteric vein and splenic vein. Type 3: Inferior mesenteric vein drains into the superior mesenteric vein. D, duodenum; IMV, inferior mesenteric vein; PVT, portal venous trunk; SMV, superior mesenteric vein; SV, splenic vein.

pancreas. The pancreaticoduodenal veins, gastrocolic trunk, and middle colic vein and the first jejunal trunk flow into the SMV at the pancreatic head region, and the SMV finally drains into the portal venous trunk (▶Fig. 11.4).[7,10]

Inferior Mesenteric Vein

The IMV is the common trunk formed by the superior rectal vein and the vein of the sigmoid colon. It receives the left colic vein. The IMV ascends lateral to the IMA and then turns right to pass just below or behind the pancreas. Finally, the IMV drains into around the confluence of the SV and SMV (▶Fig. 11.4).[7]

Gastric Veins

Left Gastric Vein: The left gastric vein (LGV) is better known as the gastric coronary vein, which runs from a right-inferior to a left-superior direction along the lesser curvature of the stomach (▶Fig. 11.5). On the axial, contrast-enhanced computed tomography (CT), the left hepatic artery is a good indicator to find the LGV because they run in parallel. There are three patterns of connection to the portal venous system: to the splenic vein (16.7%), to the portal vein (24.4%), and to the confluence of splenic and portal veins (58.9%).[7,9,11] The LGV connects to esophageal and azygos veins, a major route for development of portosystemic venous shunting.

Right Gastric Vein: The right gastric vein (RGV) is a vessel that corrects the flow of right-sided lesser curvature of the stomach. It is also called as pyloric vein and anastomoses with the LGV. The draining point of RGV is at the hilar side of the PVT opposite to the LGV confluence. This vein can cause a pseudo-filling defect in the liver by its aberrant venous flow into the hepatic parenchyma.[7]

Short Gastric Vein: The short gastric veins (SGV) are small drainage veins of the gastric fundus and splenic parenchyma.[7,9] The SGV will be dilated and form gastric varices when the SV is occluded. It is also responsible for the afferent veins for gastric varices.

Gastroepiploic Vein: The gastroepiploic vein (GEV) is divided into the right and left, and runs along the greater curvature of the stomach, parallel to the right gastroepiploic artery (GEA). The right GEV leaves the right GEA at the gastric pylorus and descends along the second portion of the duodenum, and finally drains into the gastrocolic trunk (GCT).[7,12] The GCT has several variations according to the confluence patterns of the afferent veins (▶Fig. 11.6).

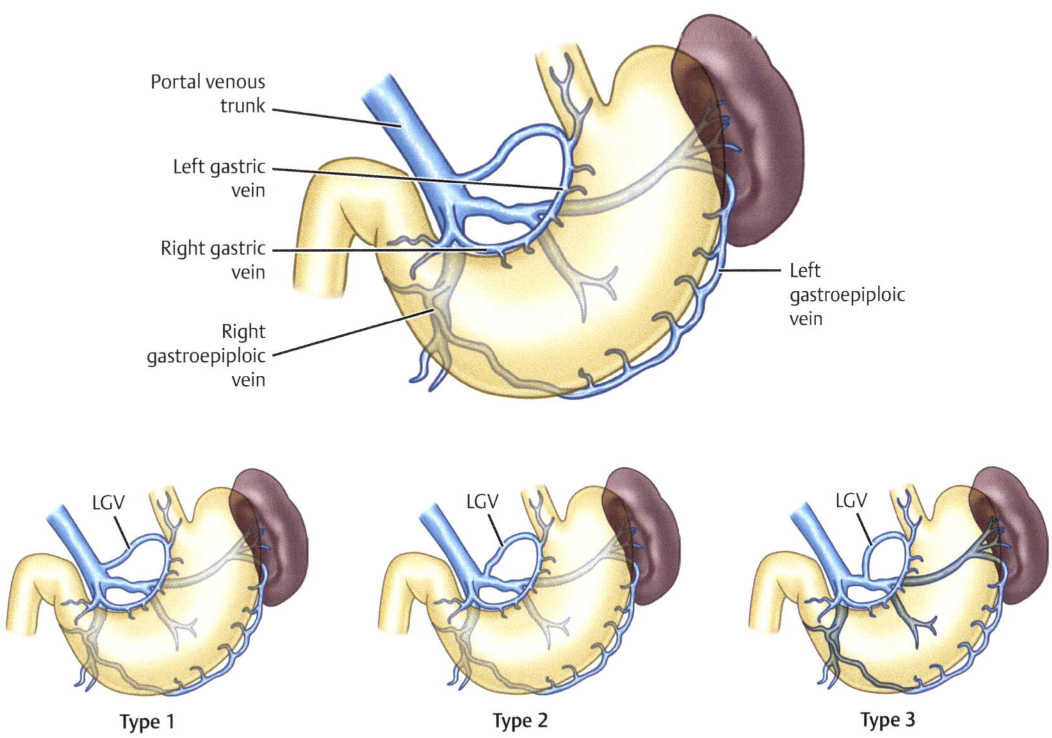

Fig. 11.5 Confluence patterns of the left gastric vein to the portal venous trunk. Type 1: Left gastric vein (LGV) directly drains into the portal venous trunk. Type 2: Left gastric vein drains into the junction of the superior mesenteric vein and splenic vein. Type 3: Left gastric vein drains into the splenic vein. LGV, left gastric vein.

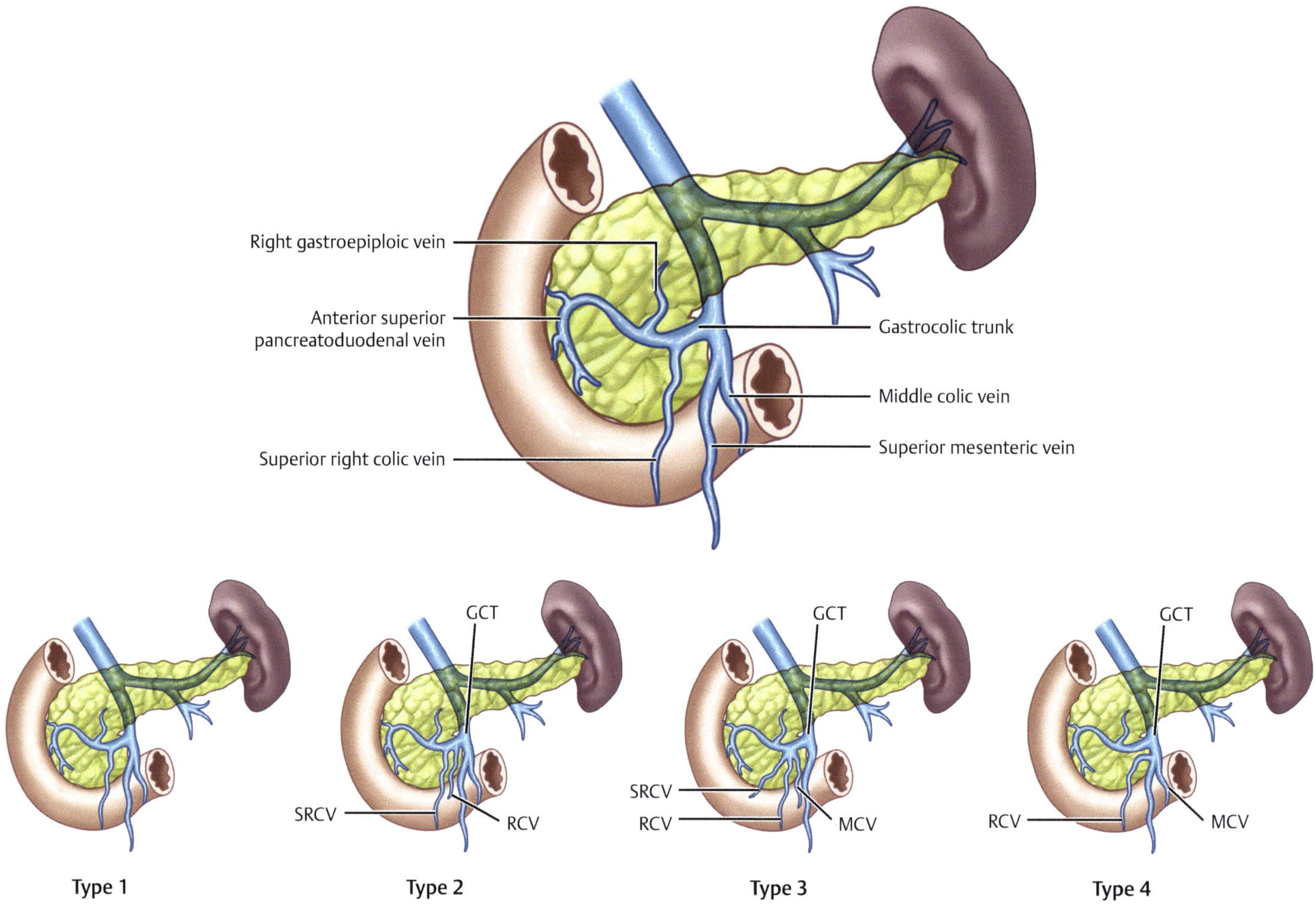

Fig. 11.6 Variation of the forming pattern of the gastrocolic trunk. Type1: Superior right colic vein (SRCV), anterior superior pancreatoduodenal vein, and right gastric epiploic vein form the gastrocolic trunk (GCT). Type 2: Right colic vein (RCV) drains into the GCT near the draining site of SRCV. Type 3: SRCV, anterior superior pancreatoduodenal vein, and middle colic vein (MCV) join and then drain into the GCT. Type 4: RCV and MCV drain into the superior mesenteric vein. ASPDV, anterior superior pancreatoduodenal vein; RGEV, right gastroepiploic vein; SMV, superior mesenteric vein.

Peripancreatic Veins

In the pancreatic head region, four peri-pancreatic veins form a venous network around the pancreatic head (▶ Fig. 11.7).[12,13] The peri-pancreatic veins and the gastrocolic trunk are easily identified on contrast-enhanced CT, which is a diagnostic key for evaluation of pancreatic cancer invasion.[12,14,15] In the pancreatic body and tail region, small pancreatic veins drain into the splenic vein at the superior part of the pancreatic body and tail, while the pancreatic veins of the inferior part drain into the inferior pancreatic vein or transverse pancreatic vein, running along the inferior margin of the pancreatic body and tail.

Gallbladder Vein

The gallbladder vein covered by the peritoneum runs parallel to the gastric artery and drains into the right branch of the portal vein or PVT. Also evident are numerous tiny gallbladder veins facing the gallbladder bed of the liver and directly entering the hepatic parenchyma.[4,16]

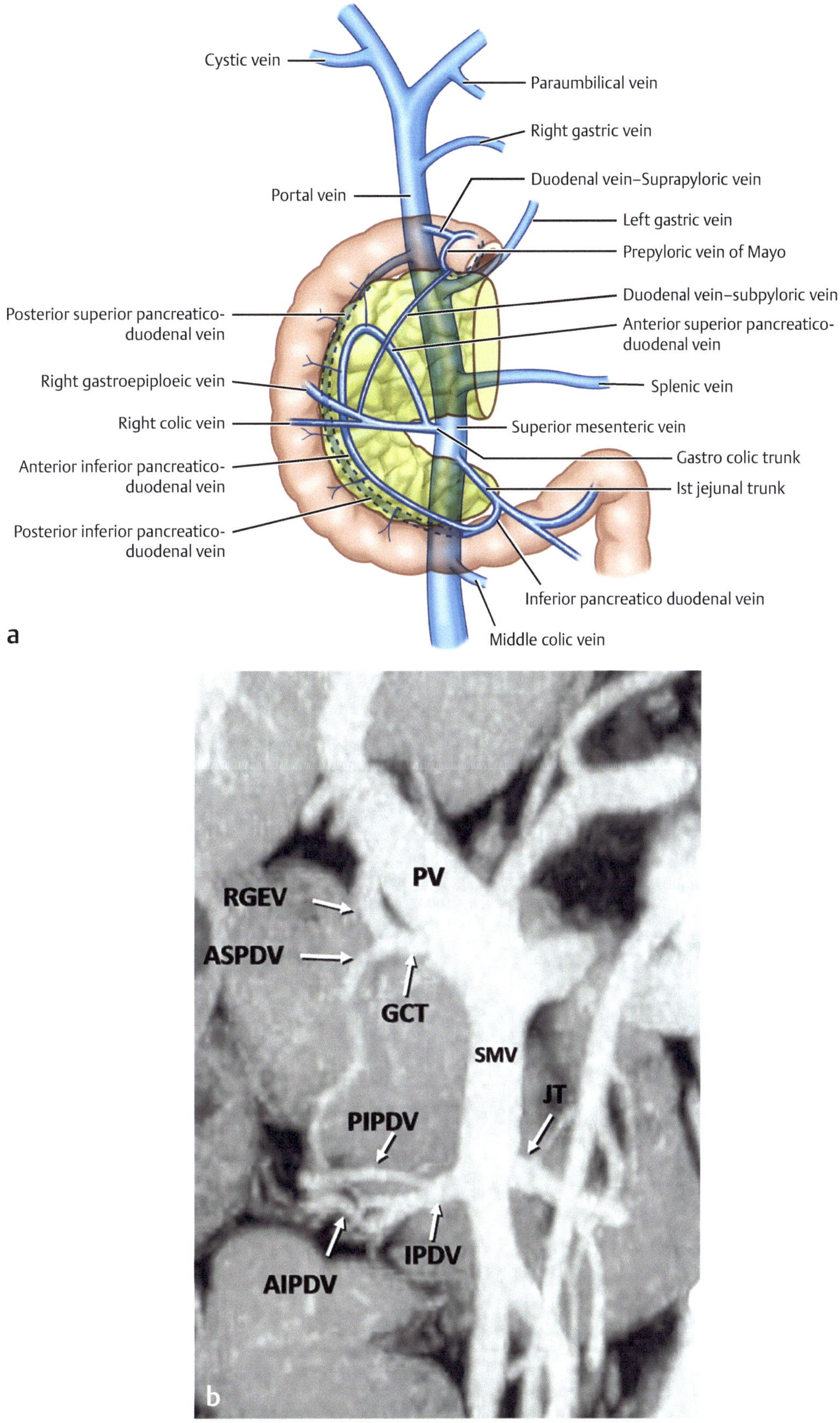

Fig. 11.7 Pancreaticoduodenal veins drain the head of the pancreas and adjacent second and third portions of the duodenum. **(a)** There are four pancreaticoduodenal veins which form an anastomosis between the superior and inferior veins. The two major tributaries of the superior mesenteric vein include the gastrocolic trunk and the first jejunal trunk which join the superior mesenteric vein roughly at the same level but on opposite sides. (Modified with permission from Sharma et al.[13]) **(b)** Coronal maximum intensity projection (MIP) image of a postcontrast multidetector row computed tomography (MDCT) showing two major venous arcades (the anterior and posterior pancreatoduodenal veins) at the surface of the pancreatic head. AIPDV, anterior inferior pancreatoduodenal vein; ASPDV, anterior superior pancreatoduodenal vein; GCT, gastrocolic trunk; IPDV, inferior pancreatoduodenal vein; JT, first jejunal trunk; PIPDV, posterior inferior pancreatoduodenal vein; PV, portal vein; RGEV, right gastroepiploic vein; SMV, superior mesenteric vein.

◆ Developmental Anomalies

Congenital Agenesis of the Portal Vein

Congenital agenesis of the portal vein (CAPV) is a rare malformation of the splanchnic venous system that is defined as a congenital loss of the portal vein accompanied by an abnormal drainage of the supramesenteric and splenic veins directly into the systemic circulation.[17] Since histologic confirmation is difficult, a patient with a large portal–systemic venous shunt is frequently misdiagnosed as CAPV on US and angiography because the intrahepatic portal vein branches are difficult to be identified. Morgan et al classified the CAPV into two types; type 1 has a complete absence of the portal venous flow, and type 2 has a patent but hypoplastic intrahepatic portal venous flow (▸Fig. 11.8). Type 1 is frequently associated with other complex anomalies, with higher occurrence of hepatic tumors.[18,19] In all patients with CAPV, a large extrahepatic portosystemic venous shunt is present. In about 50% of cases, the origin of the portal vein has a shunt with the inferior vena cava (IVC) at the hepatic segment. Other types of shuntings include the IMV–IVC, IMV–gonadal vein, and IMV–internal iliac vein shunts.

Intrahepatic Portosystemic Venous Shunt

Intrahepatic portosystemic venous shunt (IHPSVT) is divided into the congenital type and the acquired type.[17,20] The acquired type frequently shows the shunt that connects the portal vein in the umbilical portion with the epigastric veins through the medial segment of the left liver lobe. Less commonly, there is a shunt between the right posterior branch of portal vein and the right inferior phrenic vein through the hepatic parenchyma. The congenital type is caused by either the remnant of the vitelline sinusoid that forms an intrahepatic portal vein–hepatic vein shunt or the remnant of the ductus venosus that forms a shunt between the left branch of portal vein and the IVC (▸Fig. 11.9, ▸Fig. 11.10).[21] However, several anomalies other than IHPSVT can occur by the persistent vitelline sinusoid. Hereditary hemorrhagic telangiectasia (HHT;

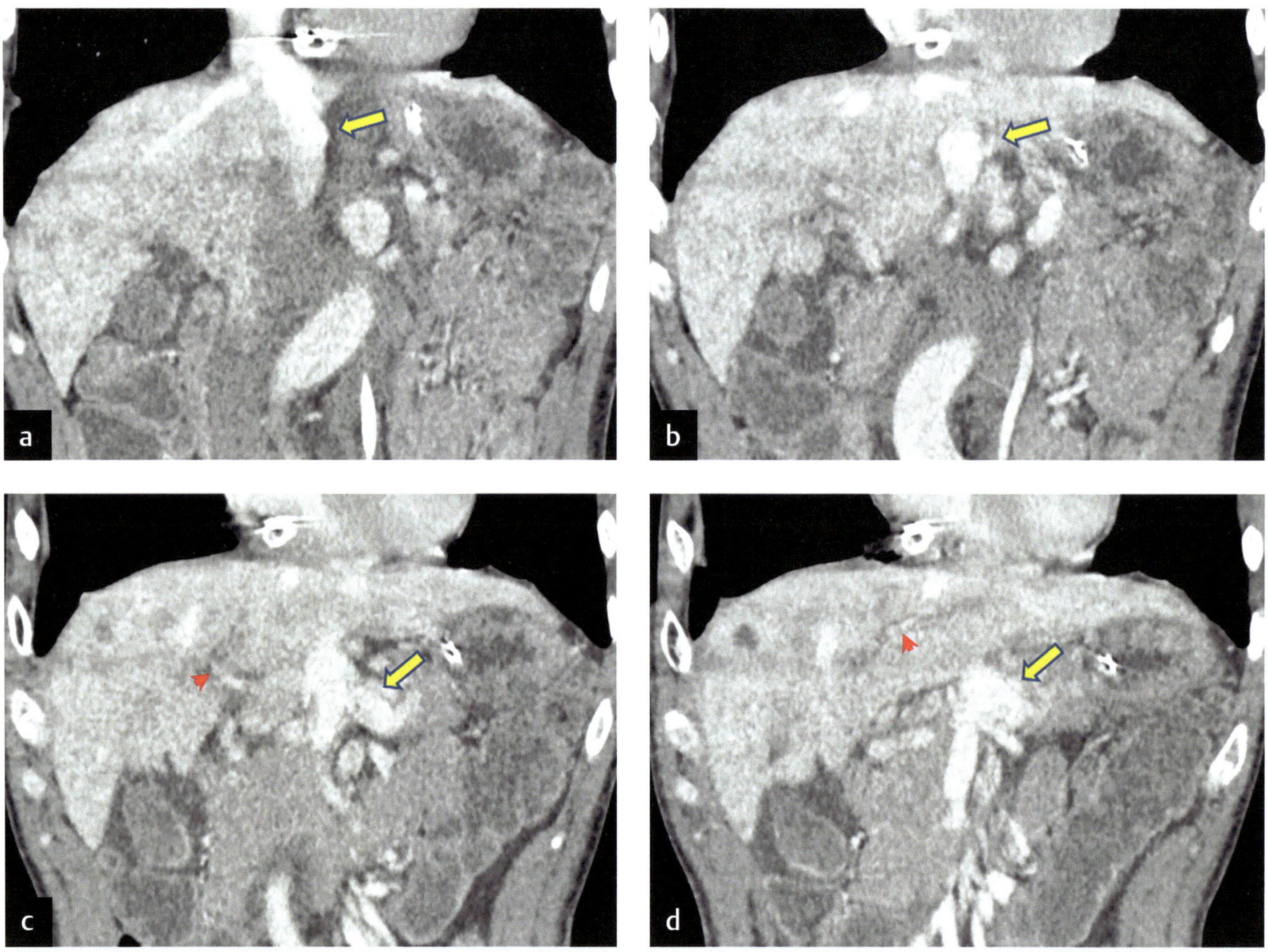

Fig. 11.8 Congenital agenesis of the portal vein. **(a–d)** Postcontrast coronal computed tomography (CT) images show hypoplasia of the intrahepatic portal veins (*arrowheads*). The hepatic portal venous trunk is directly continuous with inferior vena cava (*arrows*). This is thought to be a type 2 congenital agenesis of the portal vein based on Morgan's classification.

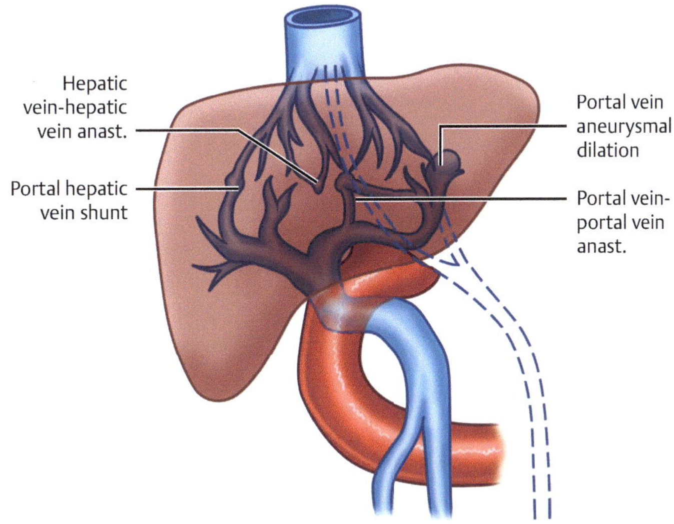

Fig. 11.9 Congenital type of intrahepatic portosystemic shunt caused by the remnant of vitelline sinusoid. Intrahepatic portosystemic shunt between the hepatic vein and the portal vein branches is formed by the remnant of vitelline sinusoid. anst., anastomosis.

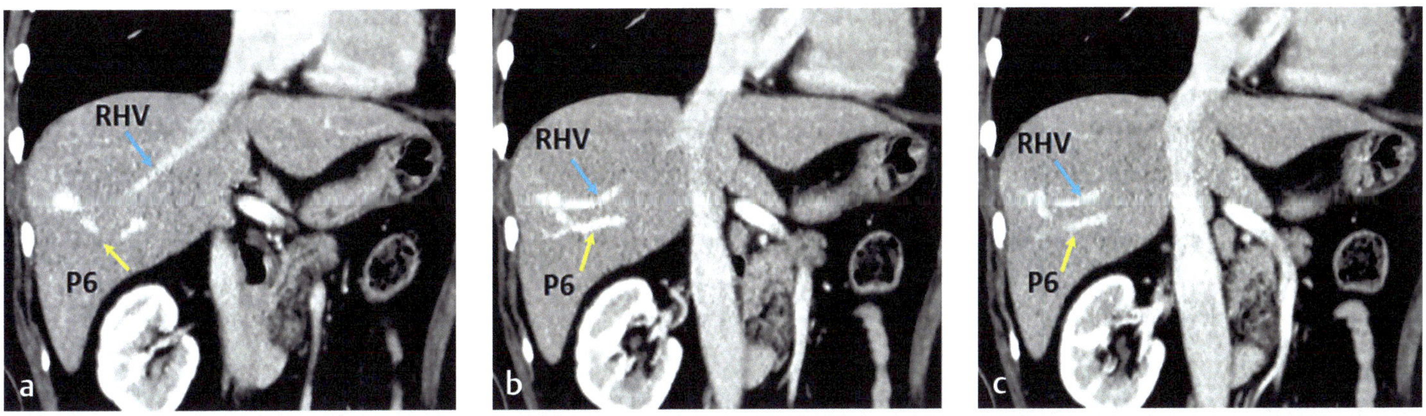

Fig. 11.10 Congenital type of intrahepatic portosystemic shunt. **(a–c)** Postcontrast coronal multidetector row computed tomography (MDCT) images show direct connection of the posterior inferior branch of the right portal vein (P6) with the right hepatic vein (RHV), suggesting an intrahepatic portosystemic shunt.

Rendu-Osler-Weber disease) resulting from the abnormality of integrin demonstrates diffuse IHPSVTs as well as hepatic artery–hepatic vein shunts or hepatic artery–portal vein shunts in the liver (▶Fig. 11.11).[22,23] This disease is also associated with pulmonary arterial–venous fistula or cerebral arterial–venous malformation, and telangiectasia of the nasopharyngeal mucosa.[1]

Extrahepatic Portosystemic Venous Shunt

Extrahepatic portosystemic venous shunt (EHPSVT) is frequently seen as the hepatofugal collateral pathways caused by portal hypertension in daily medical practice. However, there are EHPSVTs without portal hypertension, which are likely related to the persistence of the congenital anastomoses which normally regress.[24,25] As described before, EHPSVT is constantly identified when CAPV is present, but it is more common in patients without developmental anomaly of intrahepatic portal vein. In normal population, there are tiny anastomoses between the portal venous system and the IVC, including duodenal vein–IVC and colic vein–gonadal vein/renal vein anastomoses. These small connections are known the vein of Retzius.[26] Congenital EHPSVT is probably due to the persistence and enlargement of such anastomoses (▶Fig. 11.12).

Arterial–Portal Venous Shunt

Arterial–portal venous shunt (APVT) is caused by a variety of pathologic conditions such as chronic liver diseases, hepatocellular carcinoma, and hepatic injury.[17] As described before, HHT (Rendu-Osler-Weber disease) is frequently associated with diffuse and multiple APVTs as well as IHPSVT and hepatic artery–portal venous shunt (▶Fig. 11.11).[22,23] In a case without HHT, a single type of APVT is more frequently seen. The clinical symptoms mainly include ascites or intestinal hemorrhage due to secondary portal hypertension.[27] When the shunt volume is massive, cardiac failure can occur.

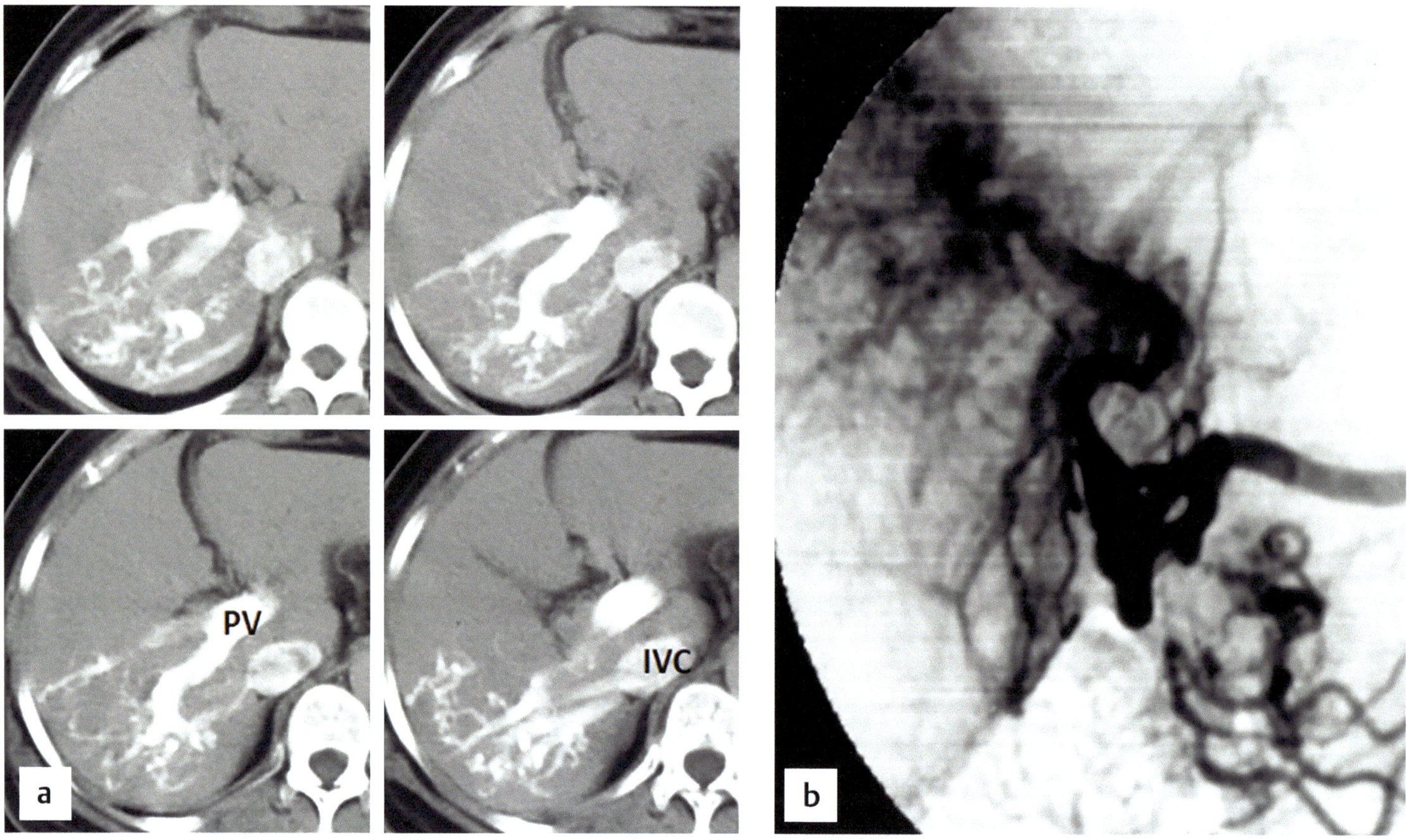

Fig. 11.11 Hereditary hemorrhagic telangiectasia with intrahepatic portosystemic shunts. **(a)** Axial computed tomography (CT) images after superior mesenteric arterial portography show diffuse and multiple intrahepatic portosystemic shunts. **(b)** Hepatic arterial angiography shows dilated and tortuous hepatic artery with peripheral arterioportal shunt and arteriovenous shunts. IVC, inferior vena cava; PV, portal vein.

Preduodenal Portal Vein

Preduodenal portal vein (PDPV) is a rare anomaly showing an abnormal vascular pathway, where the portal vein occupies an anterior position in the free edge of the lesser omentum. It is attributed to the embryonic maldevelopment of the portal venous system (▶Fig. 11.13, ▶Fig. 11.14). It is due to persistence of the preduodenal component of the caudal–ventral communicating channels.[4,28] About half of the patients with PDPV present with symptom related to the intestinal obstruction, which is mainly caused by associated anomalies such as duodenal diaphragm, intestinal malrotation, and annular pancreas. Intestinal obstruction directly induced by PDPV is extremely rare.[29] When abdominal surgery is executed for associated biliary diseases, knowledge of this portal venous anomaly is essential to avoid complications.

Portal Vein Aneurysm

The etiology of portal vein aneurysm (PVA) includes congenital causes such as diverticular formation due to incomplete regression of the distal right vitelline vein, weakened vessel wall, or anomalous portal vein branching (▶Fig. 11.15).[17] Acquired causes of PVA include liver cirrhosis, portal hypertension, trauma, surgery, and pancreatitis.[30]

Portal Annular Pancreas

Portal annular pancreas (PAP) is an asymptomatic congenital anomaly of the pancreas in which portal and/or mesenteric veins are encircled by pancreatic parenchyma (▶Fig. 11.16). PAP is caused by the developmental anomaly of the two pancreatic primordia. During embryogenesis, the pancreas is formed by two analgen originating from the endoderm in the primitive duodenum which later fuse together. In rare events, this fusion occurs to the left of the portal and/or mesenteric veins, then encircling the portal vein. PAP has several types depending on the relationship between the portal vein and the main pancreatic duct. Careful attention to this anomaly is essential in the case of pancreatic surgery.[31,32]

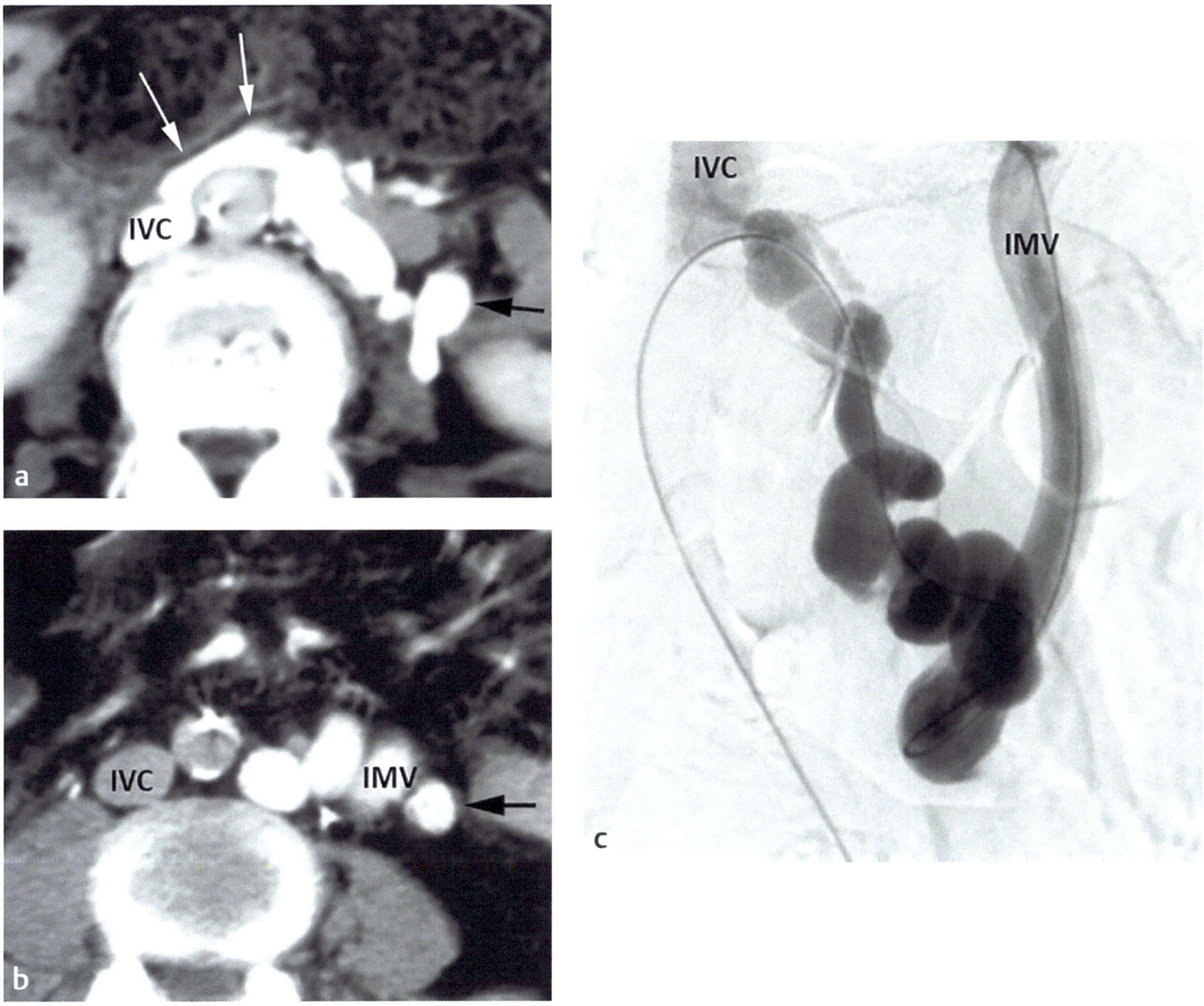

Fig. 11.12 Extrahepatic portosystemic venous shunt between the inferior mesenteric vein and the inferior vena cava. **(a, b)** Postcontrast axial computed tomography (CT) images show a large extrahepatic portosystemic venous shunt between a dilated inferior mesenteric vein (IMV) (*black arrows*) and the inferior vena cava (IVC) via the veins of Retzius (*white arrows*). **(c)** Retrograde inferior mesenteric venography shows dilated and tortuous extrahepatic portosystemic venous shunt between the IMV and IVC.

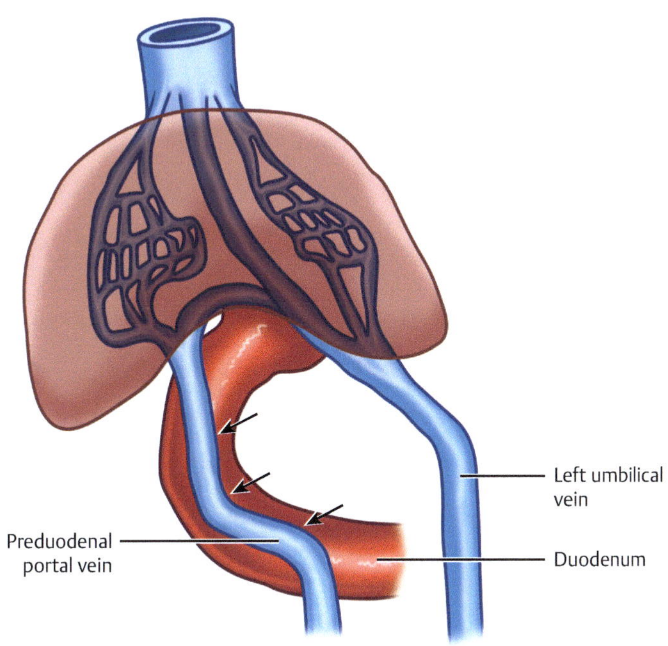

Fig. 11.13 Development of preduodenal portal vein (*arrows*). The preduodenal portal vein occurs when middle anastomosis running behind the duodenum diminishes and caudal anastomosis, which normally disappears, persists.

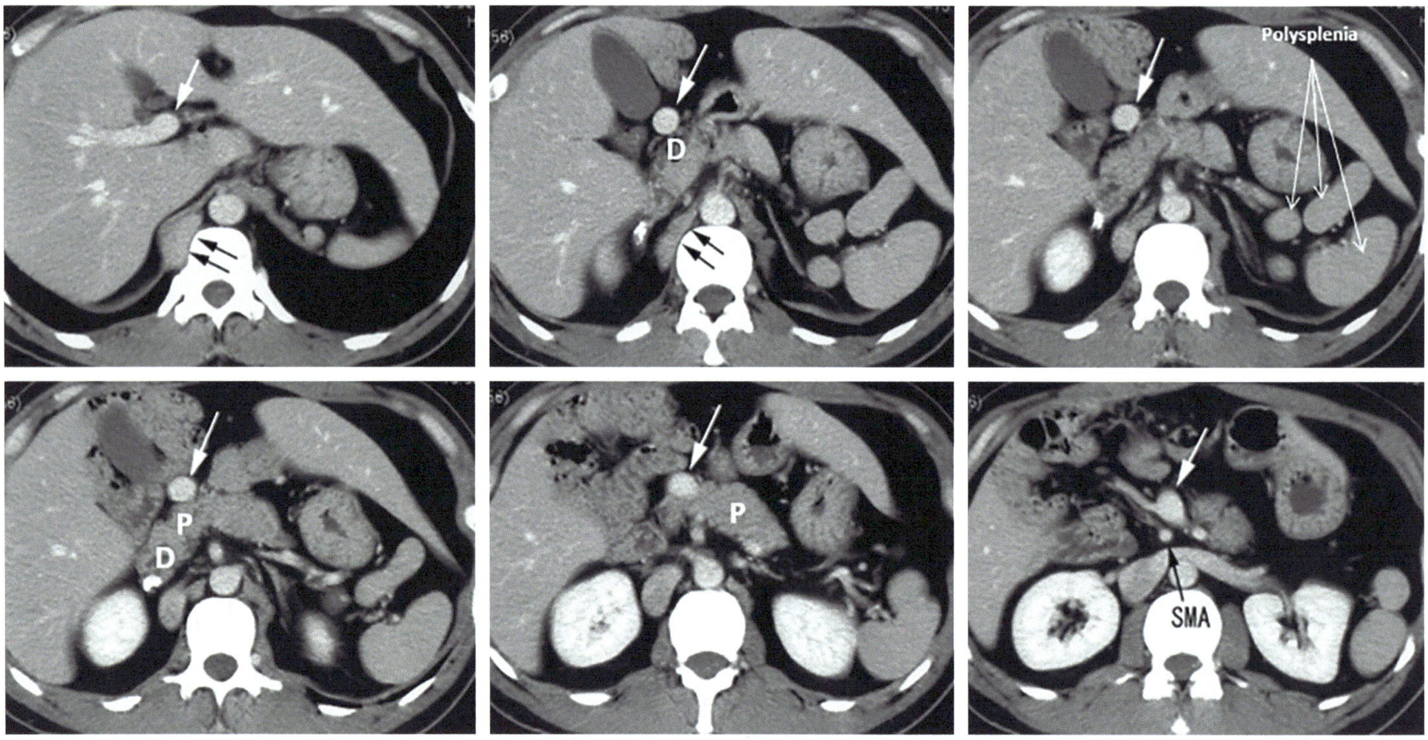

Fig. 11.14 Preduodenal portal vein. Postcontrast axial computed tomography (CT) images show the portal vein running anterior to the duodenum (D) and pancreas (P). The superior mesenteric artery (SMA) runs posterior to the portal vein (*white arrows*) and intestinal malrotation is observed. Furthermore, there are an interrupted inferior vena cava (IVC) at the hepatic segment with azygos continuation (*black arrows*) and polysplenia.

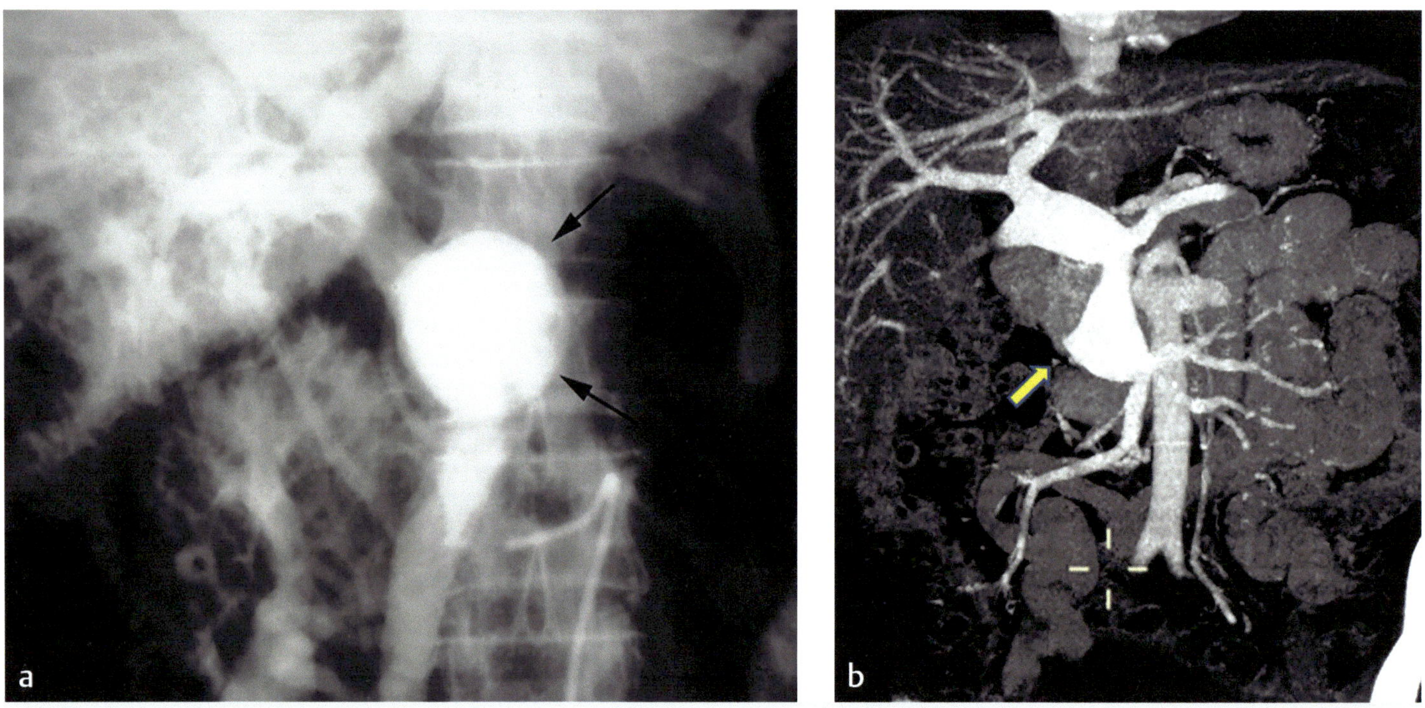

Fig. 11.15 (a) Portal vein aneurysm. Superior mesenteric arterial portography shows saccular aneurysm (*arrows*) of the portal venous trunk. It is thought to be due to remnant of the vitelline vein around the duodenum or partial remnant of the right vitelline vein. **(b)** Superior mesenteric vein aneurysm.

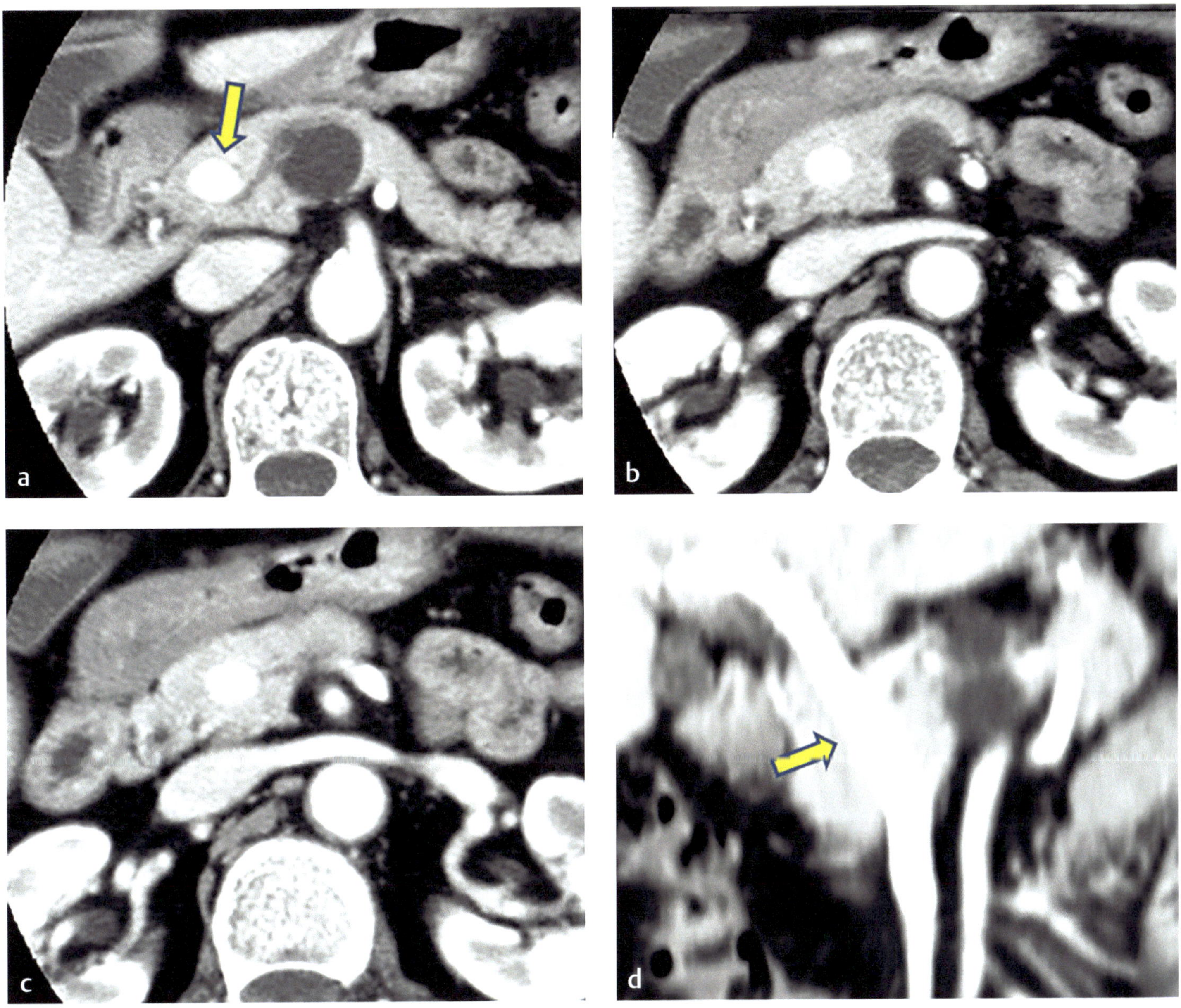

Fig. 11.16 Portal annular pancreas. **(a–c)** Postcontrast axial computed tomography (CT) images show the portal vein running in the pancreatic parenchyma (*arrow*). A cystic lesion suggestive of intraductal papillary mucinous neoplasm (IPMN) is also seen. **(d)** Coronal reconstructed CT image also shows the portal vein running in the pancreatic parenchyma.

◆ Portal Venous Diseases

Portal Vein Thrombosis

Portal vein thrombosis (PVT) is caused by various clinical conditions, including liver cirrhosis, infectious diseases, neoplasms, hypercoagulable states, and surgeries. It is reported that the occurrence of bland PVT ranges from 11.2 to 15.8% in patients with liver cirrhosis.[33,34] Acute PVT might be initially asymptomatic, but can exacerbate portal hypertension, or can result in variceal bleeding and shock. On the other hand, chronic thrombosis can be associated with gastroesophageal varices, ascites, and chronic gastric bleeding. Cavernous transformation, referring to multiple and tortuous venous collaterals at the porta hepatis is frequently seen in chronic occlusion of the portal vein[17,35] (▸ Fig. 11.17).

Portal Vein Gas

Portal vein gas (PVG) is recognized as a severe pathological condition due to mesenteric ischemia in adults or necrotizing enterocolitis in infants.[35] There are many causes of mesenteric ischemia, including arterial and venous mesenteric thrombosis, aortic dissection, hypoperfusion secondary to nonocclusive vascular disease, embolic disease, and disease processes following intestinal obstruction. On CT scan, the intrahepatic PVG can reach to the periphery of the liver. This pattern of distribution distinguishes

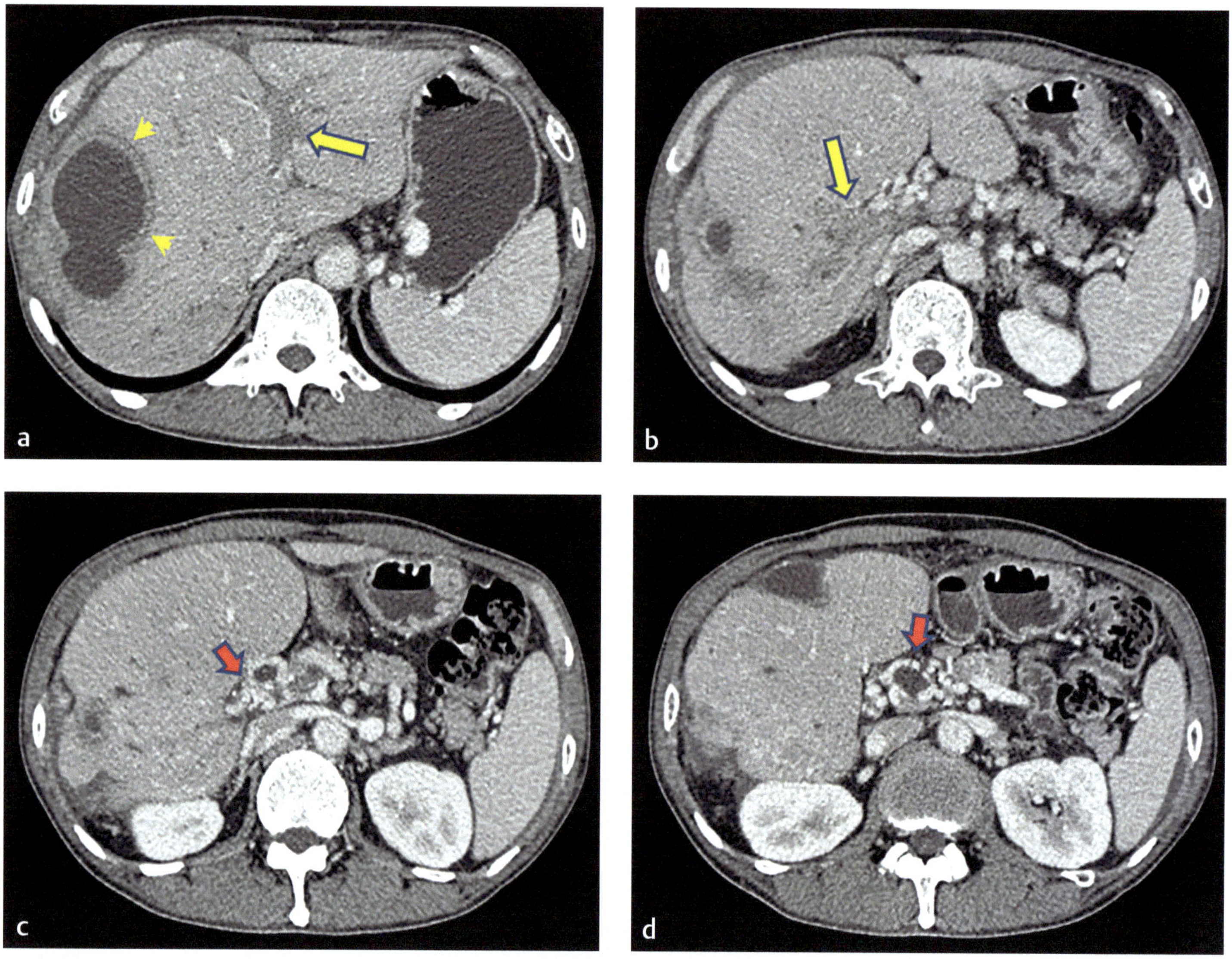

Fig. 11.17 (a–d) Portal vein thrombosis. Postcontrast axial computed tomography (CT) images show occlusion of the intrahepatic portal vein (*arrows*) due to thrombosis. Note multiple tortuous venous collaterals at the porta hepatis, the so-called "cavernous transformation" (*red arrow*). A hepatic abscess (*arrowheads*) is also identified at the right lobe of the liver.

the PVG from intrahepatic pneumobilia which remains central around the hepatic hilum.[36]

Portal Hypertension

In portal hypertension (i.e., cirrhosis) a variety of intra- and extrahepatic portosystemic collaterals develop to compensate for the large portal blood flow and increased portal pressure. Portosystemic collaterals can also develop in patients with obstruction of the portal or the splenic vein, or as a congenital communication. Common collaterals include gastroesophageal varices, paraesophageal varices, gastro- or splenorenal shunt, paraumbilical shunts, and perirectal shunts. Unusual portosystemic collateral pathways occur within the hepatic parenchyma, around the gallbladder/duodenum/pancreas, within the retroperitoneum, or through the iliac and gonadal veins (▶Fig. 11.18).

Gastric Varices

Most gastric varices arise at hepatofugal collateral pathways and drain into the systemic vein through one or two different types of portosystemic collateral drainage systems: the gastroesophageal venous system and the gastrophrenic venous system. The gastroesophageal junction venous system (▶Fig. 11.19) can form the gastric varices, esophageal varices, and paraesophageal varices draining into the azygos vein, or the superior vena cava. Gastric varices relating to the gastroesophageal venous system can be treated with endoscopic techniques. On the other hand, the gastrophrenic venous system can form isolated gastric varices (IGV) at the cardia and/or fundus of the stomach (▶Fig. 11.20). The majority of the IGVs consist of venous shunts that develop based on the potential anastomoses between the gastric vein and the left inferior phrenic vein (IPV) due to portal hypertension.[37–39] The left IPV, which has abundant anastomoses with the peridiaphragmatic and retroperitoneal veins, terminates into the left renal vein (gastrorenal shunt) (▶Fig. 11.21) or the inferior vena cava just below the diaphragm (gastrocaval shunt) (▶Fig. 11.22). Balloon-occluded retrograde transvenous obliteration (B-RTO) is an ideal treatment option for the IGVs. Gastric varices can occasionally arise from the hepatopetal collateral pathway that develops secondary to localized portal hypertension caused by splenic

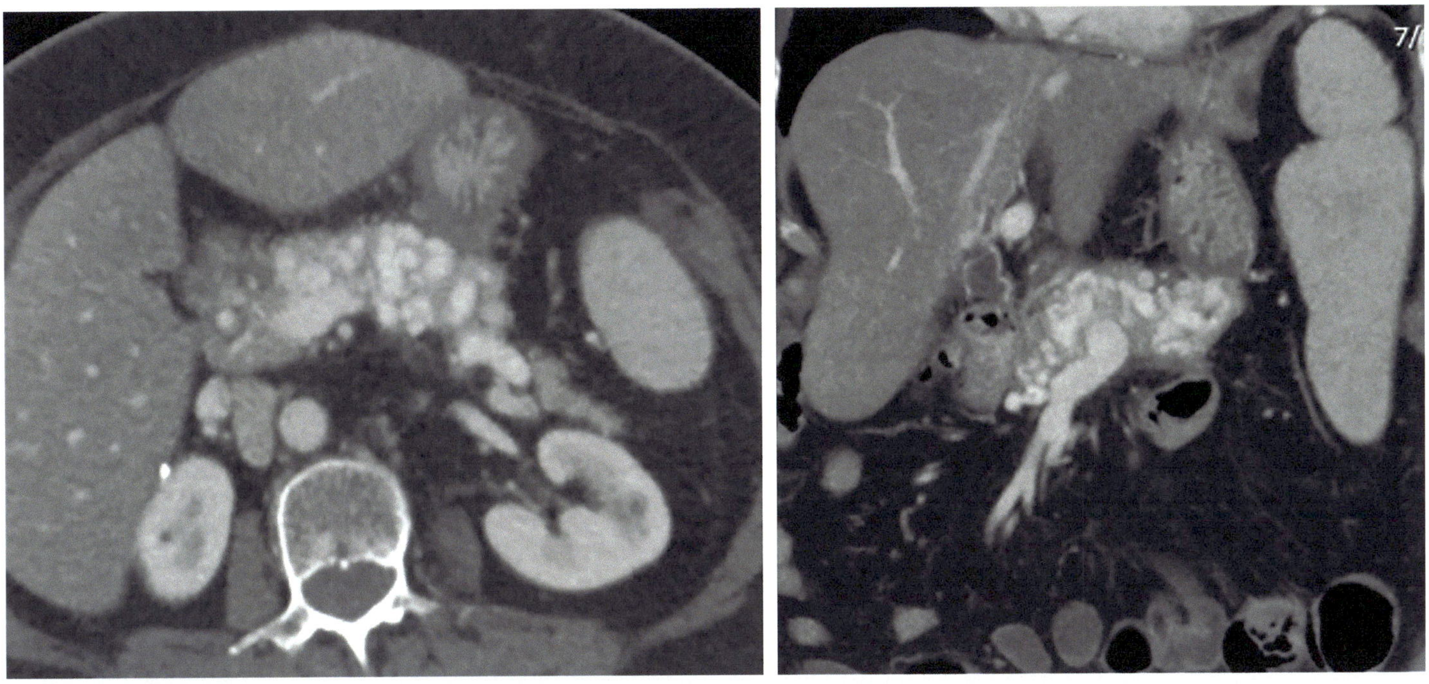

Fig. 11.18 Intrapancreatic varices in a patient with chronic splenic vein thrombosis.

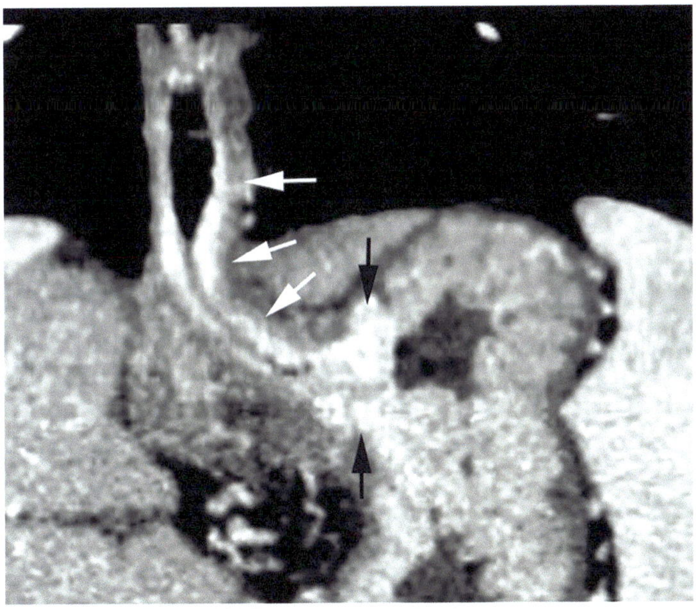

Fig. 11.19 Esophageal varices. Postcontrast coronal multidetector row computed tomography (MDCT) image shows the esophageal varices (*white arrows*) contiguous with the cardiac varices (*black arrows*).

vein occlusion. In such cases, splenectomy is often necessary for the treatment.[38]

Duodenal Varix

Duodenal varices usually develop on mesocaval collaterals between the pancreaticoduodenal vein and the right gonadal vein or the right renal capsular vein (▸Fig. 11.23). This type of mesocaval collateral can cause more complications than other portosystemic venous shunts such as gastrorenal shunt and splenorenal shunt. There are several potential anastomoses between the portal venous system and the IVC in the retroperitoneum, and between branches of the superior mesenteric veins or IMV with the IVC. These anastomoses are generally called the veins of Retzius, remnants of the anastomotic veins that develop during embryonic stage between the vitelline venous system and the posterior/subcardinal venous system (▸Fig. 11.12). It is reported that the most frequently demonstrated pathway of the veins of Retzius is the right gonadal vein followed by the right capsular vein[26,40] (▸Fig. 11.23). Rupture of duodenal varix is rare,

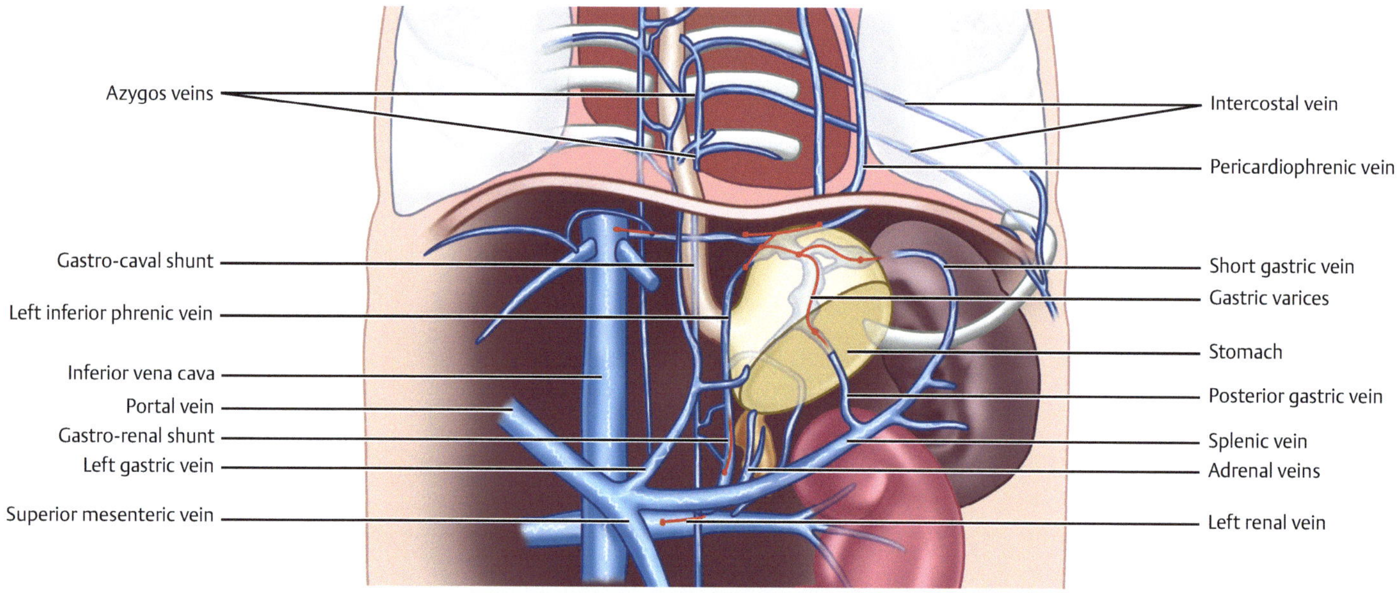

Fig. 11.20 Portosystemic shunts through isolated gastric varices. Blood flow of the isolated gastric varices at the fundus and/or cardia can terminate into the left inferior phrenic vein, thereby forming a gastrorenal shunt between the gastric varices and the left renal vein or the adrenal vein. A second form, portosystemic connection, is via a gastrocaval shunt which is formed when the left inferior renal vein drains into the inferior vena cava or left hepatic vein. The left inferior renal vein also has anastomoses with pericardiophrenic vein, subcostal and intercostal veins, and azygos vein, which can be a drainage root of gastric varices. (Modified from Kiyosue et al.[38])

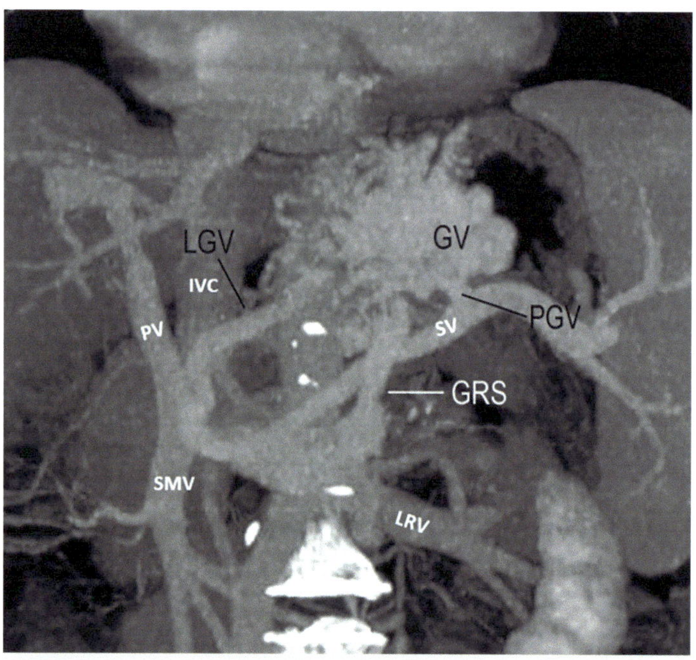

Fig. 11.21 Isolated gastric varices with gastrorenal shunt. Coronal maximum intensity projection (MIP) image on postcontrast computed tomography (CT) shows isolated gastric varices (GV) connecting with the left renal vein (LRV) through a gastrorenal shunt (GRS) formed by the left phrenic vein. Note the afferent veins of left gastric vein (LGV) and posterior gastric vein (PGV). IVC, inferior vena cava; PV, portal vein; SMV, superior mesenteric vein; SV, splenic vein.

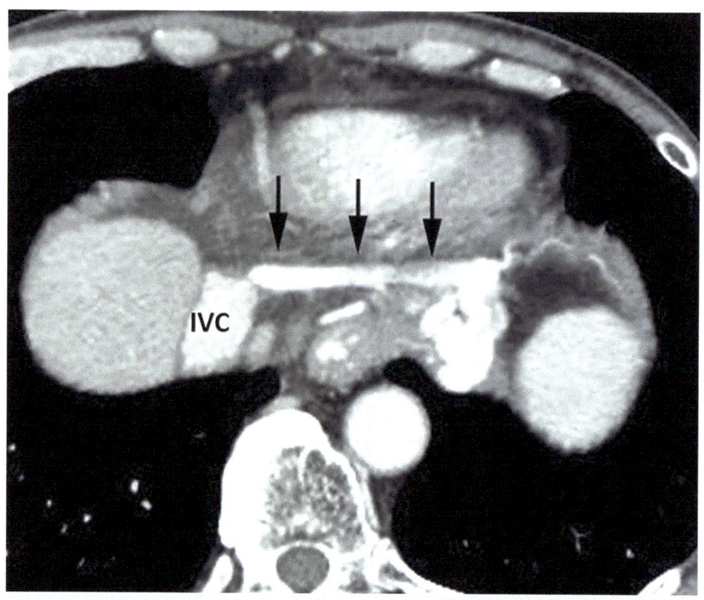

Fig. 11.22 Isolated gastric varices with gastrocaval shunt. Axial maximum intensity projection (MIP) image of postcontrast computed tomography (CT) shows isolated gastric varices at the fundus, terminating into the inferior vena cava (IVC) via a dilated left inferior phrenic vein (*arrows*).

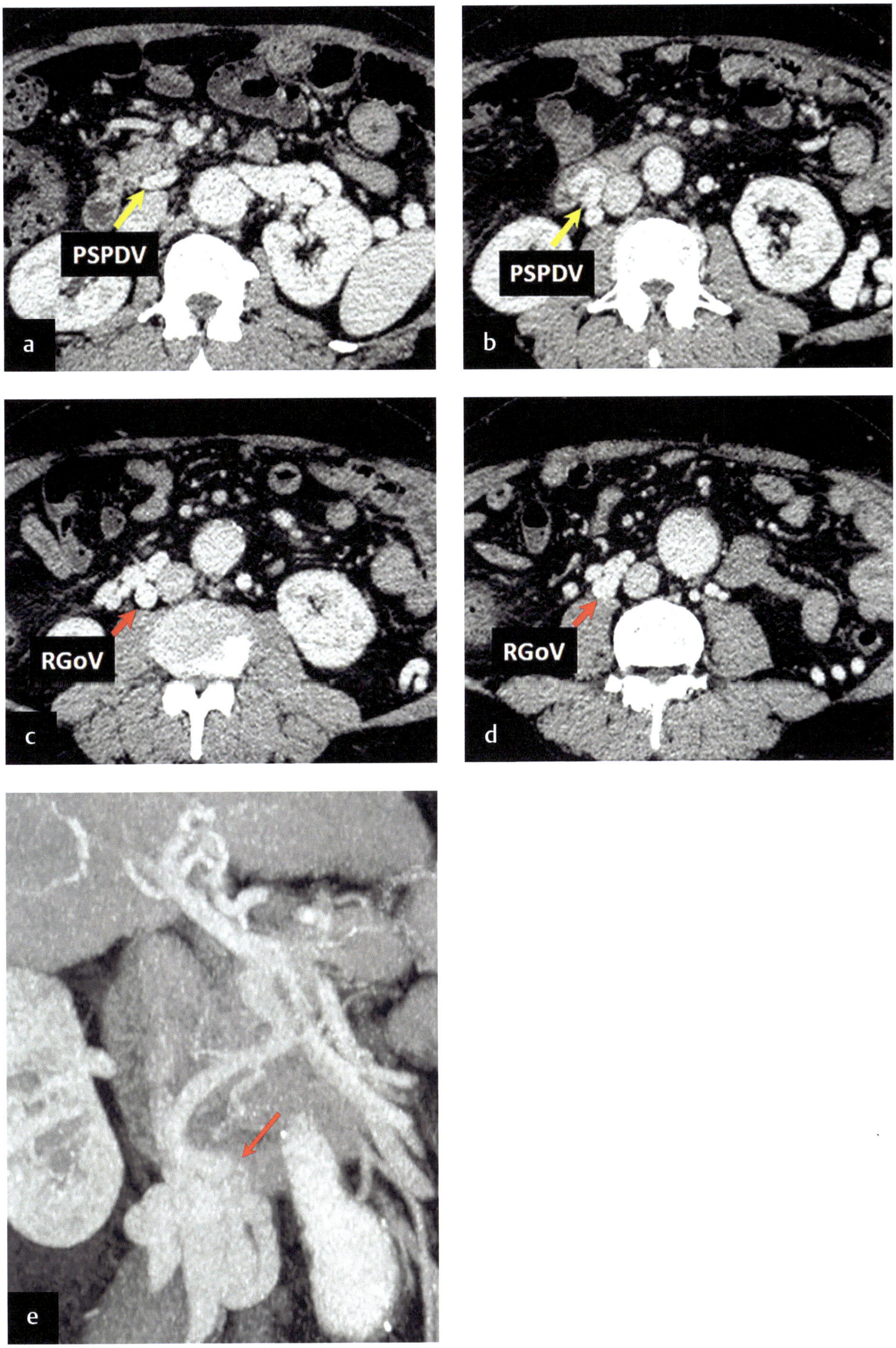

Fig. 11.23 Duodenal varices. **(a–d)** Postcontrast axial multidetector row computed tomography (MDCT) images show the duodenal varices developing between the posterior superior pancreatoduodenal vein (PSPDV) and the right gonadal vein (RGoV). **(e)** Coronal maximum intensity projection (MIP) image of a postcontrast computed tomography (CT) clearly shows the duodenal varices (*arrow*) developing on a mesocaval shunt.

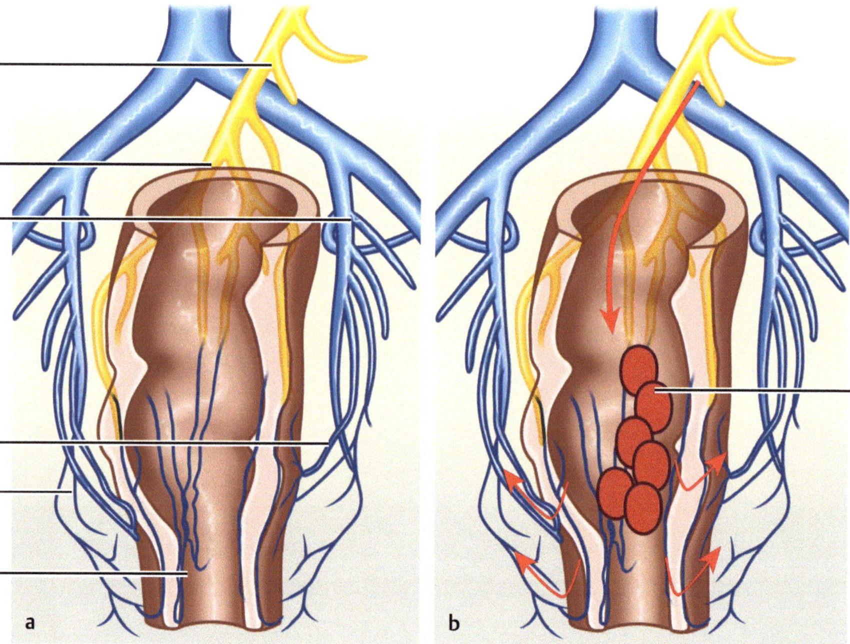

Fig. 11.24 Schematic drawing of the anatomy of the rectal veins **(a)** and rectal varices **(b)**. The venous blood from superior part of the rectal venous plexus drains via the superior rectal veins into the inferior mesenteric vein. The middle and the inferior rectal veins receive blood from the middle and inferior portions of the rectum; they join the internal iliac vein via the inferior vesical vein and the inferior rectal vein, respectively.

but when bleeding occurs, it is often life-threatening. Although the treatment of duodenal varix is still challenging, selective B-RTO is increasingly used owing to its effectiveness.[40]

Rectal Varix

Rectal varices are the most common ectopic varices, which are observed in 40% of patients with portal hypertension and in approximately 8% of variceal bleeding. The venous blood from rectum normally drains via the superior, middle, and inferior rectal veins into either the portal venous system or the internal iliac vein. The superior rectal vein drains the superior portion of the rectum and directly joins the IMV (▶Fig. 11.24). The middle and the inferior rectal veins drain into the internal iliac vein bilaterally via the inferior vesical vein and the internal pudendal vein. In cases of portal hypertension, this drainage system can be a hepatofugal collateral pathway from the IMV to the internal iliac vein, and rectal varices can arise on the portosystemic pathway. Therefore, the rectal varices are always supplied by the IMV and drain bilaterally into the internal iliac veins (▶Fig. 11.25).

Other Varices

Gonadal veins normally drain into the inferior vena cava and left renal vein. Gonadal veins are uncommon mesocaval collateral pathways between the iliac veins or the IVC and the intestinal or retroperitoneal tributaries of the superior mesenteric or splenic veins. On cross-sectional imaging, gonadal varices are demonstrated as dilated and tortuous retroperitoneal vascular structures connecting to the renal vein or the IVC (▶Fig. 11.23, ▶Fig. 11.26). Intrapancreatic varices are rare and may be seen after chronic thrombosis of the portal or splenic veins (▶Fig. 11.18).

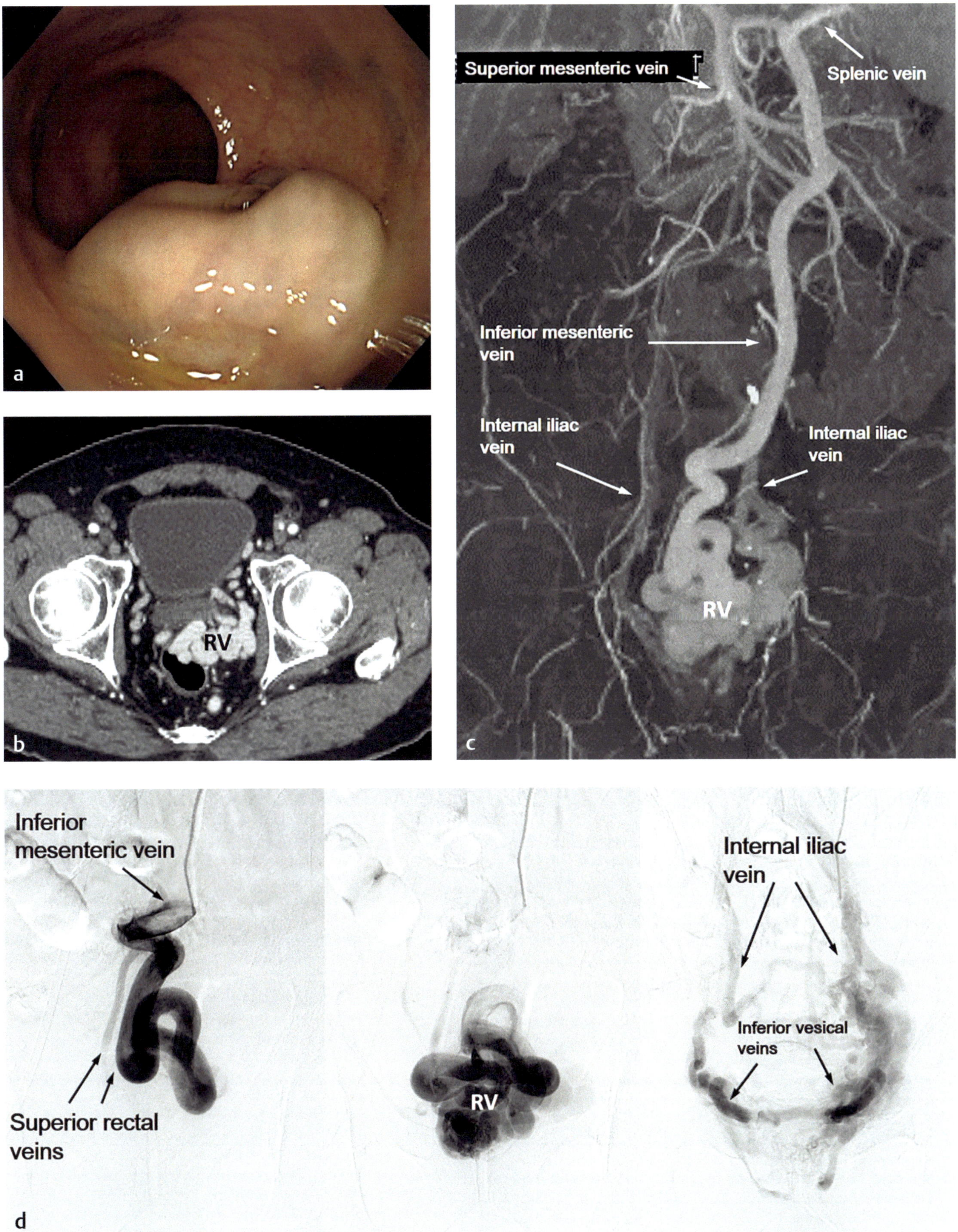

Fig. 11.25 A case of large rectal varices. **(a)** Proctoscopy shows large varices in the rectum. **(b)** Contrast-enhanced computed tomography (CT) shows large varices and dilated veins surrounding the rectum and the prostate gland. **(c)** CT angiography and **(d)** inferior mesenteric venography show large rectal varices (RV) on the portosystemic collateral pathway from the inferior mesenteric vein to the bilateral internal iliac veins.

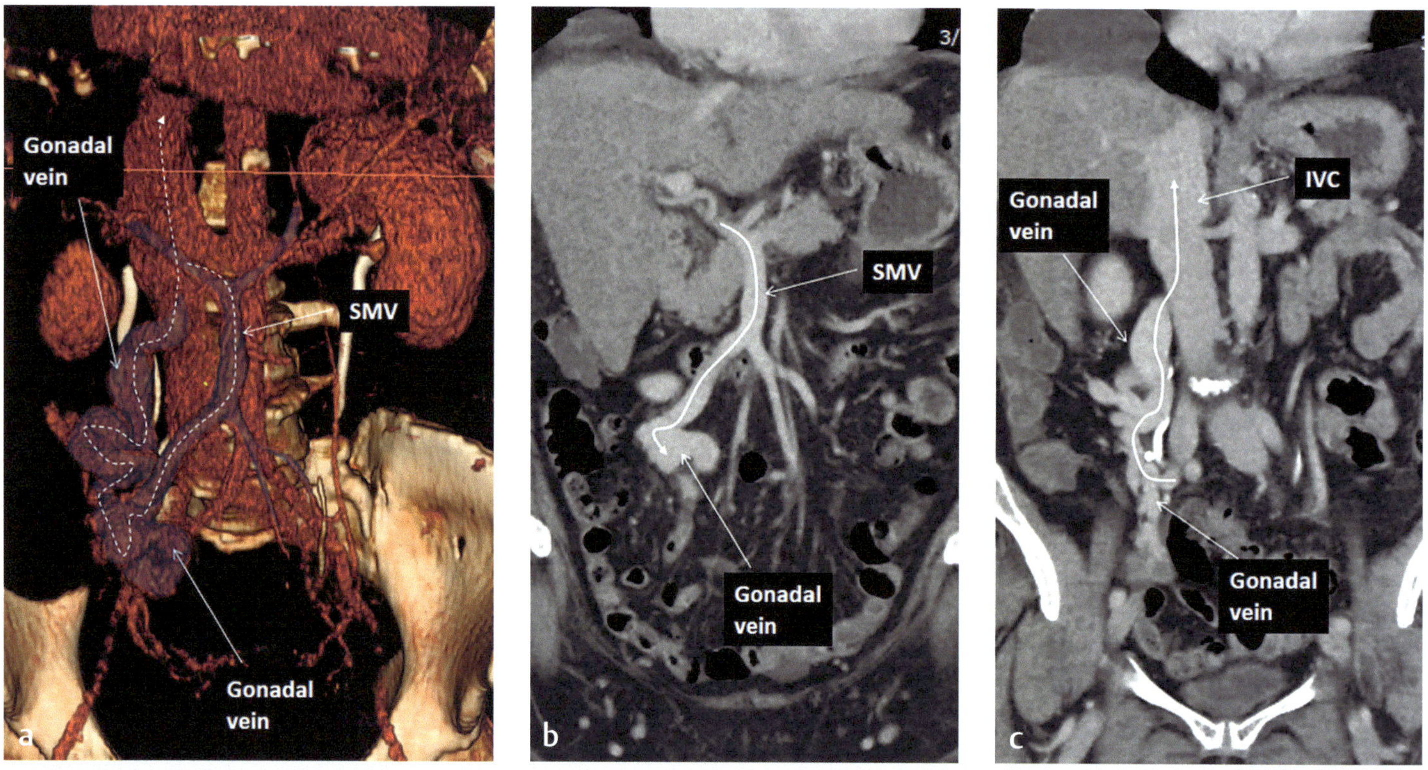

Fig. 11.26 (a–c) Coronal computed tomography (CT) images showing portosystemic shunt (*dashed long arrows*) via the superior mesenteric vein (SMV) and right gonadal vein causing varicoid enlargement of the gonadal vein in a man with liver cirrhosis. The inferior vena cava (IVC) is also markedly enlarged.

11

References

1. Moore KL, Persaud TVN. The Developing Human. 6th ed. St. Louis, Philadelphia: W.B. Saunders; 1998:349–355
2. FitzGerald MJT, FitzGerald M. Human Embryology. London: Bailliere Tindal; 1994:119–127
3. Larsen WJ. Human Embryology. 2nd ed. Edinburgh & London: Churchill Livingstone; 1997:204–211
4. Marks C. Developmental basis of the portal venous system. Am J Surg 1969; 117(5):671–681
5. Standing S. GRAY'S Anatomy: The Anatomical Basis of Clinical Practice. 40th ed. Edinburgh & London: Churchill Livingstone; 2008:1163–1175
6. Sexton CC, Zeman RK. Correlation of computed tomography, sonography, and gross anatomy of the liver. AJR Am J Roentgenol 1983;141(4):711–718
7. Douglass BE, Baggenstoss AH, Hollinshead WH. The anatomy of the portal vein and its tributaries. Surg Gynecol Obstet 1950;91(5):562–576
8. Reichardt W, Cameron R. Anatomy of the pancreatic veins. A post mortem and clinical phlebographic investigation. Acta Radiol Diagn (Stockh) 1980;21(1): 33–41
9. Gerber AB, Lev M, Goldberg SL. The surgical anatomy of the splenic vein. Am J Surg 1951;82(3):339–343
10. Jin G, Tuo H, Sugiyama M, et al. Anatomic study of the superior right colic vein: its relevance to pancreatic and colonic surgery. Am J Surg 2006;191(1):100–103
11. Gilfillan RS, Hills HL. Anatomic study of the portal vein and its main branches. Arch Surg 1950;61(3):449–461
12. Mori H, McGrath FP, Malone DE, Stevenson GW. The gastrocolic trunk and its tributaries: CT evaluation. Radiology 1992;182(3):871–877
13. Sharma M, Babu CS, Garg S, Rai P. Portal venous system and its tributaries: a radial endosonographic assessment. Endosc Ultrasound 2012;1(2):96–107
14. Mori H, Miyake H, Aikawa H, et al. Dilated posterior superior pancreaticoduodenal vein: recognition with CT and clinical significance in patients with pancreaticobiliary carcinomas. Radiology 1991;181(3):793–800
15. Yamada Y, Mori H, Kiyosue H, Matsumoto S, Hori Y, Maeda T. CT assessment of the inferior peripancreatic veins: clinical significance. AJR Am J Roentgenol 2000;174(3):677–684
16. Sugita M, Ryu M, Satake M, et al. Intrahepatic inflow areas of the drainage vein of the gallbladder: analysis by angio-CT. Surgery 2000;128(3):417–421
17. Gallego C, Velasco M, Marcuello P, Tejedor D, De Campo L, Friera A. Congenital and acquired anomalies of the portal venous system. Radiographics 2002;22(1): 141–159
18. Morgan G, Superina R. Congenital absence of the portal vein: two cases and a proposed classification system for portasystemic vascular anomalies. J Pediatr Surg 1994;29(9):1239–1241
19. Takagaki K, Kodaira M, Kuriyama S, et al. Congenital absence of the portal vein complicating hepatic tumors. Intern Med 2004;43(3):194–198
20. Mori H, Hayashi K, Fukuda T, et al. Intrahepatic portosystemic venous shunt: occurrence in patients with and without liver cirrhosis. AJR Am J Roentgenol 1987;149(4):711–714
21. Yoshimoto Y, Shimizu R, Saeki T, et al. Patent ductus venosus in children: a case report and review of the literature. J Pediatr Surg 2004;39(1):E1–E5
22. Ianora AA, Memeo M, Sabba C, Cirulli A, Rotondo A, Angelelli G. Hereditary hemorrhagic telangiectasia: multi-detector row helical CT assessment of hepatic involvement. Radiology 2004;230(1):250–259
23. Matsumoto S, Mori H, Yamada Y, Hayashida T, Hori Y, Kiyosue H. Intrahepatic porto-hepatic venous shunts in Rendu-Osler-Weber disease: imaging demonstration. Eur Radiol 2004;14(4):592–596
24. Akahoshi T, Nishizaki T, Wakasugi K, et al. Portal-systemic encephalopathy due to a congenital extrahepatic portosystemic shunt: three cases and literature review. Hepatogastroenterology 2000;47(34):1113–1116
25. Alonso-Gamarra E, Parrón M, Pérez A, Prieto C, Hierro L, López-Santamaría M. Clinical and radiologic manifestations of congenital extrahepatic portosystemic shunts: a comprehensive review. Radiographics 2011;31(3):707–722
26. Ibukuro K, Tsukiyama T, Mori K, Inoue Y. Veins of Retzius at CT during arterial portography: anatomy and clinical importance. Radiology 1998;209(3):793–800
27. Iñon AE, D'Agostino D. Portal hypertension secondary to congenital arterioportal fistula. J Pediatr Gastroenterol Nutr 1987;6(3):471–473
28. Walsh G, Williams MP. Congenital anomalies of the portal venous system—CT appearances with embryological considerations. Clin Radiol 1995;50(3): 174–176

29. Masumoto K, Teshiba R, Esumi G, et al. Duodenal stenosis resulting from a preduodenal portal vein and an operation for scoliosis. World J Gastroenterol 2009;15(31):3950–3953
30. Koc Z, Oguzkurt L, Ulusan S. Portal venous system aneurysms: imaging, clinical findings, and a possible new etiologic factor. AJR Am J Roentgenol 2007;189(5): 1023–1030
31. Harnoss JM, Harnoss JC, Diener MK, et al. Portal annular pancreas: a systematic review of a clinical challenge. Pancreas 2014;43(7):981–986
32. Narita M, Hata H, Ikai I. Portal annular pancreas: an unusual pancreatic anomaly. J Visc Surg 2016;153(2):153–155
33. Amitrano L, Guardascione MA, Brancaccio V, et al. Risk factors and clinical presentation of portal vein thrombosis in patients with liver cirrhosis. J Hepatol 2004;40(5):736–741
34. Willatt JM, Hussain HK, Adusumilli S, Marrero JA. MR imaging of hepatocellular carcinoma in the cirrhotic liver: challenges and controversies. Radiology 2008;247(2):311–330
35. Tirumani SH, Shanbhogue AK, Vikram R, Prasad SR, Menias CO. Imaging of the porta hepatis: spectrum of disease. Radiographics 2014;34(1):73–92
36. Sebastià C, Quiroga S, Espin E, Boyé R, Alvarez-Castells A, Armengol M. Portomesenteric vein gas: pathologic mechanisms, CT findings, and prognosis. Radiographics 2000;20(5):1213–1224, discussion 1224–1226
37. Kiyosue H, Mori H, Matsumoto S, Yamada Y, Hori Y, Okino Y. Transcatheter obliteration of gastric varices. Part 1. Anatomic classification. Radiographics 2003;23(4):911–920
38. Kiyosue H, Ibukuro K, Maruno M, Tanoue S, Hongo N, Mori H. Multidetector CT anatomy of drainage routes of gastric varices: a pictorial review. Radiographics 2013;33(1):87–100
39. Johns TN, Evans BB. Collateral pathways in portal hypertension. Ann Surg 1962;155:838–845
40. Okahara M, Kiyosue H, Ueda S, et al. Anatomic features and retrograde transvenous obliteration of duodenal varices associated with mesocaval collateral pathway. J Vasc Interv Radiol 2012;23(10):1339–1346

12 Peritoneal Spaces

Ali Rivandi, Corinne Deurdulian, and Farhood Saremi

◆ Introduction

Knowledge of peritoneal anatomy on imaging is important in order to identify patterns of disease spread within the abdomen and pelvis. Proper identification of the peritoneal spaces and ligaments by the radiologist helps the surgeon assess the sites of disease related to infectious, inflammatory, traumatic, or neoplastic etiologies more accurately. These disease processes often involve the peritoneal spaces and their reflections. The peritoneal reflections, mesenteries, and omenta act as disease process boundaries and play an important role in the spread of disease. Therefore, the radiologist's understanding of the peritoneal spaces and their boundaries contributes to more accurate localization of the pathologic processes and differential diagnosis based on the involved locations.

◆ Peritoneal Cavity

The peritoneum is a complexly arranged thin serous membrane comprised of parietal and visceral layers. The parietal peritoneum lines the inner abdominal wall and the visceral peritoneum envelops the abdominal organs and serves as a medium for the passage of the vessels and nerves to the organs. The peritoneal lining consists of a single layer of simple low-cuboidal epithelium called mesothelium.[1] The peritoneal cavity is a potential space between the parietal and visceral peritoneum (▶Fig. 12.1). Approximately 50 to 100 mL of serous fluid separates the two layers of the peritoneum from one another and lubricates the peritoneal surfaces.[2] The peritoneal cavity is closed in men, but in women, it communicates with the extraperitoneal pelvis through the fallopian tubes, uterus, and vagina.

The peritoneal ligaments, mesenteries, and omenta divide the peritoneum into two major compartments: the larger compartment is called the greater peritoneal sac, and a smaller compartment is called the lesser sac (or omental bursa) (▶Fig. 12.1).[3] Peritoneal ligaments are double layers of peritoneum that support structures within the peritoneal cavity; the omentum and mesentery are specifically named peritoneal ligaments. The omentum extends from the stomach to adjacent organs. The mesentery attaches the bowel wall to the posterior abdominal wall. The omenta and mesenteries contain blood vessels, nerves, lymph nodes, and variable amount of fat.

The distribution of the peritoneal fluid is determined by the peritoneal folds and recesses, the action of gravity, and the hydrostatic pressure. Fluid accumulates first in the dependent part of the peritoneal space. In a supine patient, peritoneal fluid flows toward the dependent part of the pelvic cavity (i.e., rectovaginal recess). In the abdomen, the most dependent parts are posterior subdiaphragmatic recesses and the right subhepatic space (▶Fig. 12.2). The hydrostatic pressure under the diaphragm is normally subatmospheric especially during inspiration and stimulates collection of peritoneal inflammation and metastatic disease under the diaphragm and around the liver.[3]

Embryology

The peritoneal cavity develops from the right and left pericardioperitoneal (pleuroperitoneal) canals of the intraembryonic coelom cavity. The visceral (splanchnopleuritic coelomic epithelium) and parietal (somatopleuritic coelomic epithelium) surfaces of the coelomic cavity form the visceral and parietal layers of the peritoneum. The visceral layer forms the serosal layer of the guts and covers all the peritoneal ligaments and mesenteries. With development of the transverse septum, the pleural cavity will be separated from the peritoneal cavity. The medial walls of the pleuroperitoneal canals fuse with each other to form the primitive mesentery where all vessels to the abdominal organs pass. The final mesentery is composed of two layers of the peritoneum and the vessels in between.

The position of the gut within the sagittally oriented primitive mesentery plane divides the primitive mesentery into ventral and dorsal portions that undergo further development including rotation, descents, and resorption throughout the fetal life. The liver grows into the ventral mesentery while the spleen, pancreas, and a large portion of the gut grow into the dorsal mesentery (▶Fig. 12.3).

The dorsal mesentery connects the primitive gut, from the diaphragm to the beginning of the rectum, and to the posterior abdominal wall. At the level of stomach this mesentery is called mesogastrium. This dorsal mesentery forms the greater omentum of the stomach, the small bowel mesentery of the jejunum and ileum, the transverse mesocolon of the transverse colon, and the sigmoid mesocolon of the sigmoid. The spleen and dorsal pancreas also grow in the dorsal mesentery. The dorsal pancreas serosal surface fuses with the dorsal parietal peritoneum and becomes retroperitoneal except for a small portion of its distal tail which along with the distal splenic vessels remain intraperitoneal within the splenorenal ligament. The length of splenorenal ligament is variable. Later, the embryonic dorsal mesentery of the duodenum, ascending colon, and descending colon regresses and fuses these organs to the parietal peritoneum of the posterior abdominal wall, making them secondarily retroperitoneal.

In contrast to the dorsal mesentery which is a continuous partition, the embryonic ventral mesentery is limited to the foregut (abdominal esophagus to the duodenal ampulla) and below that the two pleuroperitoneal canals communicate anterior to the gut. The growing embryonic liver in the ventral mesentery divides it into two segments. The segment extending between the liver to the anterior abdominal wall forms the falciform ligament and the segment between the liver and the stomach develops the lesser omentum.

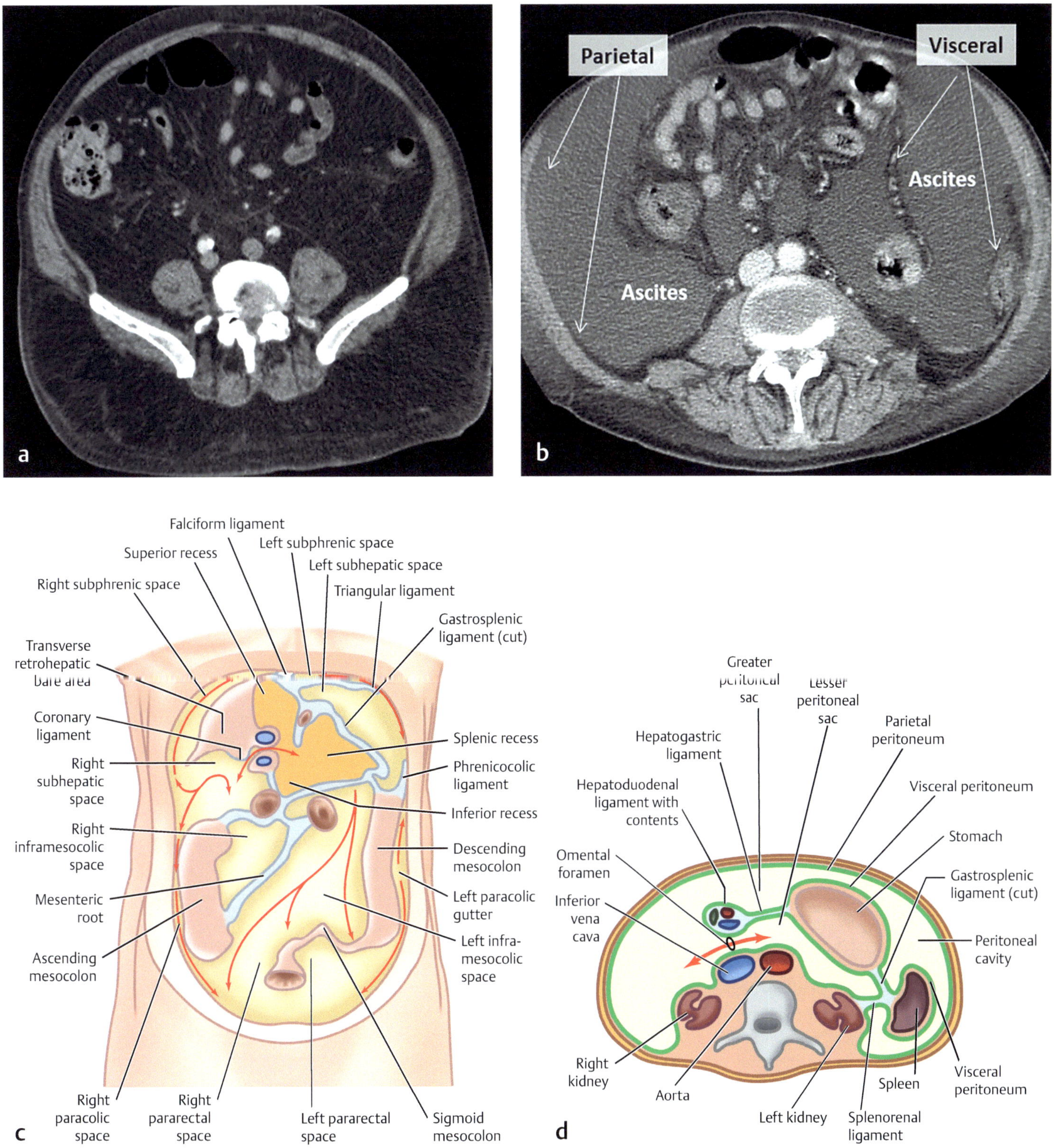

Fig. 12.1 (a) Peritoneum in a patient with large mesenteric and retroperitoneal fat. **(b)** Peritoneum in a patient with large ascites. The peritoneal cavity is a potential space between the parietal and visceral peritoneum which in normal individuals is not visible. It will be seen only if it is filled with fluid or air. **(c)** Peritoneal attachments to the posterior abdominal wall. **(d)** Cross section of abdomen at the level of foramen of Winslow.

Peritoneal Ligaments and Omenta

Suspensory Ligaments of the Liver

The liver is covered by the visceral peritoneum except at the bare area, bed of the gallbladder, and porta hepatis.[3,4] The coronary, triangular, and falciform ligaments are the suspensory ligaments of the liver formed by peritoneal reflections (see Chapter 6 "Liver"). The left and right coronary/triangular ligaments outline the bare area of the liver (▸Fig. 12.4, ▸Fig. 12.5). The left triangular ligament is formed by fusion of the superior and inferior peritoneal leaves of the left coronary ligaments. This ligament is short and does not compartmentalize the left subphrenic space. The right triangular ligament is formed by fusion of the superior and inferior leaves of the right coronary ligament on the lateral side and separates the right subphrenic space from the right subhepatic space (▸Fig. 12.4, ▸Fig. 12.5).[1] The inferior leaf of the right coronary ligament is located at the level of the right kidney, hence the name "hepatorenal ligament," and extends medially to the posterior wall of the lesser omentum.

Imaging of the triangular ligaments is difficult especially on the left side. In patients with large ascites, the right triangular ligament is visible in most cases (▸Fig. 12.5).

The falciform ligament, the remnant of the ventral part of the ventral mesentery, connects the liver to the anterior abdominal wall. It extends between the diaphragm and the umbilicus and contains the obliterated umbilical vein (the ligamentum teres) (▸Fig. 12.5). Near the liver surface, the right and left peritoneal leaves of the falciform ligament separate and merge with the anterosuperior peritoneal reflection of the right and left extensions of the coronary ligament. At its diaphragmatic attachment, the falciform ligament acts as an incomplete barrier between the right and left subphrenic spaces. It is important to know that peritoneal metastases may involve the falciform ligament and could be confused with hepatic metastases (▸Fig. 12.6).

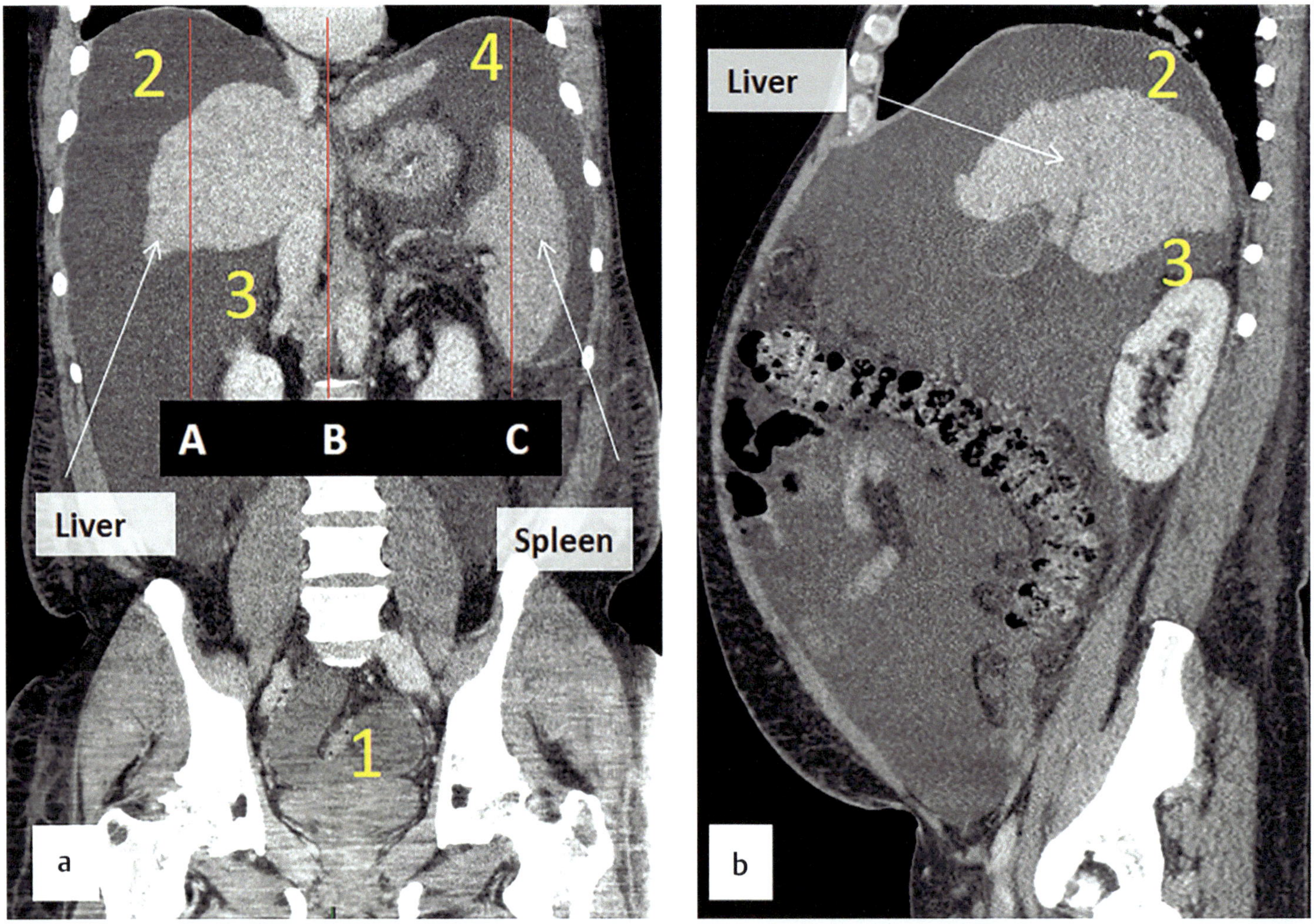

Fig. 12.2 Fluid accumulates first in the dependent part of the peritoneal space. Using a coronal reference image **(a)**, three sagittal views, **(b)** right side. **(c)** midline, and **(d)** left side, in a patient with large ascites due to cirrhosis. In a supine patient, peritoneal fluid flows toward the dependent part (#1) of the pelvic cavity (i.e., rectovesical or rectovaginal recess). In the abdomen, the most dependent parts are right posterior subphrenic recess (#2), left posterior subphrenic recesses (#4), and the right subhepatic space (#3).

(Continued)

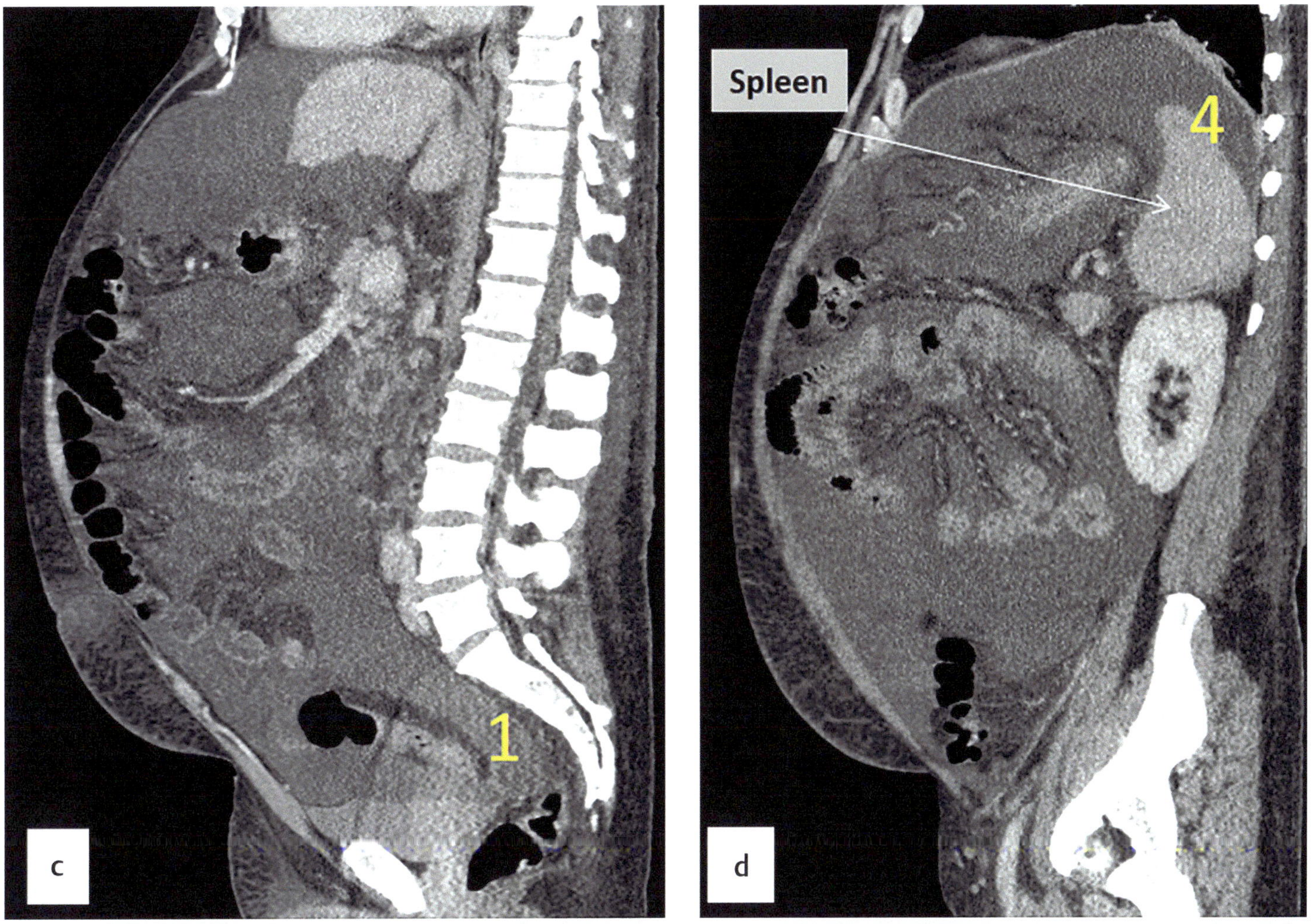

Fig. 12.2 (c, d) (*Continued*)

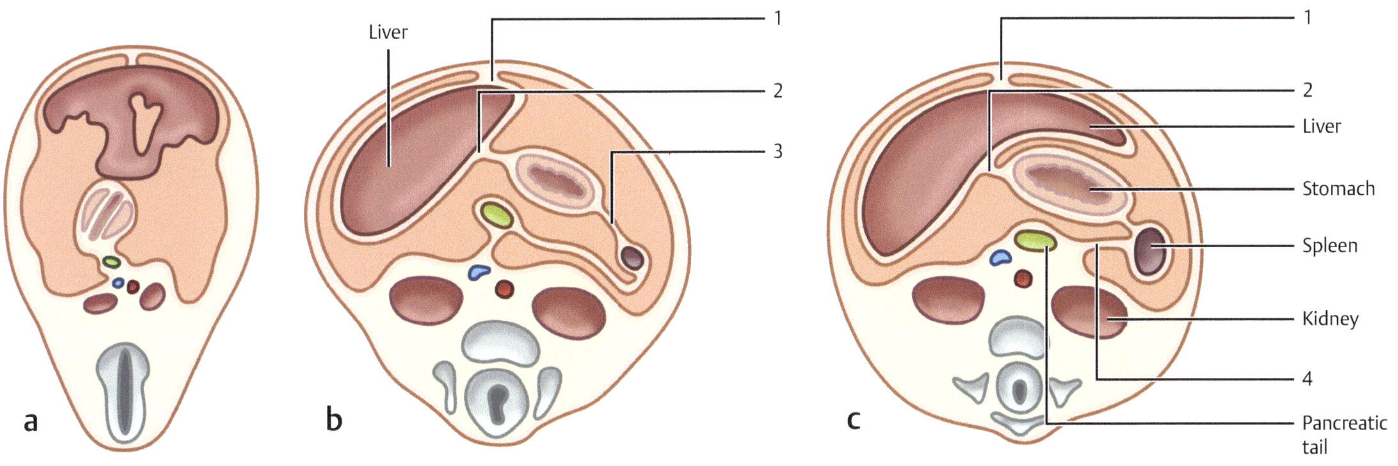

Fig. 12.3 Embryologic development of the dorsal and ventral mesentery at weeks 4 **(a)**, 5 **(b)**, and 6 **(c)** of gestation. Growing liver in the ventral mesentery divides it into two segments and growing stomach in the dorsal mesentery divides it into two segments. The anterior part of the ventral mesentery becomes the falciform ligament (1), and the posterior part of the ventral mesentery becomes the lesser omentum (2). The anterior part of the dorsal mesentery becomes the gastrosplenic ligament (3), and the posterior part of the dorsal mesentery becomes the splenorenal ligament (4). The liver (L) arises in the ventral mesentery, whereas the stomach (St), spleen (Sp), and pancreatic tail (P) develop in the dorsal mesentery. As the liver expands in the 5th and 6th weeks of gestation, the stomach and spleen are pushed to the left of peritoneal space, and the pancreatic tail fuses with the retroperitoneum. K, kidney.

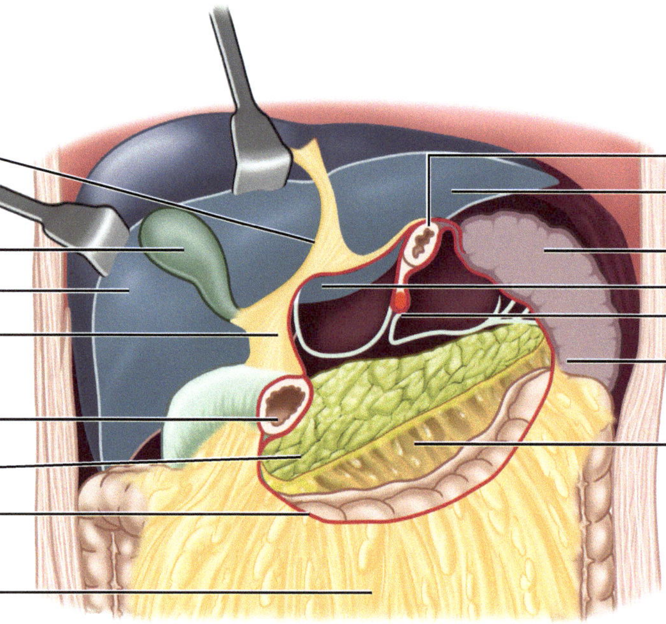

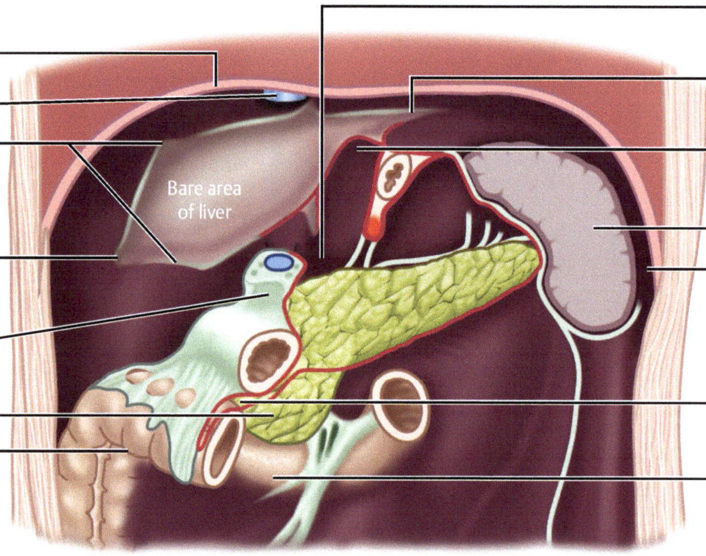

Fig. 12.4 Lesser sac. **(a, b)** Frontal views of the abdomen showing the bare area bordered by the suspensory ligaments of the liver and the lesser sac recesses. The coronary, triangular, and falciform ligaments are the suspensory ligaments of liver formed by peritoneal reflections. The lesser sac (omental bursa) is comprised of superior and inferior (including splenic) recesses that are separated by the gastropancreatic peritoneal fold containing left gastric artery and left gastric vein (coronary vein).

In patients with long-standing portal hypertension, the paraumbilical vein in the falciform ligament is often recanalized, enlarged, and communicates with the epigastric veins in the anterior abdominal wall, creating a portosystemic shunt (▶ Fig. 12.7).

Peritoneal Ligaments of the Stomach

Peritoneal ligaments of the stomach are called omentum. There are two principal omenta: the lesser omentum and the greater omentum[5] (▶ Fig. 12.8).

The development of the omentum is closely related to adjacent structures, especially the stomach and spleen. The lesser omentum arises from the dorsal portion of the ventral mesogastrium (▶ Fig. 12.1). The greater omentum develops by the folding of the dorsal mesogastrium along with rapid growth of the greater curvature of the stomach (▶ Fig. 12.9). Growth of the primitive lesser sac separates spleen from the stomach. The lesser sac becomes continuous with the peritoneal cavity by formation of the foramen of Winslow at 9th to 12th weeks of gestation.[2] Later, the inferior part of greater omentum fuses with the transverse mesocolon (▶ Fig. 12.9). This transforms the greater omentum from four layers to two layers of visceral peritoneum filled with subserosal fat and passing vessels. By the age of 5 years, most of the intestines are covered by the greater omentum.

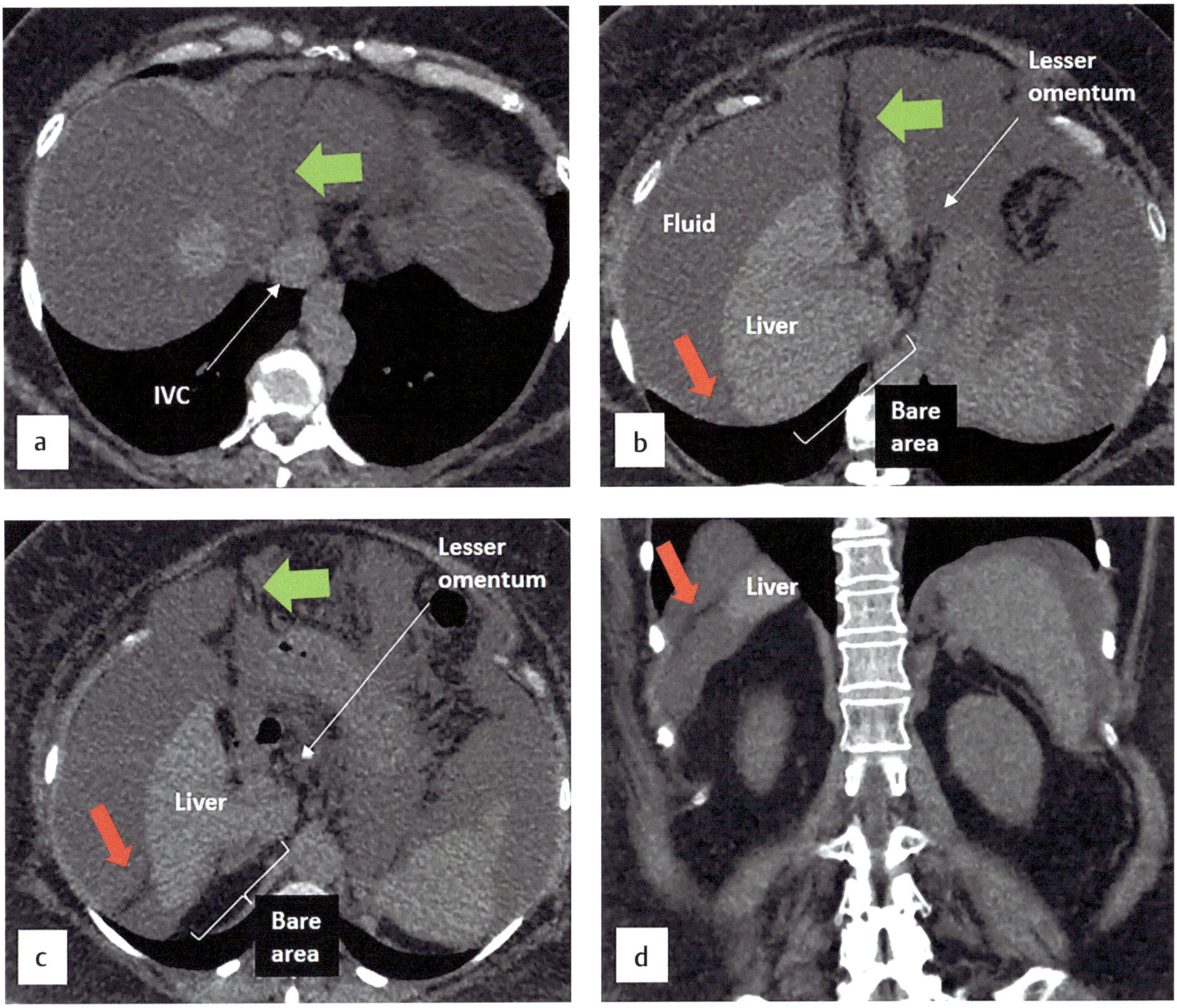

Fig. 12.5 Suspensory ligaments of the liver. **(a–c)** Axial and **(d)** coronal noncontrast computed tomography (CT) images in a patient with large ascites due to cirrhosis. The liver appears small and fibrotic. Suspensory ligaments, attaching the liver to the diaphragm, include the falciform ligament, and the coronary ligaments with the right and left triangular ligaments. These ligaments are only visible when surrounded by fluid or air. In this example the right triangular (*red arrow*) and the falciform ligament (*green arrow*) are visible as linear fat density structures between the two layers of the visceral peritoneum. The falciform ligament also contains the ligamentum teres. At the base of the falciform ligament along the liver, the hepatic veins drain into the inferior vena cava (IVC).

Lesser Omentum

The two peritoneal layers of lesser omentum form the gastrohepatic and hepatoduodenal ligaments. The gastrohepatic ligament starts from the anterior aspect of the abdominal esophagus (esophageohepatic ligament) and attaches the lesser curvature of the stomach to the inferior margin of the sulcus for the ductus venosus and the porta hepatis. It contains the coronary vein, left gastric artery, and lymph nodes in a fibrofatty stroma[6,7] (▶Fig. 12.10). The hepatoduodenal ligament is the thicker part of the lesser omentum at the level of the porta hepatis. It attaches the liver to the first part of the duodenum. The free margin of the hepatoduodenal ligament forms the roof of the epiploic foramen of Winslow which connects the lesser sac with the greater peritoneal cavity (▶Fig. 12.11). The hepatoduodenal ligament contains the portal vein, main hepatic artery, common hepatic ducts, lymphatic, nerves, and part of the cystic duct. All these structures are enclosed in a fibrous capsule known as Glisson's capsule. The portal vein, in the hepatoduodenal ligament, is behind the common hepatic duct and hepatic artery. The right hepatic artery usually passes behind the common hepatic duct. A replaced right hepatic artery passes posterior to the portal vein and common bile duct.

The hepatoduodenal ligament is surrounded by lymphatics, soft tissue, and peritoneal covering. Typical nodes in the hepatoduodenal ligament are the hilar, cystic, choledochal, retroportal,

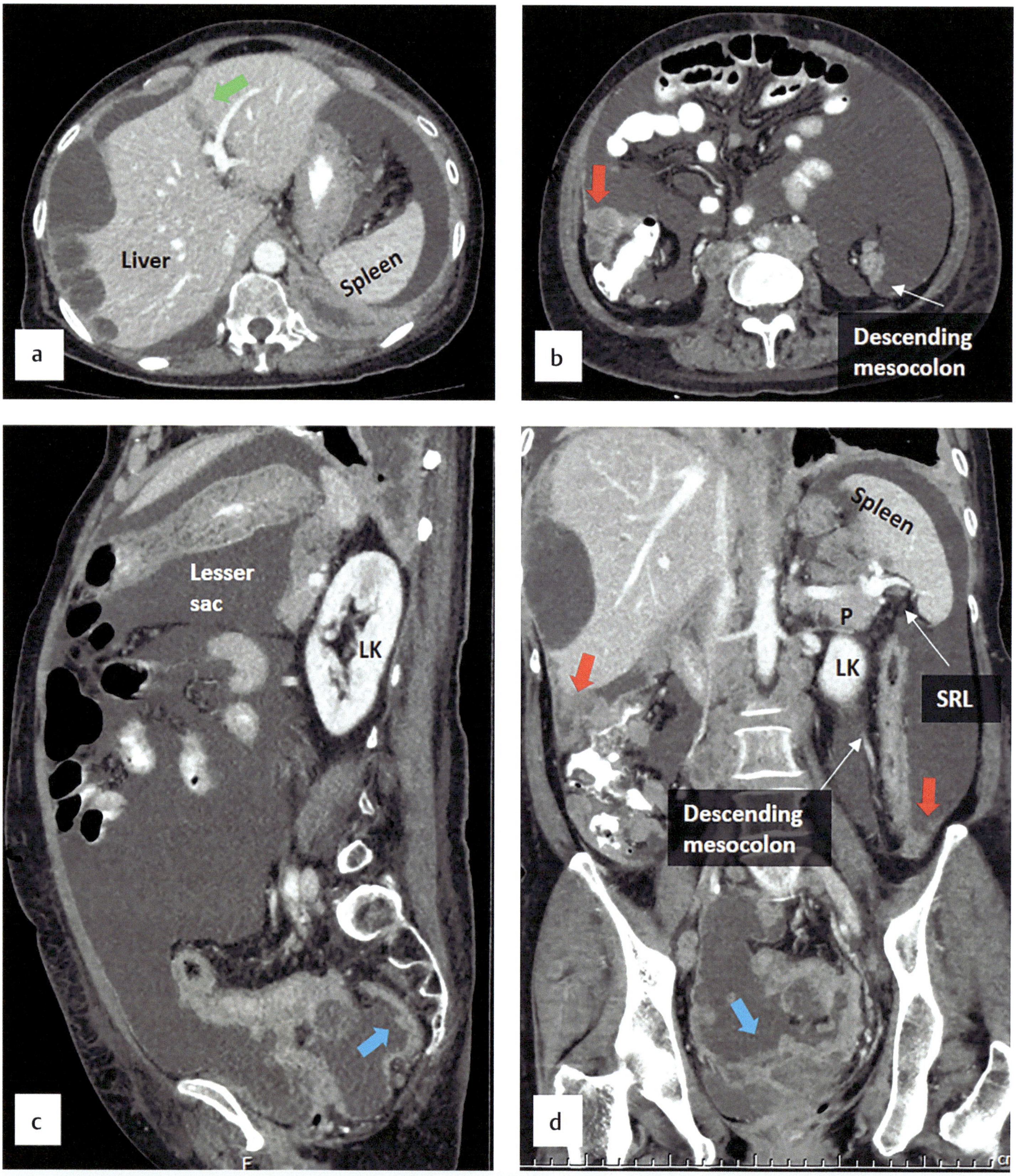

Fig. 12.6 Peritoneal carcinomatosis in a patient with ovarian tumor. **(a, b)** Axial, **(c)** sagittal, and **(d)** coronal computed tomography (CT) showing ascites and multiple peritoneal masses involving liver, umbilical fissure (*green arrow*), lesser sac, paracolic gutters (*red arrows*), and the cul de sac (*blue arrow*). LK, left kidney; P, pancreas; SRL, splenorenal ligament.

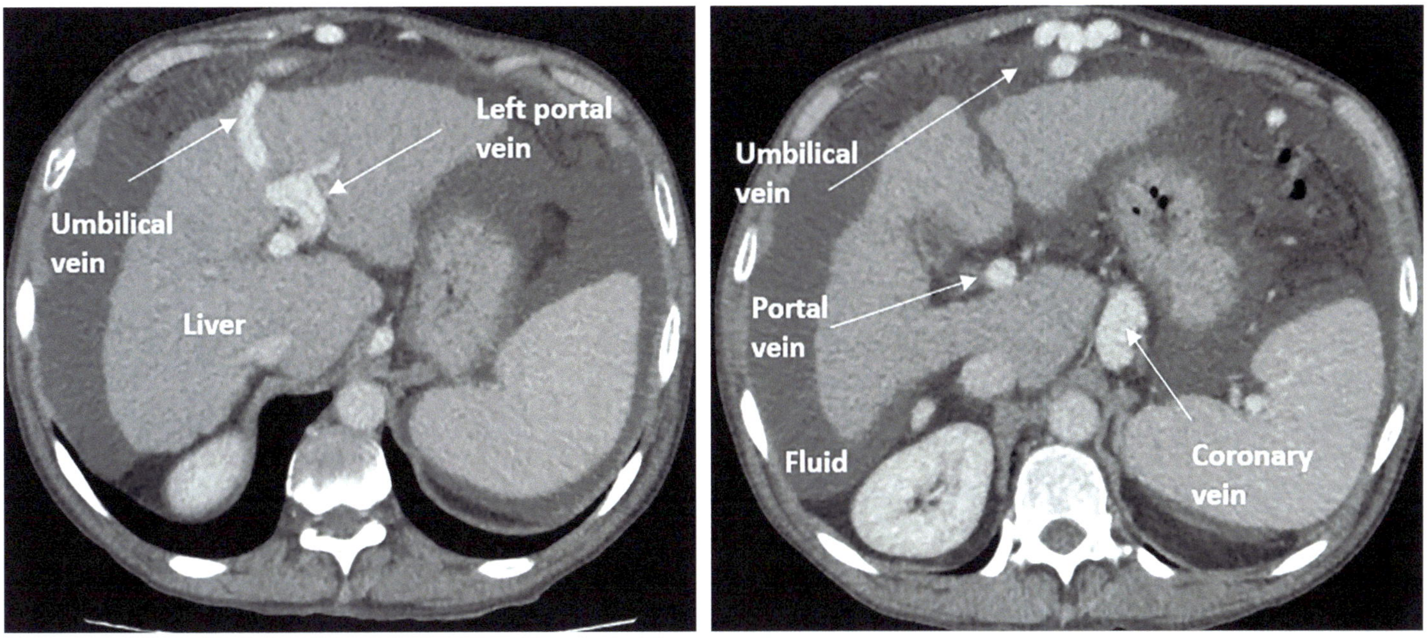

Fig. 12.7 Axial computed tomography (CT) images showing varicosities involving the recanalized umbilical vein and the coronary vein in a patient with large ascites and portal hypertension due to cirrhosis. The coronary vein is seen in the lesser omentum.

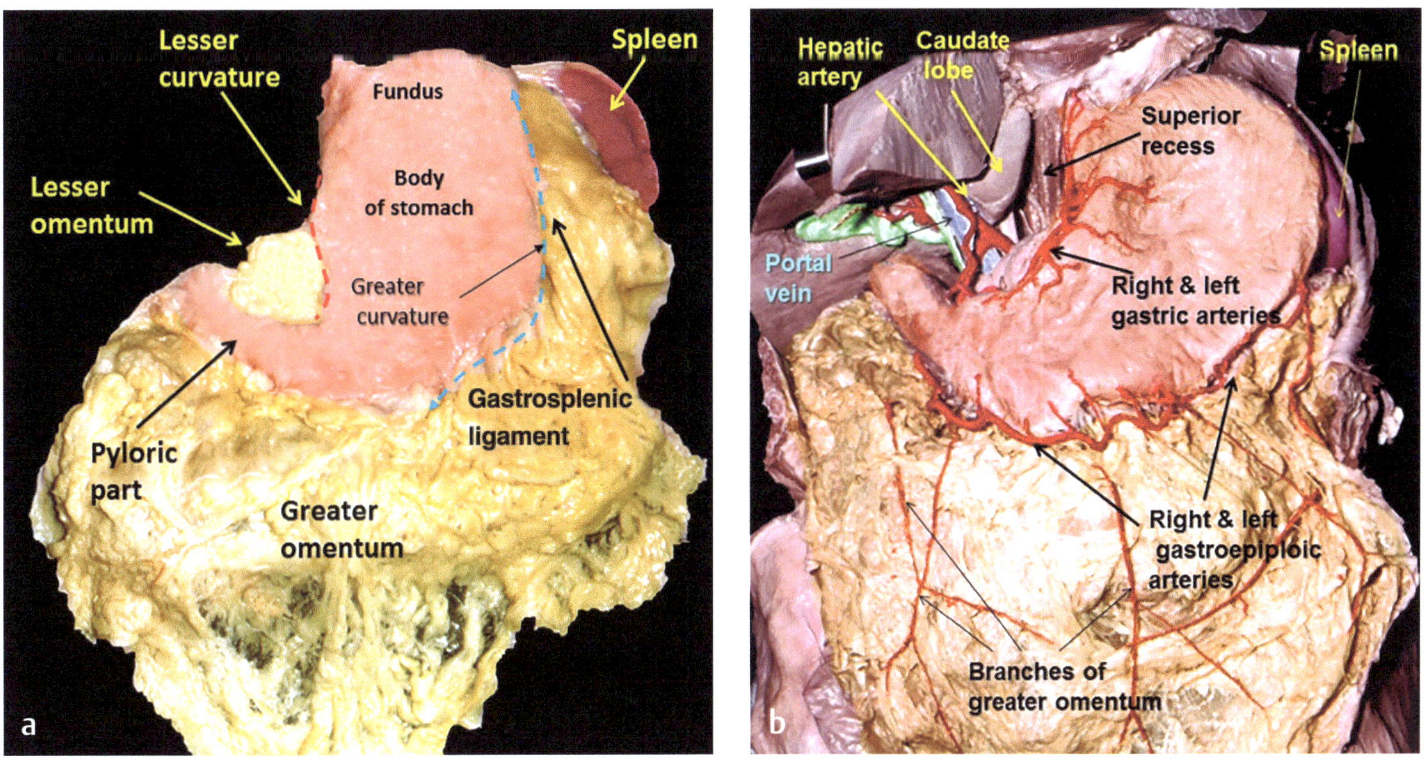

Fig. 12.8 (a, b) Anatomic extant and arterial supply of the greater and lesser omenta. On the right side of stomach, the greater omentum may extend as far as the pyloric area. On the left side of stomach, it extends along the greater curvature and forms the gastrosplenic ligament.

12

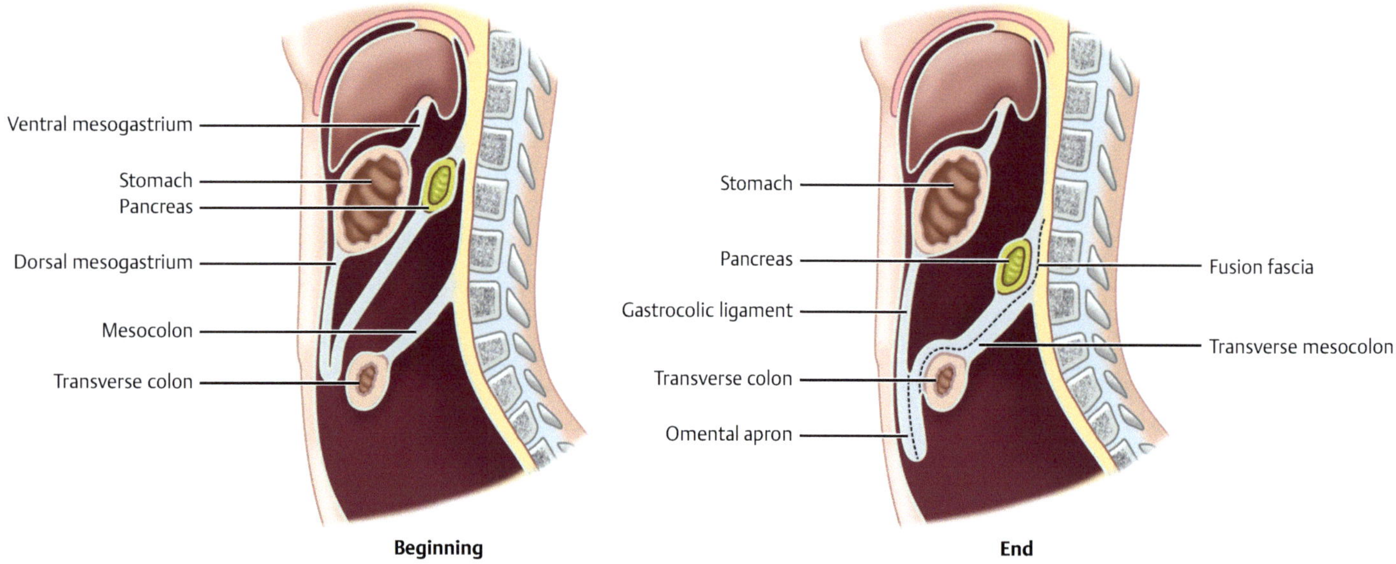

Fig. 12.9 Development of the greater omentum.

and foramen of Winslow nodes.[6] The ventral anlage of the pancreas is present in the hepatoduodenal ligament until 8 weeks of embryonic life; this explains the spread of pancreatic disease and edematous processes in this region to the liver and porta hepatis through hepatoduodenal ligament. The lesser omentum could be the first path in the local or lymphatic spread of abdominal esophagus, gastric, or biliary tumors. Gastric tumors can invade the liver through the lesser omentum.

Greater Omentum

The greater omentum also known as the gastrocolic ligament is a free-hanging mesenteric tissue apron formed by folding a double sheet of peritoneum (total of four layers) that stretches between the greater curvature of the stomach and transverse colon behind the anterior abdominal wall. It acts as a barrier against the spread of the infection into the lower abdominal cavity.[7]

The diameters of the greater omentum vary from 14 to 36 cm in length and from 20 to 46 cm in width. In obese individuals, omentum may be larger due to fat deposits. On CT scan, it appears as a band of fatty tissue of variable thickness. It becomes more visible when involved by diseases or if surrounded by ascetic fluid (▶ Fig. 12.12).

On the right side, the greater omentum may extend as far as the pyloric area/duodenum or the ascending colon to form a colohepatic ligament[2] (▶ Fig. 12.12). On the left side, it extends along the greater curvature of the stomach to the fundus and occasionally up to the left anterior aspect of intra-abdominal esophagus and forms the gastrosplenic ligament. The gastrocolic ligament is the portion between the stomach and colon which forms part of the anterior wall of the lesser sac.

The omentum is composed of a core of trabecular fatty connective tissue that carries vessels sandwiched between two layers of mesothelial cells. The arterial supply is provided by the right and left gastroepiploic arteries arising from the gastroduodenal artery and splenic artery respectively (▶ Fig. 12.8). These two arteries anastomose and give off the gastric and epiploic branches (▶ Fig. 12.8). From 5 to 13 epiploic arteries originate from the right epiploic artery and only one main vessel from the left epiploic artery.[2] Occlusion of these arteries can cause omental infarction and fat necrosis (▶ Fig. 12.13). The venous drainage of the omentum parallels the arteries and empties into the portal system. The lymphatics of the right omentum empty to the subpyloric lymph nodes and the left omentum toward the splenic nodes.

True lymph nodes seem to be absent within the omentum. On the other hand, tiny rounded, 0.5 to 3.5 mm size, opaque bodies are seen in the omentum of young adults. These milky spots are omentum-associated lymphoid tissue and may help in antibacterial defense and local immunologic response.[8] Cancer cells, particularly in ovarian cancer, tend to lodge within the milky spots, take energy from fat cells of omentum and grow, and finally form a large mass known as omental cake visible on imaging studies (▶ Fig. 12.13).

For years, surgeons have used the greater omentum to protect suture lines from intestinal anastomosis, to cure vesicovaginal fistulas by interposition of the omentum, to revascularize areas of ischemic damage, to repair defects, and for reconstructive surgeries using omental flaps on its vascular pedicle.[2]

Peritoneal Ligaments of the Spleen

There are four principal ligaments of the spleen: the gastrosplenic ligament, the colicosplenic ligament, the phrenicocolic ligament, and the splenorenal (also known as phrenicosplenic)[9] (see Chapter 7 "Spleen"). The phrenicocolic ligament is a fold of peritoneum between the splenic flexure of the colon and the diaphragm at the level of the 10th and 11th ribs serving to support the spleen.

Gastrosplenic Ligament: The gastrosplenic ligament attaches the greater curvature of the stomach to the spleen and contains short gastric and left gastroepiploic vessels (▶ Fig. 12.14). This ligament develops from superior region of the ventral part of the

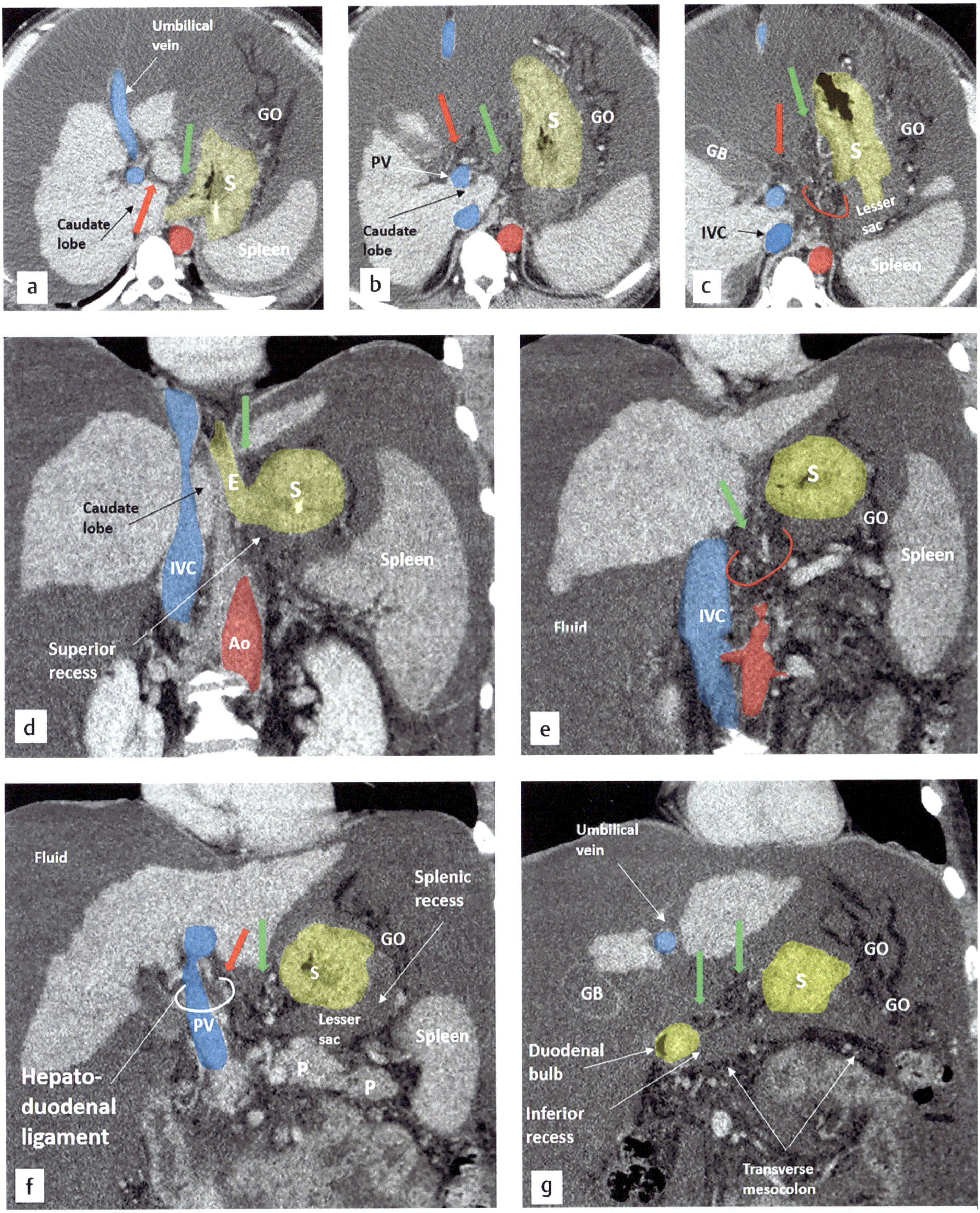

Fig. 12.10 Lesser omentum. **(a–c)** Axial computed tomography (CT). **(d–g)** Coronal CT from posterior to anterior. The two peritoneal layers of lesser omentum (*green arrows*) form the gastrohepatic (*red arrow* in **a**) and hepatoduodenal (*red arrows* in **b** and **c**) ligaments. Lesser sac recesses are shown. The gastropancreatic peritoneal fold containing fat, the left gastric artery, and the left gastric vein (coronary vein) is shown by *red rings*. The lesser sac is bordered by the stomach anteriorly, the caudate lobe of the liver medially, the pancreas posteriorly and inferiorly, the transverse mesocolon inferiorly, and the spleen laterally. Ao, aorta; E, esophagus; GB, gall bladder; GO, greater omentum; IVC, inferior vena cava; P, pancreas; PV, portal vein; S, stomach.

12

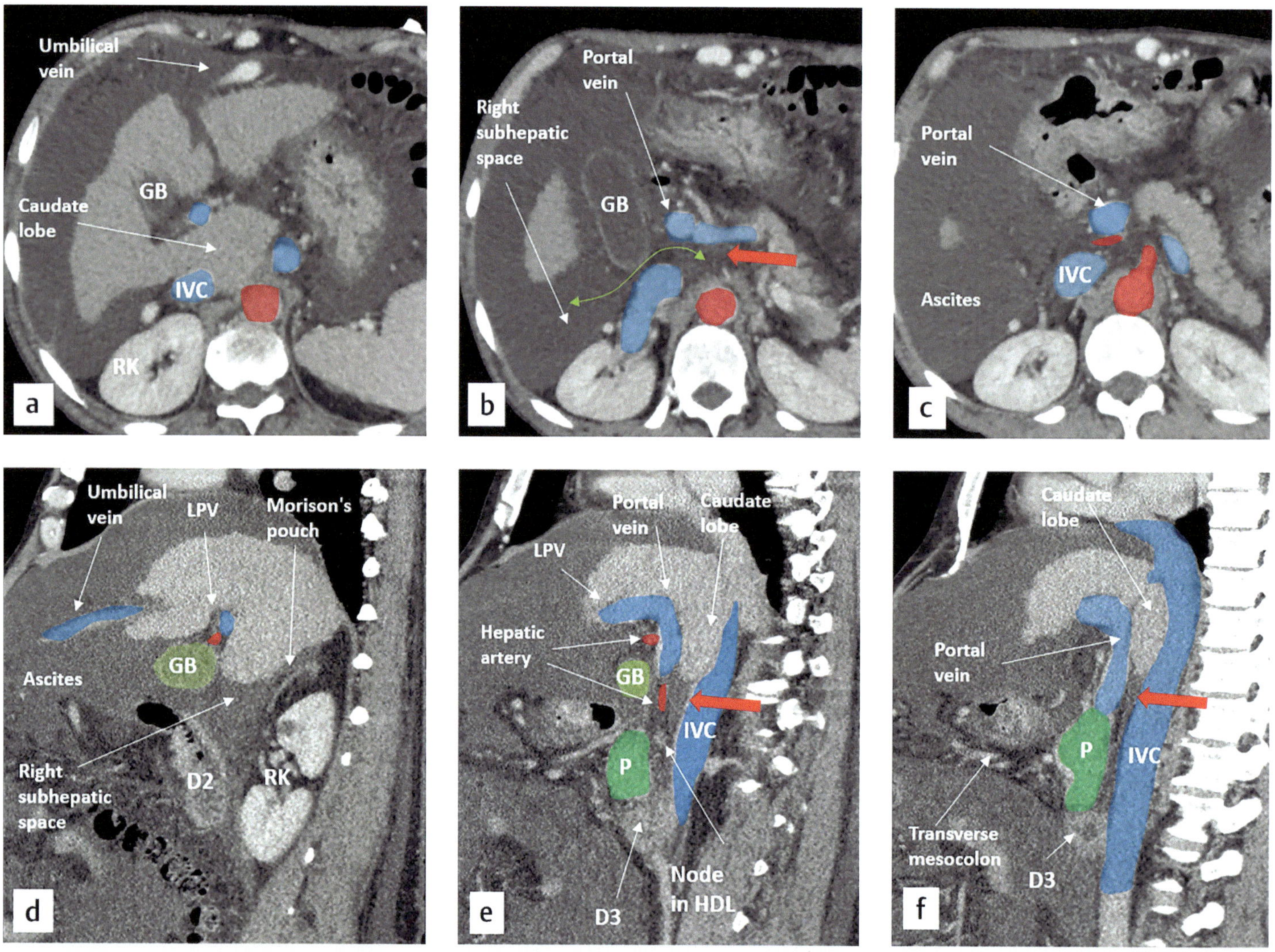

Fig. 12.11 Foramen of Winslow. **(a–c)** Axial computed tomography (CT). **(d–f)** Sagittal CT from right to left at the level of the foramen. The foramen (*red arrows*) is bordered by the portal vein in the hepatoduodenal ligament (HDL) anteriorly, the inferior vena cava posteriorly, the caudate lobe superiorly, and the origin of the hepatic artery, D2, and pancreatic head inferiorly. This region is surrounded by lymph nodes. D2, second duodenum; D3, third part duodenum; GB, gallbladder; IVC, inferior vena cava; LPV, left portal vein; p, pancreas; RK, right kidney.

dorsal mesentery. It forms a collateral route for venous flow in portal hypertension and in splenic vein thrombosis. Although it does not communicate with the peritoneal cavity, the fluid collection within this ligament is usually mistaken for fluid in the lesser sac.[4]

Splenorenal Ligament (Lienorenal Ligament): The splenorenal ligament is a peritoneal fold that extends from the left kidney to the hilum of the spleen. It develops from the most dorsal aspect of the dorsal mesentery. It contains the pancreatic tail and serves as a route for development of venous collaterals in patients with portal hypertension (▶ Fig. 12.14). Spontaneous splenorenal shunt between the splenic vein and its tributaries with the left renal vein develops through this ligament (▶ Fig. 7.11, Chapter 7 “Spleen”).

Mesenteries

According to new literature, the mesentery can now be classified as an organ due to its structural and functional features.[10] Traditionally described as multiple separate, somewhat fragmented structures, the mesentery is now considered one

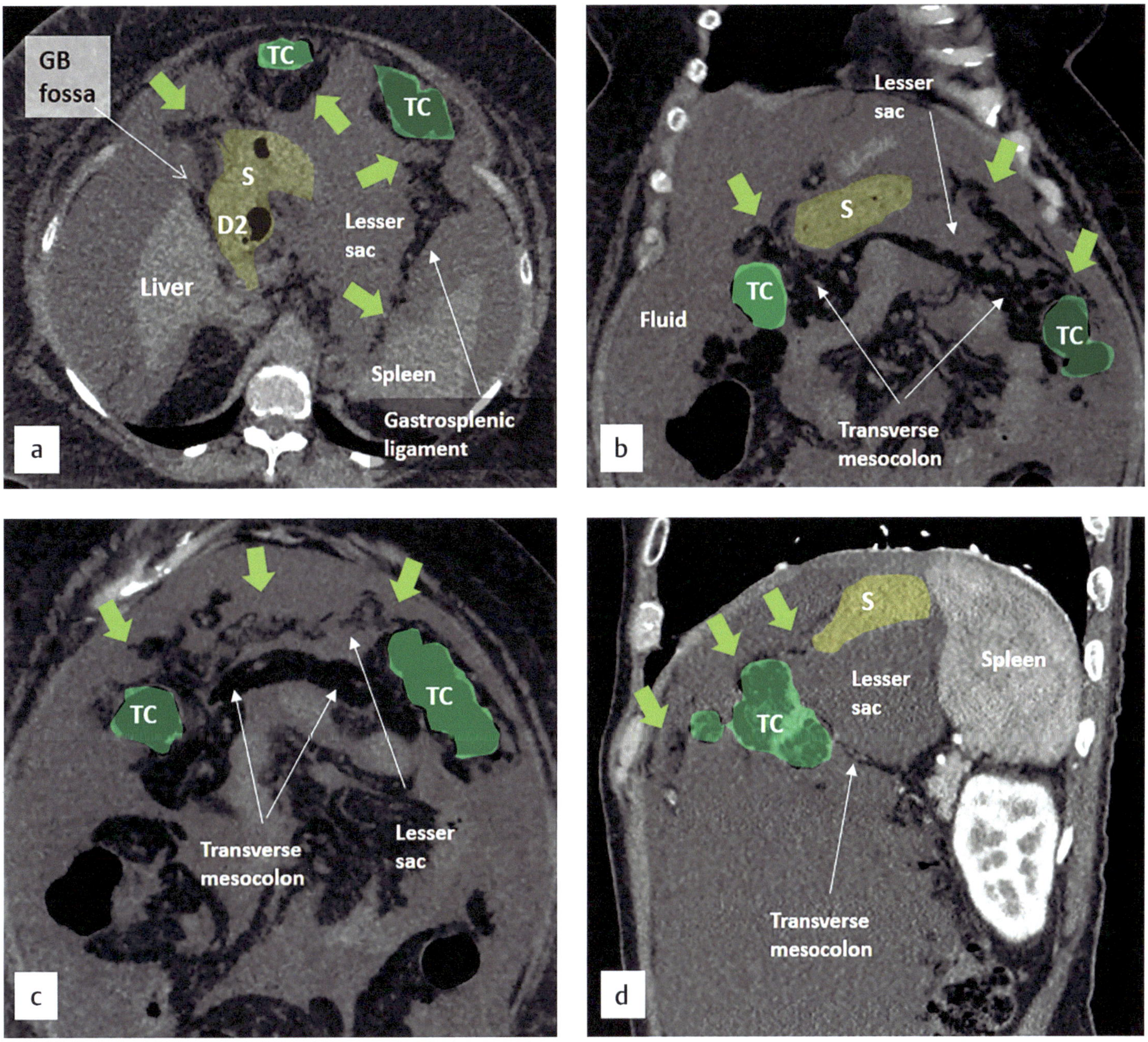

Fig. 12.12 Greater omentum shown in a patient with large ascites due to cirrhosis. Liver appears small and fibrotic. **(a)** Axial, **(b, c)** coronal, and **(d–f)** sagittal computed tomography (CT) showing the extent of the greater omentum between the two ends of the transverse colon (TC) and fusion with the transverse mesocolon at both ends (*arrows*). D2, second duodenum; GB, gallbladder; p, pancreas; S, stomach.

(Continued)

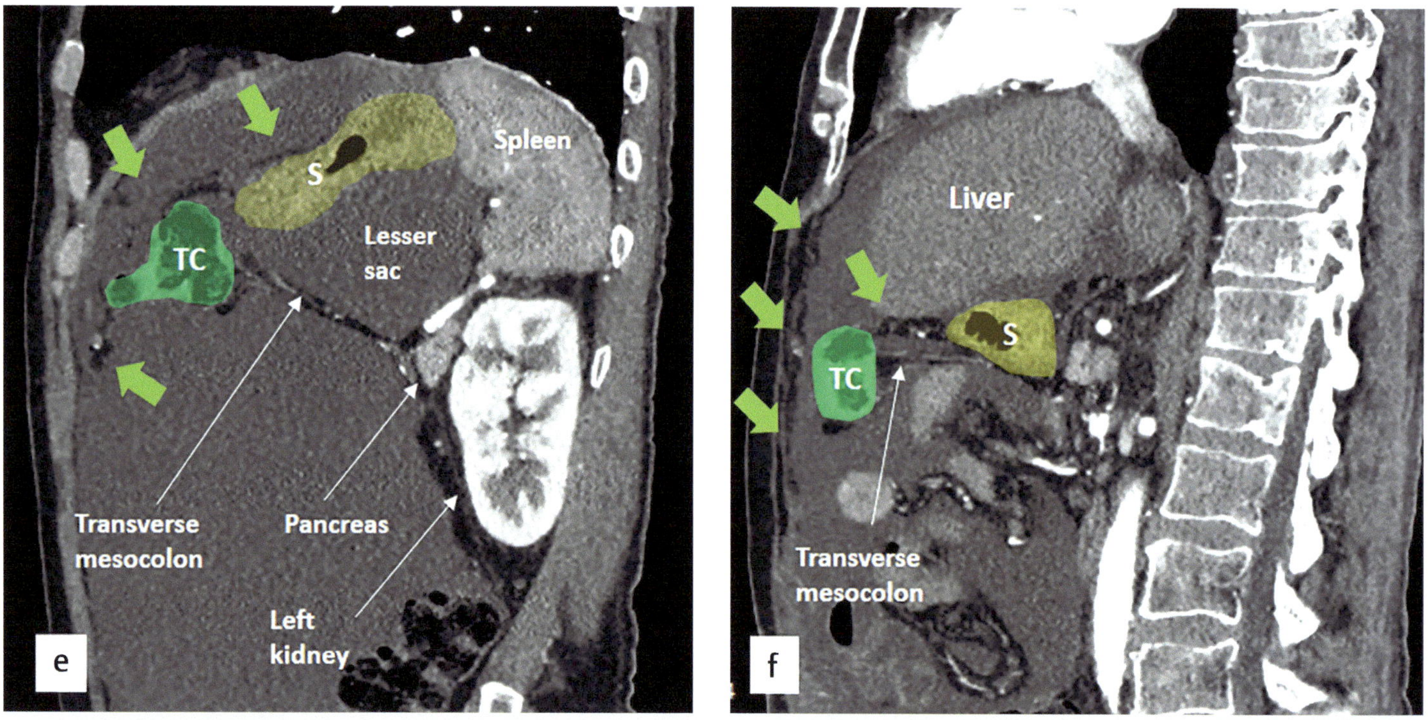

Fig. 12.12 (*Continued*) (e, f)

contiguous organ. In this chapter, the traditionally described individual components of the mesentery are described for purposes of anatomic juxtaposition.[11] The mesentery includes the transverse mesocolon, the small bowel mesentery (mesentery proper or mesenterium), and the sigmoid mesocolon. In some cases, small mesenteries are present for the cecum as well as the ascending and descending colon.

12

Transverse Mesocolon

The transverse mesocolon is a peritoneal fold that attaches the transverse colon, between the hepatic and splenic flexures, to the retroperitoneum and contains the middle colic vessels and their marginal branches[5] (▶Fig. 12.15). It is fused to the two posterior layers of the greater omentum.

The root of the transverse mesocolon crosses the descending duodenum (duodenocolic ligament), pancreatic head, and lower edge of the pancreatic body and tail (▶Fig. 12.4). It meets the root of the small bowel mesentery around the uncinate process of the pancreas. At the level of the anatomic splenic flexure, the transverse mesocolon meets the phrenicocolic and colicosplenic ligaments (▶Fig. 12.14).

It is a possible route for local extension of the duodenal, pancreatic head, colonic, and mesenteric root diseases. Pancreatic tumor extension into the mesocolon renders this cancer inoperable. The inflammatory process in severe pancreatitis may spread anteriorly to involve the transverse colon along its taenia mesocolica or laterally to involve the anatomic splenic flexure of colon.[5,12,13] Surgically created defects in the transverse mesocolon could be a route for internal hernias following Roux-en-Y gastric bypass surgery.

Small Bowel Mesentery

The small bowel mesentery is a peritoneal fold between the small bowel and the posterior abdominal wall. It spans approximately 15 cm from ligament of Treitz at the duodenojejunal flexure (near L2 level) to the ileocecal valve (▶Fig. 12.1, ▶Fig. 12.16). The root of mesentery is a bare area which is continuous posteriorly with the anterior pararenal space of the retroperitoneum and superiorly with the hepatoduodenal ligament and anteriorly with the transverse mesocolon[14] (▶Fig. 12.15, ▶Fig. 12.16). It contains the superior mesenteric vessels and their branches, nerves, lymphatic vessels, and nodes. The root of mesentery passes the aorta, the IVC, the right major psoas muscle, and the right ureter.

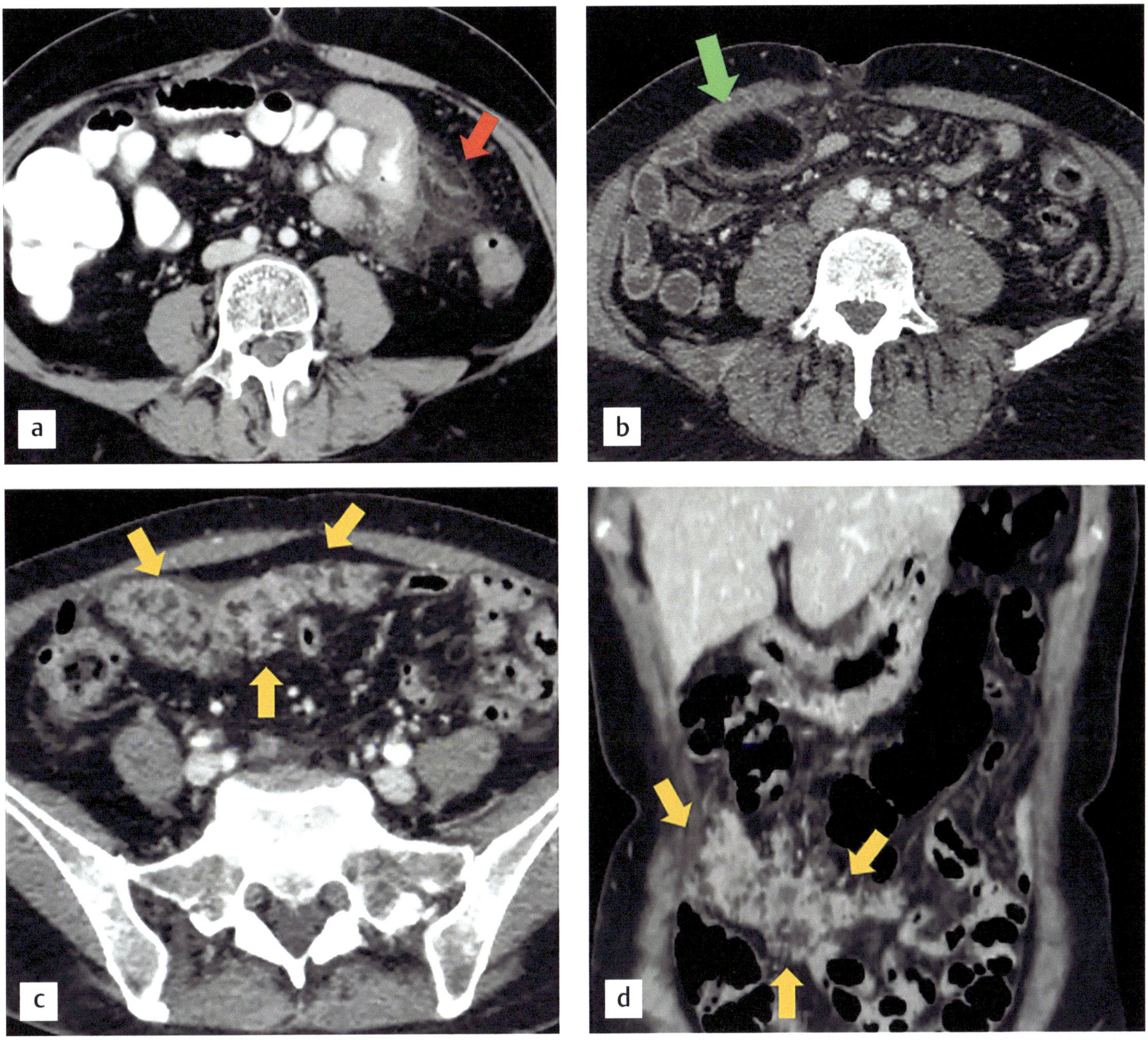

Fig. 12.13 **(a)** Omental infarct (*arrow*) in the left upper quadrant. **(b)** Omental fat necrosis (*arrow*) after surgery in the right upper quadrant. **(c)** Axial and **(d)** coronal computed tomography (CT) showing omental cake (*arrows*) containing multiple nodule due to ovarian peritoneal carcinomatosis. Cancer cells, particularly in ovarian cancer, tend to lodge within the milky spots of the omentum.

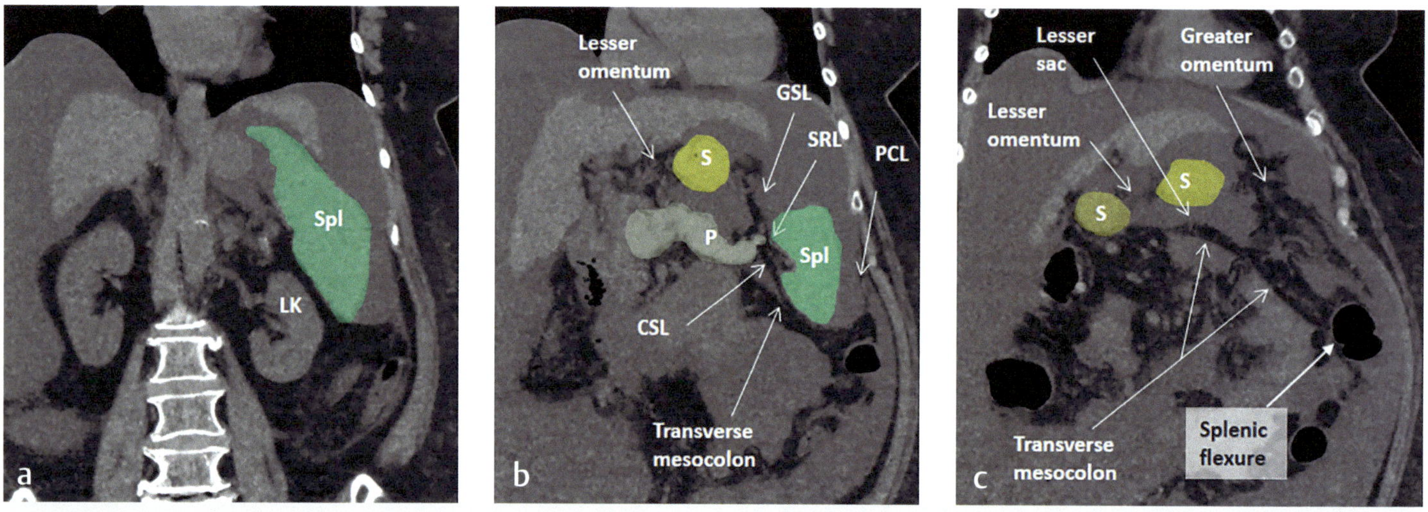

Fig. 12.14 Splenic ligaments. **(a–c)** Coronal computed tomography (CT) images from posterior to anterior are shown. There are four principal ligaments of the spleen: the gastrosplenic ligament (GSL), the colicosplenic ligament (CSL), the phrenicocolic ligament (PCL), and the splenorenal (SRL, also known as phrenicosplenic) ligament. LK, left kidney; P, pancreas; S, stomach; Spl, spleen.

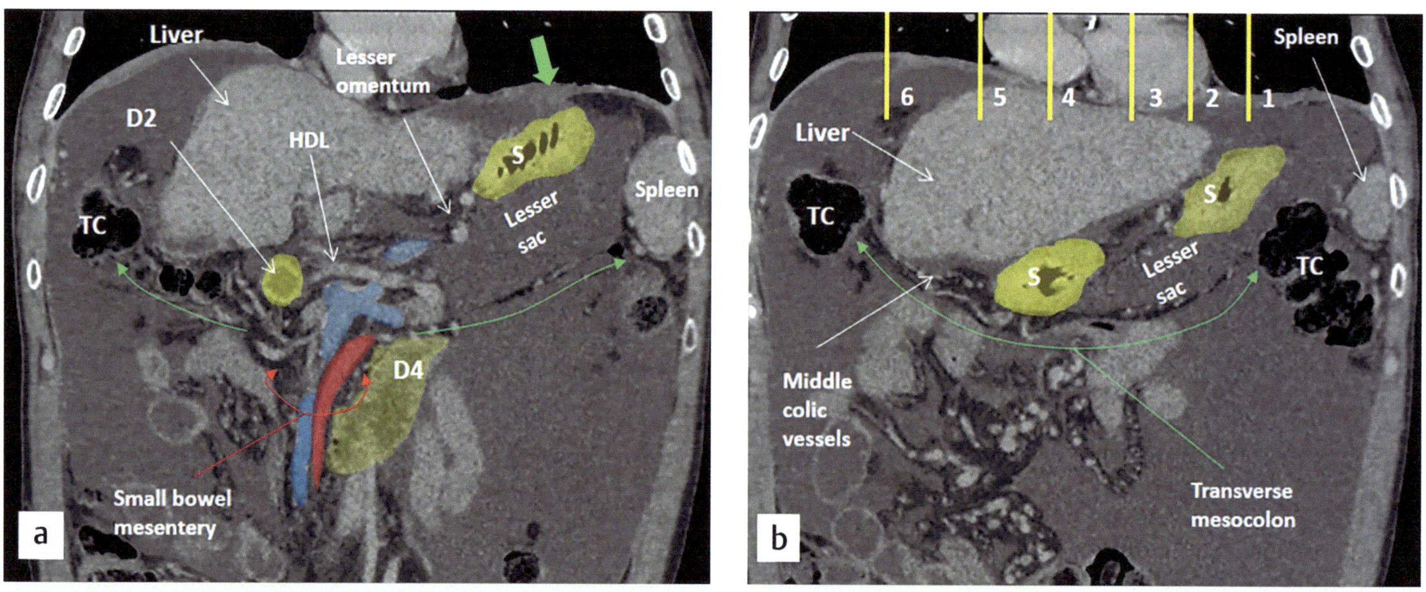

Fig. 12.15 Transverse mesocolon. **(a, b)** Coronal views. (1–6) Sagittal views from left to right as referenced on **b**. The transverse mesocolon extends between the hepatic and splenic flexures and contains the middle colic vessels and their marginal branches. It crosses the second part of duodenum (D2) and the pancreatic body and tail (P). In mid abdomen, it meets the root of the small bowel mesentery which continues along the hepatoduodenal ligament (HDL) toward the hepatic hilum incorporating all portal vein branches. The confluence of mesenteries is an important route for spread of inflammatory and malignant processes between the pancreas, transverse colon (TC), and porta hepatis. AC, ascending colon; D1–D4, first through fourth portions of duodenum; DC, descending colon; GB, gallbladder; LK, left kidney; P, pancreas; RK, right kidney; S, stomach; SRL, splenorenal.

(*Continued*)

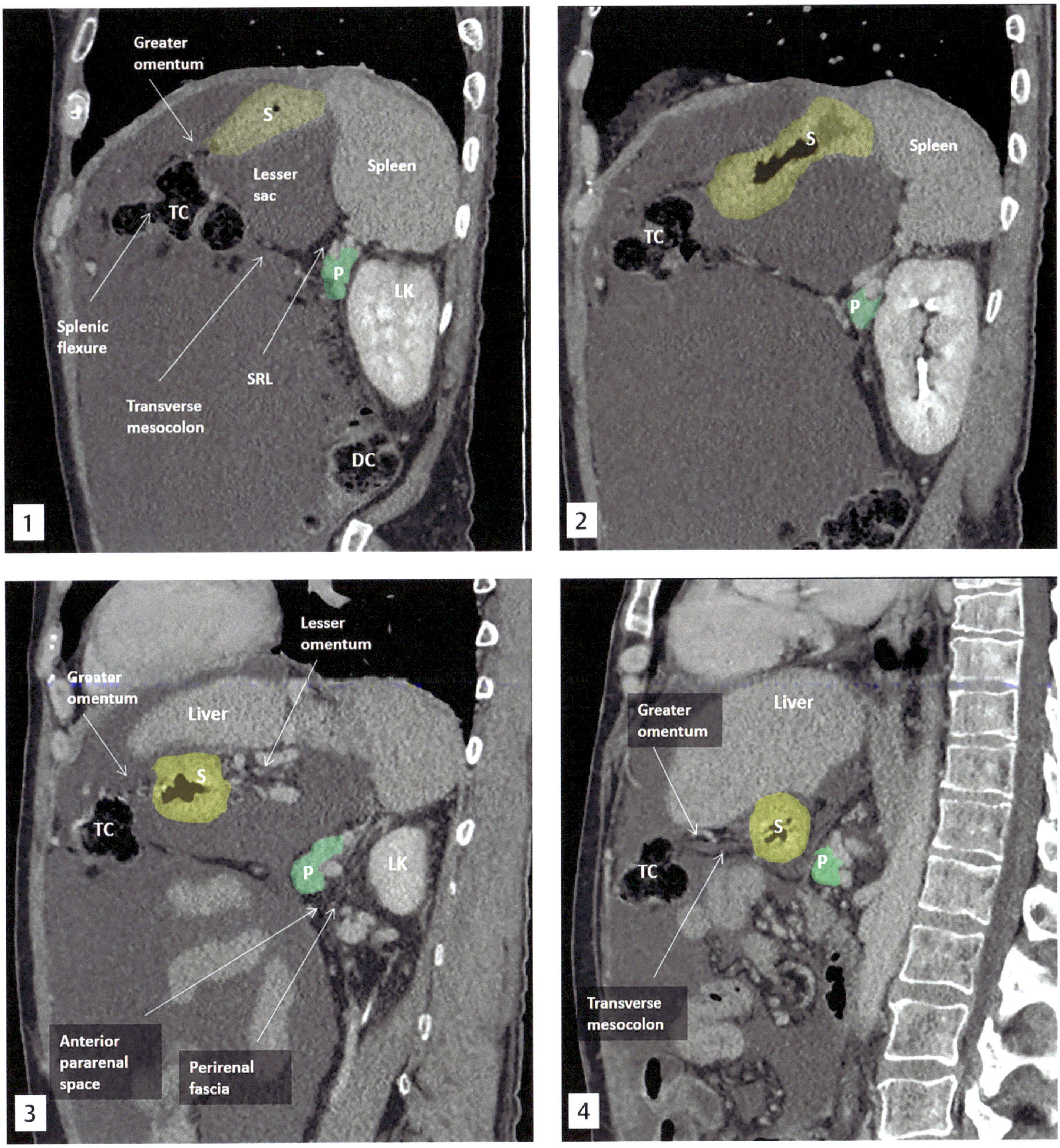

Fig. 12.15 (*Continued*)

(*Continued*)

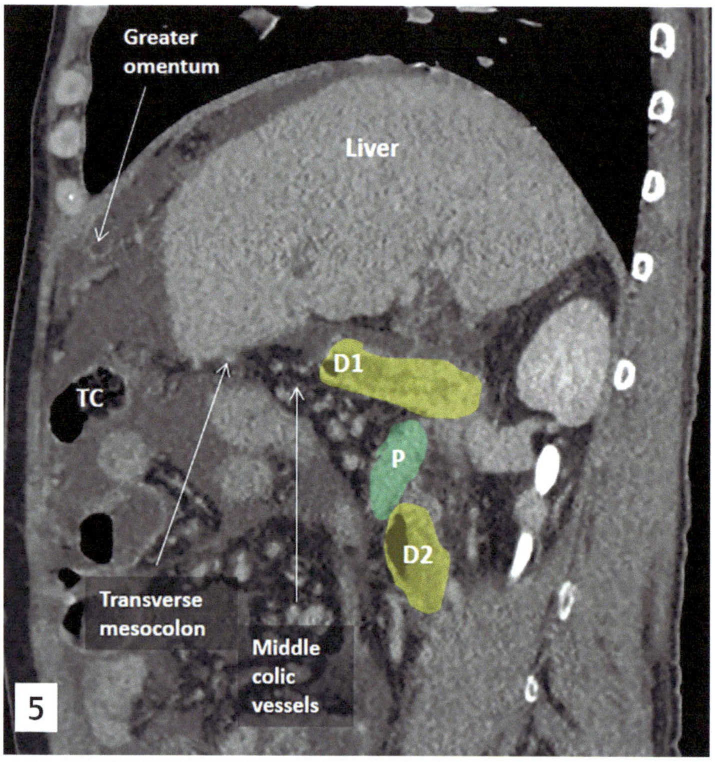

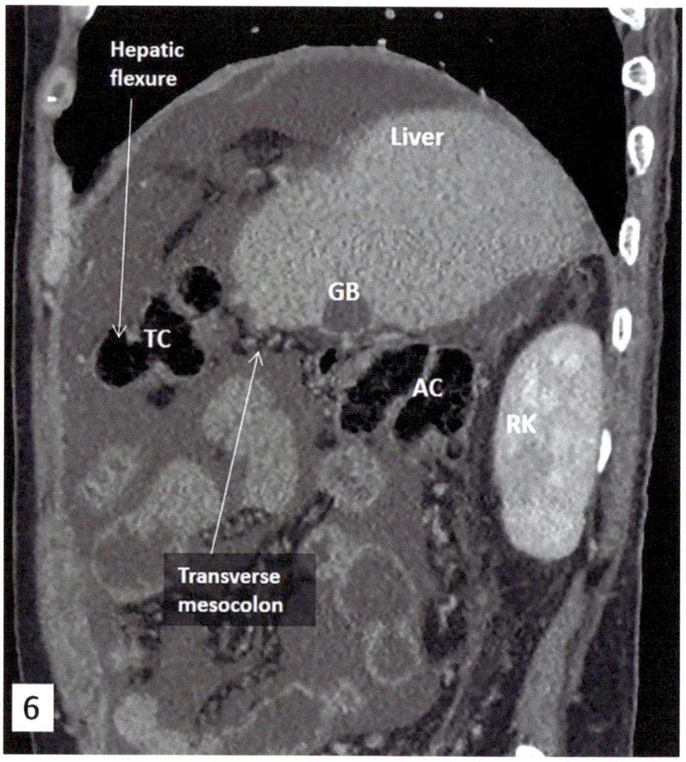

Fig. 12.15 (*Continued*)

The gastrocolic venous trunk is located at the junction of the transverse mesocolon and the root of the small bowel mesentery. This vein is formed by the union of the gastroepiploic, right colic, and anterior superior pancreaticoduodenal veins before entering the right side of the superior mesenteric vein. The celiac, superior mesenteric, and pancreatic capital plexuses are important neural element of the mesenteric root.[15]

The small bowel mesentery is common site of secondary involvement by metastatic and inflammatory/infectious conditions (i.e., pancreatitis), or direct tumor invasion. Pathological processes of the duodenum and ileocecal region can grow from one end to the other end of the mesenteric root.[14]

Cancer of the pancreatic body may spread directly from the retroperitoneum to the root of the small bowel mesentery, transverse mesocolon, hepatoduodenal ligament, and splenorenal ligament, and involve major vessels, lymphatics, and neural plexuses. Hepatobiliary and gastroduodenal cancers may also extend to the root of the small bowel mesentery (▶Fig. 12.17).

Sigmoid Mesocolon

The sigmoid mesocolon attaches the sigmoid colon to the posterior wall of the pelvis (at the level of the left common iliac artery bifurcation) and contains sigmoid and superior hemorrhoidal vessels (▶Fig. 12.18, ▶Fig. 12.19). It passes over the left ureter. The sigmoid mesocolon may be involved in pathologic processes such as diverticulitis, sigmoid colon cancer, or inflammatory bowel disease. Sigmoid mesocolon inflammatory and neoplastic process may extend to the pelvis or upper abdomen through the extraperitoneal planes (▶Fig. 12.18).

Mesocecum

At least the lower half of the cecum, if not the entire cecum, is invested by the peritoneum forming a retrocecal recess in the peritoneal cavity. A small mesentery connects the appendix to the cecum (see Chapter 5 "Colon"). The mesenteries of the cecum and appendix are in continuity with the small bowel mesentery where ileocolic vessels travel (▶Fig. 12.20, ▶Fig. 12.21). In retrocecal intraperitoneal variant, the appendix is located in the retrocecal recess (see Chapter 4 "Small Intestine" and Chapter 5 "Colon"). Inflammation of the appendix in this location may result in a retrocecal abscess which can remain contained in that location. A long mesentery predisposes the cecum to volvulus.

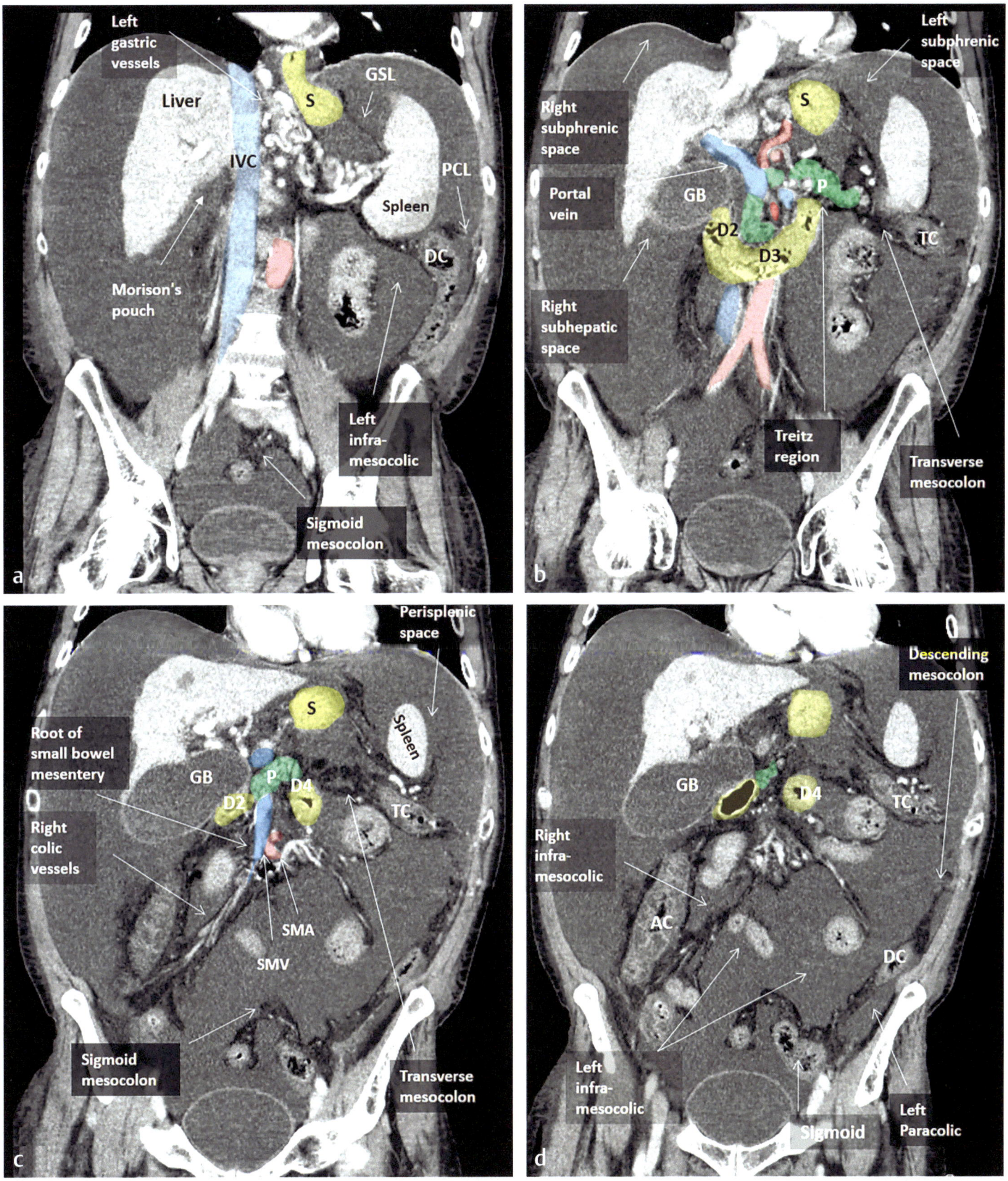

Fig. 12.16 (a–f) Coronal computed tomography (CT) views from posterior to anterior. The small bowel mesentery extends from ligament of Treitz to the ileocecal valve. The root of mesentery containing vessels and nerves continues posteriorly with the anterior pararenal space of the pancreas, superiorly with the hepatoduodenal ligament, and anteriorly with the transverse mesocolon. The small bowel mesentery with the transverse mesocolon divides the peritoneal space into three major compartments. AC, ascending colon; D1–D4, first through fourth portions of duodenum; DC, descending colon, GB, gallbladder; GSL, gastrosplenic ligament; IVC, inferior vena cava; P, pancreas; PCL, phrenicocolic ligament; S, stomach; SMA, superior mesenteric artery; SMV, superior mesenteric vein; TC, transverse colon.

(Continued)

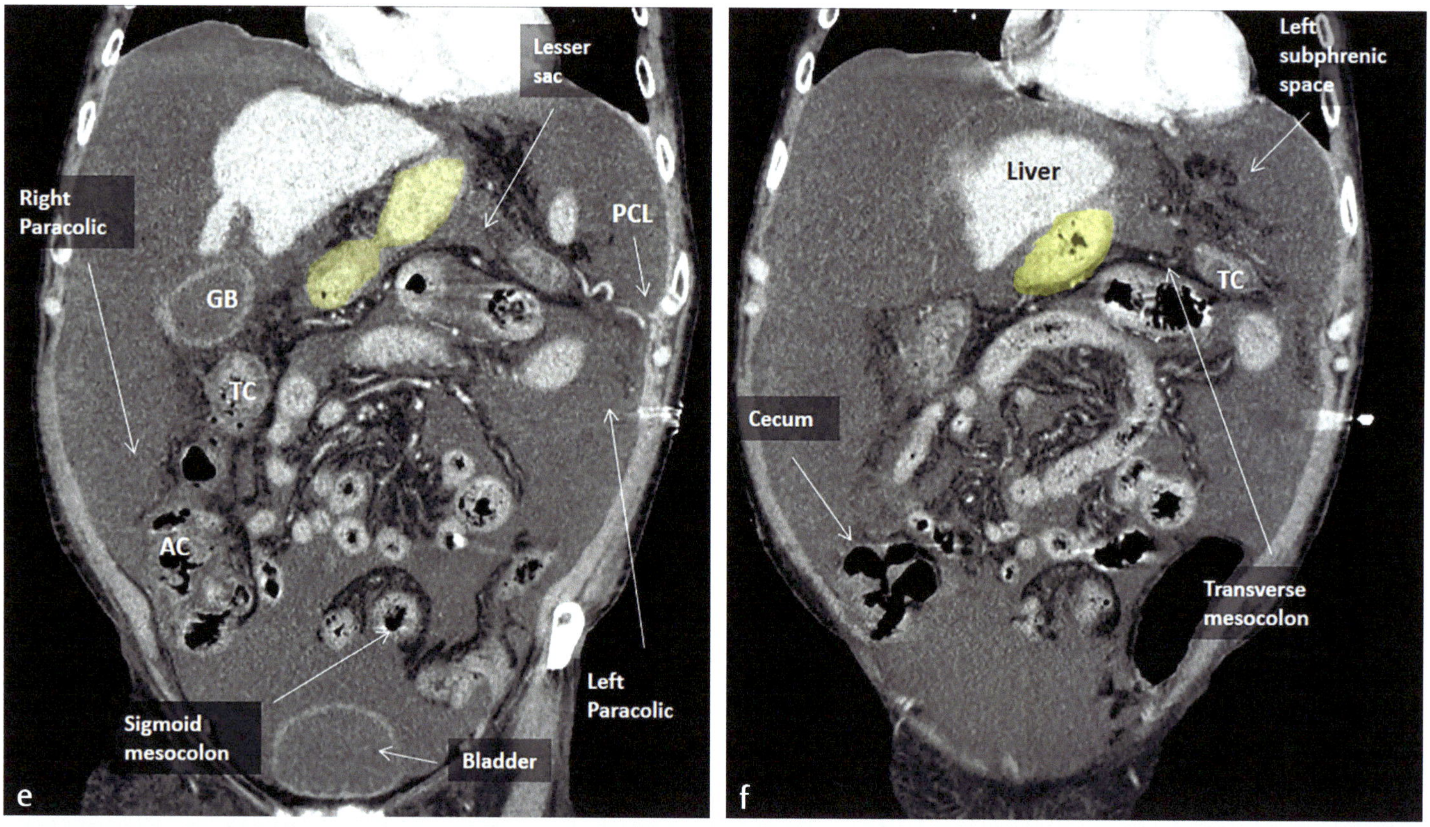

Fig. 12.16 *(Continued)*

◆ Peritoneal Spaces

12

The fold, mesenteries, and omenta divide the peritoneum into supramesocolic, inframesocolic, bilateral paracolic, and pelvic spaces[1,3] (▶Fig. 12.1, ▶Fig. 12.17, ▶Fig. 12.18). The transverse mesocolon separates the supramesocolic spaces from the inframesocolic spaces.[16] Small fluid accumulation in the peritoneal cavity is often limited to one or two spaces in the dependent parts of the peritoneum (i.e., ovarian cyst rupture, small laceration of the spleen). Large amount of fluid in severe ascites result in accumulation of fluid into multiple spaces (▶Fig. 12.17).

Supramesocolic Spaces

The area above the mesocolon is divided into the subphrenic spaces, subhepatic spaces, perisplenic space, and lesser omentum (▶Fig. 12.2, ▶Fig. 12.17).

The right triangular ligament separates the right subphrenic space from the right subhepatic space (▶Fig. 12.5). The left subhepatic space is localized between the inferior margin of the left liver lobe and the stomach (▶Fig. 12.10). The phrenicocolic ligament separates the perisplenic space from the left paracolic gutter (▶Fig. 12.14). The spread of pathologic processes from the left paracolic gutter to the left subphrenic space may be partially impeded by the phrenicocolic ligament.

Perihepatic spaces can communicate each other freely.[16] The posterior part of the right subhepatic space, also known as hepatorenal space or Morison's pouch, is the most gravity-dependent space in this region and a common site of fluid collections due to liver laceration (▶Fig. 12.2, ▶Fig. 12.11). The posterior right subhepatic space communicates freely with the right paracolic gutter.

The falciform ligament is the border between the right and the left subphrenic spaces (▶Fig. 12.1, ▶Fig. 12.10). The left subhepatic space also communicates freely with the left subphrenic and perisplenic spaces.

Lesser Sac

The lesser sac is unique peritoneal space that extends behind the stomach, anterior to the pancreas.[17] The lesser sac (omental bursa) is comprised of superior and inferior compartments that are separated by the gastropancreatic peritoneal fold containing the left gastric artery and left gastric vein (coronary vein) (▶Fig. 12.10, ▶Fig. 12.22). The superior compartment or recess is a small vertically oriented tubular shaped area bordered superiorly and laterally by the caudate lobe of the liver. Its posterior wall is only formed by the posterior parietal peritoneum (▶Fig. 12.22, ▶Fig. 12.23). The superior recess is difficult to see even in patients with large ascites. In large ascites, the superior peritoneal recess can herniate into the thorax through the hiatus of the diaphragm (▶Fig. 12.22, ▶Fig. 12.23). The inferior compartment is the larger part of the lesser sac. The inferior compartment is bounded between the stomach anteriorly, the body of pancreas

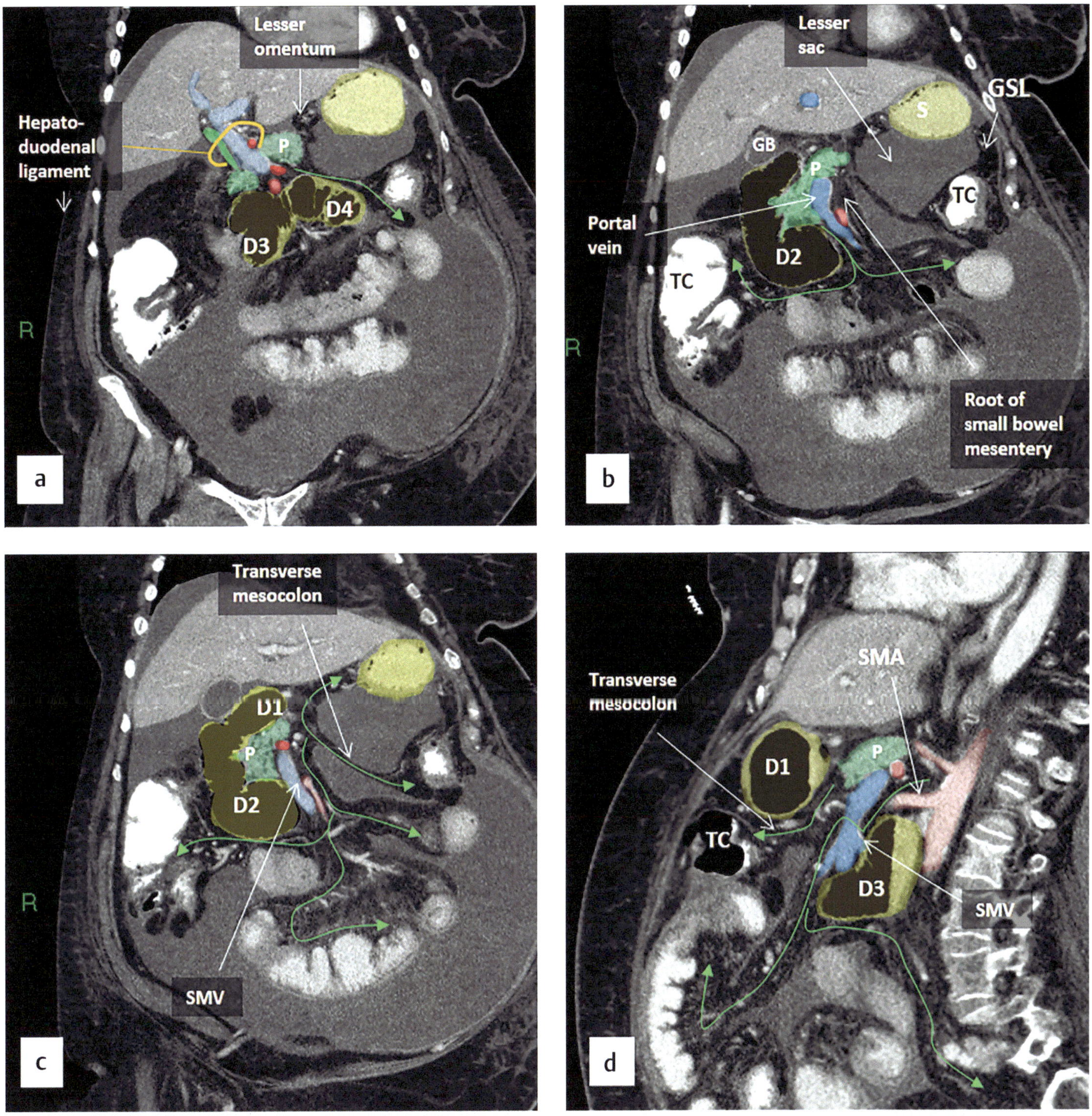

Fig. 12.17 Root of small bowel mesentery. **(a–c)** Coronal images from posterior to anterior. **(d)** Sagittal view of computed tomography (CT) scan. The root of mesentery is a subperitoneal tissue filled with fat, vessels, and lymphatics covered with the visceral peritoneum. It is continuous posteriorly with the anterior pararenal space of the retroperitoneum and superiorly with the hepatoduodenal ligament and anteriorly with the transverse mesocolon. Pathological processes of the duodenum, portal hepatis, pancreas, transverse colon, and ileocecal region can extend toward each other through this extensive subperitoneal pathway (*green arrows*). D1–D4, first through fourth portions of duodenum; GB, gallbladder; GSL, gastrosplenic ligament; p, pancreas; S, stomach; SMA, superior mesenteric artery; SMV, superior mesenteric vein; TC, transverse colon.

and the visceral side of spleen posteriorly, and the transverse mesocolon inferiorly (▶Fig. 12.10, ▶Fig. 12.17). The left side of the inferior compartment between the gastrosplenic and splenorenal ligaments is called the "*splenic recess.*" Very inferior part of the inferior compartment near the junction of the transverse mesocolon with the second part of the duodenum is called the "*inferior recess*" (▶Fig. 12.10). Anteriorly, the inferior recess may communicate with a potential space between the layers of the greater omentum.

The inferior recess of the lesser sac communicates with the right subhepatic space through the foramen of Winslow, or epiploic foramen. This foramen is bordered by the IVC posteriorly, the

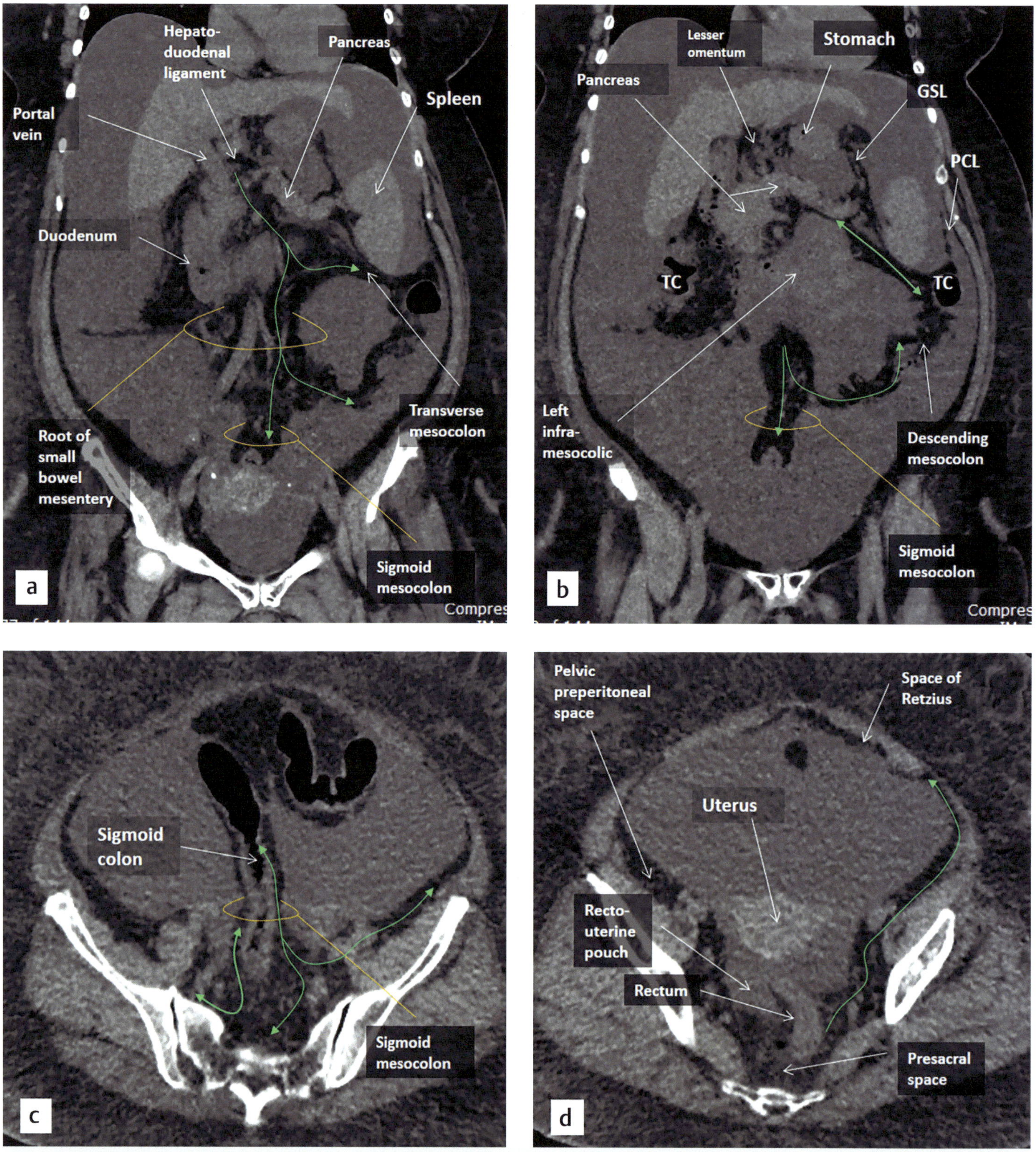

Fig. 12.18 Sigmoid mesocolon. **(a, b)** Coronal, **(c, d)** axial, and **(e, f)** sagittal noncontrast computed tomography (CT) in a patient with large ascites showing the mesenteries. The sigmoid mesocolon attaches the sigmoid colon to the posterior wall of the pelvis. All mesosigmoid is continuous with other mesenteries and with the extraperitoneal spaces, including the retroperitoneum (i.e., anterior pararenal space), preperitoneum (i.e., space of Retzius), and infraperitoneal pelvic (i.e., presacral) spaces. GSL, gastrosplenic ligament; PCL, phrenicocolic ligament; TC, transverse colon.

(*Continued*)

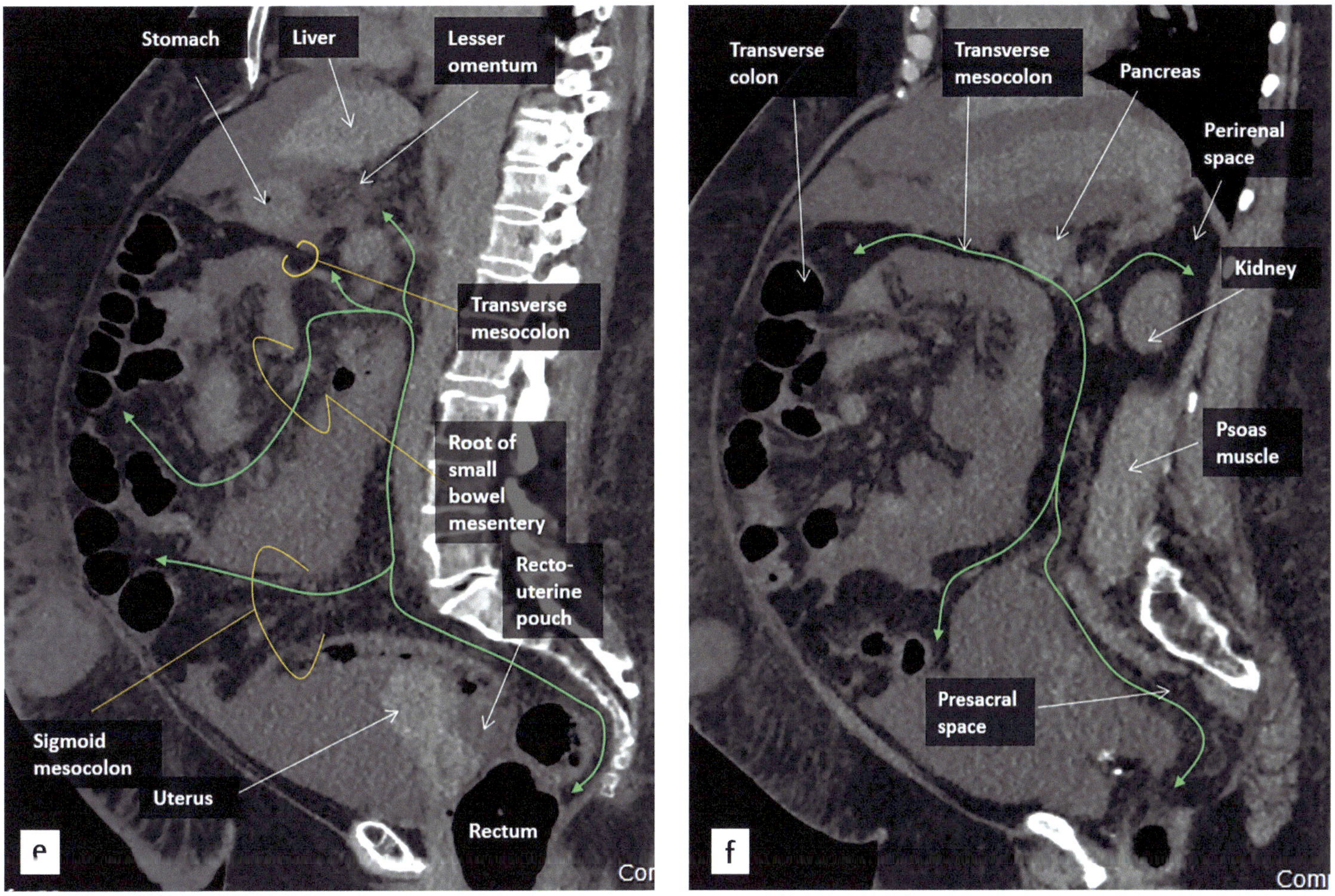

Fig. 12.18 (*Continued*) (e, f)

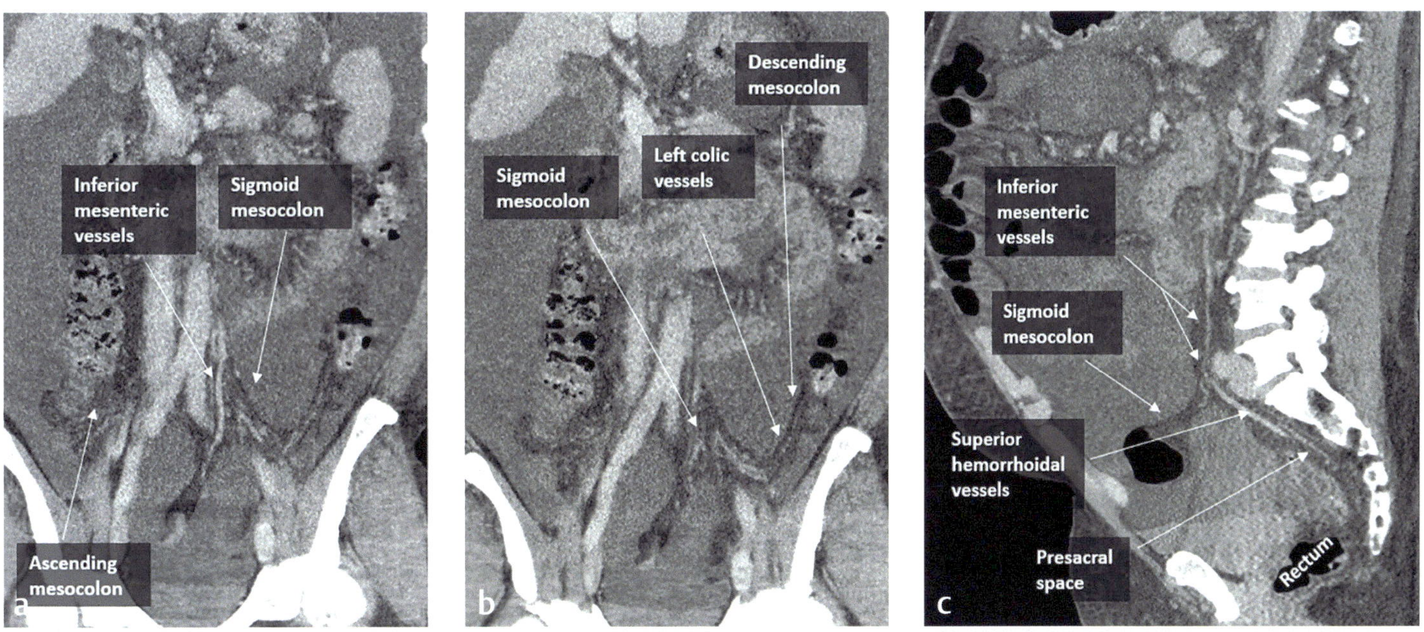

Fig. 12.19 Vessels of the sigmoid mesocolon. **(a, b)** Coronal and **(c)** sagittal contrast-enhanced computed tomography (CT) in a patient with large ascites showing the inferior mesenteric vessels traveling in the sigmoid mesocolon to reach the sigmoid. Branches of these vessels travel further into the descending mesocolon and mesorectum to reach the left colon and rectum, respectively.

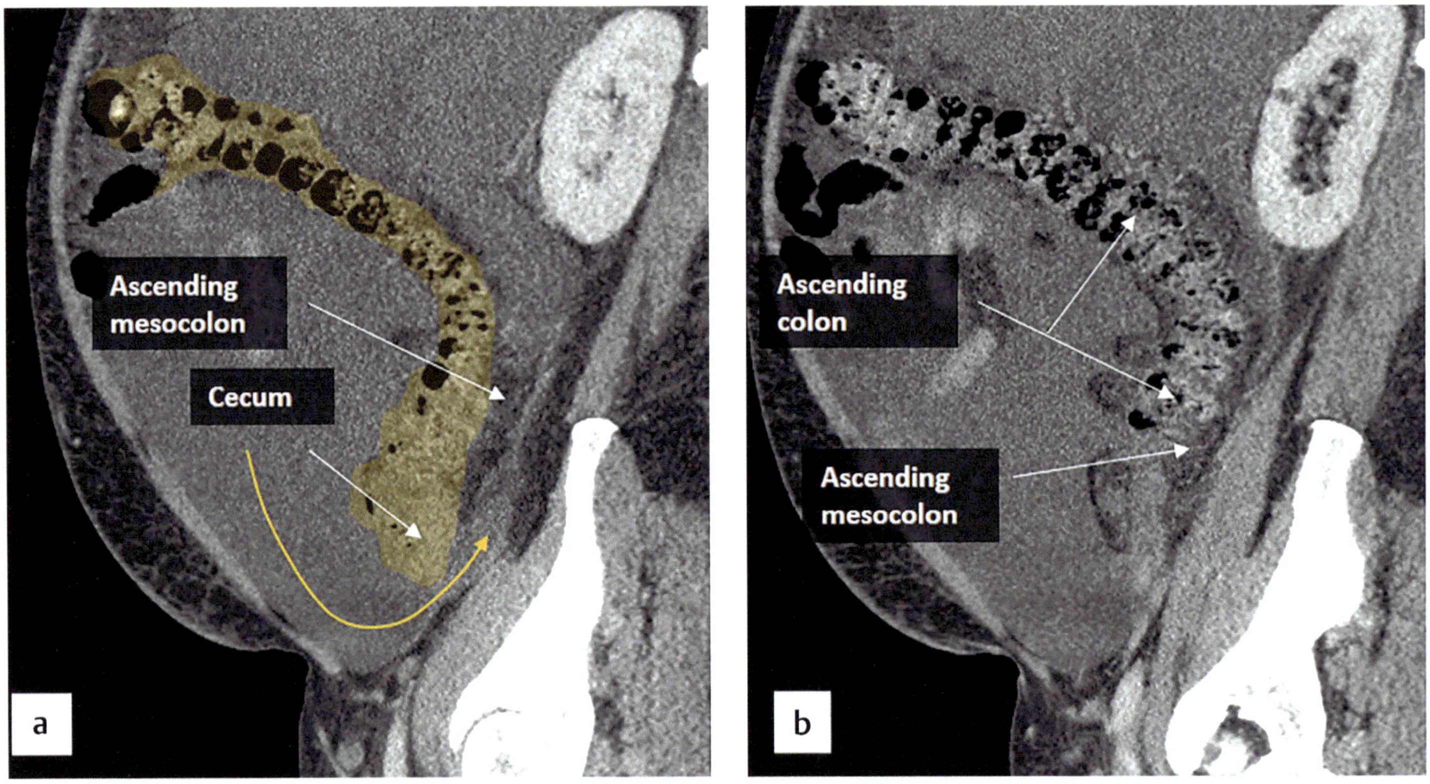

Fig. 12.20 Mesocecum. **(a, b)** Sagittal contrast-enhanced computed tomography (CT) in a patient with large ascites. The lower half of the cecum, if not the entire cecum, is invested by the peritoneum forming a retrocecal peritoneal recess (*orange arrow*), a potential site for internal hernias. The retrocecal recess is formed by the cecum anteriorly, the iliac fossa posteriorly, the right colic gutter laterally, and the mesentery medially.

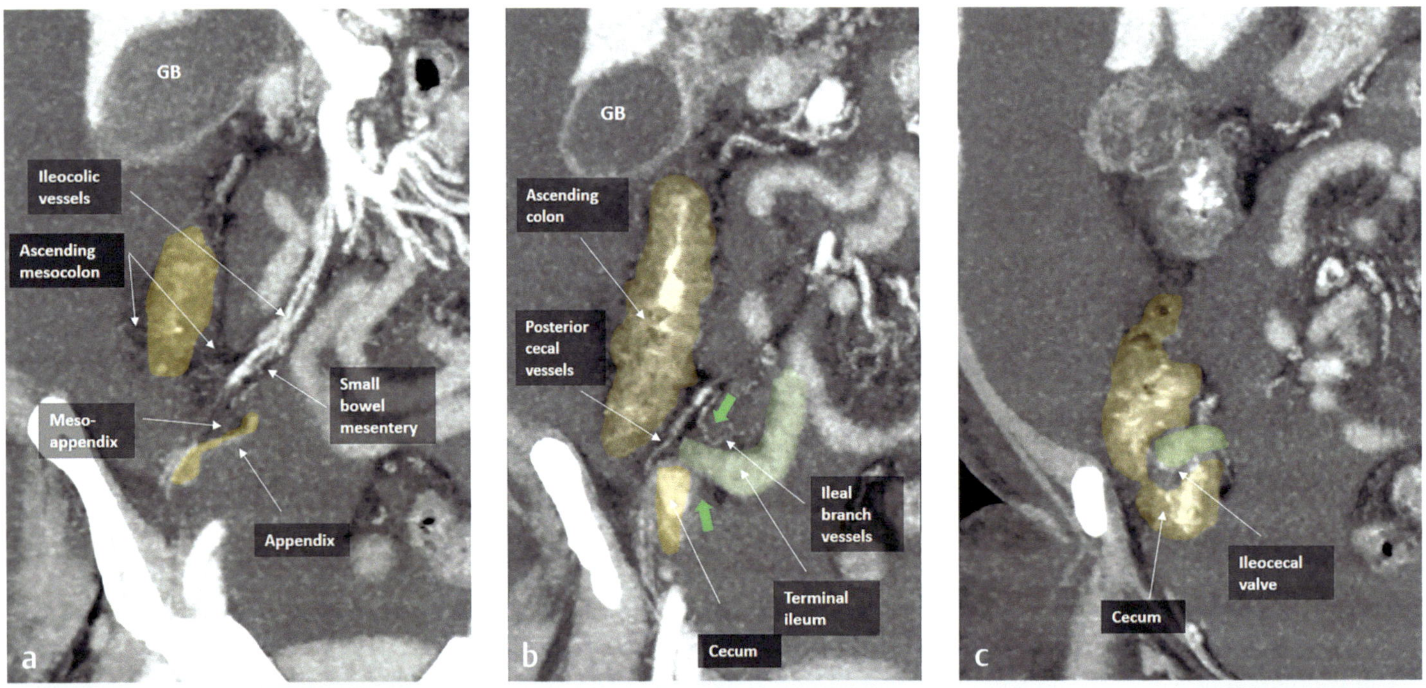

Fig. 12.21 Ileocecal region. **(a–c)** Coronal contrast-enhanced computed tomography (CT) from posterior to anterior in a patient with large ascites outlining fat-containing mesenteries. A small mesentery connects the appendix and cecum to the small bowel mesentery. Therefore, pathologies can extend from the cecum or appendix along the small bowel mesentery toward the third part of duodenum. Ileocecal recesses (superior and inferior) are shown by *green arrows*. Superior and inferior ileocecal recesses formed by a peritoneal fold originating from the terminal ileum to the cecum are potential sites for paracecal internal hernias. GB, gallbladder.

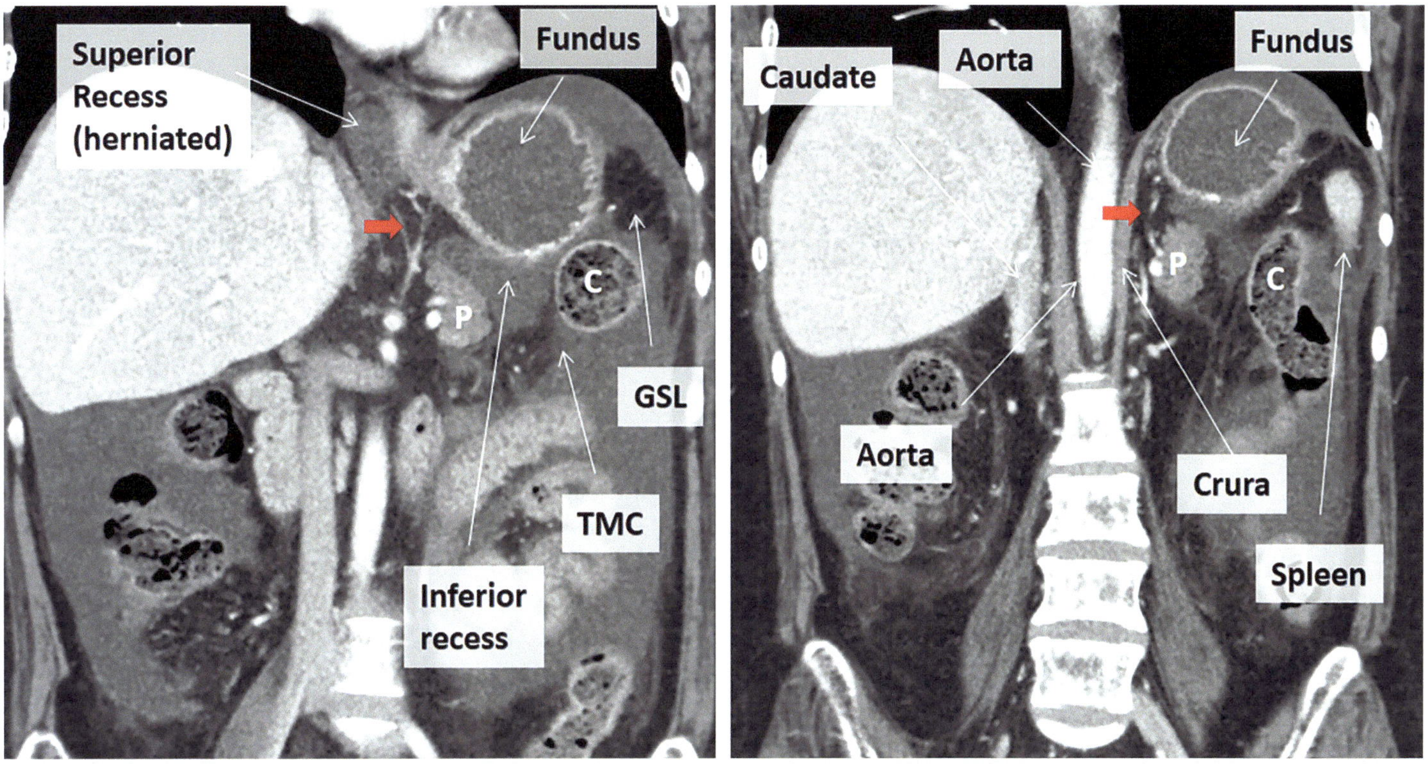

Fig. 12.22 Coronal images showing the gastropancreatic peritoneal fold containing left gastric artery and left gastric vein (*red arrows*) separates the superior from inferior lesser sac spaces. The superior recess is a small vertically oriented tubular shaped area bordered superiorly and laterally by the caudate lobe of the liver. Its posterior wall is only formed by the posterior parietal peritoneum at the level of the aorta and diaphragmatic crux. C, colon; GSL, gastrosplenic ligament; P, pancreas; TMC, transverse mesocolon.

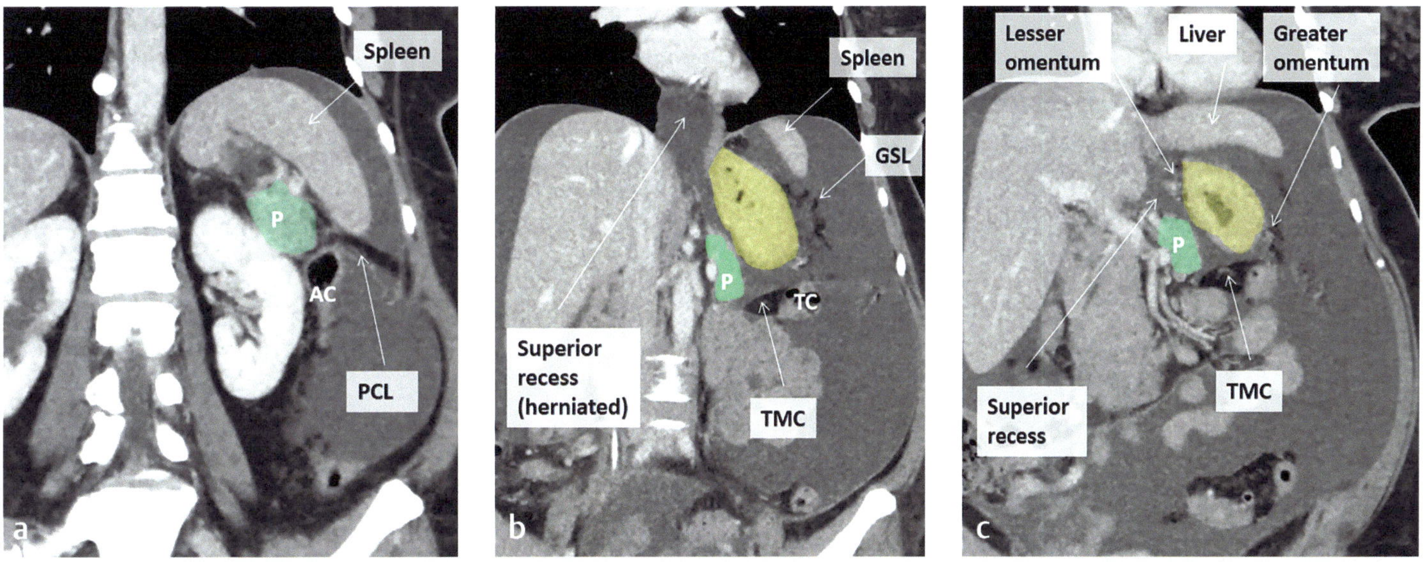

Fig. 12.23 (a–c) Coronal computed tomography (CT) images from posterior to anterior are shown. A hiatal hernia is seen containing peritoneal fluid of the lesser sac. AC, ascending colon; GSL, gastrosplenic ligament; P, pancreas; PCL, phrenicocolic ligament; TC, transverse colon; TMC, transverse mesocolon.

12

caudate lobe of the liver superiorly, the pars superior duodeni inferiorly, and the hepatoduodenal ligament anteriorly (▶ Fig. 12.11).

The lesser sac is collapsed at normal times, so only parts of its boundaries, such as the posterior gastric wall and pancreatic body, are observed on CT scans. The lesser sac is a common location for fluid collections in acute pancreatitis and perforated duodenal bulb ulcer (see Chapter 7 "Spleen," Fig. 7.19). Herniation of bowel into the lesser sac through foramen of Winslow is a rare form of internal hernia. The foramen of Winslow can be closed by inflammation of the peritoneum, enlarged regional lymph nodes, or tumor processes.

Inframesocolic Spaces

The inframesocolic cavity is divided into the right and left compartments by the root of small bowel mesentery which runs from the duodenojejunal flexure to the right lower quadrant, the ileocecal valve (▶ Fig. 12.1, ▶ Fig. 12.16).

The smaller right inframesocolic space is separated from the right paracolic gutter by the ascending colon and is bordered inferiorly by the attachment of small bowel mesentery to the cecum. The fluid collections in this space may not extend to the pelvis.[18] The larger left inframesocolic space is separated from the left paracolic gutter by the descending colon and communicates freely with the pelvis (▶ Fig. 12.16). Both spaces communicate at the duodenojejunal flexure.

Paracolic Spaces (Gutters)

The right and the left paracolic gutters are located lateral to the peritoneal reflections of the ascending and descending colon, respectively (▶ Fig. 12.1, ▶ Fig. 12.16). The right paracolic gutter communicates superiorly with the right subphrenic space. On the left side, the phrenicocolic ligament limits the left paracolic gutter connection with the left subphrenic and left subhepatic spaces[18] (▶ Fig. 12.16). However, both paracolic gutters communicate inferiorly with the pelvis.

◆ Peritoneal Recesses

Peritoneal recesses are formed by peritoneal folds caused by incomplete adhesion between the posterior parietal peritoneum and the posterior abdominal wall where vessels pass. These are formed during the 5th and the 11th weeks of gestation. Most of these recesses are around duodenum and cecum (see Chapter 4 "Small Intestine") (▶ Fig. 12.21). These recesses are of surgical importance since they may become the site of internal hernia, where a segment of intestine enters a recess and gets incarcerated or even strangulated by the peritoneal fold guarding the entrance to the recess.

Duodenal Recesses

Duodenal recesses (fossas) are formed by incomplete adhesion of the Treitz's fascia to the posterior abdominal wall during embryonic development. The duodenal recesses include the inferior (50–70%), superior (30–50%), duodenojejunal (18%), paraduodenal (2–12%), mesentericoparietal (1–3%), and retroduodenal (1–3%) recesses (▶ Fig. 12.24). Some believe that duodenal recesses are in fact serosal folds formed by the inferior mesenteric vein. Some have suggested that the inferior duodenal fold is a remnant of the mesoduodenum.[19–21]

These recesses are common sites for internal abdominal hernias (▶ Fig. 12.25). Paraduodenal hernias account for <1% of small bowel obstructions. A left side hernia is more common accounting for 75% of all hernia cases.

The inferior and superior duodenal recesses are located on the left side of the fourth part of duodenum bounded anteriorly by a duodenal fold. Their depth and width are 3 to 5 cm and 1 to 3 cm, respectively.

The paraduodenal recess (recess venosus, fossa of Landzert) is located on the left of the fourth part of the duodenum, posterior to the inferior mesenteric vein and ascending branch of the left colic artery. The inferior mesenteric vein passes behind the third portion of the duodenum before entering the splenic vein or superior mesenteric vein. A diagnostic sign of left paraduodenal hernia is displacement of the inferior mesenteric vein anteriorly (▶ Fig. 12.25).

Retroduodenal recess is located behind the third and fourth parts of duodenum and anterior to the aorta. Duodenojejunal recess is located between the duodenojejunal flexure and root of transverse mesocolon. Mesentericoparietal recess (of Waldeyer) is seen in the first part of the mesojejunum immediately behind the superior mesenteric artery, and just below the third part of the duodenum. Internal hernia through the duodenojejunal recess is rare (▶ Fig. 12.24).

◆ Pelvic Spaces

The anterior aspect of the lower abdomen and pelvic cavities are divided into several fossae by the anatomic course of the median umbilical fold (the obliterated urachus), medial umbilical folds (partially obliterated umbilical arteries), and inferior epigastric vessels (see Chapter 1: Abdominopelvic Wall).

Between the medial and median umbilical folds, the supravesical fossae are formed. Lateral to the medial umbilical folds, the inferior epigastric artery divides the anterior pelvis on each side

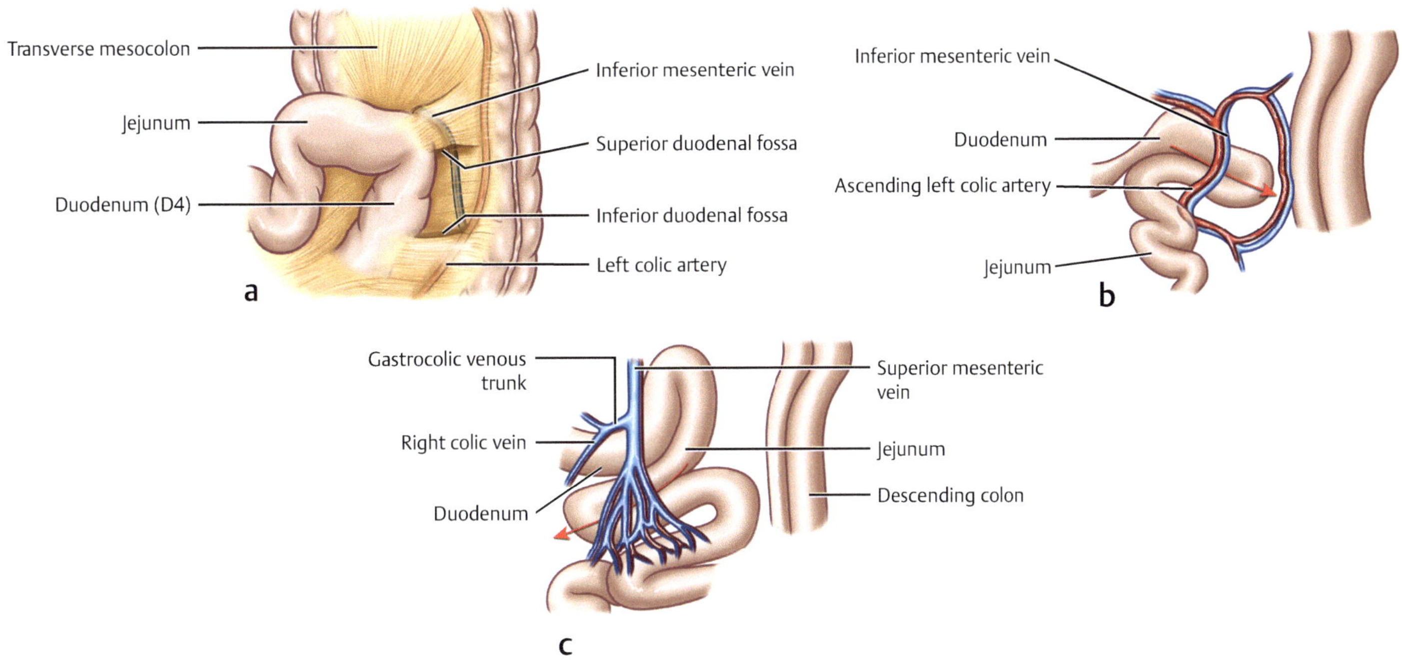

Fig. 12.24 Duodenal recesses. **(a)** The inferior and superior duodenal recesses are shown located on the left side of the fourth part of duodenum (D4) bounded anteriorly by a duodenal fold. **(b)** Left paraduodenal hernia. The jejunum enters the hernia sac behind the descending mesocolon through a left paraduodenal recess behind the inferior mesenteric vein. **(c)** Right paraduodenal hernia. The jejunum enters the mesentericoparietal fossa of Waldeyer (*arrow*). The gastrocolic venous trunk is located at the junction of the transverse mesocolon and the root of the small bowel mesentery. This vein is formed by the union of the gastroepiploic vein, the right colic vein, and the anterior superior pancreaticoduodenal veins before entering the right side of the superior mesenteric vein.

into the lateral and medial inguinal fossae. These fossae are the sites of indirect (lateral to inferior epigastric vessels) and direct (medial to inferior epigastric vessels) inguinal hernias, respectively. Lateral to the bladder on each side, the peritoneum forms a cavity called the paravesical fossa. Lateral to the rectum on each side, the pararectal fossae are formed by the peritoneal reflection. The most gravity-dependent site for fluid accumulation in men is the rectovesical pouch/space. In women, the uterus divides the rectovesical space into an anterior vesicouterine recess and a posterior rectouterine recess. The retrouterine recess, or pouch of Douglas, is the deepest (▶Fig. 12.2). When large amount of fluid accumulates in the pelvic cavity, it can spread cranially along the paracolic gutter into the subhepatic and subphrenic spaces.

◆ Pathology

The peritoneum is a common path for the spread of disease between intra-abdominal and intrapelvic organs. CT scan provides more complete evaluation of the peritoneal cavity than magnetic resonance imaging (MRI) or sonography.

Peritoneal Fluid and Air

Free intraperitoneal fluid or ascites can be caused by cirrhosis in 75% of cases, peritoneal carcinomatosis in 10%, and cardiac failure in 5%. Accumulation of blood, urine, chyle, bile, or pancreatic juice is less common. A spontaneous pneumoperitoneum is often a result of perforation of an intraperitoneal hollow viscus most commonly seen after duodenal ulcer perforation. CT is the best technique for detection and localization of a small pneumoperitoneum.

Localized Peritoneal Fluid Collection

Peritonitis is defined as diffuse inflammation of the serosal membrane. Localized peritoneal fluid collection is usually due to infection and less commonly blood. Infected peritoneal fluid will be encapsulated and form an abscess by fibrinous exudate, omentum, or the adjacent visceral organs or diaphragm. The most common sites for abscess formation are the pouch of Douglas and the subphrenic and subhepatic

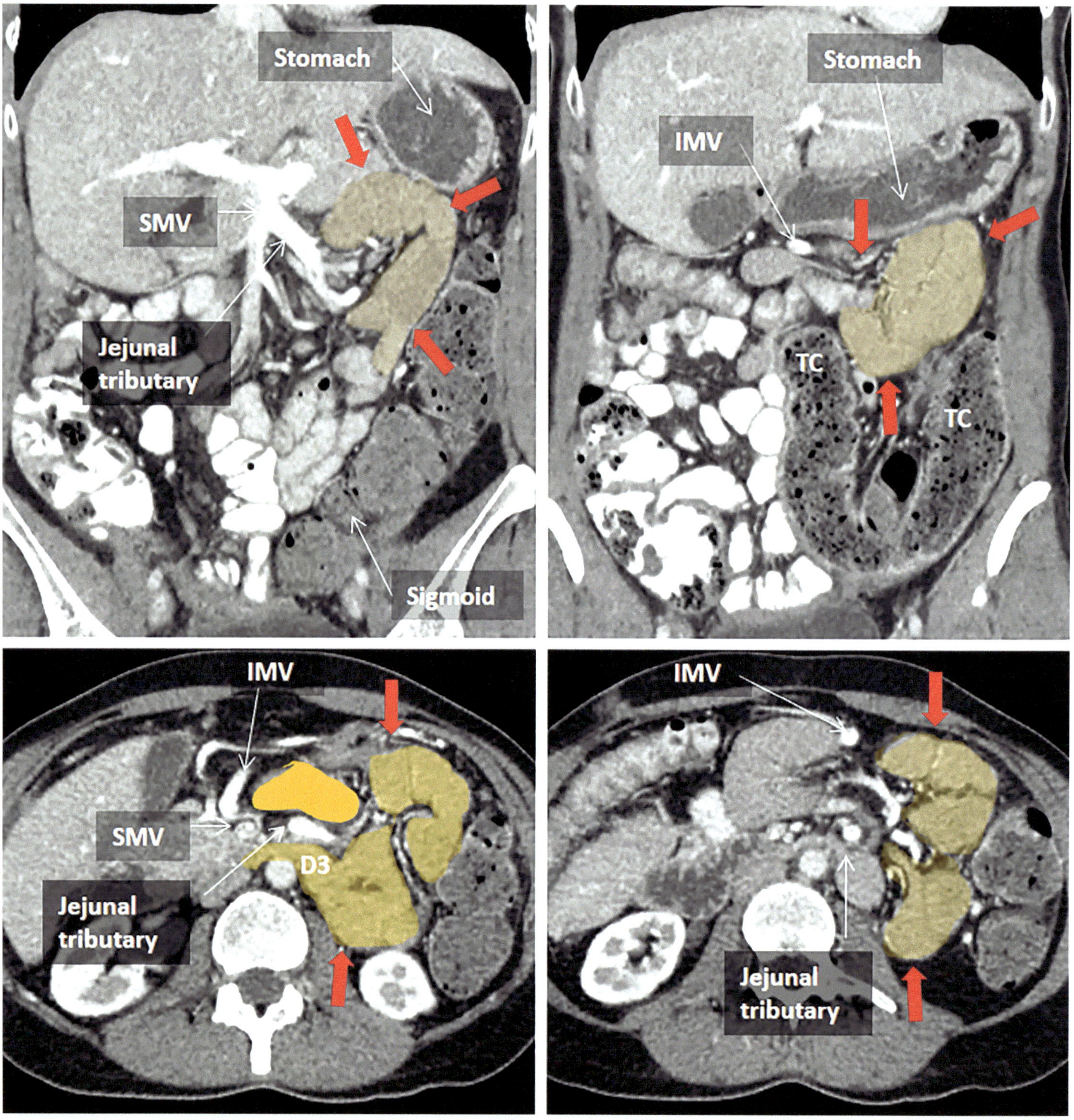

Fig. 12.25 Left paraduodenal hernia of proximal jejunal loops. The inferior mesenteric vein (IMV) was displaced anteriorly by the hernia sac (*red arrows*) and finally drained into the superior mesenteric vein (SMV). The hernia sac resides in the transverse mesocolon with the transverse colon staying inferior to the herniated loops. D3, third part of duodenum; TC, transverse colon.

spaces (▶ Fig. 12.27). Peritoneal abscess formation is most frequently seen after surgery. It may also complicate appendicitis, Crohn's disease, or sigmoid diverticulitis. Both ultrasound and CT have significant roles to play in diagnosis, needle aspiration, and percutaneous drainage of abscesses. A described earlier, inflammatory processes may travel from one location to a remote area along mesenteric and extraperitoneal planes. An example of pancreatic infection extending toward the mesenteric root, right lower quadrant, and presacral space along subperitoneal fat planes is shown in ▶ Fig. 12.28.

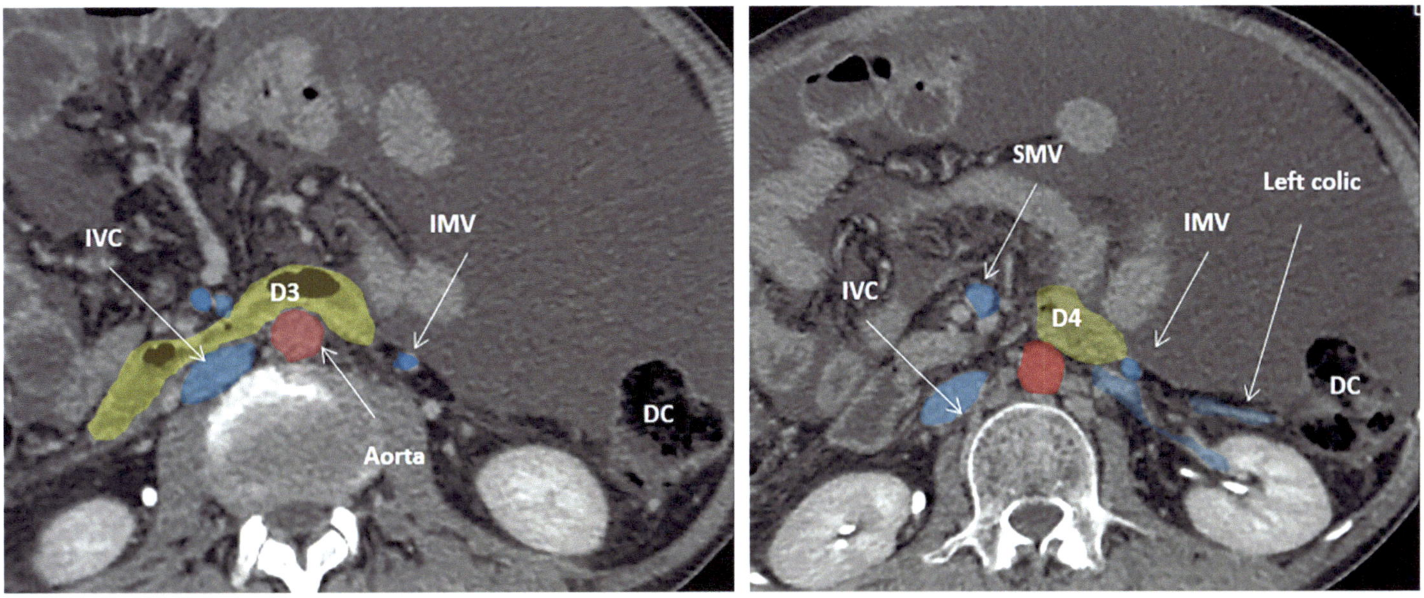

Fig. 12.26 Relationship of the third (D3) and fourth (D4) parts of the duodenum with the inferior mesenteric vein (IMV). The IMV passes behind the D3 before entering the splenic vein or SMV. DC, descending colon; IVC, inferior vena cava; SMV, superior mesenteric vein.

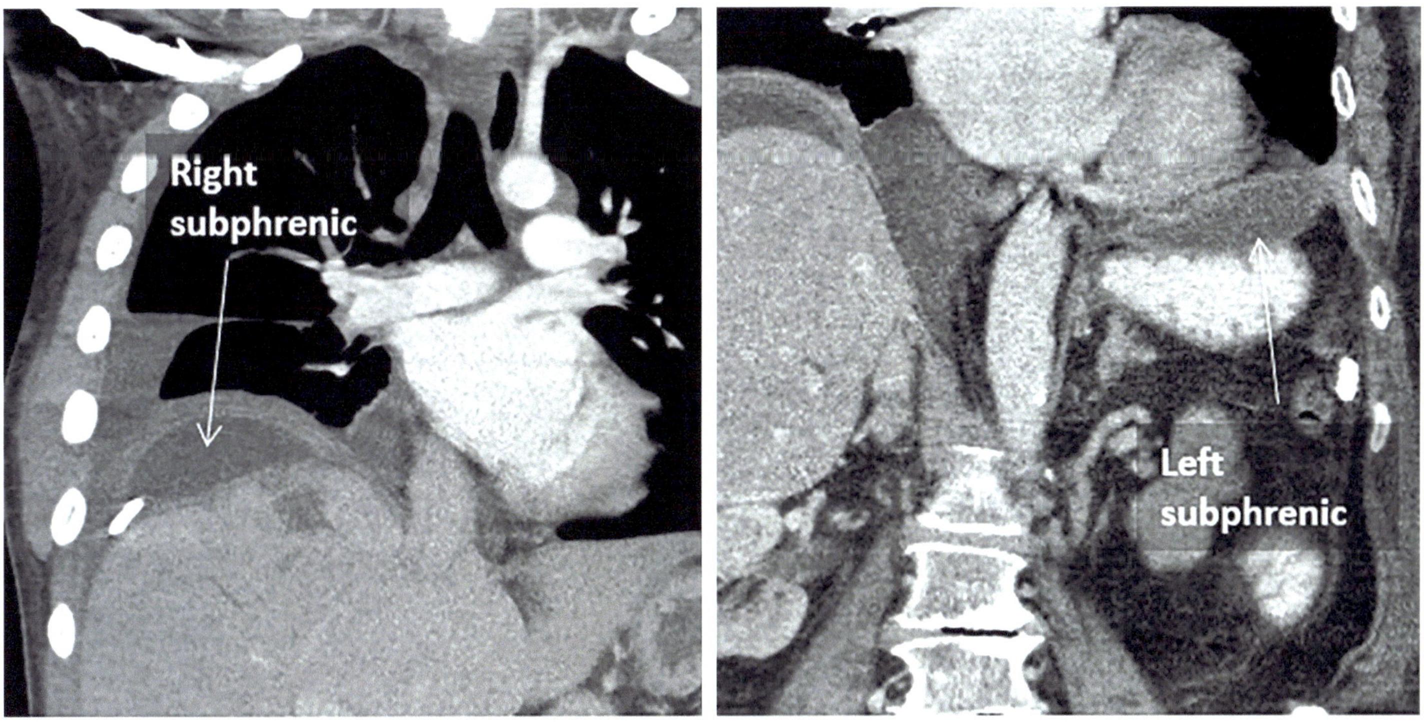

Fig. 12.27 Subphrenic abscesses.

Peritoneal Carcinomatosis

Peritoneal carcinomatosis is malignant involvement of the peritoneum and is most commonly seen in malignancies of the ovary, colon, stomach, and pancreas (▶Fig. 12.6).

Omental Infarct

Spontaneous omental infarction is a relatively uncommon disorder; however, small omental infarcts after abdominal surgery due to ligation of bleeding vessels is not uncommon. This condition causes acute abdominal pain, usually on the right side, that simulates common surgical emergencies such as cholecystitis, appendicitis, or diverticulitis. CT demonstrates a heterogeneous soft tissue density mass in the fatty omentum (▶Fig. 12.13).

Mesenteritis

Involvement of the mesentery and omentum along the subperitoneal fat occurs after surgery, trauma, autoimmune diseases, paraneoplastic process, and ischemia or infection. Edema of the mesentery (misty mesentery) is a common transient finding

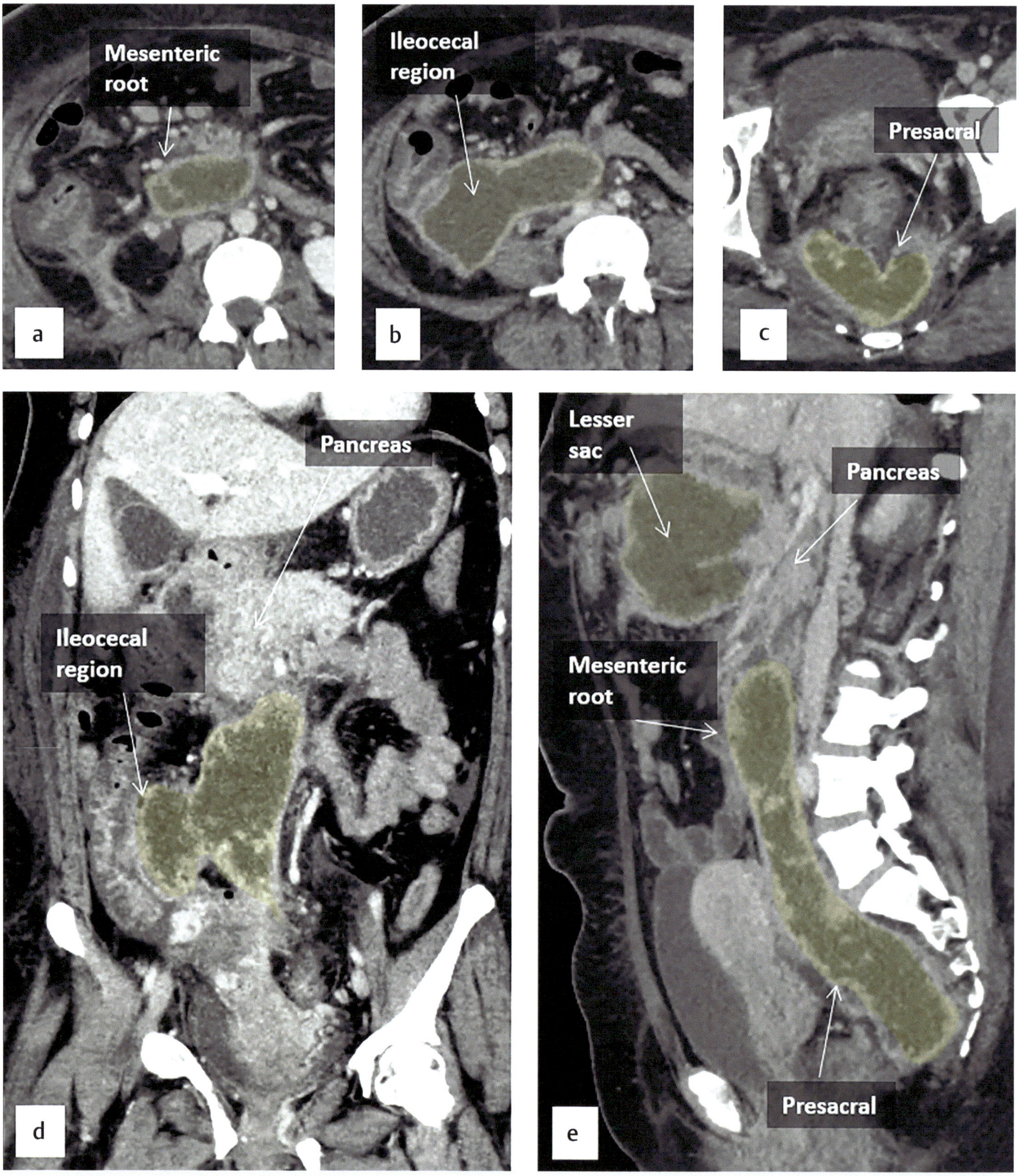

Fig. 12.28 Pancreatic abscess in the lesser sac and the small bowel mesenteric root. **(a–c)** Axial, **(d)** coronal, and **(e)** sagittal computed tomography (CT) images. The mesenteric root abscess extends toward the ileocecal region and the small bowel mesentery and from there into the pelvic presacral space along the extraperitoneal fat planes.

12

in CT scans after surgery, trauma, and following peritonitis (▶ Fig. 12.29).

Sclerosing mesenteritis also known as mesenteric lipodystrophy, mesenteric panniculitis, retractile mesenteritis, and mesenteric fibrosis is a relatively rare condition of unclear etiology. It is characterized by progressive fat necrosis, inflammation, and finally fibrosis and calcification of the abdominal mesentery.[21] Clinically, the disease presents in a patient at the age of 50 to 70 years with abdominal pain, nausea, vomiting, weight loss, and fever.

CT scan findings to diagnose sclerosing mesenteritis are increased density, mass effect, fat ring sign (halo of uninvolved fat around noses and vessels), and a pseudo-capsule around the mesenteritis (▶ Fig. 12.30). Enlargement of the mesenteric lymph nodes is a common finding in any form of mesenteritis.

Trauma

Mesenteric injuries are found in approximately 5% of patients undergoing laparotomy after blunt abdominal trauma.[22] In most cases a small hematoma or free fluid is seen with no significant clinical complication. In severe cases, mesenteric arterial disruption with bowel ischemia and active hemorrhage may necessitate immediate operation. Lesions most commonly occur near the ligament of Treitz or the ileocecal valve. The presence of a mesenteric hematoma should prompt a careful search for evidence of gastrointestinal perforation or laceration of mesenteric vasculature (▶ Fig. 12.29).

Internal Hernia

Paraduodenal hernias are the most common type of internal abdominal hernias. They are usually asymptomatic and discovered incidentally with higher frequency on the left side (▶ Fig. 12.24, ▶ Fig. 12.25). Partial or complete obstruction of the small intestine occurs in 50% of cases.[23]

Left paraduodenal hernias extend to left side of the transverse mesocolon through a peritoneal defect at the duodenojejunal junction. The inferior mesenteric vein and left colic artery are located at the neck of hernia sac. Encapsulated bowel loops are evident behind the descending mesocolon and displace the inferior mesenteric vein anteriorly. The inferior mesenteric vein is an important landmark for diagnosis. Normally it runs behind the duodenojejunal junction to join with the splenic vein or superior mesenteric vein (▶ Fig. 12.26).

Right paraduodenal hernias occur into the mesentericoparietal fossa of Waldeyer behind the root of the small bowel mesentery with the superior mesenteric artery running anterior to the neck of hernia (▶ Fig. 12.24).

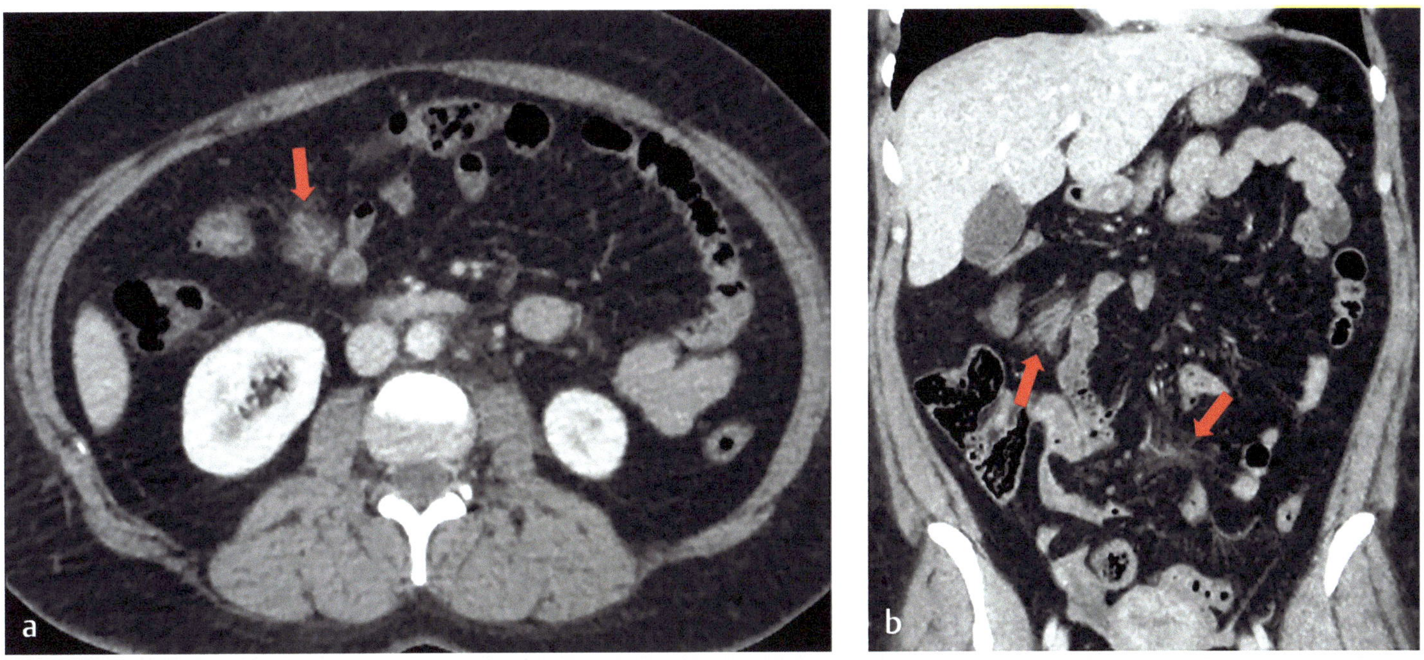

Fig. 12.29 Axial **(a)** and coronal **(b)** computed tomography (CT) images after blunt abdominal trauma showing small foci of mesenteric hematoma and edema (*arrow*).

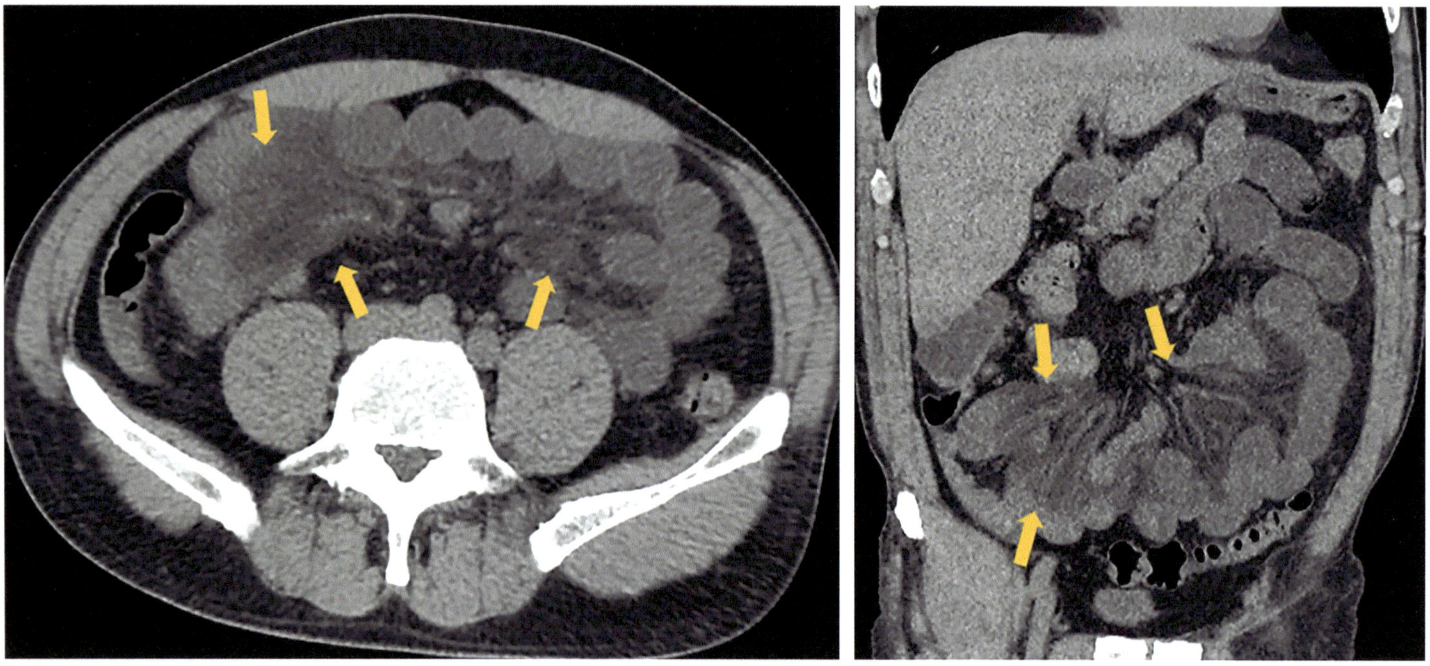

Fig. 12.30 Increased attenuation of the mesenteric fat (*arrows*) with unclear etiology in a 55-year-old man.

References

1. Tirkes T, Sandrasegaran K, Patel AA, et al. Peritoneal and retroperitoneal anatomy and its relevance for cross-sectional imaging. Radiographics 2012; 32(2):437–451
2. Liebermann-Meffert D. The greater omentum. Anatomy, embryology, and surgical applications. Surg Clin North Am 2000;80(1):275–293, xii
3. Kim S, Kim TU, Lee JW, et al. The perihepatic space: comprehensive anatomy and CT features of pathologic conditions. Radiographics 2007;27(1):129–143
4. DeMeo JH, Fulcher AS, Austin RF Jr. Anatomic CT demonstration of the peritoneal spaces, ligaments, and mesenteries: normal and pathologic processes. Radiographics 1995;15(4):755–770
5. Meyers MA, Oliphant M, Berne AS, Feldberg MA. The peritoneal ligaments and mesenteries: pathways of intraabdominal spread of disease. Radiology 1987;163(3):593–604
6. Gadzijev EM. Surgical anatomy of hepatoduodenal ligament and hepatic hilus. J Hepatobiliary Pancreat Surg 2002;9(5):531–533
7. Yoo E, Kim JH, Kim MJ, et al. Greater and lesser omenta: normal anatomy and pathologic processes. Radiographics 2007;27(3):707–720
8. Krist LF, Kerremans M, Broekhuis-Fluitsma DM, Eestermans IL, Meyer S, Beelen RH. Milky spots in the greater omentum are predominant sites of local tumour cell proliferation and accumulation in the peritoneal cavity. Cancer Immunol Immunother 1998;47(4):205–212
9. Ostermann PA, Schreiber HW, Lierse W. [The ligament system of the spleen and its significance for surgical interventions]. Langenbecks Arch Chir 1987; 371(3):207–216
10. Coffey JC, O'leary DP. Defining the mesentery as an organ and what this means for understanding its roles in digestive disorders. Expert Rev Gastroenterol Hepatol 2017;11(8):703–705
11. Coffey JC, O'Leary DP. The mesentery: structure, function, and role in disease. Lancet Gastroenterol Hepatol 2016;1(3):238–247
12. Meyers MA, Feldberg MA, Oliphant M. Grey Turner's sign and Cullen's sign in acute pancreatitis. Gastrointest Radiol 1989;14(1):31–37
13. Pannu HK, Oliphant M. The subperitoneal space and peritoneal cavity: basic concepts. Abdom Imaging 2015;40(7):2710–2722
14. Okino Y, Kiyosue H, Mori H, et al. Root of the small-bowel mesentery: correlative anatomy and CT features of pathologic conditions. Radiographics 2001;21(6):1475–1490
15. Charnsangavej C, DuBrow RA, Varma DG, Herron DH, Robinson TJ, Whitley NO. CT of the mesocolon. Part 1. Anatomic considerations. Radiographics 1993; 13(5):1035–1045
16. Healy JC, Reznek RH. The peritoneum, mesenteries and omenta: normal anatomy and pathological processes. Eur Radiol 1998;8(6):886–900
17. Dodds WJ, Foley WD, Lawson TL, Stewart ET, Taylor A. Anatomy and imaging of the lesser peritoneal sac. AJR Am J Roentgenol 1985;144(3):567–575
18. Mahmud T, Arastu M, Ellis H. Distribution of fluid within the peritoneal cavity: a cadaveric study. Clin Anat 2005;18(6):443–445
19. Peltier J, Gars DL, Page C, Yzet T, Laude M. The duodenal fossae: anatomic study and clinical correlations. Surg Radiol Anat 2005;27(4):303–307
20. Barberini F, Zani A, Ripani M, Di Nitto V, Brunone F. The complex arrangement of an "aorto-jejunal paraduodenal" fossa, as revealed by dissection of human posterior parietal peritoneum. Ann Anat 2007;189(3):299–303
21. Green MS, Chhabra R, Goyal H. Sclerosing mesenteritis: a comprehensive clinical review. Ann Transl 2018;6(17):336
22. Hawkins AE, Mirvis SE. Evaluation of bowel and mesenteric injury: role of multidetector CT. Abdom Imaging 2003;28(4):505–514
23. Takeyama N, Gokan T, Ohgiya Y, et al. CT of internal hernias. Radiographics 2005;25(4):997–1015

13 Adrenal Glands, Kidneys, Ureters, and Bladder

Kevin G. King and Farhood Saremi

◆ Introduction

In this chapter, the anatomy of the adrenal glands, kidneys, ureters, and bladder will be reviewed. All of these organs are extraperitoneal, although the bladder dome is separated from the overlying peritoneum only by the umbilicovesical fascia. Compared with other organs in the body, the kidneys and upper urinary tracts are notable for their somewhat greater degree of anatomic variation, which is attributable to their relatively complex embryonic development. Among the organs of the body, the kidneys are also notable for their high volume of blood flow relative to their size, the consequence of their important roles in hemofiltration and regulation of blood pressure. The adrenal glands are small organs which are anatomically closely associated with the kidneys, despite their dissimilar embryonic origins and functions.

◆ Adrenal Glands

The two components of the adrenal (suprarenal) gland, the cortex and medulla, are embryologically and functionally distinct. The cortex arises from the coelomic mesoderm of the urogenital ridge, and produces mineralocorticoids, glucocorticoids, and gonadocorticoids; the medulla arises from the neural crest, and produces epinephrine and norepinephrine.[1,2]

Development

The adrenal cortex develops from a proliferation of mesoderm from a groove in the coelom between the mesonephros and the gonad. The adrenal medulla arises from the ectoderm as a migration of neural crest cells into the developing adrenal in its upper lumbar location. Sympathetic ganglia and paraganglia similarly develop from this migration of neural crest cells.[10]

Anatomy

The adrenal glands weigh about 5 g each and are retroperitoneal, located in the superior part of the perirenal space within the Gerota's fascia. Transverse fibrous lamellae and a variable amount of fat separate the adrenals from the kidneys.

The normal adrenal varies in shape but typically has the shape of an arrowhead, inverted Y, or inverted V (▶ Fig. 13.1, ▶ Fig. 13.2). Each gland has a central body from which extends medial and lateral limbs. The body is typically 1 cm in thickness, and each limb 4 to 5 mm in thickness.[3,4] The glands are asymmetric in shape, with the right gland being more pyramidal and the left more semilunar.

The left adrenal gland is superomedial to the left kidney, and lateral to the left crus of the diaphragm and the left celiac ganglion. It lies posterior to the splenic vessels. The lumbar portion of the left hemidiaphragm arches posterior and superior to the gland. It is not uncommon for the lateral limb of the left adrenal to extend anterior to the superior pole of the left kidney (▶ Fig. 13.1).

The right adrenal gland is superomedial to the right kidney. The liver covers the lateral and superior aspects of the adrenal, while the lumbar portion of the right hemidiaphragm arches posterior and superior to the gland. The inferior vena cava (IVC) is anterior to the right adrenal, and the right crus of the diaphragm and the right celiac ganglion lie medial to it.

Vessels and Nerves

Although there is variation in arterial anatomy, generally the adrenal glands are supplied by three arteries (▶ Fig. 13.3): superior, middle, and inferior adrenal arteries, which arise from the inferior phrenic artery, aorta, and renal artery, respectively.[5] Adrenal venous drainage is predominantly through a single central vein, which empties to the IVC on the right, and to the left renal vein on the left[6] (▶ Fig. 13.3, ▶ Fig. 13.4). The left adrenal vein is 1 to 4 cm in length and commonly confluences with the left inferior phrenic vein to form a common trunk passing anterior to the left celiac ganglion, before emptying into the left renal vein (▶ Fig. 13.4). The right adrenal vein is short, being typically less than 1 cm in length, predisposing this vessel to injury during surgery. The central vein on the right may be duplicated or rarely triplicated. It is common to find superficial capsular veins that extend from the surface of the gland into the perirenal fat and renal capsule. Other communications include to the azygos vein, hemiazygos vein, right renal vein, or direct connection to the IVC on the right.

Lymphatic vessels drain from the adrenals medially, some to para-aortic and paracaval lymph nodes, and others directly to the cisterna chyli.[7]

Most of the innervation of the adrenal is composed of preganglionic sympathetic fibers which arise in the lower thoracic spinal cord and course to the adrenal medulla via the splanchnic nerves; in this way the catecholamine-producing medulla is analogous to a sympathetic ganglion. However, some sympathetic fibers synapse in the celiac, aorticorenal, and renal ganglia before providing the adrenal with postganglionic innervation, and additionally some afferent nerves of the adrenal have been demonstrated, with cell bodies in dorsal root ganglia and vagal sensory ganglia.[8]

Imaging for Adrenal Lesions

Adrenal lesions are described in the setting of adrenal hyperfunction, hypofunction, or normal function. Indications for evaluation of adrenal hyperfunction include hormonal abnormalities such as Cushing's syndrome (due to hypersecretion of cortisol), Conn's syndrome (due to hypersecretion of aldosterone), virilization, or feminization. Hypoadrenalism is often idiopathic, but the patient

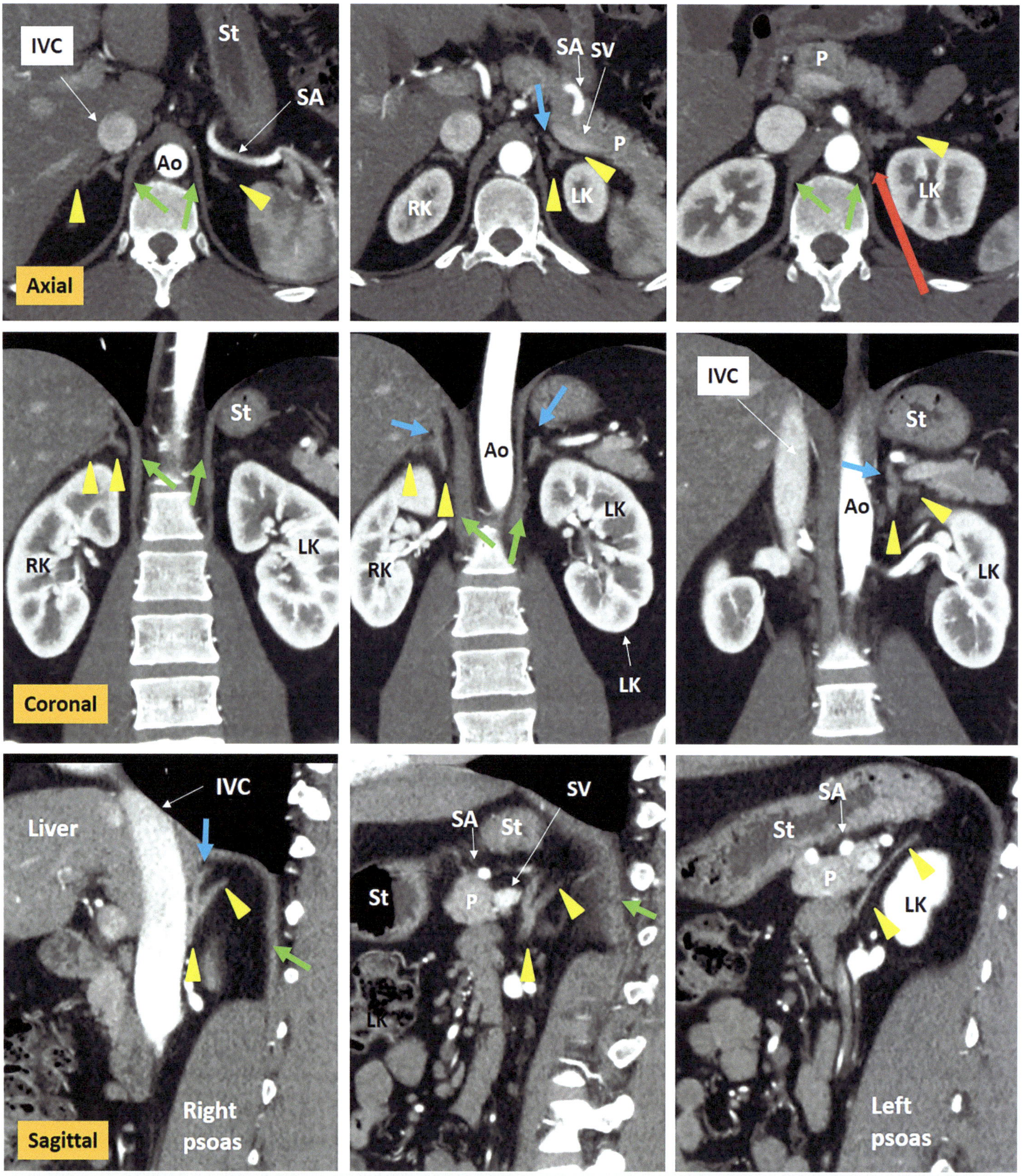

Fig. 13.1 Axial, coronal, and sagittal computed tomography (CT) images of normal adrenal glands. The body (*blue arrow*) and medial and lateral limbs (*yellow arrowheads*) of both adrenal glands are seen. The adrenals are superomedial to the kidneys (right kidney [RK] and left kidney [LK]) and lateral to the crura of the diaphragm (*green arrows*). The liver is lateral and superior and the inferior vena cava (IVC) is anterior to the right adrenal. Axial images show the left adrenal lateral to the left celiac ganglion (*long red arrow*) and the crus of the left hemidiaphragm (*green arrow*), and posterior to the splenic artery (SA) and splenic vein (SV). The LK is inferior and posterior. Images also show the lateral limb of the adrenals extending anterior to the kidneys, especially on the left. The diaphragm is superior and posterior to the adrenals. Ao, abdominal aorta; P, pancreas; St, stomach.

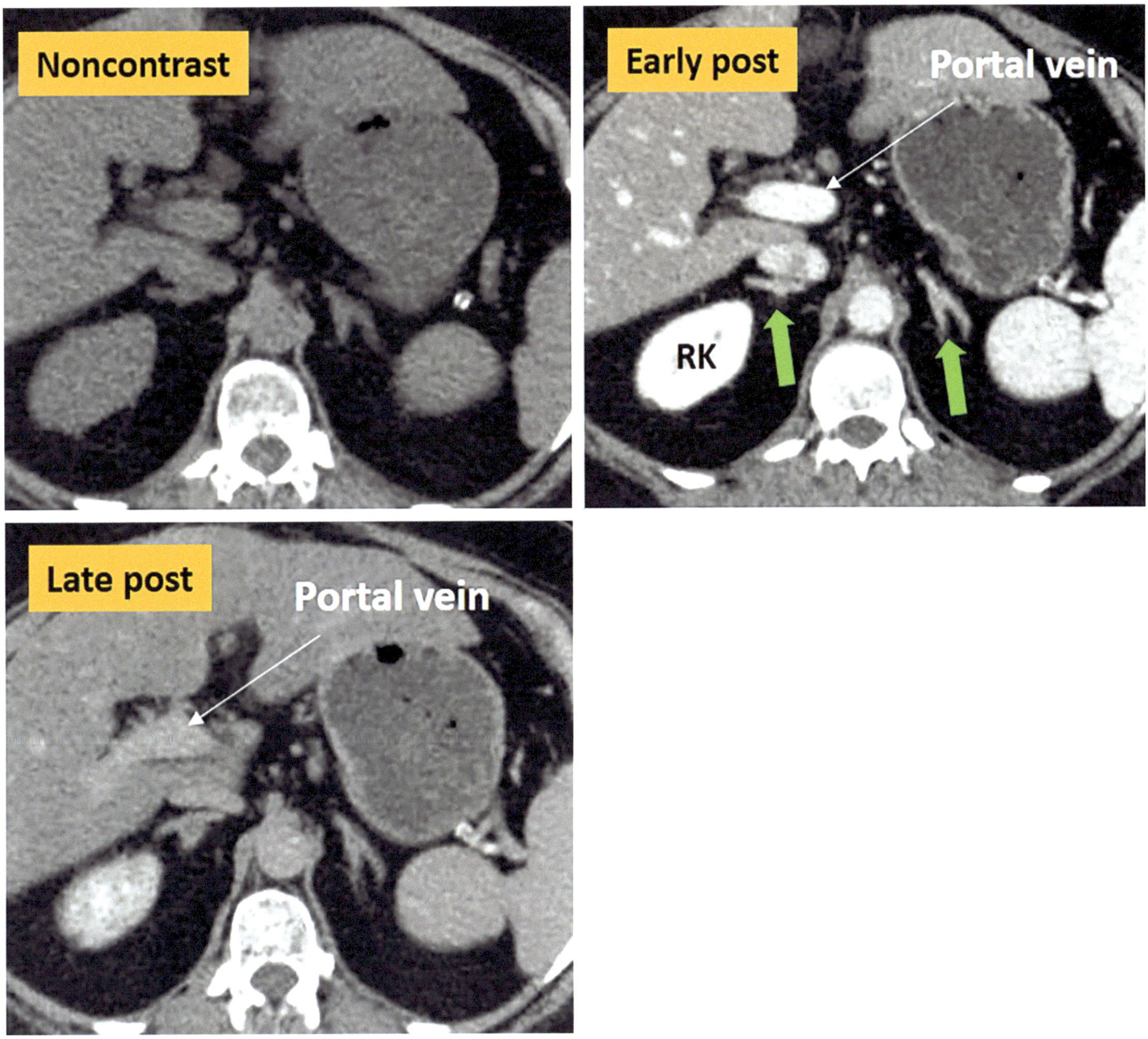

Fig. 13.2 Patten of enhancement of the normal adrenal glands. The adrenal glands enhance more in the early vascular phase (*arrows*) than the late phase. RK, right kidney.

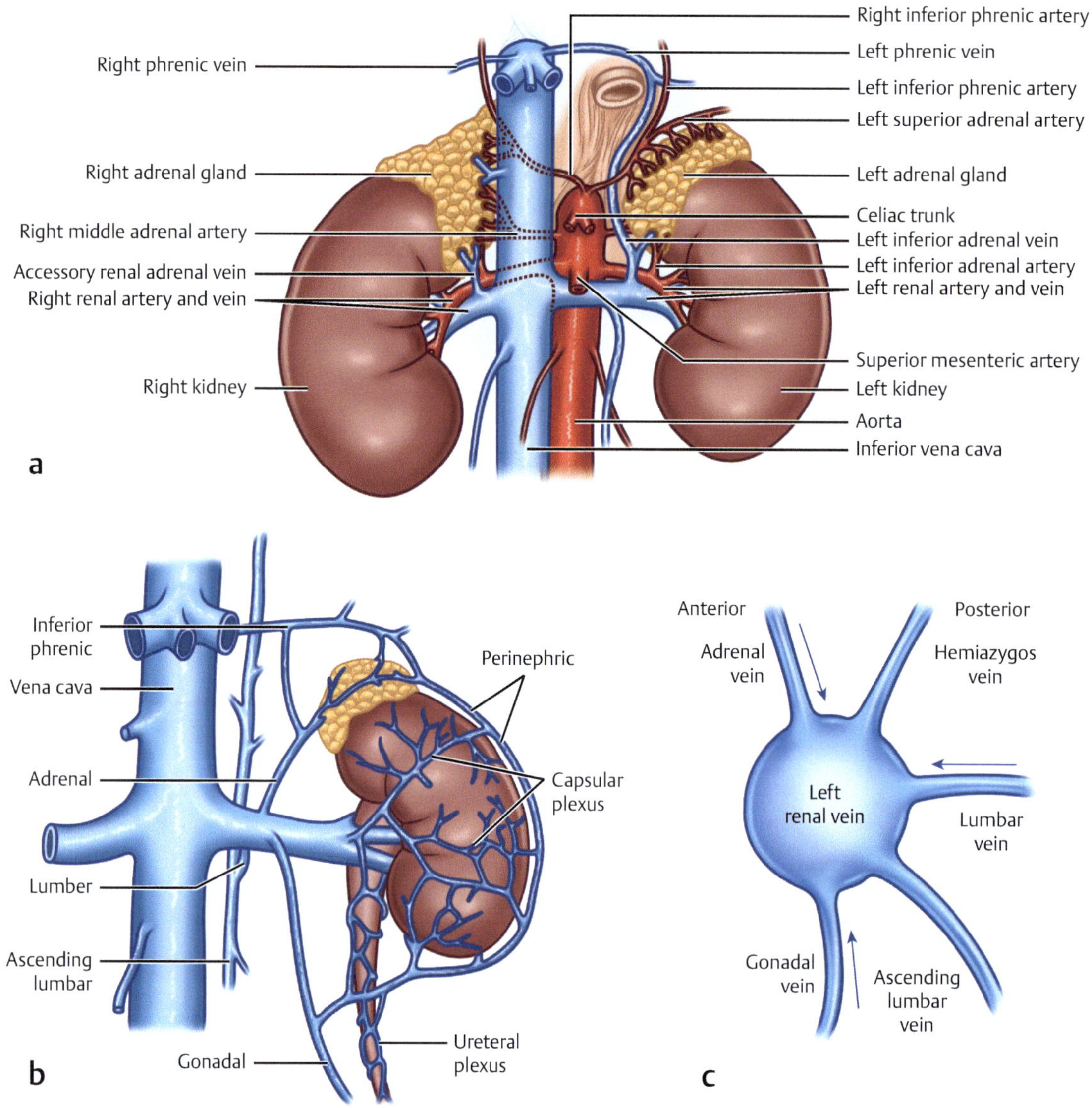

Fig. 13.3 (a) The superior, middle, and inferior adrenal arteries. Approximate locations of the adrenal veins are shown. **(b)** Frontal and **(c)** sagittal views show the left renal vein tributaries. In patients with chronic portal hypertension the left renal vein can communicate with the splenic vein through the retroperitoneal and ascending lumbar veins forming a splenorenal shunt.

13

may benefit from the exclusion of treatable etiologies. Computed tomography (CT) and magnetic resonance imaging (MRI) are the most commonly used imaging techniques to evaluate adrenal lesions. Transcatheter adrenal venography is used for venous blood sampling to evaluate whether autonomous hormone production is unilateral or bilateral[6] (▶Fig. 13.4).

The normal density of the adrenal on noncontrast CT resembles the kidney (▶Fig. 13.2). The normal signal on MRI is isointense or slightly hypointense to the liver on both T1- and T2-weighted images.[2] The normal adrenal enhances homogeneously on arterial phase contrast-enhanced imaging (▶Fig. 13.2).

Adrenal lesions discovered incidentally or in the presence of known malignancy may warrant further evaluation. Adrenal calcifications are a common incidental imaging finding, usually related to prior hemorrhage or inflammation. The most commonly encountered focal adrenal lesion is one or more nodules in a patient with normal adrenal function. Most benign adenomas have CT attenuation of less than 10 HU on precontrast images, and in many examples, have low density on delayed postcontrast images (▶Fig. 13.5). A homogeneous nodule with an attenuation of less than 10 HU on precontrast CT, or a decrease in signal intensity at opposed-phase MRI, is diagnostic for benign adenoma. Most adrenal adenomas contain large amounts of intracellular lipid. This lipid is the basis for loss of signal on opposed-phase chemical shift imaging. Drop in signal relative to spleen or liver on chemical shift imaging, which compares in- and opposed-phase gradient recalled echo (GRE) images, is the best MRI technique to discriminate adenoma from nonadenoma. Lesions that do not have these features are indeterminate and may necessitate follow-up or biopsy.

Adrenal myelolipoma is another incidentally found benign adrenal lesion (▶Fig. 13.5). It has a characteristic appearance that reflects

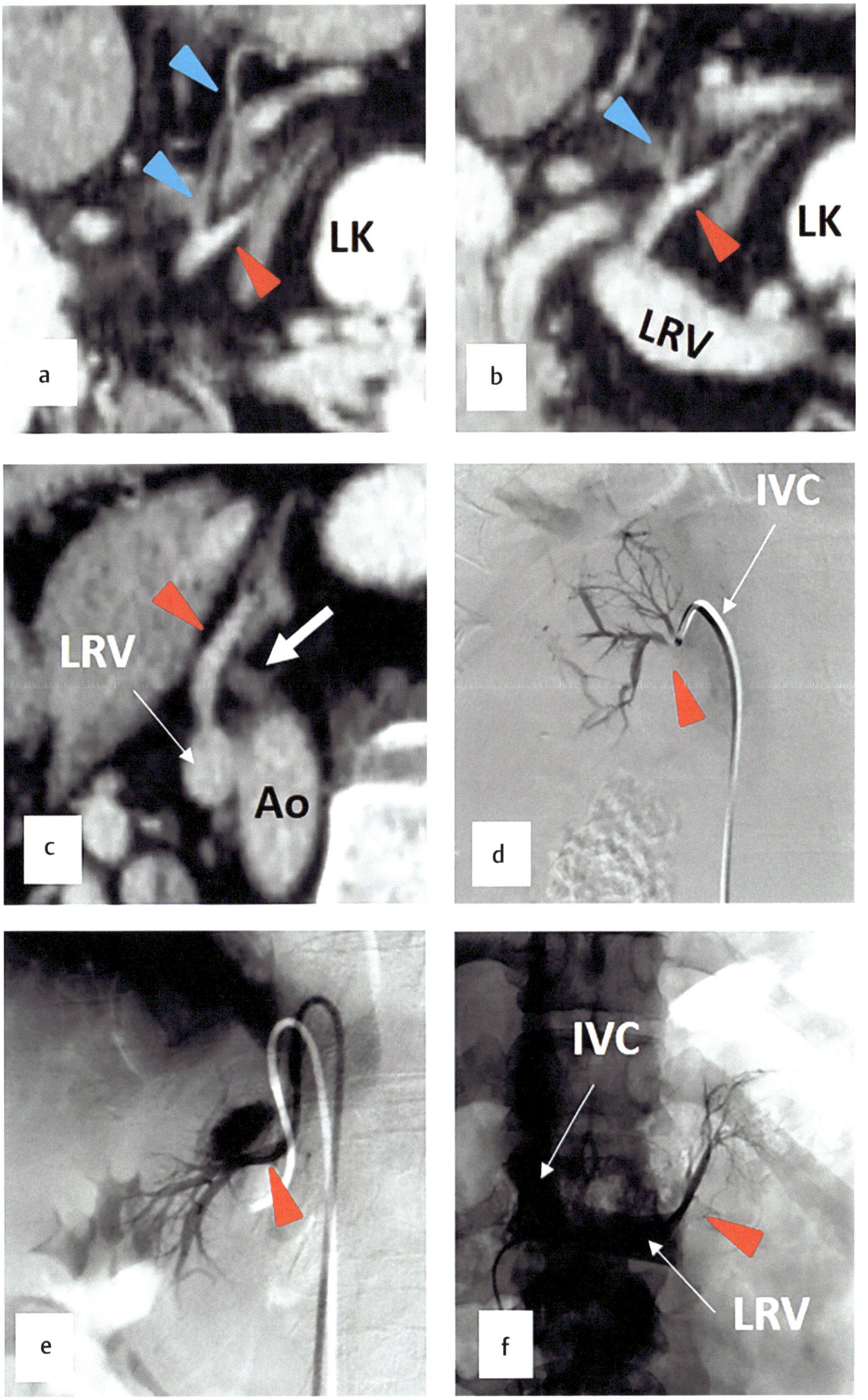

Fig. 13.4 Adrenal veins. **(a, b)** Two consecutive coronal oblique computed tomography (CT) sections showing the left inferior phrenic vein (*blue arrowhead*) joining the left adrenal vein (*red arrowhead*) to form a common trunk, before emptying into the left renal vein (LRV). **(c)** Sagittal oblique CT showing the common trunk of the left adrenal vein (*red arrowhead*) anterior to the celiac ganglion (*white arrow*) just before emptying into the left renal vein (LRV). **(d–f)** Selective adrenal venography using transcatheter injection of contrast. **(d, e)** The trunk of the right adrenal vein emptying into the inferior vena cava (IVC). The common trunk is long in **e.** **(e)** The wedged position of the catheter tip has resulted in some extravasation or staining above the common trunk. **(f)** Left adrenal vein draining into the left renal vein (LRV). Ao, abdominal aorta; LK, left kidney.

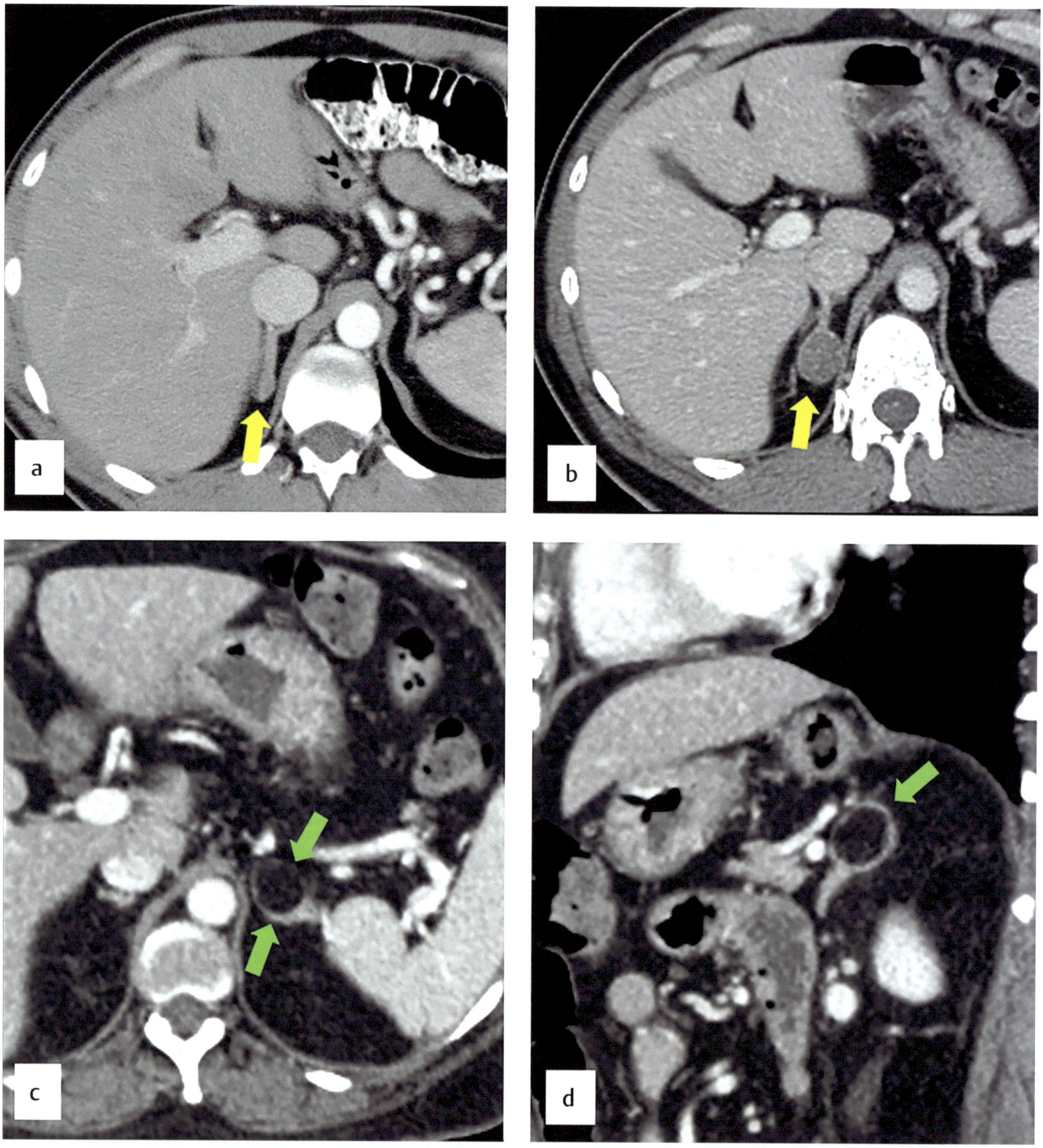

Fig. 13.5 (a, b) Enlarging benign adenoma in 4 years follow-up. Note low attenuation of the lesion on delayed postcontrast view (*arrow*). **(c)** Axial and **(d)** sagittal computed tomography (CT) showing fatty lesion of the left adrenal (*arrows*) consistent with myelolipoma.

13

the presence of macroscopic fat deposits.[9] Adrenal metastases commonly arise from lung, breast, lymphoma, or melanoma primaries. Metastatic lesions are typically larger and more heterogeneous than adenomas. Pheochromocytoma is a rare lesion which most commonly arises from the adrenal medulla, although it may arise in the retroperitoneal sympathetic ganglia, organ of Zuckerkandl, or urinary bladder. Whole-body meta-iodobenzylguanidine (MIBG) compounds labeled with radio-iodine can detect functional extra-adrenal pheochromocytomas.

Adrenal hematomas generally result from trauma, sepsis, hypotension, anticoagulation therapy, or left renal vein thrombosis. Adrenal hematomas are often associated with severe trauma, but in some cases may be the only traumatic image finding (▶ Fig. 13.6).

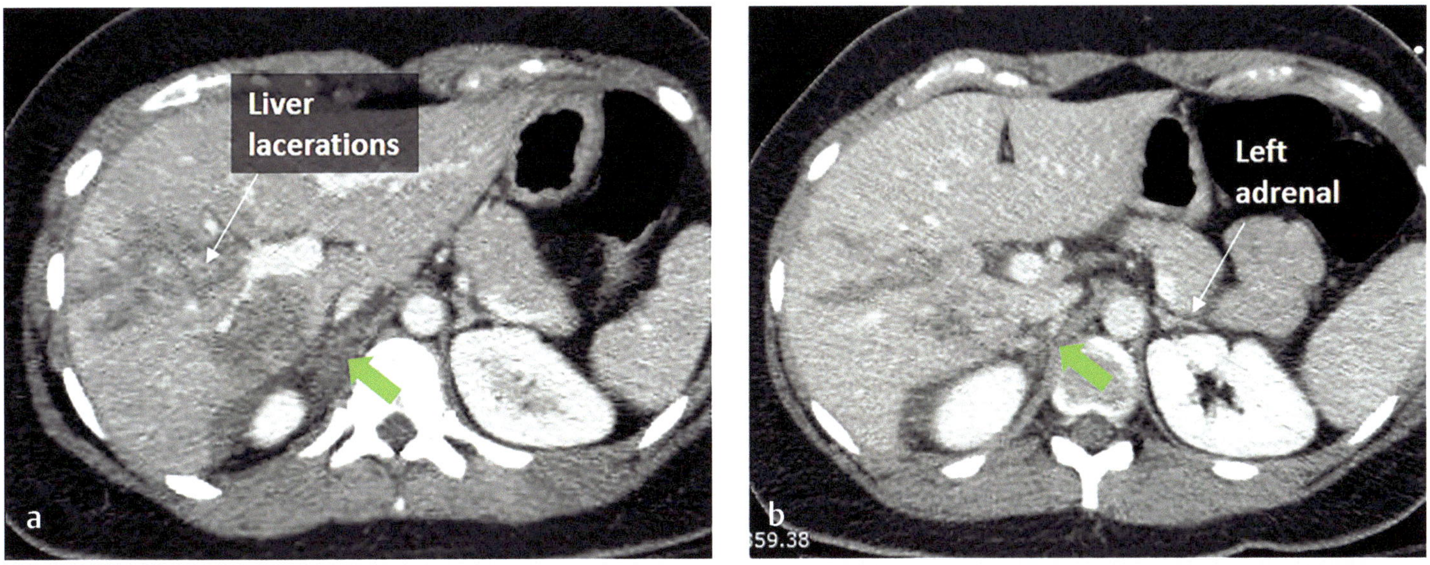

Fig. 13.6 Axial computed tomography (CT) scan after motor vehicle accident. **(a)** Severe right liver and right adrenal (*arrow*) injury. **(b)** Follow-up after 5 days showing improving lacerations (*arrow*).

◆ Kidneys

Development

Renal development occurs through several stages (▸Fig. 13.7). Around the 3rd week of gestation, a pair of comb-like formations of epithelial tubules called nephrotomes interconnected by a nephric duct appear at the cranial end on each side of the fetus. Soon, each primitive nonfunctional kidney, known as the pronephros, will be replaced by a similar series of nephrotomes, called the mesonephros, which grow progressively toward the caudal end of the embryo to connect to the cloaca. The epithelial tubules of the mesonephros, known as the mesonephric tubules, are more advanced and receive vessels from the aorta to form glomerulus-like structures that filtrate clear plasma into the nephric (mesonephric or Wolffian) ducts. Around the 5th to 6th weeks of gestation, a ureteric bud develops at the cloacal end of the mesonephric duct and grows cranially to form the future ureter. The distal end of the ureteric bud, known as the metanephric blastema, will then proliferate in nearby mesenchyme and, along with ingrowing vessels, differentiate into the renal collecting system (e.g., the renal pelvis and calyces) and the glomeruli of the definitive kidney. This nephrogenic process continues until after birth to complete formation of all nephrons. The metanephric kidneys ascend cranially from their embryonic pelvic position to reach their final location underneath the adrenal gland.

Parenchymal Anatomy

The kidneys are bean-shaped organs which, on average, are 10 to 12 cm in craniocaudal length and 130 to 150 g in weight. They are located within the perirenal space of the retroperitoneum on either side of the spine, with the right kidney generally 1 to 2 cm lower than the left kidney because of the presence of the liver. When supine, the kidneys span through T12 to L3 levels, but their position relative to the spine is not fixed; the kidneys can descend several centimeters with inspiration or upon standing. The top of the upper pole of the right kidney is generally at the level of the right 12th rib, while the upper pole of the left kidney reaches 1 to 2 cm above this.

The kidneys do not have a strictly vertical orientation, but instead their craniocaudal axis is directed from superomedial to inferolateral (▸Fig. 13.8a). The hila of the kidneys face medially, having rotated from anteriorly facing during the embryological ascension of the kidneys upwards from the pelvis. Embryological nonrotation, malrotation, and over-rotation describe hila which face anteriorly, laterally, or posteriorly, respectively.

There are several developmental anomalies that can result in abnormal renal number and location. Unilateral renal agenesis prevalence is 0.05%, and is thought to occur from failure of ureteric bud formation, or failure of interaction between the ureteric bud and the metanephros.[10] Pelvic kidney, with a prevalence of 0.08%, results when a kidney fails to ascend during normal embryological development (▸Fig. 13.8b). Crossed fused renal ectopia is rare and occurs when the two kidneys ascend on the same side, usually fused to one another (▸Fig. 13.8c). Horseshoe kidney has a prevalence of 0.25% and is characterized by an isthmus of renal parenchyma or connective tissue coursing transversely to join the lower poles of the kidneys (▸Fig. 13.8d).[11] The isthmus crosses anterior to the aorta, usually inferior to the inferior mesenteric artery, having been arrested by it during embryological ascension (▸Fig. 13.9). The horseshoe kidney typically demonstrates nonrotation, with the pelvises facing anteriorly, resulting in chronic urinary stasis and increased rate of stone formation and infection. Abnormally located or fused kidneys frequently show vascular and collecting system anomalies (see section Vasculature and Nerves).

A thin capsule surrounds the kidney, comprising connective tissue, elastic fibers, and smooth muscle fibers (▸Fig. 13.10, ▸Fig. 13.11). The renal capsule is adhered to the kidney and cannot be distinguished from the normal kidney, unless separated by

13

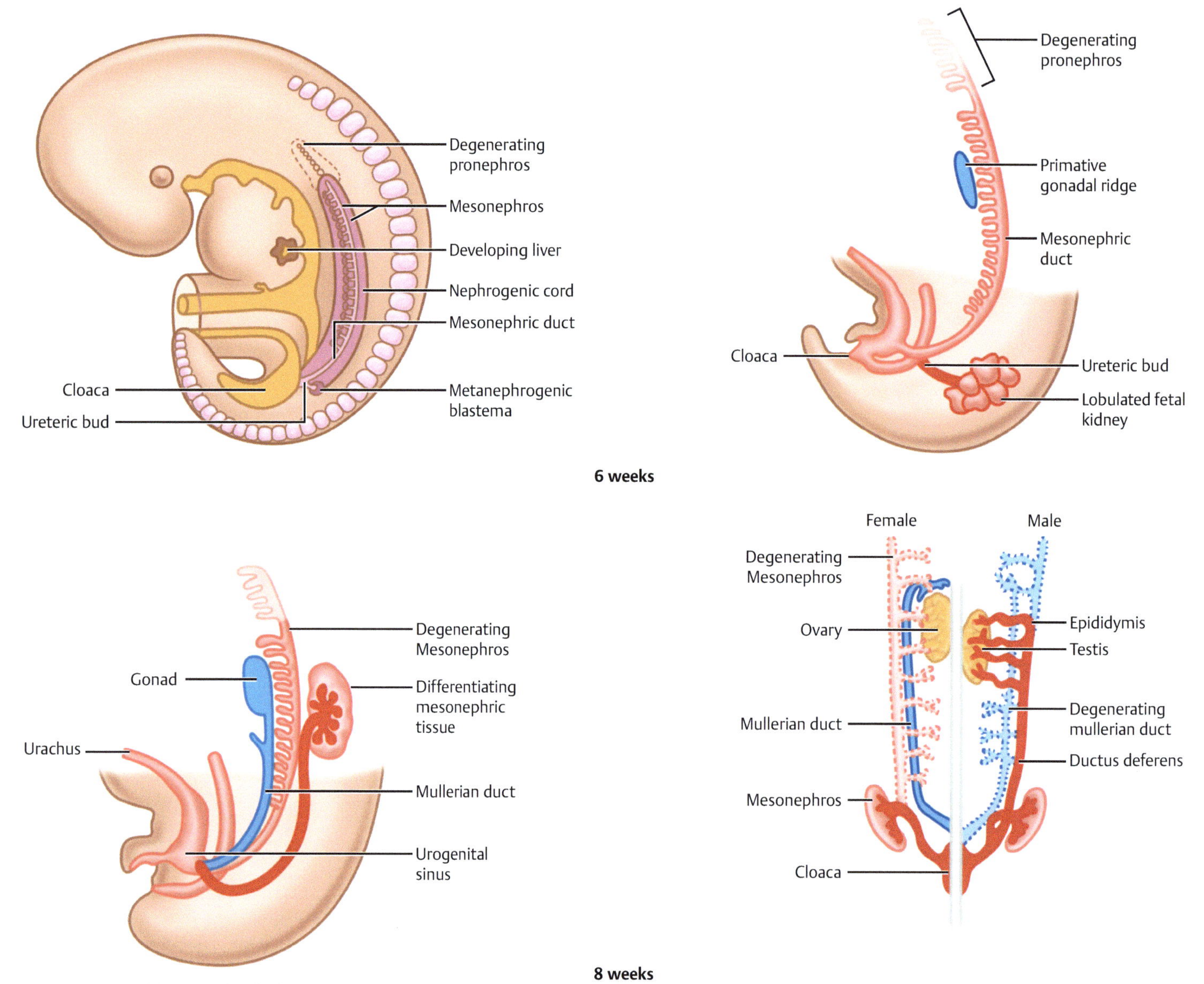

Fig. 13.7 At around 5 weeks of gestation, the primitive nonfunctional pronephros kidneys will be replaced by the mesonephros. Gonads form from a proliferation of the coelomic epithelium near the mesonephric (Wolffian) ducts. The mesonephric (Wolffian) ducts develop to form the male genital ducts, but they degenerate in the female. Instead, the paramesonephric (Müllerian) ducts form the female internal genital tract. In both sexes, the ureteric buds arise posteriorly from the distal end of mesonephric duct to form the definitive collecting system and kidneys.

blood (▶Fig. 13.11). At the center of the medial side of the kidney is the hilum, containing the renal pelvis, vessels, and nerves. Typically, the renal vein is anterior to the renal artery (▶Fig. 13.12), and the renal pelvis is posterior to the renal artery. The renal hilum leads inwards into the central renal sinus. The renal sinus is lined by the renal capsule and contains fat in the spaces surrounding the renal pelvis, vessels, and nerves (▶Fig. 13.10). The renal parenchyma can be divided into the internal medulla and the external cortex. The medulla is composed of multiple renal pyramids, which are oriented with their bases toward the periphery, and their apices toward the center of the kidney (▶Fig. 13.10, ▶Fig. 13.12). The medullary apices, also called the renal papillae, protrude into the lumen of calyces and contain converging medullary collecting ducts from which urine empties into the collecting system. The cortex covers over the bases of the pyramids, and also extends in between the pyramids, as projections of tissue called renal columns (column of Bertin). During CT or MR angiography, the normal renal cortex enhances almost simultaneously with the arterial branches, and the pyramids enhance after the renal cortex, after a few seconds' delay. Excretion of contrast into the collecting system (pyelogram) starts around 60 seconds after contrast injection, and may continue for hours. Many pathologies can change the timing of renal enhancement and excretion (▶Fig. 13.13).

Parenchymal Pathology

The kidneys are common sites of injuries after blunt trauma.[12] With CT, different categories of injures can be diagnosed including contusion, laceration, intrarenal or subcapsular hematomas, and renal collecting system injury (▶Fig. 13.11, ▶Fig. 13.14).

Renal cysts are common after the age of 50. Most cysts are asymptomatic and less than 2 cm (▶Fig. 13.15). Common solid renal lesions include angiomyolipoma and renal cell carcinoma. An enlarged column of Bertin is a common pseudotumor found on imaging studies and should not be confused with a renal mass (▶Fig. 13.15). Angiomyolipomas are benign lesions containing vascular and fatty components, often found incidentally

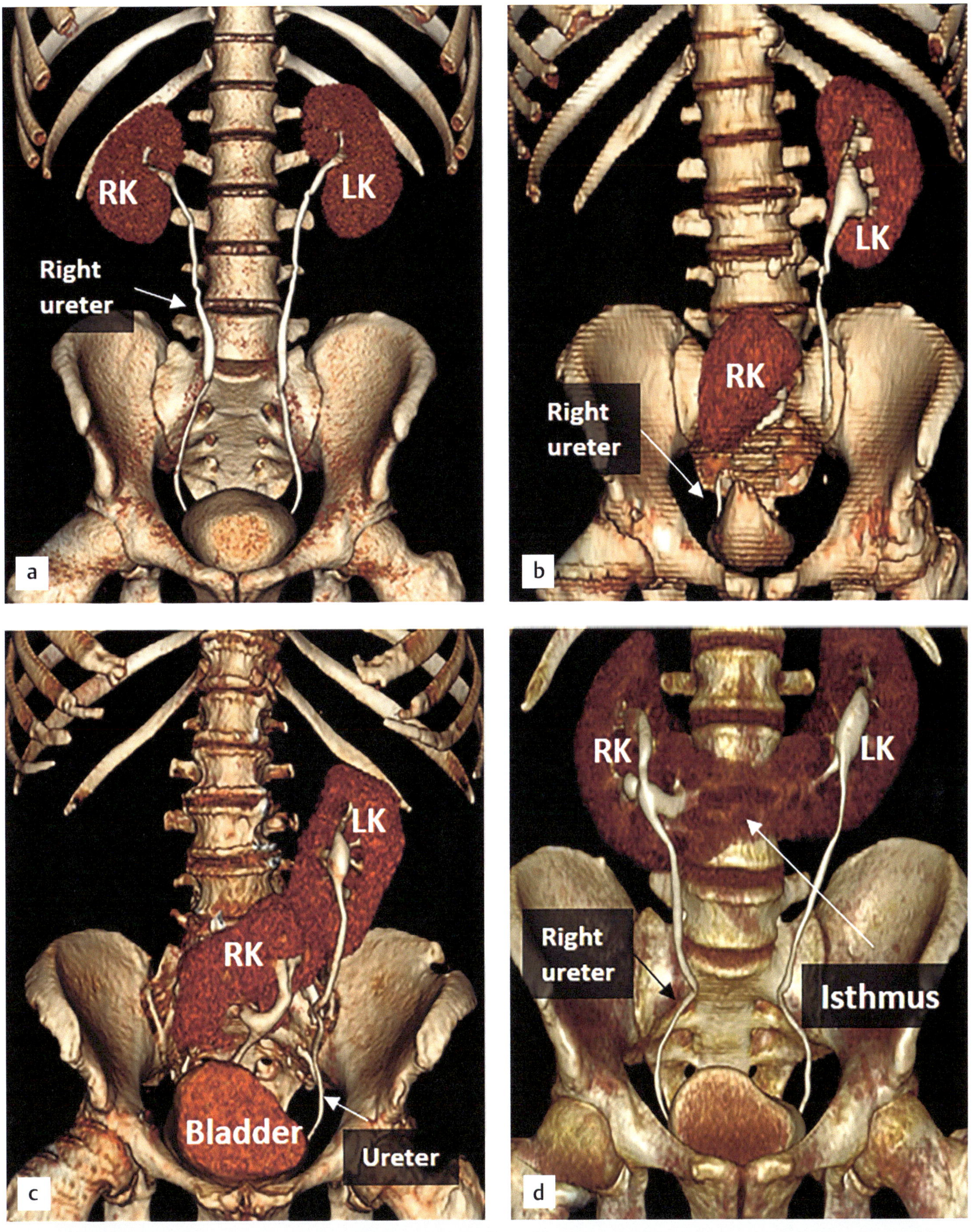

Fig. 13.8 (a) Normal position of the kidneys, with the craniocaudal axis directed from superomedial to inferolateral. Renal pelvises face medially. **(b)** Pelvic right kidney (RK). **(c)** Crossed fused right kidney. **(d)** Horseshoe kidney. Note anteriorly facing renal pelvises due to incomplete embryonic ascension and nonrotation of the kidneys.

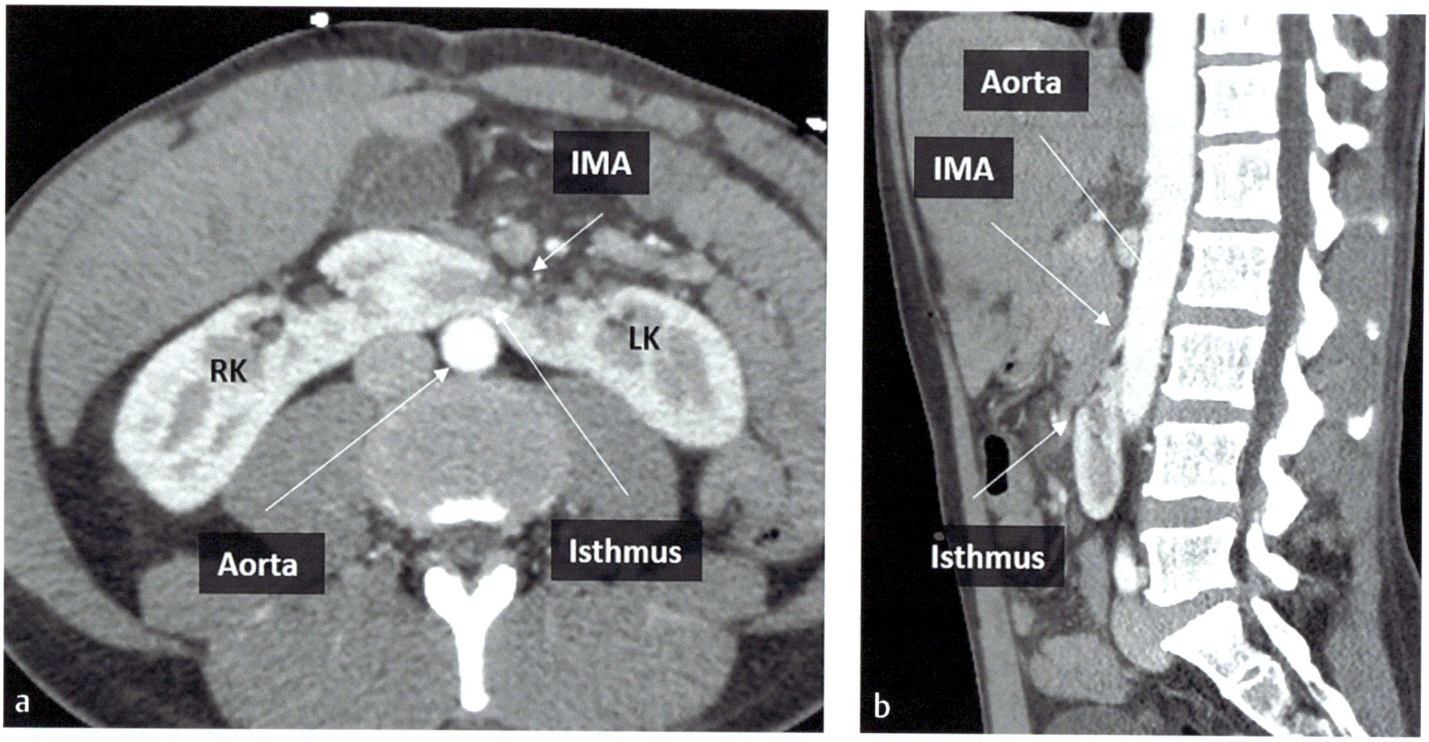

Fig. 13.9 Horseshoe kidney. **(a)** Axial and **(b)** sagittal computed tomography (CT) angiography showing the horseshoe kidney characterized by an isthmus of renal parenchyma connecting the kidneys. The isthmus crosses anterior to the aorta, and usually inferior to the inferior mesenteric artery (IMA). LK, left kidney; RK, right kidney.

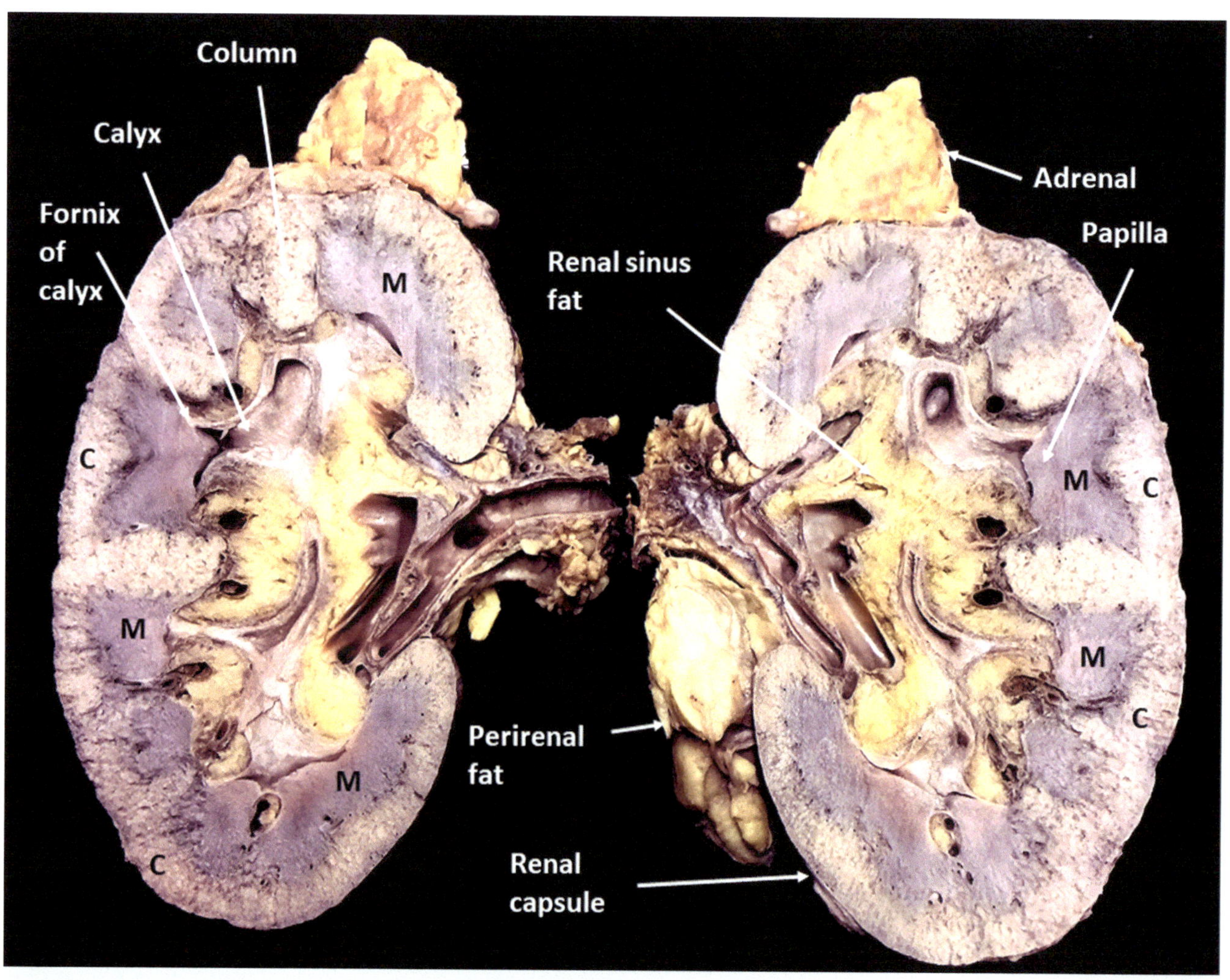

Fig. 13.10 Coronal section of a kidney specimen. C, cortex; M, medulla.

13

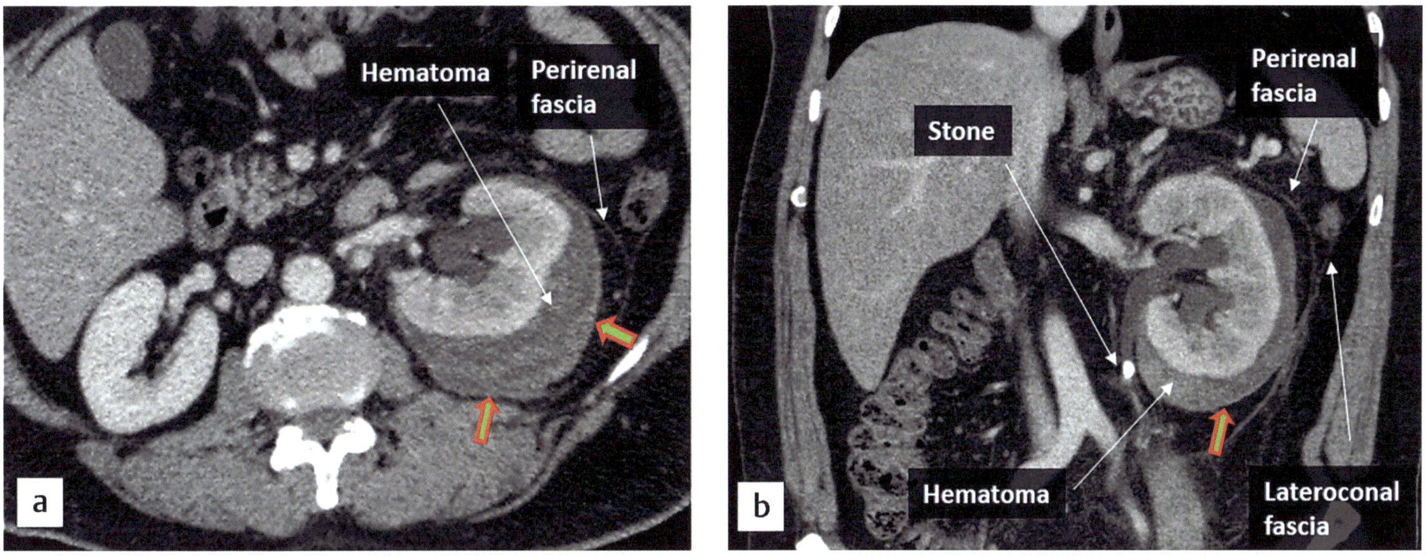

Fig. 13.11 Renal capsule. The renal capsule is usually adherent to the kidney and normally cannot be distinguished on computed tomography (CT). The presence of subcapsular blood can separate the capsule from the kidney and make it visible. **(a)** Axial CT and **(b)** coronal CT images showing left subcapsular hematoma separating the renal capsule (*arrows*) from the kidney. There is also mild hydronephrosis due to an obstructing stone in the proximal ureter.

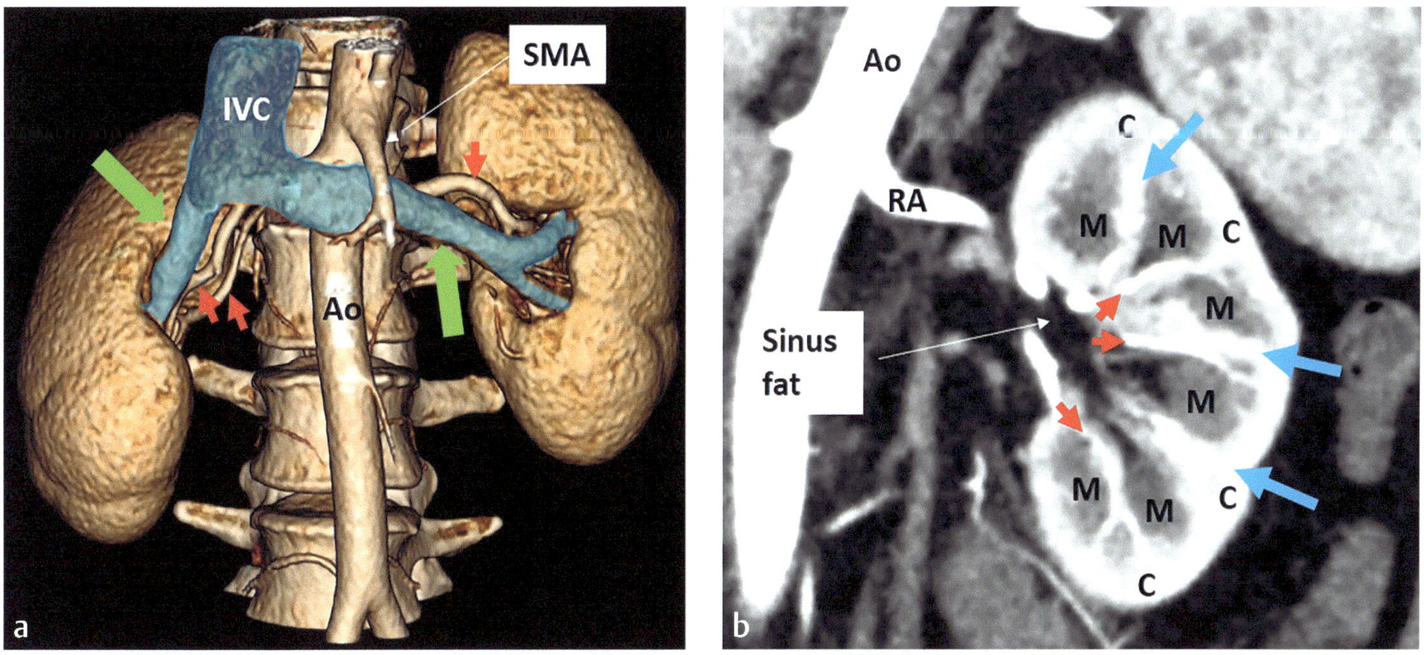

Fig. 13.12 (a) Frontal volume-rendered view of the kidneys. The renal veins (*long green arrows*) course anterior to the renal arteries (*short red arrows*). The right renal artery is duplicated in this case. The infrarenal portion of the inferior vena cava (IVC) has been removed from view. The left renal vein passes between the aorta (Ao) and the superior mesenteric artery (SMA). **(b)** Coronal computed tomography (CT) angiography of the left kidney at corticomedullary phase, showing the enhancing renal cortex (C) surrounding the medullary pyramids (M). Renal columns (*blue arrows*) are projections of cortex extending between the pyramids. The interlobar arteries (*short red arrows*) also course between the pyramids. Ao, abdominal aorta; RA, left renal artery.

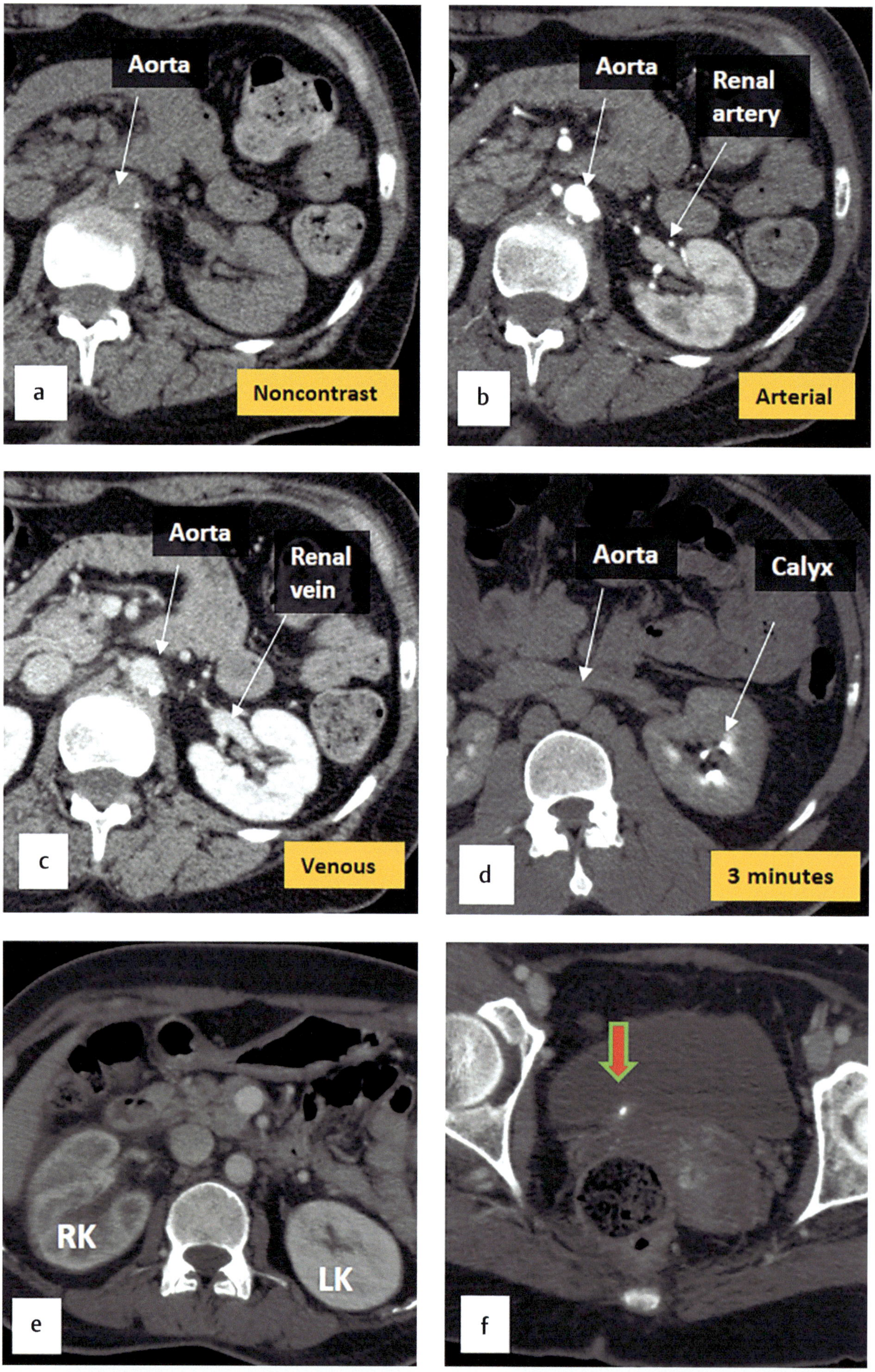

Fig. 13.13 (a–d) Axial computed tomography (CT) nephrogram images at different phases. Differentiation of the cortex from the medulla is only possible in arterial **(a)** and early venous phases. Later, the cortex and medulla become isoenhancing to one another **(b)**. Excretion of contrast into the renal collecting system is shown here 3 minutes after contrast injection **(d)**. **(e, f)** Delayed right kidney (RK) nephrogram compared to the left kidney (LK) due to some degree of ureteric obstruction by a small vesicoureteral junction stone (*arrow*).

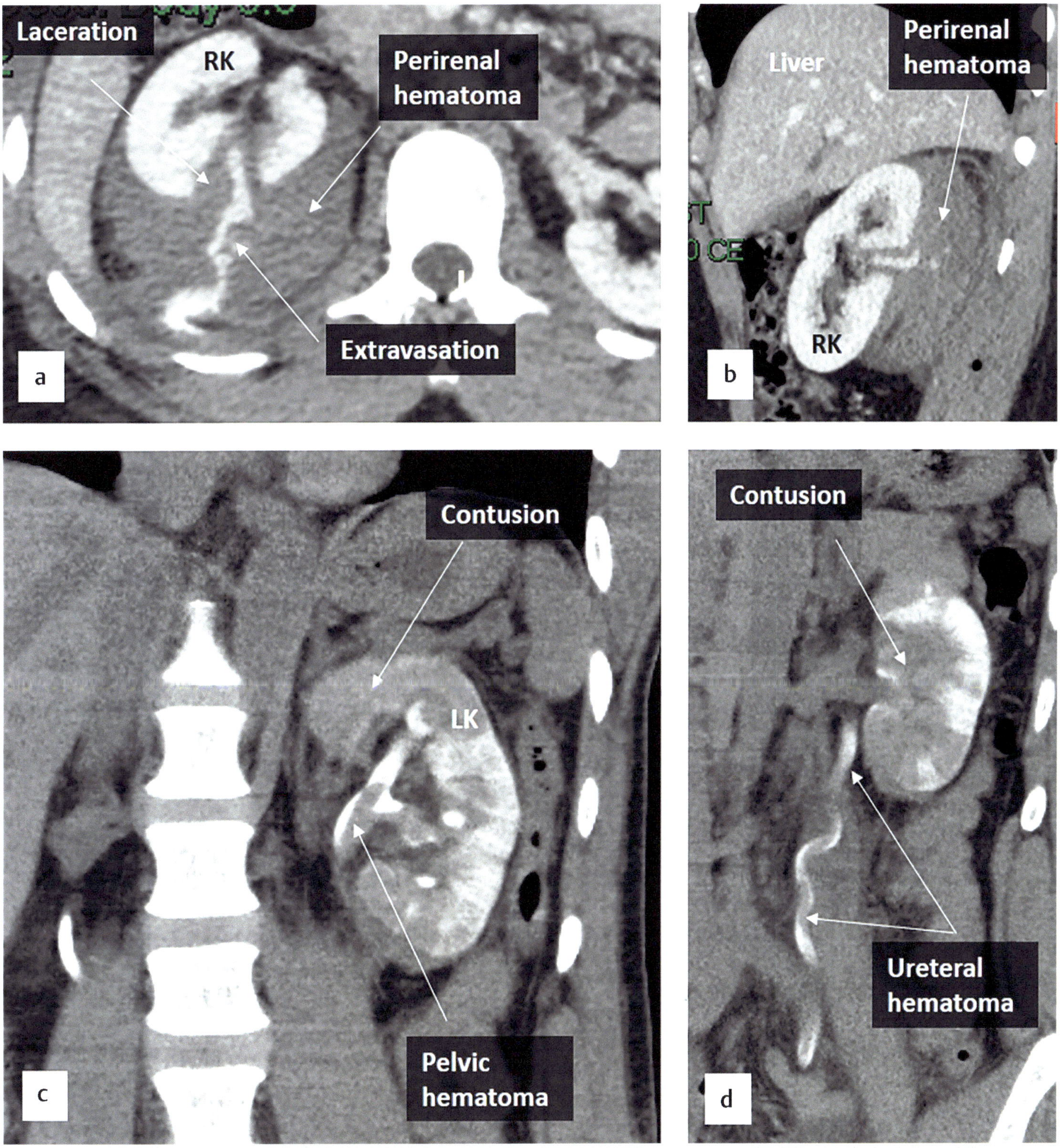

Fig. 13.14 Renal injuries. **(a)** Axial and **(b)** sagittal computed tomography (CT) showing a right kidney (RK) laceration, with a resultant large perirenal hematoma. Note active contrast extravasation into the perirenal space. **(c, d)** Coronal CT images show a severe left kidney (LK) contusion. Low attenuation hematoma surrounded by excreted contrast is seen within the left collecting system.

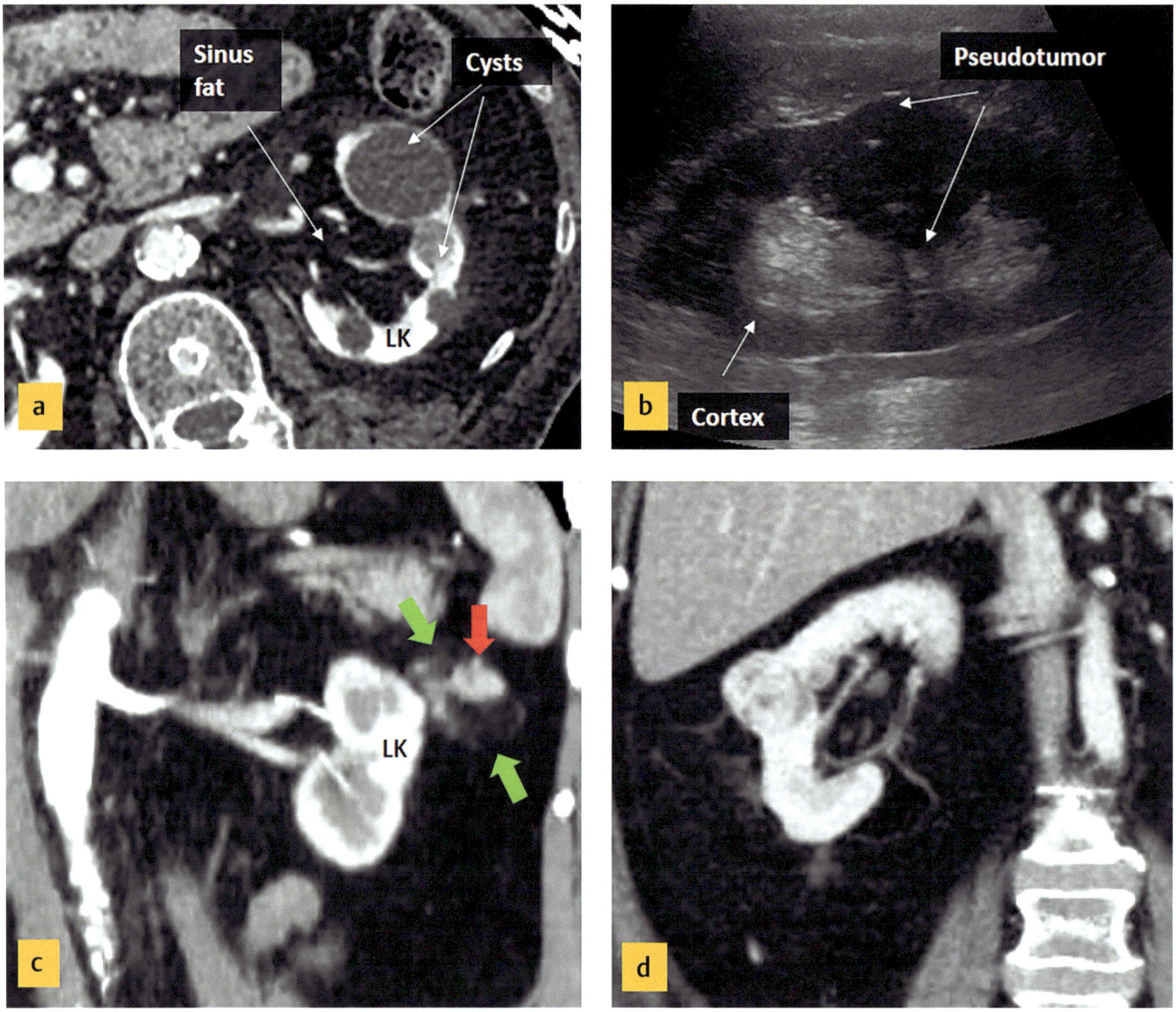

Fig. 13.15 (a) Axial computed tomography (CT) showing nonenhancing cysts in the left kidney (LK). **(b)** Sonography of the LK showing a mass-like column of Bertin. **(c)** LK exophytic angiomyolipoma containing fat (*green arrows*) and vascular (*red arrow*) elements. **(d)** Hypervascular right renal cell carcinoma on coronal postcontrast CT.

13

in the normal population, although sometimes presenting with hemorrhage when large. In patients with tuberous sclerosis, the kidneys can demonstrate a great number of angiomyolipomas. Renal cell carcinoma usually presents as a single well-defined hypervascular mass (▶ Fig. 13.15). Local invasion into the renal capsule, the renal vein, or the renal sinus, accompanied by regional lymphadenopathy, are other significant imaging findings.

Vasculature and Nerves

There is variability in the distribution and numbers of the arteries and veins of the renal macrovasculature. These common anatomical variations are mainly related to the embryonic development of the kidneys and systemic veins.[13,14] Around 5 weeks of development, the mesonephros are supplied by various numbers of lateral branches from the dorsal aorta, and drainage occurs into the subcardinal veins. With further development some arteries disappear, while the ones between the 10th thoracic and the 3rd lumbar segments (between the two mesenteric arteries) persist as gonadal, suprarenal, metanephric, and inferior diaphragmatic arteries (▶ Fig. 13.16). During superior migration of the metanephros or "future kidney" to the lumbar region, the kidneys are supplied by various arteries. One of these arteries remains as the main renal artery while the rest either disappear or persist as accessory renal arteries. The development of the renal veins from their primitive origin is also complex (▶ Fig. 13.17). The mesonephros drains to the subcardinal veins that develop during the 5th week. The two subcardinal veins connect together in the middle and with the supracardinal veins posteriorly to form the future

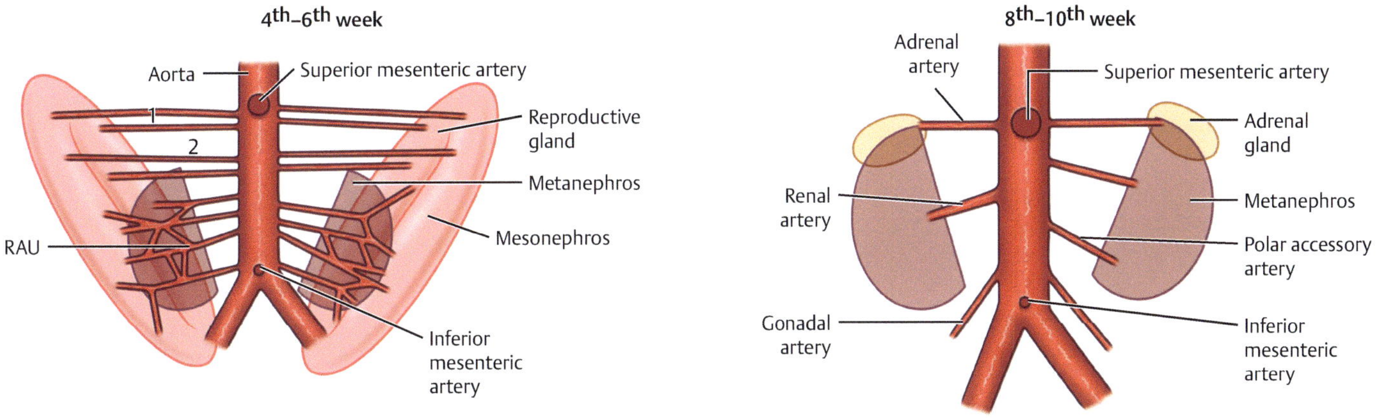

Fig. 13.16 Embryonic development of the arterial blood supply to the mesonephric (temporary) and metanephric (permanent) kidneys. The adrenal artery usually develops from the cranial group (#1) of mesonephric arteries whereas the gonadal artery forms from arteries of the caudal group (RAU). The definitive renal artery comes either from the last branch of the middle group (#2) or the first of the caudal group (RAU). Persisting mesonephric arteries form persistent accessory vessels (PAA). (Modified with permission from Arévalo Pérez et al.[13])

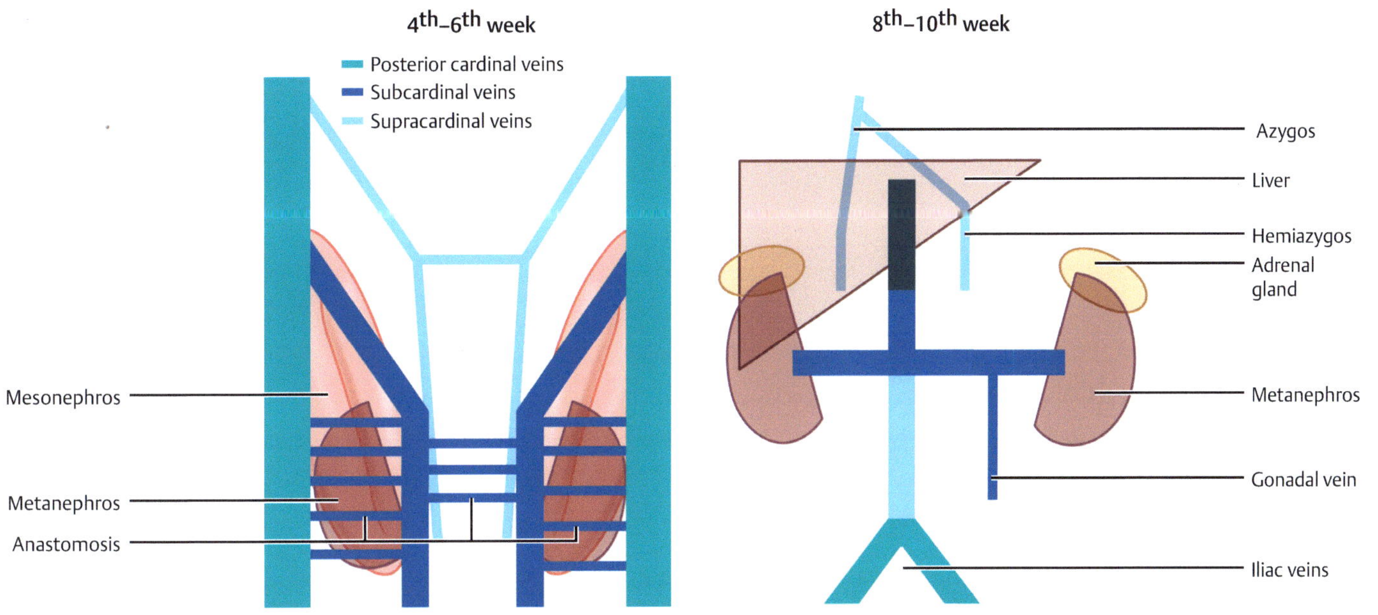

Fig. 13.17 Development of the renal venous system. There are three main venous systems that contribute to the formation of the renal veins and the inferior vena cava (IVC). The right subcardinal vein forms the suprarenal segment of the IVC, whereas the renal veins are derived from the anastomosis between the posterior cardinal and subcardinal veins. The hepatic segment forms from the vitelline vein. The azygos and hemiazygos systems are formed from the cranial part of the supracardinal vein. The caudal part of the right supracardinal vein persists as the infrarenal segment of the IVC. The left subcardinal vein regresses partially, with its distal portion remaining as the left gonadal vein. Finally, the lower segments of the posterior cardinal veins persist to form the iliac veins. (Modified with permission from Arévalo Pérez et al.[13])

IVC. They also develop a venous network that connects with the posterior cardinal veins (see Chapter 20 "Systemic Veins of the Abdomen and Pelvis"). The anastomotic branches between the posterior cardinal and subcardinal veins form the renal veins. Major portions of the posterior cardinal veins regress and the supracardinal veins transform into the lower IVC and lumbar and azygos/hemiazygos venous systems. Persistence of some of these anastomotic channels explains many of the variants found on imaging studies.

Arterial Supply

The renal vascular anatomy is an important consideration for embolization techniques, partial nephrectomy, and renal transplant surgery.[13–16] The renal arteries arise from either side of the aorta laterally, near the level of the L1–L2 interspace, 1 to 2 cm inferior to the origin of the superior mesenteric artery (▸Fig. 13.18). The abdominal aorta being left-sided, the right renal artery is longer than the left. Although the left kidney is located slightly

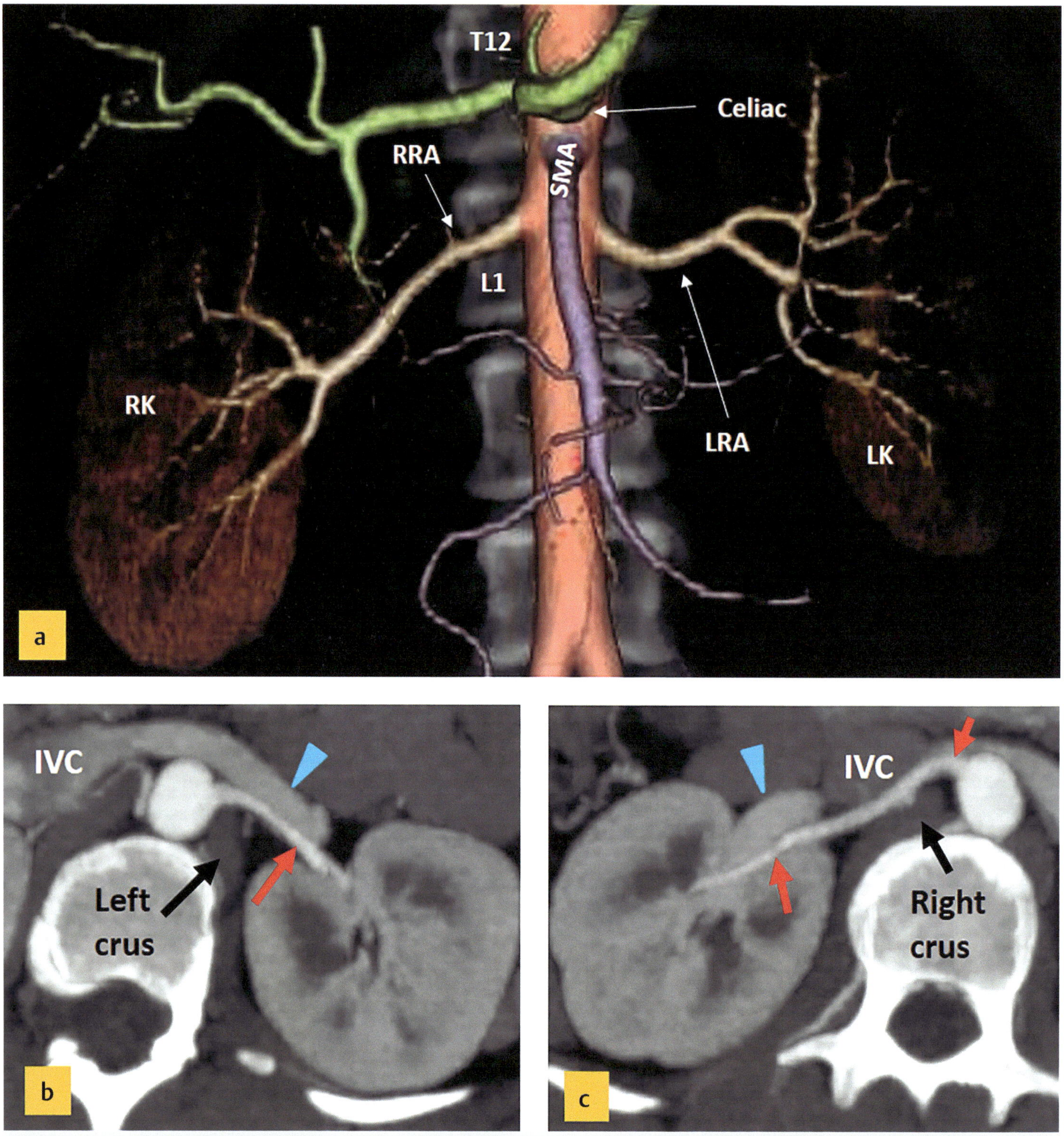

13

Fig. 13.18 Renal arteries. **(a)** The renal arteries arise from either side of the aorta laterally, near the level of the L1–L2 interspace, 1 to 2 cm inferior to the origin of the superior mesenteric artery (SMA). **(b)** The left renal artery (*red arrow*) courses between the left renal vein (*arrowhead*) and the left crus of the diaphragm. **(c)** The right renal artery is longer (*red arrows*) and courses between the IVC and the right crus of the diaphragm. Lateral to the IVC, the artery is posterior to the right renal vein (*arrowhead*). IVC, inferior vena cava; LK, left kidney; LRA, left renal artery; RK, right kidney; RRA, right renal artery.

higher in the abdomen than the right kidney, the left renal artery frequently arises slightly more inferior than the right renal artery. As a result, the left renal artery usually takes a fairly horizontal course, while the right renal artery usually has a more diagonal course inferolaterally.

The number and location of the renal arteries can vary significantly (▶Fig. 13.19). Only about 70% of kidneys are supplied by one artery, and so kidneys supplied by two or more arteries are not rare.[13] In multiple renal arteries with a separate origin, the largest vessel is considered the "main" renal artery, and the rest are called accessory or supernumerary. The main renal artery enters the kidney at the renal hilum. However, accessory arteries arising from the aorta may enter the renal hilum, may penetrate the cortex of either the upper or lower pole directly (▶Fig. 13.19), or may course around the kidney (termed as a capsular artery).

The left renal artery passes anterior to the left crus of the diaphragm, and posterior to the left renal vein (▶Fig. 13.18). The right renal artery passes anterior to the right crus of the diaphragm, and posterior to the IVC and right renal vein in 95% of cases[17] (▶Fig. 13.18). Accessory right renal arteries may course posterior or anterior to the IVC, and are consequently termed retrocaval or precaval, respectively. Accessory arteries most commonly arise from the aorta either superior or inferior to the main renal artery, and can occasionally arise from the common iliac arteries, but have also even rarely been reported to arise from superior and inferior mesenteric, celiac, middle colic, lumbar, gonadal, or middle sacral arteries.[13,18] Capsular arteries are generally small but can be large when serving as collaterals in renal artery stenosis. These arteries arise from the main renal artery but may originate from the segmental branches or the retroperitoneal arteries.

Early branching of the renal arteries (<2 cm from the aorta on the left side or retrocaval on the right side) is a common anatomical variant, especially on the right. Expert surgeons require at least 1 cm of main renal artery in order to clamp and properly anastomose the vessel. In less than 5% of patients, more than two arteries exist on the same side. The significance of the presence of more than two renal arteries is mainly for renal transplant surgery, and can be a contraindication to donation. If the third artery is a small superior polar artery less than 2 mm in diameter, it can be ligated without significant parenchymal loss.[15,19] The inferior polar artery may supply the collecting system and its ligation can cause ureteral ischemia, stenosis, or urinary tract leakage.

Renal artery branching is variable as well. Although typically the renal artery divides into segmental branches near the hilum, it may begin giving off branches anywhere along its course. The partitioning of the kidney by segmental arterial supply is not constant, but there is a prevailing pattern (▶Fig. 13.20). Typically, the renal artery divides first into posterior and anterior branches before reaching the renal hilum, and the posterior branch continues as the posterior segmental artery. The anterior branch gives rise to four segmental branches: the apical, upper, middle, and

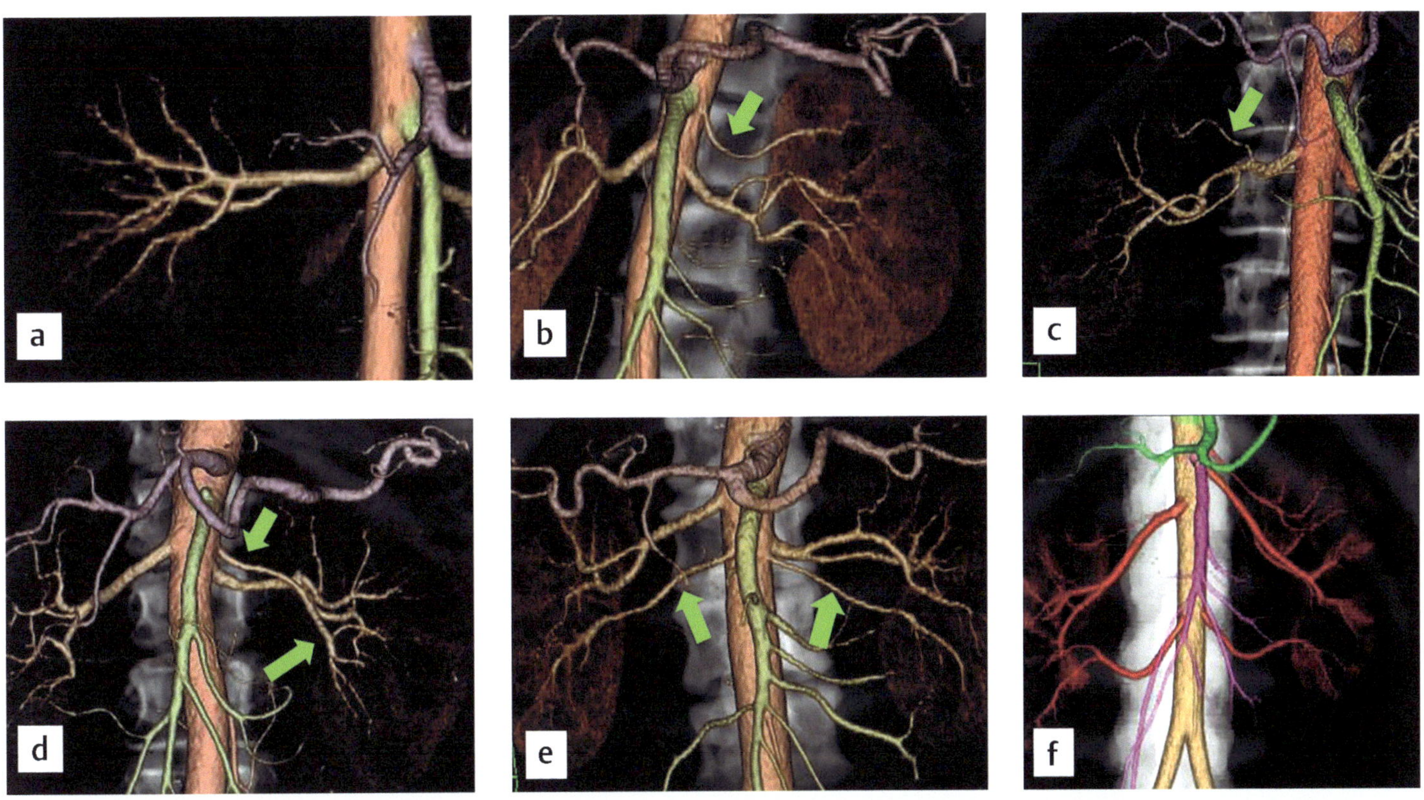

Fig. 13.19 Renal artery anatomical variations. **(a)** Classic pattern of the right kidney with two segmental branches. The anterior segment supplies apical, upper, middle, and lower segments while the posterior segment mainly feeds the posterior segment. **(b)** Accessory polar branch, arising from the aorta, supplies the left apical segment. **(c)** Polar artery arises from the right main renal artery (*arrow*). **(d)** Similar to **b** but the accessory branch supplies the inferior pole (*arrows*). **(e)** Accessory branches (*arrows*) on both sides to the lower poles. **(f)** Two renal arteries on the right and three on the left.

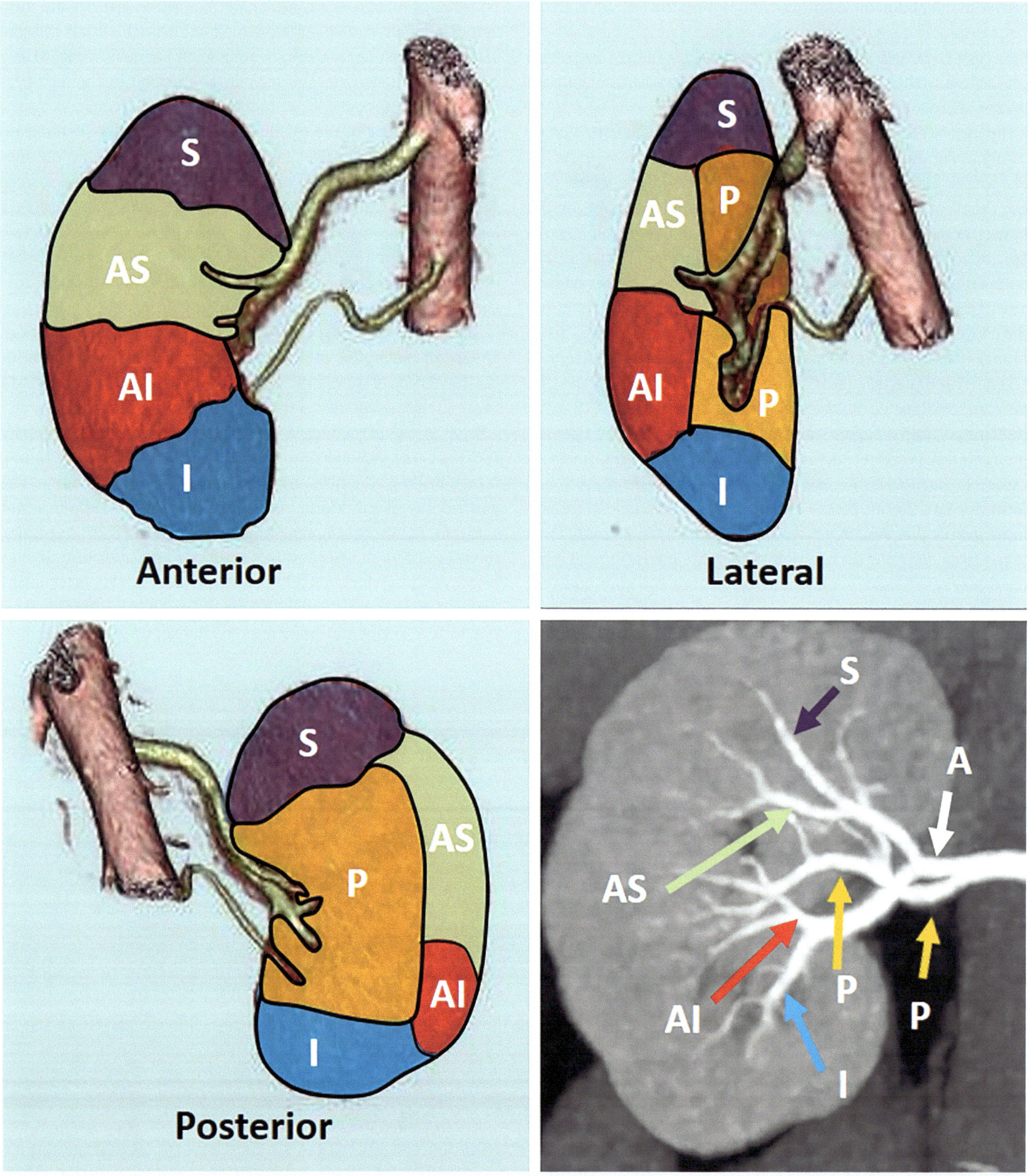

Fig. 13.20 Segmental anatomy of the kidney. The size of each segment may vary depending on the blood supply. The posterior (p) segment may extend to the superior apex. It is supplied by the posterior segmental artery or a separate accessory artery. A, anterior branch; AI, anteroinferior (middle); AS, anterosuperior (upper); I, inferior (lower); P, posterior; S, superior (apical).

lower segmental arteries (▶Fig. 13.20); alternatively, these four segments are sometimes called superior, anterosuperior, anteroinferior, and inferior.

There is no collateral arterial circulation between the segments, and each renal segment is supplied by its own segmental artery. The segmental arteries divide into lobar branches, generally one to each pyramid. The lobar arteries branch into the interlobar arteries, which course around each pyramid (▶Fig. 13.21). The interlobar arteries branch into the arcuate arteries, which course along the corticomedullary junction, parallel to the renal capsule. The interlobular arteries then branch orthogonally from the arcuate arteries, extending radially out to the cortical periphery. The interlobular branches give rise to the afferent and efferent glomerular arterioles. Occasionally, an interlobular artery can perforate the renal surface to anastomose with a capsular artery.

Renal artery pathologies are common and can be diagnosed by different imaging modalities. CT angiography plays an important role in the diagnosis of renal vascular diseases, including renal infarcts and ischemia secondary to renal artery stenosis, vascular injuries, arterial aneurysm, fibromuscular dysplasia, and arteriovenous communications (▶Fig. 13.22, ▶Fig. 13.23). Renal artery stenosis is secondary to atherosclerosis in two-thirds of patients and when severe can cause renovascular hypertension.[19] Atherosclerotic lesions usually occur at the origin or proximal third of the renal artery (▶Fig. 13.23). Fibromuscular dysplasia is the second most common cause of renal artery stenosis and usually involves main or segmental renal arteries in a young female causing a string-of-pearls appearance, with stenoses and small focal aneurysms (▶Fig. 13.22). The degree of stenosis, extent, and multiplicity of the renal artery can be diagnosed with CT or MR angiography noninvasively. Doppler examination is another modality to evaluate the hemodynamic significance of the stenosis (▶Fig. 13.23). The normal Doppler waveform of the renal artery demonstrates a rapid upstroke in systole, a low resistance profile (resistive index of <0.7) with continuous forward flow throughout the cardiac cycle, and a peak systolic velocity of <150 cm/second. Velocities greater than 200 cm/second usually indicate a >60% stenosis. High-grade stenosis results in downstream tardus parvus waveforms, characterized by delayed systolic rise and low systolic velocity.

The most common cause of renal artery aneurysms is atherosclerosis. Renal artery pseudoaneurysms are usually posttraumatic. Extrarenal renal artery aneurysms are found incidentally in 1% of imaging studies and in most cases measure less than 2 cm. Renal aneurysms account for 22% of visceral aneurysms.[19] Aneurysms that are <2 cm in diameter and partially calcified in a nonhypertensive patient can be followed by imaging for any interval growth. Renal infarctions are usually thromboembolic in nature, from an intracardiac or intravascular source, but also can rarely occur after traumatic renal artery injury (▶Fig. 13.22).

Venous and Lymphatic Drainage

The renal parenchyma is drained consecutively by the stellate, arcuate, and interlobar veins, and as opposed to the segmental arrangement of intrarenal arteries, there are extensive intrarenal anastomoses between venous tributaries (▶Fig. 13.24). In turn, ligation of the venous branches is possible without the risk of parenchymal loss. The renal veins have a more uniform anatomical pattern and are single in 92% of individuals.[20] Renal vein tributaries usually pass anterior to the renal pelvis but the incidence of retropelvic tributaries is also common, ranging from 30 to 45%.[21]

The left renal vein is approximately 8 cm long, and from the IVC it passes between the superior mesenteric artery and the aorta (▶Fig. 13.25). Continuing toward the kidney, it passes anterior to the left renal artery and posterior to the inferior mesenteric vein. It is also often posterior to the distal duodenum, although at times it may be posterior to the splenic vein and pancreatic body. The left gonadal vein usually empties into the left renal vein, approaching from below (▶Fig. 13.26) (see Chapter 15 "Male Genitourinary" and Chapter 16 "Female Genital System"). Thrombosis of the left renal vein may result in enlargement of

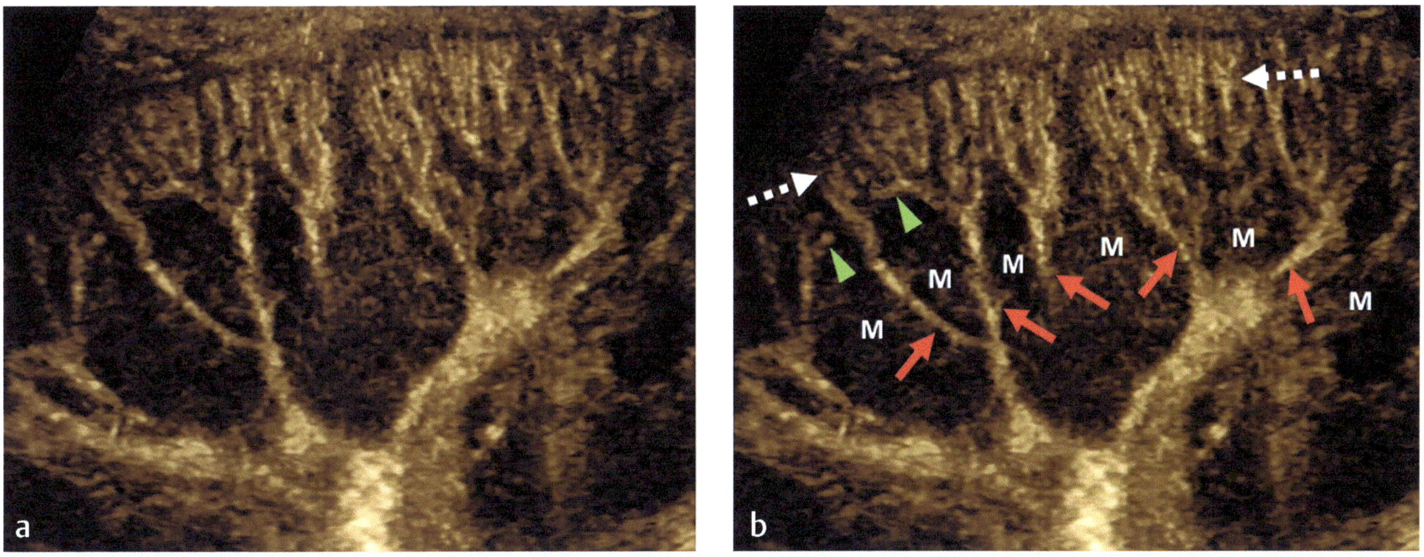

Fig. 13.21 B-flow ultrasound image of a transplant kidney (**a**, unlabeled, **b**, labeled) showing interlobar vessels (*red arrows*) coursing around the medullary pyramids (M). Some arcuate vessels are shown coursing across the bases of the pyramids at the corticomedullary junction (*green arrowheads*). Interlobular vessels are shown packed together in the cortex (between the *dashed arrows*), extending out toward the capsule.

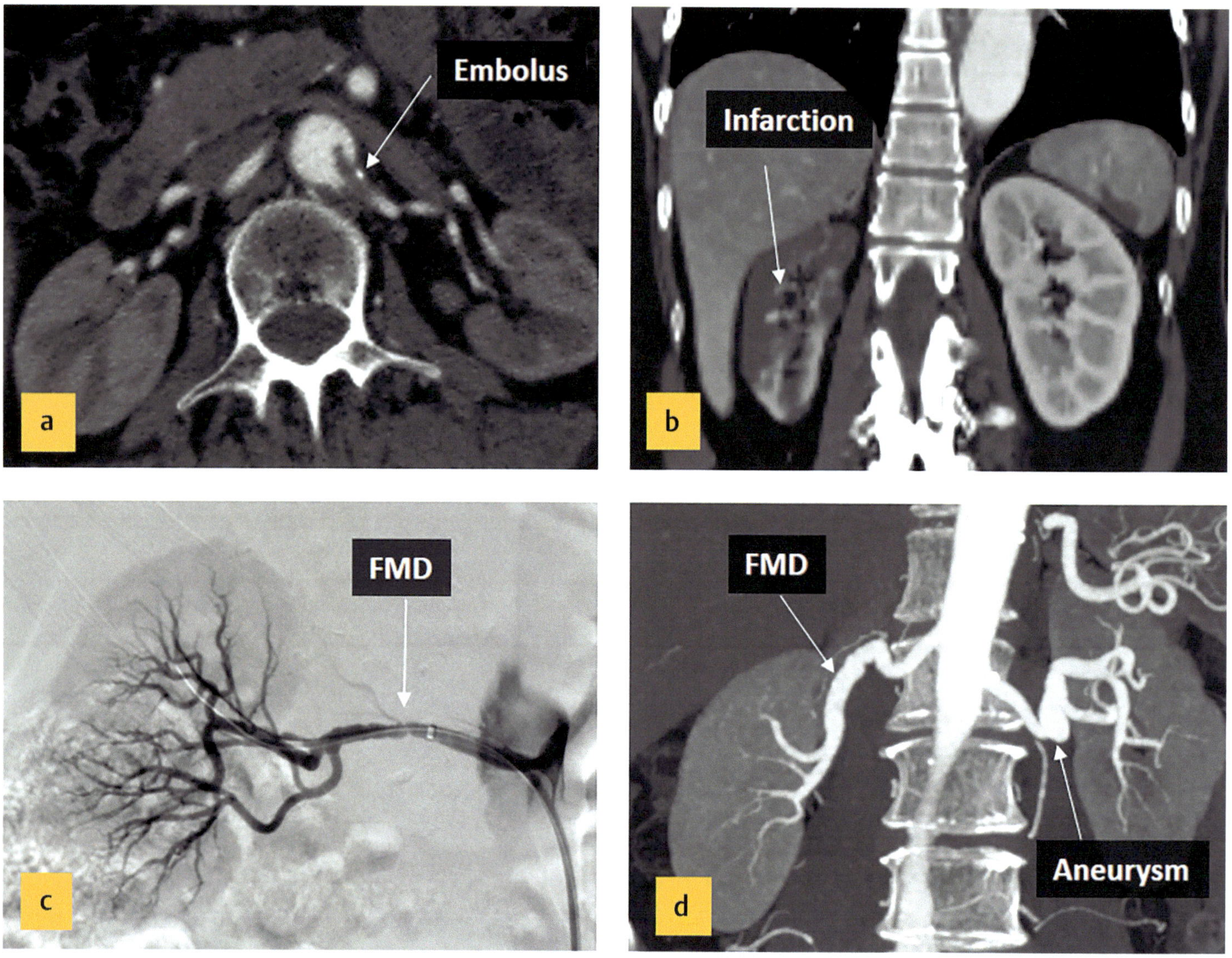

Fig. 13.22 (a) Left renal artery clot originating from the heart in a patient with atrial fibrillation. **(b)** Heterogeneous nephrogram in a small chronically infarcted right kidney. **(c, d)** Fibromuscular dysplasia (FMD) in two different patients. **(c)** Selective renal artery angiography showing typical string-of-pearls appearance of the right renal artery. **(d)** Bilateral FMD creating the string-of-pearls appearance with a stenosis and poststenotic aneurysm.

13

the left gonadal vein causing ovarian or testicular varices. From above, the left adrenal vein empties into the left renal vein, after having confluenced with the left inferior phrenic vein. The left second lumbar vein also usually empties into the left renal vein (▶Fig. 13.3, ▶Fig. 13.27) (see Chapter 20 "Systemic Veins of the Abdomen and Pelvis"). The left renal vein may connect with the ascending lumbar and hemiazygos veins in 30 to 75% of individuals.[21,22] The lumbar veins are a potential source of bleeding after left nephrectomy. Ascending lumbar veins are also a major route of retroperitoneal splenorenal shunt development in portal hypertension (see Chapter 11 "Portal Venous System" and Chapter 12 "Peritoneal Spaces").

Sometimes the left renal vein may have two moieties, with one in the normal location, and the other passing posterior to the aorta. This variant, present in 5 to 7% of patients, is termed a circumaortic left renal vein, and usually the posterior moiety is more inferior and smaller (see Chapter 20 "Systemic Veins of the Abdomen and Pelvis"). Occasionally, the orthotopic moiety is not present, and the moiety posterior to the aorta is the only left renal vein, in which case it is termed a retroaortic left renal vein. Retroaortic renal vein is a less common venous anomaly with a frequency of 2 to 5%[23] (▶Fig. 13.27).

The right renal vein is less than 2.5 cm in length, receives no extrarenal tributaries, and usually empties into the IVC with an obliquely upwards course (▶Fig. 13.18). It passes anterior to the right renal artery and posterior to the second portion of the duodenum. The right gonadal vein usually does not empty into the right renal vein, but instead into the IVC. Direct drainage of the right gonadal veins and retroperitoneal veins into the right renal vein is seen in 5% of individuals. The right adrenal vein drains into the right renal vein in 30% of cases. Duplication of the right renal vein is not rare but unusual (▶Fig. 13.27).

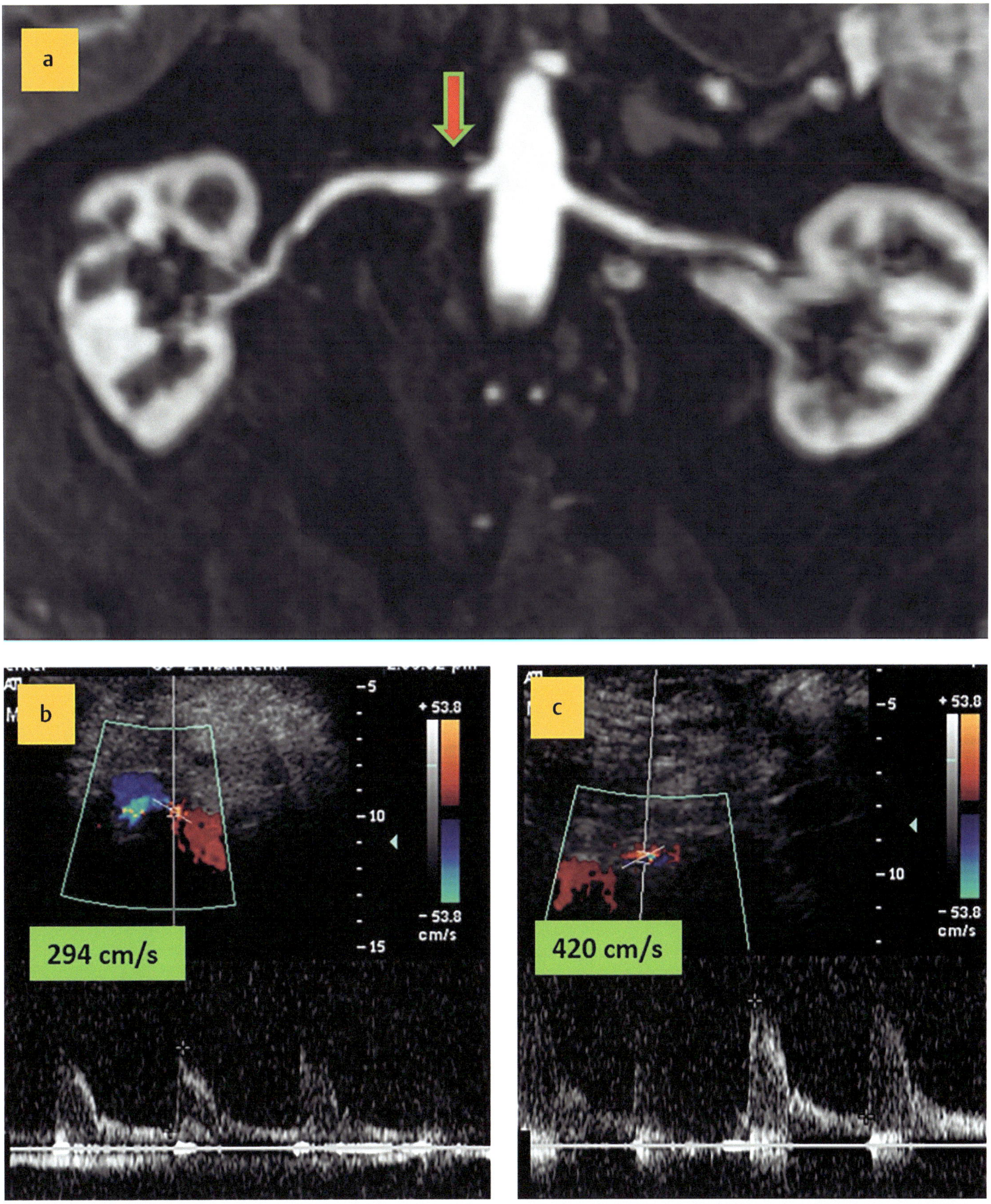

Fig. 13.23 (a) Right renal artery stenosis (*arrow*) is shown by coronal magnetic resonance (MR) angiography. **(b)** Doppler of the left renal artery origin showing mildly elevated peak systolic velocity whereas on the right **(c)** the velocity is markedly elevated.

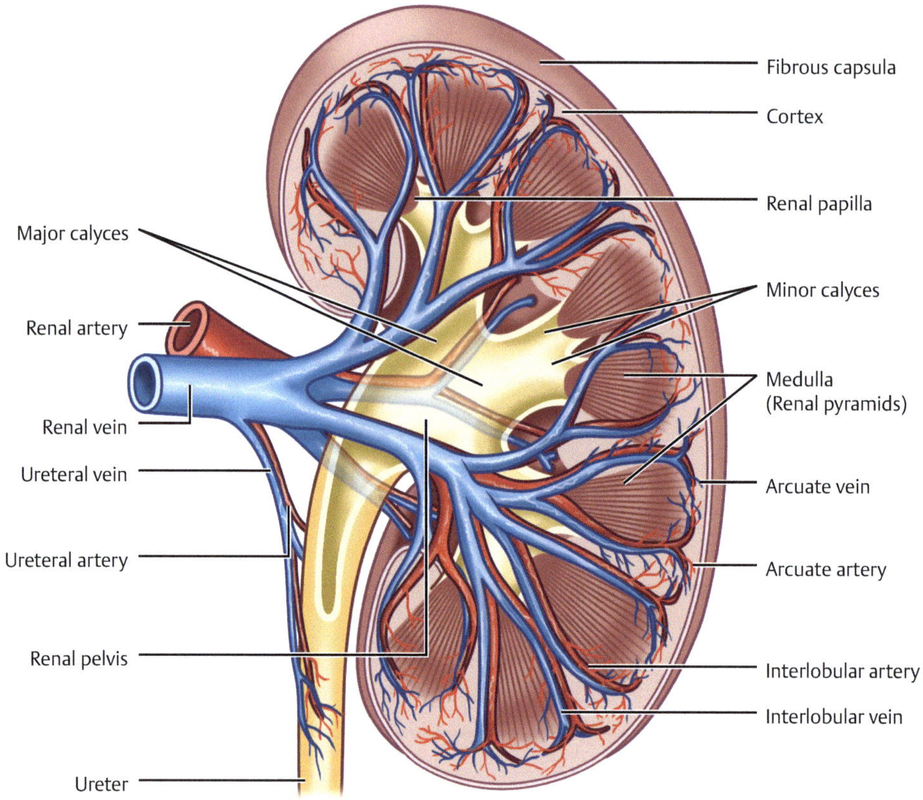

Fig. 13.24 Illustration showing the lobar, interlobar, arcuate, and interlobular vessels.

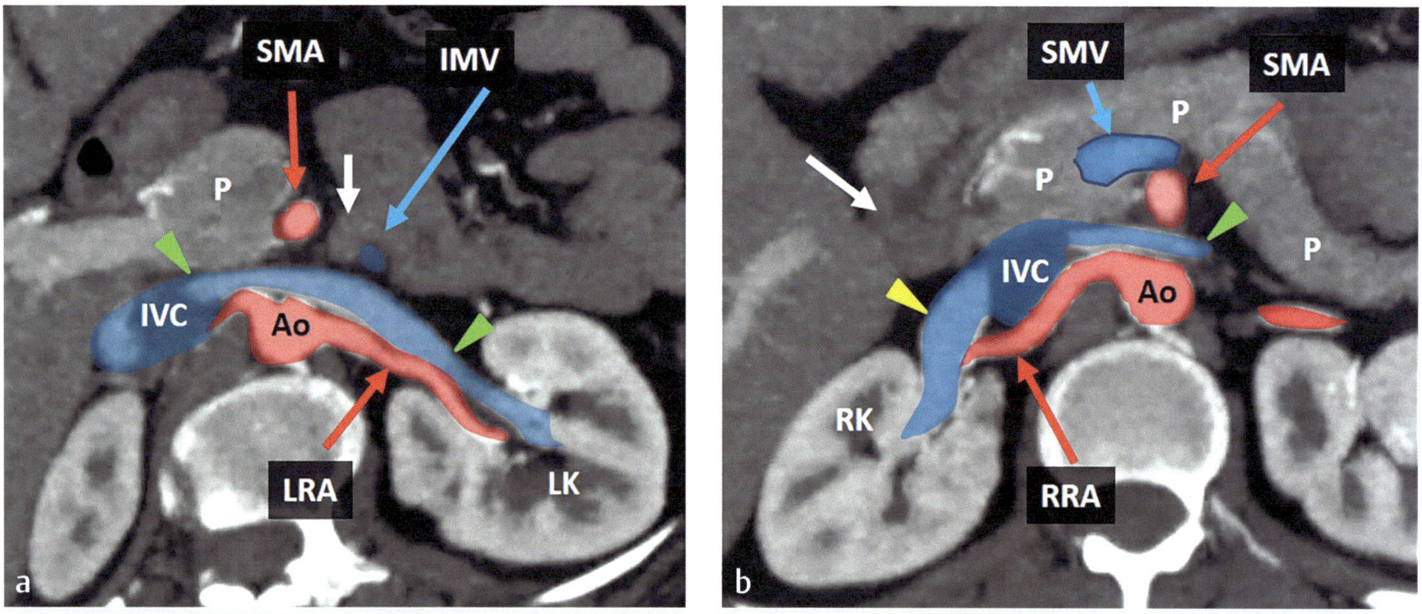

Fig. 13.25 (a) Axial oblique computed tomography (CT) showing the left renal vein (*green arrowheads*) coursing anterior to the left renal artery (LRA) and posterior to the inferior mesenteric vein (IMV) and the junction of the third/fourth portions of the duodenum (*short white arrow*). The left renal vein then passes between the aorta (Ao) and the superior mesenteric artery (SMA), before emptying into the inferior vena cava (IVC). **(b)** Axial oblique CT showing the right renal vein (*yellow arrowhead*) passing anterior to the right renal artery (RRA) and posterior to the second portion of the duodenum (*short white arrow*), before reaching the IVC. LK, left kidney; P, pancreas; RK, right kidney.

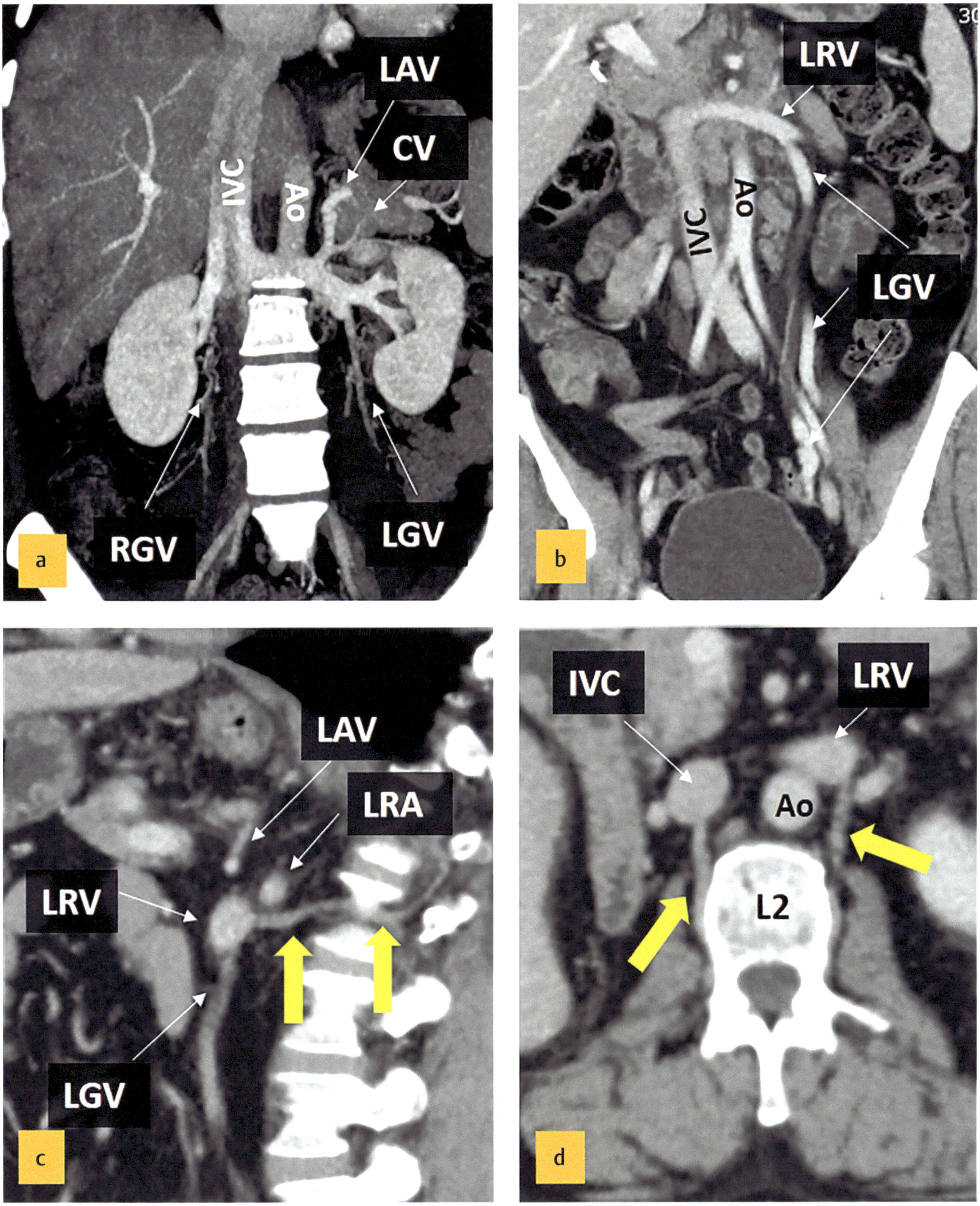

Fig. 13.26 Tributaries of the left renal vein (LRV). **(a)** Coronal computed tomography (CT) shows confluences of the left gonadal vein (LGV), left adrenal vein (LAV), and capsular vein (CV) with the LRV. **(b)** Coronal CT shows an enlarged LGV emptying into the LRV. **(c)** Sagittal CT shows the LRV cross section with its tributaries, including the LGV, LAV, and the lumbar vein (*arrow*). **(d)** Axial CT at L2 level shows the right lumbar vein entering the IVC and the left lumbar vein entering the LRV. Ao, aorta; LRA, left renal artery; RGV, right gonadal vein.

13

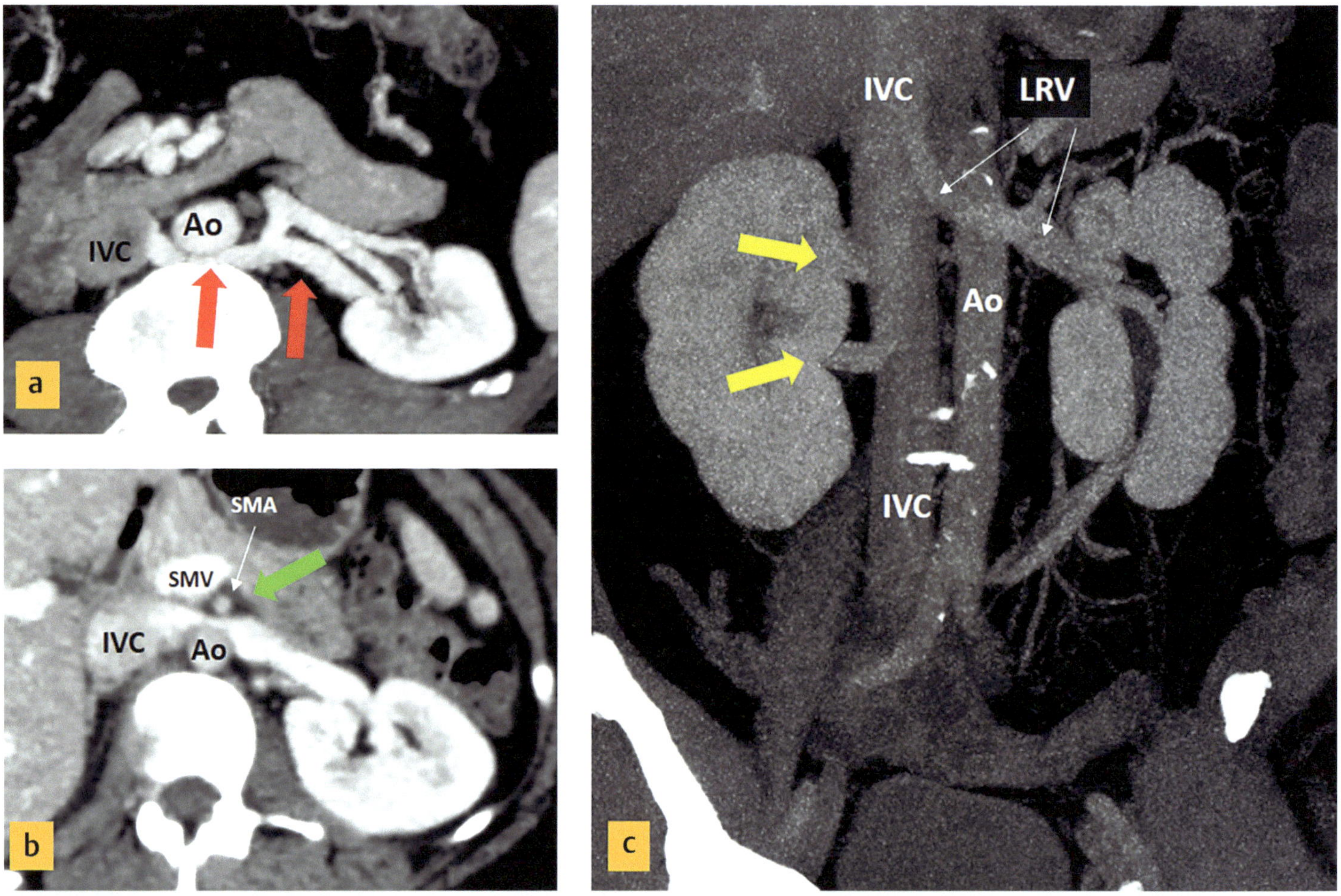

Fig. 13.27 (a) Axial computed tomography (CT) showing a retroaortic left renal vein being compressed (*arrows*) between the aorta (Ao) and the spine. **(b)** Axial CT showing a preaortic left renal vein being compressed (*arrows*) between the aorta and the superior mesenteric artery (SMA). **(a, b)** Potential examples of the so-called nutcracker phenomenon. **(c)** Coronal CT shows a duplicated right renal vein (*arrows*). This is an unusual venous variant. Ao, aorta; IVC, inferior vena cava.

13

Nutcracker phenomenon refers to compression of the left renal vein between the superior mesenteric artery and the aorta (▸Fig. 13.27). On rare occasions, compression of the retroaortic left renal vein by the aorta can also cause the nutcracker phenomenon. Nutcracker phenomenon is a common variant which is seen in 27% of asymptomatic population. Dilated gonadal and lumbar veins may be an associated finding in 50% of cases.[24] Clinical and laboratory findings reported in association with nutcracker syndrome include hematuria, proteinuria, flank pain, and pelvic congestion syndrome. Thrombosis is another occasionally encountered pathological entity of the renal veins, especially on the left side where the renal vein is longer (▸Fig. 13.28). In acute renal vein thrombosis, renal perfusion can be disturbed involving both the cortex and renal pyramids.[25] The renal veins and IVC can be invaded by tumor, for example, by renal cell carcinoma, and the level of superior extent of tumoral invasion up the IVC is an important factor for surgeons in operative planning (▸Fig. 13.28).

The drainage system of the renal lymphatics follows the intrarenal arteries and veins, eventually reaching the renal sinus.[26] From there the main lymphatic trunks follow the renal veins, to para-aortic nodes on the left, and paracaval nodes on the right.[27]

Microvasculature

Microvasculature refers to vessels less than 150 to 200 µm in diameter, including arterioles, capillaries, and venules.[28–30] The afferent arterioles, originating from interlobular arteries, usually lead to the glomerular capillaries which then form the efferent arterioles. The afferent arterioles also form a second capillary network constituted by the peritubular capillaries, which surround the renal tubules (▸Fig. 13.29). The peritubular capillaries are followed by the venous system that runs parallel to the arteriolar vessels and progressively forms the interlobular, arcuate, interlobar, and renal veins. Therefore, the distinct feature of the renal circulation is a dual capillary bed system residing in both the glomerulus and around the tubules. This dual capillary bed arrangement is important for efficient filtration, secretion, and reabsorption of minerals and removal of unwanted substances from the filtrate.

The kidneys are relatively small organs and account for 0.5% of total body weight. In contrast, renal blood flow (RBF) is high and involves 1 L/minute which is 20 to 22% of the cardiac output.[28] This high blood flow is necessary for the kidneys to maintain

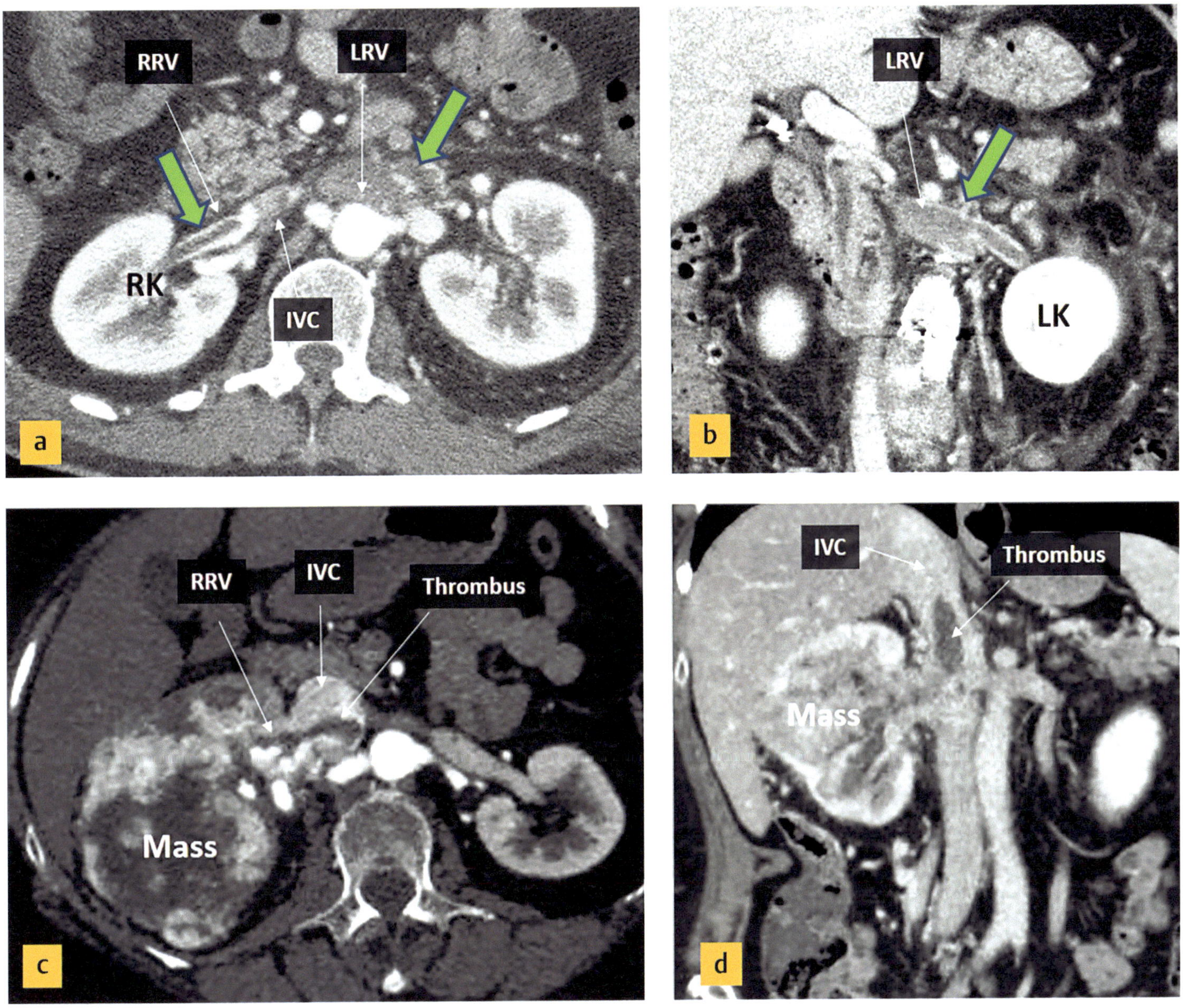

Fig. 13.28 (a) Axial computed tomography (CT) and **(b)** coronal CT angiography showing bilateral renal veins and inferior vena cava (IVC) thrombosis (*arrows*) causing pulmonary emboli (not shown). **(c)** Axial CT and **(d)** coronal CT angiography showing right renal vein (RRV) and IVC tumor thrombosis secondary to a large renal cell carcinoma of the right kidney (RK). LK, left kidney; LRV, left renal vein.

their function and meet metabolic requirements. Renal perfusion is very heterogeneous, ranging from 7 mL/minute/g tissue in the cortex to approximately 3 mL/minute/g tissue at the corticomedullary junction, 2 mL/minute/g in the outer medulla, and 0.5 mL/minute/g tissue in the inner medulla. Despite high RBF, the kidneys are relatively sensitive to hypoxia, especially in the medulla. This is because of the unique microvasculature architecture of the kidney. The oxygen tensions in the kidney vary from 70 mmHg in the cortex to 20 mmHg in the medulla.[30] The reason for this remarkable discrepancy between blood supply and oxygenation is the regionally specialized renal microvasculature. The cortex receives a large amount of blood for the filtration function, but 80% of renal oxygen consumption is used to drive Na–K-ATPase to reabsorb most of the NaCl filtered into the proximal tubule. This phenomenon, in combination with arteriovenous shunting between the descending and ascending vasa recta, further compromises oxygen supply to the medulla. It was shown by Leong et al that RBF could be reduced or increased by approximately 30%, without detectable changes in tissue pO_2 in the cortex or medulla.[31] Although an increase in RBF is accompanied by an increase in renal O_2 delivery, the renal O_2 consumption remains stable. This stability in renal O_2 consumption despite an increase in O_2 delivery (i.e., high RBF) is due to direct shunting between the precapillary arterioles and venules, causing dynamic regulation of renal parenchymal oxygenation.[31]

Renal medullary circulation is critical in maintaining water and fluid balance as well as blood pressure.[32] Although only approximately 10% of total RBF enters the medulla, deficiency

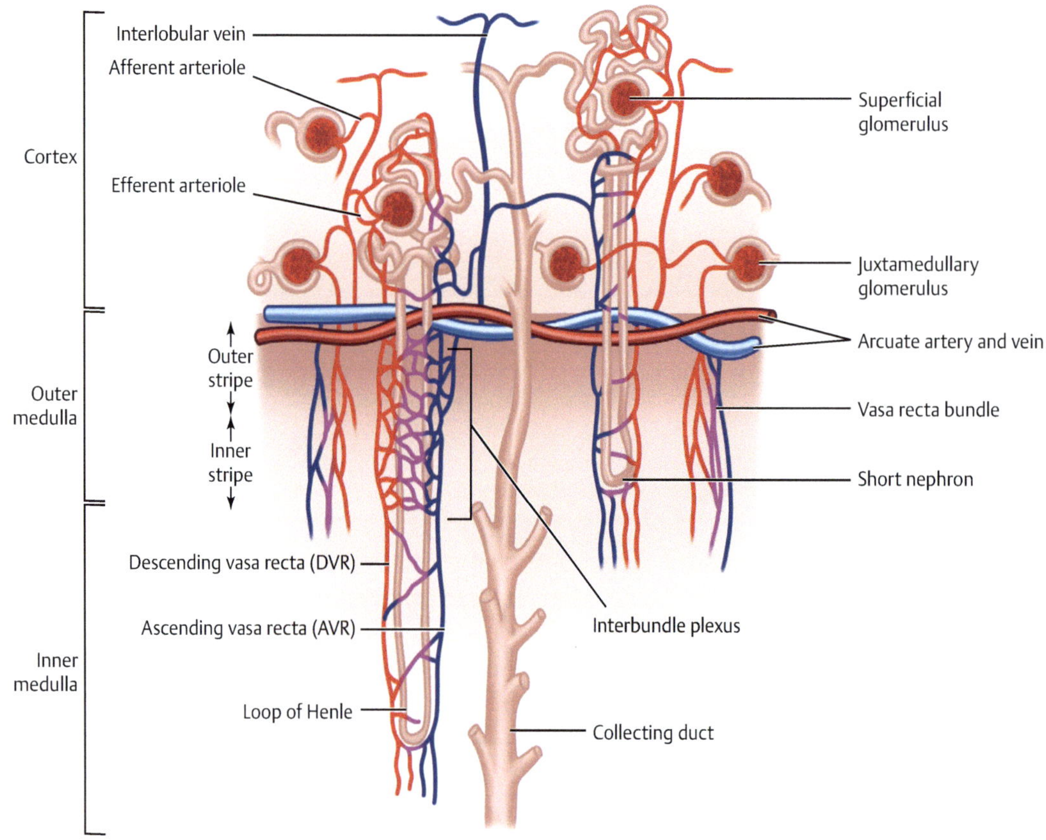

Fig. 13.29 Microcirculation of the renal cortex and medulla. The interlobular arteries, arising from the arcuate artery, ascend toward the cortical surface. The majority of blood flow to the medulla arises from juxtamedullary efferent arterioles. In the outer medulla, juxtamedullary efferent arterioles give rise to descending vasa recta (DVR) that coalesce with ascending vasa recta (AVR) and, sometimes, thin descending limbs of Henle to form vascular bundles. Vascular bundles are prominent in the inner stripe of the outer medulla. Peripheral DVRs form the interbundle capillary plexus that supplies nephrons and central DVRs perfuse the inner medulla. In the inner medulla, vascular bundles disappear and vasa recta become dispersed.

in medullary perfusion can have a significant impact on normal renal function. The descending vasa recta (DVR), originating from the juxtamedullary efferent arteriole, supply all blood flow to the renal medulla and extend to the inner medulla (▶Fig. 13.29). In the inner stripe of the outer medulla, vascular bundles formed by the DVRs move in parallel with the ascending vasa recta (AVR) and short-looped thin descending limbs of Henle. The AVR are larger and more numerous than the DVR and their endothelium is highly fenestrated, while the endothelium of the DVR is continuous with tight junctions.[33]

Vascular nephropathies account for over a third of all cases of end-stage renal disease and are more common in metabolic syndrome, diabetes, and in the presence of atherosclerosis and hypertension.[30] Renal artery stenosis may progressively induce renal tissue injury, ischemic nephropathy, and end-stage renal disease by promoting microvascular endothelial dysfunction, damage, and loss.[34]

Nervous Supply

Autonomic supply to the kidney comes from the aorticorenal ganglion, the celiac ganglion, the major splanchnic nerves, and the superior mesenteric ganglion.[35] There are also afferent nerve fibers from the kidney, with cell bodies in the T12–L3 dorsal root ganglia.[36] Nerves reach the kidney following the renal artery.

◆ Collecting System and Ureters

The arrangement of the renal collecting system is subject to some variation. Each renal papilla tends to empty into a single minor calyx (simple papilla/calyx), although two or three papillae (compound papilla) can sometimes drain into a single minor calyx. Minor calyces are funnel-shaped, surrounding the papillae that project into them (▶Fig. 13.30). Each kidney may have 8 to 14 minor calyces, and generally 4 to 5 adjacent minor calyces unite to form a major calyx, of which there are 2 to 3 per kidney. The medullary collecting ducts, also known as papillary ducts, terminate at the renal papilla and empty into the minor calyces. These papillary ducts may be seen at contrast CT as converging tubules ending in the calyces (▶Fig. 13.30). The calyces empty into the infundibula, which join to form the renal pelvis. The renal pelvis may be intrarenal or extrarenal; if extrarenal, and not confined by the space of the renal sinus, it is common for it to distend more than a typical intrarenal pelvis (▶Fig. 13.31). The extrarenal pelvis should not be confused with ureteropelvic junction (UPJ) obstruction or renal sinus cysts. The renal pelvis smoothly tapers

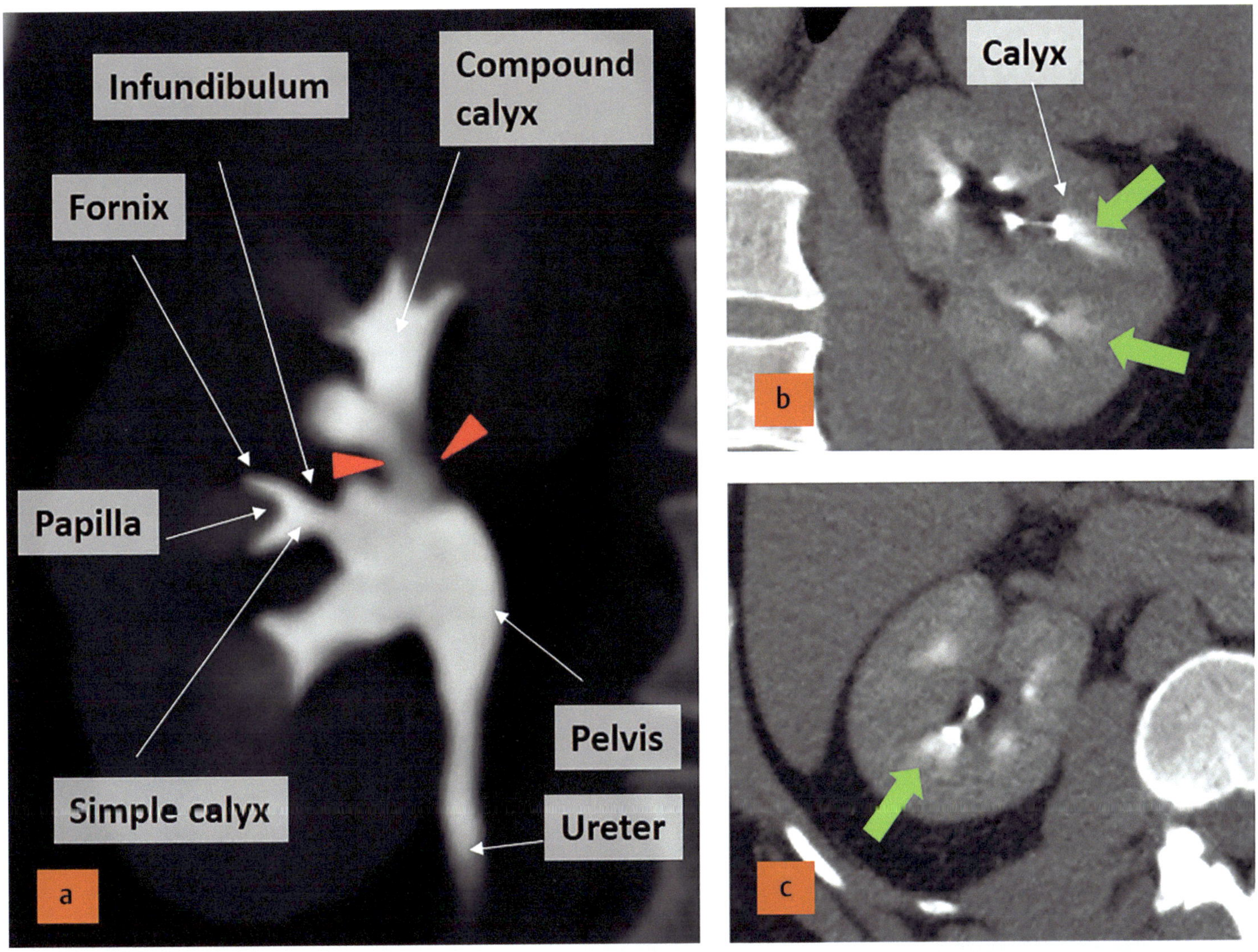

Fig. 13.30 (a) Coronal computed tomography (CT) urogram. The renal papillae protrude into minor calyces, which join to form a major calyx (*arrowheads*). The major calyces empty into the renal pelvis. Calyces can be classified as simple or compound based on papillary morphology. For compound papillae, which are found most frequently in the renal poles, the papillary ducts have round or oval orifices, which are more prone to intrarenal urine reflux. Normal calyces are cup-shaped, with a sharp free edge called the fornix. Destruction of the papilla from reflux nephropathy and papillary necrosis can change the shape of calyces. The ureteropelvic junction is medial to the renal lower pole, distal to which the upper urinary tract becomes a narrow tube, the ureter. **(b)** Coronal contrast-enhanced CT of the left kidney and **(c)** axial CT of the right kidney showing converging papillary ducts (arrows) ending in calyces.

13

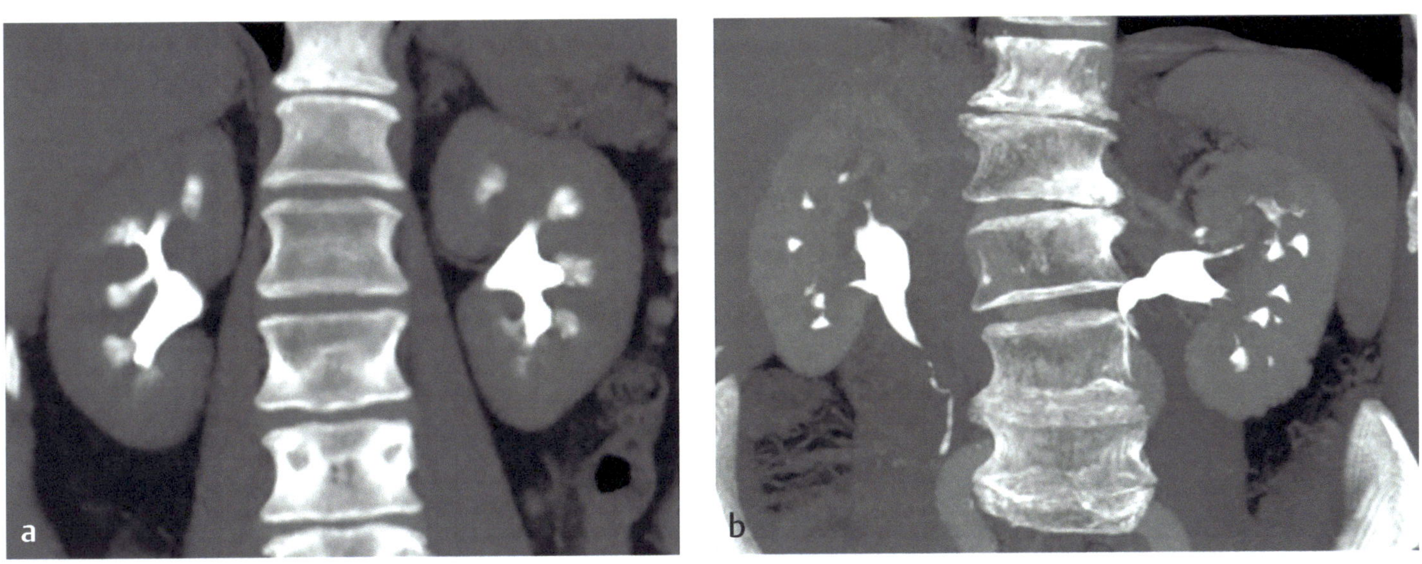

Fig. 13.31 Coronal computed tomography (CT) urogram showing **(a)** intrarenal and **(b)** extrarenal pelvises.

as it courses inferiorly to the UPJ, which is located medial to the lower pole of the kidney.

The tubular continuations of the renal pelvises are the ureters, which deliver urine to the urinary bladder (▶Fig. 13.32). The combination of the renal collecting system and ureter together is known as the upper urinary tract. The ureter is a retroperitoneal conduit that measures 25 to 30 cm in length, 3 to 4 mm in diameter, and has a muscular wall capable of periodic peristaltic contractions. Each ureter is composed of a stretchable inner cell layer of transitional urothelium which is surrounded by the lamina propria layer, inner longitudinal and outer circumferential smooth muscle layers, and an outer adventitial layer of connective tissue with numerous small blood vessels. Since the ureter is a retroperitoneal structure, it has no peritoneal serosa.

After its origin from the renal pelvis, the ureter courses down through the retroperitoneum on the anterior surface of the psoas major muscle. As the ureter courses down the psoas, the gonadal vessels cross over it anteriorly, from medial to lateral (▶Fig. 13.33, ▶Fig. 13.34).

Before entering the true pelvis, the ureter crosses over the genitofemoral nerve. The ureter enters the true pelvis by crossing over the iliac vessels near the transition from common iliac to external iliac. On the left side at this same location, the ureter courses underneath the root of the sigmoid mesocolon near the intersigmoid recess. Overstretching of the left ureter after urinary diversion surgery can cause obstruction where it passes underneath the sigmoid mesocolon. On the right side, the ureter courses underneath the root of the small bowel mesentery, just above its crossing of the iliac vessels.

In the true pelvis, the ureter courses along the pelvic sidewall between the external and internal iliac vessels, medial to the obturator nerve (▶Fig. 13.33, ▶Fig. 13.34). Then, after turning medially away from the pelvic sidewall, the ureter courses toward the base of the bladder. In the male, the ureter passes underneath

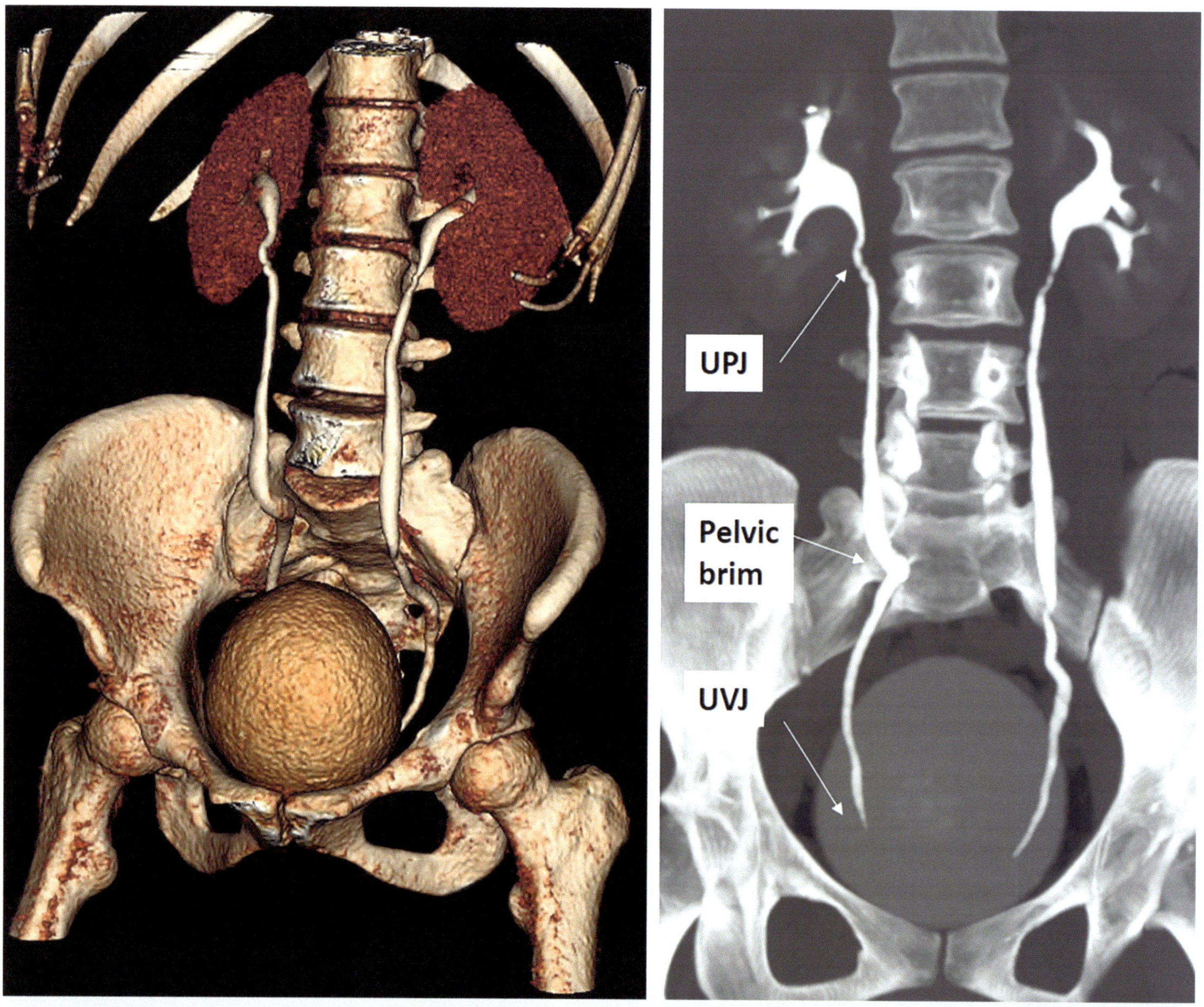

Fig. 13.32 Normal anatomic course of the ureters shown with computed tomography (CT) urography. Three areas of relative narrowing of the ureter are seen: At the ureteropelvic junction (UPJ), at the pelvic brim where the ureter crosses over the iliac vessels to enter the pelvis, and at the ureterovesical junction (UVJ).

13

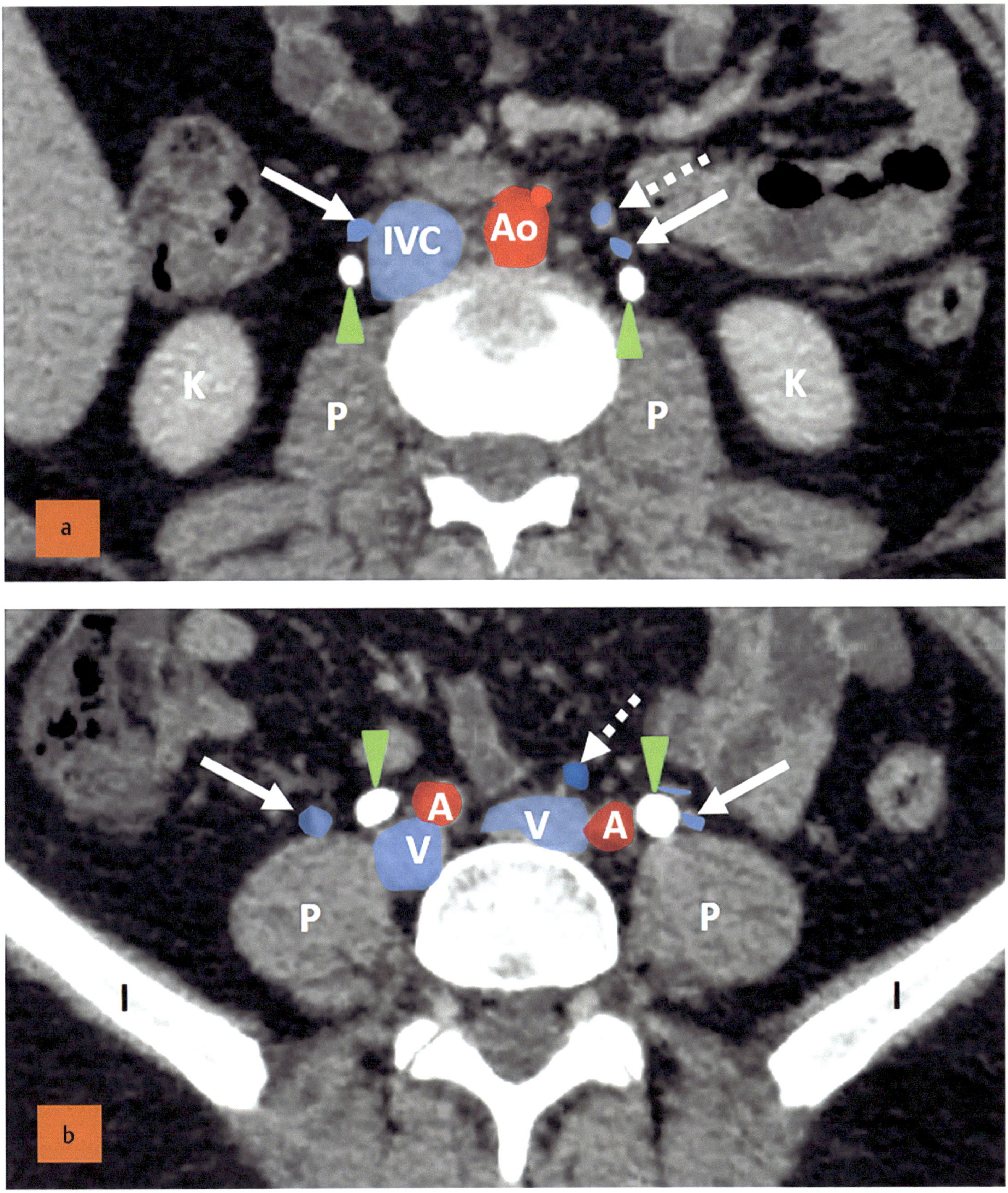

Fig. 13.33 Axial computed tomography (CT) images in urographic phase. **(a)** Superiorly, at the level of the lower poles of the kidneys (K), the ureters (*green arrowheads*) lie anterior to the psoas muscle (P). The gonadal veins (*white arrows*) are anterior and slightly medial to the ureters at this level. **(b)** Inferiorly, at the level of the top of the iliac wings (I), the ureters (*green arrowheads*) are again anterior to the psoas (P), but the gonadal veins have crossed over them and are now lateral (*white arrows*). Between the ureters are the common iliac arteries (A) and veins (V). Ao, aorta; IVC, inferior vena cava (*dashed arrow* = inferior mesenteric vein).

13

the vas deferens, and reaches the bladder just anterosuperior to the seminal vesicle. In the female, the ureter passes posterior to the ovary, forming part of the posterior boundary of the ovarian fossa, and then reaches the bladder by passing underneath the uterine artery, as this artery proceeds from the pelvic sidewall to the uterus via the cardinal ligament (see Chapter 16 "Female Genital System"). Having reached the ureterovesical junction (UVJ), the ureter passes obliquely through the wall of the bladder to reach the ureteral orifice, located at one of the angles of the bladder trigone.

Three areas of relative narrowing of the ureter are well known as urinary calculi tend to lodge at these locations: at the UPJ, at the pelvic brim where the ureter crosses over the iliac vessels to enter the true pelvis, and at the UVJ (▶ Fig. 13.32). A common example of partial ureteral obstruction, especially on the right side, occurs in the third trimester of pregnancy, during which the midureter may be compressed at the pelvic brim by the enlarged uterus.

Vasculature

Vascular supply to the ureter shows a great deal of variation. As the ureter courses from the kidneys to the bladder it receives branches from the renal, gonadal, common iliac, internal iliac, vesical, and uterine arteries. The ureter may also receive branches directly from the abdominal aorta. In general, the abdominal portion of the ureter receives its arterial branches along its medial side, while in the pelvis the ureter receives most of its arterial supply along its lateral side. Once arterial branches reach the ureter, they travel longitudinally along its adventitia, forming anastomotic networks with adjacent arterial branches. The venous

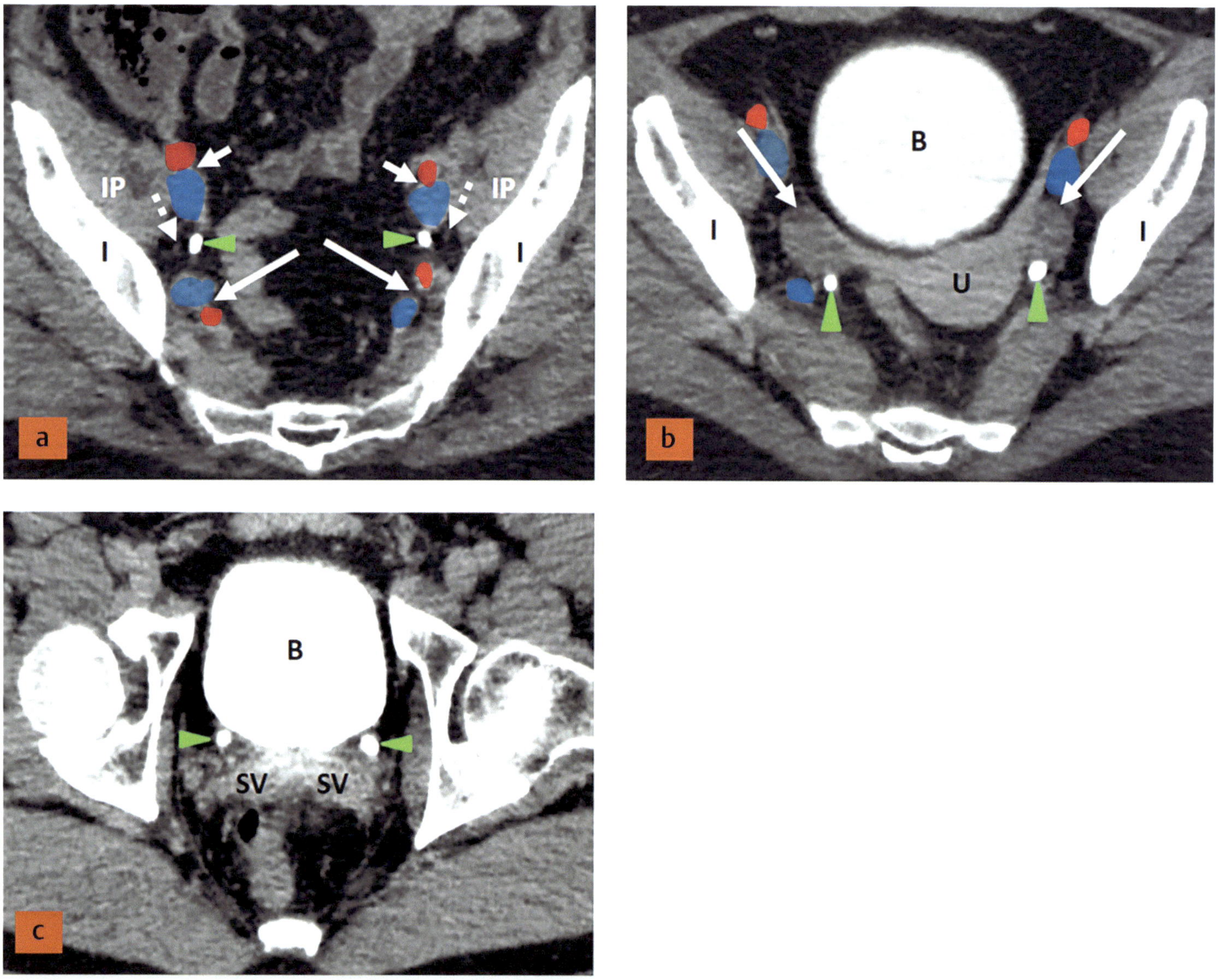

Fig. 13.34 (a) Axial computed tomography (CT) through the pelvis in urographic phase shows the ureters (*green arrowheads*) posterior to the external iliac vessels (*short white arrows*) and anterior to the internal iliac vessels (*long white arrows*). The obturator nerves (*dashed arrows*) are lateral to the ureters. **(b)** In the female, at a lower level the ureters (*arrowheads*) proceed toward the bladder (B) by passing posterior to the ovaries (*arrows*). **(c)** In the male, the ureters (*arrowheads*) reach the bladder (B) by passing anterior to the seminal vesicles (SV). I, ilium; IP, iliopsoas muscle.

drainage of the ureter tends to follow the arterial branches. As one progresses from proximal to distal along the ureter, lymphatic drainage is to the lateral aortic nodes, the common iliac nodes, and then the external/internal iliac nodes.

Function

According to the myogenic theory of ureteral peristalsis, peristalsis is initiated by pacemaker cells in the renal pelvis, and propagated anterograde through the smooth muscle of the ureteral wall.[37] The autonomic innervation of the ureter does not appear to contribute to peristalsis initiation or propagation, but instead has a modulatory role on ureteral contractility. Autonomic innervation to the ureter is via the renal, aortic, superior hypogastric, and inferior hypogastric plexuses.[38] Because of peristalsis, occasionally it can be difficult to opacify the entire length of the ureter during CT urography.

Variants

One of the most common embryological variants of the upper tracts is duplication of the collecting system.[39] In this variant, the kidney has two separate renal pelvises leading to two separate ureters termed the upper and lower moieties (▶Fig. 13.35, ▶Fig. 13.36). The two ureters may unite somewhere along their course to form one ureter (termed partial duplication) or may remain completely separate and end at two different ureteric orifices (termed complete duplication). According to the Weigert-Meyer rule, when the duplication is complete, the most common pattern is that the ureter of the lower pole moiety tends to have an orthotopic bladder insertion, but is prone to vesicoureteral reflux (▶Fig. 13.36). The ureter of the upper pole moiety tends to insert ectopically on the bladder more inferior and medial than the lower moiety, and is prone to development of a ureterocele, with resultant ureteral obstruction (▶Fig. 13.36). A ureterocele is a pouch-like bulging of the distal ureter into the wall of the urinary bladder, usually congenital in etiology. The Weigert-Meyer rule does not apply to every case of complete duplication (▶Fig. 13.36). Ectopic ureter insertion can present with urinary tract infections and incontinence. Presence of a small ureterocele in an otherwise normal ureter is not uncommon (▶Fig. 13.37).

Ureteral Pathologies

The most common causes of noncalculous filling defects within the urinary tract at CT or MR urography are blood clots and tumor.[40,41] Differentiation of clots from calcified stones is straightforward in most cases (▶Fig. 13.14). Near the UVJ, small calcified stones may be difficult to differentiate from pelvic phleboliths, a common imaging finding (▶Fig. 13.13). The presence of edema around the kidney and collecting system, as well as a delayed

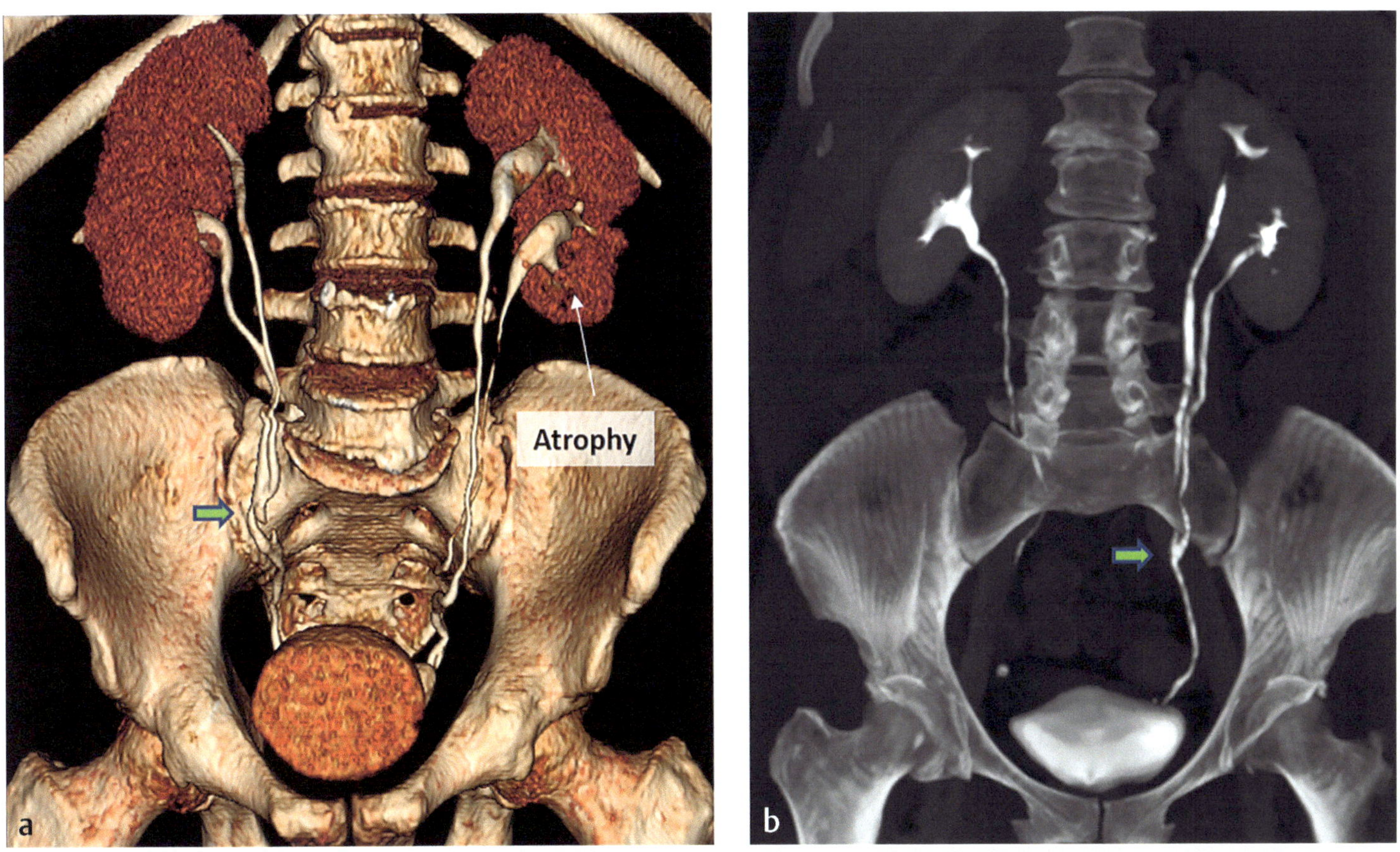

Fig. 13.35 (a) Bilateral duplicated collecting systems. Partial right (*arrow*) and complete left. Some cortical atrophy of the lower pole of the left kidney is present, likely secondary to chronic reflux nephropathy. **(b)** Left-sided partially duplicated collecting system. The ureters merge at the pelvic brim (*arrow*).

13

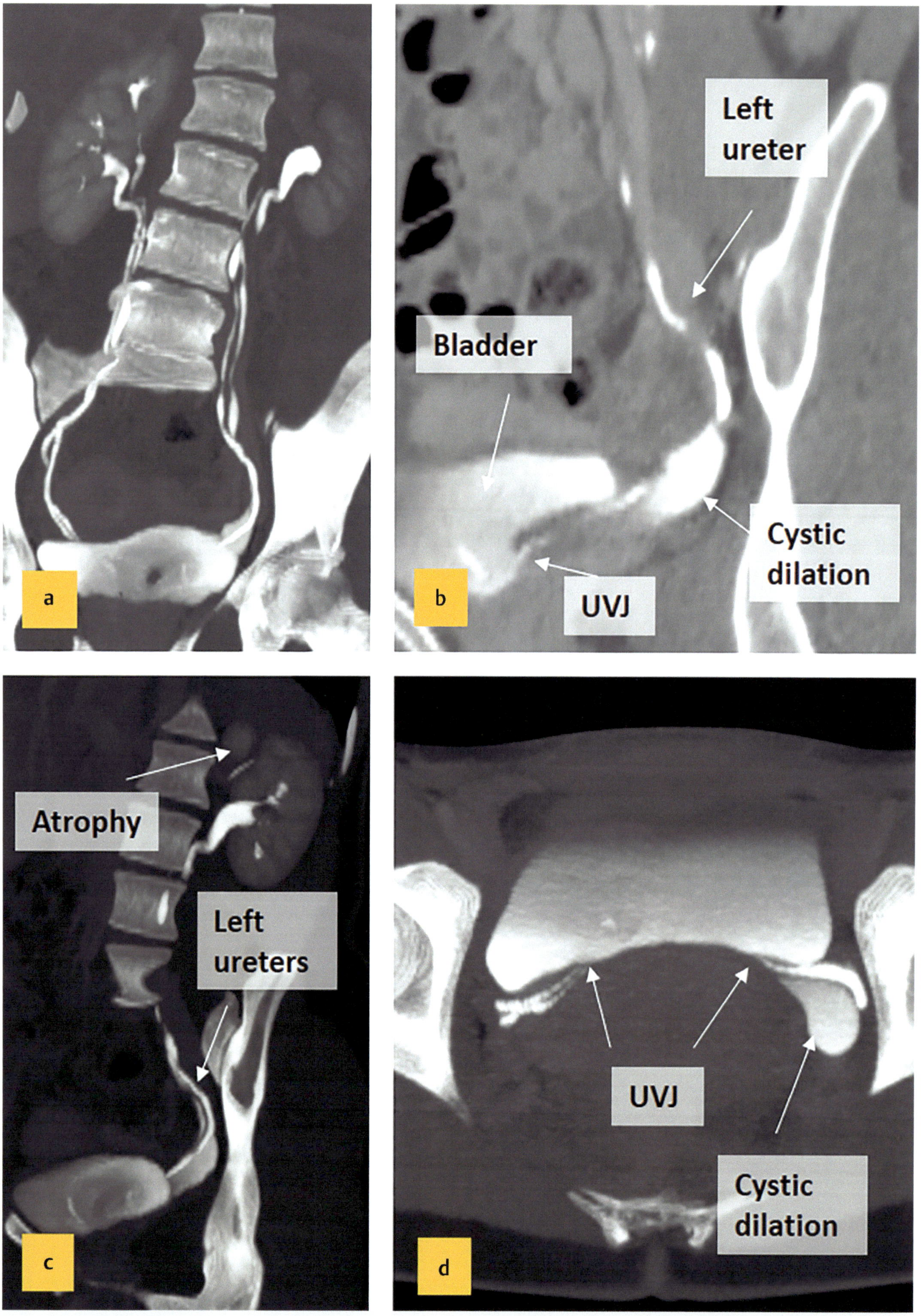

Fig. 13.36 Complete duplication of collecting system bilaterally. **(a–c)** Coronal and **(d)** axial computed tomography (CT) views. On the right, both ureters ended at the ureterovesical junction (UVJ) close to each other. On the left, the inferior moiety ureter showed a cystic dilatation in the distal part (*arrows*) whereas the upper moiety ureter entered the UVJ. The left superior moiety shows parenchymal atrophy.

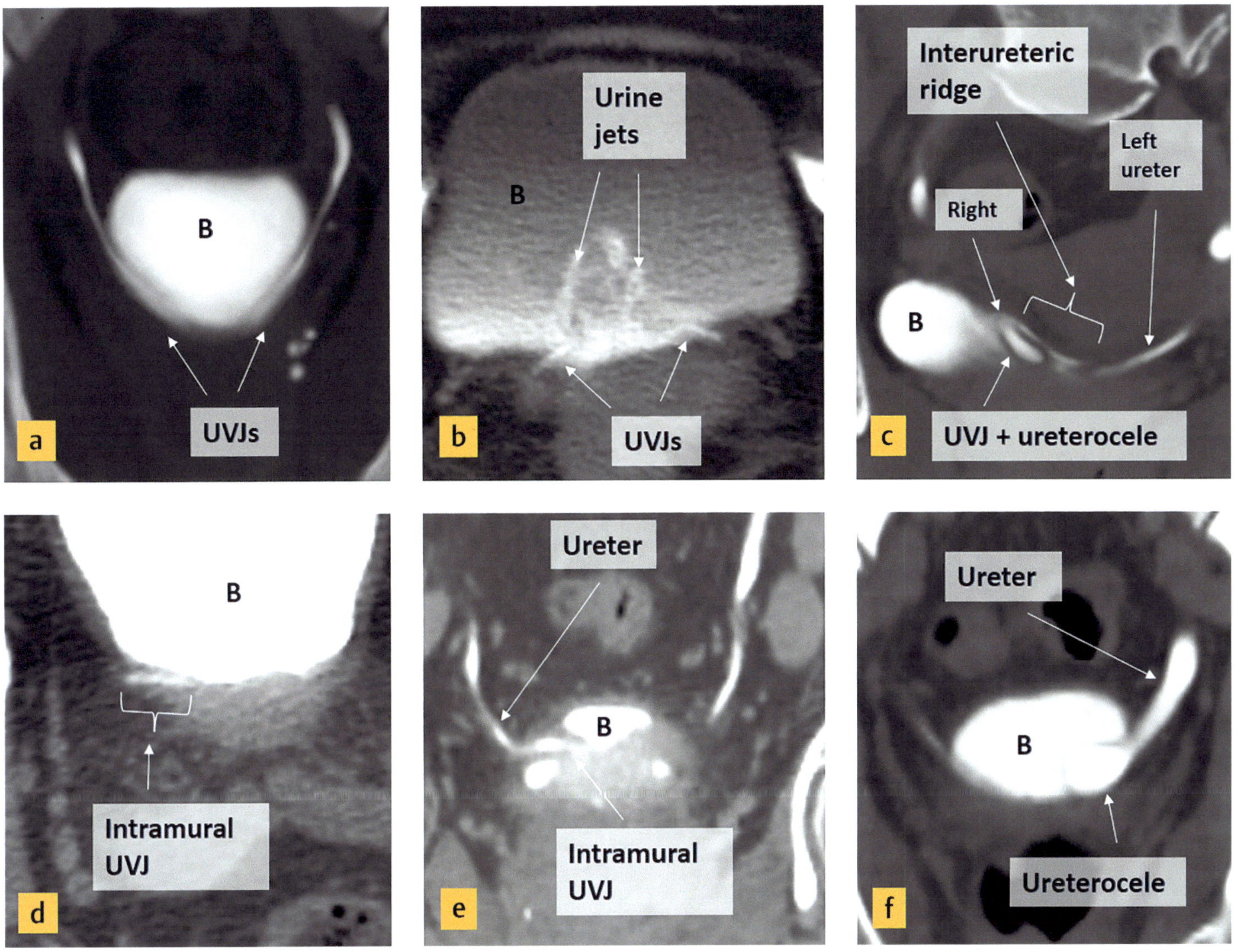

Fig. 13.37 Ureterovesical junction (UVJ) and ureterocele shown by computed tomography (CT) urogram in different patients. **(a)** Coronal view showing UVJ with tiny ureteroceles. **(b)** Ureteral jets of contrast entering the bladder lumen. **(c)** Interureteric ridge between the orifices of the ureters. A small ureterocele is seen on the right side. **(d, e)** The intramural course of the right UVJ. **(f)** A relatively large ureterocele on the left.

nephrogram, can be helpful in distinguishing acute from chronic ureteric obstruction[40] (▶Fig. 13.13).

Urothelial carcinoma is the most common neoplasia of the collecting system causing an intraluminal filling defect or focal stricture (▶Fig. 13.38). UPJ obstruction is a common etiology of benign obstruction and hydronephrosis, especially in children, and can be due to abnormal muscle or innervation, crossing vessel, or fibrotic scar.

Ureterovesical Junction and Vesicoureteral Reflux

The valve of the UVJ is responsible for protection of the low pressure upper urinary tract and kidneys from reflux of urine from the higher pressure bladder.[42,43] Normally, the distal ureter has an intramural course within the base of the bladder. The mean length of the intramural segment of the ureter at the UVJ is 22 ± 0.7 mm[42] (▶Fig. 13.37, ▶Fig. 13.39). The submucosal portion of the ureter functions as a passive valve, and appears to be the fundamental factor for the antireflux mechanism. The intramural ureter of the trigone consists of three layers: longitudinal ureteral muscle, connective tissue, and detrusor muscle. Normally, the extramural ureter consists of circular outer and longitudinal inner muscle layers; however, in the intramural segment only the longitudinally oriented muscle fibers continue forward to contribute to the superficial layer of the trigone.[43] The intramural ureter is surrounded by a cylinder of connective tissue known as Waldeyer's sheath, which allows the ureter to slide relatively freely in its transvesical channel (▶Fig. 13.39). The periureteral connective tissue is continuous with the adventitia of the ureter proximally and adventitia of the bladder distally and provides a path for distribution of edema and malignancy along the adventitia of the ureter.

At the UVJ, there is no connection between the detrusor muscle of the bladder and the longitudinal ureteral muscle surrounded by the connective tissue sheath. Independent contraction of the

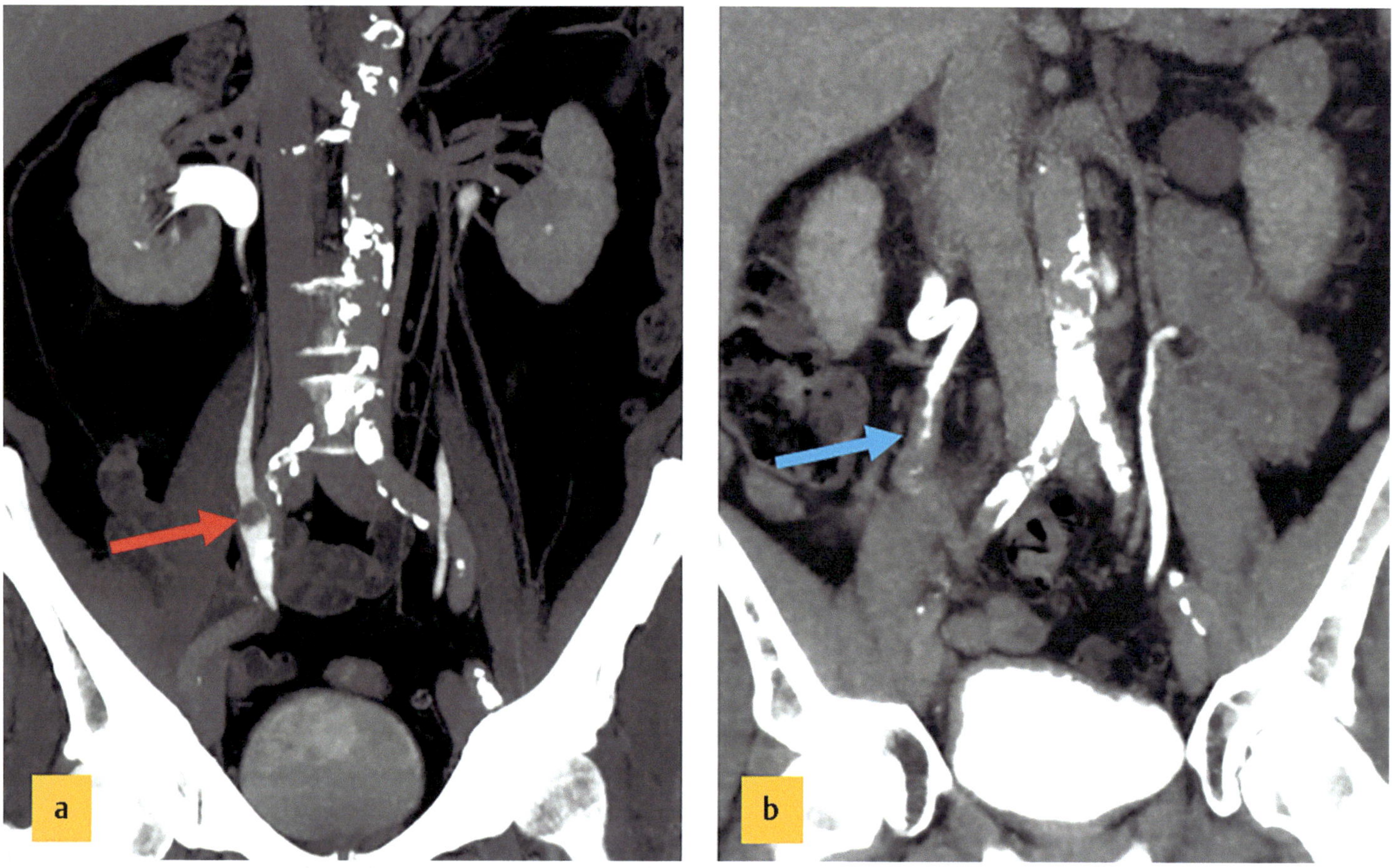

Fig. 13.38 Urothelial carcinoma of the ureter in two patients, appearing as **(a)** a polypoid intraluminal filling defect (*red arrow*) and **(b)** as segmental wall thickening causing a stricture (*blue arrow*).

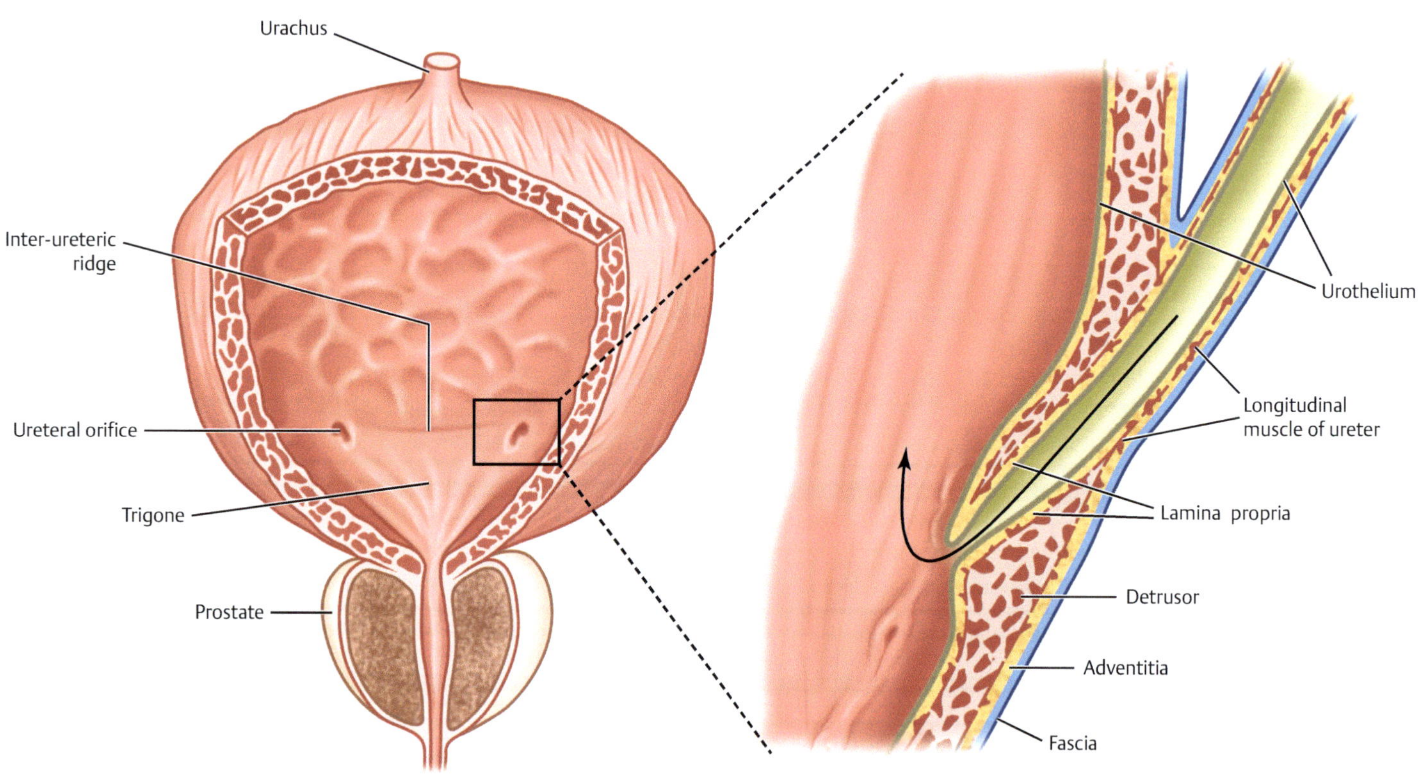

Fig. 13.39 Anatomy of the bladder trigone showing the ureteral orifices, the urethral orifice, and the interureteric ridge with cross-sectional detail of the left uretrovesical junction (UVJ). The mean length of the intramural segment of the ureter at the UVJ is 2 cm. The intramural ureter of the trigone consists of three layers, namely, longitudinal ureteral muscle, periureteral connective tissue, and detrusor muscle. The periureteral connective tissue is continuous with the adventitia of the ureter and bladder.

longitudinal musculature allows shortening and narrowing of the intramural segment, which together drives the urine bolus into the bladder and provides the ureteral antireflux mechanism. An incompetent antireflux mechanism of the UVJ in early childhood is common, especially in girls, and can result in vesicoureteral reflux (VUR). Shortness of the transmural ureteral segment relative to its diameter could be the main cause of VUR. Further maturation of the antireflux mechanism can occur in childhood, and most cases of VUR resolve spontaneously within the first 2 years of life. There is a direct relationship between the severity of VUR and the risk of recurrent urinary tract infection. In affected children, continuous VUR and associated urinary infection can lead to pyelonephritis and renal scarring with loss of parenchyma, a process called reflux nephropathy[44] (▸Fig. 13.40). VUR is four to five times more common in females than males. Children with higher grades of reflux appear more likely to have larger areas of parenchymal scarring. Renal scars could be produced by a water-hammer effect of refluxing urine. These scars are usually patchy or segmental, being more common in the upper poles with a high incidence of intrarenal reflux in compound papillae.[45,46] Based on the morphology of the papillary duct openings, the papillae are classified into two groups.[45,46] Simple papillae, usually seen in the midportion of the kidney, show collecting ducts with slit-like orifices that enter on the side of the papillae. The slit-like morphology makes the papillary outlet resistant to intrarenal backflow and renal injury, even with elevated bladder pressures. On the contrary, the compound papillae—fused papillae found in the polar regions—possess collecting ducts with large orifices that enter the concave surface of the papillae. Compound papillae are believed to be more susceptible to urine reflux and transfer of bacteria into the renal parenchyma, causing pyelonephritis and parenchymal scarring.[44] As a result of gradual pyramidal tissue loss from reflux nephropathy, the calyces become club-shaped on CT pyelography (▸Fig. 13.40). A small atrophic kidney is sometimes the end result of reflux nephropathy in early childhood.

The grade of VUR can be evaluated with voiding cystourethrography (VCUG), a technique in which contrast is administered retrograde into the bladder, and then fluoroscopy is performed to observe the bladder and ureters during voiding (▸Fig. 13.41). Other imaging findings of reflux nephropathy including renal scarring and chronic pyelonephritis can be detected with CT or MR urography or Tc 99m dimercaptosuccinic acid (DMSA).

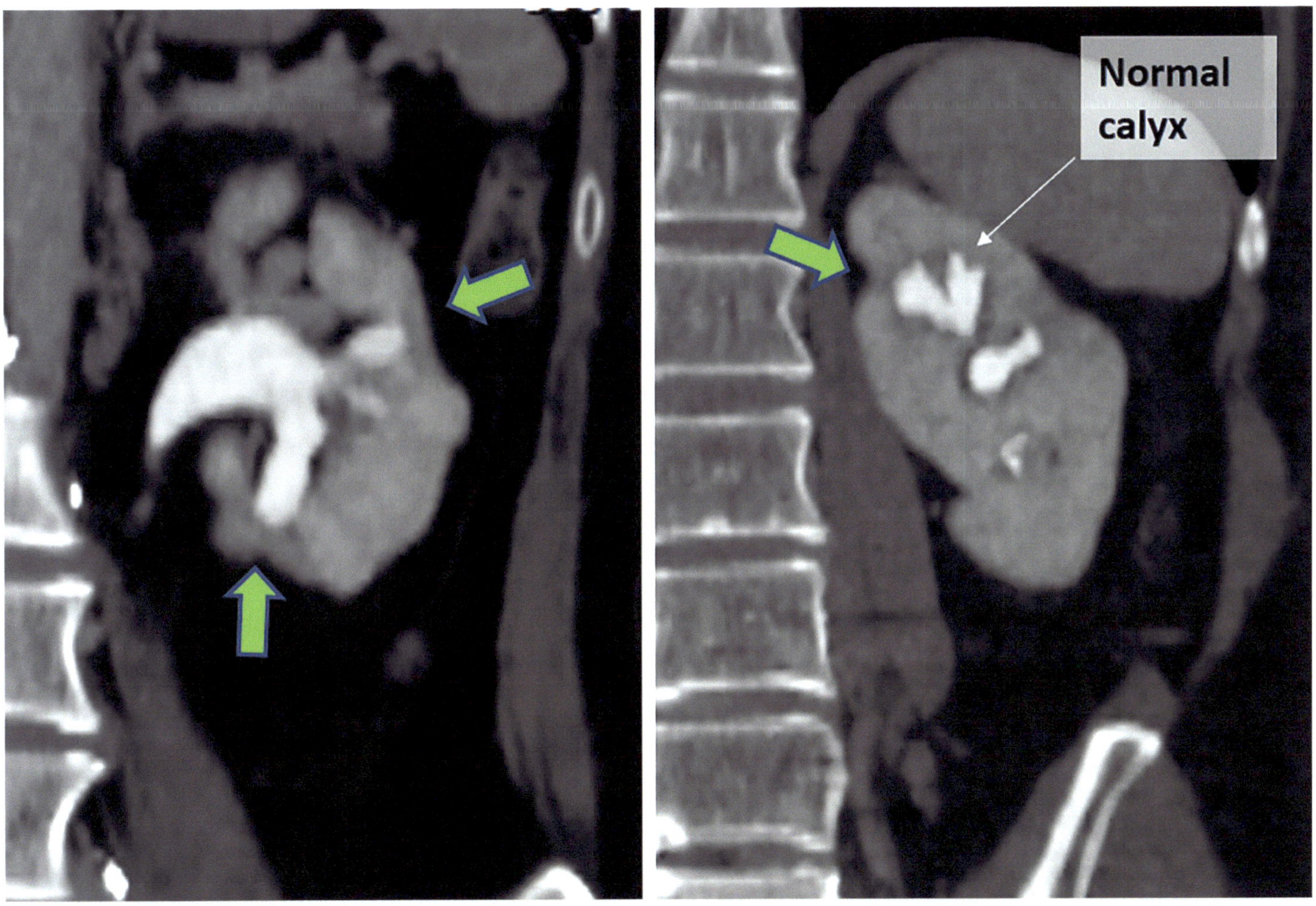

Fig. 13.40 Reflux nephropathy. Coronal views from a computed tomography (CT) urogram showing patchy areas of parenchymal loss overlying calyces of the upper and lower poles. The involved calyces have a clubbed appearance due to atrophy of the papillae. A normal calyx is shown for comparison.

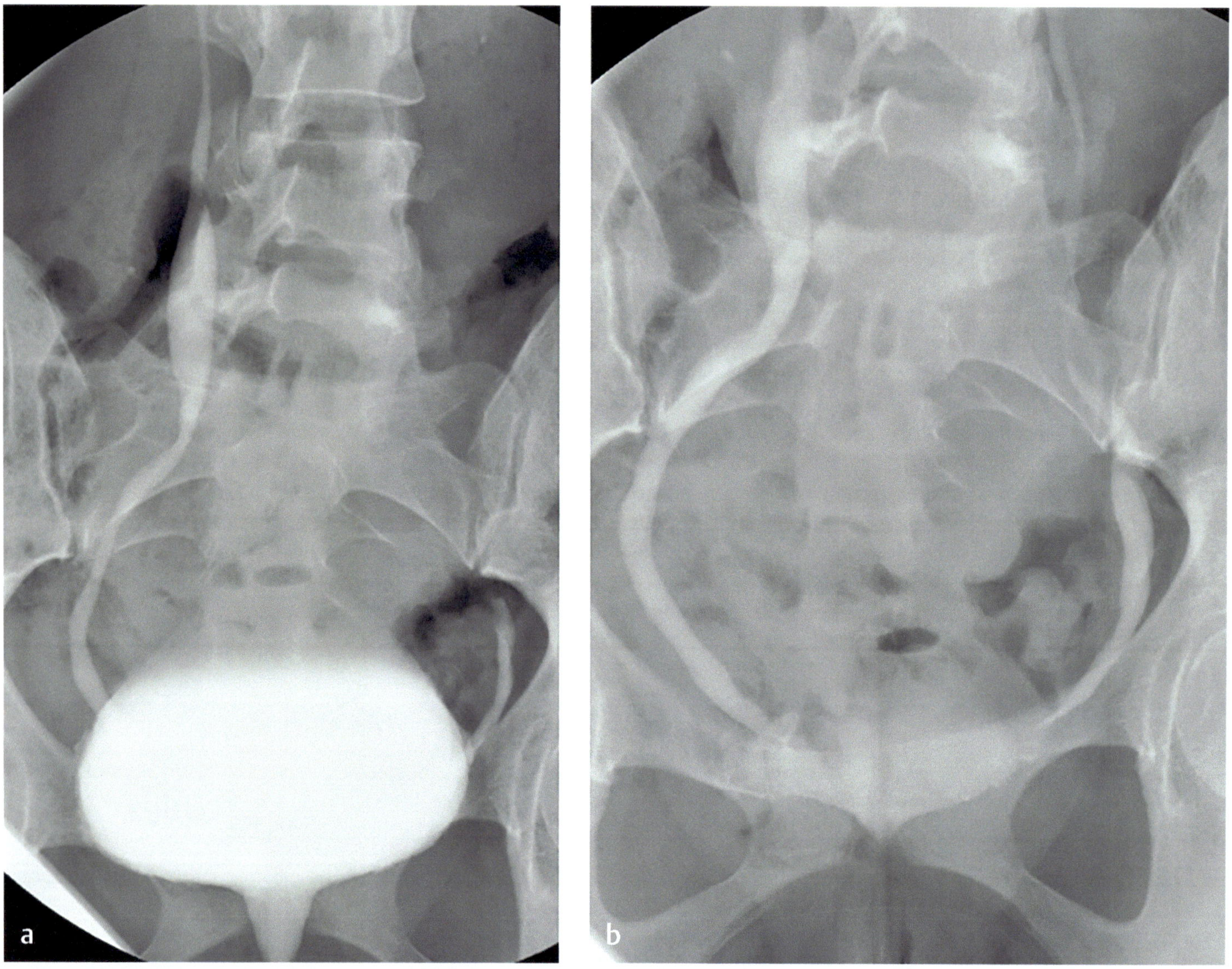

Fig. 13.41 Voiding cystourethrography (VCUG) before **(a)** and after **(b)** bladder evacuation in a 23-year-old female with recurrent urinary tract infections, showing bilateral right greater than left vesicoureteral reflux (VUR).

◆ Urinary Bladder

The bladder develops from the anterior part of the terminal hindgut, the cloaca, which is contiguous with the allantois ventrally and superiorly. The embryonic allantois is a finger-like outpouching from the caudal wall of the yolk sac. Initially, the embryonic bladder extends to the level of the umbilicus and opens into the allantois. By the 4th or 5th month of gestation, the pelvic part of the bladder remains but its apical end along with the allantois narrow to form the urachus.

The urinary bladder acts as a reservoir for urine, and as such is habituated to regular expansion and contraction. The posteroinferior wall of the bladder is termed the base, and the superior wall the dome. The median umbilical ligament (remnant of the embryological urachus) attaches to the anterosuperior aspect of the bladder (▶Fig. 13.42) creating a slightly peaked appearance to this area called the bladder apex. Inferiorly, the bladder opens to the urethra, and the portion of the bladder around the internal urethral orifice is the neck (▶Fig. 13.42).

The bladder is enveloped by the visceral pelvic fascia called the umbilicovesical fascia, which continues up superiorly around the median (remnant of the urachus) and medial (remnants of the obliterated umbilical arteries) umbilical ligaments to the umbilicus (see Chapter 14 "Extraperitoneal Space"). Fatty areolar tissue between the umbilicovesical fascia and the transversalis fascia forms the prevesical space, located both anterior to and along the sides of the bladder[18,20] (▶Fig. 13.42). Some authors refer to this entire space as the space of Retzius, while others use this eponym only to refer to the inferior portion of the prevesical space between the pubic symphysis and the bladder. The inferior border of the prevesical/Retzius space is marked by the pubovesical ligaments, which anchor the bladder neck to the pubic symphysis.

Parietal peritoneum covers over the bladder dome (▶Fig. 13.42). In females, posterior to the bladder dome the peritoneum reflects

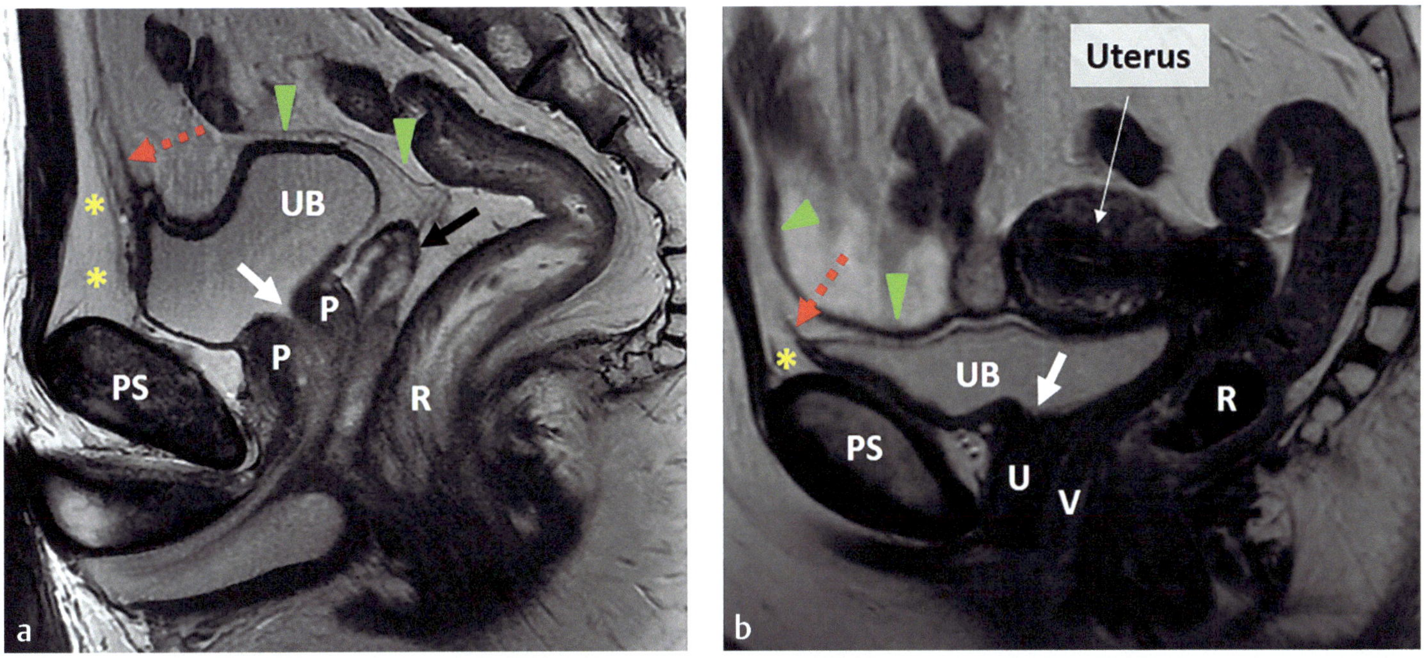

Fig. 13.42 Sagittal magnetic resonance imaging (MRI) in the male **(a)** and female **(b)** showing the urinary bladder (UB). The bladder neck is where the internal urethra orifice (*white arrow*) opens to the urethra (U). It is posterior to the lower half of the pubic symphysis (PS). In the male, the first portion of the urethra courses through the prostate (P). The apex of the bladder is attached to the medial umbilical ligament (*dashed red arrow*). The peritoneum (*green arrowheads*) covers over the dome of the bladder. The *black arrow* denotes the seminal vesicle in **a**. The prevesical space (*) is anterior to the bladder, between the transversalis fascia and the umbilicovesical fascia. R, rectum; V, vagina.

over the uterus, forming the vesicouterine pouch. In males, the peritoneum continues posteriorly to form the rectovesical pouch, before reflecting onto the rectum. Bladder rupture at the dome has the potential to create an intraperitoneal leak (which requires surgical repair), while rupture of other portions of the bladder tends to create an extraperitoneal leak (which is usually managed conservatively).[47] Extraperitoneal leaks constitute approximately 80% of all bladder ruptures.

With bladder expansion, the dome elevates upwards into the abdomen. The bladder neck remains fairly constant in position, being in communication with the urethra. The bladder neck is posterior to the mid to lower pubic symphysis, and in the male rests upon the prostate. The base of the bladder is anterior to the vagina in females, and anterior to the seminal vesicles and rectovesical pouch in males (see Chapter 18 "Pelvic Floor").

The bladder urothelium is also a transitional epithelium separated from the muscularis propria by the lamina propria. The muscularis propria, also known the detrusor muscle, is the main muscular component of the urinary bladder wall responsible for bladder contraction. The muscular fibers of the detrusor form three layers, with the outer and inner layers being longitudinally oriented and the middle layer circular. The outer surface of the detrusor is covered by the tunica adventitia, containing fat and vessels. The visceral pelvic fascia is the outermost structure. Viewed intraluminally, the bladder mucosa has a folded appearance and is without many landmarks, the exception being the trigone, an area of the posterior–inferior bladder wall where the mucosa has a smoother appearance. The trigone is an inverted triangle with the ureteral orifices at the upper angles, and the urethral orifice forming the lower apex of the triangle (▶Fig. 13.39). The horizontal line between the two ureteral orifices is called the interureteric ridge or bar, and it forms the superior boundary of the trigone (▶Fig. 13.37, ▶Fig. 13.39).

Common Pathologies

The normal bladder wall thickness is 3 to 6 mm and varies with bladder distension. In chronic bladder outlet obstruction, the bladder wall thickens and becomes trabeculated. Bladder outlet obstruction can be mechanical or neurogenic. Posterior urethral valves are the most common cause of mechanical outlet obstruction in male children, while in adults, enlargement of the prostate is the most common etiology (▶Fig. 13.43). In traumatic injury of the spinal cord involving the upper motor tracts, detrusor sphincter dyssynergia is a common cause of bladder outlet obstruction. Diverticula of the bladder wall are common in these patients (▶Fig. 13.43). Congenital bladder wall diverticula are uncommon.[48] A particular diverticulum in a paraureteral location is also known as a Hutch diverticulum. A paraureteral diverticulum is located at or adjacent to the UVJ, where Waldeyer's sheath normally seals the potential space between the intramural ureter and the bladder muscle. A paraureteral diverticulum can interfere with normal ureterovesical competence, resulting in vesicoureteric reflux. Urinary stasis within a large bladder diverticulum can lead to infection. Urothelial carcinoma is the most common type of tumor of the urinary bladder. Lesions appear as regional wall thickening or masses protruding into the bladder lumen (▶Fig. 13.44). Urothelial carcinoma is more common in the urinary bladder than in the upper urinary tracts.

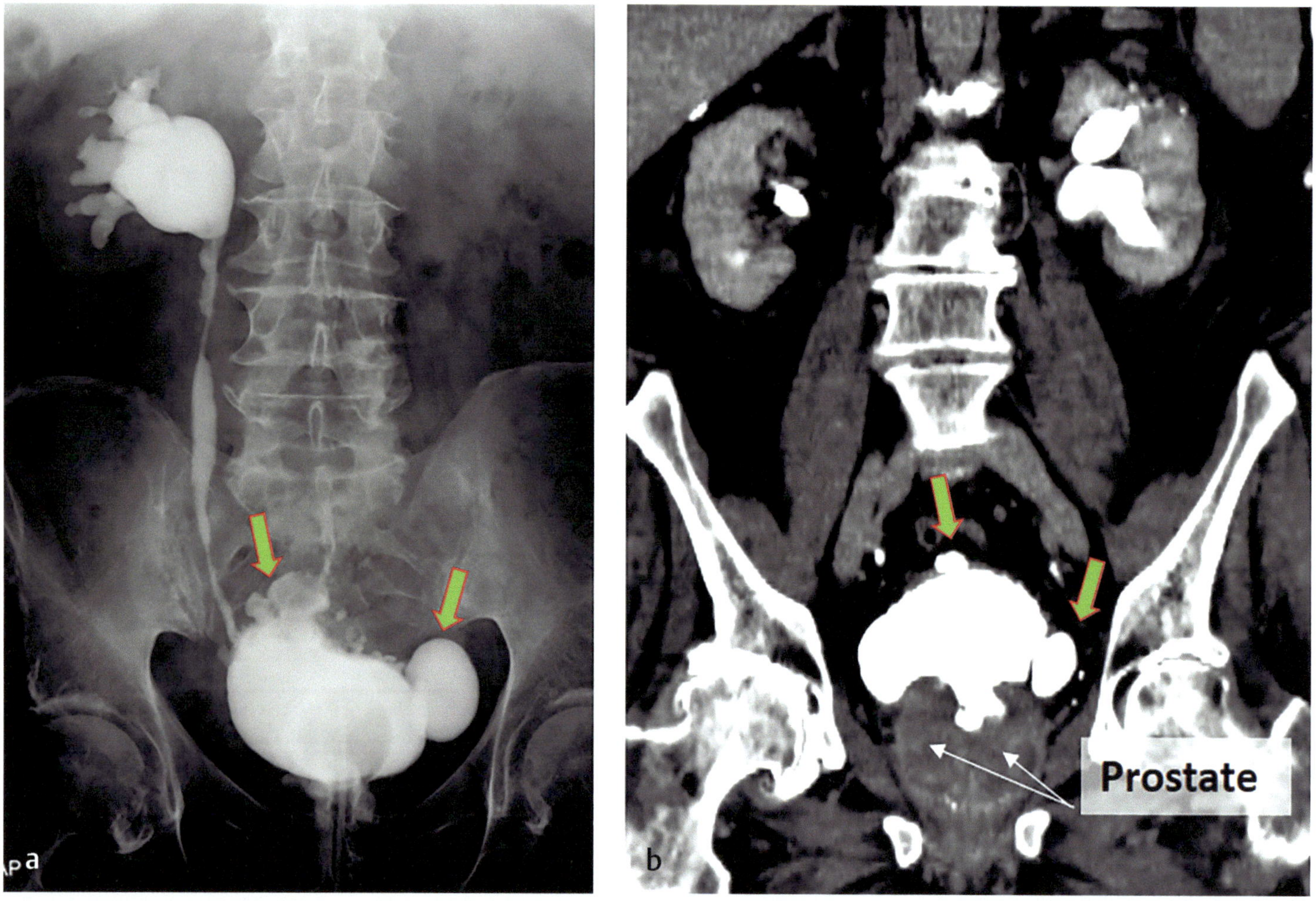

Fig. 13.43 Bladder outlet obstruction due to an enlarged prostate. **(a)** Retrograde urogram showing right hydro-ureteronephrosis and a trabeculated bladder wall with multiple diverticula (*arrows*). **(b)** Computed tomography (CT) urogram showing hydronephrosis and diverticula (*arrows*) of the bladder wall due to an enlarged prostate.

13

Vasculature and Nerves

Arterial Supply

Arterial supply to the bladder derives from the anterior division of the internal iliac artery, mainly from the superior and inferior vesical arteries, although additional branches to the bladder can also arise from the obturator, inferior gluteal, uterine, and vaginal arteries. Internal iliac arterial branching is highly variable, but the superior vesical artery tends to be one of the first branches. The inferior vesical artery tends to branch more distally, and frequently arises as a common trunk with the middle rectal artery.

Venous and Lymphatic Drainage

Venous drainage from the bladder forms a rich plexus surrounding the bladder and draining posteriorly, mostly along its inferolateral surfaces, to empty into the internal iliac veins. Most of the lymphatic drainage of the bladder is to external iliac lymph nodes, although some lymph drainage can also occur to internal iliac, obturator, and even common iliac nodes.

Nervous Supply and Neurogenic Bladder Dysfunction

The S2–S4 segments of the spinal cord give rise to the parasympathetic supply to the bladder via the pelvic splanchnic nerves. Sympathetic innervation arises from the T10–L2 segments of the spinal cord, and reaches the bladder via the hypogastric plexuses. There are also afferent fibers from the bladder, which predominantly run with the parasympathetic nerves, involved in pain and detection of distension.

Neurogenic bladder dysfunction can occur due to diseases of the central nervous system (such as spinal cord injury) or peripheral nerves (such as diabetes). There are numerous causes for the development of neurogenic bladder, ranging from spinal dysraphisms to spinal cord trauma. Urinary tract infection is one of

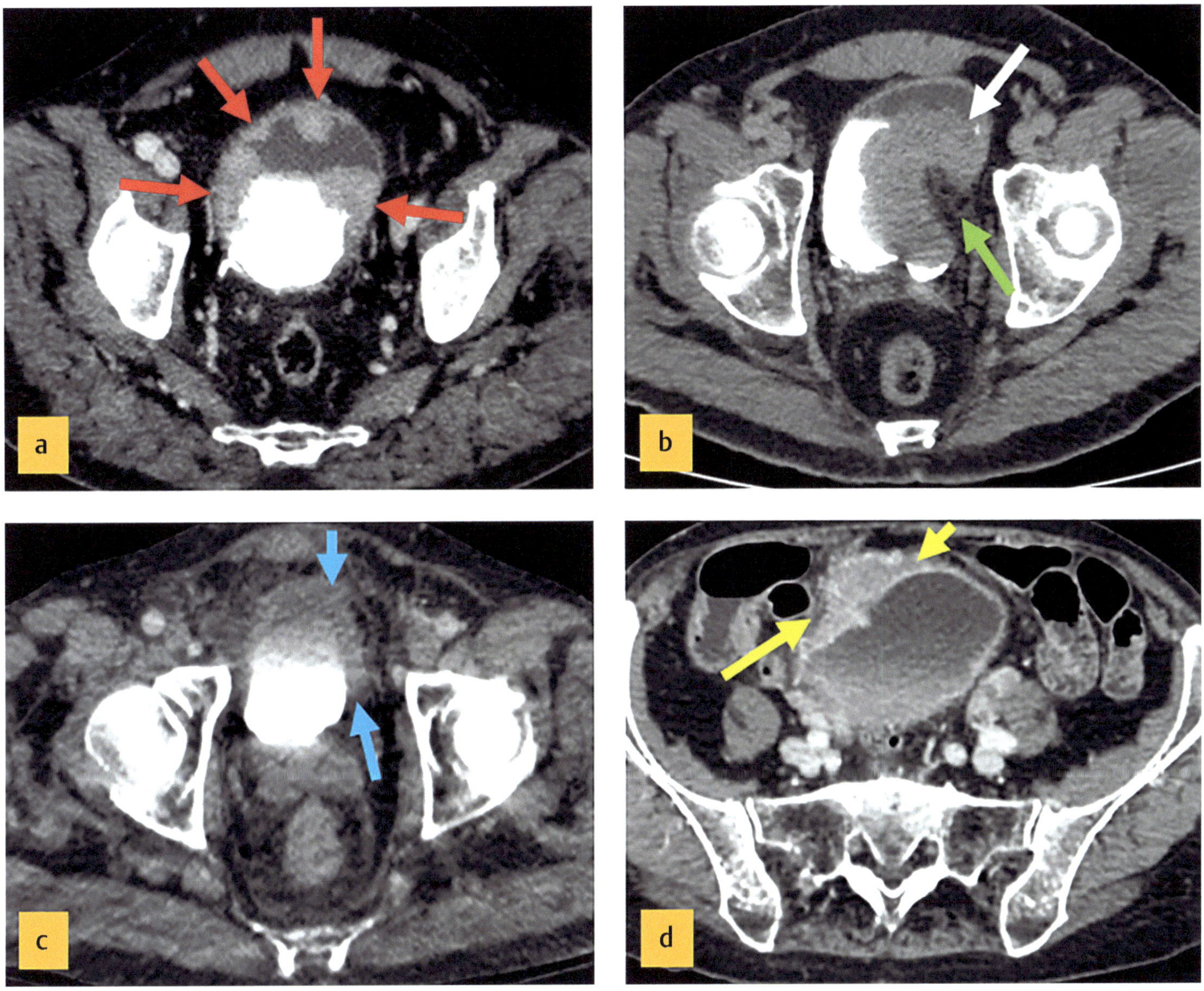

Fig. 13.44 Bladder cancer in four patients showing a variety of appearances: **(a)** Multiple masses projecting intraluminally (*red arrows*). **(b)** A single large mass (*white arrow*) also causing inwards puckering of the bladder wall (*green arrow*). **(c)** A broad area of smooth wall thickening (*blue arrows*). **(d)** Mass-like thickening of the wall showing predominantly outwards growth (*yellow arrows*).

the most common complications. Increased bladder pressure and urine retention increase the rate of VUR. Compared to primary VUR, secondary VUR in neurogenic bladder patients is less likely to spontaneously resolve and is less likely to be cured with antireflux surgery.

◆ Urachal Remnants

In late fetal life, the urachal lumen narrows and finally obliterates by fibrous tissue proliferation. The urachal ligament is usually located in the midline (median umbilical ligament) but an off-center position or merger with the obliterated umbilical arteries is not unusual (▸Fig. 13.45). It varies from 3 to 10 cm in length and from 2 to 10 mm in diameter. The ligament consists of an innermost transitional epithelium (70%) or columnar epithelium (30%) and an outer muscular layer in continuity with the detrusor muscle.[49,50] It is enveloped in the umbilicovesical fascia, along with the medial umbilical ligaments and the bladder.

Persistence of an embryonic urachal remnant is not rare and can give rise to various clinical problems (▸Fig. 13.45). Congenital urachal anomalies include patent urachus, umbilical–urachal sinus, vesicourachal diverticulum, and urachal cyst. The urachal cyst is the most common and may be seen anywhere from the bladder wall to the umbilicus. The urachal cyst can become infected or undergo malignant transformation. Although the urachus is most commonly lined by transitional epithelium, urachal carcinomas are usually adenocarcinoma.[49]

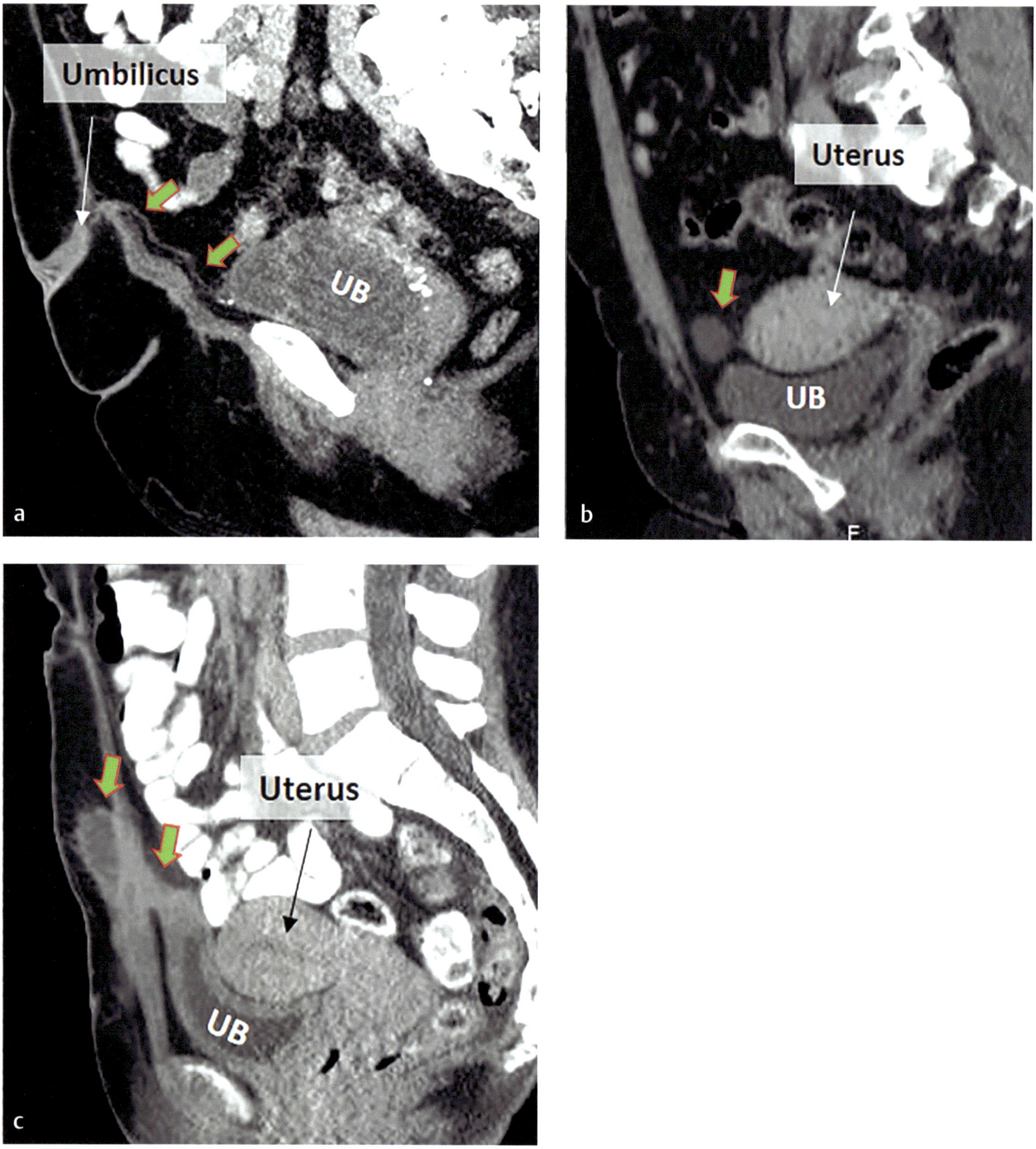

Fig. 13.45 Urachal remnants. Sagittal computed tomography (CT) images are shown. **(a)** The urachal ligament (*arrows*). **(b)** A urachal cyst (*arrow*) above the bladder apex. **(c)** Superinfected urachal remnant causing soft tissue thickening and abscess (*arrows*). UB, urinary bladder.

◆ Postsurgical Urinary Tract

Nephrectomy

Partial nephrectomy is a common procedure for removal of small tumors. Small lesions can also be obliterated by radiofrequency- or cryo-ablation techniques. Total nephrectomy is performed by a transperitoneal or retroperitoneal approach for the treatment of both benign and malignant conditions.[51] Advantages of a retroperitoneal approach include decreased risk of peritonitis or tumoral peritoneal seeding, although with the disadvantage of a smaller working space. On the other hand, advantages of the transperitoneal approach include a larger working space and easier visualization of anatomic landmarks, although this approach also requires mobilization and reflection of the colon. Following nephrectomy, the empty space will be replaced by the small or mobilized large bowel (▸ Fig. 13.46).

Renal Transplantation

In adults, most kidney allografts are placed in an extraperitoneal location in the iliac fossa.[52] The renal graft vessels are usually anastomosed with the external or common iliac vessels using the end-to-side technique (▸ Fig. 13.47). The graft artery may also be connected to the recipient internal iliac artery using an end-to-end anastomosis. In case of multiple renal arteries, they can be sometimes patched together before final anastomosis (▸ Fig. 13.48). An intravesical anastomosis is usually performed for the ureter which involves creation of a submucosal vesical tunnel near the trigone to avoid reflux. Contrast-enhanced MR or CT is used to evaluate the renal graft artery and vein and the iliac arterial system. Ultrasound with Doppler is a common technique for evaluation vessel integrity, rejection, hydronephrosis, renal artery stenosis, and perirenal fluid collection (▸ Fig. 13.49).

Urinary Diversion

Surgical resection of the bladder (cystectomy) is most commonly performed for bladder malignancy, although rarely cystectomy may be performed for benign diseases as well, such as neurogenic bladder, radiation cystitis, interstitial cystitis, and severe incontinence.[53] Once the bladder has been removed, there are many techniques for reconstituting the urinary stream. Some techniques result in an incontinent urostomy, the most common being the ileal conduit (▸ Fig. 13.50a), which is also known as the Bricker procedure.[54,55] Other reconstructions are continent, and can entail either a pouch/reservoir that can be catheterized through a stoma at the skin (such as the Indiana pouch), or the creation of a neobladder which is anastomosed to the native urethra (▸ Fig. 13.50b). Most of the different types of urinary reconstructions are fashioned using a segment of bowel.

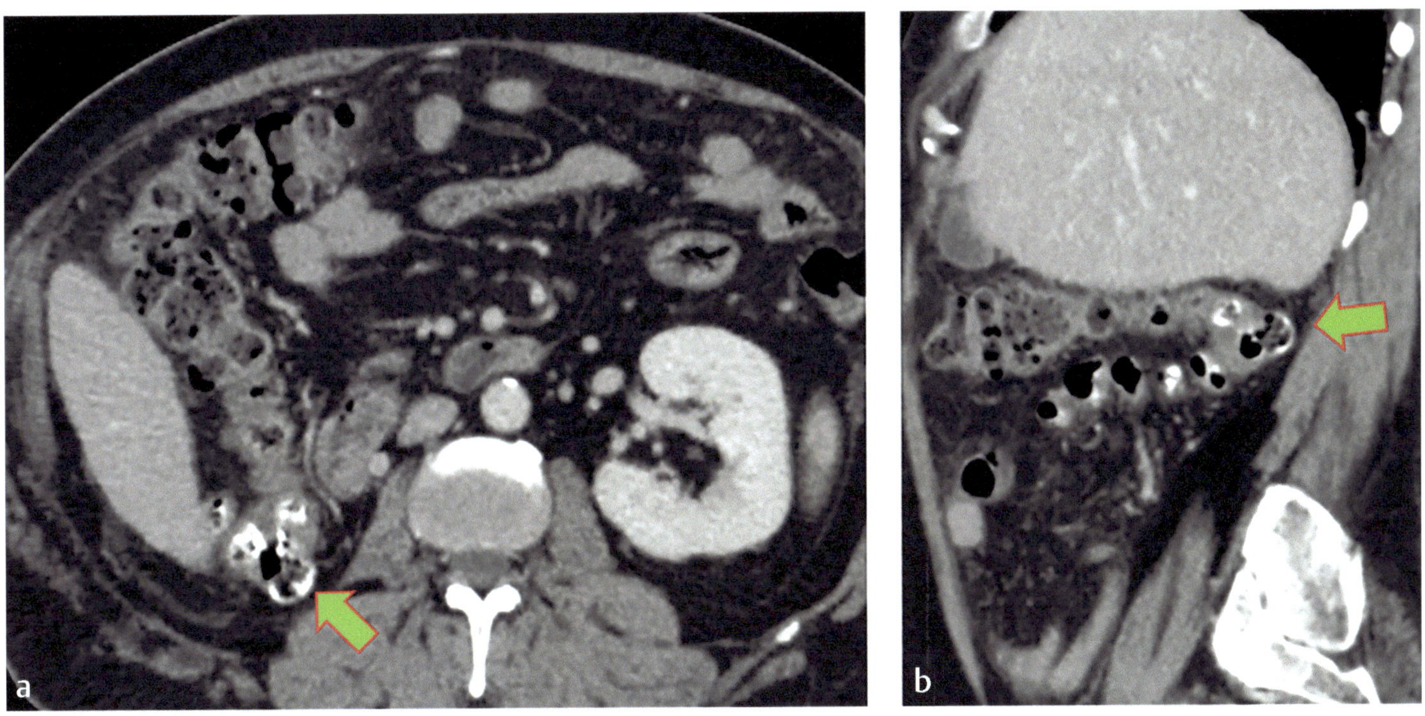

Fig. 13.46 Post right nephrectomy. **(a)** Axial and **(b)** sagittal computed tomography (CT) images in a patient post right radical nephrectomy showing migration of the hepatic flexure of the colon into the nephrectomy fossa (*arrows*).

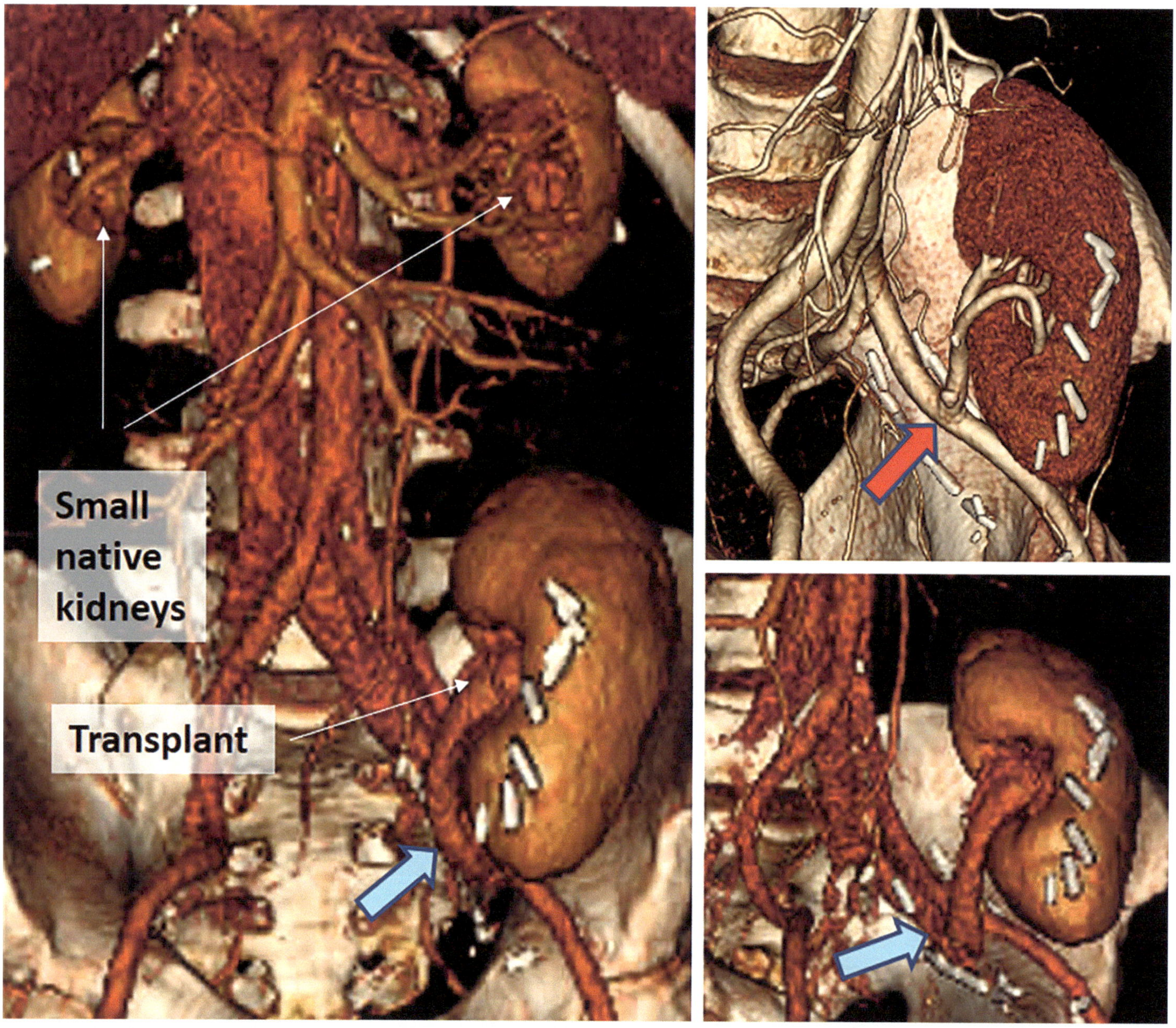

Fig. 13.47 Left iliac renal transplant. Renal artery (*red arrow*) and vein (*blue arrow*) are sutured to the external iliac artery and vein, respectively.

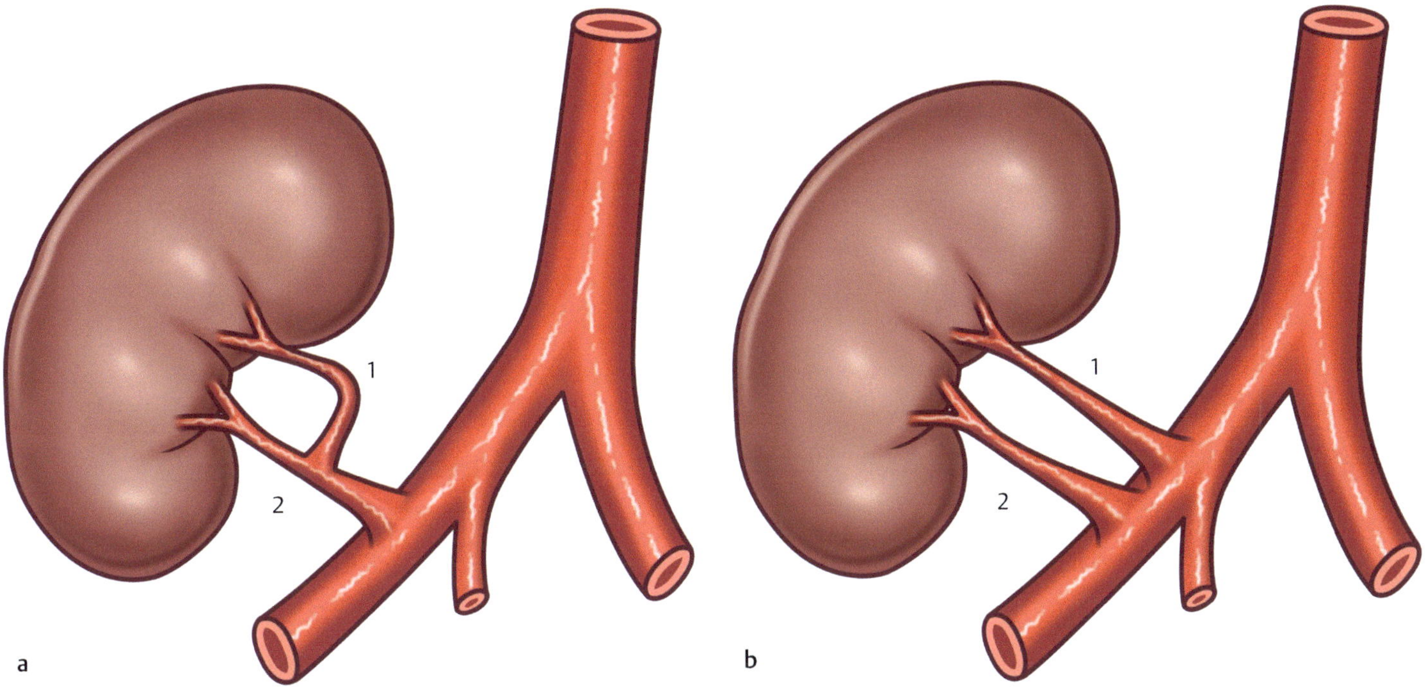

Fig. 13.48 Renal transplant. **(a, b)** End-to-side anastomosis for donors with two renal arteries.

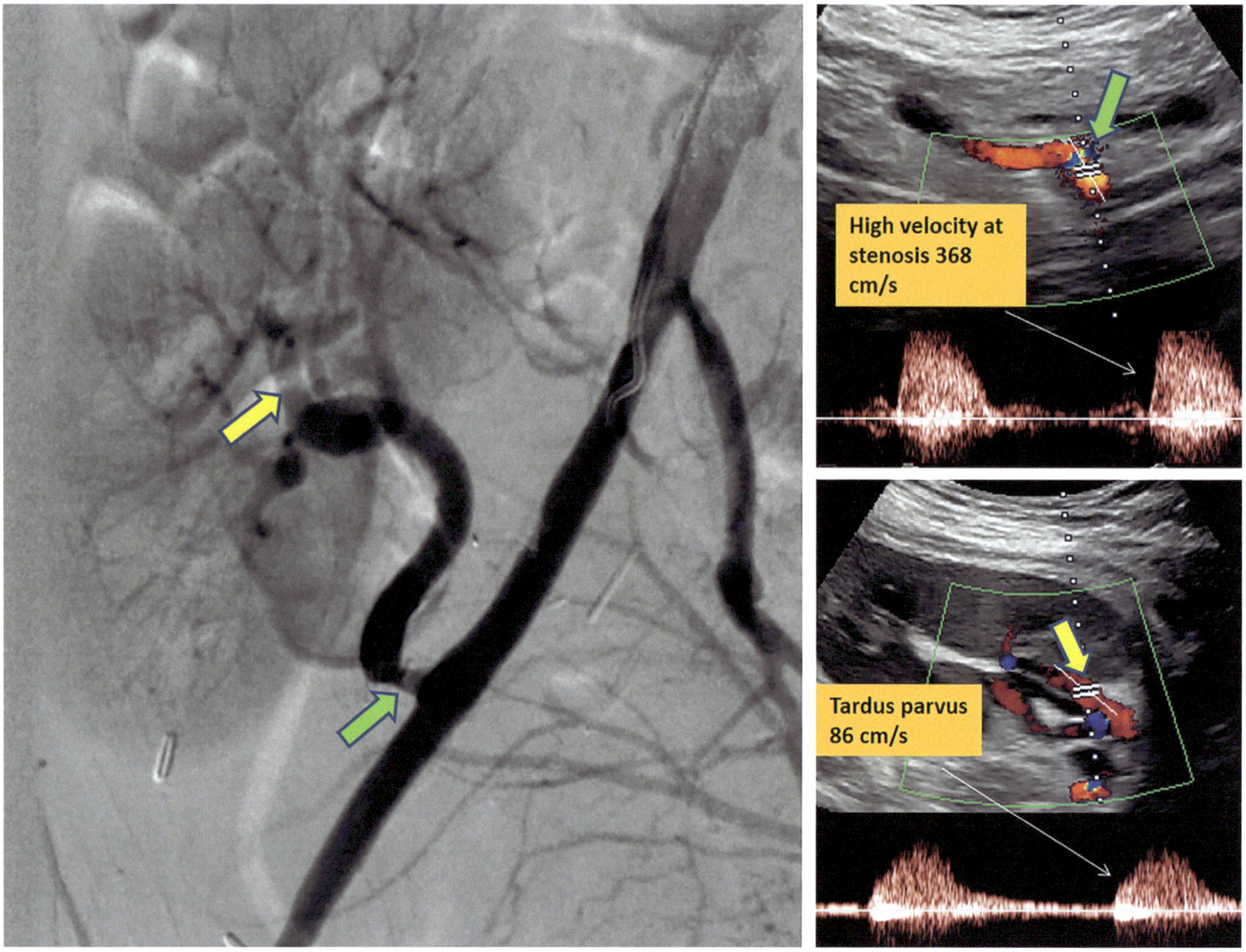

Fig. 13.49 Post-transplant renal artery anastomosis stenosis. Selective angiography and Doppler study of the right renal transplant showing markedly elevated velocity (368 cm/s) at the stenotic anastomosis (*green arrows*) of the transplant renal artery. There is also slow flow with delayed systolic acceleration (tardus parvus) waveform with peak velocity of 86 cm/s in the distal artery (*yellow arrows*).

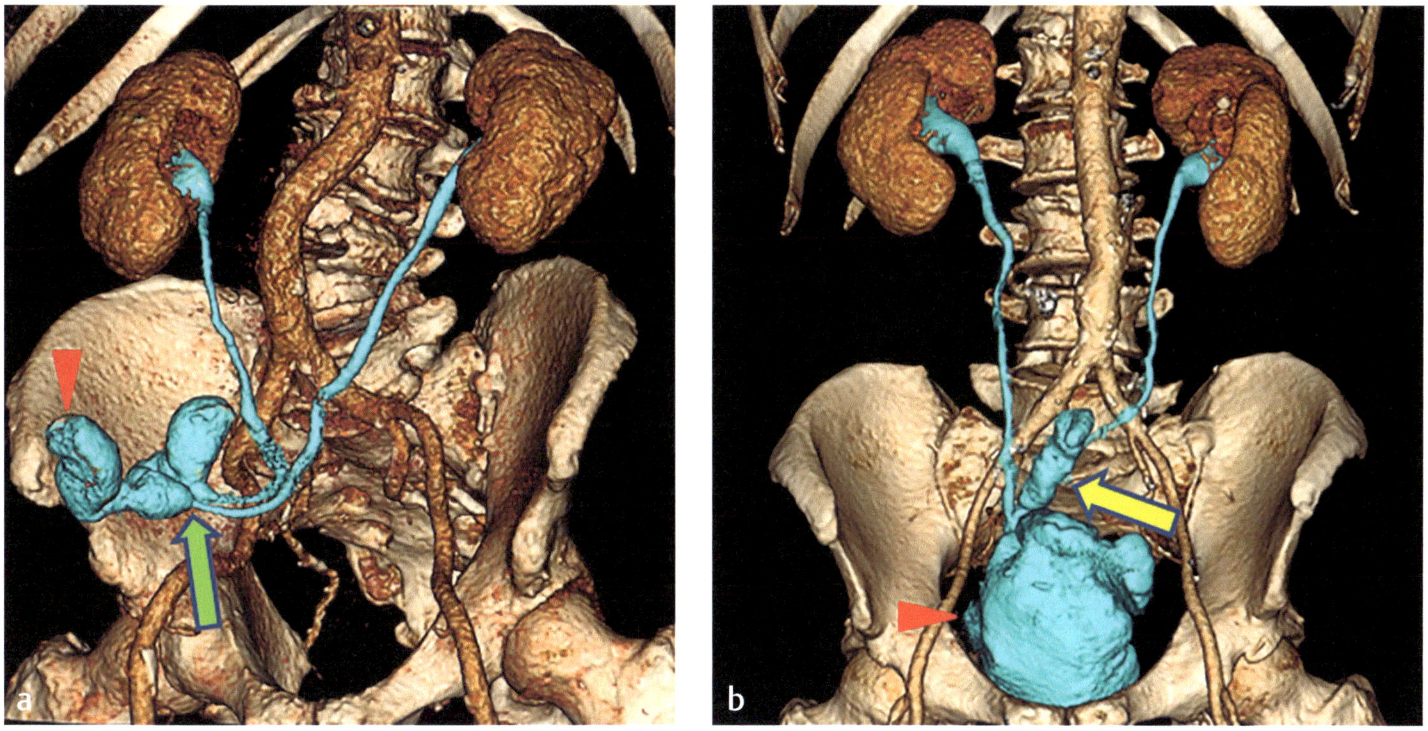

Fig. 13.50 Urinary diversion techniques. Urographic phase computed tomography (CT) in two patients showing **(a)** an ileal conduit and **(b)** a neobladder. **(a)** The image is rotated to show the ureters inserting on the posterior end of the conduit (*arrow*). The conduit courses forward to the skin forming the urostomy (*arrowhead*). **(b)** With a neobladder, the two ureters insert onto the afferent limb (*arrow*), which then empties into the neobladder (*arrowhead*) that is anastomosed to the urethra.

References

1. Barwick TD, Malhotra A, Webb JA, Savage MO, Reznek RH. Embryology of the adrenal glands and its relevance to diagnostic imaging. Clin Radiol 2005;60(9):953–959
2. Krebs TL, Wagner BJ. The adrenal gland: radiologic-pathologic correlation. Magn Reson Imaging Clin N Am 1997;5(1):127–146
3. Vincent JM, Morrison ID, Armstrong P, Reznek RH. The size of normal adrenal glands on computed tomography. Clin Radiol 1994;49(7):453–455
4. Tang YZ, Bharwani N, Micco M, Akker S, Rockall AG, Sahdev A. The prevalence of incidentally detected adrenal enlargement on CT. Clin Radiol 2014; 69(1): e37–e42
5. Kahn PC, Nickrosz LV. Selective angiography of the adrenal glands. Am J Roentgenol Radium Ther Nucl Med 1967;101(3):739–749
6. Daunt N. Adrenal vein sampling: how to make it quick, easy, and successful. Radiographics 2005;25(Suppl 1):S143–S158
7. Merklin RJ. Suprarenal gland lymphatic drainage. Am J Anat 1966;119(3):359–374
8. Parker TL, Kesse WK, Mohamed AA, Afework M. The innervation of the mammalian adrenal gland. J Anat 1993;183(Pt 2):265–276
9. Lockhart ME, Smith JK, Kenney PJ. Imaging of adrenal masses. Eur J Radiol 2002;41(2):95–112
10. Surabhi VR, Menias CO, George V, Matta E, Kaza RK, Hasapes J. MDCT and MR urogram spectrum of congenital anomalies of the kidney and urinary tract diagnosed in adulthood. AJR Am J Roentgenol 2015;205(3):W294-304
11. Glodny B, Petersen J, Hofmann KJ, et al. Kidney fusion anomalies revisited: clinical and radiological analysis of 209 cases of crossed fused ectopia and horseshoe kidney. BJU Int 2009;103(2):224–235
12. Kawashima A, Sandler CM, Corl FM, et al. Imaging of renal trauma: a comprehensive review. Radiographics 2001;21(3):557–574
13. Arévalo Pérez J, Gragera Torres F, Marín Toribio A, Koren Fernández L, Hayoun C, Daimiel Naranjo I. Angio CT assessment of anatomical variants in renal vasculature: its importance in the living donor. Insights Imaging 2013;4(2):199–211
14. Cuttino JT, Clark RL. The normal vasculature of the genitourinary tract: embryology, anatomy, and hemodynamics. In: Hillman BJ, ed. Clinical Urography. Vol. 3. Philadelphia: Saunders; 1990:2076–2091
15. Raman SS, Pojchamarnwiputh S, Muangsomboon K, Schulam PG, Gritsch HA, Lu DS. Surgically relevant normal and variant renal parenchymal and vascular anatomy in preoperative 16-MDCT evaluation of potential laparoscopic renal donors. AJR Am J Roentgenol 2007;188(1):105–114
16. Al-Katib S, Shetty M, Jafri SM, Jafri SZ. Radiologic assessment of native renal vasculature: a multimodality review. Radiographics 2017;37(1):136–156
17. Yeh BM, Coakley FV, Meng MV, Breiman RS, Stoller ML. Precaval right renal arteries: prevalence and morphologic associations at spiral CT. Radiology 2004; 230(2):429–433
18. Sebastià C, Peri L, Salvador R, et al. Multidetector CT of living renal donors: lessons learned from surgeons. Radiographics 2010;30(7):1875–1890
19. Kawashima A, Sandler CM, Ernst RD, Tamm EP, Goldman SM, Fishman EK. CT evaluation of renovascular disease. Radiographics 2000;20(5):1321–1340
20. Pollak R, Prusak BF, Mozes MF. Anatomic abnormalities of cadaver kidneys procured for purposes of transplantation. Am Surg 1986;52(5):233–235
21. Yi SQ, Ueno Y, Naito M, Ozaki N, Itoh M. The three most common variations of the left renal vein: a review and meta-analysis. Surg Radiol Anat 2012;34(9): 799–804
22. Zhu J, Zhang L, Yang Z, Zhou H, Tang G. Classification of the renal vein variations: a study with multidetector computed tomography. Surg Radiol Anat 2015; 37(6):667–675
23. Koc Z, Ulusan S, Oguzkurt L, Tokmak N. Venous variants and anomalies on routine abdominal multi-detector row CT. Eur J Radiol 2007;61(2):267–278
24. Grimm LJ, Engstrom BI, Nelson RC, Kim CY. Incidental detection of nutcracker phenomenon on multidetector CT in an asymptomatic population: prevalence and associated findings. J Comput Assist Tomogr 2013;37(3):415–418
25. Grzelak P, Kurnatowska I, Sapieha M, et al. Disturbances of kidney graft perfusion as indicators of acute renal vein thrombosis in contrast-enhanced ultrasonography. Transplant Proc 2011;43(8):3018–3020

26. Bonsib SM. Renal lymphatics, and lymphatic involvement in sinus vein invasive (pT3b) clear cell renal cell carcinoma: a study of 40 cases. Modern Pathology 2006;19(5):746–753
27. Mehta V, Mudaliar K, Ghai R, et al. Renal lymph nodes for tumor staging: appraisal of 871 nephrectomies with examination of hilar fat. Arch Pathol Lab Med 2013;137(11):1584–1590
28. Chade AR. Renovascular disease, microcirculation, and the progression of renal injury: role of angiogenesis. Am J Physiol Regul Integr Comp Physiol 2011;300(4):R783–R790
29. Nordsletten DA, Blackett S, Bentley MD, Ritman EL, Smith NP. Structural morphology of renal vasculature. Am J Physiol Heart Circ Physiol 2006;291(1):H296–H309
30. Le Dorze M, Legrand M, Payen D, Ince C. The role of the microcirculation in acute kidney injury. Curr Opin Crit Care 2009;15(6):503–508
31. Leong CL, Anderson WP, O'Connor PM, Evans RG. Evidence that renal arterial-venous oxygen shunting contributes to dynamic regulation of renal oxygenation. Am J Physiol Renal Physiol 2007;292(6):F1726–F1733
32. Pallone TL, Turner MR, Edwards A, Jamison RL. Countercurrent exchange in the renal medulla. Am J Physiol Regul Integr Comp Physiol 2003;284(5):R1153–R1175
33. Pallone TL, Edwards A, Mattson DL. Renal medullary circulation. Compr Physiol 2012;2(1):97–140
34. Thang OH, Serné EH, Grooteman MP, et al. Premature aging of the microcirculation in patients with advanced chronic kidney disease. Nephron Extra 2012;2(1):283–292
35. DiBona GF, Kopp UC. Neural control of renal function. Physiol Rev 1997;77(1):75–197
36. Abdulla MH, Johns EJ. The innervation of the kidney in renal injury and inflammation: a cause and consequence of deranged cardiovascular control. Acta Physiol (Oxf) 2017;220(4):404–416
37. Santicioli P, Maggi CA. Myogenic and neurogenic factors in the control of pyeloureteral motility and ureteral peristalsis. Pharmacol Rev 1998;50(4):683–722
38. Vernez SL, Okhunov Z, Wikenheiser J, et al. Precise characterization and 3-dimensional reconstruction of the autonomic nerve distribution of the human ureter. J Urol 2017;197(3 Pt 1):723–729
39. Privett JTJ, Jeans WD, Roylance J. The incidence and importance of renal duplication. Clin Radiol 1976;27(4):521–530
40. Leyendecker JR, Barnes CE, Zagoria RJ. MR urography: techniques and clinical applications. Radiographics 2008;28(1):23–46, discussion 46–47
41. Potenta SE, D'Agostino R, Sternberg KM, Tatsumi K, Perusse K. CT urography for evaluation of the ureter. Radiographics 2015;35(3):709–726
42. Roshani H, Dabhoiwala NF, Verbeek FJ, Kurth KH, Lamers WH. Anatomy of ureterovesical junction and distal ureter studied by endoluminal ultrasonography in vitro. J Urol 1999;161(5):1614–1619
43. Roshani H, Dabhoiwala NF, Verbeek FJ, Lamers WH. Functional anatomy of the human ureterovesical junction. Anat Rec 1996;245(4):645–651
44. Smith EA. Pyelonephritis, renal scarring, and reflux nephropathy: a pediatric urologist's perspective. Pediatr Radiol 2008;38(Suppl 1):S76–S82
45. Ransley PG, Risdon RA. The pathogenesis of reflux nephropathy. Contrib Nephrol 1979;16:90–97
46. Ransley PG, Risdon RA. Renal papillary morphology and intrarenal reflux in the young pig. Urol Res 1975;3(3):105–109
47. Gross JS, Rotenberg S, Horrow MM. Resident and fellow education feature. Bladder injury: types, mechanisms, and diagnostic imaging. Radiographics 2014;34(3):802–803
48. Blane CE, Zerin JM, Bloom DA. Bladder diverticula in children. Radiology 1994;190(3):695–697
49. Yu JS, Kim KW, Lee HJ, Lee YJ, Yoon CS, Kim MJ. Urachal remnant diseases: spectrum of CT and US findings. Radiographics 2001;21(2):451–461
50. Parada Villavicencio C, Adam SZ, Nikolaidis P, Yaghmai V, Miller FH. Imaging of the urachus: anomalies, complications, and mimics. Radiographics 2016;36(7):2049–2063
51. Bhardwaj N. Retroperitoneal versus transperitoneal approach for nephrectomy in children: anesthetic implications. J Anaesthesiol Clin Pharmacol 2015;31(1):25–26
52. Sebastià C, Quiroga S, Boyé R, Cantarell C, Fernandez-Planas M, Alvarez A. Helical CT in renal transplantation: normal findings and early and late complications. Radiographics 2001;21(5):1103–1117
53. Osborn DJ, Dmochowski RR, Kaufman MR, Milam DF, Mock S, Reynolds WS. Cystectomy with urinary diversion for benign disease: indications and outcomes. Urology 2014;83(6):1433–1437
54. Catalá V, Solà M, Samaniego J, et al. CT findings in urinary diversion after radical cystectomy: postsurgical anatomy and complications. Radiographics 2009;29(2):461–476
55. Moomjian LN, Carucci LR, Guruli G, Klausner AP. Follow the stream: imaging of urinary diversions. Radiographics 2016;36(3):688–709

14 Extraperitoneal Space

Farhood Saremi and Petros Mirilas

◆ Introduction

The extraperitoneal space is the space exterior to the peritoneal cavity. This space is filled with a variable amount of fat, connective tissue, membranes, vessels, nerves, lymphatics, and various organs.[1] In the abdomen, the extraperitoneal space is the anatomic site of the kidney, adrenal, part of duodenum, pancreas, ascending colon, and descending colon. Most intrapelvic organs are extraperitoneal. A variety of diseases, such as pancreatitis, infection, and trauma related, can cause fluid collections in the extraperitoneal space. Understanding the normal anatomy of the extraperitoneal space is the key to correct localization of pathologies and predicting the distribution of inflammatory fluid collections, extravasated blood, and solid masses in this compartment of the human body.[2,3,4] The anatomy of the extraperitoneal spaces has gained a new interest for surgical purposes.[1] Preoperative knowledge of the anatomy of this region may be important in order to identify the extraperitoneal fascia and interfascial spaces for better surgical planning. In extraperitoneal laparoscopic operations, surgeons need to dilate the potential interfascial space into a larger space for further operation. Correct placement of balloon dissectors in the exact fascial space provides a clean and clear space without significant hemorrhage after balloon dilatation.

◆ Extraperitoneal Space

The extraperitoneal space is the space between the peritoneum internally and the investing inner fascia of the muscles of the anterior, lateral, and posterior walls of the abdomen and pelvis[5,6] (▶Fig. 14.1, ▶Fig. 14.2). The transversalis fascia is the investing inner fascia of the abdominal muscles carpeting the diaphragm and transverse abdomen muscles. The parietal pelvic fascia is the investing fascia of the pelvic muscles such as the iliacus, obturator internus, and levator muscles. The transversalis and parietal pelvic fasciae both attach to the iliac crest periosteum.[6] The posterior and lateral part of the extraperitoneal space is called the retroperitoneal space. The retroperitoneal space is divided into the anterior pararenal, perirenal, and posterior pararenal spaces[5] (▶Fig. 14.1). The extraperitoneal space of the anterior abdomen (anterior extraperitoneal space) is called preperitoneal space and in the pelvis, it is called subperitoneal space[1,6] (▶Fig. 14.2). The anterior extraperitoneal space can also be subdivided into the diaphragmatic, supraumbilical, infraumbilical, prevesical, and inguinal zones (see Chapter 1 "Abdominopelvic Wall" and Chapter 10 "Diaphragm" of Volume 1). Some have arbitrarily divided the extraperitoneal space into four surgical anatomic zones: centromedial, right lateral, left lateral, anterior abdominal, and pelvic.[6] The centromedial is the anatomic location of the aorta, inferior vena cava, pancreas, and duodenum. The lateral zones are the anatomic sites of the kidneys, adrenals, ureters, ascending colon, and descending colon. The bladder, uterus, prostate, rectum, iliac vessels, and space of Retzius have been included in the pelvic zone.

◆ Extraperitoneal Fasciae

The fasciae are lamina of connective tissue approximately 1- to 2-mm thick that divide the extraperitoneal space into smaller spaces. For example, the renal fascia surrounds the kidney and forms the retroperitoneal spaces, and the umbilical prevesical fascia along with the transversalis fascia form the prevesical space.

The abdominal and pelvic fasciae are classified into the parietal, extraperitoneal, and visceral fasciae, all of which evolved from the three layers of the embryonic extraperitoneal mesenchyme called the inner, intermediate, and outer strata, respectively[6,7] (▶Fig. 14.3). The parietal fascia originates from the outer stratum and comprises of three investing fasciae, namely, superficial, intermediate, and deep which envelop the anterior and posterior abdominal muscles.[8] The deep parietal fascia in the abdomen is known as transversalis fascia. The extraperitoneal fascia and all retroperitoneal organs arise from the intermediate stratum. The extraperitoneal fascia includes an outer membranous layer and an inner fatty layer. The membranous layer forms the retroperitoneal fasciae by condensation (see below). Additionally, a continuous membranous layer exists around the abdomen and pelvic walls, and in the inguinal area. The visceral fascia arises from the inner stratum and covers the intraperitoneal organs. The pancreas, second and third portions of the duodenum, and ascending/descending colon are initially intraperitoneal. Later, the posterior peritoneal layer of these organs will be fused with the anterior renal fascia (fusion fascia) and they become retroperitoneal in the anterior pararenal space.[8] The vesical fascia of bladder (once intraperitoneal) is also a visceral fascia.

Hayes classified the fetal abdominal fasciae into three types[9]:

- Fusion fasciae formed from remnants of the peritoneum.
- Migration fasciae are formed by condensation in response to mechanical stress caused by rapid growth of adjacent structures.
- Parietal fasciae enveloping the body of wall muscles (e.g., transversalis fascia).

Most extraperitoneal fasciae are classified as migration fascia. These fascias develop around 20 weeks of gestation after the renal development or retroperitoneal fixation of the pancreas, duodenum, and colon.[10,11] It is believed that mechanical stress from the rapid growth of the retroperitoneal organs such as kidneys, pancreas, or duodenum cause thin collagen fibers to form bundles in order to reinforce the retroperitoneal organs. The perirenal fascia is a typical migration fascia which develops earlier than the retropancreatic fascia of Treitz. Cranial migration and growth of the definitive kidney are well-known processes that help to form the migration fascia around the kidneys. Therefore, in renal agenesis the renal fascia does not form on that side.[10] In ectopic kidneys, however, perirenal fascia exists. Examples of migration fascia in the pelvic are rectal and umbilical prevesical fasciae.[6]

In the early embryonic stage around the 7th to 8th weeks, kidneys are embedded in a homogeneous loose mesenchymal

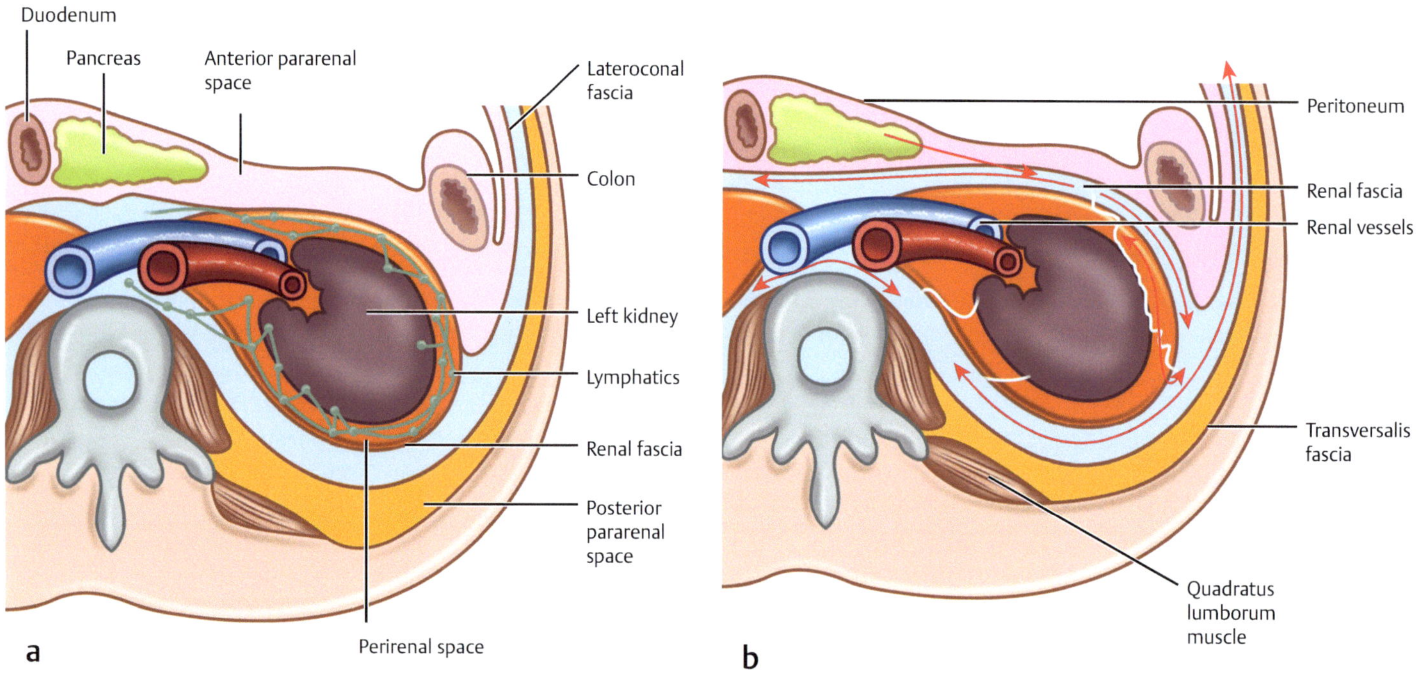

Fig. 14.1 Cross-sectional diagrams of the retroperitoneal space compartments. **(a)** Traditional tricompartmental classification of the retroperitoneum. The retroperitoneal space is divided into the anterior pararenal, perirenal, and posterior pararenal spaces. **(b)** Interfascial and perinephric pathways concept. Using this concept, inflammatory processes can easily spread between different compartments and within the perirenal fat along bridging septa (*red arrows*). Lymphatic as shown in **a** can be a potential pathway to spread inflammatory and neoplastic diseases. The perirenal lymphatics communicate with lymph nodes at renal hilum and from there with the periaortic and pericaval lymph nodes. LK, left kidney.

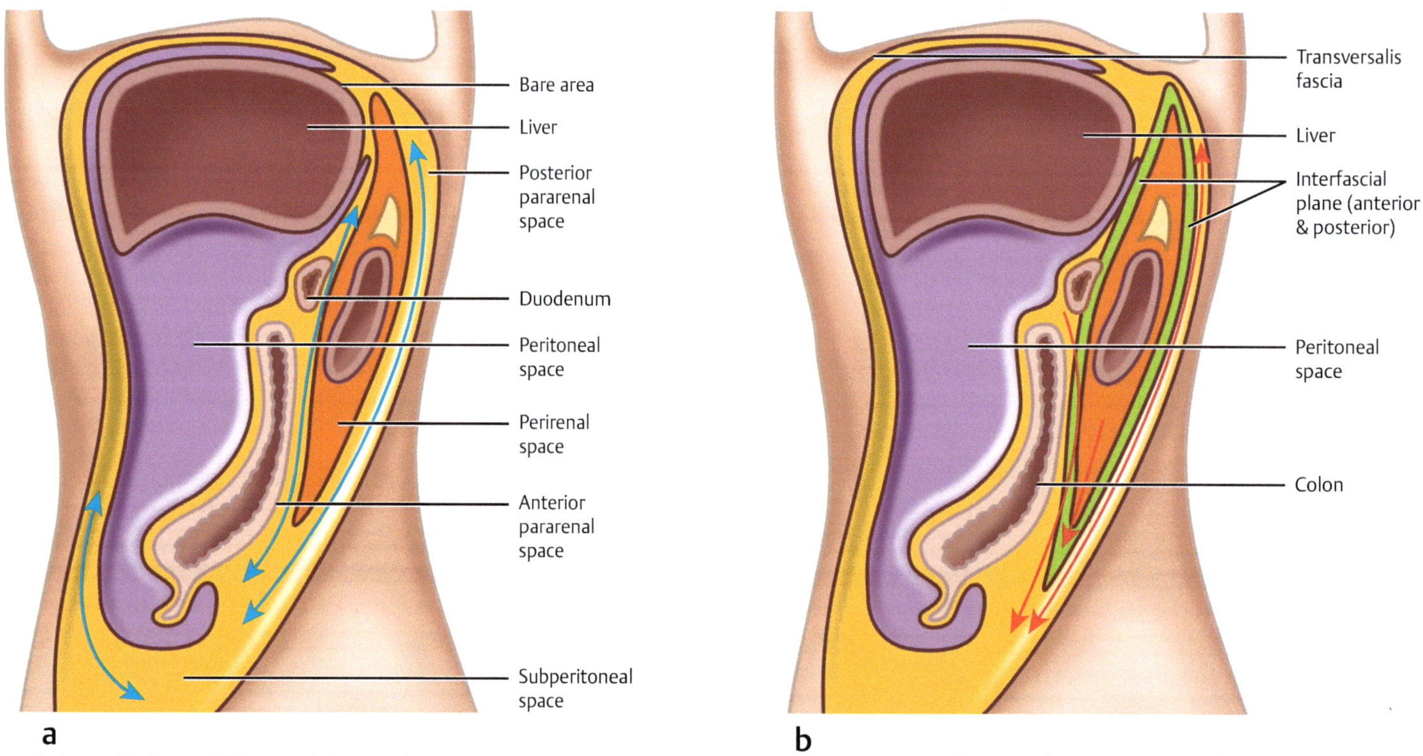

Fig. 14.2 Longitudinal diagram of the extraperitoneal space on the right side. **(a)** Traditional tricompartmental classification of the retroperitoneum. In this diagram the superior–inferior extent of the perirenal space is shown. In some cases, the superior border of the perirenal space is continuous with the bare area of the liver. The perirenal space is closed inferiorly. The anterior and posterior pararenal spaces communicate with each other below the perirenal space and both are continuous with the subperitoneal (pelvis) and preperitoneal (lateral and anterior of the abdomen) spaces (*yellow arrows*). **(b)** The concept of interfascial planes. Pancreatic and renal lesions can reach the interfascial planes and spread within into the pelvis.

14

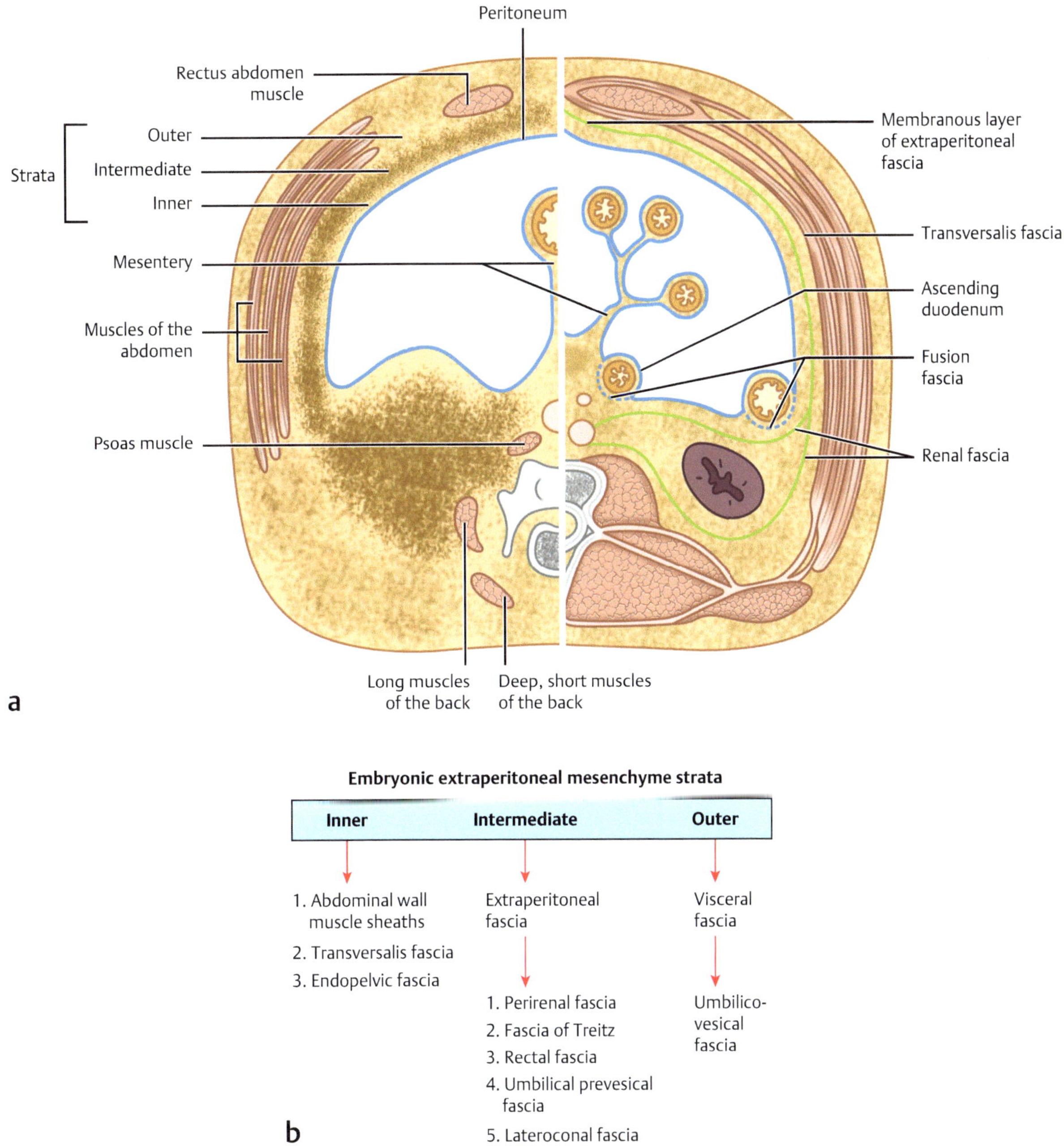

Fig. 14.3 (a) Drawing depicting a transverse section through the abdomen of human embryo (at about L1–L2). Right half section: Arrangement of primitive muscle-masses in the body wall and mesenchymal strata at 7 weeks. Left half section: Architecture of parietal and visceral abdominal fasciae at 7 months. (Modified with permission from Mirilas et al.[7]) **(b)** Layers of the embryonic extraperitoneal mesenchyme strata.

tissue with no retroperitoneal fasciae. The kidneys ascend from the pelvis to a lumbar level by the 6th week.[11] The primitive renal fascia and transversalis fascia appear around the 11th to 12th weeks and the process of development will be completed by the 23rd week.[9,11] The anterior renal fascia extends along the peritoneum and is often in part fused with the peritoneum.[11] The posterior renal fascia extends inferomedially toward the anterior aspect of the aorta and inferior vena cava.[12] However, at the level of the renal hilus, the posterior renal fascia connects with vascular sheaths of the great vessels (▶Fig. 14.4). At final stages after development of the lymph nodes, the medial continuation of the renal fascia will be penetrated by the developing lymph vessels communicating with the paraaortic and mesenteric nodes. Adipose tissue develops between the renal fascia and transversalis fascia. The inferomedial extension of the renal fascia forms the periureteric fascia (ureteral sheath).[13] The ureteral sheath runs close the hypogastric nerve in the same sheath.[12]

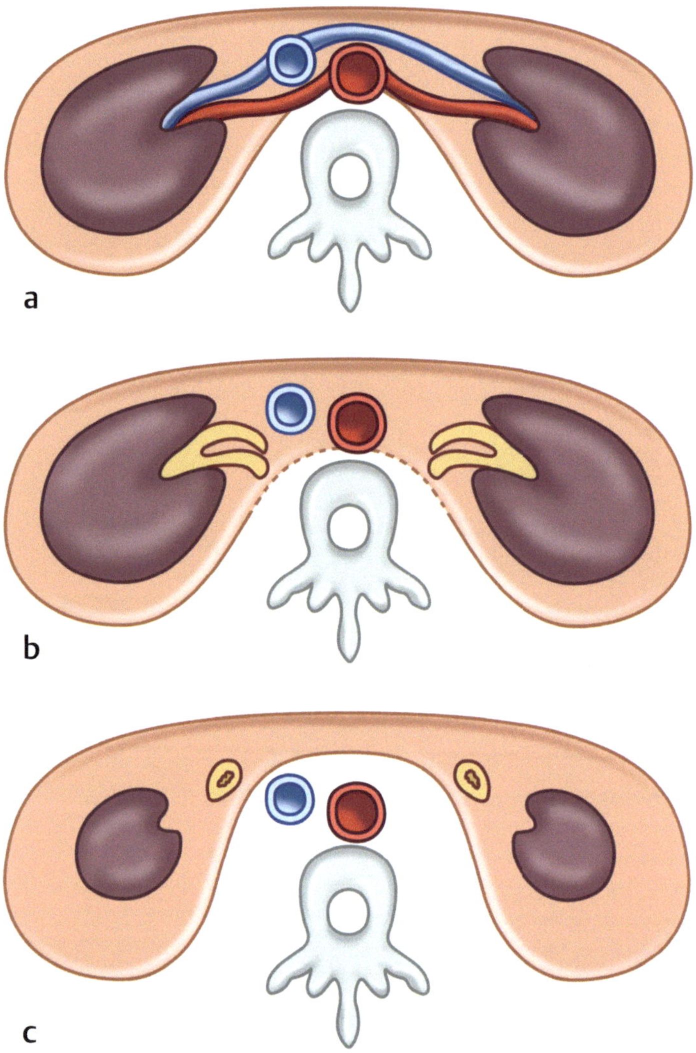

Fig. 14.4 Diagrams showing the difference in the medial continuation of the posterior renal fascia in fetuses at three levels: **(a)** At the renal vascular pedicle, the renal fascia connects with the vascular sheath of the great vessels. **(b)** The posterior fascia becomes unclear near the ureteropelvic junction. **(c)** At the level of the inferior pole of the kidney, the posterior renal fascia extends to the anterior side of the great vessels. (Modified with permission from Matsubara et al.[12])

14

The lateroconal fascia is also a migration fascia formed in the retroperitoneal loose connective tissue by mechanical compression of rapidly growing kidney (▶Fig. 14.5). It appears as a multilaminar structure between the renal fascia and transversalis fascia after 20 weeks of gestation.[12] The space is filled with fat (retroperitoneal). At 23rd to 25th weeks of fetal age, fasciae are evident on the surface of both the psoas and quadratus lumborum muscles. At this stage, the colon is attached to the retroperitoneal tissue by peritoneal fusion. After retroperitoneal fixation of the colon, the fusion fascia of the peritoneum may appear on the posterolateral side of colon by condensation of the connective tissue and in continuity with the lateroconal fascia.[12]

Transversalis Fascia

The transversalis fascia is the deep investing fascia that covers the inner surface of the diaphragm and transversus muscle. It forms the outer limit of the extraperitoneal space and is an important anatomical structure during extraperitoneal laparoscopic operations. The transversalis fascia is continuous with the investing fascia of other muscles of the inner surface of the abdominopelvic wall. Superiorly, it is continuous with the diaphragm, inferiorly with the parietal pelvic fasciae, and medially with the thoracolumbar fascia of the psoas and quadratus lumborum muscles. In the posterior wall of the inguinal canal and behind the lower rectus sheath where the transverse abdomen muscle is absent, only the transversalis fascia exists.[6,8] Behind the rectus abdominis and below the arcuate line, the transversalis fascia invests the rectus and separates it from the preperitoneal connective tissue. The transversalis fascia is continuous at the contralateral side. Thus, it forms a continuous layer that surrounds the peritoneum and extraperitoneal tissue (▶Fig. 14.6). In the pelvis, the transversalis fascia forms the internal inguinal ring and gives a fascial sling around the spermatic cord (the internal spermatic fascia). It is prolonged in front of the femoral vessels into the thigh as the anterior part of the femoral sheath.

There has been much confusion about the exact development, anatomy, and structure of the transversalis fascia. The term was introduced by Cooper in 1844.[14] Some believe that the transversalis fascia is a single layer fascia. Bilaminar structure of the fascia is also accepted by most investigators at the level of the inguinal canal near the insertion to the Cooper's ligament, with the inferior epigastric vasculature passing in between.[15] A recent investigation has shown bilayer nature of the transversalis fascia with fat in between encasing the entire retroperitoneal space, extraperitoneal space, Retzius space, and Bogros space[16] (▶Fig. 14.6). Mirilas et al believe that the so-called second layer of the transversalis fascia is in fact the membranous layer of the extraperitoneal fascia[7,17] (▶Fig. 14.3). In certain areas, this membranous layer can be very well developed (migration fascia), which is possibly related to the condensation of the extraperitoneal fascia due to the growth of adjacent organs. Examples include the perirenal, lateroconal, and umbilical prevesical fasciae.[6,17] In the lower anterior abdominal wall, the deep inferior epigastric arteries pass in the space between the transversalis fascia and the membranous layer of the extraperitoneal fascia. The median and medial umbilical ligaments (encased by the umbilicovesical fascia), vas deferens and its artery, gonadal vessels, and lower ureters are located in the preperitoneal fat between the peritoneum and the membranous layer of the extraperitoneal fascia.

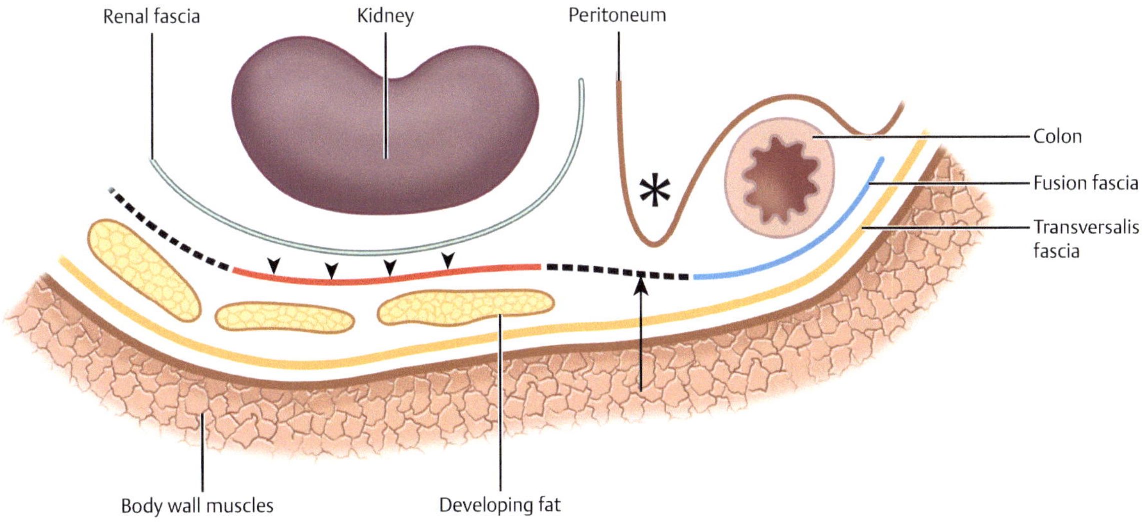

Fig. 14.5 A diagram showing the development of the lateroconal fascia (*red line*). In the connective tissue space between the renal fascia and transversalis fascia, the lateroconal fascia (*arrowheads*) secondarily develops along fatty tissue masses. When a peritoneal recess (*asterisk*) between the kidney and colon disappears, the fusion fascia (*blue line*) formed after retroperitoneal fixation of the colon is likely to connect with the lateroconal fascia (*large black arrow* shows the connection line). The fusion fascia (*blue line*) comes from the attachment of the posterior leaf of the mesentery of the colon and the peritoneum of the posterior wall (the two peritoneal leaves fuse). (Modified with permission from Matsubara et al.[12])

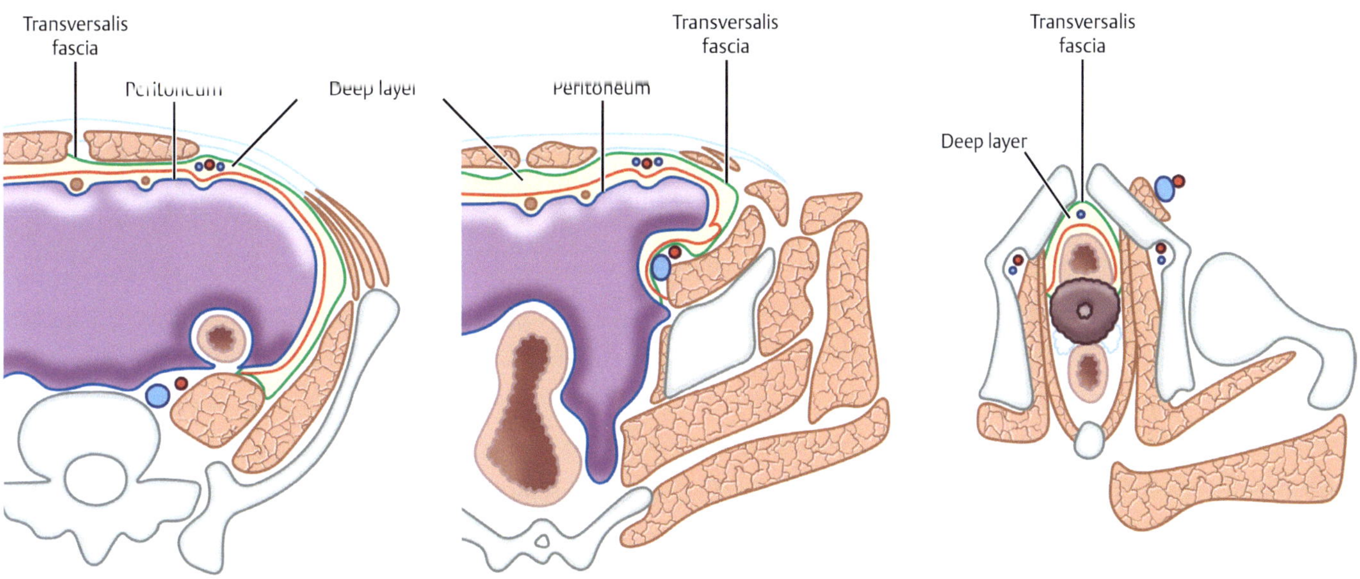

Fig. 14.6 Bilayer concept of transversalis fascia. Axial planes at three levels are shown. The superficial layer of the transversalis fascia (*green line*) attaches to the muscles and the deep layer (*red line*) is close to the peritoneum. Using this concept, the lateroconal and umbilical prevesical fasciae are the deep layer of the transversalis fascia formed later in embryonic life from compressive effect of the adjacent organs, hence the name "condensation fascia." (Modified with permission from Li et al.[16])

◆ Retroperitoneal Space

The retroperitoneum is divided into three compartments according to their demarcation by well-defined fascial planes[5] (▶Fig. 14.1):

- Anterior pararenal space is between the posterior parietal peritoneum and anterior renal fascia. It is confined laterally by the lateroconal fascia.
- Perirenal space formed by the renal fascia is the anatomic location of kidneys and adrenal glands.
- Posterior pararenal space is between the posterior renal fascia and transversalis fascia.

Perirenal Space

The perirenal spaces are inverted cone-shaped compartments that lie lateral to the lumbar spine. Each space contains the kidney, adrenal gland, upper ureter, perinephric fat, fibrous bridging septa, and a network of small perirenal vessels (arteries, veins, and lymphatics) (▶Fig. 14.1, ▶Fig. 14.2). This space is covered by the renal fascia composed of multilamellar anterior and posterior layers.[3] The anterior layer of the renal fascia (Gerota's fascia or fascia of Toldt) is a thin layer of connective tissue. The posterior layer of the renal fascia (Zuckerkandl's fascia) appears thicker possibly because of its close apposition with the lateroconal fascia. These two layers of the renal fascia appear well defined anteriorly and posteriorly, and join one another laterally; after they join, they continue more laterally with the lateroconal fascia. However, their disposition, fusion, or dehiscence along the superior, medial, and inferior borders are the sources of major differences of opinion among investigators. On the right side of the abdomen, the anterior renal fascia is closely related to the ascending colon, descending duodenum, hepatorenal peritoneal recess, hepatorenal ligament, and liver (▶Fig. 14.2). On the left side, it is related to the spleen, pancreatic tail, pancreaticorenal ligament, lower aspect of the stomach, distal transverse colon, and proximal descending colon (▶Fig. 14.7). The posterior renal fascia is clearly distinguished over its whole length. It is separated from the posterior parietal muscles (transverse abdomen, quadratum lumborum, and iliopsoas) by the posterior pararenal fat.

14

In the upper part of the renal space, the anterior and posterior layers of the renal fascia fuse one another and together attach to the inferior fascia of the diaphragm (transversalis fascia). On the right side, the anterior renal fascia may be deficient superiorly, being exposed to the bare area of the liver.[5]

The medial attachments of the anterior and posterior layers of the renal fascia are not clearly described. The posterior layer of the renal fascia may be attached to the psoas and quadratus lumborum muscles' inner investing fascia. The medial extension of the anterior layer of the renal fascia is highly controversial. Some anatomic dissections have shown that the perirenal spaces are separated from one another and do not communicate medially.[18] Some authors believe that the right and left anterior layers of the renal fasciae connect in the midline along the vascular sheath of the great vessels.[2,12,19] This creates a potential conduit between the two perirenal spaces known as the "Kneeland channel" that allows free diffusion of gas and fluid in front of the aorta and the inferior vena cava (▶Fig. 14.8). A cadaveric computed tomography (CT) study by Thornton et al after injection of iodinated contrast medium showed that the perirenal spaces communicate with each other across the midline and directly with the pelvic extraperitoneal spaces.[20] However, it is important to note that air and contrast material are diffusible and may cross through the midline even if the fascia is intact. Furthermore, the perinephric space contains multiple small bridging perinephric septa that support the kidneys and divide it into multiple small compartments.[21] These septa move in different directions and may extend between the two fasciae. Therefore, fluid collection may remain within a portion of the perirenal space or move outside along these septa[22] (▶Fig. 14.9, ▶Fig. 14.10). There is also a complex network of retroperitoneal lymphatics that can act as conduits for disease spread to or from the perinephric space.

The inferior extension of the renal fascia is also a matter of debate. Most authors also describe an inferior closure of the perirenal spaces by fusion of the anterior and posterior layers of the renal fascia that includes the ureters and gonadal vessels pass.[23,24] However, some advocate the idea that the perirenal space is potentially open and communicates inferiorly along the ureters to the pelvis. Clinical implication of this communication is that perinephric collections can potentially flow into the pelvis.[20]

Overall, although the perirenal space appears closed anatomically on all sides, thin layers of the renal fascia at the level of the hilum of the kidneys and along the ureters may easily disrupt if the pressure interior to the renal fascia is suddenly elevated; that permits extravasation of air, contrast material, blood, and inflammatory fluid outside the fascial conus.[18,23] It is important to know that in case of aortic rupture, the blood does not usually extend into the perirenal space. Instead, it preferentially dissects into the posterior pararenal space along the psoas muscle. In patients with renal atrophy, the renal capsule may appear separate from the kidney. In this case the renal capsule should not be mistaken with the renal fascia (▶Fig. 14.11).

Lateroconal Fascia

The term "lateroconal fascia" was introduced by Congdon et al.[25,26] They described that the lateroconal fascia is the result of fusion of the two layers of the renal fascia behind the ascending or descending colon and that the fascia terminates laterally by blending with the peritoneal reflection to form the paracolic gutter. However, Raptopoulos et al[27] demonstrated that the lateroconal fascia is distinct from the renal fascia and runs close to the posterior renal fascia, and this combination looks like a single-layered fascia (▶Fig. 14.8). Recently, the embryologic development of the lateroconal fascia was explained by Matsubara et al.[12] They showed that the lateroconal fascia is in fact a migration fascia originating by condensation from retroperitoneal loose connective tissue outside the renal fascia after 20 weeks of gestation. It is continuous with fusion fascia of the peritoneum on the posterolateral side of colon[12] (▶Fig. 14.5). As mentioned above, some investigators believe that the lateroconal fascia is in fact the membranous layer of the extraperitoneal fascia.[8]

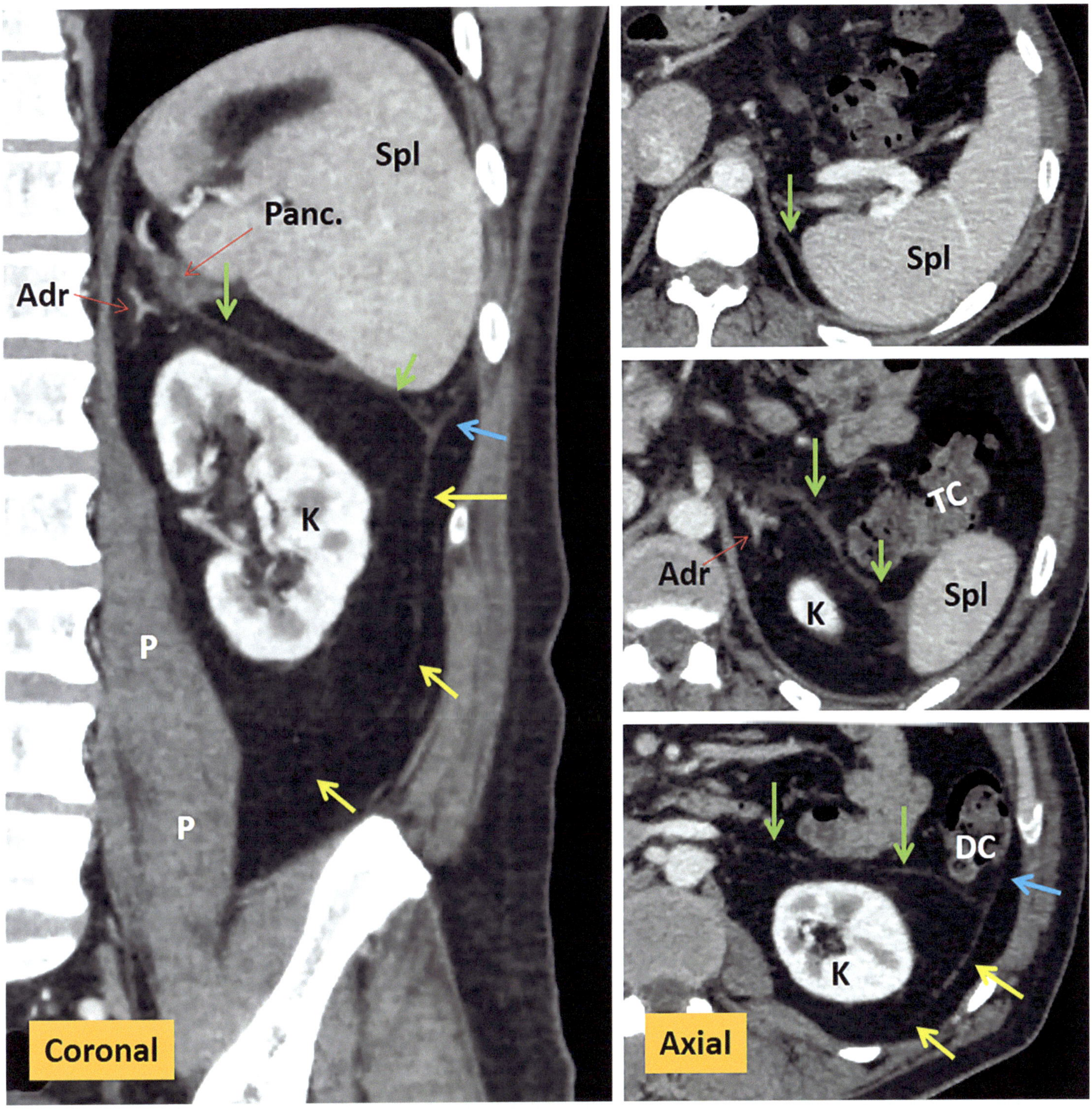

Fig. 14.7 Renal fascia. Coronal and axial computed tomography (CT) images showing the extent of the renal fascia conus. The anterior renal fascia (*green arrows*) and the posterior renal fascia (*yellow arrows*) meet the lateroconal fascia (*blue arrows*) laterally. The perirenal space is formed by the renal fascia. It contains kidney (K) and adrenal gland (Adr). The pancreas is located anterior to the anterior renal fascia within the anterior pararenal space. The spaces outside the posterior renal fascia and the lateroconal fascia are the posterior pararenal space and properitoneal fat, respectively, which are in continuity. The renal fascia is related to the spleen (Spl), pancreatic tail (Panc.), pancreaticorenal ligament, lower aspect of the stomach, distal transverse colon (TC), and proximal descending colon (DC). P, psoas muscle.

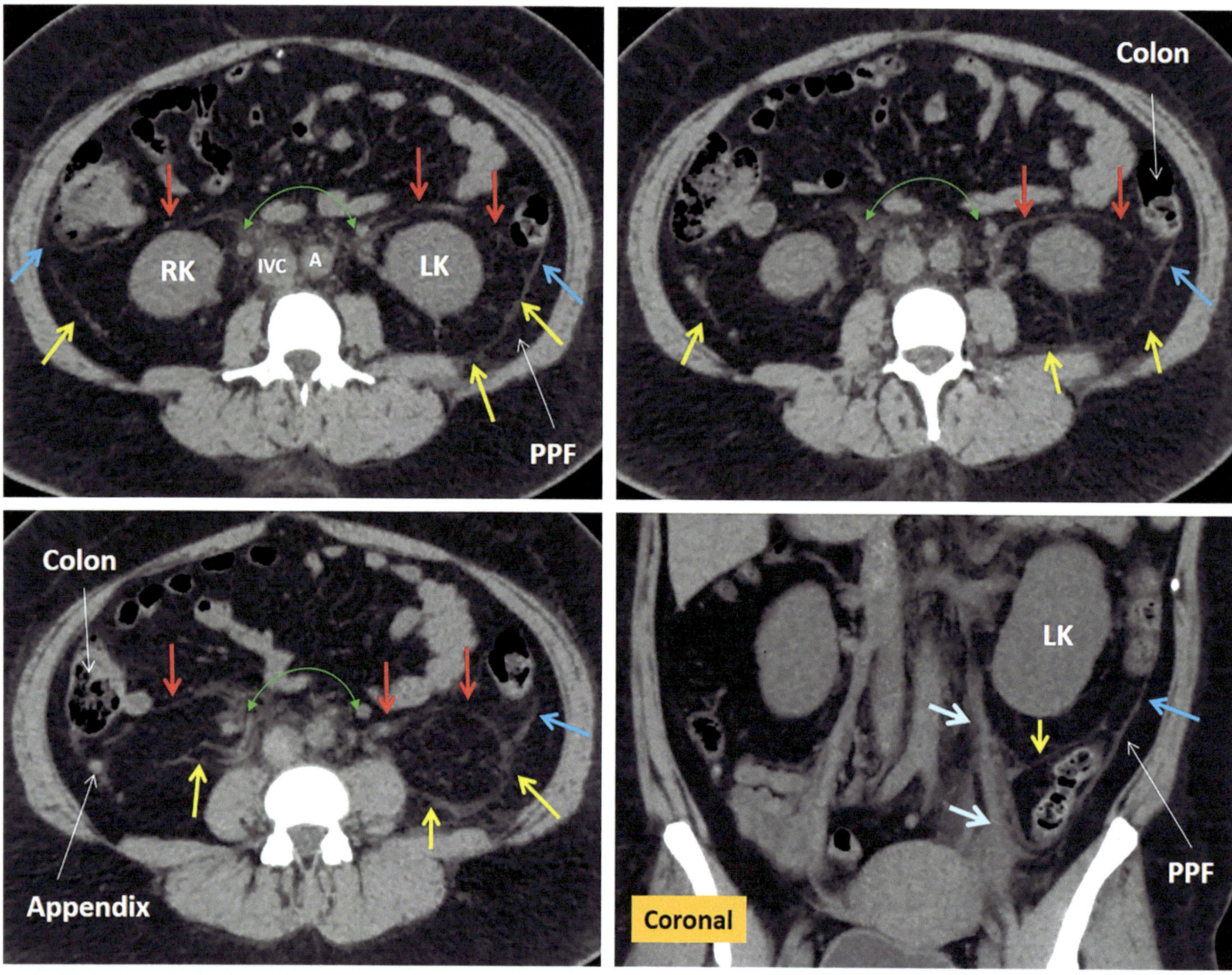

Fig. 14.8 Noncontrast, axial and coronal computed tomography (CT) scans demonstrate the renal fascia and perirenal space in a patient with recently passed left-sided stone. Stone-related edema of the perirenal fascia and anterior interfascial planes crossing the midline (*green double-headed arrows*) to the contralateral side. Edema also extends inferiorly along the left perigonadal vessel and periureteral fascial planes (*light blue arrows*). Properitoneal fat (PPF) is seen outside the posterior renal and lateroconal fasciae. A, aorta; *blue arrows*, laterocoanl fascia; IVC, inferior vena cava; LK, left kidney; *red arrows*, anterior renal fascia; RK, right kidney; *yellow arrows*, posterior renal fascia.

Anterior Pararenal Space

The anterior pararenal space is bordered anteriorly by the posterior parietal peritoneum, posteriorly by the anterior layer of the renal fascia, and laterally by the laterocoanl fascia.[5] The ascending and descending colon, duodenal loop (second and third portions), and pancreas (head and body) are included within the anterior pararenal space (▶ Fig. 14.12). This space is continuous across the midline. However, except for fluid accumulations in pancreatitis and/or retroperitoneal appendicitis, many diseases remain localized to one side.[27] In the duodenal perforation, air dissects into the right side of the anterior pararenal space.

This space is continuous with the fat tissue around the mesenteric vessels of the small bowel mesentery (▶ Fig. 14.12). It is also connected with pericholecystic, parabiliary, and periportal connective tissue spaces (▶ Fig. 14.13). Below the renal fascia, the anterior pararenal space merges with the posterior pararenal space (see the section Interfascial Planes). Anteriorly and laterally, the anterior pararenal space is continuous with the subperitoneal space (pelvis) and preperitoneal fat (abdomen)[28] (▶ Fig. 14.2).

During embryonic development, the second and third segments of the duodenum as well as the head and body of the pancreas become secondarily retroperitoneal in the anterior pararenal space. The posterior layer of embryonic mesentery of the pancreas fuses with the anterior layer of the renal fascia and the anterior interfascial plane is formed. This fusion fascia is known as fascia of Treitz. Recently, the mechanism of embryogenesis of this fascia was revised. According to Cho et al,[29] immediately after retroperitoneal fixation of the pancreas, at around 10 weeks of gestation, no distinct fascia behind the pancreas can be detected; instead, only a loose connective tissue layer is observed. Since the retropancreatic fascia emerges later, after 20 weeks of gestation, it seems that the retropancreatic fascia is one of the migration fasciae formed by mechanical compression due to development

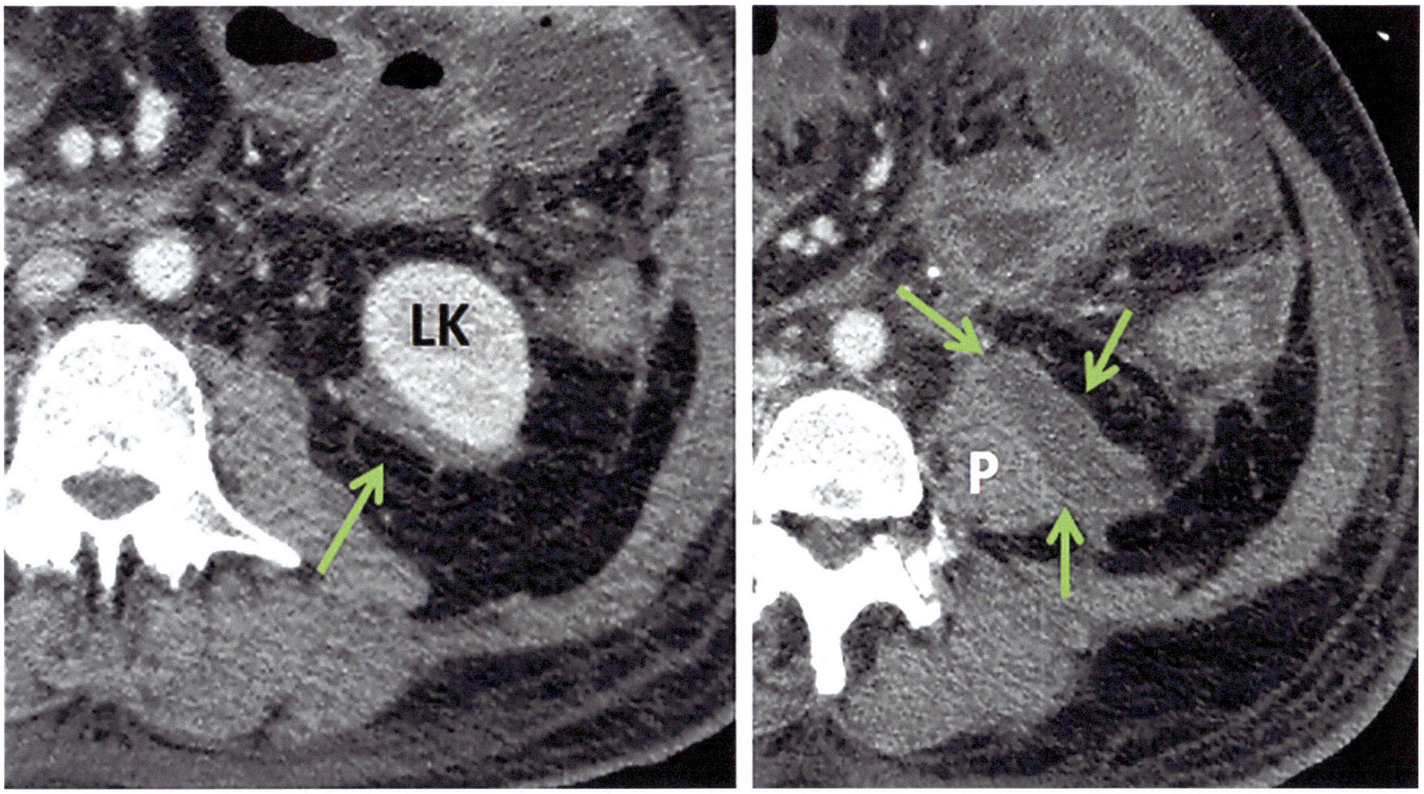

Fig. 14.9 Perirenal fluid collection is localized to a small subcompartment (*green arrows*) by the bridging septa in the perirenal space. LK, left kidney; P, psoas.

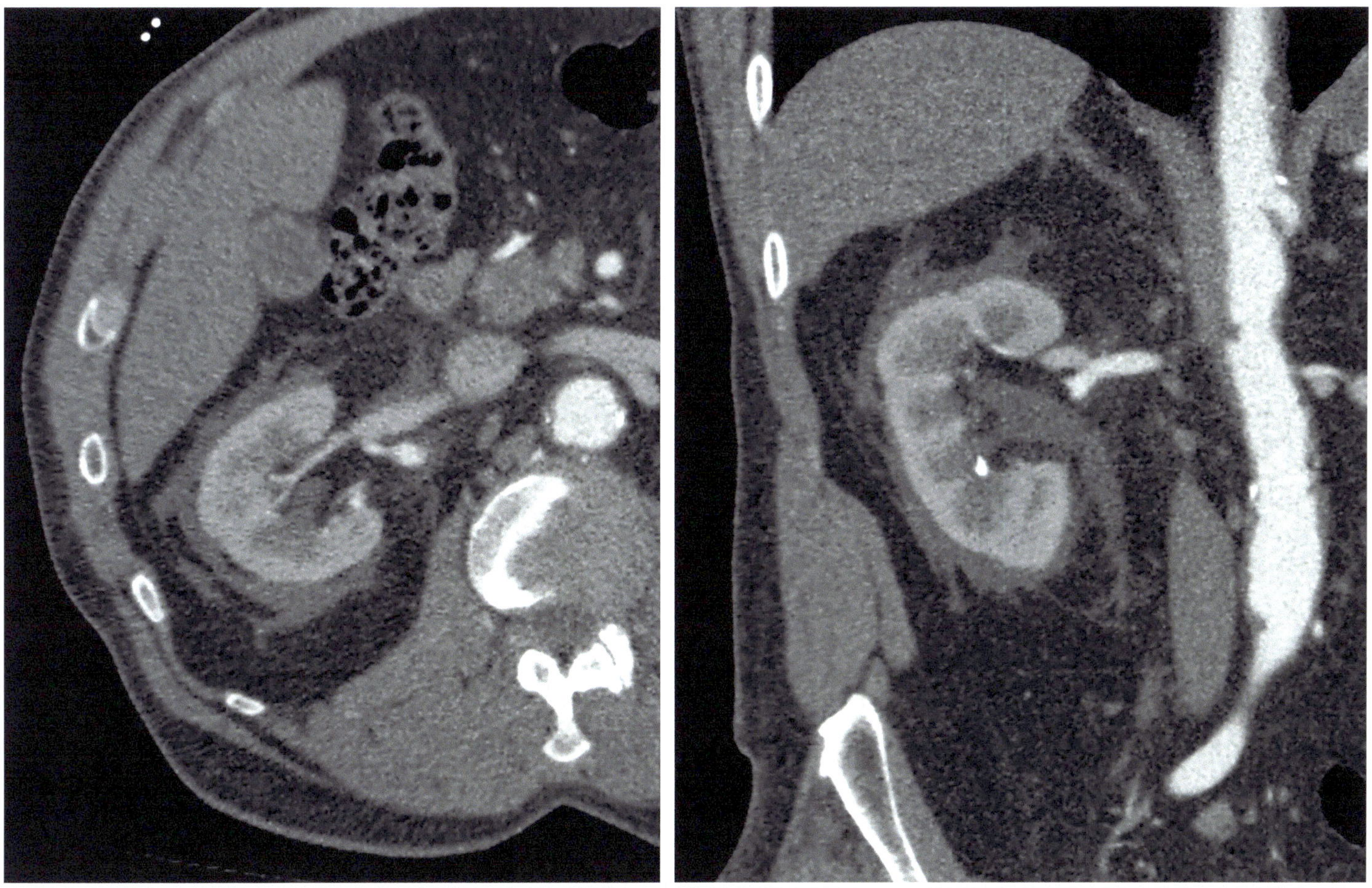

Fig. 14.10 Perirenal edema due to an obstructing right ureteral stone mainly collected within the renal fascia. The collection appears lobulated and partially compartmentalized possibly due to the bridging septa. Mild hydronephrosis and a small stone are seen.

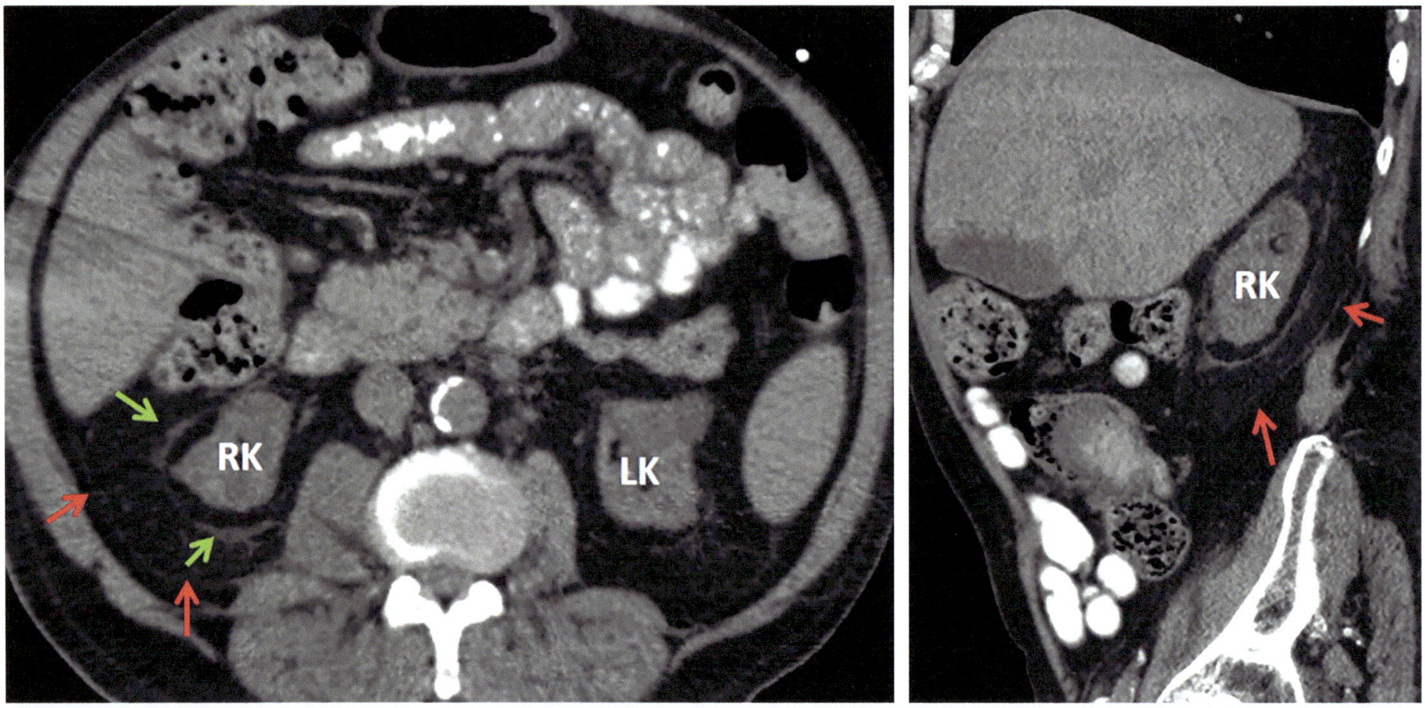

Fig. 14.11 Perirenal fat versus intracapsular fat. In this patient with atrophy of kidneys due to chronic renal failure, subcapsular fat deposition separates the renal capsule (*green arrows*) from the kidney. Another possibility for this pattern could be bridging of septa in the perirenal space. LK, left kidney; *red arrows*, renal fascia; RK, right kidney.

of the pancreas and duodenum[12] (▸Fig. 14.14). As a migration fascia, the fascia of Treitz does not usually connect to the peritoneum, but is likely to connect with other migration fasciae such as the renal fascia or ureteral sheath. This retromesenteric fascia allows surgical dissection of the pancreas and duodenum by the Kocher's maneuver which is defined as opening of the retroperitoneum at the antimesenteric—right—border of the duodenum and mobilization of duodenum together with the head of the pancreas.[30] When Kocher's maneuver is used during pancreaticoduodenectomy, the fascia adheres to the pancreatic parenchymal side, but not to the vena cava side. The pancreaticoduodenal arteries, veins, and nerves are situated on this membrane and the portal vein tributaries and superior mesenteric artery branches are covered by the fusion fascia, found on its ventral side.

14

Posterior Pararenal Space

The posterior pararenal space is the space behind the kidneys, filled with fat, bounded by the posterior renal fascia anteriorly and the transversalis fascia posteriorly (▸Fig. 14.1, ▸Fig. 14.2). The posterior pararenal space continues laterally external to the lateroconal fascia as the preperitoneal fat of the abdominal wall (▸Fig. 14.8). Superiorly, it is bordered by the diaphragm and continues as a thin subdiaphragmatic layer of fat, internal to the transversalis fascia which also invests the diaphragm. The space is open laterally toward the flank and anterior abdomen, and inferiorly toward the pelvis. The posterior pararenal spaces are in the same anatomical level with the prevesical and presacral spaces and all of them communicate one another.[4,31,32] Pathologies can reach from this space to the diaphragm superiorly and the femoral vascular sheath space and inguinal canals inferiorly[4] (▸Fig. 14.15).

Interfascial Planes

CT observations and cadaveric dissections and CT-guided injections of contrast material in cadavers suggest that rapidly developing fluid collections can accumulate and spread between the perirenal, anterior pararenal, posterior pararenal, and superioneal spaces along the so-called "interfascial planes."[2,4,11,24,31]

The concept of interfascial planes has been introduced in the radiology literature to explain the spread of retroperitoneal disease between the extraperitoneal spaces.[4,31,32,33] The interfascial planes are potential spaces between the apposed layers of the fascia that act as pathways within which retroperitoneal pathology can extend beyond the original compartment (▸Fig. 14.1). For example, pancreatitis starts in the anterior pararenal space and may extent into the pelvis or perirenal space by dissection of the pancreatic fluid along these interfascial planes[27] (▸Fig. 14.16). Using the interfascial planes' concept the severity of retroperitoneal pathologies can be stratified.[33,34]

In an embryologic study, Ishikawa et al explained that the interfascial planes are formed by the initial development of the retroperitoneal loose connective tissue followed by the development of migration fasciae.[11] The interfascial planes have been named based on their anatomic interposition.[24] The anterior interfascial (retromesenteric) plane is located anterior to the anterior renal fascia and is produced by fusion of the mesentery (fusion fascia) of the pancreas, duodenum, ascending colon, and descending colon. It is

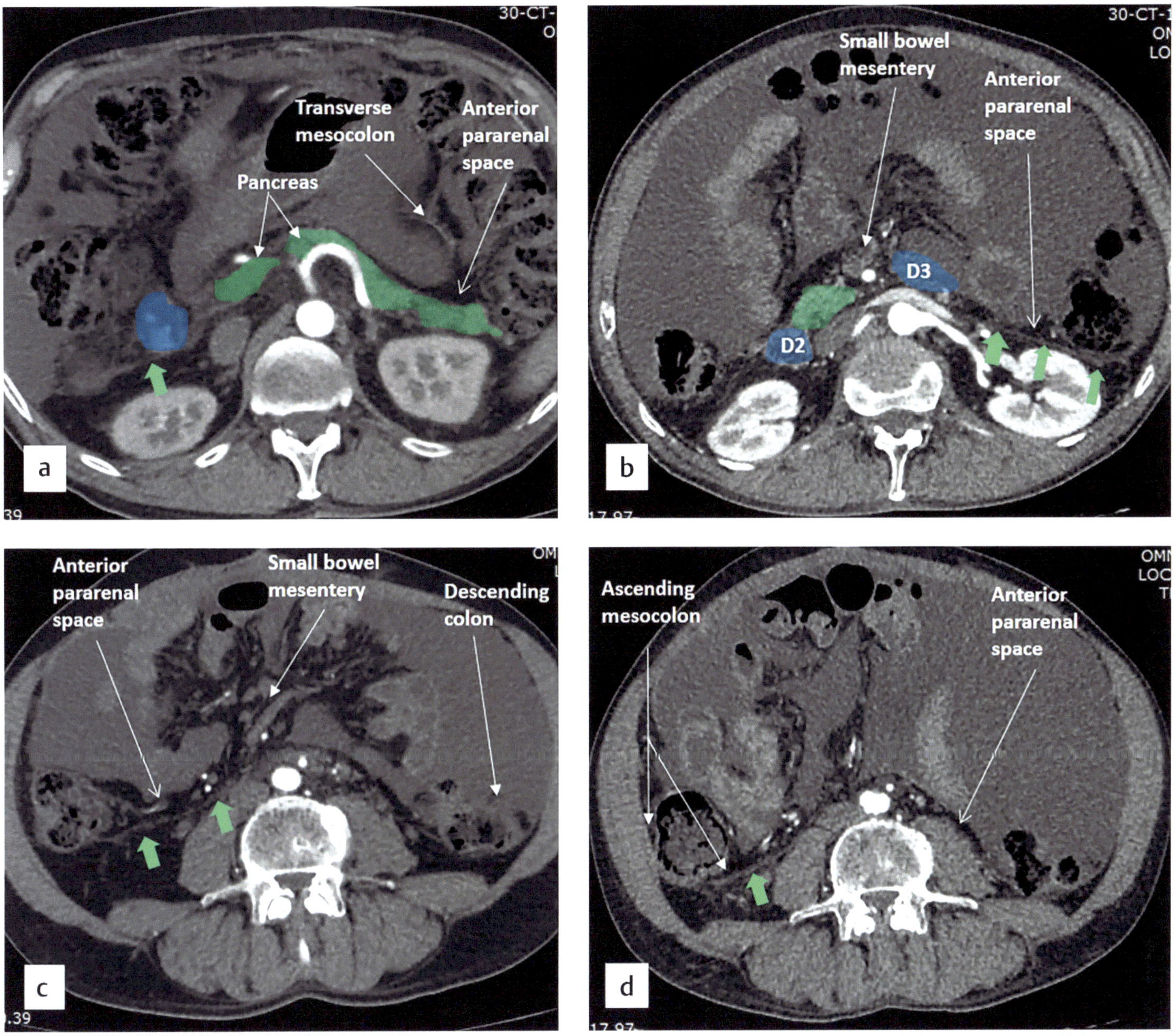

Fig. 14.12 Anterior pararenal space boundaries anterior to the Gerota's fascia (*green arrows*) are clearly shown in this patient with large ascites. Retroperitoneal compartments and mesenteric fat are outlined by fat. **(a–d)** Axial and **(e, f)** sagittal images. Note that anterior pararenal space is continuous with descending colon, ascending colon, transverse colon, and small bowel mesenteries containing all their vessels. The pancreas (*green*), second (D2) and third (D3) parts of duodenum, ascending colon, and descending colon reside in the anterior pararenal space. The mesentery of the ascending and descending colon is attached to the posterior abdominal wall and has become part of the anterior pararenal space.

(*Continued*)

continuous across midline along the retropancreatic fascia of Treitz. The posterior interfascial (retrorenal) plane is between the posterior renal fascia and the lateroconal fascia (membranous fascia).[24]

Surgico-anatomically, anterior to the renal fascia, the multilaminar anterior interfascial plane, which is continuous across the midline, permits communication between the two anterior pararenal spaces. Through the same plane the anterior pararenal space may extend to the lateroconal fascia. Thus, the lateral border of the anterior pararenal space reaches the transversalis fascia. Evidently, from thereon, the anterior pararenal space is found in the same surgical layer with the posterior pararenal space and the two may communicate. From there on, since the posterior pararenal space communicates with the prevesical and presacral spaces, it can be understood that there is no border between the pararenal and these subperitoneal pelvic spaces.[6]

In radiology, the normal interfascial planes are thin and may not be visible. However, in the presence of edema or retention of a small amount of fluid (e.g., volume overload associated with heart/renal failure, or septic shock), they become thicker and distinguishable (▶ Fig. 14.8).

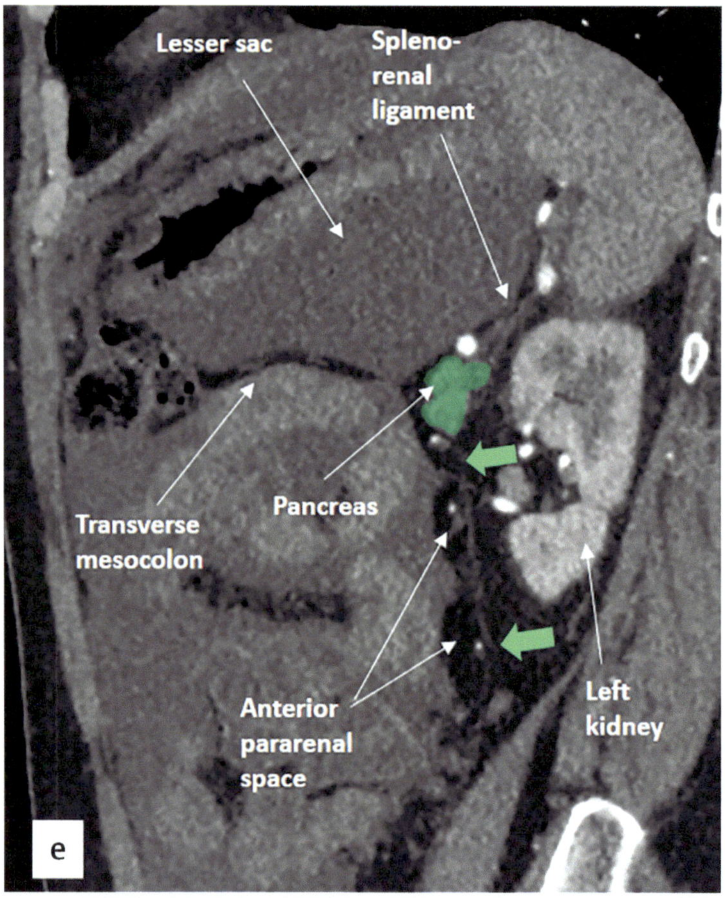

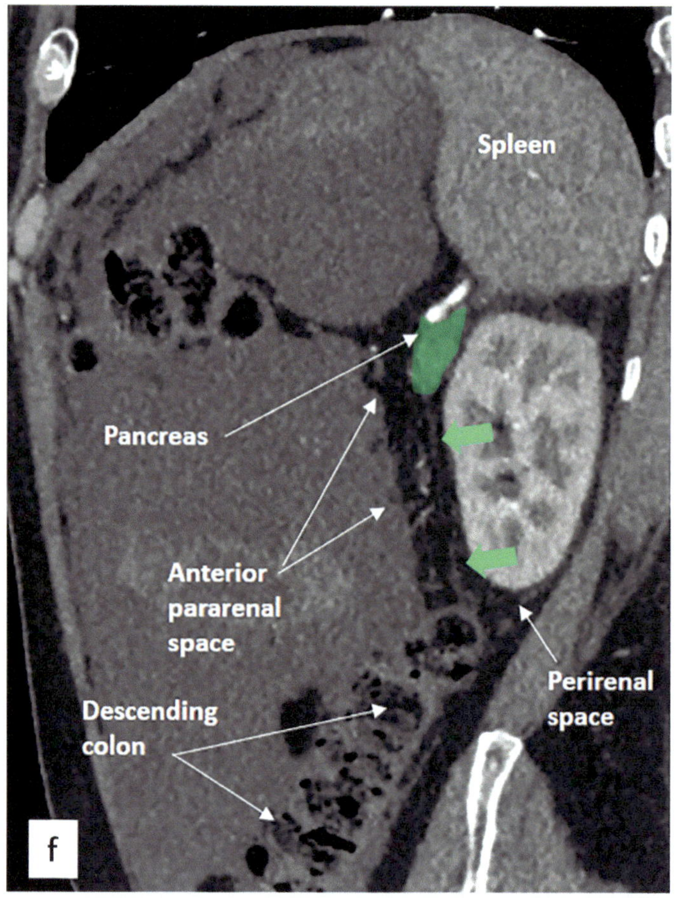

Fig. 14.12 (*Continued*) (e, f)

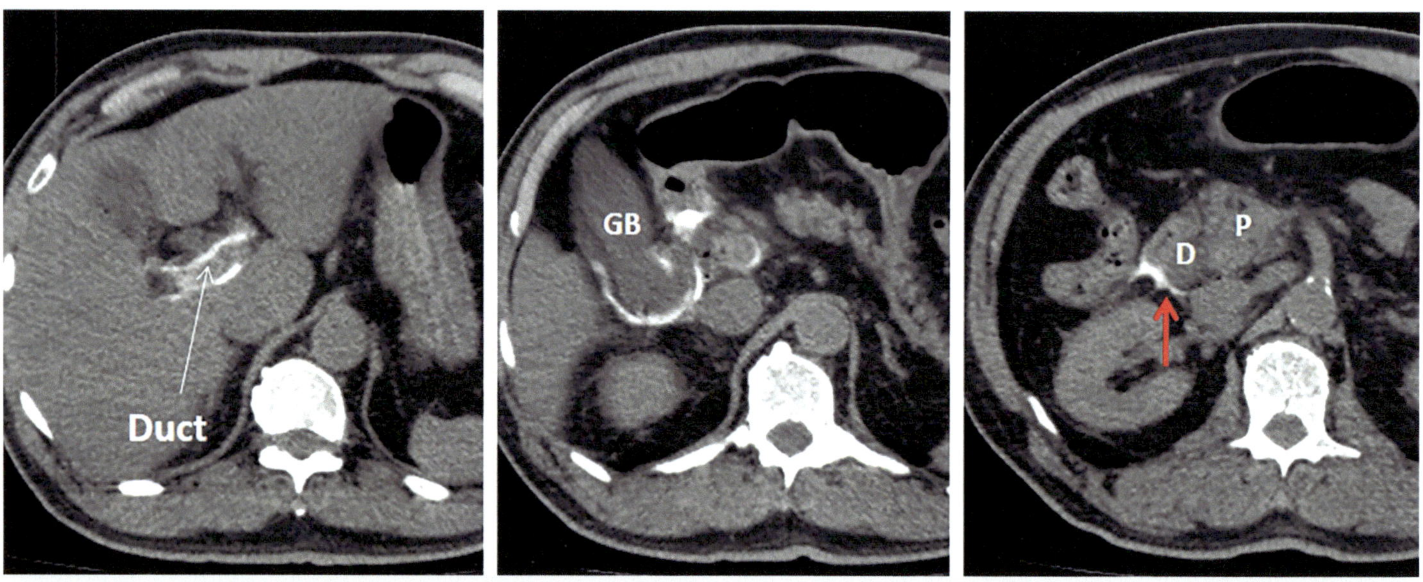

Fig. 14.13 Pancreatobiliary fascial planes. Axial computed tomography (CT) images showing contrast material extravasation after sphincterotomy for bile duct stone. Injected contrast material has entered into the anterior pararenal space and delineates the posterior margin of the second duodenum (D) (*red arrow*). Contrast material also dissects superiorly along the fascial planes in pericholecystic tissue around the gall bladder (GB) and along the extrahepatic biliary ducts. P, pancreatic head.

14

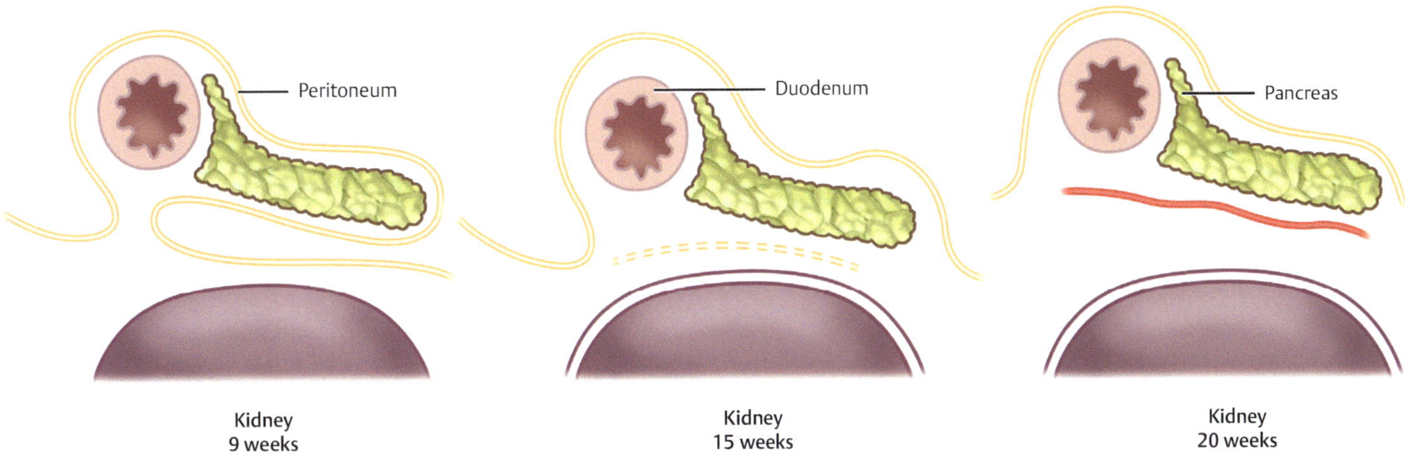

Fig. 14.14 Development of the retropancreatic fascia. At 9 weeks of gestation, the primitive peritoneum forms a deep recess behind the retro-pancreatic space to form peritoneal fusion fascia. The fusion fascia remains indistinct (*dotted line*) at 15 weeks. At 20 weeks, the retropancreatic fascia (*red band*) becomes evident especially behind the duodenum. The lateral ends of the fascia are likely to connect with other retroperitoneal migration fasciae such as the anterior renal fascia. This retromesenteric fascia allows surgical dissection of the pancreas and duodenum by the Kocher maneuver. (Modified with permission from Cho et al.[29])

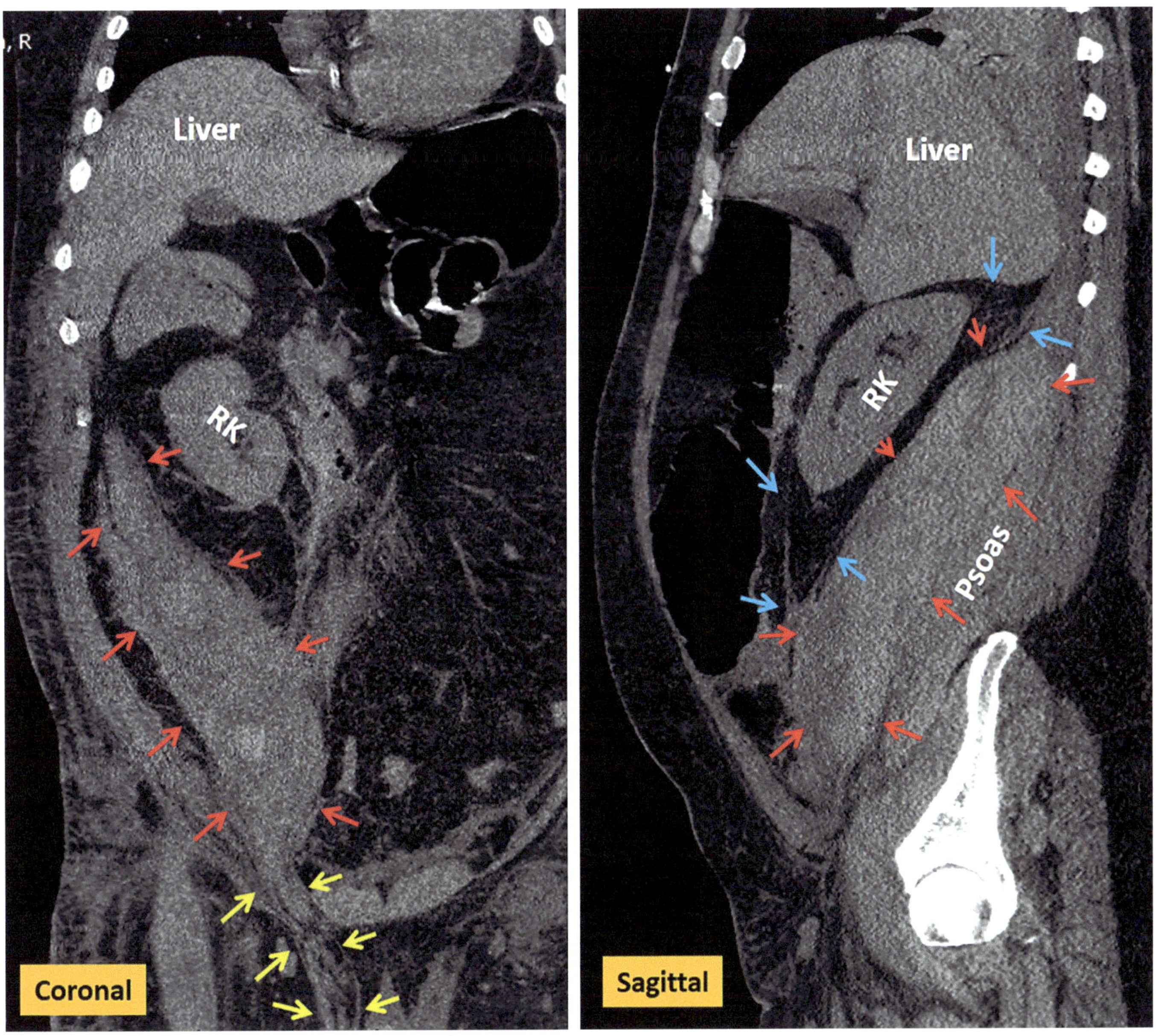

Fig. 14.15 Retroperitoneal hematoma (area between *red arrows*) formed in the right posterior pararenal space displacing the right kidney (RK) anteriorly. The perirenal fascia is demarcated by *blue arrows*. Note that hematoma is located between the psoas muscle posteriorly and the posterior renal fascia anteriorly. It is also shown on the coronal view that the hematoma extends toward the inguinal canal (between *yellow arrows*). Note that inguinal canal, obturator canal, and femoral sheath are in connection with the posterior pararenal space.

14

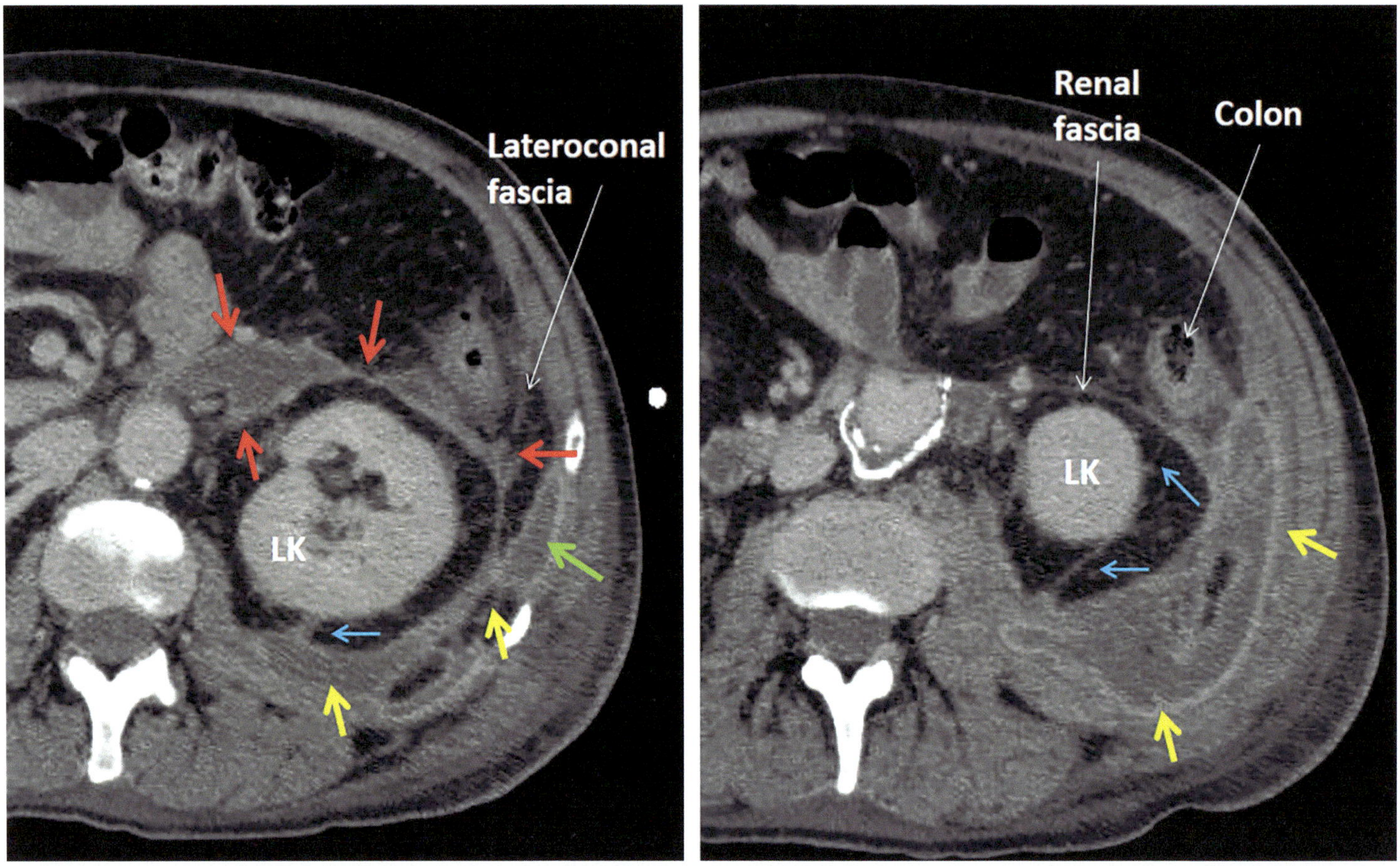

Fig. 14.16 Track of inflammatory fluid in a patient with pancreatitis. Computed tomography (CT) scan shows extension of the anterior pararenal space fluid (*red arrows*) along the lateroconal fascia and interfascial planes into posterior pararenal space (*yellow arrows*) and preperitoneal fat (*green arrow*). The perirenal space is relatively spared. This is a common phenomenon. Note edema of the abdominal wall muscle fasciae and the superficial fascia. Also note, thickened bridging septa in the perirenal space (*blue arrows*). Perinephric bridging septa provide channels between the kidney and the interfascial planes. LK, left kidney.

Subperitoneal Pelvic Space

The subperitoneal pelvic space is the extraperitoneal space below the pelvic peritoneal reflection and is covered by the endopelvic fascia. The endopelvic fasciae consist of both visceral and parietal components, with the visceral portion enveloping the pelvic viscera and the parietal portion covering the walls and floor of the pelvis. Similar to the transversalis fascia of the abdomen, the parietal investing fascia of the pelvic muscles (endopelvic) is locally named after the underlying muscle. For example, the fascia covering the inner aspect of the iliacus muscle is called iliac (iliacus) fascia and the one covering the obturator internus is named obturator fascia. The parietal pelvic fascia also covers the levator ani and piriformis muscles. It is continuous with the transversalis fascia and the fascia of the psoas and quadratus lumborum muscles. The inner aspect of the parietal fascia is filled with loose fatty connective tissue which is in continuity with posterior pararenal spaces (▶ Fig. 14.17; see also ▶ Table 14.1).

The visceral pelvic fascia is the reflection of deep parietal investing fascia that covers the pelvic organs and is firmly attached to them. The ureters, vas deferens, and vessels have their own adventitia which is continuous with the visceral fascia of the respective pelvic organs.

Thus, pelvic organs have their own visceral fascia. The urinary bladder is enveloped by the vesical fascia (the prostate by the prostatic fascia and so on). The continuation of the urinary bladder toward the umbilicus with the obliterated urachus is also ensheathed with the respective umbilicovesical fascia, which surrounds the remnants of the urachus (median umbilical ligament) and umbilical arteries (medial umbilical ligaments). It also partially covers the prostate. The ductus deferens and round ligament of the uterus traverse the umbilicovesical fascia en route to the deep inguinal ring (▶ Fig. 14.18).

In the pelvis, there is also the splanchnic (or more correctly parasplanchnic) pelvis fascia that comprises loose connective tissue. This fascia is for "packing" the continually extending

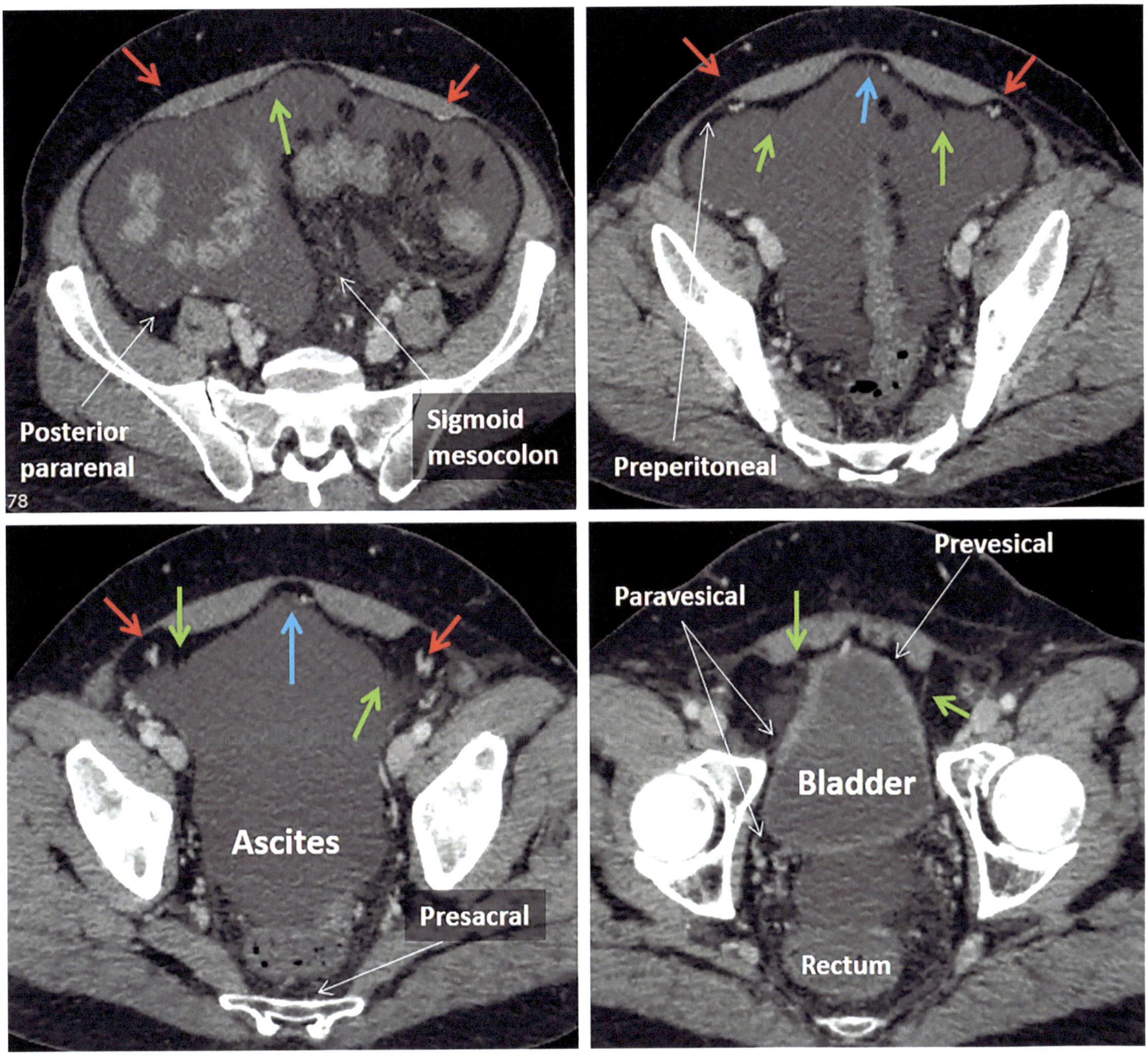

Fig. 14.17 Subperitoneal pelvic space in a patient with large ascites. Fat-containing extraperitoneal structures stand out against fluid-filled intraperitoneal space. *Arrows* point to the anatomic vessels and ligaments in the preperitoneal (prevesical) space. *Blue*, remnant of the urachus (median umbilical ligament); *green*, remnant of the umbilical arteries (medial umbilical ligaments); *red*, inferior epigastric vessels. Note continuity of the prevesical space, and paravesical and presacral spaces with each other and with the posterior pararenal space.

and shrinking pelvic organs (bladder, rectum, vagina, uterus—this in pregnancy). Thus, the perivesical fascia is the splanchnic pelvic fascia around the bladder, as is the perirectal fascia, etc. (▶ Fig. 14.18).

Similar to the renal fascia, condensation of the extraperitoneal tissue (pelvic splanchnic fascia) may be responsible for the genesis of the fascia in the parasplanchnic extraperitoneal spaces. Examples of these condensation fasciae include the umbilical prevesical fascia, rectovesical fascia in male, and rectovaginal fascia in female.

The retroperitoneal space of pelvis contains branches of the lumbar plexus, external iliac vessels, ureters, lymphatic vessels, and lymph nodes. The femoral nerve passes between the psoas and iliac muscles. The common and external iliac arteries and

Table 14.1 Extraperitoneal structures

Organs:
• ***Primarily retroperitoneal***
– Kidneys
– Adrenal glands
– Ureters
• ***Secondarily retroperitoneal***
– Pancreas
– Duodenum: descending, horizontal, and a segment of ascending parts
– Ascending and descending colon
– Part of cecum (variable)
– Rectum to the sacral flexure
Vessels:
• ***Abdominal aorta and branches***
• ***Inferior vena cava and tributaries***
• ***Ascending lumbar veins***
• ***Portal vein (before coursing in the hepatoduodenal ligament) and tributaries***
• ***Lymph nodes (sacral, lumbar, iliac, paraaortic, paracaval), cisterna chili***
Nerves:
• ***Lumbar plexus branches: iliohypogastric, ilioinguinal, genitofemoral, lateral femoral cutaneous, femoral, obturator nerves***
• ***Sympathetic trunk***
• ***Autonomic ganglia and plexuses***

veins are also covered by a sheath generated by medial extension of the iliac fascia[6] or coalescence of surrounding parietal pelvic fasciae.

Prevesical, Perivesical, and Paravesical Spaces

14

There are two fascial planes below the umbilicus and between the two medial umbilical ligaments, namely, the inner umbilicovesical fascia (visceral fascia) and the outer umbilical prevesical fascia (condensation fascia).

The umbilical prevesical fascia is a migration fascia and extends between the umbilicus and inferolateral surface of the bladder down to the pubic symphysis. It is behind the transversalis fascia in the lower anterior abdominal wall, and, as described above, is a local thickening of the membranous layer of the preperitoneal tissue. The umbilical prevesical fascia, umbilicovesical fascia, and transversalis fascia fuse with each other at the median umbilical ligaments and the umbilicus.

The space between the umbilical prevesical fascia and transversalis fascia is the retropubic or prevesical space described by Anders Adolph Retzius in 1858.[7] The space of Retzius is relatively large space situated anterior and lateral to the urinary bladder, extending from the muscular floor of the pelvis to the level of umbilicus (▶Fig. 14.18, ▶Fig. 14.19, ▶Fig. 14.20, ▶Fig. 14.21). In axial cross section, the space of Retzius has a U-shaped appearance with the open side facing posteriorly toward the lateral vesical ligaments. Anatomically, the exact posterior end of this space is at the deep transverse perineal muscles in the perineum. Anteriorly, it is bounded by the bodies of the pubic bones, medial pubic rami, and transversalis fascia along the rectus muscles to the level of the arcuate lines. The prevesical space is a common location for hematoma formation in pubic bone fractures. The source is usually from the retropubic venous plexus; however, pathologies in the posterior pararenal space can also extend into this space (▶Fig. 14.21).

The umbilical prevesical fascia delimits the perivesical space anteriorly and laterally, and the peritoneum bounds the perivesical space superiorly. It is important to realize the adherence of the umbilicovesical fascia to the bladder and urachus, and knowing that radiologically, it would be difficult to discern it separately from the bladder (▶Fig. 14.20). In the anterior abdominal wall, it would be difficult to radiologically separate the prevesical space from the perivesical space and any fluid collection in this region can be described as within the space of Retzius (▶Fig. 14.18). Only in certain cases the perivesical space may be filled with blood or contrast material, being visualized separately from the prevesical space of Retzius (▶Fig. 14.19, ▶Fig. 14.22).

Laterally, the spaces between the umbilical prevesical fascia and the parietal pelvic fascia are the paravesical spaces which are in continuation with the prevesical space, pararectal space, space of Bogros, and the femoral sheath. The paravesical space can extend along the round ligament and ductus deferens into the inguinal canal.[2] Retained fluid in the paravesical spaces can dissect superiorly into the space of Retzius and posteriorly along the sidewalls of the pelvis into the posterior pararectal space.

An aberrant obturator artery occurs in up to 50% of population and may be a hidden cause of hematoma in pubic rami fractures (▶Fig. 14.23). Normally, the obturator artery arises from the internal iliac artery, but an aberrant obturator arises from the inferior epigastric or external iliac arteries. Its recognition is important as bleeding from this artery may be unrecognized during interventional procedures.[35] They can also be injured in surgical procedures of the inguinal region.

Rectal Fascia

The rectal (mesorectal) fascia is the visceral fascia that surrounds the rectum and forms the perirectal space containing fibrofatty tissue and perirectal neurovascular structures also known as mesorectum (▶Fig. 14.21, ▶Fig. 14.24, ▶Fig. 14.25). Because of confusion with mesenterium, the term mesorectum is not listed in Terminologia Anatomica produced by the Federative International Programme for Anatomical Terminology (FIPAT). As a reminder,

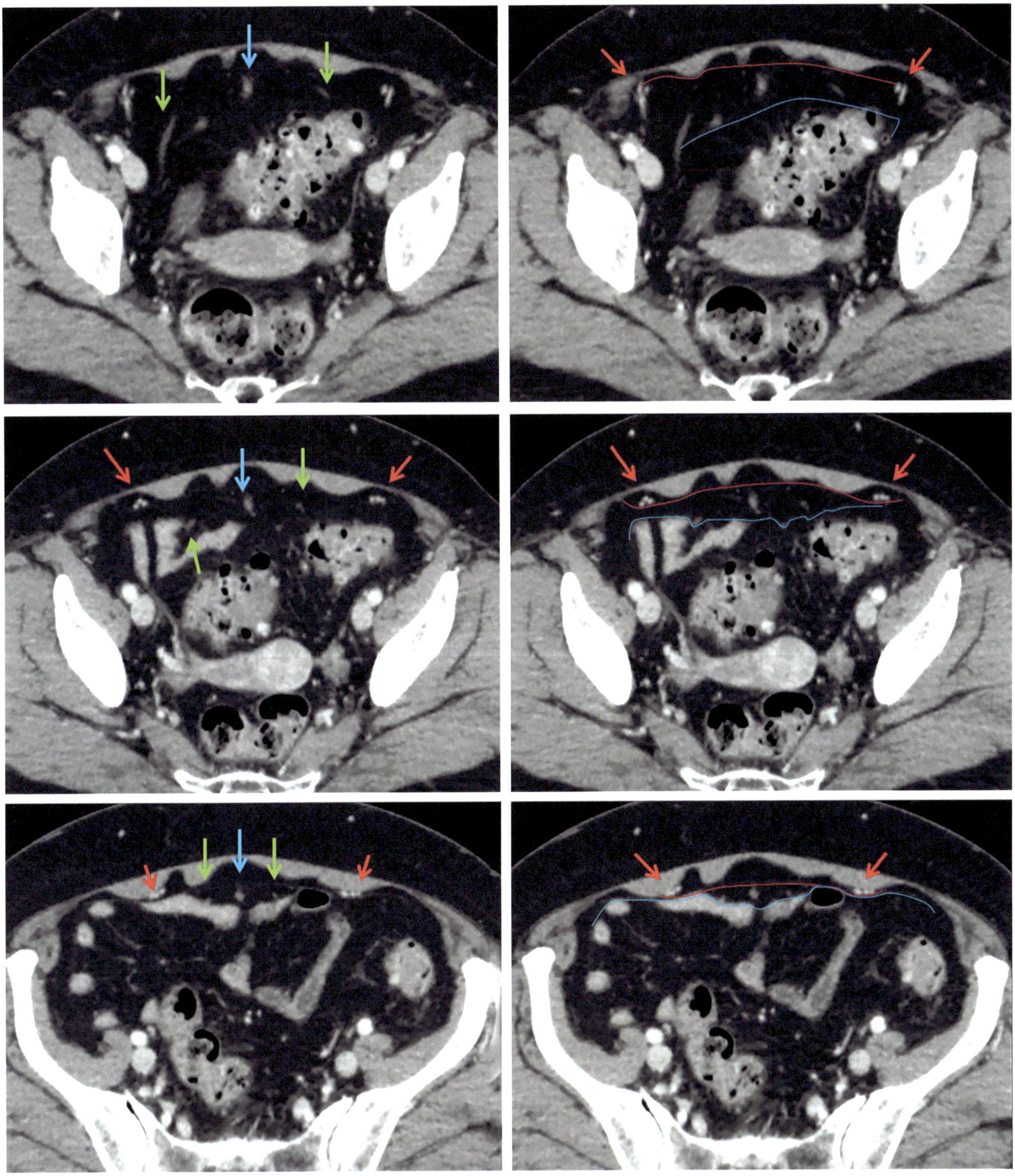

Fig. 14.18 Axial computed tomography (CT) images showing the radiologic space of Retzius in the lower anterior abdomen. The peritoneum is demarcated by the *blue lines* and the umbilical prevesical fascia by the *red lines*. Note that the inferior epigastric vessels (*red arrows*) are located anterior to the umbilical prevesical fascia, and the remnants of the urachus (*blue arrows*) and umbilical arteries (*green arrows*) are behind the umbilical prevesical fascia.

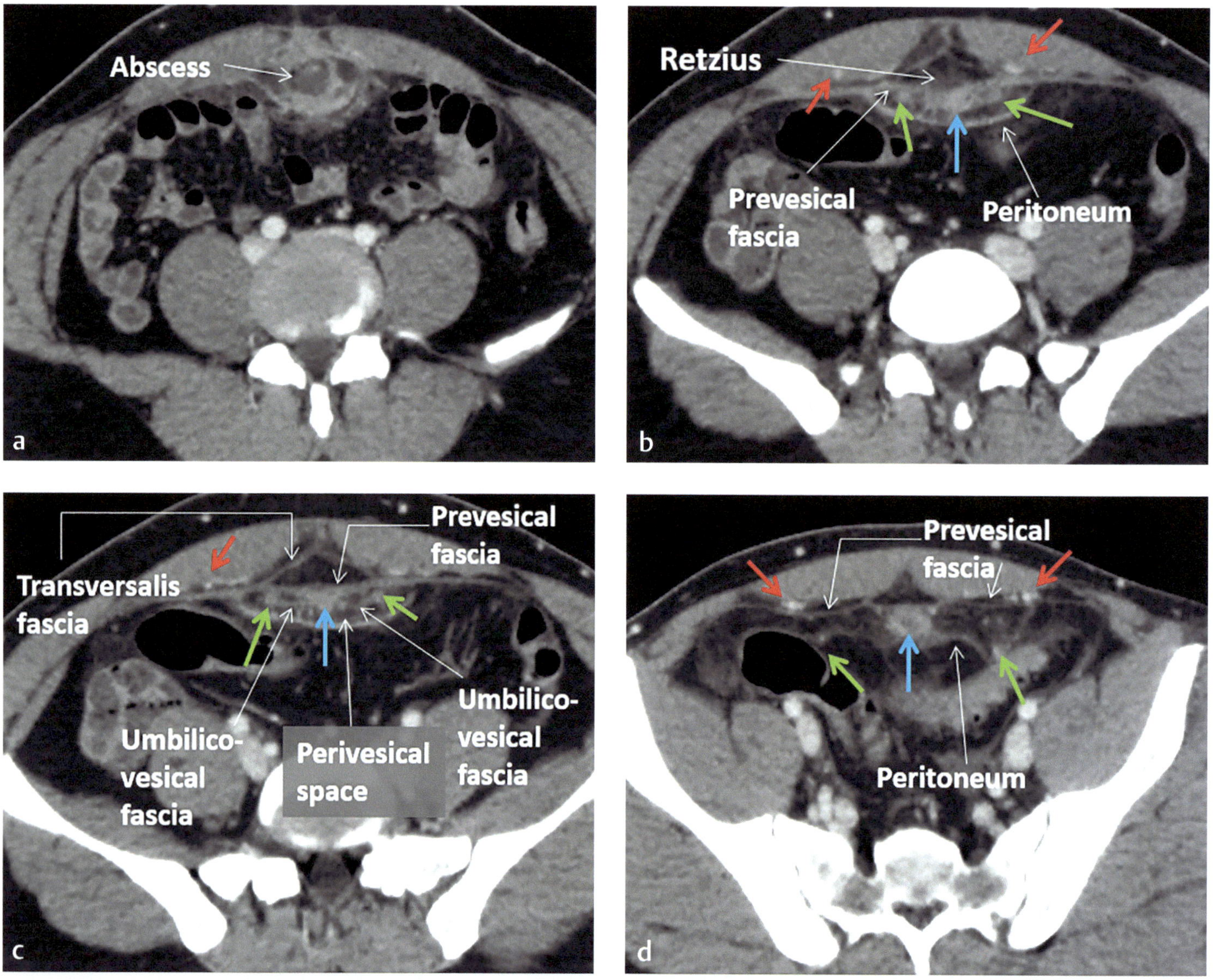

Fig. 14.19 (a–g) A case of infected urachal cyst (abscess) inferior to the umbilicus. Axial images from superior to inferior and sagittal images from midline to left show extensive edema of the fascial planes extending between the umbilicus and bladder. Fascial planes are nicely depicted. The inferior epigastric vessels (*red arrows*) are located anterior to the umbilical prevesical fascia, and the remnant of the urachus (*blue arrows*) and umbilical arteries (*green arrows*) are behind the umbilical prevesical fascia. Note the umbilicovesical fascia surrounds and connects the remnants of the urachus (median umbilical ligament) and umbilical arteries (medial umbilical ligaments). Also note that the umbilical prevesical fascia is delimited by the inferior epigastric vessels (lateral umbilical ligament). The space between the umbilical prevesical fascia and transversalis fascia is the retropubic or prevesical space of Retzius.

(*Continued*)

the mesenterium is defined as a structure composed of a double layer of peritoneum which does not apply to the endopelvic fascia of the mesorectum.[36] The superior rectal artery enters in the mesorectum from the top. Outside the rectal fascia exists the pararectal space containing loose areolar fatty tissue, full of pelvic splanchnic plexus (▶Fig. 14.24). The middle rectal artery and inferior hypogastric plexus branches travel in this interfascial plane and through the lateral rectal ligaments to enter the mesorectum. The posterior pararectal space faces the presacral space and the lateral pararectal space is continuous with the paravesical space. The pararectal space is a potential space between the rectal fascia and the presacral fascia. These two fasciae are very close, and it would be difficult to distinguish them with CT or MR unless separated by fluid, edema, or blood (▶Fig. 14.21, ▶Fig. 14.25). These interfascial spaces continue superiorly with the pararenal spaces and anteriorly with the prevesical and the extraperitoneal space of the inguinal region. The anatomy of the posterior pararectal space between the endopelvic and rectal

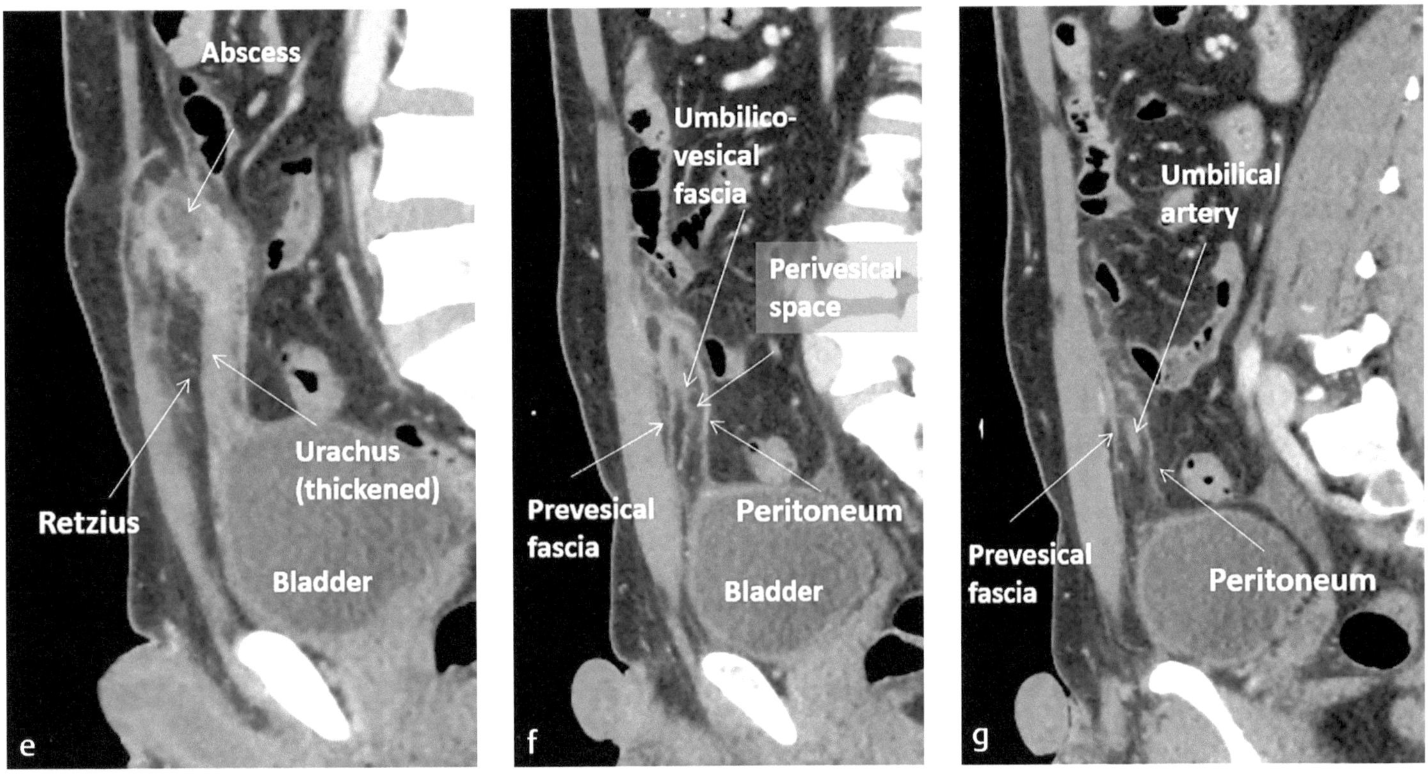

Fig. 14.19 (*Continued*) (e–g)

fasciae has been a matter of controversy especially about which fascia covers the hypogastric nerve.[37–39] This controversial interpretation is likely due to multilaminar fascial structure and interindividual variability of it in this region. The presacral fascia is a condensation fascia formed by compaction of the extraperitoneal tissue in the posterior pararectal space. It is close to the surface of sacrum and covering presacral space. The median sacral vessels, presacral venous plexus, and part of sympathetic chain reside in the presacral space. The presacral fascia laminae continue laterally to the endopelvic fascia of the levator ani and ventrolaterally to connect with Denonvilliers' fascia. The Waldeyer's fascia (also known as rectosacral, presacral, or precoccygeal fascia) is in the posterior pararectal space extending from superior to inferior between the presacral fascia and the rectal fascia.[37–39] The hypogastric nerve passes immediately posterior to the Waldeyer's fascia. The Waldeyer's fascia limits the floor of the retrorectal space (rectosacral ligament) and is an important avascular access (Holy plane) for complete excision of the mesorectum in rectal cancer surgery. It originates from the presacral fascia at the S2–S4 level and fuses with the posterior leaf of the rectal fascia.

Anteriorly, the rectal fascia is separated from the prostate by the rectoprostatic fascia of Denonvilliers, from the vagina by the rectovaginal fascia, and from the sacrum by the presacral fascia (▸ Fig. 14.25). Lateral continuation of Denonvilliers' fascia reaches the presacral fascia. These fasciae prevent spread of disease between organs. Rectal pathologies can easily penetrate into the perirectal fat (▸ Fig. 14.26).

Extraperitoneal Space of the Inguinal Region

At the level of inguinal canal, the transversalis fascia forms two laminae, anterior and posterior. Therefore, two spaces are formed: one between the peritoneum and the posterior lamina of the transversalis fascia and second one between the two facial laminae.[40,41] Both laminae are inferiorly inserted into the ligament of Cooper. Superiorly, they unite and continue upward on the upper anterior abdominal wall as "transversalis fascia."

Bogros (1786–1825) described a triangular space in the iliac region between the iliac fascia, transversalis fascia, and parietal peritoneum. This is called the retroinguinal space of Bogros which is the extraperitoneal space in continuity with the posterior pararenal space situated deep to the inguinal ligament[40,41] (see Chapter 1 "Abdominopelvic Wall"). It is limited by the fascia

14

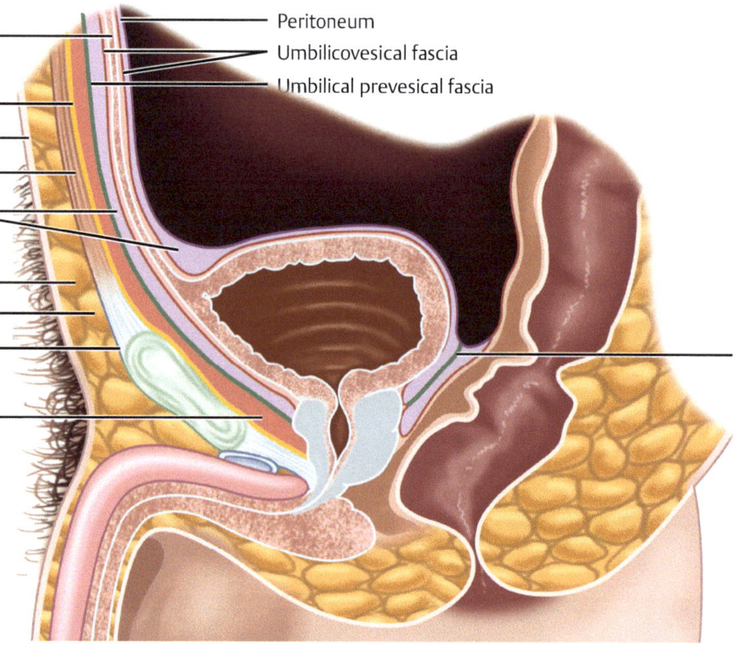

a

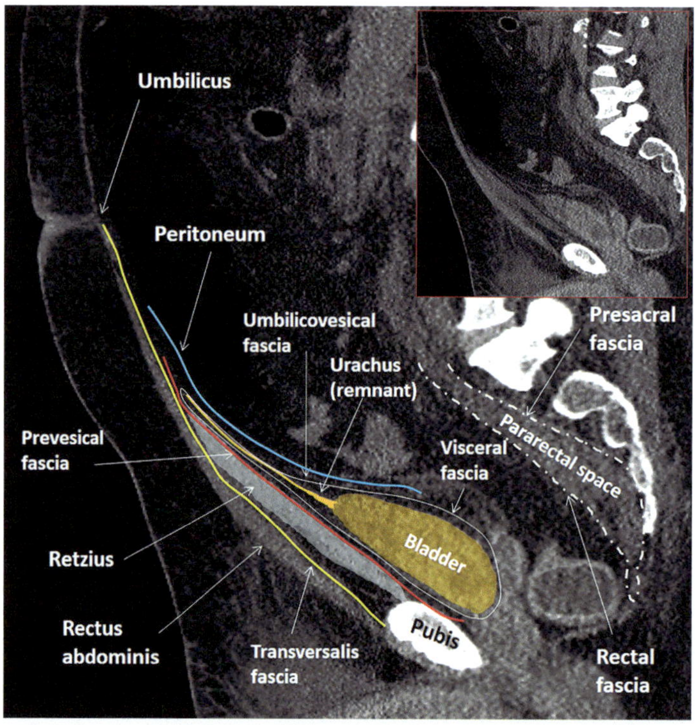

b

Fig. 14.20 Prevesical and perivesical spaces in suprapubic region. **(a)** Schematic drawing. (Modified with permission from Mirilas and Skandalakis.[6]) **(b)** Midsagittal computed tomography (CT) plane of the pelvis in a patient with hematoma and edema in the prevesical space of Retzius and posterior pararectal space. These two spaces are in connection with each other and with the pararenal spaces superiorly. Fascial planes are color coded. Inlay image is the same view without annotations. The umbilicovesical fascia is the visceral fascia that firmly envelopes the bladder, urachus, medial umbilical ligaments, and prostate. The umbilical prevesical fascia is located between the umbilicovesical fascia and the transversalis fascia. The space between the umbilical prevesical fascia and transversalis fascia is the retropubic or prevesical space of Retzius. The perivesical space is located outside the umbilicovesical fascia and delimited by the peritoneum superiorly and by the umbilical prevesical fascia anterolaterally. In general, differentiation of the prevesical and perivesical spaces by CT scan is difficult and fluid collections in these spaces are collectively addressed within the space of Retzius.

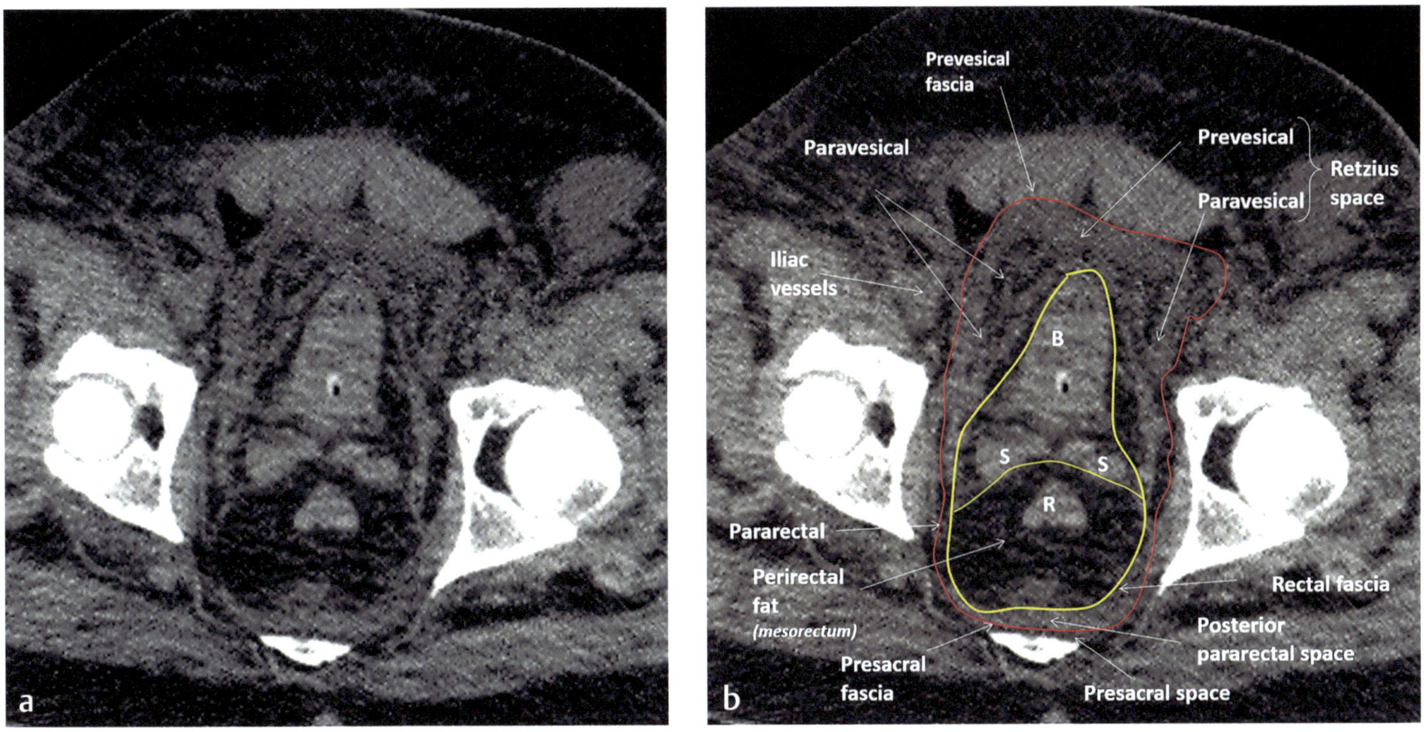

Fig. 14.21 (a, b) Axial computed tomography (CT) images in a patient with retroperitoneal hemorrhage. Hemorrhagic and edema from the right posterior pararenal space (not shown) dissected into the pelvis. There is a dental shaped accumulation of fluid along the anterolateral margins of the bladder base in the paravesical and prevesical spaces, the so-called "molar tooth sign." Blood/edema also extends into the posterior pararectal space. These spaces (within *red line*) are in direct connection with the posterior pararenal space. The perirectal and perivesical are closed perisplanchnic spaces (within *yellow lines*), hence remained clear. B, bladder; R, rectum; S, seminal vesicles.

transversalis anteriorly, the peritoneum posteriorly, and the iliac fascia laterally. In the modern concept, the space of Bogros lies between the peritoneum and the posterior lamina of the transversalis fascia.

The space of Bogros is divided into two compartments. The medial compartment contains the femoral artery and vein (femoral sheath). The lateral compartment allows for passage of the illiopsoas and femoral nerve. The retroinguinal space of Bogros is separated medially from the space of Retzius by the inferior epigastric vessels. In fact, the space of Bogros is a lateral extension of the space of Retzius.

◆ Conclusion

The extraperitoneal space is the space between the peritoneum and the inner investing fascia of the muscles of the abdomen and pelvis walls. The retroperitoneal space is the posterior and lateral part of the extraperitoneal space of abdomen. It is divided by the renal fascia into the anterior pararenal, perirenal, and posterior pararenal spaces. The extraperitoneal space of the anterior abdomen is called preperitoneal space and of the pelvis is called subperitoneal space.

In general, perisplanchnic (where retroperitoneal viscera is located) spaces are closed, except where the neurovascular pedicles are located. Anterior to the anterior layer of the renal fascia, the anterior interfascial plane permits communication between the perirenal spaces across the midline. On the other hand, the parasplanchnic extraperitoneal spaces of the abdomen (anterior and posterior pararenal spaces) and pelvis communicate with each other and across the midline. The extraperitoneal spaces of the pelvis including the prevesical, paravesical, and pararectal spaces can also communicate with each other and with the posterior pararenal space. The perirectal and perivesical spaces are closed perisplanchnic spaces.

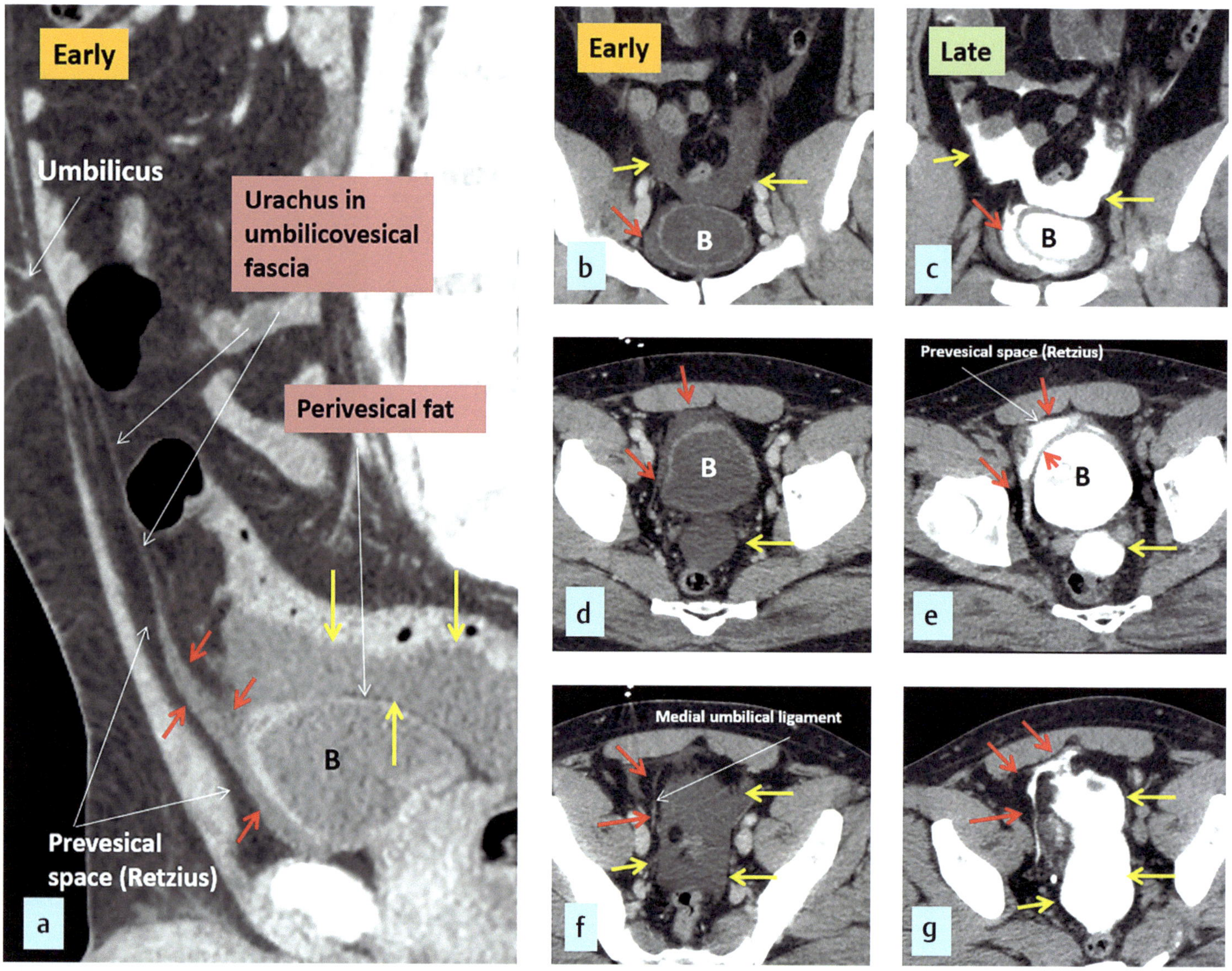

Fig. 14.22 A case of perforation of the bladder in the anterior extraperitoneal (perivesical space, *red arrows*) and intraperitoneal spaces (*yellow arrows*). **(a)** Sagittal, **(b, c)** coronal, and **(d–g)** axial computed tomography (CT) images are shown. Early and late postcontrast images are shown. Delay images showing contrast-enhanced urine extravasation from the bladder lumen into the perivesical space anterolaterally followed by intraperitoneal spillage. The fat-containing perivesical space and its contents (the bladder, urachus, and medial umbilical ligaments) are surrounded by umbilicovesical fascia and is delimited by the peritoneum posterosuperiorly and by the umbilical prevesical fascia anterolaterally. In the sagittal image, the fat-containing perivesical space is seen superior to the bladder which is interrupted anteriorly by the fluid exiting the injured bladder wall.

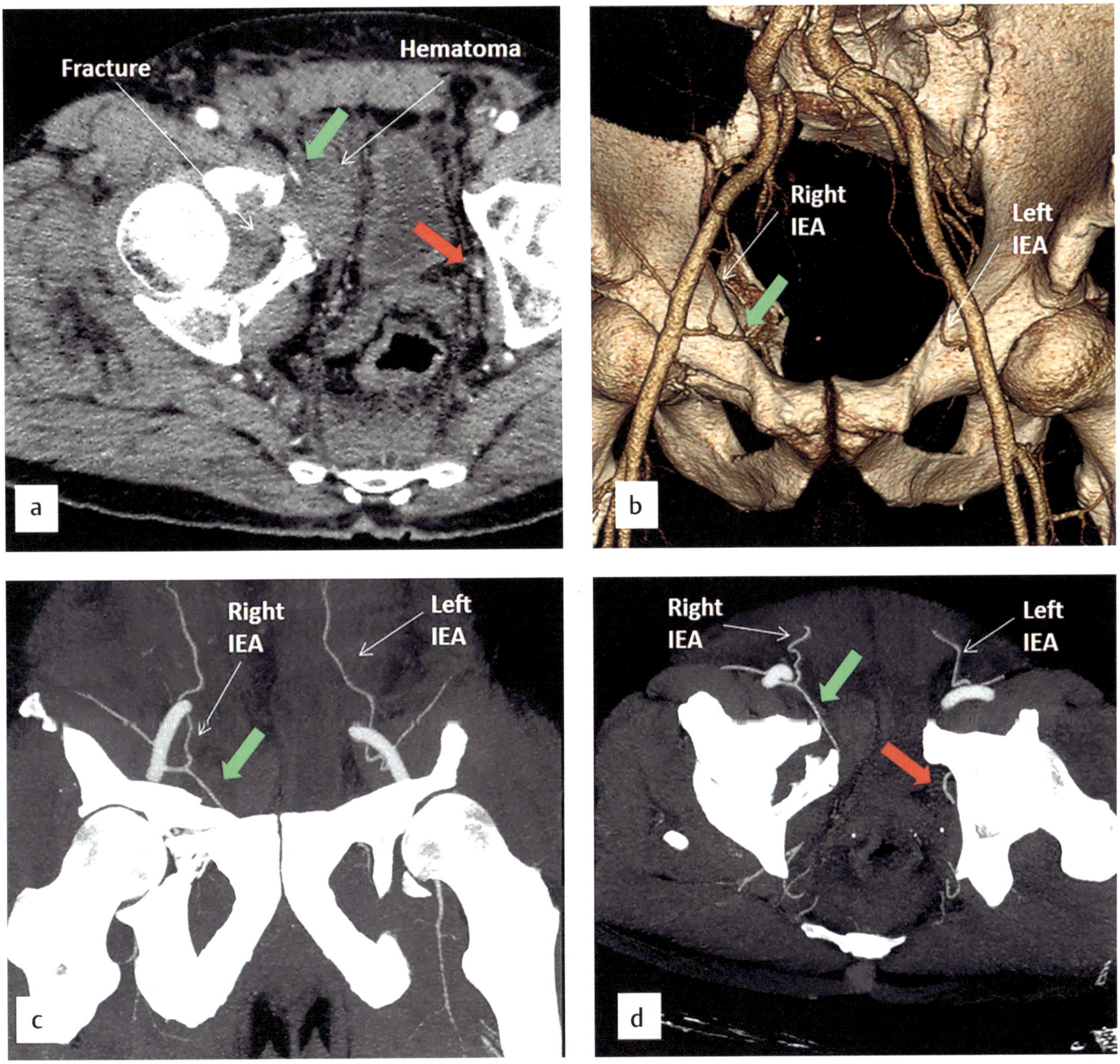

Fig. 14.23 (a–d) Injury of an aberrant obturator artery (*green arrows*) after right acetabulum fracture. The aberrant artery arises from the right inferior epigastric artery (IEA). The left obturator artery (*red arrows*) is seen in the obturator canal. Normally, the obturator artery arises from the internal iliac and passes in the obturator canal.

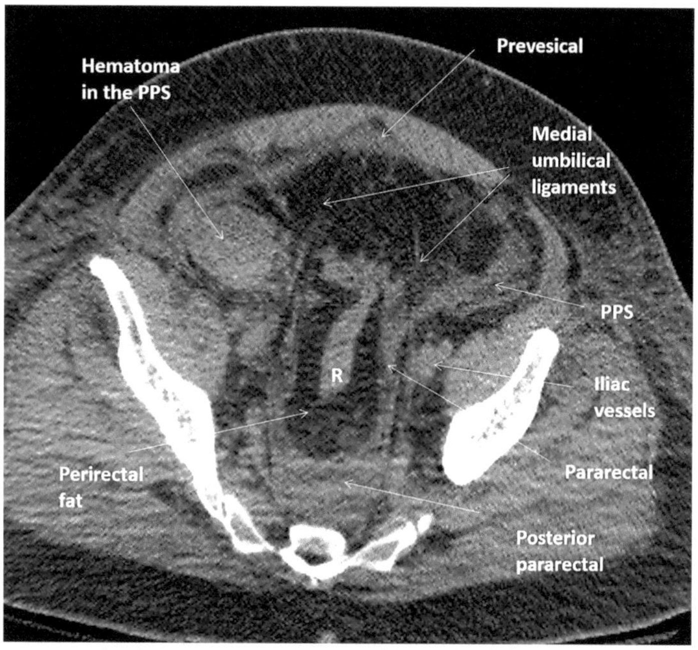

Fig. 14.24 Axial computed tomography (CT) in a patient with retroperitoneal hemorrhage. The medial umbilical ligaments are shown. These ligaments with their covering fascia (umbilical prevesical fascia) separate the perivesical space from the paravesical and pararectal spaces. The posterior pararenal space (PPS) is in continuity with the paravesical and pararectal spaces. Rectal fascia separates the perirectal space from the pararectal space. R, rectum.

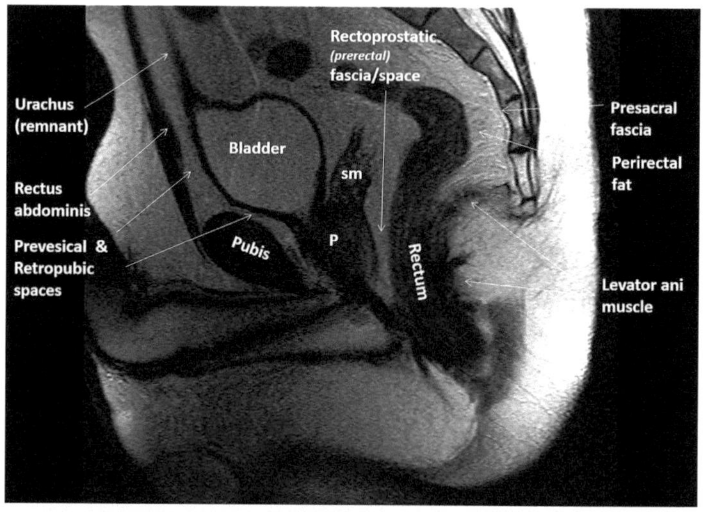

Fig. 14.25 Sagittal T1-weighted magnetic resonance imaging (MRI) of the pelvis showing large amount of fat in the prevesical space, perirectal space, and along the rectoprostatic fascia of Denonvilliers. P, prostate; SM, seminal vesicles.

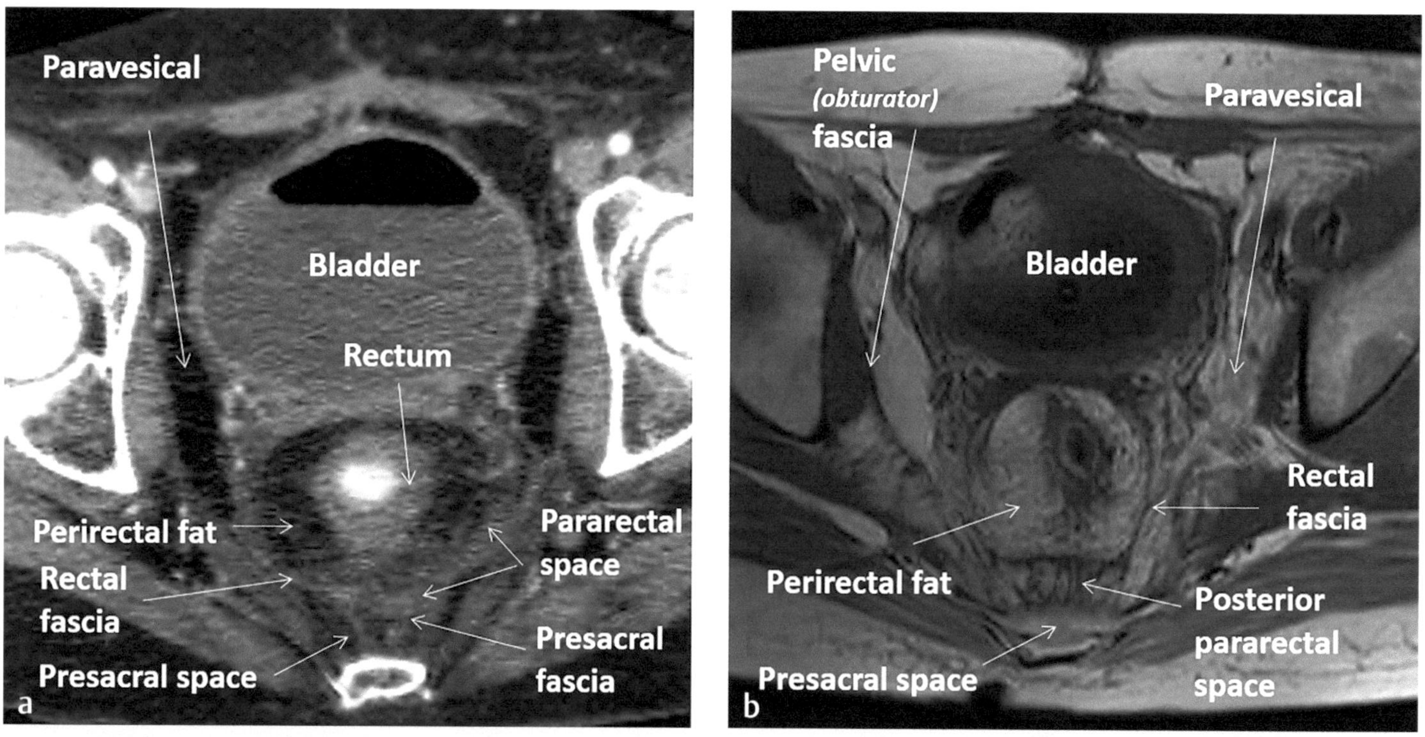

Fig. 14.26 Perirectal space containing fat is shown in two different patients with rectal cancer. **(a)** Computed tomography (CT) scan. **(b)** Magnetic resonance imaging (MRI). Increased attenuation of the perirectal fat (mesorectum) and thickened rectal fascia/pararectal space are shown related to postradiation edema and/or tumor infiltrates.

14

References

1. Kingsnorth AN, Skandalakis PN, Colborn GL, Weidman TA, Skandalakis LJ, Skandalakis JE. Embryology, anatomy, and surgical applications of the preperitoneal space. Surg Clin North Am 2000;80(1):1–24
2. Korobkin M, Silverman PM, Quint LE, Francis IR. CT of the extraperitoneal space: normal anatomy and fluid collections. AJR Am J Roentgenol 1992;159(5): 933–942
3. Gore RM, Balfe DM, Aizenstein RI, Silverman PM. The great escape: interfascial decompression planes of the retroperitoneum. AJR Am J Roentgenol 2000; 175(2):363–370
4. Aizenstein RI, Wilbur AC, O'Neil HK. Interfascial and perinephric pathways in the spread of retroperitoneal disease: refined concepts based on CT observations. AJR Am J Roentgenol 1997;168(3):639–643 Pub Gore RM, Balfe DM, Aizenstein RI, Silverman PM. The great escape: interfascial decompression planes of the retroperitoneum. AJR Am J Roentgenol 2000;175(2):363–370
5. Meyers MA, Whalen JP, Peelle K, Berne AS. Radiologic features of extraperitoneal effusions. An anatomic approach. Radiology 1972;104(2):249–257
6. Mirilas P, Skandalakis JE. Surgical anatomy of the retroperitoneal spaces part II: the architecture of the retroperitoneal space. Am Surg 2010;76(1):33–42
7. Mirilas P, Mentessidou A, Skandalakis JE. Secondary internal inguinal ring and associated surgical planes: surgical anatomy, embryology, applications. J Am Coll Surg 2008;206(3):561–570
8. Mirilas P, Skandalakis JE. Surgical anatomy of the retroperitoneal spaces—part I: embryogenesis and anatomy. Am Surg 2009;75(11):1091–1097
9. Hayes MA. Abdominopelvic fasciae. Am J Anat 1950;87(1):119–161
10. Tobin CE. The renal fascia and its relation to the transversalis fascia. Anat Rec 1944;89:295–311
11. Ishikawa K, Nakao S, Murakami G, et al. Preliminary embryological study of the radiological concept of retroperitoneal interfascial planes: what are the interfascial planes? Surg Radiol Anat 2014;36(10):1079–1087
12. Matsubara A, Murakami G, Niikura H, Kinugasa Y, Fujimiya M, Usui T. Development of the human retroperitoneal fasciae. Cells Tissues Organs 2009; 190(5):286–296
13. Mitchell GAG. The renal fascia. Br J Surg 1950;37(147):257–266
14. Cooper A. Of the anatomy of the parts concerned with inguinal hernia. In: The Anatomy and Surgical Treatment of Inguinal and Congenital Hernia. Philadelphia, PA: Lea & Blanchard; 1844:26–27
15. Read RC. Cooper's posterior lamina of transversalis fascia. Surg Gynecol Obstet 1992;174(5):426–434
16. Li G, Qian Y, Bai H, et al. Intertransversalis fascia approach in urologic laparoscopic operations. Arch Surg 2012;147(2):159–167
17. Mirilas P. "Intertransversalis" approach for laparoscopic urology: surgical anatomy concerns. Arch Surg 2012;147(10):980, author reply 980–982
18. Wolfram-Gabel R, Kahn J, Rapp E. Is the renal space closed? Clin Anat 2000;13(3):168–176
19. Kneeland JB, Auh YH, Rubenstein WA, et al. Perirenal spaces: CT evidence for communication across the midline. Radiology 1987;164(3):657–664
20. Thornton FJ, Kandiah SS, Monkhouse WS, Lee MJ. Helical CT evaluation of the perirenal space and its boundaries: a cadaveric study. Radiology 2001; 218(3):659–663
21. Kunin M. Bridging septa of the perinephric space: anatomic, pathologic, and diagnostic considerations. Radiology 1986;158(2):361–365
22. Raptopoulos V, Touliopoulos P, Lei QF, Vrachliotis TG, Marks SC Jr. Medial border of the perirenal space: CT and anatomic correlation. Radiology 1997; 205(3):777–784
23. Raptopoulos V, Lei QF, Touliopoulos P, Vrachliotis TG, Marks SC Jr. Why perirenal disease does not extend into the pelvis: the importance of closure of the cone of the renal fasciae. AJR Am J Roentgenol 1995;164(5):1179–1184
24. Molmenti EP, Balfe DM, Kanterman RY, Bennett HF. Anatomy of the retroperitoneum: observations of the distribution of pathologic fluid collections. Radiology 1996;200(1):95–103
25. Congdon ED, Edson JN. The cone of renal fascia in the adult white male. Anat Rec 1941;80:289–313
26. Congdon ED, Blumberg R, Henry W. Fasciae of fusion and elements of the fused enteric mesenteries in the human adult. Am J Anat 1942;70:251–279
27. Raptopoulos V, Kleinman PK, Marks S Jr, Snyder M, Silverman PM. Renal fascial pathway: posterior extension of pancreatic effusions within the anterior pararenal space. Radiology 1986;158(2):367–374
28. O'Connell AM, Duddy L, Lee C, Lee MJ. CT of pelvic extraperitoneal spaces: an anatomical study in cadavers. Clin Radiol 2007;62(5):432–438
29. Cho BH, Kimura W, Song CH, Fujimiya M, Murakami G; BH. An investigation of the embryologic development of the fascia used as the basis for pancreaticoduodenal mobilization. J Hepatobiliary Pancreat Surg 2009;16(6):824–831
30. Kimura W. Surgical anatomy of the pancreas for limited resection. J Hepatobiliary Pancreat Surg 2000;7(5):473–479
31. Mastromatteo JF, Mindell HJ, Mastromatteo MF, Magnant MB, Sturtevant NV, Shuman WP. Communications of the pelvic extraperitoneal spaces and their relation to the abdominal extraperitoneal spaces: helical CT cadaver study with pelvic extraperitoneal injections. Radiology 1997;202(2):523–530
32. Mindell HJ, Mastromatteo JF, Dickey KW, et al. Anatomic communications between the three retroperitoneal spaces: determination by CT-guided injections of contrast material in cadavers. AJR Am J Roentgenol 1995;164(5):1173–1178
33. Ishikawa K, Tohira H, Mizushima Y, Matsuoka T, Mizobata Y, Yokota J. Traumatic retroperitoneal hematoma spreads through the interfascial planes. J Trauma 2005;59(3):595–607, discussion 607–608
34. Ishikawa K, Idoguchi K, Tanaka H, et al. Classification of acute pancreatitis based on retroperitoneal extension: application of the concept of interfascial planes. Eur J Radiol 2006;60(3):445–452
35. Requarth JA, Miller PR. Aberrant obturator artery is a common arterial variant that may be a source of unidentified hemorrhage in pelvic fracture patients. J Trauma 2011;70(2):366–372
36. Rodríguez-Luna MR, Guarneros-Zárate JE, Tueme-Izaguirre J. Total mesorectal excision, an erroneous anatomical term for the gold standard in rectal cancer treatment. Int J Surg 2015;23(Pt A):97–100
37. Jin ZM, Peng JY, Zhu QC, Yin L. Waldeyer's fascia: anatomical location and relationship to neighboring fasciae in retrorectal space. Surg Radiol Anat 2011; 33(10):851–854
38. Kinugasa Y, Murakami G, Suzuki D, Sugihara K. Histological identification of fascial structures posterolateral to the rectum. Br J Surg 2007;94(5):620–626
39. García-Armengol J, García-Botello S, Martinez-Soriano F, Roig JV, Lledó S. Review of the anatomic concepts in relation to the retrorectal space and endopelvic fascia: Waldeyer's fascia and the rectosacral fascia. Colorectal Dis 2008;10(3):298–302
40. Bendavid R. The space of Bogros and the deep inguinal venous circulation. Surg Gynecol Obstet 1992;174(5):355–358
41. Skandalakis JE, Colborn GL, Androulakis JA, Skandalakis LJ, Pemberton LB. Embryologic and anatomic basis of inguinal herniorrhaphy. Surg Clin North Am 1993;73(4):799–836

15 Male Genitourinary

Navid Moradshahi, Mittul Gulati, and Farhood Saremi

◆ Introduction

The male reproductive tract can be divided into four different parts. The testes are where spermatogenesis occurs, the vas deferens facilitates transit of sperm, the seminal vesicles and accessory glands provide necessary substrate to produce semen, and the penis allows for deposition of semen within the female reproductive tract.

Each of these components of the male reproductive system plays an integral role in fertility and through medical imaging, potential pathologies may be identified and treated. Various modalities including ultrasound, magnetic resonance imaging (MRI), and computed tomography (CT) may be used to evaluate the male genitourinary organs with a background understanding of normal imaging anatomy.

◆ Scrotum

The scrotum contains the testes, epididymis, and vas deferens. It is covered by skin and multiple layers of fasciae, including the dartos, external spermatic (intercrural) fascia, cremasteric fascia, internal spermatic (infundibuliform) fascia, and parietal tunica vaginalis (peritoneum) (▶Fig. 15.1). The external spermatic fascia is the continuation of the aponeurosis of the external oblique muscle. The cremasteric muscle and fascia are extension of the internal oblique muscle fibers and their fascia. The internal spermatic fascia is an extension of the transversalis fascia.

The dartos is a superficial connective tissue layer containing smooth muscles that cover the scrotum. It also forms the superficial fascia of penis which envelopes the Buck's fascia of the penis. It is continuous with the membranous fascia of the perineum (Colles' fascia) and the abdomen (Scarpa's fascia). With current imaging resolution, differentiation of different layers of dartos in not possible and the skin, muscle, and Colles' fascia at the scrotal level are collectively called the dartos. The scrotum's function is to maintain the temperature of the testes at 35°C by adjusting the distance of the testicles from the abdomen. This function is primarily performed by the action of the dartos and cremasteric muscles. The scrotum is supplied by the anterior and posterior scrotal arteries. The anterior scrotal artery is a branch of the deep external pudendal artery, and posterior scrotal artery is a branch of the internal pudendal artery.

◆ Testes

Embryologically, the undifferentiated gonad is comprised of three cell types, namely, the germ cells, mesenchyme, and mesothelium. Genital ridges form clusters of mesenchyme on each side of the midline and are covered by the mesothelium. The inward growth and migration of these mesothelial cells form the primary sex cords. The germ cells develop in the wall of the yolk sac and migrate into the genital ridge, where they become part of the primary sex cords. At this point in development in the male embryo, the primary sex cords form the seminiferous tubules containing two distinct cell lines described above, the mesothelial cells and the germ cells. The mesothelial cells become the Sertoli cells, the germ cells become the spermatogonia, and the mesenchymal cells become the Leydig cells. Subsequent differentiation of the mesonephric ducts into the epididymis, vas deferens, seminal vesicles, and ejaculatory ducts is dependent upon testosterone secreted by the Leydig cells[1] (▶Fig. 15.2). The testis in the developing urogenital ridge is anchored by two ligaments, the cranial suspensory ligament and the caudal genitoinguinal ligament or gubernaculum. The gubernaculum steers the testis to the scrotum. The mechanism of testicular descent from the abdomen to the scrotum is a complex hormonally controlled process. In this process, regression of the cranial suspension ligament by testosterone occurs along with enlargement and subsequent migration (between 25 and 35 weeks of gestation) of the gubernaculum into the scrotum by hormonal control of the Leydig cell.[2] In female fetus, the cranial suspension ligament is replaced by the suspensory ligament of the ovary and the gubernaculum transforms into the round ligament of the uterus.

The adult testes are paired structures which are located within the scrotum and measure between 3 and 5 cm in length, 2 and 4 cm in width, and 3 cm in anteroposterior dimension.[3] The testes appear as homogeneous structure with medium level echoes on ultrasound (▶Fig. 15.3). The testes are the site of spermatogenesis, which occurs within structures known as the seminiferous tubules. Each testicle is located within a hemiscrotum that is separated from the contralateral side by a septum.

The arterial blood supply to the testicle comes from three arteries, testicular, vasal, and cremasteric arteries. Blood flow to the testes is mainly supplied by paired testicular arteries which originate from the abdominal aorta and predominantly terminate as upper and lower polar branches.[4] There is minor collateral flow supplied by the cremasteric artery which arises from the inferior epigastric artery and anastomoses with the lower polar branch of the testicular artery as well as the artery of the vas deferens which anastomoses with branches of the testicular artery along the mediastinum testis[4] (▶Fig. 15.4, ▶Fig. 15.5). All three arteries anastomose to allow adequate blood supply to the testis.

Venous drainage from the testes is relatively complicated, with many individual variations. The development of gonadal vein is closely related to the development of renal vein and inferior vena cava. The drainage is initially via multiple small tributaries which unite to form the paired pampiniform venous plexuses, which subsequently coalesce to form the testicular veins.[5] Small connections also exist to the pudendal and cremasteric veins. Most commonly the right testicular vein drains directly into the inferior vena cava and the left testicular vein drains into the left renal vein; however, they may demonstrate variant anatomy, including duplication (8%), atypical drainage of the left into the inferior vena cava, termination of right testicular vein in the right renal

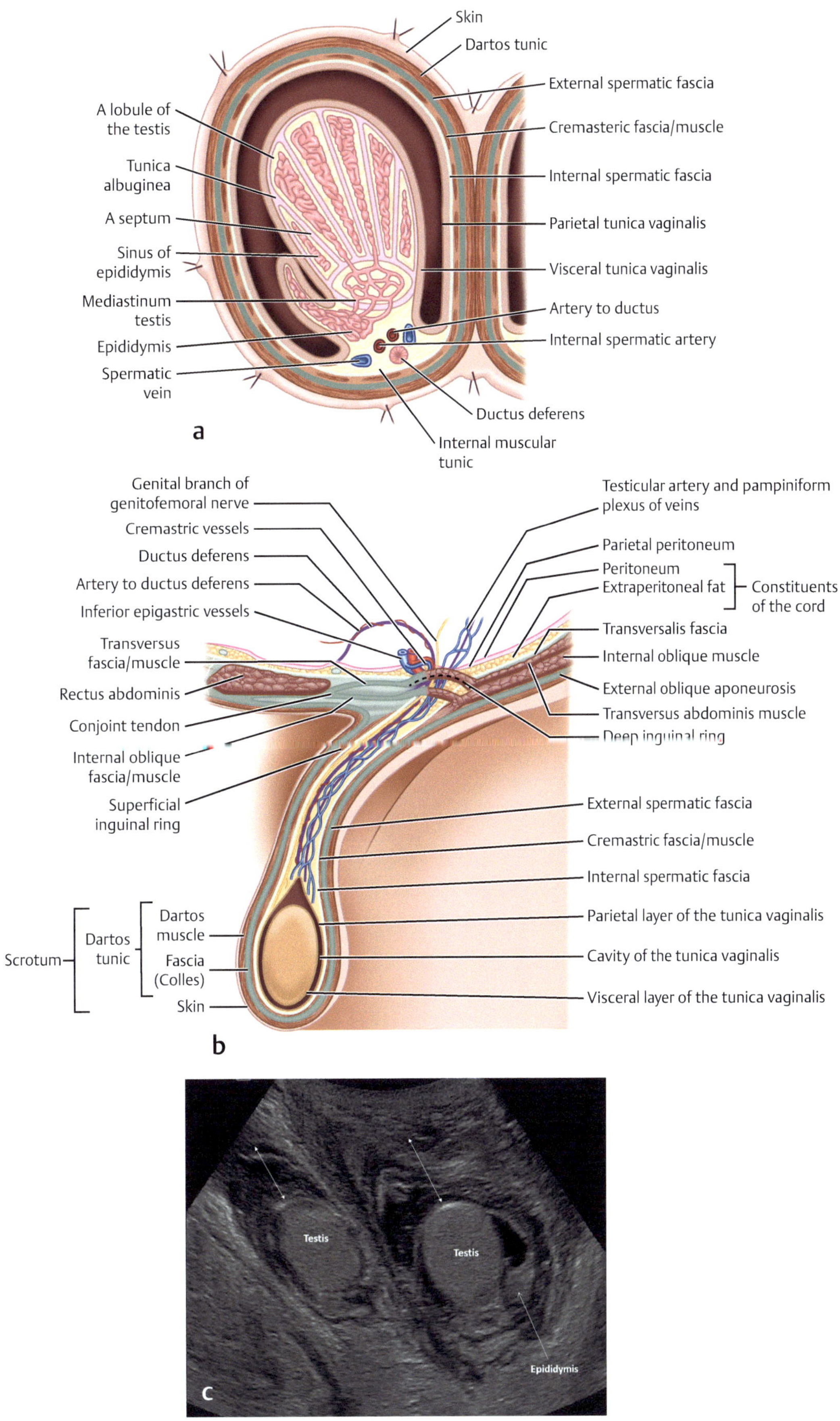

Fig. 15.1 (a, b) Diagrams showing the scrotal wall layers. **(c)** Edematous scrotal wall (*double-headed arrows*) is shown by ultrasound.

vein, and cross-communication between the left and right testicular venous systems[5,6] (▸Fig. 15.5). Duplication is more common on left side. Idiopathic venous dilation of the testicular veins (varicocele) is common (10 to 15%) and usually asymptomatic. There is a clear association between varicocele, infertility, and testicular growth arrest. Approximately 30 to 50% of males with primary infertility have varicocele. Embolization of the testicular veins is performed as a way to treat the varicoceles[7] (▸Fig. 15.6).

Seminiferous Tubules

The seminiferous tubules make up a significant portion of testicular parenchyma and contain spermatogenic cells, which eventually differentiate into mature sperm cells. Although these structures are distinguishable on the microscopic level, current imaging modalities only allow us to visualize the architecture of the testis on a macroscopic scale where individual tubules are not delineated. The parenchyma of the testis normally demonstrates a coarse, homogeneous echotexture on ultrasound with symmetric flow on color Doppler and low-resistance arterial waveforms on spectral Doppler[8] (▸Fig. 15.3).

There are connective tissue layers surrounding the testis, including the following.

Tunica Vaginalis

The tunica vaginalis is a product of embryonic development, specifically the closure of the superior portion of the processus vaginalis. This structure is composed of two mesothelium-lined layers, which surround the testis (▸Fig. 15.1). The visceral layer is intimately associated with the testis, epididymis, and gubernaculum and encapsulates most of the testis except for the most posterior portion. The parietal layer of the tunica vaginalis lies against the scrotal wall and is closely opposed to the visceral layer with only a potential

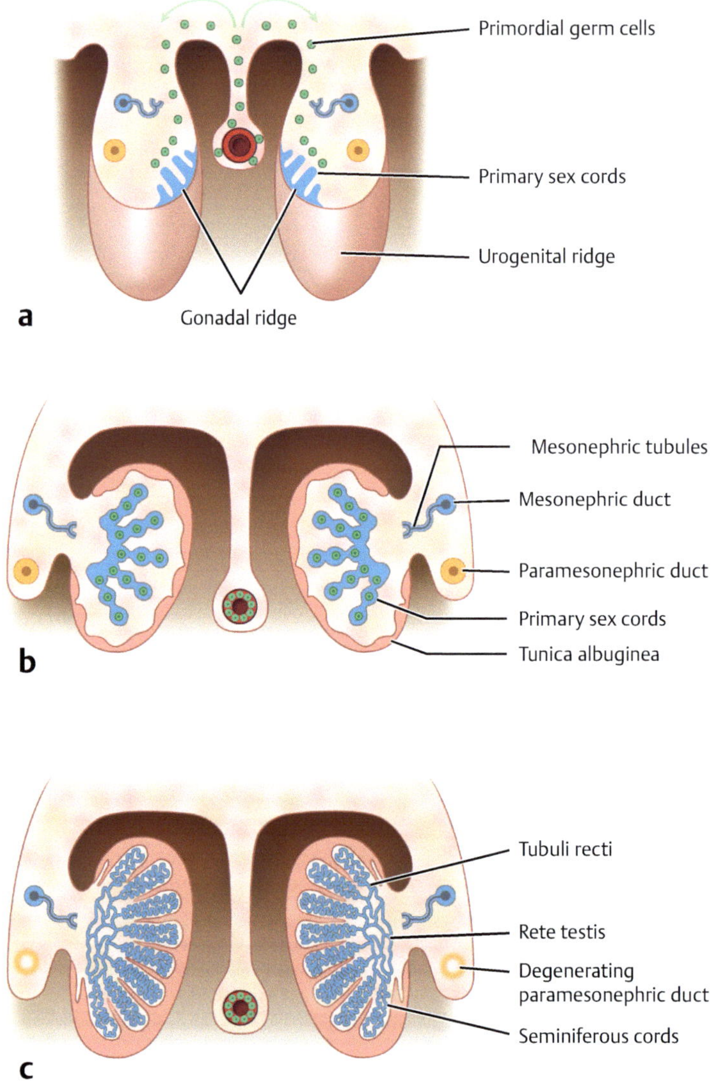

Fig. 15.2 Embryology of the testes. **(a)** The undifferentiated gonads are formed by the intermediate mesoderm within the urogenital ridges of the embryo. **(b)** Under gene control, the testes develop and testosterone produced by the Leydig cells within the testis promotes mesonephric (Wolffian) ducts growth which becomes the duct of the developing testis. **(c)** Sertoli cells within the testis produce anti-Müllerian hormone, which induces regression of the paramesonephric (Müllerian) ducts.

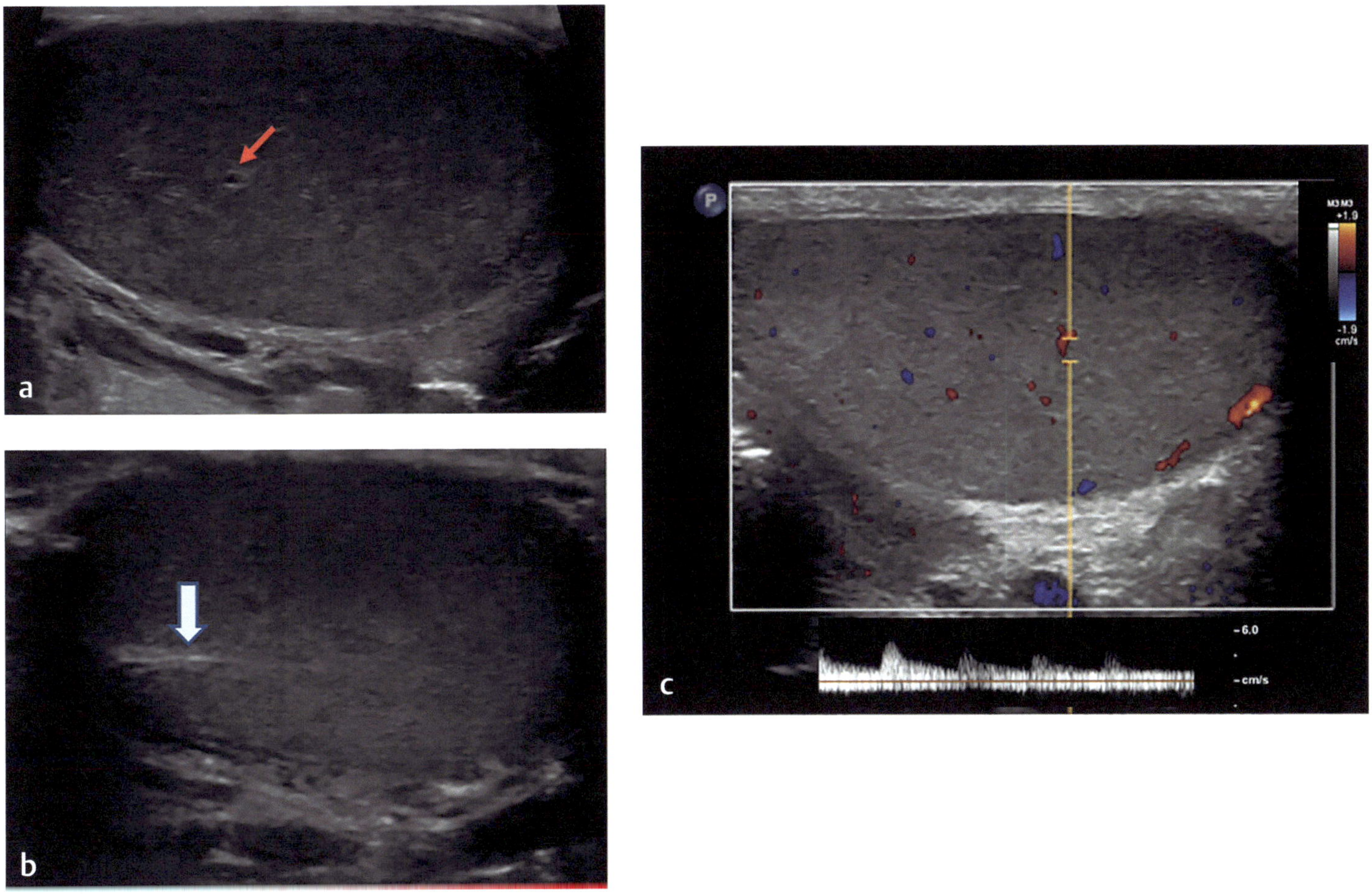

Fig. 15.3 (a) Scrotal ultrasound of the normal testis. An intratesticular artery appears as a focal hypoechoic region (*arrow*) which demonstrated expected arterial flow on color and spectral doppler (not shown). **(b)** Mediastinum testis appears as a hyperechoic linear structure (*arrow*). **(c)** Spectral Doppler ultrasound of the testis demonstrating the normal low-resistance arterial waveform.

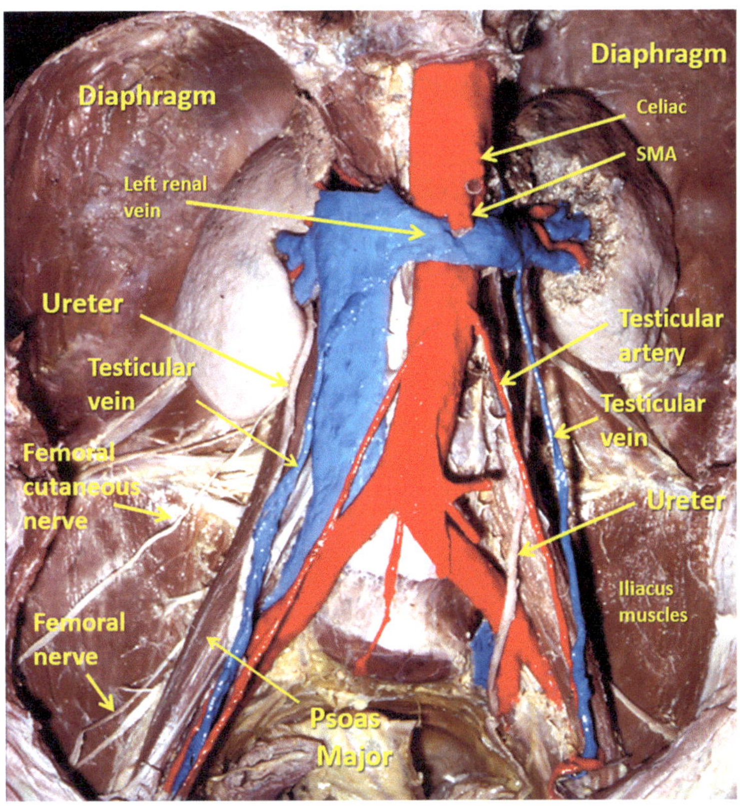

Fig. 15.4 Testicular artery and vein. SMA, superior mesenteric artery.

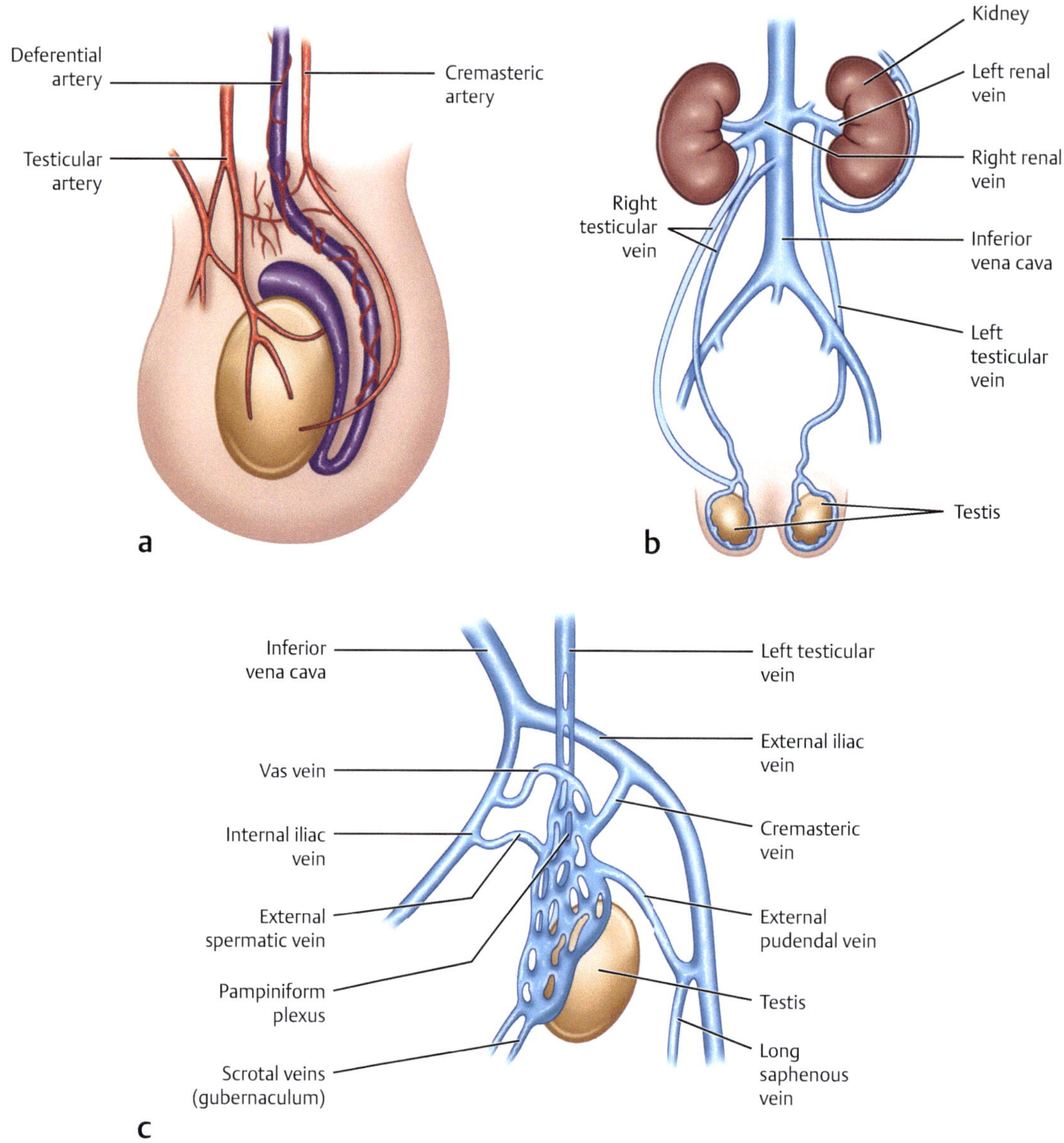

Fig. 15.5 (a) Testicular arterial anatomy. Illustration demonstrates the main arterial supply of the testis, the testicular artery. Additional minor arterial anastomoses from the cremasteric artery and deferential artery are also shown. **(b, c)** Venous drainage of the testes. Left testicular vein illustrated with most common route of drainage into the left renal vein. The right testicular vein is shown draining directly into the inferior vena cava, which is most common. Variant drainage of the right testicular vein into the right renal vein is demonstrated. (Modified from Paduch and Skoog.[7])

15

space separating the two layers[8,9] (▶Fig. 15.7a). The potential space between the visceral and parietal tunica vaginalis becomes filled with fluid when a hydrocele is present[9] (▶Fig. 15.7b, c). The collapsed layers may be identified on ultrasound as an echogenic outline of the testis; however, it is indiscernible from the tunica albuginea in the absence of pathology. Nonclosure of the processus vaginalis leads to a congenital inguinoscrotal hernia. Incomplete closure may result in a closed cystic space proximal to the testis known as "encysted hydrocele" of the cord.

Tunica Albuginea

The tunica albuginea is a fibrous capsule that encapsulates and extends into the testis (▶Fig. 15.1). This capsule lies deep to the visceral tunica vaginalis described above and has an inward reflection into the testis that forms the structure known as the mediastinum testis.[8,10] The tunica albuginea is not normally visualized under ultrasound imaging unless it is filled with fluid.

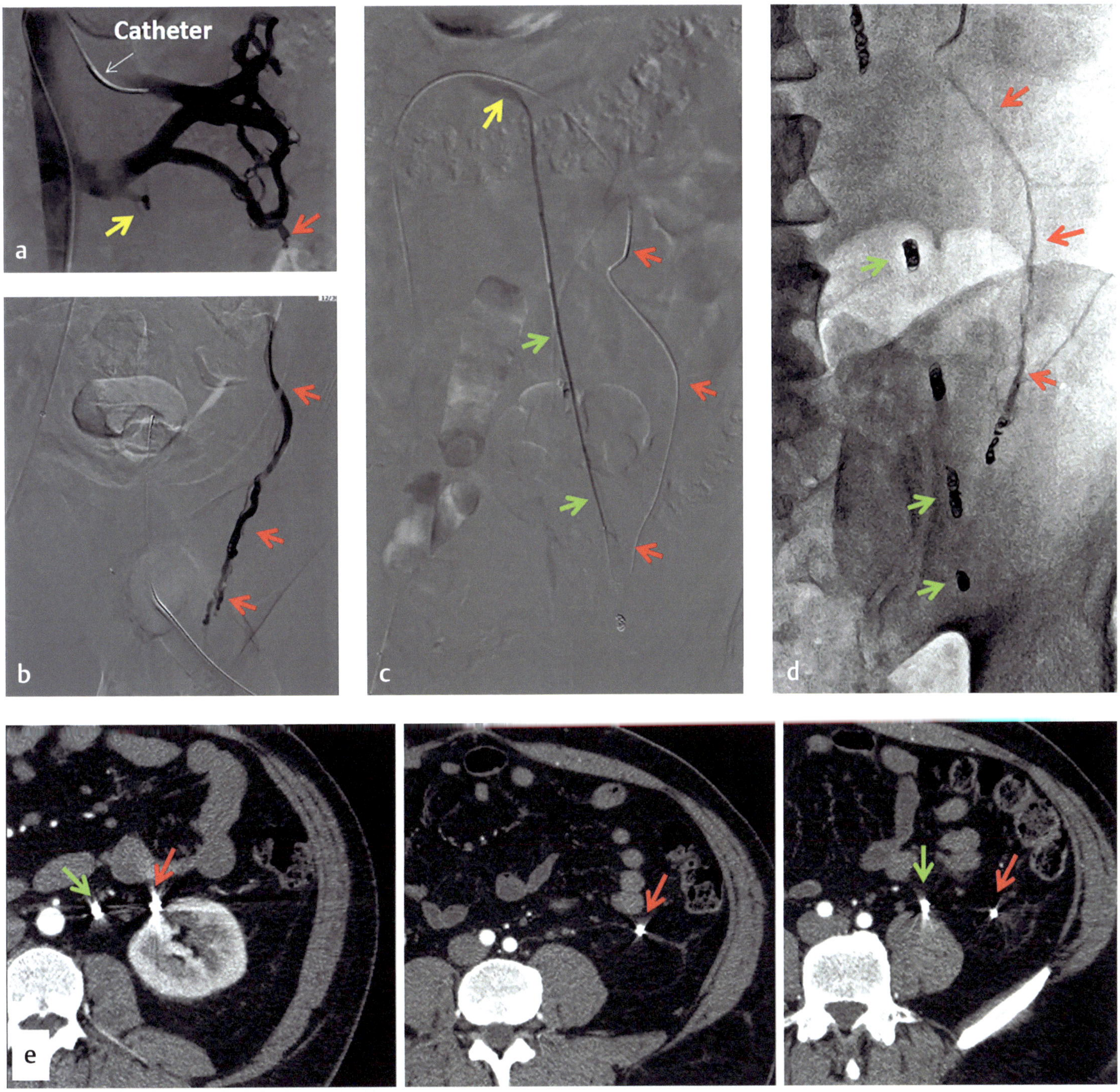

Fig. 15.6 Left renal venography in a patient with left varicocele and history of infertility. **(a)** Retrograde renal vein injection showing a circumaortic left renal drainage system. Catheter tip is seen in superior left renal vein. **(b)** Retrograde selective injection of the left spermatic vein showing drainage into the lateral aspect of the inferior left renal vein (*red arrows*). **(c)** Following advancement of the catheter in the proximal part of the left spermatic vein (*red arrows*), contrast injection showed another draining vein (*green arrows*) with an anatomic course to the medial aspect of the left inferior renal vein (*yellow arrows*). **(d)** Both veins are embolized with multiple coils. For successful treatment of varices all draining veins including collaterals should be embolized. **(e)** Computed tomography (CT) images after embolization show the anatomic location of the two left spermatic veins along the anterior renal fascia.

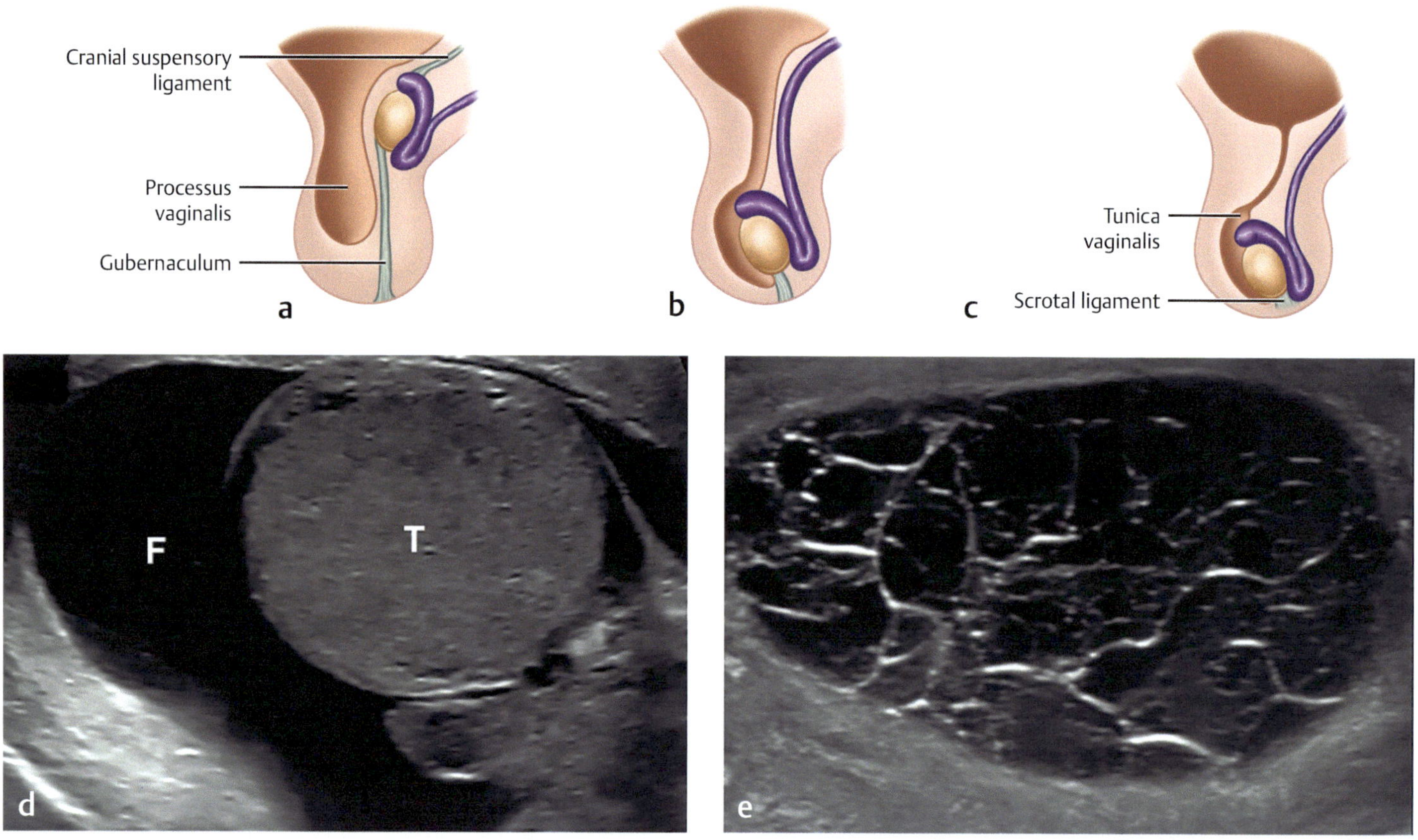

Fig. 15.7 (a–c) Normal embryologic development of the processus vaginalis, which arises as an outpouching of the parietal peritoneum at the beginning of the 3rd month of gestation. After the testis descends into the scrotum (between the 7th and 9th months of gestation), the processus vaginalis is obliterated. The gubernaculum initially enlarges to bring the testis into the scrotum and finally regresses into the scrotal ligament. **(d)** Hydrocele. Longitudinal gray-scale ultrasound demonstrates anechoic fluid (F) within the scrotum surrounding the testis (T). **(e)** Longitudinal ultrasound shows complex intrascrotal fluid with echogenic debris represented a pyocele in a patient with history of epididymo-orchitis.

Mediastinum Testis

This highly vascular connective tissue structure results from the invagination of a portion of the tunica albuginea described above and is composed of the rete testis which is a network of converged tubules that resorb fluid allowing for concentration of the sperm as they exit the seminiferous tubules. On ultrasound the mediastinum testis typically appears as a linear echogenic structure coursing within the parenchyma of the testis.[7,8] On occasion, the mediastinum testes may demonstrate hypoechoic cystic-appearing regions which represent the dilated rete testis (▶ Fig. 15.3). This condition is referred to as tubular ectasia of the rete testes, and should not be mistaken for a testicular mass lesion.[8]

Testicular cancer is most prevalent in young men with painless scrotal masses and ultrasound is the initial imaging modality of choice for evaluation. Testicular tumors will typically appear hypoechoic compared with the normal testicular parenchyma and may demonstrate additional features such as heterogeneous echotexture, calcifications, or cystic changes[7] (▶ Fig. 15.8).

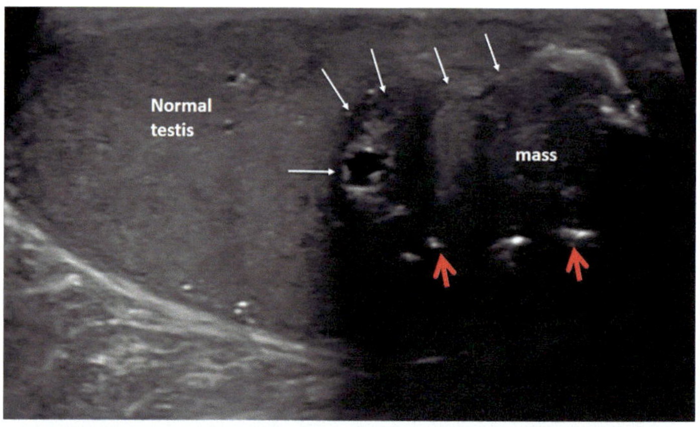

Fig. 15.8 Testicular cancer. Longitudinal gray-scale ultrasound of the testis demonstrates an ill-defined, heterogeneous solid mass (*white arrows*) within the testicular parenchyma with associated calcifications (*red arrows*). This mass demonstrated flow on color Doppler (not shown).

15

◆ Epididymis

The epididymides are intrascrotal, elongated, tightly coiled tubular structure which are positioned along the superior and posterior aspects of the testes. The epididymis drains and concentrates the spermatic fluid from the rete testis via the efferent ducts. It is divided into the superiorly located head, the epididymal body, and the inferiorly located tail. The epididymis usually appears isoechoic or hypoechoic on ultrasound relative to the testis (▶Fig. 15.9). The diameter of the epididymis varies throughout its course but measures up to 8 mm in its thickest regions.[11] The spermatic fluid concentrates and the sperm undergo maturation as they flow from the head to the tail of the epididymis. During the emission stage of ejaculation, the sperm within the epididymal tail are propelled via peristalsis into the vas deferens.

Small epididymal cysts are common in the head of the epididymis. Varicoceles are also a common imaging finding (▶Fig. 15.10). Epididymo-orchitis is a common cause of acute scrotal pain in sexually active adolescents and adults and primarily involves the epididymis in most cases[8] (▶Fig. 15.11). Ultrasound is the primary imaging modality used for evaluation and will demonstrate asymmetrically increased vascularity in the affected testis and epididymis (▶Fig. 15.11).[8] On gray-scale imaging, the testes may

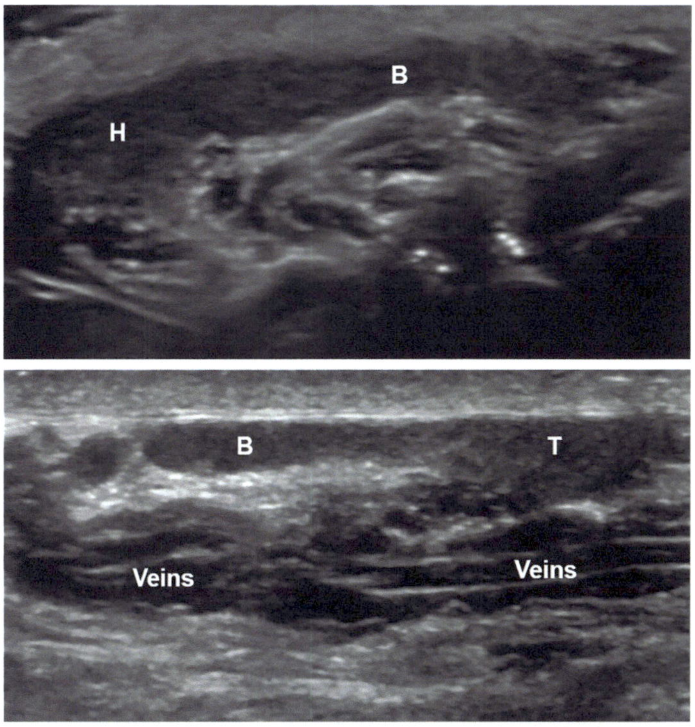

Fig. 15.9 Longitudinal ultrasound (US) scan of normal hypoechoic tubular appearance of the epididymis with head (H), body (B), and tail (T) shown. Spermatic veins are also shown.

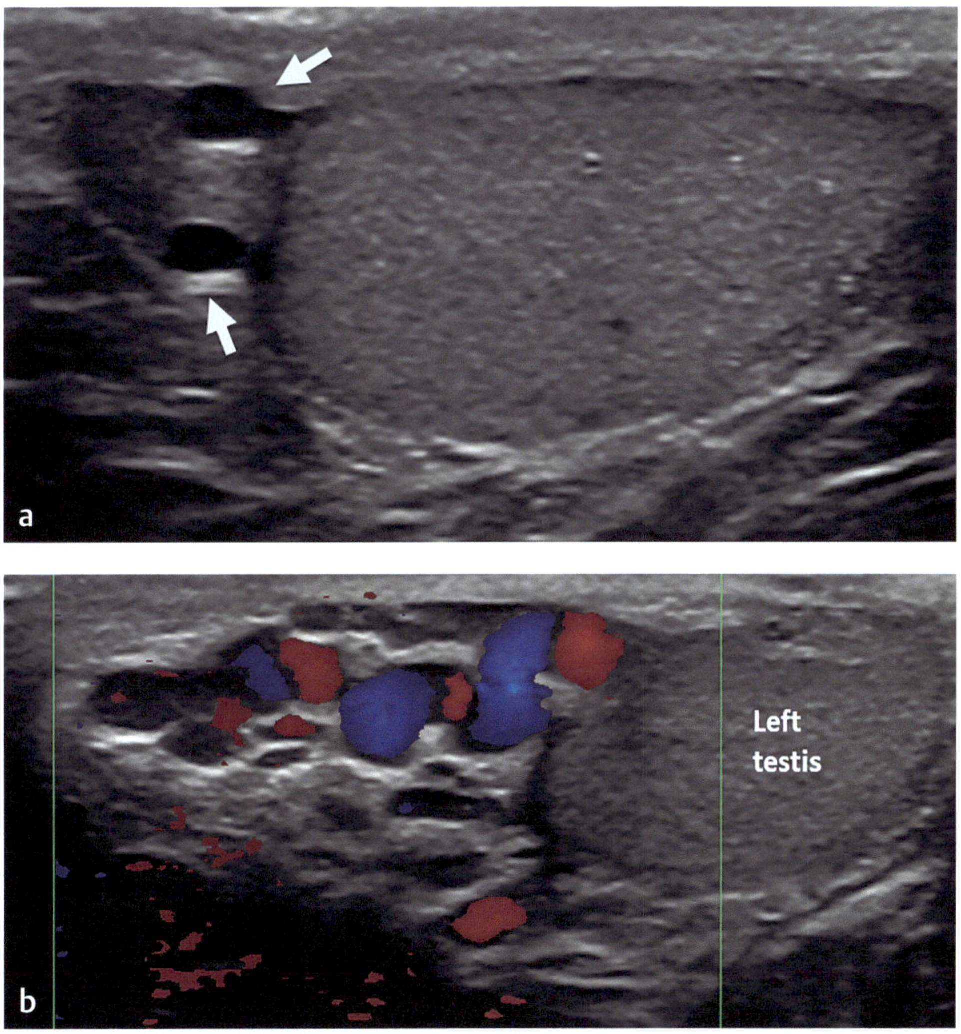

Fig. 15.10 Epididymal cyst. **(a)** The epididymal head containing several cysts (*arrows*), the largest measured 5 mm. This is a common image finding. **(b)** Left-sided varicoceles are shown by color Doppler ultrasound.

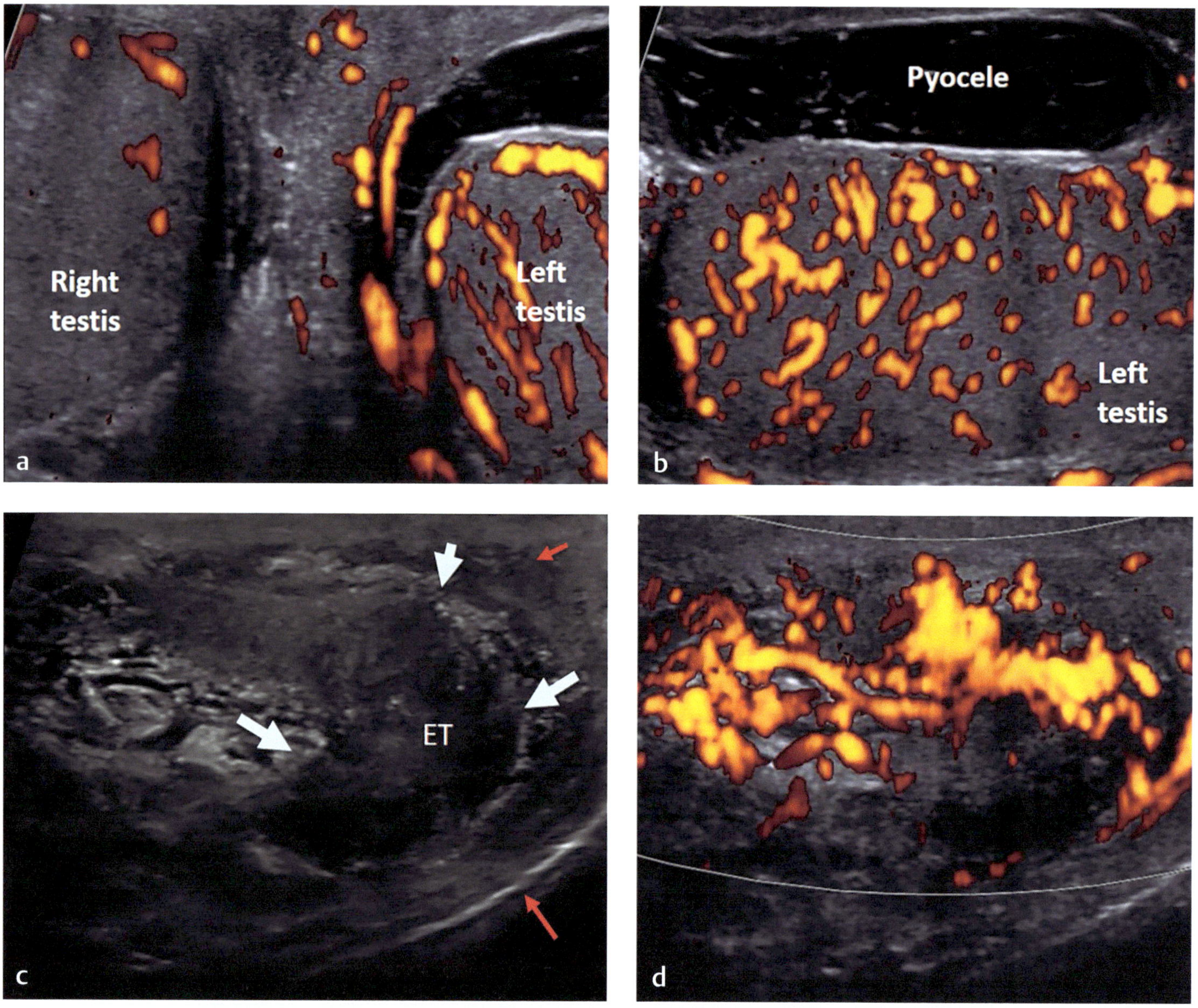

Fig. 15.11 Epididymo-orchitis. **(a, b)** Orchitis. Scrotal ultrasound with power Doppler demonstrates a normal right testis with asymmetrically increased flow in the left testis and associated complex scrotal fluid with multiple septations. An associated pyocele is present. **(c, d)** Epididymitis. Longitudinal gray-scale ultrasound of the left epididymal tail (ET) which appears markedly thickened (*blue arrows*) with overlying edema and thickening of the scrotal wall (*red arrow*). Corresponding color Doppler demonstrating markedly increased vascularity.

demonstrate heterogeneous echotexture with enlargement of the epididymis and thickening of the overlying scrotal wall.

◆ Spermatic Cord and Vas Deferens

The spermatic cord consists of the vas deferens, arteries, veins, lymphatics, nerves, and tunica vaginalis (▶Fig. 15.1). All these structures are surrounded by the external spermatic fascia, the cremasteric muscle, and the internal spermatic fascia. Three arteries running in the spermatic cord are the testicular, deferential, and cremasteric arteries (▶Fig. 15.12). The number of testicular arteries ranges from one to three. Veins are continuation of the pampiniform plexus. In each cord, there are five to eight spermatic veins. Interruption of the arterial flow and venous congestion are consequences of testicular torsion, a common clinical entity that can cause necrosis of the testis. Branches of the genitofemoral and ilioinguinal nerves also run along the spermatic cord.

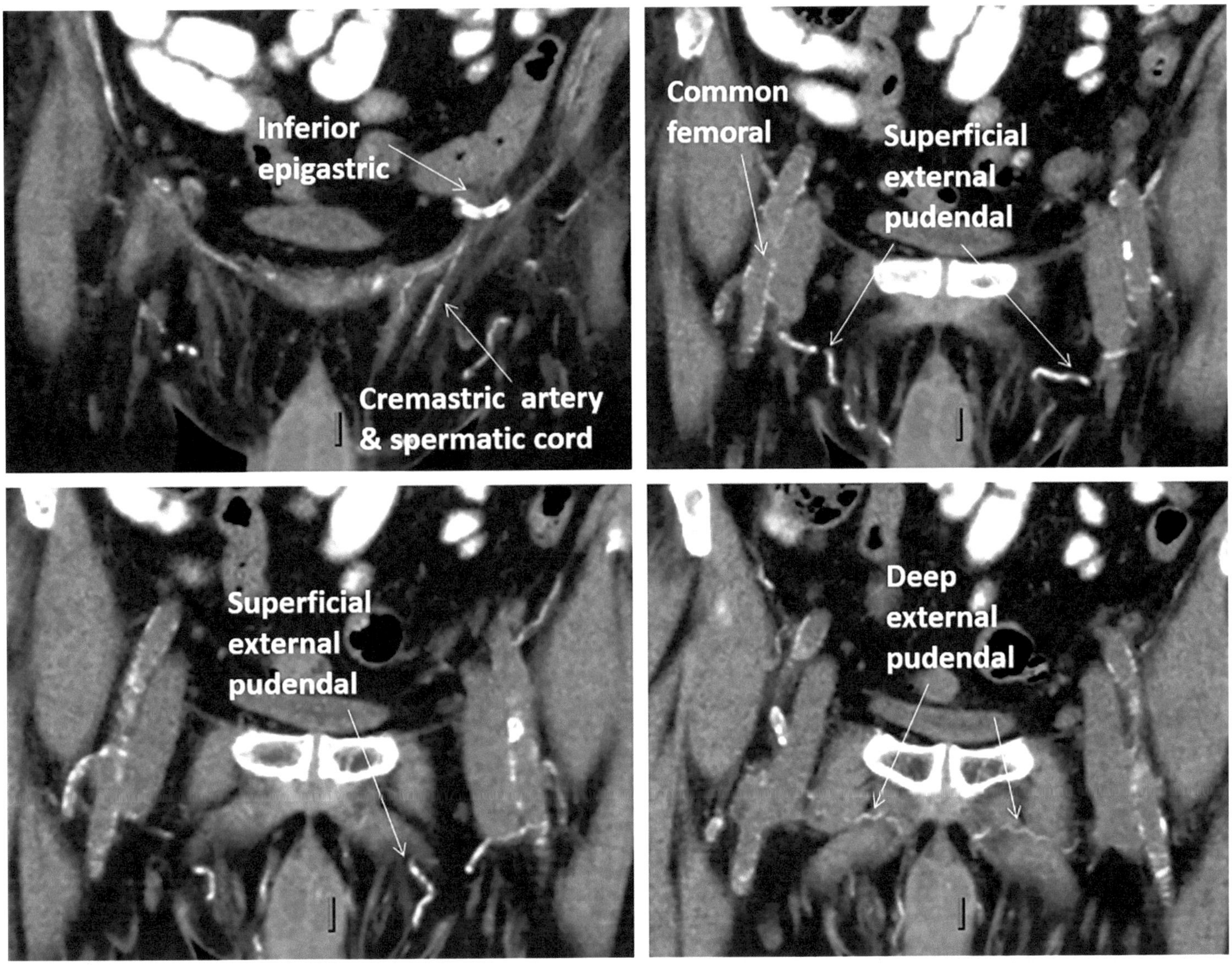

Fig. 15.12 Noncontrast, coronal computed tomography (CT) in a male patient with renal failure. Calcified arteries are shown from anterior to posterior. The cremasteric artery is a branch of the inferior epigastric artery which enters the spermatic cord. The deep and superficial external pudendal arteries are also seen arising from femoral artery to supply the penis and scrotum where they anastomose with the internal pudendal arteries.

The vas deferens (ductus deferens) originates from the mesonephric duct after development of the seminal vesicles. It is responsible for transporting sperm from the epididymis to the seminal vesicles. The proximal portion of the vas deferens is continuous with the epididymal tail and courses within the spermatic cord through the inguinal canal. The vas deferens is 30-cm long and continues on an extraperitoneal course until joining with the excretory duct of each seminal vesicle gland, at which point the ejaculatory duct is formed (▶ Fig. 15.13). The vas deferens meets the testicular vessels at the internal inguinal ring to participate in formation of the spermatic cord. The vas deferens is supplied by the artery of vas deferens, a branch of the superior vesical artery. The vas deferens appears as a low signal tubular structure on T1- and T2-weighted MRI.[12] Calcification of the vas deferens in CT images is a common finding and likely related to the calcification of the accompanying vessels (▶ Fig. 15.14). Agenesis of the vas deferens is commonly associated anomalies of the Wolffian duct derivatives, such as the epididymis and seminal vesicle.[12]

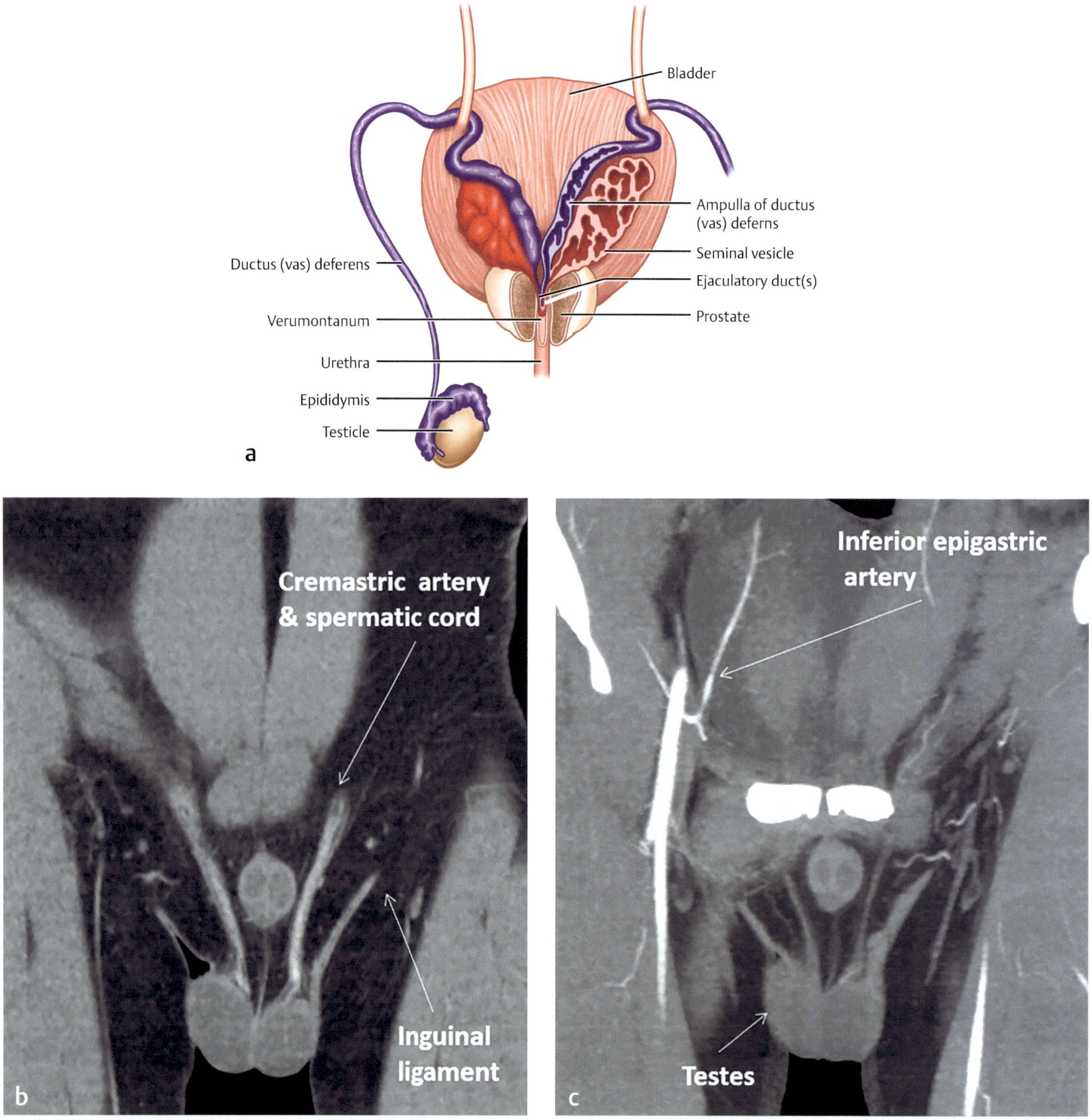

Fig. 15.13 (a) Schematic of male reproductive tract showing seminal vesicles, ejaculatory ducts, and vas deferens. **(b, c)** Postcontrast, coronal computed tomography (CT) images showing the spermatic cord outside the inguinal canal containing spermatic vessels.

◆ Seminal Vesicles

The seminal vesicles are paired structures located in the pelvis, posterior to the urinary bladder and anterior to the rectum (▶Fig. 15.13). They develop as a bulbous swelling of the distal mesonephric duct at 12 weeks of gestation.[12] Each gland is composed of a coiled, blind-ended, single tubular structure that is responsible for secreting a large proportion of the supportive fluid that will eventually become semen. The appearance of these glands on MRI is characterized by high signal intensity on T2 and low-to-intermediate signal intensity on T1 sequences[12] (▶Fig. 15.14). Although the seminal vesicles can also be visualized

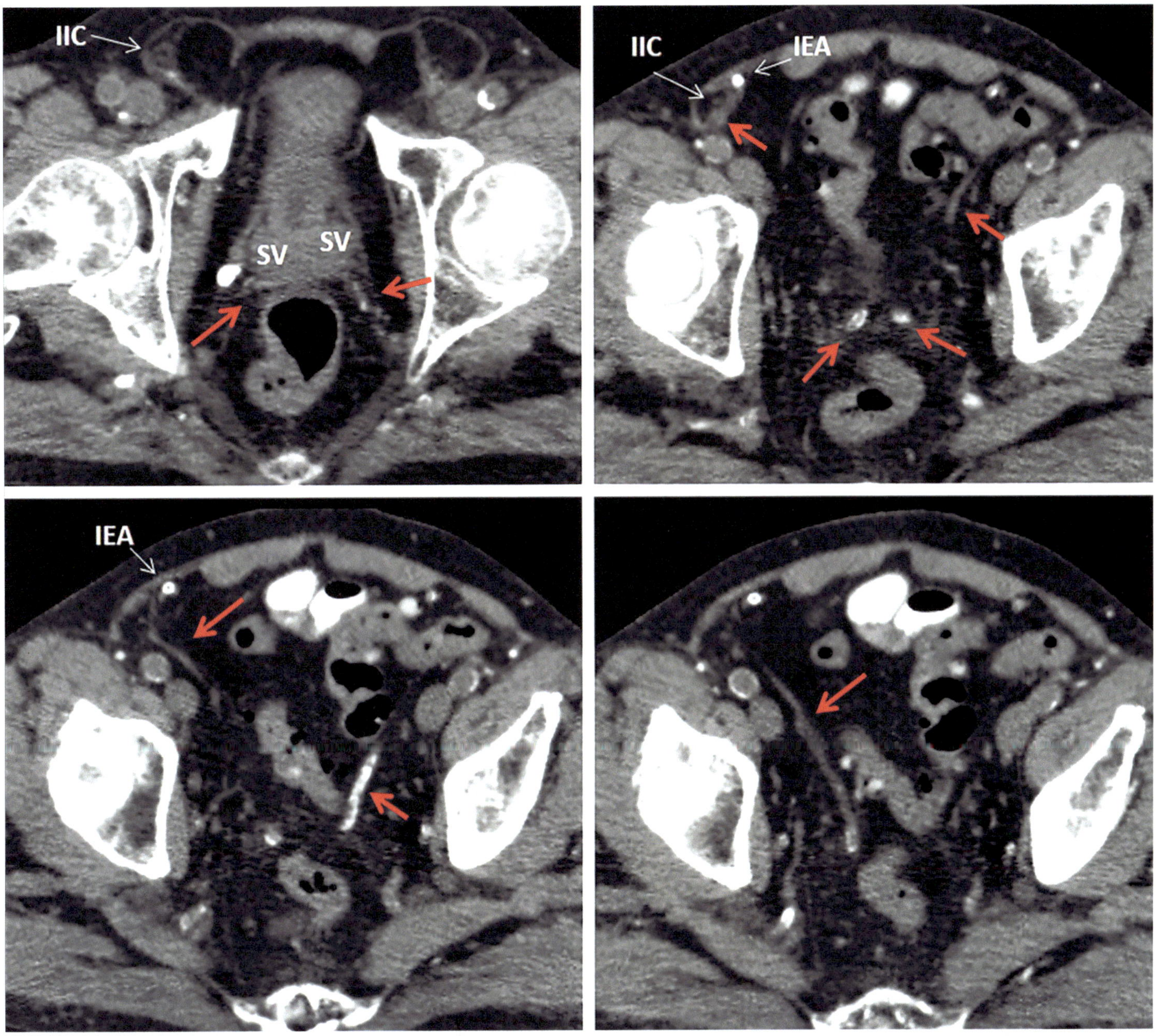

Fig. 15.14 Axial computed tomography (CT) images from inferior to superior show the anatomic course of partially calcified vas deferens (*red arrows*). The vas deferens is 30 cm long and continues on an extraperitoneal course until joining with the ejaculatory duct of each seminal vesicle (SV) gland. The vas deferens meets the testicular vessels at the internal inguinal ring to participate in formation of the spermatic cord. IEA, internal epigastric artery; IIC, internal inguinal canal.

on CT as a soft-tissue attenuation structure superior to the prostate, the underlying tubular appearance is not as well seen as on MRI[12] (▶ Fig. 15.15). The gland terminates distally into an excretory duct, which combines with its respective vas deferens to form the ejaculatory duct and subsequently propels semen into the prostatic urethra. Cystic anomalies and agenesis of the seminal vesicles are common and may coexist with the upper urinary tract anomalies because of the close embryologic association between the male reproductive and urinary systems. Cysts of the seminal vesicles are common in patients with autosomal dominant polycystic kidney disease. In agenesis of the seminal vesicle, the vas deferens is also absent (▶ Fig. 15.15c). Inflammation and post-radiation therapy changes of the seminal vesicles are common pathologic changes causing thickening and enlargement of the gland.

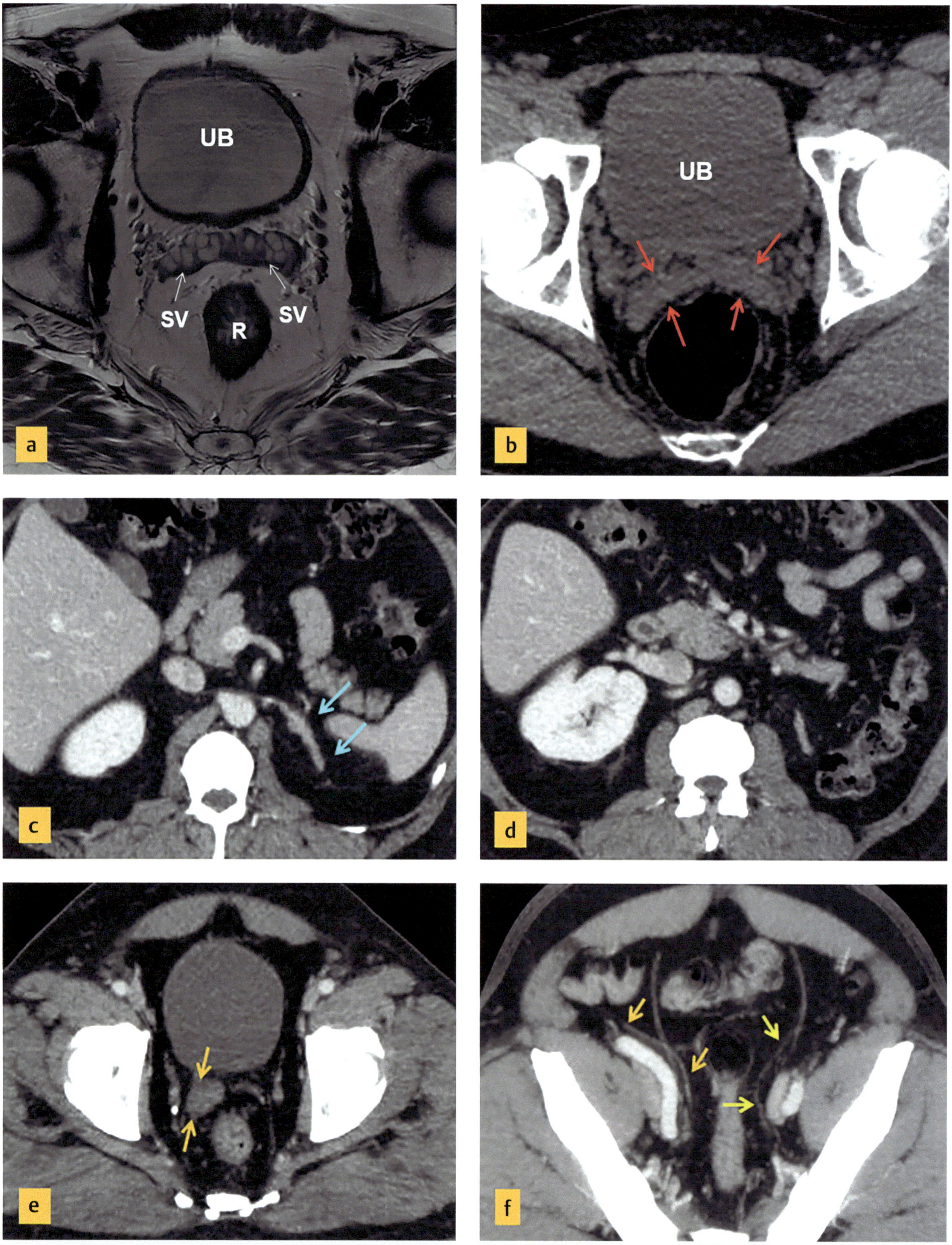

Fig. 15.15 Seminal vesicles. **(a)** Axial T2-weighted image demonstrates the paired midline tubular hyperintense structures representing the seminal vesicles (SV) located between the urinary bladder (UB) and rectum (R). **(b)** Noncontrast, axial computed tomography (CT) image demonstrates the midline soft tissue attenuation seminal vesicles (*red arrows*). Note that the internal structure of the seminal vesicle is better seen on magnetic resonance imaging (MRI). **(c–f)** Axial CT images showing agenesis of the left kidney and SV in a 61-year-old man. **(c)** Unusual linear configuration of the left adrenal (*arrows*). **(d)** Agenesis of the left kidney. **(e, f)** Normal left SV and vas deferens (*orange arrows*) but absence of the left SV and hypoplastic left vas deferens (*yellow arrows*).

◆ Prostate Gland

The prostate gland is the largest accessory gland and contributes to the formation of semen through secretion of fluid. The embryonic origin of the prostate is from the urogenital sinus as the result of endoderm invaginating into the surrounding mesenchyme with subsequent differentiation under the influence by androgens.[13] It is located inferiorly and anterior to the seminal vesicles, circumferentially surrounding the proximal part of the posterior urethra, which is thus named the prostatic urethra.

The prostate gland can be described anatomically with its base in continuity with the urinary bladder and its apex located inferiorly at the urogenital diaphragm. The midprostate is the middle one-third of the prostate gland which includes the verumontanum and the midprostatic urethra. The gland can be divided into three distinct glandular zones in the axial plane, which are best discerned using ultrasound and T2-weighted MRI as described below[14] (▸Fig. 15.16).

Peripheral Zone

This zone accounts for the largest portion of prostate volume (approximately 70%) and is the portion of the prostate that is closest to the rectum.[14] The peripheral zone surrounds the distal urethra. It is primarily derived from the urogenital sinus. On T2-weighted MRI, the peripheral zone demonstrates homogeneous high signal intensity. In this zone, carcinoma, chronic prostatitis, and postinflammatory atrophy are relatively more common than in the other zones.

Central Zone

This zone is between the peripheral and transitional zones, is immediately adjacent to the seminal vesicles at the base of prostate, and surrounds the ejaculatory ducts as they insert into the prostatic urethra. It comprises 20% of prostate. The verumontanum, a longitudinal mucosal fold, is located at this region. The central zone is derived from the Wolffian duct with low incidence of prostate cancer. The central zone demonstrates heterogeneous and relatively low signal intensity compared with the peripheral zone on T2-weighted MRI.

Transition Zone

This zone is the deepest within the prostate, surrounding the proximal prostatic urethra just superior to the verumontanum. It is the smallest part but may make up to 20% of the prostate volume.[14] The transition zone also derives from the urogenital sinus. Like the central zone, the transition zone demonstrates lower signal on T2-weighted MRI than the peripheral zone. This is the portion of the glandular tissue that enlarges due to benign prostatic hyperplasia thanks to its fibromuscular stroma. In case of benign prostatic hyperplasia because of increased ductal and acinar elements and secretions, this zone appears brighter on MR.

Anterior Fibromuscular Stroma

This region does not contain any glandular tissue and demonstrates a low-signal intensity on T2-weighted MRI.

The ratio of peripheral zone to transition/central zone tissue increases toward the apex of the prostate (▸Fig. 15.16). Denonvilliers' fascia is a thin but firm layer of connective tissue that separates the prostate and seminal vesicles from the rectum posteriorly.

MRI is optimal for evaluation of prostate zonal anatomy. However, on ultrasound, the above zonal anatomy may be distinguishable with the peripheral zone demonstrating uniform low echogenicity which is relatively hyperechoic to the central and transition zones. The sensitivity and specificity of transrectal ultrasound has historically been poor in identifying small volumes of prostate cancer and is therefore largely used in concert with MRI to guide biopsies rather than characterize the extent of cancer.[15,16]

Although the prostate gland is not surrounded by a true capsule, it is partially encapsulated by an incomplete and inseparable covering of fibromuscular tissue.[17] This incomplete capsule demonstrates low signal intensity on T2-weighted MRI, and focal loss of the T2 hypointensity may serve as an important marker for extraprostatic extension of tumor in patients with prostate cancer.

The periprostatic neurovascular bundles (NVBs) have classically been described as paired dorsolateral structures composed of the cavernous nerves as well as the arterial and venous prostatic vessels (▸Fig. 15.16, ▸Fig. 15.17). Current literature, however, suggests a more variable distribution with some studies identifying a distinct neurovascular bundle in only 52% of autopsy cases.[18] Many patients may therefore demonstrate a dispersed distribution of cavernous sympathetic and parasympathetic nerves in the dorsolateral aspect of the prostate, a finding which may impact surgical approach in patients who are candidates for nerve-sparing radical prostatectomy. Nerves also exist in the ventrolateral position to a variable degree. In patients with the classical neurovascular bundles, the structures may be seen as high-signal intensity structures on T2-weighted MRI (▸Fig. 15.16, ▸Fig. 15.17, ▸Fig. 15.18).

The arterial supply to the prostate is from the prostatic arteries, which demonstrate variable anatomic origin, but arise most commonly from the anterior division of the internal iliac artery (▸Fig. 15.17d), and less commonly from the middle rectal, the internal pudendal artery, superior vesicular artery, or obturator artery[3,15] (▸Fig. 15.17d). Two arterial pedicles are identified, superior and inferior. The superior prostatic pedicle provides the main arterial supply to the prostate, inferior bladder, and ejaculatory system. In 60% of cases, the prostate artery branches communicate with contralateral branches. The prostatic veins are responsible for venous drainage from the prostate and form the prostatic venous plexus. The prostatic venous plexus is in communication with adjacent pudendal and vesicular venous plexuses, and eventually drains into the internal iliac vein (▸Fig. 15.17c).

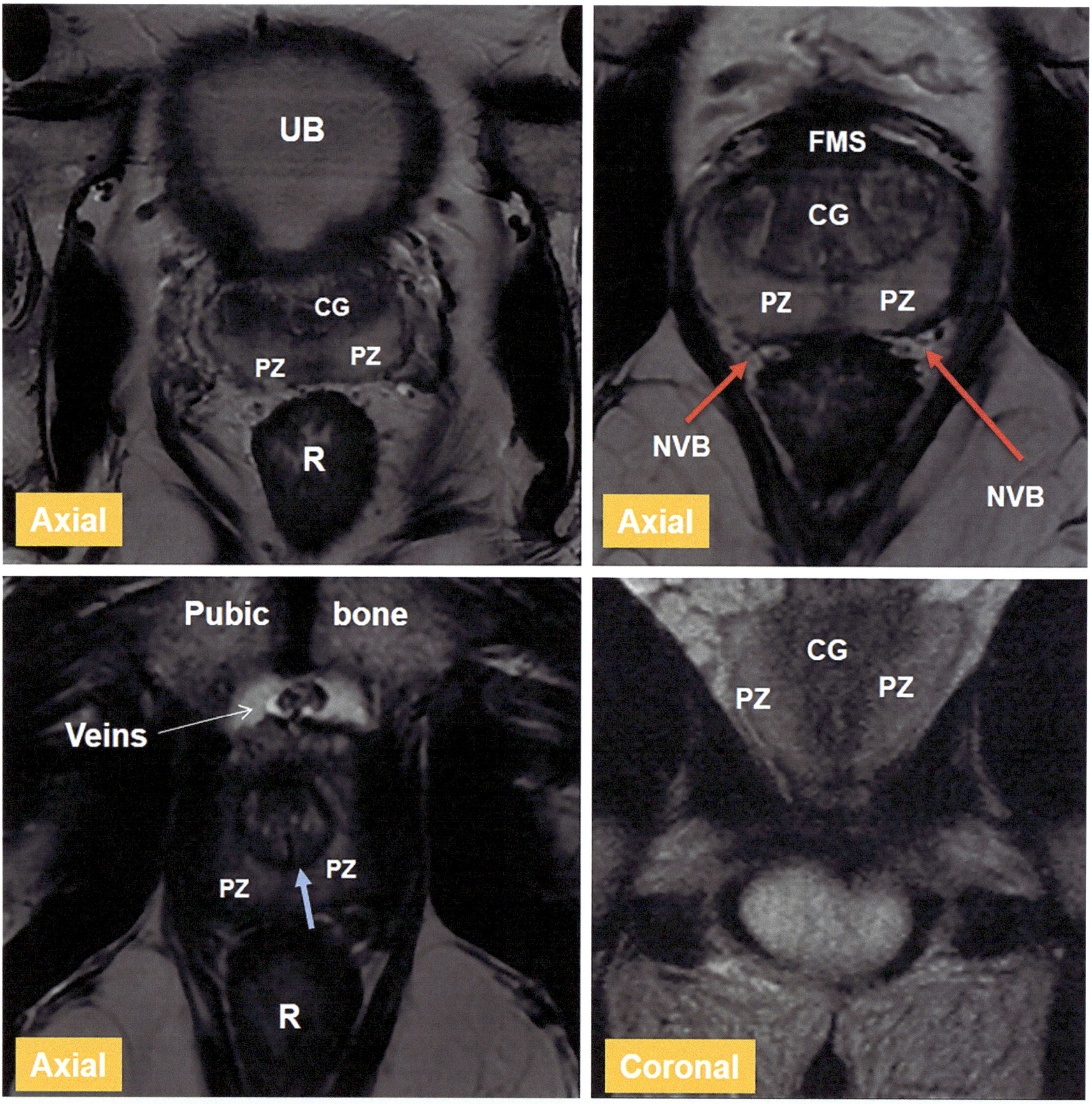

15

Fig. 15.16 Normal prostate gland. T2-weighted axial magnetic resonance (MR) image of the prostate gland at three (base, mid, and apical) levels from superior to inferior demonstrating the hyperintense peripheral zone (PZ), hypoechoic central gland (CG), and anterior fibromuscular stroma (FMS). The hyperintense neurovascular bundles are noted at mid level (*red arrow*, NVB). Retropubic veins are shown. The distal prostatic urethra is shown by *light blue arrow*. T2-weighted coronal and sagittal images show the same structures. Sagittal schematic of the prostate gland demonstrating the zonal anatomy and the relationship of these various zones to adjacent structures. R, rectum; SV, seminal vesicle; UB, urinary bladder.

(*Continued*)

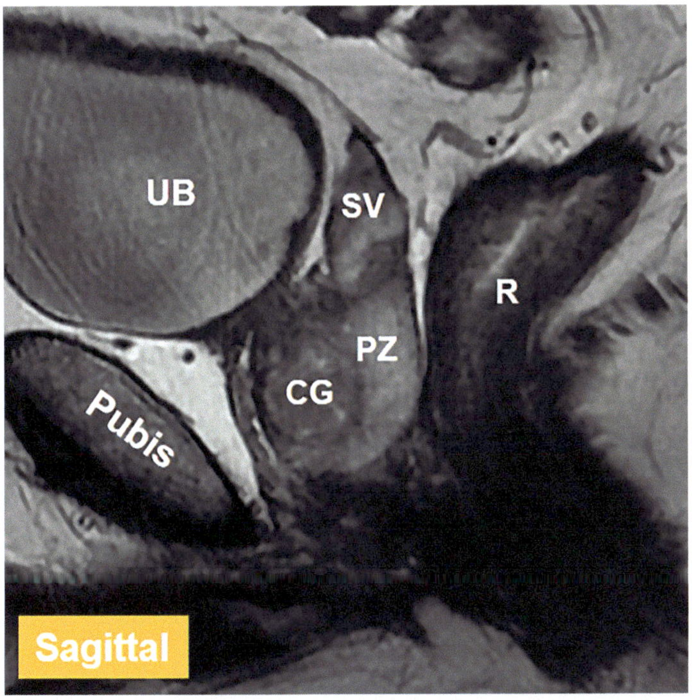

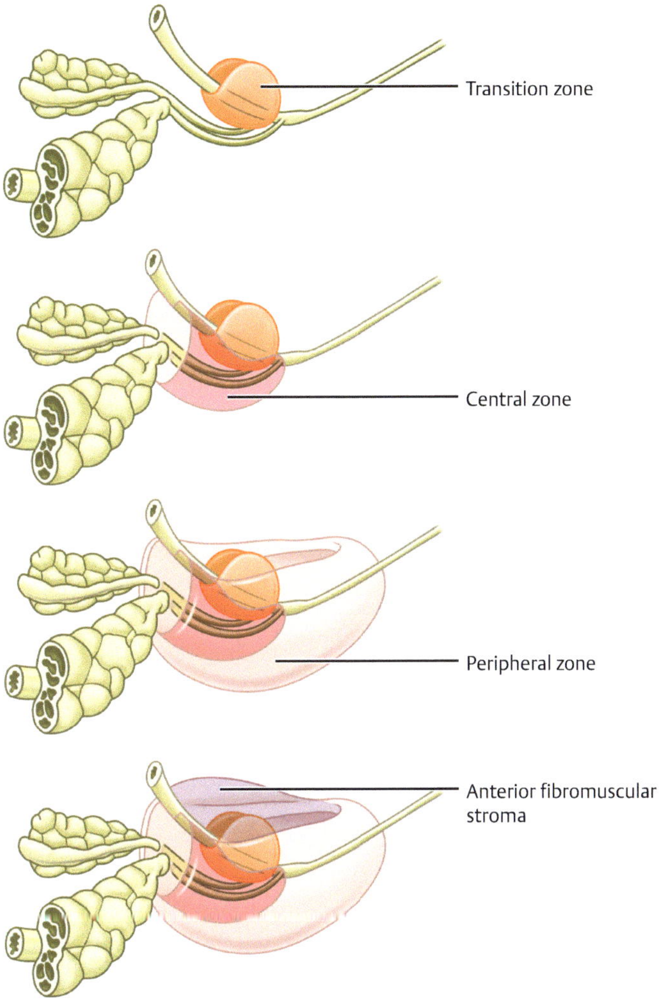

Fig. 15.16 *(Continued)*

Prostate Pathology

Benign prostatic hyperplasia is a common pathology causing urinary obstruction (▶Fig. 15.18). Treatment usually requires transurethral resection (TUR). Prostatic arterial embolization has been used as an alternative treatment for benign prostatic hyperplasia.[15] After TUR, a relatively large cavity is usually seen at the site of resection and the prostatic urethra may remain wide open. Utricle cysts are small cysts in the prostate and most often are detected in midline in young individuals (▶Fig. 15.18). Inflammation of prostate is also a relatively common benign problem. Evaluation for malignant prostate lesions is the most common indication for MRI. Prostate cancer usually appears hypointense on T2-weighted sequences. High-grade focal prostate cancers also demonstrate restriction on diffusion-weighted sequences as well as early enhancement and washout on dynamic contrast enhancement series[14,19] (▶Fig. 15.19). Invasion into surrounding tissue may significantly impact treatment decisions, and identification of anatomic landmarks including the neurovascular bundles and prostate capsule are an important aspect of staging the disease process.[19] Erectile dysfunction is one of the complications after radical retropubic prostatectomy. Preservation of the neurovascular bundles at surgery may prevent this complication.

15

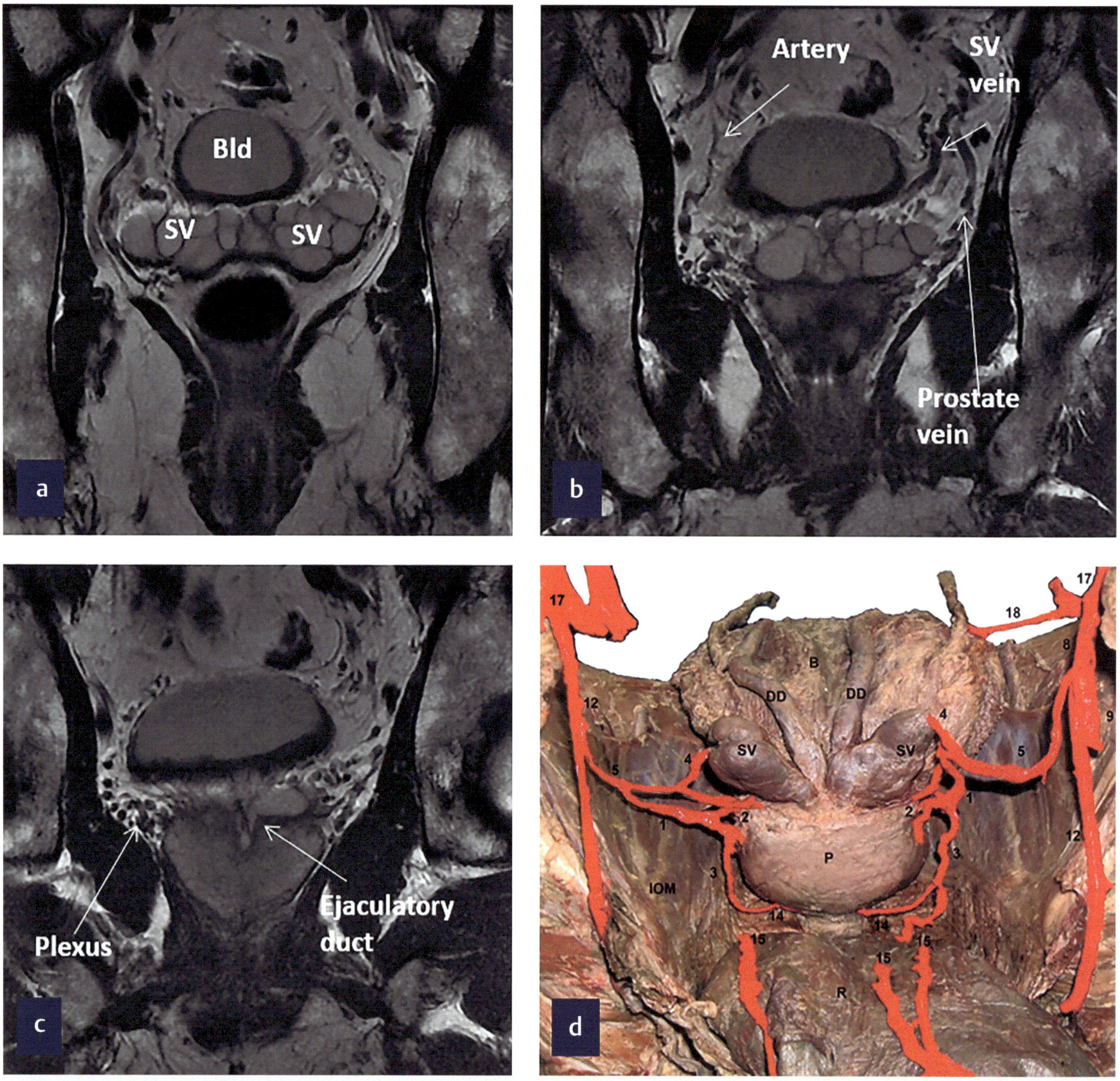

15

Fig. 15.17 (a–c) Coronal magnetic resonance (MR) images from posterior to anterior showing vessels of the bladder (bld) base, seminal vesicles (SV), and prostate. **(d)** Prostate arterial anatomy, posterior view. The left superior prostatic pedicle (1) branches from a prostatic-vesical artery (5) originating from the internal pudendal artery. On the right, the superior prostatic pedicle originates directly from the anterior trunk of the internal iliac artery (8) as a prostatic-vesical artery (5). B, bladder; DD, ductus/vas deference; P, prostate gland; SV, seminal vesicles; 1, superior prostatic pedicle; 2, medial prostatic branch; 3, lateral prostatic branch; 4, inferior vesical; 5, prostatic vesical; 8, gluteopudendal trunk; 9, obturator; 12, internal pudendal; 14, inferior prostatic pedicle; 15, superior rectal; 17, hypogastric; 18, superior vesical. (Reproduced with permission from Garcia-Monaco et al.[15])

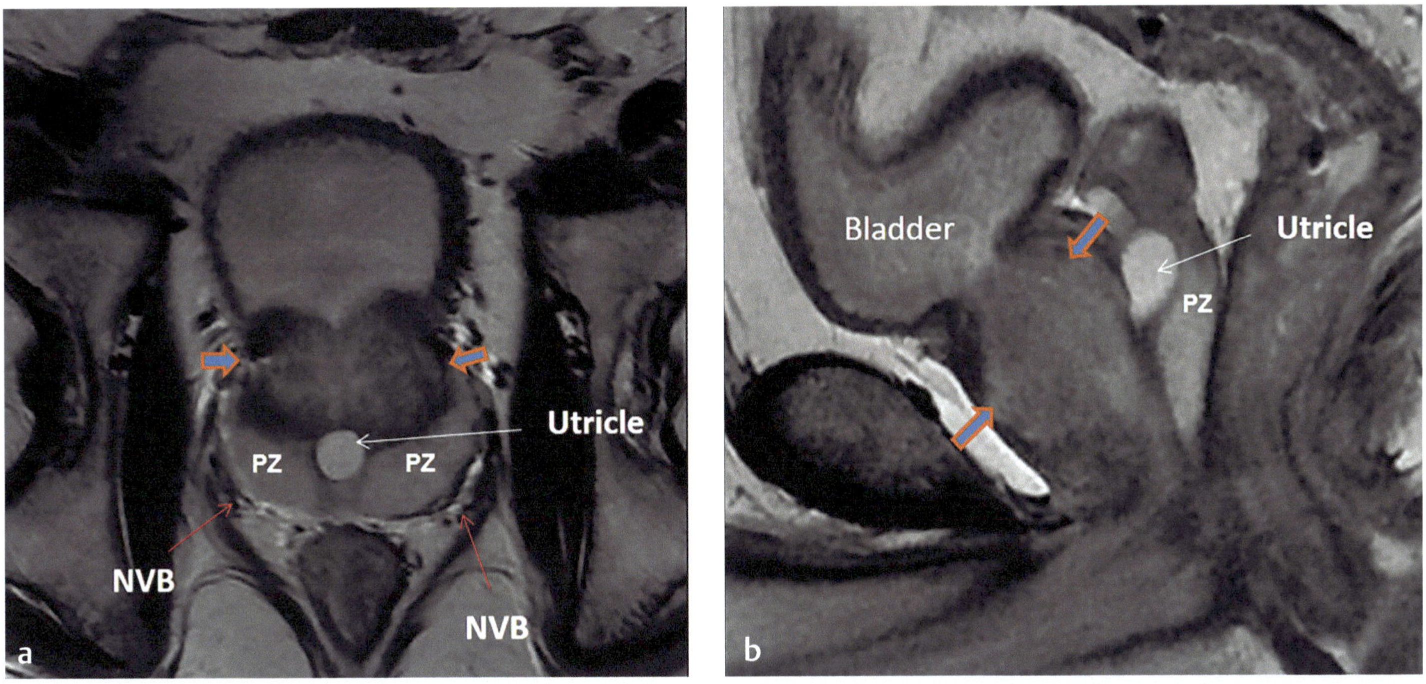

Fig. 15.18 Benign prostatic hyperplasia. **(a)** Axial and **(b)** sagittal magnetic resonance (MR) images showing the hyperplastic region (*arrows*) in the center of the prostate projecting into the bladder lumen. A prostatic utricle is seen. NVB, neurovascular bundle; PZ, peripheral zone.

◆ Penis

Differentiation of external genitalia starts in the 7th week of gestation and is completed by 16th to 17th weeks. It begins with a genital tubercle arising from mesenchyme near the embryonic cloaca. Under the influence of androgens, the mesoderm of the genital tubercle forms the phallus, with the corpora cavernosa and corpus pongiosum arising from mesenchymal condensations[13] (▸Fig. 15.20).

The penis is predominantly composed of three endothelium-lined cylindrical structures. The larger paired lateral corpora cavernosa are fused anteriorly and form two penile crura on the sides posteriorly (▸Fig. 15.21, ▸Fig. 15.22, ▸Fig. ▸15.23). The median septum between the corpora cavernosa is fenestrated allowing for communication of the cavernosal sinuses in the penis (▸Fig. 15.21). The third structure is a separate single midline corpus spongiosum which courses ventrally.

The crura of the corpora cavernosa attach proximally to the inferior pubic rami and the perineal membrane and are closely associated with the ischiocavernosus muscle. These paired structures demonstrate symmetric intermediate to high T2 and low to intermediate T1 signal on MRI, which correlates with the blood flow within their cavernous spaces and the incomplete septum which joins them (▸Fig. 15.21, ▸Fig. 15.22, ▸Fig. 15.23). The ischiocavernosus muscle on each side of the perineum extends between the inner surface of the inferior pubic ramus to the side of the corpus cavernosum and undersurface of the crus penis helping to maintain erection.

The proximal aspect of the corpus spongiosum is known as the penile bulb attaching to the lines in apposition with the perineal membrane (inferior fascia of the urogenital diaphragm). The distal end forms the glans penis. The corpus spongiosum protects the urethra from pinching during erection. The penile bulb is surrounded inferiorly by the bulbospongiosus, a superficial muscle of the perineum (▸Fig. 15.22, ▸Fig. 15.23). The bulbospongiosus muscle arises from the central point of the perineum and from the median raphe and attaches to the perineal membrane and dorsum of the penis. This muscle contributes to erection and contractions during expulsion phase of ejaculation. The penile urethra courses throughout the corpus spongiosum, which distally is continuous with the glans. The corpus spongiosum demonstrates high signal intensity on T2 and low-to-intermediate signal on T1 (▸Fig. 15.22, ▸Fig. 15.23). However, as it does not communicate with the cavernosa, the spongiosum may demonstrate distinct signal intensity characteristics.

Proximally the penis is attached to the pubic symphysis by the suspensory ligament which is an extension of the Buck's fascia. The three cavernous spaces described above are surrounded by the tunica albuginea which along with Buck's fascia attach to the undersurface of the pubic symphysis by the suspensory ligament of the penis (▸Fig. 15.23). It is composed of inner circular and outer longitudinal collagen bundles. Peyronie's disease is a wound-healing disorder in which a fibrosis of the tunica albuginea layer of the penis can lead to penile curvature during erection, penile pain, and erectile dysfunction. Fibrotic plaques may be detected by ultrasound. The Buck's fascia surrounds the tunica albuginea and separates the corpora cavernosa from the corpus spongiosum by enveloping both cavernosa dorsally and splitting to encase the spongiosum ventrally (▸Fig. 15.21, ▸Fig. 15.22, ▸Fig. 15.23). It is continuous with the deep pelvic muscle fascia and fixed distally to the corona of the glans. Superficial to Buck's fascia is dartos fascia which is continuous

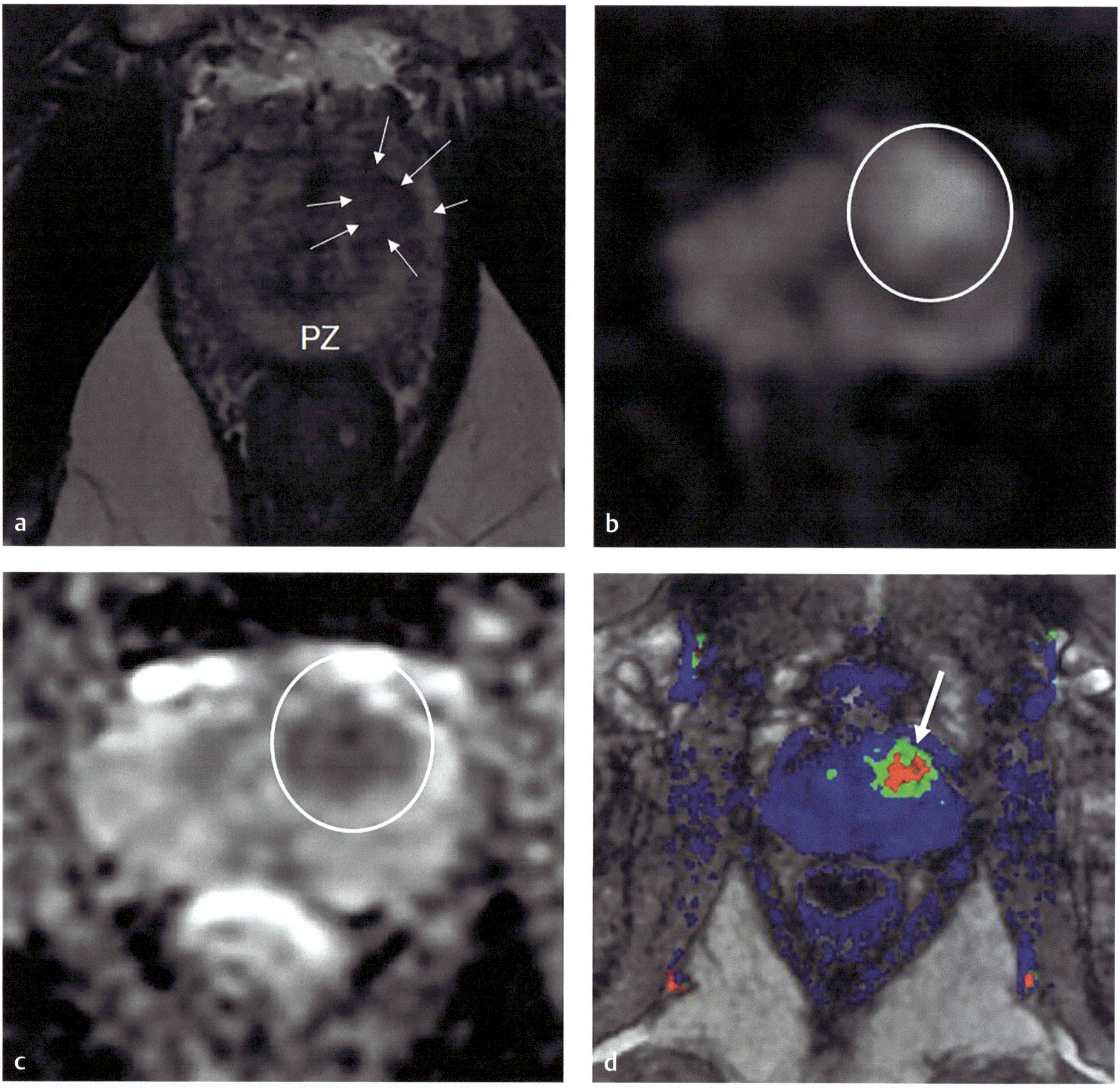

15

Fig. 15.19 Prostate cancer. Typical magnetic resonance (MR) diagnosis. **(a)** T2-weighted axial image at midgland demonstrating hypointense lesion (*arrows*) in the left anterior peripheral zone (PZ). **(b)** Axial diffusion-weighted MR demonstrating an area of diffusion restriction (*circle*) in the region of abnormality. **(c)** Axial apparent diffusion coefficient (ADC) map demonstrating region of low ADC values (*circle*) corresponding with area of diffusion restriction in **b**. **(d)** Contrast-enhanced images are overlaid on T2-weighted MR images (wash-in and washout). The color map demonstrates asymmetric focal marked enhancement in the left midgland with rapid washout (*arrow*).

with the Colles–Scapa's fascia. On imaging these two, the tunica albuginea and Buck's fascia may not be separable and all appear hypointense on T1- and T2-weighted images (▶Fig. 15.22). MRI may be utilized for evaluation of penile trauma, where it is especially useful for visualizing disruption of the tunica albuginea, which often necessitates surgical intervention[20] (▶Fig. 15.24). Although the site and extent of penile fracture may be identified on MRI, associated injury of the urethra, which commonly coexists with penile fracture, is better evaluated with retrograde urethrogram, which is further described below.

Arterial supply to the dorsal superficial skin of the penis is provided by the left and right superficial external pudendal arteries which originate from the femoral arteries and collateralize across the midline (▶Fig. 15.12). The artery may be duplicated. The superficial external pudendal arteries are important collateral pathways in occlusion of the femoral arteries. The ventral skin of

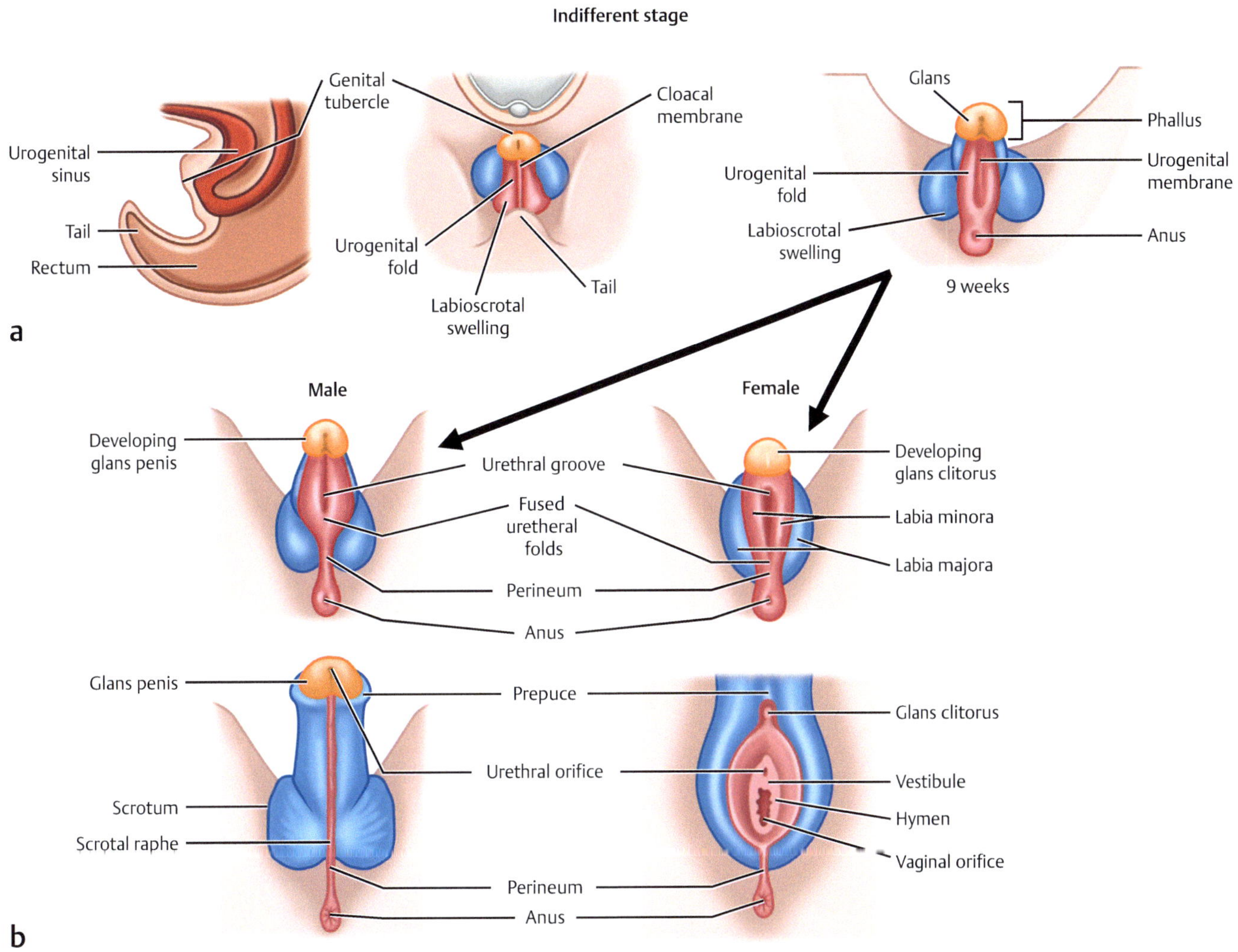

Fig. 15.20 Human urogenital embryology. **(a)** Sex undifferentiated precursors prior to 7 weeks of gestational age. **(b)** Male development in gestational weeks 7 to 17 from precursors shown in **a**. (Modified from Yiee and Baskin.[22])

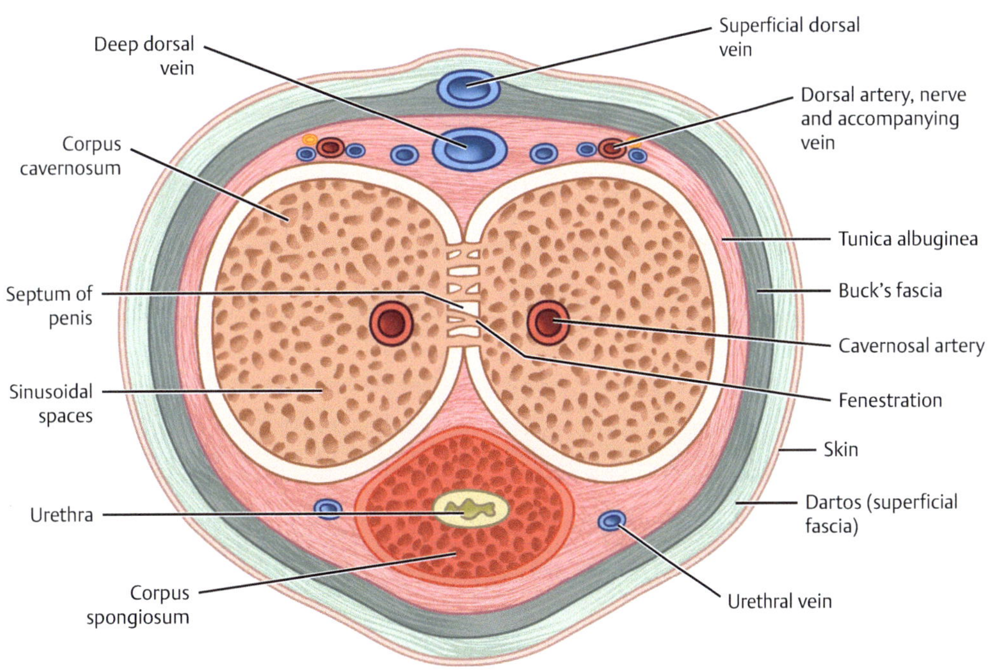

Fig. 15.21 Cross-sectional anatomy of the penis. (Modified from Kirkham.[20])

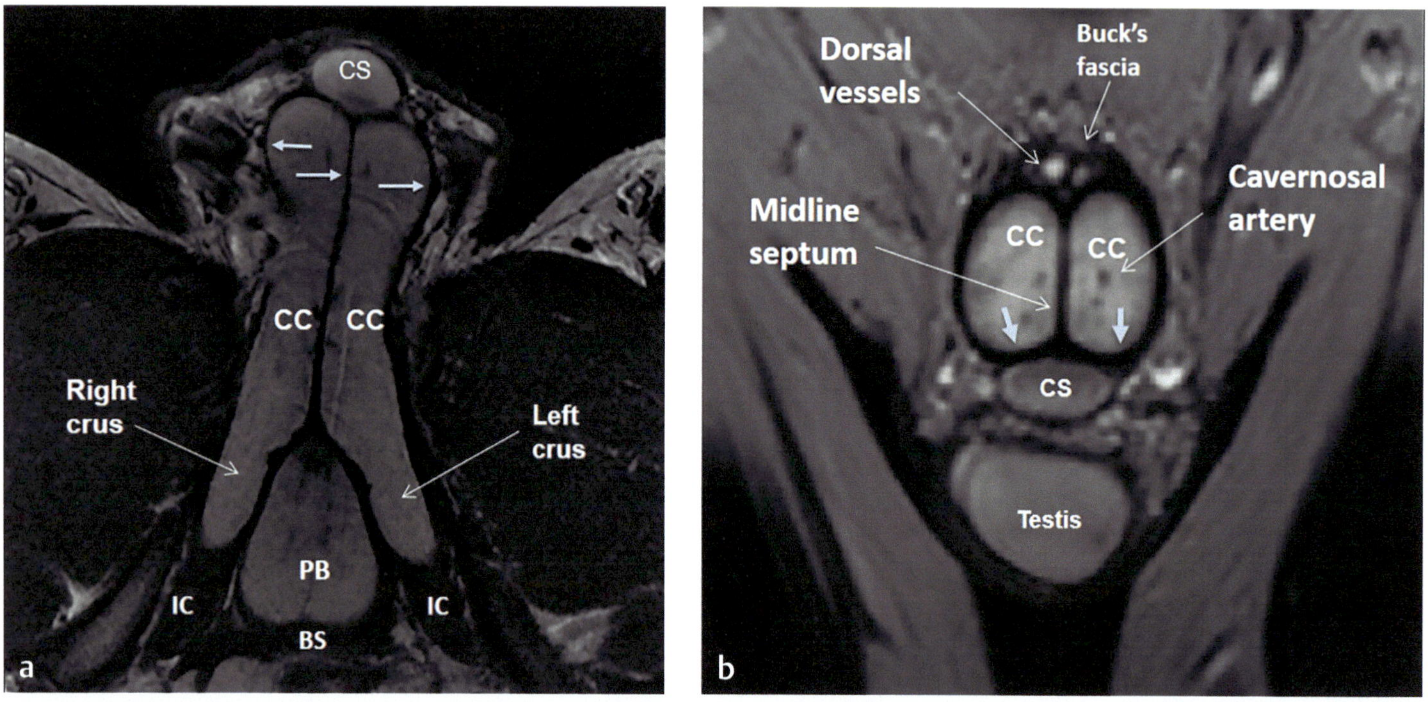

Fig. 15.22 Normal penis. **(a)** T2-weighted axial image at the base level and **(b)** coronal image at mid portion of the penis demonstrating the intermediate intensity midline penile bulb (PB) which is the most proximal portion of the corpus spongiosum (CS). The paired intermediate to high intensity corpora cavernosa (CC) are seen with their proximal crura continuous with the ischiocavernosus muscles (IC). The thin linear hypointense structure represents the tunica albuginea (*arrows*). BS, bulbospongiosus muscle.

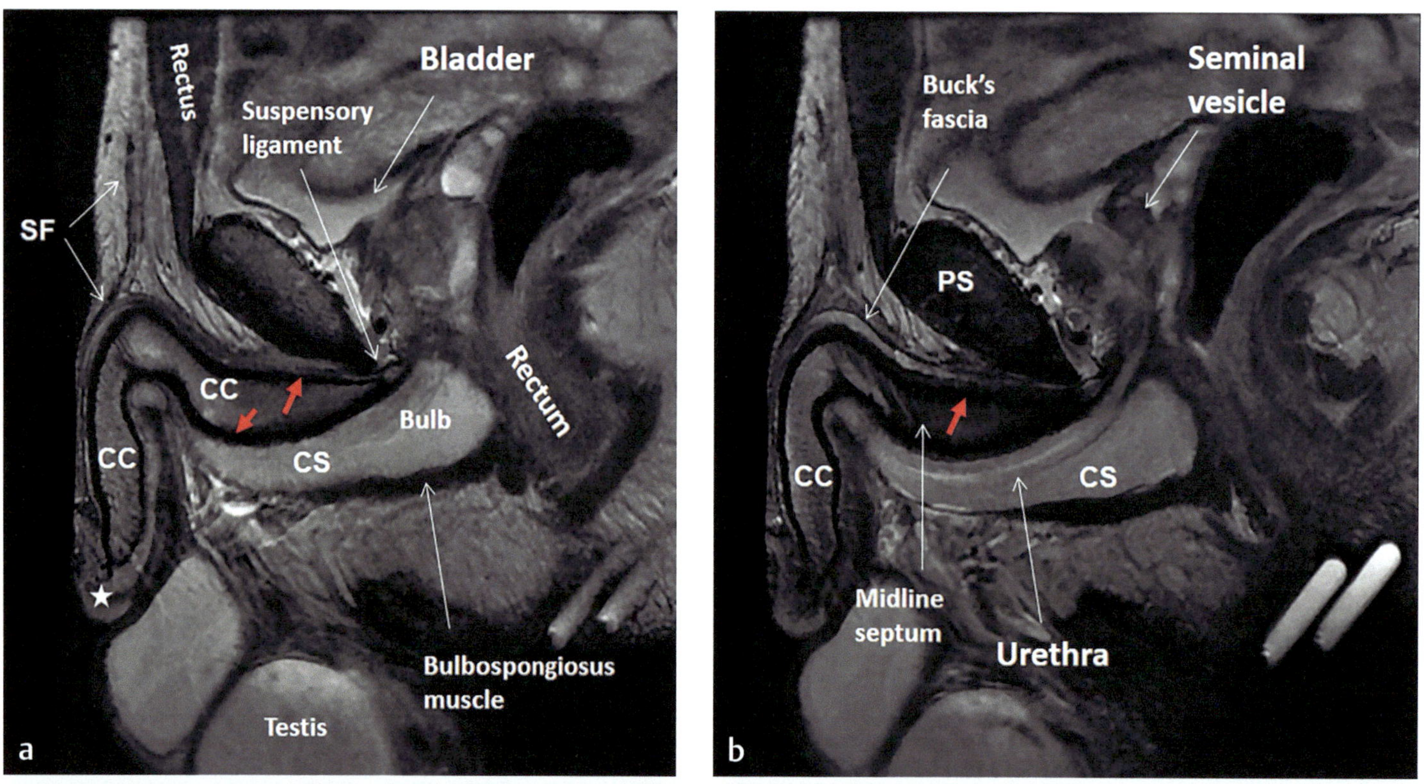

Fig. 15.23 Normal penis. T2-weighted sagittal image through the penis at paramedian **(a)** and median **(b)** levels demonstrating the hyperintense corpus cavernosum (CC) and the ventral intermediate intensity corpus spongiosum (CS). The hypointense tunica albuginea is denoted by the *arrows*. The glans penis is marked by an *asterisk*. PS, pubic symphysis; SF, superficial fascia.

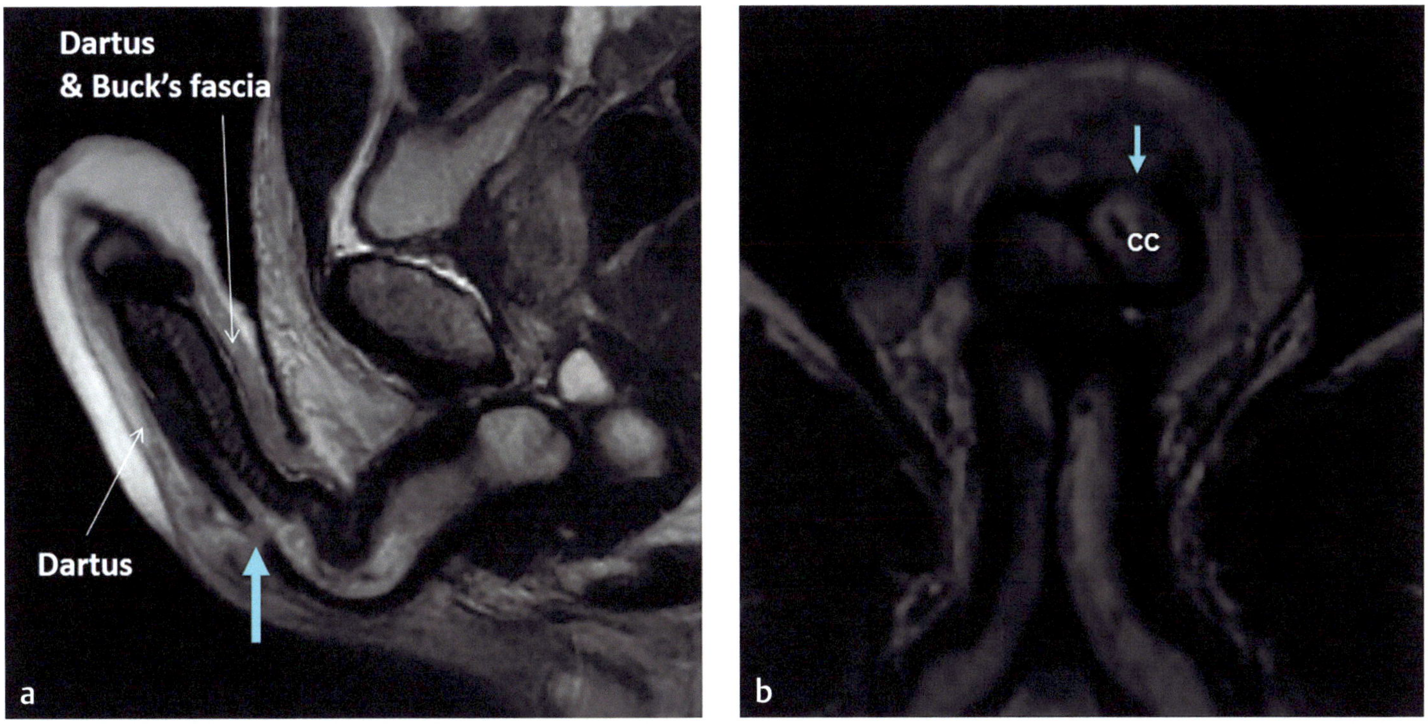

Fig. 15.24 Penile fracture. T2-weighted sagittal **(a)** and axial **(b)** images of the penis demonstrating a ventral tunica albuginea defect in the proximal third of the left corpus cavernosum (CC) (*blue arrow*).

the penis and scrotum is supplied by the posterior scrotal artery which is a branch of the deep internal pudendal artery.[3]

The deeper structures of the penis receive arterial flow by the artery of the penis which is the distal continuation of the internal pudendal artery (▶Fig. 15.25, ▶Fig. 15.26, ▶Fig. 15.27). The internal pudendal artery, after giving off its perineal branch, continues as the penile artery. It divides into three branches: the bulbourethral artery, the dorsal artery, and the deep (cavernosal) artery of the penis (▶Fig. 15.25). The deep artery of the penis may rarely arise directly from the internal iliac artery. In this situation, the dorsal artery of the penis is a branch of internal iliac artery or the inferior epigastric artery. Penile branches arising from an accessory pudendal artery are also common. The bulbourethral artery travels through the Buck's fascia to supply the bulb of the penis and the penile urethra. As the name implies, the dorsal artery travels along the dorsal aspect of the penis with multiple circumflex branches, eventually terminating in the glans. The deep artery of the penis enters the corpus cavernosum at the crus and travels the entire length of the penile shaft, branching into the helicine arteries[21] (▶Fig. 15.25). All three arteries communicate distally near the glans to provide an extensive anastomotic network. Superselective embolization of the internal pudendal artery is a common procedure to stop bleeding from the penis and perineal region.

There is only one deep dorsal vein that runs alongside the dorsal arteries and nerves in Buck's fascia above the tunica albuginea. Hence, the penile vascular system does not comply with the general rule in the body in that the number of veins is normally greater than that of the arteries[22] (▶Fig. 15.21). The deep dorsal vein receives drainage from the distal two-thirds of the corpora cavernosa via emissary veins and the corpus spongiosum via circumflex veins and finally drains to the periprostatic plexus (▶Fig. 15.28).

Penile innervation consists of the dorsal, cavernosal, and perineal nerves. The dorsal and perineal nerves arise from the pudendal nerves. The cavernosal nerves arise from the autonomic pelvic plexus. The pudendal nerve is a branch of the sacral plexus (S2–S4). The pudendal nerve gives off three branches, the inferior rectal branch, the perineal branch, and the dorsal nerve of the penis or clitoris.

Erectile Dysfunction

Male impotence is a multifactorial malfunction of the penis to achieve sufficient rigidity for sexual activity, which is estimated to affect approximately 300 million men worldwide.[23] After neurological, psychological, and endocrine disorders have been excluded by medical evaluation, radiology plays a significant role in the evaluation for atherosclerotic disease, one of the most common causes of erectile dysfunction.[24]

Adequate erection requires a balance of both increased inflow of blood as well as restricted outflow to maintain rigidity. Vascular causes of erectile dysfunction may therefore be arteriogenic, venous, or mixed in etiology. Therefore, knowledge of vascular anatomy of pelvis and penis is essential for evaluation of these patients.

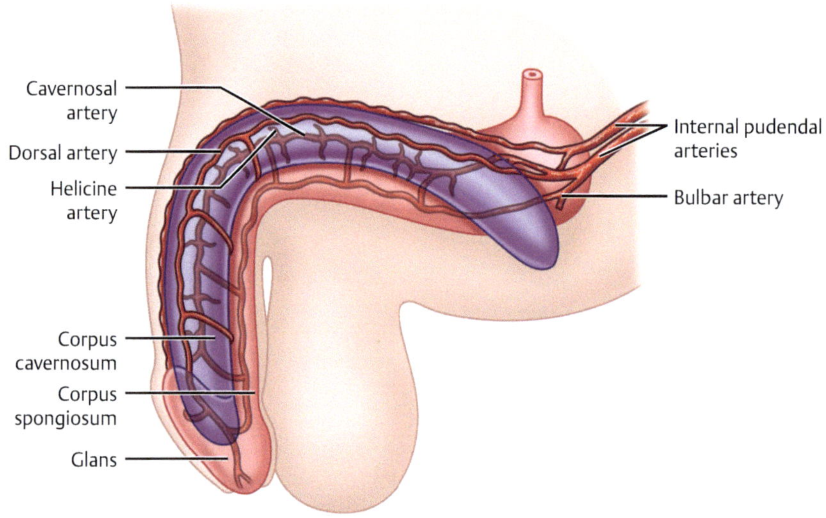

Fig. 15.25 Penile arterial anatomy. (Modified from Patel et al.[25])

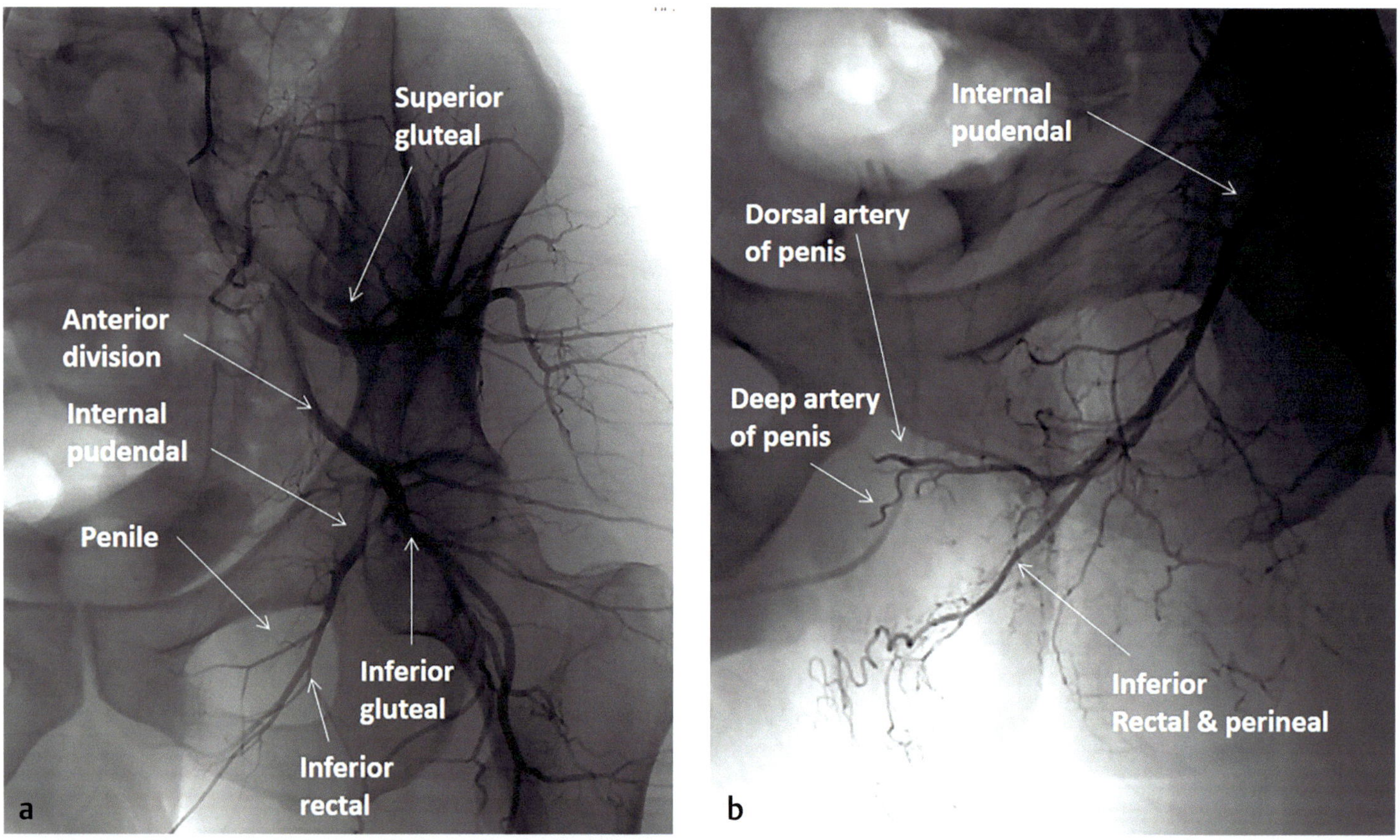

Fig. 15.26 Internal pudendal artery. **(a)** Left internal iliac artery injection. **(b)** Selective internal pudendal artery injection. The internal pudendal artery is the terminal branch of the anterior trunk of the internal iliac artery. Branches of the internal pudendal artery include inferior rectal, perineal, posterior scrotal, bulbourethral, dorsal artery of penis, and deep (cavernosal) artery of penis.

Arteriogenic erectile dysfunction affects the inflow aspect of tumescence and is secondary to atherosclerotic steno-occlusive disease affecting the arterial branches supplying the penis including the internal iliac, pudendal, cavernosal, and helicine arteries.

Normal venous outflow reduction in the erected state is mediated by the increased lacunar pressure caused by the inflow of arterial blood, which compresses the emissary veins against the rigid tunica albuginea. Impairment of the normal veno-occlusive state during erection may be secondary to venous shunts, arteriovenous malformations, or inadequate compression of the emissary and subtunical veins secondary to structural changes from old age or previous traumatic injury to the tunica albuginea.

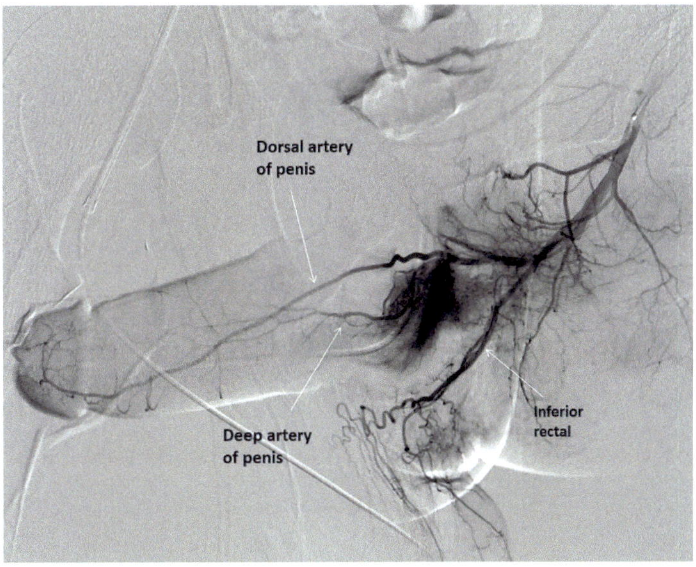

Fig. 15.27 Penile artery. Selective internal pudendal artery injection after administration of papaverine. Terminal branches including the dorsal artery of penis, deep (cavernosal) artery of penis, and inferior rectal are shown.

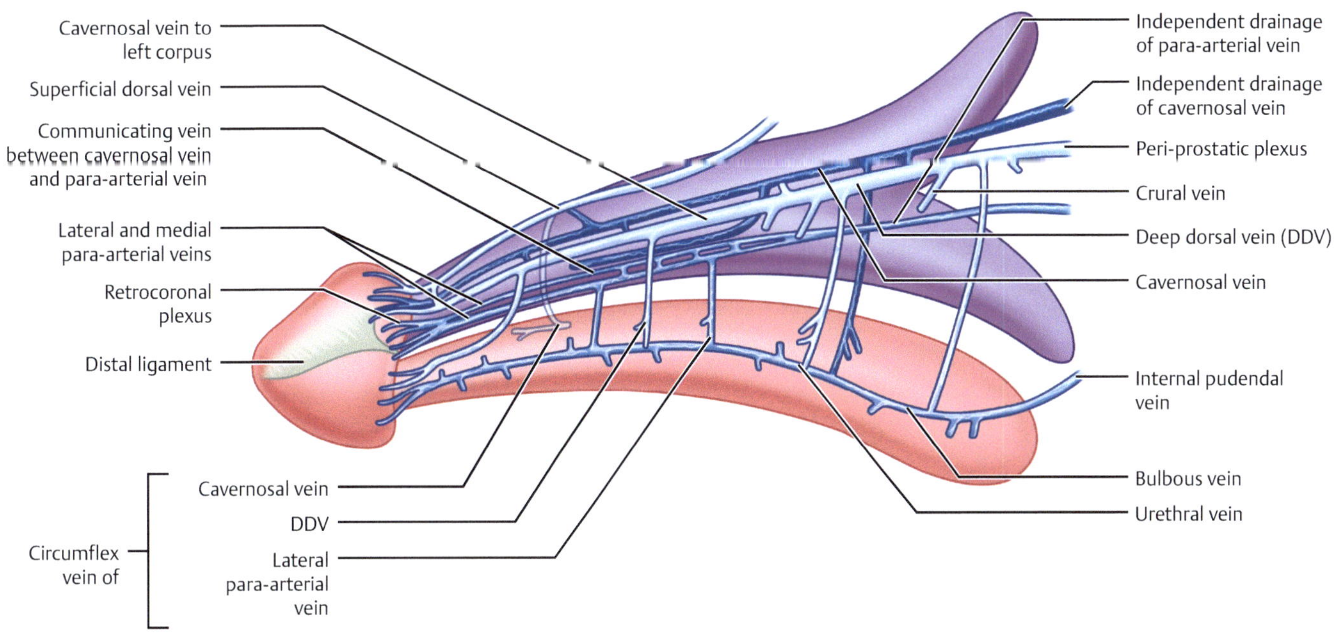

Fig. 15.28 Schematic illustration showing anatomy of the veins in the human penis. The deep dorsal vein (DDV) is consistently located in the median position and receives blood of the emissary veins from the corpora cavernosa and of the circumflex vein from the corpus spongiosum.

Color Doppler sonographic imaging is the preferred initial modality for evaluation in patients with suspected vascular etiology for erectile dysfunction. Sonographic imaging of the penis allows for evaluation of dorsal and cavernosal penile arteries as well as venous etiologies of impotence. The waveforms of the normal cavernosal arteries are multiphasic and will vary according to the phase of erection. Soon after injection of the pharmacologic agent, the systolic and especially diastolic velocities are expected to increase followed by a high resistance wave-form characterized by increased systolic velocity with narrowing of the systolic peak and eventually diastolic flow reversal; the latter represents elevated intracavernosal pressure secondary to an intact venous occlusion component. In the presence of venous incompetence, persistent, nonreversed diastolic flow may be seen with elevated end diastolic velocities.

A peak systolic velocity of less than 25 cm/s has been correlated with severe arterial disease on angiographic imaging.[25] Although literature suggests an end diastolic velocity of greater than 7 cm/s may correlate with dysfunction of the veno-occlusive mechanism, this is less reliable and may be misleading in the setting of concurrent arterial insufficiency.[26]

Angiographic evaluation for erectile dysfunction is useful as a second-line modality for documentation of steno-occlusive disease, congenital arteriovenous malformations, and anatomical variations which may contribute to impotence.[27] Superselective evaluation of the internal pudendal artery is performed with visualization of the penile artery and its branches (▶ Fig. 15.27). Endovascular treatment of arterial stenosis via percutaneous transluminal angioplasty and stent placement may be pursued.[27] Dysfunction of the veno-occlusive mechanism may conversely be treated by selective percutaneous retrograde embolization.

Priapism

Priapism is pathological persistent tumescence unrelated to sexual stimulation. It is classified into low-flow and high-flow types. Low-flow priapism is usually due to venous obstruction and ischemia. High-flow priapism is due to uncontrolled increased arterial influx in cases of fistula formation or cavernosal metastasis of a solid tumor. It most frequently develops as a result of perineal trauma (▶ Fig. 15.29).

Diagnostic methods include CT angiography, color flow Doppler ultrasound, internal pudendal arteriography, and blood gas analysis of cavernous body. CT of catheter angiography of the penis demonstrates the presence of a fistula between the right

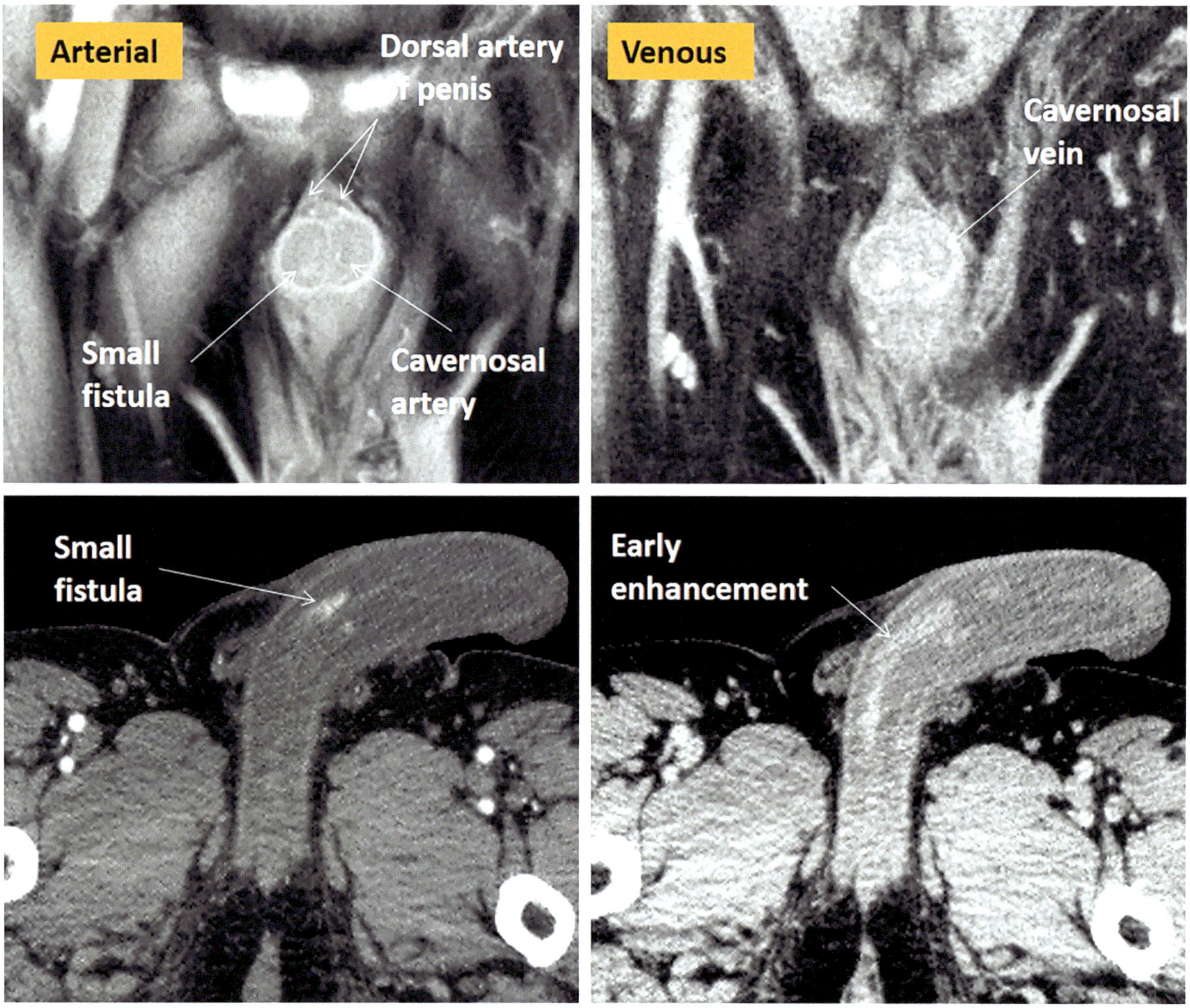

Fig. 15.29 Posttraumatic priapism. Axial (top) and coronal (bottom) computed tomography (CT) angiography images during arterial and early venous showing small arteriovenous fistula and early venous filling at the right penoscrotal junction of the corpus cavernosum consistent with high-flow priapism.

cavernous artery and the corpus cavernosum. The arterial–lacunar fistula bypasses the helicine arteries, directly extending into the cavernosal tissue and appearing as a characteristic contrast blush and turbulent high-velocity flow. The fistula can be treated with autologous clot embolization of the involved branches of the pudendal artery.

◆ Urethra

The embryologic origin of the urethra is from the urogenital sinus with the endodermal layer of the genital tubercle eventually forming the epithelium and splanchnic mesenchyme contributing to the surrounding soft tissue.

The urethra can be divided into anterior and posterior portions[28] (▶Fig. 15.30, ▶Fig. 15.31). The posterior urethra consists of the prostatic urethra, extending from the bladder neck to the verumontanum, as well as the shorter membranous urethra, measuring approximately 1 cm, extending from verumontanum to penile bulb. Some have classified the bladder neck in the posterior urethra.[22] An osseous radiographic landmark for the junction of posterior and anterior urethra is the obturator foramen. The anterior urethra can be divided into the bulbar urethra, which appears as a widened "cone" from proximal bulb to penoscrotal junction, and the penile or pendulous urethra, extending from penoscrotal junction to fossa navicularis (▶Fig. 15.30, ▶Fig. 15.31).

At the junction between the bladder and urethra, three smooth muscle fiber layers form the neck internal urethral sphincter. The inner and outer layers are longitudinal and the intermediate layer is circular fibers. The internal urethral sphincter is important for urinary continence and prevents retrograde ejaculation. Damage to this sphincter is common after prostate surgery. At the level of the membranous urethra, a layer of circular striated muscle forms the external urethral sphincter.

Retrograde urethrography (RUG) is the primary imaging modality for evaluation of urethra. Normal anatomic structures can be seen on RUG (▶Fig. 15.31). These include the verumontanum

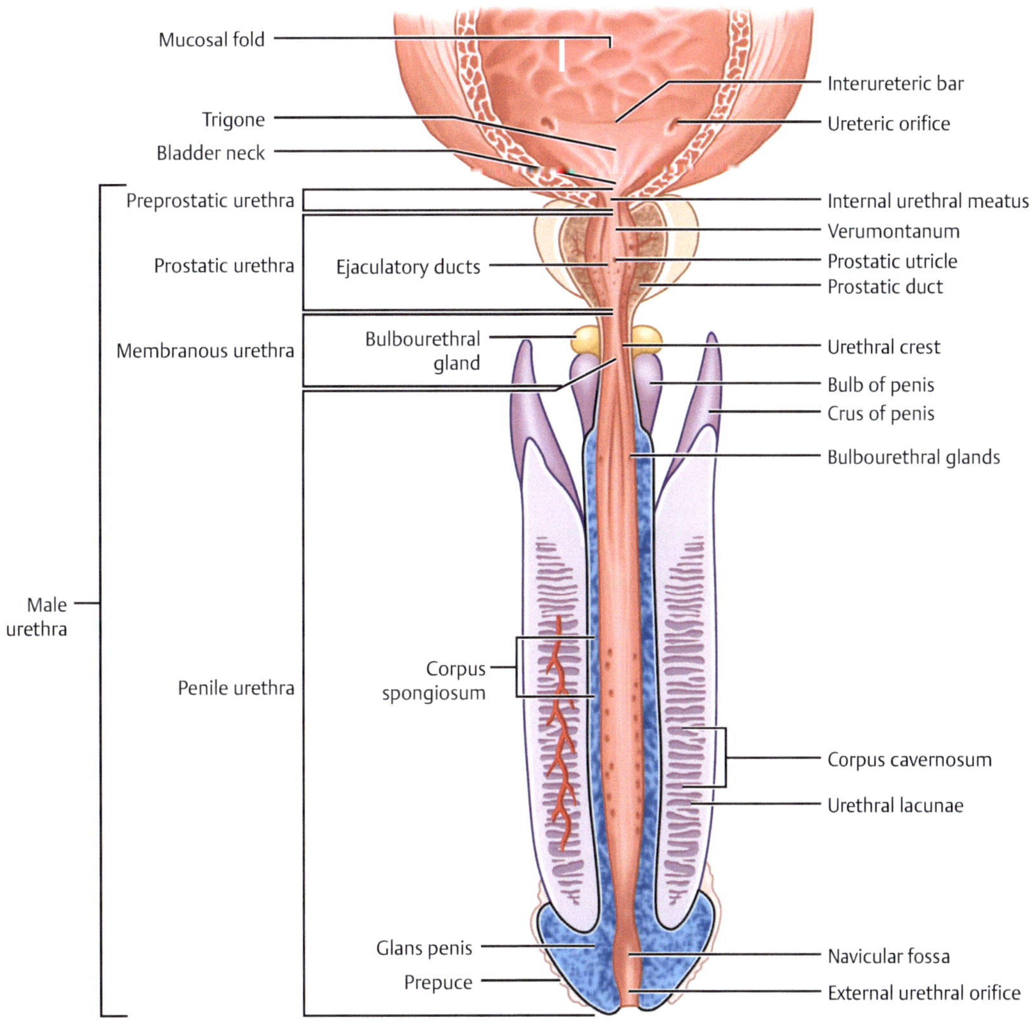

Fig. 15.30 Urethral anatomy.

15

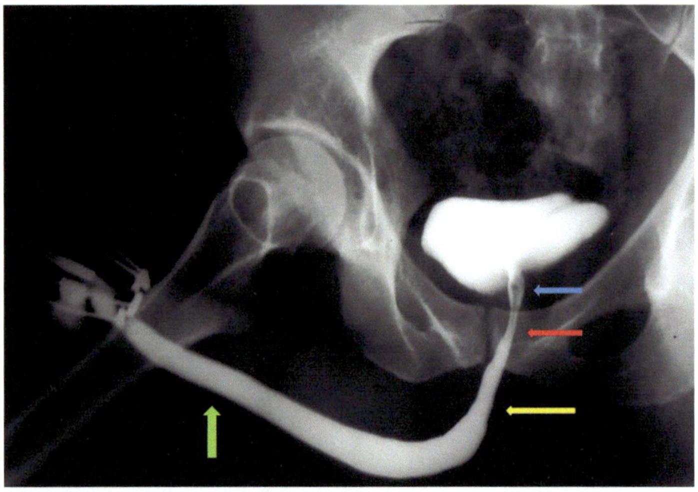

Fig. 15.31 Segments of the male urethra on retrograde urethrogram (RUG). The posterior urethra consists of the prostatic urethra, extending from the bladder base to the verumontanum, and the short membranous urethra, measuring approximately 1 cm, extending from verumontanum to penile bulb. An osseous radiographic landmark for the junction of posterior and anterior urethra is the obturator foramen. The anterior urethra consists of the bulbar urethra, which appears as a widened "cone" from the proximal bulb to penoscrotal junction, and the penile or pendulous urethra, extending from the penoscrotal junction to fossa navicularis.

(meaning "mountain ridge," also known as the seminal colliculus) on the floor of the distal prostatic urethra. The verumontanum appears as a filling defect on RUG, and should not be mistaken for a mass (▶ Fig. 15.32). The utricle is a blind duct in the prostatic urethra, flanked on both sides by the opening to the ejaculatory ducts. Cowper's glands (bulbourethral glands) are paired structures located at the level of the membranous urethra, approximately 2.5-cm long. Retrograde filling of the Cowper's glands can be seen on RUG; the glands drain via the Cowper's duct slightly distal to their origin, terminating in the midline ventral bulbar urethra. The glands of Littre, or periurethral glands, can appear as small, contrast-filled outpouchings along the penile urethra (▶ Fig. 15.31b), and are typically seen in the setting of strictures or inflammation.[29]

Infection and trauma, both often resulting in strictures seen on RUG, are the most common etiologies of male urethral pathology.[30] Infectious strictures are typically beaded and multifocal (▶ Fig. 15.33a). Posttraumatic strictures most often result from straddle injury. Strictures following saddle injury are short segment and affect the bulbar urethra (▶ Fig. 15.33b).

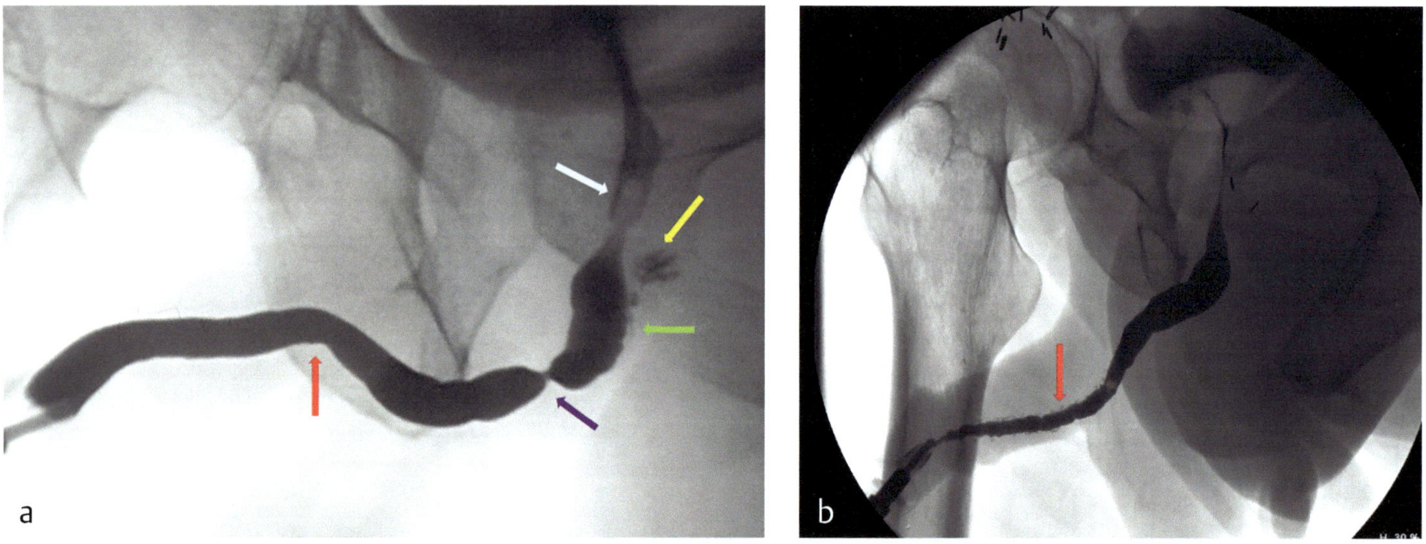

Fig. 15.32 (a) Male urethral structures on retrograde urethrogram (RUG). The verumontanum, on the floor of the distal prostatic urethra, appears as a filling defect, and should not be mistaken for a mass. Cowper's glands (bulbourethral glands), at the level of the membranous urethra, drain distally via Cowper's duct into the midline ventral bulbar urethra. **(b)** Male urethral structures on RUG. The glands of Littre are small, contrast-filled outpouchings located along the penile urethra. Visualization of Littre's glands on RUG is abnormal, and is typically seen in the setting of strictures or inflammation.

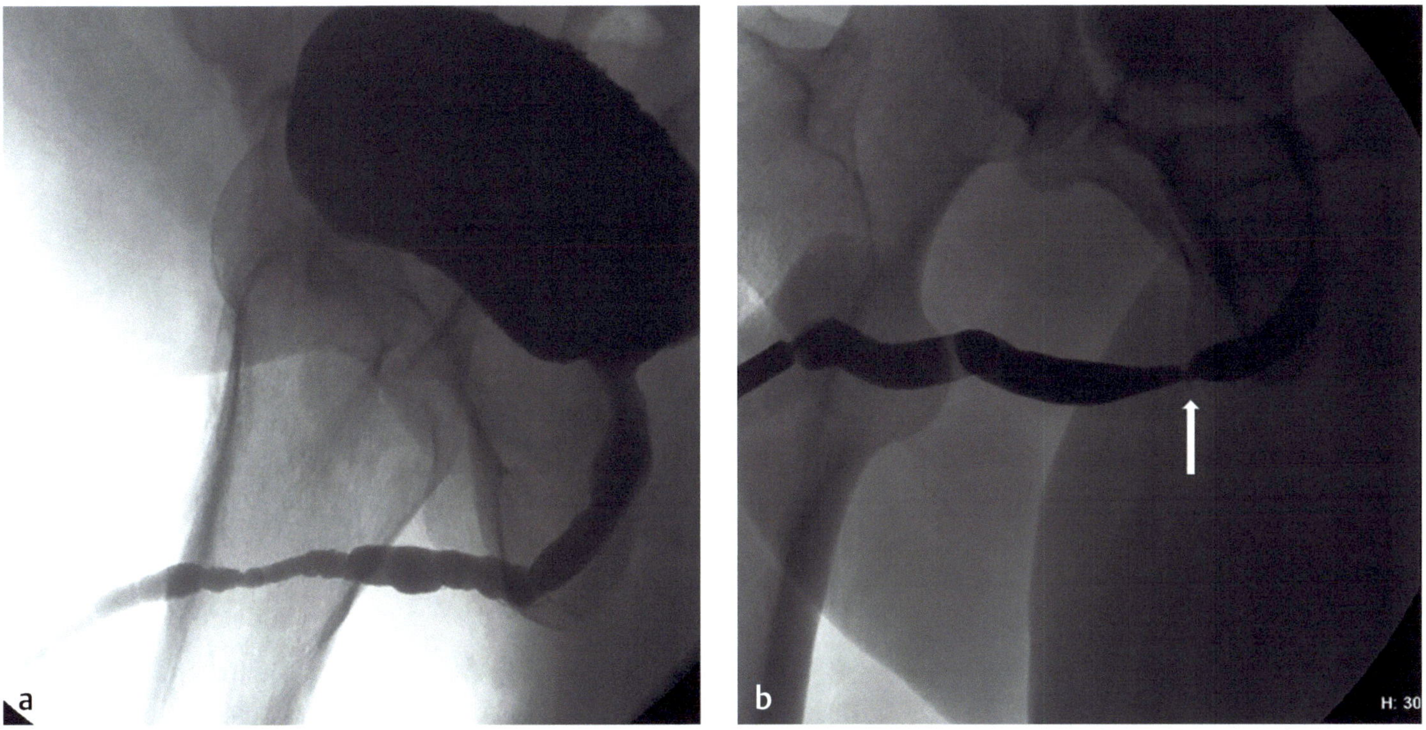

Fig. 15.33 (a) Infectious stricture on retrograde urethrogram (RUG). Typical gonococcal stricture is long and multifocal, with a beaded appearance and affects the anterior (bulbar/penile) urethra. **(b)** Posttraumatic stricture on RUG. Typical stricture following straddle injury is unifocal, short segment, and affects the bulbar urethra.

◆ Urogenital Diaphragm

The urogenital diaphragm is situated inferior to the prostate gland and anterior to the anorectal region. The urogenital diaphragm is actually the deep perineal space which is bounded between the endopelvic fascia and perineal membrane. It should not be confused with the pelvic floor muscles. The urogenital diaphragm is composed of the deep transverse muscle of the perineum and the sphincter urethrae and has extensive attachments to surrounding structures, including the perineal body, external anal sphincter, and bulbospongiosus muscle[31] (▶Fig. 15.34). The membranous urethra traverses through this structure.[32]

◆ Conclusion

In summary, various modalities including ultrasound, MRI, and CT are needed to evaluate the male genitourinary organs. A thorough understanding of normal imaging anatomy is crucial to understanding pathology in this system.

15

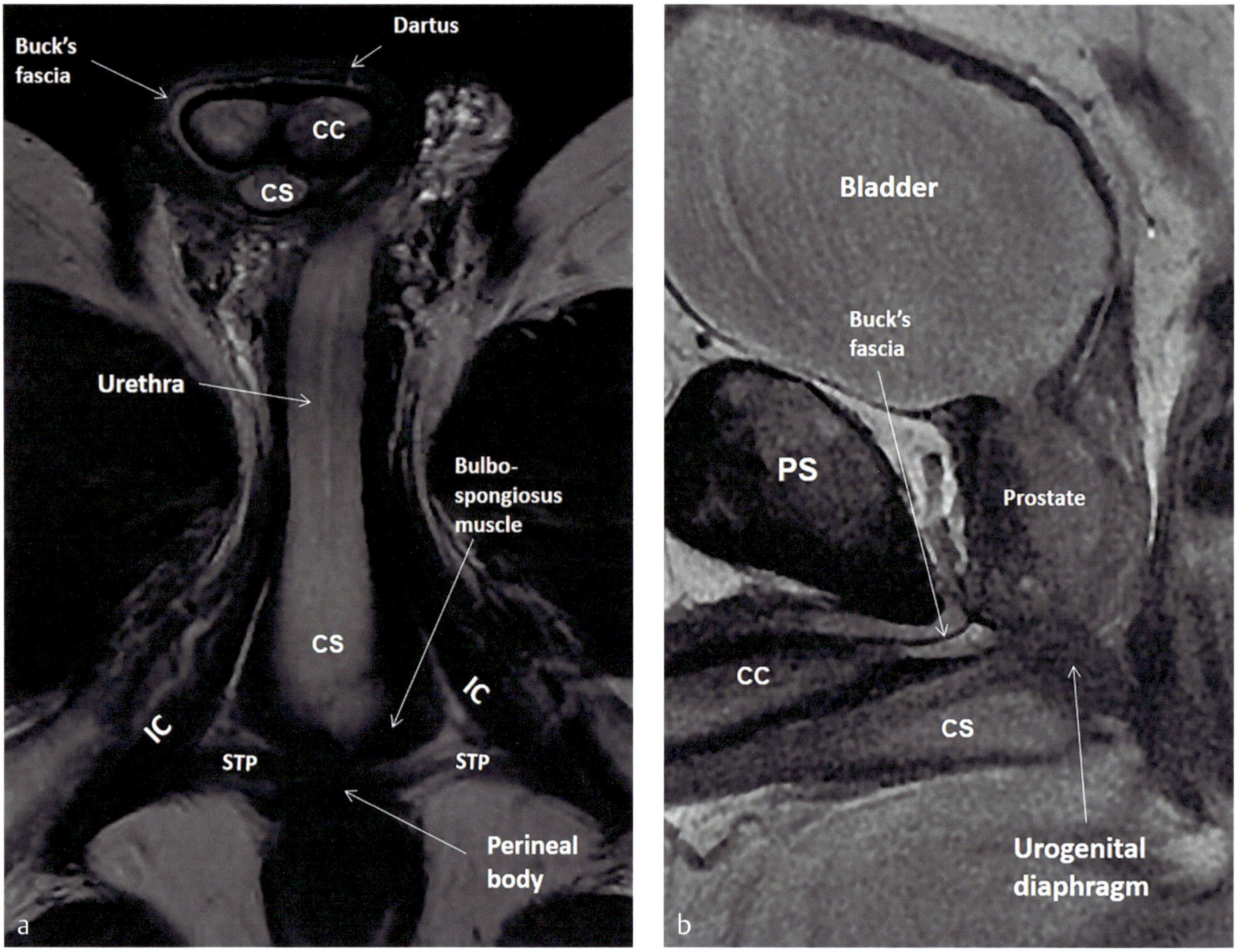

Fig. 15.34 (a) Axial magnetic resonance (MR) image showing the perineum. The perineum is divided into two triangles, anal and urogenital. The urogenital triangle is the area anterior to the superficial transverse perineal (STP) muscles that includes the external genitalia. **(b)** Urogenital diaphragm (inferior fascia of the perineal membrane). The urogenital diaphragm has extensive attachments to surrounding structures, including the perineal body, external anal sphincter, and bulbospongiosus muscle. The penile bulb and crura of the corpora cavernosa also reach to the level of the urogenital diaphragm. CC, corpus cavernosum; CS, corpus spongiosum; IC, ischiocavernosus muscles; PS, pubic symphysis.

References

1. Woodward PJ, Sohaey R, O'Donoghue MJ, Green DE. From the archives of the AFIP: tumors and tumorlike lesions of the testis: radiologic-pathologic correlation. Radiographics 2002;22(1):189–216
2. Virtanen HE, Toppari J. Embryology and physiology of testicular development and descent. Pediatr Endocrinol Rev 2014;11(Suppl 2):206–213
3. Deurdulian C, Mittelstaedt CA, Chong WK, Fielding JR. US of acute scrotal trauma: optimal technique, imaging findings, and management. Radiographics 2007;27(2):357–369
4. Mostafa T, Labib I, El-Khayat Y, El-Rahman El-Shahat A, Gadallah A. Human testicular arterial supply: gross anatomy, corrosion cast, and radiologic study. Fertil Steril 2008;90(6):2226–2230
5. Wishahi MM. Anatomy of the venous drainage of the human testis: testicular vein cast, microdissection and radiographic demonstration. A new anatomical concept. Eur Urol 1991;20(2):154–160
6. Asala S, Chaudhary SC, Masumbuko-Kahamba N, Bidmos M. Anatomical variations in the human testicular blood vessels. Ann Anat 2001;183(6):545–549
7. Paduch DA, Skoog SJ. Current management of adolescent varicocele. Rev Urol 2001;3(3):120–133
8. Dogra VS, Gottlieb RH, Oka M, Rubens DJ. Sonography of the scrotum. Radiology 2003;227(1):18–36
9. Garriga V, Serrano A, Marin A, Medrano S, Roson N, Pruna X. US of the tunica vaginalis testis: anatomic relationships and pathologic conditions. Radiographics 2009;29(7):2017–2032
10. Ragheb D, Higgins JL Jr. Ultrasonography of the scrotum: technique, anatomy, and pathologic entities. J Ultrasound Med 2002;21(2):171–185
11. Puttemans T, Delvigne A, Murillo D. Normal and variant appearances of the adult epididymis and vas deferens on high-resolution sonography. J Clin Ultrasound 2006;34(8):385–392
12. Kim B, Kawashima A, Ryu JA, Takahashi N, Hartman RP, King BF Jr. Imaging of the seminal vesicle and vas deferens. Radiographics 2009;29(4):1105–1121
13. Wilson JD, Griffin JE, Leshin M, George FW. Role of gonadal hormones in development of the sexual phenotypes. Hum Genet 1981;58(1):78–84
14. Bonekamp D, Jacobs MA, El-Khouli R, Stoianovici D, Macura KJ. Advancements in MR imaging of the prostate: from diagnosis to interventions. Radiographics 2011;31(3):677–703
15. Garcia-Monaco R, Garategui L, Kizilevsky N, Peralta O, Rodriguez P, Palacios-Jaraquemada J. Human cadaveric specimen study of the prostatic arterial anatomy: implications for arterial embolization. J Vasc Interv Radiol 2014;25(2):315–322
16. Harvey CJ, Pilcher J, Richenberg J, Patel U, Frauscher F. Applications of transrectal ultrasound in prostate cancer. Br J Radiol 2012;85(Spec No 1):S3–S17
17. Bhavsar A, Verma S. Anatomic imaging of the prostate. Biomed Res Int 2014;2014(4):728539
18. Kiyoshima K, Yokomizo A, Yoshida T, et al. Anatomical features of periprostatic tissue and its surroundings: a histological analysis of 79 radical retropubic prostatectomy specimens. Jpn J Clin Oncol 2004;34(8):463–468
19. Hricak H, Choyke PL, Eberhardt SC, Leibel SA, Scardino PT. Imaging prostate cancer: a multidisciplinary perspective. Radiology 2007;243(1):28–53
20. Kirkham A. MRI of the penis. Br J Radiol 2012;85(Spec No 1):S86–S93
21. Chapple CR, Steers WD. Practical Urology Essential Principles and Practice. New York; London: Springer; 2011
22. Yiee JH, Baskin LS. Penile embryology and anatomy. ScientificWorldJournal 2010;10:1174–1179
23. NIH Consensus Development Panel on Impotence. NIH Consensus Conference. Impotence. JAMA 1993;270(1):83–90
24. Schwartz BG, Kloner RA. Cardiology patient page: cardiovascular implications of erectile dysfunction. Circulation 2011;123(21):e609–e611
25. Patel DV, Halls J, Patel U. Investigation of erectile dysfunction. Br J Radiol 2012;85(Spec No 1):S69–S78
26. Wilkins CJ, Sriprasad S, Sidhu PS. Colour Doppler ultrasound of the penis. Clin Radiol 2003;58(7):514–523
27. Spiliopoulos S, Shaida N, Katsanos K, Krokidis M. The role of interventional radiology in the diagnosis and management of male impotence. Cardiovasc Intervent Radiol 2013;36(5):1204–1212
28. Ingram MD, Watson SG, Skippage PL, Patel U. Urethral injuries after pelvic trauma: evaluation with urethrography. Radiographics 2008;28(6):1631–1643
29. McCallum RW, Rogers JM, Alexander MW. The radiologic assessment of iatrogenic urethral injury. J Can Assoc Radiol 1985;36(2):122–126
30. Mundy AR. Management of urethral strictures. Postgrad Med J 2006;82(970):489–493
31. Kureel SN, Gupta A, Gupta RK. Surgical anatomy of urogenital diaphragm and course of its vessels in exstrophy-epispadias. Urology 2011;78(1):159–163
32. Stoker J, Halligan S, Bartram CI. Pelvic floor imaging. Radiology 2001;218(3):621–641

15

16 Female Genital System

Farhood Saremi and Mojgan Motevali

◆ Introduction

The female genital (reproductive) system consists the external genitalia (vulva) and internal organs. The internal organs include the vagina, cervix, uterus, uterine adnexa, and ovaries (▶Fig. 16.1, ▶Fig. 16.2, ▶Fig. 16.3). The uterine adnexa contain ligament, uterine tubes, and vessels as well as the ovaries in some classifications. Successful imaging evaluation of this system requires detailed knowledge of anatomy to distinguish pathologies. Ultrasonography is currently the primary modality to evaluate this system specifically for the study of uterus, endometrium (i.e., polyps, pregnancy), and ovaries. Whereas magnetic resonance imaging (MRI) and computed tomography (CT) are used more for the evaluation of the uterine myometrium (i.e., uterine contour, fibroids), ovaries, tumor staging, and study of adjacent structures. Hysterosalpingography (HSG) is used to study the fallopian tubes and endometrial cavity.

◆ Embryology

At the time of conception, the oocyte provides an X chromosome, and the sperm cell carries either an X or Y chromosome. During fertilization, an XX female or XY male determines the genetic sex. The sex-determining region gene (SRY) on the Y chromosome encodes a protein known as testis-determining factor (TDF) which induces a male phenotype. In the absence of SRY, a female phenotype develops. Notably, the genes from both X chromosomes are needed for female development. The sexual phenotype (appearance of external genitalia) is affected by hormonal expression. The female phenotype is the default, but the male

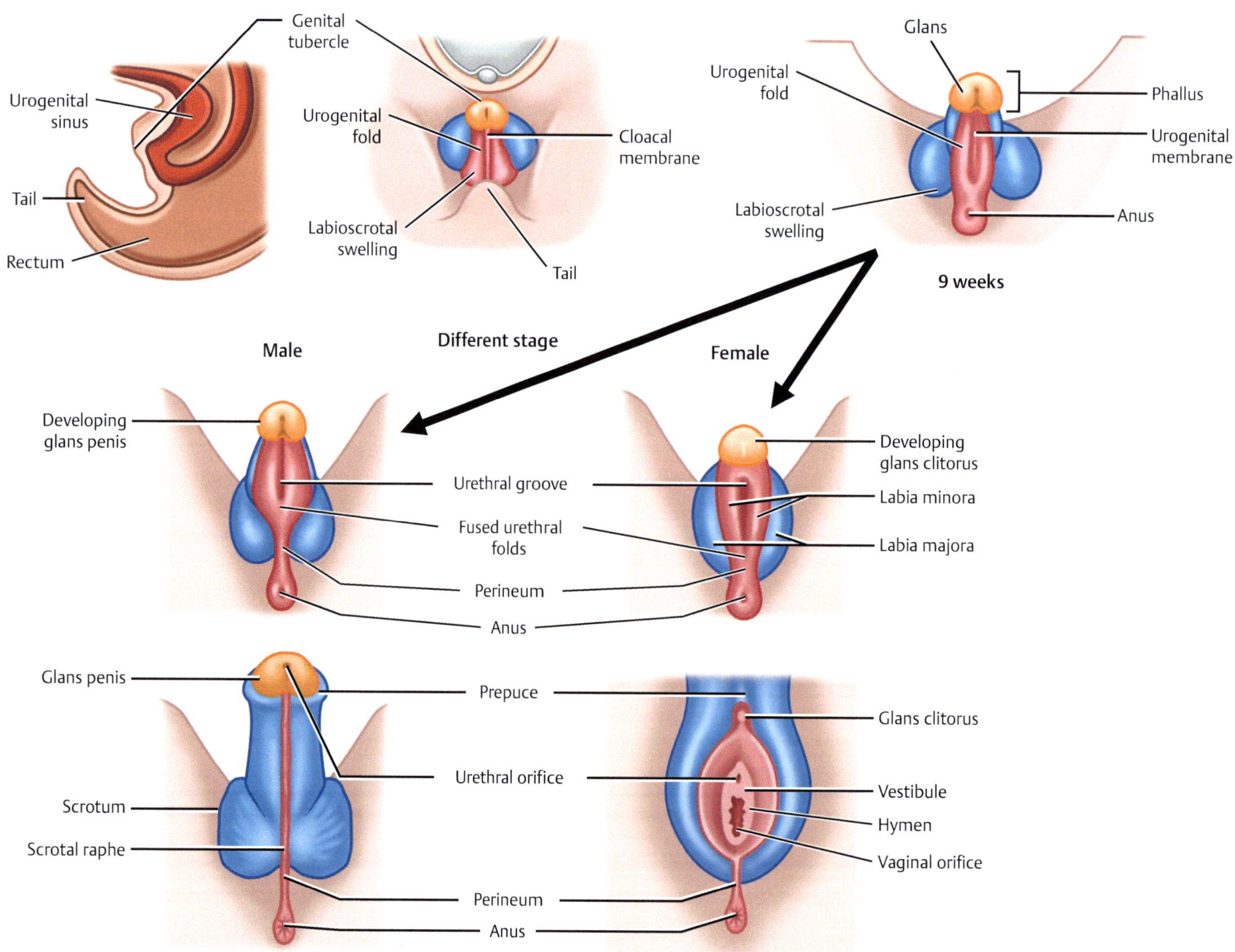

Fig. 16.1 External genitalia development. The undifferentiated external genitalia consists of five mesenchymal swellings: one genital tubercle, two urogenital (urethral) folds, and two labioscrotal folds. In female, the genital tubercle develops into the clitoris, and the urogenital and labioscrotal folds develop into the labia minora and labia majora, respectively.

16

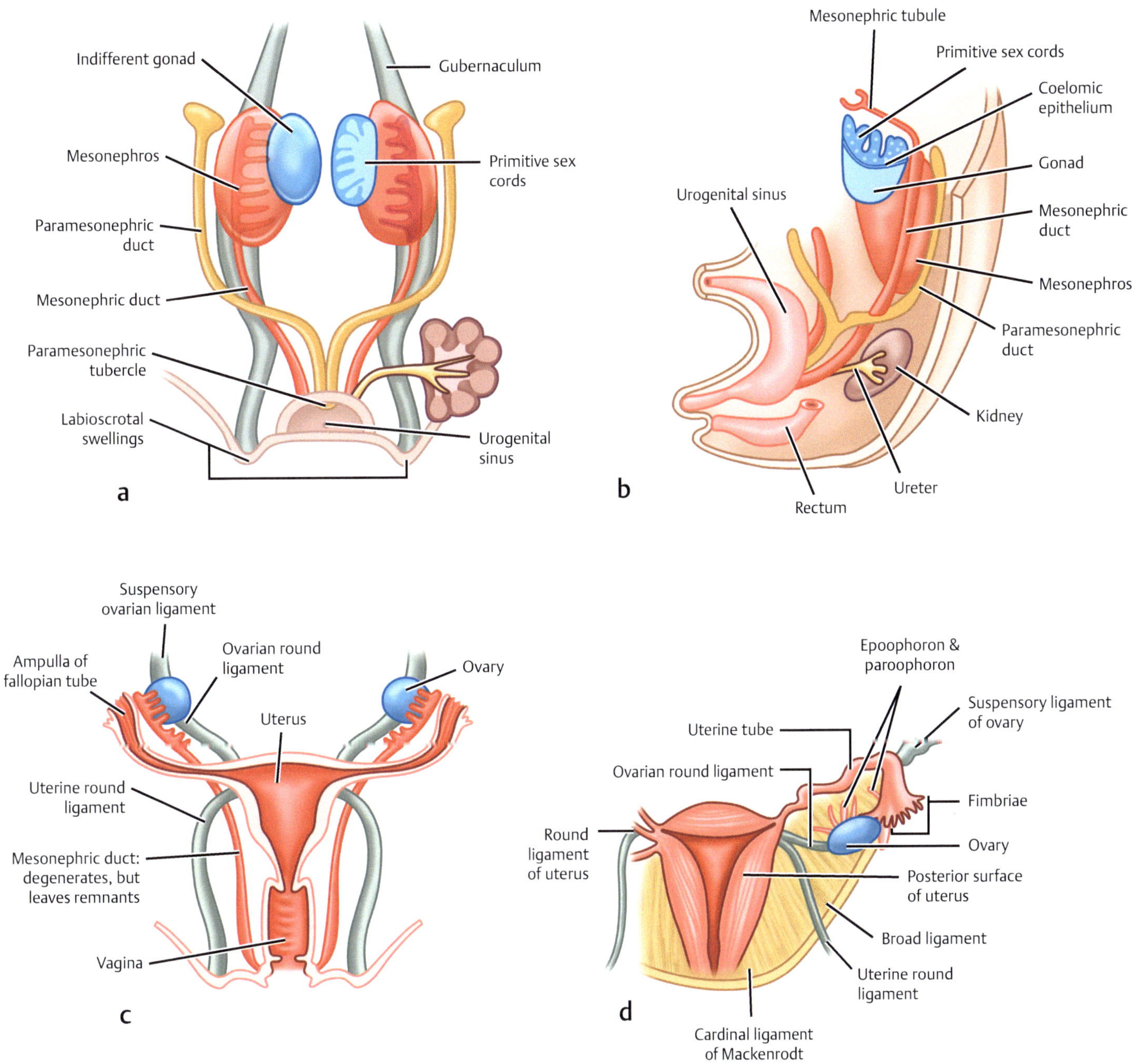

Fig. 16.2 Female internal genitalia and ligamentous development. **(a, b)** The lower portion of the paramesonephric duct fuses at the midline, forming the uterus and vagina. The upper portion of the paramesonephric duct does not fuse and forms the Fallopian tube. **(c, d)** The gubernaculum divides into three segments that form the suspensory ovarian ligament, ovarian round ligament, and uterine round ligament. The broad ligament covers the entire uterus and expands as the developing uterus descends through the pelvis, pulling epithelium and mesenchyme from the lining of the body cavity.

phenotype requires expression of the testosterone by the interstitial Leydig cells and Müllerian-inhibiting factor (MIF) by the Sertoli cells.[1,2]

External Genitalia

The external genitalia in the two sexes develop from common primordia, namely, the genital tubercle, folds, and swellings. Development of the female phenotype or the appearance of external genitalia requires two X chromosomes. Consequently, X chromosome causes the ovarian development from the primary indifferent gonad. Then ovaries determine the type of sexual differentiation that occurs in the genital ducts and external genitalia.

The external genitalia are differentiated by the 9th week of fetal life and can be completed by the 12th week. In the female, lack of testes and low level of androgen lead to the development of the lower vagina and labia. Maternal and to a lower extent fetal estradiol contribute to feminization of the genitalia.

In the 4th week of gestation, mesenchymal protuberances in the ectoderm of cloacal membrane form the external genitalia (▶ Fig. 16.1):

- A single genital tubercle becomes the clitoris.

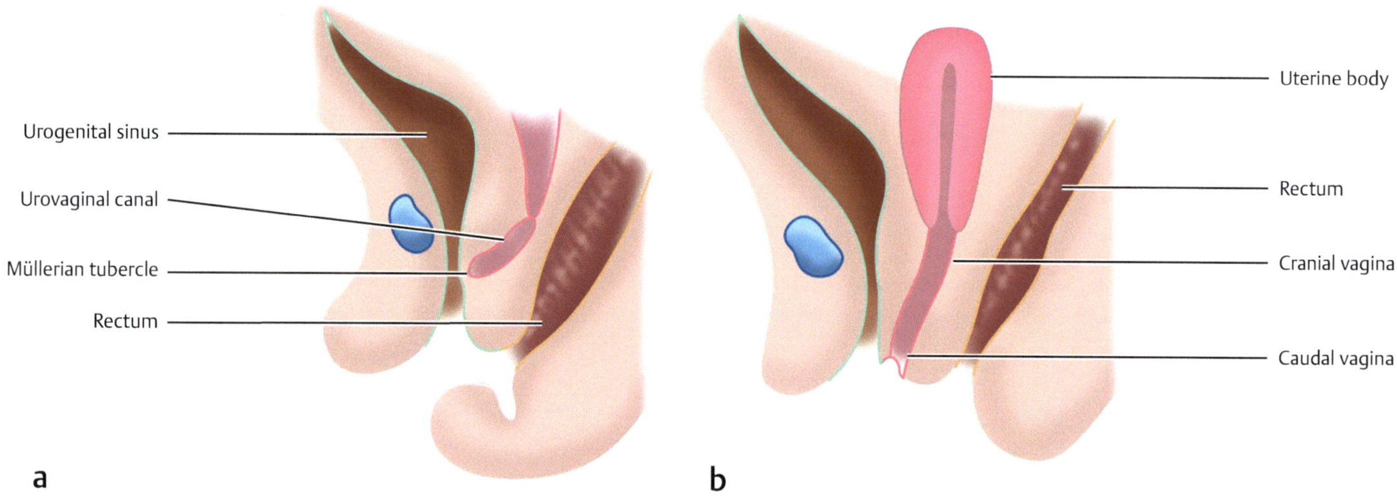

Fig. 16.3 Distal vagina formation. The caudal vagina forms due to caudal extension of the Müllerian duct. **(a)** The caudal end of the Müllerian duct inserts into the urogenital sinus (UGS) wall. The caudal end of the Müllerian duct is solidified to form the Müllerian tubercle (the caudal vaginal primordium) which extends caudally and reaches the perineum **(b)**. The vaginal primordium reverts to generate the epithelium and eventually attains canalization. Therefore, the definitive vagina is formed by the Müllerian ducts. (Modified from Cai.[4])

- Two urogenital folds that are partially fused develop the labia minora.
- Two labioscrotal folds develop the labia majora.

Genital Ducts

The paramesonephric (Müllerian) ducts form the female internal genital tract and mesonephric (Wolffian) ducts develop to form the male genital ducts[1,2] (▶Fig. 16.2). Although these two duct systems consist of different cell types, both originate from the intermediate mesoderm. The paramesonephric ducts develop after the mesonephric ducts. The cloacal membrane divides to form the urogenital orifice and the anus. The gubernaculum is a ligament which develops between the gonads and the labioscrotal swellings. In male, under the influence of the anti-Müllerian hormone (AMH), which is secreted from the Sertoli cells, testosterone and androstenedione stimulate mesonephric duct development to form the epididymis, vas deferens, and seminal vesicle, while the paramesonephric duct only remains as the rudimentary appendix testis (hydatid of Morgagni) and the prostatic utricle. In both sexes, the ureteric bud arises posteriorly from the distal end of mesonephric duct to form the definitive kidney. In female, each paramesonephric duct grows as longitudinal solid cord dorsolateral to the mesonephric tubules. Subsequently, each cord canalizes to form a lumen (▶Fig. 16.2, ▶Fig. 16.3). The proximal portion of the paramesonephric duct forms an opening into the future peritoneal cavity known as infundibulum of the uterine tube. At midline, the paramesonephric duct fuses with the contralateral duct to form a canal (uterovaginal canal) that will become the Y-shaped primary uterus and the proximal vagina (▶Fig. 16.2). The two ducts are initially separated by a septum, but later the septum vanishes to form the uterus. Vaginal development has been a long-standing controversy.[3,4] It is an old concept that the rodent vagina has a dual origin: the cranial part is derived from the Müllerian duct (Müllerian vagina) and the caudal part derived from the urogenital sinus (sinus vagina). New observations from genetic and molecular studies contradict the urogenital sinus origin of the sinus vagina. The distal end of the Wolffian and Müllerian ducts are fused to the urogenital sinus by the sinus ridges which in later stages are separated from the sinus by lateral furrows. Wolffian ducts and sinus ridges regress and the solid caudal vaginal primordium reaches the perineum. The vaginal primordium reverts to generate the epithelium and eventually attains canalization. Therefore, the definitive vagina is formed by the Müllerian ducts (▶Fig. 16.3).

Once the Müllerian ducts have fused distally in the midline, a broad transverse peritoneal fold (broad ligaments) is formed extending from the lateral sides of the fused paramesonephric ducts toward the lateral walls of the bony pelvis with the ovaries located on its posterior surface.

The gubernaculum in female forms the round ligament of the uterus and suspensory ligaments of the ovary. In female, the mesonephros degenerates entirely, although vestigial structures such as Gartner's ducts, the epoophoron, and paroophoron (rudimentary tubules in the broad ligament) are common. The epoophoron or organ of Rosenmüller is a parovarian remnant of the mesonephric tubules (homologous to the epididymis in the male) that may grow to form a cyst (▶Fig. 16.2). In males, distal mesonephros tubules give rise to the efferent ductules of the testis, epididymis, vas deferens, and seminal vesicle.

Gonads

In both sexes, the gonads arise in the gonadal ridges. The gonadal ridges are ridge-like protrusions that are formed in the 5th

16

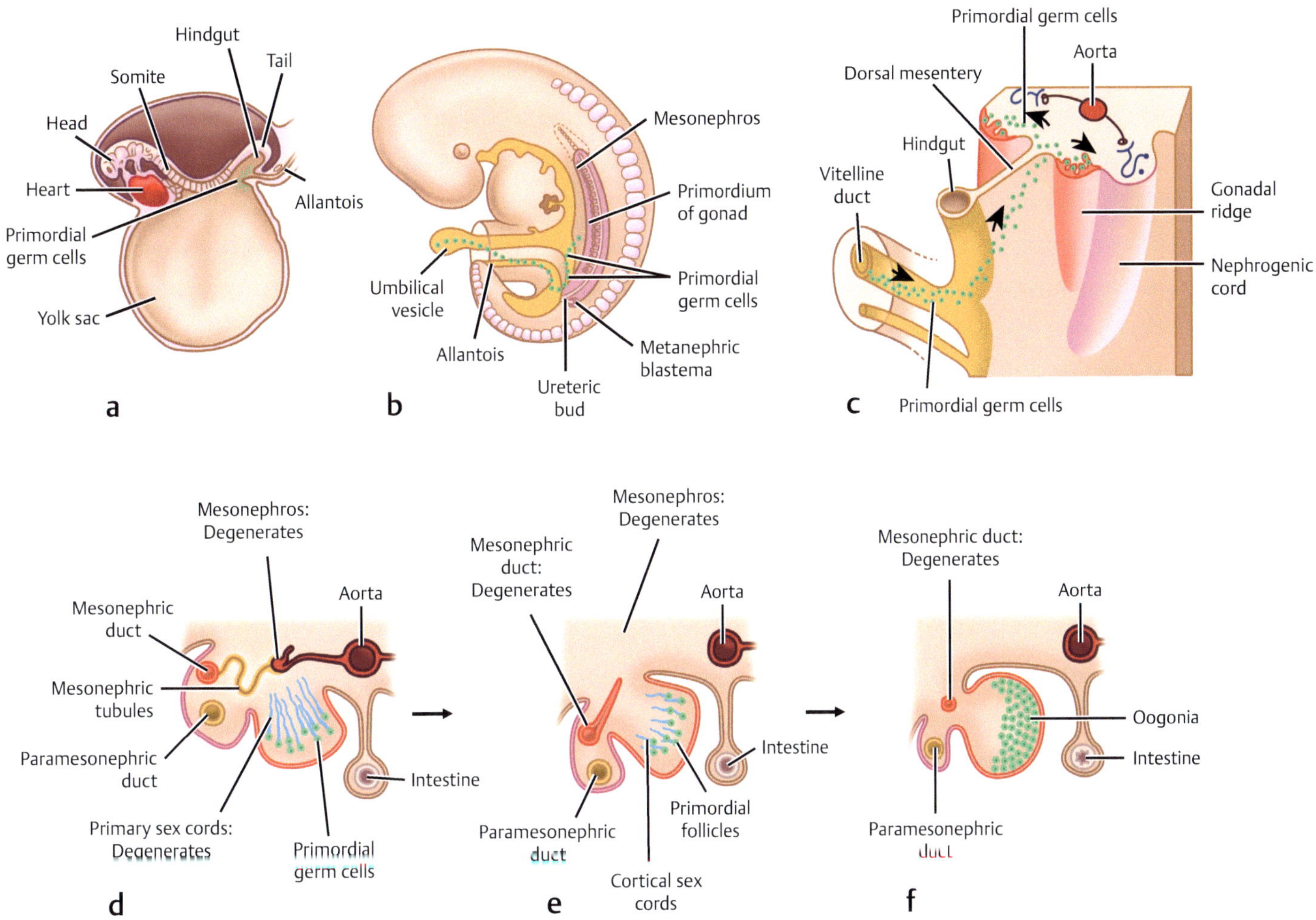

Fig. 16.4 (a–c) Development of the ovaries. For both sexes the gonads arise in the gonadal ridges. In the 6th week, the primordial germ cells immigrate from the umbilical vesicle into both gonadal ridges through the dorsal mesentery of the hindgut. **(d–f)** The primordial germ cells will be surrounded by the primary sex cords arising from the coelomic epithelial cells. These primary cords will be degenerated in female. The primordial germ cells are the precursor of the oogonia. They migrate into the cortical sex cords and form the primordial follicles and later differentiate into oogonia. As the ovaries develop, the mesonephros (nephrogenic cord) degenerates. The mesonephros is the embryonic kidney that functions between the 6th and 10th weeks of embryological life.

week through the proliferation of the coelomic epithelium and the thickening of the underlying mesenchyme. They are located medial to the nephrogenic cord and represent the primitive gonadal primordium (▶Fig. 16.3). The X chromosome has genes that contribute to ovarian development. Between the 4th and 6th weeks, primordial germ cells immigrate to the gonadal ridges through the dorsal mesentery of the hindgut (▶Fig. 16.4). The germ cells proliferate in the gonadal ridge located medial to the developing mesonephros (primary kidney). Three cell types are embedded in the gonadal ridge: the mesenchymal cells that will develop to ovarian stroma, the mesothelial cells that will transform to ovarian follicles, and the primordial germ cells which are the precursor of the oogonia.[1]

◆ Anatomy

External Genitalia

The female external genitalia or vulva includes the following structures: mons pubis, labia majora, labia minora, prepuce, frenulum, clitoris, vestibule, orifice of urethra and vagina, vestibular bulbs, Bartholin's glands, and paraurethral Skene's glands (▶Fig. 16.4, ▶Fig. 16.5, ▶Fig. 16.6, ▶Fig. 16.7, ▶Fig. 16.8, ▶Fig. 16.9).

Mons Pubis

The mons pubis is the hairy area of skin and fat tissue over the pubic symphysis. This area is flat and hairless before the puberty. With puberty the area becomes prominent with fat and covered with coarse hair.

16

Labia Majora

The labia majora (major) include two anteroposterior folds of skin that extend between the mons pubis and the perineum and cover the vulva. Each labium majora is lined externally with squamous epithelium of the skin and hair. Their internal surface contains sebaceous glands. The subcutaneous layer consists of fat and Colles' fascia. The round ligament of the uterus terminates in the labia major. The labia join anteriorly and below the mons

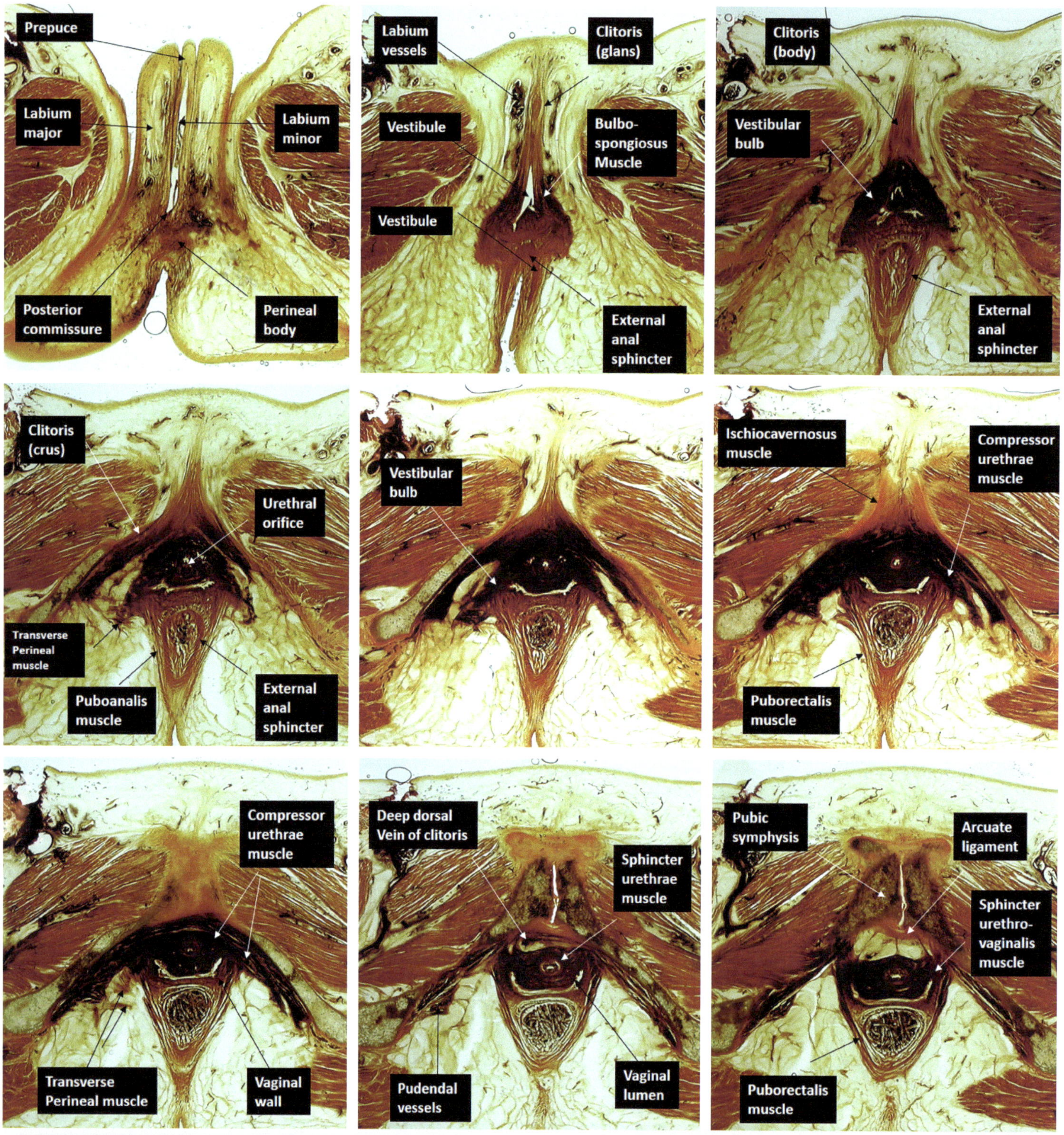

Fig. 16.5 Female external genitalia (vulvar region) are shown by axial cuts in cadaveric specimen. Note tight connection between the urethra and the anterior wall of the vagina whereas loose connective tissue exists between the rectal and vaginal walls.

pubis to form the anterior commissure and posteriorly to form the posterior commissure. The perineum is located posterior to the vulva and anterior to the anus (▸Fig. 16.1).

Labia Minora

The labia minora (minor) are two small cutaneous folds that lie between the labia majora and anterior to the hymen that enclose the vulvar vestibule and the urethra (▸Fig. 16.5). The labia minora meet anteriorly to form the prepuce which covers the clitoris. Posterior confluence of the labia below the vaginal orifice forms the posterior commissure of the labia minora or fourchette (▸Fig. 16.5). The internal surface of the labia minora also contains many sebaceous glands. Increased thickness of the labia minora occurs during sexual arousal due to increased vascularity and engorgement with blood.

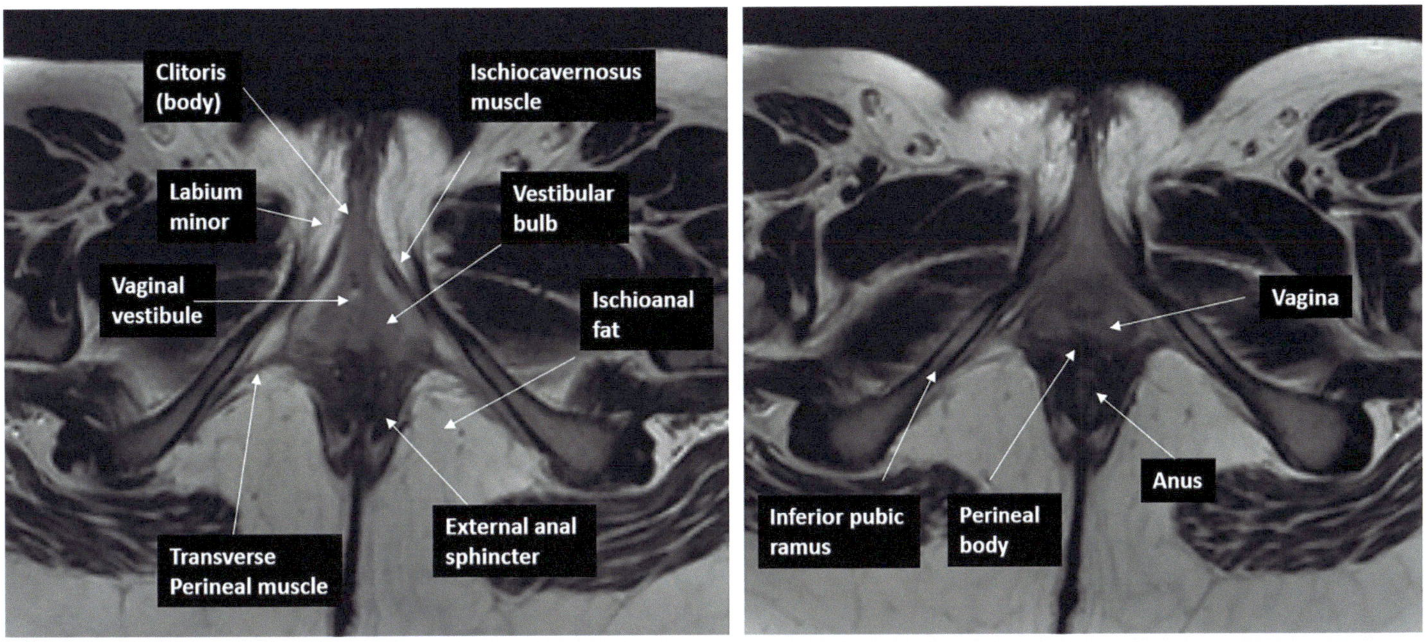

Fig. 16.6 Female perineal region shown by axial T1-weighted (T1W) magnetic resonance imaging (MRI).

Vestibule

The vestibule is the area located medial to the labia minora that extends between the clitoris and the fourchette of labia minora. It contains the external urethral orifice, the vaginal orifice, the openings of the two Bartholin's glands, and the Skene's ducts. The external urethral orifice is immediately anterior of that of the vagina and about 2.5 cm behind the clitoris (▶Fig. 16.7). The vestibular bulbs (bulbs of the vestibule) are two small piriform erectile bodies on each side of the vestibule, beneath the bulbospongiosus muscles and superficial to the inferior fascia (perineal membrane) of the urogenital diaphragm (▶Fig. 16.5, ▶Fig. 16.6, ▶Fig. 16.7, ▶Fig. 16.8, ▶Fig. 16.9, ▶Fig. 16.10). They are about 2.5 to 3 cm in length and are composed of aggregated venous plexuses. Anteriorly, the vestibular bulbs communicate at the base of clitoris and form the bulbar commissure (▶Fig. 16.8). Posteriorly, they are delimited by the Bartholin's gland. The vestibular bulbs are homologous to the bulb of penis and adjoining part of the corpus spongiosum. The vestibular bulbs cause vaginal orgasmic contractions during sexual arousal through the rhythmic contraction of the bulbocavernosus muscles. The Bartholin's glands consist of two small glands located around the vaginal orifice at the posterior end of the vestibular bulb. The vestibular glands open into the vestibule by 1- to 2-cm-long ducts situated in the inferior aspect of the cleft between the hymen and the labium minora.

Clitoris

The clitoris is located under the prepuce at the top of the labia minora (▶Fig. 16.5, ▶Fig. 16.6, ▶Fig. 16.7, ▶Fig. 16.8, ▶Fig. 16.9). It develops from the genital tubercle of the embryo and is embryologically homologous to the male glans and corpora cavernosa. Its components include paired crura (and roots), paired corpora (body), and a glans. The glans is a small round structure at the tip of clitoris containing spongy nonerectile tissue which is connected to the body. The clitoris body contains two erectile corpora cavernosa that are enveloped in a fibrous membrane with a midline incomplete septum and each is attached to the ischiopubic ramus via a crus that extends from the root of clitoris. The crus is located under the ischiocavernosus muscle. The clitoris is attached to the symphysis pubis by a midline fibrous suspensory ligament.

Vessels

The arterial blood supply of the vulva is derived from the superficial and deep external pudendal branches of the superficial femoral artery superiorly and the internal pudendal artery inferiorly (▶Fig. 16.11). Venous drainage of the vulvar skin is via external pudendal veins to the great saphenous vein. Venous drainage of the clitoris is via deep dorsal veins to the internal pudendal vein, and via superficial dorsal veins to the external pudendal and great saphenous veins.

The lymphatic channels of vulva are drained to the superficial inguinal lymph nodes. The lower part of the labia majora drains to the rectal lymphatic plexus. The clitoris is drained to the superficial inguinal, deep inguinal, and internal iliac nodes.

Innervation

The main nerve supply of the vulva is the pudendal nerve (S2, S3, and S4) through its inferior rectal and perineal branches and the dorsal nerve of the clitoris (▶Fig. 16.7, Chapter 17 "Perineum"). The sensory nerve to the anterior third of labia majora is supplied by the ilioinguinal nerve (L1), the posterior two-third is supplied by the pudendal nerve through the posterior labial branches of the perineal nerve (S3), and the lateral aspect is innervated by the perineal branch of the posterior cutaneous nerve of the thigh (S2).

Vagina

The vagina is a fibromuscular tube that extends from the cervix to the vulvar vestibule (▶Fig. 16.5, ▶Fig. 16.6, ▶Fig. 16.7, ▶Fig. 16.8,

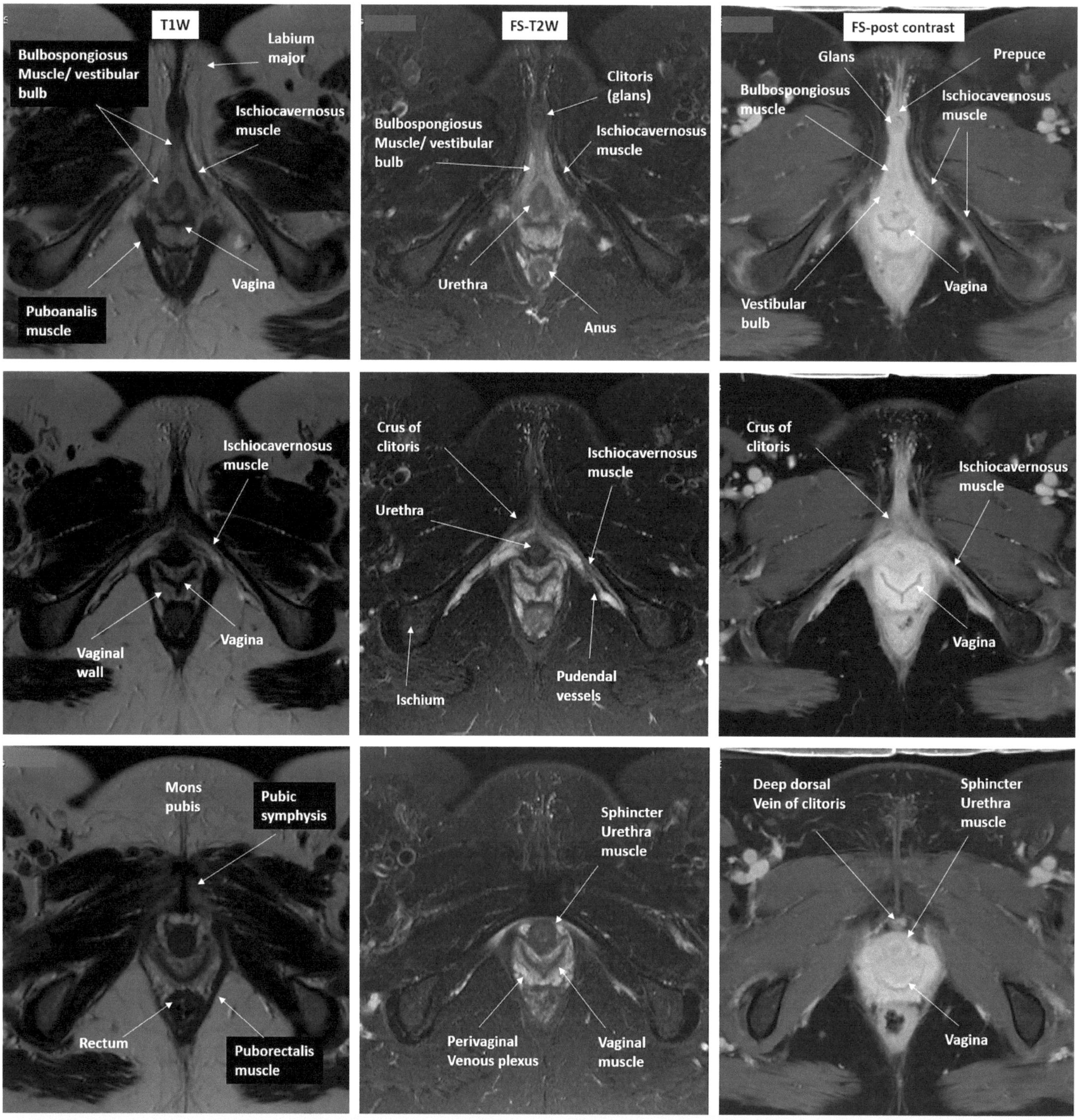

16

Fig. 16.7 Female external genitalia (vulvar region) are shown by T1-weighted (T1W), fast suppressed (FS)-T2W and postcontrast, axial magnetic resonance (MR) images. Vessels and mucosal tissues are bright on T2W and postcontrast images. The female external genitalia include mons pubis, labia major, labia minor, prepuce, frenulum, clitoris, vestibule, orifice of urethra and vagina, vestibular bulbs, Bartholin's glands, and paraurethral Skene's glands. Note vascular structures including bulb of vestibule, crura of the clitoris, and mucosa enhance along with vessels and venous plexuses.

▶Fig. 16.9). Hymen is a mucosal membrane located at the inferior orifice of vagina. There is a circumferential recess at the junction of the cervix and vagina known as fornix (▶Fig. 16.12). It is divided into anterior, posterior, and lateral fornices. The length of vagina is about 7 to 9.5 cm and it is divided into upper third, middle third, and lower third. The posterior wall is longer and ends in the posterior fornix and the anterior wall is shorter terminating in the anterior fornix[5] (▶Fig. 16.12). The vagina is located in the middle compartment of the pelvis.

The vaginal wall is composed of three layers, the mucosa, the muscularis, and the adventitia. The vaginal epithelium is derived in its entirety from the Müllerian duct epithelium. The mucosa consists of nonkeratinized stratified squamous epithelium without glands and an underlying lamina propria of connective tissue.

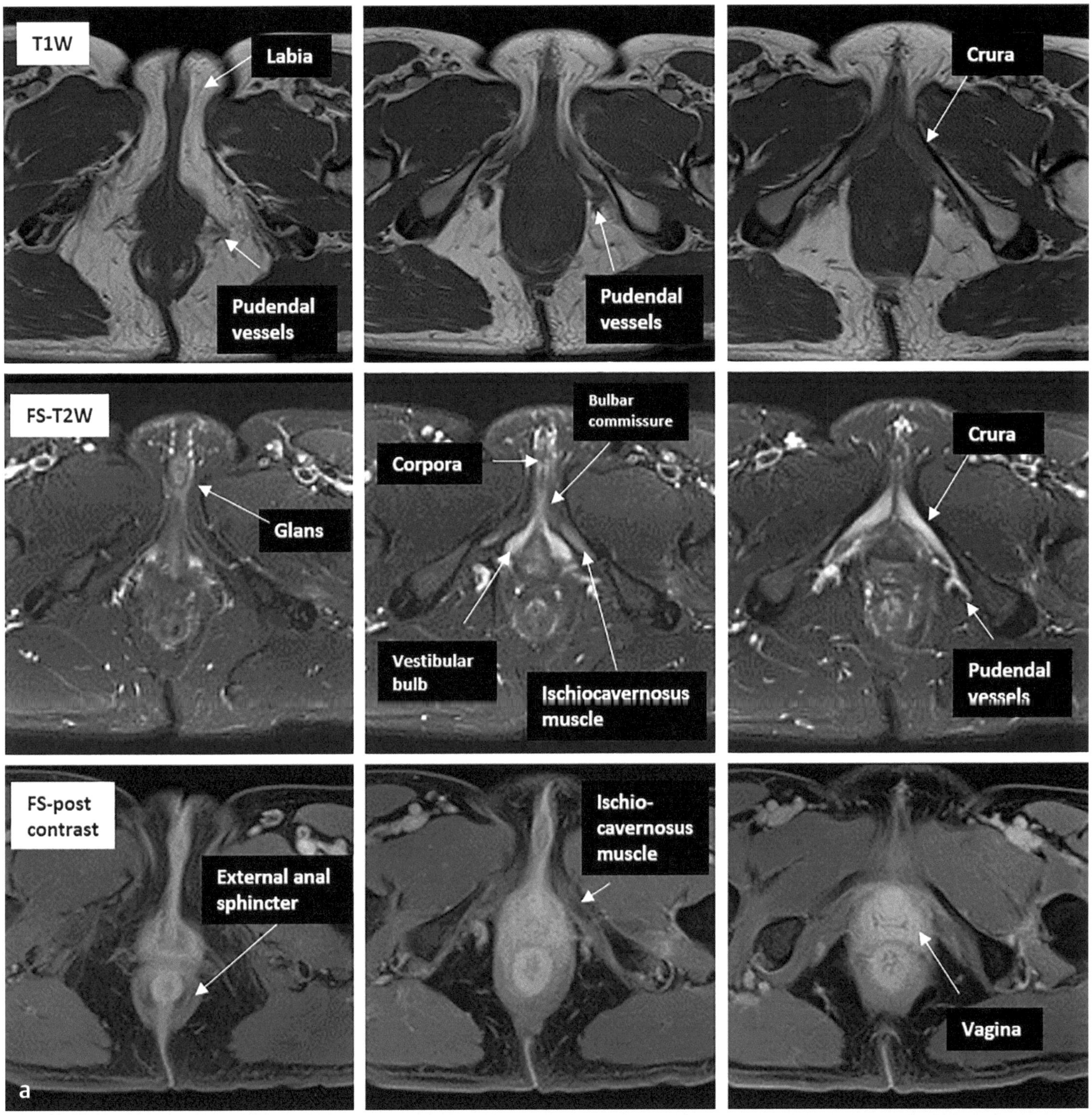

Fig. 16.8 Clitoral apparatus shown by axial **(a)** and coronal **(b)** T1-weighted (T1W), T2-weighted (T2W), fast suppressed (FS)-T2W, and postcontrast magnetic resonance (MR) images. Vessels and mucosal tissue are bright on T2W and postcontrast images. Note that glans is vascular and enhancing but it remains lower signal on T2W views compared to the rest of clitoral structures probably because of relative lack of venous lakes in it. Some believe that glans is not an erectile part of the clitoris.

(Continued)

It is lubricated by secretions from the cervical and Bartholin's glands. The muscularis layer consists of smooth muscle and connective tissues (collagen and elastin). The adventitia is continuous with the endopelvic fascia. Nerves and vasculature are concentrated in the lamina propria.

Anterior to the vagina are the urethra and neck-trigone area of the bladder, and posterior to the vagina are the anal canal and lower rectum (▶Fig. 16.10, ▶Fig. 16.12). Adipose and loose fibro-connective tissue separate the urethra from the anterior vagina. The rectouterine peritoneal pouch is located in the space between the upper third of vagina and rectum. The Denonvilliers' fascia is seen between the rectum and the middle third of vagina. The inferior third is separated from the anal canal with a fibromuscular membrane (▶Fig. 16.13).

The vagina is supported by several ligaments. In the upper one-third of the vagina, the vaginal walls are connected laterally and dorsally by the cardinal and vaginal portions of the uterosacral ligament. The lower one-third of the vagina is fused with the

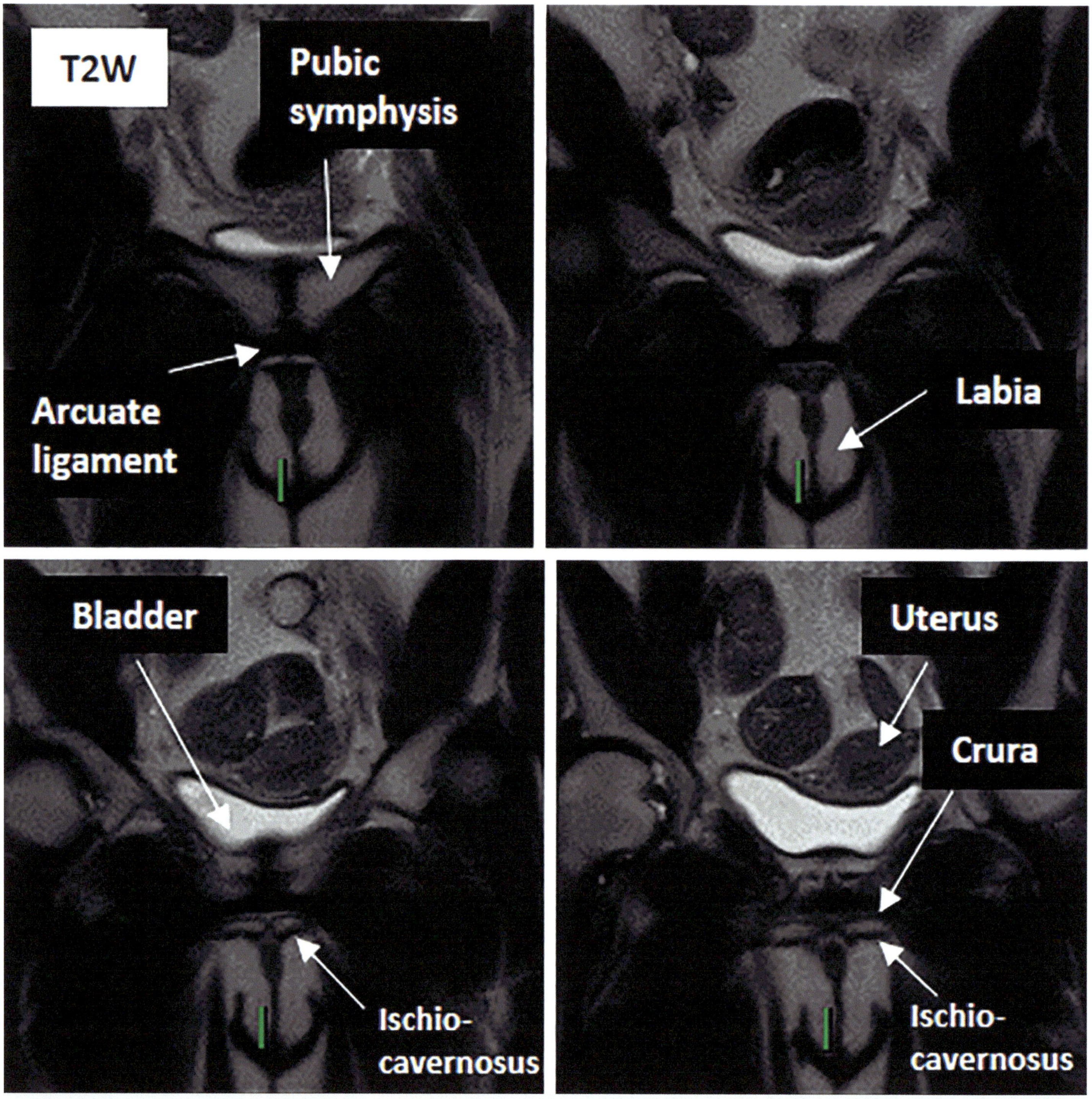

Fig. 16.8 (*Continued*)

(*Continued*)

perineal body. The midvaginal wall is connected to the endopelvic fascia that merges with the arcus tendineus fascia of the pelvis. The cardinal ligaments (lateral cervical at the base of the broad ligament of the uterus) originate from the greater sciatic foramen and insert into the lateral fascia of the cervix and adjacent vaginal wall (▶Fig. 16.14). The uterosacral ligaments originate from the sacral vertebrae and insert into the lateral vaginal fornices (▶Fig. 16.14). Vaginal support is also provided by the levator ani, and the pubocervix, pubouterine, and pubovaginal ligaments. Contraction of the levator ani muscle squeezes the midurethra, distal vagina, and rectum against the pubic bone. This action closes the vaginal orifice and constricts the urogenital hiatus through which prolapse occurs.

Imaging study of the vagina can be performed with CT, MRI, and ultrasound.[7] In axial cross section, the normal vagina is seen as a collapsed, H-shaped tubular structure with a central enhancing mucosa especially in premenopausal females (▶Fig. 16.14, ▶Fig. 16.15). The imaging appearance of the vagina changes with patient's age and the phase of the menstrual cycle. The vaginal wall and central mucosa are thickest during the midsecretory phase of the menstrual cycle. In postmenopausal women who are not undergoing hormone replacement therapy, the vaginal

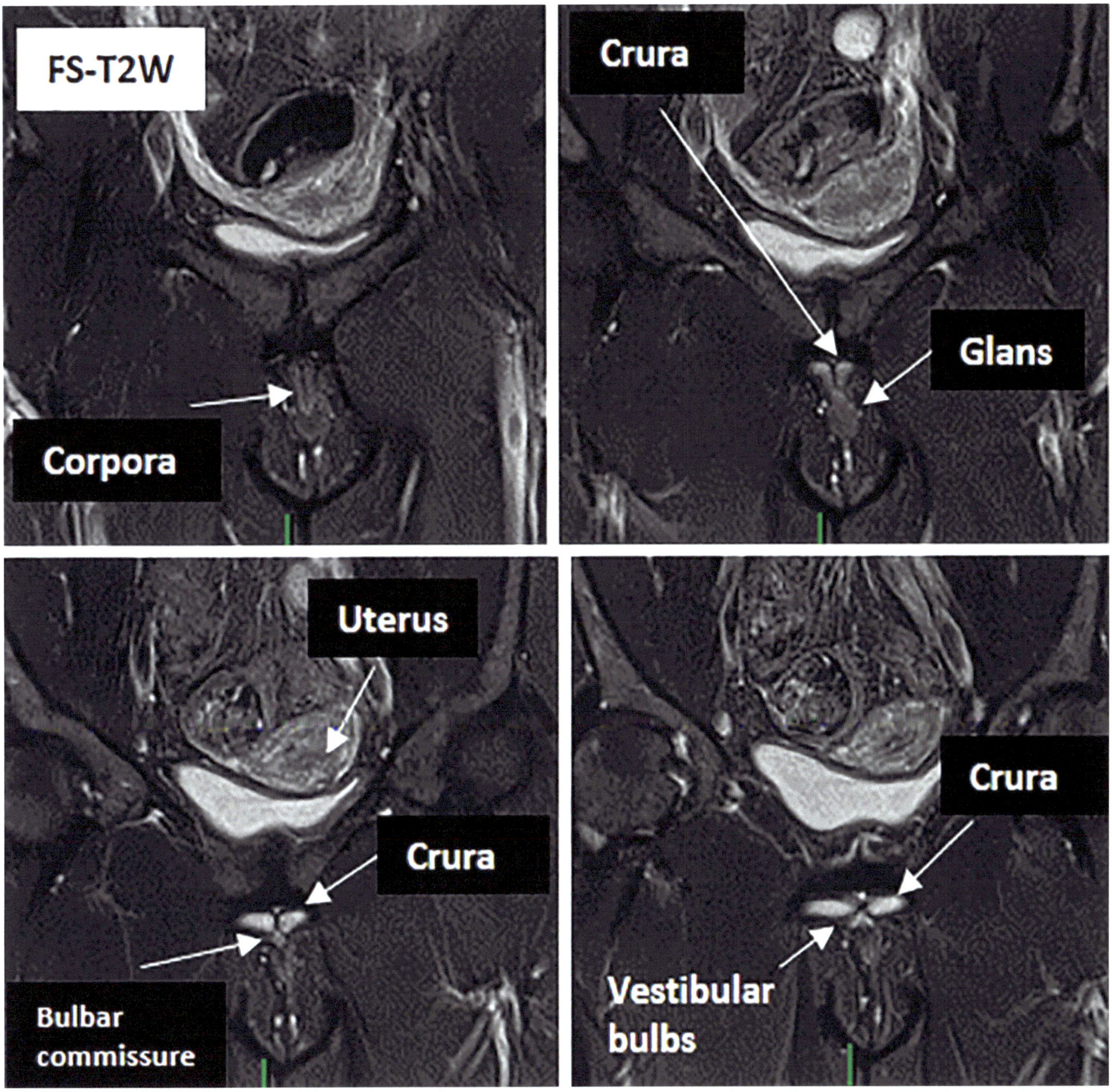

Fig. 16.8 *(Continued)*

(Continued)

wall is less distinct and the mucous layer is thinned.[5] Thickening of vaginal wall is not uncommon and in part could be due to engorgement of the perivaginal venous plexus. Engorgement of the perivaginal and parauterine venous plexuses is common in nulliparous women with ovarian vein insufficiency (▶Fig. 16.15). After hysterectomy atrophy of vagina is a common finding (▶Fig. 16.16).

Vascularity

The blood supply to vagina derives from the branches of the internal iliac artery, including the vaginal artery, the descending branch of the uterine artery, the internal pudendal artery, and the middle rectal artery. The vaginal veins on each side drain to the internal iliac veins (▶Fig. 16.17). The lymphatic drainage of the upper vagina is to the internal iliac and external iliac nodes, and the lower vagina to the superficial inguinal lymph nodes.

Innervation

The upper vagina is supplied via the pelvic plexus with input from the hypogastric plexus, prevertebral ganglia, and sacral nerves. The lower vagina is innervated by the pudendal nerve. The distal third of the anterior vaginal wall has more small

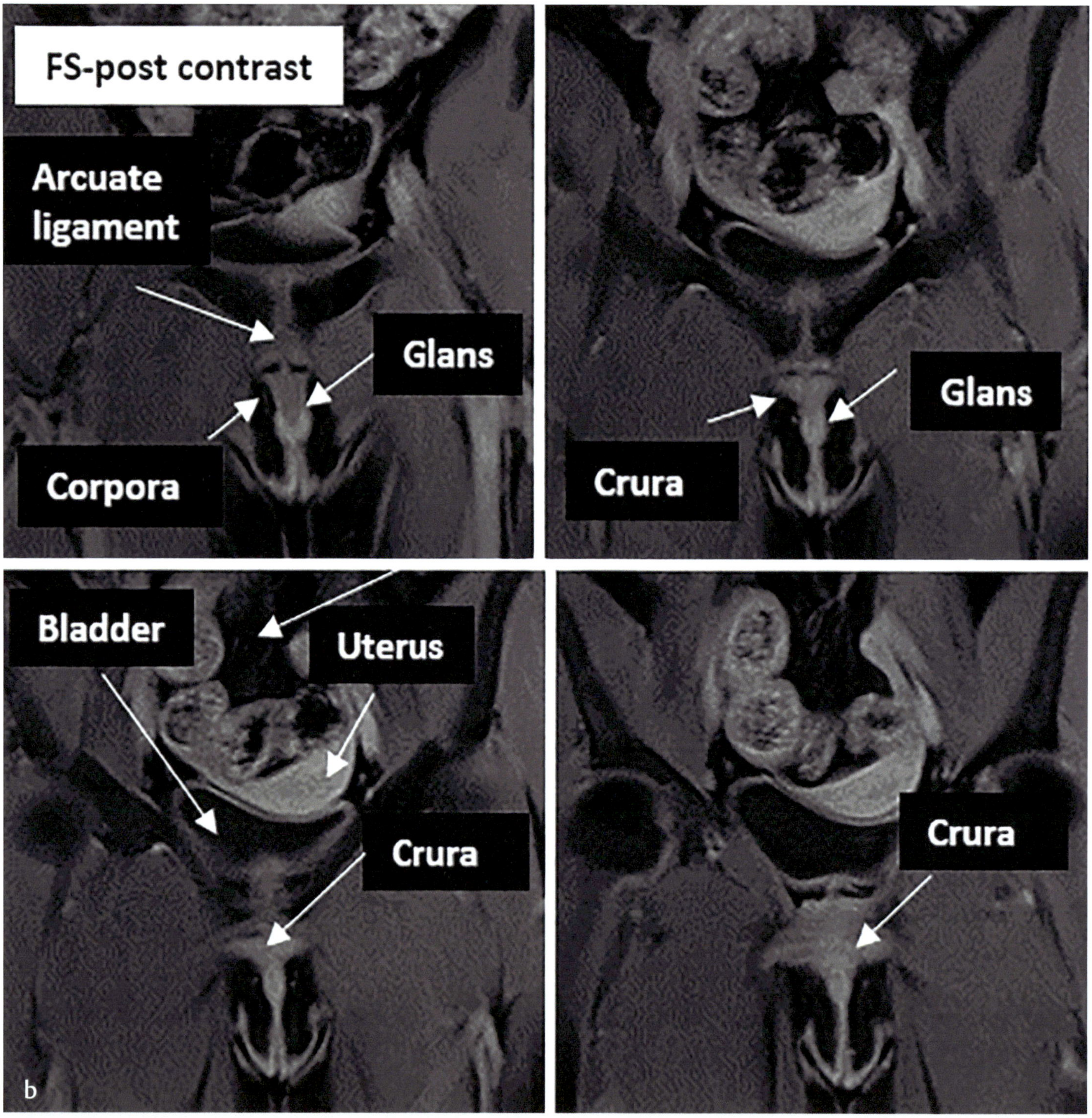

Fig. 16.8 (*Continued*) (b)

nerve fiber innervation in the lamina propria than the proximal third.[6]

Common Pathologies of the Vagina

Congenital Anomalies

Congenital anomalies of the vagina result from abnormalities in embryologic development.[3,7,8,9] As mentioned earlier, the uterus, cervix, and upper two-thirds of the vagina are formed by fusion of the caudal ends of the paramesonephric (Müllerian) ducts (▶Fig. 16.3). The lower third of the vagina develops from the bilateral sinovaginal bulbs, which arise where the Wolffian and Müllerian ducts are fused to the urogenital sinus. The Wolffian duct will regress and the sinovaginal bulbs fuse into a solid mass called the vaginal plate, which undergoes canalization in the second trimester to form the vaginal lumen. Because the developing renal-urinary system (mesonephric ducts) is closely related to the reproductive tract (Müllerian ducts), the renal, uterine, and vaginal anomalies may be seen together. Imperforate hymen is the most common congenital abnormality of the female genital tract. Incomplete canalization results in a persistent transverse septum. Transverse vaginal septum is less common than

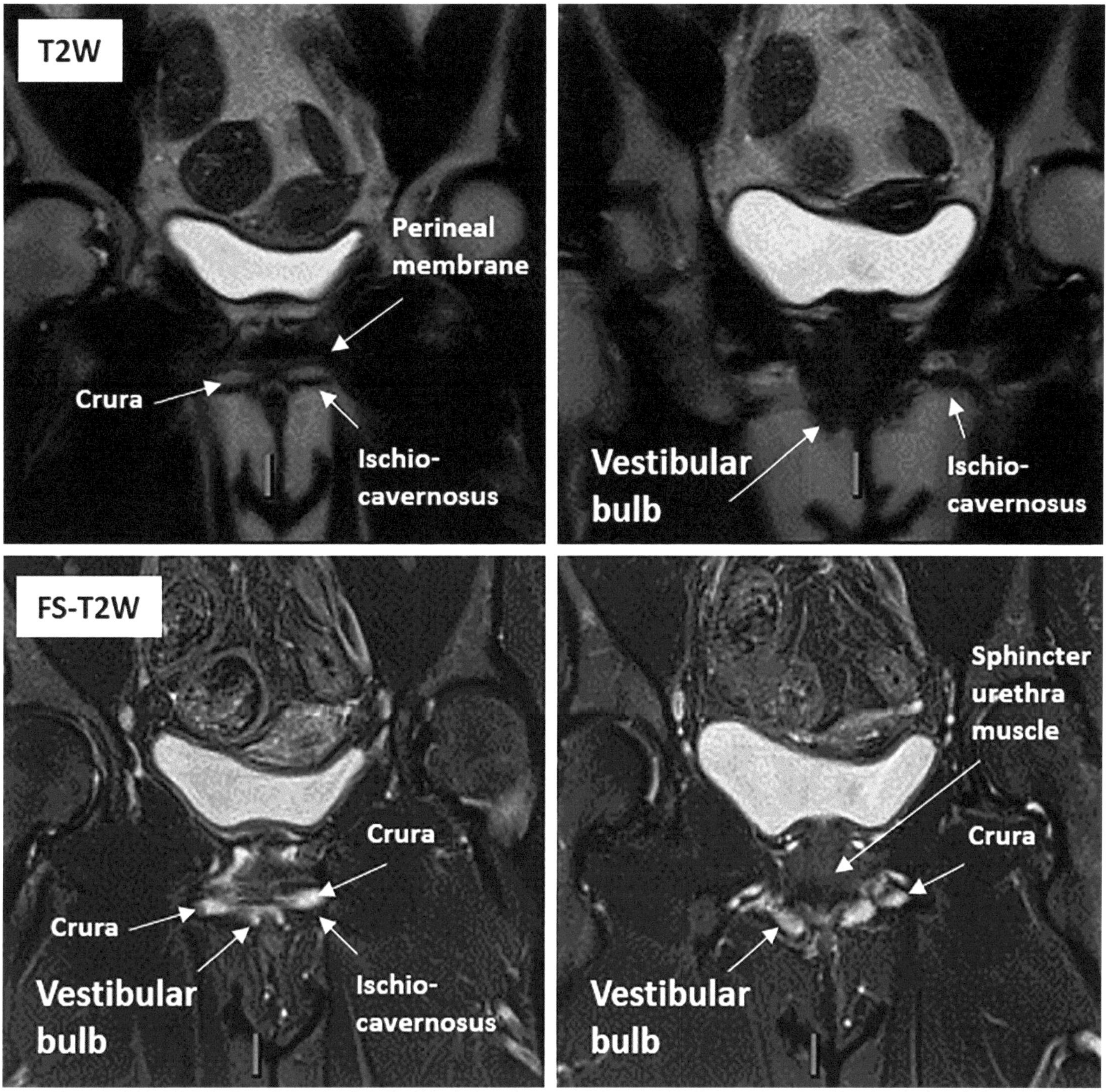

Fig. 16.9 Vestibular bulbs and urethra are shown by T2-weighted (T2W) and fast suppressed (FS)-T2W coronal magnetic resonance (MR) images. The bulbs of vestibule remain bright on T2W images. Muscles are dark. The crura of the clitoris appear bright. They are surrounded by the ischio-cavernosus muscles superficially and the peroneal membrane deeply.

imperforate hymen and both may result in complete vaginal obstruction causing amenorrhea or hematocolpos. If the fusion of the Müllerian ducts is incomplete, a longitudinal septum will develop (▶Fig. 16.18). A longitudinal septum divides the vagina into two parallel cavities. It is usually asymptomatic. The most common cause of vaginal atresia is Mayer-Rokitansky-Küster-Hauser syndrome which is characterized by congenital aplasia of the uterus and upper two-thirds of the vagina, short fallopian tubes, normal female puberty, normal ovaries, and normal 46 XX karyotype.

Vaginal Cyst

Bartholin gland cyst is cystic dilation (up to 4 cm) of the paired glands which is located in the posterolateral aspect of the vulvovaginal vestibule (▶Fig. 16.18, ▶Fig. 16.19). The cysts can be asymptomatic if infected. Bartholin gland cyst should be differentiated from other cysts. The Skene's glands are paired structures located near the external urethral meatus, with ducts draining directly into the urethral lumen. Gartner duct cysts result from incomplete regression of the Wolffian (mesonephric) ducts.[9] Bartholin gland cyst is in posterolateral aspect of

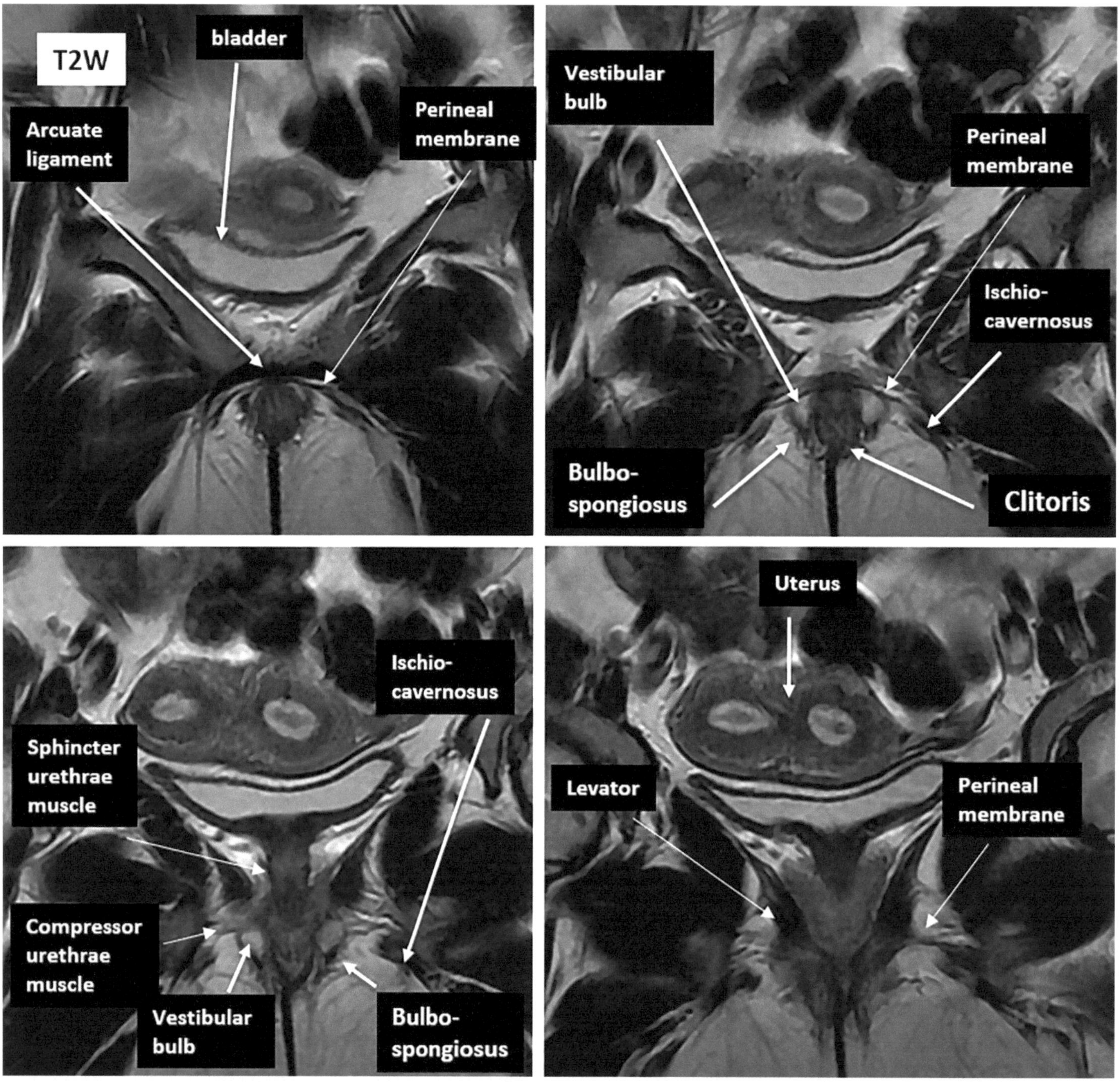

Fig. 16.10 Coronal T2-weighted (T2W) and fast suppressed (FS)-T2W magnetic resonance (MR) images at the level of the female urethra are shown. The bulbs of vestibule remain bright on T2W images. Muscles are dark. The crura of the clitoris appear bright. The bulbospongiosus muscles are seen as a linear dark structure on the surface of the vestibular bulb. The peroneal membrane is deep to the vestibular bulb and the ischiocavernosus muscles. Also shown in this patient is a bicornuate uterus.

the vestibule, below the pelvic diaphragm, whereas Gartner duct cyst is in anterolateral aspect of the upper vagina and above the pelvic diaphragm.[10]

Uterus

The uterus is a muscular, inverted, pear-shaped organ in the true pelvis. The uterus components include fundus, body, isthmus, and cervix (▶ Fig. 16.20). The fallopian tubes are attached to the superolateral part of uterine body where it is called the uterine cornua. The diameter of the uterus gradually reduces from superior to inferior, and at the junction between the body and cervix forms the isthmus. The cervix protrudes into the vaginal cuff and forms the vaginal fornix.

The normal uterine size in adult nulliparous women is approximately 7.5 cm in length, 5-cm diameter at the fundus, and 2.5-cm diameter at the cervix. The uterine cavity is 6 cm in length and up to 1.2 cm in thickness. The uterine dimensions increase in multiparous women but gradually decrease after the menopause. The normal cervix measures 2.5 to 3 cm in length and 2.5 cm in

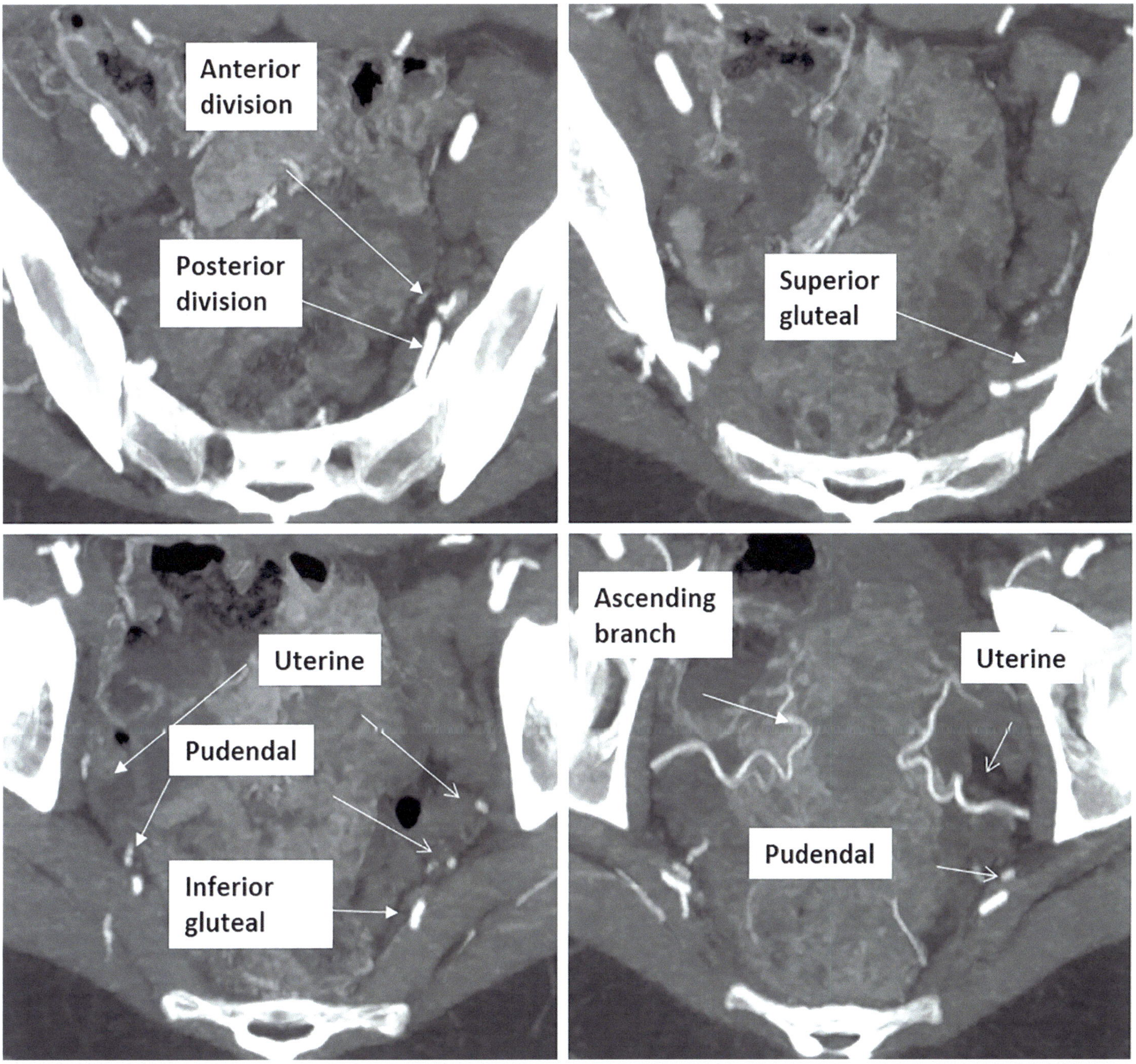

Fig. 16.11 Arterial branches to the female pelvic floor. **(a)** Axial views. **(b)** Coronal computed tomography (CT) views. Arterial supply to the pelvic floor and female organs is from branches of the anterior division of the internal iliac artery. The internal pudendal artery is the terminal branch of the anterior trunk of the internal iliac artery and passes through the pudendal (Alcock) canal to stay below the levator muscles, and then moves along the perineal membrane. The internal pudendal artery branches include dorsal and deep arteries of clitoris in female, perineal artery, posterior labial and bulbar, and inferior rectal. The superficial external pudendal (pubic) artery is one of the three pudendal arteries that run medially from the superficial femoral artery (SFA) to supply the skin of the labium major. It connects to the labial branch of the internal pudendal artery. The urethral artery may arise directly from the internal pudendal artery.

(*Continued*)

width. The uterine body is covered by the peritoneum that laterally becomes the broad ligament, anteriorly forms the vesicouterine pouch, and posteriorly forms the rectouterine or Douglas pouch (▶Fig. 16.21).

The uterus is composed of approximately 65 to 70% smooth muscle tissue. The muscular layer of uterus includes an inner longitudinal smooth muscle layer and an outer circular layer. The uterine cavity is covered by the endometrium with variable thickness depending on age and hormonal status.

Imaging of the Uterus

Ultrasound is considered the primary imaging technique for evaluation of the uterus and pregnancy-related issues. MRI and CT are usually used for staging of cancers. Hysterosalpingography is used in certain cases to assess the uterine cavity (▶Fig. 16.22). In this technique, the uterine cavity and fallopian tubes will be opacified by iodinated contrast followed by X-ray or CT scan. It can also be performed after the injection of sterile saline or water

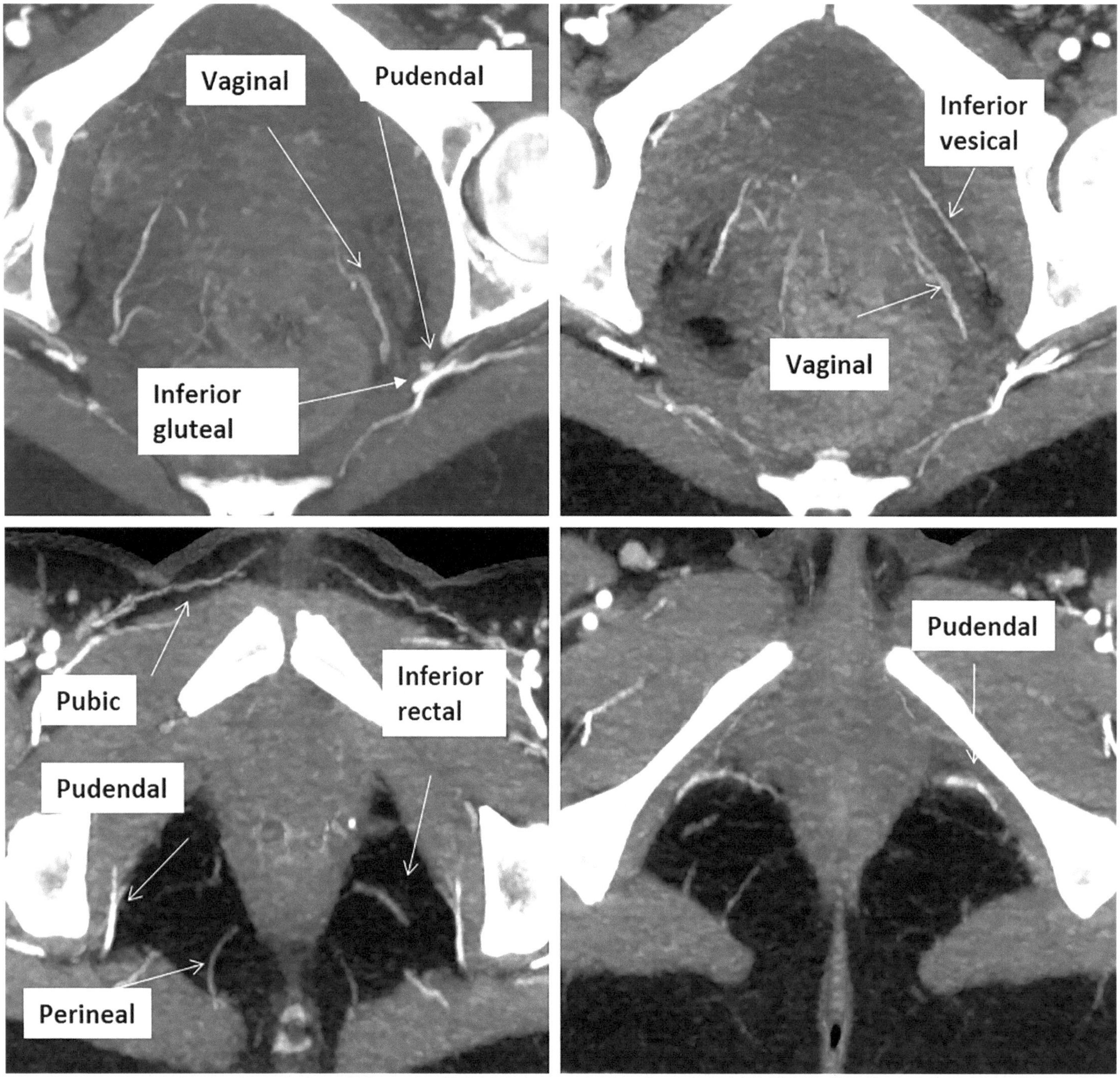

Fig. 16.11 *(Continued)*

(Continued)

into the endometrial cavity during an endovaginal ultrasound examination.[11]

Imaging appearance of the uterus varies during life. In neonate, the uterus may reach as large as 3.5 cm, the cervix is large and bulbous, and the endometrium appears echogenic containing fluid due to exposure to maternal hormones. In the next following months, the uterus and cervix become smaller and tubular. A thin endometrium is visible in 50% of girls examined at 4 to 12 months of age (▶Fig. 16.23). In infants, reflux of small amounts of urine into the vagina is seen and should not be confused with pathology. The mean uterine length can reach up to 3 to 4.5 cm for prepubertal girls, and 5 to 8 cm for pubertal girls.[12,13] A small amount of free fluid in the cul-de-sac is also considered a common physiologic finding in female of all ages (▶Fig. 16.21).

At puberty, the uterine body becomes larger and pear shaped and the ratio of the uterus to the cervix reaches 1.5:1 (▶Fig. 16.24). Before puberty, MRI shows a homogenous low to medium signal intensity uterus with indistinct zonal anatomy (▶Fig. 16.24). Later, the zonal anatomy can be appreciated on T2-weighted (T2W) MR images. The cervix is homogeneous in texture and similar to the uterine body (▶Fig. 16.24).

During reproductive years, the uterine size varies with patient's age and parity and the endometrium is typically easily visible, varying in thickness with the phase of the menstrual cycle. In

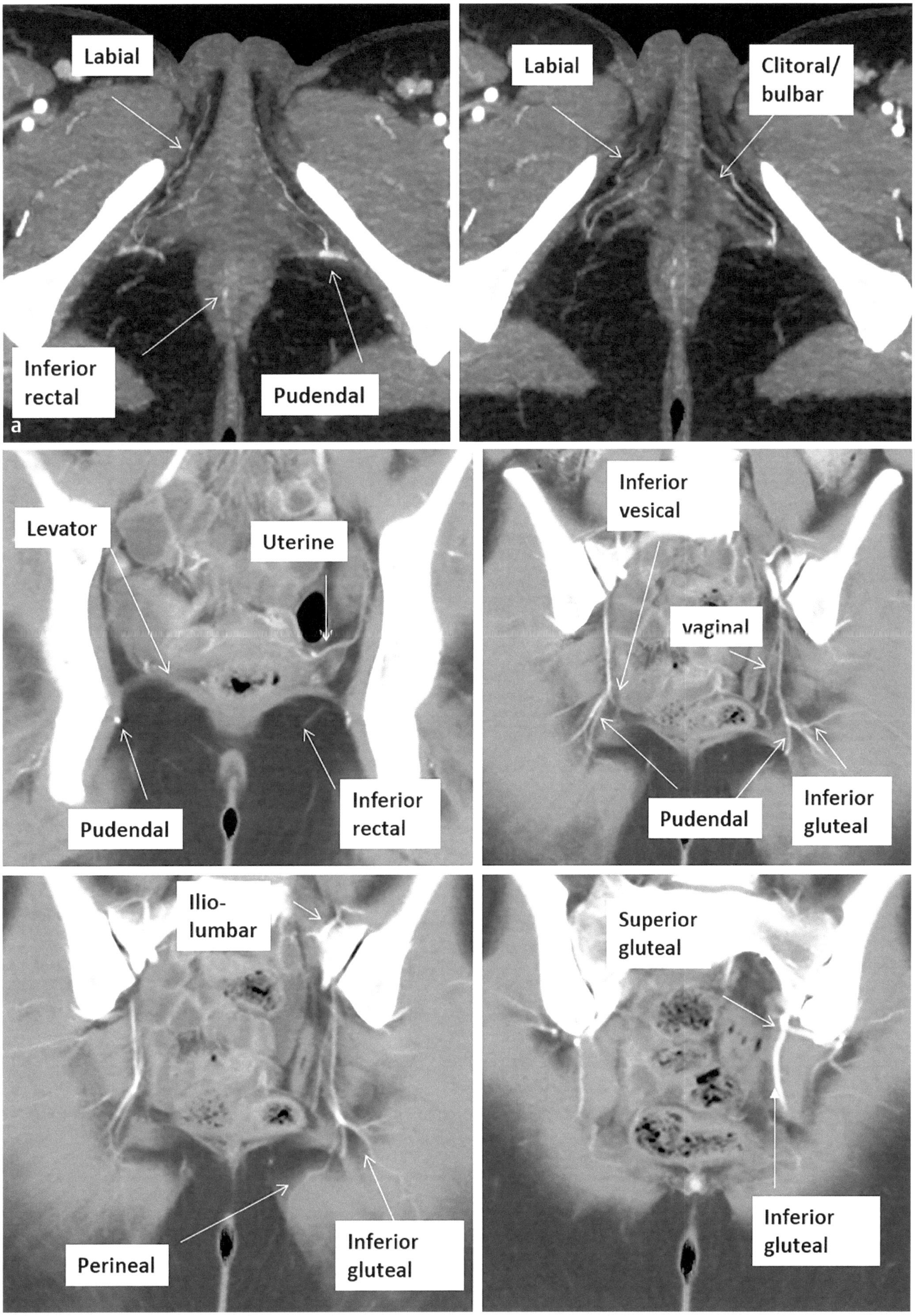

Fig. 16.11 (*Continued*)

(Continued)

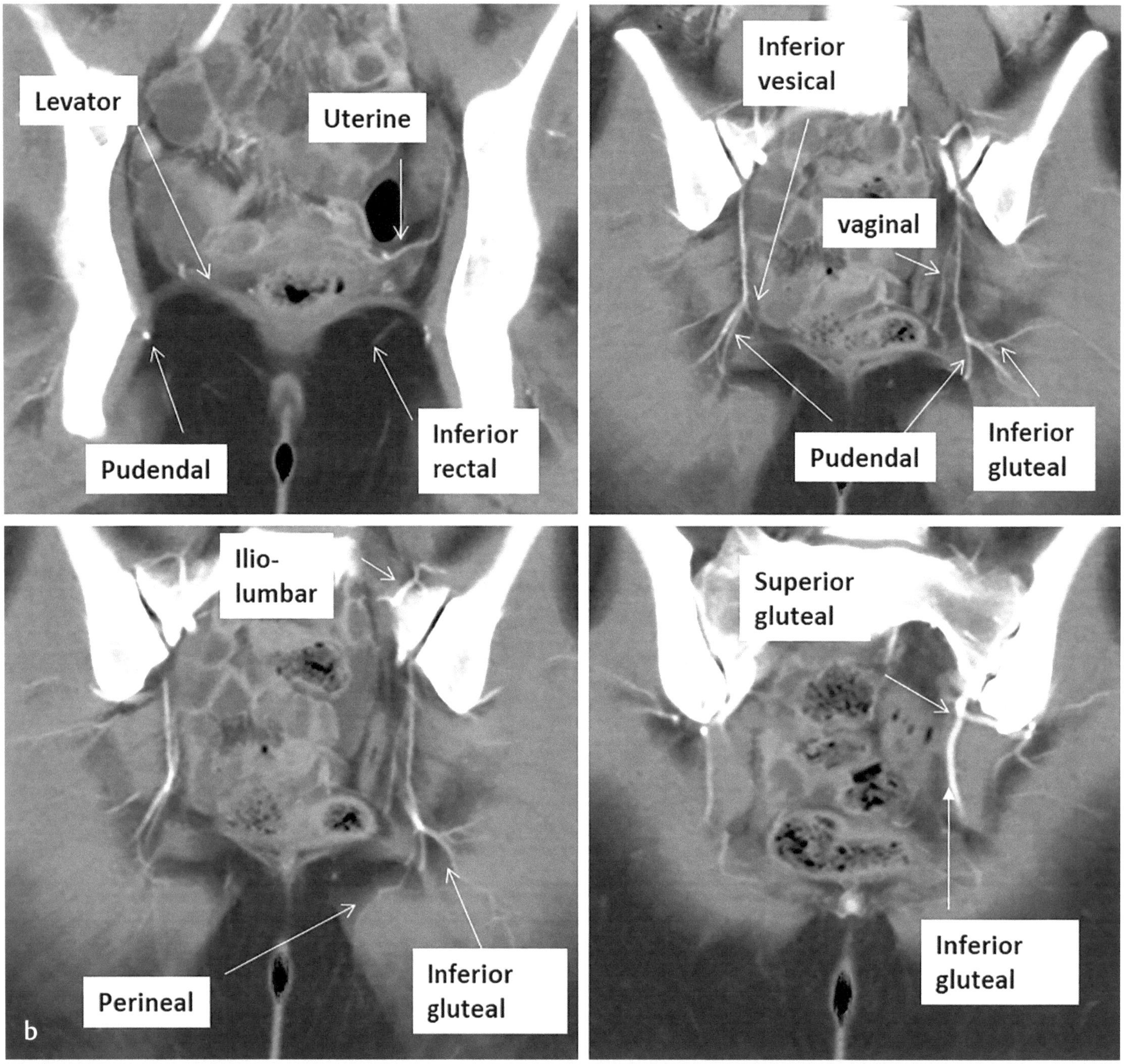

Fig. 16.11 *(Continued)* (b)

general, the multiparous uterus is larger than the nulliparous uterus. Adult uterus is pear shaped and the ratio of the uterus to the cervix is 2:1 (▶Fig. 16.24).

The outer uterine surface is typically pear shaped and the inner surface facing the endometrial canal is flat, with the body approximately twice the size of the cervix.

The normal positions of the uterus are defined in relation to the long axis angle of the uterine body to the cervix (flexion) and the long axis angle of the uterus to the vagina (version). The most common position is in anteversion or anteflexion (▶Fig. 16.22, ▶Fig. 16.24). Other uncommon positions are retroversion and retroflexion (▶Fig. 16.24, ▶Fig. 16.25). The uterus positioned in retroversion can be mistaken for a pelvic mass on axial CT images, especially if it displaces or indents the rectum. After cesarean delivery, the uterus typically is more elongated, and the myometrium is usually thinner at the site of incision (▶Fig. 16.25). The uterus endometrium and/or inner myometrium shows wave-like activity throughout the menstrual cycle which can be shown by ultrasound. Women with normal periovulatory wave-like activity have higher rates of successful fertilization compared with women with uncoordinated or absent wave activity.[14]

MRI is the best technique to show the trilaminar zonal anatomy of the uterine corpus and cervix[17,18] (▶Fig. 16.24). The innermost zone of the uterine body or cervix represents the endometrium and appears high intensity on T2W as a result of mucin-rich endometrial glands. The junctional zone or the inner myometrial

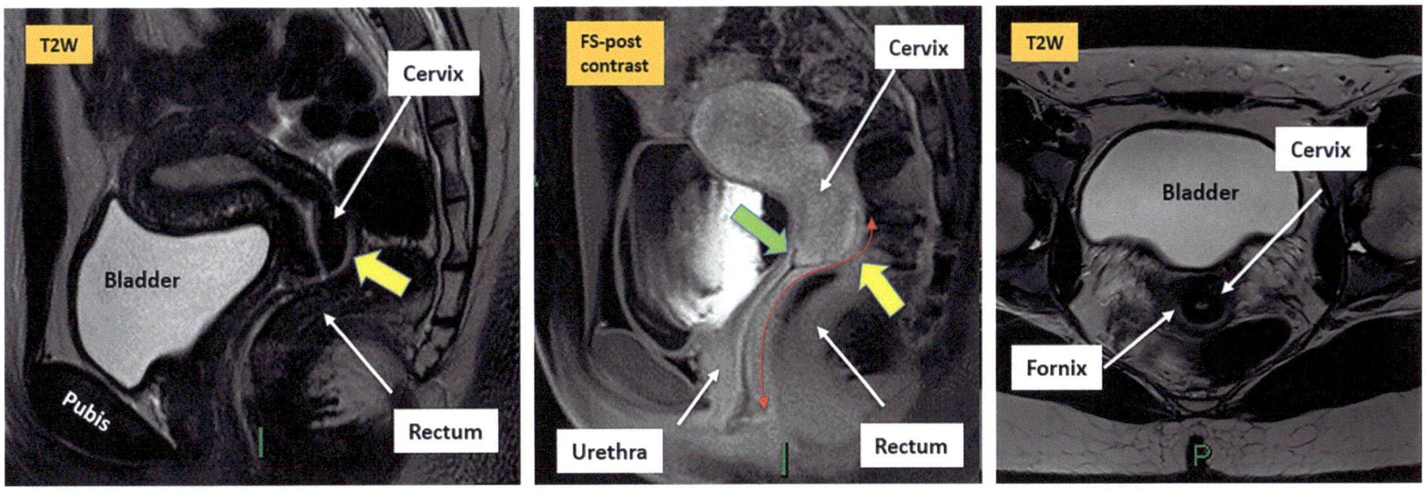

Fig. 16.12 Vaginal fornices are shown by sagittal T2-weighted (T2W), fast suppressed (FS)-post contrast, and axial T2W magnetic resonance (MR) images. The vagina is located in the middle compartment of the pelvis. The length of vagina is about 7 to 9.5 cm (*double-headed red arrow*). The posterior wall is longer and ends in the posterior fornix (*yellow arrows*) and the anterior wall is shorter terminating in the anterior fornix (*green arrow*).

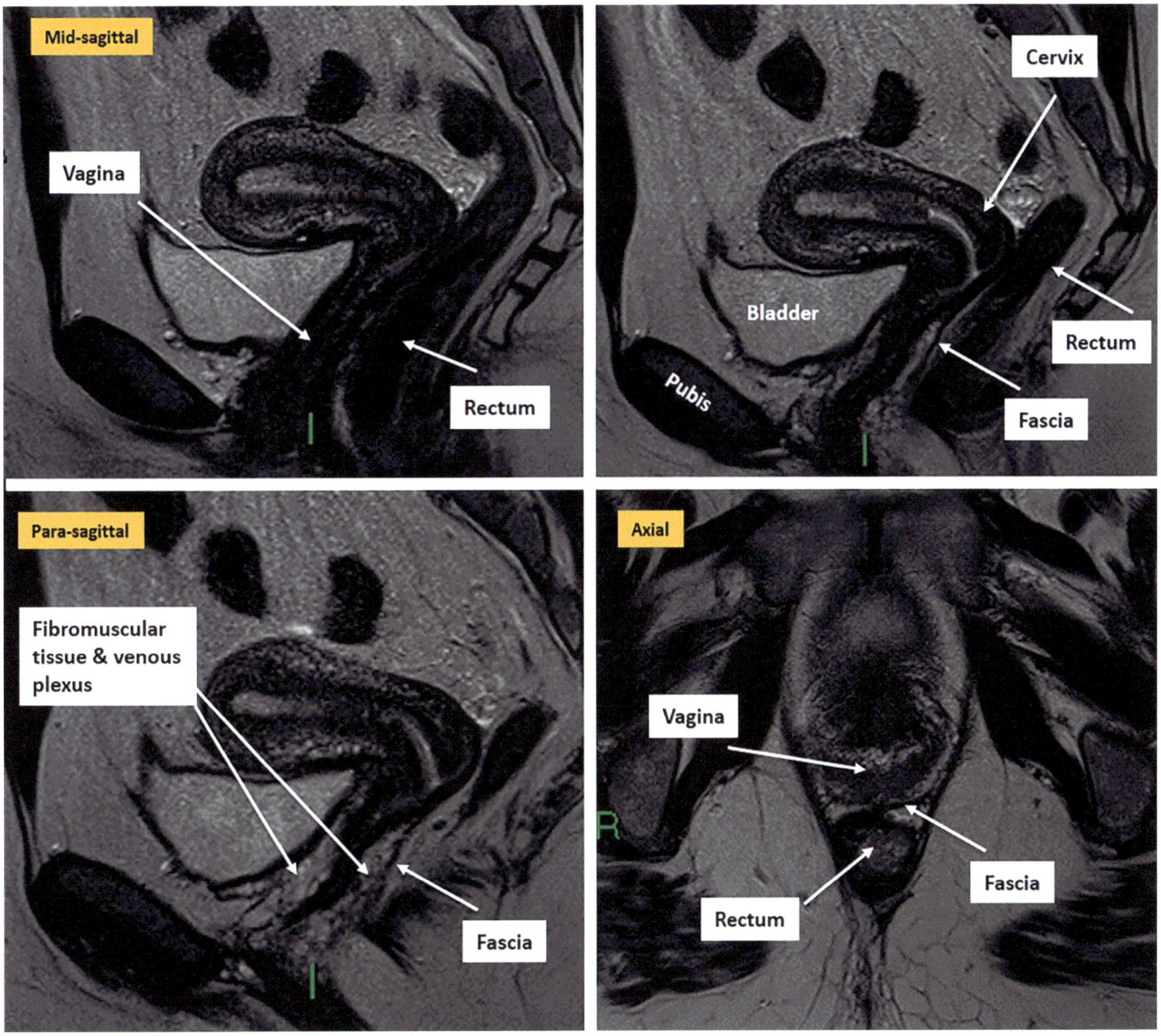

Fig. 16.13 Denonvilliers' fascia is shown by sagittal and axial T2-weighted (T2W) magnetic resonance (MR) images. The vagina is located in the middle compartment of the pelvis. The fascia is between the rectum and the middle third of vagina. The inferior third is separated from the anal canal with a vascular fibromuscular membrane.

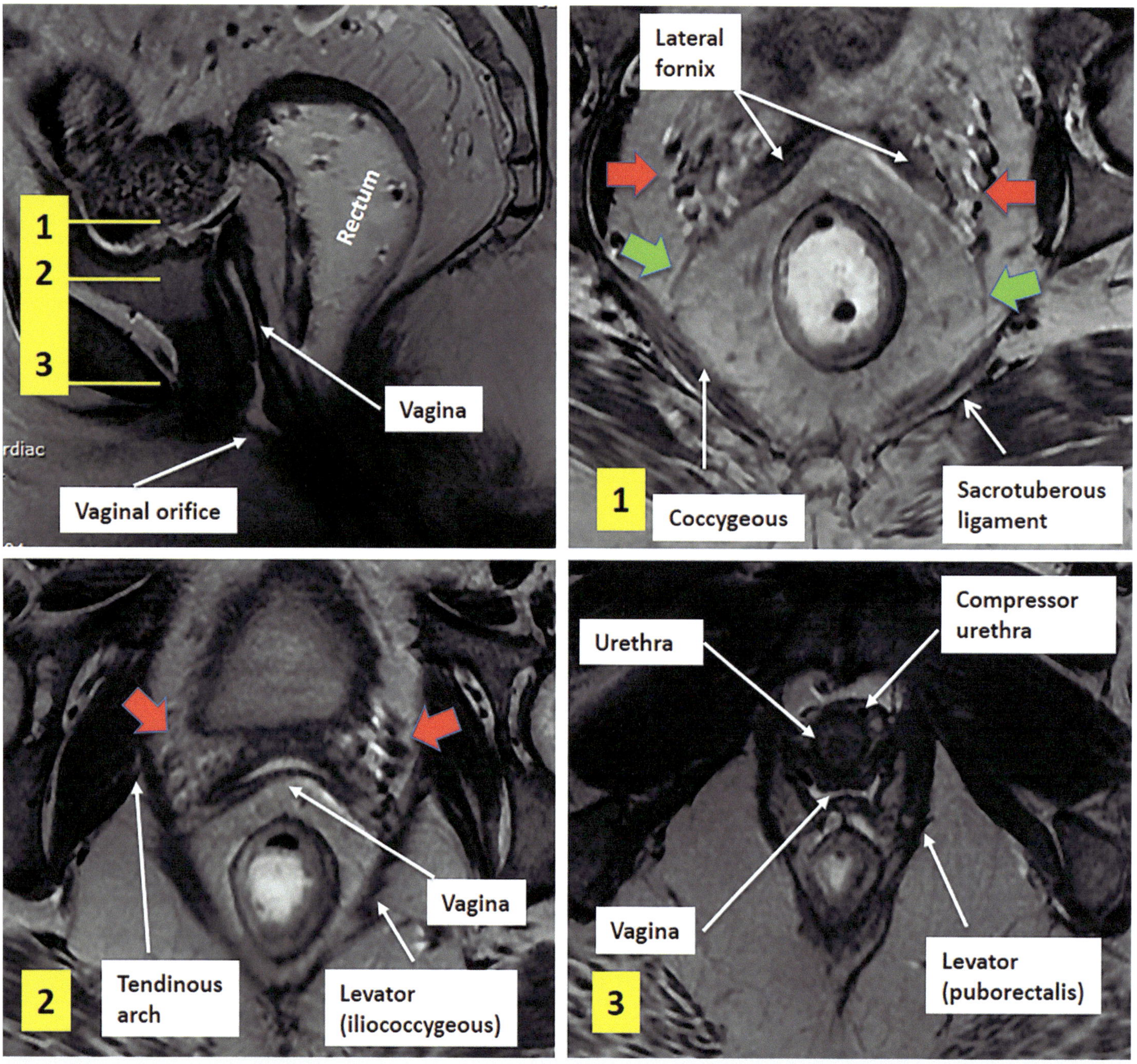

Fig. 16.14 Vaginal supports. Status post hysterectomy. Using sagittal reference, three T2-weighted (T2W) axial views are presented. The cardinal ligaments insert into the lateral fascia of the cervix and adjacent vaginal wall. Part of it has remained (*red arrows*) in this patient after hysterectomy. The uterosacral ligaments (*green arrows*) are seen originating from the sacral vertebrae and sacrotuberous ligament to insert into the lateral vaginal fornices.

layer appears low signal on T2W images due to the presence of compact smooth muscle cells and a lesser amount of extracellular matrix which corresponds to the hypoechoic subendometrial layer observed at ultrasound[16] (▶ Fig. 16.26). The normal junctional zone is usually less than 1.2-cm thick and thicker values may indicate adenomyosis.[18] Thickening of the junctional zone and myometrium during sustained contraction of the uterus should not be mistaken with adenomyosis or leiomyoma[19] (▶ Fig. 16.26). The inner myometrial layer is contiguous with the fibromuscular stroma (middle layer) in the cervix.[17] The outer myometrium demonstrates variable signal intensity depending on cellularity and vascularity. It is usually intermediate in signal intensity due to the presence of less compact smooth muscle, greater water content, and blood vessels.[29] Enlargement of venous sinuses in the subserosal surface of the outer myometrium layer is common in pelvic congestion syndrome and pregnancy (▶ Fig. 16.27).

Endometrial Cycle

The endometrium consists of a single layer of columnar epithelium and its basal lamina residing on a thick vascular stromal

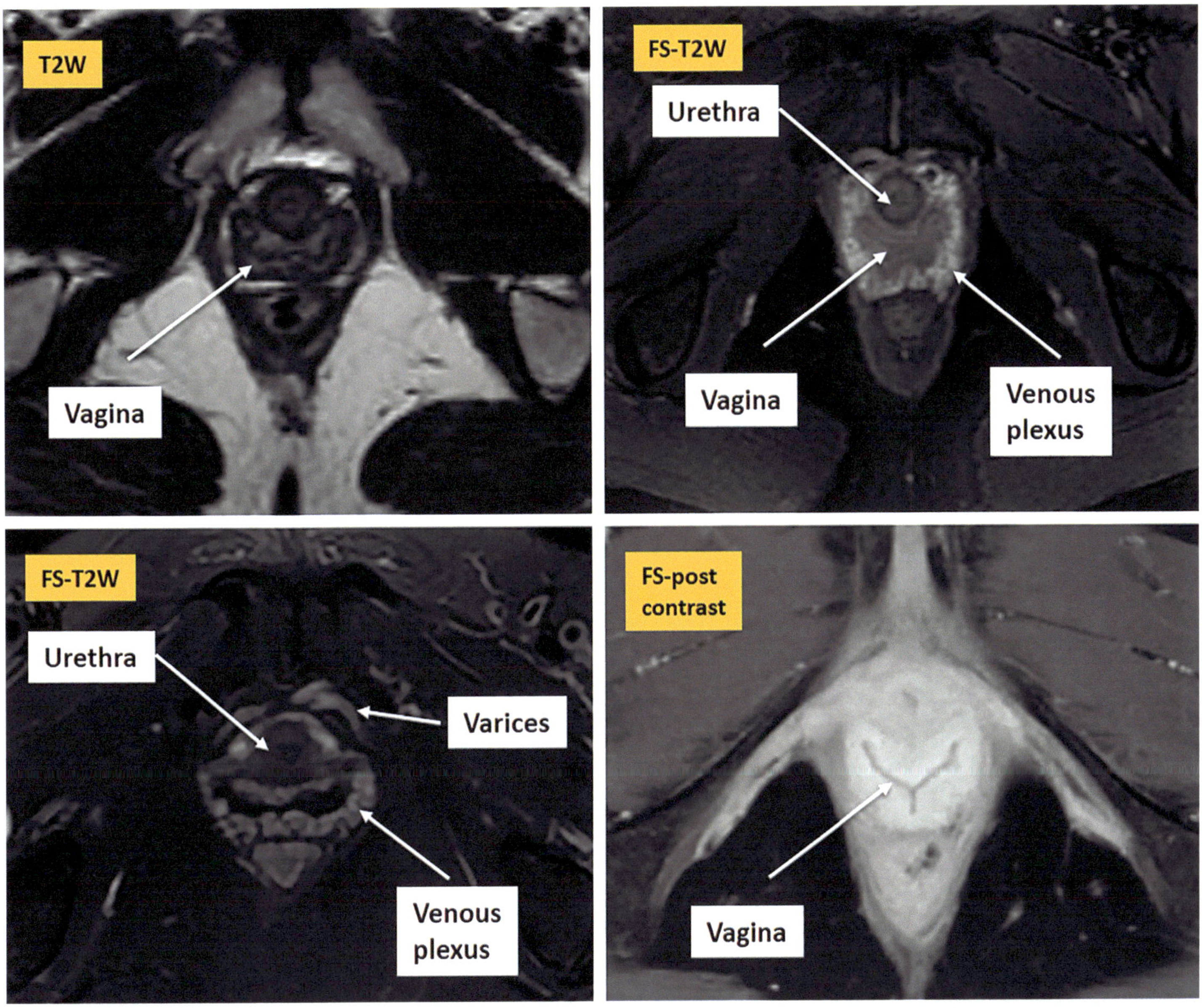

Fig. 16.15 Normal appearance of the vagina in different individuals using axial magnetic resonance (MR) images in different sequences. The normal vagina is seen as a collapsed, H-shaped tubular structure with a central enhancing mucosa. On T2-weighted (T2W) image the mucosa appears bright and muscular wall appears dark. Both structures may enhance on postcontrast images. Most of perivaginal venous plexus also varies in different individuals and in some cases venous congestion is prominent as seen in two cases.

connective tissue layer. The later contains many spiral arteries (end branches of the uterine arteries) and veins and its thickness varies during the menstrual cycle.[20,21] The normal endometrial cycle includes the following stages: (1) proliferative (follicular), (2) secretory (luteal), and (3) desquamative (menstruation) (▸Fig. 16.28). During the follicular phase of the ovarian cycle (4th to 14th day) which is governed by estrogen (from ovarian follicle), proliferation of the endometrium occurs involving the glandular epithelium, stroma, and vascular bed. During the luteal phase, along with secretion of progesterone from the corpus luteum, the secretory glands of the endometrium accumulate glycogen in their cytoplasm, and the vessels increase and become more spiral. In the absence of fertilization, corpus luteum involutes, progesterone and estrogen are reduced, and arteriolar spasm occurs leading to ischemic necrosis and shedding of the endometrial surface (functional layer). This phase is called the menstruation phase (1st to the 4th day) (▸Fig. 16.28).

16

Endometrial Thickness and Appearance

The normal endometrium appears echogenic on ultrasound. The thickness of the normal endometrium changes during menstrual cycle (▸Fig. 16.26, ▸Fig. 16.29). It is thinner during menses with a mean of 5 mm but gradually thickens during proliferative phase to reach a maximum of 12 mm a day before ovulation when the luteinizing hormone (LH) surge occurs.[13,15,22] Between days 13 and 16, the endometrium has a striated pattern, with a hypoechoic inner layer surrounded by a hyperechoic peripheral layer

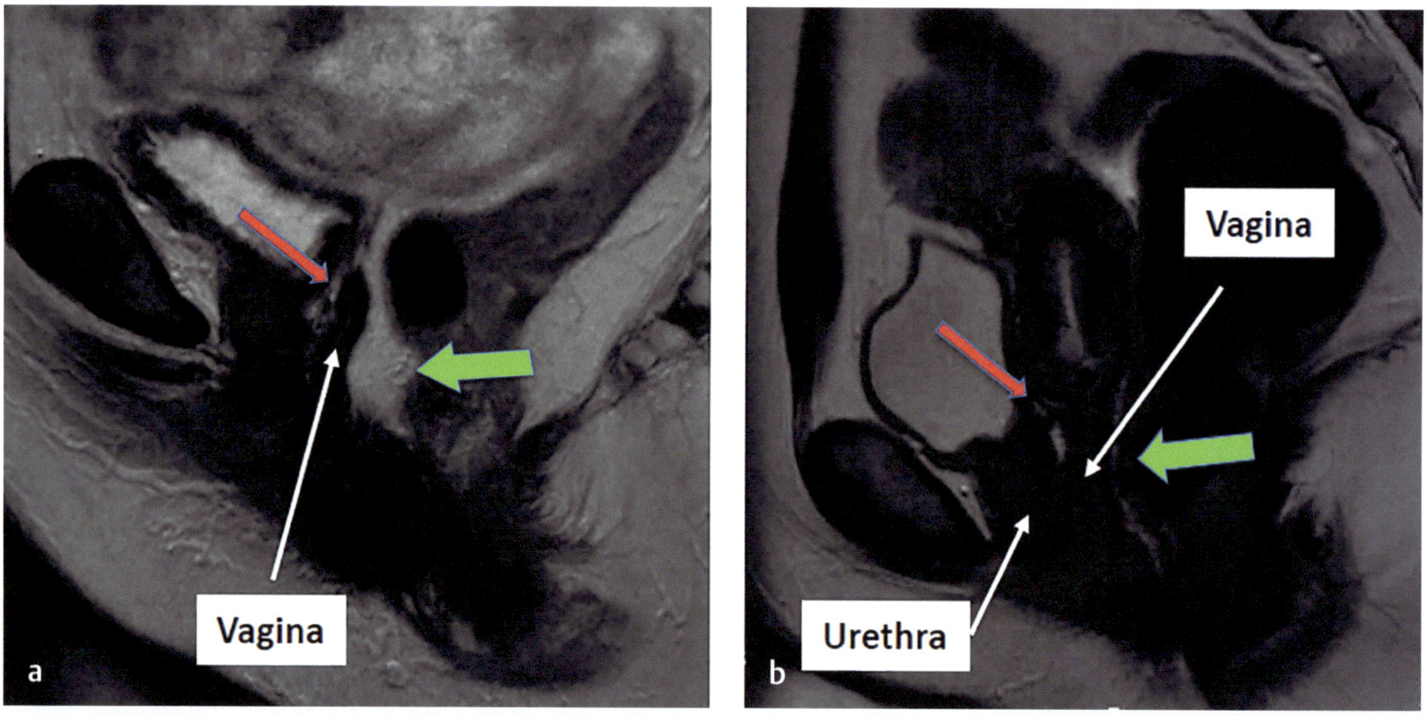

Fig. 16.16 Sagittal T2-weighted (T2W) magnetic resonance (MR) images. **(a)** Status post hysterectomy showing vaginal atrophy in a 73-year-old. Note enlarged fat plane between the rectum and vagina (*green arrow*). **(b)** Normal fat plane between the rectum and posterior wall of the vagina (*green arrow*). In normal adult female pelvis, there is tight connection between the bladder, urethra, and vagina (*red arrows*) and a looser connection between the rectum and vagina (*green arrow*). In this case there is a small fat plane between the proximal urethra and the anterior vaginal wall (*red arrows*) but there is no fat plane between the vagina and distal urethra.

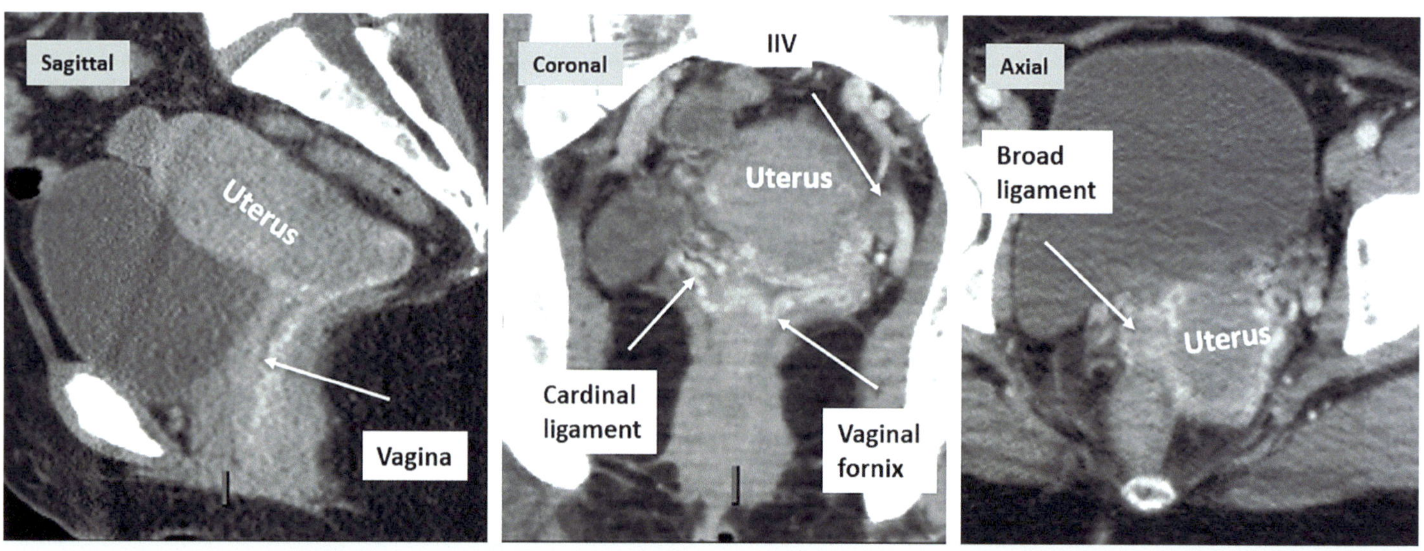

Fig. 16.17 Contrast-enhanced computed tomography (CT) images showing enlarged vaginal and parametrial veins draining into the internal iliac veins (IIV).

(▶Fig. 16.26, ▶Fig. 16.29). The thin central hyperechoic line is also seen representing the interface between the anterior and posterior endometrial surfaces. The striated pattern usually disappears 48 hours after ovulation. It is important to know that the day of the LH surge and ovulation is highly variable and may occur as early as day 8 to as late as day 20 in the menstrual cycle. A thin endometrium with lack of a striated pattern during periovulatory phase (midcycle) is abnormal and has been linked to infertility.[13]

The endometrial thickness is generally measured from the anterior to the posterior endometrial–myometrial interfaces. The hypoechoic layer peripheral to the endometrium, which represents the compact inner layer of myometrium, should not be included in endometrial thickness measurements[13] (▶Fig. 16.26). During secretory phase of the menstrual cycle which typically occurs between days 16 and 28, the striated pattern is replaced by a homogeneous echogenic and thinner endometrium[22]

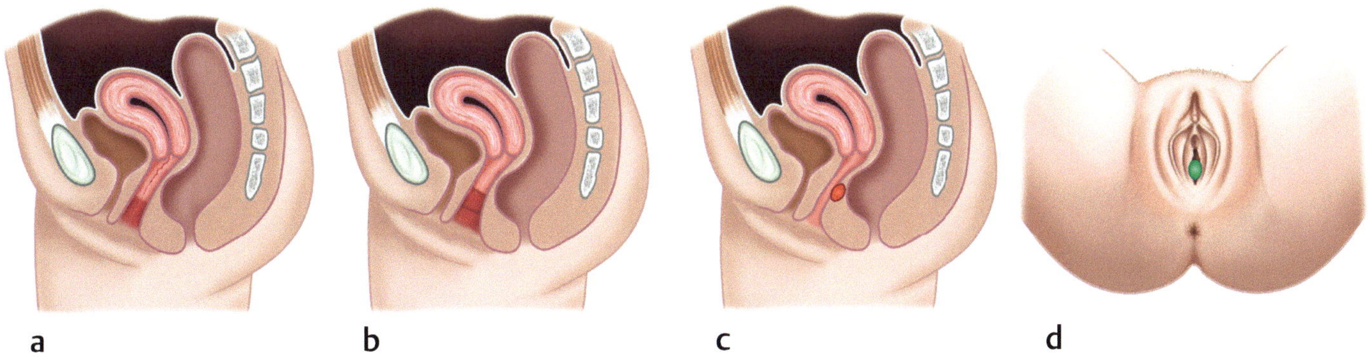

Fig. 16.18 (a) A longitudinal vaginal septum due to incomplete fusion of the Müllerian ducts. **(b)** Various possible levels for a transverse vaginal septum due to incomplete canalization of the caudal vagina. **(c, d)** Vaginal and perivaginal cysts. Drawings show typical locations of Gartner duct cysts (*red*) in the anterolateral vaginal wall above the inferior border of the pubic symphysis, Bartholin gland cysts (*blue*) in the distal posterolateral vaginal wall posterolateral to the vaginal introitus, and periurethral (Skene) glands (*green*) adjacent to the distal urethra.

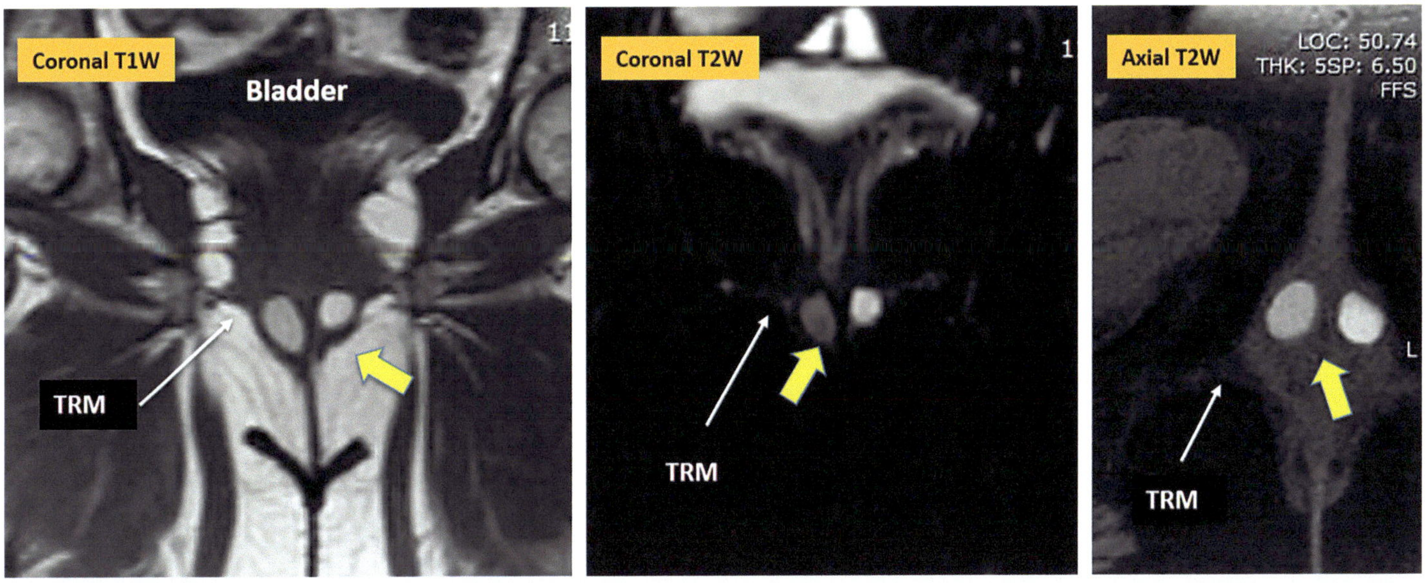

Fig. 16.19 Bartholin gland cyst. Cystic dilation of the paired glands (*arrows*) is shown in the posterolateral aspect of the vulvovaginal vestibule above the transverse perineal muscle (TRM). Cysts appear bright on both T1- and T2-weighted images.

(▸Fig. 16.29). A small amount of fluid or mucus within the endometrial cavity is a normal finding. The endometrium is generally thin throughout the menstrual cycle in women taking oral contraceptives and those with intrauterine contraceptive devices (IUCDs) (▸Fig. 16.30). Assessment of the endometrium for focal abnormalities is best performed in the early phase of the menstrual cycle when the endometrium is thicker.

Reactive thickening of the endometrium in response to a rising progesterone is common in early pregnancy. This thickening is called decidual reaction and presents as a thickened echogenic cast of endometrium (▸Fig. 16.30c–e). Decidual cast is also seen in ectopic pregnancy. As pregnancy progresses, along with invasion of the trophoblastic tissue, the imaging appearance of the endometrium changes (▸Fig. 16.30e,f).

Postpartum Uterus

Puerperium uterus is large immediately after childbirth but involutes 50% in the first 24 to 48 hours.[23] Ultrasound shows a large uterus of up to 20 cm with a heterogeneous myometrium that is highly vascular especially at the placenta insertion site. In the following days of postpartum, the uterine involution is gradual, with the uterus measuring 11-cm long at 3 weeks and 9-cm long at 6 weeks postpartum. Restoration of the endometrium to normal takes about 3 to 6 weeks. Intraendometrial cavity fluid, blood products, and small amounts of gas are common early findings and may last for 3 weeks after vaginal delivery and do not indicate postpartum complications.[24] Presence of larger amounts of fluid or soft tissue within the endometrium could be abnormal

16

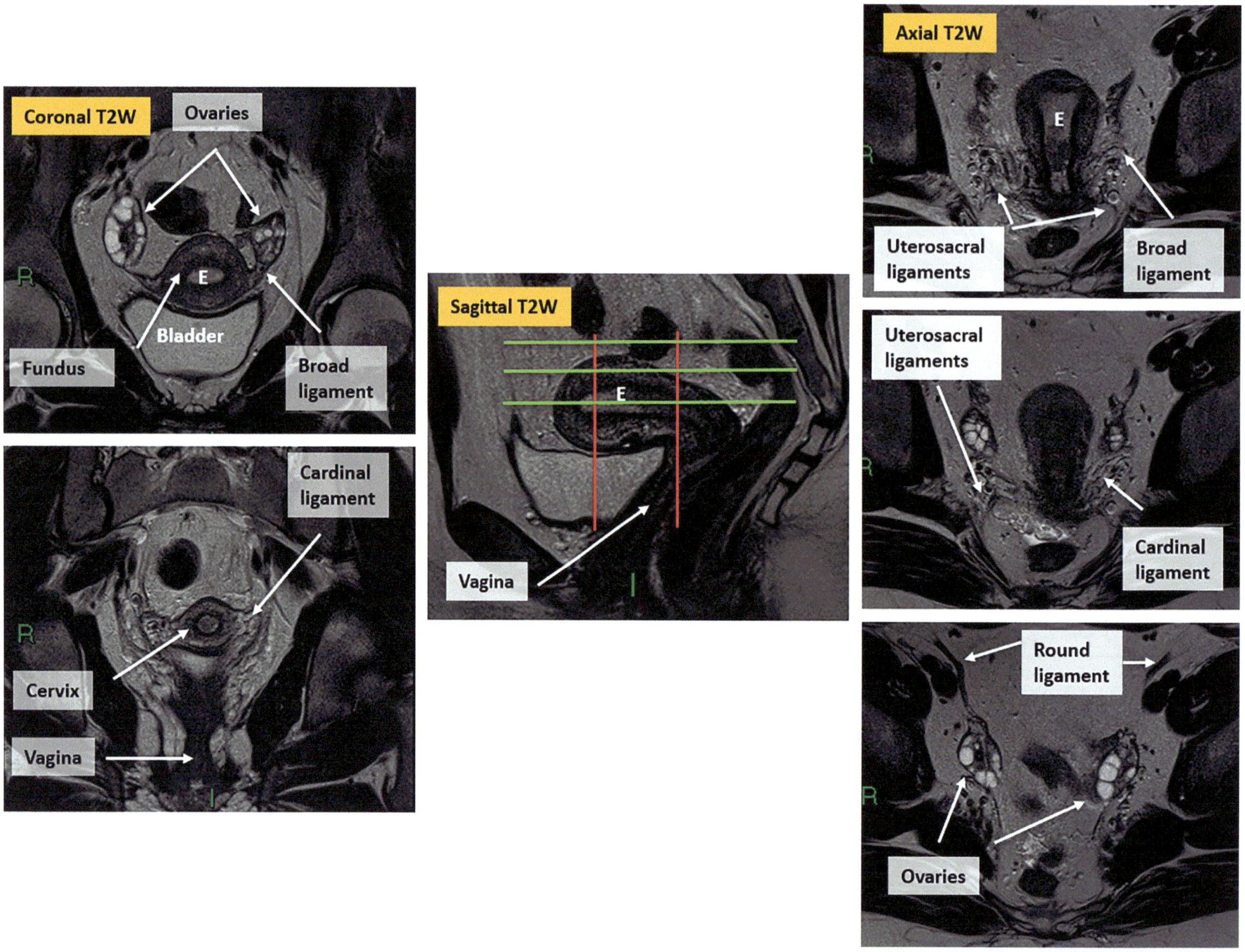

Fig. 16.20 Anatomy of the uterus and associated structures. The uterus components include fundus, body, isthmus, and cervix. The uterosacral ligaments are seen originating from the sacral vertebrae and sacrotuberous ligament to insert into the lateral vaginal fornices. E, endometrium.

and requires further studies to exclude endometritis and retained products of conception. After cesarean delivery, small free pelvic fluid and hematoma of the myometrium at incision site can be found on both CT and MR images. Diffuse myometrial, parametrial, and cervical edema as well as small foci of blood products within the endometrial cavity are common findings on T2W MR images. After approximately 6 months postpartum, the normal zonal anatomy reappears.

16

Postmenopausal Uterus

Patients are called postmenopausal when they have experienced amenorrhea for 12 consecutive months from the date of the last menstruation.[25] Generally, shrinkage in uterine size occurs within the first 10 years after menopause, with a more gradual decline thereafter.[26] The uterus sizes ranges from 3.5 to 7.5 cm in length and 1.5 to 3.5 cm in transverse diameter in patients older than 65 years.[26] The variation in size of uterus depends on patient's parity, postmenopausal period, and the presence of intrauterine myomas.

Postmenopausal myometrial is often more heterogeneous than premenopausal women. The uterine body-to-cervix ratio is approximately 1:1 (▶Fig. 16.31). The uterine zonal anatomy in postmenopausal women is usually indistinct or absent (▶Fig. 16.31). Common findings included small calcified and noncalcified myomas, vascular calcifications, and small nabothian cysts. Small amount of free peritoneal fluid is not uncommon in postmenopausal women.

The postmenopausal endometrium is typically seen as a narrow echogenic or high signal band because of the lack of estrogen stimulation measuring 1 to 2 mm in maximum diameter[13] (▶Fig. 16.31). As mentioned above, the relatively hypoechoic (low signal on MRI) band of compact inner myometrial layer (junctional zone) which surrounds the endometrium should not be included in the measurement of the endometrium. Postmenopausal patients who present with vaginal bleeding

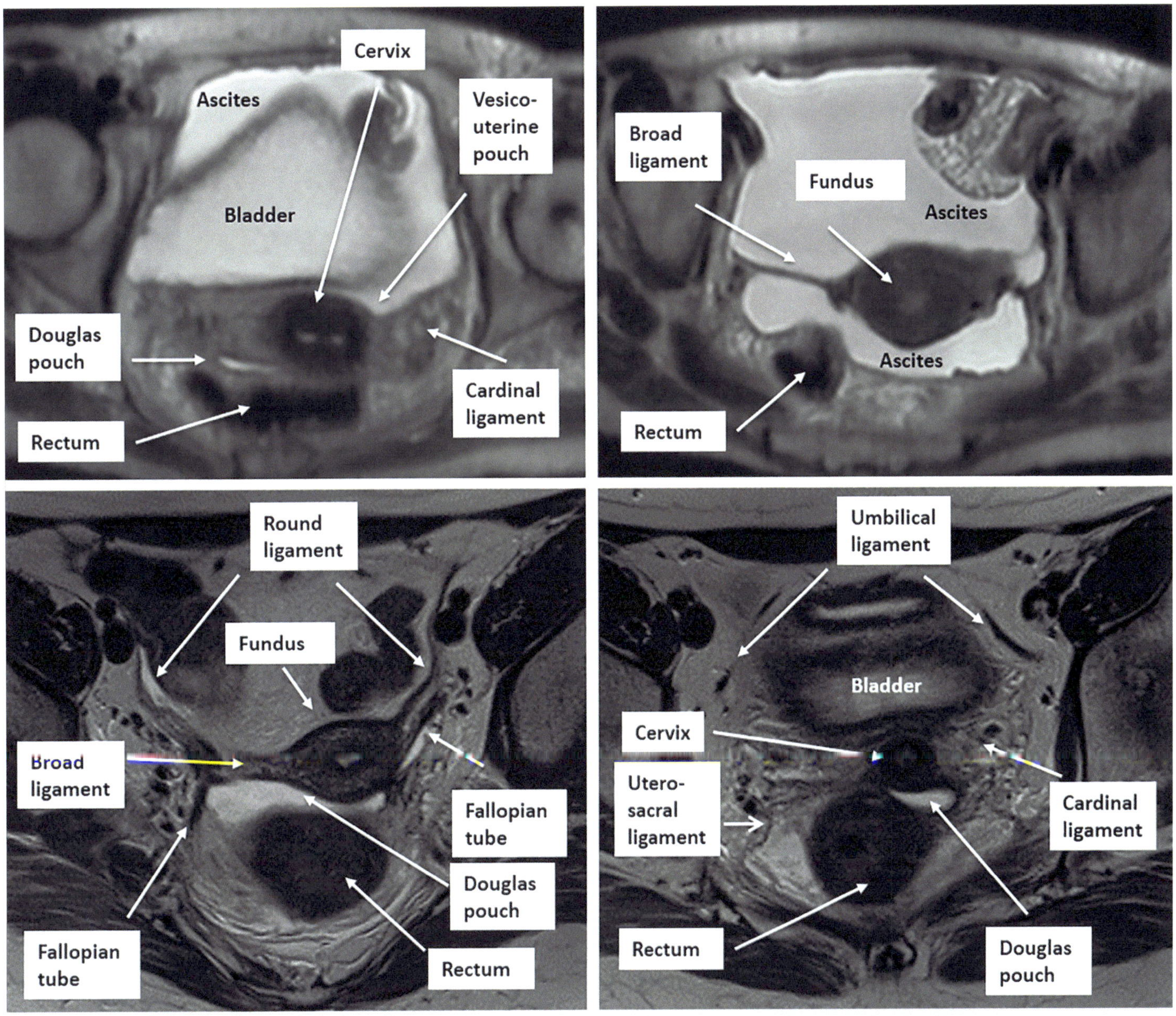

Fig. 16.21 Axial T2-weighted (T2W) images showing large ascites in a patient with ovarian cancer (*upper row*) and physiologic amount of peritoneal fluid (*lower images*). The uterine body is covered by the peritoneum that forms the broad ligament laterally, the vesicouterine pouch anteriorly, and the rectouterine or pouch of Douglas posteriorly.

and have an endometrial thickness of >5 mm should be investigated with additional testing such as endometrial biopsy to exclude endometrial carcinoma[27] (▶ Fig. 16.31c). In asymptomatic patients, sampling is suggested only when the endometrial thickness exceeds 8 mm.

Hormone replacement therapy is often used in perimenopausal patients. The use of continuous estrogen and progesterone regimens leads to endometrial atrophy not hyperplasia and thickening. Therefore, thickened endometrium in these patients should be carefully followed with ultrasound. On the other hand, unopposed estrogen therapy increases the risk of endometrial hyperplasia or carcinoma and is typically prescribed only for posthysterectomy patients.[28] Tamoxifen, commonly used for the treatment of breast cancer, is associated with increased incidence of endometrial hyperplasia (>8 mm) and polyps especially after 3 years of therapy.[29]

Cervix

Unlike the uterine body, the cervical stroma is primarily collagenous tissue and only contains 15% smooth muscle. The cervix is composed of a supravaginal portion and a lower portion termed the ectocervix which protrudes into the upper vaginal canal forming the vaginal fornices.[17,30] The opening of the ectocervix which is palpable into the vaginal canal is called the external os. The junction of the cervix with the uterine body is referred to as the isthmus. The endocervix at the isthmus narrows to 1 to 12 mm in width over a length of 1 cm to form the internal cervical os (▶ Fig. 16.32).

The endocervical mucosa is a glandular epithelium which is arranged into longitudinal ridges with interdigitating branching folds known as the plicae palmatae, sometimes visible by imaging techniques (▶ Fig. 16.22). Cyclical menstrual change of the

16

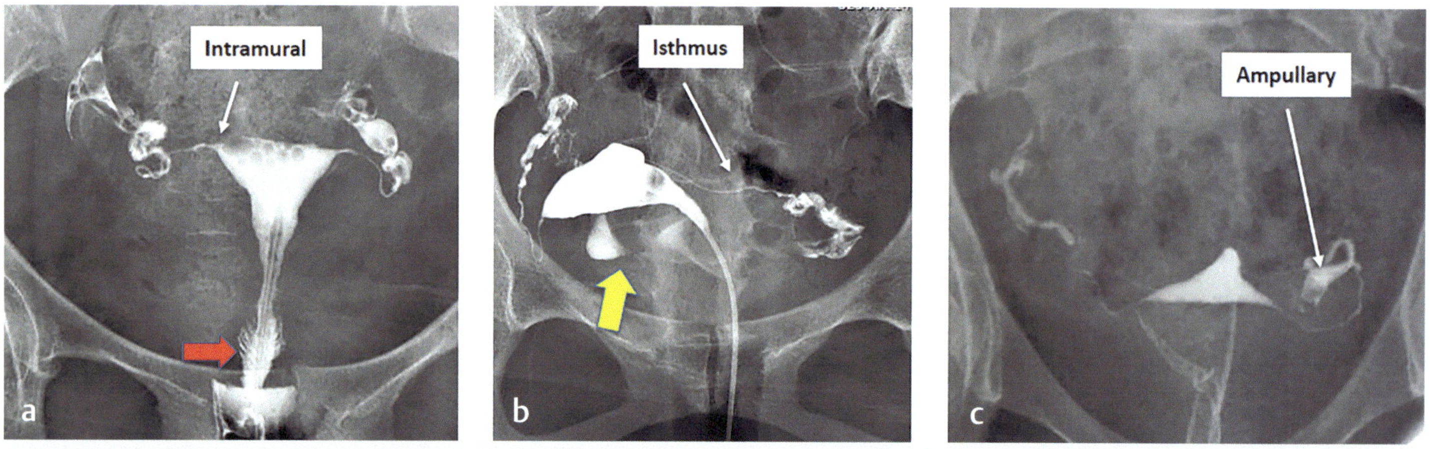

Fig. 16.22 Hysterosalpingography. Spot radiographs in different patients obtained during the filling stage of the uterus. The uterus fully distended with contrast material and both fallopian tubes are opacified. The study allows evaluation for filling defects and contour abnormalities of the uterus and patency of the tubes. A fallopian tube is divided into four anatomic segments. From medial attachment point to uterus to the distal end these segments include interstitial (intramural), isthmic, ampullary, and infundibular. **(a)** The interstitial, isthmic, and ampullary portions of both fallopian tubes. The contrast outlines multiple thin mucosal folds of the endocervix that represents the plicae palmatae (*arrow*). **(b)** Intraperitoneal contrast material spillage from the fallopian tubes (*yellow arrow*). Lateral deviation of the uterus is also seen. **(c)** Anteflexion of the uterine fundus.

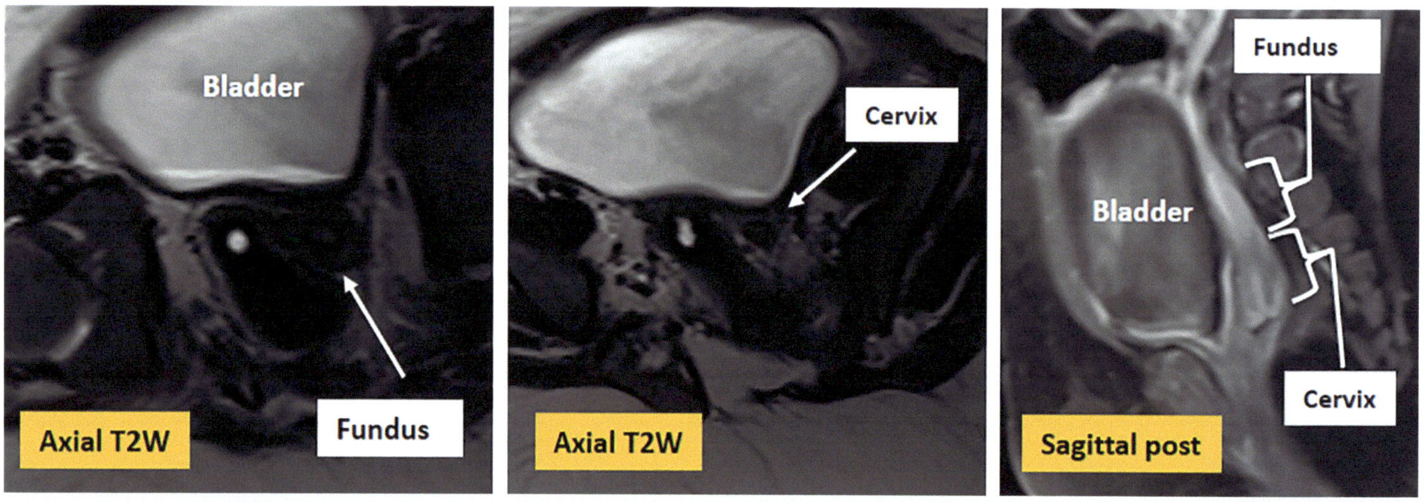

Fig. 16.23 Uterus in a 5-month-old infant. The cervix and fundus are equal in length. The cervix enhances less on postcontrast sagittal image. In neonate, the cervix is large and bulbous, and the endometrium may appear echogenic due to exposure to maternal hormones. In the next few months, the uterus and cervix become smaller and tubular. A thin endometrium is visible in 50% of girls examined at 4 to 12 months of age.

16

endocervix is less dramatic than the endometrium. The glandular columnar epithelium of the endocervix transforms into the stratified squamous epithelium (similar to the vaginal epithelium) at the ectocervix. The location of this squamocolumnar junction moves from the ectocervix surface to endocervix with age. Most carcinomas of the cervix arise in this transformation zone.[31]

At ultrasound, the endocervical canal is visible as an echogenic line that represents the interface between the anterior and posterior mucosal layers[30] (▶Fig. 16.32). A small amount of fluid or mucus may be present in the canal, especially in the periovulatory period (▶Fig. 16.32e). Occasionally, the plicae palmatae are visible. The cervix is best imaged with MRI. MRI is commonly used to characterize indeterminate ultrasound abnormalities and to stage cervical carcinoma. On T2W images, the cervix shows a distinctive trilaminar appearance with an innermost hyperintense layer measuring 3 to 8 mm that corresponds to mucosa and secretions, a middle low-signal-intensity layer representing the inner cervical fibrous stroma that measures 3 to 8 mm, and an outermost intermediate-signal-intensity layer representing the outer cervical stroma that measures 2 to 8 mm (▶Fig. 16.32). The inner and outer cervical stromal layers typically appear contiguous with the uterine junctional zone and the outer myometrium, respectively (▶Fig. 16.24, ▶Fig. 16.26, ▶Fig. 16.32). The plicae palmatae are visible at MRI in 50% of young women and should not be mistaken for a cervical septum.[32]

Nabothian cysts, which are formed by retention of mucus within the obstructed endocervical glands, are commonly seen

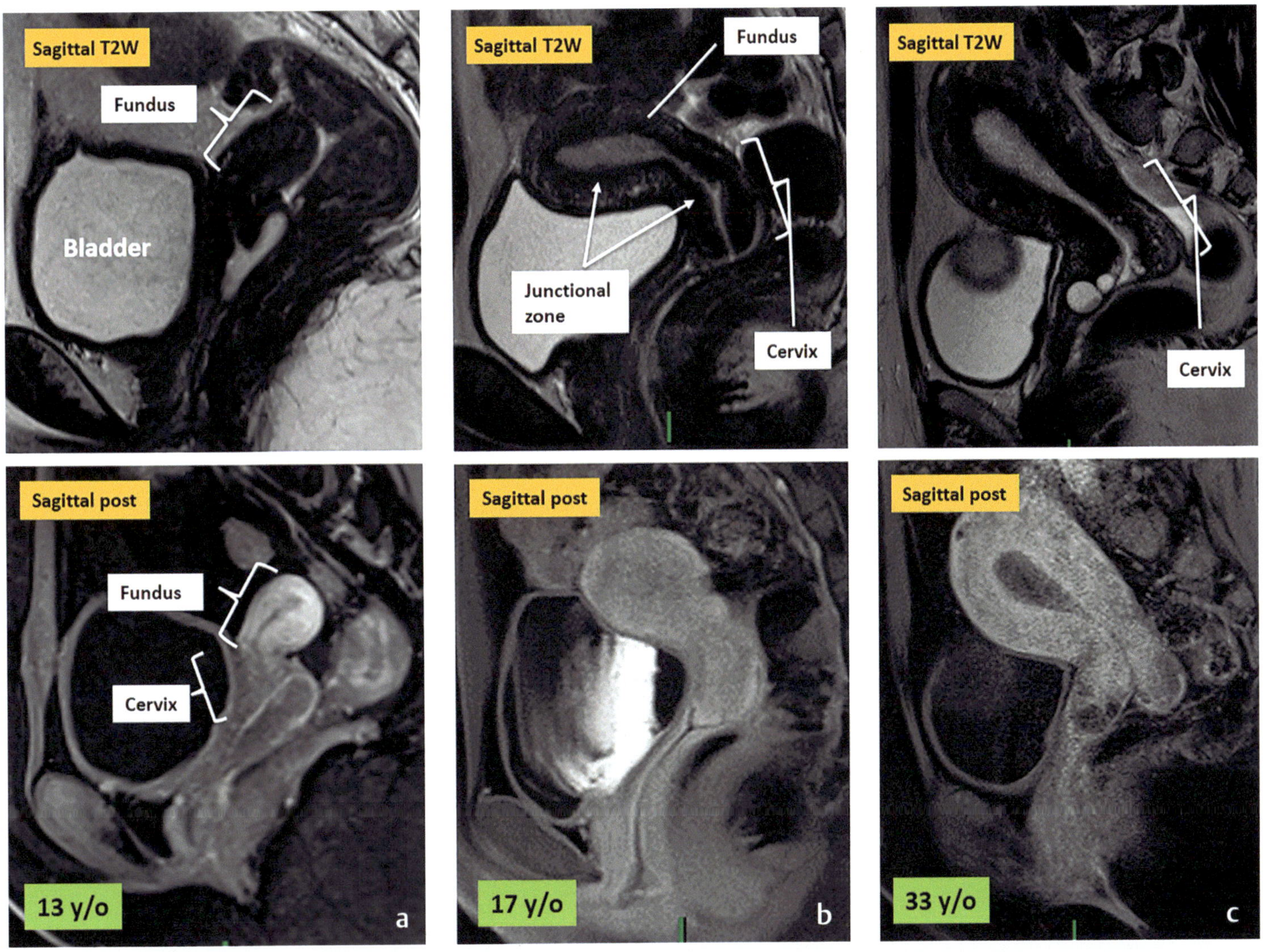

Fig. 16.24 Uterine morphology and position. **(a)** At puberty, the uterine body shows indistinct zonal anatomy and the ratio of the uterus to the cervix reaches 1.5:1. **(b)** Later, the zonal anatomy can be appreciated on T2-weighted (T2W) magnetic resonance (MR) images and the cervix is homogeneous in texture and low in signal intensity on T2W images (middle image). **(c)** Adult uterus is pear shaped and the ratio of the uterus to the cervix reaches 2:1. The most common position of the uterus is anteflexion **(b)** or anteversion **(c)**. In **a,** the uterus is retroflexed. Nabothian cysts of the cervix are shown in **c**. Trilaminar zonal anatomy of the uterine corpus and cervix is best shown in **c**. The endometrium appears high intensity on T2W due to mucin-rich endometrial glands. The junctional zone or the inner myometrial layer appears low signal on T2W due to the presence of compact smooth muscle cells and a lesser amount of extracellular matrix.

along the endocervix of adult women (▶Fig. 16.24, ▶Fig. 16.32). They typically measure less than 2 cm and contain simple fluid, but occasionally contain debris or hemorrhage.

Postcontrast Uterus and Cervix

Generally, the myometrium enhances on both arterial and venous phase images, whereas the endometrium enhances slowly with delay and appears hypointense/hypodense in comparison with the myometrium (▶Fig. 16.33). The normal cervix usually demonstrates less intense enhancement or enhances later than the uterine corpus at CT or MR because of the relatively larger proportion of fibrous tissue in the cervix[33] (▶Fig. 16.33). An intact endocervical canal and a distinct margin between the enhancing myometrium and the hypoenhancing normal cervix at the isthmus help discriminate between the normal cervix and a cervical mass.

Uterine Enhancement

Variable patterns of normal uterine enhancement related to age, parity, and menstrual status have been described.[34] Furthermore, normal variations in anatomic blood supply, timing of the contrast material injection, and cardiac output have a significant influence on uterine and cervical enhancement patterns (▶Fig. 16.27). Kaur et al[34] characterized the following three subtypes of uterine and cervical enhancement (▶Fig. 16.34): type 1, thick or thin subendometrial bandlike enhancement observed at 30 to 120 seconds after injection with or without progressive outer myometrial enhancement at delayed phase imaging predominantly in premenopausal women (▶Fig. 16.27c, ▶Fig. 16.33). The thin layer of enhancement outlining the endometrial canal could represent the junction of the basal layer of the endometrium with the underlying myometrium, whereas the thick band of enhancement might correspond to the junctional zone.[34] In type 2, either peripheral

16

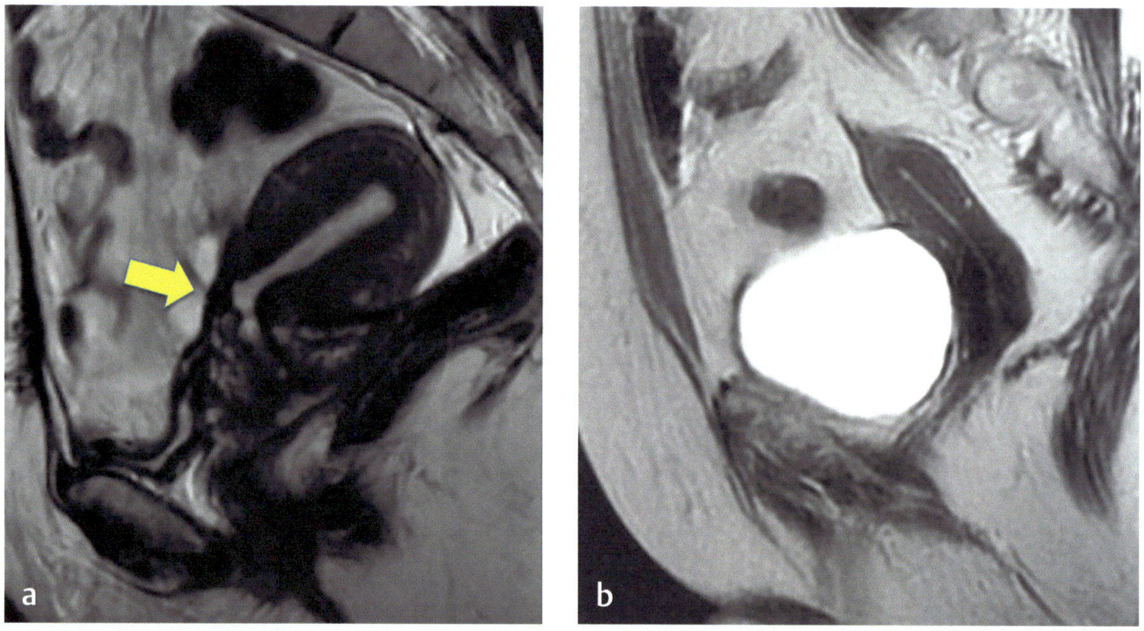

Fig. 16.25 Uterine position. Uncommon positions of the uterus are retroversion and retroflexion. **(a)** Retroflexion; also shown is the scar of cesarean section (*arrow*) in the lower uterine segment. **(b)** Mild retroversion with a distended bladder.

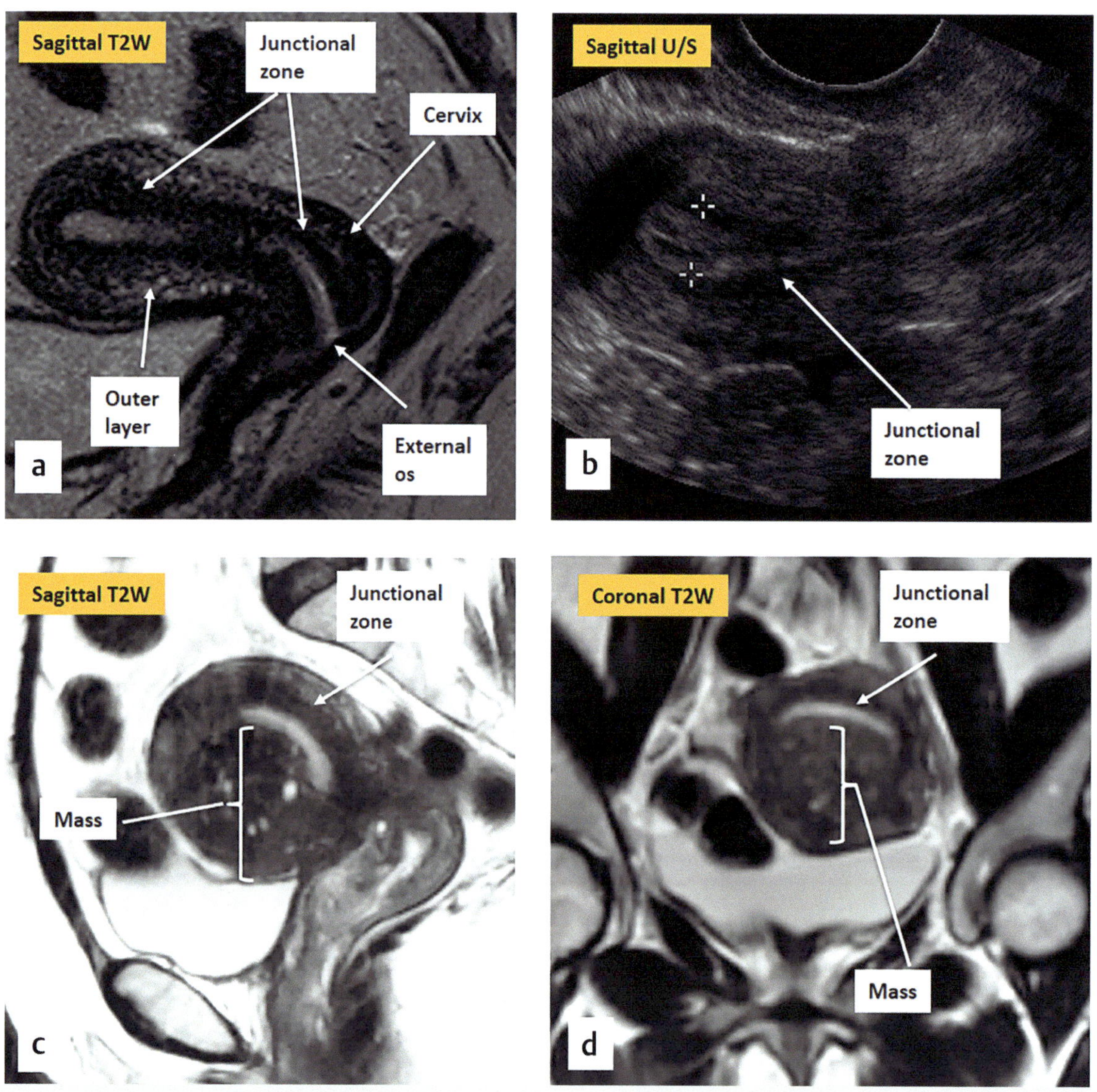

Fig. 16.26 Normal and abnormal junctional zone. **(a)** Magnetic resonance imaging (MRI) shows normal junctional zone as a low signal rim around the bright signal endometrium which corresponds to the hypoechoic subendometrial layer observed at ultrasound **(b)**. It continues with the fibromuscular stroma (middle layer) in the cervix. The normal junctional zone is usually less than 1.2 cm thick and thicker values may indicate adenomyosis. **(b)** Endovaginal ultrasound (U/S) image obtained near midcycle shows a normal striated pattern of the endometrium (between calipers). Normal endometrial measurement includes the more echogenic outer layer and more hypoechoic inner layer. The thin central hyperechoic line represents the interface between endometrial walls. **(c, d)** Mass-like adenomyosis cystic changes in posterior uterus replacing normal junctional zone.

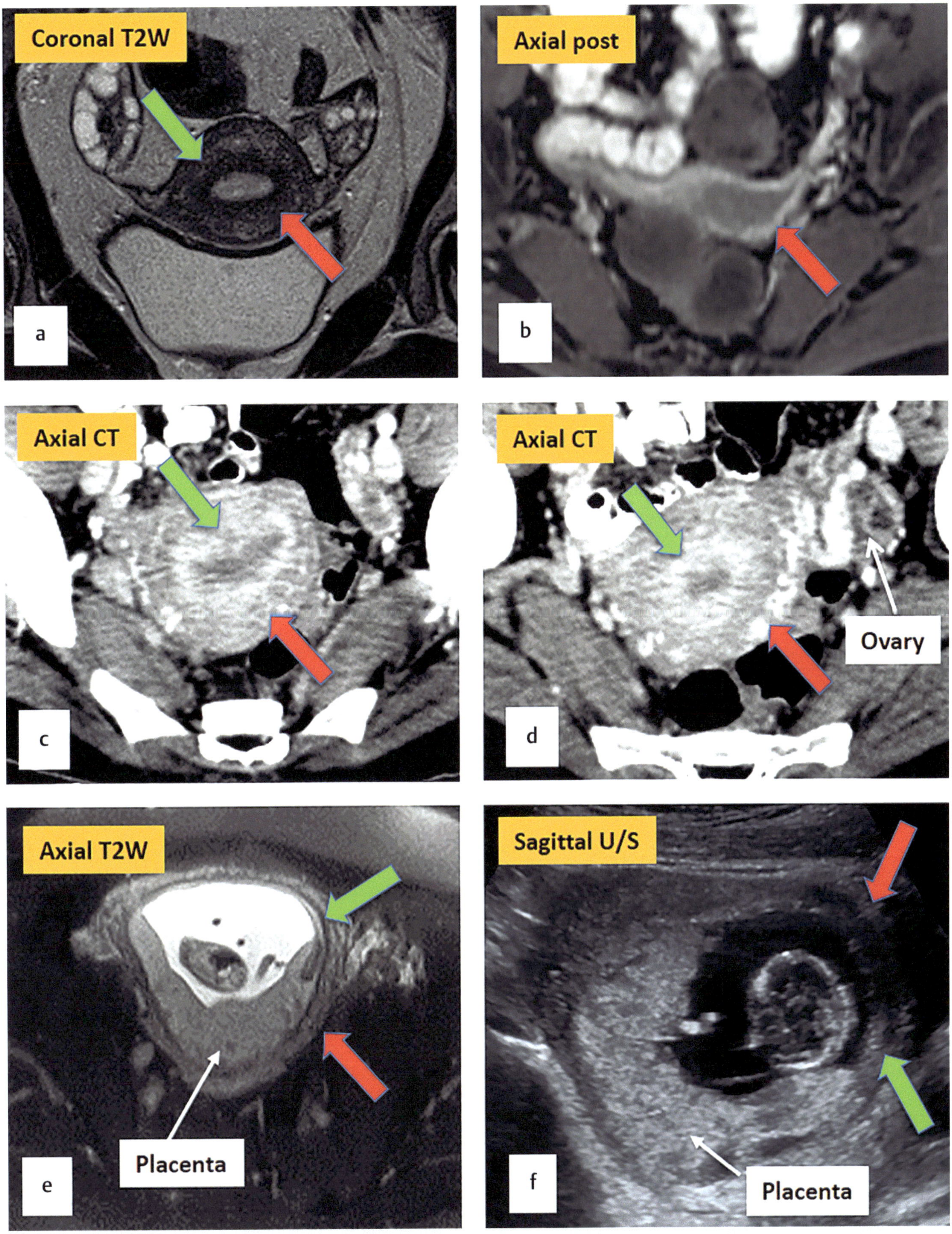

Fig. 16.27 Outer myometrium. It demonstrates variable signal intensity depending on cellularity and vascularity. **(a)** The outer myometrium is usually intermediate in signal intensity due to the presence of less compact smooth muscle, greater water content, and blood vessels. **(b–f)** Enlarged venous sinuses in the subserosal surface of the outer myometrium layer (*red arrows*) is common in multiparous women **(c, d),** pelvic congestion syndrome, and pregnancy **(e, f)**. Junctional zone (*green arrows*) appears low signal on T2-weighted (T2W) images **(a, e)** but may enhance vigorously on the arterial phase images **(c, d)**.

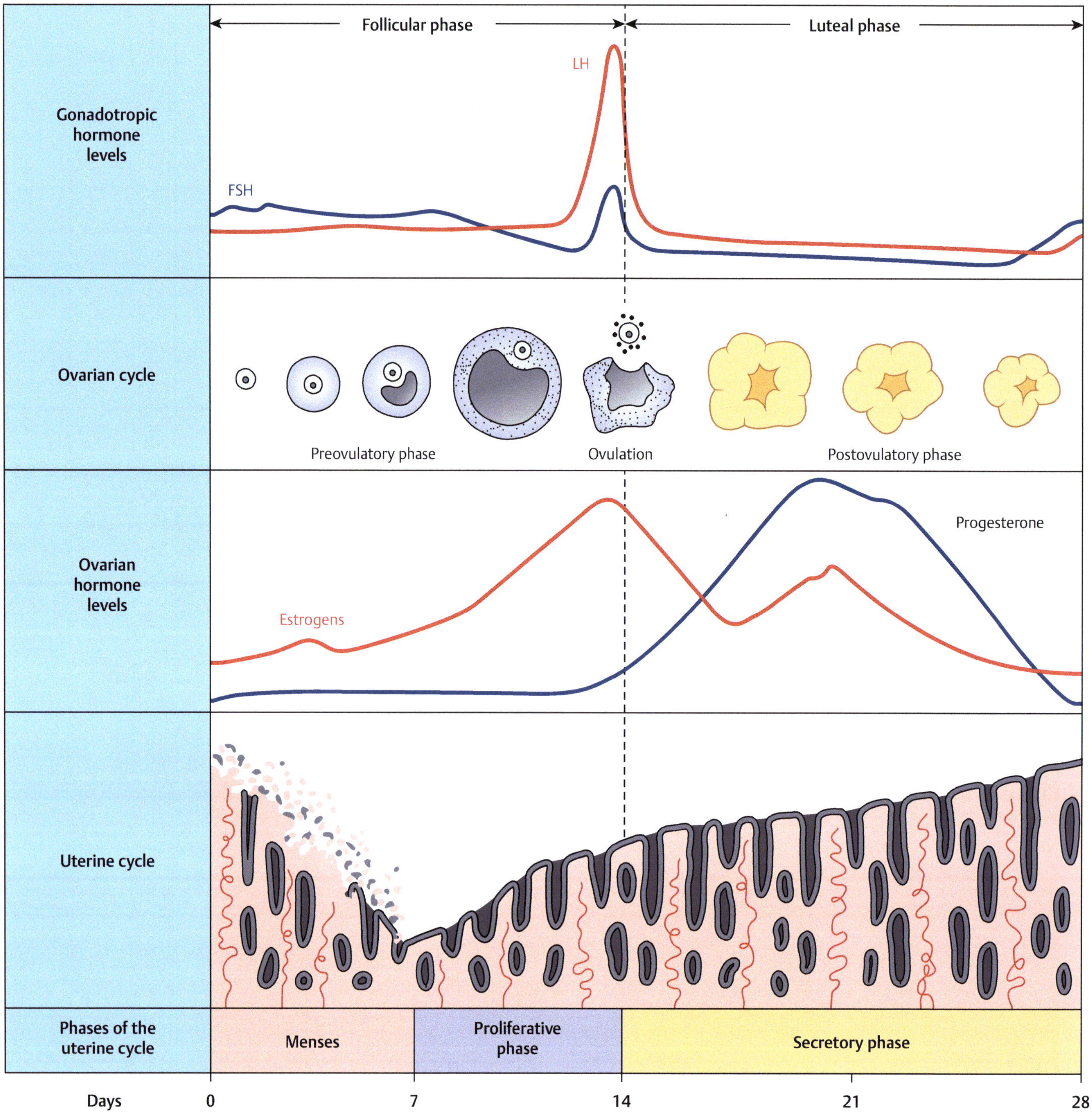

Fig. 16.28 Phases of endometrial growth and menstruation during each monthly female sexual cycle. FSH, follicle stimulating hormone; LH, luteinizing hormone.

16

or diffuse enhancement of the myometrium is seen without subendometrial enhancement (▸Fig. 16.27b, ▸Fig. 16.33). Type 2 enhancement which is observed equally in pre- and postmenopausal women might be related to the higher density of smooth muscle cells and microvessels in the periphery of the uterus. In type 3, faint diffuse myometrial enhancement is observed, exclusively in postmenopausal women and possibly due to age-related decrease in muscle mass and atheromatous process (▸Fig. 16.31).

Cervical Enhancement

The cervix also often demonstrates delayed enhancement in comparison with the uterus and may have a lower density/intensity after contrast administration (▸Fig. 16.31, ▸Fig. 16.33). The relatively greater amount of fibrous tissue in the cervix may account for these differences. In general, following MR contrast material administration, the endocervix (along with vaginal and

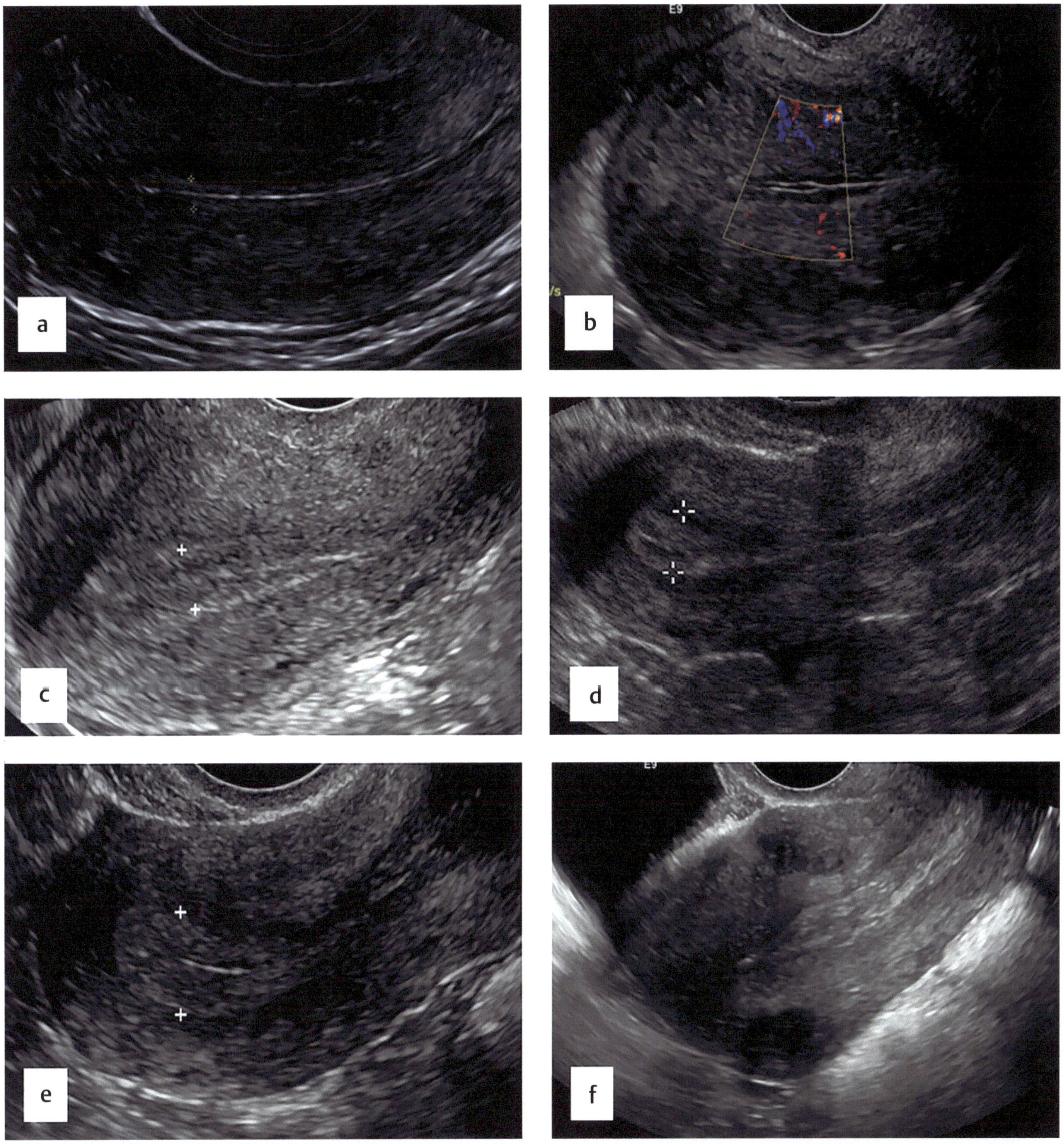

Fig. 16.29 Endometrial appearance during menstrual cycle in premenopausal women. It is thinner during menses **(a)** and early post menstruation **(b)** with a mean of 5 mm but gradually thickens during proliferative phase to reach around 12 mm a day before ovulation **(c)**. The endometrium has a striated pattern, with a hypoechoic inner layer surrounded by a hyperechoic peripheral layer. The striated pattern usually disappears 48 hours after ovulation **(d)**. During secretory phase of the menstrual cycle the striated pattern is replaced **(d)** by a homogeneous echogenic endometrium **(e)**. **(f)** Endometrial canal appears heterogenous during active bleeding.

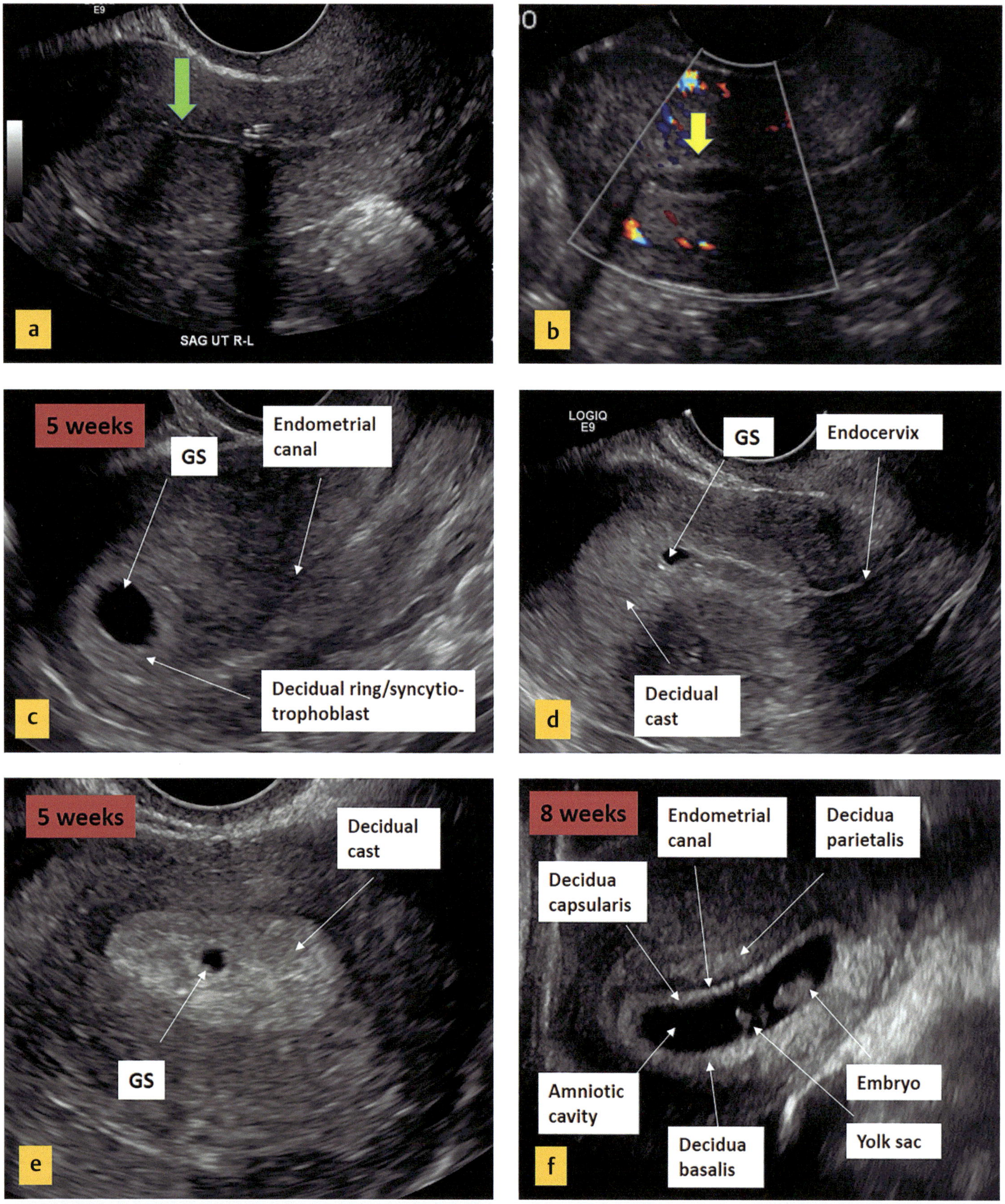

Fig. 16.30 Intrauterine contraceptive devices (IUCDs). **(a)** Decidual reaction (*arrow*) and uniform thinning of the surface endometrium by progesterone-releasing IUCDs is always found after 6 months of treatment. **(b)** Endometrium in a patient with IUCD and vaginal bleeding. The echogenic margin is the endometrium (*yellow arrow*) and hypoechoic region in the center represents blood. **(c–e)** Decidual reaction in early pregnancy after implantation. **(c)** At 5 weeks pregnancy showing a 1 cm gestational sac (GS) surrounded by the syncytio-trophoblast tissue. No significant endometrial proliferation is seen. **(d, e)** Thickening of the endometrial tissue (decidual cast) in response to pregnancy is evident. **(f)** Normal 8 weeks pregnancy. The part of the decidua that interacts with the trophoblast is called the decidua basalis.

16

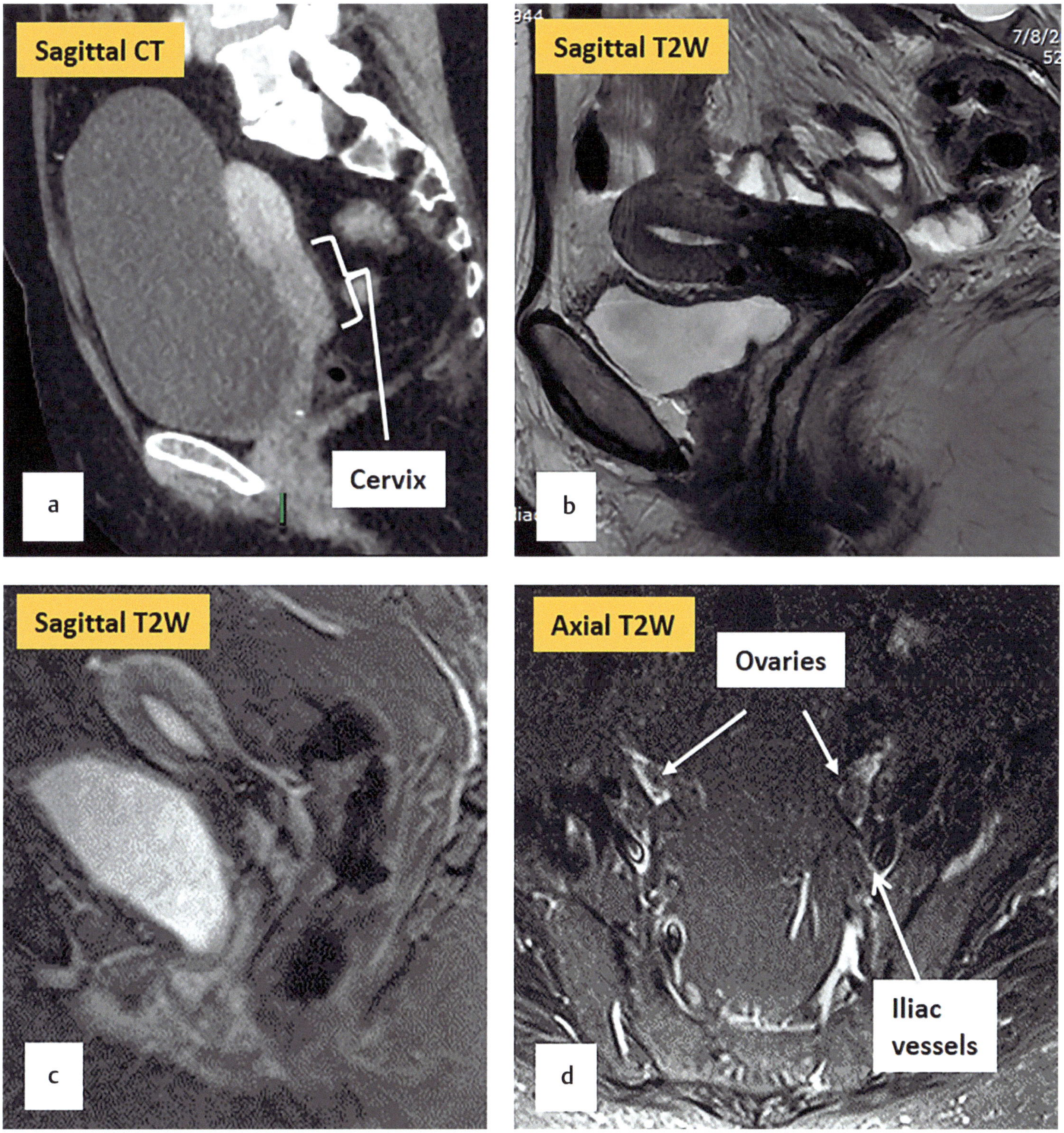

Fig. 16.31 Postmenopausal uterus. **(a)** Postcontrast computed tomography (CT) in a 65-year-old female. The uterine body-to-cervix ratio is approximately 1:1. Faint diffuse myometrial enhancement is observed (Type 3). **(b)** T2-weighted (T2W) magnetic resonance imaging (MRI) in a 72-year-old female shows indistinct transition zone. **(c, d)** Postmenopausal bleeding in a 61-year-old woman. The endometrium appears thickened and hyperplastic. Note small ovaries with no follicles adjacent to the iliac vessels.

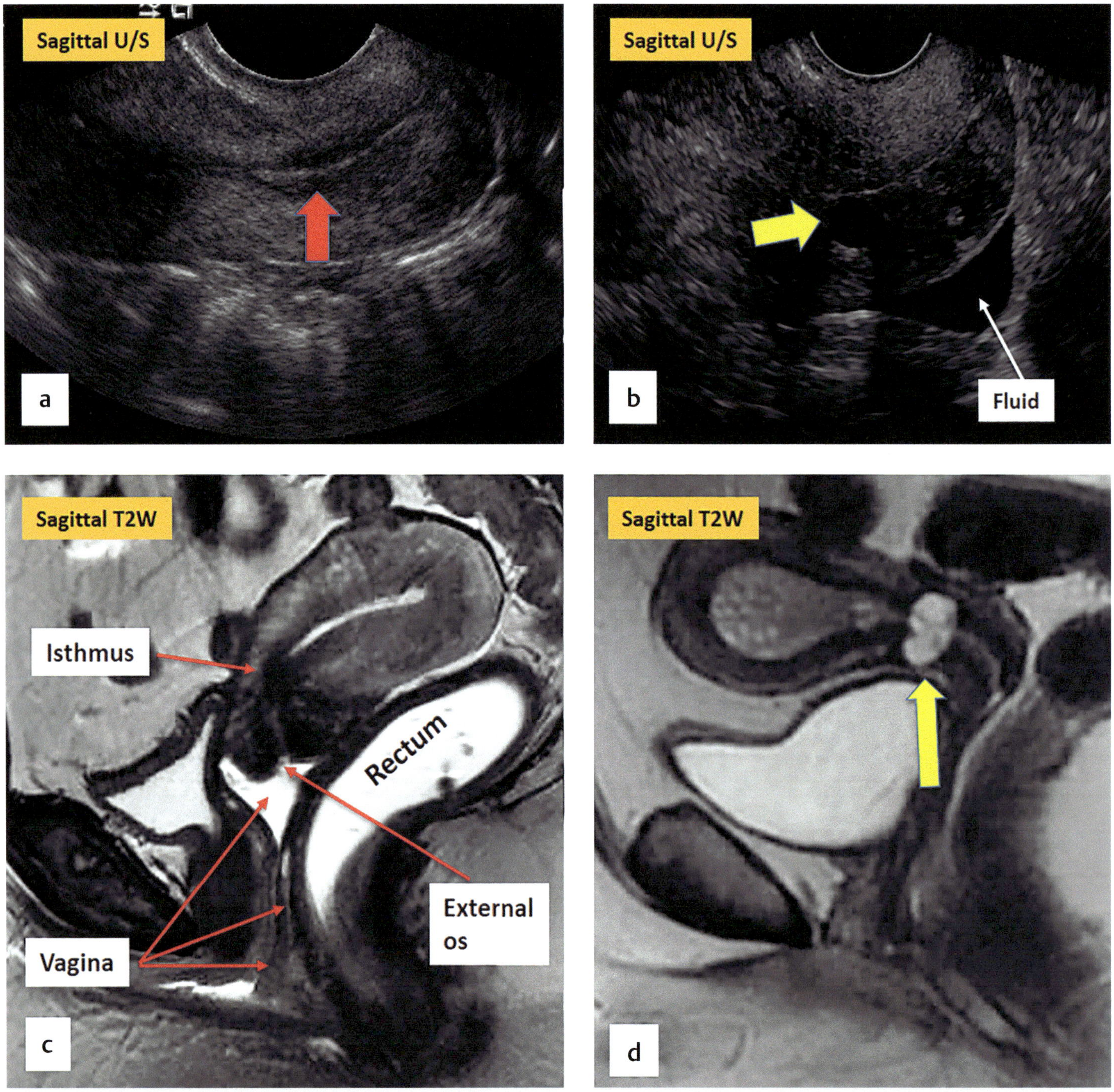

Fig. 16.32 Cervix and endocervix. **(a, b)** Sagittal ultrasound showing a thin echogenic endocervix. It is surrounded by the hypoechoic middle layer of the cervix (*arrow* in **a**). **(c–e)** On T2-weighted (T2W) images, the cervix shows a distinctive trilaminar appearance with an innermost hyperintense layer corresponding to mucosa and secretions, a middle low-signal-intensity layer representing the inner cervical fibrous stroma that is continuous with the junctional zone, and an outermost intermediate-signal-intensity layer representing the outer cervical stroma. Nabothian cysts are shown by *yellow arrows*. **(f)** Normal cervix in pregnancy. Note thickening and edema of the outer cervical stroma.

(*Continued*)

16

rectal mucosa) enhances rapidly, whereas the stroma demonstrates more gradual enhancement[33] (▶Fig. 16.33, ▶Fig. 16.35). A target-shaped zonal enhancement of the cervix has been seen on axial images. It is characterized by intense central mucosal enhancement surrounded by less enhancing inner fibromuscular stroma and an enhancing outer fibromuscular stromal layer (▶Fig. 16.36). Cervical zonal enhancement is common with a type 1 or type 2 uterine enhancement pattern, but not in type 3 uterine enhancement pattern. Higher numbers of fibroblasts and muscle cells in the inner layer may account for its relative

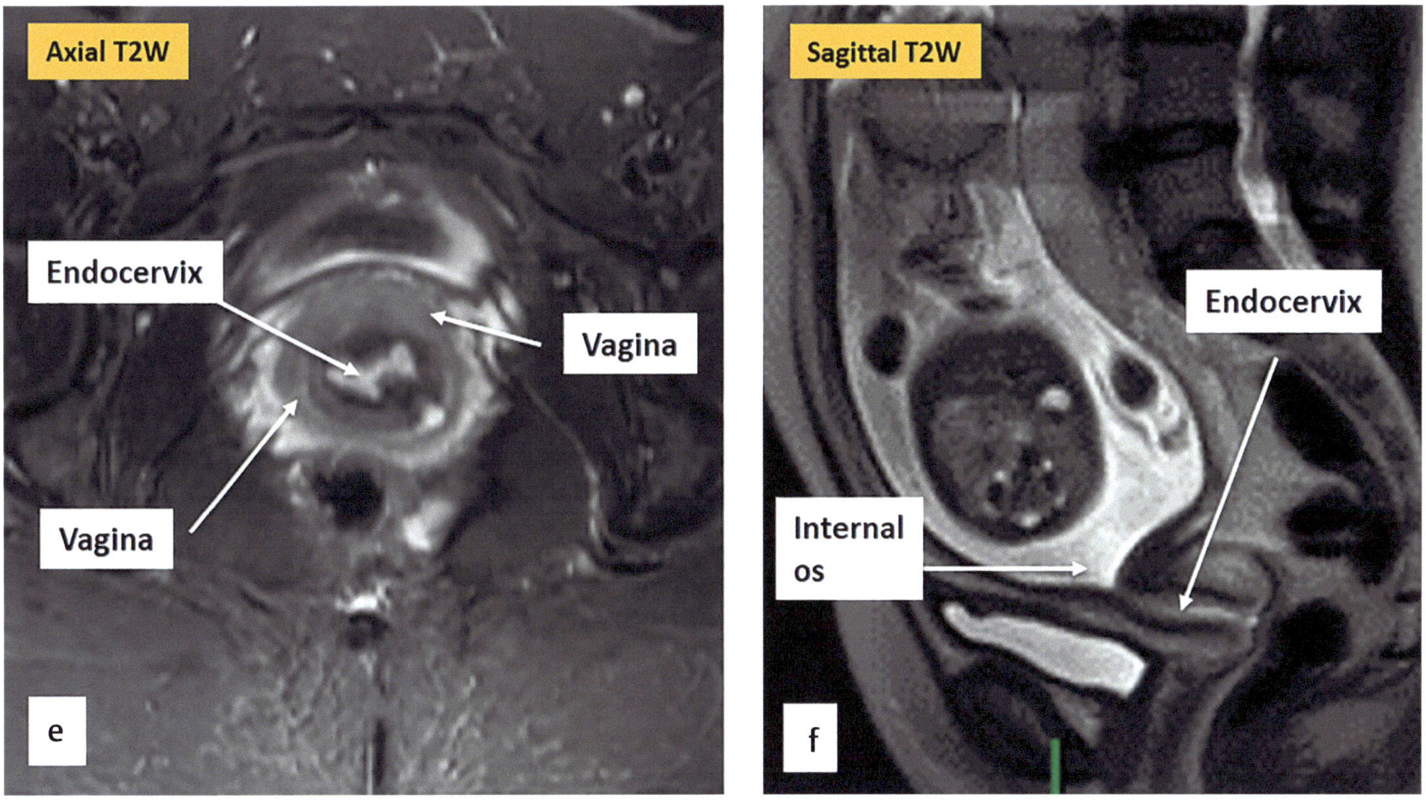

Fig. 16.32 (*Continued*) (e,f)

hypointensity in comparison with the outer stromal layer, which consists of more loosely packed cells and more prominent extracellular space.

Common Pathologies of the Uterus and Cervix

Congenital Anomalies

Mild congenital variants of the uterus are common.[35,36] In general, disruption of Müllerian duct development during embryogenesis can result in a broad and complex spectrum of congenital abnormalities which are collectively termed Müllerian duct anomalies (MDAs) (▶Fig. 16.37). Due to different embryonic origin, MDAs are not associated with anomalies of the external genitalia or ovarian development. On the other hand, MDAs are frequently associated with renal anomalies, and less commonly with vertebral and skeletal abnormalities.[9] Diagnosis of MDAs is clinically important because of the high associated risk of infertility, endometriosis, and miscarriage.

The most common congenital anomaly of the uterus is a septate uterus characterized by a midline uterocervical septum with varying length due to failure of resorption of the medial uterovaginal septum (▶Fig. 16.38). Partial fusion of the Müllerian ducts is known as bicornuate uterus. This anomaly can be complete or incomplete. In incomplete form, the two cornua communicate and form a single endometrial cavity and cervix. In complete form of the anomaly, the failed fusion extends to the cervix and sometimes the vagina (25%) (▶Fig. 16.38). The differentiation of septate uterus from bicornuate uterus is sometimes difficult. On MRI, in a septate uterus, the apex of the external fundal contour is more than 5 mm above a line drawn between the uterine ostia, whereas in a bicornuate or didelphys uterus, the apex of the external fundal contour is below or less than 5 mm above the interostial line.[36] Complete failure of fusion of the Müllerian ducts is known as uterine didelphys which is characterized by two separate uterine horns cervix and endometrial cavity. A longitudinal vaginal septum is seen in 75% of patients. Failure of one Müllerian ducts to develop and elongate results in a unicornuate uterus but the cervix is usually intact. In Müllerian aplasia or Mayer-Rokitansky-Küster-Hauser syndrome, the uterus, cervix, and upper part of the vagina are absent.

Acquired Pathologies

Common acquired pathologies include uterine fibroid (leiomyoma), endometrial hyperplasia, endometrial cancer, and cervical cancer.[37,38]

Uterine fibroids are the most common benign tumors of smooth muscle cell origin in female genital tract (▶Fig. 16.39). They are discrete nodular tumors varying in size and number and may be found intramurally, subserosally, or submucosally. The intramural form is the most common type. Small uterine fibroids are usually asymptomatic but larger ones can cause bleeding, pain,

16

and infertility, and rarely (<2%) undergo sarcomatous changes. Imaging appearance of the fibroids is different depending on degree of cellularity, fibrosis, calcification, necrosis, cystic transformation, and hemorrhage (▶Fig. 16.39). Fibroids are usually hypervascular and if symptomatic will be treated with surgery or percutaneous catheter embolization. The degree of vascularity (postcontrast enhancement) is important to predict the response of lesions to uterine arterial embolization.

Endometrial diseases including hyperplasia, polyps, and cancer are important category of diseases that can be evaluated with

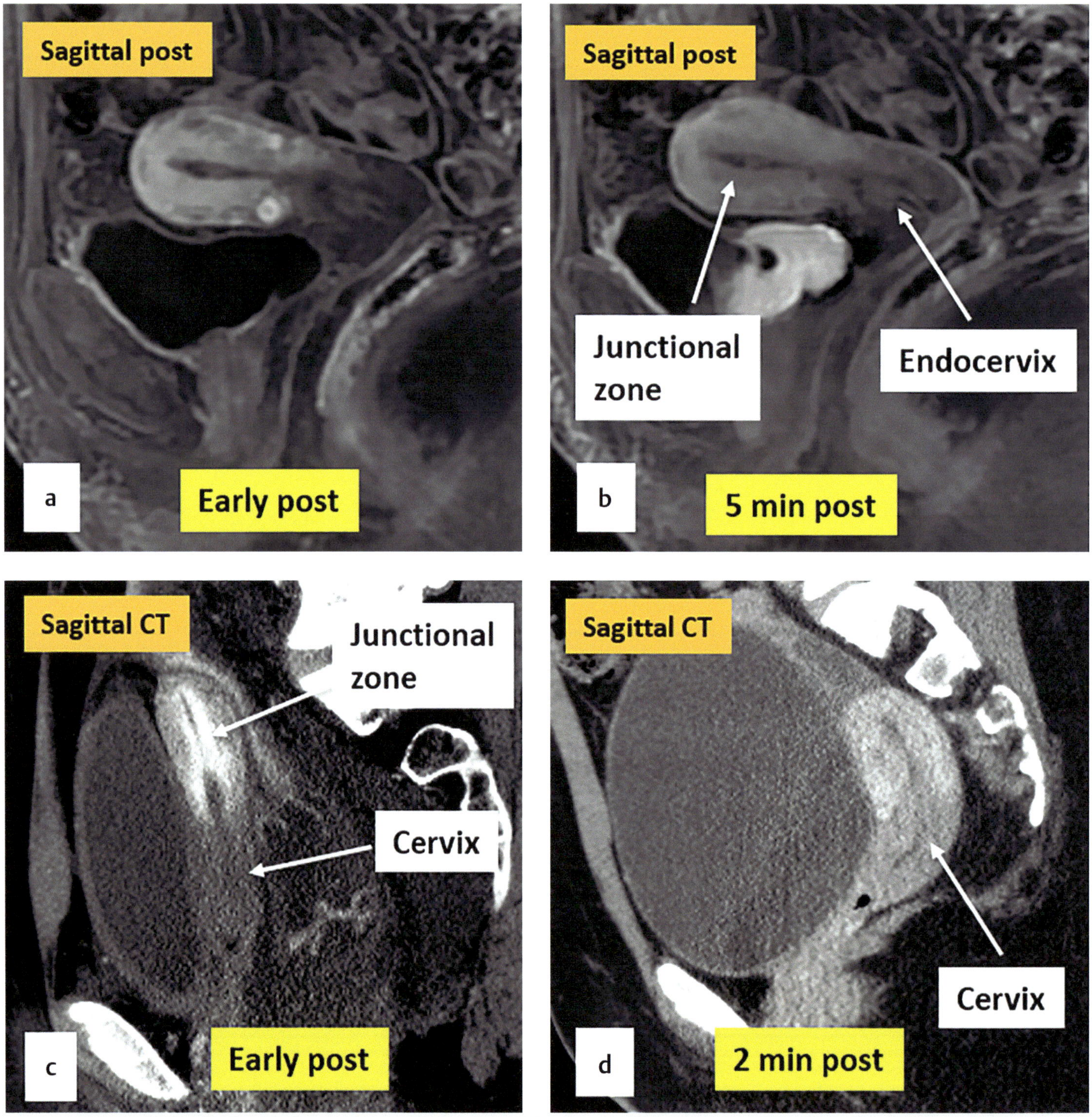

Fig. 16.33 Postcontrast uterus and cervix. **(a, c)** On magnetic resonance imaging (MRI) and computed tomography (CT), the myometrium enhances on both arterial and venous phase images, whereas the endometrium enhances less and appears hypointense/hypodense in comparison with the myometrium. The normal cervix usually demonstrates less intense enhancement or enhances later than the uterine corpus. Thick band of enhancement corresponding to the junctional zone is seen in early arterial phase images (Type 1). **(b)** Less enhancing junctional zone on delay image. **(d)** Homogenous enhancement of the myometrium. In general, after the contrast material is administered, the endocervix (along with vaginal and rectal mucosa) enhances rapidly.

16

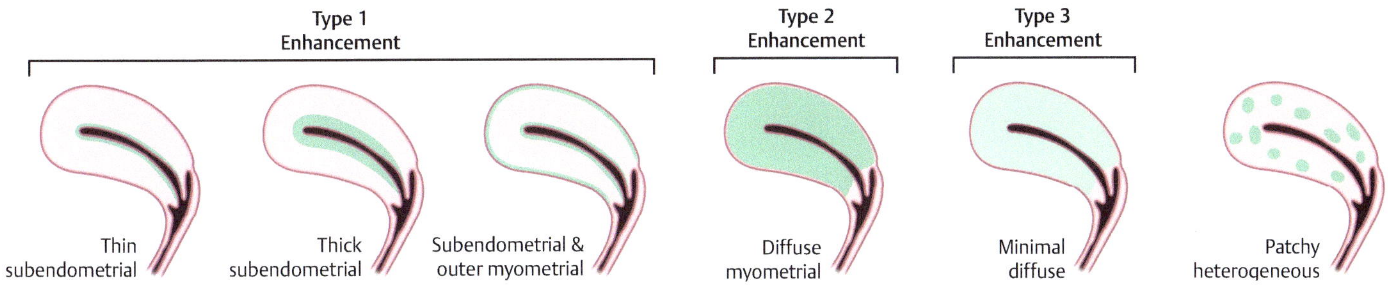

Fig. 16.34 Schematic shows the patterns of uterine enhancement that may be seen at postcontrast computed tomography (CT) or magnetic resonance imaging (MRI). (Modified with permission from Yitta et al.[33])

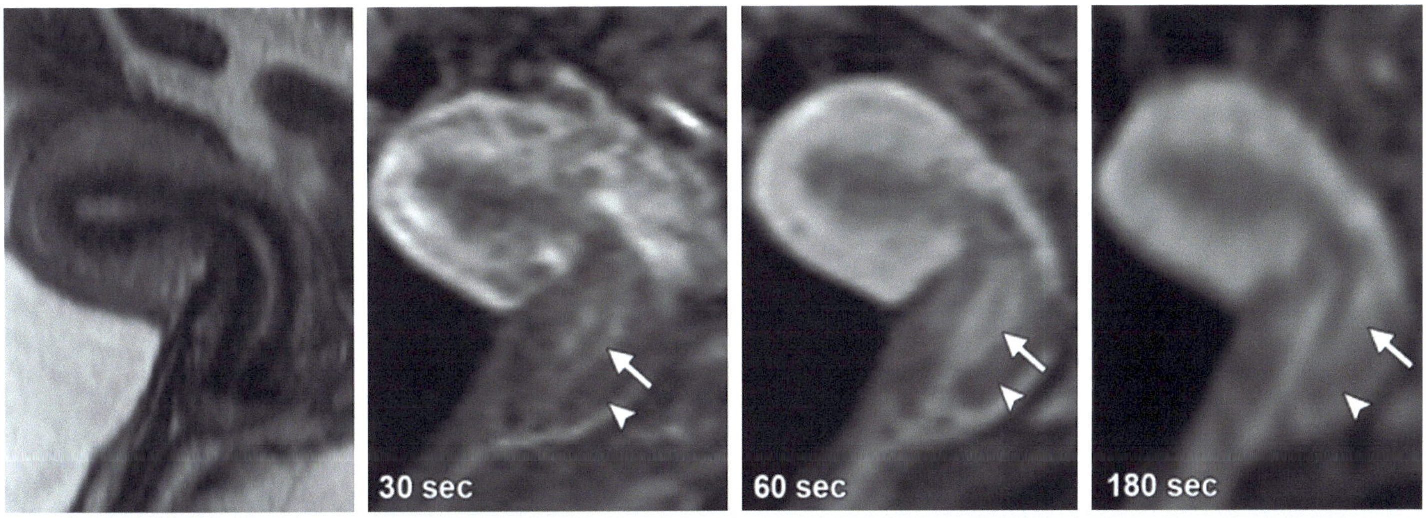

Fig. 16.35 Sagittal T2-weighted (T2W) magnetic resonance (MR) image (first image on left) and sequential T1-weighted (T1W) MR images obtained with delays of 30, 60, and 180 seconds after the administration of contrast material show normal appearances of the uterus and cervix. The myometrium appears hyperintense throughout contrast-enhanced T1W imaging, whereas the endometrium and junctional zone are relatively hypointense at 30 and 60 seconds, showing gradual enhancement over time. The endocervix (*arrows*) enhances rapidly and remains enhanced, whereas the inner and outer fibrous stromal layers (*arrowheads*) enhance gradually. (Reproduced with permission from Yitta et al.[33])

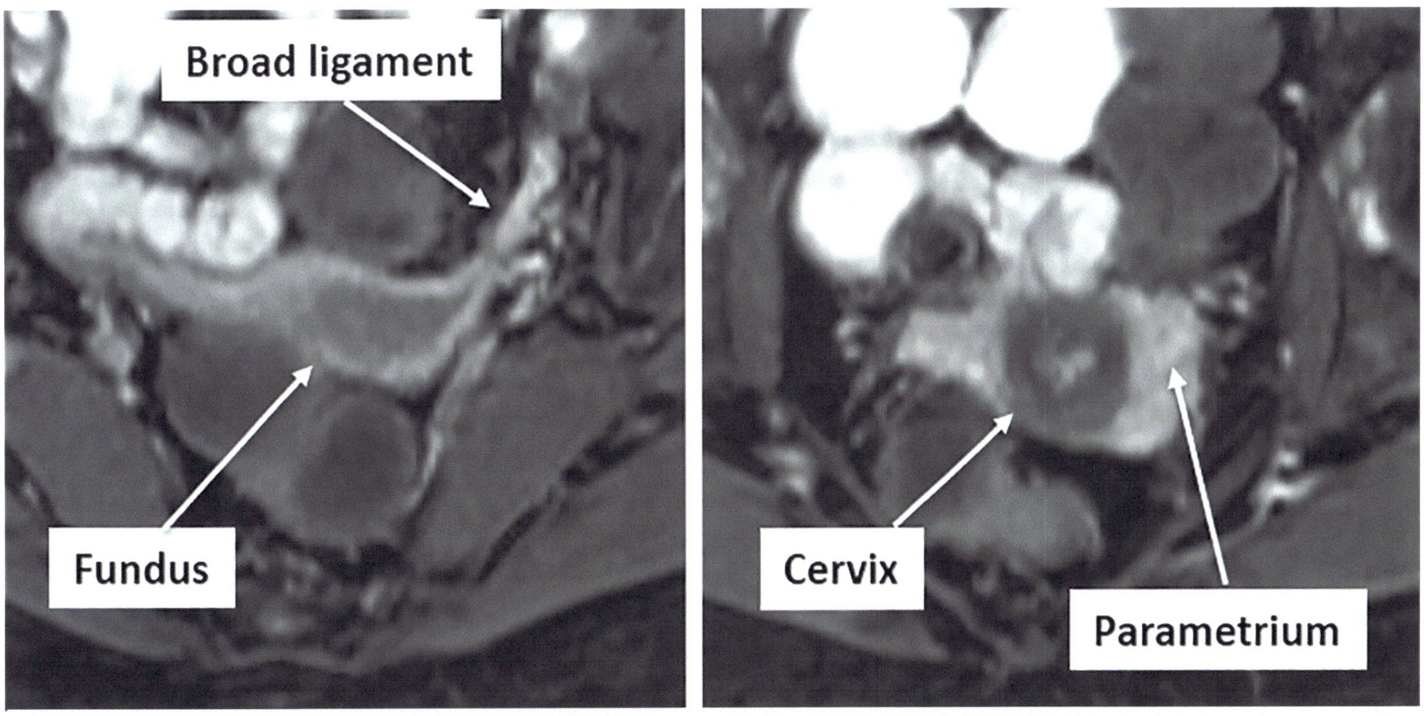

Fig. 16.36 Early, postcontrast, axial magnetic resonance (MR) images showing early enhancement of the endocervix whereas the endometrium remains unenhanced. A target-shaped zonal enhancement of the cervix is characterized by intense central mucosal enhancement surrounded by less enhancing inner fibromuscular stroma and an enhancing outer fibromuscular stromal layer. Also note type 3 peripheral enhancement of the fundus and intense enhancement of the parametrium and broad ligaments.

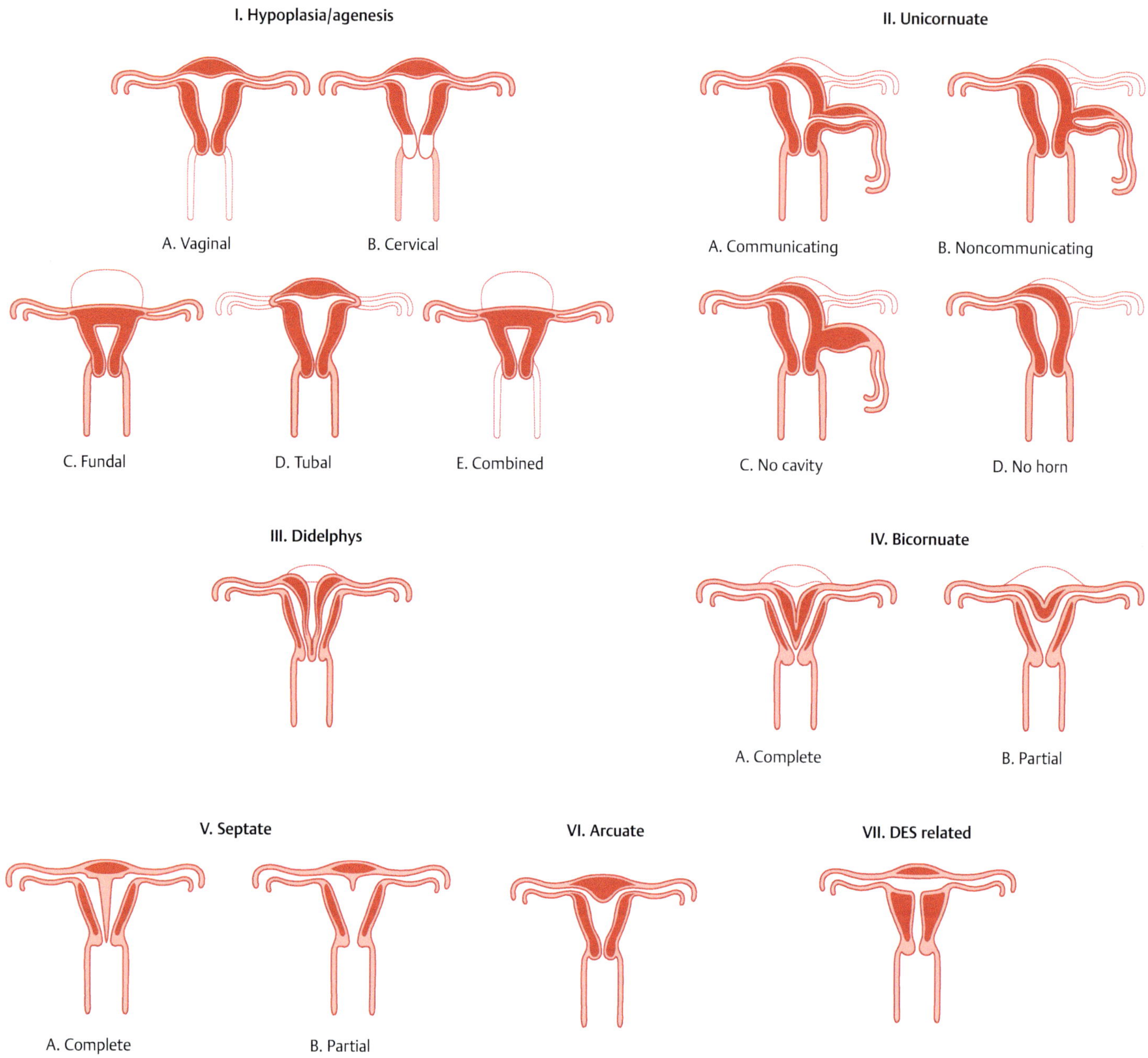

Fig. 16.37 Classification of Müllerian duct anomalies on the basis of the American Society for Reproductive Medicine system. DES, diethylstilbestrol.

ultrasound and MRI. Diffuse endometrial hyperplasia is usually a nonspecific incidental finding which is frequently seen in women who use oral progestational agents for contraception and in early pregnancy, ectopic pregnancy, or postpartum patients (▶Fig. 16.29, ▶Fig. 16.31, ▶Fig. 16.40). Endometrial polyps are polypoid lesions of endometrial glands and stroma that are associated with endometrial hyperplasia and possible malignant changes. Endometrial adenocarcinoma is the most common gynecologic malignancy in the United States. On ultrasound, the lesion usually appears as thickening of the endometrium, polypoid thickening, or heterogeneous mass. The disruption of subendometrial halo may be suggestive of endometrial invasion. MRI has a higher accuracy to diagnose myometrial invasion and cervical extension.

Cervical cancer is the leading gynecologic cancer in women and the third female malignancy in the United States. The two most common histologic subtypes of cervical cancer are squamous cell that arises from the squamocolumnar junction and adenocarcinoma originating from the endocervical cavity. Early stages of the disease are diagnosed with clinical examination, pap smear, and biopsy. CT and MRI are used for staging of the disease before surgery and to determine tumor size, parametrial invasion, pelvic sidewall invasion, adenopathy, and distant metastasis.

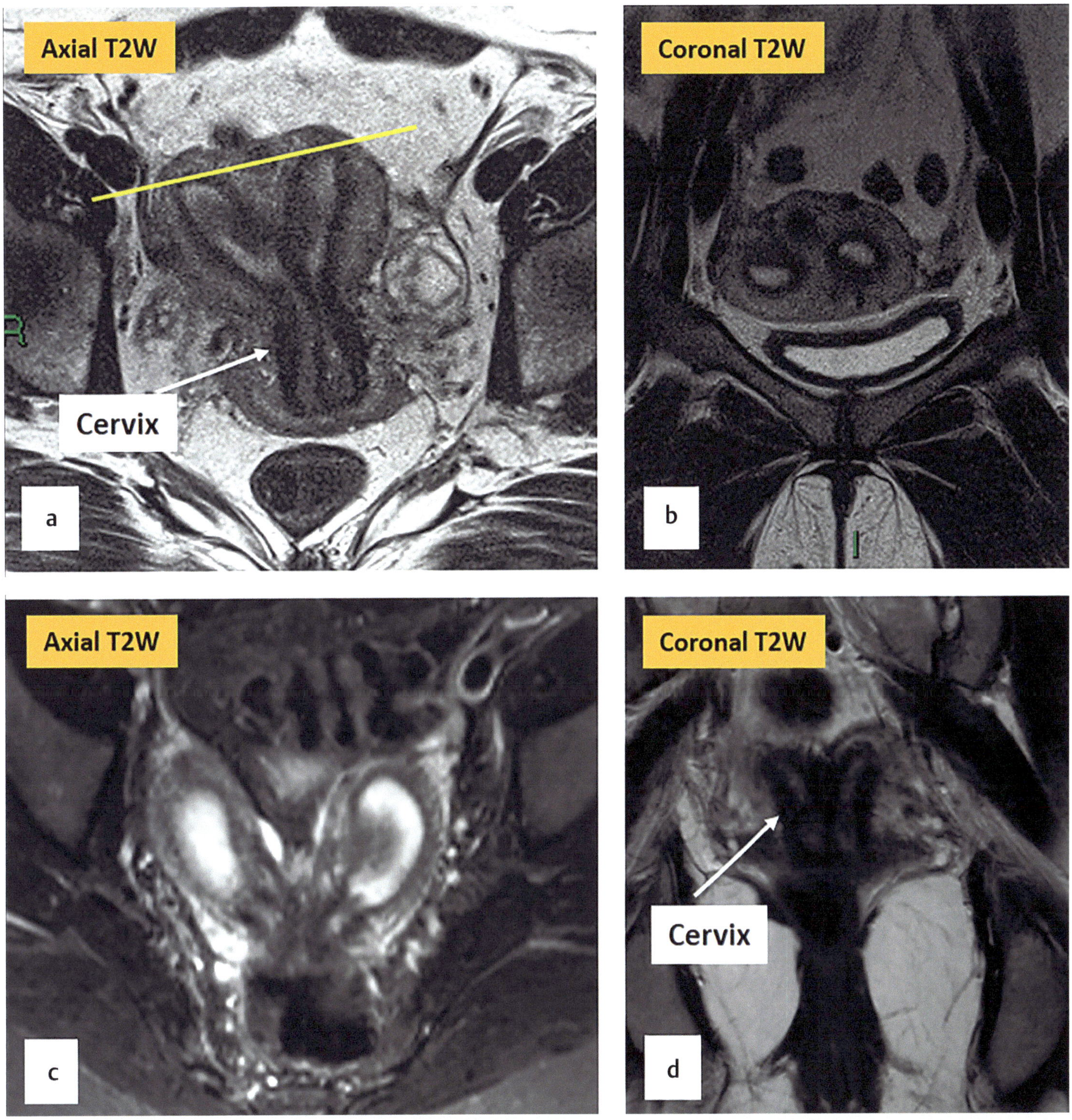

Fig. 16.38 (a, b) Septate uterus in a 34-year-old. The septum extends into the cervix. **(c, d)** Complete bicornuate uterus in a 30-year-old. The septum is seen in the endocervix **(d)**. When the apex of the fundal contour is above a *yellow line* drawn between the tubal ostia, the uterus is septate. When the apex of the fundal contour is below a line drawn between the tubal ostia, the uterus is bicornuate.

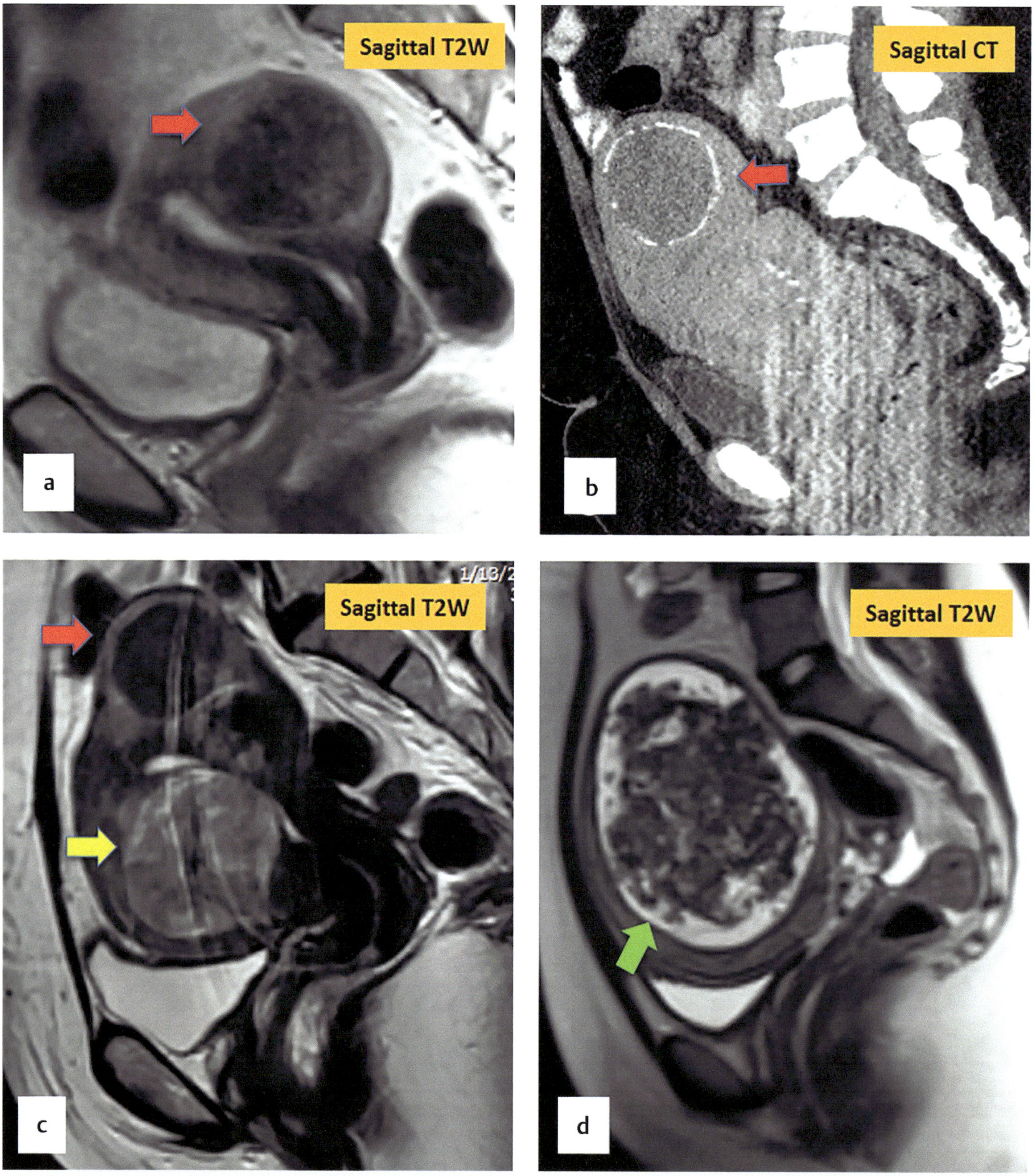

Fig. 16.39 Uterine fibroids. Uterine fibroids demonstrate different imaging appearance. **(a–c)** Typical appearance is a low signal mass on T2-weighted (T2W) image with or without calcification (*red arrows*). **(c)** Lesion may appear heterogeneously bright on T2W due to less cells and higher amount of connective tissue (*yellow arrow*). **(d)** Rarely, hemorrhage and cystic degeneration occurs (*green arrow*).

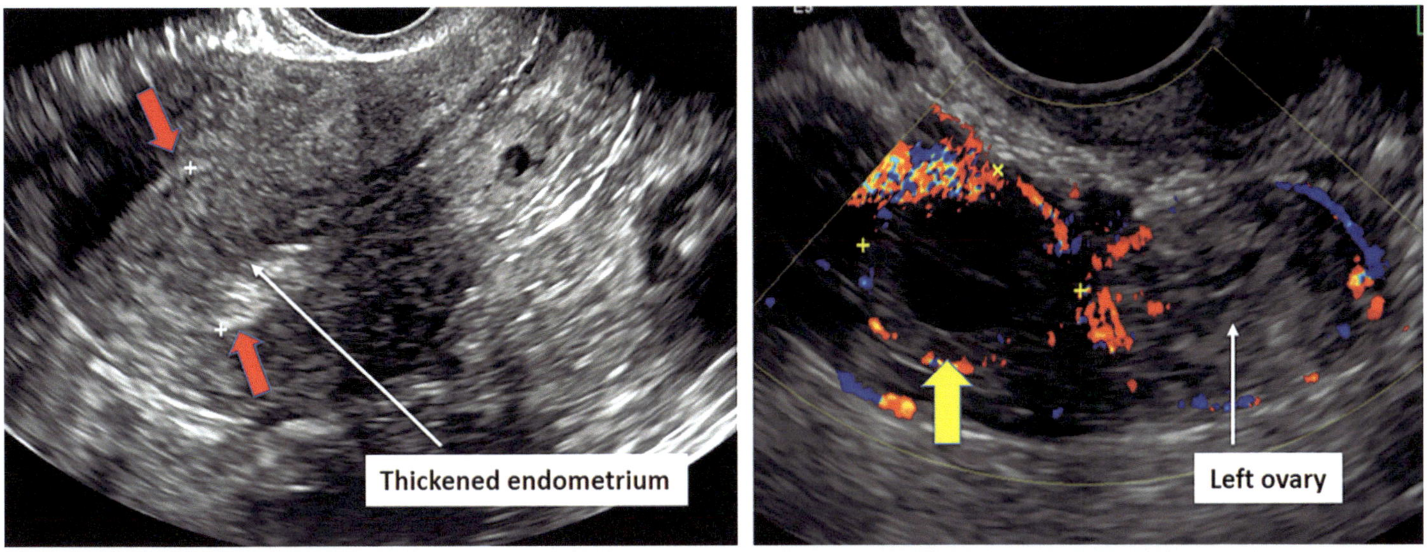

Fig. 16.40 Endometrial thickening (2 cm shown by *red arrows*) in a 42-year-old women with tubal ectopic pregnancy (*yellow arrow*).

Fallopian Tube

The fallopian tubes are derived from the paramesonephric (Müllerian) ducts at 5 weeks of intrauterine development (▶Fig. 16.2). The cranial portions of the paramesonephric ducts develop into the fallopian tubes, and the caudal portions fuse to form the uterus. The tubes are 10 to 12 cm in length and 1 to 4 mm in diameter. They extend from the ovaries to the uterine cornua (▶Fig. 16.22). Located in the superior edge of the broad ligament, each tube serves to transport the ovum from the ovary to the endometrial cavity. The peritoneal reflection draping over the salpinges forms the mesosalpinx. The fallopian tubes extend anteriorly from the posterosuperior aspect of the uterine fundus and open into the peritoneal cavity. A fallopian tube is divided into four anatomic segments. From medial attachment point to uterus to the distal end these segments are interstitial, isthmic, ampullary, and infundibular[39] (▶Fig. 16.22). The interstitial or intramural segment is located within the myometrium at the uterine cornua, above the round ligament of the uterus. The intramural portion is encased by the smooth muscle of the uterus and may appear narrowed or occluded if there is spasm of the muscle during imaging study. The isthmic segment is the narrowest and relatively long portion of the fallopian tube. Closer to the ovary, the tube widens to form the ampullary segment, which accounts for over one-half of its total length (▶Fig. 16.2). The most lateral segment is the funnel-shaped infundibulum.[40] The free edge is composed of 25 fingerlike projections, or fimbriae (▶Fig. 16.2, ▶Fig. 16.41). The fimbriae are suspended over the ovary and open in the peritoneal (coelomic) cavity, thereby capturing the ovum once it is released from the ovary.

The wall of the fallopian tube is composed of three layers: mucosa, muscularis, and serosa. Mucosal folds form fingerlike projections called plicae that increase in complexity toward the infundibulum, where they become contiguous with the fimbriae. The epithelium is composed of ciliated and nonciliated columnar cells. The fluid secreted by the epithelium progresses toward the fimbrial end of the tube and is released into the peritoneal cavity. Ciliated epithelium and mucosal plicae propel the ovum toward the uterine cavity. The muscular layer is composed of both circular and longitudinal muscle.

The normal fallopian tube is difficult to distinguish from surrounding vessels and ligaments at imaging. The normal tube is not usually visualized on cross-sectional imaging studies unless it is abnormally dilated (hydrosalpinx) or surrounded by fluid (▶Fig. 16.42). Occasionally, it can be seen as an elongated structure adjacent to the ovary (▶Fig. 16.21). In this case, the fallopian tubes are seen as paired, thin, serpentine juxtauterine structures extending either anteriorly or posteriorly into the cul-de-sac. The normal fimbriae can be seen as short fingerlike projections when surrounded by periovulatory fluid on CT, ultrasound, or when opacified by contrast at HSG. The round ligament, which is often observed on CT and MR images, can be distinguished from the fallopian tube by virtue of its anterior course toward the internal inguinal ring (▶Fig. 16.21). HSG is the radiographic study of choice for evaluation of the fallopian tubes and is used predominantly in the evaluation of infertility.[40] Radiographic images are obtained following retrograde administration of the contrast into the endometrial cavity through a catheter (▶Fig. 16.22). CT with intrauterine contrast, MR hysterography with a heavily T2W sequence, and contrast-enhanced sonohysterography have also been used for visualization of the fallopian tubes.[41]

Common Pathologies

Congenital anomalies of the fallopian tubes include aplasia, hypoplasia, accessory ostia, and congenital diverticula.[39] Common acquired diseases of the fallopian tubes are infection and ectopic pregnancy. Fallopian ligation is a common procedure to prevent pregnancy. Infections of the upper female genital tract (pelvic inflammatory disease) affecting the fallopian tube include salpingitis, pyosalpinx, and tubo-ovarian abscess which are best shown by MRI (▶Fig. 16.42). Hydrosalpinx is fluid-filled dilatation of the fallopian tube resulting from obstruction of the ampullary

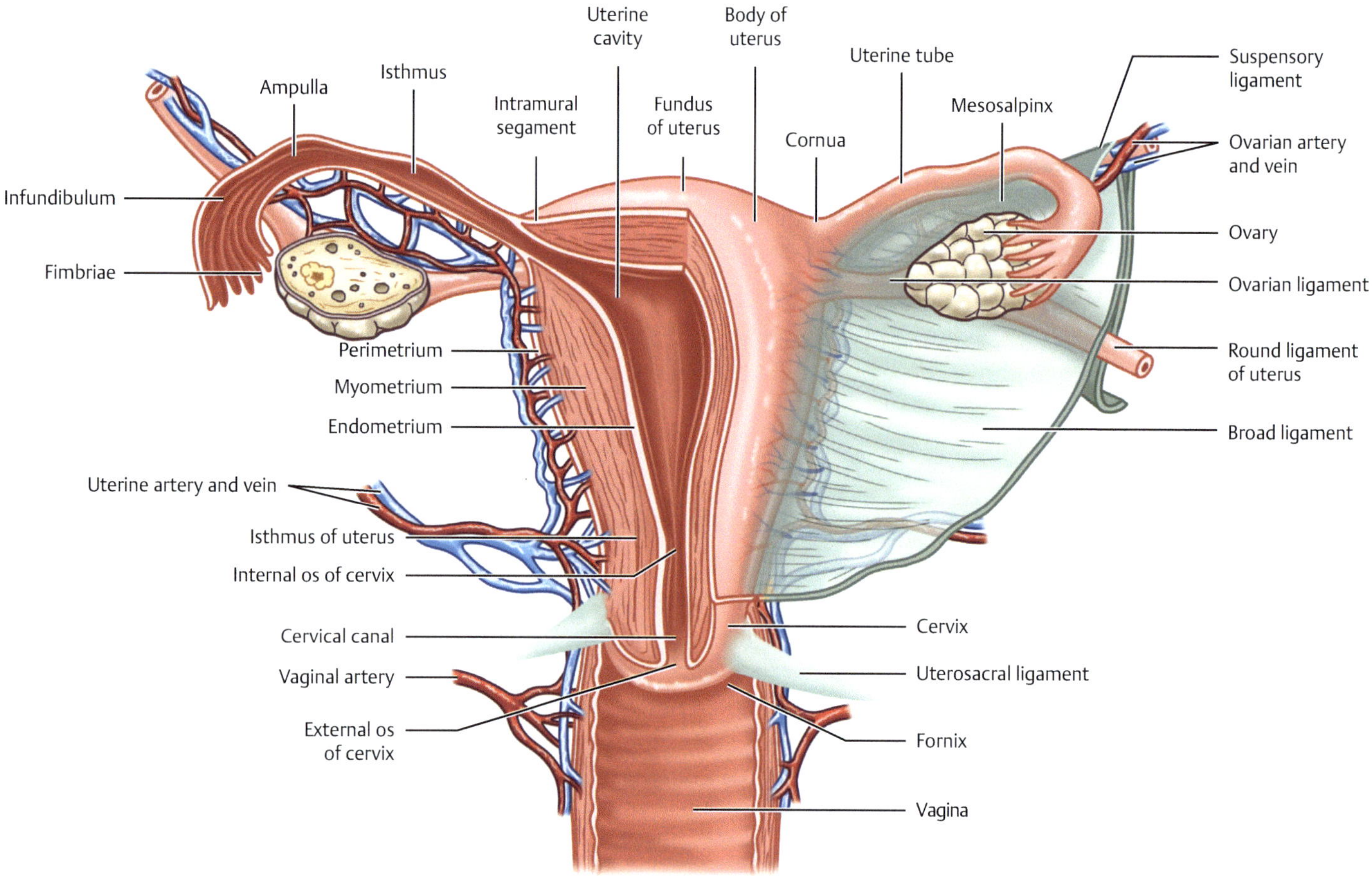

Fig. 16.41 Posterior view of the uterus and fallopian tube structures.

segment. The most common cause is adhesions from prior infections. A dilated fallopian tube is seen as a thin- or thick-walled C- or S-shaped fluid-filled tubular structure close to the uterus and ovary (▶Fig. 16.43). The presence of longitudinal tubal folds within the dilated tube differentiates hydrosalpinx from other cystic structures.[39]

Vascularity and Innervation of Uterus and Fallopian Tubes

Uterine Arteries

16

The internal iliac artery terminates at the upper border of the greater sciatic notch (60%) into two main stems, one anterior and one posterior, in most cases.[42–44] The main branches of the posterior division are the lateral sacral, iliolumbar, and superior gluteal arteries. The anterior division gives rise to the parietal (obturator, inferior gluteal, and internal pudendal) and visceral (vesical, uterovaginal, and middle rectal) vessels (▶Fig. 16.11). The uterine artery provides the dominant blood flow to the uterus; however, the ovarian artery (arising from the abdominal aorta) is frequently involved in blood supply to hypervascular lesions (i.e., large or multiple uterine fibroids) (▶Fig. 16.44, ▶Fig. 16.45, ▶Fig. 16.46, ▶Fig. 16.47). The anterior division of the internal iliac artery where the uterine artery originates is subject to numerous branching variations (▶Fig. 16.44). The uterine artery commonly (51%) arises from the anterior division of the internal iliac artery (▶Fig. 16.11, ▶Fig. 16.44). However, it may also arise as the first branch of the internal iliac artery (6%) or rarely originate directly from the ovarian artery. The uterus may be supplied by multiple small uterine branches.

Catheter, CT, and MR angiographies are common imaging methods to show the uterine arteries. On catheter angiogram, the best projection to identify the origin of the uterine artery is the contralateral anterior oblique with 20 to 30 degrees inclination[42] (▶Fig. 16.44a). The artery is enlarged in pregnant patients and those with leiomyoma (▶Fig. 16.47). It crosses approximately 2 cm from the cervix above the ureter to which it supplies a small branch. Reaching the side of the uterus near the uterocervical junction, it ascends in a tortuous manner between the two layers of the broad ligament to reach the junction of the uterine tube and uterus[33,44] (▶Fig. 16.45, ▶Fig. 16.46). The uterine artery on each side divides into three branches, ascending, transverse, and descending branches. The ascending branch of uterine artery course along the broad ligament and extends laterally to form terminal tubo-ovarian branches which anastomose with the ovarian artery branches at the ovarian hilum (▶Fig. 16.45, ▶Fig. 16.46). The descending branch of the uterine artery supplies the cervix and anastomoses with the branches of the vaginal artery to form two median longitudinal vessels known as azygos

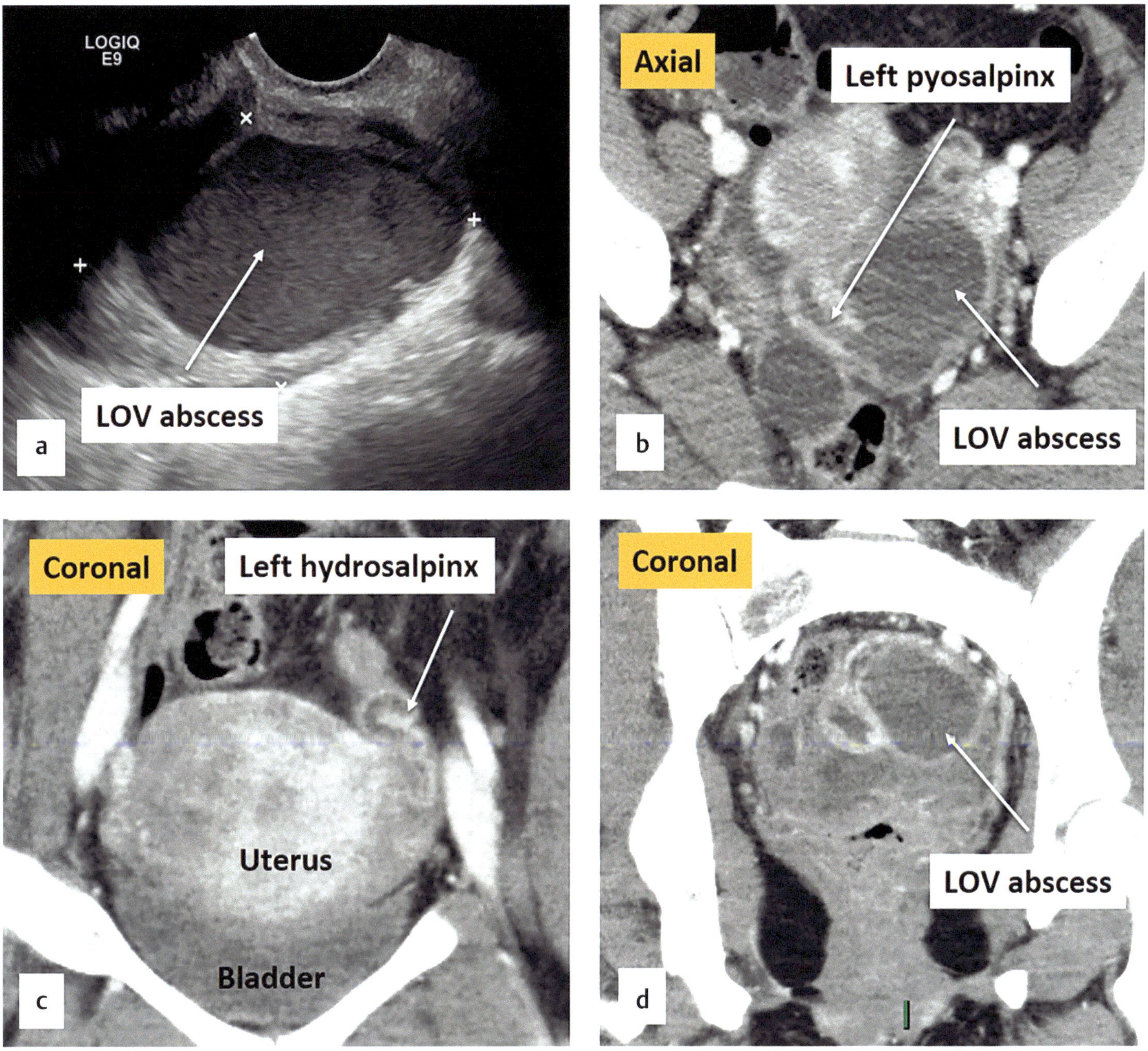

Fig. 16.42 Left tubo-ovarian abscess and hydrosalpinx in a 35-year-old patient. **(a)** Transvaginal ultrasound. **(b–d)** Contrast-enhanced computed tomography (CT). A large left side abscess is seen in the left ovary (LOV). The abscess communicates with the dilated left fallopian tube consistent with pyosalpinx.

16

arteries that descend along the anterior and posterior wall of the vagina (▶ Fig. 16.45). Multiple tortuous transverse branches, also known as the arcuate arteries, penetrate the uterine wall, divide, and run circumferentially as anterior and posterior arcuate arteries to reach the midline where they anastomose with the arcuate arteries on the contralateral side to form arcades.[45] Many helical arteriolar rami, also known as basal arteries, extend from the arcuate arteries and penetrate the endometrium[45] (▶ Fig. 16.45). These small endometrial arteries are subject to variation in size and numbers with the menstrual cycle. Their length, caliber, and tortuosity will diminish during the proliferative phase and increase during the secretory phase. It is believed that the territorial divide between the uterine and ovarian arterial supply to the uterus change during the menstrual cycle. It is shown that the cornual area of the uterus is supplied mainly from the ovarian artery in the proliferative (follicular) phase and from the uterine artery in the secretory luteal phase.[46,47]

For treatment of obstetrical hemorrhage especially from the uterine fibroids, uterine artery embolization is commonly used as an alternative technique to myomectomy and hysterectomy (▶ Fig. 16.48). During catheter embolization of the uterine artery attention to anastomotic variation of uterine artery with other

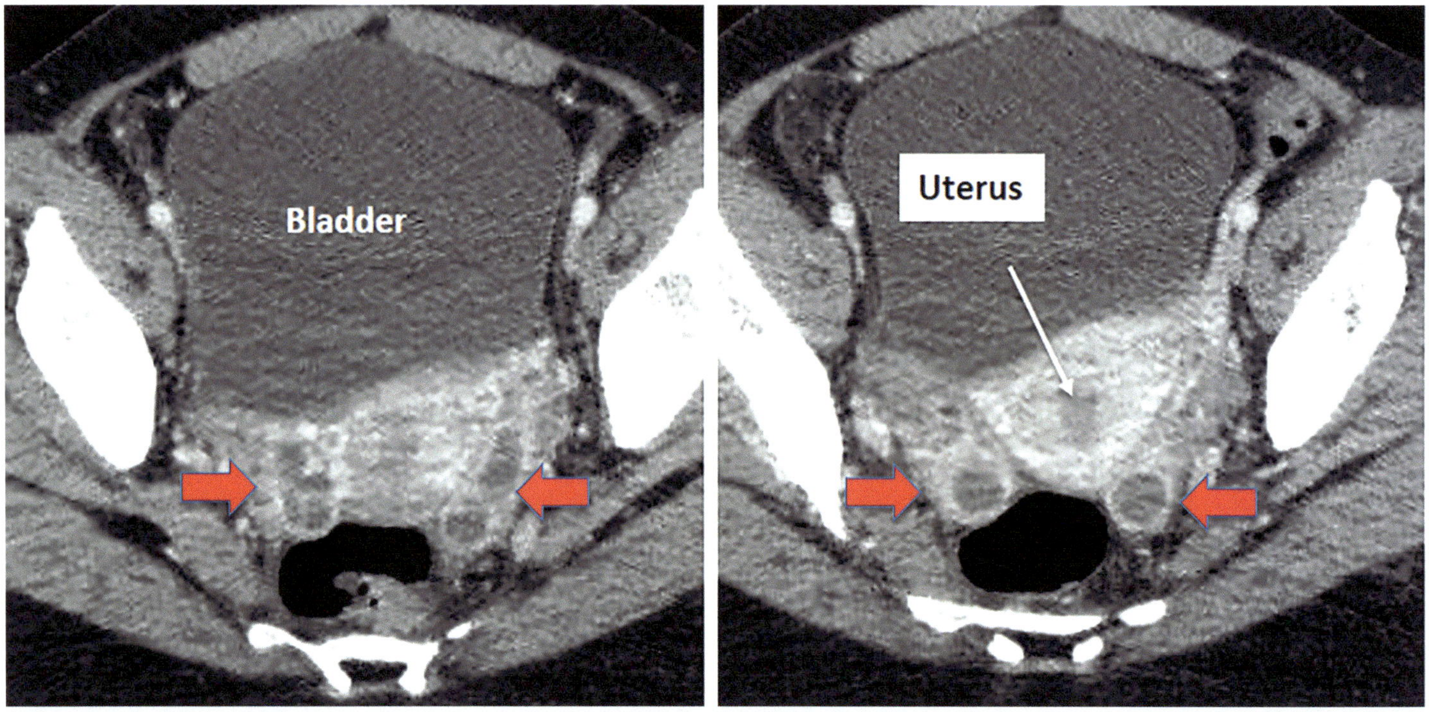

Fig. 16.43 Hydrosalpinx shown by contrast-enhanced computed tomography (CT). Fluid-filled dilatation of the fallopian tubes resulting possibly from obstruction of the ampullary segment by adhesions from prior infections are seen (*arrows*).

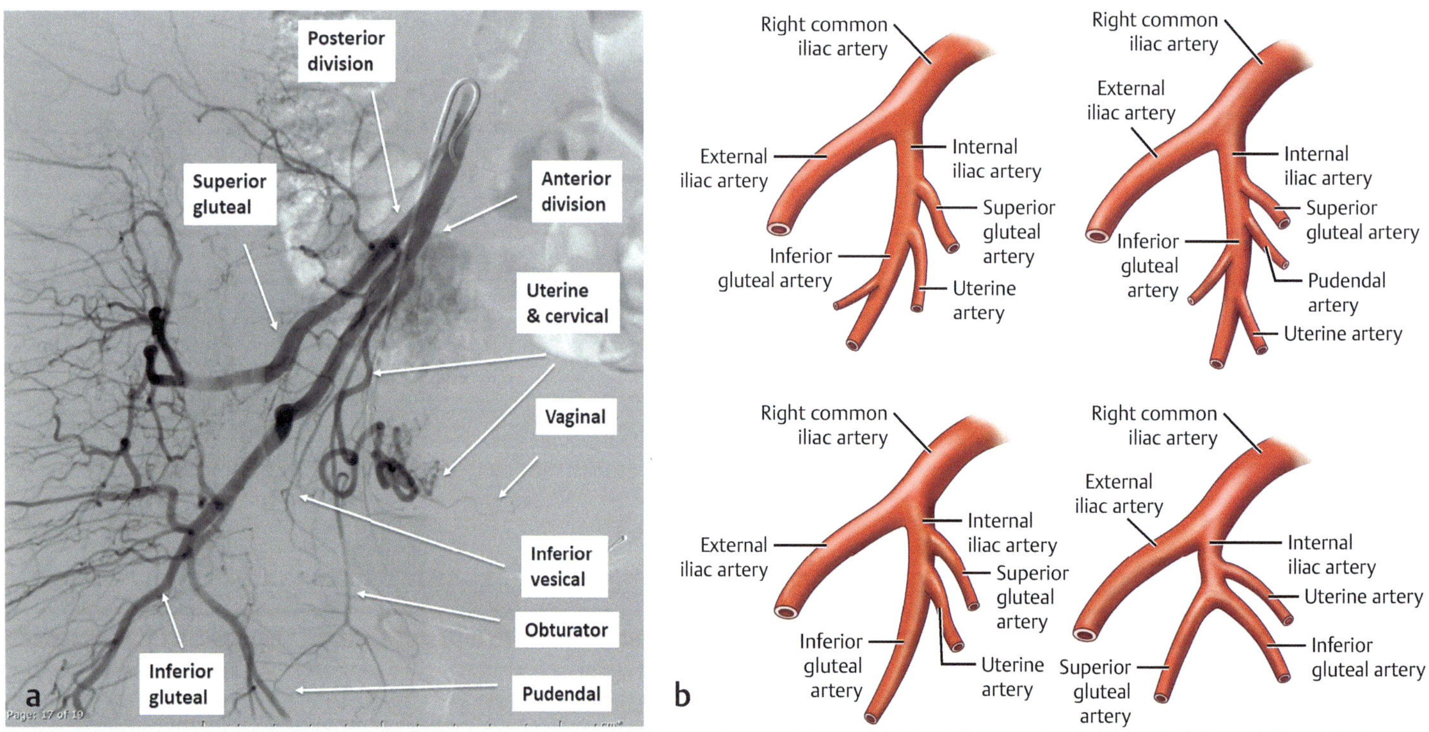

Fig. 16.44 (a) Internal iliac artery branching pattern shown by catheter angiogram. **(b)** Anatomical variations in internal iliac artery branching pattern with specific reference to uterine artery origin. Top left: In type I, the uterine artery is the first branch of the inferior gluteal artery. Top right: In type II, the uterine artery is the second or third branch of the inferior gluteal artery. Bottom left: In type III, the uterine artery, inferior gluteal artery, and superior gluteal artery are at the same level. Bottom right: In type IV, the uterine artery is the first branch proximal to the superior gluteal artery. (Reproduced with permission from Gomez-Jorge et al.[44])

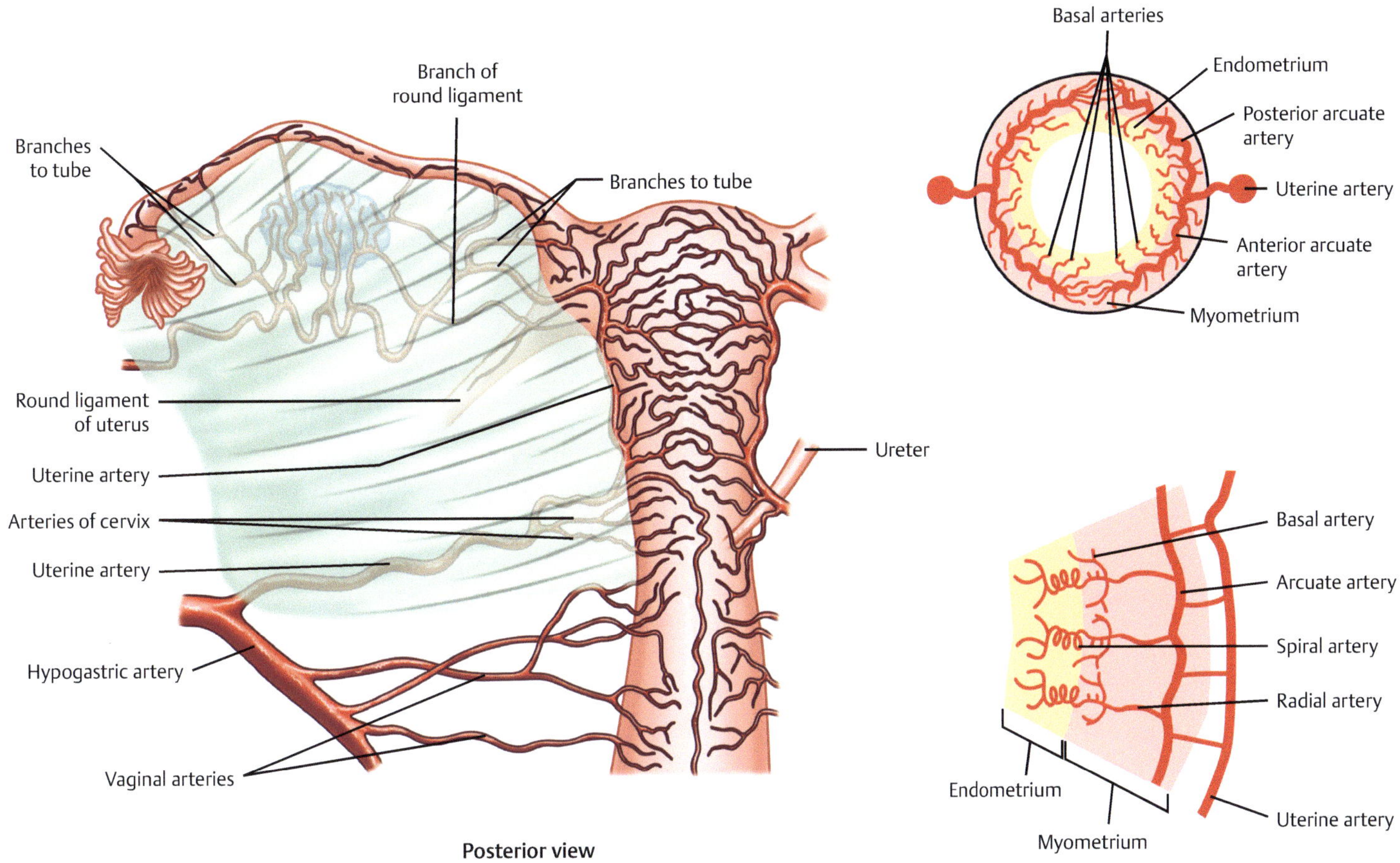

Fig. 16.45 Classic branching pattern of the uterine artery. The tortuous basal arteries, which arise from the anterior and posterior arcuate arteries to supply the endometrium, undergo cyclic changes during the menstrual cycle, with involution during the proliferative phase and growth during the secretory phase.

arterial branches including ovarian, cervical, vaginal, and vesical arteries is important. Inadvertent embolization of the ovarian and vaginal arteries has been suggested a possible cause of vaginal and ovarian dysfunctions.[48,49] Therefore, it is suggested that both uterine arteries should be embolized due to the communication branches between the two sides and that the embolization should be performed superselective distal to origin of the cervicovaginal.

Anastomoses between the uterine and ovarian arteries are demonstrated in 40% of angiography studies. These anastomoses are associated with a higher incidence of repeat intervention for treatment of hemorrhagic fibroids than in patients without anastomoses.[50] Three types of ovarian artery-to-uterine artery anastomoses has been described by Razavi et al.[48] In type I (22%), flow from the ovarian artery to the uterus is through anastomoses with the main uterine artery. In type II (4%), the ovarian artery supplies the fibroids directly. In type III (7%), the major blood supply to the ovary is from the uterine artery (▶ Fig. 16.48). Despite high prevalence of anastomoses between the ovarian artery and uterine artery, not all anastomotic connections are functional and become visible on angiography, likely due to hemodynamic factors and preferential direction of flow when a hypervascular mass exists.[51–53] In the presence of large fundal vascular fibroids, the ovarian arteries are more likely to supply the uterine mass (▶ Fig. 16.48). In other words, a dynamic balance exists between the uterine artery and the ovarian artery via ovarian artery-to-uterine artery anastomoses.

Uterine Veins

The uterine veins run parallel with the arterial branches. They form interconnecting plexuses in the broad ligament that drain into internal iliac vein (▶ Fig. 16.27c, d). The uterine venous plexus merges with the vaginal plexus inferiorly and the utero-ovarian plexus superiorly[54] (▶ Fig. 16.17). Whereas most of the venous uterine blood is collected into the iliac veins, the fundal part of the uterus drains to the ovarian veins that finally drains into the inferior vena cava on the right and the renal vein on the left (▶ Fig. 16.49). During retrograde venography of the ovarian vein especially in varicoid veins, a good visualization of the uterine veins can be obtained. Bilateral enlargement of the vaginal, parametrial, uterine, and ovarian veins is common in nulliparous women and pelvic congestion syndrome (▶ Fig. 16.49).[26] Unilateral enlargement should raise the possibility of the ovarian vein thrombosis or tumoral obstruction (▶ Fig. 16.50). Calcification of the uterine, parametrial, ovarian, and vaginal veins (phleboliths) are common in CT studies of the pelvis (▶ Fig. 16.51).

The utero-ovarian venous plexus facilitates availability of the hormones and substances secreted from the ovary or endometrium to both uterus and ovary. Aside from distribution of the

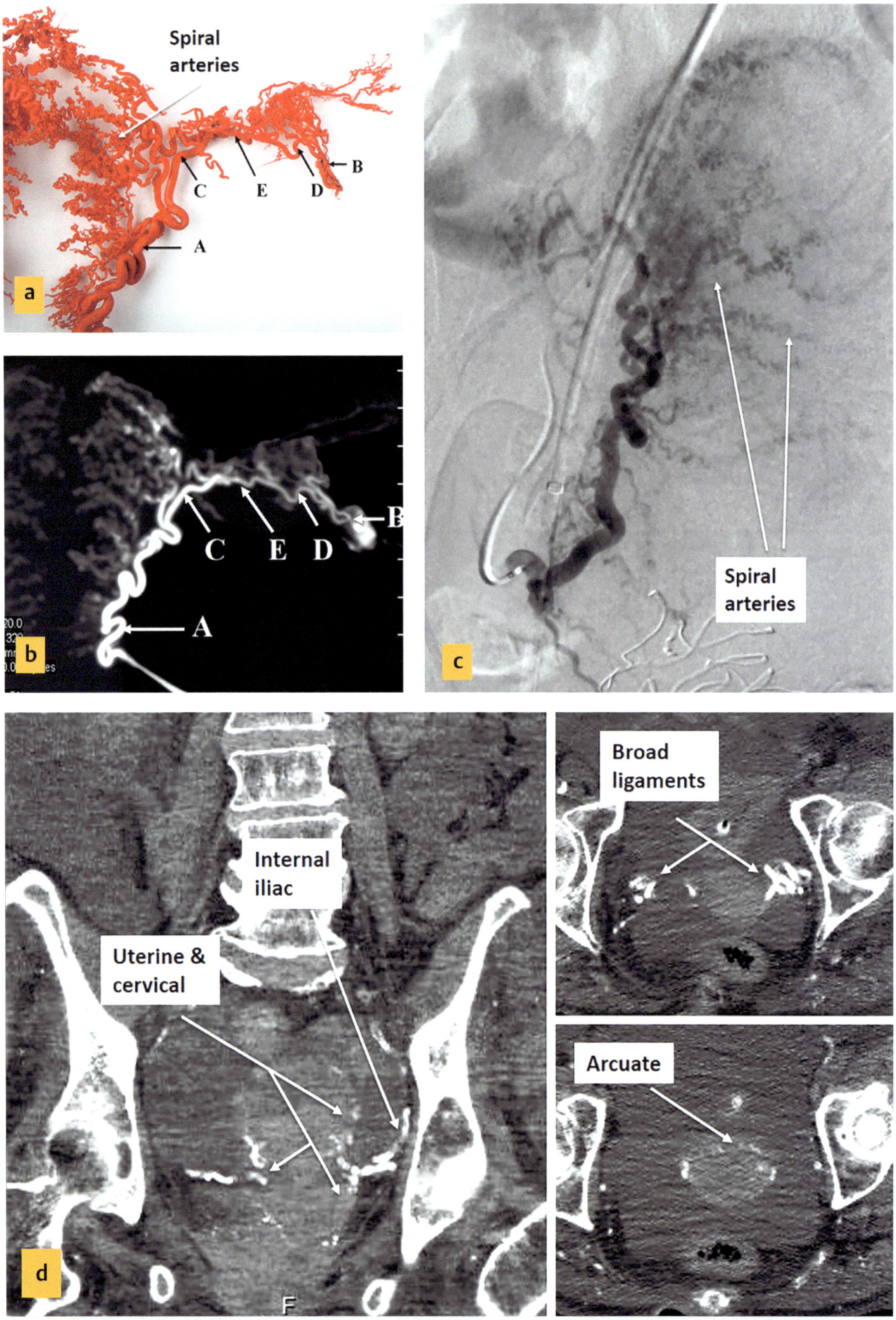

Fig. 16.46 Vascular casting specimen and radiographs of the uterine artery. **(a, b)** Left uterine artery (A), left ovarian artery (B), ovarian branch of uterine artery (C), ovarian branch of ovarian artery (D), and ovarian artery-to-uterine artery anastomoses (E). **(c)** Selective catheter angiogram of the right uterine artery showing prominent arcuate arteries. (Reproduced with permission from Ouyang et al.[49]) **(d)** Vascular cast of uterine and pelvic arteries due to extensive calcification in a patient with renal failure showing ascending, descending, and parametrial components of the uterine arteries. The uterine arteries cross the ureters at the base of the broad ligaments marking the superior margin of the cardinal ligament.

16

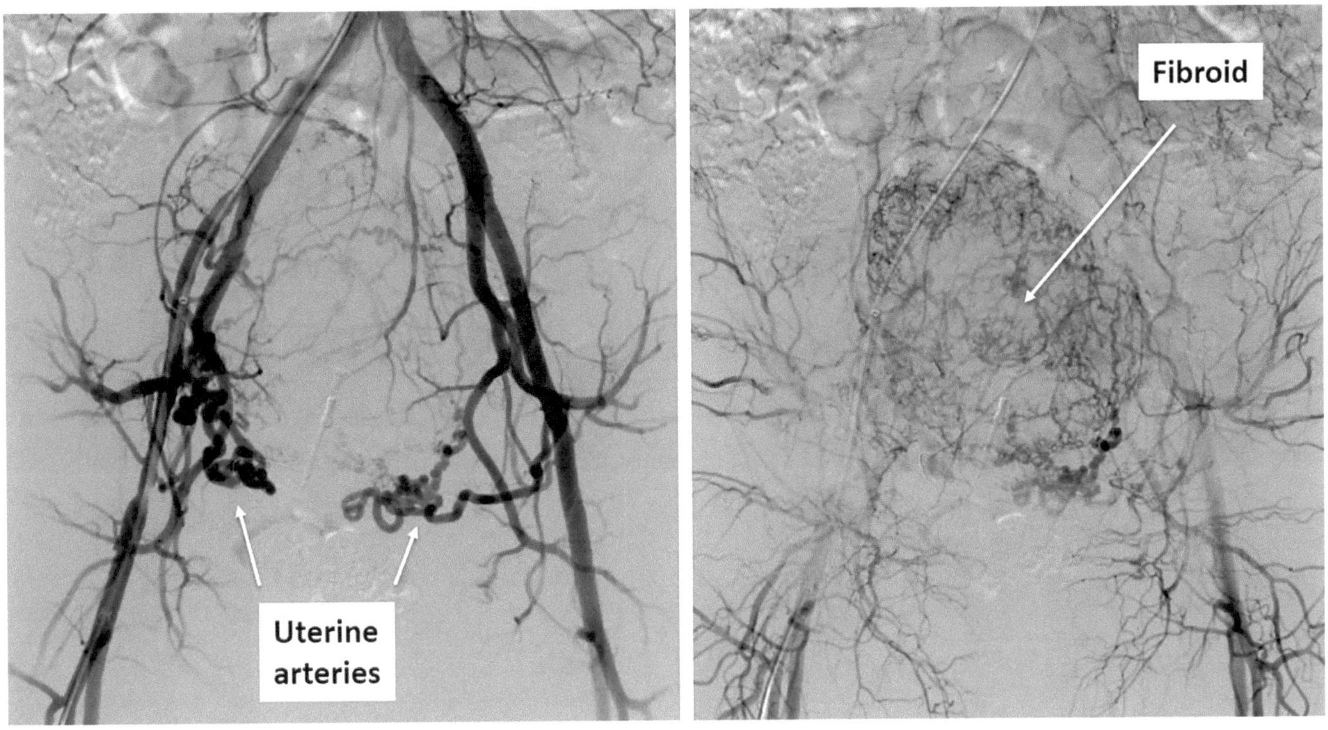

Fig. 16.47 Enlarged uterine arteries in a patient with a large hypervascular fibroid.

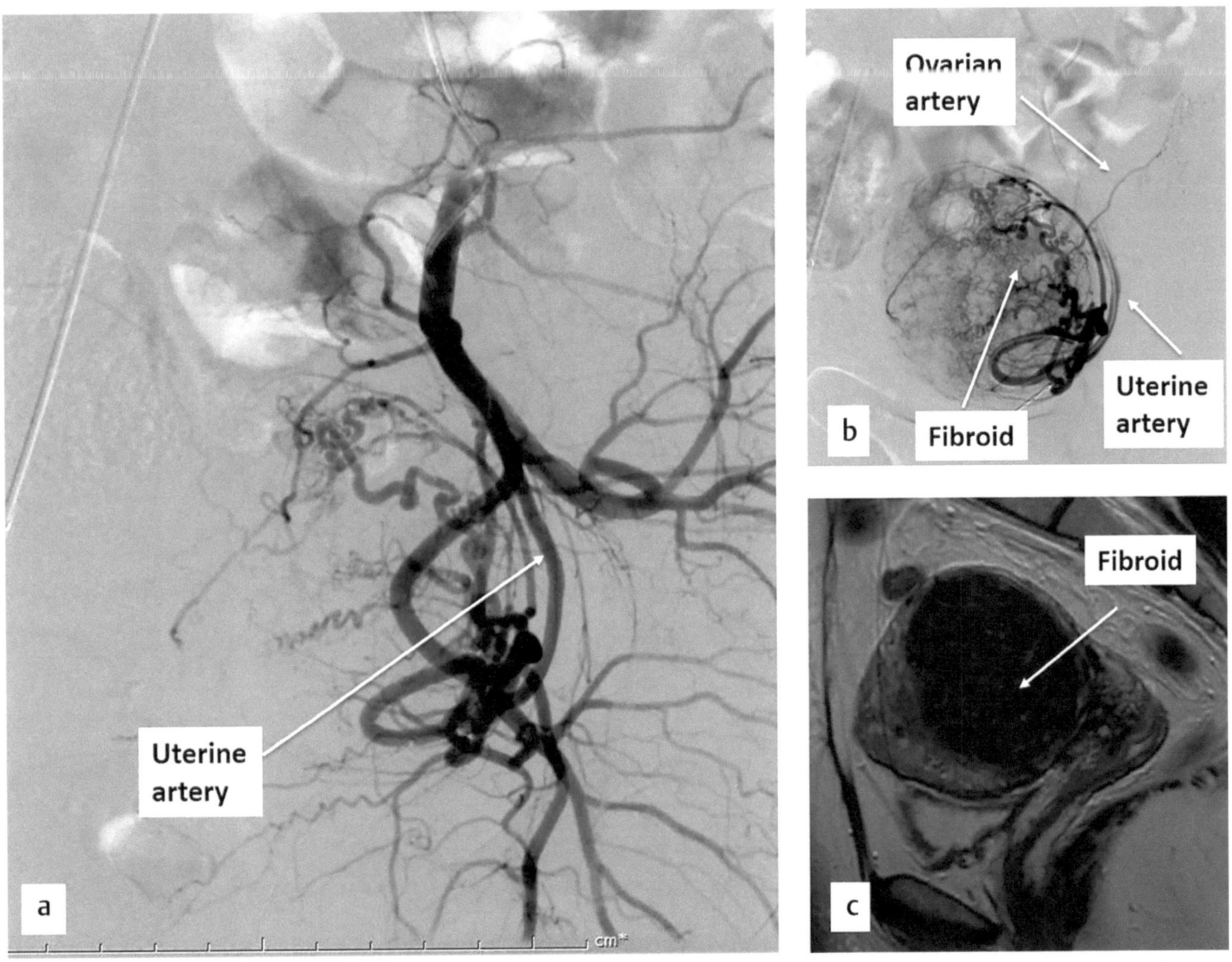

Fig. 16.48 Uterine artery embolization. **(a)** Left Internal iliac artery injection showing enlarged uterine artery. **(b)** Selective injection showing hypervascular mass which was embolized. Note visualization of an accessory left ovarian artery. **(c)** Magnetic resonance imaging (MRI) showing low signal fibroid tumor. Uterine artery embolization is commonly used as an alternative technique to myomectomy and hysterectomy. During embolization of the uterine artery attention to anastomotic connection to the ovarian artery is important. Inadvertent embolization of the ovarian arteries has been suggested a possible cause of ovarian dysfunction.

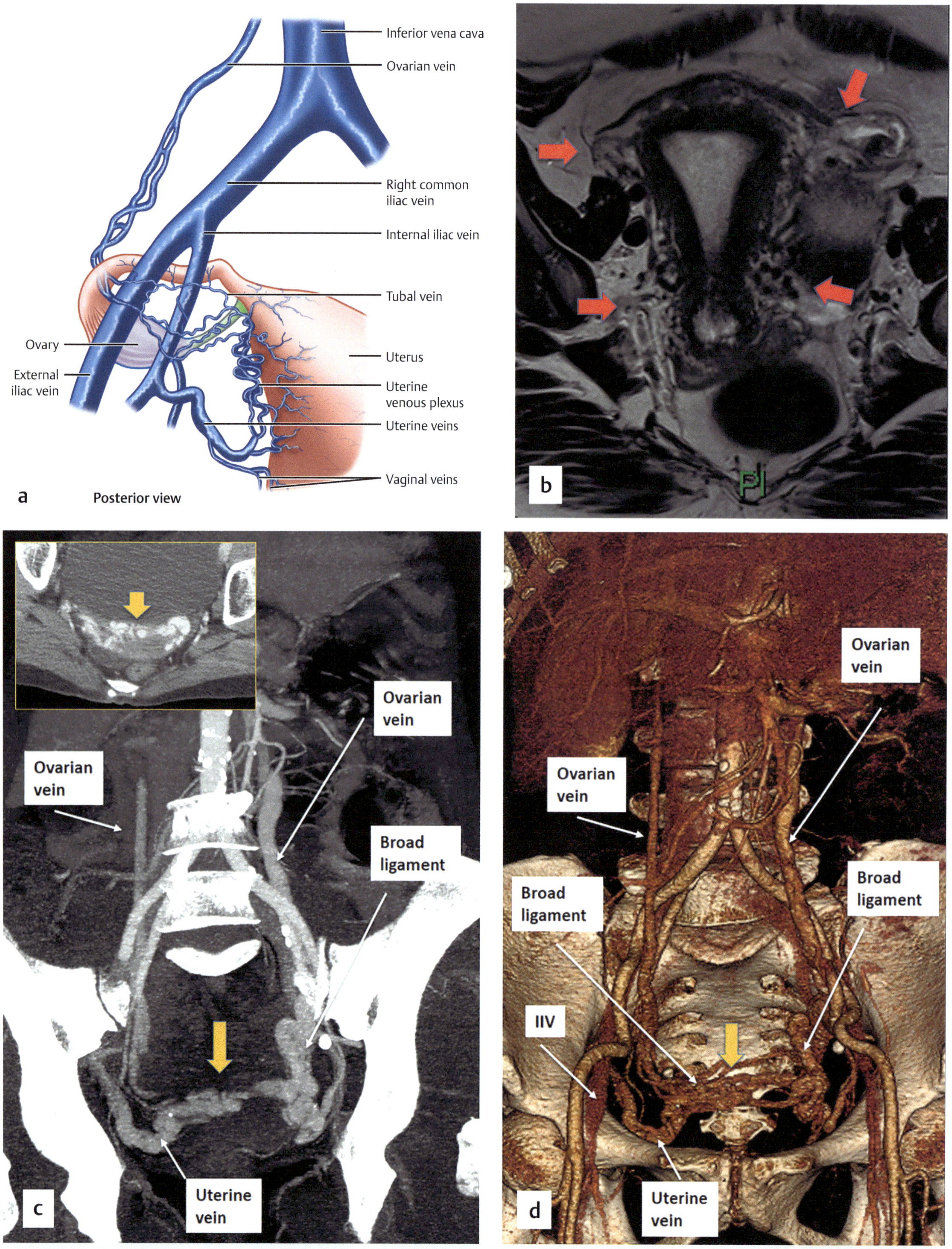

Fig. 16.49 (a) Anatomy of the uterine and ovarian veins. Uterine veins form interconnecting plexuses in the broad ligament that drain into internal iliac veins (IIV). **(b)** Coronal T2-weighted (T2W) view of the uterine body showing venous plexus of the broad and cardinal ligaments (*red arrows*). **(c, d)** Computed tomography (CT) scan shows idiopathic pelvic venous congestion involving the uterine and ovarian veins. Note collateral venous connections along the uterine surface between left and right parametrial venous plexuses (*yellow arrows*).

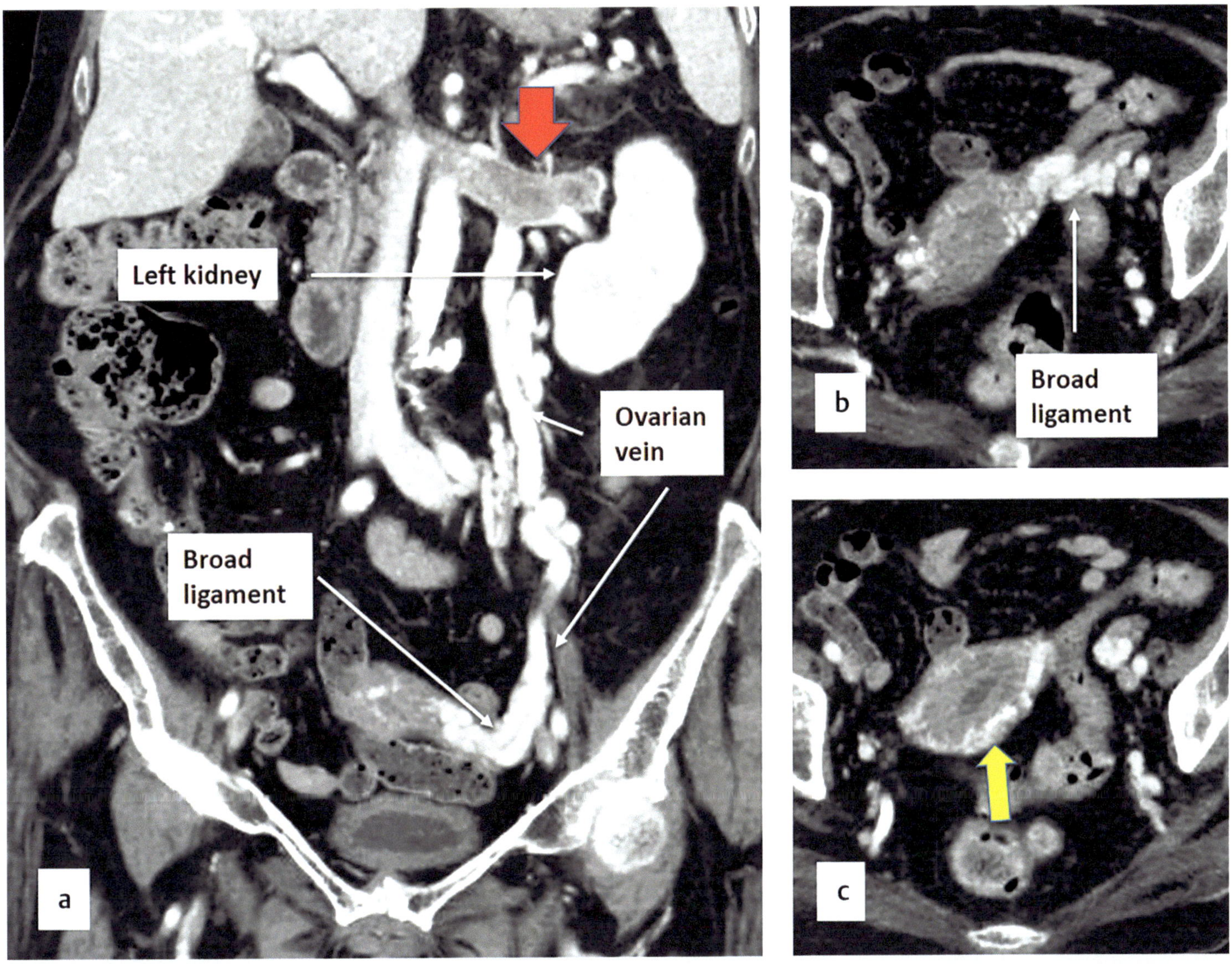

Fig. 16.50 (a–c) Unilateral enlargement of the left uterine venous plexus and left ovarian vein due to tumor thrombosis of the left renal vein (*red arrow*) shown by contrast-enhanced computed tomography (CT). Note collateral venous connections along the uterine surface between left and right parametrial venous plexuses (*yellow arrow*).

hormones, the vaginal–uterine–ovarian venous plexus has an important cooling effect on adjacent structures and during hormonal cycles that is important for normal function of genital organs.[46]

Lymphatics and Nerves

The external iliac, internal iliac, and common iliac nodes are the main lymphatic drainage of the uterus (please see Chapter 21 "Lymphatics of the Abdomen, Pelvis, and Perineum"). The nerve supply to the uterus is predominantly from the inferior hypogastric plexus.[55] The sympathetic fibers are derived from neurons in the T12 and L1 segments and produce uterine contraction and vasoconstriction. The parasympathetic fibers arising from the neurons in the S2–S3–S4 spinal segments produce vasodilation.

Uterine Tubes

The blood supply to the uterine tubes is derived from ovarian and uterine arteries. Venous drainage of the lateral two-thirds of the uterine tube is via the ovarian veins, and the medial two-thirds of the tube drain via the uterine veins. Lymph drainage is via ovarian vessels to the para-aortic nodes and uterine lymphatic channels to the internal iliac chain.

Ovaries

Ovaries are paired oval structures in the lateral aspect of the fallopian tubes, suspended by the mesovarium (▶Fig. 16.4, ▶Fig. 16.41). In embryonic and early fetal life, the ovaries are formed in the gonadal ridges medial to the mesonephric or primary kidney. Initially, the primordial germ cells migrate to the gonadal ridge and transform to oogonia (▶Fig. 16.4). The oogonia will then be surrounded by a layer of connective tissue cells (pre-granulosa cells) to form rudimentary ovarian follicles. Similar to the testes, the ovaries will then descend along the gubernaculum to the pelvic cavity. In female, the gubernaculum (caudal genital ligament) adheres to the fundus of the uterus, and thus the ovary can only descend to the level of fundus. The cranial part of the gubernaculum between the ovary and uterus remains

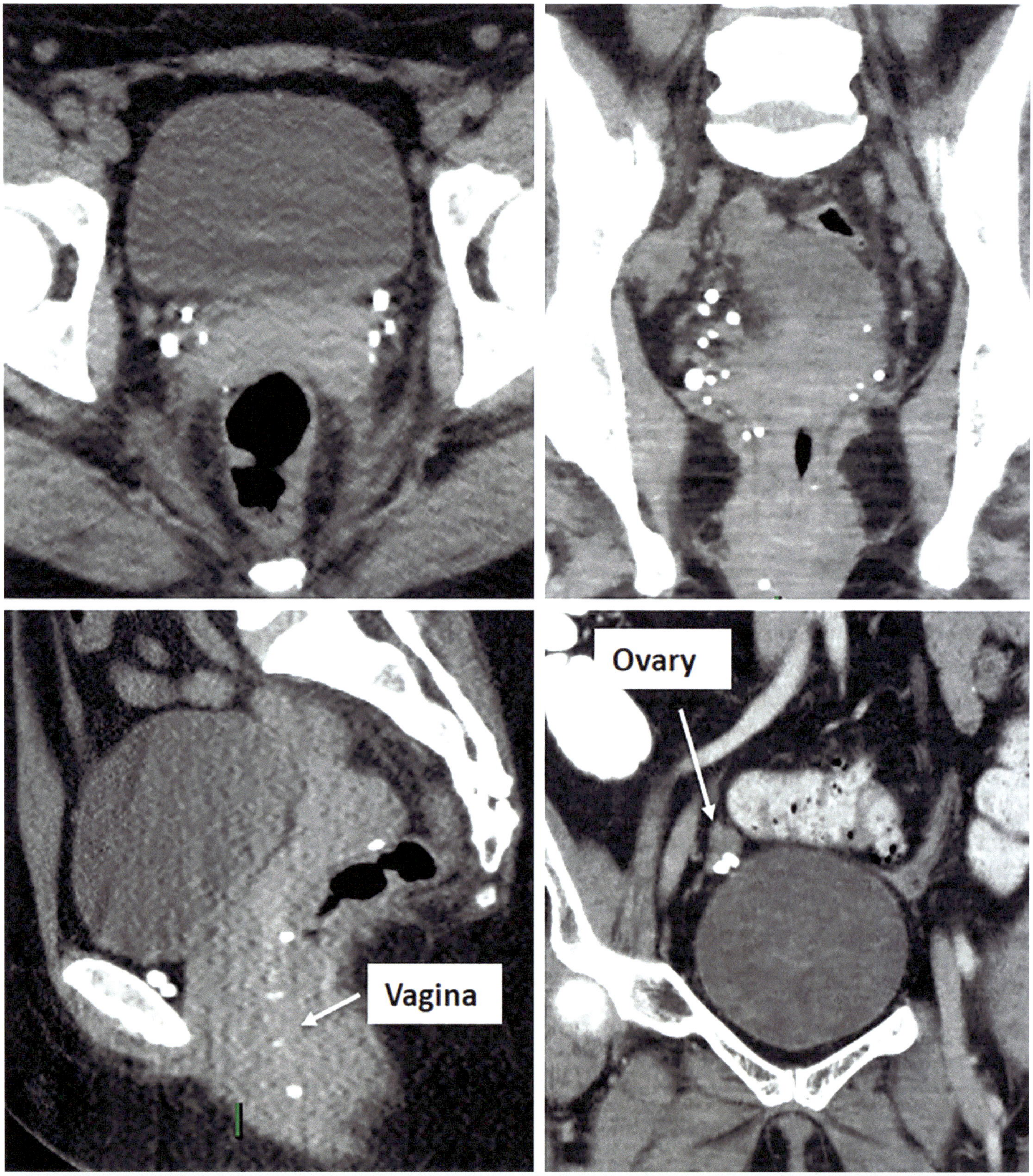

Fig. 16.51 Pelvic vein calcifications (phleboliths), possibility due to calcified thrombi, are shown by noncontrast computed tomography (CT) involving uterine, parametrial, vaginal, and ovarian veins.

as the ovarian ligament, while the lower part between the uterus and the labium majus forms the round ligament of the uterus (▶Fig. 16.2). Accessory ovarian tissue may occur in the mesovarium or in rare occasion somewhere in the pathway of the gubernaculum to the labium major.

Ovarian Physiology

During normal female cycle, gonadotropin-releasing hormone (GnRH) is released from the hypothalamus to stimulate the release of follicle stimulating hormone (FSH) from the pituitary gland. Under the effect of FSH on the ovary, 5 to 10 primordial follicles will grow but only one of the follicles becomes the dominant follicle to ovulate in the midcycle (▶Fig. 16.28). Stimulation of the follicular granulosa and theca cells releases estrogen which has a positive regulatory effect on GnRH, which in turn stimulates the pituitary gland to release LH. The LH surge in midcycle stimulates the release of the oocyte from the follicle.

Ovarian Development after Birth

The ovaries vary in size and appearance, depending on the patient's age and hormonal influences.[56,57] The ovaries descend from a superior pelvic position in neonatal period to a lower level in the pelvis adjacent to the uterus as the child ages.[57] In adults, the ovaries appear as ovoid soft-tissue structures on either side of the uterus within the mesosalpinx of the broad ligament, parallel and medial to the iliac vessels (▶Fig. 16.52).

Gonadotropin levels influence the number of individual follicles and the size of the ovary and stimulate the development of ovarian cysts at all ages. At birth, the size of ovaries may be large as large as 3.5 cm^3 but rapidly decreases to approximately 1.5 cm^3 and remains stable until age 7. It grows later and reaches about 4 cm^3 before puberty, and more than 4 cm^3 after puberty.[13,58] The ovaries are easily identified on MR and CT images in adolescents than in girls younger than 8 years. Both immature and mature ovaries can have follicles in varying stages of development. Ovarian follicles of up to 1 cm have been noted as early as 28 weeks of gestation, but detection of follicles in early childhood is very unusual.[58,59] Approximately 20% of female neonates have ovarian cysts greater than 1 cm. The development of larger physiologic ovarian cysts in prepubertal life stages is not uncommon and most of them resolve spontaneously. Cysts greater than 2.5 cm containing septa and hemorrhage should be followed with ultrasound.

The ovarian volume varies greatly in reproductive age and demonstrates variable appearance over the course of the menstrual cycle. Normal ovarian volume decreases after the age of 30 years. Mean ovarian volume is 6.6 cm^3 in women younger than 30 years, 4.8 cm^3 in 40- to 49-year-old women, and 2.6 cm^3 in 50- to 59-year-old women.[58] Developing and immature subcentimeter follicles are seen throughout the entire cycle (▶Fig. 16.52, ▶Fig. 16.53). One or more follicles will grow to a dominant diameter of 20 to 25 mm before ovulation and then rupture to release the oocyte[60] (▶Fig. 16.54). The oocyte and its supporting structures (i.e., the cumulus oöphorus) within dominant follicle may be seen in ultrasound as a curvilinear septation within the follicle (▶Fig. 16.54). After ovulation, the corpus luteum, a remnant of the mature follicle, is typically visible in the secretory phase of the menstrual cycle as a hyperechoic structure, less than 3 cm in length with low-resistance blood flow. CT scan shows a low attenuation structure with peripheral enhancement (▶Fig. 16.55). In the absence of conception, the corpus luteum becomes corpora albicans and finally becomes atretic (corpus albicans). Atretic corpus luteum may be seen in the following cycles as small complex cystic structures. In case of pregnancy the corpus luteum persists into the first trimester of pregnancy.[59] Some mature follicles may fail to ovulate and then continue growing to as large as 5 cm.[60] These physiologic follicular cysts are usually asymptomatic and appear as simple unilocular nonvascular cysts on ultrasound or MRI which involute in the subsequent cycles (▶Fig. 16.55). In women of reproductive age, cysts smaller than 5 cm are benign and do not need follow-up, those between 5 and 7 cm should be monitored yearly, and those greater than 7 cm require further evaluation with MRI or surgery if needed.[61]

On MRI, normal ovary is seen as an oval-shaped, mildly enhancing structure with the normal stromal medulla showing increased signal intensity on T2W relative to the cellular cortex[62] (▶Fig. 16.53). Normal follicles are best shown on T2W images. The corpus luteum appears as thick wall enhancing structures with variable signal intensity related to fluid and hemorrhagic products. At positron emission tomography (PET), uptake of fluorine 18 fluorodeoxyglucose (FDG) as a result of normal physiologic function in the ovaries and especially in the corpus luteum is common in women of reproductive age (▶Fig. 16.56). Endometrial uptake has also been noted at PET, most often during the ovulatory phase and during menstruation. To minimize the potential confusion with abnormalities, it is suggested to perform PET within a week before or after menses.[63]

The postmenopausal ovaries may be difficult to identify at imaging due to atrophy and the paucity or absence of follicles and cysts.[64] They typically appear as small flattened soft-tissue structures, often adjacent to the uterus or iliac vessels (▶Fig. 16.31). Identification of the ovaries may be facilitated by localizing and following the gonadal vessels into the pelvis where they end the ovary (▶Fig. 16.50). Simple ovarian cysts as large as 3 cm may be seen in some cases but most cysts are less than 1 cm. In postmenopausal years the volume of the ovary ranges from 1.2 to 5.8 cm^3. An ovary that is twice as large as the contralateral ovary should be considered abnormal.[57] Punctate calcified foci are not unusual in postmenopausal ovaries indicating dystrophic calcifications in atretic follicles or phleboliths[65] (▶Fig. 16.51).

Vasculature and Innervation of the Ovaries

The ovarian arteries originate from the anterolateral wall of the abdominal aorta below the renal arteries (▶Fig. 16.57). Each artery runs obliquely behind the peritoneum inferiorly and laterally to

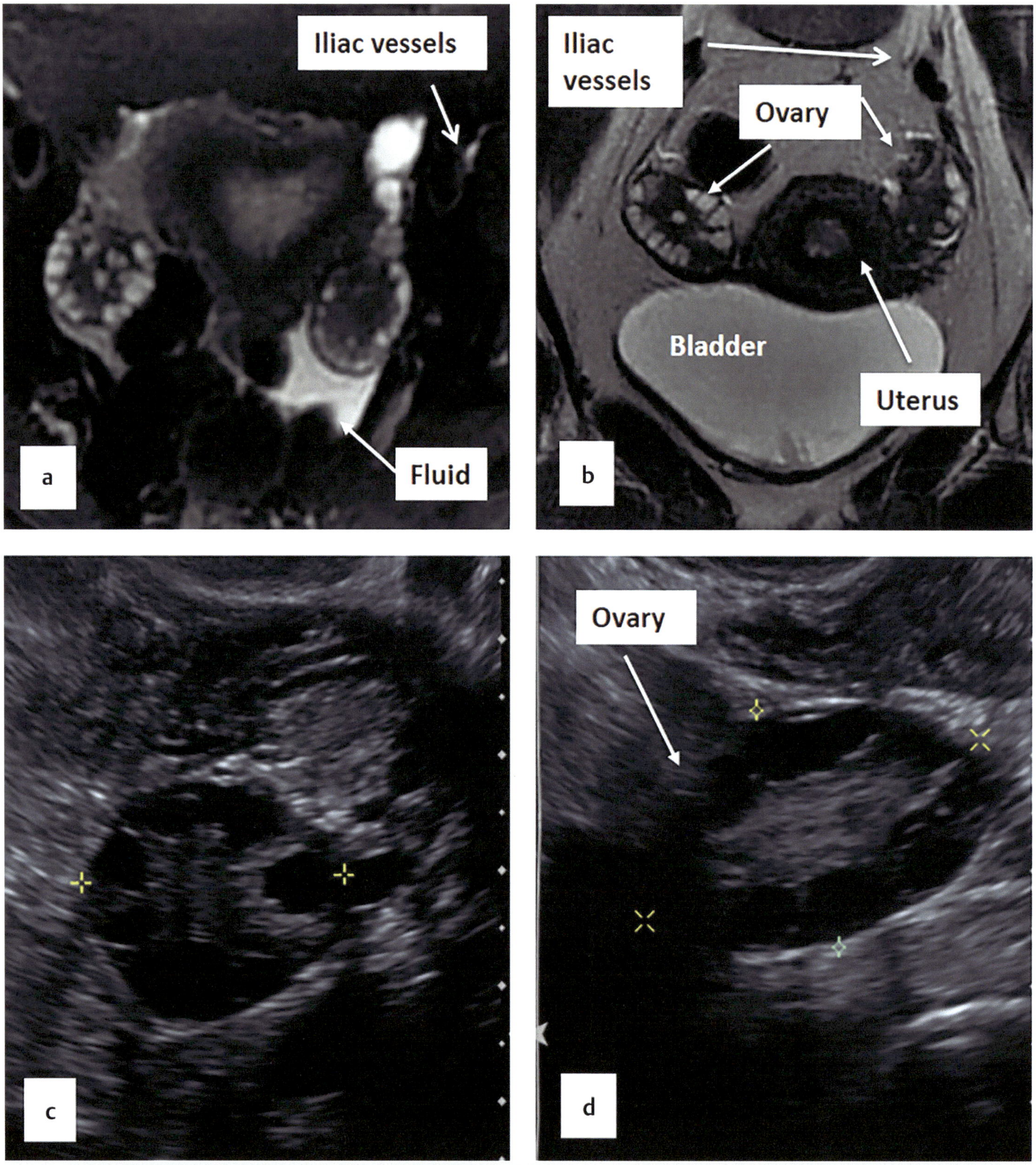

Fig. 16.52 Common appearance of normal ovaries in premenopausal stage. **(a, b)** Axial and coronal T2-weighted (T2W) magnetic resonance imaging (MRI). **(c, d)** Transvaginal ultrasound. The ovaries appear as ovoid soft-tissue structures on either side of the uterus within the mesosalpinx of the broad ligament, parallel and medial to the iliac vessels. Developing subcentimeter follicles are seen in the periphery of the ovaries.

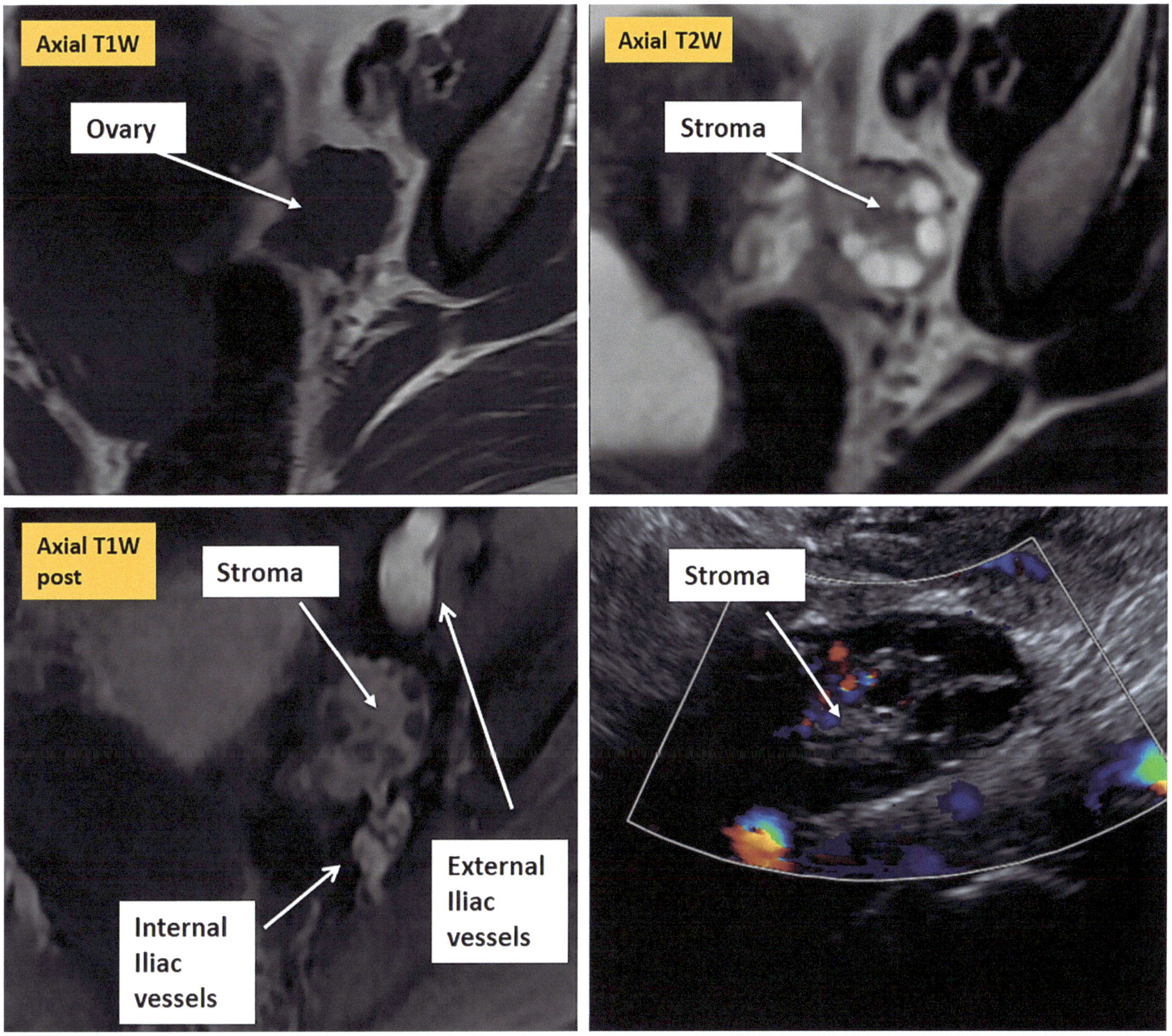

Fig. 16.53 Ovarian structure in premenopausal stage shown by magnetic resonance imaging (MRI) and ultrasound. Developing subcentimeter follicles are seen in the periphery of the ovaries. The ovarian stroma is seen as heterogeneous medium signal on T2-weighted (T2W) image which enhances on postcontrast image. Ultrasound clearly showing vessels in the stroma of the ovary.

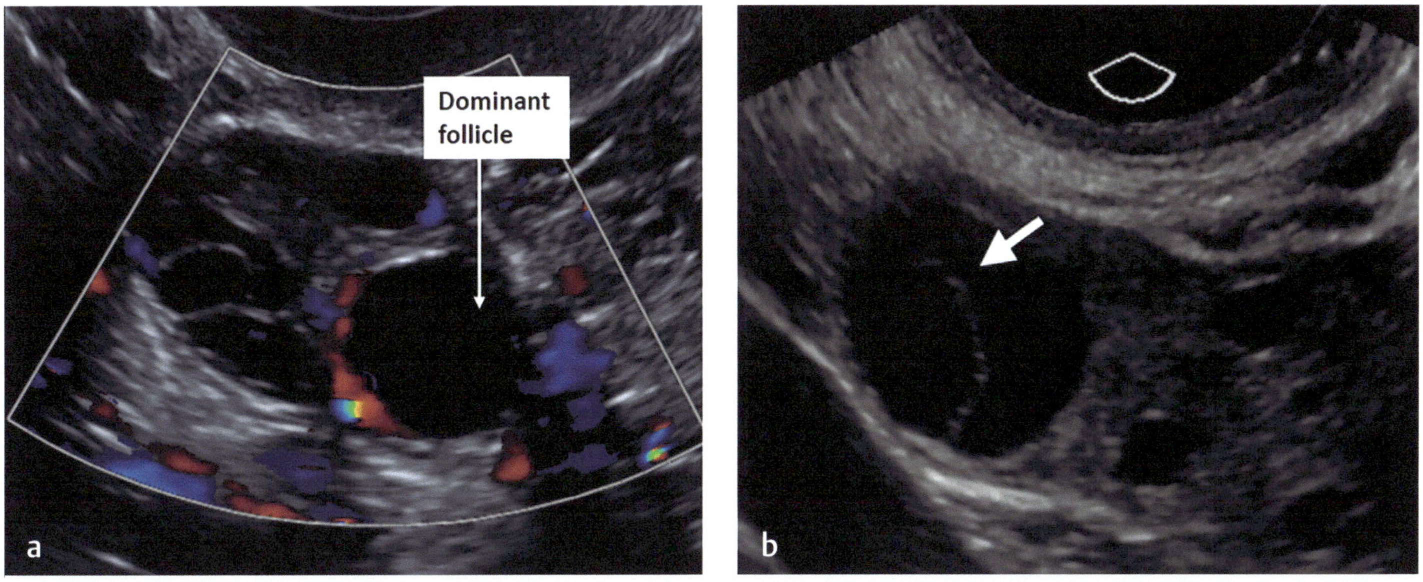

Fig. 16.54 Ovarian structure in premenopausal stage shown by ultrasound. **(a)** One or more follicles will grow to a dominant diameter of 20 to 25 mm before ovulation and then rupture to release the oocyte. **(b)** Ultrasound image shows a normal dominant follicle and cumulus oöphorus (*arrow*). (Reproduced with permission from Langer et al.[13])

16

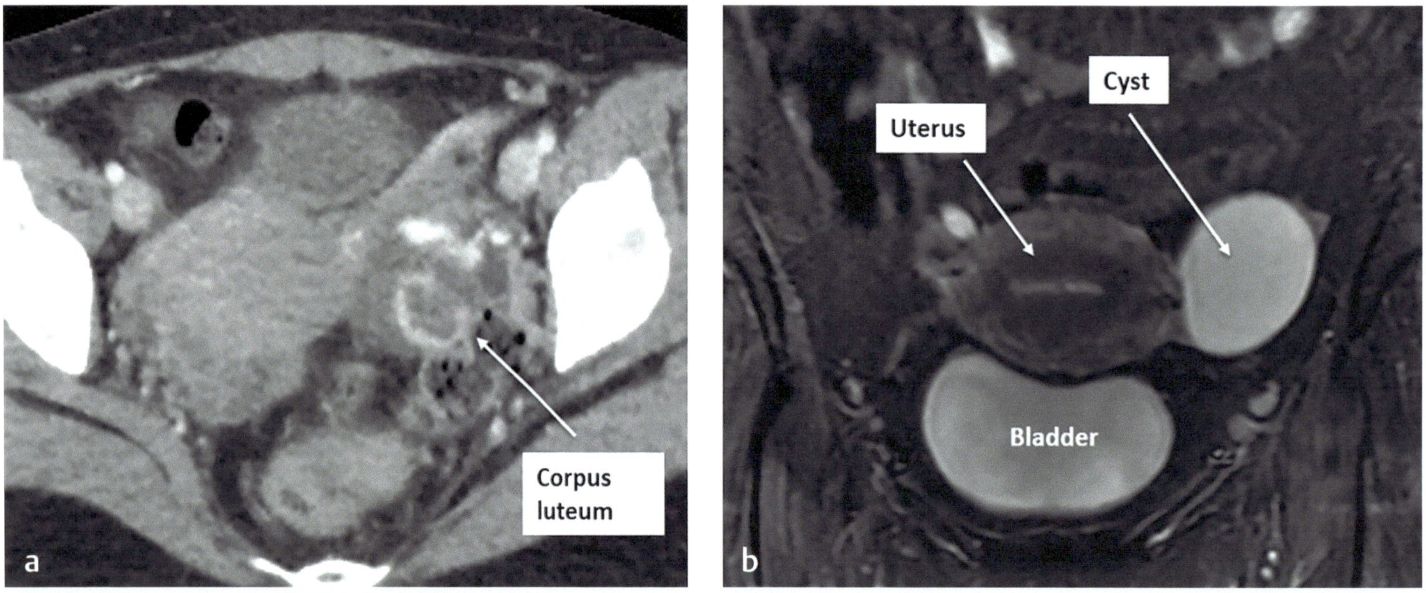

Fig. 16.55 Ovarian structure in premenopausal stage shown by computed tomography (CT) and magnetic resonance imaging (MRI). **(a)** Axial, postcontrast CT showing a corpus luteum with enhancing margin in the left ovary. **(b)** T2-weighted (T2W) axial MRI showing a left follicular cyst. In women of reproductive age, cysts smaller than 5 cm are benign and do not need follow-up.

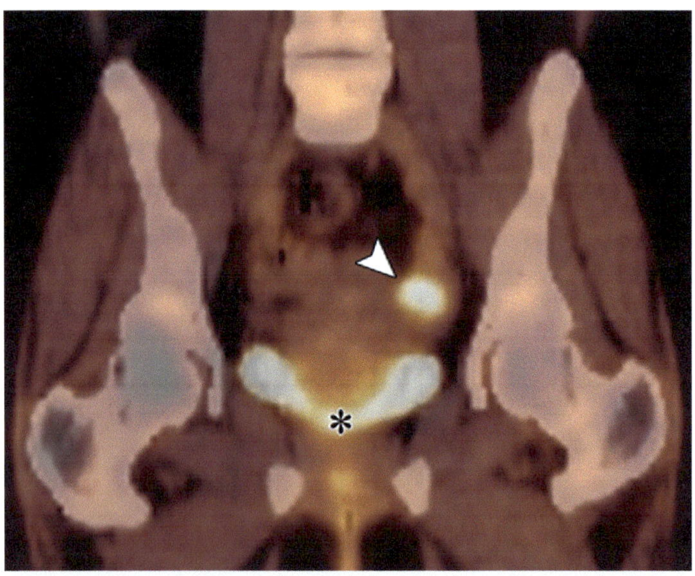

Fig. 16.56 Normal physiologic uptake in a 32-year-old woman who underwent fluorine 18 fluorodeoxyglucose (FDG) positron emission tomography–computed tomography (PET/CT). Coronal fused PET/CT images demonstrate physiologic uptake of FDG representing metabolic activity in the corpus luteum of the left ovary (*arrowhead*). Normal excretion of FDG is noted in the urinary bladder (*). (Reproduced with permission from Langer et al.[13])

16

the pelvic brim where it turns medially to enter the suspensory ligament of the ovary. It then passes through the mesovarium to reach the ovary and fallopian tubes.

Aberrant uterine supply from the ipsilateral ovarian artery is a relatively common anatomical variant in uterine blood supply. Attention to this variant is important for a successful uterine artery embolization.[66,67] Abnormality of the ovarian artery is unusual. Damage to the artery during pregnancy, cesarian section, and abdominal surgery may result in pseudoaneurysm, rupture, and hematoma (▶ Fig. 16.58). Spontaneous rupture of an ovarian artery aneurysm is a rare and a potentially life-threatening event. Diagnosis is important since it can be treated with transcatheter embolization.[68]

Small veins around the ovaries form a plexus known as pampiniform plexus which extends along the suspensory ligament toward the lateral pelvis and ascends as a single vein to drain in the right wall of the inferior vena cava on the right side or the left renal vein on the left side (▶ Fig. 16.49).

The lymphatic drainage of the ovaries is along the ovarian vessels into the superior paraaortic nodes near the origin of the renal arteries. The ovarian innervation is provided by the autonomic branches of the renal and aortic plexuses and the superior and inferior hypogastric plexuses.

Common Pathologies of the Ovarian Region

Cysts

Ovarian cysts are common and usually asymptomatic that are incidentally found on routine pelvic examination or during imaging studies. Common benign cysts include physiologic cyst, follicular cyst, corpus luteum cyst, theca lutein cyst, hemorrhagic cyst, and polycystic ovaries[69,70] (▶ Fig. 16.55). Physiologic cysts are usually less than 3 cm and will be resolved spontaneously. Follicular cysts are usually more than 3 cm, unilocular, anechoic, and thin walled with no solid enhancing component. Corpus luteum cyst appears as thick-walled unilocular cyst, often with internal hemorrhagic debris and high blood flow in the periphery (▶ Fig. 16.59). Hemorrhagic cysts are hyperintense on T1W MR images with variable signal intensity on T2W images[70] (▶ Fig. 16.59). Endometrioma is defined as a cystic lesion of endometrial stroma and most commonly occurs in the ovaries. It is characterized by a relatively thick-walled cyst containing hemorrhagic product with high concentration of iron and protein.[71]

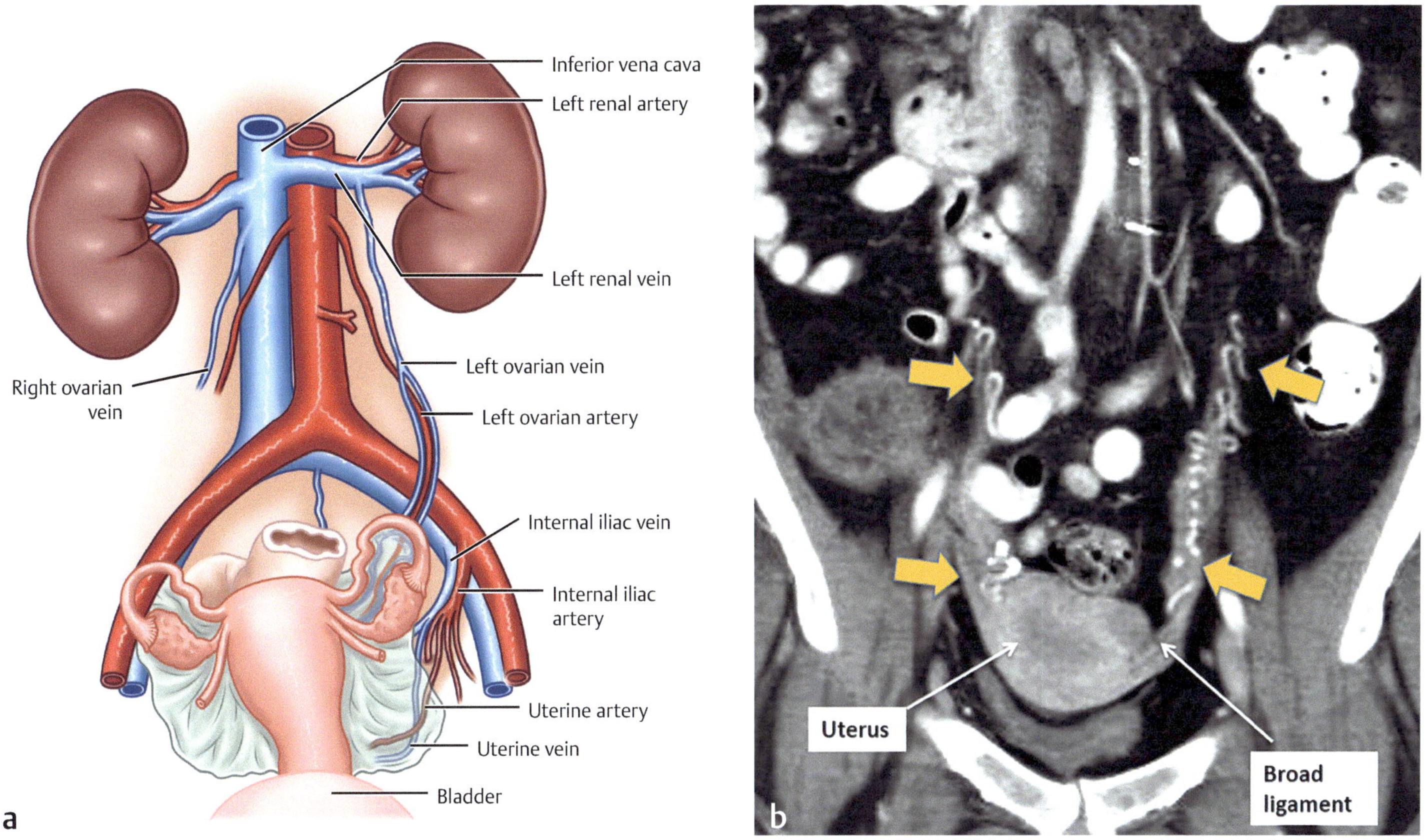

Fig. 16.57 (a) Normal ovarian vessels. **(b)** Coronal, contrast-enhanced, arterial phase computed tomography (CT) showing tortuous ovarian arteries (*arrows*) entering the broad ligaments. IVC, inferior vena cava.

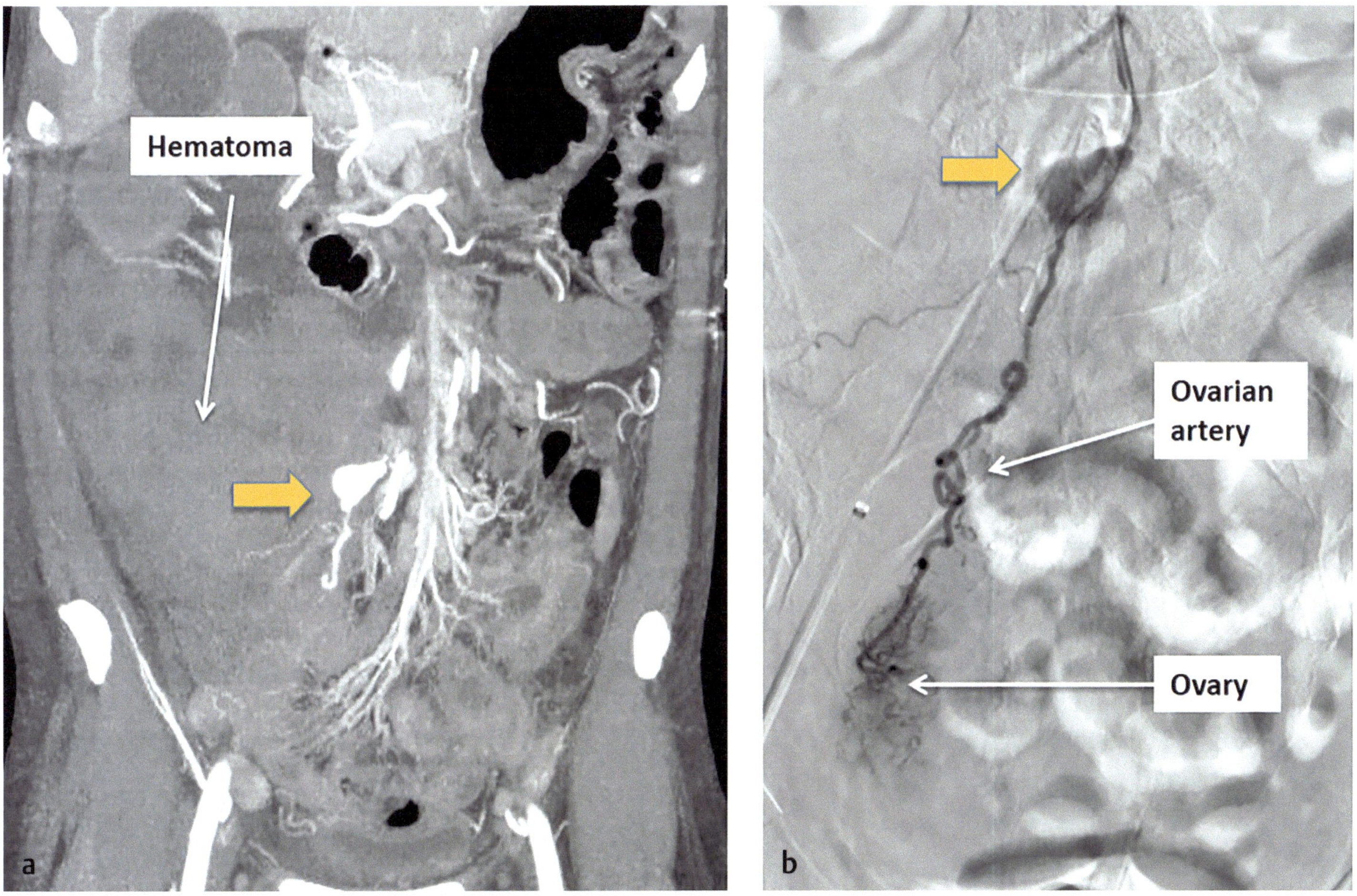

Fig. 16.58 (a) Ovarian artery pseudoaneurysm. A 15 × 12 mm pseudoaneurysm (*arrows*) is visualized emanating off the ovarian artery causing a large right retroperitoneal hematoma. **(b)** The artery was embolized.

16

Dermoid cysts are mature cystic teratoma containing fatty component; areas of acoustic shadowing and echogenic lines and dots are characteristic (▶ Fig. 16.59).

Paraovarian Cyst

Paraovarian cysts can arise from the peritoneal mesothelium of the broad ligament and less commonly from the paramesonephric or mesonephric remnants. Cysts arising from the paramesonephric duct are formed after cystic dilatation of the longest fimbria of the fallopian tube and are referred to fimbrial or hydatid cysts of Morgagni.[72,73] Paraovarian cysts are usually asymptomatic and appear as thin-walled unilocular or lobulated cysts. Cysts of Morgagni are located at the fimbrial end of the tube. Unlike ovarian cysts, paraovarian cysts do not regress over time.

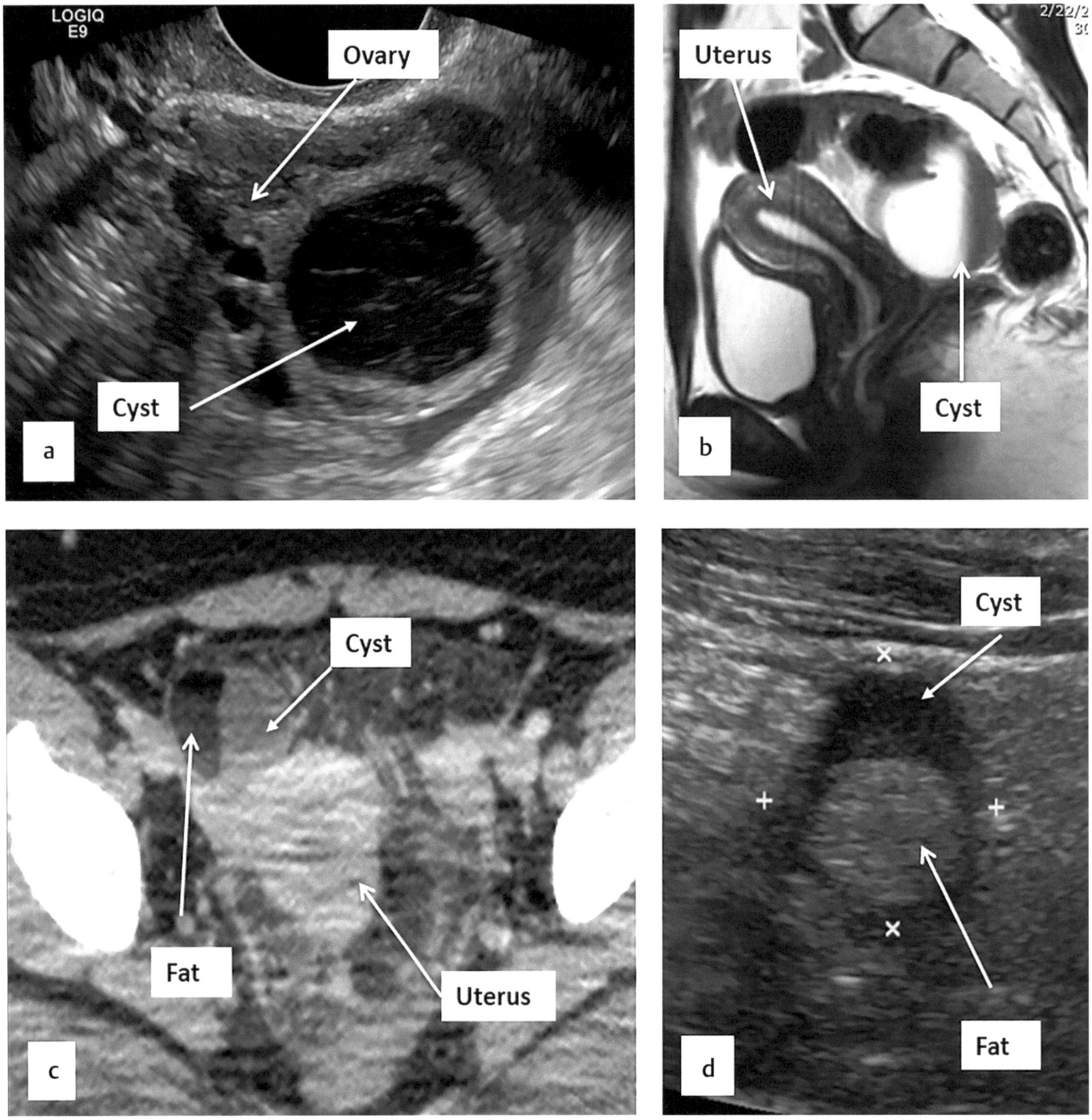

Fig. 16.59 Common ovarian complications in premenopausal stage. **(a)** Ultrasound image shows a dominant follicle with internal echoes due to hemorrhage, possibly a corpus luteum cyst. **(b)** Sagittal T2-weighted (T2W) magnetic resonance imaging (MRI) showing an ovarian cyst with fluid-fluid level due to hemorrhage. The hemorrhage is seen as dark layer in the dependent portion of the cyst. **(c, d)** Computed tomography (CT) and ultrasound show a right dermoid cyst. The fatty component of the cyst appears low signal on CT and echogenic on ultrasound.

16

Ovarian Torsion

The term refers to rotation of the ovary and portion of the fallopian tube on the supplying vascular pedicle and is a gynecologic emergency. The etiology of torsion includes ovarian mass, ovarian hypermobility, and long fallopian tube. Imaging features of torsion include enlarged ovary (>4 cm), free fluid in the pelvic cavity, abnormal ovarian position, and decrease or absence of ovarian flow.[74] Due to dual blood supply of ovary from both the ovarian and uterine arteries, the blood flow may appear normal on Doppler examination.

Pelvic Congestion Syndrome

The etiologies of pelvic venous congestion are physiologic increase in pelvic blood flow, hormonal vasodilatation, retroaortic left renal vein behind the aorta, left renal vein compression between the aorta and the superior mesenteric vein (nutcracker syndrome), compression of the right common iliac vein under the right common iliac artery, and primary valvular insufficiency of the ovarian veins[75,76] (▶Fig. 16.49). Left-sided ovarian vein congestion may be secondary to left renal vein thrombosis (▶Fig. 16.50). The ovarian veins provide drainage of the parametrium, cervix, mesosalpinx, and pampiniform plexus. These veins have an average diameter of 3 mm and possess two or three valves. Varicoid ovarian veins are generally >5 mm and demonstrate free reflux with filling of contrast material across midline on venography (▶Fig. 16.60). The pelvic venous congestion syndrome is characterized by noncyclic pelvic pain for longer than 6 months (chronic pelvic vein) usually involving multiparous patients with imaging demonstration of

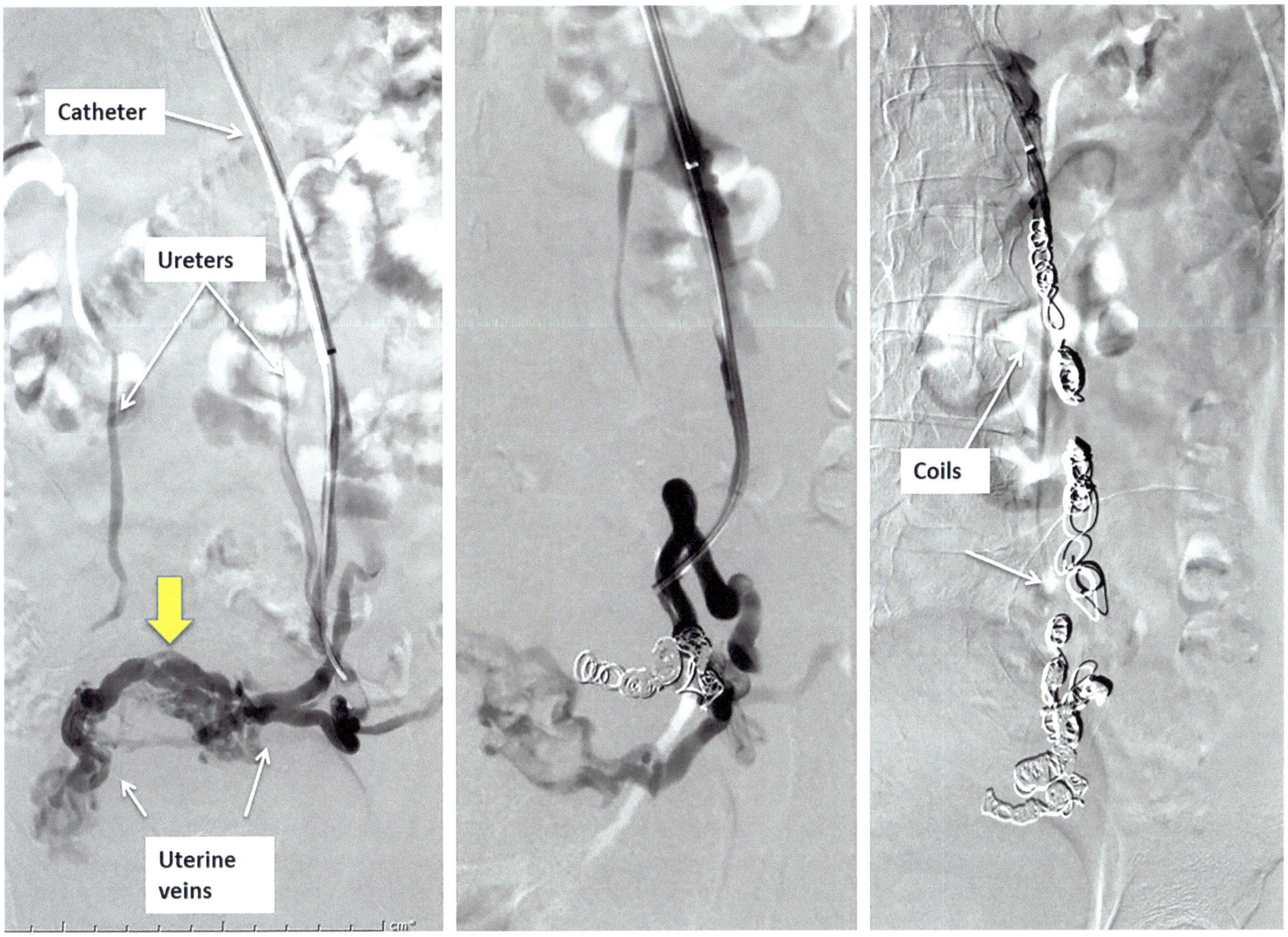

Fig. 16.60 A 48-year-old female with pelvic congestion syndrome. Left renal venography shows a dilated left ovarian vein with retrograde filling of the uterine veins. There is collateral venous connection along the uterine surface between left and right parametrial uterine venous plexuses (*yellow arrow*) filling the right internal iliac vein. After coil embolization, complete occlusion of the left ovarian vein and lack of filling of the cross pelvic collaterals were achieved.

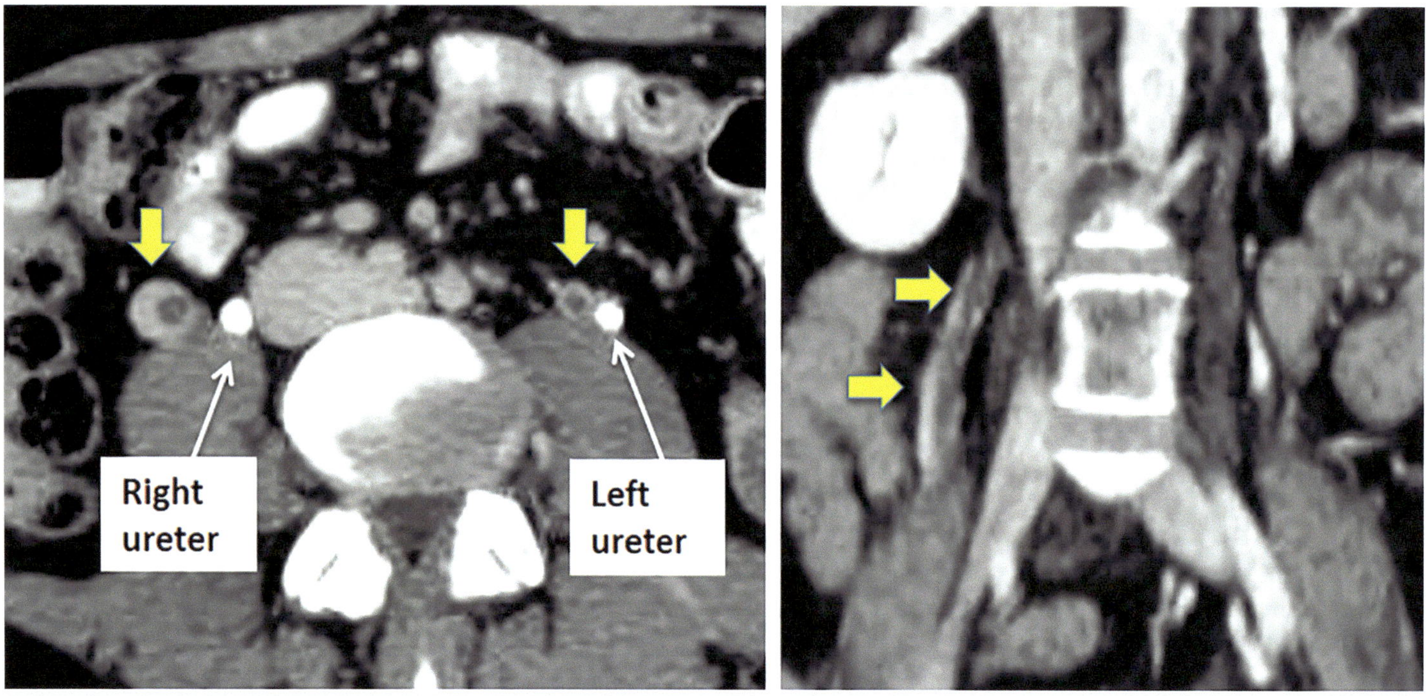

Fig. 16.61 Ovarian vein thrombosis. Axial and coronal, contrast-enhanced computed tomography (CT) showing bilateral ovarian vein thrombosis (*arrows*) in a patient with history of hysterectomy.

tortuous parauterine veins and the ovarian vein diameter of greater than 8 mm. Embolization of internal iliac vein or ovarian veins could be useful for treatment of the disease as an option to surgical hysterectomy with oophorectomy and vein ligation[77] (▶ Fig. 16.60).

Ovarian Vein Thrombosis

Ovarian vein thrombosis is usually a complication of vaginal deliveries, cesarean deliveries, and hysterectomy causing right lower quadrant pain, fever, and rarely pulmonary thromboembolism.[78] Due to anatomical and physiologic factors 80 to 90% of the thrombosis occurs in the right ovarian vein. CT and MR angiography are best techniques to show the abnormality (▶ Fig. 16.61).

◆ Postsurgical Anatomy

In reviewing images of the pelvis, a variety of foreign objects related to previous interventions may be encountered.[79] Common devices and postinterventional objects include IUCDs, tubal sterilization, varicoceles, tubal sterilization clips, and embolization coils for varicoceles, and vaginal pessary (▶ Fig. 16.62, ▶ Fig. 16.63). A pessary is a plastic or rubber ring that is inserted into the vagina to prevent uterine prolapse and cystoceles.

Intrauterine devices (IUDs) have different morphologies with the T configuration the most common in use. IUDs induce a foreign body reaction with mild inflammation and erosion in the surrounding endometrium.[80] Decidual reaction and uniform thinning of the surface endometrium by progesterone-releasing IUDs is always found after 6 months of treatment (▶ Fig. 16.62). In recent years, the Essure device (an intratubal coil) has become an increasingly popular option for tubal occlusion permanent tubal blockage[81] (▶ Fig. 16.63).

In classic cesarean section the uterus is also incised vertically allowing a large amount of space for delivering the baby. However, the procedure is associated with postsurgical complications and is not commonly used today. A lower uterine segment cesarean section is the common method in which a horizontal incision is made in the lower uterine segment. Scar of cesarean section due to myometrial discontinuity at the site of a previous incision may be seen on imaging study of the uterus (▶ **Fig. 16.25**). Scar defects are wider on retroflexed uterus than anteflexed uterus.[82] Scars are asymptomatic but sometimes can cause dysmenorrhea and chronic pelvic pain.

Hysterectomy is the most common gynecologic surgery that can be performed with either a transabdominal, laparoscopic, or transvaginal approach[83] (▶ Fig. 16.64). Total hysterectomy involves removal of the uterus and cervix. Subtotal or partial hysterectomy involves removal of the uterus but not the cervix (supracervical), and it is performed for benign causes such as fibroid uterus or dysfunctional uterine bleeding. Radical hysterectomy involves removal of the uterus, cervix, and ovaries to treat malignancies.

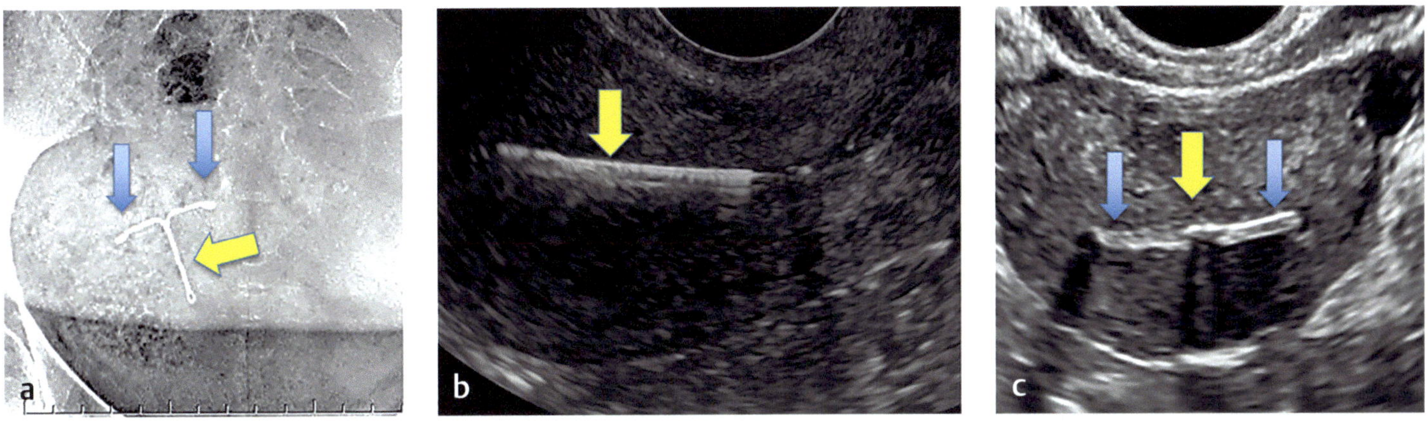

Fig. 16.62 T-shaped intrauterine device (IUD) shown by plain film and ultrasound (*arrows*). **(a)** Plain film. **(b)** Longitudinal and **(c)** axial ultrasound views of the uterus.

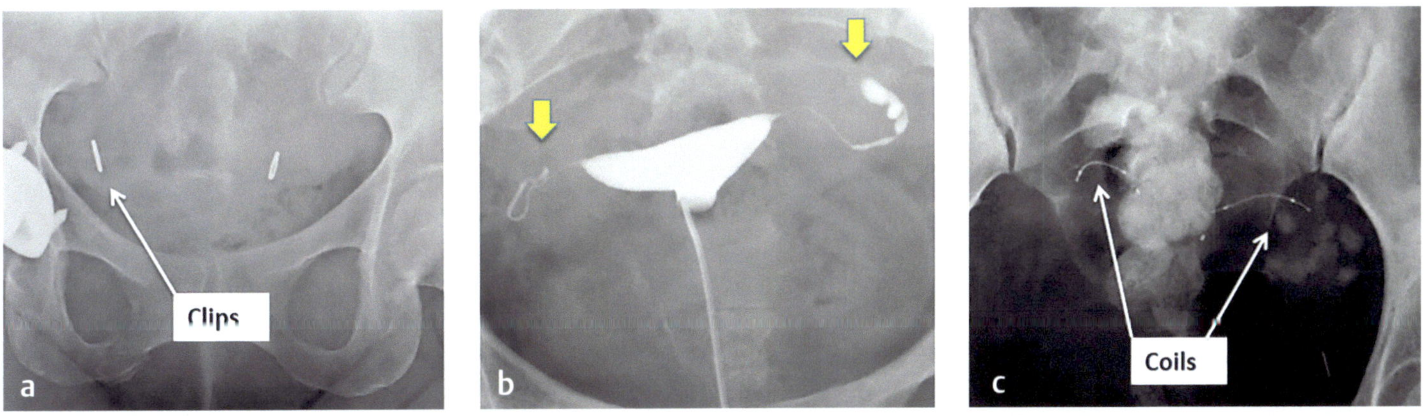

Fig. 16.63 Image findings after tubal sterilization. **(a)** Tubal surgical clips. **(b)** Tubal ligation (*arrows*) shown by hysterosalpingography. **(c)** Intratubal coils (Essure coils).

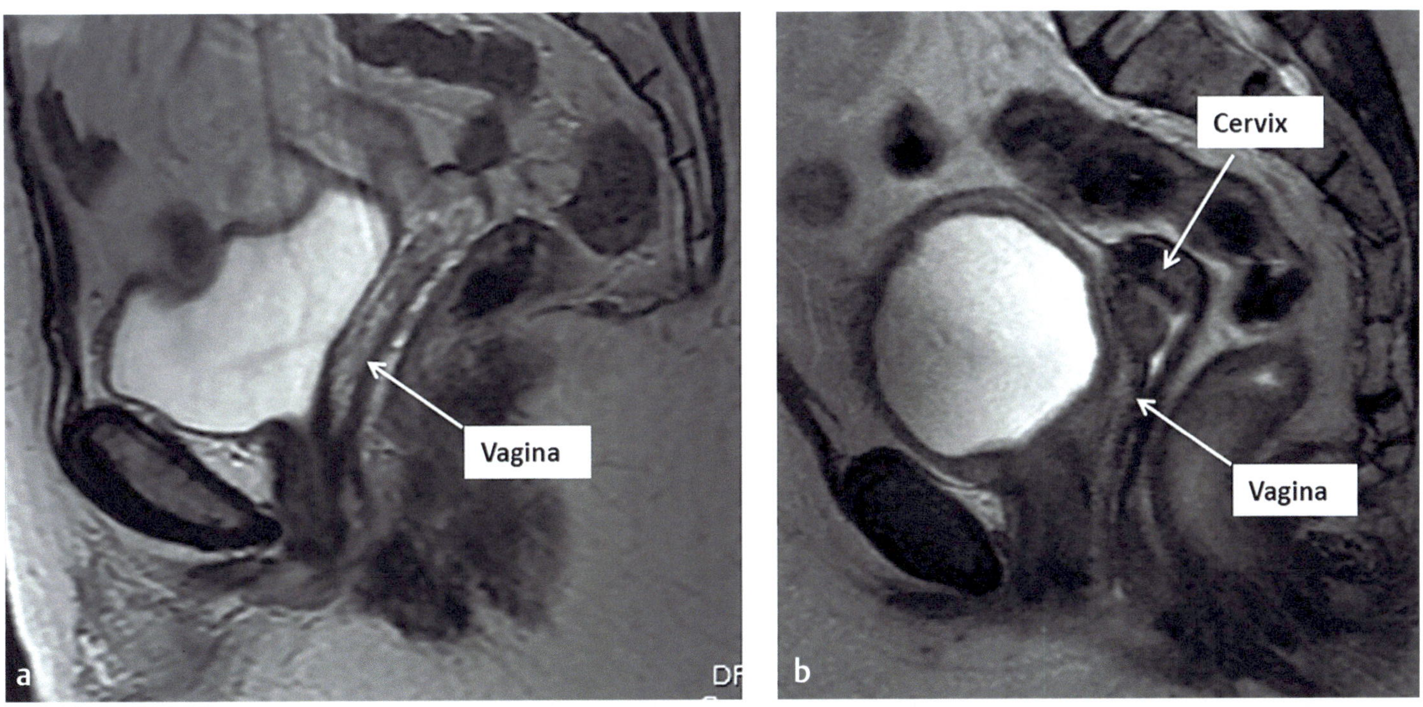

Fig. 16.64 Hysterectomy. Sagittal T2-weighted (T2W) magnetic resonance imaging (MRI) showing **(a)** total hysterectomy and **(b)** supracervical hysterectomy. Subtotal or partial hysterectomy involves removal of the uterus but not the cervix (supracervical).

◆ Supportive Ligaments

The female pelvic organs are embedded in subperitoneal connective tissue containing nerves and vessels. Taken together, these structures constitute a morphological as well as an important functional unit. The walls of the vagina, bladder, urethra, and rectum are connected and fixed by the vesicouterine, vesicovaginal, urethrovaginal, and rectovaginal septa. Pelvic ligaments are connective tissues formed by peritoneal folds or from the remnant of embryonic structures (▶Fig. 16.65, ▶Fig. 16.66). These structures are the main supportive gateway for passage of the arteries, veins, nerves, and lymphatics to the genital tract. They also act as suspensory structure to maintain the pelvic organs in their correct anatomic locations.

Broad and Cardinal Ligaments

The broad ligament is part of the lateral fold of peritoneum (lateral pelvic ligament) that runs from the lateral pelvic wall to the uterus, and encloses the uterine tube in its superior margin and suspends the ovary from its posterior aspect (▶Fig. 16.45). The uterus and the broad ligaments divide the pelvic cavity into the uterorectal pouch posteriorly and the uterovesical pouch anteriorly.

The broad ligament has three parts, the lateral mesometrium, which extends from the lateral pelvic walls to the body of the uterus, the mesosalpinx, which suspends the uterine tube in the pelvic cavity, and the mesovarium, a posterior extension of the broad ligament, which attaches to the ovary[84,85] (▶Fig. 16.45). The ovarian ligament, the round ligament of uterus, and the suspensory ligament of the ovary are included in the broad ligament. The cardinal ligament (parauterine, parametrium, lateral uterine ligament, transverse cervical ligament) forms from condensation of the base of the broad ligament and along with uterosacral ligament attach to the cervix and upper vagina medially and to the anterior trunk of the internal iliac artery (35%) or the upper border of the greater sciatic foramen (65%) laterally[84,85] (▶Fig. 16.14, ▶Fig. 16.21, ▶Fig. 16.65, ▶Fig. 16.66). The uterine arteries cross the ureters at the base of the broad ligaments marking the superior margin of the cardinal ligament (▶Fig. 16.27, ▶Fig. 16.46). Uterine vessels pass along the lateral margin of the uterus and within the broad ligaments. The ureter passes inside the cardinal ligament and crosses under the uterine artery (▶Fig. 16.65, ▶Fig. 16.67). Therefore, the ureters are located at the inferior edge of the cardinal ligament and above the uterosacral ligaments.[86]

The cardinal and broad ligaments have a complex structure consisting of vessels, nerves, and connective tissue rather than a single band of connective tissue, making imaging difficult to show their boundaries (▶Fig. 16.49). In patients with large ascites, demonstration of the broad ligaments is easy (▶Fig. 16.21).

Defect in the broad ligament, usually related to previous surgeries, may predispose it to internal herniation of organs such as colon, ovary, or small bowel through the defect. Defects occur in the broad ligament inferior or superior to the round ligament of the uterus and rarely through the mesovarium.[87]

Round Ligament

During embryonic development, the round ligament is derived from the gubernaculum. The round ligament is a fibrous cord attached to the uterine cornu anterior to the fallopian tube and passing over the pelvic inlet and external iliac vessels to reach the inguinal canal through the deep inguinal foramen and then coursing through the inguinal canal to reach the labium major (▶Fig. 16.21, ▶Fig. 16.68, ▶Fig. 16.69). It may play a role in keeping the uterus anteflexed. There is a two-layered peritoneal surface between the round ligament and the broad ligament. A pouch of parietal peritoneum may accompany the round ligament in the inguinal canal to form the "canal of nuck." It is analogous to the processus vaginalis in males and normally undergoes complete

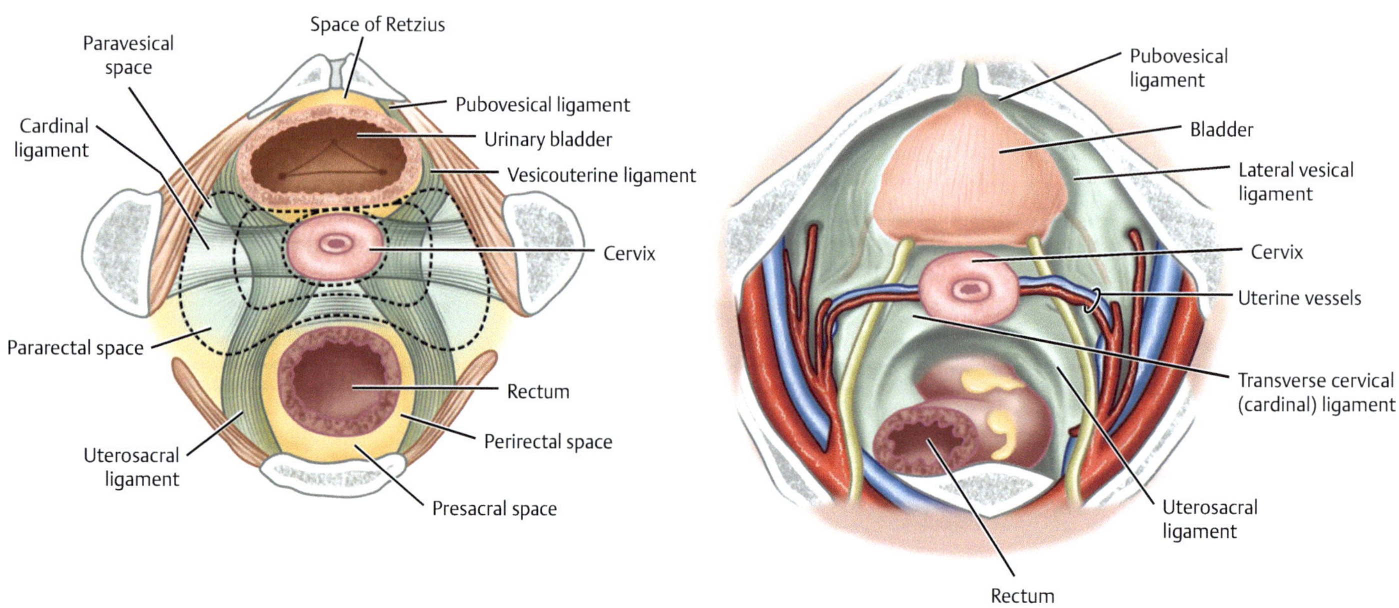

Fig. 16.65 Supportive ligaments of the pelvic organs.

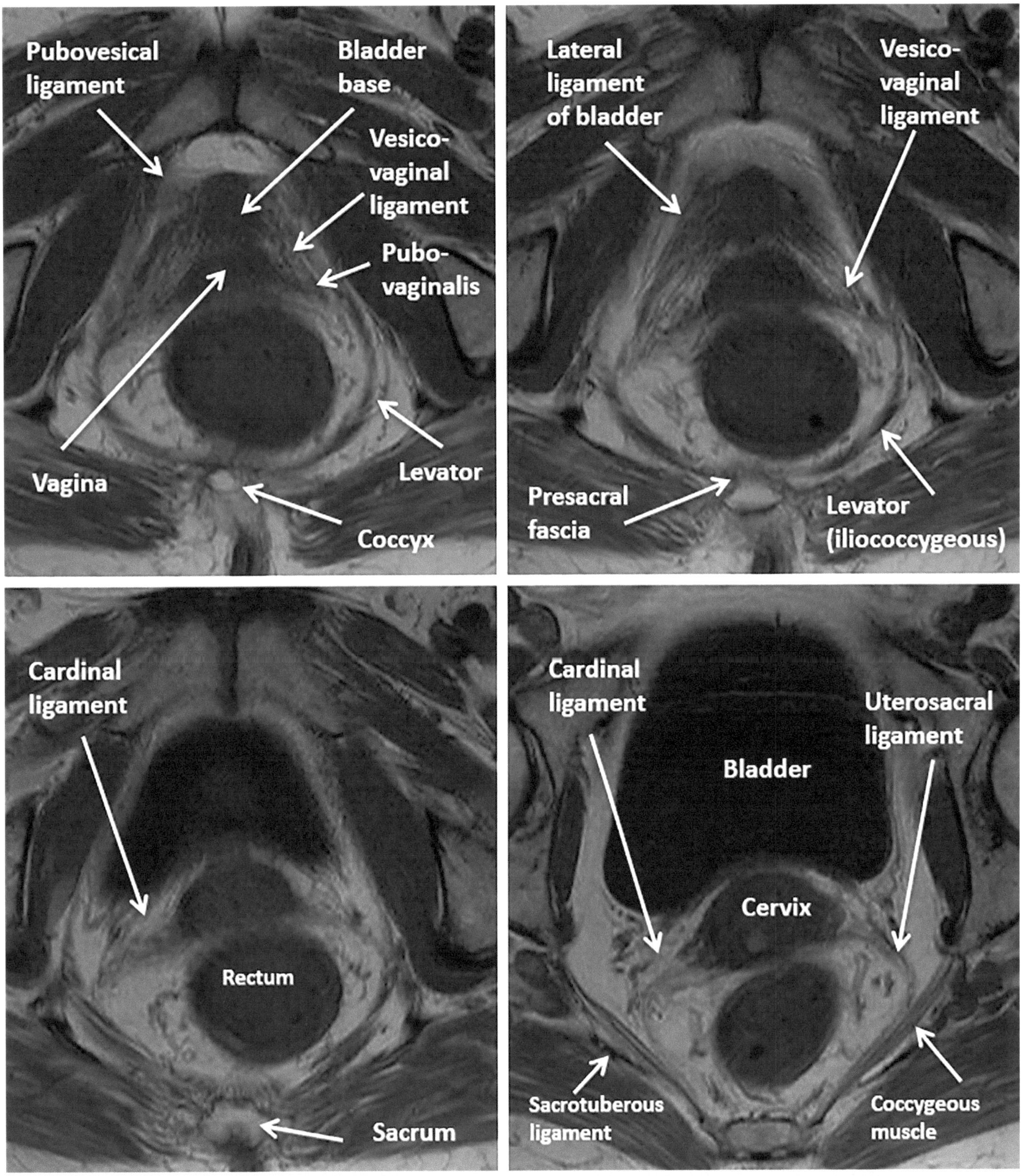

Fig. 16.66 Supportive ligaments of the pelvic organs shown by axial T1-weighted (T1W) magnetic resonance imaging (MRI).

16

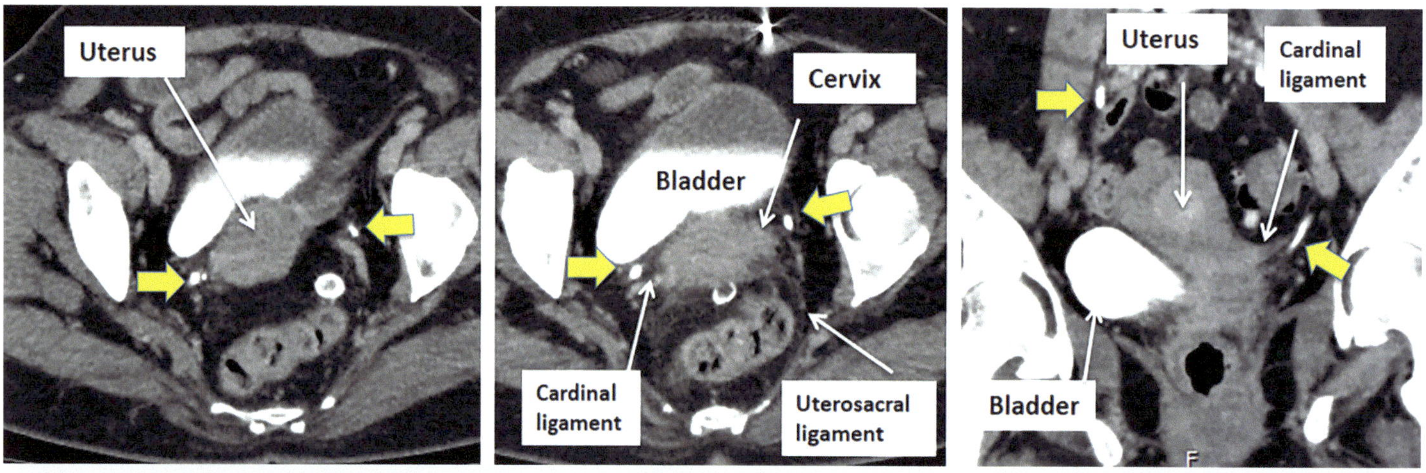

Fig. 16.67 Computed tomography (CT) urogram showing the relationship of the distal ureters (*arrows*) and the cardinal ligaments at the uterine isthmus. The ureter passes inside the cardinal ligament and crosses under the uterine artery. The ureters are located at the inferior edge of the cardinal ligament and anterior to the uterosacral ligaments.

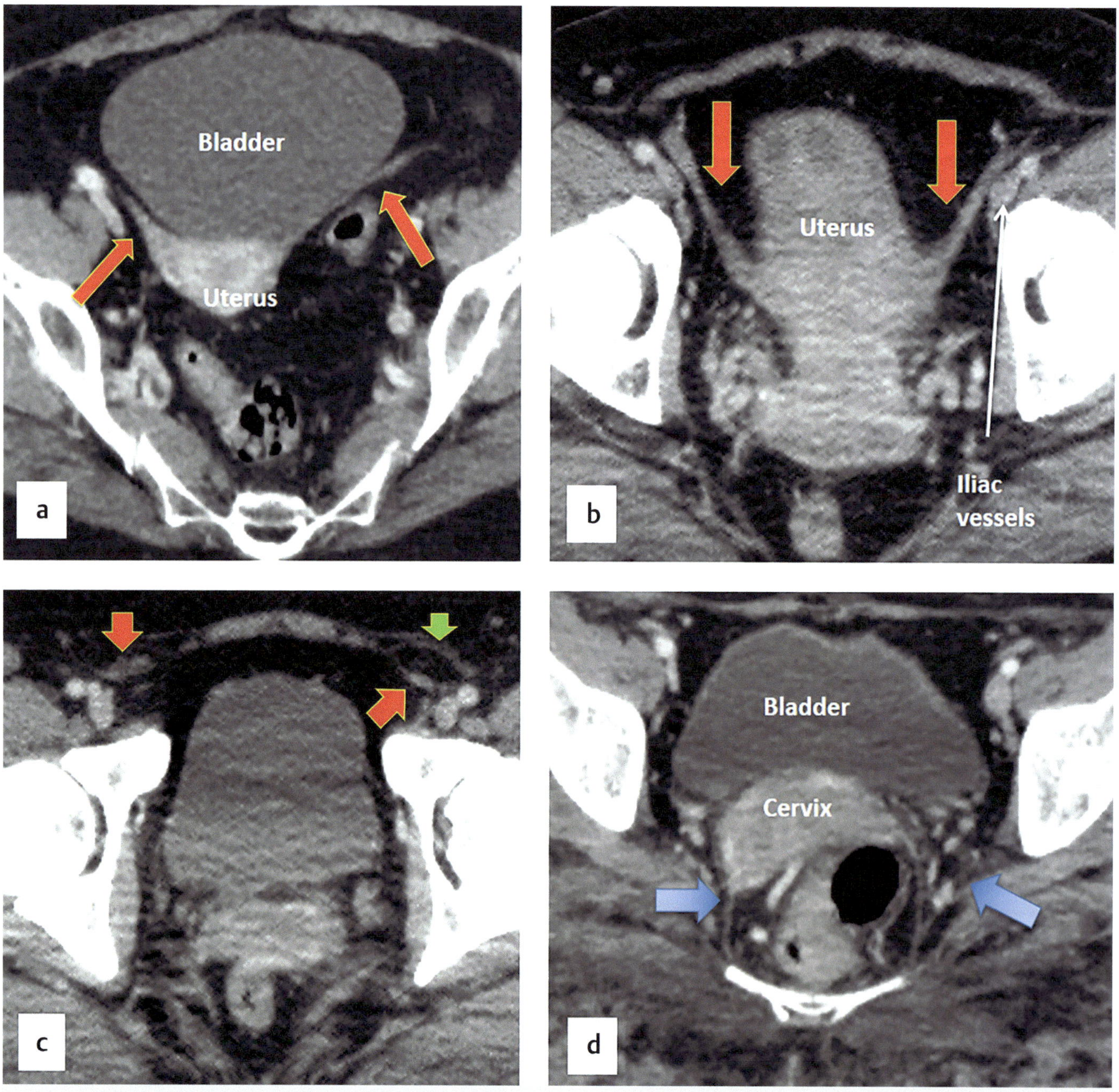

Fig. 16.68 Round (*red arrow*) and uterosacral (*blue arrow*) ligaments shown by axial computed tomography (CT) scans in different patients **(a–d)**. The round ligaments (*red arrows*) are fibrous cords attached to the uterine cornu anterior to the fallopian tube, passing over the external iliac vessels to reach the inguinal canal through the deep inguinal foramen (*green arrow*).

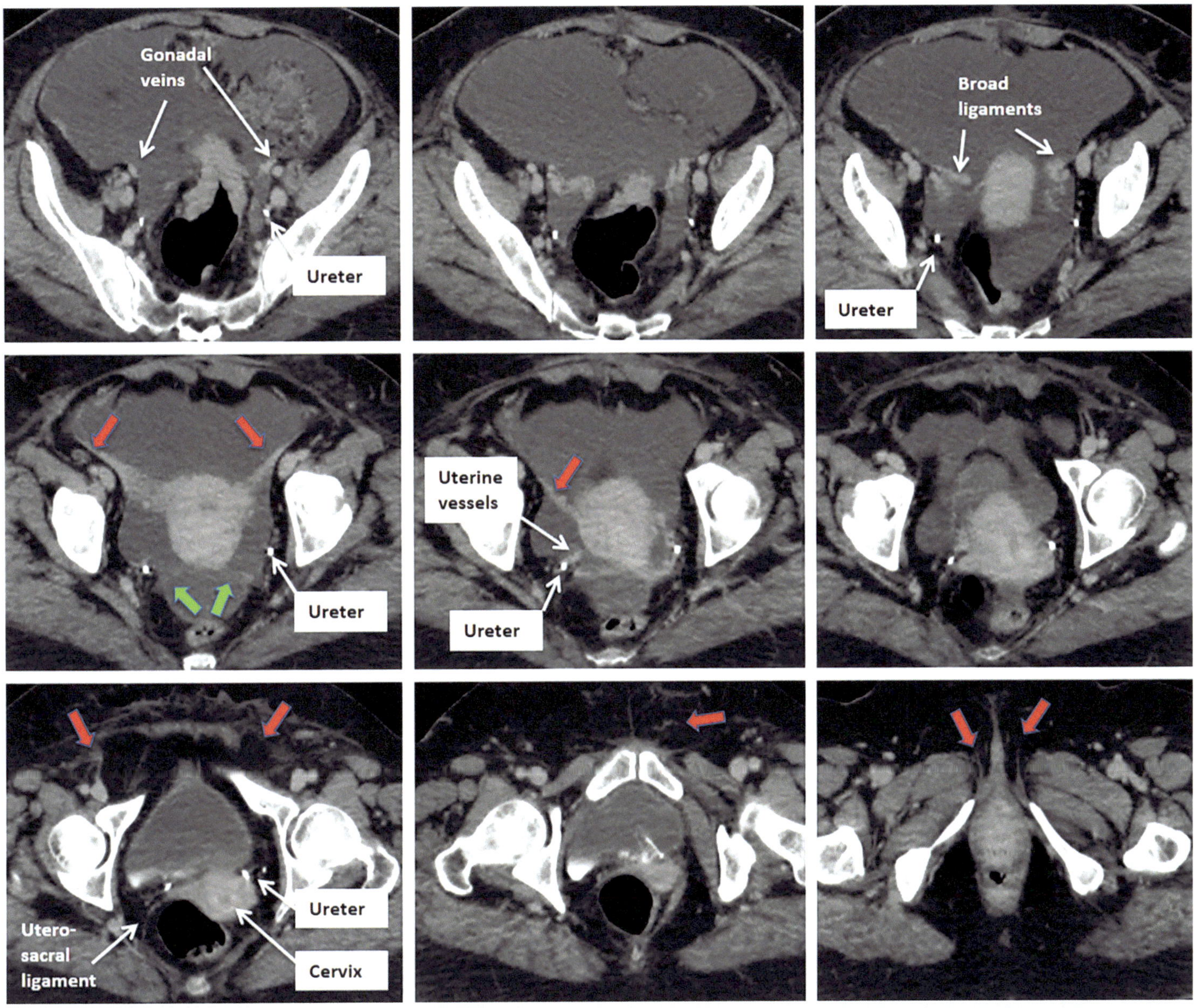

Fig. 16.69 Axial computed tomography (CT) images from superior to inferior in a patient with large ascites secondary to carcinomatosis showing pelvic ligaments. The cardinal ligaments are seen originating from the sacral lower uterine segment to insert into the pelvic fascia (*green arrows*). The ureter passes inside the cardinal ligament and crosses under the uterine artery. The right uterosacral ligament is seen originating from the sacrotuberous ligament to insert into the cervix. The round ligaments (*red arrows*) are fibrous cords attached to the uterine cornu, passing over the external iliac vessels to reach the inguinal canal through the deep inguinal foramen and then coursing through the inguinal canal to reach the labium major.

obliteration during the first year of life.[88] Incomplete obliteration of the peritoneal pouch causes indirect inguinal hernia or hydrocele of the canal of nuck in women. The round ligament artery, also known as Sampson's artery, plays a minor physiologic role.[53] This branch, which arises either from the inferior epigastric artery or from the external iliac artery, may be responsible for persistent bleeding after hysterectomy. Venous drainage of the round ligament is toward the broad ligament of the uterus. It may be enlarged in pelvic congestion syndrome and during pregnancy (▸Fig. 16.70).

Uterosacral Ligaments

The uterosacral ligaments are condensation of the visceral pelvic fascia (endopelvic fascia) reinforced by a considerable amount of fibrous tissue.[84] The ligament has a fan shape, with mean widths of 5.2 ± 0.9, 2.7 ± 1.0, and 2.0 ± 0.5 cm at the sacral, intermediate, and cervical portions, respectively. These ligaments extend from the posterolateral aspect of the cervix and vaginal dome where it is confluent with the attachment of the cardinal ligament (▸Fig. 16.65, ▸Fig. 16.66, ▸Fig. 16.68, ▸Fig. 16.69). They travel on the sides of the rectum and merge with the lateral ligament

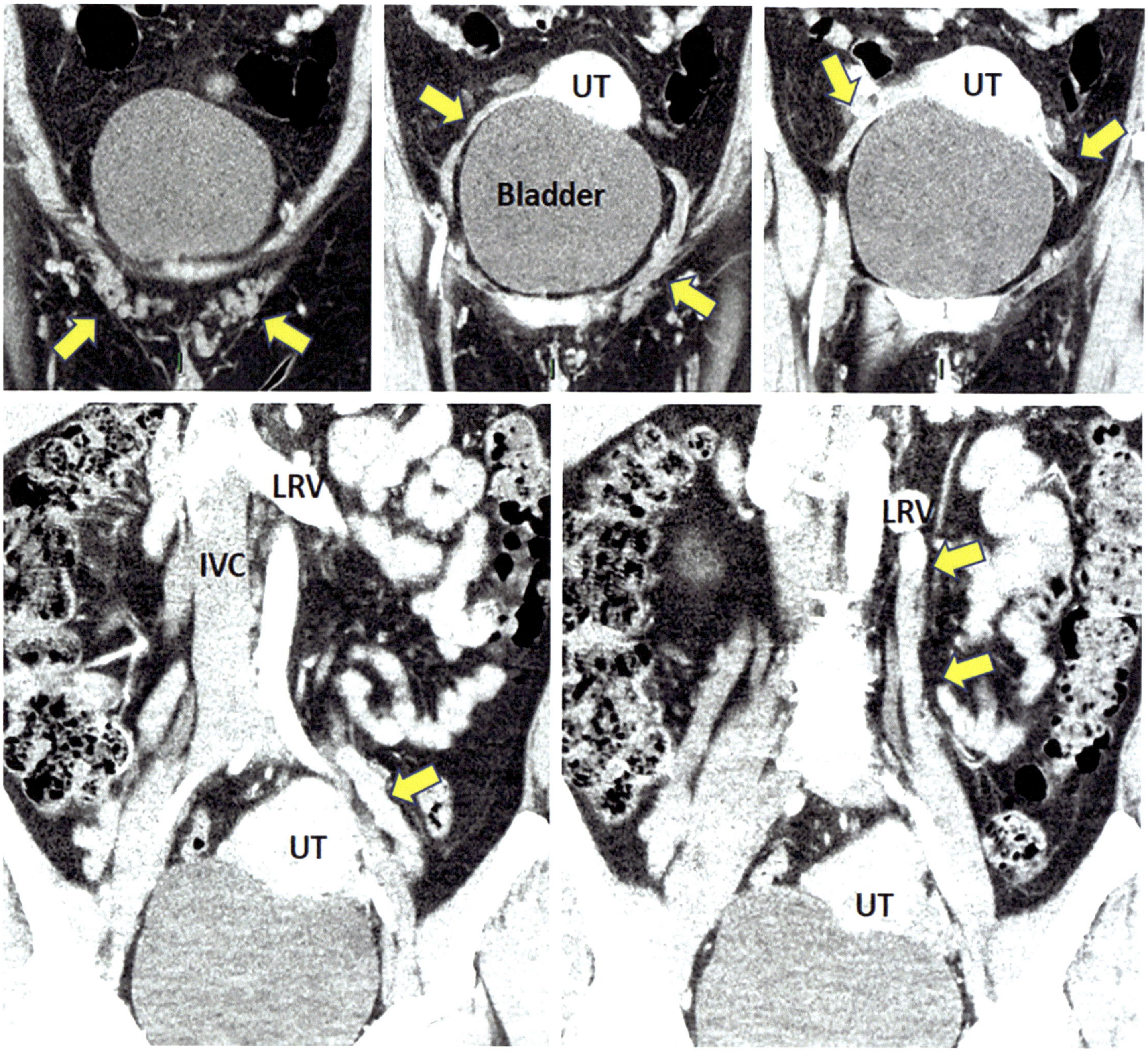

Fig. 16.70 Coronal, postcontrast computed tomography (CT) images from anterior to posterior showing vulvoperineal varicosities (*arrows*) draining along the round ligament toward the broad ligament of the uterus and joining the gonadal veins in a patient with pelvic congestion syndrome. The vulvoperineal varicosities are common findings in pelvic congestion syndrome and during pregnancy. The pelvic congestion syndrome is due to ovarian venous incompetence due to anatomic or hormonal factors. IVC, inferior vena cava; LRV, left renal vein; UT, uterus.

of the rectum before attaching to the sacrospinous ligament–coccygeus muscle complex, the anterior aspect of the sacrum from the sacrococcygeal joint to the S3, and the presacral fascia (▸Fig. 16.14, ▸Fig. 16.66). At the cervical internal os, the ureter is 1 cm lateral to the anterior/superior margin of the uterosacral ligament. At the sacral side, the ligament is close to S1–S4 and the hypogastric nerves. The ischial spine is beneath the middle portion of the ligament and the superior gluteal vein lies beneath the sacral side of the ligament. These ligaments are seen on CT or MR images[89] (▸Fig. 16.6, ▸Fig. 16.68).

Laxity and thinning of the pelvic ligaments and levator ani muscles can result in pelvic floor prolapse. Pelvic floor prolapse is defined as downward descent of the pelvic floor compartments, anterior compartment (bladder and urethra), middle compartment (uterus and vagina), and posterior compartment (rectum) (▸Fig. 16.71). Abnormality of the cardinal and uterosacral ligaments are the major factors for descent of the uterus, with the anterior vaginal wall descent being the most common type (85%) of pelvic organ prolapse.[84] Prolapse of the anterior vaginal wall is commonly seen with cystocele and prolapse of the posterior

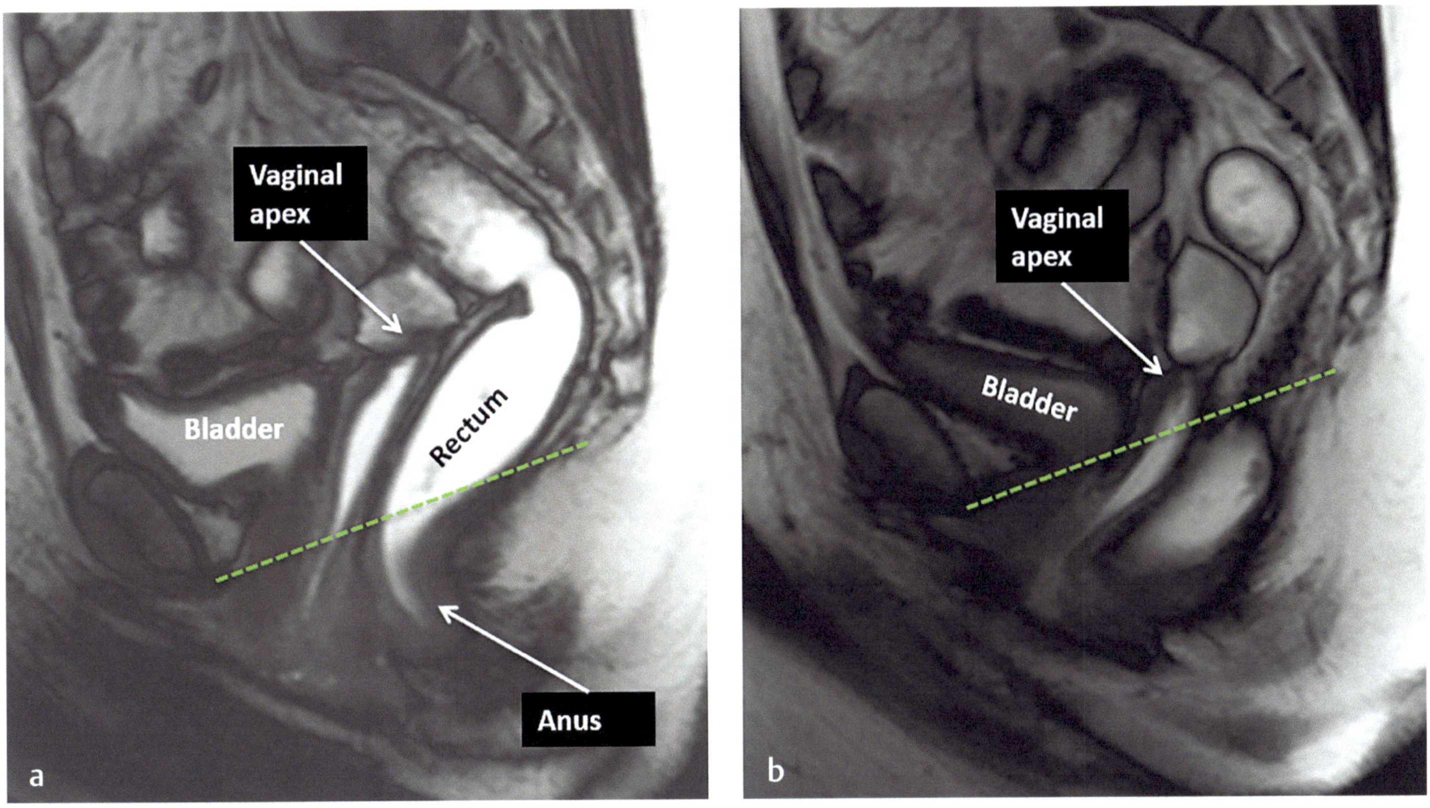

Fig. 16.71 Pelvic descent shown by dynamic magnetic resonance imaging (MRI), status post total hysterectomy. True FISP, midsagittal acquisitions before and during evacuation of rectal contrast. The uterus is surgically absent. **(a)** The pelvic organs are in normal position during rest. **(b)** On active evacuation, there is descent of pelvic organs below the pubococcygeal line. There is urethral hypermobility and the base of the bladder descends mildly below the pubococcygeal line. The apex of the vagina also descends to the pubococcygeal line. The anorectal junction descends approximately 6 cm below the pubococcygeal line.

wall of vagina is seen with rectocele. Uterosacral ligaments are important supporting structures of the pelvic organs preventing prolapse of the vagina or uterus. Uterosacral ligament suspension surgery is designed to restore support to the uterus or vaginal vault in case of prolapse. The mid part of the ligament is thicker and suitable for surgical use.

Ovarian Support

The ovaries connect to the lateral angle of the uterus by ovarian ligament, to the posterior surface of the broad ligament by the mesovarium, and to the pelvic side wall by a peritoneal fold called the suspensory (infundibulopelvic) ligament of the ovary (▶ Fig. 16.2). The ovarian ligament is continuous with the proximal part of the round ligament as both arise from the gubernaculum. The ovarian vessels pass in the suspensory ligament of the ovary.

References

1. Sajjad Y. Development of the genital ducts and external genitalia in the early human embryo. J Obstet Gynaecol Res 2010;36(5):929–937
2. Hashimoto R. Development of the human Müllerian duct in the sexually undifferentiated stage. Anat Rec A Discov Mol Cell Evol Biol 2003;272(2):514–519
3. Mauch RB, Thiedemann KU, Drews U. The vagina is formed by downgrowth of Wolffian and Müllerian ducts. Graphical reconstructions from normal and Tfm mouse embryos. Anat Embryol (Berl) 1985;172(1):75–87
4. Cai Y. Revisiting old vaginal topics: conversion of the Müllerian vagina and origin of the "sinus" vagina. Int J Dev Biol 2009;53(7):925–934
5. Walker DK, Salibian RA, Salibian AD, Belen KM, Palmer SL. Overlooked diseases of the vagina: a directed anatomic-pathologic approach for imaging assessment. Radiographics 2011;31(6):1583–1598
6. Li T, Liao Q, Zhang H, Gao X, Li X, Zhang M. Anatomic distribution of nerves and microvascular density in the human anterior vaginal wall: prospective study. PLoS One 2014;9(11):e110239
7. Parikh JH, Barton DP, Ind TE, Sohaib SA. MR imaging features of vaginal malignancies. Radiographics 2008;28(1):49–63, quiz 322
8. Wilson JD, Griffin JE, Leshin M, George FW. Role of gonadal hormones in development of the sexual phenotypes. Hum Genet 1981;58(1):78–84
9. Troiano RN, McCarthy SM. Mullerian duct anomalies: imaging and clinical issues. Radiology 2004;233(1):19–34
10. Hahn WY, Israel GM, Lee VS. MRI of female urethral and periurethral disorders. AJR Am J Roentgenol 2004;182(3):677–682
11. Shi AA, Lee SI. Radiological reasoning: algorithmic workup of abnormal vaginal bleeding with endovaginal sonography and sonohysterography. AJR Am J Roentgenol 2008;191(6, Suppl):S68–S73
12. Ziereisen F, Guissard G, Damry N, Avni EF. Sonographic imaging of the paediatric female pelvis. Eur Radiol 2005;15(7):1296–1309
13. Langer JE, Oliver ER, Lev-Toaff AS, Coleman BG. Imaging of the female pelvis through the life cycle. Radiographics 2012;32(6):1575–1597
14. van Gestel I, IJland MM, Hoogland HJ, Evers JLH. Endometrial wave-like activity in the non-pregnant uterus. Hum Reprod Update 2003;9(2):131–138
15. Randall JM, Templeton A. Transvaginal sonographic assessment of follicular and endometrial growth in spontaneous and clomiphene citrate cycles. Fertil Steril 1991;56(2):208–212
16. Togashi K, Nakai A, Sugimura K. Anatomy and physiology of the female pelvis: MR imaging revisited. J Magn Reson Imaging 2001;13(6):842–849

16

17. Scoutt LM, McCauley TR, Flynn SD, Luthringer DJ, McCarthy SM. Zonal anatomy of the cervix: correlation of MR imaging and histologic examination of hysterectomy specimens. Radiology 1993;186(1):159–162
18. Reinhold C, McCarthy S, Bret PM, et al. Diffuse adenomyosis: comparison of endovaginal US and MR imaging with histopathologic correlation. Radiology 1996;199(1):151–158
19. Fujiwara T, Togashi K, Yamaoka T, et al. Kinematics of the uterus: cine mode MR imaging. Radiographics 2004;24(1):e19
20. Nalaboff KM, Pellerito JS, Ben-Levi E. Imaging the endometrium: disease and normal variants. Radiographics 2001;21(6):1409–1424
21. Davis PC, O'Neill MJ, Yoder IC, Lee SI, Mueller PR. Sonohysterographic findings of endometrial and subendometrial conditions. Radiographics 2002;22(4): 803–816
22. Bakos O, Lundkvist O, Bergh T. Transvaginal sonographic evaluation of endometrial growth and texture in spontaneous ovulatory cycles—a descriptive study. Hum Reprod 1993;8(6):799–806
23. Wachsberg RH, Kurtz AB, Levine CD, Solomon P, Wapner RJ. Real-time ultrasonographic analysis of the normal postpartum uterus: technique, variability, and measurements. J Ultrasound Med 1994;13(3):215–221
24. Kamaya A, Ro K, Benedetti NJ, Chang PL, Desser TS. Imaging and diagnosis of postpartum complications: sonography and other imaging modalities. Ultrasound Q 2009;25(3):151–162
25. Seungdamrong A, Weiss G. Ovulation in a postmenopausal woman. Fertil Steril 2007;88(5):1438.e1–1438.e2
26. Miller EI, Thomas RH, Lines P. The atrophic postmenopausal uterus. J Clin Ultrasound 1977;5(4):261–263
27. Goldstein SR. The role of transvaginal ultrasound or endometrial biopsy in the evaluation of the menopausal endometrium. Am J Obstet Gynecol 2009;201(1):5–11
28. Smith-Bindman R, Weiss E, Feldstein V. How thick is too thick? When endometrial thickness should prompt biopsy in postmenopausal women without vaginal bleeding. Ultrasound Obstet Gynecol 2004;24(5):558–565
29. Gerber B, Krause A, Müller H, et al. Effects of adjuvant tamoxifen on the endometrium in postmenopausal women with breast cancer: a prospective long-term study using transvaginal ultrasound. J Clin Oncol 2000;18(20):3464–3470
30. Wildenberg JC, Yam BL, Langer JE, Jones LP. US of the nongravid cervix with multimodality imaging correlation: normal appearance, pathologic conditions, and diagnostic pitfalls. Radiographics 2016;36(2):596–617
31. Autier P, Coibion M, Huet F, Grivegnee AR. Transformation zone location and intraepithelial neoplasia of the cervix uteri. Br J Cancer 1996;74(3):488–490
32. El Jack AK, Siegelman ES. "Pseudoseptum" of the uterine cervix on MRI. J Magn Reson Imaging 2007;26(4):963–965
33. Yitta S, Hecht EM, Mausner EV, Bennett GL. Normal or abnormal? Demystifying uterine and cervical contrast enhancement at multidetector CT. Radiographics 2011;31(3):647–661
34. Kaur H, Loyer EM, Minami M, Charnsangavej C. Patterns of uterine enhancement with helical CT. Eur J Radiol 1998;28(3):250–255
35. Junqueira BL, Allen LM, Spitzer RF, Lucco KL, Babyn PS, Doria AS. Müllerian duct anomalies and mimics in children and adolescents: correlative intraoperative assessment with clinical imaging. Radiographics 2009;29(4):1085–1103
36. Behr SC, Courtier JL, Qayyum A. Imaging of Müllerian duct anomalies. Radiographics 2012;32(6):E233–E250
37. Okamoto Y, Tanaka YO, Nishida M, Tsunoda H, Yoshikawa H, Itai Y. MR imaging of the uterine cervix: imaging-pathologic correlation. Radiographics 2003;23(2): 425–445, quiz 534–535
38. Kier R. Nonovarian gynecologic cysts: MR imaging findings. AJR Am J Roentgenol 1992;158(6):1265–1269
39. Rezvani M, Shaaban AM. Fallopian tube disease in the nonpregnant patient. Radiographics 2011;31(2):527–548
40. Simpson WL Jr, Beitia LG, Mester J. Hysterosalpingography: a reemerging study. Radiographics 2006;26(2):419–431
41. Carrascosa PM, Capuñay C, Vallejos J, Martín López EB, Baronio M, Carrascosa JM. Virtual hysterosalpingography: a new multidetector CT technique for evaluating the female reproductive system. Radiographics 2010;30(3):643–661
42. Pelage JP, Le Dref O, Soyer P, et al. Arterial anatomy of the female genital tract: variations and relevance to transcatheter embolization of the uterus. AJR Am J Roentgenol 1999;172(4):989–994
43. Cicinelli E, Einer-Jensen N, Galantino P, Alfonso R, Nicoletti R. The vascular cast of the human uterus: from anatomy to physiology. Ann N Y Acad Sci 2004;1034:19–26
44. Gomez-Jorge J, Keyoung A, Levy EB, Spies JB. Uterine artery anatomy relevant to uterine leiomyomata embolization. Cardiovasc Intervent Radiol 2003;26(6):522–527
45. Farrer-Brown G, Beilby JO, Tarbit MH. The blood supply of the uterus. 1. Arterial vasculature. J Obstet Gynaecol Br Commonw 1970;77(8):673–681
46. Cicinelli E, Einer-Jensen N, Barba B, Luisi D, Alfonso R, Tartagni M. Blood to the cornual area of the uterus is mainly supplied from the ovarian artery in the follicular phase and from the uterine artery in the luteal phase. Hum Reprod 2004;19(4):1003–1008
47. Chen CL, Guo HX, Liu P, et al. Three-dimensional reconstruction of the uterine vascular supply through vascular casting and thin slice computed tomography scanning. Minim Invasive Ther Allied Technol 2009;18(2):98–102
48. Razavi MK, Wolanske KA, Hwang GL, Sze DY, Kee ST, Dake MD. Angiographic classification of ovarian artery-to-uterine artery anastomoses: initial observations in uterine fibroid embolization. Radiology 2002;224(3):707–712
49. Ouyang Z, Liu P, Yu Y, et al. Role of ovarian artery-to-uterine artery anastomoses in uterine artery embolization: initial anatomic and radiologic studies. Surg Radiol Anat 2012;34(8):737–741
50. Kim HS, Tsai J, Patra A, Lee JM, Griffith JG, Wallach EE. Effects of utero-ovarian anastomoses on clinical outcomes and repeat intervention rates after uterine artery embolization. J Vasc Interv Radiol 2006;17(5):783–789
51. Chrisman HB, Saker MB, Ryu RK, et al. The impact of uterine fibroid embolization on resumption of menses and ovarian function. J Vasc Interv Radiol 2000; 11(6):699–703
52. Kaump GR, Spies JB. The impact of uterine artery embolization on ovarian function. J Vasc Interv Radiol 2013;24(4):459–467
53. Chait A, Moltz A, Nelson JH Jr. The collateral arterial circulation in the pelvis. An angiographic study. Am J Roentgenol Radium Ther Nucl Med 1968; 102(2):392–400
54. Bereza T, Tomaszewski KA, Lis GJ, et al. "Venous lakes"—a corrosion cast scanning electron microscopy study of regular and myomatous human uterine blood vessels. Folia Morphol (Warsz) 2014;73(2):164–168
55. Li H, Jia J, Xiao Y, Kang L, Cui H. Anatomical basis of female pelvic cavity for nerve sparing radical hysterectomy. Surg Radiol Anat 2015;37(6):657–665
56. Cohen HL, Eisenberg P, Mandel F, Haller JO. Ovarian cysts are common in premenarchal girls: a sonographic study of 101 children 2–12 years old. AJR Am J Roentgenol 1992;159(1):89–91
57. Cohen HL, Shapiro MA, Mandel FS, Shapiro ML. Normal ovaries in neonates and infants: a sonographic study of 77 patients 1 day to 24 months old. AJR Am J Roentgenol 1993;160(3):583–586
58. Pavlik EJ, DePriest PD, Gallion HH, et al. Ovarian volume related to age. Gynecol Oncol 2000;77(3):410–412
59. Ritchie WGM. Sonographic evaluation of normal and induced ovulation. Radiology 1986;161(1):1–10
60. Zaidi J, Jurkovic D, Campbell S, Collins W, McGregor A, Tan SL. Luteinized unruptured follicle: morphology, endocrine function and blood flow changes during the menstrual cycle. Hum Reprod 1995;10(1):44–49
61. Levine D, Brown DL, Andreotti RF, et al. Management of asymptomatic ovarian and other adnexal cysts imaged at US: Society of Radiologists in Ultrasound Consensus Conference Statement. Radiology 2010;256(3):943–954
62. Outwater EK, Talerman A, Dunton C. Normal adnexa uteri specimens: anatomic basis of MR imaging features. Radiology 1996;201(3):751–755
63. Lerman H, Metser U, Grisaru D, Fishman A, Lievshitz G, Even-Sapir E. Normal and abnormal 18F-FDG endometrial and ovarian uptake in pre- and postmenopausal patients: assessment by PET/CT. J Nucl Med 2004;45(2):266–271
64. Hall DA, McCarthy KA, Kopans DB. Sonographic visualization of the normal postmenopausal ovary. J Ultrasound Med 1986;5(1):9–11
65. Muradali D, Colgan T, Hayeems E, Burns PN, Wilson SR. Echogenic ovarian foci without shadowing: are they caused by psammomatous calcifications? Radiology 2002;224(2):429–435
66. Matson M, Nicholson A, Belli AM. Anastomoses of the ovarian and uterine arteries: a potential pitfall and cause of failure of uterine embolization. Cardiovasc Intervent Radiol 2000;23(5):393–396

67. Pelage JP, Walker WJ, Le Dref O, Rymer R. Ovarian artery: angiographic appearance, embolization and relevance to uterine fibroid embolization. Cardiovasc Intervent Radiol 2003;26(3):227–233
68. Enakpene CA, Stern T, Barzallo Salazar MJ, Mukherjee P. Spontaneous rupture of an ovarian artery aneurysm: a rare postpartum complication. Case Rep Obstet Gynecol 2016;2016:1029561
69. Saini A, Dina R, McIndoe GA, Soutter WP, Gishen P, deSouza NM. Characterization of adnexal masses with MRI. AJR Am J Roentgenol 2005;184(3):1004–1009
70. Rajkotia K, Veeramani M, Macura KJ. Magnetic resonance imaging of adnexal masses. Top Magn Reson Imaging 2006;17(6):379–397
71. Togashi K, Nishimura K, Kimura I, et al. Endometrial cysts: diagnosis with MR imaging. Radiology 1991;180(1):73–78
72. Athey PA, Cooper NB. Sonographic features of parovarian cysts. AJR Am J Roentgenol 1985;144(1):83–86
73. Kim JS, Woo SK, Suh SJ, Morettin LB. Sonographic diagnosis of paraovarian cysts: value of detecting a separate ipsilateral ovary. AJR Am J Roentgenol 1995;164(6):1441–1444
74. Mashiach R, Melamed N, Gilad N, Ben-Shitrit G, Meizner I. Sonographic diagnosis of ovarian torsion: accuracy and predictive factors. J Ultrasound Med 2011;30(9):1205–1210
75. Balica AC, Nassiri N, Horne J, Egan S, Wang XK. Pelvic congestion syndrome. J Minim Invasive Gynecol 2015;22(6S):S152
76. Ignacio EA, Dua R, Sarin S, et al. Pelvic congestion syndrome: diagnosis and treatment. Semin Intervent Radiol 2008;25(4):361–368
77. Pyra K, Woźniak S, Drelich-Zbroja A, Wolski A, Jargiełło T. Evaluation of effectiveness of embolization in pelvic congestion syndrome with the new vascular occlusion device (ArtVentive EOS™): preliminary results. Cardiovasc Intervent Radiol 2016;39(8):1122–1127
78. Takeda A, Yamase Y, Koike W, Hayashi S, Imoto S, Nakamura H. Pulmonary thromboembolism as a result of ovarian vein thrombosis after laparoscopic-assisted vaginal hysterectomy for uterine myoma. J Obstet Gynaecol Res 2016; 42(6):743–747
79. Margolis M, McLennan MK. Radiology rounds. Intrauterine contraceptive device. Can Fam Physician 1993;39(2113):2113, 2283–2284
80. Sheppard BL. Endometrial morphological changes in IUD users: a review. Contraception 1987;36(1):1–10
81. Guelfguat M, Gruenberg TR, Dipoce J, Hochsztein JG. Imaging of mechanical tubal occlusion devices and potential complications. Radiographics 2012;32(6): 1659–1673
82. Wang CB, Chiu WW, Lee CY, Sun YL, Lin YH, Tseng CJ. Cesarean scar defect: correlation between Cesarean section number, defect size, clinical symptoms and uterine position. Ultrasound Obstet Gynecol 2009;34(1):85–89
83. Paspulati RM, Dalal TA. Imaging of complications following gynecologic surgery. Radiographics 2010;30(3):625–642
84. Ramanah R, Berger MB, Parratte BM, DeLancey JO. Anatomy and histology of apical support: a literature review concerning cardinal and uterosacral ligaments. Int Urogynecol J Pelvic Floor Dysfunct 2012;23(11):1483–1494
85. Yabuki Y, Sasaki H, Hatakeyama N, Murakami G. Discrepancies between classic anatomy and modern gynecologic surgery on pelvic connective tissue structure: harmonization of those concepts by collaborative cadaver dissection. Am J Obstet Gynecol 2005;193(1):7–15
86. Touboul C, Fauconnier A, Zareski E, Bouhanna P, Daraï E. The lateral infraureteral parametrium: myth or reality? Am J Obstet Gynecol 2008;199(3):242.e1–242.e6
87. Aas-Eng MK, Langebrekke A, Lieng M, Qvigstad E. Management of broad ligament defects and herniation of colon. J Minim Invasive Gynecol 2017;24(5): 713–714
88. Kono R, Terasaki H, Murakami N, Tanaka M, Takeda J, Abe T. Hydrocele of the canal of Nuck: a case report with magnetic resonance hydrography findings. Surg Case Rep 2015;1:86
89. Umek WH, Morgan DM, Ashton-Miller JA, DeLancey JO. Quantitative analysis of uterosacral ligament origin and insertion points by magnetic resonance imaging. Obstet Gynecol 2004;103(3):447–451

17 Perineum

Petros Mirilas, Ioannis Georgopoulos, and Farhood Saremi

◆ Introduction

The perineum is a narrow area located between the musculature of the gluteal and thigh areas and extends from the coccyx to the pubis. The perineum can be defined as the area below the pelvic diaphragm, which, at the same time, is the lower border of the pelvis.[1–4]

With abduction of the thighs, the perineum has a diamond-shaped configuration. This diamond is divided into two triangles by a transverse line passing just anteriorly to the ischial tuberosities and in front of the anal canal. The anterior half is the urogenital triangle, bordered by the ischiopubic rami and the pubic symphysis. The posterior half, the anal triangle, is bordered by the inferior border of the gluteus maximus muscle, the ischial tuberosities, the sacrotuberous ligaments, and the coccyx (▶ Fig. 17.1, ▶ Fig. 17.2).

◆ The Anal Region

The anal triangle is the simpler of the two parts of the perineum and is the same in both male and female. In this region, the most important structures are the terminal portion of the anal canal, the sphincter ani externus, and the ishiorectal fossa.

Anal Canal

The anal canal is the distal part of the large bowel. Currently, the anal canal is surgically considered together with the rectum as a unit known as anorectum. The anatomy of the anorectum is not within the aim of this chapter (see Chapter 5 "Colon").

Ishiorectal Fossa

The ishiorectal (or more correctly ischioanal) fossa is the space lying between the skin of the anal region below and the levator ani above and its investing fascia (inferior fascia of the pelvic diaphragm) (▶ Fig. 17.3, ▶ Fig. 17.4). The lateral wall of this fossa is the obturator internus muscle and its investing fascia. The superior and medial wall is formed by the levator ani muscle and its fascia, and the external anal sphincter. The inferior fascia of the pelvic diaphragm joins the fascia lunata medially. The fascia lunata continues laterally on the fascia of the obturator internus muscle to form the pudendal (Alcock's) canal; the latter contains the internal pudendal vessels and the pudendal nerve. The fascia lunata divides the ischiorectal fossa into an upper suprategmental space and a lower infrategmental space (▶ Fig. 17.3, ▶ Fig. 17.4).

Except for the vessels and nerve packed in the pudendal canal, the ischiorectal fossa mostly contains fat supported by numerous thin fibrous septa. In this fat pad, only the inferior rectal vessels and nerve run across the fossa to reach the anal canal. This neurovascular bundle needs attention when an abscess of the ischiorectal fossa is evacuated.

Posteriorly, each space extends beyond the anal region, above the gluteus maximus muscle, reaching the sacrotuberous ligament. Anteriorly, the fossa extends above the perineal membrane and below the levator ani toward the urogenital region as far as the so-called urogenital diaphragm (▶ Fig. 17.5, ▶ Fig. 17.6). The two ischiorectal fossae, one on either side, communicate with each other behind the anal canal between the external anal sphincter and the levator ani.

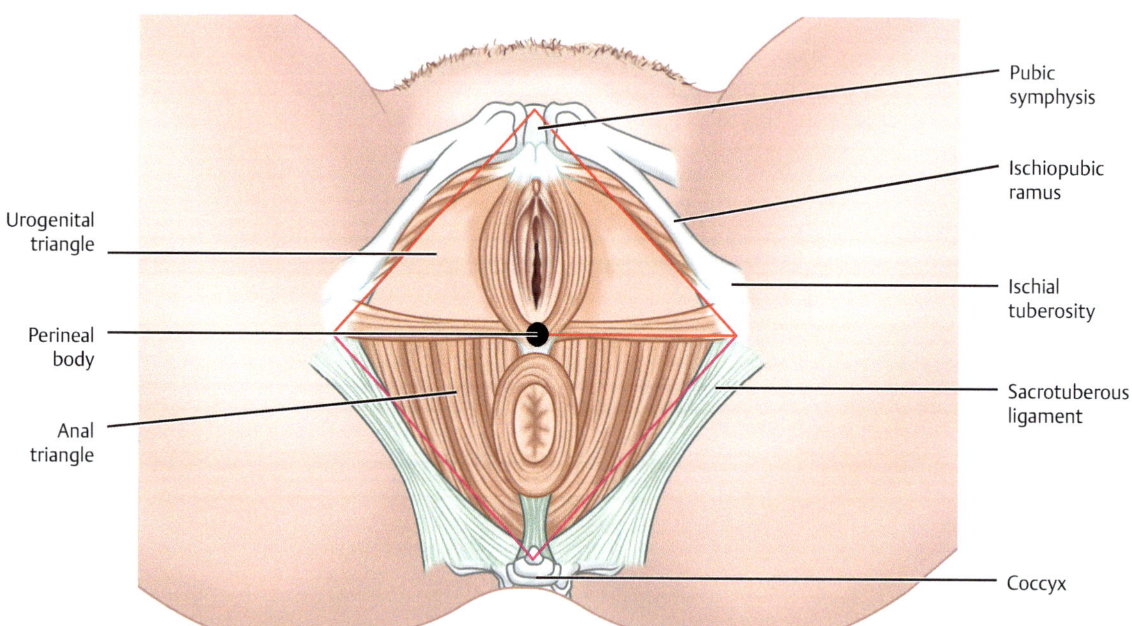

Fig. 17.1 The borders of the urogenital and anal triangle and their respective positions.

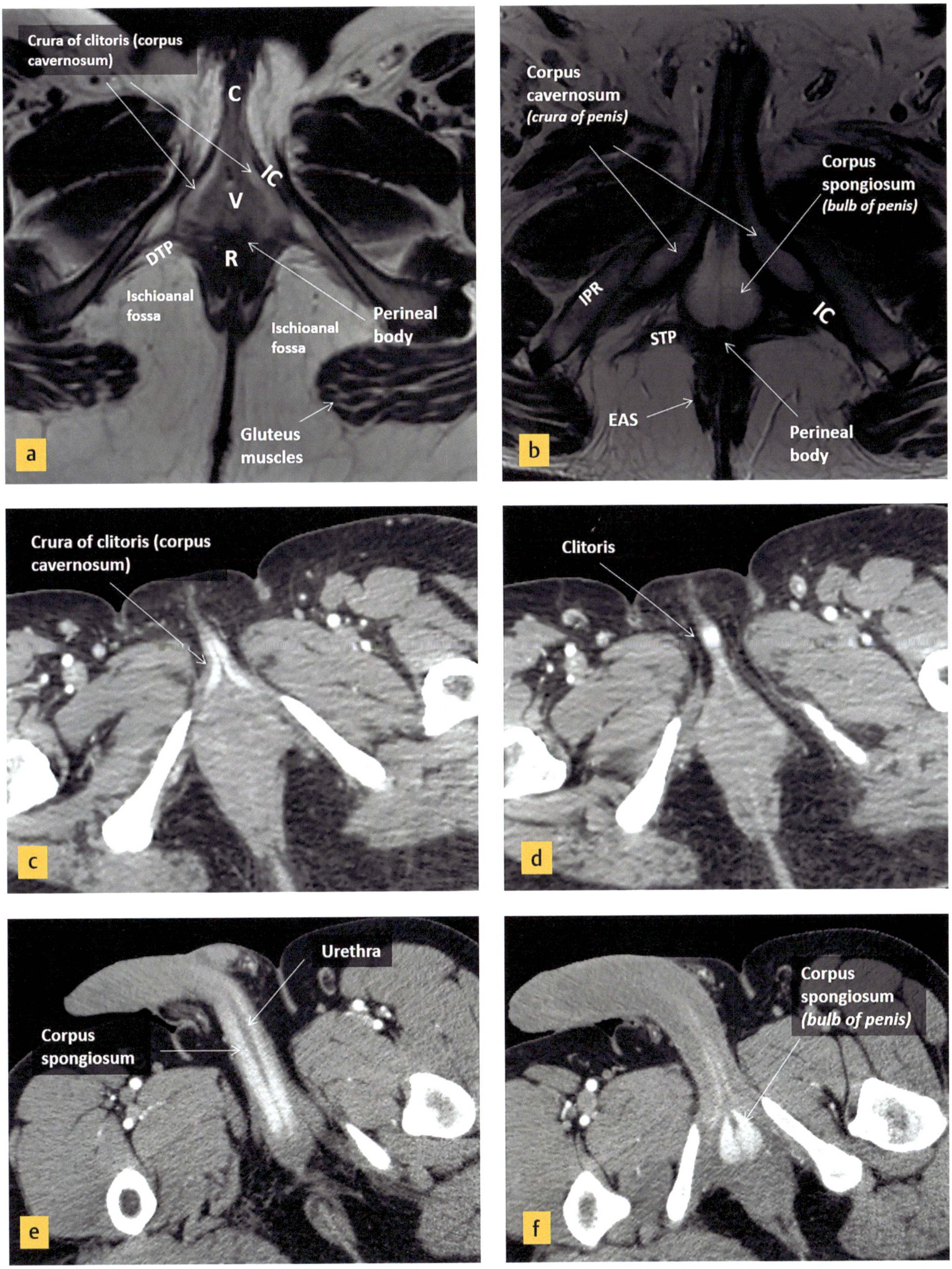

Fig. 17.2 Axial magnetic resonance (MR) images in **(a)** female and **(b)** male showing the transverse perineal (TP) and ischiocavernosus (IC) muscles. The perineum is divided into the anal and urogenital triangles by a line passing between the two ischial tuberosities. The anal triangle contains the anal orifice. The urogenital triangle includes the external genitalia. The perineal body is the meeting point of the perineal structures. Anteriorly, in front of the perineal body, the pelvic diaphragm (levator muscles) has an aperture called urogenital hiatus, which allows the passage of urethra and vagina to the perineum. **(c–f)** Axial, postcontrast computed tomography (CT) in a female **(c, d)** and male **(e, f)** showing retained enhancement of the crura of clitoris (corpus cavernosum) in female and the corpus spongiosum in male. C, body of clitoris; DTP, deep transverse muscle; EAS, external anal sphincter; IPR, inferior pubic ramus; R, rectum; STP, superficial transverse muscle; V, vagina.

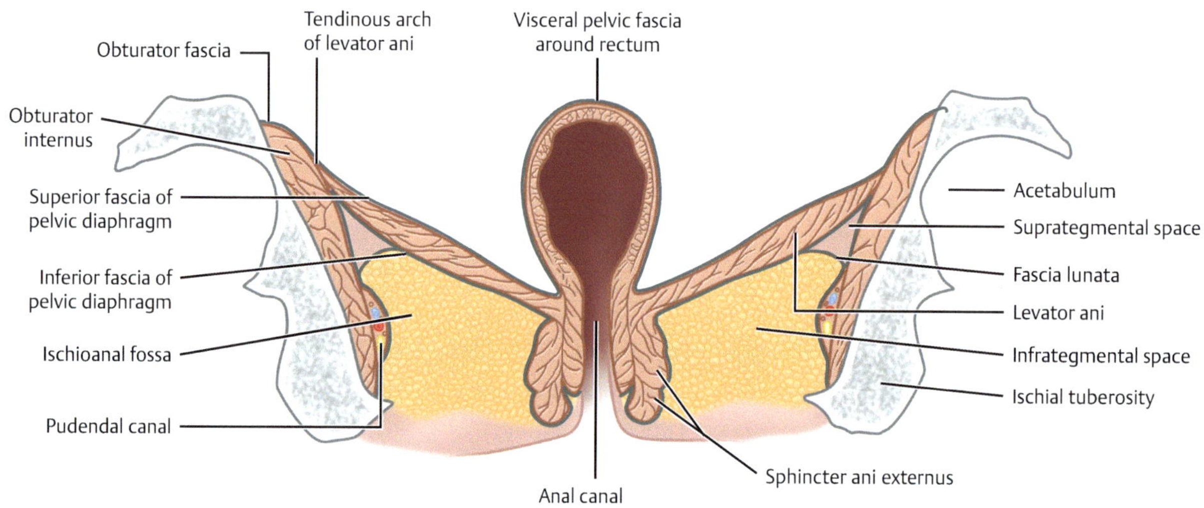

Fig. 17.3 Ischiorectal fossa, the surrounding muscles, and pelvic diaphragm.

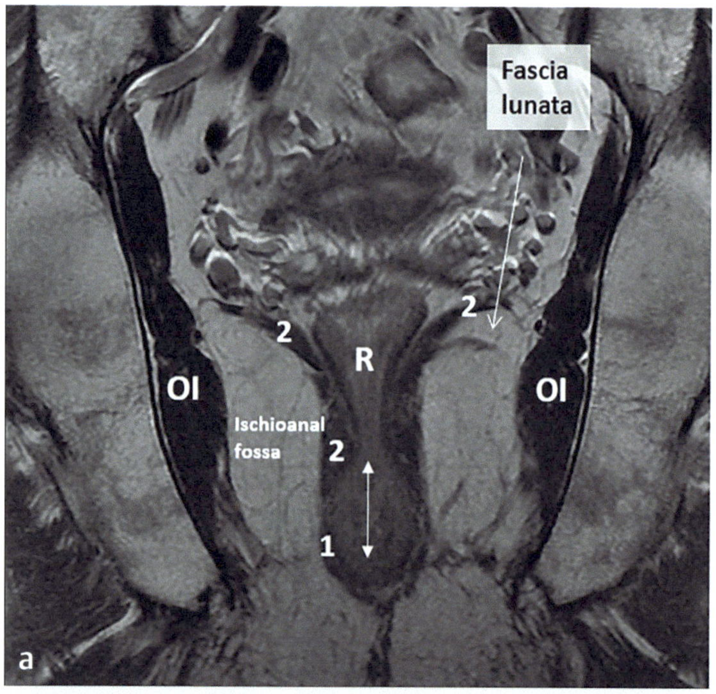

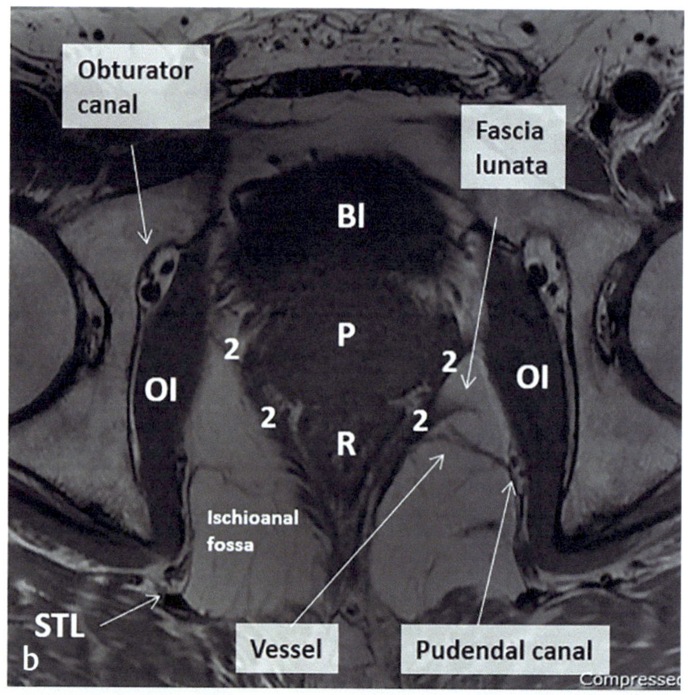

Fig. 17.4 Normal perianal anatomy. **(a)** Coronal and **(b)** axial magnetic resonance imaging (MRI). The fascia lunata divides the ischioanal fossa into an upper suprategmental space and a lower infrategmental space. 1, external anal sphincter; 2, levator ani; Bl, bladder; *double-headed arrow*, anal canal; OI, obturator internus muscle; P, prostate; R, rectum; STL, sacrotuberous ligament.

17

External Anal Sphincter

The external anal sphincter (▶ Fig. 17.4) surrounds the anal canal and lowest rectum, for about 2 cm from the levator ani to the anal verge. For most part, it surrounds the internal sphincter, which is a thickening of the circular smooth muscle. Between these two sphincters, there is a composite longitudinal muscle formed by the levator ani and the longitudinal smooth muscle of the anorectum.

The external anal sphincter is usually described to comprise three parts, a subcutaneous, a superficial, and a deep; however, this division may be just artificial and for didactic purposes. *The subcutaneous part* surrounds the lowest anal canal and the fibers of the muscle decussate behind the canal. *The superficial part* of the muscle is mainly laterally to the subcutaneous part and surrounds the upper portion of it. The shape of the muscle is elliptical, as it arises from the terminal segment of the coccyx, splits around the anal canal, and inserts into the perineal body. *The deep part* of the external sphincter, the most important part, is in contact above with the levator ani and below with the superficial part (Chapter 5 "Colon").

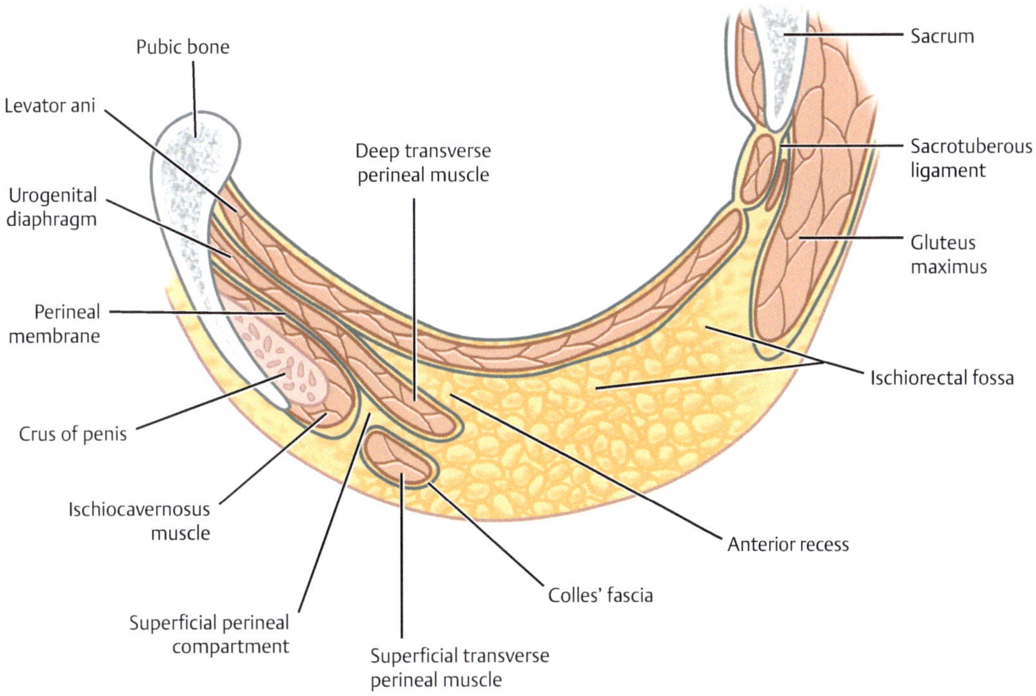

Fig. 17.5 Parasagittal section through the ishiorectal/ischioanal fossa in a male.

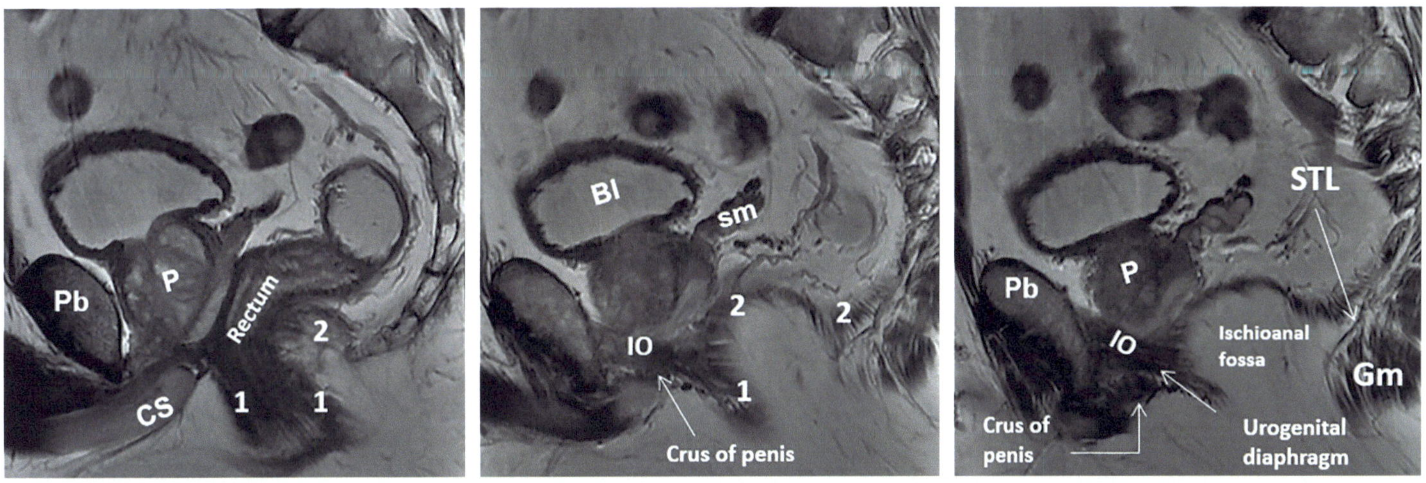

Fig. 17.6 Normal perianal anatomy. Sagittal magnetic resonance imaging (MRI) from midline to lateral show the extent of ischioanal fossa. Posteriorly, each fossa extends beyond the anal region, above the gluteus maximus (gm) muscle, reaching the sacrotuberous ligament (STL). 1, external anal sphincter; 2, levator ani; Bl, bladder; CS, corpus spongiosum; Gm, gluteus maximus; IO, internal obturator muscle; P, prostate; Pb, pubic bone; R, rectum; STL, sacrotuberous ligament.

Arteries and Veins

The internal pudendal artery is a branch of the internal iliac (hypogastric) artery and accompanies the pudendal nerve, found lateral to it, in its course in the pudendal canal. Then it gives off the inferior rectal artery (or arteries), which crosses the ischioanal fossa to supply the external sphincter muscle, the anal canal, the lower part of the rectum, and the skin around the anal opening (see Chapter 15 "Male Genitourinary").

The inferior rectal vein empties into the internal iliac (hypogastric) vein. If there is more than one vein, they join together before entering the pelvis.

Lymphatics

As a general rule, lymphatics follow the arteries (but in a retrograde flow). Thus, lymphatic drainage of the anal canal above the pectinate line goes to the pelvic nodes. Below the pectinate line, lymph flow is addressed to the inguinal nodes. Same happens for the perianal skin.

Nerves

The most important nerve of the perineum is *the pudendal nerve* (from the second, third, and fourth sacral nerve roots) (see

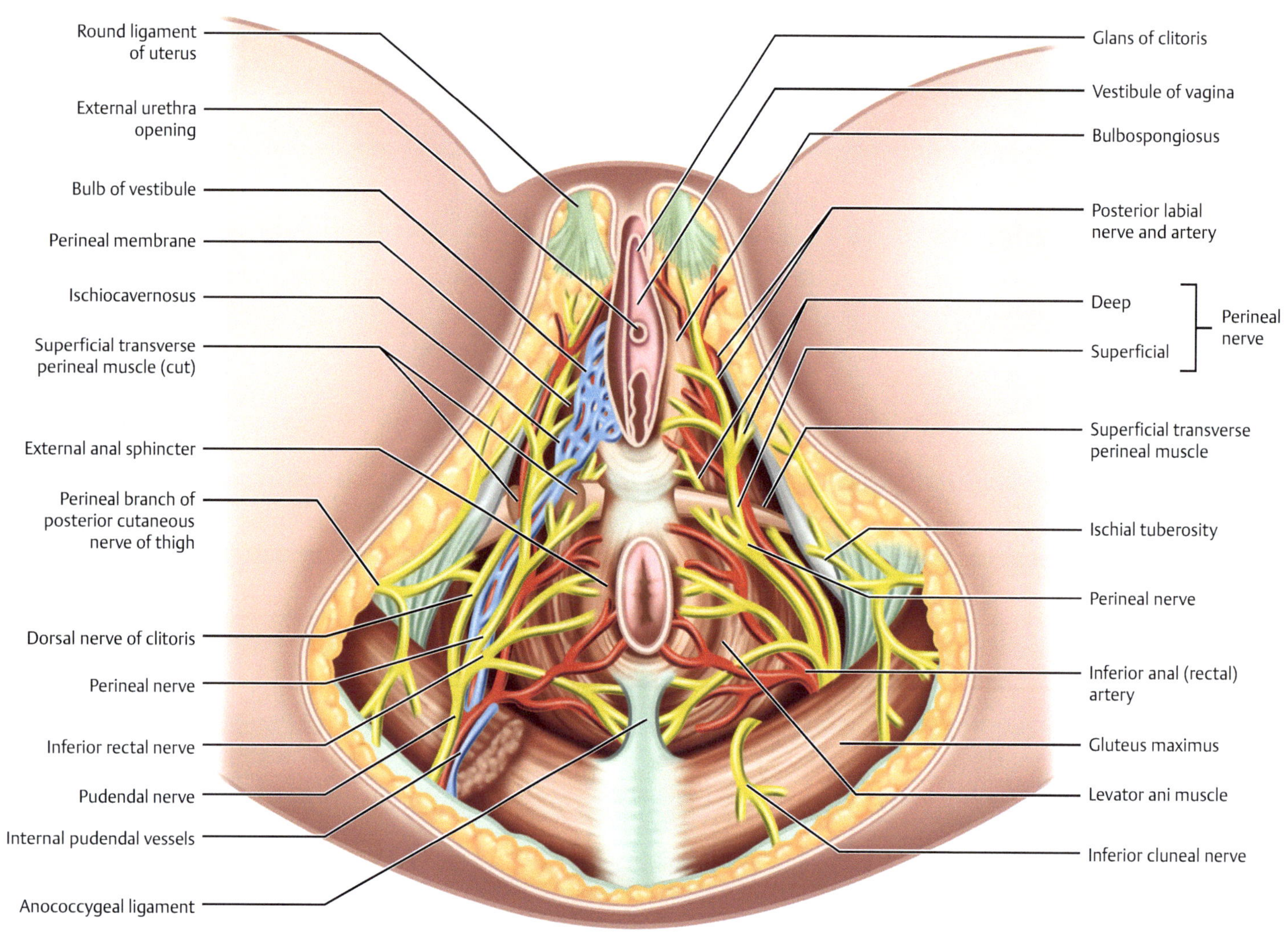

Fig. 17.7 Pudendal nerve. The pudendal nerve enters the pudendal (Alcock's) canal and branches into the inferior rectal nerve, the perineal nerve, and the dorsal nerve of the penis or clitoris.

Chapter 13 "Lumbosacral Plexus"). It enters the pudendal canal together with the homonymous artery and vein (▶Fig. 17.7). The first branch is the inferior rectal nerve, which penetrates the median wall of the pudendal canal (after a short distance, compared with the artery), passes medially through the ishiorectal fossa, and is distributed with many small branches to the sphincter ani externus, levator ani, perianal skin, and anal canal (up to the pectinate line; from there downwards, the innervation is somatic). Posteriorly in the fossa, there is also a small *perineal branch of the fourth sacral nerve* distributed to the posterior part of the external anal sphincter.

17

◆ The Urogenital Region

The structure of the urogenital triangle can be better understood if followed by layers and spaces, from superficial to deep. The reader is invited to study the text together with ▶Fig. 17.8 and ▶Fig. 17.9.

The fasciae that serve as boundaries for the spaces of the urogenital region of the perineum are the following:

- The superficial fascia of the perineum, as in the abdominal wall, comprises a superficial fatty layer and a deep membranous layer. On the penis, it is continued as superficial fascia, deprived of fat. The membranous layer in the perineum (Colles' fascia) is continuous with that of the abdominal wall (Scarpa's fascia). It is attached laterally to the ishiopubic rami and posteriorly to the posterior margin of the deep transverse perineal muscles. In the scrotum, the membranous layer is continuous with the dartos.
- The deep fascia of the perineum, as in the abdominal wall, is the fascia that invests the ischiocavernosus and bulbocavernosus muscles (Gallaudet's fascia), which forms a continuous layer attached to the ishiopubic rami (above the membranous layer) (▶Fig. 17.8, ▶Fig. 17.9).
- The perineal membrane (or triangular ligament) consists of strong fibrous tissue extending between the pubic arch and the ischiopubic rami. In the male, it is thickened anteriorly to form the transverse ligament of the perineum, in front of which passes the dorsal vein of the penis. In the female, the perineal membrane is anteriorly continuous with a pubourethral ligament—analogous to the puboprostatic ligament.

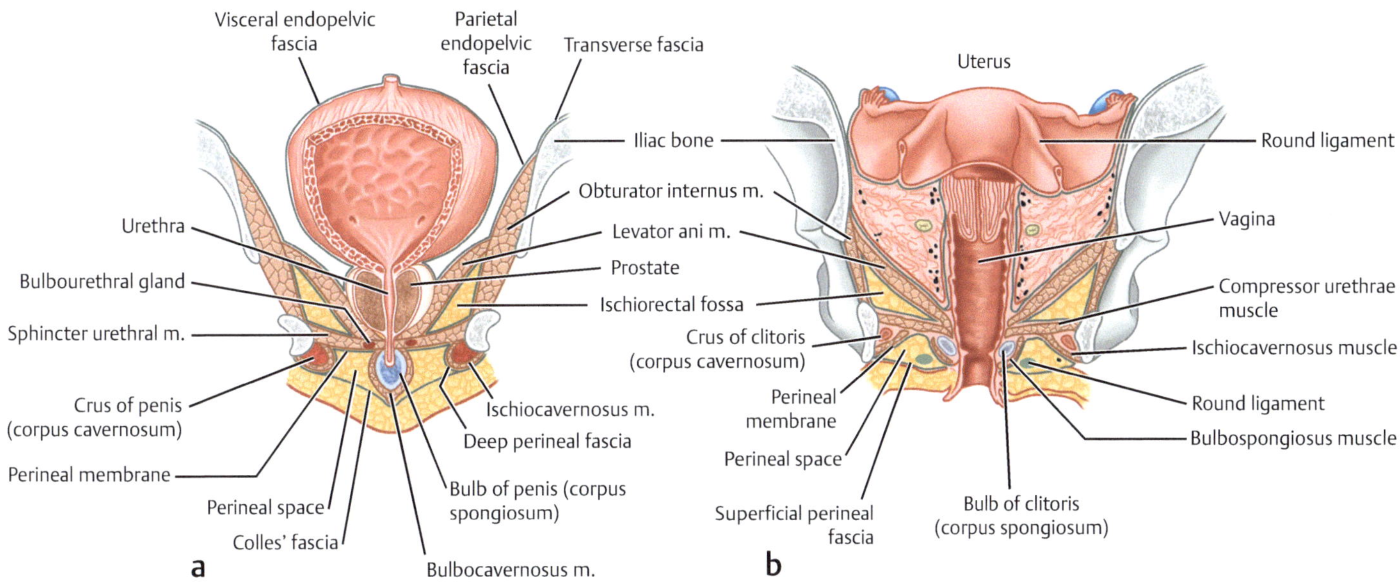

Fig. 17.8 Fasciae of the pelvis and urogenital region. **(a)** Male. Coronal section through the prostatic part of urethra. **(b)** Female. Coronal section through the vagina.

The spaces of the urogenital region are listed hereafter, together with their contents and neurovascular supply.

The Subcutaneous Perineal Pouch (or Superficial Perineal Cleft)

It is a potential space bounded by the membranous layer of the superficial fascia and the deep fascia of the perineum. There are no contents in this space.

Main innervation of this space is supplied by the posterior scrotal nerve (branch of the posterior femoral cutaneous) and the superficial perineal branches (branches of the pudendal nerve).

The Superficial Perineal Space (or Compartment or Pouch)

It is delimited by the deep perineal fascia (Gallaudet's) and the perineal membrane. It contains the superficial transverse perineal muscle, erectile tissue elements such as the corpora cavernosa of the penis or clitoris, corpora spongiosa (penile bulb or vestibular bulbs) covered by the associated muscles (bulbospongiosus and ischiocavernosus), and vestibular (Bartholin's) glands in the female, as well as branches of the internal pudendal vessels and the pudendal nerves.

Muscles in the Superficial Perineal Space

The Superficial Transverse Perineal Muscles

The superficial transverse perineal muscles (paired) arise from the ischial tuberosities and pass medially to insert into the perineal body. This muscle is usually poorly developed or even absent (▸Fig. 17.2).

The Ischiocavernosus Muscles

The ischiocavernosus muscles (paired) cover the free surfaces of the corpora cavernosa of the penis (▸Fig. 17.9). Each muscle arises from the medial side of the ischial tuberosities and the adjacent ramus of the ischium, continues forward around the crus, and inserts into it just as the two crura come together. Contraction of the ishiocavernosus may contribute to erection of the corpora cavernosa penis and clitoris through some restriction of the venous flow (▸Fig. 17.10).

The Bulbospongiosus Muscle

The bulbospongiosus muscle arises from the median raphe of the bulb of the penis and from the perineal body. The direction of fibers is upward and, with its counterpart, surrounds the corpus spongiosum and the dorsal penis as a whole (▸Fig. 17.9). In the female, the muscles are separated by the pudendal cleft and cover the vestibular bulbs (▸Fig. 17.11). In the male, it is supposed to assist in urethral compression and ejaculation, and in the female, it serves as an external sphincter for the vaginal introitus.

Erectile Tissue Elements in the Superficial Perineal Space

The Corpora Cavernosa

The corpora cavernosa in both penis and clitoris are attached posteriorly to the ischial tuberosities and the rami of the ischia; these parts of the corpora are the crura of the penis (▸Fig. 17.2). Near the pubic symphysis, the two crura are attached to the pubic bone

17

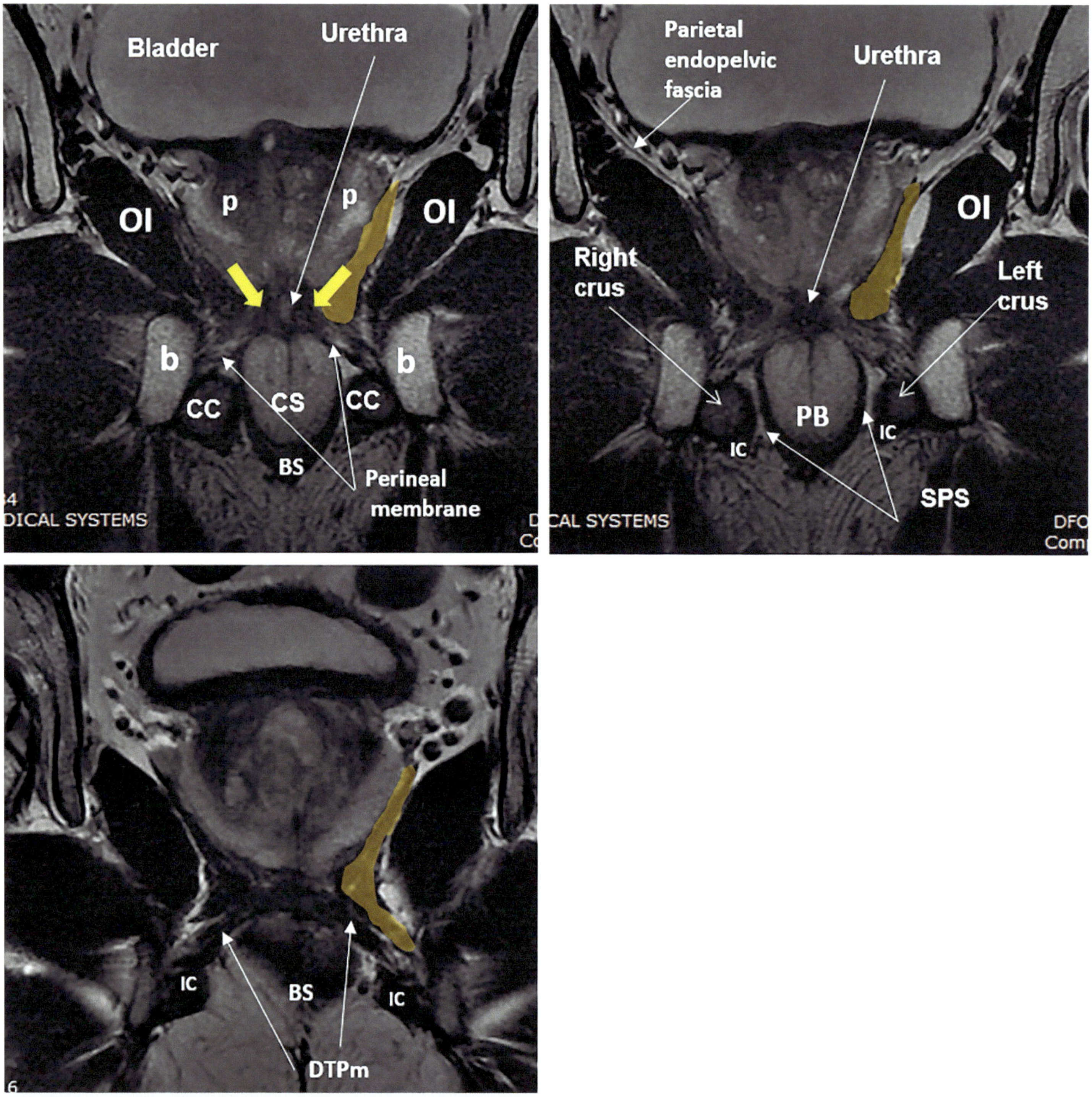

Fig. 17.9 Normal perianal anatomy. Coronal magnetic resonance imaging (MRI) in a male pelvis from anterior to posterior at the level of the prostate. The prostate (P) is enlarged. The deep perineal space or urogenital diaphragm is shown. It contains the sphincter urethrae muscle (*yellow arrows*) and the deep transverse perineal muscle (DTPm). Its inferior surface is demarcated by the perineal membrane. The superficial perineal space is delimited by the deep perineal fascia and the perineal membrane. It contains the corpora cavernosa (CC) of the penis, and corpora spongiosa (CS) or penile bulb (PB) covered by the ischiocavernosus (IC) and bulbospongiosus (BS) muscles, respectively. Parietal endopelvic fascia (transverse fascia) is clearly shown covering the inner surface of the IO muscles. Levator ani muscles are colored in orange. b, bone; OI, obturator internus muscle; P, prostate.

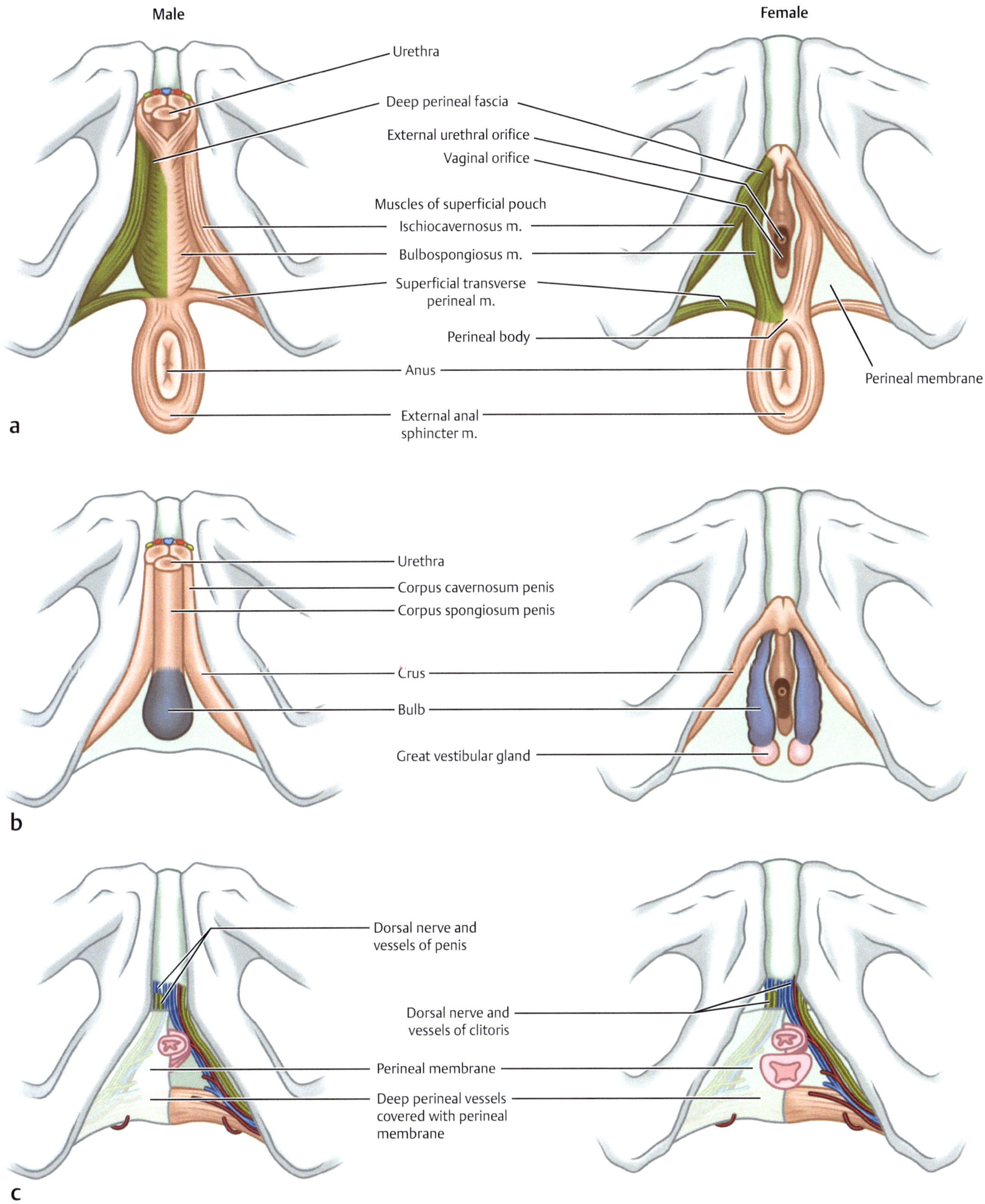

Fig. 17.10 (a–e) Inferior view of the perineal region showing boundaries of the two perineal triangles (anal and urogenital) and layers of the perineum from superficial to deep.

(*Continued*)

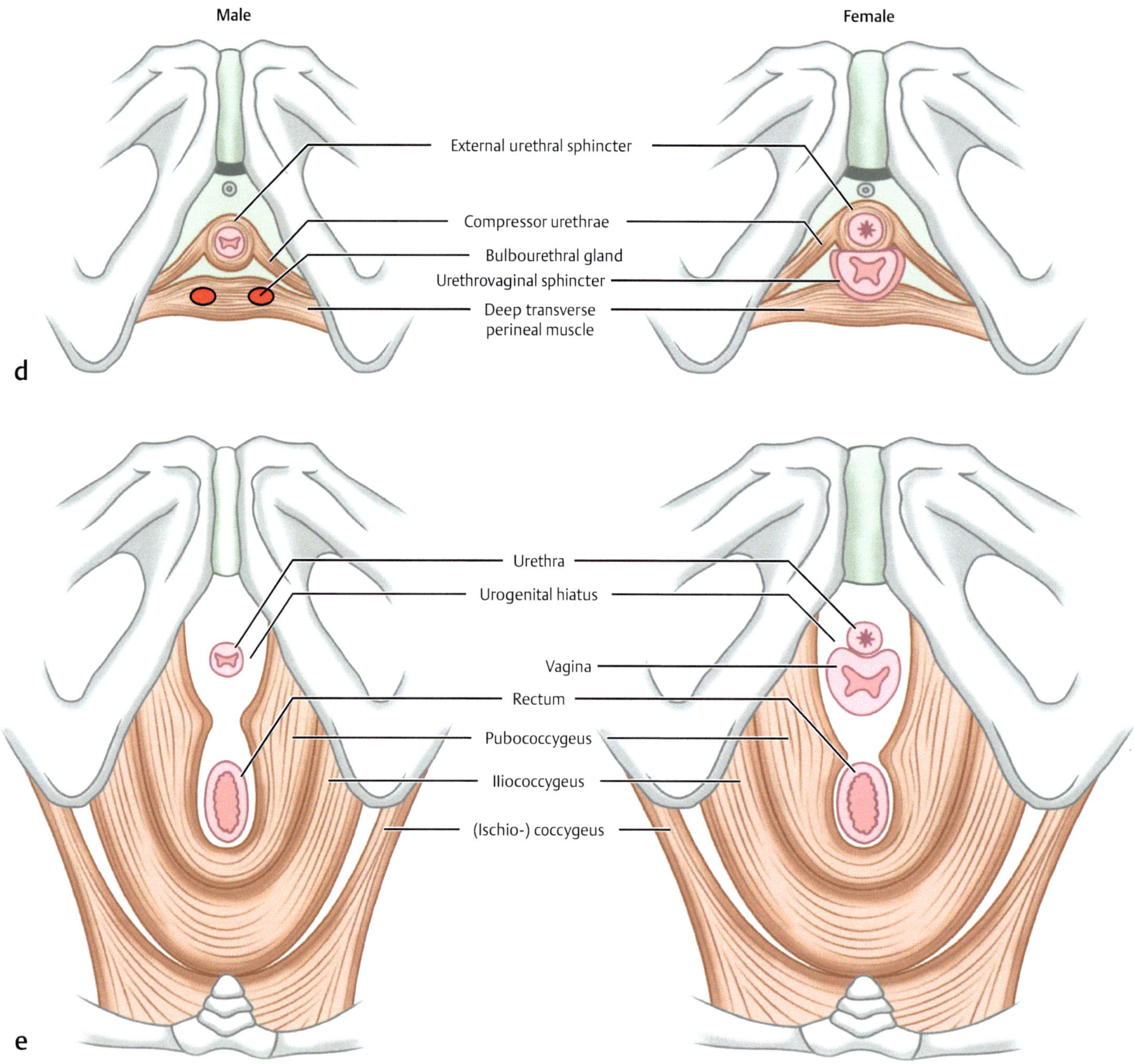

Fig. 17.10 *(Continued)* **(d,e)**

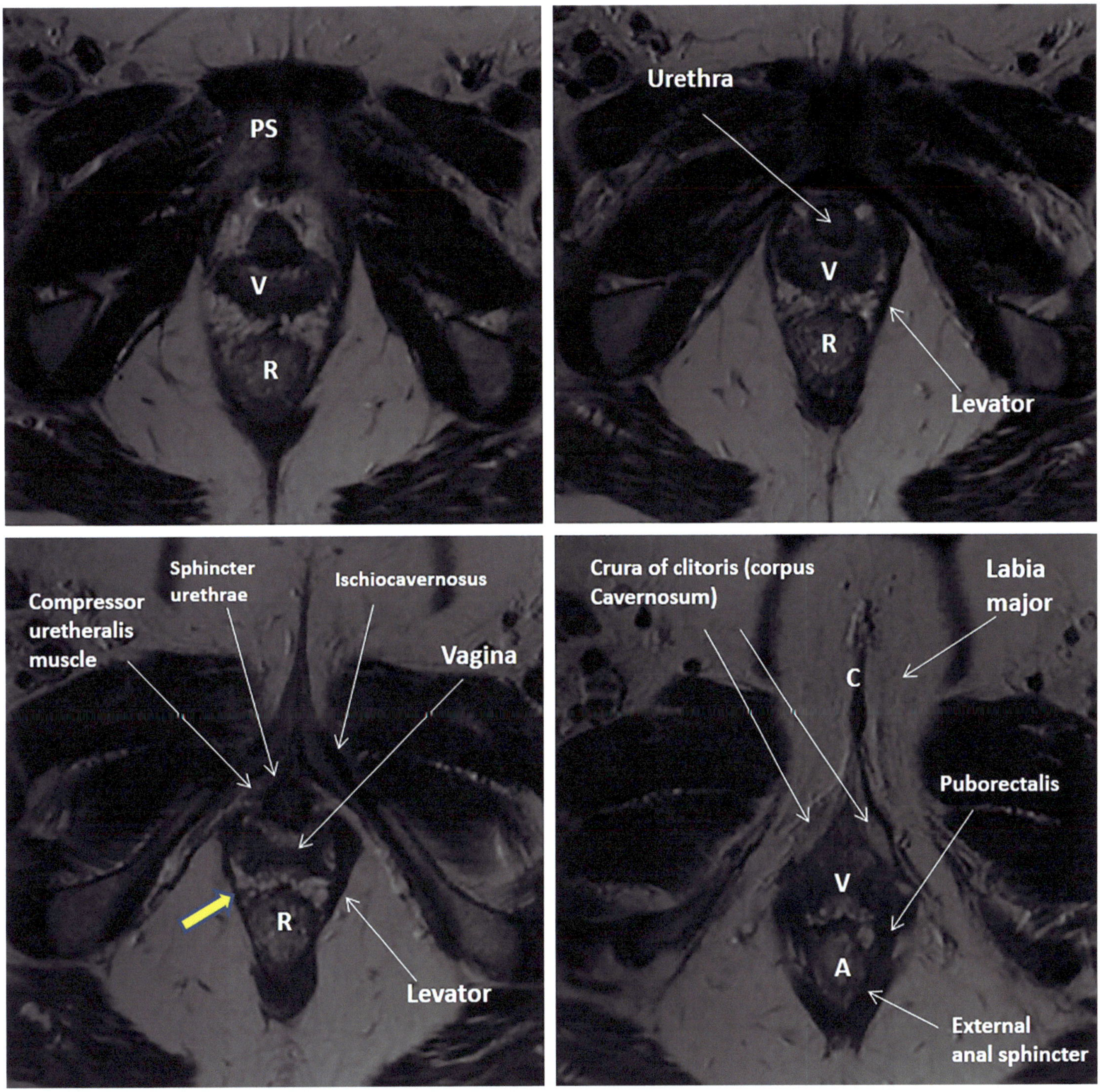

Fig. 17.11 Normal female pelvic floor anatomy. Axial T2-weighted magnetic resonance (MR) images from superior to inferior are shown. Compressor urethralis muscle which arises from the anterior wall of the urethra is seen. Note the typical H- or butterfly-like shape of the vagina and the close apposition of the puborectalis muscle to the vaginal wall. Note atrophy of the right levator muscle (*yellow arrow*). Note tight connection between the urethra and the anterior wall of the vagina whereas loose connective tissue exists between the rectal and vaginal walls. A, anal canal; C, clitoris; PS, pubic symphysis; R, rectum; V, vagina.

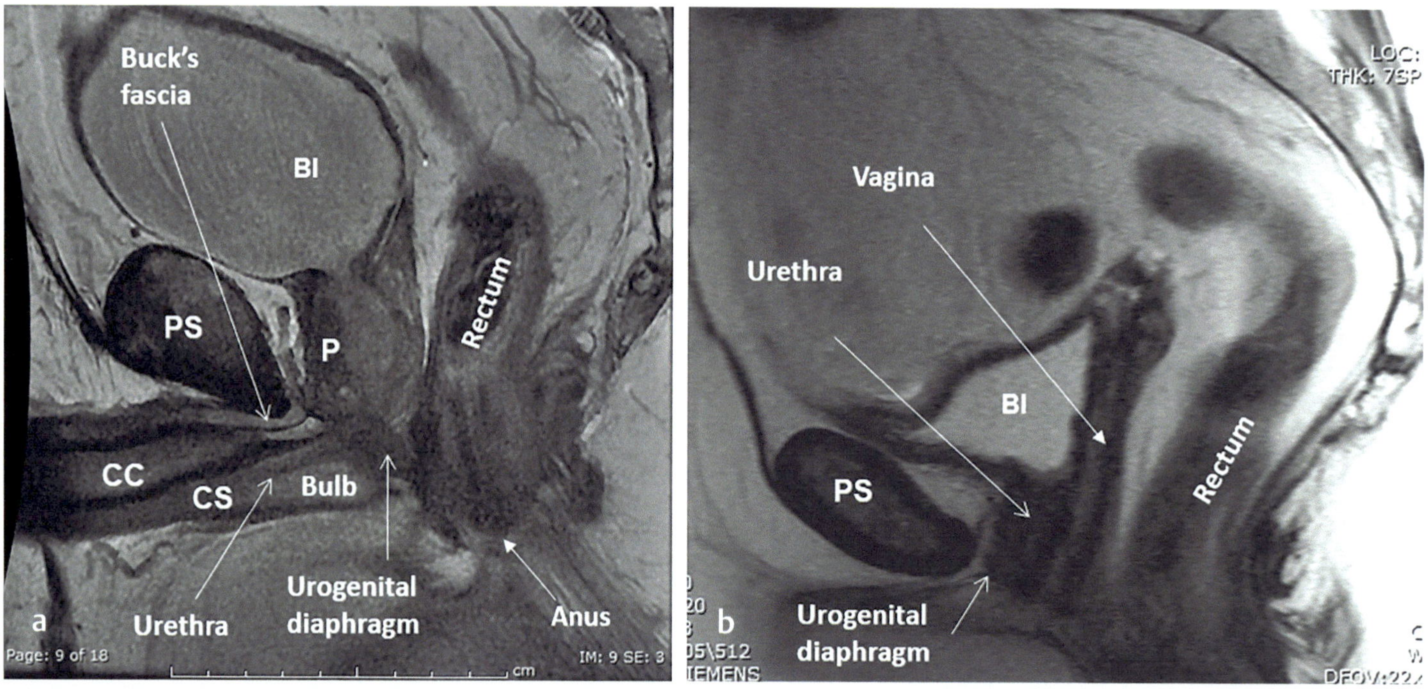

Fig. 17.12 Normal perianal anatomy. Sagittal magnetic resonance (MR) views. **(a)** Male. **(b)** Female. Bl, bladder; CS, corpus spongiosum; CC, corpora cavernosum; P, prostate; PS, pubic symphysis.

and then converge and join the corpus spongiosum of the urethra. In the female, the joined parts of the corpora form the body of the clitoris (▶Fig. 17.2) (see Chapter 16 "Female Genital System").

The Corpus Spongiosum

The corpus spongiosum of the male is the midline corpus of the penis that transmits the penile urethra. It begins as an enlarged mass named urethral bulb then gradually diminishes in diameter, stays ventral to the two corpora cavernosa, and at the top of the penis expands to form an acorn-shaped glans that caps the blunt distal parts of the corpora cavernosa (▶Fig. 17.9, ▶Fig. 17.12).

The Glans Clitoridis

The glans clitoridis is a small mass of erectile tissue, which caps the body of the clitoris, the homologue of the glans penis.

Vestibular Bulbs and Vestibular (Bartholin's) Glands

In the female, the vestibular bulbs are two paired masses of erectile tissue found on either side of the vestibule, under cover of the bulbocavernosus muscles (▶Fig. 17.11). They are thought to be the homologue of the bulb of the penis and the adjacent part of the corpus spongiosum. These masses are broader behind and become narrower anteriorly. Each vestibular bulb anteriorly sends a thin strand, the pars intermedia, which fuses with the opposite in the commissure that ends in the glans clitoris (see Chapter 16 "Female Genital System").

The greater vestibular glands are two small ovoid glands found behind or partly under cover of each vestibular bulb. Their ducts open at 5 and 7 o' clock positions, lateral to the vaginal opening. The homologue bulbourethral glands in the male lie in the deep perineal compartment.

Innervation and Blood Supply of the Superficial Perineal Space

The pudendal nerve, while in the pudendal canal, gives three branches: the inferior rectal nerve addressed exclusively to the anal region as said above (▶Fig. 17.7), the pudendal nerve, and the dorsal nerve of the penis. The dorsal nerve is deeper to the pudendal nerve. The pudendal nerve gives superficial perineal branches, the medial and lateral. The superficial perineal nerves travel from deep to superficial to reach the skin, and pass through the superficial perineal space. Although branches of the perineal nerve end in the muscles of the superficial perineal space, they appear in this space briefly during their passage. Mostly this happens at the posterior part of the space, where the superficial nerves pass around or through the transverse perineal muscle. The deep branches of the perineal nerve are not seen in the superficial perineal space.

The pudendal artery, the co-traveler of the pudendal nerve in the pudendal canal, has a similar pattern. As stated above, it gives early the inferior rectal artery that is addressed to the rectoanal region (corresponding to the homonymous nerve). After that, the pudendal artery and accompanying vein has a middle position in the pudendal canal, between the perineal and dorsal nerves. Near the transverse perineal muscle, it divides to perineal artery and artery of the penis. The perineal gives the transverse perineal artery and continues as posterior scrotal arteries (▸Fig. 17.7). The scrotal branches give twigs for the muscles of the superficial perineal space, but again, just as the homonymous nerves do, they just pass through this space. The verves and vessels of the female perineum are shown in ▸Fig. 17.7.

The Deep Perineal Space

The deep perineal space is the deeper space of the perineum. It is bounded below by the perineal membrane, but cannot be considered a closed compartment. The reason is that there is no deep, well-defined transverse boundary layer. The contents of the deep perineal space extend into the pelvis. This space is described elsewhere as "urogenital diaphragm," which is erroneous because there is no upper boundary of this "diaphragm." The term "urogenital diaphragm" was used formerly for the sphincter urethrae muscle and the deep transverse perineal muscle (▸Fig. 17.8, ▸Fig. 17.9, ▸Fig. 17.11). The former is clearly not a flat muscle as believed before (see below).

The superior fascia of the deep perineal space should not be thought as a true fascia, but a fusion of many fasciae, such as the superior fascia of the pelvic diaphragm (fascia transversalis), pubovesical, vaginal, periprostatic, visceral pelvic fascia, etc. In any case, the compartment surrounding these fasciae is open and continuous to the pelvic cavity.

Muscles in the Deep Perineal Space

The Deep Transverse Perineal Muscle

The deep transverse perineal muscle (paired), like its superficial homologue, extends from the ramus of the ischium to the perineal body. There is much variability concerning its degree of development. In this muscle rests the coupled bulbourethral glands of the male (▸Fig. 17.9).

The Sphincter Urethrae

The sphincter urethrae are usually named the external sphincter, as opposed to the internal sphincter of the bladder (sphincter vesicae); the latter is a nonstriated muscle with autonomous innervation surrounding the neck of the bladder and preprostatic urethra. The internal sphincter can be anatomically recognized only in the male. The external urethral sphincter is a skeletal muscle with differences between the two sexes.

In the male, the sphincter urethra is a vertical structure lying within the urogenital hiatus, extending from the base of the bladder and the proximal border of the prostate, traveling along the prostate, covering the membranous urethra, and resting on the perineal membrane (▸Fig. 17.9, ▸Fig. 17.11, ▸Fig. 17.13). The sphincter forms an omega-like loop around the upper subvesical portion and prostatic part of the urethra. Caudal to the prostate the fibers are circular and form a true sphincter of the urethra. The sphincter urethrae are mainly supplied by the pudendal nerve and additionally by a branch of the sacral plexus that runs on the pelvic surface of the levator ani.

In the female, there is a complex of muscles forming a urogenital sphincter, consisting of the sphincter urethrae, the compressor urethrae, and the sphincter urethrovaginalis (▸Fig. 17.11, ▸Fig. 17.14). The deeper pelvic part of the sphincter urethrae begins at the base of the bladder and surrounds the urethra in the middle-third of its length with various degrees of encirclement. The sphincter urethrae are continuous with the compressor urethrae, which are perineal muscle. The compressor urethrae arise bilaterally from the ischiopubic rami and form an arch on the anterior surface of the urethra (▸Fig. 17.15, ▸Fig. 17.16, ▸Fig. 17.17). The sphincter urethrovaginalis surrounds both the vagina and the distal urethra without crossing between them, and inserts into the perineal body. The compressor and the sphincter urethrovaginalis act as sphincters of the vagina and the urethra. They lie above the perineal membrane, in the deep perineal space.

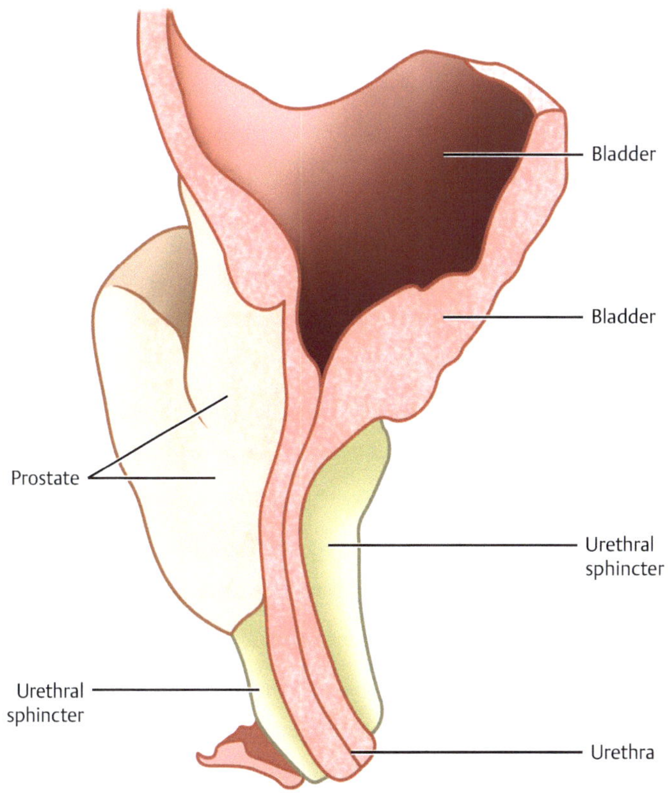

Fig. 17.13 Male urethral sphincter. Median section of bladder, urethra, prostate, and urethral sphincter showing the manner in which the urethral sphincter encircles the urethra and its extent. (Modified with permission from Mirilas and Skandalakis.[3])

17

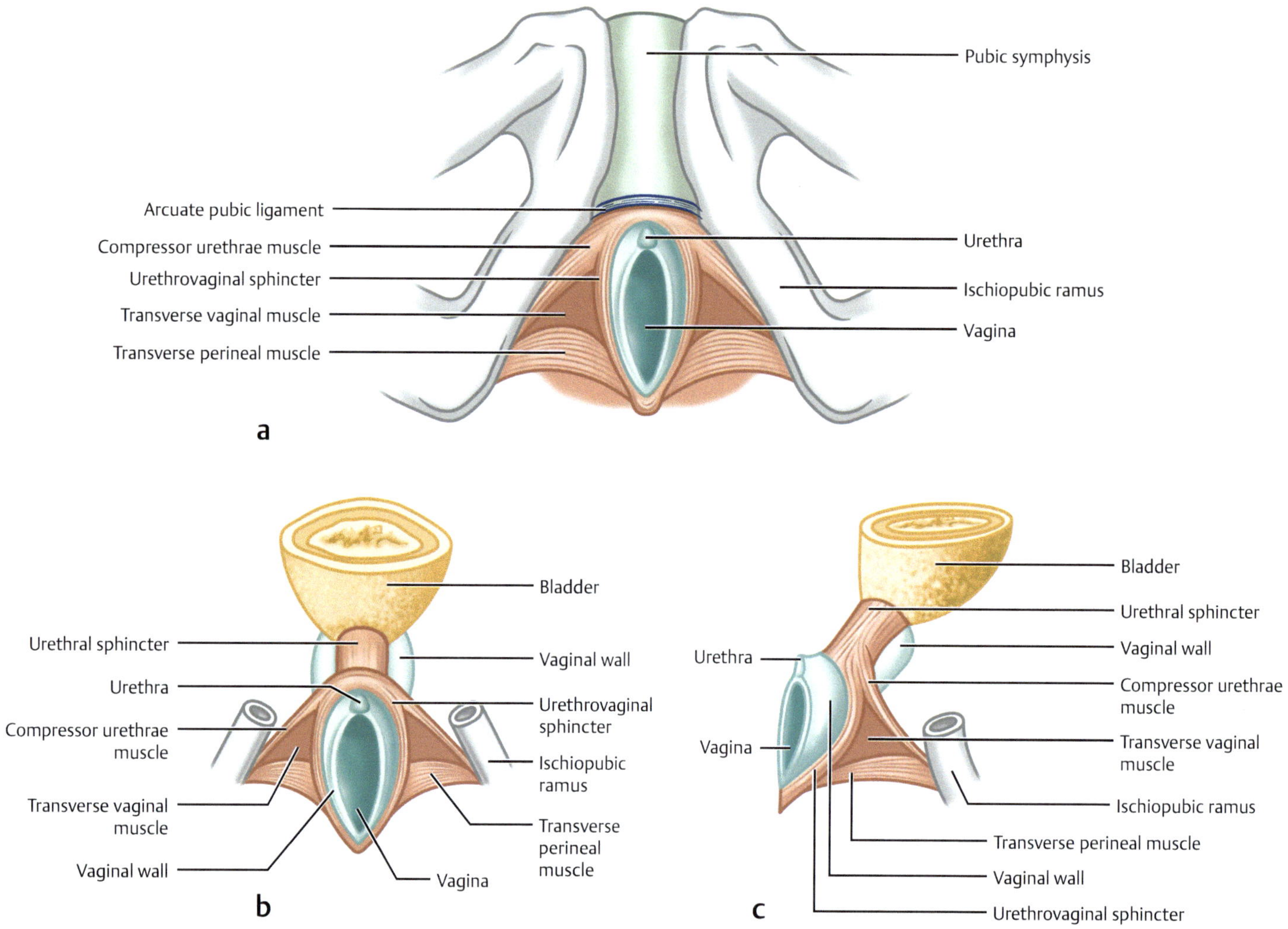

Fig. 17.14 Urogenital sphincter muscle complex in female. In the female, there is a complex of muscles forming a urogenital sphincter, consisting of the sphincter urethrae, the compressor urethrae, and the sphincter urethrovaginalis. **(a)** Perineal view of urogenital sphincter. **(b)** Urogenital sphincter musculature, bladder, and vagina wall. Bladder is cut in coronal direction, pubic symphysis is removed, and the ischial rami spread. **(c)** Oblique view.

Innervation and Blood Supply of the Deep Perineal Space

The innervation of this region is supplied by deep perineal branches of the perineal nerve. These branches leave the pudendal canal at the posterior border of the perineal membrane. The more posterior branches supply the anterior part of the sphincter, while other pierce the fasciae of this region and travel farther to supply the deep transverse perineal muscle and the sphincter urethra muscles. The bulb of the urethra is also supplied by branches of the deep perineal nerves.

The blood supply of the musculature of this space is contributed mainly by branches of the internal pudendal artery and vein. As stated above, the internal pudendal artery, already in the pudendal canal, gives the perineal artery, which continues as artery of the penis over the perineal membrane (its fellow traveler is the dorsal nerve of the penis). In the deep perineal space, this artery gives a branch for the bulb of the urethra, and then divides into the deep artery of the penis and the dorsal artery of the penis. The former enters the corpora cavernosa immediately, while the dorsal artery passes between the crus and pubic bone, to supply superficial structures of the dorsum of the penis. The veins are not that constant and usually more than one in every side but usually follow the aforementioned arteries (in a retrograde aspect).

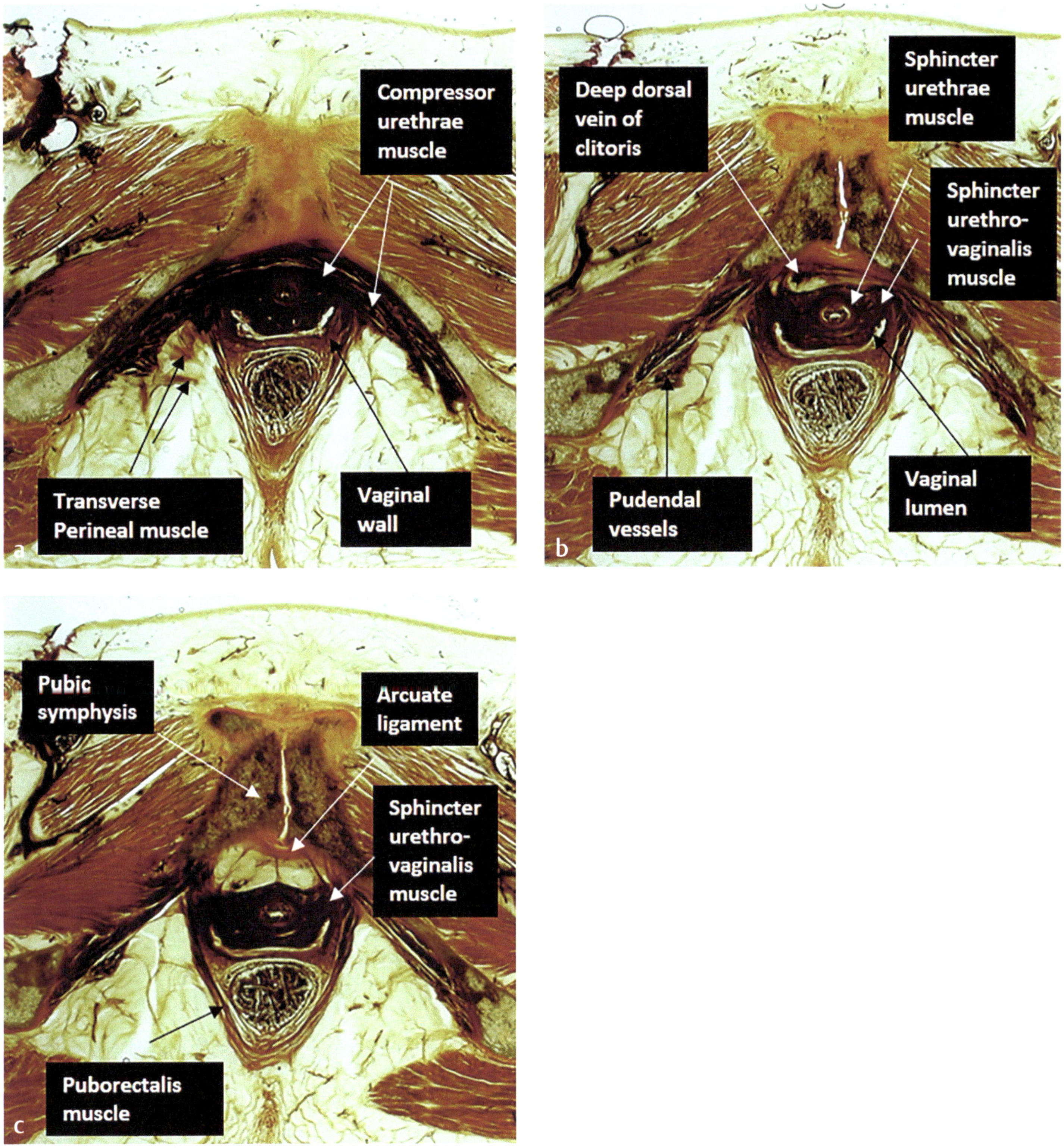

Fig. 17.15 (a–c) Female urethrovaginal region is shown by axial cuts in a cadaveric specimen. The compressor urethra arises on the sides from the ischiopubic rami and forms an arch on the anterior surface of the urethra. The sphincter urethrovaginalis surrounds both the vagina and the distal urethra. Note tight connection between the urethra and the anterior wall of the vagina whereas loose connective tissue exists between the rectal and vaginal walls.

17

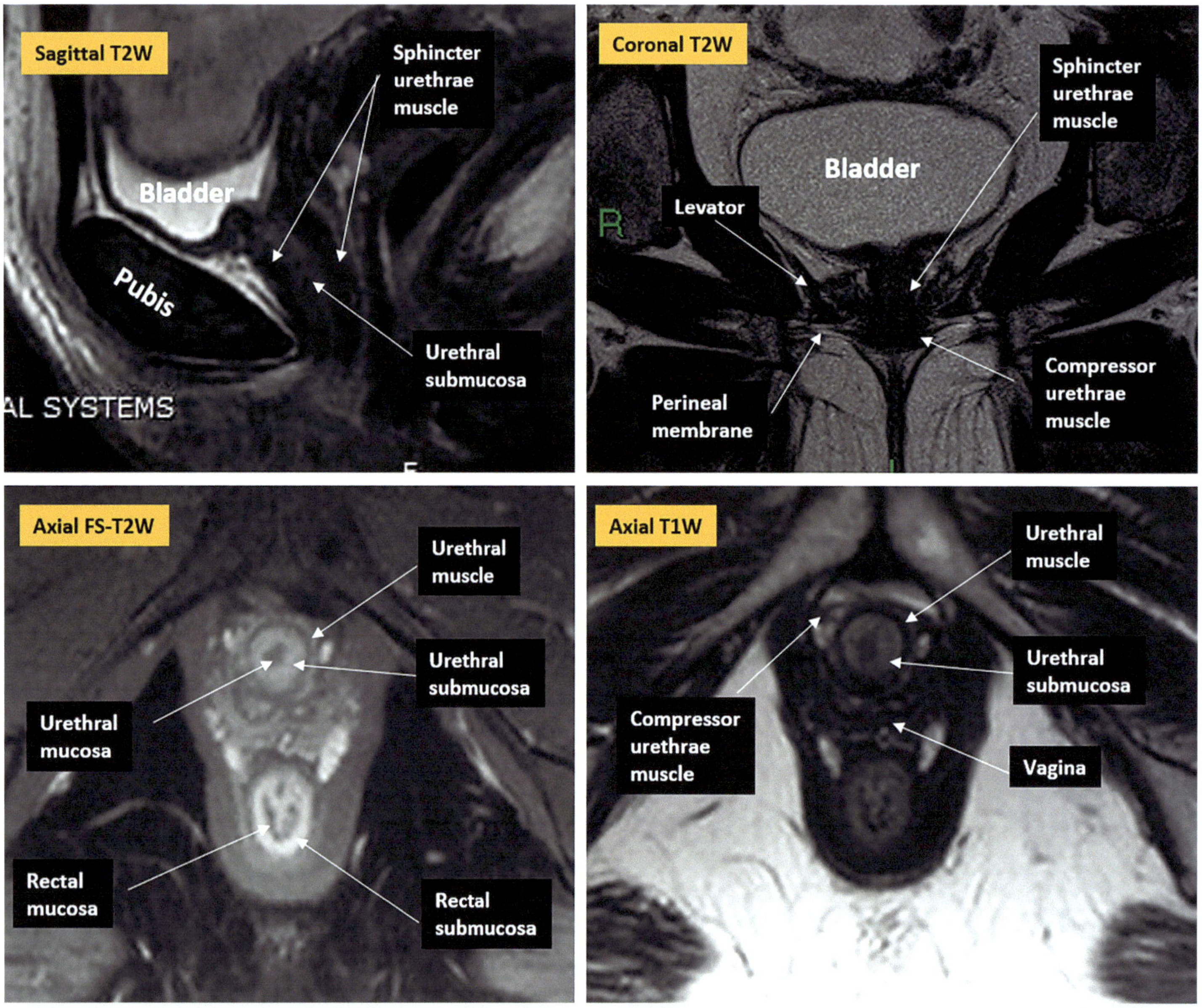

Fig. 17.16 Zonal anatomy of the normal female urethra shown by noncontrast magnetic resonance imaging (MRI). T2-weighted (T2W) MR images demonstrate a low-signal-intensity outer layer representing smooth muscle and a thicker high-signal-intensity middle layer representing vascularized connective tissue submucosal layer (also shown in the rectum). The mucosa is seen as a low-signal-intensity central area on axial images.

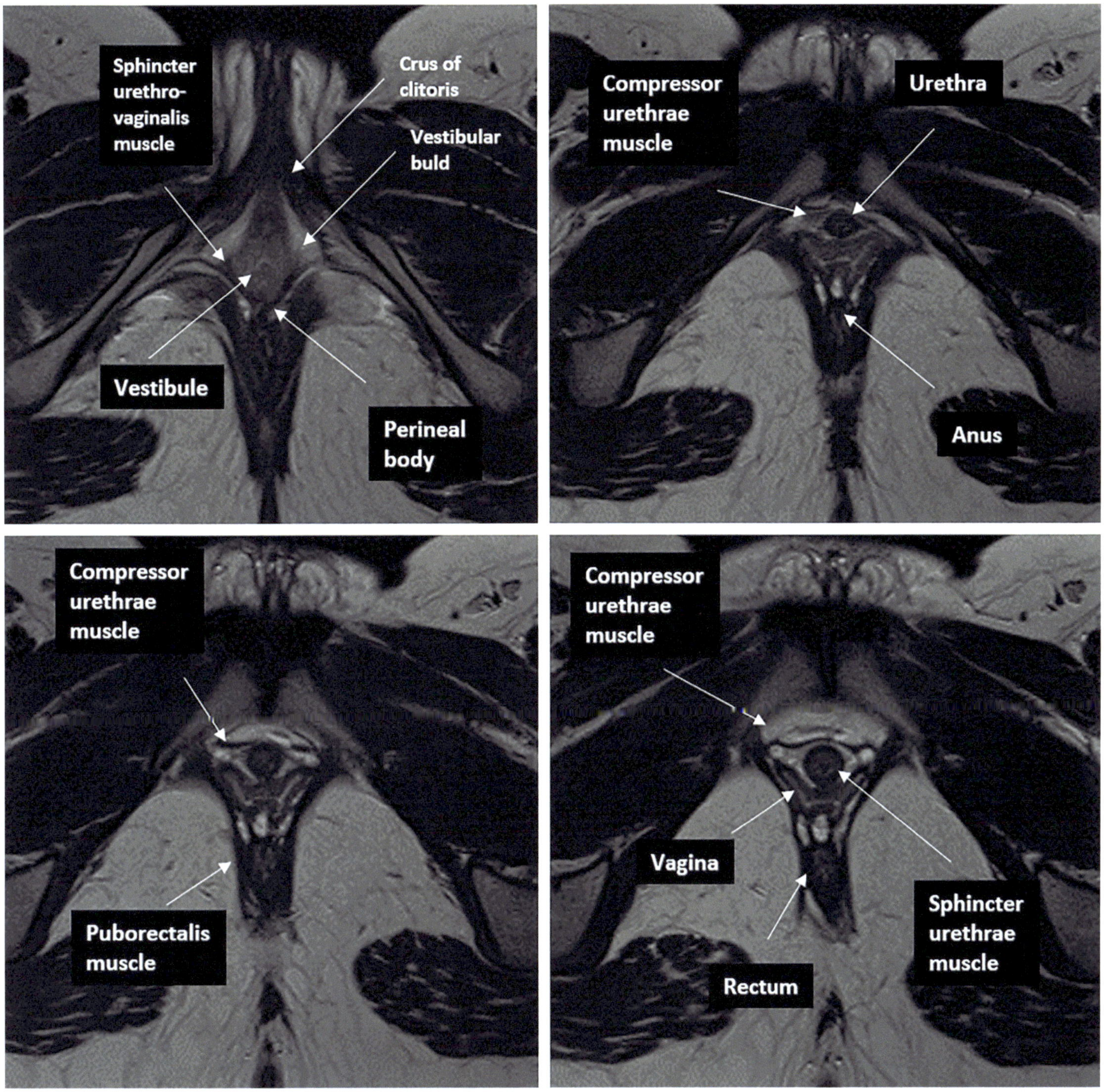

Fig. 17.17 Axial T2-weighted (T2W) magnetic resonance imaging (MRI) from inferior to superior showing urethral muscle complex in female. Complex of muscles forming the urogenital sphincter consist of the sphincter urethrae, the compressor urethrae, and the sphincter urethrovaginalis. The compressor urethrae commonly seen with MRI. It arises bilaterally from the ischiopubic rami and forms an arch on the anterior surface of the urethra. The sphincter urethrovaginalis surrounds both the vagina and the distal urethra and inserts into the perineal body.

References

1. O'Rahilly R. Gardner-Gray-O'Rahilly Anatomy: A Regional Study of Human Structure. 5th ed. Philadelphia: W. B. Saunders; 1986
2. Hollinshead WH. Anatomy for Surgeons. Vol. 2. The Thorax, Abdomen & Pelvis. New York: Hoeber & Harper; 1956
3. Mirilas P, Skandalakis JE. Urogenital diaphragm: an erroneous concept casting its shadow over the sphincter urethrae and deep perineal space. J Am Coll Surg 2004;198(2):279–290
4. Skandalakis JE, Colborn GL, Weidman TA, et al. Skandalakis' Surgical Anatomy: The Embryologic and Anatomic Basis of Modern Surgery. Vol. 2. Athens: Paschalidis Medical Publications; 2004

18 Pelvic Floor

Petros Mirilas, Eleni Paschalidou, and Farhood Saremi

◆ Introduction

The pelvic floor provides support to pelvic organs and permits multiple vital functions such as urination, defecation, and childbirth. Pelvic floor dysfunctions result in a spectrum of disorders that include urinary incontinence, defecatory dysfunction, pelvic organ prolapse, sexual dysfunction, and chronic pain syndromes. The complex anatomy of the pelvic floor has gained increased interest in the last few years as modern imaging techniques and reconstructive surgical procedures using tapes and meshes have been evolved.

◆ Definition

Many authorities describe the pelvic floor to consist entirely of the pelvic diaphragm (levator ani and coccygeus muscles). However, in this chapter, the pelvic floor is considered to include, beyond the pelvic diaphragm, all the structures that support the abdominal and pelvic organs. This support is effected by the muscles of the pelvic diaphragm and the ligaments given by the pelvic fascia. According to Ashton-Miller and DeLancey,[1] what happens with pelvic organs and pelvic floor is similar to a ship floating in a harbor while attached with ropes to the dock. Thus, the pelvic fascia (ropes) suspends the pelvic organs from above, while the pelvic floor (sea) supports them from below against the intrabdominal forces; the muscles of the pelvic floor are also responsible for the closure of the hiatus of the pelvic diaphragm. Therefore, we consider the pelvic floor as the complex structure that comprises the pelvic diaphragm, perineal membrane, muscles of deep perineal space (deep transverse perineal muscle, sphincter urethrae, compressor urethrae, sphincter urethrovaginalis, and transverse vaginal muscle), as well as the attachments of the ligaments formed by the pelvic fascia (▶Fig. 18.1). These ligaments suspend the pelvic organs and anchor them on the pelvic diaphragm; consequently, their attachments onto the pelvic diaphragm must be included in the definition of the total pelvic floor.

Other authors further include in the pelvic floor the superficial perineal pouch and the external anal sphincter. According to O'Rahilly,[2] the pelvic floor comprises all structures between the peritoneum roofing the pelvis and the pelvic diaphragm below. Wei and De Lancey[3] describe the pelvic floor with an even wider perception, as consisting of the structures between the peritoneum and the perineal skin.

◆ Pelvic Diaphragm

The pelvic diaphragm is a muscular structure formed by the levator ani and coccygeus muscles (▶Fig. 18.1). Evidently, the diaphragm includes the investing fascia covering of the upper and lower surfaces of each muscle.

Being interposed between the pelvis and the perineum, the pelvic diaphragm is the main component of the concave pelvic floor; simultaneously, it is the convex roof of the perineum. The muscles of the pelvic diaphragm arise from the pelvic surface of the ischial spine, the tendinous arch of the levator ani muscle, and the posterior surface of the body of pubis. Posteriorly, the fusion of levator ani in the midline forms the *levator plate*. Anteriorly, in the median plane in front of the perineal body, the pelvic diaphragm has an aperture called *urogenital hiatus*, which allows the passage of urogenital organs to the perineum (▶Fig. 18.2). The hiatus is delineated by the medial borders of levator ani muscles and sealed by the perineal membrane (▶Fig. 18.1).

The muscles of the pelvic diaphragm vary in thickness and strength. They contract continuously and tonically most of the time, providing a physical support to the pelvic organs against increases of the intrabdominal pressure. The entire levator ani and coccygeus muscles are innervated from branches of the ventral rami of the second to fifth sacral spinal nerves, as well as from the perineal branch of the pudendal nerve. However, recent literature suggests that the pudendal nerve may not contribute to levator ani muscle innervation.[4-6]

Traditionally and for didactic purposes, the levator ani muscle is divided into three muscles (iliococcygeus, pubococcygeus, and puborectalis) according to muscle attachments and fibers course.[1-8] However, this description is not verified in anatomy and surgery, and the boundaries of the three muscles are not clearly distinguished.[7-11] The two most commonly seen components of the levator ani using magnetic resonance (MR) images are the iliococcygeus and puborectalis muscles. The boundaries and components of the pubococcygeus are hard to define with MRI. A traditional description is given in the following sections.

Iliococcygeus Muscle

The iliococcygeus muscle is the flat, thinner, posterolateral part of the levator ani muscle with a horizontal orientation (▶Fig. 18.1, ▶Fig. 18.2, ▶Fig. 18.3, ▶Fig. 18.4, ▶Fig. 18.5). It is attached to the pelvic surface of ischial spine and posterior part of the tendinous arch of the levator ani muscle. Then it passes obliquely medially and joins the contralateral homologue. The junction is marked by a musculofibrous condensation line, the *median* or *anococcygeal (iliococcygeal) raphe* (▶Fig. 18.4). On its inferior aspect, the iliococcygeus muscle is continuous with the anococcygeal ligament. It may be poorly developed or fibrous in some areas.

Pubococcygeus Muscle

The pubococcygeus muscle originates from the posterior (pelvic) surface of the body of the pubis and the anterior part of the tendinous arch of the levator ani muscle[7] (▶Fig. 18.1, ▶Fig. 18.2, ▶Fig. 18.3). It lies medial to the iliococcygeus muscle and passes posteriorly toward the coccyx. Immediately behind the pubis, there is a small gap between the paired pubococcygeus muscles through which pass the dorsal vein of the clitoris or penis (▶Fig. 18.1, ▶Fig. 18.2). The posterior fibers of the pubococcygeus

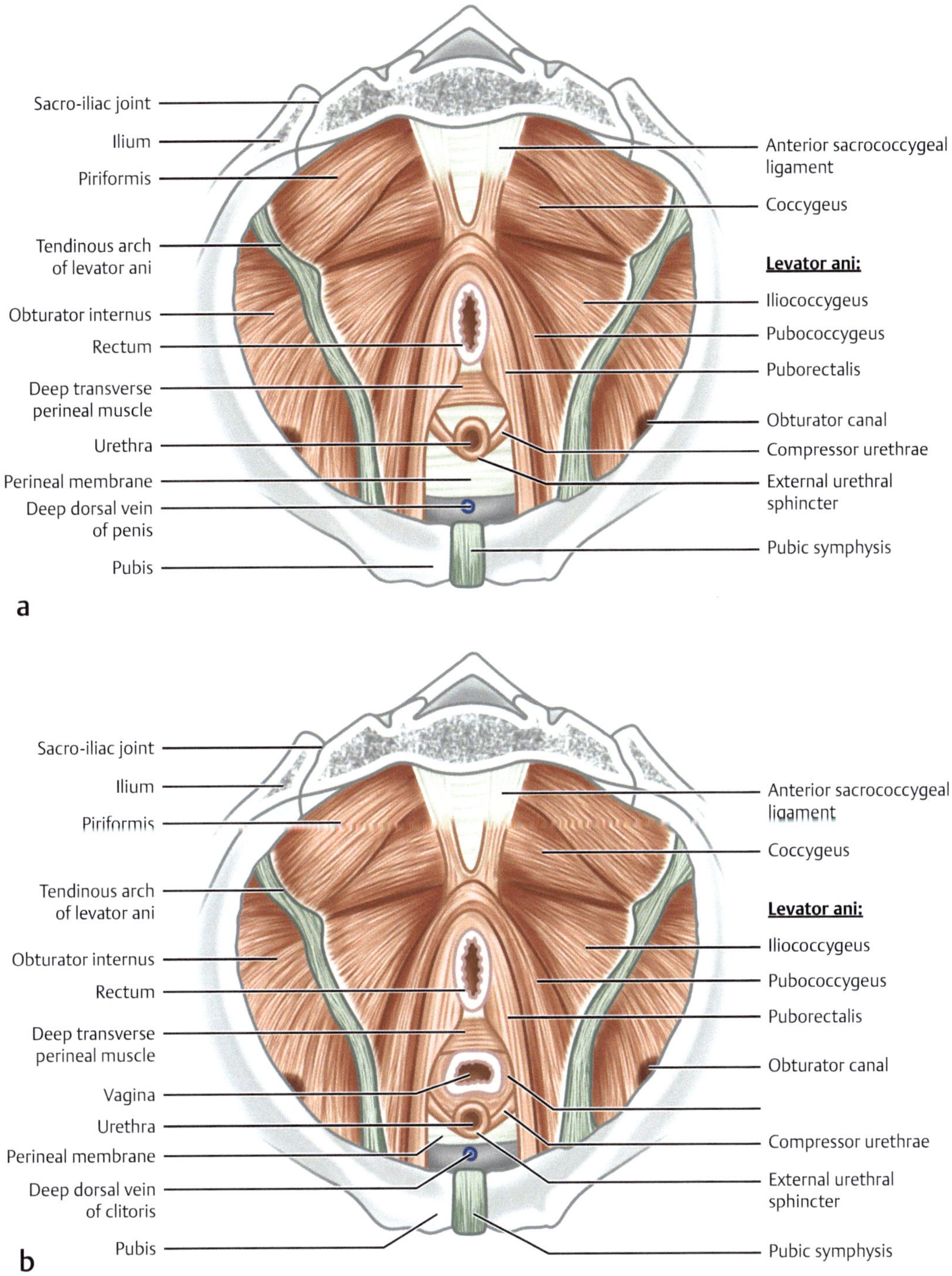

Fig. 18.1 Superior view of the pelvic floor. **(a)** Male. **(b)** Female. The pelvic floor is a complex structure that comprises the pelvic diaphragm (levator ani and coccygeus muscles), perineal membrane, muscles of deep perineal space (deep transverse perineal muscle, sphincter urethrae, compressor urethrae, sphincter urethrovaginalis, and transverse vaginal muscle), and the attachments of the ligaments formed by the pelvic fascia.

18

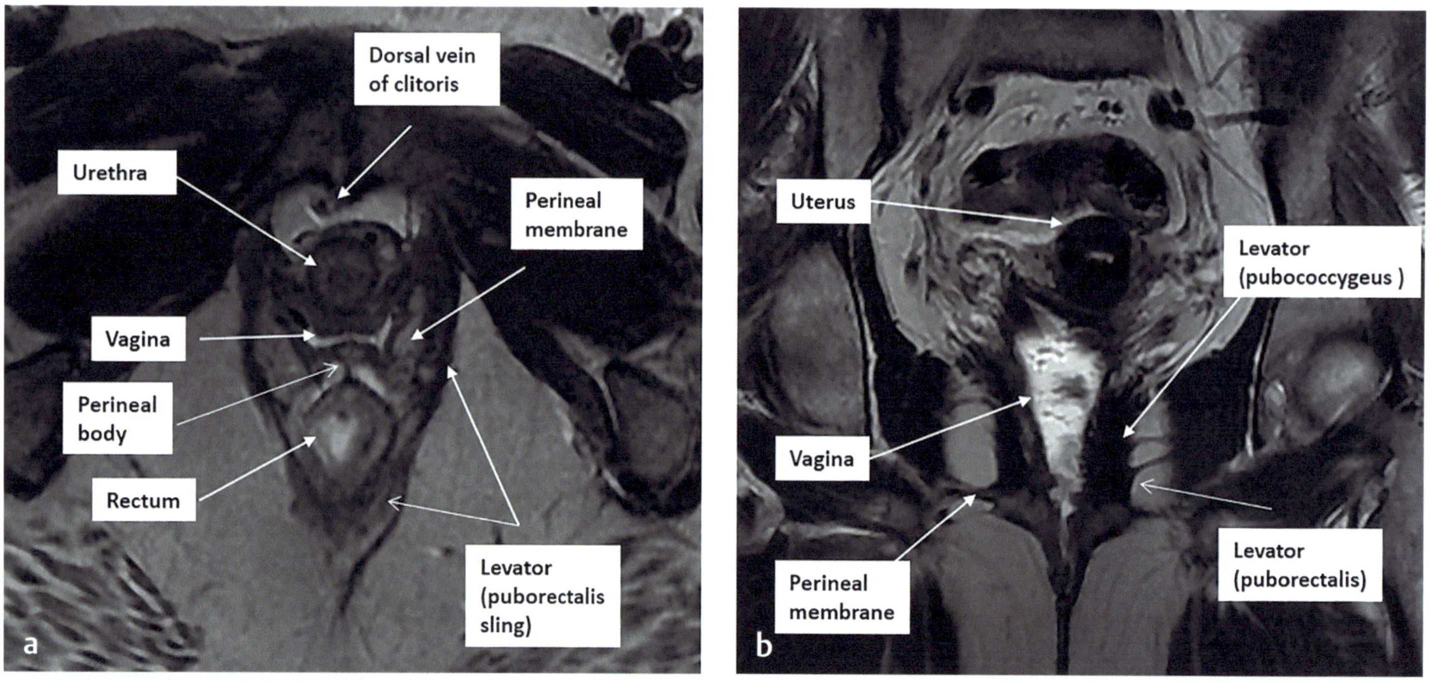

Fig. 18.2 (a) Axial and **(b)** coronal T2-weighted magnetic resonance (MR) images show the urogenital hiatus formed by the pelvic diaphragm. The pelvic diaphragm is formed by the iliococcygeus, pubococcygeus, puborectalis, and ischiococcygeus muscles. Anteriorly, in front of the perineal body, the pelvic diaphragm has an aperture called urogenital hiatus, which allows the passage of urogenital organs to the perineum. The urogenital hiatus is normally not larger than 4.5 cm on the transverse axis and not wider than 5.5 cm in the sagittal plane. In **a,** the U-shaped puborectalis sling as it inserts on the pubic bones and encloses the vagina, urethra, and anorectum is seen. Note the typical H shape of the vagina at the middle third.

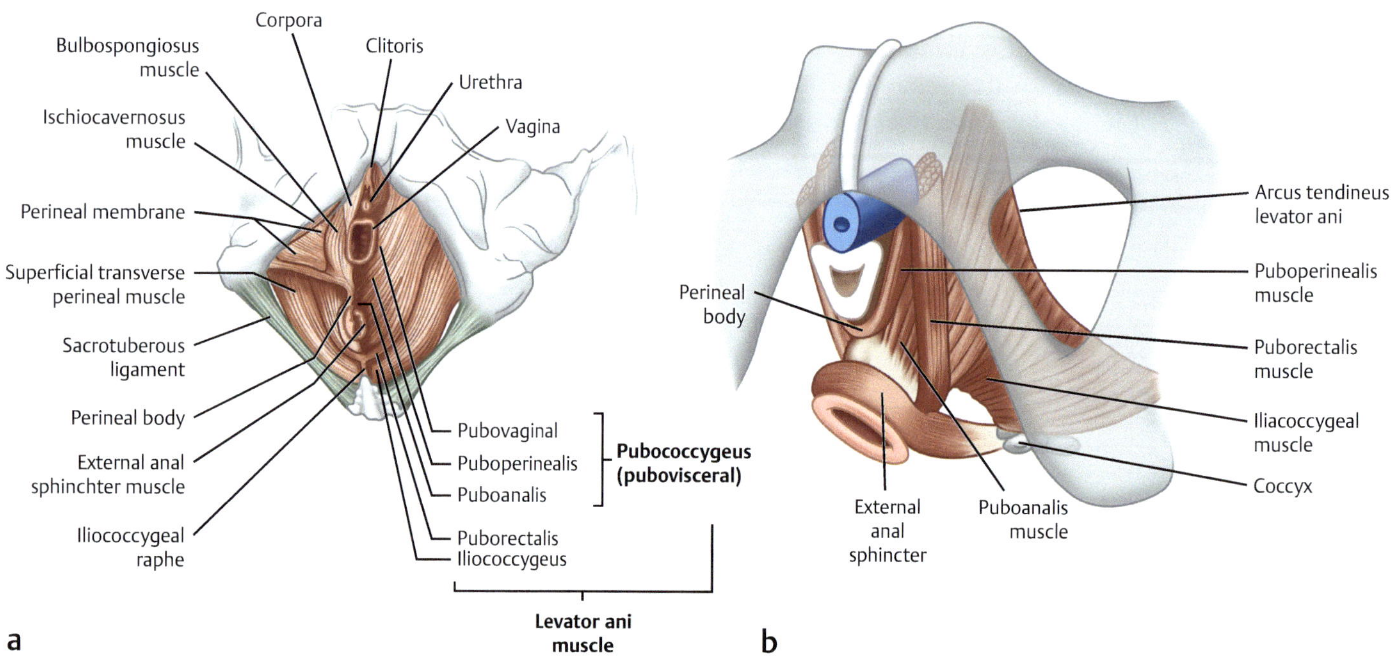

Fig. 18.3 (a) Inferior view of the pelvic floor showing components of the levator complex. (Modified with permission from Corton.[5]) **(b)** Schematic view of the levator ani muscles from below after the vulvar structures and perineal membrane have been removed showing the perineal body uniting the two ends of the puboperineal muscle. Note that the urethra and vagina have been transected just above the hymenal ring. (Modified from DeLancey.[12])

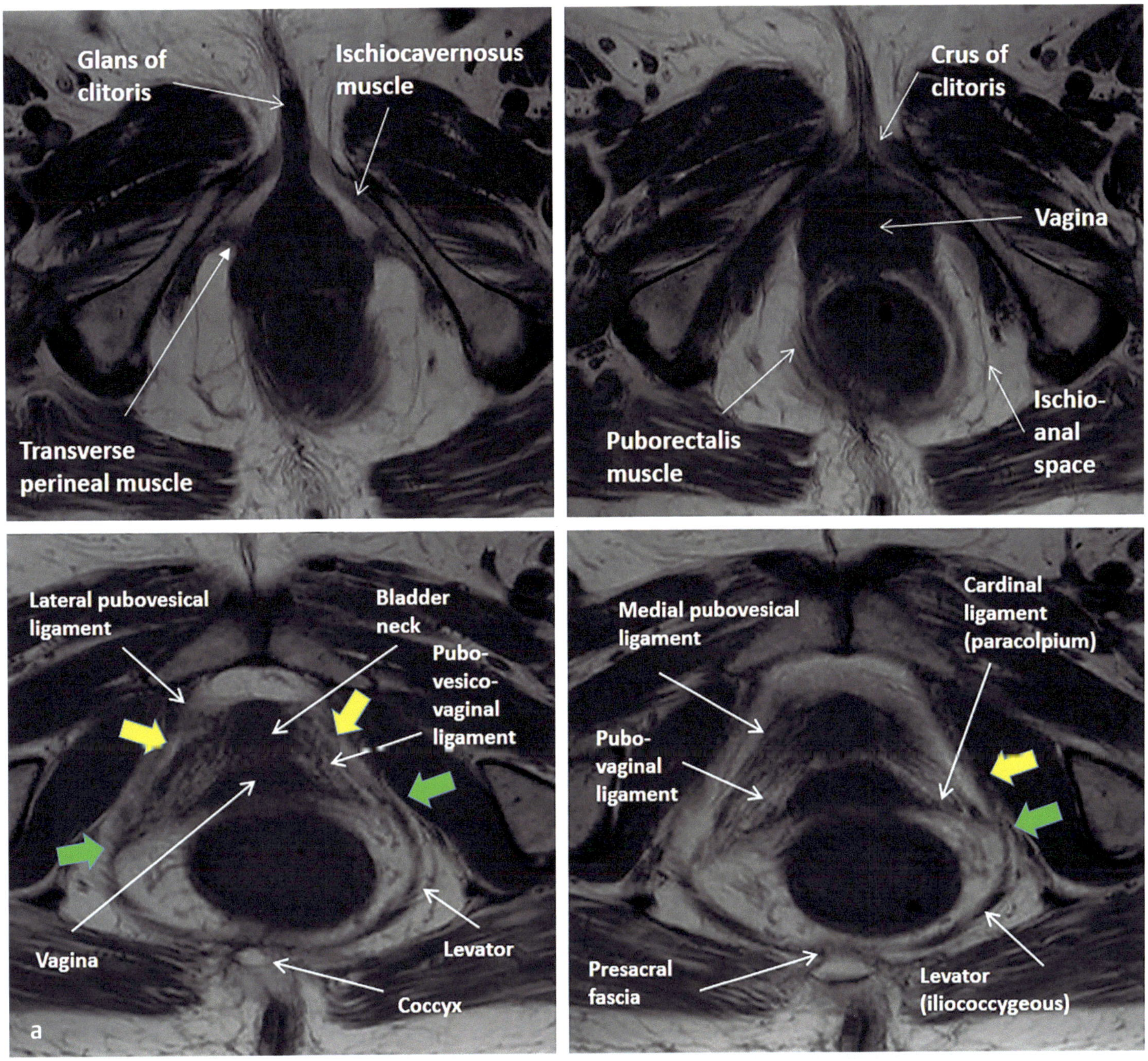

Fig. 18.4 (a, b) Supportive ligaments of the pelvic organs in a female with laxity of the pelvic floor (levator muscles) shown by axial T1-weighted (T1W) magnetic resonance imaging (MRI) from inferior to superior. The tendinous arch of the pelvic fascia is shown by *green arrows*. *Yellow arrows* denote the tendinous arch of the levator ani muscle. These two tendinous arches are close each other at their posterior thirds near the ischial spine. Part of the rectogenital fascia (Denonvilliers' fascia) is shown by *red arrows*. The presacral fascia which is partly visible in the last image is an extraperitoneal tissue band located anterior to the sacrum. The presacral space is behind it and the pararectal space anterior to it. Median sacral artery and vein are seen in this space. The sacrospinous ligament passes behind the ischiococcygeus muscle. Additional findings in this patient with mild pelvic floor dysfunction include sagging of the iliococcygeus muscle and also posterior bladder wall descent known as the "saddlebag" sign, possibly due to lateral weakening of the endopelvic fascia.

(Continued)

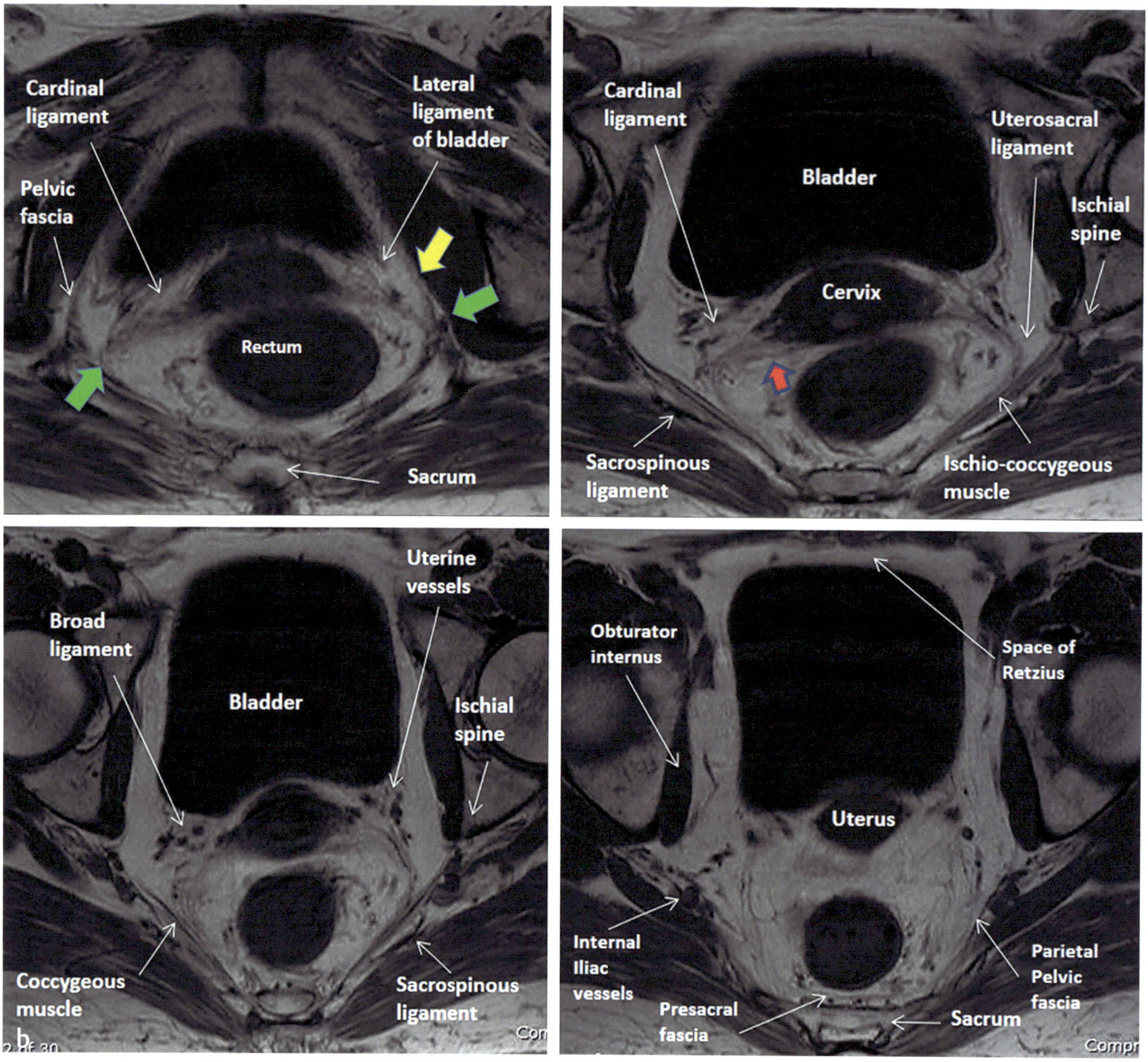

Fig. 18.4 (*Continued*) (**b**)

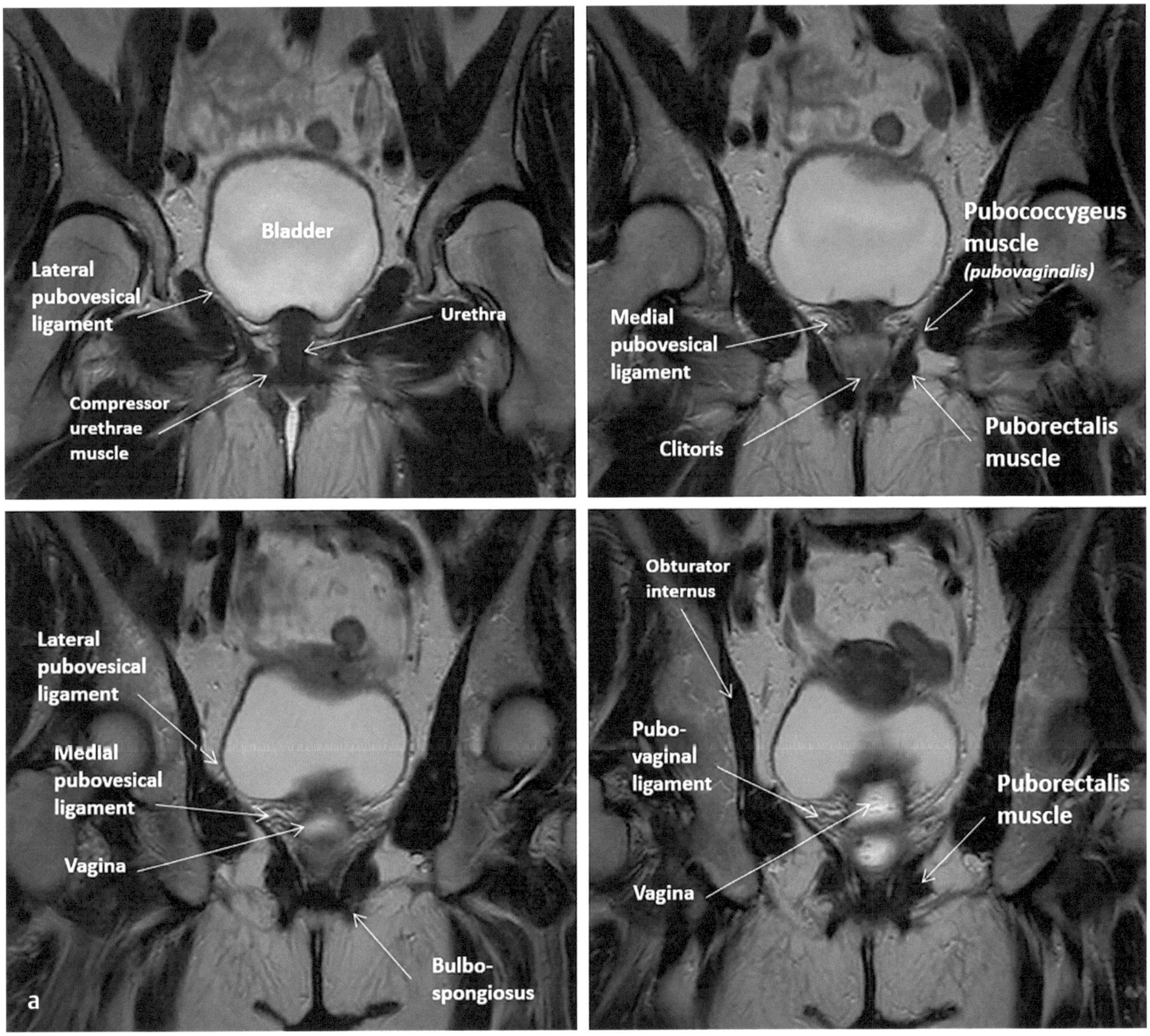

Fig. 18.5 (a, b) Supportive ligaments of the pelvic organs in a female with laxity of the pelvic floor (levator muscles) shown by coronal T2-weighted (T2W) magnetic resonance imaging (MRI) from anterior to posterior. Note sagging of the iliococcygeus muscle. The tendinous arch of the pelvic fascia is shown by *green arrows*. *Yellow arrows* denote the tendinous arch of the levator ani muscle.

(Continued)

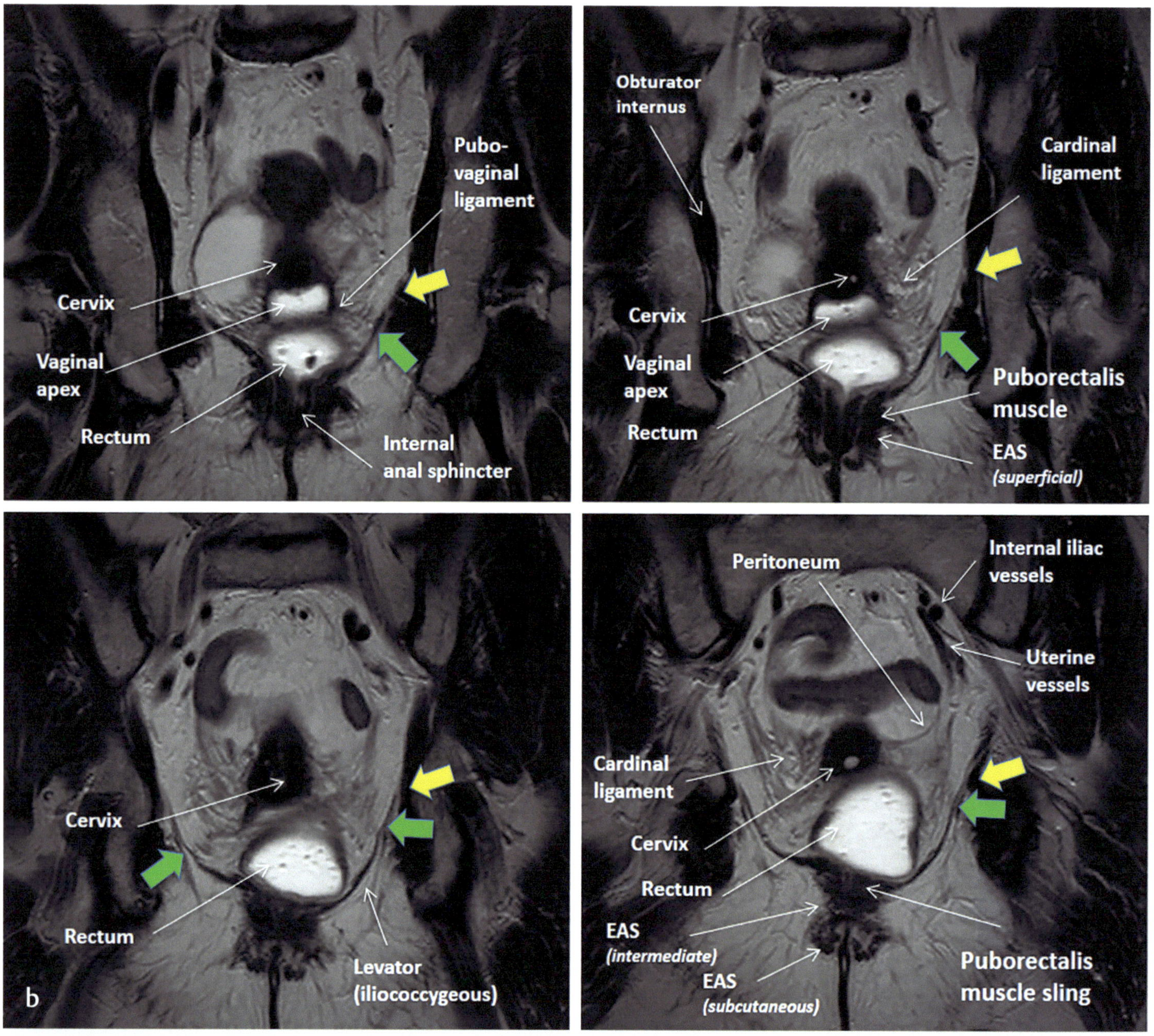

Fig. 18.5 (*Continued*) (b)

Table 18.1 Classification of the levator muscle complex

Terms and Descriptions Used in Terminolgia Anatomica		
Terminologia Anatomica	**Origin/Insertion**	**Function**
"Pubococcygeus" (recommend pubovisceral)		
Puboperinealis	Pubis/perineal boday	Tonic activity pulls perineal body ventrally toward pubis
Pubovaginalis	Pubis/vaginal wall at the level of the midurethra	Elevates vagina in region of the midurethra
Puboanalis	Pubis/intersphincteric groove between internal and external and sphincter to end in the anal skin	Inserts into the inter-sphincteric groove to elevate the anus and its attached anoderm
Puborectalis	Pubis/forms sling behind the rectum	Forms a sling behind the rectum forming the anorectal angle and closing the pelvic floor
Iliococcygeus	Tendinous arch of the levator ani/the two sides fuse in the iliococcygeal rephe	The two sides form a supportive diaphragm that spans the pelvic canal

18

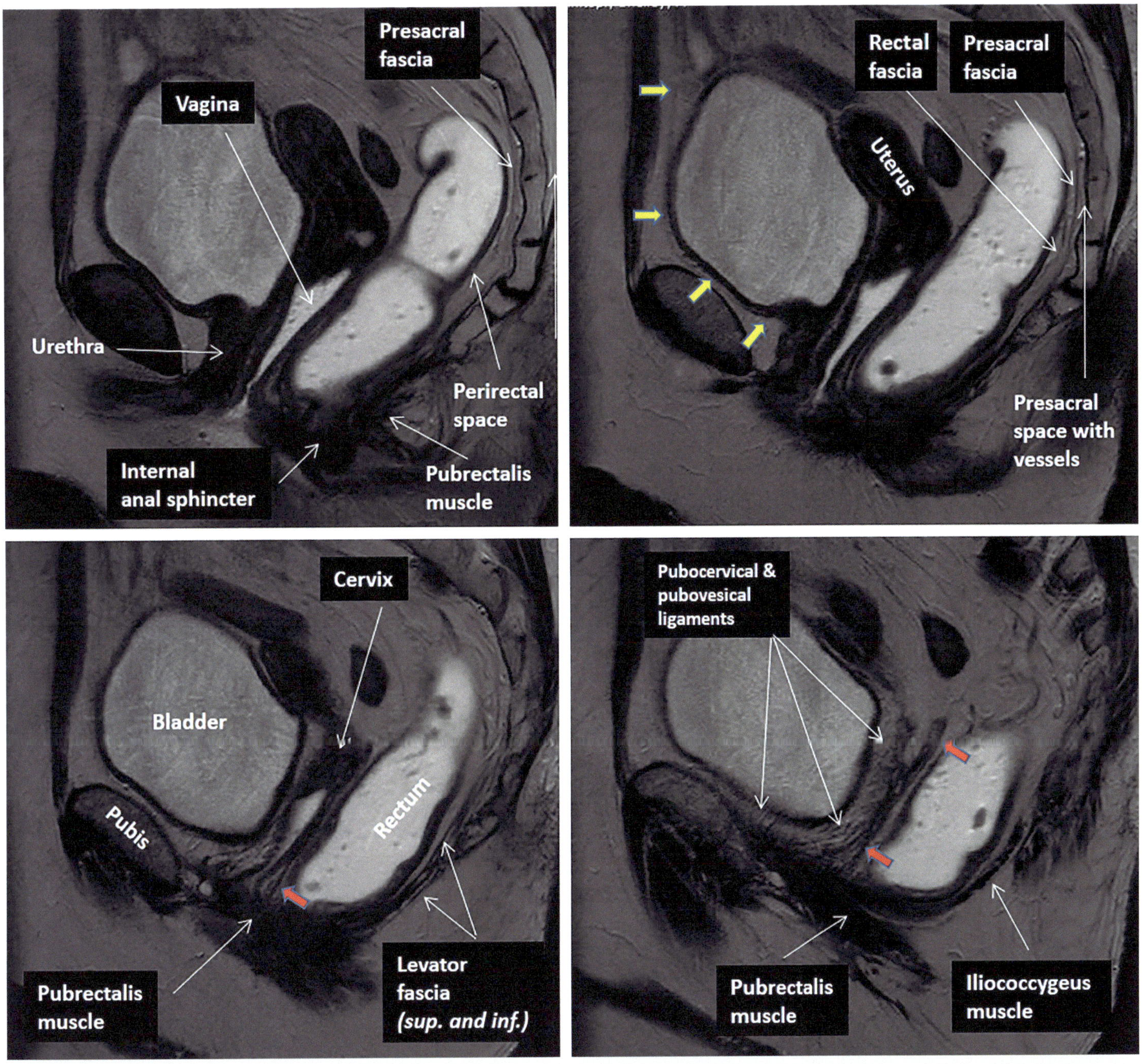

Fig. 18.6 Pelvic fasciae. From midsagittal to parasagittal magnetic resonance imaging (MRI) in a female. In the pelvis, there are three condensation fasciae: the umbilical prevesical fascia (*yellow arrows*), the rectogenital fascia (Denonvilliers' fascia) (*red arrows*), and the presacral fascia. These fasciae are continuous inferiorly with the superior fascia of the pelvic diaphragm (levator muscle) and laterally with the parietal fascia of the pelvic walls. Also shown is the pubocervical ligament which passes forwards from the vaginal vault and the uterine cervix, in a curvilinear motion around the urethra (along with the lateral pubovesical ligament), and attaches to the tendinous arch of pelvic fascia.

18

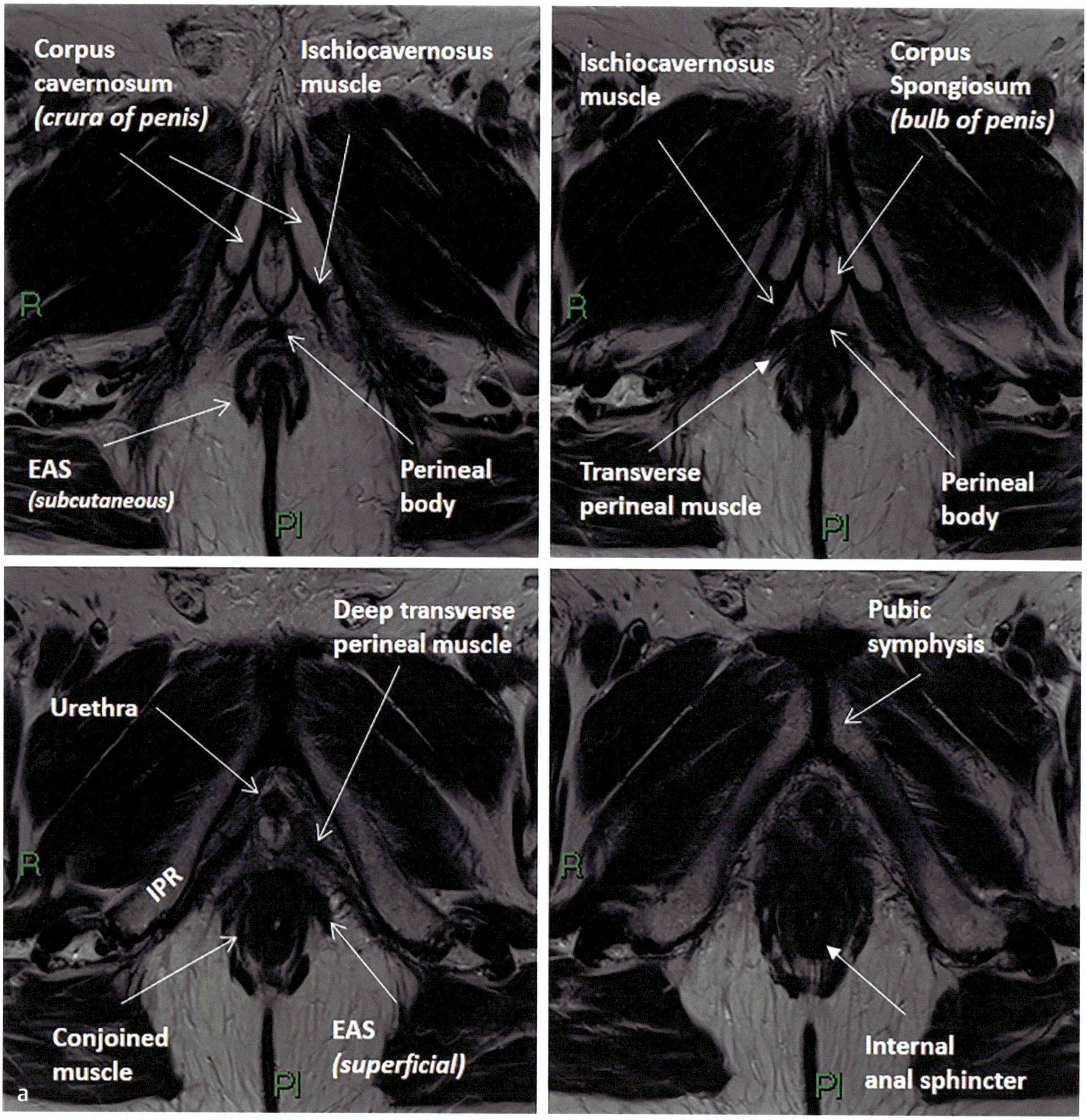

Fig. 18.7 (a–c) Supportive ligaments of the pelvic organs in a male shown by axial T2-weighted (T2W) magnetic resonance imaging (MRI) from inferior to superior. *Yellow arrow* denotes the tendinous arch of the levator ani muscle. EAS, external anal sphincter; IPR, inferior pubic ramus.

(Continued)

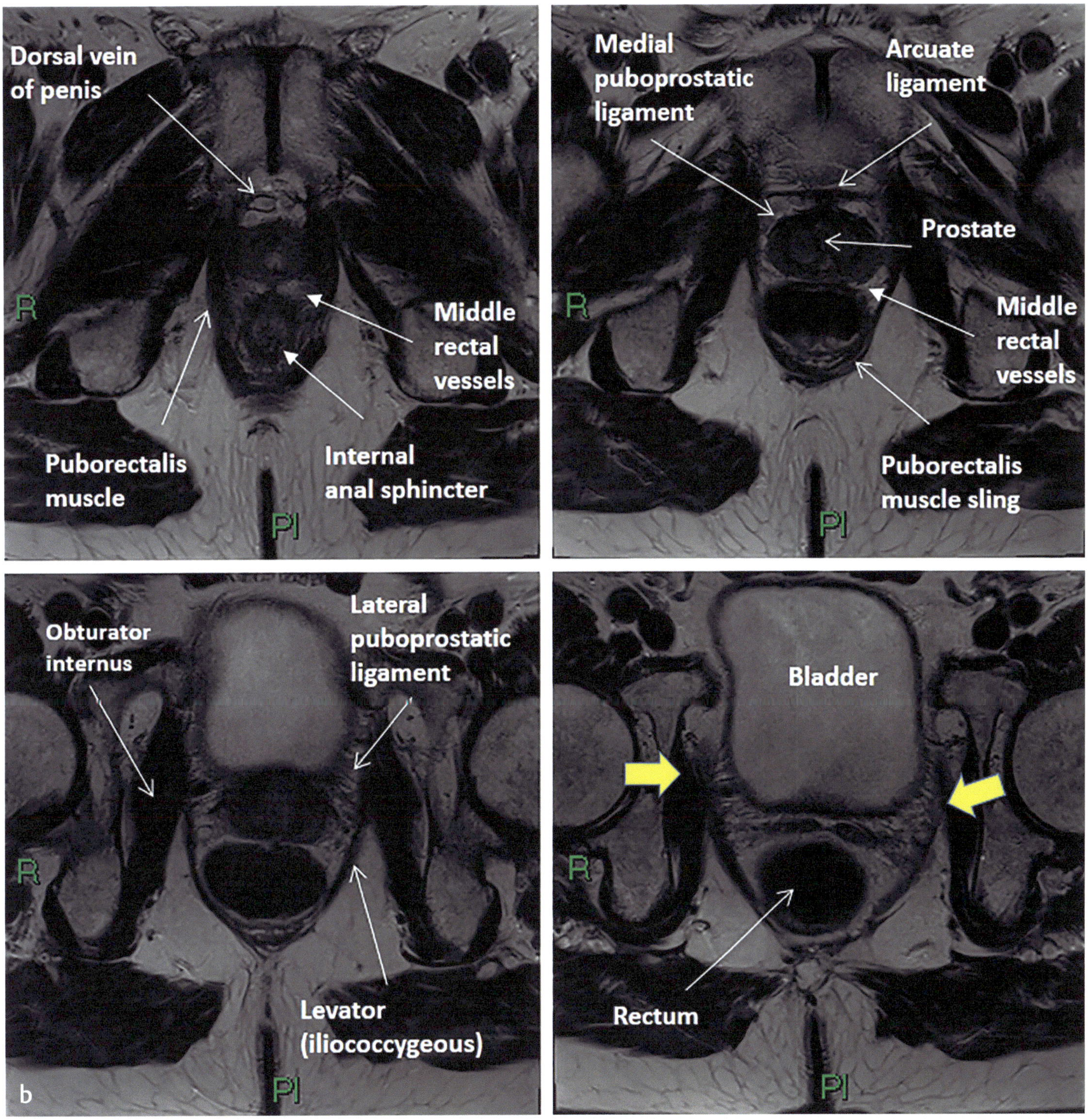

Fig. 18.7 (*Continued*) (**b**)

(*Continued*)

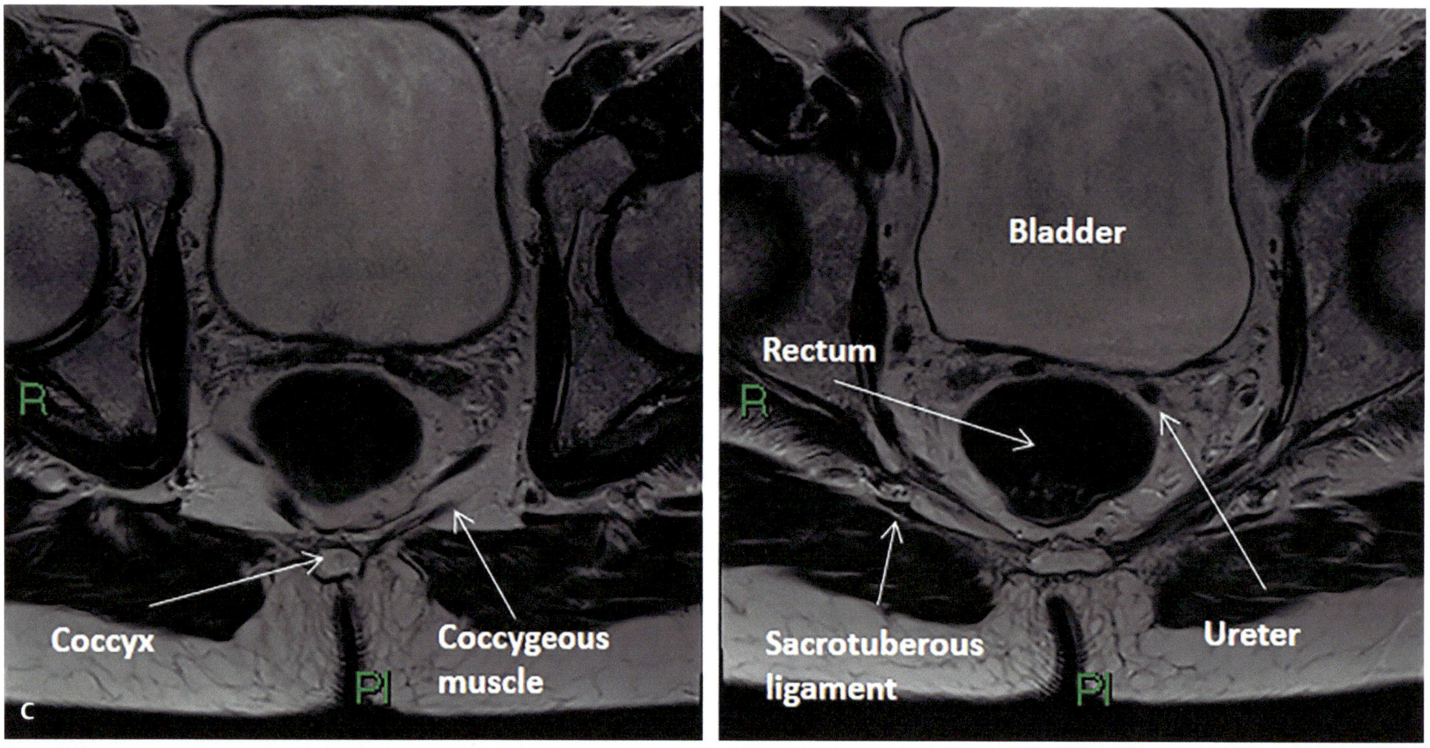

Fig. 18.7 *(Continued)* (**c**)

muscle merge with those of the contralateral homologue to from part of the levator plate. Some medial fibers merge with the pelvic organ walls and the perineal body. Various subdivisions of the medial portions of the muscle have been often described such as puboperineus, pubovaginalis or puboprostaticus (or levator prostate), puboanalis, or collectively as pubovisceralis (▸Table 18.1)[8,9] (▸Fig. 18.3, ▸Fig. 18.4).

Puborectalis Muscle

The puborectalis muscle is the medial part of the levator ani that arises from the pelvic surface of the ischiopubic rami (▸Fig. 18.1, ▸Fig. 18.2, ▸Fig. 18.3, ▸Fig. 18.4, ▸Fig. 18.5, ▸Fig. 18.6, ▸Fig. 18.7, ▸Fig. 18.8, ▸Fig. 18.9). It forms a U-shaped sling around the rectum and bounds the urogenital hiatus (▸Fig. 18.2). The puborectalis muscle lies cranial to the external anal sphincter, while some puborectalis fibers merge with it.

Coccygeus (Ischiococcygeus) Muscle

The coccygeus muscle lies immediately cranial and behind the levator ani muscle and may be tendinous in structure (▸Fig. 18.1, ▸Fig. 18.4). It originates from the ischial spine and traverses medially to attach to the lateral aspects of the inferior end of the sacrum

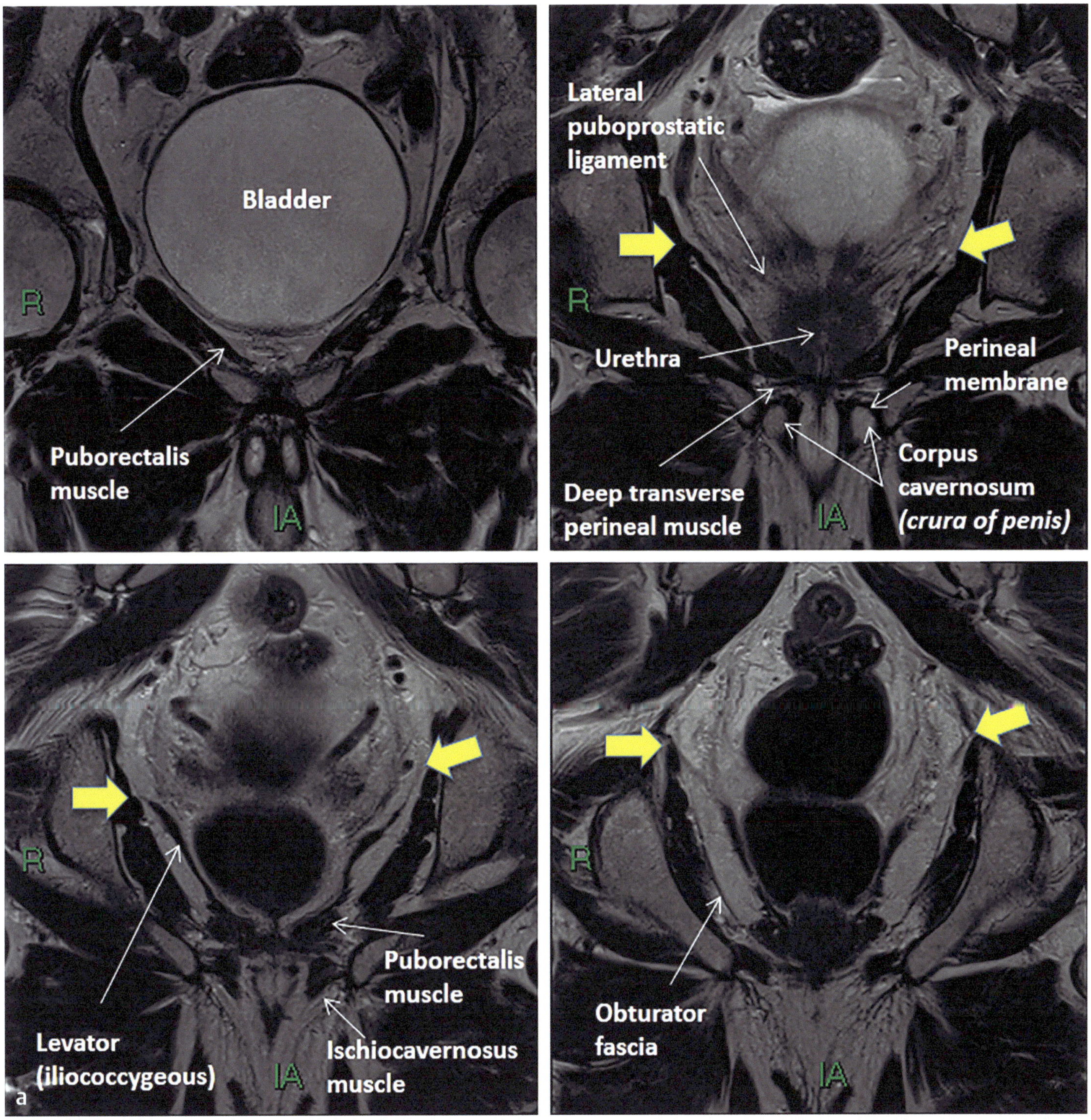

Fig. 18.8 (a, b) Supportive ligaments of the pelvic organs in a male shown by coronal T2-weighted (T2W) magnetic resonance imaging (MRI) from anterior to posterior. *Yellow arrow* denotes the tendinous arch of the levator ani muscle. EAS, external anal sphincter.

(*Continued*)

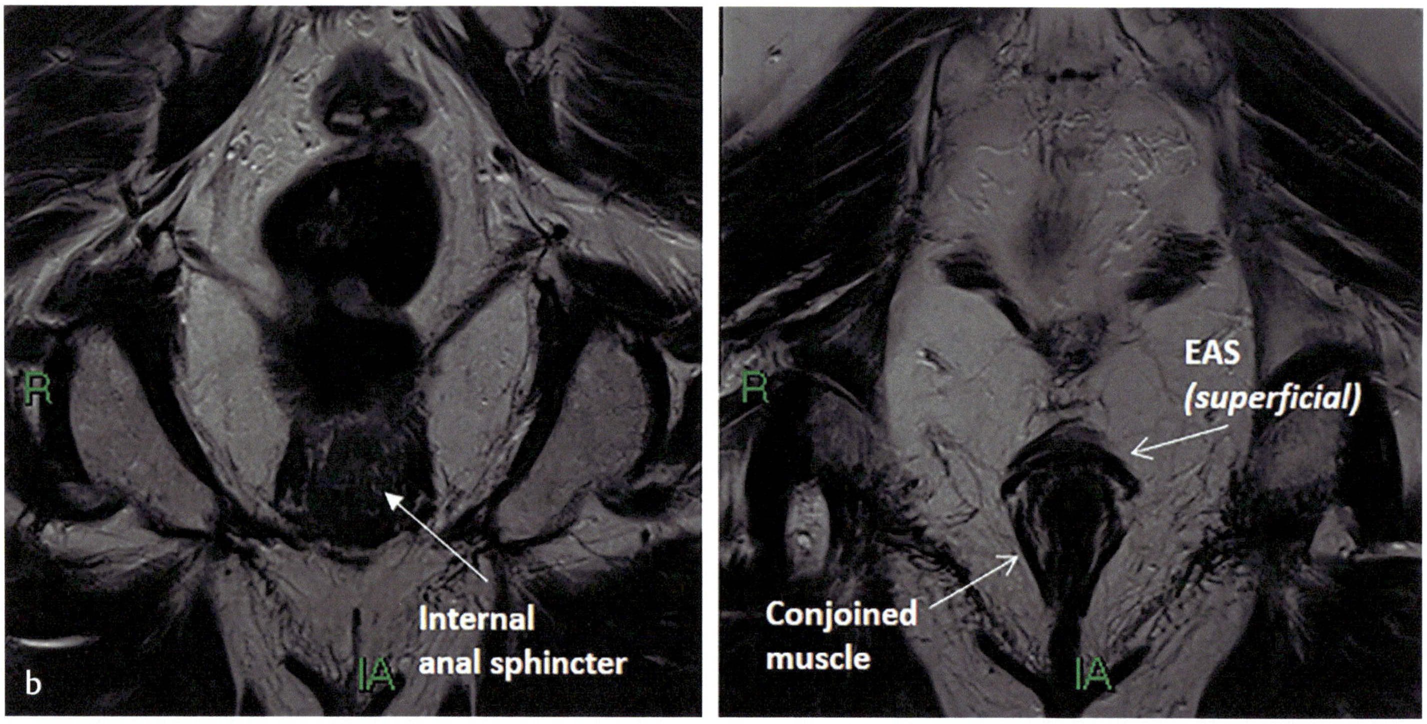

Fig. 18.8 (*Continued*) (b)

and coccyx. The coccygeus muscle lies on the upper surface of the sacrospinous ligament.

◆ Pelvis Architecture

Assessment of the architecture of the pelvis as a three-dimensional structure is crucial for understanding the anatomy and imaging of the pelvic floor. The pelvis can be regarded as an apartment with two main rooms, the urogenital compartment anteriorly and the rectal compartment posteriorly, which, respectively, contain the caudal parts of the urinary and genital tracts and the caudal digestive tract.[13] Other than the peritoneum, there is no true ceiling to separate from the abdominal cavity. The pelvic diaphragm is the floor, and all contents stand on it.

Similar to the preperitoneal and retroperitoneal spaces, in the subperitoneal pelvic space, the interval between the pelvic organs, muscles of the pelvic wall, and muscles of the pelvic floor, with the peritoneum, is filled with a variable mass of extraperitoneal connective tissue called *pelvic (or endopelvic) fascia* (▶Fig. 18.2, ▶Fig. 18.10, ▶Fig. 18.11, ▶Fig. 18.12). The pelvic fascia investing the pelvic wall and floor muscles is called parietal fascia, while the fascia on top of viscera (urinary bladder, rectum, etc.) is the visceral fascia. The endopelvic fascia fills in the space between parietal and visceral fascia and offers a packing of pelvic organs, which is important for their independent movement and function. In addition, the endopelvic fascia generates ligaments, which can be just supportive or, in addition, have the role of mesenteries—which do not exist in the extraperitoneal pelvis—and transfer vessels and nerves to respective organs (called hereafter "supplier" ligaments) (▶Fig. 18.10, ▶Fig. 18.11, ▶Fig. 18.12). The ligaments that support the pelvic organs and the fasciae separating them (e.g., Denovilliers') attach on pelvic floor. Therefore, it is important to know the traces of such ligaments and fasciae on the "floor plan" of the pelvic floor.

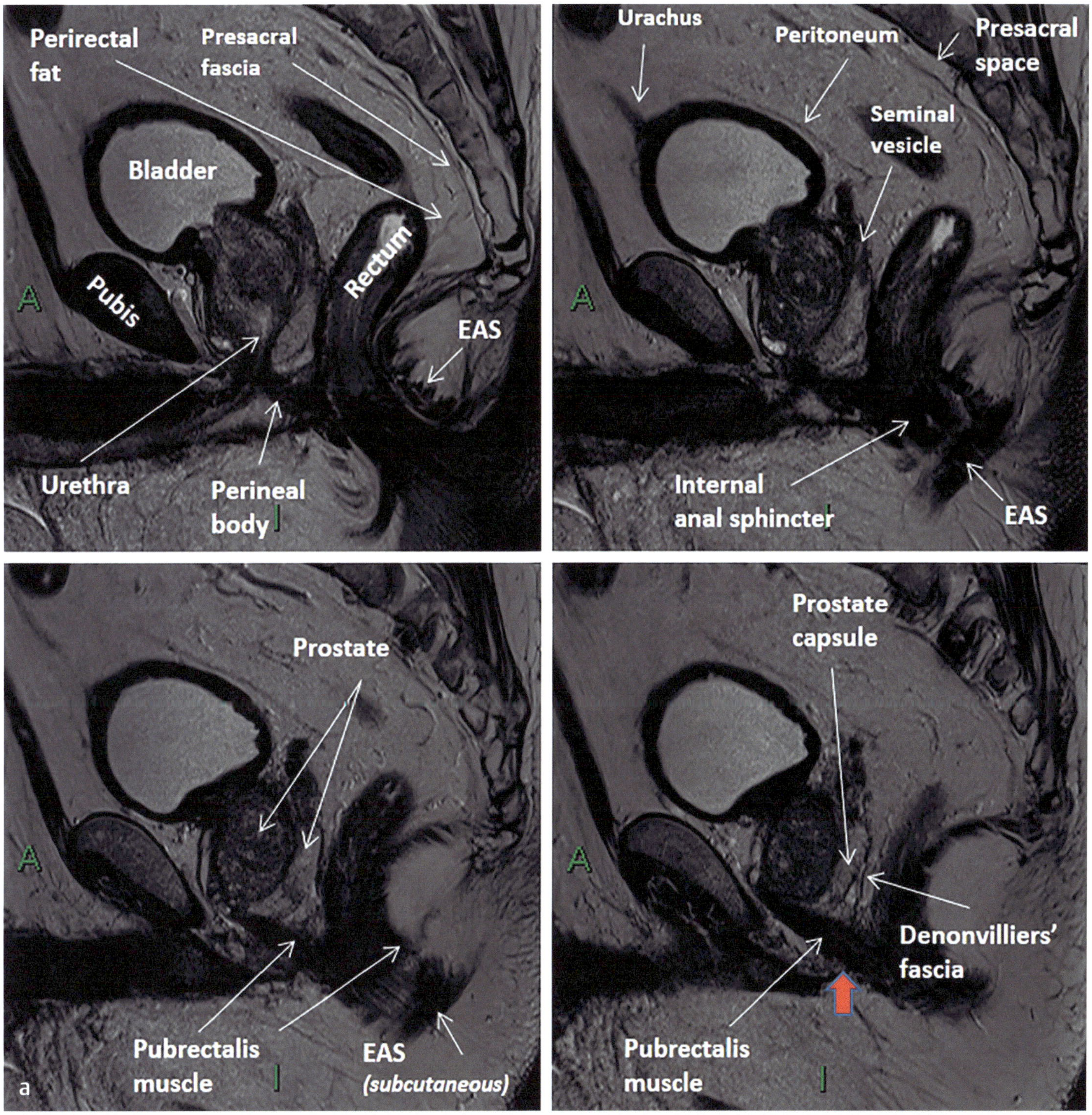

Fig. 18.9 (a, b) Pelvic fasciae. From midsagittal to parasagittal magnetic resonance imaging (MRI) in a male. The rectogenital fascia (Denonvilliers' fascia) and the presacral fascia are clearly shown. These fasciae are continuous inferiorly with the superior fascia of the pelvic diaphragm (levator muscle) and laterally with the parietal fascia of the pelvic walls. The level of the insertion of the Denonvilliers' fascia on the superior fascia of the pubrectalis muscle is shown by *red arrow*. Prevesical fascia is seen which is inseparable from the rectal fascia in this patient with large perirectal fat. EAS, external anal sphincter.

(*Continued*)

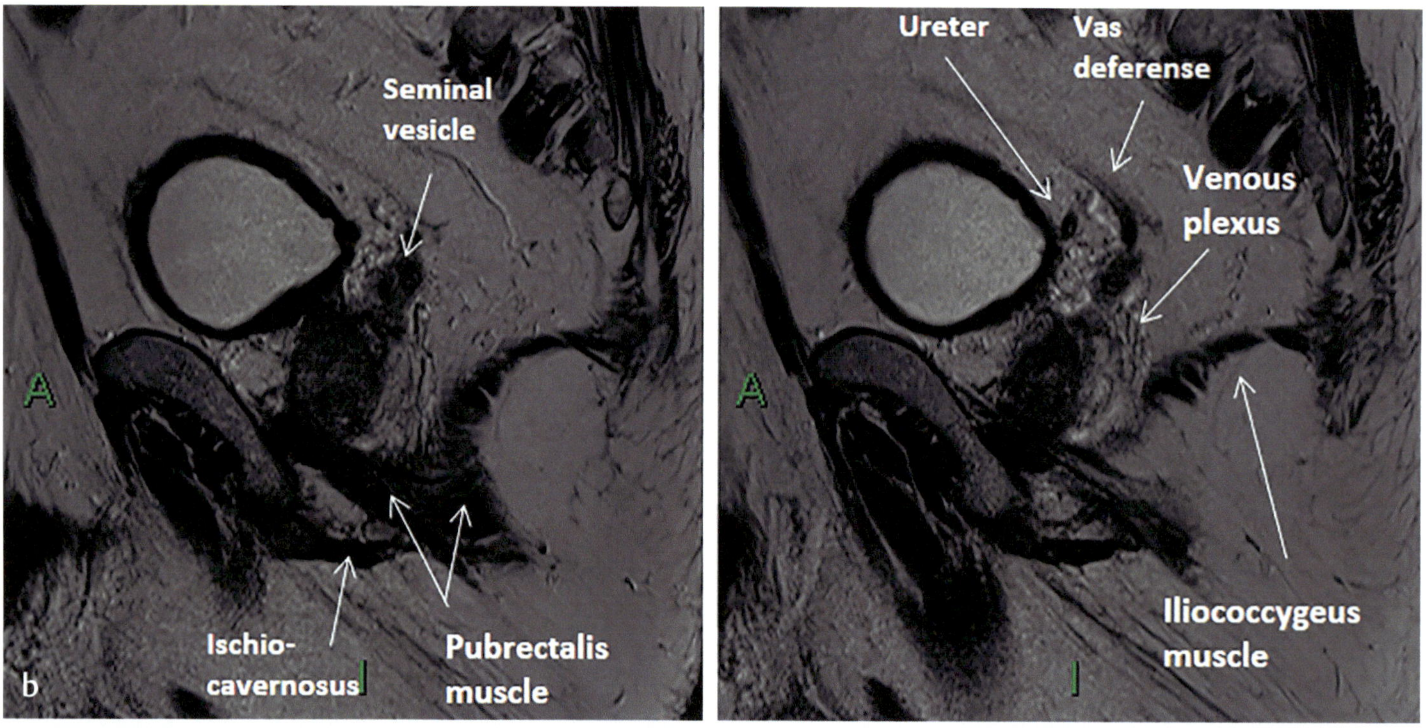

Fig. 18.9 (*Continued*) (**b**)

Peritoneum: *The False Ceiling*

The parietal peritoneum of the abdominal wall descends over the pelvic brim and covers parts of the pelvic walls and pelvic floor (▶ Fig. 18.11, ▶ Fig. 18.13). When covering the pelvic floor, it is evident that "ceiling" and floor become adjacent, with only the parietal pelvic fascia separating the two. Laterally, the peritoneum is loosely attached to the muscles of the pelvic diaphragm by extraperitoneal connective tissue containing blood vessels and nerves that supply the pelvic organs.

The parietal peritoneum reflects as visceral peritoneum onto the pelvic organs covering them to various extent (depending on their fullness) and blending with their visceral fascia. Pelvic organs are not fully covered by peritoneum, except from the uterine tubes. The visceral peritoneum reflects from the posterior surface of bladder in the male, or from the posterior vaginal fornix in the female, to cover the anterolateral surfaces of the rectum and become continuous with the parietal peritoneum of posterior abdominal wall (▶ Fig. 18.11, ▶ Fig. 18.12, ▶ Fig. 18.13). Since the pelvic organs are distendable, the distance of the visceral peritoneum from the pelvic floor varies. Mostly, the peritoneum is elevated from the pelvic floor by pelvic viscera and ligaments, but laterally, on the pelvic walls, the peritoneum may lie directly on the pelvic parietal fascia and muscles (▶ Fig. 18.12).

Visceral Pelvic Fascia: *The Shell of Pelvic Organs*

The *visceral pelvic fascia* ensheaths the pelvic organs and is firmly attached to them (e.g., vesical fascia, prostatic fascia, etc.). It becomes continuous with the parietal fascia where the organs pass through the openings of the pelvic floor. It accepts the neurovascular bundles to the respective pelvic organs carried by ligaments given from the pelvic fascia.

Condensations of Extraperitoneal Pelvic Fascia: *The Inner Walls of Pelvic Compartments and Spaces*

During embryogenesis, growth of organs may form a condensation layer in the extraperitoneal pelvic tissue due to pressure of the latter against the muscles of the pelvic floor.[6,13] In the pelvis, such condensation layers are the umbilical prevesical fascia, the

18

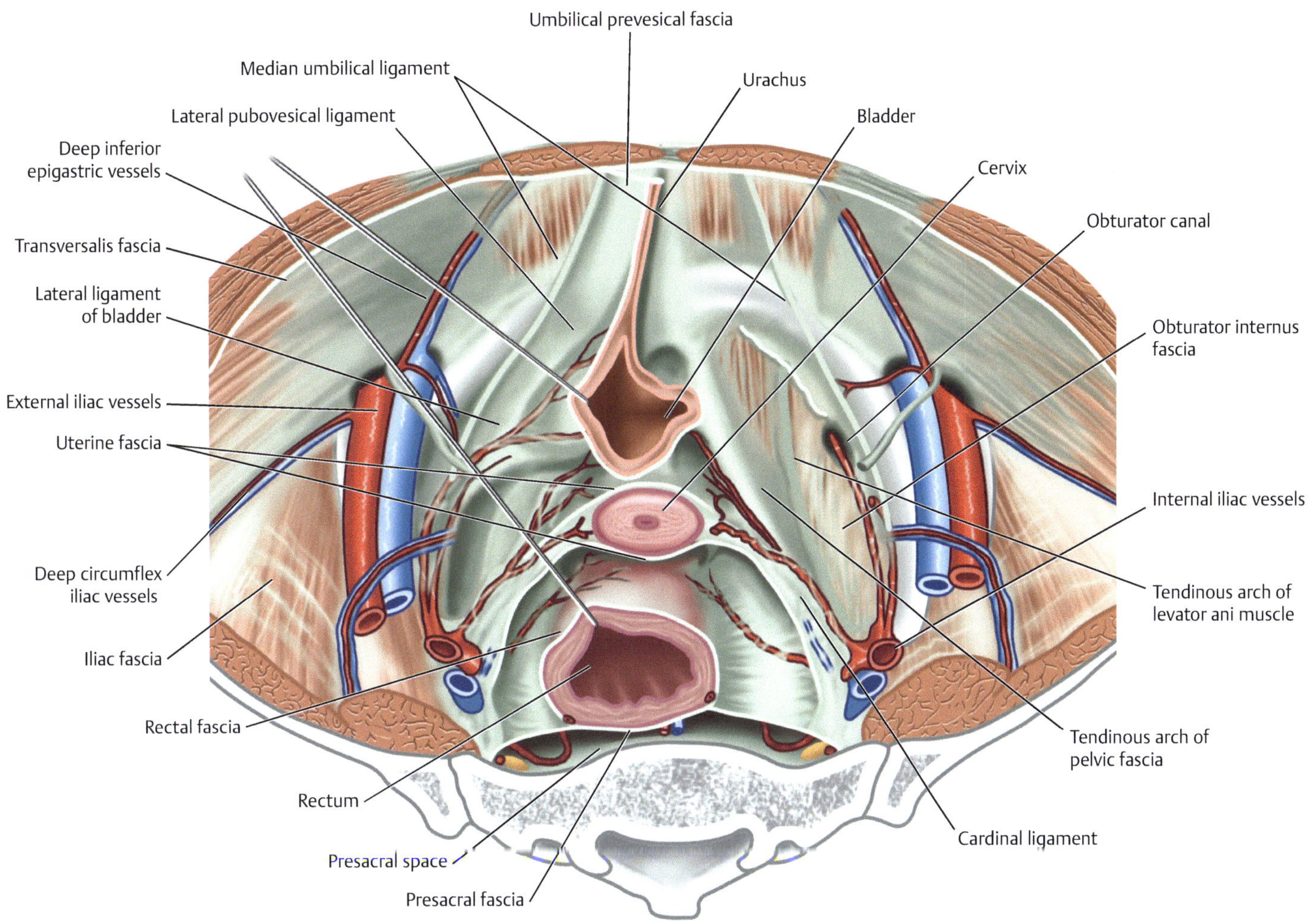

Fig. 18.10 Endopelvic fascia shown in a superior view.

rectogenital fascia (Denonvilliers' fascia), and the presacral fascia (▸Fig. 18.6, ▸Fig. 18.13, ▸Fig. 18.14). These fasciae are continuous inferiorly with the superior fascia of the pelvic diaphragm and laterally with the parietal fascia of the pelvic walls. These fasciae divide the pelvis into rooms (most prominent are the urogenital and rectal compartments), and thus can be regarded as the inner walls of the pelvis.

The *rectogenital septum (Denonvilliers' fascia)* is a single-layer fascia located between the rectum and prostate (and seminal vesicles) in men (▸Fig. 18.13, ▸Fig. 18.14), or between the rectum and vagina in women (▸Fig. 18.6, ▸Fig. 18.15). The septum extends from just below the rectovesical or rectouterine pouch up to the pelvic floor. Laterally and caudally, it fuses with the parietal fascia covering the upper surface of the pelvic diaphragm along the tendinous arch of rectogenital septum (see below, parietal pelvic fascia and its thickenings) (▸Fig. 18.13). The rectogenital septum anchors to the pelvic diaphragm anterior to the perineal body and divides the pelvis into an anterior urogenital compartment and a posterior rectal compartment (▸Fig. 18.6).

The ***umbilical prevesical fascia*** extends from the umbilicus to the urinary bladder and lies internal to the transversalis fascia and external to the umbilicovesical fascia, fusing with them at

18

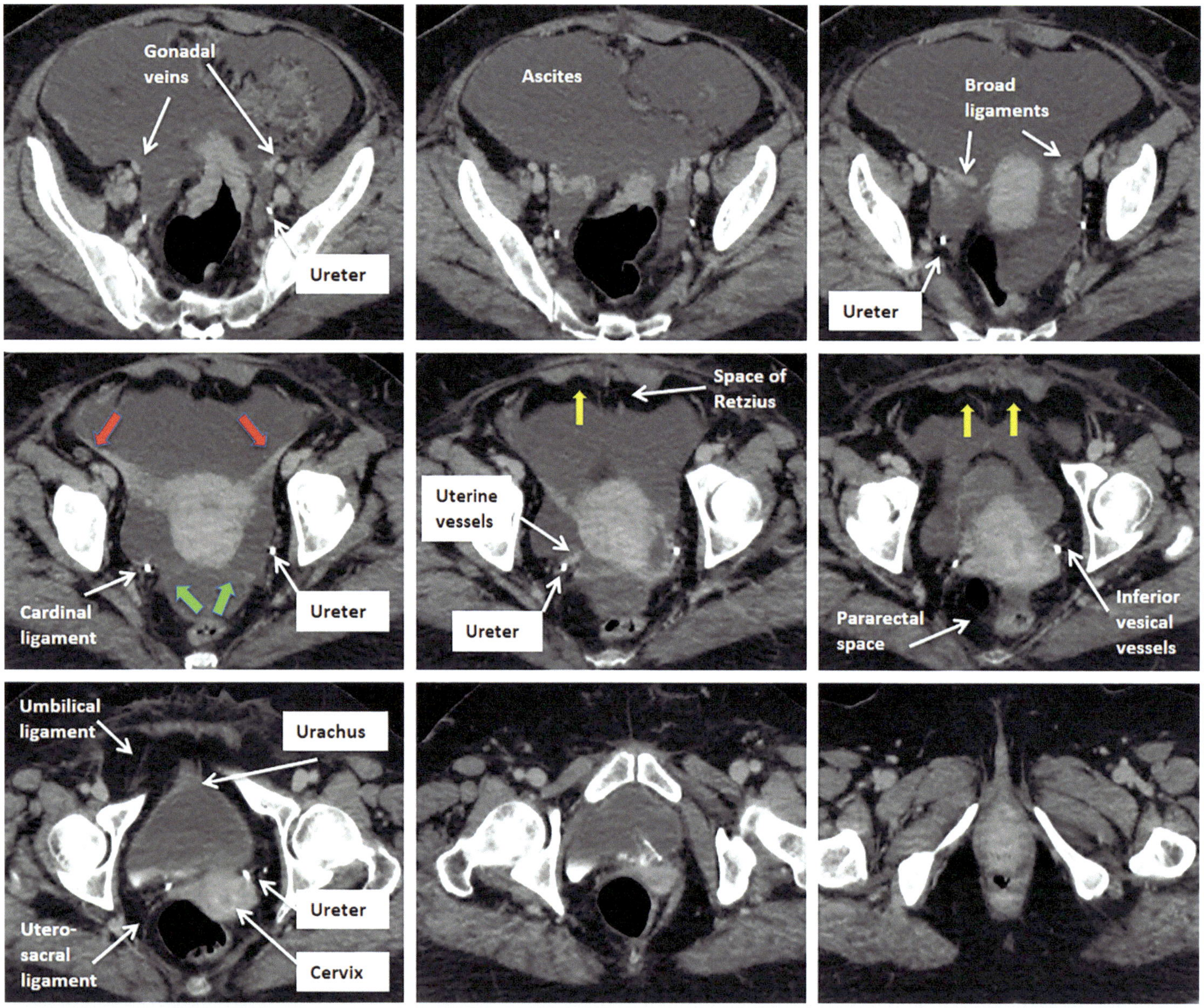

Fig. 18.11 Axial computed tomography (CT) images from superior to inferior in a patient with large ascites secondary to carcinomatosis showing the subperitoneal pelvic space including the pelvic fascia and pelvic ligaments. The interval between the pelvic organs, muscles of the pelvic wall, and muscles of the pelvic floor, with the peritoneum, is filled with a variable amount of extraperitoneal fibrofatty connective tissue called pelvic (or endopelvic) fascia. In this case, the endopelvic fascia which is filled with fat is clearly outlined by the peritoneal fluid and pelvic organs internally and pelvic bones and muscles externally. In addition, the endopelvic fascia generates ligaments, which can be just supportive, or, in addition, have the role of mesenteries that transfer vessels and nerves to respective organs. The cardinal ligaments are seen originating from the sacral lower uterine segment to insert into the pelvic fascia (*green arrows*). The ureter passes inside the cardinal ligament and crosses under the uterine artery. Note the endopelvic fascia is continuous with the space of Retzius anteriorly. The round ligaments (*red arrows*) are seen passing over the external iliac vessels to reach the inguinal canal. The umbilical prevesical fascia is shown by *yellow arrows*.

18

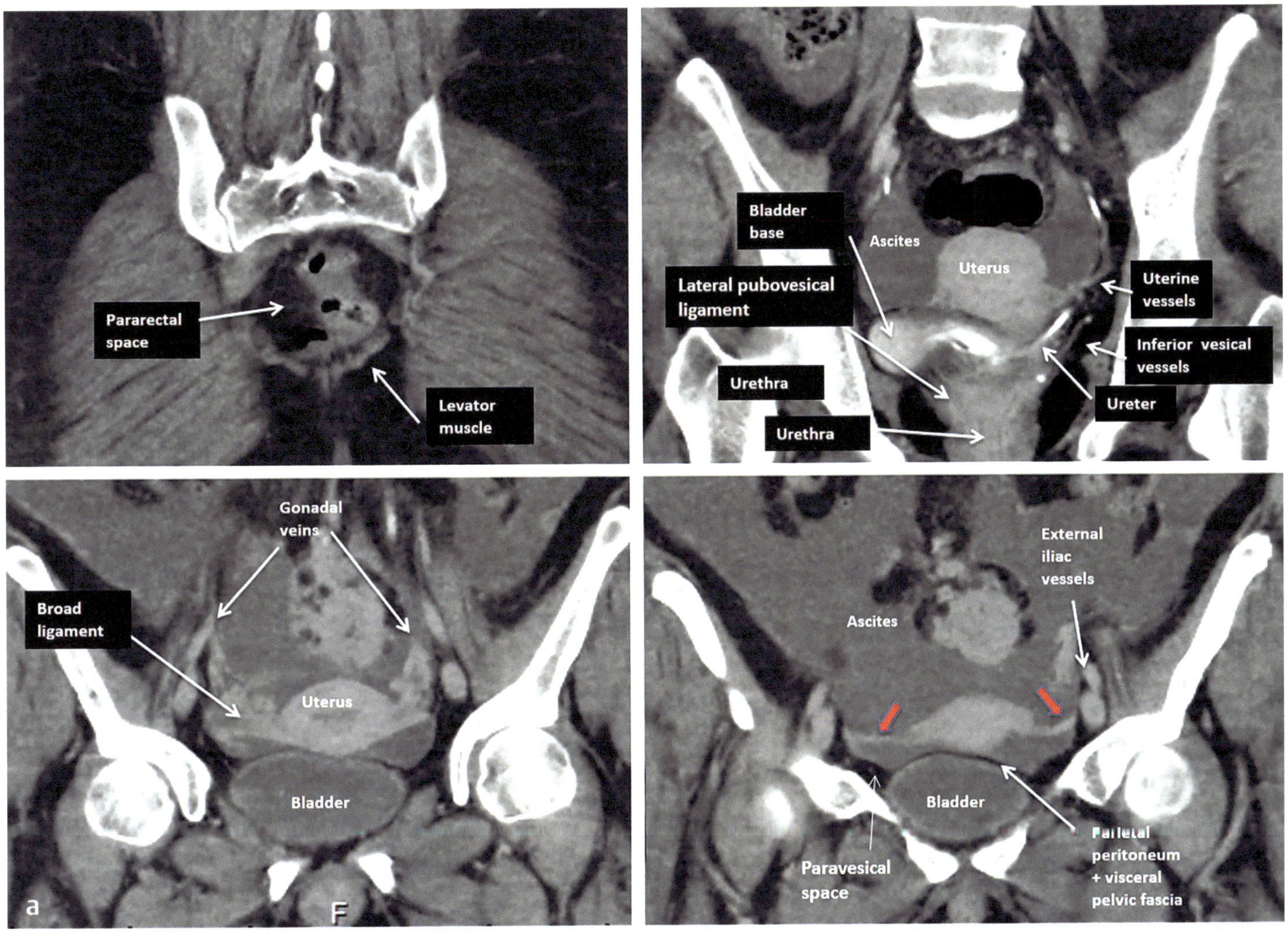

Fig. 18.12 (a, b) Coronal computed tomography (CT) images from posterior to anterior in a patient with large ascites secondary to carcinomatosis showing the subperitoneal endopelvic fascia. In this case, the endopelvic fascia which is filled with fat is clearly outlined by the peritoneal fluid and pelvic organs internally and pelvic bones and muscles externally. The left ureter is seen passing inside the cardinal ligament and crossing under the uterine artery. The round ligaments (*red arrows*) are seen passing over the external iliac vessels to reach the inguinal canal. Note the endopelvic fascia is continuous with the space of Retzius anteriorly.

(*Continued*)

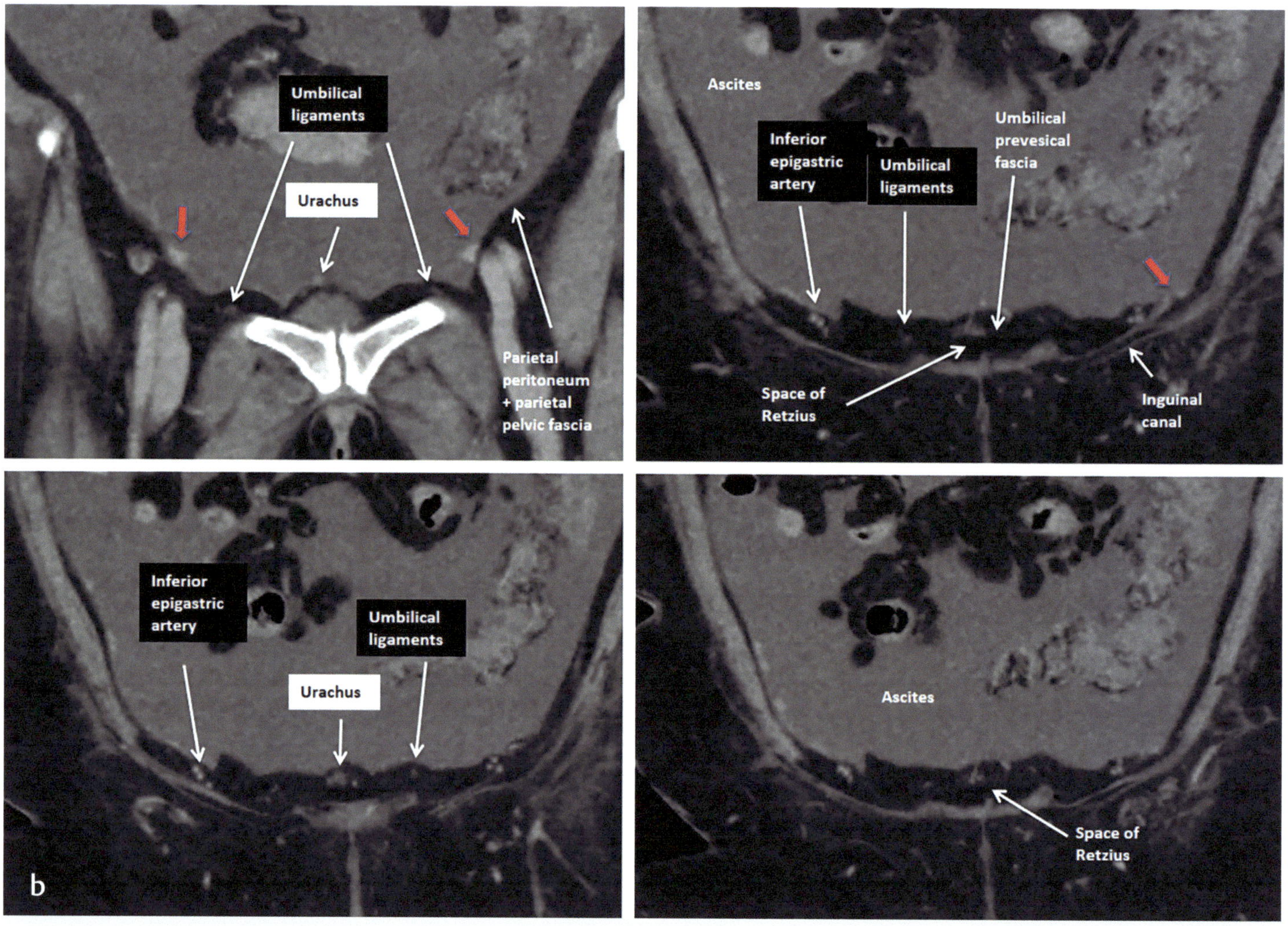

Fig. 18.12 *(Continued)* (b)

the medial umbilical ligaments and the umbilicus (see Chapter 14 "Extraperitoneal Space"). Inferiorly, the prevesical fascia fuses with the visceral fascia of the lower anterior surface of the bladder; thus, this fascia does not reach the pelvic floor (▶Fig. 18.6, ▶Fig. 18.13). The retropubic (or prevesical) space of Retzius is found between the transversalis fascia and the umbilical prevesical fascia (▶Fig. 18.11, ▶Fig. 18.12, ▶Fig. 18.13). This space also does not reach the pelvic floor. It is limited inferiorly by the medial puboprostatic ligaments and then, around the bladder, by the lateral puboprostatic and lateral ligaments of the bladder. The lateral ligaments of the bladder are fascial supports that extend backward on each side from the base of the bladder to the sacrogenital fold.

The rectum is surrounded by a variable mass of extraperitoneal tissue bordered by the rectogenital septum anteriorly and the presacral fascia posteriorly. The *presacral fascia* is an extraperitoneal tissue band located anterior to the sacral fascia (▶Fig. 18.6). A potential space, the *presacral space*, can be developed by separating these two fasciae containing the middle sacral artery and the presacral venous plexus. The presacral fascia extends between the tendinous arch of pelvic fascia on either side blending with the superior fascia of the levator ani muscle inferiorly (▶Fig. 18.10).

18

Parietal Pelvic Fascia and Its Thickenings: *The Inner Lining and Support of the Pelvic Walls and Pelvic Floor*

The parietal pelvic fascia bears several bilateral thickenings on the pelvic walls and pelvic floor. These fascial thickenings provide lines of attachment for the connective tissue that supports the pelvic organs (▶Fig. 18.10). Several such "tendinous arches" and "anchoring ligaments" are described in relation to the pelvic floor.

The *tendinous arch of the levator ani muscle* (▶Fig. 18.4, ▶Fig. 18.5, ▶Fig. 18.8, ▶Fig. 18.10) is a condensation of the obturator fascia that provides a line of origin of the levator ani muscle (the main muscle of the pelvic diaphragm). This arch is a line of fusion of the obturator fascia with the superior and inferior fascia of the pelvic diaphragm (i.e., the investing fascia of the superior and inferior surfaces of the levator ani muscle). The arch extends from the posterior surface of the pubis to the ischial spine and divides the obturator internus muscle into a superior pelvic portion and an inferior perineal portion (▶Fig. 18.8, ▶Fig. 18.10, ▶Fig. 18.16).

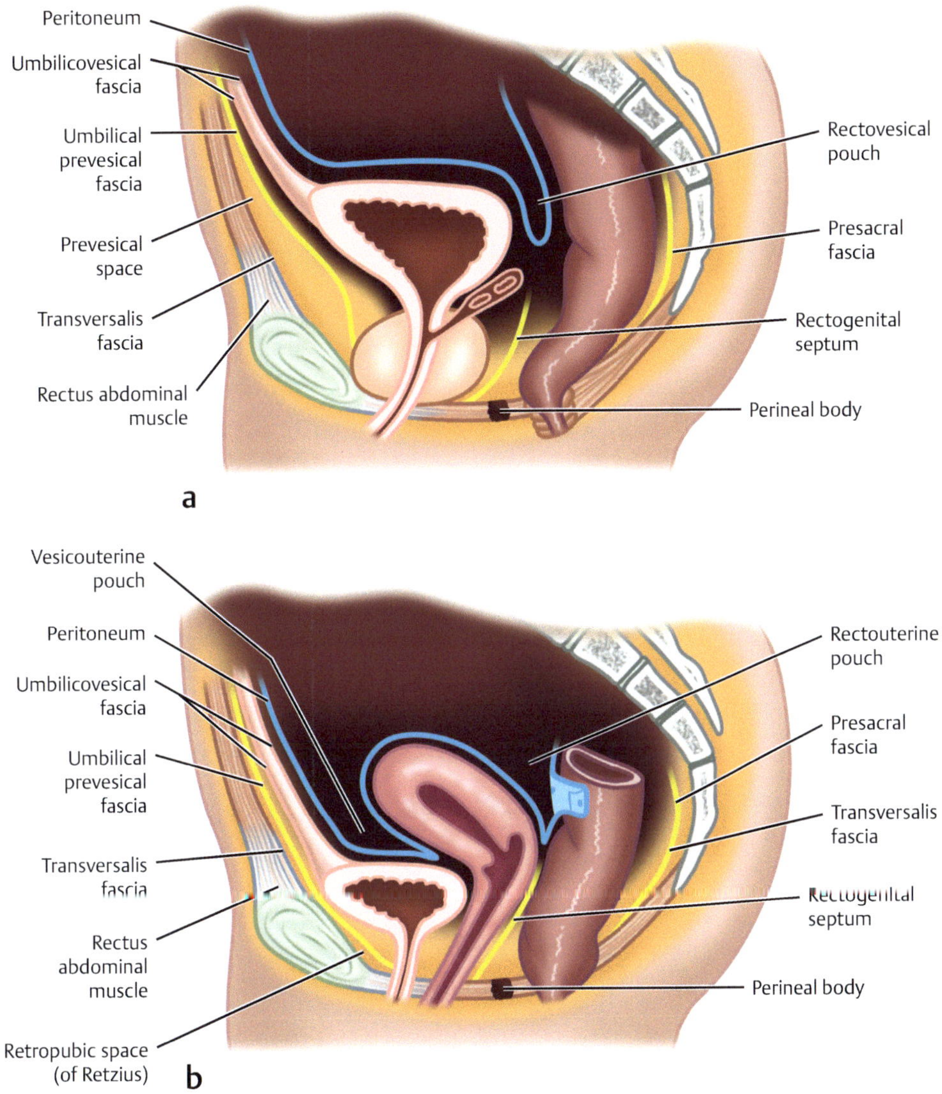

Fig. 18.13 (a) Sagittal section of the male pelvis. **(b)** Sagittal section of the female pelvis.

The *tendinous arch of the pelvic fascia* (▶Fig. 18.4, ▶Fig. 18.5, ▶Fig. 18.10, ▶Fig. 18.16) is a thickening of the superior fascia of the pelvic diaphragm. It extends from the posteroinferior aspect of the pubic symphysis to the tendinous arch of the levator ani muscle (the two arches fuse at their posterior thirds) (▶Fig. 18.16). The anterior part of the tendinous arch of the pelvic fascia blends with the medial and lateral puboprostatic or pubovesical ligaments. The lateral ligaments of the bladder and the transverse cervical ligaments fuse with parietal fascia at the tendinous arch of pelvic fascia (▶Fig. 18.7, ▶Fig. 18.8, ▶Fig. 18.16) (see below).

The *tendinous arch of rectogenital septum (Denonvilliers' fascia)* (▶Fig. 18.6, ▶Fig. 18.9, ▶Fig. 18.13) represents the lateral attachment of the rectogenital septum to the pelvic sidewall where it fuses with the parietal fascia covering the levator ani muscle. It extends from the perineal body toward the tendinous arch of pelvic fascia. The tendinous arch of rectogenital septum fuses with the posterior part of the tendinous arch of pelvic fascia near the ischial spine.

The medial and lateral puboprostatic (or pubovesical) ligaments are localized thickenings of the superior fascia of the pelvic diaphragm attaching to the tendinous arch of pelvic fascia. In the male, the ***medial puboprostatic ligaments,*** which are forward continuation of the tendinous arch of pelvic fascia, connect the pubis to the prostatic fascia (▶Fig. 18.7, ▶Fig. 18.16). Two more such ligaments, the *lateral puboprostatic ligaments,* connect the

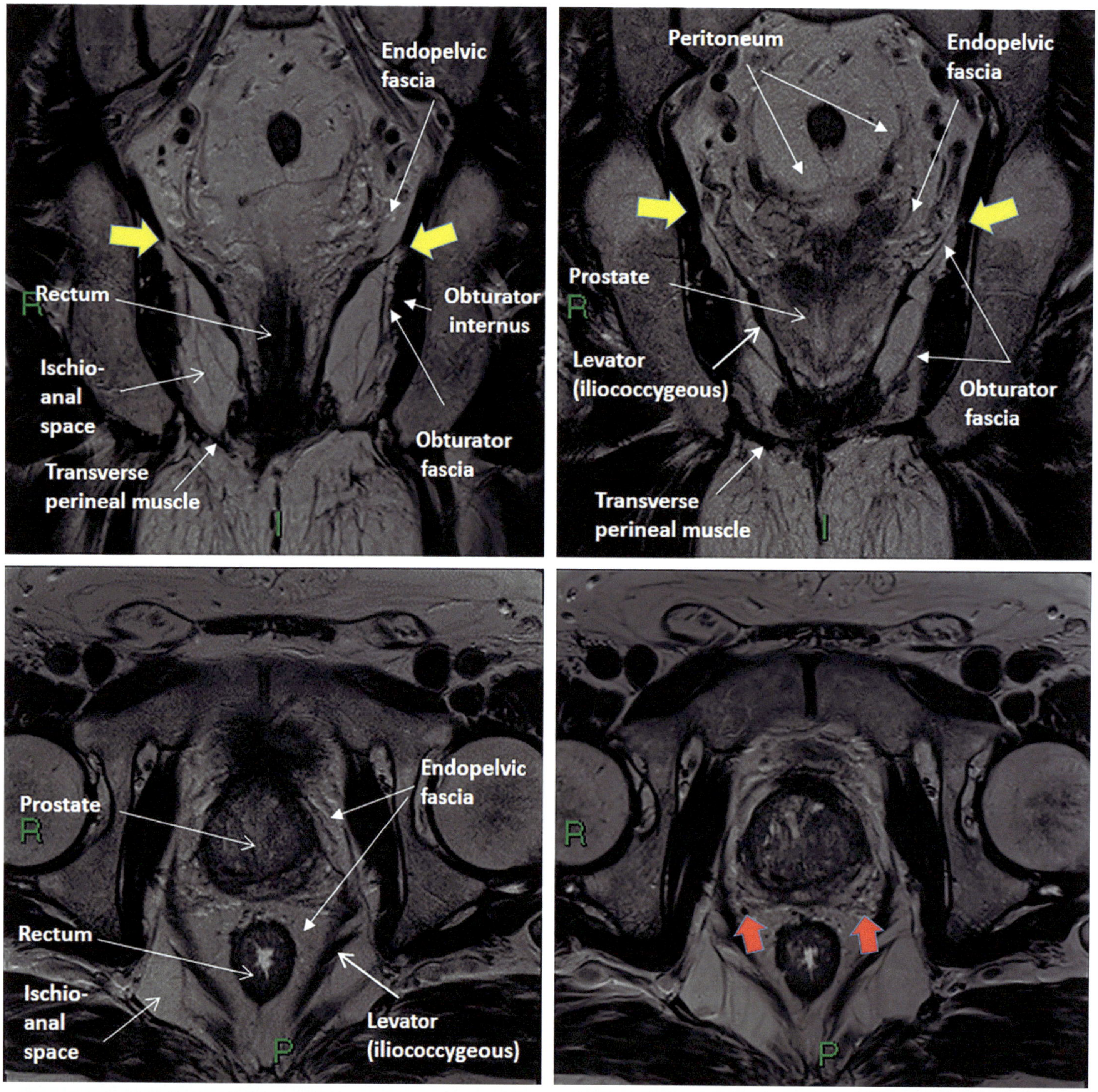

Fig. 18.14 Endopelvic fascia. Upper row. Coronal magnetic resonance imaging (MRI) showing the endopelvic fascia. The pelvic (or endopelvic) fascia is pelvic extraperitoneal connective tissue which is the space between parietal and visceral fascia and offers packing and support of pelvic organs. The tendinous arch of the levator ani muscle (*yellow arrows*) is a line of fusion of the obturator fascia with the superior and inferior fascia of the levator ani muscle, the main muscles of the pelvic diaphragm. Lower row. Axial MRI at the level of the prostate showing the Denonvilliers' fascia (*red arrows*). The rectogenital septum (Denonvilliers' fascia) is a single-layer fascia located between the rectum and prostate in men. The lateral puboprostatic ligaments connect the prostatic fascia more laterally on to the tendinous arch of pelvic fascia forming its anterior end.

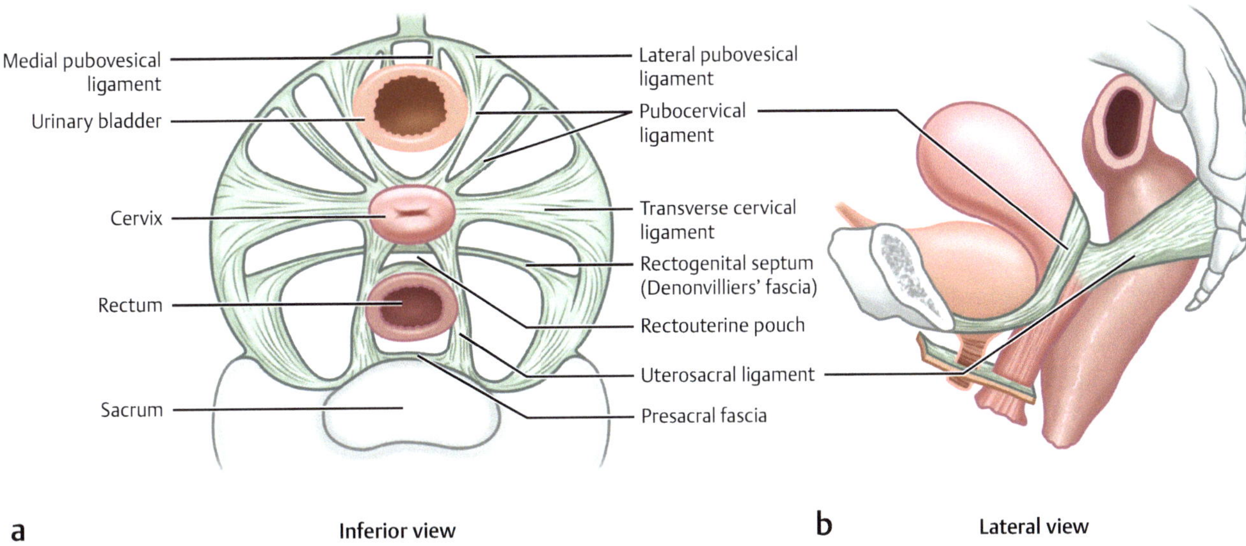

Fig. 18.15 Uterine ligaments from below **(a)** and side **(b)**. (Modified from Skandalakis et al.[13])

prostatic fascia more laterally on to the tendinous arch of pelvic fascia forming its anterior end (▶Fig. 18.16). The medial and lateral puboprostatic ligaments connect with the prostatic visceral fascia which is in continuation with the vesical fascia, and thus provide support for the bladder neck.

In the female, the bladder neck is anchored to pubic bone and tendinous arch of the pelvic fascia directly through the *medial* and *lateral pubovesical ligaments*, respectively (▶Fig. 18.4, ▶Fig. 18.6, ▶Fig. 18.15). Given that the urethra is fused with the lower two-thirds of the anterior vaginal wall, the vagina also receives support from these ligaments. The pubocervical ligament is another supporting ligament which passes forward from the vaginal vault and the uterine cervix, in a curvilinear motion around the urethra, and attaches to the tendinous arch of pelvic fascia (▶Fig. 18.6, ▶Fig. 18.15). It is described as a horizontal hummock constituting the main ventral support structure of the bladder base and neck. As already mentioned, the pubocervical ligament and the rectogenital septum originate in the tendinous arch of pelvic fascia and the tendinous arch of rectovaginal fascia, respectively, fusing with the superior fascia of the levator ani. They stretch the vagina transversely between the bladder and the rectum fusing with the anterior and posterior vaginal wall.

Pelvic Fascia "Supplier" Ligaments: *The Water Supply and the Electricity*

The major blood vessels (internal iliac) and nerves designated for pelvic organs descend extraperitoneally along the pelvic sidewalls embedded in the internal iliac vessel sheath (▶Fig. 18.5, ▶Fig. 18.7, ▶Fig. 18.8, ▶Fig. 18.10, ▶Fig. 18.16). This sheath is formed by the joining of the rectogenital, presacral, umbilical, prevesical, and parietal fasciae. The internal iliac artery provides the main vascular supply to the pelvic viscera, pelvic wall, and perineum. A complex neural network from the inferior hypogastric plexus is distributed to the pelvic viscera accompanying the visceral branches of the internal iliac artery, travelling on the wall of the arteries.

As introduced above, in the pelvis, the extraperitoneal pelvic fascia forms **"supplier ligaments,"** which provide way to neurovascular bundles while supporting the pelvic organs (▶Fig. 18.5, ▶Fig. 18.8, ▶Fig. 18.9. ▶Fig. 18.10, ▶Fig. 18.16, ▶Fig. 18.17). These ligaments originate from the sheaths of the internal iliac vessels, follow the branches of the vessels that turn medially toward the pelvic organs, and fuse with the visceral fascia of the respective organ. In their entire course from the pelvic wall to the organs, the ligaments are fastened to the superior fascia of the pelvic diaphragm at different sites (see below, separate for each ligament).

Anteriorly, the *lateral ligaments of the bladder* pass medially and anteriorly toward the base of the bladder (▶Fig. 18.8, ▶Fig. 18.16). Laterally, they blend with the superior fascia of the pelvic diaphragm at the tendinous arch of pelvic fascia forming its anterior portion. These ligaments carry the superior vesical vessels, the vesical plexus, the ureter, and part of the vas (in the male).

Posteriorly, the *lateral ligaments of the rectum* connect the visceral fascia of the dorsolateral lower portion of the rectum with the sheath of internal iliac vessels at the level of the ischial spine. Inferiorly, on either side, these ligaments attach with the pelvic floor on the superior fascia of the levator ani muscles adjacent to the rectum (▶Fig. 18.7, ▶Fig. 18.15, ▶Fig. 18.18). They carry the middle rectal vessels and autonomic nerves to the rectum (see Chapter 5 "Colon").

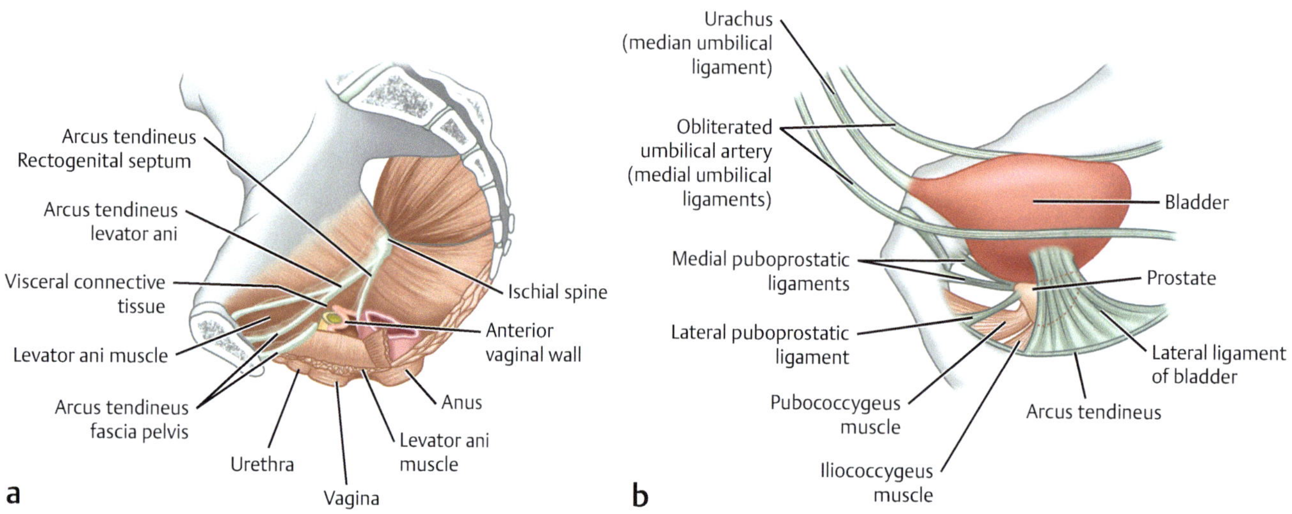

Fig. 18.16 (a) Lateral view of the pelvic floor and sidewall. Pelvic organs were transected at the level of the proximal urethra. **(b)** Diagram of the bladder and some of its ligaments.

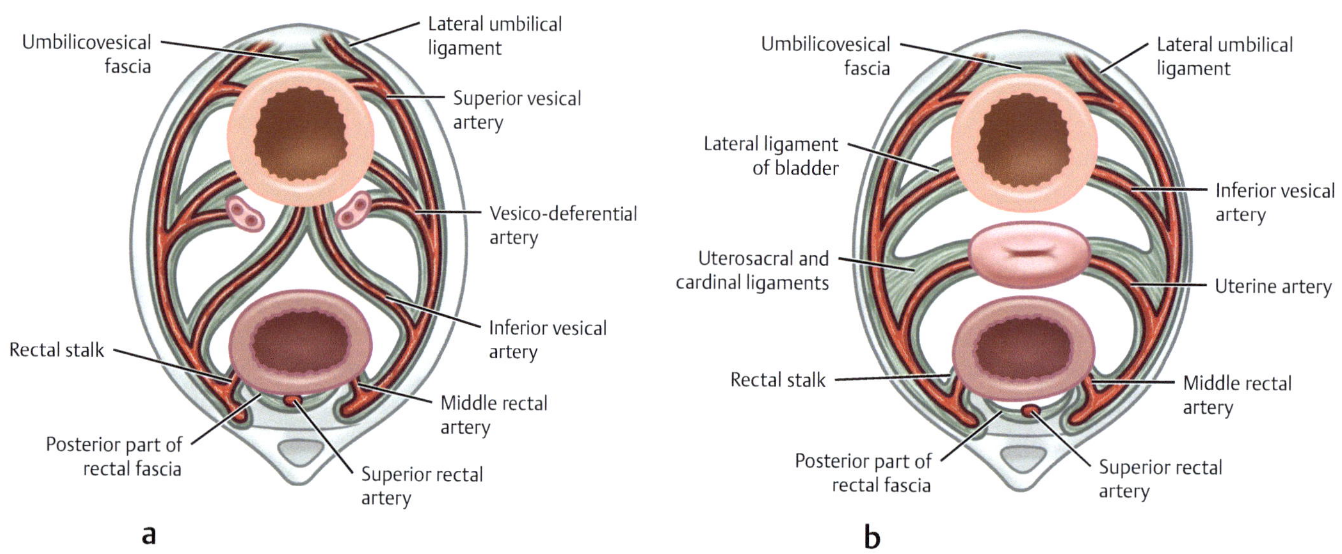

Fig. 18.17 Horizontal section of the pelvic viscera showing the pelvic ligaments in the male **(a)** and the female pelvis **(b)**. The major blood vessels (internal iliac) and nerves designated for pelvic organs descend extraperitoneally along the pelvic sidewalls embedded in the internal iliac vessel sheath. This sheath is formed by the joining of the rectogenital, presacral, umbilical, prevesical, and parietal pelvic fasciae.

In the female, the uterine vessels run in the *transverse cervical ligaments* (or cardinal ligaments or ligaments of Mackenrodt) (see Chapter 16 "Female Genital System"). They arise from the cervix and the upper part of the vagina and spread laterally, merging with the investing fascia of the lateral pelvic wall and floor at the tendinous arch of pelvic fascia; finally, they join the internal iliac sheath (▸Fig. 18.4, ▸Fig. 18.5, ▸Fig. 18.15). The transverse cervical ligaments are located in the base of the broad ligament and are responsible for the widening of the broad ligament at its base. The distal part of the ureter traverses the cardinal ligaments passing beneath the uterine artery.

The transverse cervical ligaments continue posteriorly with the *uterosacral ligaments* (▸Fig. 18.4, ▸Fig. 18.10, ▸Fig. 18.15). These ligaments originate at the posterolateral aspect of the cervix, at the level of the internal cervical os and the vaginal fornices, and extend to the sacrum posteriorly, where they blend with the parietal fascia. The uterosacral and the transverse cervical ligaments, by connecting the uterus and vagina to the pelvic floor, act together with the tendinous arch of the pelvic fascia and the muscles of the pelvic floor in order to provide support and prevent pelvic organ prolapse and dysfunction.

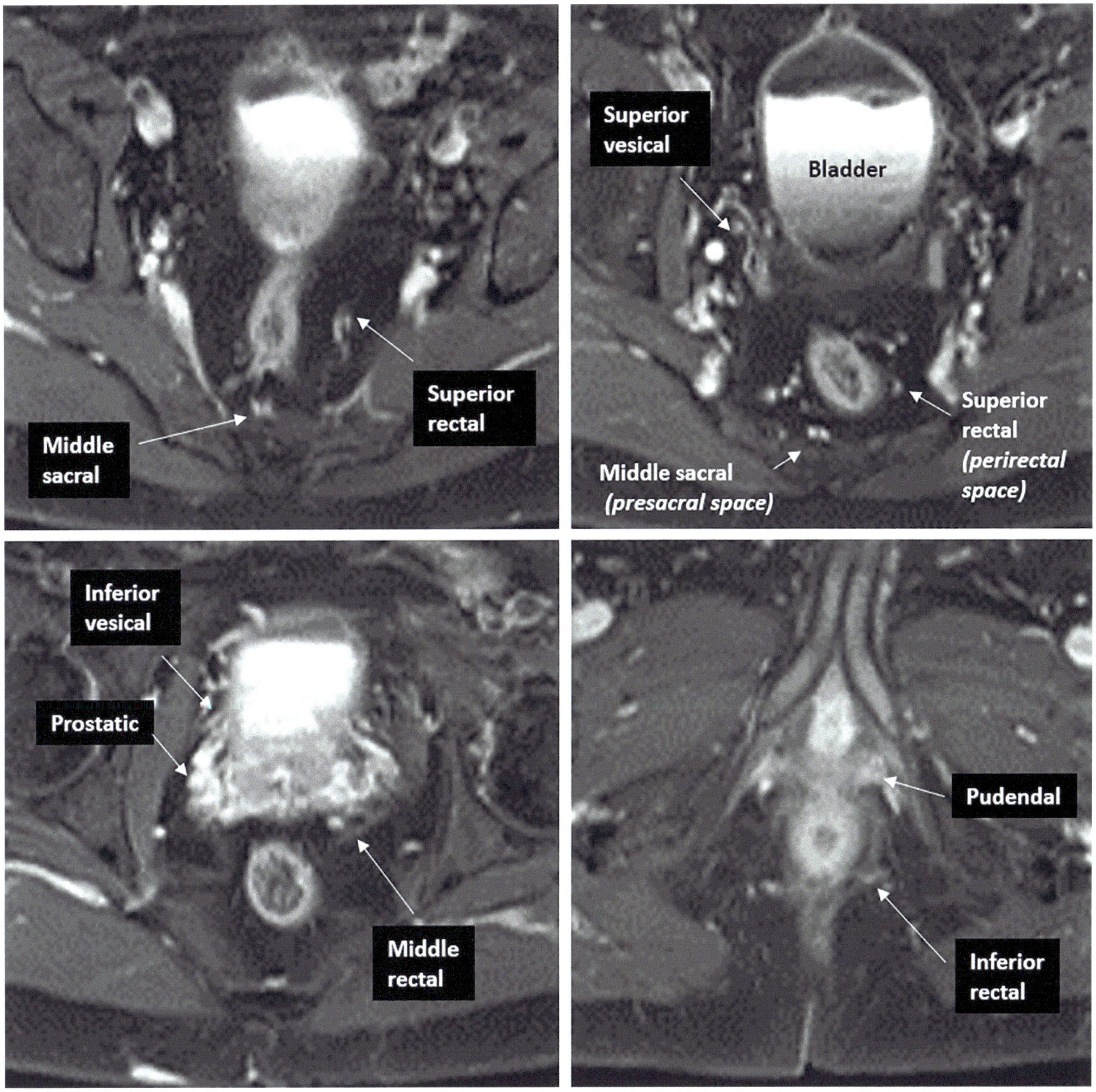

Fig. 18.18. Axial T2-weighted (T2W) magnetic resonance imaging (MRI) showing the extraperitoneal pelvic ligament vascular suppliers containing prostatic, and inferior vesical vessels and rectal neurovascular bundles.

◆ The Pelvic Floor below the Pelvic Diaphragm

As already mentioned, the perineal membrane and the muscles of deep perineal space (deep transverse perineal muscle, sphincter urethrae, compressor urethrae, sphincter urethrovaginalis, and transverse vaginal muscle) are the most caudal structures of the pelvic floor. They give support to the pelvic viscera with their multiple attachment to the surrounding structures and to the bony pelvis. Their detailed description is given in the perineum chapter.

◆ Pelvic Floor Dysfunction

Pelvic floor failure is a common disorder that affects approximately one out of four women in the United States.[14–16] Pelvic floor dysfunction may manifest with urinary stress and urge

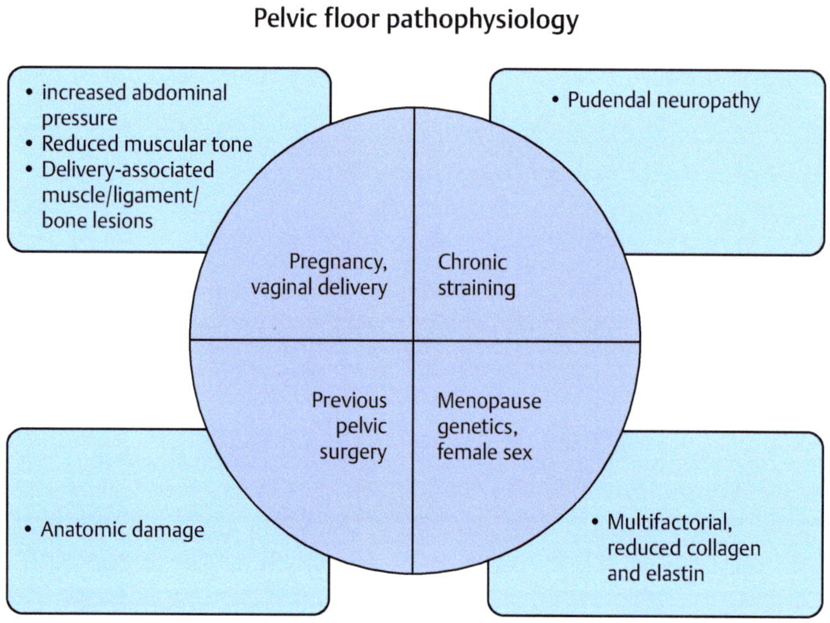

Fig. 18.19 Pathophysiology of pelvic floor failure. (Modified from Bitti et al.[15])

incontinence, vaginal problems, fecal incontinence, and rectal/vaginal/bladder prolapse. The mechanism involved in pelvic floor dysfunction is partially understood and in many cases is multifactorial (▶Fig. 18.19).

Female sex, menopause, and advanced age are the major risk factors. Obesity, pregnancy, parity, and chronic constipation cause increased abdominal pressure and reduced muscle tone. Vaginal delivery, hysterectomy, and perineal surgery can be associated with various muscular and ligamentous injury, disruption of the pubovesical and rectovaginal fasciae, and denervation of puborectalis muscle.

Evaluation of women with pelvic floor failure requires a comprehensive approach that includes clinical assessment, physiologic testing, and imaging. Knowledge of pelvic floor anatomy is essential for diagnosis of pelvic floor injury and dysfunction. MRI and dynamic MR with interactive T2-weighted, single-shot, fast spin echo sequences of the pelvic floor at rest and during straining and defecation enables real-time evaluation of patterns of dysfunction to diagnose various abnormalities, including descending perineum syndrome, rectal prolapse, cystocele, and vaginal prolapse[17,18] (▶Fig. 18.20). Prolapse can be graded in each compartment of the pelvis using MR measurements criteria (▶Fig. 18.21).

The puborectalis muscle and pudendal nerve can be damaged during vaginal delivery causing thinning, bowing, and atrophy of the muscle. Weakness and sagging of the iliococcygeus muscle and anococcygeal ligament open the urogenital hiatus and predispose to pelvic organ prolapse (▶Fig. 18.4, ▶Fig. 18.5). The attachment of the tendinous arch of the pelvic fascia to the ischial spine is commonly avulsed after vaginal deliveries. The external sphincter of the anus can also be damaged during vaginal delivery.

The endopelvic fascial support is important to maintain suspension of the urogenital organs if levator muscle support fails. Three levels of endopelvic fascial support have been described for female genital organs:

- ***Level I:*** Suspends the upper third of the vagina and cervix through the cardinal and uterosacral ligaments. Disruption of the uterosacral ligaments after total hysterectomy may cause bilateral sagging of the vagina.
- ***Level II:*** Suspends the posterior wall of bladder and the middle third of the vagina by lateral pubovesical and pubocervical ligaments and Denonvilliers' fascia. Lateral disruption endopelvic fascia at this level causes posterior bladder wall descent known as the "saddlebag" sign (▶Fig. 18.4). Other image findings include loss of H-shaped, cross-sectional configuration of the vagina and vaginal prolapse (▶Fig. 18.22). Damage of the Denonvilliers' fascia may produce a gap between the rectum and vagina where peritoneal bowel or fat can herniate, causing a peritoneocele or an enterocele (▶Fig. 18.6).
- ***Level III:*** Suspends the urethra and lower third of the vagina medial and lateral pubovesical ligaments. The urethra is fused with the lower two-thirds of the anterior vaginal wall[12] (▶Fig. 18.23). Therefore, the urethral suspension ligaments are also part of the level III endopelvic fascia support. Disruption of the urethral suspension ligaments may result in an increased space between the pubic bone and urethra, a finding called the "drooping mustache" sign.

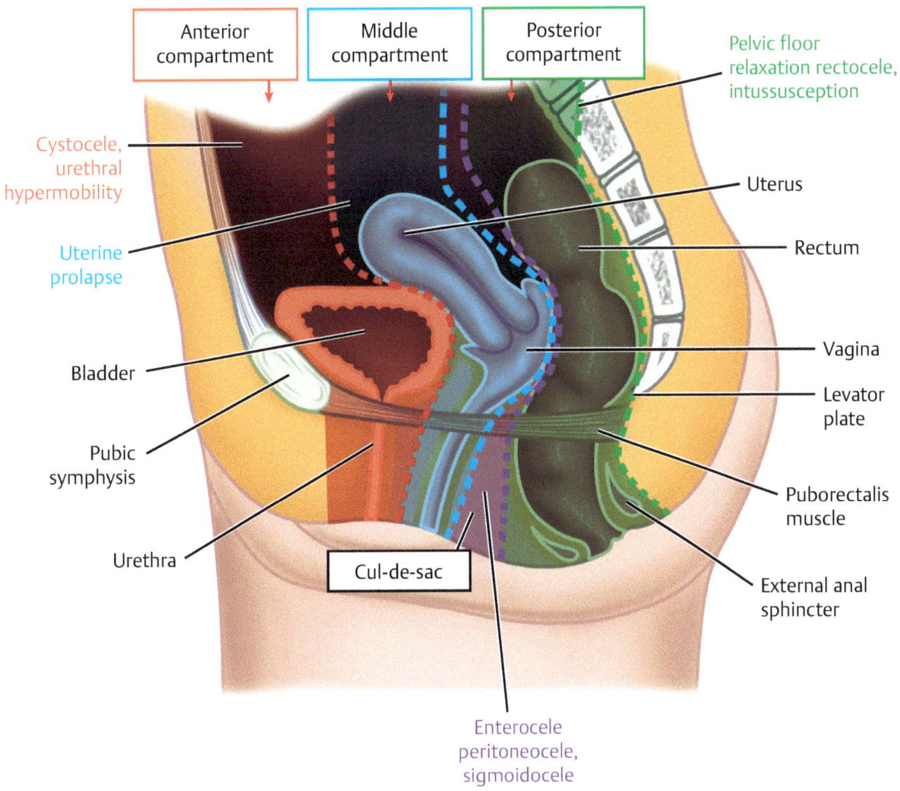

Fig. 18.20 Functional anatomic compartments of the female pelvis and pathologic conditions. The anterior compartment, containing the bladder (B) and urethra (U); the middle compartment, containing the uterus (Ut), cervix, and vagina (V); and the posterior compartment, containing the anus, anal canal, rectum (R), and sigmoid colon. A fourth "virtual" compartment called the cul-de-sac is also shown. Note that the puborectalis muscle surrounds the bladder neck, vagina, and rectum. PS, pubic symphysis. (Modified from García del Salto et al.[14])

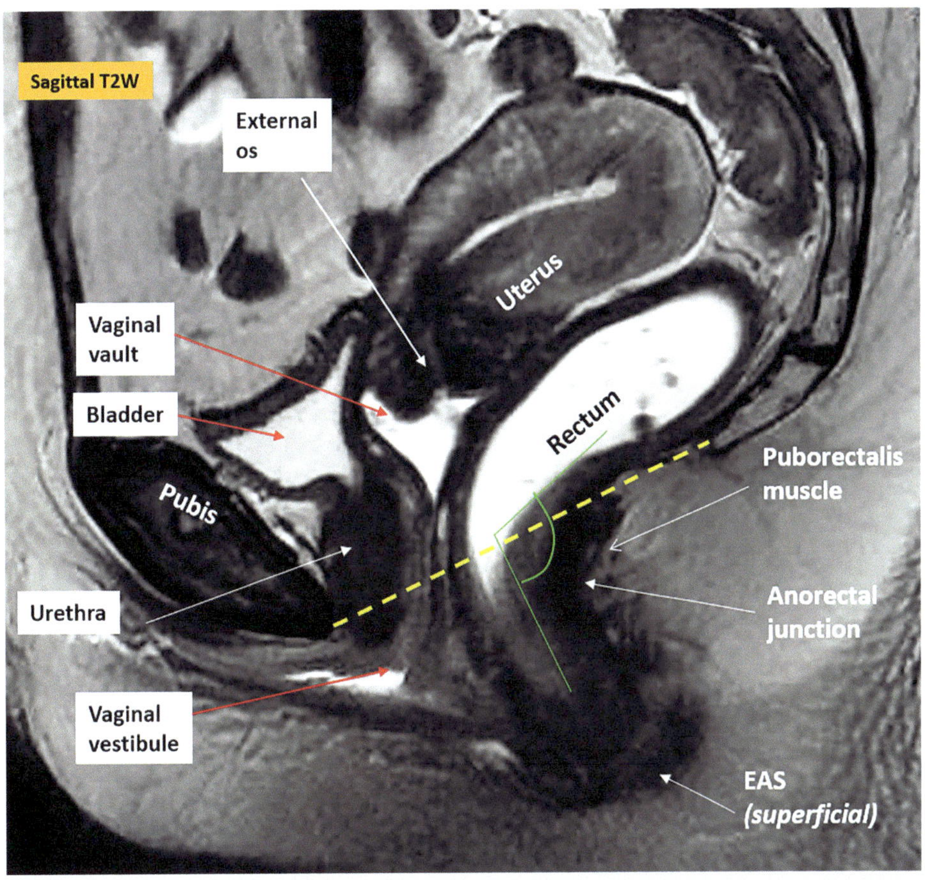

Fig. 18.21 Normal pelvic floor measurements at sagittal magnetic resonance imaging (MRI) used for grading of prolapse. The pubococcygeal line (PCL) joins the inferior border of the symphysis pubis to the last coccygeal joint in the midsagittal plane (*yellow line*) and is the reference line for most measurements. The PCL is used as the reference line to determine the extent of a cystocele, anorectal junction descent, and vaginal vault descent. Prolapse is graded as mild (<3 cm below the PCL), moderate, or severe (>6 cm). The anorectal angle is normally 90°–110° and opens during defecation. EAS, external anal sphincter.

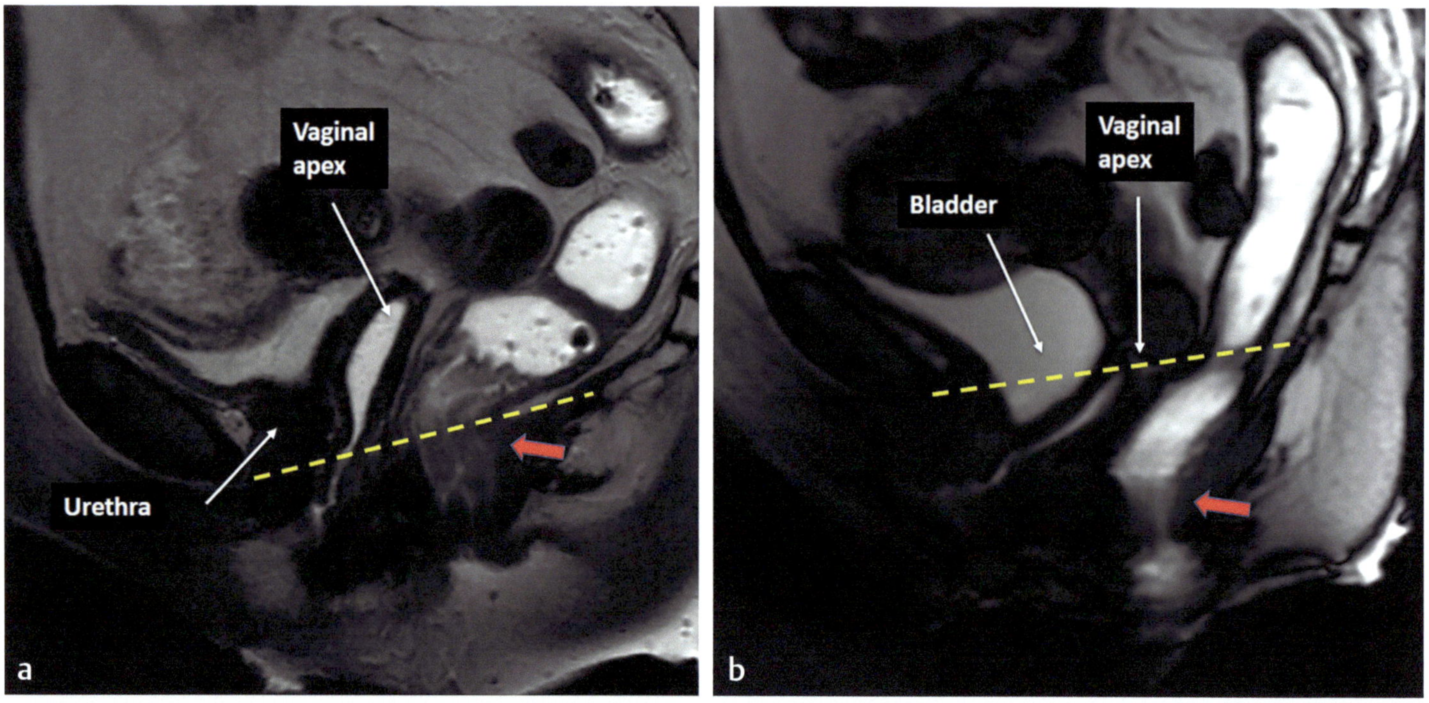

Fig. 18.22 Pelvic descent shown by dynamic magnetic resonance imaging (MRI) before **(a)** and during **(b)** evacuation of rectal contrast, status post total hysterectomy. On active evacuation, there is descent of pelvic organs below the pubococcygeal line. There is urethral hypermobility and the base of the bladder descends mildly below the pubococcygeal line. The apex of the vagina descends to the pubococcygeal line. The anorectal junction angle (*red arrows*) widens and descends approximately 6 cm below the pubococcygeal line.

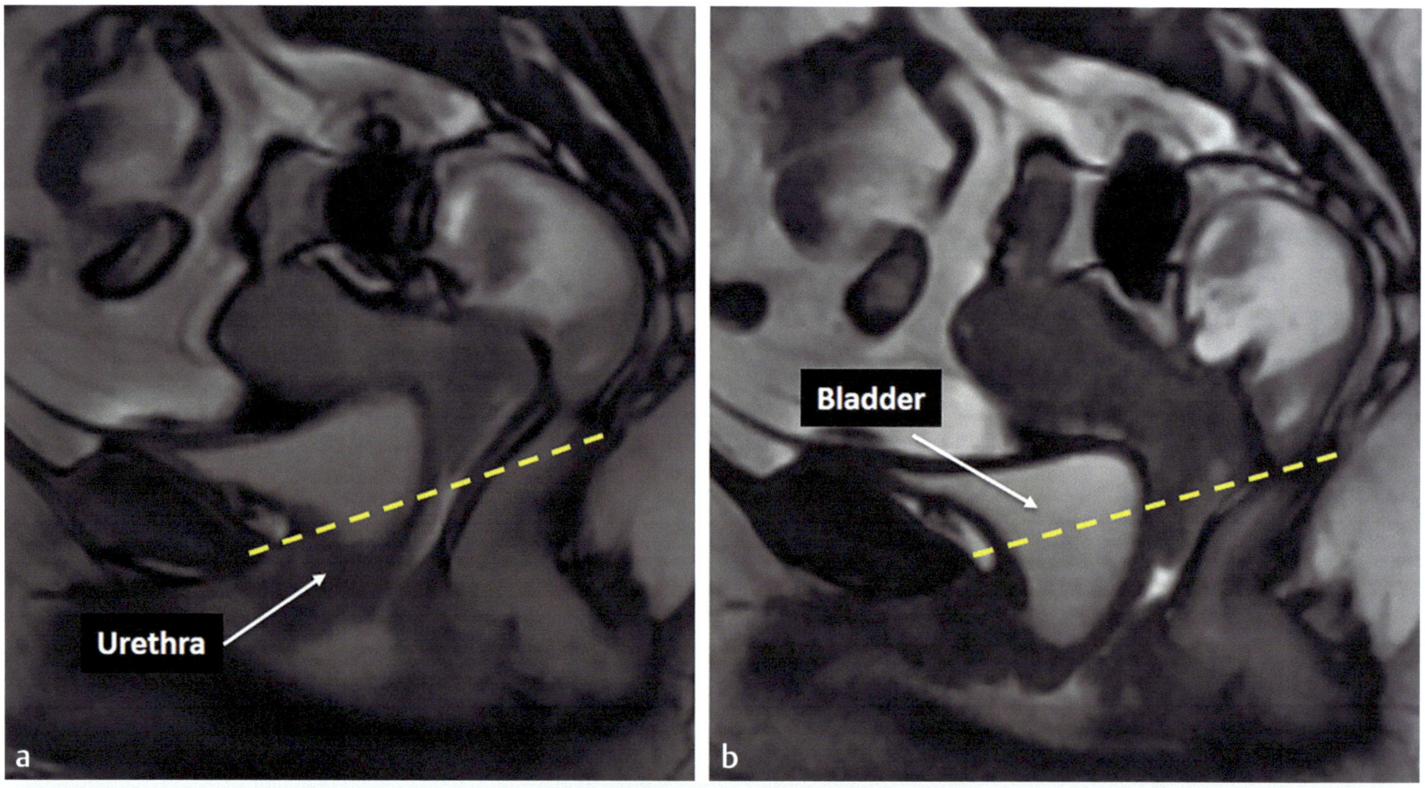

Fig. 18.23 Sagittal dynamic magnetic resonance (MR) T2-weighted sequences. The levator ani complex was intact, without evidence of muscle disruption or hernia. Moderate cystocele and urethrocele shown by sagittal dynamic magnetic resonance (MR) T2-weighted sequences at relaxation **(a)** and at maximal strain **(b)**. The levator ani complex was intact, without evidence of muscle disruption or hernia.

References

1. Ashton-Miller JA, DeLancey JOL. Functional anatomy of the female pelvic floor. Ann N Y Acad Sci 2007;1101:266–296
2. O'Rahilly R. Gardner-Gray-O'Rahilly Anatomy. A Regional Study of the Human Structure. 5th ed. Philadelphia: W.B. Saunders Co.; 1986
3. Wei JT, De Lancey JOL. Functional anatomy of the pelvic floor and lower urinary tract. Clin Obstet Gynecol 2004;47(1):3–17
4. Barber MD, Bremer RE, Thor KB, Dolber PC, Kuehl TJ, Coates KW. Innervation of the female levator ani muscles. Am J Obstet Gynecol 2002;187(1):64–71
5. Corton MM. Anatomy of pelvic floor dysfunction. Obstet Gynecol Clin North Am 2009;36(3):401–419
6. Barber MD. Contemporary views on female pelvic anatomy. Cleve Clin J Med 2005;72(Suppl 4):S3–S11
7. Hartmann D, Sarton J. Chronic pelvic floor dysfunction. Best Pract Res Clin Obstet Gynaecol 2014;28(7):977–990
8. Lawson JO. Pelvic anatomy. I. Pelvic floor muscles. Ann R Coll Surg Engl 1974;54(5):244–252
9. Kearney R, Sawhney R, DeLancey JO. Levator ani muscle anatomy evaluated by origin-insertion pairs. Obstet Gynecol 2004;104(1):168–173
10. Kim J, Betschart C, Ramanah R, Ashton-Miller JA, DeLancey JO. Anatomy of the pubovisceral muscle origin: macroscopic and microscopic findings within the injury zone. Neurourol Urodyn 2015;34(8):774–780
11. Margulies RU, Hsu Y, Kearney R, Stein T, Umek WH, DeLancey JO. Appearance of the levator ani muscle subdivisions in magnetic resonance images. Obstet Gynecol 2006;107(5):1064–1069
12. DeLancey JO. Anatomic aspects of vaginal eversion after hysterectomy. Am J Obstet Gynecol 1992;166(6 Pt 1):1717–1724, discussion 1724–1728
13. Skandalakis JE, Colborn GL, Weidman TA, et al, eds. Skandalakis' Surgical Anatomy: The Embryologic and Anatomic Basis of Modern Surgery. Athens: Paschalidis Medical Publications; 2004. Fig. 26–36
14. García del Salto L, de Miguel Criado J, Aguilera del Hoyo LF, et al. MR imaging-based assessment of the female pelvic floor. Radiographics 2014;34(5):1417–1439
15. Bitti GT, Argiolas GM, Ballicu N, et al. Pelvic floor failure: MR imaging evaluation of anatomic and functional abnormalities. Radiographics 2014;34(2):429–448
16. Nygaard I, Barber MD, Burgio KL, et al; Pelvic Floor Disorders Network. Prevalence of symptomatic pelvic floor disorders in US women. JAMA 2008;300(11):1311–1316
17. Miller JM, Brandon C, Jacobson JA, et al. MRI findings in patients considered high risk for pelvic floor injury studied serially after vaginal childbirth. AJR Am J Roentgenol 2010;195(3):786–791
18. Maglinte DD, Bartram CI, Hale DA, et al. Functional imaging of the pelvic floor. Radiology 2011;258(1):23–39

19 Abdominal Aorta and Major Branches

Norio Hongo and Hiro Kiyosue

◆ Introduction

In this chapter, the anatomy of the abdominal aorta and its major branches will be reviewed.

◆ Abdominal Aorta

Abdominal aorta extends from the aortic hiatus in the diaphragm at the level of Th11 and Th12 vertebra to the L4 level where it bifurcates into left and right common lilac arteries. From top to bottom, the celiac plexus, left renal vein, third part of the duodenum, and the root of the small bowel mesentery and sigmoid mesentery are located anterior to the abdominal aorta.[1] The mean diameter of the abdominal aorta gradually decreases from 26 mm at the hiatus to 19 mm at its bifurcation.[2] Unpaired visceral branches of the aorta including the celiac trunk, superior mesenteric artery (SMA), and inferior mesenteric artery (IMA) originate from the anterior wall of the aorta and supply the digestive tracts (▶ Fig. 19.1). Paired visceral branches of the aorta including the renal, testicular, and ovarian arteries are small branches. Parietal (segmental) arteries, including the inferior phrenic and lumbar arteries, supply the diaphragm and body wall. Direct needle angiogram of the aorta was first introduced by Dos Santos in 1929.[3] Later, this technique was replaced by catheter angiogram first introduced by Seldinger.[4] Today, most angiograms of the aorta are performed by computed tomography (CT) or magnetic resonance imaging (MRI).

Celiac Artery

The celiac artery (trunk, axis) is the first branch of the abdominal aorta at the Th12 or L1 level. It measures 1 to 2 cm in length and soon trifurcates into the common hepatic artery, the splenic artery, and the left gastric artery. The left gastric artery is usually the first branch of the celiac artery. One or both inferior phrenic arteries often originate from the ostium of the celiac

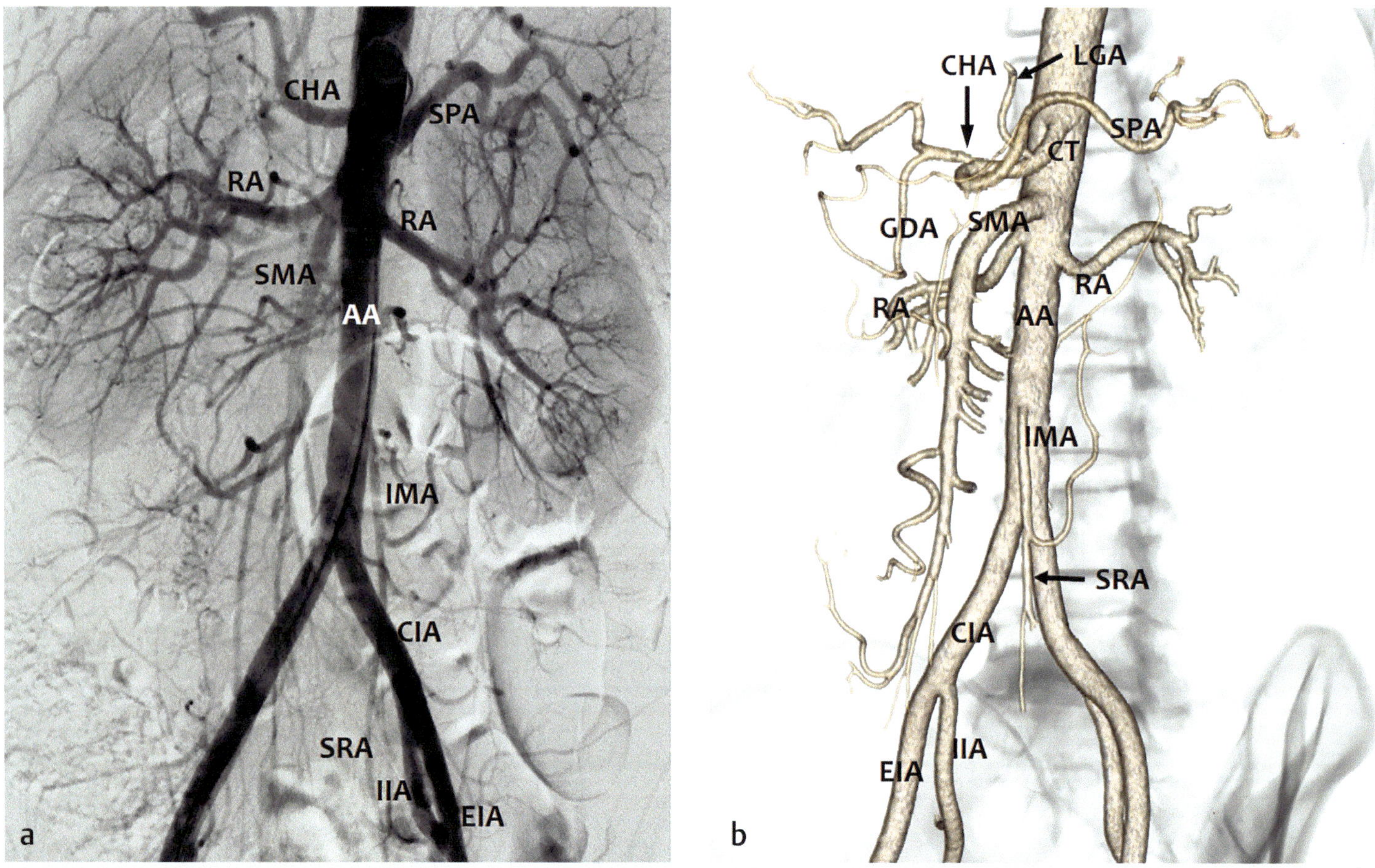

Fig. 19.1 Abdominal aorta and its major branches. **(a)** Digital subtraction angiography (DSA) and **(b)** three-dimensional computed tomography (CT) angiography of the aorta and its branches with volume rendering reconstruction. AA, abdominal aorta; CHA, common hepatic artery; CIA, common iliac artery; CT, celiac trunk; EIA, external iliac artery; GDA, gastroduodenal artery; IIA, Internal iliac artery; IMA, inferior mesenteric artery; LGA, left gastric artery; RA, renal artery; SMA, superior mesentric artery; SPA, splenic artery; SRA, superior rectal artery.

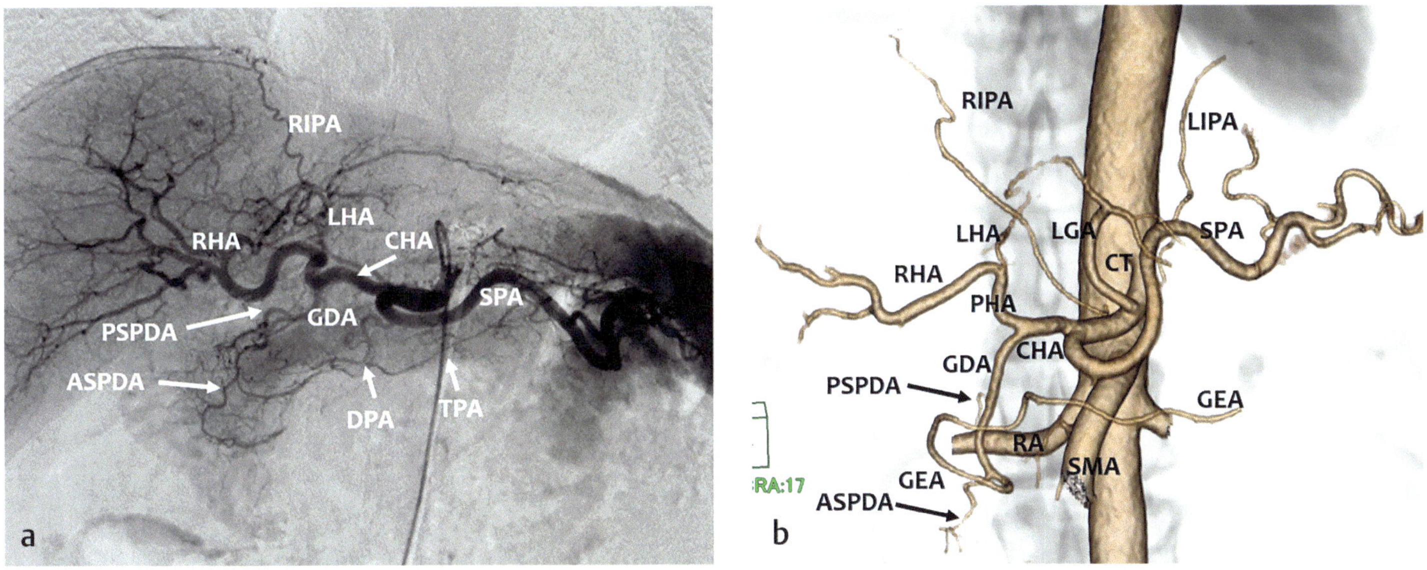

Fig. 19.2 Celiac artery and its major branches. **(a)** Digital subtraction angiography (DSA) and **(b)** three-dimensional computed tomography (CT) angiography of the aorta and its branches with volume rendering reconstruction. ASPDA, anterior superior pancreatoduodenal artery; CHA, common hepatic artery; CT, celiac trunk; DPA, doral pancreatic artery; GDA, gastroduodenal artery; GEA, gastroepiploic artery; LGA, left gastric artery; LHA, left hepatic artery; LIPA, left inferior phrenic artery; PHA, proper hepatic artery; PSPDA, posterior superior pancreatoduodenal artery; RA, renal artery; RHA, right hepatic artery; RIPA, right inferior phrenic artery; SMA, superior mesenteric artery; SPA, splenic artery; TPA, transverse pancreatic artery.

axis[1] (▶Fig. 19.2). The celiac axis is surrounded by celiac plexus, ganglion, and lymph nodes.[5] It courses under the median arcuate ligament which often compress and narrow the celiac axis (▶Fig. 19.3). There are many variations in the branching pattern of the celiac axis. Understanding these variations is essential to conduct and interpret abdominal angiography.

Splenic Artery

The splenic artery runs transversely along superior border of the pancreas and branches into the dorsal pancreatic, greater pancreatic, and pancreatic tail arteries to supply the pancreatic body and tail. The dorsal pancreatic artery gives off anastomotic branches to the pancreatic head arteries and transverse pancreatic artery (▶Fig. 19.2a). The distal splenic artery typically bifurcates into superior and inferior terminal branch at splenic hilum. The superior terminal branches are usually longer than the inferior terminal branches and supply major part of the spleen. Typically, superior polar artery originates from the distal part of splenic artery but it may originate from the superior terminal artery.[6] The short gastric and left gastroepiploic arteries often branch from splenic artery at the level of splenic hilum. An inferior polar artery often gives off the left gastroepiploic artery. The left gastroepiploic artery may also arise from the distal splenic or inferior terminal artery and run along the greater curvature of the stomach. Numerous short gastric arteries arise from the terminal splenic or left gastroepiploic artery to supply the gastric cardia and fundus.

Left Gastric Artery

The left gastric artery originates from superior aspect of the celiac artery and runs cranially into the lesser omentum to curve along the lesser curvature of the stomach (▶Fig. 19.2). Small esophageal branch supplying lower part of esophagus usually branches from the left gastric artery. The distal part of this artery anastomoses with the right gastric artery. It sometimes shares the trunk with the left hepatic artery or the inferior phrenic artery.

Common Hepatic Artery

Common hepatic artery runs to the right toward the porta hepatis and bifurcates into the gastroduodenal artery and the proper hepatic artery. The latter further divides into the right and left hepatic arteries (▶Fig. 19.2). The gastroduodenal artery gives rise to the posterior anterior and superior pancreaticoduodenal arteries before terminating to the right gastroepiploic artery which runs along the greater curvature of the stomach to anastomose with the left gastroepiploic artery. Most often the right gastric artery originates from the common hepatic or proper hepatic artery and runs along the lesser curvature of the stomach to anastomose with the left gastric artery.[5]

Superior Mesenteric Artery

The superior mesenteric artery originates from the anterior aspect the abdominal aorta caudal to the celiac axis at the level of L1 vertebral body (▶Fig. 19.4). It runs anteroinferiorly behind the pancreas body before crossing anterior to the third part of the duodenum and the left renal vein. The superior mesenteric artery runs within the small bowel mesentery. It branches into the inferior pancreaticoduodenal artery, middle colic artery, jejunal arteries, ileal arteries, and finally terminates as the ileocolic artery. The inferior pancreaticoduodenal artery is usually the first branch of the superior mesenteric artery arising as an isolated

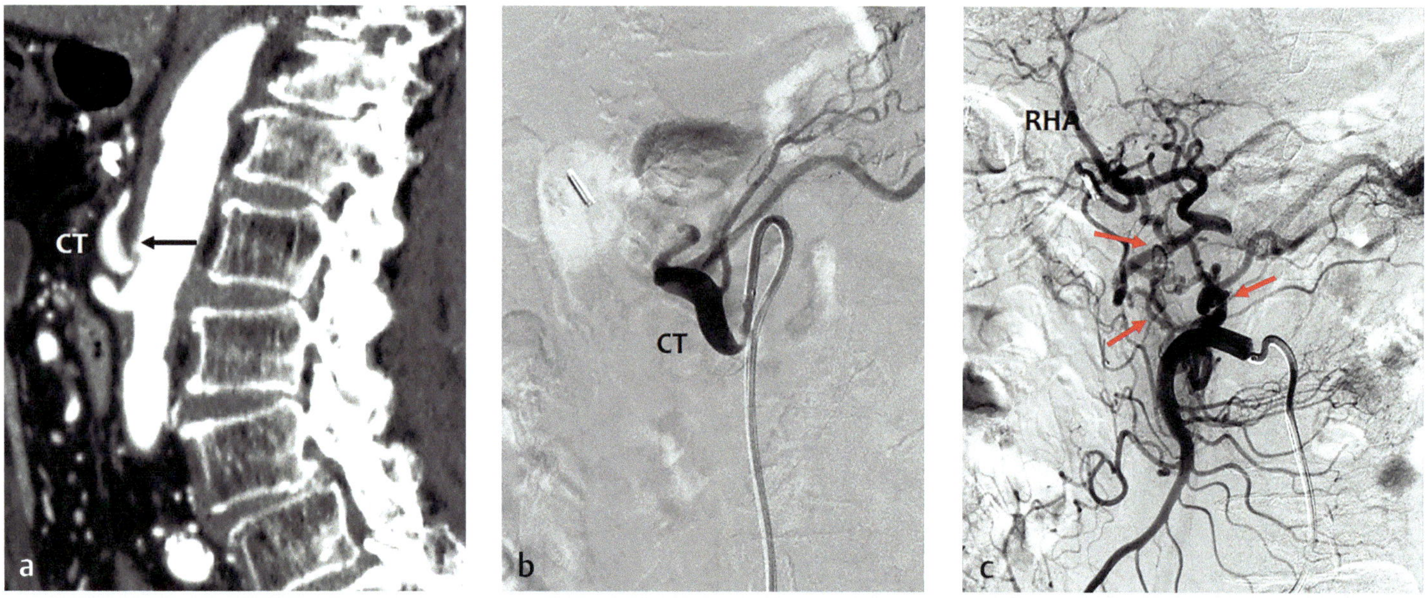

Fig. 19.3 Sagittal reconstruction of contrast-enhanced computed tomography (CT) **(a)**, digital subtraction angiography (DSA) of the celiac trunk **(b)**, and superior mesenteric artery (SMA) **(c)** of a patient with stenosis of celiac trunk (CT) due to compression by the median arcuate ligament of the diaphragm (*arrow* in **a**). DSA shows that the right hepatic artery is visualized from the superior mesenteric artery via tortuous collateral of the pancreatoduodenal arcade (*red arrows* in **c**). RHA, right hepatic artery.

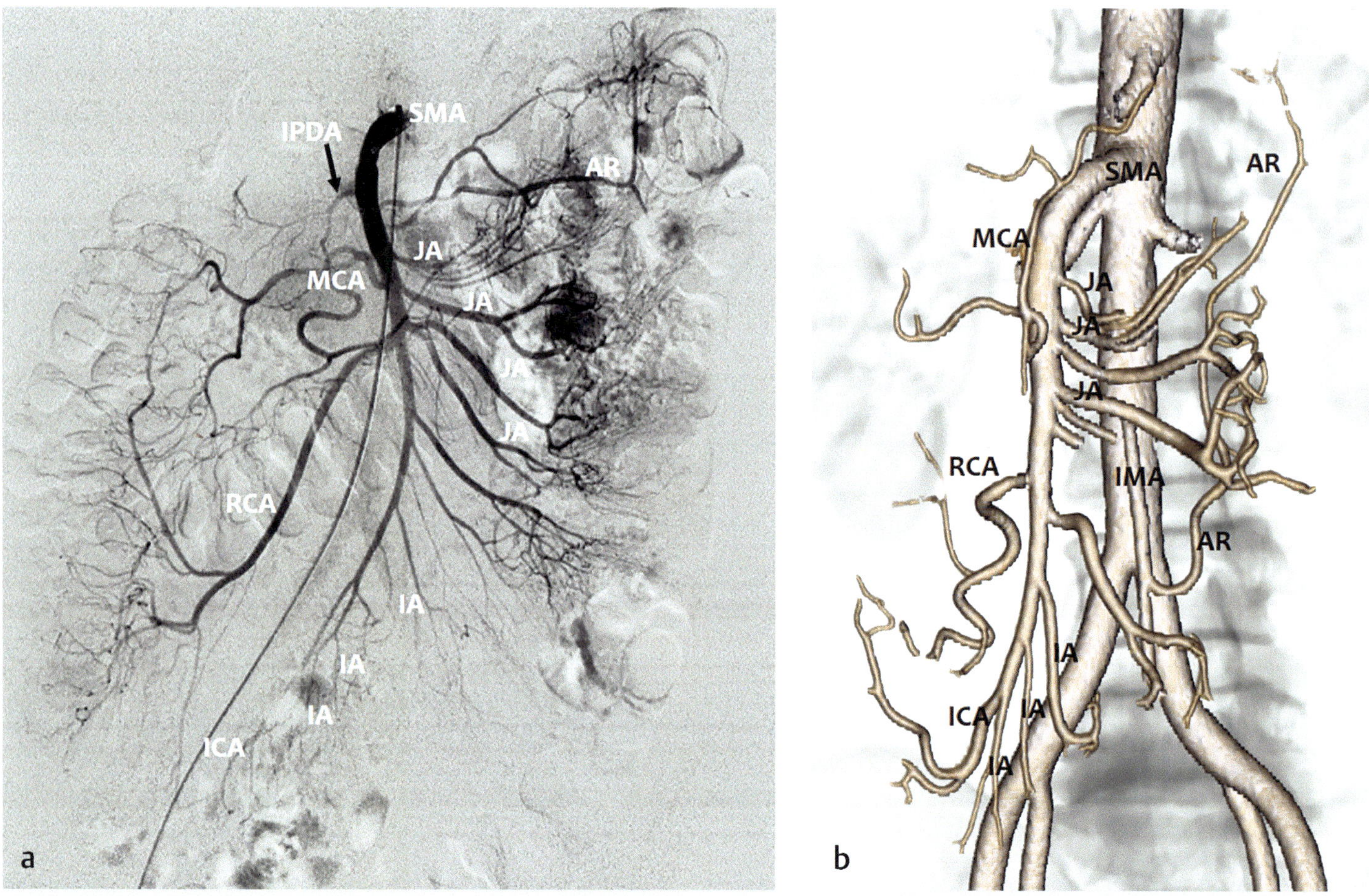

Fig. 19.4 Superior mesenteric artery and its major branches. **(a)** Digital subtraction angiography (DSA) and **(b)** three-dimensional computed tomography (CT) angiography of the aorta and its branches with volume rendering reconstruction. AR, arc of Riolan; IA, ileal artery; ICA, ileocolic artery; IMA, inferior mesenteric artery; IPDA, inferior pancreaticoduodenal artery; JA, jejunal artery; MCA, middle colic artery; RCA, right colic artery; SMA, superior mesenteric artery.

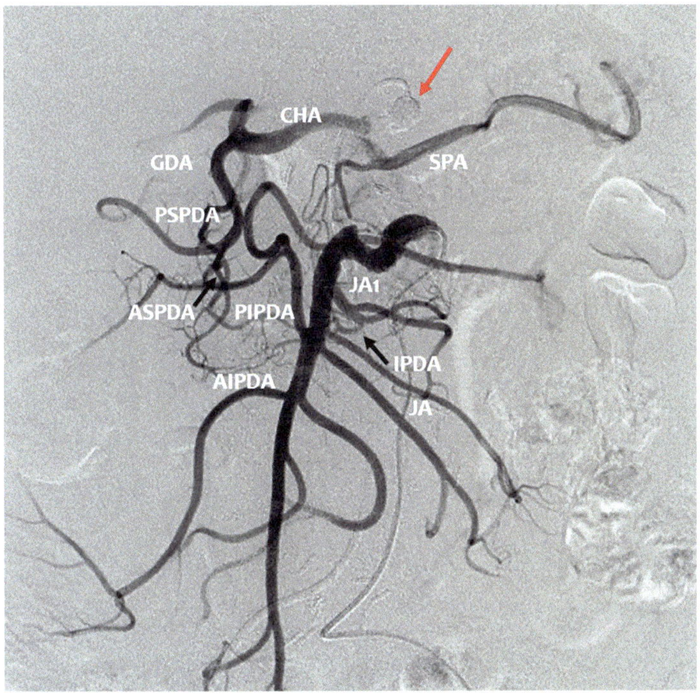

Fig. 19.5 Pancreatoduodenal arcade. Digital subtraction angiography (DSA) of the superior mesenteric artery during balloon occlusion of the celiac artery (*red arrow*) clearly depicts the pancreatoduodenal arcade. AIPDA, anterior inferior pancreatoduodenal artery; ASPDA, anterior superior pancreatoduodenal artery; CHA, common hepatic artery; GDA, gastroduodenal artery; IPDA, inferior pancreatoduodenal artery; JA, jejunal artery; JA1, first jejunal artery; PIPDA, posterior inferior pancreatoduodenal artery; PSPDA, posterior superior pancreatoduodenal artery; SPA, splenic artery.

branch or a common trunk with the first jejunal artery.[7] It anastomoses with the anterior or inferior superior pancreaticoduodenal artery to form the pancreatoduodenal arcade which supplies the head of the pancreas and the distal duodenum (▶Fig. 19.5).

Middle Colic Artery

The middle colic artery originates from the superior mesenteric artery just distal to the pancreatoduodenal artery and divides into the right and left branches. The right branch anastomoses with the right colic artery. The left branch anastomoses with the left colic artery through marginal artery and/or arc of Riolan with inferior mesenteric artery (▶Fig. 19.4). The right colic artery often arises from the middle colic artery or the ileocolic artery. It may arise directly from the superior mesenteric artery. The right colic artery anastomoses with the middle colic artery and the ileocolic artery to form the marginal artery along the ascending colon.

Inferior Mesenteric Artery

The inferior mesenteric artery takes off from the anterior aspect of the abdominal aorta at the level of L2 or L3 vertebral body (▶Fig. 19.6). It courses anteroinferiorly with an acute angle left to the abdominal aorta and within the root of sigmoid mesentery. Main branches include the left colic and sigmoidal arteries before terminating into the superior rectal artery.

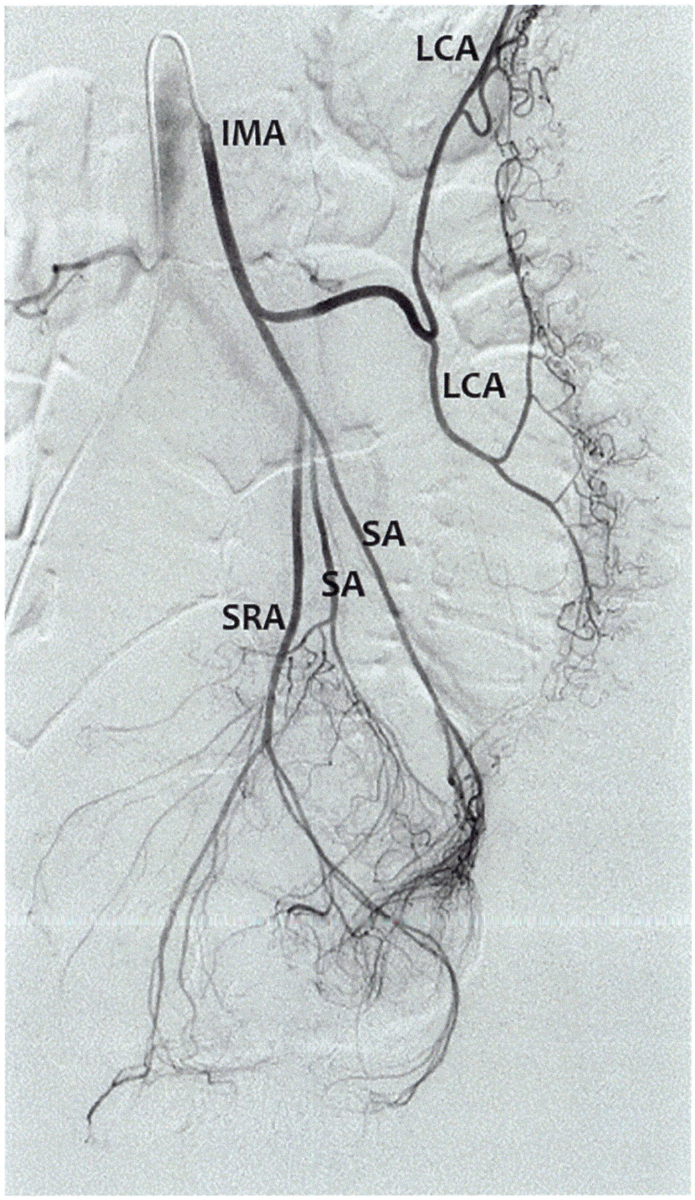

Fig. 19.6 Inferior mesenteric artery. IMA, inferior mesenteric artery; LCA, left colic artery; SA, sigmoid artery; SRA, superior rectal artery.

Renal Artery

The renal arteries rise from the lateral aspects of abdominal aorta at the level of L1 or L2 vertebral body (▶Fig. 19.7). The longer right renal artery runs posteriorly to the inferior vena cava (IVC). It usually bifurcates into the ventral and dorsal branches near the hilum, and each will be subdivided into several segmental end arteries.[5] About 25% of the people have multiple renal arteries. It is common to see accessory arteries that supply the poles of the kidneys. These polar arteries can occasionally compress the ureter.[8] The renal artery gives off to the superior capsular artery, middle capsular artery, inferior capsular artery, inferior adrenal artery, and rarely an inferior phrenic artery.

Adrenal Artery

The adrenal glands are usually supplied by the superior, middle, and inferior adrenal arteries (▶Fig. 19.7). Typically, the middle

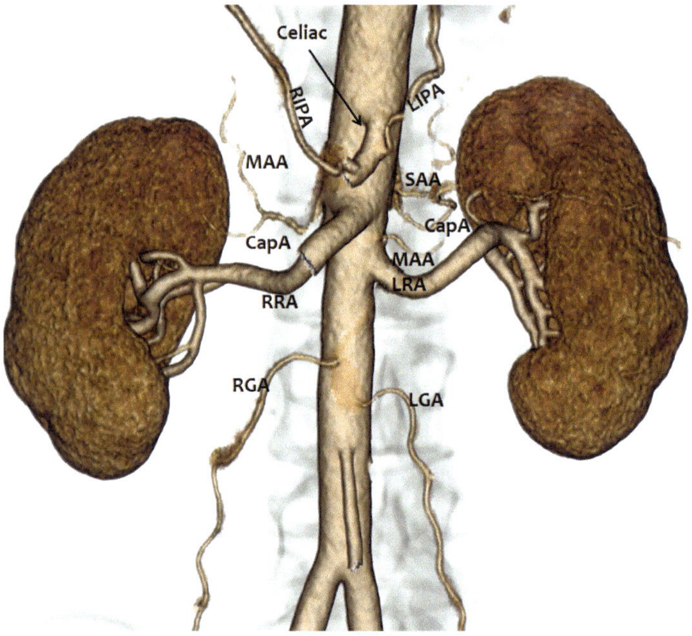

Fig. 19.7 Three-dimensional computed tomography (CT) angiography of the renal arteries, adrenal arteries, and gonadal arteries with volume rendering reconstruction. CapA, capsular artery; LGA, left gonadal artery; LIPA, left inferior phrenic artery; LRA, left renal artery; MAA, middle adrenal artery; RGA, right gonadal artery; RIPA, right inferior phrenic artery; RRA, right renal artery; SAA, superior adrenal artery.

adrenal artery arises from lateral aspect of the abdominal artery. The superior adrenal artery often arises from the ipsilateral inferior phrenic artery. The inferior adrenal artery often arises from the ipsilateral renal artery. Variation in the number and distribution of the adrenal arteries is common. About half of the individuals have a right middle adrenal artery arising from the aorta while around 90% of the cases have a left middle adrenal artery arising from the aorta.[9] When the inferior phrenic artery arises from the right renal artery, the inferior adrenal artery originates from the inferior phrenic artery.

Gonadal Artery (Testicular Artery and Ovarian Artery)

Usually, paired gonadal arteries arise anterolaterally from aorta between the level of the renal arteries and inferior mesenteric artery (▸Fig. 19.7). They descend caudally anterior to the psoas major muscle accompanied by the gonadal vein and ureter within the retroperitoneum. The right gonadal artery runs anterior to the IVC. The gonadal arteries cross the internal iliac artery at the pelvic inlet and eventually enter the mesovarium and anastomose with the uterine artery.[9] In male, the gonadal artery passes the deep inguinal ring to enter the spermatic cords.

Lumbar Artery

The lumbar arteries are four paired arteries originating from the posterolateral aspect of the abdominal aorta (▸Fig. 19.8). They run along the sides of the vertebral bodies and then posterior

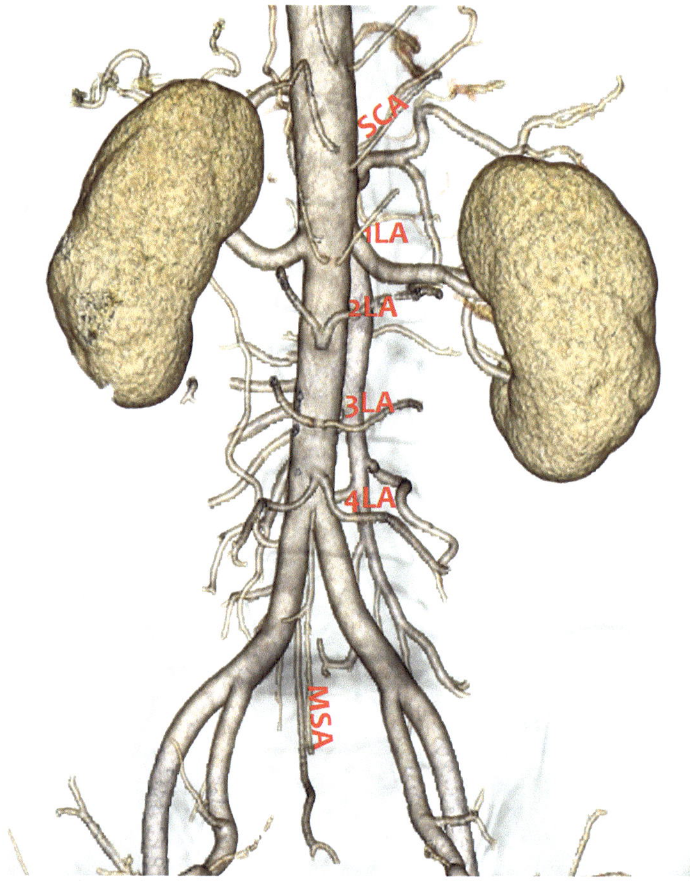

Fig. 19.8 Posterior view of the three-dimensional, volume-rendered reconstruction of computed tomography (CT) angiography of the abdominal aorta. 1LA, first lumbar artery; 2LA, second lumbar artery; 3LA, third lumbar artery; 4LA, fourth lumbar artery; MSA, middle sacral artery; SCA, subcostal artery.

to the psoas major and sympathetic trunks. They bifurcate into the ventral and dorsal branches and continue into the psoas and abdominal wall muscles. The lumbar arteries of the right side pass posterior to the inferior vena cava.[9] The fifth lumbar arteries are usually branches of the iliolumbar arteries but occasionally arise from the median sacral artery.

◆ Pelvic Arteries

The abdominal aorta bifurcates into the left and right common iliac arteries at the L4 vertebral body. The common iliac arteries descend posterolaterally and divide into the external and internal iliac arteries at the L5 or S1 vertebral level. The ureter crosses the common iliac artery or the iliac bifurcation. The middle sacral artery takes off from the posterior wall of the aorta at its bifurcation and descends anterior to the sacrum. It gives off the fifth lumbar artery and anastomoses with the lateral sacral artery.

Internal Iliac Artery

The internal iliac artery arises from the common iliac artery on sacroiliac joint, descends posteromedial to the external iliac artery, and enters the lesser pelvis to supply most of the pelvic viscera and muscles (▸Fig. 19.9, ▸Fig. 19.10). The internal iliac

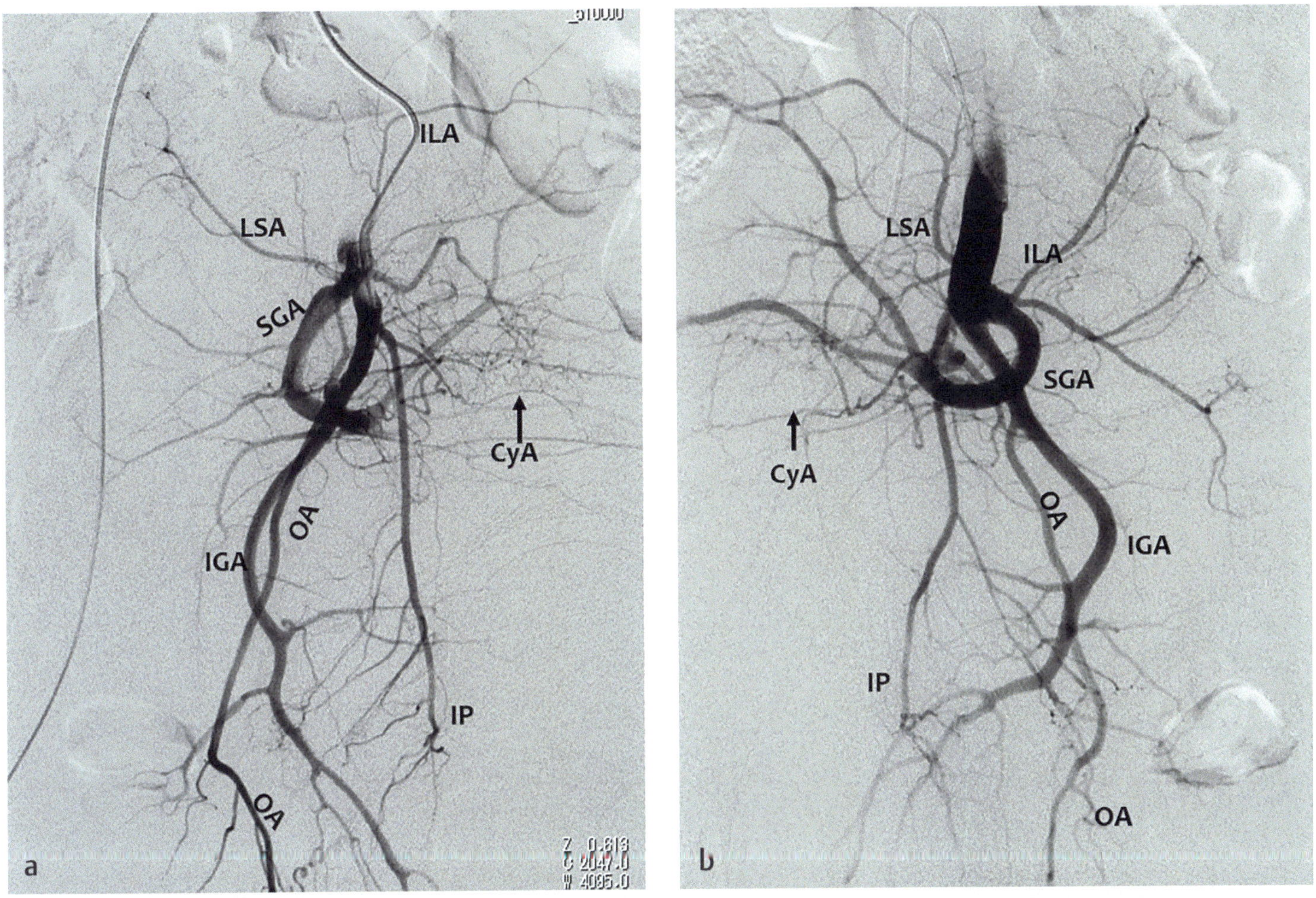

Fig. 19.9 Internal iliac artery and its major branches. Right **(a)** and left anterior oblique **(b)** projection of digital subtraction angiography (DSA) of the internal iliac artery. CyA, cystic artery; IGA, inferior gluteal artery; ILA, iliolumbar artery; IP, internal pudendal artery; LSA, lateral sacral artery; OA, obturator artery; SGA, superior gluteal artery.

usually ends at superior edge of the greater sciatic foramen and divides into the anterior and posterior division.[9] The anterior division of the internal iliac artery gives off the obturator, bladder, ureteral, inferior rectal, internal pudendal, and inferior gluteal arteries. The posterior division gives rise to the iliolumbar and lateral sacral arteries and terminates as the superior gluteal artery.[5]

Iliolumbar Artery

The iliolumbar artery originates superolaterally from the posterior division of the internal iliac artery. It runs posterior to the psoas major muscle and divides into the iliac and lumbar branches. The lumbar branch supplies the psoas major and quadratus lumborum muscles and rarely anastomose with the spinal branches, while the iliac branch supplies the iliacus muscle and the ilium and anastomoses with other branches.

Lateral Sacral Artery

Typically, the lateral sacral artery originates from the posterior division of the internal iliac artery and occasionally from the internal iliac artery. It rapidly splits into the superior and inferior branches and enter into anterior sacral foramina supplying the sacral vertebrae and the contents of the sacral canal. It passes through the corresponding dorsal foramen to reach the posterior aspect of the sacrum. It usually has multiple anastomoses with the middle sacral artery.

Superior Gluteal Artery

The superior gluteal artery forms the main trunk of posterior division of internal iliac artery and it is the largest branch of the internal iliac artery. It runs posteriorly to pierce the pelvic fascia between the lumbosacral trunk and the first sacral nerve root. It leaves the pelvis via the greater sciatic foramen above the piriformis and splits into the superficial and deep branches to supply the gluteal region. The superficial branch runs into the deep surface of gluteus maximus. The deep branch runs between the gluteus medius and the iliac bone. A lower branch travels toward the hip joint and contributes to the trochanteric anastomosis along with the inferior gluteal, medial femoral circumflex, and lateral femoral circumflex arteries.

Obturator Artery

The obturator artery arises from the anterior trunk of the internal iliac artery. Descending in deep lateral pelvis, it passes through the obturator foramen, exits the pelvis, and gives off muscular

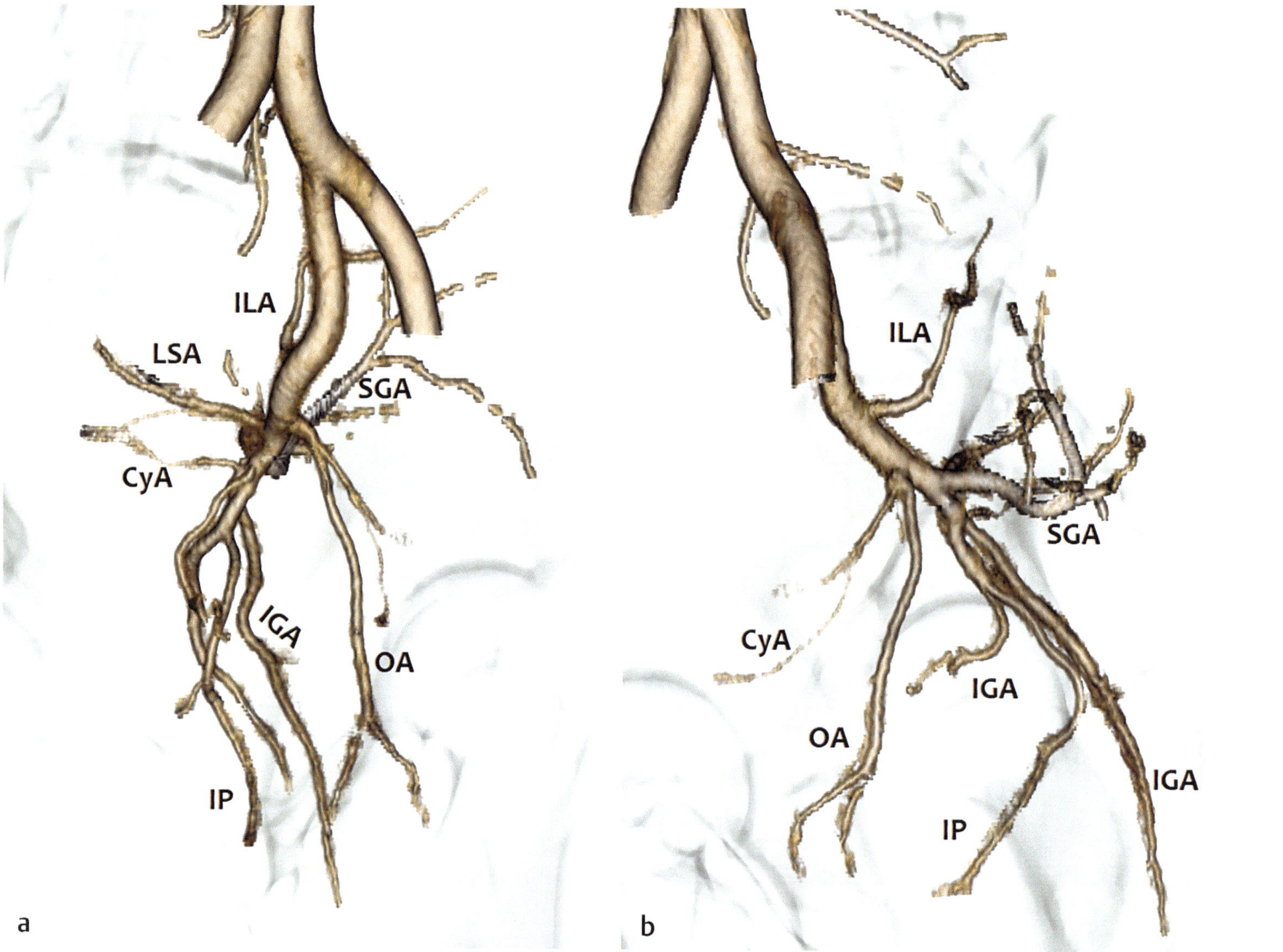

Fig. 19.10 Internal iliac artery and its major branches. Right **(a)** and left anterior oblique **(b)** projection of three-dimensional, volume-rendered reconstruction of computed tomography (CT) angiography. CyA, cystic artery; IGA, inferior gluteal artery; ILA, iliolumbar artery; IP, internal pudendal artery; LSA, lateral sacral artery; OA, obturator artery; SGA, superior gluteal artery.

branch to the obturator muscle. It then splits into the anterior and posterior branches to feed the adductor muscles and the femoral head. It has communications with the internal circumflex artery of the deep femoral artery, which often work as collaterals once the internal iliac artery is occluded. Aberrant origin of the obturator artery from the deep inferior epigastric artery is common.

Inferior Gluteal Artery

19

The inferior gluteal artery is the second-largest branch of the internal iliac artery following the superior gluteal artery and the largest terminal branch of the anterior trunk of the internal iliac artery. It descends anterior to the piriformis and posterior to the internal pudendal artery. It passes through the greater sciatic foramen below the piriformis muscle and exits the pelvis to supply the inferior part of the gluteus maximus and the dorsal femoral muscle. It also has anastomosis with the lateral and medial circumflex arteries, which work as collateral circulation in case of occlusion of the internal iliac artery. As an anatomical variation, the inferior gluteal artery and the internal pudendal artery may arise from a common trunk.

Superior Vesical Artery

The superior vesical artery is commonly the first branch of the anterior division of the internal iliac artery. It descends on lateral wall of the pelvis, then runs medially toward the upper portion of the bladder, and supplies the upper half of the bladder. The superior vesical artery is the persistent proximal part of the fetal umbilical artery. The distal part is regressed and forms the umbilical ligament.

Inferior Vesical Artery

The inferior vesical artery is a branch of the anterior division of the internal iliac artery. It originates just below the superior vesical artery coursing anteromedially to supply the seminal vesicle and prostate in male. In female, the vaginal artery replaces this artery. It may arise from the middle rectal artery.

Uterine Artery

The uterine artery usually arises from the anterior division of the internal iliac. It can originate from the ovarian artery. It courses medially along the broad ligament to the junction of the uterus and vagina and splits into the descending vaginal and ascending uterine branches.

Middle Rectal Artery

The middle rectal artery may arise from the anterior division of the internal iliac artery. Occasionally, it originates as a common trunk with the inferior vesical artery or the internal pudendal artery.

Internal Pudendal Artery

The internal pudendal artery is a branch of the anterior division of the internal iliac artery. It is the primary arterial supply of the perineum and appears larger in males than in females. It penetrates the greater sciatic foramen between the piriformis muscle and the coccygeus muscle and curves around the ischial spine to re-enter the pelvis through the lesser sciatic foramen. It terminates into the perineal artery and dorsal artery of the penis or clitoris on exiting the pudendal canal.

External Iliac Artery

The external iliac artery runs along the medial border of the psoas major muscle. After it enters the thigh under the inguinal ligament, it becomes the femoral artery. It gives off the deep circumflex iliac artery and the inferior epigastric artery at the proximity of the inguinal ligament.

The inferior epigastric artery takes off medially and superiorly in an acute angle from the external iliac artery and ascends obliquely along the lower abdominal wall. It anastomoses with the superior epigastric artery.

The deep circumflex iliac artery originates laterally from the external iliac artery and ascends toward the anterior iliac spine. Its terminal branches anastomose with the iliolumbar and superior gluteal arteries and the ascending branch of the lateral circumflex femoral artery.

◆ Collateral Pathway in Aortoiliac Occlusive Disease

Occlusive disease of the abdominal aorta or iliac artery can occur by various conditions including atherosclerotic disease, thromboembolic disease, or aortitis. Numerous potential anastomotic collaterals occur between aorta and iliac artery in chronic aortoiliac occlusive diseases. Assessment of arterial collateral pathways is clinically important for surgical planning and avoiding complications from vascular injury or ischemic events of the lower extremities.

Following collateral pathways to the lower extremities develop when the infrarenal aorta is occluded[10]:

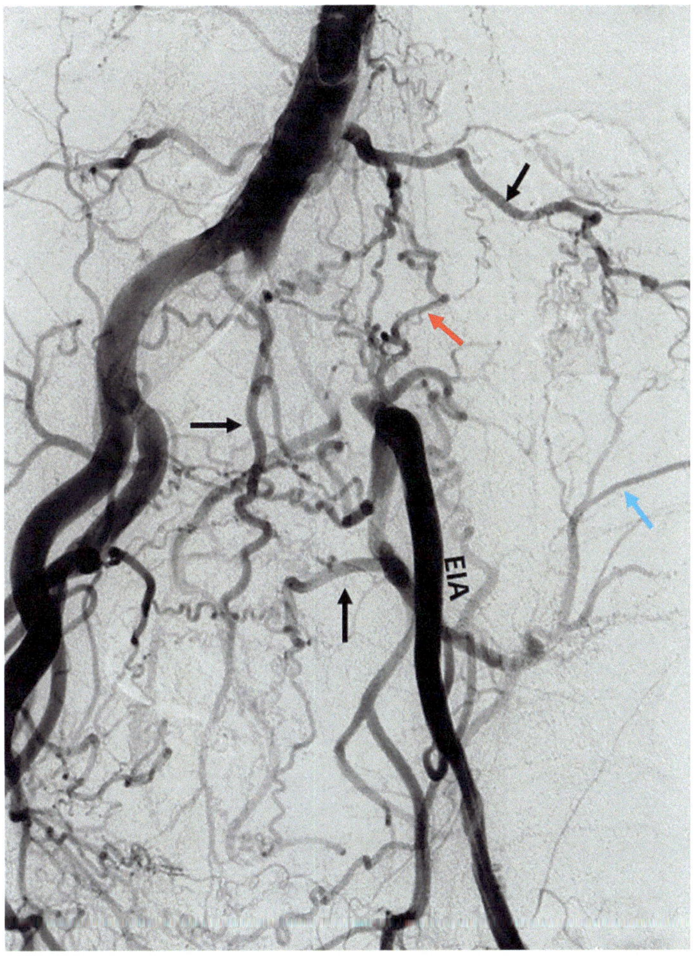

Fig. 19.11 Systemic collateral pathway in aortoiliac occlusive disease. Pelvic arteriogram of a patient with iliac occlusive disease shows the left external iliac artery (EIA) visualized through the collateral tract from the fourth lumbar artery (*black arrows*) to the superior gluteal artery (*blue arrow*) and to the iliolumbar artery (*red arrow*).

- Visceral pathway: The visceral pathway is collaterals from the superior and/or inferior mesenteric artery via the rectal arteries into the internal iliac arteries.
- Systemic pathway: In systemic pathway collaterals occur via a systemic-to-systemic anastomosis. This pathway includes the lumbar, intercostal, deep circumflex iliac, internal thoracic, inferior epigastric, and obturator arteries. Major systemic pathways reported are as follows:
  - The intercostal artery, subcostal artery, or lumbar artery to the abdominal wall arteries, iliac arteries, and distal aorta.
  - The iliolumbar artery to the superior gluteal artery and the internal iliac artery (▸ Fig. 19.11).
  - The deep circumflex iliac artery to superficial circumflex iliac artery and the external iliac artery.
- The visceral–systemic pathway: Potential anastomosis between the visceral and systemic arteries is a rare pathway to the lower extremities. Pathway via the renal or gonadal arteries to the inferior epigastric artery or other pelvic arteries has been reported.[10]
- Winslow's pathway: A relatively rare collateral pathway between the subclavian and femoral arteries via anastomosis of the internal mammary arteries to the superior and inferior epigastric arteries[11] (▸ Fig. 19.12).

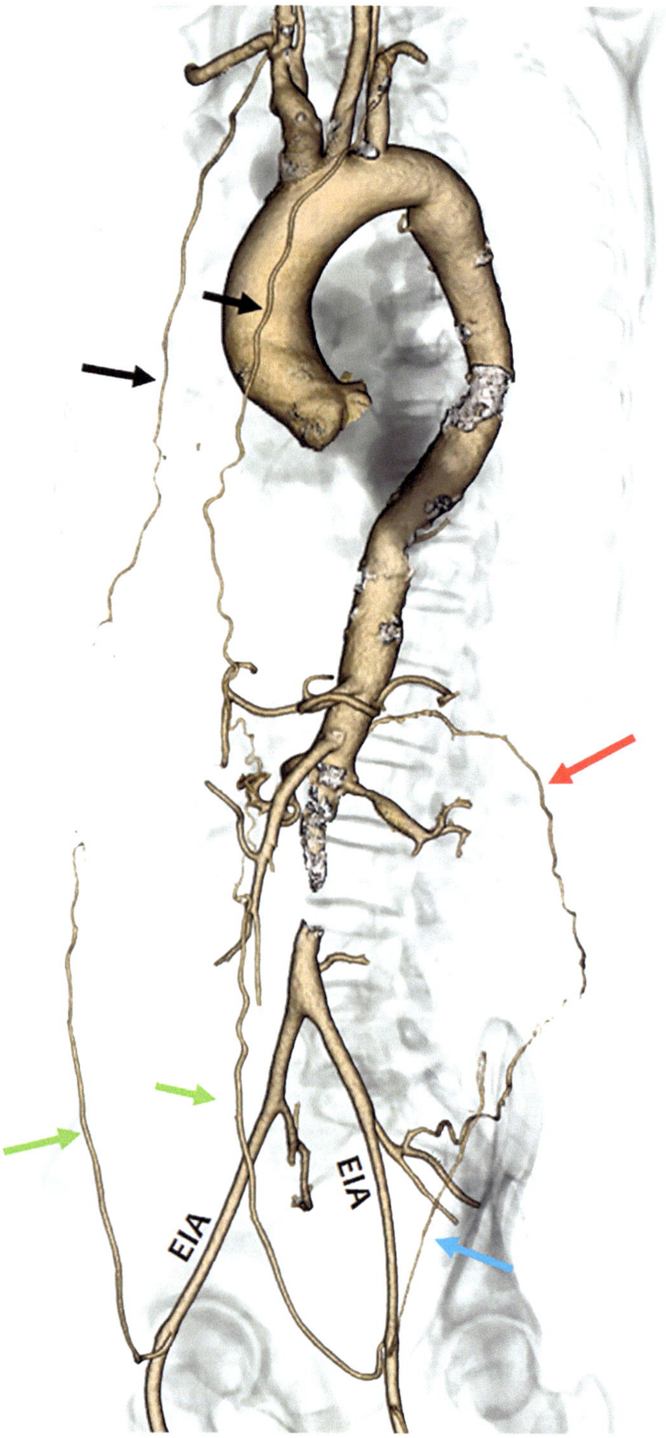

Fig. 19.12 Winslow's collateral pathway in aortoiliac occlusive disease. Pelvic arteriogram of a patient with aortic occlusive disease due to aortitis shows the external iliac arteries (EIA) visualized through the collateral tract via anastomosis of the internal mammary (*black arrows*) and inferior epigastric arteries (*green arrows*) and also the first lumbar artery (*red arrow*) with the deep circumflex iliac artery (*blue arrow*).

◆ Collateral Pathway of Type II Endoleak after EVAR

Since endovascular aortic repair (EVAR) has been widely used in the treatment of abdominal aortic aneurysm, the type II endoleak, which can be explained by retrograde sac filling from aortic tributaries, has been recognized as a cause of chronic expansion of the aneurysmal sac after EVAR (▶Fig. 19.13). The understanding of possible collateral pathway of the type II endoleak is mandatory to assess the differentiation of endoleaks and the treatment of the type II endoleak. The most common arterial collateral pathway is endoleak fed by inferior mesenteric artery through arc of Riolan or marginal artery through middle colic artery from superior mesenteric artery. Type II endoleak from lumbar arteries is commonly suppled via iliolumbar artery from internal iliac artery. Lumbar arteries are also supplied by upper lumbar artery or intercostal/ subcostal artery. Other type II endoleak from lateral sacral artery via iliolumbar artery is also observed frequently.

◆ Embryological Development

The formation and transformation of the dorsal aorta (primitive descending and abdominal aorta) begins at early period of embryologic development. The paired right and left dorsal aortas exist at the dorsal part of the embryo and the caudal part of these paired dorsal aortas unite after 1 week of gestation to form the unpaired descending and abdominal aorta of the adult.[12,13]

Many segmental arteries originate from the dorsal aortas; however, most of the segmental arteries regress or form the intercostal or lumbar arteries.[10] The segmental artery becomes the celiac artery to supply the foregut from the esophagus to the distal duodenum. The 13th segmental artery becomes the superior mesenteric artery that supplies the midgut, which corresponds to the intestinal segment between the proximal jejunum and the mid transverse colon. Lastly, the 22nd segmental artery develops into the inferior mesenteric artery to supply the hindgut from the mid transverse colon to the rectum.[7,14]

The umbilical arteries are ventral branches at the terminal part of the left and right dorsal aortas that develop even after the fusion of the paired dorsal aortas. During the 4th week, each umbilical artery acquires an anastomosis with a fifth lumbar segmental artery and loses its earliest origin. The proximal portions of the umbilical arteries become the internal iliac and superior vesical arteries and the fifth lumbar segmental arteries form the common and external iliac arteries (▶Fig. 19.14).

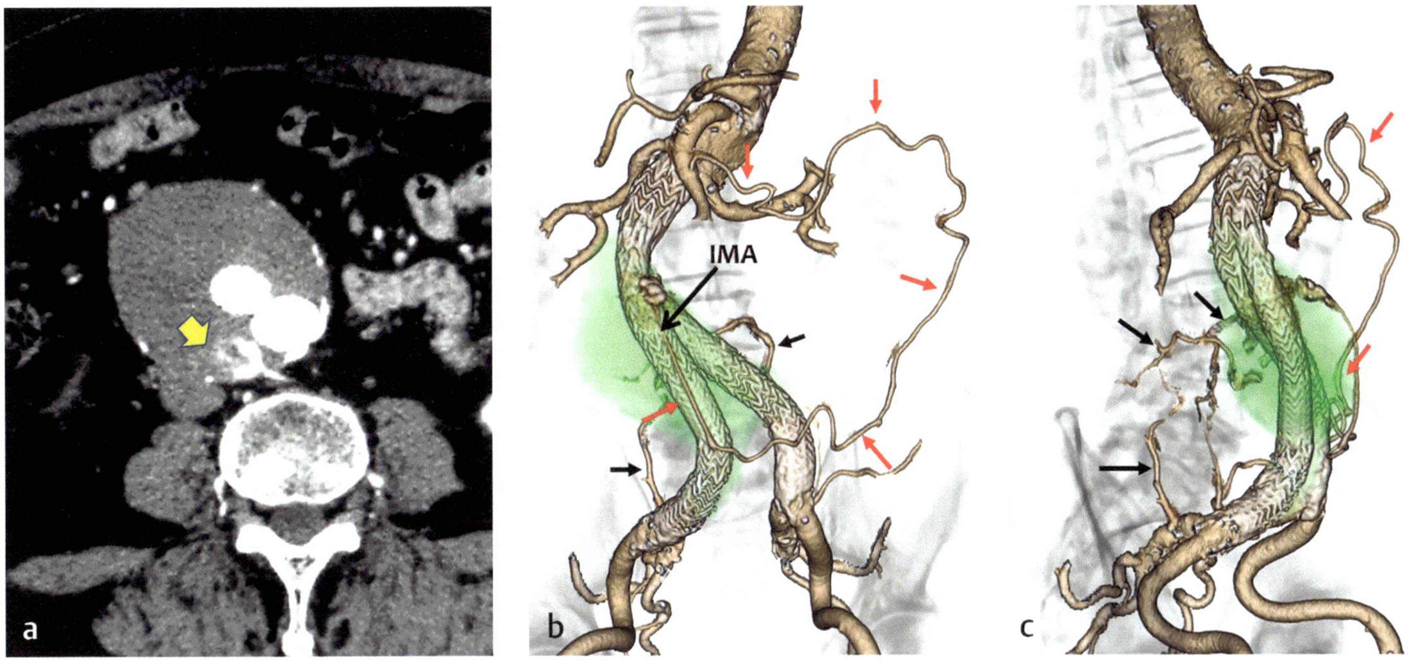

Fig. 19.13 Type II endoleak after endovascular aortic repair (EVAR) of abdominal aortic aneurysm. **(a)** Contrast-enhanced computed tomography (CT) after an EVAR shows an endoleak within the space between the aneurysmal sac and the implanted bifurcated stent-graft (*yellow arrow*). **(b)** Anterior view and **(c)** right anterior oblique view of three-dimensional volume rendering of the contrast-enhanced CT clearly visualize the type II endoleaks from the inferior mesenteric artery filled via the arc of Riolan (*red arrows*) and from the fourth lumbar arteries through iliolumbar artery (*black arrows*). IMA, inferior mesenteric artery.

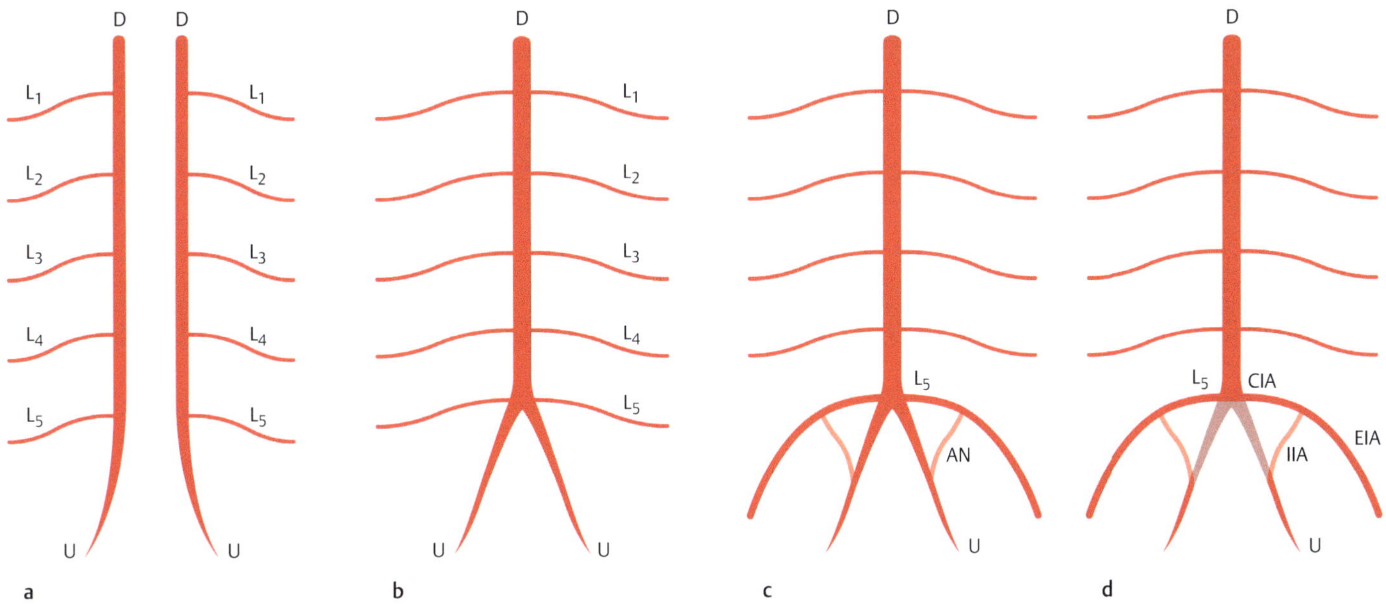

Fig. 19.14 Schematic drawing of development of the iliac artery. **(a)** Five segmental arteries (L1– L5) arise from lumbar part of paired dorsal aortas (D) which terminate into umbilical artery (U). **(b)** The paired dorsal aortas unite at the level of the fifth lumbar segmental artery after about 1 week of gestation forming the primitive abdominal aorta. **(c)** During the 4th week of gestation each umbilical artery acquires an anastomosis (AN) with a fifth lumbar segmental artery. **(d)** The umbilical artery loses its earliest origin and through anastomosis (AN) form the proximal part of the internal iliac artery (IIA) while distal part of the fifth lumbar segmental arteries form the external iliac artery (EIA). CIA, common iliac artery.

◆ Anatomical Variations and Anomalies

Celiac Artery and Superior Mesenteric Artery

There are many variations of origins of the celiac and superior mesenteric arteries. Commonly, the celiac artery and the superior mesenteric artery arise separately from the aorta, and the celiac artery gives rise to common hepatic artery. About 90% of the cases have normal aortic origin of the celiac and superior mesenteric arteries.[15] Remaining 10% of the cases have anatomical variation at the origin of the celiac and superior mesenteric arteries.[15] Hepatomesenteric trunk is seen in 4.47%, celiomesenteric trunk in 3.4% (▶Fig. 19.15), and splenomesenteric trunk in 1.2%.[16,17] There are many subtypes among these variants in regard to the origin and distribution of the common hepatic, left gastric, and splenic arteries.[18]

Tandler's hypothesis explains these variations in regard to embryonic development.[19]

Four primitive splanchnic branches arising from the abdominal aorta in early human embryos are connected by the ventral longitudinal anastomoses. Tandler[19] and Morita[20] explained that the four primitive mesenteric intestinal arteries corresponding later to the left gastric, common hepatic, splenic, and superior mesenteric arteries are connected by a longitudinal anterior anastomosis (▶Fig. 19.16). The root of the middle two to four regress and the longitudinal anastomosis persists to supply the first three

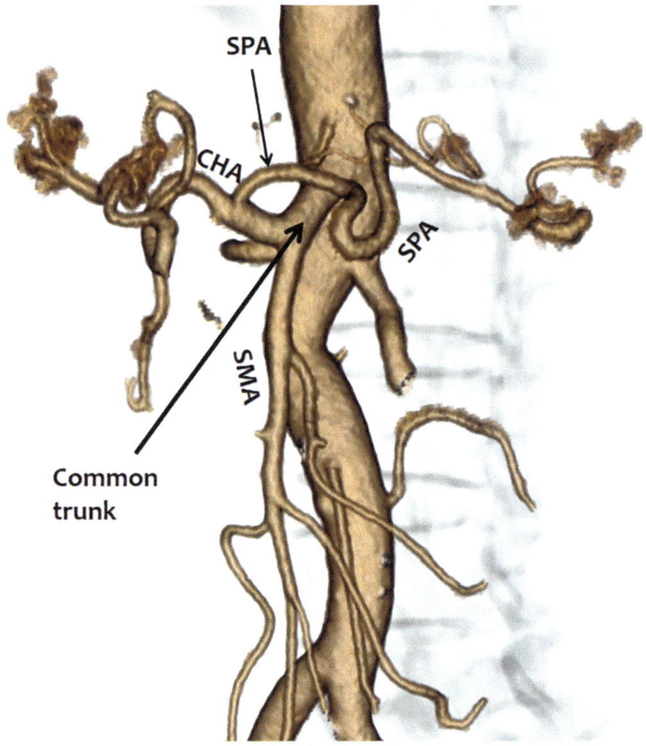

Fig. 19.15 Three-dimensional, volume-rendered reconstruction of the computed tomography (CT) angiogram in a case with common celiacomesenteric trunk. CHA, common hepatic artery; SMA, superior mesenteric artery; SPA, splenic artery.

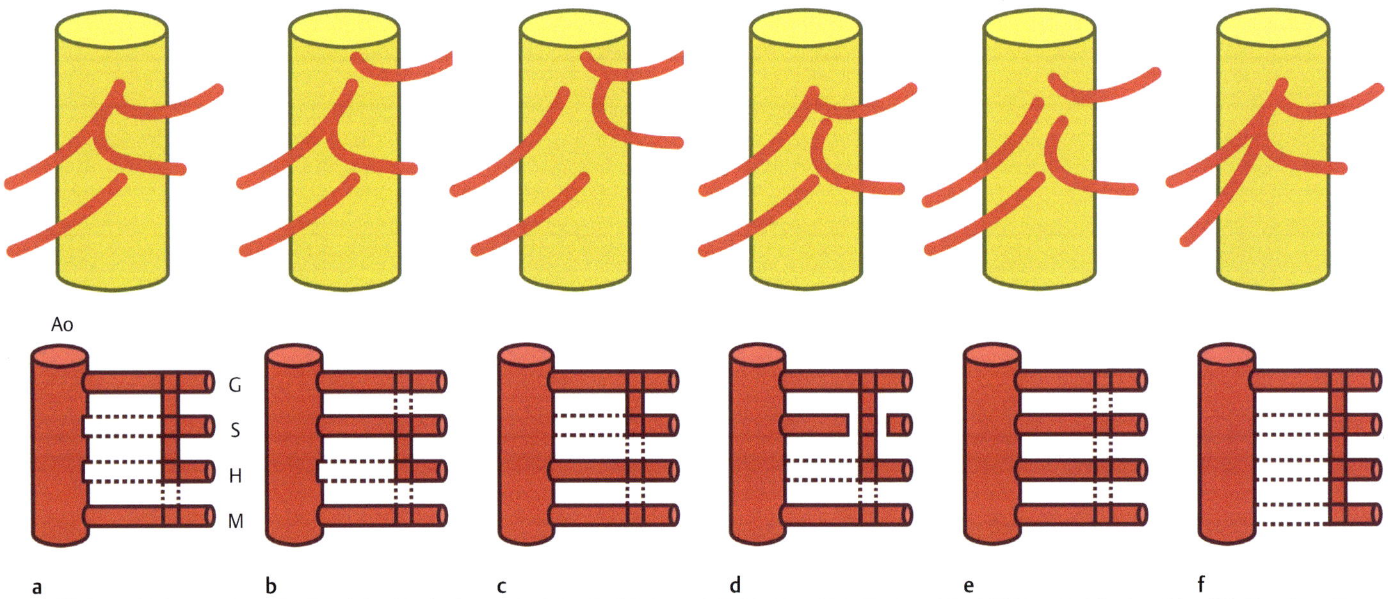

Fig. 19.16 Schematic drawing explaining variation of the origin of the celiac artery and superior mesenteric artery (SMA) on the basis of Tandler's hypothesis. Typical branching pattern **(a)** can be explained as the regression of the root of the middle two of four primitive splanchnic arteries and longitudinal anastomosis between the hepatic artery and the superior mesenteric artery. Aortic origin of the left gastric artery **(b)**, aortic origin of the hepatic artery **(c)**, aortic origin of the splenic artery **(d)**, four independent aortic origin **(e)**, and celiacomesenteric trunk **(f)** can also be explained by the hypothesis. Ao, aorta; G, gastric artery; H, hepatic artery; M, mesenteric artery; S, splenic artery.

19

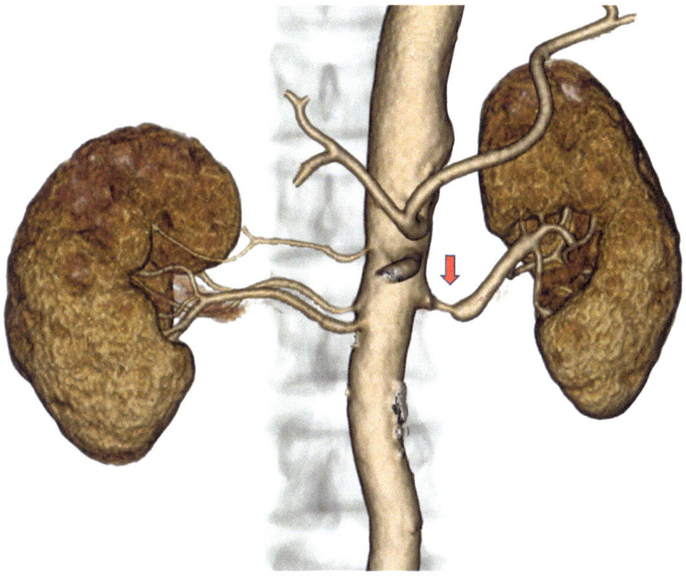

Fig. 19.17 Three-dimensional, volume-rendered reconstruction of the computed tomography (CT) angiogram in a case with multiple renal arteries. Right kidney has three renal arteries including a polar artery. Atherosclerotic stenosis (*arrow*) of the left renal artery is also noted.

roots, and the left gastric, common hepatic, and splenic arteries take off from this anastomosis. Partial remnant or regression of this primitive arterial plexus could lead to numerous anomalous variations of the celiac and superior mesenteric arteries.

Variations of Renal Artery

Variation of the origin of renal artery is very common. Both right and left renal artery usually arise at the L1 and L2 disk levels, and right renal artery tends to be higher.

Multiple renal arteries is very common as a variation, accounting for around 30% of the cases (▶ Fig. 19.17). There is embryological explanation for multiple renal arteries. During the developing process from the mesonephros to metanephros of mature kidney, mesonephros is initially located anterior to the sacrum and migrate cephalad toward lumbar portion[8] supplied by nine pairs of lateral mesonephric arteries arising from the dorsal aorta. As the mesonephros gives way to the metanephros as a permanent kidney, the middle group (from the third to fifth arteries) transform to renal arteries and the rest of the arteries diminish. Persistence of more than one renal arteries of the middle group results in multiple renal arteries.[21]

Middle Mesenteric Artery

The middle mesenteric artery is a mesenteric arterial anomaly which is seen in less than 1% of the cases. It results from incomplete regression of the primitive segmental ventral arteries from the dorsal aorta which supply the small and large bowel directly from the aorta at a level between the superior and inferior mesenteric arteries.[22] The arterial supplying area, which can vary largely in each case, can be a distal part of the small intestine and/or the proximal colon. It is also called duplicated superior mesenteric artery. The patient with this anomaly is often excluded from endovascular abdominal aortic repair using aortic stent-graft to prevent colon ischemia because the middle mesenteric artery will be usually covered by the stent-graft originating from the proximal aspect of the aortic aneurysm.[23]

References

1. Moore KL, Dalley AF, Agur AMR. Clinically Oriented Anatomy. Wolters Kluwer Health; 2013
2. Leithner C, Sinzinger H, Hohenecker J, Wicke L, Olbert F, Feigl W. Radiologic anatomy of the abdominal aorta and their large branches. Okajimas Folia Anat Jpn 1975;52(2-3):119–150
3. Dos Santos R, Lamas A, Pereira-Caldas J. Arteriografia da aorta e dos vasos abdominals. Med Contemp 1929;47:93–96
4. Seldinger SI. Catheter replacement of the needle in percutaneous arteriography; a new technique. Acta Radiol 1953;39(5):368–376
5. Standring S, Borley NR. Gray's Anatomy: The Anatomical Basis of Clinical Practice: Churchill Livingstone/Elsevier; 2008
6. Pandey SK, Bhattacharya S, Mishra RN, Shukla VK. Anatomical variations of the splenic artery and its clinical implications. Clin Anat 2004;17(6):497–502
7. Lin PH, Chaikof EL. Embryology, anatomy, and surgical exposure of the great abdominal vessels. Surg Clin North Am 2000;80(1):417–433, xiv
8. Aubert J, Koumare K. Variations of origin of the renal artery: a review covering 403 aortographies. Eur Urol 1975;1(4):182–188
9. Parson SH. Clinically Oriented Anatomy, 6th ed. Journal of Anatomy. 2009; 215(4):474
10. Hardman RL, Lopera JE, Cardan RA, Trimmer CK, Josephs SC. Common and rare collateral pathways in aortoiliac occlusive disease: a pictorial essay. AJR Am J Roentgenol 2011;197(3):W519-24
11. Prager RJ, Akin JR, Akin GC, Binder RJ. Winslow's pathway: a rare collateral channel in infrarenal aortic occlusion. AJR Am J Roentgenol 1977;128(3):485–487
12. Congdon E. Transformation of the aortic arch system during the development of the human embryo. Contrib Embryol 1922;14(68):47–110
13. Sato Y. Dorsal aorta formation: separate origins, lateral-to-medial migration, and remodeling. Dev Growth Differ 2013;55(1):113–129
14. Rana MS, Sizarov A, Christoffels VM, Moorman AF. Development of the human aortic arch system captured in an interactive three-dimensional reference model. Am J Med Genet A 2014;164A(6):1372–1383
15. Shoumura S, Emura S, Utsumi M, et al. [Anatomical study on the branches of the celiac trunk (IV). Comparison of the findings with Adachi's classification]. Kaibogaku Zasshi 1991;66(5):452–461
16. Wang Y, Cheng C, Wang L, Li R, Chen JH, Gong SG. Anatomical variations in the origins of the celiac axis and the superior mesenteric artery: MDCT angiographic findings and their probable embryological mechanisms. Eur Radiol 2014;24(8):1777–1784
17. Yi SQ, Terayama H, Naito M, et al. A common celiacomesenteric trunk, and a brief review of the literature. Ann Anat 2007;189(5):482–488
18. Adachi B. Anatomie der Japaner: Das Arterien system der Japaner. Kyoto: Verlag der Kaiserlich-Japanische Universität zu Kyoto, im kommission bei "Maruzen Company"; 1928
19. Tandler J. Zur entwickelunggeschicte des menschliden darmarterien. Anat Hefte 1904;23:189–209, 475–499
20. Morita M. Reports and conception of three anomalous cases on the area of the celiac and the superior mesenteric arteries. Igaku Kenkyu (Acta Med) 1935;9: 1993–2006
21. Keibel F, Mall FP. Manual of Human Embryology. Wentworth Press; 2016
22. Milnerowicz S, Milnerowicz A, Taboła R. A middle mesenteric artery. Surg Radiol Anat 2012;34(10):973–975
23. Falkensammer J, Hakaim AG, Paz-Fumagalli R, Oldenburg WA. Endovascular repair of an infrarenal abdominal aortic aneurysm with a dominant middle mesenteric artery: a case report. Vasc Endovascular Surg 2006;40(4):338–341

20 Systemic Veins of the Abdomen and Pelvis

Farhood Saremi, Mojgan Motevali, and Daiji Uchiyama

◆ Introduction

In this chapter, the anatomy of the inferior vena cava (IVC) and pelvic vein will be reviewed.

◆ Inferior Vena Cava

The IVC conveys blood to the right atrium from all of the structures below the diaphragm. Most of its course is within the abdomen, but a small segment lies within the pericardium in the thorax. It passes through the central tendon of diaphragm at the level of the eighth and ninth thoracic vertebrae and drains into the right atrium. The thoracic part is very short and is partly inside and partly outside the pericardial sac (▶Fig. 20.1, ▶Fig. 20.2).

It traverses the vena caval hiatus of the diaphragm at the T8 level approximately 2 cm to the right of the midline between the middle and right leaves of the central tendon. In addition to the IVC, this diaphragmatic opening also provides a passage for branches of the right phrenic nerve and pericardiophrenic vessels. It is not uncommon to see the retroperitoneal fat entering this region causing narrowing of the IVC (▶Fig. 20.3). The thoracic IVC then passes through the pericardium and connects to the posteroinferior part of the right atrium (▶Fig. 20.1). At the insertion into the right atrium, a relatively large Eustachian valve guards the orifice of the IVC anteriorly in 25% of individuals. The abdominal portion of the IVC is devoid of valves.

The abdominal IVC lies in the retroperitoneal space in the right paravertebral region. At its origin (L5 level), the IVC is related anteriorly to the right common iliac artery (▶Fig. 20.2). From below upwards, its anterior surface is crossed obliquely by the root of the small bowel mesentery including its vessels and nerves, the right gonadal artery, and the third part of the duodenum (▶Fig. 20.2). Further cranially, it lies behind the head of the pancreas and first part of the duodenum, separated from these structures by the common bile duct and portal vein. Above the duodenum, the anterior surface of the IVC is covered by the posterior peritoneum which also forms the posterior wall of the epiploic foramen and separates the IVC from the right free border of the lesser omentum and its contents. Near the diaphragm, the IVC passes behind the liver in a groove (▶Fig. 20.2).

The posterior relations of the lower IVC include the lower three lumbar vertebral bodies and their intervertebral disks, the anterior longitudinal ligament, sympathetic trunk, right third and fourth lumbar arteries, and the right psoas major. Superior to these structures, the IVC is related posteriorly to the right renal and middle suprarenal arteries, the medial part of the right suprarenal gland, the right celiac ganglion, and the right inferior phrenic arteries. The right ureter, medial border of the right kidney, second part of the duodenum, and the right lobe of the liver are all lateral to the right side of the inferior vena cava. The abdominal aorta, right crus of the diaphragm, and the caudate lobe of the liver are left-sided relations.

Embryology

By the 5th week, the sinus venosus opens into the inferoposterior of the primitive cardiac atrium and receives on each side three major veins, which contribute to the IVC development[1–4]:

- The vitelline (omphalomesenteric) veins, carrying blood from the yolk sac to the sinus venosus.
- The umbilical (placental) veins, originating in the chorionic villi and carrying oxygenated blood to the embryo.
- The common cardinal (systemic) veins draining the body of the embryo (▶Fig. 20.4).

IVC Segments

The normal IVC is composed of different components developed from different embryonic vessels which will be completed by the 8th week. Embryologically, the IVC has been divided into three segments: the hepatic, the renal, and the infrarenal (▶Fig. 20.5).

The hepatic (hepatocardiac) segment of the IVC is derived from the right vitelline vein and cranial end of the right subcardinal vein. The intraperitoneal IVC segment is likely part of the sinus venosus. The renal/suprarenal segment is from the right subcardinal vein and cranial part of the left subcardinal vein forms the left adrenal vein. The infrarenal segment of the IVC is formed by the right supracardinal vein. The renal veins are formed from coalitions of the supracardinal and subcardinal veins. Initially, two renal veins exist. The anterior vein persists to form the renal vein; however, the posterior vein usually regresses.

Vitelline Veins

The vitelline veins form a plexus around the duodenum and pass through the septum transversum to enter the sinus venosus. At approximately 6 weeks, the liver grows to form sinusoids and invades the umbilical and omphalomesenteric veins. During liver development, reduction of the left sinus channels causes blood shifting from the left side of the liver to the right, resulting in an enlargement of the right vitelline vein (right hepatocardiac channel) that ultimately forms the hepatocardiac portion of the IVC (▶Fig. 20.5, ▶Fig. 20.6). The umbilical veins and the left omphalomesenteric vein lose their connections with the sinus venosus. The vitelline venous network around the duodenum develops to the portal vein. The superior mesenteric vein, which drains the primary intestinal loop, derives from the right vitelline vein. The proximal and distal portions of the left vitelline vein disappear

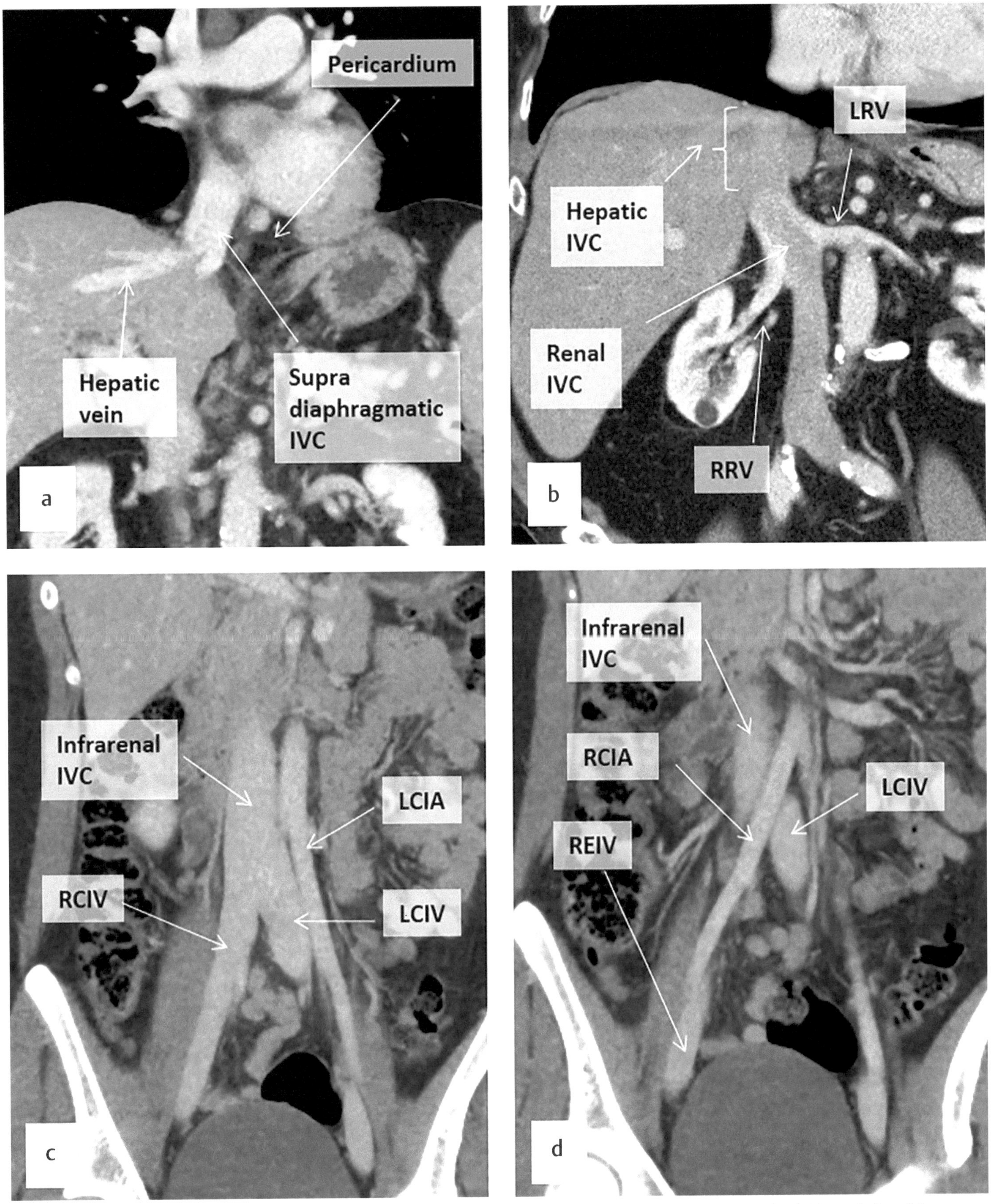

Fig. 20.1 (a–d) Coronal computed tomography (CT) images with contrast showing parts of the inferior vena cava (IVC). The majority of the supradiphragmatic IVC is intrapericardial. In **d**, note compression of the left common iliac vein (LCIV) by the right common iliac artery (RCIA). LCIA, left common iliac artery; LRV, left renal vein; RCIV, right common iliac vein; REIV, right external iliac vein; RRV, right renal vein.

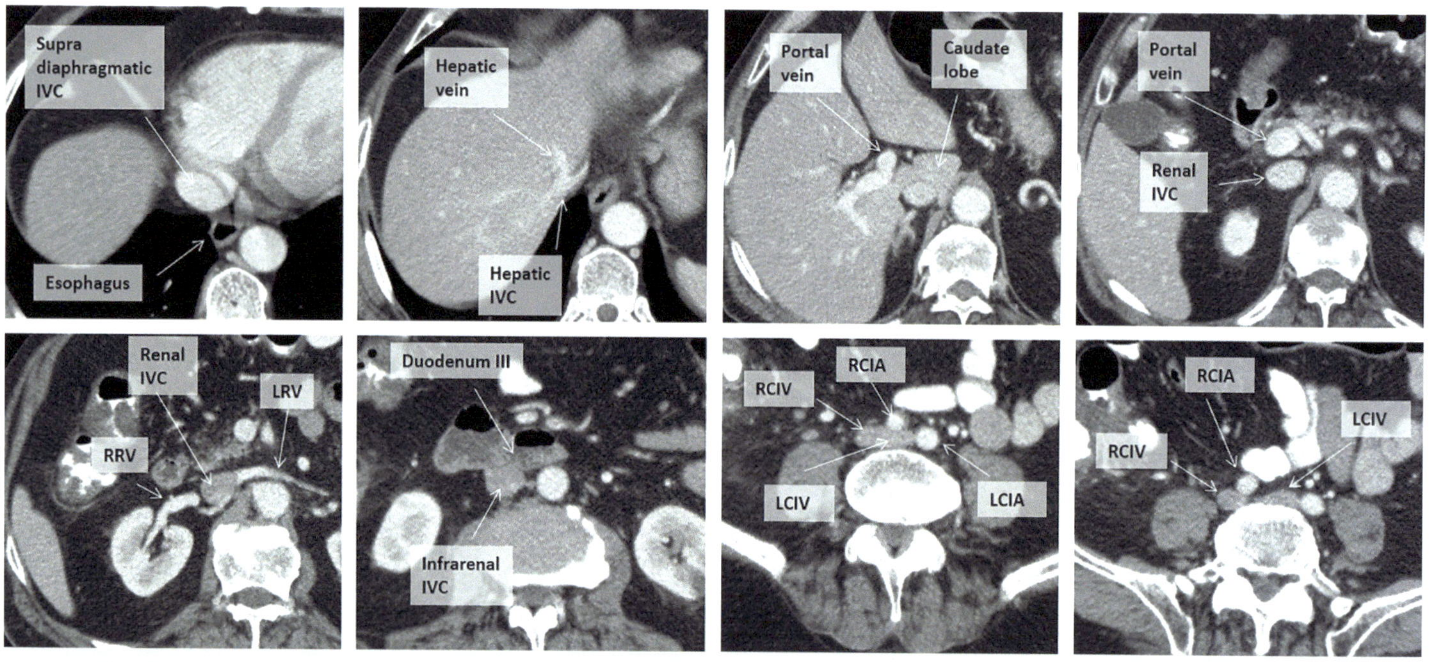

Fig. 20.2 Axial computed tomography (CT) images with contrast showing parts and relationships of the inferior vena cava (IVC). Note passage of the left common iliac vein (LCIV) behind the right common iliac artery (RCIA). LCIA, left common iliac artery; LRV, left renal vein; RCIV, right common iliac vein; RRV, right renal vein.

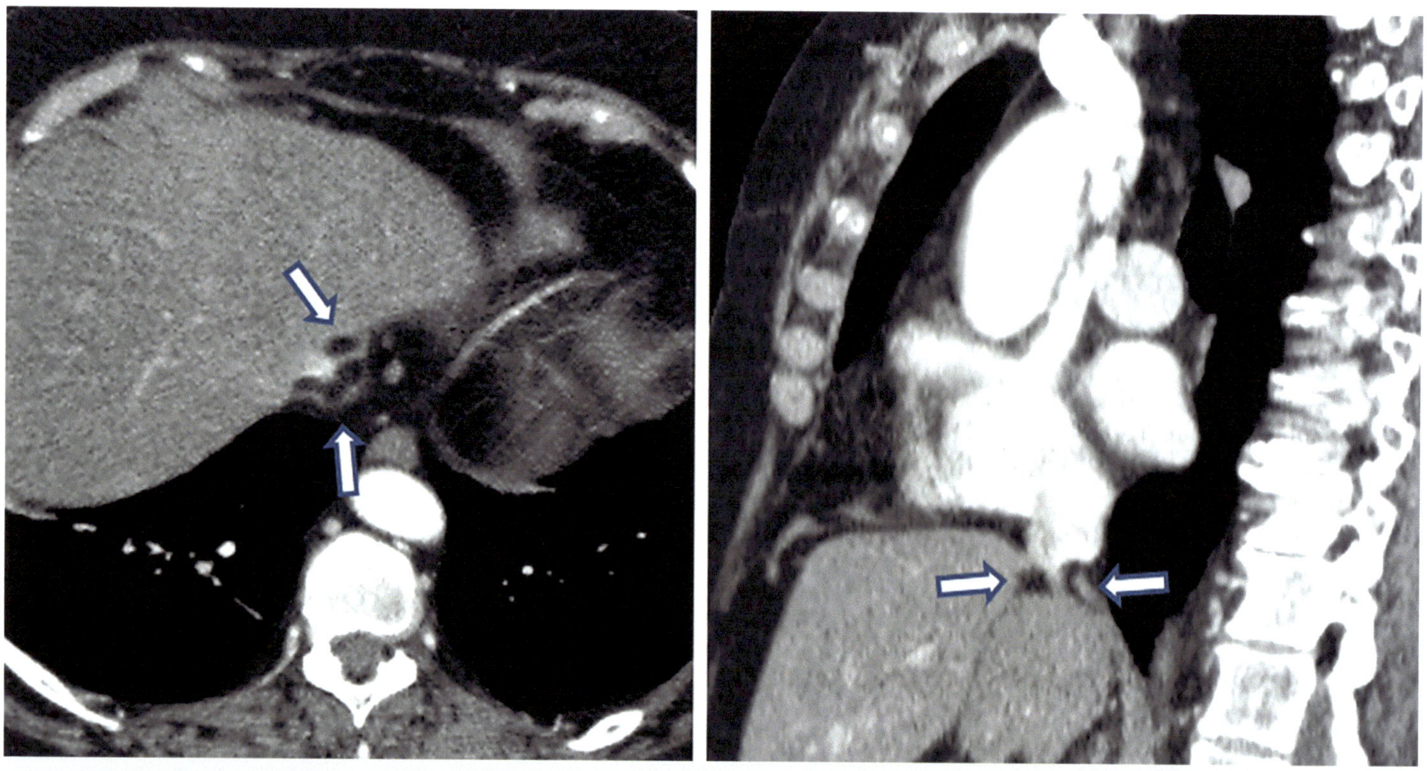

Fig. 20.3 Axial and sagittal computed tomography (CT) show accumulation of fat in the vena cava hiatus causing narrowing of the inferior vena cava in that area (*arrows*).

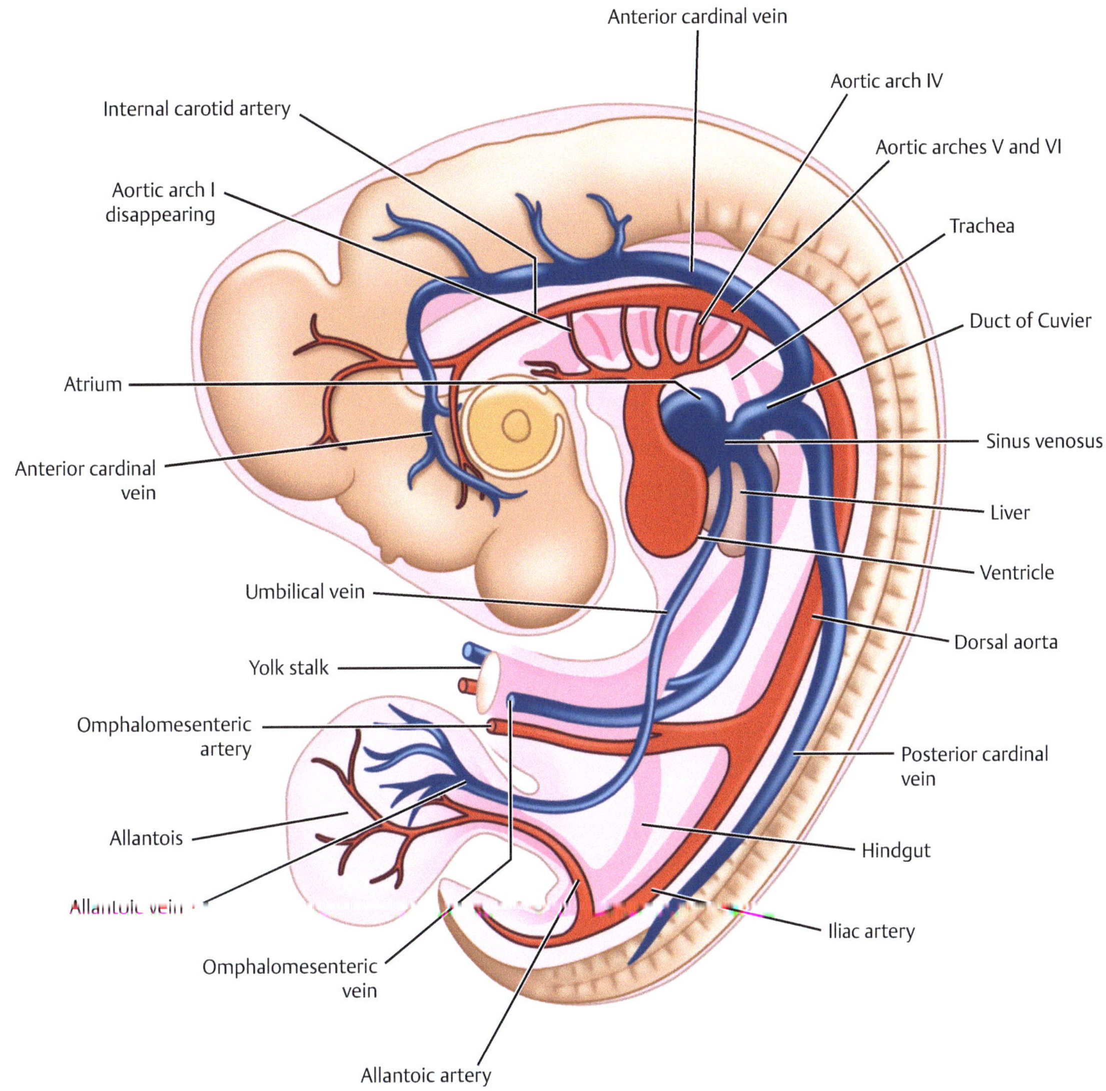

Fig. 20.4 Three pairs of venous channels during the first 4 weeks of fetal development draining into the sinus venosus. (1) The omphalomesenteric (vitelline) veins, carrying blood from the yolk sac; (2) the umbilical vein, originating in the chorionic villi and carrying oxygenated blood from the placenta; and (3) the common cardinal veins (ducts of Cuvier), draining the body of the embryo.

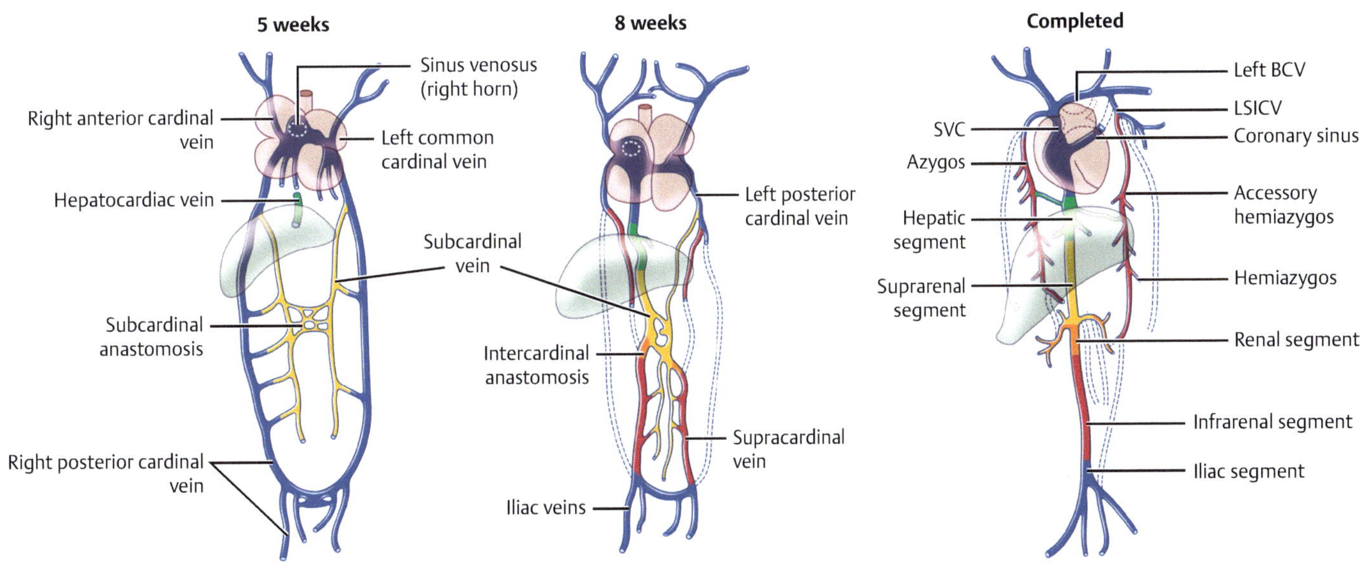

Fig. 20.5 Development of the cardinal veins. Anterior views. Acc. HAz, accessory hemiazygos; BCV, brachiocephalic vein; LSICV, left superior intercostal vein; SVC, superior vena cava.

(▶ Fig. 20.6). After these changes, the sinus venosus only receives the two common cardinal veins and the hepatocardiac portion of the IVC. Later, the left common cardinal vein regresses.

Umbilical Veins

In the 4th and 5th weeks of pregnancy, the umbilical veins pass on each side of the liver (▶ Fig. 20.6). The proximal part of both umbilical veins and the remainder of the right umbilical vein then disappear so that the left vein is the only one to carry blood from the placenta to the liver; then a direct communication forms between the left umbilical vein and the right hepatocardiac channel, the ductus venosus (▶ Fig. 20.6). After birth, the left umbilical vein and ductus venosus are obliterated and form the ligamentum teres and ligamentum venosum, respectively.

Cardinal Veins

This systemic venous network consists of the anterior cardinal veins, which drain the upper part of the embryo superior to the primitive heart and the posterior cardinal veins, which drain the lower part of the embryo inferior to the heart. The anterior and posterior veins join together and form the common cardinal vein connecting to the sinus venosus (▶ Fig. 20.5).

During the 5th to the 7th weeks, the posterior cardinal veins start to regress and three additional major veins will be formed: the subcardinal veins, which mainly drain the kidneys, the supracardinal veins, which drain the body wall by way of the intercostal veins, and the sacrocardinal veins, which drain the lower extremities (▶ Fig. 20.5). Multiple anastomoses form between the subcardinal and supracardinal veins and these veins become dominant on the right side. The anastomosis between the subcardinal veins forms the left renal vein. The rest of subcardinal veins disappear, and only their distal portions remain as the gonadal veins. The caudal ends of the subcardinal veins remain as the gonadal vessels.[3]

The anastomosis between the sacrocardinal veins forms the left common iliac vein. The right sacrocardinal vein becomes the sacrocardinal segment of the inferior vena cava.

The cranial ends of the supracardinal veins form the azygous system. The 4th to 11th right intercostal veins empty into the right supracardinal vein and with a portion of the posterior cardinal vein form the azygos vein. On the left, the 4th to 11th intercostal veins enter into the left supracardinal vein, then known as the hemiazygos vein, and empty into the azygos vein (▶ Fig. 20.5).

IVC Diameters and Flow Pattern

The normal IVC diameter is about 20 mm with 3-mm standard deviation. The vena cava diameter will be affected by respiration, intraabdominal pressure, venous flow rate, drugs, position, central venous pressure, patient stature, congestive heart failure, tricuspid regurgitation, cor pulmonale, and other pathologic conditions affecting the venous diameter.

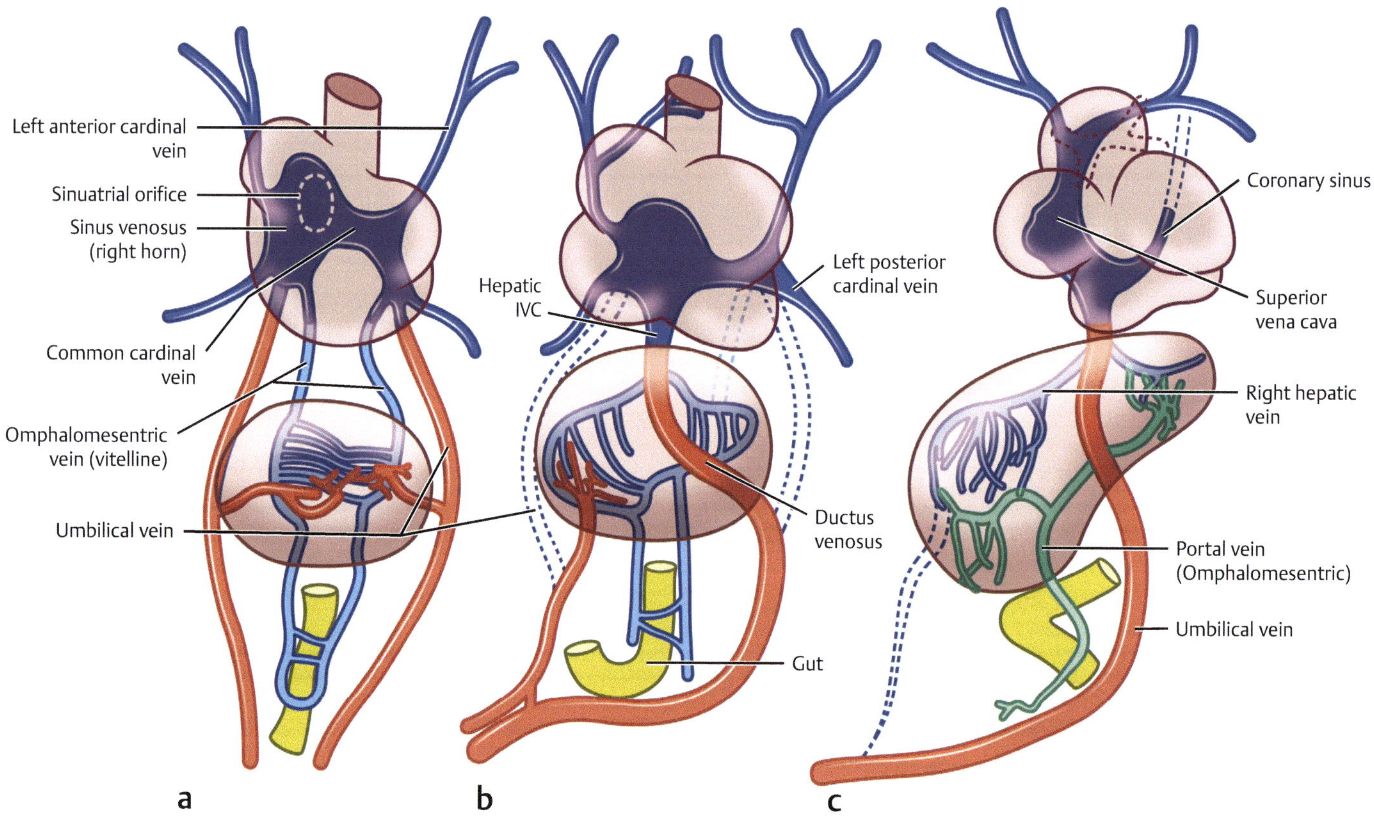

Fig. 20.6 (a–c) Development of the sinus venosus and omphalomesenteric system between 5 and 12 weeks. The timing for the cartoons is speculative. The right vitelline vein (right hepatocardiac channel) ultimately forms the hepatocardiac portion of the inferior vena cava (IVC). R, right; SVC, superior vena cava.

The IVC diameter of more than 28 mm is known as megacava. This large diameter is problematic during IVC filter deployment and may cause dislodgment of the filter.[5]

The collapsibility index of IVC is measured in percent with this formula:

$$cIVC = (Dmax - Dmin)/Dmax$$

The index is used to evaluate the central venous pressure and to decide whether fluid infusion is necessary.

Duplex ultrasound scanning is the method of choice for evaluating the venous flow. Magnetic resonance imaging (MRI) using phase contrast technique is also used to evaluate the flow in the IVC. Since there is no valve in the IVC, reverse flow is common.

IVC flow during breath holding depends on the phase of cardiac cycle. It is shown that the flow will be maximal at the time of ventricular systole (30 to 45 cm/s), forward but less stronger during diastole, and minimal or reversed at atrial systole.[6–8] During inspiration, velocity transiently increases. Reduction of flow velocity is seen with the Müller and Valsalva maneuvers with an immediate overshoot when the maneuvers end. Coughing produces a reduction of flow velocity and leg exercise increases the flow velocity. During computed tomography (CT) angiography of chest, it is common to see reflux of contrast from the right atrium into the IVC and even hepatic veins especially in patients with increased right heart pressure.[6–8] Enlargement of the IVC is a common finding in patient with right heart failure.

Anomalies

Normal IVC and its tributaries are shown in ▶Fig. 20.7. A common classification for anomalies of the IVC is based on abnormal development of each segment (▶Table 20.1).[3,4,9–13]

The most frequent developmental IVC anomalies include left IVC, duplicated IVC, and azygos continuation of the IVC. Renal vein anomalies are also included in the list of IVC anomalies. Circumaortic left renal vein (1.5–8.7%) and retroaortic left renal

Table 20.1 Classification of inferior vena caval anomalies

Anomalies of the posterior cardinal veins

- Retrocaval/circumcaval ureter

Anomalies of the subcardinal veins

- Interruption of the inferior vena cava with azygous/ hemiazygous continuation

Anomalies of the supracardinal veins

- Persistence of the left supracardinal vein—left inferior vena cava
- Persistence of both left and right supracardinal veins
- Veins double inferior vena cava

Anomalies of the renal segment

- Circumaortic venous ring
- Retroaortic renal vein
- Multiple renal veins

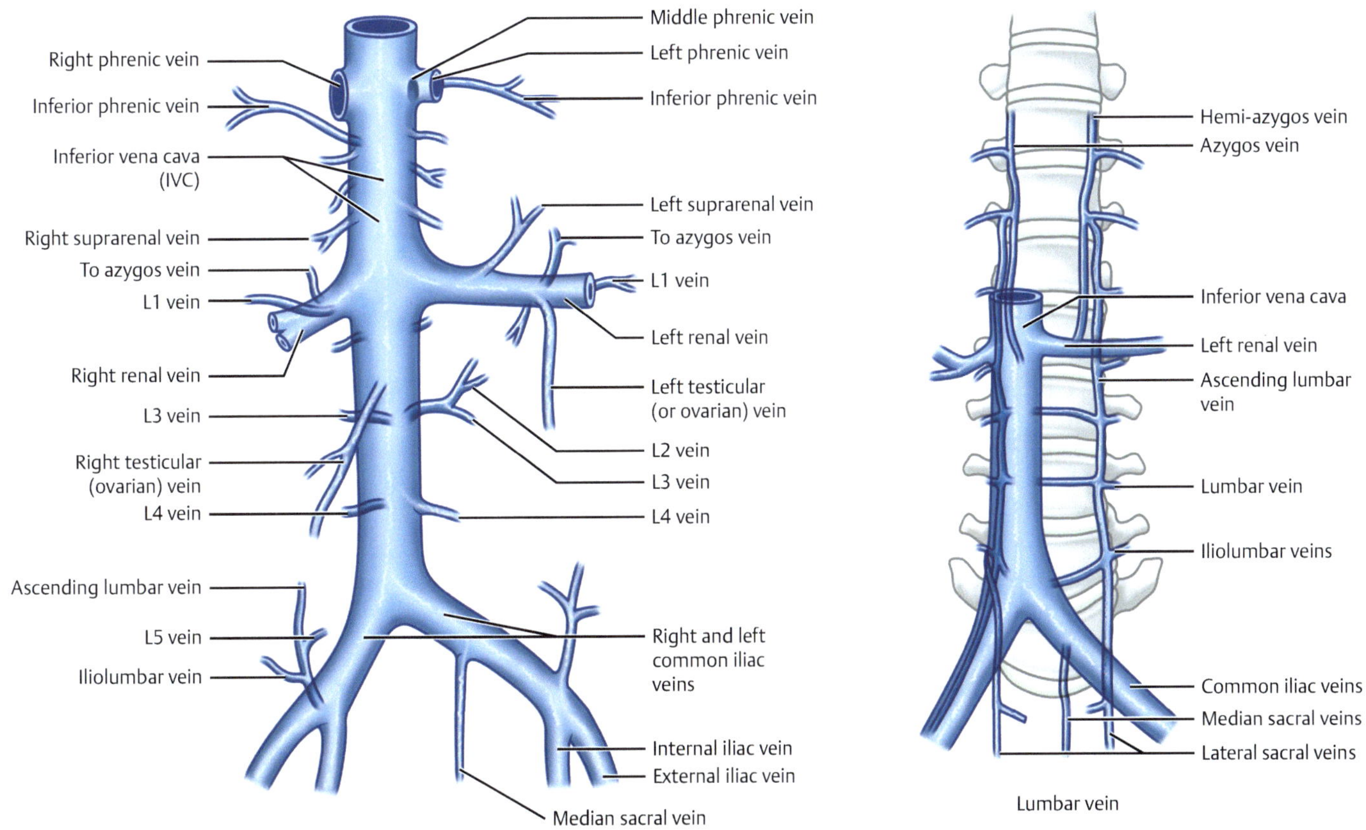

Fig. 20.7 Schematic anatomy of inferior vena cava (IVC) and its tributaries.

veins (1.2–3.3%) are two common findings in abdominal imaging studies.[14]

Multiple renal veins variant is seen in up to 15%. It is more frequent on the right side presumably due to the fact that the postsubcardinal anastomosis does not regress and participates in the formation of the renal segment of the IVC. Consequently, the dorsal renal vein may persist as well. On the contrary, the post-subcardinal anastomosis regresses on the left side and facilitates the regression of the left dorsal renal vein.

Interrupted IVC

In the embryo, the supracardinal vein empties cranially into the posterior cardinal vein (▶Fig. 20.5). The supracardinal vein has three separate segments. The lowermost portion becomes the infrarenal segment of the IVC. The uppermost portion becomes the azygos vein. The midsegment will be obliterated. Abnormal development of the subhepatic portion of the IVC results in interruption of the IVC in this region. Therefore, blood from the lower extremities and abdomen is forced to continue through the entire length of the supracardinal vein. Thus, the middle segment of the supracardinal vein does not regress and results in azygos continuation of IVC.[15]

In this anomaly, the hepatic veins remain connected to the IVC segment above the hepatic segment, which enters in the right atrium (▶Fig. 20.8, ▶Fig. 20.9, ▶Fig. 20.10). Because interrupted IVC with azygos continuation usually does not result in a physiologic abnormality, this venous abnormality does not result in clinical manifestations. Its clinical importance is due to its frequent association with the cardiac anomaly, heterotaxy syndrome, and polysplenia. In adult, it is not rare to find this anomaly in patients with normal hearts and rarely in patients with asplenia.

In rare cases, the infrahepatic segment of the IVC may continue to both right and left SVC via bilateral azygos veins. In patients

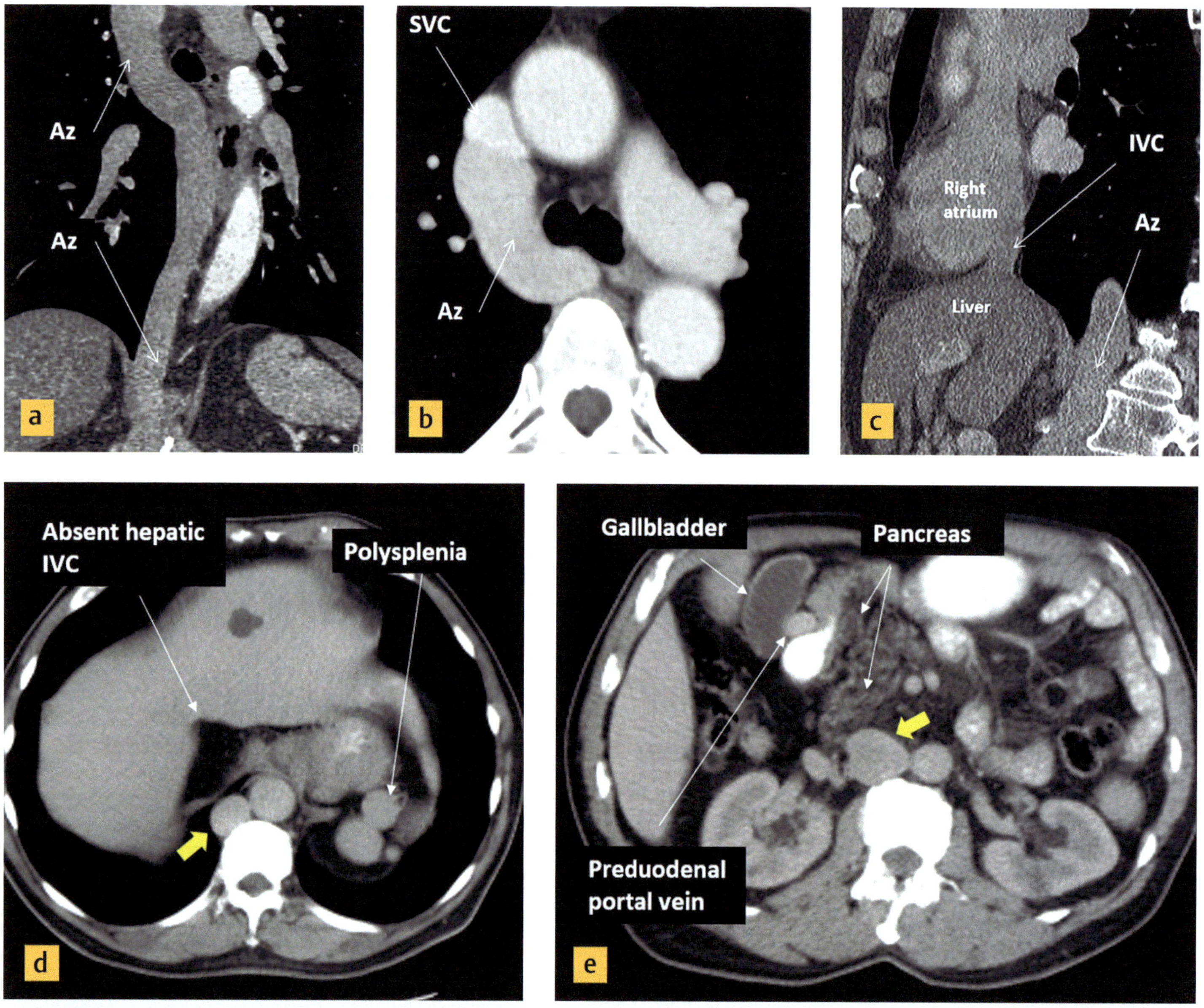

Fig. 20.8 Azygos (Az) continuation of the inferior vena cava (IVC). **(a)** Coronal, **(b)** axial, and **(c)** sagittal computed tomography (CT) images. Suprahepatic segment of the IVC is intact and connects to the right atrium. **(d, e)** Polysplenia syndrome. Azygos continuation of the IVC is present (*yellow arrows*). Absence of intrahepatic segment of the IVC due to failure of the union between the hepatic and prerenal segments during embryologic development. Evidence of polysplenia. The pancreas is mainly collected on the right side. Preduodenal course of the portal vein is seen. Findings consistent with intestinal malrotation were also present (not shown) with the third portion of the duodenum passing anterior to the pancreas. The colon was mainly on the right side and the transverse colon passed behind the mesenteric vessels. SVC, superior vena cava.

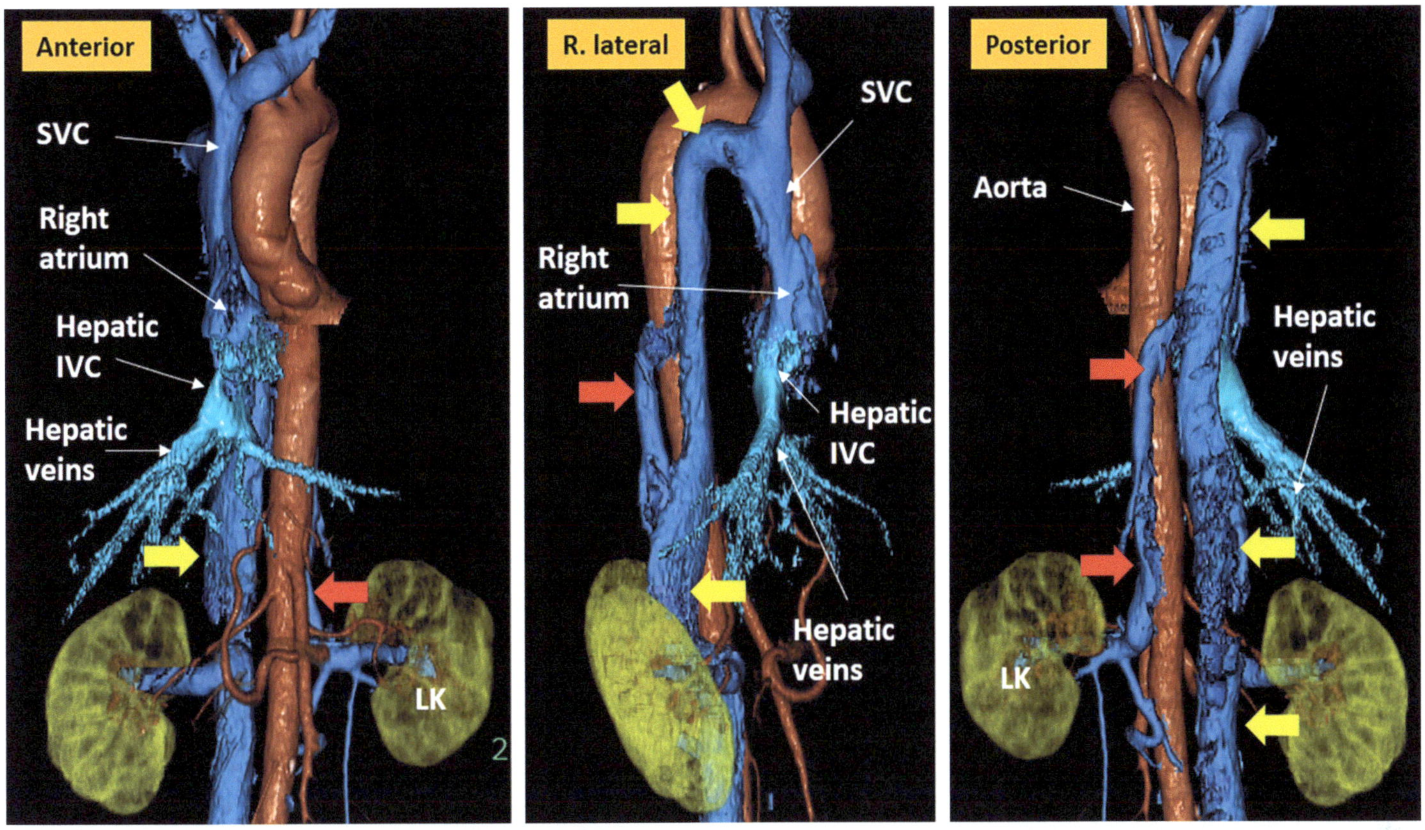

Fig. 20.9 Azygos continuation of the inferior vena cava (IVC) shown by volume-rendered computed tomography (CT). Enlarged azygos vein (*yellow arrows*) and hemiazygos vein (*red arrow*) drain into the superior vena cava (SVC). There is lack of contiguity between the prerenal segment of the IVC and the hepatic segment of IVC. The hepatic veins (*light blue*) drain directly into the right atrium. The renal portion of the IVC receives blood flow from right kidney and passes posterior to the diaphragmatic crura to enter the thorax as the azygos vein. The blood flow from left kidney (LK) passes posterior to the diaphragmatic crura to enter the thorax as the hemiazygos vein. LK, left kidney.

with situs inversus, the anomaly presents as hemiazygous continuation of a left-sided IVC.

Interrupted IVC with azygos continuation can complicate cardiac catheterization and interventional procedures such as radiofrequency catheter ablation. In patients whose associated cardiovascular anomalies require surgical therapy that involves redirection of the systemic venous return to the pulmonary arteries (bidirectional Glenn and modified Fontan procedures), awareness of the anomaly and appropriate surgical planning are important.

Absence of the Infrarenal IVC

Complete absence of the infrarenal IVC with preservation of the suprarenal segment is an extremely rare anomaly. This condition is not embryologic in origin and may be a secondary perinatal thrombosis of the IVC. Affected patients are prone to develop chronic venous insufficiency and deep vein thrombosis. Lower extremity venous return occurs via the ascending lumbar veins, which drain into the azygous–hemiazygous system.[16,17] An acquired form of this condition occurs in patient with chronic thrombosis of the IVC (▸Fig. 20.11).

Duplication

This anomaly occurs when the left sacrocardinal vein fails to lose its connection with the left subcardinal vein (▸Fig. 20.5). The left common iliac vein may or may not be present, but the left gonadal vein remains in normal condition.

The incidence of IVC duplication is up to 3% in the general population.[9,18,19] This anomaly is usually silent and clinically nonsignificant but the importance of these anomalies is during surgery or interventional procedures (▸Fig. 20.12, ▸Fig. 20.13, ▸Fig. 20.14).

A morphologic classification of this anomaly according to the size of IVCs relative to the preaortic trunk has been described.[9] Type of connection of the iliac veins to the left or right IVCs is also variable (▸Fig. 20.12, ▸Fig. 20.13). Double IVC with retroaortic left renal vein and azygos continuation of the IVC is also reported. This anomaly could be associated with retroaortic right renal vein, with hemiazygos continuation of the IVC.[20]

From a clinical point of view, in case of recurrent pulmonary embolism following placement of an IVC filter, IVC duplication could be the underlying cause. Surgeons, interventionists, and urologists should be aware of these anatomical variants to avoid hemorrhage and other complications. Misdiagnosis of the aberrant veins with lymphadenopathy should be avoided.

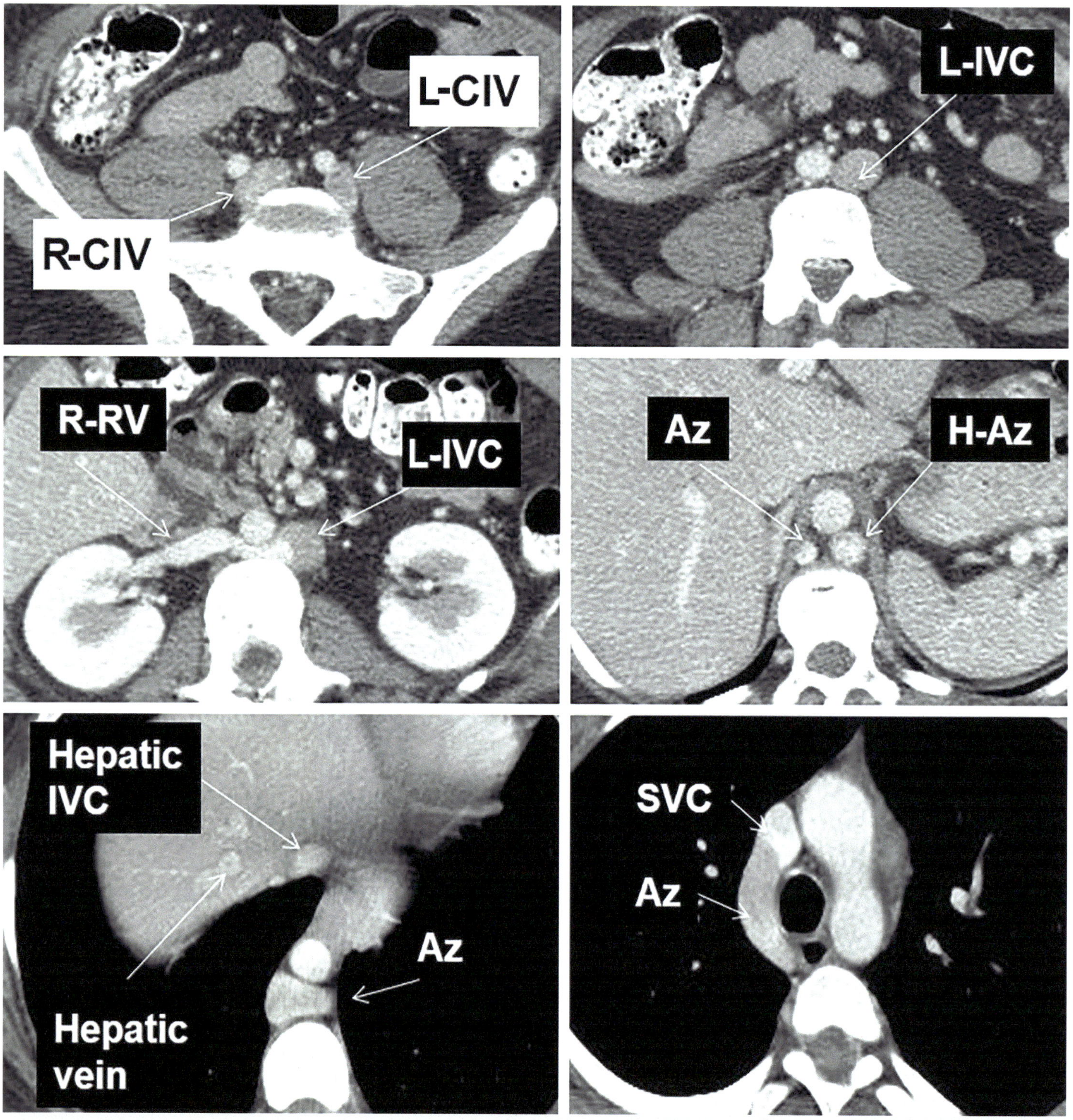

Fig. 20.10 Azygos (Az) continuation of the left inferior vena cava (L-IVC) which receives both renal veins and continues superiorly as the azygos veins. A retroaortic right renal vein (R-RV) is seen. Hepatic segment of the IVC is intact receiving the hepatic veins. H-Az, hemiazygos; L-CIV, left common iliac vein; R-CIV, right common iliac vein; SVC, superior vena cava.

Double Right IVC

It arises from the persistence of the right supracardinal and subcardinal veins with confluence into a single vessel at the level of kidney. The lateral course of the ureter with respect to the two vessels rules out a persistence of the right posterior cardinal vein.[1]

Left-Sided IVC

The prevalence of this anomaly is 0.2 to 0.5%.[3,21] Abnormal persistence of left supracardinal vein with regression of the right supracardinal vein results in this anatomic variant. The left renal vein joins to the left-sided IVC and together join the normal right IVC at the level of the renal segment of the IVC (▶ Fig. 20.15). This

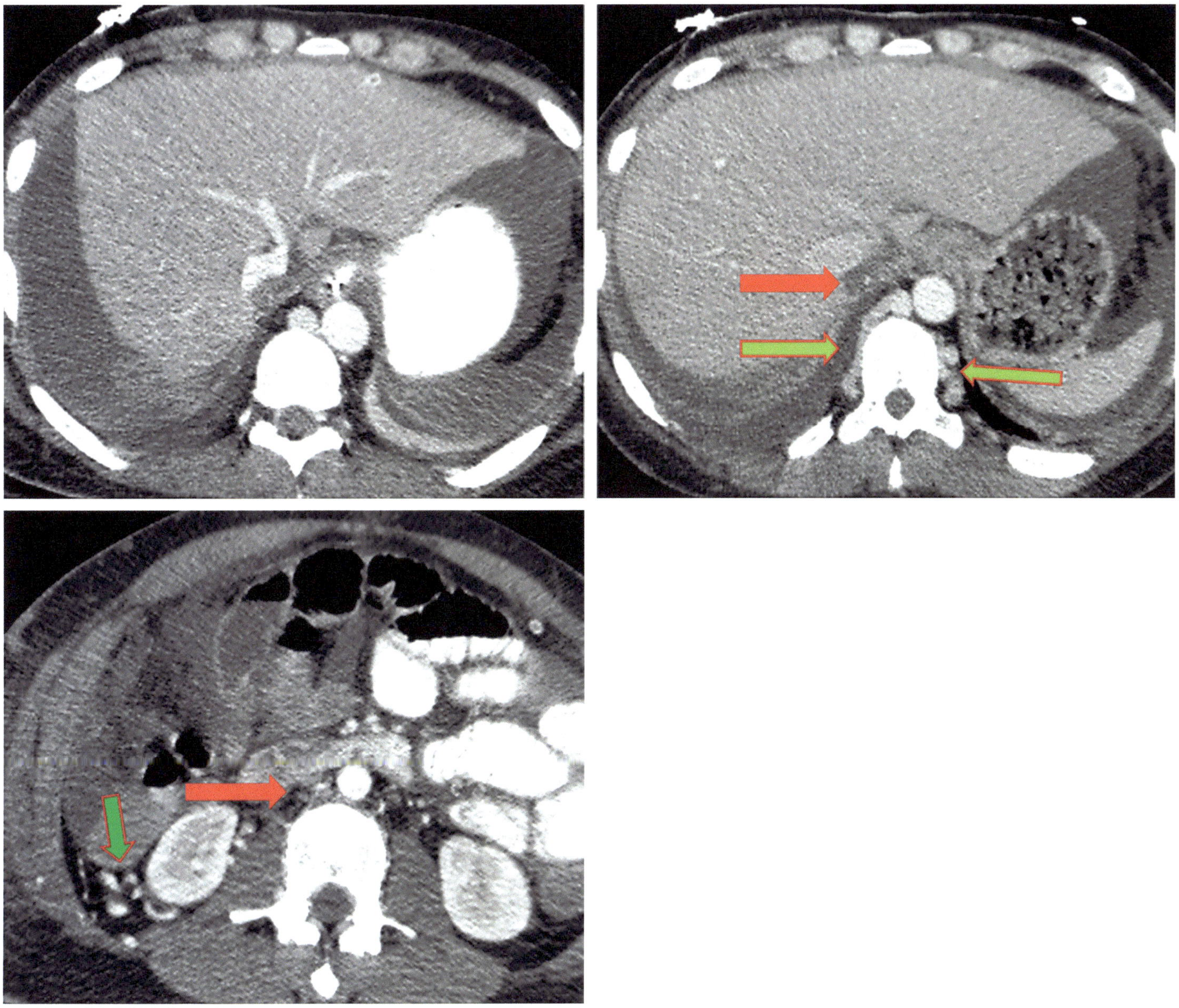

Fig. 20.11 Chronically occluded inferior vena cava (*red arrows*). Large lumbar collaterals are seen filling the azygos (*green arrows*).

anomaly has no clinical symptom, but it may be mistaken for paraaortic adenopathy or result in difficulty in performing interventional procedure, infrarenal IVC filter placement, or surgical repair of an aortic aneurysm.[22] A left-sided IVC may cause confusion between venous or arterial access line.[23]

Retroaortic Left Renal Vein

The occurrence of this anomaly is 1.7 to 3.4%. It is usually a benign isolated anomaly or associated with other IVC anomalies (▶Fig. 20.16). Retroaortic left renal vein anomaly arises from persistent intersupracardinal anastomoses passing dorsal to the aorta and regression of the intersubcardinal anastomoses, hence absence of the anterior left renal vein[24,25] (▶Fig. 20.5). In rare cases, compression of the retroaortic left renal vein by aorta (nutcracker phenomenon) can result in hematuria, periureteric varices, and hypertension.[26]

Circumaortic Left Renal Vein

Persistence of both the intersupracardinal and intersubcardinal veins results in a circumaortic renal vein, with one vein coursing anterior and the other posterior to the aorta.[25] The prevalence of this anomaly is 2.4 to 8.7% (▶Fig. 20.17). The clinical importance of an anomalous left renal vein is during nephrectomy.

Retrocaval Ureter

This anomaly is also known as circumcaval ureter (▶Fig. 20.18, ▶Fig. 20.19). The posterior cardinal vein persists and the supracardinal vein fails to develop.[2,3] The proximal ureter courses posterior to the IVC, then emerges to the right of the aorta, coming to lie anterior to the right iliac vessels (▶Fig. 20.18). It almost always occurs on the right side and is very rare on the left side. Patients with retrocaval ureter may present with ureteral obstruction or

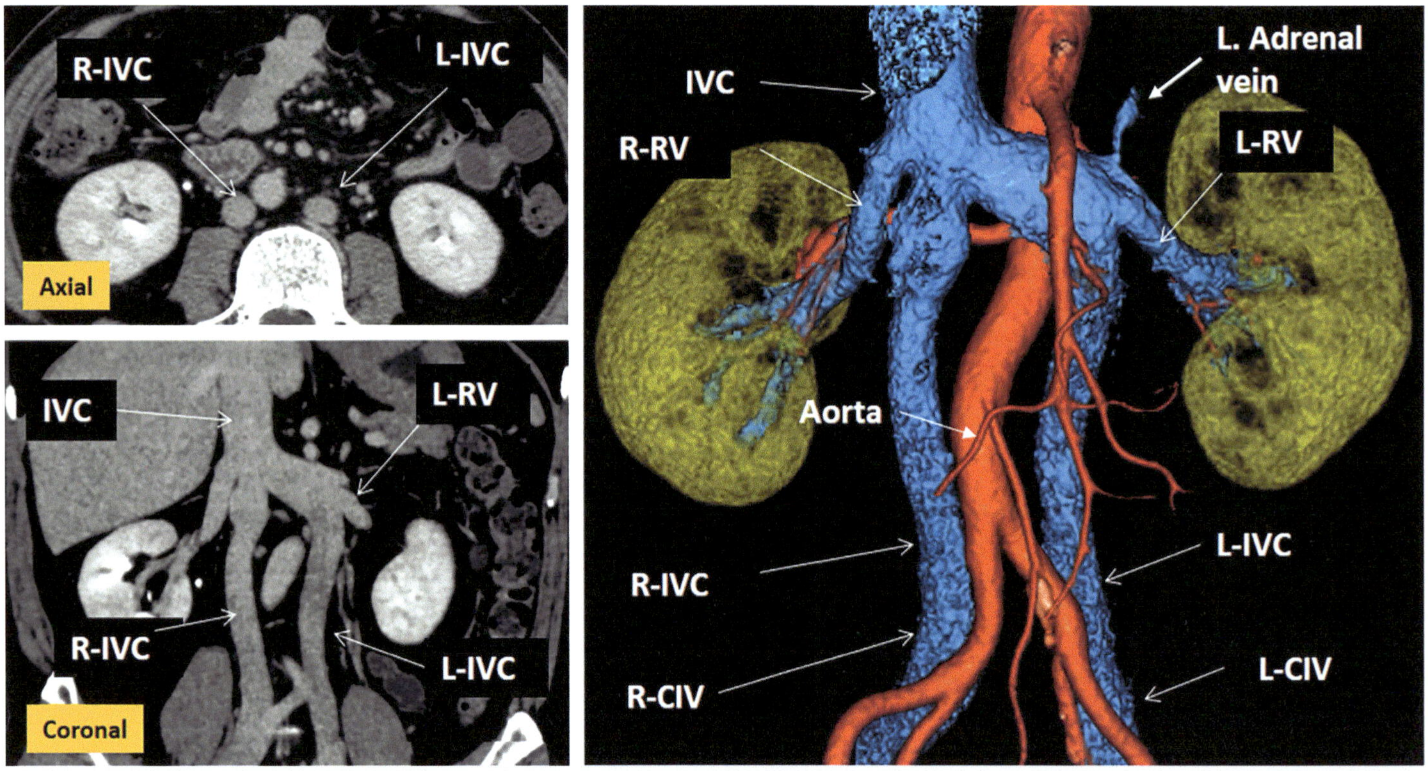

Fig. 20.12 Duplicated inferior vena cava (IVC) shown by computed tomography (CT) angiography. The right IVC (R-IVC) is same size as the left IVC (L-IVC). Right IVC is receiving the right common iliac vein (R-CIV) and the L-IVC is receiving the left common iliac vein (L-CIV). The left renal vein (L-RV) crosses anterior to the aorta in the normal fashion to join a normal prerenal IVC. R-RV, right renal vein.

recurrent urinary tract infections. Therapeutic options include surgical transfer of the ureter anterior to the IVC.[3,4]

IVC Webs

Normal IVC does not contain venous valves. IVC web is an uncommon anomaly that is formed as a result of a congenital vascular anomaly or as a complication of thrombus formation.[27] The structure and length of this anomaly varies from a complete membrane to an incomplete perforated membrane.[21] A large web could cause hepatic outflow obstruction. In congenital or acquired form of IVC web, patients may present with Budd-Chiari syndrome. Treatment of the stenosis or occlusion includes angioplasty, stent placement, or transjugular intrahepatic portosystemic shunt.

Congenital Portocaval Shunts

Occurrence of the congenital portocaval shunts can be explained by close association of the IVC and vitelline veins development. Congenital portocaval shunts can be extrahepatic or intrahepatic. The extrahepatic shunts have two subtypes. Type I shunt is a complete end-to-side shunt between the portal vein and the IVC, and the main portal vein distal to the shunt is atretic. Type II extrahepatic shunt is a partial side-to-side communication between the portal vein and the IVC. Liver transplantation is the only definitive treatment for type I shunt, whereas treatment of patients with type II shunt is ligation or coil embolization of the shunt vessel.[24]

Intrahepatic portosystemic shunts are categorized into four types. The first and most common type consists of a single large tube that connects the right portal vein to the IVC. The second type is a localized peripheral shunt in which one or more communications are found with a single hepatic segment. The third type is a portosystemic shunt through an aneurysm, and in fourth type multiple communications between peripheral portal and hepatic veins exist. A persistent ductus venosus could be considered as a fifth type of portosystemic shunt. The presumed basis for intrahepatic portosystemic shunts is a persistent communication between the vitelline veins of the omphalomesenteric system and the sinus venosus due to a focal absence of sinusoid formation.[28–30]

IVC Obstruction

Inferior vena cava obstruction is caused by a variety of pathologies including congenital absence, thrombosis, tumor invasion, external compression, retroperitoneal fibrosis, or iatrogenic (i.e., surgical ligation, IVC filter)[31–33] (▶ Fig. 20.11, ▶ Fig. 20.20, ▶ Fig. 20.21).

According to the level of obstruction three different types are seen: distal with occlusion between the hepatic veins and the right atrium, middle between the renal and hepatic veins, and proximal below the renal veins. Different collateral pathways of

20

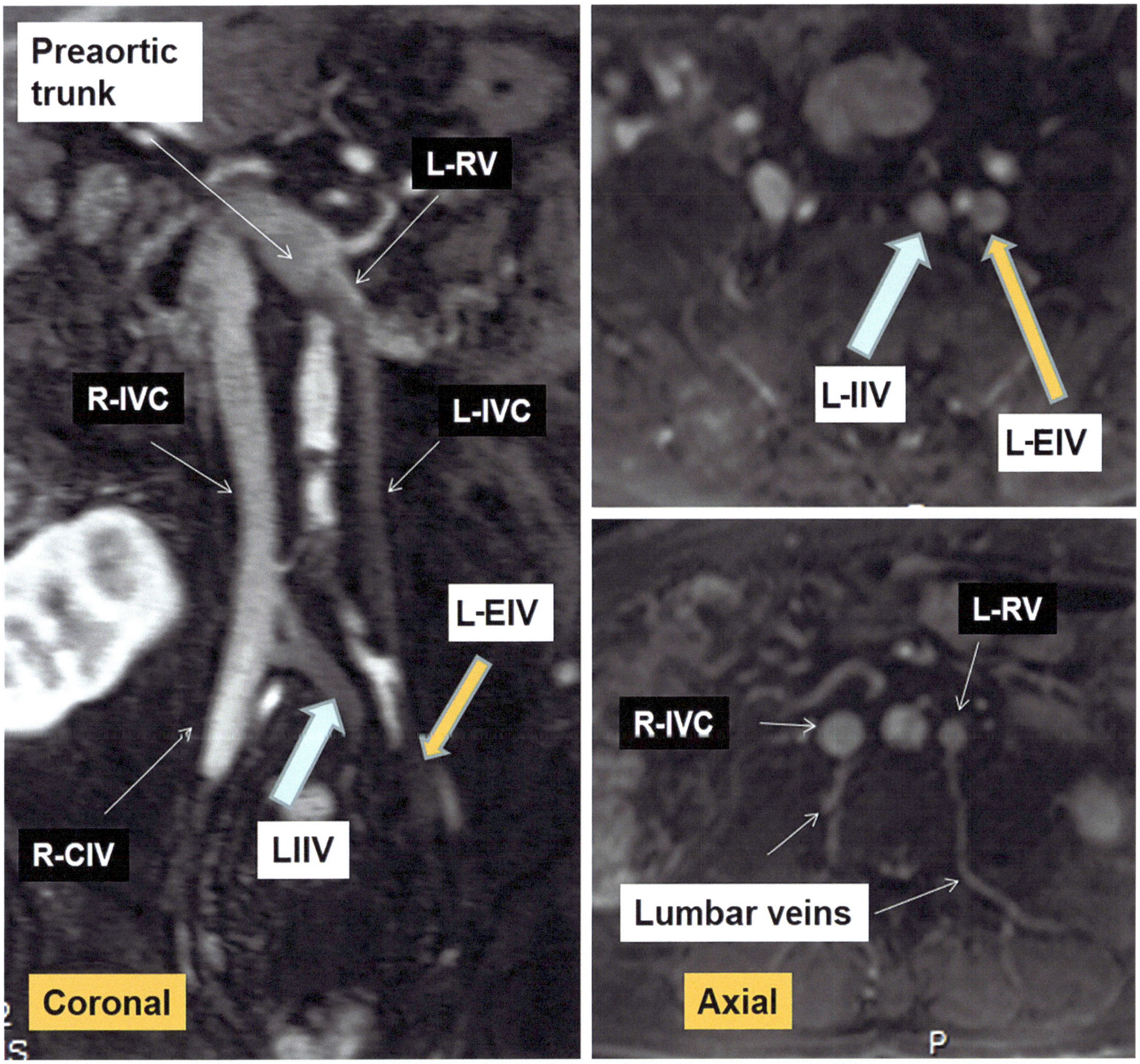

Fig. 20.13 Duplicated inferior vena cava (IVC) shown by magnetic resonance (MR) angiography. The right IVC (R-IVC) is larger than the left IVC (L-IVC) because it is receiving the right common iliac vein (R-CIV) and the left internal iliac vein (L-IIV). Whereas the L-IVC is only receiving the left external iliac vein (L-EIV). L-RV, left renal vein.

venous drainage may develop in each of these types. In proximal IVC occlusion, gonadal veins bypass the venous return from the pelvis to the patent renal and suprarenal segments of the IVC (▶Fig. 20.20). When the distal part of IVC becomes occluded, intrahepatic collaterals between the hepatic and portal veins or extrahepatic portosystemic shunts play the role. In mid IVC occlusion, collateral pathways through the azygos, hemiazygos, and lumbar plexus directs flow to the superior vena cava (▶Fig. 20.11).

The symptoms and sign of IVC occlusion are dependent on the level and duration of occlusion. In acute form of IVC occlusion, bilateral lower extremity edema and occasionally ascites will develop. In high obstruction, patients may present with Budd-Chiari syndrome. The treatment options of IVC depend on the etiology. Thrombus within the IVC is commonly managed with thrombolytic techniques with or without filter placement. For other forms of occlusion or stenosis, surgical decompression can be considered. Stenting plays a major role in relieving symptoms in this group of patients.[34–36]

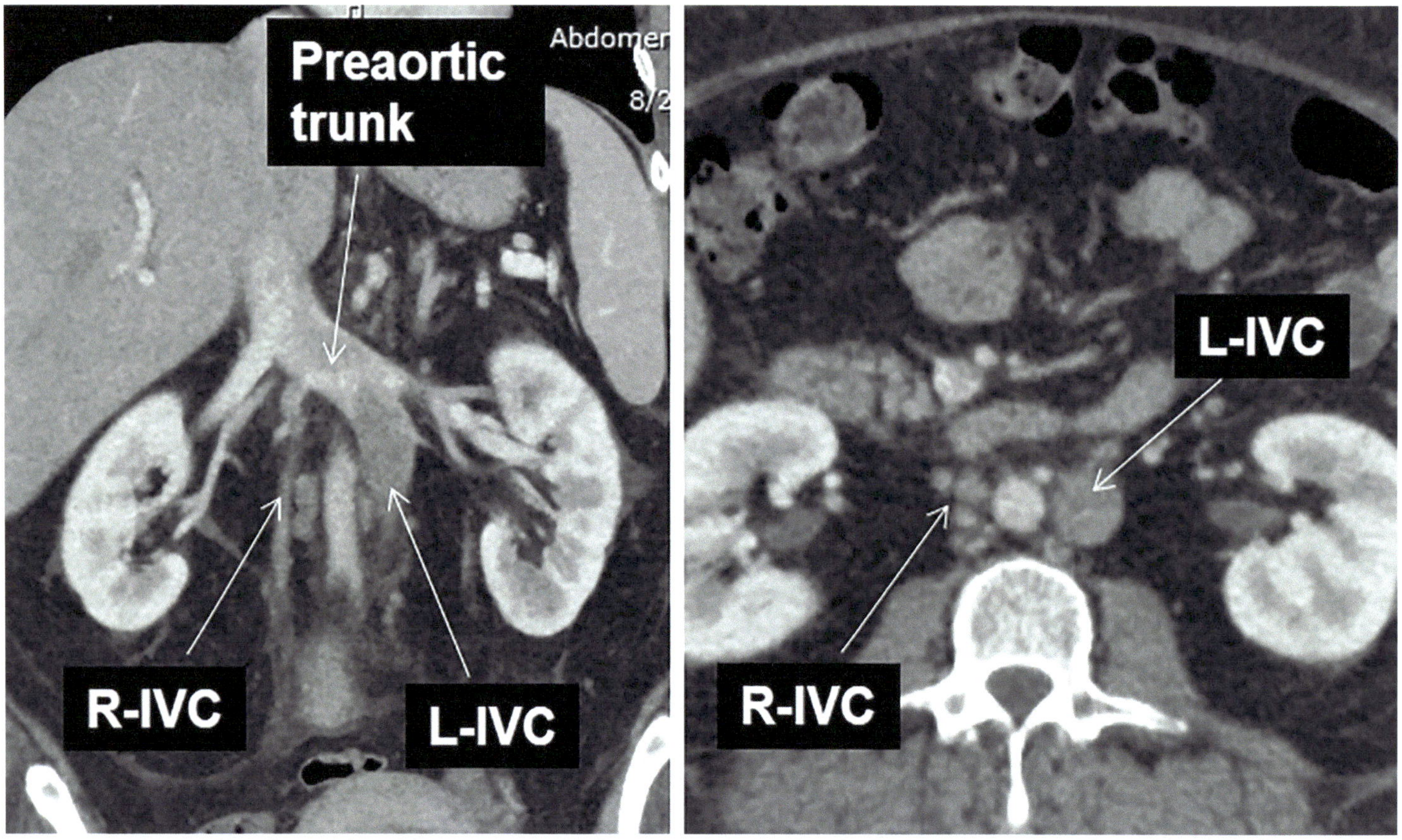

Fig. 20.14 Duplicated inferior vena cava (IVC) shown by computed tomography (CT) angiography. The right IVC (R-IVC) is atretic but the left IVC (L-IVC) is large.

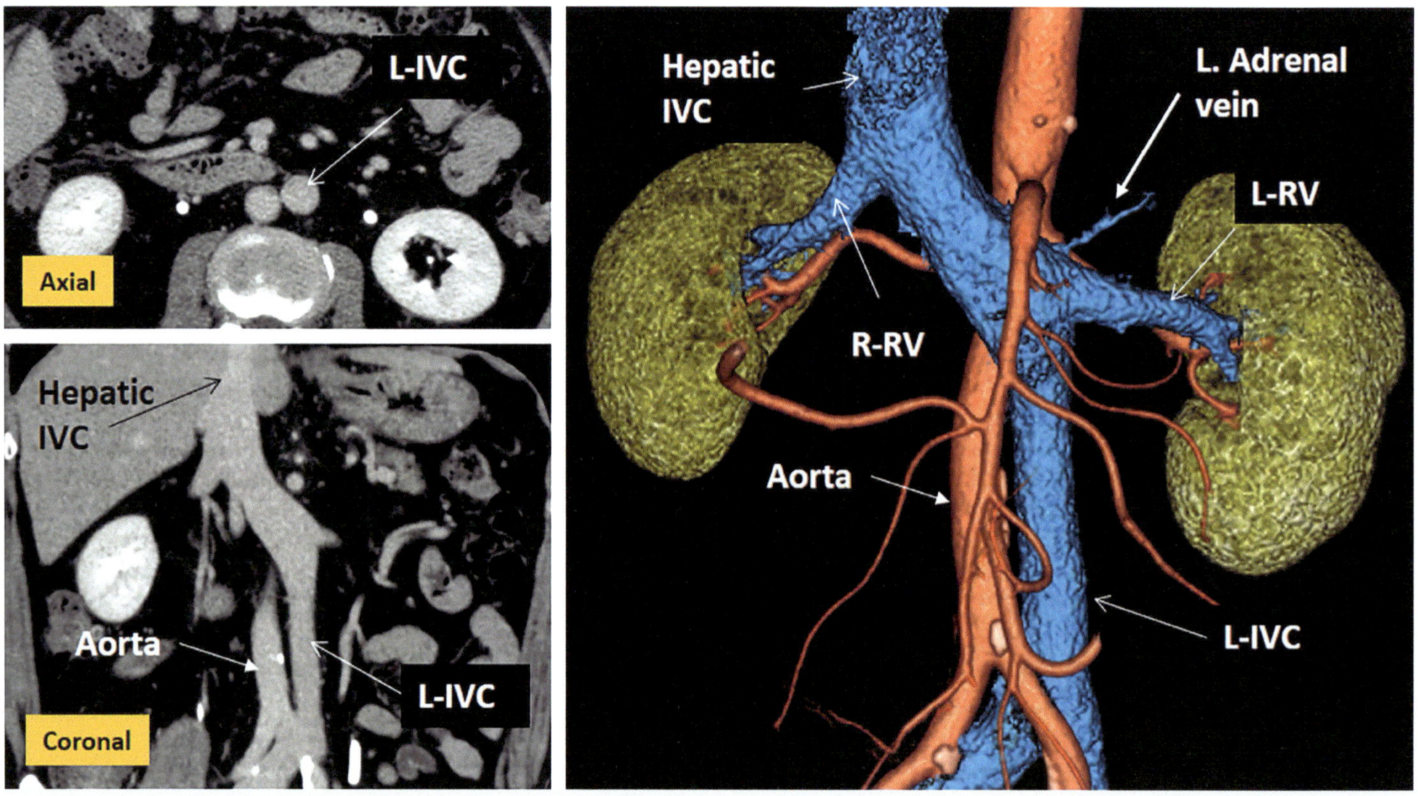

Fig. 20.15 Computed tomography (CT) images showing left inferior vena cava (L-IVC) which connects the left renal vein (L-RV) and together continue as a preaortic cava. The prehepatic segment of the IVC is on the right side which receives the right renal vein (R-RV).

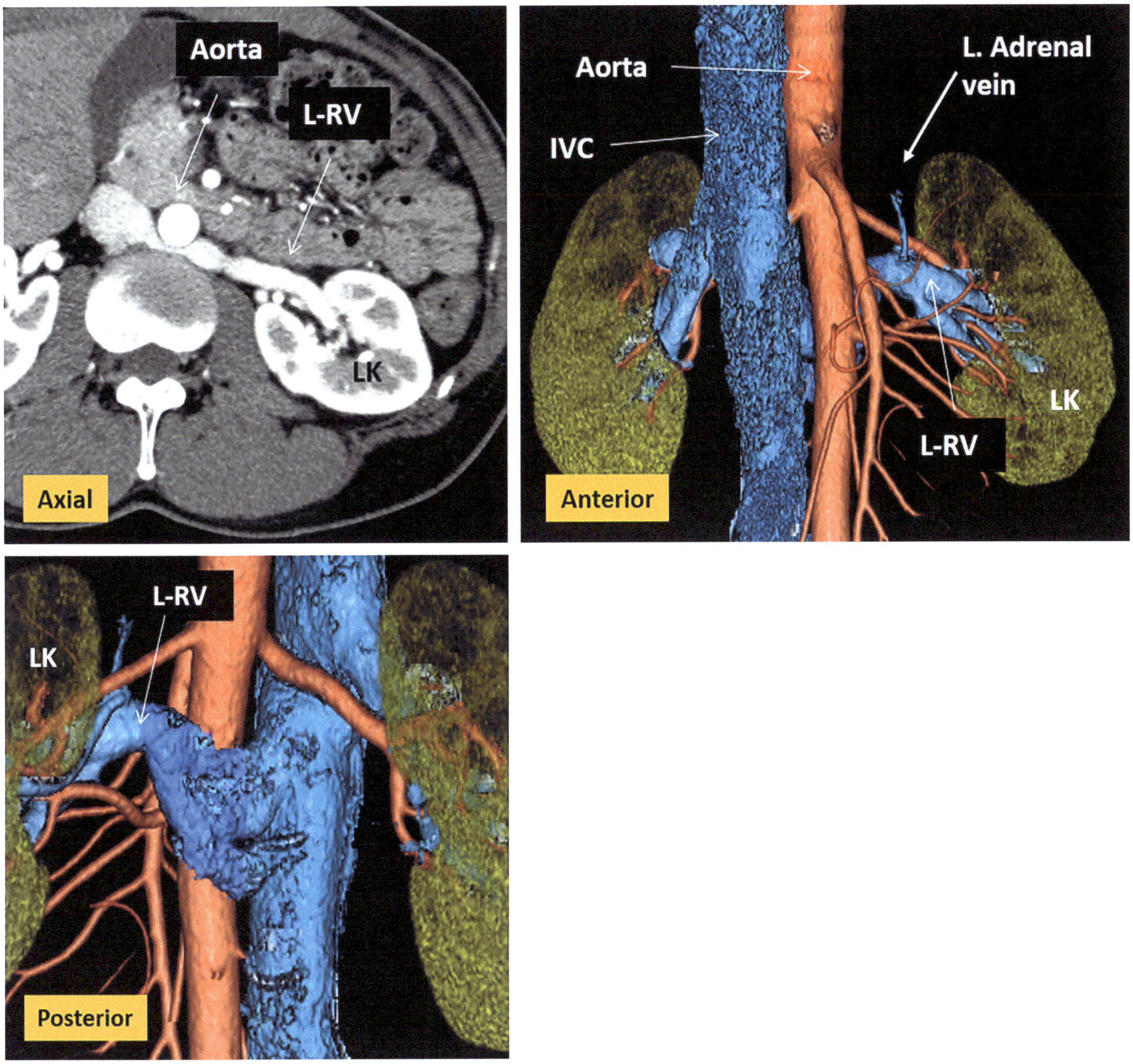

Fig. 20.16 Axial image and three-dimensional computed tomography (CT) views showing a left renal vein (L-RV) passing between the aorta and the vertebral column. IVC, inferior vena cava; LK, left kidney.

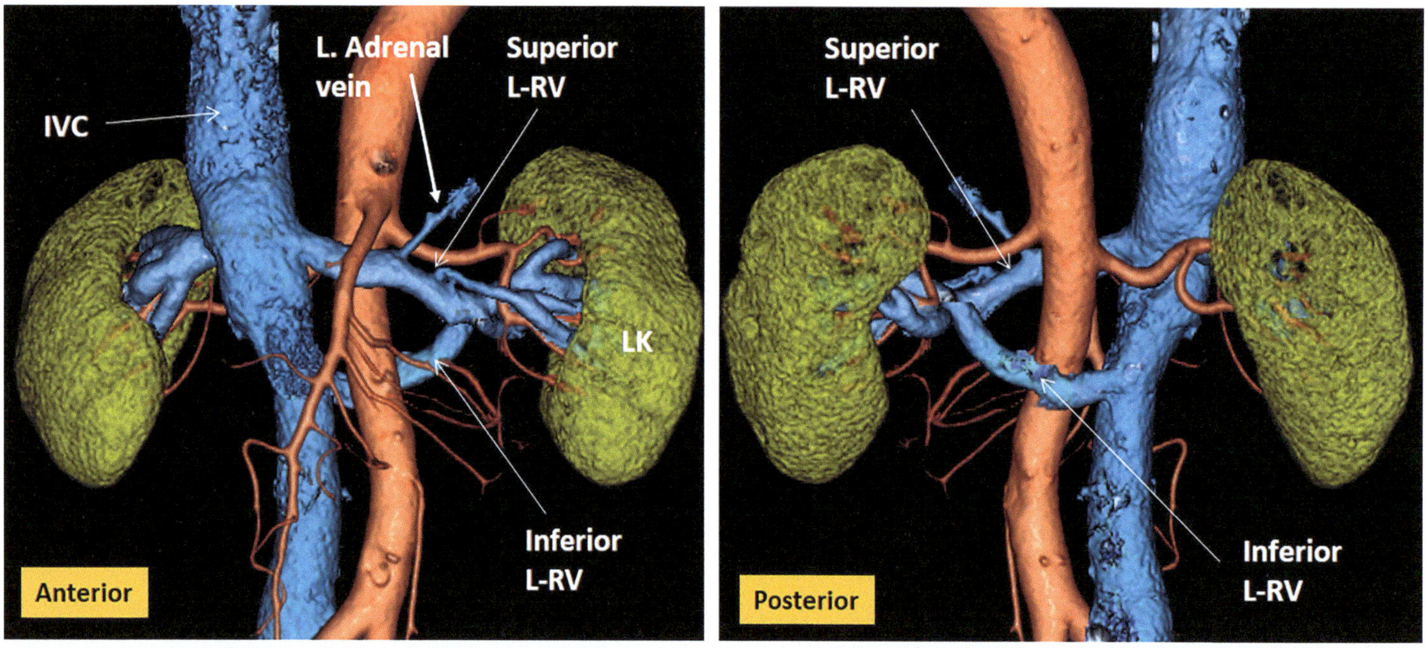

Fig. 20.17 Circumaortic left renal vein (L-RV) shown by volumetric computed tomography (CT) views. Preaortic (superior) and retroaortic (inferior) segments are seen. IVC, inferior vena cava; LK, left kidney.

Neoplasm of IVC

Primary benign neoplasm of IVC such as leiomyoma accounts for 15% of tumors arising from large veins. Leiomyoma with the IVC is usually secondary to intravascular growth of uterine leiomyoma through the gonadal veins[37] (▶Fig. 20.21). Primary IVC malignancy is extremely rare and leiomyosarcoma represents more than 75% of tumors arising from large veins.[38] On CT or MRI, the mass appears as a heterogeneously contrast material–enhancing filling defect of the IVC that may show cystic necrosis.[39] Secondary invasion of the IVC in abdominal malignancy is more common than primary IVC neoplasms. Secondary tumoral invasion of the IVC is usually due to intraluminal growth of renal cell carcinoma, adrenal carcinoma, hepatocellular carcinoma, Wilms' tumor, or primary retroperitoneal tumors. Tumors can grow from the IVC into the right atrium (▶Fig. 20.20).

Pitfalls in Imaging of the IVC

Pitfalls in imaging of the IVC are primarily due to a flow-related phenomenon or partial volume-averaging artifact during contract-enhanced CT or MRI.[40] Intraluminal filling defects occur frequently in early phase, contrast-enhanced CT of the abdomen commonly at the entrance site of renal veins into IVC due to mixing of enhanced blood in the renal veins with unenhanced venous return from the lower extremity (▶Fig. 20.1, ▶Fig. 20.2). This artifact will be eliminated on delayed venous imaging performed after 1.5 to 2 minutes of contrast injection.

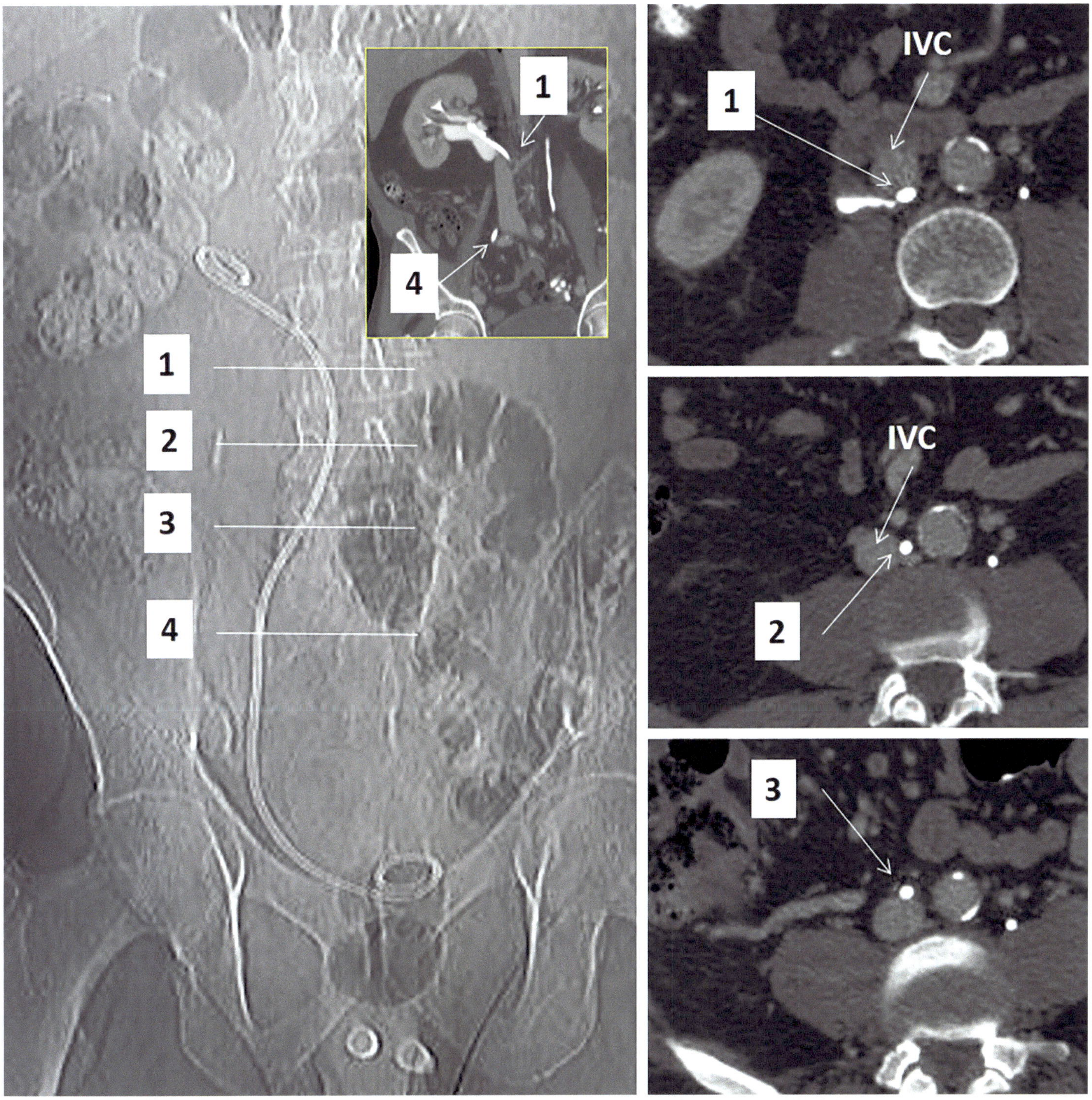

Fig. 20.18 Right retrocaval ureter. A stent is placed in the right ureter to remove obstruction. Mild right hydronephrosis was evident. IVC, inferior vena cava.

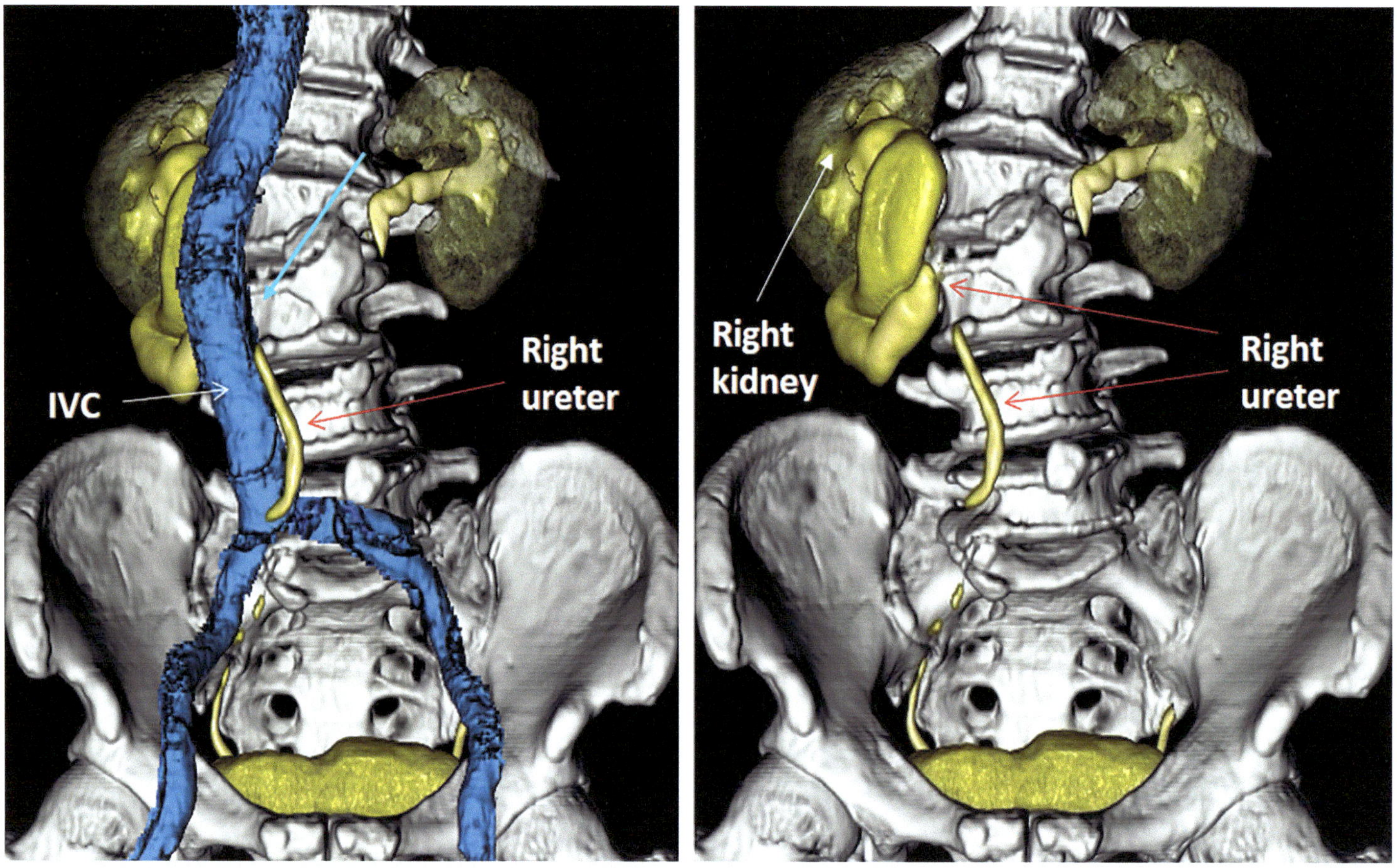

Fig. 20.19 Circumcaval ureter. Anterior three-dimensional computed tomography (CT) images showing the right ureter posterior to the inferior vena cava (IVC). There is dilatation and tortuosity of the right ureter due to obstruction causing right hydronephrosis. The ureter then courses to the left of IVC. Finally, the ureter crosses anterior to the IVC.

Early enhancement of the IVC at the confluence with the hepatic veins and right atrium is very common in first pass contrast CT angiography especially with rapid injection of above 3 mL/s through the upper extremity of jugular veins. This early enhancement is usually due to reflux of contrast from the right atrium into the IVC. It may extend into the hepatic veins especially in patients with increased right heart pressure (▶ Fig. 20.22). In low injection rates (<3 mL/s), retrograde contrast enhancement of the IVC could be CT sign of right-sided heart diseases such as tricuspid regurgitation with low sensitivity and high specificity.[41] Unusual causes include aortocaval fistula, SVC obstruction with venous collateralization, or hypervascular liver tumors with arteriovenous shunting.

IVC Filter

Presence of an IVC filter is a common finding in CT studies of the abdomen. Prophylactic IVC filter placement is a challenging issue. It is indicated in the patients with pulmonary embolism or above-knee deep venous thrombosis which cannot be anticoagulated.

20

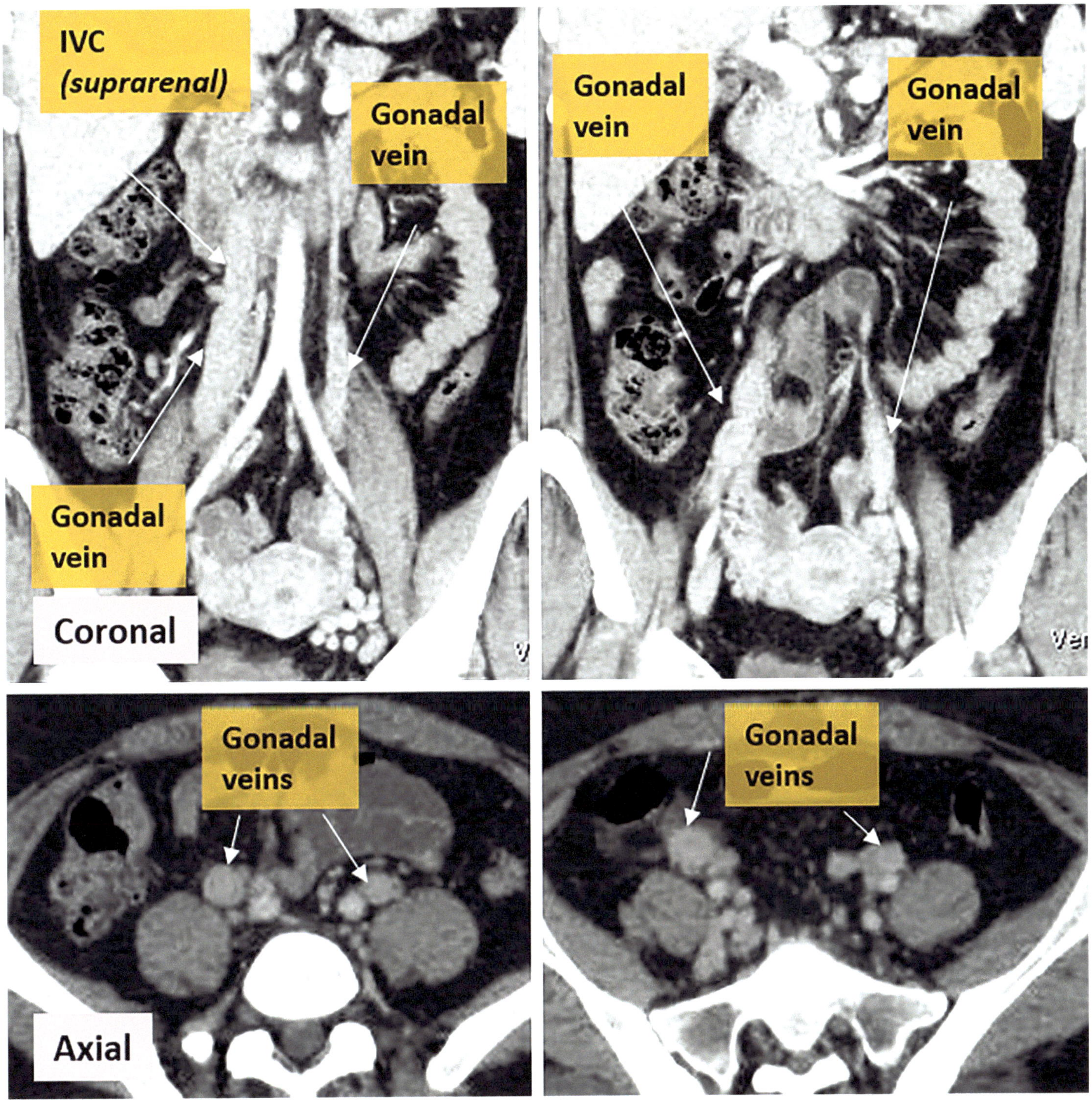

Fig. 20.20 Enlargement of the gonadal veins acting as collateral channels in a patient with chronic infrarenal thrombosis of the inferior vena cava (IVC).

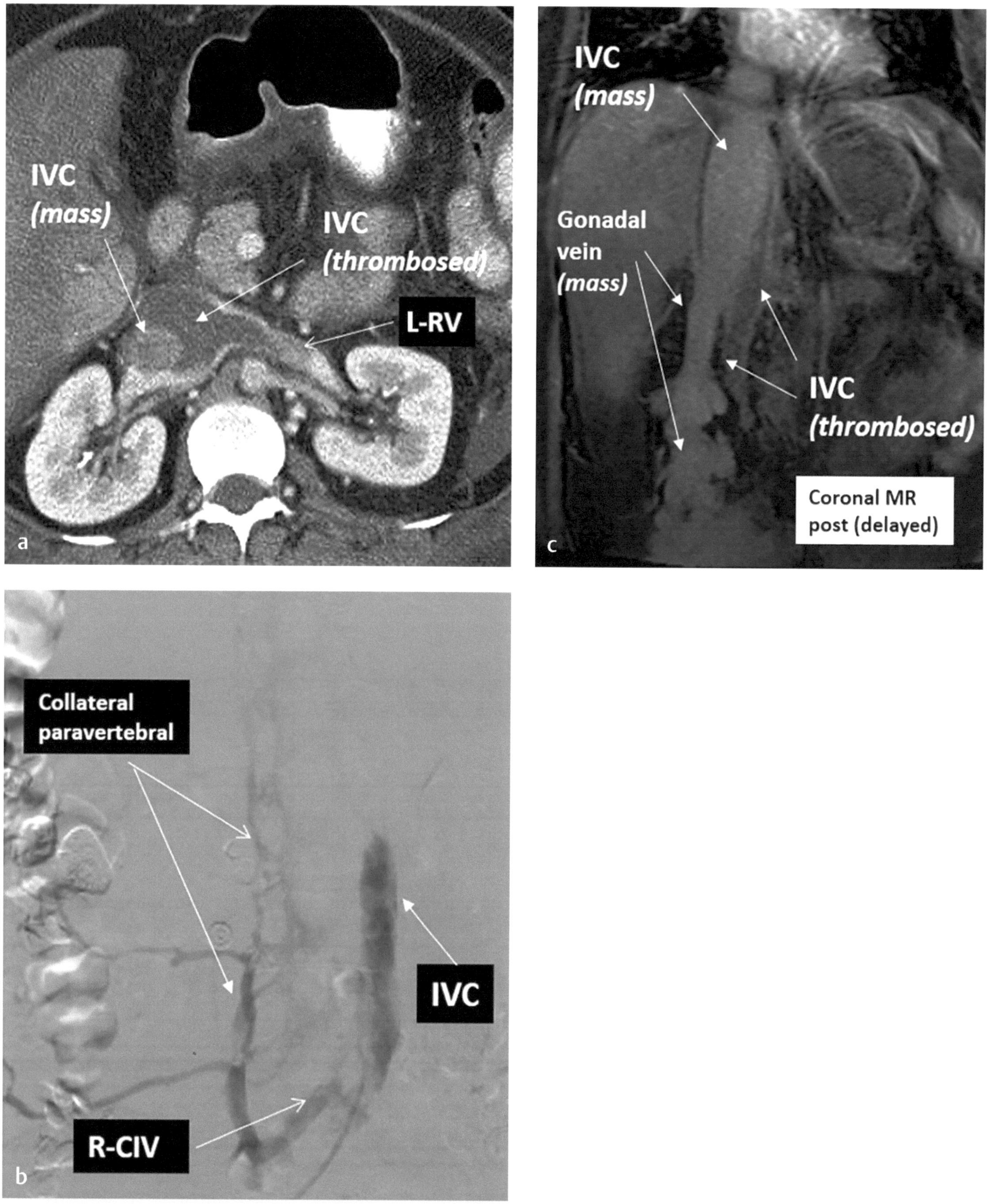

Fig. 20.21 Intravenous leiomyomatosis growing from the uterus in the right gonadal vein and distal IVC. **(a)** Postcontrast computed tomography (CT) shows extensive thrombosis of the inferior vena cava (IVC) with extension into the left renal vein (L-RV). **(b)** Coronal, postcontrast magnetic resonance (MR) shows large enhancing mass growing superiorly from the superolateral aspect of the uterus into the right gonadal vein and superior IVC. The inferior IVC is thrombosed. **(c)** Oblique view of a venogram showing occlusion of IVC with contrast filling of the paravertebral collaterals. R-CIV, right common iliac vein.

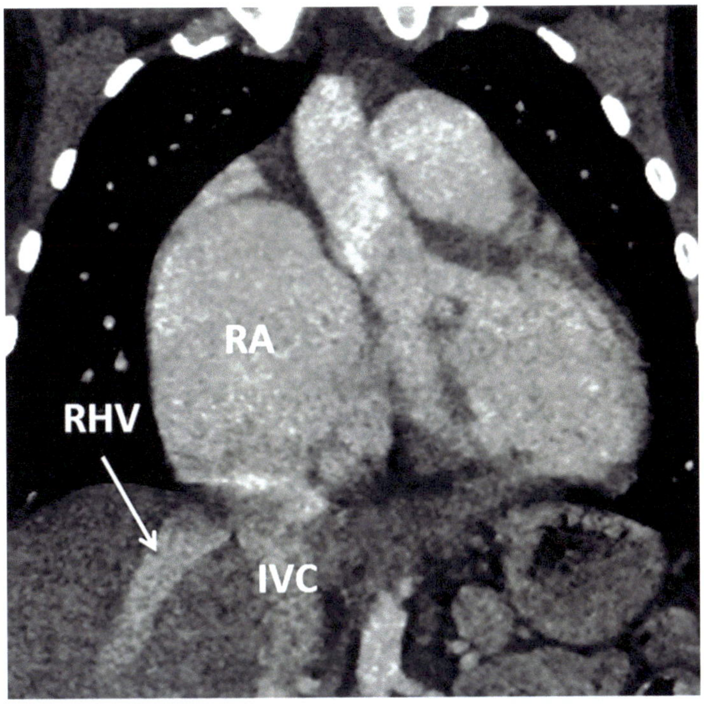

Fig. 20.22 Coronal computed tomography (CT) showing reflux of contrast into right hepatic vein (RHV) and inferior vena cava (IVC) in a patient with increased right heart pressure. RA, right atrium.

◆ Pelvic Veins

The anatomy of the iliac veins is very variable. A single internal iliac trunk usually drains into the external iliac vein to form the common iliac vein. The common iliac veins unite on the right side of the fifth lumbar vertebrae to form the IVC. The right common iliac vein ascends along a relatively straight course to the IVC, and the left common iliac vein runs transversely anterior to vertebral body and posterior to the right common iliac artery to join the right common iliac vein at an acute angle. On rare occasion, a high-riding left common iliac vein is seen running anterior to the common iliac arteries to merge the right common iliac vein near the renal veins. In this case, the infrarenal segment of the IVC is very short making it unsuitable for placement of an IVC filter (▶ Fig. 20.23).

The internal iliac vein drains both visceral and parietal tributaries that have extensive valveless interconnections. These interconnections may become important in cases of iliocaval obstruction. The visceral tributaries include hemorrhoidal, vesicoprostatic, uterine, gonadal, and vesicovaginal plexuses. The parietal tributaries include superior and inferior gluteal, sacral, sciatic, lumbar, obturator, and internal pudendal veins.

The inferior epigastric vein is formed by the union of the venae comitantes of the inferior epigastric artery and communicates superiorly with the superior epigastric vein. It joins the external

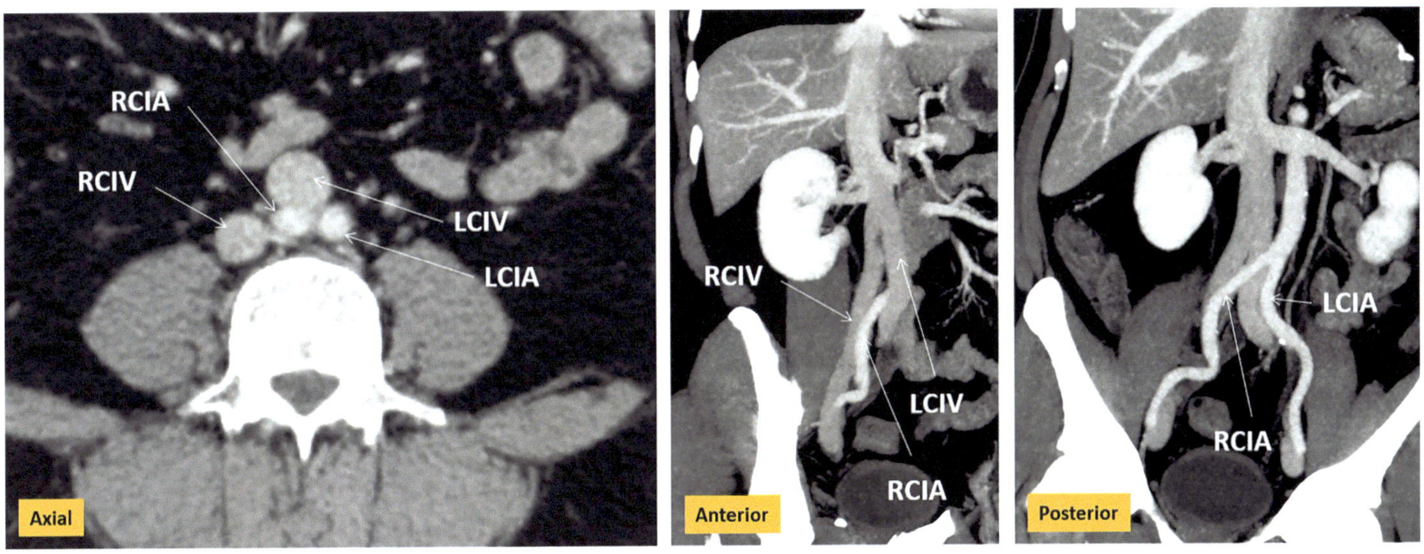

Fig. 20.23 Axial and coronal computed tomography (CT) images showing high-riding common iliac veins. The left common iliac vein (LCIV) is seen running anterior to the common iliac arteries. As a result, the infrarenal segment of the inferior vena cava (IVC) is very short making it unsuitable for placement of an IVC filter. Two renal veins are seen on the right side. LCIA, left common iliac artery; RCIA, right common iliac artery; RCIV, right common iliac vein. (Courtesy of Professor Miriam Romero, USC.)

iliac vein just above the inguinal ligament. The deep circumflex iliac vein is formed by the union of the venae comitantes of the deep circumflex iliac artery and also joins the external iliac vein just superior to the inguinal ligament.

The iliac veins rarely contain valves. Valves are present in 1% of common iliac veins, 27% of external iliac veins, and 10% of internal iliac veins.[42] Varicosities of the vulva, vagina, and posteromedial thigh as well as symptoms of the pelvic congestion syndrome are also commonly attributed to pelvic venous insufficiency and incompetent tributaries of the internal iliac vein. Since only 10% of normal internal iliac veins have valves, the role of pelvic venous incompetence is not clear.

Compression of the left common iliac vein between the right common iliac artery and the vertebrae is a common finding and usually asymptomatic.[43,44] In tortuous arteries, the left iliac vein may be compressed by the left iliac artery[45] (▶Fig. 20.24).

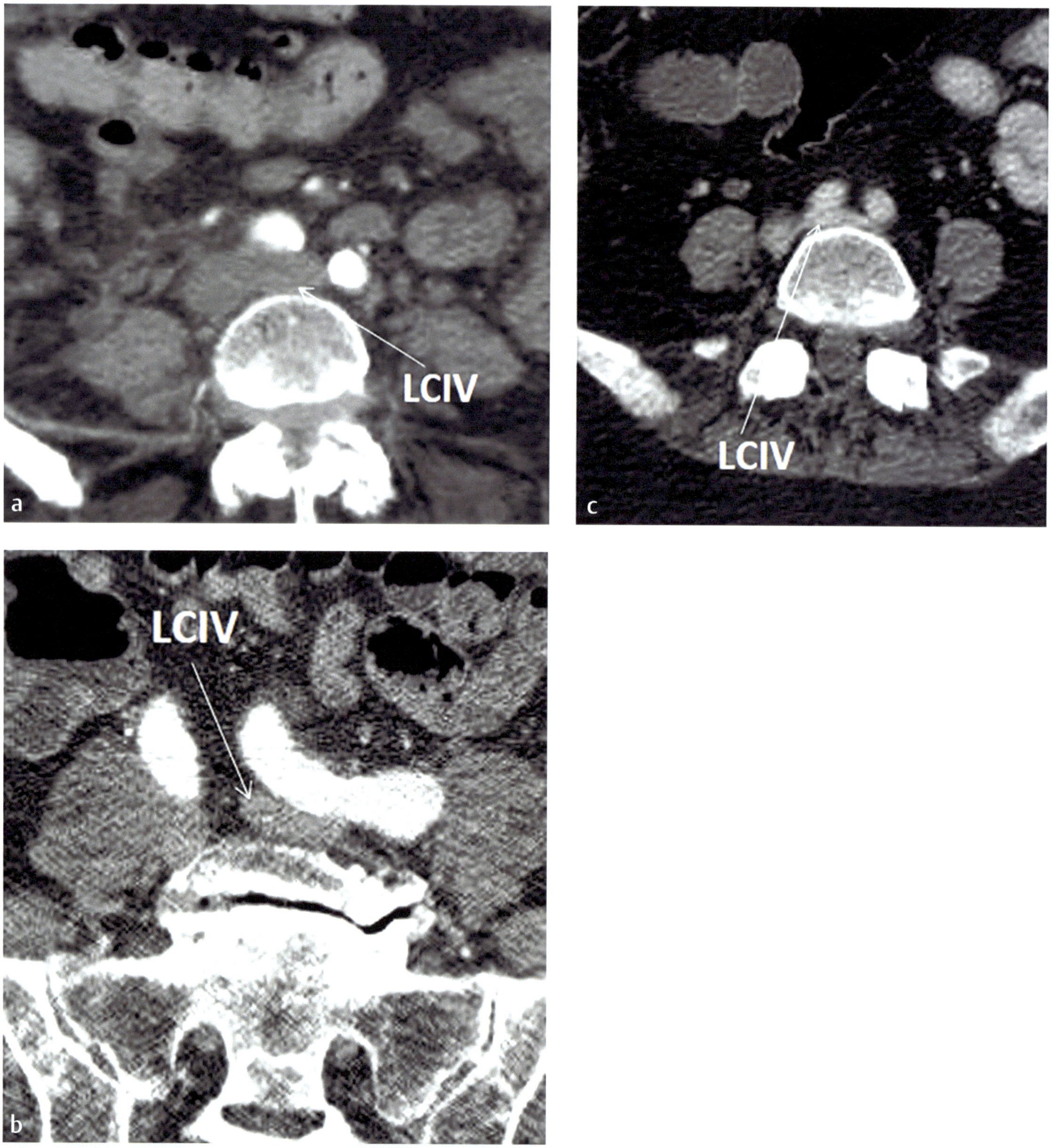

Fig. 20.24 Axial pelvic computed tomography (CT) images showing three possible compression patterns of the left common iliac vein (LCIV). **(a)** Both arteries. **(b)** Right common iliac artery. **(c)** Left common iliac artery (tortuous arteries).

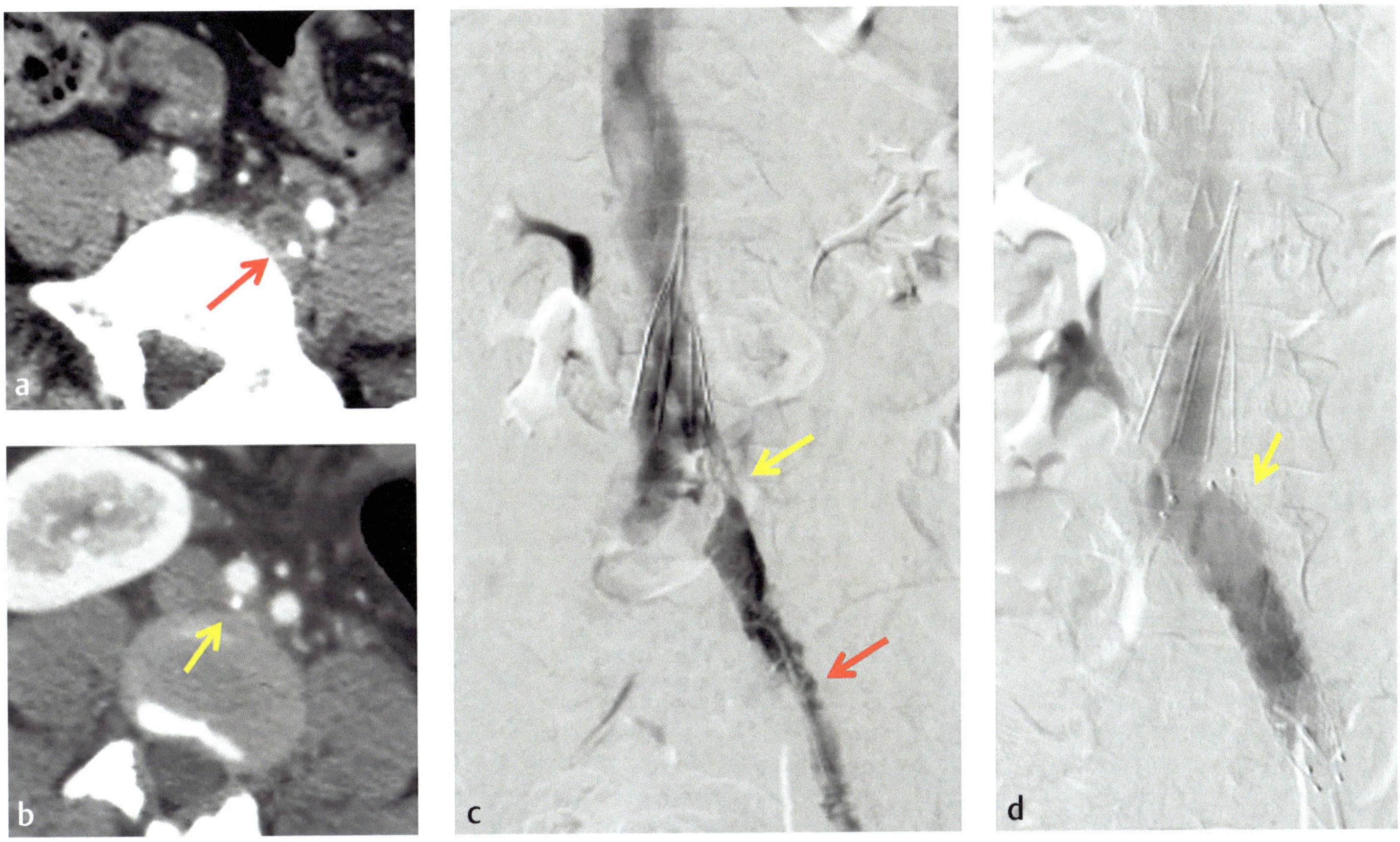

Fig. 20.25 May-Thurner syndrome. **(a, b)** Axial computed tomography (CT) angiography images showing thrombosed left iliac vein (*red arrow*) and its compression where it passes behind the right iliac artery. Thrombolysis catheter is seen within the thrombosed vein. **(c, d)** Venograms before **(c)** and after **(d)** stenting showing narrowed segment of the left iliac vein which is dilated after stent placement (*yellow arrows*). Note thrombosed vein in **c** (*red arrow*). Inferior vena cava filter is in place.

Venous compression above 50% can be detected in 25 to 35% of asymptomatic patients.[43] Chronic compression can lead to intimal hypertrophy and stenosis of the left common iliac vein, venous stasis, and thrombosis. The symptomatic cases are called iliac vein compression syndrome, also known as May-Thurner syndrome or Cockett syndrome.[46] This syndrome occurs predominantly in young to middle-aged women (▸ Fig. 20.25).

References

1. Mayo J, Gray R, St Louis E, Grosman H, McLoughlin M, Wise D. Anomalies of the inferior vena cava. AJR Am J Roentgenol 1983;140(2):339–345
2. Bass JE, Redwine MD, Kramer LA, Huynh PT, Harris JH Jr. Spectrum of congenital anomalies of the inferior vena cava: cross-sectional imaging findings. Radiographics 2000;20(3):639–652
3. Mathews R, Smith PA, Fishman EK, Marshall FF. Anomalies of the inferior vena cava and renal veins: embryologic and surgical considerations. Urology 1999; 53(5):873–880
4. Eldefrawy A, Arianayagam M, Kanagarajah P, Acosta K, Manoharan M. Anomalies of the inferior vena cava and renal veins and implications for renal surgery. Cent European J Urol 2011;64(1):4–8
5. Prince MR, Novelline RA, Athanasoulis CA, Simon M. The diameter of the inferior vena cava and its implications for the use of vena caval filters. Radiology 1983;149(3):687–689
6. Huisman TW, van den Eijnde SM, Stewart PA, Wladimiroff JW. Changes in inferior vena cava blood flow velocity and diameter during breathing movements in the human fetus. Ultrasound Obstet Gynecol 1993;3(1):26–30
7. Wexler L, Bergel DH, Gabe IT, Makin GS, Mills CJ. Velocity of blood flow in normal human venae cavae. Circ Res 1968;23(3):349–359
8. Abel FL, Waldhausen JA. Respiratory and cardiac effects on venous return. Am Heart J 1969;78(2):266–275
9. Natsis K, Apostolidis S, Noussios G, Papathanasiou E, Kyriazidou A, Vyzas V. Duplication of the inferior vena cava: anatomy, embryology and classification proposal. Anat Sci Int 2010;85(1):56–60
10. Honma S, Tokiyoshi A, Kawai K, Koizumi M, Kodama K. Left inferior vena cava with regressed right inferior vena cava. Anat Sci Int 2008;83(3):173–178
11. Babaian RJ, Johnson DE. Major venous anomalies complicating retroperitoneal surgery. South Med J 1979;72(10):1254–1258
12. Doyle AJ, Melendez MG, Simons MA. Ipsilateral duplication of the inferior vena cava. J Clin Ultrasound 1992;20(7):481–485
13. Tagliafico A, Capaccio E, Rosenberg I, Martinoli C, Derchi LE. Double right inferior vena cava associated with an anomalous venous ring encircling the right common iliac artery: report of one case with CT and US. Eur J Radiol Extra 2007;64(3):111–115
14. Kudo FA, Nishibe T, Miyazaki K, Flores J, Yasuda K. Left renal vein anomaly associated with abdominal aortic aneurysm surgery: report of a case. Surg Today 2003;33(8):609–611
15. Milledge RD. Absence of the inferior vena cava. Radiology 1965;85(5):860–865
16. Lambert M, Marboeuf P, Midulla M, et al. Inferior vena cava agenesis and deep vein thrombosis: 10 patients and review of the literature. Vasc Med 2010;15(6): 451–459
17. Salgado Ordóñez F, Gavilán Carrasco JC, Bermúdez Recio FJ, Aguilar Cuevas R, Fuentes López T, González Santos P. Absence of the inferior vena cava causing repeated deep venous thrombosis in an adult—a case report. Angiology 1998; 49(11):951–956
18. Banerjee A, Maharana S, Kumar IA, Jhansi P. Duplication of the inferior vena cava—report of a rare congenital variation. Int J Anat Var 2012;5:141–143

19. Chen H, Emura S, Nagasaki S, Kubo K-Y. Double inferior vena cava with interiliac vein: a case report and literature review. Okajimas Folia Anat Jpn 2012; 88(4):147–151
20. Haswell DM, Berrigan TJ Jr. Anomalous inferior vena cava with accessory hemiazygos continuation. Radiology 1976;119(1):51–54
21. Kandpal H, Sharma R, Gamangatti S, Srivastava DN, Vashisht S. Imaging the inferior vena cava: a road less traveled. Radiographics 2008;28(3):669–689
22. Siegfried MS, Rochester D, Bernstein JR, Miller JW. Diagnosis of inferior vena cava anomalies by computerized tomography. Comput Radiol 1983;7(2):119–123
23. Smillie RP, Shetty M, Boyer AC, Madrazo B, Jafri SZ. Imaging evaluation of the inferior vena cava. Radiographics 2015;35(2):578–592
24. Chuang VP, Mena CE, Hoskins PA. Congenital anomalies of the inferior vena cava. Review of embryogenesis and presentation of a simplified classification. Br J Radiol 1974;47(556):206–213
25. Minniti S, Visentini S, Procacci C. Congenital anomalies of the venae cavae: embryological origin, imaging features and report of three new variants. Eur Radiol 2002;12(8):2040–2055
26. Gibo M, Onitsuka H. Retroaortic left renal vein with renal vein hypertension causing hematuria. Clin Imaging 1998;22(6):422–424
27. Kew MC, Hodkinson HJ. Membranous obstruction of the inferior vena cava and its causal relation to hepatocellular carcinoma. Liver Int 2006;26(1):1–7
28. Park JH, Cha SH, Han JK, Han MC. Intrahepatic portosystemic venous shunt. AJR Am J Roentgenol 1990;155(3):527–528
29. Santamaría G, Pruna X, Serres X, Inaraja L, Zuasnabar A, Castellote A. Congenital intrahepatic portosystemic venous shunt: sonographic and magnetic resonance imaging. Eur Radiol 1996;6(1):76–78
30. Lane MJ, Jeffrey RB Jr, Katz DS. Spontaneous intrahepatic vascular shunts. AJR Am J Roentgenol 2000;174(1):125–131
31. Chang TC, Zaleski GX, Lin BH, Funaki B, Leef J. Treatment of inferior vena cava obstruction in hemodialysis patients using Wallstents: early and intermediate results. AJR Am J Roentgenol 1998;171(1):125–128
32. Giannoukas AD. Stent-grafts in the inferior vena cava. J Endovasc Ther 2011; 18(2):255
33. Hassan B, Tung K, Weeks R, Mead GM. The management of inferior vena cava obstruction complicating metastatic germ cell tumors. Cancer 1999;85(4): 912–918
34. Zamora CA, Sugimoto K, Mori T, et al. Prophylactic stenting of the inferior vena cava before transcatheter embolization of renal cell carcinomas: an alternative to filter placement. J Endovasc Ther 2004;11(1):84–88
35. Hartung O, Benmiloud F, Barthelemy P, Dubuc M, Boufi M, Alimi YS. Late results of surgical venous thrombectomy with iliocaval stenting. J Vasc Surg 2008; 47(2):381–387
36. Yamagami T, Nakamura T, Kato T, Iida S, Nishimura T. Hemodynamic changes after self-expandable metallic stent therapy for vena cava syndrome. AJR Am J Roentgenol 2002;178(3):635–639
37. Barksdale J, Abolhoda A, Saremi F. Intravenous leiomyomatosis presenting as acute Budd-Chiari syndrome. J Vasc Surg 2011;54(3):860–863
38. Hilliard NJ, Heslin MJ, Castro CY. Leiomyosarcoma of the inferior vena cava: three case reports and review of the literature. Ann Diagn Pathol 2005;9(5):259–266
39. Sessa B, Iannicelli E, Caterino S, et al. Imaging of leiomyosarcoma of the inferior vena cava: comparison of 2 cases and review of the literature. Cancer Imaging 2010;10:80–84
40. Kaufman LB, Yeh BM, Breiman RS, Joe BN, Qayyum A, Coakley FV. Inferior vena cava filling defects on CT and MRI. AJR Am J Roentgenol 2005;185(3):717–726
41. Yeh BM, Kurzman P, Foster E, Qayyum A, Joe B, Coakley F. Clinical relevance of retrograde inferior vena cava or hepatic vein opacification during contrast-enhanced CT. AJR Am J Roentgenol 2004;183(5):1227–1232
42. LePage PA, Villavicencio JL, Gomez ER, Sheridan MN, Rich NM. The valvular anatomy of the iliac venous system and its clinical implications. J Vasc Surg 1991;14(5):678–683
43. Kibbe MR, Ujiki M, Goodwin AL, Eskandari M, Yao J, Matsumura J. Iliac vein compression in an asymptomatic patient population. J Vasc Surg 2004;39(5):937–943
44. Nazzal M, El-Fedaly M, Kazan V, et al. Incidence and clinical significance of iliac vein compression. Vascular 2015;23(4):337–343
45. Moreland NC, Ujiki M, Matsumura JS, et al. Decreased incidence of left common iliac vein compression in patients with abdominal aortic aneurysms. J Vasc Surg 2006;44(3):595–600
46. May R, Thurner J. The cause of the predominantly sinistral occurrence of thrombosis of the pelvic veins. Angiology 1957;8(5):419–427

21 Lymphatics of the Abdomen, Pelvis, and Perineum

Scott A. Sylvester, Frederic J. Bertino, R. Shane Tubbs, and Farhood Saremi

◆ Introduction

A working knowledge of the abdominopelvic lymphatic anatomy from an imaging perspective is paramount for the noninvasive assessment, staging, and surgical or radiation direction for the management of male and female lymphomatous, solid organ, and urogenital malignancies. Abdominopelvic tumors, depending on their origin, follow different anatomic pathways of lymphatic drainage that may aid in staging and in identifying the most likely origin of disease, and similarly, predict the progression of spread.

The genesis of the lymphatics is not completely understood. It is believed that the lymphatic system originates from the venous system. The lymphatic system consists of lymph vessels and interposed lymph nodes that finally connects to the venous system. Mammalian lymph nodes develop in regions occupied by lymph sacs and lymph sacs will be transformed into lymphatic plexuses. The lymphatics begin with the initial lymph vessels in the periphery of the body and drain interstitial fluid from tissues and return it to the blood. This drainage is necessary for the maintenance of normal tissue fluid balance. The lymph nodes are encapsulated by dense connective tissue with blood and lymphatic vessels and nerves attached to them at specific locations (▶Fig. 21.1). The lymph nodes serve as a filtration unit and participate in immune system. The term sentinel or first lymph node is used to describe the first remote lymph node involved in a malignant neoplasm.

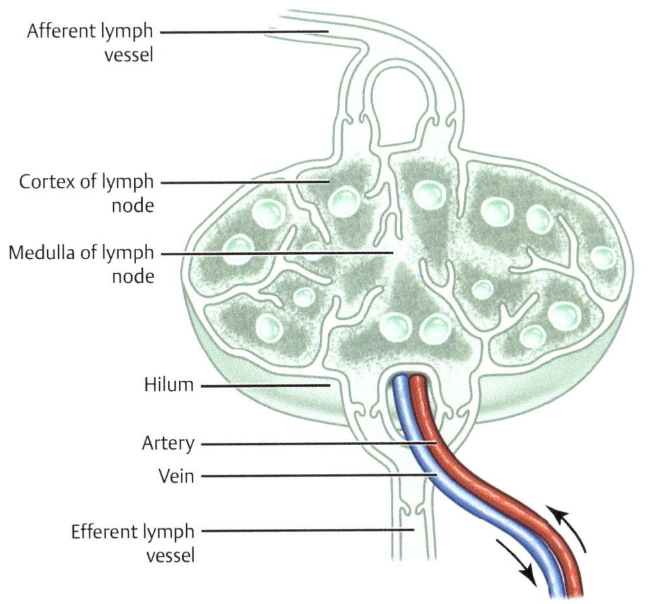

Fig. 21.1 The lymph node as a filtration unit.

◆ Major Lymph Node Groups of Pelvis and Perineum

There are several major lymph node regions and smaller subgroups within these regions that have been described in detailed anatomy when considering contouring for external-beam radiation therapy. Similarly, understanding the boundaries between lymphatic groups in the pelvis allows for appropriate staging of male and female urogenital malignancy.[1,2]

Paraaortic Lymph Nodes

Divided among seven subgroups, the paraaortic nodes are named for their location with respect to the aorta and inferior vena cava. The lateral-aortic, preaortic, and retroaortic subgroups surround the aorta directly.[2] The right lateral-aortic subgroup may be further subdivided into aortocaval, lateral-caval, precaval, and retrocaval (▶Fig. 21.2, ▶Fig. 21.3).

Common Iliac Lymph Nodes

The common iliac nodes run along the common iliac vessels in the region extending to a 7-mm diameter circumferential to the vessel. This nodal group extends posterior and lateral to the borders of the psoas muscle and vertebral body bilaterally.[1] The common iliac nodes lie in between the craniocaudal borders of the aortic bifurcation and common iliac vessels, respectively.[3]

This group may be further subdivided into three chains: lateral, medial, and middle (▶Fig. 21.4). The lateral chain extends laterally as it accompanies the lateral aspect of the common iliac artery.[2] The medial chain rests within the vascular triangle between both common iliac arteries from the aortic bifurcation inferiorly to the common iliac bifurcation into the external and internal iliac arteries.[2] The middle chain of common iliac nodes resides in the lumbosacral fossa. This region is bordered posteromedially by the inferior-most lumbar vertebrae and the superior sacral bones where anteriorly the psoas muscle joins the common iliac vessels from lateral to medial, respectively.[2,3]

External Iliac Lymph Nodes

The external iliac nodes course along the external iliac vessels and can be visualized with a similar 7-mm border extending circumferentially with an additional 10-mm anterolateral border extension along the iliopsoas muscle to incorporate the lateral external iliac node subgroup.[1]

The external iliac nodes are bound inferiorly by the inguinal ligament, and superiorly by the bifurcation of the common iliac

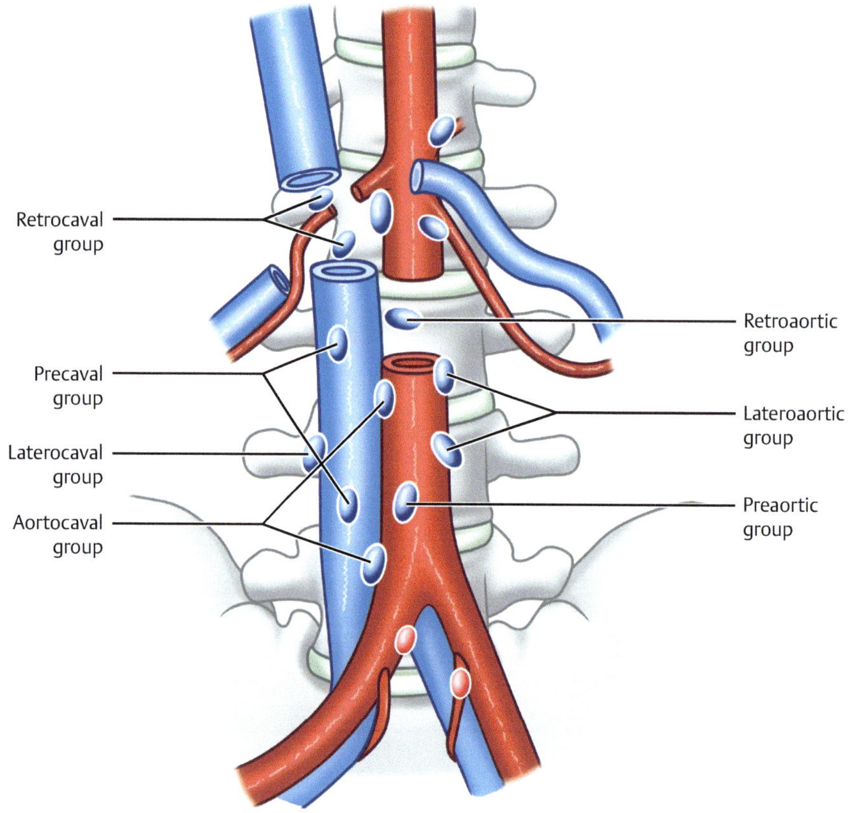

Fig. 21.2 Schematic showing the seven subgroups of paraaortic lymph nodes. The subgroups are distinguished from one another on the basis of their positions in relation to the aorta and the inferior vena cava.

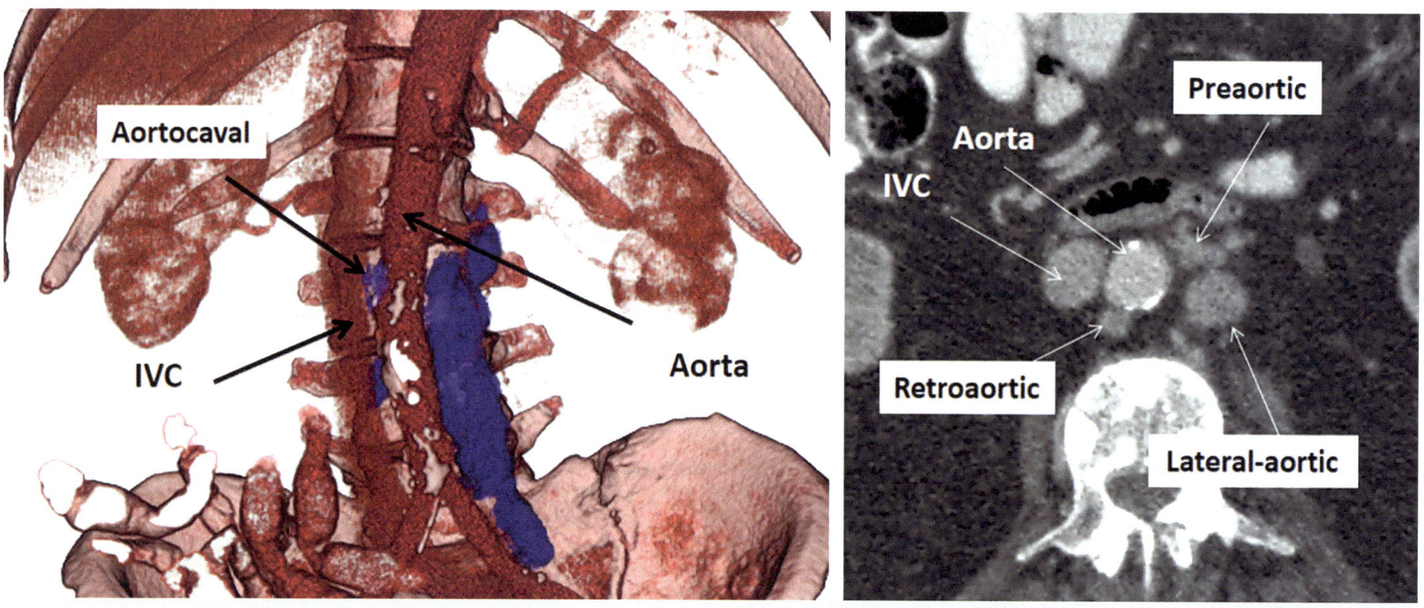

Fig. 21.3 Paraaortic lymph nodes (shown in *blue*). The lateral-aortic, preaortic, and retroaortic subgroups are shown surrounding the aorta. IVC, inferior vena cava.

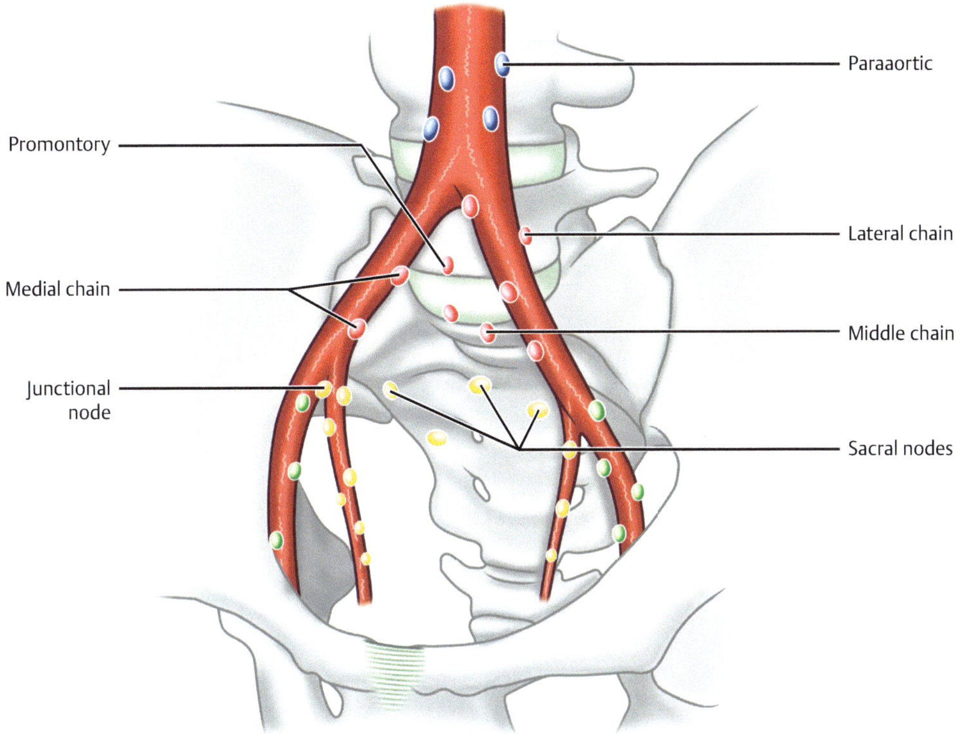

Fig. 21.4 Common iliac (*red*), internal iliac (*yellow*), and external iliac (*yellow*) lymph nodes. The common iliac nodal chain includes the lateral chain along the lateral margin of the arteries, the medial chain along the medial border of the common iliac arteries, and nodes at their junction (sacral promontory); and the middle chain, which consists of nodes within the lumbosacral fossa. The internal iliac lymph nodes accompany the branches of the internal iliac vessels. The sacral and junctional (bifurcation of the arteries) nodes are also shown.

vessels.[3] Like the common iliac group, three main chains of lymph nodes extend from the external iliac group (▶ Fig. 21.5). The lateral chain includes those nodes lying lateral to the external iliac artery. The middle chain consists of nodes seated between the external iliac artery and vein. The medial chain is situated posteromedially to the external iliac vein. The nodes included in the medial chain subgroup of the external iliac group are also referred to as the obturator nodes because of their close position to the obturator canal.[2,3] The obturator group borders the lateral pelvic side wall in a 17-mm strip and runs adjacently to the borders of the external and iliac nodal groups.[1] The external iliac group also receives lymphatic drainage from the inferior epigastric and circumflex iliac nodes, which are intimately associated with their similarly named vessels.

The pelvic ureter may drain to the external, internal, or common iliac nodes.[4] The urinary bladder contains lymphatic vessels originating at the level of the mucosal, intramuscular, and serosal layers of the organ which drain to the external iliac nodes. The seminal vesicles and vas deferens both contain lymphatic vessels that drain to the external iliac nodes. The seminal vesicles may also drain to the internal iliac nodes.

Internal Iliac (Hypogastric) Lymph Nodes

The hypogastric nodes accompany the internal iliac vessels visualized in a 7-mm diameter circumferentially with a further extension laterally to the pelvic wall.[1] As a nodal group, the internal iliac nodes lie more posteriorly in the pelvis compared to their external iliac counterparts. The nodes in the internal iliac group divide into individual subgroups each of which follows its own branch of the internal iliac artery (▶ Fig. 21.4).

The lateral sacral nodes, which join the lateral sacral arteries, may also be grouped into the internal iliac lymphatics. Some regard the presacral nodes as an independent nodal group within the pelvis, while others may regard this group as a subdivision of the internal iliac nodes.[3] Similarly, the term "hypogastric nodes" may be used synonymously with the internal iliac nodal group as a whole, or refer to only the most anterior lymph nodes in this group.[3]

The junctional nodes, which lie at the junction of the internal and external iliac groups, are particularly important as their

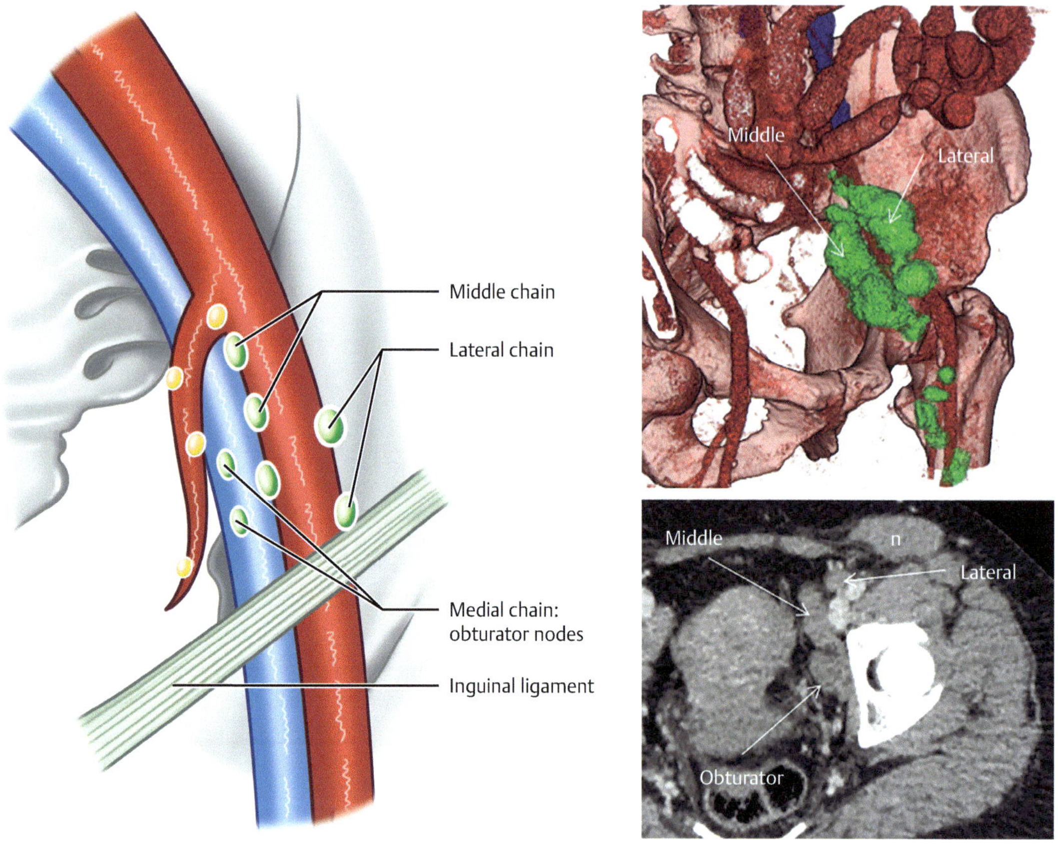

Fig. 21.5 External iliac lymph nodes. Schematic and computed tomography (CT) images show the three chains of the external iliac nodal group. The lateral chain is positioned along the lateral margin of the external iliac artery. The middle chain is located between the external iliac artery and vein. The medial chain (also known as obturator nodes) is positioned medial to the external iliac vein. n, node.

location plays a key role when identifying pathways of lymphatic dissemination of disease.[2]

The internal iliac nodes drain the pelvic ureter as previously described. Lymphatics from the superior aspect of the bladder may drain to the common or internal iliac nodes. The lymphatics of the female urethra primarily drain to the internal and external iliac nodes.

The prostatic lymphatic vessels drain mainly to the internal iliac nodes, obturator nodes, and sacral nodes. The seminal vesicles may drain to either the internal or external iliac group as previously described.

The posterior penile urethra contains vessels that mainly drain to the internal iliac nodes. Traveling with the internal pudendal artery, lymphatics from the membranous urethra may drain to any of the nodes that cross the arterial course. Like the posterior penile urethra, the primary location of membranous urethra drainage is the internal iliac nodes. The anterior urethral lymphatic vessels accompany those of the glans penis and most commonly end in the deep inguinal nodes. However, some of these vessels may drain alternatively to the superficial nodes or course the inguinal canal to the external iliac nodes. The female urethra, shorter in course, has primary drainage to the internal and external iliac nodes.[4]

Presacral Lymph Nodes

Two groups of presacral nodes are described—subaortic and mesorectal.[1] The subaortic nodes that reside over the sacral promontory are often included as part of the medial chain of lymph nodes in the common iliac group (▶ Fig. 21.4).[1] The inferior extension of the medial chain of common iliac nodes may overlap with a distinct subaortic group depending on individual anatomy.

Mesorectal nodes lie within the middle portion of an extraperitoneal space between the rectal wall inferior to the rectal-peritoneal reflection and parietal-pelvic fascia.[1,5]

Inguinal Lymph Nodes

Two major subgroups, the superficial and deep inguinal nodes, are distinguished by their anterior–posterior relationship to the inguinal ligament. As a rule, the inguinal lymph nodes reside inferior to the inguinal ligament, capped by the external iliac nodes, which lie superior to the inguinal ligament[6,7] (▶ Fig. 21.6).

The superficial inguinal group lies in the subcutaneous tissue anterior and superficial to the inguinal ligament (▶ Fig. 21.6b). This nodal group runs with the superficial femoral and saphenous veins to confluence into the common femoral vein. Sentinel nodes for the superficial subgroup rest within the junction

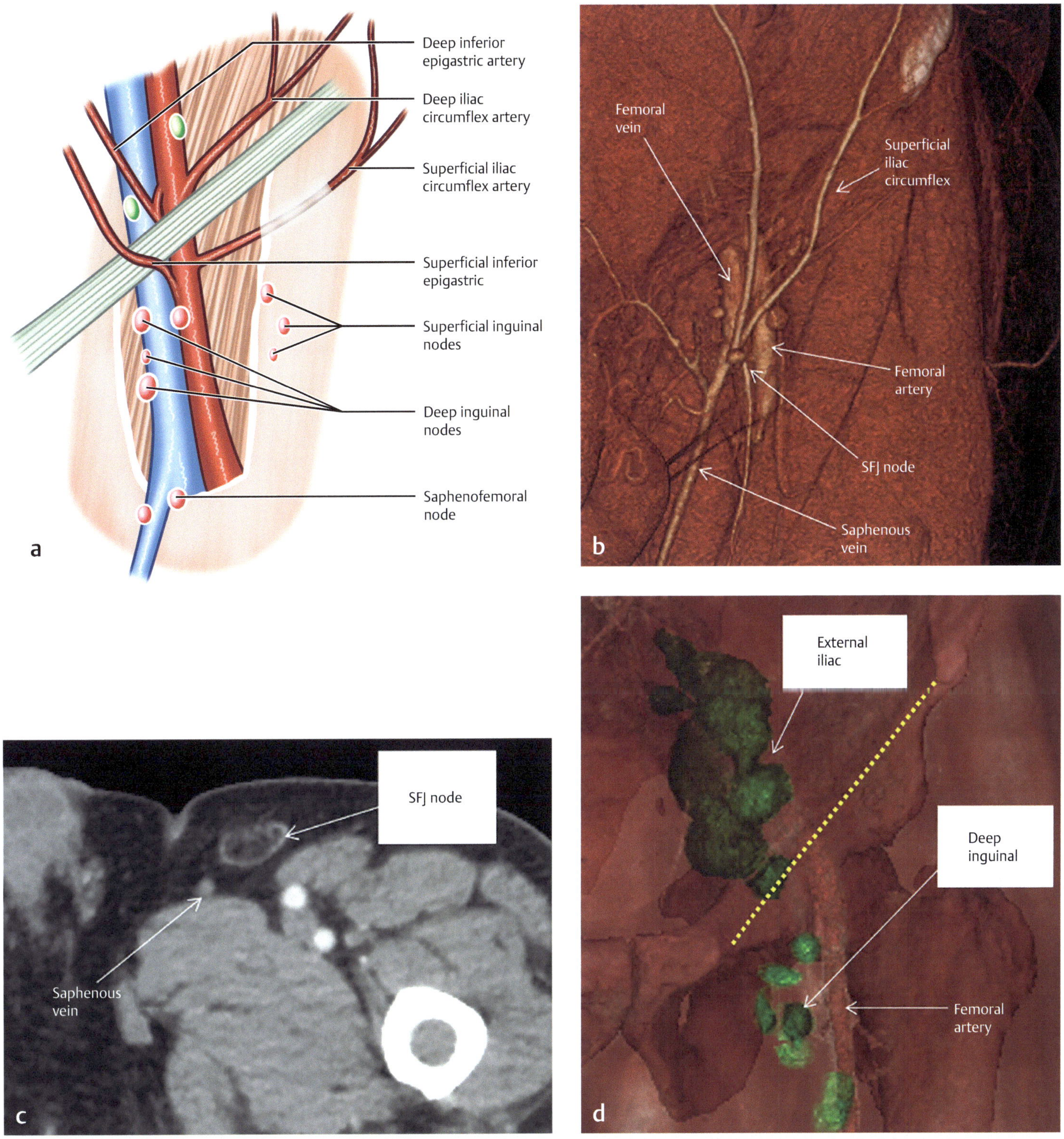

Fig. 21.6 Inguinal lymph nodes. **(a)** Schematic view. **(b)** The superficial inguinal group lies in the subcutaneous tissue anterior and superficial to the inguinal ligament. The sentinel nodes for the superficial subgroup rest at the junction between the saphenous and femoral veins (SFJ, saphenofemoral junction nodes). Metastatic tumor cells (i.e., from the penis) can ascend toward the deep inguinal and external iliac nodes. **(c)** Axial computed tomography (CT) showing an example of a normal SFJ node. **(d)** The deep inguinal group surrounds the common femoral vessels, posterior and inferior to the inguinal ligament (*dotted yellow line*).

between the saphenous and femoral veins. One of the most consistently seen lymph nodes on lymphangiography is the node of Rosenmüller, which lies medial to the common femoral vein.[6]

Perineal structures and skin contain lymphatic vessels that drain superiorly to lymphatic nodal groups in the pelvis. These lymphatics generally travel with regional blood vessels, following a similar course. The penile, clitoral and labial, vulvar structures, and perineal skin are drained by lymphatic vessels that travel with the external pudendal arteries and veins. These lymphatics first drain to the superficial inguinal nodes.[4]

The external female genitalia drain via lymphatics extending from the superficial inguinal nodes to the deep inguinal nodes along pathways lying just medial to the femoral veins. These lymphatics subsequently drain to pelvic nodes. Some lymphatic drainage of the clitoris and labia may pass directly to the internal iliac nodes. Perineal tissues and lymphatics draining the inferior part of the labia majora drain to the rectal lymphatic plexus.

The deep inguinal group surrounds the common femoral vessels, posterior and inferior to the inguinal ligament (▸ Fig. 21.6c). It is important to distinguish the deep inguinal nodes from the obturator group (the medial group of external iliac nodes) in that the deep inguinal resides inferior to the inguinal ligament, inferior epigastric artery, and circumflex iliac artery. The latter group runs superior to these vascular and ligamentous structures.[2] The deep inguinal nodes drain the anterior urethra and the glans penis (in males).[4]

Perivisceral Nodes

The perivisceral lymph nodes are not a distinct group of pelvic lymphatics. These lymph nodes surround the immediate pelvic viscera for which they are named. Those nodes, considered to be within the mesorectal (perirectal) group, may theoretically fall into the category of perivisceral nodes as these nodes also include those perirectal nodes surrounding the rectum. Perivesical and periprostatic nodes also exist which all serve to drain the immediate structure to which they are adjacent.

◆ Drainage Pathways of the Pelvic Lymphatics

Pelvic viscera drain into lymph nodes opposed to the internal and external iliac arteries, which then drain into nodes surrounding the common iliac arteries. With respect to the male anatomy, lymphatic drainage of pelvic tumors usually follows one of three main lymphatic drainage routes.[2,8,9]

Superficial Inguinal Pathway

The superficial inguinal pathway is considered the primary route of anatomic lymph drainage when studying the route of perineal tumors (including penile cancers). The saphenofemoral junctional node serves as sentinel node and from there, drainage continues to the deep inguinal and external iliac nodes (▸ Fig. 21.6).

Pelvic Pathways

Four main pelvic pathways exist that have been similarly studied for determining urogenital and pelvic tumor metastasis pathways.

Anterior Pelvic Route

This route drains mainly tissues extending from the anterior bladder wall along the (obliterated) umbilical artery remnant to the internal iliac nodes (hypogastric).

Lateral Pelvic Route

This route begins at the pelvic viscera where lymphatics subsequently drain to the medial chain of the external iliac nodes (obturator nodes). The lateral pelvic route has clinical significance as it tends to be one of the more common routes of carcinomas of the lateral aspect of the bladder and prostate adenocarcinomas.

Internal Iliac (Hypogastric) Route

Drainage begins at the pelvic organs at the visceral aspects of the lymphatic ducts of the internal iliac nodes and extends to the junctional nodes at the confluence of the external and internal iliac vessels.

Presacral Pathway

Relying mainly on the subaortic nodal plexus that lies just anteriorly to the sacrum, drainage of lower pelvic organs and the prostate may extend to the presacral space rising to the common iliac nodes. The mesorectal nodes also drain these lower pelvic structures.

Paraaortic Pathway

The paraaortic pathway is unique in that of all the pelvic lymphatic groups' lymphatic drainage bypasses all other pelvic lymphatic routes to ascend intimately with the gonadal vessels through the spermatic cord beyond the inguinal ring superiorly and anteriorly to the psoas muscles. The lymphatic pathway ends at the paraaortic and paracaval nodes at the level of the renal hilum; however, downward and retrograde spread of disease from nodal metastasis may extend to the aortic bifurcation. The paraaortic pathway primarily drains lymphatics from the testes (▸ Fig. 21.7).

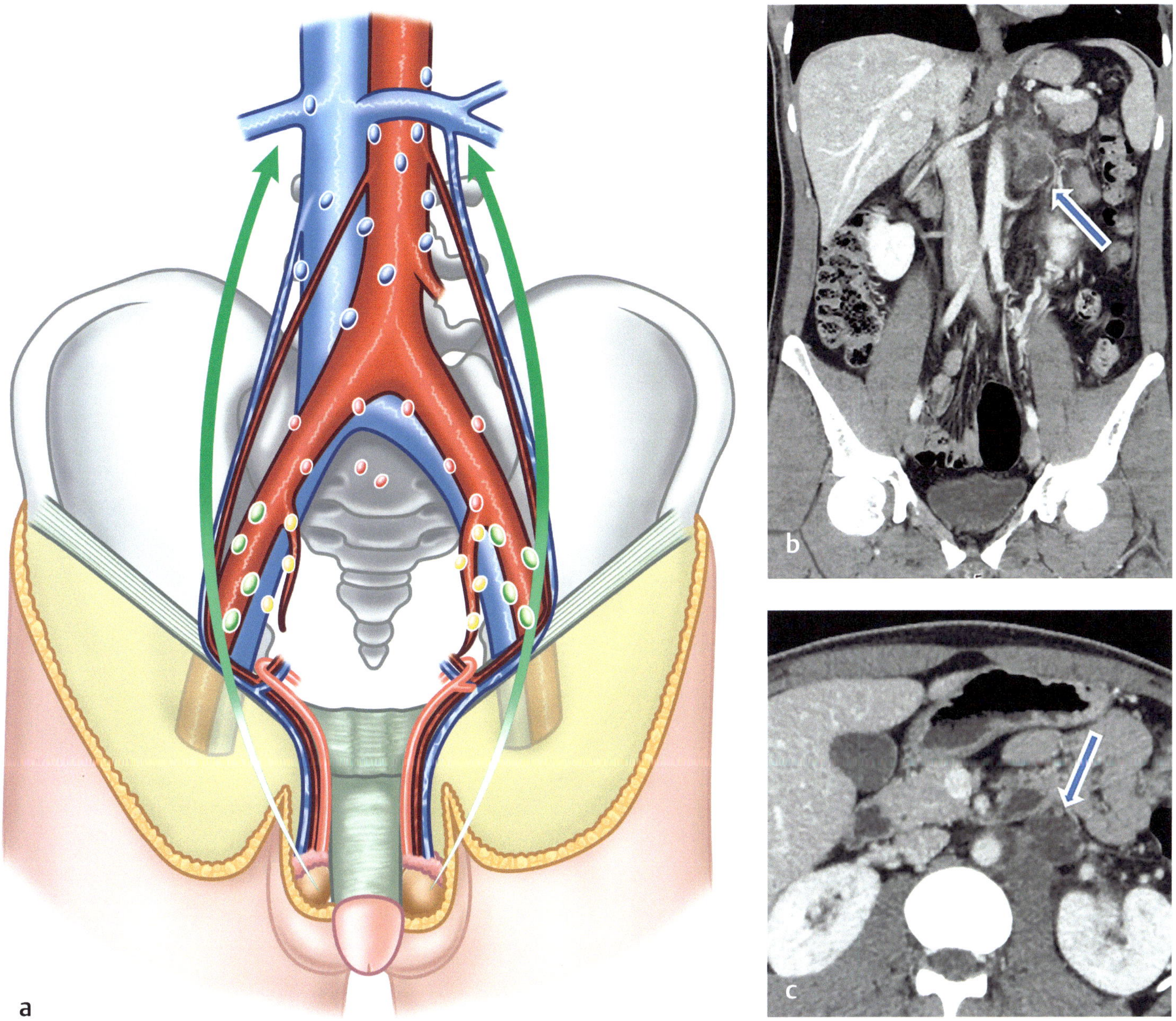

Fig. 21.7 Paraaortic pathway. **(a)** Schematic shows the paraaortic pathway of metastasis (*arrows*) by which malignant cells from testicular tumors can proceed upward through lymphatic ducts that follow the gonadal vessels to nodes (sentinel nodes) at the renal hilum where testicular veins connect to the left renal vein or inferior vena cava, completely bypassing the pelvic nodes. **(b)** Coronal and **(c)** axial computed tomography (CT) images in a patient with left-sided testicular cancer showing large confluent nodal metastasis to the left paraaortic nodes (*arrows*) at the level of the left renal hilum where the left testicular vein enters the left renal vein.

◆ Lymphatic Drainage of the Pelvis with Respect to the Pelvic Viscera

Uterus and Fallopian Tubes

The uterus drains primarily to the external iliac and internal iliac (obturator) nodes along with joining lymphatic branches of the cervix[4] (▶ Fig. 21.8). The uterine fundus and tubes may also drain to the paraaortic nodes, as they travel with common lymphatic pathways of the ovaries. Lymphatic vessels running along the round ligament to the superficial inguinal nodes drain the isthmus of the uterine tube. Fallopian tube drainage is shared by paraaortic nodes as the lymphatics course alongside ovarian vessels and uterine vessels to the internal iliac nodes. As previously mentioned, these vessels may also reach the internal iliac nodal group via the round ligament. Major lymphatic drainage of the cervix follows laterally to the iliac nodes. In addition to the lateral lymphatic trunks, smaller anterior and posterior lymphatic trunks are present. The anterior channels pass behind the bladder and terminate predominantly in the external iliac lymph nodes. The posterior lymphatic channels course in the uterosacral ligaments and connect to the lateral sacral, common iliac, and promontory lymph nodes (▶ Fig. 21.8).

Ovaries/Testes

The testes are drained from eight collecting trunks that rise superiorly with the spermatic cord and follow the gonadal (testicular) vessels (▶ Fig. 21.7). These vessels ascend over the psoas major muscles and drain to the paraaortic nodes (lateral aortic and preaortic subgroups). Like the testes, gonadal structures drain via lymphatics running along gonadal vessels (ovarian, testicular) to the paraaortic nodes to the level of the renal arteries. Ovaries, being intraabdominal, may also drain via pelvic nodes into lower paraaortic nodes and may also drain to the inguinal nodes via the round ligament.[4]

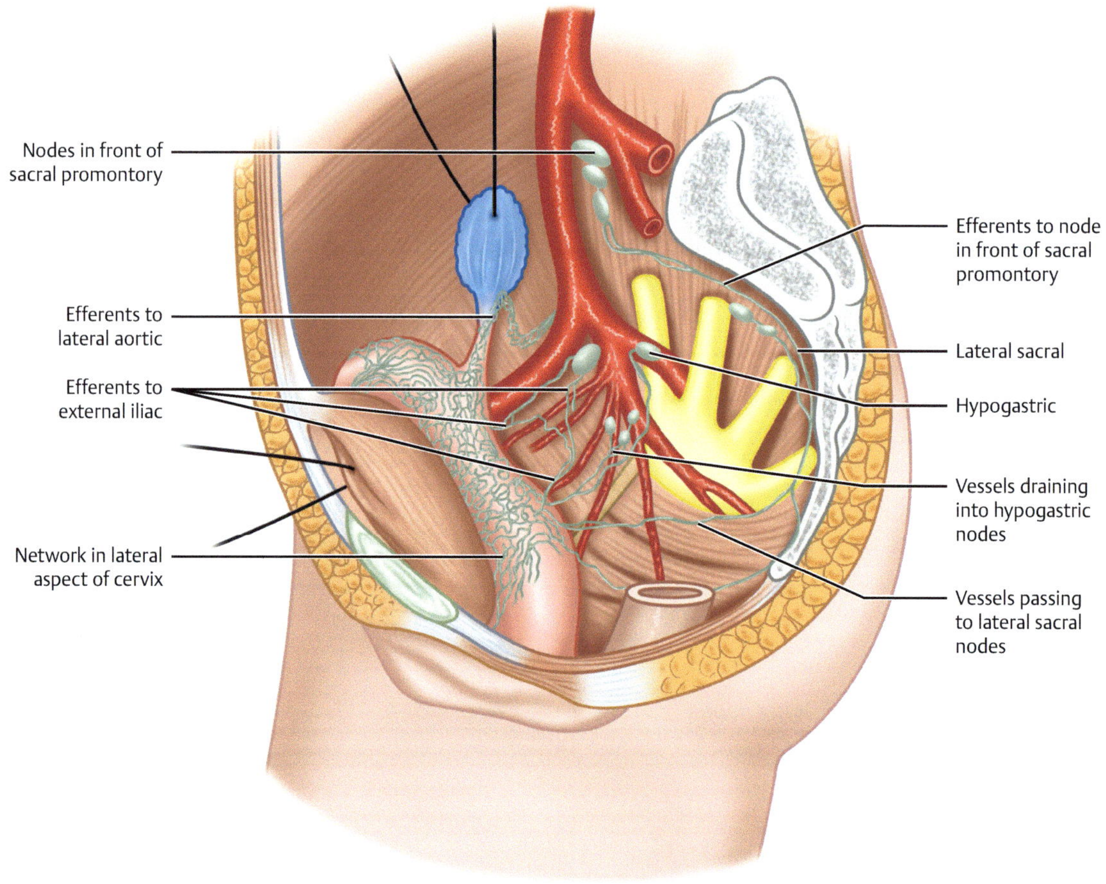

Fig. 21.8 Lymphatic of the uterus and cervix.

Prostate, Ductus Deferens, Seminal Vesicles

The external iliac nodes drain the vas deferens. The external and internal iliac nodes drain the seminal vesicles as previously described (▶Fig. 21.9). Obturator, internal iliac, and sacral nodal groups drain the prostate[4] (▶Fig. 21.9, ▶Fig. 21.10).

Penis, Vagina

The glans penis and clitoris are drained by lymphatics that drain to the deep inguinal and external iliac nodes, while lymphatics from erectile tissues (corpora cavernosa, corpora spongiosum, clitoral prepuce) drain directly to the deep inguinal lymph nodes.[4]

Vaginal lymphatic vessels join with cervical, rectal, and vulvar lymphatic vessels to common superior drainage pathways. Superior lymphatics travel with the uterine artery to the internal and external iliac nodes. Intermediate vessels run alongside the vaginal artery to the internal iliac nodes. Tissues drained inferior to the level of the hymen follow the same route as perineal skin and the majority of vulvar structures to the superficial inguinal nodes.

Bladder

Drain primarily to the external iliac nodes from three groups of draining vessels. The bladder trigone may drain to the external iliac, internal iliac, or common iliac groups[4] (▶Fig. 21.11).

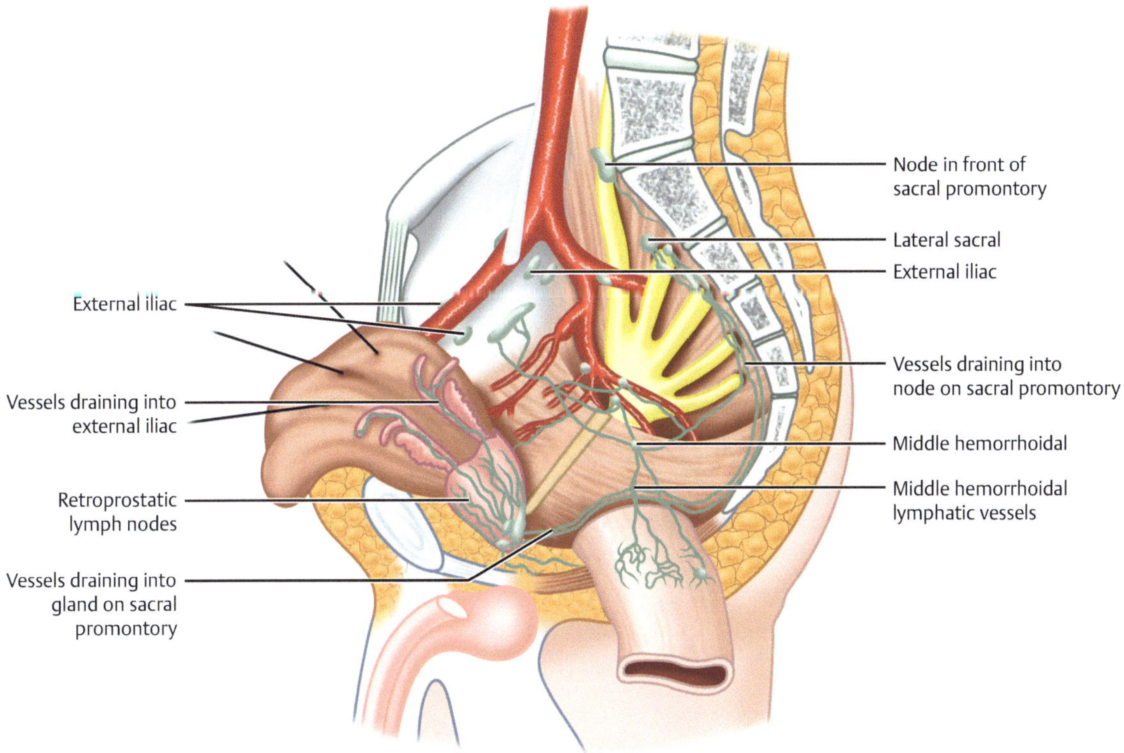

Fig. 21.9 Lymphatic of the prostate and seminal vesicles. The obturator, internal iliac, and sacral nodal groups drain the prostate. The obturator nodes in the external iliac nodal group are sentinel nodes along the lateral route, and the junctional nodes in the internal iliac nodal group are sentinel nodes along the hypogastric route.

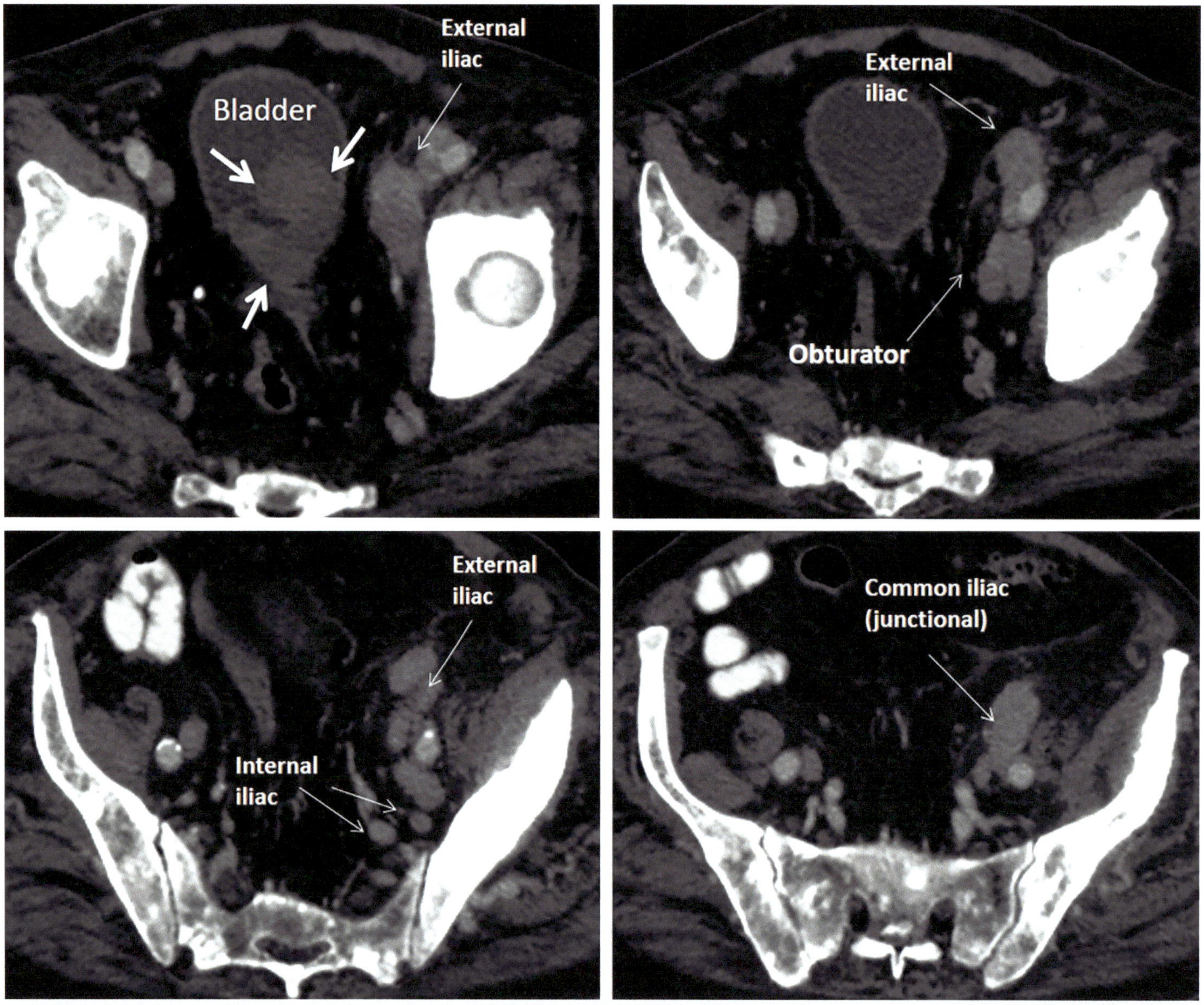

Fig. 21.10 Prostate cancer (*arrows*) with nodal and bone metastasis. Internal, external, and common iliac chains on the left side of pelvis are involved.

◆ Lymphatic Drainage of the Abdomen

The Anterior Abdominal Wall

The anterior abdominal wall drains via superficial and deep lymphatic vessels that accompany the blood vessels in the abdominal wall. Lumbar and gluteal region lymphatics run with the superficial iliac vessels while those from the infraumbilical skin accompany the superficial epigastric vessels. Superficial lymphatics drain to the superficial inguinal nodes. The parasternal nodes, axillary nodes, and pectoral nodes drain the abdominal wall in supraumbilical position. The posterior abdominal wall lymphatics will drain to the lateral aortic or retroaortic nodes. The anterior wall structures will accompany the superior epigastric vessels to the parasternal nodes. The lower deep abdominal wall will drain to the circumflex iliac, inferior epigastric, and external iliac nodes.[4]

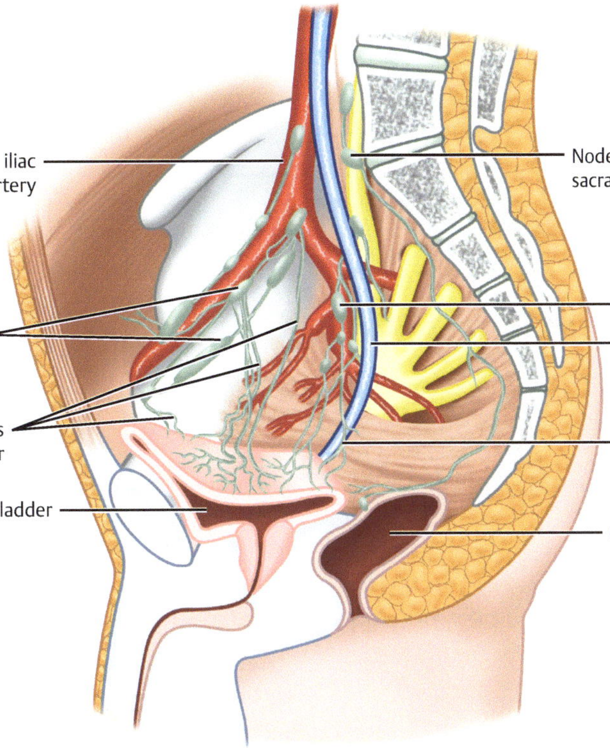

Fig. 21.11 Lymphatic of the bladder. The bladder lymphatics drain primarily to the external iliac nodes. The bladder neck drains to sacral and common iliac nodes.

The Posterior Abdominal Wall and Retroperitoneum

The lymphatics of the posterior abdominal wall drain in four regions. The lateral aortic nodes drain the upper left and right regions and the axillary lymph nodes on the same side. The lower left and right drain to the lateral and retroaortic nodes as well as to the right superficial inguinal nodes.[4]

The Lymphatic Drainage of the Abdominal Viscera

The cisterna chyli and thoracic duct almost exclusively drain the abdominal viscera, with minimal drainage occurring across the diaphragm (especially during thoracic duct obstruction). The lymph nodes of the retroperitoneum encircle the aorta forming the paraaortic nodal group, consisting of three subdivisions, the preaortic, lateral aortic, and retroaortic groups. It is difficult to approximate each subdivision on cross-sectional imaging and even on gross surgical dissection.

Preaortic Nodal Group

The preaortic group consists of the celiac nodes, superior mesenteric, and inferior mesenteric nodal groups. These nodes generally drain structures along the visceral arteries of the same name. This group covers lymphatic drainage from the inferior esophagus to anus as well as more superior abdominal organs including the liver, gallbladder, spleen, and pancreas.

Celiac Nodal Group

The celiac group drains the regional left gastric, hepatic, and pancreaticosplenic nodes while giving rise to the left and right intestinal lymphatic trunks. These nodes are positioned anterior to the aorta at the level of the celiac artery. Gastric, hepatic, and pancreaticosplenic nodes drain to the celiac node group, and are discussed with respect to their specific organs.[4]

Superior and Inferior Mesenteric Lymph Nodes

Superior and inferior mesenteric lymph nodes include lymphatics drained from the duodenojejunal flexure to the superior anal canal. The distal small intestine and entirety of the large and sigmoid intestine are drained by mesenteric, ileocolic, colonic, and pararectal nodes, which subsequently drain to the superior and inferior mesenteric groups.

Lateral Aortic Nodal Group

The lateral aortic nodes lie paramedian to the aorta on the anterior surface of the psoas major muscle and the diaphragmatic crura. Nodes in this group are also positioned anterior to the inferior vena cava and right renal vein. Superior lateral groups receive lymphatic drainage from the deep tissue of the posterior abdominal wall, adrenal glands, kidneys, ureters, gonads, fallopian tubes, and superior uterine fundus and body.

The lateral aortic group also contains the receiving nodes of the pelvic viscera and wall after converging at the level of the iliac nodes. The gonads drain directly to the lateral aortic and preaortic groups. Lumbar trunks emerge from this nodal group allowing further drainage into the confluence and thoracic duct.

Retroaortic Group

The retroaortic group of nodes drains a largely unspecific region of tissue. These nodes generally serve as peripheral nodes of the lateral groups and help to connect other lymph groups and plexuses. The paraspinal posterior abdominal wall may directly drain into this group.

◆ Cisterna Chyli and Thoracic Duct

The four main abdominal lymphatic trunks (right and left lumbar and intestinal) converge around the inferior border of the 12th thoracic vertebra to form the abdominal confluence of lymphatic trunks. This confluence can take on a number of different anatomic variations, ranging from a simple ductal structure to duplicated, triplicated, or plexiform variants with internal loculations. "Cisterna chyli" refers to a singular fusiform dilatation, which is present in approximately 60% of all individuals. The lymphatic confluence may extend inferiorly for 5 to 7 cm and lie anterolateral to the right of the lumbar vertebral bodies and disks. It lies lateral to the right of the abdominal aorta, which may lead to inadvertent chyle leaks after aortic surgery.[4]

The right lateral aortic lymph nodes and right-sided celiac and superior mesenteric preaortic groups surround the abdominal confluence of lymph trunks with branches that drain directly into this confluence. The lumbar arteries and azygos vein (right) are situated between the confluence and the vertebral column. The confluence itself receives the left and right lumbar lymph trunks and is responsible for draining the lateral aortic nodes, and subsequently, the lower extremities, the pelvic, perineal, and infraumbilical abdominal walls, pelvic viscera and gonads, and urologic and urogenital structures.

The intestinal lymph trunks receive celiac nodal and preaortic nodal group (superior and inferior mesenteric nodes) drainage. Anatomically, the abdominal gastrointestinal tract is drained completely from subdiaphragm to anus.

The thoracic duct is the diaphragm-crossing superior portion of cisterna chyli, entering the thorax posterolateral to the aorta. The duct can be subdivided into abdominal, thoracic, and cervical segments. The abdominal part of the thoracic duct originates from the superior aspect of the cisterna chyli. The thoracic duct passes through the "aortic hiatus" of the diaphragm with the aorta on its left and the azygos vein on its right. The thoracic duct and cisterna chyli relationship may possess significant anatomic variants (▶Fig. 21.12). These variations include a complete left-sided thoracic duct and cisterna chyli emptying into the left venous vasculature, a complete right-sided course with right-sided venous drainage, distal and proximal partial duplication of the thoracic duct, plexiform variant anatomy of the thoracic duct (where the duct is replaced by a plexiform channel network), and complete absence of cisterna chyli.[10]

◆ Lymphatic Drainage with Respect to the Abdominal Viscera

Abdominal Esophagus

The abdominal esophagus has lymphatics that drain to the left gastric, and left and right paracardial nodes. The posterior side of the abdominal esophagus may also drain inferiorly to the most superior paraaortic nodes.

Stomach

The stomach has regional divisions that drain to specific nodal groups that are shared with the lymphatic drainage routes with other abdominal organs. The lymphatic drainage from the stomach regionally follows the pathway of the corresponding arterial supply (▶Fig. 21.13a). The gastroesophageal junction shares lymphatic drainage with that of the lower esophagus to the superior-most paraaortic nodes, left gastric, and left and right paracardial nodes. The pylorus also drains the nodes draining the duodenum (suprapyloric nodes, infrapyloric nodes). The right and left gastroepiploic nodes drain the body and fundus of the stomach. Lymphatic metastasis from gastric cancer may involve supraclavicular lymph nodes through direct metastasis along the thoracic duct. Enlarged supraclavicular lymph node is called Virchow lymph nodes (▶Fig. 21.14). It is not exclusive of gastric cancer and may also be seen in colorectal and pelvic cancers.

Duodenum

Sharing some drainage with the stomach, the principal nodes of duodenal lymphatics are the anterior and posterior pancreaticoduodenal nodes (▶Fig. 21.13b). These pancreaticoduodenal nodes subsequently drain into the nearby pyloric groups (inferior and superior), as well as hepatic nodal groups (hepatoduodenal, common hepatic, and superior mesenteric nodes.)

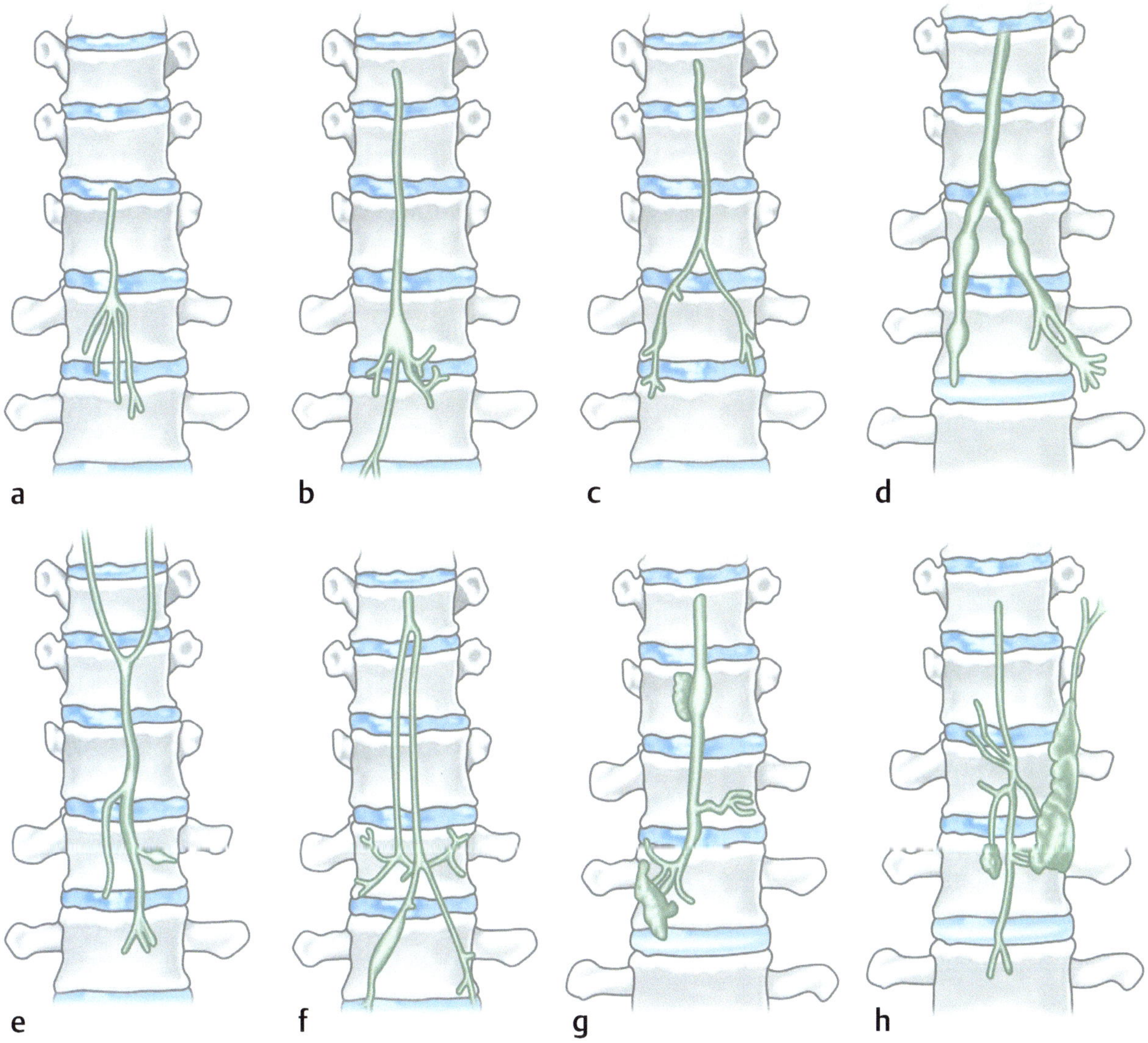

Fig. 21.12 (a–h) Anatomical variations of the cisterna chyli and abdominal segment of the thoracic duct from Anson's atlas.

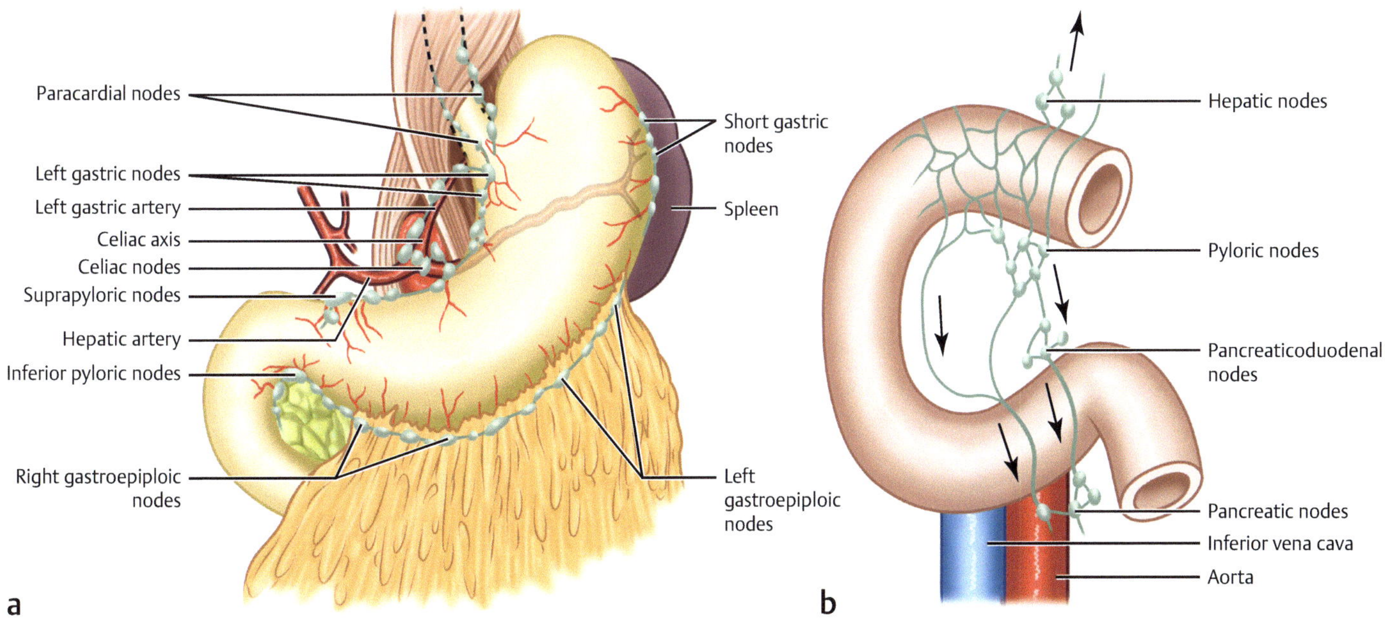

Fig. 21.13 Lymphatic drainage of the gastric and pyloric regions **(a)** and duodenum **(b)**. Ao, aorta; IVC, inferior vena cava.

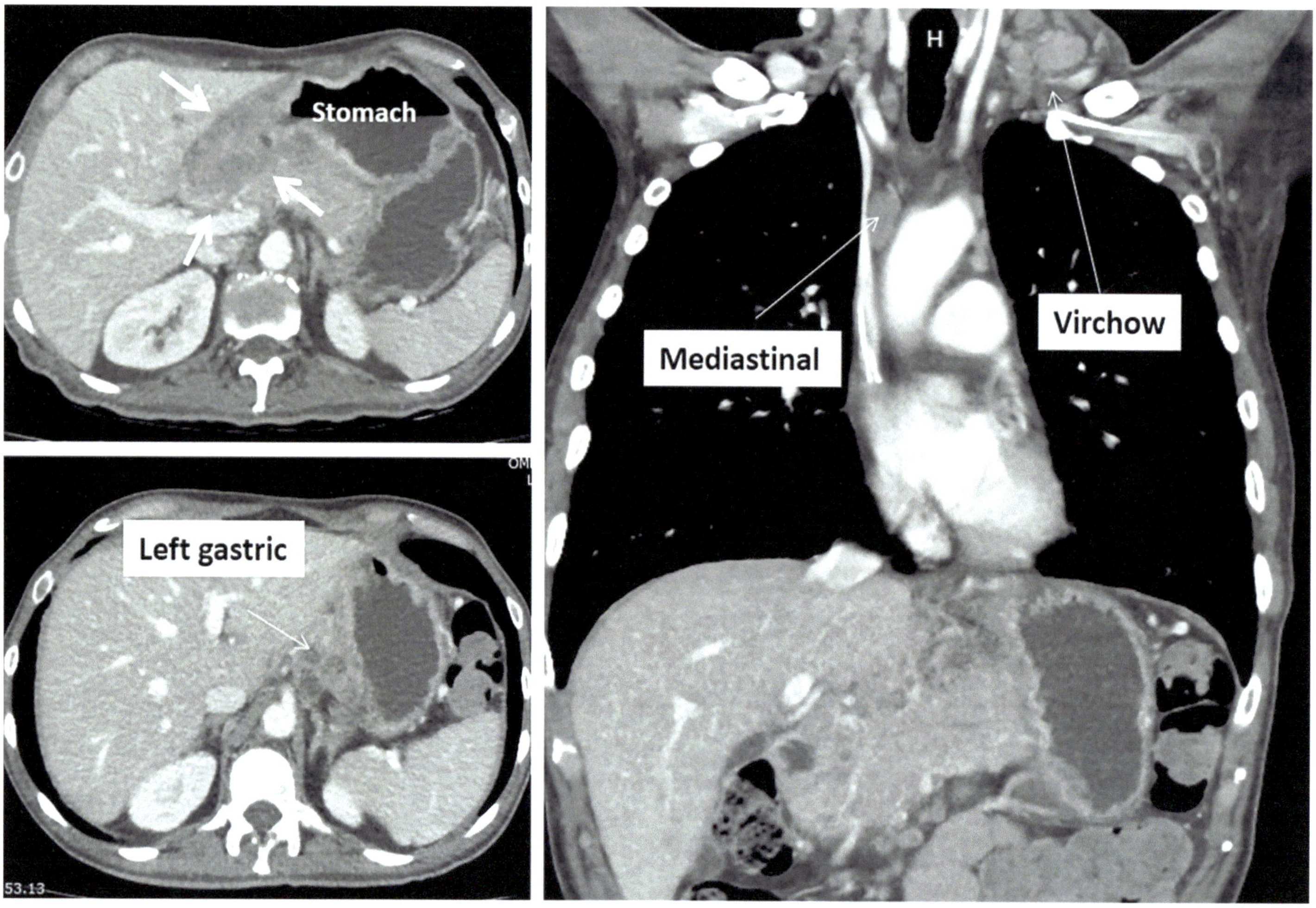

Fig. 21.14 Gastric cancer (*arrows*) involving the antrum of stomach with multiple necrotic regional and distant lymph node metastasis. The left supraclavicular (Virchow) lymph node is involved.

Jejunum and Ileum

The mucosa and muscular layers of the jejunum contain lymph vessels called lacteals that eventually coalesce as they exit the gut wall and drain to mesenteric nodes via routes following mesenteric arteries. Surrounding mesenteric nodes drain to the superficial mesenteric nodes situated superiorly at the origins of the superior mesenteric artery.

Large Intestine

As a general rule, lymphatics from the cecum, ascending colon, and proximal transverse colon drain to lymph nodes situated about the superior mesenteric artery. The distal transverse colon, descending colon, sigmoid, and rectum drain to nodes in proximity to the inferior mesenteric artery. There are four groups of colic nodes (▶Fig. 21.15).

Epicolic

The epicolic nodes are small nodularities of the serosa of the colon. These nodes may also be found in the appendices epiploicae.

Paracolic

Flank the medial borders of the ascending and descending colon and also reside on the mesenteric borders of the transverse and sigmoid colon.

Intermediate Colic

The intermediate colic or mesocolic lymph nodes accompany arterial drainage paths of the same names. These nodes communicate with more major nodal groups around the origins of the great abdominal arteries.

Preterminal (Principal) Colic

Are next to the main trunks of the great abdominal vessels (superior and inferior mesenteric arterial origins). These nodes eventually extend to the paraaortic nodes.

Ascending Colon, Cecum, Appendix

The ileocolic and right colic arteries provide a highway for the lymphatics of the ascending colon, with a predominance around

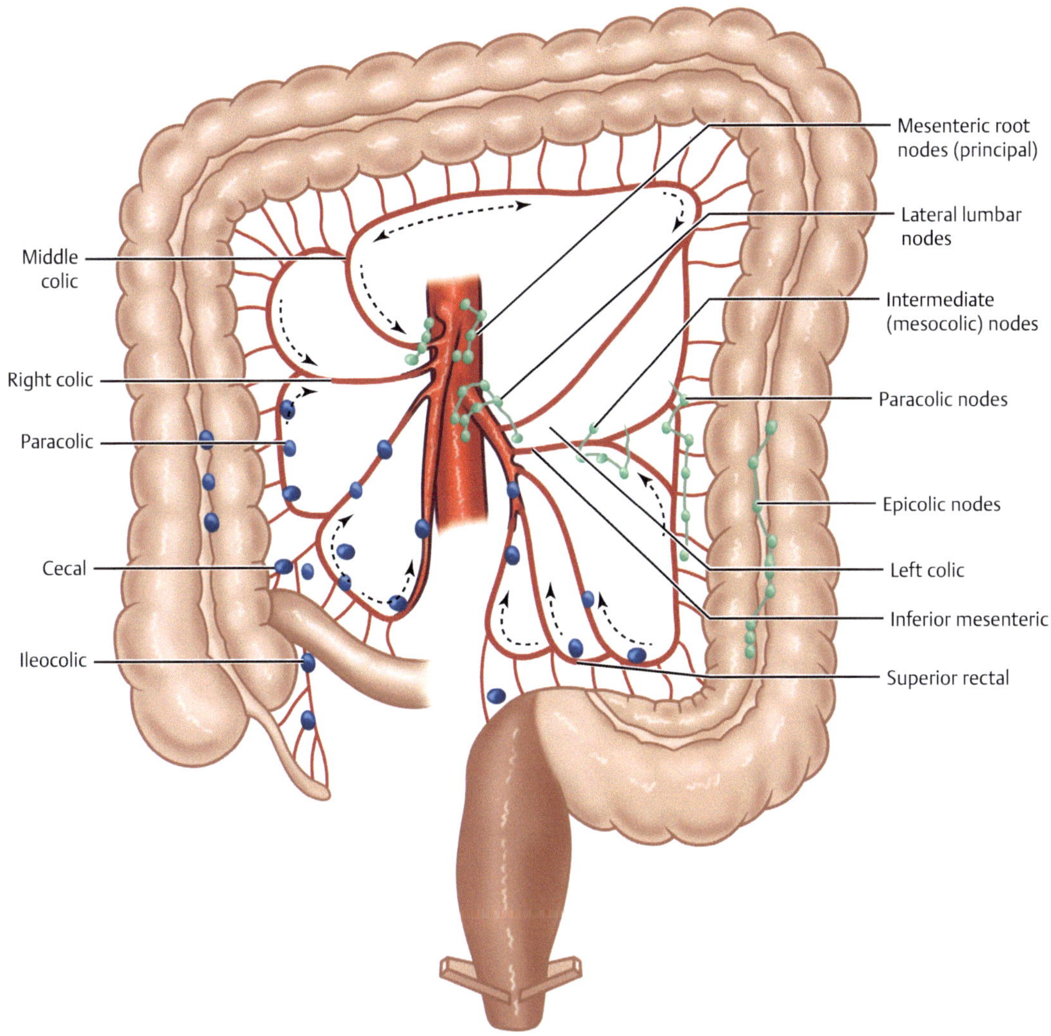

Fig. 21.15 Lymphatic of the colon.

the right colic artery as this is the desired route by lymphatics from the hepatic flexure and distal ascending colon. The ileocolic nodes at the level of the superior mesenteric artery are the ultimate drainage destination (▶Fig. 21.15, ▶Fig. 21.16). Appendiceal lymphatics are numerous and widespread throughout the appendage wall. Confluence of these lymphatic drains may enter directly into nodes with close proximity to the mesoappendix. Eventually, terminal drainage is along the route of the ileocolic chain to the inferior and superior nodes of the same name.

Transverse Colon

Lymphatics of the transverse colon follow the middle colic arteries to drain to the superior mesenteric nodes (▶Fig. 21.15). The splenic flexure is usually drained by nodes along the left colic artery (and follows drainage routes of the descending colon). Eventually, the inferior mesenteric nodes receive lymphatic drainage from the splenic flexure. There is a level of variability in the drainage from the transverse colon, largely dependent on the arterial vasculature.

Descending Colon

The left colic artery is surrounded by lymph nodes and vessels that receive drainage from the descending colon. The sigmoid colon drains to sigmoid nodes in the mesocolon, joining the course of the superior rectal and left colic arteries, and eventually drains to the inferior mesenteric nodes.

Rectum

The mesorectal nodes are held in a loose fatty tissue meshwork around the gut wall. Pararectal nodes lie within the mesorectum and epirectal nodes are situated close to the outer muscular fibers of the rectal longitudinal muscle. Drainage of the rectal nodes occurs superiorly along the course of the superior rectal artery. Intermediate groups of nodes will drain to more superior locations along the origins of the inferior mesenteric artery. Generally, the rectum and anus above the dentate line are drained by the inferior mesenteric nodes (▶Fig. 21.17, ▶Fig. 21.18).

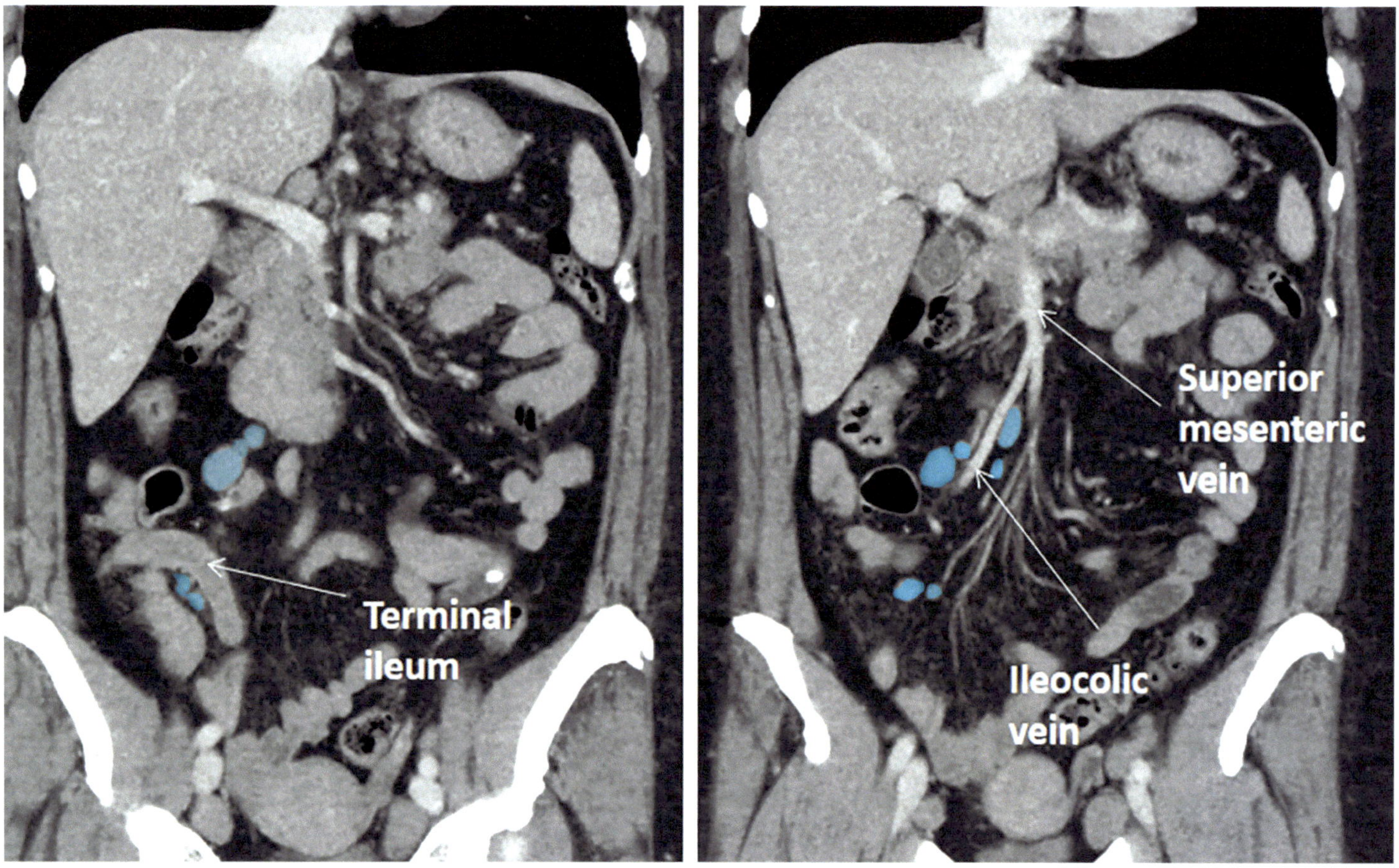

Fig. 21.16 Ileocolic mesenteric lymph nodes in lymphoma.

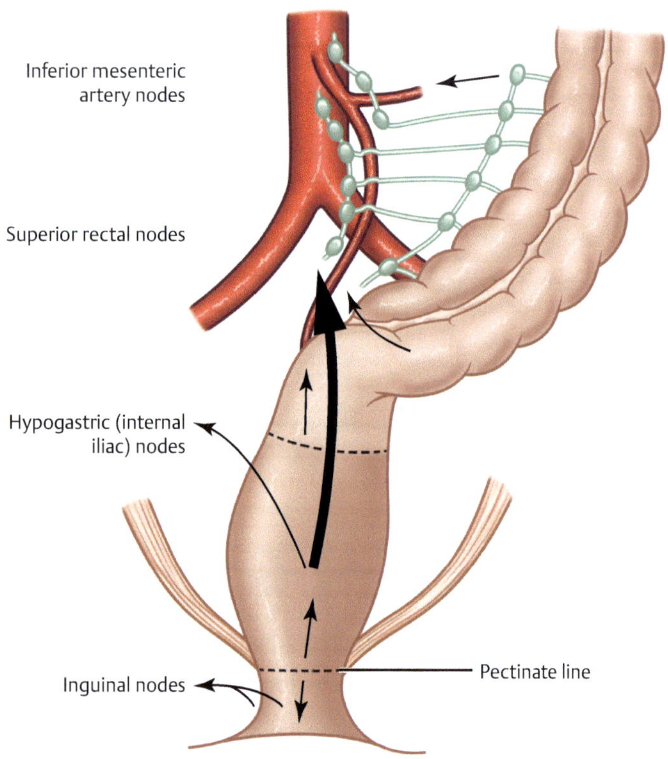

Fig. 21.17 Lymphatic drainage of the sigmoid, rectum, and anus. Below the pectinate line, lymphatic drains to inguinal nodes. Above the pectinate line, lymphatic drains to inferior mesenteric nodes.

Anus

The submucosal and intramural lymphatics drain the upper anal mucosa and internal anal sphincter along with the conjoint longitudinal coat. The perianal plexus drains more inferior anal structures (external anal sphincter and inferior mucosa). The perianal plexus drains to the external iliac lymph nodes. The puborectalis is drained by vessels eventually running to the internal iliac nodes. The mesorectal lymph nodes drain the more superior anus and lower rectum.

Liver

Superficial hepatic nodes run over the superficial surface of the liver in the subserosa. The coronary and right triangular ligaments can drain directly to the thoracic duct. The superior surface of the liver drains to the diaphragmatic nodes. Pericaval (portal) nodes drain the posterior surface, caudate lobe, and posteroinferior portion of the right lobe (▶Fig. 21.19). The hepatic nodes drain the superior, inferior, and anterior surfaces at the porta hepatis. The paracardiac nodes may drain the posterolateral left hepatic lobe. Finally, the celiac nodes may drain the right surface, and right-superior surface of the right lobe via vessels that course with the inferior phrenic artery. The superficial hepatic nodes connect with deep nodes of the liver. The deep hepatic vessels that drain the liver parenchyma largely drain to the hilar, hepatic, and portal nodal groups (▶Fig. 21.20).

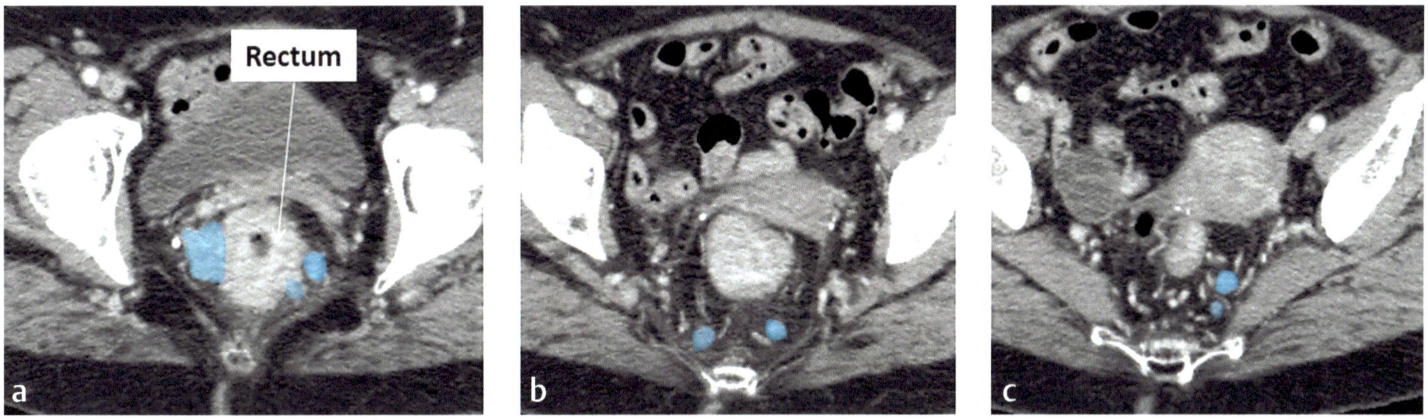

Fig. 21.18 Rectal cancer with perirectal **(a)**, retrorectal **(b)**, and superior rectal **(c)** lymph node metastasis (shown in *blue*). Pararectal (perirectal and retrorectal) nodes lie within the fatty mesorectum.

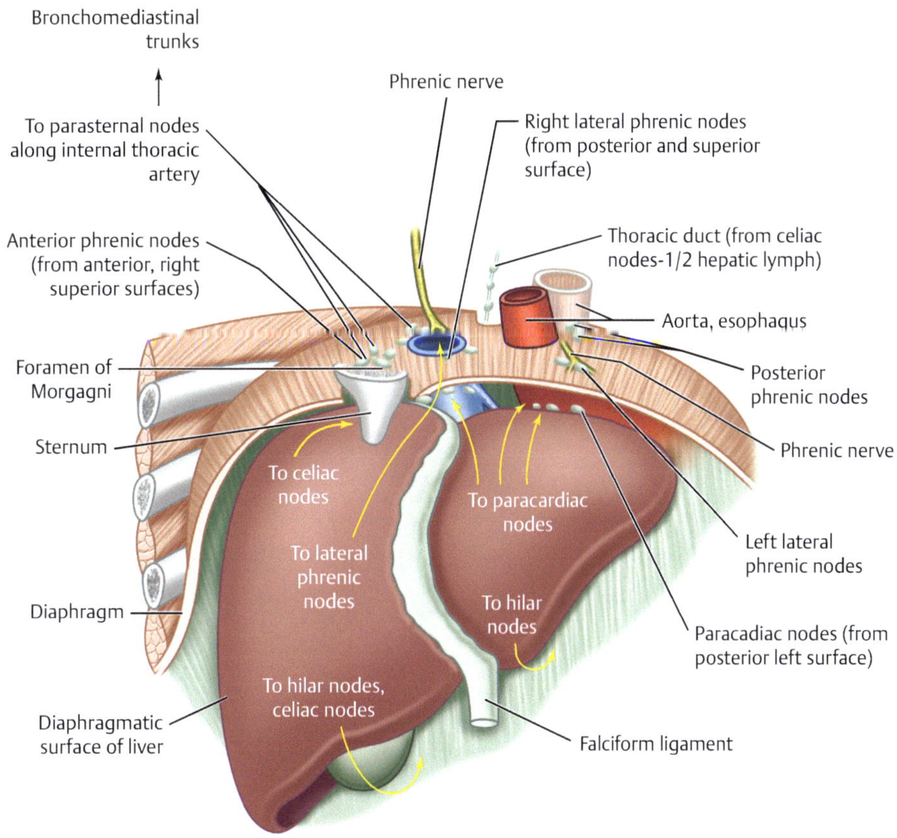

Fig. 21.19 Superficial hepatic nodes drainage.

Gallbladder/Biliary Tree

Intrahepatic lymph vessels are responsible for draining the majority of the gallbladder bordering the liver, via subserosal and submucosal lymphatic plexuses. The cystic node is a specific node draining nonhepatic bordering gallbladder. It lies above the cystic duct in Calot's triangle (▸Fig. 21.20). A secondary node drains the cystic node and some direct lymphatic channels and is located on the anterior border of the lesser omentum. The more superior bile ducts and lymphatics accompanying hepatic ducts drain into hepatic nodes within the porta hepatis. The lower bile ducts are drained by the inferior hepatic and upper pancreaticosplenic nodes.

Pancreas

The pancreas has multiple lymphatic drainage routes that are often unpredictable as the nodes may drain any part of the gland, despite location. The larger lymphatic vessels follow the arteries around the organ and drain into adjacent nodal groups (▸Fig. 21.21). The pancreatic tail and body had a predisposition to

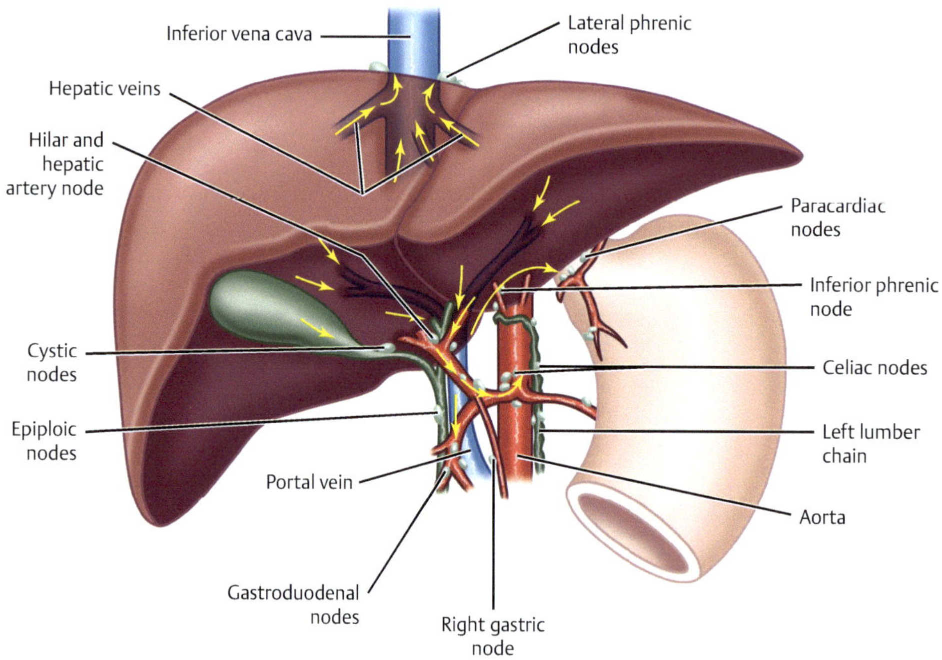

Fig. 21.20 Deep hepatic nodes drainage.

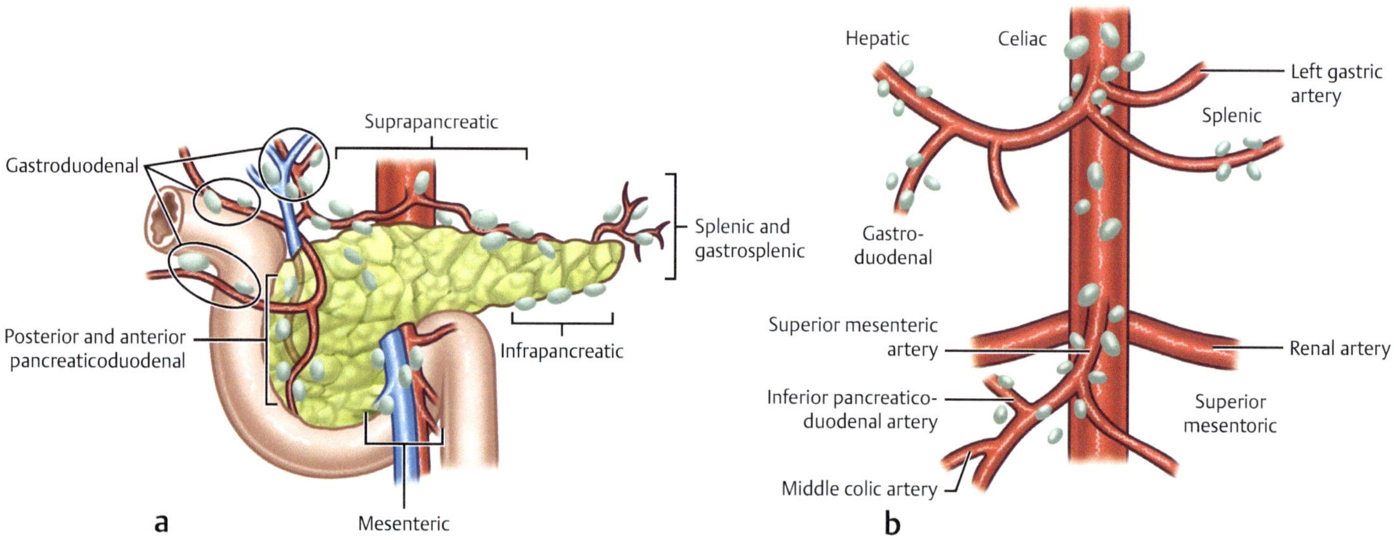

Fig. 21.21 (a, b) Lymph nodes draining the pancreas. Typical location of lymph nodes surrounding the pancreas and major vessels. The pancreas has multiple lymphatic drainage routes that are often unpredictable as the nodes may drain any part of the gland.

drain to the pancreaticosplenic nodes and preaortic nodes. The head and neck of the pancreas drain to the nodal groups of the proximal small intestine, including the preaortic nodes and celiac axis nodes. The pancreaticoduodenal arteries, superior mesenteric artery, and hepatic arteries direct lymphatic flow toward these nodal groups.

Spleen

Lymphatic vessels course via the splenic trabeculae, and subsequently along the splenic vasculature, exiting the hilum. There are nodes that exist at the hilum that receive these vessels but may also course the length of the splenic artery and drain to the celiac nodes directly.

Adrenal Glands

The paraaortic lymph nodes (lateral groups) drain the adrenal gland via lymphatics that arise as plexuses that drain both the cortex and medulla.

Kidney

Three lymphatic plexuses (around renal tubules, inferior to the renal capsule, and in the perirenal fat) give rise to major trunks that course with the renal vein toward the lateral aortic nodes (▸Fig. 21.22).

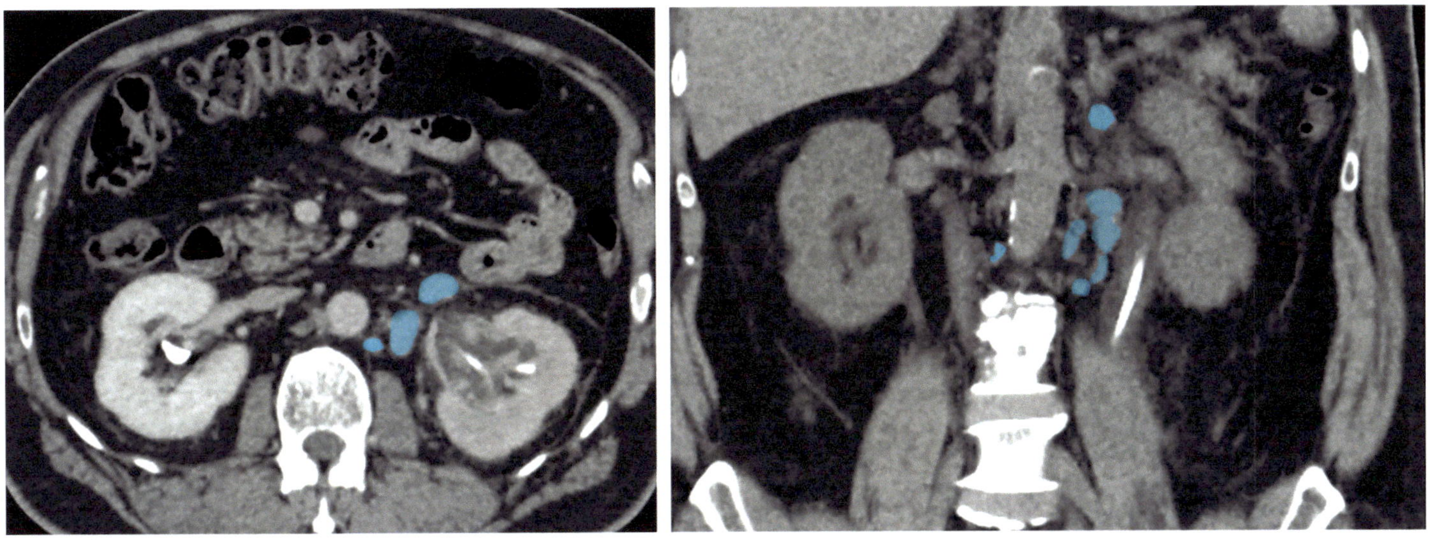

Fig. 21.22 Pyelonephritis involving left kidney. Note regional paraaortic and aortocaval enlarged lymph nodes (in *blue*).

Ureter

Because of the length of the ureter, it has a superior and inferior distribution of lymphatic drainage. A network of lymphatic plexuses begins in the muscular walls and adventitia of the ureter itself. These vessels join to create lymphatics that drain to renal collecting vessels or continue to the lateral aortic nodes at the origin of the gonadal arteries. The inferior ureter drains via pelvic nodes (common, external, and internal iliac nodes).

References

1. Taylor A, Rockall AG, Powell MEB. An atlas of the pelvic lymph node regions to aid radiotherapy target volume definition. Clin Oncol (R Coll Radiol) 2007; 19(7):542–550
2. Paño B, Sebastià C, Buñesch L, et al. Pathways of lymphatic spread in male urogenital pelvic malignancies. Radiographics 2011;31(1):135–160
3. McMahon CJ, Rofsky NM, Pedrosa I. Lymphatic metastases from pelvic tumors: anatomic classification, characterization, and staging. Radiology 2010; 254(1):31–46
4. Standring S. Gray's Anatomy: The Anatomical Basis of Clinical Practice. 40th ed. Edinburgh: Elsevier Churchill Livingstone; 2008
5. Chen N, Min PQ, Liu ZY, Wu B, Yang KQ, Lu CY. Radiologic and anatomic study of the extraperitoneal space associated with the rectum. AJR Am J Roentgenol 2010;194(3):642–652
6. Kadir S. Atlas of Normal and Variant Angiographic Anatomy. Philadelphia: W. B. Saunders; 1991:496–498, 503–515
7. Grey AC, Carrington BM, Hulse PA, Swindell R, Yates W. Magnetic resonance appearance of normal inguinal nodes. Clin Radiol 2000;55(2):124–130
8. Mao Y, Hedgire S, Prapruttam D, Harisinghani M. Imaging of pelvic lymph nodes. Curr Radiol Rep 2014;2:70
9. Vinnicombe SJ, Norman AR, Nicolson V, Husband JE. Normal pelvic lymph nodes: evaluation with CT after bipedal lymphangiography. Radiology 1995; 194(2):349–355
10. Johnson OW, Chick JFB, Chauhan NR, et al. The thoracic duct: clinical importance, anatomic variation, imaging, and embolization. Eur Radiol 2016;26(8): 2482–2493

22 Surface Anatomy and Projectional Surface Anatomy

Ali Mirjalili

◆ Introduction

Surface anatomy is a core component of clinical practice and accurate knowledge of surface anatomy is essential for safe clinical practice. To better understand the surface anatomy and projectional surface anatomy, anatomical planes are essentials.

◆ Important Anatomical Planes

The planes are a rational and convenient method of subdividing the human body using a number of vertical and horizontal lines (▶Fig. 22.1). These planes provide a common reference point when identifying the location of internal structures between clinicians[1,2]:

1. *The C6 plane* is axial plane at the vertebral level of the sixth cervical vertebra. The lower border of the cricoid cartilage, the vertebral artery entering the foramen transversarium, the origin of the inferior thyroid artery, the transition from the pharynx to the esophagus, the transition from the larynx to the trachea, and the position of the middle cervical ganglion are at the C6 plane.
2. *The sternal angle plane* is the level at which a horizontal plane through the manubriosternal joint intersects with the anterior border of the vertebral column (▶Fig. 22.2). It is located primarily at the fourth/fifth thoracic vertebral levels, and it should be noted that variability exists (▶Table 22.1). The deep structures projected at the level of this plane are: the tracheal bifurcation (or carina), convexity of the arch of the aorta, azygous vein draining to superior vena cava junction, and bifurcation of the pulmonary trunk all occur **approximately** at this plane. It should be noted that, most often, these structures listed above are located just below the sternal angle (▶Fig. 22.3).
3. *The transpyloric plane* is the level at which there is a horizontal plane midway between the suprasternal notch and the upper border of the pubic symphysis (▶Fig. 22.4). It is located primarily at the lower half of the first lumbar vertebral body and it should be noted that variability exists (▶Table 22.1). An alternative definition of the transpyloric plane in some textbooks is the level at which there is a horizontal plane midway between the xiphisternal joint and umbilicus, but this surface marking has shown to be below the true level of the plane in the majority of people. The deep structures projected at the level of this plane are: the origin of the superior mesenteric artery (SMA),

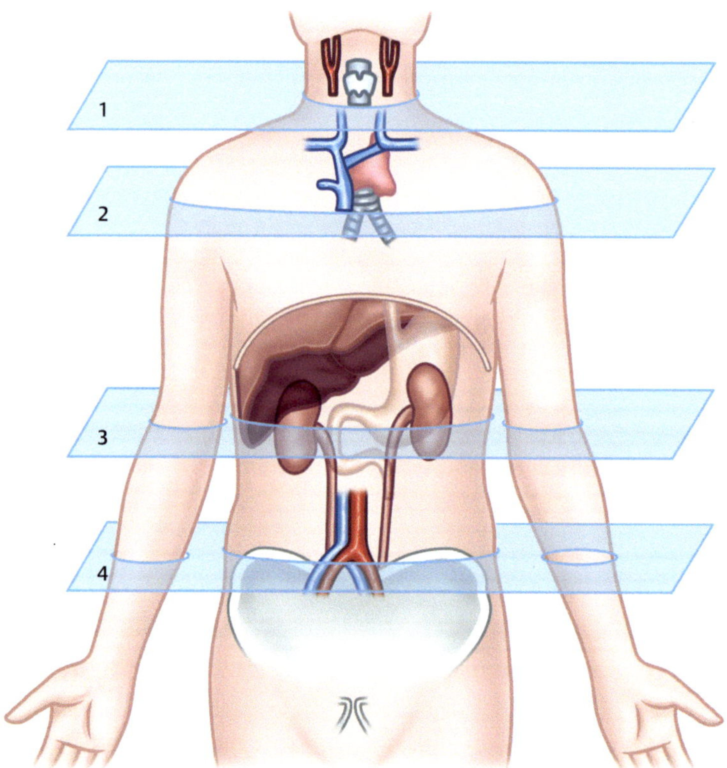

Fig. 22.1 Clinically relevant anatomical planes: 1, C6 plane; 2, sternal angle plane; 3, transpyloric plane; 4, supracristal plane. C6, sixth cervical vertebra.

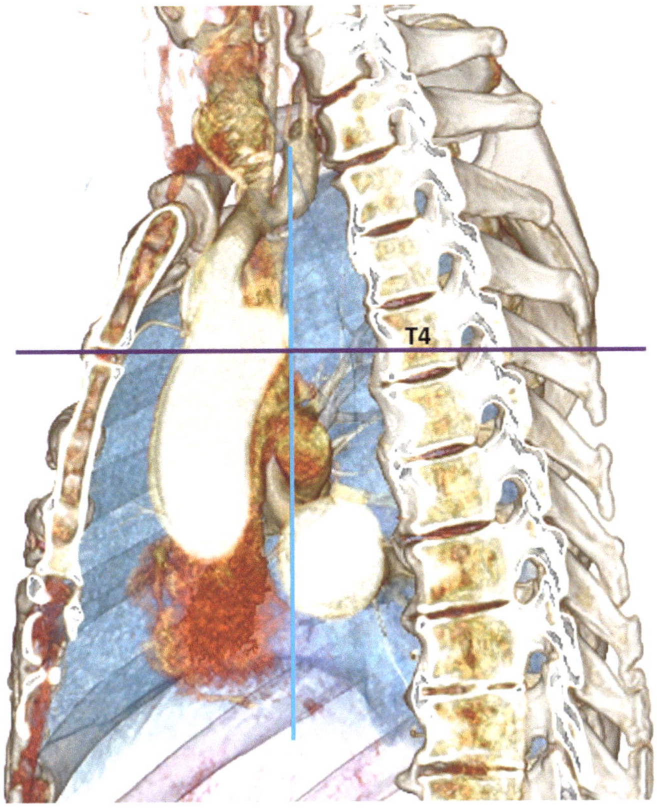

Fig. 22.2 The sternal angle plane on sagittal scan (*purple line*). It is located at the fourth/fifth thoracic vertebral levels but variability exists.

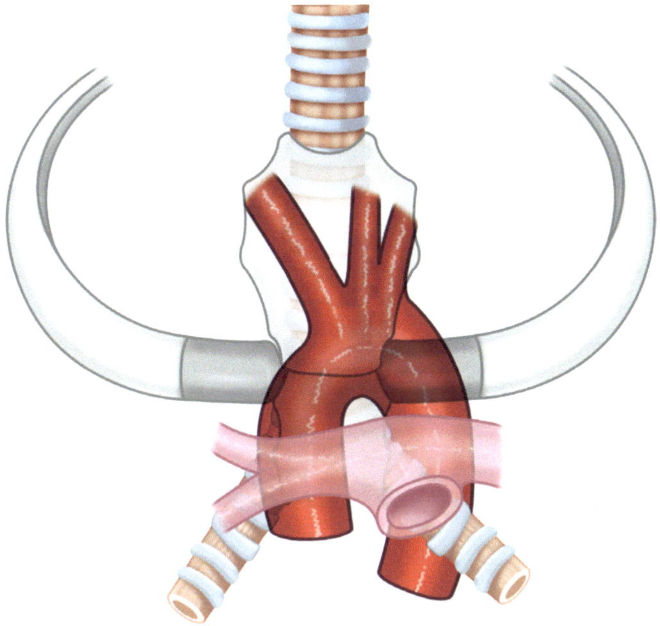

Fig. 22.3 Surface anatomy of the deep structures (arch of aorta, tracheal bifurcation, and pulmonary trunk) to the sternal angle plane. This figure shows the deep structures located just below the plane of the sternal angle.

Table 22.1 Vertebral levels of the sternal angle plane, transpyloric plane, subcostal plane, and supracristal plane with their variations

	Sternal angle plane	Transpyloric plane	Subcostal plane	Supracristal plane
Most common VL	T4/T5	L1	L2	L4
Variability	T2–T6	T12–L2	T12–L3	L3–L5

Abbreviations: L, lumbar; T, thoracic; VL, vertebral level.

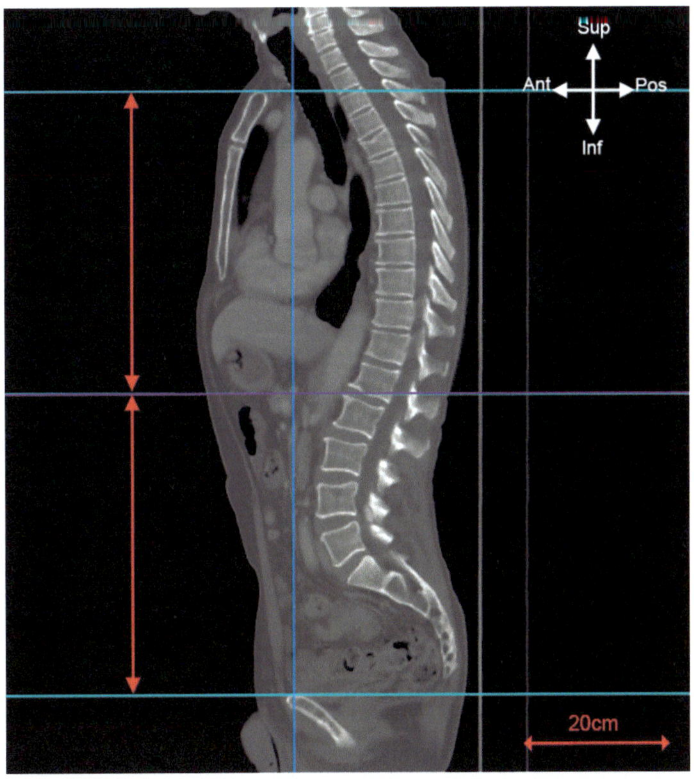

Fig. 22.4 The transpyloric plane on sagittal scan (*purple line*). It is usually located at the lower half of the first lumbar vertebral body.

the head and neck of the pancreas, the hilum of the spleen, the formation of the portal vein, and the left and right renal hila all occur approximately at this plane (▶Fig. 22.5). Note that the fundus of the gallbladder commonly lies a few centimeters below the transpyloric plane in a majority of people.

4. *The subcostal plane* is a horizontal plane that runs through the lowermost limit of the costal margins and intersects with the anterior border of the vertebral column (▶Fig. 22.6). It occurs primarily at the level of the second lumbar vertebra. It should be noted that variability exists (▶Table 22.1) and changes occur due to respiration. The deep structures projected at the level

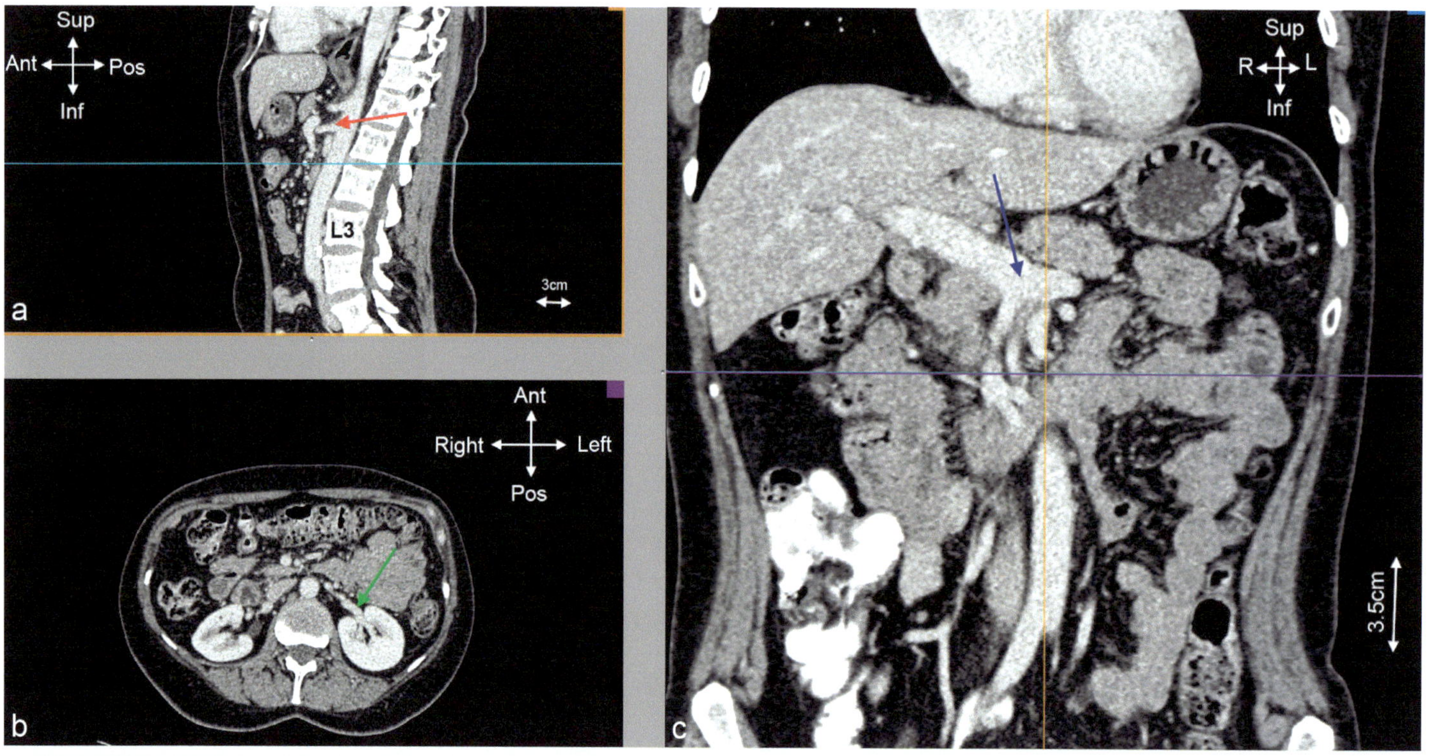

Fig. 22.5 Surface anatomy of the deep structures to the transpyloric plane. **(a)** The superior mesenteric artery (*red arrow*) above the transpyloric plane in sagittal scan, **(b)** the left renal hilum (*green arrow*) at the level of the transpyloric plane on axial scan, and **(c)** the formation of the portal vein (*blue arrow*) above the transpyloric plane in coronal section.

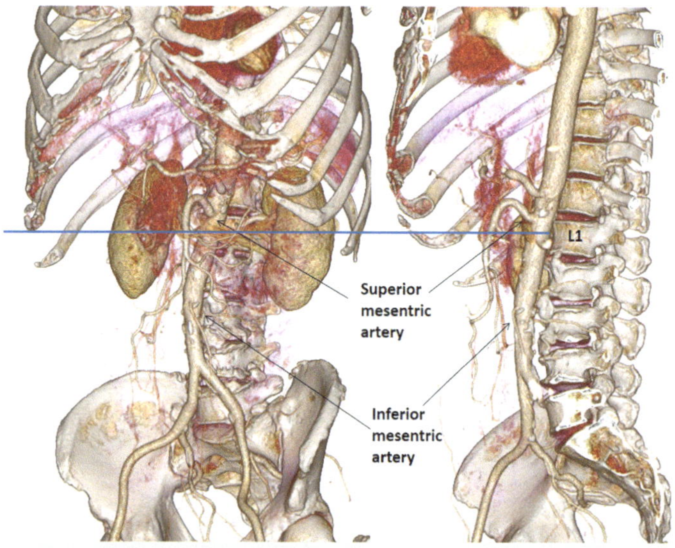

Fig. 22.6 The subcostal plane on a stacked coronal scan (*green line*). It occurs primarily at the level of first or second lumbar level where superior mesenteric artery orifice is located. The inferior mesenteric artery origin is commonly located a few centimeters below the subcostal plane.

of this plane are: the inferior mesenteric artery (IMA) is commonly located a few centimeters below the subcostal plane.

5. *The supracristal plane* is a horizontal plane, which crosses the highest points of the two iliac crests and intersects with the anterior border of the vertebral column (▸Fig. 22.7a). It is primarily at the level of the fourth lumbar vertebra, and it should be noted that variability exists (▸Table 22.1) (▸Fig. 22.7b). The supracristal plane is an important clinical landmark for performing a lumbar puncture and is most commonly at the level of the fourth lumbar vertebra. The spinal cord terminates at the lower border of the first or upper border of the second lumbar vertebra (L1/L2), and thus this plane is a reliable landmark to perform a lumbar puncture. The deep structures projected at the level of this plane are: the bifurcation of the abdominal aorta is approximately at this level.
6. *The transtubercular plane* also known as intertubercular plane is a horizontal plane, which crosses the iliac tuberosities and intersects with the anterior border of the fifth lumbar vertebra. The deep structures projected at the level of this plane are: the formation of the inferior vena cava (IVC) is approximately at this level.

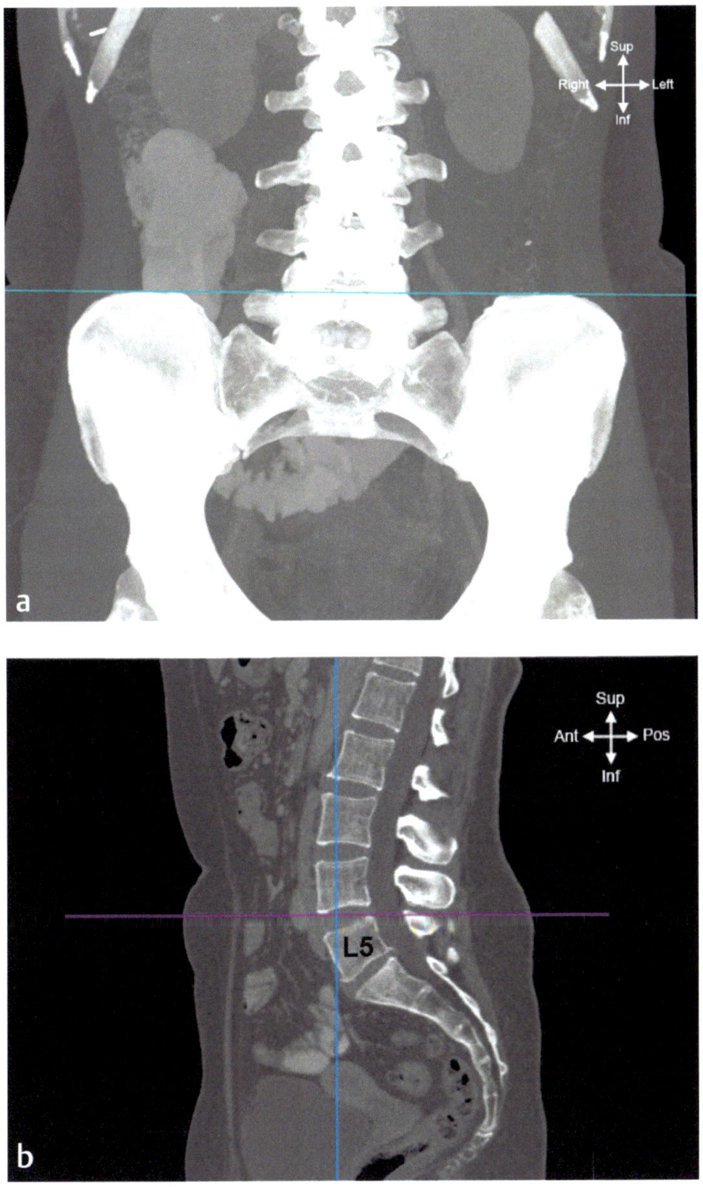

Fig. 22.7 (a) The supracristal line on a stacked coronal scan (*green line*) and **(b)** the vertebral level of the supracristal line on sagittal scan (*purple line*). The supracristal plane is an important clinical landmark for performing a lumbar puncture.

◆ Face

Facial Nerve

Some branches of the facial nerve are at risk of injury due to their superficial course and lack of protective fat padding compared to the other branches of the facial nerve. It is important to be aware of the course of these two branches to avoid injury during surgical incisions (e.g., skin cancer removal). The surface anatomy of the most commonly injured branches of the facial nerve are summarized as follows:

- ***Frontotemporal nerve***: The branches of the frontotemporal nerve can run within a triangle; the corners of the triangle are the pterion, the supraorbital notch, and the inferior rim of the orbit.

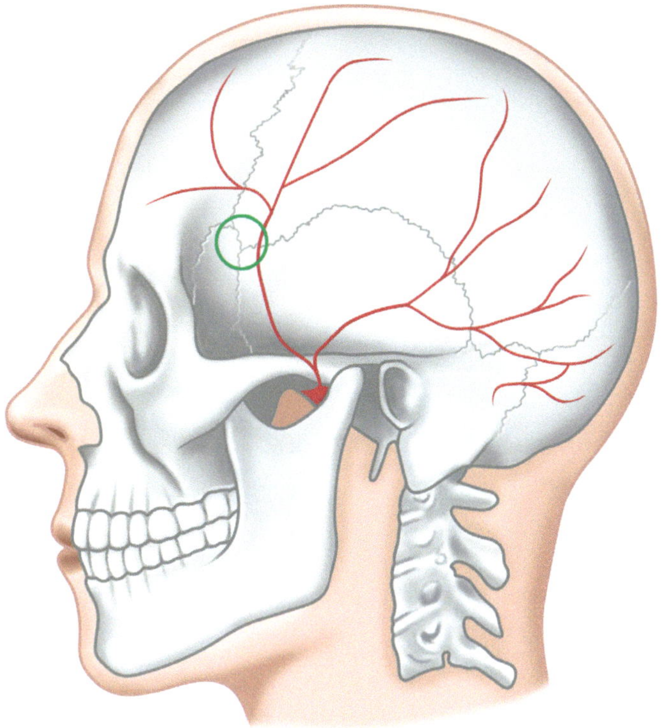

Fig. 22.8 Surface anatomy of the pterion which is located within a circle (in *green*) of radius 1 cm, 2.5 cm behind, and 1.5 cm above the posterolateral aspect of the frontozygomatic suture.

- ***Marginal mandibular nerve***: The course of the marginal mandibular nerve runs inferiorly to the inferior border of the mandible, within 3.5 cm or 2 "finger-breadths" of the lower border of the mandible.

Palpable Arteries

Facial artery can be palpated as it crosses the lower mandibular border immediately to masseter muscle and on the face approximately 1 cm lateral to the angle of the mouth. Another artery, which is also palpable, is the superficial temporal artery. It is palpable immediately anterior to the tragus. It divides into frontal and parietal branches approximately 2.5 cm above the zygomatic arch.

Pterion[3]

The pterion is a structural weak point of the skull as it is the point of union between the sphenoparietal, coronal, and squamous sutures. The pterion is located within a circle of radius 1 cm, 2.5 cm behind, and 1.5 cm above the posterolateral aspect of the frontozygomatic suture (▸ Fig. 22.8). It is a common location for skull fractures and it is of particular importance to note that the anterior division of the middle meningeal artery runs deep to the pterion. Damage to this structure may result in an extradural hemorrhage.

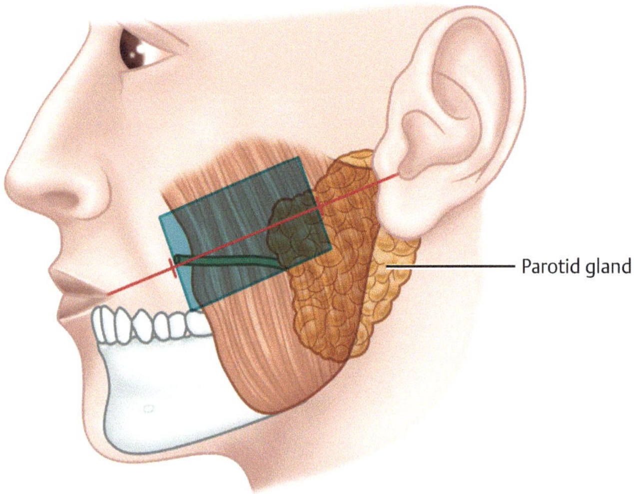

Fig. 22.9 Surface anatomy of the parotid duct: it is located in the middle half of a line between the lower border of the tragus to the commissure of the mouth running within 1.5 cm above or below this line. PG, parotid gland.

Parotid Duct[4]

The parotid duct has a superficial course in the face. It is at risk of injury by a relatively superficial lesion of the face. The parotid duct is located in the middle half of a line between the lower border of the tragus to the commissure of the mouth running within 1.5 cm above or below this line (▶Fig. 22.9).

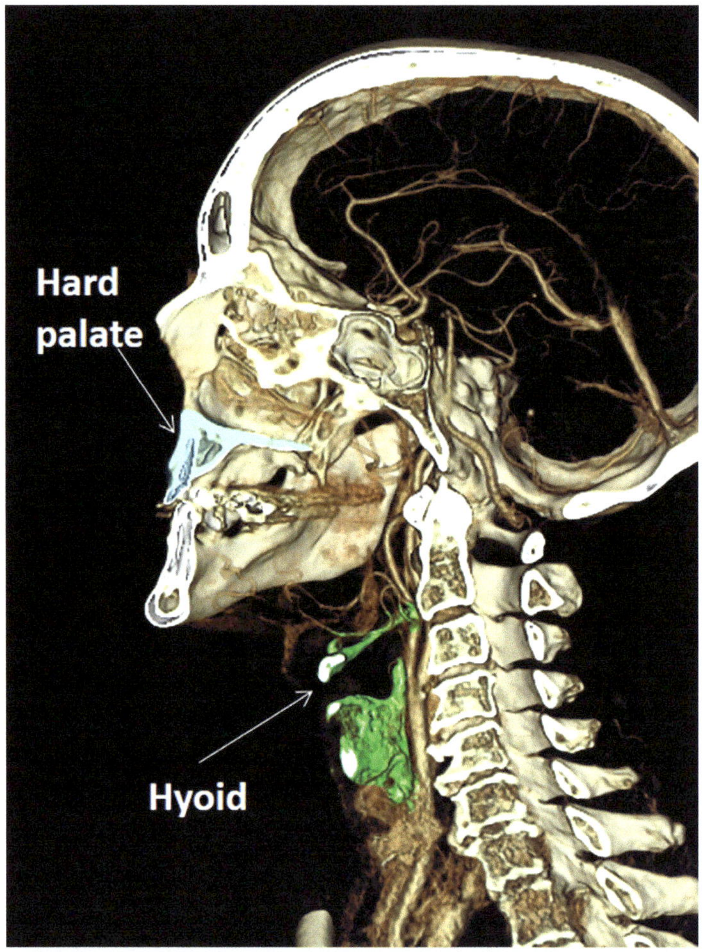

Fig. 22.10 The hyoid bone is at the level of the third cervical vertebra and the hard palate is located most commonly at the first cervical vertebral level on a sagittal scan.

◆ Neck

Vertebral Level of Clinically Important Structures of the Neck[5]

The vertebral level of the hard palate, hyoid bone, thyroid cartilage, cricoid cartilage, and bifurcation of the common carotid are clinically important and are summarized in the following.

The hard palate is located most commonly at the first cervical vertebral level and the hyoid bone at the third cervical vertebra (▶Fig. 22.10). The inferior border of the cricoid cartilage in the midline anteriorly is primarily located at the level of the seventh cervical vertebra. The bifurcation of the common carotid artery is most commonly at the level of the third cervical vertebra and can vary from the second and fifth cervical vertebrae.

Major Vessels[5]

Carotid Artery

The carotid artery runs through the entire course of the neck and may be presented by a more or less straight line passing from sternoclavicular joint to a point just behind the mandibular condyle. It bifurcates into the internal and external carotid arteries most commonly at the level of the third cervical vertebra (▶Fig. 22.11). It may be compressed against the prominent

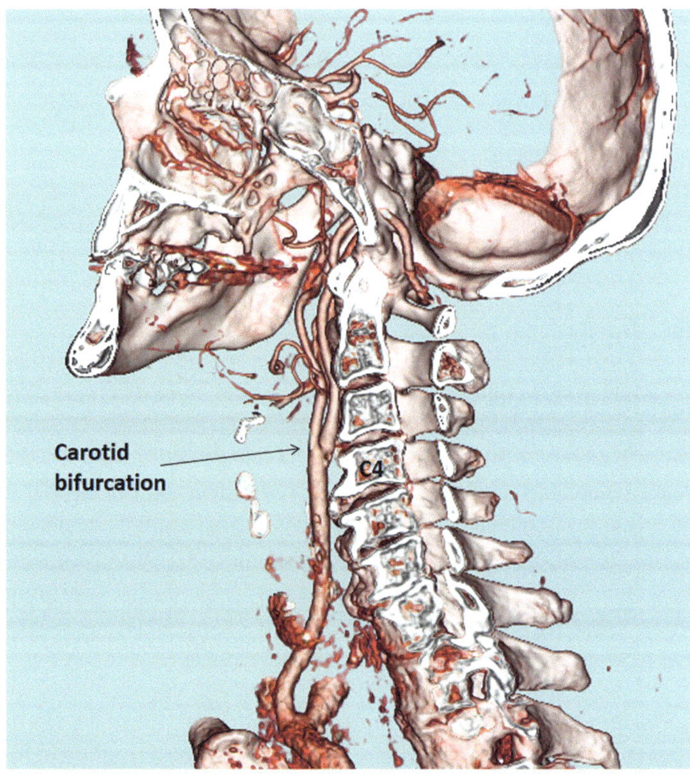

Fig. 22.11 Bifurcation of the left common carotid artery (*red arrow*) at the vertebral level of C4 on a sagittal scan.

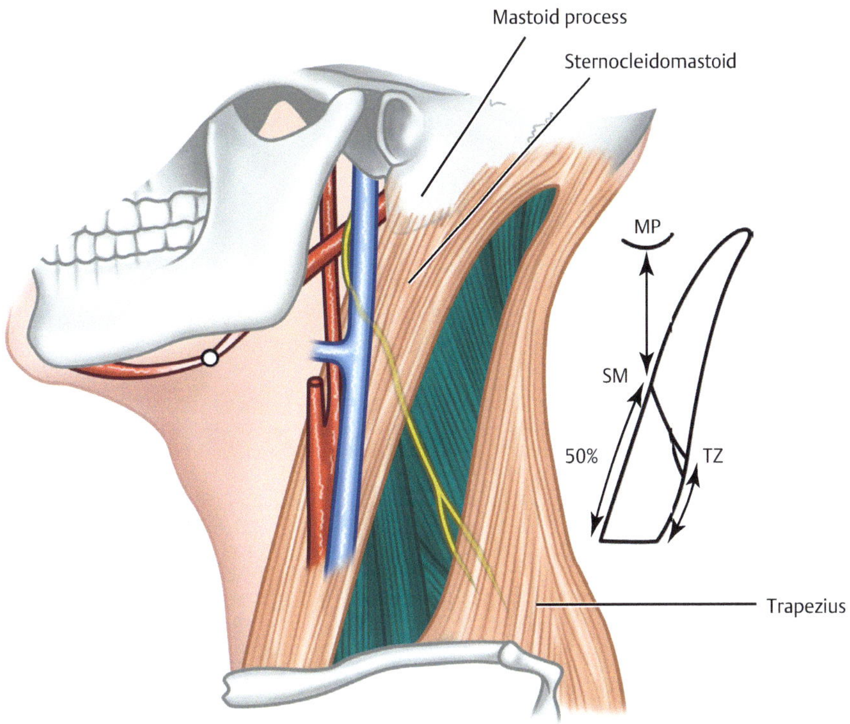

Fig. 22.12 Surface anatomy of the spinal accessory nerve (*yellow*) in the posterior triangle of the neck. MP, mastoid process; SM, sternocleidomastoid; TZ, trapezius.

transverse process of the sixth vertebra (Chassaignac's or carotid tubercle), which lies lateral to the cricoid cartilage. At this level, the pulsation may be felt under the anterior border of the sternocleidomastoid muscle.

Subclavian Artery

At the root of the neck, the subclavian artery passes through the inferior part of the palpable interscalene groove. It crosses the upper surface of the first rib behind the middle of the clavicle. It then runs upward to reach a point approximately a few centimeters (2 cm) above the clavicle deep to the posterior border of the sternocleidomastoid muscle. It is palpable where it crosses the first rib.

Subclavian Vein

The subclavian vein lies anterior to the subclavian artery and follows a similar course by passing over the first rib into the root of the neck. It joins the internal jugular vein behind the sternoclavicular joint and lies closest to the clavicle at the junction of its middle and medial thirds, approximately 7 cm from the midline.

Internal Jugular Vein

The internal jugular vein runs in the carotid sheath together with carotid artery and the vagus nerve. It sits lateral or anterolateral to the carotid artery. It has similar surface markings to the common and internal carotid artery. It may be accessed for central vein cannulation along the posterior border of the sternocleidomastoid muscle at the lower part of the neck or between the two heads of the sternocleidomastoid muscle above the sternoclavicular joint.

Spinal Accessory Nerve (SAN) In the Posterior Triangle of the Neck[6]

The SAN (cranial nerve XI) is the nerve that controls the sternocleidomastoid and trapezius muscles. The SAN runs deep to or between both heads of sternocleidomastoid, and then enters the posterior triangle approximately halfway down the posterior border of sternocleidomastoid (▶Fig. 22.12). It runs superficially in either a straight or tortuous path through the posterior triangle, approximately 3 mm below the skin surface. It finally exits the posterior triangle a few centimeters above the clavicle by piercing through the anterior border of trapezius. *Note that the course of the accessory nerve is highly variable in the posterior triangle, which makes it a commonly damaged structure during surgical procedures (e.g., posterior triangle lymph node biopsies).*

◆ Thorax

Thoracic Venous Structures[7,8]

Knowledge of the thoracic venous structures is clinically and surgically important. Emergency physicians, anesthetics, and general surgeons perform central venous catheterization in many clinical settings. Complications of this procedure are not uncommon and can be fatal. The thoracic central venous structures include: brachiocephalic veins (BCs), superior vena cava (SVC), and superior vena cava/right atrium (SVC/RA) junction. Their surface anatomy is summarized as follows.

Brachiocephalic Veins (Left and Right)

The left and right brachiocephalic veins are formed behind the ipsilateral sternoclavicular joint (▶Fig. 22.13) and are only lateral

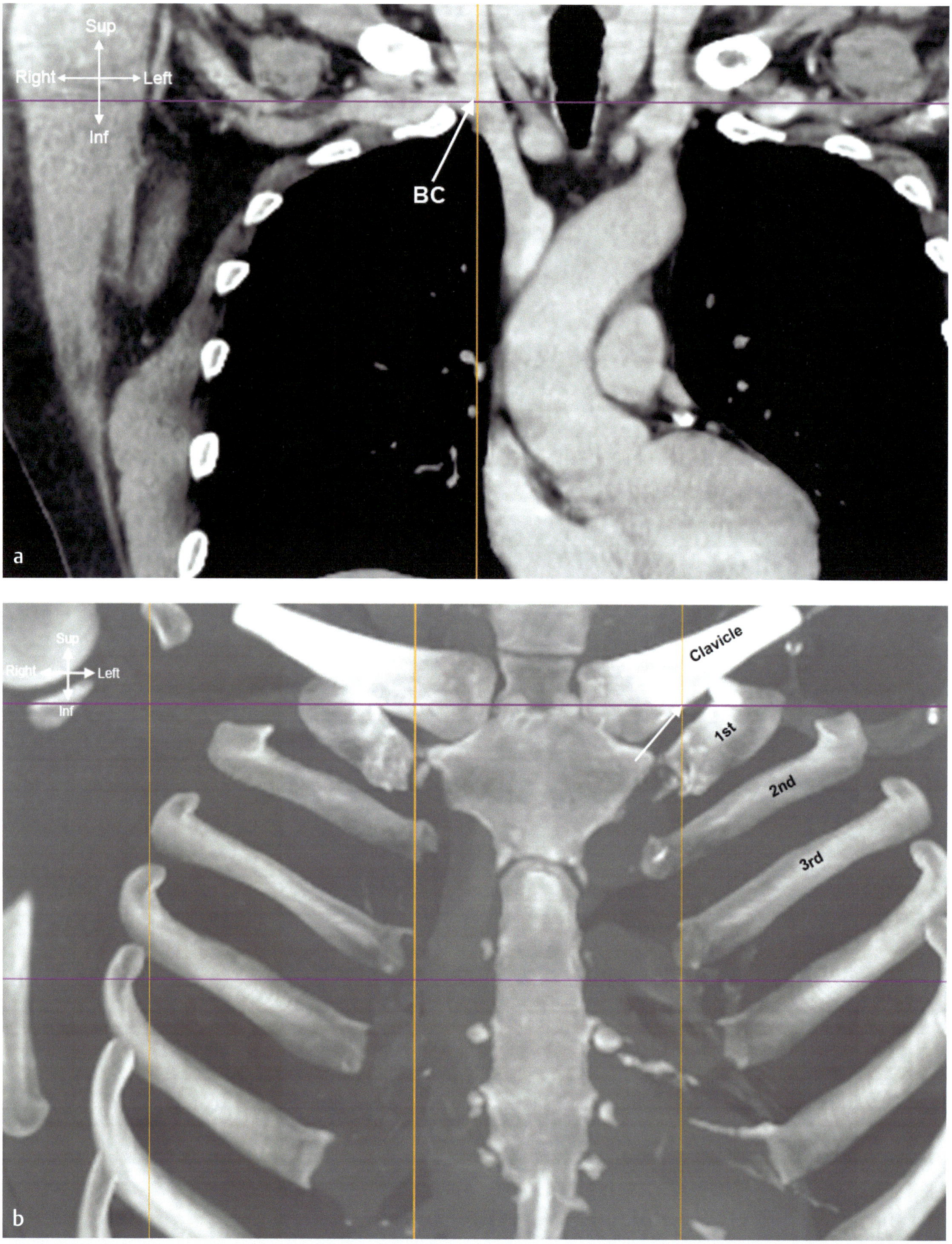

Fig. 22.13 Formation of the right brachiocephalic vein behind the ipsilateral sternoclavicular joint on a coronal scan (*white arrow*) **(a)**. Note the formation of this structure lies lateral to the sternoclavicular joint in a stacked coronal scan (*white arrow*) **(b)**. BC, brachiocephalic vein.

or superior to this joint in a small proportion of cases (about 10%) (▶Table 22.2). *It should be noted that recent evidence indicates that the right brachiocephalic vein formation is lateral to the sternoclavicular joint in the adult Chinese population* (▶Table 22.2).

Superior Vena Cava (SVC)

SVC is primarily formed behind the right second costal cartilage and is partially overlapped by the manubrium (▶Table 22.2, ▶Fig. 22.14).

SVC to Right Atrium (SVC-RA) Junction

This junction occurs most commonly at the right fourth costal cartilage (▶Table 22.2, ▶Fig. 22.15).

Breast[2]

In adult females, the base of the breast is relatively constant in position (vertically from the second/third to the sixth rib and horizontally from the lateral edge of the sternum to the midaxillary

Table 22.2 Surface marking of the central venous structures and their variations

	Left brachiocephalic vein formation	Right brachiocephalic vein formation	Superior vena cava (SVC) formation	SVC to right atrium junction
Most common location	Posterior to SCJ	Posterior to SCJ*	Right 2nd CC	Right 4th CC
Variability	Posterior, superior, or lateral to the SCJ	Posterior, superior, or lateral to the SCJ	1st CC to 3rd CC	3rd CC to 6th CC

Abbreviations: CC, costal cartilage; SCJ, sternoclavicular joint.
Note: * The right brachiocephalic vein forms lateral to the sternoclavicular joint in the adult Chinese population

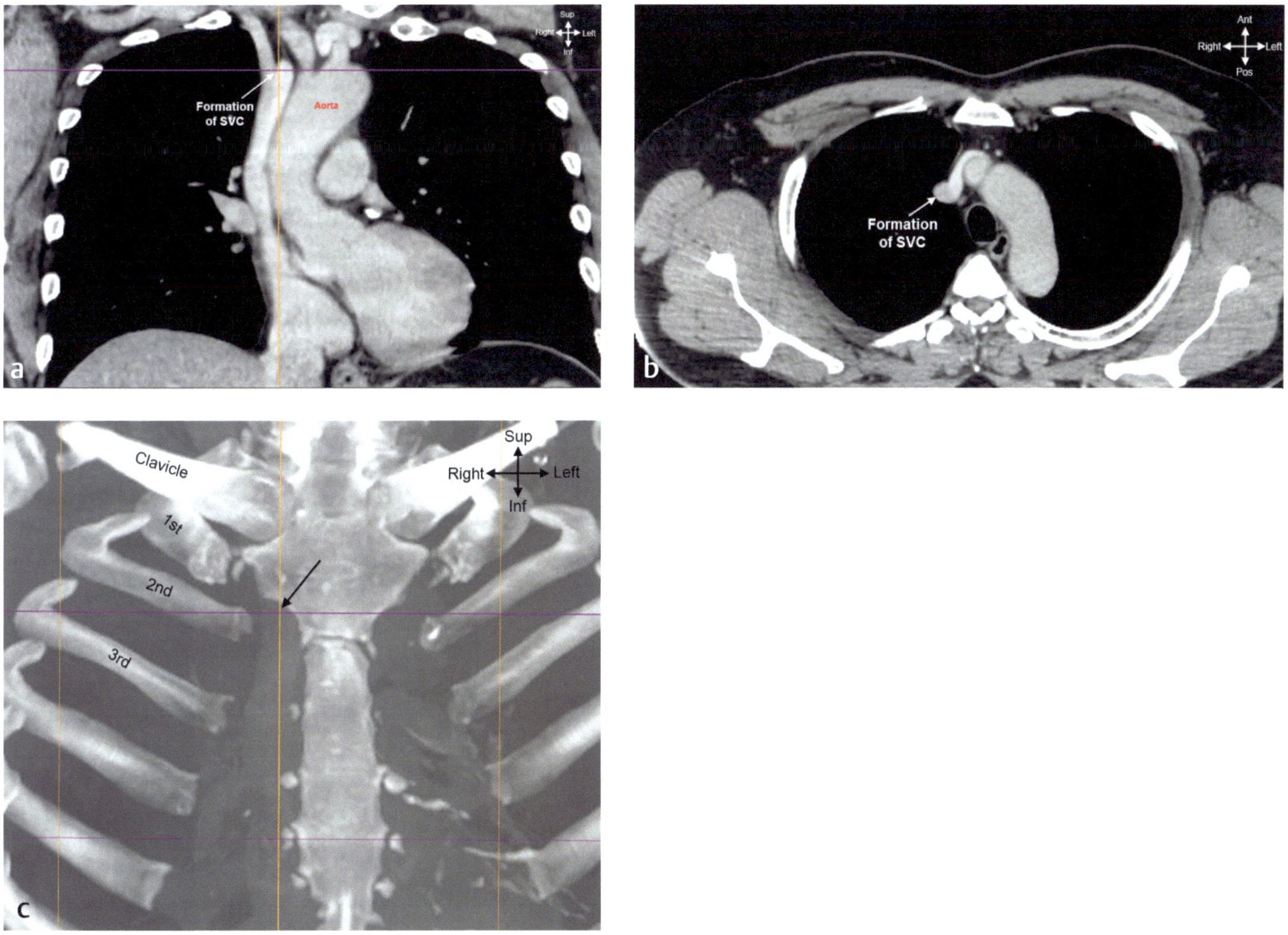

Fig. 22.14 Formation of the superior vena cava in coronal **(a)**, axial **(b)**, and stacked coronal **(c)** scans. SVC, superior vena cava.

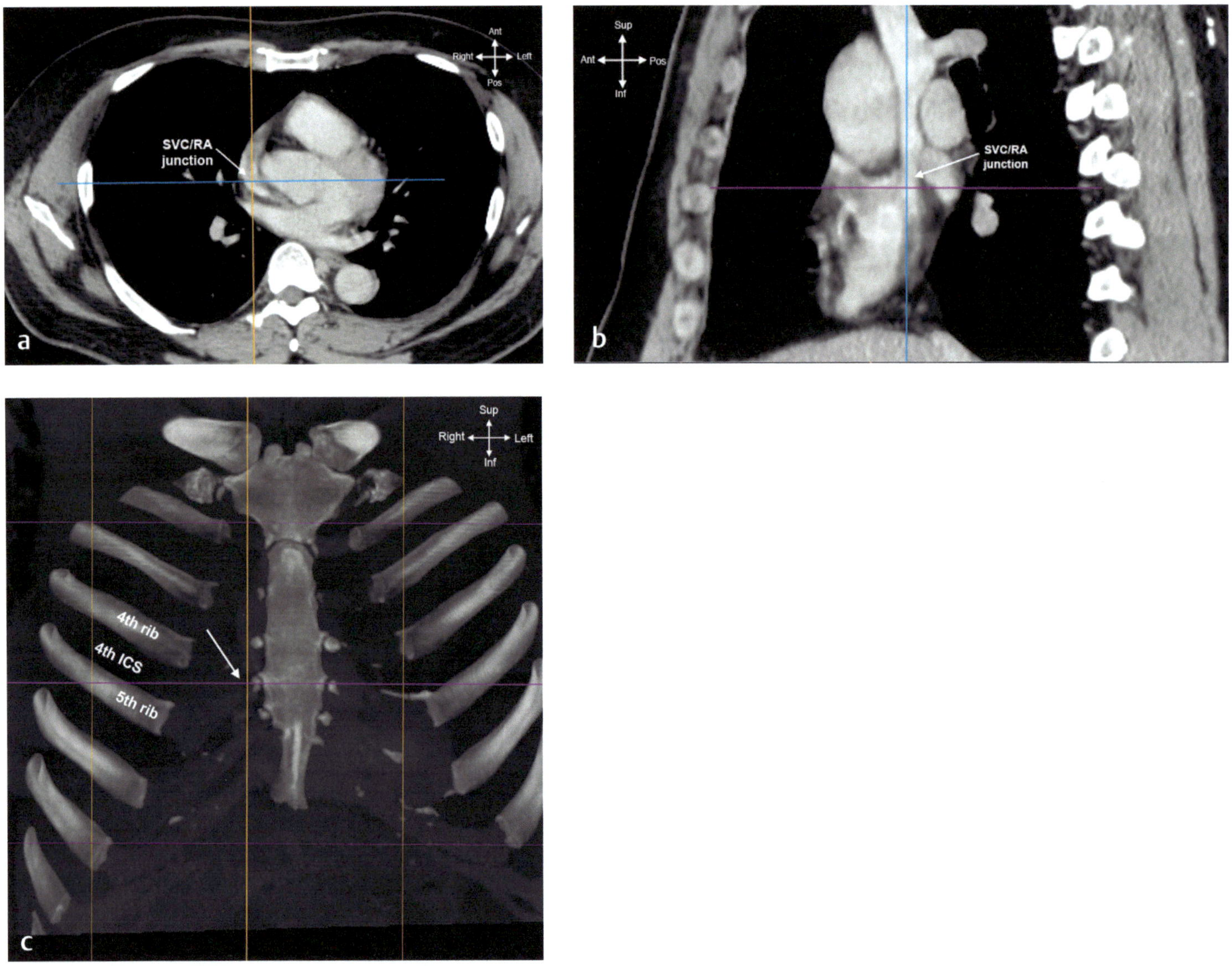

Fig. 22.15 The superior vena cava/right atrium junction in axial **(a)**, sagittal **(b)**, and stacked coronal **(c)** scans. ICS, intercostal space; SVC/RA junction, superior vena cava/right atrium junction.

line). The site of the nipple is dependent on the size and the shape of the breast. In adult males, the nipple is sited in the fourth intercostal space or the fifth rib in the midclavicular line.

Heart[2]

The heart sits anteroinferiorly and to the left in the thorax. In a clockwise direction, the borders of the heart begin at the left second costal cartilage 1 cm from the sternal edge, and then extend to the fifth left intercostal space where the cardiac apex is located. The heart border then continues to the right sixth costal cartilage, 1 to 2 cm lateral to the sternal edge. Finally, the border goes to the right third costal cartilage. *It is important to understand that the position of the heart and its borders can vary due to age, sex, stature, obesity, and posture.* The cardiac apex is located in the midclavicular line, in the fifth intercostal space, approximately 9 cm from the midline (▶ Fig. 22.16).

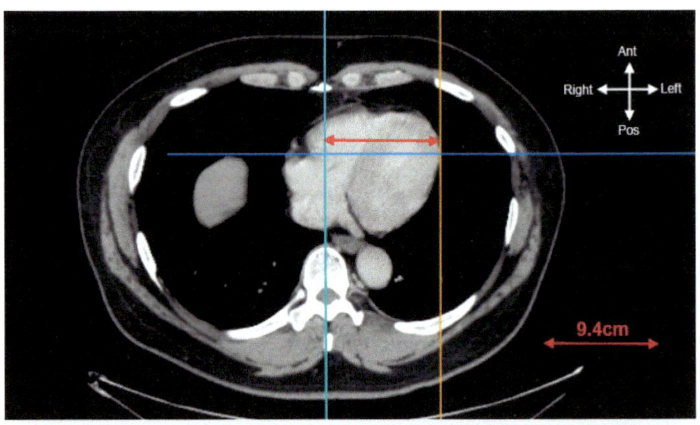

Fig. 22.16 The cardiac apex in the fifth anterior intercostal space in the midclavicular line (*orange line*), with its distance from the midline indicated (*red arrow*). Ant, anterior; Pos, posterior.

The surface anatomy of the cardiac valves is clinically important. The orifice of the pulmonary valve is represented by a horizontal line approximately 2.5-cm long that lies over the superior border of the left third costal cartilage and behind the corresponding region of the sternum. The orifice of the aortic valve is located below and to the right of the orifice of the pulmonary and is represented by a line approximately 2.5-cm long running inferolaterally and to the right from the medial end of the left third intercostal space. The orifice of the tricuspid valve is represented by a line approximately 4-cm long and starts near the midline just below the level of the right fourth costal cartilage and passes down and slightly to the right. The orifice of the mitral valve is represented by a line approximately 3-cm long and descending to the right and is at level with the left half of the sternum opposite the left fourth costal cartilage.

The sounds are best heard when the stethoscope is positioned so that it overlies the downstream flow of blood. Thus, the sites of the cardiac valve auscultation do not correspond to the surface anatomy of the valves. The tricuspid valve is best heard on the left inferior aspect of the sternum near the fifth intercostal space. The mitral valve is best heard over the apex of the heart (fifth left intercostal space in the midclavicular line). The pulmonary valve is best heard over the left second intercostal space near the sternum. The aortic valve is best heard over the right second intercostal space near the sternum.

Diaphragm and Major Structures Passing Through It[7]

The domes of the diaphragm are most commonly located at the fifth anterior intercostal space on the right (▸Fig. 22.17) and the sixth anterior intercostal space on the left at end normal inspiration. It should be noted though that both posture and respiration can affect the position of the domes of the diaphragm. Computed

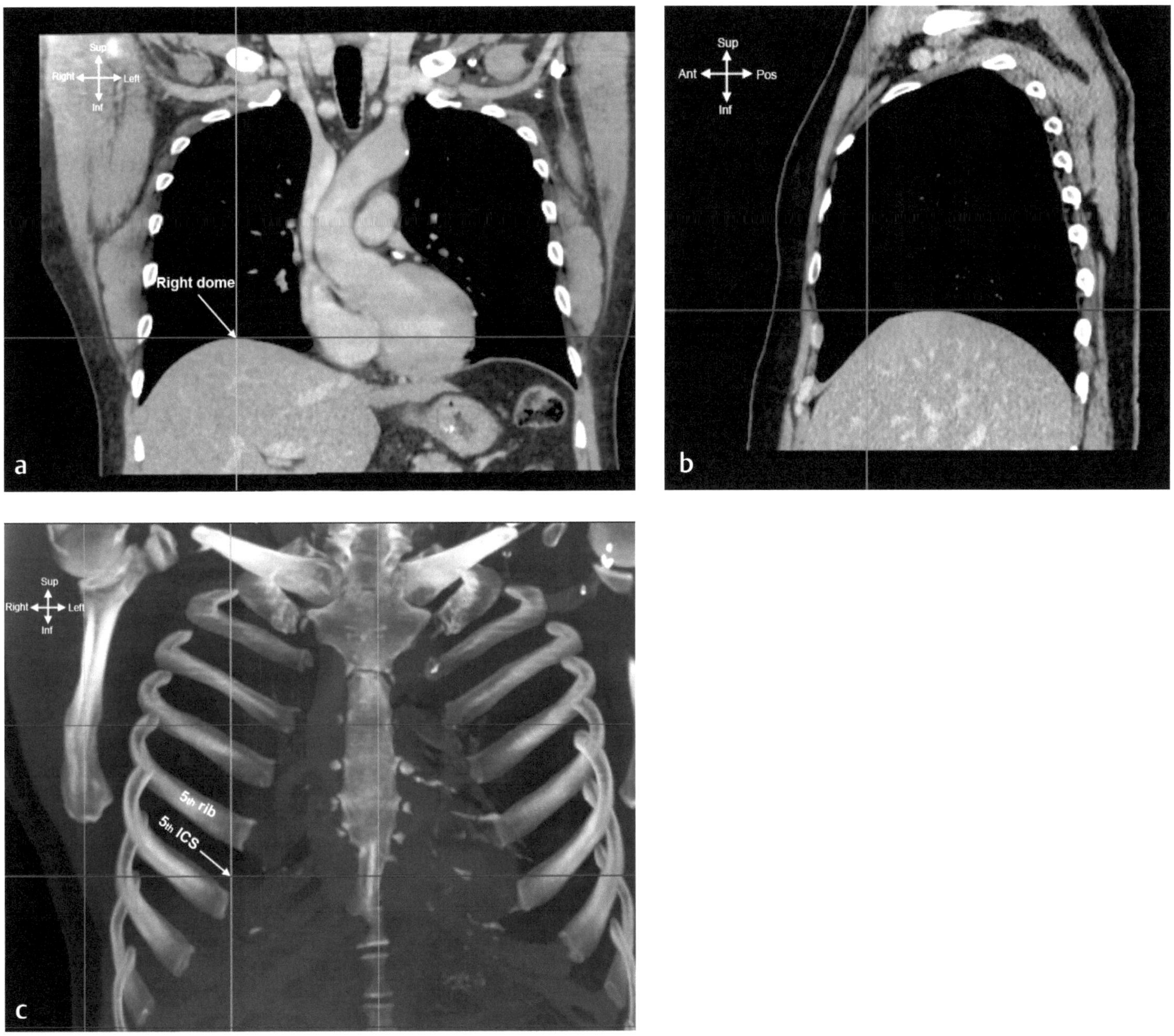

Fig. 22.17 The dome of the right hemi-diaphragm to its corresponding anterior rib level in coronal **(a)**, sagittal **(b)**, and a stacked coronal **(c)** scans. ICS, intercostal space.

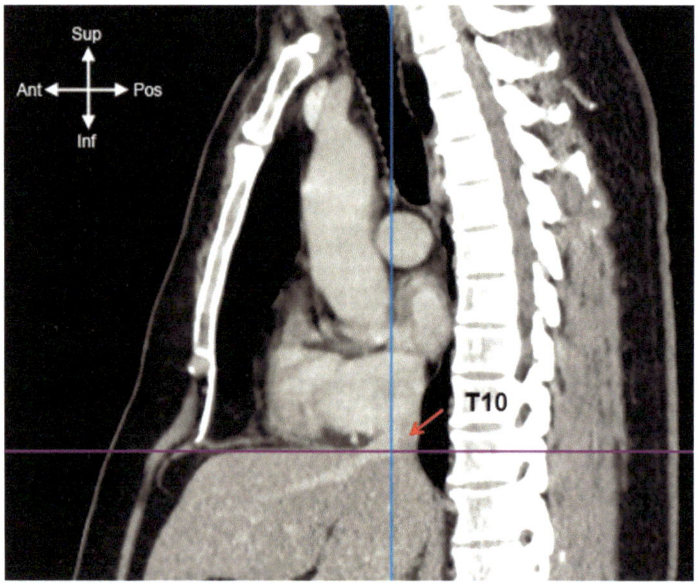

Fig. 22.18 The vertebral level at which the inferior vena cava (*red arrow*) passes through the diaphragm in a sagittal scan.

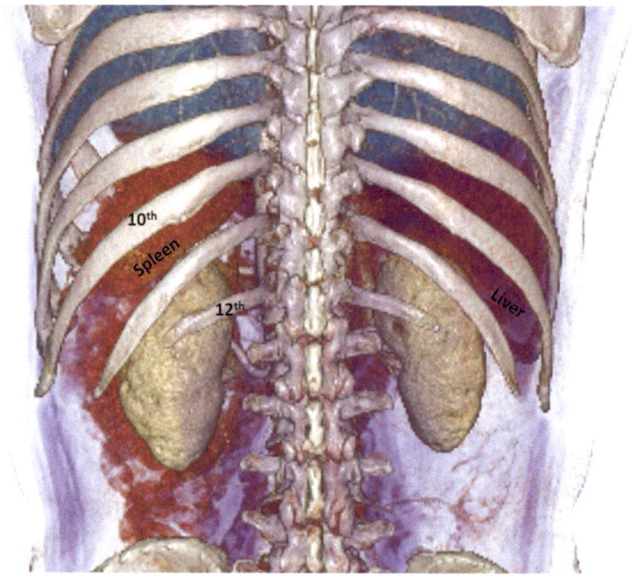

Fig. 22.19 The location of the spleen on a posterior view scan.

tomography (CT) assessment reveals that the IVC and the esophagus pass through the diaphragm at levels slightly lower than those predicted from the chest radiographs. *The IVC passes through the diaphragm at the **10th or 11th thoracic vertebra*** (▶Fig. 22.18), *the esophagus at the **11th thoracic vertebra**, and the aorta at **the 12th thoracic vertebra**.*

Lung and Its Fissures

The apex (dome) of the lung is located behind the medial third of the clavicle. The anterior lower border of the lung in the midclavicular line is most commonly located at the level of the fifth rib. The level of the inferior border of the lung adjacent to the vertebral column is most commonly at the level of T12 or below. Interlobar fissures show wide variability in their positions and shapes. Evidence indicates that the oblique fissure of the left lung is most commonly located posteriorly at the fourth rib, laterally at the sixth rib, and inferiorly in the anterior third of the hemi-diaphragm. The horizontal fissure most often originates in the middle third of the oblique fissure and then continues to terminate anteriorly at the level of the fourth rib.

◆ Abdomen

Solid Organs[8,9]

Kidneys and Ureters

The left kidney is most commonly positioned from the 12th thoracic vertebra to the 3rd lumbar vertebra, with its hilum at the 2nd lumbar vertebra (upper half). The right kidney is most commonly positioned lower than the left kidney from first lumbar vertebra to the fourth lumbar vertebra, with its hilum at the second lumbar vertebra (lower half). The left and right kidneys in adults have mean dimensions of 12 × 6 × 3 cm with an average weight of approximately 130 g.

The ureters begin approximately at the level of the transpyloric plane 5 cm from the midline and descend inferiorly, medial to the tips of the transverse processes of the lumbar vertebrae.

Spleen

The spleen is most commonly located between the 9th and 11th (▶Fig. 22.19), or 10th and 12th ribs. It is likely to span three ribs in obese individuals and commonly extends a few centimeters anterior to the midaxillary line.

Pancreas

The surface projection of the head of the pancreas lies within the duodenal curve on the right side of the second lumbar vertebra. The neck of the pancreas lies at the transpyloric plane and the body of the pancreas passes obliquely up toward the spleen.

Major Vessels[8,9]

The celiac trunk is most commonly located at the level of 12th thoracic vertebra, the superior mesenteric artery at the level of the 1st lumbar vertebra, the right and left renal arteries at the level of second lumbar vertebra, and the IMA at the level of the 3rd lumbar vertebra. The aortic bifurcation is commonly located at the level of the fourth lumbar vertebra (▶Fig. 22.20) and is a few centimeters left of the midline.

The formation of the IVC occurs most commonly at the level of fifth lumbar vertebra, a mean distance of 1.4 cm right of the midline. The formation of the portal vein occurs behind the neck of the pancreas, approximately at the level of the transpyloric plane, which is at the lower border of the first lumbar vertebra.

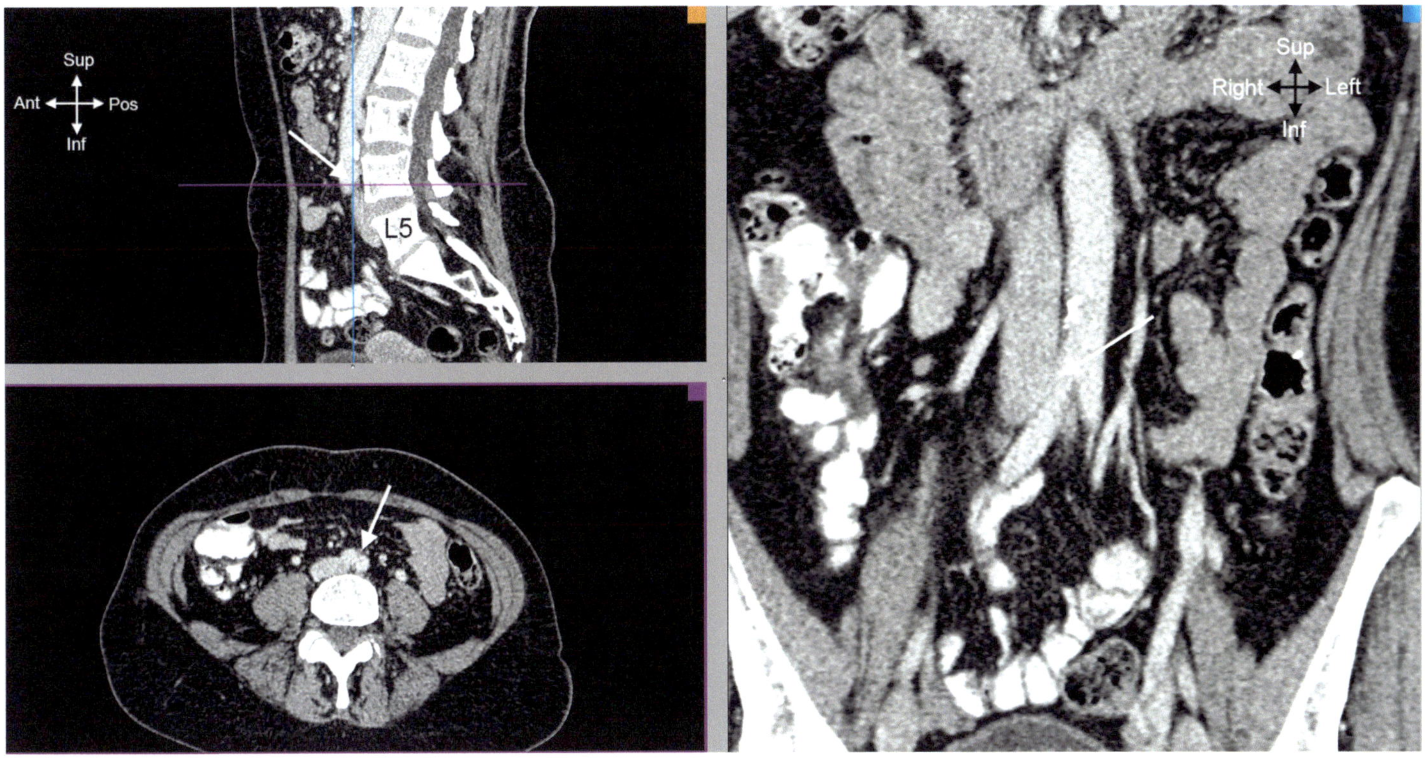

Fig. 22.20 The bifurcation of the abdominal aorta (*white arrow*) in sagittal, axial, and coronal scans. The aortic bifurcation is commonly located at the level of the fourth lumbar vertebra. L, lumbar vertebra.

Hollow Organs[8,9]

Gastroesophageal Junction (GE Junction), Duodenum, Duodenojejunal Flexure (DJ Flexure), and Appendix

Changes in the position or size and shape of the GE junction, duodenum, and DJ flexure are clinically and surgically important (e.g., hiatus hernia or intestinal malrotation). The GE junction is located most commonly at the level of 11th thoracic vertebra to the left of the midline. This junction tends to be higher in obese individuals. The first part of the duodenum is at the level of the first lumbar, the second part reaches the lower border of third lumbar vertebra, the third part crosses the aorta, IVC, and the third lumbar vertebra, and the fourth part passes upwards to the level of the lower border of the first lumbar vertebra to terminate at the DJ flexure. The appendix is located in the right lower quadrant of the abdomen, but it is highly variable in its length and position. Recent studies analyzing barium enemas in supine adults have shown that the McBurney's point (a point two-thirds of the way along a line joining the anterior superior iliac spine to the umbilicus) is not a reliable landmark for the surface anatomy of the appendix.

◆ Upper Limb

The pectoral girdle consists of the proximal humerus, clavicle, and scapula. It connects the upper limb to the axial skeleton. The scapula has bony landmarks, which are palpable and important for certain procedures such as bursal injections. These are the coracoid process inferior to the lateral clavicle, the acromion which is a shelf of bone superior to the head of the humerus and lateral to the coracoid, and finally the spine of the scapula which is a superolateral bony ridge located on the posterior border of the scapula. The deltoid tubercle of the spine of the scapula is palpable on the medial aspect of the spine. The inferior angle of the scapula is also palpable, and the triangle of auscultation lies medially to this.

The greater tubercle of the humerus is palpable inferolateral to the acromion and the lesser tubercle of the humerus is palpable lateral to the coracoid process through the deltoid muscle. On the distal aspect of the humerus, there are three important palpable bony landmarks, the medial epicondyle (the ulnar nerve runs posteriorly to this), the lateral epicondyle, and the capitulum which sits medial to the lateral epicondyle and articulates inferiorly with the radial head. Note that the cubital fossa is a triangular region located in the anterior elbow. It is bordered medially by pronator teres, laterally by brachioradialis, and superiorly (the "base") by a line between the humeral epicondyles. The roof of the cubital fossa is formed by the skin, subcutaneous fascia, deep fascia, and biceps aponeurosis. *The median cubital vein courses through the roof within the subcutaneous fascia and is commonly used for venepuncture. The median cubital vein is highly variable in its courses and three of the most common variations are shown in ▸Fig. 22.21.* Medial to lateral, the contents of the cubital fossa include the median nerve, brachial artery, and biceps tendon (the latter is palpable).

The raised prominence of the posterior elbow is the olecranon process of the ulna which articulates with the olecranon fossa between the humeral condyles. Proximally, the radial head is not

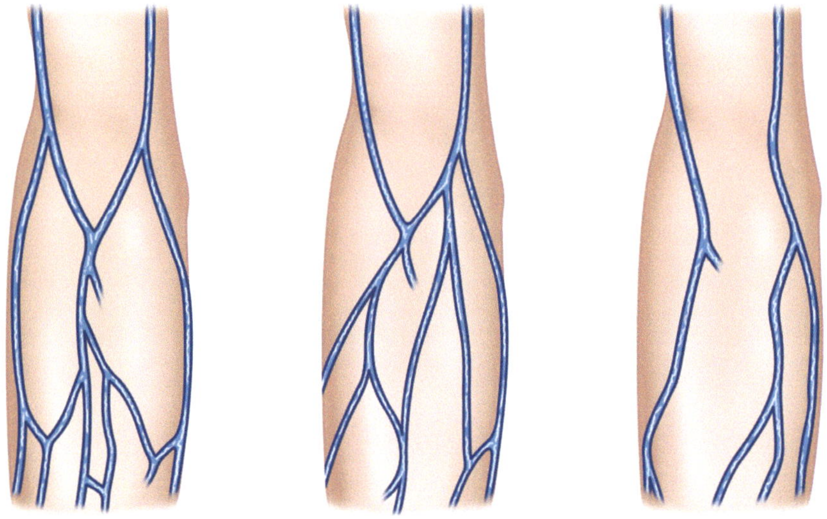

Fig. 22.21 Three of the common variations of the median cubital vein.

palpable due to overlying muscle. Distally, it is palpable on the lateral and posterior aspects of the shaft. The radial styloid process projects laterally from the distal radius and is palpable here, approximately in the "anatomical snuffbox." The "anatomical snuffbox" is a region lateral to the wrist. It is best seen when the thumb and wrist are both extended. The anatomical snuffbox is bordered anteriorly by the tendon of abductor pollicis longus, and posteriorly by the tendon of extensor pollicis longus. The floor is formed, proximal to distal, by the radial styloid process, scaphoid, trapezium attached to first carpometacarpal joint, and the base of the first metacarpal. The contents include the radial artery, cephalic vein, and superficial branch of the radial nerve. *Note that the radial artery runs deep to the tendon of abductor pollicis longus and extensor pollicis brevis, but the cephalic vein and superficial branch of the radial nerve run superficial to these tendons.*

The distal head of the ulna is rounded and palpable on the dorsomedial wrist while the ulna styloid process projects distally from the medial side of the head. Anteriorly, the two proximal and distal wrist creases are visible. The proximal wrist crease marks the line of the wrist joint, and the distal crease marks the entrance of the carpal tunnel. The carpal tunnel is formed between the U shape of the carpal bones and the flexor retinaculum. The roof (flexor retinaculum) lies between the palpable bony prominences: the pisiform and hook of the hamate medially, and the tubercles of the scaphoid and trapezium laterally. *The carpal tunnel includes nine tendons and the median nerve.*

The carpal bones are located in the proximal hand and can be palpated to serve as landmarks for important neurovascular structures. The metacarpals are palpable in the dorsal hand. The first carpometacarpal joint is palpable in the "anatomical snuffbox." The metacarpophalangeal joints are palpable on either side of the extensor tendons. The phalanges are palpable on the dorsal hand.

Knowledge of surface anatomy of the commonly injured nerves of the upper limb at primary health care level is essential. These nerves include: the axillary nerve, median nerve, and superficial branch of the radial nerve.

Axillary Nerve

The axillary nerve originates as a terminal branch of the posterior cord of the brachial plexus. Once it reaches the inferior border of the subscapularis muscle, passing into the quadrilateral space, its two fascicles split into anterior and posterior branches, which collectively innervate the deltoid muscle, teres minor, and skin sensation over the lateral shoulder. *The usual landmark to avoid injury to this nerve, which is 5 cm from the acromion, may not always be reliable. Injections made at the midpoint or below the midpoint of the deltoid muscle could be an alternative method of selecting a safe injection site.*

Median Nerve

The median nerve is formed by the contributions of the medial and lateral cords of the brachial plexus. It travels down the arm initially lateral to the brachial artery, but eventually lies medial to the brachial artery in the cubital fossa. It travels distally between flexor digitorum superficialis and flexor digitorum profundus. *At the level of the wrist joint, between the proximal and distal wrist creases, it lies just medial to the flexor carpi radialis tendon and lateral to palmaris longus or flexor digitorum superficialis (in the absence of the palmaris longus).* It gives rise to the palmar branch approximately at the level of the proximal wrist crease, which runs superficial to the flexor retinaculum. The recurrent branch comes off the median nerve 2 to 3 cm distal to the distal wrist crease, after its course within the carpal tunnel, to innervate the thenar muscle group. *Its terminal branches, the digital nerves, run on either side of the palmar aspect of the length of the digits* (lateral three and a half digits).

Superficial Branch of the Radial Nerve

This nerve emerges between the tendons of brachioradialis and extensor carpi radialis longus and then perforates the fascia of the forearm to become subcutaneous. Within the proximal two-thirds of the forearm, the superficial branch of the radial nerve has not yet emerged and is unlikely to be injured. *It is likely to be injured over the "anatomical snuffbox"* during venepuncture.

◆ Lower Limb

In the lower back, the posterior superior iliac spine is palpable and may be indicated by an overlying sacral dimple; it is at level with the second sacral vertebra (S2) which marks the usual termination of the dural sac. In the gluteal region, the gluteal crease or fold, represents the junction of the posterior thigh and buttock. It is crossed obliquely by gluteus maximus and is not produced by the lower border of this muscle. The sciatic nerve lies deep to gluteus maximus after it emerges from the pelvis just below piriformis. Its surface marking is approximately one-third of the way along a line between the posterior superior iliac spine and the ischial tuberosity, to midway between the greater trochanter and ischial tuberosity; all three bony landmarks are easily palpable (▸Fig. 22.22). The course of the sciatic nerve corresponds roughly to a line from midway between the greater trochanter and ischial tuberosity to the apex of the popliteal fossa.

Anteriorly, the inguinal ligament runs between the anterior superior iliac spine and pubic tubercle. The inguinal/groin skin crease marks the junction between the anterior abdominal wall and thigh and is several centimeters distal to the inguinal ligament. *The most reliable surface landmark for the femoral artery is the midinguinal point, midway between the anterior superior iliac spine and the pubic symphysis.* The femoral artery usually gives off its profunda femoris branch proximal to the inguinal skin crease. If the thigh is slightly flexed, abducted, and externally rotated, the course of the femoral artery corresponds roughly to the upper two-thirds of a line from the midinguinal point to the adductor tubercle. The surface marking of the great saphenous vein is described as originating from the medial marginal vein of the dorsal venous arch of the foot, passing just anterior to the medial malleolus, and running up the medial side of the leg. It lies a hands breadth behind the medial border of the patella, travels up the medial side of the thigh, and pierces the cribriform fascia covering the saphenous opening in the femoral triangle to merge with the femoral vein on its anteromedial aspect. The surface marking of the saphenofemoral junction is that its center lies in a square, extending 1 to 4 cm lateral and up to 3 cm below the pubic tubercle in >90% of adults (▸Fig. 22.23).[10]

On the lateral thigh, the iliotibial tract is visible and palpable. Around the knee, numerous structures are visible and/or

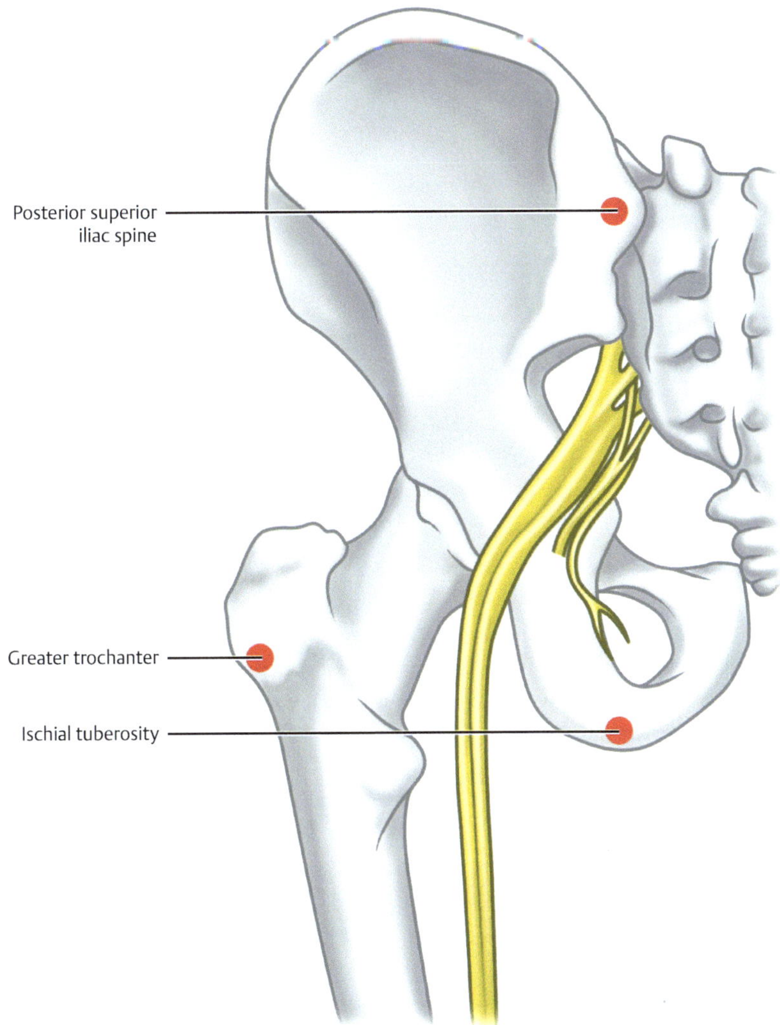

Fig. 22.22 The surface markings of the sciatic nerve which is approximately one-third of the way along a line between the posterior superior iliac spine (PSIS) and the ischial tuberosity (IT) to midway between the greater trochanter (GT) and ischial tuberosity.

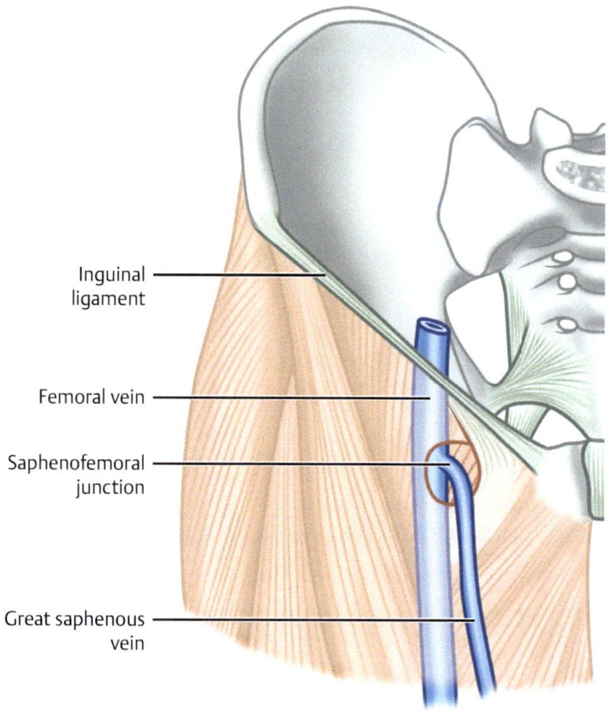

Fig. 22.23 The surface marking of the saphenofemoral junction, located in a square, extending 1 to 4 cm lateral and up to 3 cm below the pubic tubercle. FV, femoral vein; GSV, great saphenous vein; SF junction, saphenofemoral junction.

palpable: the patella and patellar tendon, the knee joint line at the junction between the tibial and femoral condyles, the boundaries of the popliteal fossa, and the head of the fibula. The common fibular nerve can be pressed against the bone as it spirals around the neck of the fibula.

At the ankle and foot, notable bony landmarks include the malleoli and the palpable tuberosity at the base of the fifth metatarsal (insertion site of fibularis brevis). Also palpable are the fibular trochlea just below the tip of the lateral malleolus and the sustentaculum tali of the calcaneus below the medial malleolus. The anterior tibial artery can be located at the ankle midway between the anterior borders of the malleoli. The dorsalis pedis artery lies immediately lateral to the tendon of extensor hallucis longus on the dorsum of the foot, and the posterior tibial artery is palpable below and behind the medial malleolus.

Nerves of the lower limb are commonly injured at primary health care level. These nerves include: the sural nerve, lateral femoral cutaneous nerve, superficial peroneal nerve, and sciatic nerve.

Sural Nerve

In the case of skin lesions in the posterior aspect of the lower limb, the calcaneal tendon is a useful landmark for the location of the sural nerve. A point around 10 cm proximal from the distal attachment of the calcaneal tendon can be used as a guide to estimate the point of intersection of the sural nerve with the lateral border of the calcaneal tendon. *For minor procedures in the proximal calf, it is unlikely that the sural nerve has formed and thus it is not likely to be injured. Note that the sural nerve may form in the popliteal fossa rarely.*

Lateral Femoral Cutaneous Nerve

This nerve pierces Colles' fascia within approximately 2 cm inferior to the anterior superior iliac spine (ASIS). *The course of the nerve as it runs between the ASIS and inguinal ligament is quite superficial and thus it is here that it is most likely to be injured.* Due to this nerve's high variability and a lack of consistent evidence, safer recommendations for injection sites (for a nerve block) cannot be made at this time.

Superficial Peroneal Nerve

This nerve originates from the common peroneal nerve at the level of the neck of the fibula. It then travels down the leg in the lateral compartment, most commonly between peroneus brevis and the intermuscular septum. As it travels more distally, it continues rising more superficially until it pierces the crural fascia above the superior extensor retinaculum of the foot. It runs superficial to the superior and inferior extensor retinacula of the foot to innervate the sensation to the dorsum of the foot (except the first web space). *Distal to the crural fascia, due to its superficial course, it is vulnerable to injury from surgical procedures or trauma.*

Sciatic Nerve

The sciatic nerve crosses a point approximately one-third of the way along a line between the posterior superior iliac spine and the ischial tuberosity. It then curves laterally and inferiorly to a point midway between the greater trochanter and ischial tuberosity. The "upper outer quadrant of the buttock" has historically been a common site for an intramuscular (IM) injection, but the sciatic nerve may be injured partly because it is not always understood that the buttock extends from the iliac crest to the gluteal fold. Iatrogenic sciatic nerve injury from misplaced IM injections continues to be a global problem. *A safer site for IM injection is the anterolateral thigh. If the buttock is used, the ventrogluteal area is a safer alternative because it has less subcutaneous fat, is well away from the sciatic nerve, and is easily accessible.*

◆ Dermatomes

A dermatome is a cutaneous area (area of skin) which is supplied by the dorsal root (sensory division) of a single spinal nerve. The dermatomes can be mapped out over the body so that certain areas of the skin are said to be supplied by a particular spinal nerve. For example, the cutaneous area at the level of the umbilicus is supplied by the spinal nerve T10 and at the level of the nipple line by T4. Note that historically the dermatome map of the human body has shown distinct distributions of each and every spinal nerve. Recent evidence-based research indicates, however, that there is overlap of nervous supply for adjacent dermatomes and that individual variability in dermatome mapping exists (▶ Fig. 22.24).

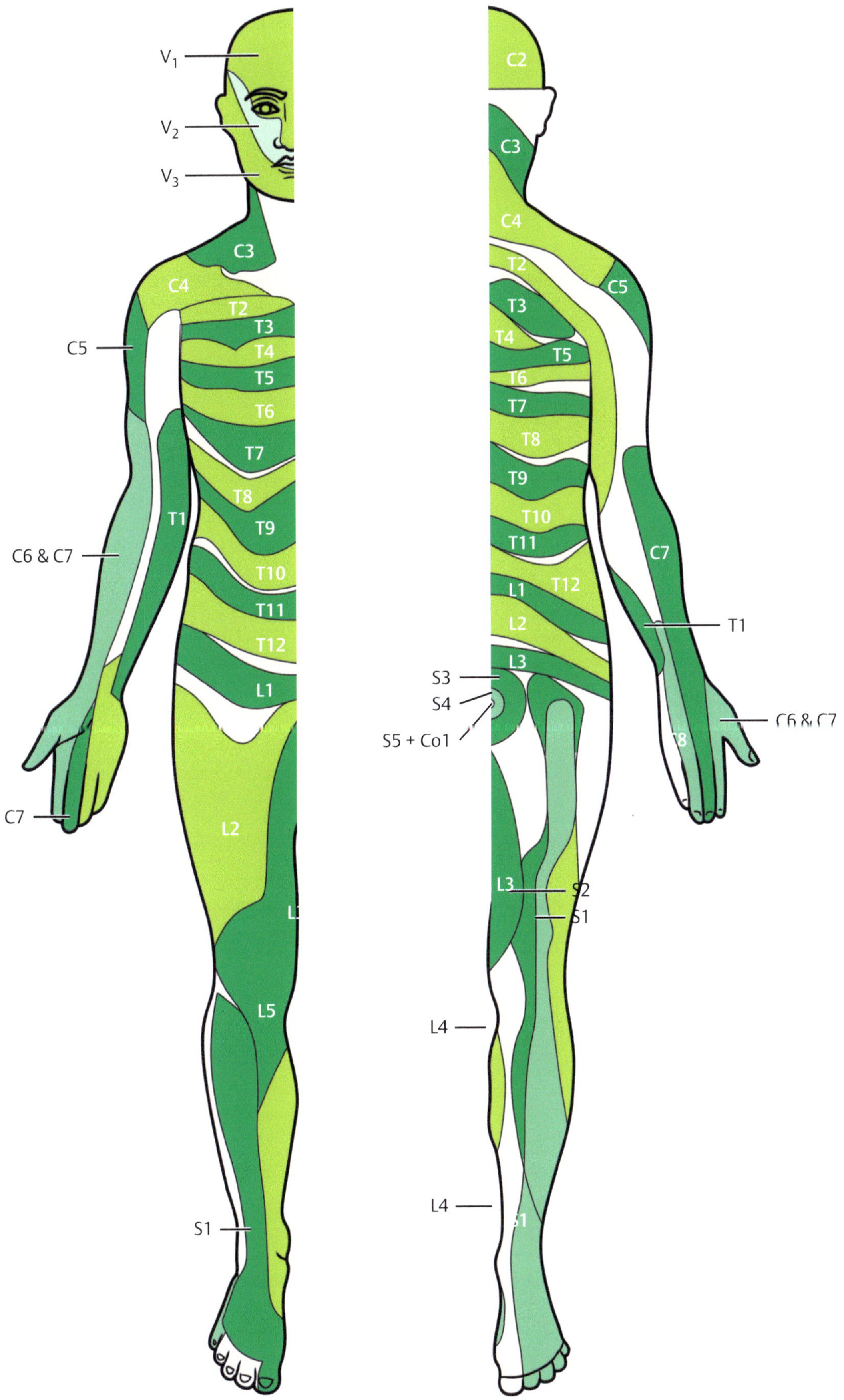

Fig. 22.24 The evidence-based dermatome map representing the most consistent tactile dermatomal areas for each spinal dorsal nerve root found in most individuals. Blank regions indicate areas of major variability and overlap.

The dermatomes have many important clinical applications. The areas of sensory loss can indicate which spinal nerves have been damaged or compressed in a radiculopathy. Knowledge of the distribution of the dermatomes is especially important for determining the level at which anesthetic is applied in anesthesiology and also for the diagnosis of many diseases which have referred pain (e.g., referred pain to the shoulders with angina).

References

1. Mirjalili SA, McFadden SL, Buckenham T, Wilson B, Stringer MD. Anatomical planes: are we teaching accurate surface anatomy? Clin Anat 2012;25(7): 819–826 Med
2. Hale SJM, Mirjalili SA, Stringer MD. Inconsistencies in surface anatomy: the need for an evidence-based reappraisal. Clin Anat 2010;23(8):922–930 PubMed
3. Ma S, Baillie LJM, Stringer MD. Reappraising the surface anatomy of the pterion and its relationship to the middle meningeal artery. Clin Anat 2012;25(3): 330–339
4. Stringer MD, Mirjalili SA, Meredith SJ, Muirhead JC. Redefining the surface anatomy of the parotid duct: an in vivo ultrasound study. Plast Reconstr Surg 2012; 130(5):1032–1037 Med
5. Mirjalili SA, McFadden SL, Buckenham T, Stringer MD. Vertebral levels of key landmarks in the neck. Clin Anat 2012;25(7):851–857
6. Mirjalili SA, Muirhead JC, Stringer MD. Ultrasound visualization of the spinal accessory nerve in vivo. J Surg Res 2012;175(1):e11–e16
7. Mirjalili SA, Hale SJM, Buckenham T, Wilson B, Stringer MD. A reappraisal of adult thoracic surface anatomy. Clin Anat 2012;25(7):827–834
8. Shen XH, Su BY, Liu JJ, et al. A reappraisal of adult thoracic and abdominal surface anatomy via CT scan in Chinese population. Clin Anat 2016;29(2):165–174
9. Mirjalili SA, McFadden SL, Buckenham T, Stringer MD. A reappraisal of adult abdominal surface anatomy. Clin Anat 2012;25(7):844–850
10. Mirjalili SA, Muirhead JC, Stringer MD. Redefining the surface anatomy of the saphenofemoral junction in vivo. Clin Anat 2014;27(6):915–919

Index

A

B

C

D

E

F

G

T

U

V

W

Z